Before you begin, review your basic study skills.

To take full advantage of the **Study Skills Workshops** that begin each chapter, you may choose to review them in the early weeks of your course. Each one includes action items, in addition to simple suggestions that can put you on a clear path to success. Below, we have included a table of contents to aid you in locating these:

Study Skills Workshop ▶ Making Homework a Pri

Attending class and taking notes are important, but they are not enough. The only way that you are really going to learn algebra is by doing your homework.

WHEN TO DO YOUR HOMEWORK: Homework should be started on the day it is assigned, when the material is fresh in your mind. It's best to break your homework sessions into 30-minute periods, allowing for short breaks in between.

HOW TO BEGIN YOUR HOMEWORK: Review your notes and the examples in your text before starting your homework assignment.

GETTING HELP WITH YOUR HOMEWORK: It's normal to have some questions when doing homework. Talk to a tutor, a classmate, or your instructor to get those questions answered.

Now Try This ▶

1. Write a one-page paper that describes *when*, *where*, and *how* you go about completing your algebra homework assignments.
2. For each problem on your next homework assignment, find an example in this book that is similar. Write the example number next to the problem.
3. Make a list of questions that you have while doing your next assignment. Then decide whom you are going to ask to get those questions answered.

Intermediate Algebra

5e

Intermediate Algebra

▶ **Alan S. Tussy**
Citrus College

▶ **R. David Gustafson**
Rock Valley College

BROOKS/COLE
CENGAGE Learning™

Australia • Brazil • Japan • Korea • Mexico • Singapore • Spain • United Kingdom • United States

Intermediate Algebra, **Fifth Edition**
Alan S. Tussy, R. David Gustafson

Publisher: Charlie Van Wagner

Senior Developmental Editor:
    Danielle Derbenti

Assistant Editor: Carrie Jones

Senior Editorial Assistant: Jennifer Cordoba

Media Editors: Heleny Wong,
    Guanglei Zhang

Marketing Manager: Laura McGinn

Marketing Assistant: Shannon Maier

Marketing Communications Manager:
    Darlene Macanan

Content Project Manager: Jennifer Risden

Design Director: Rob Hugel

Art Director: Vernon Boes

Print Buyer: Judy Inouye

Rights Acquisitions Specialist:
    Tom McDonough

Production Service: Chapter Two,
    Ellen Brownstein

Text Designer: Terri Wright

Photo Researcher: Bill Smith Group

Copy Editor: Chapter Two, Ellen Brownstein

Illustrator: Lori Heckelman

Cover Designer: Terri Wright

Cover Image: Artie Ng/www.flickr.com/
    photos/artiephotography

Compositor: Graphic World, Inc.

For product information and technology assistance, contact us at
Cengage Learning Customer & Sales Support, 1-800-354-9706.

For permission to use material from this text or product,
submit all requests online at **www.cengage.com/permissions.**
Further permissions questions can be e-mailed to
permissionrequest@cengage.com.

Library of Congress Control Number: 2010939896

Student Edition:
ISBN-13: 978-1-111-56767-5
ISBN-10: 1-111-56767-0

Loose-leaf Edition:
ISBN-13: 978-1-111-98770-1
ISBN-10: 1-111-98770-X

Brooks/Cole
20 Davis Drive
Belmont, CA 94002-3098
USA

Cengage Learning is a leading provider of customized learning solutions with office locations around the globe, including Singapore, the United Kingdom, Australia, Mexico, Brazil, and Japan. Locate your local office at **www.cengage.com/global.**

Cengage Learning products are represented in Canada by Nelson Education, Ltd.

To learn more about Brooks/Cole, visit **www.cengage.com/brookscole**
Purchase any of our products at your local college store or at our preferred online store **www.cengagebrain.com.**

Printed in the United States of America
1 2 3 4 5 6 7 15 14 13 12 11

To the mathematics instructors of Edgewood High School, the University of Redlands, and California State University at Los Angeles, whose love of their discipline and dedication to their students continue to be an inspiration to me.

—AST

To my wife, Carol,
with love and appreciation.

—RDG

ABOUT THE AUTHORS

Alan S. Tussy

Alan Tussy teaches all levels of developmental mathematics at Citrus College in Glendora, California. He has written nine math books—a paperback series and a hard-cover series. A meticulous, creative, and visionary teacher who maintains a keen focus on his students' greatest challenges, Alan Tussy is an extraordinary author, dedicated to his students' success. Alan received his Bachelor of Science degree in Mathematics from the University of Redlands and his Master of Science degree in Applied Mathematics from California State University, Los Angeles. He has taught up and down the curriculum from prealgebra to differential equations. He is currently focusing on the developmental math courses. Professor Tussy is a member of the American Mathematical Association of Two-Year Colleges.

R. David Gustafson

R. David Gustafson is Professor Emeritus of Mathematics at Rock Valley College in Illinois and coauthor of several best-selling math texts, including Gustafson/Frisk's *Beginning Algebra, Intermediate Algebra, Beginning and Intermediate Algebra: A Combined Approach, College Algebra,* and the Tussy/Gustafson developmental mathematics series. His numerous professional honors include Rock Valley Teacher of the Year and Rockford's Outstanding Educator of the Year. He earned a Master of Arts from Rockford College in Illinois, as well as a Master of Science from Northern Illinois University.

Contents

PREFACE

We are excited to present the Fifth Edition of *Intermediate Algebra*. We believe the revision process has produced an even stronger instructional experience for both students and teachers. First, we have fine-tuned several of the popular features of our series, including the *Strategy* and *Why* example structure, the problem-solving strategy, and the online homework.

Second, we have introduced some new features to further promote student understanding and success. The new *Look Alikes* problems in the *Study Sets* will help students improve their problem recognition. The *Are You Ready?* exercises that begin each section give students the opportunity to review necessary prerequisite skills before they are asked to apply them in the study of new topics. Furthermore, the additional *Campus to Careers* problems inserted in the *Study Sets* expand on the real-life applications of the chapter material.

We want to thank all of you throughout the country who have provided suggestions and input about the previous edition. Your insight has proven invaluable. Throughout this process, our fundamental belief has remained the same: Algebra is a language in its own right. And, as always, the prime objective of this textbook is to teach students how to read, write, speak, and think using the language of algebra.

New to the Fifth Edition

Sections That Begin with Review: *Are You Ready?*

Each section begins with a set of *Are You Ready?* problems. These problems review crucial prerequisite skills that students need to have mastered in order to be successful with the new topics of that section.

> **ARE YOU READY?**
>
> ▼ *The following problems review some basic skills that are needed to find the slope of a line.*
>
> 1. Evaluate: $\dfrac{4-1}{8-3}$ 2. Evaluate: $\dfrac{-10-1}{-4-(-4)}$ 3. Simplify: $\dfrac{24}{28}$
>
> 4. Fill in the blanks. _____ lines do not intersect. _____ lines intersect to form right angles.

Study Sets with More Problem-Recognition Practice: *Look Alikes*

After a poorer than expected performance on a test, students often tell their instructors, "I could do the homework each night, but when it comes to the test, I get confused." The new *Look Alikes* feature builds students' problem-recognition skills. It requires students to distinguish between similar looking problem types and then to select the correct strategy to solve the problem. Encountering such situations in the homework assignments will better prepare students for quizzes and tests.

> **Look Alikes . . .**
>
> *Solve the inequality in part a. Graph the solution set and write it in interval notation. Then use your answer to part a to determine the solution set for the inequality in part b. (No new work is necessary!) Graph the solution set and write it in interval notation.*
>
> 81. a. $12x - 33.16 \le 5.84$ b. $12x - 33.16 > 5.84$
>
> 82. a. $-\dfrac{3}{4}x > -\dfrac{21}{32}$ b. $-\dfrac{3}{4}x \le -\dfrac{21}{32}$
>
> 83. a. $3(2x + 2) > 5(x - 1) + 3x$
>   b. $3(2x + 2) < 5(x - 1) + 3x$

And Even More Problem-Recognition Practice: *Try It Yourself*

Designed to promote problem recognition, instructors and reviewers requested more *Try It Yourself* problems for the Fifth Edition. These problem types are thoroughly mixed, giving students an opportunity to practice decision making and strategy selection as they would when taking a test or quiz. With more than 80% more problems added, students will have even more opportunities to practice this essential skill.

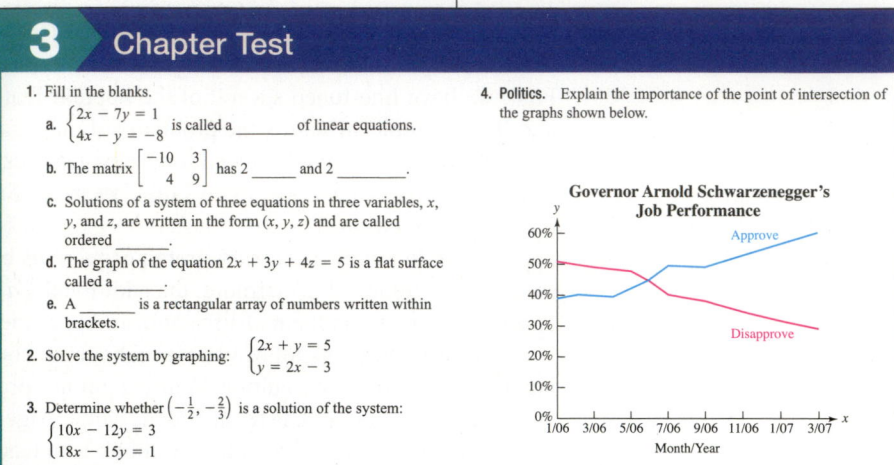

TRY IT YOURSELF

Solve each equation and inequality. For the inequalities, graph the solution set and write it using interval notation.

75. $|3x + 2| + 1 > 15$

76. $|2x - 5| - 5 > 20$

77. $6\left|\dfrac{x-2}{3}\right| \le 24$

78. $8\left|\dfrac{x-2}{3}\right| > 32$

79. $-7 = 2 - |0.3x - 3|$

80. $-1 = 1 - |0.1x + 8|$

81. $|2 - 3x| \ge -8$

82. $|-1 - 2x| > 5$

83. $|7x + 12| = |x - 6|$

84. $|8 - x| = |x + 2|$

85. $2 \ge 3|2 - 3x| + 2$

86. $7 \ge |15x - 45| + 7$

Comprehensive Test Preparation: *Chapter Tests*

Instructors often assign an end-of-chapter test for students to use as a means to study for a classroom exam. However, after taking the exam, students often remark that the classroom exam included problem types that weren't in the *Chapter Test*. To address this issue, we have made sure that each *Chapter Test* is a comprehensive collection of problems that covers *all* of the topics discussed in the chapter. As a result, the *Chapter Tests* are lengthy. If your students have time to complete a *Chapter Test,* that would be optimal. If, because of time constraints, they are unable to do so, assign an appropriate subset of problems that reflects the types of problems that the students will see on your exam. This should alleviate the discrepancy between what your students practice and what they will see on the test.

3 **Chapter Test**

1. Fill in the blanks.
   a. $\begin{cases} 2x - 7y = 1 \\ 4x - y = -8 \end{cases}$ is called a _____ of linear equations.
   b. The matrix $\begin{bmatrix} -10 & 3 \\ 4 & 9 \end{bmatrix}$ has 2 _____ and 2 _____.
   c. Solutions of a system of three equations in three variables, x, y, and z, are written in the form (x, y, z) and are called ordered _____.
   d. The graph of the equation $2x + 3y + 4z = 5$ is a flat surface called a _____.
   e. A _____ is a rectangular array of numbers written within brackets.

2. Solve the system by graphing: $\begin{cases} 2x + y = 5 \\ y = 2x - 3 \end{cases}$

3. Determine whether $\left(-\frac{1}{2}, -\frac{2}{3}\right)$ is a solution of the system:
   $\begin{cases} 10x - 12y = 3 \\ 18x - 15y = 1 \end{cases}$

4. **Politics.** Explain the importance of the point of intersection of the graphs shown below.

Governor Arnold Schwarzenegger's Job Performance

Approve

Disapprove

Month/Year

Additional Relevant, Motivating Applications in the *Examples* and *Study Sets*

We have included many new applied examples and problems that involve relevant topics such as our environment and sustainability issues, energy savings, technology and social media, and recycling. For a complete list of topics, see the *Index of Applications* following the *Preface*. To see a sampling of new topics added to each chapter, see *Content Changes by Chapter* in this *Preface*.

©Jeffrey M. Frank/Shutterstock.com

EXAMPLE 5 **Farmers Markets.** The increasing popularity of farmers markets in the United States can be modeled by the linear equation $f = 265t + 2,728$, where t represents the number of years after 2000 and f represents the number of farmers markets. (*Source:* U.S. Department of Agriculture) **a.** Graph the equation. **b.** Suppose the trend continues. Use the graph to estimate the number of farmers markets in the year 2015.

Strategy We will find three solutions of the equation, plot them on a rectangular coordinate system, and then draw a straight line passing through the points.

Why To *graph* a linear equation in two variables means to make a drawing that represents all of its solutions.

Solution **a.** The variables t and f are used in the equation. If we associate t with the horizontal axis and f with the vertical axis, then the ordered pair solutions has the form (t, f).

To graph the equation, we select three values for t, substitute them into the equation, and find each corresponding value of f. Since t represents the number of years *after* 2000, we will not select any page.

For $t = 0$ (The year 2000)
$f = 265t + 2,728$
$f = 265(0) + 2,728$
$f = 0 + 2,728$
$f = 2,728$

37. e-Readers. The perimeter of the rectangular display screen of Amazon's Kindle 3 is 16.8 inches. If the height is 1.2 inches greater than the width, find the dimensions of the screen.

©Alex Segre/Alamy

A More Precise Problem-Solving Strategy

In an effort to better describe the problem-solving strategy used in this book, we have inserted a new second step in what was formerly a five-step process. This additional step (*Assign a variable to represent an unknown value in the problem.*) better delineates the thought process students should use as they solve application problems. The six steps of the problem-solving strategy are now: *Analyze the problem*, *Assign a variable*, *Form an equation*, *Solve the equation*, *State the conclusion*, and *Check the result*.

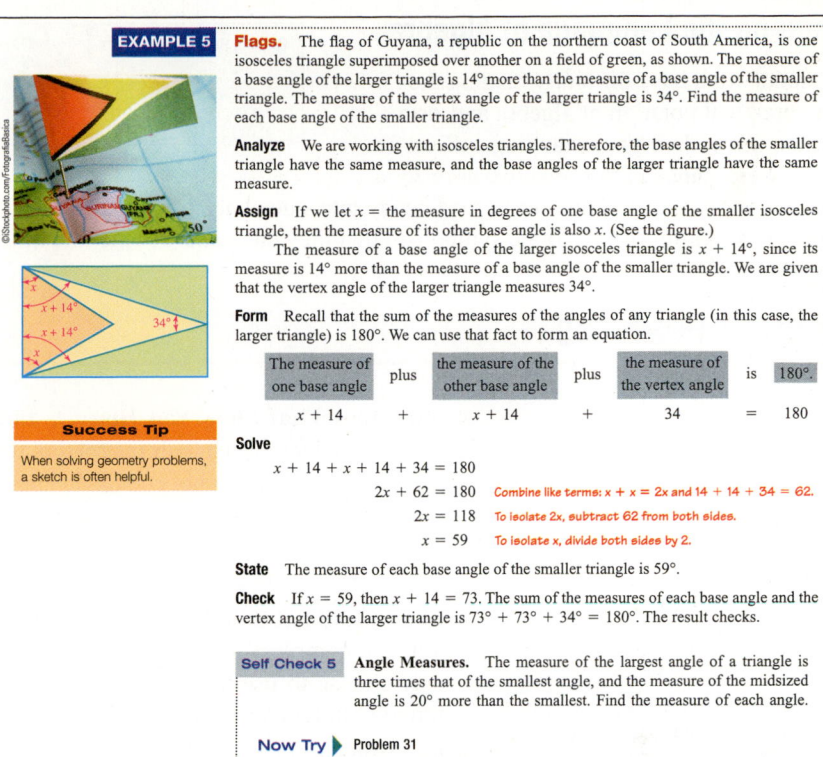

EXAMPLE 5

Flags. The flag of Guyana, a republic on the northern coast of South America, is one isosceles triangle superimposed over another on a field of green, as shown. The measure of a base angle of the larger triangle is 14° more than the measure of a base angle of the smaller triangle. The measure of the vertex angle of the larger triangle is 34°. Find the measure of each base angle of the smaller triangle.

Analyze We are working with isosceles triangles. Therefore, the base angles of the smaller triangle have the same measure, and the base angles of the larger triangle have the same measure.

Assign If we let x = the measure in degrees of one base angle of the smaller isosceles triangle, then the measure of its other base angle is also x. (See the figure.)

The measure of a base angle of the larger isosceles triangle is $x + 14°$, since its measure is 14° more than the measure of a base angle of the smaller triangle. We are given that the vertex angle of the larger triangle measures 34°.

Form Recall that the sum of the measures of the angles of any triangle (in this case, the larger triangle) is 180°. We can use that fact to form an equation.

The measure of one base angle	plus	the measure of the other base angle	plus	the measure of the vertex angle	is	180°.
$x + 14$	$+$	$x + 14$	$+$	34	$=$	180

Success Tip

When solving geometry problems, a sketch is often helpful.

Solve

$$x + 14 + x + 14 + 34 = 180$$
$$2x + 62 = 180 \quad \text{Combine like terms: } x + x = 2x \text{ and } 14 + 14 + 34 = 62.$$
$$2x = 118 \quad \text{To isolate } 2x, \text{ subtract 62 from both sides.}$$
$$x = 59 \quad \text{To isolate } x, \text{ divide both sides by 2.}$$

State The measure of each base angle of the smaller triangle is 59°.

Check If $x = 59$, then $x + 14 = 73$. The sum of the measures of each base angle and the vertex angle of the larger triangle is $73° + 73° + 34° = 180°$. The result checks.

Self Check 5 **Angle Measures.** The measure of the largest angle of a triangle is three times that of the smallest angle, and the measure of the midsized angle is 20° more than the smallest. Find the measure of each angle.

Now Try Problem 31

More Emphasis on "When Will I Use This?"

Each chapter now has three *Campus to Careers* problems that explore the mathematical connections to careers that are presented at the beginning of each chapter.

A Review of Basic Algebra

66. **from Campus to Careers**
Registered Dietitian
When discussing weight management with their patients, dietitians stress the importance of limiting calorie intake and increasing physical activity. One method dietitians use to determine if a patient has a daily calorie surplus (or deficit) is to subtract the patient's daily

99. **from Campus to Careers**
Registered Dietitian
Dietitians often calculate a patient's BMI (Body Mass Index) to screen for weight categories that may lead to health problems. BMI is a number that is calculated from one's weight and height. It is an indication of a person's total body weight that comes from fat. The formula for BMI, as it appears in dietary textbooks, is:

$$\text{BMI} = \frac{\text{weight(lb)} \cdot 703}{\text{height}^2(\text{in.}^2)}$$

57. **from Campus to Careers**
Registered Dietitian
Suppose, as registered dietitian for a school district, you must make sure that only extra lean ground beef (16% fat) is served in the cafeteria. Further suppose that the kitchen has 8 pounds of regular ground beef (30% fat) on hand. How many pounds of extra lean ground beef (12% fat) must be purchased and added to the regular ground beef to obtain a mixture that has the correct fat content?

from Campus to Careers
Registered Dietitian
One of the most important things that you can do to protect yourself from cancer, diabetes, heart disease, and stroke is to eat right. No one knows this better than dietitians. They work in hospitals, health care centers, schools, and correctional facilities, where they plan dietary programs and supervise the preparation of healthy meals. The job of dietitian requires mathematical skills such as calculating calorie intake, analyzing the nutritional content of food, and budgeting for the purchase of groceries and supplies.

Problem 65 in **Study Set 1.1**, **problem 99** in **Study Set 1.6**, and **problem 57** in **Study Set 1.8** involve situations that a registered dietitian might encounter on the job. The mathematical concepts discussed in this chapter can be used to solve those problems.

JOB TITLE:
Registered Dietitian
EDUCATION:
A bachelor's degree in foods and nutrition (or a related field) and supervised practice
JOB OUTLOOK:
Growth faster than the average, 18%–26% increase
ANNUAL EARNINGS:
The median base salary in 2009 was $53,230.
FOR MORE INFORMATION:
www.bls.gov/oco/ocos077.htm

1

Reading the Language of Algebra

Students often have difficulty reading the mathematical notation of algebra and, as a consequence, their understanding suffers. To provide assistance in this area, we have inserted notes that explain how to read newly introduced notation. Your students will appreciate the ever-present "Read as ..." statements that follow algebraic symbolism that they encounter for the first time.

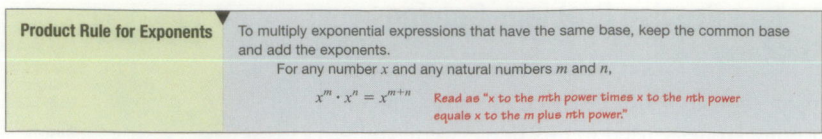

Product Rule for Exponents — To multiply exponential expressions that have the same base, keep the common base and add the exponents.

For any number x and any natural numbers m and n,

$$x^m \cdot x^n = x^{m+n}$$

Read as "x to the mth power times x to the nth power equals x to the m plus nth power."

Trusted Features

- **Examples That Show You How ... and WHY:** Why? That question is often asked by students as they watch their instructors solve problems in class and as they are working on problems at home. It's not enough to know how a problem is solved. Students gain a deeper understanding of the algebraic concepts if they know why a particular approach was taken. This instructional truth was the motivation for adding a Strategy and Why explanation to each worked example.

- **Examples That Offer Immediate Feedback:** Each worked example includes a Self Check. You can complete these on your own after reading the worked example or your instructor may choose to use them as classroom lecture examples. The Self Check answers can be found in the *Answers to Selected Exercises* in the back of the text.

- **Examples That Ask You to Work Independently:** Each worked example ends with a Now Try problem. These are the final step in the learning process. Each one is linked to a similar problem found within the *Guided Practice* sections of the *Study Sets,* offering you a smooth transition into the homework.

- **Study Sets** found in each section offer a multifaceted approach to practicing and reinforcing the concepts taught in each section. They are designed for students to build their knowledge of the section concepts methodically, from basic recall to increasingly complex problem solving, through reading, writing, and thinking mathematically.

Vocabulary—Each *Study Set* begins with the important *Vocabulary* discussed in that section. The fill-in-the-blank vocabulary problems emphasize the main concepts taught in the chapter and provide the foundation for learning and communicating the language of algebra.

Concepts—In *Concepts,* students are asked about the specific subskills and procedures necessary to successfully complete the *Guided Practice* and *Try It Yourself* problems that follow.

Notation—*Notation* problems review the new symbols introduced in a section. Often, students are asked to fill in steps of a sample solution. This strengthens their ability to read and write mathematics and prepares them for the *Guided Practice* problems by modeling solution formats.

Guided Practice—The *Guided Practice* section of each *Study Set* consistently provides 1-to-1 linking for each problem type to a single worked example or objective (*i.e.,* See Example 1, See Example 2, See Example 3, and so on). Students will appreciate this 1-to-1 linking as opposed to the all-encompassing linking statements such as See Examples 2–8 or See Example 7–11 that are found in some textbooks.

Try It Yourself—To promote problem recognition, the *Try It Yourself* problems are thoroughly mixed and are *not* linked to worked examples, giving students an opportunity to practice decision making and strategy selection as they would when taking a test or quiz.

Applications—The *Applications* provide students the opportunity to apply their newly acquired algebraic skills to relevant and interesting real-life situations.

Writing—The *Writing* problems help students build mathematical communication skills.

Review—The *Review* problems consist of randomly selected problems from previous chapters. These problems are designed to keep students' successfully mastered skills up-to-date before they move on to the next section.

Challenge Problems—The *Challenge Problems* provide students with an opportunity to stretch themselves and develop their skills beyond the basics. Instructors often find these to be useful as extra-credit problems.

- **Detailed Author Notes** that guide students along in a step-by-step process appear in the solutions to every worked example.

- **The Language of Algebra** boxes draw connections between mathematical terms and everyday references to reinforce the language of algebra thread that runs throughout the text.

- **The Notation, Success Tips, Caution,** and **Calculators** boxes offer helpful tips to reinforce correct mathematical notation, improve students' problem-solving abilities, warn students of potential pitfalls and increase clarity, and offer tips on using scientific calculators.

- **Chapter Tests,** at the end of every chapter, can be used as preparation for the class exam.

- **Cumulative Reviews** follow the end-of-chapter material and keep students' skills current before moving on to the next chapter. Each problem is linked to the associated section from which the problem came for ease of reference. The final *Cumulative Review* often is used by instructors as a final exam review.

- **Using Your Calculator** is an optional feature that is designed for instructors who want to use calculators as part of the instruction in this course. This feature introduces keystrokes and shows how scientific and graphing calculators can be used to solve problems. In the *Study Sets,* icons are used to denote problems that may be solved using a calculator.

Content Changes by Chapter

Based on feedback from colleagues and users of the Fourth Edition, the following changes have been made in an effort to further streamline and update the text.

Chapter 1

- New example and exercise applications include topics such as weight management, shoreline erosion, U.S. stock market inflows, world records, windsurfing, energy savings, leading employers, video games, and e-readers.

- New instructional features include a new worked example and new *Study Set* problems about translation from words to mathematical symbols, comparison and contrast of types of equations in one variable, a more precise definition of an identity, increased emphasis on geometric formulas, a new step was added to the problem-solving strategy (*Assign a variable.*), additional *Language of Algebra* features, a more comprehensive *Chapter Test,* and an upgraded *Group Project.*

Chapter 2

- New example and exercise applications include topics such as heart rates, farmers markets, bottled water, population, oil drilling, Alzheimer's disease, demand for nurses, and wood production.

- New instructional features include a new introduction to paired data, additional graph-reading problems, comparison and contrast of linear and nonlinear equations in two variables, more *Success Tips* and *Language of Algebra* features, a new worked example involving perpendicular lines, a more detailed explanation for finding the domain of a function, increased emphasis on writing linear equation models and linear functions, additional instruction about graphing linear functions, and a more comprehensive *Chapter Test.*

Chapter 3

- New example and exercise applications include topics such as footwear imports, area codes, investing, interior design, fashion design, physical therapy, hair-color treatments, production planning, manufacturing, nutrition, NBA centers, television vs. internet video viewing, student loans, gourmet fruit, weight training, and railroad safety.

- New instructional features include moving the application problems to the end of the chapter in Sections 3.6 and 3.7 so that any of the methods (substitution, elimination, matrices, Cramer's rule) could be used to solve the system of equations, updated graphing calculator keystrokes instructions, a more precise definition of a set of dependent equations, reduced row-echelon form and Gauss-Jordan elimination, more *Notation* and *Language of Algebra* features, additional use of tables for problem solving, and a more comprehensive *Chapter Test*.

Chapter 4

- New example and exercise applications include topics such as car engines, furnace filters, Wikipedia, national parks, comparing job offers, renting a moving truck, comparing telephone plans, earthquakes, paint spray guns, printing costs, the number of women graduating from medical schools, and the number of U.S. college graduates.

- New instructional features include a new worked example involving a linear function and an inequality, a new worked example and *Study Set* problems to find the better financial outcome, a strategy box about solving compound inequalities, a new *Caution* feature, a comparison and contrast of linear inequalities in one and two variables, more detailed explanations of graphing linear inequalities in two variables, and a more comprehensive *Chapter Test*.

Chapter 5

- New example and exercise applications include topics such as microbiology, landscaping, U.S. manufacturing, warning flares, Medicare costs, and investing.

- New instructional features include new *Language of Algebra* and *Caution* features, additional notes explaining how the notation is read, the new objective "Find function values and the domain and range of a polynomial function graphically," new worked examples and *Study Set* problems about addition, subtraction, and multiplication of polynomial functions, a more extensive discussion about discarding solutions, a new example and *Study Set* problems about quadratic functions, and a more comprehensive *Chapter Test*.

Chapter 6

- New example and exercise applications include topics such as online research, Web design, server speeds, train travel, kayaking, currency exchange, recycling, remodeling, cell phone towers, advertising budgets, wildlife management, and amusement park attendance.

- New instructional features include a more in-depth discussion of rational functions and their graphs, new examples and *Study Set* problems about addition, subtraction, multiplication, and division of polynomial functions, a screened color 1 is used when explaining how to build and simplify fractions, additional author notes explaining addition and subtraction of rational expressions graphically, new worked examples and *Study Set* problems about addition, subtraction, and multiplication of polynomial functions, a strategy box about division of polynomials, more *Success Tips,* a new worked example and *Study Set* problems about quadratic functions, and a more comprehensive *Chapter Test*.

Chapter 7

- New example and exercise applications include topics such as matting art and beach pollution.

- New instructional features include some new *Language of Algebra* and *Caution* features, a more in-depth discussion of radical functions and their graphs, additional notes explaining how the notation is read, a screened color 1 is used when explaining how to rationalize expressions, additional worked examples and *Study Set* problems about addition, subtraction, and multiplication of radical expressions, a new worked example and *Study Set* problems about solving equations containing fractional exponents, a more detailed discussion of the complex number system, and a more comprehensive *Chapter Test*.

Chapter 8

- New example and exercise applications include topics such as physics, shopping centers, women in law enforcement, crowd control, hospital emergency departments, comparing job offers, and oceanography.

- New instructional features include a new worked example and *Study Set* problems about quadratic functions, some new *Language of Algebra, Caution,* and *Success Tip* features, a new worked example and *Study Set* problems about quadratic equations with complex-number solutions, a revamped strategy for solving quadratic equations, and a more comprehensive *Chapter Test*.

Chapter 9

- New example and exercise applications include topics such as medication absorption, atmospheric carbon dioxide concentrations, book publishing, U.S. poverty rates, stocking lakes, the Richter scale, social services case loads, buying advertising time, bed bugs, cross-country skiing, weight training progress, gross domestic product of U.S. states, and human growth hormone.

- New instructional features include additional notes explaining how the notation is read, a new worked example and *Study Set* problems about evaluating sum, difference, product, quotient, and composition functions, comparison and contrast of other types of functions to exponential functions, a more detailed discussion of exponential growth and decay models and their graphs, new visuals relating exponential and logarithmic forms, a new visual explaining logarithmic function notation, a new worked example and *Study Set* problems about logarithmic growth, the former Fourth Edition Sections 9.4 and 9.6 were combined to create a new *Section 9.5: Base-e Exponential and Logarithmic Functions,* updated calculator keystrokes instructions, a revamped strategy for solving quadratic equations, more author notes and steps shown in the solutions for worked examples solving logarithmic equations, and a more comprehensive *Chapter Test*.

Chapter 10

- New example and exercise applications include topics such as civil engineering, landscaping, nuclear power, and advertising.

- New instructional features include several new *Success Tips* about determining h and k, a more in-depth discussion of the role of the center when graphing conics, a discussion of the curvature of the branches of a hyperbola, and a more comprehensive *Chapter Test*.

Chapter 11

- New example and exercise applications include topics such as union membership, manufacturing costs, and flight times.

- New instructional features include several new *Notation* and *Success Tip* features explaining notation, several new *Language of Algebra* features, a discussion of the curvature of the branches of a hyperbola, and a more comprehensive *Chapter Test*.

Student Resources

Print Ancillaries

Complete Course Notebook (1-133-36392-X)
Ann Ostberg
NEW! The *Complete Course Notebook* is your blueprint for success in your algebra course and in all your future studies! This notebook organizes and guides you through each section of your textbook. Each chapter contains a chapter readiness assessment, structured note-taking guides, questions for reflection, activities, and end-of-chapter test prep. The notebook also contains general tip sheets on note-taking, studying, assessing your test scores, and much more.

Student Workbook (1-111-98764-5)
Maria H. Andersen, *Muskegon Community College*
Get a head-start. The *Student Workbook* contains assessments, activities, and worksheets for classroom discussions, in-class activities, and group work.

Student Solutions Manual (1-111-98758-0)
Kristy Hill, *Hinds Community College*
The *Student Solutions Manual* provides worked-out solutions to the odd-numbered problems in the text.

Enhanced WebAssign: Start Smart Guide for Students (0-495-38479-8)
If your instructor has chosen to package *Enhanced WebAssign* with your text, this manual will help you get up and running quickly with the *Enhanced WebAssign* system so you can study smarter and improve your performance in class.

Electronic Ancillaries

Enhanced WebAssign (0-538-73810-3)
Get instant feedback on your homework assignments with *Enhanced WebAssign* (assigned by your instructor). This online homework system is easy to use and includes a multimedia eBook, video examples, and problem-specific tutorials.

Website www.cengagebrain.com or www.cengage.com/math/tussy
Visit us on the web and search by title to access a wealth of learning resources, including *Study Skills Workshop* materials, tutorials, final exams, chapter outlines, chapter reviews, web links, videos, flashcards, study skills handouts, and more.

Printed Access Card for CourseMate with eBook for *Intermediate Algebra,* Fifth Edition (1-133-50788-3)

Instant Access Card for CourseMate with eBook for *Intermediate Algebra,* Fifth Edition (1-133-50695-X)
The more you study, the better the results. Make the most of your study time by accessing everything you need to succeed in one place: read your textbook, take notes, review flashcards, watch videos, and take practice quizzes—online with *CourseMate*.

Acknowledgments

We want to express our gratitude to our accuracy checkers, Diane Koenig and Steve Odrich, as well as many others for their help with this project: Maria H. Andersen, Sheila Pisa, Alexander Lee, Ed Kavanaugh, Karl Hunsicker, Cathy Gong, Dave Ryba, Terry Damron, Marion Hammond, Lin Humphrey, Doug Keebaugh, Robin Carter, Tanja Rinkel, Bob Billups, Jeff Cleveland, Jo Morrison, Sheila White, Jim McClain, Paul Swatzel, Brandon Tussy, Liz Tussy, Dan Davison, Marshall Dean, Dennis Korn, Matt Greenbeck, Joyce Low, Ralph Tippins, Mohamad Trad, and the Citrus College library staff (including Barbara Rugeley) for their help with this project. Your encouragement, suggestions, and insight have been invaluable to us.

We also would like to express our thanks to the Cengage Learning editorial, marketing, production, and design staff for helping us craft this new edition: Charlie Van Wagner, Danielle Derbenti, Gordon Lee, Carrie Jones, Jennifer Cordoba, Heleny Wong, Guanglei Zhang, Maureen Ross, Sam Subity, Jennifer Risden, Vernon Boes, Terri Wright, Ellen Brownstein, Helen Walden, Lori Heckelman, and Graphic World.

Additionally, we would like to say that authoring a textbook is a tremendous undertaking. A revision of this scale would not have been possible without the thoughtful feedback and support from the colleagues listed below. Their contributions to this edition have shaped this revision in countless ways.

Alan S. Tussy
R. David Gustafson

Advisory Reviewers

Ashish Gupta, *William Paterson University*
Katrina Keating, *Diablo Valley College*
Roger Larson, *Anoka Ramsey Community College*

Lorraine Lopez, *San Antonio College*
Paul J. Vroman, *St. Louis Community College at Florissant Valley*

Reviewers of the Fourth and Fifth Editions

Andrea Adlman, *Ventura College*
Khadija Ahmed, *Monroe Community College*
Rodney Alford, *Calhoun Community College*
Maria Andersen, *Muskegon Community College*
Hamid Attarzadeh, *Jefferson Community and Technical College*
Victoria Baker, *University of Houston–Downtown*
Betty Barks, *Lansing Community College*
Scott Barnett, *Henry Ford Community College*
Susan Beane, *University of Houston–Downtown*
David Behrman, *Somerset Community College*
Chad Bemis, *Riverside Community College*
John F. Beyers, *University of Maryland University College*
Barbara Blass, *Oakland Community College*
Candace Blazek, *Anoka Ramsey Community College*
Jennifer Bluth, *Anoka Ramsey Community College*
A. Elena Bogardus, *Camden Community College*
Carilynn Bouie, *Cuyahoga Community College*
Charles A. Bower, *St. Philip's College*
Jeanne Bowman, *University of Cincinnati*
Kim Brown, *Tarrant Community College*
Kirby Bunas, *Santa Rosa Junior College*
Shawna M. Bynum, *Napa Valley College*
Kim Caldwell, *Volunteer State Community College*
Carole Carney, *Brookdale Community College*
Edythe L. Carter, *Amarillo College*
Joe Castillo, *Broward Community College*
Sandra Chandler, *Tidewater Community College*
Carol Cheshire, *Macon State College*
John Close, *Salt Lake Community College*
Chris Copple, *Northwest State Community College*
Tony Craig, *Paradise Valley Community College*
Patrick Cross, *University of Oklahoma*
Mary Deas, *Johnson County Community College*
Suzanne Doviak, *Old Dominion University*
Archie Earl, *Norfolk State University*
Melody Eldred, *State University of New York at Cobleskill*
Peter Embalabala, *Lincoln Land Community College*

Joan Evans, *Texas Southern University*
Mike Everett, *Santa Ana College*
Betsy Farber, *Bucks County Community College*
Rita Fielder, *University of Central Arkansas*
Maggie Flint, *Northeast State*
Anissa Florence, *Jefferson Community and Technical College*
Pat Foard, *South Plains College*
Nancy Forrest, *Grand Rapids Community College*
Tom Fox, *Cleveland State Community College*
Heng Fu, *Thomas Nelson Community College*
Douglas Furman, *SUNY Ulster Community College*
Abel Gage, *Skagit Valley College*
John Garlow, *Tarrant Community College–Southeast Campus*
Vicki Gearhart, *San Antonio College*
Radu Georgescu, *Prince George's Community College*
Rebecca Giles, *Jefferson State Community College*
Alketa Gjikuria, *Cecil College*
Megan Goodwin, *Anoka Ramsey Community College*
Kim Gregor, *Delaware Technical Community College–Wilmington*
Thomas Grogan, *Cincinnati State*
Sally Haas, *Angelina College*
Paula Jean Haigis, *Calhoun Community College*
Haile Kebede Haile, *Minneapolis Community and Technical College*
Kelli Jade Hammer, *Broward Community College*
Mehdi Hakim Hashemi, *Normandale Community College*
Julia Hassett, *Oakton Community College*
Jennifer Hastings, *Northeast Mississippi Community College*
Alan Hayashi, *Oxnard College*
Kristy Hill, *Hinds Community College*
Jim Hodge, *Mountain State University*
Amy Hoherz, *Johnson County Community College*
Laura Hoye, *Trident Technical College*
Becki Huffman, *Tyler Junior College*
Jeffrey Hughes, *Hinds Community College*
Vera Hu-Hyneman, *SUNY–Suffolk Community College*
Angela Jahns, *North Idaho College*

Cassandra Johnson, *Robeson Community College*

Cynthia Johnson, *Heartland Community College*

Leslie Johnson, *John C. Calhoun State Community College*

Pete Johnson, *Eastern Connecticut State University*

Shelbra Jones, *Wake Technical Community College*

Ed Kavanaugh, *Schoolcraft College*

Leonid Khazanov, *Borough of Manhattan Community College*

MC Kim, *Suffolk County Community College*

Lynette King, *Gadsden State Community College*

Mike Kirby, *Tidewater Community College*

Alex Kolesnik, *Ventura College*

Patricia Kopf, *Kellogg Community College*

Elena Kravchuk, *University of Alabama–Birmingham*

Marlene Kutesky, *Virginia Commonwealth University*

Fred Lang, *Art Institute of Washington*

Hoat Le, *San Diego Community College*

Alexander Lee, *Hinds Community College, Rankin Campus*

Wayne (Paul) Lee, *Saint Philip's College*

Richard Leedy, *Polk Community College*

Mary Legner, *Riverside Community College*

Lamar Lider-Manuel, *Seminole Community College*

Daniel Lopez, *Brookdale Community College*

Ann Loving, *J. Sargeant Reynolds Community College*

Yixia Lu, *South Suburban College*

Keith Luoma, *Augusta State University*

Julie L. Mays, *Angelina College*

Mikal McDowell, *Cedar Valley College*

Marcus McGuff, *Austin Community College*

Owen Mertens, *Missouri State University*

Susan Meshulam, *Indiana University/Purdue University Indianapolis*

James Metz, *Kapi'olani Community College*

Trudy Meyer, *El Camino College*

Pam Miller, *Phoenix College*

Molly Misko, *Gadsden State Community College*

Catherine Moushon, *Elgin Community College*

Tania Munding, *Ohlone College*

Charlie Naffziger, *Central Oregon Community College*

Oscar Neal, *Grand Rapids Community College*

Doug Nelson, *Central Oregon Community College*

Elsie Newman, *Owens Community College*

Charlotte Newsom, *Tidewater Community College*

Katrina Nichols, *Delta College*

Randy Nichols, *Delta College*

Stephen Nicoloff, *Paradise Valley Community College*

Megan Nielsen, *St. Cloud State University*

Charles Odion, *Houston Community College*

Jason Pallett, *Longview Community College*

Mary Beth Pattengale, *Sierra College*

Naeemah Payne, *Los Angeles Community College*

Fred Peskoff, *Borough of Manhattan Community College*

Sheila Pisa, *Riverside Community College–Moreno Valley*

Carol Ann Poore, *Hinds Community College*

Jill Rafael, *Sierra College*

Pamela Reed, *North Harris Montgomery Community College*

Pamelyn Reed, *Cy-Fair College*

Nancy Ressler, *Oakton Community College*

Elaine Richards, *Eastern Michigan University*

Harriette Roadman, *New River Community College*

Lilia Ruvalcaba, *Oxnard College*

Jeffrey Saikali, *San Diego Miramar College*

Fary Sami, *Harford Community College*

Emma Sargent, *Tennessee State University*

Ned Schillow, *Lehigh Carbon Community College*

Joe Sedlacek, *Kirkwood Community College*

Wendiann Sethi, *Seton Hall University*

Debra Shafer, *University of North Carolina*

Hazel Shedd, *Hinds Community College*

Patty Sheeran, *McHenry Community College*

Karen Smith, *Nicholls State University*

Christa Solheid, *Santa Ana College*

Donald Solomon, *University of Wisconsin–Milwaukee*

Frankie Solomon, *University of Houston–Downtown*

Jim Spencer, *Santa Rosa Junior College*

John Squires, *Cleveland State Community College*

Michael Stack, *South Suburban College*

Kristen Starkey, *Rose State College*

Robin Steinberg, *Pima Community College*

Kristin Stoley, *Blinn College*

Eleanor Storey, *Front Range Community College–Westminster Campus*

Teresa Sutcliffe, *Los Angeles Valley College*

Eden Thompson, *Utah Valley State College*

Cindy Thore, *Central Piedmont Community College*

Rose Toering, *Kilian Community College*

Fariheh Towfiq, *Palomar College*

James Vallade, *Monroe County Community College*

Gowribalan "Ana" Vamadeva, *University of Cincinnati*

Maggie Pasqua Viz, *Brookdale Community College*

Beverly Vredevelt, *Spokane Falls Community College*

Andreana Walker, *Calhoun Community College*

Carol Walker, *Hinds Community College*

Cynthia Wallin, *Central Virginia Community College*

John Ward, *Kentucky Community and Technical College–Jefferson Community College*

Richard Watkins, *Tidewater Community College*

Diane Williams, *Northern Kentucky University*

Antoinette Willis, *St. Philip's College*

Jackie Wing, *Angelina College*

Judith Wood, *Central Florida Community College*

Nazar Wright, *Guilford Technical Community College*

Valerie Wright, *Central Piedmont Community College*

Shishen Xie, *University of Houston–Downtown*

Catalina Yang, *Oxnard College*

Heidi Young, *Bryant and Stratton College*

Mary Young, *Brookdale Community College*

Ghidei Zedingle, *Normandale Community College*

Loris Zucca, *Kingwood College*

INDEX OF APPLICATIONS

Examples that are applications are shown with **boldface** page numbers.
Exercises that are applications are shown with lightface page numbers.

A Review of Basic Algebra

©iStockphoto.com/Kirby Hamilton

from Campus to Careers

Registered Dietitian

One of the most important things that you can do to protect yourself from cancer, diabetes, heart disease, and stroke is to eat right. No one knows this better than dietitians. They work in hospitals, health care centers, schools, and correctional facilities, where they plan dietary programs and supervise the preparation of healthy meals. The job of dietitian requires mathematical skills such as calculating calorie intake, analyzing the nutritional content of food, and budgeting for the purchase of groceries and supplies.

Problem 65 in **Study Set 1.1, problem 99** in **Study Set 1.6,** and **problem 57** in **Study Set 1.8** involve situations that a registered dietitian might encounter on the job. The mathematical concepts discussed in this chapter can be used to solve those problems.

JOB TITLE:
Registered Dietitian

EDUCATION:
A bachelor's degree in foods and nutrition (or a related field) and supervised practice

JOB OUTLOOK:
Growth faster than the average, 18%–26% increase

ANNUAL EARNINGS:
The median base salary in 2009 was $53,230.

FOR MORE INFORMATION:
www.bls.gov/oco/ocos077.htm

Starting a new course is exciting, but it also might make you a bit nervous. In order to be successful in your algebra class, you need a plan.

MAKE TIME FOR THE COURSE: As a general guideline, 2 hours of independent study time is recommended for every hour in the classroom.

KNOW WHAT IS EXPECTED: Read your instructor's syllabus thoroughly. It lists class policies about attendance, homework, tests, calculators, grading, and so on.

BUILD A SUPPORT SYSTEM: Know where to go for help. Take advantage of your instructor's office hours, your school's tutorial services, the resources that accompany this textbook, and the assistance that you can get from classmates.

Now Try This ▶

Each of the forms referred to below can be found online at: www.cengage.com/math/tussy

1. To help organize your schedule, fill out the *Weekly Planner Form*.
2. Review the class policies by completing the *Course Information Sheet*.
3. Use the *Support System Worksheet* to build your course support system.

SECTION 1.1

The Language of Algebra

OBJECTIVES

1. Write verbal and mathematical models.
2. Use equations to construct tables of data.

ARE YOU READY?

The following problems review some basic skills that are needed for this section. Answers to the Are You Ready problems are located in Appendix 2, at the back of the book.

1. What operation is indicated by each word?
   a. sum
   b. difference
   c. product
   d. quotient
2. If $m = 3d + 5$, find m if $d = 5$.
3. What is 22 increased by 10?
4. What is 6 less than 17?

Algebra is the result of contributions from many cultures over thousands of years. The word *algebra* comes from the title of the book *Al-jabr wa'l muquabalah,* written by the Arabian mathematician al-Khwarizmi around A.D. 800. Using the vocabulary and notation of algebra, we can mathematically **model** many situations in the real world. In this section, we will review some of the basic components of the language of algebra.

1 Write Verbal and Mathematical Models.

In the following rental agreement, we see that two operations need to be performed to calculate the cost of renting the banquet hall.

- First, we must *multiply* the $100-per-hour rental cost by the number of hours that the hall is to be rented.
- To that result, we must *add* the cleanup fee of $200.

Rental Agreement
ROYAL VISTA BANQUET HALL
Wedding Receptions•Dances•Reunions•Fashion Shows

Rented to_____ Date_____
Lessee's Address_____

Rental Charges
• $100 per hour
• Nonrefundable $200 cleanup fee

We can describe the process to calculate the rental cost in words using the following **verbal model:**

| The cost (in dollars) of renting the hall | is | 100 | times | the number of hours it is rented | plus | 200. |

The table below lists some key words and phrases that are often used in mathematics to indicate the operations of addition, subtraction, multiplication, and division.

Addition +	Subtraction −	Multiplication ·	Division ÷
added to	subtracted from	multiplied by	divided by
sum	difference	product	quotient
plus	minus	times	ratio
more than	less than	percent (or fraction) of	half
increased by	decreased by	twice	into
greater than	discounted	double	per
exceeds	reduced by	triple	equal parts

We can use vocabulary from the table to write a verbal model that describes how to calculate the cost of renting the banquet hall. One such model is:

The cost (in dollars) of renting the hall is the *product* of 100 and the number of hours it is rented, *increased* by 200.

We also can describe the procedure for calculating the rental cost of the banquet hall using *variables* and mathematical symbols. A **variable** is a letter that stands for a number. If we let the letter h represent the number of hours that the hall is rented, the cost to rent the hall can be represented by the notation $100h + 200$. We call $100h + 200$ an **algebraic expression,** or more simply, an **expression.**

Algebraic Expressions   Variables and/or numbers can be combined with the operations of addition, subtraction, multiplication, division, raising to a power, and finding a root to create **algebraic expressions.**

Here are some more examples of algebraic expressions.

$5a - 12$ This expression is a combination of the numbers 5 and 12, the variable a, and the operations of multiplication and subtraction.

$\dfrac{5.7 - y}{3y^3}$ This expression is a combination of the numbers 5.7 and 3, the variable y, and the operations of subtraction, division, multiplication, and raising to a power.

$\sqrt{a^2 + b^2}$ This expression is a combination of the variables a and b, and the operations of addition, raising to a power, and finding a root.

Since many problems in this course are presented in words, it is important to be able to translate those words into mathematical symbols.

EXAMPLE 1   Translate each phrase to an algebraic expression.

   a. 19 pounds less than the weight

   b. 27% of the population

   c. 6 feet more than twice the length

   d. one-half of the sum of a number and 8

Strategy   We will begin by identifying any key words or phrases.

Why   Key words and phrases can be translated to mathematical symbols.

Solution

a. **Key phrase:** *less than*   **Translation:** subtraction
If we let w represent the unknown weight (in pounds), then we need to make w 19 pounds less. The algebraic expression is: $w - 19$.

b. **Key word:** *of*   **Translation:** multiplication
If we let p represent the unknown population, then we need to multiply it by 27%. Recall that to perform a calculation involving a percent, we write the percent as a decimal (or a fraction). The algebraic expression is: $0.27p$.

c. **Key phrase:** *more than*   **Translation:** addition
   Key phrase: *twice*   **Translation:** multiplication by 2
If we let l represent the unknown length (in feet), we need to multiply it by 2 and then add 6 to that result. The algebraic expression is: $2l + 6$.

d. **Key word:** *of*   **Translation:** multiplication
   Key word: *sum*   **Translation:** addition
If we let x represent the unknown number, we need to add x and 8, and then multiply the result by $\frac{1}{2}$. The algebraic expression is: $\frac{1}{2}(x + 8)$.

CAUTION   In this answer, the sum $x + 8$ must be written within parentheses. The expression $\frac{1}{2}x + 8$, without the parentheses, is the translation of a similar, but slightly different phrase: *the sum of one-half of a number and* 8.

> **Self Check 1**   Translate each phrase to an algebraic expression:
>   **a.** 16 miles less than the distance
>   **b.** 9% of the parking spaces
>   **c.** 44 meters more than triple the height
>   **d.** one-fourth of the difference of a number and 1
>
> **Now Try** ▶   Problems 11, 13, 15, and 17

Caution

Be careful when translating subtraction. Order is important. For example, when a translation involves the phrase *less than*, note how the terms are reversed.

19 pounds less than the weight

$w \; - \; 19$

Caution

Answers to these problems may vary depending on the letters chosen to use as variables.

Success Tip

The answers to the Self Check problems are given in Appendix 2 at the back of the book.

The Language of Algebra

The equal symbol = can be represented by verbs such as:

is   are   gives   yields

The symbol ≠ is read as *"is not equal to."*

In the banquet hall example, if we let the letter *c* represent the cost to rent the hall, we can translate the verbal model to a **mathematical model.**

The cost of renting the hall	is	100	times	the number of hours it is rented	plus	200.
c	=	100	·	*h*	+	200

The statement $c = 100h + 200$ is called an *equation.* The = symbol indicates that the expressions on either side of it have the same value.

Equations   ▼   An **equation** is a mathematical sentence that contains an = symbol.

EXAMPLE 2   Translate each verbal model to a mathematical model.

   a. The distance in miles traveled by a vehicle is the product of its average rate of speed in mph and the time in hours it travels at that rate.

   b. The sale price of an item is the difference between the regular price and the discount.

Strategy   We will look for key words and phrases that indicate arithmetic operations, and we will represent any unknown quantities using variables.

Why   To translate a verbal (word) model into a mathematical model means to represent it using mathematical symbols.

Solution   **a.** Since the word *product* indicates multiplication, we have: *d* represents the distance traveled in miles, *r* the vehicle's average rate of speed in mph, and *t* the length of time traveled in hours.

$$d = rt$$

   b. Since the word *difference* indicates subtraction, we have: *s* represents the sale price of the item, *p* the regular price, and *d* the discount.

$$s = p - d$$

> **Self Check 2**   Translate to a mathematical model: The simple interest earned by a deposit is the product of the principal, the annual rate of interest, and the time.
>
> **Now Try** ▶ **Problem 39**

Many applied problems require insight and analysis to determine which mathematical operations to use when writing a verbal or mathematical model.

EXAMPLE 3   **Catering.**   It costs $16 a person to have a dinner catered. A $100 discount is given for groups of more than 200 people. Write a verbal and a mathematical model that describe the relationship between the catering cost and the number of people being served for groups larger than 200.

Strategy   We will carefully read the problem to identify any phrases that indicate an arithmetic operation and then represent any unknown quantities using variables.

Why   To write a verbal or mathematical model, we must determine what arithmetic operations are involved.

©Caro/Alamy

Solution The phrase *$16 a person* indicates multiplication by 16 and the phrase *a $100 discount* indicates subtraction. Thus, to find the catering cost c (in dollars) for groups larger than 200, we need to *multiply* the number n of people served by $16 and then *subtract* the $100 discount. A verbal model is:

> The catering cost (in dollars) is the *product* of 16 and
> the number of people served, *decreased by 100.*

The comma in the verbal model is needed to convey the correct meaning. Without it, it is unclear what is to be decreased by 100. In symbols, the mathematical model is:

$$c = 16n - 100$$

Self Check 3 **Lotteries.** After winning a lottery, three friends split the prize equally. Each person had to pay $2,000 in taxes on his or her share. Write a mathematical model that relates the amount of each person's share, after taxes, to the amount of the lottery prize.

Now Try ▶ Problem 47

2 Use Equations to Construct Tables of Data.

Equations such as $c = 100h + 200$, which express a mathematical relationship between two or more variables, are called **formulas**. Some applications require the repeated use of a formula. The results are often shown in a **two-column table**.

EXAMPLE 4 **Event Planning.** Find the cost of renting the banquet hall for 3 hours and for 4 hours. Write the results in a table.

Strategy We will substitute 3 (and then 4) for h in the equation $c = 100h + 200$ and evaluate the right side.

Why The cost c to rent the hall for h hours is given by the equation $c = 100h + 200$.

Solution First, we construct the table shown below with the appropriate column headings: h for the number of hours the hall is rented and c for the cost (in dollars) to rent the hall. Then we enter the number of hours of each rental time in the left column.

Next, we use the equation $c = 100h + 200$ to calculate the total rental cost for 3 hours and for 4 hours.

For 3 hours:

$c = 100h + 200$
$c = 100(3) + 200$ Substitute 3 for h.
$\quad = 300 + 200$ Multiply.
$\quad = 500$ Add.

For 4 hours:

$c = 100h + 200$
$c = 100(4) + 200$ Substitute 4 for h.
$\quad = 400 + 200$ Multiply.
$\quad = 600$ Add.

Finally, we enter these results in the right column of the table: $500 for a 3-hour rental and $600 for a 4-hour rental.

h	c
3	500
4	600

Self Check 4 **Event Planning.** Find the cost of renting the hall for 6 hours and for 7 hours. Write the results in a table.

Now Try ▶ Problem 51

SECTION 1.1 ▶ STUDY SET

VOCABULARY

Fill in the blanks.

1. A _____ is a letter that stands for a number.
2. Variables and/or numbers can be combined with mathematical operations to create algebraic _____.
3. An _____ is a mathematical sentence that contains an = symbol.
4. Words such as *is, was, gives,* and *yields* translate to an _____ symbol.
5. Phrases such as *increased by* and *more than* are used to indicate the operation of _____. Phrases such as *decreased by* and *less than* are used to indicate the operation of _____.
6. Equations that express a mathematical relationship between two or more variables are called _____.

CONCEPTS

7. Classify each of the following as an expression or an equation.

   a. $6x - 5$　　　　b. $P = a + b + c$

   c. $\dfrac{s + 9t}{8}$　　　d. $\sqrt{2w^2}$

8. What arithmetic operations does the expression $\dfrac{40 - 8n}{5}$ contain? What variable does it contain?

NOTATION

9. Translate each verbal model into a mathematical model.

   a. | 7 | times | the age of a dog in years | gives | the dog's equivalent human age. |

   b. | The take-home pay | will be | $2,500 | minus | any deductions. |

10.

   a. | The number of decades | is | the number of years | divided by | 10. |

   b. | The cost of dining out | equals | the cost of the meal | plus | $15 for parking. |

GUIDED PRACTICE

Translate each phrase to an algebraic expression. Answers may vary depending on the variables chosen. See Example 1.

11. 7 minutes less than the time walking
12. 50 meters less than the height
13. 54% of the enrollment
14. 12% of a number
15. 35 dollars more than twice the price
16. 0.25 ounces more than twice the weight
17. one-half of the sum of a number and 4
18. one-third of the sum of the length and width
19. three-fourths of the pressure
20. nine-sixteenths of the thickness
21. the ratio of the number of games won and games played
22. the ratio of the number of vacation days and work days
23. 950 increased by 10% of the volume
24. 100 less than triple the attendance a
25. w reduced by 500
26. 8,000 split n equal ways
27. 150 feet per m minutes
28. exceeds the cost by 25,000
29. 7 times the total of 77, h, and 88
30. decrease a number by -1
31. triple the number of waiters
32. 680 fewer than the population
33. the product of d and 4, decreased by 15
34. the quotient of the base and twice the height
35. 95% of the sum of 200 and the tonnage
36. 14 less than eleven-ninths of a number
37. one-hundredth of the distance
38. double the difference of a number and 18

Translate each verbal model into a mathematical model. Answers may vary depending on the variables chosen. See Example 2.

39. The cost each semester is the sum of $13 times the number of units taken and a student services fee of $24.
40. The yearly salary is $25,000 plus $75 times the number of years of experience.
41. The quotient of the number of clients and seventy-five gives the number of social workers needed.
42. The difference between 500 and the number of people in a theater gives the number of unsold tickets.
43. Each test score was increased by 15 points to give a new adjusted test score.
44. The weight of a super-size order of French fries is twice that of a regular-size order.
45. The product of the number of boxes of crayons in a case and 12 gives the number of crayons in a case.

46. The perimeter of an equilateral triangle can be found by tripling the length of one of its sides.

Write a mathematical model for each situation. Answers may vary depending on the variables chosen. See Example 3.

47. **Taxes.**   A married couple has decided to split the money equally when they receive their federal income tax refund. Furthermore, the husband is going to donate $75 of his share to charity. Describe the relationship between the amount of money that the husband will keep and the amount of the couple's refund.

48. **Copiers.**   A business is going to rent a copy machine. Under the rental agreement, the company is charged $105 per month and 3¢ for every copy that is made. Describe the relationship between the monthly copier expense and the number of copies made.

49. **Bottled Water.**   A driver left a production plant with 300 five-gallon bottles of drinking water on his truck. His delivery route consisted of office buildings, each of which was to receive 6 bottles of water. Describe the relationship between the number of bottles of water left on his truck and the number of stops that he has made.

50. **Collectibles.**   A woman inherited 9 antique dolls. She decided to add to her collection by purchasing two more dolls each month. Describe the relationship between the number of antique dolls in her collection and the number of months since she began to purchase them.

Use the given equation to complete each table. See Example 4.

51. $c = \dfrac{p}{12}$

Number of packages p	Cartons c
24	
72	
180	

52. $y = 100c$

Number of centuries c	Years y
1	
6	
21	

53. $n = 22.44 - K$

K	n
0	
1.01	
22.44	

54. $y = 2x + 15$

x	y
0	
15	
30	

TRY IT YOURSELF

Look Alikes . . .

Translate each phrase to an algebraic expression. Answers may vary depending on the variables chosen.

55. **a.** s subtracted from S
    b. S subtracted from s

56. **a.** the product of 4 and d, decreased by 15
    b. the product of 4 and d decreased by 15

57. **a.** the absolute value of the difference of a and 2
    b. the difference of the absolute value of a and 2

58. **a.** the quotient of a number and 6 increased by the number
    b. the quotient of a number and 6, increased by the number

59. **a.** 6 miles more than 15.5% of the altitude
    b. 15.5% of 6 miles more than the altitude

60. **a.** twice the sum of the tax and 200
    b. the sum of twice the tax and 200

61. **a.** the square of 14 less than a number
    b. 14 less than the square of a number

62. **a.** double the cube of a number
    b. the cube of double a number

APPLICATIONS

63. **Production Planning.**   Suppose r towel racks like that shown below are to be manufactured. Complete the two equations that could be used to order the necessary number of bar holders b and wood screws s.

$$b = \boxed{}\,r \qquad s = \boxed{}\,r$$

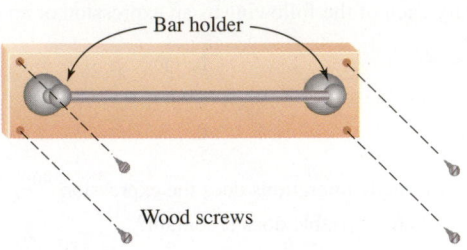

Bar holder

Wood screws

64. **Staircases.**   A builder is going to construct h new homes, each of which will have a staircase as shown. Complete the four equations that could be used to order the necessary number of balusters b, handrails r, posts p, and treads t for the entire project.

$$b = \boxed{}\,h$$
$$r = \boxed{}\,h$$
$$p = \boxed{}\,h$$
$$t = \boxed{}\,h$$

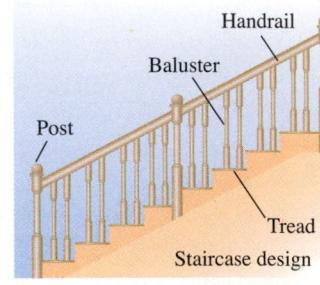

Handrail

Baluster

Post

Tread

Staircase design

65. from **Campus to Careers**

Registered Dietitian

When discussing weight management with their patients, dietitians stress the importance of limiting calorie intake and increasing physical activity. One method dietitians use to determine if a patient has a daily calorie surplus (or deficit) is to subtract the patient's daily resting metabolic rate and the calories that he/she burns during physical activity that day from the calories he/she ingests that day. Translate this verbal model into a mathematical model.

Source: conwayfitness

©iStockphoto.com/Kirby Hamilton

66. Carpet Cleaning. See the ad below.

a. Write a verbal model that states the relationship between the cost C of renting the carpet cleaning system and the number of hours h it is rented.

b. Translate the verbal model written in part (a) to a mathematical model.

c. Use your result from part (b) to complete the table.

Rent the in-home **Carpet Cleaning System**

Safe, effective   Costs only $10 an hour plus $20 for supplies

Do it yourself and save!

h	C
1	
2	
3	
4	
8	

Use the data in each table to find an equation that mathematically describes the relationship between the two quantities. (Answers may vary.)

67.

Tower height (ft)	Height of base (ft)
15.5	5.5
22	12
25.25	15.25
45.125	35.125

68.

Seasonal employees	Employees
25	75
50	100
60	110
80	130

WRITING

69. Explain the difference between an expression and an equation. Give examples.

70. Use each word below in a sentence that indicates a mathematical operation. If you are unsure of the meaning of a word, look it up in a dictionary.

quadrupled	deleted	bisected	garnished
confiscated	annexed	docked	quintupled

CHALLENGE PROBLEMS

71. Use each of the variables a, b, c, and d only once to write:

a. a sum of two differences

b. a difference of two sums

72. Fill in the blank: If $T = 16s$ and $s = \dfrac{r}{2}$, then $T = \boxed{}r$.

SECTION 1.2

The Real Numbers

OBJECTIVES

1 Define the set of natural numbers, whole numbers, and integers.

2 Define the set of rational numbers.

3 Define the set of irrational numbers.

4 Classify real numbers.

5 Graph real numbers.

6 Order the real numbers.

7 Find the opposite and the absolute value of a real number.

ARE YOU READY?

The following problems review several types of numbers that we use in everyday life.

1. Count the number of letters in the word *supercalifragilisticexpialidocious*.

2. What number represents a temperature that is 27 degrees below zero?

3. What type of number is used to express a grade point average (GPA)?

4. Suppose a recipe calls for only part of a full cup of flour. What type of number is normally used to describe such an amount?

In this course, we will work with *real numbers*. The set of real numbers is a collection of several other important sets of numbers.

1 Define the Set of Natural Numbers, Whole Numbers, and Integers.

Natural numbers are the numbers that we use for counting. To write sets, we list the **elements** (or **members**) of the set within **braces** { }.

Natural Numbers	The set of **natural numbers**, denoted by the symbol ℕ, is {1, 2, 3, 4, 5, . . . }.
	Read as "the set containing one, two, three, four, five, and so on."

The symbol $\in$ is used to indicate that an element belongs to a set. For example,

$3 \in \mathbb{N}$ *Read as "3 is an element of the set of natural numbers."*

We can write $4.5 \notin \mathbb{N}$ to indicate that 4.5 *is not an element* of the set of natural numbers. The natural numbers, together with 0, form the set of **whole numbers.**

Whole Numbers

The set of **whole numbers,** denoted by the symbol $\mathbb{W}$, is $\{0, 1, 2, 3, 4, 5, \ldots\}$.

When all the members of one set are also members of a second set, we say the first set is a **subset** of the second set. Since every natural number is also a whole number, the set of natural numbers is a subset of the set of whole numbers. We use the symbol $\subseteq$ to indicate this.

$\mathbb{N} \subseteq \mathbb{W}$ *Read as "The set of natural numbers is a subset of the set of whole numbers."*

Since the set of whole numbers contains an element that the natural numbers do not, namely 0, we can write

$\mathbb{W} \not\subseteq \mathbb{N}$ *Read as "The set of whole numbers is not a subset of the set of natural numbers."*

Two other important subsets of the whole numbers are the *prime numbers* and the *composite numbers*.

Prime Numbers and Composite Numbers

A **prime number** is a whole number greater than 1 that has only itself and 1 as factors. The first ten prime numbers are 2, 3, 5, 7, 11, 13, 17, 19, 23, and 29.

A **composite number** is a whole number, greater than 1, that is not prime. The first ten composite numbers are 4, 6, 8, 9, 10, 12, 14, 15, 16, and 18.

Recall from arithmetic that every composite number can be written as the product of prime numbers. For example,

$$6 = 2 \cdot 3, \quad 25 = 5 \cdot 5, \quad \text{and} \quad 168 = 2 \cdot 2 \cdot 2 \cdot 3 \cdot 7$$

Whole numbers are not adequate for describing many real-life situations. For example, if you write a check for more than what is in your account, the account balance will be less than zero.

We can use the **number line** below to visualize numbers less than zero. On the number line, numbers greater than 0 are to the right of 0, and they are called **positive numbers.** Numbers less than 0 are to the left of 0, and they are called **negative numbers.**

For each natural number on the number line, there is a corresponding number, called its *opposite,* to the left of 0. In the diagram, we see that 3 and -3 (negative three) are opposites, as are -5 (negative five) and 5. Note that 0 is its own opposite.

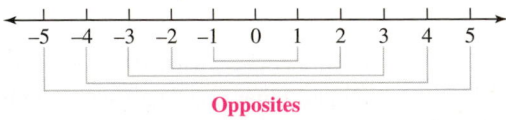

Opposites

Opposites

Two numbers that are the same distance from 0 on the number line, but on opposite sides of it, are called **opposites.**

The whole numbers, together with their opposites, form the set of *integers*.

Integers

The set of **integers,** denoted by the symbol $\mathbb{Z}$, is $\{\ldots, -4, -3, -2, -1, 0, 1, 2, 3, 4, \ldots\}$.

The Language of Algebra

The **positive integers** are:

$$1, 2, 3, 4, 5, \ldots$$

The **negative integers** are:

$$-1, -2, -3, -4, -5, \ldots$$

Integers that are divisible by 2 are called *even integers,* and integers that are not divisible by 2 are called *odd integers.*

Even integers: $\{ \ldots, -6, -4, -2, 0, 2, 4, 6, \ldots \}$

Odd integers: $\{ \ldots, -5, -3, -1, 1, 3, 5, \ldots \}$

Since every whole number is also an integer, the set of whole numbers is a subset of the set of integers. In symbols, we can write this as $\mathbb{W} \subseteq \mathbb{Z}$.

EXAMPLE 1   Determine whether each statement is true or false.   **a.** $-6 \in \mathbb{W}$ **b.** $\mathbb{Z} \not\subseteq \mathbb{N}$

Strategy   In part a, we will determine whether -6 is an element of the set of whole numbers. In part b, we will determine whether the set of integers is not a subset of the set of natural numbers.

Why   The symbol $\in$ means "is an element of" and the symbol $\not\subseteq$ means "is not a subset of."

Solution   **a.** Since -6 is not an element of the set of whole numbers, the statement $-6 \in \mathbb{W}$ is false.

b. Since the set of integers contains at least one element that does not belong to the set of natural numbers (-1 for example), the set of integers is not a subset of the set of natural numbers. Thus, the statement $\mathbb{Z} \not\subseteq \mathbb{N}$ is true.

Self Check 1   Determine whether each statement is true or false.
a. $-1.7 \notin \mathbb{Z}$ **b.** $\mathbb{Z} \subseteq \mathbb{W}$

Now Try   ▶   Problems 27 and 31

2 Define the Set of Rational Numbers.

In this course, we will work with positive and negative fractions. For example, the slope of a line might be $\frac{7}{12}$ or the rate at which a tank is draining might be represented by $-\frac{40}{3}$ gallons per minute. We also will work with mixed numbers. For instance, we might speak of $5\frac{7}{8}$ cups of flour or of a river that is $3\frac{1}{2}$ feet below flood stage $\left(-3\frac{1}{2} \text{ ft}\right)$. These fractions and mixed numbers are examples of *rational numbers.*

Rational Numbers   A **rational number** is any number that can be written in the form $\frac{a}{b}$, where a and b represent integers and $b \neq 0$.

Other examples of rational numbers are $\frac{3}{4}, \frac{25}{25}$, and $\frac{19}{6}$. To show that negative fractions are rational numbers, we can use the following fact.

Negative Fractions   Let a and b represent numbers, where $b \neq 0$,

$$-\frac{a}{b} = \frac{-a}{b} = \frac{a}{-b}$$

To illustrate this rule, consider $-\frac{40}{3}$. It is a rational number because it can be written as $\frac{-40}{3}$, or as $\frac{40}{-3}$.

Positive and negative mixed numbers such as $5\frac{7}{8}$ and $-3\frac{1}{2}$ are rational numbers because they can be expressed as fractions. For example, $5\frac{7}{8} = \frac{47}{8}$ and $-3\frac{1}{2} = -\frac{7}{2} = \frac{-7}{2}$. Any natural number, whole number, or integer can be expressed as a fraction with a denominator of 1. For example, $5 = \frac{5}{1}$, $0 = \frac{0}{1}$, and $-3 = \frac{-3}{1}$. Therefore, every natural number, whole number, and integer is also a rational number.

In this course, we also will work with decimals. **Terminating decimals** such as 0.11, 127.279, and -2.7 are rational numbers because they can be written as fractions with integer numerators and nonzero integer denominators.

$$0.11 = \frac{11}{100} \qquad 127.279 = 127\frac{279}{1,000} = \frac{127,279}{1,000} \qquad -2.7 = -2\frac{7}{10} = \frac{-27}{10}$$

Examples of **repeating decimals** are 0.333 . . . and 4.252525. . . . Any repeating decimal can be expressed as a fraction with an integer numerator and a nonzero integer denominator. For example, $0.333 \ldots = \frac{1}{3}$ and $4.252525 \ldots = 4\frac{25}{99} = \frac{421}{99}$. Since every repeating decimal can be written as a fraction, repeating decimals are also rational numbers.

Repeating decimals are often written using an **overbar.** For example,

$$0.333 \ldots = 0.\overline{3} \qquad \text{and} \qquad 4.252525 \ldots = 4.\overline{25}$$

Rational Numbers

The set of **rational numbers,** denoted by the symbol $\mathbb{Q}$, is the set of all terminating and all repeating decimals.

The set of rational numbers is too extensive to list its members in the same way that we listed the members of the natural numbers, whole numbers, and integers. Instead, we will use the following **set-builder** notation to describe it.

Rational Numbers

The set of rational numbers is

$$\left\{ \frac{a}{b} \,\middle|\, a \text{ and } b \text{ are integers, with } b \neq 0. \right\}$$

Read as "the set of all numbers of the form $\frac{a}{b}$, such that a and b are integers, with $b \neq 0$."

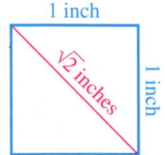

The length of a diagonal of the square is $\sqrt{2}$ inches.

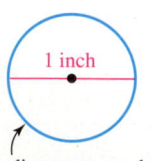

The distance around the circle is π inches.

3   Define the Set of Irrational Numbers.

Numbers that cannot be expressed as fractions with an integer numerator and a nonzero integer denominator are called **irrational numbers.** One example is $\sqrt{2}$. It can be shown that a square, with sides of length 1 inch, has a diagonal that is $\sqrt{2}$ inches long.

The number represented by the Greek letter π (pi) is another example of an irrational number. It can be shown that a circle, with a 1-inch diameter, has a circumference of π inches. Expressed in decimal form,

$$\sqrt{2} = 1.414213562 \ldots \qquad \text{and} \qquad \pi = 3.141592654 \ldots$$

These decimals neither terminate nor repeat.

Irrational Numbers

An **irrational number** is a nonterminating, nonrepeating decimal. An irrational number cannot be expressed as a fraction with an integer numerator and a nonzero integer denominator. The set of irrational numbers is denoted by the symbol $\mathbb{H}$.

Caution

Don't classify a number such as 4.12122122212222 . . . as a repeating decimal. Although it exhibits a pattern, no single block of digits repeats forever. It is a nonterminating, nonrepeating decimal—an irrational number.

Some other examples of irrational numbers are

$$\sqrt{97} = 9.848857802\ldots$$

$$-\sqrt{7} = -2.64575131\ldots \quad \textcolor{red}{\textit{This is a negative irrational number.}}$$

$$2\pi = 6.283185307\ldots \quad \textcolor{red}{\textbf{2}\boldsymbol{\pi} \textit{ means } \textbf{2} \cdot \boldsymbol{\pi}.}$$

Not all square roots are irrational numbers. When we simplify square roots such as $\sqrt{9}$, $\sqrt{36}$, and $\sqrt{400}$, it is apparent that they are rational numbers: $\sqrt{9} = 3$, $\sqrt{36} = 6$, and $\sqrt{400} = 20$.

Using Your Calculator ▶   Approximating Irrational Numbers

We can approximate the value of irrational numbers with a scientific calculator. To find the value of π, we press the $\boxed{\pi}$ key.

$\boxed{\pi}$ (You may have to use a $\boxed{\text{2nd}}$ or $\boxed{\text{Shift}}$ key first.)   `3.141592654`

We see that $\pi \approx 3.141592654$. (Read $\approx$ as "is approximately equal to.") To the nearest thousandth, $\pi \approx 3.142$.

To approximate $\sqrt{2}$, we enter 2 and press the square root key $\boxed{\sqrt{}}$.

$2 \boxed{\sqrt{}}$ `1.414213562`

We see that $\sqrt{2} \approx 1.414213562$. To the nearest hundredth, $\sqrt{2} \approx 1.41$.

To find π and $\sqrt{2}$ with a graphing calculator, we press:

$\boxed{\text{2nd}} \;\; \pi \;\; \boxed{\text{ENTER}}$

```
π
          3.141592654
```

$\boxed{\text{2nd}} \;\; \boxed{\sqrt{}} \;\; 2 \; \boxed{)} \;\; \boxed{\text{ENTER}}$

```
√(2)
          1.414213562
```

---

4 Classify Real Numbers.

The set of rational numbers together with the set of irrational numbers form the set of **real numbers.** This means that every real number can be written as either a terminating decimal, a repeating decimal, or a nonterminating, nonrepeating decimal. Thus, the set of real numbers is the set of all decimals.

The Real Numbers

A **real number** is any number that is either a rational or an irrational number. The set of real numbers is denoted by the symbol $\mathbb{R}$.

The figure on the next page shows how the sets of numbers introduced in this section are related; it also gives some specific examples of each type of number. Note that a number can belong to more than one set. For example, -6 is an integer, a rational number, and a real number.

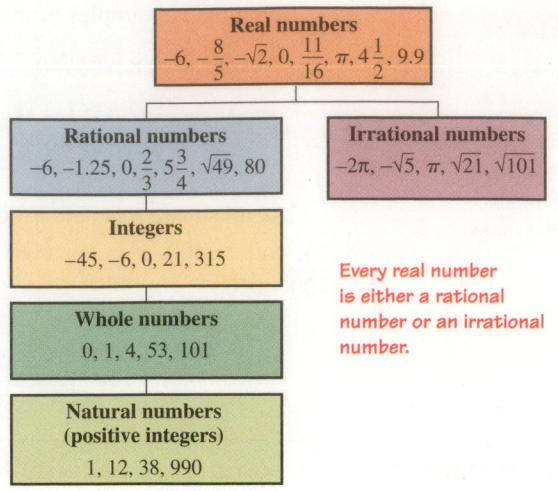

EXAMPLE 2

Classifying Real Numbers.   Which numbers in the following set are natural numbers, whole numbers, integers, rational numbers, irrational numbers, and real numbers?

$$\left\{ \frac{5}{8}, \ -0.03, \ 45, \ -9, \ \sqrt{7}, \ 5\frac{2}{3}, \ 0, \ -9.010010001\ldots, \ 0.\overline{25} \right\}$$

Strategy   We begin by scanning the given set, looking for any natural numbers. Then we scan it five more times, looking for whole numbers, for integers, for rational numbers, for irrational numbers, and, finally, for real numbers.

Why   We need to scan the given set of numbers six times, because numbers in the set can belong to more than one classification.

Solution

Natural numbers: 45   45 is a member of {1, 2, 3, 4, 5, . . .}.

Whole numbers: 45, 0   45 and 0 are members of {0, 1, 2, 3, 4, 5, . . .}.

The Language of Algebra

The prefix *ir* means *not*. For example,
irresponsible ↔ not responsible
irregular ↔ not regular
irrational ↔ not rational

Integers: 45, −9, 0   45, −9, and 0 are members of { . . . , −2, −1, 0, 1, 2, . . .}.

Rational numbers: $\frac{5}{8}$, 45, −9, $5\frac{2}{3}$, and 0 are rational numbers because each of them can be expressed as a fraction: $45 = \frac{45}{1}$, $-9 = \frac{-9}{1}$, $5\frac{2}{3} = \frac{17}{3}$, and $0 = \frac{0}{1}$. The terminating decimal $-0.03 = \frac{-3}{100}$ and the repeating decimal $0.\overline{25} = \frac{25}{99}$ are also rational numbers.

Irrational numbers: The nonterminating, nonrepeating decimals $\sqrt{7} = 2.645751311\ldots$ and $-9.010010001\ldots$ are irrational numbers.

Real numbers: $\frac{5}{8}$, −0.03, 45, −9, $\sqrt{7}$, $5\frac{2}{3}$, 0, −9.010010001 . . . , 0.$\overline{25}$

Every natural number, whole number, integer, rational number, and irrational number is a real number.

Self Check 2   Use the instructions for Example 2 with the following set:

$$\left\{ -\pi, \ -5, \ 3.4, \ \sqrt{19}, \ 1, \ \frac{16}{5}, \ 9.\overline{7} \right\}$$

Now Try   Problems 35, 37, and 39

---

5   Graph Real Numbers.

We can illustrate real numbers using a number line. To each real number, there corresponds a point on the line. Furthermore, to each point on the line, there corresponds a real number, called its **coordinate.**

EXAMPLE 3 Graph each number in the set $\left\{-\dfrac{8}{3},\ -1.1,\ 0.\overline{56},\ \dfrac{\pi}{2},\ -\sqrt{15},\ 2\sqrt{2}\right\}$ on a number line.

Strategy We locate the position of each number on the number line, draw a bold dot, and label it.

Why To *graph a number* means to make a drawing that represents the number.

Solution To help locate the graph of each number, we make some observations.

The Language of Algebra

An example of a number that is not on the real number line is $\sqrt{-4}$. It is called an **imaginary number.** We will discuss such numbers in Chapter 8.

- Expressed as a mixed number, $-\dfrac{8}{3} = -2\dfrac{2}{3}$.

- Since -1.1 is less than -1, its graph is to the left of -1.

- $0.\overline{56} \approx 0.6$

- From a calculator, $\dfrac{\pi}{2} \approx 1.6$ and $-\sqrt{15} \approx -3.9$.

- $2\sqrt{2}$ means $2 \cdot \sqrt{2}$. From a calculator, $2\sqrt{2} \approx 2.8$.

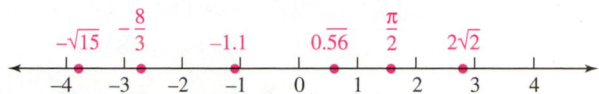

Self Check 3 Graph the numbers in the set $\left\{\pi,\ -2.\overline{1},\ \sqrt{3},\ \dfrac{11}{4},\ -0.9\right\}$ on a number line.

Now Try ▶ Problem 47

6 Order the Real Numbers.

As we move right on the number line, the values of the real numbers increase. As we move left, the values decrease. To compare real numbers, we often use one of the **inequality symbols** shown in the following table.

The Language of Algebra

If a real number x is positive, then $x > 0$. If a real number x is nonnegative, then $x \ge 0$. If a real number x is a negative number, then $x < 0$.

Symbol	Read as	Examples
$\ne$	"is not equal to"	$6 \ne 9$ and $0.33 \ne \dfrac{3}{5}$
$<$	"is less than"	$\dfrac{22}{3} < \dfrac{23}{3}$ and $-7 < -6$
$>$	"is greater than"	$19 > 5$ and $\dfrac{1}{2} > 0.3$
$\le$	"is less than or equal to"	$3.5 \le 3.\overline{5}$ and $1\dfrac{4}{5} \le 1.8$
$\ge$	"is greater than or equal to"	$29 \ge 29$ and $-15.2 \ge -16.7$

$1\dfrac{4}{5} \le 1.8$ is true because $1\dfrac{4}{5} = 1.8$.

$29 \ge 29$ is true because $29 = 29$.

The / symbol is also used with inequality symbols. For example, $\not<$ means "is not less than."

It is always possible to write an equivalent inequality with the inequality symbol pointing in the opposite direction. For example:

If $-3 < 4$, it is also true that $4 > -3$.

If $5.3 \ge 2.9$, it is also true that $2.9 \le 5.3$.

EXAMPLE 4   Use one of the symbols $>$ or $<$ to make each statement true.   **a.** $-24 \;\blacksquare\; -25$

b. $\dfrac{3}{8} \;\blacksquare\; 0.377$

Strategy   To pick the correct inequality symbol to place between a given pair of numbers, we need to determine the position of each on a number line.

Why   For any two numbers on a number line, the number to the *left* is the smaller number and the number to the *right* is the larger number.

Solution   **a.** Since -24 is to the right of -25 on the number line, $-24 > -25$.

b. If we find the **decimal equivalent** of the fraction $\dfrac{3}{8}$, we can easily compare it to 0.377. Since the fraction bar indicates division, we divide the numerator 3 by the denominator 8.

$$
\begin{array}{r}
.375 \\
8)\overline{3.000} \\
-2\,4 \\
\hline
60 \\
-56 \\
\hline
40 \\
-40 \\
\hline
0
\end{array}
$$

Since $\dfrac{3}{8} = 0.375,$ $\dfrac{3}{8} < 0.377.$

Self Check 4   Use one of the symbols $\geq$ or $\leq$ to make each statement true.   **a.** $\dfrac{2}{3} \;\blacksquare\; \dfrac{4}{3}$
b. $8\dfrac{1}{2} \;\blacksquare\; 8.4$

Now Try   Problems 59 and 63

7   Find the Opposite and the Absolute Value of a Real Number.

Two numbers that are the same distance from 0 on the number line, but on opposite sides of it, are called **opposites** or **additive inverses.** To write the opposite of a positive number, we simply insert a negative sign $-$ in front of it. For example, the opposite of 10 is -10.

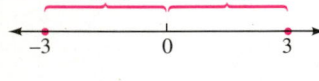

Parentheses are used to express the opposite of a negative number. For example, the opposite of -3 is written as $-(-3)$. Since -3 and 3 are the same distance from zero, the opposite of -3 is 3. In symbols, this can be written as $-(-3) = 3$.

In general, we have the following.

Opposites	The **opposite** of a number a is the number $-a$. If a is a real number, then $-(-a) = a$.

A number line can be used to measure the distance from one number to another. To express the distance that a number is from 0 on a number line, we can use absolute values.

Absolute Value	The **absolute value** of a number is its distance from 0 on the number line.

To indicate the absolute value of a number, we write the number between two vertical bars. From the number line above, we see that -3 is a distance of 3 units from 0. Thus, $|-3| = 3$. This is read as "the absolute value of negative 3 is 3." It also follows from the number line that $|3| = 3$.

The absolute value of a number can be defined more formally as follows.

Definition of Absolute Value	For any real number a, $\begin{cases} \text{If } a \geq 0, \text{ then } |a| = a \\ \text{If } a < 0, \text{ then } |a| = -a \end{cases}$

Caution

The second part of this definition is often misunderstood. Study it carefully. It indicates that the absolute value of a negative number is the opposite (or additive inverse) of the number.

The first part of this definition states that if a is a nonnegative number (that is, if $a \geq 0$), the absolute value of a is a. The second part of the definition states that if a is a negative number (that is, if $a < 0$), the absolute value of a is the opposite of a. For example, if $a = -8$, then

$$|a| = |-8| = \underbrace{-(-8)}_{\text{The opposite of } a} = 8$$

EXAMPLE 5 Evaluate each expression: **a.** $|34|$ **b.** $\left|-\dfrac{4}{5}\right|$ **c.** $|0|$ **d.** $-|-1.8|$

Strategy We need to determine the distance that the number within the vertical absolute value bars is from 0.

Why The absolute value of a number is the distance between 0 and the number on a number line.

Solution **a.** $|34| = 34$ Because 34 is a distance of 34 from 0 on a number line.

The Language of Algebra

To **evaluate** an expression means to find its value.

b. $\left|-\dfrac{4}{5}\right| = \dfrac{4}{5}$ Because $-\frac{4}{5}$ is a distance of $\frac{4}{5}$ from 0 on a number line.

c. $|0| = 0$ Because 0 is a distance of 0 from 0 on a number line.

d. The negative sign outside the absolute value bars means to find the opposite of $|-1.8|$.

$$-|-1.8| = -(1.8) \quad \text{Find } |-1.8| \text{ first to get 1.8.}$$
$$= -1.8$$

Self Check 5 Find the value of each expression. **a.** $|-9.6|$

b. $-|-12|$ **c.** $\left|\dfrac{3}{2}\right|$

Now Try ▶ Problems 75 and 79

SECTION 1.2 ▶ **STUDY SET**

VOCABULARY

Fill in the blanks.

1. The set of _____ numbers is {0, 1, 2, 3, 4, 5, . . . }, the set of _____ numbers is {1, 2, 3, 4, 5, . . . }, and the set of _____ is {. . . , −2, −1, 0, 1, 2, . . . }.

2. When all the members of one set are members of a second set, we say the first set is a _____ of the second set.

3. A _____ number is a whole number greater than 1 that has only itself and 1 as factors. A _____ number is a whole number greater than 1 that is not prime.

4. A _____ number is any number that can be written as a fraction with an integer numerator and a nonzero integer denominator.

5. _____ numbers are nonterminating, nonrepeating decimals.

6. The set of rational numbers together with the set of irrational numbers form the set of _____ numbers.

7. $>, \geq, <,$ and $\leq$ are called _____ symbols.

8. The _____ _____ of any real number is the distance between the number and zero on a number line.

CONCEPTS

9. Name two numbers that are 6 units away from −2 on the number line.

10. Show that each of the following numbers is a rational number by expressing it as a fraction with an integer numerator and a nonzero integer denominator: $7, \ -7\frac{3}{5}, \ 0.007, \ 700.1$

Determine whether each number is a repeating or a nonrepeating decimal, and whether it is a rational or an irrational number.

11. 0.090090009 . . .

12. $0.0\overline{9}$

13. 5.41414141 . . .

14. 1.414213562 . . .

15. The following diagram shows how the sets of numbers introduced in this section are related. Copy the diagram and fill in the blanks.

16. Determine whether each statement is true or false.

a. All prime numbers are odd numbers.

b. $6 \geq 6$

c. 0 is neither even nor odd.

d. Every real number is a rational number.

17. Write each statement with the inequality symbol pointing in the opposite direction.

a. $19 > 12$ b. $-6 \leq -5$

18. Fill in the blanks:

For any real number a, $\begin{cases} \text{If } a \geq 0, \text{ then } |a| = \boxed{} \\ \text{If } a < 0, \text{ then } |a| = \boxed{} \end{cases}$

NOTATION

Fill in the blanks.

19. The symbol $<$ means "___ ____ ____" and the symbol $\geq$ means "___ _____ ____ ___ _____ ___."

20. $|-2|$ is read as "the _____ value ___ -2."

21. The symbols $\{\ \}$ are called _____.

22. The symbol $\in$ is read as "is an _____ of" and the symbol $\subseteq$ is read as "is a _____ of."

23. Describe the set of rational numbers using set-builder notation.

24. What set of numbers does each symbol represent?

a. $\mathbb{N}$ b. $\mathbb{W}$

c. $\mathbb{Z}$

25. What set of numbers does each symbol represent?

a. $\mathbb{Q}$ b. $\mathbb{H}$

c. $\mathbb{R}$

26. List two other ways that the fraction $-\frac{2}{3}$ can be written.

GUIDED PRACTICE

Determine whether each statement is true or false. See Example 1.

27. $12 \in \mathbb{N}$ **28.** $9 \in \mathbb{N}$

29. $-5 \notin \mathbb{Z}$ **30.** $-55 \notin \mathbb{H}$

31. $\mathbb{R} \subseteq \mathbb{W}$ **32.** $\mathbb{W} \subseteq \mathbb{R}$

33. $\mathbb{H} \not\subseteq \mathbb{Q}$ **34.** $\mathbb{Q} \not\subseteq \mathbb{Z}$

List the elements of

$$\left\{-3, -\tfrac{8}{5}, 0, \tfrac{2}{3}, 1, \sqrt{3}, 2, \pi, 4.75, 9\,16.\overline{6}\right\}$$

that belong to the following sets. See Example 2.

35. Natural numbers   **36.** Whole numbers

37. Integers   **38.** Rational numbers

39. Irrational numbers   **40.** Real numbers

41. Even natural numbers   **42.** Odd integers

43. Prime numbers   **44.** Composite numbers

45. Odd composite numbers   **46.** Odd prime numbers

Graph each set on a number line. See Example 3.

47. $\left\{-\tfrac{5}{2}, -0.1, 2.142765\ldots, \tfrac{\pi}{3}, -\sqrt{11}, 2\sqrt{3}\right\}$

48. $\left\{2\tfrac{1}{9}, -3.821134\ldots, -\tfrac{\pi}{2}, \sqrt{15}, -0.9, \tfrac{\sqrt{2}}{2}\right\}$

49. $\left\{3.\overline{15}, \tfrac{22}{7}, 3\tfrac{1}{8}, \pi, \sqrt{10}, 3.1\right\}$

50. $\left\{-0.\overline{331}, -0.331, -\tfrac{1}{3}, -\sqrt{0.11}\right\}$

51. The set of prime numbers less than 8

52. The set of integers between -7 and 0

53. The set of odd integers between 10 and 18

54. The set of composite numbers less than 10

55. The set of positive odd integers less than 12

56. The set of negative even integers greater than -7

57. The set of even integers from -6 to 6

58. The set of odd natural numbers less than or equal to 5

Insert either a $<$ or a $>$ symbol to make a true statement. See Example 4.

59. $-9 \ \boxed{} \ -8$ **60.** $-11 \ \boxed{} \ -12$

61. $-(-5) \ \boxed{} \ -10$ **62.** $|-3| \ \boxed{} \ -(-6)$

63. $6.\overline{1} \ \boxed{} \ -(-6)$ **64.** $-6.07 \ \boxed{} \ -\tfrac{17}{6}$

65. $-7.999 \ \boxed{} \ -7.1$ **66.** $4\tfrac{1}{2} \ \boxed{} \ \tfrac{7}{2}$

67. $\tfrac{3}{5} \ \boxed{} \ 0.06$ **68.** $\tfrac{21}{50} \ \boxed{} \ 0.4$

69. $-\tfrac{11}{15} \ \boxed{} \ -0.73$ **70.** $-\tfrac{7}{30} \ \boxed{} \ -0.23$

71. $\tfrac{27}{22} \ \boxed{} \ 1.2\overline{28}$ **72.** $\tfrac{25}{990} \ \boxed{} \ 0.0\overline{26}$

73. $0.0\overline{16} \ \boxed{} \ \tfrac{2}{125}$ **74.** $1.1\overline{875} \ \boxed{} \ \tfrac{19}{16}$

Find the value of each expression. See Example 5.

75. $|20|$ **76.** $|-20|$

77. $|-5.9|$ **78.** $-|1.\overline{27}|$

79. $-|-6|$ **80.** $-|-8|$

81. $-\left|\tfrac{9}{4}\right|$ **82.** $-\left|-\tfrac{5}{16}\right|$

APPLICATIONS

83. Drafting. Express each dimension in the drawing of a bracket as a four-place decimal. Approximate when necessary.

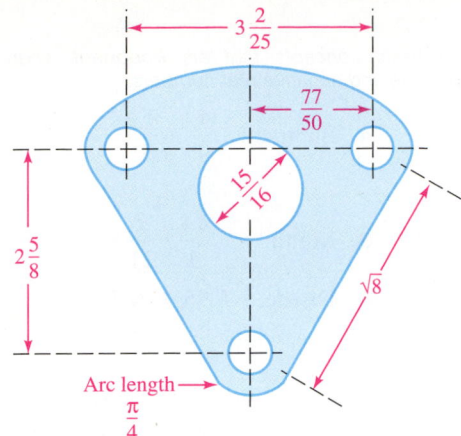

State Average Coastline Erosion/Accretion, 2006 (in meters per year)

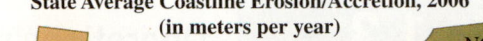
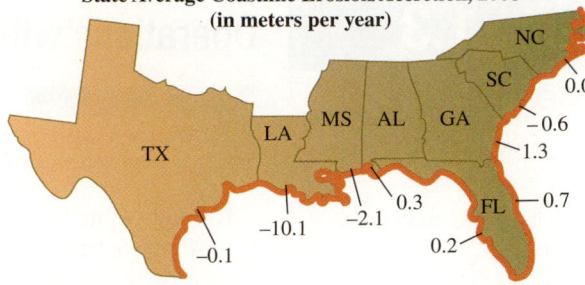

Source: U.S. Geological Survey

84. Chemistry. The pH scale is used to measure the strength of acids and bases (alkalines). It can be thought of as a number line. On the scale, graph and label each pH measurement given below.

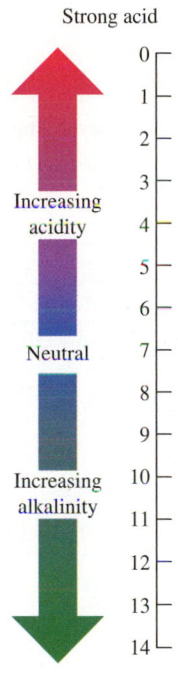

Solution	pH
Seawater	8.5
Cola	2.9
Battery acid	1.0
Milk	6.6
Blood	7.4
Ammonia	11.9
Saliva	6.1
Oven cleaner	13.2
Black coffee	5.0
Toothpaste	9.9
Tomato juice	4.1

86. Rings. The formula $C = \pi D$ gives the circumference C of a circle, where D is the length of its diameter. Find the circumference of the gold wedding band. Give an *exact* answer and then an *approximate* answer, rounded to the nearest hundredth of an inch.

WRITING

87. Explain why the whole numbers are a subset of the integers.

88. What is a real number? Give examples.

89. Explain why there are no even prime numbers greater than 2.

90. Explain why every integer is a rational number, but not every rational number is an integer.

REVIEW

91. Is $\frac{3x - 4}{2}$ an equation or an expression?

92. Translate into mathematical symbols: The weight of an object in ounces is the product of 16 and its weight in pounds.

93. Fill in the blank: A _____ is a letter that stands for a number.

Complete the table.

94. $T = x - 1.5$

x	T
3.7	
10	

85. Erosion. In the map in the next column, negative numbers indicate erosion of the shoreline and positive numbers indicate a building up (called *accretion*) of the shoreline.

  a. Which state had the greatest average shoreline accretion?

  b. Which state had the worst average shoreline erosion?

CHALLENGE PROBLEMS

95. How many integers have an absolute value that is less than 1,000?

96. Find a fraction whose decimal equivalent is $0.\overline{61}$.

97. The **trichotomy property** of real numbers states that if a and b are real numbers, then $a < b$, $a = b$, or $a > b$. Explain why this is true.

98. Let a and b represent real numbers. Which of the following statements are always true?

  a. $|a + b| = |a| + |b|$ **b.** $|a \cdot b| = |a| \cdot |b|$

  c. $|a + b| \le |a| + |b|$

SECTION 1.3

Operations with Real Numbers

OBJECTIVES

1 Add and subtract real numbers.

2 Multiply and divide real numbers.

3 Find powers and square roots of real numbers.

4 Use the order of operations rule.

5 Evaluate algebraic expressions.

ARE YOU READY?

The following problems review some basic concepts that are important when adding, subtracting, multiplying, and dividing positive and negative real numbers.

1. Find $|4|$ and $|-6|$. Which number, 4 or -6, has the larger absolute value?

2. Add: $9.37 + 2.8$

3. What is the opposite of -11?

4. Subtract: $\dfrac{4}{9} - \dfrac{1}{6}$

5. Multiply: $\dfrac{14}{3} \cdot \dfrac{2}{21}$

6. Divide: $5.95 \div 0.7$

Six operations can be performed with real numbers: addition, subtraction, multiplication, division, raising to a power, and finding a root. In this section, we will review the rules for performing these operations and discuss how to evaluate numerical expressions involving several operations.

1 Add and Subtract Real Numbers.

When two numbers are added, the result is their **sum.** The rules for adding real numbers are as follows:

Adding Two Real Numbers

To **add two positive numbers,** add them in the usual way. The final answer is positive.

To **add two negative numbers,** add their absolute values and make the final answer negative.

To **add a positive number and a negative number,** subtract the smaller absolute value from the larger.

1. If the positive number has the larger absolute value, the final answer is positive.

2. If the negative number has the larger absolute value, make the final answer negative.

EXAMPLE 1 Add: **a.** $-5 + (-3)$ **b.** $8.9 + (-5.1)$ **c.** $-\dfrac{13}{15} + \dfrac{3}{5}$ **d.** $6 + (-10) + (-1)$

Strategy We will use the rules for adding positive and negative real numbers.

Why Each sum involves signed numbers.

Solution **a.** $-5 + (-3) = -8$ Both numbers are negative. Add their absolute values, 5 and 3, to get 8, and make the final answer negative.

The Language of Algebra

Positive and negative numbers are often referred to as **signed numbers.**

b. $8.9 + (-5.1) = 3.8$ One number is positive and the other is negative. Subtract their absolute values, 5.1 from 8.9, to get 3.8. Because 8.9 has the larger absolute value, the final answer is positive.

c. $-\dfrac{13}{15} + \dfrac{3}{5} = -\dfrac{13}{15} + \dfrac{9}{15}$ Express $\frac{3}{5}$ in terms of the lowest common denominator, 15: $\frac{3}{5} = \frac{3}{5} \cdot \frac{3}{3} = \frac{9}{15}$.

$= -\dfrac{4}{15}$ Subtract the absolute values, $\frac{9}{15}$ from $\frac{13}{15}$, to get $\frac{4}{15}$, and make the final answer negative because $-\frac{13}{15}$ has the larger absolute value.

d. To add three or more real numbers, add from left to right.

$$6 + (-10) + (-1) = -4 + (-1)$$
$$= -5$$

Self Check 1 Add: **a.** $-34 + 25$ **b.** $-70.4 + (-21.2)$
c. $\frac{7}{4} + \left(-\frac{3}{2}\right)$ **d.** $-16 + 17 + (-5)$

Now Try ▶ Problems 15 and 21

When two numbers are subtracted, the result is their **difference.** To find a difference, we can change the subtraction into an equivalent addition. For example, the subtraction $7 - 4$ is equivalent to the addition $7 + (-4)$, because they have the same answer:

$$7 - 4 = 3 \qquad \text{and} \qquad 7 + (-4) = 3$$

This suggests that to subtract two numbers, we can change the sign of the number being subtracted and add.

Subtracting Two Real Numbers	To **subtract two real numbers,** add the first number to the opposite (additive inverse) of the number to be subtracted. Let a and b represent real numbers, $$a - b = a + (-b)$$

EXAMPLE 2   Subtract: **a.** $2 - 8$ **b.** $-1.3 - 5.5$ **c.** $-\dfrac{14}{3} - \left(-\dfrac{7}{3}\right)$

d. Subtract 9 from -6 **e.** $-11 - (-1) - 5$

Strategy To find each difference, we will apply the rule for subtraction: *Add the first number to the opposite of the number to be subtracted.*

Why It is easy to make an error when subtracting signed numbers. We will probably be more accurate if we write each subtraction as addition of the opposite.

Solution

The Language of Algebra

The rule for subtracting real numbers is often summarized as: *Subtraction is the same as adding the opposite.*

a. $2 - \mathbf{8} = 2 + (\mathbf{-8})$ Here, 8 is being subtracted, so we change the
                                   sign of 8 and add. Do not change the sign of 2.

the opposite

$= -6$ Use the rule for adding two numbers with unlike signs.

b. $-1.3 - \mathbf{5.5} = -1.3 + (\mathbf{-5.5})$ Change the sign of 5.5 and add.
                                          Do not change the sign of -1.3.

$= -6.8$ Use the rule for adding two numbers with like signs.

c. $-\dfrac{14}{3} - \left(-\dfrac{\mathbf{7}}{\mathbf{3}}\right) = -\dfrac{14}{3} + \dfrac{7}{3}$ Change the sign of $-\frac{7}{3}$ and add.

$= -\dfrac{7}{3}$ Use the rule for adding two numbers with unlike signs.

d. The number to be subtracted is 9. When we translate, we must reverse the order in which 9 and -6 appear in the sentence.

Subtract 9 from -6.

$-6 - 9 = -6 + (-9)$ Add the opposite of 9.

$= -15$ Use the rule for adding two numbers with like signs.

e. To subtract three or more real numbers, subtract from left to right.

$$-11 - (-1) - 5 = -10 - 5 \quad \text{Think: } -11 - (-1) = -11 + 1 = -10.$$
$$= -15 \quad \text{Think: } -10 - 5 = -10 + (-5) = -15.$$

Self Check 2 Subtract: **a.** $-15 - 4$ **b.** $-12.1 - (-7.6)$
c. $\frac{5}{9} - \frac{7}{9}$ **d.** Subtract 1 from -5 **e.** $5 - 4 - (-15)$

Now Try ▶ Problems 23 and 29

2 Multiply and Divide Real Numbers.

When two numbers are multiplied, we call the numbers **factors** and the result is their **product.** The rules for multiplying real numbers are as follows:

Multiplying Two Real Numbers	1. **With unlike signs:** To multiply a positive number and a negative number, multiply their absolute values and make the final answer negative. 2. **With like signs:** To multiply two real numbers with the same sign, multiply their absolute values. The final answer is positive.

EXAMPLE 3 Multiply: **a.** $4(-7)$ **b.** $-5.2(-3)$ **c.** $-\frac{7}{9}\left(\frac{3}{16}\right)$ **d.** $8(-2)(-3)$

Strategy We will use the rules for multiplying positive and negative real numbers.

Why Each product involves signed numbers.

Solution **a.** $4(-7) = -28$ — Multiply the absolute values, 4 and 7, to get 28. Since the signs are unlike, make the final answer negative.

b. $-5.2(-3) = 15.6$ — Multiply the absolute values, 5.2 and 3, to get 15.6. Since the signs are like, the final answer is positive.

c. $-\dfrac{7}{9}\left(\dfrac{3}{16}\right) = -\dfrac{7 \cdot 3}{9 \cdot 16}$ — Multiply the numerators and multiply the denominators. Since the signs of the factors are unlike, the final answer is negative.

$$= -\dfrac{7 \cdot \overset{1}{\cancel{3}}}{\cancel{3} \cdot 3 \cdot 16}$$ — Factor 9 as $3 \cdot 3$ and then simplify the fraction by removing the common factor of 3 in the numerator and denominator.

$$= -\dfrac{7}{48}$$ — Multiply the remaining factors in the numerator and in the denominator.

d. To multiply three or more real numbers, multiply from left to right.

$$8(-2)(-3) = -16(-3) \quad \text{Think: } 8(-2) = -16.$$
$$= 48 \quad \text{Since the factors have like signs, the final answer is positive.}$$

Self Check 3 Multiply: **a.** $(-6)(5)$ **b.** $(-4.1)(-8)$ **c.** $\left(\frac{4}{3}\right)\left(-\frac{1}{8}\right)$
d. $-4(-9)(-3)$

Now Try ▶ Problems 31 and 37

When two numbers are divided, the result is their **quotient.** In the division $\frac{x}{y} = q$, the quotient q is a number such that $y \cdot q = x$. We call $y \cdot q = x$ the **related multiplication statement** for $\frac{x}{y} = q$. We can use this relationship to find rules for dividing real numbers.

$$\frac{10}{2} = 5, \text{ because } 2(5) = 10 \qquad\qquad \frac{-10}{-2} = 5, \text{ because } -2(5) = -10$$

$$\frac{-10}{2} = -5, \text{ because } 2(-5) = -10 \qquad\qquad \frac{10}{-2} = -5, \text{ because } -2(-5) = 10$$

These results suggest the following rules for dividing real numbers. Note that they are similar to those for multiplying real numbers.

Dividing Two Real Numbers	To **divide two real numbers,** divide their absolute values. 1. The quotient of two numbers with *like* signs is positive. 2. The quotient of two numbers with *unlike* signs is negative.

EXAMPLE 4 Divide: **a.** $\dfrac{-44}{11}$ **b.** $\dfrac{-2.7}{-9}$

Strategy We will use the rules for dividing positive and negative real numbers.

Why Each quotient involves signed numbers.

Solution **a.** $\dfrac{-44}{11} = -4$ Divide the absolute values, 44 by 11, to get 4. Since the signs are unlike, make the final answer negative.

$$\begin{array}{r} .3 \\ 9\overline{)2.7} \\ \underline{-2\,7} \\ 0 \end{array}$$

b. $\dfrac{-2.7}{-9} = 0.3$ Divide the absolute values, 2.7 by 9, to get 0.3. Since the signs are like, the final answer is positive.

Self Check 4 Divide: **a.** $\dfrac{55}{-5}$ **b.** $\dfrac{-7.2}{-6}$

Now Try Problems 39 and 43

To divide two fractions, we multiply the first fraction by the **reciprocal** of the second fraction. In symbols, if a, b, c, and d are real numbers, and no denominators are 0, then

$$\frac{a}{b} \div \frac{c}{d} = \frac{a}{b} \cdot \frac{d}{c} \qquad \tfrac{d}{c} \text{ is the reciprocal of } \tfrac{c}{d}.$$

EXAMPLE 5 Divide: **a.** $\dfrac{2}{3} \div \left(-\dfrac{3}{5}\right)$ **b.** $-\dfrac{1}{2} \div (-6)$

Strategy We will multiply the first fraction by the reciprocal of the second.

Why This is the rule for dividing two fractions.

Solution **a.** $\dfrac{2}{3} \div \left(-\dfrac{3}{5}\right) = \dfrac{2}{3} \cdot \left(-\dfrac{5}{3}\right)$ Multiply by the reciprocal of $-\tfrac{3}{5}$, which is $-\tfrac{5}{3}$.

$$= -\frac{10}{9} \qquad\qquad \text{Since the factors have unlike signs, the final answer is negative.}$$

b. $-\dfrac{1}{2} \div (-6) = -\dfrac{1}{2} \cdot \left(-\dfrac{1}{6}\right)$ Multiply by the reciprocal of -6, which is $-\frac{1}{6}$.

$$= \dfrac{1}{12}$$ Since the factors have like signs, the final answer is positive.

Self Check 5 Divide: **a.** $-\dfrac{7}{8} \div \dfrac{2}{3}$ **b.** $-\dfrac{1}{10} \div (-5)$

Now Try ▶ Problem 45

3 Find Powers and Square Roots of Real Numbers.

Exponents indicate repeated multiplication. For example,

$$3^2 = 3 \cdot 3$$ Read 3^2 as "3 to the second power" or "3 squared."

$$(-9.1)^3 = (-9.1)(-9.1)(-9.1)$$ Read $(-9.1)^3$ as "-9.1 to the third power" or "-9.1 cubed."

$$\left(\dfrac{2}{3}\right)^4 = \left(\dfrac{2}{3}\right)\left(\dfrac{2}{3}\right)\left(\dfrac{2}{3}\right)\left(\dfrac{2}{3}\right)$$ Read $\left(\frac{2}{3}\right)^4$ as "$\frac{2}{3}$ to the fourth power."

These examples illustrate the following definition.

Natural-Number Exponents

A natural-number exponent indicates how many times its base is to be used as a factor. For any real number x and any natural number n,

$$\overbrace{x^n = x \cdot x \cdot x \cdot \ldots \cdot x}^{n \text{ factors of } x}$$

The exponential expression x^n is called a **power of x,** and we read it as "x to the nth power." In this expression, x is called the **base,** and n is called the **exponent.** A natural-number exponent indicates how many times the base of an exponential expression is to be used as a factor in a product.

Base $\longrightarrow x^n \longleftarrow$ Exponent

EXAMPLE 6 Find each power: **a.** $(-2)^4$ **b.** $\left(\dfrac{3}{4}\right)^2$ **c.** -0.1 cubed

Strategy We will write each exponential expression as a product of repeated factors, and then perform the multiplication. This requires that we identify the base and the exponent.

Why The exponent indicates the number of times the base is to be written as a factor.

Solution **a.** $(-2)^4 = (-2)(-2)(-2)(-2) = 16$ The base is -2. The exponent is 4.

b. $\left(\dfrac{3}{4}\right)^2 = \dfrac{3}{4}\left(\dfrac{3}{4}\right) = \dfrac{9}{16}$ The base is $\frac{3}{4}$. The exponent is 2.

c. -0.1 cubed means $(-0.1)^3$.   The base is -0.1. The exponent is 3.

$$(-0.1)^3 = (-0.1)(-0.1)(-0.1) = -0.001$$

Self Check 6 Find each power: **a.** $(-3)^3$ **b.** $(0.8)^2$ **c.** 2^4

d. $\dfrac{7}{5}$ squared

Now Try ▶ Problems 49 and 53

Using Your Calculator ▶ The Squaring and Exponential Keys

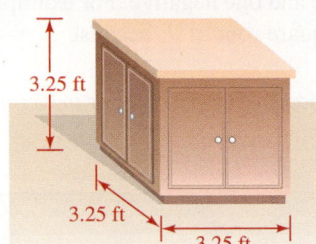

3.25 ft

3.25 ft

3.25 ft

A homeowner plans to install a cooking island in her kitchen. (See the figure.) To find the number of square feet of floor space that will be lost, we substitute 3.25 for s in the formula for the area of a square, $A = s^2$. Using the squaring key $\boxed{x^2}$ on a scientific calculator, we can find $(3.25)^2$ as follows:

3.25 $\boxed{x^2}$

> 10.5625

On a graphing calculator, we press:

3.25 $\boxed{x^2}$ $\boxed{\text{ENTER}}$

> 3.25²
> 10.5625

About 10.6 square feet of floor space will be lost.

The number of cubic feet of storage space that the cooking island will add can be found by substituting 3.25 for s in the formula for the volume of a cube, $V = s^3$. Using the exponential key $\boxed{y^x}$ ($\boxed{x^y}$ on some calculators), we can evaluate $(3.25)^3$ on a scientific calculator as follows.

3.25 $\boxed{y^x}$ 3 $\boxed{=}$

> 34.328125

On a graphing calculator, we press:

3.25 $\boxed{\wedge}$ 3 $\boxed{\text{ENTER}}$

> 3.25^3
> 34.328125

The cooking island will add about 34.3 cubic feet of storage space.

Although the expressions $(-3)^2$ and -3^2 look alike, they are not. In $(-3)^2$, the base is -3. In -3^2, the base is 3. The $-$ sign in front of 3^2 means the opposite of 3^2. When we evaluate them, we see that the results are different:

$$(-3)^2 = (-3)(-3) \qquad \text{Read as "negative} \qquad -3^2 = -(3 \cdot 3) \qquad \text{Read as "the opposite}$$
$$\text{3 squared."} \qquad\qquad\qquad\qquad \text{of the square of 3."}$$
$$= 9 \qquad\qquad\qquad\qquad\qquad\qquad\qquad = -9$$

← ———— Different results ———— →

Using Your Calculator ▶ The Parentheses and Negative Keys

To find $(-3)^2$ with a scientific calculator, use the *parentheses* keys $\boxed{(}$ $\boxed{)}$ and the *negative* key $\boxed{+/-}$. Notice that the negative key is different from the subtraction key $\boxed{-}$. To enter -3, press $\boxed{+/-}$ *after* entering 3.

$\boxed{(}$ 3 $\boxed{+/-}$ $\boxed{)}$ $\boxed{x^2}$ $\boxed{=}$

> 9

To find $(-3)^2$ with a graphing calculator, press the negative key $\boxed{(-)}$ *before* entering 3.

$\boxed{(}$ $\boxed{(-)}$ 3 $\boxed{)}$ $\boxed{x^2}$ $\boxed{\text{ENTER}}$

> (-3)²
> 9

To find -3^2 with a scientific calculator, think of the expression as $-1 \cdot 3^2$. First, find 3^2. Then press $\boxed{+/-}$, which is equivalent to multiplying 3^2 by -1.

3 $\boxed{x^2}$ $\boxed{+/-}$

> -9

A graphing calculator recognizes -3^2 as $-1 \cdot 3^2$, so we can find -3^2 by pressing:

$\boxed{(-)}$ 3 $\boxed{x^2}$ $\boxed{\text{ENTER}}$

> -3²
> -9

Since the product $3 \cdot 3$ can be denoted by the exponential expression 3^2, we say that 3 is squared. The opposite of squaring a number is called finding its **square root.**

All positive numbers have two square roots, one positive and one negative. For example, the two square roots of 9 are 3 and -3. The number 3 is a square root of 9, because $3^2 = 9$, and -3 is a square root of 9, because $(-3)^2 = 9$.

The symbol $\sqrt{}$, called a **radical symbol,** is used to represent the positive (or *principal*) square root of a number.

Principal Square Root

A number b is a square root of a if $b^2 = a$.

If $a > 0$, the expression $\sqrt{a}$ represents the **principal** (or positive) **square root** of a. The principal square root of 0 is 0: $\sqrt{0} = 0$.

The principal square root of a positive number is always positive. Although 3 and -3 are both square roots of 9, only 3 is the principal square root. The symbol $\sqrt{9}$ represents 3. To represent -3, we place a $-$ sign in front of the radical:

$$\sqrt{9} = 3 \qquad \text{and} \qquad -\sqrt{9} = -3$$

EXAMPLE 7   Find each square root: **a.** $\sqrt{121}$ **b.** $-\sqrt{49}$ **c.** $\sqrt{\dfrac{1}{4}}$ **d.** $\sqrt{0.09}$

Strategy   In each case, we will determine what positive number, when squared, produces the radicand.

Why   The symbol $\sqrt{}$ indicates that the positive square root of the number written under it should be found.

Solution   **a.** $\sqrt{121} = 11$, because $11^2 = 121$.   **b.** Since $\sqrt{49} = 7$, $-\sqrt{49} = -7$.

c. $\sqrt{\dfrac{1}{4}} = \dfrac{1}{2}$, because $\left(\dfrac{1}{2}\right)^2 = \dfrac{1}{4}$.   **d.** $\sqrt{0.09} = 0.3$, because $(0.3)^2 = 0.09$.

Self Check 7   Find each square root: **a.** $\sqrt{36}$ **b.** $-\sqrt{100}$ **c.** $\sqrt{\dfrac{4}{25}}$
d. $\sqrt{1}$ **e.** $\sqrt{0.81}$ **f.** $-\sqrt{400}$

Now Try ▶ Problems 55 and 59

---

4 Use the Order of Operations Rule.

To evaluate expressions involving several operations, we must use the following set of priority rules.

Order of Operations Rule

1. Perform all calculations within parentheses and other grouping symbols, following the order listed in steps 2–4 and working from the innermost pair to the outermost pair.

2. Evaluate all exponential expressions (powers) and roots.

3. Perform all multiplications and divisions as they occur from left to right.

4. Perform all additions and subtractions as they occur from left to right.

When all grouping symbols have been removed, repeat steps 2–4 to complete the calculations.

If a fraction is present, evaluate the expression above the bar (the *numerator*) and the expression below the bar (the *denominator*) separately. Then simplify the fraction, if possible.

EXAMPLE 8   Evaluate: **a.** $-5 + 4(-3)^2$ **b.** $-10 \div 5 - 5(3) + 6$

Strategy We will scan the expression to determine what operations need to be performed. Then we will perform those operations, one at a time, following the order of operations rule.

Why   If we don't follow the correct order of operations, the expression can have more than one value.

Solution   **a.** Although the expression contains parentheses, there are no operations to perform within the parentheses. So we proceed with steps 2, 3, and 4 of the order of operations rule.

The Language of Algebra

Sometimes, the word **simplify** is used in the place of the word **evaluate**. For instance, Example 8a could read:

  Simplify: $-5 + 4(-3)^2$

$$-5 + 4\mathbf{(-3)^2} = -5 + 4\mathbf{(9)} \qquad \text{First, evaluate the power: } (-3)^2 = 9.$$
$$= -5 + 36 \qquad \text{Multiply.}$$
$$= 31 \qquad \text{Add.}$$

b. Since the expression does not contain any powers, we perform the multiplications and divisions, working from left to right.

$$\mathbf{-10 \div 5} - 5(3) + 6 = \mathbf{-2} - 5(3) + 6 \qquad \text{Divide: } -10 \div 5 = -2.$$
$$= -2 - 15 + 6 \qquad \text{Multiply.}$$
$$= -17 + 6 \qquad \text{Working from left to right,}$$
$$\qquad\qquad\qquad\qquad \text{subtract: } -2 - 15 = -17.$$
$$= -11 \qquad \text{Add.}$$

Self Check 8   Evaluate: **a.** $-9 + 2(-4)^2$
  b. $20 \div (-5) - (-6)(-5) + (-12)$

Now Try ▶   Problems 63 and 67

Grouping symbols serve as mathematical punctuation marks. They help determine the order in which an expression is evaluated. Examples of grouping symbols are parentheses ( ), brackets [ ], absolute value bars | |, and the fraction bar —.

EXAMPLE 9   Evaluate: **a.** $3 - (4 - 8)^2$ **b.** $2 + 3[-2 - 8(4 - 3^2)]$ **c.** $|-45 + 30|(2 - 7)$

Strategy We will perform all calculations within parentheses and other grouping symbols first.

Why   This is the first step of the order of operations rule.

Solution   **a.** $3 - \mathbf{(4 - 8)^2} = 3 - \mathbf{(-4)^2}$ Perform the subtraction: $4 - 8 = -4.$
$$= 3 - 16 \qquad \text{Evaluate the power: } (-4)^2 = 16.$$
$$= -13 \qquad \text{Subtract.}$$

b. First, we work within the innermost grouping symbols, the parentheses.

$$2 + 3[-2 - 8(4 - \mathbf{3^2})] = 2 + 3[-2 - 8(4 - \mathbf{9})] \qquad \text{Find the power: } 3^2 = 9.$$
$$= 2 + 3[-2 - 8(-5)] \qquad \text{Subtract: } 4 - 9 = -5.$$

Next, we work within the outermost grouping symbols, the brackets.

$$= 2 + 3[-2 - (-40)] \qquad \text{Multiply: } 8(-5) = -40.$$
$$= 2 + 3(-2 + 40) \qquad \text{Write the subtraction as}$$
$$\qquad\qquad\qquad\qquad \text{addition of the opposite.}$$
$$\qquad\qquad\qquad\qquad \text{Since only one set of grouping}$$
$$\qquad\qquad\qquad\qquad \text{symbols was needed, write}$$
$$\qquad\qquad\qquad\qquad -2 + 40 \text{ within parentheses.}$$
$$= 2 + 3(38) \qquad \text{Add: } -2 + 40 = 38.$$
$$= 2 + 114 \qquad \text{Multiply.}$$
$$= 116 \qquad \text{Add.}$$

c. Since the absolute value bars are grouping symbols, we perform the operations within the absolute value bars and the parentheses first.

$$|-45 + 30|(2 - 7) = |-15|(-5)$$
$$= 15(-5) \quad \text{Find the absolute value: } |-15| = 15.$$
$$= -75 \quad \text{Multiply.}$$

Self Check 9   Evaluate: **a.** $(5 - 3)^3 - 40$ **b.** $-3[-2(5^3 - 3) + 4] - 1$
c. $2|-25 - (-6)(3)|$

Now Try ▶   Problems 71 and 79

Using Your **Calculator** ▶ Order of Operations

Scientific and graphing calculators are programmed to follow the rules for the order of operations. For example, when finding $3 + 2 \cdot 5$, both types of calculators give the correct answer, 13.

$3 \boxed{+} 2 \boxed{\times} 5 \boxed{=}$ 13

$3 \boxed{+} 2 \boxed{\times} 5 \boxed{\text{ENTER}}$ 3+2*5   13

Both types of calculators use parentheses keys $\boxed{(}\ \boxed{)}$ when grouping symbols are needed. To evaluate $3 - (4 - 8)^2$, we press:

$3 \boxed{-} \boxed{(} 4 \boxed{-} 8 \boxed{)} \boxed{x^2} \boxed{=}$ −13

$3 \boxed{-} \boxed{(} 4 \boxed{-} 8 \boxed{)} \boxed{x^2} \boxed{\text{ENTER}}$ 3−(4−8)²   −13

Both types of calculators require that we group the terms in the numerator together and the terms in the denominator together when calculating the value of an expression such as $\frac{200 + 120}{20 - 16}$.

$\boxed{(} 200 \boxed{+} 120 \boxed{)} \boxed{\div} \boxed{(} 20 \boxed{-} 16 \boxed{)} \boxed{=}$ 80

$\boxed{(} 200 \boxed{+} 120 \boxed{)} \boxed{\div} \boxed{(} 20 \boxed{-} 16 \boxed{)} \boxed{\text{ENTER}}$ (200+120)/(20−16)   80

If parentheses aren't used when finding $\frac{200 + 120}{20 - 16}$, you will obtain an incorrect result of 190. That is because the calculator will interpret the entry as $200 + \frac{120}{20} - 16$.

---

5 Evaluate Algebraic Expressions.

Recall that an algebraic expression is a combination of variables and numbers with the operations of arithmetic. To *evaluate* expressions, we substitute specific numbers for the variables and then apply the order of operations rule.

EXAMPLE 10   If $a = -2$, $b = 9$, and $c = -1$, evaluate: **a.** $-\frac{1}{2}a^2$ **b.** $\frac{-a\sqrt{b} + 3c^3}{c(c - b)}$

Strategy   We will replace each a, b, and c in the expression with the given value of the variable and evaluate the expression using the order of operations rule.

Why   To *evaluate an expression* means to find its numerical value, once we know the value of its variable(s).

Solution **a.** We substitute -2 for a and use the order of operations rule.

$$-\frac{1}{2}a^2 = -\frac{1}{2}(-2)^2$$

Substitute -2 for a. Write parentheses around -2 so that it is squared.

$$= -\frac{1}{2}(4)$$

Evaluate the power: $(-2)^2 = 4$.

$$= -2$$

Multiply.

Caution

When replacing a variable with its numerical value, we must often write the replacement number within parentheses to convey the proper meaning.

b.

$$\frac{-a\sqrt{b} + 3c^3}{c(c-b)} = \frac{-(-2)\sqrt{9} + 3(-1)^3}{-1(-1-9)}$$

Substitute -2 for a, 9 for b, and -1 for c.

$$= \frac{-(-2)(3) + 3(-1)}{-1(-10)}$$

In the numerator, evaluate the square root and the power: $\sqrt{9} = 3$ and $(-1)^3 = -1$. In the denominator, subtract.

$$= \frac{2(3) + 3(-1)}{-1(-10)}$$

In the numerator, simplify: $-(-2) = 2$.

$$= \frac{6 + (-3)}{10}$$

In the numerator, multiply. In the denominator, multiply.

$$= \frac{3}{10}$$

In the numerator, add.

Self Check 10 If $r = 2$, $s = -5$, and $t = 3$, evaluate: **a.** $-\frac{1}{3}s^3t$

b. $\dfrac{\sqrt{-5s}}{(s+t)r^2}$

Now Try ▶ Problems 95 and 97

Using Your Calculator ▶ **Evaluating Algebraic Expressions**

Graphing calculators can evaluate algebraic expressions. For example, to evaluate

$$\frac{-a\sqrt{b} + 3c^3}{c(c-b)}$$

(Example 10, part b) using a TI-84 Plus calculator, we first enter the values of $a = -2$, $b = 9$, and $c = -1$, using the store key $\boxed{\text{STO}}$ and the $\boxed{\text{ALPHA}}$ key. See figure (a).

$\boxed{(-)}\,2\,\boxed{\text{STO}}\,\boxed{\text{ALPHA}}\,\boxed{\text{A}}\,\boxed{\text{ALPHA}}\,\boxed{:}$ This enters $a = -2$.

$9\,\boxed{\text{STO}}\,\boxed{\text{ALPHA}}\,\boxed{\text{B}}\,\boxed{\text{ALPHA}}\,\boxed{:}$ This enters $b = 9$.

$\boxed{(-)}\,1\,\boxed{\text{STO}}\,\boxed{\text{ALPHA}}\,\boxed{\text{C}}\,\boxed{\text{ALPHA}}\,\boxed{:}$ This enters $c = -1$.

Next, enter the expression as shown in figure (b) and press $\boxed{\text{ENTER}}$ to find that the value of the expression is 0.3. To express the result as a fraction, press $\boxed{\text{MATH}}$, highlight Frac, and then press $\boxed{\text{ENTER}}$ $\boxed{\text{ENTER}}$. See figure (c).

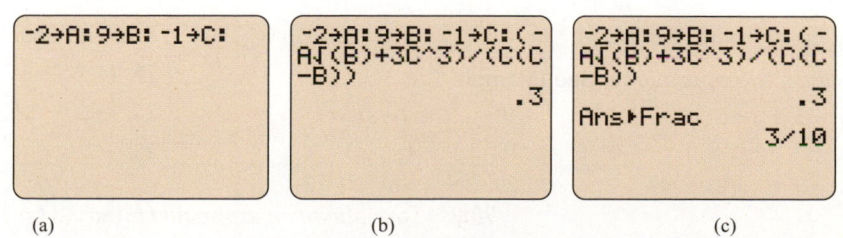

(a) (b) (c)

SECTION 1.3 .STUDY SET

VOCABULARY

Fill in the blanks.

1. When we add two numbers, the result is called the _____. When we subtract two numbers, the result is called the _____.

2. When we multiply two numbers, the result is called the _____. When we divide two numbers, the result is called the _____.

3. The _____ of $\frac{5}{9}$ is $\frac{9}{5}$.

4. In the exponential expression x^2, the _____ is x and 2 is the _____.

5. 6^2 can be read as "six _____" and 6^3 can be read as "six _____."

6. We read $\sqrt{25}$ as the _____ ____ of 25.

7. In the expression $9 + 6[22 - (6 - 1)]$, the _____ are the innermost grouping symbols, and the brackets are the _____ grouping symbols.

8. To _____ an algebraic expression, we substitute values for the variables and then apply the order of operations rule.

CONCEPTS

9. Fill in the blanks.
   a. An exponent indicates repeated _____.
   b. Subtraction is the same as adding the _____ of the number being subtracted.

10. Fill in the blanks: When multiplying signed numbers, an odd number of negative factors gives a _____ product. An even number of negative factors gives a _____ product.

11. If each of the following expressions were evaluated, what would be the *sign* of the result?
   a. $-1,763 + 1,699$
   b. $-503 - 512$
   c. $(-657)(-22)$
   d. $\dfrac{-2,744}{49}$

12. In what order should the operations be performed to evaluate $60 - (-9)^2 + 5(-1)$?

NOTATION

13. Translate each expression into symbols, and then evaluate it.
   a. Negative four, squared
   b. The opposite of the square of 4

14. What is the name of the symbol $\sqrt{}$?

GUIDED PRACTICE

Perform the operations. See Example 1.

15. $-3 + (-5)$
16. $-2 + (-8)$
17. $-7.1 + 2.8$
18. $3.1 + (-5.2)$
19. $-9 + (-8) + 4$
20. $2 + (-6) + (-3)$

21. $\dfrac{1}{2} + \left(-\dfrac{1}{3}\right)$
22. $-\dfrac{3}{4} + \left(-\dfrac{1}{5}\right)$

Perform the operations. See Example 2.

23. $-3 - 4$
24. $-11 - (-17)$
25. $-3.3 - (-3.3)$
26. $0.14 - (-0.13)$
27. Subtract $-\dfrac{3}{5}$ from $\dfrac{1}{2}$
28. Subtract $\dfrac{11}{13}$ from $\dfrac{1}{26}$
29. $-1 - 5 - (-4)$
30. $5 - (-3) - 2$

Perform the operations. See Example 3.

31. $-2(6)$
32. $-3(7)$
33. $-0.3(5)$
34. $-0.4(-0.6)$
35. $-5(6)(-2)$
36. $-9(-1)(-3)$
37. $\left(-\dfrac{3}{5}\right)\left(\dfrac{10}{7}\right)$
38. $\left(-\dfrac{6}{7}\right)\left(-\dfrac{5}{12}\right)$

Perform the operations. See Examples 4 and 5.

39. $\dfrac{-8}{4}$
40. $\dfrac{16}{-4}$
41. $\dfrac{84}{-6}$
42. $\dfrac{-78}{6}$
43. $\dfrac{-10.8}{-1.2}$
44. $\dfrac{-13.5}{-1.5}$
45. $-\dfrac{16}{5} \div \left(-\dfrac{10}{3}\right)$
46. $-\dfrac{5}{24} \div \dfrac{10}{3}$

Evaluate each expression. See Example 6.

47. 6^4
48. 2^5
49. $(-7.9)^2$
50. $(-4.6)^2$
51. -5^2
52. -8^2
53. $\left(-\dfrac{3}{5}\right)^3$
54. $\left(-\dfrac{4}{3}\right)^3$

Find each square root. See Example 7.

55. $\sqrt{64}$
56. $\sqrt{121}$
57. $-\sqrt{81}$
58. $-\sqrt{36}$
59. $-\sqrt{\dfrac{9}{16}}$
60. $-\sqrt{\dfrac{81}{49}}$
61. $\sqrt{0.04}$
62. $\sqrt{0.64}$

Evaluate each expression. See Example 8.

63. $3 - 5 \cdot 4$
64. $12 - 2 \cdot 3$
65. $4 \cdot 2^3$
66. $4 \cdot 5^3$
67. $-12 \div 3 \cdot 2$
68. $-18 \div 6 \cdot 3$
69. $7^2 - (-9)^2$
70. $4^2 - (-8)^2$

Evaluate each expression. See Example 9.

71. $(4 + 2 \cdot 3)^4$
72. $|9 - 5(1 - 8)|$
73. $\left(-3 - \sqrt{25}\right)^2$
74. $\left(-1 - \sqrt{144}\right)^2$
75. $-2|4 - 8|$
76. $|\sqrt{49} - 8(4 - 7)|$
77. $2 + 3\left(\dfrac{25}{5}\right) + (-4)$
78. $(-2)^3\left(\dfrac{-6}{-2}\right)(-1)$

79. $30 + 6[-4 - 5(6 - 4)^2]$

80. $7 - 12[7^2 - 4(2 - 5)^2]$

81. $3 - [3^3 + (3 - 1)^3]$

82. $8 - 4|-(3 \cdot 5 - 2 \cdot 6)^2|$

83. $\dfrac{1}{3}\left(\dfrac{1}{6}\right) - \left(-\dfrac{1}{3}\right)^2$

84. $\dfrac{1}{2}\left(\dfrac{1}{8}\right) + \left(-\dfrac{1}{4}\right)^2$

85. $\dfrac{-2 - 5}{-7 + (-7)}$

86. $\dfrac{-3 - (-1)}{-2 + (-2)}$

87. $\dfrac{|-25| - 2(-5)}{2^4 - 9}$

88. $\dfrac{2[-4 - 2(3 - 1)]}{3(3)(2)}$

89. $\dfrac{3[-9 + 2(7 - 3)]}{(8 - 5)(9 - 7)}$

90. $\dfrac{(6 - 5)^4 + 21}{27 - \left(\sqrt{16}\right)^2}$

Evaluate each expression for the given values. **See Example 10.**

91. $-\dfrac{2}{3}a^2$ for $a = -6$

92. $\left(-\dfrac{2}{3}a\right)^2$ for $a = -6$

93. $\dfrac{y_2 - y_1}{x_2 - x_1}$ for $x_1 = -3, x_2 = 5, y_1 = 12, y_2 = -4$

94. $P_0\left(1 + \dfrac{r}{k}\right)^{kt}$ for $P_0 = 500, r = 4, k = 2, t = 3$

95. $(x + y)(x^2 - xy + y^2)$ for $x = -4, y = 5$

96. $\dfrac{x^2}{a^2} + \dfrac{y^2}{b^2}$ for $x = -3, y = -4, a = 5, b = -5$

97. $\dfrac{-b + \sqrt{b^2 - 4ac}}{2a}$ for $a = 1, b = 2, c = -3$

98. $\dfrac{n}{2}[2a_1 + (n - 1)d]$ for $n = 50, a_1 = -4, d = 5$

99. $\sqrt{(x_2 - x_1)^2 + (y_2 - y_1)^2}$ for $x_1 = -2, x_2 = 4, y_1 = 4, y_2 = -4$

100. $\dfrac{|Ax_0 + By_0 + C|}{\sqrt{A^2 + B^2}}$ for $A = 3, B = 4, C = -5, x_0 = 2, y_0 = -1$

101. $-n(4n^2 - 27m^2)^3$ for $m = \dfrac{1}{3}$ and $n = \dfrac{1}{2}$

102. $\dfrac{-s^2 + 1 + 16r^2}{3^2 - 2}$ for $s = -10$ and $r = \dfrac{1}{4}$

TRY IT YOURSELF

Look Alikes . . .

Evaluate each expression.

103. **a.** $100 - 20 + 5$ **b.** $100 - (20 + 5)$
     c. $100 \div 20 \cdot 5$ **d.** $100 \div (20 \cdot 5)$

104. **a.** $2 \cdot 3^2$ **b.** $(2 \cdot 3)^2$

105. **a.** Subtract -3.9 from -11.2
     b. Subtract -11.2 from -3.9

106. **a.** $\left(-2 - \sqrt{64}\right)^2$ **b.** $-2 - \left(\sqrt{64}\right)^2$

APPLICATIONS

107. **The Stock Market.** In graph below, positive numbers represent new cash *inflow* into the U.S. stock market. Negative numbers represent cash *outflow* from the market. Was there a net inflow or outflow over the 9-year period from 2000 to 2009? What was it?

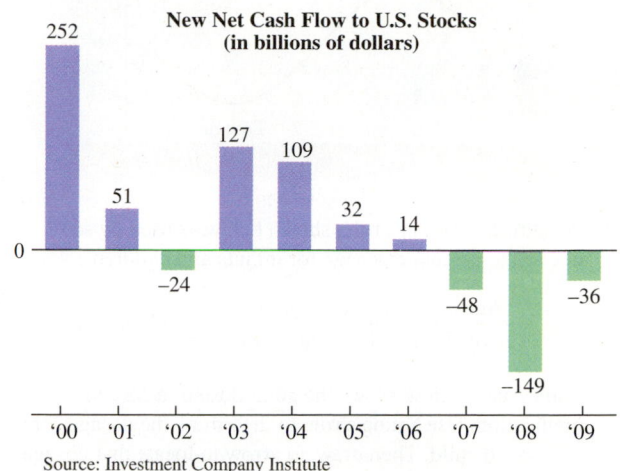

New Net Cash Flow to U.S. Stocks (in billions of dollars)

Source: Investment Company Institute

108. **Accounting.** On a financial balance sheet, debts (negative numbers) are denoted within parentheses. Assets (positive numbers) are written without parentheses. What is the 2010 fund balance for the preschool whose financial records are shown in the table?

Community Care Preschool Balance Sheet, June 2010	
Fund balances	
Classroom supplies	$ 5,889
Emergency needs	927
Holiday program	(2,928)
Insurance	1,645
Janitorial	(894)
Licensing	715
Maintenance	(6,321)
BALANCE	?

109. **Temperature Extremes.** The highest and lowest temperatures ever recorded in several cities are shown in the table. List the cities in order, from the smallest to the largest range in temperature extremes.

City	Extreme temperatures	
	Highest	Lowest
Atlanta, Georgia	105	−8
Boise, Idaho	111	−25
Helena, Montana	105	−42
New York, New York	107	−3
Omaha, Nebraska	114	−23

110. Physics. The illustration shows an example of a *standing wave*. What is the difference in the height of the crest of the wave and the depth of the trough of the wave? (m stands for meter.)

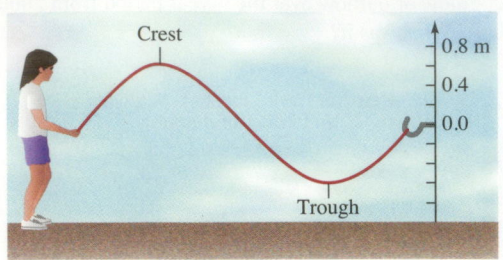

111. Pediatrics. Young's rule, shown below, is used by some doctors to calculate dosage for infants and children.

$$\frac{\text{Age of child}}{\text{Age of child} + 12}\left(\frac{\text{average}}{\text{adult dose}}\right) = \text{child's dose}$$

The syringe below shows the adult dose of a certain medication. Use Young's rule to determine the dosage for a 6-year-old child. Then draw an arrow to locate that dosage on the calibration.

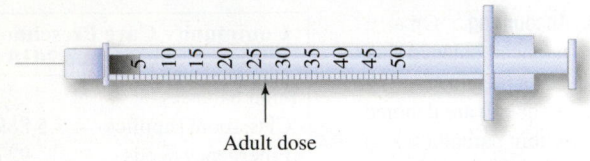

Adult dose

112. Dosages. The adult dosage of procaine penicillin is 300,000 units daily. Calculate the dosage for a 12-year-old child using Young's rule. (See Exercise 111.)

WRITING

113. Explain what the statement $x - y = x + (-y)$ means.

114. Explain why the order of operations rule is necessary.

REVIEW

115. What two numbers are a distance of 5 away from -2 on the number line?

116. Place the proper symbol ($>$ or $<$) in the blank:
$-4.6 \quad \blacksquare \quad -4.5$

117. Write the set of integers.

118. Translate into mathematical symbols: ten less than twice x.

119. True or false: The real numbers is the set of all decimals.

120. True or false: Irrational numbers are nonterminating, nonrepeating decimals.

CHALLENGE PROBLEMS

121. Insert one pair of parentheses in the expression so that its value is 0.

$$71 - 1 - 2 \cdot 5^2 + 10$$

122. Evaluate:

$$\left(\frac{\dfrac{12 \div 3 \cdot 4}{[-9^2 - 4(-1)^9(20)]^4} - \sqrt{6\sqrt{\left|-\dfrac{3}{2}(24)\right|}}}{\left|-1.5\left|\dfrac{-200}{10^2}\right|\right|^2 + \dfrac{(-2)^3}{-4 - \dfrac{4 + 2}{1 + \dfrac{8}{2 - 6}}}}\right)^5$$

SECTION 1.4

Simplifying Algebraic Expressions Using Properties of Real Numbers

OBJECTIVES

1 Identify terms, factors, and coefficients.

2 Identify and use properties of real numbers.

3 Simplify products.

4 Use the distributive property.

5 Combine like terms.

ARE YOU READY?

The following problems review some basic concepts that are important when simplifying expressions using properties of real numbers.

1. How do the expressions $9 \cdot x$ and $x \cdot 9$ differ?

2. How do the expressions $19 + (11 + x)$ and $(19 + 11) + x$ differ?

3. Evaluate $6(5 + 2)$ and $6 \cdot 5 + 6 \cdot 2$ and compare the results.

4. How do the terms $8x$ and $8y$ differ? What do they have in common?

In algebra, we frequently replace one algebraic expression with another that is equivalent and simpler in form. That process, called *simplifying an expression,* often involves the use of one or more properties of real numbers.

1 Identify Terms, Factors, and Coefficients.

Addition symbols separate algebraic expressions into parts called *terms*. For example, the expression $3x^2 + x + 4$ has three terms.

$$\underset{\substack{\text{First}\\\text{term}}}{3x^2} + \underset{\substack{\text{Second}\\\text{term}}}{x} + \underset{\substack{\text{Third}\\\text{term}}}{4}$$

In general, a **term** is a product or quotient of numbers and/or variables. A single number or variable is also a term. Examples of terms are:

$$4, \qquad y, \qquad 6r, \qquad -w^3, \qquad 3.7x^5, \qquad \frac{3}{n}, \qquad -15ab^2$$

Since subtraction can be written as addition of the opposite, the expression $6a - 5b$ has two terms.

$$6a - 5b = \underset{\substack{\text{First}\\\text{term}}}{6a} + \underset{\substack{\text{Second}\\\text{term}}}{(-5b)}$$

A term such as 9, that consists of a single number, is called a **constant term.**

The numerical factor of a term is called the **numerical coefficient** or simply the **coefficient** of the term. The coefficients of the terms of the expression $3x^2 + x + 4$ are 3, 1, and 4, respectively. The coefficients of the terms of $6a - 5b$ are 6 and -5, respectively.

2 Identify and Use Properties of Real Numbers.

The following properties of real numbers are used to simplify algebraic expressions.

Properties of Real Numbers

If a, b, and c represent real numbers, we have

The commutative properties of addition and multiplication

$$a + b = b + a \qquad ab = ba$$

The associative properties of addition and multiplication

$$(a + b) + c = a + (b + c) \qquad (ab)c = a(bc)$$

The *commutative properties* enable us to add or multiply two numbers in either order and obtain the same result. Here are two examples.

$$3 + (-5) = -2 \qquad \text{and} \qquad -5 + 3 = -2$$
$$-2.6(-8) = 20.8 \qquad \text{and} \qquad -8(-2.6) = 20.8$$

We can use the commutative properties (and other properties of real numbers) to write *equivalent expressions*. **Equivalent expressions** represent the same number. For example, $x + 3$ and $3 + x$ are equivalent expressions because for each value of x, they represent the same number. For instance, if $x = 6$, both expressions represent 9. If $x = -4$, both expressions represent -1, and so on.

$$\textbf{\textit{If x = 6}} \qquad\qquad\qquad\qquad \textbf{\textit{If x = -4}}$$

$$x + 3 = 6 + 3 \qquad 3 + x = 3 + 6 \qquad\qquad x + 3 = -4 + 3 \qquad 3 + x = 3 + (-4)$$
$$\quad = 9 \qquad\qquad\quad = 9 \qquad\qquad\qquad\qquad = -1 \qquad\qquad\qquad = -1$$

Subtraction and division are not commutative, because performing these operations in different orders will give different results. For example,

$$8 - 4 = 4 \qquad \text{but} \qquad 4 - 8 = -4$$
$$8 \div 4 = 2 \qquad \text{but} \qquad 4 \div 8 = \frac{1}{2}$$

The Language of Algebra

Associative is a form of the word *associate,* meaning to join a group. For example, the National Basketball *Association* (NBA) is a group of professional basketball teams.

The *associative properties* enable us to group the numbers in an addition or multiplication any way that we wish and get the same result. For example,

$$(\mathbf{19 + 7}) + 3 = \mathbf{26} + 3 = 29 \qquad \text{and} \qquad 19 + (\mathbf{7 + 3}) = 19 + \mathbf{10} = 29$$
$$(\mathbf{4 \cdot 2})6 = \mathbf{8} \cdot 6 = 48 \qquad \text{and} \qquad 4(\mathbf{2 \cdot 6}) = 4 \cdot \mathbf{12} = 48$$

Subtraction and division are not associative, because different groupings give different results. For example,

$$(\mathbf{8 - 4}) - 2 = \mathbf{4} - 2 = 2 \qquad \text{but} \qquad 8 - (\mathbf{4 - 2}) = 8 - \mathbf{2} = 6$$
$$(\mathbf{8 \div 4}) \div 2 = \mathbf{2} \div 2 = 1 \qquad \text{but} \qquad 8 \div (\mathbf{4 \div 2}) = 8 \div \mathbf{2} = 4$$

The real numbers 0 and 1 have special importance. For example,

$$7 + \mathbf{0} = 7, \qquad \mathbf{1}(5.4) = 5.4, \qquad \left(-\frac{7}{3}\right)\mathbf{1} = -\frac{7}{3}, \qquad \text{and} \qquad -19(\mathbf{0}) = 0$$

These examples illustrate the following properties:

Properties of 0 and 1

Additive identity: The sum of 0 and any number is the number itself.

$$0 + a = a + 0 = a$$

Multiplicative identity: The product of 1 and any number is the number itself.

$$1 \cdot a = a \cdot 1 = a$$

Multiplication property of 0: The product of any number and 0 is 0.

$$a \cdot 0 = 0 \cdot a = 0$$

If the sum of two numbers is 0, they are called **additive inverses,** or **opposites** of each other. For example, 6 and -6 are additive inverses, because $6 + (-6) = 0$.

The Additive Inverse Property

For every real number a, there exists a real number $-a$ such that

$$a + (-a) = -a + a = 0$$

If the product of two numbers is 1, the numbers are called **multiplicative inverses** or **reciprocals** of each other.

The Multiplicative Inverse Property

For every nonzero real number a, there exists a real number $\frac{1}{a}$ such that

$$a \cdot \frac{1}{a} = \frac{1}{a} \cdot a = 1$$

Caution

The reciprocal of 0 does not exist, because $\frac{1}{0}$ is undefined.

Some examples of reciprocals (multiplicative inverses) are

- 5 and $\frac{1}{5}$ are reciprocals, because $5\left(\frac{1}{5}\right) = 1$.
- $\frac{3}{2}$ and $\frac{2}{3}$ are reciprocals, because $\frac{3}{2}\left(\frac{2}{3}\right) = 1$.
- -0.25 and -4 are reciprocals, because $-0.25(-4) = 1$.

EXAMPLE 1   Complete each statement so that the indicated property is illustrated.

a. $(14 + 92) + 8 = $ _____   Associative property of addition

b. $\dfrac{7}{6} \cdot$ _____ $= \dfrac{7}{6}$ Multiplicative identity property   **c.** $x \cdot 5 = $ _____   Commutative property of multiplication

Strategy   For problems like these, it is important to have memorized the properties of real numbers by name. To fill in each blank, we will determine in what way the indicated property enables us to write an equivalent expression.

Why   We should memorize the properties of real numbers by name because their names remind us how to use them.

Solution   **a.** To *associate* means to group together. The associative property of addition enables us to group the numbers in a different way. Thus, we have

$$(14 + 92) + 8 = \underline{14 + (92 + 8)}$$ Associative property of addition

Success Tip

In part a, note that the *order* of the terms does not change.

b. The word *identical* means to be exactly the same. The multiplicative identity property indicates that if we multiply $\frac{7}{6}$ by 1, it remains the same. Thus, we have

$$\dfrac{7}{6} \cdot \underline{1} = \dfrac{7}{6}$$ Multiplicative identity property

c. To *commute* means to go back and forth. The commutative property of multiplication enables us to change the order of the factors. Thus, we have

$$x \cdot 5 = \underline{5 \cdot x}$$ Commutative property of multiplication

Self Check 1   Use the given property to complete each statement.
a. $-23 + \underline{} = 0$ Additive inverse property

b. $-16\left(\dfrac{1}{2} \cdot 7\right) = $ _____   Associative property of multiplication

Now Try ▶ Problems 23 and 27

Recall that when a number is divided by 1, the result is the number itself, and when a nonzero number is divided by itself, the result is 1.

Division Properties

Division by 1: If a represents any real number, then $\dfrac{a}{1} = a$.

Division of a number by itself: For any nonzero real number a, $\dfrac{a}{a} = 1$.

There are three possible cases to consider when discussing division involving 0.

Division with 0

Division of 0: For any nonzero real number a, $\dfrac{0}{a} = 0$.

Division by 0: For any nonzero real number a, $\dfrac{a}{0}$ is undefined.

Division of 0 *by* 0: $\dfrac{0}{0}$ is indeterminate.

To show that division of 0 by 0 is **indeterminate**, we consider $\frac{0}{0} = ?$ and its related multiplication statement $0(?) = 0$.

Multiplication fact   **Division fact**

$0(?) = 0$ $\dfrac{0}{0} = ?$

Any number multiplied by 0 gives 0.   We cannot determine this—it could be any number.

3 Simplify Products.

The commutative and associative properties of multiplication can be used to simplify certain products. For example, let's simplify $6(5x)$.

$6(5x) = 6 \cdot (5 \cdot x)$ Rewrite 5x as 5 · x.

$\quad\quad = (6 \cdot 5) \cdot x$ Use the associative property of multiplication to group 6 with 5.

$\quad\quad = 30x$ Multiply within the parentheses.

Since $6(5x) = 30x$, we say that $6(5x)$ **simplifies** to $30x$.

EXAMPLE 2   Simplify: **a.** $9(10t)$ **b.** $-5.3r(-2s)$ **c.** $-\dfrac{21}{2}a\left(\dfrac{1}{3}\right)$

Strategy   We will use the commutative and associative properties of multiplication to reorder and regroup the factors in each expression.

Why   We want to group all of the numerical factors of an expression together so that we can find their product.

Solution   **a.** $9(10t) = (9 \cdot 10)t$ Use the associative property of multiplication to regroup the factors.

$\quad\quad\quad\quad = 90t$ Multiply within the parentheses: 9 · 10 = 90.

b. $-5.3r(-2s) = [-5.3(-2)](r \cdot s)$ Use the commutative and associative properties to group the numbers and group the variables.

$\quad\quad\quad\quad\quad = 10.6rs$ Multiply within the brackets.

c. $-\dfrac{21}{2}a\left(\dfrac{1}{3}\right) = -\dfrac{21}{2}\left(\dfrac{1}{3}\right)a$ Use the commutative property of multiplication to change the order of the factors a and $\frac{1}{3}$.

$\quad\quad\quad\quad = -\dfrac{7}{2}a$ Multiply the fractions: $-\frac{21}{2} \cdot \frac{1}{3} = -\frac{21 \cdot 1}{2 \cdot 3} = -\frac{\overset{1}{\cancel{3}} \cdot 7 \cdot 1}{2 \cdot \cancel{3}} = -\frac{7}{2}.$

Self Check 2   Simplify: **a.** $14 \cdot 3s$ **b.** $-1.6b(3t)$ **c.** $-\frac{2}{3}x(-9)$

Now Try   Problems 35 and 39

4 Use the Distributive Property.

Another property that we can use to simplify algebraic expressions is the **distributive property.** To introduce it, we will evaluate $4(5 + 3)$, in two ways.

Method 1: *Use the order of operations* | **Method 2:** *Distribute the multiplication*

$4(5 + 3) = 4(8)$ $4(5 + 3) = 4(5) + 4(3)$

$\quad\quad\quad = 32$ $\quad\quad\quad\quad = 20 + 12$

$\quad\quad\quad\quad\quad\quad\quad = 32$

Each method gives a result of 32. This observation illustrates the following property.

The Distributive Property of Multiplication Over Addition	If a, b, and c represent real numbers, $a(b + c) = ab + ac$

EXAMPLE 3　Multiply:　**a.** $6(a + 9)$　　**b.** $-15(4b - 1)$　　**c.** $-(-21 - 20m)$

Strategy　We will distribute the multiplication by the factor outside the parentheses over each term within the parentheses.

Why　We cannot simplify the expression within the parentheses. To multiply, we must use the distributive property.

Solution

a. $6(a + 9) = \mathbf{6} \cdot a + \mathbf{6} \cdot 9$　　Distribute the multiplication by 6.

$\qquad\qquad = 6a + 54$　　Perform each multiplication.

The Language of Algebra

When we use the distributive property to write a product, such as $6(a + 9)$, as the sum, $6a + 54$, we say that we have **removed** or **cleared** parentheses.

b. $\mathbf{-15}(4b - 1) = \mathbf{-15}(4b) - (\mathbf{-15})(1)$　　Distribute the multiplication by −15.

$\qquad\qquad\qquad = -60b + 15$　　Multiply.

c. To use the distributive property to simplify $-(-21 - 20m)$, we interpret the $-$ symbol as a factor of -1, and proceed as follows.

Success Tip

We can use the commutative property of addition to reorder the terms of the result. It is standard practice to write answers with the variable terms first, followed by the constant term.

$-(-21 - 20m) = \mathbf{-1}(-21 - 20m)$　　Write the − sign in front of the parentheses as −1.

$\qquad\qquad\qquad = \mathbf{-1}(-21) - (\mathbf{-1})(20m)$　　Distribute the multiplication by −1.

$\qquad\qquad\qquad = 21 + 20m$　　Perform each multiplication.

$\qquad\qquad\qquad = 20m + 21$　　Write the variable term of the answer first.

Notice that the $-$ symbol in front of the parentheses changes the signs of each term within the parentheses.

Self Check 3　Multiply:　**a.** $9(r + 4)$　　**b.** $-11(-3x - 5)$　　**c.** $-(-27k + 15)$

Now Try ▶ Problems 43 and 47

A more general form of the distributive property is the **extended distributive property.**

$$a(b + c + d + e + \cdots) = ab + ac + ad + ae + \cdots$$

Since multiplication is commutative, we can write the distributive property in the following forms.

$$(b + c)a = ba + ca, \quad (b - c)a = ba - ca, \quad (b + c + d)a = ba + ca + da$$

EXAMPLE 4　Multiply:　**a.** $-0.5(7 - 5y + 6z)$　　**b.** $(8x - 3y)\dfrac{3}{2}$

Strategy　We will multiply each term within the parentheses by the factor outside the parentheses.

Why　We cannot simplify the expression within the parentheses. To multiply, we must use an extension of the distributive property.

Solution **a.** $-0.5(7 - 5y + 6z)$

$= -0.5(7) - (-0.5)(5y) + (-0.5)(6z)$ Distribute the multiplication by -0.5.

$= -3.5 + 2.5y - 3z$ Perform each multiplication.

$= 2.5y - 3z - 3.5$ Write the variable terms of the answer first.

b. $(8x - 3y)\dfrac{3}{2} = (8x)\dfrac{3}{2} - (3y)\dfrac{3}{2}$ Distribute the multiplication by $\frac{3}{2}$.

$= \dfrac{24x}{2} - \dfrac{9y}{2}$ Perform the multiplication.

$= 12x - \dfrac{9}{2}y$ Simplify the first fraction.

Self Check 4 Multiply: **a.** $0.8(-6t + 3s - 10)$

b. $(10a + 16b)\dfrac{3}{5}$

Now Try ▶ Problems 55 and 59

---

5 Combine Like Terms.

Before we can discuss methods for simplifying algebraic expressions involving addition and subtraction, we must define like and unlike terms.

Like Terms   **Like terms** are terms with exactly the same variables raised to exactly the same powers. Any constant terms in an expression are considered to be like terms. Terms that are not like terms are called **unlike terms**.

Here are some examples of like and unlike terms.

- $5x$ and $6x$ are like terms.
- $4x$ and $-17y$ are unlike terms, because they have different variables.
- $27x^2y^3$ and $-326x^2y^3$ are like terms.
- $15x^2y$ and $6xy^2$ are unlike terms, because the variables have different exponents.

Simplifying the sum or difference of like terms is called **combining like terms**. To simplify expressions containing like terms, we use the distributive property in reverse. For example,

$$5x + 6x = (5 + 6)x \qquad \text{and} \qquad 32y - 16y = (32 - 16)y$$
$$= 11x \qquad\qquad\qquad\qquad = 16y$$

These examples illustrate the following general rule.

Combining Like Terms   Like terms can be combined by adding or subtracting the coefficients of the terms and keeping the same variables with the same exponents.

EXAMPLE 5   Simplify: **a.** $-8f + 12f$ **b.** $0.6s^3 - 0.2s^3 - (-0.9s^3)$ **c.** $-\dfrac{1}{2}ab + \dfrac{1}{3}ab$
d. $16n + 8n^2 - 42n + 4n^2$

Strategy   We will use the distributive property to add (or subtract) the coefficients of the like terms.

Why   To *combine like terms* means to add or subtract the like terms in an expression.

Solution   **a.** Since $-8f$ and $12f$ are like terms with the common variable f, we can combine them by adding the coefficients of the like terms and keeping the variable f.

$$-8f + 12f = 4f \qquad \textcolor{red}{\text{Think: } (-8 + 12)f = 4f.}$$

b. $0.6s^3 - 0.2s^3 \textcolor{red}{\,-\,(-0.9s^3)} = 0.6s^3 - 0.2s^3 \textcolor{red}{+\, 0.9s^3}$ $\textcolor{red}{\text{Add the opposite of } -0.9s^3.}$

$\qquad\qquad\qquad\qquad\qquad\quad = 1.3s^3$ $\textcolor{red}{\text{Think: } (0.6 - 0.2 + 0.9)s^3 = 1.3s^3.}$

c. $-\dfrac{1}{2}ab + \dfrac{1}{3}ab = -\dfrac{1}{2} \cdot \dfrac{\textcolor{red}{3}}{\textcolor{red}{3}}ab + \dfrac{1}{3} \cdot \dfrac{\textcolor{red}{2}}{\textcolor{red}{2}}ab$ $\textcolor{red}{\text{Build each fraction into an equivalent fraction}}$
$\textcolor{red}{\text{that has the LCD 6 for its denominator.}}$

$\qquad\qquad\qquad\quad = -\dfrac{3}{6}ab + \dfrac{2}{6}ab$ $\textcolor{red}{\text{Multiply the numerators. Multiply the denominators.}}$

$\qquad\qquad\qquad\quad = -\dfrac{1}{6}ab$ $\textcolor{red}{\text{Think: } \left(-\frac{3}{6} + \frac{2}{6}\right)ab = -\frac{1}{6}ab.}$

Success Tip

We can use the commutative property of addition to reorder the terms of the result. It is standard practice to write such answers in descending powers of the variable.

d. We will combine the n-terms and combine the n^2-terms.

$$\textcolor{red}{16n + 8n^2 - 42n + 4n^2} = \textcolor{red}{-26n + 12n^2} \qquad \textcolor{red}{\text{Think: } (16 - 42)n = -26n}$$
$$\textcolor{red}{\text{and } (8 + 4)n^2 = 12n^2.}$$

$$\qquad\qquad\qquad\qquad\qquad = 12n^2 - 26n \qquad \textcolor{red}{\text{Write the terms of the result}}$$
$$\textcolor{red}{\text{in descending powers of } n.}$$

Self Check 5   Simplify: **a.** $5k + 8k$ **b.** $-600a^2 - (-800a^2) + 100a^2$
c. $\dfrac{2}{3}xy - \dfrac{3}{4}xy$ **d.** $c + 32d^2 - 19c - 20d^2$

Now Try ▶   Problems 65, 67, and 71

EXAMPLE 6   Simplify: **a.** $20b^2 - 5(3b^2 + 1) + 8$ **b.** $6\left(\dfrac{3}{2}d - \dfrac{4}{3}\right) + 6\left(\dfrac{5}{6}d\right)$

Strategy   We will use the distributive property to remove parentheses and then combine any like terms.

Why   Since we cannot simplify the expressions within the parentheses, we will perform the indicated multiplication.

Solution   **a.** $20b^2 - 5(3b^2 + 1) + 8 = 20b^2 - 15b^2 - 5 + 8$ $\textcolor{red}{\text{Distribute the multiplication by } -5.}$

$\qquad\qquad\qquad\qquad\qquad\quad = 5b^2 + 3$ $\textcolor{red}{\text{Combine like terms: } 20b^2 - 15b^2 = 5b^2 \text{ and}}$
$\textcolor{red}{-5 + 8 = 3.}$

b. $\textcolor{red}{6}\left(\dfrac{3}{2}d - \dfrac{4}{3}\right) + 6\left(\dfrac{5}{6}d\right) = \textcolor{red}{6}\left(\dfrac{3}{2}d\right) - \textcolor{red}{6}\left(\dfrac{4}{3}\right) + 6\left(\dfrac{5}{6}d\right)$ $\textcolor{red}{\text{Distribute the multiplication by 6.}}$

$\qquad\qquad\qquad\qquad\quad = \dfrac{18}{2}d - \dfrac{24}{3} + \dfrac{30}{6}d$ $\textcolor{red}{\text{Perform each multiplication.}}$

$\qquad\qquad\qquad\qquad\quad = 9d - 8 + 5d$ $\textcolor{red}{\text{Simplify each fraction.}}$

$\qquad\qquad\qquad\qquad\quad = 14d - 8$ $\textcolor{red}{\text{Combine like terms: } 9d + 5d = 14d.}$

Self Check 6 Simplify: **a.** $44a^3 - 2(10a^3 - a^2) - a^2$

b. $12\left(\frac{5}{6}r^2 + \frac{1}{4}r\right) + 12\left(\frac{2}{3}r\right)$

Now Try ▶ Problems 79 and 85

EXAMPLE 7 Simplify: $3x + 4[6x - 2(7x + 8)]$

Strategy We will simplify the expression by working from the innermost grouping symbols (the parentheses) to the outermost grouping symbols (the brackets).

Why To simplify expressions, we follow the order of operations rule.

Solution

$3x + 4[6x - \mathbf{2}(7x + 8)] = 3x + 4[6x - 14x - 16]$ Within the parentheses, distribute the multiplication by -2.

$= 3x + 4[-8x - 16]$ Combine like terms within the brackets: $6x - 14x = -8x$.

$= 3x - 32x - 64$ Distribute the multiplication by 4.

$= -29x - 64$ Combine like terms: $3x - 32x = -29x$.

Self Check 7 Simplify: $8t - 4[10t + 2(2t + 1) - 3]$

Now Try ▶ Problem 87

SECTION 1.4 ▶ STUDY SET

VOCABULARY

Fill in the blanks.

1. A _____ is a product or quotient of numbers and/or variables, such as $6r$, $-t^3$, and $\frac{44}{m}$.
2. The _____ of the term $-8c$ is -8.
3. A term, such as 9, that consists of a single number is called a _____ term.
4. To _____ expressions, we use properties of real numbers to write equivalent expressions in a less complicated form.
5. The _____ properties of real numbers involve changing *order* and the _____ properties of real numbers involve changing *grouping*.
6. We can use the _____ property to remove parentheses in the expression $2(x + 8)$.
7. _____ terms are terms with exactly the same variables raised to exactly the same powers.
8. Simplifying the sum or difference of like terms is called _____ like terms.

CONCEPTS

9. **a.** Using the variables x, y, and z, write the associative property of addition.
   b. Using the variables x and y, write the commutative property of multiplication.
   c. Using the variables r, s, and t, write the distributive property of multiplication over addition.

10. Complete each property of addition. Then give its name.
    a. $a + (-a) = $ ▢
    b. $a + 0 = $ ▢
    c. $a + b = b + $ ▢
    d. $(a + b) + c = a + $ ▢

11. Complete each property of multiplication. Then give its name.
    a. $a \cdot b = b \cdot $ ▢
    b. $(ab)c = $ ▢
    c. $0 \cdot a = $ ▢
    d. $1 \cdot a = $ ▢
    e. $a\left(\frac{1}{a}\right) = $ ▢

12. Complete each property of division.

    a. $\dfrac{a}{1} =$ ▢

    b. $\dfrac{a}{a} =$ ▢

    c. $\dfrac{0}{a} =$ ▢

    d. $\dfrac{a}{0}$ is ▢

13. a. What is the additive identity?

    b. What is the multiplicative identity?

    c. What is the additive inverse (opposite) of x?

    d. What is the multiplicative inverse (reciprocal) of x?

14. What number should be

    a. subtracted from 5 to obtain 0?

    b. added to 5 to obtain 0?

15. By what number should

    a. 5 be divided to obtain 1?

    b. 5 be multiplied to obtain 1?

16. Are the terms listed here like terms? If they are, combine them.

    a. $2x$, $6x$

    b. $-3x$, $5y$

    c. $-5xy$, $-7yz$

    d. $24t^2$, $24t^3$

NOTATION

17. In $-(x - 7)$, what does the negative sign in front of the parentheses represent?

18. Does the distributive property apply?

    a. $2(3)(5)$

    b. $2(3 \cdot 5)$

    c. $2(3x)$

    d. $2(x - 3)$

GUIDED PRACTICE

What are the terms of the expression? Give the coefficient of each term. See Objective 1.

19. $3x^3 + 11x^2 - x + 9$

20. $2y^4 - y^3 + 6y + 4$

21. $\dfrac{11}{12}a^4 - \dfrac{3}{4}b^2 + 25b$

22. $0.78m^3 - 1.55n - 0.99$

Complete each statement so that the indicated property is illustrated. See Example 1.

23. $3 + 7 =$ _____ Commutative property of addition

24. $2(5 \cdot 97) =$ _____ Associative property of multiplication

25. $3(2 + d) =$ _____ Distributive property

26. $1 \cdot y =$ _____ Commutative property of multiplication

27. $c + 0 =$ ___ Additive identity property

28. $-4(x - 2) =$ _____ Distributive property and simplifying

29. $25 \cdot \dfrac{1}{25} =$ ___ Multiplicative inverse property

30. $z + (9 - 27) =$ _____ Commutative property of addition

31. $8 + (7 + a) =$ _____ Associative property of addition

32. ___ $\cdot 3 = 3$ Multiplicative identity property

33. $(x + y)2 =$ _____ Commutative property of multiplication

34. $h + (-h) =$ ___ Additive inverse property

Multiply. See Example 2.

35. $9(8m)$

36. $12n(4)$

37. $5(-9q)$

38. $-3(2t)$

39. $\dfrac{7}{8}x(-56)$

40. $\dfrac{5}{9}r(-45)$

41. $-4(8r)(-2y)$

42. $-6s(-4t)(-1)$

Multiply. See Example 3.

43. $9(9x + 2)$

44. $7(6y + 1)$

45. $-4(-3t + 3)$

46. $-4(-5y + 3)$

47. $-(24 - d)$

48. $-(19 - w)$

49. $\dfrac{2}{3}(3s^2 - 9)$

50. $\dfrac{1}{5}(5b^3 - 15)$

51. $0.7(m + 2n)$

52. $2.5(6c - 8d)$

53. $100(0.09x + 0.02y)$

54. $100(8.36x - 2.75y)$

Multiply. See Example 4.

55. $5(9t^2 - 12t - 3)$

56. $25(2a^2 - 3a + 1)$

57. $3\left(\dfrac{4}{3}x - \dfrac{5}{3}y + \dfrac{1}{3}\right)$

58. $6\left(-\dfrac{4}{3} + \dfrac{7}{6}s + \dfrac{16}{3}t\right)$

59. $(16t + 24)\dfrac{1}{8}$

60. $(18q + 9)\dfrac{1}{9}$

61. $(y - 2)(-3)$

62. $(2t + 5)(-2)$

Simplify by combining like terms. See Example 5.

63. $3x + 15x$

64. $12y - 17y$

65. $0.7h - 3.8h$

66. $-5.7m + 5.3m$

67. $1.8x^2 - 5.1x^2 + 4.1x^2$

68. $3.7x^2 + 3.3x^2 - 1.1x^2$

69. $-8x + 5x - (-x)$

70. $-20y + 3y - (-6y)$

71. $\dfrac{2}{5}ab - \left(-\dfrac{1}{2}ab\right)$

72. $-\dfrac{3}{4}st - \dfrac{1}{3}st$

73. $\dfrac{3}{5}t + \dfrac{1}{3}t$

74. $\dfrac{3}{16}x - \dfrac{5}{4}x$

75. $-9a + 11ad - 35a + ad$

76. $-7a + 2ab - 7a + 12ab$

77. $4m - t - (-2m) + 3t$

78. $14g + h - (-g) - 8h$

Simplify. See Example 6.

79. $2x^2 + 4(3x - x^2) + 3x$

80. $3p^2 - 6(5p^2 + p) + p^2$

81. $-3(p - 2) + 2(p + 3) - 5(p - 1)$

82. $5(q + 7) - 3(q - 1) - (q + 2)$

83. $36\left(\dfrac{2}{9}x - \dfrac{3}{4}\right) + 36\left(\dfrac{1}{2}\right)$

84. $40\left(\dfrac{3}{8}y - \dfrac{1}{4}\right) + 40\left(\dfrac{4}{5}\right)$

85. $24\left(\dfrac{5}{6}y - \dfrac{9}{8}\right) - 24\left(\dfrac{3}{24}y\right)$

86. $18\left(\dfrac{11}{18}w - \dfrac{7}{2}\right) - 18\left(\dfrac{1}{9}w\right)$

Simplify. See Example 7.

87. $3[2(x + 2)] - 5[3(x - 5)] + 5x$

88. $-5[3(x - 4) - 2(x + 2)] - 7(x - 3)$

89. $2\left[6\left(\dfrac{1}{3}a + 2b\right) - 8\left(\dfrac{1}{4}a - 2b\right) + 3\right]$

90. $10\left[\dfrac{3}{5}(2s + 2t) - \dfrac{4}{5}(s - t) + 1\right]$

TRY IT YOURSELF

Simplify each expression.

91. $-(a + 2A + 1) - (a - A + 2)$
92. $3T - 2(t - T) + t$
93. $8(2cd + 7c) - 2(cd - 3c)$
94. $2tz + 5(tz - 4) - 10(8 - tz)$
95. $6.4a^2 + 11.8a - 9.2a + 5.7$
96. $9.1m^2 - 6.1m + 12.3m - 4.9$
97. $-\dfrac{7}{16}x - \dfrac{3}{4}x$ 98. $-\dfrac{5}{9}y - \dfrac{7}{18}y$
99. $-2[4(z - 9) - 6(3z - 7)] - 7(2z - 1)$
100. $9(m^3 + 3) - 5(3 - m^3) - 8(-1 - m^3)$
101. $21\left(\dfrac{6}{7}h^2 - \dfrac{15}{21}h\right) + 21\left(\dfrac{1}{3}h\right)$
102. $\dfrac{1}{12}(y - 12x) - \dfrac{1}{3}(y - 3x)$
103. $4.3(y + 9) - 8.1y$ 104. $2.1(4 + 5z) + 0.9z$
105. $3x^2 - (-2x^2) - 5x^2$ 106. $8x^3 - x^3 - (-2x^3)$
107. $19a - \{-2[4a - 2(a - 16)] - 3a\}$
108. $41m - \{-3[-2m - 7(m + 1)] - 6m\}$
109. $\dfrac{1}{2}(4a - 8) - 6[2(5a - 1) - a]$
110. $\dfrac{1}{3}(6t - 9) - 12[3(2t - 1) - t]$

Look Alikes . . .

111. **a.** $12(8n)5$ **b.** $12(8n + 5)$
112. **a.** $-3(-4t)(-2)$ **b.** $-3(-4t) - 2$
113. **a.** $6a + 6a + 6a$ **b.** $6a + 6b + 6c$
114. **a.** $9x - 2(-3x + 4)$ **b.** $9x - 2(-3x) + 4$

APPLICATIONS

115. **Parking Areas.**   Refer to the illustration below.
    a. Express the area of the entire parking lot as the product of its length and width.
    b. Express the area of the entire lot as the sum of the areas of the self-parking space and the valet parking space.
    c. Write an equation that shows that your answers to parts (a) and (b) are equal. What property of real numbers is illustrated by this example?

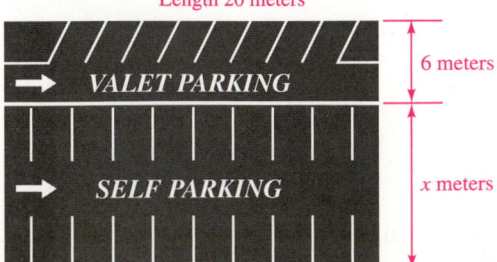

Length 20 meters

6 meters

VALET PARKING

SELF PARKING

x meters

116. **Checking Accounts.**   To find the total dollar amount of the checks entered in the register below, we could add the check amounts in the order in which they are written: $\$39 + \$75 + \$34 + \$25 + \$111 + \$16$. Write an expression with the amounts reordered and grouped in such a way that the addition is easier. Then find the sum. What properties of real numbers did you use?

Number	Date	Description of Transaction	Payment/Debit	
101	3/6	DR. OKAMOTO, DDS	$39	00
102	3/6	UNION OIL CO.	$75	00
103	3/8	STATER BROS.	$34	00
104	3/9	LITTLE LEAGUE	$25	00
105	3/11	NORDSTROM	$111	00
106	3/12	OFFICE MAX	$16	00

WRITING

117. Explain why the distributive property does not apply when simplifying $6(2 \cdot x)$.
118. In each case, explain what you can conclude about one or both of the numbers.
    a. When the two numbers are added, the result is 0.
    b. When the two numbers are subtracted, the result is 0.
    c. When the two numbers are multiplied, the result is 0.
    d. When the two numbers are divided, the result is 0.
119. What are like terms?
120. Use each of the words *commute*, *associate*, and *distribute* in a sentence in which the context is nonmathematical.

REVIEW

Evaluate each expression.

121. $\left(-\dfrac{3}{2}\right)\left(\dfrac{7}{12}\right)$ 122. $\dfrac{1}{2} - \left(-\dfrac{4}{5}\right)$
123. $-3|4 - 8| + (4 + 2 \cdot 3)^3$
124. $\left(\dfrac{-\sqrt{4^3} - 5^2}{2 \cdot 2^2 - (1^9 - 4)}\right)^3$

CHALLENGE PROBLEMS

125. Simplify: $\dfrac{x}{2} + \dfrac{x}{3} + \dfrac{x}{4} + \dfrac{x}{5} + \dfrac{x}{6}$
126. Fill in the blank:
    a. $\boxed{}(0.005x + 0.02y - 0.0003z) = 50x + 200y - 3z$
    b. $\boxed{}\left(\dfrac{x}{54} - \dfrac{1}{42}\right) = 7x - 9$
127. What two real numbers are their own reciprocals?
128. Explain how the distributive property can be used to evaluate the expression $52.713(21) + 52.713(79)$ mentally.

Solving Linear Equations Using Properties of Equality

ARE YOU READY?

The following problems review some basic skills that are needed when solving equations.

1. Simplify: $-9x - 6 + 9x$

2. Simplify: $10a + 3 - 3$

3. Simplify: $5m - 2(8m - 4)$

4. Multiply: $\dfrac{8}{3}\left(\dfrac{3}{8}x\right)$

5. Multiply: $21\left(\dfrac{6}{7}n\right)$

6. Multiply: $100 \cdot 0.27$

One of the most useful concepts in algebra is the equation. Writing and then solving an equation is a powerful problem-solving strategy. In this section, we will review some fundamental properties that are used to solve equations.

1 Determine Whether a Number Is a Solution.

An **equation** is a statement indicating that two expressions are equal. All equations contain an equal symbol $=$. An example of an equation is $7x - 3 = 4$. The equal symbol separates the equation into two parts: The expression $7x - 3$ is the **left side** and 4 is the **right side.** The letter x is the **variable** (or the **unknown**). Since the sides of an equation can be reversed, we can write $7x - 3 = 4$ or $4 = 7x - 3$.

- An equation can be true: $6 + 3 = 9$.
- An equation can be false: $2 + 4 = 7$.
- An equation can be neither true nor false. For example, $7x - 3 = 4$ is neither true nor false because we don't know what number x represents.

An equation that contains a variable is made true or false by substituting a number for the variable. A number that makes an equation true when substituted for the variable is called a **solution,** and it is said to *satisfy* the equation. The **solution set** of an equation is the set of all numbers that make the equation true.

The Language of Algebra

It is important to know the difference between an **equation** and an **expression**. An equation contains an $=$ symbol; an expression does not.

EXAMPLE 1   Check to determine whether 3 is a solution of $3x + 2 = 2x + 5$.

Strategy   We will substitute 3 for each x in the equation and evaluate the expressions on the left side and the right side separately.

Why   If a true statement results, 3 is a solution of the equation. If we obtain a false statement, 3 is not a solution.

Solution

The Language of Algebra

Read $\stackrel{?}{=}$ as "is possibly equal to."

Evaluate the expression on the left side.

$$3x + 2 = 2x + 5 \qquad \text{This is the original equation.}$$
$$3(3) + 2 \stackrel{?}{=} 2(3) + 5 \qquad \text{Substitute 3 for x.}$$
$$9 + 2 \stackrel{?}{=} 6 + 5 \qquad \text{Multiply.}$$
$$11 = 11 \qquad \text{True}$$

Evaluate the expression on the right side.

Since the resulting statement $11 = 11$ is true, 3 satisfies the equation, and we say that 3 is a solution of $3x + 2 = 2x + 5$.

Self Check 1   Is -4 a solution of $2x - 5 = 3x$?

Now Try ▶ Problem 15

2 Use Properties of Equality to Solve Equations.

Usually, we do not know the solutions of an equation—we need to find them. In this course, we will discuss how to solve many different types of equations. The easiest equations to solve are *linear equations in one variable.*

Linear Equations	A linear equation in one variable can be written in the form
	$ax + b = c$ where a, b, and c are real numbers, and $a \neq 0$.

Some examples of linear and nonlinear equations in one variable are shown below. Notice that for the linear equations the highest power on the variable is 1.

Linear equations in one variable (x)

$5x + 3 = 8$ Think: $5x^1 + 3 = 8$

$-\dfrac{9}{4}x - 2 = 0$ Think: $-\dfrac{9}{4}x^1 - 2 = 0$

Nonlinear equations in one variable (x)

$x^2 - x - 20 = 0$ The exponent on x is not 1.

$\dfrac{1}{15} + \dfrac{4}{x} = \dfrac{2}{5}$ x is in the denominator.

When solving linear equations, the objective is to *isolate* the variable on one side of the equation. This is achieved by undoing the operations performed on the variable. As we undo the operations, we produce a series of simpler equations, all having the same solutions. Such equations are called *equivalent equations.*

Equivalent Equations	Equations with the same solutions are called **equivalent equations**.

The solution of the equation $x = 2$ is obviously 2, because replacing x with 2 yields a true statement, $2 = 2$. The equation $x + 4 = 6$ also has a solution of 2. Since $x = 2$ and $x + 4 = 6$ have the same solution, they are equivalent equations.

We can use the following properties to write equivalent equations.

Properties of Equality	Adding the same number to, or subtracting the same number from, both sides of an equation does not change the solution.
	If a, b, and c are real numbers and $a = b$,
	$a + c = b + c$ **Addition property of equality**
	$a - c = b - c$ **Subtraction property of equality**
	Multiplying or dividing both sides of an equation by the same nonzero number does not change the solution.
	If a, b, and c are real numbers with $c \neq 0$, and $a = b$,
	$ca = cb$ **Multiplication property of equality**
	$\dfrac{a}{c} = \dfrac{b}{c}$ **Division property of equality**

To solve linear equations in one variable, we use properties of equality to write a series of simpler equivalent equations. In this process, we first isolate the **variable term** and then the variable itself on one side of the equation.

EXAMPLE 2   Solve: **a.** $2x - 8 = 0$ **b.** $-35.6 = 77.89 - x$

Strategy   We will use a property of equality to isolate the *variable term* on one side of the equation and then use another property to isolate the *variable*.

Why   To solve the original equation, we want to find a simpler equivalent equation of the form $x = $ **a number**, whose solution is obvious.

Solution   **a.** We note that x is multiplied by 2, and then 8 is subtracted from that product. To isolate x on the left side of the equation, we use the order of operations rule in reverse.

- To undo the subtraction of 8, we add 8 to both sides.
- To undo the multiplication by 2, we divide both sides by 2.

The Language of Algebra

Since division by 2 is the same as multiplication by $\frac{1}{2}$, we can also solve $2x = 8$ using the multiplication property of equality. To isolate x we can multiply both sides by the **multiplicative inverse** of 2, which is $\frac{1}{2}$:

$$2x = 8$$
$$\frac{1}{2} \cdot 2x = \frac{1}{2} \cdot 8$$
$$x = 4$$

$$2x - 8 = 0 \qquad \text{This is the equation to solve.}$$
$$2x - 8 + 8 = 0 + 8 \qquad \text{Use the addition property of equality:}$$
$$\text{Add 8 to both sides to isolate the variable term } 2x.$$
$$2x = 8 \qquad \text{Simplify both sides of the equation.}$$
$$\frac{2x}{2} = \frac{8}{2} \qquad \text{Use the division property of equality:}$$
$$\text{Divide both sides by 2 to isolate the variable } x.$$
$$x = 4 \qquad \text{Do the division.}$$

Check: We substitute 4 for x to verify that it satisfies the original equation.

$$2x - 8 = 0$$
$$2(4) - 8 \stackrel{?}{=} 0 \qquad \text{Substitute 4 for } x.$$
$$8 - 8 \stackrel{?}{=} 0 \qquad \text{Multiply.}$$
$$0 = 0 \qquad \text{True}$$

Since we obtain a true statement, 4 is the solution of $2x - 8 = 0$ and the solution set is $\{4\}$.

The Language of Algebra

Since subtracting 77.89 is the same as adding -77.89, we can also solve the equation using the addition property of equality. To isolate $-x$, we can add the **additive inverse** of 77.89, which is -77.89, to both sides:

$$-35.6 + (-77.89) =$$
$$77.89 - x + (-77.89)$$

b.
$$-35.6 = 77.89 - x \qquad \text{This is the equation to solve.}$$
$$-35.6 - 77.89 = 77.89 - x - 77.89 \qquad \text{Use the subtraction property of equality:}$$
$$\text{Subtract 77.89 from both sides to isolate the variable term, } -x.$$
$$-113.49 = -x \qquad \text{Simplify each side of the equation.}$$

The variable x is not yet isolated, because there is a $-$ sign in front of it. Since the term $-x$ has an understood coefficient of -1, we can write $-x$ as $-1x$. To isolate x, we can either multiply or divide both sides by -1.

$$-113.49 = -1x \qquad \text{Write } -x = -1x.$$
$$\frac{-113.49}{-1} = \frac{-1x}{-1} \qquad \text{Use the division (or multiplication) property of equality:}$$
$$\text{Divide (or multiply) both sides by } -1 \text{ to isolate } x.$$
$$113.49 = x \qquad \text{Simplify each side of the equation.}$$
$$x = 113.49 \qquad \text{If you like, you can reverse the sides of the equation}$$
$$\text{so that } x \text{ is on the left.}$$

Verify that 113.49 is the solution by checking it in the original equation.

Self Check 2   Solve: **a.** $3a + 15 = 0$ **b.** $-1.3 = -2.6 - x$

Now Try ▶ Problems 19 and 29

EXAMPLE 3 Solve: $\dfrac{3}{4}y = -7$

Strategy We will isolate y by multiplying both sides of the equation by $\frac{4}{3}$.

Why On the left side, y is multiplied by $\frac{3}{4}$. We can undo the multiplication by dividing both sides by $\frac{3}{4}$. Since division by $\frac{3}{4}$ is equivalent to multiplication by its *reciprocal*, it is easier to isolate y by multiplying both sides by $\frac{4}{3}$.

Solution

$$\frac{3}{4}y = -7 \qquad \text{This is the equation to solve.}$$

$$\frac{4}{3}\left(\frac{3}{4}y\right) = \frac{4}{3}(-7) \qquad \begin{array}{l}\text{Use the multiplication property of equality to isolate y.}\\ \text{Multiply both sides by the reciprocal of } \frac{3}{4}\text{, which is } \frac{4}{3}.\end{array}$$

$$\left(\frac{4}{3}\cdot\frac{3}{4}\right)y = \frac{4}{3}(-7) \qquad \text{Use the associative property of multiplication to regroup.}$$

$$1y = \frac{4}{3}(-7) \qquad \text{The product of a number and its reciprocal is 1: } \frac{4}{3}\cdot\frac{3}{4}=1.$$

$$y = -\frac{28}{3} \qquad \text{On the left side, 1y = y. On the right side, multiply.}$$

Notation

Variable terms with fractional coefficients can be written in two ways. For example, $\frac{3}{4}y = \frac{3y}{4}$ and $\frac{2}{3}b = \frac{2b}{3}$.

Caution

After checking a result, be careful when stating your conclusion. Here, it would be incorrect to say:

~~The solution is −7.~~

The number we were checking was $-\dfrac{28}{3}$, not −7.

Check:

$$\frac{3}{4}y = -7 \qquad \text{This is the original equation.}$$

$$\frac{3}{4}\left(-\frac{28}{3}\right) \stackrel{?}{=} -7 \qquad \text{Substitute } -\frac{28}{3} \text{ for y.}$$

$$-\frac{\overset{1}{3}\cdot\overset{1}{4}\cdot 7}{\underset{1}{4}\cdot\underset{1}{3}} \stackrel{?}{=} -7 \qquad \begin{array}{l}\text{Multiply the numerators and the denominators.}\\ \text{Factor 28 as } 4\cdot 7 \text{ and simplify by removing}\\ \text{common factors of the numerator and denominator.}\end{array}$$

$$-7 = -7 \qquad \text{True}$$

Since the resulting statement is true, the solution is $-\frac{28}{3}$ and the solution set is $\left\{-\frac{28}{3}\right\}$.

Self Check 3 Solve: $\dfrac{2}{3}b = -15$

Now Try ▶ Problem 31

The equation in Example 3 can be solved using an alternate two-step approach.

$$\frac{3}{4}y = -7 \qquad \text{This is the equation to solve.}$$

$$4\left(\frac{3}{4}y\right) = 4(-7) \qquad \text{Multiply both sides by 4 to undo the division by 4.}$$

$$3y = -28 \qquad \text{Simplify: } 4\left(\frac{3}{4}y\right) = \frac{4}{1}\left(\frac{3}{4}y\right) = \frac{\overset{1}{4}\cdot 3}{1\cdot\underset{1}{4}}y = 3y.$$

$$\frac{3y}{3} = \frac{-28}{3} \qquad \text{To isolate y, undo the multiplication by 3 by dividing both sides by 3.}$$

$$y = -\frac{28}{3} \qquad \text{This is the same result as that in Example 3.}$$

3   Simplify Expressions to Solve Equations.

To solve more complicated equations, we often need to combine like terms.

EXAMPLE 4

Solve: $4b - 7 + 2b = 1 + 2b + 8$

Strategy   We will combine like terms on each side of the equation and then eliminate $2b$ from the right side by subtracting $2b$ from both sides.

Why   To solve for b, all the terms containing b must be on the same side of the equation.

Solution

$$4b - 7 + 2b = 1 + 2b + 8 \qquad \text{This is the equation to solve.}$$
$$6b - 7 = 2b + 9 \qquad \text{Combine like terms: } 4b + 2b = 6b \text{ and } 1 + 8 = 9.$$

Success Tip

Recall that:
Subtraction undoes addition.
Addition undoes subtraction.
Division undoes multiplication.
Multiplication undoes division.

We note that terms involving b appear on both sides of the equation. To isolate b on the left side, we need to eliminate $2b$ on the right side.

$$6b - 7 = 2b + 9$$
$$6b - 7 - 2b = 2b + 9 - 2b \qquad \text{Subtract } 2b \text{ from both sides.}$$
$$4b - 7 = 9 \qquad \begin{array}{l}\text{Combine like terms on each side:} \\ 6b - 2b = 4b \text{ and } 2b - 2b = 0.\end{array}$$
$$4b - 7 + 7 = 9 + 7 \qquad \begin{array}{l}\text{To isolate the variable term } 4b, \text{ undo the} \\ \text{subtraction of 7 by adding 7 to both sides.}\end{array}$$
$$4b = 16 \qquad \text{Simplify each side of the equation.}$$
$$\frac{4b}{4} = \frac{16}{4} \qquad \begin{array}{l}\text{To isolate } b, \text{ undo the multiplication} \\ \text{by 4 by dividing both sides by 4.}\end{array}$$
$$b = 4$$

Caution

When checking solutions, always use the original equation.

Check:
$$4b - 7 + 2b = 1 + 2b + 8 \qquad \text{This is the original equation.}$$
$$4(4) - 7 + 2(4) \overset{?}{=} 1 + 2(4) + 8 \qquad \text{Substitute 4 for } b.$$
$$16 - 7 + 8 \overset{?}{=} 1 + 8 + 8 \qquad \text{Multiply.}$$
$$17 = 17 \qquad \text{True}$$

Since the resulting statement is true, the solution is 4 and the solution set is $\{4\}$.

Self Check 4   Solve: $-6t - 16 + 6t = 1 + 2t - 5$

Now Try ▶ Problem 45

EXAMPLE 5

Solve: **a.** $d - 3(d - 7) = 2(4d + 10)$ **b.** $6[x - (2 - x)] = -4(8x + 3)$

Strategy   We will use the distributive property to remove all sets of parentheses (and brackets), simplify each side of the equation by combining like terms, and isolate the variable.

Why   It's best to simplify each side of an equation before attempting to isolate the variable.

Solution   **a.**

$$d - 3(d - 7) = 2(4d + 10)$$ This is the equation to solve.

$$d - 3d + 21 = 8d + 20$$ Distribute the multiplication by -3 and by 2.

$$-2d + 21 = 8d + 20$$ Combine like terms: $d - 3d = -2d$.

$$-2d + 21 + 2d = 8d + 20 + 2d$$ To eliminate the term $-2d$ on the left side, add 2d to both sides.

$$21 = 10d + 20$$ Combine like terms on both sides: $-2d + 2d = 0$.

$$21 - 20 = 10d + 20 - 20$$ To undo the addition of 20, subtract 20 from both sides.

$$1 = 10d$$ Combine like terms on both sides: $20 - 20 = 0$.

$$\frac{1}{10} = \frac{10d}{10}$$ To isolate d, undo the multiplication by 10 by dividing both sides by 10.

$$\frac{1}{10} = d$$ Do the division.

> ### Success Tip
>
> We could have eliminated $8d$ from the right side by subtracting $8d$ from both sides:
>
> $$-2d + 21 = 8d + 20$$
> $$-2d + 21 - 8d = 8d + 20 - 8d$$
> $$-10d + 21 = 20$$
>
> However, it is usually easier to isolate the variable term on the side that will result in a *positive* coefficient, as we did.

Check: To simplify the calculations, we can use the decimal equivalent of $\frac{1}{10}$, which is 0.1, in the check.

$$d - 3(d - 7) = 2(4d + 10)$$ This is the original equation.

$$0.1 - 3(0.1 - 7) \overset{?}{=} 2[4(0.1) + 10]$$ Substitute 0.1 for d.

$$0.1 - 3(-6.9) \overset{?}{=} 2[0.4 + 10]$$

$$0.1 + 20.7 \overset{?}{=} 2(10.4)$$

$$20.8 = 20.8$$ True

Since the resulting statement is true, the solution is $\frac{1}{10}$ or 0.1.

b. To simplify the expression on the left side of the equation, we will work from the innermost grouping symbols (the parentheses) to the outermost grouping symbols (the brackets).

$$6[x - (2 - x)] = -4(8x + 3)$$ This is the equation to solve.

$$6[x - 2 + x] = -32x - 12$$ Within the brackets, distribute the multiplication by -1. On the right side, distribute the multiplication by -4.

$$6[2x - 2] = -32x - 12$$ Combine like terms within the brackets.

$$12x - 12 = -32x - 12$$ Distribute the multiplication by 6.

$$12x - 12 + 32x = -32x - 12 + 32x$$ To eliminate the term $-32x$ on the right side, add 32x to both sides.

$$44x - 12 = -12$$ Combine like terms.

$$44x - 12 + 12 = -12 + 12$$ To isolate the variable term 44x, undo the subtraction of 12 by adding 12 to both sides.

$$44x = 0$$ Simplify each side of the equation.

$$\frac{44x}{44} = \frac{0}{44}$$ To isolate the variable x, undo the multiplication by 44 by dividing both sides by 44.

$$x = 0$$ Do the division.

Verify that 0 is the solution by checking it in the original equation.

Self Check 5   Solve: **a.** $y + 9(y - 5) = 5(4y + 1)$
b. $10[h - (1 - 3h)] = -2(5h + 25)$

Now Try ▶   Problems 51 and 53

In general, we will follow these steps to solve linear equations in one variable. Not every step is needed to solve every equation.

Strategy for Solving Linear Equations in One Variable	1. **Clear the equation of fractions or decimals:** Multiply both sides by the LCD to clear fractions or multiply both sides by a power of 10 to clear decimals.
	2. **Simplify each side of the equation:** Use the distributive property to remove parentheses and combine like terms on each side.
	3. **Isolate the variable term on one side:** Add (or subtract) to get the variable term on one side of the equation and a number on the other using the addition (or subtraction) property of equality.
	4. **Isolate the variable:** Multiply (or divide) to isolate the variable using the multiplication (or division) property of equality.
	5. **Check the result:** Substitute the proposed solution for the variable in the *original* equation to see if a true statement results.

4 Clear Equations of Fractions and Decimals.

Since equations are often easier to solve when they don't contain fractions, we will use the multiplication property of equality to clear an equation of fractions before we solve it. To do so, we will multiply both sides of the equation by the least common denominator of the fractions contained within the equation.

EXAMPLE 6   Solve: $\dfrac{1}{3}(2x - 1) = \dfrac{5}{4}x + \dfrac{31}{12}$

Strategy   We will follow the steps of the equation-solving strategy.

Why   This is the best way to solve a linear equation in one variable.

Solution   **Step 1:**   We can clear the equation of fractions by multiplying both sides by the least common denominator (LCD) of $\frac{1}{3}$, $\frac{5}{4}$, and $\frac{31}{12}$, which is 12.

$$\frac{1}{3}(2x - 1) = \frac{5}{4}x + \frac{31}{12} \qquad \text{This is the equation to solve.}$$

Success Tip

Before multiplying both sides of an equation by the LCD, frame the left side and frame the right side with parentheses or brackets.

$$12\left[\frac{1}{3}(2x - 1)\right] = 12\left[\frac{5}{4}x + \frac{31}{12}\right] \qquad \begin{array}{l}\text{To eliminate the fractions, multiply} \\ \text{both sides by the LCD, 12.}\end{array}$$

$$4(2x - 1) = 12 \cdot \frac{5}{4}x + 12 \cdot \frac{31}{12} \qquad \begin{array}{l}\text{On the left side, multiply: } 12 \cdot \frac{1}{3} = 4. \text{ Don't} \\ \text{``distribute'' the 12 over } \frac{1}{3} \text{ and } (2x - 1). \text{ On the} \\ \text{right side, distribute the multiplication by 12.}\end{array}$$

$$4(2x - 1) = 15x + 31 \qquad \text{Perform the multiplication on the right side.}$$

Step 2:   We remove parentheses.

$$8x - 4 = 15x + 31 \qquad \text{Distribute the multiplication by 4.}$$

Step 3:   To get the variable term on the right side and the constant on the left side, subtract $8x$ and 31 from both sides.

$$8x - 4 \mathbf{- 8x - 31} = 15x + 31 \mathbf{- 8x - 31}$$

$$-35 = 7x \qquad \text{Simplify each side of the equation.}$$

Step 4:   To isolate the variable, undo the multiplication by 7 by dividing both sides by 7.

$$\frac{-35}{7} = \frac{7x}{7}$$ Divide both sides by 7.

$$-5 = x$$ Do the division.

Step 5:   We check by substituting -5 for x in the original equation and simplifying:

$$\frac{1}{3}(2x - 1) = \frac{5}{4}x + \frac{31}{12}$$

$$\frac{1}{3}[2(-5) - 1] \overset{?}{=} \frac{5}{4}(-5) + \frac{31}{12}$$

$$\frac{1}{3}[-11] \overset{?}{=} -\frac{25}{4} + \frac{31}{12}$$

$$-\frac{11}{3} \overset{?}{=} -\frac{75}{12} + \frac{31}{12}$$

$$-\frac{11}{3} \overset{?}{=} -\frac{44}{12}$$

$$-\frac{11}{3} = -\frac{11}{3}$$ True

Since the resulting statement is true, the solution is -5.

Self Check 6   Solve: $\frac{1}{6}(4x + 10) = \frac{1}{9}x - \frac{5}{3}$

Now Try ▶ Problem 57

EXAMPLE 7   Solve: $\frac{8}{5} - \frac{x + 9}{2} = \frac{x + 2}{5} - 4x$

Strategy   We will follow the steps of the equation-solving strategy.

Why   This is the best way to solve a linear equation in one variable.

Solution   Some of the steps used to solve an equation can be done in your head, as you will see in this example.

$$\frac{8}{5} - \frac{x + 9}{2} = \frac{x + 2}{5} - 4x$$ This is the equation to solve.

$$10\left(\frac{8}{5} - \frac{x + 9}{2}\right) = 10\left(\frac{x + 2}{5} - 4x\right)$$ To clear the equation of the fractions, multiply both sides by the LCD, 10.

$$10 \cdot \frac{8}{5} - 10 \cdot \frac{x + 9}{2} = 10 \cdot \frac{x + 2}{5} - 10 \cdot 4x$$ On each side, distribute the 10.

$$2(8) - 5(x + 9) = 2(x + 2) - 40x$$ Perform each multiplication by 10. Don't forget the parentheses.

$$16 - 5x - 45 = 2x + 4 - 40x$$ On each side, distribute.

$$-5x - 29 = -38x + 4$$ On each side, combine like terms.

$$33x = 33$$ To isolate the variable term on the right side, add 38x and 29 to both sides. Since these steps can be done mentally, we don't show them.

$$x = 1$$ To isolate x, divide both sides by 33. This step is also done mentally.

The solution is 1. Check by substituting it for x in the original equation.

Self Check 7 Solve: $\dfrac{a+3}{2} + 2a = \dfrac{3}{2} - \dfrac{a+27}{5}$

Now Try ▶ Problem 59

For more complicated equations involving decimals, we can multiply both sides of the equation by a power of 10 to clear the equation of decimals.

EXAMPLE 8 Solve: $0.04(12) + 0.01x = 0.02(12 + x)$

Strategy To clear the equation of decimals, we will multiply both sides by a carefully chosen power of 10.

Why It's easier to solve an equation that involves only integers.

Solution The equation contains the decimals 0.04, 0.01, and 0.02. Multiplying both sides by $10^2 = 100$ changes the decimals in the equation to integers.

> **Success Tip**
>
> When we write the decimals in the equation as fractions, it becomes more apparent why it is helpful to multiply both sides by the LCD, 100.
>
> $\dfrac{4}{100}(12) + \dfrac{1}{100}x = \dfrac{2}{100}(12 + x)$

$$0.04(12) + 0.01x = 0.02(12 + x) \qquad \text{This is the equation to solve.}$$
$$\mathbf{100}[0.04(12) + 0.01x] = \mathbf{100}[0.02(12 + x)] \qquad \text{To make 0.04, 0.01, and 0.02 integers, multiply both sides by 100.}$$
$$\mathbf{100} \cdot 0.04(12) + \mathbf{100} \cdot 0.01x = \mathbf{100} \cdot 0.02(12 + x) \qquad \text{On the left side, distribute the multiplication by 100.}$$
$$4(12) + 1x = 2(12 + x) \qquad \text{Perform each multiplication by 100.}$$
$$48 + x = 24 + 2x \qquad \text{Distribute.}$$
$$48 + x - 24 - x = 24 + 2x - 24 - x \qquad \text{To isolate the variable term on the right side, subtract 24 and x from both sides.}$$
$$24 = x \qquad \text{Simplify each side.}$$

Verify that 24 is the solution by substituting it for x in the original equation.

Self Check 8 Solve: $0.08x + 0.07(15,000 - x) = 1,110$

Now Try ▶ Problem 63

5 Identify Identities and Contradictions.

Each of the equations that we solved in Examples 1 through 8 had exactly one solution. However, not every equation in one variable has a single solution. Some equations are made true by *any* permissible replacement value for the variable. Such an equation is called an **identity**.

EXAMPLE 9 Solve: $-2(x - 1) - 4 = -4(1 + x) + 2x + 2$

Strategy We will follow the steps of the equation-solving strategy.

Why This is the best way to solve a linear equation in one variable.

> **Success Tip**
>
> We know the given equation is an identity because in Step 3 we see that it is equivalent to the equation $-2x - 2 = -2x - 2$, which is true for all values of x.

Solution
$$-2(x - 1) - 4 = -4(1 + x) + 2x + 2 \qquad \text{This is the equation to solve.}$$
$$-2x + 2 - 4 = -4 - 4x + 2x + 2 \qquad \text{Use the distributive property on both sides.}$$
$$-2x - 2 = -2x - 2 \qquad \text{On each side, combine like terms.}$$
$$-2x - 2 + \mathbf{2x} = -2x - 2 + \mathbf{2x} \qquad \text{To attempt to isolate the variable on one side of the equation, add 2x to both sides.}$$
$$-2 = -2 \qquad \text{True}$$

In the solution process, the terms involving x drop out. The resulting true statement indicates that the original equation is true for every permissible value of x. Therefore, *all real numbers* are solutions and this equation is an identity. Its solution set is written as {all real numbers} or using the symbol $\mathbb{R}$.

Self Check 9 Solve $3(a + 4) + 5 = 2(a - 1) + a + 19$ and give the solution set.

Now Try ▶ Problem 67

Another type of equation, called a **contradiction,** is false for all replacement values for the variable. An example follows.

EXAMPLE 10 Solve: $-6.2(-x - 1) - 4 = 4.2x - (-2x)$

Strategy We will follow the steps of the equation-solving strategy.

Why This is the best way to solve a linear equation in one variable.

Solution

$$-6.2(-x - 1) - 4 = 4.2x - (-2x) \quad \text{This is the equation to solve.}$$

$$6.2x + 6.2 - 4 = 4.2x + 2x \quad \text{On the left side, distribute. On the right side, write the subtraction as addition of the opposite.}$$

$$6.2x + 2.2 = 6.2x \quad \text{On each side, combine like terms.}$$

$$6.2x + 2.2 - \mathbf{6.2x} = 6.2x - \mathbf{6.2x} \quad \text{To attempt to isolate the variable on one side of the equation, subtract 6.2x from both sides.}$$

$$2.2 = 0 \quad \text{False}$$

The Language of Algebra

Contradiction is a form of the word *contradict,* meaning conflicting ideas. During a trial, evidence might be introduced that *contradicts* the testimony of a witness.

In the solution process, the terms involving x drop out. The resulting false statement indicates that no value for x makes the original equation true. Therefore, this equation has *no solution* and it is a contradiction. Its solution set is the **empty set,** which is written as $\{ \ \}$ or using the symbol $\varnothing$.

Self Check 10 Solve: $3(a + 4) + 2 = 2(a - 1) + a + 19$

Now Try ▶ Problem 71

SECTION 1.5 ▶ **STUDY SET**

VOCABULARY

Fill in the blanks.

1. An _____ is a statement that two expressions are equal.
2. $2x + 1 = 4$ is an example of a _____ equation in one variable.
3. A number that makes an equation true when substituted for the variable is called a _____.
4. If two equations have the same solution set, they are called _____ equations.
5. An equation that is made true by any permissible replacement value for the variable is called an _____.
6. An equation that is false for all replacement values for the variable is called a _____.

CONCEPTS

Fill in the blanks.

7. If $a = b$, then $a + c = b +$ ▢ and $a - c = b -$ ▢. Adding (or subtracting) the same number to (or from) _____ sides of an equation does not change the solution.
8. If $a = b$, then $ca =$ ▢ and $\dfrac{a}{c} = \dfrac{b}{▢}$. Multiplying (or dividing) both sides of an equation by the _____ nonzero number does not change the solution.
9. Solve each equation mentally.
   a. $x + 3 = 6$ b. $x - 3 = 6$
   c. $3x = 6$ d. $\dfrac{x}{3} = 6$

10. a. When solving $\frac{x+1}{3} - \frac{2}{15} = \frac{x-1}{5}$, why would we multiply both sides by 15?

    b. When solving $1.45x - 0.5(1 - x) = 0.7x$, why would we multiply both sides by 100?

11. a. Suppose you solve a linear equation in one variable, the variable drops out, and you obtain $8 = 8$. What is the solution set? What symbol is used to represent the solution set?

    b. Suppose you solve a linear equation in one variable, the variable drops out, and you obtain $8 = 7$. What is the solution set? What symbol is used to represent the solution set?

12. a. Simplify: $5y + 2 - 3y$

    b. Solve: $5y + 2 - 3y = 8$

    c. Evaluate $5y + 2 - 3y$ for $y = 8$.

    d. Check: Is -1 a solution of $5y + 2 - 3y = 8$?

NOTATION

Complete the solution to solve the equation. Then check the result.

13.
$$-2(x + 7) = 20$$
$$\boxed{} - 14 = 20$$
$$-2x - 14 + \boxed{} = 20 + \boxed{}$$
$$-2x = 34$$
$$\frac{-2x}{\boxed{}} = \frac{34}{\boxed{}}$$
$$x = -17$$

Check:
$$-2(x + 7) = 20$$
$$-2(\boxed{} + 7) \stackrel{?}{=} 20$$
$$-2(\boxed{}) \quad 20$$
$$\boxed{} = 20$$

The solution is $\boxed{}$.

14. Fill in the blanks to make the statements true.

    a. $-x = \boxed{} x$ **b.** $\frac{2t}{3} = \boxed{} t$

GUIDED PRACTICE

Use a check to determine whether 5 is a solution of each equation. See Example 1.

15. $3x + 2 = 17$ **16.** $7x - 2 = 53 - 5x$

17. $3(2m - 3) = 15$ **18.** $\frac{3}{5}p - 5 = -2$

Solve each equation. Check each result. See Example 2.

19. $2x - 12 = 0$ **20.** $3x - 24 = 0$

21. $8k - 2 = 13$ **22.** $3x + 1 = 3$

23. $\frac{x}{4} - 6 = 1$ **24.** $\frac{m}{3} + 10 = 8$

25. $1.6a + (-4) = 0.032$ **26.** $5.51 = 0.05y + (-9)$

27. $0.7 - 4y = 1.74$ **28.** $0.3 - 2x = -0.92$

29. $-6 - y = -13$ **30.** $-1 - h = -9$

Solve each equation. Check each result. See Example 3.

31. $\frac{2}{3}c = 10$ **32.** $\frac{9}{7}d = 81$

33. $-\frac{4}{5}s = 2$ **34.** $-\frac{9}{8}s = 3$

35. $-\frac{7}{16}w - 26 = -19$ **36.** $-\frac{5}{8}a - 20 = -10$

37. $\frac{5}{6}k - 7.5 = 7.5$ **38.** $\frac{2}{5}c - 12.2 = 1.8$

Solve each equation. Check each result. See Example 4.

39. $8m + 44 = 4m$ **40.** $9n + 36 = 6n$

41. $60t - 50 = 15t - 5$ **42.** $100s - 75 = 50s + 75$

43. $9.8 - 16r = -15.7 - r$ **44.** $15s + 8.1 - 2s = 8.1 - s$

45. $8b - 2 + b = 5 + 5b + 10$ **46.** $w + 7 + 3w = 3 + 10w + 1$

Solve each equation. Check each result. See Example 5.

47. $3(k - 4) = -36$ **48.** $4(x + 6) = 84$

49. $2(a - 5) - (3a + 1) = 0$

50. $8(3a - 5) - 4(2a + 3) = 12$

51. $9(x - 2) = -6(4 - x) + 18$

52. $3(x + 2) - 2 = -(5 + x) + x$

53. $12 + 3(x - 4) - 21 = 5[5 - 4(4 - x)]$

54. $1 + 3[-2 + 6(4 - 2x)] = -(x + 3)$

Solve each equation. Check each result. See Example 6.

55. $\frac{1}{2}(a - 2) = \frac{2}{3}a - 6$ **56.** $\frac{2}{3}(b + 3) = \frac{5}{4}b + \frac{17}{12}$

57. $\frac{1}{2}(3y + 2) - \frac{5}{8} = \frac{3}{4}y$ **58.** $-\frac{3}{4}(4c - 3) + \frac{7}{8}c = \frac{19}{16}$

Solve each equation. Check each result. See Example 7.

59. $\frac{a + 1}{3} - \frac{a - 1}{5} = \frac{8}{15}$

60. $\frac{3 + p}{3} - 4p = 1 - \frac{p + 7}{2}$

61. $\frac{2z + 3}{3} + \frac{3z - 4}{6} = \frac{z - 2}{2}$

62. $\frac{4 - t}{2} - \frac{3t}{5} = 2 + \frac{t + 1}{3}$

Solve each equation. Check each result. See Example 8.

63. $0.45 = 16.95 - 0.25(75 - 3x)$

64. $0.02x + 0.0175(15{,}000 - x) = 277.5$

65. $0.04(12) + 0.01t - 0.02(12 + t) = 0$

66. $0.25(t + 32) = 3.2 + t$

Solve each equation. If the equation is an identity or a contradiction, so indicate. See Example 9.

67. $8x + 3(2 - x) = 5x + 6$

68. $4(2 - 3t) + 6t = -6t + 8$

69. $2(x - 3) = \frac{3}{2}(x - 4) + \frac{x}{2}$

70. $y + \dfrac{1}{2} = \dfrac{5}{2}(0.2y + 1) - \dfrac{1}{2}(4 - y)$

Solve each equation. If the equation is an identity or a contradiction, so indicate. See Example 10.

71. $2x - 6 = -2x + 4(x - 2)$

72. $3(x - 4) + 6 = -2(x + 4) + 5x$

73. $-3x = -2x + 1 - (5 + x)$

74. $5(y + 2) + 7 - 3y = 2(y + 9)$

TRY IT YOURSELF

Solve each equation.

75. $\dfrac{3}{4}x - 5 = \dfrac{2}{3}x + \dfrac{1}{4}$

76. $\dfrac{3}{5}x + \dfrac{7}{10} = x - \dfrac{4}{5}$

77. $8x = x$

78. $-z = 5z$

79. $-x + 12 = -17$

80. $6 = -x + 41$

81. $\dfrac{1}{2}b - \dfrac{19}{6} = \dfrac{1}{3}b + \dfrac{5}{6}$

82. $\dfrac{1}{2}w - \dfrac{7}{6} = \dfrac{53}{6} - \dfrac{1}{3}w$

83. $a + 18 = 5a - 3 + a$

84. $4a - 21 - a = -2a - 7$

85. $-(2t - 0.71) = 0.9(1.4 - t)$

86. $-(9m - 11.13) = 7.7(6 + m)$

87. $2(2x + 1) = x + 15 + 2x$

88. $-2(x + 5) = x + 30 - 2x$

89. $\dfrac{5}{2}a - 12 = \dfrac{1}{3}a + 1$

90. $3(x - 2) + 4 = 3x - 2$

91. $\dfrac{4}{5}a = -12$

92. $4j + 12.54 = 18.12$

93. $0.06(a + 200) + 0.1a = 172$

94. $0.03x + 0.05(6,000 - x) = 280$

95. $-4[p - (3 - p)] = 3(6p - 2)$

96. $2[5(4 - a) + 2(a - 1)] = 3 - a$

97. $2(x - 2) = \dfrac{2}{3}(3x + 8) - 2$

98. $5 - \dfrac{x + 2}{3} = 7 - x$

99. $13.5y + 16.2 = 0$

100. $\dfrac{7}{3}y + 1 = 0$

101. $\dfrac{4}{5}(x + 5) = \dfrac{7}{8}(3x + 23) - 7$

102. $\dfrac{2}{3}(2x + 2) + 4 = \dfrac{1}{6}(5x + 29)$

103. $\dfrac{t - 2}{5} + 5t = \dfrac{7}{5} - \dfrac{t - 2}{2}$

104. $\dfrac{2}{3}(3m - 2) = \dfrac{3}{4}m + \dfrac{11}{12}$

105. $6 + 4t - 1 = 6 - 15t + 12t - 8$

106. $5c - 8 - 3c = 10 + 2c - 3$

Look Alikes . . .

Simplify each expression and solve each equation.

107. a. $\dfrac{1}{2}(6x + 8) - 10 - \dfrac{2}{3}(6x - 9)$

   b. $\dfrac{1}{2}(6x + 8) - 10 = -\dfrac{2}{3}(6x - 9)$

108. a. $6.31w + 9.22 + 5(7.21w - 1.13)$

   b. $6.31w + 9.22 = 5(7.21w - 1.13)$

109. a. $-4\{6x - [3(7x - 1) - x]\} + 46x$

   b. $-4\{6x - [3(7x - 1) - x]\} = 46x$

110. a. $8[4 - (5 + 6r)] - 8r - 11 + 2(4 - 12r)$

   b. $8[4 - (5 + 6r)] - 8r = -11 + 2(4 - 12r)$

WRITING

111. What does it mean to *solve an equation?*

112. Why doesn't the equation $x = x + 1$ have a real-number solution?

113. What is an identity? Give an example.

114. When solving a linear equation in one variable, the objective is to isolate the variable on one side of the equation. What does that mean?

REVIEW

Use variables to state each property of real numbers.

115. a. Commutative property of addition

   b. Associative property of multiplication

   c. Distributive property of multiplication over addition

116. a. Additive inverse property

   b. Multiplicative inverse property

117. a. Additive identity property

   b. Multiplicative identity property

118. a. Division of 0

   b. Division by 0

CHALLENGE PROBLEMS

119. Find the value of k that makes 4 a solution of the following linear equation in x.

$$k + 3x - 6 = 3kx - k + 16$$

120. Solve: $0.75(x - 5) - \dfrac{4}{5} = \dfrac{1}{6}(3x + 1) + 3.2$

Solving Formulas; Geometry

OBJECTIVES

1. Use formulas from geometry.
2. Solve for a specified variable.
3. Solve application problems using formulas.

ARE YOU READY?

The following problems review some basic skills that are needed when working with formulas.

Determine which concept (perimeter, area, volume, or circumference) should be used to find each of the following.

1. The amount of storage in a refrigerator
2. The distance the tip of an airplane propeller travels
3. The distance around a dance floor
4. The amount of floor space to carpet

A **formula** is an equation that states a mathematical relationship between two or more variables. Formulas are used in business, science, banking, and many other fields. A large collection of formulas associated with geometric figures such as squares, rectangles, triangles, circles, and cylinders can be found on the inside back cover of this textbook.

1 Use Formulas from Geometry.

The Language of Algebra

When you hear the word **peri**meter, think of the distance around the "rim" of a flat figure.

To find the **perimeter** of a plane (two-dimensional, flat) geometric figure, we find the distance around the figure by computing the sum of the lengths of the sides. Perimeter is measured in American units of inches, feet, yards, and in metric units such as millimeters, meters, and kilometers. Several perimeter formulas are shown in the margin.

EXAMPLE 1

World Records. The world's largest beach towel was displayed in Las Palmas, Spain, in June 2010. It had a perimeter of 224.68 meters and a length of 87.14 meters. Find the width of the towel. (Source: guinessworldrecords.com)

Strategy Since the towel is rectangular, we will substitute the given values in the formula $P = 2l + 2w$ and solve for w.

Why The variable w represents the unknown width of the towel.

Solution

©ANGEL MEDINA G./epa/Corbis

$P = 2l + 2w$	This is the formula for the perimeter of a rectangle.
$224.68 = 2(87.14) + 2w$	Substitute 224.68 for the perimeter P and 87.14 for the length l.
$224.68 = 174.28 + 2w$	Do the multiplication.
$50.4 = 2w$	To isolate the variable term $2w$, undo the addition of 174.28 by subtracting 174.28 from both sides.
$25.2 = w$	To isolate w, undo the multiplication by 2 by dividing both sides by 2.

The width of the towel is 25.2 meters.

Perimeter formulas

$P = 2l + 2w$ (rectangle)
$P = 4s$ (square)
$P = a + b + c$ (triangle)

Self Check 1

Flags. The largest flag that consistently flies is the flag of Brazil flown in Brasilia, the county's capital. It has a perimeter of 1,116 feet and a length of 328 feet. Find its width.

Now Try ▶ Problem 15

The **area** of a plane (two-dimensional, flat) geometric figure is the amount of surface that it encloses. Area is measured in square units, such as square inches, square feet, square yards, and square meters (written as in.^{2}, ft^2, yd^2, and m^2, respectively). Several area formulas are shown in the margin.

EXAMPLE 2

Area formulas

$A = lw$ (rectangle)
$A = s^2$ (square)
$A = \frac{1}{2}bh$ (triangle)
$A = \frac{1}{2}h(b_1 + b_2)$ (trapezoid)

Patio Covers. A triangular shade sail has an area of 187 ft^2 and the length of its base is 17 feet. Find the height.

Strategy Since the sail is triangular, we will substitute the given values in the formula $A = \frac{1}{2}bh$ and solve for h.

Why The variable h represents the unknown height of the sail.

Solution

$$A = \frac{1}{2}bh \qquad \text{This is the formula for the area of a triangle.}$$

$$187 = \frac{1}{2}(17)h \qquad \text{Substitute 187 for the area } A \text{ and 17 for the length of the base } b.$$

$$2 \cdot 187 = 2 \cdot \frac{1}{2}(17)h \qquad \text{To clear the equation of the fraction, multiply both sides by 2.}$$

$$374 = 17h \qquad \text{Do the multiplication: } 2 \cdot \frac{1}{2} = 1.$$

$$22 = h \qquad \text{To isolate } h, \text{ undo the multiplication by 17 by dividing both sides by 17.}$$

The height of the sail is 22 feet.

Self Check 2 **Triangles.** The area of a triangle is 90 yd^2. If its height is 15 yards, find the length of its base.

Now Try ▶ Problem 25

A **circle** is one of the most important geometric figures of all. A **diameter** of a circle is a chord that passes through its center. A segment drawn from the center of a circle to a point on the circle is called a **radius.** The **circumference** of a circle is the distance around it.

EXAMPLE 3

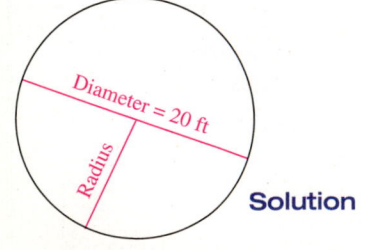

a. Find the circumference of a circle with diameter 20 feet. Round to the nearest tenth of a foot. **b.** Find the area of the circle. Round to the nearest tenth of a square foot.

Strategy We will substitute the values of D and r into the formulas $C = \pi D$ and $A = \pi r^2$ and find C and A.

Why In the formulas, the variable C represents the unknown circumference and A represents the unknown area of the circle.

Solution **a.** To find the circumference C of a circle with diameter D equal to 20 ft, we proceed as follows.

Circle formulas

$D = 2r$ (diameter)
$r = \frac{1}{2}D$ (radius)
$C = 2\pi r$ (circumference)
$C = \pi D$ (circumference)
$A = \pi r^2$ (area)

$$C = \pi D \qquad \text{This is the formula for the circumference of a circle. } \pi D \text{ means } \pi \cdot D.$$
$$C = \pi(20) \qquad \text{Substitute 20 for the diameter } D.$$
$$= 20\pi \qquad \text{The exact circumference of the circle is } 20\pi \text{ ft.}$$
$$\approx 62.83185307 \qquad \text{To use a scientific calculator to approximate the circumference,}$$

enter π ☒ 20 ═ . If you do not have a calculator, use 3.14 as an approximation of π. (Answers may vary slightly depending on which approximation of π is used.)

The circumference is exactly 20π ft. Rounded to the nearest tenth, this is 62.8 ft.

b. The radius r of the circle is one-half the diameter, or 10 feet. To find the area A of the circle, we proceed as follows.

$$A = \pi r^2 \qquad \text{This is the formula for the area of a circle. } \pi r^2 \text{ means } \pi \cdot r^2.$$

$$A = \pi(\mathbf{10})^2 \qquad \text{Substitute 10 for the radius } r.$$

$$= 100\pi \qquad \text{Evaluate the exponential expression. The exact area is } 100\pi \text{ ft}^2.$$

$$\approx 314.1592654 \qquad \text{To use a scientific calculator to approximate the area, enter 100 } \boxed{\times}\ \boxed{\pi}\ \boxed{=}.$$

The area is exactly 100π ft². To the nearest tenth, the area is 314.2 ft². *When finding area, remember to write the appropriate square units in your answer.*

Self Check 3   **Coins.**   The diameter of a U.S. penny is 0.75 inch. Find the circumference and the area (of one side) of a penny. Round to the nearest hundredth.

Now Try ▶ Problems 27 and 31

The **volume** of a three-dimensional geometric solid is the amount of space it encloses. Volume is measured in cubic units, such as cubic inches, cubic feet, and cubic meters (written as in.³, ft³, and m³, respectively). Several volume formulas are shown in the margin.

EXAMPLE 4   **Timers.**   Find the amount of sand in the hourglass.

Strategy   We will substitute the given values into the formula for the volume of a cone, $V = \frac{1}{3}\pi r^2 h$, and find V.

Why   To find the amount of sand in the hourglass, we need to find the amount of space that it occupies by finding a volume.

Solution   The sand is in the shape of a cone whose radius is one-half the diameter of the base of the hourglass and whose height is one-half the height of the hourglass. To find the amount of sand, we substitute 1 for r and 2.5 for h in the formula for the volume of a cone.

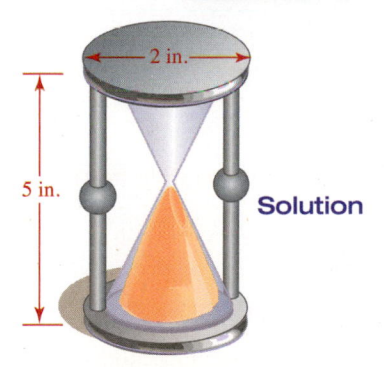

$$V = \frac{1}{3}\pi r^2 h \qquad \text{This is the formula for the volume of a cone.}$$

$$V = \frac{1}{3}\pi(\mathbf{1})^2(\mathbf{2.5}) \qquad \begin{array}{l}\text{Substitute 1 for } r, \text{ the radius of the circular base,}\\ \text{and 2.5 for } h, \text{ the height of the cone.}\end{array}$$

$$V = \frac{2.5\pi}{3} \qquad \text{The exact volume of sand is } \frac{2.5\pi}{3} \text{ in.}^3.$$

$$V \approx 2.617993878 \qquad \text{Use a calculator. Enter: 2.5 } \boxed{\times}\ \boxed{\pi}\ \boxed{\div}\ \boxed{3}\ \boxed{=}.$$

There are exactly $\frac{2.5\pi}{3}$ in.³ of sand in the hourglass. Rounded to the nearest tenth, this is 2.6 in.³. **When finding volume, remember to write the appropriate cubic units in your answer.**

Self Check 4   **Straws.**   Find the volume of a drinking straw that is 250 millimeters long with an inside diameter of 6 millimeters. Round to the nearest cubic millimeter.

Now Try ▶ Problem 35

Volume formulas

$V = lwh$ (rectangular solid)

$V = s^3$ (cube)

$V = \frac{4}{3}\pi r^3$ (sphere)

$V = \pi r^2 h$ (cylinder)

$V = \frac{1}{3}\pi r^2 h$ (cone)

$V = \frac{1}{3}Bh^*$ (pyramid)

*B represents the area of the base.

2 Solve for a Specified Variable.

Real-world applications sometimes call for a formula solved for one variable to be solved for a different variable. To **solve a formula for a specified variable** means to isolate that variable on one side of the equation, with all other variables and constants on the opposite side.

EXAMPLE 5   Solve each formula for the specified variable.

a. $V = lwh$ for w *The formula for the volume of a rectangular solid.*   **b.** $A = \dfrac{1}{2}h(b_1 + b_2)$ for b_1 *The formula for the area of a trapezoid.*

c. $v_f = v_i + at$ for t *A motion formula from physics.*   **d.** $A = P(1 + rt)$ for P *A formula from banking.*

Strategy   To solve for a specified variable, we treat it as if it were the only variable in the equation. To isolate this variable, we will use the same strategy that we used to solve linear equations in one variable. (See page 49 if you need to review the strategy.)

Why   We can solve a formula as if it were an equation in one variable because all the other variables are treated as if they were numbers (constants).

Solution   **a.**

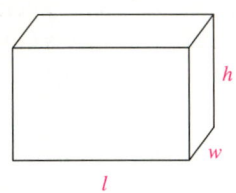
Rectangular solid

⌐— *To solve for w, we will isolate w on this side of the equation.*

$$V = lwh$$

$$\frac{V}{lh} = \frac{lwh}{lh}$$ *To isolate w, undo the multiplication by l and h by dividing both sides by lh.*

$$\frac{V}{lh} = w$$ *On the right side, remove the common factors of l and h:* $\dfrac{l w \overset{1}{\cancel{h}}}{\underset{1}{l}\,\overset{1}{\cancel{h}}} = w.$

$$w = \frac{V}{lh}$$ *Reverse the sides of the equation so that w is on the left.*

b.

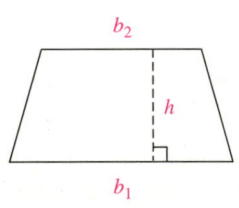
Trapezoid

—————— *To solve for b_1, we will isolate b_1 on this side of the equation.*

$$A = \frac{1}{2}h(b_1 + b_2)$$ *Read b_1 as "b-sub-one" and b_2 as "b-sub-two."*

$$2 \cdot A = 2 \cdot \frac{1}{2}h(b_1 + b_2)$$ *Multiply both sides by 2 to clear the equation of the fraction.*

$$2A = h(b_1 + b_2)$$ *Simplify each side of the equation.*

$$2A = hb_1 + hb_2$$ *Distribute the multiplication by h.*

$$2A - hb_2 = hb_1$$ *Subtract hb_2 from both sides to isolate the variable term hb_1 on the right side. This step is done mentally.*

$$\frac{2A - hb_2}{h} = \frac{hb_1}{h}$$ *To isolate b_1, undo the multiplication by h by dividing both sides by h.*

$$\frac{2A - hb_2}{h} = b_1$$ *On the right side, remove the common factor of h:* $\dfrac{\overset{1}{\cancel{h}}b_1}{\underset{1}{\cancel{h}}} = b_1.$

$$b_1 = \frac{2A - hb_2}{h}$$ *Reverse the sides of the equation so that b_1 is on the left.*

Notation

The 1 and 2 in b_1 and b_2 are called **subscripts.** This notation allows us to distinguish between the variables b_1 and b_2, while still showing that each represents the length of a base of the trapezoid.

The Language of Algebra

We say that the formula is **solved for b_1** because b_1 is alone on one side of the equation and the other side does not contain b_1.

When solving formulas for a specified variable, there is often more than one way to express the result. In this case, we could perform the division by h on the right side term-by-term: $b_1 = \frac{2A}{h} - \frac{hb_2}{h}$. After removing the common factor of h in the numerator and denominator of the second fraction, we obtain the following equivalent form of the result: $b_1 = \frac{2A}{h} - b_2$. However, do not try to simplify the result in the following way. It is incorrect because h is not a factor of the entire numerator.

$$b_1 = \frac{2A - \overset{1}{\cancel{h}}b_2}{\underset{1}{\cancel{h}}}$$

c.

To solve for t, we will isolate t on this side of the equation.

$$v_f = v_i + at$$ Read v_f as "v-sub-f and v_i as "v-sub-i."

$$v_f - v_i = at$$ To isolate the term at, subtract v_i from both sides. This step is done mentally.

$$\frac{v_f - v_i}{a} = t$$ To isolate t, undo the multiplication by a by dividing both sides by a. This step is done mentally.

$$t = \frac{v_f - v_i}{a}$$ Reverse the sides of the equation so that t is on the left.

d.

To solve for P, we will isolate P on this side of the equation.

$$A = P(1 + rt)$$

$$\frac{A}{1 + rt} = \frac{P(1 + rt)}{1 + rt}$$ To isolate P, undo the multiplication by $1 + rt$ by dividing both sides by $1 + rt$.

$$\frac{A}{1 + rt} = P$$ On the right side, remove the common factor of $1 + rt$: $\frac{P\overset{1}{(\cancel{1 + rt})}}{\underset{1}{\cancel{1 + rt}}} = P$.

$$P = \frac{A}{1 + rt}$$ Write the equation with P on the left side.

Self Check 5   **a.** Solve $I = Prt$ for r.

b. Solve $S = \frac{n}{2}(f + l)$ for f.

c. Solve $E = \dfrac{T_h - T_c}{T_h}$ for T_c.

d. Solve $W = g(h - 3t^2)$ for g.

Now Try Problems 41, 47, 51, and 57

In Chapter 2, we will work with equations in two variables, such as $6x + 5y = 35$ and $3x - 4y = 20$. It is often necessary to solve such equations for y.

EXAMPLE 6   Solve the equation $3x - 4y = 20$ for y.

Strategy   To solve for y, we will treat it as if it were the only variable in the equation and isolate y on one side of the equation.

Why   We can solve the formula as if it were an equation in one variable, y, because the other variable, x, is treated as if it were a number (constant).

Solution

To solve for y, we will isolate y on this side of the equation.

$$3x - 4y = 20$$

$$3x - 4y - 3x = 20 - 3x$$ To isolate $-4y$, subtract $3x$ from both sides.

$$-4y = 20 - 3x$$ Simplify each side of the equation.

$$\frac{-4y}{-4} = \frac{20 - 3x}{-4}$$ To isolate y, undo the multiplication by -4 by dividing both sides by -4.

$$y = \frac{20 - 3x}{-4}$$ Simplify the left side: $\dfrac{\overset{1}{\cancel{-4y}}}{\underset{1}{\cancel{-4}}} = y$.

If we reorder the terms of the numerator and perform the division by -4 term-by-term, we obtain the equivalent result

$$y = \frac{-3x}{-4} + \frac{20}{-4}$$ which simplifies to $y = \frac{3}{4}x - 5$ Divide: $\frac{20}{-4} = -5$.

Self Check 6   Solve $6x + 5y = 35$ for y.

Now Try ▶ Problem 63

---

3 Solve Application Problems Using Formulas.

EXAMPLE 7   **Melting Points.** The formula that relates a Fahrenheit temperature F to a Celsius temperature C is $F = \frac{9}{5}C + 32$. Calculate the melting points in degrees Celsius for the metals shown in the table.

Strategy   First, we will solve $F = \frac{9}{5}C + 32$ for C. Then, we will substitute each Fahrenheit temperature for F in the formula that is solved for C and calculate its corresponding Celsius temperature.

Why   It would be time consuming to substitute each melting point temperature for F in the formula $F = \frac{9}{5}C + 32$ and then repeatedly solve for C. A quicker way is to solve the formula for C first, substitute values for F, and calculate C directly.

Solution

$$F = \frac{9}{5}C + 32$$ This is the Fahrenheit temperature formula.

$$F - 32 = \frac{9}{5}C$$ To isolate the variable term $\frac{9}{5}C$, subtract 32 from both sides. This step is done mentally.

$$\frac{5}{9}(F - 32) = \frac{5}{9}\left(\frac{9}{5}C\right)$$ To isolate C, multiply both sides by the reciprocal of $\frac{9}{5}$, which is $\frac{5}{9}$.

$$\frac{5}{9}(F - 32) = C$$ Simplify the right side: $\frac{5}{9} \cdot \frac{9}{5} = 1$.

Metal	Melting point
Gold	1,948°F
Nickel	2,647°F
Aluminum	1,221°F

The Language of Algebra

In 1724, Daniel Gabriel *Fahrenheit*, a German scientist, introduced the temperature scale that bears his name. The Celsius scale was invented in 1742 by Swedish astronomer Anders *Celsius*.

To convert each melting point temperature from degrees Fahrenheit to degrees Celsius, we substitute for F in the formula $C = \frac{5}{9}(F - 32)$ and find C using a calculator.

Gold: **1,948°F**

$$C = \frac{5}{9}(1,948 - 32)$$
$$C \approx 1,064$$
Gold: 1,064°C

Nickel: **2,647°F**

$$C = \frac{5}{9}(2,647 - 32)$$
$$C \approx 1,453$$
Nickel: 1,453°C

Aluminum: **1,221°F**

$$C = \frac{5}{9}(1,221 - 32)$$
$$C \approx 661$$
Aluminum: 661°C

Self Check 7   **Boiling Water.** Water boils at 212°F. Find the boiling point temperature for water in degrees Celsius.

Now Try ▶ Problem 91

SECTION 1.6 ▶ STUDY SET

VOCABULARY

Fill in the blanks.

1. A _____ is an equation that states a mathematical relationship between two or more variables.

2. To find the _____ of a plane geometric figure, such as a rectangle or triangle, we calculate the distance around the figure. The _____ of a plane geometric figure is the amount of surface that it encloses.

3. The _____ of a three-dimensional geometric solid is the amount of space it encloses.

4. To _____ a formula for a specified variable means to isolate that variable on one side of the equation, with all other variables and constants on the opposite side.

CONCEPTS

5. Determine which concept (perimeter, circumference, area, or volume) should be used to find each of the following situations. Then determine which unit of measurement, ft, ft^2, or ft^3, would be appropriate.

   a. The amount of ground covered by a lawn

   b. The amount of storage in a safe

   c. The distance traveled by a rider on a Ferris wheel

   d. The distance around a tennis court

6. The area of a circle is exactly 54π ft^2. Approximate the area to the nearest tenth of a square foot.

7. Fill in the blanks to express the result in an equivalent form:

$$y = \frac{-5x + 8}{2}$$

$$y = \frac{-5x}{\boxed{}} + \frac{8}{\boxed{}}$$

$$y = \frac{-5}{2}x + \boxed{}$$

8. Fill in the blanks: To solve a formula for a specified variable, we treat it as if it were the _____ variable in the equation. We treat all other variables as if they were _____ (constants).

NOTATION

Complete the solution.

9. Solve $t = ad + bc$ for c.

$$t - \boxed{} = ad + bc - \boxed{}$$

$$t - ad = \boxed{}$$

$$\frac{t - ad}{\boxed{}} = \frac{bc}{\boxed{}}$$

$$\frac{t - ad}{b} = \boxed{}$$

$$c = \boxed{}$$

10. Fill in the blanks: In the notation b_1 and v_f, the number 1 and the variable f are called _____.

GUIDED PRACTICE

Find the perimeter of each figure. **See Example 1.**

11. A square with sides 2 yd long

12. A triangle with sides 1.8, 1.8, and 1.5 cm long

13. A trapezoid with parallel sides 10 in. and 15 in. long and the other two sides each 6 in. long

14. A parallelogram with two adjacent sides 50 m and 100 m long

Use a perimeter formula to find the unknown measurement. **See Example 1.**

15. **Energy Savings.** One hundred inches of foam weather stripping tape was placed around the perimeter of a rectangular window. If the length of the window is 33.5 inches, what is its width?

16. **Interior Decorating.** Sixty-six feet of crown molding was installed around a rectangular dining room. If the length of the room is 18.75 feet, what is its width?

17. **Landscaping.** The perimeter of a square flowerbed is 26 yards. Find the length of one side of the flowerbed.

18. **Advertising.** The perimeter of a triangular banner is 250 inches. If two sides of the banner are each 94.76 in. long, what is the length of the third side?

Find the area of each figure. **See Example 2.**

19. A triangle with a base that is 2.4 ft long and height 8.5 ft

20. A rectangle with sides that measure $8\frac{1}{4}$ ft and $5\frac{1}{2}$ ft

21. A square with sides 17.2 mi long

22. A trapezoid whose parallel sides measure 8 cm and 12 cm and whose height is 10.5 cm

Use an area formula to find the unknown measurement. **See Example 2.**

23. **Wrapping Presents.** There are 30 ft^2 of gift wrapping paper on a roll. When unrolled, the sheet of paper is 12 feet long. What is its width?

24. **First Aid.** A rectangular band-aid covers $2\frac{1}{2}$ in.^{2} of skin. If the width of the band-aid is $\frac{5}{8}$ in., find its length.

25. **Windsurfing.** The area of a triangular sail on a windsurfing board is 42 ft^2. If the length of the base of the sail is 7 feet, what is its height?

26. **Camping.** The area of a trapezoid-shaped canvas flap on a tent is 110 in.^{2} If the upper base of the flap is 8 in. long, and the height is 11 in., find the length of its lower base.

Find the circumference of each circle to the nearest hundredth. **See Example 3.** *(Answers may vary slightly depending on which approximation of π is used.)*

27. A circle with diameter 7.5 in.

28. A circle with diameter $6\frac{1}{4}$ m

29. A circle with radius $2\frac{1}{2}$ ft

30. A circle with radius 12.3 yd

Find the area of each circle to the nearest tenth. See Example 3. (Answers may vary slightly depending on which approximation of π is used.)

31. A circle with radius 5.7 in.
32. A circle with radius $5\frac{3}{4}$ cm
33. A circle with diameter $10\frac{1}{2}$ ft
34. A circle with diameter 12.25 m

Find the volume of each figure to the nearest hundredth. See Example 4. (Answers may vary slightly depending on which approximation of π is used.)

35. A rectangular solid with dimensions 2.51 ft, 3.71 ft, and 10.21 ft
36. A pyramid whose base is a square with each side measuring 2.57 cm and with a height of 12.32 cm
37. A sphere with radius 5.78 meters
38. A cone whose base has a radius of 5.50 in. and whose height is 8.52 in.

Solve each formula for the specified variable. See Example 5.

39. $d = rt$ for t
40. $E = mc^2$ for m
41. $V = lwh$ for h
42. $I = Prt$ for t
43. $V = \frac{1}{3}\pi r^2 h$ for h
44. $A = \frac{1}{2}bh$ for b
45. $T = W + ma$ for W
46. $V = \frac{1}{3}Bh$ for B
47. $h = 48t + \frac{1}{2}at^2$ for a
48. $H = 17 - \frac{A}{2}$ for A
49. $A = \frac{1}{2}h(b_1 + b_2)$ for b_2
50. $\bar{v} = \frac{1}{2}(v + v_0)$ for v_0
51. $l = a + (n - 1)d$ for n
52. $P = 2(l + w)$ for l
53. $P = 2(w + h + l)$ for w
54. $P = 2(w + h + l)$ for h
55. $\lambda = A(x + B)$ for A
    (λ is the letter lambda from the Greek alphabet.)
56. $S = C(1 - r)$ for C
57. $T_f = T_a(1 - F)$ for T_a
58. $S = \frac{n}{2}(f + l)$ for n
59. $l = a + (n - 1)d$ for d
60. $S = \frac{n(a + l)}{2}$ for n
61. $v = \frac{1}{t}(d_1 - d_2)$ for t
62. $A = \frac{1}{2}(b_1 + b_2)h$ for h

Solve each equation for y. See Example 6.

63. $2x - 5y = 20$
64. $5x - 6y = 12$
65. $-4x = 12 + 3y$
66. $7x - 21 = -3y$

TRY IT YOURSELF

Solve for the specified variable.

67. $y = mx + b$ for x
68. $P = 2l + 2w$ for l
69. $L = 2d + 3.25(r + R)$ for R
70. $l = \frac{a - S + Sr}{r}$ for a
71. $s = \frac{1}{2}gt^2 + vt$ for g
72. $K = \frac{Mv_0^2 + Iw^2}{2}$ for I
73. $y - y_1 = m(x - x_1)$ for x
74. $s = v_0 t - 16t^2$ for v_0
75. $G = U - TS + pV$ for S
76. $F = \frac{Gm_1m_2}{r^2}$ for m_1
77. $PV = nrt$ for r
78. $P = s_1 + s_2 + b$ for s_2
79. $E = IR + Ir$ for R
80. $Ax + By = C$ for x
81. $A = \frac{1}{3}(s_1 + s_2 + s_3)$ for s_3
82. $P_1V_1T_2 = P_2V_2T_1$ for V_2
83. $S = \frac{n}{2}[2a + (n - 1)d]$ for d
84. $x^2 = 4py$ for p
85. $d = \frac{4}{3}\pi h$ for h
86. $I_Q = \frac{100M}{C}$ for C

APPLICATIONS

87. **Floor Mats.**   Estimate the amount of plastic trim used around the floor mat shown below.

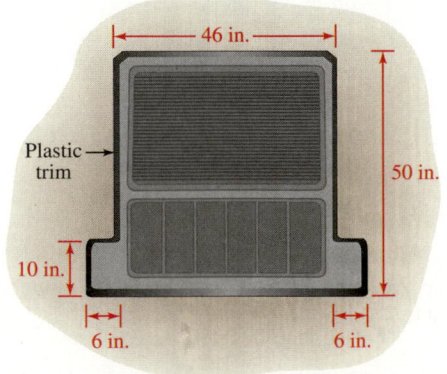

88. Aluminum Foil. Find the number of *square feet* of aluminum foil on a roll if the dimensions printed on the box are $8\frac{1}{3}$ yards $\times$ 12 inches.

89. Paper Products. When folded, the paper sheet shown in the illustration below forms an envelope. The formula

$$A = \frac{1}{2}h_1(b_1 + b_2) + b_3h_3 + \frac{1}{2}b_1h_2 + b_1b_3$$

gives the amount of paper (in square units) used in the design. Explain what each of the four terms in the formula finds. Then evaluate the formula for $b_1 = 6$, $b_2 = 2$, $b_3 = 3$, $h_1 = 2$, $h_2 = 2.5$, and $h_3 = 3$. All dimensions are in inches.

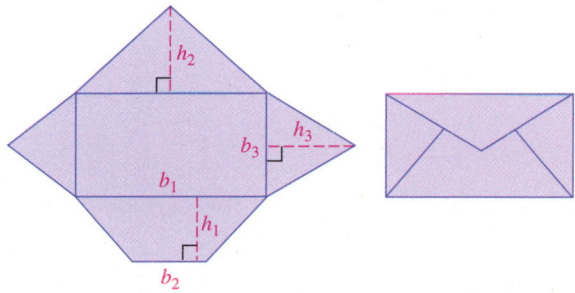

90. Hockey. A goal is scored in hockey when the puck, a vulcanized rubber disk 2.5 cm (1 in.) thick and 7.6 cm (3 in.) in diameter, is driven into the opponent's goal. Find the volume of a puck in cubic centimeters and cubic inches. Round to the nearest tenth.

91. Planets' Temperature Ranges. Solve the Fahrenheit/Celsius temperature formula $F = \frac{9}{5}C + 32$ for C. Then use your result to make the conversions for the data shown in the table. Round to the nearest degree.

Planet	High °F	Low °F	High °C	Low °C
Mercury	810	−290		
Earth	136	−129		
Mars	63	−87		

92. Ice Cream. If the two equal-sized scoops of ice cream melt completely into the cone, will they overflow the cone?

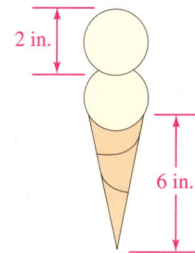

93. Wiper Design. The area cleaned by the windshield wiper shown below is given by the formula

$$A = \frac{d\pi(r_1^2 - r_2^2)}{360}$$

Solve the equation for d and use your result to find the number of degrees d the wiper arm must swing for each type of vehicle. Round to the nearest degree.

Vehicle	Area cleaned	d (deg)
Luxury car	513 in.^{2}	
Sport utility vehicle	586 in.^{2}	

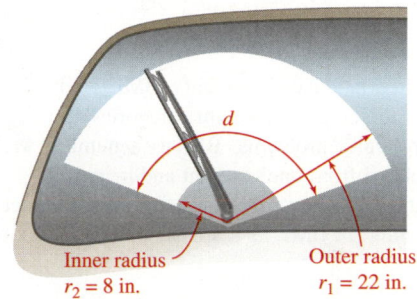

Inner radius $r_2 = 8$ in. Outer radius $r_1 = 22$ in.

94. Electronics. The illustration below is a diagram of a resistor connected to a voltage source of 60 volts. As a result, the resistor loses power in the form of heat. The power P lost when a voltage E is placed across a resistance R (in ohms) is given by the formula

$$P = \frac{E^2}{R}$$

Solve for R. If P is 4.8 watts and E is 60 volts, find R.

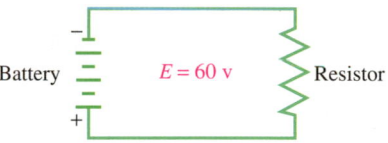

Battery $E = 60$ v   Resistor

95. Chemistry. In chemistry, the ideal gas law equation is $PV = nR(T + 273)$, where P is the pressure, V the volume, T the temperature, and n the number of moles of a gas. Solve the equation for n. Then use your result and the data from the lab notebook below to find the value of n to the nearest thousandth for trial 1 and for trial 2.

Ideal gas law Lab #1 R = 0.082 (Constant)		Betsy Kinsell Chem 1 Section A	
Data:	Pressure (Atm)	Volume (L)	Temp (°C)
Trial 1	0.900	0.250	90
Trial 2	1.250	1.560	−10

96. Investments. An amount P, invested at a simple interest rate r, will grow to an amount A in t years according to the formula $A = P(1 + rt)$. Solve for P. Suppose a man invested some money at 5.5%. If after 5 years, he had $6,693.75 on deposit, what amount did he originally invest?

97. Cost of Electricity. The cost of electricity in a city is given by the formula $C = 0.07n + 6.50$, where C is the cost in dollars and n is the number of kilowatt hours used. Solve for n. Then find the number of kilowatt hours used each month by a homeowner whose checks to pay the monthly electric bills are: $49.97, $76.50, and $125.

98. Cost of Water. A monthly water bill in a certain city is calculated by using the formula $n = \frac{5{,}000C - 17{,}500}{6}$, where n is the number of gallons used and C is the monthly cost in dollars. Solve for C and compute the bill for quantities of 500, 1,200, and 2,500 gallons.

---

99. from **Campus to Careers**

Registered Dietitian

Dietitians often calculate a patient's BMI (Body Mass Index) to screen for weight categories that may lead to health problems. BMI is a number that is calculated from one's weight and height. It is an indication of a person's total body weight that comes from fat. The formula for BMI, as it appears in dietary textbooks, is:

$$BMI = \frac{weight(lb) \cdot 703}{height^2(in.^2)}$$

Solve the formula for weight.

---

100. Surface Area. To find the amount of tin needed to make the coffee can shown in below, we use the formula for the surface area of a right circular cylinder, $A = 2\pi r^2 + 2\pi rh$. Solve the formula for h.

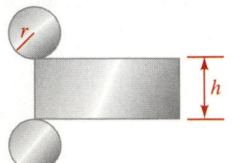 

WRITING

101. Explain the difference between what perimeter measures and what area measures.

---

102. After solving a formula for m, a student compared her answer with that at the back of the textbook. Could this problem have two different-looking answers? Explain why or why not.

Student: $m = \frac{5}{9}ar + 1$ Book: $m = \frac{5ar + 9}{9}$

103. Explain the error made below.

$$T = \frac{adx + \overset{1}{\cancel{y}}}{\underset{1}{\cancel{y}}}$$

104. A student solved $x + 5c = 3c + a$ for c. His answer was $c = \frac{3c + a - x}{5}$. Explain why the equation is not solved for c.

REVIEW

Simplify each expression.

105. $(16b + 8)\left(\dfrac{5}{4}\right) - 8b$

106. $-7(a - 3) - 5[3(a - 4) - 2(a + 2)]$

107. $-5.7pt - p + 5.1pt + 12p$

108. $\dfrac{3}{5}t - \dfrac{2}{3}t$

CHALLENGE PROBLEMS

109. Point C is the center of the largest circle in the figure. Find the area of the shaded region. Round to the nearest tenth.

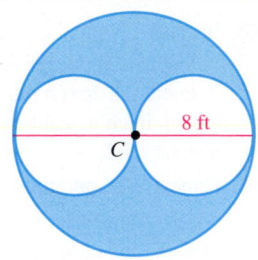

110. Solve $d_1 d_2 = f d_2 + f d_1$ for d_1.

111. Subscripts are used in other disciplines besides mathematics. In what disciplines are the following symbols used?

   a. H_2O and CO_2

   b. C_7 and G_7

   c. B_6 and B_{12}

112. Evaluate $2a_2^2 + 3a_3^3 + 4a_4^4$ for $a_2 = 2$, $a_3 = 3$, and $a_4 = 4$.

SECTION 1.7

Using Equations to Solve Problems

OBJECTIVES

1. Apply the steps of a problem-solving strategy.
2. Solve number-value problems.
3. Solve geometry problems.
4. Use formulas to solve problems.

ARE YOU READY?

The following problems review some basic skills that are needed to solve the application problems in this section.

1. Translate to symbols: *8 less than twice a number x*
2. If $x = 15$, find $6x + 3$.
3. If mouse pads cost $5.95 each, what is the cost of 10 of them?
4. What is the formula for the perimeter of a rectangle?
5. What is the sum of the measures of the angles of a triangle?
6. Solve: $2x + 10 + 53x + 9 + x = 1{,}251$

An objective of this course is to improve your problem-solving skills. In the next two sections, you will have the opportunity to do that as we discuss how to use equations to solve many different types of real-world problems.

1 Apply the Steps of a Problem-Solving Strategy.

To become a good problem solver, you need a plan to follow, such as the following six-step strategy.

Problem Solving

1. *Analyze the problem* by reading it carefully to understand the given facts. What information is given? What are you asked to find? What vocabulary is given? Often a diagram or table will help you understand the facts of the problem.
2. *Assign a variable* to represent an unknown value in the problem. This means, in most cases, to let $x =$ what you are asked to find. If there are other unknown values, represent each of them using an algebraic expression that involves the variable.
3. *Form an equation* by translating the words of the problem into mathematical symbols.
4. *Solve the equation* formed in step 3.
5. *State the conclusion clearly.* Be sure to include the units (such as feet, seconds, or pounds) in your answer.
6. *Check the result* using the original wording of the problem, not the equation that was formed in step 3.

In order to solve problems, which are usually given in words, we must translate those words into mathematical symbols. In the next example, we use translation to write an equation that mathematically models the situation.

EXAMPLE 1 **Leading Employers.** In 2009, Target and Home Depot were two of the nation's top employers. Their combined work forces totaled 616,650 people. If Home Depot employed 85,350 fewer people than Target, how many employees did each company have? (Source: money.cnn.com)

Analyze

- The phrase *combined work forces totaled 616,650* suggests that if we add the number of employees of each company, the result will be 616,650.
- The phrase *Home Depot employed 85,350 fewer people than Target* suggests that the number of employees of Home Depot can be found by subtracting 85,350 from the number of employees of Target.
- We are to find the number of employees of each company.

Assign If we let x = the number of employees of Target, then $x - 85{,}350$ = the number of employees of Home Depot.

Form We can now translate the words of the problem into an equation.

The number of employees of Target	plus	the number of employees of Home Depot	is	616,650.
x	$+$	$x - 85{,}350$	$=$	$616{,}650$

Caution

For this problem, one common mistake is to let

~~x = the number of employees of each company~~

Since Target and Home Depot have different numbers of employees, x cannot represent both unknowns.

Solve

$$x + x - 85{,}350 = 616{,}650$$
$$2x - 85{,}350 = 616{,}650 \qquad \text{Combine like terms: } x + x = 2x.$$
$$2x = 702{,}000 \qquad \text{To isolate } 2x, \text{ add } 85{,}350 \text{ to both sides.}$$
$$x = 351{,}000 \qquad \text{To isolate } x, \text{ divide both sides by 2.}$$

Since x represents the number of employees of Target, we can find the number of employees of Home Depot by evaluating $x - 85{,}350$ for $x = 351{,}000$.

$$x - 85{,}350 = \mathbf{351{,}000} - 85{,}350$$
$$= 265{,}650$$

State In 2009, Target had 351,000 employees and Home Depot had 265,650 employees.

Check Since $351{,}000 + 265{,}650 = 616{,}650$ and since 265,650 is 85,350 less than 351,000, the results check.

Self Check 1 **Package Delivery.** In 2009, Fed Ex and United Parcel Service (UPS) were two of the nation's top employers. Their combined work forces totaled 680,142 people. If UPS employed 171,858 more people than FedEx, how many employees did each company have? (Source: money.cnn.com)

Now Try ▶ Problem 11

When solving problems, diagrams are often helpful, because they allow us to visualize the facts of the problem.

EXAMPLE 2 **Triathlons.** A triathlon includes swimming, long-distance running, and cycling. The long-distance run is 11 times longer than the distance the athletes swim. The distance they cycle is 85.8 miles longer than the run. Overall, the competition covers 140.6 miles. Find the length of each part of the triathlon and round each length to the nearest tenth of a mile.

Analyze The entire triathlon course covers a distance of 140.6 miles. We note that the distance the athletes run is related to the distance that they swim, and the distance they cycle is related to the distance that they run.

Success Tip

When there is more than one unknown value in a problem, let the variable represent the unknown value upon which any other unknown values are based.

Assign If x = the distance in miles that the athletes swim, then $11x$ = the length of the long-distance run, and $11x + 85.8$ = the distance they cycle.

Form From the diagram below, we can see that the sum of the parts of the triathlon must equal the total distance covered.

Swimming x mi	Running $11x$ mi	Cycling $(11x + 85.8)$ mi

$\longleftarrow$ 140.6 m $\longrightarrow$

We can now form the equation.

The distance they swim	plus	the distance they run	plus	the distance they cycle	equals	the total length of the course.
x	$+$	$11x$	$+$	$11x + 85.8$	$=$	140.6

Solve

$$x + 11x + 11x + 85.8 = 140.6$$
$$23x + 85.8 = 140.6 \qquad \text{Combine like terms: } x + 11x + 11x = 23x.$$
$$23x = 54.8 \qquad \text{To isolate } 23x, \text{ subtract } 85.8 \text{ from both sides.}$$
$$x \approx 2.382608696 \qquad \text{To isolate } x, \text{ divide both sides by } 23. \text{ This is the distance (in miles) that the athletes swim.}$$

State To the nearest tenth, the distance the athletes swim is 2.4 miles. The distance they run is $11x$, or approximately $11(2.382608696) = 26.20869565$ miles. To the nearest tenth, that is 26.2 miles. The distance they cycle is $11x + 85.8$, or approximately $26.20869565 + 85.8 = 112.0086957$ miles. To the nearest tenth, that is 112.0 miles.

Check If we add the approximate lengths of the three parts of the triathlon, we get 140.6 miles. The results seem reasonable.

Self Check 2   **Road Trips.**   A three-part drive consists of a first part that is 1 mile longer than the second, and a third part that is 1 mile longer than 4 times the second part. If the entire drive is 17 miles, find the length of each part.

Now Try ▶ Problem 15

The wording of a problem doesn't always contain key phrases that translate directly to an equation. In such cases, an analysis of the problem will give clues that help us write an equation.

EXAMPLE 3   **Travel Promotions.**   The price of a 7-day Alaskan cruise, normally $2,752 per person, is reduced by $1.75 per person for large groups traveling together. How large a group is needed for the price to be $2,500 per person?

Analyze It is often helpful to consider some specific situations before attempting to form an equation. For a group of, say, 20 people, the price of the cruise would be reduced by 20($1.75), and each person would pay $2,752 − 20($1.75). For a group of 30 people, the price of the cruise would be reduced by 30($1.75), and each person would pay $2,752 − 30($1.75).

Assign We let x = the group size necessary for the price of the cruise to be $2,500 per person.

Form We can form the following equation:

The price of the cruise	is	$2,752	minus	the number of people in the group	times	$1.75.
2,500	$=$	2,752	$-$	x	$\cdot$	1.75

Solve

$$2,500 = 2,752 - 1.75x$$
$$2,500 - \mathbf{2{,}752} = 2,752 - 1.75x - \mathbf{2{,}752} \qquad \text{To isolate the variable term } -1.75x, \text{ subtract } 2{,}752 \text{ from both sides.}$$
$$-252 = -1.75x \qquad \text{Do the subtraction.}$$
$$144 = x \qquad \text{To isolate } x, \text{ divide both sides by } -1.75. \text{ This is the required size of the group.}$$

State If 144 people travel together, the price will be $2,500 per person.

Check For 144 people, the cruise cost of $2,752 will be reduced by 144($1.75) = $252. If we subtract, $2,752 − $252 = $2,500. The result checks.

> **Self Check 3** **Broadway Shows.** Tickets to a Broadway show normally sell for $90 per person. This price is reduced by $0.50 per person for large groups. How large a group is needed for the price to be $75 per person?
>
> **Now Try** ▶ Problem 19

2 Solve Number-Value Problems.

Some problems deal with quantities that have a value. In these problems, we must distinguish between the *number of* and the *value of* the unknown quantity. For problems such as these, we will use the relationship

$$\text{Number} \cdot \text{value} = \text{total value}$$

EXAMPLE 4 **Stocks.** A college foundation owns stock in Kodak (selling at $5 per share), Coca-Cola (selling at $59 per share), and IBM (selling at $135 per share). The foundation owns an equal number of shares of Kodak and Coca-Cola stock, but five times as many shares of IBM stock. If this collection of stocks is worth $517,300, how many shares of each stock does the foundation own?

Analyze The value of the Kodak stock plus the value of the Coca-Cola stock plus the value of the IBM stock must equal $517,300. We need to find the number of shares of each of these stocks held by the foundation.

Assign If we let x = the number of shares of Kodak stock, then x = the number of shares of Coca-Cola stock. Since the foundation owns five times as many shares of IBM stock as Kodak or Coca-Cola stock, $5x$ = the number of shares of IBM.

Form The value of the shares of each stock is the *product* of the number of shares of that stock and its per-share value, as shown in the last column of the table below.

> **Success Tip**
>
> We can let x represent the number of shares of Kodak stock *and* the number of shares of Coca-Cola stock because the foundation owns an *equal number* of shares of these stocks.

Stock	Number of shares ·	Value per share =	Total value of the stock
Kodak	x	5	$5x$
Coca-Cola	x	59	$59x$
IBM	$5x$	135	$135(5x)$

Total: $517,300

Enter this information first.

Use the information in this column to form an equation.

We can now form the equation.

The value of Kodak stock	plus	the value of Coca-Cola stock	plus	the value of IBM stock	is	the total value of all of the stock.
$5x$	$+$	$59x$	$+$	$135(5x)$	$=$	$517{,}300$

Solve

$$5x + 59x + 675x = 517{,}300 \qquad \textcolor{red}{\text{Multiply: } 135(5x) = 675x.}$$

$$739x = 517{,}300 \qquad \textcolor{red}{\text{Combine like terms on the left side:}}$$
$$\textcolor{red}{5x + 59x + 675x = 739x.}$$

$$x = 700 \qquad \textcolor{red}{\text{To isolate } x, \text{ divide both sides by 739. This is the}}$$
$$\textcolor{red}{\text{number of shares of Kodak and Coca-Cola stock.}}$$

State   The foundation owns 700 shares of Kodak, 700 shares of Coca-Cola, and $5(700) = 3{,}500$ shares of IBM.

Check   The value of 700 shares of Kodak stock is $700(\$5) = \$3{,}500$. The value of 700 shares of Coca-Cola is $700(\$59) = \$41{,}300$. The value of 3,500 shares of IBM is $3{,}500(\$135) = \$472{,}500$. The sum is $\$3{,}500 + \$41{,}300 + \$472{,}500 = \$517{,}300$. The results check.

Self Check 4   **Electronic Readers.**   A college bought some Apple iPads (at $345 each), twice as many Amazon Kindles (at $180 each), and half as many Barnes & Noble Nooks (at $150 each) as iPads. If the purchase totaled $15,600, how many of each reader did they buy?

Now Try ▶ Problem 23

---

3 Solve Geometry Problems.

Sometimes we can use a geometric fact or formula to solve a problem. The following illustrations show two important types of geometric figures: angles and triangles. A **right angle** is an angle whose measure is 90°. A **straight angle** is an angle whose measure is 180°. An **acute angle** is an angle whose measure is greater than 0° and less than 90°. An angle whose measure is greater than 90° and less than 180° is called an **obtuse angle.**

If the sum of two angles equals 90°, the angles are called **complementary,** and each angle is called the **complement** of the other. If the sum of two angles equals 180°, the angles are called **supplementary,** and each angle is the **supplement** of the other.

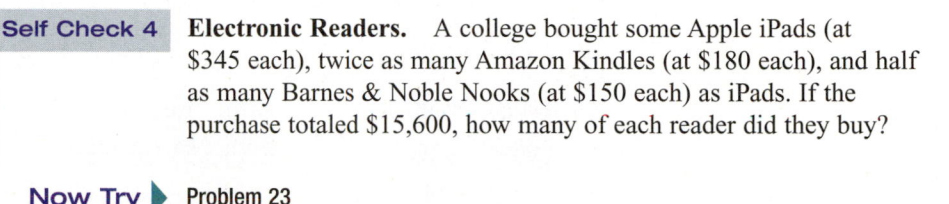

A **right triangle** is a triangle with one right angle. An **isosceles triangle** is a triangle with two sides of equal measure that meet to form the **vertex angle.** The angles opposite the equal sides, called the **base angles,** are also equal. An **equilateral triangle** is a triangle with three equal sides and three equal angles. It is important to note that *the sum of the angle measures of any triangle is 180°.*

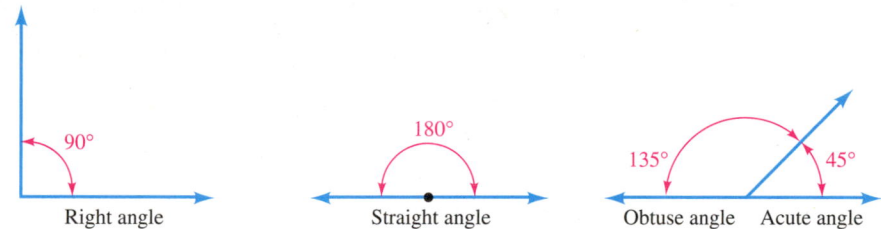

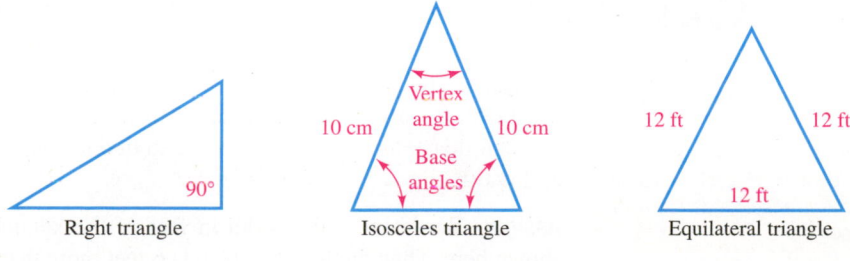

EXAMPLE 5

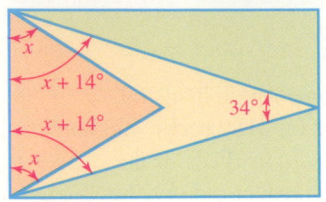

Flags. The flag of Guyana, a republic on the northern coast of South America, is one isosceles triangle superimposed over another on a field of green, as shown. The measure of a base angle of the larger triangle is 14° more than the measure of a base angle of the smaller triangle. The measure of the vertex angle of the larger triangle is 34°. Find the measure of each base angle of the smaller triangle.

Analyze We are working with isosceles triangles. Therefore, the base angles of the smaller triangle have the same measure, and the base angles of the larger triangle have the same measure.

Assign If we let x = the measure in degrees of one base angle of the smaller isosceles triangle, then the measure of its other base angle is also x. (See the figure.)

The measure of a base angle of the larger isosceles triangle is $x + 14°$, since its measure is 14° more than the measure of a base angle of the smaller triangle. We are given that the vertex angle of the larger triangle measures 34°.

Form Recall that the sum of the measures of the angles of any triangle (in this case, the larger triangle) is 180°. We can use that fact to form an equation.

The measure of one base angle	plus	the measure of the other base angle	plus	the measure of the vertex angle	is	180°.
$x + 14$	$+$	$x + 14$	$+$	34	$=$	180

Solve

$$x + 14 + x + 14 + 34 = 180$$
$$2x + 62 = 180 \qquad \text{\color{red}Combine like terms: } x + x = 2x \text{ and } 14 + 14 + 34 = 62.$$
$$2x = 118 \qquad \text{\color{red}To isolate } 2x, \text{ subtract 62 from both sides.}$$
$$x = 59 \qquad \text{\color{red}To isolate } x, \text{ divide both sides by 2.}$$

State The measure of each base angle of the smaller triangle is 59°.

Check If $x = 59$, then $x + 14 = 73$. The sum of the measures of each base angle and the vertex angle of the larger triangle is $73° + 73° + 34° = 180°$. The result checks.

Success Tip

When solving geometry problems, a sketch is often helpful.

Self Check 5 **Angle Measures.** The measure of the largest angle of a triangle is three times that of the smallest angle, and the measure of the midsized angle is 20° more than the smallest. Find the measure of each angle.

Now Try ▶ Problem 31

4 Use Formulas to Solve Problems.

When preparing to write an equation to solve a problem, the given facts of the problem often suggest a formula that we can use to model the situation mathematically.

EXAMPLE 6

The Language of Algebra

Dimensions are measurements of length and width. We might speak of the *dimensions* of a dance floor or a TV screen.

Kennels. A man has a 50-foot roll of fencing to make a rectangular kennel. If he wants the kennel to be 6 feet longer than it is wide, find its dimensions.

Analyze The perimeter P of the rectangular kennel is 50 feet. Recall that the formula for the perimeter of a rectangle is $P = 2l + 2w$. We need to find its length and width.

Assign We let w = the width in feet of the kennel shown here. Then the length, which is 6 feet more than the width, is represented by the expression $w + 6$.

Form   We can now form the equation by substituting 50 for the perimeter P and $w + 6$ for the length l in the formula for the perimeter of a rectangle.

$$P = 2\textbf{\textit{l}} + 2w$$
$$\textbf{50} = 2(\textbf{w + 6}) + 2w$$

Solve

$50 = 2(w + 6) + 2w$	
$50 = 2w + 12 + 2w$	On the right side, distribute the multiplication by 2.
$50 = 4w + 12$	Combine like terms: 2w + 2w = 4w.
$38 = 4w$	To isolate the variable term 4w, subtract 12 from both sides.
$9.5 = w$	To isolate w, divide both sides by 4.

State   The width of the kennel is 9.5 feet. The length is 6 feet more than this, or 15.5 feet.

Check   If a rectangle has a width of 9.5 feet and a length of 15.5 feet, its length is 6 feet more than its width, and the perimeter is 2(9.5) feet + 2(15.5) feet = 50 feet. The results check.

Self Check 6   **Crime Scene.**   Police used 400 feet of yellow tape to fence off a rectangular lot for an investigation. If the width is 50 feet less than the length, find the dimensions of the lot.

Now Try ▶ Problem 37

SECTION **1.7**   STUDY SET

VOCABULARY

Fill in the blanks.

1. An _____ angle has a measure of more than 0° and less than 90°.
2. A _____ angle is an angle whose measure is 90°.
3. If the sum of the measures of two angles equals 90°, the angles are called _____ angles.
4. If the sum of the measures of two angles equals 180°, the angles are called _____ angles.
5. If a triangle has a right angle, it is called a _____ triangle.
6. If a triangle has two sides with equal measures, it is called an _____ triangle.
7. The sum of the measures of the _____ of a triangle is 180°.
8. An _____ triangle has three sides of equal length and three angles of equal measure.

CONCEPTS

9. The unit used to measure the intensity of sound is called the *decibel.* Use the comments in the right column to write algebraic expressions that represent the decibel level for a vacuum cleaner, a circular saw, and whispering.

Activity	Decibels	Compared to conversation
Conversation	d	—
Vacuum cleaner		15 decibels more
Circular saw		10 decibels less than twice
Whispering		10 decibels less than half

10. The following table shows the four types of problems an instructor put on a history test.
    a. Complete the table.
    b. Write an equation that could be used to find x.

Type of question	Number	·	Value	=	Total value
Multiple choice	x		5		
True/false	$3x$		2		
Essay	$x - 2$		10		
Fill-in	x		5		

Total: 110 points

APPLICATIONS

11. **Airplanes.**   Together, a Delta B747 and a Delta B777 seat 681 passengers. If the B777 seats 125 less people than the B747, how many passengers does each seat? (Source: deltaskymag.com)

12. **Denzel.**   As of October 2010, Denzel Washington's three top domestic grossing films, *American Gangster, Remember the Titans,* and *The Pelican Brief,* had earned a total of $346.7 million. If *American Gangster* earned $14.5 million more than *Remember the Titans,* and if *Remember the Titans* earned $14.9 million more than *The Pelican Brief,* how much did each film earn as of that date? (Source: boxofficemojo.com)

13. Statue of Liberty. From the foundation of the large base on which it sits to the top of the torch, the Statue of Liberty National Monument measures 305 feet. The base is 3 feet taller than the statue. Find the height of the base and the height of the statue.

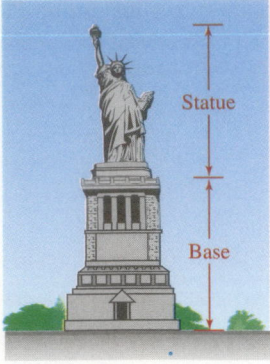

14. Woodworking. A carpenter saws a board that is 22 feet long into two pieces. One piece is to be 1 foot longer than twice the length of the shorter piece. Find the length of each piece.

15. Flutes. When it is assembled, a flute is 29 inches long. The middle piece is 4 inches less than twice as long as the first piece. The last piece is two-thirds as long as the first piece. Find the length of each piece of the flute.

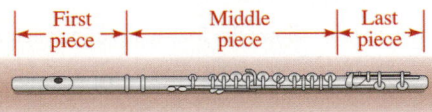

16. Commercials. For the typical "one-hour" prime-time television slot, the number of minutes of commercials is $\frac{3}{7}$ of the number of minutes of the actual program. Determine how many minutes of the program are shown in that one hour.

17. Kitchen Drawers. A woodworker wants to put two partitions crosswise in a drawer that is 28 inches deep, as shown below. He wants to place the partitions so that the spaces created increase by 3 inches from front to back. If the thickness of each partition is $\frac{1}{2}$ inch, how far from the front end should he place the first partition?

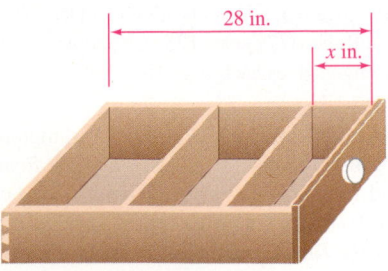

18. Shelving. A carpenter wants to put four shelves on an 8-foot wall so that the five spaces created decrease by 6 inches as we move up the wall. (See below.) If the thickness of each shelf is $\frac{3}{4}$ inch, how far will the bottom shelf be from the floor?

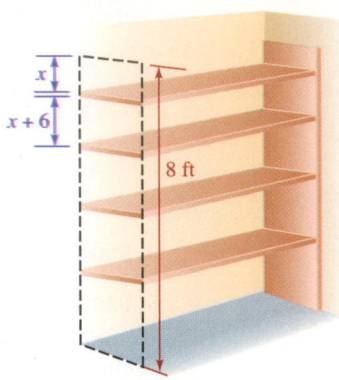

19. Spring Tours. A group of junior high students will be touring Washington, D.C. Their chaperons will have the $1,810 cost of the tour reduced by $15.50 for each student they personally supervise. How many students will a chaperon have to supervise so that his or her cost to take the tour will be $1,500?

20. Machining. Each pass through a lumber plane shaves off 0.015 inch of thickness from a board. How many times must a board, originally 0.875 inch thick, be run through the planer if a board of thickness 0.74 inch is desired?

21. Moving. To help move his furniture, a man rents a truck for $41.50 per day plus 35¢ per mile. If he has budgeted $150 for transportation expenses, how many miles will he be able to drive the truck if the move takes 1 day?

22. Salaries. A student working for a delivery company earns $57.50 per day plus $4.75 for each package she delivers. How many deliveries must she make each day to earn $200 a day?

23. IRAs. In an Individual Retirement Account (IRA) valued at $49,050, a couple has twice as many shares of stock in Big Bank Corporation as in Safe Savings and Loan. If Big Bank sells for $115 per share and Safe Savings sells for $97 per share, how many shares of each does the couple own?

24. Selling Calculators. Last month, a bookstore ran the following ad. Sales of $5,370 were generated, with 15 more graphing calculators sold than scientific calculators. How many of each type of calculator did the bookstore sell?

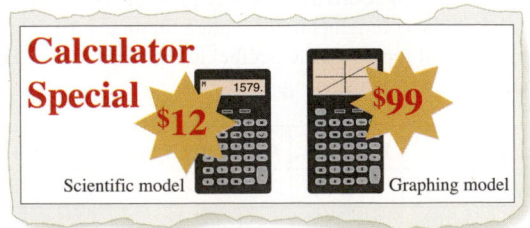

25. **Pension Funds.** A pension fund owns 2,000 fewer shares in mutual stock funds than mutual bond funds. Currently, the stock funds sell for $12 per share, and the bond funds sell for $15 per share. How many shares of each does the pension fund own if their total value is $165,000?

26. **Selling Seed.** A seed company sells two grades of grass seed. A 100-pound bag of a mixture of rye and Kentucky bluegrass sells for $245, and a 100-pound bag of bluegrass sells for $347. How many bags of each are sold in a week when the receipts for 19 bags are $5,369?

27. **Nursing.** The illustration shows the angle a needle should make with the skin when administering a certain type of injection. Find the measure of both angles labeled.

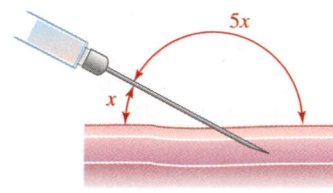

28. **Supplementary Angles.** Refer to the illustration and find x. Then find the measure of each angle.

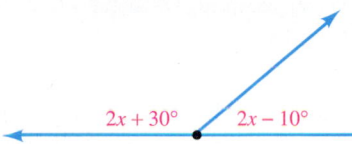

29. **The Tower of Pisa.** Because of soft soil and a shallow foundation, the Leaning Tower of Pisa in Italy is not vertical. Engineers predict that if the indicated angle gets to be 7°, the walls will not be able to support the structure and it will come crumbling down.

This angle is 0.4° more than 15 times the other angle.

 a. How many degrees from vertical is the tower now?

 b. How many more degrees of lean must occur to cause the predicted collapse?

30. **The iPad Building.** The sleek design of Apple's iPod has inspired a 23-floor residential and office building in Dubai, called the iPAD. It closely resembles a gigantic iPod sitting at an angle in its charger as shown below. How many degrees from the vertical is the building?

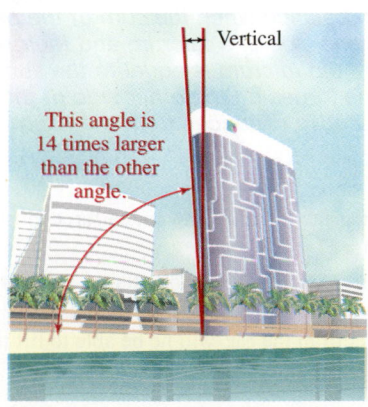

This angle is 14 times larger than the other angle.

31. **Stepstools.** In the illustration, the measure of ∠2 (angle 2) is 10° larger than the measure of ∠1. The measure of ∠3 is 10° larger than the measure of ∠2. Find each angle measure.

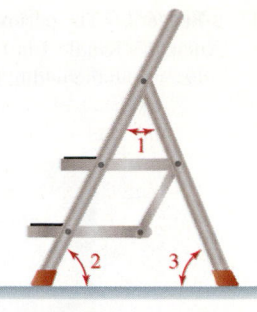

32. **Isosceles Triangles.** Find the measure of one base angle of an isosceles triangle if the measure of the vertex angle is 101°.

33. **Geometry.** In the illustration, lines r and s are cut by a third line l to form ∠1 (angle 1) and ∠2. When lines r and s are parallel, ∠1 and ∠2 are supplementary. If ∠1 = $x + 50°$, ∠2 = $2x - 20°$, and lines r and s are parallel, find x.

34. **Geometry.** In the illustration, $r \| s$ (read as "line r is parallel to line s"), and $a = 103$. Find b, c, and d. (*Hint:* See Problem 33.)

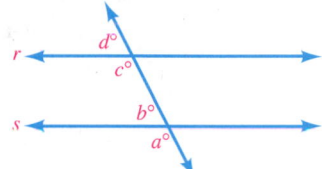

35. **Vertical Angles.** When two lines intersect, four angles are formed. Angles that are side-by-side, such as ∠1 (angle 1) and ∠2, are called **adjacent angles.** Angles that are nonadjacent, such as ∠1 and ∠3 or ∠2 and ∠4, are called **vertical angles.** From geometry, we know that if two lines intersect, vertical angles have the same measure. If ∠1 = $3x + 10°$ and ∠3 = $5x - 10°$, find x.

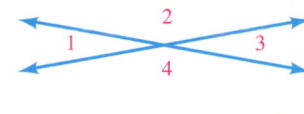

36. **Geometry.** In the illustration, $r \| s$ (read as "line r is parallel to line s"), and $b = 137$. Find a and c. (*Hint:* See Problems 33 and 35.)

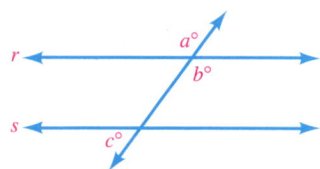

37. e-Readers. The perimeter of the rectangular display screen of Amazon's Kindle 3 is 16.8 inches. If the height is 1.2 inches greater than the width, find the dimensions of the screen.

©Alex Segre/Alamy

38. U.S. Currency. The perimeter of a one-dollar bill is 17.5 inches and the length is 0.92 in. more than twice the width. Find the dimensions of a one-dollar bill.

39. Swimming Pools. A woman wants to enclose the pool shown and have a walkway of uniform width all the way around. How wide will the walkway be if the woman uses 180 feet of fencing?

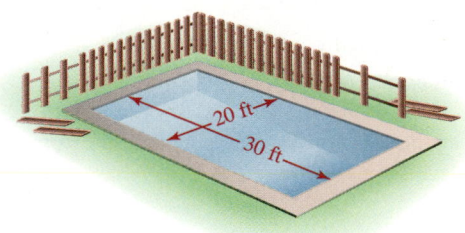
20 ft
30 ft

40. Quilting. Throughout history, most artists and designers have felt that *golden rectangles* with a length 1.618 times as long as their width have the most visually attractive shape. A woman is planning to make a quilt in the shape of a golden rectangle. She has exactly 22 feet of a special lace that she plans to sew around the edge of the quilt. What should the length and width of the quilt be? Round both answers up to the nearest hundredth.

41. Ranching. A farmer has 624 feet of fencing to enclose a pasture. Because a river runs along one side, fencing will be needed on only three sides. Find the dimensions of the pasture if its length is double its width.

Length
Width

42. Fencing. A man has 150 feet of fencing to build the two-part pen shown below. If one part is a square and the other a rectangle, find the outside dimensions of the pen.

Square
Rectangle
x ft
$(x + 5)$ ft

43. Solar Heating. One solar panel in the illustration is to be 3 feet wider than the other. To be equally efficient, they must have the same area. Find the width of each.

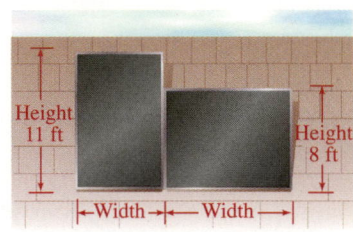
Height 11 ft
Height 8 ft
Width
Width

44. Height of a Triangle. If the height of a triangle with a base of 8 inches is tripled, its area is increased by 96 square inches. Find the height of the triangle.

WRITING

45. Briefly explain what should be accomplished in each of the steps (*analyze, assign, form, solve, state,* and *check*) of the problem-solving strategy used in this section.

46. Write a problem about a triangle that can be modeled by the following equation.

$$x + 2x + x + 10 = 180$$

REVIEW

47. When expressed as a decimal, is $\frac{7}{9}$ a terminating or repeating decimal?

48. Solve: $x + 20 = 4x - 1 + 2x$

49. Write the set of integers.

50. Solve $T - R = ma$ for R.

CHALLENGE PROBLEMS

In Problems 51–54, use the following property of levers: A lever will be in balance when the sum of the products of the forces on one side of a fulcrum and their respective distances from the fulcrum is equal to the sum of the products of the forces on the other side of the fulcrum and their respective distances from the fulcrum.

51. Moving a Stone. A woman uses a 10-foot bar to lift a 210-pound stone. If she places another rock 3 feet from the stone to act as the fulcrum, how much force must she exert to move the stone?

52. Lifting a Car. A 350-pound football player brags that he can lift a 2,500-pound car. If he uses a 12-foot bar with the fulcrum placed 3 feet from the car, will he be able to lift the car?

53. Balancing a Lever. Forces are applied to a lever as indicated in the illustration. Find x, the distance of the smallest force from the fulcrum.

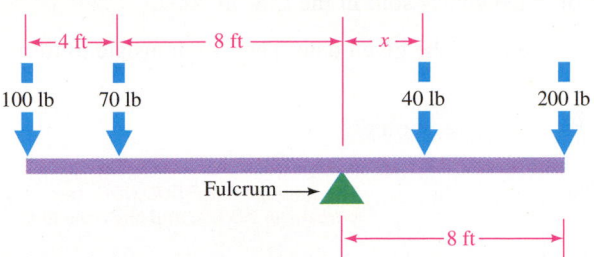

Fulcrum →

54. Balancing a Seesaw. Jim and Bob sit at opposite ends of an 18-foot seesaw, with the fulcrum at its center. Jim weighs 160 pounds, and Bob weighs 200 pounds. Kim sits 4 feet in front of Jim, and the seesaw balances. How much does Kim weigh?

SECTION 1.8

OBJECTIVES

1 Solve percent problems.

2 Solve investment problems.

3 Solve uniform motion problems.

4 Solve mixture problems.

More about Problem Solving

ARE YOU READY?

The following problems review some basic skills that are needed to solve the application problems in this section.

1. a. Write 27.8% as a decimal.

   b. Write 0.8356 as a percent.

2. Find the amount of interest earned by $15,000 invested at a 2% annual simple interest rate for 1 year.

3. At 55 miles per hour, how far will a car travel in 4 hours?

4. A pharmacist has 2 liters of a 30% alcohol solution. How many liters of pure alcohol are in the solution?

5. At $5.55 per pound, what is the value of 6 pounds of roast beef?

6. A couple invested x of their $60,000 lottery winnings in bonds. How much do they have left to invest in stocks?

In this section, we will use equations again as we solve a variety of problems.

1 Solve Percent Problems.

The Language of Algebra

The names of the parts of a percent sentence are:

  5   is   50%   of   10.
amount **percent**   **base**

They are related by the formula:

Amount = percent · base

Percents are often used to present numeric information. **Percent** means parts per one hundred. One method to solve percent problems is to use the given facts to write a **percent sentence** of the form:

☐ is ☐ % of ☐ ?

We enter the appropriate numbers in two of the blanks and the words "what number" or "what percent" in the remaining blank. Then we translate the sentence to mathematical symbols and solve the resulting equation.

EXAMPLE 1 **Video Games.** Refer to the following **circle graph**. Determine the total number of video games sold in the United States in 2009.

VIDEO GAME CATEGORIES BY UNITS SOLD IN THE U.S., 2009

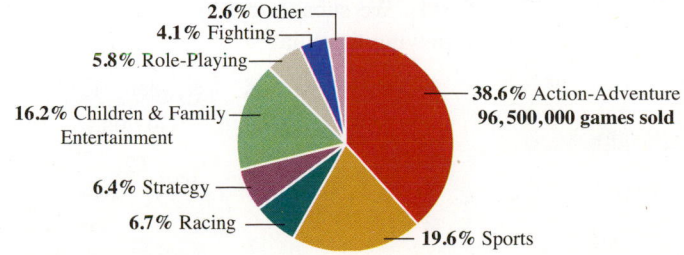

Source: The NPD Group

Analyze In the graph, we see that 96,500,000 action-adventure video games were 38.6% of the total number of video games sold.

Assign Let $x =$ the total number of video games sold in the U.S. in 2009.

Form First, we write a percent sentence using the given data. Then we translate to form a percent equation.

96,500,000	is	38.6%	of	what number?
↓	↓	↓	↓	↓
96,500,000	=	38.6%	·	x

The amount is 96,500,000, the percent is 38.6%, and the base is x.

Solve To find the total number of video games sold in 2009, we solve for x.

$$96,500,000 = 0.386x$$ Write 38.6% as a decimal: 38.6% = 0.386.

$$\frac{96,500,000}{0.386} = \frac{0.386x}{0.386}$$ To isolate x, divide both sides by 0.386.

$$250,000,000 = x$$ Do the division using a calculator.

State There were a total of 250,000,000 video games sold in the United States in 2009.

Check If a total of 250,000,000 video games were sold, then the 96,500,000 action-adventure video games sold were $\frac{96,500,000}{250,000,000} = 0.386$ or 38.6% of the games sold. The result checks.

Self Check 1   **Video Games.** Use the result from Example 1 to find the number of racing video games sold in the U.S. in 2009.

Now Try ▶ Problem 23

When the regular price of merchandise is reduced, the amount of reduction is called **markdown** (or discount).

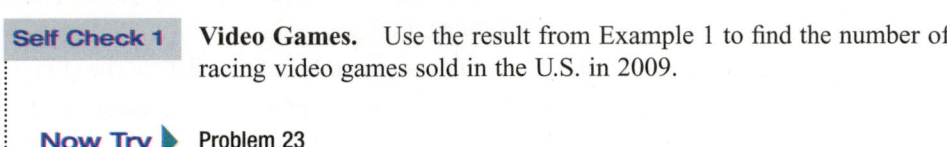

Sale price	=	regular price	−	markdown

Usually, the markdown is expressed as a percent of the regular price.

Markdown	=	percent of markdown	·	regular price

EXAMPLE 2   **Wedding Gowns.** At a bridal shop, a wedding gown that normally sells for $397.98 is on sale for $265.32. Find the percent of markdown.

Analyze In this case, $265.32 is the sale price, $397.98 is the regular price, and the markdown is the *product* of $397.98 and the percent of markdown.

Assign Let $r =$ the percent of markdown, expressed as a decimal.

Form We substitute $265.32 for the sale price and $397.98 for the regular price in the following formula:

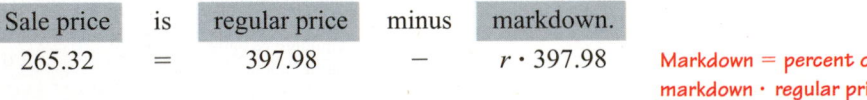

Sale price	is	regular price	minus	markdown.
265.32	=	397.98	−	$r \cdot 397.98$

Markdown = percent of markdown · regular price.

©iStockphoto.com/Gene Chutka

Solve

$$265.32 = 397.98 - 397.98r \quad \text{Rewrite } r \cdot 397.98 \text{ as } 397.98r.$$

$$-132.66 = -397.98r \quad \text{Subtract 397.98 from both sides.}$$

$$\frac{-132.66}{-397.98} = r \quad \text{To isolate } r, \text{ divide both sides by } -397.98.$$

$$0.333333\ldots = r \quad \text{Do the division using a calculator.}$$

$$33.3333\ldots\% = r \quad \text{Write the decimal as a percent. Multiply } 0.333333\ldots$$
$$\text{by 100 by moving the decimal point two places to}$$
$$\text{the right and insert a \% sign.}$$

State The percent of markdown on the wedding gown is 33.3333 . . . % or $33\frac{1}{3}\%$.

Check The markdown is $33\frac{1}{3}\%$ of \$397.98, or \$132.66. The sale price is \$397.98 − \$132.66, or \$265.32. The result checks.

> **Self Check 2**   **Seasonal Markdowns.** At an after-season sale, a winter coat that normally sells for \$279 is on sale for \$112. Find the percent of markdown.
>
> **Now Try ▶** Problem 27

Percents are often used to describe how a quantity has changed. To describe such changes, we use **percent of increase** or **percent of decrease.**

EXAMPLE 3   **Entertainment.** Use the following data to determine the percent of increase in the number of indoor movie theater screens in the United States from 1990 to 2009. Round to the nearest one percent.

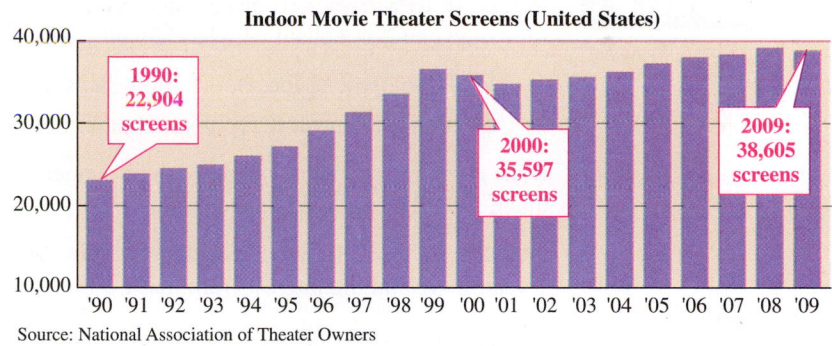

Indoor Movie Theater Screens (United States)

1990: 22,904 screens

2000: 35,597 screens

2009: 38,605 screens

Source: National Association of Theater Owners

Analyze To find the percent of increase, we first find the *amount of increase* by subtracting the number of screens in 1990 from the number in 2009.

$$38,605 - 22,904 = 15,701$$

Then we find what percent of the original 22,904 screens the 15,701 increase represents.

Assign Let x = the unknown percent.

Form We write a percent sentence using the given data. Then we translate to form a percent equation.

15,701	is	what percent	of	22,904?
15,701	=	x	⋅	22,904

The amount is 15,701, the percent is x, and the base is 22,904.

> **Caution**
>
> Always find the percent of increase (or decrease) with respect to the *original* amount.

Solve

$$15{,}701 = 22{,}904x \qquad \text{Write } x \cdot 22{,}904 \text{ as } 22{,}904x.$$

$$\frac{15{,}701}{\mathbf{22{,}904}} = \frac{22{,}904x}{\mathbf{22{,}904}} \qquad \text{To isolate } x, \text{ divide both sides by } 22{,}904.$$

$$0.685513447\ldots \approx x \qquad \text{Do the division using a calculator.}$$

$$68.5513447\ldots\% \approx x \qquad \text{Write the decimal as a percent.}$$

$$69\% \approx x \qquad \text{Round to the nearest one percent.}$$

> **Success Tip**
>
> One way to check an answer is using *estimation*. If your estimate and your answer are close, you can be reasonably sure that your answer is correct.

State   There was a 69% increase in the number of movie screens in the United States from 1990 to 2009.

Check   A 50% increase from 22,904 screens would be approximately 11,000 additional screens. It seems reasonable that 15,701 more screens would be a 69% increase.

> **Self Check 3**   **Entertainment.** In 2008, there were 38,934 indoor movie theater screens in the U.S. Determine the percent decrease from 2008 to 2009. Round to the nearest one-hundredth of one percent.
>
> **Now Try** ▶ Problem 33

2   Solve Investment Problems.

The money an investment earns is called **interest. Simple interest** is calculated by the formula $\boldsymbol{I = Prt}$, where I is the interest earned, P is the principal (amount invested), r is the annual interest rate, and t is the length of time the principal is invested.

EXAMPLE 4   **Interest Income.**   To protect against a major loss, a financial analyst suggested the following plan for a client who has $50,000 to invest for 1 year.

1. Alco Development, Inc. Builds mini-malls. High yield: 12% per year. Risky!
2. Certificate of deposit (CD). Insured, safe. Low yield: 4.5% annual interest.

If the client puts some money in each investment and wants to earn $3,600 in interest, how much should be invested at each rate?

Analyze   In this case, we are working with two investments made at two different rates for 1 year. If we add the interest from the two investments, the sum should equal $3,600.

> **Caution**
>
> Don't forget to move the decimal point in 4.5% two places to the left to express it as a decimal: 4.5% = 0.045.

Assign   We let $x =$ the number of dollars invested at 12%. If $\$x$ is invested at 12%, there is $\$(50{,}000 - x)$ to invest at 4.5%. These facts are entered in the table below.

Form   The interest earned by each investment is the product of the *principal, rate,* and *time,* as shown in the last column of the table.

	P · r · $t =$			I	
Alco Development, Inc.	x	0.12	1	$0.12x$	← This is $x \cdot 0.12 \cdot 1$.
Certificate of deposit	$50{,}000 - x$	0.045	1	$0.045(50{,}000 - x)$	← This is $(50{,}000 - x) \cdot 0.045 \cdot 1$.

Total: $3,600

Enter this information first.   Use the information in this column to form an equation.

The sum of the two amounts of interest should equal $3,600. We now translate the words into an equation.

The interest earned at 12%	plus	the interest earned at 4.5%	equals	the total interest earned.
$0.12x$	$+$	$0.045(50,000 - x)$	$=$	$3,600$

Solve

$$0.12x + 0.045(50,000 - x) = 3,600$$

$$1,000[0.12x + 0.045(50,000 - x)] = 1,000(3,600) \qquad \text{To clear the equation of decimals, multiply both sides by 1,000.}$$

$$120x + 45(50,000 - x) = 3,600,000 \qquad \text{Distribute the multiplication by 1,000 and simplify both sides.}$$

$$120x + 2,250,000 - 45x = 3,600,000 \qquad \text{Distribute 45.}$$

$$75x + 2,250,000 = 3,600,000 \qquad \text{Combine like terms.}$$

$$75x = 1,350,000 \qquad \text{To isolate 75x, subtract 2,250,000 from both sides.}$$

$$x = 18,000 \qquad \text{To isolate x, divide both sides by 75.}$$

> **Caution**
>
> On the left side of the equation, do not incorrectly distribute the multiplication by 1,000 over addition **and** multiplication.

State　$18,000 should be invested at 12% and $(50,000 - 18,000) = $32,000 should be invested at 4.5%.

Check　The annual interest on $18,000 is 0.12($18,000) = $2,160. The interest earned on $32,000 is 0.045($32,000) = $1,440. The total interest is $2,160 + $1,440 = $3,600. The results check.

Self Check 4　**Interest Income.**　A college student invested a total of $12,000, some at 6% and the rest at 9%. If the total interest earned in 1 year was $945, find the amount invested at each rate.

Now Try ▶　Problem 35

3　Solve Uniform Motion Problems.

Problems that involve an object traveling at a constant rate for a specified period of time over a certain distance are called **uniform motion** problems. To solve these problems, we use the formula *d* = *rt*, where *d* is distance, *r* is rate, and *t* is time.

EXAMPLE 5　**Travel Time.**　After a stay on her grandparents' farm, a girl is to return home, 385 miles away. To split up the drive, the parents and grandparents start at the same time and drive toward each other, planning to meet somewhere along the way. If the parents travel at an average rate of 60 mph and the grandparents at 50 mph, how long will it take them to meet?

Analyze　The vehicles are traveling toward each other as shown in the figure on the next page. We know the rates the cars are traveling (60 mph and 50 mph). We also know that they will travel for the same amount of time.

Assign　Let *t* = the time in hours that each vehicle travels. This information is entered in the table.

> **Caution**
>
> When using *d* = *rt*, make sure the units are consistent. For example, if the rate is given in miles per hour, the time must be expressed in hours.

Form   The distance traveled by the parents and the grandparents is the product of the *rate* and the *time*, as shown in the last column of the table.

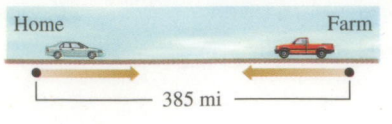

	$r \cdot t =$		d
Parents	60	t	$60t$
Grandparents	50	t	$50t$

} Multiply r · t to obtain an expression for each distance traveled.

Total: 385 mi.

Enter this information first.   Use the information in this column to form an equation.

We now translate the words of the problem into an equation.

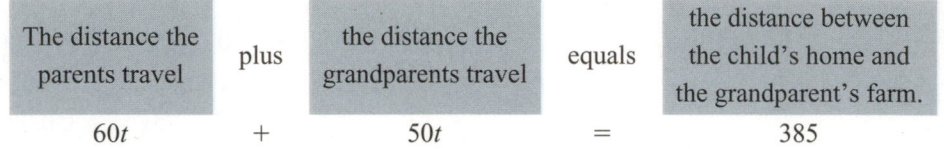

The distance the parents travel	plus	the distance the grandparents travel	equals	the distance between the child's home and the grandparent's farm.
$60t$	$+$	$50t$	$=$	385

Solve

$$60t + 50t = 385$$
$$110t = 385 \quad \text{Combine like terms: } 60t + 50t = 110t.$$
$$t = 3.5 \quad \text{To isolate } t, \text{ divide both sides by 110.}$$

State   The parents and grandparents will meet in $3\frac{1}{2}$ hours.

Check   The parents travel $3.5(60) = 210$ miles. The grandparents travel $3.5(50) = 175$ miles. The total distance traveled is $210 + 175 = 385$ miles. The result checks.

> **Self Check 5**   **Travel Time.**   Two cars leave at the same time from two towns that are 345 miles apart. The first car is traveling at a rate of 65 mph, and the other car is traveling at 50 mph. If the cars are traveling toward each other, how long will it take the two cars to meet?
>
> **Now Try** ▶ Problem 41

4 Solve Mixture Problems.

We now discuss two types of mixture problems. In the first example, a **dry mixture** of a specified value is created from two differently priced ingredients (or components). The value of each of its ingredients and the value of the dry mixture is given by

Amount · price = total value

EXAMPLE 6   **Snacks.**   The owner of a produce store notices that 20 pounds of gourmet cashews did not sell because of their high price of $12 per pound. The owner decides to mix peanuts with the cashews to lower the price per pound. If peanuts sell for $3 per pound, how many pounds of peanuts must be mixed with the cashews to make a mixture that could be sold for $6 per pound?

Analyze   We need to determine how many pounds of peanuts to mix with 20 pounds of cashews to obtain a mixture that could be sold for $6 per pound.

Assign   Let $x =$ the number of pounds of peanuts to be used. Then $20 + x =$ the number of pounds in the mixture. We enter this in the table.

Form   The total value for each ingredient and the mixture is the product of the *amount* and the *price*, as shown in the last column of the table.

	Amount	· Price	= Total value
Cashews	20	12	240
Peanuts	x	3	$3x$
Mixture	$20 + x$	6	$6(20 + x)$

Multiply amount · price three times to fill in this column.

Enter this information first.

Use the information in this column to form an equation.

We can now form the equation.

The value of the cashews	plus	the value of the peanuts	equals	the value of the mixture.
240	+	$3x$	=	$6(20 + x)$

Solve

$240 + 3x = 6(20 + x)$

$240 + 3x = 120 + 6x$ Distribute the multiplication by 6.

$120 = 3x$ To isolate the variable term, subtract 3x and 120 from both sides.

$40 = x$ To isolate x, divide both sides by 3.

State   The owner should mix 40 pounds of peanuts with the 20 pounds of cashews.

Check   The cashews are valued at $12(20) = \$240$, and the peanuts are valued at $3(40) = \$120$. The mixture is valued at $6(60) = \$360$. Since the value of the cashews plus the value of the peanuts equals the value of the mixture, the result checks.

Self Check 6   **Coffee Blends.**   A store sells regular coffee for $8 a pound and gourmet coffee for $14 a pound. To get rid of 40 pounds of the gourmet coffee, a shopkeeper makes a blend to put on sale at $10 a pound. How many pounds of the regular coffee should he use?

Now Try ▶ Problem 49

In the next example, a **liquid mixture** of a desired strength is to be made from two solutions with different strengths (concentrations).

EXAMPLE 7   **Milk Production.**   Owners of a dairy find that milk with a 2% butterfat content is their best seller. Suppose the dairy has large quantities of whole milk having a 4% butterfat content and milk having a 1% butterfat content. How much of each type of milk should be mixed to obtain 120 gallons of milk that is 2% butterfat?

Analyze   We are to find the amount of 4% milk to mix with 1% milk to get 120 gallons of a milk that has a 2% butterfat content. The amount of pure butterfat in each solution is given by:

Amount of solution · strength of the solution = amount of pure butterfat

Assign   Refer to the figure on the next page. If we let g = the number of gallons of the 4% milk used in the mixture, then $120 - g$ = the number of gallons of the 1% milk needed to obtain the desired strength.

Form   The amount of butterfat in a tank is the product of the *amount* of milk in the tank and its *strength*. In the first tank shown in the figure, 4% of the g gallons, or $0.04g$ gallons, is butterfat. In the second tank, 1% of the $(120 - g)$ gallons, or $0.01(120 - g)$ gallons, is butterfat. Upon mixing, the third tank will have $0.02(120)$ gallons of butterfat in it. These results are recorded in the last column of the table.

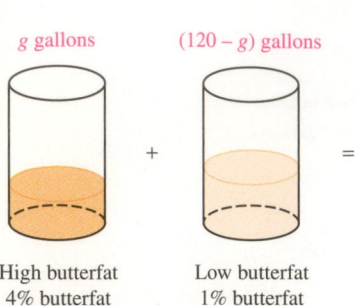

g gallons   (120 − g) gallons   120 gallons

High butterfat   Low butterfat   Mixture
4% butterfat   1% butterfat   2% butterfat

	Amount	**· Strength =**	**Amount of pure butterfat**
High butterfat	g	0.04	$0.04g$
Low butterfat	$120 - g$	0.01	$0.01(120 - g)$
Mixture	120	0.02	$0.02(120)$

⎫
⎬ Multiply three times to fill in this column.
⎭

Enter this information first.

↑
Use the information in this column to form an equation.

We now translate the words of the problem into an equation.

The amount of butterfat in g gallons of 4% milk	plus	the amount of butterfat in $(120 - g)$ gallons of 1% milk	equals	the amount of butterfat in 120 gallons of the mixture.
$0.04(g)$	$+$	$0.01(120 - g)$	$=$	$0.02(120)$

Solve

$$0.04(g) + 0.01(120 - g) = 0.02(120)$$

$$4(g) + 1(120 - g) = 2(120)$$ Multiply both sides by 100 to clear the equation of decimals. This is done mentally.

$$4g + 120 - g = 240$$ Simplify.

$$3g + 120 = 240$$ Combine like terms: $4g - g = 3g$.

$$3g = 120$$ Subtract 120 from both sides.

$$g = 40$$ To isolate g, divide both sides by 3.

State   40 gallons of 4% milk and $120 - 40 = 80$ gallons of 1% milk should be mixed to get 120 gallons of milk with a 2% butterfat content.

Check   The 40 gallons of 4% milk contains $0.04(40) = 1.6$ gallons of butterfat, and 80 gallons of 1% milk contains $0.01(80) = 0.8$ gallons of butterfat—a total of $1.6 + 0.8 = 2.4$ gallons of butterfat. The 120 gallons of the 2% mixture contains 2.4 gallons of butterfat. The results check.

Self Check 7   **Mixing Antifreeze.**   A 50% antifreeze solution is mixed with a 25% antifreeze solution to obtain 30 liters of a 30% solution. How much of each concentration should be used?

Now Try ▶   Problem 59

SECTION 1.8   **STUDY SET**

VOCABULARY

Fill in the blanks.

1. _____ means parts per one hundred.

2. In the statement, "10 is 20% of 50," 10 is the _____, and 50 is the _____.

3. When the regular price of an item is reduced, the amount of reduction is called the _____.

4. When an investment is made, the amount of money invested is called the _____. The money an investment earns is called _____.

CONCEPTS

5. One method to solve applied percent problems is to use the given facts to write a percent sentence. What is the basic form of a percent sentence?

6.

> Total Paid Circulation: *People* Magazine
> **2007:** 3,618,718   **2008:** 3,691,819

Source: *The World Almanac*, 2010

a. Find the *amount* of increase in circulation of *People* magazine.

b. Fill in the percent sentence that can be used to find the percent of increase in circulation.

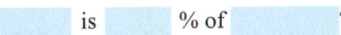 is ____ % of _____ ?

7. Complete each formula used in this section.

a. $I =$ ____

b. $d =$ ____

c. Total value = amount · _____

d. Amount pure = amount · _____

8. A student inherited $5,000 and deposits x dollars in American Savings. Write an algebraic expression that represents the amount of money left to deposit in a City Mutual account.

$5,000

American Savings   City Mutual

9. Solution 1 is poured into solution 2. Write an algebraic expression that represents the number of ounces in the mixture.

Solution 1
20 ounces

Solution 2
x ounces

10. Cashews were mixed with p pounds of peanuts to make 100 pounds of a mixture. Write an algebraic expression that represents the number of pounds of cashews that were used.

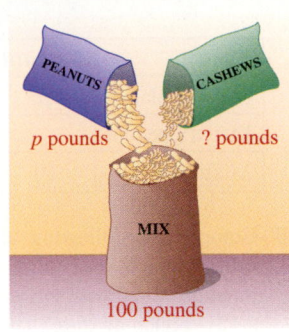

PEANUTS   CASHEWS

p pounds   ? pounds

MIX

100 pounds

11. Complete the following table for each investment. Then write an equation that could be used to solve this investment problem.

	Principal ·	Rate ·	Time =	Interest
Bonds	x	0.055	1	
Stocks	$10,850 - x$	0.07	1	

Total: $1,205

12. The following table shows how a retired teacher invested a total of $8,000 in two accounts for 1 year. The investments earned $290 in interest. Complete the table. Then write an equation that could be used to solve this investment problem.

	P ·	r ·	$t =$	I
S & L	x	0.03		
Credit Union		0.04		

Total:

13. A banker invested the *same* amount in two money-making opportunities for 1 year and earned $3,300 in interest. Complete the table below. Then write an equation that could be used to solve this investment problem.

	P ·	r ·	$t =$	I
Cattle futures	x	0.15		
Soybeans		0.18		

Total: $3,300

14. A husband and wife drive in opposite directions to work. Their drives last the same amount of time and their workplaces are 40 miles apart. Complete the table. Then write an equation that could be used to solve this uniform motion problem.

	r ·	$t =$	d
Husband	35	t	
Wife	45		

Total:

15. A 50-pound mixture of plain and peanut M&M's is to be made. Complete the table. Then write an equation that could be used to solve this mixture problem.

	Amount	·	Price	= Total Value
M&M's plain	p		7.45	
M&M's peanut			8.25	
Mixture	50		7.75	

16. Complete the following table that could be used to solve this problem: How many pints of punch from the orange cooler must be mixed with the entire contents of the blue cooler to get a 12% punch mixture? Then write an equation that could be used to solve this liquid mixture problem.

	Amount	·	Strength	= Pure concentrate
Too strong	20			
Too weak			0.10	
Mixture				

20 pints of punch, 20% concentrate x pints of punch, 10% concentrate

NOTATION

17. **a.** Write 2.5% as a decimal.
    b. Write 0.06 as a percent.
18. Multiply: $100(0.67x)$

GUIDED PRACTICE

Translate each statement into mathematical symbols. Do not solve. See Example 1.

19. What number is 5% of 10.56?
20. 16 is what percent of 55?
21. 32.5 is 74% of what number?
22. What is 83.5% of 245?

APPLICATIONS

23. **Energy.** In 2010, the United States alone accounted for 20.2% of the world's total energy consumption, using 103.3 quadrillion British thermal units (Btu). What was the world's energy consumption in 2010? Round to the nearest quadrillion. (Source: U.S. Energy Information Administration)

24. **Computers.** In 2010, 78 million, or 68.7%, of U.S. households had personal computers with an Internet connection. How many U.S. households were there in 2010? Round to the nearest million. (Source: U.S. Census)

25. **Boating Accidents.** According to the Insurance Information Institute, 6.5% of the 4,730 recreational boating accidents in 2009 involved alcohol use. How many boating accidents were alcohol-related? Round up to the nearest one accident. (Source: U.S. Coast Guard)

26. **Basketball.** The following data is for the 2009 NCAA men's college basketball season. What was the average number of three-point shots made per game? Round to the nearest hundredth.

THREE-POINT SHOTS
Average number of attempts per game: 18.33
Percent made: 34.40% (Souce: NCAA)

27. **Buying Appliances.** Use the following ad to find the percent of markdown of the sale.

28. **Buying Furniture.** A bedroom set regularly sells for $983. If it is on sale for $737.25, what is the percent of markdown?

29. **Flea Markets.** A vendor sells tool chests at a flea market for $65. If she makes a profit of 30% on each unit sold, what does she pay the manufacturer for each tool chest? (*Hint:* The retail price = the wholesale price + the markup.)

30. **Bookstores.** A bookstore sells a textbook for $39.20. If the bookstore makes a profit of 40% on each sale, what does the bookstore pay the publisher for each book? (*Hint:* The retail price = the wholesale price + the markup.)

31. **Improving Horsepower.** The following graph shows how the installation of a special computer chip increases the horsepower of a truck. Find the percent of increase in horsepower for the engine running at 4,000 revolutions per minute (rpm) and round to the nearest tenth of one percent.

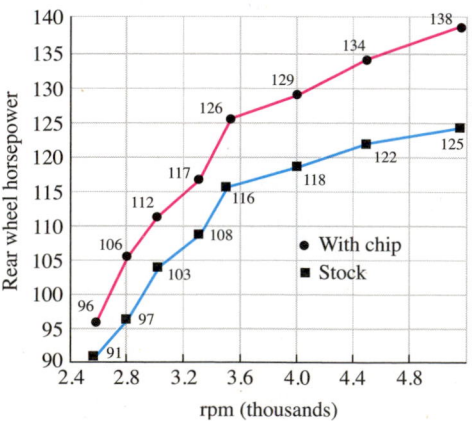

32. Greenhouse Gases. The U.S. energy-related carbon dioxide emissions in 2008 were 5,802 million metric tons. In 2007, that figure was 5,967 million metric tons. Find the percent of decrease and round to the nearest tenth of one percent. (Source: Energy Information Administration)

33. Broadway Shows. Complete the table to find the percent of increase or decrease in attendance at Broadway shows for each season compared with the previous season. Round to the nearest tenth of one percent.

Season	Broadway attendance	% of increase or decrease
2006–07	12.31 million	—
2007–08	12.27 million	
2008–09	12.15 million	

Source: LiveBroadway.com

34. Drive-Ins. The number of drive-in movie theaters in the United States peaked in 1958. Since then, the numbers have steadily declined. Determine the percent of decrease in the number of drive-ins. Round to the nearest one percent.

Number of Drive-ins
1958: 4,063
2010: 374

Source: United Drive-in Theater Owners Association

35. Highest Rates. Based on the information in the table, a woman invested $12,000, some in an account paying the highest rate and the rest in an account paying the second highest rate. How much was invested in each account if the interest from both investments is $1,060 per year?

First Republic Savings and Loan	
Account	**Rate**
NOW	5.5%
Savings	7.5%
Money market	8.0%
Checking	4.0%
5-year CD	9.0%

36. Entrepreneurs. Last year, a women's professional organization made two small-business loans totaling $28,000 to young women beginning their own businesses. The money was lent at 7% and 10% simple interest rates. If the annual income the organization received from these loans was $2,560, what was each loan amount?

37. Inheritances. Paula used some of the money that she received from an inheritance to invest in a certificate of deposit paying 7% annual interest and the rest of the money in a promising biotech company offering an annual return of 10%. She invested twice as much in the 10% investment as she did in the 7% investment. Her combined annual income from the two investments was $4,050.

  a. How much did she invest in each account?

  b. How much did she inherit?

38. Tax Returns. On a federal income tax form, Schedule B, a taxpayer forgot to write in the amount of interest income he earned for the year. From what is written on the form, determine the amount of interest earned from each investment and the amount he invested in stocks.

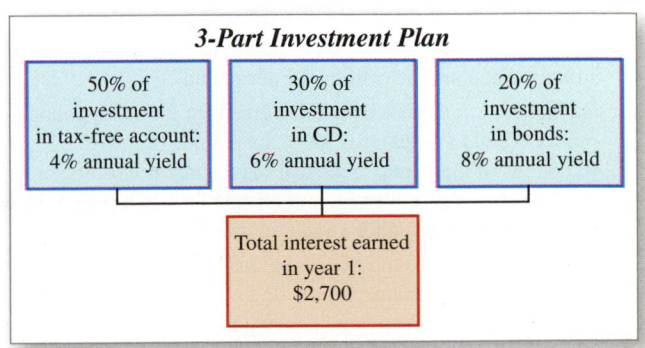

39. Money-Laundering. Use the evidence compiled by investigators to determine how much money a suspect deposited in the Cayman Islands bank.

  • On 6/1/09, the suspect electronically transferred $300,000 to a Swiss bank account paying an 8% annual yield.

  • That same day, the suspect opened another account in a Cayman Islands bank that offered a 5% annual yield.

  • A document dated 6/3/10 was seized during a raid of the suspect's home. It stated, "The total interest earned in one year from the two overseas accounts was 7.25% of the total amount deposited."

40. Financial Presentation. A financial planner showed her client the following investment plan. Find the total amount the client will have to invest to earn $2,700 in interest.

41. **Travel Times.**   A man called his wife to tell her that they needed to switch vehicles so he could use the family van to pick up some building materials after work. The wife left their home, traveling toward his office in their van at 35 mph. At the same time, the husband left his office in his car, traveling toward their home at 45 mph. If his office is 20 miles from their home, how long will it take them to meet so they can switch vehicles?

42. **Air Traffic Control.**   An airplane leaves Los Angeles bound for Caracas, Venezuela, flying at an average rate of 500 mph. At the same time, another airplane leaves Caracas bound for Los Angeles, averaging 550 mph. If the airports are 3,675 miles apart, when will the air traffic controllers have to make the pilots aware that the planes are passing each other?

43. **Cycling.**   A cyclist leaves his training base for a morning workout, riding at the rate of 18 mph. One hour later, his support staff leaves the base in a car going 45 mph in the same direction. How long will it take the support staff to catch up with the cyclist?

44. **Marathons.**   Two marathon runners leave the starting gate, one running 12 mph and the other 10 mph. If they maintain the pace, how long will it take for them to be one-quarter of a mile apart?

45. **Radio Communications.**   At 2 P.M., two military convoys leave Eagle River, Wisconsin, one headed north and one headed south. The convoy headed north averages 50 mph, and the convoy headed south averages 40 mph. They will lose radio contact when the distance between them is more than 135 miles. When will this occur?

46. **Search and Rescue.**   Two search-and-rescue teams leave base camp at the same time, looking for a lost child. The first team, on horseback, heads north at 3 mph, and the other team, on foot, heads south at 1.5 mph. How long will it take them to search a distance of 18 miles between them?

47. **Jet Skiing.**   A jet ski can go 12 mph in still water. If a rider goes upstream for 3 hours against a current of 4 mph, how long will it take the rider to return? (*Hint:* Upstream speed is $(12 - 4)$ mph; how far can the rider go in 3 hours?)

48. **Physical Fitness.**   For her workout, Sarah walks north at the rate of 3 mph and returns at the rate of 4 mph. How many miles does she walk if the round trip takes 3.5 hours?

49. **Mixing Candy.**   How many pounds of red licorice bits that sell for $1.90 per pound should be mixed with 5 pounds of lemon gumdrops that sell for $2.20 per pound to make a candy mixture that could be sold for $2 per pound?

50. **Tea Blends.**   A store sells regular green tea for $16 a pound and an exotic loose leaf tea for $28 a pound. To get rid of 40 pounds of the exotic loose leaf tea that are not selling, a shopkeeper makes a blend to put on sale for $20 a pound. How many pounds of green tea should he use?

51. **Health Foods.**   A pound of dried pineapple bits sells for $6.19, a pound of dried banana chips sells for $4.19, and a pound of raisins sells for $2.39 a pound. Two pounds of raisins are to be mixed with equal amounts of pineapple and banana to create a trail mix that will sell for $4.19 a pound. How many pounds of pineapple and banana chips should be used?

52. **Metallurgy.**   A 1-ounce commemorative coin is to be made of a combination of pure gold, costing $380 an ounce, and a gold alloy that costs $140 an ounce. If the cost of the coin is to be $200, and 500 are to be minted, how many ounces of gold and gold alloy are needed to make the coins?

53. **Gardening.**   A wholesaler of premium organic planting mix notices that the retail garden centers are not buying her product because of its high price of $1.57 per cubic foot. She decides to mix sawdust with the planting mix to lower the price per cubic foot. If the wholesaler can buy the sawdust for $0.10 per cubic foot, how many cubic feet of each must be mixed to have 6,000 cubic feet of planting mix that could be sold to retailers for $1.08 per cubic foot?

54. **Bronze.**   A pound of tin is worth $1 more than a pound of copper. Four pounds of tin are mixed with 6 pounds of copper to make bronze that sells for $3.65 per pound. How much is a pound of tin worth?

55. **Mixing Paint.**   The colors on the paint chip card below are created by adding different amounts of red tint to a white latex base. How many gallons of Fruit Shake should be mixed with 1 gallon of Watermelon to obtain Mellow Coral?

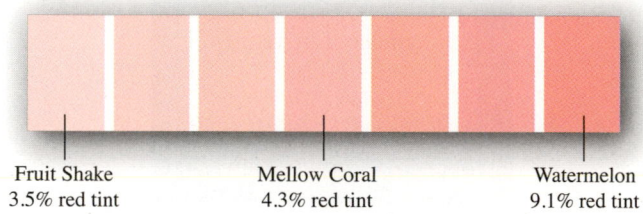

Fruit Shake   Mellow Coral   Watermelon
3.5% red tint   4.3% red tint   9.1% red tint

56. **Insect repellant.**   The active ingredient used in most insect repellents is known as DEET (N-Diethyl-meta-toluamide). How many ounces of a 1.25% DEET solution and how many ounces of a 5% DEET solution should be mixed to produce 10 ounces of a 2.5% DEET solution? (*Hint:* Express the results as mixed numbers.)

57. ▶ from **Campus to Careers**
    Registered Dietitian
    Suppose, as registered dietitian for a school district, you must make sure that only extra lean ground beef (16% fat) is served in the cafeteria. Further suppose that the kitchen has 8 pounds of regular ground beef (30% fat) on hand. How many pounds of extra lean ground beef (12% fat) must be purchased and added to the regular ground beef to obtain a mixture that has the correct fat content?

58. **Pharmacists.**   How many liters of a 1% glucose solution should a pharmacist mix with 0.5 liter of a 5% glucose solution to obtain a 2% glucose solution?

59. **Dairy Foods.**   Cream is approximately 22% butterfat. How many gallons of cream must be mixed with milk testing at 2% butterfat to get 20 gallons of milk containing 4% butterfat?

60. Flood Damage. One website recommends a 6% chlorine bleach–water solution to remove mildew. A chemical lab has 3% and 15% chlorine bleach–water solutions in stock. How many gallons of each should be mixed to obtain 100 gallons of the mildew spray?

61. Diluting Solutions. How much water should be added to 20 ounces of a 15% solution of alcohol to dilute it to a 10% alcohol solution?

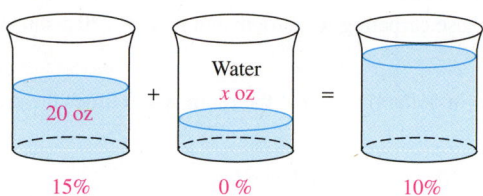

62. Evaporation. The beaker shown below contains a 2% saltwater solution.

a. How much water must be boiled away to increase the concentration of the salt solution from 2% to 3%?

b. Where on the beaker would the new water level be? (mL means milliliter.)

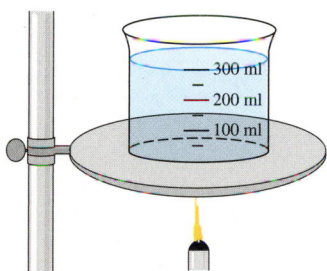

63. If a car travels at 60 mph for 30 minutes, explain why the distance traveled is not $60 \cdot 30 = 1,800$ miles.

64. If a mixture is to be made from solutions with concentrations of 12% and 30%, can the mixture have a concentration less than 12%? Can the mixture have a concentration greater than 30%? Explain.

65. Write a dry mixture problem that can be modeled by the following equation.

$$4x + 7(40 - x) = 5(40)$$

66. Write a liquid mixture problem that can be modeled by the following equation.

$$0.02(50) + 0.07x = 0.05(50 + x)$$

67. Solve the following problem. Then explain why the solution does not make sense.

Adult tickets cost $4 and student tickets cost $2. Sales of 71 tickets bring in $245. How many of each were sold?

68. Write a uniform-motion problem using the facts entered in the table.

	r	$\cdot$ t $=$	d
West	8	t	$8t$
East	6	t	$6t$

Total: 24 mi.

REVIEW

Solve each equation.

69. $9x = 6x$

70. $7a + 2 = 12 - 4(a - 3)$

71. $\dfrac{8(y - 5)}{3} = 2(y - 4)$

72. $\dfrac{t - 1}{3} = \dfrac{t + 2}{6} + 2$

CHALLENGE PROBLEMS

73. Raising Grades. A student had a score of 70% on a test that contained 30 questions. To improve his score, the instructor agreed to let him work 15 *additional* questions. How many of those must he get right to raise his grade to 80%?

74. Financial Planning. A retired teacher has a choice of two investment plans:

- An insured fund that pays 11% interest
- A risky investment that pays a 13% return

If the same amount invested at the higher rate would generate an extra $150 per year than it would at the lower rate, how much does the teacher have to invest?

1 ▶ Summary & Review

SECTION 1.1 ▶ The Language of Algebra

DEFINITIONS AND CONCEPTS	EXAMPLES
In this course, we will mathematically **model** real-life situations using verbal sentences, mathematical statements (in symbols), and tables.	Suppose we want to purchase carpeting for a room and it costs $20 a square yard. *Verbal model:* The cost c (in dollars) to carpet the room is $20 times the area a of the room (in square yards). *Mathematical model:* $c = 20a$ *Table model:*

Table model:

a	c
10	200
20	400
30	600
40	800

DEFINITIONS AND CONCEPTS	EXAMPLES
A **variable** is a letter (or symbol) that stands for a number.	Variables: x, t, a, and m
Algebraic expressions contain numbers, variables, and mathematical operations such as addition, subtraction, multiplication, division, powers, or roots.	Expressions: $5y + 7$, $\dfrac{12 - x}{5}$, and $8a(b - 3)$
An **equation** is a mathematical sentence that contains an $=$ symbol. The $=$ symbol indicates that the expressions on either side of it have the same value.	Equations: $3x + 4 = 8$, $\dfrac{t}{9} = 12$, and $I = Prt$

REVIEW EXERCISES

1. Translate each verbal model into a mathematical model.

    a. The cost C (in dollars) to rent t tables is $15 more than the product of $2 and t.

    b. A rectangle has an area of 25 in.^{2}. The length of the rectangle is the quotient of its area and its width.

    c. The waiting period for a business license is now 3 weeks less than it used to be.

2. To determine the cooking time for prime rib, a cookbook suggests using the equation $T = 30p$, where T is the cooking time in minutes and p is the weight of the prime rib in pounds. Use this equation to complete the table.

p	T
6.0	
6.5	
7.0	
7.5	
8.0	

Translate each phrase to an algebraic expression. Answers may vary depending on the variable chosen.

3. 43 ounces less than the weight

4. one-third of the difference of a number and 10

SECTION 1.2 ▶ The Real Numbers

DEFINITIONS AND CONCEPTS	EXAMPLES
Important sets of numbers	
The set of **natural numbers:**	$\mathbb{N} = \{1, 2, 3, 4, 5, 6, 7, 8, 9, 10, \ldots\}$
The set of **whole numbers:**	$\mathbb{W} = \{0, 1, 2, 3, 4, 5, 6, 7, 8, 9, 10, \ldots\}$
The set of **integers:**	$\mathbb{Z} = \{\ldots, -4, -3, -2, -1, 0, 1, 2, 3, 4, \ldots\}$
The set of **prime numbers:**	$\{2, 3, 5, 7, 11, 13, 17, 19, 23, 29, \ldots\}$
The set of **composite numbers:**	$\{4, 6, 8, 9, 10, 12, 14, 15, 16, 18, \ldots\}$
The set of **rational numbers,** denoted by the symbol $\mathbb{Q}$, contains the numbers that can be written as fractions in the form $\frac{a}{b}$, where a and b are integers and $b \neq 0$. **Terminating** and **repeating decimals** can be expressed as fractions and are, therefore, rational numbers.	Rational numbers: $$-6, \quad -3.1, \quad 0, \quad \frac{11}{12}, \quad 9\frac{4}{5}, \quad 87, \quad 0.25 = \frac{1}{4}, \quad \text{and} \quad -0.\overline{6} = -\frac{2}{3}$$
The set of **irrational numbers,** denoted by the symbol $\mathbb{H}$, contains the nonterminating, nonrepeating decimals. An irrational number cannot be expressed as a fraction with an integer numerator and a nonzero integer denominator.	Irrational numbers: $$\sqrt{2}, \quad \pi, \quad -\frac{\sqrt{11}}{4}, \quad 73.050050005\ldots, \quad \text{and} \quad -32\pi$$
The set of **real numbers,** denoted by the symbol $\mathbb{R}$, contains the numbers that are either a rational or an irrational number.	All of the numbers previously listed are real numbers.
Every real number corresponds to a point on the **number line,** and every point on the number line corresponds to exactly one real number.	Graph: $\left\{-2, -0.75, 1\frac{3}{4}, \pi\right\}$
Symbols used with sets	
The symbol $\in$ is used to indicate that an **element** belongs to a set.	$6 \in \mathbb{N}$ 6 is an element of the set of natural numbers.
	$3 \notin \mathbb{H}$ 3 is not an element of the set of irrational numbers.
When all the members of one set are also members of a second set, we say the first set is a **subset** of the second set. The symbol $\subseteq$ is used to indicate that one set is a **subset** of another set.	$\mathbb{N} \subseteq \mathbb{W}$ The set of natural numbers is a subset of the set of whole numbers.
	$\mathbb{Q} \not\subseteq \mathbb{Z}$ The set of rational numbers is not a subset of the set of integers.
Inequality symbols	
$\neq$ is not equal to	$3 \neq 5$
$<$ is less than $\leq$ is less than or equal to	$0.23 < 0.24$ $8 \leq 12$ and $-9 \leq -9$
$>$ is greater than $\geq$ is greater than or equal to	$\frac{3}{4} > \frac{2}{3}$ $-5 \geq -6$ and $8 \geq 8$
Two numbers are called **opposites** if they are the same distance from 0 on the number line but are on opposite sides of it.	Opposites: 3 and -3
The **opposite of a number** a is $-a$. $$-(-a) = a$$	$-(-11) = 11$ and $-(-20) = 20$

The **absolute value** of a number is the distance on the number line between the number and 0. For any real number x: $\begin{cases} \text{If } x \geq 0, \text{ then } \lvert x \rvert = x \\ \text{If } x < 0, \text{ then } \lvert x \rvert = -x \end{cases}$	Find the value of each expression: $\lvert 7.2 \rvert = 7.2 \qquad \lvert -1 \rvert = 1 \qquad -\left\lvert -\dfrac{13}{5} \right\rvert = -\dfrac{13}{5}$

REVIEW EXERCISES

List the numbers in

$$\left\{ -5, 0, -\sqrt{3}, 2.4, 7, -\tfrac{2}{3}, -3.\overline{6}, \pi, \tfrac{15}{4}, 0.13242368\ldots \right\}$$

that belong to the following sets.

5. a. Natural numbers   **b.** Whole numbers

6. a. Integers   **b.** Rational numbers

7. a. Irrational numbers   **b.** Real numbers

8. a. Negative numbers   **b.** Positive numbers

9. a. Prime numbers   **b.** Composite numbers

10. a. Even integers   **b.** Odd integers

11. Graph the prime numbers between 20 and 30 on the number line.

12. Graph the set $\left\{ 2.75, 2.\overline{3}, \sqrt{7}, \tfrac{8}{3}, \tfrac{3\pi}{4} \right\}$ on the number line.

Determine whether each statement is true or false.

13. a. $0 \in \mathbb{N}$ **b.** $\mathbb{W} \subseteq \mathbb{Z}$

14. a. $-5 \notin \mathbb{R}$ **b.** $\mathbb{Q} \not\subseteq \mathbb{H}$

15. Use one of the symbols $>$ or $<$ to make each statement true.

    a. $-16 \;\blacksquare\; -17$ **b.** $-(-1.8) \;\blacksquare\; 2\tfrac{1}{2}$

16. Determine whether each statement is true or false.

    a. $23.000001 \geq 23.1$ **b.** $-11 \leq -11$

Find the value of each expression.

17. $\lvert -18 \rvert$ **18.** $-\lvert -6.26 \rvert$

SECTION 1.3 ▶ Operations with Real Numbers

DEFINITIONS AND CONCEPTS	EXAMPLES
Adding real numbers: **To add two positive numbers,** add them in the usual way. The final answer is positive. **To add two negative numbers,** add their absolute values and make the final answer negative.	Add: $3 + 5 = 8$ Add: $-3 + (-5) = -8$ $5.2 + 7.3 = 12.5$ $-5.2 + (-7.3) = -12.5$
To add a positive number and a negative number, subtract the smaller absolute value from the larger. **1.** If the positive number has the larger absolute value, make the final answer positive. **2.** If the negative number has the larger absolute value, make the final answer negative.	Add: $5 + (-3) = 2$ Add: $6 + (-20) = -14$ $-5.2 + 7 = 1.8$ $-9 + 3 = -6$
To **subtract two real numbers,** add the first to the opposite of the number to be subtracted. For any real numbers a and b, $$a - b = a + (-b)$$	Subtract: $4 - 7 = 4 + (-7) = -3$ $-1 - (-2) = -1 + 2 = 1$ $6 - (-8) = 6 + 8 = 14$ $-3 - 4 = -3 + (-4) = -7$
Multiplying and dividing real numbers: With **unlike signs,** multiply (or divide) their absolute values and make the final answer negative. With **like signs,** multiply (or divide) their absolute values and make the final answer positive.	Multiply: $-5(7) = -35$ Divide: $\dfrac{-12}{4} = -3$ Multiply: $-6(-8) = 48$ Divide: $\dfrac{-25}{-5} = 5$

x^n is a *power of x*. x is the **base,** and n is the **exponent.** An exponent represents repeated multiplication. $$\overset{\text{\textcolor{blue}{n factors of x}}}{x^n = \overbrace{x \cdot x \cdot x \cdot \,\cdots\, \cdot x}}$$	In 4^3, the base is 4 and 3 is the exponent. $$4^3 = 4 \cdot 4 \cdot 4 \qquad (-3)^4 = (-3)(-3)(-3)(-3)$$
A number b is a **square root** of a if $b^2 = a$. If $a > 0$, $\sqrt{a}$ represents the **principal** (positive) **square root** of a.	Find each square root: $\sqrt{16} = 4$ **Because $4^2 = 16$.** $\sqrt{144} = 12$ **Because $12^2 = 144$.**
Order of operations rule: 1. Perform all calculations within parentheses in the order listed in steps 2–4, working from the innermost pair to the outermost pair. 2. Evaluate all powers and roots. 3. Perform all multiplications and divisions, working from left to right. 4. Perform all additions and subtractions, working from left to right. 5. When all grouping symbols have been removed, repeat steps 2–4 to finish the calculation. If a fraction is present, evaluate the numerator and denominator separately, and then simplify the fraction, if possible.	Evaluate: $\begin{aligned} &3 + 2[-4 - 7(\mathbf{5-2})^2] \\ &= 3 + 2[-4 - 7(\mathbf{3})^2] \quad \textbf{Work within the parentheses.} \\ &= 3 + 2[-4 - 7(9)] \quad \textbf{Evaluate the power.} \\ &= 3 + 2(-4 - 63) \quad \textbf{Do the multiplication.} \\ &= 3 + 2(-67) \quad \textbf{Do the subtraction with parentheses.} \\ &= 3 - 134 \quad \textbf{Do the multiplication.} \\ &= -131 \quad \textbf{Do the subtraction.} \end{aligned}$
To **evaluate an algebraic expression,** substitute the values for the variables and then use the order of operations rule.	Evaluate $\dfrac{a+b}{2(b-a)}$ for $a = 3$ and $b = -5$. $\dfrac{\mathbf{a+b}}{\mathbf{2(b-a)}} = \dfrac{\mathbf{3+(-5)}}{\mathbf{2(-5-3)}}$ **Substitute 3 for a and -5 for b.** $= \dfrac{-2}{2(-8)}$ **Evaluate numerator and denominator separately.** $= \dfrac{-2}{-16}$ **Multiply in the denominator.** $= \dfrac{1}{8}$ **Simplify the fraction.**

REVIEW EXERCISES

Perform the operations.

19. $-13 + (-14)$

20. $-70.5 + 80.6$

21. $-\dfrac{1}{2} - \dfrac{1}{4}$

22. $-6 - (-8)$

23. $(-4.2)(-3.0)$

24. $-\dfrac{1}{10} \cdot \dfrac{5}{16}$

25. $\dfrac{-2.2}{-11}$

26. $-\dfrac{9}{8} \div 21$

27. $15 - 25 - 23$

28. $-3.5 + (-7.1) + 4.9$

29. $-3(-5)(-8)$

30. $-1(-1)(-1)(-1)$

Evaluate each expression.

31. $(-3)^5$

32. $\left(-\dfrac{2}{9}\right)^2$

33. 0.4 cubed

34. -5^2

Evaluate each expression.

35. $\sqrt{4}$

36. $-\sqrt{100}$

37. $\sqrt{\dfrac{9}{25}}$

38. $\sqrt{0.64}$

Evaluate each expression.

39. $-6 + 2(-5)^2$

40. $\dfrac{-20}{4} - (-3)(-2)\left(-\sqrt{1}\right)$

41. $4 - (5-9)^2$

42. $4 + 6[-1 - 5(25 - 3^3)]$

43. $2|-1.3 + (-2.7)|$

44. $\dfrac{(7-6)^4 + 32}{36 - (\sqrt{16}+1)^2}$

45. $(-10)^3\left(\dfrac{-6}{-2}\right)(-1)$

46. $-(-2 \cdot 4)^2 \div 8 \cdot 2$

Evaluate the algebraic expression for the given values of the variables.

47. $(x+y)(x^2 - xy + y^2)$ for $x = -2$ and $y = 4$

48. $\dfrac{-b - \sqrt{b^2 - 4ac}}{2a}$ for $a = 2$, $b = -3$, and $c = -2$

SECTION 1.4 ▶ Simplifying Algebraic Expressions Using Properties of Real Numbers

DEFINITIONS AND CONCEPTS	EXAMPLES
A **term** is a product or quotient of numbers and/or variables. A single number or variable is also a term. A term that is a number is called a **constant term**. The numerical factor of a term is called its **coefficient**.	Terms: z, $-7t$, $3.7x^7$, $\dfrac{3}{5}y$, $25ab^2c^3$, and 6 (a constant) The coefficients of the terms above are: 1, -7, 3.7, $\dfrac{3}{5}$, 25, and 6
Properties of real numbers The **commutative properties** enable us to add or multiply two numbers in either order and obtain the same result. $a + b = b + a$ Commutative property of addition $ab = ba$ Commutative property of multiplication The **associative properties** enable us to group the numbers in an addition or multiplication any way that we wish and get the same result. $(a + b) + c = a + (b + c)$ Associative property of addition $(ab)c = a(bc)$ Associative property of multiplication	$3 + (-5) = -5 + 3$ Reorder the addends. $3(-5) = -5(3)$ Reorder the factors. $(-5 + 7) + 4 = -5 + (7 + 4)$ Regroup the addends. $(-5 \cdot 7)4 = -5(7 \cdot 4)$ Regroup the factors.
0 is the **additive identity**: $a + 0 = a$ and $0 + a = a$ 1 is the **multiplicative identity**: $1 \cdot a = a$ and $a \cdot 1 = a$ **Multiplication property of 0:** $a \cdot 0 = 0$ and $0 \cdot a = 0$ The **additive inverse property**: $a + (-a) = 0$ and $-a + a = 0$ The **multiplicative inverse property**: $a \cdot \dfrac{1}{a} = 1$ and $\dfrac{1}{a} \cdot a = 1$ $(a \neq 0)$	$5 + 0 = 5$ The sum of a real number and 0 is the number. $1 \cdot 22 = 22$ The product of any real number and 1 is the number. $-1.78 \cdot 0 = 0$ The product of any real number and 0 is 0. $9 + (-9) = 0$ The sum of a real number and its opposite is 0. $7 \cdot \dfrac{1}{7} = 1$ The product of a real number and its reciprocal is 1.
Division properties of real numbers: $\dfrac{a}{1} = a$ $\dfrac{a}{a} = 1$ $\dfrac{0}{a} = 0$ $\dfrac{a}{0}$ is undefined $\dfrac{0}{0}$ is indeterminate	Divide: $\dfrac{6.8}{1} = 6.8$ $\dfrac{3}{3} = 1$ $\dfrac{0}{15} = 0$ $\dfrac{-2.03}{0}$ is undefined

The **distributive property** and its related forms can be used to remove parentheses. $a(b + c) = ab + ac \qquad a(b - c) = ab - ac$ $(b + c)a = ba + ca \qquad (b - c)a = ba - ca$ The **extended distributive property:** $a(b + c + d + e + \cdots) =$ $\qquad\qquad ab + ac + ad + ae + \cdots$	Multiply: $3(a + 4) = 3a + 3 \cdot 4 \qquad (x - 5)x = x \cdot x - 5 \cdot x$ $\qquad\quad = 3a + 12 \qquad\qquad\qquad = x^2 - 5x$ $-4(2a + b - 7) = -4(2a) + (-4)b - (-4)(7)$ $\qquad\qquad\qquad\quad = -8a - 4b + 28$
Terms with exactly the same variables raised to exactly the same powers are called **like terms.**	Like terms: $6x$ and $-7x$ $\qquad -4a^2b$ and $8a^2b$ Unlike terms: $6x$ and $-7y$ $\qquad -4a^2b$ and $8a^3b$
Simplifying the sum or difference of like terms is called **combining like terms.** Like terms can be combined by adding or subtracting the coefficients of the terms and keeping the same variables with the same exponents.	Simplify: $\qquad\qquad\qquad 4a + 2a = 6a \qquad$ Think: $(4 + 2)a = 6a$. $5p^2 + p - p^2 - 9p = 4p^2 - 8p \qquad$ Think: $(5 - 1)p^2 = 4p^2$ and $(1 - 9)p = -8p$. $2(k - 1) - 3(k + 2) = 2k - 2 - 3k - 6 \qquad$ Distribute. $\qquad\qquad\qquad\qquad = -k - 8 \qquad$ Think: $(2 - 3)k = -k$ and $(-2 - 6) = -8$.

REVIEW EXERCISES

Fill in the blanks by applying the indicated property of the real numbers.

49. $3(x + 7) = $ _____ (Distributive property and simplify)

50. $t \cdot 5 = $ ___ (Commutative property of multiplication)

51. $-x + x = $ ___ (Additive inverse property)

52. $(27 + 1) + 99 = $ _____ (Associative property of addition)

53. $\frac{1}{8} \cdot 8 = $ ___ (Multiplicative inverse property)

54. $0 + m = $ ___ (Additive identity property)

55. ___ $\cdot \, 9.87 = 9.87$ (Multiplicative identity property)

56. $5(-9)(0)(2{,}345) = $ ___ (Multiplication property of 0)

57. $(-3 \cdot 5)2 = $ _____ (Associative property of multiplication)

58. $(t + z) \cdot t = $ _____ (Commutative property of addition)

59. Perform each division.

    a. $\dfrac{102}{102}$ **b.** $\dfrac{-25}{1}$

60. Perform each division, if possible.

    a. $\dfrac{0}{6}$ **b.** $\dfrac{5.88}{0}$

Multiply.

61. $8(9x + 6)$

62. $-(6y - 2)$

63. $(3x - 2y)(1.2)$

64. $\dfrac{3}{4}(8c^2 - 4c + 1)$

Simplify each expression.

65. $8(6k)$

66. $(-7.5x)(-10y)$

67. $-9(-3p)(-7)$

68. $15a + 7 + 30a + 9$

69. $3g^2 + g^2 - 3g^2 - g^2$

70. $-m + 4(m - 12n) - (-8n)$

71. $\dfrac{7}{5}x - \dfrac{3}{4}x$

72. $21.45l + (-45.99l)$

73. $4[-2(a^3 - a^2) - 2(3a^2 - 6a^3)]$

74. $\dfrac{3}{4}(2h + 9) - \dfrac{5}{4}(h - 1)$

SECTION 1.5 ▶ Solving Linear Equations Using Properties of Equality

DEFINITIONS AND CONCEPTS	EXAMPLES
A **linear equation in one variable** is an equation that can be written in the form $ax + b = c$, where $a \neq 0$.	Linear equations: Notice that the highest power on the variable is 1. $$5y - 2 = 12 \qquad \frac{11}{6}t = 7 \qquad 4b - 7 + 2b = 1 + 2b + 8$$
A number that makes an equation a true statement when substituted for the variable is called a **solution** of the equation. The **solution set** of an equation is the set of all numbers that make the equation true.	Determine whether 2 is a solution of $x + 4 = 3x$. *Check:* $\quad x + 4 = 3x$ $\quad 2 + 4 \overset{?}{=} 3(2)$ Substitute 2 for each x. $\qquad\quad 6 = 6$ True Since the resulting statement is true, 2 is a solution of $x + 4 = 3x$.
Equations with the same solution set are called **equivalent equations.**	$x + 1 = 9$ and $x = 8$ are equivalent equations because they have the same solution, 8.
To **solve an equation,** isolate the variable on one side of the equation by undoing the operations performed on it using properties of equality. **Addition (subtraction) property of equality:** If the same number is added to (or subtracted from) both sides of an equation, the result is an equivalent equation. **Multiplication (division) property of equality:** If both sides of an equation are multiplied (or divided) by the same nonzero number, the result is an equivalent equation.	If $a = b$, then $$a + c = b + c \qquad\qquad a - c = b - c$$ $$ca = cb \quad (c \neq 0) \qquad\qquad \frac{a}{c} = \frac{b}{c} \quad (c \neq 0)$$
Strategy for Solving Linear Equations in One Variable 1. Clear the equation of fractions or decimals. 2. Simplify each side of the equation by removing all sets of parentheses and combining like terms. 3. Isolate the variable term on one side of the equation. 4. Isolate the variable. 5. Check the result in the original equation.	Solve: $\quad \dfrac{x-1}{6} + x = \dfrac{2}{3} - \dfrac{x+2}{6}$ $$6\left(\frac{x-1}{6} + x\right) = 6\left(\frac{2}{3} - \frac{x+2}{6}\right)$$ Multiply both sides by 6 to clear the fractions. $$6 \cdot \frac{x-1}{6} + 6 \cdot x = 6 \cdot \frac{2}{3} - 6 \cdot \frac{x+2}{6}$$ Distribute 6 on both sides. $$x - 1 + 6x = 4 - (x + 2)$$ Simplify. Don't forget the parentheses. $$x - 1 + 6x = 4 - x - 2$$ Remove parentheses. $$7x - 1 = 2 - x$$ Combine like terms on each side. $$7x - 1 + x = 2 - x + x$$ To eliminate $-x$ on the right side, add x to both sides. $$8x - 1 = 2$$ Combine like terms on each side. $$8x - 1 + 1 = 2 + 1$$ To isolate the variable term 8x, add 1 to both sides. $$8x = 3$$ Simplify each side. $$\frac{8x}{8} = \frac{3}{8}$$ Isolate the variable x by dividing both sides by 8. $$x = \frac{3}{8}$$ The solution is $\frac{3}{8}$ and the solution set is $\left\{\frac{3}{8}\right\}$. Check this result to verify that it satisfies the *original* equation.

An equation that is satisfied by every number for which both sides are defined is called an **identity**.	When we solve $x + 5 + x = 2x + 5$, the variables drop out and we obtain a true statement $5 = 5$. All real numbers are solutions. The solution set is the set of real numbers written as $\mathbb{R}$.
A **contradiction** is an equation that is never true.	When we solve $y + 2 = y$, the variables drop out and we obtain a false statement $2 = 0$. The equation has no solutions. The solution set contains no elements and can be written as the **empty set** $\{\ \ \}$ or the **null set** $\varnothing$.

REVIEW EXERCISES

Determine whether -6 *is a solution of each equation.*

75. $6 - x = 2x + 24$ **76.** $\dfrac{5}{3}(x - 3) = -12$

Solve each equation and check the result.

77. $\dfrac{x}{5} = -45$ **78.** $t - 3.67 = 4.23$

79. $0.0035 = 0.25g$ **80.** $0 = x + 4$

81. $11 - 5x = -1$

82. $-3x - 7 + x = 6x + 20 - 5x$

83. $-4(y - 1) + (-3) = 25$

84. $5 + 3[2 - 13(x - 1)] = 17 - 18x$

85. $\dfrac{8}{3}(x - 5) = \dfrac{2}{5}(x - 4)$ **86.** $\dfrac{3y}{4} - 14 = -\dfrac{y}{3} - 1$

87. $-k = -0.06$ **88.** $\dfrac{5}{4}p = -10$

89. $\dfrac{4t + 1}{3} - \dfrac{t + 5}{6} = \dfrac{t - 3}{6}$

90. $33.9 - 0.5(75 - 3x) = 0.9$

Solve each equation. If the equation is an identity or a contradiction, so state.

91. $2(x - 6) = 10 + 2x$

92. $-5x + 2x - 1 = -(3x + 1)$

SECTION 1.6 ▶ Solving Formulas; Geometry

DEFINITIONS AND CONCEPTS	EXAMPLES
A **formula** is an equation that states a mathematical relationship between two or more variables.	Turn to the inside back cover for a complete list of geometric formulas.
The **perimeter** of a plane geometric figure is the distance around it. The **area** of a plane geometric figure is the amount of surface that it encloses. The **volume** of a three-dimensional geometric figure is the amount of space it encloses.	Find the volume of a cone whose base has a radius of 6 cm and whose height is 8 cm. $V = \dfrac{1}{3}\pi r^2 h$ *This is the formula for the volume of a cone.* $V = \dfrac{1}{3}\pi(6)^2(8)$ *Substitute 6 for r and 8 for h.* $V = \dfrac{1}{3}\pi(288)$ *Evaluate $(6)^2(8)$.* $V = 96\pi$ *Multiply $\frac{1}{3}$ and 288.* The exact volume is 96π cm^3. Rounded to the nearest tenth of a cubic centimeter, the approximate volume is 301.6 cm^3.
To **solve a formula for a specified variable** means to isolate that variable on one side of the equation, with all other variables and constants on the opposite side.	Solve $\ \ F = \dfrac{mMg}{r^2}\ $ for M. $Fr^2 = mMg$ *Multiply both sides by r^2.* $\dfrac{Fr^2}{mg} = M$ *To isolate M, divide both sides by mg.* $M = \dfrac{Fr^2}{mg}$ *Write M on the left side.*

93. Find the perimeter of a trapezoid whose parallel sides measure 10.5 feet and 12.5 feet and whose nonparallel sides measure 3.5 feet and 4.5 feet.

94. Find the circumference and the area of a circle with a diameter of 17 centimeters. Round to the nearest hundredth.

95. Find the volume of a sphere with a radius of 7.5 meters. Round to the nearest hundredth.

96. Highway Safety Cones.

   a. Find the area covered by the square rubber base if its sides are 10 inches long.

   b. The safety cone is centered atop the base, as shown. Give its exact volume and its approximate volume, rounded to the nearest tenth.

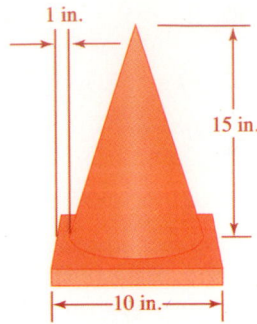

1 in.

15 in.

10 in.

Solve each formula for the specified variable.

97. $V = \frac{1}{3}\pi r^2 h$ for h

98. $K = \frac{M v_0^2 + I w^2}{2}$ for M

99. $l = a + (n-1)d$ for d

100. $9x - 5y = 35$ for y

SECTION 1.7 ▶ Using Equations to Solve Problems

DEFINITIONS AND CONCEPTS	EXAMPLES

Problem-solving strategy:

1. Analyze the problem.

2. Assign a variable.

3. Form an equation.

4. Solve the equation.

5. State the conclusion.

6. Check the result.

Some problem-solving hints

Diagrams are often helpful in solving application problems. See page 66.

For problems that deal with quantities that have a value (see page 68), use the relationship:

   Number · value = total value

It is often helpful to list the facts of a number-value problem in a **table**.

Sometimes a **geometric fact** or formula can be used to solve a problem. See pages 69–70.

The given facts of a problem often suggest a **formula** that can be used to model the situation mathematically. See page 70.

Billionaires. In 2010, *Forbes* magazine ranked Bill Gates and Warren Buffett as the two richest Americans. Their combined net worth was estimated to be $100 billion, with Gates the wealthier by $6 billion. Find the net worth of each man.

Analyze The phrase *combined net worth* suggests addition. If Gates was the wealthier, then his net worth was $6 billion *more than* Buffett's.

Assign Let x = Buffett's net worth in billions of dollars. Since Gates is the wealthier, $x + 6$ = Gates's net worth in billions of dollars.

Form We can use the words of the problem to form an equation.

Buffett's net worth	plus	Gates's net worth	equals	$100 billion.
x	$+$	$x + 6$	$=$	100

Solve

$2x + 6 = 100$ Combine like terms.

$2x = 94$ To isolate $2x$, subtract 6 from both sides.

$x = 47$ To isolate x, divide both sides by 2.

To find Gates's net worth we evaluate $x + 6$ for $x = 47$.

$x + 6 = 47 + 6 = 53$

State In 2010, Warren Buffett's net worth was $47 billion and Bill Gates's net worth was $53 billion.

Check The sum of $47 billion and $53 billion is $100 billion, and $53 billion is $6 billion more than $47 billion. The results check.

REVIEW EXERCISES

101. Airports. The world's two busiest airports are Hartsfield Atlanta International and Chicago O'Hare International. Together they served 159.3 million passengers in 2008, with Atlanta handling 20.7 million more than O'Hare. How many passengers did each airport serve? (Source: *The World Almanac*, 2010)

102. Tuition. A private school reduces the monthly tuition cost of $245 by $5 per child if a family has more than one child attending the school. Write an algebraic expression that gives the discounted monthly tuition cost per child for a family having c children.

103. Warehousing. A large warehouse stores 150 more computers than printers. The monthly storage cost for a computer is $2.50 and a printer is $1.50. If storage for the computers and printers is $2,775 per month, how many printers are in the warehouse?

104. Cable TV. A 186-foot television cable is to be cut into four pieces. Find the length of each piece if each successive piece is 3 feet longer than the previous one.

105. Tooling. The illustration shows the angle at which a drill is to be held when drilling a hole into a piece of aluminum. Find the measures of both labeled angles.

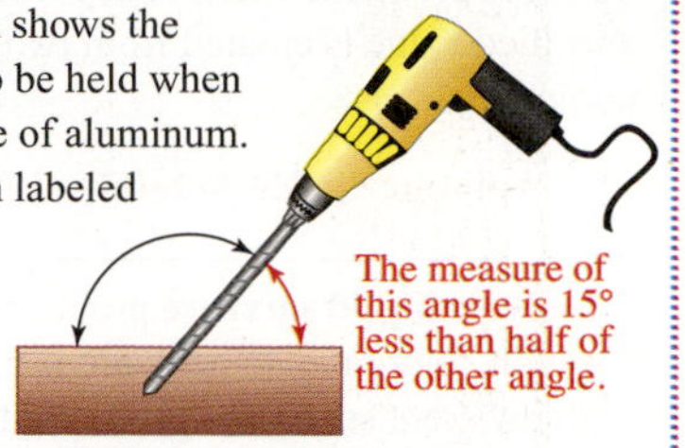

The measure of this angle is 15° less than half of the other angle.

106. Collectibles. In North America, most new movie releases are advertised using a poster size commonly referred to as a *one-sheet*. A one-sheet movie poster is rectangular, has a perimeter of 134 inches, and its length is 13 inches longer than its width. Find the dimensions of a one-sheet.

SECTION 1.8 ▶ More about Problem Solving

DEFINITIONS AND CONCEPTS	EXAMPLES
Percent means parts per one hundred.	$7\% = \dfrac{7}{100} = 0.07 \qquad 125\% = \dfrac{125}{100} = 1.25$
One method to solve percent problems is to use the given facts to write a **percent sentence** of the form: ⬚ is ⬚ % of ⬚ ? Then, translate the sentence to mathematical symbols and solve the resulting **percent equation**. Always find the **percent of increase** (or decrease) with respect to the *original* amount. Consumer formulas: Sale price = regular price − markdown Markdown = percent of markdown · regular price	**Taxes.** In Texas, $31.25 in state sales tax is charged on a $500 purchase. What is the Texas state sales tax rate? 31.25 is what percent of 500? **The percent sentence** 31.25 = x · 500 **The percent equation** $31.25 = 500x$ $\dfrac{31.25}{500} = \dfrac{500x}{500}$ To isolate x, divide both sides by 500. $0.0625 = x$ Do the division. $6.25\% = x$ Change the decimal to a percent. The Texas state sales tax rate is 6.25%.
To solve **investment** problems involving simple interest, use the formula $I = Prt$ Interest = principal · rate · time	See page 78 for an example.
To solve **uniform motion** problems, use the formula $d = rt$ Distance = rate · time	See page 79 for an example.

To solve problems where a **dry mixture** of a specified value is created from two differently priced components, use **Amount · price = total value**	See page 80 for an example.
To solve a **liquid mixture** problem, where a desired strength solution is to be made from two solutions with different strengths (concentrations), use **Amount of · strength of = amount of** **solution the solution pure ingredient**	See page 81 for an example.

REVIEW EXERCISES

107. Groundhog Day. According to groundhog.org, the weather-predicting groundhog has emerged from his burrow and seen his shadow 99 times, which is about 80% of the years on record. How many groundhog days does this website have on record?

108. Early Registration. An early bird discount lowers the registration fee for a financial seminar from $550 to $375. Find the percent of markdown. Round to the nearest percent.

109. Pickup Sales. Determine the percent of decrease in the number of Ford F-Series pickups sold in 2008 compared with 2007. Round to the nearest tenth of one percent.

110. Car Sales. Determine the percent of increase in the number of Toyota Camrys sold in 2008 compared with 2007. Round to the nearest tenth of one percent.

The Two Top-Selling Vehicles in the U.S.

2008
1. Ford F-Series P/U   235,924
2. Toyota Camry   198,309

2007
1. Ford F-Series P/U   290,282
2. Toyota Camry   193,900

Source: theautochannel.com

111. Investments. Sally has $25,000 to invest. She invests some money at 10% interest and the rest at 9%. If her total annual income from these two investments is $2,430, how much does she invest at each rate?

112. Paparazzi. A celebrity leaves a nightclub in his car and travels at 1 mile per minute (60 mph) trying to avoid a tabloid photographer. One minute later, the photographer leaves the nightclub on his motorcycle, traveling at 1.5 miles per minute (90 mph) in pursuit of the celebrity. How long will it take the photographer to catch up with the celebrity?

113. Pest Control. How much of a 4% pesticide solution must be added to 20 gallons of a 12% pesticide solution to dilute it to a 10% solution?

114. Coffee. Mild coffee that sells for $7.50 per pound is to be mixed with a robust coffee that sells for $8.40 per pound to make 90 pounds of a mixture that will be sold for $7.90 per pound. How many pounds of each type of coffee should be used?

1 ▶ CHAPTER TEST

1. Fill in the blanks.

  a. For any nonzero real number a, $\frac{a}{0}$ is _____.

  b. $>$, $\geq$, $<$, and $\leq$ are called _____ symbols.

  c. $9x^2$ and $7x^2$ are _____ _____ because they have the same variable raised to exactly the same power.

  d. To _____ an equation means to find all of the values of the variable that make the equation true.

  e. The _____ property of _____ says that adding the same number to both sides of an equation does not change the solution.

2. Translate each verbal model into a mathematical model.

  a. Each test score T was increased by 10 points to give a new adjusted test score s.

  b. The area A of a triangle is the product of one-half the length of the base b and the height h.

3. Translate each phrase to an algebraic expression.

  a. 64% of the population

  b. 15 dollars more than twice the price

  c. 27 less than seven-eighths of a number

  d. triple the difference of a number and 1

4. Consider the set: $\left\{-2, \pi, 0, -3\frac{3}{4}, 9.2, \frac{14}{5}, 5, -\sqrt{7}\right\}$

   a. Which numbers are integers?

   b. Which numbers are rational numbers?

   c. Which numbers are irrational numbers?

   d. Which numbers are real numbers?

5. Determine whether each statement is true or false.

   a. $-6 \in \mathbb{Z}$
   b. $76 \notin \mathbb{N}$
   c. $\mathbb{W} \subseteq \mathbb{R}$
   d. $\mathbb{H} \not\subseteq \mathbb{Q}$

Graph each set on a number line.

6. $\left\{\frac{7}{6}, \frac{\pi}{2}, 1.8234503\ldots, \sqrt{3}, 1.\overline{91}\right\}$

7. The set of prime numbers less than 12

8. Determine whether each statement is true or false.

   a. $-|-2.78| > -(-2.71)$
   b. $(-3)^2 \le -3^2$

Evaluate each expression.

9. $-\frac{5}{3} \div \left(-\frac{25}{4}\right)$

10. $\dfrac{2|-4 - 2(3 - 1)|}{-3\left(\sqrt{9}\right)(-2)}$

11. $10 - 3[5^2 - 6(-1 - 1)^3]$

12. Evaluate the expression for $a = 2$, $b = -3$, and $c = 4$.

    $$\frac{(-3b + c)^2 - 17a}{-b + a^2bc}$$

13. **Pediatrics.** Some doctors use Young's rule in calculating dosage for infants and children.

    $$\frac{\text{Age of child}}{\text{Age of child} + 12}\left(\begin{array}{c}\text{average}\\ \text{adult dose}\end{array}\right) = \text{child's dose}$$

    The adult dose of Achromycin is 250 milligrams (mg). What is the dose for an 8-year-old child?

Determine which property of real numbers justifies each statement.

14. a. $3 + 5 = 5 + 3$

    b. $x(yz) = (xy)z$

    c. $-17 + 17 = 0$

    d. $\frac{1}{2} \cdot 1 = \frac{1}{2}$

Simplify each expression.

15. $11.1n^2 - 7.3n + 15.1n - 9.8$

16. $-5(9s)(-2t)$

17. $-7(c - 4) - 5[3(c - 4) - 2(c + 2)]$

18. $\frac{2}{9}(x + 45y) - \frac{1}{4}(x - 24y)$

Solve each equation.

19. $9(x + 4) + 4 - 8x = 4(x - 5) + x$

20. $\dfrac{m - 1}{5} = \dfrac{2m - 3}{3} - 2$

21. $6 - (x - 3) - 5x = 3[1 - 2(x + 2)]$

22. $\dfrac{1}{2}r - \dfrac{7}{6} = -\dfrac{1}{3}r + \dfrac{53}{6}$

23. Use a check to determine whether 6.7 is a solution of $1.6y + (-3) = y + 1.02$.

24. Solve $P = L + \dfrac{s}{f}i$ for i.

25. Solve $y - y_1 = m(x - x_1)$ for x_1.

26. **Crop Circles.** In 1992, two Hungarian high school students were charged for the damage that they caused in creating a 36-meter diameter crop circle in a wheat field. Find the area covered by the crop circle. Round to the nearest square meter.

27. **Hand Tools.** With each pass that a craftsman makes with a sander over a piece of fiberglass, he removes 0.03125 inch of thickness. If the fiberglass was originally 0.9375 inch thick, how many passes are needed to obtain the desired thickness of 0.6875 inch?

28. **Rentals.** The owners of an apartment building rent equal numbers of 1- and 2-bedroom units. The monthly rent for a 1-bedroom is $950, and a 2-bedroom is $1,200. If the total monthly income is $53,750, how many of each type of unit are there?

©iStockphoto.com/ivanastar

29. **Isosceles Triangles.** The measure of a base angle of an isosceles triangle is 5° more than eight times the measure of the vertex angle. Find the measure of each angle of the triangle.

30. **Calculators.** The viewing window of a calculator has a perimeter of 26 centimeters and is 5 centimeters longer than it is wide. Find the dimensions of the window.

31. **Hybrids.** Use the data below to determine the percent of decrease in sales of hybrid vehicles from 2008 to 2009. Round to the nearest percent.

U.S. Hybrid Vehicle Sales	
2008: 315,513	2009: 290,272

Source: hybridcars.com

32. **Air Traffic Control.** An airliner leaves Berlin, Germany, headed for Montreal, Canada, flying at an average speed of 450 mph. At the same time, an airliner leaves Montreal headed for Berlin, averaging 500 mph. If the airports are 3,800 miles apart, when will the air traffic controllers have to make the pilots aware that the planes are passing each other?

33. **Investing.** An investment club invested part of $10,000 at 9% annual interest and the rest at 8%. If the annual income from these investments was $860, how much was invested at 8%?

34. **Rental Cars.** While waiting for his car to be repaired, a man rents a car for $17 per day and 33 cents per mile. His insurance company will pay up to $200 of the rental fee. If he needs the car for four days, how many miles of driving will his policy cover?

35. **Mixing Alloys.** How many ounces of a 40% copper alloy must be mixed with 10 ounces of a 10% copper alloy to obtain an alloy that is 25% copper?

36. **Men's Cologne.** How many ounces of *Skin Soother* men's cologne (unit price: $2.40 per ounce) must be mixed with *Cool Sport* men's cologne (unit price: $1.60 per ounce) to make 8 ounces of a mixture having a unit price of $1.90 per ounce?

Group Project

WRITING FRACTIONS AS DECIMALS

Overview: This is a good activity to try at the beginning of the course. You can become acquainted with other students in your class while you review the process for finding decimal equivalents of fractions.

Instructions: Form groups of 6 students. Select one person from your group to record the group's responses on the questionnaire. Express the results in fraction form and in decimal form.

What fraction (decimal) of the students in your group . . .	Fraction	Decimal
■ have the letter *a* in their first names?		
■ have a birthday in January or February?		
■ work full-time or part-time?		
■ have ever been on television?		
■ have downloaded music in the last week?		
■ log into Facebook at least once a day?		
■ send 30 or more text messages a day?		
■ have downloaded at least 10 applications to their cell phones?		

Graphs, Equations of Lines, and Functions

2

©iStockphoto.com/Jacob Wackerhausen

from Campus to Careers

Certified Fitness Instructor

Because of our busy schedules, many of us have difficulty making exercise a part of our daily routines. A certified fitness instructor can often provide the motivation, discipline, and instruction that a person needs to get and stay in shape. Fitness instructors plan and lead classes, weigh and measure clients, analyze records and graphs, and perform assessment and testing related to weight training and cardiovascular exercise.

Problem 56 in **Study Set 2.1, problem 87** in **Study Set 2.4,** and **problem 113** in **Study Set 2.5** involve situations that a Certified Fitness Instructor might encounter on the job. The mathematical concepts discussed in this chapter can be used to solve those problems.

JOB TITLE:
Certified Fitness Instructor
EDUCATION:
An increasing number of employers are requiring a bachelor's degree in a health-related field. Also, some level of training certification is often required.

JOB OUTLOOK:
Good because of rapid growth in the fitness industry.

ANNUAL EARNINGS:
$53,323 is the median annual salary.

FOR MORE INFORMATION:
www.acefitness.org

Many students think that there are two types of people—those who are good at math and those who are not—and that this cannot be changed. This isn't true! Here are some suggestions that can increase your chances for success in algebra.

DISCOVER YOUR LEARNING STYLE: Are you a visual, verbal, or audio learner? Knowing this will help you determine how best to study.

GET THE MOST OUT OF THE TEXTBOOK: This book and the software that comes with it contain many student support features. Are you taking advantage of them?

TAKE GOOD NOTES: Are your class notes complete so that they are helpful when doing your homework and studying for tests?

Now Try This ▶

1. To determine what type of learner you are, take the *Learning Style Survey* found online at http://www.metamath.com/multiple/multiple_choice_questions.html. Then, write a one-page paper explaining what you learned from the survey results and how you will use the information to help you succeed in the class.

2. To learn more about the student support features of this book, take the *Textbook Tour* found online at: www.cengage.com/math/tussy.

3. Rewrite a set of your class notes to make them more readable and to clarify the concepts and examples covered. Be sure to write your notes in outline form. Fill in any information you didn't have time to copy down in class and complete any phrases or sentence fragments.

SECTION 2.1

Graphs

OBJECTIVES

1. Plot ordered pairs and determine the coordinates of a point.
2. Graph paired data.
3. Read graphs.
4. Find the midpoint of a line segment.

ARE YOU READY?

▼ *The following problems review some basic skills that are needed when graphing ordered pairs.*

1. Graph each number in the set $\left\{\dfrac{5}{3}, -2, 0, 3, -3.75\right\}$ on a number line.

2. **a.** What number is 6 units to the right of 0 on a number line?

   b. What number is 4.5 units to the left of 0 on a number line?

3. List the first four Roman numerals.

4. Write $\dfrac{7}{2}$ and $-\dfrac{16}{5}$ in mixed-number form.

In this section, we will show how numerical relationships can be described by mathematical pictures called *graphs*. We will draw the graphs on a *rectangular coordinate system*.

1 Plot Ordered Pairs and Determine the Coordinates of a Point.

The idea of associating an ordered pair of numbers with points on a grid is attributed to the 17th-century French mathematician René Descartes. Such a grid is called a **rectangular coordinate system**, or **Cartesian coordinate system** after its inventor.

The Language of Algebra

A rectangular coordinate system is a **grid**—a network of uniformly spaced perpendicular lines. At times, some large U.S. cities have such horrible traffic congestion that vehicles can barely move, if at all. The condition is called *gridlock*.

In general, a rectangular coordinate system is formed by two intersecting perpendicular number lines, as shown in the figure. The horizontal number line is usually called the **x-axis.** The vertical number line is usually called the **y-axis.**

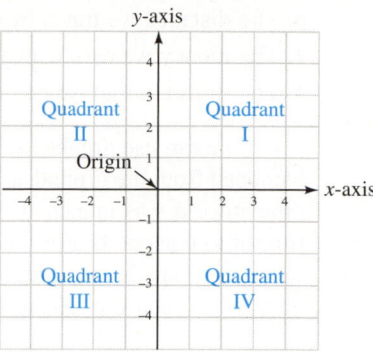

Points in quadrant **II** have a negative x- and positive y-coordinate.

Points in quadrant **I** have a positive x- and positive y-coordinate.

Points in quadrant **III** have a negative x- and negative y-coordinate.

Points in quadrant **IV** have a positive x- and negative y-coordinate.

The Language of Algebra

The prefix **quad** means four, as in quadrilateral (4 sides), quadraphonic sound (4 speakers), and quadruple (4 times).

The positive direction on the *x*-axis is to the right, and the positive direction on the *y*-axis is upward. If no scale is indicated on the axes, we assume that the axes are scaled in units of 1.

The point where the axes intersect is called the **origin.** This is the 0 point on each axis. The two axes form a **coordinate plane** and divide it into four regions called **quadrants,** which are numbered using Roman numerals.

Every point on a coordinate plane can be identified by an **ordered pair** of real numbers *x* and *y*, written as (x, y). The first number in the pair is the **x-coordinate,** and the second number is the **y-coordinate.** Together, the numbers are called the **coordinates** of the point. Some examples are $(6, -4)$, $(2, 3)$, and $(-4, 6)$.

$$(6, -4)$$

The x-coordinate is listed first.

The y-coordinate is listed second.

Read as "the point six, negative four" or as "the ordered pair six, negative four."

The process of locating a point in the coordinate plane is called **graphing** or **plotting** the point. Below, we use red arrows to graph the point $(6, -4)$. Since the *x*-coordinate, 6, is positive, we begin at the origin and move 6 units to the *right* along the *x*-axis. Since the *y*-coordinate, -4, is negative, we then move *down* 4 units, and draw a dot. This locates the point $(6, -4)$, which lies in quadrant IV.

In the figure, blue arrows are used to show how to plot $(-4, 6)$. We start at the origin, move 4 units to the *left* along the *x*-axis, and then 6 units *up* and draw a dot. This locates the point $(-4, 6)$, which lies in quadrant II.

Caution

Note that the point $(-4, 6)$ is not the same as the point $(6, -4)$. This illustrates that the order of the coordinates of a point is important.

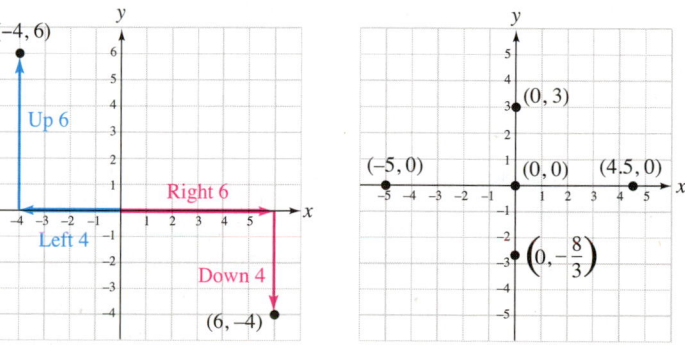

In the figure on the right above, we see that the points $(-5, 0)$, $(0, 0)$, and $(4.5, 0)$ all lie on the *x*-axis. In fact, every point with a *y*-coordinate of 0 will lie on the *x*-axis. We also see that the points $\left(0, -\frac{8}{3}\right)$, $(0, 0)$, and $(0, 3)$ all lie on the *y*-axis. In fact, every point with an *x*-coordinate of 0 will lie on the *y*-axis. Note that the coordinates of the origin are $(0, 0)$. A point may lie in one of the four quadrants or it may lie on one of the axes, in which case the point is not considered to be in any quadrant.

2 Graph Paired Data.

Every day, we deal with quantities that are related:

- The distance we travel by car depends on how fast we are going.
- The money we earn depends on the number of hours we work.
- The sales tax that we pay depends on the price of the item purchased.

We can use graphs to visualize such relationships. For example, suppose a toy rocket is launched from the ground, as shown below. Obviously, the height of the rocket depends on the time since it was launched. Suppose we know the height of the rocket at 1-second intervals from 0 to 6 seconds after launch. We can list that information in a table as **paired data** and write each pair in the form (time, height). The ordered pairs in the table can then be plotted on a rectangular coordinate system and a smooth curve drawn through them.

Time (seconds)	Height of rocket (feet)		
0	0	→	(0, 0)
1	80	→	(1, 80)
2	128	→	(2, 128)
3	144	→	(3, 144)
4	128	→	(4, 128)
5	80	→	(5, 80)
6	0	→	(6, 0)

x-coordinate   y-coordinate   The data in the table can be expressed as ordered pairs.

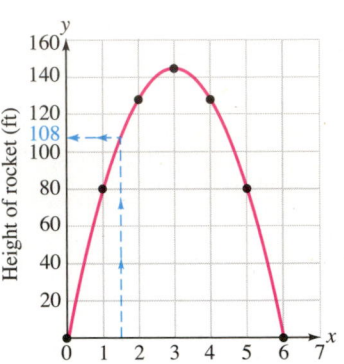

*The graph shows the height of the rocket in relation to the time since it was launched. It does not show the path of the rocket.

From the graph, we can see that the height of the rocket increases as the time increases from 0 to 3 seconds. Then the height decreases until the rocket hits the ground in 6 seconds. We also can use the graph to make observations about the height of the rocket at other times. For example, the dashed blue lines on the graph show that in 1.5 seconds, the height of the rocket will be approximately 108 feet.

3 Read Graphs.

Since graphs are becoming an increasingly popular way to present information, the ability to read and interpret them is becoming ever more important.

EXAMPLE 1   **Water Management.**   The graph below shows the water level of a reservoir before, during, and after a storm. On the x-axis, 0 represents the day the storm began. On the y-axis, 0 represents the normal water level that operators try to maintain.

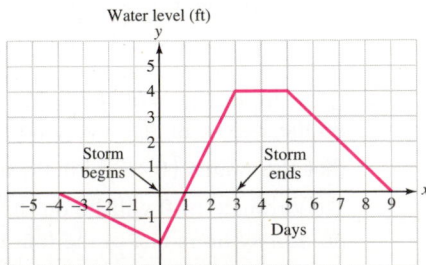

a. In anticipation of the storm, operators released water to lower the level of the reservoir. By how many feet was the water lowered prior to the storm?

b. After the storm ended, on what day did the water level begin to fall?

c. When was the water level 2 feet above normal?

Strategy   We will use ordered pairs to describe the situations mentioned in parts (a), (b), and (c).

Why   The coordinates of specific points on the graph can be used to answer the given questions.

Solution   **a.** The graph starts at the point $(-4, 0)$. This means that 4 days before the storm began, the water was at the normal level. If we look below 0 on the y-axis, we see that the point $(0, -2)$ is on the graph. So the day the storm began, the water level was lowered 2 feet.

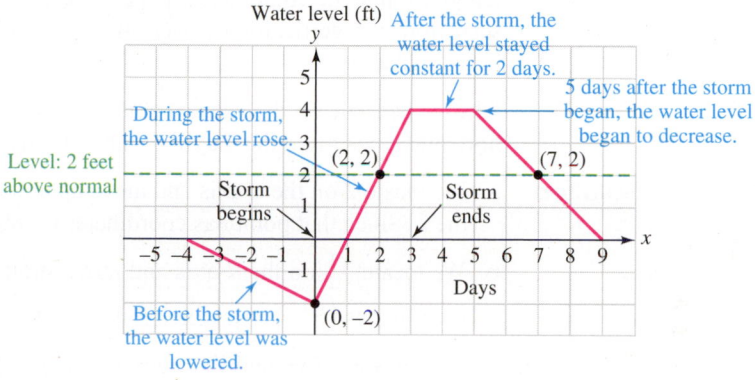

b. If we look at the x-axis, we see that the storm lasted 3 days. From the third to the fifth day, the water level remained constant, 4 feet above normal, as indicated by the horizontal line segment between $(3, 4)$ and $(5, 4)$. The graph does not begin to decrease until day 5.

c. We can draw a horizontal line passing through 2 on the y-axis. This line intersects the graph in two places—at the points $(2, 2)$ and $(7, 2)$. This means that 2 days and 7 days after the storm began, the water level was 2 feet above normal.

Self Check 1   **Water Management.**   Refer to the graph for Example 1.
a. When was the water at the normal level?

b. By how many feet did the water level rise during the storm?
c. After the storm ended and the water level began to fall, how long did it take for the water level to return to normal?

Now Try ▶ Problem 37

The graph on the right shows the cost of renting ski equipment for different periods of time. The horizontal axis is labeled with the variable d. This reinforces the fact that it is associated with the number of *days* the equipment is rented. The vertical axis is labeled with the variable c, for *cost*. In this case, ordered pairs on the graph have the form (d, c). For example, the point $(3, 100)$ on the graph tells us that the cost of renting the equipment for 3 days is \$100. The cost of renting the equipment for more than 4 and up to 5 days is \$140. We call this type of graph a **step graph**.

The point at the end of each step indicates the rental cost for 1, 2, 3, 4, 5, and 6 days. Each open circle indicates that that point is not on the graph.

Cost of Renting a Ski Equipment Package

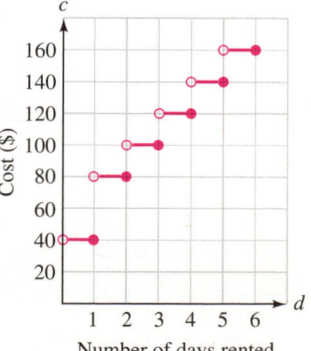

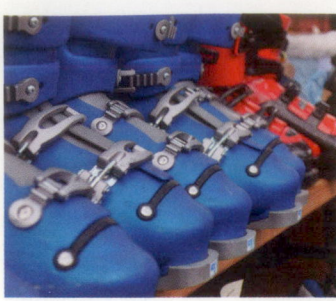

EXAMPLE 2

Ski Rental Packages.   Use the graph on the right to answer the questions about the rental cost of a ski equipment package (skis, boots, and poles).
a. Find the cost of renting the equipment for 2 days.
b. Find the cost of renting the equipment for $5\frac{1}{2}$ days.   **c.** How long can you rent the equipment if you have budgeted $120 for the rental?   **d.** Is the cost of renting the equipment the same each day?

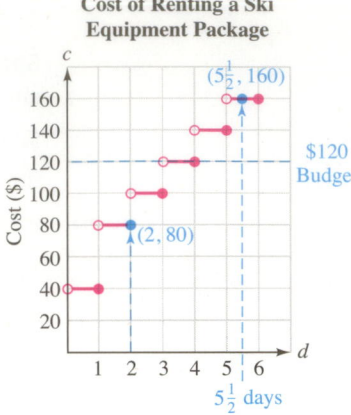

Cost of Renting a Ski Equipment Package

Strategy   To answer questions about the rental costs, we will scan from the horizontal axis, up and over, to the vertical axis. To answer questions about length of time the equipment can be rented, we will scan from the vertical axis, over and down, to the horizontal axis.

Why   The scale on the vertical axis gives the cost to rent the equipment. The scale on the horizontal axis gives the length of time the equipment is rented.

Solution

a. We locate 2 on the d-axis and move up to locate the point on the graph directly above the 2. Since that point has coordinates $(2, 80)$, a 2-day rental costs $80.

b. We locate $5\frac{1}{2}$ on the d-axis and move straight up to locate the point on the graph with coordinates $\left(5\frac{1}{2}, 160\right)$, which indicates that a $5\frac{1}{2}$-day rental would cost $160.

c. We draw a horizontal line through the point labeled 120 on the c-axis. Since this line intersects one of the steps of the graph, we can look down to the d-axis to find the d-values that correspond to a c-value of 120. We see that the equipment can be rented for more than 3 and up to 4 days for $120.

d. The cost each day is not the same. If we look at how the c-coordinates change, we see that the first-day rental fee is $40. The second day, the cost increases another $40. However, the third day, and all succeeding days, the cost increases $20 per day.

Self Check 2   **Ski Rental Packages.**   Use the graph in Example 2 to find the cost of renting the ski equipment for   **a.** 5 days   **b.** $2\frac{1}{2}$ days

Now Try ▶ Problem 39

---

4 **Find the Midpoint of a Line Segment.**

If point M in the figure on the right lies midway between point P and point Q, it is called the **midpoint** of line segment PQ. We call the points P and Q, the **endpoints** of the segment.

To distinguish between the coordinates of the endpoints of a line segment, we can use *subscript notation*. In the figure, the point P with coordinates (x_1, y_1) is read as "point P with coordinates x sub 1 and y sub 1," and the point Q with coordinates (x_2, y_2) is read as "point Q with coordinates x sub 2 and y sub 2."

To find the coordinates of point M, we find the average of the x-coordinates and the average of the y-coordinates of points P and Q. Using subscript notation, we can write the midpoint formula in the following way.

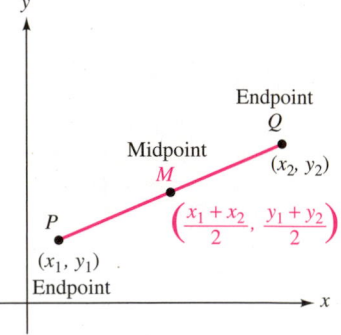

The Midpoint Formula	The **midpoint** of a line segment with endpoints (x_1, y_1) and (x_2, y_2) is the point with coordinates $$\left(\frac{x_1 + x_2}{2}, \frac{y_1 + y_2}{2}\right)$$

EXAMPLE 3 Find the midpoint of the line segment with endpoints $(-2, 5)$ and $(4, -2)$.

Strategy To find the coordinates of the midpoint, we find the average of the x-coordinates and the average of the y-coordinates of the endpoints.

Why This is what is called for by the expressions $\frac{x_1 + x_2}{2}$ and $\frac{y_1 + y_2}{2}$ of the midpoint formula.

Solution We can let $(x_1, y_1) = (-2, 5)$ and $(x_2, y_2) = (4, -2)$. After substituting these values into the expressions for the x- and y-coordinates in the midpoint formula, we evaluate each expression to find the coordinates of the midpoint.

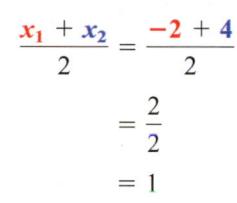

Find the x-coordinate of the midpoint:

$$\frac{x_1 + x_2}{2} = \frac{-2 + 4}{2}$$
$$= \frac{2}{2}$$
$$= 1$$

Find the y-coordinate of the midpoint:

$$\frac{y_1 + y_2}{2} = \frac{5 + (-2)}{2}$$
$$= \frac{3}{2}$$

Thus, the midpoint is $\left(1, \frac{3}{2}\right)$.

Self Check 3 Find the midpoint of the line segment with endpoints $(-1, 8)$ and $(5, 2)$.

Now Try ▶ Problem 41

EXAMPLE 4 The midpoint of the line segment joining $(-5, -3)$ and a point Q is the point $(-1, 2)$. Find the coordinates of point Q.

Strategy As in Example 3, we will use the midpoint formula to find the unknown coordinates. However, this time, we need to find x_2 and y_2.

Why We want to find the coordinates of one of the endpoints.

Solution We can let $(x_1, y_1) = (-5, -3)$ and $(x_M, y_M) = (-1, 2)$, where x_M represents the x-coordinate and y_M represents the y-coordinate of the midpoint. To find the coordinates of point Q, we substitute for x_1, x_M, y_1, and y_M in the expressions for the coordinates in the midpoint formula and solve the resulting equations for x_2 and y_2.

$$x_M = \frac{x_1 + x_2}{2}$$
$$-1 = \frac{-5 + x_2}{2} \qquad \text{Substitute.}$$
$$-2 = -5 + x_2 \qquad \text{Multiply both sides by 2.}$$
$$3 = x_2 \qquad \text{Add 5 to both sides.}$$

$$y_M = \frac{y_1 + y_2}{2} \qquad \text{Read } x_M \text{ as "x sub M" and } y_M \text{ as "y sub M."}$$
$$2 = \frac{-3 + y_2}{2} \qquad \text{Substitute.}$$
$$4 = -3 + y_2 \qquad \text{Multiply both sides by 2.}$$
$$7 = y_2 \qquad \text{Add 3 to both sides.}$$

Since $x_2 = 3$ and $y_2 = 7$, the coordinates of point Q are $(3, 7)$.

> **Self Check 4** The midpoint of the line segment joining $(-5, -3)$ and a point P is the point $(-2, 5)$. Find the coordinates of point P.
>
> **Now Try** ▶ Problem 49

SECTION 2.1 ▶ **STUDY SET**

VOCABULARY

Fill in the blanks.

1. The pair of numbers $(6, -2)$ is called an _____ pair.
2. In the ordered pair $(22, 29)$, the ___-coordinate is 29.
3. The point $(0, 0)$ is the _____.
4. The x- and y-axes divide the coordinate plane into four regions called _____.
5. Ordered pairs of numbers can be graphed on a _____ coordinate system.
6. The process of locating a point on a coordinate plane is called _____ the point.
7. If a point is midway between two points P and Q, it is called the _____ of segment PQ.
8. If a line segment joins points P and Q, points P and Q are called _____ of the segment.

CONCEPTS

Fill in the blanks.

9. To plot $(6, -3.5)$, we start at the _____ and move 6 units to the _____ and then 3.5 units _____.
10. To plot $(-6, 2)$, we start at the _____ and move 6 units to the _____ and then 2 units ___.
11. In which quadrant do points with a negative x-coordinate and a positive y-coordinate lie?
12. In which quadrant do points with a positive x-coordinate and a negative y-coordinate lie?

NOTATION

13. Do these ordered pairs name the same point?

$$\left(5.25, -\tfrac{3}{2}\right), \left(5\tfrac{1}{4}, -1.5\right), \left(\tfrac{21}{4}, -1\tfrac{1}{2}\right)$$

14. For the ordered pair (t, d), which variable is associated with the horizontal axis?
15. What type of letter is used to label points?
16. Fill in the blank: The expression x_1 is read as "___ _____ ___".
17. Explain the difference between x^2 and x_2.

18. Fill in the blanks: The x-coordinate of the midpoint of the line segment joining (x_1, y_1) and (x_2, y_2) is [____] and the y-coordinate is [____].

GUIDED PRACTICE

Plot each point on a rectangular coordinate system and name the quadrant or axis in which the point lies. See Objective 1.

19. $(4, 3)$
20. $(-2, 1)$
21. $(3.5, -2)$
22. $(-2.5, -3)$
23. $(5, 0)$
24. $(-4, 0)$
25. $\left(0, -\dfrac{8}{3}\right)$
26. $\left(0, \dfrac{10}{3}\right)$

Give the coordinates of each point. See Objective 1.

27. A
28. B
29. C
30. D
31. E
32. F
33. G
34. H

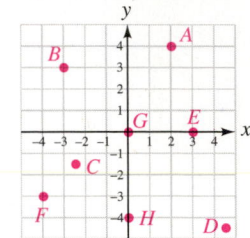

Use the information in the table to draw each graph. See Objective 2.

35. **Water Pressure.**   A tub was filled with water from a faucet. The table shows the number of gallons of water in the tub at 1-minute intervals. Plot the ordered pairs in the table on a rectangular coordinate system and then draw a line through the points.

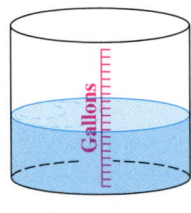

Time (min)	Water in tub (gal)
0	0
1	8
2	16
3	24
4	32

36. **Trampolines.**   The table on the next page shows the distance a girl is from the ground (in relation to time) as she bounds into the air and back down to the trampoline. Plot the ordered pairs in the table on a rectangular coordinate system and then draw a smooth curve through the points.

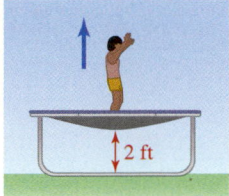

Time (sec)	Height (ft)
0	2
0.25	9
0.5	14
1.0	18
1.5	14
1.75	9
2.0	2

Use the graph to answer each question. **See Example 1.**

37. Submarines. The graph in the following illustration shows the depths of a submarine at certain times.

a. Where is the sub 2 hours after launch?

b. What is the sub doing as t increases from 2 to 3?

c. For how long does the sub travel at a depth of 1,000 feet?

d. How large an ascent does the sub begin to make 6 hours after launch?

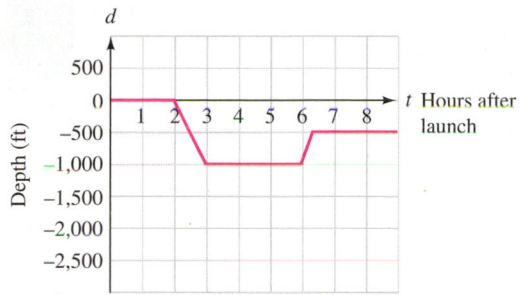

38. Airplanes. The following graph shows the altitudes of a plane at certain times.

a. Where is the plane when $t = 0$?

b. What is the plane doing as t increases from 1 to 2?

c. For how long does the airplane travel at an altitude of 5,000 feet?

d. How much of a descent does the plane begin to make 4 hours after take-off?

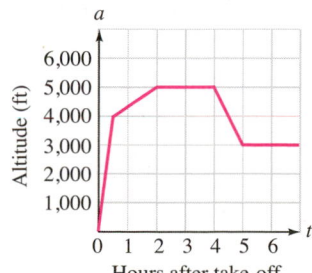

Use each graph to answer each question. **See Example 2.**

39. Video Rentals. The charges for renting a video are shown in the graph in the next column.

a. Find the 1-day rental charge.

b. Find the charge if the video is kept for 1 week.

c. How long can you rent the video if you have budgeted $5 for the rental?

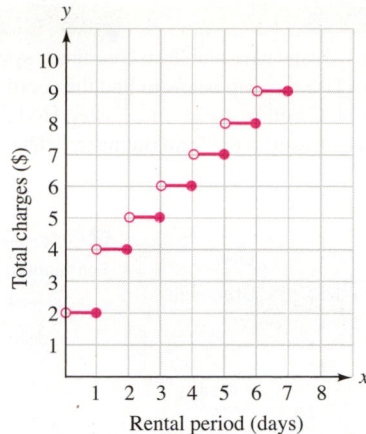

40. Taxis. The following graph gives the fares charged for rides up to 1 mile in length by a taxicab company.

a. What is the fare for a $\frac{1}{2}$-mile ride?

b. Is the fare the same for each $\frac{1}{8}$ mile traveled?

c. What is a fare for a $\frac{7}{10}$-mile long ride?

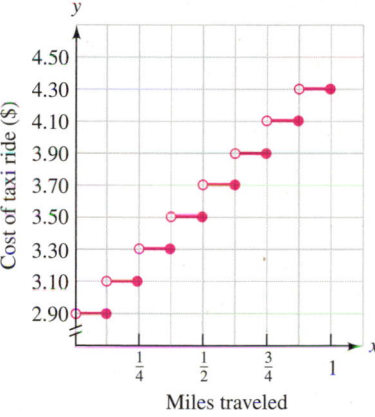

The symbol ⌇ indicates a break in the labeling of the vertical axis. The break enables us to omit a large portion of the grid that would not be used.

Find the midpoint of the line segment with the given endpoints. **See Example 3.**

41. $(0, 0), (6, 8)$ **42.** $(10, 12), (0, 0)$

43. $(6, 8), (12, 16)$ **44.** $(10, 4), (2, -2)$

45. $(-2, -8), (3, -8)$ **46.** $(-5, -2), (7, 3)$

47. $(7, 1), (-10, 4)$ **48.** $(-4, -3), (4, -8)$

Solve each problem. **See Example 4.**

49. If $(-2, 3)$ is the midpoint of segment PQ and the coordinates of P are $(-8, 5)$, find the coordinates of Q.

50. If $(6, -5)$ is the midpoint of segment PQ and the coordinates of Q are $(-5, -8)$, find the coordinates of P.

51. If $(-7, -3)$ is the midpoint of segment QP and the coordinates of Q are $(6, -3)$, find the coordinates of P.

52. If $\left(\frac{1}{2}, -2\right)$ is the midpoint of segment QP and the coordinates of P are $\left(-\frac{5}{2}, 5\right)$, find the coordinates of Q.

APPLICATIONS

53. Road Maps.   Maps have a built-in coordinate system to help locate cities. Use the map below to find the coordinates of these cities in South Carolina: Jonesville, Easley, Hodges, and Union. Express each answer in the form (number, letter).

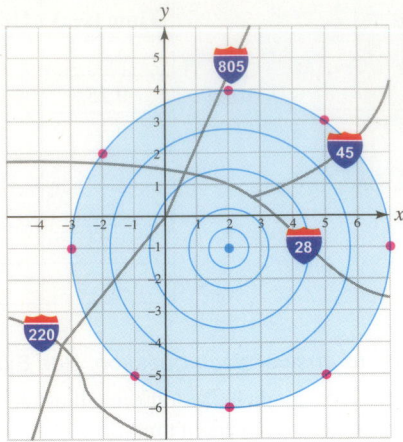

54. Hurricanes.   A coordinate system that designates the location of places on the surface of Earth uses a series of latitude and longitude lines, as shown in the illustration.

a. In 2005, Hurricane Katrina devastated New Orleans. If we list longitude first, what are the coordinates of New Orleans, expressed as an ordered pair?

b. In August 1992, Hurricane Andrew destroyed Homestead, Florida. Estimate the coordinates of Homestead.

c. Estimate the coordinates of where Hurricane Andrew hit Louisiana.

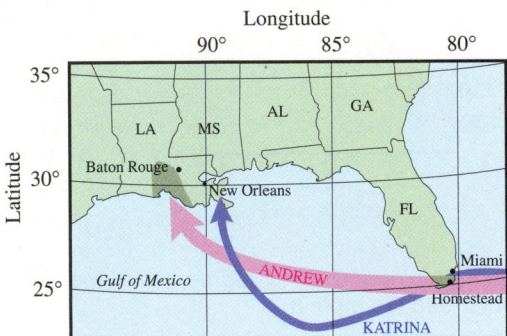

55. Earthquakes.   The graph in the next column shows the area damaged by an earthquake.

a. Find the coordinates of the *epicenter* (the source of the quake).

b. Was damage done at the point (4, 5)?

c. Was damage done at the point $(-1, -4)$?

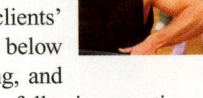

56. ▶ from **Campus to Careers**

Certified Fitness Instructor

Fitness instructors often keep records of their clients' exercise sessions to monitor the clients' progress toward physical fitness. The graph below gives the heart rate of a man before, during, and after an aerobic workout. Use it to answer the following questions.

a. What was his heart rate before beginning the workout?

b. After beginning his workout, how long did it take him to reach his training-zone heart rate?

c. What was his heart rate half an hour after beginning the workout?

d. For how long did he work out in the training zone?

e. Estimate the two times his heart rate was 100 beats per minute.

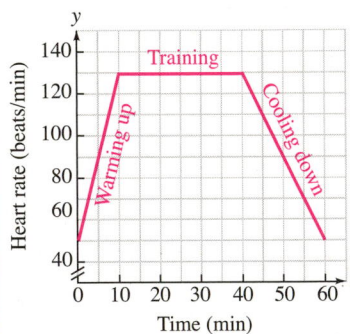

The ≉ symbol indicates a break in the labeling of the vertical axis.

The break enables us to omit a large portion of the grid that would not be used.

57. Geography. The illustration below shows a cross-sectional profile of the Sierra Nevada mountain range.

a. Estimate the coordinates of blue oak, sagebrush scrub, and tundra using an ordered pair of the form (distance, elevation).

b. The *tree line* is the highest elevation at which trees grow. Estimate the tree line for this mountain range.

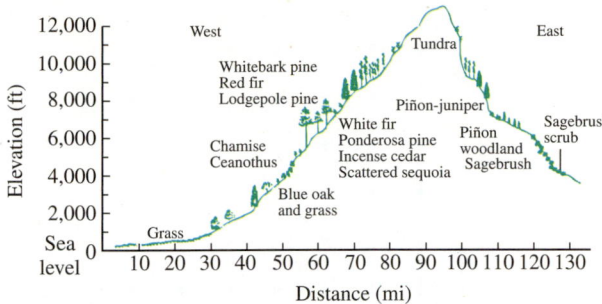

58. Long-Distance Running. Refer to the graph below that describes a two-person race.

a. Which runner ran faster at the start of the race?

b. Which runner stopped to rest first?

c. Which runner dropped his watch and had to go back and get it?

d. At which of these elapsed times (A, B, C, D) was runner 1 stopped and runner 2 running?

e. Describe what was happening at time D.

f. Which runner won the race?

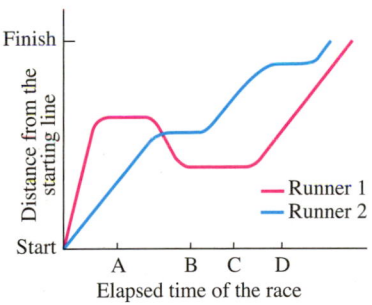

59. Campus Parking. Match each daily parking description below with the graph in the next column that best illustrates it. The parking lot holds 500 cars.

a. On Mondays, the parking lot is full by noon. It's impossible to find a parking space until late in the afternoon.

b. On Tuesdays, the lot never gets more than half-full.

c. On Wednesdays, the lot fills up quickly. It empties out around lunchtime but then it fills up fast when the evening classes begin.

d. On Thursdays, it is easy to park in the lot until the evening classes begin. Then it is almost impossible to find a parking space.

e. On Fridays, it is very difficult to find a parking space, unless you arrive early. However, the lot clears out by noon.

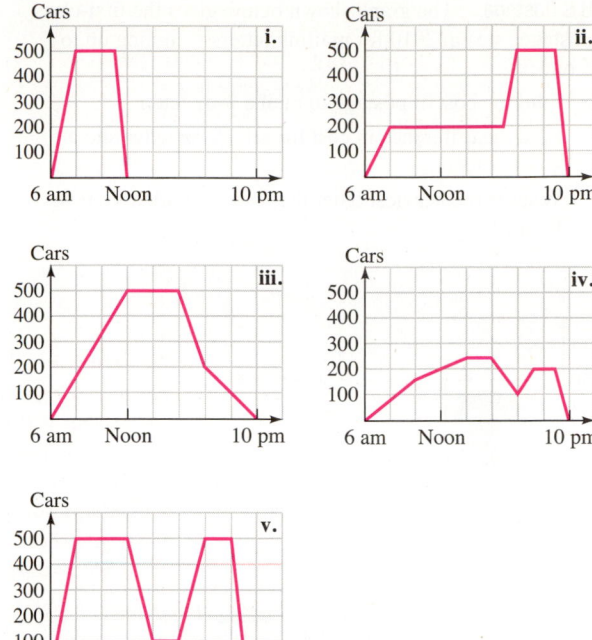

60. Match each description with the graph below that best illustrates it.

a. You make popcorn in a microwave. The number of pops per second depends on the time since you started the microwave.

b. You jump off a diving board. Your position in realation to the surface of the water depends on the time since you jumped.

c. You mow your lawn when it reaches a certain height. The height of the grass depends on the time since you last mowed it.

d. You pour water into a glass filled with ice. The temperature of the water depends on the time since you poured it.

e. You run out of gas on a highway and coast to a stop. Your car's speed depends on the time since you ran out of gas.

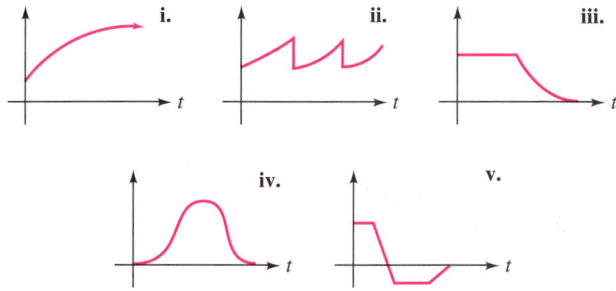

61. U.S Postage. The graph shown below gives the first-class postage rates in 2010 for mailing letters weighing up to 5 ounces.

a. Find the cost of postage to mail a 3-oz letter.

b. Find the difference in cost for a 1.75-oz letter and a 4.75-oz letter.

c. What is the heaviest letter that can be mailed first class for $1?

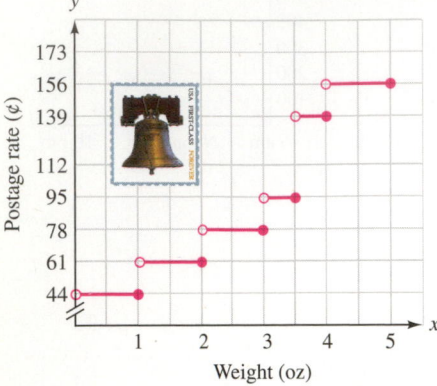

The symbol ⌇ indicates a break in the labeling of the vertical axis. The break enables us to omit a large portion of the grid that would not be used.

62. Roast Turkey. Guidelines that appear on the label of a frozen turkey are listed in the table. Draw a step graph that illustrates these instructions.

Size	Time thawing in refrigerator
10 lb to just under 18 lb	3 days
18 lb to just under 22 lb	4 days
22 lb to just under 24 lb	5 days
24 lb to just under 30 lb	6 days

63. Multicultural Studies. Social scientists use the following diagram to classify cultures. The amount of group/family loyalty in a culture is measured on the horizontal *group* axis. The amount of social mobility is measured on the vertical *social grid* axis. In the diagram, four cultures are classified. In which culture, R, S, T, or U, would you expect that

a. anyone can grow up to be president, and parents expect their children to get out on their own as soon as possible?

b. only the upper class attends college, and people must marry within their own social class?

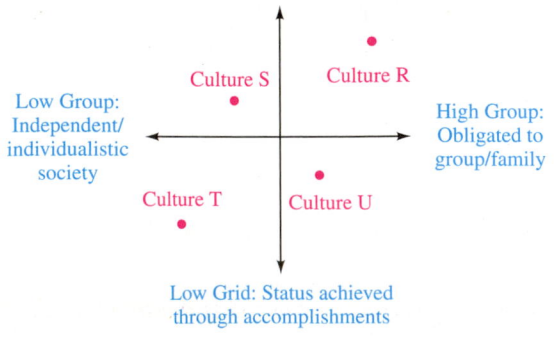

64. Psychology. The result of a personal profile test taken by an employee is plotted as an ordered pair on the grid below. The test shows whether the employee is more task oriented or people oriented. From the result, would you expect the employee to *agree* or *disagree* with each of the following statements?

a. Completing a project is almost an obsession with me, and I cannot be content until I am finished.

b. Even if I'm in a hurry while running errands, I will stop to talk with a friend.

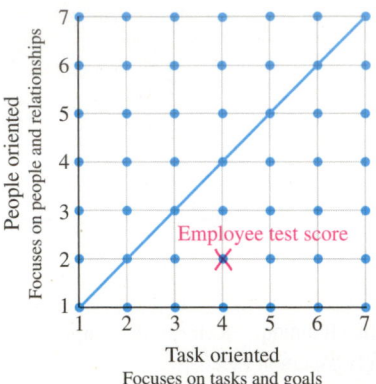

WRITING

65. Explain how to plot the point $(-2, 5)$.

66. Explain why the coordinates of the origin are $(0, 0)$.

67. Explain why the point $(-1, 6)$ is not the same as the point $(6, -1)$.

68. Explain this diagram.

$$
\begin{array}{c|c}
\text{II} & \text{I} \\
(-, +) & (+, +) \\
\hline
\text{III} & \text{IV} \\
(-, -) & (+, -)
\end{array}
$$

69. Use the Internet to perform a search of the name René Descartes. After reading about him, explain how a fly on his bedroom ceiling provided the inspiration for the concept of a rectangular coordinate system.

70. What does the open circle in the graph indicate?

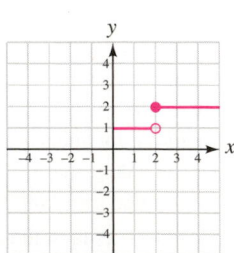

REVIEW

Evaluate each expression.

71. $-5^2 - 5 - 5(-5)$

72. $\dfrac{1}{3}\left(\dfrac{1}{6}\right) - \left(-\dfrac{1}{3}\right)^2$

73. $\dfrac{|-25| - 2(-5)}{2^4 - 9}$

74. $\dfrac{3[-9 + 2(7 - 3)]}{(8 - 5)(9 - 7)}$

75. Solve $P = 2l + 2w$ for w.

76. Solve $T_f = T_a(1 - F)$ for T_a.

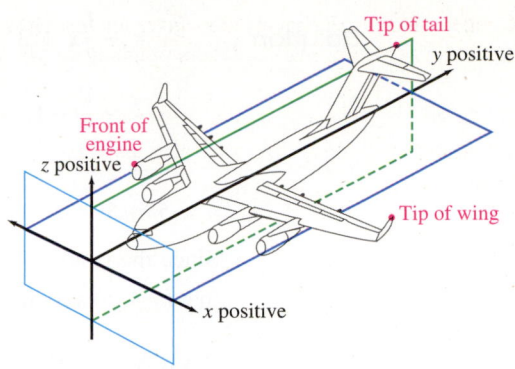

Tip of tail
y positive

Front of engine
z positive

Tip of wing

x positive

CHALLENGE PROBLEMS

77. Find the coordinates of the three points that divide the line segment joining (a, b) and (c, d) into four equal parts.

78. Airplanes. Engineers use a coordinate system with 3 axes, as shown on the right. Any point on the airplane can be described by an *ordered triple* of the form (x, y, z). The coordinates of three points on the plane are $(0, 181, 56)$, $(-46, 48, 19)$, and $(84, 94, 24)$. Which highlighted part of the plane corresponds with which ordered triple?

SECTION 2.2

Graphing Linear Equations in Two Variables

OBJECTIVES

1 Determine whether an ordered pair is a solution of an equation.

2 Find a solution of an equation in two variables.

3 Graph linear equations by plotting points.

4 Graph linear equations by finding intercepts.

5 Graph horizontal and vertical lines.

6 Use linear models to solve applied problems.

ARE YOU READY?

The following problems review some basic skills that are needed when graphing linear equations.

1. Is $4(2) - 3(-1) = 10$ a true or false statement?

2. Solve: $5(3) + 3y = 6$

3. Graph the points $(2, 0)$, $(-4, 0)$, $(0, 1)$, and $(0, -3)$.

4. What is the x-coordinate of any point that lies on the y-axis?

5. Solve: $8(0) - 5y = 10$

6. Draw a horizontal line. Draw a vertical line.

In this section, we will discuss equations that contain two variables. These equations are used to describe algebraic relationships between two quantities.

1 Determine Whether an Ordered Pair Is a Solution of an Equation.

We will now extend our equation-solving skills to find solutions of equations in two variables. To begin, let's consider $y = -\frac{1}{2}x + 3$, an equation in x and y. In general, **a solution of an equation in two variables** is an ordered pair of numbers that make a true statement when substituted into the equation.

EXAMPLE 1   Is $(-4, 5)$ a solution of $y = -\frac{1}{2}x + 3$?

Strategy   We will substitute -4 for x and 5 for y and see whether the resulting equation is true.

Why   An ordered pair is a solution of $y = -\frac{1}{2}x + 3$ if replacing the variables with the values of the ordered pair results in a true statement.

Solution $y = -\dfrac{1}{2}x + 3$ This is the given equation.

$5 \overset{?}{=} -\dfrac{1}{2}(-4) + 3$ Substitute 5 for y and −4 for x.

$5 \overset{?}{=} 2 + 3$ Evaluate the right side. Do the multiplication: $-\frac{1}{2}(-4) = 2$.

$5 = 5$ True

Since the result $5 = 5$ is a true statement, $(-4, 5)$ is a solution of $y = -\frac{1}{2}x + 3$. We say that $(-4, 5)$ **satisfies** the equation.

> **Self Check 1**   Is $(4, -1)$ a solution of $y = -\frac{1}{2}x + 3$?
>
> **Now Try** ▶ Problem 17

2 Find a Solution of an Equation in Two Variables.

To find a solution of an equation in two variables, we can select a number for one of the variables and find the corresponding value of the other variable. For example, to find a solution of $y = -\frac{1}{2}x + 3$, we can select a value for x, say 4, and find the corresponding value of y.

$y = -\dfrac{1}{2}x + 3$ This is the given equation.

$y = -\dfrac{1}{2}(4) + 3$ Substitute 4 for x.

$y = -2 + 3$ Evaluate the right side. Do the multiplication: $-\frac{1}{2}(4) = -2$.

$y = 1$ This is the y-coordinate of the solution.

Thus, $(4, 1)$ is a solution of $y = -\frac{1}{2}x + 3$.

Since we can choose any real number for x, and since any choice for x will give a corresponding value of y, the equation $y = -\frac{1}{2}x + 3$ has infinitely many solutions. It would be impossible to list all of the solutions. Instead, we can draw a mathematical picture of the solutions, called the *graph of the equation*.

3 Graph Linear Equations by Plotting Points.

Equations in two variables can be graphed in several ways. If an equation in x and y is solved for y, we can graph it by selecting values for x and calculating the corresponding values of y.

EXAMPLE 2   Graph: $y = -\dfrac{1}{2}x + 3$

Strategy We will find three solutions of the equation, plot them on a rectangular coordinate system, and then draw a straight line passing through the points.

Why To *graph* an equation in two variables means to make a drawing that represents all of its solutions.

Solution

To find three solutions of this equation, we select three values for x that will make the calculations easy. Then we find each corresponding value of y. For example, if x is -2, we have

$$y = -\frac{1}{2}x + 3 \qquad \text{This is the equation to graph.}$$

$$y = -\frac{1}{2}(-2) + 3 \qquad \text{Substitute } -2 \text{ for } x.$$

$$y = 1 + 3 \qquad \text{Evaluate the right side. Do the multiplication: } -\frac{1}{2}(-2) = 1.$$

$$y = 4 \qquad \text{This is the } y\text{-coordinate of the solution.}$$

Thus, $(-2, 4)$ is a solution. In a similar manner, we find corresponding y-values for x-values of 0 and 2 and enter the solutions in the table below.

When we plot the ordered-pair solutions on a rectangular coordinate system, we see that they lie in a straight line. Using a straightedge or ruler, we then draw a straight line through the points because the graph of any solution of $y = -\frac{1}{2}x + 3$ will lie on this line. Furthermore, every point of this line represents a solution. We call the line the **graph of the equation.** It represents all of the solutions of $y = -\frac{1}{2}x + 3$. The arrowheads drawn on either end of the line indicate that the solutions continue indefinitely in both directions, beyond what we can see on the coordinate grid. It is standard practice to write the equation of the line next to its graph.

$$y = -\frac{1}{2}x + 3$$

x	y	(x, y)
-2	4	$(-2, 4)$
0	3	$(0, 3)$
2	2	$(2, 2)$

Select x. Find y. Plot (x, y).
A table of solutions

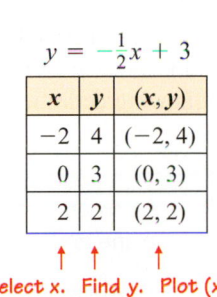
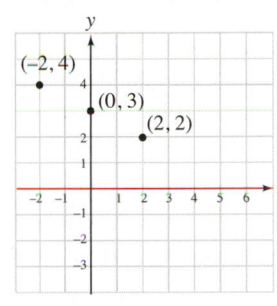

Plot the ordered pairs.

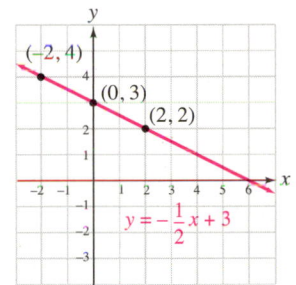

Draw a straight line through the points. This is the *graph of the equation*.

Self Check 2 Graph: $y = \frac{1}{3}x + 1$

Now Try ▶ Problems 21 and 25

The equation from Example 2, $y = -\frac{1}{2}x + 3$, is called a *linear equation in two variables*.

Standard (General) Form of a Linear Equation

A linear equation in two variables is an equation that can be written in the form

$$Ax + By = C$$

where A, B, and C are real numbers and A and B are not both 0. This form is called **standard form.***

---

*In some textbooks, the definition of the standard form of a linear equation in two variables contains additional requirements, such as: A, B, and C are integers, $A > 0$, and the greatest common factor of A, B, and C is 1.

Success Tip

The exponent on each variable of a linear equation in two variables is an understood 1. For example, $2x - 5y = 10$ can be thought of as $2x^1 - 5y^1 = 10$.

Some other examples of linear equations are

$$y = 4x - 7, \qquad 2x - 5y = 10, \qquad y = 3, \qquad \text{and} \qquad x = 2$$

Some examples of equations in two variables that are **not linear** are shown below. You will see later in this course (and in more advanced courses) that the graphs of these equations are not straight lines.

$$y = x^2 + 3, \qquad y = \sqrt{x}, \qquad y = 4^x, \qquad x^2 + y^2 = 25, \qquad \text{and} \qquad y = \frac{1}{x}$$

The graph of every linear equation in two variables is a line. We can use the following **point-plotting method** to graph linear equations solved for y.

Graphing Linear Equations Solved for y by Plotting Points

1. Find three ordered pairs that are solutions of the equation by selecting three values for x and calculating the corresponding values of y.
2. Plot the solutions on a rectangular coordinate system.
3. Draw a straight line passing through the points. If the points do not lie on a line, check your calculations.

4 Graph Linear Equations by Finding Intercepts.

In Example 2, the graph of $y = -\frac{1}{2}x + 3$ was a straight line that intersected the y-axis at the point (0, 3), called the y-intercept, and it intersected the x-axis at the point (6, 0), called the x-intercept. In general, we have the following definitions.

Intercepts of a Line

The **y-intercept** of a line is the point $(0, b)$, where the line intersects the y-axis. To find b, substitute 0 for x in the equation of the line and solve for y.

The **x-intercept** of a line is the point $(a, 0)$, where the line intersects the x-axis. To find a, substitute 0 for y in the equation of the line and solve for x.

Plotting the x- and y-intercepts of a graph and drawing a line through them is called the **intercept method of graphing a line.** This method is useful when graphing linear equations written in the standard form $Ax + By = C$.

EXAMPLE 3 Graph $2x - 5y = 10$ by finding the intercepts.

Strategy We will let $x = 0$ to find the y-intercept of the graph and then let $y = 0$ to find the x-intercept.

Why Since two points determine a line, the y-intercept and x-intercept are enough information to graph this linear equation.

Solution To find the y-intercept, we substitute 0 for x and solve for y:

$$\begin{aligned} 2x - 5y &= 10 && \text{This is the equation to graph.} \\ 2(0) - 5y &= 10 && \text{Substitute 0 for x.} \\ -5y &= 10 && \text{Simplify the left side.} \\ y &= -2 && \text{To isolate y, divide both sides by } -5. \end{aligned}$$

The y-intercept is the point $(0, -2)$. To find the x-intercept, we substitute 0 for y and solve for x:

$$2x - 5y = 10 \qquad \text{This is the equation to graph.}$$
$$2x - 5(0) = 10 \qquad \text{Substitute 0 for y.}$$
$$2x = 10 \qquad \text{Simplify the left side.}$$
$$x = 5 \qquad \text{To isolate x, divide both sides by 2.}$$

The Language of Algebra

For any two points, exactly one line passes through them. We say two points **determine** a line.

The x-intercept is the point $(5, 0)$.

Although two points provide enough information to draw the graph of the equation, it is a good idea to find and plot a third point as a check. If the three points do not lie on a line, then at least one of them is in error.

To find the coordinates of a third point, we can substitute any convenient number (such as -5) for x and solve for y:

$$2x - 5y = 10 \qquad \text{This is the equation to graph.}$$
$$2(-5) - 5y = 10 \qquad \text{Substitute -5 for x.}$$
$$-10 - 5y = 10 \qquad \text{On the left side, multiply: } 2(-5) = -10.$$
$$-5y = 20 \qquad \text{Add 10 to both sides.}$$
$$y = -4 \qquad \text{To isolate y, divide both sides by -5.}$$

The line will also pass through the point $(-5, -4)$. We plot the intercepts and the check point, draw a straight line through them, and label the line as $2x - 5y = 10$.

The Language of Algebra

Be careful with pronunciation: The point where a line **intersects** the x- or y-axis is called an **intercept**.

$$2x - 5y = 10$$

x	y	(x, y)	
0	-2	$(0, -2)$	← y-intercept
5	0	$(5, 0)$	← x-intercept
-5	-4	$(-5, -4)$	← Check point

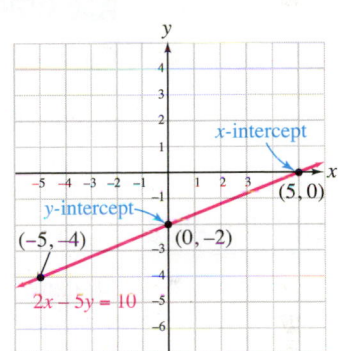

Self Check 3   Graph $5x + 15y = -15$ by finding the y- and x-intercepts.

Now Try ▶ Problem 33

Using Your **Calculator** ▶   Generating Tables of Solutions

If an equation in x and y is solved for y, we can use a graphing calculator to generate a table of solutions. The instructions in this discussion are for a TI-84 Plus graphing calculator. For specific details about other brands, please consult the owner's manual.

To construct a table of solutions for $2x - 5y = 10$, we first solve for y.

$$2x - 5y = 10$$
$$-5y = -2x + 10 \qquad \text{Subtract 2x from both sides.}$$
$$y = \frac{2}{5}x - 2 \qquad \text{Divide both sides by -5 and simplify.}$$

To enter $y = \frac{2}{5}x - 2$, we press $\boxed{Y =}$ and enter (2/5)x − 2, as shown in figure (a). (Ignore the subscript 1 on y; it is not relevant at this time.)

To enter the x-values that are to appear in the table, we press $\boxed{2nd}$ $\boxed{TBLSET}$ and enter the first value for x on the line labeled TblStart =. In figure (b), −5 has been entered on this line. Other values for x that are to appear in the table are determined by setting an **increment value** on the line labeled ΔTbl =. Figure (b) shows that an increment of 1 was entered. This means that each x-value in the table will be 1 unit larger than the previous x-value.

The final step is to press the keys $\boxed{2nd}$ $\boxed{TABLE}$. This displays a table of solutions, as shown in figure (c).

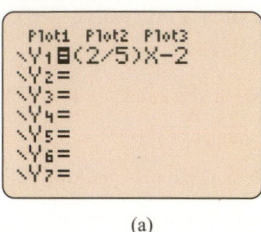   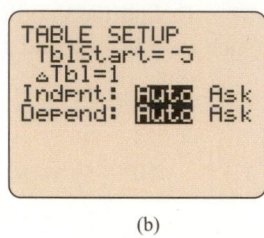   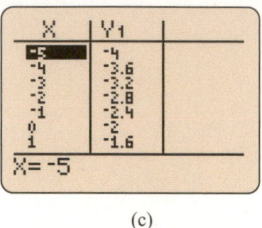

(a)   (b)   (c)

5 Graph Horizontal and Vertical Lines.

Equations such as $y = 3$ and $x = -2$ are linear equations, because they can be written in the standard form $Ax + By = C$. For example, $y = 3$ is equivalent to $0x + 1y = 3$ and $x = -2$ is equivalent to $1x + 0y = -2$. We can graph these types of equations using point-plotting.

EXAMPLE 4   Graph the equations on the same rectangular coordinate system: **a.** $y = 4$ **b.** $x = -2$

Strategy   To find three ordered-pair solutions of $y = 4$ to plot, we will select three values for x and use 4 for y each time. To find three ordered-pair solutions of $x = -2$ to plot, we must select −2 for x each time.

Why   The first equation requires that $y = 4$ and the second equation requires that $x = -2$.

Solution   **a.** We can write the equation in standard form as $0x + y = 4$. Since the coefficient of x is 0, the numbers selected for x have no effect on y. The value of y is always 4.

After plotting the ordered pairs shown in the table, and drawing a straight line through them, we see that the graph is a horizontal line, parallel to the x-axis, with a y-intercept of (0, 4). The line has no x-intercept.

b. We can write the equation in standard form as $x + 0y = -2$. Since the coefficient of y is 0, the value of y has no effect on x.

After plotting the ordered pairs shown in the table, and drawing a straight line through them, we see that the graph is a vertical line, parallel to the y-axis, with an x-intercept of (−2, 0). The line has no y-intercept.

$y = 4$

x	y	(x, y)
−3	4	(−3, 4)
0	4	(0, 4)
2	4	(2, 4)

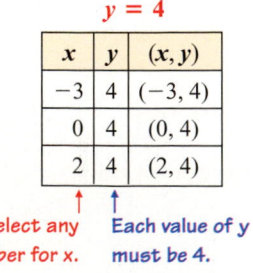

Select any number for x.   Each value of y must be 4.

$x = -2$

x	y	(x, y)
−2	−3	(−2, −3)
−2	0	(−2, 0)
−2	2	(−2, 2)

Each value of x we select must be −2.   Any number for y can be used.

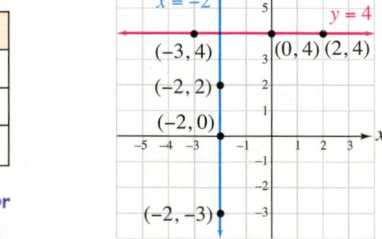

Self Check 4   Graph the equations on the same rectangular coordinate system:
a. $x = 4$ **b.** $y = -3$

Now Try ▶ Problems 41 and 43

The results of Example 4 suggest the following facts.

Equations of Horizontal and Vertical Lines

The graph of $y = b$ is a horizontal line with y-intercept $(0, b)$.	The graph of $x = a$ is a vertical line with x-intercept $(a, 0)$.	The graph of $y = 0$ is the x-axis. The graph of $x = 0$ is the y-axis.
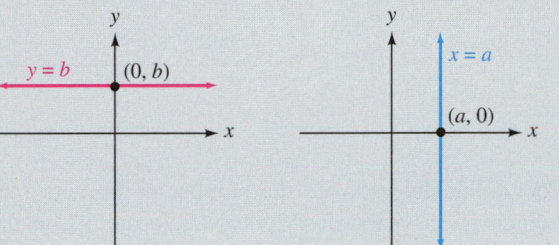		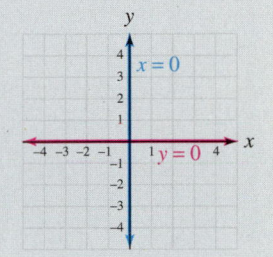

---

6 Use Linear Models to Solve Applied Problems.

In the next examples, we will see how linear equations can model real-life situations. **Linear models,** as they are called, often are written in variables other than x and y. We must make the appropriate changes when labeling the table of solutions and the graph of the equation. We can use linear models to make observations about what has occurred in the past and what might occur in the future.

EXAMPLE 5

Farmers Markets. The increasing popularity of farmers markets in the United States can be modeled by the linear equation $f = 265t + 2{,}728$, where t represents the number of years after 2000 and f represents the number of farmers markets. (*Source:* U.S. Department of Agriculture) **a.** Graph the equation.   **b.** Suppose the trend continues. Use the graph to estimate the number of farmers markets in the year 2015.

Strategy We will find three solutions of the equation, plot them on a rectangular coordinate system, and then draw a straight line passing through the points.

Why To *graph* a linear equation in two variables means to make a drawing that represents all of its solutions.

Solution   **a.** The variables t and f are used in the equation. If we associate t with the horizontal axis and f with the vertical axis, then the ordered pair solutions has the form (t, f).

To graph the equation, we select three values for t, substitute them into the equation, and find each corresponding value of f. Since t represents the number of years *after* 2000, we will not select any negative values for t. The results are listed in the table on the next page.

For $t = 0$ (The year 2000)	*For $t = 2$* (The year 2002)	*For $t = 5$* (The year 2005)
$f = 265\boldsymbol{t} + 2{,}728$	$f = 265\boldsymbol{t} + 2{,}728$	$f = 265\boldsymbol{t} + 2{,}728$
$f = 265(\boldsymbol{0}) + 2{,}728$	$f = 265(\boldsymbol{2}) + 2{,}728$	$f = 265(\boldsymbol{5}) + 2{,}728$
$f = 0 + 2{,}728$	$f = 530 + 2{,}728$	$f = 1{,}325 + 2{,}728$
$f = 2{,}728$	$f = 3{,}258$	$f = 4{,}053$

The pairs (0, 2,728), (2, 3,258), and (5, 4,053) are plotted, and a straight line is drawn through them to give the graph of the equation.

$f = 265t + 2{,}728$

t	f	(t, f)
0	2,728	(0, 2,728)
2	3,258	(2, 3,258)
5	4,053	(5, 4,053)

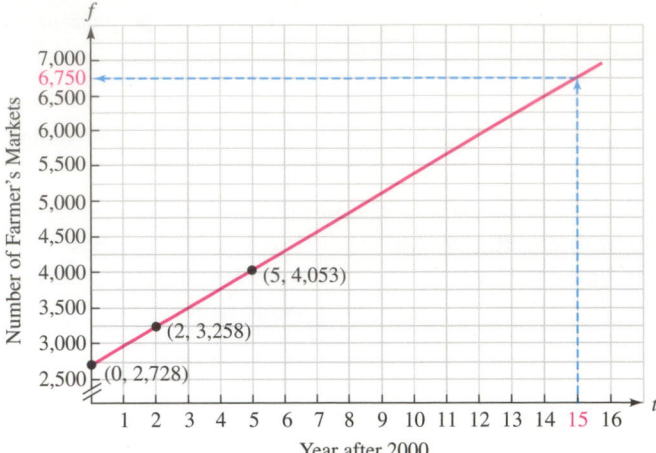

b. To graphically estimate the number of farmers markets that there will be in 2015, (which is 15 years after 2000), we locate 15 on the horizontal axis. Then we move upward and over (as shown in blue) to estimate a reading of 6,750 on the vertical axis. This means that if the current trend continues, in the year 2015, the number of farmer markets in the United States will be approximately 6,750.

Self Check 5   Use the graph in Example 5 to estimate the number of farmers markets in 2010.

Now Try ▶ Problem 77

For tax purposes, many businesses use the equation of a line to find the declining value of aging equipment. This method is called **straight-line depreciation.**

EXAMPLE 6

Depreciation.   A copy machine that was purchased for $6,750 is expected to depreciate according to the **straight-line depreciation equation** $y = -950x + 6{,}750$, where y is the value of the copier after x years of use. When will the copier have no value?

Strategy   To find when the copier will have no value, we will substitute 0 for y in the equation $y = -950x + 6{,}750$ and solve for x.

Why   The variable y represents the value of the copier. When the copier has no value, y will be equal to 0.

Solution

$$y = -950x + 6{,}750 \quad \text{This is the straight-line depreciation model.}$$
$$0 = -950x + 6{,}750 \quad \text{Substitute 0 for y.}$$
$$-6{,}750 = -950x \quad \text{To isolate the variable term } -950x, \text{ subtract 6,750 from both sides.}$$
$$7.105263158 \approx x \quad \text{To isolate x, divide both sides by } -950. \text{ Use a calculator.}$$

The copier will have no value after it has been in use for approximately 7.1 years.

The equation $y = -950x + 6{,}750$ is graphed on the next page using the intercept method. Important information can be obtained from the intercepts of the graph.

The y-intercept of the graph is (0, 6,750). This indicates that the purchase price of the copier was $6,750.

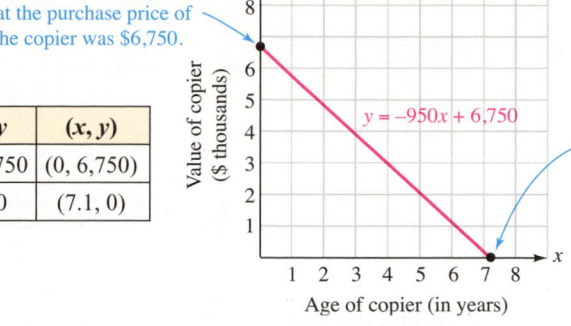

$y = -950x + 6,750$

The x-intercept of the graph is approximately (7.1, 0). This indicates that the value of the copier will be $0 in about 7.1 years.

Value of copier ($ thousands)

Age of copier (in years)

Success Tip

When the copier is new, it has been in use 0 years. In that case, x is 0. When the copier has no value, y is 0.

x	y	(x, y)
0	6,750	(0, 6,750)
7.1	0	(7.1, 0)

Self Check 6   **Depreciation.**   **a.** Use the equation $y = -950x + 6,750$ to determine when the copier will be worth $3,900.   **b.** Use the graph in Example 6 to determine when the copier will be worth $2,000.

Now Try ▶ Problem 83

Using Your Calculator ▶ Graphing Lines

We have graphed linear equations by finding solutions, plotting points, and drawing lines through those points. Graphing is often easier using a graphing calculator.

Window settings

Graphing calculators have a window to display graphs. To see the proper picture of a graph, we must decide on the minimum and maximum values for the x- and y-coordinates. A window with standard settings of

$$\text{Xmin} = -10 \qquad \text{Xmax} = 10 \qquad \text{Ymin} = -10 \qquad \text{Ymax} = 10$$

will produce a graph where the values of x and the values of y are between -10 and 10, inclusive. We can use the notation $[-10, 10]$ to describe such intervals.

Graphing lines

To graph $5x - 2y = 4$, we must first solve the equation for y.

$$y = \frac{5}{2}x - 2 \qquad \text{\textcolor{red}{Subtract 5x from both sides and then divide both sides by }} -2.$$

Next, we press $\boxed{Y =}$ and enter the right side of the equation after the symbol $Y_1 =$. See figure (a). We then press the $\boxed{\text{GRAPH}}$ key to get the graph shown in figure (b). To show more detail, we can change the window settings to $[-2, 5]$ for x and $[-4, 5]$ for y by pressing $\boxed{\text{WINDOW}}$ and entering -2 for Xmin, 5 for Xmax, -4 for Ymin, and 5 for Ymax. See figure (c).

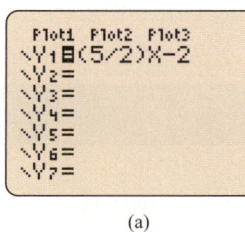   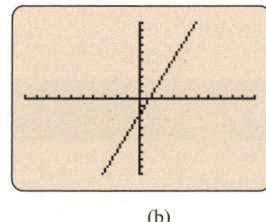   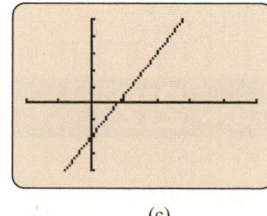

     (a)   (b)   (c)

Finding the coordinates of a point on the graph

If we reenter the standard window settings of $[-10, 10]$ for x and for y, press $\boxed{\text{GRAPH}}$, and press the $\boxed{\text{TRACE}}$ key, we get the display shown in figure (d), on the next page. The y-intercept of the graph is highlighted by the flashing cursor, and the x- and y-coordinates of that point are given at the bottom of the screen. We can use the $\boxed{\blacktriangleright}$ and $\boxed{\blacktriangleleft}$ keys to move the cursor along the line to find the coordinates of any point on the line. After pressing the ▶ key 12 times, we will get the display in figure (e).

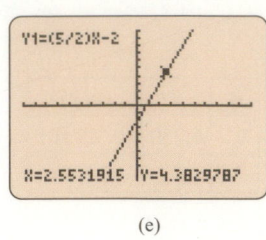

(d)   (e)   (f)

To find the y-coordinate of any point on the line, given its x-coordinate, we press [2nd] [CALC] and select the value option. We enter the x-coordinate of the point and press [ENTER]. The y-coordinate is then displayed. In figure (f), 1.5 was entered for the x-coordinate, and its corresponding y-coordinate, 1.75, was found.

The table feature, discussed on page 118, gives us a third way of finding the coordinates of a point on the line.

Determining the x-intercepts of a graph

To determine the x-intercept of the graph of $y = \frac{5}{2}x - 2$, we can use the zero option, found under the CALC menu. (Be sure to reenter the standard window settings for x and y before using CALC.) After we enter left and right bounds and a guess, as shown in figure (g), the cursor automatically moves to the x-intercept of the graph when we press [ENTER]. Figure (h) shows how the coordinates of the x-intercept are then displayed at the bottom of the screen.

We can also use the trace and zoom features to determine the x-intercept of the graph of $y = \frac{5}{2}x - 2$. After graphing the equation using the standard window settings, we press [TRACE]. Then we move the cursor along the line toward the x-intercept until we arrive at a point with the coordinates shown in figure (i). To get better results, we press [ZOOM], select the zoom in option, and press [ENTER] to get a magnified picture. We press [TRACE] again and move the cursor to the point with coordinates shown in figure (j). Since the y-coordinate is nearly 0, this point is nearly the x-intercept. We can achieve better results with more zooms and traces.

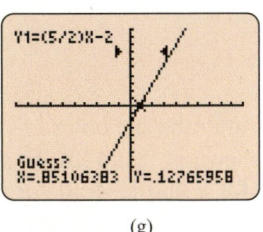

(g)   (h)   (i)   (j)

VOCABULARY

Fill in the blanks.

1. A solution of an equation in two variables is an _____ _____ of numbers that make a true statement when substituted into the equation.

2. The graph of $y = 2x + 1$ is the graph of all points (x, y) on the rectangular coordinate system whose coordinates _____ the equation.

3. The equation $y = -6x - 3$ is a _____ equation in two variables.

4. The _____ form of a linear equation in two variables is $Ax + By = C$, where not both A and B are 0.

5. The point where the graph of a linear equation intersects the y-axis is called the _____ and the point where it intersects the x-axis is called the _____.

6. The graph of any equation of the form $x = a$ is a _____ line. The graph of any equation of the form $y = b$ is a _____ line.

CONCEPTS

7. Use the graph on the right to determine three solutions of $2x + 3y = 9$.

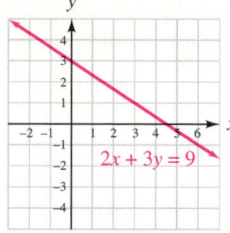

8. Consider the equation $2x + 4y = 8$. How many variables does it contain? How many solutions does it have?

9. Fill in the blanks: The exponent on each variable of a linear equation is an understood ▢. For example, $4x + 7y = 3$ can be thought of as $4x^{▢} + 7y^{▢} = 3$.

10. A table of solutions for a linear equation is given on the right. From the table, determine the x-intercept and the y-intercept of the graph of the equation.

x	y	(x, y)
-6	0	$(-6, 0)$
-2	2	$(-2, 2)$
0	3	$(0, 3)$

11. Refer to the graph on the right.

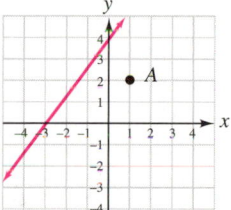

   a. What is the x-intercept and what is the y-intercept of the line?

   b. If the coordinates of point A are substituted into the equation of the line that is graphed here, will a true or a false statement result?

12. Consider the linear equation $6x - 4y = -12$.

   a. Find the x-intercept of its graph.

   b. Find the y-intercept of its graph.

   c. Does its graph pass through $(2, 6)$?

13. A graphing calculator display is shown below. It is a table of solutions for which one of the following linear equations?

$$y = -2x - 1, \quad y = -3x - 1, \quad \text{or} \quad y = -4x - 1$$

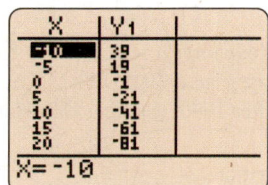

14. The graphing calculator displays in the next column show the graph of

$$y = -2x - \frac{5}{4}.$$

   a. In figure (a), what important feature of the line is highlighted by the cursor?

   b. In figure (b), what important feature of the line is highlighted by the cursor?

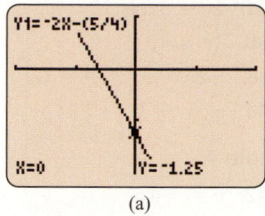   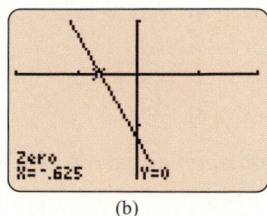

(a)   (b)

NOTATION

15. a. The graph of the equation $x = 0$ is which axis?

   b. The graph of the equation $y = 0$ is which axis?

16. A linear equation in two variables is an equation that can be written in the standard form $Ax + By = C$. The equation $x - 5y = 4$ is written in standard form. Determine A, B, and C.

GUIDED PRACTICE

Determine whether each ordered pair is a solution of the given equation. See Example 1.

17. $y = -5x - 2$

   a. $(-1, 3)$ **b.** $(3, -13)$

18. $2x - 5y = 9$

   a. $(-4, 2)$ **b.** $(2, -1)$

19. $y - 6x = 10$

   a. $\left(\frac{1}{6}, 11\right)$ **b.** $(-2.1, -0.6)$

20. $3y - 5x = 30$

   a. $(-2.3, -7.5)$ **b.** $\left(-\frac{23}{5}, \frac{7}{3}\right)$

Complete each table of solutions and use the results to graph the equation. See Example 2.

21. $y = -x + 4$

x	y
-1	
0	
2	

22. $y = x - 2$

x	y
-2	
0	
4	

23. $y = -\frac{1}{3}x - 1$

x	y
-3	
0	
3	

24. $y = -\frac{1}{2}x + \frac{5}{2}$

x	y
-1	
3	
5	

Graph each equation. See Example 2.

25. $y = \frac{x}{4} - 1$ **26.** $y = -\frac{x}{4} + 2$

27. $y = -3x + 2$ **28.** $y = 2x - 3$

29. $y = 3 - x$ **30.** $y = 5 - x$

31. $y = x$ **32.** $y = -2x$

Graph each equation using the intercept method. Label the intercepts on each graph. See Example 3.

33. $3x + 4y = 12$

34. $4x - 3y = 12$

35. $3y = 6x - 9$

36. $2x = 4y - 10$

37. $3x + 4y - 8 = 0$

38. $-2y - 3x + 9 = 0$

39. $3x = 4y - 11$

40. $-5x + 3y = 11$

Graph each equation. See Example 4.

41. $x = 3$

42. $y = -4$

43. $y = -\dfrac{1}{2}$

44. $x = \dfrac{4}{3}$

Write each equation in $y = b$ or $x = a$ form by solving for y or x. Then graph it. See Example 4.

45. $y - 2 = 0$

46. $x + 1 = 0$

47. $-2x + 3 = 11$

48. $-3y + 2 = 5$

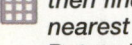 *Use a graphing calculator to graph each equation, and then find the x-coordinate of the x-intercept to the nearest hundredth. See Using Your Calculator: Determining the x-intercepts of a Graph.*

49. $y = 3.7x - 4.5$

50. $y = \dfrac{3}{5}x + \dfrac{5}{4}$

51. $1.5x - 3y = 7$

52. $0.3x + y = 7.5$

TRY IT YOURSELF

Graph each equation.

53. $5x - 4y = 13$

54. $3x - 4y = 11$

55. $y = \dfrac{5}{6}x - 5$

56. $y = \dfrac{2}{3}x - 2$

57. $x + 2y = -2$

58. $2x + 4y = -8$

59. $y = \dfrac{5}{2}$

60. $x = 0$

61. $y = 4x$

62. $y = x$

63. $x = 50 - 5y$

64. $3x = -150 - 5y$

65. $y = -\dfrac{1}{2}x$

66. $y = \dfrac{3}{4}x$

67. $y = 1.5x - 4$

68. $y = 0.5x + 3$

69. $3x + 5y = 0$

70. $4x + 3y = 0$

71. $x = -\dfrac{5}{3}$

72. $y = 0$

73. $y - 3x = -\dfrac{4}{3}$

74. $y - 2x = -\dfrac{9}{8}$

75. $4x - 5y = 0$

76. $7x + 3y = 0$

APPLICATIONS

77. **Square Footage.** The equation $s = 27t + 1,925$ approximates the median number of square feet s in a new single-family home, t years after 1995. (Source: U.S. Department of Commerce)

   a. Graph the equation.

   b. What information can be obtained from the s-intercept of the graph?

   c. From the graph, estimate the median number of square feet that a new single-family home had in 2008.

78. **Market Share.** The automobile industry classifies light-duty vehicles as cars, vans, SUVs, and pickup trucks. For sales of light-duty vehicles in the United States, the percent market share m of cars only is approximated by the equation $m = -\dfrac{4}{3}t + 69$, where t is the number of years after 1990. (Source: Environmental Protection Agency)

   a. Graph the equation.

   b. What information can be obtained from the m-intercept of the graph?

   c. Suppose the current trend continues. From the graph, estimate the market share m (in percent) of the sales of cars in 2020.

79. **Union Membership.** The equation $p = -0.25t + 11.7$ approximates the percent p of those employed in the private sector who were union members, t years after 1990. (Source: unionstats.gsu.edu)

   a. Graph the equation.

   b. What information can be obtained from the p-intercept of the graph?

   c. Suppose the current trend continues. From the graph, estimate the percent of people working in the private sector who will be union members in 2015.

80. **Farming.** The equation $a = -3.1t + 945$ gives the approximate number of acres a of farmland (in millions) in the United States, t years after 2000. (Source: U.S. Department of Agriculture)

   a. Graph the equation.

   b. What information can be obtained from the a-intercept of the graph?

   c. Suppose the current trend continues. From the graph, estimate the number of acres of farmland in the year 2020.

81. **Living Longer.** The equation $a = 0.16t + 73.7$ is a linear model that approximates the average life expectancy a (in years) in the United States, t years after 1980. (Source: World Bank)

   a. Graph the equation.

   b. What information can be obtained from the a-intercept of the graph?

   c. Suppose the current trend continues. From the graph, estimate the average life expectancy in the United States in 2030.

82. **Hourly Pay.** The equation $p = 0.4t + 10.5$ approximates the average hourly pay p (in dollars) of a U.S. manufacturing worker, t years after 1990. (Source: U.S. Bureau of Labor Statistics)

   a. Graph the equation.

   b. What information can be obtained from the p-intercept of the graph?

   c. Suppose the current trend continues. From the graph, estimate the average hourly pay for manufacturing workers in 2014.

83. **Depreciation.** The graph on the right shows how the value of a computer decreased over the age of the computer. What information can be obtained from the *x*-intercept? The *y*-intercept?

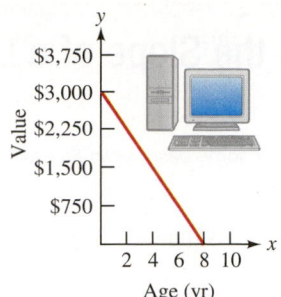

Value: $3,750, $3,000, $2,250, $1,500, $750

2 4 6 8 10

Age (yr)

84. **Spray Bottles.** The following graph shows the amount *A* of sore throat medication (in ounces) that remains in the bottle after the spray trigger has been pushed down a total of *n* times. What information can be obtained from the *n*-intercept? The *A*-intercept?

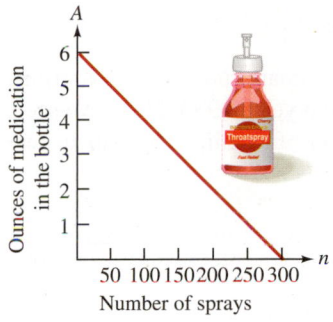

Ounces of medication in the bottle

6 5 4 3 2 1

50 100 150 200 250 300

Number of sprays

85. **Car Depreciation.** A car purchased for $17,000 is expected to lose value according to the straight-line depreciation model $y = -1,360x + 17,000$. When will the car have no value?

86. **Accounting.** A carpet company purchased a new loom for $124,000. For income tax purposes, company accountants will use the straight-line depreciation equation $y = -15,500x + 124,000$ to describe the declining value of the loom.

   a. When will the value of the loom be one-half of its purchase price?

   b. When will the loom have no value?

87. **Demand Equation.** The number of microwave ovens that consumers buy depends on price. The higher the price, the fewer microwaves people will buy. The equation that relates price to the number of microwaves sold at that price is called a **demand equation.** If the demand equation for a 2-cubic-foot countertop microwave is approximated by $p = -\frac{1}{10}q + 170$, where *p* is the price and *q* is the number of microwaves sold at that price, how many of that model microwave will be sold at a price of $150?

88. **Supply Equation.** The number of microwave ovens that manufacturers produce depends on price. The higher the price, the more microwaves manufacturers will produce. The equation that relates price to the number of microwaves produced at that price is called a **supply equation.** If the supply equation for a 2-cubic-foot countertop microwave is $p = \frac{1}{10}q + 130$, where *p* is the price and *q* is the number of microwaves produced for sale at that price, how many of that model microwave will be produced if the price is $150?

WRITING

89. Explain how to graph a line using the intercept method.

90. When graphing a line by plotting points, why is it a good practice to find three solutions instead of two?

91. What does it mean when we say that a linear equation in two variables has infinitely many solutions? Give an example.

92. On a quiz, a student was asked to graph the lines $x = -3$ and $y = 2$ on the same rectangular coordinate system. His answer is shown on the right. Explain his error.

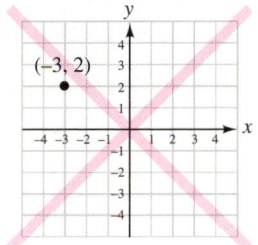

(−3, 2)

REVIEW

93. Write the set whose elements are the prime numbers between 10 and 30.

94. Write the set whose elements are the first ten composite numbers.

95. In what quadrant does the point $(-2, -3)$ lie?

96. What is the formula that gives the area of a circle?

97. Simplify: $-4(-20s)(-6)$

98. Approximate π to the nearest hundredth.

99. Simplify: $-4[-2(-3x - 8)]$

100. Simplify: $\frac{1}{3}b + \frac{1}{3}b + \frac{1}{3}b$

CHALLENGE PROBLEMS

Graph each equation.

101. $\frac{1}{5}x = 6 - \frac{3}{10}y$

102. $\frac{x}{2} - \frac{y}{3} - 4 = 0$

103. $y = \frac{x}{|x|}$

104. $y = \frac{x - 3}{|x - 3|}$

SECTION 2.3

OBJECTIVES

1 Calculate an average rate of change.

2 Find the slope of a line from its graph.

3 Find the slope of a line given two points.

4 Find the slope of horizontal and vertical lines.

5 Solve applications of slope.

6 Determine whether lines are parallel or perpendicular using slope.

Rate of Change and the Slope of a Line

ARE YOU READY?

The following problems review some basic skills that are needed to find the slope of a line.

1. Evaluate: $\dfrac{4-1}{8-3}$ **2.** Evaluate: $\dfrac{-10-1}{-4-(-4)}$ **3.** Simplify: $\dfrac{24}{28}$

4. Fill in the blanks. _____ lines do not intersect. _____ lines intersect to form right angles.

Our world is one of constant change. In this section, we will describe the change in one quantity with respect to the change in another by finding an *average rate of change*.

1 Calculate an Average Rate of Change.

The following line graphs model the approximate number of morning and evening newspapers published in the United States for the years 1990–2008. We see that the number of morning newspapers increased and the number of evening newspapers decreased over this time span.

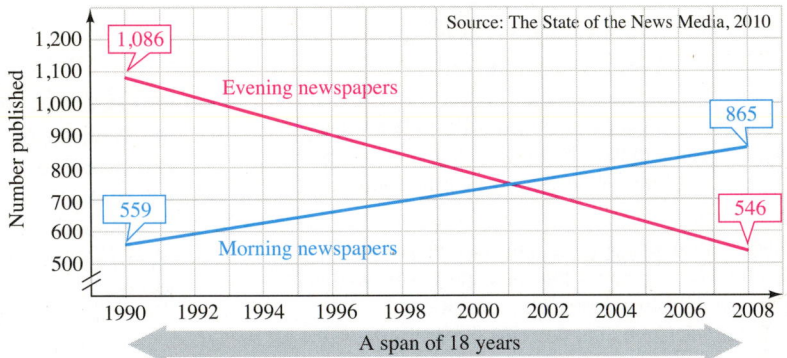

If we want to know the rate at which the number of morning newspapers increased or the rate at which the number of evening newspapers decreased, we must find an average rate of change. To find an average rate of change, we find the *ratio* of the change in the number of newspapers to the length of time in which that change took place and attach the appropriate units.

Ratios and Rates	A **ratio** is a comparison of two numbers using a quotient. In symbols, if a and b are two numbers, the ratio of a to b is $\dfrac{a}{b}$. Ratios that are used to compare quantities with different units are called **rates**.

The Language of Algebra

Ratios are used in many settings. Mortgage companies calculate *debt-to-income ratios* for loan applicants. Bicyclists choose *gear ratios* that produce a smooth ride uphill and downhill.

In the graph, we see that in 1990, the number of morning newspapers published was 559. In 2008, the number grew to 865. This is a change of $865 - 559$ or 306 over an 18-year time span. So we have

$$\begin{aligned}\text{Average rate} \atop \text{of change} &= \frac{\text{change in number of morning newspapers}}{\text{change in time}} \\ &= \frac{306 \text{ newspapers}}{18 \text{ years}}\end{aligned}$$

A rate of change is a ratio that includes units.

Subtract the number of morning newspapers in 1990 from the number in 2008: $865 - 559 = 306$.

Success Tip

Note that the numerator of the rate of change ratio contains units associated with the vertical axis of the graph. The denominator contains units associated with the horizontal axis.

$$= \frac{\overset{1}{\cancel{18} \cdot 17 \text{ newspapers}}}{\underset{1}{\cancel{18} \text{ years}}}$$

Factor 306 as 18 · 17 and simplify: $\frac{18}{18} = 1$. We could also just simply divide: $306 \div 18 = 17$.

$$= \frac{17 \text{ newspapers}}{1 \text{ year}}$$

The number of morning newspapers published in the United States increased, on average, at a rate of 17 newspapers per year from 1990 through 2008. This can be written as 17 newspapers/year.

An average rate of change also can be negative. This can be illustrated by finding the average rate of change in the number of evening newspapers published in the United States from 1990 through 2008.

$$\frac{\text{Average rate}}{\text{of change}} = \frac{\text{change in number of evening newspapers}}{\text{change in time}}$$

Success Tip

In general, to find the change in a quantity, we subtract the earlier value from the later value. The negative result here indicates a decrease in the number of evening newspapers from 1990 to 2008.

$$= \frac{-540 \text{ newspapers}}{18 \text{ years}}$$

Subtract the number of evening newspapers in 1990 from the number in 2008: $546 - 1{,}086 = -540$.

$$= \frac{\overset{-1}{\cancel{-18} \cdot 30 \text{ newspapers}}}{\underset{1}{\cancel{18} \text{ years}}}$$

Factor -540 as $-18 \cdot 30$ and simplify: $\frac{-18}{18} = -1$. We could also just simply divide: $-540 \div 18 = -30$.

$$= \frac{-30 \text{ newspapers}}{1 \text{ year}}$$

The number of evening newspapers being published changed at a rate of -30 newspapers/year. That is, on average, there were 30 fewer per year, every year, from 1990 through 2008.

2 Find the Slope of a Line from Its Graph.

The Language of Algebra

m is used to denote the slope of a line. Many historians credit this to the fact that it is the first letter of the French word **monter**, meaning to ascend or to climb.

In the newspaper example, we measured the steepness of the lines in the graph to determine the average rates of change. In doing so, we found the *slope* of each line. The **slope of a line** is a ratio that compares the vertical change to the corresponding horizontal change as we move along the line from one point to another.

To determine the slope of a line (usually denoted by the letter m) from its graph, we first pick two points on the line. Then we write the ratio of the vertical change, called the **rise,** to the corresponding horizontal change, called the **run,** as we move from one point to the other.

$$m = \frac{\text{vertical change}}{\text{horizontal change}} = \frac{\text{rise}}{\text{run}}$$

EXAMPLE 1 Find the slope of the line graphed in figure (a) below.

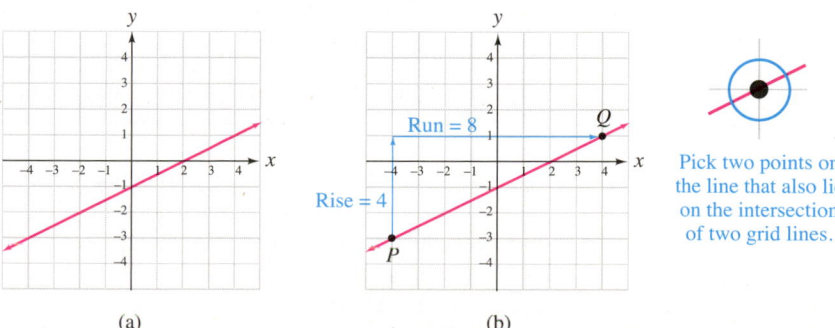

(a)   (b)

Pick two points on the line that also lie on the intersection of two grid lines.

Strategy We will pick two points on the line, construct a slope triangle, and find the rise and run. Then we will write the ratio of rise to run and simplify the result, if possible.

Why The slope of a line is the ratio of the rise to the run.

Solution   We begin by choosing two points on the line, P and Q, as shown in figure (b). One way to move from P to Q is to start at P, move upward, a rise of 4 grid squares, and then to the right, a run of 8 grid squares, to reach Q. These steps create a right triangle called a **slope triangle**.

$$m = \frac{\text{rise}}{\text{run}} = \frac{4}{8} = \frac{1}{2} \qquad \textcolor{red}{\text{Simplify the fraction. The result is positive.}}$$

The slope of the line is $\frac{1}{2}$.

The two-step process to move from P to Q can be reversed. Starting at P, we can move to the right, a run of 8; and then upward, a rise of 4, to reach Q. With this approach, the slope triangle is below the line. When we form the ratio to find the slope, we get the same result as before:

$$m = \frac{\text{rise}}{\text{run}} = \frac{4}{8} = \frac{1}{2}$$

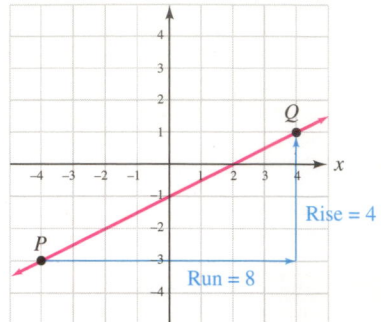

Self Check 1   Find the slope of the line shown above using two points different from those used in the solution of Example 1.

Now Try ▶ Problems 17 and 21

The identical answers from Example 1 and its Self Check illustrate that **the same value will be obtained no matter which two points on a line are used to find its slope.**

3 Find the Slope of a Line Given Two Points.

We can use the graphic method for finding slope to develop a slope formula. To begin, we select points P and Q on the line shown in the figure on the right. To distinguish between the coordinates of these two points, we use **subscript notation.** Point P has coordinates (x_1, y_1) and point Q has coordinates (x_2, y_2).

As we move from point P to point Q, the rise is the difference of the y-coordinates: $y_2 - y_1$. The run is the difference of the x-coordinates: $x_2 - x_1$. Since the slope is the ratio $\frac{\text{rise}}{\text{run}}$, we have the following formula for calculating slope.

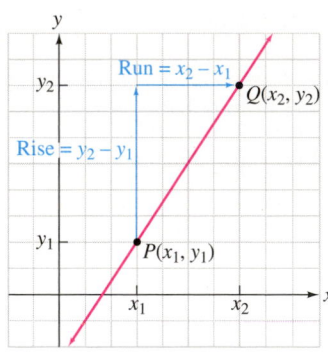

Slope of a Line ▼

The **slope** m of a line passing through points (x_1, y_1) and (x_2, y_2) is

$$m = \frac{\text{vertical change}}{\text{horizontal change}} = \frac{\text{rise}}{\text{run}} = \frac{\text{change in } y}{\text{change in } x} = \frac{y_2 - y_1}{x_2 - x_1} \quad \text{if } x_2 \neq x_1$$

Another notation that we use to define slope involves the symbol Δ, which is the letter *delta* from the Greek alphabet. If the change in y is represented by Δy (read as "delta y") and the change in x is represented by Δx (read as "delta x"), then:

$$m = \frac{\Delta y}{\Delta x} \quad \text{where } \Delta x \neq 0$$

The graph on the right shows all of the notation associated with the concept of slope of a line.

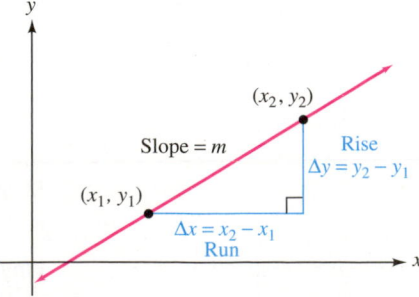

The slope formula is a valuable tool because it enables us to calculate the slope of a line without having to view its graph.

EXAMPLE 2   Find the slope of the line passing through $(-2, 4)$ and $(3, -4)$.

Strategy   We will use the slope formula to find the slope.

Why   We know the coordinates of two points on the line.

Solution   We can let $(x_1, y_1) = (-2, 4)$ and $(x_2, y_2) = (3, -4)$. Then we have

$$m = \frac{y_2 - y_1}{x_2 - x_1} \qquad \text{This is the slope formula.}$$

$$m = \frac{-4 - 4}{3 - (-2)} \qquad \text{Substitute } -4 \text{ for } y_2, 4 \text{ for } y_1, 3 \text{ for } x_2, \text{ and } -2 \text{ for } x_1.$$

$$m = -\frac{8}{5} \qquad \begin{array}{l}\text{Do the subtraction. Write } \frac{-8}{5} \text{ with the } - \text{ sign in front}\\ \text{of the fraction. The result is negative.}\end{array}$$

The slope of the line is $-\frac{8}{5}$.

When calculating slope, it doesn't matter which point we call (x_1, y_1) and which point we call (x_2, y_2). We will obtain the same result if we let $(x_1, y_1) = (3, -4)$ and $(x_2, y_2) = (-2, 4)$.

$$m = \frac{y_2 - y_1}{x_2 - x_1} = \frac{4 - (-4)}{-2 - 3} = \frac{8}{-5} = -\frac{8}{5}$$

Self Check 2   Find the slope of the line passing through $(-3, 6)$ and $(4, -8)$.

Now Try   Problems 27 and 31

> **Success Tip**
>
> When using the slope formula, you may find it helpful to begin your solution by labeling the coordinates of the points in this way:
>
> (x_1, y_1) (x_2, y_2)
> ↓ ↓   ↓ ↓
> $(-2, 4)$ $(3, -4)$

CAUTION   When using the slope formula, we must be careful to subtract the y-coordinates and the x-coordinates in the same order. For instance, in Example 2 with $(x_1, y_1) = (-2, 4)$ and $(x_2, y_2) = (3, -4)$, it would be incorrect to write either of the following:

This is $y_2 - y_1$. The subtraction is not in the same order.

$$m = \frac{-4 - 4}{-2 - 3} = \frac{-8}{-5} = \frac{8}{5}$$

This is $x_1 - x_2$.

This is $y_1 - y_2$. The subtraction is not in the same order.

$$m = \frac{4 - (-4)}{3 - (-2)} = \frac{8}{5}$$

This is $x_2 - x_1$.

4 Find the Slope of Horizontal and Vertical Lines.

If (x_1, y_1) and (x_2, y_2) are distinct points on the horizontal line shown on the left, then $y_1 = y_2$, and the numerator of the fraction

$$\frac{y_2 - y_1}{x_2 - x_1} \qquad \text{On a horizontal line, } x_2 \neq x_1.$$

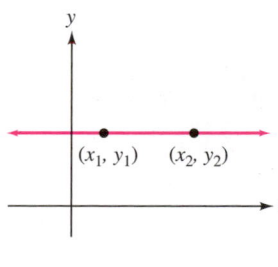

is 0. Thus, the value of the fraction is 0, and the slope of the horizontal line is 0.

If (x_1, y_1) and (x_2, y_2) are distinct points on the vertical line to the left, then $x_1 = x_2$, and the denominator of the fraction

$$\frac{y_2 - y_1}{x_2 - x_1} \qquad \text{On a vertical line, } y_2 \neq y_1.$$

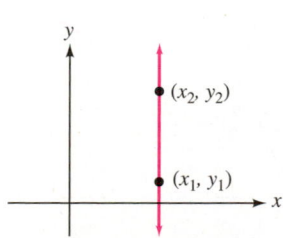

is 0. Since the denominator of a fraction cannot be 0, a vertical line has no defined slope.

Slopes of Horizontal and Vertical Lines	Horizontal lines (lines with an equation of the form $y = b$) have a slope of 0.
	Vertical lines (lines with equations of the form $x = a$) have no defined slope.

To classify the slope of a line as positive or negative, follow the line from left to right, as you would read a sentence in a book. If a line rises, its slope is positive. If a line falls, its slope is negative. If a line is horizontal, its slope is 0. If a line is vertical, it has undefined slope.

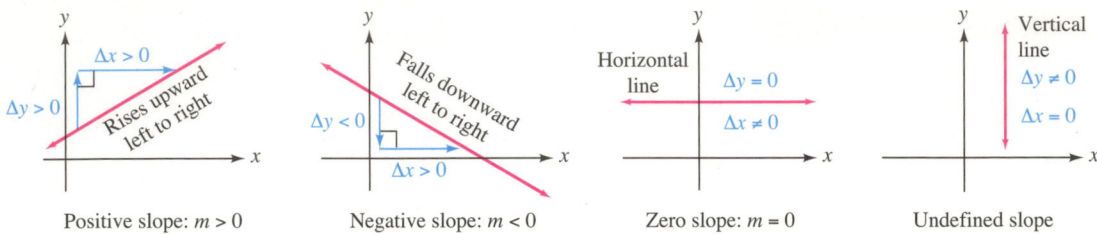

Positive slope: $m > 0$ — Negative slope: $m < 0$ — Zero slope: $m = 0$ — Undefined slope

5 Solve Applications of Slope.

The concept of slope has many applications. For example, architects use slope when designing ramps and determining the pitch of roofs. Truckers must be aware of the slope, or **grade,** of a road. Mountain resorts rate the difficulty level of ski runs by the degree of steepness.

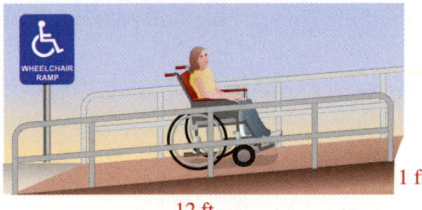

The Americans with Disabilities Act provides a guideline for the steepness of a ramp. The maximum slope for a wheelchair ramp is 1 foot of rise for every 12 feet of run: $m = \frac{1}{12}$.

The grade of an incline is its slope expressed as a percent. A 6% grade means a vertical change of 6 feet for every horizontal change of 100 feet: $m = \frac{6}{100}$, which simplifies to $\frac{3}{50}$. A grade is always expressed as a positive percent.

EXAMPLE 3 **Building Stairs.** The slope of a staircase is defined to be the ratio of the total rise to the total run, as shown in the illustration. Find the slope of the staircase.

Strategy We will express the total rise and the total run in terms of the same units and form their ratio.

Why The slope is a ratio that compares two quantities with the same units.

Solution Since the design has eight 7-inch risers, the total rise is $8 \cdot 7 = 56$ inches. The total run is 8 feet, or $8 \cdot 12 = 96$ inches. With the rise and run expressed in the same units, we can form their ratio.

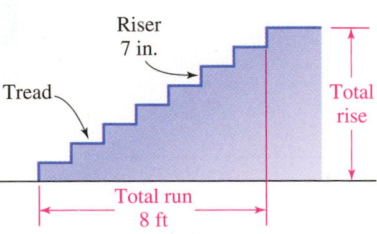

$$m = \frac{\text{total rise}}{\text{total run}} = \frac{56}{96} = \frac{7}{12}$$ Simplify the fraction: $\frac{56}{96} = \frac{7 \cdot \cancel{8}}{\cancel{8} \cdot 12} = \frac{7}{12}$.

The slope of the staircase is $\frac{7}{12}$.

Self Check 3 **Stairs.** Find the slope of the staircase if the riser height is changed to 6.5 inches.

Now Try ▶ Problem 57

EXAMPLE 4

Downhill Skiing. It takes a skier 25 minutes to complete the course shown in the illustration below. Find her average rate of descent in feet per minute.

Strategy We will describe the skier's positions on the course as ordered pairs of the form (time, elevation). Then we will use the slope formula to calculate the average rate of descent.

Why We use the slope formula because we know the coordinates of the skier at the beginning and at the end of the course.

Solution The skier's position can be described as an ordered pair of the form (time, elevation), or more simply, (t, E). To find the average rate of descent, we will calculate the slope of the line passing through the points $(t_1, E_1) = (0, 12{,}000)$ and $(t_2, E_2) = (25, 8{,}500)$ and attach the appropriate units to the numerator and denominator.

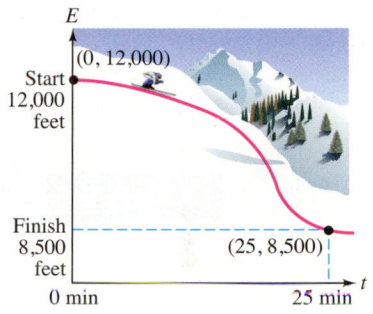

$$\text{Rate of descent} = \frac{(E_2 - E_1)\ \text{feet}}{(t_2 - t_1)\ \text{minutes}}$$

This is the slope formula adapted to a coordinate system with ordered pairs of the form (t, E).

$$= \frac{(8{,}500 - 12{,}000)\ \text{feet}}{(25 - 0)\ \text{minutes}}$$

Substitute 8,500 for E_2, 12,000 for E_1, 25 for t_2, and 0 for t_1.

$$= \frac{-3{,}500\ \text{feet}}{25\ \text{minutes}}$$

Do the subtractions. The negative number indicates a loss of elevation.

$$= -140\ \text{ft per min}$$

Do the division.

The skier's average rate of descent was 140 feet per minute. (The word *descent* itself implies a loss of elevation during the ski run. Therefore, the negative symbol $-$ need not be written in front of 140.)

Self Check 4 **Downhill Skiing.** Find the average rate of descent if the skier completes the course in 20 minutes.

Now Try ▶ Problem 63

6 Determine Whether Lines Are Parallel or Perpendicular Using Slope.

To see a relationship between parallel lines and their slopes, we refer to the parallel lines l_1 and l_2 shown on the next page, with slopes of m_1 and m_2, respectively. Because right triangles *ABC* and *DEF* are similar, it follows that

$$m_1 = \frac{\Delta y \text{ of } l_1}{\Delta x \text{ of } l_1}$$

Read l_1 as "line l sub 1."

$$m_1 = \frac{\Delta y \text{ of } l_2}{\Delta x \text{ of } l_2}$$

Since the triangles are similar, corresponding sides of △ABC and △DEF are proportional: $\frac{CB}{BA} = \frac{FE}{ED}$.

$$m_1 = m_2$$

Thus, if two nonvertical lines are parallel, they have the same slope. It is also true that when two different lines have the same slope, they are parallel.

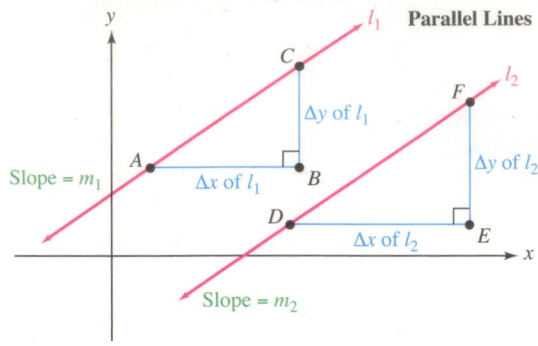

Parallel Lines

Slopes of Parallel Lines

Nonvertical parallel lines have the same slope, and different lines having the same slope are parallel.

EXAMPLE 5   Determine whether the line that passes through the points $(-6, 2)$ and $(3, -1)$ is parallel to a line with a slope of $-\dfrac{1}{3}$.

Strategy   We will compare the slopes of the lines.

Why   If the slopes are equal, the lines are parallel. If the slopes are not equal, the lines are not parallel.

Solution   We can use the slope formula to find the slope of the line that passes through $(-6, 2)$ and $(3, -1)$. We will substitute -1 for y_2, 2 for y_1, 3 for x_2, and -6 for x_1.

$$m = \frac{y_2 - y_1}{x_2 - x_1} = \frac{-1 - 2}{3 - (-6)} = \frac{-3}{9} = -\frac{1}{3} \qquad \begin{array}{l}\text{Simplify the fraction.}\\ \text{Write the } - \text{ symbol in front.}\end{array}$$

Both lines have slope $-\dfrac{1}{3}$, and therefore they are parallel.

Self Check 5   Determine whether the line that passes through the points $(4, -8)$ and $(1, -2)$ is parallel to a line with slope 2.

Now Try ▶ Problems 41 and 49

The two lines shown in the figure below meet at right angles and are called **perpendicular lines.** Each of the four angles that are formed has a measure of 90°.

The product of the slopes of two (nonvertical) perpendicular lines is -1. For example, the perpendicular lines shown in the figure have slopes of $\dfrac{3}{2}$ and $-\dfrac{2}{3}$. If we find the product of their slopes, we have

$$\frac{3}{2}\left(-\frac{2}{3}\right) = -\frac{6}{6} = -1$$

Two numbers whose product is -1, such as $\dfrac{3}{2}$ and $-\dfrac{2}{3}$, are called **negative reciprocals.** The phrase *negative reciprocal* can be used to relate perpendicular lines and their slopes.

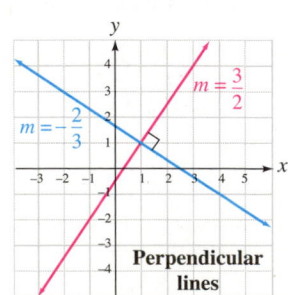

Perpendicular lines

Slopes of Perpendicular Lines

If two nonvertical lines are perpendicular, their slopes are negative reciprocals.

If the slopes of two lines are negative reciprocals, the lines are perpendicular.

We also can state the fact given above symbolically: If the slopes of two nonvertical lines are m_1 and m_2, then the lines are perpendicular if

$$m_1 \cdot m_2 = -1 \qquad \text{or} \qquad m_2 = -\frac{1}{m_1}$$

Because a horizontal line is perpendicular to a vertical line, a line with a slope of 0 is perpendicular to a line with no defined slope.

EXAMPLE 6 Are the lines l_1 and l_2 shown in the figure perpendicular?

Strategy We will compare the slopes of the two lines.

Why If the slopes are negative reciprocals, the lines are perpendicular. If the slopes are not negative reciprocals, the lines are not perpendicular.

Solution We find the slope of each line and see whether they are negative reciprocals.

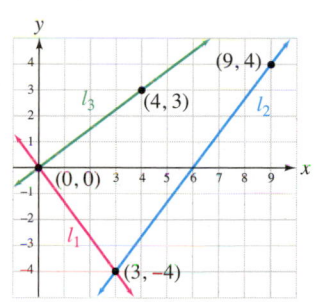

For line l_1

$$m = \frac{y_2 - y_1}{x_2 - x_1}$$

$$= \frac{-4 - 0}{3 - 0} \qquad \text{Use } (3, -4) \text{ and } (0, 0) \text{ on } l_1.$$

$$= -\frac{4}{3} \qquad \text{This is the slope of } l_1.$$

For line l_2

$$m = \frac{y_2 - y_1}{x_2 - x_1}$$

$$= \frac{4 - (-4)}{9 - 3} \qquad \text{Use } (3, -4) \text{ and } (9, 4) \text{ on } l_2.$$

$$= \frac{8}{6}$$

$$= \frac{4}{3} \qquad \text{This is the slope of } l_2.$$

Since their slopes are not negative reciprocals $\left(-\frac{4}{3} \cdot \frac{4}{3} \neq -1\right)$, the lines are not perpendicular.

Self Check 6 Is l_1 perpendicular to l_3?

Now Try ▶ Problems 43 and 51

SECTION 2.3 ▶ STUDY SET

VOCABULARY

Fill in the blanks.

1. A _____ is a comparison of two numbers using a quotient. In symbols, the ratio of a to b is $\frac{a}{b}$.

2. Ratios that are used to compare quantities with different units are called _____.

3. An average _____ of _____ describes how much one quantity changes with respect to another.

4. _____ is defined as the change in y divided by the change in x.

5. The _____ in x (written Δx) is the horizontal run of the line between two points on the line, and the change in y (written Δy) is the vertical _____ of the line between two points on the line.

6. $\frac{7}{8}$ and $-\frac{8}{7}$ are _____ reciprocals.

CONCEPTS

7. Halloween. A couple kept a graphical record of the number of trick-or-treaters who came to their door on Halloween night for the years 1990–2010. See the graph below. Fill in the blanks.

  a. Number of trick-or-treaters in 1990: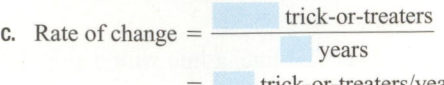

      Number of trick-or-treaters in 2010: ▢

  b. Change in the number of trick-or-treaters from 1990 to 2010: ▢

      Time span: ▢ years

  c. Rate of change $= \dfrac{\boxed{} \text{ trick-or-treaters}}{\boxed{} \text{ years}}$

      $= \boxed{}$ trick-or-treaters/year

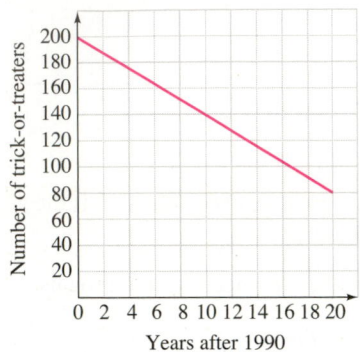

8. Fill in the blanks.

  a. $m = \dfrac{\boxed{}}{\text{horizontal change}} = \dfrac{\text{rise}}{\boxed{}} = \dfrac{\text{change in } y}{\boxed{}} = \dfrac{\boxed{}}{\Delta x}$

  b. $m = \dfrac{y_2 - y_1}{\boxed{}}$

9. Refer to the slope triangle shown on the graph.

  a. Find the rise.

  b. Find the run.

  c. Find the value of $\dfrac{\text{rise}}{\text{run}}$.

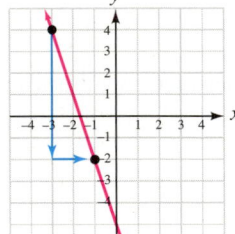

10. Refer to the slope triangle shown on the graph.

  a. Find Δy.

  b. Find Δx.

  c. Find $\dfrac{\Delta y}{\Delta x}$.

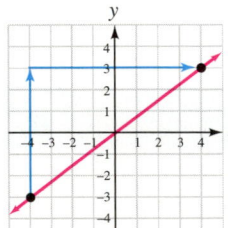

11. Refer to the graph.

  a. Which line is horizontal? Find its slope.

  b. Which line is vertical? Find its slope.

  c. Which line has a positive slope? What is it?

  d. Which line has a negative slope? What is it?

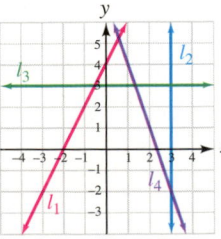

12. Fill in the blanks: _____ lines have the same slope and the slopes of _____ lines are negative reciprocals.

13. Refer to the graph below.

  a. Find the slopes of lines l_1 and l_2. Are they parallel?

  b. Find the slopes of lines l_2 and l_3. Are they perpendicular?

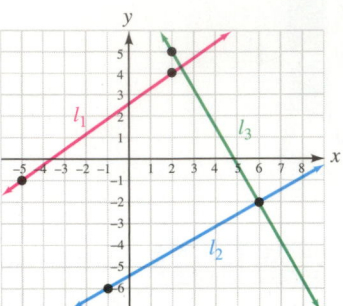

14. a. Find the slope of the line graphed in figure (a).

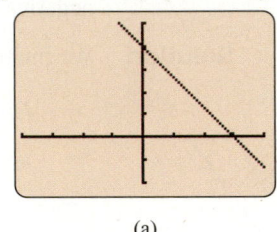

(a)

  b. A table of solutions for a linear equation is shown in figure (b). Find the slope of the graph of the equation.

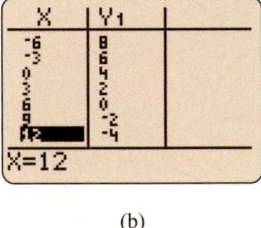

(b)

NOTATION

15. The rate of change of 10 ft/yr is read as "10 feet ____ year."

16. Fill in the blanks: The symbol Δ is the letter _____ from the _____ alphabet.

GUIDED PRACTICE

Find the slope of each line. **See Example 1.**

17.

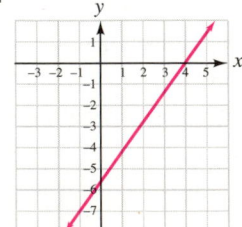

18.

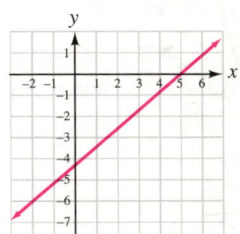

19.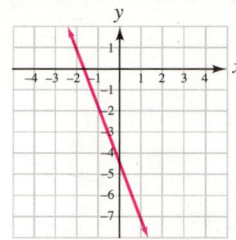

20.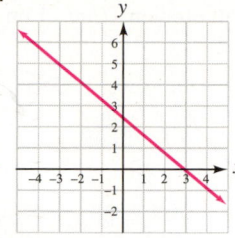

21.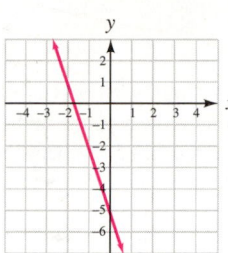

22.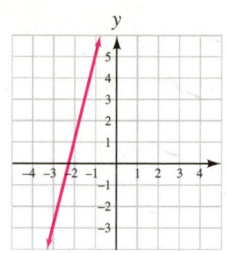

23.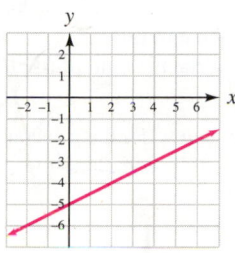

24.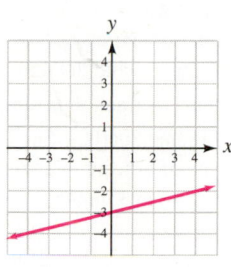

Find the slope of the line that passes through the given points, if possible. **See Example 2.**

25. $(0, 0), (3, 9)$

26. $(-5, -8), (3, 8)$

27. $(-1, 8), (6, 1)$

28. $(3, 4), (2, 7)$

29. $(3, -1), (-6, 2)$

30. $(0, -8), (-5, 0)$

31. $(28, 50), (7, 17)$

32. $(-7, -2), (70, 40)$

33. $(7, 5), (-9, 5)$

34. $(2, -8), (3, -8)$

35. $(-7, -5), (-7, -2)$

36. $(3, -5), (3, 14)$

37. $\left(\dfrac{1}{4}, \dfrac{9}{2}\right), \left(-\dfrac{3}{4}, 0\right)$

38. $\left(\dfrac{1}{8}, \dfrac{3}{4}\right), \left(\dfrac{3}{8}, -\dfrac{1}{4}\right)$

39. $(0.7, -0.6), (-0.9, 0.2)$

40. $(-1.2, 8.6), (-1.1, 7.6)$

Determine whether the line that passes through the two given points is parallel, perpendicular, or neither to a line with a slope of −2. **See Examples 5 and 6.**

41. $(3, 4), (4, 2)$

42. $(6, 4), (8, 5)$

43. $(-2, 1), (6, 5)$

44. $(3, 4), (-3, -5)$

45. $(5, 4), (6, 6)$

46. $(-2, 3), (4, -9)$

47. $(3.2, 12.3), (6.2, 6.3)$

48. $\left(\dfrac{2}{3}, \dfrac{3}{4}\right), \left(\dfrac{1}{2}, \dfrac{2}{3}\right)$

Determine whether the lines are parallel, perpendicular, or neither. **See Examples 5 and 6.**

49. A line passing through $(-3, -2)$ and $(4, 5)$

   A line passing through $(-1, 0)$ and $(6, 7)$

50. A line passing through $(-3, -1)$ and $(-3, 4)$

   A line passing through $(-4, -2)$ and $(3, -2)$

51. A line passing through $(-2, -2)$ and $(-5, 4)$

   A line passing through $(-6, 2)$ and $(-2, 4)$

52. A line passing through $(-5, 4)$ and $(5, 4)$

   A line passing through $(-6, -2)$ and $(3, -2)$

APPLICATIONS

53. U.S. Music Sales. The following line graph models the approximate number of CDs that were shipped for sale in the United States from 1980 through 2008.

   a. Find the rate of increase in the number of CDs shipped from 1990 to 2000.

   b. Find the rate of decrease in the number of CDs shipped from 2000 to 2008.

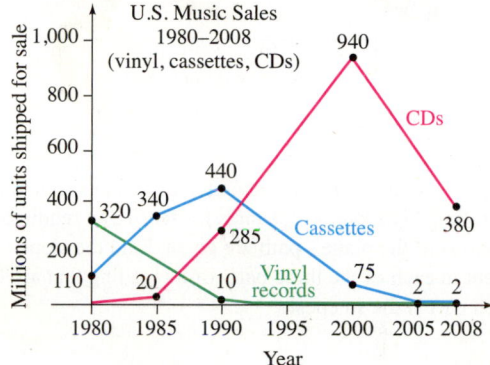

Source: World Almanac 2010

54. Music History. The line graphs in problem 53 model the approximate number of vinyl records and cassettes that were shipped for sale in the United States for the years 1980 through 2008.

   a. Find the rate of decrease in the number of vinyl records shipped from 1980 to 1990.

   b. Find the rate of decrease in the number of cassettes shipped from 1990 to 2000.

55. Global Warming. The following graph models the prediction by the National Center for Atmospheric Research of a future average global temperature rise caused by greenhouse gas emissions. Find the predicted average rate of change of temperature for this time span.

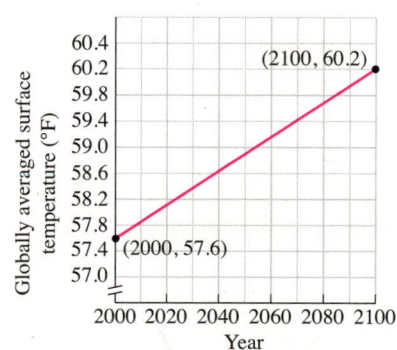

Source: Science Museum of the National Academy of Sciences

56. Air Pressure. Air pressure, measured in units called **Pascals** (Pa), decreases with altitude. Find the rate of change in Pascals for the fastest and the slowest decreasing steps of the following graph. (km stands for kilometers.)

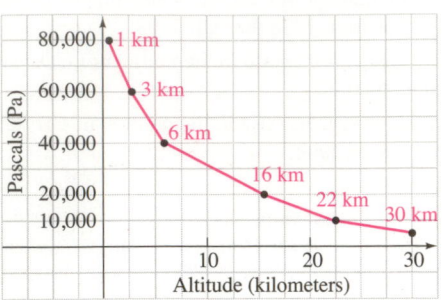

Based on data from *The Blue Planet* (Wiley, 1995)

57. Flying. A jet descends in a stairstep pattern, as shown in the illustration. (The symbol ′ stands for feet.) The required elevations of the plane's path are given. Find the slope of the descent in each of the three parts of its landing that are labeled. Which part is the steepest?

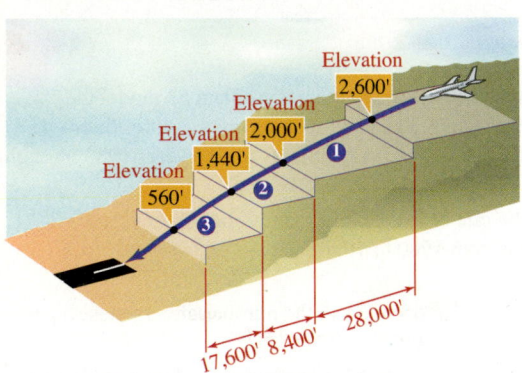

Based on data from *Los Angeles Times* (August 7, 1997), p. A8

58. Maps. Topographic maps have contour lines that connect points of equal elevation on a mountain. The vertical distance between contour lines in the illustration below is 50 feet. Find the slope of the west face and the slope of the east face of the mountain peak.

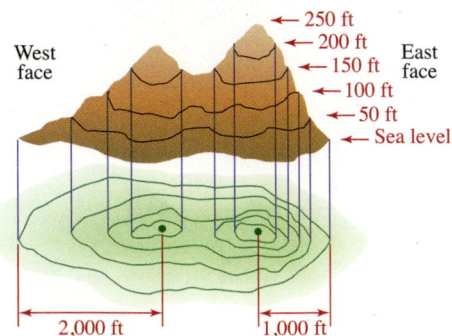

59. Steep Grades. Find the grade of the road shown in the illustration. (*Hint:* 1 mi = 5,280 ft.)

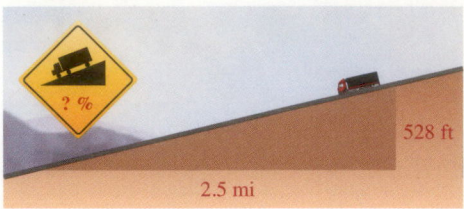

60. Railroads. The Saluda Grade in Polk County, North Carolina, is the steepest standard-gauge mainline railway grade in the United States. At one point, the grade reaches 5.1% between the towns of Melrose and Saluda. Explain what a 5.1% grade means.

61. Staircases. Common practice among American architects for interior staircases has been to make the unit rise about $7\frac{1}{2}$ inches and the unit run 9 inches. Write the ratio of rise to run as a fraction in simplest form.

62. Extreme Sports. See the illustration below. Find the average rate of descent of a street luge "pilot" if he completes the course in 3 minutes.

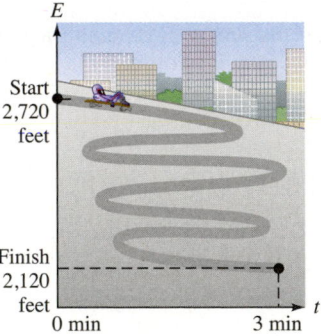

63. Bike Racing. Find the average rate of descent of a mountain biker if she completes the course in 6 minutes.

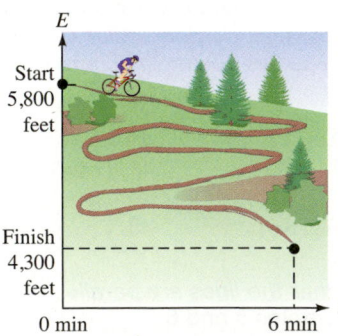

64. Computers. The price of computers has been dropping for the past ten years. If a desktop PC cost $5,700 10 years ago, and the same computing power cost $400 2 years ago, find the rate of decrease per year. (Assume a straight-line model.)

65. Skiing. The men's giant slalom course shown in the illustration is longer than the women's course. Does this mean that the men's course is steeper? Use the concept of the slope of a line to explain.

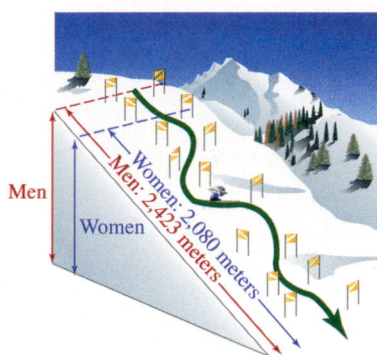

66. Decks. See the illustration below. Find the slopes of the cross-brace and the supports. Is the cross-brace perpendicular to either support?

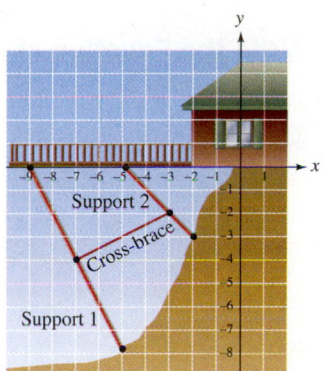

WRITING

67. Bottled Water. Refer to the graph below. Write a paragraph that explains how the consumption of bottled water has changed. Use the words *increasing* and *decreasing*, and the phrase *rate of change* in your written answer.

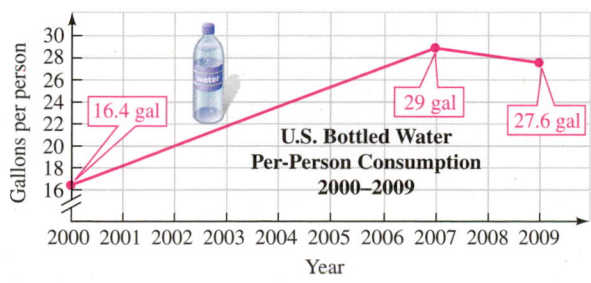

U.S. Bottled Water Per-Person Consumption 2000–2009

Source: beveragemarketing.com

68. a. When is a rate of change negative? Give an example.

    b. When is a rate of change 0? Give an example.

69. Explain why a vertical line has no defined slope.

70. Explain how to determine from their slopes whether two lines are parallel, perpendicular, or neither.

REVIEW

71. Candy. A candy maker wants to make a 60-pound mixture of two candies to sell for $2 per pound. If black licorice bits sell for $1.90 per pound and orange gumdrops sell for $2.20 per pound, how many pounds of each should be used?

72. Retirement. A nurse wants to supplement her pension with investment interest. If she invests $28,000 at 6% interest, how much would she have to invest at 7% to achieve a goal of $3,500 per year in supplemental income?

CHALLENGE PROBLEMS

73. The two lines graphed in the illustration are parallel. Find x and y.

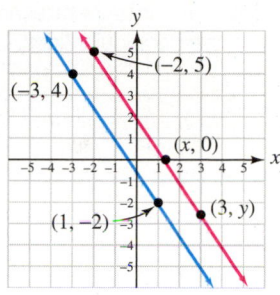

74. The line passing through $(1, 3)$ and $(-2, 7)$ is perpendicular to the line passing through points $(4, b)$ and $(8, -1)$. Without graphing, find b.

75. Find the slope of the line that passes through (a, b) and $(-b, -a)$. Assume $a \neq 0$ and $b \neq 0$.

76. Refer to the graph. Show that the line through the midpoints of sides AC and AB of $\triangle ABC$ is parallel to side BC.

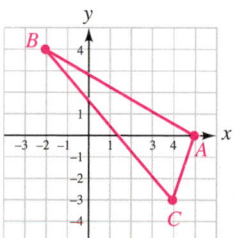

SECTION 2.4

OBJECTIVES

1. Use the slope–intercept form to write the equation of a line.
2. Write a linear equation model.
3. Use the point–slope form to write the equation of a line.
4. Write a linear depreciation equation.
5. Use slope as an aid when graphing.
6. Recognize parallel and perpendicular lines.

Writing Equations of Lines

ARE YOU READY?

The following problems review some basic skills that are needed when working with equations of lines.

1. a. Identify each term in the expression $5x - 8$.

   b. What is the coefficient of the first term?

2. Solve for y: $4x + 5y = 20$

3. On what axis does the point $(0, 3)$ lie?

4. Find the slope of the line that passes through $(-1, 0)$ and $(-10, -6)$.

5. Solve $y + 1 = 2(x - 11)$ for y.

6. Add: $-\dfrac{3}{8} + 6$

We have seen that linear relationships can be presented in graphs. In this section, we will write equations to model linear relationships.

1 Use the Slope–Intercept Form to Write the Equation of a Line.

To develop one form of the equation of a line, consider line l in the figure with y-intercept $(0, b)$ and let (x, y) be another point on the line. If we let $(x_1, y_1) = (0, b)$ and $(x_2, y_2) = (x, y)$, the slope of line l is given by the slope formula:

$$m = \frac{y - b}{x - 0}$$ Substitute y for y_2, b for y_1, x for x_2, and 0 for x_1 in $m = \frac{y_2 - y_1}{x_2 - x_1}$.

$$m = \frac{y - b}{x}$$ Simplify the denominator.

To solve the equation for y, we clear it of the fraction by multiplying both sides by x.

$$mx = y - b$$

$$mx + b = y$$ To isolate y, add b to both sides.

$$y = mx + b$$ Write the equation with y on the left side.

Because this equation displays the slope m and the y-coordinate b of the y-intercept, it is called the **slope–intercept form** of the equation of a line.

Slope–Intercept Form	The equation of the line with slope m and y-intercept $(0, b)$ is $$y = mx + b$$

Note that when an equation of a line is written in slope–intercept form, it is solved for y. Furthermore, the coefficient of the x-term is the line's slope and the constant term gives the y-coordinate of the y-intercept.

$$y = \boldsymbol{m}x + \boldsymbol{b}$$

Slope ⟶   ⟵ y-intercept: $(0, b)$

Linear equation	Equation written in slope–intercept form	Slope	y-intercept
$y = 4x - 3$	$y = 4x + (-3)$ or $y = 4x - 3$	4	$(0, -3)$
$y = -\dfrac{5}{6}x$	$y = -\dfrac{5}{6}x + 0$	$-\dfrac{5}{6}$	$(0, 0)$
$y = 7.5 - x$	$y = -x + 7.5$	-1	$(0, 7.5)$

EXAMPLE 1

a. Write an equation of the line with slope -1 and y-intercept $(0, 7)$.
b. Write an equation of the line graphed on the left.

Strategy In each case, we will use the slope–intercept form, $y = mx + b$, to write an equation of the line.

Why We are given (or we can easily find) the slope and the y-intercept of each line.

Solution

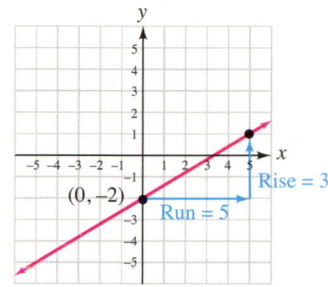

a. If the slope is -1 and the y-intercept is $(0, 7)$, then $m = -1$ and $b = 7$.

$y = mx + b$ This is the slope–intercept form.
$y = -1x + 7$ Substitute -1 for m and 7 for b.
$y = -x + 7$ Simplify: $-1x = -x$.

The equation of the line with slope -1 and y-intercept $(0, 7)$ is $y = -x + 7$.

b. In the figure on the left, we see that the y-intercept of the line is $(0, -2)$ and the slope of the line is $\frac{3}{5}$. When we substitute $\frac{3}{5}$ for m and -2 for b into the slope–intercept form $y = mx + b$, we obtain an equation of the line: $y = \frac{3}{5}x - 2$.

Self Check 1   **a.** Write an equation of the line with slope 1 and y-intercept $(0, -12)$.

b. Write an equation of the line whose graph is shown.

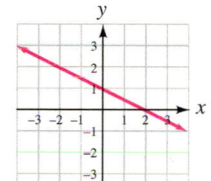

Now Try ▶ Problems 15 and 19

EXAMPLE 2

Use the slope–intercept form to write an equation of the line that has slope 4 and passes through $(5, 9)$.

Strategy Since the slope of the line is given as $m = 4$, the slope–intercept equation of the line has the form $y = 4x + b$. To find b, we will substitute 5 for x and 9 for y in $y = 4x + b$ and solve for b.

Why If the point $(5, 9)$ lies on the line, it is a solution of the equation and its coordinates must satisfy the equation of the line.

Solution

$y = mx + b$ This is the slope–intercept form.
$y = 4x + b$ Substitute 4 for m. We do not know b.

To find b, we use the fact that the point $(5, 9)$ lies on the line and its coordinates, therefore, satisfy the equation.

$9 = 4(5) + b$ Substitute 9 for y and 5 for x.
$9 = 20 + b$ Do the multiplication.
$-11 = b$ To isolate b, subtract 20 from both sides.

Because $m = 4$ and $b = -11$, the equation is $y = 4x - 11$.

> **Self Check 2**   Use the slope–intercept form to write an equation of the line that has slope -6 and passes through $(-1, 8)$.
>
> **Now Try** ▶ Problem 23

2   Write a linear equation model.

The slope–intercept form of the equation of a line can be used to model linear relationships. However, we often will use variables other than x and y to describe the quantities involved.

EXAMPLE 3   **School Supplies.**   Each turn of the handle of a pencil sharpener shaves off 0.05 inch from a 7.25-inch-long pencil.

a. Write a linear equation that gives the new length L of the pencil after the sharpener handle has been turned t times.

b. How long is the pencil after the sharpener handle has been turned 20 times?

Strategy   We will read the problem carefully, looking for information about the slope and the y-intercept of the graph of the equation.

Why   If we can determine the slope and y-intercept of the graph of the line, we can use the slope–intercept form to write an equation to model the situation.

Solution   a. Since the length L of the pencil depends on the number of turns t of the handle, the linear equation that models this situation will have the form $L = mt + b$. We need to determine m and b.

- The length of the pencil decreases as the handle is turned. This rate of change, -0.05 inch per turn, is the slope of the graph of the equation. Thus, $m = -0.05$.

- Before any turns of the handle are made (when $t = 0$), the length of the pencil is 7.25 inches. Written as an ordered pair of the form (t, L), we have $(0, 7.25)$. When graphed, this would be the L-intercept of the graph. Thus, $b = 7.25$.

Substituting for m and b, we have the linear equation that models this situation.

$$L = -0.05t + 7.25$$

This is the slope–intercept form using the variables L and t.

The slope is the rate of change of the length of the pencil.   The intercept is the original length of the pencil.

b. To find the pencil's length after the handle is turned 20 times, we proceed as follows:

$L = -0.05t + 7.25$	This is the linear model.
$L = -0.05(20) + 7.25$	Substitute 20 for t, the number of turns.
$L = -1 + 7.25$	Multiply: $-0.05(20) = -1$.
$L = 6.25$	Do the addition.

If the sharpener handle is turned 20 times, the pencil will be 6.25 inches long.

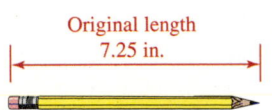

Original length
7.25 in.

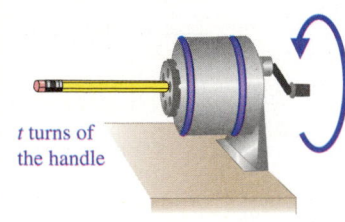

t turns of the handle

New length
L in.

> **The Language of Algebra**
>
> To determine the slope m for modeling problems like this, look for phrases that describe a **rate of change**, such as:
> - \$5 per mile
> - 7° every 2 minutes
> - 1 quart for each 8 ft²

> **Self Check 3**   **Manicures.**   Fingernails grow at an average rate of 0.12 inch a month. When a woman has her nails manicured, they are trimmed to a length of 0.75 inch.   a. Write a linear equation that gives the length L of her fingernails m months after a manicure.
> b. If she does not get them trimmed, how long will her fingernails be in one year?
>
> **Now Try** ▶ Problem 88

3 Use the Point–Slope Form to Write the Equation of a Line.

To develop another form of the equation of a line, we consider line l in the figure below that has slope m and passes through (x_1, y_1). If (x, y) is a second point on the line, the slope of the line is given by the slope formula:

$$m = \frac{y - y_1}{x - x_1} \qquad \textcolor{red}{\text{Substitute } x \text{ for } x_2 \text{ and } y \text{ for } y_2.}$$

If we multiply both sides by $x - x_1$, and reverse the sides of the equation, we have

$$y - y_1 = m(x - x_1)$$

Because this equation displays the coordinates of the point (x_1, y_1) on the line and the slope m of the line, it is called the **point–slope form** of the equation of a line.

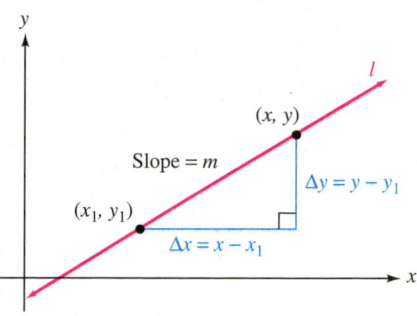

Point–Slope Form	The equation of the line that passes through (x_1, y_1) and with slope m is
	$$y - y_1 = m(x - x_1)$$

EXAMPLE 4

Find an equation of the line that has slope $-\dfrac{2}{3}$ and passes through $(-4, 5)$. Write the equation in slope–intercept form.

Strategy We will use the point–slope form to write the equation of the line.

Why We are given the slope of the line and the coordinates of a point that it passes through.

Solution

$$y - y_1 = m(x - x_1) \qquad \textcolor{red}{\text{This is the point–slope form.}}$$

$$y - 5 = -\frac{2}{3}[x - (-4)] \qquad \textcolor{red}{\text{Substitute } -\tfrac{2}{3} \text{ for } m, -4 \text{ for } x_1, \text{ and } 5 \text{ for } y_1.}$$

$$y - 5 = -\frac{2}{3}(x + 4) \qquad \textcolor{red}{\begin{array}{l}\text{Simplify the expression within the brackets.}\\ \text{This equation is in point–slope form.}\end{array}}$$

Success Tip

This is the same type of problem as that in Example 2. Here, it is solved using a different approach. We begin with point–slope form, and then write the result in slope–intercept form.

To write the equation in slope–intercept form, we solve for y.

$$y - 5 = -\frac{2}{3}x - \frac{8}{3} \qquad \textcolor{red}{\text{Distribute the multiplication by } -\tfrac{2}{3}: -\tfrac{2}{3} \cdot 4 = -\tfrac{8}{3}.}$$

$$y = -\frac{2}{3}x - \frac{8}{3} + 5 \qquad \textcolor{red}{\text{To isolate } y, \text{ add 5 to both sides.}}$$

$$y = -\frac{2}{3}x - \frac{8}{3} + \frac{5}{1} \cdot \frac{3}{3} \qquad \textcolor{red}{\begin{array}{l}\text{To prepare to add } -\tfrac{8}{3} \text{ and 5, build } \tfrac{5}{1} \text{ to produce an}\\ \text{equivalent fraction with denominator 3.}\end{array}}$$

$$y = -\frac{2}{3}x - \frac{8}{3} + \frac{15}{3} \qquad \textcolor{red}{\begin{array}{l}\text{Multiply the numerators.}\\ \text{Multiply the denominators.}\end{array}}$$

$$y = -\frac{2}{3}x + \frac{7}{3} \qquad \textcolor{red}{\begin{array}{l}\text{Add } -8 + 15, \text{ and write the sum, 7, over the common}\\ \text{denominator 3. This is the slope–intercept form for the}\\ \text{equation of the line.}\end{array}}$$

Self Check 4 Find an equation of the line that has slope $\dfrac{5}{4}$ and passes through $(2, -6)$. Write the equation in slope–intercept form.

Now Try ▶ Problems 27 and 31

EXAMPLE 5   Find an equation of the line that passes through $(-5, 4)$ and $(8, -6)$. Write the equation in slope–intercept form.

Strategy   We will use the point–slope form to write an equation of the line.

Why   Since we know the coordinates of two points on the line, we can calculate the slope of the line and solve the problem using the method of Example 4.

Solution   First we find the slope of the line by substituting -6 for y_2, 4 for y_1, 8 for x_2, and -5 for x_1 in the slope formula.

$$m = \frac{y_2 - y_1}{x_2 - x_1} = \frac{-6 - 4}{8 - (-5)} = -\frac{10}{13}$$

Since the line passes through $(-5, 4)$ and $(8, -6)$, we can choose either point and substitute its coordinates into the point–slope form. If we select $(-5, 4)$, we substitute -5 for x_1, 4 for y_1, and $-\frac{10}{13}$ for m and proceed as follows.

$$y - y_1 = m(x - x_1) \qquad \text{This is the point–slope form.}$$

$$y - 4 = -\frac{10}{13}[x - (-5)] \qquad \text{Substitute } -\tfrac{10}{13} \text{ for } m, \ -5 \text{ for } x_1, \text{ and } 4 \text{ for } y_1.$$

$$y - 4 = -\frac{10}{13}(x + 5) \qquad \begin{array}{l}\text{Simplify the expression within the brackets.}\\ \text{This equation is in point–slope form.}\end{array}$$

To write this equation in slope–intercept form, we solve for y.

$$y - 4 = -\frac{10}{13}x - \frac{50}{13} \qquad \text{Distribute the multiplication by } -\tfrac{10}{13} : -\tfrac{10}{13} \cdot 5 = -\tfrac{50}{13}$$

$$y = -\frac{10}{13}x - \frac{50}{13} + 4 \qquad \text{To isolate } y, \text{ add 4 to both sides.}$$

$$y = -\frac{10}{13}x - \frac{50}{13} + \frac{52}{13} \qquad \text{Express 4 as a fraction with denominator 13: } \tfrac{4}{1} \cdot \tfrac{13}{13} = \tfrac{52}{13}.$$

$$y = -\frac{10}{13}x + \frac{2}{13} \qquad \begin{array}{l}\text{Add } -50 + 52, \text{ and write the sum, 2, over the common}\\ \text{denominator 13. This is the slope–intercept form for the}\\ \text{equation of the line.}\end{array}$$

Success Tip

Here, either of the given points can be used as (x_1, y_1) when writing the point–slope equation. Looking ahead, we usually choose the point whose coordinates will make the computations the easiest.

Self Check 5   Find an equation of the line that passes through $(-2, 5)$ and $(4, -3)$. Write the equation in slope–intercept form.

Now Try ▶   Problem 35

---

4 Write a linear depreciation equation.

Linear models can be used to describe certain types of financial gain or loss. For example, **straight-line depreciation** is used when aging equipment declines in value and **straight-line appreciation** is used when property or collectibles increase in value.

EXAMPLE 6   **Accounting.**   After purchasing a new drill press, a machine shop owner had his accountant prepare a depreciation worksheet for tax purposes. See the illustration.
   a. Assuming straight-line depreciation, write an equation that gives the value v of the drill press after x years of use.
   b. Find the value of the drill press after $2\frac{1}{2}$ years of use.
   c. What is the economic meaning of the v-intercept of the line?
   d. What is the economic meaning of the slope of the line?

Depreciation Worksheet

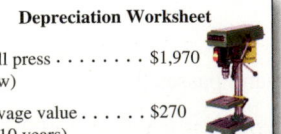

Drill press $1,970
(new)

Salvage value $270
(in 10 years)

Strategy   We will read the problem with the hope that we can find information about the slope of the line and the coordinates of a point (or points) that lie on the line.

Why   If we know the slope of the line and the coordinates of a point (or points) that lie on the line, we can write its equation using slope–intercept form or point–slope form.

Solution   **a.** The facts presented in the worksheet can be expressed as ordered pairs of the form

$$(x, v)$$

Number of years of use ⎯↑ ↑⎯ Value of the drill press

- When purchased, the new $1,970 drill press had been used 0 years: (0, 1,970).
- After 10 years of use, the value of the drill press will be $270: (10, 270).

A sketch showing these ordered pairs and the line of depreciation is helpful in visualizing the situation. Since we know two points that lie on the line, we can write its equation. As we saw in Example 5, the first step is to find the slope of the line. If we write the slope formula using the variables x and v, and let $(x_1, v_1) = (0, 1,970)$ and $(x_2, v_2) = (10, 270)$, we have

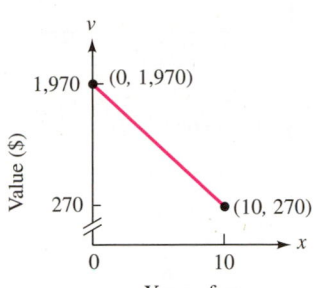

$$m = \frac{v_2 - v_1}{x_2 - x_1} = \frac{270 - 1,970}{10 - 0} = \frac{-1,700}{10} = -170$$

This negative result indicates that the drill press loses $170 in value each year.

To find the equation of the line, we substitute -170 for m, 0 for x_1, and 1,970 for v_1 in the point–slope form and simplify.

$$v - v_1 = m(x - x_1) \qquad \text{This is the point–slope form using the variables x and v.}$$
$$v - 1,970 = -170(x - 0) \qquad \text{Substitute.}$$
$$v = -170x + 1,970 \qquad \text{To isolate v, add 1,970 to both sides.}$$

The value v of the drill press after x years of use is given by the straight-line depreciation model $v = -170x + 1,970$.

b. To find the value of the drill press after $2\frac{1}{2}$ years of use, we substitute 2.5 for x in the depreciation equation and find v.

$$v = -170x + 1,970 \qquad \text{This is the straight-line depreciation equation.}$$
$$v = -170(2.5) + 1,970 \qquad \text{Substitute.}$$
$$v = -425 + 1,970$$
$$v = 1,545$$

In $2\frac{1}{2}$ years, the drill press will be worth $1,545.

c. From the sketch, we see that the v-intercept of the graph of the depreciation line is (0, 1,970). This gives the original cost of the drill press, $1,970.

d. Each year, the value of the drill press decreases by $170, because the slope of the line is -170. The slope of the line is the **annual depreciation rate.**

Self Check 6   **Depreciation.**   Find the value of the drill press in Example 6 after 4 years.

Now Try ▶ Problem 97

The Language of Algebra

In this problem, the value of the drill press **depreciates**. The value of an item can also **appreciate**, which means to increase in value over time. Certain types of art, antiques, and jewelry appreciate quickly.

---

5 Use Slope as an Aid When Graphing.

If we know the slope and the y-intercept of a line, we can graph the line without constructing a table of solutions.

EXAMPLE 7   Find the slope and the y-intercept of the line with the equation $2x + 3y = -9$ and graph the line.

Strategy   We will write the equation in slope–intercept form $(y = mx + b)$, plot the y-intercept, and use the slope to determine a second point on the line.

Why   Once we locate two points on the line, we can draw the graph of the line.

Solution

$$2x + 3y = -9 \qquad \text{The given equation is in standard form.}$$

$$3y = -2x - 9 \qquad \text{To isolate the term 3y, subtract 2x from both sides.}$$

$$\frac{3y}{3} = \frac{-2x}{3} - \frac{9}{3} \qquad \text{To isolate y, divide both sides by 3.}$$

$$y = -\frac{2}{3}x - 3 \qquad \text{Simplify both sides. We see that } m = -\frac{2}{3} \text{ and } b = -3.$$

Caution
When using the y-intercept and the slope to graph a line, remember to draw the slope triangle from the y-intercept, *not* from the origin.

Success Tip
To increase accuracy, slope triangles in a staircase pattern can be drawn to locate several points on the line.

The slope of the line is $-\frac{2}{3}$, which can be expressed as $\frac{-2}{3}$. After plotting the y-intercept, $(0, -3)$, we move 2 units downward (rise) and then 3 units to the right (run). This locates a second point on the line, $(3, -5)$. From this point, we move another 2 units downward and 3 units to the right to locate a third point on the line, $(6, -7)$. Then we draw a line through the points to obtain the graph shown in the figure.

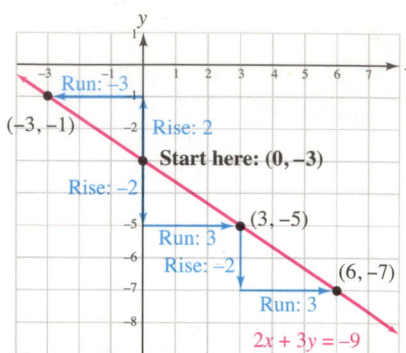

An alternate way to find another point on the line is to write the slope in the form $\frac{2}{-3}$. As before, we begin at the y-intercept $(0, -3)$. Since the rise is positive, we move 2 units *upward,* and since the run is negative, we then move 3 units to the *left.* We arrive at $(-3, -1)$, another point on the graph of $y = -\frac{2}{3}x - 3$.

Self Check 7 Find the slope and the y-intercept of the line with the equation $3x - 2y = -4$ and graph the line.

Now Try Problems 41 and 46

6 Recognize Parallel and Perpendicular Lines.

Recall these two facts from Section 2.3:

- Different lines having the same slope are parallel.
- If the slopes of two lines are negative reciprocals, the lines are perpendicular.

EXAMPLE 8

a. Show that the lines represented by $2y = -3x + 12$ and $6x + 4y = 7$ are parallel.

b. Show that the lines represented by $3x - 9y = 10$ and $-3x = y + 5$ are perpendicular.

Strategy We will write each equation in slope–intercept form and compare their slopes.

Why If the slopes are equal, the lines are parallel. If the slopes are negative reciprocals, the lines are perpendicular.

Solution

a. We solve each equation for y to see whether the lines are distinct (different) and whether their slopes are equal.

$$2y = -3x + 12$$
$$y = -\frac{3}{2}x + 6 \qquad m = -\frac{3}{2} \text{ and } b = 6$$

$$6x + 4y = 7$$
$$4y = -6x + 7$$
$$y = -\frac{3}{2}x + \frac{7}{4} \qquad m = -\frac{3}{2} \text{ and } b = \frac{7}{4}$$

Since the values of b in these equations are different $\left(6 \text{ and } \frac{7}{4}\right)$, the lines have different y-intercepts and are distinct. Since the slope of each line is $-\frac{3}{2}$, they are parallel.

b. We solve each equation for y to see whether the slopes of their straight-line graphs are negative reciprocals.

$$3x - 9y = 10 \qquad\qquad\qquad\qquad -3x = y + 5$$
$$-9y = -3x + 10 \qquad\qquad\qquad -3x - 5 = y$$
$$y = \frac{1}{3}x - \frac{10}{9} \quad m = \tfrac{1}{3} \text{ and } b = -\tfrac{10}{9} \qquad y = -3x - 5 \quad m = -3 \text{ and } b = -5$$

Since the slopes are negative reciprocals $\left(\frac{1}{3} \text{ and } -3\right)$, the lines are perpendicular.

> **Self Check 8**   **a.** Are the lines represented by $3x - 2y = 4$ and $2x = 5y + 5$ parallel?   **b.** Are the lines represented by $3x + 2y = 6$ and $2x - 3y = 6$ perpendicular?
>
> **Now Try** ▶ Problems 51 and 53

EXAMPLE 9   **a.** Find an equation of the line that passes through $(-2, 5)$ and is parallel to the line $y = 8x - 3$. Write the equation in slope–intercept form.   **b.** Find an equation of the line that passes through $(6, -4)$ and is perpendicular to the line $3x + y = 2$. Write the equation in slope–intercept form.

Strategy   We will use the point–slope form to write an equation of the line.

Why   We know a point that the line passes through. We can use the fact that the lines are parallel or perpendicular to determine the unknown slope of the desired line.

Solution   **a.** Since the equation is solved for y, the slope of the line represented by $y = 8x - 3$ is the coefficient of x. Thus, the slope of the given line is **8**. Since the desired equation is to have a graph that is parallel to the graph of $y = 8x - 3$, its slope must also be 8.

$$y - y_1 = m(x - x_1) \qquad \text{This is the point–slope form.}$$
$$y - 5 = 8[x - (-2)] \qquad \text{Substitute 5 for } y_1, 8 \text{ for } m, \text{ and } -2 \text{ for } x_1.$$
$$y - 5 = 8(x + 2) \qquad \text{Simplify within the brackets. This equation is in point–slope form.}$$
$$y - 5 = 8x + 16 \qquad \text{Distribute the multiplication by 8.}$$
$$y = 8x + 21 \qquad \text{To write the equation in slope–intercept form, isolate } y \text{ by adding 5 to both sides.}$$

b. To find the slope of the given line, we must first solve for y to find the equation in slope–intercept form.

$$3x + y = 2$$
$$y = -3x + 2 \qquad \text{The slope of the given line is } -3.$$

The desired line is to have a graph that is perpendicular to the graph of $y = -3x + 2$. Therefore, their slopes must be negative reciprocals. It follows that the slope of the desired line must be $\frac{1}{3}$.

$$y - y_1 = m(x - x_1) \qquad \text{This is point–slope form.}$$
$$y - (-4) = \frac{1}{3}(x - 6) \qquad \text{Substitute 6 for } x_1, -4 \text{ for } y_1, \text{ and } \tfrac{1}{3} \text{ for } m.$$
$$y + 4 = \frac{1}{3}x - 2 \qquad \text{Distribute the multiplication by } \tfrac{1}{3}: \tfrac{1}{3}(-6) = -2.$$
$$y = \frac{1}{3}x - 6 \qquad \text{To write the equation in slope–intercept form, isolate } y \text{ by subtracting 4 from both sides.}$$

Self Check 9   **a.** Find an equation of the line that is parallel to the line $y = 8x - 3$ and passes through the $(0, 0)$. Write the equation in slope–intercept form.
b. Find an equation of the line that passes through $(8, -2)$ that is perpendicular to the line $2x + 3y = 9$. Write the equation in slope–intercept form.

Now Try ▶ Problems 59 and 63

When you are asked to *write the equation of a line,* determine what you know about the graph of the line: its slope, its y-intercept, points it passes through, and so on. Then substitute the appropriate numbers into one of the following forms of a linear equation.

Forms of a Linear Equation		
Standard form	$Ax + By = C$	A and B cannot both be 0.
Slope–intercept form	$y = mx + b$	The slope is m. The y-intercept is (0, b).
Point–slope form	$y - y_1 = m(x - x_1)$	The slope is m. The line passes through (x_1, y_1).
A horizontal line	$y = b$	The slope is 0. The y-intercept is (0, b).
A vertical line	$x = a$	There is no defined slope. The x-intercept is (a, 0).

SECTION 2.4   STUDY SET

VOCABULARY

Fill in the blanks.

1. The _____ form of the equation of a line is $y = mx + b$.

2. The point–slope form of the equation of a line is _____.

CONCEPTS

3. Find the slope and y-intercept of the graph of the equation $y = -\frac{2}{3}x + 1$.

4. Find the slope of the graph of each equation, if possible.
 a. $y = -x$ **b.** $x = -3$

5. Find the y-intercept of the graph of each equation.
 a. $y = 2x$ **b.** $x = -3$

6. For the line in the illustration, find its slope and y-intercept. Then write the equation of the line in slope–intercept form by filling in the blanks: $y = \boxed{\ }\, x + \boxed{\ }$.

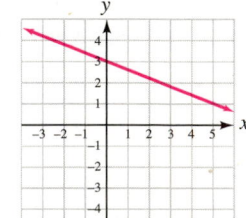

7. a. Find the slope of the graph of $y - 3 = -\frac{2}{3}(x + 1)$.

 b. What point does the equation indicate the line will pass through?

8. Find the slope of the line in the illustration that passes through the point $(-2, -3)$ and write its equation in point–slope form by filling in the blanks:

$y + \boxed{\ } = \boxed{\ }\,(x + \boxed{\ })$.

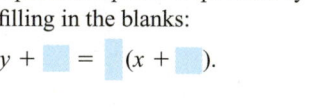

9. Consider the following situation: *In the cooling process, the temperature of a 75° liquid is lowered 5° every two minutes.* If you were to write a linear equation in slope–intercept form to model this situation, what would you use as m and what would you use as b?

10. Depreciation. Use the data in the illustration below to write two ordered pairs of the form (number of years in use, value of the copier).

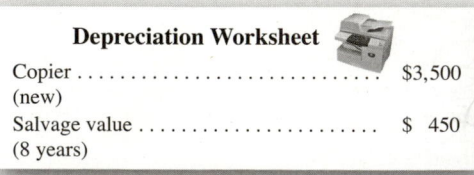

Depreciation Worksheet
Copier $3,500
(new)
Salvage value . . . . $ 450
(8 years)

11. **a.** What is the slope of a line that is parallel to the line represented by $y = 8x - 10$?

    b. What is the slope of a line that is perpendicular to the line represented by $y = \frac{3}{4}x - 6$?

12. **a.** The graphs of two lines and their equations are shown below. Are the lines perpendicular? Explain.

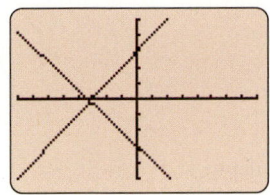 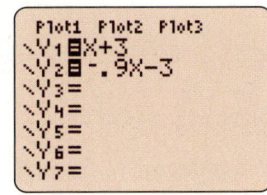

    b. The graphs of two lines and their equations are shown below. Are the lines parallel? Explain.

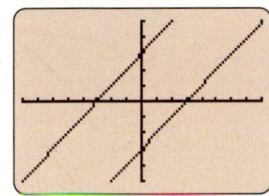

NOTATION

Complete each solution.

13. Write the following equation in slope–intercept form.

$$y + 2 = \frac{1}{3}(x + 3) \qquad \text{\color{red}Point–slope form}$$

$$y + 2 = \boxed{} + \boxed{}$$

$$y + 2 - \boxed{} = \frac{1}{3}x + 1 - \boxed{}$$

$$y = \frac{1}{3}x - \boxed{}$$

$$m = \boxed{} \quad \text{and} \quad b = \boxed{}$$

14. Write an equation of the line that has slope -2 and passes through the point $(3, 1)$.

$$y - y_1 = \boxed{}(x - x_1) \qquad \text{\color{red}Slope–intercept form}$$

$$y - \boxed{} = -2(x - \boxed{})$$

$$y - 1 = \boxed{} + 6$$

$$y = -2x + \boxed{}$$

GUIDED PRACTICE

Use the slope–intercept form to write an equation of the line with the given slope and y-intercept. See Example 1.

15. Slope 3; y-intercept $(0, 6)$

16. Slope -2; y-intercept $(0, 11)$

17. Slope $-\frac{2}{3}$; y-intercept $\left(0, -\frac{7}{3}\right)$

18. Slope $\frac{5}{7}$; y-intercept $\left(0, \frac{1}{4}\right)$

Find the slope and y-intercept of each line graphed below. Then use the slope–intercept form to write an equation of the line. See Example 1.

19.
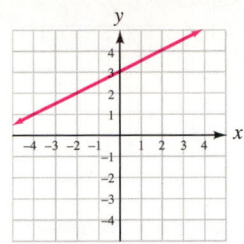

20.
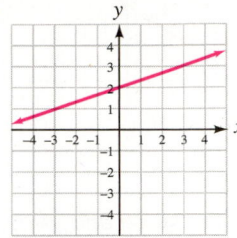

21.
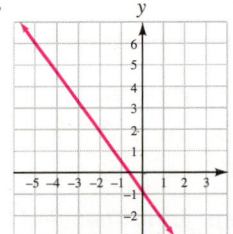

22.
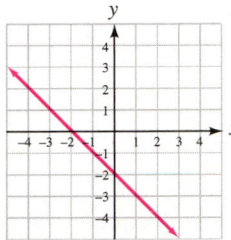

Use the slope–intercept form to write an equation of the line that has the given slope and passes through the given point. See Example 2.

23. Slope 7; passes through $(-7, 5)$

24. Slope 3; passes through $(-2, -5)$

25. Slope -9; passes through $(2, -4)$

26. Slope -10; passes through the origin

Use the point–slope form to write an equation of the line with the given properties or the given graph. Leave the answer in point–slope form. See Example 4.

27. Slope 10; passes through $(1, 7)$

28. Slope -8; passes through $(3, -2)$

29.
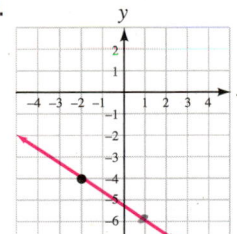

30.
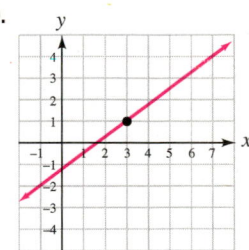

Use the point–slope form to write an equation of the line with the given properties. Then write each equation in slope–intercept form. See Example 4.

31. Slope 5; passes through $(4, -5)$

32. Slope -7; passes through $(-3, -7)$

33. Slope -9; passes through $(-3.5, 2.7)$

34. Slope 4; passes through $(7.2, -3.7)$

Use the point–slope form to write an equation of the line passing through the two given points. Then write each equation in slope–intercept form. See Example 5.

35. $(6, 8)$ and $(2, 10)$

36. $(-4, 5)$ and $(2, -6)$

37. 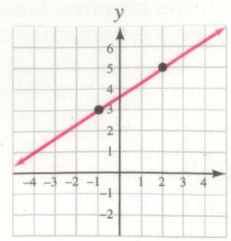   **38.**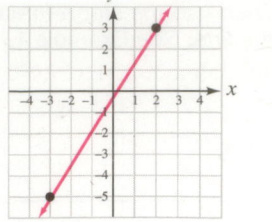

Find the slope and y-intercept and use them to draw the graph of the line. See Example 7.

39. $y = x - 1$ **40.** $y = -x + 2$

41. $y = -\dfrac{5}{4}x - 3$ **42.** $y = \dfrac{2}{3}x + 2$

43. $5y - 8x = 30$ **44.** $-2x + 4y = 12$

45. $4y - 3 = -3x - 11$ **46.** $7x + 3y + 15 = 0$

Write each equation in slope–intercept form. Then find the slope and the y-intercept of the line determined by the equation. See Example 8.

47. $3x - 2y = 8$ **48.** $-2x - 4y = -12$

49. $-2(x + 3y) = 5$ **50.** $5(2x - 3y) = 4$

Determine whether the graphs of each pair of equations are parallel, perpendicular, or neither. See Example 8.

51. $y = 3x + 4, \; y = 3x - 7$

52. $y = 4x - 13, \; y = \dfrac{1}{4}x + 13$

53. $x + y = 2, \; y = x + 5$

54. $x = y + 2, \; y = x + 3$

55. $3x + 6y = 1, \; y = \dfrac{1}{2}x$

56. $2x + 3y = 9, \; 3x - 2y = 5$

57. $y = 3, \; x = 4$

58. $y = -3, \; y = -7$

Find an equation of the line that passes through the given point and is parallel to the given line. Write the equation in slope–intercept form. See Example 9.

59. $(2, 5), \; y = 4x + 8$ **60.** $(-6, 3), \; y = -3x - 12$

61. $\left(\dfrac{2}{3}, \dfrac{1}{4}\right), \; y = 3x - 2$ **62.** $\left(\dfrac{4}{5}, -\dfrac{2}{3}\right), \; y = -5x - \dfrac{1}{2}$

Find an equation of the line that passes through the given point and is perpendicular to the given line. Write the equation in slope–intercept form. See Example 9.

63. $(-40, 10), \; y = 4x - 7$

64. $(-18, -54), \; y = -\dfrac{1}{3}x + 4$

65. $(4, -2), \; 7x = 9y - 8$

66. $(1, -5), \; 4x = -3y + 20$

Write an equation in slope–intercept form of the line with the given table of solutions, given properties, or given graph.

67.

x	y
3	4
-3	-10

68.

x	y
2	8
6	-8

69. Slope $\dfrac{4}{3}$, passes through $(5, 9)$

70. Slope $-\dfrac{7}{5}$, passes through $(-6, 0)$

71. Passes through $(2, 5)$, perpendicular to $4x - y = 7$

72. Passes through $(-6, 3)$, perpendicular to $y + 3x = -12$

73. 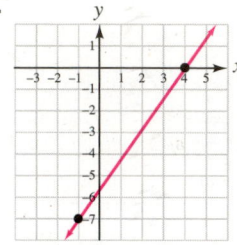   **74.**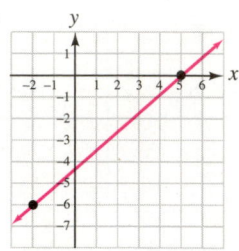

75. Passes through $(-1, -1)$ and $(4, 4)$

76. Passes through $(-5, 5)$ and $(9, -9)$

77. Passes through $(0, 0)$, parallel to $y = 4x - 7$

78. Passes through $(0, 0)$, parallel to $y = -\dfrac{1}{3}x + 4$

79. Slope -3, passes through $(4, -6)$

80. Slope 4, passes through $(-1, 16)$

81. Passes through $\left(\dfrac{2}{3}, \dfrac{1}{4}\right)$, perpendicular to $y = 3x - 2$

82. Passes through $\left(\dfrac{4}{5}, -\dfrac{2}{3}\right)$, perpendicular to $y = -5x - \dfrac{1}{2}$

83. 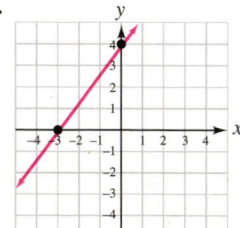   **84.**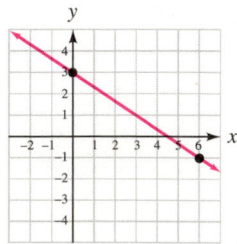

85. Passes through $(4, -2)$, parallel to $x = \dfrac{5}{4}y - 2$

86. Passes through $(1, -5)$, parallel to $x = -\dfrac{3}{4}y + 5$

APPLICATIONS

87.

from Campus to Careers

Certified Fitness Instructor

Suppose you are a fitness instructor and want to determine the number of calories a client burns during a workout. From exercise tables, you find that during the first part of the workout (aerobics) she will burn 220 calories. During the optional second part of the workout (swimming), she will burn 7.8 calories per minute.

a. Write a linear model in slope–intercept form that gives the total number of calories c that the client burns if she concludes a workout with m minutes of swimming.

b. Many fitness instructors recommend that a client burn 300 calories per exercise session to lose weight. How many minutes of swimming should the client perform to satisfy this requirement?

88. Cable TV. Since 1990, when the average monthly price for basic cable TV programming in the United States was approximately $16, the cost has risen by about $1.60 a year. (Source: Kagan Research)

a. Write an equation in slope–intercept form to predict cable TV costs in the future. Use t to represent time in years after 1990 and C to represent the average basic monthly cost.

b. If the equation in part (a) were graphed, what would be the meaning of the C-intercept and the slope of the line?

c. Use the model to predict the average monthly price for cable TV in the year 2020.

89. Alaskan Oil. According to the U.S. Energy Information Administration, in 1990, about 650,000,000 barrels of crude oil were produced from drilling in Alaska. Since then, the annual production has decreased by about 22,000,000 per year.

a. Write a linear model in slope–intercept form that gives the number of barrels b of crude oil produced in Alaska t years after 1990.

b. If the equation were graphed, what would be the meaning of the b-intercept and the slope of the line?

90. College Costs. According to the *College Board*, in 1990, the average tuition and fees at a private college were about $8,050 a year. Since then, the annual cost has increased by about $900 per year.

a. Write a linear model in slope–intercept form that gives the cost c to attend a private college t years after 1990.

b. Use the model to predict what the average tuition and fees at a private college will be in the year 2050.

91. Criminology. City growth and the number of burglaries for a certain city are related by a linear equation. Records show that 575 burglaries were reported in a year when the local population was 77,000 and that the rate of increase in the number of burglaries was 1 for every 100 new residents.

a. Using the variables p for population and B for burglaries, write an equation (in slope–intercept form) that police can use to predict future burglary statistics.

b. How many burglaries can be expected when the population reaches 110,000?

92. Fire Protection. City growth and the number of fires for a certain city are related by a linear equation. Records show that 113 fires occurred in a year when the local population was 150,000 and that the rate of increase in the number of fires was 1 for every 1,000 new residents.

a. Using the variables p for population and F for fires, write an equation (in slope–intercept form) that the fire department can use to predict future fire statistics.

b. How many fires can be expected when the population reaches 200,000?

93. Russia. According to the World Bank, since 1995, the population of Russia has decreased by about 505,000 people per year. In 2000, the Russian population was about 146,000,000.

a. Using the variables p for population and t for the number of years after 1995, write an equation in slope–intercept form that can be used to predict the future population of Russia.

b. What does the equation predict that Russian population will be in 2015?

94. Real Estate.

a. Use the information given in the description below of the property to write a straight-line appreciation equation for the house.

b. What will be the predicted value of the home when it is 25 years old?

Vacation Home
$254,000
Only 2 years old
1,635 sq. ft

• Great investment property!
• Expected to appreciate $4,000/yr

95. Undersea Diving. The illustration below shows that the pressure p that divers experience is related to the depth d of the dive. A linear model can be used to describe this relationship.

a. Write the linear model in slope–intercept form.

b. Scuba divers can safely dive to depths of 250 feet. What pressure do they experience? Round to the nearest tenth.

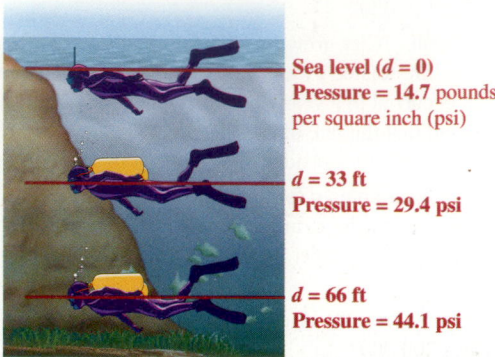

Sea level ($d = 0$)
Pressure = 14.7 pounds per square inch (psi)

$d = 33$ ft
Pressure = 29.4 psi

$d = 66$ ft
Pressure = 44.1 psi

96. Windchill. A combination of cold and wind makes a person feel colder than the actual temperature. The table shows what temperatures of 35°F and 15°F feel like when a 15-mph wind is blowing. The relationship between the actual temperature and the windchill temperature can be modeled with a linear equation.

a. Write the equation that models this relationship. Answer in slope–intercept form.

b. What information is given by the y-intercept of the graph of the equation found in part (a)?

15-mph wind	
Actual temperature x	**Windchill temperature y**
35°F	25°F
15°F	0°F

97. Craigslist. Find the straight-line depreciation equation for the TV in the following ad found on Craigslist (an online website featuring free classified advertisements).

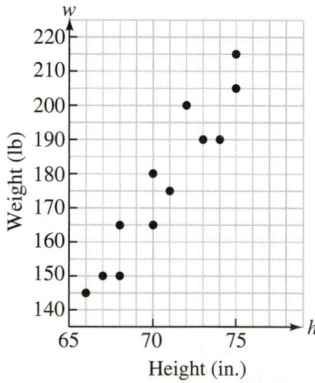

For Sale: 3-year-old 46-inch TV, with matrix surround sound & picture within picture, remote. $1,750 new. Asking $800. Call 875-5555. Ask for Mike.

98. Salvage Values. A truck was purchased for $19,984. Its salvage value at the end of 8 years is expected to be $1,600. Find the straight-line depreciation equation.

99. Paintings. According to Wikipedia, the estimated value of the *Mona Lisa* was about $100 million in 1960. By the year 2000, its value had increased to approximately $600 million.

©Dennis Hallinan/Alamy

a. Write a straight-line appreciation equation for the painting, where y represents the approximate value of the painting in millions of dollars and x represents the number of years after 1960.

b. When does the straight-line appreciation equation predict that the painting will be worth one billion dollars?

100. Real Estate. A house purchased for $275,000 is expected to be worth $282,000 in 2 years. Find the straight-line appreciation equation for the house if it continues to appreciate at the same rate. What will the house be worth in 10 years?

101. Regression Equations. The following **scatter diagram** shows the relationship between the height h and weight w of a random sample of twelve men. The data are plotted as ordered pairs in the form (h, w).

a. Draw a line through (68, 155) and (73, 195) and then determine its equation. Write the equation in slope–intercept form. This equation is called a **regression equation.** In statistics, there are exact methods to find the equation that best fits the data points. However, these methods are beyond the scope of this book. In this case, we simply draw the regression line "by eye."

b. Use the equation to predict the weight of a 72-inch-tall man.

102. Regression Equations. See problem 101. The scatter diagram on the next page shows the performance of a rat in a maze.

a. Draw a line through (1, 10) and (19, 1). Write its equation using the variables t and E. In psychology, this equation is called the **learning curve** for the rat.

b. What does the slope of the line tell us?

c. What information does the *t*-intercept of the graph give?

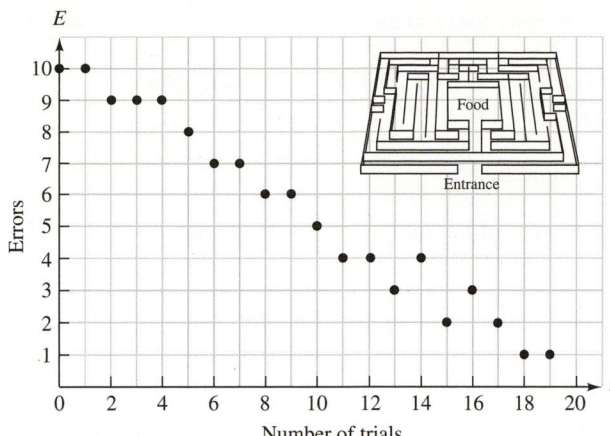

Number of trials

103. Cigarette Smoking. See problem 101. The regression equation $c = -10t + 632$ graphed below estimates the number of cigarettes c (in billions) consumed in the United States t years after 1980.

a. What is the c-intercept of the line? What information about cigarette consumption does it give?

b. What is the slope of the line? What information about the change in cigarette consumption does it give?

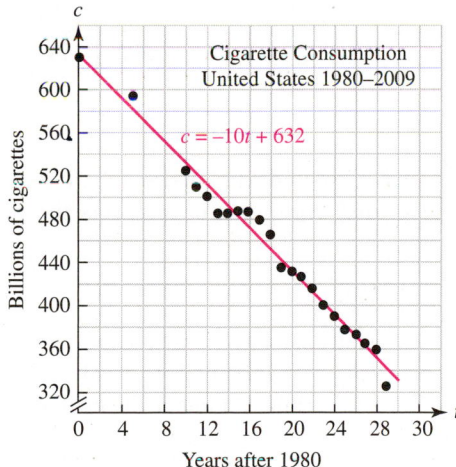

104. Drafting. The illustration in the next column shows a computer-generated drawing of an airplane part. When the designer clicks the mouse on a line on the drawing, the computer finds the equation of the line. Use a calculator to determine whether the angle where the weld is to be made is a right angle.

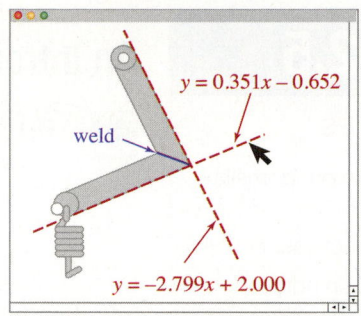

WRITING

105. Explain how to find the equation of a line passing through two given points.

106. Explain what m, x_1, and y_1 represent in the point–slope form of the equation of a line.

107. A student was asked to determine the slope of the graph of the line $y = 6x - 4$. If his answer is $m = 6x$, explain his error.

108. Linear relationships between two quantities can be described by an equation or a graph. Which do you think is the more informative? Why?

109. a. Suppose you know the slope of a line. Is that enough information about the line to write its equation? Explain.

   b. Suppose you know the coordinates of a point on a line. Is that enough information about the line to write its equation? Explain.

110. Do the equations $y - 2 = 3(x - 2)$ and $y = 3x - 4$ describe the same line? Explain.

REVIEW

111. Investments. Equal amounts are invested at 6%, 7%, and 8% annual interest. The three investments yield a total of $2,037 annual interest. Find the total amount of money invested.

112. Medications. A doctor prescribes an ointment that is 2% hydrocortisone. A pharmacist has 1% and 5% concentrations in stock. How many ounces of each should the pharmacist use to make a 1-ounce tube?

CHALLENGE PROBLEMS

113. If the graph of $y = mx + b$ passes through quadrants I, II, and IV, what do we know about the constants m and b?

114. The graph of $Ax + By = C$ passes only through quadrants I and IV. What do we know about the constants A, B, and C?

SECTION 2.5

OBJECTIVES

1 Define relation, domain, and range.

2 Identify functions.

3 Use function notation.

4 Find the domain of a function.

5 Graph linear functions.

6 Write equations of linear functions.

An Introduction to Functions

ARE YOU READY?

The following problems review some basic skills that are needed when working with functions.

1. Which of the ordered pairs in the following set have the same *x*-coordinate: $\{(3, 5), (2, 9), (6, -7), (-1, 5), (3, 0), (1, 9)\}$?

2. Substitute 8 for *x* in $y = \frac{1}{2}x + 3$ and find *y*.

3. What are the slope and the *y*-intercept of the line described by the equation $y = 3x - 8$?

4. What is the slope of a line perpendicular to the graph of the line that is described by the equation $y = \frac{2}{3}x + 1$?

The concept of a *function* is one of the most important ideas in all of mathematics. To introduce this topic, we will begin with a table that might be seen on television or printed in a newspaper.

1 Define Relation, Domain, and Range.

©Chris Howey/Shutterstock.com

The following table shows the number of women serving in the U.S. House of Representatives for several recent sessions of Congress.

Women in the U.S. House of Representatives							
Session of Congress	105th	106th	107th	108th	109th	110th	111th
Number of Women Representatives	54	56	59	59	68	71	78

Source: womenincongress.house.gov

We can display the data in the table as a set of ordered pairs, where the **first component** represents the session of Congress and the **second component** represents the number of women representatives serving during that session:

$(105, 54), \quad (106, 56), \quad (107, 59), \quad (108, 59), \quad (109, 68), \quad (110, 71), \quad (111, 78)\}$

Sets of ordered pairs like this are called **relations.** The set of all first components is called the **domain of the relation,** and the set of all second components is called the **range of the relation.** A relation may consist of a finite (countable) number of ordered pairs or an infinite (unlimited) number of ordered pairs.

EXAMPLE 1   Find the domain and range of the relation: $\{(3, 2), (5, -7), (-8, 2), (9, 0)\}$

Strategy   We will identify the first components and the second components of the ordered pairs.

Why   The set of all first components is the domain of the relation, and the set of all second components is the range.

Solution   The first components of the ordered pairs are highlighted in red, and the second components are highlighted in blue: $\{(\mathbf{3}, 2), (\mathbf{5}, -7), (\mathbf{-8}, 2), (\mathbf{9}, 0)\}$. When listing the elements of the domain and range, they are usually written in increasing order, and if a value is repeated, it is listed only once.

The domain of the relation is $\{-8, 3, 5, 9\}$ and the range of the relation is $\{-7, 0, 2\}$.

Self Check 1   Find the domain and range of the relation:
$$\{(5, 6), (-12, 4), (8, 6), (-6, -6), (5, 4)\}$$

Now Try ▶ Problem 19

2 Identify Functions.

The relation in Example 1 was defined by a set of ordered pairs. Relations also can be defined using an **arrow** or **mapping diagram**. The data from the U.S. House of Representatives example is presented on the right in that form. Relations are also often defined using **two-column tables**.

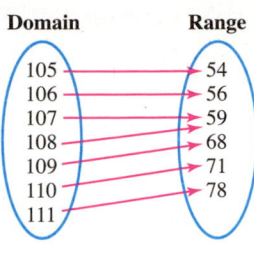

Notice that to each session of Congress, there corresponds exactly one number of women representatives. That is, to each member of the domain, there corresponds exactly one member of the range. Relations that have this characteristic are called *functions*.

Function, Domain, Range	A function is a set of ordered pairs (a relation) in which to each first component there corresponds exactly one second component. The set of first components is called the **domain of the function,** and the set of second components is called the **range of the function.**

Since we often will work with sets of ordered pairs of the form (x, y), it is helpful to define a function using the variables x and y.

y Is a Function of x	Given a relation in x and y, if to each value of x in the domain there corresponds exactly one value of y in the range, then y is said to be a function of x.

In the previous definition, since y depends on x, we call x the **independent variable** and y the **dependent variable**. The set of all possible values that can be used for the independent variable is the **domain** of the function, and the set of all values of the dependent variable is the **range** of the function.

EXAMPLE 2   Determine whether the relation defines y to be a function of x.

a.

x	y
5	4
7	6
11	10

b.

x	y
8	2
1	4
8	3
9	9

c. $\{(-2, 3), (-1, 3), (0, 3), (1, 3)\}$

Strategy   We will determine whether there is more than one value of y that corresponds to a single value of x.

Why   If to any x-value there corresponds more than one y-value, then y is not a function of x.

Solution **a.** The arrow diagram defines a function because to each value of x there corresponds exactly one value of y.

- $5 \rightarrow 4$ To the x-value 5, there corresponds exactly one y-value, 4.
- $7 \rightarrow 6$ To the x-value 7, there corresponds exactly one y-value, 6.
- $11 \rightarrow 10$ To the x-value 11, there corresponds exactly one y-value, 10.

b. The table does not define a function, because to the x-value 8 there corresponds more than one y-value.

- In the first row, to the x-value 8, there corresponds the y-value 2.
- In the third row, to the same x-value 8, there corresponds a different y-value, 3.

When the correspondence in the table is written as a set of ordered pairs, it is apparent that the relation does not define a function:

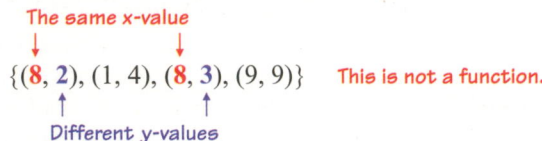

The same x-value

$\{(\mathbf{8}, \mathbf{2}), (1, 4), (\mathbf{8}, \mathbf{3}), (9, 9)\}$ This is not a function.

Different y-values

c. Since to each value of x, there corresponds exactly one value of y, the set of ordered pairs defines y to be a function of x.

- $(-2, 3)$ To the x-value -2, there corresponds exactly one y-value, 3.
- $(-1, 3)$ To the x-value -1, there corresponds exactly one y-value, 3.
- $(0, 3)$ To the x-value 0, there corresponds exactly one y-value, 3.
- $(1, 3)$ To the x-value 1, there corresponds exactly one y-value, 3.

In this case, the same y-value, 3, corresponds to each x-value.

The results from parts (b) and (c) illustrate an important fact: *Two different ordered pairs of a function can have the same y-value, but they cannot have the same x-value.*

Self Check 2   Determine whether the relation defines y to be a function of x.

a.

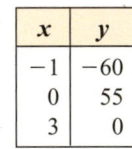

b.

x	y
-1	-60
0	55
3	0

c. $\{(4, -1), (9, 2), (16, 15), (4, 4)\}$
d. $\{(9,5), (10,5), (11,5), (12,5)\}$

Now Try ▶ Problems 23, 27, and 31

A function also can be defined by an equation. For example, $y = \frac{1}{2}x + 3$ sets up a rule in which to each value of x there corresponds exactly one value of y. To find the y-value (called an **output**) that corresponds to the x-value 4 (called an **input**), we substitute 4 for x and evaluate the right side of the equation.

$$y = \frac{1}{2}x + 3$$

$$y = \frac{1}{2}(\mathbf{4}) + 3 \quad \text{Substitute 4 for x. The input is 4.}$$

$$y = 2 + 3 \quad \text{Do the multiplication: } \frac{1}{2}(4) = 2.$$

$$y = 5 \quad \text{Do the addition. This is the output.}$$

For the function defined by $y = \frac{1}{2}x + 3$, a y-value of 5 corresponds to an x-value of 4.

Not all equations define functions, as we will see in the next example.

EXAMPLE 3   Determine whether each equation defines y to be a function of x **a.** $y = 2x - 5$
b. $y^2 = x$

Strategy   In each case, we will determine whether there is more than one value of y that corresponds to a single value of x.

Why   If to any x-value there corresponds more than one y-value, then y is not a function of x.

Solution   **a.** To find the output value y that corresponds to an input value x, we multiply x by 2 and subtract 5. Since this arithmetic gives one result, to each value of x there corresponds exactly one value of y. Thus, $y = 2x - 5$ defines y to be a function of x.

b. The equation $y^2 = x$ does not define y to be a function of x, because more than one value of y corresponds to a single value of x. For example, if x is 16, then the equation becomes $y^2 = 16$ and y can be either 4 or -4. This is because $4^2 = 16$ and $(-4)^2 = 16$.

x	y
16	4
16	-4

Self Check 3   Determine whether each equation defines y to be a function of x.

a. $y = -x + 1$ **b.** $\left| \dfrac{1}{2}y \right| = x$

Now Try ▶   Problems 35 and 39

---

3   Use Function Notation.

A special notation is used to name functions that are defined by equations.

Function Notation	The notation $y = f(x)$ indicates that y is a function of x.

If y is a function of x, the symbols y and $f(x)$ are interchangeable. In Example 3a, we saw that $y = 2x - 5$ defines y to be a function of x. To write this equation using function notation, we replace y with $f(x)$ to get $f(x) = 2x - 5$. This is read as "f of x is equal to $2x$ minus 5." In this notation, the parentheses do not indicate multiplication.

This variable represents the input.

$$f(x) = \underline{2x - 5}$$

This is the name of the function.   This expression shows how to obtain an output from a given input.

Function notation provides a compact way of representing the output value that corresponds to some input value. For example, if $f(x) = 2x - 5$, the value that corresponds to $x = 6$ is represented by $f(6)$.

$$f(x) = 2x - 5$$
$$f(6) = 2(6) - 5 \quad \text{Substitute 6 for each x. (The input is 6.)}$$
$$= 12 - 5 \quad \text{Do the multiplication.}$$
$$= 7 \quad \text{Do the subtraction. This is the output.}$$

Thus, $f(6) = 7$.

We read this result as "f of 6 equals 7." It means that when $x = 6$, $f(x)$ or y is 7. The output 7 is called a **function value.** The input, 6, and output, 7, also can be written as an ordered pair of the form $(6, 7)$.

To see why function notation is useful, we consider two sentences that ask you to do the same thing:

1. If $y = 2x - 5$, find the value of y when x is 6.
2. If $f(x) = 2x - 5$, find $f(6)$.

Statement 2, which uses $f(x)$ notation, is more compact.

The letter f used in the notation $f(x)$ represents the word *function*. However, other letters are often used to name functions. For example, in $g(x) = -5x + 15$, the name of the function is g and in $h(x) = x^3 - 7x + 9$, the name of the function is h.

Letters other than x can be used to represent the input of a function. Examples are

$$f(a) = 4a, \ g(t) = t^2 - 2t, \ \text{and} \ h(n) = -\frac{n^2 + 2}{2}.$$

EXAMPLE 4   Let $f(x) = 4x + 3$ and $g(t) = t^2 - 2t$. Find: **a.** $f(3)$ **b.** $f(-1)$
c. $f(r + 1)$ **d.** $g(-2.4)$

Strategy   We will substitute 3, -1, and $r + 1$ for each x in $f(x) = 4x + 3$ and evaluate the right side. We will substitute -2.4 for each t in $g(t) = t^2 - 2t$ and evaluate the right side.

Why   Whatever appears within the parentheses in $f(\ )$ is to be substituted for each x in $f(x) = 4x + 3$. Whatever appears within the parentheses in $g(\ )$ is to be substituted for each t in $g(t) = t^2 - 2t$.

Solution   **a.** To find $f(3)$, we replace each x with 3.   **b.** To find $f(-1)$, we replace each x with -1:

$$f(x) = 4x + 3 \qquad\qquad f(x) = 4x + 3$$
$$f(3) = 4(3) + 3 \qquad\qquad f(-1) = 4(-1) + 3$$
$$= 12 + 3 \qquad\qquad\qquad = -4 + 3$$
$$= 15 \qquad\qquad\qquad\qquad = -1$$

Thus, $f(3) = 15$ and the corresponding ordered pair is $(3, 15)$.

Thus, $f(-1) = -1$ and the corresponding ordered pair is $(-1, -1)$.

> **Notation**
>
> Note how function notation and ordered pair notation are related:
>
> $f(3) = 15$
> $x \qquad y$
> $(3, \quad 15)$

c. To find $f(r + 1)$, we replace each x with $r + 1$:

$$f(x) = 4x + 3$$
$$f(r + 1) = 4(r + 1) + 3$$
$$= 4r + 4 + 3$$
$$= 4r + 7$$

d. To find $g(-2.4)$, we replace each t with -2.4:

$$g(t) = t^2 - 2t \qquad \text{Read as "g of t."}$$
$$g(-2.4) = (-2.4)^2 - 2(-2.4)$$
$$= 5.76 + 4.8$$
$$= 10.56$$

> **The Language of Algebra**
>
> In part (d), if we use
> $g(t) = t^2 - 2t$ or $g(x) = x^2 - 2x$
> or $g(a) = a^2 - 2a$, the results are
> the same: $g(-2.4) = 10.56$. The
> independent variable is really just
> a place holder, and for this reason,
> the letter that is used is often
> referred to as a **dummy variable**.

Self Check 4   If $f(x) = -2x - 1$ and $h(n) = -\dfrac{n^2 + 2n}{2}$, find: **a.** $f(2)$
b. $f(-3)$ **c.** $f(-t)$ **d.** $h(-0.6)$

Now Try ▶   Problems 47, 67, and 71

In the next example, we are asked to find the input of a function when we are given the corresponding output.

EXAMPLE 5   Let $f(x) = 5x + 4$. For what value of x is $f(x) = -26$?

Strategy   We will substitute -26 for $f(x)$ and solve for x.

Why   In the equation, there are two unknowns, x and $f(x)$. If we replace $f(x)$ with -26, we can use our equation-solving skills to find x.

Solution $f(x) = 5x + 4$ This is the given function.

$-26 = 5x + 4$ Substitute -26 for f(x).

$-30 = 5x$ To isolate the variable term 5x, subtract 4 from both sides.

$-6 = x$ To isolate x, divide both sides by 5.

We have found that $f(x) = -26$ when $x = -6$. To check this result, we can substitute -6 for x and verify that $f(-6) = -26$.

$f(x) = 5x + 4$ This is the given function.

$f(-6) = 5(-6) + 4$ Substitute -6 for x.

$= -30 + 4$ Do the multiplication.

$= -26$ This is the desired output.

Self Check 5   Let $f(x) = 7x + 200$. For what value of x is $f(x) = 11$?

Now Try ▶   Problem 75

4   Find the Domain of a Function.

We can think of a function as a machine that takes some input x and turns it into some output $f(x)$, as shown in figure (a). The machine shown in figure (b) turns the input -6 into the output -11. The set of numbers that we put into the machine is the *domain* of the function, and the set of numbers that comes out is the *range*.

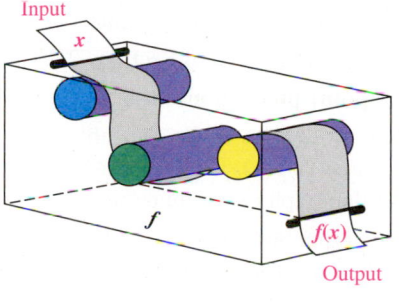

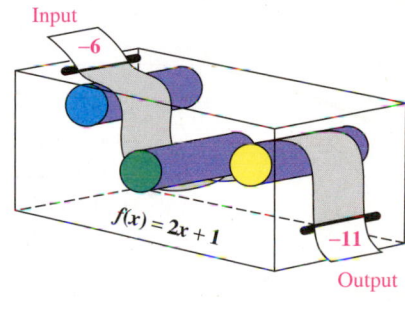

(a)   (b)

Success Tip

To find the domain of a function defined by an equation, we must identify all the values of the input variable that produce a real-number output.

EXAMPLE 6   Find the domain of each function: **a.** $f(x) = 3x + 1$ **b.** $f(x) = \dfrac{x}{3x - 6}$

Strategy   We will ask, "What values of x are permissible replacements for x in $3x + 1$ and $\dfrac{x}{3x - 6}$?"

Why   These values of x form the domain of the function.

Solution   **a.** We will be able to evaluate $3x + 1$ for any value of x that is a real number. Thus, the domain of the function is *the set of real numbers,* which can be represented by the symbol $\mathbb{R}$.

The Language of Algebra

The last two letters in the word *domain* help us remember that it is the set of all *inputs* of a function.

b. Since division by 0 is undefined, we will not be able to evaluate $\dfrac{x}{3x - 6}$ for any number x that makes the denominator equal to 0. To find such x-values, we set the denominator equal to 0 and solve for x.

$3x - 6 = 0$

$3x = 6$ To isolate the variable term 3x, add 6 to both sides.

$x = 2$ To isolate the variable, divide both sides by 3.

Success Tip

Notice that to determine the domain of $f(x) = \dfrac{x}{3x - 6}$, we don't have to examine the numerator. It can be any value, including 0.

We have found that 2 must be excluded from the domain of the function because it makes the denominator 0. However, all other real numbers are permissible replacements for x. Thus, the domain of the function *is the set of all real numbers except* 2.

Self Check 6   Find the domain of each function: **a.** $f(x) = 2x - 6$

   b. $f(x) = \dfrac{2}{-4x - 12}$

Now Try ▶ Problem 79

---

5 Graph Linear Functions.

In this course, we will study several important "families" of functions. We will begin here with the most basic family, *linear functions*.

Linear Functions	A **linear function** is a function that can be defined by an equation of the form $f(x) = mx + b$, where m and b are real numbers. The graph of a linear function is a straight line with slope m and y-intercept $(0, b)$.

Some examples of linear functions are:

$$f(x) = \frac{1}{2}x + 3, \qquad f(x) = -6x - 10, \qquad \text{and} \qquad f(x) = x$$

The input-output pairs that a linear function such as $f(x) = \frac{1}{2}x + 3$ generates can be plotted on a rectangular coordinate system to get the **graph of the function.** Since the symbols y and $f(x)$ are interchangeable, we can graph $f(x) = \frac{1}{2}x + 3$ as we would $y = \frac{1}{2}x + 3$, using the methods of Sections 2.2 and 2.4.

To use the **point-plotting method** to graph $f(x) = \frac{1}{2}x + 3$, we begin by constructing a table of function values. To make the table, we select several values for x, and find the corresponding values of $f(x)$. Then we plot the ordered pairs and draw a straight line through the points to get the graph of the function, as shown on the next page.

To use the **slope–intercept method** to graph $f(x) = \frac{1}{2}x + 3$, we begin by identifying m and b in the equation. Then we plot the y-intercept and use the slope to determine a second point on the line. Finally, we draw a line through the points to obtain a graph like that shown on the next page.

$$f(x) = \frac{1}{2}x + 3$$

Slope ⌐   ⌐ y-intercept: (0, 3)

To use the **intercept method** to graph $f(x) = \frac{1}{2}x + 3$, we begin by identifying b in the equation to determine the y-intercept. Then we let y, which in this case is $f(x)$, equal 0 and solve for x to find the x-intercept.

Find the y-intercept: Identify b	*Find the x-intercept: Let $f(x) = 0$*
$$f(x) = \frac{1}{2}x + 3$$ ↑ The y-intercept is (0, 3).	$$f(x) = \frac{1}{2}x + 3$$ $$0 = \frac{1}{2}x + 3$$ $$-3 = \frac{1}{2}x$$ $$-6 = x \qquad \text{The x-intercept is } (-6,0).$$

When we plot the intercepts $(0, 3)$ and $(-6, 0)$ and draw a straight line through them, we obtain a graph like that on the next page.

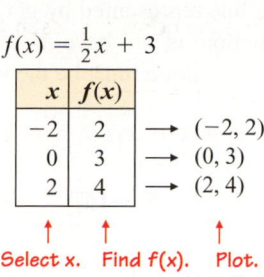

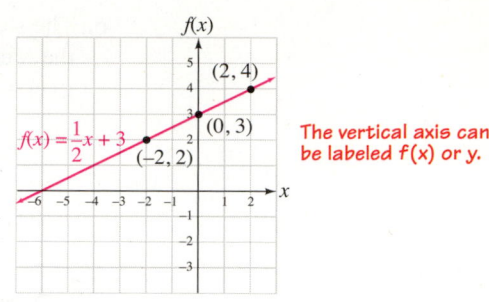

$f(x) = \frac{1}{2}x + 3$

x	$f(x)$
-2	2
0	3
2	4

$\longrightarrow (-2, 2)$
$\longrightarrow (0, 3)$
$\longrightarrow (2, 4)$

↑ Select x.   ↑ Find f(x).   ↑ Plot.

Notation

A table of function values is similar to a table of solutions, except that the second column is usually labeled $f(x)$ instead of y.

x	$f(x)$

x	y

The vertical axis can be labeled f(x) or y.

The most basic linear function is $f(x) = x$. It is called the **identity function** because it assigns each real number to itself. The graph of the identity function is a line with slope 1 and y-intercept $(0, 0)$, as shown below in figure (a).

A linear function defined by $f(x) = b$ is called a **constant function,** because for any input x, the output is the constant b. The graph of a constant function is a horizontal line. The graph of $f(x) = 2$ is shown below in figure (b).

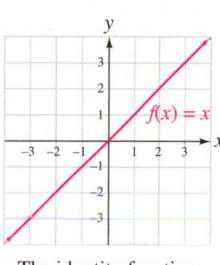

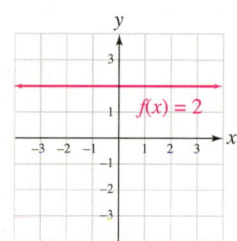

The identity function

(a)

A constant function

(b)

6 Write Equations of Linear Functions.

The slope–intercept and point–slope forms for the equation of a line that we studied in Section 2.4 can be adapted to write equations of linear functions.

EXAMPLE 7

a. Write an equation for the linear function whose graph has slope $-\frac{3}{4}$ and y-intercept $(0, 2)$.

b. Write an equation for the linear function whose graph passes through $(6, 4)$ with slope 5.

c. Write an equation for the linear function whose graph passes through $(-16, -12)$ and is perpendicular to the graph of $g(x) = -8x - 1$.

Strategy We will use either the slope–intercept or point–slope form to write each equation.

Why Writing equations of linear functions is similar to writing equations of linear equations in two variables, except that we replace y with $f(x)$.

Solution **a.** If the slope is $-\frac{3}{4}$ and the y-intercept is $(0, 2)$, then $m = -\frac{3}{4}$ and $b = 2$.

$f(x) = \boldsymbol{m}x + \boldsymbol{b}$ This is the slope–intercept form in which y is replaced with f(x).

$f(x) = -\dfrac{3}{4}x + 2$ Substitute $-\frac{3}{4}$ for m and 2 for b.

b. If the slope is 5 and the graph passes through $(6, 4)$, then $m = 5$ and $(x_1, y_1) = (6, 4)$.

$f(x) - \boldsymbol{y_1} = \boldsymbol{m}(x - \boldsymbol{x_1})$ This is the point–slope form in which y is replaced with f(x).

$f(x) - 4 = 5(x - 6)$ Substitute 4 for y_1, 5 for m, and 6 for x_1.

$f(x) - 4 = 5x - 30$ Distribute the multiplication by 5.

$f(x) = 5x - 26$ To isolate f(x), add 4 to both sides.

c. The slope of the line represented by $g(x) = -8x - 1$ is the coefficient of x: -8. Since the desired function is to have a graph that is perpendicular to the graph of $g(x) = -8x - 1$, its slope must be the negative reciprocal of -8, which is $\frac{1}{8}$.

$$f(x) - y_1 = m(x - x_1)$$ This is the point-slope form in which y is replaced with f(x).

$$f(x) - (-12) = \frac{1}{8}[x - (-16)]$$ Substitute −12 for y₁, ⅛ for m, and −16 for x₁.

$$f(x) + 12 = \frac{1}{8}(x + 16)$$ Simplify each side.

$$f(x) + 12 = \frac{1}{8}x + 2$$ Distribute the multiplication by ⅛: ⅛ · 16 = 2.

$$f(x) = \frac{1}{8}x - 10$$ To isolate f(x), subtract 12 from both sides.

Self Check 7   **a.** Write an equation for the linear function whose graph has slope $-\frac{1}{5}$ and y-intercept $(0, 9)$.   **b.** Write an equation for the linear function whose graph passes through $(7, 1)$ with slope 2.   **c.** Write an equation for the linear function whose graph passes through $(-20, -4)$ and is perpendicular to the graph of $g(x) = -10x$.

Now Try ▶ Problems 95, 97, and 103

EXAMPLE 8

Alzheimer's Disease. The graph on the right is from a recent Alzheimer's Association report. It shows how the number of people in the United States with Alzheimer's disease is expected to increase steadily at a constant rate. The report estimated there were 4,800,000 people with Alzheimer's in 2005 and 5,100,000 with the disease in 2010. (Source: Medill Reports)
a. Write the linear function that models this situation. **b.** Use the function to estimate the number of people who will have Alzheimer's disease in 2025, assuming the trend continues.

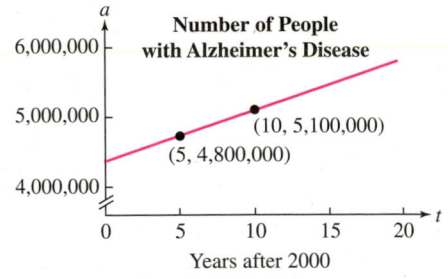
Number of People with Alzheimer's Disease
(10, 5,100,000)
(5, 4,800,000)
Years after 2000

The Language of Algebra

The phrase **constant rate** indicates that a linear function should be used to model the situation.

Strategy We will use the point–slope form to write a linear function.

Why We are given two points that lie on the graph of the function.

Solution Let a represent the approximate number of people in the U.S. with Alzheimer's disease and t represent the number of years after 2000. Two ordered pairs of the form (t, a) that lie on the graph of the function are:

- $(5, 4,800,000)$ Since the year 2005 is 5 years after 2000, t = 5.
- $(10, 5,100,00)$ Since the year 2010 is 10 years after 2000, t = 10.

Since we know two points that lie on the graph, we can write its equation. First, we find the slope of the line. If we write the slope formula using the variables t and a, and let $(t_1, a_1) = (5, 4,800,000)$ and $(t_2, a_2) = (10, 5,100,000)$, we have:

$$m = \frac{a_2 - a_1}{t_2 - t_1} = \frac{5,100,000 - 4,800,000}{10 - 5} = \frac{300,000}{5} = 60,000$$

The result indicates the number of people with Alzheimer's disease will increase at a rate of 60,000 per year.

a. To find the linear function, we substitute 60,000 for m, 5 for t_1, and 4,800,000 for a_1 in the point–slope form and simplify.

$$a(t) - a_1 = m(t - t_1)$$

This is the point–slope form using the variable t and replacing a with a(t).

$$a(t) - 4{,}800{,}000 = 60{,}000(t - 5)$$

Substitute for m, t₁, and a₁.

$$a(t) - 4{,}800{,}000 = 60{,}000t - 300{,}000$$

Distribute the multiplication by 60,000.

$$a(t) = 60{,}000t + 4{,}500{,}000$$

To isolate a(t), add 4,800,000 to both sides.

The approximate number of people with Alzheimer's disease t years after 2000 is given by $a(t) = 60{,}000t + 4{,}500{,}000$.

b. To estimate the number of people who will have Alzheimer's disease in 2025, which is 25 years after 2000, we find $a(25)$.

$$a(t) = 60{,}000t + 4{,}500{,}000$$

This is the linear function model.

$$a(25) = 60{,}000(25) + 4{,}500{,}000$$

Substitute 25 for t.

$$= 1{,}500{,}000 + 4{,}500{,}000$$

Do the multiplication.

$$= 6{,}000{,}000$$

Do the addition.

In 2025, there will be approximately 6,000,000 people in the U.S. with Alzheimer's disease.

Self Check 8   **Energy.**   The world's annual energy consumption can be modeled by a linear function. In 2004, the world consumed about 430 quadrillion Btu. By the year 2006, that number had increased to about 446 quadrillion Btu. (Source: Energy Information Administration)
a. Let t be the number of years after 2000 and E be the amount of energy (in quadrillion Btu). Write a linear function $E(t)$ to model the situation.   **b.** Predict the world's energy consumption in 2030, if the trend continues.

Now Try ▶ Problem 111

Using Your **Calculator** ▶ Evaluating Functions

We can use a graphing calculator to find function values. For example, suppose the linear function $f(c) = 12c - 18$ gives the daily income earned by a cosmetologist from serving c customers.

To find the income she earns for different numbers of customers, we first graph the income function $f(c) = 12c - 18$ as $y = 12x - 18$, using window settings of $[0, 10]$ for x and $[0, 100]$ for y to obtain figure (a). To find her income when she serves seven customers, we trace and move the cursor until the x-coordinate on the screen is nearly 7, as in figure (b). From the screen, we see that her income is about $66.26.

To find her income when she serves nine customers, we trace and move the cursor until the x-coordinate is nearly 9, as in figure (c). From the screen, we see that her income is about $90.51.

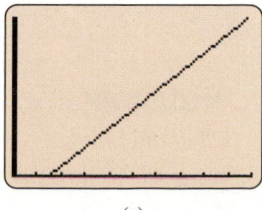   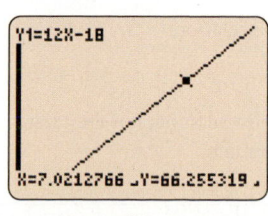   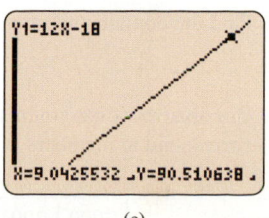

(a)   (b)   (c)

With some graphing calculator models, we can evaluate a function by entering function notation. To find $f(15)$, the income earned by the cosmetologist if she serves 15 customers, we use the following steps on a TI-84 Plus calculator.

With $f(c) = 12c - 18$ entered as $Y_1 = 12x - 18$, we call up the home screen by pressing $\boxed{\text{2nd}}$ $\boxed{\text{QUIT}}$. Then we enter $\boxed{\text{VARS}}$ $\boxed{\blacktriangleright}$ $\boxed{1}$ $\boxed{\text{ENTER}}$. The symbolism Y_1 will be displayed. See figure (a). Next, we enter the input value 15 within parentheses, as shown in figure (b), and press $\boxed{\text{ENTER}}$. In figure (c) we see that $Y_1(15) = 162$. That is, $f(15) = 162$. The cosmetologist will earn \$162 if she serves 15 customers in one day.

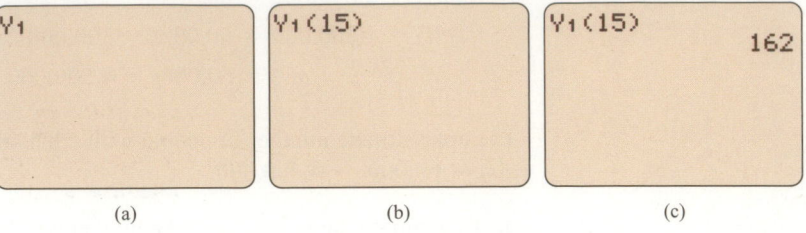

(a)   (b)   (c)

SECTION 2.5 ▸ STUDY SET

VOCABULARY

Fill in the blanks.

1. A set of ordered pairs is called a _____. The set of all first components of the ordered pairs is called the _____ and the set of all second components is called the _____.

2. A _____ is a set of ordered pairs (a relation) in which to each first component there corresponds exactly one second component.

3. Given a relation in x and y, if to each value of x in the domain there corresponds exactly one value of y in the range, y is said to be a _____ of x. We call x the independent _____ and y the _____ variable.

4. For a function, the set of all possible values that can be used for the independent variable is called the _____. The set of all values of the dependent variable is called the _____.

5. We call $f(x) = 2x + 1$ a _____ function.

6. We call $f(x) = x$ the _____ function because it assigns each real number to itself. We call $f(x) = 2$ a _____ function, because for any input x, the output is always 2.

CONCEPTS

7. **U.S. Recycling.** The following table gives the approximate number of aluminum cans (in billions) collected each year for the years 2000–2006.

   a. Display the data in the table as a relation, that is, as a set of ordered pairs.

   b. Find the domain and range of the relation.

   c. Use an arrow diagram to show how members of the range correspond to members of the domain.

Year	2000	2001	2002	2003	2004	2005	2006
Billions of aluminum cans	63	56	54	50	52	51	51

Source: Aluminum Association of America, U.S. Dept. of Commerce

8. For the given input, what value will the function machine output?

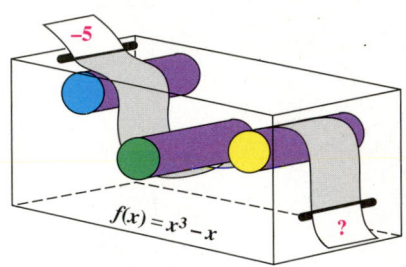

9. Explain why -4 is not in the domain of $f(x) = \frac{1}{x + 4}$.

10. Consider the linear function $y = -\frac{4}{5}x + 3$.

    a. What is the slope of its graph?

    b. What is the y-intercept of its graph?

11. Consider the linear function $f(x) = -6x - 4$.

    a. What is the y-intercept of its graph?

    b. What is the x-intercept of its graph?

12. The graphing calculator display shows a table of values for a function f. Fill in the blanks:

    $f(-1) = $ ▢ $f(3) = $ ▢

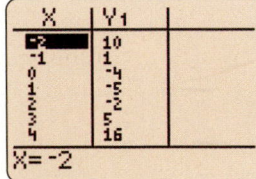

NOTATION

Fill in the blanks.

13. a. We read $f(x) = 5x - 6$ as "f ▢ x is equal to $5x$ minus 6."

    b. We read $g(t) = t + 9$ as "g ▢ t is equal to t plus 9."

14. This variable represents the ▢ .

$$f(x) = 2x - 5$$

This is the ▢   Use this expression of the function.   to find the ▢ .

15. Fill in the blank so that the following statements are equivalent:
- If $y = 5x + 1$, find the value of y when $x = 8$.
- If $f(x) = 5x + 1$, find ▢ .

16. If $f(2) = 7$, the input 2 and the output 7 can be written as the ordered pair (▢ , ▢).

17. When graphing $f(x) = -x + 5$, the vertical axis of the rectangular coordinate system can be labeled ▢ or ▢ .

18. To write the slope–intercept form $y = mx + b$ and the point–slope form $y - y_1 = m(x - x_1)$ using function notation, we simply replace y with ▢ .

GUIDED PRACTICE

Find the domain and range of each relation.
See Example 1.

19. $\{(7, -1), (-1, -11), (-5, 3), (8, -6)\}$

20. $\{(15, -3), (0, 0), (4, 6), (-3, -8)\}$

21. $\{(0, 1), (-23, 35), (7, 1)\}$
22. $\{(1, -12), (-6, 8), (5, 8), (0, 0), (1, 4)\}$

Determine whether the relation defines y to be a function of x. If it does not, find two ordered pairs where more than one value of y corresponds to a single value of x. See Example 2.

23.   **24.**

25.   **26.**

27. $\{(3, 4), (3, -4), (4, 3), (4, -3)\}$

28. $\{(-1, 1), (-3, 1), (-5, 1), (-7, 1), (-9, 1)\}$
29. $\{(-2, 7), (-1, 10), (0, 13), (1, 16)\}$
30. $\{(-2, 4), (-3, 8), (-3, 12), (-4, 16)\}$

31.

x	y
1	7
2	15
3	23
4	16
5	8

32.

x	y
30	2
30	4
30	6
30	8
30	10

33.

x	y
-4	6
-1	0
0	-3
2	4
-1	2

34.

x	y
1	1
2	1
3	1
4	1

Determine whether each equation defines y to be a function of x. If it does not, find two ordered pairs where more than one value of y corresponds to a single value of x. See Example 3.

35. $y = 2x + 3$ **36.** $y = 4x - 1$
37. $y = 4x^2$ **38.** $y^2 = x$
39. $y^4 = x$ **40.** $y = \dfrac{1}{x}$
41. $xy = 9$ **42.** $y = |x|$
43. $y = \dfrac{1}{x^2}$
44. $x + 1 = |y|$
45. $x = |y|$
46. $xy = -4$

Find f(3) and f(−1). See Example 4.

47. $f(x) = 3x$ **48.** $f(x) = -4x$
49. $f(x) = 2x - 3$ **50.** $f(x) = 3x - 5$

Find g(2) and g(3). See Example 4.

51. $g(x) = x^2 - 10$ **52.** $g(x) = x^2 - 2$
53. $g(x) = -x^3 + x$ **54.** $g(x) = x^3 - x$
55. $g(x) = (x + 1)^2$ **56.** $g(x) = (x - 3)^2$
57. $g(x) = 2x^2 - x + 1$ **58.** $g(x) = 5x^2 + 2x + 2$

Find h(5) and h(−2). See Example 4.

59. $h(x) = |x| + 2$ **60.** $h(x) = |x| - 5$
61. $h(x) = \dfrac{1}{x + 3}$ **62.** $h(x) = \dfrac{3}{x - 4}$
63. $h(x) = \dfrac{x}{x - 3}$ **64.** $h(x) = \dfrac{x}{x^2 + 2}$
65. $h(x) = \dfrac{x^2 + 2x - 35}{x^2 + 5x + 6}$ **66.** $h(x) = \dfrac{x^2 + x - 2}{x^2 - 5x}$

Complete each table. See Example 4.

67. $f(t) = |t - 2|$

t	$f(t)$
-1.7	
0.9	
5.4	

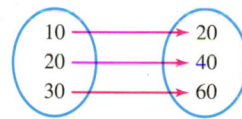

68. $f(r) = -2r^2 + 1$

Input	Output
-1.7	
0.9	
5.4	

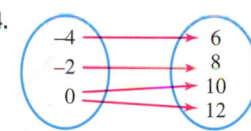

69. $g(a) = a^3$

Input	Output
$-\frac{3}{4}$	
$\frac{1}{6}$	
$\frac{5}{2}$	

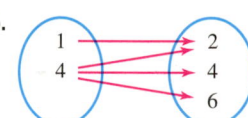

70. $g(b) = 2\left(-b - \frac{1}{4}\right)$

b	$g(b)$
$-\frac{3}{4}$	
$\frac{1}{6}$	
$\frac{5}{2}$	

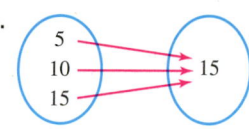

Find g(w) and g(w + 1). See Example 4.

71. $g(x) = 2x$

72. $g(x) = -3x$

73. $g(x) = 3x - 5$

74. $g(x) = 2x - 7$

Let f(x) = −2x + 5. For what value of x does function f have the given value? See Example 5.

75. $f(x) = 5$ **76.** $f(x) = -7$

Let f(x) = $\frac{3}{2}$x − 2. For what value of x does function f have the given value? See Example 5.

77. $f(x) = -\frac{1}{2}$ **78.** $f(x) = \frac{2}{3}$

Find the domain of each function. See Example 6.

79. a. $h(x) = 3x + 6$

  b. $f(x) = \frac{1}{x - 4}$

80. a. $g(x) = |x - 7|$

  b. $f(x) = \frac{5}{x + 1}$

81. a. $f(x) = x^2$

  b. $s(x) = \frac{9}{2x + 1}$

82. a. $h(x) = x^3$

  b. $t(x) = \frac{15}{1 - 3x}$

Graph each function. See Objective 5.

83. $f(x) = 2x - 1$ **84.** $f(x) = -x + 2$

85. $f(x) = -\frac{3}{2}x - 3$ **86.** $f(x) = \frac{2}{3}x - 2$

87. $f(x) = x$ **88.** $f(x) = -x$

89. $f(x) = -4$ **90.** $f(x) = 2$

91. $g(x) = 0.75x$ **92.** $g(x) = -0.25x$

93. $s(x) = \frac{7}{8}x + 2$ **94.** $s(x) = -\frac{4}{5}x + 3$

Write an equation for a linear function whose graph has the given characteristics. See Example 7.

95. Slope 5, y-intercept $(0, -3)$ **96.** Slope 2, y-intercept $(0, 11)$

97. Slope $\frac{1}{5}$, passes through $(10, 1)$

98. Slope $\frac{1}{4}$, passes through $(8, 1)$

99. Passes through $(1, 7)$ and $(-2, 1)$

100. Passes through $(-2, 2)$ and $(2, -8)$

101. Passes through $(3, 0)$, parallel to the graph of $g(x) = -\frac{2}{3}x - 4$

102. Passes through $(2, 20)$, parallel to the graph of $g(x) = 8x + 1$

103. Passes through $(1, 2)$, perpendicular to the graph of $g(x) = -\frac{x}{6} + 1$

104. Passes through $(54, 0)$, perpendicular to the graph of $g(x) = 9x + 5$

105. Horizontal, passes through $(-8, 12)$

106. Horizontal, passes through $(9, -32)$

APPLICATIONS

107. Decongestants.   The temperature in degrees Celsius that is equivalent to a temperature in degrees Fahrenheit is given by the linear function $C(F) = \frac{5}{9}(F - 32)$. Refer to the label from a bottle of decongestant shown below. Use this function to find the low and high temperature extremes, in degrees Celsius, in which the bottle should be stored.

> **DIRECTIONS:** Adults and children 12 years of age and over: Two teaspoons every 4 hours. DO NOT EXCEED 6 DOSES IN A 24-HOUR PERIOD. Store at a controlled room temperature between 68°F and 77°F.

108. Body Temperatures.   The temperature in degrees Fahrenheit that is equivalent to a temperature in degrees Celsius is given by the linear function $F(C) = \frac{9}{5}C + 32$. Convert each temperature in the following excerpt from *The Good Housekeeping Family Health and Medical Guide* to degrees Fahrenheit. (Round to the nearest degree.)

> *In disease, the temperature of the human body may vary from about 32.2°C to 43.3°C for a time, but there is grave danger to life should it drop and remain below 35°C or rise and remain at or above 41°C.*

109. Concessionaires.   A baseball club pays a vendor $125 per game for selling bags of peanuts for $4.75 each.

  a. Write a linear function that describes the profit the vendor makes for the baseball club during a game if she sells *b* bags of peanuts.

  b. Find the profit the baseball club will make if the vendor sells 110 bags of peanuts during a game.

110. Home Construction.   In a proposal to some clients, a housing contractor listed the following costs:

Fees, permits, miscellaneous	$12,000
Construction, per square foot	$95

  a. Write a linear function that the clients could use to determine the cost of building a home having *f* square feet.

  b. Find the cost to build a home having 1,950 square feet.

111. Nurses.   The demand for full-time registered nurses in the United States can be modeled by a linear function. In 2005, approximately 2,175,500 nurses were needed. By the year 2015, that number is expected to increase to about 2,586,500. (Source: National Center for Health Workforce Analysis)

  a. Let *t* be the number of years after 2000 and *N* be the number of full-time registered nurses needed in the U.S. Write a linear function $N(t)$ to model the demand for nurses.

  b. Use your answer to part a to predict the number of full-time registered nurses that will be needed in 2025, if the trend continues.

112. Wood Production. The total world wood production can be modeled by a linear function. In 1960, approximately 2,400 million cubic feet of wood were produced. Since then, the amount of increase has been approximately 25.5 million cubic feet per year. (Source: Earth Policy Institute)

a. Let t be the number of years after 1960 and W be the number of million cubic feet of wood produced. Write a linear function $W(t)$ to model the production of wood.

b. Use your answer to part a to estimate how many million cubic feet of wood the world produced in 2010.

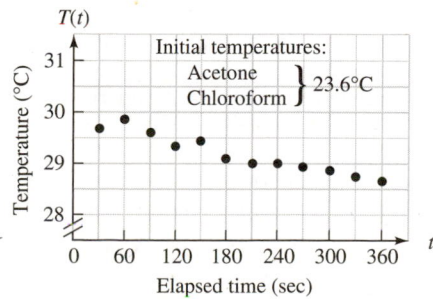

113. ▶ from **Campus to Careers**

Certified Fitness Instructor

When fitness instructors prescribe exercise workouts for elderly patients, they must take into account age-related loss of lung function. Studies show that the percent of remaining breathing capacity for someone over 30 years old can be modeled by a linear function. (Source: alsearsmd.com)

a. At 35 years of age, approximately 90% of maximal breathing capacity remains and at 55 years of age, approximately 66% of maximal breathing capacity remains. Let a be the age of a patient and L be the percent of her maximal breathing capacity that remains. Write a linear function $L(a)$ to model this situation.

b. Use your answer to part a to estimate the percent of maximal breathing capacity that remains in an 80-year-old.

114. Chemical Reactions. When students mixed solutions of acetone and chloroform, they found that heat was generated. However, as time passed, the mixture cooled down. The graph shows data points of the form (time, temperature) taken by the students.

a. The linear function $T(t) = -\frac{t}{240} + 30$ models the relationship between the elapsed time t since the solutions were combined and the temperature $T(t)$ of the mixture. Graph the function.

b. Predict the temperature of the mixture immediately after the two solutions are combined.

c. Is $T(180)$ more or less than the temperature recorded by the students for $t = 300$?

115. Taxes. The function

$$T(a) = 837.50 + 0.15(a - 8,375)$$

(where a is adjusted gross income) is a model of the instructions given on the first line of the following tax rate Schedule X.

a. Find $T(25,000)$ and interpret the result.

b. Write a function that models the second line on Schedule X.

Schedule X–Use if your filing status is **Single**			**2010**
If your adjusted gross income is: Over —	But not over —	Your tax is	of the amount over —
$ 8,375	$34,000	$ 837.50 + 15%	$ 8,375
$34,000	$82,400	$4,681.25 + 25%	$34,000

116. Archery. The area of a circle with a diameter of length d is given by the function $A(d) = \pi\left(\frac{d}{2}\right)^2$.

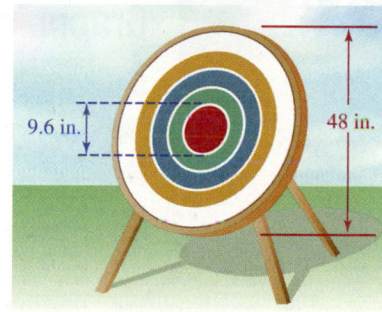

a. Find the area of the archery target to the nearest tenth of a square inch.

b. Find the area of the bull's eye to the nearest tenth of a square inch.

WRITING

117. Explain why 8 is not in the domain of the function $f(x) = \dfrac{5x - 7}{x - 8}$.

118. Explain why we can think of a function as a machine.

119. Consider the function defined by $y = 6x + 4$. Why do you think x is called the *independent* variable and y the *dependent* variable?

120. A website selling nutritional supplements contains the following sentence: "Health is a *function* of proper nutrition." Explain what this statement means.

Solve each equation. If the equation is an identity or a contradiction, so indicate.

121. $-2(t + 4) + 5t + 1 = 3(t - 4) + 7$

122. $\frac{3}{2}(a - 4) = 2(a - 3) - \frac{a}{2}$

CHALLENGE PROBLEMS

123. Let $f(x) = 4x + 6$, function g be defined by $\{(4, 6), (6, 8), (8, 4)\}$, and function h be defined by the arrow diagram below. Find $\dfrac{f(8) + g(8)}{h(8)}$.

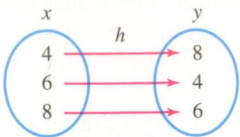

124. Find the domain of $f(x) = \dfrac{1}{5[9(x - 2) - 6(x - 3) + 3]}$.

Graphs of Functions

OBJECTIVES

1. Find function values graphically.

2. Find the domain and range of a function graphically.

3. Graph nonlinear functions.

4. Translate graphs of functions.

5. Reflect graphs of functions.

6. Use the vertical line test.

ARE YOU READY?

The following problems review some basic skills that are needed when graphing functions.

1. Fill in the blanks: $f(3) = 9$ corresponds to the ordered pair (☐ , ☐).
2. If $f(x) = x^2$, find $f(-1)$ and $f(1)$.
3. If $f(x) = x^3$, find $f(-2)$ and $f(2)$.
4. If $f(x) = |x|$, find $f(-4)$ and $f(4)$.

Since a graph is often the best way to describe a function, we need to know how to construct and interpret their graphs.

1 Find Function Values Graphically.

From the graph of a function, we can determine function values. In general, the value of $f(a)$ is given by the y-coordinate of a point on the graph of function f with x-coordinate a.

EXAMPLE 1   Refer to the graph of function f in figure (a) on the next page.   **a.** Find $f(-3)$ **b.** Find the value of x for which $f(x) = -2$.

Strategy   In each case, we will use the information provided by the function notation to locate a specific point on the graph and determine its x- and y-coordinates.

Why   Once we locate the specific point, one of its coordinates will equal the value that we are asked to find.

Solution   **a.** To find $f(-3)$, we need to find the y-coordinate of the point on the graph of f whose x-coordinate is -3. If we draw a vertical line through -3 on the x-axis, as shown in figure (b), the line intersects the graph of f at $(-3, 5)$. Therefore, 5 corresponds to -3, and it follows that $f(-3) = 5$.

b. To find the input value x that has an output value $f(x) = -2$, we draw a horizontal line through -2 on the y-axis, as shown in figure (c) and note that it intersects the graph of f at $(4, -2)$. Since -2 corresponds to 4, it follows that $f(x) = -2$ if $x = 4$.

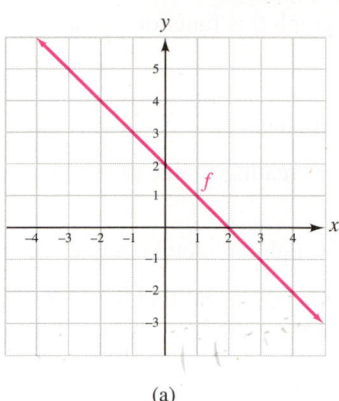

(a)

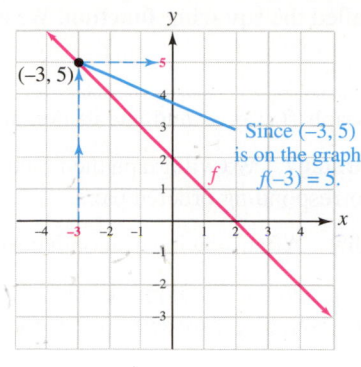

(b)

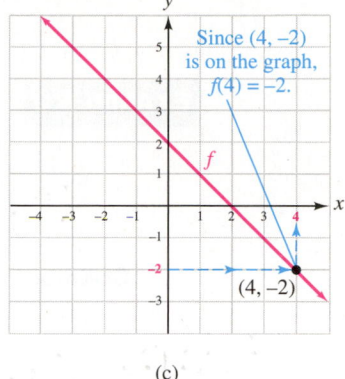

(c)

Self Check 1   Refer to the graph of function g on the right.
a. Find $g(-3)$.   **b.** Find the x-value for which $g(x) = 2$.

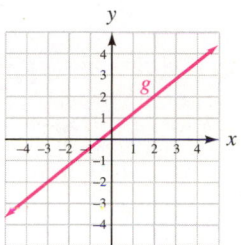

Now Try ▶ Problem 17

2 Find the Domain and Range of a Function Graphically.

We can find the domain and range of a function from its graph. For example, to find the domain of the linear function graphed in figure (a), we *project* the graph onto the x-axis. Because the graph of the function extends indefinitely to the left and to the right, the projection includes all the real numbers. Therefore, the domain of the function is the set of real numbers.

To find the range of the same linear function, we project the graph onto the y-axis, as shown in figure (b). Because the graph of the function extends indefinitely upward and downward, the projection includes all the real numbers. Therefore, the range of the function is the set of real numbers.

The Language of Algebra

Think of the **projection** of a graph on an axis as the "shadow" that the graph makes on the axis.

Project the graph onto the x-axis.

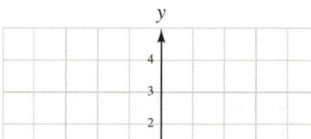
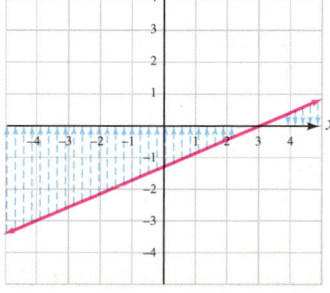

Domain: all real numbers

(a)

Project the graph onto the y-axis.

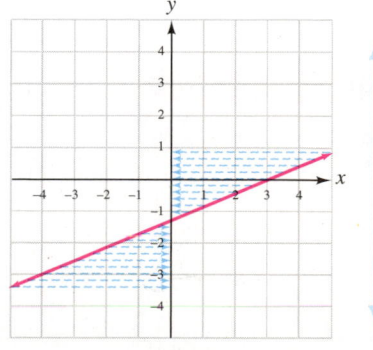

Range: all real numbers

(b)

3 Graph Nonlinear Functions.

We have seen that the graph of a linear function is a line. We will now consider several examples of **nonlinear functions** whose graphs are not lines. We will begin with $f(x) = x^2$, called the **squaring function**. We can graph this function using the **point-plotting method**.

EXAMPLE 2   Graph $f(x) = x^2$ and find its domain and range.

Strategy   We will graph the function by creating a table of function values and plotting the corresponding ordered pairs.

Why   After drawing a smooth curve through the plotted points, we will have the graph.

Solution   To graph the function, we select several x-values and find the corresponding values of $f(x)$. For example, if we select -3 for x, we have

$$f(x) = x^2 \qquad \text{This is the function to graph.}$$
$$f(-3) = (-3)^2 \qquad \text{Substitute } -3 \text{ for each } x.$$
$$= 9$$

Since $f(-3) = 9$, the ordered pair $(-3, 9)$ lies on the graph of f. In a similar manner, we find the corresponding values of $f(x)$ for six other x-values and list the ordered pairs in the table of values. Then we plot the points and draw a smooth curve through them to get the graph, called a **parabola.**

$$f(x) = x^2$$

x	$f(x)$	
-3	9	$\longrightarrow (-3, 9)$
-2	4	$\longrightarrow (-2, 4)$
-1	1	$\longrightarrow (-1, 1)$
0	0	$\longrightarrow (0, 0)$
1	1	$\longrightarrow (1, 1)$
2	4	$\longrightarrow (2, 4)$
3	9	$\longrightarrow (3, 9)$

↑ Select x.   ↑ Find $f(x)$.   ↑ Plot the point.

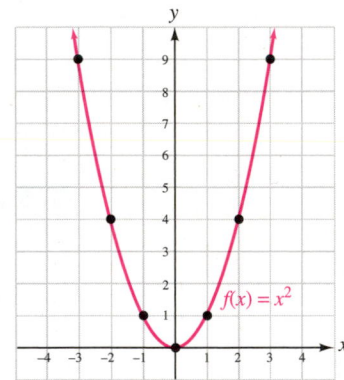

> **The Language of Algebra**
>
> The cuplike shape of a **parabola** has many real-life applications. For example, a satellite TV dish is called a **parabolic** dish.

Because the graph extends indefinitely to the left and to the right, the projection of the graph onto the x-axis includes all the real numbers. See figure (a). This means that the domain of the squaring function is the set of real numbers.

> **The Language of Algebra**
>
> The set of **nonnegative real numbers** is the set of real numbers greater than or equal to 0.

Project the graph onto the x-axis.   Project the graph onto the y-axis.

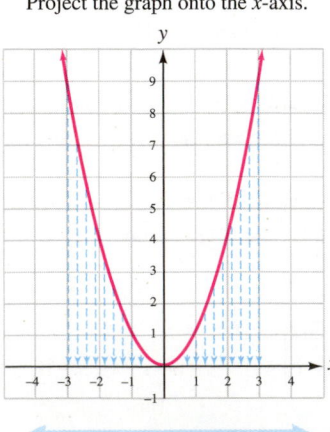

Domain: all real numbers

Range: nonnegative real numbers

(a)   (b)

Because the graph extends upward indefinitely from the point $(0, 0)$, the projection of the graph on the y-axis includes only positive real numbers and 0. See figure (b) on the previous page. This means that the range of the squaring function is the set of nonnegative real numbers.

Self Check 2 Graph $g(x) = x^2 - 2$ by plotting points. Then find its domain and range. Compare the graph to the graph of $f(x) = x^2$.

Now Try Problem 29

Another important nonlinear function is $f(x) = x^3$, called the **cubing function.**

EXAMPLE 3 Graph $f(x) = x^3$ and find its domain and range.

Strategy We will graph the function by creating a table of function values and plotting the corresponding ordered pairs.

Why After drawing a smooth curve through the plotted points, we will have the graph.

Solution To graph the function, we select several values for x and find the corresponding values of $f(x)$. For example, if we select -2 for x, we have

$$f(x) = x^3 \qquad \text{This is the function to graph.}$$
$$f(-2) = (-2)^3 \qquad \text{Substitute } -2 \text{ for each } x.$$
$$= -8$$

Since $f(-2) = -8$, the ordered pair $(-2, -8)$ lies on the graph of f. In a similar manner, we find the corresponding values of $f(x)$ for four other x-values and list the ordered pairs in the table. Then we plot the points and draw a smooth curve through them to get the graph.

Success Tip

To graph a linear function, it is recommended that you find three points on the line to draw its graph. Because the graphs of nonlinear functions are more complicated, more work is required. You need to find a sufficient number of points on the graph so that its entire shape is revealed. At least five well-chosen input values are recommended.

$$f(x) = x^3$$

x	$f(x)$		
-2	-8	→	$(-2, -8)$
-1	-1	→	$(-1, -1)$
0	0	→	$(0, 0)$
1	1	→	$(1, 1)$
2	8	→	$(2, 8)$

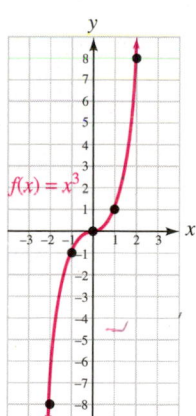

Because the graph of the function extends indefinitely to the left and to the right, the projection includes all the real numbers. Therefore, the domain of the cubing function is the set of real numbers.

Because the graph of the function extends indefinitely upward and downward, the projection includes all the real numbers. Therefore, the range of the cubing function is the set of real numbers.

Self Check 3 Graph $g(x) = x^3 + 1$ by plotting points. Then find its domain and range. Compare the graph to the graph of $f(x) = x^3$.

Now Try Problem 31

A third nonlinear function is $f(x) = |x|$, called the **absolute value function**.

EXAMPLE 4   Graph $f(x) = |x|$ and find its domain and range.

Strategy   We will graph the function by creating a table of function values and plotting the corresponding ordered pairs.

Why   After drawing straight lines through the plotted points, we will have the graph.

Solution   To graph the function, we select several x-values and find the corresponding values for $f(x)$. For example, if we choose -3 for x, we have

$$f(x) = |x| \qquad \text{This is the function to graph.}$$
$$f(-3) = |-3| \qquad \text{Substitute } -3 \text{ for each } x.$$
$$= 3$$

Since $f(-3) = 3$, the ordered pair $(-3, 3)$ lies on the graph of f. In a similar manner, we find the corresponding values of $f(x)$ for six other x-values and list the ordered pairs in the table. Then we plot the points and connect them to get the following V-shaped graph.

Success Tip

To determine the entire shape of the graph, several positive and negative values, along with 0, were selected as x-values when constructing the table of values.

$f(x) = |x|$

x	$f(x)$	
-3	3	$\longrightarrow (-3, 3)$
-2	2	$\longrightarrow (-2, 2)$
-1	1	$\longrightarrow (-1, 1)$
0	0	$\longrightarrow (0, 0)$
1	1	$\longrightarrow (1, 1)$
2	2	$\longrightarrow (2, 2)$
3	3	$\longrightarrow (3, 3)$

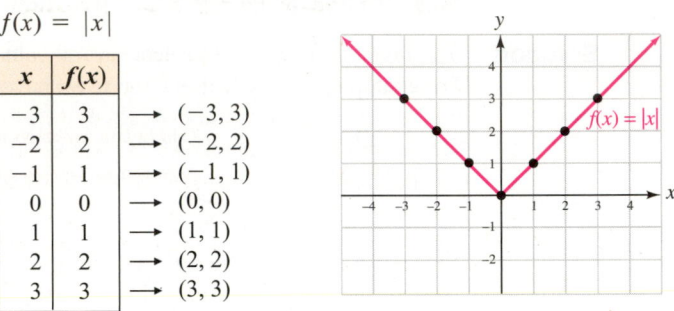

Because the graph extends indefinitely to the left and to the right, the projection of the graph onto the x-axis includes all the real numbers. Thus, the domain of the absolute value function is the set of real numbers.

Because the graph extends upward indefinitely from the point $(0, 0)$, the projection of the graph on the y-axis includes only positive real numbers and 0. Thus, the range of the absolute value function is the set of nonnegative real numbers.

Self Check 4   Graph $g(x) = |x - 2|$ by plotting points. Then find its domain and range. Compare the graph to the graph of $f(x) = |x|$.

Now Try ▶   Problem 33

Using Your Calculator ▶ Graphing Functions

We can graph nonlinear functions with a graphing calculator. For example, to graph $f(x) = x^2$ in a standard window of $[-10, 10]$ for x and $[-10, 10]$ for y, we first press $\boxed{Y =}$. Then we enter the function by typing $x \wedge 2$ (or x followed by $\boxed{x^2}$), and press the $\boxed{GRAPH}$ key. We will obtain the graph shown in figure (a).

To graph $f(x) = x^3$, we enter the function by typing $x \wedge 3$ and then press the $\boxed{GRAPH}$ key to obtain the graph in figure (b). To graph $f(x) = |x|$, we enter the function by selecting abs(from the NUM option within the MATH menu, typing x, and pressing the $\boxed{GRAPH}$ key to obtain the graph in figure (c).

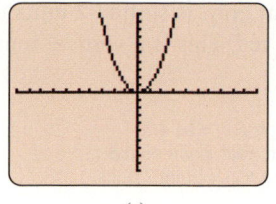
(a)

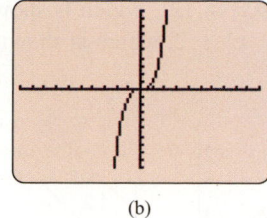
(b)

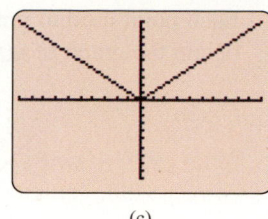
(c)

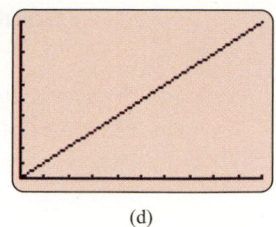
(d)

When using a graphing calculator, we must be sure that the viewing window does not show a misleading graph. For example, if we graph $f(x) = |x|$ in the window [0, 10] for x and [0, 10] for y, we will obtain a misleading graph that looks like a line. See figure (d). This is not correct. The proper graph is the V-shaped graph shown in figure (c). One of the challenges of using graphing calculators is finding an appropriate viewing window.

4 Translate Graphs of Functions.

Examples 2, 3, and 4 and their Self Checks suggest that the graphs of different functions may be identical except for their positions in the coordinate plane. For example, the figure on the right shows the graph of $f(x) = x^2 + k$ for three different values of k. If $k = 0$, we get the graph of $f(x) = x^2$, shown in red. If $k = 3$, we get the graph of $f(x) = x^2 + 3$, shown in blue. Note that it is identical to the graph of $f(x) = x^2$ except that it is shifted 3 units upward. If $k = -4$, we get the graph of $f(x) = x^2 - 4$, shown in green. It is identical to the graph of $f(x) = x^2$ except that it is shifted 4 units downward. These shifts are called **vertical translations.**

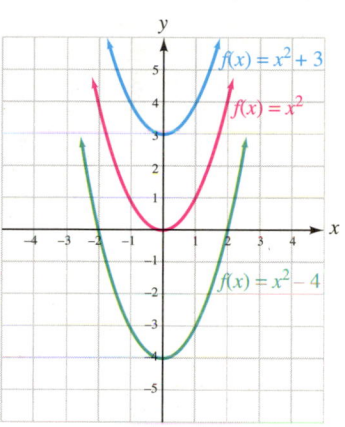

In general, we can make these observations.

Vertical Translations

If f is a function and k represents a positive number, then

- The graph of $y = f(x) + k$ is identical to the graph of $y = f(x)$ except that it is translated k units upward.

- The graph of $y = f(x) - k$ is identical to the graph of $y = f(x)$ except that it is translated k units downward.

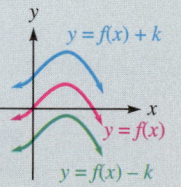

EXAMPLE 5   Graph: $g(x) = |x| + 2$

Strategy   We will graph $g(x) = |x| + 2$ by translating (shifting) the graph of $f(x) = |x|$ upward 2 units.

Why   The addition of 2 in $g(x) = |x| + 2$ causes a vertical shift of the graph of the absolute value function 2 units upward.

Solution   Each point used to graph $f(x) = |x|$, which is shown in gray, is shifted 2 units upward to obtain the graph of $g(x) = |x| + 2$, which is shown in red. This is a vertical translation.

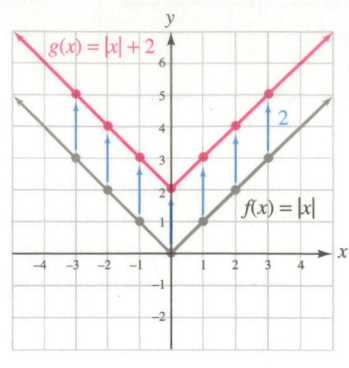

To graph $g(x) = |x| + 2$, translate each point on the graph of $f(x) = |x|$ up 2 units.

Self Check 5   Graph: $g(x) = |x| - 3$

Now Try ▶   Problem 37

The figure on the right shows the graph of $f(x) = (x + h)^2$ for three different values of h. If $h = 0$, we get the graph of $f(x) = x^2$, shown in red. The graph of $f(x) = (x - 3)^2$ shown in green is identical to the graph of $f(x) = x^2$ except that it is shifted 3 units to the right. The graph of $f(x) = (x + 2)^2$ shown in blue is identical to the graph of $f(x) = x^2$ except that it is shifted 2 units to the left. These shifts are called **horizontal translations.**

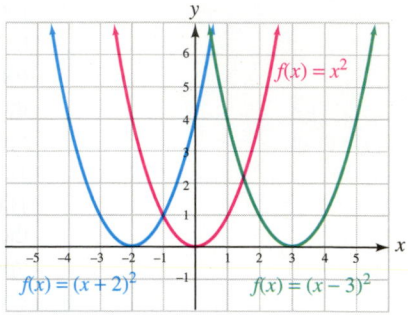

In general, we can make these observations.

Horizontal Translations

▼   If f is a function and h is a positive number, then

- The graph of $y = f(x - h)$ is identical to the graph of $y = f(x)$ except that it is translated h units to the right.

- The graph of $y = f(x + h)$ is identical to the graph of $y = f(x)$ except that it is translated h units to the left.

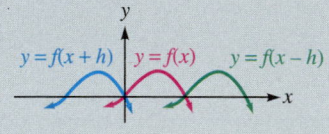

EXAMPLE 6   Graph: $g(x) = (x + 3)^3$

Success Tip

To determine the direction of the horizontal translation, find the value of x that makes the expression within the parentheses, in this case, $x + 3$, equal to 0. Since -3 makes $x + 3 = 0$, the translation is 3 units to the left.

Strategy   We will graph $g(x) = (x + 3)^3$ by translating (shifting) the graph of $f(x) = x^3$ to the left 3 units.

Why   The addition of 3 to x in $g(x) = (x + \mathbf{3})^3$ causes a horizontal shift of the graph of the cubing function 3 units to the left.

To graph $g(x) = (x + 3)^3$, translate each point on the graph of $f(x) = x^3$ to the left 3 units.

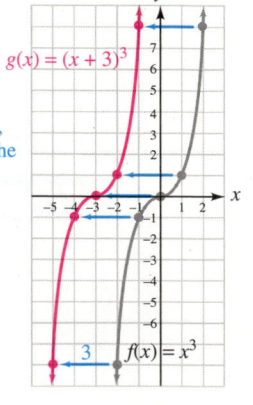

Solution   Each point used to graph $f(x) = x^3$, which is shown in gray, is shifted 3 units to the left to obtain the graph of $g(x) = (x + 3)^3$, which is shown in red. This is a horizontal translation.

Self Check 6   Graph: $g(x) = (x - 2)^2$

Now Try ▶ Problem 47

The graphs of some functions involve horizontal and vertical translations.

EXAMPLE 7   Graph: $g(x) = (x - 5)^2 - 2$

Strategy   To graph $g(x) = (x - 5)^2 - 2$, we will perform two translations by shifting the graph of $f(x) = x^2$ to the right 5 units and then 2 units downward.

Why   The subtraction of 5 from x in $g(x) = (x - \mathbf{5})^2 - 2$ causes a horizontal shift of the graph of the squaring function 5 units to the right. The subtraction of 2 causes a vertical shift of the graph 2 units downward.

Solution   Each point used to graph $f(x) = x^2$, which is shown in gray, is shifted 5 units to the right and 2 units downward to obtain the graph of $g(x) = (x - 5)^2 - 2$, which is shown in red. This is a horizontal and vertical translation.

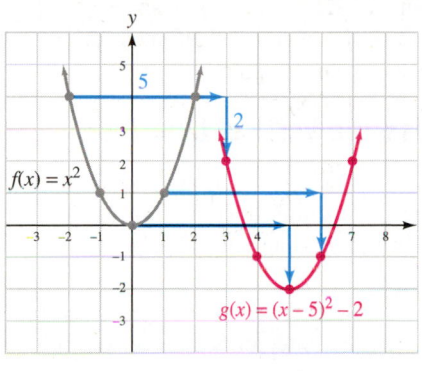

To graph $g(x) = (x - 5)^2 - 2$, translate each point on the graph of $f(x) = x^2$ to the right 5 units and then 2 units downward.

Self Check 7   Graph: $g(x) = |x + 2| - 3$

Now Try ▶ Problem 53

5 Reflect Graphs of Functions.

The following figure shows a table of values for $f(x) = x^2$ and for $g(x) = -x^2$. We note that for a given value of x, the corresponding y-value in the tables are opposites. When graphed, we see that the $-$ sign in $g(x) = -x^2$ has the effect of flipping the graph of $f(x) = x^2$ over the x-axis so that the parabola opens downward. We say that the graph of $g(x) = -x^2$ is a **reflection** of the graph of $f(x) = x^2$ about the x-axis.

$f(x) = x^2$

x	$f(x)$	
-2	4	→ $(-2, 4)$
-1	1	→ $(-1, 1)$
0	0	→ $(0, 0)$
1	1	→ $(1, 1)$
2	4	→ $(2, 4)$

$g(x) = -x^2$

x	$g(x)$	
-2	-4	→ $(-2, -4)$
-1	-1	→ $(-1, -1)$
0	0	→ $(0, 0)$
1	-1	→ $(1, -1)$
2	-4	→ $(2, -4)$

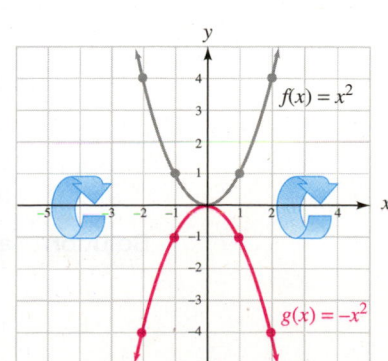

Reflection of a Graph ▼   The graph of $y = -f(x)$ is the graph of $y = f(x)$ reflected about the x-axis.

EXAMPLE 8   Graph: $g(x) = -x^3$

Strategy   We will graph $g(x) = -x^3$ by reflecting the graph of $f(x) = x^3$ about the x-axis.

Why   Because of the $-$ sign in $g(x) = -x^3$, the y-coordinate of each point on the graph of function g is the opposite of the y-coordinate of the corresponding point on the graph $f(x) = x^3$.

Solution   To graph $g(x) = -x^3$, we use the graph of $f(x) = x^3$ from Example 3. First, we reflect the portion of the graph of $f(x) = x^3$ in quadrant I to quadrant IV, as shown. Then we reflect the portion of the graph of $f(x) = x^3$ in quadrant III to quadrant II.

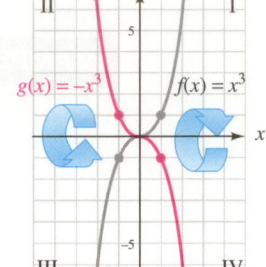

Self Check 8   Graph: $g(x) = -|x|$

Now Try ▶ Problem 63

---

6 Use the Vertical Line Test.

Some graphs define functions and some do not. If a vertical line intersects a graph more than once, the graph does not represent a function, because to one value of x there would correspond more than one value of y.

The Vertical Line Test ▼   If a vertical line intersects a graph in more than one point, the graph is not the graph of a function.

EXAMPLE 9   Determine whether the graph in figure (a) and figure (c) is the graph of a function.

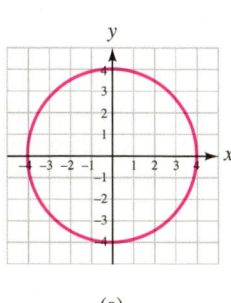

(a)

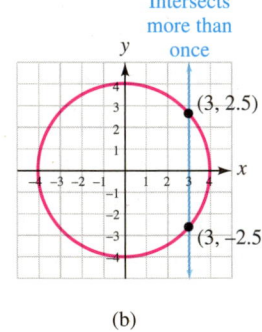

(b)

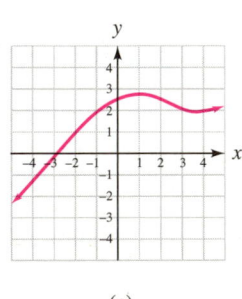

(c)

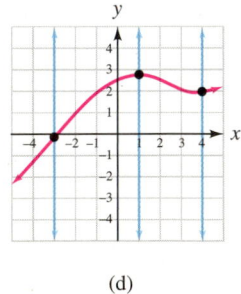

(d)

Strategy   We will check to see whether any vertical lines intersect the graph more than once.

Why   If any vertical line intersects the graph more than once, it is not the graph of a function.

Solution   **a.** Refer to figure (b) above. The graph shown in red is not the graph of a function because a vertical line intersects the graph more than once. The points of intersection of the graph and the vertical line indicate that two values of y (2.5 and -2.5) correspond to the x-value 3.

x	y
3	2.5
3	-2.5

b. Refer to figure (d) on the previous page. The graph shown in red is the graph of a function, because no vertical line intersects the graph more than once.

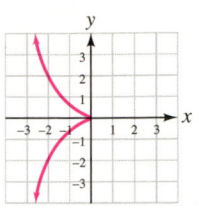

Self Check 9 Determine whether the following graph is the graph of a function.

Now Try ▶ Problems 65 and 67

SECTION 2.6 ▶ STUDY SET

VOCABULARY

Fill in the blanks.

1. Functions whose graphs are not lines are called _____ functions.
2. The graph of $f(x) = x^2$ is a cuplike shape called a _____.
3. The set of _____ real numbers is the set of real numbers greater than or equal to 0.
4. A shift of the graph of a function upward or downward is called a vertical _____.

CONCEPTS

5. Graph each basic function by plotting points and give its name.

   a. $f(x) = x^2$
   b. $f(x) = x^3$
   c. $f(x) = |x|$

6. Complete each sentence about finding function values graphically.

   a. To find $f(-3)$, we find the y-coordinate of the point on the graph whose x-coordinate is ▢.
   b. To find the value of x for which $f(x) = -2$, we find the x-coordinate of the point(s) on the graph whose y-coordinate is ▢.

7. Suppose for a function f that $f(5) = 9$. What corresponding ordered pair will be on the graph of function f?

8. Fill in the blanks. The illustration shows the projection of the graph of function f on the _____. We see that the _____ of f is the set of real numbers less than or equal to 0.

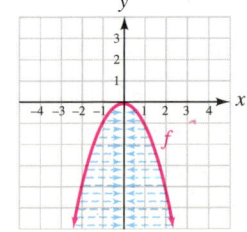

9. Consider the graph of the function f on the right.

   a. Label each arrow in the illustration with the appropriate term: *domain* or *range*.
   b. Give the domain and range of f.

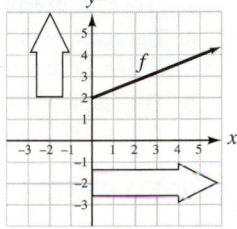

10. The graph of $f(x) = x^2 + k$ for three values of k is shown on the right. Find the value of k for
    a. the blue graph
    b. the red graph
    c. the green graph

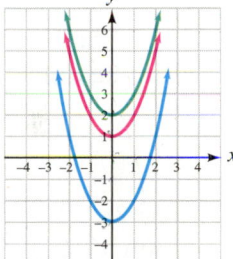

11. The graph of $f(x) = |x + h|$ for three values of h is shown. Find the value of h for
    a. the blue graph
    b. the red graph
    c. the green graph

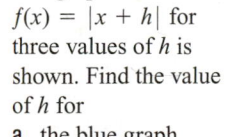
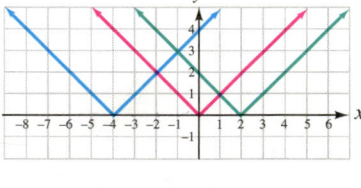

12. a. Translate each point plotted on the graph below to the left 5 units and then up 1 unit.   b. Translate each point plotted on the graph below to the right 4 units and then down 3 units.

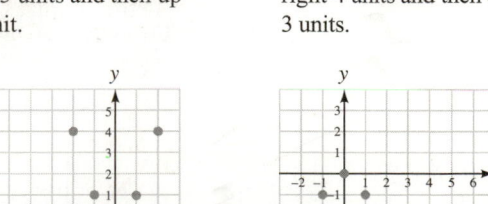
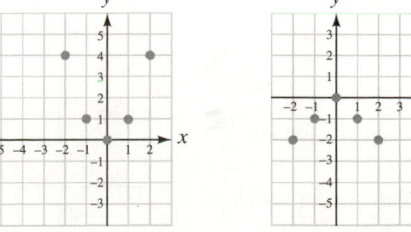

13. a. Give the coordinates of the points where the given vertical line intersects the graph in red.

   b. Is this the graph of a function? Explain.

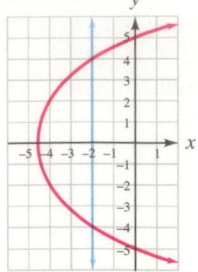

NOTATION

Fill in the blanks.

14. Fill in the blanks. The _____ line test: If a vertical line intersects a graph in more than one point, the graph is not the graph of a _____.

15. a. The graph of $f(x) = (x + 4)^3$ is the same as the graph of $f(x) = x^3$ except that it is shifted ___ units to the ____.

   b. The graph of $f(x) = x^3 + 4$ is the same as the graph of $f(x) = x^3$ except that it is shifted ___ units ____.

16. a. The graph of $f(x) = |x| - 5$ is the same as the graph of $f(x) = |x|$ except that it is shifted ___ units _____.

   b. The graph of $f(x) = |x - 5|$ is the same as the graph of $f(x) = |x|$ except that it is shifted ___ units to the ____.

GUIDED PRACTICE

Refer to the given graph to find each value. **See Example 1.**

17. a. $f(-2)$

   b. $f(0)$

   c. The value of x for which $f(x) = 4$.

   d. The value of x for which $f(x) = -2$.

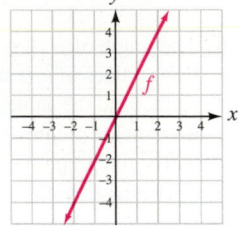

18. a. $g(-2)$

   b. $g(0)$

   c. The value of x for which $g(x) = 3$.

   d. The values of x for which $g(x) = -1$.

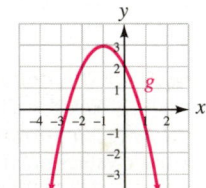

19. a. $s(-3)$

   b. $s(3)$

   c. The values of x for which $s(x) = 0$.

   d. The values of x for which $s(x) = 3$.

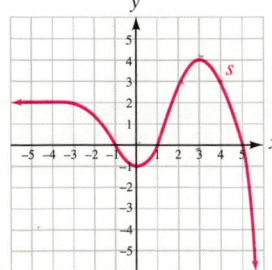

20. a. $h(-3)$

   b. $h(4)$

   c. The values of x for which $h(x) = 1$.

   d. The value of x for which $h(x) = 0$.

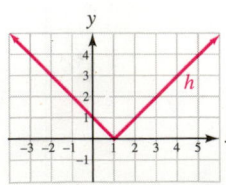

Find the domain and range of each function. **See Objective 2 and Example 2.**

21.

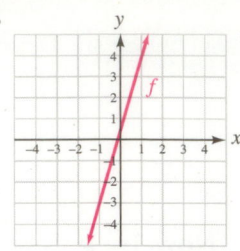

22.

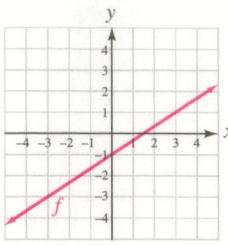

23.

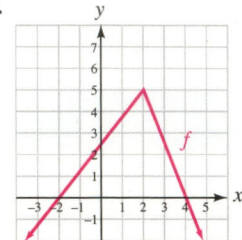

24.

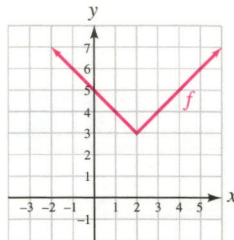

25.

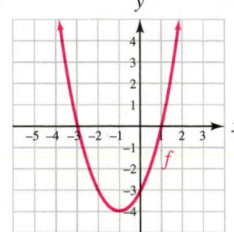

26.

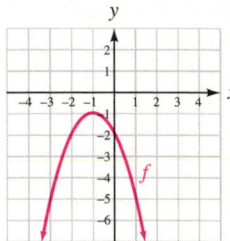

27.

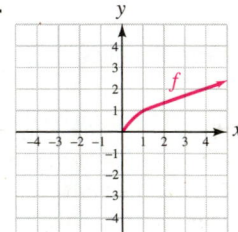

28.

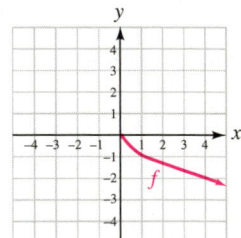

Graph each function by creating a table of function values and plotting points. Give the domain and range of the function. **See Examples 2, 3, and 4.**

29. $f(x) = x^2 + 2$

30. $f(x) = x^2 - 4$

31. $f(x) = x^3 - 3$

32. $f(x) = x^3 + 2$

33. $f(x) = |x - 1|$

34. $f(x) = |x + 4|$

35. $f(x) = (x + 4)^2$

36. $f(x) = (x - 1)^3$

For each of the following functions, first sketch the graph of its associated function, $f(x) = x^2$, $f(x) = x^3$, or $f(x) = |x|$. Then draw the graph of function g using a translation and give its domain and range. See Examples 5 and 6.

37. $g(x) = |x| - 2$

38. $g(x) = |x + 2|$

39. $g(x) = (x + 1)^3$

40. $g(x) = x^3 + 5$

41. $g(x) = x^2 - 3$

42. $g(x) = (x - 6)^2$

43. $g(x) = (x - 4)^3$

44. $g(x) = |x| + 1$

45. $g(x) = x^3 + 4$

46. $g(x) = x^2 - 5$

47. $g(x) = (x + 4)^2$

48. $g(x) = (x - 1)^3$

For each of the following functions, first sketch the graph of its associated function, $f(x) = x^2$, $f(x) = x^3$, or $f(x) = |x|$. Then draw the graph of function g using translations and/or a reflection. See Examples 7 and 8.

49. $g(x) = |x - 2| - 1$

50. $g(x) = (x + 2)^2 - 1$

51. $g(x) = (x + 1)^3 - 2$

52. $g(x) = |x + 4| + 3$

53. $g(x) = (x - 2)^2 + 4$

54. $g(x) = (x - 4)^2 + 3$

55. $g(x) = |x + 3| + 5$

56. $g(x) = (x - 3)^2 - 2$

57. $g(x) = -x^3$

58. $g(x) = -|x|$

59. $g(x) = -x^2$

60. $g(x) = -(x + 1)^2$

61. $g(x) = -|x + 5|$

62. $g(x) = -(x + 4)^3$

63. $g(x) = -x^2 + 3$

64. $g(x) = -|x| - 4$

Determine whether each graph is the graph of a function. If it is not, find two ordered pairs where more than one value of y corresponds to a single value of x. See Example 9.

65.

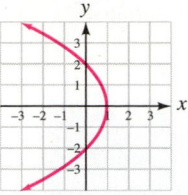

66.

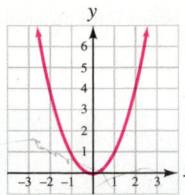

67.

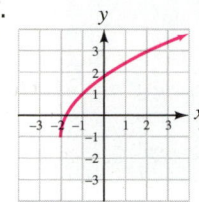

68.

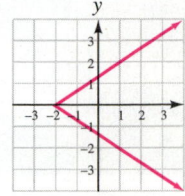

69.

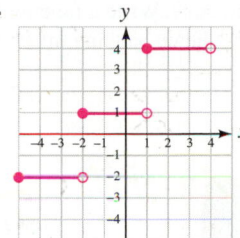

70.

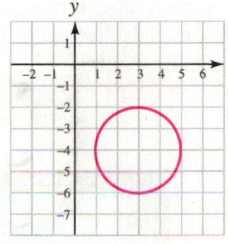

71.

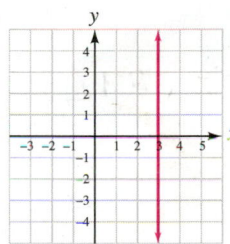

72.

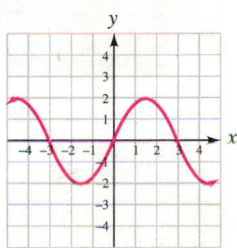

Graph each function using window settings of $[-4, 4]$ for x and $[-4, 4]$ for y. The graph is not what it appears to be. Pick a better viewing window and find a better representation of the true graph. See Using Your Calculator: Graphing Functions.

73. $f(x) = x^2 + 8$

74. $f(x) = x^3 - 8$

75. $f(x) = |x + 5|$

76. $f(x) = |x - 5|$

77. $f(x) = (x - 6)^2$

78. $f(x) = (x + 9)^2$

79. $f(x) = x^3 + 8$

80. $f(x) = x^3 - 12$

APPLICATIONS

81. Optics.   See the illustration. The **law of reflection** states that the angle of reflection is equal to the angle of incidence. What function studied in this section models the path of the reflected light beam with an angle of incidence measuring 45°?

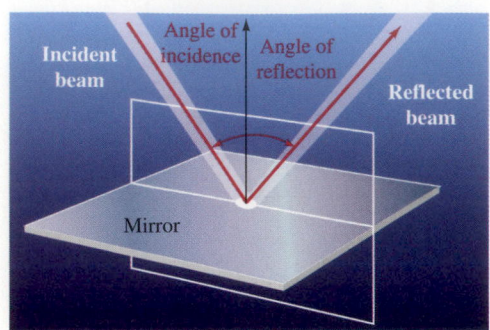

82. Billiards.   In the illustration, a rectangular coordinate system has been superimposed over a billiard table. Write a function that models the path of the ball that is shown banking off of the cushion.

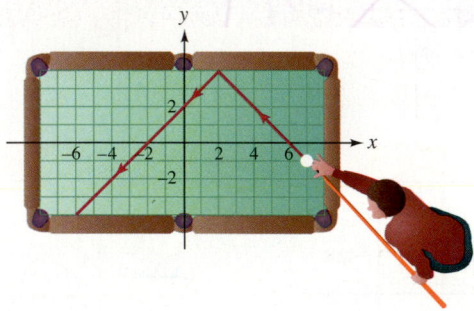

83. Center of Gravity.   See the illustration. As a diver performs a $1\frac{1}{2}$-somersault in the tuck position, her center of gravity follows a path that can be described by a graph shape studied in this section. What graph shape is that?

84. Earth's Atmosphere.   The illustration in the next column shows a graph of the temperatures of the atmosphere at various altitudes above Earth's surface. The temperature is expressed in degrees Kelvin, a scale widely used in scientific work.

a. Estimate the coordinates of three points on the graph that have an x-coordinate of 200.

b. Explain why this is not the graph of a function.

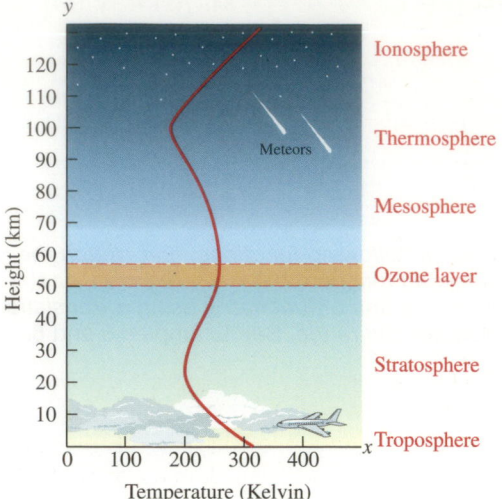

WRITING

85. Explain how to graph a function by plotting points.

86. Explain how to *project* the graph of a function onto the x-axis. Give an example.

87. What does it mean to translate a graph vertically? What does it mean to translate a graph horizontally?

88. What does it mean to reflect the graph of a function about the x-axis?

89. A student was asked to determine whether the graph shown below is the graph of a function. What is wrong with the following reasoning?

> *When I draw a vertical line through the graph, it intersects the graph only once. By the vertical line test, this is the graph of a function.*

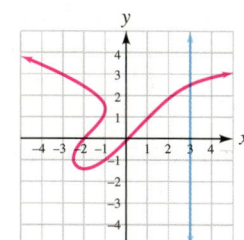

90. Explain why the graph of $g(x) = (x - 2)^2$ is two units to the right of the graph of $f(x) = x^2$.

REVIEW

Solve each formula for the indicated variable.

91. $T - W = ma$ for W

92. $a + (n - 1)d = l$ for n

93. $s = \frac{1}{2}gt^2 + vt$ for g

94. $e = mc^2$ for m

CHALLENGE PROBLEMS

Graph each function.

95. $f(x) = \begin{cases} |x| & \text{for } x \geq 0 \\ x^3 & \text{for } x < 0 \end{cases}$

96. $f(x) = \begin{cases} x^2 & \text{for } x \geq 0 \\ |x| & \text{for } x < 0 \end{cases}$

Find the domain and range of each function.

97. **a.**

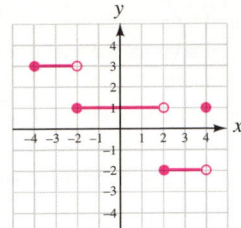

b.

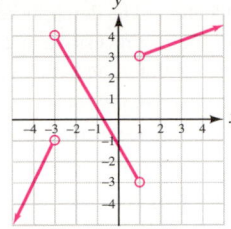

98. Light. Light beams coming from a bulb are reflected outward by a parabolic mirror as parallel rays.

a. The cross-section of a parabolic mirror is given by the function $f(x) = x^2$ for the following values of x: $-0.7, -0.6, -0.5, -0.4, -0.3, -0.2, -0.1, 0, 0.1, 0.2, 0.3, 0.4, 0.5, 0.6, 0.7$. Sketch the parabolic mirror using the following grid.

b. From the lightbulb filament at $(0, 0.25)$, draw a line segment representing a beam of light that strikes the mirror at $(-0.4, 0.16)$ and then reflects outward, parallel to the y-axis.

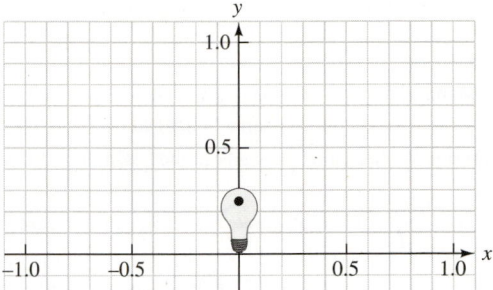

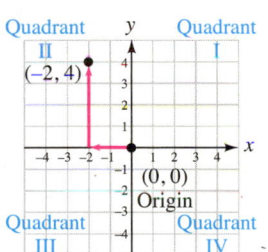

2 Summary & Review

SECTION 2.1 ▶ Graphs

DEFINITIONS AND CONCEPTS	EXAMPLES
A **rectangular coordinate** system is formed by two intersecting perpendicular number lines called the **x-axis** and the **y-axis.** The x- and y-axes divide the coordinate plane into four quadrants. The process of locating a point in the coordinate plane is called **plotting** or **graphing** that point.	Plot the point $(-2, 4)$. We begin at the origin, and move 2 units to the left along the x-axis and then 4 units up, and draw a dot.
Graphs can be used to visualize relationships between two quantities.	Refer to the **line graph** on page 104. We see that on day 6, the water level was 3 feet above normal. Refer to the **step graph** on page 105. We see that the cost to rent the ski equipment for 5 days is $140.
The **midpoint** of a line segment with endpoints (x_1, y_1) and (x_2, y_2) is the point with coordinates $$\left(\frac{x_1 + x_2}{2}, \frac{y_1 + y_2}{2} \right)$$	Find the midpoint of the segment joining $(-3, 7)$ and $(5, -8)$. We let $(x_1, y_1) = (-3, 7)$ and $(x_2, y_2) = (5, -8)$ and substitute the coordinates into the midpoint formula. $$\left(\frac{x_1 + x_2}{2}, \frac{y_1 + y_2}{2} \right) = \left(\frac{-3 + 5}{2}, \frac{7 + (-8)}{2} \right)$$ $$= \left(1, -\frac{1}{2} \right) \qquad \text{This is the midpoint.}$$

REVIEW EXERCISES

Plot each point on a rectangular coordinate system.

1. a. $(0, 3)$ **b.** $(-2, -4)$

c. $\left(\dfrac{5}{2}, -1.75\right)$ **d.** the origin

e. $(2.5, 0)$

2. Water Management. The given graph shows how the height of the water in a flood control channel changed over a 7-day period.

a. Describe the height of the water at the beginning of day 2.

b. By how much did the water level increase or decrease from day 4 to day 5?

c. During what time period did the water level stay the same?

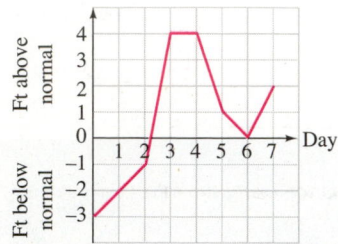

3. Auctions. The dollar increments used by an auctioneer during the bidding process depend on what initial price the auctioneer began with for the item. See the step graph.

a. What increments are used by the auctioneer if the bidding on an item began at $150?

b. If the first bid on an item being auctioned is $750, what will be the next price asked for by the auctioneer?

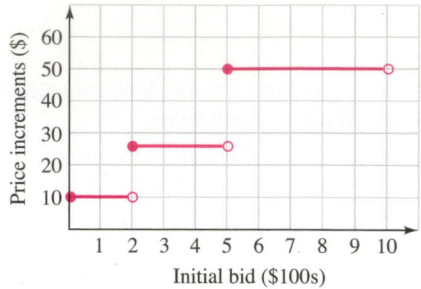

4. Find the midpoint of the segment joining $(8, -2)$ and $(6, -4)$.

SECTION 2.2 ▶ Graphing Linear Equations in Two Variables

DEFINITIONS AND CONCEPTS	EXAMPLES
A **solution of an equation in two variables** is an ordered pair of numbers that makes the equation a true statement when the pair is substituted for the variables.	Determine whether $(1, -5)$ is a solution of $3x - y = 8$. $3x - y = 8$ We substitute the coordinates into the given equation. $3(1) - (-5) \overset{?}{=} 8$ Substitute 1 for x and −5 for y. $3 + 5 \overset{?}{=} 8$ Evaluate the left side. $8 = 8$ True Since the result is true, $(1, -5)$ is a solution of the equation $3x - y = 8$.
The **standard form of a linear equation** in two variables is $Ax + By = C$, where A, B, and C are real numbers and A and B are not both zero. The graph of a linear equation in two variables is a line.	Linear equations: $y = \dfrac{2}{3}x - 5 \qquad 3x + 4y = -8 \qquad y = -\dfrac{5}{2} \qquad x = 3$
To graph linear equations solved for y by plotting points: **1.** Find three ordered pairs that are solutions of the equation by selecting three values for x and calculating the corresponding values of y. **2.** Plot the solutions on a rectangular coordinate system. **3.** Draw a straight line passing through the points. If the points do not lie on a line, check your calculations.	To graph $y = \dfrac{2}{3}x - 5$, we construct a table of solutions, plot the points, and draw the line. $y = \dfrac{2}{3}x - 5$ | x	y	(x, y)
---	---	---
-3	-7	$(-3, -7)$
0	-5	$(0, -5)$
6	-1	$(6, -1)$	 ↑ Select x. ↑ Find y. ↑ Plot (x, y). 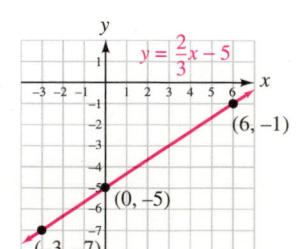

To find the **y-intercept** of a line, substitute 0 for x in the equation and solve for y. To find the **x-intercept** of a line, substitute 0 for y in the equation and solve for x.

Plotting the x- and y-intercepts of a graph and drawing a line through them is called the **intercept method for graphing a line.**

Use the y- and x-intercepts to graph $3x + 4y = -8$.

y-intercept: x = 0
$3x + 4y = -8$
$3(0) + 4y = -8$
$4y = -8$
$y = -2$

The y-intercept is $(0, -2)$.

x-intercept: y = 0
$3x + 4y = -8$
$3x + 4(0) = -8$
$3x = -8$
$x = -\dfrac{8}{3}$

The x-intercept is $\left(-\dfrac{8}{3}, 0\right)$.

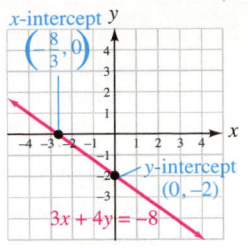

The graph of $x = a$ is a **vertical line** with x-intercept $(a, 0)$.

The graph of $y = b$ is a **horizontal line** with y-intercept $(0, b)$.

Graph: $x = 3$

Graph: $y = -\dfrac{5}{2}$

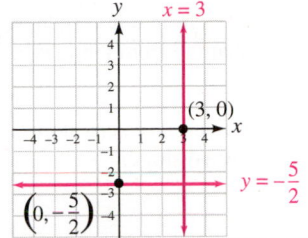

REVIEW EXERCISES

5. Is $(3, -6)$ a solution of $y = -5x + 9$?

6. The graph of a linear equation is shown in the illustration.

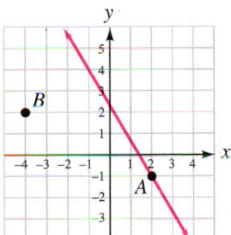

   a. If the coordinates of point A are substituted in the equation, will a true or false statement result?

   b. If the coordinates of point B are substituted in the equation, will a true or false statement result?

Complete each table.

7. $y = -3x$

x	y
-3	
0	
3	

8. $y = \dfrac{1}{2}x - \dfrac{5}{2}$

x	y
-3	
0	
3	

Graph each equation by plotting points.

9. $y = 3x + 4$

10. $y = -\dfrac{1}{3}x - 1$

Graph each equation using the intercept method.

11. $2x + y = 4$

12. $3x - 4y - 8 = 0$

Graph each equation.

13. $y = 4$

14. $x = -2$

15. Accounting. A recording studio purchased a new electronic console for \$82,800. For income tax purposes, company accountants will use the straight-line depreciation equation $y = -13,800x + 82,800$ to describe the declining value of the console.

   a. What will be the value of the console in 2 years?

   b. When will the console have no value?

16. Recycling. It takes more aluminum cans to weigh one pound than it used to because manufacturers continue to use thinner materials. The equation $n = 0.36t + 24.81$ gives the approximate number n of empty aluminum cans needed to weigh one pound, where t is the number of years since 1980. Graph the equation. (Source: The Aluminum Association)

   a. What information can be obtained from the n-intercept of the graph?

   b. From the graph, estimate the number of cans it took to weigh one pound in 2006.

SECTION 2.3 ▶ Rate of Change and the Slope of a Line

DEFINITIONS AND CONCEPTS	EXAMPLES
The **slope** m of a line is a ratio that compares the vertical and horizontal change as we move along the line from one point to another. $$m = \frac{\text{vertical change}}{\text{horizontal change}} = \frac{\text{rise}}{\text{run}} = \frac{\Delta y}{\Delta x}$$	Find the slope of the line. Pick two points on the line. Then create a **slope triangle** by moving up 5 grid squares and to the right 7 grid squares. $$m = \frac{\text{rise}}{\text{run}} = \frac{5}{7}$$ 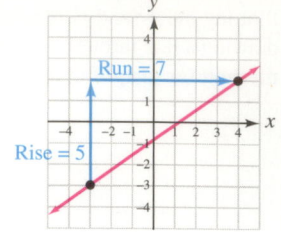
Lines that rise from left to right have a **positive slope,** and lines that fall from left to right have a **negative slope.** Horizontal lines have **zero slope** and vertical lines have **undefined slope.**	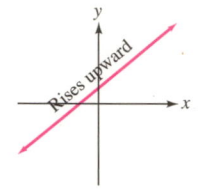 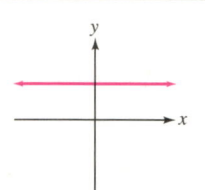 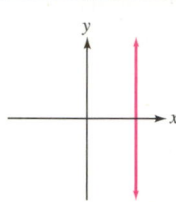
We also can find the slope of a line using the **slope formula:** $$m = \frac{y_2 - y_1}{x_2 - x_1} \qquad \text{if } x_1 \neq x_2$$	Find the slope of the line passing through $(-5, -2)$ and $(7, -14)$. $$m = \frac{y_2 - y_1}{x_2 - x_1} = \frac{-14 - (-2)}{7 - (-5)} = \frac{-12}{12} = -1$$
Parallel lines have the same slope. The slopes of two nonvertical **perpendicular lines** are negative reciprocals. The product of their slopes is -1.	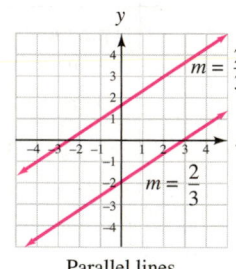 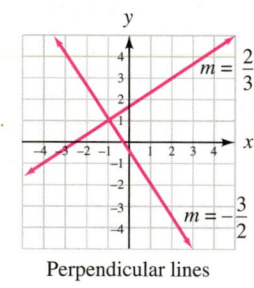 Parallel lines Perpendicular lines

REVIEW EXERCISES

17. Find the slope of lines l_1 and l_2.

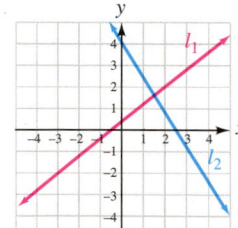

18. **Drug Stores.** The graph models the approximate number of Walgreens Drug Stores for the years 2001–2010. Find the rate of increase in the number of stores.

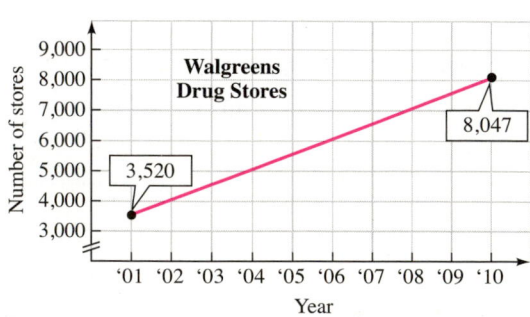

Source: walgreens.com

Find the slope of the line passing through the given points.

19. (2, 5) and (5, 8)   **20.** (3, −2) and (−6, 12)

21. (−2, 4) and (8, 4)   **22.** (−5, −4) and (−5, 8)

23. Determine whether a line that passes through (−2, 1) and (6, 5) is parallel or perpendicular to a line with slope −2.

24. Determine whether the lines are parallel, perpendicular, or neither:

A line passing through (7, 5) and (−1, 0)

A line passing through (2.2, 3.7) and (10.2, 8.7)

25. Ladders.   A health and safety manual recommends that a ladder should be set at *"a slope of 4 vertical to 1 horizontal"* for maximum stability. Refer to the illustration. How far away from the wall should the base of the ladder be positioned so that it is set at the proper slope?

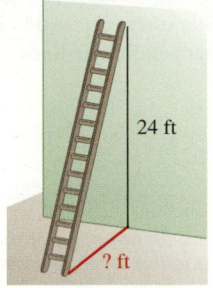

26. San Francisco.   According to the city Bureau of Engineering, the steepest street in San Francisco is 22nd Street between Church and Vicksburg. For a run of 200 feet, the street rises 63 feet. Find the grade of the street.

SECTION 2.4 ▶ **Writing Equations of Lines**

DEFINITIONS AND CONCEPTS	EXAMPLES
If a linear equation is written in **slope–intercept** form $$y = mx + b$$ the graph of the equation is a line with slope m and y-intercept $(0, b)$.	The equation of the line with slope $\frac{2}{3}$ and y-intercept $(0, 7)$ is: $$y = \frac{2}{3}x + 7$$
Slope can be used as an aid in graphing.	Find the slope and y-intercept of the line whose equation is $4x + 3y = 6$. Then graph the line. To find the slope and y-intercept, we solve the equation for y. $4x + 3y = 6$ This is the equation to graph. $3y = -4x + 6$ Subtract 4x from both sides. $y = -\frac{4}{3}x + 2$ To isolate y, divide both sides by 3. $m = -\frac{4}{3}$ and $b = 2$. The slope of the line is $-\frac{4}{3}$ and the y-intercept is $(0, 2)$. To graph the line, plot the y-intercept and then use the rise and run components of the slope to locate other points on the graph. 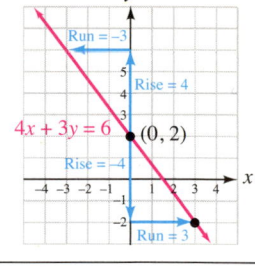
If a line with slope m passes through the point with coordinates (x_1, y_1), the equation of the line in **point–slope form** is $$y - y_1 = m(x - x_1)$$	Find an equation of the line with slope −5 that passes through (−1, 3). Write the equation in slope–intercept form. We substitute the slope and the coordinates of the point into the point–slope form. $y - y_1 = m(x - x_1)$ This is point–slope form. $y - 3 = -5[x - (-1)]$ Substitute. $y - 3 = -5(x + 1)$ Simplify within the brackets. $y - 3 = -5x - 5$ Distribute. $y = -5x - 2$ To isolate y, add 3 to both sides. This is slope–intercept form.

| Slopes can be used to identify **parallel** and **perpendicular lines.** | The graphs of the lines $y = 7x + 5$ and $y = 7x - 3$ are parallel because each line has slope 7.

The graphs of the lines $y = 6x + 1$ and $y = -\frac{1}{6}x$ are perpendicular because their slopes are 6 and $-\frac{1}{6}$, which are negative reciprocals. |

REVIEW EXERCISES

Find an equation of the line with the given properties. Write the equation in slope–intercept form.

27. Slope 3; passes through $(-8, 5)$

28. Passes through $(-2, 4)$ and $(6, -9)$

29. Passes through $(-3, -5)$; parallel to the graph of $3x - 2y = 7$

30. Passes through $(-3, -5)$; perpendicular to the graph of $3x - 2y = 7$

31. Write $3x + 4y = -12$ in slope–intercept form. Give the slope and y-intercept of the graph of the equation. Then use this information to graph the line.

32. Find the slope and the y-intercept of the line shown. Then write an equation of the line.

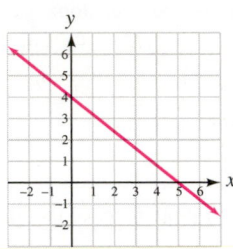

33. a. Write the equation of the x-axis
   b. Write the equation of the y-axis

34. Determine whether the graphs of $y = x + 15$ and $x + y = 4$ are parallel, perpendicular, or neither.

35. Business Growth.   City growth and the number of business licenses issued by a certain city are related by a linear equation. Records show that 250 licenses had been issued when the local population was 21,000, and that the rate of increase in the number of licenses issued was 1 for every 150 new residents. Use the variables p for population and L for the number of business licenses to write an equation (in slope–intercept form) that city officials can use to predict future business growth.

36. Depreciation.   A manufacturing company purchased a new diamond-tipped saw blade for $8,700 and will depreciate it on a straight-line basis over the next 5 years. At the end of its useful life, it will be sold for scrap for $100.

   a. Write a depreciation equation for the saw blade using the variables x and y.

   b. If the depreciation equation is graphed, explain the significance of the y-intercept.

SECTION 2.5 ▶ An Introduction to Functions

DEFINITIONS AND CONCEPTS	EXAMPLES
A **relation** is a set of ordered pairs. The set of first components is called the **domain** of the relation and the set of second components is called the **range.**	The relation $\{(2, 5), (7, -3), (4, 6)\}$ has domain $\{2, 4, 7\}$ and range $\{-3, 5, 6\}$.
A **function** is a set of ordered pairs (a relation) in which to each first component there corresponds exactly one second component. Since we often work with sets of ordered pairs of the form (x, y), it is helpful to define a function using the variables x and y: **y is a function of x:** Given a relation in x and y, if to each value of x in the domain there corresponds exactly one value of y in the range, then y is said to be a function of x. Since y depends on x, we call x the **independent variable** and y the **dependent variable.**	The relation $\{(2, 5), (7, -3), (4, 6)\}$ defines a function because to each first component there corresponds exactly one second component. The relation $\{(-9, 1), (3, 8), (0, 0), (3, 24)\}$ *does not* define a function because to the first component 3 there corresponds two second components, 8 and 24. The arrow diagram *does not* define y as a function of x because to the x-value 4 there corresponds more than one y-value: 2 and 6. The table defines a function and illustrates an important point: *Two different ordered pairs of a function can have the same y-value.* 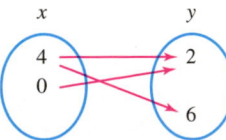

A function can be defined by an equation, however not all equations in two variables define functions.	The equation $y = 2x - 7$ defines y as a function of x because to each value of x there corresponds exactly one value of y. The equation $x = \lvert y \rvert$ does not define y as a function of x because more than one value of y corresponds to a single value of x. If x is 2, for example, the equation becomes $2 = \lvert y \rvert$ and y can be either 2 or -2.
The **function notation** $y = f(x)$ indicates that the variable y is a function of x. It is read as "f of x." Think of a function as a machine that takes some **input** x and turns it into some **output** $f(x)$, called a **function value**.	If $f(x) = 2x + 1$, find $f(-2)$ and $f(n + 1)$. $f(x) = 2x + 1$　　　　　$f(x) = 2x + 1$ $f(-2) = 2(-2) + 1$　　$f(n + 1) = 2(n + 1) + 1$ $\qquad = -4 + 1$　　　　　$\qquad = 2n + 2 + 1$ $\qquad = -3$　　　　　　　$\qquad = 2n + 3$ Thus, $f(-2) = -3$.　　　Thus, $f(n + 1) = 2n + 3$.
The input-output pairs that a function generates can be written as ordered pairs and plotted on a rectangular coordinate system to give the **graph of the function.** A **linear function** is a function that can be defined by an equation of the form $f(x) = mx + b$. The graph of a linear function is a straight line. Linear functions can be graphed using the **point-plotting method**, the **intercept method,** or the **slope–intercept method.**	To graph the linear function $f(x) = -4x - 2$, we make a table of values, plot the points, and draw the graph. 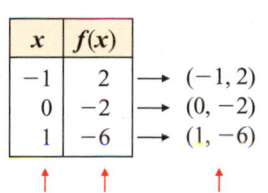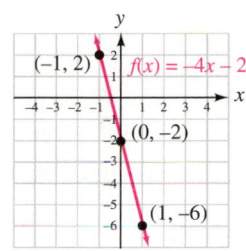 Select x.　Find f(x).　Plot the point.
The **domain** of a function is the set of input values. The **range** is the set of output values. To find the domain of a function defined by an equation, we must identify all the values of the input variable that produce a real-number output.	Find the domain of $f(x) = \dfrac{10}{x + 3}$. The number -3 cannot be substituted for x, because that would make the denominator equal to 0. Since any real number except -3 can be substituted for x, the domain is *the set of all real numbers except -3.*
To write equations of linear functions, we can use: **slope–intercept form:** $f(x) = mx + b$ **point–slope form:** $f(x) - y_1 = m(x - x_1)$	Write an equation for the linear function whose graph has slope -4 and passes through $(1, 8)$. $f(x) - y_1 = m(x - x_1)$　　This is point–slope form. $f(x) - 8 = -4(x - 1)$　　Substitute for m, y₁, and x₁. $f(x) - 8 = -4x + 4$　　　Distribute. $f(x) = -4x + 12$　　　　Solve for f(x).

37. Find the domain and the range of the relation:
$\{(-4, 0), (5, 16), (2, -2), (-1, -2)\}$

38. Fill in the blanks.

　a. A _____ is a set of ordered pairs (a relation) in which to each first component there corresponds exactly one second component.

　b. Given a relation in x and y, if to each value of x in the domain there corresponds exactly one value of y in the range, y is said to be a _____ of x. We call x the independent _____ and y the _____ variable.

Determine whether the relation defines y as a function of x. If it does not, find two ordered pairs where more than one value of y corresponds to a single value of x.

39. a.

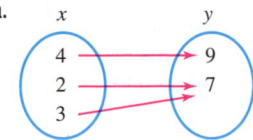

　b.

x	y
-1	8
0	5
4	1
-1	9

　c. $\{(14, 6), (-1, 14), (6, 0), (-3, 8)\}$

40. For the given input, what value will the function machine output?

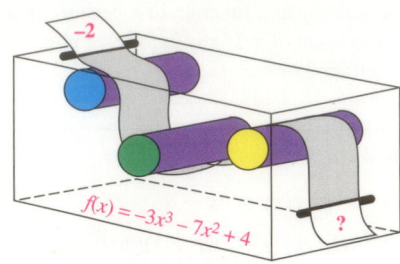

$f(x) = -3x^3 - 7x^2 + 4$

Determine whether each equation defines y as a function of x. If it does not, find two ordered pairs where more than one value of y corresponds to a single value of x.

41. $y = 6x - 4$

42. $y = 4 - x^2$

43. $y^2 = x$

44. $|y| = x + 1$

Let $f(x) = 3x + 2$ and $g(a) = \frac{a^2 - 4a + 4}{2}$. Find each function value.

45. $f(-3)$

46. $g(8)$

47. $g(-2)$

48. $f(t + 2)$

49. Let $f(x) = -5x + 7$. For what value of x is $f(x) = -8$?

50. Let $g(t) = \frac{3}{4}t - 1$. For what value of t is $g(t) = 0$?

Find the domain of each function.

51. $f(x) = 4x - 1$

52. $s(t) = t^2 + 1$

53. $h(x) = \dfrac{4}{2 - x}$

54. $g(b) = \dfrac{12}{5b + 25}$

55. What are the slope and the y-intercept of the graph of $g(x) = -2x - 16$?

56. Graph: $f(x) = \frac{2}{3}x - 2$

Write an equation for a linear function whose graph has the given characteristics.

57. Slope: $\dfrac{9}{10}$, y-intercept: $\left(0, \dfrac{7}{8}\right)$

58. Slope: $\dfrac{1}{5}$, passes through $(10, 1)$

59. Horizontal, passes through $(-8, 12)$

60. Passes through $(2, 5)$, parallel to the graph of $g(x) = 4x - 7$.

61. Passes through $(-6, 3)$, perpendicular to the graph of $g(x) = -3x - 12$.

62. Electricity. The relationship between the electrical resistance R of a coil of wire and its temperature t can be modeled with a linear function.

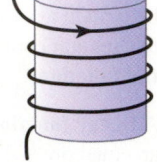

   a. Use the following data in the table to write a function that describes this relationship.

   b. Use the answer to part (a) to predict the resistance if the temperature of the coil of wire is 100° Celsius.

t **(in degrees Celsius)**	10	30
R **(in milliohms)**	5.25	5.65

SECTION 2.6 ▶ **Graphs of Functions**

DEFINITIONS AND CONCEPTS	EXAMPLES
From the graph of a function, we can determine function values. For the example, the value of $f(2)$ is given by the y-coordinate of a point on the graph of f with x-coordinate 2.	To find $f(2)$ from the graph, draw a vertical line through 2 on the x-axis. It intersects the graph at $(2, 3)$. Therefore, 3 corresponds to 2 and it follows that $f(2) = 3$. 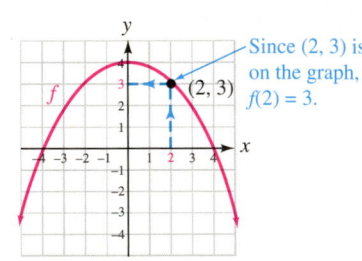
These three basic functions are used so often in algebra that you should memorize their names and their graphs. They are called **nonlinear functions** because their graphs are not lines.	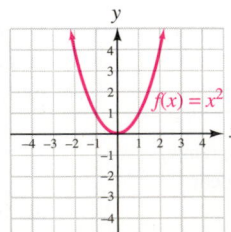 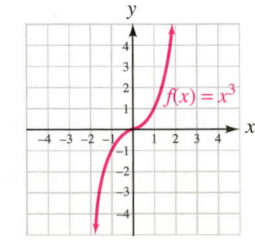 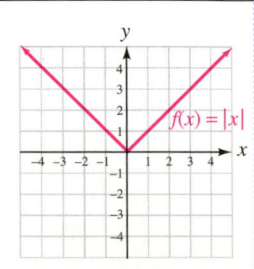 The squaring function The cubing function The absolute value function

We can find the domain and range of a function from its graph. The **domain of a function** is the projection of its graph onto the *x*-axis. **The range of a function** is the projection of its graph onto the *y*-axis.

Find the domain and range of function *f*.

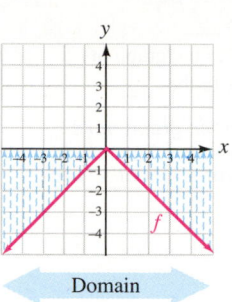

Domain

D: The set of real numbers

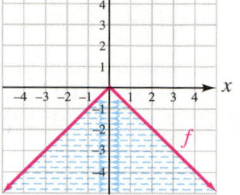

Range

R: The set of nonpositive real numbers

A **vertical translation** shifts a graph upward or downward. A **horizontal translation** shifts a graph left or right. A **reflection** "flips" a graph about the *x*-axis.

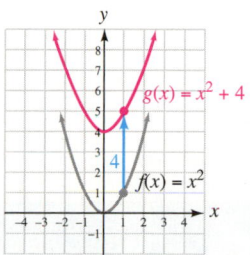

To graph $g(x) = x^2 + 4$, translate each point on the graph of $f(x) = x^2$ up 4 units.

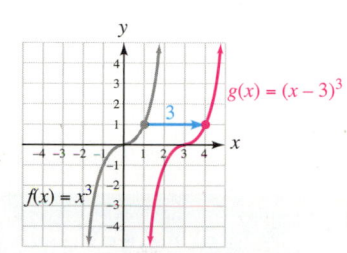

To graph $g(x) = (x - 3)^3$, translate each point on the graph of $f(x) = x^3$ to the right 3 units.

The **vertical line test:** If a vertical line intersects a graph in more than one point, the graph is not the graph of a function.

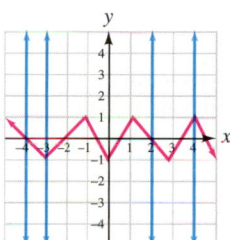

The graph of a function

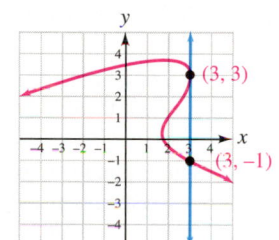

Not the graph of a function

63. Use the graph to find each value.

   a. $f(-2)$

   b. $f(3)$

   c. The value of *x* for which $f(x) = 0$.

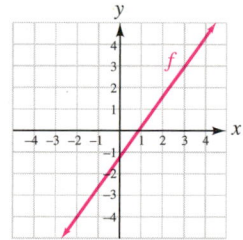

64. Use the graph to find each value.

   a. $g(0)$

   b. $g(-3)$

   c. The value(s) of *x* for which $g(x) = -4$.

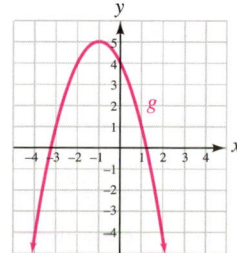

Give the domain and range of each function graphed below.

65.

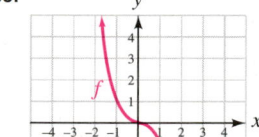

66.

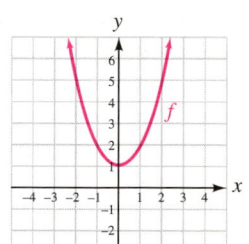

67. Graph $f(x) = |x + 2|$ by creating a table of function values and plotting points. Give the domain and range of the function.

68. Fill in the blanks.

   a. The graph of $f(x) = x^2 + 6$ is the same as the graph of
      $f(x) = x^2$ except that it is shifted ___ units ___.

   b. The graph of $f(x) = (x + 6)^2$ is the same as the graph of
      $f(x) = x^2$ except that it is shifted ___ units to the ____.

*For each of the following functions, first sketch the graph of its
associated function, $f(x) = x^2$, $f(x) = x^3$, or $f(x) = |x|$. Then draw
the graph of function g using a translation and/or a reflection and
give its domain and range.*

69. $g(x) = x^2 - 3$

70. $g(x) = |x - 4|$

71. $g(x) = (x - 2)^3 + 1$

72. $g(x) = -x^3$

*Determine whether each graph is the graph of a function. If
it is not, find two ordered pairs where more than one value of
y corresponds to a single value of x.*

73.
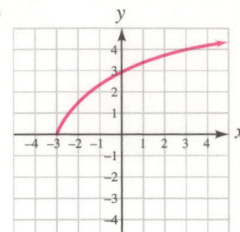

74.
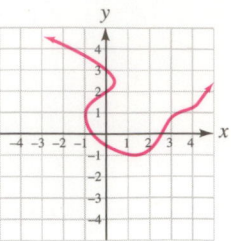

2 CHAPTER TEST

1. Fill in the blanks.

   a. Ordered pairs of numbers can be graphed on a _____
      coordinate system.

   b. A _____ is a set of ordered pairs (a relation) in which to
      each first component there corresponds exactly one second
      component.

   c. Given a relation in x and y, if to each value of x in the
      domain there corresponds exactly one value of y in the range,
      y is said to be a _____ of x. We call x the independent
      _____ and y the _____ variable.

   d. For a function, the set of all possible values that can be used
      for the independent variable is called the _____. The set of
      all values of the dependent variable is called the _____.

   e. A shift of the graph of a function to the left or right is called
      a horizontal _____.

2. **Sleepiness.** Refer to the following graph. It shows how the
   sleepiness/alertness levels of a college student change during the
   hours of the day that he is awake.

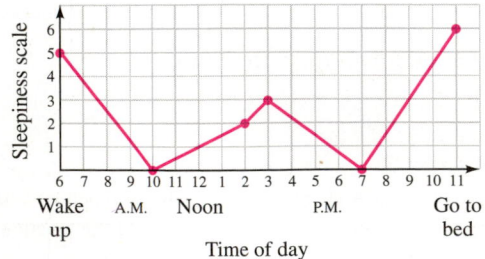

   a. At what time is the student the sleepiest?

   b. When is the student most alert?

   c. After awaking, how long does it take the student to reach
      maximum alertness?

   d. During what time periods is the student getting
      sleepy?

   e. During the middle of the day, when is the student the least
      alert?

3. **Rentals.** Use the information in the graph below to answer the
   following questions.

   a. Find the cost of renting the trailer for 2 days.

   b. Find the cost of renting the trailer for $5\frac{1}{2}$ days.

   c. For how long can you rent the trailer if you have $50?

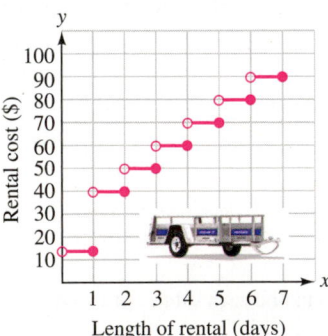

4. Find the coordinates of the midpoint of the line segment joining
   $(-2, -5)$ and $(7, -11)$.

5. Graph: $y = -\frac{2}{3}x - 2$

6. Find the x- and y-intercepts of the graph of $2x - 3y = 10$. Then
   graph the equation.

7. Graph: $y = -2$

8. Find the rate of change of the temperature for the period of time shown in the graph below.

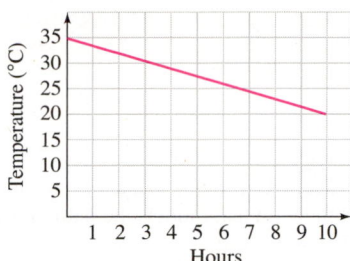

Find the slope of each line, if possible.

9. The line through $(-2, 4)$ and $(6, 8)$

10. The graph of $2x - 3y = 8$

11. a. The graph of $x = 12$

b. The graph of $y = 12$

12. Write an equation of the line graphed here. Give the answer in slope–intercept form.

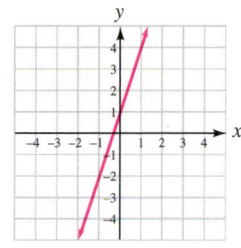

13. Find an equation of the line that passes through $(-2, 6)$ and $(-4, -10)$. Write the equation in slope–intercept form.

14. Find the slope and the y-intercept of the graph of $4x - 25 = -5y$. Then graph the line.

15. Determine whether the graphs of $y + 4x = 12$ and $y = \frac{1}{4}x + 3$ are parallel, perpendicular, or neither.

16. Find an equation of the line that passes through the origin and is parallel to the graph of $y = -\frac{3}{2}x - 7$. Write the equation in slope–intercept form.

17. Accounting. After purchasing a new color copier, a business owner had his accountant prepare a depreciation worksheet for tax purposes. (See the illustration below.)

a. Assuming straight-line depreciation, write an equation that gives the value v of the copier after x years of use.

b. If the depreciation equation is graphed, explain the significance of its v-intercept.

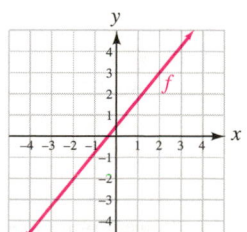

Depreciation Worksheet

Color copier $4,000
(new)

Salvage value $400
(in 6 years)

18. Pets. The relationship between the number n of dog licenses issued by a certain city and its population p can be modeled with a linear equation. When the population was 20,000, the number of licenses issued was 2,500. The rate of increase in the number of licenses issued is 11 for every 200 new residents. Find the linear equation model. Answer in slope–intercept form.

19. Determine whether the relation defines y to be a function of x. If it does not, explain why.

a.

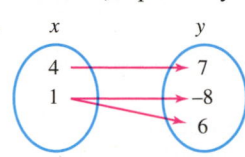

b.

x	y
-3	4
4	-3
1	5
5	1

c. $\left\{\left(0, \frac{1}{2}\right), \left(-10, \frac{1}{2}\right), \left(-20, \frac{1}{2}\right), \left(-30, \frac{1}{2}\right)\right\}$

d. $|y| = x$

20. Find the domain of $f(x) = \frac{15}{12 - 2x}$.

21. Let $f(x) = -\frac{4}{5}x - 12$. For what value of x is $f(x) = 4$?

22. Determine the slope and y-intercept of the graph of $f(x) = 8x - 9$.

23. Write an equation for the linear function whose graph passes through $(2, 0)$ and is perpendicular to the graph of $g(x) = \frac{4}{5}x + \frac{1}{9}$.

24. Vehicle Performance. The average fleet-wide performance of light-duty trucks has made constant improvement since 1980. For example, the time for the average 1980 model light-duty truck to accelerate from 0 to 60 mph was about 14.5 seconds. For 2005 models, that time was only 10.5 seconds. (Source: United States EPA)

a. Let m be the model year of a light-duty truck and T be the time in seconds for it to accelerate from 0 to 60 mph. Write a linear function $T(m)$ to model the situation.

b. Use your answer to part a to predict the time it will take the average light-duty truck to accelerate from 0 to 60 mph in 2015, if the trend continues.

Let $f(x) = 3x + 1$ and $g(t) = t^2 - 2t + 1$. Find each value.

25. $f(3)$

26. $g(-6)$

27. $g\left(\frac{1}{4}\right)$

28. $f(r + 8)$

Refer to the graph of function f here.

29. Find $f(-2)$.

30. Find the value of x for which $f(x) = 3$.

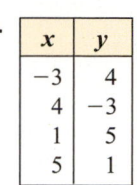

Determine whether each graph is the graph of a function.
If it is not, explain why.

31.

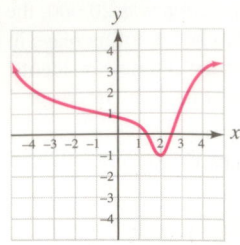

32.

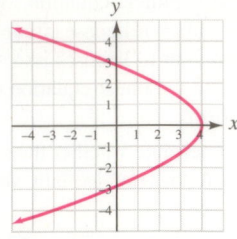

33. Graph $f(x) = |x| + 3$ by creating a table of function values and plotting points. Give the domain and range of the function.

34. Draw the graph of $g(x) = (x - 4)^3 + 1$ using a translation by first sketching the graph of its associated function. Then give the domain and range of function g.

35. Give the domain and range of function f graphed here.

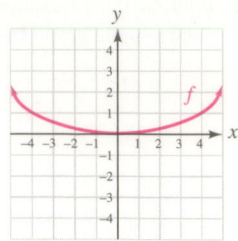

36. Drinking Fountains. Use the concepts of translation and reflection to write a function that models the path of the stream of water shown in the illustration.

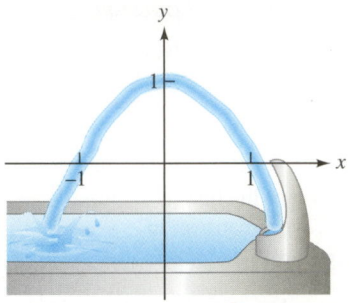

Group Project

Measuring Slope

Overview: This hands-on activity will give you a better understanding of slope.

Instructions: Form groups of 2 or 3 students. Use a ruler and a level to find the slopes of ramps or inclines on campus by measuring $\frac{\text{rise}}{\text{run}}$, as shown in the illustration. Record your results in a table, listing the slopes in order from smallest to largest.

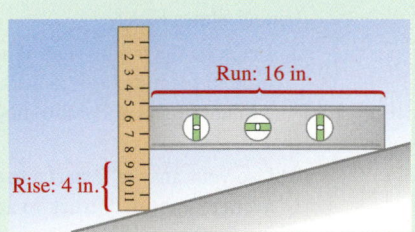

Object/location	Slope		
Ramp outside the cafeteria	$\dfrac{\text{Rise}}{\text{Run}}$	$= \dfrac{4 \text{ in.}}{16 \text{ in.}}$	$= \dfrac{1}{4}$

CUMULATIVE REVIEW ▶▶ Chapters 1–2

1. List the elements of $\{-2,\ 0,\ 1,\ 2,\ \frac{13}{12},\ 6,\ 7,\ \sqrt{5},\ \pi\}$ that belong to the following sets. [Section 1.2]

   a. Natural numbers

   b. Whole numbers

   c. Rational numbers

   d. Irrational numbers

   e. Negative numbers

   f. Real numbers

   g. Prime numbers

   h. Even numbers

2. Graph each element of the following set of numbers on a number line. [Section 1.2]

$$\left\{-1\tfrac{1}{2},\ -\pi,\ -\tfrac{35}{8},\ -0.333\ldots,\ 3,\ 4.25,\ \sqrt{2}\right\}$$

Evaluate each expression.

3. $\dfrac{|-5| + |-3|}{-|4|}$ [Section 1.3]

4. $7 - 12[7^2 - 4(2 - 5)^2]$ [Section 1.3]

5. $-\dfrac{16}{5} \div \left(-\dfrac{10}{3}\right)$ [Section 1.3]

6. $\dfrac{(9-8)^4+21}{3^3-\left(\sqrt{16}\right)^2}$ [Section 1.3]

Evaluate each expression for x = 2 and y = −3.

7. $-y-5xy$ [Section 1.3]

8. $\dfrac{x^2-y^2}{2x+y}$ [Section 1.3]

9. Determine which property of real numbers justifies each statement. [Section 1.4]

   a. $(a+b)+c=a+(b+c)$

   b. $3(x+y)=3x+3y$

   c. $(a+b)+c=c+(a+b)$

   d. $(ab)c=a(bc)$

10. a. What is the additive inverse (opposite) of 6? [Section 1.4]

   b. What is the multiplicative inverse (reciprocal) of $-\frac{1}{8}$? [Section 1.4]

Simplify each expression.

11. $-7s(-4t)(-1)$ [Section 1.4]

12. $40\left(\dfrac{3}{8}y-\dfrac{1}{4}\right)+40\left(\dfrac{4}{5}\right)$ [Section 1.4]

13. $-\dfrac{3}{4}s-\dfrac{1}{3}s$ [Section 1.4]

14. $-5[3(x-4)-2(x+2)]-7(x-3)$ [Section 1.4]

Solve each equation, if possible.

15. $-\dfrac{9}{8}s=3$ [Section 1.5]

16. $4(y-3)+4=-3(y+5)$ [Section 1.5]

17. $2x-\dfrac{3(x-2)}{2}=7-\dfrac{x-3}{3}$ [Section 1.5]

18. $0.04(24)+0.02x=0.04(12+x)$ [Section 1.5]

19. $-3x=-2x+1-(5+x)$ [Section 1.5]

20. $2[5(4-a)+2(a-1)]=3-a$ [Section 1.5]

Solve each formula for the specified variable.

21. $-Tx+3By=c$ for B [Section 1.6]

22. $A=\dfrac{1}{2}h(b_1+b_2)$ for h [Section 1.6]

23. **Collecting Signatures.** A student working for a political campaign earns $45 per day plus $1.25 for each signature on a petition she obtains. How many signatures must she collect to earn $250 a day? [Section 1.7]

24. **Isosceles Triangles.** The measure of each base angle of an isosceles triangle is 10° less than twice the measure of the vertex angle. Find the measure of each angle of the triangle. [Section 1.7]

25. **Investments.** A woman invested part of $20,000 at 6% and the rest at 7%. If the annual interest earned was $1,260, how much did she invest at 6%? [Section 1.8]

26. **Driving Rates.** John drove to a distant city in 5 hours. When he returned, there was less traffic, and the trip took only 3 hours. If he drove 26 mph faster on the return trip, how fast did he drive each way? [Section 1.8]

27. Find the midpoint of a line segment with endpoints $(-5,-2),(7,3)$. [Section 2.1]

28. Determine whether $\left(-1,\frac{25}{3}\right)$ is a solution of $3y-5x=30$. [Section 2.2]

29. Graph: $7x-3y=6$ [Section 2.2]

30. Find the slope of the line graphed here. [Section 2.3]

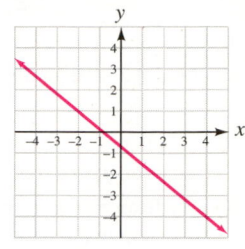

31. Find an equation of the line passing through $(-2,5)$ and $(8,-9)$. Write the equation in slope–intercept form. [Section 2.4]

32. Find an equation of the line passing through $(-2,3)$ and parallel to the graph of $3x+y=8$. Write the equation in slope–intercept form. [Section 2.4]

33. **Traffic Signals.** City growth and the number of traffic signals for a certain city are related by a linear equation. Records show that there were 50 traffic signals when the local population was 25,000 and that the rate of increase in the number of traffic signals was 1 for every 1,000 new residents.

   a. Using the variables p for population and T for traffic signals, write an equation (in slope–intercept form) that the transportation department can use to predict future traffic signal needs. [Section 2.4]

   b. How many traffic signals can be expected when the population reaches 35,000? [Section 2.4]

34. Fill in the blank: Given a relation in x and y, if to each value of x in the domain there corresponds exactly one value of y in the range, then y is said to be a _____ of x. [Section 2.5]

Refer to the following graph of function f.

35. a. Find $f(1)$. [Section 2.5]

   b. Find the value of x for which $f(x)=1$. [Section 2.5]

36. Find the domain and range of the function.

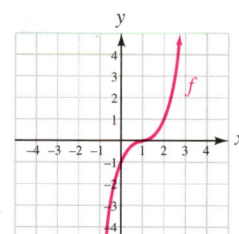

37. Find the domain and range of the relation $\{(1,-12),(-6,7),(5,8),(0,7),(0,4)\}$. Is it a function? [Section 2.5]

38. Does $y^2=x$ define y as a function of x? If it does not, find two ordered pairs where more than one value of y corresponds to a single value of x. [Section 2.5]

39. Let $h(x)=-\frac{1}{5}x-12$. For what value of x is $h(x)=0$? [Section 2.5]

40. Determine whether the graph represents a function. [Section 2.6]

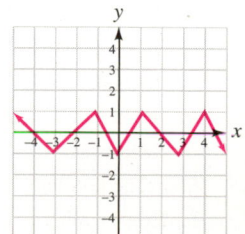

Let $f(x) = 3x^2 + 2$ and $g(x) = -2x - 1$. Find each function value.

41. $f(-1)$ [Section 2.5]

42. $g(0)$ [Section 2.5]

43. $g(-2)$ [Section 2.5]

44. $f(-r)$ [Section 2.5]

45. Find the domain of the function $f(x) = \frac{5}{x+1}$. [Section 2.5]

46. What are the slope and y-intercept of the graph of the linear function $f(x) = 6x + 15$? [Section 2.5]

47. Write an equation for the linear function whose graph is horizontal and passes through $\left(\frac{3}{2}, -\frac{7}{8}\right)$. [Section 2.5]

48. Driving. The total number of miles driven each day in the United States on all roads by all vehicles since the year 1980 can be modeled by a linear function. In 1990, approximately 5.76 billion vehicle miles were driven daily. In 2005, approximately 8.31 billion vehicle miles were driven daily. (Source: FHWA Highway Statistics) [Section 2.5]

   a. Let t be the number of years after 1980 and M be the number of billion vehicle miles driven daily. Write a linear function $M(t)$.

   b. Use your answer to part a to estimate how many billion vehicle miles will be driven daily in 2020.

49. Graph $f(x) = -x^2 + 1$ by plotting points. Then give the domain and range of the function. [Section 2.6]

50. First sketch the graph of the basic function associated with $g(x) = |x - 3| - 4$. Then draw the graph of function g using a translation. Give the domain and range of function g. [Section 2.6]

Systems of Equations

3

> from **Campus to Careers**

Fashion Designer

Fashion designers help create the billions of clothing articles, shoes, and accessories purchased every year by consumers. Fashion design relies heavily on mathematical skills, including knowledge of lines, angles, curves, and measurement. Designers also use mathematics in the manufacturing and marketing parts of the industry as they calculate labor costs and determine the markups and markdowns involved in retail pricing.

Problem 65 in **Study Set 3.1**, **problem 17** in **Study Set 3.6**, and **problem 11** in **Study Set 3.7** involve situations that a fashion designer might encounter on the job. The mathematical concepts discussed in this chapter can be used to solve those problems.

JOB TITLE:
Fashion Designer

EDUCATION:
Employers usually seek designers with a 2-year or 4-year degree who are knowledgeable about textiles, fabrics, ornaments, and fashion trends.

JOB OUTLOOK:
The best opportunities will be in designing clothing sold in department stores and retail chains.

ANNUAL EARNINGS:
The average annual income is $74,100.

FOR MORE INFORMATION:
www.bls.gov/oco/ocos291.htm

Taking a math test doesn't have to be an unpleasant experience. Here are some suggestions that can make it more enjoyable and also improve your score.

PREPARING FOR THE TEST: Begin studying several days before the test rather than cramming your studying into one marathon session the night before.

TAKING THE TEST: Follow a test-taking strategy so that you can maximize your score by using the testing time wisely.

EVALUATING YOUR PERFORMANCE: After your graded test is returned, classify the types of errors that you made on the test so that you do not make them again.

Now Try This ▶

1. Write a study session plan that explains how you will prepare on each of the four days before the test, as well as on test day. For some suggestions, see *Preparing for a Test**.

2. Develop your own test-taking strategy by answering the survey questions found in *How to Take a Math Test**.

3. Use the outline found In *Analyzing Your Test Results** to classify the errors that you made on your most recent test.

* Found online at: www.cengage.com/math/tussy

SECTION 3.1

Solving Systems of Equations by Graphing

OBJECTIVES

1 Determine whether an ordered pair is a solution of a system.

2 Solve systems of linear equations by graphing.

3 Use graphing to identify inconsistent systems and dependent equations.

4 Solve equations graphically.

ARE YOU READY?

▼ *The following problems review some basic skills that are needed when solving systems of equations by graphing.*

1. Is $(-1, 4)$ a solution of $y = 3x - 1$?
2. Use the slope and the y-intercept to graph $y = -4x + 2$.
3. Graph $3x + 4y = 12$ by finding the x- and y-intercepts.
4. Without graphing, determine whether the graphs of $y = \dfrac{1}{3}x - 2$ and $x - 3y = 2$ are parallel, perpendicular, or neither.

The red line in the graph on the next page shows the cost for a company to produce a given number of skateboards. The blue line shows the revenue the company will receive for selling a given number of those skateboards. The graph offers the company important financial information.

- *For the portion of the graph where the red line is above the blue line:* The production costs exceed the revenue earned when fewer than 400 skateboards are sold. In this case, the company loses money.
- *For the portion of the graph where the blue line is above the red line:* The revenue earned exceeds the production costs when more than 400 skateboards are sold. In this case, the company makes a profit.
- Production costs equal revenue earned if exactly 400 skateboards are sold. This fact is indicated by the point of intersection of the two lines, (400, 20,000), which is called the **break-even point.**

©Brooke Whatnall/Shutterstock.com

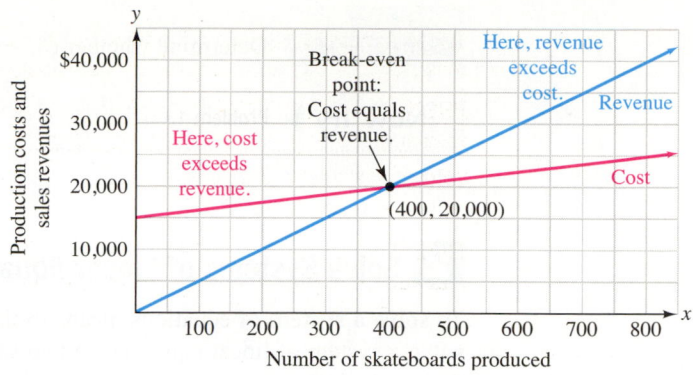

From Chapter 2, we know that the lines in the graph that show the cost and the revenue can be modeled by linear equations in two variables. Together, such a set of equations is called a *system of equations*.

In general, when two equations with the same variables are considered simultaneously (at the same time), we say that they form a **system of equations.** We will use a left brace { when writing a system of equations. An example is

$$\begin{cases} 2x + 5y = -1 \\ x - y = -4 \end{cases}$$ Read as "the system of equations $2x + 5y = -1$ and $x - y = -4$."

1   Determine Whether an Ordered Pair Is a Solution of a System.

A **solution of a system** of equations in two variables is an ordered pair that satisfies both equations of the system.

EXAMPLE 1   Determine whether $(-3, 1)$ is a solution of each system of equations.

a. $\begin{cases} 2x + 5y = -1 \\ x - y = -4 \end{cases}$ **b.** $\begin{cases} 5y = 2 - x \\ y = 3x \end{cases}$

Strategy   We will substitute the x- and y-coordinates of $(-3, 1)$ for the corresponding variables in both equations of the system.

Why   If both equations are satisfied (made true) by the x- and y-coordinates, the ordered pair is a solution of the system.

Solution   **a.** To determine whether $(-3, 1)$ is a solution, we substitute -3 for x and 1 for y in each equation.

Check:

$2x + 5y = -1$ **First equation**	$x - y = -4$ **Second equation**
$2(\mathbf{-3}) + 5(\mathbf{1}) \overset{?}{=} -1$	$\mathbf{-3} - \mathbf{1} \overset{?}{=} -4$
$-6 + 5 \overset{?}{=} -1$	$-4 = -4$ **True**
$-1 = -1$ **True**	

Since $(-3, 1)$ satisfies both equations, it is a solution of the system.

b. We substitute -3 for x and 1 for y in each equation in the second system.

Check:

$5y = 2 - x$ **First equation**	$y = 3x$ **Second equation**
$5(\mathbf{1}) \overset{?}{=} 2 - (\mathbf{-3})$	$\mathbf{1} \overset{?}{=} 3(\mathbf{-3})$
$5 \overset{?}{=} 2 + 3$	$1 = -9$ **False**
$5 = 5$ **True**	

Although $(-3, 1)$ satisfies the first equation, it does not satisfy the second. Because it does not satisfy both equations, $(-3, 1)$ is *not a solution* of the system.

2 Solve Systems of Linear Equations by Graphing.

To **solve a system** of equations means to find all of the solutions of the system. One way to solve a system of linear equations in two variables is to graph each equation and find where the graphs intersect.

The Graphing Method	▼
	1. Carefully graph each equation on the same rectangular coordinate system.
	2. If the lines intersect, determine the coordinates of the point of intersection of the graphs. That ordered pair is the solution of the system.
	3. If the graphs have no point in common, the system has no solution.
	4. Check the proposed solution in each equation of the original system.

A system of two linear equations can have exactly one solution, no solution, or infinitely many solutions. When a system of equations (as in Example 2) has at least one solution, the system is called a **consistent system.**

EXAMPLE 2   Solve the system by graphing: $\begin{cases} x + 2y = 4 \\ 2x - y = 3 \end{cases}$

Strategy   We will graph both equations on the same coordinate system.

Why   The graph of a linear equation is a picture of its solutions. If both equations are graphed on the same coordinate system, we can see whether they have any common solutions.

Solution   The intercept method is a convenient way to graph equations such as $x + 2y = 4$ and $2x - y = 3$, because they are in standard $Ax + By = C$ form.

$x + 2y = 4$

x	y	(x, y)
4	0	$(4, 0)$
0	2	$(0, 2)$
-2	3	$(-2, 3)$

$2x - y = 3$

x	y	(x, y)
$\frac{3}{2}$	0	$\left(\frac{3}{2}, 0\right)$
0	-3	$(0, -3)$
-1	-5	$(-1, -5)$

Although infinitely many ordered pairs (x, y) satisfy $x + 2y = 4$, and infinitely many ordered pairs (x, y) satisfy $2x - y = 3$, only the coordinates of the point where the graphs intersect satisfy both equations. From the graph, it appears that the intersection point has coordinates $(2, 1)$. To verify that it is the solution, we substitute 2 for x and 1 for y in both equations and show that $(2, 1)$ satisfies each one.

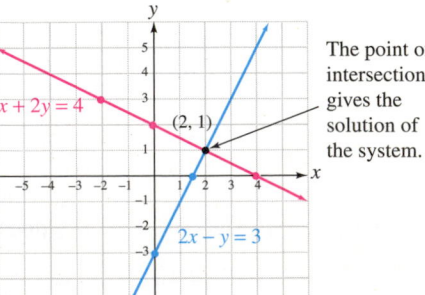

The point of intersection gives the solution of the system.

Check:

$$x + 2y = 4 \quad \text{First equation}$$
$$2 + 2(1) \stackrel{?}{=} 4$$
$$2 + 2 \stackrel{?}{=} 4$$
$$4 = 4 \quad \text{True}$$

$$2x - y = 3 \quad \text{Second equation}$$
$$2(2) - 1 \stackrel{?}{=} 3$$
$$4 - 1 \stackrel{?}{=} 3$$
$$3 = 3 \quad \text{True}$$

Since (2, 1) makes both equations true, it is the solution of the system. The solution set is {(2, 1)}.

> **Self Check 2** Solve the system by graphing: $\begin{cases} x - 3y = -5 \\ 2x + y = 4 \end{cases}$
>
> **Now Try** ▶ Problem 21

3 Use Graphing to Identify Inconsistent Systems and Dependent Equations.

When a system has no solution (as in Example 3), it is called an **inconsistent system.**

EXAMPLE 3 Solve the system $\begin{cases} 2x + 3y = 6 \\ 4x + 6y = 24 \end{cases}$ by graphing, if possible.

Strategy We will graph both equations on the same coordinate system.

Why If both equations are graphed on the same coordinate system, we can see whether they have any common solutions.

Solution Using the intercept method, we graph both equations on one set of coordinate axes, as shown on the right.

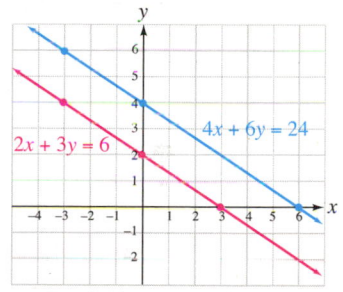

$2x + 3y = 6$

x	y	(x, y)
3	0	(3, 0)
0	2	(0, 2)
−3	4	(−3, 4)

$4x + 6y = 24$

x	y	(x, y)
6	0	(6, 0)
0	4	(0, 4)
−3	6	(−3, 6)

Parallel lines — no solution

> **Caution**
>
> A common error is to graph the parallel lines, but forget to answer with the words
>
> **no solution**

In this example, the graphs are parallel, because the slopes of the two lines are equal and they have different y-intercepts. We can see that the slope of each line is $-\frac{2}{3}$ by writing the equations in slope–intercept form. To do that, we solve each for y.

$$2x + 3y = 6 \quad \text{First equation}$$
$$3y = -2x + 6$$
$$y = -\frac{2}{3}x + 2 \quad \text{Divide both sides by 3 and simplify.}$$

$$4x + 6y = 24 \quad \text{Second equation}$$
$$6y = -4x + 24$$
$$y = -\frac{2}{3}x + 4 \quad \text{Divide both sides by 6 and simplify.}$$

Because the lines are parallel, there is no point of intersection. Such a system has *no solution* and it is called an **inconsistent system.** The solution set is the empty set, which is written ∅.

> **Self Check 3** Solve the system $\begin{cases} 3y - 2x = 6 \\ 2x - 3y = 6 \end{cases}$ by graphing, if possible.
>
> **Now Try** ▶ Problem 25

When the equations of a system have different graphs (as in Examples 2 and 3), the equations are called **independent equations.**

EXAMPLE 4   Solve the system by graphing: $\begin{cases} y = \dfrac{1}{2}x + 2 \\ 2x + 8 = 4y \end{cases}$

Strategy   We will graph both equations on the same coordinate system.

Why   If both equations are graphed on the same coordinate system, we can see whether they have any common solutions.

Solution   We graph each equation on one set of coordinate axes, as shown below.

Graph using the slope and y-intercept.

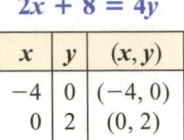

$$y = \frac{1}{2}x + 2$$

$$m = \frac{1}{2}$$

y-intercept: $(0, 2)$

Graph using the intercept method.

$$2x + 8 = 4y$$

x	y	(x, y)
-4	0	$(-4, 0)$
0	2	$(0, 2)$
2	3	$(2, 3)$

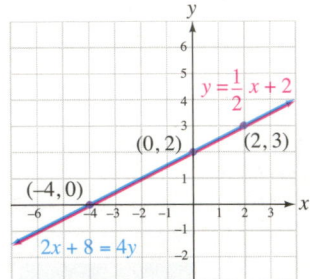

The same line — infinitely many solutions

The Language of Algebra

Here the graphs of the lines **coincide.** That is, they occupy the same location. To illustrate this concept, think of a clock. At noon and midnight, the hands of the clock *coincide.*

The graphs appear to be identical. We can verify this by writing the second equation in slope–intercept form and observing that it is the same as the first equation.

$$y = \frac{1}{2}x + 2 \quad \text{First equation}$$

$$2x + 8 = 4y \quad \text{Second equation}$$

$$\frac{2x}{4} + \frac{8}{4} = \frac{4y}{4} \quad \text{Divide both sides by 4.}$$

$$\frac{1}{2}x + 2 = y$$

We see that the equations of the system are equivalent. Because $y = \frac{1}{2}x + 2$ and $2x + 8 = 4y$ are different forms of the same equation, they are called **dependent equations.** Since the graphs are the same line, they have infinitely many points in common. All of the points that lie on the common line are solutions because the coordinates of each of those points satisfy both equations of the system. In cases like this, we say that there are *infinitely many solutions.* The solution set can be written using **set-builder notation** as

Caution

A common error is to graph the identical lines, but forget to answer with the words *infinitely many solutions* and the set-builder notation.

$$\left\{ (x, y) \mid y = \frac{1}{2}x + 2 \right\} \quad \begin{array}{l} \text{Read as, "the set of all ordered pairs } (x, y), \\ \text{such that } y = \frac{1}{2}x + 2.\text{"} \end{array}$$

We also can express the solution set using the second equation of the system in the set-builder notation: $\{(x, y) \mid 2x + 8 = 4y\}$.

Some instructors prefer that the set-builder notation use an equation in standard form with coefficients that are integers having no common factor other than 1. Such an equation that is equivalent to $y = \frac{1}{2}x + 2$ and $2x + 8 = 4y$ is $x - 2y = -4$. The set-builder notation solution for this example could, therefore, be written as $\{(x, y) \mid x - 2y = -4\}$.

From the graph, it appears that three of the infinitely many solutions are $(-4, 0)$, $(0, 2)$, and $(2, 3)$. Check each of them to verify that both equations of the system are satisfied.

Self Check 4   Solve the system by graphing: $\begin{cases} 2x - y = 4 \\ y = 2x - 4 \end{cases}$

Now Try ▶   Problem 27

We now summarize the possibilities that can occur when two linear equations, each with two variables, are graphed.

Solving a System of Equations by the Graphing Method

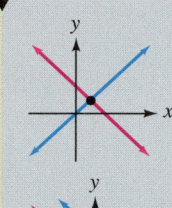

If the lines are different and intersect, the equations are independent, and the system is consistent. **One solution exists.** It is the point of intersection.

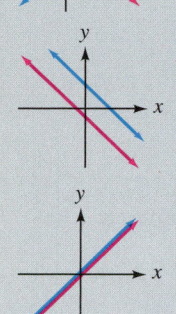

If the lines are different and parallel, the equations are independent, and the system is inconsistent. **No solution exists.**

If the lines are identical, the equations are dependent, and the system is consistent. **Infinitely many solutions exist.** Any point on the line is a solution.

If each equation in one system is equivalent to a corresponding equation in another system, the systems are called **equivalent systems.**

EXAMPLE 5   Solve the system by graphing: $\begin{cases} \dfrac{3}{2}x - y = \dfrac{5}{2} \\ \dfrac{1}{8}y = 1 - \dfrac{x}{4} \end{cases}$

Strategy   We will use the multiplication property of equality to clear both equations of fractions and solve the resulting equivalent system by graphing.

Why   It is usually easier to solve systems of equations that do not contain fractions.

Solution   We multiply both sides of $\frac{3}{2}x - y = \frac{5}{2}$ by 2 to eliminate the fractions and obtain the equation $3x - 2y = 5$. We multiply both sides of $\frac{1}{8}y = 1 - \frac{x}{4}$ by 8 to eliminate the fractions and obtain the equation $y = 8 - 2x$.

The original system   *An equivalent system*

$\begin{cases} \dfrac{3}{2}x - y = \dfrac{5}{2} \end{cases}$ →(Multiply by 2)→ $2\left(\dfrac{3}{2}x - y\right) = 2\left(\dfrac{5}{2}\right)$ →(Simplify)→ $\begin{cases} 3x - 2y = 5 \end{cases}$

$\begin{cases} \dfrac{1}{8}y = 1 - \dfrac{x}{4} \end{cases}$ →(Multiply by 8)→ $8\left(\dfrac{1}{8}y\right) = 8\left(1 - \dfrac{x}{4}\right)$ →(Simplify)→ $\begin{cases} y = 8 - 2x \end{cases}$

Since the new system is equivalent to the original system, they have the same solution. If we graph the equations of the new system, it appears that the point where the lines intersect is (3, 2).

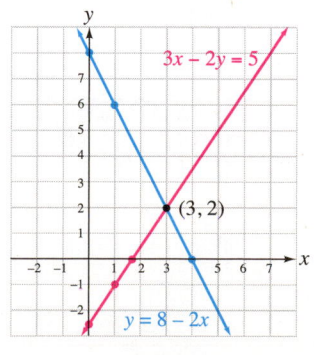

Graph using the intercept method.

$3x - 2y = 5$

x	y	(x, y)
$\frac{5}{3}$	0	$\left(\frac{5}{3}, 0\right)$
0	$-\frac{5}{2}$	$\left(0, -\frac{5}{2}\right)$
1	-1	$(1, -1)$

Graph using the slope and y-intercept.

$y = 8 - 2x$

Slope = -2
y-intercept: (0, 8)

Caution

When checking the solution of a system of equations, always substitute the values of the variables into the original equations.

To verify that (3, 2) is the solution, we substitute 3 for x and 2 for y in each equation of the original system.

Check: $\dfrac{3}{2}x - y = \dfrac{5}{2}$ First equation $\dfrac{1}{8}y = 1 - \dfrac{x}{4}$ Second equation

$\dfrac{3}{2}(3) - 2 \stackrel{?}{=} \dfrac{5}{2}$ $\dfrac{1}{8}(2) \stackrel{?}{=} 1 - \dfrac{3}{4}$

$\dfrac{9}{2} - \dfrac{4}{2} \stackrel{?}{=} \dfrac{5}{2}$ $\dfrac{2}{8} \stackrel{?}{=} \dfrac{1}{4}$

$\dfrac{5}{2} = \dfrac{5}{2}$ True $\dfrac{1}{4} = \dfrac{1}{4}$ True

Self Check 5   Solve the system by graphing: $\begin{cases} \dfrac{1}{2}x + \dfrac{1}{2}y = -1 \\ \dfrac{1}{3}x - \dfrac{1}{2}y = -4 \end{cases}$

Now Try ▶ Problem 29

Using Your Calculator ▶ **Solving Systems by Graphing**

The TRACE and INTERSECT features found on most graphing calculators enable us to get very good approximations of solutions of systems of two linear equations. To illustrate this, consider the following system. To graph the equations, we must first solve each equation for y.

To solve the system $\begin{cases} 3x + 2y = 12 \\ 2x - 3y = 12 \end{cases}$ $\xrightarrow[\text{Solve for } y]{\text{Solve for } y}$ $\begin{cases} y = -\dfrac{3}{2}x + 6 \\ y = \dfrac{2}{3}x - 4 \end{cases}$

Next, we press $\boxed{Y=}$ and enter the right side of each equation of the equivalent system after the symbols Y_1 and Y_2, as shown in figure (a) below. Then we press $\boxed{\text{GRAPH}}$. If we use window settings of $[-10, 10]$ for x and for y, the graphs of the equations will look like those in figure (b). If we zoom in on the intersection point of the two lines and trace, we will get an approximate solution like the one shown in figure (c). To get better results, we can do more zooms. We would then find that, to the nearest hundredth, the solution is $(4.63, -0.94)$. Verify that this is reasonable.

A more efficient method for finding the intersection point of two lines uses the INTERSECT feature. With this feature, the cursor automatically highlights the intersection point, and the x- and y-coordinates are displayed. To locate INTERSECT, press $\boxed{\text{2nd}}$, $\boxed{\text{CALC}}$, 5, followed by $\boxed{\text{ENTER}}$. The result is a graph similar to figure (d). The display shows the approximate coordinates of the point of intersection.

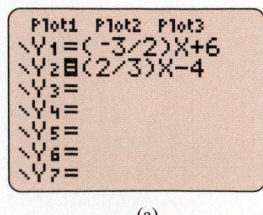   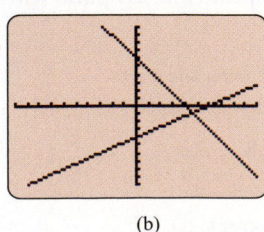   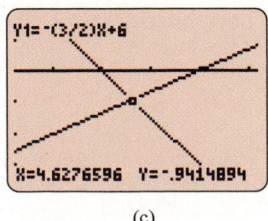   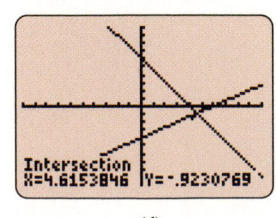

       (a)   (b)   (c)   (d)

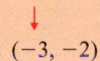

4 Solve Equations Graphically.

The graphing method of this section can be used to solve linear equations in one variable.

EXAMPLE 6

Solve the equation $2x + 4 = -2$ graphically.

Strategy To find the solution of $2x + 4 = -2$, we will set y equal to the left side of this equation, and y equal to the right side of the equation, and solve the system $\begin{cases} y = 2x + 4 \\ y = -2 \end{cases}$ by graphing.

Why To solve $2x + 4 = -2$, we need to find the value of x that makes $2x + 4$ equal -2. That x-value will be equal to the x-coordinate of the point of intersection of the graphs of $y = 2x + 4$ and $y = -2$.

Solution

The graphs of $y = 2x + 4$ and $y = -2$ are shown here. The point of intersection of the lines appears to be $(-3, -2)$. This indicates that if $x = -3$, the expression $2x + 4$ equals -2. So the solution of $2x + 4 = -2$ is -3. We can verify this result by checking:

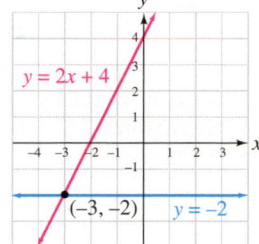

$$2x + 4 = -2 \quad \text{\color{red}This is the equation to solve.}$$
$$2(-3) + 4 \stackrel{?}{=} -2 \quad \text{\color{red}Substitute.}$$
$$-6 + 4 \stackrel{?}{=} -2 \quad \text{\color{red}Evaluate the left side.}$$
$$-2 = -2 \quad \text{\color{red}True}$$

Success Tip

Note that it is the first coordinate of the point of intersection of the lines that is the solution of the equation $2x + 4 = -2$.

↓

$(-3, -2)$

Self Check 6 Solve $2x + 4 = 2$ graphically.

Now Try ▶ Problem 41

Using Your Calculator ▶ **Solving Equations Graphically**

To solve $2(x - 3) + 3 = 7$ with a graphing calculator, we graph the left side of the equation and the right side of the equation in the same window by entering

$$Y_1 = 2(x - 3) + 3 \qquad \text{and} \qquad Y_2 = 7$$

Figure (a) shows the graphs generated using settings of $[-10, 10]$ for x and for y. The coordinates of the point of intersection of the graphs can be determined using the INTERSECT feature. In figure (b), we see that the point of intersection is $(5, 7)$, which indicates that 5 is a solution of $2(x - 3) + 3 = 7$.

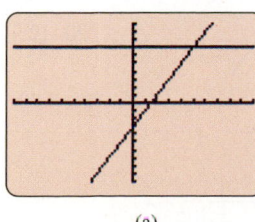
(a)

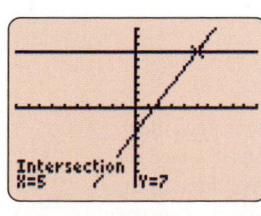
(b)

SECTION 3.1   **STUDY SET**

VOCABULARY

Fill in the blanks.

1. $\begin{cases} x - 2y = 4 \\ 2x - y = 3 \end{cases}$ is called a _____ of linear equations.

2. When a system of equations has at least one solution, it is called a _____ system. If a system has no solutions, it is called an _____ system.

3. If two equations have different graphs, they are called _____ equations. Two equations with the same graph are called _____ equations.

4. When solving a system of two linear equations by the graphing method, we look for the point of _____ of the two lines.

CONCEPTS

5. Refer to the illustration. Decide whether a true or a false statement would be obtained

   a. when the coordinates of Point A are substituted into the equation for line l_1.

   b. when the coordinates of Point B are substituted into the equation for line l_1.

   c. when the coordinates of Point C are substituted into the equation for line l_1.

   d. when the coordinates of Point C are substituted into the equation for line l_2.

6. Refer to the illustration.

   a. How many ordered pairs satisfy the equation $3x + y = 3$? Name three.

   b. How many ordered pairs satisfy the equation $\frac{2}{3}x - y = -3$? Name three.

   c. How many ordered pairs satisfy both equations? Name it or them.

7. a. How many solutions does the system of equations graphed on the right have? Are the equations dependent or independent?

   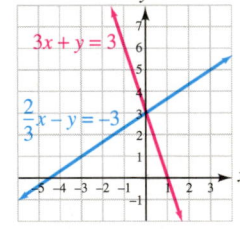

   b. How many solutions does the system of equations graphed on the right have? Give three of the solutions. Is the system consistent or inconsistent?

   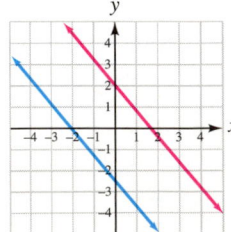

   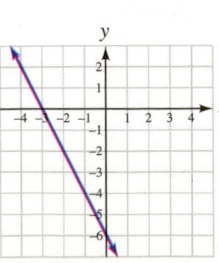

8. Estimate the solution of the system of linear equations shown in the following display.

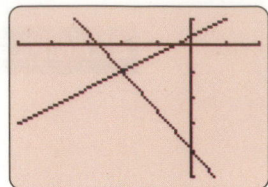

Use the graphs in the illustration below to solve each equation.

9. $-3x + 2 = x - 2$

10. $-x - 4 = x - 2$

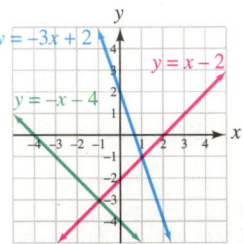

NOTATION

Fill in the blanks.

11. The symbol { is called a left _____. It is used when writing a system of equations.

12. We read the set-builder notation $\{(x, y) \mid 3x - 5y = 1\}$ as "the _____ of all ordered pairs (x, y) _____ that $3x - 5y = 1$."

GUIDED PRACTICE

Determine whether the ordered pair is a solution of the system of equations. **See Example 1.**

13. $(-4, 3)$; $\begin{cases} 4x - y = -19 \\ 3x + 2y = -6 \end{cases}$

14. $(-1, 2)$; $\begin{cases} 3x - y = -5 \\ x - y = -4 \end{cases}$

15. $(2, -3)$; $\begin{cases} y + 2 = \frac{1}{2}x \\ 3x + 2y = 0 \end{cases}$

16. $(1, 2)$; $\begin{cases} 2x - y = 0 \\ y = \frac{1}{2}x + \frac{3}{2} \end{cases}$

17. $\left(\frac{1}{2}, \frac{1}{3}\right)$; $\begin{cases} 2x + 3y = 2 \\ 4x - 9y = 1 \end{cases}$

18. $\left(-\frac{3}{4}, \frac{2}{3}\right)$; $\begin{cases} 4x + 3y = -1 \\ 4x - 3y = -5 \end{cases}$

19. $(-0.2, 0.5)$; $\begin{cases} 2x + 5y = 2.1 \\ 5x + y = -0.5 \end{cases}$

20. $(2.1, -3.2)$; $\begin{cases} x + y = -1.1 \\ 2x - 3y = 13.8 \end{cases}$

Solve each system by graphing. See Example 2.

21. $\begin{cases} x + y = 6 \\ x - y = 2 \end{cases}$

22. $\begin{cases} x - y = 4 \\ 2x + y = 5 \end{cases}$

23. $\begin{cases} y = -2x + 1 \\ x - 2y = -7 \end{cases}$

24. $\begin{cases} 3x - y = -3 \\ y = -2x - 7 \end{cases}$

Solve each system by graphing, if possible. If a system is inconsistent or if the equations are dependent, state this. See Examples 3 and 4.

25. $\begin{cases} 3x - 3y = 4 \\ x - y = 4 \end{cases}$

26. $\begin{cases} 5x + 2y = 6 \\ -10x - 4y = -12 \end{cases}$

27. $\begin{cases} x = 3 - 2y \\ 2x + 4y = 6 \end{cases}$

28. $\begin{cases} 3x = 5 - 2y \\ 3x + 2y = 7 \end{cases}$

Solve each system by graphing. See Example 5.

29. $\begin{cases} \dfrac{1}{6}x = \dfrac{1}{3}y + \dfrac{1}{2} \\ y = x \end{cases}$

30. $\begin{cases} x = y + 3 \\ \dfrac{1}{4}x - \dfrac{1}{6}y = \dfrac{1}{3} \end{cases}$

31. $\begin{cases} \dfrac{1}{3}x - \dfrac{7}{6}y = \dfrac{1}{2} \\ \dfrac{1}{5}y = \dfrac{1}{3}x + \dfrac{7}{15} \end{cases}$

32. $\begin{cases} \dfrac{3}{5}x + \dfrac{1}{4}y = -\dfrac{11}{10} \\ \dfrac{1}{8}x = \dfrac{13}{24} + \dfrac{1}{3}y \end{cases}$

Solve each equation graphically. See Example 6.

33. $2x + 1 = 5$

34. $3x + 5 = -4$

35. $-2x + 8 = 3x - 7$

36. $2x - 3 = 3x - 3$

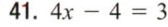 *Use a graphing calculator to solve each system. Give all answers to the nearest hundredth. See Using Your Calculator: Solving Systems by Graphing.*

37. $\begin{cases} y = 3.2x - 1.5 \\ y = -2.7x - 3.7 \end{cases}$

38. $\begin{cases} y = -0.45x + 5 \\ y = 5.55x - 13.7 \end{cases}$

39. $\begin{cases} 1.7x + 2.3y = 3.2 \\ y = 0.25x + 8.95 \end{cases}$

40. $\begin{cases} 2.75x = 12.9y - 3.79 \\ 7.1x - y = 35.76 \end{cases}$

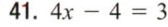 *Use a graphing calculator to solve each equation.*

41. $4x - 4 = 3x$

42. $4(x - 3) - x = x - 6$

43. $11x + 6(3 - x) = 3$

44. $2x + 4 = 12 - 2x$

Solve each system by graphing, if possible. If a system is inconsistent or if the equations are dependent, state this. (Hint: Several coordinates of points of intersection are fractions.)

45. $\begin{cases} y = -\dfrac{5}{2}x + \dfrac{1}{2} \\ 2x - \dfrac{3}{2}y = 5 \end{cases}$

46. $\begin{cases} y = -\dfrac{5}{6}x + 2 \\ \dfrac{5}{2}x + 3y = 6 \end{cases}$

47. $\begin{cases} x + y = 0 \\ y = 2x - 6 \end{cases}$

48. $\begin{cases} 3x + y = 3 \\ 3x + 2y = 0 \end{cases}$

49. $\begin{cases} y = 3 \\ x = 2 \end{cases}$

50. $\begin{cases} 2x + 3y = -15 \\ 2x + y = -9 \end{cases}$

51. $\begin{cases} x = \dfrac{11 - 2y}{3} \\ y = \dfrac{11 - 6x}{4} \end{cases}$

52. $\begin{cases} x = \dfrac{1 - 3y}{4} \\ y = \dfrac{12 + 3x}{2} \end{cases}$

53. $\begin{cases} 4x - 3y = 5 \\ y = -2x \end{cases}$

54. $\begin{cases} 2x + 2y = -1 \\ 3x + 4y = 0 \end{cases}$

55. $\begin{cases} x = 13 - 4y \\ 3x = 4 + 2y \end{cases}$

56. $\begin{cases} 3x = 7 - 2y \\ 2x = 2 + 4y \end{cases}$

57. $\begin{cases} x = 2 \\ y = -\dfrac{1}{2}x + 2 \end{cases}$

58. $\begin{cases} y = -2 \\ y = \dfrac{2}{3}x - \dfrac{4}{3} \end{cases}$

59. $\begin{cases} x = \dfrac{5}{2}y - 2 \\ x - \dfrac{5}{3}y + \dfrac{1}{3} = 0 \end{cases}$

60. $\begin{cases} 2x = 5y - 11 \\ 3x = 2y \end{cases}$

61. $\begin{cases} x + 3y = 6 \\ y = -\dfrac{1}{3}x + 2 \end{cases}$

62. $\begin{cases} 2x - y = -4 \\ 2y = 4x - 6 \end{cases}$

63. $\begin{cases} x = -\dfrac{3}{2}y \\ 2x = 3y - 4 \end{cases}$

64. $\begin{cases} 4x = 3y - 1 \\ 3y = 4 - 8x \end{cases}$

65.

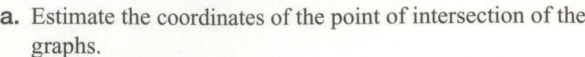

from **Campus to Careers**

Fashion Designer

One of the line graphs below gives the percent share of the U.S. footwear market for shoes produced in the United States. The other line gives the percent share of the U.S. footwear market for imports.

a. Estimate the coordinates of the point of intersection of the graphs.

b. What important percent-of-the-market information does your answer to part a give?

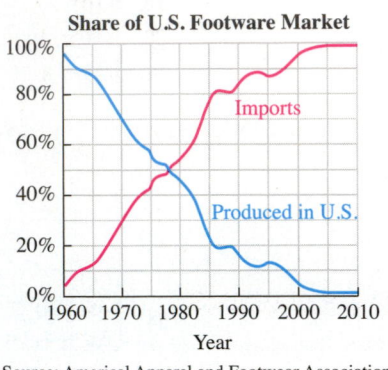

Share of U.S. Footware Market

Source: Americal Apparel and Footwear Association

66. Business.   Estimate the break-even point (where cost = revenue) on the graph in the illustration. Explain why it is called the *break-even point*.

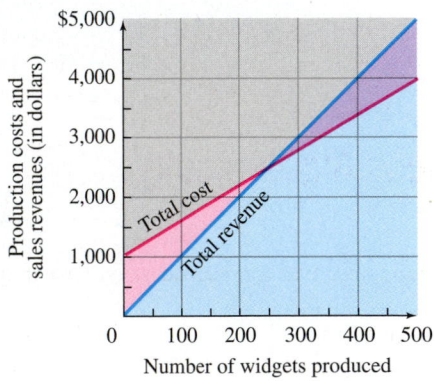

67. Hearing Tests.   See the illustration in the next column. At what frequency and decibel level were the hearing test results the same for the left and right ear? Write your answer as an ordered pair.

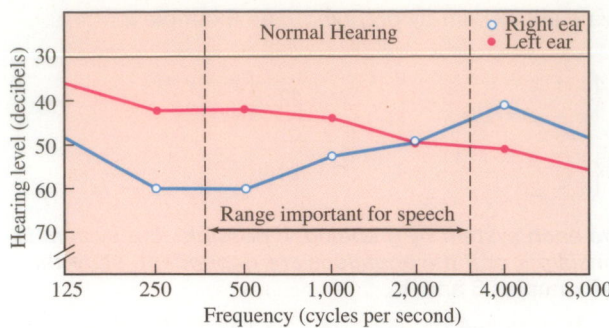

68. The Internet.

The graph below shows the growing importance of the Internet in the daily lives of Americans. Determine when the time spent on the following activities was the same. Approximately how many hours per year were spent on each?

a. Internet and reading magazines

b. Internet and reading newspapers

c. Internet and reading books

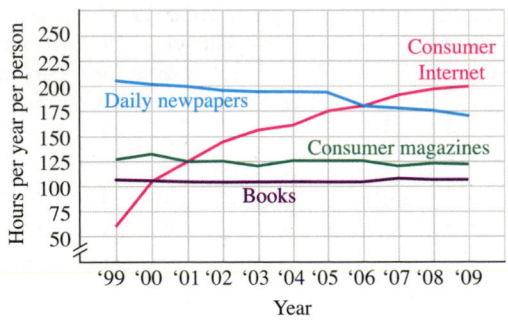

Source: Vernois Suhler Stevenson

69. Law of Supply and Demand.   The demand function, graphed below, describes the relationship between the price x of a certain camera and the demand for the camera.

a. The supply function, $S(x) = \frac{25}{4}x - 525$, describes the relationship between the price x of the camera and the number of cameras the manufacturer is willing to supply. Graph this function in the illustration.

b. For what price will the supply of cameras equal the demand?

c. As the price of the camera is increased, what happens to supply and what happens to demand?

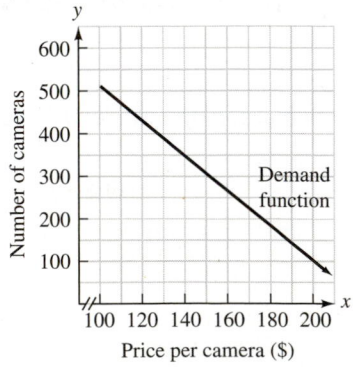

70. Cost and Revenue. The function $C(x) = 200x + 400$ gives the cost for a college to offer x sections of an introductory class in CPR (cardiopulmonary resuscitation). The function $R(x) = 280x$ gives the amount of revenue the college brings in when offering x sections of CPR.

   a. Find the break-even point (where cost = revenue) by graphing each function on the same coordinate system.

   b. How many sections does the college need to offer to make a profit on the CPR training course?

71. Navigation. The paths of two ships are tracked on the same coordinate system. One ship is following a path described by the equation $2x + 3y = 6$, and the other is following a path described by the equation $y = \frac{2}{3}x - 3$.

   a. Graph both equations on the same coordinate system. Is there a possibility of a collision?

   b. Is a collision a certainty?

72. Air Traffic Control. Two airplanes, flying at the same altitude, are tracked using the same coordinate system on a radar screen. One plane is following a path described by the equation $y = 0.4x - 2$, and the other is following a path described by the equation $2x = 5y + 7$. Graph both equations on the same coordinate system. Is there a possibility of a collision?

WRITING

73. Suppose the solution of a system of two linear equations is $\left(\frac{14}{5}, -\frac{8}{3}\right)$. Knowing this, explain any drawbacks you might encounter when solving the system by the graphing method.

74. Can a system of two linear equations have exactly two solutions? Why or why not?

75. Suppose the graphs of the two linear equations of a system are the same line. What is wrong with this statement? *The system has infinitely many solutions. Any ordered pair is a solution of the system.*

76. a. Without graphing, how can you tell that the graphs of $y = 2x + 1$ and $y = 3x + 2$ intersect?

   b. Without graphing, how can you tell that the graphs of $y = 2x + 1$ and $y = 2x + 2$ do not intersect?

   c. Without graphing, how can you tell that the graphs of $y = 2x + 3$ and $2y = 4x + 6$ are the same line?

REVIEW

Let $f(x) = -x^3 + 2x - 2$ *and* $g(x) = \frac{2 - x}{9 + x}$ *and find each value.*

77. $f(-1)$ **78.** $f(10)$

79. $g(2)$ **80.** $g(-20)$

81. $f(t)$ **82.** $g(s + 1)$

83. Find the domain of the function $f(x) = \dfrac{1}{3x + 6}$.

84. Find the domain of the function $f(x) = |x|$.

CHALLENGE PROBLEMS

85. Write a dependent system of equations with a solution of $(-5, 2)$.

86. Write a system of *three* independent linear equations in two variables that has the solution $(-5, 2)$.

SECTION 3.2

Solving Systems of Equations Algebraically

OBJECTIVES

1 Solve systems of linear equations by substitution.

2 Solve systems of linear equations by the elimination (addition) method.

3 Use substitution and elimination (addition) to identify inconsistent systems and dependent equations.

4 Determine the most efficient method to use to solve a linear system.

ARE YOU READY?

The following problems review some basic skills that are needed when solving systems of equations algebraically.

1. In $6x + y = 9$, what is the coefficient of y?

2. Solve $5x - y = -4$ for y.

3. Substitute 4 for x in $y = -2x - 1$ and find y.

4. In $8x - 3y = 12$, what is the *opposite* of the coefficient of y?

5. Substitute -3 for y in $5x + 6y = 2$ and find x.

6. Multiply both sides of the equation $7x - y = 9$ by -4.

The graphing method enables us to visualize the process of solving systems of equations. However, it can be difficult to determine the exact coordinates of the point of intersection. We now discuss two other methods we can use to find the exact solutions of systems of equations.

1 Solve Systems of Linear Equations by Substitution.

The substitution method works well for solving systems where one equation is solved, or can be easily solved, for one of the variables. To solve a system of two linear equations in x and y by the substitution method, we can follow these steps.

The Substitution Method	
	1. Solve one of the equations for either x or y—preferably a variable with a coefficient of 1 or -1. If this is already done, go to step 2. (We call this equation the **substitution equation**.)
	2. Substitute the expression for x or for y obtained in step 1 into the other equation and solve that equation.
	3. Substitute the value of the variable found in step 2 into the substitution equation to find the value of the remaining variable.
	4. Check the proposed solution in each equation of the original system. Write the solution as an ordered pair.

EXAMPLE 1 Solve the system by substitution: $\begin{cases} 4x + y = 13 \\ -2x + 3y = -17 \end{cases}$

Strategy We will use the substitution method. Since the system does not contain an equation solved for x or y, we must choose an equation and solve it for x or y. It is easiest to solve for y in the first equation, because y has a coefficient of 1.

Why Solving $4x + y = 13$ for x or solving $-2x + 3y = -17$ for x or y would involve working with cumbersome fractions.

Solution **Step 1:** We solve the first equation for y, because y has a coefficient of 1.

$$4x + y = 13$$
$$4x + y - 4x = -4x + 13 \qquad \text{To isolate y, subtract 4x from both sides.}$$
$$y = -4x + 13 \qquad \text{This is the substitution equation.}$$
$$\text{It could also be written: } y = 13 - 4x.$$

The Language of Algebra

Since substitution involves algebra and not graphing, it is called an **algebraic** method for solving a system.

Because y and $-4x + 13$ are equal, we can substitute $-4x + 13$ for y in the second equation of the system.

$$y = \boxed{-4x + 13} \qquad -2x + 3y = -17$$

Success Tip

Throughout the course, we have been substituting numbers for variables. With this method, we substitute a *variable expression for a variable*. The objective is to use an appropriate substitution to obtain *one* equation in *one* variable.

Step 2: We then substitute $-4x + 13$ for y in the second equation to eliminate the variable y from that equation. The result will be an equation containing only one variable, x.

$$-2x + 3y = -17 \qquad \text{This is the second equation of the system.}$$
$$-2x + 3(-4x + 13) = -17 \qquad \text{Substitute } -4x + 13 \text{ for y. Write the parentheses so}$$
$$\text{that the multiplication by 3 is distributed over both}$$
$$\text{terms of } -4x + 13.$$
$$-2x - 12x + 39 = -17 \qquad \text{Distribute the multiplication by 3.}$$
$$-14x + 39 = -17 \qquad \text{Combine like terms.}$$
$$-14x = -56 \qquad \text{Subtract 39 from both sides.}$$
$$x = 4 \qquad \text{To solve for x, divide both sides by } -14.$$
$$\text{This is the x-value of the solution.}$$

The Language of Algebra

The phrase **back-substitute** can also be used to describe step 3 of the substitution method. To find y, we *back-substitute* 4 for x in the equation $y = -4x + 13$.

Step 3: To find y, we substitute 4 for x in the substitution equation and evaluate the right side.

$y = -4x + 13$ This is the substitution equation.

$y = -4(4) + 13$ Substitute 4 for x.

$y = -16 + 13$ Multiply: $-4(4) = -16$.

$y = -3$ This is the y-value of the solution.

Step 4: To verify that $(4, -3)$ satisfies both equations, we substitute 4 for x and -3 for y into each equation of the original system and simplify.

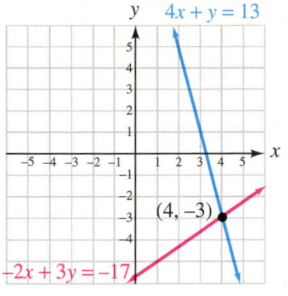

Check:

$4x + y = 13$ First equation

$4(4) + (-3) \stackrel{?}{=} 13$

$16 - 3 \stackrel{?}{=} 13$

$13 = 13$ True

$-2x + 3y = -17$ Second equation

$-2(4) + 3(-3) \stackrel{?}{=} -17$

$-8 - 9 \stackrel{?}{=} -17$

$-17 = -17$ True

Since $(4, -3)$ satisfies both equations of the system, it is the solution of the system. The solution set is $\{(4, -3)\}$. The graphs of the equations of the system help to verify this—they appear to intersect at $(4, -3)$, as shown on the left.

Self Check 1   Solve the system by substitution: $\begin{cases} x + 3y = 9 \\ 2x - y = -10 \end{cases}$

Now Try ▶ Problem 17

EXAMPLE 2   Solve the system by substitution: $\begin{cases} \dfrac{2}{9}x - \dfrac{2}{9}y = \dfrac{2}{3} \\ 0.1x = 0.2 - 0.1y \end{cases}$

Strategy   We will find an equivalent system without fractions or decimals and use the substitution method to solve it.

Why   It's usually easier to solve a system of equations that involves only integers.

Solution   To clear the first equation of fractions, we multiply both sides by 9, which is the least common denominator of the fractions in the equation. To clear the second equation of the decimals, we multiply both sides by 10.

The original system   *An equivalent system*

$\begin{cases} \dfrac{2}{9}x - \dfrac{2}{9}y = \dfrac{2}{3} \\ 0.1x = 0.2 - 0.1y \end{cases}$ Multiply by 9 $\longrightarrow$ $9\left(\dfrac{2}{9}x - \dfrac{2}{9}y\right) = 9\left(\dfrac{2}{3}\right)$ Simplify $\longrightarrow$ $\begin{cases} 2x - 2y = 6 \\ x = 2 - y \end{cases}$

Multiply by 10 $\longrightarrow$ $10(0.1x) = 10(0.2 - 0.1y)$ Simplify

These results form the following equivalent system, which has the same solution as the original one. We number the equations (1) and (2) to help describe how the system is solved using the substitution method.

(1) $\begin{cases} 2x - 2y = 6 \\ x = 2 - y \end{cases}$

(2)   This is the substitution equation.

Since the variable x is isolated in equation 2, we will substitute $2 - y$ for x in equation 1. This step will eliminate x from equation 1, leaving an equation containing only one variable, y. We then solve for y.

$$2\mathbf{x} - 2y = 6 \qquad \text{This is equation 1 of the equivalent system.}$$
$$2(\mathbf{2 - y}) - 2y = 6 \qquad \text{Substitute } 2 - y \text{ for } x. \text{ Don't forget the parentheses.}$$
$$4 - 2y - 2y = 6 \qquad \text{Distribute the multiplication by 2.}$$
$$-4y = 2 \qquad \text{Combine like terms and subtract 4 from both sides.}$$
$$y = -\frac{1}{2} \qquad \text{To solve for y, divide both sides by } -4 \text{ and simplify the fraction } -\frac{2}{4}. \text{ This is the } y\text{-value of the solution.}$$

We can find x by substituting $-\frac{1}{2}$ for y in equation 2 and simplifying:

$$x = 2 - \mathbf{y} \qquad \text{This is the substitution equation.}$$
$$x = 2 - \left(-\frac{\mathbf{1}}{\mathbf{2}}\right) \qquad \text{Substitute } -\frac{1}{2} \text{ for y.}$$
$$x = 2 + \frac{1}{2} \qquad \text{Write the subtraction as the addition of the opposite.}$$
$$x = \frac{5}{2} \qquad \text{Do the addition: } 2 + \frac{1}{2} = \frac{4}{2} + \frac{1}{2} = \frac{5}{2}. \text{ This is the } x\text{-value of the solution.}$$

The solution is $\left(\frac{5}{2}, -\frac{1}{2}\right)$ and the solution set is $\left\{\left(\frac{5}{2}, -\frac{1}{2}\right)\right\}$. Verify that this result satisfies both equations of the original system.

Self Check 2   Solve the system by substitution: $\begin{cases} \dfrac{x}{8} + \dfrac{y}{4} = \dfrac{1}{2} \\ 0.01y = -0.02x + 0.04 \end{cases}$

Now Try ▶   Problem 19

2 Solve Systems of Linear Equations by the Elimination (Addition) Method.

The substitution method for solving a system of equations can be difficult to use if none of the variables has a coefficient of 1 or -1. This is the case for the system

$$\begin{cases} 5x + 3y = 2 \\ 2x - 3y = -16 \end{cases}$$

Solving either equation for x or y involves working with cumbersome fractions. For example, if we solve the first equation for x, the resulting substitution equation is $x = \dfrac{2 - 3y}{5}$. Fortunately, we can solve systems like this one using an easier method called the **elimination** or the **addition method.**

The elimination (addition) method for solving a system is based on the **addition property of equality:** *When equal quantities are added to both sides of an equation, the results are equal.* In symbols, if $A = B$ and $C = D$, then adding the left sides and the right sides of these equations, we have $A + C = B + D$. This procedure is called *adding the equations.*

Add the terms on the left sides.

$$\begin{array}{c} A = B \\ C = D \\ \hline A + C = B + D \end{array}$$

Add the terms on the right sides.

With the elimination method, we add the equations in a way that will eliminate the terms involving one of the variables.

To solve a system of linear equations in x and y by the elimination (addition) method, we can use the following steps.

The Elimination (Addition) Method

1. Write both equations of the system in standard form: $Ax + By = C$.

2. If necessary, multiply one or both of the equations by a nonzero number chosen to make the coefficients of x (or the coefficients of y) opposites.

3. Add the equations to eliminate the terms involving x (or y).

4. Solve the equation resulting from step 3.

5. Find the value of the remaining variable by substituting the solution found in step 4 into any equation containing both variables. Or, repeat steps 2–4 to eliminate the other variable.

6. Check the proposed solution in each equation of the original system. Write the solution as an ordered pair.

EXAMPLE 3

Solve the system by elimination: $\begin{cases} 5x + 3y = 2 \\ 2x - 3y = -16 \end{cases}$

Strategy Since the coefficients of the y-terms are opposites, we will add the left and right sides of the given equations to eliminate y.

Why When we add the equations in this way, the result will be an equation that contains only one variable, x.

Solution

Step 1: Since both equations are already written in standard $Ax + By = C$ form, we can move to the next step.

Step 2: Because the coefficients of the terms $3y$ in the first equation and $-3y$ in the second equation are opposites, we can move to the next step.

Step 3: We can add the left sides and add the right sides of the given equations as shown below.

$$\begin{aligned} 5x + 3y &= 2 \\ \underline{2x - 3y} &= \underline{-16} \\ 7x \quad\;\; &= -14 \end{aligned}$$

To add the equations, add the like terms, column by column.

$2 + (-16) = -14$
$3y + (-3y) = 0$
$5x + 2x = 7x$

Step 4: Because the sum of the terms $3y$ and $-3y$ is 0, we say that the variable y has been *eliminated*. Since the resulting equation has only one variable, we can solve it for x.

$$7x = -14$$
$$x = -2 \qquad \text{To solve for x, divide both sides by 7. This is the x-value of the solution.}$$

Step 5: To find the y-value of the solution, substitute -2 for x in either equation of the original system. If we use the first equation of the system, we have:

$$\begin{aligned} 5x + 3y &= 2 && \text{This is the first equation of the system.} \\ 5(-2) + 3y &= 2 && \text{Substitute } -2 \text{ for x.} \\ -10 + 3y &= 2 && \text{Multiply: } 5(-2) = -10. \\ 3y &= 12 && \text{Add 10 to both sides.} \\ y &= 4 && \text{To solve for y, divide both sides by 3.} \\ & && \text{This is the y-value of the solution.} \end{aligned}$$

Notation

When a linear equation is written in standard $Ax + By = C$ form, the *variable terms* are on the left side of the equal symbol and the *constant* term is on the right side.

The Language of Algebra

In the equations $5x + 3y = 2$ and $2x - 3y = -16$, the coefficients of the y-terms, 3 and -3, are **opposites** or **additive inverses**.

Caution

When using the substitution or elimination method to solve a system of equations, a common error is to find the value of one of the variables, say x, and forget to find the value of the other. Remember that a solution of a linear system of two equations is an ordered pair (x, y).

Step 6: Now we check the proposed solution $(-2, 4)$ in the equations of the original system.

Check:

$5x + 3y = 2$	First equation	$2x - 3y = -16$	Second equation
$5(-2) + 3(4) \overset{?}{=} 2$		$2(-2) - 3(4) \overset{?}{=} -16$	
$-10 + 12 \overset{?}{=} 2$		$-4 - 12 \overset{?}{=} -16$	
$2 = 2$	True	$-16 = -16$	True

Since $(-2, 4)$ satisfies both equations, it is the solution. The solution set is written $\{(-2, 4)\}$. The graphs of the equations of the system help to verify this—they appear to intersect at $(-2, 4)$, as shown on the left.

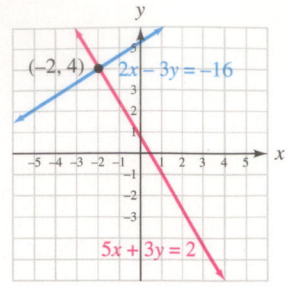

Self Check 3 Solve the system by elimination: $\begin{cases} 6x - 7y = 8 \\ 2x + 7y = -16 \end{cases}$

Now Try ▶ Problem 23

For many systems, we cannot immediately eliminate a variable by adding. In such cases, we will use the multiplication property of equality to create coefficients that are opposites.

EXAMPLE 4 Solve the system by elimination: $\begin{cases} 9a + 5b = 16 \\ 3a - 4b = -6 \end{cases}$

Strategy To use the elimination method to solve this system, we will multiply both sides of the second equation by -3 and add the equations to eliminate the terms involving the variable a.

Why This will give one equation involving only the variable b.

Solution

Step 1: Since both equations are written in standard form, we can move to the next step.

Step 2: The coefficient of a in the first equation is 9. If we multiply both sides of the second equation by -3, the coefficient of a in that equation will be -9. Then the coefficients of a will differ only in sign.

$$\begin{cases} 9a + 5b = 16 \\ 3a - 4b = -6 \end{cases} \xrightarrow[\text{Multiply by } -3]{\text{Unchanged}} \begin{array}{l} 9a + 5b = 16 \\ -3(3a - 4b) = -3(-6) \end{array} \xrightarrow[\text{Simplify}]{\text{Unchanged}} \begin{cases} 9a + 5b = 16 \\ -9a + 12b = 18 \end{cases}$$

Step 3: When the resulting equations are added, the terms involving a drop out (or are eliminated), and we get an equation that contains only the variable b. We then proceed by solving for b.

$$\begin{array}{r} 9a + 5b = 16 \\ \underline{-9a + 12b = 18} \\ 17b = 34 \end{array}$$

Add the like terms, column by column: $9a + (-9a) = 0$, $5b + 12b = 17b$, and $16 + 18 = 34$.

Step 4: Since the resulting equation has only one variable, we can solve it for b.

$$17b = 34$$
$$b = 2 \qquad \text{To solve for } b, \text{ divide both sides by 17. This is the } b\text{-value of the solution.}$$

Step 5: To find a, we can substitute 2 for b in either of the equations of the original system, or in $-9a + 12b = 18$. The calculations will involve fewer negative numbers if we use $9a + 5b = 16$.

$$9a + 5b = 16 \quad \text{This is the first equation of the original system.}$$
$$9a + 5(2) = 16 \quad \text{Substitute 2 for } b.$$
$$9a + 10 = 16$$
$$9a = 6 \quad \text{To isolate the variable term, subtract 10 from both sides.}$$
$$a = \frac{2}{3} \quad \text{To solve for } a, \text{ divide both sides by 9 and simplify the fraction } \frac{6}{9}.$$
This is the a-value of the solution.

Step 6: Verify that the solution is $\left(\frac{2}{3}, 2\right)$ by substituting $\frac{2}{3}$ for a and 2 for b in the original equations.

Notation

Unless told otherwise, list the values of the variables of a solution in *alphabetical* order. Here, the equations of the system involve the variables a and b, so we write the solution in the form (a, b).

Self Check 4 Solve the system by elimination: $\begin{cases} 15a - 2b = -9 \\ 5a - 5b = -29 \end{cases}$

Now Try Problems 25

Sometimes we must apply the multiplication property of equality to both equations to create coefficients of one variable that are opposites.

EXAMPLE 5 Solve the system by elimination: $\begin{cases} 4x = 3(2 + y) \\ 3(x - 10) = -2y \end{cases}$

Strategy We will write each equation in standard form $Ax + By = C$ and use the elimination (addition) method to solve the resulting equivalent system.

Why In their current form, the equations do not contain terms with coefficients that are opposites.

Solution To write each equation in standard form, we use the distributive property to remove the parentheses and then rearrange the terms.

The first equation	The second equation
$4x = 3(2 + y)$	$3(x - 10) = -2y$
$4x = 6 + 3y$	$3x - 30 = -2y$
$4x - 3y = 6$	$3x + 2y = 30$ This is $Ax + By = C$ form.

We proceed by solving the following equivalent system:

(1) $\begin{cases} 4x - 3y = 6 \\ 3x + 2y = 30 \end{cases}$
(2)

Since the coefficients of y have *opposite signs,* we choose to eliminate y. To make the y-terms drop out when we add the equations, we multiply both sides of equation 1 by 2 and both sides of equation 2 by 3. See the justification for this approach in the Success Tip at the left.

Success Tip

We create the term $-6y$ from $-3y$ and the term $6y$ from $2y$. Note that the *least common multiple* of 2 and 3 is 6:

2, 4, **6**, 8, 10, 12, 14, . . .
3, **6**, 9, 12, 15, 18, . . .

$\begin{cases} 4x - 3y = 6 \\ 3x + 2y = 30 \end{cases}$ $\xrightarrow[\text{Multiply by 3}]{\text{Multiply by 2}}$ $\begin{cases} 2(4x - 3y) = 2(6) \\ 3(3x + 2y) = 3(30) \end{cases}$ $\xrightarrow{\text{Simplify}}$ $\begin{cases} 8x - 6y = 12 \\ 9x + 6y = 90 \end{cases}$

When the results are added, the *y*-terms drop out and we can solve for *x*.

$$8x - 6y = 12 \qquad \text{Add like terms, column by column:}$$
$$\underline{9x + 6y = 90} \qquad 8x + 9x = 17x, \ -6y + 6y = 0, \text{ and } 12 + 90 = 102.$$
$$17x \qquad = 102 \qquad \text{This equation has only one variable, } x.$$
$$x = 6 \qquad \text{To solve for } x, \text{ divide both sides by 17.}$$

To find *y*, we can substitute 6 for *x* in any equation involving both variables. If we substitute 6 for *x* in equation 2, we get

$$3x + 2y = 30 \qquad \text{This is equation 2.}$$
$$3(6) + 2y = 30 \qquad \text{Substitute 6 for } x.$$
$$18 + 2y = 30 \qquad \text{Perform the multiplication.}$$
$$2y = 12 \qquad \text{Subtract 18 from both sides.}$$
$$y = 6 \qquad \text{Divide both sides by 2.}$$

The solution is (6, 6). Check this result by substituting 6 for *x* and 6 for *y* in the equations of the original system.

> **Success Tip**
>
> With this method, it doesn't matter which variable is eliminated. We could have created terms of $12x$ and $-12x$ to eliminate *x*. We will get the same solution, (6, 6).

Self Check 5   Solve the system by elimination: $\begin{cases} 4(2x - y) = 20 \\ 3(x - 2) = -5y - 18 \end{cases}$

Now Try ▶   Problem 29

---

3   Use Substitution and Elimination (Addition) to Identify Inconsistent Systems and Dependent Equations.

We have solved inconsistent systems and systems of dependent equations by graphing. We also can solve these systems using the substitution and elimination methods.

EXAMPLE 6   Solve the system: $\begin{cases} y = 2x + 4 \\ 8x - 4y = 7 \end{cases}$

Strategy   We will use the substitution method to solve this system.

Why   The substitution method works well when one of the equations of the system (in this case, $y = 2x + 4$) is solved for a variable.

Solution   Since the first equation is solved for *y*, we will use the substitution method.

$$y = 2x + 4 \qquad \text{This is the substitution equation.}$$
$$8x - 4y = 7 \qquad \text{This is the second equation of the system.}$$
$$8x - 4(2x + 4) = 7 \qquad \text{Substitute } 2x + 4 \text{ for } y.$$

Now we can try to solve this equation for *x*:

$$8x - 8x - 16 = 7 \qquad \text{Distribute the multiplication by } -4.$$
$$-16 = 7 \qquad \text{Simplify the left side: } 8x - 8x = 0.$$

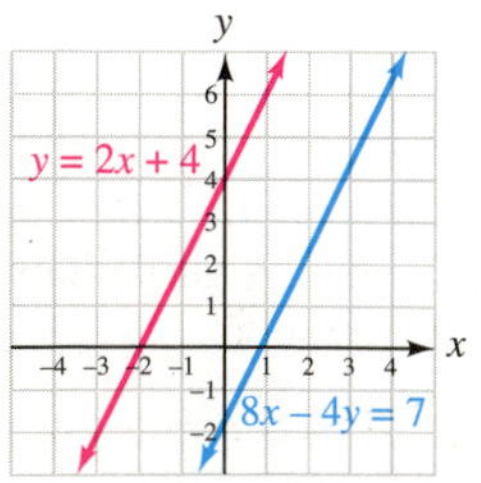

Here, the terms involving *x* drop out, and we get $-16 = 7$. This false statement indicates that the system has *no solution* and is, therefore, inconsistent. The solution set is ∅. The graphs of the equations of the system help to verify this—they appear to be parallel lines, as shown on the left.

Self Check 6   Solve the system: $\begin{cases} x = -2.5y + 8 \\ y = -0.4x + 2 \end{cases}$

Now Try ▶   Problem 33

EXAMPLE 7   Solve the system: $\begin{cases} 4x + 6y = 12 \\ -2x - 3y = -6 \end{cases}$

Strategy   Since both equations are written in standard form, we will use the elimination method.

Why   Since no variable has a coefficient 1 or -1, it would be difficult to solve this system using substitution.

Solution   We can copy the first equation and multiply both sides of the second equation by 2 to get

$$
\begin{array}{rcl}
4x + 6y &=& 12 \\
-4x - 6y &=& -12 \\
\hline
0 &=& 0
\end{array}
$$

When we add like terms, column by column, the result is $0x + 0y = 0$, which simplifies to $0 = 0$.

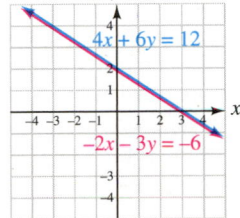

Here, both the x- and y-terms drop out. The resulting true statement $0 = 0$ indicates that the equations are dependent and that the system has *infinitely many solutions*. The solution set is written using set-builder notation as $\{(x, y) \mid 4x + 6y = 12\}$ and is read as "the set of all ordered pairs (x, y) such that $4x + 6y = 12$."

Note that the equations of the system are dependent equations, because when the second equation is multiplied by -2, it becomes the first equation. The graphs of these equations are, therefore, the same line. To find some of the infinitely many solutions of the system, we can substitute 0, 3, and -3 for x in either equation to obtain $(0, 2)$, $(3, 0)$, and $(-3, 4)$.

Self Check 7   Solve the system: $\begin{cases} 2x - 5y = 19 \\ -\dfrac{2}{5}x + y = -\dfrac{19}{5} \end{cases}$

Now Try ▶ Problem 35

Examples 6 and 7 illustrate the following facts.

Inconsistent Systems and Dependent Equations

When solving a system of two linear equations in two variables using substitution or elimination (addition):

1. If the variables drop out and a true statement (identity) is obtained, the system has an infinite number of solutions. The equations are dependent and the system is consistent.

2. If the variables drop out and a false statement (contradiction) is obtained, the system has no solution and is inconsistent.

---

4 Determine the Most Efficient Method to Use to Solve a Linear System.

If no method is specified for solving a particular linear system, the following guidelines can be helpful in determining whether to use graphing, substitution, or elimination.

1. If you want to show trends and see the point that two graphs have in common, use the **graphing method.** However, this method can be lengthy and is not exact.

2. If one of the equations is solved for one of the variables, or easily solved for one of the variables, use the **substitution method.**

3. If both equations are in standard $Ax + By = C$ form, and no variable has a coefficient of 1 or -1, use the **elimination (addition) method.**

4. If the coefficient of one of the variables is 1 or -1, you have a choice. You can write each equation in standard form ($Ax + By = C$) and use elimination, or you can solve for the variable with coefficient 1 or -1 and use substitution.

Here are some examples of suggested approaches:

$$\begin{cases} 5x + 6y = 1 \\ y = x - 4 \end{cases} \qquad \begin{cases} a - 16 = 5b \\ 4a - 8b = -11 \end{cases} \qquad \begin{cases} 3x + 2y = 17 \\ 7x + 5y = -33 \end{cases} \qquad \begin{cases} 3x - y = -1 \\ 9x + 8y = 2 \end{cases}$$

  Substitution   Substitution   Elimination   Elimination or substitution

SECTION 3.2 STUDY SET

VOCABULARY

Fill in the blanks.

1. $Ax + By = C$ is the _____ form of a linear equation.

2. In the equation $x + 3y = -1$, the x-term has an understood _____ of 1.

3. When we add the two equations of the system $\begin{cases} x + y = 5 \\ x - y = -3 \end{cases}$, the y-terms are _____.

4. To solve $\begin{cases} y = 3x + 1 \\ x + y = 4 \end{cases}$, we can _____ $3x + 1$ for y in the second equation.

CONCEPTS

5. If the system $\begin{cases} 4x - 3y = 7 \\ 3x - y = 6 \end{cases}$ is to be solved using the substitution method, what variable in what equation would it be easier to solve for?

6. Given the equation $3x + y = -4$,
   a. solve for x.
   b. solve for y.
   c. Which variable was easier to solve for? Explain why.

7. If the system $\begin{cases} 4x - 3y = 7 \\ 3x - 2y = 6 \end{cases}$ is to be solved using the elimination (addition) method, by what number should each equation be multiplied if
   a. the x-terms are to drop out?
   b. the y-terms are to drop out?

8. Consider the system: $\begin{cases} \dfrac{2}{3}x - \dfrac{y}{6} = \dfrac{16}{9} \\ 0.03x + 0.02y = 0.03 \end{cases}$
   a. What algebraic step should be performed to clear the first equation of fractions?
   b. What algebraic step should be performed to clear the second equation of decimals?

9. Can the system $\begin{cases} 2x + 5y = 7 \\ 4x - 3y = 16 \end{cases}$ be solved more easily by the substitution or the elimination method?

10. The substitution method was used to solve three systems of linear equations. The results after y was eliminated and the remaining equation was solved for x are listed below. Match each result with one of the possible graphs shown.
    a. $-2 = 3$ b. $x = 3$ c. $3 = 3$

    Possible graphs
    i.   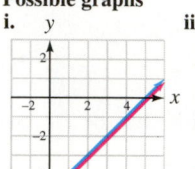   ii.   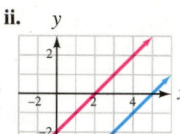   iii.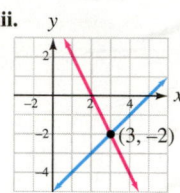
                                                                              $(3,-2)$

NOTATION

11. For the following system, write each equation in standard $Ax + By = C$ form.
    a. $\begin{cases} 4y = 8 - 7x \\ 3x - y = 2(x + 4) \end{cases} \longrightarrow \begin{cases} \quad\quad \\ \quad\quad \end{cases}$

    b. For the following system, clear the equations of any fractions or decimals.
    $\begin{cases} \dfrac{x}{5} + \dfrac{y}{10} = \dfrac{6}{5} \\ 0.3x - 0.9y = 17 \end{cases} \longrightarrow \begin{cases} \quad\quad \\ \quad\quad \end{cases}$

12. Fill in the blanks: We read $\{(x, y) \mid x - 5y = 9\}$ as "the set of all _____ pairs (x, y), _____ _____ $x - 5y = 9$."

GUIDED PRACTICE

Solve each system by substitution. **See Examples 1 and 2.**

13. $\begin{cases} y = 3x \\ x + y = 8 \end{cases}$

14. $\begin{cases} y = x + 2 \\ x + 2y = 16 \end{cases}$

15. $\begin{cases} x = 2 + y \\ 2x + y = 13 \end{cases}$

16. $\begin{cases} x = -5 + y \\ 3x - 2y = -7 \end{cases}$

17. $\begin{cases} x + 2y = 6 \\ 3x - y = -10 \end{cases}$

18. $\begin{cases} 2x - y = -21 \\ 4x + 5y = 7 \end{cases}$

19. $\begin{cases} 0.3a + 0.1b = 0.5 \\ \dfrac{4}{3}a + \dfrac{1}{3}b = 3 \end{cases}$

20. $\begin{cases} 0.9p + 0.2q = 1.2 \\ \dfrac{2}{3}p + \dfrac{1}{9}q = 1 \end{cases}$

Solve each system by elimination. See Examples 3 and 4.

21. $\begin{cases} x - y = 7 \\ x + y = 11 \end{cases}$

22. $\begin{cases} a + b = 5 \\ a - b = 11 \end{cases}$

23. $\begin{cases} 2s + 3t = -8 \\ 2s - 3t = -8 \end{cases}$

24. $\begin{cases} x + 2y = -21 \\ x - 2y = 11 \end{cases}$

25. $\begin{cases} 5x + 2y = 11 \\ 7x + 6y = 9 \end{cases}$

26. $\begin{cases} 3x + 4y = -24 \\ 5x + 12y = -72 \end{cases}$

27. $\begin{cases} 5x + 3y = 72 \\ 3x + 5y = 56 \end{cases}$

28. $\begin{cases} 2x + 3y = 31 \\ 3x + 2y = 39 \end{cases}$

Solve each system by elimination. See Example 5.

29. $\begin{cases} 2(a + b) = 94 \\ 4(a - 9) = 3b - 23 \end{cases}$

30. $\begin{cases} c - 2d = 29 \\ 2(c - 5) = d - 21 \end{cases}$

31. $\begin{cases} 2(x + y) + 1 = 0 \\ 3x + 4y = 0 \end{cases}$

32. $\begin{cases} 5x + 3y = -7 \\ 3(x - y) - 7 = 0 \end{cases}$

Solve each system by any method, if possible. If a system is inconsistent or if the equations are dependent, state this. See Examples 6 and 7.

33. $\begin{cases} 2(a + b) = a + 12 \\ a = 14 - 2b \end{cases}$

34. $\begin{cases} 3c = 2(6 - d) \\ 2d = 3(4 - c) \end{cases}$

35. $\begin{cases} 2x - \dfrac{5}{2} = y \\ 0.04x - 0.02y = 0.05 \end{cases}$

36. $\begin{cases} \dfrac{3}{2}x + 2 = y \\ 0.6x - 0.4y = -0.4 \end{cases}$

TRY IT YOURSELF

Solve each system by any method, if possible. If a system is inconsistent or if the equations are dependent, state this.

37. $\begin{cases} 3x - 4y = 9 \\ x + 2y = 8 \end{cases}$

38. $\begin{cases} 3x - 2y = -10 \\ 6x + 5y = 25 \end{cases}$

39. $\begin{cases} 4(x - 2) = -9y \\ 2(x - 3y) = -3 \end{cases}$

40. $\begin{cases} 2(2x + 3y) = 5 \\ 8x = 3(1 + 3y) \end{cases}$

41. $\begin{cases} 0.16x - 0.08y = 0.32 \\ 2x - 4 = y \end{cases}$

42. $\begin{cases} x = \dfrac{3}{2}y + 5 \\ 2x - 3y = 8 \end{cases}$

43. $\begin{cases} x = \dfrac{2}{3}y \\ y = 4x + 50 \end{cases}$

44. $\begin{cases} y = -2x - 165 \\ x = \dfrac{2}{3}y + 5 \end{cases}$

45. $\begin{cases} \dfrac{m - n}{5} + \dfrac{m + n}{2} = 6 \\ \dfrac{m - n}{2} - \dfrac{m + n}{4} = 3 \end{cases}$

46. $\begin{cases} \dfrac{r - 2}{5} + \dfrac{s + 3}{2} = 5 \\ \dfrac{r + 3}{2} + \dfrac{s - 2}{3} = 6 \end{cases}$

47. $\begin{cases} 0.5x + 0.5y = 6 \\ \dfrac{x}{2} - \dfrac{y}{2} = -2 \end{cases}$

48. $\begin{cases} \dfrac{x}{2} - \dfrac{y}{3} = -4 \\ 0.009x + 0.002y = 0 \end{cases}$

49. $\begin{cases} \dfrac{3}{4}x + \dfrac{2}{3}y = 7 \\ \dfrac{3}{5}x - \dfrac{1}{2}y = 18 \end{cases}$

50. $\begin{cases} \dfrac{2}{3}x - \dfrac{1}{4}y = -8 \\ \dfrac{1}{2}x - \dfrac{3}{8}y = -9 \end{cases}$

51. $\begin{cases} \dfrac{x}{4} = 1 + \dfrac{y}{5} \\ x = \dfrac{4}{5}(y + 10) \end{cases}$

52. $\begin{cases} \dfrac{5}{3}x + 2 = 2(y + 6) \\ 3y + 5 = \dfrac{5}{2}(x - 4) \end{cases}$

53. $\begin{cases} \dfrac{3}{2}x - \dfrac{2}{3}y = 0 \\ \dfrac{3}{4}x + \dfrac{4}{3}y = \dfrac{5}{2} \end{cases}$

54. $\begin{cases} \dfrac{3}{5}x + \dfrac{5}{3}y = 2 \\ \dfrac{6}{5}x - \dfrac{5}{3}y = 1 \end{cases}$

55. $\begin{cases} 12x - 5y - 21 = 0 \\ \dfrac{3}{4}x + \dfrac{2}{3}y = -\dfrac{13}{8} \end{cases}$

56. $\begin{cases} 4y + 5x - 7 = 0 \\ \dfrac{10}{7}x - \dfrac{4}{9}y = \dfrac{17}{21} \end{cases}$

57. $\begin{cases} y = -2.2x + 3.5 \\ y = -1.8x + 2.4 \end{cases}$

58. $\begin{cases} y = -1.6x - 1.6 \\ y = 2.4x + 4.8 \end{cases}$

59. $\begin{cases} 0.05a - 0.03b = 0.24 \\ 0.003a + 0.005b = 0.028 \end{cases}$

60. $\begin{cases} 0.06c - 0.03d = -0.03 \\ 0.004c + 0.005d = -0.009 \end{cases}$

Solve each system. To do so, substitute a for $\frac{1}{x}$ and b for $\frac{1}{y}$ and solve for a and b. Then find x and y using the fact that $a = \frac{1}{x}$ and $b = \frac{1}{y}$.

61. $\begin{cases} \dfrac{1}{x} + \dfrac{1}{y} = \dfrac{5}{6} \\ \dfrac{1}{x} - \dfrac{1}{y} = \dfrac{1}{6} \end{cases}$

62. $\begin{cases} \dfrac{1}{x} + \dfrac{1}{y} = \dfrac{9}{20} \\ \dfrac{1}{x} - \dfrac{1}{y} = \dfrac{1}{20} \end{cases}$

63. $\begin{cases} \dfrac{1}{x} + \dfrac{2}{y} = -1 \\ \dfrac{2}{x} - \dfrac{1}{y} = -7 \end{cases}$

64. $\begin{cases} \dfrac{3}{x} - \dfrac{2}{y} = -30 \\ \dfrac{2}{x} - \dfrac{3}{y} = -30 \end{cases}$

APPLICATIONS

65. Area Codes. The entire state of Alaska (except for the small community of Hyder) has just one telephone area code. The same is true for the state of Hawaii. If we let A = the area code of Alaska and H = the area code of Hawaii, we can find A and H by solving the system:

$$\begin{cases} A + H = 1{,}715 \\ A - H = 99 \end{cases}$$

What is the area code of each state?

66. Driving Big Rigs. The two angle measures shown in the illustration of the jack-knifed trailer can be found by solving the system:

$$\begin{cases} x + y = 180 \\ y = 4x + 5 \end{cases}$$

Find x and y.

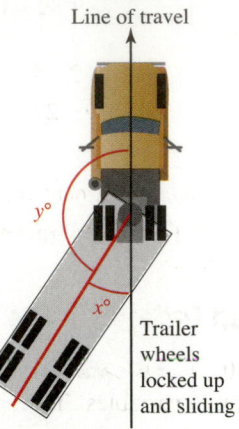

Line of travel

$y°$

$x°$

Trailer wheels locked up and sliding

WRITING

67. Which method would you use to solve the system $\begin{cases} 4x + 6y = 5 \\ 8x - 3y = 3 \end{cases}$? Explain.

68. Which method would you use to solve the system $\begin{cases} x - 2y = 2 \\ 2x + 3y = 11 \end{cases}$? Explain.

69. Why is one of the methods for solving systems that is discussed in this section called the *elimination method*?

70. When using the elimination (addition) method, how can you tell whether

  a. a system of linear equations has no solution?

  b. a system of linear equations has infinitely many solutions?

71. If the elimination (addition) method is to be used to solve this system, what is wrong with the form in which it is written?

$$\begin{cases} 3x - 5y = 10 \\ 8 + 5x = 7y \end{cases}$$

72. Construct a table like that shown below, listing one advantage and one disadvantage for each of the methods that can be used to solve a system of two linear equations in two variables.

Method	Advantage	Disadvantage
Graphing		
Substitution		
Elimination		

REVIEW

Find the slope of each line.

73.

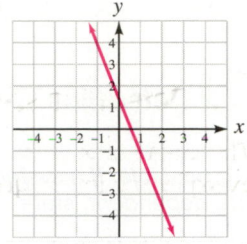

74.

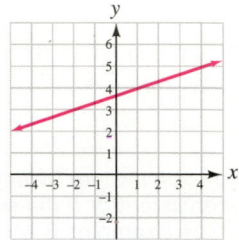

75. The line that passes through $(0, -8)$ and $(-5, 0)$

76. The line with equation $y = -3x + 4$

77. The line with equation $4x - 3y = -3$

78. The line with equation $y = 3$

CHALLENGE PROBLEMS

79. If the solution of the system $\begin{cases} Ax + By = -2 \\ Bx - Ay = -26 \end{cases}$ is $(-3, 5)$, what are the values of the constants A and B?

80. Solve: $\begin{cases} 2ab - 3cd = 1 \\ 3ab - 2cd = 1 \end{cases}$ for a and c. Assume that b and d are nonzero constants.

Solving Systems of Equations in Three Variables

ARE YOU READY?

The following problems review some basic skills that are needed when solving systems of equations in three variables.

1. What is the coefficient of each term on the left side of the equation $x - 4y + 5z = 2$?

2. For $6x + y + 2z = 36$, if $x = 5$ and $y = -2$, what is z?

3. Write the equation $7x + 5y = z + 9$ so that all three variable terms are on the left side.

4. Solve the system by elimination:
$$\begin{cases} 5x + 2y = -5 \\ -6x - 2y = 10 \end{cases}$$

In previous sections, we solved systems of linear equations in two variables. We will now extend this discussion to consider systems of linear equations in *three* variables.

1 Determine Whether an Ordered Triple Is a Solution of a System.

The equation $x - 5y + 7z = 10$, where each variable is raised to the first power, is an example of a linear equation in three variables. In general, we have the following definition.

Standard Form	A **linear equation in three variables** is an equation that can be written in the form $$Ax + By + Cz = D$$ where A, B, C, and D are real numbers and A, B, and C are not all 0.

A solution of a linear equation in three variables is an **ordered triple** of numbers of the form (x, y, z) whose coordinates satisfy the equation. For example, $(2, 0, 1)$ is a solution of $x + y + z = 3$ because a true statement results when we substitute 2 for x, 0 for y, and 1 for z: $2 + 0 + 1 = 3$.

A **solution of a system of three linear equations** in three variables is an ordered triple that satisfies each equation of the system.

EXAMPLE 1 Determine whether $(-4, 2, 5)$ is a solution of the system:
$$\begin{cases} 2x + 3y + 4z = 18 \\ 3x + 4y + z = 1 \\ x + y + 3z = 13 \end{cases}$$

Strategy We will substitute the x-, y-, and z-coordinates of $(-4, 2, 5)$ for the corresponding variables in each equation of the system.

Why If each equation is satisfied by the x-, y-, and z-coordinates, the ordered triple is a solution of the system.

Solution We substitute -4 for x, 2 for y, and 5 for z in each equation.

The first equation

$$2x + 3y + 4z = 18$$
$$2(-4) + 3(2) + 4(5) \stackrel{?}{=} 18$$
$$-8 + 6 + 20 \stackrel{?}{=} 18$$
$$18 = 18 \quad \text{True}$$

The second equation

$$3x + 4y + z = 1$$
$$3(-4) + 4(2) + 5 \stackrel{?}{=} 1$$
$$-12 + 8 + 5 \stackrel{?}{=} 1$$
$$1 = 1 \quad \text{True}$$

The third equation

$$x + y + 3z = 13$$
$$-4 + 2 + 3(5) \stackrel{?}{=} 13$$
$$-4 + 2 + 15 \stackrel{?}{=} 13$$
$$13 = 13 \quad \text{True}$$

Since $(-4, 2, 5)$ satisfies each equation, it is a solution of the system.

Self Check 1   Is $(6, -3, 1)$ a solution of: $\begin{cases} x - y + z = 10 \\ x + 4y - z = -7 \\ 3x - y + 4z = 24 \end{cases}$

Now Try ▶ Problem 11

The Language of Algebra

Recall that when a system of equations has at least one solution, the system is called a **consistent** system, and if a system has no solution, the system is called **inconsistent**.

The graph of an equation of the form $Ax + By + Cz = D$ can be drawn on a coordinate system with three axes. The graph of such an equation is a flat surface called a **plane.** A system of three linear equations with three variables is consistent or inconsistent, depending on how the three planes corresponding to the three equations intersect. The following illustration shows some of the possibilities. Just as in the case of two variables, a system of three linear equations in three variables can have exactly one solution, no solution, or infinitely many solutions, as shown below.

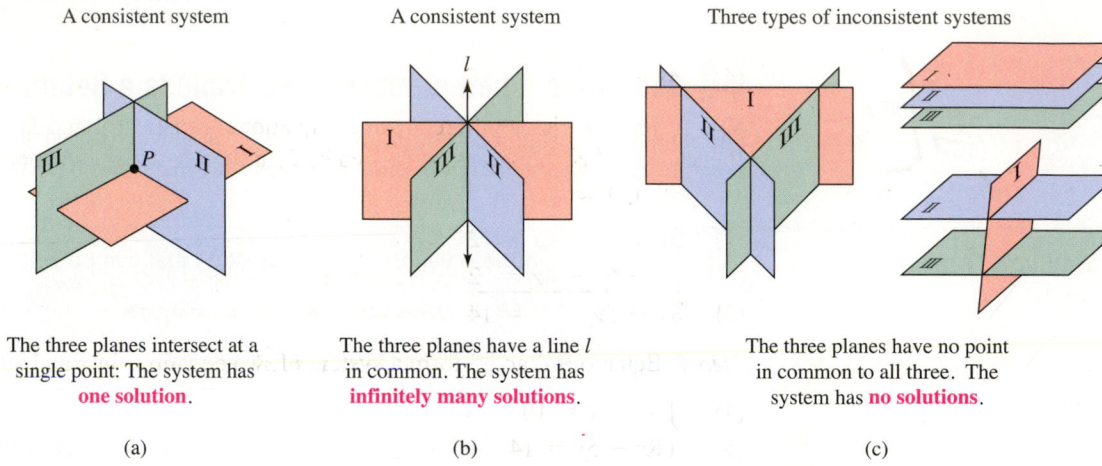

A consistent system   A consistent system   Three types of inconsistent systems

The three planes intersect at a   The three planes have a line l The three planes have no point
single point: The system has   in common. The system has   in common to all three. The
one solution.   **infinitely many solutions**.   system has **no solutions**.

(a)   (b)   (c)

2 Solve Systems of Three Linear Equations in Three Variables.

To **solve a system of three linear equations** in three variables means to find all of the solutions of the system. Solving such a system by graphing is not practical because it requires a three-dimensional coordinate system.

The substitution method is useful to solve systems of three equations where one or more equations have only two variables. However, the best way to solve systems of three linear equations in three variables is usually the elimination method.

Solving a System of Three Linear Equations by Elimination

1. Write each equation in standard form $Ax + By + Cz = D$ and clear any decimals or fractions.

2. Pick any two equations and eliminate a variable.

3. Pick a different pair of equations and eliminate the same variable as in step 1.

4. Solve the resulting pair of two equations in two variables.

5. To find the value of the third variable, substitute the values of the two variables found in step 4 into any equation containing all three variables and solve the equation.

6. Check the proposed solution in all three of the original equations. Write the solution as an ordered triple.

EXAMPLE 2

Solve the system: $\begin{cases} 2x + y + 4z = 12 \\ x + 2y + 2z = 9 \\ 3x - 3y - 2z = 1 \end{cases}$

Strategy Since the coefficients of the z-terms are opposites in the second and third equations, we will add the left and right sides of those equations to eliminate z. Then we will choose another pair of equations and eliminate z again.

Why The result will be a system of two equations in x and y that we can solve by elimination.

Solution

Step 1: We can skip step 1 because each equation is written in standard form and there are no fractions or decimals to clear. We will number each equation and move to step 2.

Notation

We number the equations (1), (2), and (3) to help describe how the system is solved using the elimination method.

(1) $\begin{cases} 2x + y + 4z = 12 \\ (2) \quad x + 2y + 2z = 9 \\ (3) \quad 3x - 3y - 2z = 1 \end{cases}$

Step 2: If we pick equations 2 and 3 and add them, the variable z is eliminated.

(2) $\quad x + 2y + 2z = \ \ 9$
(3) $\quad 3x - 3y - 2z = \ \ 1$
(4) $\quad 4x - \ y \qquad = 10$ This equation does not contain z.

Step 3: We now pick a different pair of equations (equations 1 and 3) and eliminate z again. If each side of equation 3 is multiplied by 2, and the resulting equation is added to equation 1, z is eliminated.

(1) $\quad 2x + \ y + 4z = 12$
$\qquad 6x - 6y - 4z = \ \ 2$ This is 2(3x − 3y − 2z) = 2(1).
(5) $\quad 8x - 5y \qquad = 14$ This equation does not contain z.

Step 4: Equations 4 and 5 form a system of two equations in x and y.

Success Tip

With this method, we use elimination to reduce a system of three equations in three variables to a system of two equations in two variables.

(4) $\begin{cases} 4x - y = 10 \\ (5) \quad 8x - 5y = 14 \end{cases}$

To solve this system, we multiply equation 4 by -5 and add the resulting equation to equation 5 to eliminate y.

$\qquad -20x + 5y = -50$ This is −5(4x − y) = −5(10).
(5) $\qquad \underline{\ \ 8x - 5y = \ \ 14}$
$\qquad -12x \qquad = -36$
$\qquad\qquad x = 3$ Divide both sides by −12. This is the x-value of the solution.

To find y, we substitute 3 for x in any equation containing x and y (such as equation 5) and solve for y:

(5) $\quad 8x - 5y = 14$
$\quad 8(3) - 5y = 14$ Substitute 3 for x.
$\quad 24 - 5y = 14$ Simplify.
$\qquad -5y = -10$ Subtract 24 from both sides.
$\qquad\quad y = 2$ Divide both sides by −5. This is the y-value of the solution.

Step 5: To find z, we substitute 3 for x and 2 for y in any equation containing x, y, and z (such as equation 1) and solve for z:

(1) $\quad 2x + y + 4z = 12$
$\quad 2(3) + 2 + 4z = 12$ Substitute 3 for x and 2 for y.
$\qquad\quad 8 + 4z = 12$ Simplify.
$\qquad\qquad 4z = 4$ Subtract 8 from both sides.
$\qquad\qquad z = 1$ Divide both sides by 4. This is the z-value of the solution.

Step 6: To verify that the solution is $(3, 2, 1)$, we substitute 3 for x, 2 for y, and 1 for z in the three equations of the original system. The solution set is written as $\{(3, 2, 1)\}$. Since this system has a solution, it is a consistent system.

Self Check 2 Solve the system: $\begin{cases} 2x - 3y + 2z = -7 \\ x + 4y - z = 10 \\ 3x + 2y + z = 4 \end{cases}$

Now Try ▶ Problem 15

---

3 Solve Systems of Equations with Missing Variable Terms.

When one or more of the equations of a system is missing a variable term, the elimination of a variable that is normally performed in step 2 of the solution process can be skipped.

EXAMPLE 3 Solve the system: $\begin{cases} 3x = 6 - 2y + z \\ -y - 2z = -8 - x \\ x = 1 - 2z \end{cases}$

Strategy Since the third equation does not contain the variable y, we will work with the first and second equations to obtain another equation that does not contain y.

Why Then we can use the elimination method to solve the resulting system of two equations in x and z.

Solution **Step 1:** We use the addition property of equality to write each equation in the standard form $Ax + By + Cz = D$ and number each equation.

(1) $\begin{cases} 3x + 2y - z = 6 \quad$ Add 2y and subtract z from both sides of $3x = 6 - 2y + z$.
(2) $\phantom{\{} x - y - 2z = -8 \quad$ Add x to both sides of $-y - 2z = -8 - x$.
(3) $\phantom{\{} x + 2z = 1 \quad$ Add 2z to both sides of $x = 1 - 2z$.

Step 2: Since equation 3 does not have a y-term, we can skip to step 3, where we will find another equation that does not contain a y-term.

Step 3: If each side of equation 2 is multiplied by 2 and the resulting equation is added to equation 1, y is eliminated.

(1) $\quad 3x + 2y - z = 6$
$\quad\quad \underline{2x - 2y - 4z = -16} \quad$ This is $2(x - y - 2z) = 2(-8)$.
(4) $\quad 5x - 5z = -10$

Step 4: Equations 3 and 4 form a system of two equations in x and z:

(3) $\begin{cases} x + 2z = 1 \\ 5x - 5z = -10 \end{cases}$
(4)

To solve this system, we multiply equation 3 by -5 and add the resulting equation to equation 4 to eliminate x:

$\quad -5x - 10z = -5 \quad$ This is $-5(x + 2z) = -5(1)$.
(4) $\quad \underline{5x - 5z = -10}$
$\quad\quad\quad -15z = -15$
$\quad\quad\quad\quad z = 1 \quad$ Divide both sides by -15. This is the z-value of the solution.

To find x, we substitute 1 for z in equation 3.

(3) $\quad x + 2z = 1$
$\quad x + 2(1) = 1 \quad$ Substitute 1 for z.
$\quad\quad x + 2 = 1 \quad$ Multiply.
$\quad\quad\quad x = -1 \quad$ Subtract 2 from both sides.

Success Tip

We don't have to find the values of the variables in alphabetical order. In step 2, choose the variable that is the easiest to eliminate. In this example, the value of z is found first.

Step 5: To find y, we substitute -1 for x and 1 for z in equation 1:

(1) $3x + 2y - z = 6$

       $3(-1) + 2y - 1 = 6$ Substitute -1 for x and 1 for z.

           $-3 + 2y - 1 = 6$ Multiply.

                    $2y = 10$ Simplify and add 4 to both sides.

                    $y = 5$ Divide both sides by 2.

The solution of the system is $(-1, 5, 1)$ and the solution set is $\{(-1, 5, 1)\}$.

Step 6: Check the proposed solution in all three of the original equations.

Self Check 3   Solve the system: $\begin{cases} x + 2y = 1 + z \\ 2x = 3 + y - z \\ x + z = 3 \end{cases}$

Now Try ▶ Problem 23

EXAMPLE 4   Solve the system: $\begin{cases} x - y + 4z = -30 & \textbf{(1)} \\ x + 2y = 200 & \textbf{(2)} \\ y + z = 30 & \textbf{(3)} \end{cases}$

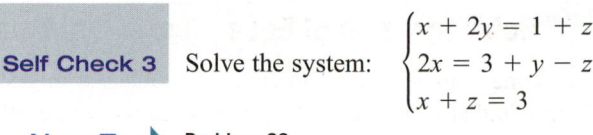

Strategy   Since the second equation does not contain the variable z, we will work with the first and third equations to obtain another equation that does not contain z.

Why   Then we can use the elimination method to solve the resulting system of two equations in x and y.

Solution   *Step 1:* Each of the equations is written in standard form, however, two of the equations are missing variable terms.

Step 2: Since equation 2 does not have a z-term, we can skip to step 3, where we will find another equation that does not contain a z-term.

Step 3: If each side of equation 3 is multiplied by -4 and the resulting equation is added to the equation 1, z is eliminated.

(1) $x - y \;\;\;\; + 4z = -30$

      $\underline{-4y - 4z = -120}$ This is $-4(y + z) = -4(30)$.

(4) $x - 5y \;\;\;\;\;\;\;\;\;\; = -150$

Step 4: Equations 2 and 4 form a system of two equations in x and y.

(2) $\begin{cases} x + 2y = 200 \\ x - 5y = -150 \end{cases}$
(4)

To solve this system, we multiply equation 4 by -1 and add the resulting equation to equation 2 to eliminate x.

(2) $x + 2y = 200$

      $\underline{-x + 5y = 150}$ This is $-1(x - 5y) = -1(-150)$.

               $7y = 350$

                 $y = 50$ To find y, divide both sides by 7.

We then find the value of x to complete step 4 and the value of z (step 5) in the same way as in Examples 2 and 3.

Step 6: Use a check to verify that the solution is $(100, 50, -20)$.

Self Check 4   Solve the system: $\begin{cases} x + y + 3z = 35 \\ x + 3y = -20 \\ 2y + z = -35 \end{cases}$

Now Try ▶ Problem 27

In the following example, we will solve the same system from Example 4 using a different approach.

EXAMPLE 5   Use substitution to solve the system: $\begin{cases} x - y + 4z = -30 \\ x + 2y = 200 \\ y + z = 30 \end{cases}$

Strategy   We will solve the second equation for x and the third equation for z. This creates two substitution equations. Then we will substitute the results for x and for z in the first equation.

Why   These substitutions will produce one equation in the variable y.

Solution   The method we will use is much like solving systems of two equations by substitution. However, we must find two substitution equations, instead of one.

$$\begin{cases} x - y + 4z = -30 \\ x + 2y = 200 \\ y + z = 30 \end{cases} \quad \xrightarrow[\text{Solve for } x]{} \quad \begin{cases} x - y + 4z = -30 & \textbf{(1)} \\ x = \boxed{200 - 2y} & \textbf{(2)} \\ z = \boxed{30 - y} & \textbf{(3)} \end{cases}$$

Since the variable x is isolated in equation 2, we will substitute $200 - 2y$ for x in equation 1, and since the variable z is isolated in equation 3, we will substitute $30 - y$ for z in equation 1. These substitutions will eliminate x and z from equation 1, leaving an equation in one variable, y.

$x - y + 4z = -30$	This is equation 1.
$200 - 2y - y + 4(30 - y) = -30$	Substitute $200 - 2y$ for x and $30 - y$ for z.
$200 - 2y - y + 120 - 4y = -30$	Distribute the multiplication by 4.
$320 - 7y = -30$	On the left side, combine like terms.
$-7y = -350$	Subtract 320 from both sides.
$y = 50$	To solve for y, divide both sides by -7.

> **Success Tip**
>
> Just like with systems of two equations, the objective is to use the appropriate substitutions to obtain one equation in one variable.

As expected, this is the same value for y that we obtained using elimination in Example 4. We can substitute 50 for y in equation 2 to find that $x = 100$ and 50 for y in equation 3 to find that $z = -20$. Using elimination or substitution, we find that the solution is $(100, 50, -20)$.

Self Check 5   Use substitution to solve the system: $\begin{cases} x + y - 4z = -54 \\ x - y = -6 \\ 3y + z = 12 \end{cases}$

Now Try ▶ Problem 31

4 Identify Inconsistent Systems and Dependent Equations.

We have seen that a system of three linear equations in three variables represents three planes. If the planes have no common point of intersection, the system is said to be **inconsistent** with no solution. Illustrations of these types of inconsistent systems are shown in figure c on page 218.

EXAMPLE 6   Solve the system: $\begin{cases} 2a + b - 3c = -3 & \textbf{(1)} \\ 3a - 2b + 4c = 2 & \textbf{(2)} \\ 4a + 2b - 6c = -7 & \textbf{(3)} \end{cases}$

Strategy   Since the coefficients of the b-terms are opposites in the second and third equations, we will add the left and right sides of those equations to eliminate b. Then we will choose another pair of equations and eliminate b again.

Why   The result will be a system of two equations in a and c that we can attempt to solve by elimination.

Solution   We add equations 2 and 3 of the system to eliminate b.

$$\begin{array}{ll} \textbf{(2)} & 3a - 2b + 4c = 2 \\ \textbf{(3)} & \underline{4a + 2b - 6c = -7} \\ \textbf{(4)} & 7a \quad\quad - 2c = -5 \end{array}$$

We can multiply both sides of equation 1 by 2 and add the resulting equation to equation 2 to eliminate b again:

$$\begin{array}{ll} & 4a + 2b - 6c = -6 \quad\quad \textcolor{red}{\text{This is } 2(2a + b - 3c) = 2(-3).} \\ \textbf{(2)} & \underline{3a - 2b + 4c = 2} \\ \textbf{(5)} & 7a \quad\quad - 2c = -4 \end{array}$$

Equations 4 and 5 form a system in a and c.

$$\begin{array}{l} \textbf{(4)} \\ \textbf{(5)} \end{array} \begin{cases} 7a - 2c = -5 \\ 7a - 2c = -4 \end{cases}$$

If we multiply both sides of equation 5 by -1 and add the result to equation 4, the terms involving a and c are both eliminated.

$$\begin{array}{ll} \textbf{(4)} & 7a - 2c = -5 \\ & \underline{-7a + 2c = 4} \quad\quad \textcolor{red}{\text{This is } -1(7a - 2c) = -1(-4).} \\ & \quad\quad\quad 0 = -1 \end{array}$$

Success Tip

If you obtain a false statement at any time in the solution process, you need not proceed. Such an outcome indicates that the system has no solution.

Note that in the solution process, all three variables have been eliminated. The false statement $0 = -1$ indicates that the system has *no solution* and is, therefore, inconsistent. The solution set is $\varnothing$.

Self Check 6   Solve the system: $\begin{cases} 2a + b - 3c = 8 \\ 3a - 2b + 4c = 10 \\ 4a + 2b - 6c = -5 \end{cases}$

Now Try ▶ Problem 35

When the equations in a system of *two* equations with *two* variables are dependent, the system has infinitely many solutions. In such cases, we classify the system as consistent. This is not always true for systems of three equations with three variables. In fact, a system can have dependent equations* and still be inconsistent. The following illustration shows the different possibilities.

A consistent system

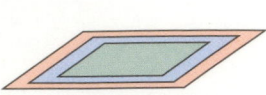

When three planes coincide, the equations are dependent, and there are **infinitely many solutions**.

(a)

A consistent system

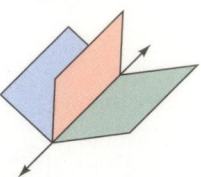

When three planes intersect in a common line, the equations are dependent, and there are **infinitely many solutions**.

(b)

An inconsistent system

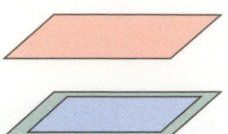

When two planes coincide and are parallel to a third plane, the system is inconsistent, and there are **no solutions**.

(c)

*A set of equations is dependent if at least one equation can be expressed as a sum of multiples of other equations in that set.

EXAMPLE 7　Solve the system: $\begin{cases} \dfrac{4}{5}x - y + z = \dfrac{53}{5} \\ x - 2y - z = 8 \\ 0.2x - 0.3y + 0.1z = 2.3 \end{cases}$

Strategy　We will find an equivalent system without fractions or decimals and use elimination to solve it.

Why　It's easier to solve a system of equations that involves only integers.

Solution　To clear equation 1 of fractions, we multiply both sides by the LCD of the fractions, which is 5. To clear equation 3 of decimals, we multiply both sides by 10.

(1) $\begin{cases} \dfrac{4}{5}x - y + z = \dfrac{53}{5} \end{cases}$ $\xrightarrow{\text{Multiply by 5}}$ $\begin{cases} 4x - 5y + 5z = 53 \end{cases}$ **(4)**

(2) $\begin{cases} x - 2y - z = 8 \end{cases}$ $\xrightarrow{\text{Unchanged}}$ $\begin{cases} x - 2y - z = 8 \end{cases}$ **(2)**

(3) $\begin{cases} 0.2x - 0.3y + 0.1z = 2.3 \end{cases}$ $\xrightarrow{\text{Multiply by 10}}$ $\begin{cases} 2x - 3y + z = 23 \end{cases}$ **(5)**

If each side of equation 2 is multiplied by 5 and the resulting equation is added to equation 4, the variable z is eliminated.

(4)　$4x - 5y + 5z = 53$

$\underline{\ 5x - 10y - 5z = 40}$　This is $5(x - 2y - z) = 5(8)$.

(6)　$9x - 15y = 93$

If we add equations 2 and 5, the variable z is eliminated again.

(2)　$x - 2y - z = 8$

(5)　$\underline{2x - 3y + z = 23}$

(7)　$3x - 5y = 31$

Equations 6 and 7 form a system in x and y. If each side of equation 7 is multiplied by -3 and the resulting equation is added to equation 6, both x and y are eliminated.

(6)　$9x - 15y = 93$

$\underline{-9x + 15y = -93}$　This is $-3(3x - 5y) = -3(31)$.

$0 = 0$

Note that in the solution process, all three variables are eliminated. The resulting true statement, $0 = 0$, indicates that we are working with a set of dependent equations and that the system has an infinite number of solutions.

Success Tip

If you obtain a true statement at any time in the solution process, you need not proceed. Such an outcome indicates that the system contains dependent equations.

Self Check 7 Solve the system:
$$\begin{cases} x - 2y - z = 1 \\ x + \dfrac{4}{3}y + z = \dfrac{5}{3} \\ 0.02x + 0.01y + 0.01z = 0.03 \end{cases}$$

Now Try ▶ Problem 37

SECTION 3.3 ▶ STUDY SET

VOCABULARY

Fill in the blanks.

1. $\begin{cases} 2x + y - 3z = 0 \\ 3x - y + 4z = 5 \\ 4x + 2y - 6z = 0 \end{cases}$ is called a _____ of three linear
   equations in three variables. Each equation is written in
   _____ $Ax + By + Cz = D$ form.

2. If the first two equations of the system in Exercise 1 are added,
   the variable y is _____.

3. Solutions of a system of three equations in three variables, x, y,
   and z, are written in the form (x, y, z) and are called ordered
   _____.

4. The graph of the equation $2x + 3y + 4z = 5$ is a flat surface
   called a _____.

5. When three planes coincide, the equations of the system are
   _____, and there are infinitely many solutions.

6. When three planes intersect in a line, the system will have
   _____ many solutions.

CONCEPTS

7. For each graph of a system of three equations, determine
   whether the solution set contains one solution, infinitely many
   solutions, or no solution.

   a. 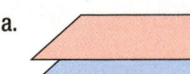 b.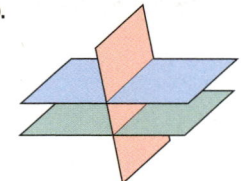

8. Consider the system: (1) $\begin{cases} -2x + y + 4z = 3 \\ x - y + 2z = 1 \\ x + y - 3z = 2 \end{cases}$ (2) (3)

   a. What is the result if equation 1 and equation 2 are added?

   b. What is the result if equation 2 and equation 3 are added?

   c. What variable was eliminated in the steps performed in parts
      (a) and (b)?

NOTATION

9. For the following system, clear the equations of any fractions
   or decimals and write each equation in $Ax + By + Cz = D$
   form.

   $\begin{cases} x + y = 3 - 4z \\ 0.7x - 0.2y + 0.8z = 1.5 \\ \dfrac{x}{2} + \dfrac{y}{3} - \dfrac{z}{6} = \dfrac{2}{3} \end{cases}$ ⟶ $\begin{cases} \\ \\ \end{cases}$

10. What is the purpose of the numbers shown in red in front of the
    equations below?

    (1) $\begin{cases} x + y - z = 6 \\ 2x - y + z = 3 \\ 5x + 3y - z = -2 \end{cases}$
    (2)
    (3)

GUIDED PRACTICE

*Determine whether the ordered triple is a solution of the
system. See Example 1.*

11. $(2, 1, 1)$
    $\begin{cases} x - y + z = 2 \\ 2x + y - z = 4 \\ 2x - 3y + z = 2 \end{cases}$

12. $(-3, 2, -1)$
    $\begin{cases} 3x + y - z = -6 \\ 2x + 2y + 3z = -1 \\ x + y + 2z = 1 \end{cases}$

13. $(6, -7, -5)$
    $\begin{cases} 3x - 2y - z = 37 \\ x - 3y = 27 \\ 2x + 7y + 2z = -48 \end{cases}$

14. $(-4, 0, 9)$
    $\begin{cases} x + 2y - 3z = -31 \\ 2x + 6z = 46 \\ 3x - y = -12 \end{cases}$

Solve each system. See Example 2.

15. $\begin{cases} x + y + z = 4 \\ 2x + y - z = 1 \\ 2x - 3y + z = 1 \end{cases}$

16. $\begin{cases} x + y + z = 4 \\ x - y + z = 2 \\ x - y - 2z = -1 \end{cases}$

17. $\begin{cases} 3x + 2y - 5z = 3 \\ 4x - 2y - 3z = -10 \\ 5x - 2y - 2z = -11 \end{cases}$

18. $\begin{cases} 5x + 4y + 2z = -2 \\ 3x + 4y - 3z = -27 \\ 2x - 4y - 7z = -23 \end{cases}$

19. $\begin{cases} 2x + 6y + 3z = 9 \\ 5x - 3y - 5z = 3 \\ 4x + 3y + 2z = 15 \end{cases}$

20. $\begin{cases} 4x - 3y + 5z = 23 \\ 2x - 5y - 3z = 13 \\ -4x - 6y + 7z = 7 \end{cases}$

21. $\begin{cases} 4x - 5y - 8z = -52 \\ 2x - 3y - 4z = -26 \\ 3x + 7y + 8z = 31 \end{cases}$

22. $\begin{cases} 2x + 6y + 3z = -20 \\ 5x - 3y - 5z = 47 \\ 4x + 3y + 2z = 4 \end{cases}$

43. $\begin{cases} b + 2c = 7 - a \\ a + c = 2(4 - b) \\ 2a + b + c = 9 \end{cases}$

44. $\begin{cases} 0.02a = 0.02 - 0.03b - 0.01c \\ 4a + 6b + 2c - 5 = 0 \\ a + c = 3 + 2b \end{cases}$

Solve each system. See Example 3.

23. $\begin{cases} 3x + 3z = 6 - 4y \\ 7x - 5z = 46 + 2y \\ 4x = 31 - z \end{cases}$

24. $\begin{cases} 5x + 6z = 4y - 21 \\ 9x + 2y = 3z - 47 \\ 3x + y = -19 \end{cases}$

45. $\begin{cases} 2x + y - z = 1 \\ x + 2y + 2z = 2 \\ 4x + 5y + 3z = 3 \end{cases}$

25. $\begin{cases} 2x + z = -2 + y \\ 8x - 3y = -2 \\ 6x - 2y + 3z = -4 \end{cases}$

26. $\begin{cases} 3y + z = -1 \\ -x + 2z = -9 + 6y \\ 9y + 3z = -9 + 2x \end{cases}$

46. $\begin{cases} 2x + 2y + 3z = 10 \\ 3x + y - z = 0 \\ x + y + 2z = 6 \end{cases}$

Solve each system using elimination. See Example 4.

27. $\begin{cases} x + y + 3z = 35 \\ -x - 3y = 20 \\ 2y + z = -35 \end{cases}$

28. $\begin{cases} x + 2y + 3z = 11 \\ 5x - y = 13 \\ 2x - 3z = -11 \end{cases}$

47. $\begin{cases} 0.4x + 0.3z = 0.4 \\ 2y - 6z = -1 \\ 4(2x + y) = 9 - 3z \end{cases}$

48. $\begin{cases} a + b + c = 180 \\ \dfrac{a}{4} + \dfrac{b}{2} + \dfrac{c}{3} = 60 \\ 2b + 3c - 330 = 0 \end{cases}$

29. $\begin{cases} 3x + 2y - z = 7 \\ 6x - 3y = -2 \\ 3y - 2z = 8 \end{cases}$

30. $\begin{cases} 2x + y = 4 \\ -x - 2y + 8z = 7 \\ -y + 4z = 5 \end{cases}$

49. $\begin{cases} r + s + 4t = 3 \\ 3r + 7t = 0 \\ 3s + 5t = 0 \end{cases}$

50. $\begin{cases} x - y = 3 \\ 2x - y + z = 1 \\ x + z = -2 \end{cases}$

Solve each system using substitution. See Example 5.

31. $\begin{cases} r + s - 3t = 21 \\ r + 4s = 9 \\ 5s + t = -4 \end{cases}$

32. $\begin{cases} r - s + 6t = 12 \\ r + 6s = -28 \\ 7s + t = -26 \end{cases}$

51. $\begin{cases} 0.5a + 0.3b = 2.2 \\ 1.2c - 8.5b = -24.4 \\ 3.3c + 1.3a = 29 \end{cases}$

52. $\begin{cases} 4a - 3b = 1 \\ 6a - 8c = 1 \\ 2b - 4c = 0 \end{cases}$

33. $\begin{cases} x - 8z = -30 \\ 3x + y - 4z = 5 \\ y + 7z = 30 \end{cases}$

34. $\begin{cases} x + 6z = -36 \\ 5x + 3y - 2z = -20 \\ y + 4z = -20 \end{cases}$

53. $\begin{cases} 2x + 3y = 6 - 4z \\ 2x = 3y + 4z - 4 \\ 4x + 6y + 8z = 12 \end{cases}$

54. $\begin{cases} -x + 5y - 7z = 0 \\ 4x + y - z = 0 \\ x + y - 4z = 0 \end{cases}$

Solve each system. If a system is inconsistent or if the equations are dependent, state this. See Examples 6 and 7.

35. $\begin{cases} 7a + 9b - 2c = -5 \\ 5a + 14b - c = -11 \\ 2a - 5b - c = 3 \end{cases}$

36. $\begin{cases} 3x + 4y + z = 10 \\ x - 2y + z = -3 \\ 2x + y + z = 5 \end{cases}$

55. $\begin{cases} a + b = 2 + c \\ a = 3 + b - c \\ -a + b + c - 4 = 0 \end{cases}$

56. $\begin{cases} 0.1x - 0.3y + 0.4z = 0.2 \\ 2x + y + 2z = 3 \\ 4x - 5y + 10z = 7 \end{cases}$

37. $\begin{cases} 7x - y - z = 10 \\ x - 3y + z = 2 \\ x + 2y - z = 1 \end{cases}$

38. $\begin{cases} 2a - b + c = 6 \\ -5a - 2b - 4c = -30 \\ a + b + c = 8 \end{cases}$

57. $\begin{cases} x + \dfrac{1}{3}y + z = 13 \\ \dfrac{1}{2}x - y + \dfrac{1}{3}z = -2 \\ x + \dfrac{1}{2}y - \dfrac{1}{3}z = 2 \end{cases}$

58. $\begin{cases} x - \dfrac{1}{5}y - z = 9 \\ \dfrac{1}{4}x + \dfrac{1}{5}y - \dfrac{1}{2}z = 5 \\ 2x + y + \dfrac{1}{6}z = 12 \end{cases}$

TRY IT YOURSELF

Solve each system, if possible. If a system is inconsistent or if the equations are dependent, state this.

39. $\begin{cases} 2a + 3b - 2c = 18 \\ 5a - 6b + c = 21 \\ 4b - 2c - 6 = 0 \end{cases}$

40. $\begin{cases} r - s + t = 4 \\ r + 2s - t = -1 \\ r + s - 3t = -2 \end{cases}$

41. $\begin{cases} 2x + 2y - z = 2 \\ x + 3z - 24 = 0 \\ y = 7 - 4z \end{cases}$

42. $\begin{cases} r - 3t = -11 \\ r + s + t = 13 \\ s - 4t = -12 \end{cases}$

APPLICATIONS

59. Graphs of Systems. Explain how each of the following pictures is an example of the graph of a system of three equations. Then describe the solution, if there is any.

a.

b.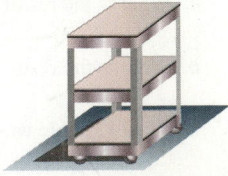

c.

d.

60. Zoology. An X-ray of a mouse revealed a cancerous tumor located at the intersection of the coronal, sagittal, and transverse planes. From this description, would you expect the tumor to be at the base of the tail, on the back, in the stomach, on the tip of the right ear, or in the mouth of the mouse?

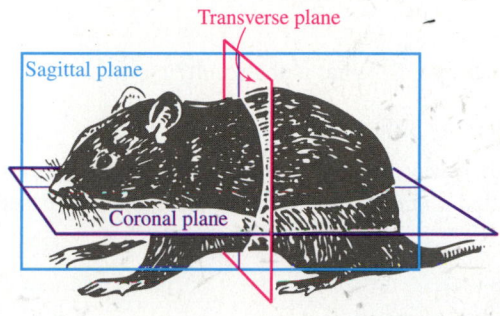

61. NBA Records. The three highest one-game point totals by one player in a National Basketball Association game are shown below. Solve the following system to find x, y, and z.

$$\begin{cases} x + y + z = 259 \\ x - y = 19 \\ x - z = 22 \end{cases}$$

Pts	Player, team	Date
x	Wilt Chamberlain, Philadelphia	3/2/1962
y	Kobe Bryant, Los Angeles	1/22/2006
z	Wilt Chamberlain, Philadelphia	12/8/1961

62. Bicycle Frames. The angle measures of the triangular part of the bicycle frame shown can be found by solving the following system. Find x, y, and z.

$$\begin{cases} x + y + z = 180 \\ x + y = 120 \\ y + z = 135 \end{cases}$$

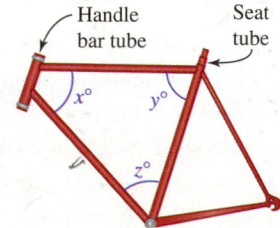

Handle bar tube   Seat tube

WRITING

63. Explain how a system of three equations in three variables can be reduced to a system of two equations in two variables.

64. What makes a system of three equations with three variables inconsistent?

65. What does the graph of a linear equation in three variables such as $2x - 3y + 9z = 10$ look like?

66. What situation discussed in this section looks like two walls of a room and the floor meeting in a corner?

REVIEW

Graph each of the basic functions.

67. $f(x) = |x|$

68. $g(x) = x^2$

69. $h(x) = x^3$

70. $S(x) = x$

CHALLENGE PROBLEMS

Solve each system.

71.
$$\begin{cases} w + x + y + z = 3 \\ w - x + y + z = 1 \\ w + x - y + z = 1 \\ w + x + y - z = 3 \end{cases}$$

72.
$$\begin{cases} \dfrac{1}{x} + \dfrac{1}{y} + \dfrac{1}{z} = 3 \\ \dfrac{2}{x} + \dfrac{1}{y} - \dfrac{1}{z} = 0 \\ \dfrac{1}{x} - \dfrac{2}{y} + \dfrac{4}{z} = 21 \end{cases}$$

73.
$$\begin{cases} 4a + b + 2c - 3d = -16 \\ 3a - 3b + c - 4d = -20 \\ a - 2b - 5c - d = 4 \\ 5a + 4b + 3c - d = -10 \end{cases}$$

74.
$$\begin{cases} a + c + 2d = -4 \\ b - 2c = 1 \\ a + 2b - c = -2 \\ 2a + b + 3c - 2d = -4 \end{cases}$$

SECTION 3.4

SECTION 3.4 — Solving Systems of Equations Using Matrices

OBJECTIVES

1 Define a matrix and determine its order.

2 Write the augmented matrix for a system.

3 Perform elementary row operations on matrices.

4 Use matrices to solve a system of two equations.

5 Use matrices to solve a system of three equations.

6 Use matrices to identify inconsistent systems and dependent equations.

ARE YOU READY?

The following problems review some basic skills that are needed when solving systems of equations using matrices.

1. Consider the system $\begin{cases} 3x - 7y = 14 \\ 5x - y = -9 \end{cases}$. What are the coefficients of the variable terms on the left side of:
   a. the first equation?   b. the second equation?

2. Multiply -2 times 4 and add the result to 9. What is the answer?

3. Multiply -18 by $-\dfrac{1}{18}$. What is the result?

4. Multiply both sides of $2x - 6y = 5$ by -2. What is the result?

In this section, we will discuss another way to solve systems of linear equations. This technique uses a mathematical tool called a *matrix* in a series of steps that are based on the elimination (addition) method.

1 Define a Matrix and Determine Its Order.

Another way to solve systems of equations involves rectangular arrays of numbers called *matrices* (plural of matrix).

Matrices	A **matrix** is any rectangular array of numbers arranged in rows and columns, written within brackets.

The Language of Algebra

An **array** is an orderly arrangement. For example, a jewelry store might display an impressive *array* of gemstones.

The **rows** of a matrix are horizontal and the **columns** are vertical. A pair of brackets is used to write a matrix. Matrices are often named using capital letters. Two examples of matrices are:

$$A = \begin{bmatrix} 1 & -3 & 8 \\ 2 & 5 & -1 \end{bmatrix} \begin{matrix} \leftarrow \text{Row 1} \\ \leftarrow \text{Row 2} \end{matrix} \qquad B = \begin{bmatrix} 1 & 4 & -2 & -4 \\ 6 & -2 & 6 & 1 \\ 3 & 8 & -3 & 12 \end{bmatrix} \begin{matrix} \leftarrow \text{Row 1} \\ \leftarrow \text{Row 2} \\ \leftarrow \text{Row 3} \end{matrix}$$

Column 1   Column 2   Column 3   Column 1   Column 2   Column 3   Column 4

Each number in a matrix is called an **element** or an **entry** of the matrix. A matrix with m rows and n columns has **order** $m \times n$, which is read as "m by n." Because matrix A has two rows and three columns, its order is 2×3. The order of matrix B is 3×4 because it has three rows and four columns.

2 Write the Augmented Matrix for a System.

To show how to use matrices to solve systems of linear equations, consider the following system (written in standard form) and the corresponding **augmented matrix** to its right.

Caution

The equations of a system must be written in standard form before the corresponding augmented matrix can be written.

A system of two linear equations

$$\begin{cases} x - y = 4 \\ 2x + y = 5 \end{cases}$$

Its augmented matrix

$$\begin{bmatrix} 1 & -1 & \vdots & 4 \\ 2 & 1 & \vdots & 5 \end{bmatrix}$$

Each row of the augmented matrix represents one equation of the system. The first two columns of the augmented matrix are determined by the coefficients of x and y in the equations of the system. The last column is determined by the constants in the equations. A dashed vertical bar is written in place of the equal symbols.

$$\begin{bmatrix} 1 & -1 & \vdots & 4 \\ 2 & 1 & \vdots & 5 \end{bmatrix}$$ This row represents the equation $x - y = 4$.
This row represents the equation $2x + y = 5$.

Coefficients   Coefficients   Constants
of x of y

EXAMPLE 1   Represent each system using an augmented matrix:

a. $\begin{cases} 3x + y = 11 \\ x - 8y = 0 \end{cases}$ **b.** $\begin{cases} 2a + b - 3c = -3 \\ 9a + 4c = 2 \\ a - b - 6c = -7 \end{cases}$

Strategy   We will write the coefficients of the variables and the constants from each equation in rows to form a matrix. The coefficients are written to the left of a vertical dashed line and constants to the right.

Why   In an augmented matrix, each row represents one equation of the system.

Solution   Since the equations of each system are written in standard form, we can easily write the corresponding augmented matrices.

a. $\begin{cases} 3x + y = 11 & \leftrightarrow \\ x - 8y = 0 & \leftrightarrow \end{cases} \begin{bmatrix} 3 & 1 & \vdots & 11 \\ 1 & -8 & \vdots & 0 \end{bmatrix}$

b. $\begin{cases} 2a + b - 3c = -3 & \leftrightarrow \\ 9a + 4c = 2 & \leftrightarrow \\ a - b - 6c = -7 & \leftrightarrow \end{cases} \begin{bmatrix} 2 & 1 & -3 & \vdots & -3 \\ 9 & 0 & 4 & \vdots & 2 \\ 1 & -1 & -6 & \vdots & -7 \end{bmatrix}$ In the second row, 0 is entered as the coefficient of the missing b-term.

Self Check 1   Represent each system using an augmented matrix:

a. $\begin{cases} 2x - 4y = 9 \\ 5x - y = -2 \end{cases}$ **b.** $\begin{cases} a + b - c = -4 \\ -2b + 7c = 0 \\ 10a + 8b - 4c = 5 \end{cases}$

Now Try ▶ Problem 13

---

3 **Perform Elementary Row Operations on Matrices.**

To solve a system of linear equations using matrices, we transform the entries to the left of the dashed line in the augmented matrix into an equivalent matrix that has 1's down its **main diagonal** and 0's in all the remaining entries directly above and below this diagonal. A matrix written in this form is said to be in **reduced row-echelon form.** We can easily determine the solution of the associated system of equations when an augmented matrix is written in this form.

Reduced row-echelon form

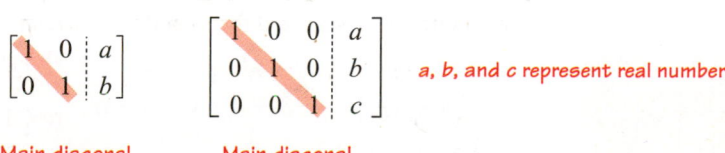

$$\begin{bmatrix} 1 & 0 & \vdots & a \\ 0 & 1 & \vdots & b \end{bmatrix} \qquad \begin{bmatrix} 1 & 0 & 0 & \vdots & a \\ 0 & 1 & 0 & \vdots & b \\ 0 & 0 & 1 & \vdots & c \end{bmatrix}$$ a, b, and c represent real numbers.

Main diagonal   Main diagonal

To write an augmented matrix in reduced row-echelon form, we use three operations called **elementary row operations.**

Elementary Row Operations

Type 1: Any two rows of a matrix can be interchanged.
Type 2: Any row of a matrix can be multiplied by a nonzero constant.
Type 3: Any row of a matrix can be changed by adding a nonzero constant multiple of another row to it.

None of these row operations affect the solution of a given system of equations. The changes to the augmented matrix produce equivalent matrices that correspond to systems with the same solution.

- A type 1 row operation corresponds to interchanging two equations of the system.
- A type 2 row operation corresponds to multiplying both sides of an equation by a nonzero constant.
- A type 3 row operation corresponds to adding a nonzero multiple of one equation to another.

EXAMPLE 2   Perform the following elementary row operations.

$$A = \begin{bmatrix} 2 & 4 & | & -3 \\ 1 & -8 & | & 0 \end{bmatrix} \qquad B = \begin{bmatrix} 1 & -1 & | & 2 \\ 4 & -8 & | & 0 \end{bmatrix} \qquad C = \begin{bmatrix} 2 & 1 & -8 & | & 4 \\ 0 & 1 & 4 & | & -2 \\ 0 & 0 & -6 & | & 24 \end{bmatrix}$$

a. Type 1: Interchange rows 1 and 2 of matrix A.

b. Type 2: Multiply row 3 of matrix C by $-\frac{1}{6}$.

c. Type 3: To the numbers in row 2 of matrix B, add the results of multiplying each number in row 1 by -4.

Strategy   We will perform elementary row operations on each matrix as if we were performing those operations on the equations of a system.

Why   The rows of an augmented matrix correspond to the equations of a system.

Solution   **a.** Interchanging rows 1 and 2 of matrix $A = \begin{bmatrix} 2 & 4 & | & -3 \\ 1 & -8 & | & 0 \end{bmatrix}$ gives $\begin{bmatrix} 1 & -8 & | & 0 \\ 2 & 4 & | & -3 \end{bmatrix}$.

We can represent the instruction to interchange rows 1 and 2 with the symbol $R_1 \leftrightarrow R_2$.

b. We multiply each number in row 3 of matrix $C = \begin{bmatrix} 2 & 1 & -8 & | & 4 \\ 0 & 1 & 4 & | & -2 \\ 0 & 0 & -6 & | & 24 \end{bmatrix}$ by $-\frac{1}{6}$. Note that rows 1 and 2 remain unchanged.

$$\begin{bmatrix} 2 & 1 & -8 & | & 4 \\ 0 & 1 & 4 & | & -2 \\ 0 & 0 & 1 & | & -4 \end{bmatrix}$$

We can represent the instruction to multiply the third row by $-\frac{1}{6}$ with the symbol $-\frac{1}{6}R_3$.

c. We multiply each number from the first row of matrix $B = \begin{bmatrix} 1 & -1 & | & 2 \\ 4 & -8 & | & 0 \end{bmatrix}$ by -4 to get

$$-4 \qquad 4 \qquad -8 \qquad \text{This is } -4R_1.$$

We then add these numbers to the entries in row 2 of matrix B. (Note that row 1 remains unchanged.)

$$\begin{bmatrix} 1 & -1 & | & 2 \\ 4 + (-4) & -8 + 4 & | & 0 + (-8) \end{bmatrix}$$

This procedure is represented by $-4R_1 + R_2$, which means "Multiply row 1 by -4 and add the result to row 2."

After simplifying the bottom row, we have the matrix $\begin{bmatrix} 1 & -1 & | & 2 \\ 0 & -4 & | & -8 \end{bmatrix}$.

4 Use Matrices to Solve a System of Two Equations.

We can solve a system of two linear equations using a series of elementary row operations on the augmented matrix to produce a simpler, equivalent matrix. This process is called **Gauss-Jordan elimination.**

EXAMPLE 3 Use Gauss-Jordan elimination to solve the system: $\begin{cases} 2x + y = 5 \\ x - y = 4 \end{cases}$

Strategy We will represent the system with an augmented matrix and use a series of elementary-row operations to produce an equivalent matrix in reduced row-echelon form.

Why The values of the variables x and y that solve the original system of equations will be in the last column of the reduced row-echelon form matrix.

Solution We can represent the system with the following augmented matrix:

$$\left[\begin{array}{cc|c} 2 & 1 & 5 \\ 1 & -1 & 4 \end{array}\right]$$

The Language of Algebra

This matrix method of solving a system of linear equations is called **Gauss-Jordan elimination,** after the German mathematicians Carl Friedrich Gauss (1777–1855) and Wilhelm Jordan (1842–1899).

First, we want to get a 1 in the top row of the first column where the shaded 2 is. This can be done by applying a type 1 row operation and interchanging rows 1 and 2.

$$\left[\begin{array}{cc|c} 1 & -1 & 4 \\ 2 & 1 & 5 \end{array}\right] \quad R_1 \leftrightarrow R_2$$

To get a 0 in the first column where the shaded 2 is, we use a type 3 row operation and multiply each entry in row 1 by -2 to get

$$-2 \quad 2 \quad -8$$

and add these numbers to the entries in row 2.

$$\left[\begin{array}{cc|c} 1 & -1 & 4 \\ 2 + (-2) & 1 + 2 & 5 + (-8) \end{array}\right] \quad \text{This is } -2R_1 + R_2.$$

Success Tip

Follow this order in getting 1's and 0's in the proper positions of the augmented matrix.

Step 1: $\left[\begin{array}{cc|c} 1 & \blacksquare & \blacksquare \\ \blacksquare & \blacksquare & \blacksquare \end{array}\right]$

Step 2: $\left[\begin{array}{cc|c} 1 & \blacksquare & \blacksquare \\ 0 & \blacksquare & \blacksquare \end{array}\right]$

Step 3: $\left[\begin{array}{cc|c} 1 & \blacksquare & \blacksquare \\ 0 & 1 & \blacksquare \end{array}\right]$

Step 4: $\left[\begin{array}{cc|c} 1 & 0 & \blacksquare \\ 0 & 1 & \blacksquare \end{array}\right]$

After simplifying the bottom row, we have

$$\left[\begin{array}{cc|c} 1 & -1 & 4 \\ 0 & 3 & -3 \end{array}\right]$$

To get a 1 in the bottom row of the second column where the shaded 3 is, we use a type 2 row operation and multiply row 2 by $\frac{1}{3}$.

$$\left[\begin{array}{cc|c} 1 & -1 & 4 \\ 0 & 1 & -1 \end{array}\right] \quad \frac{1}{3}R_2$$

To get a 0 in the first row where the shaded -1 is, we use a type 3 row operation and add to those numbers the entries in row 2 to get

$$\left[\begin{array}{cc|c} 1 + 0 & -1 + 1 & 4 + (-1) \\ 0 & 1 & -1 \end{array}\right] \quad \text{This is } R_1 + R_2.$$

After simplifying the top row, we obtain an equivalent matrix in the desired reduced row-echelon form.

$$\left[\begin{array}{cc|c} 1 & 0 & 3 \\ 0 & 1 & -1 \end{array}\right]$$

This augmented matrix represents a simpler system of equations that is equivalent to the original system:

$$\begin{cases} 1x + 0y = 3 \\ 0x + 1y = -1 \end{cases}$$

Writing the equations without the coefficients, and dropping the $0y$ and $0x$ terms, we have

$$\begin{cases} x = 3 \\ y = -1 \end{cases}$$

Thus, the solution of the given system is $(3, -1)$ and the solution set is $\{(3, -1)\}$. Verify that this ordered pair satisfies both equations in the original system.

Self Check 3   Use Gauss-Jordan elimination to solve the system: $\begin{cases} 2x + y = -9 \\ x - 3y = 13 \end{cases}$

Now Try ▶ Problem 27

When a system of linear equations has one solution, we can use the following steps to solve it.

Solving Systems of Linear Equations Using Gauss-Jordan Elimination

1. Write an augmented matrix for the system.
2. Use elementary row operations to transform the augmented matrix into a matrix in reduced row-echelon form with 1's down its main diagonal and 0's directly above and below the 1's.
3. When step 2 is complete, write the resulting equivalent system to find the solution.
4. Check the proposed solution in the equations of the original system.

5 Use Matrices to Solve a System of Three Equations.

To show how to use Gauss-Jordan elimination to solve systems of three linear equations in three variables, consider the following system (written in standard form) and the corresponding augmented matrix to its right.

A system of three linear equations   *Its augmented matrix*

$$\begin{cases} 3x + y + 5z = 8 \\ 2x + 3y - z = 6 \\ x + 2y + 2z = 10 \end{cases} \qquad \left[\begin{array}{ccc|c} 3 & 1 & 5 & 8 \\ 2 & 3 & -1 & 6 \\ 1 & 2 & 2 & 10 \end{array}\right]$$

To solve the 3×3 system of equations, we transform the augmented matrix into a matrix in reduced row-echelon form with 1's down its main diagonal and 0's directly above and below its main diagonal.

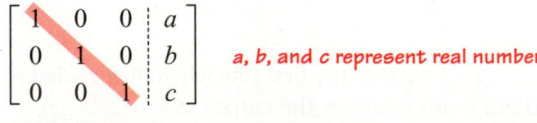

$$\left[\begin{array}{ccc|c} 1 & 0 & 0 & a \\ 0 & 1 & 0 & b \\ 0 & 0 & 1 & c \end{array}\right] \qquad \textit{a, b, and c represent real numbers.}$$

Main diagonal

EXAMPLE 4   Use Gauss-Jordan elimination to solve the system:

$$\begin{cases} 3x + y + 5z = 8 \\ 2x + 3y - z = 6 \\ x + 2y + 2z = 10 \end{cases}$$

Strategy   We will represent the system with an augmented matrix and use a series of elementary row operations to produce an equivalent matrix in reduced row-echelon form.

Why   The values of the variables x, y, and z that solve the original system of equations will be in the last column of the reduced row-echelon form matrix.

Solution   This system can be represented by the augmented matrix

$$\left[\begin{array}{ccc|c} 3 & 1 & 5 & 8 \\ 2 & 3 & -1 & 6 \\ 1 & 2 & 2 & 10 \end{array}\right]$$

To get a 1 in the first column where the shaded 3 is, we perform a type 1 row operation by interchanging rows 1 and 3.

$$\left[\begin{array}{ccc|c} 1 & 2 & 2 & 10 \\ 2 & 3 & -1 & 6 \\ 3 & 1 & 5 & 8 \end{array}\right] \quad R_1 \leftrightarrow R_3$$

To get a 0 where the shaded 2 is, we perform a type 3 row operation by multiplying each entry in row 1 by -2 to get: $\mathbf{-2 \quad -4 \quad -4 \quad -20}$. Then we add these numbers to the entries in row 2. And to get a 0 where the shaded 3 is, we perform another type 3 row operation by multiplying the entries in row 1 by -3 and adding the results to row 3.

$$\left[\begin{array}{ccc|c} 1 & 2 & 2 & 10 \\ 0 & -1 & -5 & -14 \\ 0 & -5 & -1 & -22 \end{array}\right] \quad \begin{array}{l} \text{This is } -2R_1 + R_2. \\ -3R_1 + R_3 \end{array}$$

To get a 1 where the shaded -1 is, we perform a type 2 row operation by multiplying row 2 by -1.

$$\left[\begin{array}{ccc|c} 1 & 2 & 2 & 10 \\ 0 & 1 & 5 & 14 \\ 0 & -5 & -1 & -22 \end{array}\right] \quad -1R_2$$

To get a 0 where the shaded 2 is, we perform a type 3 row operation by multiplying the entries in row 2 by -2 and adding the results to row 1. And to get a 0 where the shaded -5 is, we multiply the entries in row 2 by 5 and add the results to row 3.

$$\left[\begin{array}{ccc|c} 1 & 0 & -8 & -18 \\ 0 & 1 & 5 & 14 \\ 0 & 0 & 24 & 48 \end{array}\right] \quad \begin{array}{l} -2R_2 + R_1 \\ 5R_2 + R_3 \end{array}$$

To get a 1 where the shaded 24 is, we perform a type 2 row operation by multiplying row 3 by $\frac{1}{24}$.

$$\left[\begin{array}{ccc|c} 1 & 0 & -8 & -18 \\ 0 & 1 & 5 & 14 \\ 0 & 0 & 1 & 2 \end{array}\right] \quad \frac{1}{24}R_3$$

To get a 0 where the shaded 5 is, we multiply the entries in row 3 by -5 and add the results to row 2. And to get a 0 where the shaded -8 is, we multiply the entries in row 3 by 8 and add the results to row 1.

$$\left[\begin{array}{ccc|c} 1 & 0 & 0 & -2 \\ 0 & 1 & 0 & 4 \\ 0 & 0 & 1 & 2 \end{array}\right]$$

$8R_3 + R_1$
$-5R_3 + R_2$ This is the desired reduced row-echelon form.

This augmented matrix represents a simpler system of equations that is equivalent to the original system:

$$\begin{cases} 1x + 0y + 0z = -2 \\ 0x + 1y + 0z = 4 \\ 0x + 0y + 1z = 2 \end{cases}$$ which can be written without the coefficients of 0 and 1 as $$\begin{cases} x = -2 \\ y = 4 \\ z = 2 \end{cases}$$

Thus, the solution of the given system is $(-2, 4, 2)$ and the solution set is $\{(-2, 4, 2)\}$. Verify that this ordered triple satisfies each equation of the original system.

Self Check 4   Use Gauss-Jordan elimination to solve: $$\begin{cases} 2x - y + z = 5 \\ x + y - z = -2 \\ -x + 2y + 2z = 1 \end{cases}$$

Now Try ▶ Problem 29

6 Use Matrices to Identify Inconsistent Systems and Dependent Equations.

In the next example, we will see how to recognize inconsistent systems and systems of dependent equations when matrices are used to solve them.

EXAMPLE 5   Use Gauss-Jordan elimination to solve the system:

a. $\begin{cases} x + y = -1 \\ -3x - 3y = -5 \end{cases}$ **b.** $\begin{cases} 2x - y = 4 \\ -6x + 3y = -12 \end{cases}$

Strategy   We will represent the system with an augmented matrix and attempt to use Gauss-Jordan elimination to produce an equivalent matrix in reduced row-echelon form.

Why   It is easy to recognize an inconsistent system or dependent equations during that process.

Solution   **a.** The system $\begin{cases} x + y = -1 \\ -3x - 3y = -5 \end{cases}$ can be represented by the augmented matrix

$$\left[\begin{array}{cc|c} 1 & 1 & -1 \\ -3 & -3 & -5 \end{array}\right]$$

Since the matrix has a 1 in the top row of the first column, we proceed to get a 0 where the shaded -3 is by multiplying row 1 by 3 and adding the results to row 2.

$$\left[\begin{array}{cc|c} 1 & 1 & -1 \\ 0 & 0 & -8 \end{array}\right]$$ $3R_1 + R_2$

The second row of this augmented matrix represents the equation $0 = -8$. This false statement indicates that the system is inconsistent and has no solution. The solution set is $\varnothing$.

b. The system $\begin{cases} 2x - y = 4 \\ -6x + 3y = -12 \end{cases}$ can be represented by the augmented matrix

$$\begin{bmatrix} \boxed{2} & -1 & \vdots & 4 \\ -6 & 3 & \vdots & -12 \end{bmatrix}$$

To get a 1 where the shaded 2 is, we perform a type 2 row operation by multiplying row 1 by $\frac{1}{2}$.

$$\begin{bmatrix} 1 & -\dfrac{1}{2} & \vdots & 2 \\ \boxed{-6} & 3 & \vdots & -12 \end{bmatrix} \quad \frac{1}{2}R_1$$

To get a 0 where the shaded -6 is, we perform a type 3 row operation by multiplying the entries in row 1 by 6 and adding the results to the entries in row 2.

$$\begin{bmatrix} 1 & -\dfrac{1}{2} & \vdots & 2 \\ 0 & 0 & \vdots & 0 \end{bmatrix} \quad 6R_1 + R_2$$

The second row of this augmented matrix represents the equation $0 = 0$. This true statement indicates that the equations are dependent and that the system has infinitely many solutions. The solution set is $\{(x, y) \mid 2x - y = 4\}$.

Self Check 5   Use Gauss-Jordan elimination to solve the system:

a. $\begin{cases} 4x - 8y = 9 \\ x - 2y = -5 \end{cases}$

b. $\begin{cases} x - 3y = 6 \\ -4x + 12y = -24 \end{cases}$

Now Try ▶ Problems 33 and 35

Using Your Calculator ▶ **Augmented Matrices**

Systems of linear equations like those found in this section can be solved on a graphing calculator by entering their corresponding augmented matrices. For example, the augmented matrix for the system in Example 4 is shown below in figure (a). Note that it has been defined to be matrix "A." See your owner's manual for the specific instructions on how to enter the elements of a matrix.

We then use the reduced row-echelon matrix command (on a TI-84, locate **B:rref** by pressing MATRIX, arrowing to MATH, and scrolling down) to put the matrix in reduced row-echelon form, as shown in figure (b). The solution of the given system, $(-2, 4, 2)$, can be read from the last column of the display.

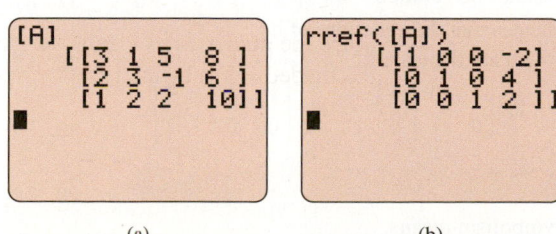

(a)   (b)

SECTION 3.4 ▶ STUDY SET

VOCABULARY

Fill in the blanks.

1. A _____ is a rectangular array of numbers written within brackets.

2. Each number in a matrix is called an _____ or entry of the matrix.

3. If the order of a matrix is 3×4, it has 3 _____ and 4 _____. We read 3×4 as "3 ___ 4."

4. A matrix that represents the equations of a system is called an _____ matrix.

5. Elementary _____ operations can be used on an augmented matrix to produce a simpler equivalent matrix that gives the solution of a system. This process is called _____-_____ elimination.

6. The matrix $\begin{bmatrix} 1 & 0 & \vdots & -2 \\ 0 & 1 & \vdots & 4 \end{bmatrix}$, with 1's down its main _____ and 0's directly above and below it, is in reduced row-_____ form.

CONCEPTS

7. For each matrix, determine the number of rows and the number of columns.

   a. $\begin{bmatrix} 4 & 6 & \vdots & -1 \\ 1 & 9 & \vdots & -3 \end{bmatrix}$

   b. $\begin{bmatrix} 1 & -2 & 3 & \vdots & 1 \\ 0 & 1 & 6 & \vdots & 4 \\ 0 & 0 & 1 & \vdots & \frac{1}{3} \end{bmatrix}$

8. Fill in the blanks to complete each elementary row operation:

   a. Type 1: Any two rows of a matrix can be _____.

   b. Type 2: Any row of a matrix can be _____ by a nonzero constant.

   c. Type 3: Any row of a matrix can be changed by _____ a nonzero constant multiple of another row to it.

9. Gauss-Jordan elimination was used to solve a system. The final augmented matrix is shown. Fill in the blanks.

   a. $\begin{bmatrix} 1 & 0 & \vdots & 10 \\ 0 & 1 & \vdots & 6 \end{bmatrix}$ represents $\begin{cases} \quad = 10 \\ \quad = 6 \end{cases}$

   The solution of the system is (,).

   b. $\begin{bmatrix} 1 & 0 & 0 & \vdots & -16 \\ 0 & 1 & 0 & \vdots & 8 \\ 0 & 0 & 1 & \vdots & 4 \end{bmatrix}$ represents $\begin{cases} \quad = -16 \\ y = 8 \\ \quad = 4 \end{cases}$

   The solution of the system is (, ,).

10. a. Which matrix shown below indicates that its corresponding system of equations has no solution?

    b. Which matrix indicates that the equations of its corresponding system are dependent?

    i. $\begin{bmatrix} 1 & 2 & \vdots & -4 \\ 0 & 0 & \vdots & 0 \end{bmatrix}$ ii. $\begin{bmatrix} 1 & 0 & \vdots & 6 \\ 0 & 1 & \vdots & 0 \end{bmatrix}$ iii. $\begin{bmatrix} 1 & 2 & \vdots & -4 \\ 0 & 0 & \vdots & 2 \end{bmatrix}$

NOTATION

11. Explain what each symbolism means.

    a. $R_1 \leftrightarrow R_2$

    b. $\frac{1}{2} R_1$

    c. $6R_2 + R_3$

12. Complete the solution to solve the system using matrices.

$$\begin{cases} 4x - y = 14 \\ x + y = 6 \end{cases}$$

$\begin{bmatrix} 4 & \blacksquare & \vdots & 14 \\ 1 & 1 & \vdots & 6 \end{bmatrix}$ *This is the augmented matrix.*

$\begin{bmatrix} \blacksquare & 1 & \vdots & 6 \\ 4 & -1 & \vdots & 14 \end{bmatrix}$ $R_1 \leftrightarrow \blacksquare$

$\begin{bmatrix} 1 & 1 & \vdots & 6 \\ 0 & \blacksquare & \vdots & -10 \end{bmatrix}$ $-4R_1 + \blacksquare$

$\begin{bmatrix} 1 & 1 & \vdots & 6 \\ 0 & \blacksquare & \vdots & 2 \end{bmatrix}$ $-\frac{1}{5}R_2$

$\begin{bmatrix} 1 & 0 & \vdots & 4 \\ 0 & 1 & \vdots & 2 \end{bmatrix}$ $-R_2 + \blacksquare$

This augmented matrix represents the equivalent system

$$\begin{cases} x = 4 \\ \blacksquare = 2 \end{cases}$$

The solution is (, 2).

GUIDED PRACTICE

Represent each system using an augmented matrix. See Example 1.

13. $\begin{cases} x + 2y = 6 \\ 3x - y = -10 \end{cases}$

14. $\begin{cases} x + y + z = 4 \\ 2x + y - z = 1 \\ 2x - 3y = 1 \end{cases}$

For each augmented matrix, give the system of equations that it represents.

15. $\begin{bmatrix} 1 & 6 & \vdots & 7 \\ 0 & 1 & \vdots & 4 \end{bmatrix}$

16. $\begin{bmatrix} 1 & -2 & 9 & \vdots & 1 \\ 0 & 1 & 4 & \vdots & 0 \\ 0 & 0 & 1 & \vdots & -7 \end{bmatrix}$

Perform each of the following elementary row operations on the augumented matrix $\begin{bmatrix} -3 & 1 & \vdots & -6 \\ 1 & -4 & \vdots & 4 \end{bmatrix}$. *See Example 2.*

17. $R_1 \leftrightarrow R_2$

18. $5R_2$

19. $-\frac{1}{3}R_1$

20. $3R_2 + R_1$

Perform each of the following elementary row operations

on the augmented matrix $\begin{bmatrix} 3 & 6 & -9 & | & 0 \\ 1 & 5 & -2 & | & 1 \\ -2 & 2 & -2 & | & 5 \end{bmatrix}$. *See Example 2.*

21. $R_2 \leftrightarrow R_3$

22. $-\frac{1}{2}R_3$

23. $-R_1 + R_2$

24. $2R_2 + R_3$

Use matrices to solve each system of equations. See Example 3.

25. $\begin{cases} x + y = 2 \\ x - y = 0 \end{cases}$

26. $\begin{cases} x + y = 3 \\ x - y = -1 \end{cases}$

27. $\begin{cases} 2x + y = 1 \\ x + 2y = -4 \end{cases}$

28. $\begin{cases} 5x - 4y = 10 \\ x - 7y = 2 \end{cases}$

Use matrices to solve each system of equations. See Example 4.

29. $\begin{cases} x + y + z = 6 \\ x + 2y + z = 8 \\ x + y + 2z = 7 \end{cases}$

30. $\begin{cases} x + y + z = 6 \\ x + 2y + z = 8 \\ x + y + 2z = 9 \end{cases}$

31. $\begin{cases} 3x + y - 3z = 5 \\ x - 2y + 4z = 10 \\ x + y + z = 13 \end{cases}$

32. $\begin{cases} 2x + y - 3z = -1 \\ 3x - 2y - z = -5 \\ x - 3y - 2z = -12 \end{cases}$

Use matrices to solve each system of equations. If the equations of a system are dependent or if a system is inconsistent, state this. See Example 5.

33. $\begin{cases} x - 3y = 9 \\ -2x + 6y = 18 \end{cases}$

34. $\begin{cases} -6x + 12y = 10 \\ 2x - 4y = 8 \end{cases}$

35. $\begin{cases} -4x - 4y = -12 \\ x + y = 3 \end{cases}$

36. $\begin{cases} 5x - 15y = 10 \\ x - 3y = 2 \end{cases}$

TRY IT YOURSELF

Use matrices to solve each system of equations. If the equations of a system are dependent or if a system is inconsistent, state this.

37. $\begin{cases} 2x + 3y - z = -8 \\ x - y - z = -2 \\ -4x + 3y + z = 6 \end{cases}$

38. $\begin{cases} 2a + b + 3c = 3 \\ -2a - b + c = 5 \\ 4a - 2b + 2c = 2 \end{cases}$

39. $\begin{cases} 2x - y = -1 \\ x - 2y = 1 \end{cases}$

40. $\begin{cases} 2x - y = 0 \\ x + y = 3 \end{cases}$

41. $\begin{cases} 3x + 4y = -12 \\ 9x - 2y = 6 \end{cases}$

42. $\begin{cases} 2x - 3y = 16 \\ -4x + y = -22 \end{cases}$

43. $\begin{cases} 2x + y - z = 1 \\ x + 2y + 2z = 2 \\ 4x + 5y + 3z = 3 \end{cases}$

44. $\begin{cases} x - y = 1 \\ 2x - z = 0 \\ 2y - z = -2 \end{cases}$

45. $\begin{cases} 8x - 2y = 4 \\ 4x - y = 2 \end{cases}$

46. $\begin{cases} 9x - 3y = 6 \\ 3x - y = 8 \end{cases}$

47. $\begin{cases} 2x + y - 2z = 6 \\ 4x - y + z = -1 \\ 6x - 2y + 3z = -5 \end{cases}$

48. $\begin{cases} 2x - 3y + 3z = 14 \\ 3x + 3y - z = 2 \\ -2x + 6y + 5z = 9 \end{cases}$

49. $\begin{cases} 6x + y - z = -2 \\ x + 2y + z = 5 \\ 5y - z = 2 \end{cases}$

50. $\begin{cases} 2x + 3y - 2z = 18 \\ 5x - 6y + z = 21 \\ 4y - 2z = 6 \end{cases}$

51. $\begin{cases} 5x + 3y = 4 \\ 3y - 4z = 4 \\ x + z = 1 \end{cases}$

52. $\begin{cases} y + 2z = -2 \\ x + y = 1 \\ 2x - z = 0 \end{cases}$

APPLICATIONS

53. Digital Photography. A digital camera stores the black and white photograph shown on the right as a 512×512 matrix. Each element of the matrix corresponds to a small dot of grey scale shading, called a *pixel*, in the picture. How many elements does a 512×512 matrix have?

© Cengage Learning/Heinle Image Resource Bank

54. Digital Imaging. A scanner stores a black and white photograph as a matrix that has a total of 307,200 elements. If the matrix has 480 rows, how many columns does it have?

55. Complementary Angles. The following system can be used to find the angle measures (in degrees) of the two complementary angles shown in the illustration below. Solve the system using matrices to find the measure of each angle. (*Hint:* Each equation must be in standard $Ax + By = C$ form.)

$$\begin{cases} x + y = 90 \\ y = x + 46 \end{cases}$$

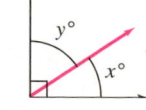

56. Supplementary Angles. The following system can be used to find the angle measures (in degrees) of the two supplementary angles shown in the illustration below. Solve the system using matrices to find the measure of each angle. (*Hint:* Each equation must be in standard $Ax + By = C$ form.)

$$\begin{cases} x + y = 180 \\ y = x - 14 \end{cases}$$

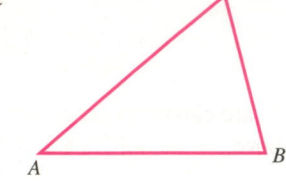

57. Triangles. The following system can be used to find the measures (in degrees) of $\angle A$, $\angle B$, and $\angle C$ shown in the illustration below. Solve the system using matrices to find the measure of each angle of the triangle. (*Hint:* Each equation must be in standard $Ax + By = C$ form.)

$$\begin{cases} A + B + C = 180 \\ B = A + 25 \\ C = 2A - 5 \end{cases}$$

58. Triangles. The following system can be used to find the measures (in degrees) of $\angle A$, $\angle B$, and $\angle C$ shown in the illustration below. Solve the system using matrices to find the measure of each angle of the triangle. (*Hint:* Each equation must be in standard $Ax + By = C$ form.)

$$\begin{cases} A + B + C = 180 \\ A = B - 10 \\ B = C - 10 \end{cases}$$

WRITING

59. For the system $\begin{cases} 2x - 3y = 5 \\ 4x + 8 = y \end{cases}$, explain what is wrong with writing its corresponding augmented matrix as $\begin{bmatrix} 2 & -3 & | & 5 \\ 4 & 8 & | & 1 \end{bmatrix}$. How should it be written?

60. Explain what is meant by the phrase *reduced row-echelon form*. Give an example of how it was used in this section.

61. Explain how a type 3 row operation is similar to the elimination (addition) method of solving a system of equations.

62. If the system represented by the augmented matrix at the right has no solution, what do you know about k? Explain your answer.
$$\begin{bmatrix} 1 & 1 & 0 & | & 1 \\ 0 & 0 & 1 & | & 2 \\ 0 & 0 & 0 & | & k \end{bmatrix}$$

REVIEW

63. What is the formula used to find the slope of a line, given two points on the line?

64. What is the form of the equation of a horizontal line? Of a vertical line?

65. What is the point–slope form of the equation of a line?

66. What is the slope–intercept form of the equation of a line?

CHALLENGE PROBLEMS

Use matrices to solve the system.

67. $\begin{cases} x^2 + y^2 + z^2 = 14 \\ 2x^2 + 3y^2 - 2z^2 = -7 \\ x^2 - 5y^2 + z^2 = 8 \end{cases}$

68. $\begin{cases} w + x + y + z = 0 \\ w - 2x + y - 3z = -3 \\ 2w + 3x + y - 2z = -1 \\ 2w - 2x - 2y + z = -12 \end{cases}$

SECTION 3.5

Solving Systems of Equations Using Determinants

OBJECTIVES

1. Evaluate 2×2 and 3×3 determinants.

2. Use Cramer's rule to solve systems of two equations.

3. Use Cramer's rule to solve systems of three equations.

ARE YOU READY?

The following problems review some basic skills that are needed when solving systems of equations using determinants.

1. How many rows and columns does the matrix $\begin{bmatrix} 2 & -4 & 9 \\ 1 & 0 & 4 \\ -6 & 5 & 11 \end{bmatrix}$ have?

2. What numbers lie on the main diagonal of the matrix $\begin{bmatrix} 3 & 5 \\ 6 & -1 \end{bmatrix}$?

3. Evaluate: $6(-8) - (-2)(-3)$

4. Evaluate: $6[3 - (-1)] - 9[7 - (-8)] + 4(-2 - 1)$

In this section, we will discuss another method for solving systems of linear equations. With this method, called *Cramer's rule,* we work with combinations of the coefficients and the constants of the equations written as *determinants.*

1 Evaluate 2 × 2 and 3 × 3 Determinants.

An idea related to the concept of matrix is the **determinant.** A determinant is a number that is associated with a **square matrix,** a matrix that has the same number of rows and columns. For any square matrix A, the symbol $|A|$ represents the determinant of A. To write a determinant, we put the elements of a square matrix between two vertical lines.

<table>
<tr><td>

$$\begin{bmatrix} 3 & 2 \\ 6 & 9 \end{bmatrix}$$

Matrix
</td><td>

$$\begin{vmatrix} 3 & 2 \\ 6 & 9 \end{vmatrix}$$

Determinant
</td><td>

$$\begin{bmatrix} 9 & 3 & -2 \\ 0 & 1 & 8 \\ 1 & 2 & 4 \end{bmatrix}$$

Matrix
</td><td>

$$\begin{vmatrix} 9 & 3 & -2 \\ 0 & 1 & 8 \\ 1 & 2 & 4 \end{vmatrix}$$

Determinant
</td></tr>
</table>

Like matrices, determinants are classified according to the number of rows and columns they contain. The determinant above, on the left, is a 2 × 2 determinant. The other is a 3 × 3 determinant.

The determinant of a 2 × 2 matrix is the number that is equal to the product of the numbers on the main diagonal minus the product of the numbers on the other diagonal.

$$\begin{vmatrix} a & b \\ c & d \end{vmatrix}$$
Main diagonal

$$\begin{vmatrix} a & b \\ c & d \end{vmatrix}$$
Other diagonal

Notation

Brackets are used to write matrices and vertical bars are used to write determinants.

$$\begin{bmatrix} 2 & 1 \\ 8 & -3 \end{bmatrix}$$
Matrix

$$\begin{vmatrix} 7 & -6 \\ 4 & 0 \end{vmatrix}$$
Determinant

Value of a 2 × 2 Determinant

If a, b, c, and d are numbers, the **determinant** of the matrix $\begin{bmatrix} a & b \\ c & d \end{bmatrix}$ is

$$\begin{vmatrix} a & b \\ c & d \end{vmatrix} = ad - bc$$

EXAMPLE 1 Evaluate each determinant: **a.** $\begin{vmatrix} 3 & 2 \\ 6 & 9 \end{vmatrix}$ **b.** $\begin{vmatrix} -20 & 1 \\ -8 & 4 \end{vmatrix}$

Strategy We will find the product of the numbers on the main diagonal and the product of the numbers along the other diagonal and subtract the results.

Why The value of a determinant of the form $\begin{vmatrix} a & b \\ c & d \end{vmatrix}$ is $ad - bc$.

Solution To evaluate these 2 × 2 determinants, we proceed as follows:

a.

This is always minus.

$$\begin{vmatrix} 3 & 2 \\ 6 & 9 \end{vmatrix} = 3(9) - 2(6) = 27 - 12 = 15$$

b.

$$\begin{vmatrix} -20 & 1 \\ -8 & 4 \end{vmatrix} = -20(4) - 1(-8) = -80 - (-8) = -80 + 8 = -72$$

Caution

Although the symbols used to write a determinant are a pair of vertical lines just like the vertical lines used for absolute value, the two concepts are unrelated. In part (b), notice that the value of the determinant is negative.

Self Check 1 Evaluate: $\begin{vmatrix} 4 & -3 \\ 2 & 1 \end{vmatrix}$

Now Try ▶ Problems 15 and 17

A 3×3 determinant is evaluated by **expanding by minors.** The following definition shows how we can evaluate a 3×3 determinant by expanding by minors along the first row.

Value of a 3×3 Determinant

$$\begin{vmatrix} a_1 & b_1 & c_1 \\ a_2 & b_2 & c_2 \\ a_3 & b_3 & c_3 \end{vmatrix} = a_1 \overbrace{\begin{vmatrix} b_2 & c_2 \\ b_3 & c_3 \end{vmatrix}}^{\text{Minor of } a_1} - b_1 \overbrace{\begin{vmatrix} a_2 & c_2 \\ a_3 & c_3 \end{vmatrix}}^{\text{Minor of } b_1} + c_1 \overbrace{\begin{vmatrix} a_2 & b_2 \\ a_3 & b_3 \end{vmatrix}}^{\text{Minor of } c_1}$$

The Language of Algebra

The 2×2 determinants in the definition box above are called **minors** of the elements in the 3×3 determinant.

To find the minor of a_1, we cross out the elements of the determinant that are in the same row and column as a_1:

$$\begin{vmatrix} a_1 & b_1 & c_1 \\ a_2 & b_2 & c_2 \\ a_3 & b_3 & c_3 \end{vmatrix} \quad \text{The minor of } a_1 \text{ is } \begin{vmatrix} b_2 & c_2 \\ b_3 & c_3 \end{vmatrix}.$$

To find the minor of b_1, we cross out the elements of the determinant that are in the same row and column as b_1:

$$\begin{vmatrix} a_1 & b_1 & c_1 \\ a_2 & b_2 & c_2 \\ a_3 & b_3 & c_3 \end{vmatrix} \quad \text{The minor of } b_1 \text{ is } \begin{vmatrix} a_2 & c_2 \\ a_3 & c_3 \end{vmatrix}.$$

To find the minor of c_1, we cross out the elements of the determinant that are in the same row and column as c_1:

$$\begin{vmatrix} a_1 & b_1 & c_1 \\ a_2 & b_2 & c_2 \\ a_3 & b_3 & c_3 \end{vmatrix} \quad \text{The minor of } c_1 \text{ is } \begin{vmatrix} a_2 & b_2 \\ a_3 & b_3 \end{vmatrix}.$$

EXAMPLE 2   Evaluate the determinant: $\begin{vmatrix} 1 & 3 & -2 \\ 2 & 0 & 3 \\ 1 & 2 & 3 \end{vmatrix}$

Strategy   We will expand the determinant along the first row using the numbers in the first row and their corresponding minors.

Why   We can then evaluate the resulting 2×2 determinants and simplify.

Solution   To evaluate this 3×3 determinant, we can use the first row and expand the determinant by minors:

$$\begin{vmatrix} 1 & 3 & -2 \\ 2 & 0 & 3 \\ 1 & 2 & 3 \end{vmatrix} = 1 \overbrace{\begin{vmatrix} 0 & 3 \\ 2 & 3 \end{vmatrix}}^{\text{Minor of } 1} - 3 \overbrace{\begin{vmatrix} 2 & 3 \\ 1 & 3 \end{vmatrix}}^{\text{Minor of } 3} + (-2) \overbrace{\begin{vmatrix} 2 & 0 \\ 1 & 2 \end{vmatrix}}^{\text{Minor of } -2}$$

$= 1(0 - 6) - 3(6 - 3) - 2(4 - 0)$ Evaluate each 2×2 determinant.

$= 1(-6) - 3(3) - 2(4)$ Do each subtraction within parentheses.

$= -6 - 9 - 8$ Do each multiplication.

$= -23$ Do the subtraction.

Self Check 2

Evaluate: $\begin{vmatrix} 2 & 3 & -1 \\ 0 & 2 & 4 \\ -2 & 5 & 6 \end{vmatrix}$

Now Try ▶ Problem 27

We can evaluate a 3×3 determinant by expanding it by the minors of any row or column. We will get the same value. To determine the signs between the terms of the expansion of a 3×3 determinant, we use the following array of signs.

Array of Signs for a 3 × 3 Determinant	$\begin{array}{ccc} + & - & + \\ - & + & - \\ + & - & + \end{array}$ *This array of signs is often called the checkerboard pattern.*

To remember the sign pattern, note that there is a $+$ sign in the upper left position and that the signs alternate for all of the positions that follow.

EXAMPLE 3   Evaluate the determinant $\begin{vmatrix} 1 & 3 & -2 \\ 2 & 0 & 3 \\ 1 & 2 & 3 \end{vmatrix}$ by expanding by the minors of the middle column. (This is the determinant of Example 2.)

Strategy   We will expand the determinant using the numbers in the middle column and their corresponding minors. We will use the sign pattern $- + -$ between the terms of the expansion.

Why   We can then evaluate the resulting 2×2 determinants and simplify.

Solution   To evaluate the determinant, we proceed as follows:

$$\begin{vmatrix} 1 & 3 & -2 \\ 2 & 0 & 3 \\ 1 & 2 & 3 \end{vmatrix} = -3\begin{vmatrix} 2 & 3 \\ 1 & 3 \end{vmatrix} + 0\begin{vmatrix} 1 & -2 \\ 1 & 3 \end{vmatrix} - 2\begin{vmatrix} 1 & -2 \\ 2 & 3 \end{vmatrix}$$

Use the middle column of the checkerboard pattern: $\begin{array}{ccc} + & - & + \\ - & + & - \\ + & - & + \end{array}$

Minor of 3, Minor of 0, Minor of 2

$$= -3(6 - 3) + 0 - 2[3 - (-4)] \qquad \text{Evaluate each } 2 \times 2 \text{ determinant.}$$
$$= -3(3) + 0 - 2(7) \qquad \text{Do each subtraction within parentheses.}$$
$$= -9 + 0 - 14 \qquad \text{Do each multiplication.}$$
$$= -23 \qquad \text{Add and subtract.}$$

As expected, we get the same value as in Example 2.

Success Tip

When evaluating a determinant, expanding along a row or column that contains 0's can simplify the calculations.

Self Check 3   Evaluate $\begin{vmatrix} 2 & 3 & -1 \\ 0 & 2 & 4 \\ -2 & 5 & 6 \end{vmatrix}$ by expanding by the minors of the first column.

Now Try ▶ Problem 31

Using Your Calculator ▶ Evaluating Determinants

It is possible to use a graphing calculator to evaluate determinants. For example, to evaluate the determinant in Example 3, we first enter the matrix by pressing the MATRIX key, selecting EDIT, and pressing the ENTER key. Next, we enter the dimensions and the elements of the matrix to get figure (a). We then press 2nd QUIT to clear the screen, press MATRIX, select MATH, and press 1 to get figure (b). We then press MATRIX, select NAMES, press 1, and press) and ENTER to get the value of the determinant. Figure (c) shows that the value of the determinant is −23.

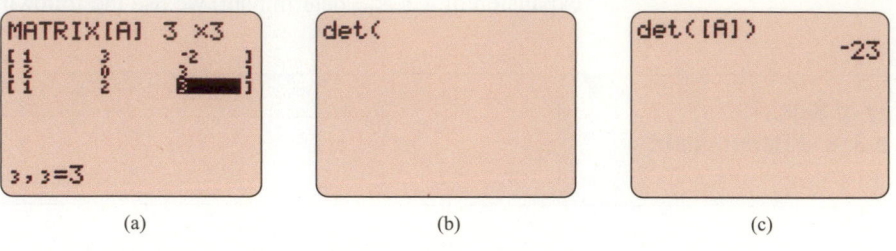

| (a) | (b) | (c) |

2 Use Cramer's Rule to Solve Systems of Two Equations.

The method of using determinants to solve systems of linear equations is called **Cramer's rule,** named after the 18th-century Swiss mathematician Gabriel Cramer. To develop Cramer's rule, we consider the system

$$\begin{cases} ax + by = e \\ cx + dy = f \end{cases}$$

where x and y are variables and a, b, c, d, e, and f are constants.

If we multiply both sides of the first equation by d and multiply both sides of the second equation by $-b$, we can add the equations and eliminate y:

$$\begin{array}{ll} a\mathbf{d}x + b\mathbf{d}y = e\mathbf{d} & \text{This is } \mathbf{d}(ax + by) = \mathbf{d}(e). \\ \underline{-\mathbf{b}cx - \mathbf{b}dy = -\mathbf{b}f} & \text{This is } -\mathbf{b}(cx + dy) = -\mathbf{b}(f). \\ adx - bcx = ed - bf & \end{array}$$

To solve for x, we use the distributive property to write $adx - bcx$ as $(ad - bc)x$ on the left side and divide each side by $ad - bc$:

$$(ad - bc)x = ed - bf$$

$$x = \frac{ed - bf}{ad - bc} \quad \text{where } ad - bc \neq 0$$

We can find y in a similar way. After eliminating the variable x, we get

$$y = \frac{af - ec}{ad - bc} \quad \text{where } ad - bc \neq 0$$

Note that the denominator for both x and y is

$$\begin{vmatrix} a & b \\ c & d \end{vmatrix} = ad - bc$$

The numerators can be expressed as determinants also:

$$x = \frac{ed - bf}{ad - bc} = \frac{\begin{vmatrix} e & b \\ f & d \end{vmatrix}}{\begin{vmatrix} a & b \\ c & d \end{vmatrix}} \quad \text{and} \quad y = \frac{af - ec}{ad - bc} = \frac{\begin{vmatrix} a & e \\ c & f \end{vmatrix}}{\begin{vmatrix} a & b \\ c & d \end{vmatrix}}$$

Cramer's Rule for Systems of Two Equations in Two Variables

The solution of the system $\begin{cases} ax + by = e \\ cx + dy = f \end{cases}$ is given by $x = \dfrac{D_x}{D}$ and $y = \dfrac{D_y}{D}$, where

$$D = \begin{vmatrix} a & b \\ c & d \end{vmatrix}$$ Use only the coefficients of the variables.

$$D_x = \begin{vmatrix} e & b \\ f & d \end{vmatrix}$$ Replace the x-term coefficients with the constants.

$$D_y = \begin{vmatrix} a & e \\ c & f \end{vmatrix}$$ Replace the y-term coefficients with the constants.

If every determinant is 0, the system is consistent, but the equations are dependent.

If $D = 0$ and D_x or D_y is nonzero, the system is inconsistent. If $D \neq 0$, the system is consistent, and the equations are independent.

The following observations are helpful when memorizing the three determinants of Cramer's rule.

- The denominator determinant, D, is formed by using the coefficients a, b, c, and d of the variables in the equations.

$$D = \begin{vmatrix} a & b \\ c & d \end{vmatrix}$$

x-term coefficients   y-term coefficients

- The numerator determinants, D_x and D_y, are the same as the denominator determinant, D, except that the column of coefficients of the variable for which we are solving is replaced with the column of constants e and f.

$$D_x = \begin{vmatrix} e & b \\ f & d \end{vmatrix} \qquad\qquad D_y = \begin{vmatrix} a & e \\ c & f \end{vmatrix}$$

Replace the x-term coefficients with the constants.   Replace the y-term coefficients with the constants.

EXAMPLE 4   Use Cramer's rule to solve the system: $\begin{cases} 4x - 3y = 6 \\ -2x + 5y = 4 \end{cases}$

Strategy   We will evaluate three determinants, D, D_x, and D_y.

Why   The x-value of the solution of the system is the quotient of D_x and D and the y-value of the solution is the quotient of two determinants, D_y and D.

Solution   By Cramer's rule, $x = \dfrac{D_x}{D}$ and $y = \dfrac{D_y}{D}$. It is best to find D first because if $D = 0$, the system is inconsistent or has dependent equations. The denominator determinant D is made up of the coefficients of x and y:

Success Tip

To find D, D_x, and D_y, we must evaluate three 2×2 determinants. The values of D_x and D_y cannot be obtained from the calculations used to find D.

$$D = \begin{vmatrix} 4 & -3 \\ -2 & 5 \end{vmatrix} = 4(5) - (-3)(-2) = 20 - 6 = 14$$

We form the numerator determinant D_x from D by replacing its first column (the coefficients of x) with the column of constants (**6** and **4**).

$$D_x = \begin{vmatrix} 6 & -3 \\ 4 & 5 \end{vmatrix} = 6(5) - (-3)(4) = 30 + 12 = 42$$

We form the numerator determinant D_y from D by replacing the second column (the coefficients of y) with the column of constants (**6** and **4**).

$$D_y = \begin{vmatrix} 4 & 6 \\ -2 & 4 \end{vmatrix} = 4(4) - 6(-2) = 16 + 12 = 28$$

Thus, $x = \dfrac{D_x}{D} = \dfrac{42}{14} = 3$ and $y = \dfrac{D_y}{D} = \dfrac{28}{14} = 2$. The solution is $(3, 2)$ and the solution set is $\{(3, 2)\}$. Verify that $(3, 2)$ satisfies both equations of the given system.

Self Check 4 Use Cramer's rule to solve the system: $\begin{cases} 2x - 3y = -16 \\ 3x + 5y = 14 \end{cases}$

Now Try ▶ Problem 39

In the next example, we will see how to recognize inconsistent systems when Cramer's rule is used to solve them.

EXAMPLE 5 Use Cramer's rule to solve: $\begin{cases} 7x = 8 - 4y \\ 2y = 3 - \frac{7}{2}x \end{cases}$

Strategy We will evaluate three determinants, D, D_x, and D_y.

Why The x-value of the solution of the system is the quotient of D_x and D, and the y-value of the solution is the quotient of two determinants, D_y and D.

Solution Before we can form the required determinants, the equations of the system must be written in standard $Ax + By = C$ form.

$$\begin{cases} 7x = 8 - 4y \\ 2y = 3 - \frac{7}{2}x \end{cases} \xrightarrow[\text{Multiply both sides by 2 and add 7x to both sides.}]{\text{Add 4y to both sides.}} \begin{cases} 7x + 4y = 8 \\ 7x + 4y = 6 \end{cases}$$

When we attempt to use Cramer's rule to solve this system for x, we obtain

$$D = \begin{vmatrix} 7 & 4 \\ 7 & 4 \end{vmatrix} = 28 - 28 = 0 \quad \text{and} \quad D_x = \begin{vmatrix} 8 & 4 \\ 6 & 4 \end{vmatrix} = 32 - 24 = 8$$

Thus,

$$x = \frac{8}{0}, \text{ which is undefined.}$$

Since the denominator determinant D is 0 and the numerator determinant D_x is not 0, the system is inconsistent. It has no solution and the solution set is $\varnothing$.

We can see directly from the system that it is inconsistent. For any values of x and y, it is impossible that 7 times x plus 4 times y could be both 8 and 6.

Self Check 5 Use Cramer's rule to solve the system: $\begin{cases} 3x = 8 - 4y \\ y = \frac{5}{2} - \frac{3}{4}x \end{cases}$

Now Try ▶ Problem 47

3 Use Cramer's Rule to Solve Systems of Three Equations.

Cramer's rule can be extended to solve systems of three linear equations with three variables.

Cramer's Rule for Three Equations in Three Variables

The solution of the system $\begin{cases} ax + by + cz = j \\ dx + ey + fz = k \\ gx + hy + iz = l \end{cases}$ is given by

$$x = \frac{D_x}{D}, \qquad y = \frac{D_y}{D}, \quad \text{and} \quad z = \frac{D_z}{D}, \qquad \text{where}$$

$$D = \begin{vmatrix} a & b & c \\ d & e & f \\ g & h & i \end{vmatrix} \quad \text{Use only the coefficients of the variables.} \qquad D_x = \begin{vmatrix} j & b & c \\ k & e & f \\ l & h & i \end{vmatrix} \quad \text{Replace the x-term coefficients with the constants.}$$

$$D_y = \begin{vmatrix} a & j & c \\ d & k & f \\ g & l & i \end{vmatrix} \quad \text{Replace the y-term coefficients with the constants.} \qquad D_z = \begin{vmatrix} a & b & j \\ d & e & k \\ g & h & l \end{vmatrix} \quad \text{Replace the z-term coefficients with the constants.}$$

If every determinant is 0, the system is consistent, but the equations are dependent.

If $D = 0$ and D_x or D_y or D_z is nonzero, the system is inconsistent. If $D \neq 0$, the system is consistent, and the equations are independent.

EXAMPLE 6 Use Cramer's rule to solve the system: $\begin{cases} 2x + y + 4z = 12 \\ x + 2y + 2z = 9 \\ 3x - 3y - 2z = 1 \end{cases}$

Strategy We will evaluate four determinants, D, D_x, D_y, and D_z.

Why The x-value of the solution of the system is the quotient of D_x and D, the y-value of the solution is the quotient of two determinants, D_y and D, and the z-value of the solution is the quotient of D_z and D.

Solution The denominator determinant D is the determinant formed by the coefficients of the variables. The numerator determinants, D_x, D_y, and D_z, are formed by replacing the coefficients of the variable being solved for by the column of constants. We will evaluate each determinant by expanding by minors about the first row:

$$D = \begin{vmatrix} 2 & 1 & 4 \\ 1 & 2 & 2 \\ 3 & -3 & -2 \end{vmatrix} = 2\begin{vmatrix} 2 & 2 \\ -3 & -2 \end{vmatrix} - 1\begin{vmatrix} 1 & 2 \\ 3 & -2 \end{vmatrix} + 4\begin{vmatrix} 1 & 2 \\ 3 & -3 \end{vmatrix} = 2(2) - 1(-8) + 4(-9) = -24$$

$$D_x = \begin{vmatrix} 12 & 1 & 4 \\ 9 & 2 & 2 \\ 1 & -3 & -2 \end{vmatrix} = 12\begin{vmatrix} 2 & 2 \\ -3 & -2 \end{vmatrix} - 1\begin{vmatrix} 9 & 2 \\ 1 & -2 \end{vmatrix} + 4\begin{vmatrix} 9 & 2 \\ 1 & -3 \end{vmatrix} = 12(2) - 1(-20) + 4(-29) = -72$$

$$D_y = \begin{vmatrix} 2 & 12 & 4 \\ 1 & 9 & 2 \\ 3 & 1 & -2 \end{vmatrix} = 2\begin{vmatrix} 9 & 2 \\ 1 & -2 \end{vmatrix} - 12\begin{vmatrix} 1 & 2 \\ 3 & -2 \end{vmatrix} + 4\begin{vmatrix} 1 & 9 \\ 3 & 1 \end{vmatrix} = 2(-20) - 12(-8) + 4(-26) = -48$$

$$D_z = \begin{vmatrix} 2 & 1 & 12 \\ 1 & 2 & 9 \\ 3 & -3 & 1 \end{vmatrix} = 2\begin{vmatrix} 2 & 9 \\ -3 & 1 \end{vmatrix} - 1\begin{vmatrix} 1 & 9 \\ 3 & 1 \end{vmatrix} + 12\begin{vmatrix} 1 & 2 \\ 3 & -3 \end{vmatrix} = 2(29) - 1(-26) + 12(-9) = -24$$

Thus,

$$x = \frac{D_x}{D} = \frac{-72}{-24} = 3$$

$$y = \frac{D_y}{D} = \frac{-48}{-24} = 2$$

$$z = \frac{D_z}{D} = \frac{-24}{-24} = 1$$

The solution is $(3, 2, 1)$ and the solution set is $\{(3, 2, 1)\}$. Verify that $(3, 2, 1)$ satisfies the three original equations.

Self Check 6 Use Cramer's rule to solve the system: $\begin{cases} x + y + 2z = 6 \\ 2x - y + z = 9 \\ x + y - 2z = -6 \end{cases}$

Now Try ▶ Problem 51

SECTION 3.5 ▶ STUDY SET

VOCABULARY

Fill in the blanks.

1. $\begin{vmatrix} 4 & 9 \\ -6 & 1 \end{vmatrix}$ is a _____. The numbers 4 and 1 lie along its main _____.

2. A determinant is number that is associated with a _____ matrix.

3. The _____ of b_1 in $\begin{vmatrix} a_1 & b_1 & c_1 \\ a_2 & b_2 & c_2 \\ a_3 & b_3 & c_3 \end{vmatrix}$ is $\begin{vmatrix} a_2 & c_2 \\ a_3 & c_3 \end{vmatrix}$.

4. _____ rule uses determinants to solve systems of linear equations.

CONCEPTS

Fill in the blanks.

5. $\begin{vmatrix} a & b \\ c & d \end{vmatrix} = \boxed{} - \boxed{}$

6. To find the minor of 5, we cross out the elements of the determinant that are in the same row and column as $\boxed{}$.

$$\begin{vmatrix} 3 & 5 & 1 \\ 6 & -2 & 2 \\ 8 & -1 & 4 \end{vmatrix}$$

7. In evaluating the determinant below, about what row or column was it expanded?

$$\begin{vmatrix} 5 & 1 & -1 \\ 8 & 7 & 4 \\ 9 & 7 & 6 \end{vmatrix} = -1 \begin{vmatrix} 8 & 7 \\ 9 & 7 \end{vmatrix} - 4 \begin{vmatrix} 5 & 1 \\ 9 & 7 \end{vmatrix} + 6 \begin{vmatrix} 5 & 1 \\ 8 & 7 \end{vmatrix}$$

8. What is the denominator determinant D for the system $\begin{cases} 3x + 4y = 7 \\ 2x - 3y = 5 \end{cases}$?

9. What is the denominator determinant D for the system $\begin{cases} x + 2y = -8 \\ 3x + y - z = -2 \\ 8x + 4y - z = 6 \end{cases}$?

10. For the system $\begin{cases} 3x + 2y = 1 \\ 4x - y = 3 \end{cases}$, $D_x = -7$, $D_y = 5$, and $D = -11$. Find the solution of the system.

11. For the system $\begin{cases} 2x + 3y - z = -8 \\ x - y - z = -2 \\ -4x + 3y + z = 6 \end{cases}$, $D_x = -28$, $D_y = -14$, $D_z = 14$, and $D = 14$. Find the solution of the system.

12. Fill in the blank. If the denominator determinant D for a system of equations is 0, the equations of the system are dependent or the system is _____.

NOTATION

Complete the evaluation of each determinant.

13. $\begin{vmatrix} 5 & -2 \\ -2 & 6 \end{vmatrix} = 5(\boxed{}) - (-2)(-2)$

$$= \boxed{} - 4$$

$$= 26$$

14. $\begin{vmatrix} 2 & 1 & 3 \\ 3 & 4 & 2 \\ 1 & 5 & 3 \end{vmatrix} = 2 \begin{vmatrix} 4 & \blacksquare \\ 5 & 3 \end{vmatrix} - 1 \begin{vmatrix} 3 & 2 \\ \blacksquare & 3 \end{vmatrix} + 3 \begin{vmatrix} 3 & 4 \\ 1 & \blacksquare \end{vmatrix}$

$= 2(\blacksquare - 10) - 1(9 - \blacksquare) + 3(15 - \blacksquare)$

$= 2(2) - 1(\ ) + \blacksquare(11)$

$= 4 - 7 + \blacksquare$

$= 30$

GUIDED PRACTICE

Evaluate each determinant. **See Example 1.**

15. $\begin{vmatrix} 2 & 3 \\ 2 & 5 \end{vmatrix}$

16. $\begin{vmatrix} 3 & 2 \\ 2 & 4 \end{vmatrix}$

17. $\begin{vmatrix} -9 & 7 \\ 4 & -2 \end{vmatrix}$

18. $\begin{vmatrix} -1 & 2 \\ 3 & -4 \end{vmatrix}$

19. $\begin{vmatrix} 5 & 20 \\ 10 & 6 \end{vmatrix}$

20. $\begin{vmatrix} 10 & 15 \\ 15 & 5 \end{vmatrix}$

21. $\begin{vmatrix} -6 & -2 \\ 15 & 4 \end{vmatrix}$

22. $\begin{vmatrix} 3 & -2 \\ 12 & -8 \end{vmatrix}$

23. $\begin{vmatrix} -9 & -1 \\ -10 & -5 \end{vmatrix}$

24. $\begin{vmatrix} -7 & -7 \\ -6 & -4 \end{vmatrix}$

25. $\begin{vmatrix} 8 & 8 \\ -9 & -9 \end{vmatrix}$

26. $\begin{vmatrix} 20 & -3 \\ 20 & -3 \end{vmatrix}$

Evaluate each determinant. **See Examples 2 and 3.**

27. $\begin{vmatrix} 3 & 2 & 1 \\ 4 & 1 & 2 \\ 5 & 3 & 1 \end{vmatrix}$

28. $\begin{vmatrix} 6 & 2 & 3 \\ 1 & 5 & 4 \\ 2 & 3 & 5 \end{vmatrix}$

29. $\begin{vmatrix} 1 & -2 & 3 \\ -2 & 1 & 1 \\ -3 & -2 & 1 \end{vmatrix}$

30. $\begin{vmatrix} 1 & 1 & 2 \\ 2 & 1 & -2 \\ 3 & 1 & 3 \end{vmatrix}$

31. $\begin{vmatrix} -2 & 5 & 1 \\ 0 & 3 & 4 \\ -1 & 2 & 6 \end{vmatrix}$

32. $\begin{vmatrix} 4 & -1 & 2 \\ 6 & -1 & 0 \\ 1 & -3 & 4 \end{vmatrix}$

33. $\begin{vmatrix} 1 & -4 & 1 \\ 3 & 0 & -2 \\ 3 & 1 & -2 \end{vmatrix}$

34. $\begin{vmatrix} 8 & -3 & 1 \\ 1 & 0 & 2 \\ 3 & -9 & 4 \end{vmatrix}$

35. $\begin{vmatrix} 1 & 2 & 1 \\ -3 & 7 & 3 \\ -4 & 3 & -5 \end{vmatrix}$

36. $\begin{vmatrix} 1 & 4 & 7 \\ 2 & 5 & 8 \\ 3 & 6 & 9 \end{vmatrix}$

37. $\begin{vmatrix} 1 & 2 & 0 \\ 0 & 1 & 2 \\ 0 & 0 & 1 \end{vmatrix}$

38. $\begin{vmatrix} 1 & 0 & 1 \\ 0 & 1 & 0 \\ 1 & 1 & 1 \end{vmatrix}$

Use Cramer's rule to solve each system of equations. **See Example 4.**

39. $\begin{cases} x + y = 6 \\ x - y = 2 \end{cases}$

40. $\begin{cases} x - y = 4 \\ 2x + y = 5 \end{cases}$

41. $\begin{cases} x + 2y = -21 \\ x - 2y = 11 \end{cases}$

42. $\begin{cases} 5x + 2y = 11 \\ 7x + 6y = 9 \end{cases}$

43. $\begin{cases} 3x - 4y = 9 \\ x + 2y = 8 \end{cases}$

44. $\begin{cases} 2x + 2y = -1 \\ 3x + 4y = 0 \end{cases}$

45. $\begin{cases} 2x + 3y = 31 \\ 3x + 2y = 39 \end{cases}$

46. $\begin{cases} 5x + 3y = 72 \\ 3x + 5y = 56 \end{cases}$

Use Cramer's rule to solve each system of equations. If a system is inconsistent or if the equations are dependent, so indicate. **See Example 5.**

47. $\begin{cases} 3x + 2y = 11 \\ 6x + 4y = 11 \end{cases}$

48. $\begin{cases} 5x - 4y = 20 \\ 10x - 8y = 30 \end{cases}$

49. $\begin{cases} \dfrac{5}{6}x = 2 - y \\ 10x + 12y = 24 \end{cases}$

50. $\begin{cases} 16x - 8y = 32 \\ x - 2 = \dfrac{y}{2} \end{cases}$

Use Cramer's rule to solve each system of equations. **See Example 6.**

51. $\begin{cases} x + y + z = 4 \\ x + y - z = 0 \\ x - y + z = 2 \end{cases}$

52. $\begin{cases} x + y + z = 4 \\ x - y + z = 2 \\ x - y - z = 0 \end{cases}$

53. $\begin{cases} 3x + 2y - z = -8 \\ 2x - y + 7z = 10 \\ 2x + 2y - 3z = -10 \end{cases}$

54. $\begin{cases} x + 2y + 2z = 10 \\ 2x + y + 2z = 9 \\ 2x + 2y + z = 1 \end{cases}$

TRY IT YOURSELF

Use Cramer's rule to solve each system of equations. If a system is inconsistent or if the equations are dependent, so indicate.

55. $\begin{cases} 2x + y + z = 5 \\ x - 2y + 3z = 10 \\ x + y - 4z = -3 \end{cases}$

56. $\begin{cases} x + y + 2z = 7 \\ x + 2y + z = 8 \\ 2x + y + z = 9 \end{cases}$

57. $\begin{cases} y = \dfrac{-2x + 1}{3} \\ 3x - 2y = 8 \end{cases}$

58. $\begin{cases} 2x + 3y = -1 \\ x = \dfrac{y - 9}{4} \end{cases}$

59. $\begin{cases} 4x - 3y = 1 \\ 6x - 8z = 1 \\ 2y - 4z = 0 \end{cases}$

60. $\begin{cases} 4x + 3z = 4 \\ 2y - 6z = -1 \\ 8x + 4y + 3z = 9 \end{cases}$

61. $\begin{cases} 2x + y - z - 1 = 0 \\ x + 2y + 2z - 2 = 0 \\ 4x + 5y + 3z - 3 = 0 \end{cases}$

62. $\begin{cases} 2x - y + 4z + 2 = 0 \\ 5x + 8y + 7z = -8 \\ x + 3y + z + 3 = 0 \end{cases}$

63. $\begin{cases} 3x - 16 = 5y \\ -3x + 5y - 33 = 0 \end{cases}$

64. $\begin{cases} 2x + 5y - 13 = 0 \\ -2x + 13 = 5y \end{cases}$

65. $\begin{cases} x + y = 1 \\ \dfrac{1}{2}y + z = \dfrac{5}{2} \\ x - z = -3 \end{cases}$

66. $\begin{cases} \dfrac{1}{2}x + y + z + \dfrac{3}{2} = 0 \\ x + \dfrac{1}{2}y + z - \dfrac{1}{2} = 0 \\ x + y + \dfrac{1}{2}z + \dfrac{1}{2} = 0 \end{cases}$

67. $\begin{cases} 2x + 3y = 0 \\ 4x - 6y = -4 \end{cases}$ **68.** $\begin{cases} 4x - 3y = -1 \\ 8x + 3y = 4 \end{cases}$

69. $\begin{cases} 2x + 3y + 4z = 6 \\ 2x - 3y - 4z = -4 \\ 4x + 6y + 8z = 12 \end{cases}$ **70.** $\begin{cases} x - 3y + 4z - 2 = 0 \\ 2x + y + 2z - 3 = 0 \\ 4x - 5y + 10z - 7 = 0 \end{cases}$

 Use a calculator with matrix capabilities. Evaluate each determinant. See Using Your Calculator: **Evaluating Determinants.**

71. $\begin{vmatrix} 25 & -36 & 44 \\ -11 & 21 & 54 \\ 37 & -31 & 19 \end{vmatrix}$ **72.** $\begin{vmatrix} 13 & -27 & 62 \\ -38 & 27 & -52 \\ 10 & -300 & 42 \end{vmatrix}$

73. $\begin{vmatrix} -280 & 191 & -356 \\ -211 & -102 & -422 \\ 400 & -213 & -333 \end{vmatrix}$ **74.** $\begin{vmatrix} 4.1 & 2.2 & -3.3 \\ 2.7 & -5.9 & 6.8 \\ 2.3 & 5.3 & 0.6 \end{vmatrix}$

APPLICATIONS

75. Signaling. Solve the following system using Cramer's rule to find the measures (in degrees) of the angles shown in the illustration. (*Hint:* Each equation must be written in standard $Ax + By = C$ form.)

$$\begin{cases} 2x + y = 180 \\ y = x + 30 \end{cases}$$

76. Investing. Solve the following system using Cramer's rule to find the amount invested in each stock. (*Hint:* Each equation must be written in standard $Ax + By + Cz = D$ form.)

$$\begin{cases} x + y + z = 20{,}000 \\ 0.10x + 0.05y + 0.06z = 1{,}320 \\ x = \dfrac{1}{3}(y + z) \end{cases}$$

Stock	Amount Invested	Rate of return
HiTech	x	10%
SaveTel	y	5%
OilCo	z	6%

WRITING

77. Explain the difference between a matrix and a determinant. Give an example of each.

78. When evaluating $\begin{vmatrix} 4 & -1 & 2 \\ 6 & -1 & 0 \\ 1 & -3 & 4 \end{vmatrix}$, why is it helpful to expand by the minors of the numbers in the third column?

79. Explain how to find the minor of an element of a determinant.

80. Explain how to find x when solving a system of three linear equations in x, y, and z by Cramer's rule. Use the words *coefficients* and *constants* in your explanation.

81. Explain how the following checkerboard pattern is used when evaluating a 3×3 determinant.

$$\begin{matrix} + & - & + \\ - & + & - \\ + & - & + \end{matrix}$$

82. Briefly describe each of the five methods of this chapter that can be used to solve a system of two linear equations in two variables.

REVIEW

83. Are the lines described by $y = 2x - 7$ and $x - 2y = 7$ perpendicular?

84. Are the lines described by $y = 2x - 7$ and $2x - y = 10$ parallel?

85. How are the graphs of $f(x) = x^2$ and $g(x) = x^2 - 2$ related?

86. Is the graph of a circle the graph of a function?

87. The graph of a line passes through $(0, -3)$. Is this the x-intercept or the y-intercept of the line?

88. What is the name of the function $f(x) = |x|$?

89. For the function $y = 2x^2 + 6x + 1$, what is the independent variable and what is the dependent variable?

90. If $f(x) = x^3 - x$, what is $f(-1)$?

CHALLENGE PROBLEMS

91. Show that $\begin{vmatrix} x & y & 1 \\ -2 & 3 & 1 \\ 3 & 5 & 1 \end{vmatrix} = 0$ represents the equation of the line passing through $(-2, 3)$ and $(3, 5)$.

92. Show that $\dfrac{1}{2} \begin{vmatrix} 0 & 0 & 1 \\ 3 & 0 & 1 \\ 0 & 4 & 1 \end{vmatrix}$ represents the area of the triangle with vertices at $(0, 0)$, $(3, 0)$, and $(0, 4)$.

Problem Solving Using Systems of Two Equations

OBJECTIVES

1 Assign variables to two unknowns.

2 Use systems to solve geometry problems.

3 Use systems to solve number-value problems.

4 Use systems to find the break point.

5 Use systems to solve interest, uniform motion, and mixture problems.

ARE YOU READY?

The following problems review some basic skills that are needed when solving application problems using systems of two equations.

1. Translate to mathematical symbols: 150 less than the weight w.

2. At \$260 per ounce, what is the value of four ounces of Chanel No.5 perfume?

3. At 12 miles per hour, how far will a bicyclist travel in 6 hours?

4. An 80 pound bag of quick-set concrete is 40% sand by weight. How many pounds of sand are in the bag?

In Chapter 1, we solved applied problems involving two unknown quantities by modeling the situation with an equation in one variable. It's often easier to solve such problems using a two-variable approach. We write two equations in two variables to model the situation, and then we use the methods of this chapter to solve the system formed by the pair of equations.

1 Assign Variables to Two Unknowns.

The following steps are helpful when solving problems involving two unknown quantities.

Problem-Solving Strategy

1. **Analyze the problem** by reading it carefully to understand the given facts. What information is given? What are you asked to find? What vocabulary is given? Often a diagram or table will help you understand the facts of the problem.

2. **Assign variables** to represent unknown values in the problem. This means, in most cases, to let $x =$ one of the unknowns that you are asked to find, and $y =$ the other unknown.

3. **Form a system of equations** by translating the words of the problem into mathematical symbols.

4. **Solve the system** of equations using graphing, substitution, elimination, matrices, or Cramer's rule.

5. **State the conclusion** clearly. Be sure to include the units.

6. **Check the results** using the words of the problem, not the equations that were formed in step 3.

EXAMPLE 1

Pets. In 2010, there were an estimated 172 million dogs and cats owned in the United States. If the number of dogs was 16 million less than the number of cats, how many dogs and how many cats were owned in the United States that year? (Source: Humane Society)

Analyze

- In 2010, the total number of dogs and cats was 172 million.
- The number of dogs was 16 million less than the number of cats.
- Find the number of dogs and the number of cats that year.

Assign We will let $x =$ the number of dogs (in millions) and $y =$ the number of cats (in millions).

Form   We can translate the words of the problem into two equations.

The number of dogs	plus	the number of cats	was	172 million.
x	$+$	y	$=$	172

The number of dogs	was	16 million	less than	the number of cats.
x	$=$	y	$-$	16

The resulting system is: $\begin{cases} x + y = 172 \\ x = y - 16 \end{cases}$

Solve   Since the second equation is solved for x, we will use substitution to solve the system.

$$x + y = 172 \qquad \text{\color{red}This is the first equation of the system.}$$
$$y - 16 + y = 172 \qquad \text{\color{red}Substitute } y - 16 \text{ for } x.$$
$$2y - 16 = 172 \qquad \text{\color{red}Combine like terms.}$$
$$2y = 188 \qquad \text{\color{red}Add 16 to both sides.}$$
$$y = 94 \qquad \text{\color{red}Divide both sides by 2. This is the number of cats (in millions).}$$

To find x, we substitute 94 for y in the second equation of the system.

$$x = y - 16$$
$$x = 94 - 16 \qquad \text{\color{red}Substitute 94 for } y.$$
$$x = 78 \qquad \text{\color{red}This is the number of dogs (in millions).}$$

State   In 2010, the number of dogs owned in the United States was 78 million and the number of cats was 94 million.

Check   Since 78 million + 94 million = 172 million, and 78 million is 16 million less than 94 million, the results check.

> **Self Check 1**   **Pet Fish.**   In 2010, there were an estimated 183 million freshwater and saltwater fish owned as pets in the United States. The number of freshwater fish was 7 million more than 15 times the number of saltwater fish. How many freshwater fish and how many saltwater fish were owned as pets that year? (Source: Humane Society)
>
> **Now Try ▶ Problem 13**

Caution

If two variables are used to represent two unknown quantities, we must form a system of two equations to find the unknowns.

Success Tip

If two variables are used to represent two unknowns, a two-part check must be performed to verify the results.

2 Use Systems to Solve Geometry Problems.

Sometimes we can use geometric facts or formulas to solve application problems.

EXAMPLE 2   **Parallelograms.**   Refer to parallelogram *ABCD*. Find the unknown degree measures represented by *x* and *y*.

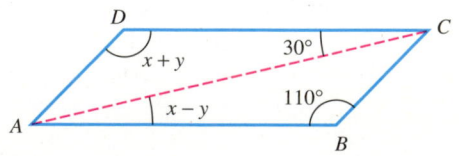

Analyze To solve this problem, we will use two facts about parallelograms.

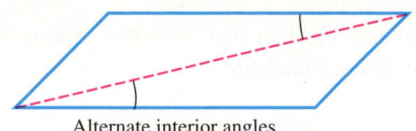

Alternate interior angles

When a diagonal intersects two parallel sides of a parallelogram, pairs of alternate interior angles have the same measure.

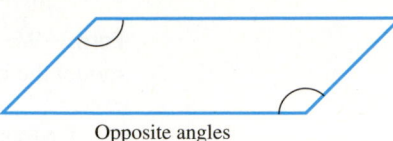

Opposite angles

Opposite angles of a parallelogram have the same measure.

Assign The variable expressions $x + y$ and $x - y$ were given in the illustration on the previous page.

Form In the figure, $\angle BAC$ and $\angle DCA$ are alternate interior angles and therefore have the same measure. Thus, $x - y = 30$. Since $\angle B$ and $\angle D$ in the figure are opposite angles of the parallelogram, $x + y = 110$.

The resulting system is: $\begin{cases} x - y = 30 \\ x + y = 110 \end{cases}$

Solve Since the coefficients of the terms $-y$ and y are opposites, we will use elimination to solve the system.

$$
\begin{aligned}
x - y &= 30 \\
\underline{x + y} &= \underline{110} \\
2x &= 140 \qquad \text{Add the equations. The } y\text{-terms drop out.} \\
x &= 70 \qquad \text{To solve for } x \text{, divide both sides by 2.}
\end{aligned}
$$

We can substitute 70 for x in the second equation and solve for y.

$$
\begin{aligned}
x + y &= 110 \\
70 + y &= 110 \qquad \text{Substitute 70 for } x. \\
y &= 40 \qquad \text{To solve for } y \text{, subtract 70 from both sides.}
\end{aligned}
$$

State Thus, $x = 70°$ and $y = 40°$.

Check Since $70° - 40° = 30°$ and $70° + 40° = 110°$, the results check.

Self Check 2 **Geometry.** In a right triangle, the sum of the two acute angles is 90°. If one of the acute angles is 36° more than the other, find the measure of the two angles.

Now Try ▶ Problem 19

3 Use Systems to Solve Number-Value Problems.

| **EXAMPLE 3** | **Wedding Pictures.** A professional photographer offers two different packages for wedding pictures. Use the information in the illustration to determine the cost of one 8×10-inch photograph and the cost of one 5×7-inch photograph. |

Wedding Pictures

Package 1
8 - 8 x 10's
12 - 5 x 7's
Only $133.00

Package 2
6 - 8 x 10's
22 - 5 x 7's
Only $168.00

Analyze

- Eight 8×10 and twelve 5×7 pictures cost $133.
- Six 8×10 and twenty-two 5×7 pictures cost $168.
- Find the cost of one 8×10 photograph and the cost of one 5×7 photograph.

Assign   Let x = the cost of one 8×10 photograph (in dollars), and let y = the cost of one 5×7 photograph (in dollars).

Form   We can use the fact that **Number · value = total value** to construct tables that model the cost of each package.

Package 1

Size of photo	Number	· Value =	Total value
8×10	8	x	$8x$
5×7	12	y	$12y$

Total: $133

↑
One equation comes
from this column.

Package 2

Size of photo	Number	· Value =	Total value
8×10	6	x	$6x$
5×7	22	y	$22y$

Total: $168

↑
The second equation
comes from this column.

From the *Total value column* of the first table:

The cost of eight 8×10 photographs	plus	the cost of twelve 5×7 photographs	is	the cost of the first package.
$8x$	$+$	$12y$	$=$	133

From the *Total value column* of the second table:

The cost of six 8×10 photographs	plus	the cost of twenty-two 5×7 photographs	is	the cost of the second package.
$6x$	$+$	$22y$	$=$	168

The resulting system is:
$$\begin{cases} 8x + 12y = 133 & \textbf{(1)} \\ 6x + 22y = 168 & \textbf{(2)} \end{cases}$$

Solve   We will use elimination to solve this system. To make the x-terms drop out, we multiply both sides of equation 1 by 3. Then we multiply both sides of equation 2 by -4, add the resulting equations, and solve for y:

$$24x + 36y = 399 \quad \text{This is } 3(8x + 12y) = 3(133).$$
$$\underline{-24x - 88y = -672} \quad \text{This is } -4(6x + 22y) = -4(168).$$
$$-52y = -273 \quad \text{Add the terms, column by column. The x-terms drop out.}$$
$$y = 5.25 \quad \text{Divide both sides by } -52. \text{ This is the cost of one } 5 \times 7 \text{ photograph.}$$

To find x, we substitute 5.25 for y in equation 1 and solve for x:

$$8x + 12y = 133$$
$$8x + 12(5.25) = 133 \quad \text{Substitute 5.25 for y.}$$
$$8x + 63 = 133 \quad \text{Do the multiplication.}$$
$$8x = 70 \quad \text{Subtract 63 from both sides.}$$
$$x = 8.75 \quad \text{Divide both sides by 8. This is the cost of one } 8 \times 10 \text{ photograph.}$$

State   The cost of one 8×10 photo is $8.75, and the cost of one 5×7 photo is $5.25.

Check   If the first package contains eight 8×10 and twelve 5×7 photographs, the value of the package is $8(\$8.75) + 12(\$5.25) = \$70 + \$63 = \$133$. If the second package contains six 8×10 and twenty-two 5×7 photographs, the value of the package is $6(\$8.75) + 22(\$5.25) = \$52.50 + \$115.50 = \$168$. The results check.

Self Check 3   **Theater Snacks.**   At a movie theater, 1 large popcorn and 2 medium drinks cost $10.50. Another package offers 1 large popcorn and 3 medium drinks for $13.00. Find the cost of one large popcorn and 1 medium drink.

Now Try ▶ Problem 25

4 Use Systems to Find the Break Point.

Running a machine involves both **setup costs** and **unit costs**. Setup costs include the cost of preparing a machine to do a certain job. The costs to make one item are unit costs. They depend on the number of items to be manufactured, including costs of raw materials and labor.

EXAMPLE 4   **Manufacturing.**   The setup cost of a machine that mills brass plates is $750. After setup, it costs $0.25 to mill each plate. Management is considering the purchase of a larger machine that can produce a plate at a cost of $0.20 per plate. If the setup cost of the larger machine is $1,200, how many plates would the company have to produce to make the purchase worthwhile?

Analyze   We need to find the number of plates (called the **break point**) that will cost equal amounts to produce using either machine.

Assign   We can let c = the cost (in dollars) of milling p plates.

Form   If we call the machine currently in use machine 1 and the new, larger machine 2, we can form two equations.

The cost of making p plates using machine 1	equals	the setup cost of machine 1	plus	the cost per plate of machine 1	times	the number of plates p to be made.
c	=	750	+	0.25	·	p

The cost of making p plates using machine 2	equals	the setup cost of machine 2	plus	the cost per plate of machine 2	times	the number of plates p to be made.
c	=	1,200	+	0.20	·	p

To find the break point, we must solve the system $\begin{cases} c = 750 + 0.25p \\ c = 1,200 + 0.20p \end{cases}$

Solve   Since the costs are equal for the break point, we can use substitution to solve the system.

$\begin{cases} c = \boxed{750 + 0.25p} \\ c = 1,200 + 0.20p \end{cases}$ We could multiply both sides of the equations by 100 to eliminate the decimals. However, in this case, it is easier to work with the equations as they are.

$750 + 0.25p = 1,200 + 0.20p$ Substitute $750 + 0.25p$ for c in the second equation.

$0.25p = 450 + 0.20p$ Subtract 750 from both sides.

$0.05p = 450$ Subtract $0.20p$ from both sides.

$p = 9,000$ Divide both sides by 0.05.

State   Since the cost will be the same on either machine when 9,000 plates are milled, 9,000 is the break point.

Check   We can check the result by substituting 9,000 for p in each equation of the system and verifying that 3,000 is the value of c in both cases.

We can graph the two equations using the point-plotting method to illustrate the break point.

Machine 1
$c = 750 + 0.25p$

p	c
0	750
1,000	1,000
3,000	1,500

Machine 2
$c = 1,200 + 0.20p$

p	c
0	1,200
4,000	2,000
12,000	3,600

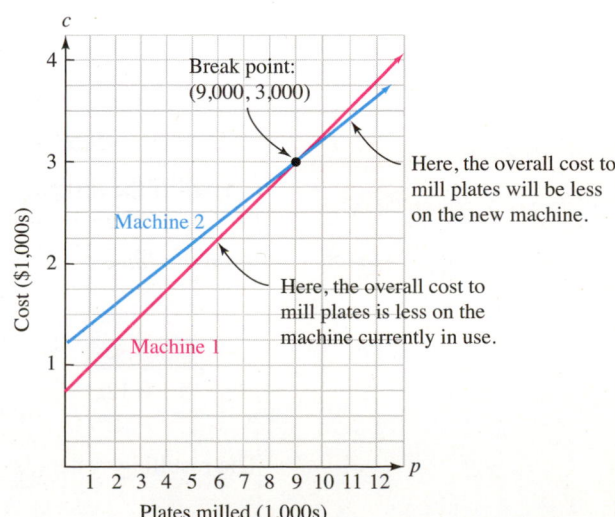

Self Check 4 **Break Point.** If in Example 4, the setup cost of the smaller machine is $600 with a $.30 cost to mill each plate, and the larger machine's setup is $1,000 with $0.22 to mill each plate, what would the break point be?

Now Try ▶ Problem 31

5 Use Systems to Solve Interest, Uniform Motion, and Mixture Problems.

To compare one-variable and two-variable approaches, we will solve the investment problem of Example 4 in Section 1.8 using two variables to find the unknown investment amounts.

EXAMPLE 5 | **Interest Income.** To protect against a major loss, a financial analyst suggested the following plan for a client who has $50,000 to invest for 1 year.

1. Alco Development, Inc. Builds mini-malls. High yield: 12% per year. Risky!
2. Certificate of deposit (CD). Insured, safe. Low yield: 4.5% annual interest.

If the client puts some money in each investment and wants to earn $3,600 in interest, how much should be invested at each rate?

Analyze A total of $50,000 is invested at two different rates for 1 year. The total interest earned is $3,600.

Assign Let $x =$ the number of dollars invested at 12% and $y =$ the number of dollars invested at 4.5%.

Form We can construct a table to help us to form two equations that model the situation. Figure (a) below shows the first stage in constructing a table. The table is completed in figure (b) below using the formula $I = Prt$ to determine that $x invested at 12% for 1 year will earn $0.12x and $y invested at 4.5% for 1 year will earn $0.045y.

> **Success Tip**
>
> With a one-variable approach, we let $x =$ the number of dollars invested at 12% and $50,000 - x =$ the number of dollars invested at 4.5%. With a two-variable approach, again we let $x =$ the number of dollars invested at 12%, but then we let $y =$ the number of dollars invested at 4.5%.

	P ·	r ·	$t =$	I
Alco	x	0.12	1	
CD	y	0.045	1	
Total	50,000			3,600

↑ Enter this information first.

(a)

	P ·	r ·	$t =$	I
Alco	x	0.12	1	**0.12x**
CD	y	0.045	1	**0.045y**
Total	50,000			3,600

↑ One equation comes from this column.　↑ The second equation comes from this column.

(b)

From the *Principal (P) column* of the table:

The amount invested in Alco	plus	the amount invested in the CD	is	$50,000.
x	$+$	y	$=$	50,000

From the *Interest (I) column* of the table:

The interest earned at 12%	plus	the interest earned at 4.5%	equals	the total interest earned.
$0.12x$	$+$	$0.045y$	$=$	3,600

> **Caution**
>
> Don't forget to express 12% and 4.5% as decimals.
> $12\% = 0.12$ $4.5\% = 0.045$

The resulting system is:
$$\begin{cases} x + y = 50,000 & \textbf{(1)} \\ 0.12x + 0.045y = 3,600 & \textbf{(2)} \end{cases}$$

Solve   To solve this system by substitution, we can solve equation 1 for y:

$$x + y = 50{,}000$$

$$y = 50{,}000 - x \quad \text{This is the substitution equation.}$$

Then we substitute $50{,}000 - x$ for y in equation 2 and solve for x.

$$0.12x + 0.045y = 3{,}600$$

$$0.12x + 0.045(\mathbf{50{,}000 - x}) = 3{,}600 \quad \text{Substitute } 50{,}000 - x \text{ for } y.$$

$$120x + 45(50{,}000 - x) = 3{,}600{,}000 \quad \text{To clear the equation of the decimals, multiply both sides by 1,000.}$$

$$120x + 2{,}250{,}000 - 45x = 3{,}600{,}000 \quad \text{Distribute the multiplication by 45.}$$

$$75x = 1{,}350{,}000 \quad \text{Combine like terms and subtract 2,250,000 from both sides.}$$

$$x = 18{,}000 \quad \text{Divide both sides by 75. This is the amount that should be invested at 12\%.}$$

To find y, we can substitute $18{,}000$ for x in the substitution equation:

$$y = 50{,}000 - \mathbf{x}$$

$$y = 50{,}000 - \mathbf{18{,}000} \quad \text{Substitute 18,000 for } x.$$

$$y = 32{,}000 \quad \text{This is the amount that should be invested at 4.5\%.}$$

State $18{,}000 should be invested at 12% and $32{,}000 should be invested at 4.5%.

Check   We note that $18{,}000 plus $32{,}000 equals the required amount of money invested. The annual interest on $18{,}000 is $0.12(\$18{,}000) = \$2{,}160$. The interest earned on $32{,}000 is $0.045(\$32{,}000) = \$1{,}440$. The total interest is $\$2{,}160 + \$1{,}440 = \$3{,}600$. The results check.

Self Check 5   **Investing.**   A student has $15{,}000 to invest, some at 8% and the rest at 3.5%. If the annual interest earned from the two accounts is $840, how much was invested in each account?

Now Try ▶ Problem 39

EXAMPLE 6

Blimps.   *The Spirit of America,* one of the Goodyear blimps, flew 175 miles in 5 hours with the wind. The return trip took 7 hours flying against the wind. Find the speed of the blimp in still air and the speed of the wind.

Analyze   Traveling with the wind, the speed of the blimp will be faster than it would be in still air. Traveling against the wind, the speed of the blimp will be slower than it would be in still air.

Tailwind
w mph

Headwind
w mph

Assign   Let s = the speed of the blimp (in mph) in still air and w = the speed of the wind (in mph). Then the speed of the blimp flying with the wind is $s + w$ and the speed of the blimp flying against the wind is $s - w$.

Form   We can construct a table to help us form two equations that model the situation. First, we enter the rates and times for both trips. Then we use the formula $d = rt$, to find that $5(s + w)$ represents the distance traveled with the wind and $7(s - w)$ represents the distance traveled against the wind. This information is shown in the last column of the table.

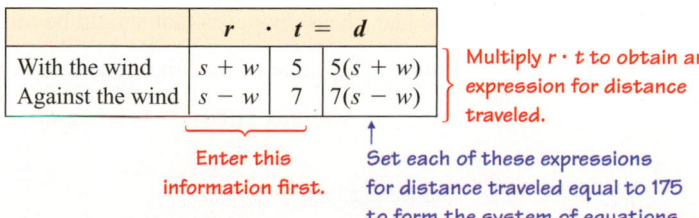

	r	$\cdot$ t =	d
With the wind	$s + w$	5	$5(s + w)$
Against the wind	$s - w$	7	$7(s - w)$

Multiply $r \cdot t$ to obtain an expression for distance traveled.

Enter this information first.

Set each of these expressions for distance traveled equal to 175 to form the system of equations.

Since each trip is 175 miles long, the information in the *Distance (d) column* of the table can be used to form two equations in two variables. To write each equation in standard form, we use the distributive property.

$$\begin{cases} 5(s + w) = 175 \xrightarrow{\text{Distribute}} \\ 7(s - w) = 175 \xrightarrow{\text{Distribute}} \end{cases} \begin{cases} 5s + 5w = 175 \quad \textbf{(1)} \\ 7s - 7w = 175 \quad \textbf{(2)} \end{cases}$$

Solve   Since the coefficients of w have opposite signs, we will eliminate w. To do this, we will create terms of $35w$ and $-35w$ by multiplying both sides of the equation 1 by 7 and both sides of the equation 2 by 5.

$$\begin{array}{ll} 35s + 35w = 1{,}225 & \text{This is } 7(5s + 5w) = 7(175). \\ \underline{35s - 35w = 875} & \text{This is } 5(7s - 7w) = 5(175). \\ 70s \quad\quad = 2{,}100 & \\ \quad\quad s = 30 & \text{Divide both sides by 70. This is the speed of the blimp in still air.} \end{array}$$

To find w, we will substitute 30 for s in equation 1.

$$\begin{array}{ll} 5s + 5w = 175 & \\ 5(\textbf{30}) + 5w = 175 & \text{Substitute 30 for } s. \\ 150 + 5w = 175 & \text{Do the multiplication.} \\ 5w = 25 & \text{Subtract 150 from both sides.} \\ w = 5 & \text{Divide both sides by 5. This is the speed of the wind.} \end{array}$$

State   The speed of the blimp in still air is 30 mph and the speed of the wind is 5 mph.

Check   The blimp's speed traveling with a 5-mph wind will be $30 + 5 = 35$ mph. In 5 hours, it will travel $35 \cdot 5 = 175$ miles. The blimp's speed traveling against a 5-mph wind will be $30 - 5 = 25$ mph. In 7 hours, it will travel $25 \cdot 7 = 175$ miles. The results check.

Self Check 6   **Boating.**   A boat traveled 63 miles downstream in 3 hours. The return trip took 7 hours going upstream. Find the speed of the boat in still water and the speed of the current.

Now Try   Problem 45

The liquid and dry mixture problems that we studied in Section 1.8 can be solved with a two-variable approach.

EXAMPLE 7   **Popcorn.**   A tin of jalapeño-flavored popcorn sells for $36, while the same size tin of cheddar cheese-flavored popcorn sells for $24. How many tins of each type of popcorn should be used to create 10 tins of a jalapeño-cheddar mix that can be sold for $27 per tin?

Analyze   We will use a two-variable approach to solve this dry mixture problem.

Assign   Let $x =$ the number of tins of jalapeño popcorn and $y =$ the number of tins of cheddar cheese popcorn that should be mixed.

Form   The value of the mixture and the value of each of its components are given by

Amount · price = total value

©Kim Karpeles/Alamy

Thus, the value of x tins of jalapeño popcorn is $36x$ and the value of y tins of cheddar cheese popcorn is $24y$. The sum of these values is also equal to the total value of the final mixture that is $10 \cdot \$27$ or $\$270$. This information is shown in the table.

	Amount	· Price	= Total value
Jalapeño popcorn	x	36	$36x$
Cheddar cheese popcorn	y	24	$24y$
Mixture	10	27	10(27)

} Multiply Amount · Price three times to fill in this column.

↑ One equation comes from this column.   ↑ A second equation comes from this column.

From the *Amount column* of the table:

The number of tins of jalapeño popcorn	plus	the number of tins of cheddar cheese popcorn	equals	10.
x	$+$	y	$=$	10

From the *Total value column* of the table:

The value of the jalapeño popcorn	plus	the value of the cheddar cheese popcorn	equals	the value of the mixture.
$36x$	$+$	$24y$	$=$	10(27)

The resulting system is: $\begin{cases} x + y = 10 \\ 36x + 24y = 270 \end{cases}$ Multiply: 10(27) = 270.

Solve   We will use elimination to solve this system. To make the y-terms drop out, we multiply both sides of the first equation of the system by -24 and add the resulting equations to solve for x:

$$-24x - 24y = -240$$ This is $-24(x + y) = -24(10)$.
$$\underline{36x + 24y = 270}$$
$$12x = 30$$ Add like terms, column by column.

$$x = \frac{30}{12} = 2.5$$ Divide both sides by 12 and simplify. This is the number of tins of jalapeño popcorn needed.

To find y, we substitute 2.5 for x in the first equation of the system and solve for y:

$$x + y = 10$$
$$2.5 + y = 10$$ Substitute 2.5 for x.
$$y = 7.5$$ Subtract 2.5 from both sides. This is the number of tins of cheddar cheese popcorn needed.

State   To obtain 10 tins of jalapeño-cheddar popcorn, 2.5 tins of japaleño and 7.5 tins of cheddar cheese popcorn should be combined.

Check   When 2.5 tins and 7.5 tins are combined, the result is 10 tins. The 2.5 tins of jalapeño popcorn are valued at 2.5($36) = $90 and the 7.5 tins of cheddar cheese popcorn are valued at 7.5($24) = $180. The sum of those values, $90 + $180 = $270, is the same as the value of the mixture, 10($27) = $270. The results check.

Self Check 7   **Snacks.**   How many pounds of peanuts (selling for $4 per pound) and how many pounds of M&M's (selling for $3.52 per pound) must be combined to get a 12-pound mixture worth $3.75 per pound?

Now Try ▶   Problem 51

Success Tip

Determine the similarities and the differences of the one-variable and two-variable approaches by reviewing the dry mixture problem of Example 6 in Section 1.8.

EXAMPLE 8   **Water Treatment.**   A technician determines that 100 fluid ounces of a 15% muriatic acid solution needs to be added to the water in a swimming pool to kill a growth of algae. If the technician has 5% and 20% muriatic solutions on hand, how many ounces of each must be combined to create the 15% solution?

Analyze   We will use a two-variable approach to solve this liquid mixture problem. We need to find the number of ounces of a 5% solution and the number of ounces of a 20% solution that must be combined to obtain 100 ounces of a 15% solution.

Assign   Let x = the number of ounces of the 5% solution and let y = the number of ounces of the 20% solution that are to be mixed.

Form   The amount of pure muriatic acid in each solution is given by

$$\textbf{Amount of solution} \cdot \textbf{strength} = \textbf{amount of pure muriatic acid}$$

Thus, the amount of muriatic acid in the 5% solution is $0.05x$ ounces, and the amount of muriatic acid in the 20% solution is $0.20y$ ounces. The sum of these amounts is also the amount of muriatic acid in the final mixture, which is 15% of 100 ounces or $0.15(100)$. This information is shown in the table.

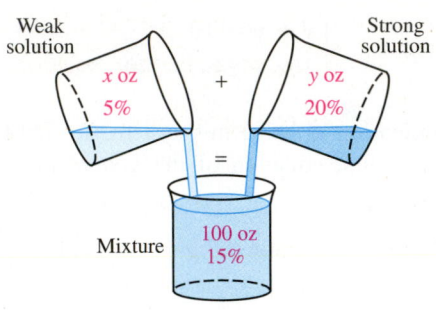

	Amount ·	Strength =	Amount of muriatic acid
Weak	x	0.05	$0.05x$
Strong	y	0.20	$0.20y$
Mixture	100	0.15	$0.15(100)$

↑ One equation comes from the information in this column.

↑ Another equation comes from the information in this column.

Success Tip

Determine the similarities and the differences of the one-variable and two-variable approaches by reviewing the liquid mixture problem of Example 7, Section 1.8.

From the *Amount column* of the table:

The number of ounces of 5% solution	plus	the number of ounces of 20% solution	equals	the total number of ounces in the 15% mixture.
x	$+$	y	$=$	100

From the *Amount of muriatic acid column* of the table:

The number of ounces of acid in the 5% solution	plus	the number of ounces of acid in the 20% solution	is	the number of ounces of acid in the 15% mixture.
$0.05x$	$+$	$0.20y$	$=$	$0.15(100)$

Caution

Don't forget to express 5%, 20%, and 15% as decimals.

$5\% = 0.05 \qquad 20\% = 0.20$

$15\% = 0.15$

The resulting system is: $\begin{cases} x + y = 100 & \textbf{(1)} \\ 0.05x + 0.20y = 15 & \textbf{(2)} \end{cases}$ Multiply: 0.15(100) = 15.

Solve   To solve this system by substitution, we can solve the first equation for y:

$$x + y = 100$$

$$y = 100 - x \qquad \text{Subtract } x \text{ from both sides. This is the substitution equation.}$$

Then we substitute $100 - x$ for y in equation 2 and solve for x.

$$0.05x + 0.20\mathbf{y} = 15$$

$$0.05x + 0.20(\mathbf{100 - x}) = 15 \qquad \text{Substitute } 100 - x \text{ for } y.$$

$$5x + 20(100 - x) = 1{,}500 \qquad \text{To clear the equation of decimals,}$$
$$\text{multiply both sides by 100.}$$

$$5x + 2{,}000 - 20x = 1{,}500 \qquad \text{Distribute the multiplication by 20.}$$

$$-15x = -500 \qquad \text{Combine like terms and subtract 2,000 from both sides.}$$

$$x = \frac{-500}{-15} \qquad \text{To solve for x, divide both sides by } -15.$$

$$x = \frac{100}{3} \qquad \text{Simplify. This is the number of ounces}$$
$$\text{of the 5\% acid solution that is needed.}$$

To find y, we can substitute $\frac{100}{3}$ for x in the substitution equation:

$$y = 100 - \mathbf{x}$$

$$y = 100 - \frac{\mathbf{100}}{\mathbf{3}} \qquad \text{Substitute } \tfrac{100}{3} \text{ for x.}$$

$$y = \frac{200}{3} \qquad \text{To subtract, think: } \tfrac{300}{3} - \tfrac{100}{3}. \text{ This is the number}$$
$$\text{of ounces of the 20\% acid solution that is needed.}$$

State   To obtain 100 ounces of a 15% solution, the technician must mix $\frac{100}{3}$ or $33\frac{1}{3}$ ounces of the 5% solution with $\frac{200}{3}$ or $66\frac{2}{3}$ ounces of the 20% solution.

Notation

For application problems, it is often more meaningful to present the results as mixed numbers $\left(\text{such as } 33\frac{1}{3} \text{ and } 66\frac{2}{3}\right)$ than as improper fractions $\left(\frac{100}{3} \text{ and } \frac{200}{3}\right)$.

Check   We note that $33\frac{1}{3}$ ounces of solution plus $66\frac{2}{3}$ ounces of solution equals the required 100 ounces of solution. The $33\frac{1}{3} = \frac{100}{3}$ ounces of 5% solution contains $0.05\left(\frac{100}{3}\right) = \frac{5}{3}$ ounces of muriatic acid, and the $66\frac{2}{3} = \frac{200}{3}$ ounces of 20% solution contains $0.20\left(\frac{200}{3}\right) = \frac{40}{3}$ ounces of muriatic acid—a total of $\frac{5}{3} + \frac{40}{3} = \frac{45}{3}$ or 15 ounces of muriatic acid. The 100 ounces of the 15% mixture contains $0.15(100) = 15$ ounces of acid. The results check.

Self Check 8   **Nutrition.**   A nutritionist suggested that 20 fluid ounces of 4% butterfat milk should be included in an underweight toddler's diet. If the mother has cream that is approximately 22% butterfat and 2% butterfat milk, how many ounces of each must be combined to create the 4% butterfat milk solution?

Now Try ▶ Problem 57

VOCABULARY

Fill in the blanks.

1. A parallelogram is a four-sided figure with two pairs of _____ sides.

2. Suppose a hammer can be manufactured in two different ways. The number of hammers that will cost equal amounts to produce either way is called the _____ point.

CONCEPTS

3. **a.** Refer to the parallelogram below. Find the measure of $\angle SRU$ and $\angle TSU$.

   b. Fill in the blank: $\angle SUR$ and $\angle TSU$ are called _____ _____ angles.

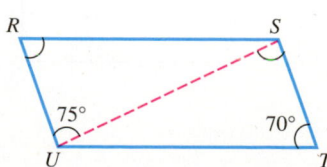

4. A company charges a $75 setup fee plus $5.25 per shirt to silkscreen a design on specialty t-shirts. Write an equation that gives the cost of purchasing x shirts.

5. In still water, a person swims at the rate of x mph. Find the speed of the swimmer for each of the following situations.

   a. With the current

   b. Against the current

6. a. Write an expression that represents the total value of x ounces of ginseng tea that costs $32 per pound.

   b. Write an expression that represents the amount of hydrochloric acid in x gallons of a 3% hydrochloric acid solution.

NOTATION

7. Write each percent as a decimal.

   a. 6%   b. 4.8%   c. $13\frac{1}{2}$%

8. What is the formula that finds

   a. Simple interest   b. Distance traveled

GUIDED PRACTICE

Complete each table and write a system of two equations that can be used to find x and y. DO NOT SOLVE THE SYSTEM. See Examples 5–8.

9. **Investments.** A total of $25,000 was invested in two accounts for 1 year and earned a total of $1,050 in interest.

	Principal ·	Rate ·	Time =	Interest
Township Bank	x	0.05	1	
Ameritech Savings	y	0.04	1	
Total				

10. **Physical Fitness.** A jogger and cyclist started at the same point and traveled for 2 hours in opposite directions until they were 42 miles apart. The cyclist traveled 10 mph faster than the jogger.

	Rate ·	Time =	Distance
Jogger	x	2	
Cyclist	y	2	
Total			

11. **Candy.** A company combines dark chocolate (selling for $13.90 per pound) with white chocolate (selling for $5.10 per pound) to get 6 pounds of a mixture that will sell for $10.25 per pound.

	Amount ·	Value =	Total value
Dark chocolate	x	13.90	
White chocolate	y	5.10	
Mixture		10.25	

12. **Disinfectants.** A 1% bleach solution is to be mixed with a 5% bleach solution to obtain 15 ounces of a 3% bleach solution.

	Amount ·	Strength =	Amount of bleach
Weak	x	0.01	
Strong	y	0.05	
Mix		0.03	

APPLICATIONS

Write a system of two equations in two variables to solve each problem.

13. **Electronics.** In the illustration, two resistors in the voltage divider circuit have a total resistance of 1,375 ohms. To provide the required voltage, R_1 must be 125 ohms greater than R_2. Find both resistances.

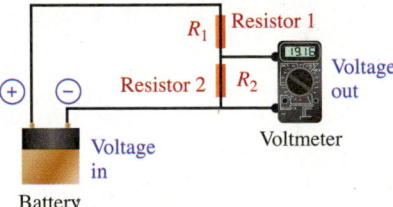

14. **Desserts.** A slice of Mrs. Smith's apple pie and one scoop of Häagen-Dazs vanilla bean ice cream totals 600 calories. The pie has 20 more calories than the ice cream. Find the number of calories in each.

15. **Area Codes.** The entire state of Montana has just one telephone area code. The same is true for Idaho. The sum of their area codes is 614 and the difference is 198, and Montana has the numerically larger one. Find the area code of each of these states.

16. Hiking. The Pacific Crest Trail runs from the U.S. border with Mexico to its border with Canada. The Appalachian Trail extends between Springer Mountain in Georgia and Mount Katahdin in Maine. The sum of the lengths of the trails is 4,824 miles, the difference is 476 miles, and the Pacific Crest Trail is the longer. Find the length of each trail.

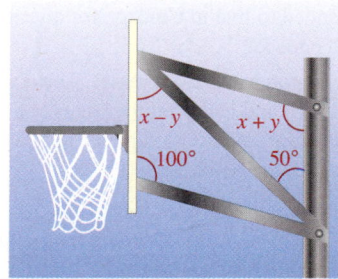

17. from **Campus to Careers**

Fashion Designer

In 2009, there was a combined total of 4,046 Gap and Aéropostale clothing stores worldwide. The number of Gap stores was $3\frac{1}{4}$ times more than the number of Aéropostale stores. How many Gap stores and how many Aéropostale stores were there that year? (Source: wikinvest.com)

18. Avalanches. For the 2009–2010 snow season, the total number of avalanche fatalities in the United States and Canada was 48. If the number in the United States was three times greater than the number in Canada, how many avalanche fatalities were there in each country? (Source: Avalanche.org)

19. Bracing. The bracing of a basketball backboard forms a parallelogram. Find the unknown degree measures represented by x and y.

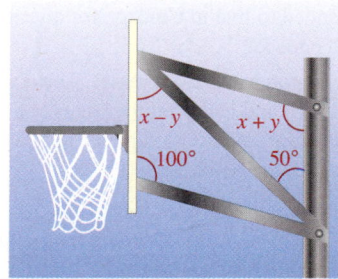

20. Decorative Mirrors. An interior designer is ordering a mirror that is in the shape of a parallelogram. Find the unknown degree measures represented by x and y.

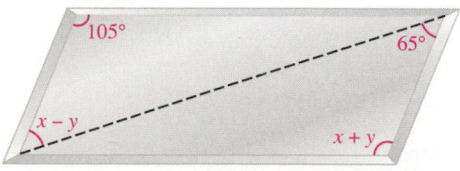

21. Traffic Signals. In the illustration, braces A and B are perpendicular. Find the values of x and y.

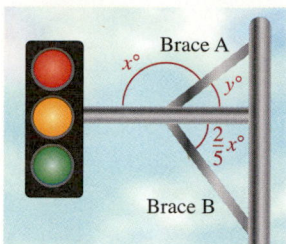

22. Physical Therapy. After an elbow injury, a volleyball player has restricted movement of her arm. Her range of motion (the measure of $\angle 1$) is 28° less than the measure of $\angle 2$. Find the measure of each angle.

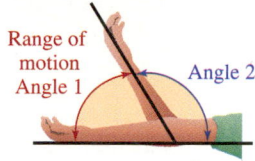

23. Fencing a Field. The perimeter of a rectangular field is surrounded by 72 meters of fencing. If the field is partitioned into two parts as shown, a total of 88 meters of fencing is required. Find the dimensions of the field.

24. New York City. The triangular-shaped Flatiron Building in Manhattan has a perimeter of 499 feet at its base. It is bordered on each side by a street. The 5th Avenue front of the building is 198 feet long. The Broadway front is 43 feet more than twice as long as the East 22nd Street front. Find the length of the Broadway front and East 22nd Street front. (Source: New York Public Library)

25. **Advertising.** Use the information in the ad to find the cost of a 15-second and a 30-second radio commercial on radio station KLIZ.

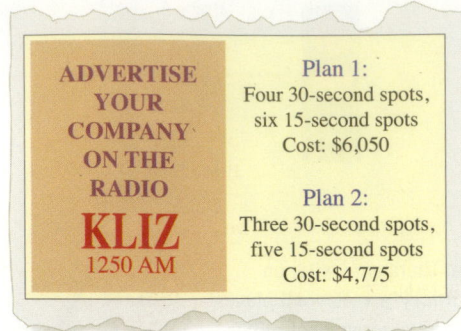

ADVERTISE YOUR COMPANY ON THE RADIO
KLIZ
1250 AM

Plan 1:
Four 30-second spots,
six 15-second spots
Cost: $6,050

Plan 2:
Three 30-second spots,
five 15-second spots
Cost: $4,775

26. **Temporary Help.** A law firm hired several workers to help finish a large project. From the following billing records, determine the daily fee charged by the employment agency for a clerk-typist and for a computer programmer. (*Hint:* Count the number of last names in each type of job.)

TEMPORARY EMPLOYMENT, INC.

Billed to: Archer Law Offices

Day	Position/Employee Name	Total cost
Mon. 3/22	*Clerk-typists:* K. Amad, B. Tran, S. Smith *Programmers:* T. Lee, C. Knox	$685
Tues. 3/23	*Clerk-typists:* K. Amad, B. Tran, S. Smith, W. Morada *Programmers:* T. Lee, C. Knox, B. Morales	$975

27. **Production Planning.** A manufacturer builds racing bikes and mountain bikes, with the per-unit manufacturing costs shown in the table. The company has budgeted $26,150 for materials and $31,800 for labor. How many bicycles of each type can be built?

Model	Cost of materials	Cost of labor
Racing	$110	$120
Mountain	$140	$180

28. **Farming.** A farmer keeps some animals on a strict diet. Each animal is to receive 15 grams of protein and 7.5 grams of carbohydrates. The farmer uses two food mixes, with nutrients as shown in the table. How many grams of each mix should be used to provide the correct nutrients for each animal?

Mix	Protein	Carbohydrates
Mix *A*	12%	9%
Mix *B*	15%	5%

29. **Summer Concerts.** According to *StubHub.com,* in 2009, two tickets to a Jonas Brothers concert and two tickets to an Elton John concert cost, on average, a total of $560. At those prices, four tickets to see the Jonas Brothers and two tickets to see Elton John cost $806. What was the average cost of a Jonas Brothers ticket and an Elton John ticket in 2009?

30. **Measurement.** A *furlong* is a measure of distance and is used to express the length of certain horse races. A *fathom* is a measure of distance and is used to express depths of water. Four furlongs and five fathoms is a total of 2,670 feet. Five furlongs and four fathoms is a total of 3,324 feet. Find the length of a furlong and a fathom.

31. **Making Tires.** A company has two molds to form tires. One mold has a setup cost of $1,000 and the other has a setup cost of $3,000. The cost to make each tire with the first mold is $15, and the cost to make each tire with the second mold is $10.

   a. Find the break point.

   b. Check your result by graphing both equations on the coordinate system below.

   c. If a production run of 500 tires is planned, determine which mold should be used.

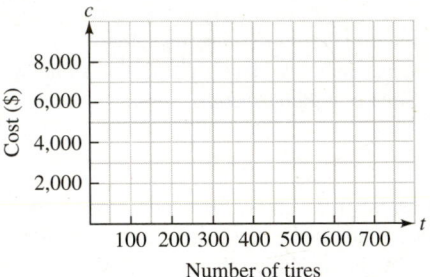

32. **Choosing a Furnace.** A high-efficiency 90+ furnace can be purchased for $2,250 and costs an average of $824 per year to operate in Chicago, Illinois. An 80+ furnace can be purchased for only $1,710, but it costs $932 per year to operate.

   a. Find the break point.

   b. If you intended to live in Chicago for 4 years, which furnace would you choose?

33. **Publishing.** A printer has two presses. The older press has a setup cost of $210 and can print the pages of a certain book for $5.98. The newer press has a setup cost of $350 and can print the pages of the same book for $5.95.

   a. Find the break point.

   b. If the publisher has advance orders for 5,100 copies of the book, which press should be used?

In Problems 34–37 recall that the money a business spends to produce a product (or service) is called its cost *and the money that it takes in from the sales of a product (or service) is called the* revenue. *In business and economics, it is important to determine the value at which costs equal the revenue, called the* break-even point.

34. **Cosmetology.** A beauty shop specializing in hair color treatments has fixed costs of $2,310 per month. The owner estimates that the cost for each treatment is $22, which covers labor, chemicals, and electricity.

   a. If her shop can give as many hair color treatments as she wants at a price of $64 each, how many must be given each month to break even? (*Hint:* To break even, revenue = costs.)

   b. How many hair color treatments must be given each month to make a profit?

35. **Recording Companies.** An executive invested a total of $105,000 to start a recording company to produce reissues of classic jazz. Each release will be a boxed set of CDs that will retail for $45 per set.

   a. If each set can be produced for $28.20, how many sets must be sold for the executive to break even? (*Hint:* To break even, revenue = costs.)

   b. How many sets must be sold for the executive to make a profit?

36. **Production Planning.** A paint manufacturer can choose between two processes for making house paint, with monthly costs as shown in the table. Assume that the paint sells for $36 per gallon.

Process	Fixed costs	Unit cost (per gallon)
A	$32,500	$26
B	$80,600	$10

   a. How many gallons of paint must be sold per month for the manufacturer to break even if process A is used to produce the paint? (*Hint:* To break even, revenue = costs.)

   b. How many gallons of paint must be sold per month for the manufacturer to break even if process B is used to produce the paint?

   c. If expected sales are 3,500 gallons per month, which process should the company use?

37. **Manufacturing.** A manufacturer of automobile water pumps is considering retooling for one of two manufacturing processes, with monthly fixed costs and unit costs as indicated in the table. Each water pump can be sold for $50.

Process	Fixed costs	Unit cost
A	$12,390	$29
B	$20,460	$17

   a. How many water pumps must be sold per month for the manufacturer to break even if process A is used to produce the pumps? (*Hint:* To break even, revenue = costs.)

   b. How many water pumps must be sold per month for the manufacturer to break even if process B is used to produce the pumps? (*Hint:* To break even, revenue = costs.)

   c. If expected sales are 550 water pumps per month, which process should be used?

38. **Salary Options.** A sales clerk can choose from two salary plans:

   1. a straight 7% commission
   2. $150 + 2% commission

   How much would the clerk have to sell for each plan to produce the same monthly paycheck?

39. **Investment Clubs.** Part of $8,000 was invested by an investment club at 10% interest and the rest at 12%. If the annual interest income from these investments is $900, how much was invested at each rate?

40. **Retirement Income.** A retired couple invested part of $12,000 at 6% interest and the rest at 7.5%. If their annual income from these investments is $810, how much was invested at each rate?

41. **Investing.** A woman invested some money in a credit union paying 5% annual simple interest and three times as much in a money market account paying 4.25% annual simple interest. If she earned $1,420 interest in one year, how much did she invest in each account?

42. **IRA Accounts.** A teacher started an Individual Retirement Account (IRA) by investing a total of $30,500 in two municipal bond funds, one paying 5% annual interest and the other paying $5\frac{1}{2}$% annual interest. At the end of the year, the funds earned a total of $1,615 in interest. How much did the teacher invest in each fund?

43. **Credit Cards.** A couple stopped using their VISA credit card charging 1.5% per month interest and their Robinsons-May credit card charging 1.75% per month interest because they had built up a combined debt of $16,500 on the two cards. For one month, they made no purchases or payments on the accounts. If the total amount of interest the credit cards accumulated during the month was $259.25, what amount did they owe on each card when they stopped using them?

44. **Loans.** A student had a car loan charging 0.75% interest per month and a tuition loan charging 0.5% interest per month. How much did he owe on each account if he paid a total of $95.50 monthly interest on a total debt of $14,200?

45. **Aviation.** The jet stream is a wind current that flows across the United States from west to east. Flying with the jet stream, an airplane flew 2,700 miles in 4.5 hours. Against the same wind, the return trip took 6 hours. Find the speed of the plane in still air and the speed of the jet stream.

46. **Salmon.** It takes a salmon 40 minutes to swim 10,000 feet upstream and 8 minutes to swim that same portion of a river downstream. Find the speed of the salmon in still water and the speed of the current.

47. Airport Walkways. A man walks at a steady pace as he steps onto a moving walkway. It takes him 40 seconds to reach the end, 320 feet away. If he walks at the same rate against the flow of the walkway, it would take him 80 seconds to reach the end. Find his rate of walking and the rate of the moving walkway.

48. Snowmobiling. A man rode a snowmobile at the rate of 20 mph and then skied cross country at the rate of 4 mph. During the 6-hour trip, he traveled 48 miles. How long did he snowmobile, and how long did he ski?

49. Jet Skis. A Jet Ski rider can travel 10 miles against the current of the lower Mississippi River in $\frac{1}{2}$ hour and make the return trip with the current in $\frac{1}{3}$ hour. Find the speed of the Jet Ski in still water and the speed of the current.

50. Rollerblading. An in-line skater headed west at the rate of 6 mph. One hour later, a moped rider left the same spot and headed west on the same road at 30 mph. How long will it take the moped rider to catch the skater?

51. Mixing Candy. How many pounds of each candy shown in the illustration must be mixed to obtain 60 pounds of candy that would be worth $4 per pound?

52. Gasoline. A truck owner drove his pickup to a service station to fill the nearly empty 24-gallon gas tank. If the truck runs on 89-octane gasoline, but the station only sells 87-octane and 93-octane gas, how many gallons of each should be pumped to fill the tank with an 89-octane blend?

53. Mixing Coffee. How many pounds of regular coffee (selling for $4 per pound) and how many pounds of Kona coffee (selling for $11.50 per pound) must be combined to get 20 pounds of a mixture worth $6 per pound?

54. Room Freshener. A florist sells mixtures of dried, fragrant plant material that provides a gentle natural scent for houses. She wants to mix lilac (that sells for $18.25 a pound) with lavender (that sells for $12.25 a pound) to create 30 pounds of a blend that sells for $15 a pound. How many pounds of each should the florist use?

55. Confetti. How many pounds of $14.50-per-pound small flake confetti should be mixed with $24.50-per-pound mylar confetti stars to obtain one ton of a confetti mix that would be worth $20 per pound? (*Hint:* How many pounds equal one ton?)

56. Bath Salts. The owner of a kitchen and bath store wants to combine $7.92-per-pound eucalyptus bath salts with 88¢-per-pound Epsom salt to create 40 pounds of a $2.64-per-pound bath salt mix. How many pounds of each should be used?

57. Antifreeze. How many pints of a 10% antifreeze solution and how many pints of a 40% antifreeze solution must be mixed to obtain 24 pints of a 30% solution?

58. Mixing Solutions. How many ounces of the two alcohol solutions must be mixed to obtain 100 ounces of a 12.2% solution?

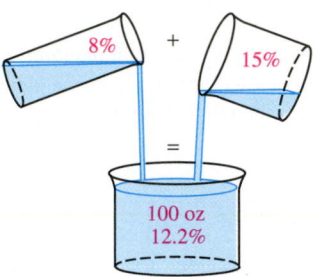

59. Dermatology. Tests of an antibacterial face wash cream showed that a mixture containing 0.3% Triclosan (active ingredient) gave the best results. How many grams of cream from each tube shown in the illustration should be used to make an equal-size tube of the 0.3% cream?

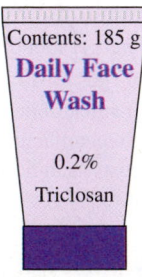 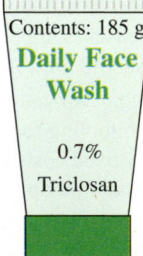

60. Salads. A chef wants to make 1 gallon (128 ounces) of a 50% vinegar-to-oil salad dressing. He only has pure vinegar and a mild 4% vinegar-to-oil salad dressing on hand. How many ounces of each should he mix to make the desired dressing?

WRITING

61. Write a problem that can be solved by solving the system:
$$\begin{cases} x + y = 36 \\ \$1.29x + \$2.29y = \$72.44 \end{cases}$$

62. Write a problem to fit the information given in the table.

	Ounces · Strength	=	Amount of insecticide
Weak	x	0.02	0.02x
Strong	y	0.10	0.10y
Mixture	80	0.07	0.07(80)

63. What is a *break point*? Give an example.

64. A woman paid $219 for two blouses and four pairs of pants. If we let x = the cost of a blouse and y = the cost of a pair of pants, an equation modeling the purchase is $2x + 4y = 219$. Explain why there is not enough information to determine the cost of a blouse or the cost of a pair of pants.

65. To solve mixture problems, do you prefer the one-variable or two-variable solution strategy? Explain why.

66. Write a system of two equations in two variables to attempt to solve the following problem. Then explain why the problem has no solution.

How many gallons of a 20% salt solution and how many gallons of a 30% salt solution should be mixed to obtain 10 gallons of a 50% salt solution?

REVIEW

Fill in the blanks.

67. A _____ number is any number that can be written as a fraction with an integer numerator and a nonzero integer denominator.

68. The _____ of the term $-8c$ is -8.

69. An equation that is true for all values of its variable is called an _____.

70. The _____ of a three-dimensional geometric solid is the amount of space it encloses.

71. If a triangle has exactly two sides with equal measures, it is called an _____ triangle.

72. The _____ of $\frac{5}{9}$ is $\frac{9}{5}$.

CHALLENGE PROBLEMS

73. Managing an Apartment. The manager of an apartment complex is also a tenant. He pays only three-fourths of the monthly rent that each of the remaining 5 tenants pays. Each month, a total of $5,520 in rent is paid by the 6 occupants. How much rent does the manager pay?

74. Digits Problem. The sum of the digits of a two-digit number is 10. If we interchange the digits, then the new number formed is 54 less than the original. Find the original number.

Problem Solving Using Systems of Three Equations

OBJECTIVES

1 Assign variables to three unknowns.

2 Use systems to solve curve-fitting problems.

ARE YOU READY?

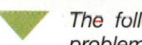

The following problems review some basic skills that are needed when solving application problems using systems of three equations.

1. Translate to mathematical symbols: 75 more than the number n.

2. Write an algebraic expression that represents the value of x coats, if each coat has a value of $150.

3. Write an algebraic expression that represents the value (in cents) of n nickels.

4. Solve the system $\begin{cases} x - y + 6z = 12 \\ x + 6y = -28 \\ 7y + z = -26 \end{cases}$ by substitution.

In this section, we will solve application problems involving three unknown quantities.

1 Assign Variables to Three Unknowns.

The six step problem-solving strategy that was used to solve application problems involving two unknowns in Section 3.6 can be extended to situations involving three unknowns. In the Assign step, we assign *three* variables to represent *three* unknown quantities. In the Form and Solve steps, we form a system of three equations in three variables and solve it using either elimination, substitution, matrices, or Cramer's rule.

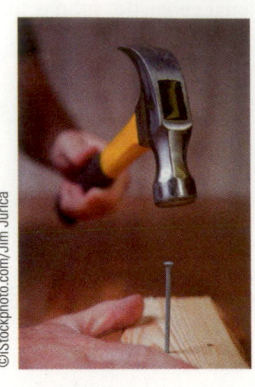

EXAMPLE 1

Tool Manufacturing. A company makes three types of hammers, which are marketed as "good," "better," and "best." The cost of manufacturing each type of hammer is $4, $6, and $7, respectively, and the hammers sell for $6, $9, and $12. Each day, the cost of manufacturing 100 hammers is $520, and the daily revenue from their sale is $810. How many hammers of each type are manufactured per day?

Analyze   We need to find how many of each type of hammer are manufactured daily. Since there are three unknowns, we must write three equations to find them.

Assign   Let x = the number of good hammers
        y = the number of better hammers
        z = the number of best hammers

Form   We can organize the facts of the problem in a table, as shown below. To fill in the "Cost to manufacture" column, we multiply the number of good hammers, x, by the cost to manufacture a good hammer, $4, to find that it costs $4 \cdot x$, or $4x$, to manufacture the good hammers. Similarly, we use multiplication to find that the manufacturing cost for y better hammers is $6y$ and for z best hammers is $7z$. The total manufacturing cost of $520 is entered at the bottom of that column.

   To complete the "Revenue received" column, we again use multiplication. The revenue received from the sale of x good hammers, sold for $6 per hammer, is $6 \cdot x$, or $6x$. Similarly, the revenue received from the sale of y better hammers is $9y$, and the revenue received from the sale of z best hammers is $12z$. The total revenue received of $810 is entered at the bottom of that column.

   We can use the facts of the problem to write three equations.

Type of hammer	Number	Cost to manufacture	Revenue received
Good	x	$4x$	$6x$
Better	y	$6y$	$9y$
Best	z	$7z$	$12z$

        Total: 100   Total: $520   Total: $810
              ↑   ↑   ↑
One equation   A second equation   A third equation
comes from   comes from   comes from
this column.   this column.   this column.

The resulting system is: $\begin{cases} x + y + z = 100 & \textbf{(1)} \\ 4x + 6y + 7z = 520 & \textbf{(2)} \\ 6x + 9y + 12z = 810 & \textbf{(3)} \end{cases}$

Solve   We will use the elimination method to solve this system of three equations in three variables. If we multiply equation 1 by -7 and add the result to equation 2, we get

$\qquad -7x - 7y - 7z = -700$ This is $-7(x + y + z) = -7(100)$.

(2) $\qquad \underline{4x + 6y + 7z = 520}$

(4) $\qquad -3x - y = -180$

If we multiply equation 1 by -12 and add the result to equation 3, we get

$\qquad -12x - 12y - 12z = -1{,}200$ This is $-12(x + y + z) = -12(100)$.

(3) $\qquad \underline{6x + 9y + 12z = 810}$

(5) $\qquad -6x - 3y = -390$

We can multiply equation 4 by -3 and add it to equation 5 to eliminate y.

$$9x + 3y = 540 \qquad \text{This is } -3(-3x - y) = -3(-180).$$
$$(5) \quad \underline{-6x - 3y = -390}$$
$$3x \qquad = 150$$
$$x = 50 \qquad \text{To solve for x, divide both sides by 3.}$$
$$\text{This is the number of good hammers manufactured.}$$

To find y, we substitute 50 for x in equation 4:

$$-3x - y = -180$$
$$-3(50) - y = -180 \qquad \text{Substitute 50 for x.}$$
$$-150 - y = -180$$
$$-y = -30 \qquad \text{Add 150 to both sides.}$$
$$y = 30 \qquad \text{To solve for y, divide both sides by } -1.$$
$$\text{This is the number of better hammers manufactured.}$$

To find z, we substitute 50 for x and 30 for y in equation 1:

$$x + y + z = 100$$
$$50 + 30 + z = 100$$
$$z = 20 \qquad \text{To solve for z, subtract 80 from both sides.}$$
$$\text{This is the number of best hammers manufactured.}$$

Success Tip

If three variables are used to represent three unknowns, then three equations must be formed, and a three-part check must be performed to verify the results.

State Each day, the company manufactures 50 good hammers, 30 better hammers, and 20 best hammers.

Check If the company manufactures **50** good hammers, **30** better hammers, and **20** best hammers each day, that is a total of **50** + **30** + **20** = 100 hammers. The cost of manufacturing the three types of hammers is $4(**50**) + $6(**30**) + $7(**20**) = $200 + $180 + $140 or $520. The revenue from the sale of the hammers is $6(**50**) + $9(**30**) + $12(**20**) = $300 + $270 + $240 or $810. The results check.

Self Check 1 **Computer Storage.** A manufacturer of memory cards makes 1-GB, 2-GB, and 4-GB storage size cards. The cost of manufaturing each is $2, $3, and $5, respectively. Each day the cost of manufacturing 500 cards is $1,500. The cards sell for $15, $20, and $30, respectively, with a daily revenue of $10,000. How many memory cards of each type are manufactured?

Now Try Problem 7

EXAMPLE 2 **The Olympics.** The three countries that won the most medals in the 2008 summer Olympic games were the United States, China, and Russia, in that order. Together they won a total of 282 medals, with the U.S. medal count 10 more than China's and China's medal count 28 more than Russia's. Find the number of medals won by each country. (Source: sportsillustrated.cnn.com)

Analyze We need to find how many medals the United States, China, and Russia won. Since there are three unknowns, we must write three equations to find them.

Assign Let x = the number of medals won by the United States
y = the number of medals won by China
z = the number of medals won by Russia

Form   We can use the facts of the problem to write three equations.

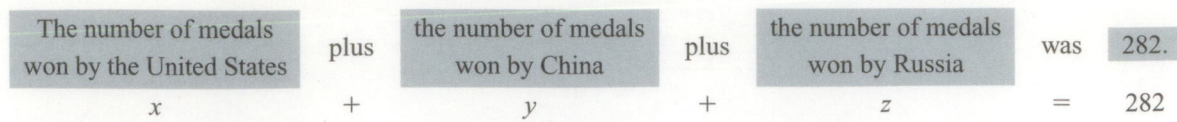

The number of medals won by the United States	plus	the number of medals won by China	plus	the number of medals won by Russia	was	282.
x	$+$	y	$+$	z	$=$	282

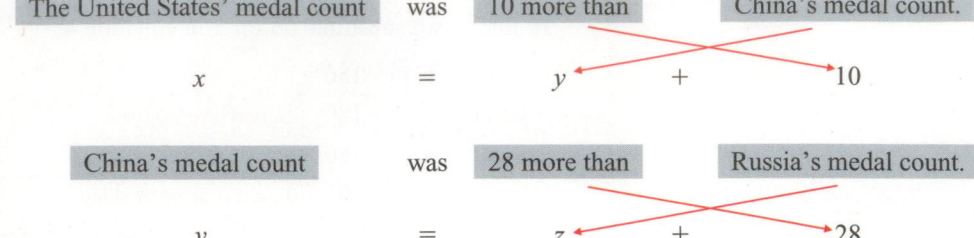

The United States' medal count	was	10 more than	China's medal count.
x	$=$	y	$+$	10

China's medal count	was	28 more than	Russia's medal count.
y	$=$	z	$+$	28

Solve   We can use substitution to solve the resulting system of three equations. If we solve equation 3 for z, then the resulting equation 4 and equation 2 can serve as substitution equations.

(1) $\begin{cases} x + y + z = 282 \\ (2)\ x = y + 10 \\ (3)\ y = z + 28 \end{cases}$

Unchanged
Unchanged
Solve for z

$\begin{cases} x + y + z = 282 \quad (1) \\ x = \boxed{y + 10} \quad (2) \\ z = \boxed{y - 28} \quad (4) \end{cases}$

When we substitute for x and z in equation 1, we obtain an equation in one variable, y.

$x + y + z = 282$ This is equation 1.

$y + 10 + y + y - 28 = 282$ Substitute $y + 10$ for x and $y - 28$ for z.

$3y - 18 = 282$ On the left side, combine like terms.

$3y = 300$ Add 18 to both sides.

$y = 100$ To solve for y, divide both sides by 3.
            This is the number of medals won by China.

To find x, we substitute 100 for y in equation 2. To find z, we substitute 100 for y in equation 4.

$x = y + 10$	This is equation 2.	$z = y - 28$	This is equation 4.
$x = 100 + 10$		$z = 100 - 28$	
$x = 110$	This is the U.S. medal count.	$z = 72$	Russia's medal count

State   In the 2008 summer Olympics, the United States won 110 medals, China won 100, and Russia won 72.

Check   The sum of $110 + 100 + 72$ is 282. Furthermore, 110 is 10 more than 100, and 100 is 28 more than 72. The results check.

Self Check 2   **The Winter Olympics.**   The three countries that are at the top of the list for the most all-time medal wins are Norway, the United States, and Austria, in that order. Together they have won a total of 618 medals, with Norway's medal count 70 more than the U.S., and the U.S. medal count 31 more than Austria's. Find the number of medals won by each country. (Source: nationmaster.com)

Now Try ▶   Problem 13

2 Use Systems to Solve Curve-Fitting Problems.

The process of determining an equation whose graph contains given points is called **curve fitting.**

EXAMPLE 3   The equation of a parabola opening upward or downward is of the form $y = ax^2 + bx + c$. Find the equation of the parabola graphed on the left by determining the values of a, b, and c.

Strategy   We will substitute the x- and y-coordinates of three points that lie on the graph into the equation $y = ax^2 + bx + c$. This will produce a system of three equations in three variables that we can solve to find a, b, and c.

Why   Once we know a, b, and c, we can write the equation.

Solution   Since the parabola passes through the points $(-1, 5)$, $(1, 1)$, and $(2, 2)$, each pair of coordinates must satisfy the equation $y = ax^2 + bx + c$. If we substitute each pair into $y = ax^2 + bx + c$, we will get a system of three equations in three variables.

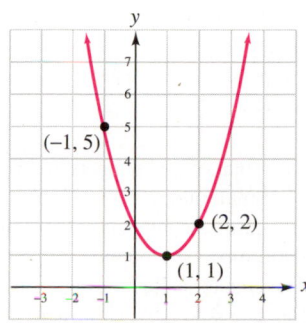

Substitute $(-1, 5)$	Substitute $(1, 1)$	Substitute $(2, 2)$
$y = ax^2 + bx + c$	$y = ax^2 + bx + c$	$y = ax^2 + bx + c$
$5 = a(-1)^2 + b(-1) + c$	$1 = a(1)^2 + b(1) + c$	$2 = a(2)^2 + b(2) + c$
$5 = a - b + c$	$1 = a + b + c$	$2 = 4a + 2b + c$
This is equation 1.	This is equation 2.	This is equation 3.

The three equations above give the system, which we can solve to find a, b, and c.

$$
\begin{array}{ll}
(1) & \\
(2) & \\
(3) &
\end{array}
\begin{cases}
a - b + c = 5 \\
a + b + c = 1 \\
4a + 2b + c = 2
\end{cases}
$$

Success Tip

If a point lies on the graph of an equation, it is a solution of the equation, and the coordinates of the point satisfy the equation.

If we add equations 1 and 2, we obtain

$$
\begin{array}{l}
\quad a - b + c = 5 \\
\underline{\quad a + b + c = 1} \\
(4) \ 2a + 2c = 6
\end{array}
$$

If we multiply equation 1 by 2 and add the result to equation 3, we get

$$
\begin{array}{l}
\quad\quad 2a - 2b + 2c = 10 \\
(3) \ \underline{4a + 2b + c = 2} \\
(5) \ 6a + 3c = 12
\end{array}
$$

We can then divide both sides of equation 4 by 2 to get equation 6 and divide both sides of equation 5 by 3 to get equation 7. We now have the system

$$
\begin{array}{ll}
(6) & \\
(7) &
\end{array}
\begin{cases}
a + c = 3 \\
2a + c = 4
\end{cases}
$$

To eliminate c, we multiply equation 6 by -1 and add the result to equation 7. We get

$$
\begin{array}{ll}
\quad -a - c = -3 & \text{This is } -1(a + c) = -1(3). \\
\underline{\quad 2a + c = 4} & \\
\quada = 1 &
\end{array}
$$

To find c, we can substitute 1 for a in equation 6 and find that $c = 2$. To find b, we can substitute 1 for a and 2 for c in equation 2 and find that $b = -2$.

After we substitute these values of a, b, and c into the equation $y = ax^2 + bx + c$, we have the equation of the parabola.

$$
\begin{array}{l}
y = ax^2 + bx + c \\
y = 1x^2 - 2x + 2 \\
y = x^2 - 2x + 2 \quad \text{This is the equation of the parabola graphed above.}
\end{array}
$$

> **Self Check 3** Find the equation of the parabola, $y = ax^2 + bx + c$, that passes through $(1, 6)$, $(-4, 1)$, and $(-3, -2)$.
>
> **Now Try** ▶ Problem 27

SECTION 3.7 ▶ STUDY SET

VOCABULARY

Fill in the blanks.

1. If a point lies on the graph of an equation, it is a solution of the equation, and the coordinates of the point _____ the equation.

2. The process of determining an equation whose graph contains given points is called curve _____.

CONCEPTS

Write a system of three equations in three variables that models the situation. Do not solve the system.

3. **Desserts.** A bakery makes three kinds of pies: chocolate cream, which sells for $5; apple, which sells for $6; and cherry, which sells for $7. The cost to make the pies is $2, $3, and $4, respectively. Let x = the number of chocolate cream pies made daily, y = the number of apple pies made daily, and z = the number of cherry pies made daily.

   - Each day, the bakery makes 50 pies.
   - Each day, the revenue from the sale of the pies is $295.
   - Each day, the cost to make the pies is $145.

4. **Fast Foods.** Let x = the number of calories in a Big Mac hamburger, y = the number of calories in a small order of French fries, and z = the number of calories in a medium Coca-Cola.

   - The total number of calories in a Big Mac hamburger, a small order of French fries, and a medium Coke is 1,000.
   - The number of calories in a Big Mac is 260 more than in a small order of French fries.
   - The number of calories in a small order of French fries is 40 more than in a medium Coke. (Source: McDonald's USA)

5. What equation results when the coordinates of the point $(2, -3)$ are substituted into $y = ax^2 + bx + c$?

6. The equation $y = 5x^2 - 6x + 1$ is written in the form $y = ax^2 + bx + c$. What are a, b, and c?

APPLICATIONS

7. **Making Statues.** An artist makes three types of ceramic statues (large, medium, and small) at a monthly cost of $650 for 180 statues. The manufacturing costs for the three types are $5, $4, and $3. If the statues sell for $20, $12, and $9, respectively, how many of each type should be made to produce $2,100 in monthly revenue?

8. **Puppets.** A toy company makes a total of 500 puppets in three sizes during a production run. The small puppets cost $5 to make and sell for $8 each, the standard-size puppets cost $10 to make and sell for $16 each, and the super-size puppets cost $15 to make and sell for $25. The total cost to make the puppets is $4,750 and the revenue from their sale is $7,700. How many small, standard, and super-size puppets are made during a production run?

9. **Nutrition.** A dietician is to design a meal using Foods A, B, and C that will provide a patient with exactly 14 grams of fat, 13 grams of carbohydrates, and 9 grams of protein.

   - Each ounce of Food A contains 2 grams of fat, 3 grams of carbohydrates, and 2 grams of protein.
   - Each ounce of Food B contains 3 grams of fat, 2 grams of carbohydrates, and 1 gram of protein.
   - Each ounce of Food C contains 1 gram of fat, 1 gram of carbohydrates, and 2 grams of protein.

   a. Complete the following table and then form a system of three equations that could be used to determine how many ounces of each food should be used in the meal.

Name of food	Number of ounces used	Grams of fat	Grams of carbohydrates	Grams of protein
A	a	$2a$	$3a$	
B	b	$3b$		b
C	c		c	$2c$
		Total: 14	Total:	Total: 9

   b. Solve the system from part a.

10. **Nutritional Planning.**   One ounce of each of three foods has the vitamin and mineral content shown in the table. How many ounces of each must be used to provide exactly 22 milligrams (mg) of niacin, 12 mg of zinc, and 20 mg of vitamin C?

Milligrams per ounce in each food type

Food	Niacin	Zinc	Vitamin C
A	1 mg	1 mg	2 mg
B	2 mg	1 mg	1 mg
C	2 mg	1 mg	2 mg

 11. ▶ from **Campus to Careers**

Fashion Designer

A clothing manufacturer makes coats, shirts, and slacks. The time required for cutting, sewing, and packaging each item is shown in the table. How many of each should be made to use all available labor hours?

	Coats	Shirts	Slacks	Time available
Cutting	20 min	15 min	10 min	115 hr
Sewing	60 min	30 min	24 min	280 hr
Packaging	5 min	12 min	6 min	65 hr

12. **Sculpting.**   A wood sculptor carves three types of statues with a chainsaw. The number of hours required for carving, sanding, and painting a totem pole, a bear, and a deer are shown in the table. How many of each should be produced to use all available labor hours?

	Totem pole	Bear	Deer	Time available
Carving	2 hr	2 hr	1 hr	14 hr
Sanding	1 hr	2 hr	2 hr	15 hr
Painting	3 hr	2 hr	2 hr	21 hr

13. **NFL Records.**   Jerry Rice, who played the majority of his career with the San Francisco 49ers and the Oakland Raiders, holds the all-time record for touchdown (TD) passes caught. Here are some interesting facts about this feat.

- He caught 30 more TD passes from Steve Young than he did from Joe Montana.
- He caught 39 more TD passes from Joe Montana than he did from Rich Gannon.
- He caught a total of 156 TD passes from Young, Montana, and Gannon.

Determine the number of touchdown passes Rice has caught from Young, from Montana, and from Gannon.

14. **Hot Dogs.**   In 10 minutes, the top three finishers in the 2010 Nathan's Hot Dog Eating Contest consumed a total of 136 hot dogs. The winner, Joey Chestnut, ate 9 more hot dogs than the runner-up, Tim Janus. Pat Bertoletti finished a distant third, 8 hot dogs behind Janus. How many hot dogs did each person eat? (Source: nathansfamous.com)

15. **Earth's Atmosphere.**   Use the information in the circle graph to determine what percent of Earth's atmosphere is nitrogen, is oxygen, and is other gases.

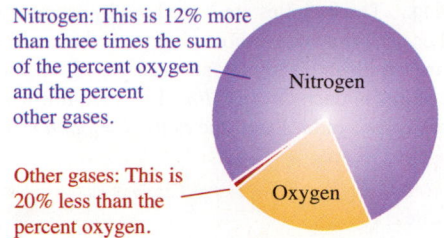

Nitrogen: This is 12% more than three times the sum of the percent oxygen and the percent other gases.

Other gases: This is 20% less than the percent oxygen.

16. **Deceased Celebrities.**   Between October 2009 and October 2010, the estates of Michael Jackson, Elvis Presley, and J.R.R. Tolkien (author of *Lord of the Rings*) earned a total of $385 million. Together, the Presley and Tolkien estates earned $165 million less than the Jackson estate. The Jackson estate earned 5.5 times as much as the Tolkien estate. Use this information to label each bar on the graph below. (Source: Forbes.com)

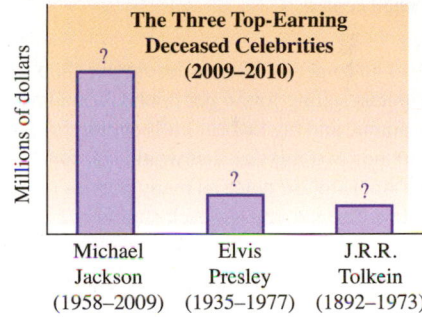

17. **Triangles.**   The sum of the measures of the angles of any triangle is 180°. In $\triangle ABC$, $\angle A$ measures 100° less than the sum of the measures of $\angle B$ and $\angle C$, and the measure of $\angle C$ is 40° less than twice the measure of $\angle B$. Find the measure of each angle of the triangle.

18. **Quadrilaterals.**   A quadrilateral is a four-sided polygon. The sum of the measures of the angles of any quadrilateral is 360°. In the illustration, the measures of $\angle A$ and $\angle B$ are the same. The measure of $\angle C$ is 20° greater than the measure of $\angle A$, and the measure of $\angle D$ is 60° less than $\angle B$. Find the measure of $\angle A$, $\angle B$, $\angle C$, and $\angle D$.

19. **TV History.**   *X-Files, Will & Grace,* and *Seinfeld* are three of the most popular television shows of all time. The total number of episodes of these three shows is 575. There are 21 more episodes of *X-Files* than *Seinfeld,* and the difference between the number of episodes of *Will & Grace* and *Seinfeld* is 14. Find the number of episodes of each show.

20. Traffic Lights. At a traffic light, one cycle through green-yellow-red lasts for 80 seconds. The green light is on eight times longer than the yellow light, and the red light is on eleven times longer than the yellow light. For how long is each colored light on during one cycle?

21. Ice Skating. Three circles are traced out by a figure skater during her performance as shown below. If the centers of the circles are the given distances apart (10 yd, 14 yd, and 18 yd), find the radius of each circle. (*Hint:* Each red line segment is composed of two radii. Label the radii r_1, r_2, and r_3.)

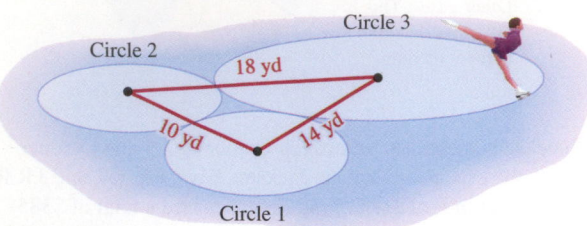

22. NBA Centers. Together, Shaquille O'Neil (Boston Celtics) and Dwight Howard (Orlando Magic) weigh 590 pounds. Together O'Neil and Yao Ming (Houston Rockets) weigh 635 pounds. Together, Howard and Yao Ming weigh 575 pounds. Find the weight of each center.

23. Potpourri. The owner of a home decorating shop wants to mix dried rose petals selling for $6 per pound, dried lavender selling for $5 per pound, and buckwheat hulls selling for $4 per pound to get 10 pounds of a mixture that would sell for $5.50 per pound. She wants to use twice as many pounds of rose petals as lavender. How many pounds of each should she use?

24. Mixing Nuts. The owner of a candy store wants to mix some peanuts worth $3 per pound, some cashews worth $9 per pound, and some Brazil nuts worth $9 per pound to get 50 pounds of a mixture that will sell for $6 per pound. She uses 15 fewer pounds of cashews than peanuts. How many pounds of each did she use?

25. Piggy Banks. When a child breaks open her piggy bank, she finds a total of 64 coins, consisting of nickels, dimes, and quarters. The total value of the coins is $6. If the nickels were dimes, and the dimes were nickels, the value of the coins would be $5. How many nickels, dimes, and quarters were in the piggy bank?

26. Theater Seating. The illustration shows the cash receipts and the ticket prices from two sold-out Sunday performances of a play. Find the number of seats in each of the three sections of the 800-seat theater.

Sunday Ticket Receipts	
Matinee	$13,000
Evening	$23,000

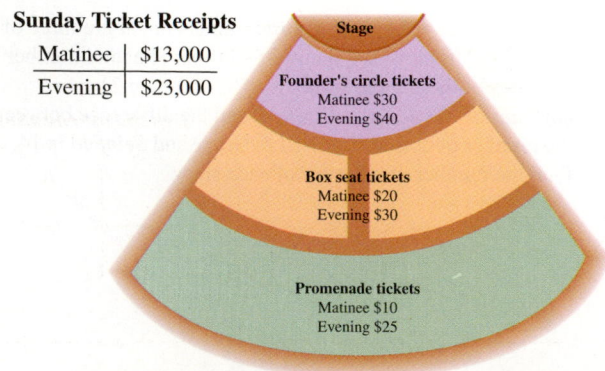

27. Astronomy. Comets have elliptical orbits, but the orbits of some comets are so large that they look much like a parabola. Find an equation of the form $y = ax^2 + bx + c$ for the parabola that closely describes the orbit of the comet shown in the illustration.

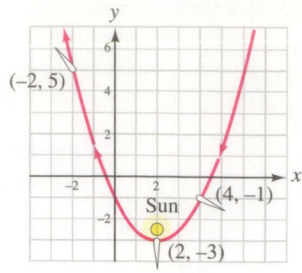

28. Curve Fitting. Find an equation of the form $y = ax^2 + bx + c$ for the parabola shown in the illustration.

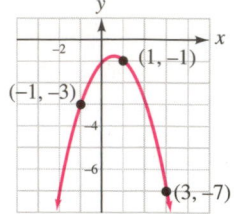

29. Walkways. A circular sidewalk is to be constructed in a city park. The walk is to pass by three particular areas of the park, as shown in the illustration. If an equation of a circle is of the form $x^2 + y^2 + Cx + Dy + E = 0$, find an equation that describes the path of the sidewalk by determining C, D, and E.

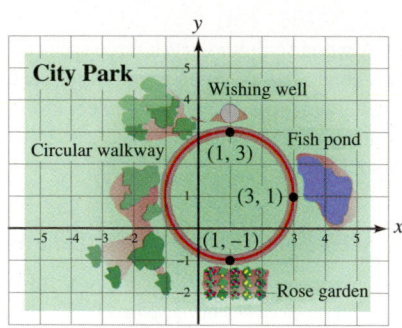

30. Curve Fitting. The equation of a circle is of the form $x^2 + y^2 + Cx + Dy + E = 0$. Find an equation of the circle shown in the illustration by determining C, D, and E.

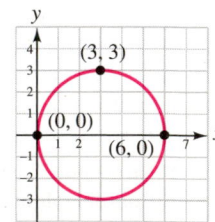

WRITING

31. Explain why the following problem does not give enough information to answer the question: The sum of three integers is 48. If the first integer is doubled, the sum is 60. Find the integers.

32. Write an application problem that can be solved using a system of three equations in three variables.

REVIEW

Determine whether each equation defines y to be a function of x. If it does not, find two ordered pairs where more than one value of y corresponds to a single value of x.

33. $y = \dfrac{1}{x}$

34. $y^4 = x$

35. $xy = 9$

36. $y = |x|$

37. $x + 1 = |y|$

38. $y = \dfrac{1}{x^2}$

39. $y^2 = x$

40. $x = |y|$

CHALLENGE PROBLEMS

41. Digits Problems.   The sum of the digits of a three-digit number is 8. Twice the hundreds digit plus the tens digit is equal to the ones digit. If the digits of the number are reversed, the new number is 82 more than twice the original number. What is the three-digit number?

42. Purchasing Pets.   A pet store owner spent $100 to buy 100 animals. He bought at least one iguana, one guinea pig, and one mouse, but no other kinds of animals. If an iguana cost $10.00, a guinea pig cost $3.00, and a mouse cost $0.50, how many of each did he buy?

3 Summary & Review

SECTION 3.1 ▶ Solving Systems of Equations by Graphing

DEFINITIONS AND CONCEPTS	EXAMPLES
When two equations are considered at the same time, we say that they form a **system of equations.** A **solution of a system** of equations in two variables is an ordered pair that satisfies both equations of the system.	The ordered pair $(2, -3)$ is a solution of the system $\begin{cases} x + y = -1 \\ x - 2y = 8 \end{cases}$ because its coordinates, $x = 2$ and $y = -3$, satisfy both equations. $x + y = -1$ **First equation.** $x - 2y = 8$ **Second equation.** $2 + (-3) \overset{?}{=} -1$ $2 - 2(-3) \overset{?}{=} 8$ $-1 = -1$ **True** $8 = 8$ **True**
To **solve a system graphically:** 1. Graph each equation on the same rectangular coordinate system. 2. Determine the coordinates of the point where the graphs intersect. That ordered pair is the solution. 3. If the graphs have no point in common, the system has no solution. 4. Check the proposed solution in each equation of the original system.	To solve the system $\begin{cases} y = -x - 1 \\ x - 2y = 8 \end{cases}$ by graphing, we graph each equation as shown on the right. The graphs appear to intersect at the point $(2, -3)$. The check shown above verifies that $(2, -3)$ is the solution of the system. 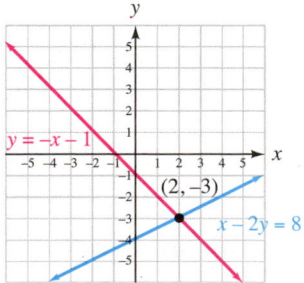
A system of equations that has at least one solution is called a **consistent system.** If the graphs are parallel lines, the system has no solution, and it is called an **inconsistent system.** Equations with different graphs are called **independent equations.** If the graphs are the same line, the system has infinitely many solutions. The equations are called **dependent equations.**	Since the system shown above has a solution, it is a *consistent system.* Since the graphs are different, the equations are *independent.* *Consistent system* *Inconsistent system* 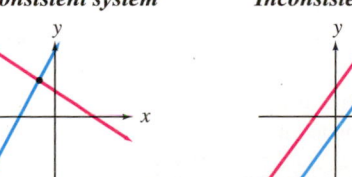 *Dependent equations* 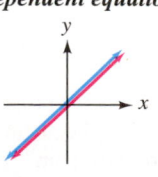

REVIEW EXERCISES

Determine whether the ordered pair is a solution of the system of equations.

1. $\left(-1, \frac{1}{2}\right)$, $\begin{cases} x + 2y = 0 \\ x + 4y = 1 \end{cases}$ **2.** $(13, 23)$, $\begin{cases} 3a - 2b + 7 = 0 \\ -2a + b = -4 \end{cases}$

3. See the illustration.

a. Give three points that satisfy the equation $2x + y = 5$.

b. Give three points that satisfy the equation $y = x - 4$.

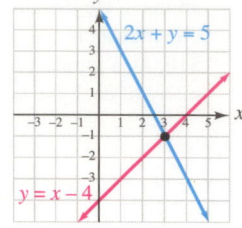

c. Find the solution of: $\begin{cases} 2x + y = 5 \\ y = x - 4 \end{cases}$

4. Video Viewing.   The graph below by the Diffusion Group (TDG) predicts American video viewing habits in the future. In what year does the graph predict that the weekly amount of time spent viewing live broadcast television will be the same as that spent viewing Internet video? Approximately how many hours of weekly viewing of each type will that be?

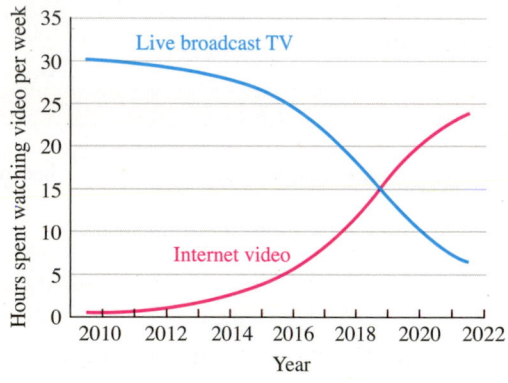

Solve each system by the graphing method, if possible. If a system is inconsistent or if the equations are dependent, state this.

5. $\begin{cases} 2x + y = 11 \\ -x + 2y = 7 \end{cases}$ **6.** $\begin{cases} y = -\frac{3}{2}x \\ 2x - 3y + 13 = 0 \end{cases}$

7. $\begin{cases} \frac{1}{2}x + \frac{1}{3}y = 2 \\ y = 6 - \frac{3}{2}x \end{cases}$ **8.** $\begin{cases} \frac{x}{3} - \frac{y}{2} = 1 \\ 6x - 9y = 3 \end{cases}$

Use the graphs in the illustration to solve each equation. Check each answer.

9. $2(2 - x) + x = x$

10. $2(2 - x) + x = 5$

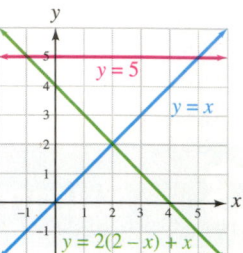

SECTION 3.2 ▶ Solving Systems of Equations Algebraically

DEFINITIONS AND CONCEPTS	EXAMPLES

To solve a system of two linear equations in x and y by the **substitution method**:

1. Solve one equation for either x or y. This is called the **substitution equation.**

2. Substitute the resulting expression for that variable into the other equation and solve it.

3. Substitute the value of the variable found in step 2 into the substitution equation and solve that equation.

4. Check the proposed solution in each of the original equations. Write the solution as an ordered pair.

Use substitution to solve the system: $\begin{cases} x + y = -1 \\ x - 2y = 8 \end{cases}$

Step 1: We will solve $x + y = -1$ for y.

$y = -x - 1$ This is the substitution equation.

Step 2: We substitute $-x - 1$ for y in the second equation and solve for x.

$x - 2(\mathbf{-x - 1}) = 8$

$x + 2x + 2 = 8$ Distribute.

$3x + 2 = 8$ Combine like terms.

$3x = 6$ Subtract 2 from both sides.

$x = 2$ Divide both sides by 3.

If in step 2 the variable drops out and a false statement results, the system has **no solution.** If a true statement results, the system has **infinitely many solutions** and we can use **set-builder notation** to write the solution set.

Step 3: We substitute 2 for x in the substitution equation and solve for y.

$$y = -\textbf{\textit{x}} - 1$$
$$y = -\textbf{2} - 1$$
$$y = -3$$

Step 4: The solution is $(2, -3)$. Verify this by checking it in each of the original equations.

To solve a system of two linear equations in x and y by the **elimination (addition) method:**

1. Write both equations of the system in standard form: $Ax + By = C$.

2. If necessary, multiply one or both of the equations by a nonzero number chosen to make the coefficients of x (or the coefficients of y) opposites.

3. Add the equations to eliminate the terms involving x (or y).

4. Solve the equation resulting from step 3.

5. Find the value of the remaining variable by substituting the solution found in step 4 into any equation containing both variables. Or, repeat steps 2–4 to eliminate the other variable.

6. Check the proposed solution in each equation of the original system. Write the solution as an ordered pair.

If in step 3 both variables drop out and a false statement results, the system has **no solution.** If a true statement results, the system has **infinitely many solutions** and we can use **set-builder notation** to write the solution set.

Use elimination to solve the system: $\begin{cases} x + y = -1 \\ x - 2y = 8 \end{cases}$

Step 1: Since both equations are in standard form, we move to step 2.

Step 2: We can multiply both sides of the first equation by 2 to get the coefficients of y to be opposites.

$$(1) \begin{cases} x + y = -1 \\ (2) \quad x - 2y = 8 \end{cases} \xrightarrow[\text{Unchanged}]{\text{Multiply by 2}} \begin{cases} 2x + 2y = -2 \ (3) \\ x - 2y = 8 \quad (2) \end{cases}$$

Step 3: We add equations 3 and 2 to eliminate y.

$$\begin{array}{r} 2x + 2y = -2 \\ \underline{x - 2y = 8} \quad \text{Add like terms, column-by-column.} \\ 3x = 6 \end{array}$$

Step 4: Since the resulting equation has only one variable, we can solve it for x.

$$3x = 6$$
$$x = 2 \quad \text{Divide both sides by 3.}$$

Step 5: To find y, we can substitute 2 for x in equation 1.

$$\textbf{\textit{x}} + y = -1$$
$$\textbf{2} + y = -1$$
$$y = -3 \quad \text{Subtract 2 from both sides.}$$

Step 6: The solution is $(2, -3)$. Verify this by checking it in each of the original equations.

REVIEW EXERCISES

Solve each system using the substitution method.

11. $\begin{cases} x = y - 4 \\ 2x + 3y = 7 \end{cases}$

12. $\begin{cases} y = 2x + 5 \\ 3x - 5y = -4 \end{cases}$

Solve each system using the elimination (addition) method.

13. $\begin{cases} x - 2y = 11 \\ x + 2y = -21 \end{cases}$

14. $\begin{cases} 4a + 5b = -9 \\ 6a = 3b - 3 \end{cases}$

Solve each system by any method, if possible. If a system is inconsistent or if the equations are dependent, state this.

15. $\begin{cases} \dfrac{1}{2}a - \dfrac{3}{8}b = -9 \\ \dfrac{2}{3}a - \dfrac{1}{4}b = -8 \end{cases}$

16. $\begin{cases} y = \dfrac{x - 3}{2} \\ x = \dfrac{2y + 7}{2} \end{cases}$

17. $\begin{cases} 4x = 8y + 5 \\ 8x = 1 - 2y \end{cases}$

18. $\begin{cases} x + 3y = -2 \\ -2(x + 3y) = 4 \end{cases}$

19. $\begin{cases} 0.07x = 0.05 + 0.09y \\ 7x - 9y = 8 \end{cases}$

20. $\begin{cases} 0.1x + 0.2y = 1.1 \\ 2x - y = 2 \end{cases}$

21. Estimate the solution of the system $\begin{cases} y = -\dfrac{2}{3}x \\ 2x - 3y = -4 \end{cases}$ from the graphs in the illustration. Then solve the system algebraically.

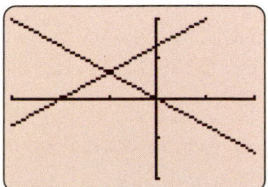

22. Give one advantage and one disadvantage for each of the following methods for solving a system of two linear equations in two variables.

   a. Graphing method   **b.** Substitution method

   c. Elimination method

SECTION 3.3 ▶ Solving Systems of Equations In Three Variables

DEFINITIONS AND CONCEPTS	EXAMPLES

The graph of an equation of the form $Ax + By + Cz = D$ is a flat surface called a **plane.**

A **solution of a system of three linear equations** in three variables is an **ordered triple** that satisfies each equation of the system.

The ordered triple $(4, 0, -3)$ is a solution of $\begin{cases} x + y - z = 7 \\ x - y + z = 1 \\ 2x + y + z = 5 \end{cases}$ because its

coordinates, $x = 4$, $y = 0$, and $z = -3$, satisfy each equation:

$$x + y - z = 7 \qquad\qquad x - y + z = 1 \qquad\qquad 2x + y + z = 5$$

$$4 + 0 - (-3) \overset{?}{=} 7 \qquad 4 - 0 + (-3) \overset{?}{=} 1 \qquad 2(4) + 0 + (-3) \overset{?}{=} 5$$

$$7 = 7 \qquad\qquad\qquad 1 = 1 \qquad\qquad\qquad 5 = 5$$

$$\text{True} \qquad\qquad\qquad \text{True} \qquad\qquad\qquad \text{True}$$

To **solve a system of three linear equations** by elimination:

1. Write each equation in standard form $Ax + By + Cz = D$ and clear any decimals or fractions.

2. Pick any two equations and eliminate a variable.

3. Pick a different pair of equations and eliminate the same variable as in step 2.

4. Solve the resulting pair of two equations in two variables.

5. To find the value of the third variable, substitute the values of the two variables found in step 4 into any equation containing all three variables and solve the equation.

6. Check the proposed solution in all three of the original equations. Write the solution as an ordered triple.

If at any time in the elimination process the variables drop out and a false statement results, the system has **no solution.** If a true statement results, the system has **infinitely many solutions.**

Solve the system: $\begin{cases} x + 2y - z = 1 \quad \textbf{(1)} \\ 2x - y + z = 6 \quad \textbf{(2)} \\ x + 3y - z = 2 \quad \textbf{(3)} \end{cases}$

Step 1: Each equation is written in standard form.

Step 2: To eliminate z, we add equations 1 and 2.

(1) $\quad x + 2y - z = 1$

(2) $\quad \underline{2x - y + z = 6}$

(4) $\quad 3x + y \qquad = 7$

Step 3: To eliminate z again, we add equations 2 and 3.

(2) $\quad 2x - y + z = 6$

(3) $\quad \underline{x + 3y - z = 2}$

(5) $\quad 3x + 2y \qquad = 8$

Step 4: Equations 4 and 5 form a system of two equations in x and y. To solve this system, we multiply equation 4 by -1 and add the resulting equation to equation 5 to eliminate x.

$$-3x - y = -7 \qquad \text{This is } -1(3x + y) = -1(7).$$

(5) $\quad \underline{3x + 2y = \quad 8}$

$$\qquad\qquad y = \quad 1$$

To find x, we substitute 1 for y in any equation containing x and y (such as equation 4) and solve for x:

$$3x + y = 7 \qquad \text{This is equation 4.}$$

$$3x + 1 = 7 \qquad \text{Substitute 1 for y.}$$

$$x = 2 \qquad \text{Solve for x.}$$

Step 5: To find z, we substitute 2 for x and 1 for y in any equation containing x, y, and z (such as equation 2) and solve for z:

$$2x - y + z = 6 \qquad \text{This is equation 2.}$$

$$2(2) - 1 + z = 6 \qquad \text{Substitute for x and y.}$$

$$4 - 1 + z = 6$$

$$z = 3 \qquad \text{Solve for z.}$$

Step 6: The solution is $(2, 1, 3)$. Verify this by checking it in each of the original equations.

REVIEW EXERCISES

23. Determine whether $(2, -1, 1)$ is a solution of the system:
$$\begin{cases} x - y + z = 4 \\ x + 2y - z = -1 \\ x + y - 3z = -1 \end{cases}$$

24. A system of three linear equations in three variables is graphed on the right. Does the system have a solution? If so, how many solutions does it have?

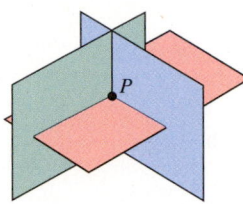

Solve each system, if possible. If a system is inconsistent or if the equations are dependent, state this.

25. $\begin{cases} x - 2y + 3z = -7 \\ -x + 3y + 2z = -8 \\ 2x - y - z = 7 \end{cases}$

26. $\begin{cases} x + y + z = 4 \\ x - 2y - z = 1 \\ 2x - y - 2z = -1 \end{cases}$

27. $\begin{cases} x + y - z = -3 \\ x + z = 2 \\ 2x - y = 3 - 2z \end{cases}$

28. $\begin{cases} b - 4c = 2 \\ a - b + 2c = 1 \\ 2a - 2b = -2 - 5c \end{cases}$

29. $\begin{cases} x + 2z = 10 \\ 3x + 2y - 3z = 8 \\ y + 4z = 6 \end{cases}$

30. $\begin{cases} x + 3y + z = 14 \\ x - 5y = -19 \\ 3y + z = 13 \end{cases}$

31. $\begin{cases} 2x + 3y + z = -5 \\ -x + 2y - z = -6 \\ 3x + y + 2z = 4 \end{cases}$

32. $\begin{cases} 3x + 3y + 6z = -6 \\ -x - y - 2z = 2 \\ 2x + 2y + 4z = -4 \end{cases}$

SECTION 3.4 ▶ **Solving Systems of Equations Using Matrices**

DEFINITIONS AND CONCEPTS	EXAMPLES
A **matrix** is a rectangular array of numbers. Each number in a matrix is called an **element** or an **entry** of the matrix. A matrix with m rows and n columns has **order** $m \times n$.	A 2×3 matrix: $\begin{bmatrix} 2 & -7 & 5 \\ -3 & 4 & 1 \end{bmatrix}$ A 3×3 matrix: $\begin{bmatrix} 5 & -3 & 12 \\ 4 & 7 & -5 \\ 1 & -4 & 2 \end{bmatrix}$
A system of linear equations can be represented by an **augmented matrix.** Each row of the augmented matrix represents one equation of the system.	The system of equations $\begin{cases} 3x + 5y = 12 \\ 2x - 7y = -5 \end{cases}$ can be represented by the augmented matrix $\begin{bmatrix} 3 & 5 & \vdots & 12 \\ 2 & -7 & \vdots & -5 \end{bmatrix}$.
Systems of linear equations can be solved by using Gauss-Jordan elimination and **elementary row operations:** **1.** Any two rows can be interchanged. **2.** Any row can be multiplied by a nonzero constant. **3.** Any row can be changed by adding a nonzero constant multiple of another row to it. To solve a system of two linear equations in two unknowns using matrices, we transform the augmented matrix into an equivalent matrix that has 1's down its main diagonal and 0's directly above and below the 1's. A matrix written in this form is said to be in **reduced row-echelon form.** $\begin{bmatrix} 1 & 0 & \vdots & a \\ 0 & 1 & \vdots & b \end{bmatrix}$ *a and b represent real numbers.* Main diagonal Matrices also can be used to solve systems of three linear equations containing three variables.	Solve the system using matrices: $\begin{cases} 2x - 3y = 0 \\ x + 2y = 7 \end{cases}$ This system is represented by the augmented matrix $\begin{bmatrix} 2 & -3 & \vdots & 0 \\ 1 & 2 & \vdots & 7 \end{bmatrix}$ We can write this matrix in reduced row-echelon form by performing the following elementary row operations $\begin{bmatrix} 1 & 2 & \vdots & 7 \\ 2 & -3 & \vdots & 0 \end{bmatrix}$ Interchange row 1 and row 2. In symbols: $R_1 \leftrightarrow R_2$. $\begin{bmatrix} 1 & 2 & \vdots & 7 \\ 0 & -7 & \vdots & -14 \end{bmatrix}$ Multiply row 1 by -2 and add to row 2: $-2R_1 + R_2$. $\begin{bmatrix} 1 & 2 & \vdots & 7 \\ 0 & 1 & \vdots & 2 \end{bmatrix}$ Multiply row 2 by $-\frac{1}{7}$. In symbols: $-\frac{1}{7}R_2$. $\begin{bmatrix} 1 & 0 & \vdots & 3 \\ 0 & 1 & \vdots & 2 \end{bmatrix}$ Multiply row 2 by -2 and add to row 1: $-2R_2 + R_1$. This augmented matrix represents the equivalent system: $\begin{cases} x = 3 \\ y = 2 \end{cases}$ Thus, the solution of the given system is $(3, 2)$. The solution set is $\{(3, 2)\}$.

Represent each system of equations using an augmented matrix.

33. $\begin{cases} 5x + 4y = 3 \\ x - y = -3 \end{cases}$

34. $\begin{cases} x + 2y + 3z = 6 \\ x - 3y - z = 4 \\ 6x + y - 2z = -1 \end{cases}$

35. Perform each of the following elementary row operations on the augmented matrix $\begin{bmatrix} 6 & 12 & | & -6 \\ 1 & 3 & | & -2 \end{bmatrix}$.

  a. $R_1 \leftrightarrow R_2$

  b. $\frac{1}{6} R_1$

  c. $-6R_2 + R_1$

36. Perform each of the following elementary row operations on the augmented matrix: $\begin{bmatrix} 2 & -1 & 1 & | & 3 \\ 1 & 1 & 0 & | & -1 \\ 3 & -1 & -2 & | & 7 \end{bmatrix}$

  a. $R_1 \leftrightarrow R_2$

  b. $3R_2$

  c. $-2R_2 + R_1$

Solve each system using matrices, if possible. If a system is inconsistent or if the equations are dependent, state this.

37. $\begin{cases} x - y = 4 \\ 3x + 7y = -18 \end{cases}$

38. $\begin{cases} x + 2y - 3z = 5 \\ x + y + z = 0 \\ 3x + 4y + 2z = -1 \end{cases}$

39. $\begin{cases} 16x - 8y = 32 \\ -2x + y = -4 \end{cases}$

40. $\begin{cases} x + 2y - z = 4 \\ x + 3y + 4z = 1 \\ 2x + 4y - 2z = 3 \end{cases}$

SECTION 3.5 ▶ **Solving Systems of Equations Using Determinants**

DEFINITIONS AND CONCEPTS	EXAMPLES
A determinant is a number that is associated with a **square matrix,** a matrix that has the same number of rows and columns.	A 2 × 2 determinant: $\begin{vmatrix} 3 & -3 \\ -4 & 5 \end{vmatrix}$ A 3 × 3 determinant: $\begin{vmatrix} 3 & 8 & 3 \\ 7 & 2 & 2 \\ 1 & 5 & 1 \end{vmatrix}$
To **evaluate** a 2 × 2 determinant: $\begin{vmatrix} a & b \\ c & d \end{vmatrix} = ad - bc$	Evaluate: $\begin{vmatrix} 3 & -3 \\ -4 & 5 \end{vmatrix} = 3(5) - (-3)(-4) = 15 - 12 = 3$
To evaluate a 3 × 3 determinant, we **expand it by minors** along any row or column.	Evaluate: $\begin{vmatrix} 3 & 8 & 3 \\ 7 & 2 & 2 \\ 1 & 5 & 1 \end{vmatrix} = 3\begin{vmatrix} 2 & 2 \\ 5 & 1 \end{vmatrix} - 8\begin{vmatrix} 7 & 2 \\ 1 & 1 \end{vmatrix} + 3\begin{vmatrix} 7 & 2 \\ 1 & 5 \end{vmatrix}$ $= 3(-8) - 8(5) + 3(33)$ $= -24 - 40 + 99$ $= 35$

Cramer's rule can be used to solve systems of two linear equations in two variables.	Use Cramer's rule to solve: $\begin{cases} 2x - 3y = 0 \\ x + 2y = 7 \end{cases}$
Cramer's rule can be extended to solve systems of **three linear equations** with three variables.	The denominator determinant is D: $\begin{vmatrix} 2 & -3 \\ 1 & 2 \end{vmatrix} = 4 - (-3) = 7$

The numerator determinant for x is D_x: $\begin{vmatrix} 0 & -3 \\ 7 & 2 \end{vmatrix} = 0 - (-3)(7) = 21$

The numerator determinant for y is D_y: $\begin{vmatrix} 2 & 0 \\ 1 & 7 \end{vmatrix} = 14 - 0 = 14$

Thus, we have:

$$x = \frac{D_x}{D} = \frac{21}{7} = 3 \quad \text{and} \quad y = \frac{D_y}{D} = \frac{14}{7} = 2$$

Thus, the solution of the given system is $(3, 2)$. The solution set is $\{(3, 2)\}$.

REVIEW EXERCISES

Evaluate each determinant.

41. $\begin{vmatrix} 2 & 3 \\ -4 & 3 \end{vmatrix}$

42. $\begin{vmatrix} -3 & -4 \\ 5 & -6 \end{vmatrix}$

43. $\begin{vmatrix} -1 & 2 & -1 \\ 2 & -1 & 3 \\ 1 & -2 & 2 \end{vmatrix}$

44. $\begin{vmatrix} 3 & -2 & 2 \\ 1 & -2 & -2 \\ 2 & 1 & -1 \end{vmatrix}$

Use Cramer's rule to solve each system, if possible. If a system is inconsistent or if the equations are dependent, state this.

45. $\begin{cases} 3x + 4y = 10 \\ 2x - 3y = 1 \end{cases}$

46. $\begin{cases} -6x - 4y = -6 \\ 3x + 2y = 5 \end{cases}$

47. $\begin{cases} x + 2y + z = 0 \\ 2x + y + z = 3 \\ x + y + 2z = 5 \end{cases}$

48. $\begin{cases} 2x + 3y + z = 2 \\ x + 3y + 2z = 7 \\ x - y - z = -7 \end{cases}$

SECTION 3.6 ▶ **Problem Solving Using Systems of Two Equations**

DEFINITIONS AND CONCEPTS	EXAMPLES

To solve problems involving two unknown quantities, we can use the following **problem-solving strategy:**

1. **Analyze** the facts of the problem.

2. **Assign** different variables to represent the two unknown quantities.

3. **Form** two equations involving those variables.

4. **Solve** the system of equations by graphing, substitution, elimination, using matrices, or using Cramer's rule.

5. **State** the conclusion.

6. **Check** the result in the words of the problem.

Retirement Income. A retired office manager invested $10,000 in two accounts, one paying 5% annual interest and the other 6%. If the interest earned for the first year was $540, how much did she invest at each rate?

Analyze A total of $10,000 is invested at two different rates for 1 year. The total interest earned was $540.

Assign Let $x =$ the number of dollars invested at 5% and $y =$ the number of dollars invested at 6%.

Form We can use the formula $I = Prt$ to determine that the interest earned on the 5% investment is $\$0.05x$ and the interest earned on the 6% investment is $\$0.06y$. This information is shown in the table.

	P $\cdot$	r $\cdot$	t $=$	I
First account	x	0.05	1	$0.05x$
Second account	y	0.06	1	$0.06y$
Total	$10,000			$540

↑ One equation comes from this column. ↑ A second equation comes from this column.

The resulting system is: $\begin{cases} x + y = 10,000 \\ 0.05x + 0.06y = 540 \end{cases}$

Solve   To solve by elimination we can multiply both sides of the first equation by −6, multiply the second equation by 100, and add the equations.

$$-6x - 6y = -60,000 \quad \text{This is } -6(x + y) = -6(10,000).$$
$$\underline{5x + 6y = 54,000} \quad \text{This is } 100(0.05x + 0.06y) = 100(540).$$
$$-x = -6,000$$
$$x = 6,000$$

To find y, we substitute 6,000 for x in the first equation of the system.

$$x + y = 10,000$$
$$6,000 + y = 10,000$$
$$y = 4,000$$

State   The amount invested at 5% was $6,000 and the amount invested at 6% was $4,000.

Check   The sum of $6,000 and $4,000 is $10,000.

$6,000 invested at 5% for 1 year would earn $6,000 · 0.05 = $300.
$4,000 invested at 6% for 1 year would earn $4,000 · 0.06 = $240.
The total interest earned is $300 + $240 = $540. The results check.

REVIEW EXERCISES

Write a system of two equations in two variables to solve each problem.

49. Maps.   Refer to the illustration. The distance between Austin and Houston is 4 miles less than twice the distance between Austin and San Antonio. The round trip from Houston to Austin to San Antonio and back to Houston is 442 miles. Determine the mileages between Austin and Houston and between Austin and San Antonio.

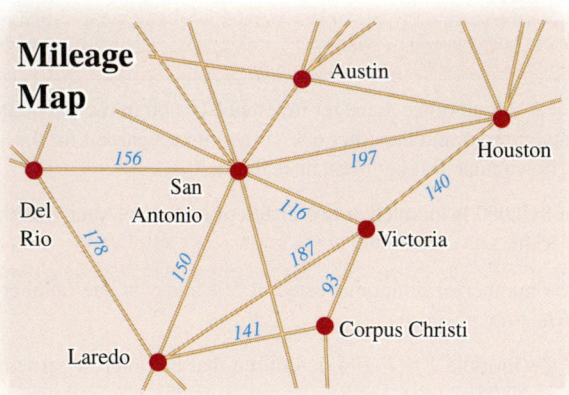

Mileage Map

50. Riverboats.   A Mississippi riverboat travels 30 miles downstream in three hours and then makes the return trip upstream in five hours. Find the speed of the riverboat in still water and the speed of the current.

51. Mixing Solutions.   How many fluid ounces of 6% sucrose solution must be mixed with 18% sucrose solution to make 750 ounces of a 10% sucrose solution?

52. Investing.   One year, a couple invested a total of $10,000 in two projects. The first investment, a mini-mall, made a 6% profit. The other investment, a skateboard park, made a 12% profit. If their investments made $960, how much was invested at each rate?

53. Cooking.   Two teaspoons and five tablespoons is a total of 85 milliliters of liquid. Five teaspoons and two tablespoons is a total of 55 milliliters of liquid. Find the number of milliliters in one teaspoon and the number of milliliters in one tablespoon.

54. Break Points.   A bottling company is considering purchasing a new piece of equipment for their production line. The machine they currently use has a setup cost of $250 and a cost of $0.04 per bottle. The new machine has a setup cost of $600 and a cost of $0.02 per bottle. Find the break point.

SECTION 3.7 ▸ Problem Solving Using Systems of Three Equations

DEFINITIONS AND CONCEPTS	EXAMPLES

DEFINITIONS AND CONCEPTS

Problems that involve **three unknown quantities** can be solved using a strategy similar to that for solving problems involving two unknowns.

Elimination, substitution, matrices, or Cramer's rule can be used to solve the system of equations.

EXAMPLES

Batteries.   A hardware store sells three types of batteries: AA size for $1 each, C size for $1.50 each, and D size for $2.00 each. One Saturday, the store sold 25 batteries for a total of $34. If the number of C batteries that were sold was four less than the number of AA batteries that were sold, how many of each size battery were sold?

Analyze   To find the three unknowns we will write a system of three equations in three variables.

Assign   Let A = the number of AA batteries sold, C = the number of C batteries sold, and D = the number of D batteries sold.

Form   The information can be shown in a table.

Battery	Number ·	Value =	Total value
AA size	A	1	A
C size	C	1.50	1.50C
D size	D	2	2D
Total	25		$34

One equation comes from this column.   Another equation comes from this column.

$$\begin{cases} A + C + D = 25 \\ 1A + 1.50C + 2D = 34 \\ C = A - 4 \end{cases}$$

The total number of batteries sold was 25.
The total value of the batteries sold was $34.
The number of C batteries sold was 4 less than AA batteries sold.

If we multiply the second equation by 10 to clear the decimal and write the third equation in standard form, we have the system:

$$\begin{matrix}(1)\\(2)\\(3)\end{matrix} \begin{cases} A + C + D = 25 \\ 10A + 15C + 20D = 340 \\ -A + C = -4 \end{cases}$$

Solve   Since equation 3 does not contain a D-term, we will find another equation that does not contain a D-term. If each side of equation 1 is multiplied by -20 and the resulting equation is added to the equation 2, D is eliminated, and we obtain

$$-20A - 20C - 20D = -500 \quad \text{This is } -20(A + C + D) = -20(25).$$
$$(2) \quad \underline{10A + 15C + 20D = \quad 340}$$
$$(4) \quad -10A - 5C \quad\quad = -160$$

Equations 3 and 4 form a system of two equations in A and C that can be solved in the usual manner. (The remaining work is left to the reader.)

$$\begin{cases} -10A - 5C = -160 \\ -A + C = -4 \end{cases}$$

State   There were 12 AA batteries, 8 C batteries, and 5 D batteries sold.

Check   Verify that these results are correct by checking them in the words of the problem.

55. Teddy Bears.   A toy company produces three sizes of teddy bears. Each day, the total cost to produce the bears is $850, the total time needed to stuff them is 480 minutes, and the total time needed to sew them is 1,260 minutes. Use the information in the table to determine how many of each type of teddy bear are produced daily.

Size of teddy bear	Production cost	Stuffing time	Sewing time
Small	$3	2 min	6 min
Medium	$5	3 min	8 min
Large	$10	5 min	12 min

56. Financial Planning.   A financial planner invested $22,000 in three accounts, paying 5%, 6%, and 7% annual interest. She invested $2,000 more at 6% than at 5%. If the total interest earned in one year was $1,370, how much was invested at each rate?

57. Ballistics.   The path of a thrown object is a parabola with an equation of $y = ax^2 + bx + c$. The parabola passes through the points $(0, 0)$, $(8, 12)$, and $(12, 15)$. Find a, b, and c.

58. Veterinary Medicine.   The daily requirements of a balanced diet for an animal are shown in the nutritional pyramid. The number of grams per cup of nutrients in three food mixes are shown in the table. How many cups of each mix should be used to meet the daily requirements for protein, carbohydrates, and essential fatty acids in the animal's diet?

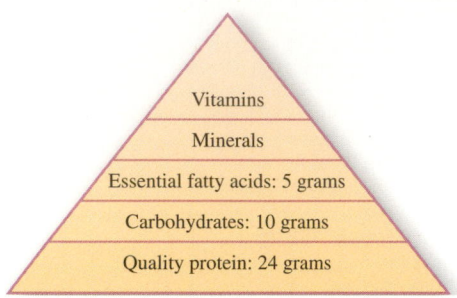

Vitamins

Minerals

Essential fatty acids: 5 grams

Carbohydrates: 10 grams

Quality protein: 24 grams

Grams per cup			
	Protein	Carbohydrates	Fatty acids
Mix A	5	2	1
Mix B	6	3	2
Mix C	8	3	1

3 ► Chapter Test

1. Fill in the blanks.

a. $\begin{cases} 2x - 7y = 1 \\ 4x - y = -8 \end{cases}$ is called a _____ of linear equations.

b. The matrix $\begin{bmatrix} -10 & 3 \\ 4 & 9 \end{bmatrix}$ has 2 _____ and 2 _____.

c. Solutions of a system of three equations in three variables, x, y, and z, are written in the form (x, y, z) and are called ordered _____.

d. The graph of the equation $2x + 3y + 4z = 5$ is a flat surface called a _____.

e. A _____ is a rectangular array of numbers written within brackets.

2. Solve the system by graphing: $\begin{cases} 2x + y = 5 \\ y = 2x - 3 \end{cases}$

3. Determine whether $\left(-\frac{1}{2}, -\frac{2}{3}\right)$ is a solution of the system:
$\begin{cases} 10x - 12y = 3 \\ 18x - 15y = 1 \end{cases}$

4. Politics.   Explain the importance of the point of intersection of the graphs shown below.

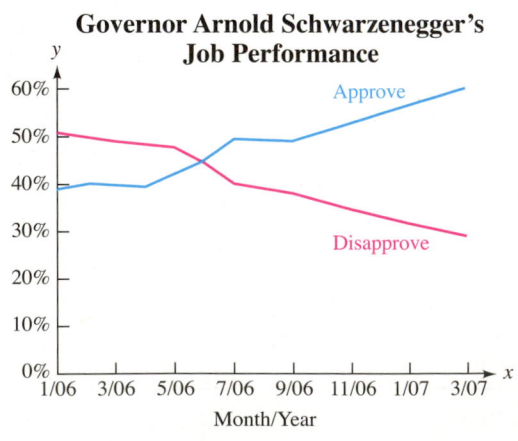

Governor Arnold Schwarzenegger's Job Performance

5. Use the vocabulary of this chapter to describe each system of two linear equations in two variables graphed below. Does the system have a solution (or solutions)?

a.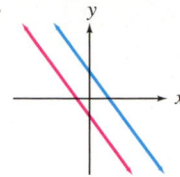

b.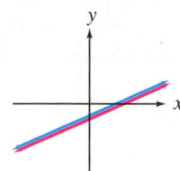

6. Use the graphs in the illustration to solve $3(x - 2) - 2(-2 + x) = 1$.

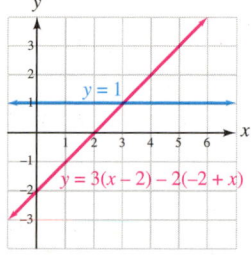

7. Use substitution to solve the system: $\begin{cases} 2x - 4y = 14 \\ x + 2y = 7 \end{cases}$

8. Use elimination (addition) to solve the system:
$$\begin{cases} 2c + 3d = -5 \\ 3c - 2d = 12 \end{cases}$$

Solve each system by any method, if possible. If a system is inconsistent or if the equations are dependent, state this.

9. $\begin{cases} 3(x + y) = x - 3 \\ -y = \dfrac{2x + 3}{3} \end{cases}$

10. $\begin{cases} 0.6x + 0.5y = 1.2 \\ x - \dfrac{4}{9}y + \dfrac{5}{9} = 0 \end{cases}$

Write a system of two equations in two variables to solve each problem.

11. Traffic Signs. In the sign, find x and y, if y is 15 more than x.

12. Antifreeze. How much of a 40% antifreeze solution must a mechanic mix with an 80% antifreeze solution if 20 gallons of a 50% antifreeze solution are needed?

13. Break Points. A metal stamping plant is considering purchasing a new piece of equipment. The machine they currently use has a setup cost of $1,775 and a cost of $5.75 per impression. The new machine has a setup cost of $3,975 and a cost of $4.15 per impression. Find the break point.

14. Parallelograms. Refer to parallelogram *ABCD* at the right. Find the unknown degree measures represented by x and y.

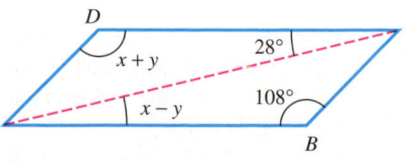

15. Student Loans. A student took out two loans to pay for $8,500 of college expenses. One of the loans was at 2.5% simple annual interest and the other at 4% simple annual interest. After one year, he owed a total of $265 in interest. What was the amount of each loan?

16. Gourmet Fruit. A fruit pack that contains 6 Royal Riviera pears and 4 Honey Crisp apples costs $25.50. Another pack with 4 pears and 10 apples costs $31.30. Find the cost of one pear and the cost of one apple.

17. Determine whether $\left(-1, -\frac{1}{2}, 5\right)$ is a solution of:
$$\begin{cases} x - 2y + z = 5 \\ 2x + 4y = -4 \\ -6y + 4z = 22 \end{cases}$$

18. A system of three linear equations in three variables is graphed on the right. Does the system have a solution? If so, how many solutions does it have?

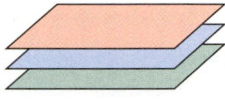

19. Solve the system: $\begin{cases} x + y + z = 4 \\ x + y - z = 6 \\ 2x - 3y + z = -1 \end{cases}$

20. Solve the system: $\begin{cases} z - 2y = 1 \\ x + y + z = 1 \\ x + 5y = 4 \end{cases}$

21. Movie Tickets. The receipts for one showing of a movie were $410 for an audience of 100 people. The ticket prices are given in the table. If twice as many children's tickets as general admission tickets were purchased, how many of each type of ticket were sold?

Ticket prices	
Children	$3
General admission	$6
Seniors	$5

22. Weight Training. Refer to the following illustrations. Determine the weight of the bar, the weight of one large plate, and the weight of one small plate.

Total weight: 155 lb

Total weight: 245 lb

Total weight: 375 lb

23. Let $A = \begin{bmatrix} 1 & 7 & | & -3 \\ 3 & -1 & | & 13 \end{bmatrix}$. Write the matrix obtained when the elementary row operations $-3R_1 + R_2$ are performed on matrix A.

Use matrices to solve each system, if possible. If a system is inconsistent or if the equations are dependent, state this.

24. $\begin{cases} x + y = 4 \\ 2x - y = 2 \end{cases}$

25. $\begin{cases} x - 3y + 2z = 1 \\ x - 2y + 3z = 5 \\ 2x - 6y + 4z = 3 \end{cases}$

26. $\begin{cases} a + 2b + 2c = 10 \\ 2a + 3b - c = 6 \\ 3a + b + 5c = 8 \end{cases}$

Evaluate each determinant.

27. $\begin{vmatrix} 2 & -3 \\ -4 & 5 \end{vmatrix}$

28. $\begin{vmatrix} 1 & 2 & 0 \\ 2 & 0 & 3 \\ 1 & -2 & 2 \end{vmatrix}$

29. Use Cramer's rule to solve the system:
$$\begin{cases} x - y = -6 \\ 3x + y = -6 \end{cases}$$

30. Solve the following system for z only, using Cramer's rule.

Group Project

$\begin{cases} x + y + z = 4 \\ x + y - z = 6 \\ 2x - 3y + z = -1 \end{cases}$

Methods of Solution

Overview: In this activity, you will explore the advantages and disadvantages of several methods for solving a system of linear equations.

Instructions: Form groups of 5 students. Have each member of your group solve the system

$$\begin{cases} x - y = 4 \\ 2x + y = 5 \end{cases}$$

in a different way. The methods to use are graphing, substitution, elimination, matrices, and Cramer's rule. Have each person briefly explain his or her method of solution to the group. After everyone has presented a solution, discuss the advantages and drawbacks of each method. Then rank the five

CUMULATIVE REVIEW ▶▶ Chapters 1–3

methods, from most desirable to least desirable, for this system.

1. Complete the illustration by labeling the rational numbers, irrational numbers, integers, and whole numbers. [Section 1.2]

Real numbers

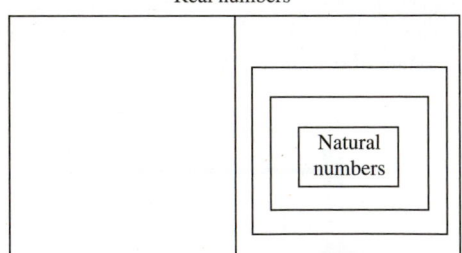

Natural numbers

2. Insert either a $<$ or a $>$ symbol to make a true statement. [Section 1.2]

    a. -5.96 ◻ -5.95 **b.** $-(-1)$ ◻ $-|-18|$

Evaluate each expression for a = −3 and b = −5.

3. $-|b| - ab^2$

    [Section 1.3]

4. $\dfrac{14 + 2[2a - (b - a)]}{-b - 2}$

    [Section 1.3]

Simplify each expression.

5. $40\left(\dfrac{3}{8}m - \dfrac{1}{4}\right) + 40\left(\dfrac{4}{5}\right)$ [Section 1.4]

6. $3[2(a + 2)] - 5[3(a - 5)] + 5a$ [Section 1.4]

Solve each equation, if possible. If an equation is an identity or a contradiction, state this.

7. $\dfrac{3}{4}x + 1.5 = -19.5$ [Section 1.5]

8. $1 + 3[-2 + 6(4 - 2x)] = -(x + 3)$ [Section 1.5]

9. $\dfrac{x + 7}{3} = \dfrac{x - 2}{5} - \dfrac{x}{15} + \dfrac{7}{3}$ [Section 1.5]

10. $3p - 6 = 4(p - 2) + 2 - p$ [Section 1.5]

Solve each equation for the specified variable.

11. $\lambda = Ax + AB$ for B [Section 1.6]

12. $v = \dfrac{d_1 - d_2}{t}$ for d_2 [Section 1.6]

13. **Antique Shows.** A traveling antique show will be on the road for 17 weeks, visiting three cities. They will be in Los Angeles for 2 weeks longer than they will be in Las Vegas. Their stay in Dallas will be 1 week less than twice that in Las Vegas. How many weeks will they be in each city? [Section 1.7]

14. U.S. Federal Budget. The proposed federal spending for the fiscal year 2009 was $3,518 billion. The illustration shows how a typical dollar of the budget was to be spent. Determine the total amount to be spent on Social Security, Medicare, and Medicaid. Round to the nearest billion dollars. [Section 1.8]

Social Security: 20¢   Defense: 23¢   Other entitlements: 17¢   Medicare/Medicaid: 19¢   TARP: Interest: 5¢   4¢ Other spending: 12¢

15. Commuting. Use the following facts to determine a commuter's average speed when she drives to work.

- If she drives her car, it takes a quarter of an hour to get to work.
- If she rides the bus, it takes half an hour to get to work.
- When she drives, her average speed is 10 miles per hour faster than that of the bus. [Section 1.8]

16. Dried Fruits. Dried apple slices cost $4.60 per pound, and dried banana chips sell for $3.40 per pound. How many pounds of each should be used to create a 10-pound mixture that sells for $4 per pound? [Section 1.8]

Graph each equation.

17. $3x = 4y - 11$
[Section 2.2]

18. $y = -4$
[Section 2.2]

19. Find the slope of the line, shown on the right. [Section 2.3]

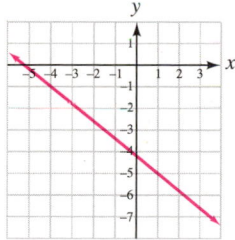

20. Determine whether the line passing through $(-3, -1)$ and $(-3, 4)$ and the line passing through $(-4, -2)$ and $(5, -2)$ are parallel, perpendicular, or neither. [Section 2.3]

21. Find an equation of the line that passes through $(4, 5)$ and is parallel to the graph of $y = -3x$. Write the equation in slope–intercept form. [Section 2.4]

22. Collectibles. A collector buys the Hummel figurine shown in the illustration, anticipating that it will be worth $650 in 20 years. Assuming straight-line appreciation, write an equation that gives the value v of the figurine x years after it is purchased. [Section 2.4]

Price: $300.00

© Christa Knijff/Alamy

If $f(x) = -x^2 - \dfrac{x}{2}$, find each of the following.

23. $f(10)$ [Section 2.5]

24. $f(r)$ [Section 2.5]

25. We can think of a function as a machine. (See the illustration.) Write a function that turns the given input into the given output. [Section 2.5]

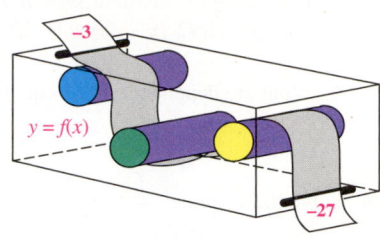

$y = f(x)$

26. Does the table define y as a function of x? [Section 2.5]

x	y
-2	5
-1	2
0	2
5	5

27. Railroad Safety. There has been a downward trend in the number of highway–railroad-crossing accidents in the United States since 1990. For example, the number of such accidents in 1995 was about 4,500 and the number in 2005 was about 2,800. (Source: Federal Railroad Administration) [Section 2.5]

a. Let t be the number of years since 1990 and A be the number of highway–railroad-crossing accidents in the U.S. Write a linear function $A(t)$ to model the situation.

b. Use your answer to part a to predict the number of highway–railroad-crossing accidents in 2015, if the trend continues.

28. Use the graph of function f to find each of the following. [Section 2.6]

a. $f(-1)$

b. The value of x for which $f(x) = 3$

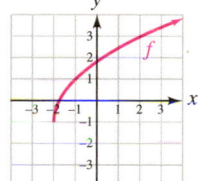

29. Determine whether the graph on the right is the graph of a function. Explain why or why not. [Section 2.6]

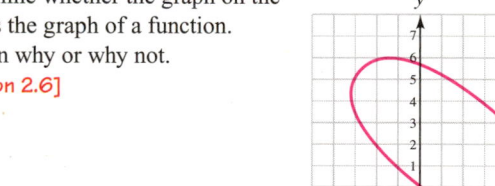

30. Graph the function $f(x) = x^2$. Then draw the graph of the associated function $g(x) = (x + 4)^2$ using a translation. Give the domain and range of function g. [Section 2.6]

31. The graph of function g on the right is a vertical translation of the graph of $f(x) = |x|$. Write the equation that describes function g. [Section 2.6]

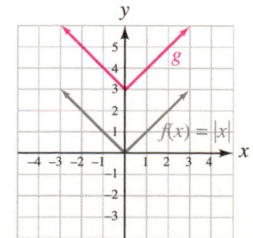

$f(x) = |x|$

32. Solve the system by graphing: $\begin{cases} 3x - y = -3 \\ y = -2x - 7 \end{cases}$ [Section 3.1]

33. a. What is an inconsistent system of two linear equations in two variables? [Section 3.1]

    b. What are dependent linear equations in two variables? [Section 3.1]

34. Solve the system: $\begin{cases} y = \dfrac{-2x + 1}{3} \\ 3x - 2y = 8 \end{cases}$ [Section 3.2]

35. Solve the system: $\begin{cases} -x + 3y + 2z = 5 \\ 3x + 2y + z = -1 \\ 2x - y + 3z = 4 \end{cases}$ [Section 3.3]

36. Use matrices to solve the system: $\begin{cases} x - y = 5 \\ 2x - 5y = 1 \end{cases}$

[Section 3.4]

37. Evaluate each determinant. [Section 3.5]

    a. $\begin{vmatrix} 6 & 2 \\ -2 & 6 \end{vmatrix}$

    b. $\begin{vmatrix} 2 & 1 & -3 \\ -2 & 2 & 4 \\ 1 & -2 & 2 \end{vmatrix}$

38. Use Cramer's rule to solve the system: $\begin{cases} x + 2y = 6 \\ x - y = 4 \end{cases}$

[Section 3.5]

39. Inventories. The table shows an end-of-the-year inventory report for a warehouse that supplies electronics stores. If the warehouse stocks two models of cordless telephones, one valued at $67 and the other at $100, how many of each model of phone did the warehouse have at the time of the inventory? [Section 3.6]

Item	Number	Merchandise value
Cordless phones	360	$29,400

40. Investing. A woman wants to earn $2,200 the first year that she invests $30,000 in three certificates of deposit. (See the table.) She wants to invest five times as much in the 36-month CD as in the 12-month CD. How much should she invest in each CD? [Section 3.7]

Type of CD	Annual rate of return
12-month	6%
24-month	7%
36-month	8%

Inequalities

4

from Campus to Careers

Heating, Ventilation, and Air Conditioning Technician

HVAC technicians make sure that we are warm in the winter and cool in the summer. They install, maintain, and repair heating and cooling systems in residential, commercial, and industrial buildings. HVAC technicians constantly work with numbers as they take measurements, read blueprints, and prepare work estimates. The installation instructions and diagrams they follow require strong mathematical skills in algebra and geometry.

Problem 91 in **Study Set 4.1, problem 83** in **Study Set 4.2,** and **problem 51** in **Study Set 4.5** involve situations that a Heating, Ventilation, and Air Conditioning Technician might encounter on the job. The mathematical concepts discussed in this chapter can be used to solve those problems.

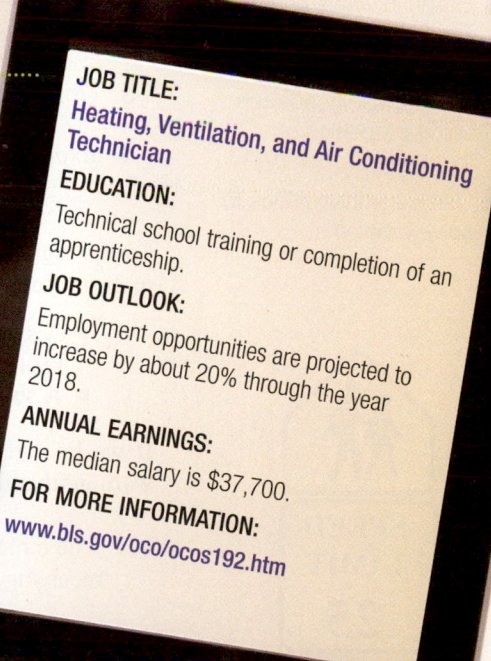

JOB TITLE:
Heating, Ventilation, and Air Conditioning Technician

EDUCATION:
Technical school training or completion of an apprenticeship.

JOB OUTLOOK:
Employment opportunities are projected to increase by about 20% through the year 2018.

ANNUAL EARNINGS:
The median salary is $37,700.

FOR MORE INFORMATION:
www.bls.gov/oco/ocos192.htm

Attending class and taking notes are important, but they are not enough. The only way that you are really going to learn algebra is by doing your homework.

WHEN TO DO YOUR HOMEWORK: Homework should be started on the day it is assigned, when the material is fresh in your mind. It's best to break your homework sessions into 30-minute periods, allowing for short breaks in between.

HOW TO BEGIN YOUR HOMEWORK: Review your notes and the examples in your text before starting your homework assignment.

GETTING HELP WITH YOUR HOMEWORK: It's normal to have some questions when doing homework. Talk to a tutor, a classmate, or your instructor to get those questions answered.

Now Try This ▶

1. Write a one-page paper that describes *when, where,* and *how* you go about completing your algebra homework assignments.
2. For each problem on your next homework assignment, find an example in this book that is similar. Write the example number next to the problem.
3. Make a list of questions that you have while doing your next assignment. Then decide whom you are going to ask to get those questions answered.

SECTION 4.1

Solving Linear Inequalities in One Variable

OBJECTIVES

1. Read and interpret inequality symbols.
2. Graph intervals and use interval and set-builder notation.
3. Solve linear inequalities using properties of inequality.
4. Use linear inequalities to solve problems.

ARE YOU READY?

▼ *The following problems review some basic skills that are needed to solve inequalities.*

1. Fill in the blank: The symbol $>$ means   "___ ___ ___."
2. Is $-11 > -10$ a true or false statement?
3. Graph each number in the set $\left\{ -3.8, -\dfrac{2}{5}, 0.6, \dfrac{9}{4} \right\}$ on a number line.
4. Express the fact that $6 > 0$ using an $<$ symbol.

Traffic signs like the one shown here often appear in front of schools. From the sign, a motorist knows that

- A speed *greater than* 25 miles per hour (mph) breaks the law and could possibly result in a ticket for speeding.
- A speed *less than or equal to* 25 mph is within the posted speed limit.

If we let s represent the motorist's speed, we can use *inequality symbols* to describe the warning that the traffic sign gives to drivers.

The motorist is in danger of receiving a speeding ticket if	The motorist is observing the posted speed limit if
$s > 25$	$s \le 25$
↑	↑
is greater than	**is less than or equal to**

The symbols $>$ and $\le$ are two types of inequality symbols that we will work with in this chapter.

1   Read and Interpret Inequality Symbols.

Inequalities are statements indicating that two expressions are unequal and they contain one or more of the following symbols.

Inequality symbol	Meaning	Example
$\neq$	is not equal to	$7 \neq 9$
$<$	is less than	$3 < 5$
$>$	is greater than	$52 > 38$
$\leq$	is less than or equal to	$19 \leq 30$
$\geq$	is greater than or equal to	$4 \geq 4$

By definition, $a \leq b$ is true if either $a < b$ or $a = b$ is true, and $a \geq b$ is true if either $a > b$ or $a = b$ is true. Therefore, $19 \leq 30$ is true because $19 < 30$ and $4 \geq 4$ is true because $4 = 4$.

In this section, we will solve **linear inequalities** in one variable. A linear *inequality* in one variable is similar to a linear *equation* in one variable except that the equal symbol is replaced with an inequality symbol.

Success Tip

To distinguish between the inequality symbols $<$ and $>$, remember that each one points to the smaller of the two numbers involved.

$$3 < 5 \qquad 52 > 38$$

Points to the smaller number

Linear Inequalities

A **linear inequality** in one variable (say, x) is any inequality that can be written in one of the following forms, where a, b, and c represent real numbers and $a \neq 0$.

$$ax + b < c \qquad ax + b \leq c \qquad ax + b > c \qquad ax + b \geq c$$

Some examples of linear inequalities in one variable are

$$x > -5, \qquad x \leq 7, \qquad 6x - 27 < 9, \qquad \text{and} \qquad \frac{2}{3}(x + 2) \geq \frac{4}{5}(x - 3)$$

Most of the linear inequalities that we will solve have infinitely many solutions. To represent such solutions, we can use number line graphs and two special types of notation.

2   Graph Intervals and Use Interval and Set-Builder Notation.

The graph of a set of real numbers that is a portion of a number line is called an **interval.** The following graph represents all real numbers that are greater than -5. This interval contains numbers that satisfy the inequality $x > -5$, such as -4.99, -1.8, 0, $2\frac{3}{4}$, π, and $1,070$. The left **parenthesis** at -5 indicates that -5 is not included in the interval.

We can also express this interval in **interval notation** as $(-5, \infty)$, where ∞ (read as **positive infinity**) indicates that the interval extends indefinitely to the right. The left **parenthesis** is used to show that the endpoint -5 is not included.

Set-builder notation is another way of describing the set of real numbers graphed in the figure above. With this notation, the condition for membership in the set is described using a variable. For example, the set of real numbers greater than -5 is written in set-builder notation as

$$\{x \mid x > -5\}$$

The set of all real numbers x   such that   x is greater than -5

Notation

The symbols ∞ and $-\infty$ do not represent numbers. Instead, ∞ indicates that an interval extends indefinitely to the right and $-\infty$ indicates that an interval extends indefinitely to the left. Note that a parenthesis rather than a bracket is written next to an infinity symbol:

$$(-\infty, 7] \qquad (-5, \infty)$$

The interval shown in the following figure is the graph of the real numbers less than or equal to 7. It contains the numbers that satisfy the inequality $x \leq 7$. The right **bracket** at 7 indicates that 7 is included in the interval. To express this interval in interval notation, we write $(-\infty, 7]$, where $-\infty$ (read as **negative infinity**) indicates that the interval extends indefinitely to the left. The bracket is used to show that 7 is included in the interval. To describe the interval using set-builder notation, we write $\{x \mid x \leq 7\}$.

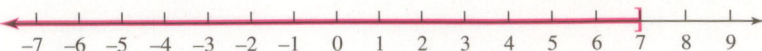

If an interval extends forever in one direction, as in the previous examples, it is called an **unbounded interval.** The following chart illustrates the various types of unbounded intervals and shows how they are described using set-builder notation and a graph.

Unbounded Intervals	
$\{x \mid x > a\}$ represents the interval (a, ∞).	
$\{x \mid x \geq a\}$ represents the interval $[a, \infty)$.	
$\{x \mid x < a\}$ represents the interval $(-\infty, a)$.	
$\{x \mid x \leq a\}$ represents the interval $(-\infty, a]$.	
The interval $(-\infty, \infty)$ includes all real numbers x. The graph of this interval is the entire number line.	

Notation

Parentheses are used to exclude endpoints. Brackets are used to include endpoints.

When graphing intervals, an open circle can be used to show that a point is not included in a graph, and a solid circle can be used to show that a point is included. For example,

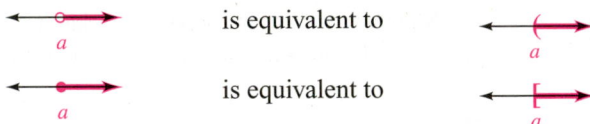

We will use parentheses and brackets when graphing intervals because they are consistent with interval notation.

EXAMPLE 1   Represent each set using a graph, interval notation, and set-builder notation.
a. The set of real numbers greater than or equal to 8
b. The set of real numbers less than -3

Strategy   To graph the interval on a number line, we will determine whether the endpoint is in the graph. Then we will determine whether the real numbers to the right or the left of the endpoint should be shaded.

Why   We draw the graph first because the corresponding interval notation and set-builder notation follow directly from it.

Solution   **a.** The set of real numbers *greater than or equal to* 8 is graphed on the left. Because equality with 8 is allowed, a bracket is used to include the endpoint 8. The numbers to the right of 8 are shaded because they are greater than 8. The interval is written as $[8, \infty)$ and the set-builder notation is written as $\{x \mid x \geq 8\}$.

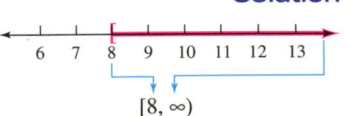

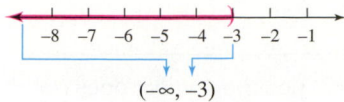

$(-\infty, -3)$

b. The set of real numbers *less than* -3 is graphed on the left. Since the elements of the set must be strictly less than -3, we exclude -3 using a parenthesis and shade the real numbers to its left. The interval notation is written as $(-\infty, -3)$ and the set-builder notation is written as $\{x \mid x < -3\}$.

> **Self Check 1** Represent each set using a graph, interval notation, and set-builder notation.
> **a.** The set of real numbers greater than 1
> **b.** The set of real numbers less than or equal to -2
>
> **Now Try** ▶ Problems 17 and 19

3 Solve Linear Inequalities Using Properties of Inequality.

To **solve a linear inequality** means to find the values of its variable that make the inequality true. The set of all solutions of an inequality is called its **solution set.** We will use the following properties to solve inequalities in one variable.

Addition and Subtraction Properties of Inequality	Adding the same number to, or subtracting the same number from, both sides of an inequality does not change the solution.
	For any real numbers a, b, and c,
	$$\text{If } a < b, \quad \text{then} \quad a + c < b + c \quad \text{and} \quad a - c < b - c.$$
	Similar statements can be made for the symbols $\le$, $>$, or $\ge$.

As with equations, there are properties for multiplying and dividing both sides of an inequality by the same number. To develop the **multiplication property of inequality,** we consider the true statement $2 < 5$. If both sides are multiplied by a positive number, such as 3, another true inequality results.

$$2 < 5$$
$$\mathbf{3 \cdot 2} < \mathbf{3 \cdot 5} \qquad \text{Multiply both sides by 3.}$$
$$6 < 15 \qquad \text{This is a true inequality.}$$

However, if we multiply both sides of $2 < 5$ by a negative number, such as -3, the direction of the inequality symbol must be reversed to produce another true inequality.

$$2 < 5$$
$$\mathbf{-3 \cdot 2} > \mathbf{-3 \cdot 5} \qquad \text{Multiply both sides by the negative number } -3 \text{ and change } < \text{ to } >.$$
$$-6 > -15 \qquad \text{This is a true inequality.}$$

Dividing both sides of a true inequality by the same positive number leads to another true inequality. However, dividing both sides of a true inequality by the same negative number requires that the direction of the inequality symbol be reversed to produce another true inequality.

$$6 > -4 \qquad \text{This is a true inequality.}$$
$$\frac{6}{-2} < \frac{-4}{-2} \qquad \text{Divide both sides by the negative number } -2 \text{ and change } > \text{ to } <.$$
$$-3 < 2 \qquad \text{This is a true inequality.}$$

These examples illustrate the multiplication and division properties of inequality.

Multiplication and Division Properties of Inequality	Multiplying or dividing both sides of an inequality by the same positive number does not change the solutions. For any real numbers a, b, and c, where c is **positive**, $\quad$ If $a < b$, $\quad$ then $\quad ac < bc$ $\quad$ and $\quad \dfrac{a}{c} < \dfrac{b}{c}$. If we multiply or divide both sides of an inequality by a negative number, the direction of the inequality symbol must be reversed for the inequalities to have the same solutions. For any real numbers a, b, and c, where c is **negative**, $\quad$ If $a < b$, $\quad$ then $\quad ac > bc$ $\quad$ and $\quad \dfrac{a}{c} > \dfrac{b}{c}$. Similar statements can be made for the symbols $\leq$, $>$, or $\geq$.

After applying one of the properties of inequality, the resulting inequality is equivalent to the original one. Like equivalent equations, **equivalent inequalities** have the same solution set.

EXAMPLE 2 Solve: $6x - 27 < 9$. Graph the solution set and write it using interval notation.

Strategy We will use properties of inequality to isolate the variable on one side.

Why Once we have obtained an equivalent inequality of the form $x < a$, with the variable isolated on one side, the solution set is obvious.

Solution We use the same steps to solve inequalities as we used to solve equations.

$$6x - 27 < 9 \qquad \text{\color{red}This is the inequality to solve.}$$

$$6x - 27 \mathbf{\color{red}{+\ 27}} < 9 \mathbf{\color{red}{+\ 27}} \qquad \text{\color{red}To isolate the variable term 6x, undo the subtraction}$$
$$\text{\color{red}of 27 on the left side by adding 27 to both sides.}$$

$$6x < 36 \qquad \text{\color{red}Simplify each side.}$$

$$\frac{6x}{\mathbf{\color{red}6}} < \frac{36}{\mathbf{\color{red}6}} \qquad \text{\color{red}To isolate x on the left side, undo the multiplication}$$
$$\text{\color{red}by 6 by dividing both sides by 6.}$$

$$x < 6 \qquad \text{\color{red}Simplify each side.}$$

The Language of Algebra

Because $<$ requires one number to be strictly less than another number and $>$ requires one number to be strictly greater than another number, $<$ and $>$ are called **strict inequalities**.

The graph of the solution set is shown on the right. It can be written in interval notation as $(-\infty, 6)$ and in set-builder notation as $\{x \mid x < 6\}$.

$$5 \quad 6 \quad 7$$

Success Tip

The solution of an inequality can be represented by a graph, using interval notation, or set-builder notation. Check with your instructor to see his or her preferred way (or ways) to present such solutions.

Since the solution set contains infinitely many real numbers, we cannot check all of them to see whether they satisfy the original inequality. However, as an informal check, we can pick one number in the graph, near the endpoint, such as 5, and see whether it satisfies the inequality. We also can pick one number not in the graph, but near the endpoint, such as 7, and see whether it fails to satisfy the inequality.

Check a value in the graph: $x = 5$	*Check a value not in the graph:* $x = 7$
$6\mathbf{\color{red}x} - 27 < 9$	$6\mathbf{\color{red}x} - 27 < 9$
$6(\mathbf{\color{red}5}) - 27 \overset{?}{<} 9 \quad \text{\color{red}Substitute 5 for x.}$	$6(\mathbf{\color{red}7}) - 27 \overset{?}{<} 9 \quad \text{\color{red}Substitute 7 for x.}$
$30 - 27 \overset{?}{<} 9$	$42 - 27 \overset{?}{<} 9$
$3 < 9 \quad \text{\color{red}True}$	$15 < 9 \quad \text{\color{red}False}$

Since 5 satisfies $6x - 27 < 9$ and 7 does not, the solution set appears to be correct.

Self Check 2 Solve: $8x + 4 < -44$. Graph the solution set and write it using interval notation.

Now Try ▶ Problem 21

EXAMPLE 3

Solve: $2x - 9 - 10x \leq 3 + 4x + 12$. Graph the solution set and write it using interval notation.

Strategy We will combine the like terms on each side of the inequality and use properties of inequality to isolate the variable on one side.

Why Once we have obtained an equivalent inequality, with the variable isolated on one side, the solution set is obvious.

Solution

$2x - 9 - 10x \leq 3 + 4x + 12$	This is the inequality to solve.
$-8x - 9 \leq 4x + 15$	Combine like terms: $2x - 10x = -8x$ and $3 + 12 = 15$.
$-8x - 9 - \mathbf{4x} \leq 4x + 15 - \mathbf{4x}$	To eliminate $4x$ on the right, subtract $4x$ from both sides.
$-12x - 9 \leq 15$	Simplify each side.
$-12x - 9 + \mathbf{9} \leq 15 + \mathbf{9}$	To isolate the variable term $-12x$, undo the subtraction of 9 on the left side by adding 9 to both sides.
$-12x \leq 24$	Simplify each side.
$\dfrac{-12x}{\mathbf{-12}} \geq \dfrac{24}{\mathbf{-12}}$	To isolate x, undo the multiplication by -12 by dividing both sides by -12. Because of the division by a negative number, reverse the $\leq$ symbol.
$x \geq -2$	Simplify each side.

Caution

A common error is to reverse the inequality symbol one step too late. It must be reversed on this line, when the division by -12 takes place. →

The graph of the solution set is shown on the right. It can be written in interval notation as $[-2, \infty)$ and in set-builder notation as $\{x \mid x \geq -2\}$.

$$\overset{}{\underset{-3 \quad -2 \quad -1}{\longleftarrow\!\!\!\!\!\!\!\!\vdash\!\!\!\!-\!\!\!\!-\!\!\!\!}}$$

Self Check 3 Solve: $x + 4 - 5x \leq 1 + 2x - 15$. Graph the solution set and write it using interval notation.

Now Try ▶ Problem 31

EXAMPLE 4

Solve: $-7 > \dfrac{16}{15}t + 1$. Graph the solution set and write it using interval notation.

Strategy We will use properties of inequality to isolate the variable on one side.

Why Once we have obtained an equivalent inequality, with the variable isolated on one side, the solution set is obvious.

Solution

Caution

When solving inequalities, the variable can end up on the right side. This causes some students trouble when graphing the solution set. It is helpful to write such an inequality in an equivalent form with the variable on the left and then draw the graph.

$-7 > \dfrac{16}{15}t + 1$	This is the inequality to solve.
$-8 > \dfrac{16}{15}t$	Subtract 1 from both sides.
$\dfrac{\mathbf{15}}{\mathbf{16}}(-8) > \dfrac{\mathbf{15}}{\mathbf{16}}\left(\dfrac{16}{15}t\right)$	To undo the multiplication by $\frac{16}{15}$ and isolate t, multiply both sides by the reciprocal, which is $\frac{15}{16}$.
$-\dfrac{15}{2} > t$	Simplify each side.

We can write an equivalent inequality with the variable on the left side.

$t < -\dfrac{15}{2}$	If $-\frac{15}{2}$ is greater than t, then t must be less than $-\frac{15}{2}$.

The graph of the solution set is shown on the right. It can be written in interval notation as $\left(-\infty, -\dfrac{15}{2}\right)$ and in set-builder notation as $\left\{t \mid t < -\dfrac{15}{2}\right\}$.

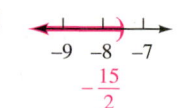

$$\underset{-9 \quad -8 \quad -7}{\longleftarrow\!\!\!\!\!\!\!\!\longleftarrow}$$
$$-\dfrac{15}{2}$$

Self Check 4 Solve: $7 < \dfrac{10}{9}s + 2$. Graph the solution set and write it using interval notation.

Now Try ▶ Problem 33

EXAMPLE 5   Solve: $\frac{2}{3}(x + 2) > \frac{4}{5}(x - 3)$. Graph the solution set and write it using interval notation.

Strategy   We will clear the inequality of fractions by multiplying both sides by the LCD of $\frac{2}{3}$ and $\frac{4}{5}$.

Why   It's easier to solve an inequality that involves only integers.

Solution   The LCD of $\frac{2}{3}$ and $\frac{4}{5}$ is 15.

$$\frac{2}{3}(x + 2) > \frac{4}{5}(x - 3)$$ This is the inequality to solve.

$$15 \cdot \frac{2}{3}(x + 2) > 15 \cdot \frac{4}{5}(x - 3)$$ Multiply both sides by 15.

$$10(x + 2) > 12(x - 3)$$ Simplify: $15 \cdot \frac{2}{3} = 10$ and $15 \cdot \frac{4}{5} = 12$.

$$10x + 20 > 12x - 36$$ Distribute the multiplication by 10 and 12.

$$-2x + 20 > -36$$ To eliminate 12x on the right side, subtract 12x from both sides.

$$-2x > -56$$ To isolate the variable term $-2x$, undo the addition of 20 by subtracting 20 from both sides.

$$\frac{-2x}{-2} < \frac{-56}{-2}$$ To isolate x, undo the multiplication by -2 by dividing both sides by -2 and reverse the $>$ symbol.

$$x < 28$$ Simplify each side.

Success Tip

You may find it easier to graph the solution set first and then use the symbols that occur in that form of the answer to write the interval notation.

The graph of the solution set is shown on the right. It can be written in interval notation as $(-\infty, 28)$ and in set-builder notation as $\{x \mid x < 28\}$.

$$\xleftarrow{\hspace{1cm}} \quad \underset{27 \quad 28 \quad 29}{\longmapsto}$$

Self Check 5   Solve: $\frac{3}{2}(x + 2) \le \frac{3}{5}(x - 3)$. Graph the solution set and write it using interval notation.

Now Try ▶ Problem 37

When solving equations, we have seen that some are true for all real numbers while others have no solution. Similar situations can occur when solving inequalities. Some inequalities are made true by *any* permissible replacement value for the variable. Such inequalities are called **identities.** Another type of inequality, called a **contradiction,** is false for all replacement values for the variable.

EXAMPLE 6   Solve each inequality. Graph the solution set and write it using interval notation.

a. $\dfrac{3a - 4}{-5} > \dfrac{3a + 15}{-5}$ **b.** $1 - 2a \ge 2(1 - a)$

Strategy   We will use properties of inequality to isolate the variable on one side.

Why   Once we have obtained an equivalent inequality, with the variable isolated on one side, the solution set is obvious.

Solution   **a.** $\dfrac{3a - 4}{-5} > \dfrac{3a + 15}{-5}$ This is the inequality to solve.

$$-5\left(\frac{3a - 4}{-5}\right) < -5\left(\frac{3a + 15}{-5}\right)$$ To clear the inequality of fractions, multiply both sides by -5. Since we are multiplying both sides by a negative number, reverse the direction of the inequality.

$$3a - 4 < 3a + 15$$ Simplify each side.

$$3a - 4 - 3a < 3a + 15 - 3a$$ Subtract 3a from both sides.

$$-4 < 15$$ This is a true statement.

In the solution process, the terms involving a drop out. The resulting true statement indicates that the original inequality is true for every permissible value of a. Therefore, *all real numbers* are solutions and this inequality is an identity. Its solution set is the set of real numbers written as $(-\infty, \infty)$ or using the symbol $\mathbb{R}$, and its graph is as shown.

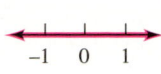

b.

$1 - 2a \geq 2(1 - a)$	This is the inequality to solve.
$1 - 2a \geq 2 - 2a$	Distribute the multiplication by 2.
$1 - 2a + 2a \geq 2 - 2a + 2a$	Add 2a to both sides.
$1 \geq 2$	This is a false statement.

In the solution process, the terms involving a drop out. The resulting false statement indicates that no value for a makes the original inequality true. Therefore, this inequality has *no solution* and it is a contradiction. Its solution set is the empty set, which is written $\{\ \}$ or using the symbol $\varnothing$.

Self Check 6 Solve each inequality. Graph the solution set and write it using interval notation. **a.** $-8n + 10 \geq 1 - 2(4n - 2)$
b. $\dfrac{4d - 5}{-10} > \dfrac{2(2d - 3)}{-10}$

Now Try ▶ Problems 41 and 43

Using Your Calculator ▶ Solving Linear Inequalities in One Variable

There are several ways to solve linear inequalities graphically. For example, to solve $3(2x - 9) < 9$, we can subtract 9 from both sides and solve the equivalent inequality $3(2x - 9) - 9 < 0$. Using standard window settings of $[-10, 10]$ for x and $[-10, 10]$ for y, we graph $y = 3(2x - 9) - 9$ and then use TRACE. Moving the cursor closer and closer to the x-axis, as shown in figure (a), we see that the graph is below the x-axis for x-values in the interval $(-\infty, 6)$. This interval is the solution, because in this interval, $3(2x - 9) - 9 < 0$.

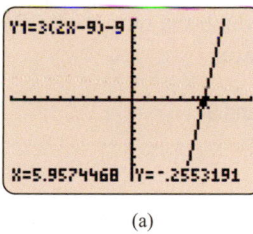
(a)

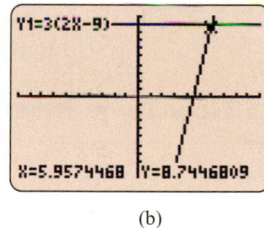
(b)

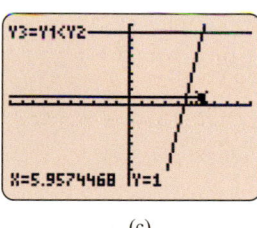
(c)

Another way to solve $3(2x - 9) < 9$ is to graph $y = 3(2x - 9)$ and $y = 9$. We can then trace to see that the graph of $y = 3(2x - 9)$ is below the graph of $y = 9$ for x-values in the interval $(-\infty, 6)$. See figure (b). This interval is the solution, because in this interval, $3(2x - 9) < 9$.

A third approach is to enter and then graph

$Y_1 = 3(2x - 9)$

$Y_2 = 9$

$Y_3 = Y_1 < Y_2$ To do this, use the VARS key. Consult your owner's manual for the specific directions.

The graphs of $y = 3(2x - 9)$, $y = 9$, and a horizontal line 1 unit above the x-axis will be displayed, as shown in figure (c). In the TRACE mode, we then move the cursor to the rightmost endpoint of the horizontal line to determine that the interval $(-\infty, 6)$ is the solution of $3(2x - 9) < 9$.

4   Use Linear Inequalities to Solve Problems.

In some application problems, an inequality, rather than an equation, is needed to find the solution. This is the case when we are asked to determine when one quantity *is more* (or *is less*) than another. Other phrases that call for an inequality are listed in the table below.

The statement	Translates to		The statement	Translates to
a does not exceed *b*.	$a \leq b$		*a* will exceed *b*.	$a > b$
a is at most *b*.	$a \leq b$		*a* is at least *b*.	$a \geq b$
a is no more than *b*.	$a \leq b$		*a* is not less than *b*.	$a \geq b$

EXAMPLE 7   **Car Engines.**   The horsepower $h(r)$ produced by the engine in a Porsche 911 can be approximated by the linear function $h(r) = 0.06r - 36$, where r is the engine speed in revolutions per minute (rpm). Use an inequality to determine the rpm at which the output of the engine is at least 300 horsepower.   (Source: ultimatecarpage.com)

Strategy   Since r represents the rpm of the engine and $h(r)$ the horsepower it produces, we need to find the values of r for which $h(r) \geq 300$.

Why   The phrase *at least* means to be greater than or equal to and, therefore, translates to the inequality symbol $\geq$.

Solution

$$h(r) \geq 300 \qquad \text{We want the horsepower } h(r) \text{ to be greater than or equal to 300.}$$

$$0.06r - 36 \geq 300 \qquad \text{Substitute } 0.06r - 36 \text{ for } h(r).$$

$$0.06r \geq 336 \qquad \text{To isolate the variable term } 0.06r, \text{ add 36 to both sides.}$$

$$\frac{0.06r}{0.06} \geq \frac{336}{0.06} \qquad \text{To isolate the variable } r, \text{ divide both sides by 0.06.}$$

$$r \geq 5{,}600 \qquad \text{Do the division.}$$

The output of the engine exceeds 300 horsepower when the engine speed is 5,600 rpm or greater.

Self Check 7   **Car Engines.**   Refer to Example 7. Use an inequality to determine the rpm at which the output of a Porsche engine is no more than 240 horsepower.

Now Try ▶ Problems 77 and 91

For more complicated application problems involving inequalities, we can adapt the six-step problem-solving strategy used in earlier chapters to find the solution. In such cases, Steps 3 and 4 of the strategy are as follows:

   3. Form an inequality by translating the words of the problem into mathematical symbols.
   4. Solve the inequality.

In the next example, an inequality is used to compare two rental plans to determine which one is financially better.

EXAMPLE 8   **Renting RVs.**   *Adventure RVs* charges $2,300 and $0.28 per mile to rent a 36-foot recreational vehicle for one week. *Roadmaster RV's* weekly charge for the same model is $1,900 and $0.36 per mile. For what range of miles driven is *Adventure RV's* rental plan better?

Analyze   *Adventure RV's* rental plan will be the better deal if it costs less than *Roadmaster RV's* rental plan. The phrase *less than* indicates an inequality should be used to model the situation.

Assign   Let m = the number of miles the RV is driven during the week it is rented.

Form   We can use the following formula to write algebraic expressions that represent the cost to rent the RV from each company:

The cost to rent the RV for a week	is	the basic fee	+	the cost per mile	·	the number of miles driven.

Thus,

Adventure RV's cost = 2,300 + 0.28m and   *Roadmaster RV's* cost = 1,900 + 0.36m

We want to find the range of miles that can be driven during the week so that:

The cost to rent the RV from *Adventure RVs*	is less than	the cost to rent the RV from *Roadmaster RVs.*
2,300 + 0.28m	<	1,900 + 0.36m

Solve $2,300 + 0.28m < 1,900 + 0.36m$

$$2,300 < 1,900 + 0.08m$$

To eliminate 0.28m on the left side, subtract 0.28m from both sides: 0.36m − 0.28m = 0.08m.

$$400 < 0.08m$$

To isolate the variable term 0.08m on the right side, subtract 1,900 from both sides: 2,300 − 1,900 = 400.

$$\frac{400}{0.08} < \frac{0.08m}{0.08}$$

To isolate m on the right side, undo the multiplication by 0.08 by dividing both sides by 0.08.

$$5,000 < m$$

Do the division.

$$m > 5,000$$

Write an equivalent inequality with the variable on the left side.

State   *Adventure RVs* will be the better deal if the RV is going to be driven more than 5,000 miles during the week.

Check   We can check the result by calculating each RV rental for a number of miles driven that is slightly greater than 5,000, say 5,010.

Adventure RVs cost = 2,300 + 0.28m	*Roadmaster RVs* cost = 1,900 + 0.36m
= 2,300 + 0.28(**5,010**)	− 1,900 + 0.36(**5,010**)
= 3,702.80	= 3,703.60

Since *Adventure RVs* rental cost of $3,702.80 is less than *Roadmaster RVs* rental cost of $3,703.60, the result seems reasonable.

Self Check 8   **Renting Cars.**   *Great Value Car Rental* charges $12 a day and $0.15 per mile to rent a Ford Fusion. *Be Thrifty Car Rental's* daily charge for the same car is $15 and $0.12 per mile. If a businessman wants to rent the car for one day, for what range of miles driven is the *Be Thrifty* rental plan better?

Now Try ▸ Problem 95

SECTION 4.1 ▶ STUDY SET

VOCABULARY

Fill in the blanks.

1. $<$, $>$, $\leq$, and $\geq$ are _____ symbols.

2. $3x + 2 \geq 7$ is an example of a _____ inequality in one variable.

3. The graph of a set of real numbers that is a portion of a number line is called an _____.

4. In $(-\infty, 5)$, the right _____ is used to show that 5 is not included in the interval. In $[12, \infty)$, the left _____ is used to show that 12 is included in the interval.

5. We read the set-_____ notation $\{x \mid x < 1\}$ as "the set of all real numbers x _____ _____ x is less than 1."

6. To _____ an inequality means to find all values of the variable that make the inequality true.

CONCEPTS

7. Which of the following are inequalities?

$$6 - x = 8 \qquad 5 + a \qquad 7t - 5 > 4 \qquad \frac{x}{2} \leq -1$$

8. Perform each step listed below on the inequality $4 > -2$ and give the resulting true inequality.

   a. Add 2 to both sides.   b. Subtract 4 from both sides.

   c. Multiply both sides by 4.   d. Divide both sides by -2.

9. Use a check to determine whether each number is a solution of $3x + 6 \leq 6$.

   a. 0   b. $\frac{2}{3}$

   c. -10 d. 1.5

10. The solution set of a linear inequality in x is graphed on the right. Determine whether a true or false statement results when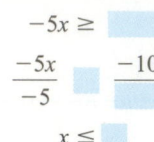

   a. -4 is substituted for x.

   b. -3 is substituted for x.

   c. 0 is substituted for x.

11. a. Suppose that when solving a linear inequality, the variable drops out, and the result is $6 \leq 10$. Write the solution set in interval notation and graph it.

   b. Suppose that when solving a linear inequality, the variable drops out, and the result is $7 < -1$. What symbol is used to represent the solution set?

12. Insert the correct symbol, $<$, $\leq$, $>$, or $\geq$, in each blank.

   a. As many as 16 people were seriously injured: The number of people seriously injured ___ 16.

   b. There were no fewer than 8 references to taxes in the speech: The number of tax references ___ 8.

   c. The weight w of the roast is at most 8 pounds: w ___ 8.

   d. The temperature t exceeded 100°: t ___ 100.

NOTATION

Complete the solution to solve the inequality.

13. $-5x - 1 \geq -11$

$$-5x \geq \boxed{}$$

$$\frac{-5x}{-5} \boxed{} \frac{-10}{-5}$$

$$x \leq \boxed{}$$

The solution set is $\left(\boxed{}, 2\right]$. Using set-builder notation, it is $\{x \mid \boxed{}\}$.

14. Match each interval with its graph.

   a. $(-\infty, -1]$ i. (number line with 0, 1, 2)

   b. $(-\infty, 1)$ ii. (number line with -2, -1, 0)

   c. $[-1, \infty)$ iii. (number line with -2, -1, 0)

15. Fill in the blank: If $-10 > x$ then x ___ -10.

16. Fill in the blank: ∞ is a symbol representing positive _____.

GUIDED PRACTICE

Represent each set using a graph, interval notation, and set-builder notation. **See Example 1.**

17. The set of real numbers less than 14

18. The set of real numbers greater than 6

19. The set of real numbers greater than or equal to -2

20. The set of real numbers less than or equal to -7

Solve each inequality. Graph the solution set and write it using interval notation. **See Example 2.**

21. $x + 4 < 5$ 22. $x - 5 > 2$

23. $3x > -9$ 24. $4x < -36$

25. $2x - 7 \geq -29$ 26. $6x + 8 \leq -16$

27. $9a + 11 \leq 29$ 28. $3b - 26 \geq 4$

Solve each inequality. Graph the solution set and write it using interval notation. **See Example 3.**

29. $2x + 4 + 6x > 2 - 3x + 2$

30. $5x + 6 + 2x \geq 2 - x + 4$

31. $t + 1 - 3t \geq t - 20$

32. $a + 4 - 10a > a - 16$

Solve each inequality. Graph the solution set and write it using interval notation. See Example 4.

33. $4 \leq \dfrac{9}{10}x + 1$

34. $2 \leq \dfrac{9}{4}x + 8$

35. $-3 > \dfrac{7}{8}x - 1$

36. $-10 > \dfrac{11}{2}x - 6$

Solve each inequality. Graph the solution set and write it using interval notation. See Example 5.

37. $\dfrac{3}{4}(x - 3) < \dfrac{1}{3}(x - 4)$

38. $\dfrac{5}{16}(x + 1) \geq \dfrac{1}{4}(x - 3)$

39. $\dfrac{2}{5}(3 - 2n) \geq \dfrac{3}{8}(2 - 3n)$

40. $\dfrac{1}{5}(1 - n) < \dfrac{1}{3}(2 - n)$

Solve each inequality. Graph the solution set and write it using interval notation, if possible. See Example 6.

41. $2(5x - 6) > 4x - 15 + 6x$

42. $3(4x - 2) > 14x - 7 - 2x$

43. $\dfrac{5x + 2}{-4} > \dfrac{5x + 1}{-4}$

44. $\dfrac{7 - n}{-6} > \dfrac{1 - n}{-6}$

TRY IT YOURSELF

Solve each inequality. Graph the solution set and write it using interval notation.

45. $-5t + 3 \leq 5$

46. $-9t + 6 \geq 16$

47. $\dfrac{2}{5} > \dfrac{4}{5}x$

48. $\dfrac{5}{9} < \dfrac{11}{9}x$

49. $-0.6x \leq -36$

50. $-0.2x > -8$

51. $7 < \dfrac{5}{3}a - 3$

52. $5 > \dfrac{7}{2}a - 9$

53. $-7y + 5 > -5y - 1$

54. $-2s - 105 \leq -7s - 205$

55. $\dfrac{6 - d}{-2} \leq -6$

56. $\dfrac{9 - 3b}{-8} < 3$

57. $0.4x + 0.4 \leq 0.1x + 0.85$

58. $0.05 + 0.8x \leq 0.5x - 0.7$

59. $3(z - 2) \leq 2(z + 7)$

60. $5(3 + z) > -3(z + 3)$

61. $\dfrac{3b + 7}{3} \leq \dfrac{2b - 9}{2}$

62. $-\dfrac{5x}{4} > \dfrac{3 - 5x}{4}$

63. $\dfrac{x - 7}{2} - \dfrac{x - 1}{5} \geq -\dfrac{x}{4}$

64. $\dfrac{3a + 1}{3} - \dfrac{4 - 3a}{5} \geq -\dfrac{1}{15}$

65. $\dfrac{1}{2}x + 6 \geq 4 + 2x$

66. $\dfrac{1}{3}x + 1 < 4 + 5x$

67. $5(2n + 2) - n > 3n - 3(1 - 2n)$

68. $-1 + 4(y - 1) + 2y \leq \dfrac{1}{2}(12y - 30) + 15$

69. $\dfrac{1}{2}y + 2 \geq \dfrac{1}{3}y - 4$

70. $\dfrac{1}{4}x - \dfrac{1}{3} \leq x + 2$

71. $-11(2 - b) < 4(2b + 2)$

72. $-9(h - 3) + 2h \leq 8(4 - h)$

73. $\dfrac{2}{3}x + \dfrac{3}{2}(x - 5) \leq x$

74. $\dfrac{5}{9}(x + 3) - \dfrac{4}{3}(x - 3) \geq x - 1$

75. $5[3t - (t - 4)] - 11 \leq -12(t - 6) - (-t)$

76. $2 - 2[3h - (7 - h)] > 6[-(19 + h) - (1 - h)]$

77. Let $f(x) = \dfrac{1}{2}x - \dfrac{2}{3}$ and $g(x) = x + \dfrac{4}{3}$. Find all values of x for which $f(x) < g(x)$.

78. Let $s(x) = \dfrac{1}{4}x - \dfrac{1}{2}$ and $g(x) = \dfrac{1}{2}x - \dfrac{2}{3}$. Find all values of x for which $s(x) \geq g(x)$.

79. Let $y_1 = 0.7x - 0.15$ and $y_2 = x + 0.3$. Find all values of x for which y_2 exceeds y_1.

80. Let $y_1 = 0.8x - 1.1$ and $y_2 = 3.1 - 0.6x$. Find all values of x for which y_1 does not exceed y_2.

Look Alikes . . .

Solve the inequality in part a. Graph the solution set and write it in interval notation. Then use your answer to part a to determine the solution set for the inequality in part b. (No new work is necessary!) Graph the solution set and write it in interval notation.

81. a. $12x - 33.16 \leq 5.84$ **b.** $12x - 33.16 > 5.84$

82. a. $-\dfrac{3}{4}x > -\dfrac{21}{32}$ **b.** $-\dfrac{3}{4}x \leq -\dfrac{21}{32}$

83. a. $3(2x + 2) > 5(x - 1) + 3x$
b. $3(2x + 2) < 5(x - 1) + 3x$

84. a. $\dfrac{x - 3}{2} \leq \dfrac{1}{2} - \dfrac{x - 5}{4}$ **b.** $\dfrac{x - 3}{2} > \dfrac{1}{2} - \dfrac{x - 5}{4}$

Use a graphing calculator to solve each inequality. Write the solution set using interval notation. See Using Your Calculator: Solving Linear Inequalities in One Variable.

85. $2x + 3 < 5$ **86.** $5x + 2 \geq 4x - 2$

87. $3x - 2 > 4$ **88.** $3x - 4 \leq 2x + 4$

APPLICATIONS

89. Real Estate. Refer to the graph below. For which regions of the country were the following inequalities true in August, 2010?

a. Median sales price $<$ U.S. median price

b. Median sales price $\geq$ U.S. median price

August, 2010: Median Price of Existing Single-Family Homes

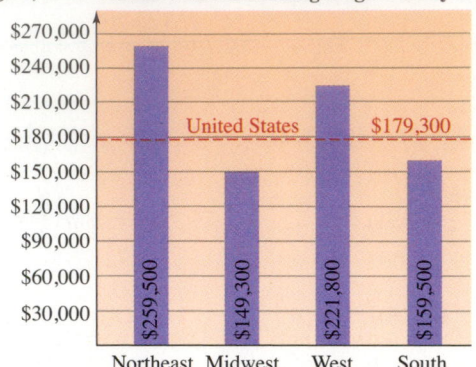

Source: National Association of Realtors

90. Public Education. For which years shown in the graph below is the inequality true?

a. Grade 4 enrollment $\geq$ Grade 1 enrollment

b. Projected grade 1 enrollment $>$ Projected grade 4 enrollment

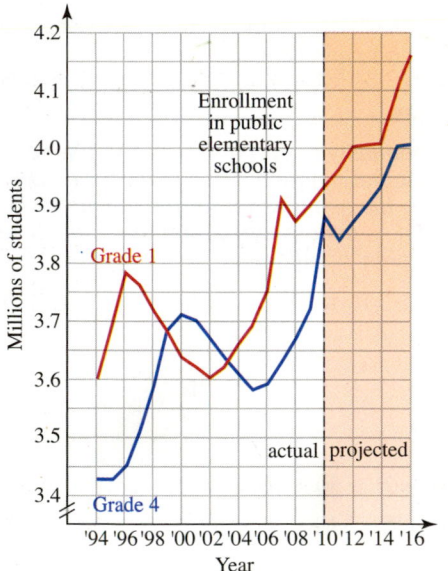

Source: National Center for Education Statistics

In problems 91–104, write and then solve an inequality to answer each question.

91.
from Campus to Careers

Heating, Ventilation, and Air Conditioning Technician

The percent of the air-borne particles in a room that a Climatec® furnace filter can remove is approximated by the linear function $p(m) = \frac{6}{5}m$, where m is the time in minutes that the furnace has been operating. Use an inequality to determine the time after which at least 60% of the air-borne particles in the room will have been removed. (Source: climatec.com)

92. Wikipedia. The number of articles $a(t)$ in millions in the English-language edition of Wikipedia can be approximated by the function $a(t) = 0.54t + 0.5$, where t is the number of years since 2005. Use an inequality to determine those years for which the number of articles surpassed 2.5 million. (Source: Wikipedia article: *Size of Wikipedia*)

The Wikipedia logo and trademark are used by permission from the Wikipedia Foundation.

93. National Parks. The number of visitors $v(t)$ to U.S. national parks can be approximated by the function $v(t) = -100{,}000t + 3{,}600{,}000$, where t is the number of years after 1990. Use an inequality to determine those years for which the number of visitors fell below 2,400,000. (Source: National Park Service Stats)

94. Medicare. The number of workers $w(t)$ for each Medicare beneficiary (each person receiving Medicare benefits) is approximated by the function $w(t) = -0.05t + 4$, where t is the number of years after 2000. Use an inequality to determine those years for which there will be less than 2 workers for each Medicare beneficiary. (Source: Kaiser Family Foundation)

95. Moving Day. *Valley Truck Rentals* charges $25.50 per day and $0.75 per mile to rent a 14-foot truck. *Nationwide Truck Rentals'* daily charge for the same vehicle is $36.75 and $0.60 per mile. If the truck is rented for one day, for what range of miles driven is *Nationwide's* plan better?

96. Telephone Service. A telephone company offers two long-distance calling plans.

■ Plan 1: $17 per month and 2¢ per minute

■ Plan 2: $8 per month and 8¢ per minute

For how many minutes of long distance calls would Plan 2 save the caller money?

97. Business Losses. It costs a company $1,400 to set up the necessary machinery and then $5 each to manufacture bath towels. They sell the towels for $8.50 each. Up to this point, the company has lost money on the towels. What is the greatest number of towels they could have sold for this to happen?

98. Job Offers. A company has offered a newly hired salesperson her choice of compensation packages:

- Package 1: $2,500 salary per month and a 5% sales commission
- Package 2: $3,500 salary per month and a 3% sales commission

What amount of sales per month would the salesperson have to make so that Package 1 is more profitable for her?

99. Averaging Grades. A student has scores of 70, 77, and 85 on three government exams. Use an inequality to determine the score she needs on a fourth exam to give her an average of 80 or better.

100. Video Game Systems. A student who can afford to spend up to $1,000 sees the ad shown in the illustration and decides to buy the video game system. Use an inequality to find the greatest number of video games that she can also purchase. (Disregard sales tax.)

VIDEO GAME SYSTEM
only $449⁹⁹

YOUR FAVORITE GAMES
only $45⁹⁹ each

101. Work Schedules. A student works two part-time jobs. He earns $8 an hour for working at the college library and $15 an hour for construction work. To save time for study, he limits his work to 25 hours a week. If he enjoys the work at the library more, what is the greatest number of hours he can work at the library and still earn at least $300 a week?

102. Scheduling Equipment. An excavating company charges $300 an hour for the use of a backhoe and $500 an hour for the use of a bulldozer. (Part of an hour counts as a full hour.) The company employs one operator for 40 hours per week to operate the machinery. If the company wants to bring in at least $18,500 each week from equipment rental, how many hours per week can it schedule the operator to use a backhoe?

103. Fundraising. A school PTA wants to rent a dunking tank for its annual school fundraising carnival. The cost is $85.00 for the first three hours and then $19.50 for each additional hour or part thereof. How long can the tank be rented if up to $185 is budgeted for this expense?

104. Investments. If a woman has invested $10,000 at 8% annual interest, how much more must she invest at 9% so that her annual income will exceed $1,250?

105. How are the methods for solving linear equations and linear inequalities similar? How are they different?

106. Explain how the symbol ∞ is used in this section. Is ∞ a real number?

107. Explain what is wrong with the following statement:

When solving inequalities involving negative numbers, the direction of the inequality symbol must be reversed.

108. In each case, determine what is wrong with the interval notation.
  a. $(\infty, -3)$ **b.** $[-\infty, -3)$

109. Geometry. The **triangle inequality** states an important relationship between the sides of any triangle:

$$\text{The sum of the lengths of two sides of a triangle} > \text{the length of the third side.}$$

Use the triangle inequality to explain why the dimensions of the triangle shown here must be mislabeled.

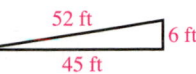
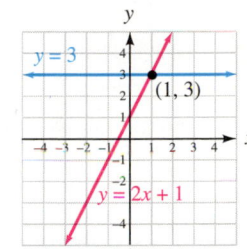

110. Explain how to use the graph to solve $2x + 1 < 3$.

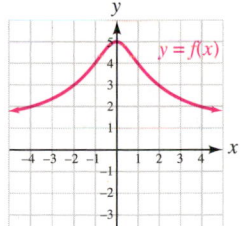

Use the graph of the function to find $f(-1)$, $f(0)$, and $f(2)$.

111.

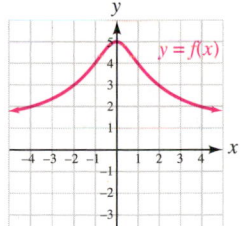

112.

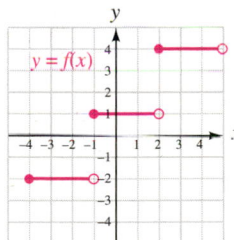

113. In the illustration, which of the following are true?
  i. $-a \geq 0$ **ii.** $a - b < 0$ **iii.** $b - a < 0$ **iv.** $ab > 0$

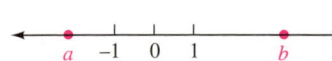

114. Consider the following "solution" of the inequality $\frac{1}{3} > \frac{1}{x}$, where it appears that the solution set is the interval $(3, \infty)$.

$$\frac{1}{3} > \frac{1}{x}$$

$$3x\left(\frac{1}{3}\right) > 3x\left(\frac{1}{x}\right)$$

$$x > 3$$

a. Show that $x = -1$ makes the original inequality true.

b. If $x = -1$ makes the original inequality true, there must be an error in the solution. Where is it?

115. Medical Plans.   A college provides its employees with a choice of the two medical plans shown in the following table. For what hospital bill dollar amount is Plan 2 better for the employee than Plan 1? (*Hint:* The cost to the employee includes both the deductible payment and the employee's coinsurance payment.)

Plan 1	Plan 2
Employee pays $100	Employee pays $200
Plan pays 70% of the rest	Plan pays 80% of the rest

116. Medical Plans.   To save costs, the college in Exercise 115 raised the employee deductible, as shown in the following table. For what hospital bill dollar amount is Plan 2 better for the employee than Plan 1? (*Hint:* The cost to the employee includes both the deductible payment and the employee's coinsurance payment.)

Plan 1	Plan 2
Employee pays $200	Employee pays $400
Plan pays 70% of the rest	Plan pays 80% of the rest

SECTION 4.2

Solving Compound Inequalities

OBJECTIVES

1 Find the intersection and the union of two sets.

2 Solve compound inequalities containing the word *and*.

3 Solve double linear inequalities.

4 Solve compound inequalities containing the word *or*.

ARE YOU READY?

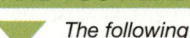

The following problems review some basic skills that are needed to solve inequalities.

1. Let $A = \{-6, 1, 2, 3, 4\}$ and $B = \{0, 3, 4, 5, 6\}$. What numbers do sets A and B have in common?

2. Solve $6x + 8 + x \geq 5(x - 3) + 9$. Graph the solution set and write it in interval notation.

3. Consider the statements $x > 4$ and $x < 8$. Does the number 7 make both inequalities true?

4. Graph the solution set for $x > 2$ and the solution set for $x \leq -1$ on the same number line.

A label on a first-aid cream warns the user about the temperature at which the medication should be stored. A careful reading reveals that the storage instructions consist of two parts:

The storage temperature
should be at least 59°F

and

the storage temperature
should be at most 77°F

First Aid Cream

Store at 59° to 77° F

When the word *and* or the word *or* is used to connect pairs of inequalities, we call the statement a **compound inequality**. To solve compound inequalities, we need to know how to find the *intersection* and *union* of two sets.

1 Find the Intersection and the Union of Two Sets.

Just as operations such as addition and multiplication are performed on real numbers, operations also can be performed on sets. The operation of intersection of two sets produces a new third set that consists of all of the elements that the two given sets have in common.

The Intersection of Two Sets	The **intersection of set A and set B**, written $A \cap B$, is the set of all elements that are common to set A and set B.

The operation of union of two sets produces a third set that is a combination of all of the elements of the two given sets.

The Union of Two Sets	The **union of set A and set B**, written $A \cup B$, is the set of elements that belong to set A or set B or both.

The Language of Algebra

The **union** of two sets is the collection of elements that belong to either set. The concept is similar to that of a family *reunion*, which brings together the members of several families.

Venn diagrams can be used to illustrate the intersection and union of sets. The area shown in purple in figure (a) represents $A \cap B$ and the area shown in both shades of red in figure (b) represents $A \cup B$.

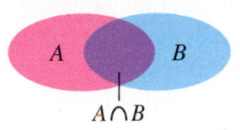

$A \cap B$

Read as "A intersect B."

(a)

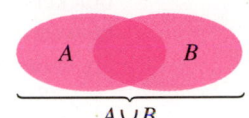

$A \cup B$

Read as "A union B."

(b)

EXAMPLE 1

Let $A = \{0, 1, 2, 3, 4, 5, 6\}$ and $B = \{-4, -2, 0, 2, 4\}$.
a. Find $A \cap B$.   **b.** Find $A \cup B$.

Strategy In part (a), we will find the elements that sets A and B have in common, and in part (b), we will find the elements that are in one set, or the other set, or both.

Why The symbol $\cap$ means intersection, and the symbol $\cup$ means union.

Solution

The Language of Algebra

The **intersection** of two sets is the collection of elements that they have in common. When two streets cross, we call the area of pavement that they have in common an *intersection*.

a. Since the numbers 0, 2, and 4 are common to both sets A and B, we have

$A \cap B = \{0, 2, 4\}$ Since the intersection is, itself, a set, braces are used.

b. Since the numbers in either or both sets are $-4, -2, 0, 1, 2, 3, 4, 5$, and 6, we have

$A \cup B = \{-4, -2, 0, 1, 2, 3, 4, 5, 6\}$ 0, 2, and 4 are not listed twice.

Self Check 1   Let $C = \{8, 9, 10, 11\}$ and $D = \{3, 6, 9, 12, 15\}$.
a. Find $C \cap D$.   **b.** Find $C \cup D$.

Now Try ▶ Problems 17 and 21

2 Solve Compound Inequalities Containing the Word *And*.

When two inequalities are joined with the word *and*, we call the statement a **compound inequality.** Some examples are

The Language of Algebra

Compound means composed of two or more parts, as in *compound* inequalities, chemical *compounds*, and *compound* sentences.

$$x \geq -3 \qquad \text{and} \qquad x \leq 6$$

$$\frac{x}{2} + 1 > 0 \qquad \text{and} \qquad 2x - 3 < 5$$

$$x + 3 \leq 2x - 1 \qquad \text{and} \qquad 3x - 2 < 5x - 4$$

The **solution set of a compound inequality containing the word** *and* includes all numbers that make both of the inequalities true. That is, it is the intersection of their solution sets. We can find the solution set of the compound inequality $x \geq -3$ and $x \leq 6$, for example, by graphing the solution sets of each inequality on the same number line and looking for the numbers common to both graphs.

In the following figure, the graph of the solution set of $x \geq -3$ is shown in red, and the graph of the solution set of $x \leq 6$ is shown in blue.

Preliminary work to determine the graph of the solution set

The figure below shows the graph of the solution of the compound inequality $x \geq -3$ and $x \leq 6$. The purple shaded interval, where the red and blue graphs intersect (or overlap), represents the real numbers that are common to the graphs of $x \geq -3$ and $x \leq 6$.

The graph of the solution set

The solution set of $x \geq -3$ and $x \leq 6$ is the **bounded interval** $[-3, 6]$, where the brackets indicate that the endpoints, -3 and 6, are included. It represents all real numbers between -3 and 6, including -3 and 6. Intervals such as this, which contain both endpoints, are called **closed intervals.**

Since the solution set of $x \geq -3$ and $x \leq 6$ is the intersection of the solution sets of the two inequalities, we can write

$$[-3, \infty) \cap (-\infty, 6] = [-3, 6]$$

The solution set of the compound inequality $x \geq -3$ and $x \leq 6$ can be expressed in several ways:

1. As a graph:
   $$\xleftarrow{\hspace{1em}} \underset{-3 \quad\ \ 0 \quad\ \ 6}{[\rule{2em}{0pt}]} \xrightarrow{\hspace{1em}}$$

2. In words: all real numbers from -3 to 6

3. In interval notation: $[-3, 6]$

4. Using set-builder notation: $\{x \mid x \geq -3 \ \text{and} \ x \leq 6\}$

EXAMPLE 2   Solve $\dfrac{x}{2} + 1 > 0$ and $2x - 3 < 5$. Graph the solution set and write it using interval notation and set-builder notation.

Strategy   We will solve each inequality separately. Then we will graph the two solution sets on the same number line and determine their intersection.

Why   The solution set of a compound inequality containing the word *and* is the intersection of the solution sets of the two inequalities.

Solution   In each case, we can use properties of inequality to isolate the variable on one side of the inequality.

$$\frac{x}{2} + 1 > 0 \qquad \text{and} \qquad 2x - 3 < 5 \qquad \text{\textcolor{red}{This is the compound inequality to solve.}}$$

$$\frac{x}{2} > -1 \qquad\qquad\qquad 2x < 8$$

$$x > -2 \qquad\qquad\qquad x < 4$$

When graphing on a number line, $(-2, 4)$ represents an *interval*. When graphing on a rectangular coordinate system, $(-2, 4)$ is an *ordered pair* that gives the coordinates of a point.

Next, we graph the solutions of each inequality on the same number line and determine their intersection.

Preliminary work to determine the graph of the solution set

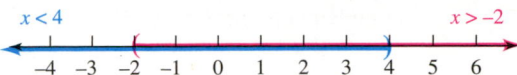

We see that the intersection of the graphs is the set of all real numbers between -2 and 4. The solution set of the compound inequality is the interval $(-2, 4)$, whose graph is shown below. This bounded interval, which does not include either endpoint, is called an **open interval.** Written using set-builder notation, the solution set is $\{x \mid x > -2 \text{ and } x < 4\}$.

The graph of the solution set

Self Check 2   Solve $3x > -18$ and $\frac{x}{5} - 1 \leq 1$. Graph the solution set and write it using interval notation.

Now Try ▶ **Problem 27**

The solution of the compound inequality in the Self Check of Example 2 is the interval $(-6, 10]$. A bounded interval such as this, which includes only one endpoint, is called a **half-open interval.** The following chart shows the various types of bounded intervals, along with the inequalities and interval notation that describe them.

Intervals			
	Open intervals	The interval (a, b) includes all real numbers x such that $a < x < b$.	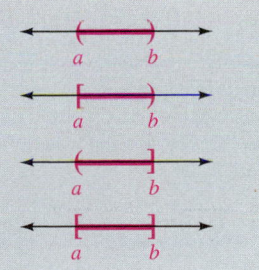
	Half-open intervals	The interval $[a, b)$ includes all real numbers x such that $a \leq x < b$.	
		The interval $(a, b]$ includes all real numbers x such that $a < x \leq b$.	
	Closed intervals	The interval $[a, b]$ includes all real numbers x such that $a \leq x \leq b$.	

EXAMPLE 3   Solve $x + 3 \leq 2x - 1$ and $3x - 2 < 5x - 4$. Graph the solution set and write it using interval notation and set-builder notation.

Strategy   We will solve each inequality separately. Then we will graph the two solution sets on the same number line and determine their intersection.

Why   The solution set of a compound inequality containing the word *and* is the intersection of the solution sets of the two inequalities.

Solution   In each case, we can use properties of inequality to isolate the variable on one side.

$$
\begin{array}{ll}
x + 3 \leq 2x - 1 \quad \text{and} & 3x - 2 < 5x - 4 \qquad \text{\textcolor{red}{This is the compound inequality to solve.}} \\
\quad\quad 4 \leq x & \quad\quad 2 < 2x \\
\quad\quad x \geq 4 & \quad\quad 1 < x \\
& \quad\quad x > 1
\end{array}
$$

The graph of $x \geq 4$ is shown below in red and the graph of $x > 1$ is shown below in blue.

Preliminary work to determine the graph of the solution set

Only those values of x where $x \geq 4$ and $x > 1$ are in the solution set of the compound inequality. Since all numbers greater than or equal to 4 are also greater than 1, the solutions are the numbers x where $x \geq 4$. The solution set is the interval $[4, \infty)$, whose graph is shown below. Written using set-builder notation, the solution set is $\{x \mid x \geq 4\}$.

The graph of the solution set

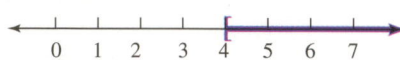

Self Check 3   Solve $2x + 3 < 4x + 2$ and $3x + 1 < 5x + 3$. Graph the solution set and write it using interval notation.

Now Try ▶ Problem 29

EXAMPLE 4   Solve $x - 1 > -3$ and $2x < -8$, if possible.

Strategy   We will solve each inequality separately. Then we will graph the two solution sets on the same number line and determine their intersection, if any.

Why   The solution set of a compound inequality containing the word *and* is the intersection of the solution sets of the two inequalities.

Solution   In each case, we can use properties of inequality to isolate the variable on one side.

$$x - 1 > -3 \quad \text{and} \quad 2x < -8 \qquad \textcolor{red}{\text{This is the compound inequality to solve.}}$$
$$x > -2 \qquad\qquad\qquad x < -4$$

> **Notation**
>
> The graphs of two linear inequalities can intersect at a single point, as shown below. The interval notation used to describe this point of intersection is $[3, 3]$.
>
>

The graphs of the solution sets shown below do not intersect. Since there are no numbers that make both parts of the original compound inequality true, $x - 1 > -3$ and $2x < -8$ has no solution.

Preliminary work to determine the graph of the solution set

The solution set of the compound inequality is the empty set, which can be written as $\varnothing$. Since there is no solution, a graph is not needed.

Self Check 4   Solve $2x - 3 < x - 2$ and $0 < x - 3.5$, if possible.

Now Try ▶ Problem 31

---

3 Solve Double Linear Inequalities.

Inequalities that contain exactly two inequality symbols are called **double inequalities**. An example is

$$-3 \leq 2x + 5 < 7 \qquad \textcolor{red}{\text{Read as "}-3 \text{ is less than or equal to } 2x + 5 \text{ and } 2x + 5 \text{ is less than 7."}}$$

Any double linear inequality can be written as a compound inequality containing the word *and*. In general, the following is true.

Double Linear Inequalities	The compound inequality $c < x < d$ is equivalent to $c < x$ and $x < d$.

Thus, the double inequality $-3 \le 2x + 5 < 7$ is a shorter form for the compound inequality

$$-3 \le 2x + 5 \quad \text{and} \quad 2x + 5 < 7$$

As you will see in Example 5, it is easier to solve the double inequality because it enables us to solve both inequalities at once.

EXAMPLE 5 Solve $-3 \le 2x + 5 < 7$. Graph the solution set and write it using interval notation and set-builder notation.

Strategy We will solve the double inequality by applying properties of inequality to *all three of its parts* to isolate x in the middle.

Why This double inequality $-3 \le 2x + 5 < 7$ means that $-3 \le 2x + 5$ and $2x + 5 < 7$. We can solve it more easily by leaving it in its original form.

Solution The objective is to isolate x in the middle of the double inequality.

$$-3 \le 2x + 5 < 7 \qquad \text{This is the double inequality to solve.}$$

$$-3 - 5 \le 2x + 5 - 5 < 7 - 5 \qquad \text{To undo the addition of 5, subtract}$$
$$\text{5 from all three parts.}$$

$$-8 \le 2x < 2 \qquad \text{Perform the subtractions.}$$

$$\frac{-8}{2} \le \frac{2x}{2} < \frac{2}{2} \qquad \text{To isolate } x, \text{ undo the multiplication by 2}$$
$$\text{by dividing all three parts by 2.}$$

$$-4 \le x < 1 \qquad \text{Perform the divisions.}$$

The solution set of the double linear inequality is the half-open interval $[-4, 1)$, whose graph is shown below. Written using set-builder notation, the solution set is $\{x \mid -4 \le x < 1\}$.

Self Check 5 Solve $-5 \le 3x - 8 \le 7$. Graph the solution set and write it using interval notation.

Now Try ▶ Problem 35

CAUTION When multiplying or dividing all three parts of a double inequality by a negative number, don't forget to reverse the direction of both inequalities. As an example, we solve $-15 < -5x \le 25$.

$$-15 < -5x \le 25$$

$$\frac{-15}{-5} > \frac{-5x}{-5} \ge \frac{25}{-5} \qquad \text{Divide all three parts by } -5 \text{ to isolate } x \text{ in}$$
$$\text{the middle. Reverse both inequality signs.}$$

$$3 > x \ge -5 \qquad \text{Perform the divisions.}$$

$$-5 \le x < 3 \qquad \text{Write an equivalent double inequality with}$$
$$\text{the smaller number, } -5, \text{ on the left.}$$

4 Solve Compound Inequalities Containing the Word *Or.*

A warning on the water temperature gauge of a commercial dishwasher cautions the operator to shut down the unit if

The water temperature
goes below 140°

or

The water temperature
goes above 160°

When two inequalities are joined with the word *or,* we also call the statement a compound inequality. Some examples are

$$x < 140 \quad \text{or} \quad x > 160$$
$$x \le -3 \quad \text{or} \quad x \ge 2$$
$$\frac{x}{3} > \frac{2}{3} \quad \text{or} \quad -(x - 2) > 3$$

The **solution set of a compound inequality containing the word** *or* includes all numbers that make one or the other or both inequalities true. That is, it is the union of their solution sets. We can find the solution set of $x \le -3$ or $x \ge 2$, for example, by drawing the graphs of each inequality on the same number line.

In the following figure, the graph of the solution set of $x \le -3$ is shown in red, and the graph of the solution set of $x \ge 2$ is shown in blue.

Preliminary work to determine the graph of the solution set

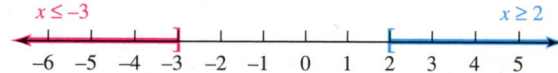

The figure below shows the graph of the solution set of $x \le -3$ or $x \ge 2$. This graph is a union of the graph of $x \le -3$ with the graph of $x \ge 2$.

The graph of the solution set

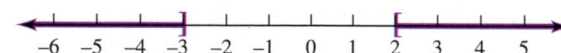

For the compound inequality $x \le -3$ or $x \ge 2$, we can write the solution set as the union of two intervals:

$$(-\infty, -3] \cup [2, \infty)$$

We can express the solution set of the compound inequality $x \le -3$ or $x \ge 2$ in several ways:

1. As a graph:
   -3   0   2

2. In words: all real numbers less than or equal to -3 *or* greater than or equal to 2

3. As the union of two intervals: $(-\infty, -3] \cup [2, \infty)$

4. Using set-builder notation: $\{x \mid x \le -3 \text{ or } x \ge 2\}$

EXAMPLE 6   Solve $\frac{x}{3} > \frac{2}{3}$ or $-(x - 2) > 3$. Graph the solution set and write it using interval notation and set-builder notation.

Strategy   We will solve each inequality separately. Then we will graph the two solution sets on the same number line to show their union.

Why   The solution set of a compound inequality containing the word *or* is the union of the solution sets of the two inequalities.

Solution   To solve each inequality, we proceed as follows:

$$\frac{x}{3} > \frac{2}{3} \quad \text{or} \quad -(x-2) > 3 \qquad \text{{\small\color{red}This is the compound inequality to solve.}}$$

$$x > 2 \qquad\qquad -x + 2 > 3$$
$$-x > 1$$
$$x < -1$$

The Language of Algebra
The meaning of the word **or** in a compound inequality differs from our everyday use of the word. For example, when we say, "I will go shopping today *or* tomorrow," we mean that we will go one day or the other, but *not* both. With compound inequalities, *or* includes one possibility, or the other, or both.

Next, we graph the solutions of each inequality on the same number line and determine their union.

Preliminary work to determine the graph of the solution set

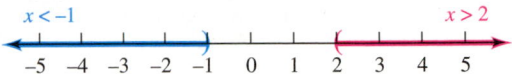

The union of the two solution sets consists of all real numbers less than -1 or greater than 2. The solution set of the compound inequality is the union of two intervals: $(-\infty, -1) \cup (2, \infty)$. Its graph appears below. Written using set-builder notation, the solution set is $\{x \mid x > 2 \text{ or } x < -1\}$.

The graph of the solution set

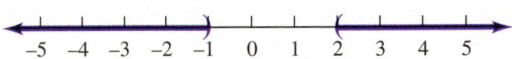

Self Check 6   Solve $\frac{x}{2} > 2$ or $-3(x-2) > 0$. Graph the solution set and write it using interval notation.

Now Try ▶   Problem 41

EXAMPLE 7   Solve $x + 3 \geq -3$ or $-x > 0$. Graph the solution set and write it using interval notation and set-builder notation.

Strategy   We will solve each inequality separately. Then we will graph the two solution sets on the same number line to show their union.

Why   The solution set of a compound inequality containing the word *or* is the union of the solution sets of the two inequalities.

Solution   To solve each inequality, we proceed as follows:

$$x + 3 \geq -3 \quad \text{or} \quad -x > 0 \qquad \text{{\small\color{red}This is the compound inequality to solve.}}$$
$$x \geq -6 \qquad\qquad x < 0$$

We graph the solution set of each inequality on the same number line and determine their union.

Preliminary work to determine the graph of the solution set

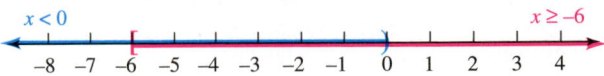

Since the entire number line is shaded, all real numbers satisfy the original compound inequality and the solution set is denoted as $(-\infty, \infty)$ or $\mathbb{R}$. Its graph is shown below. Written using set-builder notation, the solution set is $\{x \mid x \text{ is a real number}\}$.

The graph of the solution set

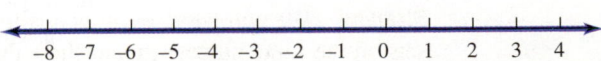

> **Self Check 7** Solve $x - 1 < 5$ or $-2x \leq 10$. Graph the solution set and write it using interval notation.
>
> **Now Try** ▶ Problem 43

Solving Compound Inequalities

▼

1. Solve each inequality separately and graph their solution sets in different colors on the same number line.

2. If the inequalities are connected with the word *and*, find the intersection of the two solution sets. If the inequalities are connected with the word *or*, find the union of the two solution sets.

3. Write the solution set of the compound inequality using interval notation or set-builder notation, and graph it on a new number line.

SECTION 4.2 ▶ STUDY SET

VOCABULARY

Fill in the blanks.

1. The _____ of two sets is the set of elements that are common to both sets and the _____ of two sets is the set of elements that are in one set, or the other, or both.

2. $x \geq 3$ and $x < 4$ is a _____ inequality.

3. $-6 < x + 1 \leq 1$ is a _____ linear inequality.

4. (2, 8) is an example of an open _____, [−4, 0] is an example of a _____ interval, and (0, 9] is an example of a half-_____ interval.

CONCEPTS

Fill in the blanks.

5. a. The solution set of a compound inequality containing the word *and* includes all numbers that make _____ inequalities true.

   b. The solution set of a compound inequality containing the word *or* includes all numbers that make _____, or the other, or _____ inequalities true.

6. The double inequality $4 < 3x + 5 \leq 15$ is equivalent to $4 < 3x + 5$ _____ $3x + 5 \leq 15$.

7. a. When solving a compound inequality containing the word *and*, the solution set is the _____ of the solution sets of the inequalities.

   b. When solving a compound inequality containing the word *or*, the solution set is the _____ of the solution sets of the inequalities.

8. When multiplying or dividing all three parts of a double inequality by a negative number, the direction of both inequality symbols must be _____.

9. Use a check to determine whether −3 is a solution of the compound inequality.

   a. $\dfrac{x}{3} + 1 \geq 0$ and $2x - 3 < -10$

   b. $2x \leq 0$ or $-3x < -5$

10. Use a check to determine whether −3 is a solution of the double linear inequality.

    a. $-1 < -3x + 4 < 12$

    b. $-1 < -3x + 4 < 14$

11. Use interval notation, if possible, to describe the intersection of each pair of graphs.

    a.

    b.

    c.

12. Use interval notation to describe the union of each pair of graphs.

    a.

    b.

    c.

NOTATION

13. Fill in the blanks: We read ∪ as _____ and ∩ as _____.

14. Match each interval with its corresponding graph.

    a. [2, 3)   i.

    b. (2, 3)   ii.

    c. [2, 3]   iii.

15. What set is represented by the interval notation $(-\infty, \infty)$? Graph it.

16. a. Graph: $(-\infty, 2) \cup [3, \infty)$

    b. Graph: $(-\infty, 3) \cap [-2, \infty)$

GUIDED PRACTICE

Let $A = \{0, 1, 2, 3, 4, 5, 6\}$, $B = \{4, 6, 8, 10\}$, $C = \{-3, -1, 0, 1, 2\}$, and $D = \{-3, 1, 2, 5, 8\}$. Find each set. See Example 1.

17. $A \cap B$

18. $A \cap D$

19. $C \cap D$

20. $B \cap C$

21. $B \cup C$

22. $A \cup C$

23. $A \cup D$

24. $C \cup D$

Solve each compound inequality, if possible. Graph the solution set (if one exists) and write it using interval notation. See Examples 2–4.

25. $x > -2$ and $x \leq 5$

26. $x \leq -4$ and $x \geq -7$

27. $2x - 1 > 3$ and $x + 8 \leq 11$

28. $5x - 3 \geq 2$ and $6 \geq 4x - 3$

29. $6x + 1 < 5x - 3$ and $\frac{x}{2} + 9 \leq 6$

30. $\frac{2}{3}x + 1 > -9$ and $\frac{3}{4}x - 1 > -10$

31. $x + 2 < -\frac{1}{3}x$ and $-6x < 9x$

32. $\frac{3}{2}x + \frac{1}{5} < 5$ and $2x + 1 > 9$

Solve each double inequality. Graph the solution set and write it using interval notation. See Example 5.

33. $4 \leq x + 3 \leq 7$

34. $-5.3 \leq x - 2.3 \leq -1.3$

35. $0.9 < 2x - 0.7 < 1.5$

36. $7 < 3x - 2 < 25$

Solve each compound inequality. Graph the solution set and write it using interval notation. See Examples 6 and 7.

37. $x \leq -2$ or $x > 6$

38. $x \geq -1$ or $x \leq -3$

39. $x - 3 < -4$ or $-x + 2 < 0$

40. $4x < -12$ or $\frac{x}{2} > 4$

41. $3x + 2 < 8$ or $2x - 3 > 11$

42. $3x + 4 < -2$ or $3x + 4 > 10$

43. $2x > x + 3$ or $\frac{x}{8} + 1 < \frac{13}{8}$

44. $2(x + 2) < x - 11$ or $-\frac{x}{5} < 20$

TRY IT YOURSELF

Solve each compound inequality, if possible. Graph the solution set (if one exists) and write it using interval notation.

45. $-4(x + 2) \geq 12$ or $3x + 8 < 11$

46. $4.5x - 1 < -10$ or $6 - 2x \geq 12$

47. $2.2x < -19.8$ and $-4x < 40$

48. $\frac{1}{2}x \leq 2$ and $0.75x \geq -6$

49. $-2 < -b + 3 < 5$

50. $2 < -t - 2 < 9$

51. $4.5x - 2 > 2.5$ or $\frac{1}{2}x \leq 1$

52. $0 < x$ or $3x - 5 > 4x - 7$

53. $5(x - 2) \geq 0$ and $-3x < 9$

54. $x - 1 \leq 2(x + 2)$ and $x \leq 2x - 5$

55. $-x < -2x$ and $3x > 2x$

56. $-\frac{x}{4} > -2.5$ and $9x > 2(4x + 5)$

57. $-6 < -3(x - 4) \leq 24$

58. $-4 \leq -2(x + 8) < 8$

59. $2x + 1 \geq 5$ and $-3(x + 1) \geq -9$

60. $2(-2) \leq 3x - 1$ and $3x - 1 \leq -1 - 3$

61. $\frac{4.5x - 12}{2} < x$ or $-15.3 > -3(x - 1.4)$

62. $y + 0.52 < 1.05y$ or $9.8 - 15y > -15.7$

63. $\frac{x}{0.7} + 5 > 4$ and $-4.8 \leq \frac{3x}{-0.125}$

64. $5(x + 1) \leq 4(x + 3)$ and $x + 12 < -3$

65. $-24 < \frac{3}{2}x - 6 \leq -15$

66. $-4 > \frac{2}{3}x - 2 > -6$

67. $\frac{x}{3} - \frac{x}{4} > \frac{1}{6}$ or $\frac{x}{2} + \frac{2}{3} \leq \frac{3}{4}$

68. $\frac{a}{2} + \frac{7}{4} > 5$ or $\frac{3}{8} + \frac{a}{3} \leq \frac{5}{12}$

69. $0 \leq \frac{4 - x}{3} \leq 2$

70. $-2 \leq \frac{5 - 3x}{2} \leq 2$

71. $x \leq 6 - \frac{1}{2}x$ and $\frac{1}{2}x + 1 \geq 3$

72. $3\left(x + \frac{2}{3}\right) \leq -7$ and $2(x + 2) \geq -2$

73. $-6 < f(x) \leq 0$ where $f(x) = 3x - 9$

74. $-3 < f(x) < 7$ where $f(x) = \frac{2}{3}x - \frac{1}{3}$

75. Let $f(x) = 5x + 14$ and $g(x) = 2x + 8$. Find all values of x for which $f(x) > 29$ and $g(x) < 20$.

76. Let $f(x) = x - 2$. Find all values of x for which $f(x) > 5$ or $f(x) < -1$.

Look Alikes . . .

Solve the inequality in part a. Graph the solution set and write it in interval notation. Then use your work from part a to determine the solution set for the compound inequality in part b. (No new work is necessary!) Graph the solution set and write it in interval notation.

77. a. $3x - 2 \geq 4$ and $x + 6 \geq 12$
  b. $3x - 2 \geq 4$ or $x + 6 \geq 12$

78. a. $x + 2 \le 10$ or $x - 3 \ge 2$

   b. $x + 2 \le 10$ and $x - 3 \ge 2$

79. a. $2x + 1 \le 7$ and $3x + 5 \ge 23$

   b. $2x + 1 \le 7$ or $3x + 5 \ge 23$

80. a. $7 \le 4x + 1 < 23$

   b. $7 < 4x + 1 \le 23$

APPLICATIONS

81. Baby Furniture. Refer to the illustration. A company manufactures various sizes of play yard cribs having perimeters between 128 and 192 inches, inclusive.

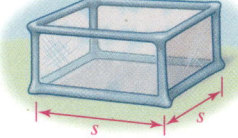

a. Complete the double inequality that describes the range of the perimeters of the play yard shown.

$$\boxed{} \le 4s \le \boxed{}$$

b. Solve the double inequality to find the range of the side lengths of the play yard.

82. Trucking. The distance that a truck can travel in 8 hours, at a constant rate of r mph, is given by $8r$. A trucker wants to travel at least 350 miles, and company regulations don't allow him to exceed 450 miles in one 8-hour shift.

a. Complete the double inequality that describes the mileage range of the truck.

$$\boxed{} \le 8r \le \boxed{}$$

b. Solve the double inequality to find the range of the average rate (speed) of the truck for the 8-hour trip.

83. from **Campus to Careers**

Heating, Ventilation, and Air Conditioning Technician

During business hours, as shown in figure (a), the *Temp range* control in an office is set at 5. This means that the heater comes on when the room temperature gets 5 degrees below the *Temp setting* and the air conditioner comes on when the room temperature gets 5 degrees above the *Temp setting*.

a. Use interval notation to describe the temperature range when neither the heater nor the air conditioner will come on during business hours.

b. After business hours, the *Temp range* setting is changed to save energy. See figure (b). Use interval notation to describe the after-business-hours temperature range when neither the heater nor the air conditioner will come on.

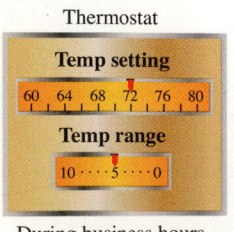
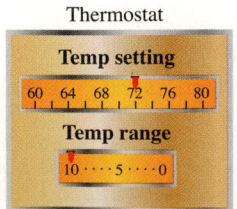

During business hours (a)   After business hours (b)

84. Treating Fevers. Use the flow chart to determine what action should be taken for a 13-month-old child who has had a 99.8° temperature for 3 days and is not suffering any other symptoms. T represents the child's temperature, A the child's age in months, and S the number of hours the child has experienced the symptoms.

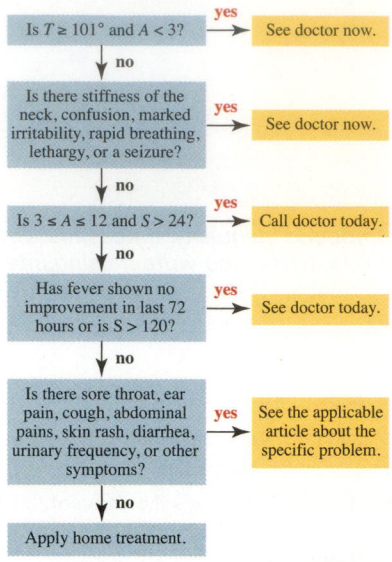

Based on information from *Take Care of Yourself* (Addison-Wesley, 1993)

85. U.S. Health Care. Refer to the following graph. Let P represent the percent of children covered by private insurance, M the percent covered by Medicaid, and N the percent not covered. For what years are the following true?

a. $P \ge 63$ and $M \ge 26$

b. $P > 60$ or $M \ge 29$

c. $M \ge 29$ and $N \le 10$

d. $M \ge 30$ or $N < 9.2$

U.S. Health Care Coverage for People Under 18 Years of Age (in percent)

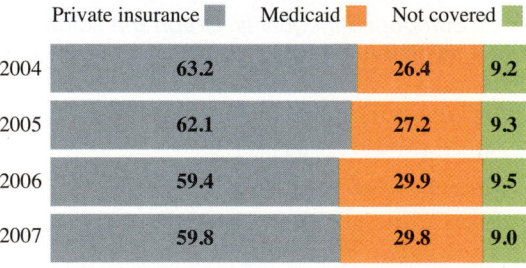

	Private insurance	Medicaid	Not covered
2004	63.2	26.4	9.2
2005	62.1	27.2	9.3
2006	59.4	29.9	9.5
2007	59.8	29.8	9.0

Source: U.S. Department of Health and Human Services

(The small percent of people covered each year by some nonstandard program is not shown.)

86. Polls. For each response to the poll question shown below, the *margin of error* is +/− (read as "plus or minus") 2.8%. This means that for the statistical methods used to do the polling, the actual response could be as much as 2.8 points more or 2.8 points less than shown. Use interval notation to describe the possible interval (in percent) for each response.

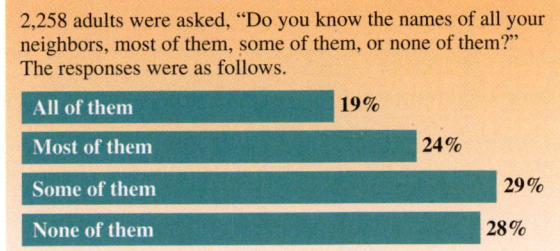

2,258 adults were asked, "Do you know the names of all your neighbors, most of them, some of them, or none of them?" The responses were as follows.

All of them	19%
Most of them	24%
Some of them	29%
None of them	28%

87. Street Intersections. Refer to figure (a) below.
  a. Shade the area that represents the intersection of the two streets shown in the illustration.
  b. Shade the area that represents the union of the two streets.

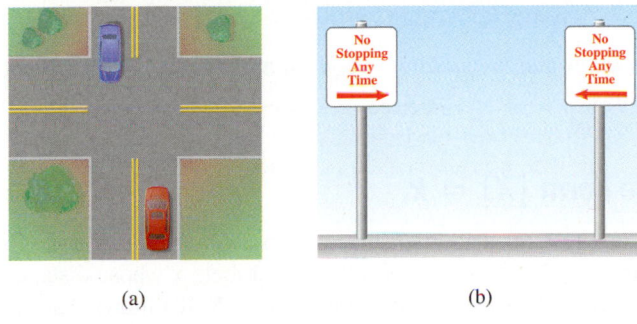

(a)   (b)

88. Traffic Signs. The pair of signs shown in figure (b) above are a real-life example of which concept discussed in this section?

WRITING

89. Explain how to find the union and how to find the intersection of $(-\infty, 5)$ and $(-2, \infty)$ graphically.

90. Explain why the double inequality $2 < x < 8$ can be written in the equivalent form $2 < x$ and $x < 8$.

91. Explain the meaning of the notation $(-1, 2)$ for each type of graph.

a.
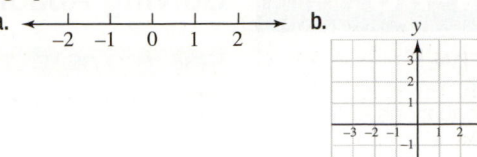

b.

92. The meaning of the word *or* in a compound inequality differs from our everyday use of the word. Explain the difference.

93. Describe each set in words.
  a. $(-3, 3)$ **b.** $[7, 12]$
  c. $(-\infty, 5] \cup (6, \infty)$

94. What is incorrect about the double inequality $3 < -3x + 4 < -3$?

REVIEW

95. Earthquakes. In 2008, the number of recorded earthquakes in the United States was 3,618. In 2009, the total was 4,264. What was the percent increase in the number of earthquakes from 2008 to 2009? Round to the nearest percent. (Source: U.S. Geological Survey)

96. Earthquakes. In 2005, the total number of recorded earthquakes in the United States was 3,685. In 2006, the total was 2,783. What was the percent decrease in the number of earthquakes from 2005 to 2006? Round to the nearest percent. (Source: U.S. Geological Survey)

CHALLENGE PROBLEMS

Solve each compound inequality. Graph the solution set and write it in interval notation.

97. $-5 < \dfrac{x + 2}{-2} < 0$ or $2x + 10 \geq 30$

98. $-2 \leq \dfrac{x - 4}{3} \leq 0$ and $\dfrac{x - 5}{2} \geq -3$

99. $x - 12 < 4x < 2x + 16$

100. $6(x - 3) \leq 3(3x + 2) \leq 4(2x + 3)$

SECTION 4.3

Solving Absolute Value Equations and Inequalities

OBJECTIVES

1 Solve equations of the form $|X| = k$.

2 Solve equations with two absolute values.

3 Solve inequalities of the form $|X| < k$.

4 Solve inequalities of the form $|X| > k$.

ARE YOU READY?

The following problems review some basic skills that are needed to solve absolute value equations and inequalities.

1. Find each absolute value.
   a. $|12|$ b. $|-7.5|$

2. Tell whether each statement is true or false.
   a. $|-3| \geq 2$ b. $|-26| < -27$

3. Solve $3x + 6 \leq -3$ or $3x + 6 \geq 9$. Graph the solution set and write it using interval notation.

4. Solve $-8 < 2x + 8 < 16$. Graph the solution set and write it using interval notation.

Many quantities studied in mathematics, science, and engineering are expressed as positive numbers. To guarantee that a quantity is positive, we often use absolute value. In this section, we will consider equations and inequalities involving the absolute value of an algebraic expression. Some examples are

$$|3x - 2| = 5, \qquad |2x - 3| < 9, \qquad \text{and} \qquad \left|\frac{3 - x}{5}\right| \geq 6$$

To solve these *absolute value equations* and *inequalities,* we write and then solve equivalent compound equations and inequalities.

1 Solve Equations of the Form $|X| = k$.

Recall that the absolute value of a real number is its distance from 0 on a number line. To solve the **absolute value equation** $|x| = 5$, we must find all real numbers x whose distance from 0 on the number line is 5. There are two such numbers: 5 and -5. It follows that the solutions of $|x| = 5$ are 5 and -5 and the solution set is $\{5, -5\}$.

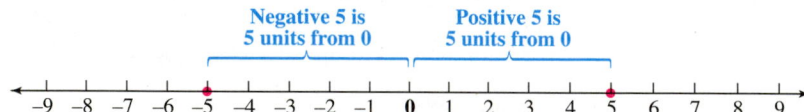

The results from this example suggest the following approach for solving absolute value equations.

Solving Absolute Value Equations	For any positive number k and any algebraic expression X: To solve $|X| = k$, solve the equivalent compound equation $$X = k \qquad \text{or} \qquad X = -k$$

The statement $X = k$ or $X = -k$ is called a **compound equation** because it consists of two equations joined with the word *or*.

EXAMPLE 1   Solve: **a.** $|s| = 0.003$ **b.** $|3x - 2| = 5$ **c.** $|10 - x| = -40$

Strategy   To solve the first two equations, we will write and then solve an equivalent compound equation. We will solve the third equation by inspection.

Why   All three of the equations are of the form $|X| = k$. However, the standard method for solving absolute value equations cannot be applied to $|10 - x| = -40$ because k is negative.

Solution   **a.** The absolute value equation $|s| = 0.003$ is equivalent to the compound equation

The Language of Algebra

When we say that the absolute value equation and a compound equation are **equivalent,** we mean that they have the same solution(s).

$$s = 0.003 \quad \text{or} \quad s = -0.003$$

Therefore, the solutions of $|s| = 0.003$ are 0.003 and -0.003, and the solution set is $\{0.003, -0.003\}$.

b. The equation-solving method used in part a can be extended to equations where the expression within absolute value bars is more complicated than a single variable. The absolute value equation $|3x - 2| = 5$ is equivalent to the compound equation

$$3x - 2 = 5 \quad \text{or} \quad 3x - 2 = -5$$

Now we solve each equation for x:

$$
\begin{array}{lll}
3x - 2 = 5 & \quad \text{or} \quad & 3x - 2 = -5 \\
3x = 7 & & 3x = -3 \\
x = \dfrac{7}{3} & & x = -1
\end{array}
$$

Caution

When you see the word *solve,* you probably think of steps such as combining like terms, distributing, or doing something to both sides. However, to solve absolute value equations of the form $|X| = k$, those familiar steps are not used until an equivalent compound equation is written.

The results must be checked separately to see whether each of them produces a true statement. We substitute $\frac{7}{3}$ for x and then -1 for x in the original equation.

$$
\text{Check:} \qquad \text{For } x = \frac{7}{3} \qquad\qquad\qquad \text{For } x = -1
$$

$$
\begin{array}{ll}
|3x - 2| = 5 & \qquad\qquad |3x - 2| = 5 \\
\left|3\left(\dfrac{7}{3}\right) - 2\right| \stackrel{?}{=} 5 & \qquad\qquad |3(-1) - 2| \stackrel{?}{=} 5 \\
|7 - 2| \stackrel{?}{=} 5 & \qquad\qquad |-3 - 2| \stackrel{?}{=} 5 \\
|5| \stackrel{?}{=} 5 & \qquad\qquad |-5| \stackrel{?}{=} 5 \\
5 = 5 \quad \text{True} & \qquad\qquad 5 = 5 \quad \text{True}
\end{array}
$$

The resulting true statements indicate that the equation has two solutions: $\frac{7}{3}$ and -1. The solution set is $\left\{\frac{7}{3}, -1\right\}$.

c. Since an absolute value can never be negative, there are no real numbers x that make $|10 - x| = -40$ true. The equation has no solution and the solution set is $\varnothing$.

Self Check 1   Solve: **a.** $|x| = \frac{1}{2}$ **b.** $|2x - 3| = 7$

c. $\left|\frac{x}{4} - 1\right| = -3$

Now Try   Problems 19, 23, and 29

When solving absolute value equations (or inequalities), we must **isolate the absolute value expression on one side** before writing the equivalent compound statement.

EXAMPLE 2   Solve: $\left|\dfrac{2}{3}x + 3\right| + 4 = 10$

Strategy   We will first isolate $\left|\frac{2}{3}x + 3\right|$ on the left side of the equation and then write and solve an equivalent compound equation.

Why   After isolating the absolute value expression on the left, the resulting equation will have the desired form $|X| = k$.

Solution

$$\left|\frac{2}{3}x + 3\right| + 4 = 10 \qquad \text{This is the equation to solve.}$$

$$\left|\frac{2}{3}x + 3\right| = 6 \qquad \text{To isolate the absolute value expression, subtract 4 from both sides. The resulting equation is in the form } |X| = k.$$

With the absolute value now isolated, we can solve $\left|\frac{2}{3}x + 3\right| = 6$ by writing and solving an equivalent compound equation:

$$\frac{2}{3}x + 3 = 6 \qquad \text{or} \qquad \frac{2}{3}x + 3 = -6$$

Now we solve each equation for x:

$$
\begin{array}{ccc}
\frac{2}{3}x + 3 = 6 & \text{or} & \frac{2}{3}x + 3 = -6 \\[6pt]
\frac{2}{3}x = 3 & & \frac{2}{3}x = -9 \\[6pt]
2x = 9 & & 2x = -27 \\[6pt]
x = \frac{9}{2} & & x = -\frac{27}{2}
\end{array}
$$

Verify that both $\frac{9}{2}$ and $-\frac{27}{2}$ are solutions by substituting them into the original equation.

Self Check 2 Solve: $|0.4x - 2| - 0.6 = 0.4$

Now Try ▶ Problem 33

EXAMPLE 3 Solve: $3\left|\frac{1}{2}x - 5\right| - 4 = -4$

Strategy We will first isolate $\left|\frac{1}{2}x - 5\right|$ on the left side of the equation and then write and solve an equivalent compound equation.

Why After isolating the absolute value expression, the resulting equation will have the desired form $|X| = k$.

Solution

$$3\left|\frac{1}{2}x - 5\right| - 4 = -4 \qquad \text{This is the equation to solve.}$$

$$3\left|\frac{1}{2}x - 5\right| = 0 \qquad \text{To isolate the absolute value expression } \left|\frac{1}{2}x - 5\right|, \text{ we first add 4 to both sides.}$$

$$\left|\frac{1}{2}x - 5\right| = 0 \qquad \text{To complete the process to isolate the absolute value expression, divide both sides by 3. The resulting equation is in the form } |X| = k.$$

Since 0 is the only number whose absolute value is 0, the expression $\frac{1}{2}x - 5$ must be 0.

$$\frac{1}{2}x - 5 = 0 \qquad \text{Set the expression within the absolute value bars equal to 0 and solve for } x.$$

$$\frac{1}{2}x = 5 \qquad \text{To isolate the variable term } \frac{1}{2}x, \text{ add 5 to both sides.}$$

$$x = 10 \qquad \text{To isolate } x, \text{ multiply both sides by 2.}$$

The solution is 10 and the solution set is $\{10\}$. Verify that it satisfies the original equation.

Self Check 3   Solve: $-5\left|\frac{2}{3}x + 4\right| + 1 = 1$

Now Try   Problem 39

In Section 2.6 we discussed absolute value functions and their graphs. If we are given an output of an absolute value function, we can work in reverse to find the corresponding input(s).

EXAMPLE 4   Let $f(x) = |x + 4|$. For what value(s) of x is $f(x) = 20$?

Strategy   We will substitute 20 for $f(x)$ and solve for x.

Why   In the equation, there are two unknowns, x and $f(x)$. If we replace $f(x)$ with 20, we can solve the resulting absolute value equation for x.

Solution

$f(x) = |x + 4|$ This is the given function.

$20 = |x + 4|$ Substitute 20 for f(x).

$|x + 4| = 20$ Rewrite the equation so that the absolute value expression is on the left side.

To solve $|x + 4| = 20$, we write and then solve an equivalent compound equation:

$x + 4 = 20$ or $x + 4 = -20$

Now we solve each equation for x:

$x + 4 = 20$ or $x + 4 = -20$
$x = 16$ | $x = -24$

The values of x for which $f(x) = 20$ are 16 and -24. To check, find $f(16)$ and $f(-24)$, and verify that the result is 20 in each case.

Self Check 4   Let $f(x) = |x + 4|$. For what value(s) of x is $f(x) = 11$?

Now Try   Problem 71

2   Solve Equations with Two Absolute Values.

Equations can contain two absolute value expressions. To develop a strategy to solve them, consider the following four true statements.

$|3| = |3|$ or $|-3| = |-3|$ or $|3| = |-3|$ or $|-3| = |3|$

The numbers are the same.   The numbers are the same.   The numbers are opposites.   The numbers are opposites.

These four possible cases are really just two cases: *Two absolute value expressions are equal when the expressions within the absolute value bars are equal to or opposites of each other.* This observation suggests the following approach for solving equations having two absolute value expressions.

Solving Equations with Two Absolute Values

For any algebraic expressions X and Y:

To solve $|X| = |Y|$, solve the compound equation $X = Y$ or $X = -Y$.

EXAMPLE 5 Solve: $|5x + 3| = |3x + 25|$

Strategy To solve this equation, we will write and then solve an equivalent compound equation.

Why We can use this approach because the equation is of the form $|X| = |Y|$.

Solution The equation $|5x + 3| = |3x + 25|$, with the two absolute value expressions, is equivalent to the following compound equation:

The expressions within the absolute value symbols are equal

The expressions within the absolute value symbols are opposites

$$5x + 3 = 3x + 25 \qquad \text{or} \qquad 5x + 3 = -(3x + 25)$$
$$2x = 22 \qquad\qquad\qquad\qquad 5x + 3 = -3x - 25 \quad \text{Solve each equation.}$$
$$x = 11 \qquad\qquad\qquad\qquad\qquad 8x = -28$$
$$x = -\frac{28}{8}$$
$$x = -\frac{7}{2} \quad \text{Simplify the fraction.}$$

> **Caution**
>
> Don't forget to use parentheses to write the opposite of expressions that have more than one term.
>
> | Expression | Opposite |
> |---|---|
> | $3x + 25$ | $-(3x + 25)$ |

Verify that both solutions, 11 and $-\frac{7}{2}$, check by substituting them into the original equation. The solution set is $\left\{11, -\frac{7}{2}\right\}$.

Self Check 5 Solve: $|2x - 3| = |4x + 9|$

Now Try ▶ Problem 47

3 Solve Inequalities of the Form $|X| < k$.

To solve the **absolute value inequality** $|x| < 5$, we must find all real numbers x whose distance from 0 on the number line is less than 5. From the graph, we see that there are many such numbers. For example, -4.999, -3, -2.4, $-1\frac{7}{8}$, $-\frac{3}{4}$, 0, 1, 2.8, 3.001, and 4.999 all meet this requirement. We conclude that the solution set is all numbers between -5 and 5, which can be written in interval notation as $(-5, 5)$.

The real numbers in this interval are less than 5 units from 0

Since x is between -5 and 5, it follows that $|x| < 5$ is equivalent to $-5 < x < 5$. This observation suggests the following approach for solving absolute value inequalities of the form $|X| < k$ and $|X| \leq k$.

Solving $\|X\| < k$ **and** $\|X\| \leq k$	For any positive number k and any algebraic expression X:
	To solve $\|X\| < k$, solve the equivalent double inequality $-k < X < k$.
	To solve $\|X\| \leq k$, solve the equivalent double inequality $-k \leq X \leq k$.

EXAMPLE 6

Solve $|2x - 3| < 9$ and graph the solution set.

Strategy To solve this absolute value inequality, we will write and solve an equivalent double inequality.

Why We can use this approach because the inequality is of the form $|X| < k$, and k is positive.

Solution The absolute value inequality $|2x - 3| < 9$ is equivalent to the double inequality

$$-9 < 2x - 3 < 9$$

which we can solve for x:

$$-9 < 2x - 3 < 9$$
$$-6 < 2x < 12 \qquad \text{To isolate the variable term 2x, add 3 to all three parts.}$$
$$-3 < x < 6 \qquad \text{To isolate x, divide all parts by 2.}$$

Any number between -3 and 6 is in the solution set, which can be written as $\{x \mid -3 < x < 6\}$. This is the interval $(-3, 6)$; its graph is shown on the right.

Self Check 6 Solve $|3x + 2| < 4$ and graph the solution set.

Now Try ▶ Problems 55 and 59

Because it is related to distance, absolute value can be used to describe the amount of error involved when measurements are taken.

EXAMPLE 7

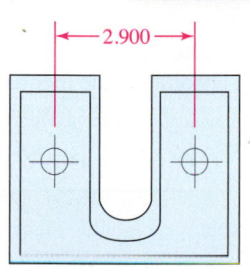

Unless otherwise specified, dimensions are in inches.
Tolerances ±0.015

Tolerances. When manufactured parts are inspected by a quality control engineer, they are classified as acceptable if each dimension falls within a given **tolerance range** of the dimensions listed on the blueprint. For the bracket shown in the margin, the distance between the two drilled holes is given as 2.900 inches. Because the tolerance is ± 0.015 inch, this distance can be as much as 0.015 inch longer or 0.015 inch shorter, and the part will be considered acceptable. The acceptable distance d between holes can be represented by the absolute value inequality $|d - 2.900| \le 0.015$. Solve the inequality and explain the result.

Strategy To solve $|d - 2.900| \le 0.015$, we will write and solve an equivalent double inequality.

Why We can use this approach because the inequality is of the form $|X| \le k$, and k is positive.

Solution The absolute value inequality $|d - 2.900| \le 0.015$ is equivalent to the double inequality

$$-0.015 \le d - 2.900 \le 0.015$$

which we can solve for d:

$$-0.015 \le d - 2.900 \le 0.015$$
$$2.885 \le d \le 2.915 \qquad \text{To isolate d, add 2.900 to all three parts.}$$

The solution set is the interval $[2.885, 2.915]$. This means that the distance between the two holes should be between 2.885 and 2.915 inches, inclusive. If the distance is less than 2.885 inches or more than 2.915 inches, the part should be rejected.

Self Check 7 **Tolerances.** Refer to Example 7. Find the tolerance range if the tolerance is ± 0.0015.

Now Try ▶ Problem 105

EXAMPLE 8 Solve: $|4x - 5| < -2$

Strategy We will solve this inequality by inspection.

Why The inequality $|4x - 5| < -2$ is of the form $|X| < k$. However, the standard method for solving such inequalities cannot be used because k (in this case, -2) is not positive.

Solution Since $|4x - 5|$ is always greater than or equal to 0 for any real number x, this absolute value inequality has no solution. The solution set is $\varnothing$.

> **Self Check 8** Solve: $|6x + 24| < -51$
>
> **Now Try** ▶ Problem 61

4 Solve Inequalities of the Form $|X| > k$.

To solve the absolute value inequality $|x| > 5$, we must find all real numbers x whose distance from 0 on the number line is greater than 5. From the following graph, we see that there are many such numbers. For example, -5.001, -6, -7.5, and $-8\frac{3}{8}$, as well as 5.001, 6.2, 7, 8, and $9\frac{1}{2}$ all meet this requirement. We conclude that the solution set is all numbers less than -5 or greater than 5, which can be written as the union of two intervals: $(-\infty, -5) \cup (5, \infty)$.

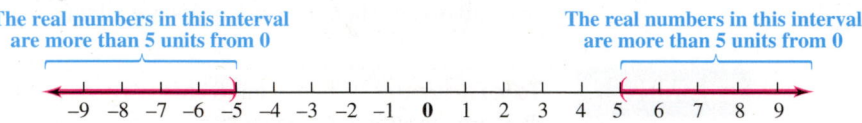

The real numbers in this interval are more than 5 units from 0

The real numbers in this interval are more than 5 units from 0

Since x is less than -5 or greater than 5, it follows that $|x| > 5$ is equivalent to $x < -5$ or $x > 5$. This observation suggests the following approach for solving absolute value inequalities of the form $|X| > k$ and $|X| \geq k$.

Solving $|X| > k$ and $|X| \geq k$

For any positive number k and any algebraic expression X:

To solve $|X| > k$, solve the equivalent compound inequality $X < -k$ or $X > k$.

To solve $|X| \geq k$, solve the equivalent compound inequality $X \leq -k$ or $X \geq k$.

EXAMPLE 9 Solve $\left|\dfrac{3 - x}{5}\right| \geq 6$ and graph the solution set.

Strategy To solve this absolute value inequality, we will write and solve an equivalent compound inequality.

Why We can use this approach because the inequality is of the form $|X| \geq k$, and k is positive.

Solution The absolute value inequality $\left|\dfrac{3 - x}{5}\right| \geq 6$ is equivalent to the compound inequality

$$\frac{3 - x}{5} \leq -6 \quad \text{or} \quad \frac{3 - x}{5} \geq 6$$

Now we solve each inequality for x:

$$\frac{3-x}{5} \le -6 \qquad \text{or} \qquad \frac{3-x}{5} \ge 6$$

$$3 - x \le -30 \qquad\qquad 3 - x \ge 30 \qquad \text{\color{red}{To clear the fraction, multiply both sides by 5.}}$$

$$-x \le -33 \qquad\qquad -x \ge 27 \qquad \text{\color{red}{To isolate the variable term} } -x\text{,}$$
$$\text{\color{red}{subtract 3 from both sides.}}$$

$$x \ge 33 \qquad\qquad\qquad x \le -27 \qquad \text{\color{red}{To isolate x, divide both sides by -1 and reverse}}$$
$$\text{\color{red}{the direction of the inequality symbol.}}$$

The solution set is the union of two intervals: $(-\infty, -27] \cup [33, \infty)$. Using set-builder notation, the solution set is written as $\{x \mid x \le -27 \text{ or } x \ge 33\}$. Its graph appears on the right.

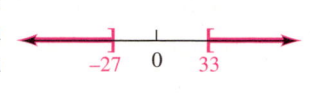

Self Check 9 Solve $\left|\dfrac{2-x}{4}\right| \ge 1$ and graph the solution set.

Now Try ▶ Problems 63 and 65

EXAMPLE 10 Solve $6 < \left|\dfrac{2}{3}x - 2\right| - 3$ and graph the solution set.

Strategy We will first write the inequality in an equivalent form with the absolute value on the left side.

Why It's usually easier to solve an absolute value inequality if the absolute value appears on the left side of the inequality.

Solution

$$6 < \left|\frac{2}{3}x - 2\right| - 3 \qquad \text{\color{red}{This is the inequality to solve.}}$$

$$\left|\frac{2}{3}x - 2\right| - 3 > 6 \qquad \text{\color{red}{Write the inequality with the absolute value on the left side.}}$$

$$\left|\frac{2}{3}x - 2\right| > 9 \qquad \text{\color{red}{Add 3 to both sides to isolate the absolute value expression.}}$$

After isolating the absolute value expression on the left side, the resulting inequality has the form $|X| > k$. To solve this absolute value inequality, we write and solve an equivalent compound inequality:

$$\frac{2}{3}x - 2 < -9 \qquad \text{or} \qquad \frac{2}{3}x - 2 > 9$$

$$\frac{2}{3}x < -7 \qquad\qquad \frac{2}{3}x > 11 \qquad \text{\color{red}{Add 2 to both sides.}}$$

$$2x < -21 \qquad\qquad 2x > 33 \qquad \text{\color{red}{Multiply both sides by 3.}}$$

$$x < -\frac{21}{2} \qquad\qquad x > \frac{33}{2} \qquad \text{\color{red}{To isolate x, divide both sides by 2.}}$$

The solution set is the union of two intervals: $\left(-\infty, -\dfrac{21}{2}\right) \cup \left(\dfrac{33}{2}, \infty\right)$. Its graph appears on the right.

Using set-builder notation, the solution set can be written as $\left\{x \mid x < -\dfrac{21}{2} \text{ or } x > \dfrac{33}{2}\right\}$.

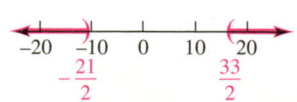

Self Check 10   Solve $3 < \left|\frac{3}{4}x + 2\right| - 1$ and graph the solution set.

Now Try   Problem 67

EXAMPLE 11   Solve $\left|\dfrac{x}{8} - 1\right| \geq -4$ and graph the solution set.

Strategy   We will solve this inequality by inspection.

Why   The inequality $\left|\frac{x}{8} - 1\right| \geq -4$ is of the form $|x| \geq k$. However, the standard method for solving such inequalities cannot be used because k is not positive.

Solution   Since $\left|\frac{x}{8} - 1\right|$ is always greater than or equal to 0 for any real number x it will also be greater than or equal to -4 for any real number x. Therefore, this absolute value inequality is true for all real numbers. The solution set is the interval $(-\infty, \infty)$ or $\mathbb{R}$. Its graph appears on the right.

Self Check 11   Solve $|-x - 9| > -0.5$ and graph the solution set.

Now Try   Problem 69

The following summary shows how we can interpret absolute value in three ways. Assume $k > 0$.

Geometric description	*Graphic description*	*Algebraic description*
1. $\|x\| = k$ means that x is k units from 0 on the number line.	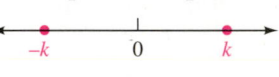	$\|x\| = k$ is equivalent to $x = k$ or $x = -k$.
2. $\|x\| < k$ means that x is less than k units from 0 on the number line.	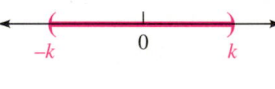	$\|x\| < k$ is equivalent to $-k < x < k$.
3. $\|x\| > k$ means that x is more than k units from 0 on the number line.	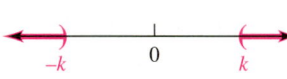	$\|x\| > k$ is equivalent to $x > k$ or $x < -k$.

Using Your Calculator ▶ Solving Absolute Value Equations and Inequalities

We can solve absolute value equations and inequalities with a graphing calculator. For example, to solve $|2x - 3| = 9$, we graph the equations $y = |2x - 3|$ and $y = 9$ on the same coordinate system, as shown in the figure. The equation $|2x - 3| = 9$ will be true for all x-coordinates of points that lie on *both* graphs. Using the TRACE or the INTERSECT feature, we can see that the graphs intersect at the points $(-3, 9)$ and $(6, 9)$. Thus, the solutions of the absolute value equation are -3 and 6.

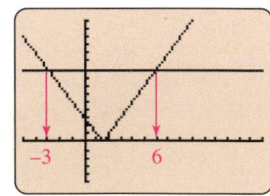

The inequality $|2x - 3| < 9$ will be true for all x-coordinates of points that lie on the graph of $y = |2x - 3|$ and *below* the graph of $y = 9$. We see that these values of x are between -3 and 6. Thus, the solution set is the interval $(-3, 6)$.

The inequality $|2x - 3| > 9$ will be true for all x-coordinates of points that lie on the graph of $y = |2x - 3|$ and *above* the graph of $y = 9$. We see that these values of x are less than -3 or greater than 6. Thus, the solution set is the union of two intervals: $(-\infty, -3) \cup (6, \infty)$.

VOCABULARY

Fill in the blanks.

1. The _____ _____ of a number is its distance from 0 on a number line.
2. $|2x - 1| = 10$ is an absolute value _____ and $|2x - 1| > 10$ is an absolute value _____.
3. To _____ the absolute value in $|3 - x| - 4 = 5$, we add 4 to both sides.
4. When we say that the absolute value equation and a compound equation are equivalent, we mean that they have the same _____.
5. When two equations are joined by the word *or*, such as $x + 1 = 5$ or $x + 1 = -5$, we call the statement a _____ equation.
6. $f(x) = |6x - 2|$ is called an absolute value _____.

CONCEPTS

Fill in the blanks.

7. To solve absolute value equations and inequalities, we write and solve equivalent _____ equations and inequalities.
8. Two absolute value expressions are equal when the expressions within the absolute value bars are equal to or _____ of each other.
9. Consider the following real numbers:
$-3, -2.01, -2, -1.99, -1, 0, 1, 1.99, 2, 2.01, 3$
   a. Which of them make $|x| = 2$ true?
   b. Which of them make $|x| < 2$ true?
   c. Which of them make $|x| > 2$ true?
10. Determine whether -3 is a solution of the given equation or inequality.
    a. $|x - 1| = 4$ b. $|x - 1| > 4$
    c. $|x - 1| \leq 4$ d. $|5 - x| = |x + 12|$
11. For each absolute value equation, write an equivalent compound equation.
    a. $|x - 7| = 8$ is equivalent to

    $x - 7 = \boxed{}$ or $x - 7 = \boxed{}$

    b. $|x + 10| = |x - 3|$ is equivalent to

    $x + 10 = \boxed{}$ or $x + 10 = \boxed{}$

12. For each absolute value inequality, write an equivalent compound inequality.
    a. $|x + 5| < 1$ is equivalent to

    $\boxed{} < x + 5 < \boxed{}$

    b. $|x - 6| \geq 3$ is equivalent to

    $x - 6 \leq \boxed{}$ or $x - 6 \geq \boxed{}$

13. For each absolute value equation or inequality, write an equivalent compound equation or inequality.
    a. $|x| = 8$ b. $|x| \geq 8$
    c. $|x| \leq 8$ d. $|5x - 1| = |x + 3|$
14. Perform the necessary steps to isolate the absolute value expression on one side of the equation. *Do not solve.*
    a. $|3x + 2| - 7 = -5$
    b. $6 + 2|5x - 19| \leq 40$
15. Determine the solution set of each absolute value equation or inequality by inspection. (No work is necessary.) Your answer should be either *all real numbers* or *no solution*.
    a. $|7x + 6| = -8$ b. $|7x + 6| \leq -8$
    c. $|7x + 6| \geq -8$
16. Write the inequality $10 > |16x - 3|$ in an equivalent form with the absolute value expression on the left side.

NOTATION

17. Match each equation or inequality with its graph.
    a. $|x| = 1$ i.
    b. $|x| > 1$ ii.
    c. $|x| < 1$ iii.

18. Describe the set graphed below using interval notation.

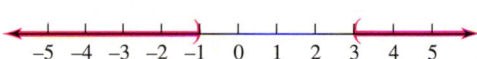

GUIDED PRACTICE

Solve each equation. See Example 1.

19. $|x| = 23$ 20. $|x| = 90$

21. $|x - 5| = 8$ 22. $|x - 7| = 4$

23. $|3x + 2| = 16$ 24. $|5x - 3| = 22$

25. $\left|\dfrac{x}{5}\right| = 10$ 26. $\left|\dfrac{x}{7}\right| = 2$

27. $|2x + 3.6| = 9.8$ 28. $|4x - 24.8| = 32.4$

29. $\left|\dfrac{7}{2}x + 3\right| = -5$ 30. $|x - 2.1| = -16.3$

Solve each equation. See Example 2.

31. $|x - 3| - 19 = 3$ **32.** $|x - 10| + 30 = 50$

33. $|3x - 7| + 8 = 22$ **34.** $|6x - 3| + 7 = 28$

35. $|3 - 4x| + 1 = 6$ **36.** $|8 - 5x| - 8 = 10$

37. $\left|\frac{7}{8}x + 5\right| - 2 = 7$ **38.** $\left|\frac{3}{4}x + 4\right| - 5 = 11$

Solve each equation. See Example 3.

39. $\left|\frac{1}{5}x + 2\right| - 8 = -8$ **40.** $\left|\frac{1}{9}x + 4\right| + 25 = 25$

41. $2|3x + 24| = 0$ **42.** $8\left|\frac{2x}{3} + 10\right| = 0$

43. $-5|2x - 9| + 14 = 14$ **44.** $-10|16x + 4| - 3 = -3$

45. $6 - 3|10x + 5| = 6$ **46.** $15 - |12x + 12| = 15$

Solve each equation. See Example 5.

47. $|5x - 12| = |4x - 16|$ **48.** $|4x - 7| = |3x - 21|$

49. $|10x| = |x - 18|$ **50.** $|6x| = |x + 45|$

51. $|2 - x| = |3x + 2|$ **52.** $|4x + 3| = |9 - 2x|$

53. $|5x - 7| = |4(x + 1)|$ **54.** $|2x + 1| = |3(x + 1)|$

Solve each inequality. Graph the solution set and write it using interval notation. See Examples 6 and 8.

55. $|x| < 4$ **56.** $|x| < 9$

57. $|x + 9| \le 12$ **58.** $|x - 8| \le 12$

59. $|3x - 2| < 10$ **60.** $|4 - 3x| \le 13$

61. $|5x - 12| < -5$ **62.** $|3x + 2| \le -3$

Solve each inequality. Graph the solution set and write it using interval notation. See Examples 9–11.

63. $|x| > 3$ **64.** $|x| > 7$

65. $|x - 12| > 24$ **66.** $|x + 5| \ge 7$

67. $0 \le |5x - 1| - 2$ **68.** $0 \le |6x - 3| - 5$

69. $|4x + 3| \ge -5$ **70.** $|7x + 2| \ge -8$

See Examples 4, 6, and 9.

71. Let $f(x) = |x + 3|$. For what value(s) of x is $f(x) = 3$?

72. Let $g(x) = |2 - x|$. For what value(s) of x is $g(x) = 2$?

73. Let $f(x) = |2(x - 1) + 4|$. For what value(s) of x is $f(x) < 4$?

74. Let $h(x) = \left|\frac{x}{5} - \frac{1}{2}\right|$. For what value(s) of x is $h(x) > \frac{9}{10}$?

TRY IT YOURSELF

Solve each equation and inequality. For the inequalities, graph the solution set and write it using interval notation.

75. $|3x + 2| + 1 > 15$ **76.** $|2x - 5| - 5 > 20$

77. $6\left|\frac{x - 2}{3}\right| \le 24$ **78.** $8\left|\frac{x - 2}{3}\right| > 32$

79. $-7 = 2 - |0.3x - 3|$ **80.** $-1 = 1 - |0.1x + 8|$

81. $|2 - 3x| \ge -8$ **82.** $|-1 - 2x| > 5$

83. $|7x + 12| = |x - 6|$ **84.** $|8 - x| = |x + 2|$

85. $2 \ge 3|2 - 3x| + 2$ **86.** $7 \ge |15x - 45| + 7$

87. $-14 = |x - 3|$ **88.** $-75 = |x + 4|$

89. $\frac{6}{5} = \left|\frac{3x}{5} + \frac{x}{2}\right|$ **90.** $\frac{11}{12} = \left|\frac{x}{3} - \frac{3x}{4}\right|$

91. $-|2x - 3| < -7$ **92.** $-|3x + 1| < -8$

93. $|0.5x + 1| < -23$ **94.** $15 \ge 7 - |1.4x + 9|$

Look Alikes . . .

95. **a.** $\frac{x}{10} - 1 = 1$ **b.** $\left|\frac{x}{10} - 1\right| = 1$

     c. $\frac{x}{10} - 1 > 1$ **d.** $\left|\frac{x}{10} - 1\right| > 1$

96. **a.** $4x - 5 = 15$ **b.** $|4x - 5| = 15$

     c. $4x - 5 \le 15$ **d.** $|4x - 5| \le 15$

97. **a.** $0.9 - 0.3x = 8.4$ **b.** $|0.9 - 0.3x| = 8.4$

     c. $0.9 - 0.3x > 8.4$ **d.** $|0.9 - 0.3x| > 8.4$

98. **a.** $8(x - 4) = 6x - 44$

     b. $|8(x - 4)| = |6x - 44|$

Solve the absolute value inequality in part a. Graph the solution set and write it in interval notation. Then use your work from part a to determine the solution set for the absolute value inequality in part b. (No new work is necessary!) Graph the solution set and write it in interval notation.

99. **a.** $|8x - 40| \le 16$ **b.** $|8x - 40| \ge 16$

100. a. $0 \le |14 - 27x|$ **b.** $0 > |14 - 27x|$

101. a. $\left|\dfrac{4x - 4}{3}\right| - 1 > 11$

    b. $\left|\dfrac{4x - 4}{3}\right| - 1 \le 11$

102. a. $\left|-\dfrac{1}{2}x - 3\right| + 2 > 7$

    b. $\left|-\dfrac{1}{2}x - 3\right| + 2 < 7$

APPLICATIONS

103. Temperature Ranges. The temperatures on a sunny summer day satisfied the inequality $|t - 78°| \le 8°$, where t is a temperature in degrees Fahrenheit. Solve this inequality and express the range of temperatures as a double inequality.

104. Operating Temperatures. A car CD player has an operating temperature of $|t - 40°| < 80°$, where t is a temperature in degrees Fahrenheit. Solve the inequality and express this range of temperatures as an interval.

105. Auto Mechanics. On most cars, the bottoms of the front wheels are closer together than the tops, creating a *camber angle*. This lessens road shock to the steering system. (See the illustration.) The specifications for a certain car state that the camber angle c of its wheels should be $0.6° \pm 0.5°$.

    a. Express the range with an inequality containing absolute value symbols.

    b. Solve the inequality and express this range of camber angles as an interval.

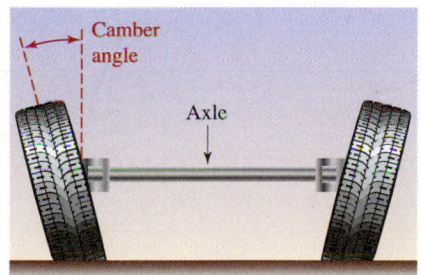

106. Steel Production. A sheet of steel is to be 0.250 inch thick with a tolerance of 0.025 inch.

    a. Express this specification with an inequality containing absolute value symbols, using x to represent the thickness of a sheet of steel.

    b. Solve the inequality and express the range of thickness as an interval.

107. Error Analysis. In a lab, students measured the percent of copper p in a sample of copper sulfate. The students know that copper sulfate is actually 25.46% copper by mass. They are to compare their results to the actual value and find the amount of *experimental error*. Which measurements shown in the illustration satisfy the absolute value inequality $|p - 25.46| \le 1.00$?

Lab 4	Section A
Title:	
"Percent copper (Cu) in copper sulfate (CuSO$_4\cdot$5H$_2$O)"	
Results	
	% Copper
Trial #1:	22.91%
Trial #2:	26.45%
Trial #3:	26.49%
Trial #4:	24.76%

108. Error Analysis. See Exercise 107. Which measurements satisfy the absolute value inequality $|p - 25.46| > 1.00$?

WRITING

109. Explain the error.

Solve: $|x| + 2 = 6$

$\cancel{x + 2 = 6}$ or $\cancel{x + 2 = -6}$

$\cancel{x = 4}$ | $\cancel{x = -8}$

110. Explain why the equation $|x - 4| = -5$ has no solution.

111. Explain the differences between the solution sets of $|x| = 8$, $|x| < 8$, and $|x| > 8$.

112. Explain how to use the graph in the illustration to solve the following.

    a. $|x - 2| = 3$

    b. $|x - 2| \le 3$

    c. $|x - 2| \ge 3$

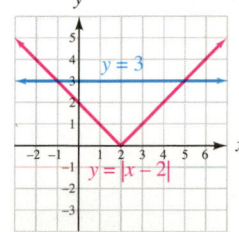

REVIEW

113. Railroad Crossing. The warning sign in the illustration is to be painted on the street in front of a railroad crossing. If y is 30° more than twice x, find x and y.

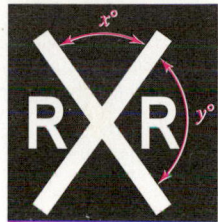

114. Solve:

$$\begin{cases} 2a + 3b + 2c = 7 \\ a + 2b - c = 4 \\ 3a - b + c = 5 \end{cases}$$

CHALLENGE PROBLEMS

115. a. For what values of k does $|x| + k = 0$ have exactly two solutions?

    b. For what values of k does $|x| + k = 0$ have exactly one solution?

116. Solve: $2^{|2x-3|} = 64$

Linear Inequalities in Two Variables

OBJECTIVES

1 Graph linear inequalities.

2 Graph inequalities with a boundary through the origin.

3 Graph inequalities having horizontal and vertical boundary lines.

4 Solve applied problems involving linear inequalities in two variables.

ARE YOU READY?

The following problems review some basic skills that are needed when graphing linear inequalities.

1. True or false: $4(-3) + 10 > -2$ **2.** True or false: $-6 \le -6$

3. Graph: $4x - 5y = 20$

4. Determine whether each of the following points lies *above, below,* or *on* the line graphed in problem 3.

    a. $(2, -5)$ **b.** $(0, 0)$ **c.** $(0, -4)$

In the first three sections of this chapter, we worked with linear inequalities in one variable. Some examples are

$$x \ge -7, \qquad 10 < 2x - 6, \qquad \text{and} \qquad 5(3 + z) > -3(z + 3)$$

These inequalities have infinitely many solutions. Recall that when their solutions are graphed on a number line, we obtain an interval. For example, the solution set of $10 < 2x + 6$ is the interval $(2, \infty)$. The graph of the solution set is shown below.

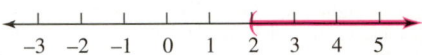

In this section, we will discuss **linear inequalities in two variables.** Some examples are:

$$y > 3x + 2, \qquad 2x - 3y \le 6, \qquad \text{and} \qquad y < 2x$$

Linear Inequalities in Two Variables	A linear inequality in x and y is any inequality that can be written in the form $$Ax + By < C \quad \text{or} \quad Ax + By > C \quad \text{or} \quad Ax + By \le C \quad \text{or} \quad Ax + By \ge C$$ where A, B, and C are real numbers and A and B are not both 0.

A **solution of a linear inequality in two variables** is an ordered pair whose coordinates satisfy the inequality. For example, $(-2, 4)$ is a solution of $y > 3x + 2$ because a true statement results when we substitute -2 for x and 4 for y. The ordered pair $(5, 1)$ is not a solution of $y > 3x + 2$ because a false statement results when we substitute 5 for x and 1 for y.

Notation

The symbol $\overset{?}{>}$ is read as "is possibly greater than."

$y > 3x + 2$	$y > 3x + 2$
$4 \overset{?}{>} 3(-2) + 2$ **Substitute.**	$1 \overset{?}{>} 3(5) + 2$ **Substitute.**
$4 \overset{?}{>} -6 + 2$ **Multiply.**	$1 \overset{?}{>} 15 + 2$ **Multiply.**
$4 > -4$ **True**	$1 > 17$ **False**

We can graph the solutions of linear inequalities in two variables on a rectangular coordinate system.

1 Graph Linear Inequalities.

The **graph of a linear inequality** in x and y is the graph of all ordered pairs (x, y) whose coordinates *satisfy* the inequality. In general, such graphs are regions bounded by a line.

EXAMPLE 1

Graph: $y > 3x + 2$

Strategy   We will graph the **related equation** $y = 3x + 2$ to establish a **boundary line** between two regions of the coordinate plane. Then we will determine which region contains points whose coordinates satisfy the given inequality.

Why   The graph of a linear inequality in two variables is a region of the coordinate plane on one side of a boundary line.

Solution

To graph the boundary line, $y = 3x + 2$, we can use the fact that the equation is in slope–intercept form and that $m = 3 = \frac{3}{1}$ and $b = 2$. Since the symbol $>$ does *not* include an equal symbol, the points on the graph of $y = 3x + 2$ are not part of the graph of $y > 3x + 2$. Therefore, the boundary line should be dashed, as shown in part (a) of the illustration below. Note that the boundary line divides the coordinate plane into two **half-planes.**

To find which half-plane is the graph of $y > 3x + 2$, we can substitute the coordinates of any point in either half-plane. We will choose the origin as the test point because its coordinates, $(0, 0)$, make the calculations easy. We substitute 0 for x and 0 for y into the inequality and simplify the right side.

Check the test point $(0, 0)$:

$y > 3x + 2$ This is the inequality to graph.

$0 \stackrel{?}{>} 3(0) + 2$ Substitute 0 for y and 0 for x. Read $\stackrel{?}{>}$ as "is possibly greater than."

$0 > 2$ False

Since $0 > 2$ is a false statement, the point $(0, 0)$ does not satisfy the inequality. This indicates that it is not on the side of the dashed line we want to shade. Instead, we shade in red the region on the other side of the boundary line. The graph of the solution set of $y > 3x + 2$ is the half-plane to the left of the dashed line, as shown in part (b).

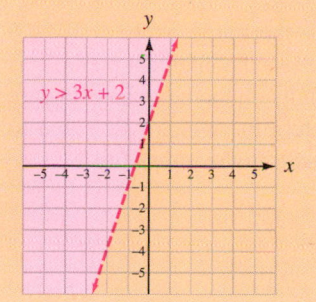

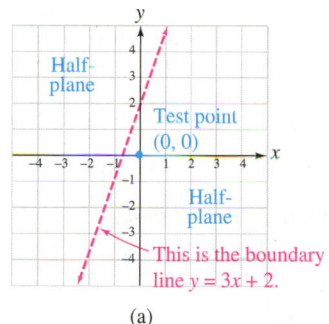

(a)

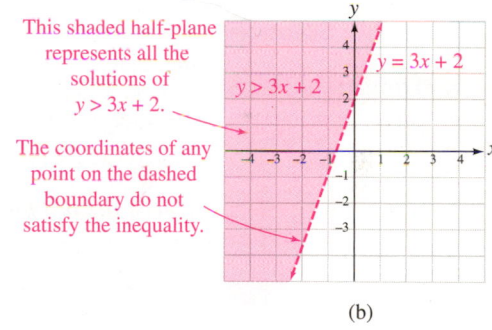

(b)

Since the solution set contains infinitely many ordered pairs, we cannot check all of them to see whether they satisfy $y > 3x + 2$. However, as an informal check, we can pick one point in the graph that is near the boundary, such as $(-1, 1)$, and see whether it satisfies the inequality. We also can pick one point not in the graph but near the boundary, such as $(1, 3)$, and see whether it fails to satisfy the inequality.

For $(-1, 1)$, in the shaded region:

$y > 3x + 2$

$1 \stackrel{?}{>} 3(-1) + 2$ Substitute for x and y.

$1 \stackrel{?}{>} -3 + 2$

$1 > -1$ True

For $(1, 3)$, not in the shaded region:

$y > 3x + 2$

$3 \stackrel{?}{>} 3(1) + 2$ Substitute for x and y.

$3 \stackrel{?}{>} 3 + 2$

$3 > 5$ False

Since $(-1, 1)$ satisfies $y > 3x + 2$ and $(1, 3)$ does not, the graph appears to be correct.

Self Check 1   Graph: $y > 2x - 4$

Now Try ▶   Problem 11

The previous example suggests the following **test-point method** to graph linear inequalities in two variables.

Graphing Linear Inequalities in Two Variables	1. Replace the inequality symbol with an equal symbol $=$ and graph the boundary line of the region. If the original inequality allows the possibility of equality (the symbol is either $\leq$ or $\geq$), draw the boundary line as a solid line. If equality is not allowed ($<$ or $>$), draw the boundary line as a dashed line.
	2. Pick a test point that is on one side of the boundary line. (Use the origin if possible.) Replace x and y in the inequality with the coordinates of that point. If a true statement results, shade the side that contains that point. If a false statement results, shade the other side of the boundary.

EXAMPLE 2   Graph: $2x - 3y \leq 6$

Strategy   We will graph the related equation $2x - 3y = 6$ to establish a boundary line between two regions of the coordinate plane. Then we will determine which region contains points that satisfy the given inequality.

Why   The graph of a linear inequality in two variables is a region of the coordinate plane on one side of a boundary line.

Solution   Since the inequality $2x - 3y \leq 6$ includes an equal symbol, the graph of $2x - 3y \leq 6$ includes the graph of $2x - 3y = 6$. To graph $2x - 3y = 6$, we use the intercept method, as shown in part (a) of the illustration below. To show that the points on the boundary line are solutions of $2x - 3y \leq 6$, we draw it as a solid line. To determine which half-plane to shade, we check to see whether the coordinates of the origin satisfy the inequality.

Notation

The inequality $2x - 3y \leq 6$ means
$$2x - 3y = 6$$
or
$$2x - 3y < 6$$

Check the test point **(0, 0):**

$$2x - 3y \leq 6 \quad \text{This is the inequality to graph.}$$
$$2(0) - 3(0) \overset{?}{\leq} 6 \quad \text{Substitute 0 for } x \text{ and 0 for } y. \text{ Read } \overset{?}{\leq} \text{ as}$$
$$\text{"is possibly less than or equal to."}$$
$$0 \leq 6 \quad \text{True}$$

Success Tip

Draw a solid boundary line if the inequality has $\leq$ or $\geq$. Draw a dashed line if the inequality has $<$ or $>$.

The coordinates of the origin satisfy the inequality. In fact, the coordinates of every point on the same side of the boundary line as the origin satisfy the inequality. We then shade that half-plane in red to complete the graph of $2x - 3y \leq 6$, shown in figure (b).

A table of solutions to graph the boundary

x	y	(x, y)
0	-2	$(0, -2)$
3	0	$(3, 0)$
1	$-\dfrac{4}{3}$	$\left(1, -\dfrac{4}{3}\right)$

Let $x = 0$ and find y.
Let $y = 0$ and find x.
As a check, let $x = 1$ and find y.

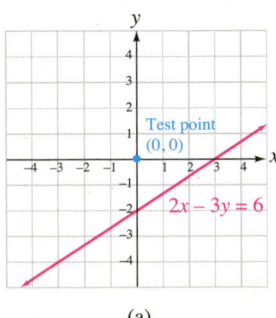

(a)

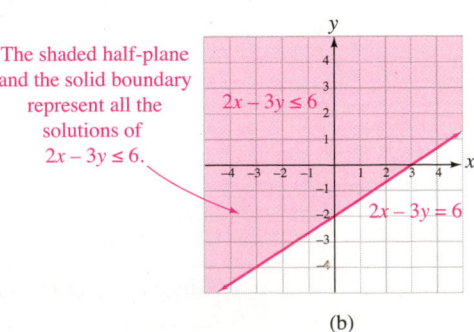

The shaded half-plane and the solid boundary represent all the solutions of $2x - 3y \leq 6$.

(b)

Self Check 2   Graph: $3x - 2y \geq 12$

Now Try ▶   Problem 15

2 Graph Inequalities with a Boundary through the Origin.

EXAMPLE 3   Graph: $y < 2x$

Strategy   We will graph the related equation $y = 2x$ to establish the boundary line between two regions of the coordinate plane. Then we will determine which region contains points whose coordinates satisfy the given inequality.

Why   The graph of a linear inequality in two variables is a region of the coordinate plane on one side of a boundary line.

Solution   To graph $y = 2x$, we can use the fact that the equation is in slope–intercept form and that $m = 2 = \frac{2}{1}$ and $b = 0$, or we can use the point-plotting method, as shown below. Since the symbol $<$ does not include an equal symbol, the points on the graph of $y = 2x$ are not on the graph of $y < 2x$. We draw the boundary line as a dashed line to show this, as in figure (a) below.

To determine which half-plane to shade, we substitute the coordinates of a point that lies on one side of the boundary line into $y < 2x$. We cannot use the origin as a test point, because the boundary line passes through the origin. However, we can choose a different point, say, $(2, 0)$, because it does not lie on the boundary. To see whether it satisfies $y < 2x$, we substitue 2 for x and 0 for y in the inequality, and simplify the right side.

> **Success Tip**
>
> The origin $(0, 0)$ is usually a smart choice for a test point because calculations involving 0 are usually easy. If the origin is on the boundary, choose a test point not on the boundary that has one coordinate that is 0, such as $(0, 1)$ or $(2, 0)$.

Check the test point $(2, 0)$:

$y < 2x$ This is the inequality to graph.

$0 \overset{?}{<} 2(2)$ Substitute 2 for x and 0 for y.

$0 < 4$ True

Since $0 < 4$ is a true statement, the point $(2, 0)$ satisfies the inequality and is in the graph of $y < 2x$. We then shade in red the half-plane containing $(2, 0)$, as shown in figure (b).

A table of solutions to graph the boundary
$y = 2x$

x	y	(x, y)
0	0	(0, 0)
−1	−2	(−1, −2)
1	2	(1, 2)

Select three values for x and find the corresponding values for y.

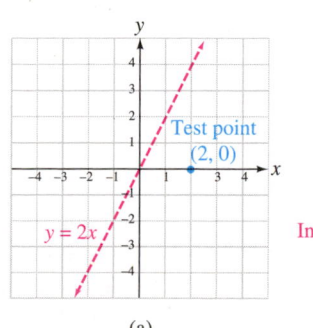

(a)

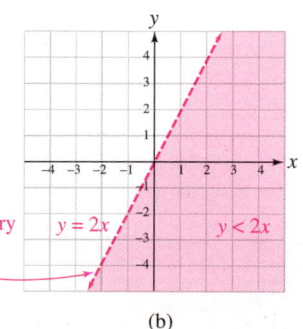

In this case, the boundary is not included.

(b)

Self Check 3   Graph: $y > -x$

Now Try ▶   Problem 21

3 Graph Inequalities Having Horizontal and Vertical Boundary Lines.

Recall that the graph of $x = a$ is a vertical line with x-intercept at $(a, 0)$, and the graph of $y = b$ is a horizontal line with y-intercept at $(0, b)$.

EXAMPLE 4   Graph: $x \geq -1$

Strategy   We will graph the related equation $x = -1$ to establish the boundary line between two regions of the coordinate plane. Then we will determine which region contains points whose coordinates satisfy the given inequality.

Why   The graph of a linear inequality in two variables is a region of the coordinate plane on one side of a boundary line.

Solution   The graph of the boundary $x = -1$ is a vertical line passing through $(-1, 0)$. Because $x \geq -1$ contains an $\geq$ symbol, draw the boundary as a solid line to show that it is part of the solution. See figure (a).

In this case, we need not pick a test point. The inequality $x \geq -1$ is satisfied by points with an x-coordinate greater than or equal to -1. Points satisfying this condition lie to the right of the boundary. We shade that half-plane in red, as shown in figure (b), to complete the graph of $x \geq -1$.

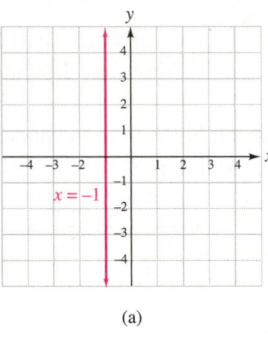

(a)

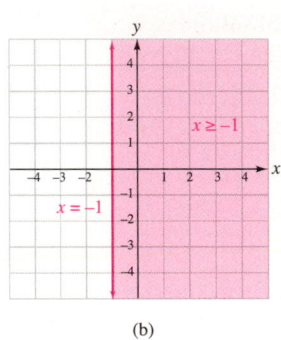

(b)

Self Check 4   Graph: $y < 4$

Now Try ▶ Problem 23

4 Solve Applied Problems Involving Linear Inequalities in Two Variables.

When solving applied problems, phrases such as *at least, at most,* and *should not exceed* indicate that an inequality should be used.

©iStockphoto.com/asiseeit

EXAMPLE 5   **Social Security.**   Retirees, ages 62–65, can earn as much as $13,560 and still receive their full Social Security benefits. If their annual earnings exceed $13,560, their benefits are reduced. A 64-year-old retired woman receiving Social Security works two part-time jobs: one at the library, paying $565 per week, and another at a pet store, paying $452 per week. Write an inequality that represents the number of weeks the woman can work at each job during the year without losing any of her benefits. Then find three possible combinations of weeks that she can work at those jobs.

Analyze   We need to find the various combinations of weeks that she can work at the library and at the pet store so that her annual income is less than or equal to $13,560.

Assign   Let $x =$ the number of weeks she works at the library and $y =$ the number of weeks she works at the pet store.

Form Now we translate the words of the problem to an inequality in two variables.

The weekly rate on the library job	·	the weeks worked on the library job	plus	the weekly rate on the pet store job	·	the weeks worked on the pet store job	should not exceed	$13,560.
$565	·	x	+	$452	·	y	$\leq$	$13,560

Solve The graph of the inequality $565x + 452y \leq 13{,}560$ is shown below in part (a). Any point in the shaded region represents a possible combination of weeks that the woman can work at each job without losing her benefits.

State If the woman works only full weeks (represented by whole numbers), the black highlighted points in part (b) show three acceptable work combinations.

(4, 20):   4 weeks at the library, 20 weeks at the pet store

(12, 15):   12 weeks at the library, 15 weeks at the pet store

(16, 5):   16 weeks at the library, 5 weeks at the pet store

A table of solutions to graph the boundary
$565x + 452y = 13{,}560$

x	y	(x, y)
0	30	(0, 30)
24	0	(24, 0)

Let $x = 0$ and find y.
Let $y = 0$ and find x.

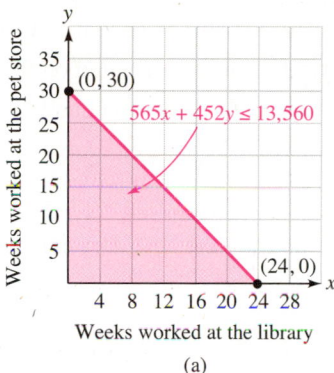

(a)

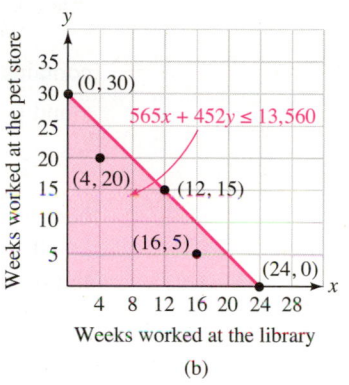

(b)

Check As an informal check, we will consider the earnings represented by a point that lies in the graph and a point that does not. For (12, 15), a point in the graph, she will earn

$565(12) + $452(15) = $6,780 + $6,780   This represents 12 weeks at the library and 15 weeks at the pet store.

= $13,560   The amount does not exceed $13,560.

For (20, 10), a point not in the graph, she will earn

$565(20) + $452(10) = $11,300 + $4,520   This represents 20 weeks at the library and 10 weeks at the pet store.

= $15,820   The amount exceeds $13,560.

From these results, it appears that the graph is correct.

Self Check 5   **Babysitting.** One babysitter charges $10 per hour and another charges $12 per hour. Marie can afford no more than $180 per month for babysitting. Write an inequality representing the number of hours Marie can hire the first babysitter (x) and the number of hours that she can hire the second babysitter (y). Then find three possible combinations of hours that she can afford.

Now Try ▶ Problem 53

Using Your Calculator ▶ Graphing Inequalities

Some graphing calculators (such as the TI-84) have a graph style icon in the $y =$ editor. Some of the different graph styles are

\	line	A straight line or curved graph is shown.	\Y$_1$ =
◥	above	Shading covers the area above a graph.	◥Y$_1$ =
◣	below	Shading covers the area below a graph.	◣Y$_1$ =

We can change the icon in front of Y$_1$ by placing the cursor on it and pressing the [ENTER] key.

To graph $2x - 3y \le 6$ of Example 2, we first write it in an equivalent form, with y isolated on the left side.

$$2x - 3y \le 6$$
$$-3y \le -2x + 6 \qquad \text{Subtract 2x from both sides.}$$
$$y \ge \frac{2}{3}x - 2 \qquad \begin{array}{l}\text{Divide both sides by } -3. \text{ Change the}\\ \text{direction of the inequality symbol.}\end{array}$$

We then change the graph style icon to above (◥), because the inequality $y \ge \frac{2}{3}x - 2$ contains an $\ge$ symbol and y is isolated on the left. Using window settings of $[-10, 10]$ for x and $[-10, 10]$ for y, we enter the boundary equation $y = \frac{2}{3}x - 2$. See figure (a). Finally, we press the [GRAPH] key to get figure (b).

To graph $y < 2x$ from Example 3, we change the graph style icon to below (◣), because the inequality contains an $<$ symbol. Using window settings of $[-10, 10]$ for x and $[-10, 10]$ for y, we enter the boundary equation $y = 2x$ and press the [GRAPH] key to get figure (c).

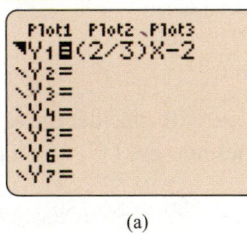
(a)

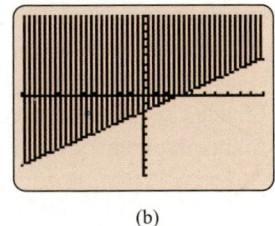
(b)

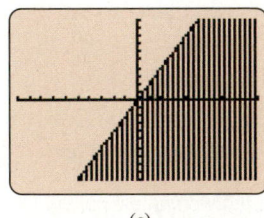
(c)

If your calculator does not have a graph style icon, you can graph linear inequalities with a SHADE feature. Graphing calculators do not distinguish between solid and dashed lines to show whether the edge of a region is included in the graph.

SECTION 4.4 ▶ STUDY SET

VOCABULARY

Fill in the blanks.

1. $4x - 2y \ge -8$ is an example of a _____ inequality in ____ variables.

2. Graphs of linear inequalities are half-_____.

3. The graph of a linear inequality in two variables is a region of the coordinate plane on one side of a _____ line.

4. A _____ of a linear inequality in two variables is an ordered pair whose coordinates satisfy the inequality.

CONCEPTS

5. Check to determine whether each ordered pair is a solution of $3x - 2y \ge 5$.

   a. $(3, 1)$ **b.** $(0, 3)$

   c. $(-1, -4)$ **d.** $\left(1, \frac{1}{2}\right)$

6. The graph of a linear inequality in two variables is shown. Tell whether each point satisfies the inequality.
   a. $(-1, 4)$
   b. $(3, -2)$
   c. $(0, 0)$
   d. $(-3, -3)$

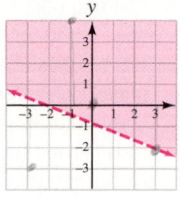

7. a. To graph the inequality $y > 3x - 1$, we begin by graphing the boundary line $y = 3x - 1$. What is the slope m of the line? What is its y-intercept?
   b. To graph the inequality $2x + 3y \leq -6$, we begin by graphing the boundary line $2x + 3y = -6$. What are its x- and y-intercepts?

8. **Zoos.** To determine the allowable number of juvenile chimpanzees x and adult chimpanzees y that can live in an enclosure, a zookeeper refers to the illustration. Can 7 juvenile and 4 adult chimps be kept in the enclosure?

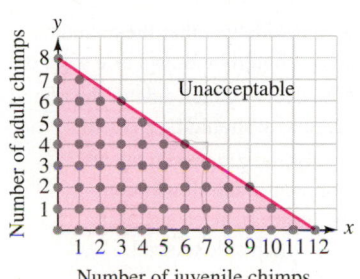

Number of juvenile chimps

NOTATION

9. Tell whether the graph of each inequality includes the boundary line. In each case, would the boundary be a solid or a dashed line?
   a. $y < 3x - 1$ b. $2x + 3y \geq -6$
   c. $y \leq -10$ d. $x > 1$

10. Match each inequality or compound inequality with the correct graph of its solution set.
    a. $2x + 4 \geq 8$ b. $x \leq 2$ or $x \geq 4$
    c. $x \leq 4$ and $x \geq 2$ d. $2x + 4y \geq 8$

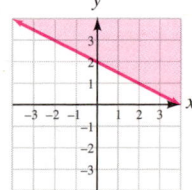

GUIDED PRACTICE

Graph each inequality. See Examples 1 and 2.

11. $y > x + 1$ 12. $y < 2x - 1$

13. $y \geq -\dfrac{3}{2}x + 1$ 14. $y < \dfrac{x}{3} - 1$

15. $2x + y \leq 6$ 16. $x - 2y \geq 4$

17. $3x + 5y > -9$ 18. $5x + 3y < 10$

Graph each inequality. See Example 3.

19. $y \geq x$ 20. $y \leq 2x$

21. $y < -\dfrac{x}{2}$ 22. $y > \dfrac{x}{3}$

Graph each inequality. See Example 4.

23. $x < 4$ 24. $y \geq -2$

25. $y < 0$ 26. $x \geq 0$

TRY IT YOURSELF

Graph each inequality.

27. $3x \geq -y + 3$ 28. $2x \leq -3y - 12$

29. $3x + y > 2 + x$ 30. $3x - y > 6 + y$

31. $y \geq \dfrac{8}{3}$ 32. $x < -3.75$

33. $y + 4x \geq 0$ 34. $y \geq -\dfrac{5}{4}x$

35. $\dfrac{x}{2} + \dfrac{y}{2} \leq 2$ 36. $\dfrac{x}{3} - \dfrac{y}{2} \geq 1$

37. $y - 4.5 < 0$ 38. $x - 3 \geq 0$

39. $x < -\dfrac{1}{2}y$ 40. $x < \dfrac{1}{6}y$

41. $0.3x + 0.4y \geq -1.2$ 42. $0.8x - 0.3y < 2.4$

Look Alikes . . .

Graph the given inequality in part a. Then use your answer to part a to help you quickly graph the associated inequality in part b. (Hint: If you spot the relationship between the inequalities, the graph in part b can be completed without having to use the test-point method.)

43. a. $5x - 3y < -15$ b. $5x - 3y \geq -15$

44. a. $y \leq -\dfrac{2}{3}x + 2$ b. $y > -\dfrac{2}{3}x + 2$

45. a. $y + 2x \geq 0$ b. $y + 2x < 0$

46. a. $y > \dfrac{1}{4}x$ b. $y \leq \dfrac{1}{4}x$

Use a graphing calculator to graph each inequality. See Using Your Calculator: Graphing Inequalities.

47. $y < 0.27x - 1$ 48. $y > -3.5x + 2.7$

49. $y \geq -2.37x + 1.5$ 50. $y \leq 3.37x - 1.7$

APPLICATIONS

51. **The Korean War.** After World War II, the 38th parallel of north latitude was established as the boundary between North Korea and South Korea. In the illustration, shade the region of the Korean Peninsula south of the 38th parallel.

52. Geography. A region of the continental United States is shaded in the following map.

    a. What is the boundary that separates the shaded and unshaded regions?

    b. In words, describe the shaded area with respect to the boundary.

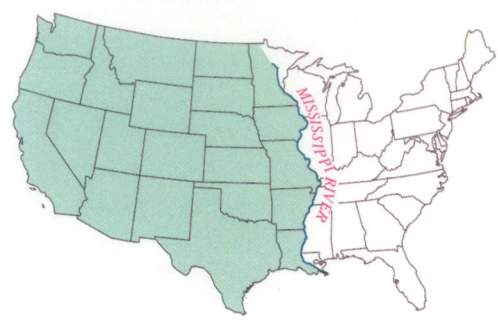

53. Restaurant Seating. As part of a remodeling project, a restaurant owner will be installing new booths that seat 4 persons, and new tables that seat 6 persons. The overall seating must conform to the sign shown below. Write an inequality that describes the possible combinations of the number of booths (x) and the number of tables (y) that the owner can install. Graph the inequality and give three ordered-pair solutions.

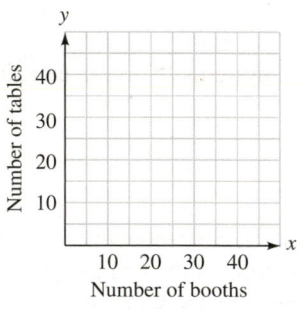

54. Gardening. During an Arbor Day sale, a garden store sold more than $2,000 worth of 6-foot maple and 5-foot pine trees. A 6-foot maple sold for $100 and a 5-foot pine sold for $125. Write an inequality that describes the possible combinations of the number of the maple trees (x) and the number of pine trees (y) that could have been sold during the sale. Graph the inequality and give three ordered-pair solutions.

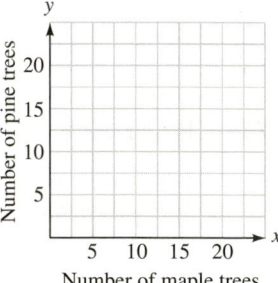

55. Sporting Goods. A sporting goods manufacturer allocates at least 1,200 units of time per day to make fishing rods and reels. It takes 10 units of time to make a rod and 15 units of time to make a reel. Write an inequality that describes the possible combinations of the number of units of time devoted to make rods (x) and the number of units of time devoted to make reels (y). Graph the inequality and give three ordered-pair solutions.

56. Housekeeping. One housekeeper charges $12 per hour, and another charges $9 per hour. Sarah, who can afford no more than $54 per month to clean her house, decides to hire the housekeepers. Write an inequality that describes the possible combinations of the number of hours that she can hire the first housekeeper (x) and the number of hours that she can hire the second housekeeper (y). Graph the inequality and give three ordered-pair solutions.

WRITING

57. Explain how to decide whether the boundary of the graph of a linear inequality should be drawn as a solid or a dashed line.

58. Explain how to decide which side of the boundary of the graph of a linear inequality should be shaded.

59. The boundary for the graph of a linear inequality is shown in the illustration. Why can't the origin be used as a test point to decide which side to shade?

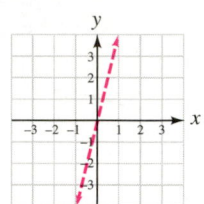

60. Explain the difference between the graph of the solution set of $x + 2 > 6$, an inequality in *one variable,* and the graph of $x + 2y > 6$, an inequality in *two variables.*

REVIEW

Decide whether the ordered pair $(-4, 3)$ is a solution of the system of linear equations.

61. $\begin{cases} 4x - y = -19 \\ 3x + 2y = -6 \end{cases}$

62. $\begin{cases} y = 2x + 11 \\ \dfrac{x}{2} + y = 0 \end{cases}$

Solve each system of equations.

63. $\begin{cases} x = \dfrac{2}{3}y \\ y = 4x + 5 \end{cases}$

64. $\begin{cases} x - \dfrac{y}{2} = -2 \\ 0.01x + 0.02y = 0.03 \end{cases}$

CHALLENGE PROBLEMS

65. Can an inequality in two variables be an identity, one that is satisfied by all pairs (x, y)? If so, give an example.

66. Can an inequality in two variables have no solutions? If so, give an example.

Find the equation of the boundary line. Then give the inequality whose graph is shown.

67. a.

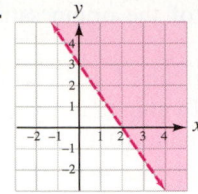

b.

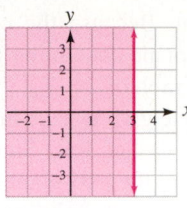

68. a.

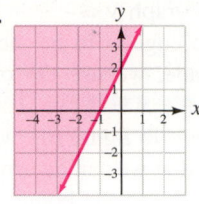

b.

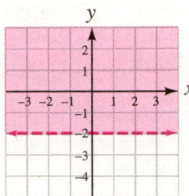

SECTION 4.5

OBJECTIVES

1. Solve systems of linear inequalities.
2. Graph compound inequalities.
3. Solve problems involving systems of linear inequalities.

Systems of Linear Inequalities

ARE YOU READY?

The following problems review some basic skills that are needed when graphing systems of linear inequalities.

1. Check to determine whether $(-3, 5)$ is a solution of $2x - 3y \leq -25$.
2. Is the graph of the equation $x = 4$ on a rectangular coordinate system a vertical or horizontal line?
3. Graph: $x - 2y \geq 4$
4. What inequality symbol is indicated by the phrase *at least*?

We have discussed how to solve systems of linear equations by the graphing method. For example, to solve

$$\begin{cases} y = -x + 1 \\ 2x - y = 2 \end{cases}$$

we graph both equations on the same rectangular coordinate system and then find the coordinates of the point of intersection of the straight lines. We will now discuss how to solve **systems of linear inequalities** graphically, such as

$$\begin{cases} y \leq -x + 1 \\ 2x - y > 2 \end{cases}$$

1 Solve Systems of Linear Inequalities.

When we graph the solution of a linear inequality in x and y, the result is a half-plane. To solve a system of linear inequalities, we graph each of the inequalities on one set of coordinate axes and find the intersection of the shaded half-planes.

EXAMPLE 1 Graph the solution set of: $\begin{cases} y \leq -x + 1 \\ 2x - y > 2 \end{cases}$

Strategy We will graph the solutions of $y \leq -x + 1$ in one color and the solutions of $2x - y > 2$ in another color on the same rectangular coordinate system.

Why We then can see where the graphs of the two inequalities intersect. This is the solution set of the system.

Solution

To graph $y \leq -x + 1$, we graph the boundary $y = -x + 1$, as shown in figure (a). Since the edge is to be included, we draw it as a solid line. To determine which half-plane to shade, we use the origin as a test point. Because the coordinates of the origin satisfy $y \leq -x + 1$, we shade (in red) the half-plane containing the origin.

In figure (b), we superimpose the graph of $2x - y > 2$ on the graph of $y \leq -x + 1$ so that we can determine the points that the graphs have in common. To graph $2x - y > 2$, we graph the boundary $2x - y = 2$ as a dashed line. Since the test point $(0, 0)$ does not satisfy $2x - y > 2$, we then shade (in blue) the half-plane that does not contain $(0, 0)$.

The area that is shaded twice represents the solutions of the given system. Any point in the doubly shaded region in purple (including the purple portion of the solid boundary) has coordinates that satisfy both inequalities.

$$y = -x + 1$$
$$m = -1 = \frac{-1}{1} \qquad \text{y-intercept: } (0, 1)$$

$$2x - y = 2$$

x	y	(x, y)
0	-2	$(0, -2)$
1	0	$(1, 0)$

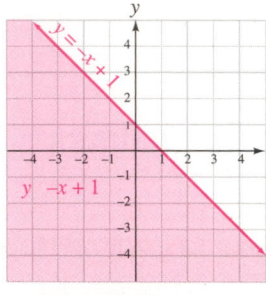

The graph of $y \leq -x + 1$ is shaded in red.

(a)

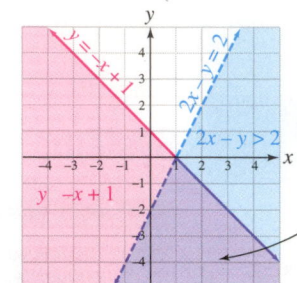

The graph of $2x - y > 2$ is shaded in blue. It is drawn over the graph of $y \leq -x + 1$.

(b)

The solutions of the system are shaded in purple. The purple region is the intersection or overlap of the red and blue regions.

Since there are infinitely many solutions, we cannot check each of them. However, as an informal check, we can select several points near the boundaries that lie in the doubly shaded region and show that their coordinates satisfy both inequalities of the system. For example, to check $(0, -3)$, we have:

$y \leq -x + 1$	The first inequality.	$2x - y > 2$	The second inequality.
$-3 \overset{?}{\leq} -(0) + 1$	Substitute.	$2(0) - (-3) \overset{?}{>} 2$	Substitute.
$-3 \leq 1$	True	$3 > 2$	True

Self Check 1   Graph the solution set of: $\begin{cases} x + y \geq 1 \\ 2x - y < 2 \end{cases}$

Now Try ▶ Problem 9

Using Your Calculator ▶ **Solving Systems of Inequalities**

To solve the system of Example 1 with a graphing calculator, we can use window settings of $x = [-10, 10]$ and $y = [-10, 10]$. To graph $y \leq -x + 1$, we enter the boundary equation $y = -x + 1$ and change the graph style icon to below (◣). See figure (a). To graph $2x - y > 2$, we first write it in equivalent form as $y < 2x - 2$. Then we enter the boundary equation $y = 2x - 2$ and change the graph style icon to below (◣). See figure (a). Finally, we press $\boxed{\text{GRAPH}}$ to obtain figure (b).

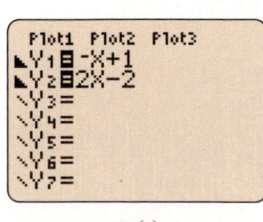

(a)

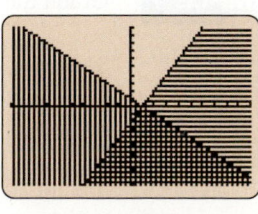

(b)

In general, to solve systems of linear inequalities, we will follow these steps.

Solving Systems of Linear Inequalities	1. Graph each inequality on the same rectangular coordinate system.
	2. Use shading to highlight the intersection of the graphs (the region where the graphs overlap). The points in this region are the solutions of the system.
	3. As an informal check, pick a point from the region and verify that its coordinates satisfy each inequality of the original system.

EXAMPLE 2   Graph the solution set of: $\begin{cases} x \geq 1 \\ y \geq x \\ 4x + 5y < 20 \end{cases}$

Strategy   We will graph the solutions of $x \geq 1$, $y \geq x$, and $4x + 5y < 20$ on the same coordinate system.

Why   We then can see where the graphs of the three inequalities intersect. This is the solution set of the system.

Solution   We will find the graph of the solution set of the system in stages using three graphs.

The graph of $x \geq 1$ includes the points that lie on the graph of $x = 1$ and to the right, as shown in red in figure (a) below.

Figure (b) below shows the graph of $x \geq 1$ and the graph of $y \geq x$. The graph of $y \geq x$, in blue, includes the points that lie on the graph of the boundary $y = x$ and above it.

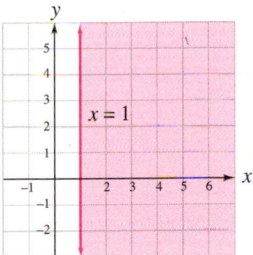

The graph of $x \geq 1$ is shaded in red.

(a)

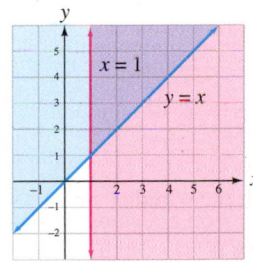

The graph of $y \geq x$ is shaded in blue.

(b)

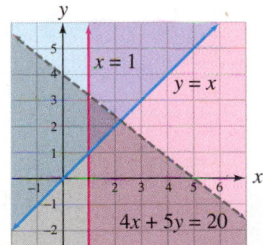

The graph of $4x + 5y < 20$ is shaded in gray.

(c)

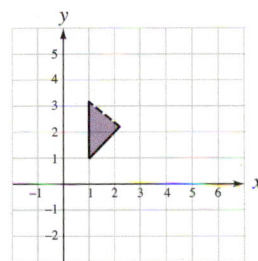

This is the graph of the solution of the system.

(d)

Figure (c) above shows the graphs of $x \geq 1$, $y \geq x$, and $4x + 5y < 20$. The graph of $4x + 5y < 20$ includes the points that lie below the graph of the boundary $4x + 5y = 20$.

The graph of the solution of the system includes the points that lie within the shaded triangle together with the points on the two sides of the triangle that are drawn with solid line segments, as shown in figure (d).

Check: Pick a point in the shaded region, such as $(1.5, 2)$, and show that it satisfies each inequality of the system.

Self Check 2   Graph the solution set of: $\begin{cases} x \geq 0 \\ y \leq 0 \\ y \geq -2 \end{cases}$

Now Try ▶ Problem 13

2 Graph Compound Inequalities.

We have graphed the solution set of double linear inequalities, such as $2 < x \le 5$, on a number line. In the next example, we will graph the solution set of $2 < x \le 5$ in the context of two variables.

EXAMPLE 3   Graph $2 < x \le 5$ on a rectangular coordinate system.

Strategy   We will write the double inequality $2 < x \le 5$ as an equivalent system of inequalities. Then we will graph the system.

Why   The compound inequality $2 < x \le 5$ is equivalent to $2 < x$ and $x \le 5$.

Solution   The compound inequality $2 < x \le 5$ is equivalent to the following system of two linear inequalities:

$$\begin{cases} 2 < x \\ x \le 5 \end{cases}$$

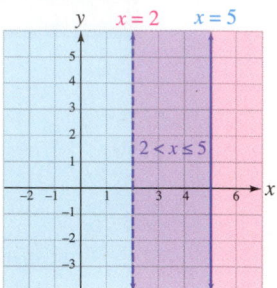

The graph of $2 < x$, shown in the figure, is the half-plane to the right of the vertical line $x = 2$. The graph of $x \le 5$, shown in the figure, includes the line $x = 5$ and the half-plane to its left. The graph of $2 < x \le 5$ will contain all points in the plane that satisfy the inequalities $2 < x$ and $x \le 5$ simultaneously. These points are in the purple-shaded region.

> **Self Check 3**   Graph $-2 \le y < 3$ on the rectangular coordinate plane.
>
> **Now Try** ▶ Problem 17

Success Tip

Colored pencils often are used to graph systems of inequalities. A standard pencil also can be used. Just draw different patterns of lines instead of shading.

EXAMPLE 4   Graph $y \ge -1$ or $y < -3$ on a rectangular coordinate system.

Strategy   We will graph the inequalities $y \ge -1$ and $y < -3$ on the same rectangular coordinate system and determine their union.

Why   When solving a compound inequality containing the word *or*, the solution set is the union of the solution sets of the two inequalities.

Solution   Since the operation of union combines the graphs, we can draw them in the same color. The graph of $y \ge -1$, shown in red, includes the line $y = -1$ and the half plane above it. The graph of $y < -3$, also shown in the figure in red, is the half-plane below the line $y = -3$. The graph of $y \ge -1$ or $y < -3$ contains all the points in the plane that satisfy one or the other inequalities.

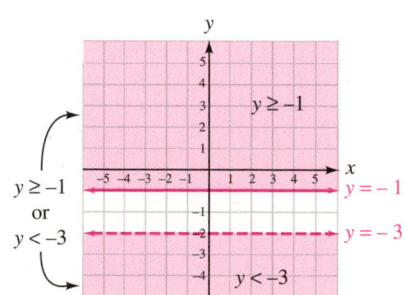

> **Self Check 4**   Graph $x \le -2$ or $x > 3$ on a rectangular coordinate system.
>
> **Now Try** ▶ Problem 19

3 Solve Problems Involving Systems of Linear Inequalities.

EXAMPLE 5

Landscaping. A homeowner has a budget of $300 to $600 for trees and bushes to landscape his yard. After shopping, he finds that trees cost $150 each and bushes cost $75 each. What combinations of trees and bushes can he afford to buy?

Analyze We must find the number of trees and bushes that the homeowner can afford. This suggests we should use two variables. We know that he is willing to spend *at least* $300 and *at most* $600 for trees and bushes. These phrases suggest that we should write two inequalities that model the situation.

Assign Let x = the number of trees purchased and y = the number of bushes purchased.

Form Using the given information, we can form the following system of linear inequalities.

The cost of a tree	times	the number of trees purchased	plus	the cost of a bush	times	the number of bushes purchased	should be at least	$300.
$150	·	x	+	$75	·	y	≥	$300

The cost of a tree	times	the number of trees purchased	plus	the cost of a bush	times	the number of bushes purchased	should be at most	$600.
$150	·	x	+	$75	·	y	≤	$600

Solve We graph the system

$$\begin{cases} 150x + 75y \geq 300 \\ 150x + 75y \leq 600 \end{cases}$$

as shown on the left. The coordinates of each point highlighted in the graph give a possible combination of trees (x) and bushes (y) that can be purchased. Because the homeowner cannot buy a negative number of trees or bushes, we graph the system for $x \geq 0$ and $y \geq 0$.

State The possible combinations of trees and bushes that can be purchased are given by

(0, 4), (0, 5), (0, 6), (0, 7), (0, 8),
(1, 2), (1, 3), (1, 4), (1, 5), (1, 6),   *The ordered pair (1, 6), for example, indicates that*
(2, 0), (2, 1), (2, 2), (2, 3), (2, 4),   *the homeowner can afford 1 tree and 6 bushes.*
(3, 0), (3, 1), (3, 2), (4, 0)

Only these points can be used, because the homeowner cannot buy a portion of a tree or a bush.

Check Check some of the ordered pairs to verify that they satisfy both inequalities.

Self Check 5   **Buying Media.** An electronics store sells CDs for $10 and DVDs for $20. Laurie wants to spend at least $100 but no more than $200 on ($x$) CDs and ($y$) DVDs. What combinations of CDs and DVDs can she afford to buy?

Now Try ▶ Problem 49

VOCABULARY

Fill in the blanks.

1. $\begin{cases} x + y \leq 2 \\ x - 3y > 10 \end{cases}$ is a system of linear _____ in two variables.

2. To solve a system of inequalities by graphing, we graph each inequality. The solution is the region where the graphs overlap or _____.

3. To determine which half-plane to shade when graphing a linear inequality, we see whether the coordinates of a test _____ satisfy the inequality.

4. When we graph a system of two linear inequalities, any point in the doubly shaded region has coordinates that _____ both inequalities.

CONCEPTS

5. Check to determine whether each point satisfies the following system of linear inequalities: $\begin{cases} x + y \leq 2 \\ x - 3y > 10 \end{cases}$

   a. $(2, -3)$ **b.** $(12, -1)$

   c. $(0, -3)$ **d.** $(-0.5, -5)$

6. **a.** Check to determine whether $(-3, 10)$ satisfies the compound inequality $-5 < x \leq 8$ in the rectangular coordinate system.

   b. Check whether $(-3, 3)$ satisfies the compound inequality $y \leq 0$ or $y > 4$ in the rectangular coordinate system.

7. In the illustration, the solution of one linear inequality is shaded in red, and the solution of a second is shaded in blue. Determine whether a true or false statement results if the coordinates of the given point are substituted into the given inequality.

   a. A, inequality 1
   b. A, inequality 2
   c. B, inequality 1
   d. B, inequality 2
   e. C, inequality 1
   f. C, inequality 2

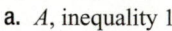

8. Match each equation, inequality, or system with the graph of its solution shown below.

   a. $2x + y = 2$

   b. $2x + y \geq 2$

   c. $\begin{cases} 2x + y = 2 \\ 2x - y = 2 \end{cases}$

   d. $\begin{cases} 2x + y \geq 2 \\ 2x - y \leq 2 \end{cases}$

   i.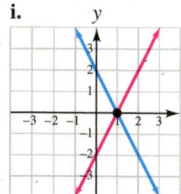

   ii.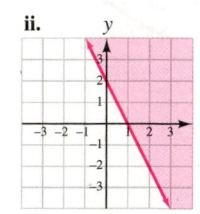

   iii.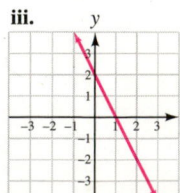

   iv.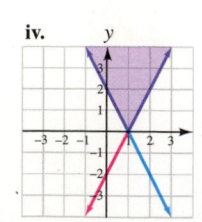

GUIDED PRACTICE

Graph the solution set of each system of inequalities. See Example 1.

9. $\begin{cases} y < 3x + 2 \\ 2x + y < 3 \end{cases}$ 10. $\begin{cases} y \leq x - 2 \\ 2x - y \leq -1 \end{cases}$

11. $\begin{cases} 3x + y \leq 1 \\ -x + 2y \geq 6 \end{cases}$ 12. $\begin{cases} 3x + 2y > 6 \\ x + 3y \leq 2 \end{cases}$

Graph the solution set of each system of inequalities. See Example 2.

13. $\begin{cases} y - x \leq 2 \\ y > -2 \\ x < 2 \end{cases}$ 14. $\begin{cases} x - y < 4 \\ y \leq 0 \\ x \geq 0 \end{cases}$

15. $\begin{cases} 2x + 3y \leq 6 \\ 3x + y \leq 1 \\ x \leq 0 \end{cases}$ 16. $\begin{cases} 2x + y \leq 2 \\ y \geq x \\ x \geq 0 \end{cases}$

Graph each inequality on a rectangular coordinate system. See Examples 3 and 4.

17. $-2 \leq x < 0$ 18. $-3 < y \leq -1$

19. $y < -2$ or $y > 3$ 20. $x \leq 1$ or $x > 5$

TRY IT YOURSELF

Graph the solution set of each system of inequalities on a rectangular coordinate system.

21. $\begin{cases} 2x < 3y \\ 2x + 3y \geq 12 \end{cases}$ 22. $\begin{cases} x > 2 - y \\ x - y < -2 \end{cases}$

23. $\begin{cases} x > 0 \\ y > 0 \end{cases}$ 24. $\begin{cases} x \leq 0 \\ y < 0 \end{cases}$

25. $\begin{cases} y \geq x \\ y \leq \dfrac{1}{3}x + 1 \\ x > -3 \end{cases}$ 26. $\begin{cases} x - y \leq 6 \\ x + 2y \leq 6 \\ x \geq 0 \end{cases}$

27. $5 > y \geq 2$ 28. $0 \geq x > -4$

29. $\begin{cases} x + y < 2 \\ x + y \leq 1 \end{cases}$ 30. $\begin{cases} x + 2y < 3 \\ 2x + 4y < 8 \end{cases}$

31. $\begin{cases} y < -\dfrac{3}{2}x - 3 \\ 3x + 2y \geq 2 \end{cases}$ 32. $\begin{cases} y > \dfrac{1}{4}x + 3 \\ x - 4y > 4 \end{cases}$

33. $\begin{cases} 3y - 5x < 0 \\ 5x - 3y \geq -12 \end{cases}$ 34. $\begin{cases} x + y < 2 \\ x \leq 1 - y \end{cases}$

35. $-x \leq 1$ or $x \geq 2$ 36. $y < 1$ or $y \leq -5$

37. $\begin{cases} x < 1 \\ x > -1 \\ x - y + 4 \geq 0 \\ y - x \geq -4 \end{cases}$ 38. $\begin{cases} x \geq 0 \\ y \geq 0 \\ 9x + 3y \leq 18 \\ 3x + 6y \leq 18 \end{cases}$

39. $\begin{cases} 2x - 3y \leq 3 \\ 3y \leq 2x - 3 \end{cases}$ 40. $\begin{cases} x \leq 3 \\ x \geq 3 \end{cases}$

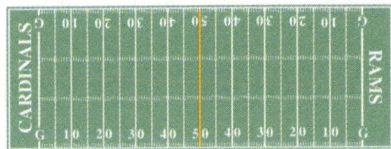

Use a graphing calculator to solve each system. **See Using Your Calculator: Solving Systems of Inequalities.**

41. $\begin{cases} y < 3x + 2 \\ y < -2x + 3 \end{cases}$

42. $\begin{cases} y > -x + 2 \\ y < -x + 4 \end{cases}$

43. $\begin{cases} 2x + y \geq 6 \\ y \leq 2(2x - 3) \end{cases}$

44. $\begin{cases} 3x + y < -2 \\ y > 3(1 - x) \end{cases}$

APPLICATIONS

45. Football. In 2009, the Arizona Cardinals scored either a touchdown or a field goal 69.8% of the time when their offense was in the *red zone*. This was the best record in the NFL. Refer to the illustration below. If *x* represents the yard line the football is on, the Cardinals' red zone is the area on the left end of the field that can be described by the following system. Shade the red zone.

$$\begin{cases} x > 0 \\ x \leq 20 \end{cases}$$

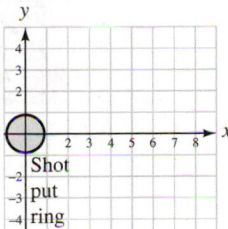

← Cardinals moving this direction

46. Track and Field. In the shot put, the solid metal ball must land in a marked sector for it to be a fair throw. In the illustration, graph the system of inequalities that describes the region in which a shot must land.

$$\begin{cases} y \leq \dfrac{3}{8}x \\ y \geq -\dfrac{3}{8}x \\ x \geq 1 \end{cases}$$

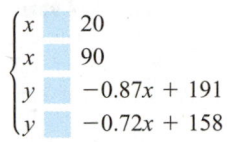

47. No-Fly Zones. After the first Gulf War, U.S. and Allied forces enforced northern and southern "no-fly" zones over Iraq. Iraqi aircraft was prohibited from flying in this air space. If *y* represents the north latitude parallel measurement, the no-fly zones can be described by

$$y \geq 36 \ \text{ or } \ y \leq 33$$

On the map, shade the regions of Iraq over which there was a no-fly zone.

48. Cardiovascular Fitness. The graph in the illustration shows the range of pulse rates that persons ages 20–90 should maintain during aerobic exercise to get the most benefit from the training. The shaded region "Effective Training Heart Rate Zone" can be described by a system of linear inequalities. Determine what inequality symbol should be inserted in each blank.

$$\begin{cases} x \ \underline{\quad} \ 20 \\ x \ \underline{\quad} \ 90 \\ y \ \underline{\quad} \ -0.87x + 191 \\ y \ \underline{\quad} \ -0.72x + 158 \end{cases}$$

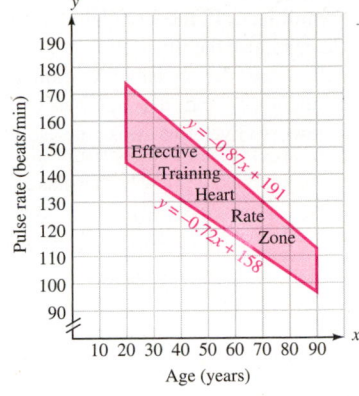

49. Compact Discs. A music store has compact discs on sale for either $10 or $15. A customer wants to spend at least $30 but no more than $60 on CDs at the store. Use the illustration to graph a system of inequalities that will show the possible combinations of the number of $10 CDs (*x*) and the number of $15 CDs (*y*) that the customer can buy. Give two possible solutions.

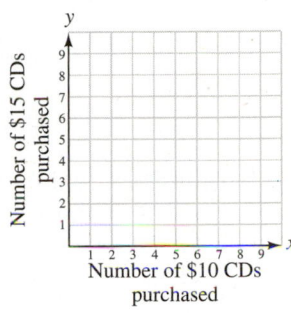

50. Boat Sales. Dry Boat Works wholesales aluminum boats for $800 and fiberglass boats for $600. Northland Marina wants to order at least $2,400 worth but no more than $4,800 worth of boats from them. Use the illustration to graph a system of inequalities that will show the possible combinations of the number of aluminum boats (*x*) and the number of fiberglass boats (*y*) that can be ordered. Give two possible solutions.

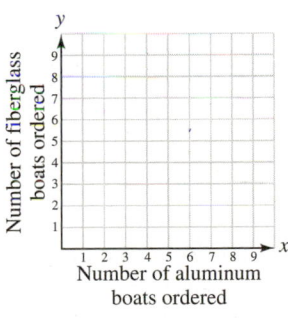

from Campus to Careers

Heating, Ventilation, and Air Conditioning Technician

A heating/air conditioning company wants to order no more than $2,000 worth of electronic air cleaners and humidifiers from a wholesaler that charges $500 for air cleaners and $200 for humidifiers. If the company wants more humidifiers than air cleaners, use the illustration to graph a system of inequalities that will show the possible combinations of the number of air cleaners (x) and the number of humidifiers (y) that can be ordered. Give two possible solutions.

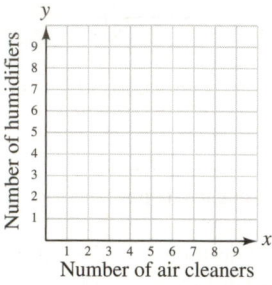

52. Furniture Sales. A distributor wholesales desk chairs for $150 and side chairs for $100. Best Furniture wants to order no more than $900 worth of chairs, including more side chairs than desk chairs, from that distributor. Use the illustration to graph a system of inequalities that will show the possible combinations of the number of desk chairs (x) and the number of side chairs (y) that can be ordered. Give two possible solutions.

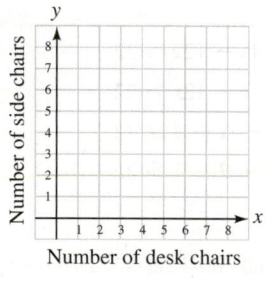

WRITING

53. Explain how to solve a system of two linear inequalities graphically.

54. Explain how a system of two linear inequalities might have no solution.

55. A student graphed the following system as shown. Explain how to informally check the result.

$$\begin{cases} -x + 3y > 0 \\ x + 3y < 3 \end{cases}$$

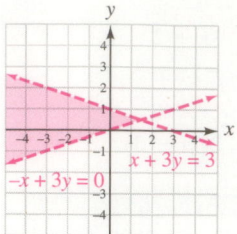

56. Describe the result when $-3 \le x < 4$ is graphed on a number line. Describe the result when $-3 \le x < 4$ is graphed on the rectangular coordinate plane.

REVIEW

Use the given conditions to determine in which quadrant of a rectangular coordinate system each point (x, y) is located.

57. $x > 0$ and $y < 0$ **58.** $x < 0$ and $y < 0$

59. $x < 0$ and $y > 0$ **60.** $x > 0$ and $y > 0$

CHALLENGE PROBLEMS

61. Write a compound inequality whose graph is shown in figure (a) below.

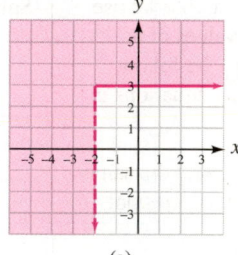   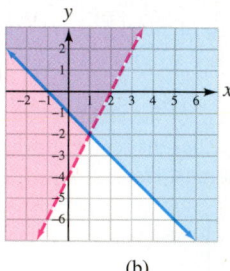

(a)   (b)

62. Write a system of linear inequalities in two variables whose graph is shown in figure (b) above.

63. The solution of a system of inequalities in two variables is *bounded* if it is possible to draw a circle around the solution.

 a. Can the solution of two linear inequalities be bounded?

 b. Can the solution of three linear inequalities be bounded?

64. The solution of $\begin{cases} y \ge |x| \\ y \le k \end{cases}$ has an area of 25. Find k.

4 Summary & Review

DEFINITIONS AND CONCEPTS	EXAMPLES
Inequalities are statements that contain one or more **inequality symbols.**	$5 \neq 7$ 5 is not equal to 7. $-1 < 9$ −1 is less than 9. $8 > 5$ 8 is greater than 5. $3 \leq 6$ 3 is less than or equal to 6. $4 \geq 4$ 4 is greater than or equal to 4.
A **linear inequality in one variable** is any inequality that can be written in the form $ax + b < c$, where a, b, and c represent real numbers and $a \neq 0$. The inequality symbols $>$, $\leq$, and $\geq$ can also be used.	Linear inequalities in one variable: $2x + 3 < 6$ $8x > 2x - 7$ $5(x + 9) \leq 6(x - 6)$
The **solution of a linear inequality** is a number that satisfies the inequality.	5 is a solution of $12 < 3x + 11$ because it satisfies the inequality. $12 < 3x + 11$ $12 \overset{?}{<} 3(5) + 11$ Substitute 5 for x. $12 \overset{?}{<} 15 + 11$ Multiply. $12 < 26$ True

To **solve a linear inequality** in one variable we use **properties of inequality** to find the values of its variable that make the inequality true. We follow steps similar to those for solving linear equations in one variable.
 	
 Adding the same number to, or subtracting the same number from, both sides of an inequality does not change the solutions.
 	
 Multiplying or dividing both sides of an inequality by the same positive number does not change the solutions.
 	
 If we multiply or divide both sides of an inequality by a negative number, the direction of the **inequality symbol must be reversed** for the inequalities to have the same solutions.
 	
 The set of all solutions of an inequality is called its **solution set.**
	Solve: $7(x + 1) - 12x > 22$
	 $\qquad 7x + 7 - 12x > 22$ Distribute the multiplication by 7.
	 $\qquad\qquad -5x + 7 > 22$ Combine like terms.
	 $\qquad -5x + 7 - 7 > 22 - 7$ Subtract 7 from both sides.
	 $\qquad\qquad\qquad -5x > 15$
	 $\qquad\qquad \dfrac{-5x}{-5} < \dfrac{15}{-5}$ Divide both sides by −5 and reverse
	                  the direction of the inequality symbol.
	 $\qquad\qquad\qquad\quad x < -3$
	 The solution set is:
	
	     Graph Interval notation Set-builder notation
	                  $(-\infty, -3)$ $\{x \,\vert\, x < -3\}$
	     $-4 \ -3 \ -2$

In some application problems, an inequality, rather than an equation, is needed to find the solution. This is the case when we are asked to determine when one quantity *is more* (or *is less*) than another. For application problems involving inequalities, we can adapt the **six-step problem-solving strategy** used in earlier chapters to find the solution. Study Example 8 in Section 4.1.	*a is at most b* means $a \leq b$. *a is at least b* means $a \geq b$. *a is no more than b* means $a \leq b$. *a is no less than b* means $a \geq b$. *a does not exceed b* means $a \leq b$. *a exceeds b* means $a > b$. *a is between b and c* means $b < a < c$.

REVIEW EXERCISES

1. Check to determine whether -2 is a solution of each inequality.

   a. $3x - 6 < x - 10$ b. $\frac{x}{2} - 3 \geq 4(x + 1)$

2. Represent the set of real numbers greater than or equal to -5 with a graph, using interval notation, and using set-builder notation.

Solve each inequality. Graph the solution set and write it using interval notation and set-builder notation.

3. $5(x - 2) \leq 5$

4. $0.3x - 0.4 \geq 1.2 - 0.1x$

5. $-16 < -\frac{4}{5}t$

6. $\frac{7}{4}(x + 3) < \frac{3}{8}(x - 3)$

7. $7 - [6t - 5(t - 3)] > 2(t - 3) - 3(t + 1)$

8. $\frac{2b + 7}{2} \leq \frac{3b - 1}{3}$

9. Let $f(x) = \frac{x}{2} - \frac{5}{6}$ and $g(x) = x + \frac{7}{8}$. Find all values of x for which $f(x) < g(x)$.

10. Let $f(x) = 6(4x - 1) - 11(2x - 1)$. Find all values of x for which $f(x)$ is no more than 6.

11. **Paint Spray Guns.**   The conventional spray guns used by many auto body shops are inefficient and wasteful. The annual dollar savings $s(g)$ on paint purchases and disposal costs that shops can amass by using a new high-volume low-pressure (HVLP) spray gun is approximated by $s(g) = 1,700g$, where g is the number of gallons of paint used per week. Use an inequality to determine the number of gallons of paint that a shop must use per week to have an annual savings that exceeds $21,250. (Source: epa.gov)

12. **Printing Costs.**   For business cards, Speedy Printers charges a set-up fee of $45 and 2¢ per card. Vista Printing charges a set-up fee of $24 and 3¢ per card. Use an inequality to determine the number of business cards for which Speedy Printers is more economical.

13. **Investments.**   A woman has invested $10,000 at 6% annual interest. How much more must she invest at 7% so that her annual income is at least $2,000?

14. **Ice Skating.**   For the free-skating portion of a competition, an ice skater received scores of 5.3, 4.8, 4.7, 4.9, and 5.1 from the first five judges. What score must she receive from the sixth and final judge to average better than 5.0 for her performance?

15. **Lawyers.**   A lawyer earns $200 an hour for telephone consultations and $300 an hour for office consultations with clients. To save time for court appearances, she limits her consulting to 15 hours a week. What is the greatest number of hours that she can spend on the phone and still earn at least $4,000 in consulting fees a week?

16. Explain how to use the graphs of $y = 1$ and $y = x - 3$ to solve $x - 3 \leq 1$.

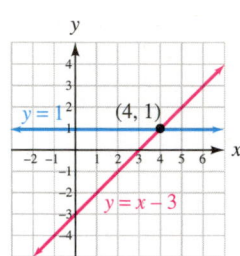

SECTION 4.2 ▶ Solving Compound Inequalities

DEFINITIONS AND CONCEPTS	EXAMPLES
The **intersection** of two sets A and B, written $A \cap B$, is the set of all elements that are common to set A and set B.	Let $A = \{-2, 0, 3, 5\}$ and $B = \{-3, 0, 5, 7\}$. $A \cap B = \{0, 5\}$ The intersection contains the elements that the sets have in common.
The **union** of two sets A and B, written $A \cup B$, is the set of all elements that are in set A, set B, or both.	$A \cup B = \{-3, -2, 0, 3, 5, 7\}$ The union contains the elements that are in one or the other set, or both.
When the word *and* or the word *or* is used to connect pairs of inequalities, we call the statement a **compound inequality**. The solution set of a **compound inequality containing the word *and*** includes all numbers that make both of the inequalities true. That is, it is the intersection of their solution sets.	Solve: $2x - 1 \leq 5$ and $5x + 1 > 4$ We solve each inequality separately. Then we graph the two solution sets (one in red, the other in blue) on the same number line and determine their intersection.

Solve: $2x - 1 \leq 5$ and $5x + 1 > 4$

$2x - 1 \leq 5$	and	$5x + 1 > 4$	***Preliminary Work:***
$2x \leq 6$		$5x > 3$	
$x \leq 3$		$x > \frac{3}{5}$	

The purple-shaded interval shown below is where the red and blue graphs overlap. Thus, the solution set is:

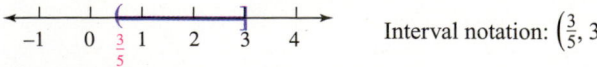

Interval notation: $\left(\frac{3}{5}, 3\right]$

Inequalities that contain exactly two inequality symbols are called **double inequalities.** Any double linear inequality can be written as a compound inequality containing the word *and.* For example:

$$c < x < d \text{ is equivalent to } c < x \text{ and } x < d$$

Solve: $-7 \le 3x - 1 < 5$

We apply properties of inequality to *all three of its parts* to isolate x in the middle.

$$-7 \le 3x - 1 < 5$$
$$-7 + 1 \le 3x - 1 + 1 < 5 + 1 \quad \text{Add 1 to all three parts.}$$
$$-6 \le 3x < 6$$
$$\frac{-6}{3} \le \frac{3x}{3} < \frac{6}{3} \quad \text{Divide each part by 3.}$$
$$-2 \le x < 2$$

The solution set is:

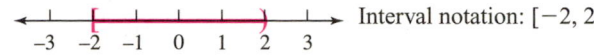

Interval notation: $[-2, 2)$

The solution set of a **compound inequality containing the word *or*** includes all numbers that make one or the other, or both, inequalities true. That is, it is the union of their solution sets.

Solve: $2x - 1 > 5$ or $-(5x - 7) \ge 2$

We solve each inequality separately. Then we graph the two solution sets on the same number line to show their union.

$$\begin{array}{ll} 2x - 1 > 5 & \quad -(5x - 7) \ge 2 \\ 2x > 6 & \quad -5x + 7 \ge 2 \\ x > 3 & \quad -5x \ge -5 \\ & \quad x \le 1 \end{array}$$

The solution set is:

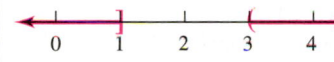

Interval notation: $(-\infty, 1] \cup (3, \infty)$
This is the union of two intervals.

REVIEW EXERCISES

Let $A = \{-6, -3, 0, 3, 6\}$ and $B = \{-5, -3, 3, 8\}$.

17. Find $A \cap B$.   **18.** Find $A \cup B$.

Check to determine whether -4 is a solution of the compound inequality.

19. $x < 0$ and $x > -5$

20. $x + 3 < -3x - 1$ and $4x - 3 > 3x$

Graph each set.

21. $(-3, 3) \cup [1, 6]$ **22.** $(-\infty, 2] \cap [1, 4)$

Solve each compound inequality. Graph the solution set and write it using interval notation.

23. $-2x > 8$ and $x + 4 \ge -6$

24. $5(x + 2) \le 4(x + 1)$ and $11 + x < 0$

25. $\frac{2}{5}x - 2 < -\frac{4}{5}$ and $-\frac{x}{3} < -1$

26. $4\left(x - \frac{1}{4}\right) \le 3x - 1$ and $x \ge 0$

Solve each double inequality. Graph the solution set and write it using interval notation.

27. $3 < 3x + 4 < 10$ **28.** $-2 \le \frac{5 - x}{2} \le 2$

Check to determine whether -4 is a solution of the compound inequality.

29. $x < 1.6$ or $x > -3.9$

30. $x + 1 < 2x - 1$ or $4x - 3 > 3x$

Solve each compound inequality. Graph the solution set and write it using interval notation.

31. $x + 1 < -4$ or $x - 4 > 0$

32. $\frac{x}{2} + 3 > -2$ or $4 - x > 4$

33. Rugs. A manufacturer makes a line of decorator rugs that are 4 feet wide and of varying lengths x (in feet). The floor area covered by the rugs ranges from 17 ft^2 to 25 ft^2. Write and then solve a double linear inequality to find the range of the lengths of the rugs.

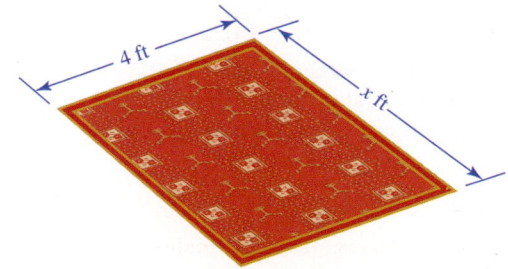

34. Match each word in Column I with *two* associated items in Column II.

Column I	Column II
a. or	**i.** $\cap$
	ii. $\cup$
b. and	**iii.** intersection
	iv. union

35. Let $f(x) = \frac{5}{4}x - 140$. Find all values of x for which $f(x) < -48$ or $f(x) \geq 32$.

36. Let $f(x) = 3x - 5$. Find all values of x for which $-4 \geq f(x) > -12$.

SECTION 4.3 ▶ Solving Absolute Value Equations and Inequalities

DEFINITIONS AND CONCEPTS	EXAMPLES
To **solve absolute value equations** of the form $\|X\| = k$, where $k > 0$, solve the equivalent **compound equation** $$X = k \quad \text{or} \quad X = -k$$ If k is negative, then $\|X\| = k$ has no solution. Recall that **equivalent equations** have the same solutions.	Solve: $\|2x + 1\| = 7$ This absolute value equation is equivalent to the following compound equation: $$\begin{array}{ccc} 2x + 1 = \mathbf{7} & \text{or} & 2x + 1 = \mathbf{-7} \\ 2x = 6 & & 2x = -8 \\ x = 3 & & x = -4 \end{array}$$ This equation has two solutions: 3 and -4. The solution set is $\{-4, 3\}$.
Since absolute value expresses distance, the absolute value of a number is always positive or zero, but never negative.	Solve: $\|4x - 5\| = -3$ Since an absolute value can never be negative, there are no real numbers x that make $\|4x - 5\| = -3$ true. The equation has no solution and the solution set is $\varnothing$.
To **solve absolute value equations** of the form $\|X\| = \|Y\|$, solve the compound equation $$X = Y \quad \text{or} \quad X = -Y$$ The expressions within the absolute value symbols are equal, or they are opposites.	Solve: $\|3x - 2\| = \|2x + 4\|$ This equation is equivalent to the following compound equation: $$\begin{array}{ccc} 3x - 2 = 2x + 4 & \text{or} & 3x - 2 = \mathbf{-}(2x + 4) \\ x - 2 = 4 & & 3x - 2 = -2x - 4 \\ x = 6 & & 5x - 2 = -4 \\ & & 5x = -2 \\ & & x = -\frac{2}{5} \end{array}$$ This equation has two solutions: 6 and $-\frac{2}{5}$. The solution set is $\left\{-\frac{2}{5}, 6\right\}$.
To **solve absolute value inequalities** of the form $\|X\| < k$, where $k > 0$, solve the equivalent double inequality $-k < X < k$. Use a similar approach to solve $\|X\| \leq k$. When solving absolute value equations or inequalities, **isolate the absolute value expression** on one side **before** writing the equivalent compound statement.	Solve: $\|4x - 3\| - 2 < 7$ $$\|4x - 3\| - 2 \mathbf{+ 2} < 7 \mathbf{+ 2} \quad \text{To isolate the absolute value, add 2 to both sides.}$$ $$\|4x - 3\| < 9$$ The resulting inequality is equivalent to the following double inequality: $$-9 < 4x - 3 < 9$$ $$-6 < 4x < 12 \quad \text{Add 3 to all three parts.}$$ $$-\frac{3}{2} < x < 3 \quad \text{Divide each part by 4 and simplify.}$$ The solution set is: Interval notation: $\left(-\frac{3}{2}, 3\right)$ Set builder: $\left\{x \mid -\frac{3}{2} < x < 3\right\}$

To **solve absolute value inequalities** of the form $|X| \geq k$, where $k > 0$, solve the equivalent compound inequality $X \leq -k$ or $X \geq k$.

Use a similar approach to solve $|X| > k$.

Solve: $|3x + 1| \geq 7$

This inequality is equivalent to the following compound inequality:

$$3x + 1 \leq -7 \quad \text{or} \quad 3x + 1 \geq 7$$
$$3x \leq -8 \qquad\qquad 3x \geq 6$$
$$x \leq -\frac{8}{3} \qquad\qquad x \geq 2$$

Interval notation: $\left(-\infty, -\frac{8}{3}\right] \cup [2, \infty)$

This is the union of two intervals.

Set builder: $\left\{ x \mid x \leq -\frac{8}{3} \text{ or } x > 2 \right\}$

REVIEW EXERCISES

Solve each absolute value equation.

37. $|4x| = 8$

38. $2|3x + 1| - 1 = 19$

39. $\left|\frac{3}{2}x - 4\right| - 10 = -1$

40. $\left|\frac{2 - x}{3}\right| = -4$

41. $|-4(2x - 6)| = 0$

42. $\left|\frac{3}{8} + \frac{x}{3}\right| = \frac{5}{12}$

43. $|3x + 2| = |2x - 3|$

44. $\left|\frac{2(1 - x) + 1}{2}\right| = \left|\frac{3x - 2}{3}\right|$

Solve each absolute value inequality. Graph the solution set and write it using interval notation.

45. $|x| \leq 3$

46. $|2x + 7| < 3$

47. $2|5 - 3x| \leq 28$

48. $\left|\frac{2}{3}x + 14\right| + 6 < 6$

49. $|x| > 1$

50. $\left|\frac{1 - 5x}{3}\right| \geq 7$

51. $|3x - 8| - 4 > 0$

52. $\left|\frac{3}{2}x - 14\right| \geq 0$

53. Produce.   Before packing, freshly picked tomatoes are weighed on the scale shown. Tomatoes having a weight w (in ounces) that falls within the highlighted range are sold to grocery stores.

a. Complete the following absolute value inequality that expresses the acceptable weight range:
$|w - \boxed{}| \leq \boxed{}$

b. Solve the inequality from part (a) and express the acceptable weight range using interval notation.

54. Let $f(x) = \frac{1}{3}|6x| - 1$. For what value(s) of x is $f(x) = 5$?

55. Let $f(x) = 2|3(x + 4)| + 1.5$. Find all values of x for which $f(x) = 25.5$.

56. Let $f(x) = |7 - x|$. Find all values of x for which $f(x) < 5$.

57. Explain why $|0.04x - 8.8| < -2$ has no solution.

58. Explain why the solution set of $\left|\frac{3x}{50} + \frac{1}{45}\right| \geq -\frac{4}{5}$ is the set of all real numbers.

SECTION 4.4 ▶ Linear Inequalities In Two Variables

DEFINITIONS AND CONCEPTS	EXAMPLES
The graph of a **linear inequality in x and y** is the graph of all ordered pairs (x, y) whose coordinates satisfy the inequality.	Graph: $y < 3x + 2$

The graph of a **linear inequality in x and y** is the graph of all ordered pairs (x, y) whose coordinates satisfy the inequality.

To **graph a linear inequality in x and y,** graph the **boundary line.** Draw a solid boundary line if the inequality is $\leq$ or $\geq$. Draw a dashed line if the inequality is $<$ or $>$.

Then use a **test point** to decide which side of the boundary should be shaded. If the inequality is satisfied, shade the **half-plane** that contains the test point. If the inequality is not satisfied, shade the half-plane on other side of the boundary.

Graph: $y < 3x + 2$

To find the boundary line, we graph $y = 3x + 2$. Since the inequality symbol $<$ does not include an $=$ symbol, points on the boundary line are not included in the graph, and we draw a dashed boundary line.

To find which half-plane is the graph of $y < 3x + 2$, we choose the origin $(0, 0)$ as the test point and see whether its coordinates satisfy the inequality.

$$y < 3x + 2$$
$$0 \overset{?}{<} 3(0) + 2 \quad \text{Substitute 0 for x and 0 for y.}$$
$$0 < 2 \quad \text{True}$$

Since the coordinates of the origin satisfy the inequality, the origin is in the graph. In fact, the coordinates of every point on the same side of the boundary line as the origin satisfy the inequality. We then shade that half-plane in red to complete the graph.

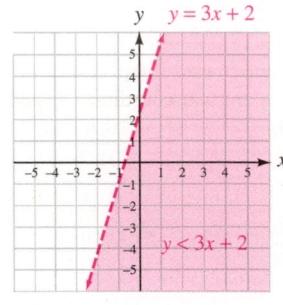

REVIEW EXERCISES

59. Check to determine whether $(-1, -4)$ is a solution of the linear inequality $6x - 4y \geq 15$.

60. Does the graph of $6x - 4y \geq 15$ include the boundary line?

Graph each inequality in the rectangular coordinate system.

61. $2x + 3y > 6$

62. $y \leq 4 - x$

63. $y < \dfrac{1}{2}x$

64. $x \geq -\dfrac{3}{2}$

65. Find the equation of the boundary line. Then give the inequality whose graph is shown.

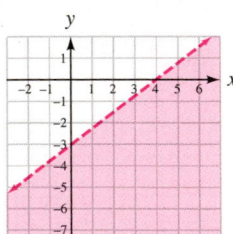

66. Concert Tickets. Tickets to a concert cost $6 for reserved seats and $4 for general admission. If receipts must be at least $10,200 to meet expenses, find an inequality that shows the possible combinations of the number of reserved seats (x) and the number of general admission tickets (y) that the box office can sell. Then graph the inequality for nonnegative values of x and y and give three ordered pairs that satisfy the inequality.

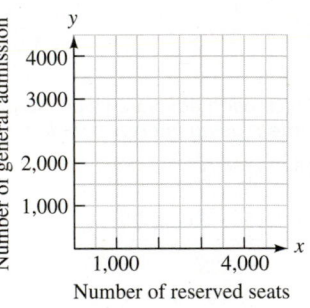

SECTION 4.5 ▶ Systems of Linear Inequalities

DEFINITIONS AND CONCEPTS	EXAMPLES
To **solve a system of linear inequalities,** graph each of the inequalities on the same rectangular coordinate system and look for the intersection of the shaded half-planes. The area that is shaded twice represents the solutions of the given system.	Graph the solution set of: $\begin{cases} y < x - 1 \\ 2x + y \geq 2 \end{cases}$ The graph of $y < x - 1$ is shaded in red. The graph of $2x + y \geq 2$ is shaded in blue. Any point in the doubly shaded region has coordinates that satisfy both inequalities. 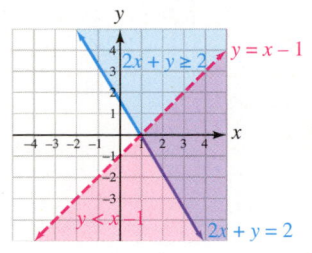 $y = x - 1$ $2x + y = 2$ $m = 1$ $b = -1$ <table><tr><th>x</th><th>y</th><th>(x, y)</th></tr><tr><td>0</td><td>2</td><td>(0, 2)</td></tr><tr><td>1</td><td>0</td><td>(1, 0)</td></tr></table>
Compound inequalities can be graphed in the rectangular coordinate system.	Graph: $-2 < x \leq 4$ Graph: $y \geq 1$ or $y < -3$ The graph of this double inequality contains all points in the plane that satisfy the inequalities $-2 < x$ and $x \leq 4$. These points are in the purple-shaded region of figure (a) below. The graph of this compound inequality contains all the points in the plane that satisfy one or the other inequalities. They are shaded in red in figure (b) below. 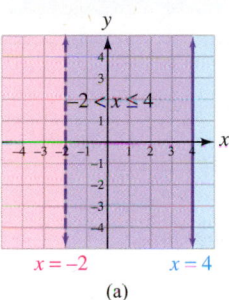 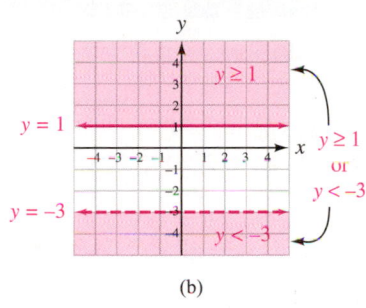 (a) (b)

REVIEW EXERCISES

67. Check to determine whether $(1, -2)$ is a solution of the system of linear inequalities $\begin{cases} y \leq -x + 1 \\ 2x - y > 2 \end{cases}$.

68. In the illustration, the solution of one linear inequality is shaded in red, and the solution of a second is shaded in blue. Determine whether a true or false statement results if the coordinates of the given point are substituted into the given inequality.

  a. *A*, inequality 1   **b.** *A*, inequality 2
  c. *B*, inequality 1   **d.** *B*, inequality 2
  e. *C*, inequality 1   **f.** *C*, inequality 2

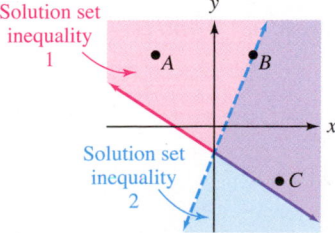

Graph the solution set of each system of inequalities.

69. $\begin{cases} y \geq x + 1 \\ 3x + 2y < 6 \end{cases}$ **70.** $\begin{cases} x - y < 3 \\ y \leq 0 \\ x \geq 0 \end{cases}$

Graph each compound inequality in the rectangular coordinate system.

71. $-2 < x < 4$ **72.** $y \leq -2$ or $y > 1$

73. Inventory. A men's clothing store carries shirts that sell for \$20 and some that sell for \$30. The store manager wants to stock at least \$300 but no more than \$600 in shirts at the store. Graph a system of inequalities that will show the possible combinations of the number of \$20 shirts (x) and the number of \$30 shirts (y) that the store can stock. Give two possible solutions.

74. Petroleum Exploration. Organic matter converts to oil and gas within a specific range of temperature and depth called the *petroleum window*. The petroleum window shown can be described by a system of linear inequalities, where x is the temperature in °C of the soil at a depth of y meters. Determine what inequality symbol should be inserted in each blank.

$$\begin{cases} x \quad\rule{1cm}{0.4pt}\quad 35 \\ x \quad\rule{1cm}{0.4pt}\quad 130 \\ y \quad\rule{1cm}{0.4pt}\quad -56x + 280 \\ y \quad\rule{1cm}{0.4pt}\quad -18x + 90 \end{cases}$$

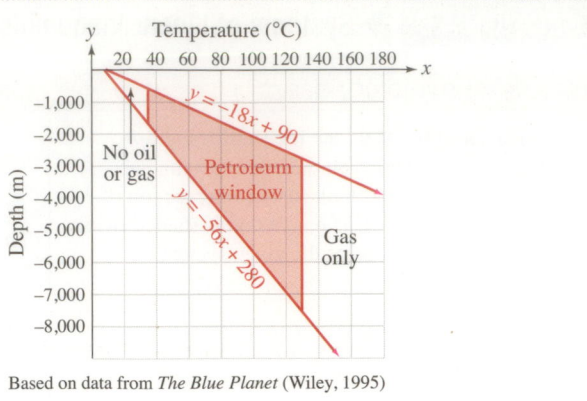

Based on data from *The Blue Planet* (Wiley, 1995)

---

4 ▶ Chapter Test

1. Fill in the blanks.

   a. $<$, $>$, $\leq$, and $\geq$ are _____ symbols.

   b. ∞ is a symbol representing _____.

   c. $x + 1 > 2$ or $2x - 3 \leq 8$ is a _____ inequality.

   d. We read $\cup$ as _____ and $\cap$ as _____.

   e. $\begin{cases} x + y > 10 \\ 3x - 2y \geq 4 \end{cases}$ is a _____ of linear inequalities in _____ variables.

2. Check to determine whether -2 is a solution of the inequality $3(x - 2) \leq 2(x + 7)$.

Solve each inequality. Graph the solution set and write it using interval notation.

3. $\dfrac{2}{3}t - 1 > 7$

4. $-2(2x + 3) \geq 14$

5. $\dfrac{x}{4} - \dfrac{1}{3} > \dfrac{5}{6} + \dfrac{x}{3}$

6. $4 - 4[3t - 2(3 - t)] \leq -15t - (5t - 28)$

7. **Averaging Grades.** Use the information from the gradebook to determine what score Karen Nelson-Sims needs on the fifth exam so that her exam average exceeds 80.

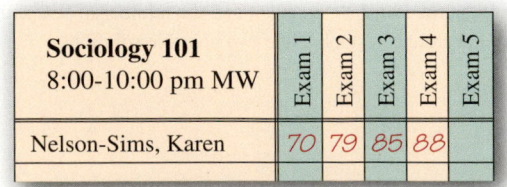

8. **Insurance Coverage.** A fire damage restoration crew charges $175 for the first hour of cleanup and $80 for each additional hour or part thereof. For how long can the crew work at a home with smoke and water damage if the homeowner's insurance policy will only cover up to $1,000 of the cleanup cost?

9. **Medical Schools.** The number of women graduating annually from U.S. medical schools $w(t)$ can be approximated by the function $w(t) = 150t + 5,200$, where t is the number of years after 1990. Use an inequality to determine those years for which the number of women graduating from medical school was at most 6,550. (Source: Wikipedia article: AAMC)

10. **Job Offers.** A recent graduate of a culinary school has job offers from two local bakeries.

    ■ Corner Bakery: $15.50 an hour but employee must purchase $135 worth of uniforms to start

    ■ Main Street Bakery: $12.80 an hour, uniforms are provided by employer

    How many hours must be worked at the Corner Bakery for it to be at least as profitable as working at Main Street Bakery?

11. Determine whether 4 is a solution of the compound inequality $x + 6 \geq 10$ and $3x - 8 > 4$.

Let $A = \{-4, 0, 7, 8, 9, 11\}$ and $B = \{-5, -4, 0, 10, 11\}$.

12. Find $A \cap B$.

13. Find $A \cup B$.

14. Graph each set.

    a. $(-3, 6) \cup [5, \infty)$

    b. $[-2, 7] \cap (-\infty, 1)$

Solve each compound inequality. Give the result in interval notation, if possible, and graph the solution set.

15. $3x \geq -2x + 5$ and $7 \geq 4x - 2$

16. $3x < -9$ or $-\dfrac{x}{4} < -2$

17. $-2 < \dfrac{x - 4}{3} < 4$

18. $\dfrac{4}{5}(x + 1) > 1$ and $-(0.3x + 1.5) > 2.9 - 0.2x$

19. Let $f(x) = 0.08x + 6.48$. Find all values of x for which $f(x) > 24.8$ and $-72.8 \leq f(x)$.

20. Match each inequality or compound inequality with the graph of its solution set.

   a. $x > 1$ **b.** $x > 1$ or $x < -1$

   c. $-1 < x < 1$ **d.** $x < 1$

   i.

           −3 −2 −1 0 1 2 3

   ii.

           −3 −2 −1 0 1 2 3

   iii.

           −3 −2 −1 0 1 2 3

   iv.

           −3 −2 −1 0 1 2 3

Solve each equation.

21. $|4 - 3x| = 19$

22. $|3x + 4| = |x + 12|$

23. $10 = 4\left|\dfrac{3x}{8} - \dfrac{3x}{2}\right| + 6$

24. $|16x| = -16$

25. $5|20 - 2x| = 0$

26. Aluminum Production. A sheet of aluminum is to be 0.0625 inch thick, with a tolerance of 0.0015 inch. Write and then solve an absolute value inequality that describes this specification, using x to represent the thickness of a sheet of aluminum.

Solve each inequality. Graph the solution set and write it using interval notation.

27. $|x + 3| \leq 4$

28. $\left|\dfrac{x - 2}{2}\right| > 5.5$

29. $|4 - 2x| + 48 > 50$

30. $2|3(x - 2)| \leq 4$

31. $|4.5x - 0.9| \geq -0.7$

32. $-160 \geq |4x|$

33. Let $f(x) = |2x + 9|$. For what values of x is $f(x) < 3$?

34. Let $f(x) = \left|12 - \dfrac{3}{5}x\right|$ and $g(x) < 9$. Find all values of x for which $f(x) < g(x)$.

Graph each solution set on a rectangular coordinate system.

35. $5x + 3y \leq 10$

36. $y < x$

37. $\begin{cases} 2x - 3y \geq 6 \\ y < -x + 1 \end{cases}$

38. $-2 \leq y < 5$

39. $x < -3$ or $x \geq 4$

40. Accounting. On average, it takes an accountant 1 hour to complete a simple tax return and 3 hours to complete a complicated return. If the accountant wants to work less than 9 hours per day, find an inequality that shows the possible combinations of the number of simple returns (x) and the number of complicated returns (y) that can be completed each day. Then graph the inequality and give three ordered pairs that satisfy it.

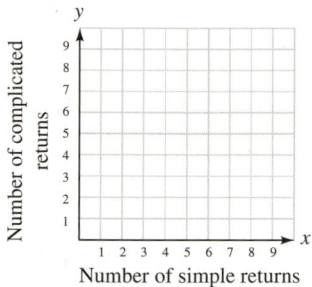

Number of simple returns

41. Two linear inequalities are graphed on the same coordinate axes in the illustration. The solution set of the first inequality is shaded in red, and the solution set of the second in blue.

   a. Determine from the graph whether $(3, -4)$ is a solution of either inequality.

   b. Is $(3, -4)$ a solution of the system of two linear inequalities? Explain your answer.

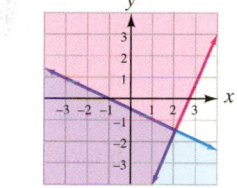

42. Indoor Climates. The general zone of comfort acceptable to most people when working in an office can be described by a system of linear inequalities where x is the dry bulb temperature and y is the percent relative humidity. See the illustration. Determine what inequality symbol should be inserted in each blank.

$$\begin{cases} y \quad\underline{}\quad 60 \\ y \quad\underline{}\quad 27 \\ y \quad\underline{}\quad -11x + 852 \\ y \quad\underline{}\quad -5x + 445 \end{cases}$$

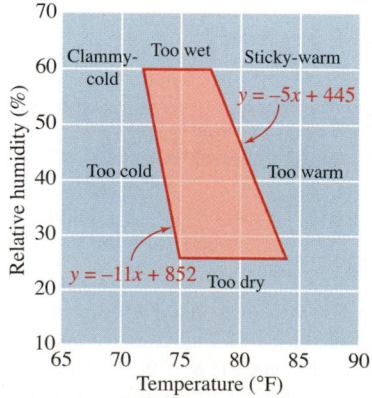

43. Match each equation, inequality, or system with the graph of its solution set.

a. $x - 2y = -2$ **b.** $x - 2y \leq -2$

c. $\begin{cases} x - 2y = -2 \\ 2x + 3y = 6 \end{cases}$ **d.** $\begin{cases} x - 2y \leq -2 \\ 2x + 3y \leq 6 \end{cases}$

i.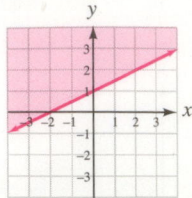

ii.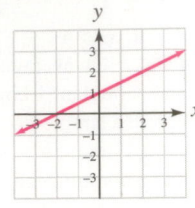

iii.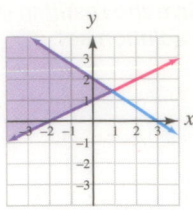

iv.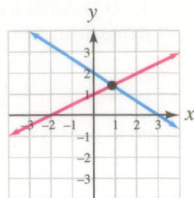

44. Can the statement
$$x < 3 \quad \text{or} \quad x \geq 10$$
be written as $10 \leq x < 3$? Explain why or why not.

Group Project

Venn Diagrams

Overview: In this activity, we will discuss several of the fundamental concepts of what is known as *set theory*.

Instructions: *Venn diagrams* are a convenient way to visualize relationships between sets and operations on sets. They were invented by the English mathematician John Venn (1834–1923). To draw a Venn diagram, we begin with a large rectangle, called the *universal set*. Ovals or circles are then drawn in the interior of the rectangle to represent subsets of the universal set.

Form groups of 2 or 3 students. Study the following figures, which illustrate three set operations: union, intersection, and complement.

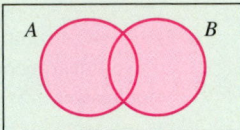

$A \cup B$
The shaded region is the *union* of set A and set B.

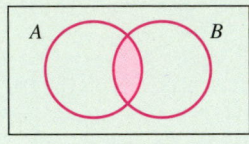

$A \cap B$
The shaded region is the *intersection* of set A and set B.

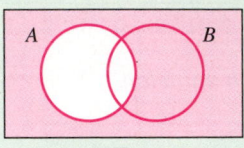

$\overline{A}$
The shaded region is the complement of set A.

For each of the following exercises, sketch the following blank Venn diagram and then shade the indicated region.

1. $A \cup B$
2. $A \cap B$
3. $A \cap C$
4. $A \cup C$
5. $A \cup B \cup C$
6. $A \cap B \cap C$
7. $(B \cup C) \cap A$
8. $C \cup (A \cap B)$
9. $\overline{A}$
10. $\overline{B} \cup \overline{C}$
11. $\overline{B} \cap \overline{C}$
12. $\overline{A \cup B}$

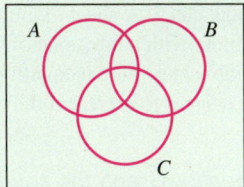

CUMULATIVE REVIEW ▶▶ Chapters 1–4

1. Fill in the blanks: The set of _____ numbers together with the set of _____ numbers form the set of real numbers. [Section 1.2]

2. Hardware. The thread profile of a screw is determined by the distance between threads. This distance, indicated by the letter p, is known as the *pitch*. If $p = 0.125$, find each of the other dimensions labeled in the illustration. [Section 1.3]

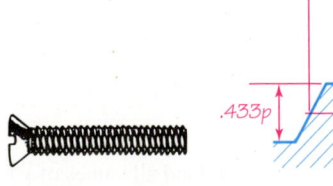

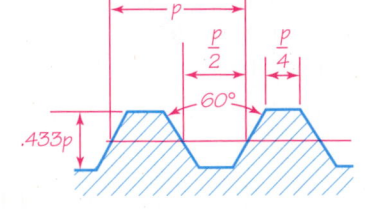

Evaluate each expression for x = 2 and y = −4.

3. $|x| - xy$ [Section 1.3]

4. $\dfrac{x^2 - y^2}{3x + y}$ [Section 1.3]

5. Use the given property to complete each statement. [Section 1.4]

   a. $9 + x = \underline{}$ Commutative property of addition

   b. $6(10n) = \underline{}$ Associative property of multiplication

6. a. What is the additive inverse (opposite) of x? [Section 1.4]

   b. What is the multiplicative inverse (reciprocal) of x? [Section 1.4]

Simplify each expression.

7. $-(a + 2) - (a - b)$ [Section 1.4]

8. $36\left(\dfrac{2}{9}t - \dfrac{3}{4}\right) + 36\left(\dfrac{1}{2}\right)$ [Section 1.4]

Solve each equation, if possible.

9. $6(x - 1) = 2(x + 3)$ [Section 1.5]

10. $\dfrac{5b}{2} - 10 = \dfrac{b}{3} + 3$ [Section 1.5]

11. $2a - 5 = -2a + 4(a - 2) + 1$ [Section 1.5]

12. $\dfrac{2z + 3}{3} + \dfrac{3z - 4}{6} = \dfrac{z - 2}{2}$ [Section 1.5]

13. Solve $l = a + (n - 1)d$ for d. [Section 1.5]

14. Plastic Wrap. Estimate the number of *square feet* of plastic wrap on a roll if the dimensions printed on the box describe the roll as 205 feet long by $11\frac{3}{4}$ inches wide. [Section 1.6]

15. Investments. Find the amount of money that was invested at $8\frac{7}{8}\%$ if it earned \$1,775 in simple interest in one year. [Section 1.6]

16. Marathons. Two marathon runners leave the starting gate at the same time, one running 12 mph and the other 10 mph. If they maintain the pace, how long will it take for them to be one-half of a mile apart? [Section 1.8]

17. Find the slope of the line that passes through $(0, -8)$ and $(-5, 0)$. [Section 2.3]

18. Prisons. The graph in the next column shows the growth of the U.S. prison population from 1980 to 2010.

   a. Find the rate of change in the prison population from 2000 to 2005. [Section 2.3]

   b. During what five-year period was the rate of change in the U.S. prison population the greatest? Find the rate of change. [Section 2.3]

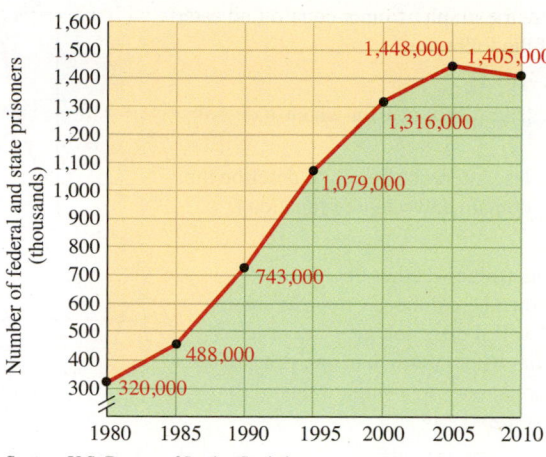

Source: U.S. Bureau of Justice Statistics

19. Determine whether the lines represented by the equations $3x = y + 4$ and $y = 3(x - 4) - 1$ are parallel, perpendicular, or neither. [Section 2.4]

20. Find an equation of the line that passes through $(-2, 3)$ and is perpendicular to the graph of $3x + y = 8$. Write the equation in slope–intercept form. [Section 2.4]

21. Find the domain and range of the relation: $\{(5, 6), (-12, 4), (8, 6), (-6, -6), (5, 4)\}$ [Section 2.5]

22. Does the equation $x = |y|$ define y to be a function of x? [Section 2.5]

Let $f(x) = 3x^2 - x$. Find each of the following.

23. $f(-2)$ [Section 2.5]   **24.** $f(t)$ [Section 2.5]

25. Aging. The graph below shows the effects of aging on cardiac output (the amount of blood that the heart can pump in one minute). [Section 2.5]

   a. Write the linear function that models this situation.

   b. Use your answer to part a to determine the cardiac output at age 90.

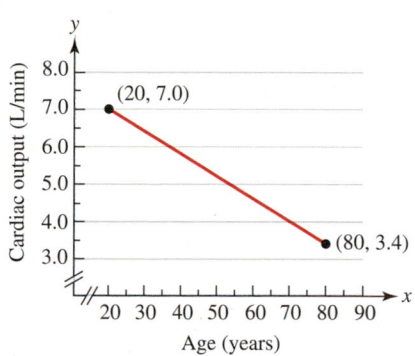

Based on data from *Cardiopulmonary Anatomy and Physiology, Essentials for Respiratory Care*, 2nd ed. (Delmar Publishers, 1994)

26. Use the graph of function f to find each of the following. [Section 2.5]

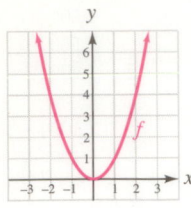

   a. $f(-1)$

   b. The values of x for which $f(x) = 4$

   c. Use interval notation to describe the domain and range of function f.

27. Determine whether the relation shown here defines y to be a function of x. If it does not, give two ordered pairs where more than one value of y corresponds to a single value of x. [Section 2.5]

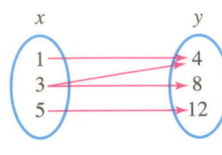

28. Determine whether the equation $y = x^3 + 1$ defines y to be a function of x. If it does not, give two ordered pairs where more than one value of y corresponds to a single value of x. [Section 2.5]

29. First sketch the graph of the basic function associated with $g(x) = |x| - 2$. Then sketch the graph of g on the same coordinate system using a translation. Describe the domain and range of function g in words. [Section 2.6]

30. Is the graph at the right the graph of a function? [Section 2.6]

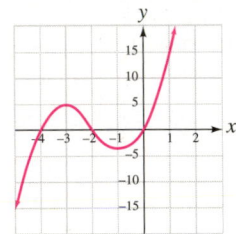

31. Use graphing to solve: $\begin{cases} 2x + y = 5 \\ x - 2y = 0 \end{cases}$ [Section 3.1]

32. College Graduates. The following illustration shows the percent of all college degrees awarded in the United States by gender for the years 1970 through 2017 (projected). Estimate the point of intersection of the red and blue line graphs. Explain its significance. [Section 3.1]

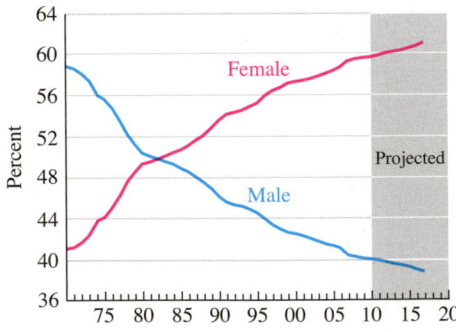

Source: *Dept. of Education*

33. Use elimination to solve: $\begin{cases} \dfrac{x}{10} + \dfrac{y}{5} = \dfrac{1}{2} \\ \dfrac{x}{2} - \dfrac{y}{5} = \dfrac{13}{10} \end{cases}$ [Section 3.2]

34. Use substitution to solve: $\begin{cases} 3x = 4 - y \\ 4x - 3y = -1 + 2x \end{cases}$ [Section 3.2]

35. Solve: $\begin{cases} x + y + z = 1 \\ 2x - y - z = -4 \\ x - 2y + z = 4 \end{cases}$ [Section 3.3]

36. Use matrices to solve the system. $\begin{cases} 4x - 3y = -1 \\ 3x + 4y = -7 \end{cases}$ [Section 3.4]

37. Evaluate the determinant: $\begin{vmatrix} -9 & 7 \\ 4 & -2 \end{vmatrix}$ [Section 3.5]

38. Use Cramer's rule to solve the system.

$\begin{cases} 5x + 2y = 11 \\ 7x + 6y = 9 \end{cases}$ [Section 3.5]

39. Entrepreneurs. A person invests \$18,375 to set up a small business producing a piece of computer software that will sell for \$29.95. If each piece can be produced for \$5.45, how many pieces must be sold to break even? [Section 3.6]

40. Ticket Sales. Tickets for a concert cost \$5, \$3, and \$2. Twice as many \$5 tickets were sold as \$2 tickets. The receipts for 750 tickets were \$2,625. How many tickets were sold at each price? [Section 3.7]

Solve each inequality. Graph the solution set and write it using interval notation.

41. $-3(x - 4) \geq x - 32$ [Section 4.1]

42. $-8 < -3x + 1 < 10$ [Section 4.2]

Solve each compound inequality. Graph the solution set and write it using interval notation.

43. $3x + 2 < 8$ or $2x - 3 > 11$ [Section 4.2]

44. $5x - 3 \geq 2$ and $6 \geq 4x - 3$ [Section 4.2]

45. Solve: $2|4x - 3| + 1 = 19$ [Section 4.3]

46. Let $f(x) = |2x - 1|$ and $g(x) = |3x + 4|$. Find all values of x for which $f(x) = g(x)$. [Section 4.3]

Solve. Graph the solution set and write it using interval notation.

47. $|3x - 2| \leq 4$ [Section 4.3]

48. Let $f(x) = |2x + 3| - 1$. Find all values of x for which $f(x) > 4$. [Section 4.3]

Graph the solution set.

49. $2x - 3y \leq 12$ [Section 4.4]   **50.** $\begin{cases} y < x + 2 \\ 3x + y \leq 6 \end{cases}$ [Section 4.5]

Exponents, Polynomials, and Polynomial Functions

5

©Leslie Harris/Photolibrary

from Campus to Careers

Landscape Architect

Whether it's a community park, a college campus, or simply someone's backyard, landscape architects are skilled at creating outdoor areas that are both functional and beautiful. They use algebra and geometry to prepare working drawings, design scale models, and estimate costs. Throughout the planning and construction phases, they make calculations to find everything from drainage slopes and sunlight angles to walkway elevations.

Problem 123 in **Study Set 5.1, problem 91** in **Study Set 5.3,** and **problem 97** in **Study Set 5.9** involve situations that a landscape architect might encounter on the job. The mathematical concepts discussed in this chapter can be used to solve those problems.

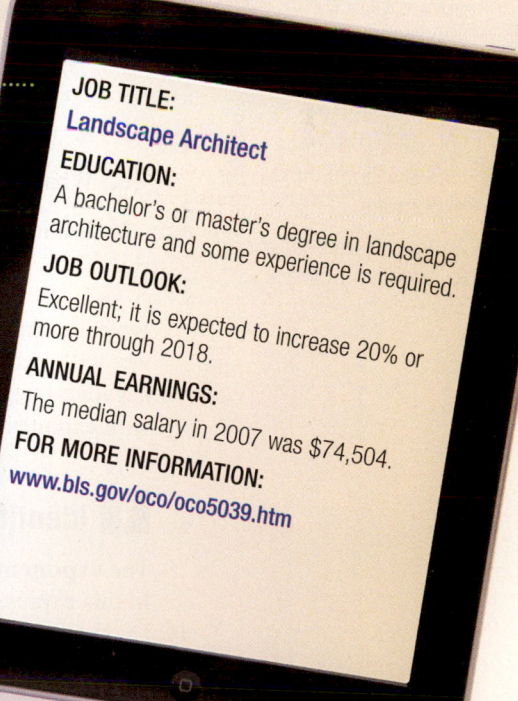

JOB TITLE:
Landscape Architect

EDUCATION:
A bachelor's or master's degree in landscape architecture and some experience is required.

JOB OUTLOOK:
Excellent; it is expected to increase 20% or more through 2018.

ANNUAL EARNINGS:
The median salary in 2007 was $74,504.

FOR MORE INFORMATION:
www.bls.gov/oco/oco5039.htm

It's not uncommon for students' enthusiasm to lessen toward the middle of the term. Sometimes, their effort and attendance begin to slip. Realize that missing even one class can have a great effect on your grade. Being tardy takes its toll as well. If you are just a few minutes late, or miss an entire class, you risk getting behind. So, keep the following tips in mind.

ARRIVE ON TIME, OR A LITTLE EARLY: When you arrive, get out your note-taking materials and homework. Identify any questions that you plan to ask your instructor once the class starts.

IF YOU MUST MISS A CLASS: Get a set of notes, the homework assignments, and any handouts that the instructor may have provided for the day(s) that you missed.

STUDY THE MATERIAL YOU MISSED: Take advantage of the online resources that are available with this textbook, such as video examples and problem-specific tutorials. Watch the explanations of the material from the section(s) that you missed.

Now Try This ▶

1. Plan ahead! List five possible situations that could cause you to be late to class or miss a class. (Some examples are: parking/traffic delays, lack of a babysitter, oversleeping, and job responsibilities.) What can you do ahead of time so that these situations won't cause you to be tardy or absent?

2. Watch one section on the video series that accompanies this book. Take notes as you watch the explanations.

SECTION 5.1

OBJECTIVES

1. Identify bases and exponents.
2. Use the product and power rules for exponents.
3. Use the zero and the negative-integer exponent rules.
4. Use the quotient rule for exponents.
5. Simplify quotients raised to negative powers.

Exponents

ARE YOU READY?

The following problems review some basic skills that are needed when working with exponents.

1. Evaluate: **a.** $6 + 6 + 6$ **b.** $6 \cdot 6 \cdot 6$

2. Evaluate: **a.** 2^5 **b.** $2 \cdot 5$

3. Write the expression $x \cdot x \cdot x \cdot x \cdot x \cdot x$ in an equivalent form using an exponent.

4. Translate to mathematical symbols:

   a. y squared   **b.** 7 cubed

5. What is the reciprocal of 8?

6. Fill in the blank: $9m^4$ means $9 \ \boxed{} \ m^4$.

We have evaluated exponential expressions having natural-number exponents. In this section, we will extend the definition of exponent to include zero exponents, as in 3^0, and negative-integer exponents, as in 3^{-2}. We also will develop several rules that simplify work with exponents.

1 Identify Bases and Exponents.

The **exponential expression** x^n is called a *power of x*, and we read it as "*x* to the *n*th power." In this expression, x is called the **base,** and n is called the **exponent.**

Base $\longrightarrow x^n \longleftarrow$ Exponent

Exponents provide a way to write products of *repeated factors* in compact form.

Natural-Number Exponents ▼	A natural-number exponent indicates how many times its base is to be used as a factor. For any number x and any natural number n,

$$x^n = \overbrace{x \cdot x \cdot x \cdot \,\cdots\, \cdot x}^{n \text{ factors of } x}$$

The base of an exponential expression can be a number, a variable, or a combination of numbers and variables. Some examples are:

$$10^5 = 10 \cdot 10 \cdot 10 \cdot 10 \cdot 10$$ The base is 10 and the exponent is 5. Read as "10 to the fifth power" or simply as "10 to the fifth."

$$y^2 = y \cdot y$$ The base is y and the exponent is 2. Read as "y squared."

$$(-2s)^3 = (-2s)(-2s)(-2s)$$ Because of the parentheses, the base is $-2s$. The exponent is 3. Read as "negative 2s raised to the third power" or "negative 2s cubed."

$$-8^4 = -(8 \cdot 8 \cdot 8 \cdot 8)$$ Since the $-$ sign is not written within parentheses, the base is 8. The exponent is 4. Read as "the opposite (or the negative) of 8 to the fourth power."

$$(x - 7)^2 = (x - 7)(x - 7)$$ Because of the parentheses, $x - 7$ is the base. The exponent is 2. Read as "the quantity x minus 7 raised to the second power."

Notation

Bases that contain a $-$ sign *must* be written within parentheses.

$$(-2s)^3 \leftarrow \text{Exponent}$$
Base

An **exponent of 1** means the base is used as a factor 1 time. For example, $4^1 = 4$ and $x^1 = x$. When an exponent is 1, it is usually not written.

2 Use the Product and Power Rules for Exponents.

Several rules for exponents come directly from the definition of exponent. To develop the first rule, we consider $x^5 \cdot x^3$, the product of two exponential expressions having the same base. Since x^5 means that x is to be used as a factor five times, and since x^3 means that x is to be used as a factor three times, $x^5 \cdot x^3$ means that x will be used as a factor eight times.

$$x^5 x^3 = \overbrace{x \cdot x \cdot x \cdot x \cdot x}^{5 \text{ factors of } x} \cdot \overbrace{x \cdot x \cdot x}^{3 \text{ factors of } x} = \overbrace{x \cdot x \cdot x \cdot x \cdot x \cdot x \cdot x \cdot x}^{8 \text{ factors of } x} = x^8$$

In general,

$$x^m x^n = \overbrace{x \cdot x \cdot x \cdot \,\cdots\, \cdot x}^{m \text{ factors of } x} \cdot \overbrace{x \cdot x \cdot x \cdot \,\cdots\, \cdot x}^{n \text{ factors of } x} = \overbrace{x \cdot x \cdot x \cdot x \cdot \,\cdots\, \cdot x}^{m + n \text{ factors of } x} = x^{m+n}$$

This result is called the *product rule for exponents*.

Product Rule for Exponents ▼	To multiply exponential expressions with the same base, keep the common base and add the exponents. For any real number x and any natural numbers m and n,

$$x^m \cdot x^n = x^{m+n}$$ Read as "x to the mth power times x to the nth power equals x to the m plus nth power."

EXAMPLE 1   Simplify each expression: **a.** $x^{11}x^5$ **b.** y^5y^4y **c.** $a^2b^3a^3b^2$ **d.** $-8x^{14}(x^{13})$

Strategy   Since there are products of the form $x^m \cdot x^n$ in each of these expressions, we will use the product rule for exponents and keep the common base and add the exponents to simplify them.

Why   We use the product rule to multiply exponential expressions with the same base.

Solution

The Language of Algebra

Here, the instruction **Simplify** means to write an equivalent expression where each base occurs only once.

a. $x^{11}x^5 = x^{11+5}$ Keep the common base x.   **b.** $y^5y^4y = (y^5y^4)y$ Group the factors.

$\phantom{x^{11}x^5}= x^{16}$ Add the exponents. $= y^9y^1$ Write y as y¹.

$= y^{10}$

c. $a^2b^3a^3b^2 = a^2a^3b^3b^2$ Reorder the factors.   **d.** $-8x^{14}(x^{13}) = -8(x^{14}x^{13})$ Regroup

$= a^5b^5$ $\phantom{-8x^{14}(x^{13})}= -8x^{27}$ the factors.

Self Check 1   Simplify each expression: **a.** $2^3 2^5$ **b.** $k \cdot k^4$
c. $a^2b^3a^3b^4$ **d.** $-8a^{40}(a^{24})$

Now Try   Problems 19, 21, and 25

CAUTION   Here are examples of two common errors associated with the product rule:

$$3^2 \cdot 3^4 \neq 9^6 \qquad\qquad 2^3 \cdot 5^2 \neq 10^5$$

Do not multiply the common bases. Keep   The power rule does not apply
the common base and add exponents to get 3⁶.   because the bases are not the same.

To develop another rule for exponents, we consider $(x^4)^3$, which means x^4 cubed.

The Language of Algebra

An exponential expression raised to a power, such as $(x^4)^3$, is called a **power of a power**.

$$(x^4)^3 = \overbrace{x^4 \cdot x^4 \cdot x^4}^{} = \underbrace{x \cdot x \cdot x \cdot x \cdot x \cdot x \cdot x \cdot x \cdot x \cdot x \cdot x \cdot x}_{} = x^{12}$$

In general, we have

$$(x^m)^n = \overbrace{x^m \cdot x^m \cdot x^m \cdot \cdots \cdot x^m}^{n \text{ factors of } x^m} = \underbrace{x \cdot x \cdot x \cdot x \cdot x \cdot \cdots \cdot x}_{mn \text{ factors of } x} = x^{mn}$$

This result is called the *power rule for exponents*.

Power Rule for Exponents   To raise an exponential expression to a power, keep the base and multiply the exponents.
For any real number x and any natural numbers m and n,

$$(x^m)^n = x^{m \cdot n} = x^{mn}$$ Read as "x to the mth power raised to the nth power
equals x to the mnth power."

EXAMPLE 2   Simplify each expression: **a.** $(3^2)^3$ **b.** $(x^{11})^5$ **c.** $(x^2x^3)^6$ **d.** $(x^2)^4(x^3)^2$

Strategy   Since there are powers of the form $(x^m)^n$ in each of these expressions, we will use the power rule for exponents and keep the base and multiply the exponents to simplify them.

Why   We use the power rule to raise an exponential expression to a power.

Solution

The Language of Algebra

Here, the instruction **Simplify** means to write an equivalent expression such that:
- No powers are raised to powers
- No parentheses appear
- Each base occurs only once

a. $(3^2)^3 = 3^{2 \cdot 3}$ Keep the base 3.

               Multiply the exponents.

      $= 3^6$

      $= 729$

b. $(x^{11})^5 = x^{11 \cdot 5}$

           $= x^{55}$

c. $(x^2 x^3)^6 = (x^5)^6$ Within the parentheses,

                  keep the common base x.

                  Add the exponents.

        $= x^{30}$ Keep the base x.

                  Multiply the exponents.

d. $(x^2)^4 (x^3)^2 = x^8 x^6$

              $= x^{14}$

Self Check 2   Simplify each expression: **a.** $(a^5)^8$ **b.** $(6^3)^5$ **c.** $(a^4 a^3)^3$ **d.** $(a^3)^3 (a^2)^3$

Now Try ▶   Problems 27, 31, and 33

To develop a third rule for exponents, we consider $(3x)^2$, which means $3x$ squared.

$$(3x)^2 = (3x)(3x) = 3 \cdot 3 \cdot x \cdot x = 3^2 x^2 = 9x^2$$

In general, we have

The Language of Algebra

Read $(xy)^n$ as "the quantity of xy, raised to the nth power."

$$(xy)^n = \overbrace{(xy)(xy)(xy) \cdot \cdots \cdot (xy)}^{n \text{ factors of } xy} = \overbrace{xxx \cdot \cdots \cdot x}^{n \text{ factors of } x} \cdot \overbrace{yyy \cdot \cdots \cdot y}^{n \text{ factors of } y} = x^n y^n$$

To develop a fourth rule for exponents, we consider $\left(\frac{x}{3}\right)^3$, which means $\frac{x}{3}$ cubed.

$$\left(\frac{x}{3}\right)^3 = \frac{x}{3} \cdot \frac{x}{3} \cdot \frac{x}{3} = \frac{x \cdot x \cdot x}{3 \cdot 3 \cdot 3} = \frac{x^3}{3^3} = \frac{x^3}{27}$$

In general, we have

The Language of Algebra

Read $\left(\frac{x}{y}\right)^n$ as "the quantity of x divided by y, raised to the nth power."

$$\left(\frac{x}{y}\right)^n = \overbrace{\left(\frac{x}{y}\right)\left(\frac{x}{y}\right)\left(\frac{x}{y}\right) \cdot \cdots \cdot \left(\frac{x}{y}\right)}^{n \text{ factors of } \frac{x}{y}} = \frac{\overbrace{xxx \cdot \cdots \cdot x}^{n \text{ factors of } x}}{\underbrace{yyy \cdot \cdots \cdot y}_{n \text{ factors of } y}} = \frac{x^n}{y^n} \quad \text{where } y \neq 0$$

The previous results are called the *power of a product* and the *power of a quotient rules.* As we work with many types of quotients, we will assume that no denominators are 0.

Powers of a Product and a Quotient Rules

To raise a product to a power, raise each factor of the product to that power. To raise a quotient to a power, raise the numerator and denominator to that power.

For any real numbers x and y, and any natural number n,

$$(xy)^n = x^n y^n \quad \text{and} \quad \left(\frac{x}{y}\right)^n = \frac{x^n}{y^n} \quad \text{where } y \neq 0$$

EXAMPLE 3   Simplify each expression: **a.** $(x^2 y)^3$ **b.** $(2y^4)^5$ **c.** $\left(\frac{x}{y^2}\right)^4$ **d.** $\left(\frac{-6x^3}{5y^4}\right)^2$

Strategy   Since these expressions have the form $(xy)^n$ and $\left(\frac{x}{y}\right)^n$, we will use the power of a product rule and the power of a quotient rule to simplify them.

Why   We use the power of a product rule to raise a product to a power, and the power of a quotient rule is used to raise a quotient to a power.

Solution   **a.** $(x^2y)^3 = (x^2)^3y^3$ Raise each factor of the
product x^2y to the 3rd power.

$= x^6y^3$

b. $(2y^4)^5 = (2)^5(y^4)^5$
$= 32y^{20}$

Caution

Never multiply an exponent and a base, or an exponent and a factor of the base! For example, in part b:

$(2y^4)^5 \neq 10y^{20}$

c. $\left(\dfrac{x}{y^2}\right)^4 = \dfrac{x^4}{(y^2)^4}$ Raise the numerator and
denominator to the 4th power.

$= \dfrac{x^4}{y^8}$

d. $\left(\dfrac{-6x^3}{5y^4}\right)^2 = \dfrac{(-6)^2(x^3)^2}{5^2(y^4)^2}$

$= \dfrac{36x^6}{25y^8}$

Self Check 3   Simplify each expression: **a.** $(a^4b^5)^2$ **b.** $\left(\dfrac{-6a^5}{b^7}\right)^3$
c. $(-2d^5)^4$

Now Try ▶ Problems 35 and 41

3 Use the Zero and Negative-Integer Exponent Rules.

To develop the definition of a zero exponent, we consider the expression $x^0 \cdot x^n$, where x is not 0. By the product rule,

The Language of Algebra

Because 0^0 is undefined, it is called an **indeterminate form**.

$$x^0 \cdot x^n = x^{0+n} = x^n = 1x^n$$

For the product rule to hold true for 0 exponents, $x^0 \cdot x^n$ must equal $\mathbf{1}x^n$. Comparing factors (as shown with the colored arrows), it follows that $x^0 = 1$. This result suggests the following definition.

Zero Exponents	A nonzero base raised to the 0 power is 1. For any nonzero base x, $$x^0 = 1 \quad \text{Read as "x to the zero power equals 1."}$$

EXAMPLE 4   Simplify each expression: **a.** 5^0 **b.** $(5x)^0$ **c.** $5x^0$ **d.** $-(5cd)^0$ **e.** $5a^0b$

Strategy   Since there are factors of the form x^0 in each of these expressions, we will use the zero exponent rule to simplify them.

Why   Any nonzero base raised to the 0 power is 1.

Solution   **a.** $5^0 = 1$ The base is 5 and the exponent is 0.

b. $(5x)^0 = 1$ The base is 5x and the exponent is 0.

The Language of Algebra

Here, the instruction **Simplify** means to write an equivalent expression such that no zero exponent appears.

c. $5x^0 = 5 \cdot x^0 = 5 \cdot 1 = 5$ The base is x and the exponent is 0.

d. $-(5cd)^0 = -1$ The base is 5cd and the exponent is 0.

e. $5a^0b = 5 \cdot a^0 \cdot b = 5 \cdot 1 \cdot b = 5b$ The base is a and the exponent is 0.

Self Check 4   Simplify each expression: **a.** 2^0 **b.** $2xy^0$ **c.** $-(xy)^0$

Now Try ▶ Problems 43 and 47

To develop the definition of a negative-integer exponent, we consider the expression $x^{-n} \cdot x^n$, where $x \neq 0$. By the product rule,

$$x^{-n} \cdot x^n = x^{-n+n} = x^0 = 1$$

Since the product is 1, x^{-n} and x^n must be reciprocals. It is also true that $\frac{1}{x^n}$ and x^n are reciprocals and their product is 1.

$$x^{-n} \cdot x^n = 1 \qquad \frac{1}{x^n} \cdot x^n = 1$$

Comparing factors (as shown with the colored arrows), it follows that x^{-n} must equal $\frac{1}{x^n}$. This result suggests the following definition.

Negative Exponents

For any nonzero real number x and any integer n,

$$x^{-n} = \frac{1}{x^n}$$

Read as "x to the negative nth power equals 1 over x to the nth power."

In words, x^{-n} is the reciprocal of x^n.

From the definition, we see that another way to write x^{-n} is to write its reciprocal and change the sign of the exponent. For example,

$$3^{-2} = \frac{1}{3^2}$$

Write the reciprocal of 3^{-2}, which is $\frac{1}{3^{-2}}$. Then change the sign of the exponent.

$$= \frac{1}{9}$$

EXAMPLE 5

Simplify each expression. Write answers using positive exponents.
a. 4^{-3} **b.** $(-2)^{-5}$ **c.** $7m^{-8}$ **d.** $-n^{-4}$

Strategy Since there are factors of the form x^{-n} in each of these expressions, we will use the negative-integer exponent rule to write equivalent expressions with positive exponents.

Why This rule enables us to rid these expressions of negative exponents by writing the reciprocal of the base and changing the sign of the exponent.

Solution

a. $4^{-3} = \frac{1}{4^3}$ Write the reciprocal of 4^{-3} and change the sign of the exponent.

$$= \frac{1}{64} \quad \text{Evaluate: } 4^3 = 64.$$

b. $(-2)^{-5} = \frac{1}{(-2)^5}$

$$= -\frac{1}{32}$$

c. $7m^{-8} = 7 \cdot m^{-8}$ Since there are no parentheses, the base is m.

$$= 7 \cdot \frac{1}{m^8} \quad \text{Write the reciprocal of } m^{-8} \text{ and change the sign of the exponent.}$$

$$= \frac{7}{m^8}$$

d. $-n^{-4} = -1 \cdot n^{-4}$

$$= -1 \cdot \frac{1}{n^4}$$

$$= -\frac{1}{n^4}$$

Caution

A negative exponent itself does not make the simplified expression negative. It indicates a reciprocal. Avoid common mistakes:

$4^{-3} = -64$ $4^{-3} = -\frac{1}{64}$

Self Check 5   Simplify each expression. Write answers using positive exponents.

a. 8^{-2} **b.** $(-3)^{-3}$ **c.** $12h^{-9}$ **d.** $-c^{-1}$

Now Try ▶ Problems 51 and 55

The rules for exponents involving products and powers are also true for negative exponents.

EXAMPLE 6   Simplify each expression. Write answers using positive exponents.
a. $x^{-5}x^3$ **b.** $(x^{-3})^{-2}$

Strategy   We will use the product rule for exponents to simplify the first expression, and power rule for exponents to simplify the second.

Why   The first expression has the form $x^m \cdot x^n$ and the second has the form $(x^m)^n$.

Solution   **a.** $x^{-5}x^3 = x^{-5+3}$ Keep the common base x   **b.** $(x^{-3})^{-2} = x^{(-3)(-2)}$ Keep the base x and
and add the exponents.   multiply the exponents.

$= x^{-2}$ $= x^6$

$= \dfrac{1}{x^2}$

Self Check 6   Simplify each expression. Write answers using positive exponents only.
a. $a^{-7}a^3$ **b.** $(a^{-5})^{-3}$

Now Try ▶   Problems 59 and 61

Negative exponents can appear in the numerator and/or the denominator of a fraction. To develop rules to apply to such situations, consider the following example.

$$\frac{x^{-4}}{y^{-3}} = \frac{\dfrac{1}{x^4}}{\dfrac{1}{y^3}} = \frac{1}{x^4} \cdot \frac{y^3}{1} = \frac{y^3}{x^4}$$

We can obtain this result in a simpler way. Beginning with $\dfrac{x^{-4}}{y^{-3}}$, move x^{-4} to the denominator and change the sign of its exponent. Then move y^{-3} to the numerator and change the sign of its exponent.

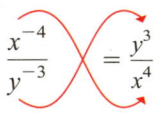 $\dfrac{x^{-4}}{y^{-3}} = \dfrac{y^3}{x^4}$ This example illustrates the following rules.

Rules for Changing from Negative to Positive Exponents	A factor can be moved from the denominator to the numerator or from the numerator to the denominator of a fraction if the sign of its exponent is changed.
	For any nonzero real numbers x and y, and any integers m and n,
	$\dfrac{1}{x^{-n}} = x^n$ and $\dfrac{x^{-m}}{y^{-n}} = \dfrac{y^n}{x^m}$

These rules streamline the process when simplifying fractions involving negative exponents.

EXAMPLE 7 Simplify each expression. Write answers using positive exponents. **a.** $\dfrac{1}{c^{-10}}$ **b.** $\dfrac{2^{-3}}{3^{-4}}$ **c.** $\dfrac{-6s^{-2}}{t^{-9}}$

Strategy Since these expressions have the form $\dfrac{1}{x^{-n}}$ or $\dfrac{x^{-m}}{y^{-n}}$, we will use the rule for changing exponents from negative to positive to write equivalent expressions with positive exponents only.

Why This rule enables us to rid these expressions of negative exponents by moving factors with negative exponents to the other side of the fraction bar and changing the sign of the exponent to positive.

Solution **a.** $\dfrac{1}{c^{-10}} = c^{10}$ Move c^{-10} to the numerator and change the sign of the exponent.

b. $\dfrac{2^{-3}}{3^{-4}} = \dfrac{3^4}{2^3}$ Move 2^{-3} to the denominator and change the sign of the exponent.
Move 3^{-4} to the numerator and change the sign of the exponent.

$\quad\quad = \dfrac{81}{8}$ Evaluate 3^4 and 2^3.

c. $\dfrac{-6s^{-2}}{t^{-9}} = -\dfrac{6t^9}{s^2}$ Since $-6s^{-2}$ has no parentheses, s is the base. Move only s^{-2} to the denominator and change the sign of the exponent. Do not move -6.
Move t^{-9} to the numerator and change the sign of the exponent.

> **Caution**
>
> In part c, a common error is to mistake the $-$ sign in -6 for a negative exponent. It is not an exponent. The -6 should not be moved to the denominator and have its sign changed.

Self Check 7 Simplify each expression. Write answers using positive exponents.
a. $\dfrac{1}{t^{-9}}$ **b.** $\dfrac{5^{-2}}{4^{-3}}$ **c.** $-\dfrac{h^{-6}}{8r^{-7}}$

Now Try ▶ Problems 63 and 69

---

4 Use the Quotient Rule for Exponents.

To develop a rule for dividing exponential expressions, we proceed as follows:

$$\frac{x^m}{x^n} = x^m\left(\frac{1}{x^n}\right) = x^m x^{-n} = x^{m+(-n)} = x^{m-n}$$

This result is called the *quotient rule for exponents.*

> **Quotient Rule for Exponents**
>
> To divide exponential expressions with the same base, keep the common base and subtract the exponents.
>
> For any nonzero number x and any integers m and n,
>
> $$\frac{x^m}{x^n} = x^{m-n}$$ Read as "x to the mth power divided by x to the nth power equals x to the m minus nth power."

EXAMPLE 8 Simplify each expression. Write answers using positive exponents. **a.** $\dfrac{a^5}{a^3}$ **b.** $\dfrac{2x^{-5}}{x^{11}}$

Strategy Since these expressions have the form $\dfrac{x^m}{x^n}$, we will use the quotient rule for exponents to simplify them.

Why We use the quotient rule to divide exponential expressions with the same base.

Solution

a. $\dfrac{a^5}{a^3} = a^{5-3}$ Keep the common base a.
Subtract the exponents.

$= a^2$

b. $\dfrac{2x^{-5}}{x^{11}} = 2x^{-5-11}$

$= 2x^{-16}$

$= \dfrac{2}{x^{16}}$

Success Tip

Since the rules for exponents can be applied in different orders, there are often equally valid ways to simplify an expression. For example, we can also simplify $\frac{2x^{-5}}{x^{11}}$ by moving x^{-5} to the denominator and changing the sign of the exponent.

$$\dfrac{2x^{-5}}{x^{11}} = \dfrac{2}{x^{11}x^5} = \dfrac{2}{x^{16}}$$

Self Check 8   Simplify each expression. Write answers using positive exponents.

a. $\dfrac{b^7}{b^5}$ **b.** $\dfrac{3b^{-3}}{b^3}$

Now Try   Problem 73

Remember, when simplifying expressions, *exponents interact only with other exponents* using the operations of addition, subtraction, or multiplication.

EXAMPLE 9   Simplify each expression. Write answers using positive exponents.

a. $\dfrac{x^4 x^3}{x^{-5}}$ **b.** $\dfrac{(x^2)^3}{(x^3)^2}$ **c.** $\dfrac{x^2 y^3}{7xy^4}$ **d.** $\left(\dfrac{2a^{-2}b^3}{3a^5b^4}\right)^3$

Strategy   To simplify these expressions, we must use more than one rule for exponents.

Why   The expressions involve products, powers, and quotients of exponential expressions with the same base as well as negative exponents.

Solution

a. $\dfrac{x^4 x^3}{x^{-5}} = \dfrac{x^7}{x^{-5}}$

$= x^{7-(-5)}$

$= x^{12}$

b. $\dfrac{(x^2)^3}{(x^3)^2} = \dfrac{x^6}{x^6}$

$= x^{6-6}$

$= x^0$

$= 1$

Success Tip

When more than one rule for exponents is involved in a simplification, more than one approach can often be used. In Example 9a, we obtain the same result with this alternate approach:

$$\dfrac{x^4 x^3}{x^{-5}} = x^4 x^3 x^5$$
$$= x^{4+3+5}$$
$$= x^{12}$$

c. $\dfrac{x^2 y^3}{7xy^4} = \dfrac{x^{2-1}y^{3-4}}{7}$

$= \dfrac{xy^{-1}}{7}$

$= \dfrac{x}{7y}$

d. $\left(\dfrac{2a^{-2}b^3}{3a^5b^4}\right)^3 = \left(\dfrac{2a^{-2-5}b^{3-4}}{3}\right)^3$

$= \left(\dfrac{2a^{-7}b^{-1}}{3}\right)^3$

$= \left(\dfrac{2}{3a^7 b}\right)^3$

$= \dfrac{8}{27a^{21}b^3}$

Self Check 9   Simplify each expression. Write answers using positive exponents.

a. $\dfrac{(a^{-2})^3}{(a^2)^{-3}}$ **b.** $\left(\dfrac{a^{-2}b^5}{5b^8}\right)^{-3}$

Now Try   Problem 77

5 Simplify Quotients Raised to Negative Powers.

To develop the final rule for exponents, we consider the following simplification:

The exponent is the opposite of -4.

$$\left(\dfrac{2}{3}\right)^{-4} = \dfrac{1}{\left(\frac{2}{3}\right)^4} = \dfrac{1}{\frac{2^4}{3^4}} = 1 \div \dfrac{2^4}{3^4} = 1 \cdot \dfrac{3^4}{2^4} = \dfrac{3^4}{2^4} = \left(\dfrac{3}{2}\right)^4$$

The base is the reciprocal of $\frac{2}{3}$.

This process can be streamlined using the following rule.

Rule for Negative Exponents and Reciprocals

A fraction raised to a power is equal to the reciprocal of the fraction raised to the opposite power.

For any nonzero real numbers x and y, and any integer n,

$$\left(\frac{x}{y}\right)^{-n} = \left(\frac{y}{x}\right)^{n}$$

EXAMPLE 10

Simplify each expression. Write answers using positive exponents.

a. $\left(\frac{2}{3}\right)^{-4}$ b. $\left(\frac{y^2}{x^3}\right)^{-3}$ c. $\left(\frac{a^{-2}b^3}{a^2a^3b^4}\right)^{-3}$ d. $\left(\frac{2x^2}{3y^{-3}}\right)^{-4}$

Strategy To simplify these expressions, we must use more than one rule for exponents.

Why The expressions involve fractions to negative powers as well as products, powers, and quotients of exponential expressions with the same base.

Solution

a. $\left(\frac{2}{3}\right)^{-4} = \left(\frac{3}{2}\right)^{4}$ Write the reciprocal of $\frac{2}{3}$ and change the exponent to 4.

$= \frac{3^4}{2^4}$

$= \frac{81}{16}$

b. $\left(\frac{y^2}{x^3}\right)^{-3} = \left(\frac{x^3}{y^2}\right)^{3}$

$= \frac{x^9}{y^6}$

The Language of Algebra

Here, the instruction **Simplify** means to write an equivalent expression such that:
- No powers are raised to powers
- No parentheses appear
- Each base occurs only once
- No zero or negative exponents appear

c. $\left(\frac{a^{-2}b^3}{a^2a^3b^4}\right)^{-3} = \left(\frac{a^2a^3b^4}{a^{-2}b^3}\right)^{3}$ Write the reciprocal of the fraction and change the exponent to 3.

$= \left(\frac{a^5b^4}{a^{-2}b^3}\right)^{3}$

$= (a^{5-(-2)}b^{4-3})^3$

$= (a^7b)^3$

$= a^{21}b^3$

d. $\left(\frac{2x^2}{3y^{-3}}\right)^{-4} = \left(\frac{3y^{-3}}{2x^2}\right)^{4}$

$= \frac{3^4y^{-12}}{2^4x^8}$

$= \frac{81}{16x^8y^{12}}$

Self Check 10 Simplify $\left(\frac{3a^3}{2b^{-2}}\right)^{-5}$ and write the answer using positive exponents.

Now Try Problems 79 and 81

We summarize the rules for exponents as follows. They are used so often in this course that you need to memorize them.

Summary of Exponent Rules

If m and n represent integers and there are no divisions by 0, then

Product rule $x^m \cdot x^n = x^{m+n}$

Power rule $(x^m)^n = x^{mn}$

Power of a product $(xy)^n = x^ny^n$

Quotient rule $\frac{x^m}{x^n} = x^{m-n}$

Power of a quotient $\left(\frac{x}{y}\right)^n = \frac{x^n}{y^n}$

Exponents of 0 and 1 $x^0 = 1$ and $x^1 = x$

Negative exponent $x^{-n} = \frac{1}{x^n}$

Negative exponents appearing in fractions $\frac{1}{x^{-n}} = x^n$ $\frac{x^{-m}}{y^{-n}} = \frac{y^n}{x^m}$ $\left(\frac{x}{y}\right)^{-n} = \left(\frac{y}{x}\right)^n$

VOCABULARY

Fill in the blanks.

1. Expressions such as x^4, 10^3, and $(5t)^2$ are called _____ expressions.

2. In the exponential expression x^n, the _____ is x and n is called the _____.

3. The expression x^4 represents a repeated multiplication where x is to be written as a _____ four times.

4. $3^4 \cdot 3^8$ is a _____ of exponential expressions with the same base, and $\dfrac{x^4}{x^2}$ is a _____ of exponential expressions with the same base.

5. $(h^3)^7$ is a _____ of an exponential expression.

6. In the expression 5^{-1}, the exponent is a _____ integer.

CONCEPTS

Complete the rules for exponents. Assume that there are no divisions by 0.

7. a. $x^m x^n = $ ▢   b. $(x^m)^n = $ ▢   c. $(xy)^n = $ ▢

   d. $\left(\dfrac{x}{y}\right)^n = $ ▢   e. $x^0 = $ ▢   f. $x^{-n} = $ ▢

   g. $\dfrac{x^m}{x^n} = $ ▢   h. $\left(\dfrac{x}{y}\right)^{-n} = \left(\dfrac{▢}{▢}\right)$

   i. $\dfrac{x^{-m}}{y^{-n}} = $ ▢

8. a. To multiply exponential expressions with the same base, such as $x^3 \cdot x^8$, keep the common base x and _____ the exponents.
   b. To divide exponential expressions with the same base, such as $\dfrac{a^9}{a^4}$, keep the common base a and _____ the exponents.

9. To raise an exponential expression to a power, such as $(x^9)^4$, keep the base x and _____ the exponents.

10. a. To raise a product to a power, such as $(4t^2)^6$, raise each _____ of the product to that power.
    b. To raise a quotient to a power, such as $\left(\dfrac{a}{b}\right)^8$, raise the numerator and the _____ to that power.

11. a. Any nonzero base raised to the 0 power is ___.
    b. Another way to write x^{-n} is to write its _____ and change the sign of the exponent.

12. a. A factor can be moved from the denominator to the numerator or from the numerator to the denominator of a fraction if the _____ of its exponent is changed.
    b. A fraction raised to a power is equal to the reciprocal of the fraction raised to the _____ power.

NOTATION

Complete each simplification.

13. $\dfrac{x^5 x^4}{x^{-2}} = \dfrac{x^{▢}}{x^{-2}}$

    $= x^{9-▢}$

    $= x^{▢}$

14. $\left(\dfrac{a^{-4}}{a^3}\right)^2 = (a^{-4-▢})^2$

    $= (a^{▢})^2$

    $= a^{▢}$

    $= \dfrac{▢}{a^{14}}$

GUIDED PRACTICE

Identify the base and the exponent. See Objective 1.

15. a. $(6x)^3$ 16. a. -7^2
    b. $-x^5$ b. $(-t)^4$
    c. $2b^6$ c. $12a^2$

17. a. $\left(\dfrac{n}{4}\right)^3$ 18. a. $\left(\dfrac{3x}{y}\right)^0$
    b. $(m-8)^6$ b. $(t+7)^8$
    c. 10^0 c. -9^0

Simplify each expression. See Example 1.

19. $x^2 x^3$ 20. $y^3 y^4$

21. $y^3 y^7 y^2$ 22. $x^2 x^3 x^5$

23. $-6t^8(t^{15})$ 24. $-9n^{20}(n^{14})$

25. $aba^3 b^4$ 26. $x^2 y^3 x^3 y^2$

Simplify each expression. See Example 2.

27. $(2^3)^2$ 28. $(4^2)^3$

29. $(x^4)^7$ 30. $(y^7)^5$

31. $(r^8 r^3)^5$ 32. $(w^5 w^2)^4$

33. $(g^4)^5(g^2)^6$ 34. $(s^5)^2(s^3)^8$

Simplify each expression. See Example 3.

35. $(x^5 y)^4$ 36. $(a^6 b)^5$

37. $(4m^7)^3$ 38. $(9t^8)^2$

39. $\left(\dfrac{m^{10}}{n}\right)^8$ 40. $\left(\dfrac{c}{d^9}\right)^4$

41. $\left(\dfrac{3a^{16}}{7n^{11}}\right)^2$ 42. $\left(\dfrac{5t^8}{3m^3}\right)^3$

Simplify each expression. See Example 4.

43. 8^0 44. 1^0

45. $(-6t)^0$ 46. $(-9s)^0$

47. $60h^0$ 48. $75b^0$

49. $-3s^0 t$ 50. $-9mn^0$

Simplify each expression. Write answers using positive exponents. See Example 5.

51. 5^{-2} 52. 3^{-4}

53. $(-3)^{-3}$ 54. $(-4)^{-3}$

55. $8x^{-9}$ 56. $12y^{-5}$

57. $-h^{-1}$ 58. $-w^{-2}$

Simplify each expression. Write answers using positive exponents. See Example 6.

59. $m^{-4} \cdot m^{-6}$ 60. $n^{-9} \cdot n^{-2}$

61. $(s^2)^{-3}$ 62. $(-t^2)^{-5}$

Simplify each expression. Write answers using positive exponents. See Example 7.

63. $\dfrac{1}{a^{-5}}$ 64. $\dfrac{3}{b^{-5}}$

65. $\dfrac{1}{7^{-2}}$ 66. $\dfrac{1}{4^{-3}}$

67. $\dfrac{2^{-4}}{1^{-10}}$ **68.** $\dfrac{3^{-4}}{1^{-9}}$

69. $-\dfrac{t^{-6}}{12p^{-8}}$ **70.** $-\dfrac{x^{-9}}{20k^{-7}}$

Simplify each expression. Write answers using positive exponents. See Example 8.

71. $\dfrac{m^{15}}{m^3}$ **72.** $\dfrac{n^{18}}{n^6}$

73. $\dfrac{33y^{-2}}{y^{10}}$ **74.** $\dfrac{24k^{-4}}{k^8}$

Simplify each expression. Write answers using positive exponents. See Example 9.

75. $\dfrac{t^4 t^9}{t^{-1}}$ **76.** $\dfrac{r^6 r^2}{r^{-7}}$

77. $\dfrac{m^3 n^2}{6mn^{11}}$ **78.** $\dfrac{a^{11} b^7}{100a^{15} b}$

Simplify each expression. Write answers using positive exponents. See Example 10.

79. $\left(\dfrac{2}{3}\right)^{-2}$ **80.** $\left(\dfrac{4}{5}\right)^{-3}$

81. $\left(\dfrac{g^{20}}{t^{30}}\right)^{-4}$ **82.** $\left(\dfrac{y^8}{z^6}\right)^{-2}$

TRY IT YOURSELF

Simplify each expression. Write answers using positive exponents.

83. $(2a^2 a^3 b^0)^4$ **84.** $(3bb^2 b^3 c^0)^4$

85. $\left(\dfrac{4a^{-2}b}{3ab^{-3}}\right)^3$ **86.** $\left(\dfrac{2ab^{-3}}{3a^{-2}b^2}\right)^2$

87. $-\dfrac{8t^{-3} \cdot t^{-11}}{t^{-14}}$ **88.** $-\dfrac{4x^{-9} \cdot x^{-3}}{x^{-12}}$

89. $(-x^8)^2 y^4 x^3 x^0$ **90.** $(-x^2)^5 y^7 y^3 x^{-2} y^0$

91. $\left(\dfrac{x^{-5}}{x^2}\right)^{-4}\left(\dfrac{x^7}{x^{-8}}\right)^3$ **92.** $\left(\dfrac{a^{-3}}{a^6}\right)^{-3}\left(\dfrac{a^6}{a^{-2}}\right)^6$

93. $5^2 r^{-5} (r^6)^3$ **94.** $8^2 d^{-8} (d^9)^2$

95. $m^{-4} \cdot m^2 \cdot m^{-8}$ **96.** $n^{-9} \cdot n^5 \cdot n^{-7}$

97. $\left(\dfrac{4a^2 b^3 z^{-4}}{3a^{-2} b^{-1} z^3}\right)^{-3}$ **98.** $\left(\dfrac{-3pqr^{-4}}{2p^2 q^{-3} r^2}\right)^{-2}$

99. $\dfrac{(4c^{-3}d)^0}{25(d+3)^0}$ **100.** $\dfrac{b^0}{2(a+b)^0}$

101. $\dfrac{(3x^2)^{-2}}{x^3 x^{-4} x^0}$ **102.** $\dfrac{y^{-3} y^{-4} y^0}{(2y^{-2})^3}$

103. $\left(\dfrac{3(d^{-1})^{-5}}{8(d^{-4})^{-2}}\right)^{-2}$ **104.** $\left(\dfrac{(c^{-2})^{-4}}{15(c^{-3})^{-7}}\right)^{-1}$

105. $\dfrac{(-3cd^2)^3 (c^{-1} d^{-3})^3}{(c^3 d)^5}$ **106.** $\dfrac{(c^3 t^{-4})^2 (2c^4 t^4)^{-2}}{(c^2 t^5)^{-3}}$

Look Alikes . . .

107. a. $x^4 \cdot x^4$ **b.** $(x^4)^4$ **c.** $x^4 + x^4$

108. a. $m^3 m^2 m^4$ **b.** $(m^3 m^2)^4$ **c.** $\dfrac{m^3 m^2}{m^4}$

109. a. $(-2t^3 t)^5$ **b.** $-2t^3 t^5$ **c.** $(2t^3 t)^{-5}$

110. a. $\left(\dfrac{c^2}{d^3}\right)^4$ **b.** $\left(\dfrac{c^2}{d^3}\right)^{-4}$ **c.** $\left(\dfrac{-c^2}{d^3}\right)^4$

111. a. 6^{-1} **b.** -6^{-1} **c.** $-(-6)^{-1}$

112. a. 7^{-2} **b.** -7^{-2} **c.** $-(-7)^{-2}$

113. a. $8xy^{-2}$ **b.** $(8xy)^{-2}$ **c.** $-8^{-2}xy$

114. a. $\left(\dfrac{a}{2b}\right)^{-1}$ **b.** $\dfrac{a^{-1}}{2b}$ **c.** $-\dfrac{a^{-1}}{2b^{-1}}$

Use a calculator to verify that each statement is true by showing that the values on either side of the equation are equal.

115. $(3.68)^0 = 1$ **116.** $(2.1)^4 (2.1)^3 = (2.1)^7$

117. $\left(\dfrac{5.4}{2.7}\right)^{-4} = \left(\dfrac{2.7}{5.4}\right)^4$ **118.** $(7.23)^{-3} = \dfrac{1}{(7.23)^3}$

APPLICATIONS

119. License Plates. The number of different Ohio license plates of the form three letters followed by four digits, as shown in the illustration, is $26 \cdot 26 \cdot 26 \cdot 10 \cdot 10 \cdot 10 \cdot 10$. Write this expression using exponents. Then evaluate it.

120. Astronomy. See the illustration. The distance d, in miles, of the nth planet from the sun is given by the formula

$$d = 9,275,200[3(2^{n-2}) + 4]$$

a. What is the value of n for the planet Earth? $n = 3$
b. Find the distance of Earth and the distance of Mercury from the sun.

121. Biology. Refer to the illustrations below where the sizes of a bacterium and a molecule are expressed as fractions of a meter. Fill in each exponent to express the fraction as a power of 10.

Bacterium

 $\dfrac{1}{1,000,000}$ meter $= 10^{\boxed{}}$ meter

Molecule

 $\dfrac{1}{1,000,000,000}$ meter $= 10^{\boxed{}}$ meter
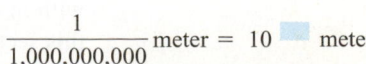

122. Physics. Albert Einstein's work in the area of special relativity resulted in the observation that the total energy E of a body is equal to its total mass m times the square of the speed of light c: $E = mc^2$. Identify the base and exponent on the right side of this famous equation.

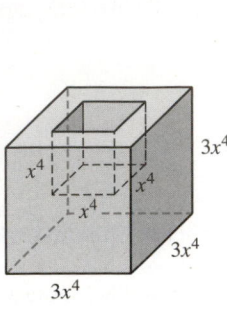

123. from **Campus to Careers**

Landscape Architect

Write an algebraic expression that represents the volume of concrete needed to make the planter shown below. The dimensions are in inches. $26x^{12}$ in.^{3}

124. Geometry. A rectangular solid is shown on the right.

a. Find the area of its base.

b. Find its volume.

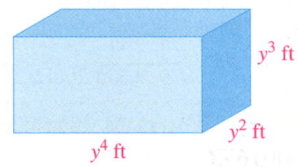

y^3 ft

y^2 ft

y^4 ft

WRITING

125. Explain how you would help a friend understand that 3^{-2} is not equal to -9.

126. Explain how you would help a friend understand that 4^0 is not equal to 0.

127. Explain how an exponential expression with a negative exponent can be expressed as an equivalent expression with a positive exponent. Give an example.

128. Explain the error in the following solution.

Write $-8ab^{-3}$ using positive exponents only.

$$-8ab^{-3} = \frac{a}{8b^3}$$

REVIEW

Solve each inequality. Graph the solution set and write it using interval notation.

129. $-9x + 5 \geq 15$

130. $\frac{1}{4}p - \frac{1}{3} \leq p + 2$

CHALLENGE PROBLEMS

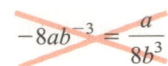

Evaluate each expression.

131. $(2^{-1} + 3^{-1} - 4^{-1})^{-1}$ **132.** $(3^{-1} + 4^{-1})^{-2}$

Simplify each expression. Assume there are no divisions by 0.

133. $\dfrac{8^{5a}(8^{6a})^5}{8^{-2a} \cdot 8^a \cdot 8^{4a}}$ **134.** $\left(\dfrac{(y^{5x})^2(y^{4x})^4}{(y^{2x} \cdot y^x)^{-3}}\right)^{-2}$

SECTION 5.2 Scientific Notation

OBJECTIVES

1. Write numbers in scientific notation.

2. Convert from scientific notation to standard notation.

3. Perform calculations with scientific notation.

ARE YOU READY?

The following problems review some basic skills that are needed when working with scientific notation.

1. Evaluate: 10^4

2. Multiply: $1,000,000 \cdot 9.63$

3. Evaluate: 10^{-3}

4. Multiply: $0.01 \cdot 4.31$

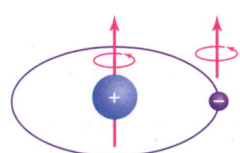

Hydrogen atom

Very large and very small numbers occur in science and other disciplines. For example, the star nearest Earth (excluding the sun) is Proxima Centauri, about 24,793,000,000,000 (read as "24 trillion, 793 billion") miles away. The mass of a hydrogen atom is approximately 0.00000000000000000000000001673 (read as "1,673 octillionths") of a gram.

These numbers, written in **standard** or **decimal notation,** are difficult to read and cumbersome to work with in calculations because they contain many zeros. In this section, we will discuss a notation that enables us to express such numbers in a more manageable form.

Atlas Image Mosaic Courtesy of 2MASS/UMass/IPAC-Caltech/NASA/NSF

©Leslie Harris/Photolibrary

1 Write Numbers in Scientific Notation.

Scientific notation provides a compact way of writing very large or very small numbers.

Scientific Notation	A positive number is written in **scientific notation** when it is written in the form $N \times 10^n$, where $1 \le N < 10$ and n is an integer.

Notation

A raised dot · is sometimes used when writing scientific notation.

$3.67 \times 10^6 = 3.67 \cdot 10^6$

Some examples of numbers written in scientific notation are

3.67×10^6 2.2×10^{-4} 9.875×10^{22}

$N = 3.67$ $N = 2.2$ $N = 9.875$

$n = 6$ $n = -4$ $n = 22$

Every positive number written in scientific notation is the product of a decimal number that is at least 1, but less than 10, and a power of 10.

A decimal that is at least 1, but less than 10 called N

An integer exponent called n

 $. \ \ \times 10$

To write numbers in scientific notation, you need to be familiar with **powers of 10,** like those listed in the table below.

Power of 10	10^4	10^3	10^2	10^1	10^0	10^{-1}	10^{-2}	10^{-3}	10^{-4}
Value	10,000	1,000	100	10	1	$\frac{1}{10} = 0.1$	$\frac{1}{100} = 0.01$	$\frac{1}{1,000} = 0.001$	$\frac{1}{10,000} = 0.0001$

EXAMPLE 1

Write each number in scientific notation: **a.** 24,793,000,000,000   **b.** 0.000000000000000000000001673

Strategy   We will write each number as the product of a number between 1 and 10 and an integer power of 10.

Why   Numbers written in scientific notation have the form $N \times 10^n$ where $1 \le N < 10$ and n is an integer.

Solution

a. The number 2.4793 is between 1 and 10. To get 24,793,000,000,000, the decimal point in 2.4793 must be moved 13 places to the *right*.

2.4,793,000,000,000.   Start with a decimal point (shown in red)
⌣⌣⌣⌣⌣⌣⌣⌣⌣⌣⌣⌣⌣↗   to the right of the first nonzero digit, which is 2.
   13 places

Success Tip

The key to writing numbers in scientific notation is to be able to determine N (the decimal that is at least 1, but less than 10) and n (the integer exponent in the power of 10).

We can move the red decimal point 13 places to the right by multiplying 2.4793 by 10^{13}.

$24,793,000,000,000 = 2.4793 \times 10^{13}$ This is the distance (in miles) from Proxima Centauri to Earth. $N = 2.4793$ and $n = 13$.

b. The number 1.673 is between 1 and 10. To get 0.000000000000000000000001673, the decimal point in 1.673 must be moved 24 places to the *left*.

0.000000000000000000000001.673   Start with a decimal point (shown in red) to
↖⌣⌣⌣⌣⌣⌣⌣⌣⌣⌣⌣⌣⌣⌣⌣⌣⌣   the right of the first nonzero digit, which is 1.
        24 places

Notation

When writing numbers in scientific notation, keep the negative exponents. Don't apply the negative exponent rule.

1.673×10^{-24} $1.673 \times \dfrac{1}{10^{24}}$

We can move the red decimal point 24 places to the left by multiplying 1.673 by 10^{-24}.

$0.000000000000000000000001673 = 1.673 \times 10^{-24}$ This is the mass (in grams) of a hydrogen atom. $N = 1.673$ and $n = -24$.

Self Check 1   Write each italicized number in scientific notation.   **a.** In 2008, the country earning the most money from tourism was the United States, *$110,100,000,000*. (Source: World Tourism Organization)
b. DNA molecules contain and transmit the information that allows cells to reproduce. They are only *0.000000002* meter wide.

Now Try ▶ Problems 13 and 15

When a number is written in scientific notation, the first factor must be at least 1, but less than 10.

EXAMPLE 2   Write each number in scientific notation: **a.** 47.2×10^{30} **b.** 0.063×10^{-2}

Strategy   To write 47.2×10^{30} in scientific notation, we will write the first factor, 47.2, in scientific notation and then multiply the powers of 10. We will answer part (b) in a similar way.

Why   This approach is necessary because neither 47.2×10^{30} nor 0.063×10^{-2} are written in scientific notation—the first factors (47.2 and 0.063) are not between 1 and 10.

Solution   **a.** $\mathbf{47.2 \times 10^{30}} = (\mathbf{4.72 \times 10^1}) \times 10^{30}$ Write 47.2 in scientific notation.

$= 4.72 \times (10^1 \times 10^{30})$ Group the powers of 10 together.

$= 4.72 \times 10^{31}$ Use the product rule for exponents: $10^1 \times 10^{30} = 10^{1+30} = 10^{31}$. $N = 4.72$ and $n = 31$.

b. $\mathbf{0.063 \times 10^{-2}} = (\mathbf{6.3 \times 10^{-2}}) \times 10^{-2}$ Write 0.063 in scientific notation.

$= 6.3 \times (10^{-2} \times 10^{-2})$ Group the powers of 10 together.

$= 6.3 \times 10^{-4}$ $10^{-2} \times 10^{-2} = 10^{-2+(-2)} = 10^{-4}$. $N = 6.3$ and $n = -4$.

> **Caution**
>
> At first glance, the following numbers might look like they are written in scientific notation, but they are not. In each case, the first factor is not a number between 1 and 10.
>
> 47.2×10^{30} 0.063×10^{-2}

Self Check 2   Write each number in scientific notation: **a.** 17.3×10^2
b. 0.0045×10^{-3}

Now Try ▶ Problems 21 and 25

2   Convert from Scientific Notation to Standard Notation.

Each of the following numbers is written in scientific and standard notation. In each case, the exponent gives the number of places that the decimal point moves, and the sign of the exponent indicates the direction that it moves:

> **Success Tip**
>
> Since $10^0 = 1$, scientific notation involving 10^0 is easily simplified. For example,
>
> $4.8 \times \mathbf{10^0} = 4.8 \times \mathbf{1} = 4.8$

$5.32 \times 10^4 = 5.3\,2\,0\,0.$ $6.45 \times 10^7 = 6.4\,5\,0\,0\,0\,0\,0.$
4 places to the right   7 places to the right

$2.37 \times 10^{-4} = 0.0\,0\,0\,2.3\,7$ $9.234 \times 10^{-2} = 0.0\,9.2\,3\,4$
4 places to the left   2 places to the left

$4.8 \times 10^0 = 4.8$
No movement of the decimal point

These results suggest the following steps for changing a number written in scientific notation to standard notation.

Converting from Scientific to Standard Notation

1. If the exponent is positive, move the decimal point the same number of places to the right as the exponent.

2. If the exponent is negative, move the decimal point the same number of places to the left as the absolute value of the exponent.

EXAMPLE 3   Convert to standard notation: **a.** 8.706×10^5 **b.** 1.1×10^{-3}

Strategy   In each case, we need to identify n, the exponent on the power of 10, and consider its sign.

Why   The exponent gives the number of decimal places that we should move the decimal point in the number N. The sign of the exponent indicates whether it should be moved to the right or the left.

Solution   **a.** Here, $N = 8.706$ and $n = 5$. Since the exponent in 10^5 is 5, the red decimal point shown below moves 5 places to the right. (Multiplication by 10^5, which is 100,000, moves the decimal point 5 places to the right.)

$$8.706 \times 10^5 = 8.7\,0\,6\,0\,0. = 870,600$$

5 places to the right

b. Here, $N = 1.1$ and $n = -3$. Since the exponent in 10^{-3} is -3, the red decimal point shown below moves 3 places to the left. (Multiplication by 10^{-3}, which is 0.001, moves the decimal point 3 places to the left.)

$$1.1 \times 10^{-3} = 0.0\,0\,1\,.\,1 = 0.0011$$

3 places to the left

Self Check 3   Convert each number written in scientific notation to standard notation.
   a. In 2010, the world's forest areas were estimated to occupy 9.88×10^9 acres. (Source: United Nations FAO)
   b. The average distance between molecules of air in a room is 3.937×10^{-7} inch.

Now Try ▶   Problems 33 and 39

The results from the previous examples suggest the following forms to use when converting numbers from standard to scientific notation.

For real numbers between 0 and 1:   ▨ $\times 10^{\text{negative integer}}$
For real numbers at least 1, but less than 10:   ▨ $\times 10^0$
For real numbers greater than or equal to 10:   ▨ $\times 10^{\text{positive integer}}$

3 Perform Calculations with Scientific Notation.

Another advantage of scientific notation becomes apparent when we evaluate products or quotients that involve very large or small numbers. If we express those numbers in scientific notation, we can use rules for exponents to make the calculations easier. To multiply two numbers written in scientific notation, use the following rule:

$$(a \times 10^m)(b \times 10^n) = (a \cdot b) \times 10^{m+n}$$

EXAMPLE 4   **Astronomy.**   The galaxy in which we live is called the *Milky Way*. It has a diameter of approximately 100,000 light years. (A light year is the distance light travels in a vacuum in one year: 9.46×10^{15} meters.) Find the diameter of the Milky Way in meters.

Strategy   To find the diameter of the Milky Way, we will convert the number of light years, 100,000, to scientific notation and multiply it by the number of meters per light year, 9.46×10^{15}.

Why   When the numbers to be multiplied (100,000 and 9.46×10^{15}) are written in scientific notation, we can use the product rule for exponents to simplify the calculation.

©Viktor Malyshchyts/Shutterstock.com

Solution   After writing 100,000 in scientific notation as 1.0×10^5, we perform the arithmetic on the decimals and the exponential expressions separately.

$$100,000(9.46 \times 10^{15})$$

$$= (1.0 \times 10^5)(9.46 \times 10^{15})$$ Multiply the number of light years by the number of meters per light year.

$$= (1.0 \cdot 9.46) \times (10^5 \cdot 10^{15})$$ Use the commutative and associative properties of multiplication to group the decimal factors together and the powers of 10 together.

$$= 9.46 \times 10^{5+15}$$ Perform the multiplication: $1.0 \cdot 9.46 = 9.46$. For the powers of 10, keep the base and add the exponents.

$$= 9.46 \times 10^{20}$$ Do the addition: $5 + 15 = 20$.

The Milky Way Galaxy is about 9.46×10^{20} meters in diameter.

> **Self Check 4**   **Astronomy.**   A light year is 5.88×10^{12} miles. Find the diameter of the Milky Way in miles.
>
> **Now Try** ▶ Problems 47 and 49

To divide two numbers written in scientific notation, use the following rule:

$$\frac{a \times 10^m}{b \times 10^n} = \frac{a}{b} \times 10^{m-n}$$

EXAMPLE 5   **World Oil Reserves/Demand.**   According to estimates in the *World Fact Book,* there were 1,350,000,000,000 (read as "1 trillion, 350 billion") barrels of crude oil reserves in the ground at the start of 2010. At that time, world demand was 31,200,000,000 (read as "31 billion, 200 million") barrels per year according to the Energy Information Administration. If annual demand as of 2010 remains the same and if no new oil discoveries are made, when will the world's oil supply run out?

Strategy   To find the number of years of crude oil that remains, we will convert the number of barrels in reserves to scientific notation and divide it by the annual number of barrels of demand, also written in scientific notation.

Why   When the numbers to be divided are written in scientific notation, we can use the quotient rule for exponents to simplify the calculation.

©Tatagatta/Shutterstock.com

Solution   First, we write 1,350,000,000,000 in scientific notation as 1.35×10^{12} and 31,200,000,000 as 3.12×10^{10}. Then we perform the arithmetic on the decimals and the exponential expressions separately.

$$\frac{1,350,000,000,000}{31,200,000,000} = \frac{\mathbf{1.35} \times \mathbf{10^{12}}}{\mathbf{3.12} \times \mathbf{10^{10}}}$$ Divide the number of barrels in reserve by the number of barrels used each year.

$$= \frac{\mathbf{1.35}}{\mathbf{3.12}} \times \frac{\mathbf{10^{12}}}{\mathbf{10^{10}}}$$ Separate the factors to divide the decimals and divide the powers of 10.

$$\approx 0.43 \times 10^{12-10}$$ Perform each division: $\frac{1.35}{3.12} \approx 0.43$. For the powers of 10, keep the base and subtract the exponents.

$$\approx 0.43 \times 10^2$$ This result is not in scientific notation: $N = 0.43$.

$$\approx (4.3 \times 10^{-1}) \times 10^2$$ Write 0.43 in scientific notation.

$$\approx 4.3 \times (10^{-1} \times 10^2)$$ Group the powers of 10 together.

$$\approx 4.3 \times 10^1$$ Add the exponents.

$$\approx 43$$ To write 0.43×10^1 in standard notation, move the decimal point 1 place to the right.

According to industry estimates, as of 2010, there were 43 years of crude oil reserves left if the demand remained constant. Under those conditions, the world's crude oil supply will run out in the year 2053.

> **Self Check 5**   **U.S. Farming.**   The total amount of rice produced in 2009 was 22,000,000,000 pounds on 3,140,000 acres planted. Find the number of pounds of rice produced per acre that year.
>
> **Now Try** ▶ Problems 55 and 59

EXAMPLE 6   Use scientific notation to evaluate: $\dfrac{(0.00000064)(24,000,000,000)}{(400,000,000)(0.0000000012)}$

Strategy   After writing each number in scientific notation, we will perform the arithmetic on the decimals and the exponential expressions separately.

Why   When the numbers to be multiplied and divided are written in scientific notation, we can use the product and quotient rules for exponents to simplify the calculation.

Solution

$$\frac{(0.00000064)(24,000,000,000)}{(400,000,000)(0.0000000012)} = \frac{(6.4 \times 10^{-7})(2.4 \times 10^{10})}{(4.0 \times 10^{8})(1.2 \times 10^{-9})} \quad \text{Convert each number to scientific notation.}$$

$$= \frac{(6.4)(2.4)}{(4)(1.2)} \times \frac{10^{-7}10^{10}}{10^{8}10^{-9}} \quad \text{Separate the decimal factors and the power of 10 factors.}$$

$$= \frac{15.36}{4.8} \times \frac{10^{3}}{10^{-1}} \quad \text{Simplify: } 10^{-7}10^{10} = 10^{-7+10} = 10^{3} \text{ and } 10^{8}10^{-9} = 10^{8+(-9)} = 10^{-1}.$$

$$= 3.2 \times 10^{3-(-1)} \quad \text{Use the quotient rule for exponents.}$$

$$= 3.2 \times 10^{4} \quad \text{Do the subtraction.}$$

The result is 3.2×10^{4}. In standard notation, this is 32,000.

> **Self Check 6**   Use scientific notation to evaluate: $\dfrac{(320)(25,000)}{0.00004}$
>
> **Now Try** ▶ Problem 69

Using Your Calculator ▶ **Using Scientific Notation**

Scientific and graphing calculators often give answers in scientific notation. For example, if we use a calculator to find 301.2^{8}, the display will read

| 6.77391496¹⁹ | *On a scientific calculator* |

| 301.2 ∧ 8 |
| 6.773914961E19   *On a graphing calculator* |

In either case, the answer is given in scientific notation and means $6.77391496 \times 10^{19}$.

Numbers can be entered into a calculator in scientific notation. For example, to enter 24,000,000,000 (which is 2.4×10^{10} in scientific notation), we enter these numbers and press these keys:

2.4 | EXP | 10   *On some scientific calculators*

2.4 | EE | 10   *On a graphing calculator and on most scientific calculators*

To use a scientific calculator to evaluate

$$\frac{(24,000,000,000)(0.00000006495)}{0.00000004824}$$

we enter each number in scientific notation, because each number has too many digits to be entered directly. In scientific notation, the three numbers are

$$2.4 \times 10^{10} \qquad 6.495 \times 10^{-8} \qquad 4.824 \times 10^{-8}$$

Using a scientific calculator, we enter these numbers and press these keys:

2.4 EE 10 × 6.495 EE 8 +/− ÷ 4.824 EE 8 +/− =

The display will read 3.231343284 ^{10}. In standard notation, the answer is 32,313,432,840. The steps are similar on a graphing calculator.

SECTION 5.2 STUDY SET

VOCABULARY

Fill in the blanks.

1. 7.4×10^6 is written in _____ notation and 7,400,000 is written in _____ notation.
2. 10^{-3}, 10^0, 10^1, and 10^4 are _____ of 10.

CONCEPTS

Fill in the blanks.

3. A positive number is written in scientific notation when it is written in the form $N \times$ ____, where $1 \le N < 10$ and n is an _____.

4. Insert > or <: 5.3×10^2 ____ 5.3×10^{-2}

5. For each number written in scientific notation form $N \times 10^n$, identify N and n.
   a. 2.316×10^{54}
   b. 1.07×10^{-21}

6. a. To change 6.31×10^{-4} to standard notation, we move the decimal point four places to the ____.
   b. To change 9.7×10^3 to standard notation, we move the decimal point three places to the ____.

NOTATION

7. a. Explain why the number 60.22×10^{22} is not written in scientific notation.
   b. Explain why the number 0.6022×10^{24} is not written in scientific notation.

8. Determine what type of integer *exponent* (positive, negative, or 0) must be used when writing each of the three categories of real numbers in scientific notation.
   a. For real numbers between 0 and 1, such as 0.000071:
      ■ × 10^{integer}
   b. For real numbers at least 1, but less than 10, such as 6.45:
      ■ × 10
   c. For real numbers greater than or equal to 10, such as 3,300,000,000:
      ■ × 10^{integer}

GUIDED PRACTICE

Write each number in scientific notation. See Example 1.

9. 3,900
10. 1,700
11. 0.0078
12. 0.068
13. 173,000,000,000,000
14. 89,800,000,000
15. 0.0000096
16. 0.000000046
17. 0.00000000203
18. 0.0000000000301
19. 50,160,000,000,000,000
20. 220,000,000,000,000,000

Write each number in scientific notation. See Example 2.

21. 23.65×10^6
22. 75.6×10^5
23. 90.09×10^{-11}
24. 20.08×10^{-13}
25. 0.0317×10^{-2}
26. 0.0012×10^{-3}
27. 0.0527×10^5
28. 0.0298×10^3
29. 323×10^5
30. 689×10^9
31. $6,000 \times 10^{-7}$
32. 765×10^{-5}

Write each number in standard notation. See Example 3.

33. 2.7×10^4
34. 7.2×10^3
35. 3.23×10^{-3}
36. 6.48×10^{-2}
37. 7.96×10^8
38. 9.67×10^9
39. 3.5×10^{-7}
40. 4.12×10^{-10}
41. 5.23×10^0
42. 8.67×10^0
43. 8.0×10^{13}
44. 4.0×10^{14}

Multiply. Give all answers in scientific notation.
See Example 4.

45. $(1.3 \times 10^4)(2.0 \times 10^5)$ **46.** $(3.0 \times 10^8)(2.2 \times 10^3)$

47. $(7.9 \times 10^5)(2.3 \times 10^6)$ **48.** $(6.1 \times 10^8)(3.9 \times 10^5)$

49. $(9.1 \times 10^{-5})(5.5 \times 10^{12})$ **50.** $(8.4 \times 10^{-13})(4.8 \times 10^9)$

51. $(9.0 \times 10^{-1})(8.0 \times 10^{-6})$ **52.** $(8.1 \times 10^{-4})(2.4 \times 10^{-15})$

Divide. Give all answers in scientific notation.
See Example 5.

53. $\dfrac{8.6 \times 10^{15}}{2.0 \times 10^6}$ **54.** $\dfrac{9.6 \times 10^{20}}{3.0 \times 10^{10}}$

55. $\dfrac{2.193 \times 10^{32}}{4.3 \times 10^{20}}$ **56.** $\dfrac{1.107 \times 10^{16}}{4.1 \times 10^2}$

57. $\dfrac{2.686 \times 10^{10}}{7.9 \times 10^{-7}}$ **58.** $\dfrac{4.216 \times 10^{31}}{6.8 \times 10^{-14}}$

59. $\dfrac{4.2 \times 10^{-12}}{8.4 \times 10^{-5}}$ **60.** $\dfrac{1.21 \times 10^{-17}}{1.1 \times 10^{-2}}$

Write each number in scientific notation and perform the operations. Give all answers in scientific notation and in standard notation. See Example 6.

61. $\dfrac{4,500,000,000,000}{0.0002}$

62. $\dfrac{6,150,000,000}{0.003}$

63. $\dfrac{0.00000128}{0.0004}$

64. $\dfrac{0.000000000117}{0.00039}$

65. $(89,000,000,000)(4,500,000,000)$

66. $(0.000000061)(3,500,000,000)$

67. $\dfrac{(640,000)(2,700,000)}{120,000}$

68. $\dfrac{(220,000)(0.000009)}{0.00033}$

69. $\dfrac{(15,000,000)(7,000,000,000)}{25,000,000}$

70. $\dfrac{(4,900,000)(2,700)}{63,000}$

71. $\dfrac{(0.0000000039)(0.00095)}{(0.0195)(4,000)}$

72. $\dfrac{(0.00024)(96,000,000)}{(640,000,000)(0.025)}$

APPLICATIONS

73. Five-Card Poker. The odds against being dealt the hand shown in the illustration are about 2.6×10^6 to 1. Express the odds using standard notation.

74. Energy. See the graph below. Express each of the following using scientific notation.

   a. U.S. energy consumption (94 quadrillion, 600 trillion Btu)

   b. U.S. energy production (72 quadrillion, 900 trillion Btu)

   c. The difference in 2009 consumption and production

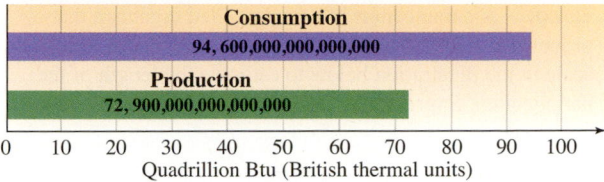

2009 U.S. Energy Consumption and Production
(petroleum, natural gas, coal, hydroelectric, nuclear, geothermal, solar, wind)

Source: Energy Information Administration, United States Department of Energy

75. Atoms. A hydrogen atom is so small that a single drop of water contains more than a million million billion hydrogen atoms. Express this number in scientific notation.

76. Astronomy. The American Physical Society recently honored first-year graduate student Gwen Bell for coming up with what it considers the most accurate estimate of the mass of the Milky Way. In pounds, her estimate is a 3 with 42 zeros after it. Express this number in scientific notation.

77. National Debt. As of January 2011, the U.S. national debt was approximately $13,950,000,000,000. The estimated population of the United States at that time was approximately 310,000,000. Express each person's share of the debt in scientific and standard notation. (Source: www.census.gov; www.usdebtclock.org)

78. Warp Speed. In the series *Star Trek,* the *U.S.S. Enterprise* traveled at warp speeds. To convert a warp speed, W, to an equivalent velocity in miles per second, v, we can use the equation $v = W^3 c$ where c is the speed of light, 1.86×10^5 miles per second. Find the velocity of a spacecraft traveling at warp 2.

79. Atoms. A simple model of a helium atom is shown. If a proton has a mass of 1.7×10^{-24} grams, and if the mass of an electron is only about $\frac{1}{2,000}$ that of a proton, find the mass of an electron.

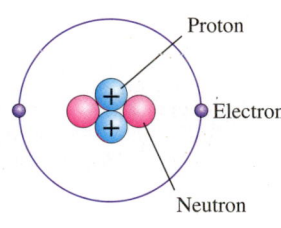

80. Oceans. The mass of the Earth's oceans is only about $\frac{1}{4,400}$ that of the Earth. If the mass of the Earth is 6.578×10^{21} tons, find the mass of the oceans.

81. Light Years. Light travels about 300,000,000 meters per second. A *light year* is the distance that light can travel in one year. Estimate the number of meters in one light year.

82. Aquariums. Express the volume of the fish tank in scientific notation.

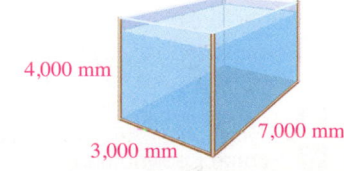

83. The Big Dipper. One star in the Big Dipper is named Merak. It is approximately 4.65×10^{14} miles from Earth.

a. If light travels about 1.86×10^5 miles/sec, how many seconds does it take light emitted from Merak to reach the Earth? (*Hint:* Use the formula $t = \frac{d}{r}$.)

b. Convert your result from part a to years.

Merak

84. Biology. A paramecium is a single-celled organism that propels itself with hair-like projections called *cilia*. Use the scale in the illustration below to estimate the length of the paramecium. Express the result in scientific and in standard notation.

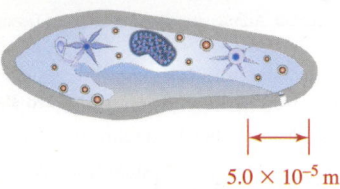

5.0×10^{-5} m

85. Comets. On March 23, 1997, Comet Hale-Bopp made its closest approach to Earth, coming within 1.3 *astronomical units*. One astronomical unit (AU) is the distance from the Earth to the sun—about 9.3×10^7 miles. Express this distance in miles, using scientific notation.

86. Diamonds. The approximate number of atoms of carbon in a $\frac{1}{2}$-carat diamond is given by the following expression. Find the number of carbon atoms in scientific and in standard notation.

$$\frac{6.0 \times 10^{23}}{1.2 \times 10^2}$$

WRITING

87. Explain how to change a number from standard notation to scientific notation.

88. Explain how to change a number from scientific notation to standard notation.

89. Explain why 9.99×10^n represents a number less than 1 but greater than 0 if n is a negative integer.

90. Explain the advantages of writing very large and very small numbers in scientific notation.

91. a. To multiply a number by 10^3, we move the decimal point. Which way, and how far?

b. To multiply a number by 10^{-3}, we move the decimal point. Which way, and how far?

92. Explain why 437.9×10^{23} is not written in scientific notation.

REVIEW

Solve each compound inequality. Graph the solution set and write it using interval notation.

93. $4x \geq -x + 5$ and $6 \geq 4x - 3$

94. $15 > 2x - 7 > 9$

95. $3x + 2 < 8$ or $2x - 3 > 11$

96. $-4(x + 2) \geq 12$ or $3x + 8 < 11$

CHALLENGE PROBLEMS

97. What is the reciprocal of the opposite of 2.5×10^{-24}? Write the result in scientific notation.

98. Solve: $(1.1 \times 10^{-16})x - (1.2 \times 10^{10}) = (6.5 \times 10^{10})$ and write the solution in scientific notation.

SECTION 5.3

Polynomials and Polynomial Functions

OBJECTIVES

1 Define and classify polynomials.

2 Evaluate polynomial functions.

3 Find function values and the domain and range of polynomial functions graphically.

4 Simplify polynomials by combining like terms.

5 Add polynomials.

6 Subtract polynomials.

ARE YOU READY?

The following problems review some basic skills that are needed when working with polynomials and polynomial functions.

1. a. How many terms does the expression $4x^2 - 8x + 1$ have?

b. What is the coefficient of the second term?

2. Fill in the blanks: The equation $f(x) = x - 3$ defines a _____, because to each value of x in the _____, there corresponds exactly one value of $f(x)$ in the range.

3. Let $f(x) = 7x + 11$. Find $f(-3)$.

4. Combine like terms: $7.4a^3 + 2.8a^3$

5. Combine like terms: $6t^2u - (-11t^2u)$

6. Simplify: $-(3b^2 - b + 24)$

In arithmetic, we add, subtract, multiply, divide, and find powers of real numbers. In algebra, we perform these operations on algebraic expressions called *polynomials*.

1 Define and Classify Polynomials.

Recall from Chapter 1 that a **term** is a product or quotient of numbers and/or variables. A single number or variable is also a term. Examples of terms are:

$$4, \quad y, \quad 6r, \quad -w^3, \quad 3.7x^5, \quad \text{and} \quad -15ab^2$$

If a term contains only a number, such as 4, it is called a **constant term,** or simply a **constant.**

The **numerical coefficient,** or simply the **coefficient,** is the numerical factor of a term. For example, the coefficient of $6r$ is 6 and the coefficient of $-15ab^2$ is -15. The coefficient of a constant term is that constant.

Polynomials	A **polynomial** is a single term or the sum of terms in which all variables have whole-number exponents. No variable appears in a denominator.

Here are some examples of polynomials. Note that polynomials are expressions, not equations.

$$5x + 3, \quad 4n^2 - 6n - 8, \quad p^3 + 3p^2q + 3pq^2 + q^3, \quad \text{and} \quad -\frac{5}{2}rs^2t^4$$

The Language of Algebra

The prefix **poly** means many. A *poly*gon is a many-sided figure and *poly*unsaturated fats are molecules having many strong chemical bonds.

The polynomial $5x + 3$ is the sum of two terms, $5x$ and 3, and we say it is a **polynomial in one variable, *x*.** Since $4n^2 - 6n - 8$ can be written as the sum $4n^2 + (-6n) + (-8)$, it has three terms, $4n^2$, $-6n$, and -8. It is written in **descending powers of *n*,** because the exponents on n decrease from left to right. When a polynomial is written in descending powers, the first term, in this case $4n^2$, is called the **leading term.** The coefficient of the leading term, in this case 4, is called the **leading coefficient.**

A polynomial can have more than one variable. For example, $p^3 + 3p^2q + 3pq^2 + q^3$ is a **polynomial in two variables,** *p* and *q*. It has four terms and is written in descending powers of *p* and **ascending powers** of *q*. Its leading term is p^3 and its leading coefficient is 1.

The polynomial $-\frac{5}{2}rs^2t^4$ is a **polynomial in three variables,** *r*, *s*, and *t*. It has one term.

CAUTION The following expressions are not polynomials:

$$\frac{2x}{x^2 + 1}, \quad 6y^{1/2} - 8, \quad \text{and} \quad c^{-3} + 2c + 24$$

The Language of Algebra

The prefix **mono** means one; Jay Leno begins the *Tonight Show* with a monologue. The prefix **bi** means two, as in bilingual or bifocals. The prefix **tri** means three, as in triangle or the *Star Wars Trilogy.*

The first expression is a quotient with a variable in the denominator. The last two contain variables with exponents that are not whole numbers.

Polynomials can be classified according to their number of terms. A polynomial with one term is called a **monomial,** a polynomial with two terms is called a **binomial,** and a polynomial with three terms is called a **trinomial.** Polynomials with four or more terms have no special names.

Polynomials			
Monomials	**Binomials**	**Trinomials**	**No special name**
$2x^3$	$2x + 5$	$2x^2 + 4x + 3$	$t^4 - t^3 + t^2 + t + 1$
a^2b	$-\frac{17}{2}x^4 - \frac{3}{5}x$	$3mn^3 - m^2n^3 + 7n$	$1.6 - 3.2d + 1.7d^2 + d^5$
$3x^3y^5z^2$	$3.2x^{13}y^5z^3 + 4.7x^3yz$	$-12x^5y^2 + 13x^4y^3 - 7x^3y^3$	$mn^2 - mn + m - 2n$

Polynomials and their terms can be classified according to the exponents on their variables.

Degree of a Term of a Polynomial	The **degree of a term** of a polynomial in one variable is the value of the exponent on the variable. If a polynomial is in more than one variable, the **degree of a term** is the sum of the exponents on the variables in that term. The **degree of a nonzero constant** is 0. The constant 0 has no defined degree.

Think of the *degree or a term* as the number of variable factors in that term.

- $15x^4$ has degree **4**. x^4 represents four variable factors: $x \cdot x \cdot x \cdot x$.
- $-2.9y^2$ has degree **2**. y^2 represents two variable factors: $y \cdot y$.
- $\frac{7}{8}m^8n^6$ has degree **14**.   Because $8 + 6 = 14$.
- 6 has degree **0** since it can be written as $6x^0$.   There is no variable factor.

We determine the *degree of a polynomial* by considering the degrees of each of its terms.

Degree of a Polynomial	The **degree of a polynomial** is the same as the highest degree of any term of the polynomial.

EXAMPLE 1   Use the vocabulary of this section to describe each polynomial: **a.** $x^4 - 2x^2 + 4$
b. $\frac{1}{5}y^3 - y$ **c.** $-2.7m^{12} - 4.5m^{10}n^4 + 9.1m^8n^6 + mn^9$

Strategy   First, we will identify the variable(s) in each polynomial and determine whether it is written in ascending or descending powers. Then we will count its number of terms and determine the degree of each term.

Why   The number of terms determines the type of polynomial. The highest degree of any term of the polynomial determines its degree.

Solution   **a.** $x^4 - 2x^2 + 4$ is a polynomial in one variable that is written in descending powers of x. Because it has three terms, x^4, $-2x^2$, and 4, it is a trinomial. The leading term is x^4, and thus, the leading coefficient is 1. Because the highest degree of any of its terms is 4, this trinomial is of degree 4.

The Language of Algebra

To **descend** means to move from higher to lower. When we write $x^4 - 2x^2 + 4$ as $x^4 + 0x^3 - 2x^2 + 0x^1 + 4x^0$, we more clearly see the *descending* powers of x.

Term	Coefficient	Degree
x^4	1	**4**
$-2x^2$	-2	2
4	4	0

Degree of the polynomial: **4**
Type of polynomial: Trinomial

b. $\frac{1}{5}y^3 - y$ is a polynomial in one variable. It is written in descending powers of y. Since it has two terms, $\frac{1}{5}y^3$ and $-y$, it is a binomial. The leading term is $\frac{1}{5}y^3$, and thus, the leading coefficient is $\frac{1}{5}$. The highest degree of any of its terms is 3, so this binomial is of degree 3.

Term	Coefficient	Degree
$\frac{1}{5}y^3$	$\frac{1}{5}$	**3**
$-y$	-1	1

Degree of the polynomial: **3**
Type of polynomial: Binomial

The Language of Algebra

The word **degree** is also used in other disciplines for classification. For example, doctors speak of second- and third-*degree* burns.

c. $-2.7m^{12} - 4.5m^{10}n^4 + 9.1m^8n^6 + mn^9$ is   a polynomial in two variables, m and n. It is written in descending powers of m and ascending powers of n. It has four terms and, therefore, has no special name. The leading term is $-2.7m^{12}$ and, thus, a leading coefficient is -2.7. The highest degree of any term is 14, so this polynomial is of degree 14.

Term	Coefficient	Degree
$-2.7m^{12}$	-2.7	12
$-4.5m^{10}n^4$	-4.5	**14**
$9.1m^8n^6$	9.1	**14**
mn^9	1	10

Degree of the polynomial: **14**
Type of polynomial: No special name

Describe the polynomial $-5.6p^6q + p^5q^3$.

Now Try ▶ Problems 19 and 21

2 Evaluate Polynomial Functions.

We have seen that **linear functions** are defined by equations of the form $f(x) = mx + b$. Some examples of linear functions are

$$f(x) = 3x + 1, \qquad g(x) = -\frac{1}{2}x - 1, \qquad \text{and} \qquad h(x) = 5x$$

In each case, the right side of the equation is a polynomial. For this reason, linear functions are members of a larger class of functions known as *polynomial functions.*

Polynomial Functions	A **polynomial function** is a function whose equation is defined by a polynomial in one variable.

Another example of a polynomial function is $f(x) = x^2 + 6x - 8$. This is a second-degree polynomial function, called a **quadratic function.** Quadratic functions are of the form $f(x) = ax^2 + bx + c$, where $a \neq 0$.

An example of a third-degree polynomial function is $f(x) = x^3 - 3x^2 - 9x + 2$. Third-degree polynomial functions, also called **cubic functions,** are of the form $f(x) = ax^3 + bx^2 + cx + d$, where $a \neq 0$.

Some functions that are *not* polynomial functions are:

$$f(x) = |x|, \qquad f(x) = \frac{9}{x - 1}, \qquad \text{and} \qquad f(x) = x^{-3}$$

Polynomial functions can be used to model many real-life situations. If we are given a polynomial function model, we can learn more about the situation by evaluating the function at specific values.

To **evaluate a polynomial function** at a specific value, we replace the variable in the defining equation with that value, called the input. Then we simplify the resulting expression to find the output.

EXAMPLE 2 **Rocketry.** If a toy rocket is shot straight up with an initial velocity of 128 feet per second, its height, in feet, t seconds after being launched is approximated by the polynomial function $h(t) = -16t^2 + 128t$. Find the height of the rocket:

a. 2 seconds after being launched. **b.** 7.9 seconds after being launched.

Strategy We will find $h(2)$ and $h(7.9)$.

Why The notation $h(2)$ represents the height of the rocket 2 seconds after being launched and $h(7.9)$ represents the height of the rocket 7.9 seconds after being launched.

©Peter Barrett/Shutterstock.com

Solution **a.** To find the height of the rocket 2 seconds after being launched, we find $h(2)$ as follows:

$$h(t) = -16t^2 + 128t \qquad \text{\color{red}{This is the given function.}}$$
$$h(2) = -16(2)^2 + 128(2) \qquad \text{\color{red}{Substitute 2 for each } t. \text{ (The input is 2.)}}$$
$$= -16(4) + 256 \qquad \text{\color{red}{Evaluate the right side.}}$$
$$= -64 + 256 \qquad \text{\color{red}{Do the multiplication.}}$$
$$= 192 \qquad \text{\color{red}{The output is 192.}}$$

We have found that $h(2) = 192$. Thus, 2 seconds after it is launched, the height of the rocket is 192 feet.

b. To find the height of the rocket 7.9 seconds after it is launched, we find $h(7.9)$ as follows:

$$h(t) = -16t^2 + 128t \qquad \text{\color{red}This is the given function.}$$
$$h(\mathbf{7.9}) = -16(\mathbf{7.9})^2 + 128(\mathbf{7.9}) \qquad \text{\color{red}Substitute 7.9 for each } t. \text{ (The input is 7.9.)}$$
$$= -16(62.41) + 1{,}011.2 \qquad \text{\color{red}Evaluate the right side.}$$
$$= -998.56 + 1{,}011.2 \qquad \text{\color{red}Do the multiplication.}$$
$$= 12.64 \qquad \text{\color{red}The output is 12.64. That means } h(7.9) = 12.64.$$

At 7.9 seconds, the height of the rocket is 12.64 feet. It has almost fallen back to Earth.

> **Self Check 2** **Rocketry.** Find the height of the rocket 4 seconds after it is launched.
>
> **Now Try** ▶ Problems 25 and 85

3 Find Function Values and the Domain and Range of Polynomial Functions Graphically.

The graphs of polynomial functions of degree 1, such as $f(x) = 4x - 1$ or $f(x) = \frac{1}{2}x + 3$, are straight lines. (See Section 2.5.) The graphs of polynomial functions of degree 2, such as $f(x) = x^2$ or $f(x) = x^2 - 2$, are parabolas. (See Section 2.6.) The graphs of polynomial functions of degree 3 or higher are often more complicated. Such graphs are always *smooth* and *continuous*. That is, they consist of only rounded curves with no sharp corners, and there are no breaks.

We can obtain important information about a polynomial function from its graph.

EXAMPLE 3 Refer to the graph of polynomial function $f(x) = x^3 - 3x^2 - 9x + 2$ in figure (a) below. **a.** Find $f(2)$. **b.** Find any values of x for which $f(x) = -25$. **c.** Find the domain and range of f.

Strategy For parts a and b, we will use the information provided by the function notation to locate a specific point on the graph and determine its x- and y-coordinates. For part c, we will project the graph onto each axis.

Why Once we locate the specific point, one of its coordinates will equal the value that we are asked to find. Projecting the graph onto the x-axis gives the domain and projecting it onto the y-axis gives the range.

Solution **a.** To find $f(2)$, we need to find the y-coordinate of the point on the graph of f whose x-coordinate is 2. If we draw a vertical line downward, from 2 on the x-axis, as shown in figure (b), the line intersects the graph of f at $(2, -20)$. Therefore, -20 corresponds to 2, and it follows that $f(2) = -20$.

b. To find any input value x that has an output value $f(x) = -25$, we draw a horizontal line through -25 on the y-axis, as shown in figure (c), and note that it intersects the graph of f at $(-3, -25)$ and $(3, -25)$. Therefore, $f(-3) = -25$ and $f(3) = -25$. It follows that the values of x for which $f(x) = -25$ are -3 and 3.

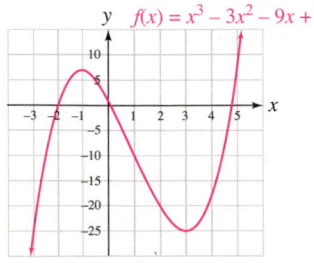

(a)

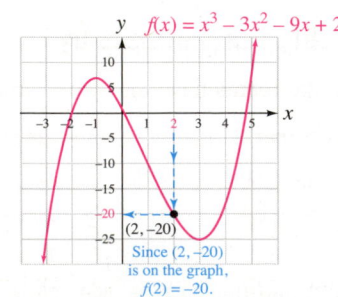

(b)

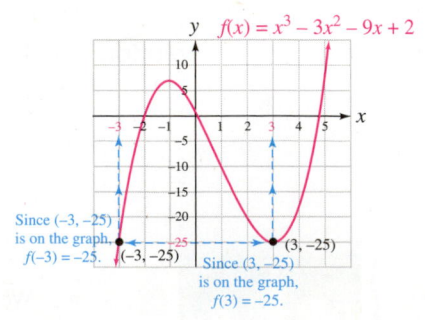

(c)

c. To find the domain of $f(x) = x^3 - 3x^2 - 9x + 2$, we project its graph onto the x-axis as shown in figure (d) below. Because the graph extends indefinitely to the left and right, the projection includes all real numbers. Therefore, the domain of the function is the set of real numbers, which can be written in interval notation as $(-\infty, \infty)$.

To determine the range of the same polynomial function, we project the graph onto the y-axis, as shown in figure (e) below. Because the graph of the function extends indefinitely upward and downward, the projection includes all real numbers. Therefore the range of the function is the set of real numbers, written $(-\infty, \infty)$.

Success Tip

The graphs of many polynomial functions of degree 3 and higher have "peaks" and "valleys" as shown here.

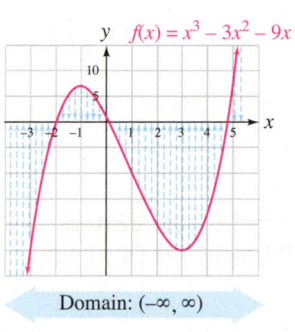

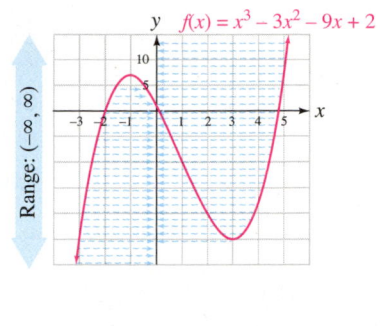

(d)

(e)

Self Check 3   Refer to the graph of function f in Example 3. Estimate each of the following: **a.** $f(1)$ **b.** Any x-values for which $f(x) = 0$

Now Try ▶ Problems 13 and 15

Using Your Calculator ▶ **Graphing Polynomial Functions**

We can graph polynomial functions with a graphing calculator. For example, to graph $f(x) = x^3 - 3x^2 - 9x + 2$ from Example 3, we enter the right side of the function notation as shown in figure (a). Using window settings of $[-8, 8]$ for x and $[-50, 50]$ for y, we get the graph shown in figure (b).

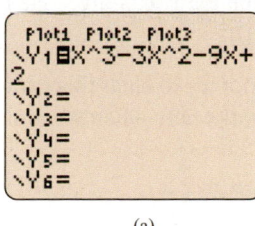

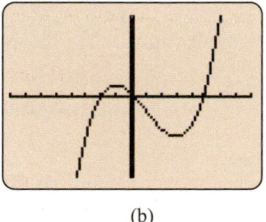

(a)

(b)

---

4   Simplify Polynomials by Combining Like Terms.

Recall that **like terms** have the same variables with the same exponents:

Like terms	*Unlike terms*	
$-7x$ and $15x$	$-7x$ and $15a$	Different variables.
$4y^3$ and $16y^3$	$4y^3$ and $16y^2$	Different exponents on the same variable.
$\dfrac{1}{2}xy^2$ and $-\dfrac{1}{3}xy^2$	$\dfrac{1}{2}xy^2$ and $-\dfrac{1}{3}x^2y$	Different exponents on different variables.

The Language of Algebra

Simplifying the sum or difference of like terms is called **combining like terms.**

Also recall that to **combine like terms,** we combine their coefficients and keep the same variables with the same exponents. For example,

$$4y + 5y = (4 + 5)y \qquad 8x^2 - x^2 = (8 - 1)x^2$$
$$= 9y \qquad\qquad\qquad = 7x^2$$

Polynomials with like terms can be simplified by combining like terms.

EXAMPLE 4

Simplify each polynomial by combining like terms.

a. $4x^4 + 81x^4$ **b.** $17x^2y^2 + 2x^2y - 6x^2y^2$

c. $r - 3r^2 - 4r^2 + 8r^2$ **d.** $\dfrac{3}{5}ab + \dfrac{4}{3}a - 7 + \dfrac{1}{2}ab - \dfrac{1}{6}a + 4$

Strategy   We will use the distributive property in reverse to add (or subtract) the coefficients of the like terms. We will keep the same variables raised to the same powers.

Why   To *combine like terms* means to add or subtract the like terms in an expression.

Solution

a. $4x^4 + 81x^4 = 85x^4$ Think: $(4 + 81)x^4 = 85x^4$.

b. The first and third terms of the given trinomial are like terms.

$$17x^2y^2 + 2x^2y - 6x^2y^2 = 11x^2y^2 + 2x^2y \quad \text{Think: } (17 - 6)x^2y^2 = 11x^2y^2.$$

Caution

When combining like terms, simply combine their coefficients. The exponents on the variables *stay the same.* Don't incorrectly add the exponents.

c. The last three terms of the given polynomial are like terms.

$$r - 3r^2 - 4r^2 + 8r^2 = r + r^2 \quad \text{Think: } (-3 - 4 + 8)r^2 = 1r^2 = r^2.$$
$$= r^2 + r \quad \text{Write the result in descending powers of } r.$$

d. The first and fourth terms are like terms, the second and fifth terms are like terms, and the third and sixth terms are like terms.

$$\frac{3}{5}ab + \frac{4}{3}a - 7 + \frac{1}{2}ab - \frac{1}{6}a + 4 \qquad \text{This polynomial has six terms.}$$

$$= \left(\frac{3}{5} + \frac{1}{2}\right)ab + \left(\frac{4}{3} - \frac{1}{6}\right)a - 7 + 4 \qquad \text{Combine like terms.}$$

$$= \left(\frac{6}{10} + \frac{5}{10}\right)ab + \left(\frac{8}{6} - \frac{1}{6}\right)a - 7 + 4 \qquad \text{To add and subtract the fractions, build equivalent fractions: } \frac{3}{5} \cdot \frac{2}{2} = \frac{6}{10}, \frac{1}{2} \cdot \frac{5}{5} = \frac{5}{10}, \text{ and } \frac{4}{3} \cdot \frac{2}{2} = \frac{8}{6}.$$

$$= \frac{11}{10}ab + \frac{7}{6}a - 3 \qquad \text{Do the addition and the subtraction.}$$

CAUTION   Do not try to clear this expression of fractions by multiplying it by the LCD 30. That strategy works only when we multiply *both sides of an equation* by the LCD.

$$30\left(\frac{3}{5}ab + \frac{4}{3}a - 7 + \frac{1}{2}ab - \frac{1}{6}a + 4\right)$$

Self Check 4   Simplify each polynomial by combining like terms.

     a. $6m^4 + 3m^4$ **b.** $17s^3t + 3s^2t - 6s^3t$

     c. $x - 19x^2 + 22x^2 - x^2$

     d. $\frac{7}{8}rs + \frac{7}{9}r + 1 - \frac{3}{4}rs + \frac{4}{3}r - 2$

Now Try ▶   Problems 29 and 35

5   Add Polynomials.

When adding polynomials horizontally, each polynomial is usually enclosed within parentheses. For example, $(3x^2 - 2x + 4) + (2x^2 + 4x - 3)$ is the sum of two trinomials.

| Adding Polynomials | To add polynomials, drop the parentheses and combine their like terms. |

EXAMPLE 5 Add: **a.** $(3x^2 - 2x + 4) + (2x^2 + 4x - 3)$

b. $(-5x^3y^2 - 4x^2y^3) + (2x^3y^2 + x^3y + 5x^2y^3)$ **c.** $\left(\frac{1}{4}m^4 + \frac{1}{2}m^3\right) + \left(\frac{3}{4}m^4 - \frac{7}{3}m^3\right)$

Strategy We will drop the parentheses and combine like terms.

Why To add polynomials means to combine their like terms.

Solution **a.** $(3x^2 - 2x + 4) + (2x^2 + 4x - 3)$ This is the sum of two trinomials.

$= 3x^2 - 2x + 4 + 2x^2 + 4x - 3$ Drop the parentheses.

$= 5x^2 + 2x + 1$ Combine like terms.

Notation

When performing operations on polynomials, it is standard practice to write the terms of a result in descending powers of one variable because that makes them easy to compare.

b. $(-5x^3y^2 - 4x^2y^3) + (2x^3y^2 + x^3y + 5x^2y^3)$ This is the sum of a binomial and a trinomial.

$= -5x^3y^2 - 4x^2y^3 + 2x^3y^2 + x^3y + 5x^2y^3$ Drop the parentheses.

$= -3x^3y^2 + x^3y + x^2y^3$ Combine like terms.

c. $\left(\frac{1}{4}m^4 + \frac{1}{2}m^3\right) + \left(\frac{3}{4}m^4 - \frac{7}{3}m^3\right)$ This is the sum of two binomials.

$= \frac{1}{4}m^4 + \frac{1}{2}m^3 + \frac{3}{4}m^4 - \frac{7}{3}m^3$ Drop the parentheses.

$= \frac{1}{4}m^4 + \frac{3}{6}m^3 + \frac{3}{4}m^4 - \frac{14}{6}m^3$ To add and subtract the fractions, build equivalent fractions $\frac{1}{2} \cdot \frac{3}{3} = \frac{3}{6}$ and $\frac{7}{3} \cdot \frac{2}{2} = \frac{14}{6}$.

$= \frac{4}{4}m^4 - \frac{11}{6}m^3$ Combine like terms.

$= m^4 - \frac{11}{6}m^3$ Simplify: $\frac{4}{4}m^4 = 1m^4 = m^4$.

Self Check 5 Add: **a.** $(2a^2 - 3a + 5) + (5a^2 + 4a - 2)$
b. $(-6a^2b^3 - 5a^3b^2) + (3a^2b^3 + 2a^3b^2 + ab^2)$
c. $\left(\frac{5}{8}t^5 - \frac{1}{6}t^4\right) + \left(\frac{3}{8}t^5 + \frac{7}{4}t^4\right)$

Now Try ▶ Problems 37, 39, and 41

The first two additions in Example 5 can be done by aligning the like terms vertically and combining them column by column.

$\begin{array}{r} 3x^2 - 2x + 4 \\ + \ 2x^2 + 4x - 3 \\ \hline 5x^2 + 2x + 1 \end{array}$ $\begin{array}{r} -5x^3y^2 \qquad - 4x^2y^3 \\ + \ \ 2x^3y^2 + x^3y + 5x^2y^3 \\ \hline -3x^3y^2 + x^3y + x^2y^3 \end{array}$

6 Subtract Polynomials.

Recall from Chapter 1 that we can use the distributive property to find the opposite of several terms enclosed within parentheses. For example, we consider $-(2x^3 - 3x^2)$.

$-(2x^3 - 3x^2) = -1(2x^3 - 3x^2)$ Replace the − symbol in front of the parentheses with −1.

$= -2x^3 + 3x^2$ Use the distributive property to remove parentheses.

This result illustrates that we can remove a $-$ sign preceding parentheses by dropping the $-$ sign and the parentheses and *changing the sign of every term within the parentheses.* This observation suggests a way to subtract polynomials.

| **Subtracting Polynomials** | To subtract two polynomials, change the signs of the terms of the polynomial being subtracted, drop the parentheses, and combine like terms. |

EXAMPLE 6   Subtract: **a.** $(8x^3 + 2x^2) - (2x^3 - 3x^2)$

b. $(3.1rt^2 + 4.3r^2t^2) - (8.7rt^2 - 4.3r^2t^2 + 5.9r^3t^2)$

Strategy   In each case, we will change the signs of the terms of the polynomial being subtracted, drop the parentheses, and combine like terms.

Why   We can remove a $-$ sign preceding parentheses by dropping the $-$ sign and the parentheses and changing the sign of every term within the parentheses.

Solution

a. $(8x^3 + 2x^2) - (2x^3 - 3x^2)$ This is the difference of two binomials.

$= 8x^3 + 2x^2 - 2x^3 + 3x^2$ Change the sign of each term of $2x^3 - 3x^2$ and drop the parentheses.

$= 6x^3 + 5x^2$ Combine like terms.

Caution

A common error is to change the sign of only the first term of the polynomial being subtracted and to forget to change the signs of its other terms.

b. $(3.1rt^2 + 4.3r^2t^2) - (8.7rt^2 - 4.3r^2t^2 + 5.9r^3t^2)$ This is the difference of a binomial and a trinomial.

$= 3.1rt^2 + 4.3r^2t^2 - 8.7rt^2 + 4.3r^2t^2 - 5.9r^3t^2$ Change the signs of the terms of the polynomial being subtracted.

$= -5.6rt^2 + 8.6r^2t^2 - 5.9r^3t^2$ Combine like terms.

$= -5.9r^3t^2 + 8.6r^2t^2 - 5.6rt^2$ Write the terms of the result in descending powers of r.

Self Check 6   Subtract: **a.** $(9m^4 + 16m^2) - (12m^4 - 18m^2)$

b. $(6.4a^2b^3 - 2.7a^2b^2) - (-2.5a^2b^3 + 8.1a^2b^2)$

Now Try ▶   Problems 45 and 49

Just as real numbers have opposites, so do polynomials. To find the opposite of a polynomial, multiply each of its terms by -1. This changes the sign of each term of the polynomial.

A polynomial   *Its opposite*

$2x^2 - 4x + 5$ →(Multiply by -1)→ $-(2x^2 - 4x + 5)$ or $-2x^2 + 4x - 5$

To subtract polynomials in vertical form, we add the opposite of the polynomial that is being subtracted.

$$\begin{array}{r} 8x^3 + 2x^2 \\ - (2x^3 - 3x^2) \end{array}$$ →Change signs and add→ $$\begin{array}{r} 8x^3 + 2x^2 \\ + \; -2x^3 + 3x^2 \\ \hline 6x^3 + 5x^2 \end{array}$$ This is the opposite of $2x^3 - 3x^2$.

EXAMPLE 7   Subtract $16x^4y - 9x^3y$ from the sum of $x^4y + 10x^3y$ and $7x^4y - 8x^3y$.

Strategy   First, we will translate the words of the problem into mathematical symbols. Then we will perform the indicated operations.

Why   The words of the problem contain the key phrases *subtract from* and *sum*.

Solution Since $16x^4y - 9x^3y$ is to be subtracted from the sum, the order must be reversed when we translate to mathematical symbols.

The Language of Algebra

Recall that we must be careful of the order when translating English phrases containing the words **"subtract from."** For example, to subtract a from b, we write $b - a$.

Subtract $16x^4y - 9x^3y$ from   the sum of $x^4y + 10x^3y$ and $7x^4y - 8x^3y$.

$$[(x^4y + 10x^3y) + (7x^4y - 8x^3y)] - (16x^4y - 9x^3y)$$

Use brackets [ ] to enclose the sum. Use parentheses to enclose the difference.

Next, we remove the grouping symbols to obtain

$$= x^4y + 10x^3y + 7x^4y - 8x^3y - 16x^4y + 9x^3y$$

Change the sign of each term within $(16x^4y - 9x^3y)$ and drop the parentheses.

$$= -8x^4y + 11x^3y \qquad \text{Combine like terms.}$$

Self Check 7 Subtract $15mn^3 - 8mn^2$ from the sum of $14mn^3 - 9mn^2$ and $3mn^3 + 6mn^2$.

Now Try ▶ Problem 57

SECTION **5.3**   STUDY SET

VOCABULARY

Fill in the blanks.

1. A _____ is the sum of one or more algebraic terms whose variables have whole-number exponents.

2. $x^3 - 8x^2 - x + 9$ is a polynomial in _____ variable, and is written in _____ powers of x. The polynomial $m^3 + 10m^2n - n^2$ is in _____ variables and is written in _____ powers of n.

3. For the polynomial $7x^2 - 5x - 12$, the _____ term is $7x^2$, and the leading _____ is 7. The _____ term is -12.

4. A _____ is a polynomial with one term. A _____ is a polynomial with two terms. A _____ is a polynomial with three terms.

5. The _____ of the term $6x^5$ is 5 because x appears as a factor 5 times: $6 \cdot x \cdot x \cdot x \cdot x \cdot x$.

6. A second-degree polynomial function is also called a _____ function. A third-degree polynomial function is also called a _____ function.

7. Terms having the same variables with the same exponents are called _____ terms.

8. The _____ of $y^2 + 4y - 2$ is $-y^2 - 4y + 2$.

CONCEPTS

9. Determine whether each expression is a polynomial.

   a. $\dfrac{3}{x^2} + \dfrac{4}{x} + 2$ **b.** $16a^2 - 25b^2$

   c. $y^{-2} - 5y^{-1}$ **d.** $t^4 + t^3 + t^2 + t + 1$

10. Classify each polynomial as a monomial, a binomial, a trinomial, or none of these.

   a. $cdy^2 + 4cd - 3c$ **b.** $3 - 5x^2$

   c. $-\dfrac{15}{16}z^{18}$ **d.** $\dfrac{3}{5}x^4 - \dfrac{2}{5}x^3 + \dfrac{3}{5}x - 1$

11. Fill in the blank so that the term has degree 6.

   a. $-15x^{\blacksquare}$ **b.** $\dfrac{9}{8}ab^{\blacksquare}$

12. Decide whether the terms are like or unlike terms. If they are like terms, add them.

   a. $12x$, $5x$ **b.** $9u^2$, $10u^2$

   c. $6x^2y^3$, $6x^2y^2$ **d.** $-27x^6y^4z$, $8x^6y^4z^2$

13. Use the graph of function f to find each of the following:

   a. $f(1)$

   b. $f(-3)$

   c. The values of x for which $f(x) = 0$.

   d. The domain and range of f.

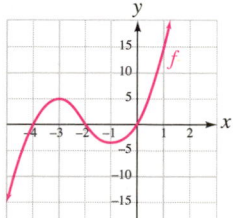

14. Use the graph of function f to find each of the following

a. $f(-1)$

b. $f(0)$

c. The values of x for which $f(x) = 3$.

d. The domain and range of f.

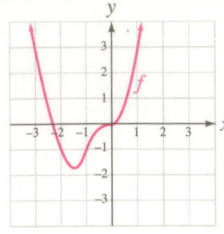

15. Use the given graph of function g to find each of the following:

a. $g(1)$

b. $g(-4)$

c. The values of x for which $g(x) = 4$.

d. The domain and range of g.

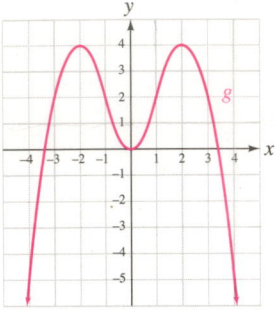

16. Use the given graph of function h to find each of the following.

a. $h(0.5)$

b. $h(-0.5)$

c. The values of x for which $h(x) = -1.5$.

d. The domain and range of h.

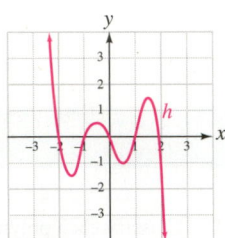

NOTATION

Complete the evaluation.

17. If $h(t) = -t^3 - t^2 + 2t + 1$, find $h(3)$.

$$h(\boxed{}) = -(\boxed{})^3 - (\boxed{})^2 + 2(3) + 1$$
$$= \boxed{} - 9 + 6 + 1$$
$$= \boxed{}$$

18. a. Write $3x - 2x^4 + 7 - 5x^2$ with the exponents on x in descending order.

b. Write $x^3y^2 + x^2y^3 - 2x^3y + x^7y^6 - 3x^6$ with the exponents on y in ascending order.

GUIDED PRACTICE

In problems 19–22, complete a table like that shown below for each given polynomial. See Example 1.

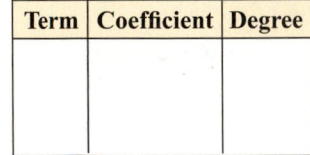

Term	Coefficient	Degree

19. $25x^2 - x + 4$

20. $6a^5 + a^3 - 9a^2 + 7a$

21. $5a^7b^4 - 33a^2b$

22. $\frac{3}{4}u^4 + \frac{5}{8}u^2v^2 - \frac{1}{2}v^4$

Find each function value. See Example 2.

23. Let $f(x) = 5x^4 - 2x^2 + 9x + 5$.

a. $f(-1)$ b. $f(2)$

24. Let $g(t) = -t^4 + 2t^3 - 5t + 8$.

a. $g(-2)$ b. $g(3)$

25. Let $h(t) = \frac{1}{4}t^2 - \frac{5}{8}t$.

a. $h(-4)$ b. $h(8)$

26. Let $s(x) = 2.5x^3 - 3.6x^2 + 1.1x$.

a. $s(-10)$ b. $s(10)$

Use a graphing calculator to graph each polynomial function. Use window settings of $[-4, 6]$ for x and $[-5, 5]$ for y.

27. $f(x) = 2.75x^2 - 4.7x + 1.5$

28. $f(x) = 0.37x^3 - 1.4x + 1.5$

Simplify each polynomial. See Example 5.

29. $15x^2 + 4x - 5x + 5x^2 + 9$

30. $8m^3 - 5m^2 - 5 + 3m^2 - 7m^3$

31. $-7y^2 - 11y^3 - 4 + y^3 - y^2$

32. $2y^3 - 3y^4 - 5 + 4y^4 - 3y^3$

33. $1.4ab^2 - 4.8ab - 0.2a + 5.4ab + 3.7ab^2$

34. $8.7c^4d + 5.4cd - 7.9cd + 1.8 - 6.3c^4d$

35. $\frac{9}{4}rst^2 - \frac{5}{3}rst - \frac{1}{2}rst^2 + \frac{5}{6}rst$

36. $\frac{1}{9}m^4np - \frac{5}{2}mnp - \frac{2}{3}m^4np + \frac{3}{4}mnp$

Add. See Example 6.

37. $(3x^2 + 2x + 1) + (2x^2 - 7x + 5)$

38. $(2a^2 + 4a - 7) + (3a^2 - a - 2)$

39. $(9p^2q^2 + p - q) + (-p^2q^2 - p - q + 8)$

40. $(6a^2x^3 - 2ax^2 + 3a^3) + (-4a^2x^3 - 2a^3)$

41. $\left(\frac{1}{5}h^6 + \frac{3}{4}h^2\right) + \left(\frac{2}{3}h^6 - \frac{1}{12}h^2\right)$

42. $\left(\frac{9}{16}n^8 - \frac{5}{6}n^4\right) + \left(\frac{1}{4}n^8 + \frac{7}{3}n^4\right)$

43. $\begin{array}{r} 6a^4 + 9a^2 + a \\ + 2a^4 - 13a^2 + a \\ \hline \end{array}$

44. $\begin{array}{r} 13b^3 - 11b^2 + 7b \\ + 2b^3 + 9b^2 + 8b \\ \hline \end{array}$

Subtract. See Example 7.

45. $(6x^3 + 3x - 2) - (2x^3 + 3x^2 + 5)$

46. $(x^2 - 3x + 8) - (3x^2 + x + 3)$

47. $(2m^2n^2 + 2m - n) - (-2m^2n^2 - 2m + n)$

48. $(2x^2y^3 + 6xy + 5y^2) - (-4x^2y^3 - 7xy + 2y^2 + y)$

49. $(7.1y^3 + 4.9y^2 + 0.1y) - (-8.4y^3 - 0.1y)$

50. $(-8.9p^3 - 2.4p) - (2.1p^3 + 0.8p^2 - p)$

51. $\left(\dfrac{5}{2}w^3 + \dfrac{1}{4}w^2 + \dfrac{3}{5}\right) - \left(\dfrac{1}{3}w^3 + \dfrac{1}{2}w^2 - \dfrac{1}{5}\right)$

52. $\left(\dfrac{3}{8}t^9 - \dfrac{1}{10}t^3 - \dfrac{13}{4}t\right) - \left(\dfrac{1}{8}t^9 - \dfrac{1}{2}t^3 - \dfrac{5}{4}t\right)$

53.
$$\begin{array}{r} 3x^2 - 4x + 17 \\ -\ (2x^2 + 4x -\ \ 5) \\ \hline \end{array}$$

54.
$$\begin{array}{r} 7y^2 -\ \ 4y + 3 \\ -\ (3y^2 + 10y - 5) \\ \hline \end{array}$$

55.
$$\begin{array}{r} 4x^3 \qquad\ \ + 6a \\ -\ (4x^3 - 2x^2 -\ \ a) \\ \hline \end{array}$$

56.
$$\begin{array}{r} -2a^3 \qquad\ \ - 7b \\ -\ (-2a^3 + 3a^2 - 6b) \\ \hline \end{array}$$

Perform the indicated operations. See Example 8.

57. Subtract $3x^2y^3 + 4xy^2 - 3x^2$ from the sum of $-2x^2y^3 - xy^2 + 7x^2$ and $5x^2y^3 + 3xy^2 - x^2$.

58. Subtract $8m^3n^3 + 2m^2n - n^2$ from the sum of $m^2n + mn^2 + 2n^2$ and $2m^3n^3 - mn^2 + 9n^2$.

59. Find the sum when the difference of $2x^2 - 4x + 3$ and $8x^2 + 5x - 3$ is added to $-2x^2 + 7x - 4$.

60. Find the sum when the difference of $7x^3 - 4x$ and $x^2 + 2$ is added to $x^2 + 3x + 5$.

TRY IT YOURSELF

Perform the indicated operations.

61. $(2.8b^2 + 1.2bc + 4.2c^2) + (5.1b^2 - 7.6bc - 3.9c^2)$

62. $(4.4t^2 - 2.9bt + 1.5b^2) + (5.6t^2 + 1.2bt - 3.3b^2)$

63. $(3x^2 + 4x - 3) + (2x^2 - 3x - 1) - (x^2 + x + 7)$

64. $(-2x^2 + 6x + 5) - (-4x^2 - 7x + 2) - (4x^2 + 10x + 5)$

65.
$$\begin{array}{r} 3x^3 - 2x^2 + 4x -\ \ 3 \\ -2x^3 + 3x^2 + 3x -\ \ 2 \\ +\ 5x^3 - 7x^2 + 7x - 12 \\ \hline \end{array}$$

66.
$$\begin{array}{r} 7a^3 \qquad\ \ + 3a +\ \ 7 \\ -2a^3 + 4a^2 \qquad\ \ - 13 \\ +\ 3a^3 - 3a^2 + 4a +\ \ 5 \\ \hline \end{array}$$

67. Find the difference when $(ay^3 - 2ay^2 + 2a)$ is subtracted from the sum of $(3ay^3 + ay^2)$ and $(-ay^3 + 6ay - 3a)$.

68. Find the difference when $(-3mn^3 - 4mn + 7m)$ is subtracted from the sum of $(2mn^2 + 3mn - 7m)$ and $(-4mn^3 - 2mn - 3m)$.

69. $(0.2xy^7 + 0.8xy^5) - (0.5xy^7 - 0.6xy^5 + 0.2xy)$

70. $(5.8g^3t + 0.9g^2t) - (1.3g^3t - 7.6g^2t + 9.9gt)$

71.
$$\begin{array}{r} -5y^3 +\ \ 4y^2 - 11y +\ \ 3 \\ -\ (-2y^3 - 14y^2 + 17y - 32) \\ \hline \end{array}$$

72.
$$\begin{array}{r} 17x^4 -\ \ 3x^2 - 65x - 12 \\ -\ (23x^4 + 14x^2 +\ \ 3x - 23) \\ \hline \end{array}$$

73. $(1 - 2x - x^2 + 4x^3) + (x^3 - 5x^2 + x + 8)$

74. $(3 - 2z - z^2 + 7z^3) + (3z^3 - z^2 + z + 1)$

75. Subtract $(k^3 - 5k)$ from $(6k^4 + 2k^2 - 16)$.

76. Subtract $(-m^4 + 9m^2)$ from $(m^3 + 2m - 4)$.

77. Let $f(x) = x^2 + 2x$ and $g(x) = 4x^2 - 2x - 1$. Find $f(x) - g(x)$.

78. Let $f(x) = 8x^2 - 7x$ and $g(x) = -10x^2 + 6x + 4$. Find $f(x) - g(x)$.

79. Let $f(x) = \dfrac{1}{3}x^6 - \dfrac{1}{6}x^4 - \dfrac{4}{3}x^2$ and $g(x) = -\dfrac{1}{6}x^6 - \dfrac{1}{2}x^4 + \dfrac{5}{6}x^2$. Find $f(x) + g(x)$.

80. Let $f(x) = -\dfrac{2}{5}x^5 + \dfrac{2}{5}x^3 - \dfrac{2}{5}x$ and $g(x) = \dfrac{1}{3}x^5 + \dfrac{1}{3}x^3 - \dfrac{1}{3}x$. Find $f(x) - g(x)$.

Look Alikes . . .

81. a.
$$\begin{array}{r} 3d^3 - 4d^2 - 3d + 5 \\ +\ 11d^3 \qquad\ \ - 8d - 2 \\ \hline \end{array}$$
b.
$$\begin{array}{r} 3d^3 - 4d^2 - 3d + 5 \\ -\ (11d^3 \qquad\ \ - 8d - 2) \\ \hline \end{array}$$

82. a.
$$\begin{array}{r} 9x^5 - 7x^3 +\ \ x - 1 \\ +\ 6x^5 - 7x^3 + 3x + 1 \\ \hline \end{array}$$
b.
$$\begin{array}{r} 9x^5 - 7x^3 +\ \ x - 1 \\ -\ (6x^5 - 7x^3 + 3x - 1) \\ \hline \end{array}$$

83. a. $(-8m^2 - 3m) + (-11m^2 + 6m + 10)$

   b. $(-8m^2 - 3m) - (-11m^2 + 6m + 10)$

84. a. $(10 - ab - 3a^2b) + (4 - 6ab)$

   b. $(10 - ab - 3a^2b) - (4 - 6ab)$

APPLICATIONS

85. Juggling. During a performance, a juggler tosses one ball straight upward while continuing to juggle three others. The height $f(t)$, in feet, of the ball is given by the polynomial function $f(t) = -16t^2 + 22t + 4$, where t is the time in seconds since the ball was thrown. Find the height of the ball 1 second after it is tossed upward.

86. Stopping Distances. The number of feet that a car travels before stopping depends on the driver's reaction time and the braking distance. For one driver, the stopping distance $d(v)$, in feet, is given by the polynomial function $d(v) = 0.04v^2 + 0.9v$, where v is the velocity of the car in mph. Find the stopping distance at 60 mph.

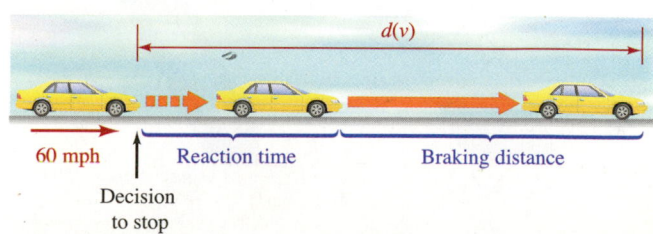

60 mph · Reaction time · Braking distance · $d(v)$ · Decision to stop

87. Storage Tanks. The volume $V(r)$ of the gasoline storage tank, in cubic feet, is given by the polynomial function $V(r) = 4.2r^3 + 37.7r^2$, where r is the radius in feet of the cylindrical part of the tank. What is the capacity of the tank if its radius is 4 feet?

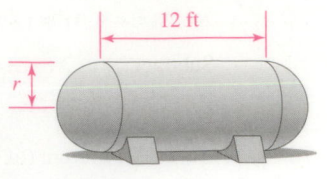

12 ft

88. Roller Coasters. The polynomial function $f(x) = 0.001x^3 - 0.12x^2 + 3.6x + 10$ models the path of a portion of the track of a roller coaster. Use the function equation to find the height of the track for $x = 0, 20, 40,$ and 60.

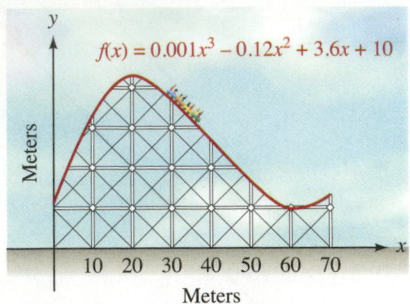

$f(x) = 0.001x^3 - 0.12x^2 + 3.6x + 10$

y

Meters

10 20 30 40 50 60 70

x

Meters

89. Packaging. To make boxes, a manufacturer cuts equal-sized squares from each corner of the 10 in. × 12 in. piece of cardboard shown below and then folds up the sides. The polynomial function $f(x) = 4x^3 - 44x^2 + 120x$ gives the volume (in cubic inches) of the resulting box when a square with sides x inches long is cut from each corner. Find the volume of a box if 3-inch squares are cut out.

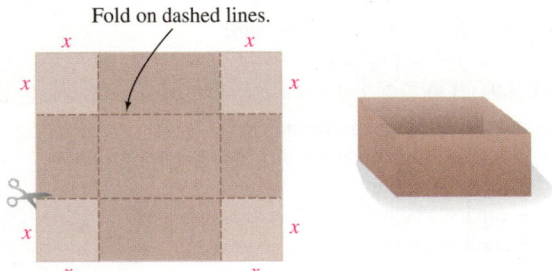

Fold on dashed lines.

x x

x x

x x

x x

90. Rain Gutters. A rectangular sheet of metal will be used to make a rain gutter by bending up its sides, as shown. If the ends are covered, the capacity $f(x)$ of the gutter is a polynomial function of x: $f(x) = -240x^2 + 1{,}440x$. Find the capacity of the gutter if x is 3 inches.

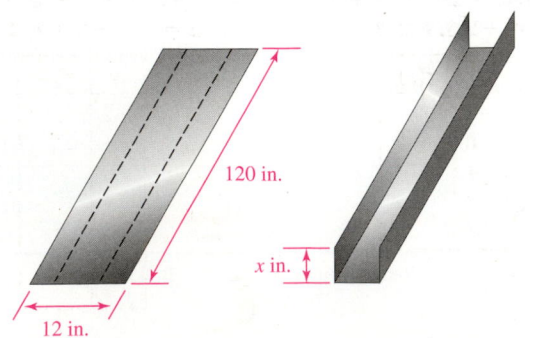

120 in.

x in.

12 in.

91. *from* **Campus to Careers**

Landscape Architect

A landscape architect has designed a set of pillars to line the driveway entrance to a college library, as shown below. The volume of a cylinder is given by the formula $V = \pi r^2 h$ and the volume of a cone is given by $V = \frac{1}{3}\pi r^2 h$. Find a formula for the volume of one pillar.

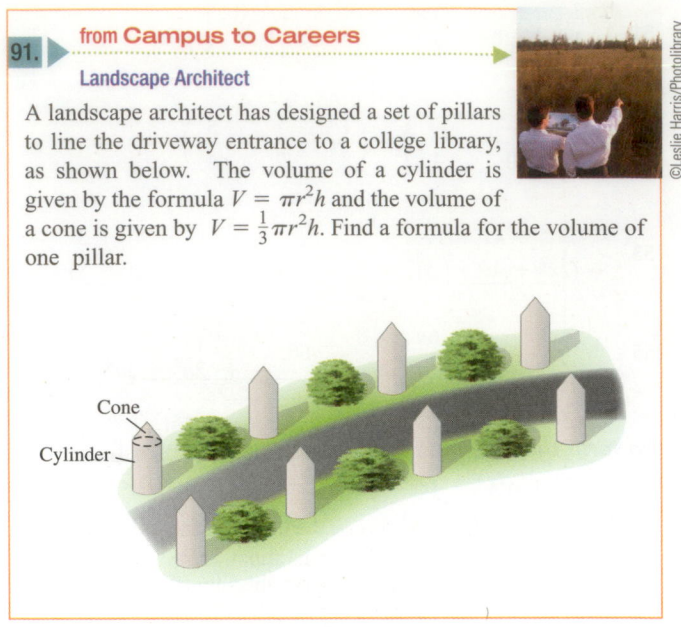

Cone

Cylinder

92. Customer Service. A software service hotline has found that on Mondays, the polynomial function $C(t) = -0.0625t^4 + t^3 - 6t^2 + 16t$ approximates the number of callers to the hotline at any one time. Here, t represents the time, in hours, since the hotline opened at 8:00 A.M. How many service technicians should be on duty on Mondays at noon if the company doesn't want any callers to the hotline waiting to be helped by a technician?

93. Labor Statistics. The polynomial function that is graphed below approximates the number of manufacturing jobs (in millions) in the United States, where x is the number of years after 2000. Use the graph to answer the following questions.

a. Estimate $J(9)$. Explain what the result means.

b. Estimate the value of x for which $J(x) = 14.5$. Explain what the result means.

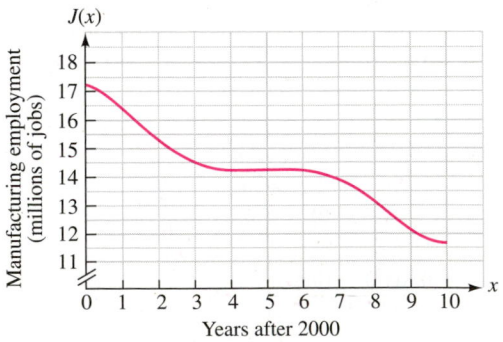

$J(x)$

Manufacturing employment (millions of jobs)

18
17
16
15
14
13
12
11

0 1 2 3 4 5 6 7 8 9 10

x

Years after 2000

(Source: Bureau of Labor Statistics)

94. Transportation Engineering. The polynomial function A that is graphed on the next page approximates the number of accidents per mile in one year on a 4-lane interstate, where x is the average daily traffic in number of vehicles. Use the graph to answer the following questions.

a. Estimate $A(20{,}000)$. Explain what the result means.

b. Estimate the value of x for which $A(x) = 2$. Explain what the result means.

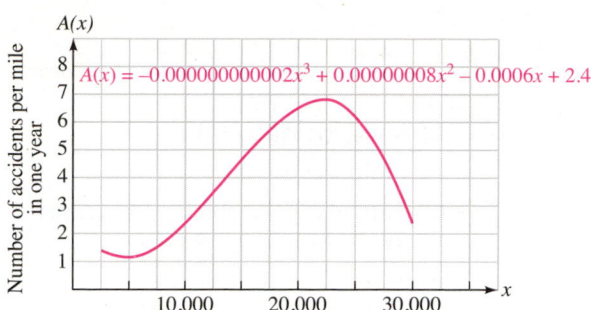

$A(x) = -0.000000000002x^3 + 0.00000008x^2 - 0.0006x + 2.45$

Source: Highway Safety Manual, Colorado Department of Transportation

95. Calculus. In the advanced mathematics course called Calculus, an important polynomial function is

$$f(x) = 1 + x + \frac{x^2}{2} + \frac{x^3}{6} + \frac{x^4}{24}$$

Refer to the polynomial on the right side of the equation.

a. How many terms does the polynomial have?

b. What is the degree of the polynomial?

c. What is the coefficient of the fourth term?

d. Find $f(1)$.

96. Business Expenses. A company purchased two vehicles for its sales force to use. The following functions give the respective values of the vehicles after x years.

Toyota Camry LE: $T(x) = -2,500x + 21,175$

Ford Explorer XLT: $F(x) = -2,900x + 31,190$

a. Find one polynomial function V that will give the combined value of both cars after x years.

b. Use your answer in part (a) to find the combined value of the two cars after 3 years.

WRITING

97. Explain why the terms $5x^2y$ and $5xy^2$ are not like terms.

98. Explain why the range of the polynomial function graphed below is not $(-\infty, \infty)$.

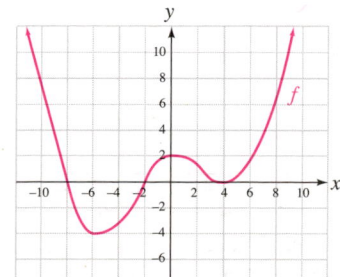

99. Explain the error in the following solution.

Subtract $2x - 3$ from $3x + 4$.

$(2x - 3) - (3x + 4) = 2x - 3 - 3x - 4$

$= -x - 7$

100. Explain why $f(x) = \dfrac{1}{x + 1}$ is not a polynomial function.

101. Use the word *descending* in a sentence in which the context is not mathematical. Do the same for the word *ascending*.

102. Look up the meaning of the prefix *poly* in a dictionary. Why do you think the name *polynomial* was given to expressions such as $x^3 - x^2 + 2x + 15$?

REVIEW

Solve each inequality. Graph the solution set and write it using interval notation.

103. $|x| \leq 5$

104. $|x| > 7$

105. $|x - 4| < 5$

106. $|2x + 1| \geq 7$

CHALLENGE PROBLEMS

107. What polynomial should be subtracted from $5x^3y - 5xy + 2x$ to obtain the polynomial $8x^3y - 7xy + 11x$?

108. Find two trinomials such that their sum is a binomial and their difference is a monomial.

Construct a table of values for each polynomial function using the given values for x. Then graph the function and find its domain and range.

109.
$f(x) = 2x^3 - 3x^2 - 11x + 6$
$x = -3, -2, -1, 0, 1, 2, 3, 4$

x	$f(x)$
-3	
-2	
-1	
0	
1	
2	
3	
4	

110.
$f(x) = -x^3 - x^2 + 6x$
$x = -4, -3, -2, -1, 0, 1, 2, 3$

x	$f(x)$
-4	
-3	
-2	
-1	
0	
1	
2	
3	

111. $f(x) = 2x^2 - 4x + 2$
$x = -1, 0, 1, 2, 3$

x	$f(x)$
-1	
0	
1	
2	
3	

112. $f(x) = -x^2 + 2x + 6$
$x = -2, -1, 0, 1, 2, 3, 4$

x	$f(x)$
-2	
-1	
0	
1	
2	
3	
4	

SECTION 5.4

OBJECTIVES

1 Multiply monomials.

2 Multiply a polynomial by a monomial.

3 Multiply two binomials.

4 Multiply any two polynomials.

5 Multiply three polynomials.

6 Find special products.

7 Use multiplication to simplify expressions.

Multiplying Polynomials

ARE YOU READY?

The following problems review some basic skills that are needed when multiplying polynomials.

1. Multiply: $9 \cdot 10a$

2. Multiply: $5(2x - 3)$

3. Multiply: $(4y + 8)7$

4. Multiply: $x^8 \cdot x^6$

5. Multiply: $-2(3)(5)$

6. Simplify by combining like terms: $9x^2 + 6x - 3x + 2$

In this section, we discuss the procedures used to multiply polynomials. These procedures involve the application of several concepts introduced earlier, such as the commutative and associative properties of multiplication, the rules for exponents, and the distributive property. We begin with the simplest case of polynomial multiplication, multiplying two monomials.

1 Multiply Monomials.

Success Tip

In this section, you will see that every polynomial multiplication is basically a series of monomial multiplications. Then, if possible, we combine like terms.

To find the product of two monomials, such as $6x^4$ and $5x^3$, we use the commutative and associative properties of multiplication to reorder and regroup the factors.

$$(6x^4)(5x^3) = (6 \cdot 5)(x^4 \cdot x^3)$$ Group the coefficients together and the variables together.

$$= 30x^7$$ Multiply: $6 \cdot 5 = 30$. Add exponents: $x^4 \cdot x^3 = x^{4+3} = x^7$.

This example suggests the following rule.

Multiplying Monomials	To multiply two monomials, multiply the numerical factors (the coefficients) and then multiply the variable factors.

EXAMPLE 1   Multiply: **a.** $(3x^2)(6x^3)$ **b.** $(25s^4t^3)\left(\frac{1}{5}st^3\right)$

Strategy   We will multiply the numerical factors and then multiply the variable factors.

Why   The commutative and associative properties of multiplication enable us to reorder and regroup the factors.

Solution   In each case, we will *multiply* the coefficients and *add* the exponents of the like bases.

a. $(3x^2)(6x^3) = 18x^5$ Think: $3 \cdot 6 = 18$ and $x^2 \cdot x^3 = x^{2+3} = x^5$.

b. $(25s^4t^3)\left(\frac{1}{5}st^3\right) = 5s^5t^6$ Think: $25 \cdot \frac{1}{5} = 5$, $s^4 \cdot s = s^{4+1} = s^5$, and $t^3 \cdot t^3 = t^{3+3} = t^6$.

Self Check 1   Multiply: **a.** $(2y^3)(4y^2)$ **b.** $(36m^4n)\left(\frac{1}{9}m^8n^6\right)$

Now Try ▶ Problems 11 and 13

2 Multiply a Polynomial by a Monomial.

Recall the **extended distributive property** from Section 1.4.

$$a(b + c + d + e + \dots) = ab + ac + ad + ae + \dots$$

This extension suggests the following rule for multiplying a monomial and a polynomial.

Multiplying Polynomials by Monomials	To multiply a monomial and a polynomial, multiply each term of the polynomial by the monomial.

EXAMPLE 2   Multiply: **a.** $3x^2(6x^6 + 4x^2)$ **b.** $-2ab^2(3a^3b - 2a^2b + 4b^2)$
c. $(xy^3z^2 - 2x^2yz)5x^3y^2z^5$

Strategy   To find each product, we will multiply each term of the polynomial by the monomial.

Why   We use the distributive property (and extensions of it) to multiply a monomial and a polynomial.

Solution   **a.** $3x^2(6x^6 + 4x^2) = 3x^2(6x^6) + 3x^2(4x^2)$ Distribute the multiplication by $3x^2$.
$$= 18x^8 + 12x^4 \qquad \text{Multiply the monomials.}$$

Since $18x^8$ and $12x^4$ are not like terms, we cannot add them.

b. $-2ab^2(3a^3b - 2a^2b + 4b^2)$
$$= -2ab^2 \cdot 3a^3b - (-2ab^2)2a^2b + (-2ab^2)4b^2 \qquad \text{Distribute } -2ab^2.$$
$$= -6a^4b^3 + 4a^3b^3 - 8ab^4 \qquad \text{Multiply the monomials.}$$

Success Tip

Since multiplication is commutative, the distributive property can be expressed as:
$(b + c)a = ba + ca.$

c. $(xy^3z^2 - 2x^2yz)5x^3y^2z^5 = (xy^3z^2)5x^3y^2z^5 - (2x^2yz)5x^3y^2z^5$ Distribute $5x^3y^2z^5$.
$$= 5x^4y^5z^7 - 10x^5y^3z^6 \qquad \text{Multiply the monomials.}$$

Self Check 2   Multiply: **a.** $2a^2(a^6 - 5a^3)$
b. $-8cd^2(4c^5d^2 + 8c^4d - d^5)$
c. $(m^9n^4p^4 - 9m^2n^3p)6mn^6p^3$

Now Try ▶   Problems 15, 19, and 21

3   Multiply Two Binomials.

The distributive property also can be used to multiply two binomials. For example, to multiply $3x + 2$ and $4x + 9$, we think of $3x + 2$ as a single quantity and distribute it over each term of $4x + 9$.

$$(3x + 2)(4x + 9) = (3x + 2)4x + (3x + 2)9$$

$$= (3x + 2)4x + (3x + 2)9 \qquad \text{The distributive property must be used two more times.}$$

$$= (3x)4x + (2)4x + (3x)9 + (2)9 \qquad \text{Distribute the multiplication by 4x and by 9.}$$

$$= 12x^2 + 8x + 27x + 18 \qquad \text{Multiply the monomials.}$$

$$= 12x^2 + 35x + 18 \qquad \text{Combine like terms: } 8x + 27x = 35x.$$

In the third line of the solution, each term of $3x + 2$ has been multiplied by each term of $4x + 9$. This example suggests the following rule.

Multiplying Two Binomials	To multiply two binomials, multiply each term of one binomial by each term of the other binomial, and then combine like terms.

A shortcut method, called the **FOIL method,** can be used to multiply each term of one binomial by each term of the other. It is a form of the distributive property. FOIL is an acronym for **F**irst terms, **O**uter terms, **I**nner terms, **L**ast terms. To use the FOIL method to multiply $3x + 2$ by $4x + 9$, we

1. Multiply the **F**irst terms $3x$ and $4x$ to obtain $12x^2$,
2. Multiply the **O**uter terms $3x$ and 9 to obtain $27x$,
3. Multiply the **I**nner terms 2 and $4x$ to obtain $8x$, and
4. Multiply the **L**ast terms 2 and 9 to obtain 18.

Then we simplify the resulting polynomial by combining like terms, if possible.

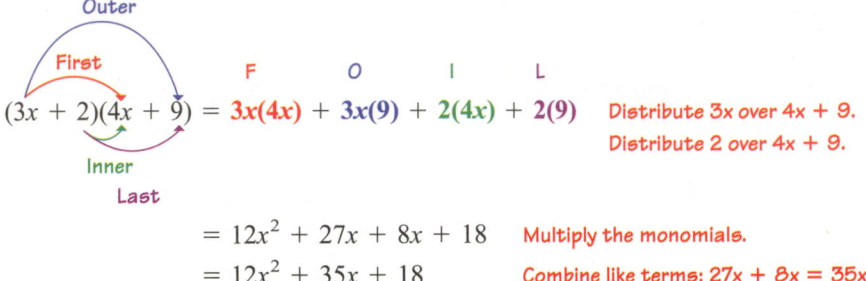

$$(3x + 2)(4x + 9) = \textbf{3x(4x)} + \textbf{3x(9)} + \textbf{2(4x)} + \textbf{2(9)} \quad \text{Distribute } 3x \text{ over } 4x + 9.$$
$$\text{Distribute } 2 \text{ over } 4x + 9.$$

$$= 12x^2 + 27x + 8x + 18 \quad \text{Multiply the monomials.}$$
$$= 12x^2 + 35x + 18 \quad \text{Combine like terms: } 27x + 8x = 35x.$$

EXAMPLE 3 Multiply: **a.** $(2x - 3)(3x + 2)$ **b.** $(7y^2 - 4)(2y^2 - 1)$

c. $\left(6ab^2 - \dfrac{1}{3}\right)\left(3ab^2 + \dfrac{5}{6}\right)$

Strategy We will use the FOIL method to multiply the binomials.

Why In each case we are to find the product of two binomials, and the FOIL method is a shortcut for multiplying two binomials.

Solution **a.** $(2x - 3)(3x + 2) = 2x(3x) + 2x(2) - 3(3x) - 3(2)$
$$= 6x^2 + 4x - 9x - 6 \quad \text{Multiply the monomials.}$$
$$= 6x^2 - 5x - 6 \quad \text{Combine like terms: } 4x - 9x = -5x.$$

b. $(7y^2 - 4)(2y^2 - 1) = 7y^2(2y^2) + 7y^2(-1) - 4(2y^2) - 4(-1)$
$$= 14y^4 - 7y^2 - 8y^2 + 4 \quad \text{Multiply the monomials.}$$
$$= 14y^4 - 15y^2 + 4 \quad \text{Combine like terms: } -7y^2 - 8y^2 = -15y^2.$$

c. $\left(6ab^2 - \dfrac{1}{3}\right)\left(3ab^2 + \dfrac{5}{6}\right) = 6ab^2(3ab^2) + 6ab^2\left(\dfrac{5}{6}\right) - \dfrac{1}{3}(3ab^2) - \dfrac{1}{3}\left(\dfrac{5}{6}\right)$

$$= 18a^2b^4 + 5ab^2 - ab^2 - \dfrac{5}{18} \quad \text{Multiply the monomials.}$$

$$= 18a^2b^4 + 4ab^2 - \dfrac{5}{18} \quad \begin{array}{l} \text{Combine like terms:} \\ 5ab^2 - ab^2 = 4ab^2. \end{array}$$

Self Check 3 Multiply: **a.** $(3a + 4)(2a - 9)$
b. $(4c^4 + 5)(3c^4 + 7)$
c. $\left(10wz^2 - \dfrac{1}{5}\right)\left(5wz^2 + \dfrac{3}{10}\right)$

Now Try ▶ Problems 23, 27, and 29

4   Multiply Any Two Polynomials.

To develop a general rule for multiplying any two polynomials, we will find the product of $x + 4$ and $x^2 + 2x + 1$. In the solution, the distributive property is used four times.

$$(x + 4)(x^2 + 2x + 1) = (x + 4)x^2 + (x + 4)2x + (x + 4)1$$ Distribute $(x + 4)$.

$$= (x + 4)x^2 + (x + 4)2x + (x + 4)1$$ The distributive property must be used 3 more times.

$$= (x)x^2 + (4)x^2 + (x)2x + (4)2x + (x)1 + (4)1$$ Distribute x^2, $2x$, and 1.

$$= x^3 + 4x^2 + 2x^2 + 8x + x + 4$$ Multiply the monomials.

$$= x^3 + 6x^2 + 9x + 4$$ Combine like terms: $4x^2 + 2x^2 = 6x^2$ and $8x + x = 9x$.

In the third line of the solution above, each term of $x^2 + 2x + 1$ has been multiplied by each term of $x + 4$. This result suggests the following rule.

Multiplying Two Polynomials	To multiply two polynomials, multiply each term of one polynomial by each term of the other polynomial, and then combine like terms.

EXAMPLE 4   Multiply: $(2a + b)(3a^2 - 4ab - b^2)$

Strategy   We will multiply each term of the trinomial, $3a^2 - 4ab - b^2$, by each term of the binomial, $2a + b$.

Why   To multiply two polynomials, we must multiply each term of one polynomial by each term of the other polynomial.

Solution   If we multiply each term of a three-term polynomial by each term of a two-term polynomial, there will be $3 \cdot 2 = 6$ monomial multiplications to perform.

Success Tip

The FOIL method cannot be applied here, only to products of two binomials.

$$(2a + b)(3a^2 - 4ab - b^2)$$

$$= 2a(3a^2) + 2a(-4ab) + 2a(-b^2) + b(3a^2) + b(-4ab) + b(-b^2)$$

$$= 6a^3 - 8a^2b - 2ab^2 + 3a^2b - 4ab^2 - b^3$$ Multiply the monomials.

$$= 6a^3 - 5a^2b - 6ab^2 - b^3$$ Combine like terms.

Self Check 4   Multiply: $(3y^2 + y)(4y^2 - 2y + 5)$

Now Try ▶   Problem 31

It is often convenient to multiply polynomials using a vertical form.

EXAMPLE 5   Use vertical form to multiply: $(6y^3 - 5y + 4)(4y^2 - 3)$

Strategy   We will write one polynomial underneath the other and multiply each term of the upper polynomial by each term of the lower polynomial. Then we will combine like terms column-by-column.

Why   *Vertical form* means to use an approach similar to that used in arithmetic to multiply two whole numbers.

Solution

With this method, it is often necessary to leave a space for a missing term to align like terms vertically.

Multiply:

$$\begin{array}{r} 6y^3 - 5y + 4 \\ 4y^2 - 3 \\ \hline -18y^3 \qquad + 15y - 12 \\ 24y^5 - 20y^3 + 16y^2 \\ \hline 24y^5 - 38y^3 + 16y^2 + 15y - 12 \end{array}$$

Multiply $6y^3 - 5y + 4$ by -3.

Multiply $6y^3 - 5y + 4$ by $4y^2$.

Leave a space for any missing powers of y. In each column, combine like terms.

The two polynomials written below the first horizontal line, $-18y^3 + 15y - 12$ and $24y^5 - 20y^3 + 16y^2$, are called **partial products.**

Self Check 5   Multiply:

$$\begin{array}{r} 3x^2 + 2x - 5 \\ 2x^2 + 1 \end{array}$$

Now Try ▶ Problem 37

EXAMPLE 6   **Scrapbooks.**   An 8.5-inch by 11-inch scrapbook page has a border of uniform width surrounding a rectangular birth announcement. Write a polynomial that represents the area of the birth announcement.

Strategy   We will write polynomials that represent the length and the width of the rectangular announcement and find their product.

Why   The area of a rectangle is the product of its length and width.

Solution   Let x represent the width of the border, in inches, as shown in the figure on the left. Then the length of the birth announcement is $11 - x - x$ or $(11 - 2x)$ inches and the width is $8.5 - x - x$ or $(8.5 - 2x)$ inches. The area of the announcement is the product of its length and width:

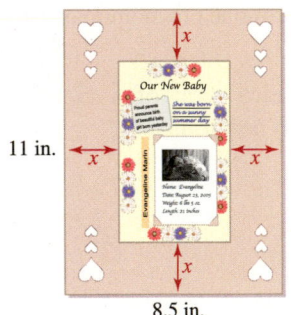

11 in.

8.5 in.

$$A = lw$$ This is the formula for the area of a rectangle.

$$A = (11 - 2x)(8.5 - 2x)$$ Substitute for l and w.

$$A = 11(8.5) - 11(2x) - 2x(8.5) - 2x(-2x)$$ Use the FOIL method.

$$A = 93.5 - 22x - 17x + 4x^2$$ Multiply the monomials.

$$A = 93.5 - 39x + 4x^2$$ Combine like terms.

The polynomial $93.5 - 39x + 4x^2$ or $4x^2 - 39x + 93.5$ represents the area of the birth announcement in square inches.

Self Check 6   **Tapestries.**   A 36-inch by 24-inch tapestry has a border of uniform width x inches surrounding a rectangular center panel. Write a polynomial that represents the area of the center panel.

Now Try ▶ Problem 117

---

5   Multiply Three Polynomials.

Multiplying Three Polynomials

To multiply three polynomials, multiply *any* two of them, and then multiply that result by the third polynomial.

EXAMPLE 7   Multiply: $5cd(c + 6d)(3c - 8d)$

Strategy   We will find the product of $c + 6d$ and $3c - 8d$ and then multiply that result by $5cd$.

Why   It is wise to perform the most difficult multiplication first. (In this case, this is the product of two binomials). Save the simpler multiplication by $5cd$ for last.

Solution

$$5cd(c + 6d)(3c - 8d) = 5cd(3c^2 - 8cd + 18cd - 48d^2)$$ Use the FOIL method to find $(c + 6d)(3c - 8d)$.

$$= 5cd(3c^2 + 10cd - 48d^2)$$ Combine like terms: $-8cd + 18cd = 10cd$.

$$= 15c^3d + 50c^2d^2 - 240cd^3$$ Distribute the multiplication by $5cd$.

Self Check 7   Multiply: $-2r(r - 2s)(5r - 4s)$

Now Try ▶ Problem 39

6 Find Special Products.

To find the square of a binomial, we can use the FOIL method. For example, to find $(x + y)^2$ and $(x - y)^2$, we proceed as follows.

$$(x + y)^2 = (x + y)(x + y) \qquad\qquad (x - y)^2 = (x - y)(x - y)$$
$$= x^2 + xy + xy + y^2 \qquad\qquad\quad = x^2 - xy - xy + y^2$$
$$= x^2 + 2xy + y^2 \qquad\qquad\qquad\quad = x^2 - 2xy + y^2$$

In each case, we see that the square of the binomial is the square of its first term, twice the product of its two terms, and the square of its last term.

The figure shows how $(x + y)^2$ can be found geometrically.

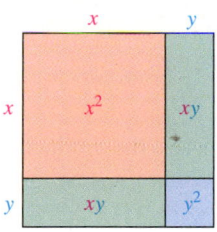

The area of the largest square is the product of its length and width: $(x + y)(x + y) = (x + y)^2$.

The area of the largest square is also the sum of its four parts: $x^2 + xy + xy + y^2 = x^2 + 2xy + y^2$.

Thus, $(x + y)^2 = x^2 + 2xy + y^2$.

Another common binomial product is the product of the sum and difference of the same two terms. An example of such a product is $(x + y)(x - y)$. To find this product, we use the FOIL method.

$$(x + y)(x - y) = x^2 - xy + xy - y^2$$
$$= x^2 - y^2 \qquad\qquad \text{Combine like terms: } -xy + xy = 0.$$

We see that the product of the sum and the difference of the same two terms is the square of the first term minus the square of the second term.

The results from these three examples suggest the following **special-product rules.**

Special-Product Rules

$$(A + B)^2 = A^2 + 2AB + B^2 \qquad \text{The square of a sum.}$$
$$(A - B)^2 = A^2 - 2AB + B^2 \qquad \text{The square of a difference.}$$
$$(A + B)(A - B) = A^2 - B^2 \qquad \text{The product of the sum and difference of two terms.}$$

CAUTION   Remember that the square of a binomial is a *trinomial*. A common error when squaring a binomial is to forget the middle term of the product. For example,

$$(x + y)^2 \neq x^2 + y^2 \qquad \text{and} \qquad (x - y)^2 \neq x^2 - y^2$$

Missing 2xy   Missing −2xy ─── └─ Should be + symbol

Also remember that the product $(x + y)(x - y)$ is the binomial $x^2 - y^2$. And since $(x + y)(x - y) = (x - y)(x + y)$ by the commutative property of multiplication,

$$(x - y)(x + y) = x^2 - y^2$$

EXAMPLE 8   Multiply: **a.** $(5c + 3d)^2$ **b.** $\left(\dfrac{1}{2}a^4 - b^2\right)^2$ **c.** $(0.2m^3 + 2.5n)(0.2m^3 - 2.5n)$

Strategy   To find the product of each pair of binomials, we will use a special-product rule.

Why   This approach is faster than using the FOIL method.

Solution   **a.** To find $(5c + 3d)^2$ using the square of a sum rule, we begin by noting that the first term of the binomial is $5c$ and the last term is $3d$.

The Language of Algebra

When squaring a binomial, the result is called a **perfect-square trinomial.** For example,

$$(t + 9)^2 = t^2 + 18t + 81$$

Perfect-square trinomial

The square of   Twice the product   The square of
the first term, 5c   of both terms   the last term, 3d

$$(5c + 3d)^2 = \quad (5c)^2 \quad + \quad 2(5c)(3d) \quad + \quad (3d)^2$$
$$= 25c^2 + 30cd + 9d^2$$

b. To find $\left(\dfrac{1}{2}a^4 - b^2\right)^2$ using the square of a difference rule, we begin by noting that the first term of the binomial is $\dfrac{1}{2}a^4$ and the last term is $-b^2$.

The square of   Twice the product   The square of
the first term, $\frac{1}{2}a^4$ of both terms   the last term, $-b^2$

$$\left(\frac{1}{2}a^4 - b^2\right)^2 = \quad \left(\frac{1}{2}a^4\right)^2 \quad + \quad 2\left(\frac{1}{2}a^4\right)(-b^2) \quad + \quad (-b^2)^2$$
$$= \frac{1}{4}a^8 - a^4b^2 + b^4$$

Success Tip

We can use the FOIL method to find each of the special products. However, these forms occur so often, it is worthwhile to learn the special-product rules.

c. $(0.2m^3 + 2.5n)(0.2m^3 - 2.5n)$ is the product of the sum and the difference of the same two terms: $0.2m^3$ and $2.5n$. Using a special-product rule, we proceed as follows.

The square of the   The square of the
first term, $0.2m^3$ second term, 2.5n

$$(0.2m^3 + 2.5n)(0.2m^3 - 2.5n) = \quad (0.2m^3)^2 \quad - \quad (2.5n)^2$$
$$= 0.04m^6 - 6.25n^2$$

Self Check 8   Multiply: **a.** $(8r + 2s)^2$
b. $\left(\dfrac{1}{3}a^3 - b^6\right)^2$
c. $(0.4x + 1.2y^4)(0.4x - 1.2y^4)$

Now Try ▶   Problems 43, 47, and 51

The special-product rules can be helpful in finding more complicated powers and products.

EXAMPLE 9 Multiply: **a.** $[(5x + y) + 4]^2$ **b.** $[12 + (c - d)][12 - (c - d)]$ **c.** $(a + b)^3$

Strategy For each problem, we will use a special-product rule to simplify the calculations.

Why Using a special-product rule often produces the result more quickly than multiplying each term of one polynomial by each term of the other.

Solution **a.** It is helpful to think of $(5x + y)$ as the first term and 4 as the second term of the square of a sum. We can then match the given expression to a special-product rule.

$$(\ A \ + B)^2 = \ A^2 \ + 2 \ A \ \ B \ + B^2$$
$$[(5x + y) + 4]^2 = (5x + y)^2 + 2(5x + y)(4) + 4^2$$
$$= 25x^2 + 10xy + y^2 + 8(5x + y) + 16 \quad \text{Square } (5x + y). \text{ Simplify.}$$
$$= 25x^2 + 10xy + y^2 + 40x + 8y + 16$$

b. It is helpful to think of 12 as the first term and $(c - d)$ as the second term of the sum and difference of the same two terms. We can then match the given expression to a special-product rule.

$$(A + \ B \ )(A - \ B \ ) = A^2 - \ B^2$$
$$[12 + (c - d)][12 - (c - d)] = 12^2 - (c - d)^2$$
$$= 144 - (c^2 - 2cd + d^2) \quad \text{Square 12 and } (c - d).$$
$$= 144 - c^2 + 2cd - d^2$$

c. Since $(a + b)^3$ can be written as $(a + b)(a + b)^2$, we can use a special-product rule to quickly find $(a + b)^2$.

$$(a + b)^3 = (a + b)(a + b)^2 \quad \text{Write } (a + b)^3 \text{ in an equivalent form.}$$
$$= (a + b)(a^2 + 2ab + b^2) \quad \text{Square } (a + b).$$
$$= a^3 + 2a^2b + ab^2 + a^2b + 2ab^2 + b^3 \quad \begin{array}{l}\text{Multiply each term of} \\ a^2 + 2ab + b^2 \text{ by each} \\ \text{term of } a + b.\end{array}$$
$$= a^3 + 3a^2b + 3ab^2 + b^3 \quad \text{Combine like terms.}$$

Self Check 9 Multiply: **a.** $[(s + 4t) + 2]^2$
b. $[9 + (m - n)][9 - (m - n)]$
c. $(4x - y)^3$

Now Try Problems 55, 59, and 61

Polynomial multiplication is also encountered when evaluating functions.

EXAMPLE 10 If $f(x) = x^2 + 9x - 5$, find $f(a + 4)$.

Strategy To find $f(a + 4)$, we will substitute $a + 4$ for each x in $f(x) = x^2 + 9x - 5$ and simplify the right side.

Why Whatever expression appears within the parentheses in $f(\ )$ is to be substituted for each x in $f(x) = x^2 + 9x - 5$.

Solution
$$f(x) = x^2 + 9x - 5$$
$$f(a + 4) = (a + 4)^2 + 9(a + 4) - 5 \quad \text{Substitute } a + 4 \text{ for } x.$$
$$= a^2 + 8a + 16 + 9a + 36 - 5 \quad \text{Use a special-product rule to find } (a + 4)^2.$$
$$= a^2 + 17a + 47 \quad \text{Combine like terms.}$$

Self Check 10   If $f(x) = x^2 - 6x + 1$, find $f(a - 8)$.

Now Try ▶ Problem 65

7   Use Multiplication to Simplify Expressions.

The procedures discussed in this section are helpful when we simplify algebraic expressions that involve the multiplication of polynomials.

EXAMPLE 11   Simplify: $(5x - 4)^2 - (x - 7)(x + 1)$

Strategy   We will use a special-product rule to find $(5x - 4)^2$ and the FOIL method to find $(x - 7)(x + 1)$. Then we will combine like terms.

Why   To simplify expressions, we follow the order of operations rule. We find powers and perform the multiplication before performing the subtraction.

Solution

$$(5x - 4)^2 - (x - 7)(x + 1) = 25x^2 - 40x + 16 - (x^2 - 6x - 7)$$

Don't forget to write parentheses around $x^2 - 6x - 7$.

$$= 25x^2 - 40x + 16 - x^2 + 6x + 7$$
$$= 24x^2 - 34x + 23$$

Combine like terms.

Self Check 11   Simplify: $(y - 7)(y + 7) - (4y + 3)^2$

Now Try ▶ Problem 67

SECTION 5.4 ▶ STUDY SET

VOCABULARY

Fill in the blanks.

1. The expression $(2x^3)(3x^4)$ is the product of two _____ and the expression $(x + 4)(x - 5)$ is the product of two _____.

2. $(x + 4)^2$ is the _____ of a sum and $(m - 9)^2$ is the square of a _____. The expression $(b + 1)(b - 1)$ is the product of the _____ and difference of two terms.

3. FOIL is an acronym for _____ terms, _____ terms, _____ terms, and _____ terms.

4. Since $x^2 + 16x + 64$ is the square of $x + 8$, it is called a _____-square trinomial. (*Hint:* This is in a Language of Algebra box.)

CONCEPTS

Fill in the blanks.

5. a. To multiply a monomial by a monomial, we multiply the numerical _____ and then multiply the variable factors.
   b. To multiply a polynomial by a monomial, we multiply each _____ of the polynomial by the monomial.
   c. To multiply a polynomial by a polynomial, we multiply each _____ of one polynomial by each term of the other polynomial.

6. The square of a binomial is the _____ of its first term, _____ the product of its two terms, plus the _____ of its last term.

   $$(x + y)^2 = (x + y)(x + y) = \rule{3cm}{0.4cm}$$

   $$(x - y)^2 = (x - y)(x - y) = \rule{3cm}{0.4cm}$$

7. The product of the sum and difference of the same two terms is the _____ of the first term minus the _____ of the second term.

   $$(x + y)(x - y) = \rule{3cm}{0.4cm}$$

8. Consider $(2x + 4)(4x - 3)$. The first terms are ☐ and ☐, the outer terms are ☐ and ☐, the inner terms are ☐ and ☐, and the last terms are ☐ and ☐.

9. Perform the indicated operation.
   a. $(4b - 1) + (2b - 1)$
   b. $(4b - 1) - (2b - 1)$
   c. $(4b - 1)(2b - 1)$

10. To find $-2a^2b(a^2 + b^2)(a^2 - b^2)$, which two polynomials would you multiply first? Explain why.

GUIDED PRACTICE

Multiply. See Example 1.

11. $(2a^2)(3a^5)$

12. $(3x^2)(3x^9)$

13. $\left(\frac{1}{6}g^4h^5\right)(36g^7h^{11})$

14. $\left(\frac{1}{16}r^9s^{10}\right)(32r^2s^{10})$

Multiply. See Example 2.

15. $10x^5(3x^5 + 2x^2)$

16. $5a^4(6a^7 + 2a^2)$

17. $-2d^4(3d^3 - 3d^2 + 2d)$

18. $-3a^8(4a^4 + 3a^3 - 4a^2)$

19. $7rs^3t(r^2 + s^2 - t^2)$

20. $3x^2yz(x^2 - 2y + 3z^2)$

21. $(7x^6y^3z - 4x^2yz^4)4xy^6z^3$

22. $(8abc^{20} - 6a^2bc^2)5ab^8c^9$

Multiply. See Example 3.

23. $(7t - 2)(2t + 3)$

24. $(5p + 3)(3p - 4)$

25. $(3y^3 - 4)(2y^3 - 1)$

26. $(2m^5 - 7)(3m^5 - 1)$

27. $\left(4b + \frac{1}{2}\right)\left(8b - \frac{3}{4}\right)$

28. $\left(6x + \frac{1}{3}\right)\left(6x - \frac{1}{6}\right)$

29. $(9b^3c - c)(3b^2 - bc)$

30. $(h^5k - k)(4h^3 - hk)$

Multiply. See Example 4.

31. $(3y + 1)(2y^2 + 3y + 2)$

32. $(a + 2)(3a^2 + 4a - 2)$

33. $(x - y)(x^2 + xy + y^2)$

34. $(x + y)(x^2 - xy + y^2)$

Use vertical form to multiply the polynomials. See Example 5.

35. $3a^2 + 4a - 2$
    $\underline{ 2a + 3}$

36. $5y^2 - y - 2$
    $\underline{ 2y - 1}$

37. $2x^2 - 10x + 14$
    $\underline{ x^2 + 2x - 3}$

38. $4x^2 - 12x + 4$
    $\underline{ x^2 - 2x + 7}$

Multiply. See Example 7.

39. $6p^2(3p - 4)(p + 3)$

40. $4a(2a + 3)(3a - 2)$

41. $4my(3m - y)(2m - y)$

42. $-3hz(2h - z)(3h - z)$

Use a special product formula to find each product. See Example 8.

43. $(2a + b)^2$

44. $(a - 2b)^2$

45. $(5r^2 + 6)^2$

46. $(6p^2 - 3)^2$

47. $\left(\frac{1}{4}b - 2\right)^2$

48. $\left(\frac{2}{3}y - 7\right)^2$

49. $(9ab^2 - 4)^2$

50. $(2yz^2 + 5)^2$

51. $(5y + 2.4)(5y - 2.4)$

52. $(9t + 0.7)(9t - 0.7)$

53. $\left(x^2y - \frac{6}{5}\right)\left(x^2y + \frac{6}{5}\right)$

54. $\left(a^4b - \frac{1}{2}c\right)\left(a^4b + \frac{1}{2}c\right)$

Multiply. See Example 9.

55. $[(6a + b) + 4]^2$

56. $[(3m + n) + 10]^2$

57. $[7 - (x - y)]^2$

58. $[12 - (r - s)]^2$

59. $[5 + (2n + p)][5 - (2n + p)]$

60. $[4 + (w + 3z)][4 - (w + 3z)]$

61. $(3x - 2)^3$

62. $(p - 2q)^3$

Evaluate each function. See Example 10.

63. If $f(x) = x^2 - 8x + 2$, find $f(b + 1)$.

64. If $f(x) = x^2 - 6x + 10$, find $f(n + 2)$.

65. If $f(x) = x^2 + 4x - 9$, find $f(a - 6)$.

66. If $f(x) = x^2 + 12x - 8$, find $f(c - 3)$.

Simplify each expression. See Example 11.

67. $(7x - 1)^2 - (x - 4)(x + 3)$

68. $(9x - 2)^2 - (x + 9)(x - 7)$

69. $(3x - 4)^2 - (2x + 3)^2$

70. $(3y + 1)^2 + (2y - 4)^2$

Use a calculator to help find each product.

71. $(3.21x - 7.85)(2.87x + 4.59)$

72. $(7.44y + 56.7)(-2.1y - 67.3)$

73. $(-17.3y + 4.35)^2$

74. $(-0.31x + 29.3)(-0.31x - 29.3)$

TRY IT YOURSELF

Perform the indicated operations.

75. $(2a - b)(4a^2 + 2ab + b^2)$

76. $(x - 3y)(x^2 + 3xy + 9y^2)$

77. $(2b + 3)(5b - 1) - (b + 2)^2$

78. $(x + 3)^2 - (2x - 1)(4x + 2)$

79. $(11m^2 + 3n^3)(5m + 2n^2)$

80. $(50m^4 - 3n^4)(2m + 2n^3)$

81. $(-5s^2tu)(-3s^4t^4u)$

82. $(-2a^3b^2c^5)(-3ab^4c^2)$

83. $(a + b + c)(2a - b - 2c)$

84. $(x + 2y + 3z)^2$

85. $(a + b)(a - b)(a - 3b)$

86. $(x - y)(x + 2y)(x - 2y)$

87. $(4k - 13)^2$

88. $(5k - 6)^2$

89. $\left(\dfrac{1}{2}x - 4\right)\left(\dfrac{1}{2}x + 16\right)$

90. $\left(9h^2 - \dfrac{2}{3}\right)\left(3h^2 + \dfrac{2}{3}\right)$

91. $(-3ab^2)(5ab)$

92. $(-2m^2n)(4mn^3)$

93. $(y^3 + 2)(y^3 - 2)$

94. $(y^4 + 3)(y^4 - 3)$

95. $(3a^3 + 4c)^2$

96. $(2y^5 + 5z)^2$

97. $\dfrac{1}{2}m^2n(m^2 + n^3)(-12mn)$

98. $\dfrac{1}{3}a^2b^3(3a^3 + b^4)(-6b)$

99. $(2x^2y^3z^5)(4xy^5z)(-5y^6z^6)$

100. $(4a^2bc^6)(-5a^3b^2c^2)(6a^4c^4)$

101. $(5s - t^3)^3$

102. $(3a - 2b)^3$

103. $[(3d + f) + 2]^2$

104. $[8 - (a - 5b)]^2$

105. Let $f(t) = 0.4t - 3$ and $g(t) = 0.5t - 3$. Find $f(t) \cdot g(t)$.

106. Let $f(d) = 0.7d - 2$ and $g(d) = 0.1d + 3$. Find $f(d) \cdot g(d)$.

107. Let $s(x) = 5x^5 + 6$ and $t(x) = 5x^5 - 6$. Find $s(x) \cdot t(x)$.

108. Let $g(x) = 3x^3 - 0.8$ and $h(x) = 3x^3 + 0.8$. Find $g(x) \cdot h(x)$.

Look Alikes . . .

109. a. $(-3ab^2)(5ab)$ **b.** $(-3a + b^2)(5a + b)$

110. a. $\left(\dfrac{1}{2}n + 2\right) + \left(\dfrac{1}{4}n - \dfrac{1}{2}\right)$ **b.** $\left(\dfrac{1}{2}n + 2\right)\left(\dfrac{1}{4}n - \dfrac{1}{2}\right)$

111. a. $(2x^2 + 5) + (3x^2 + 2x + 1)$

   b. $(2x^2 + 5)(3x^2 + 2x + 1)$

112. a. $3y + (y^7 + y) + (y^3 + y)$ **b.** $3y(y^7 + y)(y^3 + y)$

APPLICATIONS

113. Geometry. Write a polynomial that represents the area of the figure.

a.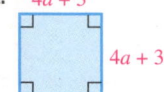

b.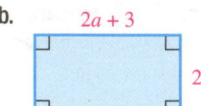

114. Geometry. Write a polynomial that represents the area of the figure.

a. 4a + 3 / 4a + 3   **b.** 2a + 3 / 2a − 3

115. The Yellow Pages. Refer to the illustration below.

a. Describe the area occupied by the ads for movers by using a product of two binomials.

b. Describe the area occupied by the ad for Budget Moving Co. by using a product. Then perform the multiplication.

c. Describe the area occupied by the ad for Snyder Movers by using a product. Then perform the multiplication.

d. Explain why your answer to part (a) is equal to the sum of your answers to parts (b) and (c). What special product does this exercise illustrate?

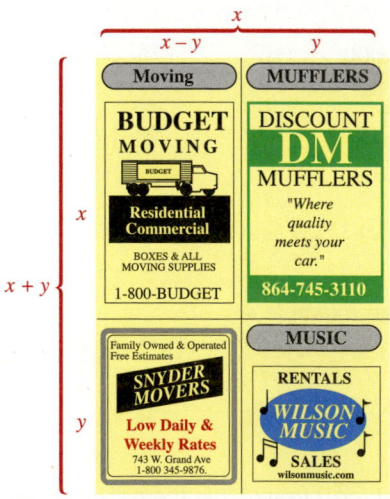

116. Helicopter Pads. To determine the amount of fluorescent paint needed to paint the circular ring on the landing pad design shown in the illustration on the next page, painters must find its area. The area of the ring is given by the expression $\pi(R + r)(R - r)$.

a. Find the product $\pi(R + r)(R - r)$.

b. If $R = 25$ feet and $r = 20$ feet, find the area to be painted. Round to the nearest tenth.

c. If a quart of fluorescent paint covers 65 ft^2, how many quarts will be needed to paint the ring?

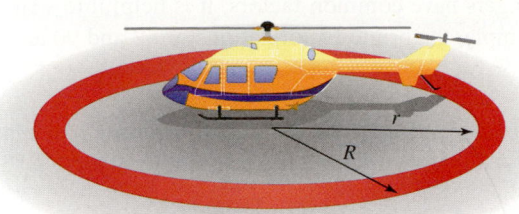

117. **Quilts.** A 36-inch by 46-inch baby crib quilt has a border of uniform width x inches surrounding a rectangular center panel. Write a polynomial that represents the area of the center panel.

46 in.

36 in.

118. **Gift Boxes.** The corners of a 12-in. by 12-in. piece of cardboard are creased, folded inward, and glued to make a gift box. (See the illustration.) Write a polynomial that gives the volume of the resulting box.

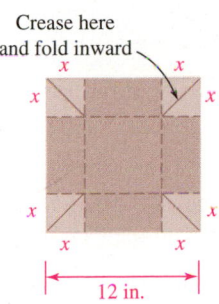

Crease here and fold inward

x x

x x

x x

x x

12 in.

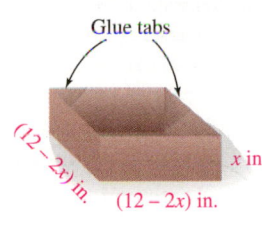

Glue tabs

$(12 - 2x)$ in.

x in.

$(12 - 2x)$ in.

WRITING

119. Explain how to use the FOIL method.
120. Explain how you would multiply two trinomials.
121. On a test, when asked to find $(x - y)^2$, a student answered $x^2 - y^2$. What error did the student make?
122. Describe each expression in words:

$$(x + y)^2 \qquad (x - y)^2 \qquad (x + y)(x - y)$$

REVIEW

Graph the solution set of each inequality or system of inequalities on a rectangular coordinate system.

123. $2x + y \leq 2$
124. $x \geq 2$
125. $\begin{cases} y - 2 < 3x \\ y + 2x < 3 \end{cases}$
126. $\begin{cases} y < 0 \\ x < 0 \end{cases}$

CHALLENGE PROBLEMS

Multiply. Write all answers without negative exponents.

127. $ab^{-2}c^{-3}(a^{-4}bc^3 + a^{-3}b^4c^3)$
128. $(5x^{-4} - 4y^2)(5x^2 - 4y^{-4})$

Multiply. Assume n is a natural number.

129. $a^{2n}(a^n + a^{2n})$
130. $(a^{3n} - b^{3n})(a^{3n} + b^{3n})$

131. If $f(x) = x^2 - 4x - 7$, find $f(a + h) - f(a)$.
132. If $f(x) = x^3 + x$, find $f(a + h) - f(a)$.

SECTION 5.5

The Greatest Common Factor and Factoring by Grouping

OBJECTIVES

1. Find the greatest common factor of a list of terms.
2. Factor out the greatest common factor.
3. Factor by grouping.
4. Use factoring to solve formulas for a specified variable.

ARE YOU READY?

The following problems review some basic skills that are needed when factoring expressions.

1. Find the prime factorization of 108.
2. Write $t \cdot t \cdot t \cdot t \cdot t$ in an equivalent form using exponents.
3. Simplify: $-(h - 9)$
4. How many terms does the expression $3n^3 + n^2 + 4n + 8$ have?
5. Multiply: $6b(b^3 + 2b)$
6. Multiply and simplify: $m(m + 2) - 5(m + 2)$

In Section 5.4, we discussed ways of multiplying polynomials. In this section, we will discuss the reverse process—*factoring* polynomials. When factoring a polynomial, the first step is to determine whether its terms have any common factors.

1 Find the Greatest Common Factor of a List of Terms.

To determine whether two or more integers have common factors, it is helpful to write them as products of prime numbers. For example, the prime factorizations of 42 and 90 are:

$$42 = \mathbf{2} \cdot \mathbf{3} \cdot 7 \qquad 90 = \mathbf{2} \cdot \mathbf{3} \cdot 3 \cdot 5$$

The highlighting shows that 42 and 90 have one factor of 2 and one factor of 3 in common. To find their *greatest common factor (GCF)*, we multiply the common factors: $2 \cdot 3 = 6$. Thus, the GCF of 42 and 90 is 6.

The Greatest Common Factor (GCF)	The **greatest common factor (GCF)** of a list of integers is the largest common factor of those integers.

Recall from arithmetic that the factors of a number divide the number exactly, leaving no remainder. Therefore, the greatest common factor of two or more integers is the largest natural number that divides each of the integers exactly.

To find the greatest common factor of a list of terms, we can use the following approach.

Strategy for Finding the GCF	1. Write each coefficient as a product of prime factors.
	2. Identify the numerical and variable factors common to each term.
	3. Multiply the common numerical and variable factors identified in step 2 to obtain the GCF. If there are no common factors, the GCF is 1.

EXAMPLE 1   Find the GCF of $6a^2b^3c$, $9a^3b^2c$, and $18a^4c^3$.

Strategy   We will prime factor each coefficient of each term in the list. Then we will identify the numerical and variable factors common to each term and find their product.

Why   The product of the common factors is the GCF of the terms in the list.

Solution   We begin by factoring each term.

$$6a^2b^3c = 2 \cdot \mathbf{3} \cdot \mathbf{a} \cdot \mathbf{a} \cdot b \cdot b \cdot b \cdot \mathbf{c} \qquad \text{This can be written as } 2 \cdot 3 \cdot a^2 \cdot b^3 \cdot c.$$
$$9a^3b^2c = \mathbf{3} \cdot 3 \cdot \mathbf{a} \cdot \mathbf{a} \cdot a \cdot b \cdot b \cdot \mathbf{c} \qquad \text{This can be written as } 3^2 \cdot a^3 \cdot b^2 \cdot c.$$
$$18a^4c^3 = 2 \cdot \mathbf{3} \cdot 3 \cdot \mathbf{a} \cdot \mathbf{a} \cdot a \cdot a \cdot \mathbf{c} \cdot c \cdot c \qquad \text{This can be written as } 2 \cdot 3^2 \cdot a^4 \cdot c^3.$$

Success Tip

The exponent on any variable in a GCF is the smallest exponent that appears on that variable in all of the terms under consideration.

Since each term has one factor of 3, two factors of a, and one factor of c in common, the GCF is

$$\mathbf{3} \cdot \mathbf{a} \cdot \mathbf{a} \cdot \mathbf{c} = 3a^2c$$

Self Check 1   Find the GCF of $24x^2y^3$, $15x^3y$, and $18x^2y^2$.

Now Try Problems 11 and 17

The concept of greatest common factor is used to factor polynomials.

2 Factor Out the Greatest Common Factor.

We have seen that the distributive property provides a method for multiplying a polynomial by a monomial. For example, to multiply $3x^2 + 4$ by $2x^3$, we proceed as follows:

Multiplication

$$2x^3(3x^2 + 4) = 2x^3 \cdot 3x^2 + 2x^3 \cdot 4$$
$$= 6x^5 + 8x^3$$

In this section, we reverse the previous steps and determine what factors were multiplied to obtain $6x^5 + 8x^3$. We call that process *factoring the polynomial*.

Factoring

$$6x^5 + 8x^3 = \mathbf{2x^3} \cdot 3x^2 + \mathbf{2x^3} \cdot 4 \qquad \text{Write } 6x^5 \text{ and } 8x^3 \text{ as the product of}$$
$$\text{the GCF, } 2x^3, \text{ and one other factor.}$$

$$= \mathbf{2x^3}(3x^2 + 4) \qquad \text{Write an expression so that the multiplication}$$
$$\text{by } 2x^3 \text{ distributes over the terms } 3x^2 \text{ and } 4.$$

Since $2x^3$ is the GCF of the terms of $6x^5 + 8x^3$, this method is called **factoring out the greatest common factor.** When we factor a polynomial, we write a sum of terms as a product of factors.

$$\underbrace{6x^5 + 8x^3}_{\text{A sum of terms}} = \underbrace{2x^3(3x^2 + 4)}_{\text{A product of factors}}$$

EXAMPLE 2   Factor: **a.** $16y^2 + 24y$ **b.** $25a^3b - 15ab^3$ **c.** $3xy^2z^3 + 6xyz^3 + 3xz^2$

Strategy   We will determine the GCF of the terms of the polynomial. Then we will write each term of the polynomial as the product of the GCF and one other factor.

Why   We can then use the distributive property to factor out the GCF.

Solution   **a.** Since the GCF of $16y^2$ and $24y$ is $8y$, we write $16y^2$ and $24y$ as the product of $8y$ and one other factor and proceed as follows:

$$16y^2 + 24y = \mathbf{8y} \cdot 2y + \mathbf{8y} \cdot 3 \qquad \text{This step can be done mentally.}$$

$$= \mathbf{8y}(2y + 3) \qquad \text{Factor out the GCF, 8y.}$$

> **Success Tip**
>
> Always verify a factorization by performing the indicated multiplication. The result should be the original polynomial.

To check, we multiply: $8y(2y + 3) = 8y \cdot 2y + 8y \cdot 3 = 16y^2 + 24y$. Since we obtain the original polynomial, $16y^2 + 24y$, the factorization is correct.

CAUTION   Remember to factor out the greatest common factor, not just a common factor. If we factored out $4y$ in the previous example, we would get

$$16y^2 + 24y = 4y(\mathbf{4y + 6})$$

However, the terms in red within parentheses have a common factor of 2, indicating that the factoring is not complete.

b. We begin by factoring each term of $25a^3b - 15ab^3$:

$$25a^3b = \mathbf{5} \cdot 5 \cdot \mathbf{a} \cdot a \cdot a \cdot \mathbf{b}$$
$$-15ab^3 = -1 \cdot \mathbf{5} \cdot 3 \cdot \mathbf{a} \cdot \mathbf{b} \cdot b \cdot b$$

> **Success Tip**
>
> On the game show *Jeopardy!*, answers are revealed and contestants respond with the appropriate questions. Factoring is similar. Answers to multiplications are given. You are to respond by telling what factors were multiplied.

Since each term has one factor of 5, one factor of a, and one factor of b in common, and there are no other common factors, $5ab$ is the GCF of the two terms. We write $25a^3b$ and $15ab^3$ as the product of the GCF, $5ab$, and one other factor and proceed as follows:

$$25a^3b - 15ab^3 = \mathbf{5ab} \cdot 5a^2 - \mathbf{5ab} \cdot 3b^2 \qquad \text{This step can be done mentally.}$$

$$= \mathbf{5ab}(5a^2 - 3b^2) \qquad \text{Factor out the GCF, 5ab.}$$

To check, we multiply: $5ab(5a^2 - 3b^2) = 25a^3b - 15ab^3$.

c. We begin by factoring each term of $3xy^2z^3 + 6xyz^3 + 3xz^2$:

$$3xy^2z^3 = \mathbf{3} \cdot \mathbf{x} \cdot y \cdot y \cdot \mathbf{z} \cdot z \cdot z$$
$$6xyz^3 = 2 \cdot \mathbf{3} \cdot \mathbf{x} \cdot y \cdot \mathbf{z} \cdot z \cdot z$$
$$3xz^2 = \mathbf{3} \cdot \mathbf{x} \cdot \mathbf{z} \cdot z$$

Since each term has one factor of 3, one factor of x, and two factors of z in common, and because there are no other common factors, $3xz^2$ is the GCF of the three terms. We write each term as the product of the GCF, $3xz^2$, and one other factor and proceed as follows:

$$3xy^2z^3 + 6xyz^3 + 3xz^2 = \mathbf{3xz^2} \cdot y^2z + \mathbf{3xz^2} \cdot 2yz + \mathbf{3xz^2} \cdot \mathbf{1}$$

This step can be done mentally.

$$= \mathbf{3xz^2}(y^2z + 2yz + \mathbf{1})$$

Don't forget to write the + 1.

Check the factorization using multiplication.

Self Check 2 Factor: **a.** $30t^2 + 20t$ **b.** $9x^4y^2 - 12x^3y^3$
  c. $4a^4b^2 + 6a^3b^2 + 2a^2b$

Now Try Problems 23, 29, and 33

A polynomial that cannot be factored is called a **prime polynomial** or an **irreducible polynomial.**

EXAMPLE 3 Factor: $9x + 16$

Strategy First, we will determine the GCF of the terms of the polynomial.

Why If the terms have no common factors (other than 1), this polynomial does not factor.

Solution We factor each term of $9x + 16$:

$$9x = 3 \cdot 3 \cdot x \qquad 16 = 2 \cdot 2 \cdot 2 \cdot 2$$

Since there are no common factors other than 1, this polynomial cannot be factored. It is a prime polynomial.

Self Check 3 Factor: $8a + 27$

Now Try Problem 31

EXAMPLE 4 Factor out -1 from $-n^3 + 2n^2 - 8$.

Strategy We will write each term of the polynomial as the product of -1 and one other factor.

Why We can then use the distributive property to factor out the -1.

Solution First, we write each term of the polynomial as the product of -1 and another factor. Then we factor out the common factor, -1.

$$-n^3 + 2n^2 - 8 = (\mathbf{-1})n^3 + (\mathbf{-1})(-2n^2) + (\mathbf{-1})8$$

This can be done mentally.

$$= \mathbf{-1}(n^3 - 2n^2 + 8)$$

Factor out -1.

$$= -(n^3 - 2n^2 + 8)$$

The 1 need not be written.

Self Check 4 Factor out -1 from $-b^4 - 3b^2 + 2$.

Now Try Problem 37

When asked to factor a polynomial whose leading coefficient is negative, we factor out the *opposite of the GCF.*

EXAMPLE 5   Factor out the opposite of the GCF from $-6u^2v^3 + 8u^3v^2$.

Strategy   We will determine the GCF of the terms of the polynomial. Then we will write each term as the product of the opposite of the GCF and one other factor.

Why   We can use the distributive property to factor out the opposite of the GCF.

Solution   Because the GCF of the two terms is $2u^2v^2$, the opposite of the GCF is $-2u^2v^2$. To factor out $-2u^2v^2$, we proceed as follows:

$$-6u^2v^3 + 8u^3v^2 = \mathbf{-2u^2v^2} \cdot 3v - (\mathbf{-2u^2v^2})4u$$ This can be done mentally.

$$= \mathbf{-2u^2v^2}(3v - 4u)$$ The leading coefficient of the polynomial within the parentheses is positive.

> **Success Tip**
>
> It is standard practice to factor in such a way that the leading coefficient of the polynomial within the parentheses is positive.

Self Check 5   Factor out the opposite of the GCF from $-8a^2b^2 - 12ab^3$.

Now Try ▶   Problem 43

A common factor can have more than one term.

EXAMPLE 6   Factor: **a.** $x(x + 1) + y(x + 1)$ **b.** $a(x - y + z) - b(x - y + z) + 3(x - y + z)$

Strategy   We will identify the terms of the expression and find their GCF.

Why   We can then use the distributive property to factor out the GCF.

Solution   **a.**   The expression is the sum of two terms: $\underline{x(x + 1)} + \underline{y(x + 1)}$

The first term   The second term

The GCF of the terms is the binomial $x + 1$, which can be factored out.

$$x(\mathbf{x + 1}) + y(\mathbf{x + 1})$$

$$= (\mathbf{x + 1})(x + y)$$ Factor out the GCF, x + 1.
Caution: Don't write $(x + 1)^2(x + y)$.

> **The Language of Algebra**
>
> In part a, we say that the terms have a **common binomial factor**, $(x + 1)$.

b.   We can factor out the GCF of the three terms, which is $(x - y + z)$.

$$a(\mathbf{x - y + z}) - b(\mathbf{x - y + z}) + 3(\mathbf{x - y + z})$$

$$= (\mathbf{x - y + z})(a - b + 3)$$

Self Check 6   Factor: **a.** $c(y^2 + 1) + d(y^2 + 1)$
b. $x(a + b - c) - y(a + b - c)$

Now Try ▶   Problems 47 and 51

3   Factor by Grouping.

Although the terms of many polynomials don't have a common factor, other than 1, it is possible to factor some of them by arranging their terms in convenient groups. This method is called **factoring by grouping.**

EXAMPLE 7   Factor: $2m - 2n + mn - n^2$

Strategy   We will factor out a common factor from the first two terms and a common factor from the last two terms.

Why   This will produce a common binomial factor that can then be factored out.

Solution

If we group the terms as highlighted below, the first two terms have a common factor of 2 and the last two terms have a common factor of n. When we factor out the common factor from each group, a common factor of $m - n$ appears.

$$\boxed{2m - 2n} + \boxed{mn - n^2} = 2(\boldsymbol{m - n}) + n(\boldsymbol{m - n})$$
Factor out 2 from $2m - 2n$ and n from $mn - n^2$. Don't forget the + sign.

$$= (\boldsymbol{m - n})(2 + n)$$
Factor out the common binomial factor, $m - n$.

We can check the factorization by multiplying:

$$(m - n)(2 + n) = 2m + mn - 2n - n^2$$
$$= 2m - 2n + mn - n^2$$
Rearrange the terms to get the original polynomial.

Self Check 7 Factor: $7r - 7s + rs - s^2$

Now Try ▶ Problems 55 and 59

The factoring method used in the previous example can be summarized as follows.

Factoring by Grouping

1. Group the terms of the polynomial so that each group has a common factor.
2. Factor out the common factor from each group.
3. Factor out the resulting common factor. If there is no common factor, regroup the terms of the polynomial and repeat steps 2 and 3.

By the multiplication property of 1, we know that 1 is a factor of every term. We can use this fact to factor certain polynomials by grouping.

EXAMPLE 8 Factor: **a.** $y^3 + 3y^2 + y + 3$ **b.** $x^2 - bx - x + b$

Strategy We will follow the steps for factoring a polynomial by grouping.

Why Since the terms of the polynomials do not have a common factor (other than 1), the only option is to attempt to factor them by grouping.

Solution

a. The first two terms, y^3 and $3y^2$, have a common factor of y^2. The only common factor of the last two terms, y and 3, is 1.

$$\boxed{y^3 + 3y^2} + \boxed{y + 3} = y^2(\boldsymbol{y + 3}) + 1(\boldsymbol{y + 3})$$
Factor out y^2 from $y^3 + 3y^2$.
Factor out 1 from $y + 3$.

$$= (\boldsymbol{y + 3})(y^2 + 1)$$
Factor out the common binomial factor, $y + 3$.

b. Since x is a common factor of the first two terms, we can factor it out and proceed as follows.

$$\boxed{x^2 - bx} - x + b = x(x - b) - x + b$$
Factor out x from $x^2 - bx$.

When factoring four terms by grouping, if the coefficient of the third term is negative, we often factor out a negative coefficient from the last two terms. If we factor -1 from

Success Tip

When we factor out -1 from the last two terms,

$x^2 - bx$ $-x + b$

$= x(x-b) - 1(x-b)$

the signs of those terms change within the parentheses. The binomials within both sets of parentheses are then identical.

$-x + b$, a common binomial factor $x - b$ appears within the second set of parentheses, which we can factor out.

$$x^2 - bx - x + b = x(x-b) - 1(x-b)$$

To factor the first two terms, we factor out x. To factor the last two terms, we factor out -1.

$$= (x-b)(x-1)$$

Factor out the common factor, $x - b$.

Self Check 8 Factor: **a.** $x^5 + 17x^4 + x + 17$ **b.** $a^2 - ab - a + b$

Now Try Problems 63 and 65

EXAMPLE 9 Factor: $5x^3 - 8 + 10x^2 - 4x$

Strategy We will follow the steps for factoring by grouping.

Why Since the four terms of the polynomial do not have a common factor (other than 1), we will attempt to factor it by grouping.

Solution Since the first two terms do not have a common factor, we cannot factor the polynomial in its current form.

$$5x^3 - 8 \quad + \quad 10x^2 - 4x$$

No common factor (other than 1)   GCF = 2x

Success Tip

An equivalent factorization, $(5x^2 - 4)(x + 2)$, results if the terms are arranged as $5x^3 - 4x + 10x^2 - 8$ or as $5x^3 - 4x - 8 + 10x^2$.

We will write the polynomial in descending powers of x, and attempt to factor by grouping again.

$$5x^3 - 8 + 10x^2 - 4x = 5x^3 + 10x^2 - 4x - 8$$

Reorder the terms and regroup.

$$= 5x^2(x + 2) - 4(x + 2)$$

Factor $5x^2$ from $5x^3 + 10x^2$ and -4 from $-4x - 8$.

$$= (x + 2)(5x^2 - 4)$$

Factor out the GCF, $x + 2$.

Self Check 9 Factor: $y^3 - 6 + 3y^2 - 2y$

Now Try Problem 67

The instruction "Factor" means for you to factor the given expression completely. Each factor of a completely factored expression will be prime. To **factor a polynomial completely**, it is often necessary to factor more than once. When factoring a polynomial, *always look for a common factor first*.

EXAMPLE 10 Factor: $3x^3y - 4x^2y^2 - 6x^2y + 8xy^2$

Strategy Since all four terms have a common factor of xy, we factor it out first. Then we will attempt to factor the resulting polynomial by grouping.

Why Factoring out the GCF first makes factoring by any method easier.

Solution We begin by factoring out the common factor xy.

$$3x^3y - 4x^2y^2 - 6x^2y + 8xy^2 = xy(3x^2 - 4xy - 6x + 8y)$$

Then we factor the four-term polynomial within the parentheses using factoring by grouping.

$$3x^3y - 4x^2y^2 - 6x^2y + 8xy^2$$
$$= xy(3x^2 - 4xy - 6x + 8y)$$
$$= xy[x(3x - 4y) - 2(3x - 4y)] \qquad \text{Factor } x \text{ from } 3x^2 - 4xy \text{ and } -2 \text{ from } -6x + 8y.$$
$$\text{Brackets [ ] are used to enclose the factoring by grouping steps.}$$
$$= xy[(3x - 4y)(x - 2)] \qquad \text{Factor out } 3x - 4y.$$
$$= xy(3x - 4y)(x - 2) \qquad \text{The brackets [ ] are not needed any longer.}$$

Because xy, $3x - 4y$, and $x - 2$ are prime, no further factoring can be done. The factorization is complete. We say the polynomial is *factored completely*.

Self Check 10   Factor: $\;3a^3b + 3a^2b - 2a^2b^2 - 2ab^2$

Now Try ▶   Problem 71

4   Use Factoring to Solve Formulas for a Specified Variable.

Factoring is often required to solve a formula for one of its variables.

EXAMPLE 11

Electronics.   The formula $r_1r_2 = rr_2 + rr_1$ is used in electronics to relate the combined resistance, r, of two resistors wired in parallel. The variable r_1 represents the resistance of the first resistor, and the variable r_2 represents the resistance of the second. Solve for r_2.

Strategy   To isolate r_2 on one side of the equation, we will get all the terms involving r_2 on the left side and all the terms not involving r_2 on the right side.

Why   To *solve a formula for a specified variable* means to isolate that variable on one side of the equation, with all other variables and constants on the opposite side.

Solution

We want to isolate this variable on one side of the equation.

$$r_1r_2 = rr_2 + rr_1$$

$$r_1r_2 - rr_2 = rr_1 \qquad \text{To eliminate } rr_2 \text{ on the right side, subtract } rr_2 \text{ from both sides.}$$

$$r_2(r_1 - r) = rr_1 \qquad \text{On the left side, factor out the GCF } r_2 \text{ from } r_1r_2 - rr_2.$$

$$\frac{r_2(r_1 - r)}{r_1 - r} = \frac{rr_1}{r_1 - r} \qquad \begin{array}{l}\text{To isolate } r_2 \text{ on the left side, undo the multiplication} \\ \text{by } r_1 - r \text{ by dividing both sides by } r_1 - r.\end{array}$$

$$r_2 = \frac{rr_1}{r_1 - r} \qquad \begin{array}{l}\text{Simplify the left side by removing the common} \\ \text{factor } r_1 - r \text{ from the numerator and denominator.}\end{array}$$

Self Check 11   Solve $A = p + prt$ for p.

Now Try ▶   Problems 75 and 79

SECTION 5.5 ▶ STUDY SET

VOCABULARY

Fill in the blanks.

1. When we write $2x + 4$ as $2(x + 2)$, we say that we have _____ $2x + 4$.
2. When we factor a polynomial, we write a sum of terms as a _____ of factors.
3. The abbreviation GCF stands for _____ _____ _____.
4. If a polynomial cannot be factored, it is called a _____ polynomial or an irreducible polynomial.
5. The terms $t(t - 5)$ and $9(t - 5)$ have the common _____ factor $t - 5$.
6. To factor $ab + 6a + 2b + 12$ by _____, we begin by writing $a(b + 6) + 2(b + 6)$.

CONCEPTS

7. The prime factorizations of three terms are shown. Find their GCF.

$$12x^2y^3 = 2 \cdot 2 \cdot 3 \cdot x \cdot x \cdot y \cdot y \cdot y$$
$$18xy^4 = 2 \cdot 3 \cdot 3 \cdot x \cdot y \cdot y \cdot y \cdot y$$
$$126x^3y^2 = 2 \cdot 3 \cdot 3 \cdot 7 \cdot x \cdot x \cdot x \cdot y \cdot y$$

8. Fill in the blanks to complete each factorization.
   a. $8x^3 + 6x^2 + 2x = 2x(4x^2 + 3x + \boxed{})$
   b. $-9x + 5 = -(9x \boxed{} 5)$
9. Consider the polynomial $5t - 25 + st - 5s$.
   a. How many terms does the polynomial have?
   b. Is there a common factor of all the terms, other than 1?
   c. What is the GCF of the first two terms and what is the GCF of the last two terms?
10. Check to determine whether each factorization is correct.
    a. $9a^4 + 15a^2 = 3a^2(3a^2 + 5)$
    b. $3x^3 + 2x^2 + 6x + 4 = (3x + 1)(x^2 + 4)$

GUIDED PRACTICE

Find the GCF of each list of terms. See Example 1.

11. $14x$, $24x^2$
12. $20a^2$, $35a$
13. $45a^5b$, $30a^4$
14. $36x^3y^2$, $27xy$
15. $16y^4$, $40y^2$, $24y^3$
16. $16m^4$, $40m^6$, $28m^3$
17. $24r^3s^5$, $36r^8s^4$, $48r^4s^4$
18. $28m^6n^6$, $56m^4n^5$, $42m^7n^6$
19. $18x^4y^3z^9$, $54xy^2z^6$, $36xy^5z^8$
20. $56x^2y^6z^7$, $24x^3y^7z^9$, $40x^2y^5z^3$
21. $10(c - d)$, $c(c - d)$
22. $16(x + y)$, $7x(x + y)$

Factor, if possible. See Examples 2 and 3.

23. $24x + 16$
24. $24y - 30$
25. $3y^3 + 5y^2 - 9y$
26. $2x^3 - 6x^2 + 11x$

27. $45a^2 - 9a$
28. $21x^2 + 7x$
29. $15x^2y - 10x^2y^2$
30. $13b^2c^3 - 26b^2c$
31. $14r + 15$
32. $100m + 33$
33. $27a^9b^4c^5 - 12a^7b^4c^2 + 3a^6b^4c^2$
34. $25t^{10}u^2v^5 - 10t^7u^2v^4 - 55t^6u^2v^4$

Factor -1 from each polynomial. See Example 4.

35. $-a - b$
36. $-2x - y$
37. $-5xy + y - 4$
38. $-7m - 12n + 16$

Factor each polynomial by factoring out the opposite of the GCF. See Example 5.

39. $-8a - 16$
40. $-6b - 30$
41. $-6x^2 - 3xy$
42. $-15y^3 - 25y^2$
43. $-18a^2b + 12ab^2$
44. $-21t^5 - 28t^3$
45. $-8a^4c^8 + 28a^3c^8 - 20a^2c^9$
46. $-30x^{10}y + 24x^9y^2 - 60x^8y^2$

Factor. See Example 6.

47. $4(x + y) + t(x + y)$
48. $5(a - b) - t(a - b)$
49. $(a - b)r - (a - b)s + (a - b)t$
50. $(x + y)u + (x + y)v - (x + y)z$
51. $3(m + n + p) + x(m + n + p)$
52. $x(x - y - z) + y(x - y - z)$
53. $3x(x + 7)^2 - 2(x + 7)^2$
54. $15x^2(x - y)^3 - 11(x - y)^3$

Factor by grouping. See Example 7.

55. $ax + 8x + ay + 8y$
56. $cd + 2c + 5d + 10$
57. $3c - c^2 + 3d - cd$
58. $x^2 - xy + 4x - 4y$
59. $a^2 + ab - 4a - 4b$
60. $7u + uv - 7v - v^2$
61. $t^3 - 3t^2 - 7t + 21$
62. $b^3 - 4b^2 - 3b + 12$

Factor by grouping. See Example 8.

63. $x^2 + xy + x + y$
64. $cd + d^2 + c + d$
65. $1 - n - m + mn$
66. $1 - c - 3t + 3ct$

Rearrange the terms and factor by grouping.
See Example 9.

67. $y^3 - 12 + 3y - 4y^2$ **68.** $h^3 - 8 + h - 8h^2$

69. $st + rv + sv + rt$ **70.** $2tx + 3dy + ty + 6dx$

Factor by grouping. Factor out the GCF first.
See Example 10.

71. $28a^3b^3c + 14a^3c - 4b^3c - 2c$
72. $12x^3z + 12xy^2z - 8x^2yz - 8y^3z$
73. $mpx + mqx + npx + nqx$
74. $abd - abe + acd - ace$

Solve for the specified variable or expression.
See Example 11.

75. $d_1d_2 = fd_2 + fd_1$ for f **76.** $2g = ch + dh$ for h

77. $b^2x^2 + a^2y^2 = a^2b^2$ for a^2 **78.** $d_1d_2 = fd_2 + fd_1$ for d_1

79. $rx - ty = by$ for y **80.** $Ar - c^2d = dm^2$ for d

81. $S(1 - r) = a - lr$ for r **82.** $Sn = (n - 2)180$ for n

TRY IT YOURSELF

Factor.

83. $-63u^3 + 28u^2$ **84.** $-56x^4 - 72x^3$

85. $a^3b^2 - 3 + a^3 - 3b^2$ **86.** $2ax^2 - 4 + a - 8x^2$

87. $45x^{10}y^3 - 63x^7y^7 + 81x^{10}y^{10}$ **88.** $48u^6v^6 - 16u^4v^4 - 3u^6v^3$

89. $ar^2t - br^2t + as^2t - bs^2t$ **90.** $2a^2x^2 - 4x^2 + 10a^2 - 20$

91. $16a^3b^3 - 27c^3d^3$ **92.** $2x^2 + x - 9y^2$

93. $\frac{3}{5}ax^4 + \frac{1}{5}bx^2 - \frac{4}{5}ax^3$ **94.** $\frac{3}{2}t^2y^4 - \frac{1}{2}ty^4 - \frac{5}{2}ry^3$

95. $24x^{50} - 18x^{40} - 42x^{30}$ **96.** $54c^{21} - 90c^{20} + 72c^{19}$

97. $a(2x + y) - b(2x + y) + c(2x + y)$
98. $b(a - b + c) - c(a - b + c)$
99. $45y^{12} + 30y^{10} + 25y^8 - 5y^6$
100. $2a^4 + 4a^8 + 6a^{12} + 8a^{16}$
101. $a^2x + bx - a^2 - b$
102. $x^2y - ax - xy + a$
103. $ab + b^2 + bc + ac + bc + c^2$
104. $x^2 + xy + xz + xy + y^2 + zy$

Look Alikes . . .

105. a. $5a^3 + 6a^2 + 15a + 18$ **b.** $3a^3 + 6a^2 + 15a + 18$

106. a. $2mn - 3m - 4n + 6$ **b.** $2mn - 3m + 4n + 6$

107. Geometric Formulas. The expression that represents the area of the part of the figure below that is shaded red is $\frac{1}{2}b_1h$. The expression that represents the area of the part of the figure that is shaded blue is $\frac{1}{2}b_2h$. Add these two expressions and factor the result. What important formula from geometry do you obtain?

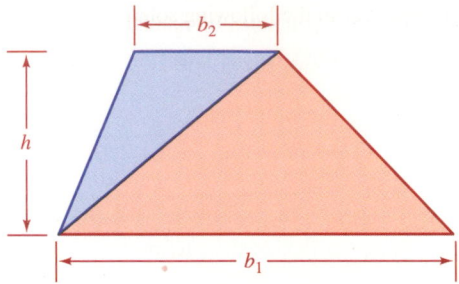

108. Packaging. The amount of cardboard needed to make the cereal box shown below can be found by computing the area A, which is given by the formula $A = 2wh + 4wl + 2lh$, where w is the width, h the height, and l the length. Solve the equation for the width.

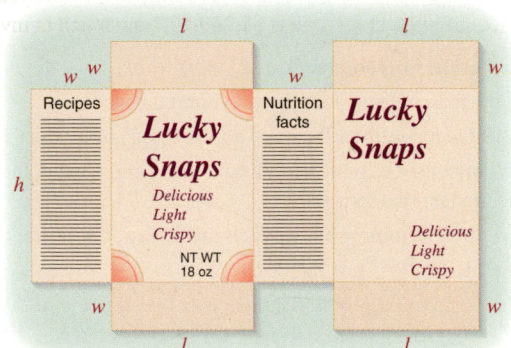

109. Landscaping. The combined area of the portions of a square lot that the sprinkler doesn't reach is given by $4r^2 - \pi r^2$, where r is the radius of the circular spray. Factor this expression.

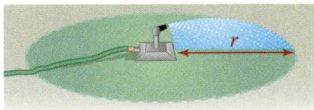

110. Crayons. The amount of colored wax used to make the crayon shown below can be found by computing its volume using the formula $V = \pi r^2 h_1 + \frac{1}{3}\pi r^2 h_2$. Factor the expression on the right side of this equation.

WRITING

111. Explain the error.

Factor: $30z^3 - 20z^2 = 5z^2(6z - 4)$

112. How are factorizations of polynomials checked? Give an example.

113. Explain why the following factorization is not complete. Then finish the solution.

$ax + ay + x + y = a(x + y) + x + y$

114. What is a prime polynomial? Give an example.

115. Explain the error in the following solution.

Solve for r_1: $r_1r_2 = rr_2 + rr_1$

$$\frac{r_1r_2}{r_2} = \frac{rr_2 + rr_1}{r_2}$$

$$r_1 = \frac{rr_2 + rr_1}{r_2}$$

116. Explain this diagram.

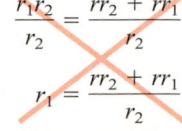

Multiplication
$6st^3(3t + 2) = 18st^4 + 12st^3$
Factoring

REVIEW

117. Investments. Equal amounts are invested in each of three accounts paying 7%, 8%, and 10.5% annually. If one year's combined interest income is $1,249.50, how much is invested in each account?

118. Search and Rescue. Two search-and-rescue teams leave base at the same time looking for a lost boy. The first team, on foot, heads north at 2 mph and the other, on horseback, south at 4 mph. How long will it take them to search a distance of 21 miles between them?

CHALLENGE PROBLEMS

Factor.

119. $2n^4p - 2n^2 - n^3p^2 + np + 2mn^3p - 2mn$

120. $a^2c^3 + ac^2 + a^3c^2 - 2a^2bc^2 - 2bc^2 + c^3$

121. $6(x^3 - 7x + 1)^2 - 3(x^3 - 7x + 1)^3$

122. $4x^2(x^2 + 1)^2 + 2x^2(x^2 + 1)^3$
123. $x^2(a - b) + 83(b - a)$
124. $ax_1 + bx_1 + cx_1 - ax_2 - bx_2 - cx_2 + ax_3 + bx_3 + cx_3$

Factor out the specified factor.

125. x^2 from $x^{n+2} + x^{n+3}$
126. y^n from $2y^{n+2} - 3y^{n+3}$
127. t^{-3} from $t^5 + 4t^{-6}$
128. $7x^{-3n}$ from $21x^{6n} + 7x^{3n} + 14$

SECTION 5.6

Factoring Trinomials

OBJECTIVES

1 Factor perfect-square trinomials.

2 Factor trinomials of the form $x^2 + bx + c$.

3 Factor trinomials of the form $ax^2 + bx + c$.

4 Use substitution to factor trinomials.

5 Use the grouping method to factor trinomials.

ARE YOU READY?

The following problems review some basic skills that are needed when factoring trinomials.

1. In $x^2 - 6x - 27$, what is the coefficient of the leading term?

2. Multiply: $(x + 8)(x - 6)$

3. Multiply: $(2a + 9)^2$

4. $x^2 - 4xy - 5y^2$ is a polynomial in how many variables?

5. Find two integers whose product is 10 and whose sum is 7.

6. Find two integers whose product is -18 and whose sum is 3.

We will now discuss techniques for factoring trinomials. These techniques are based on the fact that the product of two binomials is often a trinomial. With that observation in mind, we begin the study of trinomial factoring by considering two special products.

1 Factor Perfect-Square Trinomials.

Recall that trinomials that are squares of a binomial are called **perfect-square trinomials**. They can be factored using the following special-product rules from Section 5.4.

| **Factoring Perfect-Square Trinomials** | $A^2 + 2AB + B^2 = (A + B)^2$ $A^2 - 2AB + B^2 = (A - B)^2$ | *Each of these trinomials factors as the square of a binomial.* |

EXAMPLE 1   Factor: **a.** $n^2 + 20n + 100$ **b.** $9a^4 - 30a^2b^2 + 25b^4$

Strategy   The terms of these trinomials do not have a common factor (other than 1). We will determine whether each is a perfect-square trinomial.

Why   If it is, we can factor it using a special-product rule.

Solution   **a.** To factor $n^2 + 20n + 100$, we note that it is a perfect-square trinomial because

- The first term n^2 is the square of **n**.
- The last term 100 is the square of **10**: $10^2 = 100$.
- The middle term $20n$ is twice the product of **n** and **10**: $2(\textbf{n})(\textbf{10}) = 20n$.

To find the factorization, we match $n^2 + 20n + 100$ to the proper special-product rule.

$$A^2 + 2\ A\ \ B + B^2 = (A + \ B)^2$$
$$n^2 + 20n + 100 = n^2 + 2 \cdot \textbf{n} \cdot \textbf{10} + \textbf{10}^2 = (\textbf{n} + \textbf{10})^2$$

Thus, $n^2 + 20n + 100 = (n + 10)^2$. We can check the factorization by multiplying.

Check: $(n + 10)^2 = (n + 10)(n + 10)$
$\qquad\qquad\qquad\ = n^2 + 20n + 100$ *This is the original trinomial.*

b. The trinomial $9a^4 - 30a^2b^2 + 25b^4$ is a perfect-square trinomial because

- The first term $9a^4$ is the square of **$3a^2$**: $(3a^2)^2 = 9a^4$.
- The last term $25b^4$ is the square of **$-5b^2$**: $(-5b^2)^2 = 25b^4$.
- The middle term $-30a^2b^2$ is twice the product of **$3a^2$** and **$-5b^2$**: $2(\textbf{3}\textbf{a}^2)(\textbf{−5}\textbf{b}^2) = -30a^2b^2$.

We can match the trinomial to a special-product rule to find the factorization.

$$9a^4 - 30a^2b^2 + 25b^4 = (\textbf{3}\textbf{a}^2)^2 + 2 \cdot \textbf{3}\textbf{a}^2 \cdot (\textbf{−5}\textbf{b}^2) + (\textbf{−5}\textbf{b}^2)^2 = (\textbf{3}\textbf{a}^2 − \textbf{5}\textbf{b}^2)^2$$

Thus, $9a^4 - 30a^2b^2 + 25b^4 = (3a^2 - 5b^2)^2$. Check by multiplying.

Self Check 1   Factor: **a.** $m^2 + 12m + 36$ **b.** $49b^4 - 28b^2c^2 + 4c^4$

Now Try ▶ Problems 19 and 23

2 Factor Trinomials of the Form $x^2 + bx + c$.

To begin the discussion of trinomial factoring, we first consider trinomials of the form $x^2 + bx + c$, such as

$$x^2 + 10x + 24, \qquad x^2 + 14x + 24, \qquad t^2 - 7t + 12, \quad \text{and} \quad y^2 + 2xy - 15x^2$$

In each case, the **leading coefficient** (the coefficient of the squared variable) is 1.

To develop a method for factoring such trinomials, we will find the product of $x + 6$ and $x + 4$ and make some observations about the result.

$$(x + 6)(x + 4) = x \cdot x + 4x + 6x + 6 \cdot 4 \qquad \text{Multiply the binomials.}$$
$$= x^2 + 10x + 24$$

First term · Middle term · Last term

The result is a trinomial, where

- The first term, x^2, is the product of x and x.
- The last term, 24, is the product of 6 and 4.
- The coefficient of the middle term, 10, is the sum of 6 and 4.

These observations suggest a strategy for factoring trinomials with a leading coefficient of 1.

Factoring Trinomials Whose Leading Coefficient Is 1

To factor a trinomial of the form $x^2 + bx + c$, find two numbers whose product is c and whose sum is b.

1. If c is positive, the numbers have the same sign.

2. If c is negative, the numbers have different signs.

Then write the trinomial as a product of two binomials. You can check by multiplying.

$$x^2 + bx + c = \left(x \,\boxed{}\right)\left(x \,\boxed{}\right)$$

The product of these numbers must be c and their sum must be b.

EXAMPLE 2   Factor: $x^2 + 11x + 24$

Strategy We will assume that $x^2 + 11x + 24$ is the product of two binomials. We must find the terms of each binomial.

Why Since the terms of $x^2 + 11x + 24$ do not have a common factor (other than 1), the only option is to try to factor it as the product of two binomials.

Solution We represent the binomials using two sets of parentheses. Since the first term of the trinomial is x^2, we enter x and x as the first terms of the binomial factors.

$$x^2 + 11x + 24 = \left(x \,\boxed{}\right)\left(x \,\boxed{}\right) \qquad \text{Because } x \cdot x \text{ will give } x^2.$$

The second terms of the binomials must be two integers whose product is 24 and whose sum is 11. Since the integers must have a positive product and a positive sum, we only consider pairs of positive integer factors of 24. All such possible integer pairs are listed in the table.

Positive factors of 24	Sum of the positive factors of 24
$1 \cdot 24 = 24$	$1 + 24 = 25$
$2 \cdot 12 = 24$	$2 + 12 = 14$
$3 \cdot 8 = 24$	$3 + 8 = 11$
$6 \cdot 4 = 24$	$6 + 4 = 10$

← This is the one to choose.

The third row of the table contains the correct pair of integers 3 and 8, whose product is 24 and whose sum is 11. To complete the factorization, we enter 3 and 8 as the second terms of the binomial factors.

$$x^2 + 11x + 24 = (x + 3)(x + 8)$$

We can check this factorization by multiplying:

$$(x + 3)(x + 8) = x^2 + 8x + 3x + 24$$
$$= x^2 + 11x + 24 \qquad \text{This is the original trinomial.}$$

By the commutative property of multiplication, the order of binomial factors in a factorization does not matter. Thus, we can also write $x^2 + 11x + 24 = (x + 8)(x + 3)$.

Self Check 2   Factor: $a^2 + 13a + 40$

Now Try ▶ Problem 27

EXAMPLE 3   Factor: $-8t^2 + t^4 + 12$

Strategy   We will write the terms of the trinomial in descending powers of t.

Why   It is easier to factor a trinomial if its terms are written in descending powers of the variable.

Solution   We can use the commutative property of addition to reorder the terms:

$$-8t^2 + t^4 + 12 = t^4 - 8t^2 + 12$$

Since the first term of the trinomial is t^4, the first term of each binomial factor must be t^2.

$$t^4 - 8t^2 + 12 = \left(t^2 \;\boxed{}\right)\left(t^2 \;\boxed{}\right) \qquad \text{Because } t^2 \cdot t^2 \text{ will give } t^4.$$

The second terms of the binomials must be two integers whose product is 12 and whose sum is -8. Since the integers must have a positive product and a negative sum, we consider only pairs of negative integer factors of 12. The possible pairs are listed in the table.

Factors of 12	Sum of factors
$-1(-12) = 12$	$-1 + (-12) = -13$
$\mathbf{-2(-6) = 12}$	$\mathbf{-2 + (-6) = -8}$
$-3(-4) = 12$	$-3 + (-4) = -7$

The second row of the table contains the correct pair of integers -2 and -6, whose product is 12 and whose sum is -8. To complete the factorization, we enter -2 and -6 as the second terms of the binomial factors.

$$t^4 - 8t^2 + 12 = (t^2 - 2)(t^2 - 6) \qquad \text{Check the result using multiplication.}$$

Self Check 3   Factor: $-9m^2 + m^4 + 18$

Now Try ▶ Problem 33

EXAMPLE 4   Factor: $2x^2y^2 + 4xy^3 - 30y^4$

Strategy   We will factor out the GCF of $2y^2$ and factor the resulting trinomial.

Why   The first step in factoring any polynomial is to factor out the GCF. Factoring out the GCF first makes factoring easier.

Solution   Each term of the trinomial has a common factor of $2y^2$, which we can factor out.

$$2x^2y^2 + 4xy^3 - 30y^4 = 2y^2(x^2 + 2xy - 15y^2)$$

Next, we factor $x^2 + 2xy - 15y^2$, which is a trinomial in two variables, x and y. Since its first term is x^2, the first term of each binomial factor must be x. Since its third term contains y^2, the last term of each binomial factor must contain y. We need to determine the coefficient of each y-term.

$$2y^2(x^2 + 2xy - 15y^2) = 2y^2\left(x \;\boxed{}\; y\right)\left(x \;\boxed{}\; y\right) \qquad \begin{array}{l}\text{Because } x \cdot x \text{ will give } x^2 \\ \text{and } y \cdot y \text{ will give } y^2.\end{array}$$

Since the integers **5** and **−3** have a product of −15 and a sum of 2, they are entered as the coefficients of the second terms of the binomial factors.

Factors of −15	Sum of factors
$1(-15) = -15$	$1 + (-15) = -14$
$3(-5) = -15$	$3 + (-5) = -2$
$5(-3) = -15$	**$5 + (-3) = 2$**
$15(-1) = -15$	$15 + (-1) = 14$

$$2y^2(x^2 + 2xy - 15y^2) = 2y^2(x + 5y)(x - 3y)$$

We can check by multiplying.

Self Check 4   Factor: $3a^2b^2 + 6ab^3 - 105b^4$

Now Try ▶ Problem 41

3 Factor Trinomials of the Form $ax^2 + bx + c$.

There are more combinations of factors to consider when factoring trinomials with leading coefficients other than 1. We can use two methods to factor such trinomials. With the first method, called **trial-and-check**, we make educated guesses and then check them using multiplication. The correct factorization is determined through a process of elimination. The second method is an extention of **factoring by grouping.**

EXAMPLE 5   Factor: $5x^2 + 7x + 2$

Strategy   We will assume that this trinomial is the product of two binomials. To find their terms, we will make educated guesses and then check them using multiplication.

Why   Since the terms of the trinomial do not have a common factor (other than 1), the only option is to try to factor it as the product of two binomials.

Solution   We represent the binomial factors using two sets of parentheses.

$$5x^2 + 7x + 2 = (\boxed{}\ \boxed{})(\boxed{}\ \boxed{})$$

Since the first term of the trinomial $5x^2 + 7x + 2$ is $5x^2$, the first terms of the binomial factors must be $5x$ and x.

$$5x^2 + 7x + 2 = (5x\ \boxed{})(x\ \boxed{})$$ Because $5x \cdot x$ will give $5x^2$.

The second terms of the binomials must be two integers whose product is 2.

$$5x^2 + 7x + 2 = (5x\ \boxed{})(x\ \boxed{})$$

Because the coefficients of the terms of $5x^2 + 7x + 2$ are positive, we consider only pairs of positive integer factors of 2. Since there is only one such pair, $1 \cdot 2$, we can use 1 and 2 as the second terms of the binomials, or we can reverse the order and enter 2 and 1.

Product of the outer terms: 10x

$(5x + 1)(x + 2)$ Multiply and add to find the middle term: $10x + x = 11x$.

Product of the inner terms: x

Product of the outer terms: 5x

$(5x + 2)(x + 1)$ Multiply and add to find the middle term: $5x + 2x = 7x$.

Product of the inner terms: 2x

The second possibility shown in blue is correct, because it gives a middle term of $7x$. Thus,

$$5x^2 + 7x + 2 = (5x + 2)(x + 1)$$

We can check this result by multiplication:

$$(5x + 2)(x + 1) = 5x^2 + 5x + 2x + 2$$
$$= 5x^2 + 7x + 2$$ This is the original trinomial.

Self Check 5 Factor: $5w^2 + 8w + 3$

Now Try ▶ Problem 45

EXAMPLE 6 Factor: $3p^2 - 4p - 4$

Strategy We will assume that this trinomial is the product of two binomials. To find their terms, we will make educated guesses and then check them using multiplication.

Why Since the terms of the trinomial do not have a common factor (other than 1), the only option is to try to factor it as the product of two binomials.

Solution To factor the trinomial, we note that the first terms of the binomial factors must be $3p$ and p to give the first term of $3p^2$.

$$3p^2 - 4p - 4 = \left(3p\;\boxed{}\right)\left(p\;\boxed{}\right) \quad \text{Because } 3p \cdot p \text{ will give } 3p^2.$$

The second terms of the binomials must be two integers whose product is -4. There are three such pairs: $1(-4)$, $-1(4)$, and $-2(2)$. When these pairs are entered, and then reversed, as second terms of the binomials, there are six possibilities to consider.

For 1 and −4: $(3p + 1)(p - 4)$ or $(3p - 4)(p + 1)$

$-12p + p = -11p$ $3p + (-4p) = -p$

For −1 and 4: $(3p - 1)(p + 4)$ or $(3p + 4)(p - 1)$

$12p + (-p) = 11p$ $-3p + 4p = p$

For −2 and 2: $(3p - 2)(p + 2)$ or $(3p + 2)(p - 2)$

$6p + (-2p) = 4p$ $-6p + 2p = -4p$

Of these possibilities, only the one in blue gives the required middle term of $-4p$. Thus,

$$3p^2 - 4p - 4 = (3p + 2)(p - 2)$$

Self Check 6 Factor: $2q^2 - 17q - 9$

Now Try ▶ Problem 49

EXAMPLE 7 Factor: $7t^2 - 15t + 11$

Strategy We will assume that the trinomial is the product of two binomials and find the terms of the binomials.

Why Since the terms of the trinomial do not have a common factor (other than 1), the only option is to try to factor it as the product of two binomials.

Solution To try to factor the trinomial, we note that the first terms of the binomial factors should be $7t$ and t to give the first term of $7t^2$.

$$7t^2 - 15t + 11 = \left(7t\ \boxed{}\right)\left(t\ \boxed{}\right)$$

Because $7t \cdot t$ will give $7t^2$.

The second terms of the binomials must be two integers whose product is 11. Since the last term of $7t^2 - 15t + 11$ is positive and the coefficient of the middle term is negative, we consider only negative integer factors of the last term. There is only one such pair, $-1(-11)$. When this pair is entered, and then reversed, as second terms of the binomials, there are two possibilities to consider.

$$\overset{-77t}{(7t - 1)(t - 11)} \quad -77t + (-t) = -78t \qquad \overset{-7t}{(7t - 11)(t - 1)} \quad -7t + (-11t) = -18t$$

$$\underset{-t}{} \qquad\qquad\qquad\qquad \underset{-11t}{}$$

Since neither of the possible factorizations gives the correct middle term of $-15t$, the trinomial $7t^2 - 15t + 11$ does not factor. It is a prime trinomial.

The Language of Algebra

When a trinomial is not factorable using only integers, we say it is **prime** and that it does not factor **over the integers.**

Self Check 7 Factor: $5a^2 - 8a + 2$

Now Try ▶ Problem 53

Because guesswork is often necessary, it is difficult to give specific rules for factoring trinomials with leading coefficients other than 1. However, the following hints are helpful when using the **trial-and-check method.**

Factoring Trinomials with Leading Coefficients Other Than 1

To factor trinomials with leading coefficients other than 1:

1. Factor out any GCF (including -1 if that is necessary to make a positive in a trinomial of the form $ax^2 + bx + c$).

2. Write the trinomial as a product of two binomials. The coefficients of the first terms of each binomial factor must be factors of a, and the last terms must be factors of c.

The product of these numbers must be a.

$$ax^2 + bx + c = \left(\boxed{}x\ \boxed{}\right)\left(\boxed{}x\ \boxed{}\right)$$

The product of these numbers must be c.

3. If c is positive, the signs within the binomial factors match the sign of b. If c is negative, the signs within the binomial factors are opposites.

4. Try combinations of first terms and second terms until you find the one that gives the correct middle term. If no combination works, the trinomial is prime.

5. Check by multiplying.

EXAMPLE 8 Factor: $-15x^2 + 25xy + 60y^2$ GCF: 5 $-5(3x^2 - 5xy + 12y^2)$

Strategy We will factor out the opposite of the GCF first. Then we will factor the resulting trinomial.

Why It is easier to factor trinomials that have a positive leading coefficient.

Solution   Factor out -5 from each term.

$$-15x^2 + 25xy + 60y^2 = -5(3x^2 - 5xy - 12y^2)$$ The opposite of the GCF is −5.

To factor $3x^2 - 5xy - 12y^2$, we examine its terms.

- Since the first term is $3x^2$, the first terms of the binomial factors must be $3x$ and x.
- Since the sign of the first term of the trinomial is positive and the sign of the last term is negative, the signs within the binomial factors will be opposites.
- Since the last term of the trinomial contains y^2, the second terms of the binomial factors must contain y.

$$-5(3x^2 - 5xy - 12y^2) = -5\left(3x \boxed{} y\right)\left(x \boxed{} y\right)$$

The product of the last terms must be $-12y^2$, and the sum of the product of the outer terms and the product of the inner terms must be $-5xy$.

$$-5(3x^2 - 5xy - 12y^2) = -5\left(3x \boxed{} y\right)\left(x \boxed{} y\right)$$

Outer + Inner = −5xy

Since $1(-12)$, $2(-6)$, $3(-4)$, $-1(12)$, $-2(6)$, and $-3(4)$ all give a product of -12, there are 12 combinations to consider.

$(3x + 1y)(x - 12y)$	$(3x - 12y)(x + 1y)$	← 3x − 12y has a common factor 3.
$(3x + 2y)(x - 6y)$	$(3x - 6y)(x + 2y)$	← 3x − 6y has a common factor 3.
$(3x + 3y)(x - 4y)$	$(3x - 4y)(x + 3y)$	← 3x + 3y has a common factor 3.
$(3x - 1y)(x + 12y)$	$(3x + 12y)(x - 1y)$	← 3x + 12y has a common factor 3.
$(3x - 2y)(x + 6y)$	$(3x + 6y)(x - 2y)$	← 3x + 6y has a common factor 3.
$(3x - 3y)(x + 4y)$	$(3x + 4y)(x - 3y)$	← This is the one to choose.

The combinations in blue cannot work, because one of the factors has a common factor. This implies that $3x^2 - 5xy - 12y^2$ would have a common factor, which it doesn't.

     After mentally trying the remaining combinations, we find that only $(3x + 4y)(x - 3y)$ gives the proper middle term of $-5xy$. Thus,

$$-15x^2 + 25xy + 60y^2 = -5(3x^2 - 5xy - 12y^2)$$
$$= -5(3x + 4y)(x - 3y)$$

> **Caution**
>
> When factoring a trinomial, be sure to factor it completely. Always check to see whether any of the factors of your result can be factored further.

> **Success Tip**
>
> If the terms of a trinomial do not have a common factor, the terms of each of its binomial factors will not have a common factor.

Self Check 8   Factor: $-6x^2 - 57xy - 72y^2$

Now Try ▶   Problem 61

EXAMPLE 9   Factor: $6y^3 + 13x^2y^3 + 6x^4y^3$

Strategy   We write the expression in descending powers of x.

Why   It is easier to factor a trinomial if its terms are written in descending powers of one variable.

Solution   We write the expression in descending powers of x and factor out the greatest common factor, y^3.

$$6y^3 + 13x^2y^3 + 6x^4y^3 = 6x^4y^3 + 13x^2y^3 + 6y^3$$
$$= y^3(6x^4 + 13x^2 + 6)$$

To factor $6x^4 + 13x^2 + 6$, we examine its terms.

- Since the first term is $6x^4$, the first terms of the binomial factors must be either $2x^2$ and $3x^2$ or x^2 and $6x^2$.

$$6x^4 + 13x^2 + 6 = \left(2x^2 \,\boxed{}\right)\left(3x^2 \,\boxed{}\right) \text{ or } \left(x^2 \,\boxed{}\right)\left(6x^2 \,\boxed{}\right)$$

- Since the signs of the middle term and the last term of the trinomial are positive, the signs within each binomial factor will be positive.

- Since the product of the last terms of the binomial factors must be 6, we must find two numbers whose product is 6 that will lead to a middle term of $13x^2$.

After trying some combinations, we find the one that works.

$$6x^4y^3 + 13x^2y^3 + 6y^3 = y^3(6x^4 + 13x^2 + 6)$$
$$= y^3(2x^2 + 3)(3x^2 + 2)$$

Self Check 9   Factor: $4b + 11a^2b + 6a^4b$

Now Try ▶ Problem 65

4   Use Substitution to Factor Trinomials.

For more complicated expressions, especially those involving a quantity within parentheses, a substitution sometimes helps to simplify the factoring process.

EXAMPLE 10   Factor: $(x + y)^2 + 7(x + y) + 12$

Strategy   We will use a substitution where we will replace each expression $x + y$ with the variable z and factor the resulting trinomial.

Why   The resulting trinomial will be easier to factor because it will be in only one variable, z.

Solution   If we use the substitution $z = x + y$, we obtain

$$(x + y)^2 + 7(x + y) + 12 = z^2 + 7z + 12 \quad \text{Replace x + y with z.}$$
$$= (z + 4)(z + 3) \quad \text{Factor the trinomial.}$$

To find the factorization of $(x + y)^2 + 7(x + y) + 12$, we substitute $x + y$ for each z in the expression $(z + 4)(z + 3)$.

$$(z + 4)(z + 3) = (x + y + 4)(x + y + 3)$$

Thus, $(x + y)^2 + 7(x + y) + 12 = (x + y + 4)(x + y + 3)$.

Self Check 10   Factor: $(a + b)^2 - 3(a + b) - 10$

Now Try ▶ Problem 73

5   Use the Grouping Method to Factor Trinomials.

Another way to factor trinomials is to write them as equivalent four-termed polynomials and factor by grouping. For example, to factor $2x^2 + 5x + 3$ in this way, we proceed as follows.

First, we identify the values of a, b, and c and find the product ac, called the *key number*: $ac = 2(3) = 6$.

$$\left.\begin{array}{ccc} ax^2 & + bx & + c \\ \downarrow & \downarrow & \downarrow \\ 2x^2 & + 5x & + 3 \end{array}\right\} \; a = 2, b = 5, \text{ and } c = 3$$

Next, we find two integers whose product is $ac = 6$ and whose sum is $b = 5$. Since the integers must have a positive product and a positive sum, we consider only positive factors of 6.

The second row of the table contains the correct pair of integers 2 and 3, whose product is 6 and whose sum is 5.

$ac = 2(3) = 6$	$b = 5$
Positive factors of 6	**Sum the factors of 6**
$1 \cdot 6 = 6$	$1 + 6 = 7$
$2 \cdot 3 = 6$	$2 + 3 = 5$

Now we express the middle term, $5x$, of the trinomial as the *sum of two terms,* using the integers 2 and 3 found in the previous step as coefficients of the two terms.

$$2x^2 + 5x + 3 = 2x^2 + 2x + 3x + 3 \quad \text{Express 5x as 2x + 3x.}$$

We can then factor the equivalent four-termed polynomial by grouping:

$$2x^2 + 2x + 3x + 3 = 2x(x + 1) + 3(x + 1) \quad \text{Factor 2x out of } 2x^2 + 2x \text{ and 3 out of 3x + 3.}$$

$$= (x + 1)(2x + 3) \quad \text{Factor out the GCF, x + 1.}$$

Remember to check the factorization by multiplying.

Factoring by grouping is especially useful when the leading coefficient, a, and the constant term, c, have many factors.

Factoring Trinomials by Grouping

To factor a trinomial by grouping:

1. Factor out any GCF (including -1 if that is necessary to make the coefficient a positive in a trinomial of the form $ax^2 + bx + c$).

2. Identify a, b, and c, and find the key number ac.

3. Find two integers whose product is the key number and whose sum is b.

4. Express the middle term, bx, as the sum (or difference) of two terms. Enter the two numbers found in step 2 as coefficients of x in the form shown below. Then factor the equivalent four-term polynomial by grouping.

$$ax^2 + \boxed{}\, x + \boxed{}\, x + c$$

The product of these numbers must be ac and their sum must be b.

5. Check the factorization by multiplying.

EXAMPLE 11   Factor by grouping: **a.** $x^2 + 8x + 15$ **b.** $10x^2 + 13xy - 3y^2$

Strategy   In each case, we will express the middle term of the trinomial as the sum of two terms.

Why   We want to produce an equivalent four-termed polynomial that can be factored by grouping.

Solution   **a.** Since $x^2 + 8x + 15 = 1x^2 + 8x + 15$, we identify a as 1, b as 8, and c as 15. The key number is $ac = 1(15) = 15$. We must find two integers whose product is the key number, 15, and whose sum is $b = 8$. Since the integers must have a positive product and a positive sum, we consider only positive factors of 15.

$ac = 1(15) = 15$	$b = 8$
Positive factors of 15	**Sum the factors of 15**
$1 \cdot 15 = 15$	$1 + 15 = 16$
$3 \cdot 5 = 15$	$3 + 5 = 8$

The second row of the table contains the correct pair of integers 3 and 5, whose product is 15 and whose sum is 8.

Express the middle term, $8x$, of the trinomial as the *sum of two terms,* using the integers 3 and 5 as coefficients of the two terms and factor the equivalent four-term polynomial by grouping:

$$x^2 + \mathbf{8x} + 15 = x^2 + \mathbf{3x + 5x} + 15 \qquad \text{Express 8x as 3x + 5x.}$$

$$x^2 + 3x + 5x + 15 = x(x + 3) + 5(x + 3) \qquad \begin{array}{l}\text{Factor x out of } x^2 + 3x \text{ and}\\ \text{5 out of 5x + 15.}\end{array}$$

$$= (x + 3)(x + 5) \qquad \text{Factor out the GCF, x + 3.}$$

Check the factorization by multiplying.

b. In $10x^2 + 13xy - 3y^2$, we have $a = 10$, $b = 13$, and $c = -3$. The key number is $ac = 10(-3) = -30$. We must find a factorization of -30 in which the sum of the factors is $b = 13$. Since the factors must have a negative product, their signs must be different. The possible factor pairs are listed in the table.

The seventh row contains the correct pair of numbers 15 and -2, whose product is -30 and whose sum is 13. They serve as the coefficients of the two terms, $15xy$ and $-2xy$, that we use to represent the middle term, $13xy$, of the trinomial. We can then factor the resulting four-term polynomial by grouping.

$$ac = 10(-3) = -30 \qquad\qquad b = 13$$

It is wise to follow an order when listing the factors in the table so that you don't skip the correct combination. Here, the first factors 1, 2, 3, 5, 6, 10, 15, and 30 are listed from least to greatest.

Factors of -30	Sum of factors of -30
$1(-30) = -30$	$1 + (-30) = -29$
$2(-15) = -30$	$2 + (-15) = -13$
$3(-10) = -30$	$3 + (-10) = -7$
$5(-6) = -30$	$5 + (-6) = -1$
$6(-5) = -30$	$6 + (-5) = 1$
$10(-3) = -30$	$10 + (-3) = 7$
$\mathbf{15(-2) = -30}$	$\mathbf{15 + (-2) = 13}$
$30(-1) = -30$	$30 + (-1) = 29$

Notation

In Example 11b, the middle term, $13xy$, may be expressed as $15xy - 2xy$ or as $-2xy + 15xy$ when using factoring by grouping. The resulting factorizations will be equivalent.

$$10x^2 + \mathbf{13xy} - 3y^2 = 10x^2 + \mathbf{15xy - 2xy} - 3y^2 \qquad \text{Express 13xy as 15xy − 2xy.}$$

$$= 5x(2x + 3y) - y(2x + 3y) \qquad \begin{array}{l}\text{Factor out 5x from } 10x^2 + 15xy.\\ \text{Factor out −y from } -2xy - 3y^2.\end{array}$$

$$= (2x + 3y)(5x - y) \qquad \text{Factor out the GCF, 2x + 3y.}$$

Thus, $10x^2 + 13xy - 3y^2 = (2x + 3y)(5x - y)$. Check by multiplying.

Self Check 11 Factor by grouping: **a.** $m^2 + 13m + 42$
b. $15a^2 + 17ab - 4b^2$

Now Try ▶ Problems 29, 45, 49, and 57

EXAMPLE 12 Factor by grouping: $12x^5 - 17x^4 + 6x^3$

Strategy We will factor out the GCF of x^3 and factor the resulting trinomial using the grouping method.

Why The first step in factoring any polynomial is to factor out the GCF.

Solution The GCF of the three terms of the trinomial is x^3.

$$12x^5 - 17x^4 + 6x^3 = \mathbf{x^3}(12x^2 - 17x + 6)$$

To factor $12x^2 - 17x + 6$, we must find two integers whose product is $12(6) = 72$ and whose sum is -17. Two such numbers are -8 and -9. They serve as the coefficients of $-8x$ and $-9x$, the two terms that we use to represent the middle term, $-17x$, of the trinomial.

	$ac = 12(6) = 72$	$b = -17$
	Factors	**Sum**
	$-8(-9) = 72$	$-8 + (-9) = -17$

$$12x^2 - 17x + 6 = 12x^2 - 8x - 9x + 6 \qquad \text{Express } -17x \text{ as } -8x - 9x.$$
$$\qquad\qquad\qquad\qquad\qquad\qquad (-9x - 8x \text{ could also be used.})$$
$$= 4x(3x - 2) - 3(3x - 2) \qquad \text{Factor out } 4x \text{ and factor out } -3.$$
$$= (3x - 2)(4x - 3) \qquad \text{Factor out } 3x - 2.$$

The complete factorization of the original trinomial is

$$12x^5 - 17x^4 + 6x^3 = x^3(3x - 2)(4x - 3) \qquad \text{Don't forget to write the GCF, } x^3.$$

Check the factorization by multiplying.

Self Check 12 Factor by grouping: $21a^4 - 13a^3 + 2a^2$

Now Try ▶ Problem 65

SECTION 5.6 ▶ STUDY SET

VOCABULARY

Fill in the blanks.

1. Since $y^2 + 2y + 1 = (y + 1)^2$, we call $y^2 + 2y + 1$ a _____-square trinomial.

2. The statement $x^2 - x - 12 = (x - 4)(x + 3)$ shows that the trinomial $x^2 - x - 12$ factors into the product of two _____.

3. The _____ coefficient of the trinomial $x^2 - 3x + 2$ is 1, the _____ of the middle term is -3, and the last term is ___. The trinomial is written in _____ powers of x.

4. A trinomial is factored _____ when no factor can be factored further. A _____ polynomial cannot be factored by using only integers.

CONCEPTS

5. Consider $3x^2 - x + 16$. What is the sign of the
   a. First term?
   b. Middle term?
   c. Last term?

6. Find two integers whose
   a. Product is 10 and whose sum is 7.
   b. Product is 20 and whose sum is -9.
   c. Product is -6 and whose sum is 1.
   d. Product is -32 and whose sum is -4.

7. Complete the table and the sentence below it.

Factors of 8	Sum of the factors of 8
$1(8) = 8$	
$2(4) = 8$	
$-1(-8) = 8$	
$-2(-4) = 8$	

The numbers -1 and -8 are two integers whose _____ is 8 and whose _____ is -9.

8. Fill in the blanks. When factoring a trinomial, we write it in _____ powers of the variable. Then we factor out any _____ (including -1 if that is necessary to make the leading coefficient _____).

9. Use a check to determine whether $(3t - 1)(5t - 6)$ is the correct factorization of $15t^2 - 19t + 6$.

10. Complete the key number table and the sentence below.

Key number: $ac = 12$

Negative factors of 12	Sum of factors of 12
$-1(-12) = 12$	
	$-2 + (-6) = -8$
$-3(-4) = 12$	

The numbers -3 and -4 are two integers whose _____ is 12 and whose _____ is -7.

Complete each factorization.

11. $x^2 - 6x + 8 = (x - 4)(x \;\square\; 2)$

12. $x^2 - 3x - 18 = (x - 6)(x \;\square\; 3)$

13. $2a^2 + 9a + 4 = (2a \;\square\; 1)(a + 4)$

14. $6p^2 - 5p - 4 = (3p \;\square\; 4)(2p + 1)$

NOTATION

15. The trinomial $4m^2 - 4m + 1$ is written in the form $ax^2 + bx + c$. Identify a, b, and c.

16. Use the substitution $x = a + b$ to rewrite the trinomial $6(a + b)^2 - 17(a + b) - 3$.

GUIDED PRACTICE

Use a special-product rule to factor each perfect-square trinomial. See Example 1.

17. $a^2 + 18a + 81$

18. $b^2 + 16b + 64$

19. $4y^2 + 28y + 49$

20. $25x^2 + 60x + 36$

21. $y^4 - 10y^2 + 25$

22. $a^4 - 14a^2 + 49$

23. $9b^4 - 12b^2c^2 + 4c^4$

24. $4a^4 - 12a^2b^2 + 9b^4$

Factor each trinomial. See Examples 2 and 3 or Example 11.

25. $x^2 + 5x + 6$

26. $y^2 + 8y + 15$

27. $t^2 + 14t + 48$

28. $m^2 + 13m + 36$

29. $x^2 - 16x + 55$

30. $c^2 - 24c + 44$

31. $y^2 - 17y + 72$

32. $t^2 - 11t + 28$

33. $-13y^2 + 30 + y^4$

34. $-13b^2 + 42 + b^4$

35. $50 + 15x^2 + x^4$

36. $27 + 12d^2 + d^4$

Factor each trinomial. Factor out the GCF first. See Example 4 or Example 11.

37. $3x^2 + 12x - 63$

38. $2y^2 + 4y - 48$

39. $15x^2 + 45x + 30$

40. $9y^2 + 90y + 81$

41. $2x^2y - 12xy^2 - 14y^3$

42. $5a^2b + 20ab^2 - 25b^3$

43. $3s^2t^2 + 18st^3 - 48t^4$

44. $4h^2k^2 - 12hk^3 - 160k^4$

Factor each trinomial, if possible. See Examples 5–7 or Example 11.

45. $5x^2 + 13x + 6$

46. $5x^2 + 18x + 9$

47. $7a^2 + 12a + 5$

48. $7a^2 + 36a + 5$

49. $3r^2 + 13r - 10$

50. $3r^2 - r - 10$

51. $11y^2 + 32y - 3$

52. $2y^2 - 9y - 18$

53. $2t^2 + 9t - 6$

54. $2t^2 - t - 9$

55. $13a^2 + 2t + 1$

56. $7z^2 + 8z + 2$

57. $8a^2 - 18ay + 7y^2$

58. $6n^2 - 17ny + 5y^2$

59. $7g^2 - 12gh - 4h^2$

60. $9q^2 + 25qt - 6t^2$

Factor each trinomial. Factor out the opposite of the GCF first. See Example 8 or Example 12.

61. $-18p^2 + 14pq + 4q^2$

62. $-6m^2 + 3mn + 3n^2$

63. $-30x^2 + 25xy + 20y^2$

64. $-72y^2 + 12yz + 40z^2$

Factor each trinomial. Factor out the GCF first. See Example 9 or Example 12.

65. $9a^2b^4 + 15a^2b^2 + 4a^2$

66. $9d^2h^4 + 37d^2h^2 + 4d^2$

67. $8b^4c^8 + 14b^2c^8 - 15c^8$

68. $8t^5u^4 - 30t^5u^2 - 27t^5$

69. $10b^6 - 19b^4 + 6b^2$

70. $12x^6 - 4x^4 - 21x^2$

71. $9m^7 + 6m^5 - 8m^3$

72. $15s^7 + 19s^5 - 10s^3$

Use a substitution to help factor each expression. See Example 10.

73. $(x + a)^2 + 2(x + a) + 1$

74. $(a + b)^2 - 2(a + b) + 1$

75. $(a + b)^2 - 2(a + b) - 24$

76. $(x - y)^2 + 3(x - y) - 10$

77. $14(q - r)^2 - 17(q - r) - 6$

78. $8(h + s)^2 + 34(h + s) + 35$

79. $16(s + t)^2 - 6(s + t) - 27$

80. $15(w + z)^2 - 31(w + z) + 10$

TRY IT YOURSELF

Factor each expression, if possible. Factor out any GCF first (including −1 if the leading coefficient is negative).

81. $32 - a^2 + 4a$

82. $15 - x^2 - 2x$

83. $6z^2 + 17z + 12$

84. $10x^2 + 19x + 6$

85. $25y^2 - 10y + 1$

86. $49m^2 - 14m + 1$

87. $64h^6 + 24h^5 - 4h^4$

88. $27x^2yz + 90xyz - 72yz$

89. $5x^2 + 4x + 1$

90. $3 + 4a^2 + 20a$

91. $b^4x^2 - 12b^2x^2 + 35x^2$

92. $c^3x^4 + 11c^3x^2 - 42c^3$

93. $-3a^2 + ab + 2b^2$

94. $-2x^2 + 3xy + 5y^2$

95. $56a^2 + 42a - 70$ **96.** $150b^2 + 40b - 20$

97. $6(t + w)^2 + 11(t + w) - 10$

98. $4(x - y)^2 + 13(x - y) + 3$

99. $12y^6 + 23y^3 + 10$ **100.** $5m^8 + 29m^4 - 42$

101. $6a^2(m + n) + 13a(m + n) - 15(m + n)$

102. $15n^2(q - r) - 17n(q - r) - 18(q - r)$

103. $20a^2 - 60ab + 45b^2$ **104.** $-12x^2 - 27 + 36x$

105. $-3a^2x^2 + 15a^2x - 18a^2$ **106.** $-2bc^2y^2 - 16bc^2y + 40bc^2$

107. $25m^8 - 60m^4n + 36n^2$ **108.** $49s^6 + 84s^3n^2 + 36n^4$

Look Alikes . . .

Factor the polynomial in part a. Then use your answer to part a to find the remaining factorizations. (No new work is necessary!)

109. a. $x^2 - 5x + 6$ **110. a.** $2x^2 + 3x - 9$

b. $x^4 - 5x^2 + 6$ **b.** $2x^4 + 3x^2 - 9$

c. $x^6 - 5x^3 + 6$ **c.** $2x^6 + 3x^3 - 9$

111. a. $3x^2 - 11x + 8$ **112. a.** $3a^2 + 12a + 12$

b. $3x^2 - 11xy + 8y^2$ **b.** $3a^2 + 12ab + 12b^2$

APPLICATIONS

113. Checkers.   The area of a square checkerboard is represented by the polynomial $25x^2 - 40x + 16$. Use factoring to find an expression that represents the length of a side.

114. Ice.   The surface area of a cubical block of ice is represented by the polynomial $6x^2 + 36x + 54$. Use factoring to find an expression that represents the length of an edge of the block. (*Hint:* The surface area of a cube is 6 times the surface area of one face.)

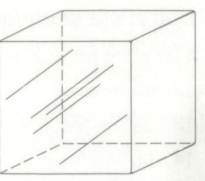

WRITING

115. Explain the error.

Factor:

$$2x^2 - 4x - 6 = \cancel{(2x + 2)(x - 3)}$$

116. How do you know when a polynomial has been factored completely?

117. How was substitution used in this section?

118. How does one determine whether a trinomial is a perfect-square trinomial?

REVIEW

119. If $f(x) = |2x - 1|$, find $f(-2)$.

120. If $g(x) = 2x^2 - 1$, find $g(-2)$.

121. Let $f(x) = -2x + 5$. For what value of x does $f(x) = -7$?

122. Let $f(x) = \frac{3}{2}x - 2$. For what value of x does $f(x) = \frac{2}{3}$?

CHALLENGE PROBLEMS

123. What are the only integer values of b for which $9m^2 + bm - 1$ can be factored?

124. Find the missing factors:
$17y^2 + 1,496y - 11,305 = ?(y - 7)?$

Factor. Assume that n is a natural number.

125. $x^{2n} + 2x^n + 1$ **126.** $2a^{6n} - 3a^{3n} - 2$

127. $x^{4n} + 2x^{2n}y^{2n} + y^{4n}$ **128.** $6x^{2n} + 7x^n - 3$

SECTION 5.7

OBJECTIVES

1 Factor the difference of two squares.

2 Factor the sum and difference of two cubes.

The Difference of Two Squares; the Sum and Difference of Two Cubes

ARE YOU READY?

The following problems review some basic skills that are needed when factoring certain types of binomials.

1. Multiply: $(n + 9)(n - 9)$

2. Simplify: $(8d)^2$

3. Evaluate: **a.** 3^3 **b.** 6^3

4. Simplify: $(4a)^3$

5. Multiply: $(b + 4)(b^2 - 4b + 16)$

6. Multiply: $(2a - 1)(4a^2 + 2a + 1)$

We will now discuss some special rules of factoring. These rules are applied to polynomials that can be written as the difference of two squares or as the sum or difference of two cubes. To use these factoring methods, we must first be able to recognize such polynomials. We begin with a discussion that will help you recognize polynomials with terms that are *perfect squares*.

1 Factor the Difference of Two Squares.

The Language of Algebra

The expression $A^2 - B^2$ is a **difference of two squares,** whereas $(A - B)^2$ is the **square of a difference.** They are not equivalent because $(A - B)^2 \neq A^2 - B^2$.

Recall the special-product rule for multiplying the sum and difference of the same two terms:

$$(A + B)(A - B) = A^2 - B^2$$

The binomial $A^2 - B^2$ is called a **difference of two squares,** because A^2 is the square of A and B^2 is the square of B. If we reverse this rule, we obtain a method for factoring a difference of two squares.

Factoring a Difference of Two Squares

To factor the square of a First quantity minus the square of a Last quantity, multiply the First plus the Last by the First minus the Last.

$$F^2 - L^2 = (F + L)(F - L)$$

To factor the difference of two squares, it is helpful to know the first twenty **perfect-square integers.** The number 400, for example, is a perfect square, because $400 = 20^2$.

$1 = 1^2$	$25 = 5^2$	$81 = 9^2$	$169 = 13^2$	$289 = 17^2$
$4 = 2^2$	$36 = 6^2$	$100 = 10^2$	$196 = 14^2$	$324 = 18^2$
$9 = 3^2$	$49 = 7^2$	$121 = 11^2$	$225 = 15^2$	$361 = 19^2$
$16 = 4^2$	$64 = 8^2$	$144 = 12^2$	$256 = 16^2$	$400 = 20^2$

EXAMPLE 1

Factor: $49x^2 - 16$

Strategy   The terms of this binomial do not have a common factor (other than 1). The only option is to attempt to factor it as a difference of two squares.

Why   If a binomial is a difference of two squares, we can factor it using a special product.

Solution $49x^2 - 16$ is the difference of two squares because it can be written as $(7x)^2 - (4)^2$. We can match it to the rule for factoring a difference of two squares to find the factorization.

$$F^2 - L^2 = (F + L)(F - L)$$
$$(7x)^2 - (4)^2 = (7x + 4)(7x - 4)$$

Therefore, $49x^2 - 16 = (7x + 4)(7x - 4)$. We can verify this result using multiplication.

$$(7x + 4)(7x - 4) = 49x^2 - 28x + 28x - 16$$
$$= 49x^2 - 16 \quad \text{\color{red}This is the original binomial.}$$

Self Check 1 Factor: $81p^2 - 25$

Now Try ▶ Problems 11 and 17

Expressions such as a^4 and x^6y^8 are also perfect squares, because they can be written as the square of another quantity:

$$a^4 = (a^2)^2, \qquad 81b^4 = (9b^2)^2, \qquad \text{and} \qquad x^6y^8 = (x^3y^4)^2$$

EXAMPLE 2 Factor: $64a^4 - 25b^2$

Strategy The terms of this binomial do not have a common factor (other than 1). The only option is to attempt to factor it as a difference of two squares.

Why If a binomial is a difference of two squares, we can factor it using a special-product rule.

Solution We can write $64a^4 - 25b^2$ in the form $(8a^2)^2 - (5b)^2$ and use the rule for factoring the difference of two squares.

$$\mathbf{F}^2 \ - \ \mathbf{L}^2 \ = (\ \mathbf{F} \ + \ \mathbf{L})(\ \mathbf{F} \ - \ \mathbf{L})$$
$$(\mathbf{8a^2})^2 - (\mathbf{5b})^2 = (\mathbf{8a^2 + 5b})(\mathbf{8a^2 - 5b})$$

Therefore, $64a^4 - 25b^2 = (8a^2 + 5b)(8a^2 - 5b)$.

Self Check 2 Factor: $36r^4 - s^2$

Now Try ▶ Problems 21 and 25

EXAMPLE 3 Factor: $x^4 - 1$

Strategy The terms of $x^4 - 1$ do not have a common factor (other than 1). To factor this binomial, we will write it in a form that shows it is a difference of two squares.

Why We can then use a special-product rule to factor it.

Solution Because the binomial is the difference of the squares of x^2 and 1, it factors into the sum of x^2 and 1 and the difference of x^2 and 1.

$$x^4 - 1 = (\mathbf{x^2})^2 - (\mathbf{1})^2$$
$$= \underline{(\mathbf{x^2 + 1})}(\mathbf{x^2 - 1})$$

\qquad\qquad\qquad\qquad {\color{red}A prime polynomial}

The factor $x^2 + 1$ is the sum of two quantities and is prime. However, the factor $x^2 - 1$ is the difference of two squares and can be factored as $(x + 1)(x - 1)$. Thus,

$$x^4 - 1 = (x^2 + 1)(\mathbf{x^2 - 1})$$
$$= (x^2 + 1)(\mathbf{x + 1})(\mathbf{x - 1}) \quad \text{\color{red}Don't forget to write $(x^2 + 1)$.}$$

> **Self Check 3** Factor: $a^4 - 81$
>
> **Now Try** ▶ Problem 29

At times, a **substitution** can be helpful in simplifying the factorization process.

EXAMPLE 4 Factor: $(x + y)^4 - z^4$

Strategy We will use a substitution to factor this difference of two squares.

Why For more complicated expressions, especially those involving a quantity within parentheses, a substitution often helps simplify the factoring process.

Solution If we use the substitution $a = x + y$, we obtain

$$(x + y)^4 - z^4 = a^4 - z^4 \qquad \text{Replace } x + y \text{ with } a.$$
$$= (a^2 + z^2)(a^2 - z^2) \qquad \text{Factor the difference of two squares.}$$
$$= (a^2 + z^2)(a + z)(a - z) \qquad \text{Factor } a^2 - z^2.\ a^2 + z^2 \text{ is prime.}$$

To find the factorization of $(x + y)^4 - z^4$, we "reverse" substitute $x + y$ for each a in the expression $(a^2 + z^2)(a + z)(a - z)$.

$$(a^2 + z^2)(a + z)(a - z) = [(x + y)^2 + z^2](x + y + z)(x + y - z)$$

Thus, $(x + y)^4 - z^4 = [(x + y)^2 + z^2](x + y + z)(x + y - z)$.

If we square the binomial within the brackets, we have

$$(x + y)^4 - z^4 = [x^2 + 2xy + y^2 + z^2](x + y + z)(x + y - z)$$

> **Caution**
>
> A common error after making a substitution is to forget to "undo" or "reverse" it. In this example, if you are to factor an expression in x, y, and z, your answer must involve only x, y, and z. It should not contain the variable a.

> **Caution**
>
> When factoring a polynomial, be sure to factor it completely. Always check to see whether any of the factors of your result can be factored further.

> **Self Check 4** Factor: $(a - b)^4 - c^4$
>
> **Now Try** ▶ Problem 35

When possible, we always factor out a common factor before factoring the difference of two squares. The factoring process is easier when all common factors are factored out first.

EXAMPLE 5 Factor: $2x^4y - 32y$

Strategy We will factor out the GCF of $2y$ and factor the resulting difference of two squares.

Why The first step in factoring any polynomial is to factor out the GCF.

Solution

$$2x^4y - 32y = 2y(x^4 - 16) \qquad \text{Factor out the GCF, which is } 2y.$$
$$= 2y(x^2 + 4)(x^2 - 4) \qquad \text{Factor } x^4 - 16.$$
$$= 2y(x^2 + 4)(x + 2)(x - 2) \qquad \text{Factor } x^2 - 4.\ x^2 + 4 \text{ is prime.}$$

> **Success Tip**
>
> Remember that a **difference of two squares** is a binomial. Each term is a square and the terms have different signs. The powers of the variables in the terms must be even.

> **Self Check 5** Factor: $3a^4 - 3$
>
> **Now Try** ▶ Problems 37 and 43

To factor some expressions, we need to use some creative grouping to begin the process.

EXAMPLE 6   Factor: **a.** $x^2 - y^2 + x - y$ **b.** $x^2 + 6x + 9 - z^2$

Strategy   The terms of each expression do not have a common factor (other than 1) and traditional factoring by grouping will not work. Instead, in part a, we will group only the first two terms of the polynomial and in part b we will group the first three terms.

Why   Hopefully, those steps will produce equivalent expressions that can be factored.

Solution   We group the first two terms, factor them as a difference of two squares, and look for a common factor.

a. $\boxed{x^2 - y^2} + x - y = \boxed{(x + y)(x - y)} + (x - y)$ Factor $x^2 - y^2$. The terms of the resulting expression have a common binomial factor, $x - y$, that can be factored out later.

$= (x + y)(x - y) + \mathbf{1}(x - y)$ Factor out 1 from $x - y$.

$= (x - y)(x + y + 1)$ Factor out the GCF, $x - y$. Don't forget the 1.

b. We group the first three terms and factor that trinomial to get:

$\boxed{x^2 + 6x + 9} - z^2 = \mathbf{(x + 3)(x + 3)} - z^2$ $x^2 + 6x + 9$ is a perfect-square trinomial.

$= (x + 3)^2 - z^2$ Write $(x + 3)(x + 3)$ as $(x + 3)^2$. The expression that results is a difference of two squares.

$= (x + 3 + z)(x + 3 - z)$ Factor the difference of two squares.

> **Success Tip**
>
> We could use a substitution to factor $(x + 3)^2 - z^2$.

Self Check 6   Factor: **a.** $a^2 - b^2 + a + b$
b. $a^2 + 4a + 4 - b^2$

Now Try ▶ Problems 45 and 49

2 Factor the Sum and Difference of Two Cubes.

The number 64 is called a perfect cube, because $4^3 = 64$. To factor the sum or difference of two cubes, it is helpful to know the first ten **perfect-cube integers:**

$1 = 1^3$	$27 = 3^3$	$125 = 5^3$	$343 = 7^3$	$729 = 9^3$
$8 = 2^3$	$64 = 4^3$	$216 = 6^3$	$512 = 8^3$	$1,000 = 10^3$

Expressions such as b^6 and $64x^9y^{12}$ are also perfect cubes, because they can be written as the cube of another quantity:

$$b^6 = (b^2)^3 \qquad \text{and} \qquad 64x^9y^{12} = (4x^3y^4)^3$$

To find rules for factoring the sum of two cubes and the difference of two cubes, we need to find the products shown below. Note that each term of the trinomial is multiplied by each term of the binomial.

$$(x + y)(x^2 - xy + y^2) = x^3 - x^2y + xy^2 + x^2y - xy^2 + y^3$$
$$= x^3 + y^3 \qquad \text{Combine like terms.}$$

$$(x - y)(x^2 + xy + y^2) = x^3 + x^2y + xy^2 - x^2y - xy^2 - y^3$$
$$= x^3 - y^3 \qquad \text{Combine like terms.}$$

We have found that

$$x^3 + y^3 = (x + y)(x^2 - xy + y^2) \qquad \text{and} \qquad x^3 - y^3 = (x - y)(x^2 + xy + y^2)$$

The binomial $x^3 + y^3$ is called the **sum of two cubes,** because x^3 represents the cube of x, y^3 represents the cube of y, and $x^3 + y^3$ represents the sum of these cubes. Similarly, $x^3 - y^3$ is called the **difference of two cubes.**

These results justify the rules for factoring the **sum and difference of two cubes.** They are easier to remember if we think of a sum (or a difference) of two cubes as the cube of a **F**irst quantity plus (or minus) the cube of the **L**ast quantity.

The Language of Algebra

The expression $x^3 + y^3$ is a **sum of two cubes,** whereas $(x + y)^3$ is the **cube of a sum.** If you expand $(x + y)^3$, you will see that they are not equivalent.

Factoring the Sum and Difference of Two Cubes

To factor the cube of a First quantity plus the cube of a Last quantity, multiply the First plus the Last by the First squared, minus the First times the Last, plus the Last squared.

$$F^3 + L^3 = (F + L)(F^2 - FL + L^2)$$

To factor the cube of a First quantity minus the cube of a Last quantity, multiply the First minus the Last by the First squared, plus the First times the Last, plus the Last squared.

$$F^3 - L^3 = (F - L)(F^2 + FL + L^2)$$

EXAMPLE 7

Factor: $a^3 + 8$

Strategy We will write the binomial in a form that shows it is the sum of two cubes.

Why We can then use the rule for factoring the sum of two cubes.

Solution Since $a^3 + 8$ can be written as $a^3 + 2^3$, it is a sum of two cubes, which factors as follows:

$$\mathbf{F}^3 + \mathbf{L}^3 = (\mathbf{F} + \mathbf{L})(\mathbf{F}^2 - \mathbf{FL} + \mathbf{L}^2)$$
$$a^3 + 2^3 = (a + 2)(a^2 - a2 + 2^2)$$
$$= (a + 2)(a^2 - 2a + 4) \quad \cdot \quad a^2 - 2a + 4 \text{ does not factor, it is prime.}$$

Therefore, $a^3 + 8 = (a + 2)(a^2 - 2a + 4)$.

We can check by multiplying.

$$(a + 2)(a^2 - 2a + 4) = a^3 - 2a^2 + 4a + 2a^2 - 4a + 8$$
$$= a^3 + 8 \quad \text{This is the original binomial.}$$

Caution

In Example 7, a common error is to try to factor $a^2 - 2a + 4$. It is not a perfect-square trinomial, because the middle term needs to be $-4a$. Furthermore, it cannot be factored by any of the methods of Section 5.6. It is prime.

Self Check 7 Factor: $p^3 + 27$

Now Try ▶ Problem 53

You should memorize the rules for factoring the sum and the difference of two cubes. Note that each has the form

(a binomial)(a trinomial)

and that there is a relationship between the signs that appear in these forms.

Success Tip

An easy way to remember the sign patterns within the binomial and trinomial of these factoring forms is with the word "SOAP."

Same, **O**pposite, **A**lways, **P**lus

The Sum of Cubes

The same sign

$$F^3 + L^3 = (F + L)(F^2 - FL + L^2)$$

Opposite signs   Always plus

The Difference of Cubes

The same sign

$$F^3 - L^3 = (F - L)(F^2 + FL + L^2)$$

Opposite signs   Always plus

EXAMPLE 8   Factor: $27a^3 - 64b^6$

Strategy   We will write the binomial in a form that shows it is the difference of two cubes.

Why   We can then use the rule for factoring the difference of two cubes.

Solution   Since $27a^3 - 64b^6$ can be written as $(3a)^3 - (4b^2)^3$, it is a difference of two cubes, which factors as follows:

$$F^3 - L^3 = (F - L)(F^2 + F\ L + L^2)$$
$$(3a)^3 - (4b^2)^3 = (3a - 4b^2)[(3a)^2 + (3a)(4b^2) + (4b^2)^2]$$
$$= (3a - 4b^2)(9a^2 + 12ab^2 + 16b^4)$$

Thus, $27a^3 - 64b^6 = (3a - 4b^2)(9a^2 + 12ab^2 + 16b^4)$.   Multiply to check.

Self Check 8   Factor: $8c^6 - 125d^3$

Now Try ▶ Problem 57

EXAMPLE 9   Factor: $a^3 - (c + d)^3$

Strategy   To factor this expression, we will use the rule for factoring the difference of two cubes.

Why   The terms a^3 and $(c + d)^3$ are perfect cubes.

Solution
$$F^3 - L^3 = (F - L)(F^2 + F\ L + L^2)$$
$$a^3 - (c + d)^3 = [a - (c + d)][a^2 + a(c + d) + (c + d)^2]$$

Success Tip

We could also use the substitution $x = c + d$, factor $a^3 - x^3$ as $(a - x)(a^2 + ax + x^2)$, and then replace each x with $c + d$.

Now we simplify the expressions within both sets of brackets. Thus,

$$a^3 - (c + d)^3 = (a - c - d)(a^2 + ac + ad + c^2 + 2cd + d^2)$$

Self Check 9   Factor: $(p + q)^3 - r^3$

Now Try ▶ Problem 61

EXAMPLE 10   Factor: $x^6 - 64$

Strategy   This binomial is both the difference of two squares and the difference of two cubes. We will write it in a form that shows it is a difference of two squares to begin the factoring process.

Why   It is easier to factor it as the difference of two squares first.

Solution
$$x^6 - 64 = (x^3)^2 - 8^2$$
$$= (x^3 + 8)(x^3 - 8)$$

Each of these binomial factors can be factored further. The first is the sum of two cubes and the second is the difference of two cubes. Thus,

$$x^6 - 64 = (x + 2)(x^2 - 2x + 4)(x - 2)(x^2 + 2x + 4)$$

Self Check 10   Factor: $1 - x^6$

Now Try ▶ Problem 65

EXAMPLE 11   Factor: $2a^5 + 250a^2$

Strategy   We will factor out the GCF $2a^2$ and factor the resulting sum of two cubes.

Why   The first step in factoring any polynomial is to factor out the GCF.

Solution   We first factor out the common factor $2a^2$ to obtain

$$2a^5 + 250a^2 = 2a^2(a^3 + 125)$$

Then we factor $a^3 + 125$ as the sum of two cubes to obtain

$$2a^5 + 250a^2 = 2a^2(a + 5)(a^2 - 5a + 25) \qquad a^2 - 5a + 25 \text{ is prime.}$$

Self Check 11   Factor: $3x^5 + 24x^2$

Now Try ▶   Problem 69

SECTION 5.7   STUDY SET

VOCABULARY

Fill in the blanks.

1. When the polynomial $4x^2 - 25$ is written as $(2x)^2 - (5)^2$, we see that it is the difference of two _____.

2. When the polynomial $8x^3 + 125$ is written as $(2x)^3 + (5)^3$, we see that it is the sum of two _____.

CONCEPTS

3. a. Write the first ten perfect-square integers.

    b. Write the first ten perfect-cube integers.

4. a. Use multiplication to verify that the sum of two squares $x^2 + 25$ does not factor as $(x + 5)(x + 5)$.

    b. Use multiplication to verify that the difference of two squares $x^2 - 25$ factors as $(x + 5)(x - 5)$.

5. Complete each factorization.

    a. $F^2 - L^2 = (F + L)()$

    b. $F^3 + L^3 = (F + L)()$

    c. $F^3 - L^3 = (F - L)()$

6. Factor each binomial.

    a. $5p^2 + 20$

    b. $5p^2 - 20$

    c. $5p^3 + 20$

    d. $5p^3 + 40$

NOTATION

7. Give an example of each.

    a. a difference of two squares

    b. a square of a difference

    c. a sum of two squares

    d. a sum of two cubes

    e. a cube of a sum

8. Fill in the blanks.

    a. $36y^2 - 49m^4 = ()^2 - ()^2$

    b. $125h^3 - 27k^6 = ()^3 - ()^3$

GUIDED PRACTICE

Factor. **See Example 1.**

9. $x^2 - 16$

10. $y^2 - 49$

11. $9y^2 - 64$

12. $16x^2 - 81$

13. $144 - c^2$

14. $25 - t^2$

15. $100m^2 - 1$

16. $144x^2 - 1$

17. $81a^2 - 49b^2$

18. $64r^2 - 121s^2$

19. $x^2 + 25$

20. $a^2 + 36$

Factor each difference of two squares. **See Example 2.**

21. $9r^4 - 121s^2$

22. $81a^4 - 16b^2$

23. $16t^2 - 25w^4$

24. $9r^2 - 25s^4$

25. $100r^2s^4 - t^4$

26. $400x^2z^4 - a^4$

27. $36x^4y^2 - 49z^6$

28. $4a^2b^4 - 9d^6$

Factor completely. **See Example 3.**

29. $x^4 - y^4$

30. $16n^4 - 1$

31. $16a^4 - 81b^4$

32. $81m^4 - 256n^4$

Factor. See Example 4.

33. $(x + y)^2 - z^2$

34. $a^2 - (b - c)^2$

35. $(r - s)^2 - t^4$

36. $(m + n)^2 - p^4$

Factor each expression. Factor out any GCF first. See Example 5.

37. $2x^2 - 288$

38. $8x^2 - 72$

39. $3x^3 - 243x$

40. $2x^3 - 32x$

41. $5ab^4 - 5a$

42. $3ac^4 - 243a$

43. $64b - 4b^5$

44. $1{,}250n - 2n^5$

Factor by first grouping the appropriate terms. See Example 6.

45. $c^2 - d^2 + c + d$

46. $s^2 - t^2 + s - t$

47. $a^2 - b^2 + 2a - 2b$

48. $m^2 - n^2 + 3m + 3n$

49. $x^2 + 12x + 36 - y^2$

50. $x^2 - 6x + 9 - 4y^2$

51. $x^2 - 2x + 1 - 9z^2$

52. $x^2 + 10x + 25 - 16z^2$

Factor each sum of cubes. See Example 7.

53. $a^3 + 125$

54. $b^3 + 64$

55. $8r^3 + s^3$

56. $27t^3 + u^3$

Factor each difference of cubes. See Example 8.

57. $64t^6 - 27v^3$

58. $125m^3 - x^6$

59. $x^3 - 216y^6$

60. $8c^6 - 343w^3$

Factor. See Example 9.

61. $(a - b)^3 + 27$

62. $(b - c)^3 - 1{,}000$

63. $64 - (a + b)^3$

64. $1 - (x + y)^3$

Factor each expression completely. Factor a difference of two squares first. See Example 10.

65. $x^6 - 1$

66. $x^6 - y^6$

67. $x^{12} - y^6$

68. $a^{12} - 64$

Factor each sum or difference of cubes. Factor out the GCF first. See Example 11.

69. $5x^3 + 625$

70. $2x^3 - 128$

71. $4x^5 - 256x^2$

72. $2x^6 + 54x^3$

TRY IT YOURSELF

Factor each expression.

73. $64a^3 - 125b^6$

74. $8x^6 - 27y^3$

75. $288b^2 - 2b^6$

76. $98x - 2x^5$

77. $x^2 - y^2 + 8x + 8y$

78. $5m - 5n + m^2 - n^2$

79. $x^9 + y^9$

80. $x^6 + y^6$

81. $144a^2t^2 - 169b^6$

82. $25x^6 - 81y^2z^2$

83. $100a^2 + 9b^2$

84. $25s^4 + 16t^2$

85. $81c^4d^4 - 16t^4$

86. $256x^4 - 81y^4$

87. $128u^2v^3 - 2t^3u^2$

88. $56rs^2t^3 + 7rs^2v^6$

89. $y^2 - (2x - t)^2$

90. $(15 - r)^2 - s^2$

91. $x^2 + 20x + 100 - 9z^2$

92. $49a^2 - b^2 - 14b - 49$

93. $(c - d)^3 + 216$

94. $1 - (x + y)^3$

95. $\dfrac{1}{36} - y^4$

96. $\dfrac{4}{81} - m^4$

97. $m^6 - 64$

98. $1 - y^6$

99. $(a + b)x^3 + 27(a + b)$

100. $(c - d)r^3 - (c - d)s^3$

101. $x^9 - y^{12}z^{15}$

102. $r^{12} + s^{18}t^{24}$

Look Alikes . . .

103. a. $q^2 - 64$

b. $q^3 - 64$

104. a. $a^2 - b^2$

b. $a^3 - b^3$

105. a. $d^2 - 25$

b. $d^3 - 125$

106. a. $m^2 + 27$

b. $m^3 + 27$

Factor the expression in part a. Then use your answer from part a to give the factorization of the expression in part b. (No new work is necessary!)

107. a. $a^6 - b^3$

b. $a^6 + b^3$

108. a. $c^3 - \dfrac{1}{8}$

b. $c^3 + \dfrac{1}{8}$

109. a. $125m^3 + 8n^3$ **b.** $125m^3 - 8n^3$

110. a. $w^3 + 0.001$ **b.** $w^3 - 0.001$

APPLICATIONS

111. Candy. To find the amount of chocolate used in the outer coating of the malted-milk ball shown, we can find the volume V of the chocolate shell using the formula

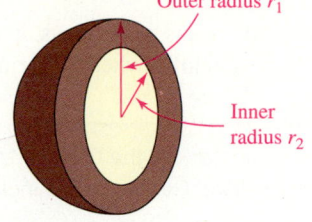
Outer radius r_1
Inner radius r_2

$$V = \frac{4}{3}\pi r_1^3 - \frac{4}{3}\pi r_2^3$$

Factor the expression on the right side of the formula.

112. Movie Stunts. The function that gives the distance a stuntwoman is above the ground t seconds after she falls over the side of a 144-foot tall building is $h(t) = 144 - 16t^2$. Factor the right side.

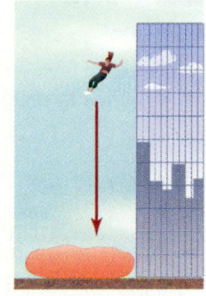

WRITING

113. Explain how the patterns used to factor the sum and difference of two cubes are similar and how they differ.

114. Explain why the factorization is not complete.

Factor: $1 - t^8 = (1 + t^4)(1 - t^4)$

115. Explain the error.

Factor: $4g^2 - 16 = \cancel{(2g + 4)(2g - 4)}$

116. When asked to factor $81t^2 - 16$, one student answered $(9t - 4)(9t + 4)$, and another answered $(9t + 4)(9t - 4)$. Explain why both students are correct.

REVIEW

Graph the line with the given characteristics.

117. Passing through $(-2, -1)$; slope $-\dfrac{2}{3}$

118. y-intercept $(0, -4)$; slope 3

119. Write the equation of line l shown on the right.

120. Write the equation of line r shown on the right.

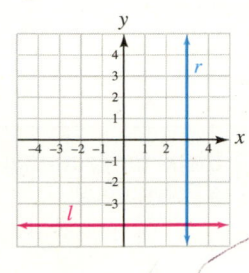

CHALLENGE PROBLEMS

Factor. Assume all variables represent natural numbers.

121. $4x^{2n} - 9y^{2n}$

122. $25 - x^{6n}$

123. $a^{3b} - c^{3b}$

124. $27x^{3n} + y^{3n}$

125. Factor: $x^{32} - y^{32}$

126. Find the error in this proof that $2 = 1$.

$$x = y$$
$$x^2 = xy$$
$$x^2 - y^2 = xy - y^2$$
$$(x + y)(x - y) = y(x - y)$$
$$\frac{(x + y)\cancel{(x - y)}}{\cancel{(x - y)}} = \frac{y\cancel{(x - y)}}{\cancel{(x - y)}}$$
$$x + y = y$$
$$y + y = y$$
$$2y = y$$
$$\frac{2y}{y} = \frac{y}{y}$$
$$2 = 1$$

SECTION 5.8

Summary of Factoring Techniques

OBJECTIVE

1 Factor random polynomials.

ARE YOU READY?

The following problems review some basic skills that are needed when factoring polynomials.

1. How many terms does each expression have?
  a. $12x^2y^2z^3 - 2xy^2z^3 - 4y^2z^3$ **b.** $x^3 + 5x^2 + 6x + x^2y + 5xy + 6y$

2. Do any of the factors in $(x^3 + y^3)(x^2 - y^2)$ factor further?

3. Multiply: $3a(a + 5)(a - 14)$

4. What is the greatest common factor of the terms of $32m^3n^2 - 24mn^3$?

The factoring methods discussed so far will be used in the remaining chapters to simplify expressions and solve equations. In such cases, we must determine the factoring method—it will not be specified. This section will give you practice in selecting the appropriate factoring method to use given a randomly chosen polynomial.

1 Factor Random Polynomials.

Recall that **to factor a polynomial** means to express it as a product of two (or more) polynomials. The following strategy is helpful when factoring polynomials.

Steps for Factoring a Polynomial	1. Is there a common factor? If so, factor out the GCF, or the opposite of the GCF, so that the leading coefficient is positive. Remember to include it in your final answer.
	2. How many terms does the polynomial have?
	If it has **two terms,** look for the following problem types:
	a. The difference of two squares
	b. The sum of two cubes
	c. The difference of two cubes
	If it has **three terms,** look for the following problem types:
	a. A perfect-square trinomial
	b. If the trinomial is not a perfect square, use the trial-and-check method or the grouping method.
	If it has **four or more terms,** try to factor by grouping.
	3. Can any factors be factored further? If so, factor them completely.
	4. Does the factorization check? Check by multiplying.

For more complicated expressions, a substitution sometimes helps to simplify the factoring process.

EXAMPLE 1 Factor: $12x^2y^2z^3 - 2xy^2z^3 - 4y^2z^3$

Strategy We will answer the four questions listed in the *Steps for Factoring a Polynomial.*

Why The answers will help us determine which factoring techniques to use.

Solution *Is there a common factor?* Yes. Factor out the greatest common factor $2y^2z^3$.

$$12x^2y^2z^3 - 2xy^2z^3 - 4y^2z^3 = 2y^2z^3(6x^2 - x - 2)$$

> **The Language of Algebra**
>
> Remember that the instruction to factor means to **factor completely**. A polynomial is factored completely when no factor can be factored further.

How many terms does it have? The polynomial within the parentheses has three terms. We can factor $6x^2 - x - 2$, using the trial-and-check method or the key number method, to get

$$12x^2y^2z^3 - 2xy^2z^3 - 4y^2z^3 = 2y^2z^3(6x^2 - x - 2)$$
$$= 2y^2z^3(3x - 2)(2x + 1)$$

Factor the trinomial. Don't forget to write the GCF from the first step.

Is it factored completely? Yes. Since each of the individual factors is prime, the factorization is complete.

Does it check? To check, we multiply.

$$2y^2z^3(3x - 2)(2x + 1) = 2y^2z^3(6x^2 - x - 2)$$
$$= 12x^2y^2z^3 - 2xy^2z^3 - 4y^2z^3$$

Multiply the binomials first.

Distribute the multiplication by $2y^2z^3$.

Since we obtain the original polynomial, the factorization is correct.

Self Check 1 Factor: $30a^2b^3c - 27ab^3c + 6b^3c$

Now Try ▶ Problem 13

EXAMPLE 2   Factor: $48a^4c^3 - 3b^4c^3$

Strategy We will answer the four questions listed in the *Steps for Factoring a Polynomial*.

Why The answers will help us determine which factoring techniques to use.

Solution *Is there a common factor?* Yes. Factor out the greatest common factor $3c^3$.

$$48a^4c^3 - 3b^4c^3 = 3c^3(16a^4 - b^4)$$

How many terms does it have? The polynomial within the parentheses, $16a^4 - b^4$, has two terms. It is the difference of two squares and factors as $(4a^2 + b^2)(4a^2 - b^2)$.

$$48a^4c^3 - 3b^4c^3 = 3c^3(\mathbf{16a^4 - b^4})$$
$$= 3c^3(\mathbf{4a^2 + b^2})(\mathbf{4a^2 - b^2})$$

> **Success Tip**
>
> Recall that if a pair of parentheses is produced in the factoring process, always see if the resulting expression within the parentheses can be factored further.

Is it factored completely? No. The binomial $4a^2 + b^2$ is the sum of two squares and is prime. However, $4a^2 - b^2$ is the difference of two squares and factors as $(2a + b)(2a - b)$.

$$48a^4c^3 - 3b^4c^3 = 3c^3(16a^4 - b^4)$$
$$= 3c^3(4a^2 + b^2)(\mathbf{4a^2 - b^2})$$
$$= 3c^3(4a^2 + b^2)(\mathbf{2a + b})(\mathbf{2a - b})$$

Since each of the individual factors is prime, the factorization is now complete.

Does it check? Multiply to verify that this factorization is correct.

Self Check 2 Factor: $3p^4r^3 - 3q^4r^3$

Now Try ▶ Problem 19

EXAMPLE 3   Factor: $x^5y + x^2y^4 - x^3y^3 - y^6$

Strategy We will answer the four questions listed in the *Steps for Factoring a Polynomial*.

Why The answers will help us determine which factoring techniques to use.

Solution *Is there a common factor?* Yes. Factor out the greatest common factor of y.

$$x^5y + x^2y^4 - x^3y^3 - y^6 = y(x^5 + x^2y^3 - x^3y^2 - y^5)$$

> **Success Tip**
>
> Something as simple as counting the number of terms that a polynomial has is very important when determining how to factor it.

How many terms does it have? The polynomial $x^5 + x^2y^3 - x^3y^2 - y^5$ has four terms. We try factoring by grouping to obtain

$$x^5y + x^2y^4 - x^3y^3 - y^6$$
$$= y(\; x^5 + x^2y^3 \;\; - x^3y^2 - y^5 \;)\qquad \text{Factor out the common factor } y.$$
$$= y[x^2(x^3 + y^3) - y^2(x^3 + y^3)]\qquad \text{Factor by grouping. Use brackets.}$$
$$= y(x^3 + y^3)(x^2 - y^2)\qquad \text{Factor out } x^3 + y^3. \text{ Drop the brackets.}$$

Is it factored completely? No. We can factor $x^3 + y^3$ (the sum of two cubes) and $x^2 - y^2$ (the difference of two squares) to obtain

$$x^5y + x^2y^4 - x^3y^3 - y^6 = y(x + y)(x^2 - xy + y^2)(x + y)(x - y)$$

Because each of the individual factors is prime, the factorization is complete.

Does it check? Multiply to verify that this factorization is correct.

> **Self Check 3**   Factor: $9a^5 - 9a^3b^2 + 9a^2b^3 - 9b^5$
>
> **Now Try** ▶   Problem 31

EXAMPLE 4   Factor: $x^3 + 5x^2 + 6x + x^2y + 5xy + 6y$

Strategy   We will answer the four questions listed in the *Steps for Factoring a Polynomial.*

Why   The answers will help us determine which factoring techniques to use.

Solution   *Is there a common factor?* No. There is no common factor (other than 1).

How many terms does it have? Since there are more than three terms, we try factoring by grouping. We can factor x from the first three terms and y from the last three terms.

> **The Language of Algebra**
>
> It is important to learn the names of all the factoring methods we have studied. Educational research has found that words play an important role in concept formation.

$$x^3 + 5x^2 + 6x \; + \; x^2y + 5xy + 6y$$
$$= x(x^2 + 5x + 6) + y(x^2 + 5x + 6)$$
$$= (x^2 + 5x + 6)(x + y) \quad \text{Factor out } x^2 + 5x + 6.$$

Is it factored completely? No. We can factor the trinomial $x^2 + 5x + 6$ to obtain

$$x^3 + 5x^2 + 6x + x^2y + 5xy + 6y = (x + 3)(x + 2)(x + y)$$

Since each of the individual factors is prime, the factorization is now complete.

Does it check? Multiply to verify that the factorization is correct.

> **Self Check 4**   Factor: $a^3 - 5a^2 + 6a + a^2b - 5ab + 6b$
>
> **Now Try** ▶   Problem 47

EXAMPLE 5   Factor: $4x^4 + 4x^3 + x^2 + 2x + 1$

Strategy   We will answer the four questions listed in the *Steps for Factoring a Polynomial.*

Why   The answers will help us determine which factoring techniques to use.

Solution   *Is there a common factor?* There is no common factor (other than 1).

How many terms does it have? Since there are more than three terms, we try factoring by grouping. We can factor x^2 from the first three terms, and group the last two terms together.

$$4x^4 + 4x^3 + x^2 \; + \; 2x + 1 = x^2(4x^2 + 4x + 1) + (2x + 1)$$

Is it factored completely? No. We recognize $4x^2 + 4x + 1$ as a perfect-square trinomial, because $4x^2 = (2x)^2$, $1 = (1)^2$, and $4x = 2 \cdot 2x \cdot 1$. Therefore, it factors as $(2x + 1)(2x + 1)$.

$$4x^4 + 4x^3 + x^2 + 2x + 1 = x^2(4x^2 + 4x + 1) + (2x + 1)$$
$$= x^2(2x + 1)(2x + 1) + (2x + 1)$$

Finally, we factor out the common factor $2x + 1$.

$$4x^4 + 4x^3 + x^2 + 2x + 1 = x^2(4x^2 + 4x + 1) + (2x + 1)$$
$$= x^2(2x + 1)(2x + 1) + (2x + 1)$$
$$= (2x + 1)[x^2(2x + 1) + 1]$$
$$= (2x + 1)(2x^3 + x^2 + 1) \quad \text{Within the brackets, distribute the multiplication by } x^2.$$

Since each of the individual factors is prime, the factorization is complete.

Does it check? Multiply to verify that this factorization is correct.

Self Check 5   Factor: $a^4 - a^3 - 2a^2 + a - 2$

Now Try ▶ Problem 59

SECTION 5.8   **STUDY SET**

VOCABULARY

Fill in the blanks.

1. Each factor of a completely factored expression will be _____.
2. When we factor a polynomial, we write a sum of terms as an equivalent product of _____.
3. $x^3 + y^3$ is called a sum of two _____ and $x^3 - y^3$ is called a difference of two _____.
4. $x^2 - y^2$ is called a _____ of two squares.

CONCEPTS

Fill in the blanks.

5. In any factoring problem, always factor out any _____ factors first.
6. If a polynomial has two terms, check to see whether the problem type is the _____ of two squares, the sum of two _____, or the _____ of two cubes.
7. If a polynomial has three terms, try to factor it as a _____.
8. If a polynomial has four or more terms, try factoring it by _____.
9. Explain how to verify that $y^2z^3(x + 6)(x + 1)$ is the factored form of $x^2y^2z^3 + 7xy^2z^3 + 6y^2z^3$.

10. Why is the polynomial $x^2 + 6$ classified as prime?

NOTATION

Complete each factorization.

11. $18a^3b + 3a^2b^2 - 6ab^3 = \boxed{}(6a^2 + ab - 2b^2)$

$\qquad\qquad = 3ab(3a + \boxed{})(\boxed{} - b)$

12. $2x^4 - 1{,}250 = 2(\boxed{})$

$\qquad\qquad = 2(\boxed{})(x^2 - 25)$

$\qquad\qquad = 2(x^2 + 25)(x + 5)(\boxed{})$

TRY IT YOURSELF

Factor each expression completely. If an expression is prime, so indicate.

13. $4a^2bc + 4abc - 120bc$
14. $8x^3y^4 - 27y$
15. $-3x^2y - 6xy^2 + 12xy$
16. $xy - ty + xs^2 - ts^2$

17. $y^3(y^2 - 1) - 27(y^2 - 1)$
18. $b^2c + b^2 + bcd + bd$
19. $36x^4 - 36$
20. $27x^9 - y^3$
21. $16c^2g^2 + h^4$
22. $12x^2 + 14x - 6$
23. $-14x + 8 + 6x^2$
24. $-13m^2 + m^4 + 36$
25. $4x^2y^2 + 4xy^2 + y^2$
26. $x^3 + a^6y^3$
27. $4x^2y^2z^2 - 26x^2y^2z^3$
28. $-2x^3 + 54$
29. $9a^6 - 48a^4 + 64a^2$
30. $-4m^7 + 36m^4 - 81m$
31. $6a^5 - 6a^3b^2 - 6a^2b^3 + 6b^5$

32. $2(x + y)^2 + (x + y) - 3$
33. $(x - y)^3 + 125$
34. $625x^4 - 256y^4$
35. $2(a - b)^2 + 5(a - b) + 3$
36. $5x + 4y + 25x^2 - 16y^2$
37. $6x^2 - 63 - 13x$
38. $a^4b^2 - 20a^2b^2 + 64b^2$
39. $-17x^2 + 16 + x^4$
40. $x^2 + 6x + 9 - y^2$
41. $x^2 + 10x + 25 - y^8$
42. $4x^2 + 4x + 1 - 4y^2$
43. $9x^2 - 6x + 1 - 25y^2$
44. $x^2 - y^2 - 2y - 1$
45. $a^2 - b^2 + 4b - 4$
46. $60q^2r^2s^4 + 78qr^2s^4 - 18r^2s^4$
47. $ax^2 - 2axy + ay^2 - x^2 + 2xy - y^2$
48. $32x^{10} + 48x^9 + 18x^8$
49. $\dfrac{81}{16}x^4 - y^{40}$
50. $\dfrac{d^2x^2}{2} - \dfrac{f^2x^2}{2} - \dfrac{c^2d^2}{2} + \dfrac{c^2f^2}{2}$
51. $16m^{16} - 16$
52. $8(4 - a^2) - x^3(4 - a^2)$
53. $9y^5 + 6y^4 + y^3 + 3y + 1$
54. $25m^4 - 10m^3 + m^2 + 5m - 1$

55. $x^3 - xy^2 - 4x^2 + 4y^2$

56. $m^2n - 9n + 9m^2 - 81$

57. $c^3 - 4a^2c + 4abc - b^2c$

58. $4t^3 - s^2t - 6stz - 9tz^2$

59. $9x^4 + 6x^3 + x^2 + 3x + 1$

60. $z^3(y^2 - 4) + 8(y^2 - 4)$

61. $(2x - 1)^2 + 4(2x - 1) + 4$

62. $(3z + 2)^2 - 12(3z + 2) + 36$

63. $a^2 + b^2 + 25$

64. $4x^3y^3 + 256$

APPLICATIONS

65. **Graphic Design.**   The logo for a marketing firm consists of small green squares (each with side length x inches) positioned within a large square (side length 1 inch). Write a polynomial that represents the number of square inches of white space in the logo and then factor it.

Jones-
Kennedy
INCORPORATED

66. **Hardware.**   Write a polynomial that represents the volume of the metal spacer shown here. Then factor the polynomial.

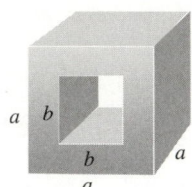

WRITING

67. What is your strategy for factoring a polynomial?

68. For the factorization below, explain why the polynomial is not factored completely.

$$48a^4c^3 - 3b^4c^3 = 3c^3(16a^4 - b^4)$$

REVIEW

Evaluate each determinant.

69. $\begin{vmatrix} -6 & -2 \\ 15 & 4 \end{vmatrix}$

70. $\begin{vmatrix} 3 & -2 \\ 12 & -8 \end{vmatrix}$

71. $\begin{vmatrix} -1 & 2 & 1 \\ 2 & 1 & -3 \\ 1 & 1 & 1 \end{vmatrix}$

72. $\begin{vmatrix} 1 & 0 & 1 \\ 0 & 1 & 0 \\ 1 & 1 & 1 \end{vmatrix}$

CHALLENGE PROBLEMS

73. Factor: $x^4 + x^2 + 1$ (*Hint:* Add and subtract x^2.)

74. Factor: $x^4 + 7x^2 + 16$ (*Hint:* Add and subtract x^2.)

Factor. Assume that n is a natural number.

75. $2a^{2n} + 2a^n - 24$

76. $ma^{2n} - mb^{2n}$

77. $54a^{3n} + 16b^{3n}$

78. $-a^{2n} - 2a^nb^n - b^{2n}$

79. $12m^{4n} + 10m^{2n} + 2$

80. $nx^{4n} + 2nx^{2n}y^{2n} + ny^{4n}$

SECTION 5.9

Solving Equations by Factoring

OBJECTIVES

1. Solve quadratic equations using the zero-factor property.

2. Solve higher-degree polynomial equations by factoring.

3. Use quadratic equations to solve problems.

ARE YOU READY?

The following problems review some basic skills that are needed when solving quadratic equations.

1. Fill in the blank: $8 \cdot \boxed{} = 0$

2. Solve: $x + 4 = 0$

3. Solve: $8x = 0$

4. Factor: $x^2 - x - 6$

5. Factor: $3n^2 - n - 2$

6. Factor: $x^3 - x^2 - 121x + 121$

We have previously solved linear equations in one variable such as $3x - 1 = 5$ and $10y + 9 = y + 4$. These equations are also called **polynomial equations** because they involve two polynomials that are set equal to each other. Some other examples of polynomial equations are

$$3x^2 = -6x, \qquad 6x^3 - x^2 = 2x, \qquad \text{and} \qquad x^4 + 4 - 5x^2 = 0$$

In this section, we will discuss a method for solving polynomial equations like these.

1   Solve Quadratic Equations Using the Zero-Factor Property.

A second-degree polynomial equation in one variable is called a *quadratic equation.*

Quadratic Equations	A **quadratic equation** is an equation that can be written in the **standard form** $$ax^2 + bx + c = 0$$ where a, b, and c represent real numbers and $a \neq 0$.

Here are some examples of quadratic equations. Only the last one is written in standard form.

$$2x^2 + 6x = 15 \qquad y^2 = -16y \qquad 4m^2 - 7m + 2 = 0$$

A **solution of a quadratic equation** is a value of the variable that makes the equation true. To *solve a quadratic equation* means to find all of its solutions. Many quadratic equations can be solved by factoring and using the zero-factor property.

The Zero-Factor Property	When the product of two real numbers is 0, at least one of them is 0. If $ab = 0$, then $a = 0$ or $b = 0$. This property also applies to three or more factors.

EXAMPLE 1   Solve: $x^2 + 5x + 6 = 0$

Strategy   To solve this equation, we will factor the trinomial on the left side and use the zero-factor property.

Why   To use the zero-factor property, we need one side of the equation to be factored completely and the other side to be 0.

Solution
$$x^2 + 5x + 6 = 0 \quad \text{This is the equation to solve.}$$
$$(x + 3)(x + 2) = 0 \quad \text{Factor the left side.}$$

Success Tip

When you see the word *solve* in this example, you probably think of steps from Chapter 1 such as combining like terms, distributing, or doing something to both sides. However, to solve this quadratic equation, we begin by factoring $x^2 + 5x + 6$.

Since the product of $x + 3$ and $x + 2$ is 0, at least one of the factors must be 0. Thus, we can set each factor equal to 0 and solve each resulting linear equation for x:

$$x + 3 = 0 \qquad \text{or} \qquad x + 2 = 0$$
$$x = -3 \qquad | \qquad x = -2$$

To check these solutions, we substitute -3 and -2 for x in the original equation and verify that each number satisfies the equation.

$$
\begin{array}{lll}
\textbf{\textit{Check:}} & x^2 + 5x + 6 = 0 & \qquad \text{or} \qquad x^2 + 5x + 6 = 0 \\
& (-3)^2 + 5(-3) + 6 \stackrel{?}{=} 0 & \qquad (-2)^2 + 5(-2) + 6 \stackrel{?}{=} 0 \\
& 9 - 15 + 6 \stackrel{?}{=} 0 & \qquad 4 - 10 + 6 \stackrel{?}{=} 0 \\
& 0 = 0 \quad \text{True} & \qquad 0 = 0 \quad \text{True}
\end{array}
$$

Both -3 and -2 are solutions, because they satisfy the equation. The solution set is $\{-3, -2\}$.

Self Check 1   Solve: $x^2 - 4x - 45 = 0$

Now Try ▶ Problems 15 and 19

The previous example suggests the following strategy to solve quadratic equations by factoring.

The Factoring Method for Solving a Quadratic Equation

1. Write the equation in standard form: $ax^2 + bx + c = 0$.
2. Factor the polynomial.
3. Use the zero-factor property to set each factor equal to zero.
4. Solve each resulting equation.
5. Check the results in the original equation.

EXAMPLE 2   Solve: **a.** $3x^2 = 6x$ **b.** $9x^2 = 25$

Strategy   To solve each equation, we will get 0 on the right side, factor the resulting binomial on the left side, and use the zero-factor property.

Why   To use the zero-factor property, we need one side of the equation to be factored completely and the other side to be 0.

Solution   **a.** We begin by writing the equation in standard $ax^2 + bx + c = 0$ form.

$$3x^2 = 6x \quad \text{This is the equation to solve.}$$
$$3x^2 - 6x = 0 \quad \text{To get 0 on the right side, subtract 6x from both sides.}$$

To solve the equation, we factor the left side, set each factor equal to 0, and solve each resulting equation for x.

$$3x^2 - 6x = 0$$
$$3x(x - 2) = 0 \quad \text{Factor out the GCF, 3x.}$$
$$3x = 0 \quad \text{or} \quad x - 2 = 0 \quad \text{By the zero-factor property, at least one of the factors must be equal to zero.}$$
$$x = 0 \quad \quad \quad x = 2 \quad \text{Solve each equation.}$$

The solutions are 0 and 2. Check each one in the original equation.

b.
$$9x^2 = 25 \quad \text{This is the equation to solve.}$$
$$9x^2 - 25 = 0 \quad \text{To get 0 on the right side, subtract 25 from both sides.}$$
$$(3x + 5)(3x - 5) = 0 \quad \text{Factor the difference of two squares.}$$
$$3x + 5 = 0 \quad \text{or} \quad 3x - 5 = 0 \quad \text{Set each factor equal to 0.}$$
$$3x = -5 \quad \quad 3x = 5 \quad \text{Solve each equation.}$$
$$x = -\frac{5}{3} \quad \quad x = \frac{5}{3}$$

The solutions are $-\frac{5}{3}$ and $\frac{5}{3}$. Check each one in the original equation.

Self Check 2   Solve: **a.** $4p^2 = 12p$ **b.** $16a^2 = 49$

Now Try ▶ Problems 25 and 29

Caution

A creative, but incorrect, approach is to divide both sides of $3x^2 = 6x$ by $3x$.

$$\frac{3x^2}{3x} = \frac{6x}{3x}$$

You will obtain $x = 2$. However, you will lose the second solution, 0.

Notation

$3x^2 - 6x = 0$ is a quadratic equation in standard form, $ax^2 + bx + c = 0$ where $a = 3$, $b = -6$, and $c = 0$. Although $9x^2 - 25 = 0$ is missing an x-term, it is also a quadratic equation in standard form, where $a = 9$, $b = 0$, and $c = -25$.

EXAMPLE 3   Solve: $x = \frac{6}{5} - \frac{6}{5}x^2$

Strategy   We will multiply both sides of the equation by 5 to clear it of fractions and use factoring to solve the resulting quadratic equation.

Why   It is easier to factor a polynomial that contains integers than one that contains fractions.

Solution   We clear the equation of fractions and proceed as follows:

$$x = \frac{6}{5} - \frac{6}{5}x^2$$ This is the equation to solve.

$$5(x) = 5\left(\frac{6}{5} - \frac{6}{5}x^2\right)$$ Multiply both sides by 5.

$$5x = 5 \cdot \frac{6}{5} - 5 \cdot \frac{6}{5}x^2$$ Distribute the multiplication by 5.

$$5x = 6 - 6x^2$$ Simplify.

To use factoring to solve this quadratic equation, one side of the equation must be 0. Since it is easier to factor a second-degree polynomial if the coefficient of the squared term is positive, we add $6x^2$ to both sides and subtract 6 from both sides to obtain

$$6x^2 + 5x - 6 = 0$$

$$(3x - 2)(2x + 3) = 0$$ Factor the trinomial.

$$3x - 2 = 0 \quad \text{or} \quad 2x + 3 = 0$$ Set each factor equal to 0 and solve for x.

$$3x = 2 \quad\quad\quad 2x = -3$$ Solve each linear equation.

$$x = \frac{2}{3} \quad\quad\quad x = -\frac{3}{2}$$

The solutions are $\frac{2}{3}$ and $-\frac{3}{2}$. Check them in the original equation.

Self Check 3   Solve: $x = \frac{6}{7}x^2 - \frac{3}{7}$

Now Try ▶   Problem 31

The Language of Algebra

Quadratic equations involve the square of a variable, not the fourth power as "quad" might suggest. Why the inconsistency? A closer look at the origin of the word **quadratic** reveals that it comes from the Latin word quadratus, meaning square.

CAUTION   To solve a quadratic equation by factoring, set the quadratic polynomial equal to 0 before factoring and using the zero-factor property. Don't make the following error:

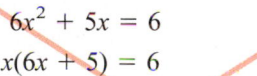

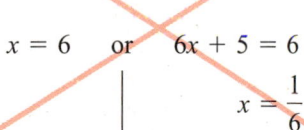

       $x = 6 \quad \text{or} \quad 6x + 5 = 6$ If the product of two numbers is 6, neither number need be 6. For example, $2 \cdot 3 = 6$.

$$x = \frac{1}{6}$$

Neither solution checks.

Many equations that don't appear to be quadratic can be written in standard form and then solved by factoring.

EXAMPLE 4   Solve: $(3x + 1)^2 = 6(3x - 1) + 3$

Strategy   We will square the binomial on the left side and distribute and combine like terms on the right side.

Why   We want to write an equivalent equation in standard form and use factoring to solve it.

Solution $(3x + 1)^2 = 6(3x - 1) + 3$ This is the equation to solve.

$$9x^2 + 6x + 1 = 18x - 6 + 3$$ Square $3x + 1$ and distribute the multiplication by 6.

$$9x^2 + 6x + 1 = 18x - 3$$ Combine like terms: $-6 + 3 = 3$.

$$9x^2 - 12x + 4 = 0$$ To get 0 on the right side, subtract 18x from and add 3 to both sides.

$$(3x - 2)(3x - 2) = 0$$ Factor the perfect-square trinomial.

$$3x - 2 = 0 \quad \text{or} \quad 3x - 2 = 0 \qquad \textcolor{red}{\text{Set each factor equal to 0.}}$$
$$x = \frac{2}{3} \qquad\qquad\qquad x = \frac{2}{3} \qquad \textcolor{red}{\text{Solve each equation.}}$$

We see that the two solutions are the same. We call $\frac{2}{3}$ a **repeated solution.** Check by substituting it into the original equation.

Self Check 4   Solve: $(2x + 1)^2 = -8(2x + 1) - 16$

Now Try ▶   Problem 39

Using Your Calculator ▶ Solving Quadratic Equations

To solve a quadratic equation such as $x^2 + 4x - 5 = \textcolor{red}{0}$ with a graphing calculator, we can use window settings of $[-10, 10]$ for x and $[-10, 10]$ for y and graph the quadratic function $\textcolor{red}{y = x^2 + 4x - 5}$, as shown in figure (a). We can then trace to find the x-coordinates of the x-intercepts of the parabola. See figures (b) and (c). For better results, we can zoom in. Since these are the numbers x that make $y = 0$, they are the solutions of the equation.

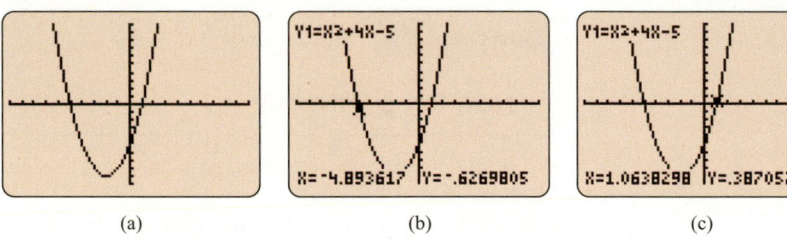

(a)   (b)   (c)

We can also find the x-intercepts of the graph of $y = x^2 + 4x - 5$ by using the ZERO feature found on most graphing calculators (Press $\boxed{\text{2nd}}$, CALC, 2). Figures (d) and (e) show how this feature locates the x-intercept and displays its coordinates.

From the displays, we can conclude that -5 and 1 are solutions of $x^2 + 4x - 5 = 0$.

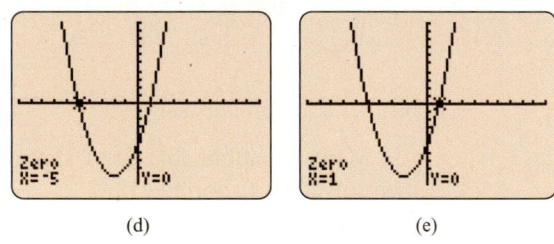

(d)   (e)

EXAMPLE 5   Let $f(x) = x^2 - 71$. For what value(s) of x is $f(x) = 10$?

Strategy   We will substitute 10 for $f(x)$ and solve for x.

Why   In the equation, there are two unknowns, x and $f(x)$. If we replace $f(x)$ with 10, we can use equation-solving techniques to find x.

Solution
$$\textcolor{red}{f(x)} = x^2 - 71$$
$$\textcolor{red}{10} = x^2 - 71 \qquad \textcolor{red}{\text{Substitute 10 for } f(x).}$$

To solve this quadratic equation, we first write it in standard form. Then we factor the difference of two squares on the right side, set each factor equal to 0, and solve each resulting equation.

$$0 = x^2 - 81 \qquad \text{\color{red}{To get 0 on the left side, subtract 10 from both sides.}}$$
$$0 = (x + 9)(x - 9) \qquad \text{\color{red}{Factor the difference of two squares.}}$$
$$x + 9 = 0 \qquad \text{or} \qquad x - 9 = 0 \qquad \text{\color{red}{Set each factor equal to 0.}}$$
$$x = -9 \qquad \qquad x = 9 \qquad \text{\color{red}{Solve each equation.}}$$

If x is -9 or 9, then $f(x) = 10$. Check these results by substituting -9 and 9 for x in $f(x) = x^2 - 71$.

Self Check 5   Let $g(x) = x^2 - 21$. For what value(s) of x is $g(x) = 100$?

Now Try ▶ Problem 43

2   Solve Higher-Degree Polynomial Equations by Factoring.

Some equations involving polynomials with degrees higher than 2 can be solved by factoring. In such cases, we use the extension of the zero-factor property: When the product of two *or more* real numbers is 0, at least one of them is 0.

EXAMPLE 6   Solve: $6x^3 - x^2 = 2x$

Strategy   This equation is not quadratic, because it contains a term involving x^3. To solve it, we will get 0 on the right side, factor the polynomial on the left side, and use the zero-factor property.

Why   To use the zero-factor property, we need one side of the equation to be factored completely and the other side to be 0.

Solution

$$6x^3 - x^2 = 2x \qquad \text{\color{red}{This is the equation to solve.}}$$
$$6x^3 - x^2 - 2x = 0 \qquad \text{\color{red}{To get 0 on the right side, subtract 2x from both sides.}}$$

The Language of Algebra

Since the highest degree of any term in $6x^3 - x^2 = 2x$ is 3, it is called a **third-degree** polynomial equation. Note that it has three solutions.

Next, we factor x from each term of the polynomial on the left side and proceed as follows:

$$6x^3 - x^2 - 2x = 0$$
$$x(6x^2 - x - 2) = 0 \qquad \text{\color{red}{Factor out the GCF, x.}}$$
$$x(3x - 2)(2x + 1) = 0 \qquad \text{\color{red}{Factor } 6x^2 - x - 2.}$$
$$x = 0 \qquad \text{or} \qquad 3x - 2 = 0 \qquad \text{or} \qquad 2x + 1 = 0 \qquad \text{\color{red}{Set each of the three factors equal to 0.}}$$
$$x = \frac{2}{3} \qquad \qquad x = -\frac{1}{2} \qquad \text{\color{red}{Solve each equation.}}$$

The solutions are 0, $\frac{2}{3}$, and $-\frac{1}{2}$. Check each one in the original equation.

Self Check 6   Solve: $5x^3 + 13x^2 = 6x$

Now Try ▶ Problem 47

EXAMPLE 7   Solve: **a.** $-x^4 + 5x^2 - 4 = 0$ **b.** $x^3 - x^2 - 9x + 9 = 0$

Strategy   To solve each equation, we will factor the polynomial on the left side and use the zero-factor property.

Why   To use the zero-factor property, we need one side of the equation to be factored completely and the other side to be 0.

Solution   **a.** To make the leading coefficient positive, we multiply both sides of the equation by -1. Then we factor the resulting trinomial and set the factors equal to 0.

$$-x^4 + 5x^2 - 4 = 0 \qquad \text{This is the equation to solve.}$$
$$\mathbf{-1(}-x^4 + 5x^2 - 4\mathbf{)} = \mathbf{-1(}0\mathbf{)} \qquad \text{Multiply both sides by } -1.$$
$$x^4 - 5x^2 + 4 = 0 \qquad \text{This is a fourth-degree polynomial equation.}$$
$$(x^2 - 1)(x^2 - 4) = 0 \qquad \text{Factor the trinomial.}$$
$$(x + 1)(x - 1)(x + 2)(x - 2) = 0 \qquad \text{Factor each difference of two squares.}$$

$$x + 1 = 0 \quad \text{or} \quad x - 1 = 0 \quad \text{or} \quad x + 2 = 0 \quad \text{or} \quad x - 2 = 0$$
$$x = -1 \quad | \quad x = 1 \quad | \quad x = -2 \quad | \quad x = 2$$

The solutions are $-1, 1, -2,$ and 2. Check each one in the original equation.

b. We can using grouping to factor the polynomial on the left side.

$$x^3 - x^2 - 9x + 9 = 0 \qquad \text{This is the equation to solve.}$$
$$x^2\mathbf{(}x - 1\mathbf{)} - 9\mathbf{(}x - 1\mathbf{)} = 0 \qquad \text{Factor } x^2 \text{ from } x^3 - x^2 \text{ and } -9 \text{ from } -9x + 9.$$
$$\mathbf{(}x - 1\mathbf{)}(x^2 - 9) = 0 \qquad \text{Factor out the GCF, } x - 1.$$
$$(x - 1)(x + 3)(x - 3) = 0 \qquad \text{Factor the difference of two squares.}$$

$$x - 1 = 0 \quad \text{or} \quad x + 3 = 0 \quad \text{or} \quad x - 3 = 0 \qquad \text{Set each factor equal to 0.}$$
$$x = 1 \quad | \quad x = -3 \quad | \quad x = 3 \qquad \text{Solve each equation.}$$

The solutions are $1, -3,$ and 3. Check each one in the original equation.

Self Check 7   Solve: **a.** $-a^4 - 36 + 13a^2 = 0$
b. $x^3 + x^2 - 100x - 100 = 0$

Now Try   Problems 51 and 53

Using Your Calculator ▶ Solving Equations

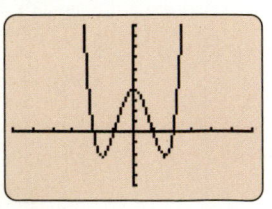

To solve $x^4 - 5x^2 + 4 = \mathbf{0}$ with a graphing calculator, we can use window settings of $[-6, 6]$ for x and $[-5, 10]$ for y and graph the polynomial function $\mathbf{y} = x^4 - 5x^2 + 4$ as shown in the figure. We can then read the values of x that make $y = 0$. They are $x = -2, -1, 1,$ and 2. If the x-coordinates of the x-intercepts were not obvious, we could approximate their values by using TRACE and ZOOM or by using the ZERO feature.

---

3 Use Quadratic Equations to Solve Problems.

We can use the six-step problem-solving strategy and the factoring method for solving quadratic equations to solve many types of application problems.

EXAMPLE 8 **Stained Glass.** The length of the base of a stained glass window is 3 times its height, and its area is 96 square feet. Find the length of its base and its height.

Analyze The formula that gives the area of a triangle is $A = \frac{1}{2}bh$, where b is the length of the base and h the height.

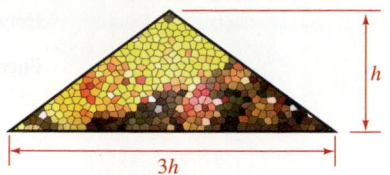

Assign Let h = the height of the window, in feet. Then $3h$ = the length of the base, in feet.

Form To form an equation in terms of h, we can substitute $3h$ for b and 96 for A in the formula for the area of a triangle.

$$A = \frac{1}{2}bh$$

$$96 = \frac{1}{2}(3h)h \qquad \text{\textcolor{red}{Substitute 96 for A, the area, and 3h for b, the length of the base.}}$$

$$96 = \frac{3}{2}h^2 \qquad \text{\textcolor{red}{Simplify the right side.}}$$

Solve To solve this quadratic equation, we will get 0 on the left side, factor the resulting binomial on the right side, and use the zero-factor property.

$$96 = \frac{3}{2}h^2$$

$$192 = 3h^2 \qquad \text{\textcolor{red}{To clear the equation of the fraction, multiply both sides by 2.}}$$

$$64 = h^2 \qquad \text{\textcolor{red}{Divide both sides by 3.}}$$

$$0 = h^2 - 64 \qquad \text{\textcolor{red}{To get 0 on the left side, subtract 64 from both sides.}}$$

$$0 = (h + 8)(h - 8) \qquad \text{\textcolor{red}{Factor the difference of two squares.}}$$

$h + 8 = 0 \qquad \text{or} \qquad h - 8 = 0 \qquad$ \textcolor{red}{Set each factor equal to 0.}

$\cancel{h = -8} \qquad\qquad h = 8 \qquad$ \textcolor{red}{Solve each linear equation.}

The Language of Algebra

When solving applications, any solution of a quadratic equation that doesn't make sense in the real-world should be **discarded** or **rejected**. In this case, the height h of a window cannot be a negative number of feet, so we discard $h = -8$.

State The solutions of the equation are -8 and 8. Since h represents the height of the window, and the height cannot be negative, we discard -8. Thus, the height of the stained glass window is 8 feet, and the length of its base is 3(8), or 24 feet.

Check The area of a triangle with a base of 24 feet and a height of 8 feet is: $\frac{1}{2}(24)(8) = 12(8) = 96$. The results check.

Self Check 8 **Landscaping.** A rectangular garden is 3 feet longer than it is wide. If the total area of the garden is 40 square feet, find the dimensions of the garden.

Now Try ▶ Problem 93

EXAMPLE 9 **Recreation.** A rectangular spa, 5 feet wide and 6 feet long, is surrounded by decking of uniform width, as shown in the illustration on the next page. If the total area of the deck is 60 ft^2, how wide is the decking?

Analyze Since the dimensions of the rectangular spa are 5 feet by 6 feet, its surface area is $6 \cdot 5 = 30$ ft^2. The decking has an area of 60 ft^2.

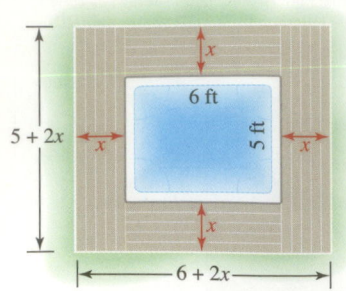

Assign Let x = the width of the decking, in feet. Then the width of the outer rectangle (the pool and the decking) is $x + 5 + x$ or $(5 + 2x)$ feet, and the length of the outer rectangle is $x + 6 + x$ or $(6 + 2x)$ feet. The area of the outer rectangle is the product of its length and width: $(6 + 2x)(5 + 2x)$ ft^2.

Form We can use subtraction to form an equation.

The area of the outer rectangle	minus	the area of the spa	equals	the area of the decking.
$(6 + 2x)(5 + 2x)$	$-$	30	$=$	60

Solve

$$(6 + 2x)(5 + 2x) - 30 = 60$$

$$30 + 12x + 10x + 4x^2 - 30 = 60 \quad \text{\color{red}Multiply the binomials on the left side.}$$

$$4x^2 + 22x = 60 \quad \text{\color{red}Simplify the left side.}$$

$$4x^2 + 22x - 60 = 0 \quad \text{\color{red}Subtract 60 from both sides to get 0 on the right side.}$$

$$2(2x^2 + 11x - 30) = 0 \quad \text{\color{red}Factor out the GCF, 2.}$$

$$2(2x + 15)(x - 2) = 0 \quad \text{\color{red}Factor the trinomial.}$$

$$2x + 15 = 0 \quad \text{or} \quad x - 2 = 0 \quad \text{\color{red}Since 2 cannot equal 0, discard that possibility. Set each of the remaining factors that contain a variable equal to 0.}$$

$$2x = -15 \qquad\qquad x = 2 \quad \text{\color{red}Solve each equation.}$$

$$x = \cancel{-\frac{15}{2}}$$

> **Success Tip**
>
> An alternate way to solve the equation $4x^2 + 22x - 60 = 0$ is to divide both sides by the GCF of the terms on its left side. When we divide both sides by the GCF, which is 2, we obtain
>
> $$\frac{4x^2}{2} + \frac{22x}{2} - \frac{60}{2} = \frac{0}{2}$$
>
> $$2x^2 + 11x - 30 = 0$$

State The solutions of the equation are $-\frac{15}{2}$ and 2. Since x represents the width of the decking, and the width cannot be negative, we discard $-\frac{15}{2}$. Thus, the width of the decking is 2 feet.

Check The illustration shows that if the decking is 2 feet wide, the total area of the decking is $18 + 12 + 18 + 12 = 60$ ft^2. The result checks.

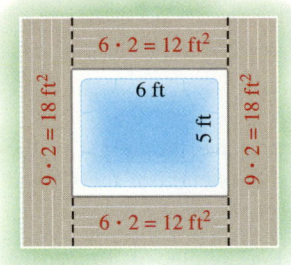

Self Check 9 **Frames.** A flat wood frame of uniform width, with outside dimensions of 6 inches by 9 inches, surrounds a rectangular photograph. If 28 in.^{2} of the photograph is visible, how wide is the material from which the frame is made?

Now Try ▶ Problem 97

A function defined by a polynomial in one variable of degree 2 is called a **quadratic function.** Some examples of quadratic functions are:

$$h(t) = -16t^2 + 128t \qquad \text{and} \qquad h(t) = -16t^2 + 64t + 80$$

Quadratic functions can be used to describe the height of falling objects or objects that have been propelled upward.

EXAMPLE 10 **Ballistics.** The height in feet $h(t)$ of an object launched straight up into the air from the ground with an initial velocity of 64 feet per second is given by the quadratic function $h(t) = -16t^2 + 64t$, where t represents the number of seconds since the object was launched. How long after launch will the object strike the ground?

Analyze When an object strikes the ground, its height is 0 feet.

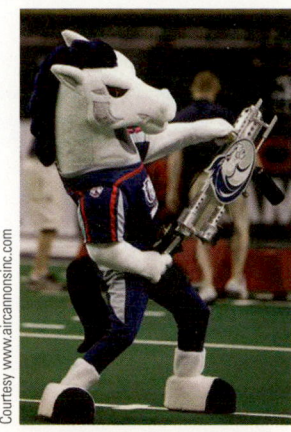

Courtesy www.aircannonsinc.com

Assign   In $h(t) = -16t^2 + 64t$, the variable t represents the unknown length of time after the launch for the object to strike the ground.

Form   In the function, we set $h(t)$, the height of the object, equal to 0. The result is a quadratic equation in the variable t.

$$h(t) = -16t^2 + 64t$$
$$0 = -16t^2 + 64t$$

Solve   To solve this quadratic equation, we will use the factoring method.

$$0 = -16t^2 + 64t$$
$$0 = -16t(t - 4)$$ Factor out $-16t$ from both terms on the right.
$$-16t = 0 \quad \text{or} \quad t - 4 = 0$$ Set each factor equal to 0.
$$t = 0 \quad | \quad t = 4$$ Solve each linear equation.

State   When t is 0, the object's height above the ground is 0 feet, because it has not been released. When t is 4, the height is again 0 feet, and the object has returned to the ground. The solution is 4 seconds.

Check   If we substitute 4 for t in $h(t) = -16t^2 + 64t$, we have

$$h(4) = -16(4)^2 + 64(4)$$
$$= -256 + 256$$
$$= 0$$

The result checks.

Success Tip

When solving equations of this type, it is helpful to determine if the coefficient of the t-term has 16 as a factor. If it does, factor out -16 from each term.

$$0 = -16t^2 + \mathbf{64}t$$
↑
This is $16 \cdot 4$.

Self Check 10

Rocketry.   The height in feet $h(t)$ of a toy rocket launched straight up into the air from the ground with an initial velocity of 128 feet per second is approximated by the quadratic function $h(t) = -16t^2 + 128t$, where t represents the number of seconds since the rocket was launched. How long after launch will the rocket strike the ground?

Now Try ▶ Problem 103

SECTION 5.9 STUDY SET

VOCABULARY

Fill in the blanks.

1. A _____ equation is any equation that can be written in the form $ax^2 + bx + c = 0$, where $a \neq 0$.

2. To _____ a quadratic equation means to find all the values of the variable that make the equation true.

3. To write the quadratic equation $x^2 - 3x = 15$ in _____ form, we subtract 15 from both sides.

4. $2x^2 - 4x = 0$, $3x^3 - x^2 - 6x = 0$, and $x^4 - 5x^2 + 4 = 0$ are examples of _____ equations.

CONCEPTS

5. Determine whether each equation is a quadratic equation.
   a. $w^2 + 7w + 12 = 0$ b. $6t + 11 = 0$
   c. $x(x + 3) = -2$ d. $k^3 - 4k^2 + k - 15 = 0$

6. Write each equation in standard form.
   a. $x^2 - x = 6$
   b. $x(x + 9) = -1$

7. a. If the product of two numbers is 0, what must be true about at least one of the numbers?
   b. Fill in the blanks: By the _____-factor property, if $ab = 0$, then $a =$ ▨ or $b =$ ▨.

8. Use the zero-factor property to solve each equation.
   a. $(x - 3)(x + 5) = 0$
   b. $6x(x - 1)(2x + 5) = 0$

9. Use a check to determine whether 4 and -5 are solutions of $a^2 - 9a + 20 = 0$.

10. What *first step* should be performed to solve each quadratic equation?
    a. $x^2 + 24 = -11x$
    b. $x^2 + x + \dfrac{1}{4} = 0$
    c. $-2x^2 + 7x + 4 = 0$
    d. $m(m + 3) = 2$

11. a. Write an expression that represents the width of the outer rectangle.
   b. Write an expression that represents the length of the outer rectangle.

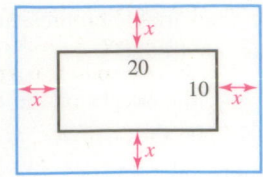

12. a. Use the graph in figure (a) below to solve $x^2 - 2x - 3 = 0$.

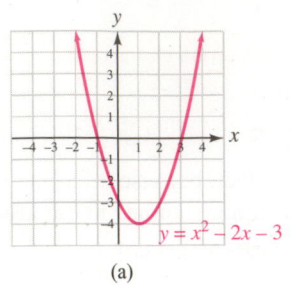

(a)

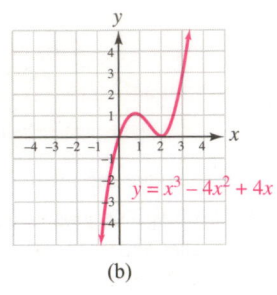

$y = x^3 - 4x^2 + 4x$

(b)

   b. Use the graph in figure (b) above to solve $x^3 - 4x^2 + 4x = 0$.

NOTATION

Complete each solution to solve each equation.

13. Solve: $y^2 - 2y - 8 = 0$

$(y - 4)() = 0$

$ = 0$ or $y + 2 = 0$

$y = 4$ $\Big|$ $y = $

14. Solve: $2x^2 - 3x - 1 = 1$

$2x^2 - 3x - 2 = $

$()(x - 2) = 0$

$2x + 1 = $ or $ = 0$

$2x = -1$ $\Big|$ $x = 2$

$x = $

GUIDED PRACTICE

Solve each equation. See Example 1.

15. $z^2 + 8z + 15 = 0$ **16.** $w^2 + 7w + 12 = 0$

17. $x^2 + 6x + 8 = 0$ **18.** $x^2 + 9x + 20 = 0$

19. $2x^2 - 3x + 1 = 0$ **20.** $2x^2 - x - 15 = 0$

21. $3m^2 + 10m + 3 = 0$ **22.** $3r^2 + 7r + 2 = 0$

Solve each equation. See Example 2.

23. $x^2 + x = 0$ **24.** $x^2 + 5x = 0$
25. $4x^2 = 8x$ **26.** $3x^2 = 9x$
27. $y^2 - 16 = 0$ **28.** $y^2 - 25 = 0$
29. $16y^2 = 9$ **30.** $81y^2 = 25$

Solve each equation by first clearing it of fractions. See Example 3.

31. $\dfrac{3a^2}{2} = \dfrac{1}{2} - a$ **32.** $x^2 + 1 = \dfrac{5}{2}x$

33. $x^2 - \dfrac{2}{5} = -\dfrac{9}{5}x$ **34.** $\dfrac{4}{5}x^2 + x = \dfrac{6}{5}$

35. $\dfrac{8}{3}a^2 = 1 - \dfrac{10}{3}a$ **36.** $\dfrac{5}{6}z^2 = 1 - \dfrac{13}{6}z$

37. $\dfrac{3}{16}m^2 - \dfrac{27}{16} = 0$ **38.** $\dfrac{2}{7}n^2 - \dfrac{128}{7} = 0$

Solve each equation. See Example 4.

39. $(x + 7)^2 = -2(x + 7) - 1$
40. $2(7x + 18) - 1 = (x + 6)^2$
41. $(m + 4)(2m + 3) - 22 = 10m$
42. $(d - 2)(d + 1) - d = 1$

See Example 5.

43. Let $f(x) = x^2 - 3x + 3$. For what value(s) of x is $f(x) = 1$?
44. Let $f(x) = 6x^2 + 5x + 2$. For what value(s) of x is $f(x) = 6$?
45. Let $f(x) = x^3 - 6x^2 + 8x + 2$. For what value(s) of x is $f(x) = 2$?
46. Let $f(x) = x^3 - 2x^2 - 8x + 10$. For what value(s) of x is $f(x) = 10$?

Solve each equation. See Example 6.

47. $x^3 - 4x^2 = 21x$ **48.** $x^3 + 8x^2 = 9x$

49. $y^3 - 49y = 0$ **50.** $2z^3 - 200z = 0$

Solve each equation. See Example 7.

51. $-z^4 + 37z^2 - 36 = 0$ **52.** $-y^4 + 10y^2 - 9 = 0$

53. $x^3 - 3x^2 - 4x + 12 = 0$ **54.** $x^3 + 4x^2 - 25x - 100 = 0$

Use a graphing calculator to find the solutions of each equation, if one exists. If an answer is not exact, give the answer to the nearest hundredth.

55. $2x^2 - 7x + 4 = 0$ **56.** $x^2 - 4x + 7 = 0$

57. $-3x^3 - 2x^2 + 5 = 0$ **58.** $-2x^3 - 3x - 5 = 0$

TRY IT YOURSELF

Solve each equation.

59. $b(6b - 7) = 10$ **60.** $2y(4y + 3) = 9$

61. $x^3 + x^2 = 0$ **62.** $2x^4 + 8x^3 = 0$

63. $\dfrac{x^2}{5} - \dfrac{4}{5} = -\dfrac{3}{5}x$ **64.** $\dfrac{x^2}{9} = \dfrac{8}{9}x - \dfrac{7}{9}$

65. $a^3 - 2a^2 - 16a + 32 = 0$ **66.** $b^3 - 5b^2 - 9b + 45 = 0$

67. $6y^2 = 25y$ **68.** $15x^2 = 7x$

69. $a^3 - 18a^2 = -81a$

70. $y^3 + 22y^2 = -121y$

71. $7t^2 - 2t - 5 = 0$

72. $15x^2 + 22x - 5 = 0$

73. $3x^3 + 3x^2 = 12(x + 1)$

74. $9(y + 4) = y^3 + 4y^2$

75. $64t^2 - 81 = 0$

76. $36t^2 - 1 = 0$

77. $\dfrac{x^2(6x + 37)}{35} = x$

78. $x^2 = -\dfrac{4x^3(3x + 5)}{3}$

79. $0 = d^2 - 5d - 66$

80. $0 = t^2 - 2t - 63$

81. $n(3n - 4) = (n - 6)^2 + 11n - 1$

82. $s(2s + 7) = (s + 1)^2 + 71 - s$

83. $(x - 5.5)(x + 3) = 0$

84. $(2y + 11)(y - 4) = 0$

85. $-x^4 + 34x^2 - 225 = 0$

86. $-x^4 + 26x^2 - 25 = 0$

87. Let $f(t) = 3t^2 - 2t + 2$ and $g(t) = 8t^2 + t$. Find all values of t such that $f(t) = g(t)$.

88. Let $f(x) = 2x^3 + 16x^2$ and $g(x) = -30x$. Find all values of x such that $f(x) = g(x)$.

89. Let $s(x) = (x + 11)(x + 3)$ and $t(x) = (x + 5)(2x + 1)$. Find all values of x such that $s(x) = t(x)$.

90. Let $f(x) = (x - 4)(x + 4)$ and $g(x) = 6x$. Find all values of x such that $f(x) = g(x)$.

APPLICATIONS

91. Integer Problems.   The product of two positive consecutive even integers is 288. Find the integers. (*Hint:* Let x = the smaller even integer and $x + 2$ = the larger even integer.)

92. Integer Problems.   The product of two positive consecutive odd integers is 143. Find the integers. (*Hint:* Let x = the smaller odd integer and $x + 2$ = the larger odd integer.)

93. Kitchens.   An efficient *kitchen work triangle* design minimizes the number of steps that the cook must take between the refrigerator, range, and sink during meal preparation and cleanup. The length of the base of the kitchen work triangle shown in the illustration is twice its height, and its area is 16 square feet. Find the length of its base and its height.

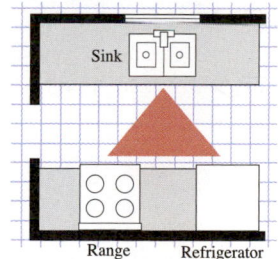

94. Weatherizing a House.   Refer to the illustration. It took 44 square feet of Tyvek house wrap to cover a trapezoid-shaped part of a house. Find the height of the trapezoid if the length of its shorter base is the same as its height. (The formula for the area of a trapezoid is $A = \frac{1}{2}h(b_1 + b_2)$, where h is its height, and b_1 and b_2 are the lengths of the two bases.)

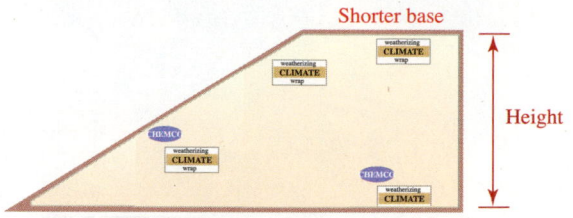

Shorter base

Height

Longer base: 18 ft

95. Cooking.   A griddle has a cooking surface of 160 square inches and its length is 6 inches longer than its width. Find its length and width.

$w + 6$

w

96. Winter Recreation.   The length of a rectangular ice-skating rink is 20 meters greater than twice its width. Find the width.

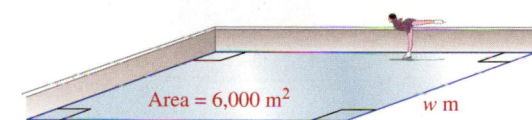

Area = 6,000 m²

w m

97. from **Campus to Careers**

Landscape Architect

Suppose you are a landscape architect designing a water fountain feature for a community park. Building codes require that the 6-foot by 8-foot fountain pool be surrounded by a uniform-width concrete walkway with a surface area of 120 square feet. How wide should the walkway be?

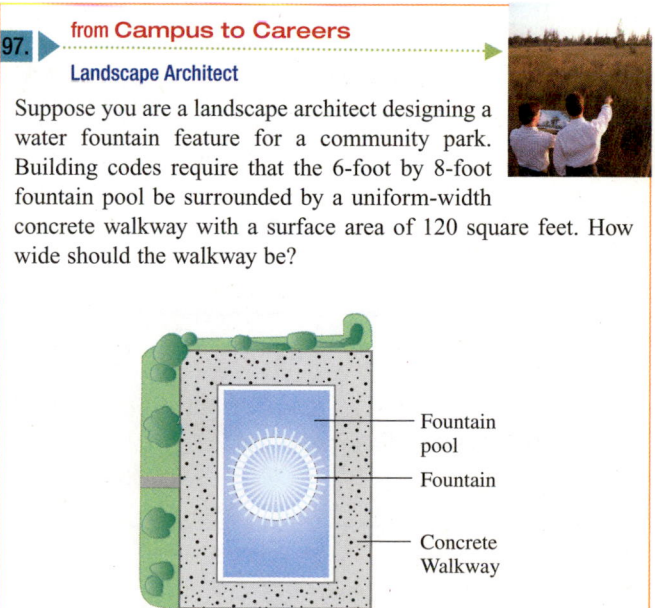

Fountain pool

Fountain

Concrete Walkway

©Leslie Harris/Photolibrary

98. Framing Posters. A green cardboard mat of uniform width was used in framing the 34 inch by 22 inch poster shown here. If 116 square inches of matting was used, what is the width of the mat?

99. Frames. A flat wooden frame of uniform width, with outside dimensions of 8 inches by 10 inches, surrounds a rectangular photograph. If 15 in.^{2} of the photograph is visible, how wide is the material from which the frame is made?

100. Fine Arts. An artist intends to paint a 45-square-foot rectangular mural on a wall that is 13 feet long and 9 feet high, as shown below. If the artist wants to leave a border of uniform width, how wide should the border be? (*Hint:* Let w represent the width of the border. Then $13 - 2w$ represents the length of the mural.)

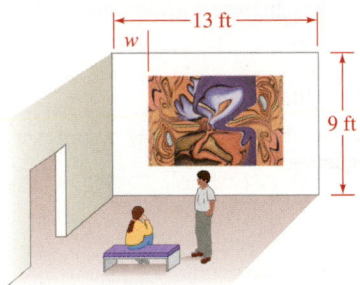

101. Architecture. The rectangular floor space outlined in black below is twice as long as it is wide. Room dividers partition it into a dining room and a den. If the den contains 560 square feet, find the dimensions of the rectangular floor space. (*Hint:* It is helpful to write an algebraic expression that represents the length of the den.)

102. Baseball. In 2007, Tim Lincecum of the San Francisco Giants consistently threw fastballs clocked at 97 mph. This is approximately 144 feet per second. If he could throw the baseball vertically into the air with this velocity, the quadratic function $h(t) = -16t^2 + 144t$ would approximate the height in feet $h(t)$ of the ball t seconds after being thrown. How long would it take for a baseball thrown in this manner to fall to the ground?

103. Burp Guns. A Ping-Pong ball is shot vertically upward from a height of 4 feet. If we neglect air resistance, the quadratic function $h(t) = -16t^2 + 63t + 4$ approximates the height in feet $h(t)$ of the ball t seconds after being shot. How long after being shot will the Ping-Pong ball hit the ground?

104. Slingshots. The quadratic function $h(t) = -16t^2 + 128t$ approximates the height in feet $h(t)$ of a stone t seconds after being shot upward into the air. At what times will a stone, shot vertically upward, be 192 feet above the ground?

105. Bungee Jumping. See the illustration below. The quadratic function $h(t) = -16t^2 + 212$ approximates the distance a bungee jumper is from the ground for the free-fall portion of the jump, t seconds after leaping off a bridge. We can find the number of seconds it takes the jumper to reach the point in the fall where the 64-foot bungee cord starts to stretch by substituting 148 for $h(t)$ and solving for t. Find t.

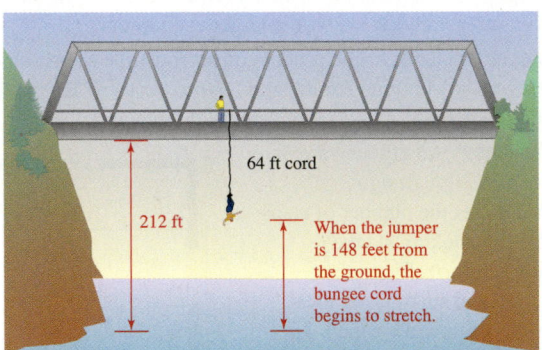

106. Ballistics. The muzzle velocity of a cannon is 480 feet per second. If a cannonball is fired vertically, the quadratic function $h(t) = -16t^2 + 480t$ approximates the height in feet $h(t)$ of the cannonball t seconds after firing. At what times will it be at a height of 3,344 feet?

107. Forensic Medicine. The kinetic energy E of a moving object is given by $E = \frac{1}{2}mv^2$, where m is the mass of the object (in kilograms) and v is the object's velocity (in meters per second). Kinetic energy is measured in joules. By measuring the damage done to a victim who has been struck by a 3-kilogram club, a pathologist finds that the energy at impact was 54 joules. Find the velocity of the club at impact.

108. Traffic Accidents. Investigators at a traffic accident used the function $d(v) = 0.04v^2 + 0.8v$, where v is the velocity of the car (in mph) and $d(v)$ is the stopping distance of the car (in feet), to reconstruct the events leading up to a collision. From physical evidence, it was concluded that it took one car 32 feet to stop. At what velocity was the car traveling prior to the accident?

109. Break-Even Point. The cost for a guitar maker to handcraft x guitars is given by the function $C(x) = \frac{1}{8}x^2 - x + 6$. The revenue taken in with the sale of x guitars is given by the function $R(x) = \frac{1}{4}x^2$. Find the number of guitars that must be sold so that the cost equals the revenue.

110. Revenue. Over the years, the manager of a store has found that the number of scented candles x she can sell in a month depends on the price p according to the formula $x = 200 - 10p$. At what price should she sell the candles if she needs to bring in \$750 in revenue a month? (*Hint:* Revenue = price · number sold = px.)

WRITING

111. Explain the zero-factor property.

112. In the work shown below, explain why the student has not solved for x.

Solve: $x^2 + x - 6 = 0$

$$x^2 + x = 6$$
$$x = 6 - x^2$$

113. Explain what is wrong with the following solution.

Solve: $x^2 = x$

$$\frac{x^2}{x} = \frac{x}{x}$$
$$x = 1$$

114. Explain what is wrong with the following solution.

Solve: $x^2 - x = 6$

$$x(x - 1) = 6$$
$$x = 6 \quad \text{or} \quad x - 1 = 6$$
$$x = 7$$

115. The following graphs of two polynomial functions $f(x) = 2x^3 - 8x$ and $f(x) = 2x(x + 2)(x - 2)$ appear to be the same. After examining their equations, explain why we know that they are identical graphs.

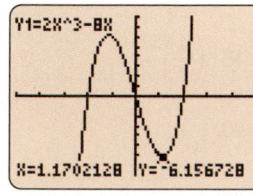 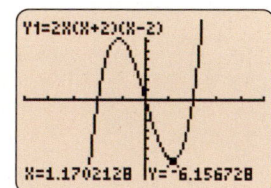

116. Explain why the x-coordinate of the x-intercept in the following graph of $y = 8x^2 + 10x - 3$ is a solution of $8x^2 + 10x - 3 = 0$.

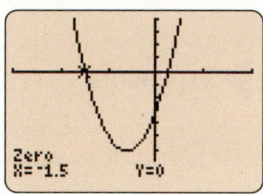

REVIEW

117. Aluminum Foil. Find the number of square feet of aluminum foil on a roll if it has dimensions of $8\frac{1}{3}$ yards × 12 inches.

118. Hockey. A hockey puck is a hard rubber disk 2.5 cm (1 in.) thick and 7.6 cm (3 in.) in diameter. Find the volume of a puck in cubic centimeters and cubic inches. Round to the nearest tenth.

CHALLENGE PROBLEMS

119. Find a quadratic equation with solutions $\frac{1}{4}$ and $-\frac{4}{3}$.

120. Find a polynomial equation of degree 3 with the solutions $-3, -2,$ and 3.

121. Solve: $\dfrac{r^2}{105} + \dfrac{r}{140} = \dfrac{1}{42}$

122. Solve: $\dfrac{a^3}{65} - \dfrac{a^2}{30} - \dfrac{a}{78} = 0$

123. Explain why $5x^2 + 3x + 2 = 2x(x + 1) + 3x^2$ is not a quadratic equation.

124. Find the domain of the function defined by:

$$f(x) = \frac{3}{x^2 - x - 6}$$

5 ▶ Summary & Review

DEFINITIONS AND CONCEPTS	EXAMPLES

An **exponent** indicates repeated multiplication. It tells how many times the **base** is to be used as a factor. If n is a natural number,

n factors of x

$$x^n = x \cdot x \cdot x \cdot x \cdots \cdot x$$

where x is the **base** and n the **exponent.**

Identify the base and the exponent in each expression.

$2^6 = 2 \cdot 2 \cdot 2 \cdot 2 \cdot 2 \cdot 2$ 2 is the base, 6 is the exponent

$p^5 = p \cdot p \cdot p \cdot p \cdot p$ p is the base, 5 is the exponent

$(-ab)^3 = (-ab)(-ab)(-ab)$ $-ab$ is the base, 3 is the exponent

$7c^4 = 7 \cdot c \cdot c \cdot c \cdot c$ c is the base, 4 is the exponent

Rules for Exponents: If m and n represent integers and there are no divisions by 0, then

Product rule:

$$x^m x^n = x^{m+n}$$

Quotient rule:

$$\frac{x^m}{x^n} = x^{m-n}$$

Power rule:

$$(x^m)^n = x^{mn}$$

Power of a product:

$$(xy)^m = x^m y^m$$

Power of a quotient:

$$\left(\frac{x}{y}\right)^n = \frac{x^n}{y^n}$$

Zero exponent:

$$x^0 = 1$$

Negative exponent:

$$x^{-n} = \frac{1}{x^n}$$

Exponent of 1:

$$x^1 = x$$

Negative exponents appearing in fractions:

$$\frac{1}{x^{-n}} = x^n \qquad \frac{x^{-m}}{y^{-n}} = \frac{y^n}{x^m}$$

$$\left(\frac{x}{y}\right)^{-n} = \left(\frac{y}{x}\right)^n$$

Simplify each expression:

$$x^2 \cdot x^3 = x^{2+3} = x^5 \qquad \frac{m^9}{m^3} = m^{9-3} = m^6$$

$$(r^4)^5 = r^{4 \cdot 5} = r^{20} \qquad (2y)^5 = 2^5 y^5 = 32y^5$$

$$\left(\frac{a}{3}\right)^4 = \frac{a^4}{3^4} = \frac{a^4}{81} \qquad 5^0 = 1$$

$$2^{-3} = \frac{1}{2^3} = \frac{1}{8} \qquad 7^1 = 7$$

$$\frac{1}{10^{-2}} = 10^2 = 100 \qquad \frac{4^{-2}}{m^{-3}} = \frac{m^3}{4^2} = \frac{m^3}{16}$$

$$\left(\frac{7}{a}\right)^{-2} = \left(\frac{a}{7}\right)^2 = \frac{a^2}{7^2} = \frac{a^2}{49}$$

REVIEW EXERCISES

Evaluate each expression.

1. 3^5

2. -2^5

3. $(-4)^3$

4. $\left(\frac{2}{3}\right)^2$

Simplify each expression. Write answers using positive exponents.

5. $x^4 \cdot x^2$

6. $m^{-3} n^{-4} m^6 n^{-1}$

7. $\dfrac{(4m^6)^3}{m^2}$

8. $(-t^2)^2 (t^3)^3$

9. $(3x^2 y^3)^2$

10. $\left(\dfrac{x^4}{b}\right)^4$

11. $-10x^0$

12. $\dfrac{1}{5h^{-12}}$

13. $\left(\dfrac{2a}{b}\right)^{-1}$

14. $\dfrac{x}{5x^{-4}}$

15. $(3x^{-3})^{-2}$

16. $\dfrac{2x^{-4} x^3}{9}$

17. $-\left(\dfrac{c^{-3}}{c^{-5}}\right)^5$

18. $\left(\dfrac{4}{5}\right)^{-2}$

19. $\dfrac{3^{-3}}{4^{-2}}$

20. $\dfrac{1}{4^{-3}}$

21. $\left(\dfrac{s^7}{s^{-8}}\right)^3 \left(\dfrac{s^{-5}}{s^2}\right)^{-4}$

22. $\left(\dfrac{-2a^4 b c^{-6}}{a^{-3} b^2 c^{-5}}\right)^{-3}$

SECTION 5.2 ▶ Scientific Notation

DEFINITIONS AND CONCEPTS	EXAMPLES

Scientific notation is a compact way of writing large and small numbers.
 	
A positive number is written in **scientific notation** when it is written in the form
 	
    $N \times 10^n$, where $1 \le N < 10$ and n is an integer.
	Write each number in scientific notation.
	
	    $98{,}100 = 9.81 \times 10^4$ and $0.0043 = 4.3 \times 10^{-3}$
	    4 decimal places *3 decimal places*
	
	Write each number in **standard notation.**
	
	    $6.42 \times 10^5 = 642{,}000$ and $7.7 \times 10^{-6} = 0.0000077$
	      5 decimal places *6 decimal places*

To **multiply or divide with very small or very large numbers,** write the numbers in scientific notation and perform the arithmetic on the decimals and the powers of 10 separately.	Use scientific notation to perform the calculation: $$\frac{(5{,}400{,}000{,}000)(0.0066)}{(0.000000022)(64{,}800)} = \frac{(5.4 \times 10^9)(6.6 \times 10^{-3})}{(2.2 \times 10^{-8})(6.48 \times 10^4)}$$ $$= \frac{(5.4)(6.6)}{(2.2)(6.48)} \times \frac{10^9 10^{-3}}{10^{-8} 10^4}$$ $$= 2.5 \times 10^{9+(-3)-(-8)-4}$$ $$= 2.5 \times 10^{10}$$

REVIEW EXERCISES

Write each number in scientific notation.

23. 19,300,000,000

24. 0.00000002735

Write each number in standard notation.

25. 7.277×10^7

26. 8.3×10^{-9}

Write each number in scientific notation and perform the operations. Give the answers in scientific notation.

27. **Speed of Light.** Light travels at about 300,000 kilometers per second. If the average distance from the sun to the planet Mars is approximately 228,000,000 kilometers, how long does it take light from the sun to reach Mars?

28. **Protons.** If the mass of 1 proton is 0.0000000000000000000000167248 gram, find the mass of 1 million protons.

Write each number in scientific notation and perform the operations. Give the answers in scientific notation.

29. $\dfrac{(616{,}000{,}000)(0.000009)}{(0.00066)}$

30. $\dfrac{0.0000000495}{(33{,}000)(800{,}000{,}000)}$

SECTION 5.3 ▶ Polynomials and Polynomial Functions

DEFINITIONS AND CONCEPTS	EXAMPLES

A **polynomial** is a single term or the sum of terms in which all variables have whole-number exponents. No variable appears in a denominator.	Examples of polynomials:
	
	    6, $-2mn^3$, $r^2 - 14s^4$, $3x^2 + x - 8$
	
	Not polynomials:
	
	    $x^3 - x^{-2}$, $6ab^4 - \dfrac{1}{b} + 5a$

A **monomial** is a polynomial with one term. A **binomial** is a polynomial with two terms. A **trinomial** is a polynomial with three terms.	Monomials: $17m$ $-5x^2y$ Binomials: $3y + 2$ $3p^4 - 2q^3$ Trinomials: $3t^2 + 5t - 6$ $8r^4s^3 - 6r^3s^2 + 4r^2s$

	Term	Degree of the term
The **degree of a term** of a polynomial in one variable is the value of the exponent on the variable. The **degree of a term** that has more than one variable is the sum of the exponents on the variables. The **degree of a nonzero constant** is 0.	$9d^3$	3
	$-1.6r^7s^4$	$7 + 4 = 11$
	19	0

	Polynomial	Degree of the polynomial
The **degree of a polynomial** is equal to the highest degree of any term of the polynomial.	$8a^4 - 3a^2 + 5a - 9$	4
	$4x^4y + 11x^3y^2 - x^2y^6$	$2 + 6 = 8$

To **evaluate a polynomial function,** we replace the variable in the defining equation with its value, called the **input.** Then we simplify to find the **output.**	Let $f(x) = x^3 - 3x^2 - 9x + 2$. Find $f(2)$. $f(x) = x^3 - 3x^2 - 9x + 2$ $f(2) = (2)^3 - 3(2)^2 - 9(2) + 2$ Substitute 2 for x. $\qquad = 8 - 3(4) - 18 + 2$ $\qquad = -20$
To **simplify a polynomial,** combine like terms.	Simplify the polynomial $8a^5 - 2a^4 + 6a^5 + a^2$. $8a^5 - 2a^4 + 6a^5 + a^2 = 14a^5 - 2a^4 + a^2$
To **add polynomials,** drop the parentheses and combine like terms (terms having the same variables with the same exponents).	Add: $(3x^2 - 2x + 7) + (2x^2 - 17) = 3x^2 - 2x + 7 + 2x^2 - 17$ $\qquad\qquad\qquad\qquad\qquad = 5x^2 - 2x - 10$
To **subtract polynomials,** change the signs of the terms of the polynomial being subtracted, drop the parentheses, and combine like terms.	Subtract: $(3x^2 - 2x + 7) - (2x^2 - 17) = 3x^2 - 2x + 7 - 2x^2 + 17$ $\qquad\qquad\qquad\qquad\qquad = x^2 - 2x + 24$

REVIEW EXERCISES

Determine whether each expression is a polynomial.

31. $\dfrac{2x^2}{x+1}$

32. $-5x^3 + x^2 - 5x - 4$

33. $2.8y^{15} - y^{10} + y^8 - \dfrac{3}{2}y^6$

34. $x^{-3} + x^{-2} - x^{-1} - 1$

Classify each polynomial as a monomial, binomial, trinomial, or none of these. Then determine the degree of the polynomial.

35. $x^2 - 8$

36. $-15a^3b$

37. $x^4 + x^3 - x^2 + x - 4$

38. $9x^2y + 13x^3y^2 + 8x^4y^4$

39. Let $f(x) = x^3 - 5x^2 - 2x - 5$. Find $f(3)$.

40. Squirt Guns. The volume of the reservoir on top of the squirt gun is given by the polynomial function $V(r) = 4.19r^3 + 25.13r^2$, where r is the radius in inches. Find $V(2)$ to the nearest cubic inch.

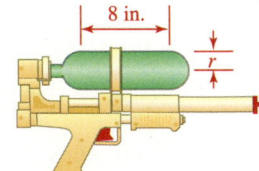

Simplify each polynomial.

41. $9t^3 - 5t^2 - 5t + 3t^2 - 7t^3$

42. $\dfrac{5}{4}ab^2c - \dfrac{5}{3}abc - \dfrac{1}{2}ab^2c + \dfrac{5}{6}abc$

Perform each operation.

43. $(2x^2y^3 - 5x^2y + 9y) + (x^2y^3 - 3x^2y - y)$

44. $(8m^2 - 7m) - (-11m^2 + 6m + 9)$

45. $\left(\dfrac{2}{3}s^6 - \dfrac{1}{6}s^4\right) + \left(-\dfrac{1}{6}s^6 - \dfrac{3}{2}s^4\right)$

46.
$$-10k^4 - 4k^3 + 5k^2 - k + 1$$
$$-\ \underline{(-16k^4 + 2k^3 - 4k^2 - k + 3)}$$

47. Subtract $6c^2d^2 + 4c^2d - 5cd^2$ from the sum of $-c^2d^2 + 5c^2d - 10cd^2$ and $11c^2d^2 - c^2d + 9cd^2$.

48. Let $f(x) = x^5 + 3x^2 + 2x$ and $g(x) = 4x^5 + 2x^2 - 5x$. Find $f(x) + g(x)$.

49. Let $s(t) = 1.7t^3 - 0.5t^2 + 0.4t - 0.3$ and $h(t) = 0.5t^3 - 0.9t^2 - 0.8t + 1.1$. Find $s(t) - h(t)$.

50. Use the graph of function f to find each of the following.

  a. $f(0)$

  b. The values of x for which $f(x) = 0$

  c. Write the domain and range of f in interval notation.

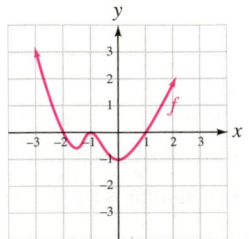

SECTION 5.4 ▶ **Multiplying Polynomials**

DEFINITIONS AND CONCEPTS	EXAMPLES
To **multiply two monomials,** multiply their numerical factors and multiply their variable factors.	Multiply: $(3x^2y)(4x^3y^2) = (3 \cdot 4)(x^2 \cdot x^3)(y \cdot y^2)$ Group the coefficients together and the variables with like bases together. $= 12x^5y^3$
To **multiply a polynomial by a monomial,** multiply each term of the polynomial by the monomial.	Multiply: $5m^2n^3(2m - 3n^2) = 5m^2n^3(2m) - 5m^2n^3(3n^2)$ Distribute. $= 10m^3n^3 - 15m^2n^5$ Multiply the monomials.

To **multiply two binomials,** multiply each term of one binomial by each term of the other binomial.
 	
The **FOIL method** can be used to multiply two binomials.
 	
  F: First   **O:** Outer   **I:** Inner   **L:** Last
	Multiply:
	
	                  F O I L
	$(7x + 4)(3x + 5) = 7x \cdot 3x + 7x \cdot 5 + 4 \cdot 3x + 4 \cdot 5$
	
	$= 21x^2 + 35x + 12x + 20$ Multiply the monomials.
	
	$= 21x^2 + 47x + 20$ Combine like terms.

To **multiply two polynomials,** multiply each term of one polynomial by each term of the other polynomial.	Multiply: $(3x + 2)(x^2 - 3x + 6)$ $= 3x(x^2) + (3x)(-3x) + 3x(6) + 2(x^2) + 2(-3x) + 2(6)$ $= 3x^3 - 9x^2 + 18x + 2x^2 - 6x + 12$ Multiply the monomials. $= 3x^3 - 7x^2 + 12x + 12$ Combine like terms.
When finding the **product of three polynomials,** begin by multiplying any two of them, and then multiply that result by the third polynomial.	Multiply: $-4x^3(x - 5)(x - 1) = -4x^3(x^2 - x - 5x + 5)$ Multiply the binomials. $= -4x^3(x^2 - 6x + 5)$ $= -4x^5 + 24x^4 - 20x^3$ Distribute.

Special products: It is worthwhile to learn these forms.
 	
The square of a binomial sum:
  $(A + B)^2 = A^2 + 2AB + B^2$
 	
The square of a binomial difference:
  $(A - B)^2 = A^2 - 2AB + B^2$
 	
The sum and difference of two terms:
  $(A + B)(A - B) = A^2 - B^2$
	Multiply: $(2x + 3)^2 = (2x)^2 + 2(2x)(3) + 3^2$
	                $= 4x^2 + 12x + 9$
	
	Multiply: $(3x - y)^2 = (3x)^2 - 2(3x)(y) + y^2$
	                $= 9x^2 - 6xy + y^2$
	
	Multiply: $(5p + 3q)(5p - 3q) = (5p)^2 - (3q)^2$
	                        $= 25p^2 - 9q^2$

REVIEW EXERCISES

Find each product.

51. $(8a^2)\left(-\dfrac{1}{2}a\right)$

52. $(-3xy^2z)(-2xz^3)(xz)$

53. $2xy^2(x^3y - 4xy^5)$

54. $-a^2b(-a^2 - 2ab + b^2)$

55. $(3x^2 + 2)(2x - 4)$

56. $(5at - 6)^2$

57. $(7c^4d^3 - d)(7c^4d^3 + d)$

58. $(5x^2 - 4x)(3x^2 - 2x + 10)$

59. $(r + s)(r - s)(r - 3s)$

60. $\left(3c - \dfrac{3}{4}\right)^2$

61. $[5 - (a - b)]^2$

62. $(2x - y - 2z)(x + y + z)$

63. Simplify: $(4a + 1)^2 + (5a - 2)^2$

64. Let $f(x) = x^2 + 5x - 1$. Find $f(b - 3)$.

65. Let $f(x) = 0.5x - 0.4$ and $g(x) = 0.1x + 2.1$. Find $f(x) \cdot g(x)$.

66. Let $s(x) = x + 3$ and $h(x) = x^2 - 3x + 9$. Find $s(x) \cdot h(x)$.

67. Ice Chests.   Write a polynomial that represents

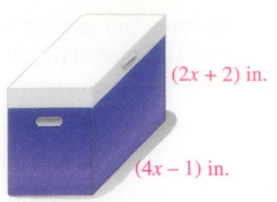

   a. The perimeter of the base of the ice chest.

   b. The area of the base of the ice chest.

   c. The volume of the ice chest.

(2x + 2) in.

(4x − 1) in.

2x in.

68. Geometry.   The length, width, and height of the rectangular solid shown here are consecutive integers.

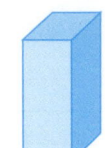

   a. Write a polynomial function that gives the volume of the solid.

   b. What is the volume of the solid if the shortest dimension is 5 inches?

SECTION 5.5 ▶ **The Greatest Common Factor and Factoring by Grouping**

DEFINITIONS AND CONCEPTS	EXAMPLES
Factoring is multiplication reversed. To **factor a polynomial** means to express it as a product of two (or more) polynomials.	Multiplication: Given the factors, we find a polynomial. ⟶ $$4x(x + 9) = 4x^2 + 36x$$ ⟵ Factoring: Given a polynomial, we find the factors.
To find the **greatest common factor, GCF,** of a list of terms 1. Write each coefficient as a product of prime factors. 2. Identify the numerical and variable factors common to each term. 3. Multiply the common numerical and variable factors identified in step 2 to obtain the GCF. If there are no common factors, the GCF is 1.	Find the GCF of $14a^4$, $35a^3$, and $56a^2$. $\left.\begin{array}{l} 14a^4 = 2 \cdot \mathbf{7} \cdot \mathbf{a} \cdot \mathbf{a} \cdot a \cdot a \\ 35a^3 = 5 \cdot \mathbf{7} \cdot \mathbf{a} \cdot \mathbf{a} \cdot a \\ 56a^2 = 2 \cdot 2 \cdot 2 \cdot \mathbf{7} \cdot \mathbf{a} \cdot \mathbf{a} \end{array}\right\}$ GCF $= \mathbf{7} \cdot \mathbf{a} \cdot \mathbf{a} = 7a^2$
When we **factor a polynomial,** we write *a sum of terms as a product of factors.* The first step of factoring a polynomial is to see whether the terms of the polynomial have a common factor. If they do, **factor out the GCF.**	Factor: $14a^4 + 35a^3 - 56a^2 = \mathbf{7a^2}(2a^2 + 5a - 8)$ Factor out the GCF, $\mathbf{7a^2}$. Use multiplication to check the factorization: $7a^2(2a^2 + 5a - 8) = 14a^4 + 35a^3 - 56a^2$ This is the original polynomial.
A polynomial that cannot be factored is a **prime polynomial.**	Since $2x + 13$ cannot be factored using integer coefficients, it is a prime binomial.

If a polynomial has four terms, try **factoring by grouping.**	Factor: $ax - 2x + 3a - 6$
1. Group the terms of the polynomial so that the first two terms have a common factor and the last two terms have a common factor.	$ax - 2x + 3a - 6 = \boxed{ax - 2x} + \boxed{3a - 6}$ Group the terms. $= x(a - 2) + 3(a - 2)$ Factor x from ax − 2x. Factor 3 from 3a − 6.
2. Factor out the common factor from each group.	$= (a - 2)(x + 3)$ Factor out the GCF, a − 2.
3. Factor out the resulting common binomial factor. If there is no common binomial factor, regroup the terms of the polynomial and repeat steps 2 and 3.	

REVIEW EXERCISES

Find the GCF of each list.

69. 42, 36, 54   **70.** $6x^2y^5,\ 15xy^3$

Factor, if possible.

71. $4x^4 + 8$

72. $\dfrac{3x^3}{5} - \dfrac{6x^2}{5} + \dfrac{x}{5}$

73. $6x - 11$

74. $7a^4b^2 + 49a^3b$

75. $5x^2(x + y) - 15x^3(x + y)$

76. $27x^3y^3z^3 + 81x^4y^5z^2 - 90x^2y^3z^7$

Factor −1 from each binomial.

77. $-x - 9$ **78.** $4r - 7$

Factor out the opposite of the greatest common factor.

79. $-7b^3 + 14c$

80. $-49a^3b^2(a - b)^4 + 63a^2b^4(a - b)^3$

Factor by grouping.

81. $xy + 2y + 4x + 8$

82. $r^2y - ar - ry + a + r - 1$

83. $t^3 - 9 + t - 9t^2$

84. $1 - x - 3z + 3xz$

85. Solve $m_1m_2 = mm_2 + mm_1$ for m_1.

86. **Geometry.** The formula for the surface area of a cylinder is $A = 2\pi r^2 + 2\pi rh$. Write the formula with the right side in factored form.

SECTION 5.6 ▶ Factoring Trinomials

DEFINITIONS AND CONCEPTS	EXAMPLES
Trinomials that are squares of a binomial are called **perfect-square trinomials.** We can factor perfect-square trinomials by applying the special-product rules in reverse. $A^2 + 2AB + B^2 = (A + B)^2$ $A^2 - 2AB + B^2 = (A - B)^2$	Factor: $x^2 + 22x + 121$ and $s^2 - 18st + 81t^2$ Match each trinomial to a special-product form in the left column. $x^2 + 22x + 121 = x^2 + 2 \cdot x \cdot 11 + 11^2 = (x + 11)^2$ $s^2 - 18st + 81t^2 = s^2 - 2 \cdot s \cdot 9t + (9t)^2 = (s - 9t)^2$
Many trinomials factor as the product of two binomials. To **factor a trinomial** of the form $x^2 + bx + c$, whose **leading coefficient is 1,** find two integers whose product is c and whose sum is b. $x^2 + bx + c = (x\ \boxed{})(x\ \boxed{})$ The product of these numbers must be c and their sum must be b. Use the FOIL method to check the factorization.	Factor: $p^2 + 14p + 45$ Here $c = 45$ and $b = 14$. We must find two integers whose product is 45 and whose sum is 14. Since $5 \cdot 9 = 45$ and $5 + 9 = 14$, two such numbers are **5** and **9**, and we have $p^2 + 14p + 45 = (p + 5)(p + 9)$ ***Check:*** $(p + 5)(p + 9) = p^2 + 9p + 5p + 45 = p^2 + 14p + 45$

We can use the **trial-and-check method** to factor trinomials with **leading coefficients other than 1.** Write the trinomial as the product of two binomials and determine four integers.

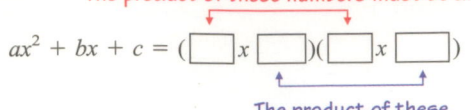

The product of these numbers must be a.

$$ax^2 + bx + c = (\boxed{}x\,\boxed{})(\boxed{}x\,\boxed{})$$

The product of these numbers must be c.

Use the FOIL method to check.

Factor: $2x^2 - 5x - 12$

Since the first term is $2x^2$, the first terms of the binomial factors must be $2x$ and x.

$$(2x\,\boxed{})(x\,\boxed{})$$ Because $2x \cdot x$ will give $2x^2$.

The second terms of the binomials must be two integers whose product is -12. There are six such pairs:

$$1(-12),\ 2(-6),\ \mathbf{3(-4)},\ 4(-3),\ 6(-2),\ \text{and}\ 12(-1)$$

The pair in blue gives the correct middle term, $-5x$, when we use the FOIL method to check:

Outer: $-8x$

$$(2x + 3)(x - 4) \qquad -8x + 3x = -5x$$

Inner: $3x$

Thus, $2x^2 - 5x - 12 = (2x + 3)(x - 4)$.

To factor $ax^2 + bx + c$ by **grouping,** write it as an equivalent four-term polynomial:

$$ax^2 + \boxed{}x + \boxed{}x + c$$

The product of these numbers must be ac, and their sum must be b.

Then factor the four-term polynomial by grouping. Use the FOIL method to check.

Factor by grouping: $2x^2 - 5x - 12$

We must find two integers whose product is $ac = 2(-12) = -24$ and whose sum is $b = -5$. Two such numbers are -8 and 3. They serve as the coefficients of $-8x$ and $3x$, the two terms that we use to represent the middle term, $-5x$, of the trinomial.

$$2x^2 - 5x - 12 = 2x^2 - 8x + 3x - 12 \qquad \text{Express } -5x \text{ as } -8x + 3x.$$
$$= 2x(x - 4) + 3(x - 4)$$
$$= (x - 4)(2x + 3) \qquad \text{Factor out } (x - 4).$$

The GCF should always be factored out first. A trinomial is **factored completely** when no factor can be factored further.

Factor: $10x^4 - 50x^3 + 60x^2$

$$10x^4 - 50x^3 + 60x^2 = 10x^2(x^2 - 5x + 6) \qquad \text{Factor out the GCF, } 10x^2.$$
$$= 10x^2(x - 3)(x - 2) \qquad \text{Factor the trinomial.}$$

REVIEW EXERCISES

Factor.

87. $x^2 + 10x + 25$

88. $49a^6 + 84a^3b^2 + 36b^4$

95. $y^8 + y^7 - 2y^6$

96. $27r^2st + 90rst - 72st$

89. $y^2 - 21y + 20$

90. $z^2 + 30 - 11z$

97. $6t^2(r + s) + 13t(r + s) - 15(r + s)$

91. $28 - x^2 - 3x$

92. $a^2 - 24b^2 - 5ab$

98. $v^4 - 13v^2 + 42$

99. $w^8 - w^4 - 90$

93. $4a^2 - 5a + 1$

94. $3b^2 + 2b + 1$

100. Use a substitution to factor $(s + t)^2 - 2(s + t) + 1$.

SECTION 5.7 ▶ The Difference of Two Squares; the Sum and Difference of Two Cubes

DEFINITIONS AND CONCEPTS	EXAMPLES
The **difference of two squares:** To factor the square of a First quantity minus the square of a Last quantity, multiply the First plus the Last by the First minus the Last. $F^2 - L^2 = (F + L)(F - L)$	Factor: $x^2y^2 - 100$ $x^2y^2 - 100 = (xy)^2 - 10^2 \qquad$ This is a difference of two squares. $= (xy + 10)(xy - 10)$

In general, the **sum of two squares** (with no common factor other than 1) cannot be factored using real numbers.	$x^2 + 100$ and $36y^2 + 49z^4$ are prime polynomials.
The **sum of two cubes:** To factor the cube of a First quantity plus the cube of a Last quantity, multiply the First plus the Last by the First squared, minus the First times the Last, plus the Last squared. $$F^3 + L^3 = (F + L)(F^2 - FL + L^2)$$	Factor: $y^3 + 27z^6$ $y^3 + 27z^6 = y^3 + (3z^2)^3$ **This is a sum of two cubes.** $= (y + 3z^2)[y^2 - y \cdot 3z^2 + (3z^2)^2]$ $= (y + 3z^2)(y^2 - 3yz^2 + 9z^4)$
The **difference of two cubes:** To factor the cube of a First quantity minus the cube of a Last quantity, multiply the First minus the Last by the First squared, plus the First times the Last, plus the Last squared. $$F^3 - L^3 = (F - L)(F^2 + FL + L^2)$$	Factor: $125s^3 - 64$ $125s^3 - 64 = (5s)^3 - 4^3$ **This is a difference of two cubes.** $= (5s - 4)[(5s)^2 + 5s \cdot 4 + 4^2]$ $= (5s - 4)(25s^2 + 20s + 16)$

REVIEW EXERCISES

Factor.

101. $z^2 - 16$

102. $x^2y^4 - 64z^6$

103. $a^2b^2 + c^2$

104. $c^2 - (a + b)^2$

105. $10m^6 - 160m^2$

106. $m^2 - n^2 + m + n$

107. $32a^4c - 162b^4c$

108. $k^2 + 2k + 1 - 9m^2$

109. $t^3 + 64$

110. $8a^3 - 125b^9$

111. $4d^7 + 4d^4$

112. $(b + c)^3 + 27$

SECTION 5.8   Summary of Factoring Techniques

DEFINITIONS AND CONCEPTS	EXAMPLES

To factor a random polynomial, use this **factoring strategy:**
 	
1. Factor out all common factors.
 	
2. How many terms does the polynomial have? If it has *two terms,* check to see whether it is
  a. The difference of two squares
  b. The sum of two cubes
  c. The difference of two cubes
  If it has *three terms,* try to factor it as
  a. A perfect-square trinomial or
  b. A general trinomial
  If it has *four or more terms,* try factoring it by grouping.
 	
3. Continue until each individual factor is prime.
 	
4. Check the results by multiplying.
	Factor: $-4t^2x^5 + 4t^2x^3y^2 - 4t^2x^2y^3 + 4t^2y^5$
	
	We first factor out the opposite of the GCF, $-4t^2$.
	
	$-4t^2x^5 + 4t^2x^3y^2 - 4t^2x^2y^3 + 4t^2y^5$
	$= -4t^2(\; x^5 - x^3y^2 \; + \; x^2y^3 - y^5 \;)$
	
	Since the expression within parentheses has four terms, we try to factor it by grouping.
	
	$= -4t^2[x^3(x^2 - y^2) + y^3(x^2 - y^2)]$
	$= -4t^2(x^2 - y^2)(x^3 + y^3)$
	
	Finally, we factor the difference of two squares and the sum of two cubes.
	
	$= -4t^2(x + y)(x - y)(x + y)(x^2 - xy + y^2)$
	
	Since each factor is prime, the factoring is complete. Check the result by multiplication.

Factor.

113. $4q^2rs + 4qrst - 120rst^2$

114. $2(m + n)^2 + (m + n) - 3$

115. $z^2 - 4 + zx - 2x$

116. $m^4 + 16n^2$

117. $x^2 + 4x + 4 - 4p^4$

118. $y^2 + 3y + 2 + 2x + xy$

119. $4a^3b^3c^2 + 256c^2$

120. $-13a^2 + 36 + a^4$

121. $4x^4 + 12x^3 + 9x^2 + 2x + 3$

122. Spanish Roof Tiles. The amount of clay used to make a roof tile is given by

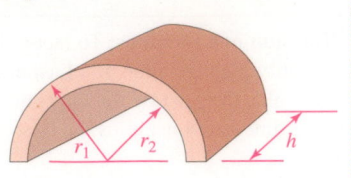

$$V = \frac{\pi}{2}r_1^2h - \frac{\pi}{2}r_2^2h$$

Factor the right side of the formula.

SECTION 5.9	▶	Solving Equations by Factoring

DEFINITIONS AND CONCEPTS	EXAMPLES
A **quadratic equation** is an equation that can be written in the **standard form** $ax^2 + bx + c = 0$, where a, b, and c are real numbers and $a \neq 0$.	Examples of quadratic equations are: $8x^2 + 64x = 0$, $25y^2 - 16 = 0$, and $5m^2 - 13m = 6$
The **zero-factor property:** If a and b are real numbers, then If $ab = 0$, then $a = 0$ or $b = 0$.	If $(x + 1)(x - 9) = 0$, then $x + 1 = 0$ or $x - 9 = 0$.

To **solve a quadratic equation by factoring:**
 	
1. Write the equation in standard form $ax^2 + bx + c = 0$.
 	
2. Factor the polynomial.
 	
3. Use the zero-factor property to set each factor equal to zero.
 	
4. Solve each resulting equation.
 	
5. Check each solution.
	Solve: $6x^2 = 5x + 6$
	
	   $6x^2 - 5x - 6 = 0$ *Write the equation in standard form.*
	
	   $(2x - 3)(3x + 2) = 0$ *Factor the trinomial.*
	
	   $2x - 3 = 0$ or $3x + 2 = 0$ *Set each factor equal to 0.*
	
	   $x = \frac{3}{2}$ $x = -\frac{2}{3}$ *Solve each equation.*
	
	The solutions are $\frac{3}{2}$ and $-\frac{2}{3}$. Check both in the original equation.

Some **polynomial equations of higher degree** can be solved by factoring. We use the extension of the zero-factor property: When the product of two *or more* real numbers is 0, at least one of them is 0.	Solve: $x^3 - 2x^2 - 8x = 0$
	
	   $x(x^2 - 2x - 8) = 0$ *Factor out the GCF, x.*
	
	   $x(x - 4)(x + 2) = 0$ *Factor the trinomial.*
	
	   $x = 0$ or $x - 4 = 0$ or $x + 2 = 0$ *Set each factor equal to 0.*
	
	           $x = 4$ $x = -2$ *Solve each equation.*
	
	The solutions are 0, 4, and -2. Check each in the original equation.

REVIEW EXERCISES

Solve each equation by factoring.

123. $4x^2 - 3x = 0$ **124.** $x^2 = 36$

125. $12x^2 = 5 - 4x$ **126.** $-d^4 + 10d^2 - 9 = 0$

127. $t^2(15t - 2) = 8t$ **128.** $u^3 = \frac{1}{3}u(19u + 14)$

129. $(y + 7)^2 + 8 = -2(y + 7) + 7$

130. $x^3 + 7x^2 - x - 7 = 0$

131. Let $f(x) = x^2 - 4x + 2$. For what value(s) of x is $f(x) = -1$?

132. Let $f(x) = 9x^2 - 14x$ and $g(x) = x + 6$. Find all values of x for which $f(x) = g(x)$.

133. Warning Flares. The height in feet $h(t)$ of a flare launched straight up into the air from the ground with an initial velocity of 112 feet per second is approximated by the quadratic function $h(t) = -16t^2 + 112t$, where t represents the number of seconds since the flare was launched. How long after launch will the flare strike the ground?

134. Pyramids. The volume of a pyramid is given by the formula $V = \frac{1}{3}lwh$, where l is the length and w is the width of its rectangular base and h is its height. The volume of the pyramid is 210 cubic meters. Find the dimensions of its rectangular base if one edge of the base is 3 meters longer than the other, and the height of the pyramid is 9 meters.

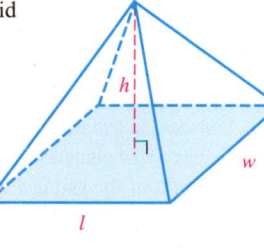

135. Gardening. A family wants to increase the size of their vegetable garden by 400 square feet by clearing the weeds off of a uniform-width border around the existing garden. How wide should the border be if the rectangular-shaped garden is currently 10 feet wide and 20 feet long?

136. Use the graph of $f(x) = 2x^2 - x - 1$, shown in the illustration, to estimate the solutions of $2x^2 - x - 1 = 0$.

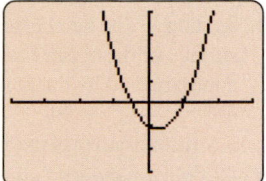

5 ▶ Chapter Test

1. Fill in the blanks.

  a. Since $x^2 + 16x + 64$ is the square of $x + 8$, it is called a _____-square trinomial.

  b. When we factor a polynomial, we write a sum of terms as a _____ of factors.

  c. The statement $x^2 - x - 12 = (x - 4)(x + 3)$ shows that the trinomial $x^2 - x - 12$ factors as the product of two _____.

  d. A _____ equation is any equation that can be written in the form $ax^2 + bx + c = 0$, where $a \neq 0$.

  e. $x^2 - y^2$ is called a _____ of two squares and $x^3 + y^3$ is called a sum of two _____.

  f. The _____ _____ factor of $12x^3$ and $8x^5$ is $4x^3$.

2. a. Write an algebraic expression that represents the area of the base of the cube.

  b. Write an algebraic expression that represents the volume of the cube.

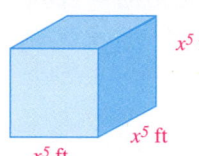

Simplify each expression. Write answers using positive exponents. Assume that no denominators are zero.

3. $9^{-1}a^{-5}m^3(a^5m^{-4})^2$ **4.** $\left(\dfrac{-2x^2y^3}{5}\right)^3$

5. $\dfrac{(-4b^2)^3(b^{-3})^3}{b^5}$ **6.** $\left(\dfrac{3m^2n^3}{m^4n^{-2}}\right)^{-2}$

7. Use scientific notation to find the quotient:

$$\frac{3{,}190{,}000{,}000{,}000}{0.00022(5{,}000{,}000{,}000)}$$

Express the answer in scientific notation and standard notation.

8. Speed of Light. Light travels 1.86×10^5 miles per second. How far does it travel in a minute? Express the answer in scientific notation.

Find the coefficient of each term and the degree of the polynomial.

9. a. $3x^3 - 4x^5 - 3x^2 - \dfrac{5}{3}$

b. $8x^5y^3 - x^8y^2 + x^9y^4 - 6x^2y^5 + 4$

10. Boating. The height (in feet) of a warning flare from the surface of the ocean t seconds after being shot into the air is approximated by the polynomial function $h(t) = -16t^2 + 80t + 10$. What is the height of the flare 2.5 seconds after being fired?

11. Use the graph of function f to find each of the following.

a. $f(4)$

b. The values of x for which $f(x) = 2$.

c. The domain and range of f.

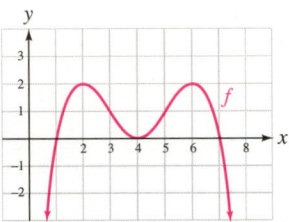

12. Let $f(x) = 2.39x^3 + 4.17x + 9.89$ and $g(x) = 1.24x^3 - 1.33x - 2.11$. Find $f(x) - g(x)$.

Perform the operations.

13. $(-2x^2y^3 + 6xy + 5y^2) - (-4x^2y^3 - 7xy + 2y^2)$

14. $\left(\dfrac{1}{2}x^5 + \dfrac{1}{3}x^2\right) + \left(\dfrac{3}{2}x^5 - \dfrac{1}{5}x^2\right)$

15. $(a^2yz^4)(2ay^5z)(-6ay^6z^7)$

16. $-5a^2b(3ab^3 - 2ab^4)$

17. $(3y^5 + 1)(2y^2 + 3y + 2)$

18. $(0.6d - 2)(0.1d + 3)$

19. $[6 + (m - n)]^2$

20. $2s(4s + 5t)(4s - 5t)$

21. Simplify: $(4t - 3)^2 - (t + 1)(t - 4)$

22. If $f(x) = x^2 - 3x + 6$, find: $f(c + 1)$

23. Let $f(x) = x^2 + 7x + 49$ and $g(x) = x - 7$. Find $f(x) \cdot g(x)$.

24. Perform the indicated operation.

a. $(4a + 3b) + (4a - b)$

b. $(4a + 3b)(4a - b)$

Factor.

25. $12a^3b^2c - 3a^2b^2c^2 + 6abc^3$

26. $hk + bz + hz + bk$

27. $x^2 - x - 30$

28. $y^4 - 81$

29. $-7x^4 + 13x^3 + 2x^2$

30. $s^4 - 13s^2 + 36$

31. $25m^2 - 40mn + 16n^2$

32. $16b^2 + 25$

33. $5x^3 + 625$

34. $64a^3 - 125b^6$

35. $(x - y)^2 + 3(x - y) - 10$

36. $6b^2 + bc - 2c^2$

37. $a^2 - b^2 + a + b$

38. $n^2 - 6n + 9 - 4m^2$

39. $a^6 - 1$

40. Solve for v: $v_1v_3 - v_3v = v_1v$

Solve each equation.

41. $5m^2 = 25m$

42. $\dfrac{2}{5}x^2 + \dfrac{3}{5}x - 7 = 0$

43. $(3x + 1)^2 + 2x = x^2 + 2(x + 5)$

44. $x^3 + 8x^2 - 9x = 0$

45. $x^3 - 16x = 16 - x^2$

46. Sailing. The triangular mainsail shown in the illustration has an area of 14 square meters. The height of the sail is 1 meter less than twice the length of the base. Find the length of its base and its height.

Mainsail

47. Product Reliability. As part of a television commercial, a piece of luggage was tossed upward from a 96-foot-tall tower to prove its durability. When it hit the ground, the luggage did not pop open. The height (in feet) of the luggage, t seconds after being thrown from the tower, is given by the function $h(t) = -16t^2 + 16t + 96$. How many seconds after being thrown did it hit the ground?

48. Statues. A bronze statue is mounted on a 5 foot by 10 foot rectangular concrete slab. The slab is surrounded by a flowerbed border of uniform width that has a total area of 54 square feet. Find the width of the border.

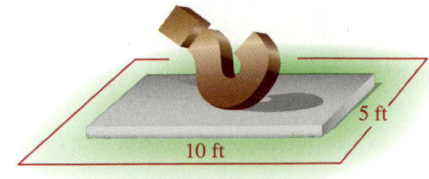

5 ft

10 ft

Group Project

A Factoring Mind Map

Overview: This activity will improve your ability to factor polynomials.

Instructions: Form groups of 2 or 3 students.

A *flowchart* is a diagram that illustrates the steps of a particular process. When completed, the flowchart shown below can be used to identify the type(s) of factoring necessary for any given polynomial having two or more terms.

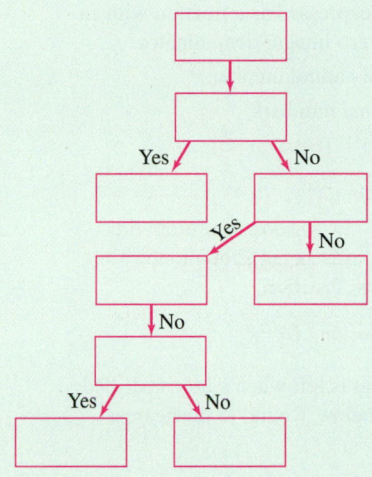

Redraw a larger version of the flowchart and write the correct statement from the list below in each box.

> Can the polynomial be factored by grouping?

> Does it factor as a perfect-square trinomial?

> Can the trinomial be factored
> 1. using the trial-and-check method?
> 2. by grouping?

> Does the polynomial have exactly two terms?

> Does it factor as a
> 1. difference of two squares?
> 2. sum of two cubes?
> 3. difference of two cubes?

> Can you find two integers whose product is the last term and whose sum is the coefficient of the middle term?

> Does the polynomial have exactly 3 terms?

> Is the leading coefficient 1?

> 1. Write the polynomial in descending powers of a variable.
> 2. Factor out the GCF.
> 3. If necessary, factor out -1 so that the leading coefficient is positive.

Use the completed flowchart to help you factor each of the following polynomials.

1. $-3x^2 + 21x - 36$ 2. $rt + 2r + st + 2s$ 3. $46w - 6 + 16w^2$
4. $v^3 - 8$ 5. $x^2 - 121y^2$ 6. $25y^2 - 20y + 4$

CUMULATIVE REVIEW Chapters 1–5

1. Determine whether each of the following statements is true or false. [Section 1.2]

   a. Every integer is a real number.

   b. Rational numbers are nonterminating, nonrepeating decimals.

   c. An irrational number can be expressed as a fraction with an integer numerator and a nonzero integer denominator.

   d. 0 is a whole number but not a natural number.

   e. All natural numbers are rational numbers.

2. Evaluate: $-3|(3^2 \cdot 5 - 2^3 \cdot 6)^2|$ [Section 1.3]

3. Simplify: $-\dfrac{7}{16}x - \dfrac{3}{4}x$ [Section 1.4]

4. Solve: $\dfrac{x+2}{5} - 4x = \dfrac{8}{5} - \dfrac{x+9}{2}$ [Section 1.5]

5. Solve $F = \dfrac{9}{5}C + 32$ for C. [Section 1.6]

6. **Machining.** Find the volume that is left when a hole is drilled through the metal block shown below. Round to the nearest hundredth. [Section 1.7]

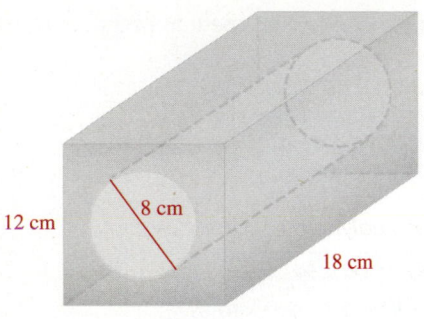

12 cm 8 cm 18 cm 12 cm

7. **Air Traffic Control.** An airplane leaves Los Angeles bound for Caracas, Venezuela, flying at an average rate of 500 mph. At the same time, another airplane leaves Caracas bound for Los Angeles, averaging 550 mph. If the airports are 3,675 miles apart, when will the air traffic controllers have to make the pilots aware that the planes are passing each other? [Section 1.8]

8. **Housekeeping.** To clean a kitchen floor, a maid mixed a concentrated cleaner with water to get the 15% solution shown below. If this mixture was too harsh on her hands, how much water would she have to add to the bucket to make it a 10% solution? (*Hint:* The strength of water is 0%.) [Section 1.8]

9. Determine whether $\left(\frac{1}{6}, 11\right)$ is a solution of $y - 6x = 10$. [Section 2.2]

10. Find the x- and y-intercepts of the graph of $2x - 5y = 10$. [Section 2.2]

11. **Health Care.** The line graph below shows the approximate projected outlays for the Medicare program. Find the rate of change of projected outlays for the years 2011 through 2019. [Section 2.4]

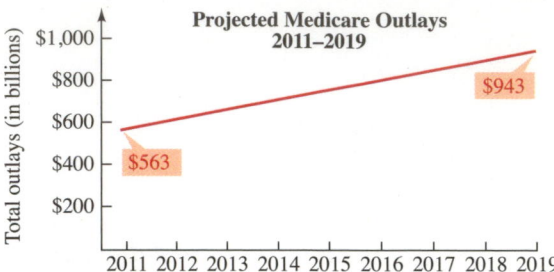

Projected Medicare Outlays 2011–2019

Total outlays (in billions)

$1,000
$800
$600
$400 $563
$200

$943

2011 2012 2013 2014 2015 2016 2017 2018 2019

Source: Kaiser Family Foundation

12. What is the slope of the graph of the line $y = 7$? [Section 2.4]

Write an equation of the line with the given properties. Express the answer in slope–intercept form.

13. Perpendicular to a line with slope $m = -\frac{1}{8}$, passing through $(-2, 5)$ [Section 2.4]

14. Passing through $(-5, 4)$ and $(8, -6)$ [Section 2.4]

15. If $h(x) = x^4 - x^3 - x^2 - x - 1$, find $h(1)$. [Section 2.5]

16. **Copier Costs.** A lease agreement for a copier charges the user a $95 monthly service fee and $2\frac{1}{2}$¢ per copy. Write a linear function that gives the cost per month to lease the copier, where x represents the total number of copies made during the month. [Section 2.5]

17. Graph $g(x) = x^2 + 2$ by translating the graph of $f(x) = x^2$. Then give the domain and range of function g. [Section 2.6]

18. Refer to the graph of function h. [Section 2.6]

    a. Find $h(5)$.   b. Find $h(0)$.

    c. Find the values of x for which $h(x) = 3$.

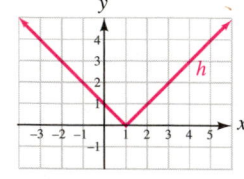

19. Determine whether the graph on the right is the graph of a function. If it is not, find two ordered pairs where more than one value of y corresponds to a single value of x. [Section 2.6]

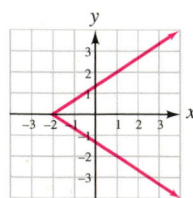

20. Solve the system by graphing. [Section 3.1]

$$\begin{cases} y = -2x + 1 \\ x - 2y = -7 \end{cases}$$

21. Solve: $\begin{cases} 4x + 6y = 5 \\ 8x = 3(1 + 3y) \end{cases}$ [Section 3.2]

22. Solve: $\begin{cases} 3x + 2y - z = -8 \\ 2x - y + 7z = 10 \\ 2x + 2y - 3z = -10 \end{cases}$ [Section 3.3]

23. Solve the system using matrices. [Section 3.4]

$$\begin{cases} 2x + y = 1 \\ x + 2y = -4 \end{cases}$$

24. Evaluate: $\begin{vmatrix} 2 & -3 & 4 \\ -1 & 2 & 4 \\ 3 & -3 & 1 \end{vmatrix}$ [Section 3.5]

25. Investments. A financial planner invested part of $50,000 at 7.5% simple annual interest and the rest at 6% simple annual interest. The total interest earned the first year was $3,240. Find the amount invested at each rate. [Section 3.6]

26. Geometry. The sum of the measure of the first angle and twice the measure of the second angle is 72° more than the measure of the third angle. If the measure of the third angle is reduced by 26°, the result is the measure of the second angle. Find the measure of each angle. [Section 3.7]

Solve each inequality. Give the solution set in interval notation and then graph it.

27. $-9(t - 3) + 2t \le 8(4 - t)$ [Section 4.1]

28. $-6 \le \dfrac{1}{3}h + 1 < 0$ [Section 4.2]

29. $|m + 5| \ge 7$ [Section 4.3]

30. $4.5x - 1 < -10$ or $6 - 2x \ge 12$ [Section 4.2]

31. Grocery Shopping.
Let $x =$ the number of items in the grocery bag shown on the right. Which statement below mathematically describes the number of items this type of bag can hold?
[Section 4.2]

  i. $x > 10$ or $x < 15$ **ii.** $x < 10$ and $x > 15$

  iii. $10 \le x \le 15$ **iv.** $x > 10$ and $x < 15$

32. Graph the solution set:

$$\begin{cases} x - y < 4 \\ y \le 0 \\ x \ge 0 \end{cases}$$ [Section 4.5]

33. Simplify $\left(\dfrac{-3a^4b^2}{-9a^5b^{-2}} \right)^{-2}$. Write the answer without using negative exponents. [Section 5.1]

34. Write 9.0895×10^{-8} in standard notation. [Section 5.2]

35. Let $f(x) = 0.4x^3 + 0.6x$ and $g(x) = -0.3x^3 + 0.2x$. Find $f(x) - g(x)$. [Section 5.3]

36. Multiply: $(4a^2b - 3c^3)(9a^2b - 2c^3)$ [Section 5.4]

37. Multiply: $(2a - b)(4a^2 + 2ab + b^2)$ [Section 5.4]

38. Simplify: $(3k + 1)^2 + (2k - 4)^2$ [Section 5.4]

Factor each polynomial completely.

39. $x^2 + 4y - xy - 4x$ [Section 5.5]

40. $6s^4 - 216s^2$ [Section 5.7]

41. $8x^6 + 125y^3$ [Section 5.7]

42. $-3a^2 + ab + 2b^2$ [Section 5.6]

43. $a^3 + 5a^2 + 6a + a^2b + 5ab + 6b$ [Section 5.8]

44. Solve: $x^2 = \dfrac{1}{2}(x + 1)$ [Section 5.9]

45. Solve $m_1m_2 = m_2g + m_1g$ for m_1. [Section 5.9]

46. Communications. The number of telephone connections $C(n)$ that can be made among n telephones is given by the quadratic function $C(n) = \dfrac{1}{2}(n^2 - n)$. How many telephones are needed to make 66 connections? [Section 5.9]

Rational Expressions and Equations

6

©Ronen/Shutterstock.com

from Campus to Careers

Webmaster

If you use the Internet, then you have seen first-hand what webmasters do. They design and maintain websites for individuals and companies on the World Wide Web. The job of webmaster requires excellent computer and technical skills. A background in business, art, and design is also helpful. Since webmasters are often called on to be troubleshooters when technical difficulties arise, those considering entering the field are encouraged to study math to strengthen their problem-solving abilities.

Problem 99 in **Study Set 6.1, problem 25** in **Study Set 6.8,** and **problem 83 in Study Set 6.9** involve situations that a webmaster might encounter on the job. The mathematical concepts discussed in this chapter can be used to solve those problems.

JOB TITLE:
Webmaster
EDUCATION:
Many webmasters have college degrees. However, some have only a year or two of college training.
JOB OUTLOOK:
Excellent—job opportunities are expected to increase between 18% to 26% through the year 2014.
ANNUAL EARNINGS:
Average salary $73,830
FOR MORE INFORMATION:
http://www.bls.gov/k12/computers05.htm

Reading an algebra textbook is different from reading a newspaper or a novel. Here are two ways that you should be reading this textbook.

SKIMMING FOR AN OVERVIEW: This is a quick way to look at material just *before* it is covered in class. It helps you become familiar with the new vocabulary and notation that will be used by your instructor in the lecture. It lays a foundation.

READING FOR UNDERSTANDING: This in-depth type of reading is done more slowly, with a pencil and paper at hand. Don't skip anything—every word counts! You should do just as much writing as you do reading. Highlight the important points and work each example. If you become confused, stop and reread the material until you understand it.

Now Try This ▶

Choose a section from this chapter and . . .

1. Quickly skim it. Write down any terms in boldface type and the titles of any properties, definitions, or strategies that are given in the colored boxes.
2. Work each *Self Check* problem. Your solutions should look like those in the *Examples*. Include author notes (the sentences in red to the right of each step of a solution).

SECTION 6.1

Rational Functions and Simplifying Rational Expressions

OBJECTIVES

1. Define rational expressions and rational functions.
2. Evaluate rational functions.
3. Find the domain of a rational function.
4. Recognize the graphs of rational functions.
5. Simplify rational expressions.
6. Simplify rational expressions that have factors that are opposites.

ARE YOU READY?

▼ *The following problems review some basic skills that are needed when simplifying rational expressions.*

1. Evaluate: **a.** $\dfrac{0}{10}$ **b.** $\dfrac{10}{0}$ **c.** $\dfrac{9}{-9}$

2. Simplify: $\dfrac{36}{28}$

3. Factor: $12a^5 - 16a^4$

4. Factor: $x^2 - 100$

5. Factor: $5n^2 + 23n - 10$

6. Factor: $2x^3 + x^2 - 14x - 7$

We have used linear and polynomial functions to model many real-world situations. In this section, we introduce another family of functions known as *rational functions*.

1 Define Rational Expressions and Rational Functions.

Fractions that are the quotient of two integers, with the denominator not 0, are *rational numbers*. For example, $\frac{7}{8}$ and $-\frac{15}{4}$ are rational numbers. Fractions that are the quotient of two polynomials, again with the denominator not 0, are called *rational expressions*.

Rational Expressions	A **rational expression** is an expression of the form $\frac{A}{B}$, where A and B are polynomials and B does not equal 0.

Some examples of rational expressions are

$$\frac{3x}{x-7}, \qquad \frac{8yz^4}{6y^2z^2}, \qquad \frac{5m+n}{m^2+4mn+4n^2}, \qquad \text{and} \qquad \frac{6a^2-13a+6}{a^3+3a^2+a-2}$$

The Language of Algebra	The rational expression $\frac{3x}{x-7}$ is the quotient of the monomial $3x$ and the binomial $x - 7$.

The Language of Algebra

Rational functions get their name from the fact that their defining equation contains a *ratio* (fraction) of two polynomials.

The rational expression $\frac{3x}{x-7}$ is the quotient of the monomial $3x$ and the binomial $x - 7$. It is in one variable, x. The rational expression $\frac{5m+n}{m^2 + 4mn + 4n^2}$ is the quotient of the binomial $5m + n$ and the trinomial $m^2 + 4mn + 4n^2$. It is in two variables, m and n.

Rational expressions in one variable are used to define *rational functions*.

Rational Functions	A **rational function** is a function whose equation is defined by a rational expression in one variable, where the value of the polynomial in the denominator is never zero.

Caution

Since division by 0 is undefined, the value of a polynomial in the denominator of a rational expression cannot be 0. For example, x cannot be -8 in the rational expression $\frac{1}{x+8}$, because the value of the denominator would be 0.

Two examples of rational functions are

$$f(x) = \frac{1}{x+8}$$

 The rational expression $\frac{1}{x+8}$ that defines function f is in one variable, x.

$$s(t) = \frac{5t^2 + t}{t^2 + 6t - 1}$$

 The rational expression $\frac{5t^2 + t}{t^2 + 6t - 1}$ that defines function s is in one variable, t.

2 Evaluate Rational Functions.

Rational functions can be used to model many types of real-world situations.

EXAMPLE 1

Internet Research. The cost of subscribing to an online research network is $6 a month plus $1.50 per hour of access time. The rational function $c(n) = \frac{1.50n + 6}{n}$ gives the average hourly cost $c(n)$ of using the network for n hours a month. Find the average hourly cost for

a. a student who used the network for 2 hours in a month.

b. an instructor who used the network for 18 hours in a month.

Strategy We will find $c(2)$ and $c(18)$.

Why The notation $c(2)$ represents the average hourly cost for using the network for 2 hours in a month and $c(18)$ represents the average hourly cost for using the network for 18 hours in a month.

Solution **a.** To find the average hourly cost the student paid for 2 hours of access time in a month, we find $c(2)$.

$$c(\mathbf{2}) = \frac{1.50(\mathbf{2}) + 6}{\mathbf{2}} = 4.5$$ *Substitute 2 for n and evaluate the right side.*

The student paid $4.50 per hour to use the online research network for 2 hours.

b. To find the average hourly cost for 18 hours of access time in a month, we find $c(18)$.

$$c(\mathbf{18}) = \frac{1.50(\mathbf{18}) + 6}{\mathbf{18}} = 1.833333333$$ *Substitute 18 for n and evaluate the right side.*

The instructor paid approximately $1.83 per hour to use the online research network for 18 hours.

Self Check 1 Find the average hourly cost when the online research network is used for 100 hours in a month.

Now Try ▶ Problem 97

3 Find the Domain of a Rational Function.

Recall that the **domain** of a function is the set of all permissible input values for the variable. Since division by 0 is undefined, any input values that make the denominator 0 in a rational function must be excluded from the domain of the function.

EXAMPLE 2

Find the domain of the function: $f(x) = \dfrac{3x + 2}{x^2 + x - 6}$

Strategy We will set $x^2 + x - 6$ equal to 0 and solve for x.

Why We don't need to examine the numerator of the rational expression; it can be any value, including 0. The domain of the function includes all real numbers, except those that make the *denominator equal to* 0.

Solution

$$x^2 + x - 6 = 0 \qquad \text{Set the denominator equal to 0.}$$
$$(x + 3)(x - 2) = 0 \qquad \text{Factor the trinomial.}$$
$$x + 3 = 0 \quad \text{or} \quad x - 2 = 0 \qquad \text{Set each factor equal to 0.}$$
$$x = -3 \qquad\qquad x = 2 \qquad \text{Solve each linear equation.}$$

The Language of Algebra

Two other ways that Example 2 could be phrased are:

- **State the restrictions** on the variable.
- Find the values of x for which the function is **undefined**.

Thus, the domain of the function is the set of all real numbers except -3 and 2. Using set-builder notation we can describe the domain as $\{x \mid x$ is a real number and $x \neq -3, x \neq 2\}$. In interval notation, the domain is $(-\infty, -3) \cup (-3, 2) \cup (2, \infty)$.

To check the answers, substitute -3 and 2 for x in $x^2 + x - 6$ and verify that the result is 0 for each.

Self Check 2 Find the domain of: $f(x) = \dfrac{x^2 + 1}{x^2 - 49}$

Now Try ▶ Problem 25

4 Recognize the Graphs of Rational Functions.

The simplest of all rational functions, $f(x) = \dfrac{1}{x}$, is called the **reciprocal function.** We can use the point-plotting method to graph it. Because of the complex shape of the graph, we must select a large number of x-values, including positive and negative fractions, to determine how it looks. After plotting all of the ordered pairs, we draw a smooth curve through the points that form a "branch" on the left and a smooth curve through the points that form a "branch" on the right. Since 0 is not in the domain of the function, there is no point on the graph for $x = 0$.

$f(x) = \dfrac{1}{x}$

x	$f(x)$	
-4	$-\dfrac{1}{4}$	$\rightarrow \left(-4, -\dfrac{1}{4}\right)$
-3	$-\dfrac{1}{3}$	$\rightarrow \left(-3, -\dfrac{1}{3}\right)$
-2	$-\dfrac{1}{2}$	$\rightarrow \left(-2, -\dfrac{1}{2}\right)$
-1	-1	$\rightarrow (-1, -1)$
$-\dfrac{1}{2}$	-2	$\rightarrow \left(-\dfrac{1}{2}, -2\right)$
$-\dfrac{1}{3}$	-3	$\rightarrow \left(-\dfrac{1}{3}, -3\right)$
$-\dfrac{1}{4}$	-4	$\rightarrow \left(-\dfrac{1}{4}, -4\right)$

↑ Select x.   ↑ Find the reciprocal.   ↑ Plot the point.

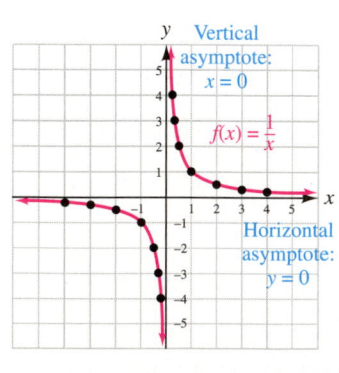

$f(x) = \dfrac{1}{x}$

x	$f(x)$	
4	$\dfrac{1}{4}$	$\rightarrow \left(4, \dfrac{1}{4}\right)$
3	$\dfrac{1}{3}$	$\rightarrow \left(3, \dfrac{1}{3}\right)$
2	$\dfrac{1}{2}$	$\rightarrow \left(2, \dfrac{1}{2}\right)$
1	1	$\rightarrow (1, 1)$
$\dfrac{1}{2}$	2	$\rightarrow \left(\dfrac{1}{2}, 2\right)$
$\dfrac{1}{3}$	3	$\rightarrow \left(\dfrac{1}{3}, 3\right)$
$\dfrac{1}{4}$	4	$\rightarrow \left(\dfrac{1}{4}, 4\right)$

↑ Select x.   ↑ Find the reciprocal.   ↑ Plot the point.

When a graph approaches a line, we call the line an **asymptote.** In this case, as x gets smaller and approaches 0, the graph of $f(x) = \frac{1}{x}$ approaches the y-axis. We say that the y-axis is a **vertical asymptote** of the graph.

On the far left and far right, the graph gets steadily closer to the x-axis. We say that the x-axis is a **horizontal asymptote** of the graph.

Using Your Calculator ▶ **Graphing Rational Functions**

We can graph rational functions using a graphing calculator. However, rational functions usually have a value (or values) of x that are excluded from the domain, and that can cause accuracy problems.

As an example, consider the rational function in Example 2, $f(x) = \frac{3x + 2}{x^2 + x - 6}$. Recall, $x = 2$ and $x = -3$ are not in the domain of this function. If we use window settings of $[-10, 10]$ for x and $[-10, 10]$ for y and graph the function in the **connected mode,** we obtain the graph in figure (a) below.

The display shows two vertical lines on the screen where asymptotes should appear. This is because graphing calculators draw graphs by connecting dots whose x-coordinates are close together. Often, when two such points straddle a vertical asymptote and their y-coordinates are far apart, the calculator draws a line between them.

If, instead, the calculator is set in **dot mode,** the vertical lines will not appear. See figure (b). The graph looks more like one that we would draw by hand.

Both graphs suggest that the domain of the function is $(-\infty, -3) \cup (-3, 2) \cup (2, \infty)$ and that $x = -3$ and $x = 2$ are vertical asymptotes. This can be confirmed using the Table feature, the Calculate Value feature, or the Trace feature.

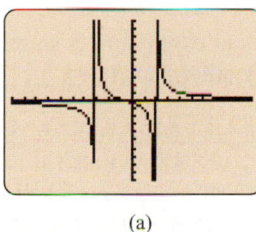 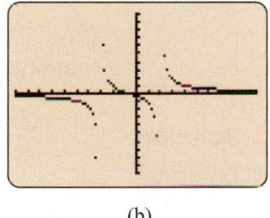

(a)   (b)

From figure (a), we can also see that

■ As x increases to the right of 2, the values of y decrease and approach the line $y = 0$.
■ As x decreases to the left of -3, the values of y increase and approach the line $y = 0$.

The line $y = 0$ (the x-axis) is a horizontal asymptote. Graphing calculators do not draw lines that appear to be horizontal asymptotes.

---

5 **Simplify Rational Expressions.**

When working with rational expressions, we will use some familiar rules from arithmetic.

Properties of Fractions

If a, b, c, d, and k represent real numbers, and if there are no divisions by 0, then

1. $\dfrac{a}{b} = \dfrac{c}{d}$ if and only if $ad = bc$

2. $\dfrac{a}{1} = a$ and $\dfrac{a}{a} = 1$

3. $\dfrac{ak}{bk} = \dfrac{a}{b} \cdot \dfrac{k}{k} = \dfrac{a}{b}$

4. $-\dfrac{a}{b} = \dfrac{-a}{b} = \dfrac{a}{-b}$

Property 3 is true because any number times 1 is that number.

$$\frac{ak}{bk} = \frac{a}{b} \cdot \frac{k}{k} = \frac{a}{b} \cdot 1 = \frac{a}{b} \quad \text{where} \quad b \neq 0 \quad \text{and} \quad k \neq 0$$

To streamline this process, we can replace $\frac{k}{k}$ in $\frac{ak}{bk}$ with the equivalent fraction $\frac{1}{1}$.

$$\frac{ak}{bk} = \frac{a\overset{1}{\cancel{k}}}{b\underset{1}{\cancel{k}}} = \frac{a}{b} \qquad \frac{k}{k} = \frac{1}{1} = 1$$

We say that we have simplified $\frac{ak}{bk}$ by *removing a factor equal to 1*.

To **simplify a rational expression** means to write it so that the numerator and denominator have no common factors other than 1.

Simplifying Rational Expressions	1. Factor the numerator and denominator completely to determine their common factors.
	2. Remove factors equal to 1 by replacing each pair of factors common to the numerator and denominator with the equivalent fraction $\frac{1}{1}$.
	3. Multiply the remaining factors in the numerator and in the denominator.

EXAMPLE 3 Simplify: $\dfrac{8yz^4}{6y^2z^2}$

Strategy We will begin by writing the numerator and denominator in factored form. Then we will remove any factors common to the numerator and denominator.

Why The rational expression is simplified when the numerator and denominator have no common factors other than 1.

Solution

$$\frac{8yz^4}{6y^2z^2} = \frac{2 \cdot 2 \cdot 2 \cdot y \cdot z \cdot z \cdot z \cdot z}{2 \cdot 3 \cdot y \cdot y \cdot z \cdot z}$$

To prepare to simplify the rational expression, factor $8yz^4$ and $6y^2z^2$ completely.

$$= \frac{2 \cdot 2 \cdot 2 \cdot \overset{1}{\cancel{y}} \cdot \overset{1}{\cancel{z}} \cdot \overset{1}{\cancel{z}} \cdot z \cdot z}{2 \cdot 3 \cdot \underset{1}{\cancel{y}} \cdot y \cdot \underset{1}{\cancel{z}} \cdot \underset{1}{\cancel{z}}}$$

Replace $\frac{2}{2}$, $\frac{y}{y}$, and $\frac{z}{z}$ with $\frac{1}{1}$. This removes the factor $\frac{2 \cdot y \cdot z \cdot z}{2 \cdot y \cdot z \cdot z} = 1$.

$$= \frac{4z^2}{3y}$$

Multiply the remaining factors in the numerator.
Multiply the remaining factors in the denominator.

Since $\frac{8yz^4}{6y^2z^2}$ is undefined for $y = 0$ and $z = 0$, the expressions $\frac{8yz^4}{6y^2z^2}$ and $\frac{4z^2}{3y}$ are equal only if $y \neq 0$ and $z \neq 0$. That is,

$$\frac{8yz^4}{6y^2z^2} = \frac{4z^2}{3y} \qquad \text{provided } y \neq 0 \text{ and } z \neq 0.$$

An alternate approach is to use the rules for exponents to simplify the rational expressions that are the quotient of two monomials.

$$\frac{8yz^4}{6y^2z^2} = \frac{\overset{1}{\cancel{2}} \cdot 2 \cdot 2 \cdot y^{1-2}z^{4-2}}{\underset{1}{\cancel{2}} \cdot 3} = \frac{4y^{-1}z^2}{3} = \frac{4z^2}{3y}$$

To divide exponential expressions with the same base, keep the base and subtract the exponents.

Self Check 3 Simplify: $\dfrac{12a^4b^2}{20ab^4}$

Now Try ▶ Problem 33

To simplify rational expressions, we often make use of the factoring methods discussed in Chapter 5.

EXAMPLE 4   Simplify: **a.** $\dfrac{6x^2}{3x^4 - 9x^3}$ **b.** $\dfrac{x^2 - 16}{2x^2 + 8x}$

Strategy   We will begin by factoring the numerator and denominator. Then we will remove any factors common to the numerator and denominator.

Why   We need to make sure that the numerator and denominator have no common factors other than 1. If that is the case, then the rational expression is simplified.

Solution   **a.** $\dfrac{6x^2}{3x^4 - 9x^3} = \dfrac{2 \cdot 3 \cdot x \cdot x}{3x^3(x - 3)}$

To prepare to simplify, factor the numerator. In the denominator, factor out the GCF, $3x^3$.

$$= \dfrac{\overset{1}{2} \cdot \overset{1}{\cancel{3}} \cdot \overset{1}{\cancel{x}} \cdot \cancel{x}}{\underset{1}{\cancel{3}} \cdot \underset{1}{\cancel{x}} \cdot \underset{1}{\cancel{x}} \cdot x \cdot (x - 3)}$$

Remove the factors common to the numerator and denominator.

$$= \dfrac{2}{x(x - 3)}$$

It is not necessary to perform the multiplication $x(x - 3)$ in the result. For later work, It is usually more convenient to leave the denominator in factored form.

b. In the numerator, to prepare to simplify, we factor the difference of two squares. In the denominator, we factor out the GCF, $2x$.

$$\dfrac{x^2 - 16}{2x^2 + 8x} = \dfrac{\overset{1}{\cancel{(x + 4)}}(x - 4)}{2x\underset{1}{\cancel{(x + 4)}}}$$

Remove the binomial factor $x + 4$ that is common to the numerator and denominator.

$$= \dfrac{x - 4}{2x}$$

This rational expression does not simplify further.

Self Check 4   Simplify: **a.** $\dfrac{28x^4}{7x^5 - 14x^4}$ **b.** $\dfrac{x^2 - 9}{5x^2 - 15x}$

Now Try ▶   Problems 35 and 39

When a rational expression is simplified, the result is an **equivalent expression.** In Example 4b, for instance, this means that $\dfrac{x^2 - 16}{2x^2 + 8x}$ and $\dfrac{x - 4}{2x}$ have the same value for **all** values of x, except those that make either denominator 0. We can use that fact to perform an informal check of our work. If we let $x = \mathbf{1}$, for example, we see that the original rational expression and the simplified expression have the same value, $-\dfrac{3}{2}$.

The original rational expression

$$\dfrac{x^2 - 16}{2x^2 + 8x} = \dfrac{(\mathbf{1})^2 - 16}{2(\mathbf{1})^2 + 8(\mathbf{1})}$$

$$= \dfrac{1 - 16}{2 + 8}$$

$$= -\dfrac{15}{10}$$

$$= -\dfrac{3}{2}$$

The resulting simplified expression

$$\dfrac{x - 4}{2x} = \dfrac{\mathbf{1} - 4}{2(\mathbf{1})}$$

$$= -\dfrac{3}{2}$$ Same result

If the results are different, an error has been made, and the problem should be reworked.

Using Your Calculator ▶ Checking an Algebraic Simplification

After simplifying an expression, we can use a scientific calculator to check the answer. One way to check whether $\dfrac{x^2 - 16}{2x^2 + 8x} = \dfrac{x - 4}{2x}$ is correct in Example 4b is to evaluate $\dfrac{x^2 - 16}{2x^2 + 8x}$ and $\dfrac{x - 4}{2x}$ for a value of x (say, **2**). The expressions should give identical results.

For $\dfrac{x^2 - 16}{2x^2 + 8x}$: $\boxed{(}\,\boxed{2}\,\boxed{x^2}\,\boxed{-}\,16\,\boxed{)}\,\boxed{\div}\,\boxed{(}\,\boxed{2}\,\boxed{\times}\,\boxed{2}\,\boxed{x^2}\,\boxed{+}\,8\,\boxed{\times}\,\boxed{2}\,\boxed{)}\,\boxed{=}$ -0.5

For $\dfrac{x - 4}{2x}$: $\boxed{(}\,\boxed{2}\,\boxed{-}\,4\,\boxed{)}\,\boxed{\div}\,\boxed{(}\,2\,\boxed{\times}\,\boxed{2}\,\boxed{)}\,\boxed{=}$ -0.5

The results of the evaluations are indeed the same. Evaluate the expressions for several other values of x. If the results differ for any given value, the original expression was not simplified correctly.

We also can use a graphing calculator to show that the simplification in Example 4b is correct. We enter the functions $f(x) = \dfrac{x^2 - 16}{2x^2 + 8x}$ and $g(x) = \dfrac{x - 4}{2x}$ as Y_1 and Y_2, respectively. See figure (a) below. Then select the TABLE feature. Reading across the table, the values of Y_1 and Y_2 should be the same for each value of x as shown in figure (b). Note for $x = -4$ the Y_1 value says error while the Y_2 value is 1. This happens as a result of removing the common factor $x + 4$ in the simplification.

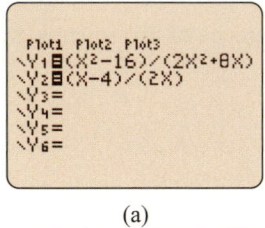

(a)

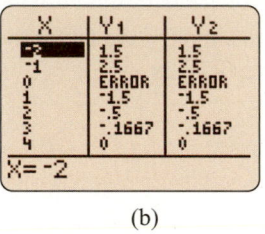

(b)

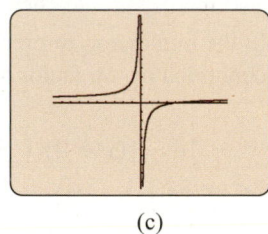

(c)

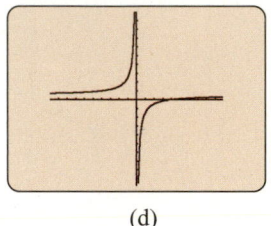

(d)

A third method to check the simplification informally is to compare the graphs of $f(x) = \dfrac{x^2 - 16}{2x^2 + 8x}$, shown in figure (c) above, and $g(x) = \dfrac{x - 4}{2x}$, shown in figure (d) above. Since the graphs appear to be the same, we can conclude that the simplification is probably correct.

EXAMPLE 5   Simplify: **a.** $\dfrac{x^2 - 10x + 25}{8x - 40}$ **b.** $\dfrac{6a^2 - 13a + 6}{3a^2 + a - 2}$

Strategy   We will begin by factoring the numerator and denominator completely. Then we will remove any factors common to the numerator and denominator.

Why   We need to make sure that the numerator and denominator have no common factors other than 1. If that is the case, then the rational expression is simplified.

Solution   **a.** To prepare to simplify, we factor the perfect-square trinomial in the numerator. In the denominator, we factor out the GCF, 8. Then we remove the common factor, $x - 5$.

$$\frac{x^2 - 10x + 25}{8x - 40} = \frac{\overset{1}{\cancel{(x - 5)}}(x - 5)}{8\underset{1}{\cancel{(x - 5)}}}$$ Remove a factor equal to 1: $\dfrac{x-5}{x-5} = 1$.

$$= \frac{x - 5}{8}$$

b. We factor the trinomials in the numerator and the denominator and then remove the common factor, $3a - 2$.

$$\frac{6a^2 - 13a + 6}{3a^2 + a - 2} = \frac{\overset{1}{\cancel{(3a - 2)}}(2a - 3)}{\underset{1}{\cancel{(3a - 2)}}(a + 1)} \qquad \text{Remove a factor equal to 1: } \frac{3a - 2}{3a - 2} = 1.$$

$$= \frac{2a - 3}{a + 1} \qquad \text{This expression does not simplify further.}$$

Self Check 5   Simplify: **a.** $\dfrac{x^2 - 6x + 9}{6x - 18}$ **b.** $\dfrac{2b^2 + 7b - 15}{2b^2 + 13b + 15}$

Now Try ▶ Problems 43 and 47

| **EXAMPLE 6** | Simplify the function given by: $f(x) = \dfrac{x^3 - 8}{x^3 - 2x^2 + 3x - 6}$. List any restrictions on the domain. |

Success Tip

When simplifying functions, it is necessary to give any restrictions on the domain.

Strategy   We will begin by factoring the numerator and denominator completely. Then we will remove any factors common to the numerator and denominator.

Why   We need to make sure that the numerator and denominator have no common factors other than 1. If that is the case, then the rational expression that defines the function is simplified.

Solution

Caution

When simplifying rational expressions, we can remove only the factors common to the entire numerator and denominator. It is incorrect to remove terms common to the numerator and denominator.

$$\frac{\overset{1}{\cancel{x^3}} - 8}{\underset{1}{\cancel{x^3}} - 2x^2 + 3x - 6}$$

x^3 is a term of $x^3 - 8$ and a term of $x^3 - 2x^2 + 3x - 6$.

$$f(x) = \frac{x^3 - 8}{x^3 - 2x^2 + 3x - 6}$$

$$= \frac{(x - 2)(x^2 + 2x + 4)}{x^2(x - 2) + 3(x - 2)} \qquad \text{In the numerator, factor the sum of two cubes. In the denominator, begin the process of factoring by grouping.}$$

$$= \frac{(x - 2)(x^2 + 2x + 4)}{(x - 2)(x^2 + 3)} \qquad \text{In the denominator, complete the factoring by grouping. It is now apparent that the function is undefined for } x = 2.$$

$$= \frac{\overset{1}{\cancel{(x - 2)}}(x^2 + 2x + 4)}{\underset{1}{\cancel{(x - 2)}}(x^2 + 3)} \qquad \text{Remove the factor common to the numerator and denominator: } \frac{x - 2}{x - 2} = 1.$$

$$= \frac{x^2 + 2x + 4}{x^2 + 3} \qquad \text{This expression does not simplify further.}$$

The simplified form is $f(x) = \dfrac{x^2 + 2x + 4}{x^2 + 3}$, provided $x \neq 2$.

Self Check 6   Simplify the function given by: $h(a) = \dfrac{a^3 - 1}{a^3 - a^2 + 6a - 6}$.
List any restrictions on the domain.

Now Try ▶ Problem 55

Sometimes we will encounter rational expressions that are already in simplified form. For example, to attempt to simplify the following expression, we factor the numerator and denominator as shown:

$$\frac{x^2 + xa + 2x + 2a}{x^2 + x - 6} = \frac{x(x + a) + 2(x + a)}{(x - 2)(x + 3)} \qquad \text{In the numerator, begin the process of factoring by grouping. In the denominator, factor the trinomial.}$$

$$= \frac{(x + a)(x + 2)}{(x - 2)(x + 3)} \qquad \text{In the numerator, complete the factoring by grouping.}$$

Since there are no common factors in the numerator and denominator, the rational expression is in *simplified form.*

6 Simplify Rational Expressions That Have Factors That Are Opposites.

If the terms of two polynomials are the same, except that they are opposite in sign, the polynomials are **opposites.** For example, $b - a$ and $a - b$ are opposites.

To simplify $\frac{b - a}{a - b}$, the quotient of opposites, we factor -1 from the numerator and remove any factors common to both the numerator and the denominator:

$$\frac{b - a}{a - b} = \frac{-a + b}{a - b} \qquad \text{\color{red}{Rewrite the numerator.}}$$

$$= \frac{\overset{1}{-(a - b)}}{\underset{1}{(a - b)}} \qquad \text{\color{red}{Factor out -1 from each term in the numerator and remove the common factor $a - b$.}}$$

$$= \frac{-1}{1}$$

$$= -1$$

In general, we have the following principle.

The Quotient of Opposites	The quotient of any nonzero polynomial and its opposite is -1.

EXAMPLE 7 Simplify: $\dfrac{3x^2 - 10xy - 8y^2}{4y^2 - xy}$

Strategy We will begin by factoring the numerator and denominator. Then we look for common factors, or factors that are opposites, and remove them.

Why We need to make sure that the numerator and denominator have no common factors other than 1. If that is the case, then the rational expression is simplified.

Solution We factor the numerator and denominator. Because $x - 4y$ and $4y - x$ are opposites, their quotient is -1.

$$\frac{3x^2 - 10xy - 8y^2}{4y^2 - xy} = \frac{(3x + 2y)\overset{-1}{(x - 4y)}}{\underset{1}{y(4y - x)}} \qquad \text{\color{red}{Since $x - 4y$ and $4y - x$ are opposites,}}$$
$$\text{\color{red}{simplify by replacing $\frac{x - 4y}{4y - x}$ with the}}$$
$$\text{\color{red}{equivalent fraction $\frac{-1}{1} = -1$.}}$$

$$= \frac{-(3x + 2y)}{y}$$

This result also can be written as $-\dfrac{3x + 2y}{y}$ or $\dfrac{-3x - 2y}{y}$.

Self Check 7 Simplify: $\dfrac{2a^2 - 3ab - 9b^2}{3b^2 - ab}$

Now Try ▶ Problem 59

SECTION **6.1**   STUDY SET

VOCABULARY

Fill in the blanks.

1. A quotient of two polynomials, such as $\frac{x^2 + x}{x^2 - 3x}$, is called a _____ expression.

2. A _____ function, such as $f(x) = \frac{x - 7}{x^2 - x - 6}$, is a function whose equation is defined by a rational expression in one variable.

3. In the rational expression $\frac{(x + 2)(3x - 1)}{(x + 2)(4x + 2)}$, the binomial $x + 2$ is a common _____ of the numerator and the denominator.

4. To _____ a rational expression, we remove factors common to the numerator and denominator.

5. Because of the division by 0, the expression $\frac{8}{0}$ is _____.

6. The binomials $x - 15$ and $15 - x$ are called _____, because their terms are the same, except that they are opposite in sign.

7. The _____ of a function is the set of all permissible input values for the variable.

8. The rational function $f(x) = \frac{9x}{x - 10}$ is _____ for $x = 10$. In other words, there is a _____ on the domain of the function: $x \neq 10$.

CONCEPTS

9. The graph of rational function f is shown below. Find each of the following.

   a. $f(1)$
   b. $f(4)$
   c. The value of x for which $f(x) = -2$
   d. The value of x for which $f(x) = 1$

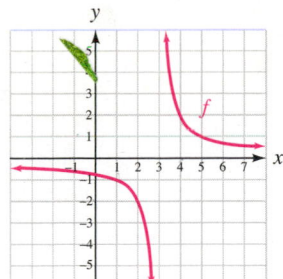

10. Fill in the blanks to simplify $\frac{x - y}{y - x}$.

$$\frac{x - y}{y - x} = \frac{-y + \boxed{}}{y - x} = \frac{\boxed{}(y - x)}{(y - x)} = \boxed{}$$

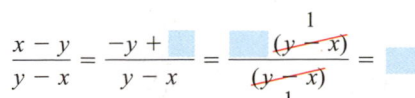

11. Simplify each expression.

   a. $\frac{3 \cdot 5 \cdot x \cdot y \cdot y}{5 \cdot 7 \cdot x \cdot x \cdot x \cdot y}$
   b. $\frac{(x + 8)(x - 3)}{(x + 2)(x + 8)}$

   c. $\frac{a^3(a - 9)}{(9 - a)(9 + a)}$

12. Simplify each rational expression, if possible.

   a. $\frac{x + 8}{x}$
   b. $\frac{3a^2 + 23}{a^2}$

13. Match each function with the correct graph shown below.

   a. $f(x) = 2$
   b. $f(x) = x$
   c. $f(x) = x^2$
   d. $f(x) = x^3$
   e. $f(x) = |x|$
   f. $f(x) = \frac{1}{x}$

   i 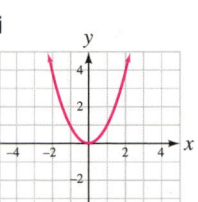 ii 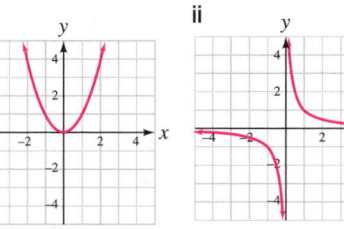 iii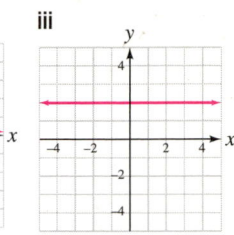

   iv 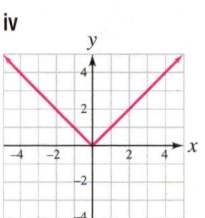 v 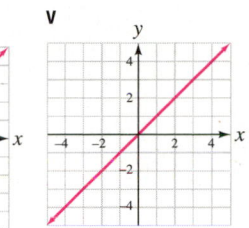 vi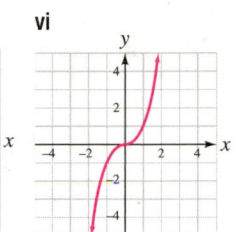

14. For what value(s) of x is each function undefined?

   a. $f(x) = \frac{x - 7}{x}$
   b. $f(x) = \frac{x + 1}{x - 3}$

   c. $f(x) = \frac{x^2 - 2}{x(x + 8)}$
   d. $f(x) = \frac{8x}{(x - 1)(x + 1)}$

15. Let $f(x) = \frac{2x + 1}{x^2 + 3x - 4}$. Find

   a. $f(0)$
   b. $f(2)$
   c. $f(1)$

16. Fill in the blank: The rational expressions $\frac{x + 2}{x^2 - 4}$ and $\frac{1}{x - 2}$ are equivalent. They have the _____ value for all values of x, except those that make either denominator 0.

NOTATION

17. A student checks his answers with those in the back of his textbook. Determine whether they are equivalent.

Answer	Book's answer	Equivalent?
$\frac{-3}{x + 3}$	$-\frac{3}{x + 3}$	
$\frac{-x + 4}{6x + 1}$	$\frac{-(x - 4)}{6x + 1}$	
$\frac{x + 7}{(x - 4)(x + 2)}$	$\frac{x + 7}{(x + 2)(x - 4)}$	
$\frac{-x - 4}{x + 4}$	$\frac{4 - x}{x - 4}$	
$\frac{a - 3b}{2b - a}$	$\frac{3b - a}{a - 2b}$	

18. What numbers are not included in each set of real numbers represented using interval notation?

a. $(-\infty, 4) \cup (4, \infty)$

b. $(-\infty, -8) \cup (-8, 0) \cup (0, \infty)$

GUIDED PRACTICE

Find the domain of each rational function. Express your answer in words and using interval notation. See Example 2.

19. $f(x) = \dfrac{2}{x}$

20. $f(x) = \dfrac{8}{x - 1}$

21. $f(x) = \dfrac{2x}{x + 2}$

22. $f(x) = \dfrac{2x + 1}{x^2 - 2x}$

23. $f(x) = \dfrac{3x - 1}{x - x^2}$

24. $f(x) = \dfrac{x^2 + 36}{x^2 - 36}$

25. $f(x) = \dfrac{x^2 + 3x + 2}{x^2 - x - 56}$

26. $f(x) = \dfrac{2x^2 - 3x - 2}{x^2 + 2x - 24}$

Simplify each rational expression. See Example 3.

27. $\dfrac{12a^3}{18a}$

28. $\dfrac{25b^4}{55b}$

29. $\dfrac{15a^2}{25a^8}$

30. $\dfrac{12x}{16x^7}$

31. $\dfrac{27st}{36st^2}$

32. $\dfrac{49xy^2}{21xy}$

33. $\dfrac{24x^3y^{10}}{18x^4y^3}$

34. $\dfrac{15a^5b^4}{21a^8b^3}$

Simplify each rational expression. See Example 4.

35. $\dfrac{4x^2}{2x^3 - 12x^2}$

36. $\dfrac{15y^2}{5y^3 + 15y^2}$

37. $\dfrac{24n^4}{16n^4 + 24n^3}$

38. $\dfrac{18m^4}{36m^4 - 9m^3}$

39. $\dfrac{2x + 18}{x^2 - 81}$

40. $\dfrac{6x - 12}{x^2 - 4}$

41. $\dfrac{4a^2 - 25}{20a - 50}$

42. $\dfrac{9b^2 - 16}{21b + 28}$

Simplify each rational expression. See Example 5.

43. $\dfrac{5x^2 - 10x}{x^2 - 4x + 4}$

44. $\dfrac{x^2 + 6x + 9}{2x^2 + 6x}$

45. $\dfrac{x^2 + 2x + 1}{x^2 + 4x + 3}$

46. $\dfrac{y^2 - 4y + 4}{y^2 - 8y + 12}$

47. $\dfrac{3d^2 + 13d + 4}{3d^2 + 7d + 2}$

48. $\dfrac{10r^2 + 17r + 3}{2r^2 + 17r + 21}$

49. $\dfrac{2h^2 + 9h - 5}{4h^2 - 4h + 1}$

50. $\dfrac{6x^2 + x - 2}{8x^2 + 2x - 3}$

Simplify each function. List any restrictions on the domain. See Example 6.

51. $f(x) = \dfrac{x^2 - 1}{x^2 + 5x - 6}$

52. $f(x) = \dfrac{x^2 + 6x - 16}{x^2 - 4}$

53. $f(x) = \dfrac{9x^2 + 81x}{x^3 + 9x^2}$

54. $f(x) = \dfrac{5x^2 + 50x}{x^5 + 10x^4}$

55. $g(x) = \dfrac{x^3 + 27}{x^3 + 3x^2 + 4x + 12}$

56. $g(x) = \dfrac{x^3 + 64}{x^3 + 4x^2 + 3x + 12}$

57. $s(a) = \dfrac{a^3 - a^2 - 6a + 6}{a^3 - 1}$

58. $h(t) = \dfrac{t^3 - 5t^2 - 5t + 25}{t^3 - 125}$

Simplify each rational expression. See Example 7.

59. $\dfrac{3m^2 - 2mn - n^2}{mn - m^2}$

60. $\dfrac{5s^2 - 4st - t^2}{st - s^2}$

61. $\dfrac{b^2 - a^2}{a - b}$

62. $\dfrac{d^2 - 16c^2}{4c - d}$

63. $\dfrac{4 - x^2}{x^2 - x - 2}$

64. $\dfrac{x^2 - 2x - 15}{25 - x^2}$

65. $\dfrac{20x^3 - 20x^4}{x^2 - 2x + 1}$

66. $\dfrac{16m^5 - 2m^6}{m^2 - 16m + 64}$

Use a graphing calculator to graph each rational function. From the graph, determine any vertical asymptotes. See Using Your Calculator: Graphing Rational Functions.

67. $f(x) = \dfrac{x}{x - 2}$

68. $f(x) = \dfrac{x + 2}{x}$

69. $f(x) = \dfrac{x + 1}{x^2 - 4}$

70. $f(x) = \dfrac{x - 2}{x^2 - 3x - 4}$

TRY IT YOURSELF

Simplify each expression. If an expression cannot be simplified, write "Does not simplify."

71. $\dfrac{x^2 + x - 30}{3x^2 - 3x - 60}$

72. $\dfrac{4x^2 + 24x + 32}{16x^2 + 8x - 48}$

73. $\dfrac{a^2 - 4}{a^3 - 8}$

74. $\dfrac{x^3 - 27}{3x^2 - 8x - 3}$

75. $\dfrac{m^3 - mn^2}{mn^2 + m^2n - 2m^3}$

76. $\dfrac{a^3 - ab^2}{ab^2 - 4a^2b + 3a^3}$

77. $\dfrac{sx + 4s - 3x - 12}{sx + 4s + 6x + 24}$

78. $\dfrac{ax + by + ay + bx}{a + b}$

79. $\dfrac{2x^2 - 3x - 9}{2x^2 + 3x - 9}$

80. $\dfrac{6x^2 - 7x - 5}{2x^2 + 5x + 2}$

81. $\dfrac{3x + 6y}{x + 2y}$

82. $\dfrac{y - xy}{xy - x}$

83. $\dfrac{x^4 + 3x^3 + 9x^2}{x^3 - 27}$

84. $\dfrac{x^3 + 8}{x^4 - 2x^3 + 4x^2}$

85. $\dfrac{2x^2 + 2x - 12}{x^3 + 3x^2 - 4x - 12}$

86. $\dfrac{3x^2 - 3y^2}{x^2 + 2y + 2x + yx}$

87. $\dfrac{4x^2 + 8x + 3}{6 + x - 2x^2}$

88. $\dfrac{6x^2 + 13x + 6}{6 - 5x - 6x^2}$

89. $\dfrac{x^2 - 6x + 9}{81 - x^4}$

90. $\dfrac{y^2 - 2y + 1}{1 - y^4}$

91. $\dfrac{16p^3q^2}{24pq^8}$

92. $\dfrac{30a^3b^{15}}{18a^9b^{10}}$

93. $\dfrac{t^3 - 5t^2 + 6t}{9t - t^3}$

94. $\dfrac{a^4 - 27a}{36a - 4a^3}$

APPLICATIONS

95. **Environmental Cleanup.** Suppose the cost (in dollars) of removing $p\%$ of the pollution in a river is given by the rational function
$$f(p) = \dfrac{50{,}000p}{100 - p} \quad \text{where } 0 \le p < 100$$
Find the cost of removing each percent of pollution.
a. 50%
b. 80%

96. **Directory Costs.** The average (mean) cost for a service club to publish a directory of its members is given by the rational function
$$f(x) = \dfrac{1.25x + 700}{x}$$
where x is the number of directories printed. Find the average cost per directory if
a. 500 directories are printed.
b. 2,000 directories are printed.

97. **Utility Costs.** An electric company charges $7.50 per month plus 9¢ for each kilowatt hour (kwh) of electricity used.
a. Find a linear function that gives the total cost of n kwh of electricity. (*Hint:* See Example 1.)
b. Find a rational function that gives the average cost per kwh when using n kwh.
c. Find the average cost per kwh when 775 kwh are used.

98. **Filling a Pool.** The rational function
$$f(t) = \dfrac{t^2 + 3t}{2t + 3}$$
gives the number of hours it would take two pipes, working together, to fill a pool that the larger pipe (working alone) could fill in t hours and the smaller pipe (working alone) could fill in $t + 3$ hours.
a. If the smaller pipe could fill a pool in 7 hours, how long would it take both pipes to fill the pool?
b. If the larger pipe could fill a pool in 8 hours, how long would it take both pipes to fill the pool?

99. ▶ from **Campus to Careers**

Webmaster

The rational function
$$f(t) = \dfrac{t^2 + 2t}{2t + 2}$$
gives the number of days it would take two webpage designers, working together, to design a standard website for a business that designer 1 (working alone) could complete in t days and designer 2 (working alone) could complete in $t + 2$ days.
a. If designer 1 could complete the website in 15 days, how long would it take both designers working together?
b. If designer 2 could complete a website in 20 days, how long would it take both designers working together?

100. **Retention Study.** After learning a list of words, two subjects were tested over a 28-day period to see what percent of the list they remembered. In both cases, their percent recall could be modeled by rational functions, as shown in the graph below.
a. Use the graphs to complete the table.

Days since learning	0	1	2	4	7	14	28
% recall—subject 1							
% recall—subject 2							

b. After 28 days, which subject had the better recall?

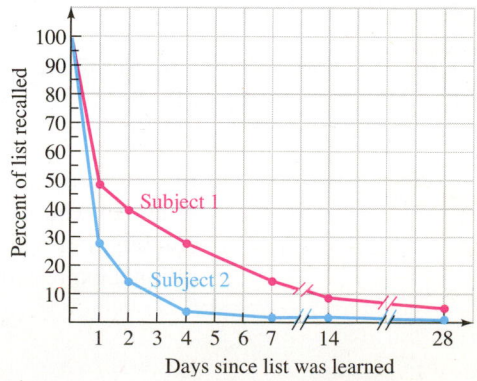

WRITING

101. a. In $\dfrac{(x + 5)\overset{1}{(x - 5)}}{x\underset{1}{(x - 5)}}$, what do the slashes show?

b. In $\dfrac{(x - 3)\overset{-1}{(x - 7)}}{(x + 3)\underset{1}{(7 - x)}}$, what do the slashes show?

102. What does it mean when we say that $\dfrac{3x - 12}{3x + 15}$ and $\dfrac{x - 4}{x + 5}$ are equivalent expressions?

103. A student simplified $\dfrac{6x^2 - 7x - 5}{2x^2 + 5x + 2}$ and obtained $\dfrac{3x - 5}{x - 2}$. As a check, she graphed $Y_1 = \dfrac{6x^2 - 7x - 5}{2x^2 + 5x + 2}$ (shown below on the left) and $Y_2 = \dfrac{3x - 5}{x - 2}$ (shown below on the right). What conclusion can be drawn from the graphs? Explain your answer.

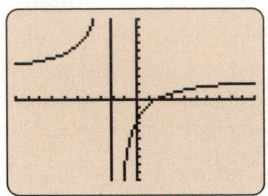 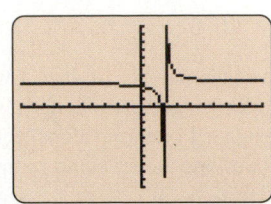

104. Simplify: $\dfrac{6x^2 + x - 2}{8x^2 + 2x - 3}$. Then explain how the table of values for $Y_1 = \dfrac{6x^2 + x - 2}{8x^2 + 2x - 3}$ and $Y_2 = \dfrac{3x + 2}{4x + 3}$ shown on the right can be used to check your result.

X	Y1	Y2
-2	.8	.8
-1	1	1
0	.66667	.66667
1	.71429	.71429
2	.72727	.72727
3	.73333	.73333
4	.73684	.73684

X= -2

REVIEW

Perform each operation.

105. $(a^2 - 4a - 3)(a - 2)$

106. $(3c^2 + 5c) + (7 - c^2 - 5c)$

107. $-3mn^2(m^3 - 7mn - 2m^2)$

108. $(4u^2 + z^2 - 3u^2z^2) - (u^3 + 3z^2 - 3u^2z^2)$

CHALLENGE PROBLEMS

Simplify each expression.

109. $\dfrac{x^{32} - 1}{x^{16} - 1}$

110. $\dfrac{20m^2(m^2 - 1) - 47m(1 - m^2) + 24(m^2 - 1)}{4m^2 - m - 3}$

111. $\dfrac{a^6 - 64}{(a^2 + 2a + 4)(a^2 - 2a + 4)}$

112. $\dfrac{(p + q)^3 + 64}{(p + q)^2 - 16}$

Graph each rational function. Show the vertical asymptote as a dashed line and label it.

113. $f(x) = \dfrac{1}{x - 1}$

114. $f(x) = \dfrac{1}{x + 4}$

SECTION 6.2

Multiplying and Dividing Rational Expressions

OBJECTIVES

1 Multiply rational expressions.

2 Find powers of rational expressions.

3 Divide rational expressions.

4 Perform mixed operations.

ARE YOU READY?

The following problems review some basic skills that are needed when multiplying and dividing rational expressions.

1. Multiply: $\dfrac{3}{8} \cdot \dfrac{1}{7}$

2. What is the reciprocal of $\dfrac{2}{3}$?

3. Divide: $\dfrac{21}{25} \div \dfrac{7}{15}$

4. Factor: $x^3 - x^4$

5. Simplify: $\dfrac{x(x + 5)(x - 7)}{4x(x - 5)(x - 7)}$

6. Simplify: $\dfrac{x^3 - 125}{2x - 10}$

In this section, we review the rules for multiplying and dividing arithmetic fractions—fractions whose numerators and denominators are integers. Then we use these rules, in combination with the simplification skills learned in Section 6.1, to multiply and divide rational expressions.

1 Multiply Rational Expressions.

Recall that to multiply fractions, we multiply the numerators and multiply the denominators. Then we simplify the result, if possible. For example,

$$\frac{3}{5} \cdot \frac{2}{7} = \frac{3 \cdot 2}{5 \cdot 7} \qquad\qquad \frac{4}{7} \cdot \frac{5}{8} = \frac{4 \cdot 5}{7 \cdot 8}$$

$$= \frac{6}{35} \qquad\qquad = \frac{\overset{1}{\cancel{2}} \cdot \overset{1}{\cancel{2}} \cdot 5}{7 \cdot \underset{1}{\cancel{2}} \cdot \underset{1}{\cancel{2}} \cdot 2}$$

Factor 4 as 2 · 2. Factor 8 as 2 · 2 · 2. Then simplify.

$$= \frac{5}{14}$$

We use the same procedure to multiply rational expressions.

Multiplying Rational Expressions	To multiply two rational expressions, multiply their numerators and their denominators. Then, if possible, factor and simplify.
	For any two rational expressions, $\frac{A}{B}$ and $\frac{C}{D}$,
	$$\frac{A}{B} \cdot \frac{C}{D} = \frac{AC}{BD}$$

EXAMPLE 1   Multiply: $\dfrac{25a^3}{17b} \cdot \dfrac{b}{5a}$

Strategy   To find the product, we will use the rule for multiplying rational expressions. In the process, we must be prepared to factor the numerators and denominators so that any common factors can be removed.

Why   We want to give the result in simplified form.

Solution

$$\frac{25a^3}{17b} \cdot \frac{b}{5a} = \frac{25a^3 \cdot b}{17b \cdot 5a}$$

Multiply the numerators.
Multiply the denominators.

It is obvious that the numerator and denominator of $\frac{25a^3 \cdot b}{17b \cdot 5a}$ have several common factors, such as 5, a, and b. These common factors become more apparent when we factor the numerator and denominator completely.

$$\frac{25a^3 \cdot b}{17b \cdot 5a} = \frac{5 \cdot 5 \cdot a \cdot a \cdot a \cdot b}{17 \cdot b \cdot 5 \cdot a}$$

To prepare to simplify, factor $25a^3$.

$$= \frac{\overset{1}{\cancel{5}} \cdot 5 \cdot \overset{1}{\cancel{a}} \cdot a \cdot a \cdot \overset{1}{\cancel{b}}}{17 \cdot \underset{1}{\cancel{b}} \cdot \underset{1}{\cancel{5}} \cdot \underset{1}{\cancel{a}}}$$

Simplify by replacing $\frac{5}{5}, \frac{a}{a}$, and $\frac{b}{b}$ with the equivalent fraction $\frac{1}{1}$. This removes the factor $\frac{5 \cdot a \cdot b}{5 \cdot a \cdot b} = 1$.

$$= \frac{5a^2}{17}$$

Multiply the remaining factors in the numerator: $1 \cdot 5 \cdot 1 \cdot a \cdot a \cdot 1 = 5a^2$.
Multiply the remaining factors in the denominator: $17 \cdot 1 \cdot 1 \cdot 1 = 17$.

Success Tip

We also could use the rules for exponents to simplify the product:

$$\frac{25a^3 \cdot b}{17b \cdot 5a} = \frac{\overset{1}{\cancel{5}} \cdot 5 \cdot a^{3-1} \cdot b^{1-1}}{17 \cdot \underset{1}{\cancel{5}}}$$

$$= \frac{5a^2 b^0}{17}$$

$$= \frac{5a^2}{17}$$

Self Check 1   Multiply: $\dfrac{x^7}{16y} \cdot \dfrac{24y}{17x^3}$

Now Try ▶ Problem 17

EXAMPLE 2   Multiply: **a.** $\dfrac{x^2 - 6x + 9}{20x} \cdot \dfrac{5x^2}{6x - 18}$ **b.** $\dfrac{x^2 - x - 6}{x^2 - 4} \cdot \dfrac{x^2 + x - 6}{x^2 - 9}$

Strategy   To find the product, we will use the rule for multiplying rational expressions. In the process, we must be prepared to factor the numerators and denominators so that any common factors can be removed.

Why   We want to give the result in simplified form.

Solution   **a.** $\dfrac{x^2 - 6x + 9}{20x} \cdot \dfrac{5x^2}{6x - 18} = \dfrac{(x^2 - 6x + 9)5x^2}{20x(6x - 18)}$ Multiply the numerators.
Multiply the denominators.

$= \dfrac{(x - 3)(x - 3)5xx}{4 \cdot 5 \cdot x \cdot 6(x - 3)}$ To prepare to simplify, factor the numerator and factor the denominator.

$= \dfrac{\overset{1}{\cancel{(x - 3)}}(x - 3)\overset{11}{\cancel{5}}xx}{4 \cdot \underset{1}{\cancel{5}} \cdot \underset{1}{\cancel{x}} \cdot 6\underset{1}{\cancel{(x - 3)}}}$ Simplify by removing common factors of the numerator and denominator.

$= \dfrac{x(x - 3)}{24}$ Multiply the remaining monomial factors in the numerator: $1 \cdot 1 \cdot 1 \cdot x = x$.
Multiply the remaining factors in the denominator: $4 \cdot 1 \cdot 1 \cdot 6 \cdot 1 = 24$.

b. $\dfrac{x^2 - x - 6}{x^2 - 4} \cdot \dfrac{x^2 + x - 6}{x^2 - 9} = \dfrac{(x^2 - x - 6)(x^2 + x - 6)}{(x^2 - 4)(x^2 - 9)}$ Multiply the numerators.
Multiply the denominators.

$= \dfrac{(x - 3)(x + 2)(x + 3)(x - 2)}{(x + 2)(x - 2)(x + 3)(x - 3)}$ To prepare to simplify, factor the polynomials.

$= \dfrac{\overset{1}{\cancel{(x - 3)}}\overset{1}{\cancel{(x + 2)}}\overset{1}{\cancel{(x + 3)}}\overset{1}{\cancel{(x - 2)}}}{\underset{1}{\cancel{(x + 2)}}\underset{1}{\cancel{(x - 2)}}\underset{1}{\cancel{(x + 3)}}\underset{1}{\cancel{(x - 3)}}}$ Simplify by removing common factors of the numerator and denominator.

$= 1$

Self Check 2   Multiply: **a.** $\dfrac{a^2 + 6a + 9}{18a} \cdot \dfrac{3a^3}{7a + 21}$

b. $\dfrac{a^2 + a - 56}{a^2 - 49} \cdot \dfrac{a^2 - a - 56}{a^2 - 64}$

Now Try ▶   Problems 19 and 23

Caution

When multiplying rational expressions, always write the result in simplest form by removing any factors common to the numerator and denominator.

Notation

In part a, we could distribute in the numerator and write the result as $\frac{x^2 - 3x}{24}$. Check with your instructor to see which form of the result he or she prefers.

Caution

Note that when all of the factors of the numerator and denominator are removed as shown, the result is 1 and not 0.

Using Your Calculator ▶   **Checking an Algebraic Simplification**

We can check the simplification in Example 2(a) by graphing the functions $f(x) = \left(\dfrac{x^2 - 6x + 9}{20x}\right)\left(\dfrac{5x^2}{6x - 18}\right)$, shown in figure (a), and $g(x) = \dfrac{x(x - 3)}{24}$, shown in figure (b), and observing that the graphs are the same, except that 0 and 3 are not included in the domain of the function f.

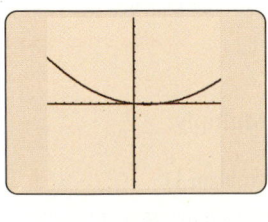

(a)   (b)

We can use the split-screen G-T (graph, table) mode to check the result of a multiplication. To set the split-screen feature on a graphing calculator, press $\boxed{\text{MODE}}$, press ▼ seven times, press ▶ twice, then press $\boxed{\text{ENTER}}$. If we enter $Y_3 = Y_1 - Y_2$, use the cursor to highlight the = sign as shown in figure (c), and then press $\boxed{\text{GRAPH}}$, we get the display shown in figure (d). The zeros under the Y_3 column indicate that the value of $\left(\frac{x^2 - 6x + 9}{20x}\right)\left(\frac{5x^2}{6x - 18}\right)$ and the value of $\frac{x(x - 3)}{24}$ are the same for different values of x. (The error message is given because when $x = 0$ and $x = 3$, $\left(\frac{x^2 - 6x + 9}{20x}\right)\left(\frac{5x^2}{6x - 18}\right)$ is undefined.)

The graph of $Y_3 = Y_1 - Y_2$ is difficult to see because it lies on the x-axis. The graph indicates that for all x-values (except those that make the rational expressions undefined), $Y_3 = 0$, or more specifically, $\left(\frac{x^2 - 6x + 9}{20x}\right)\left(\frac{5x^2}{6x - 18}\right) = \frac{x(x - 3)}{24}$.

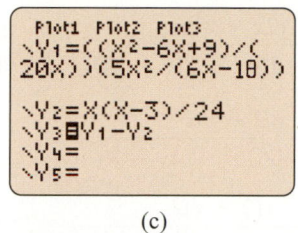

(c)

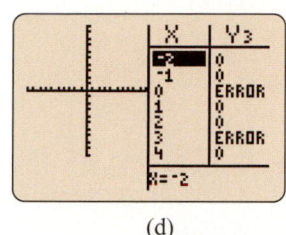

(d)

EXAMPLE 3

Multiply: $\dfrac{6x^2 + 5xy - 4y^2}{2x^2 + 5xy + 3y^2} \cdot \dfrac{8x^2 + 6xy - 9y^2}{12x^2 + 7xy - 12y^2}$

Strategy To find the product, we will use the rule for multiplying rational expressions. In the process, we must be prepared to factor the numerators and denominators so that any common factors can be removed.

Why We want to give the result in simplified form.

Solution

$\dfrac{6x^2 + 5xy - 4y^2}{2x^2 + 5xy + 3y^2} \cdot \dfrac{8x^2 + 6xy - 9y^2}{12x^2 + 7xy - 12y^2}$

$= \dfrac{(6x^2 + 5xy - 4y^2)(8x^2 + 6xy - 9y^2)}{(2x^2 + 5xy + 3y^2)(12x^2 + 7xy - 12y^2)}$ Multiply the numerators.
Multiply the denominators.

$= \dfrac{(3x + 4y)(2x - y)(4x - 3y)(2x + 3y)}{(2x + 3y)(x + y)(3x + 4y)(4x - 3y)}$ To prepare to simplify, factor the trinomials.

$= \dfrac{\overset{1}{\cancel{(3x + 4y)}}(2x - y)\overset{1}{\cancel{(4x - 3y)}}\overset{1}{\cancel{(2x + 3y)}}}{\underset{1}{\cancel{(2x + 3y)}}(x + y)\underset{1}{\cancel{(3x + 4y)}}\underset{1}{\cancel{(4x - 3y)}}}$ Simplify by removing common factors of the numerator and denominator.

$= \dfrac{2x - y}{x + y}$ Multiply the remaining factors in the numerator: $1 \cdot (2x - y) \cdot 1 \cdot 1 = 2x - y$. Multiply the remaining factors in the denominator: $1 \cdot (x + y) \cdot 1 \cdot 1 = x + y$.

Self Check 3 Multiply: $\dfrac{2a^2 + 5ab - 12b^2}{2a^2 + 11ab + 12b^2} \cdot \dfrac{2a^2 - 3ab - 9b^2}{2a^2 - ab - 3b^2}$

Now Try ▶ Problems 29 and 31

EXAMPLE 4 Multiply: $(2x - x^2) \cdot \dfrac{x}{5x^3 - 10x^2 + 20x - 40}$

Strategy We will write $2x - x^2$ as a rational expression with denominator 1. (Remember, any number divided by 1 remains unchanged.) Then we will use the rule for multiplying rational expressions.

Why Writing $2x - x^2$ as $\frac{2x - x^2}{1}$ is helpful during the multiplication process when we multiply numerators and multiply denominators.

Solution $(2x - x^2) \cdot \dfrac{x}{5x^3 - 10x^2 + 20x - 40}$

$= \dfrac{2x - x^2}{1} \cdot \dfrac{x}{5x^3 - 10x^2 + 20x - 40}$ Write $2x - x^2$ as $\frac{2x - x^2}{1}$.

$= \dfrac{(2x - x^2)x}{1(5x^3 - 10x^2 + 20x - 40)}$ Multiply the numerators.
Multiply the denominators.

$= \dfrac{x(2 - x)x}{1 \cdot 5(x^3 - 2x^2 + 4x - 8)}$ To prepare to simplify, factor out x in the numerator and factor out 5 in the denominator.

$= \dfrac{x(2 - x)x}{1 \cdot 5[x^2(x - 2) + 4(x - 2)]}$ In the denominator, begin factoring by grouping. Perform the factoring within the brackets.

$= \dfrac{x(2 - x)x}{1 \cdot 5(x - 2)(x^2 + 4)}$ In the denominator, complete the factoring by grouping. The brackets [ ] are no longer needed.

$= \dfrac{x(2 - \overset{-1}{\cancel{x)}}x}{1 \cdot 5\underset{1}{(\cancel{x - 2})}(x^2 + 4)}$ Simplify. Recall that the quotient of any nonzero quantity and its opposite is -1: $\frac{2 - x}{x - 2} = -1$.

$= \dfrac{-x^2}{5(x^2 + 4)}$ Multiply the remaining factors in the numerator: $x \cdot (-1) \cdot x = -x^2$. Multiply the remaining monomial factors in the denominator: $1 \cdot 5 \cdot 1 = 5$.

Since the $-$ sign can be written in front of the fraction, this result can be expressed as

$$-\dfrac{x^2}{5(x^2 + 4)}$$

Self Check 4 Multiply: $\dfrac{x}{8x^3 - 32x^2 + 8x - 32}(4x - x^2)$

Now Try ▶ Problem 39

2 Find Powers of Rational Expressions.

EXAMPLE 5 Find: $\left(\dfrac{x^2 + x - 1}{2x + 3}\right)^2$

Strategy We will find the product $\left(\frac{x^2 + x - 1}{2x + 3}\right)\left(\frac{x^2 + x - 1}{2x + 3}\right)$ using the rule for multiplying rational expressions.

Why The exponent 2 means the base, $\frac{x^2 + x - 1}{2x + 3}$, should be written as a factor two times.

Solution

$$\left(\frac{x^2 + x - 1}{2x + 3}\right)^2 = \left(\frac{x^2 + x - 1}{2x + 3}\right)\left(\frac{x^2 + x - 1}{2x + 3}\right)$$

$$= \frac{(x^2 + x - 1)(x^2 + x - 1)}{(2x + 3)(2x + 3)} \qquad \text{Multiply the numerators. Multiply the denominators.}$$

$$= \frac{x^4 + 2x^3 - x^2 - 2x + 1}{4x^2 + 12x + 9} \qquad \text{Use a special-product rule to find the denominator.}$$

Self Check 5 Find: $\left(\dfrac{x + 5}{x^2 - 6x}\right)^2$

Now Try ▶ Problem 43

3 Divide Rational Expressions.

Recall that one number is called the **reciprocal** of another if their product is 1. To find the reciprocal of a fraction, we invert its numerator and denominator. We have seen that to divide fractions, we multiply the first fraction by the reciprocal of the second fraction. Then we simplify the result, if possible.

$$\frac{3}{5} \div \frac{8}{9} = \frac{3}{5} \cdot \frac{9}{8} \qquad\qquad \frac{4}{7} \div \frac{2}{21} = \frac{4}{7} \cdot \frac{21}{2}$$

$$= \frac{3 \cdot 9}{5 \cdot 8} \qquad\qquad\qquad = \frac{4 \cdot 21}{7 \cdot 2}$$

$$= \frac{27}{40} \qquad\qquad\qquad = \frac{\overset{1}{2} \cdot 2 \cdot 3 \cdot \overset{1}{7}}{\underset{1}{7} \cdot \underset{1}{2}} \qquad \text{Factor 4 as } 2 \cdot 2. \text{ Factor } 21 \text{ as } 3 \cdot 7. \text{ Then simplify.}$$

$$= 6$$

We use the same procedure to divide rational expressions.

Dividing Rational Expressions

To divide two rational expressions, multiply the first by the reciprocal of the second. Then, if possible, factor and simplify.

For any two rational expressions, $\dfrac{A}{B}$ and $\dfrac{C}{D}$, where $\dfrac{C}{D} \neq 0$,

$$\frac{A}{B} \div \frac{C}{D} = \frac{A}{B} \cdot \frac{D}{C} = \frac{AD}{BC}$$

EXAMPLE 6 Divide: $\dfrac{6}{y^{19}z^{28}} \div \dfrac{20}{yz^{32}}$

Strategy We will use the rule for dividing rational expressions. After multiplying by the reciprocal, we will use rules for exponents to simplify the result.

Why We want to give the result in simplified form.

Solution

$$\frac{6}{y^{19}z^{28}} \div \frac{20}{yz^{32}} = \frac{6}{y^{19}z^{28}} \cdot \frac{yz^{32}}{20}$$

Multiply the first rational expression by the reciprocal of the second.

$$= \frac{6yz^{32}}{20y^{19}z^{28}}$$

Multiply the numerators. Multiply the denominators.

$$= \frac{2 \cdot 3 \cdot y^{1-19}z^{32-28}}{2 \cdot 10}$$

To prepare to simplify, factor 6 and 20. To divide exponential expressions with the same base, keep the base and subtract the exponents.

$$= \frac{\overset{1}{\cancel{2}} \cdot 3 \cdot y^{-18}z^4}{\underset{1}{\cancel{2}} \cdot 10}$$

Remove the common factor of 2. Simplify the exponents.

$$= \frac{3z^4}{10y^{18}}$$

Write the result without the negative exponent.

Self Check 6 Divide: $\frac{8a^{43}}{t^{21}} \div \frac{28a^{37}}{t^{18}}$

Now Try Problem 51

EXAMPLE 7 Divide: $\frac{x^3 + 8}{4x + 4} \div \frac{x^2 - 2x + 4}{2x^2 - 2}$

Strategy To find the quotient, we will use the rule for dividing rational expressions. After multiplying by the reciprocal, we will factor each polynomial that is not prime and remove any common factors of the numerator and denominator.

Why We want to give the result in simplified form.

Solution

$$\frac{x^3 + 8}{4x + 4} \div \frac{x^2 - 2x + 4}{2x^2 - 2}$$

$$= \frac{x^3 + 8}{4x + 4} \cdot \frac{2x^2 - 2}{x^2 - 2x + 4}$$

Multiply the first rational expression by the reciprocal of the second.

$$= \frac{(x^3 + 8)(2x^2 - 2)}{(4x + 4)(x^2 - 2x + 4)}$$

Multiply the numerators. Multiply the denominators.

$$= \frac{(x + 2)(\overset{1}{\cancel{x^2 - 2x + 4}})2(\overset{1}{\cancel{x + 1}})(x - 1)}{2 \cdot \underset{1}{\cancel{2}}(\underset{1}{\cancel{x + 1}})(\underset{1}{\cancel{x^2 - 2x + 4}})}$$

To prepare to simplify, factor completely. Note that $x^2 - 2x + 4$ is prime. Then simplify.

$$= \frac{(x + 2)(x - 1)}{2}$$

Multiply the remaining monomial factors in the numerator: $1 \cdot 1 \cdot 1 = 1$. Multiply the remaining factors in the denominator: $2 \cdot 1 \cdot 1 \cdot 1 = 2$.

Self Check 7 Divide: $\frac{x^3 - 8}{9x - 9} \div \frac{x^2 + 2x + 4}{3x^2 - 3x}$

Now Try Problem 55

EXAMPLE 8 Divide: $\frac{b^3 - 4b}{x - 1} \div (b - 2)$

Strategy We will begin by writing $b - 2$ as a rational expression by inserting a denominator 1. Then we will use the rule for dividing rational expressions.

Why Writing $b - 2$ over 1 is helpful when we invert its numerator and denominator to find its reciprocal.

Solution $\dfrac{b^3 - 4b}{x - 1} \div (b - 2) = \dfrac{b^3 - 4b}{x - 1} \div \dfrac{b - 2}{1}$ Write $b - 2$ as a fraction with a denominator of 1.

$$= \dfrac{b^3 - 4b}{x - 1} \cdot \dfrac{1}{b - 2}$$ Multiply the first rational expression by the reciprocal of the second.

$$= \dfrac{b^3 - 4b}{(x - 1)(b - 2)}$$ Multiply the numerators. Multiply the denominators.

$$= \dfrac{b(b + 2)(\overset{1}{\cancel{b - 2}})}{(x - 1)(\underset{1}{\cancel{b - 2}})}$$ To prepare to simplify, factor $b^3 - 4b$. Then remove the common factor: $b - 2$.

$$= \dfrac{b(b + 2)}{x - 1}$$ Multiply the remaining monomial factors in the numerator: $b \cdot 1 = b$. Multiply the remaining factors in the denominator.

Self Check 8 Divide: $\dfrac{m^4 - 9m^2}{a^2 - 3a} \div (m^2 + 3m)$

Now Try ▶ Problem 63

4 Perform Mixed Operations.

EXAMPLE 9 Simplify: $\dfrac{x^2 + 2x - 3}{6x^2 + 5x + 1} \div \dfrac{2x^2 - 2}{2x^2 - 5x - 3} \cdot \dfrac{6x^2 + 4x - 2}{x^2 - 2x - 3}$

Strategy   We will consider the division first by multiplying the first rational expression by the reciprocal of the second. Then we will find the product of the three rational expressions.

Why   By the rules for the order of operations, we must perform division and multiplication in order from left to right.

Solution   Since multiplications and divisions are done in order from left to right, we begin by focusing on the division. We introduce grouping symbols to emphasize this. To divide the expressions within the parentheses, we invert $\dfrac{2x^2 - 2}{2x^2 - 5x - 3}$ and multiply.

$$\left(\dfrac{x^2 + 2x - 3}{6x^2 + 5x + 1} \div \dfrac{2x^2 - 2}{2x^2 - 5x - 3}\right)\dfrac{6x^2 + 4x - 2}{x^2 - 2x - 3} = \left(\dfrac{x^2 + 2x - 3}{6x^2 + 5x + 1} \cdot \dfrac{2x^2 - 5x - 3}{2x^2 - 2}\right)\dfrac{6x^2 + 4x - 2}{x^2 - 2x - 3}$$

Next, we multiply the three rational expressions and simplify the result.

$$= \dfrac{(x^2 + 2x - 3)(2x^2 - 5x - 3)(6x^2 + 4x - 2)}{(6x^2 + 5x + 1)(2x^2 - 2)(x^2 - 2x - 3)}$$

$$= \dfrac{(x + 3)(\overset{1}{\cancel{x - 1}})(\overset{1}{\cancel{2x + 1}})(\overset{1}{\cancel{x - 3}})2(3x - 1)(\overset{1}{\cancel{x + 1}})}{(3x + 1)(\underset{1}{\cancel{2x + 1}})2(x + 1)(\underset{1}{\cancel{x - 1}})(\underset{1}{\cancel{x - 3}})(\underset{1}{\cancel{x + 1}})}$$ To prepare to simplify, factor each polynomial completely. Then remove the common factors.

$$= \dfrac{(x + 3)(3x - 1)}{(3x + 1)(x + 1)}$$ Multiply the remaining factors in the numerator. Multiply the remaining factors in the denominator:

Self Check 9 Simplify: $\dfrac{x^2 - 25}{4x^2 + 12x + 9} \div \dfrac{x^2 - 5x}{3x - 1} \cdot \dfrac{2x + 3}{3x^2 + 14x - 5}$

Now Try ▶ Problem 67

SECTION 6.2 STUDY SET

VOCABULARY

Fill in the blanks.

1. $\dfrac{a^2 - 9}{a^2 - 49} \cdot \dfrac{a - 7}{a + 3}$ is the product of two _____ expressions.

2. The _____ of $\dfrac{a + 3}{a + 7}$ is $\dfrac{a + 7}{a + 3}$.

3. To find the reciprocal of a rational expression, we _____ its numerator and denominator.

4. To simplify a rational expression, remove any factors _____ to the numerator and denominator.

CONCEPTS

Fill in the blanks.

5. To multiply rational expressions, multiply their _____ and multiply their _____ . In symbols, we have:

$$\frac{A}{B} \cdot \frac{C}{D} = \boxed{}$$

6. To divide two rational expressions, multiply the first by the _____ of the second. In symbols, we have:

$$\frac{A}{B} \div \frac{C}{D} = \frac{A}{B} \cdot \boxed{} = \boxed{}$$

NOTATION

Complete each solution.

7. $\dfrac{x^2 + 3x}{5x - 25} \cdot \dfrac{x - 5}{x + 3} = \dfrac{(x^2 + 3x)\boxed{}}{(5x - 25)\boxed{}}$

$$= \frac{\boxed{}(x + 3)(x - 5)}{(x - 5)(x + 3)}$$

$$= \frac{\boxed{}}{5}$$

8. $\dfrac{x^2 - x - 6}{4x^2 + 16x} \div \dfrac{x - 3}{x + 4} = \dfrac{x^2 - x - 6}{4x^2 + 16x} \cdot \dfrac{\boxed{}}{\boxed{}}$

$$= \frac{\boxed{}(x + 2)(x + 4)}{\boxed{}(x + 4)(x - 3)}$$

$$= \frac{x + 2}{\boxed{}}$$

9. A student checks her answers with those in the back of her textbook. Determine whether they are equivalent.

Student's answer	Book's answer	Equivalent?
$\dfrac{-x^{10}}{y^2}$	$-\dfrac{x^{10}}{y^2}$	
$\dfrac{x - 3}{x + 3}$	$\dfrac{3 - x}{x + 3}$	
$\dfrac{a + b}{(2 - x)(c + d)}$	$-\dfrac{a + b}{(x - 2)(c + d)}$	

10. a. Write $5x^2 + 35x$ as a fraction.

    b. What is the reciprocal of $5x^2 + 35x$?

GUIDED PRACTICE

Multiply, and then simplify, if possible. See Objective 1.

11. $\dfrac{3}{4} \cdot \dfrac{11}{3}$

12. $\dfrac{13}{6} \cdot \dfrac{6}{21}$

13. $\dfrac{15}{24} \cdot \dfrac{16}{25}$

14. $\dfrac{49}{36} \cdot \dfrac{18}{35}$

Multiply, and then simplify, if possible. See Example 1.

15. $\dfrac{3a}{10} \cdot \dfrac{2}{15a^4}$

16. $\dfrac{4p}{21} \cdot \dfrac{7}{12p^6}$

17. $\dfrac{12x^{61}}{7y^{15}} \cdot \dfrac{y}{8x^{27}}$

18. $\dfrac{b^{65}}{27a^2} \cdot \dfrac{18a^{41}}{5b^{90}}$

Multiply, and then simplify, if possible. See Example 2.

19. $\dfrac{y^2 + 6y + 9}{15y} \cdot \dfrac{3y^2}{2y + 6}$

20. $\dfrac{3p^2}{6p + 24} \cdot \dfrac{p^2 - 16}{6p}$

21. $\dfrac{x^2 + x - 6}{5x} \cdot \dfrac{5x - 10}{x + 3}$

22. $\dfrac{z^2 + 4z - 5}{25z - 25} \cdot \dfrac{5z}{z + 5}$

23. $\dfrac{x^2 + 2x + 1}{9x^3} \cdot \dfrac{2x^2 - 2x}{2x^2 - 2}$

24. $\dfrac{a^4 + 6a^3}{a^2 - 16} \cdot \dfrac{3a - 12}{3a + 18}$

25. $\dfrac{x^3 + 3x^2 - 3x - 9}{x} \cdot \dfrac{1}{x^2 + 3x}$

26. $\dfrac{m^2 + 7m^2 - 7m - 49}{m^3} \cdot \dfrac{1}{m^2 + 7m}$

Multiply, and then simplify, if possible. See Example 3.

27. $\dfrac{2x^2 - x - 3}{x^2 - 1} \cdot \dfrac{x^2 + x - 2}{2x^2 + x - 6}$

28. $\dfrac{2p^2 - 5p - 3}{p^2 - 9} \cdot \dfrac{2p^2 + 5p - 3}{2p^2 + 5p + 2}$

29. $\dfrac{3t^2 - t - 2}{6t^2 - 5t - 6} \cdot \dfrac{4t^2 - 9}{2t^2 + 5t + 3}$

30. $\dfrac{9x^2 + 3x - 20}{3x^2 - 7x + 4} \cdot \dfrac{3x^2 - 5x + 2}{9x^2 + 18x + 5}$

31. $\dfrac{x^2 + 4xy + 4y^2}{2x^2 + 4xy} \cdot \dfrac{3x - 6y}{x^2 - 4y^2}$

32. $\dfrac{x^2 - y^2}{xy} \cdot \dfrac{x^2}{x^2 + 2xy + y^2}$

33. $\dfrac{3a^2 + 7ab + 2b^2}{a^2 + 2ab} \cdot \dfrac{a^2 - ab}{3a^2 + ab}$

34. $\dfrac{a^2 + 3ab + 2b^2}{a^2 - 3ab - 4b^2} \cdot \dfrac{a^2 - 4ab}{ab^2 + 2b^3}$

Multiply, and then simplify, if possible. See Example 4.

35. $15x\left(\dfrac{x + 1}{15x}\right)$

36. $30t\left(\dfrac{t - 7}{30t}\right)$

37. $12y\left(\dfrac{y+8}{6y}\right)$

38. $16x\left(\dfrac{3x+8}{4x}\right)$

39. $(6a - a^2) \cdot \dfrac{a^3}{2a^3 - 12a^2 + 6a - 36}$

40. $(10n - n^2) \cdot \dfrac{n^6}{n^4 - 10n^3 - 2n^2 + 20n}$

41. $(x^2 + x - 2cx - 2c) \cdot \dfrac{x^2 + 3x + 2}{4c^2 - x^2}$

42. $(2ax - 10x + a - 5) \cdot \dfrac{x}{2x^2 + x}$

Find each power. See Example 5.

43. $\left(\dfrac{x-3}{x^2+4}\right)^2$

44. $\left(\dfrac{2t^2 + t}{t-1}\right)^2$

45. $\left(\dfrac{2m^2 - m - 3}{x^2 - 1}\right)^2$

46. $\left(\dfrac{k^4 + 3k}{x^2 - x + 1}\right)^2$

Divide, and then simplify, if possible. See Objective 3.

47. $\dfrac{6}{11} \div \dfrac{36}{55}$

48. $\dfrac{17}{12} \div \dfrac{34}{3}$

49. $\dfrac{12}{5} \div \dfrac{24}{45}$

50. $\dfrac{18}{7} \div \dfrac{54}{35}$

Divide, and then simplify, if possible. See Example 6.

51. $\dfrac{22x^3}{y^2} \div \dfrac{33x^9}{y^7}$

52. $\dfrac{24a^6}{b} \div \dfrac{64a^9}{b^2}$

53. $\dfrac{pq^{29}}{50} \div \dfrac{p^{10}q^{38}}{15}$

54. $\dfrac{s^{41}t^3}{12} \div \dfrac{s^{41}t^{52}}{144}$

Divide, and then simplify, if possible. See Example 7.

55. $\dfrac{x^{12}}{x^3 - 8} \div \dfrac{x^2}{x^2 - 2x}$

56. $\dfrac{x^9}{x^3 + 125} \div \dfrac{x^4}{x^2 + 5x}$

57. $\dfrac{a^2 - 18a + 81}{a^{40}} \div \dfrac{(a - 9)^3}{a^{37}}$

58. $\dfrac{m^2 + 16m + 64}{m^{50}} \div \dfrac{(m + 8)^4}{m^{33}}$

59. $\dfrac{3n^2 + 5n - 2}{12n^2 - 13n + 3} \div \dfrac{n^2 + 3n + 2}{4n^2 + 5n - 6}$

60. $\dfrac{8y^2 - 14y - 15}{6y^2 - 11y - 10} \div \dfrac{4y^2 - 9y - 9}{3y^2 - 7y - 6}$

61. $\dfrac{5cd + d^2}{6d^2} \div \dfrac{125c^3 + d^3}{6c + 6d}$

62. $\dfrac{6m - 8n}{9m^3} \div \dfrac{27m^3 - 64n^3}{9m + 9n}$

Divide, and then simplify, if possible. See Example 8.

63. $\dfrac{y^3 - 9y}{y + 2} \div (y - 3)$

64. $\dfrac{x - 2}{x} \div (x^2 - 4)$

65. $(x + 1) \div \dfrac{x^2 + 2x + 1}{2}$

66. $(y + 4) \div \dfrac{y^2 + 8y + 16}{ab}$

Perform each operation and simplify, if possible. See Example 9.

67. $\dfrac{6a^2 - 7a - 3}{a^2 - 1} \div \dfrac{4a^2 - 12a + 9}{a^2 - 1} \cdot \dfrac{2a^2 - a - 3}{3a^2 - 2a - 1}$

68. $\dfrac{x^2 - x - 12}{x^2 + x - 2} \div \dfrac{x^2 - 6x + 8}{x^2 - 3x - 10} \cdot \dfrac{x^2 - 3x + 2}{x^2 - 2x - 15}$

69. $\dfrac{2x^2 - 2x - 4}{x^2 + 2x - 8} \cdot \dfrac{3x^2 + 15x}{x + 1} \div \dfrac{4x^2 - 100}{x^2 - x - 20}$

70. $\dfrac{4a^2 - 10a + 6}{a^4 - 3a^3} \div \dfrac{3 - 2a}{2a^3} \cdot \dfrac{a - 3}{2a - 2}$

TRY IT YOURSELF

Perform the operations and simplify.

71. $\dfrac{x^2 - 6x + 9}{4 - x^2} \div \dfrac{x^2 - 9}{x^2 - 8x + 12}$

72. $\dfrac{x^3 + 1}{4} \div \dfrac{x + 1}{2}$

73. $\dfrac{2x^2 - 2x - 12}{x^2 - 4} \cdot \dfrac{x^2 - x - 2}{x^3 - 9x}$

74. $\dfrac{x^2 + 2x - 35}{12x^3} \cdot \dfrac{x^2 + 4x - 21}{x^2 - 3x}$

75. $\dfrac{p^3 - q^3}{p^2 - q^2} \cdot \dfrac{q^2 + pq}{p^3 + p^2q + pq^2}$

76. $\dfrac{x^2 - 4}{2b - bx} \div \dfrac{x^2 + 4x + 4}{2b + bx}$

77. $\dfrac{10r^2s}{6rs^2} \cdot \dfrac{3r^3}{2rs}$

78. $\dfrac{3a^3b}{25cd^3} \cdot \dfrac{5cd^2}{6ab}$

79. $10(h - 9)\dfrac{h - 3}{9 - h}$

80. $r(r - 25)\dfrac{r + 4}{r - 25}$

81. $\dfrac{2x^2 + 5xy + 3y^2}{3x^2 - 5xy + 2y^2} \div \dfrac{2x^2 + xy - 3y^2}{3x^2 - 5xy + 2y^2}$

82. $\dfrac{2p^2 - 5pq - 3q^2}{p^2 - 9q^2} \div \dfrac{2p^2 + 5pq + 2q^2}{2p^2 + 5pq - 3q^2}$

83. $(4x^2 - 9) \div \dfrac{2x^2 + 5x + 3}{x + 2} \cdot \dfrac{1}{2x - 3}$

84. $(4x + 12) \div \dfrac{2x - 6}{x^2} \cdot \dfrac{x - 3}{2}$

85. $\dfrac{x^3 - 3x^2 - 25x + 75}{x^3 - 27} \cdot \dfrac{2x^3 + 6x^2 + 18x}{x^2 + 10x + 25}$

86. $\dfrac{x^2 + 3x + xy + 3y}{x^2 - 9} \cdot \dfrac{3 - x}{x^3 + 3x^2}$

87. Let $f(x) = \dfrac{x^2 + x - 6}{x^2 - 6x + 9}$ and $g(x) = \dfrac{x^2 - 9}{x^2 - 4}$.
   Find $f(x) \cdot g(x)$.

88. Let $g(s) = \dfrac{s^2 - 5s + 6}{s^2 - 10s + 16}$ and $h(s) = \dfrac{s^2 - 6s - 16}{s^2 + 2s}$.

Find $g(s) \cdot h(s)$.

89. Let $s(x) = \dfrac{x^2 - 16}{x^2 - 25}$ and $f(x) = \dfrac{5x + 20}{10x^2 - 50x}$.

Find $s(x) \div f(x)$.

90. Let $g(x) = \dfrac{x^2 - 9}{x^2 - 49}$ and $h(x) = \dfrac{9x^2 + 27x}{3x + 21}$.

Find $g(x) \div h(x)$.

Look Alikes . . .

91. a. $\dfrac{t^2 + 9t + 20}{9t + 36} \cdot \dfrac{9t + 45}{t + 4}$

b. $\dfrac{t^2 + 9t + 20}{9t + 36} \div \dfrac{9t + 45}{t + 4}$

92. a. $\dfrac{a^2 - 5a + 6}{2a - 4} \cdot \dfrac{2a - 6}{a - 2}$

b. $\dfrac{a^2 - 5a + 6}{2a - 4} \div \dfrac{2a - 6}{a - 2}$

93. a. $\dfrac{4r - 8}{5r} \div \dfrac{5r - 10}{4r^2}$

b. $\dfrac{4r - 8}{5r} \cdot \dfrac{5r - 10}{4r^2}$

94. a. $\dfrac{3d^2 + 6}{4} \div \dfrac{4d^2 + 8}{3}$

b. $\dfrac{3d^2 + 6}{4} \cdot \dfrac{4d^2 + 8}{3}$

APPLICATIONS

95. Physics Experiments. The following table contains data from a physics experiment. Complete the table.

Trial	Rate (m/sec)	Time (sec)	Distance (m)
1	$\dfrac{k_1^2 + 3k_1 + 2}{k_1 - 3}$	$\dfrac{k_1^2 - 3k_1}{k_1 + 1}$	
2	$\dfrac{k_2^2 + 6k_2 + 5}{k_2 + 1}$		$k_2^2 + 11k_2 + 30$

96. Geometry. Find a simplified rational expression that represents the volume of the rectangular solid shown here.

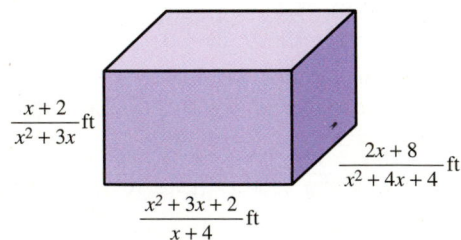

$\dfrac{x + 2}{x^2 + 3x}$ ft

$\dfrac{2x + 8}{x^2 + 4x + 4}$ ft

$\dfrac{x^2 + 3x + 2}{x + 4}$ ft

WRITING

97. Explain how to multiply two rational expressions.

98. Write some comments to the student who wrote the following solution, explaining the error.

$$\frac{x^2 + x - 2}{x^2 - 4} \cdot \frac{x - 2}{x - 1} = \frac{\cancel{(x + 2)}\cancel{(x - 1)}\cancel{(x - 2)}}{\cancel{(x + 2)}\cancel{(x - 2)}\cancel{(x - 1)}}$$

$$= 0$$

99. The graph of $Y_3 = Y_1 - Y_2$, where

$$Y_1 = \frac{2x^2 - 5x - 3}{x^2 - 9} \cdot \frac{2x^2 + 5x - 3}{2x^2 + 5x + 2} \quad \text{and} \quad Y_2 = \frac{2x - 1}{x + 2}$$

is shown. Explain how the graph and table can be used to verify that

$$\frac{2x^2 - 5x - 3}{x^2 - 9} \cdot \frac{2x^2 + 5x - 3}{2x^2 + 5x + 2} = \frac{2x - 1}{x + 2}$$

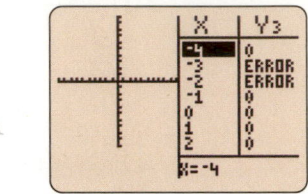

100. A student obtained an answer of $\dfrac{x + 3}{x + 7}$ after performing $\dfrac{x^2 - 9}{x^2 - 49} \div \dfrac{x + 3}{x + 7}$. As a check, he graphed $Y_3 = Y_1 - Y_2$, where

$$Y_1 = \left(\frac{x^2 - 9}{x^2 - 49}\right) \div \left(\frac{x + 3}{x + 7}\right) \quad \text{and} \quad Y_2 = \frac{x + 3}{x + 7}$$

The graph is shown. Explain what conclusion can be drawn from the graph and the table.

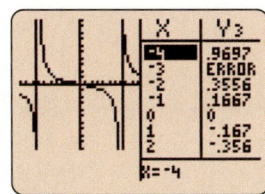

REVIEW

Complete the rules for exponents. Assume that there are no divisions by 0.

101. a. $x^m x^n = \boxed{}$ **b.** $(x^m)^n = \boxed{}$

c. $(xy)^n = \boxed{}$ **d.** $\left(\dfrac{x}{y}\right)^n = \boxed{}$

e. $x^0 = \boxed{}$ **f.** $x^{-n} = \boxed{}$

102. a. $\dfrac{x^m}{x^n} = \boxed{}$ **b.** $\left(\dfrac{x}{y}\right)^{-n} = \left(\boxed{}\right)$

c. $\dfrac{x^{-m}}{y^{-n}} = \dfrac{\boxed{}}{\boxed{}}$ **d.** $x^1 = \boxed{}$

CHALLENGE PROBLEMS

Insert either a multiplication symbol $\cdot$ or a division symbol $\div$ in each blank to make a true statement.

103. $\dfrac{x^2}{y} \;\boxed{}\; \dfrac{x}{y^2} \;\boxed{}\; \dfrac{x^2}{y^2} = \dfrac{x^3}{y}$

104. $\dfrac{x^2}{y} \;\boxed{}\; \dfrac{x}{y^2} \;\boxed{}\; \dfrac{x^2}{y^2} = \dfrac{y^3}{x}$

SECTION 6.3

Adding and Subtracting Rational Expressions

OBJECTIVES

1. Add and subtract rational expressions with like denominators.
2. Find the least common denominator.
3. Add and subtract rational expressions with unlike denominators.
4. Add and subtract rational expressions that have denominators that are opposites.
5. Perform mixed operations.

ARE YOU READY?

The following problems review some basic skills that are needed when adding and subtracting rational expressions.

1. Add: $\dfrac{5}{11} + \dfrac{3}{11}$

2. Subtract: $\dfrac{2}{3} - \dfrac{1}{5}$

3. Simplify: $x^2 + 9x - (6x - 4)$

4. Simplify: $\dfrac{x + 9}{x + 9}$

5. Simplify: $\dfrac{2a + 14}{(a + 7)(a - 2)}$

6. Factor: $5x^2 - 5$

The methods used to add and subtract rational expressions are based on the rules for adding and subtracting arithmetic fractions. In this section, we will add and subtract rational expressions with *like* and *unlike* denominators.

1 Add and Subtract Rational Expressions with Like Denominators.

To add or subtract fractions with a common denominator, we add or subtract their numerators and write the sum or difference over the common denominator. For example,

$$\frac{3}{7} + \frac{2}{7} = \frac{3 + 2}{7} \qquad \frac{3}{7} - \frac{2}{7} = \frac{3 - 2}{7}$$

$$= \frac{5}{7} \qquad\qquad = \frac{1}{7}$$

The Language of Algebra

We can describe $\frac{3}{7}$ and $\frac{2}{7}$ as having the **same** denominator, **common** denominators, or **like** denominators.

We use the same procedure to add and subtract rational expressions with like denominators.

Adding and Subtracting Rational Expressions That Have the Same Denominator

To add (or subtract) rational expressions that have the same denominator, add (or subtract) their numerators and write the sum (or difference) over the common denominator. Then, if possible, factor and simplify.

If $\frac{A}{D}$ and $\frac{B}{D}$ are rational expressions,

$$\frac{A}{D} + \frac{B}{D} = \frac{A + B}{D} \qquad \text{and} \qquad \frac{A}{D} - \frac{B}{D} = \frac{A - B}{D}$$

EXAMPLE 1

Perform the operations: **a.** $\dfrac{4}{3x} + \dfrac{7}{3x}$ **b.** $\dfrac{a^2}{a^2 - 1} - \dfrac{a}{a^2 - 1}$

Strategy In part (a), we will add the numerators and write the sum over the common denominator. In part (b), we will subtract the numerators and write the difference over the common denominator. Then, if possible, we will factor and simplify.

Why These are the rules for adding and subtracting rational expressions that have the *same* denominator.

Solution **a.** $\dfrac{4}{3x} + \dfrac{7}{3x} = \dfrac{4 + 7}{3x}$ Add the numerators. Write the sum over the common denominator, 3x.

$$= \frac{11}{3x} \qquad \text{Since 11 and 3 have no common factors (other than 1), the result does not simplify.}$$

b. $\dfrac{a^2}{a^2 - 1} - \dfrac{a}{a^2 - 1} = \dfrac{a^2 - a}{a^2 - 1}$ Subtract the numerators. Write the difference over the common denominator, $a^2 - 1$.

$= \dfrac{a(a - 1)}{(a + 1)(a - 1)}$ To prepare to simplify the result, factor the numerator and denominator.

$= \dfrac{a(\cancel{a - 1})}{(a + 1)(\cancel{a - 1})}$ Remove the common factor $a - 1$ in the numerator and denominator.

$= \dfrac{a}{a + 1}$ Multiply the remaining factors in the numerator: $a \cdot 1 = a$. Multiply the remaining factors in the denominator: $(a + 1) \cdot 1 = a + 1$.

Self Check 1   Perform the operations: **a.** $\dfrac{17}{22t} + \dfrac{13}{22t}$

b. $\dfrac{a^2}{a^2 - 2a} - \dfrac{4}{a^2 - 2a}$

Now Try ▶ Problems 19 and 25

Using Your Calculator ▶ **Checking Algebra**

We can check the subtraction in Example 1(b) by replacing each a with x and graphing the rational functions $f(x) = \dfrac{x^2}{x^2 - 1} - \dfrac{x}{x^2 - 1}$, shown in figure (a), and $g(x) = \dfrac{x}{x + 1}$, shown in figure (b), and observing that the graphs are the same. Note that -1 and 1 are not in the domain of the first function and that -1 is not in the domain of the second function.

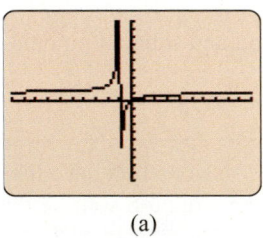   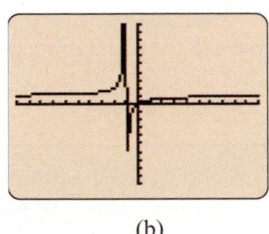   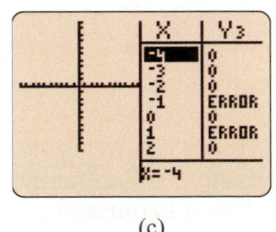

(a)   (b)   (c)

Figure (c) shows the display when the G-T mode is used to check the simplification. Here, $Y_3 = Y_1 - Y_2$, where $Y_1 = \dfrac{x^2}{x^2 - 1} - \dfrac{x}{x^2 - 1}$ and $Y_2 = \dfrac{x}{x + 1}$.

---

2 Find the Least Common Denominator.

When adding or subtracting rational expressions with unlike denominators, it is easiest if we write the rational expressions in terms of the smallest common denominator possible, called the *least* (or lowest) *common denominator* (LCD). To find the least common denominator of several rational expressions, we follow these steps.

Finding the LCD

1. Factor each denominator completely.

2. The LCD is a product that uses each different factor obtained in step 1 the greatest number of times it appears in any one factorization.

EXAMPLE 2   Find the LCD of: **a.** $\dfrac{5a}{24b}$ and $\dfrac{11a}{18b^2}$ **b.** $\dfrac{1}{x^2 - 12x + 36}$ and $\dfrac{3 - x}{x^2 - 6x}$

Strategy   We begin by factoring the denominator of each rational expression completely.

Why   Since the LCD must contain the factors of each denominator, we need to write each denominator in factored form.

Solution   **a.** We write each denominator as the product of prime numbers and variables.

$$24b = \mathbf{2 \cdot 2 \cdot 2} \cdot 3 \cdot b = \mathbf{2^3} \cdot 3 \cdot b$$
$$18b^2 = 2 \cdot \mathbf{3 \cdot 3} \cdot b \cdot b = 2 \cdot \mathbf{3^2} \cdot \mathbf{b^2}$$

To find the LCD, we form a product using each of these factors the greatest number of times it appears in any one factorization.

The greatest number of times the factor 2 appears is three times.
The greatest number of times the factor 3 appears is twice.
The greatest number of times the factor b appears is twice.

$$\text{LCD} = \mathbf{2 \cdot 2 \cdot 2} \cdot \mathbf{3 \cdot 3} \cdot \mathbf{b \cdot b} = 72b^2$$

b. We factor each denominator completely:

$$x^2 - 12x + 36 = (x - 6)(x - 6) = (x - 6)^2$$
$$x^2 - 6x = x(x - 6)$$

To find the LCD, we form a product using the highest power of each of the factors:

The greatest number of times the factor x appears is once.
The greatest number of times the factor $x - 6$ appears is twice.

$$\text{LCD} = x(x-6)^2$$

Self Check 2   Find the LCD of: **a.** $\dfrac{5x}{28z^3}$ and $\dfrac{1}{21z}$

b. $\dfrac{a - 1}{a^2 - 25}$ and $\dfrac{3 - a^2}{a^2 + 7a + 10}$

Now Try ▶ Problems 27 and 31

3   Add and Subtract Rational Expressions with Unlike Denominators.

Recall from arithmetic that writing a fraction as an equivalent fraction with a larger denominator is called **building the fraction.** For example, to write $\frac{3}{5}$ as an equivalent fraction with a denominator of 35, we multiply it by 1 in the form of $\frac{7}{7}$. It is important to note that multiplying $\frac{3}{5}$ by $\frac{7}{7}$ changes its appearance, but not its value, because we are multiplying it by 1.

$$\frac{3}{5} = \frac{3}{5} \cdot \frac{7}{7} = \frac{21}{35}$$ Multiply the numerators.
Multiply the denominators.

To add and subtract rational expressions with different denominators, we write them as equivalent expressions having a common denominator. To do so, we use a procedure similar to that shown above called *building a rational expression.*

Building Rational Expressions	To **build a rational expression**, multiply it by 1 in the form of $\dfrac{c}{c}$, where c is any nonzero number or expression.

The following steps summarize how to add or subtract rational expressions with different denominators.

Adding and Subtracting Rational Expressions with Unlike Denominators	1. Find the LCD. 2. Write each rational expression as an equivalent expression with the LCD as the denominator. To do so, build each rational expression using a form of 1 that involves any factor(s) needed to obtain the LCD. 3. Add or subtract the numerators and write the sum or difference over the LCD. 4. Simplify the resulting rational expression, if possible.

EXAMPLE 3   Add: $\dfrac{5a}{24b} + \dfrac{11a}{18b^2}$

Strategy   In Example 2, we saw that the LCD of these rational expressions is $72b^2$. We will multiply each one by the appropriate form of 1 to build it into an equivalent rational expression with a denominator of $72b^2$. Then we will use the rule for adding rational expressions with like denominators.

Why   Since the denominators are different, we cannot add these rational expressions in their present form.

Solution   We will begin by asking, "By what must we multiply each given denominator to get the common denominator, $72b^2$?"

We need to multiply the denominator of $\frac{5a}{24b}$ by $3b$ to obtain a denominator of $72b^2$. It follows that $\frac{3b}{3b}$ should be the form of 1 that should be used to build an equivalent rational expression.

We need to multiply the denominator of $\frac{11a}{18b^2}$ by 4 to obtain a denominator of $72b^2$. It follows that $\frac{4}{4}$ is the form of 1 that should be used to build an equivalent rational expression.

Caution

Don't make the mistake of removing the b's in the result. This is incorrect because b is not a factor of the *entire* numerator.

$$\dfrac{\overset{1}{\cancel{15ab + 44a}}}{\underset{1}{\cancel{72b^2}}}$$

$$\dfrac{5a}{24b} + \dfrac{11a}{18b^2} = \dfrac{5a}{24b} \cdot \dfrac{3b}{3b} + \dfrac{11a}{18b^2} \cdot \dfrac{4}{4}$$

Build the rational expressions so that each has a denominator of $72b^2$.

$$= \dfrac{15ab}{72b^2} + \dfrac{44a}{72b^2}$$

Multiply the numerators. Multiply the denominators. Note that the rational expressions now have like denominators.

$$= \dfrac{15ab + 44a}{72b^2}$$

Add the numerators. Write the sum over the common denominator, $72b^2$. The result does not simplify.

Self Check 3   Add: $\dfrac{5x}{28z^3} + \dfrac{1}{21z}$

Now Try ▶ Problem 37

EXAMPLE 4   Subtract: $\dfrac{4x}{x+2} - \dfrac{7x}{x-2}$

Strategy   The LCD for the rational expressions is $(x+2)(x-2)$. We will multiply each one by the appropriate form of 1 to build it into an equivalent rational expression with a denominator of $(x+2)(x-2)$.

Why   Since the denominators are different, we cannot subtract these rational expressions in their present form.

Solution

$$\frac{4x}{x+2}-\frac{7x}{x-2}$$

$$=\frac{4x}{x+2}\cdot\frac{x-2}{x-2}-\frac{7x}{x-2}\cdot\frac{x+2}{x+2}$$

Build each rational expression.

Multiply the numerators using the distributive property to prepare to combine like terms later.

Don't multiply the denominators. Leave them in factored form to possibly simplify the result later.

$$=\frac{4x^2-8x}{(x+2)(x-2)}-\frac{7x^2+14x}{(x+2)(x-2)}$$

Multiply the numerators. The denominators are now the same.

This numerator is written within parentheses to make sure that we subtract both of its terms.

$$=\frac{(4x^2-8x)-(7x^2+14x)}{(x+2)(x-2)}$$

Subtract the numerators. Write the difference over the common denominator, $(x+2)(x-2)$.

$$=\frac{4x^2-8x-7x^2-14x}{(x+2)(x-2)}$$

To subtract in the numerator, change the signs of $7x^2$ and $14x$, and drop the parentheses.

$$=\frac{-3x^2-22x}{(x+2)(x-2)}$$

Combine like terms in the numerator: $4x^2-7x^2=-3x^2$ and $-8x-14x=-22x$.

Notation

The numerator of the result may be written in two forms:

Not factored	Factored
$\frac{-3x^2-22x}{(x+2)(x-2)}$	$\frac{-x(3x+22)}{(x+2)(x-2)}$

Ask your instructor which form he or she prefers.

If the common factor $-x$ is factored out of the terms in the numerator, this result can be written in two other equivalent forms.

$$\frac{-3x^2-22x}{(x+2)(x-2)}=\frac{-x(3x+22)}{(x+2)(x-2)}=-\frac{x(3x+22)}{(x+2)(x-2)}$$

The result does not simplify.

Self Check 4 Subtract: $\frac{3a}{a+3}-\frac{5a}{a-3}$

Now Try Problems 39 and 43

EXAMPLE 5 Add: $3+\dfrac{7}{x-15}$

Strategy We will begin by writing 3 as $\frac{3}{1}$.

Why Then we can multiply $\frac{3}{1}$ by the appropriate form of 1 to build it into an equivalent rational expression with a denominator of $x-15$.

Solution

$$3+\frac{7}{x-15}=\frac{3}{1}+\frac{7}{x-15}$$

$3=\frac{3}{1}$

Success Tip

Since the denominator of $\dfrac{7}{x-15}$ is the LCD, we do not need to build it by multiplying it by a form of 1. The step below is unnecessary.

$$\frac{7}{x-15}\cdot\frac{1}{1}$$

$$=\frac{3}{1}\cdot\frac{x-15}{x-15}+\frac{7}{x-15}$$

Build $\frac{3}{1}$ to a rational expression with a denominator of $x-15$.

$$=\frac{3x-45}{x-15}+\frac{7}{x-15}$$

Distribute the multiplication by 3. The denominators are now the same.

$$=\frac{3x-45+7}{x-15}$$

Add the numerators. Write the difference over the common denominator, $x-15$.

$$=\frac{3x-38}{x-15}$$

Combine like terms in the numerator. The result does not simplify.

Self Check 5 Add: $6+\dfrac{5y}{6-y}$

Now Try Problem 47

EXAMPLE 6 Subtract: $\dfrac{x+1}{x^2-2x+1} - \dfrac{x-4}{x^2-1}$

Strategy We will factor each denominator, find the LCD, and build the rational expressions so each one has the LCD as its denominator.

Why Since the denominators are different, we cannot subtract these rational expressions in their present form.

Solution We factor each denominator to find the LCD:

$$x^2 - 2x + 1 = (x-1)(x-1) = (x-1)^2 \quad \begin{array}{l}\text{The greatest number of times} \\ x-1 \text{ appears is twice.}\end{array}$$
$$x^2 - 1 = (x+1)(x-1) \quad \begin{array}{l}\text{The greatest number of times} \\ x+1 \text{ appears is once.}\end{array}$$

The LCD is $(x-1)^2(x+1)$ or $(x-1)(x-1)(x+1)$.

We now write each rational expression with its denominator in factored form. Then we multiply each numerator and denominator by the missing factor, so that each rational expression has a denominator of $(x-1)(x-1)(x+1)$.

$$\dfrac{x+1}{x^2-2x+1} - \dfrac{x-4}{x^2-1}$$

$$= \dfrac{x+1}{(x-1)(x-1)} - \dfrac{x-4}{(x+1)(x-1)} \qquad \begin{array}{l}\text{Write each denominator} \\ \text{in factored form.}\end{array}$$

$$= \dfrac{x+1}{(x-1)(x-1)} \cdot \boxed{\dfrac{x+1}{x+1}} - \dfrac{x-4}{(x+1)(x-1)} \cdot \boxed{\dfrac{x-1}{x-1}} \qquad \begin{array}{l}\text{Build each rational} \\ \text{expression.}\end{array}$$

Multiply the numerators using the FOIL method to prepare to combine like terms. →

Don't multiply the denominators. → Leave them in factored form to possibly simplify the result later.

$$= \dfrac{x^2+2x+1}{(x-1)(x-1)(x+1)} - \dfrac{x^2-5x+4}{(x-1)(x-1)(x+1)} \qquad \begin{array}{l}\text{Multiply the numerators.} \\ \text{The denominators are} \\ \text{now the same.}\end{array}$$

$$= \dfrac{x^2+2x+1-(x^2-5x+4)}{(x-1)(x-1)(x+1)} \qquad \begin{array}{l}\text{Subtract the numerators. Write the difference} \\ \text{over the common denominator. Don't forget the} \\ \text{parentheses shown in blue.}\end{array}$$

$$= \dfrac{x^2+2x+1-x^2+5x-4}{(x-1)(x-1)(x+1)} \qquad \begin{array}{l}\text{To subtract in the numerator, change the signs} \\ \text{of } x^2, -5x, \text{ and 4, and drop the parentheses.}\end{array}$$

$$= \dfrac{7x-3}{(x-1)(x-1)(x+1)} \qquad \begin{array}{l}\text{Combine like terms in the numerator.} \\ \text{The result does not simplify.}\end{array}$$

Self Check 6 Subtract: $\dfrac{a+2}{a^2-4a+4} - \dfrac{a-3}{a^2-4}$

Now Try ▶ Problem 53

4 Add and Subtract Rational Expressions That Have Denominators That Are Opposites.

We can use the following fact to add or subtract rational expressions whose denominators are opposites.

| Multiplying by −1 | When a polynomial is multiplied by −1, the result is its opposite. |

EXAMPLE 7   Add: $\dfrac{x}{x-y} + \dfrac{y}{y-x}$

Strategy   Since the denominators are opposites, either one can serve as the LCD. If we choose $x - y$, we can multiply $\dfrac{y}{y-x}$ by $\dfrac{-1}{-1}$ to build it into an equivalent rational expression with the denominator $x - y$.

Why   When $y - x$ is multiplied by -1 , the subtraction is reversed, and the result is $x - y$.

Solution

$$\dfrac{x}{x-y} + \dfrac{y}{y-x} = \dfrac{x}{x-y} + \dfrac{y}{y-x} \cdot \dfrac{-1}{-1}$$ Build $\dfrac{y}{y-x}$ so that it has a denominator of $x - y$.

$$= \dfrac{x}{x-y} + \dfrac{-y}{-y+x}$$ Multiply the numerators. Multiply the denominators.

$$= \dfrac{x}{x-y} + \dfrac{-y}{x-y}$$ Write the second denominator, $-y + x$, as $x - y$. The rational expressions now have a common denominator.

$$= \dfrac{x-y}{x-y}$$ Add the numerators. Write the result over the common denominator, $x - y$.

$$= 1$$ Simplify the result.

> **Success Tip**
>
> Since $\dfrac{-1}{-1}$ is a form of 1, multiplying a rational expression by $\dfrac{-1}{-1}$ produces an equivalent rational expression.

Self Check 7   Add: $\dfrac{2a}{a-b} + \dfrac{b}{b-a}$

Now Try ▶ Problem 59

5 Perform Mixed Operations.

EXAMPLE 8   Perform the operations: $\dfrac{2x}{x^2-4} - \dfrac{1}{x^2-3x+2} + \dfrac{x+1}{x^2+x-2}$

Strategy   We will factor each denominator, find the LCD, and build the rational expressions so each one has the LCD as its denominator.

Why   Since the denominators are different, we cannot add or subtract these rational expressions in their present form.

Solution   We factor each denominator to find the LCD and note that the greatest number of times each factor appears is once.

$$\left.\begin{array}{l} x^2 - 4 = (x-2)(x+2) \\ x^2 - 3x + 2 = (x-2)(x-1) \\ x^2 + x - 2 = (x-1)(x+2) \end{array}\right\} \quad \text{LCD} = (x-2)(x+2)(x-1)$$

We then write each rational expression as an equivalent rational expression with the LCD as its denominator and do the subtraction and addition.

$$\dfrac{2x}{x^2-4} - \dfrac{1}{x^2-3x+2} + \dfrac{x+1}{x^2+x-2}$$

$$= \dfrac{2x}{(x-2)(x+2)} - \dfrac{1}{(x-2)(x-1)} + \dfrac{x+1}{(x-1)(x+2)}$$ Factor the denominators.

Multiply the numerators using the distributive property or the FOIL method to prepare to combine like terms.

$$= \frac{2x}{(x-2)(x+2)} \cdot \frac{x-1}{x-1} - \frac{1}{(x-2)(x-1)} \cdot \frac{x+2}{x+2} + \frac{x+1}{(x-1)(x+2)} \cdot \frac{x-2}{x-2}$$

Don't multiply the denominators. Leave them in factored form to possibly simplify the result later.

$$= \frac{2x^2 - 2x}{(x-2)(x+2)(x-1)} - \frac{x+2}{(x-2)(x-1)(x+2)} + \frac{x^2 - x - 2}{(x-1)(x+2)(x-2)}$$

Multiply the numerators.

$$= \frac{2x^2 - 2x - (x+2) + x^2 - x - 2}{(x+2)(x-2)(x-1)}$$

Write the sum and difference over the common denominator. Don't forget the parentheses in blue.

$$= \frac{2x^2 - 2x - x - 2 + x^2 - x - 2}{(x+2)(x-2)(x-1)}$$

In the numerator, change the signs of x and 2, and drop the parentheses.

Caution

Always write the result in simplest form by removing any factors common to the numerator and denominator.

$$= \frac{3x^2 - 4x - 4}{(x+2)(x-2)(x-1)}$$

In the numerator, combine like terms.

$$= \frac{(3x+2)(x-2)}{(x+2)(x-2)(x-1)}$$

Factor the trinomial and simplify the result by removing the common factor, x − 2.

$$= \frac{3x+2}{(x+2)(x-1)}$$

Self Check 8 Perform the operations: $\dfrac{5a}{a^2 - 25} - \dfrac{7}{a-5} + \dfrac{2}{a+5}$

Now Try ▶ Problem 65

SECTION 6.3 STUDY SET

VOCABULARY

Fill in the blanks.

1. The rational expressions $\frac{7}{6n}$ and $\frac{n+1}{6n}$ have a common _____ of 6n.

2. The least _____ _____ of $\frac{x-8}{x+6}$ and $\frac{6-5x}{x}$ is x(x + 6).

3. To _____ a rational expression, we multiply it by a form of 1. For example, $\frac{2}{n^2} \cdot \frac{8}{8} = \frac{16}{8n^2}$.

4. The polynomials x − y and y − x are _____ because their terms are the same but opposite in sign.

CONCEPTS

Fill in the blanks.

5. To add or subtract rational expressions that have the same denominator, add or subtract the _____, and write the sum or difference over the common _____.

   In symbols, if $\frac{A}{D}$ and $\frac{B}{D}$ are rational expressions,

   $$\frac{A}{D} + \frac{B}{D} = \boxed{}{D} \qquad \text{and} \qquad \frac{A}{D} - \frac{B}{D} = \boxed{}{D}$$

6. To find the least common denominator of several rational expressions, _____ each denominator completely. The LCD is a product that uses each different factor the _____ number of times it appears in any one factorization.

7. When a number is multiplied by ▢, its value does not change.

8. $\dfrac{x+3}{x+3} = \boxed{}$

9. Consider the following two procedures.

   i. $\dfrac{x^2 - 2x}{x^2 + 4x - 12} = \dfrac{x(x-2)}{(x+6)(x-2)} = \dfrac{x}{x+6}$

   ii. $\dfrac{x}{x+6} = \dfrac{x}{x+6} \cdot \dfrac{x-2}{x-2} = \dfrac{x^2 - 2x}{(x+6)(x-2)}$

   a. In which of these procedures are we *building* a rational expression? For what type of problem is this procedure often necessary?

   b. What name is used to describe the other procedure?

10. The LCD for $\dfrac{2x+1}{x^2 + 5x + 6}$ and $\dfrac{3x}{x^2 - 4}$ is

    LCD = (x + 2)(x + 3)(x − 2)

    If we want to subtract these rational expressions, what form of 1 should be used:

    a. to build $\frac{2x+1}{x^2+5x+6}$?   b. to build $\frac{3x}{x^2-4}$?

11. Consider the following factorizations of the denominators of two rational expressions:

$$(x - 2)(x - 2) \quad \text{and} \quad 3(x - 2)$$

   a. What is the greatest number of times the factor 3 appears in any one factorization?

   b. What is the greatest number of times the factor $x - 2$ appears in any one factorization?

   c. What is the LCD of the rational expressions?

12. The factorizations of the denominators of two rational expressions follow. Find the LCD.

$$\left. \begin{array}{l} 2 \cdot 3 \cdot a \cdot a \cdot a \\ 2 \cdot 3 \cdot 3 \cdot a \cdot a \end{array} \right\} \; \text{LCD} = \boxed{}$$

13. Factor each denominator completely.

   a. $\dfrac{17}{40x^2}$

   b. $\dfrac{x + 25}{2x^2 - 6x}$

   c. $\dfrac{n^2 + 3n - 4}{n^2 - 64}$

14. By what must $y - 4$ be multiplied to obtain $4 - y$?

NOTATION

Complete each solution.

15. $\dfrac{6x - 1}{3x - 1} + \dfrac{3x - 2}{3x - 1} = \dfrac{6x - 1 + \boxed{}}{3x - 1}$

$$= \dfrac{9x - \boxed{}}{3x - 1}$$

$$= \dfrac{3(\boxed{})}{3x - 1}$$

$$= \boxed{}$$

16. $\dfrac{x^2 + 3x}{x - 1} - \dfrac{2x - 1}{x - 1} = \dfrac{x^2 + 3x - (\boxed{})}{x - 1}$

$$= \dfrac{x^2 + 3x - 2x \boxed{} 1}{x - 1}$$

$$= \dfrac{x^2 + \boxed{} + \boxed{}}{x - 1}$$

17. $\dfrac{8}{3v} - \dfrac{1}{4v^2} = \dfrac{8}{3v} \cdot \dfrac{\boxed{}}{\boxed{}} - \dfrac{1}{4v^2} \cdot \dfrac{\boxed{}}{\boxed{}} \quad \text{LCD} = \boxed{}$

$$= \dfrac{\boxed{}}{12v^2} - \dfrac{3}{\boxed{}}$$

$$= \dfrac{32v - 3}{\boxed{}}$$

18. Fill in the blanks:

$$\dfrac{-2x^2 - 4x}{(x + 7)(x - 9)} = \dfrac{\boxed{}(x + 2)}{(x + 7)(x - 9)} = \dfrac{2x(x + 2)}{(x + 7)(x - 9)}$$

GUIDED PRACTICE

Add or subtract, and then simplify, if possible. See Example 1.

19. $\dfrac{8}{3x} + \dfrac{5}{3x}$

20. $\dfrac{3}{4y} + \dfrac{8}{4y}$

21. $\dfrac{t}{4r} + \dfrac{t}{4r}$

22. $\dfrac{16x}{3z^2} - \dfrac{x}{3z^2}$

23. $\dfrac{4y}{y - 4} - \dfrac{16}{y - 4}$

24. $\dfrac{3x}{2x + 2} + \dfrac{x + 4}{2x + 2}$

25. $\dfrac{3x}{x^2 - 9} - \dfrac{9}{x^2 - 9}$

26. $\dfrac{9x}{x^2 - 1} - \dfrac{9}{x^2 - 1}$

The denominators of several fractions are given. Find the LCD. See Example 2.

27. $12xy, \; 18x^2y$

28. $15ab^2, \; 27a^2b$

29. $x^2 + 3x, \; x^2 - 9$

30. $3y^2 - 6y, \; 3y(y - 4)$

31. $x^3 + 27, \; x^2 + 6x + 9$

32. $x^3 - 8, \; x^2 - 4x + 4$

33. $2x^2 + 5x + 3, \; 4x^2 + 12x + 9, \; x^2 + 2x + 1$

34. $2x^2 + 5x + 3, \; 4x^2 + 12x + 9, \; 4x + 6$

Perform the operations and simplify the result when possible. See Example 3.

35. $\dfrac{11}{5m} - \dfrac{5}{6m}$

36. $\dfrac{5}{9s} - \dfrac{1}{4s}$

37. $\dfrac{3}{4ab^2} - \dfrac{5}{2a^2b}$

38. $\dfrac{1}{5xy^3} - \dfrac{2}{15x^2y}$

Add or subtract, and then simplify, if possible. See Example 4.

39. $\dfrac{3}{x + 2} + \dfrac{5}{x - 4}$

40. $\dfrac{6}{a + 4} - \dfrac{2}{a + 3}$

41. $\dfrac{6x}{x + 3} - \dfrac{4x}{x - 3}$

42. $\dfrac{t}{t + 2} + \dfrac{8}{t - 2}$

43. $\dfrac{a - 1}{a + 4} + \dfrac{a + 3}{a - 5}$

44. $\dfrac{n - 3}{n + 7} + \dfrac{n + 3}{n - 6}$

45. $\dfrac{n + 2}{n - 4} - \dfrac{n + 5}{n + 4}$

46. $\dfrac{b - 3}{b + 4} - \dfrac{b + 2}{b - 4}$

Add or subtract, and then simplify, if possible. See Example 5.

47. $4 + \dfrac{1}{x - 2}$

48. $2 - \dfrac{1}{x + 1}$

49. $x + \dfrac{4x}{7x - 3}$

50. $x - \dfrac{3x}{3x - 2}$

Perform the operations and simplify the result when possible. See Example 6.

51. $\dfrac{1}{x + 3} + \dfrac{2}{x^2 + 4x + 3}$

52. $\dfrac{4}{y^2 + 8y + 12} + \dfrac{1}{y + 6}$

53. $\dfrac{m}{m^2 + 9m + 20} - \dfrac{4}{m^2 + 7m + 12}$

54. $\dfrac{t}{t^2 + 5t + 6} - \dfrac{2}{t^2 + 3t + 2}$

55. $\dfrac{x}{x^2 + 5x + 6} + \dfrac{x}{x^2 - 4}$

56. $\dfrac{2a}{a^2 - 2a - 8} + \dfrac{3}{a^2 - 5a + 4}$

57. $\dfrac{x + 2}{6x - 42} - \dfrac{x - 3}{5x - 35}$

58. $\dfrac{x - 1}{4x - 24} - \dfrac{3x - 2}{5x - 30}$

Add or subtract, and then simplify, if possible. See Example 7.

59. $\dfrac{5x}{x - 3} + \dfrac{4x}{3 - x}$

60. $\dfrac{8x}{x - 4} - \dfrac{10x}{4 - x}$

61. $\dfrac{9m}{m - n} - \dfrac{2}{n - m}$

62. $\dfrac{3s}{s - x} + \dfrac{1}{x - s}$

Perform the operations and simplify the result when possible. See Example 8.

63. $\dfrac{5x}{x + 1} + \dfrac{3}{x + 1} - \dfrac{2x}{x + 1}$

64. $\dfrac{4}{a + 4} - \dfrac{2\overline{a}}{a + 4} + \dfrac{3a}{a + 4}$

65. $\dfrac{1}{x + y} - \dfrac{1}{x - y} + \dfrac{2y}{x^2 - y^2}$

66. $\dfrac{a}{a - b} + \dfrac{b}{a + b} - \dfrac{a^2 + b^2}{a^2 - b^2}$

67. $\dfrac{3x}{2x - 1} + \dfrac{x + 1}{3x + 2} - \dfrac{2}{6x^2 + x - 2}$

68. $\dfrac{2}{x - 2} + \dfrac{3}{x + 2} - \dfrac{x - 1}{x^2 - 4}$

69. $\dfrac{8}{x^2 - 9} + \dfrac{2}{x - 3} - \dfrac{6}{x}$

70. $\dfrac{x}{x^2 - 4} - \dfrac{x}{x + 2} + \dfrac{2}{x}$

TRY IT YOURSELF

Perform the operations and simplify the result when possible.

71. $\dfrac{7}{2b} - \dfrac{11}{3a}$

72. $\dfrac{5}{3n} - \dfrac{7}{4m}$

73. $\dfrac{s + 7}{s + 3} - \dfrac{s - 3}{s + 7}$

74. $\dfrac{t + 5}{t - 5} - \dfrac{t - 5}{t + 5}$

75. $\dfrac{x - y}{2} + \dfrac{x + y}{3}$

76. $\dfrac{a + b}{3} + \dfrac{a - b}{7}$

77. $\dfrac{3x^2 + 3x}{x^2 - 5x + 6} - \dfrac{3x^2 - 3x + 12}{x^2 - 5x + 6}$

78. $\dfrac{2m^2 - 7}{m^4 - 9} + \dfrac{4 - m^2}{m^4 - 9}$

79. $\dfrac{a^2 + ab}{a^3 - b^3} - \dfrac{b^2}{b^3 - a^3}$

80. $\dfrac{y^2 - 3xy}{x^3 - y^3} - \dfrac{x^2 + 4xy}{y^3 - x^3}$

81. $2x + 3 + \dfrac{1}{x + 1}$

82. $x + 1 + \dfrac{1}{x - 1}$

83. $\dfrac{d}{d^2 + 11d + 30} - \dfrac{5}{d^2 + 9d + 20}$

84. $\dfrac{t}{t^2 + 9t + 20} - \dfrac{4}{t^2 + 7t + 12}$

85. $\dfrac{3}{x + 1} - \dfrac{2}{x - 1} + \dfrac{x + 3}{x^2 - 1}$

86. $\dfrac{7n^2}{m - n} + \dfrac{3m}{n - m} - \dfrac{3m^2 - n}{m^2 - 2mn + n^2}$

87. $\dfrac{8}{9y^2} + \dfrac{1}{6y^4}$

88. $\dfrac{5}{6a^3} + \dfrac{7}{8a^2}$

Perform the operations and simplify the result when possible. Be careful to apply the correct method, because these problems involve addition, subtraction, multiplication, and division of rational expressions.

89. $\dfrac{6}{b^2 - 9} \cdot \dfrac{b + 3}{2b + 4}$

90. $\dfrac{3a^2 - 22a + 7}{a - a^2} \cdot \dfrac{8a^2 - 8a}{a^2 + a - 56}$

91. $\dfrac{4a}{a - 5} + a$

92. $\dfrac{10z}{z + 4} + z$

93. $\dfrac{2a + 1}{3a - 2} - \dfrac{a - 4}{2 - 3a}$

94. $\dfrac{2x + 1}{x^4 - 81} + \dfrac{2 - x}{x^4 - 81}$

95. $\dfrac{x^2 + x}{3x - 15} \div \dfrac{(x + 1)^2}{6x - 30}$

96. $\dfrac{m}{m^2 + 5m + 6} - \dfrac{2}{m^2 + 3m + 2}$

97. $\dfrac{z^2 - 9}{z^2 + 4z + 3} \div \dfrac{z^2 - 3z}{(z + 1)^2}$

98. $\dfrac{1}{m + 1} + \dfrac{1}{m - 1} + \dfrac{2}{m^2 - 1}$

99. $\dfrac{27p^4}{35q} \div \dfrac{9p}{21q}$

100. $\dfrac{12t}{25s^5} \div \dfrac{10t}{15s^2}$

101. $\dfrac{6}{5d^2 - 5d} - \dfrac{3}{5d - 5}$

102. $\dfrac{9}{2r^2 - 2r} - \dfrac{5}{2r - 2}$

103. $\dfrac{s^3 t}{4s^2 - 9t^2} \cdot \dfrac{4s^2 - 12st + 9t^2}{s^3 t^2}$

104. $\dfrac{25x^2 - 40xy + 16y^2}{x^2y^4} \cdot \dfrac{xy^4}{25x^2 - 16y^2}$

105. $\dfrac{4}{x^2 - 2x - 3} - \dfrac{x}{3x^2 - 7x - 6}$

106. $\dfrac{x + 3}{2x^2 - 5x + 2} - \dfrac{3x - 1}{x^2 - x - 2}$

107. Let $f(x) = \dfrac{3x}{x^2 - 25}$ and $g(x) = \dfrac{4}{x + 5}$.
Find $f(x) + g(x)$.

108. Let $f(x) = \dfrac{5}{2x - 4}$ and $g(x) = \dfrac{3}{2 - x}$.
Find $f(x) - g(x)$.

Look Alikes . . .

109. **a.** $\dfrac{x - 1}{x + 2} + \dfrac{x + 2}{x - 1}$ **b.** $\dfrac{x - 1}{x + 2} \cdot \dfrac{x + 2}{x - 1}$

110. **a.** $\dfrac{n + 2}{4} - \dfrac{8}{4n + 8}$ **b.** $\dfrac{n + 2}{4} \cdot \dfrac{8}{4n + 8}$

APPLICATIONS

111. **Drafting.** Among the tools used in drafting are the 45°–45°–90° and the 30°–60°–90° triangles shown below. Find the perimeter of each triangle. Express each result as a single rational expression.

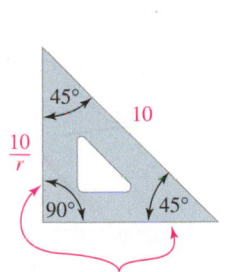 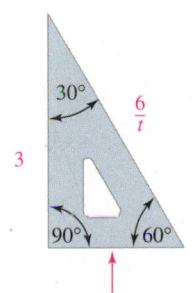

For a 45°–45°–90° triangle, these two sides are the same length.   For a 30°–60°–90° triangle, this side is half as long as the hypotenuse.

112. **The Amazon.** In Brazil, when the Amazon River is at low stage, the rate of flow is about 5 mph. Suppose that a river guide can canoe in still water at a rate of r mph.

   a. Complete the table to find rational expressions that represent the time it would take the guide to canoe 3 miles downriver and to canoe 3 miles upriver on the Amazon.

	Rate (mph)	Time (hr)	Distance (ml)
Downriver	$r + 5$		3
Upriver	$r - 5$		3

   b. Find the difference in the times for the trips upriver and downriver. Express the result as a single rational expression.

WRITING

113. Explain how to find the least common denominator of a set of rational expressions.

114. Add the rational expressions by expressing them in terms of a common denominator $24b^3$. (*Note:* This is not the LCD.) An extra step has to be performed to obtain the correct result because the lowest common denominator was not used. What was the step?

$$\frac{r}{4b^2} + \frac{s}{6b}$$

115. Write some comments to the student who wrote the following solution, explaining his misunderstanding.

Multiply: $\dfrac{1}{x} \cdot \dfrac{3}{2} = \dfrac{1}{x} \cdot \dfrac{2}{2} \cdot \dfrac{3}{2} \cdot \dfrac{x}{x}$

$$= \frac{2}{2x} \cdot \frac{3x}{2x}$$

$$= \frac{6x}{4x^2}$$

116. Write some comments to the student who wrote the following solution, pointing out where she made an error.

Subtract: $\dfrac{1}{x} - \dfrac{x + 1}{x} = \dfrac{1 - x + 1}{x}$

$$= \frac{2 - x}{x}$$

REVIEW

Solve each equation.

117. $a(a - 6) = -9$

118. $x^2 - \dfrac{1}{2}(x + 1) = 0$

119. $y^3 + y^2 = 0$

120. $5x^2 = 6 - 13x$

CHALLENGE PROBLEMS

121. Find two rational expressions, each with denominator $x^2 + 5x + 6$, such that their sum is $\dfrac{1}{x + 2}$.

122. Add: $x^{-1} + x^{-2} + x^{-3} + x^{-4} + x^{-5}$

Perform the operations and simplify the result when possible.

123. $\left(\dfrac{3}{x - 3} - \dfrac{1}{x}\right) \div \dfrac{12x + 18}{x^3 - 9x}$

124. $\dfrac{13x + 39}{4x^2 + 24x + 36} \div \left(\dfrac{7}{3x + 9} - \dfrac{5}{4x + 12}\right)$

125. $\left(\dfrac{3x}{x + 1} - \dfrac{6}{x^2 - 1} + \dfrac{4}{x - 1}\right)\left(\dfrac{x^3 - 1}{9x^2 - 4}\right)$

126. $\left(\dfrac{3}{a + 2b} - \dfrac{2b}{a^2 + 2ab}\right) \div \left(\dfrac{3b}{a^2 + 2ab} + \dfrac{5}{a}\right)$

Simplifying Complex Fractions

OBJECTIVES

1 Simplify complex fractions using division.

2 Simplify complex fractions using the LCD.

ARE YOU READY?

The following problems review some basic skills that are needed when simplifying complex fractions.

1. What operation is indicated by the fraction bar in $\dfrac{64}{16}$?

2. Divide: $\dfrac{2}{21} \div \dfrac{8}{27}$

3. What is the LCD of $\dfrac{5}{a-2}$ and $\dfrac{3}{a+1}$?

4. Multiply: $20a\left(\dfrac{3a}{5} - \dfrac{1}{a}\right)$

A rational expression whose numerator and/or denominator contain rational expressions is called a **complex rational expression** or, more simply, a **complex fraction.** The expression above the main fraction bar of a complex fraction is the numerator, and the expression below the main fraction bar is the denominator. Two examples are:

$$\dfrac{\dfrac{3a}{b}}{\dfrac{6ac}{b^2}} \qquad \leftarrow \quad \text{Numerator} \quad \rightarrow \qquad \dfrac{2x + \dfrac{1}{y}}{\dfrac{1}{x} - \dfrac{1}{y}}$$

$\leftarrow$ Main fraction bar $\rightarrow$

$\leftarrow$ Denominator $\rightarrow$

In this section, we will discuss two methods for simplifying complex fractions. To **simplify a complex fraction** means to write it in the form $\dfrac{A}{B}$, where A and B are polynomials that have no common factors.

1 Simplify Complex Fractions Using Division.

One method for simplifying complex fractions uses the fact that the main fraction bar indicates division.

Simplifying Complex Fractions	**Method 1: Using Division**
	1. Add or subtract in the numerator and/or denominator so that the numerator is a single rational expression and the denominator is a single rational expression.
	2. Perform the indicated division by multiplying the numerator of the complex fraction by the reciprocal of the denominator.
	3. Simplify the result, if possible.

EXAMPLE 1

Use Method 1 to simplify: $\dfrac{\dfrac{3a}{b}}{\dfrac{6ac}{b^2}}$

Strategy We will perform the division indicated by the main fraction bar using the procedure for dividing rational expressions from Section 6.2.

Why We can skip the first step of Method 1 and immediately divide because the numerator and the denominator of the complex fraction are already single rational expressions.

Solution

$$\frac{\dfrac{3a}{b}}{\dfrac{6ac}{b^2}} = \frac{3a}{b} \div \frac{6ac}{b^2}$$ Write the division indicated by the main fraction bar using a ÷ symbol.

$$= \frac{3a}{b} \cdot \frac{b^2}{6ac}$$ To divide two rational expressions, multiply the first by the reciprocal of the second.

$$= \frac{3a \cdot b^2}{b \cdot 6ac}$$ Multiply the numerators. Multiply the denominators.

$$= \frac{\overset{1}{\cancel{3}} \cdot \overset{1}{\cancel{a}} \cdot \overset{1}{\cancel{b}} \cdot b}{\underset{1}{\cancel{b}} \cdot 2 \cdot \underset{1}{\cancel{3}} \cdot \underset{1}{\cancel{a}} \cdot c}$$ Factor the numerator and denominator. Then simplify by removing common factors of the numerator and denominator.

$$= \frac{b}{2c}$$ Multiply the remaining factors in the numerator. Multiply the remaining factors in the denominator.

> ### Success Tip
> Method 1 works well when a complex fraction is written, or can be easily written, as a quotient of two single rational expressions.

Self Check 1   Use Method 1 to simplify: $\dfrac{\dfrac{7x^4}{8y^5}}{\dfrac{21x^3}{20y}}$

Now Try ▶ Problem 13

2 Simplify Complex Fractions Using the LCD.

A second method for simplifying complex fractions uses the concepts of LCD and multiplication by a form of 1. The multiplication by 1 produces a simpler, equivalent expression, which will not contain rational expressions in its numerator or denominator.

> **Simplifying Complex Fractions**
>
> **Method 2: Multiplying by the LCD**
> 1. Find the LCD of all rational expressions within the complex fraction.
> 2. Multiply the complex fraction by 1 in the form $\frac{\text{LCD}}{\text{LCD}}$.
> 3. Perform the operations in the numerator and denominator. No rational expressions should remain within the complex fraction.
> 4. Simplify the result, if possible.

EXAMPLE 2   Use Method 2 to simplify: $\dfrac{\dfrac{3a}{b}}{\dfrac{6ac}{b^2}}$

Strategy   We will use Method 2 to rework Example 1. We can eliminate the rational expressions in the numerator and denominator of the complex fraction by multiplying it by 1, written in the form $\frac{b^2}{b^2}$.

Why   We use $\frac{b^2}{b^2}$ because b^2 is the LCD of $\frac{3a}{b}$ and $\frac{6ac}{b^2}$.

Solution

$$\frac{\dfrac{3a}{b}}{\dfrac{6ac}{b^2}} = \frac{\dfrac{3a}{b}}{\dfrac{6ac}{b^2}} \cdot \frac{b^2}{b^2}$$

Multiply the complex fraction by a form of 1: $\frac{b^2}{b^2} = 1$.

$$= \frac{\dfrac{3ab^2}{b}}{\dfrac{6acb^2}{b^2}}$$

Multiply the numerators: $\frac{3a}{b} \cdot b^2$.

Multiply the denominators: $\frac{6ac}{b^2} \cdot b^2$.

$$= \frac{3ab}{6ac}$$

Simplify the numerator of the complex fraction (highlighted in blue).
Simplify the denominator of the complex fraction (highlighted in red).

$$= \frac{b}{2c}$$

Simplify the resulting rational expression. This is the same result that was obtained using Method 1.

Self Check 2 Use Method 2 to simplify: $\dfrac{\dfrac{7x^4}{8y^5}}{\dfrac{21x_3}{20y}}$

Now Try ▶ Problem 13

EXAMPLE 3 Simplify: $\dfrac{\dfrac{2}{x} + 5}{x + 3}$

Strategy We will simplify the complex fraction using both Method 1 and Method 2.

Why We want to show that the result is the same using either method.

Solution *Method 1*
We add in the numerator of the complex fraction make it a single rational expression.

$$\frac{\dfrac{2}{x} + 5}{x + 3} = \frac{\dfrac{2}{x} + \dfrac{5}{1} \cdot \dfrac{x}{x}}{\dfrac{x + 3}{1}}$$

Write 5 as $\frac{5}{1}$ and $x + 3$ as $\frac{x+3}{1}$. Build $\frac{5}{1}$ to have the LCD x (highlighted in blue).

$$= \frac{\dfrac{2}{x} + \dfrac{5x}{x}}{\dfrac{x + 3}{1}}$$

Multiply $\frac{5}{1}$ and $\frac{x}{x}$ (highlighted in blue).

$$= \frac{\dfrac{2 + 5x}{x}}{\dfrac{x + 3}{1}}$$

Add $\frac{2}{x}$ and $\frac{5x}{x}$ to get $\frac{2+5x}{x}$. The numerator of the complex fraction is now a single rational expression.

$$= \frac{2 + 5x}{x} \div \frac{x + 3}{1}$$

Write the division indicated by the main fraction bar using a ÷ symbol.

$$= \frac{2 + 5x}{x} \cdot \frac{1}{x + 3}$$

To divide the rational expressions, multiply the first by the reciprocal of $\frac{x+3}{1}$.

$$= \frac{2 + 5x}{x^2 + 3x}$$

Multiply the numerators and multiply the denominators. This result does not simplify further.

Method 2

The LCD of all rational expressions in the complex fraction is x.

$$\frac{\dfrac{2}{x}+5}{x+3}=\frac{\dfrac{2}{x}+5}{x+3}\cdot\frac{x}{x}$$

Multiply the complex fraction by 1 in the form $\frac{LCD}{LCD}$. In this case, that is $\frac{x}{x}$.

$$=\frac{\left(\dfrac{2}{x}+5\right)x}{(x+3)x}$$

Multiply the numerators.
Multiply the denominators.

$$=\frac{\dfrac{2}{x}\cdot x+5\cdot x}{x\cdot x+3\cdot x}$$

← In the numerator, distribute the multiplication by x.

← In the denominator, distribute the multiplication by x.

$$=\frac{2+5x}{x^2+3x}$$

Perform each of the four multiplications by x. Notice that no rational expression remains within the complex fraction. This result does not simplify further.

Notation

We could also write the result as:

$$\frac{5x+2}{x^2+3x}\ \text{ or }\ \frac{5x+2}{x(x+3)}$$

Check with your instructor to see which form of the result he or she prefers.

Self Check 3 Simplify: $\dfrac{\dfrac{3}{m}+2}{m+2}$

Now Try ▶ Problem 19

Using Your **Calculator** ▶ Checking Algebra

A check of the simplification done in Example 3 can be performed using a scientific calculator. If

$$\frac{\dfrac{2}{x}+5}{x+3}=\frac{2+5x}{x^2+3x}$$

the expressions on each side will have identical values when evaluated for a given value of x (say, $x = 4$). To evaluate the expression on the left side, we enter these numbers and press these keys.

$$\boxed{(}\ \boxed{2}\ \boxed{\div}\ \boxed{4}\ \boxed{+}\ \boxed{5}\ \boxed{)}\ \boxed{\div}\ \boxed{(}\ \boxed{4}\ \boxed{+}\ \boxed{3}\ \boxed{)}\ \boxed{=}\qquad\boxed{0.785714286}$$

To evaluate the expression on the right side, we enter these numbers and press these keys.

$$\boxed{(}\ \boxed{2}\ \boxed{+}\ \boxed{5}\ \boxed{\times}\ \boxed{4}\ \boxed{)}\ \boxed{\div}\ \boxed{(}\ \boxed{4}\ \boxed{x^2}\ \boxed{+}\ \boxed{3}\ \boxed{\times}\ \boxed{4}\ \boxed{)}\ \boxed{=}\qquad\boxed{0.785714286}$$

The results are the same, so it appears that the simplification is correct. We say "appears" because checking for only a single value of x is not definitive. The expressions should yield identical values when evaluated for any value of x for which the fractions are defined.

We can also check the simplification in Example 3 by graphing the functions $f(x)=\dfrac{\dfrac{2}{x}+5}{3+x}$ shown in figure (a) and $g(x)=\dfrac{2+5x}{x^2+3x}$ shown in figure (b) and observing that the graphs are the same. The window settings are $[-10, 10]$ for x and for y.

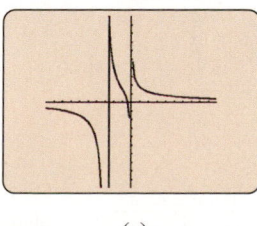

(a)   (b)

EXAMPLE 4

Simplify: $\dfrac{\dfrac{2}{x}+\dfrac{3}{y}}{\dfrac{3}{x}-\dfrac{4}{y}}$

Strategy   We will simplify the complex fraction using both Method 1 and Method 2.

Why   We want to show that using either method, the result is the same.

Solution

Method 1

To write the numerator and denominator of the complex fraction as single rational expressions, we add the rational expressions in the numerator and subtract the rational expressions in the denominator.

$$\frac{\dfrac{2}{x}+\dfrac{3}{y}}{\dfrac{3}{x}-\dfrac{4}{y}} = \frac{\dfrac{2}{x}\cdot\dfrac{y}{y}+\dfrac{3}{y}\cdot\dfrac{x}{x}}{\dfrac{3}{x}\cdot\dfrac{y}{y}-\dfrac{4}{y}\cdot\dfrac{x}{x}}$$

The LCD for the numerator is xy. Build each rational expression so that each has a denominator of xy.

The LCD for the denominator is xy. Build each rational expression so that each has a denominator of xy.

$$= \frac{\dfrac{2y}{xy}+\dfrac{3x}{xy}}{\dfrac{3y}{xy}-\dfrac{4x}{xy}}$$

Multiply the numerators.
Multiply the denominators.

$$= \frac{\dfrac{2y+3x}{xy}}{\dfrac{3y-4x}{xy}}$$

Add the rational expressions in the numerator and subtract the rational expressions in the denominator.

$$= \frac{2y+3x}{xy}\div\frac{3y-4x}{xy}$$

Write the division indicated by the main fraction bar using a ÷ symbol.

$$= \frac{2y+3x}{xy}\cdot\frac{xy}{3y-4x}$$

To divide the rational expressions, multiply the first by the reciprocal of $\frac{3y-4x}{xy}$.

$$= \frac{(2y+3x)\cdot \overset{1}{\cancel{x}}\cdot\overset{1}{\cancel{y}}}{\underset{1}{\cancel{x}}\cdot\underset{1}{\cancel{y}}\cdot(3y-4x)}$$

Multiply the rational expressions and simplify the result.

$$= \frac{2y+3x}{3y-4x}$$

Multiply the remaining factors in the numerator.
Multiply the remaining factors in the denominator.

Method 2

The LCD of the fractions appearing in the complex fraction is *xy*. We multiply the complex fraction by 1 in the form of $\frac{\text{LCD}}{\text{LCD}}$.

$$\frac{\dfrac{2}{x}+\dfrac{3}{y}}{\dfrac{3}{x}-\dfrac{4}{y}} = \frac{\dfrac{2}{x}+\dfrac{3}{y}}{\dfrac{3}{x}-\dfrac{4}{y}}\cdot\frac{xy}{xy}$$

Multiply the complex fraction by a form of 1: $\frac{xy}{xy}$.

$$= \frac{\left(\dfrac{2}{x}+\dfrac{3}{y}\right)xy}{\left(\dfrac{3}{x}-\dfrac{4}{y}\right)xy}$$

Multiply the numerators.
Multiply the denominators.

Success Tip

Notice that with this method, *separate* LCD's are found for the numerator and the denominator of the complex fraction.

Success Tip

Method 2 works well when the complex fraction has sums and/or differences in the numerator or denominator.

Success Tip

With Method 2, each term of the numerator and each term of the denominator of the complex fraction is multiplied by the LCD. Arrows can be helpful in showing this.

$$= \frac{\dfrac{2}{x} \cdot xy + \dfrac{3}{y} \cdot xy}{\dfrac{3}{x} \cdot xy - \dfrac{4}{y} \cdot xy}$$

In the numerator, distribute the multiplication by xy, as shown by the arrows.

In the denominator, distribute the multiplication by xy, as shown by the arrows.

$$= \frac{2y + 3x}{3y - 4x}$$

Perform each of the four multiplications by xy. This result does not simplify further.

Self Check 4 Simplify: $\dfrac{\dfrac{5}{s} - \dfrac{2}{t}}{\dfrac{7}{s} + \dfrac{3}{t}}$

Now Try ▶ Problem 23

If negative exponents occur, we write an equivalent expression involving positive exponents and simplify using Method 1 or Method 2.

EXAMPLE 5 Simplify: $\dfrac{x^{-1} + y^{-1}}{x^{-2} - y^{-2}}$

Strategy We will use the rule for exponents $a^{-m} = \dfrac{1}{a^m}$ to write each term of the numerator and denominator without negative exponents.

Why This will produce familiar rational expressions in the numerator and denominator of the complex fraction.

Solution We write the complex fraction without using negative exponents and use Method 2 to simplify.

Caution

Recall that a *factor* can be moved from the numerator to the denominator if the sign of its exponent is changed. However, in this case, x^{-1}, y^{-1}, x^{-2}, and y^{-2} are *terms* and, therefore, they cannot be moved "across" the fraction bar. Note:

$$\frac{x^{-1} + y^{-1}}{x^{-2} - y^{-2}} \ne \frac{x^2 + y^2}{x - y}$$

Success Tip

Notice that with this method, all the rational expressions in the complex fraction are considered to find one universal LCD.

$$\frac{x^{-1} + y^{-1}}{x^{-2} - y^{-2}} = \frac{\dfrac{1}{x} + \dfrac{1}{y}}{\dfrac{1}{x^2} - \dfrac{1}{y^2}}$$

Use the rule for negative exponents to write x^{-1}, y^{-1}, x^{-2}, and y^{-2} using positive exponents. See the Caution box.

$$= \frac{\dfrac{1}{x} + \dfrac{1}{y}}{\dfrac{1}{x^2} - \dfrac{1}{y^2}} \cdot \frac{x^2 y^2}{x^2 y^2}$$

The LCD of all rational expressions in the complex fraction is x^2y^2. Multiply the complex fraction by $\frac{LCD}{LCD}$.

$$= \frac{\left(\dfrac{1}{x} + \dfrac{1}{y}\right) x^2 y^2}{\left(\dfrac{1}{x^2} - \dfrac{1}{y^2}\right) x^2 y^2}$$

Multiply the numerators. Multiply the denominators.

$$= \frac{\dfrac{1}{x} \cdot x^2 y^2 + \dfrac{1}{y} \cdot x^2 y^2}{\dfrac{1}{x^2} \cdot x^2 y^2 - \dfrac{1}{y^2} \cdot x^2 y^2}$$

In the numerator, distribute the multiplication by x^2y^2, as shown by the arrows.

In the denominator, distribute the multiplication by x^2y^2, as shown by the arrows.

$$= \frac{xy^2 + x^2 y}{y^2 - x^2}$$

Perform each of the four multiplications by x^2y^2.

$$= \frac{xy(\overset{1}{\cancel{y + x}})}{(\cancel{y + x})(y - x)}$$

To prepare to simplify, factor the numerator and denominator.

$$= \frac{xy}{y - x}$$

Simplify the rational expression by removing the factor $y + x$ that is common to the numerator and denominator.

Self Check 5 Simplify: $\dfrac{x^{-2} + y^{-2}}{x^{-1} - y^{-1}}$

Now Try ▶ Problem 31

EXAMPLE 6 Simplify: $\dfrac{\dfrac{1}{a^2 - 3a + 2}}{\dfrac{3}{a - 2} - \dfrac{2}{a - 1}}$

Strategy We will use Method 2 to simplify.

Why Method 2 works well when the complex fraction has sums and/or differences in the numerator or denominator.

Solution To find the LCD for all the fractions appearing in the complex fraction, we must factor $a^2 - 3a + 2$.

$$\dfrac{\dfrac{1}{a^2 - 3a + 2}}{\dfrac{3}{a - 2} - \dfrac{2}{a - 1}} = \dfrac{\dfrac{1}{(a - 2)(a - 1)}}{\dfrac{3}{a - 2} - \dfrac{2}{a - 1}}$$ Factor the trinomial $a^2 - 3a + 2$.

The LCD of the fractions in the numerator and denominator of the complex fraction is $(a - 2)(a - 1)$. We multiply the numerator and the denominator by the LCD.

$$= \dfrac{\dfrac{1}{(a - 2)(a - 1)}}{\dfrac{3}{a - 2} - \dfrac{2}{a - 1}} \cdot \dfrac{(a-2)(a-1)}{(a-2)(a-1)}$$

$$= \dfrac{\left(\dfrac{1}{(a - 2)(a - 1)}\right)(a-2)(a-1)}{\left(\dfrac{3}{a - 2} - \dfrac{2}{a - 1}\right)(a-2)(a-1)}$$ Multiply the numerators.
Multiply the denominators.

$$= \dfrac{\dfrac{(a - 2)(a - 1)}{(a - 2)(a - 1)}}{\dfrac{3(a - 2)(a - 1)}{a - 2} - \dfrac{2(a - 2)(a - 1)}{a - 1}}$$ Perform the mutiplication in the numerator.

In the denominator, distribute the multiplication by the LCD, $(a - 2)(a - 1)$, as shown by the arrows.

> **The Language of Algebra**
>
> After multiplying a complex fraction by $\frac{\text{LCD}}{\text{LCD}}$ and performing the multiplications, the numerator and denominator of the complex fraction will be **cleared of fractions.**

$$= \dfrac{1}{3(a - 1) - 2(a - 2)}$$ Simplify each of the three rational expressions highlighted in blue.

$$= \dfrac{1}{3a - 3 - 2a + 4}$$ In the denominator, distribute the multiplication by 3 and by −2.

$$= \dfrac{1}{a + 1}$$ In the denominator, combine like terms: $3a - 2a = a$ and $-3 + 4 = 1$.

Self Check 6 Simplify: $\dfrac{\dfrac{b}{b + 4} + \dfrac{2}{b + 3}}{\dfrac{b}{b^2 + 7b + 12}}$

Now Try ▶ Problem 35

SECTION 6.4 ▶ STUDY SET

VOCABULARY

Fill in the blanks.

1. $\dfrac{\dfrac{x}{y} + \dfrac{1}{x}}{\dfrac{1}{y} + \dfrac{2}{x}}$ and $\dfrac{\dfrac{5a^2}{b}}{\dfrac{b}{2a^3}}$ are examples of complex _____ expressions, or more simply, _____ fractions.

2. To _____ a complex fraction means to express it in the form $\frac{A}{B}$, where A and B are polynomials with no common factors.

CONCEPTS

3. To simplify the following complex fraction, it is multiplied by what form of 1?

$$\frac{\dfrac{4}{t^2} + \dfrac{b}{t}}{\dfrac{3b}{t}} = \frac{\dfrac{4}{t^2} + \dfrac{b}{t}}{\dfrac{3b}{t}} \cdot \frac{t^2}{t^2}$$

4. Determine the LCD of the rational expressions appearing in each complex fraction.

a. $\dfrac{1 + \dfrac{4}{c}}{\dfrac{2}{c} + c}$

b. $\dfrac{\dfrac{6}{m^2} + \dfrac{1}{2m}}{\dfrac{m^2 - 1}{4}}$

c. $\dfrac{\dfrac{p}{p + 2} + \dfrac{12}{p + 3}}{\dfrac{p - 1}{p^2 + 5p + 6}}$

d. $\dfrac{2 + \dfrac{3}{x + 1}}{\dfrac{1}{x} + x + x^2}$

Fill in the blanks.

5. Method 1: To simplify a complex fraction, write its numerator and denominator as _____ rational expressions. Then perform the indicated _____ by multiplying the numerator of the complex fraction by the _____ of the denominator.

6. Method 2: To simplify a complex fraction, find the LCD of ____ rational expressions within the complex fraction. Multiply the complex fraction by 1 in the form ▢ .

NOTATION

Complete each solution to simplify the rational expression.

7. $\dfrac{\dfrac{5m^2}{6}}{\dfrac{25m}{3}} = \dfrac{5m^2}{6}$ ▢ $\dfrac{25m}{3}$ **Use Method 1**

$= \dfrac{5m^2}{6} \cdot \dfrac{▢}{▢}$

$= \dfrac{5 \cdot ▢ \cdot m \cdot ▢}{2 \cdot ▢ \cdot 5 \cdot 5 \cdot ▢}$

$= \dfrac{m}{▢}$

8. $\dfrac{\dfrac{2}{a} - \dfrac{1}{b}}{\dfrac{5}{a} + \dfrac{3}{b}} = \dfrac{\dfrac{2}{a} - \dfrac{1}{b}}{\dfrac{5}{a} + \dfrac{3}{b}} \cdot \dfrac{▢}{▢}$ **Use Method 2**

$= \dfrac{\left(\dfrac{2}{a} - \dfrac{1}{b}\right)▢}{\left(\dfrac{5}{a} + \dfrac{3}{b}\right)▢}$

$= \dfrac{\dfrac{2}{a} \cdot ▢ - \dfrac{1}{b} \cdot ▢}{\dfrac{5}{a} \cdot ▢ + \dfrac{3}{b} \cdot ▢}$

$= \dfrac{2b - ▢}{▢ + 3a}$

9. a. Fill in the blank: The expression $\dfrac{\dfrac{a}{b}}{\dfrac{c}{d}}$ is equivalent to $\dfrac{a}{b}$ ▢ $\dfrac{c}{d}$.

   b. What is the numerator and what is the denominator of the following complex fraction?

$$\frac{6 - k - \dfrac{5}{k}}{k^2 - 9}$$

10. A student checks her answers with those in the back of her textbook. Determine whether they are equivalent.

Student's answer	Book's answer	Equivalent?
$\dfrac{3 + 2t}{t^2 + 2t}$	$\dfrac{2t + 3}{t(t + 2)}$	
$\dfrac{5 - 3x^2}{x + x^2}$	$\dfrac{3x^2 - 5}{x^2 + x}$	
$\dfrac{3xy(y + x)}{(2y - x)(2y + 3x)}$	$\dfrac{3xy^2 + 3x^2y}{(2y + x)(2y - 3x)}$	

GUIDED PRACTICE

Simplify each complex fraction. See Examples 1 and 2.

11. $\dfrac{\dfrac{a^6}{2}}{\dfrac{3a}{4}}$

12. $\dfrac{\dfrac{3b^7}{4}}{\dfrac{b^9}{2}}$

13. $\dfrac{\dfrac{20x}{y}}{\dfrac{36x}{y^2}}$

14. $\dfrac{\dfrac{5t^4}{9x^2}}{\dfrac{2t}{18x}}$

15. $\dfrac{\dfrac{18x^5}{35y^2}}{\dfrac{2x^8}{21y^5}}$

16. $\dfrac{\dfrac{32m}{45n^{10}}}{\dfrac{8m^7}{3n^{11}}}$

17. $\dfrac{\dfrac{16c}{77d^4}}{\dfrac{28c^7}{55d^0}}$

18. $\dfrac{\dfrac{8m^{10}}{27n^0}}{\dfrac{32m}{63n^9}}$

Simplify each complex fraction. See Example 3.

19. $\dfrac{\dfrac{3}{a} - 2}{a - 3}$

20. $\dfrac{\dfrac{5}{t} - 5}{t - 5}$

21. $\dfrac{4p - \dfrac{4}{p}}{12 - \dfrac{4}{p}}$

22. $\dfrac{3s - \dfrac{3}{s}}{6 + \dfrac{3}{s}}$

Simplify each complex fraction. See Example 4.

23. $\dfrac{\dfrac{y}{x} - \dfrac{x}{y}}{\dfrac{1}{x} + \dfrac{1}{y}}$ 24. $\dfrac{\dfrac{y}{x} - \dfrac{x}{y}}{\dfrac{1}{y} - \dfrac{1}{x}}$ 25. $\dfrac{\dfrac{1}{a} - \dfrac{1}{b}}{\dfrac{a}{b} - \dfrac{b}{a}}$ 26. $\dfrac{\dfrac{1}{a} + \dfrac{1}{b}}{\dfrac{a}{b} - \dfrac{b}{a}}$

27. $\dfrac{\dfrac{2}{a^2} + \dfrac{1}{a}}{\dfrac{2}{a} + \dfrac{1}{a^2}}$ 28. $\dfrac{\dfrac{3}{y^2} - \dfrac{4}{y}}{\dfrac{1}{y} + \dfrac{15}{y^2}}$

29. $\dfrac{\dfrac{3}{b^2} - \dfrac{4}{b} + 1}{1 - \dfrac{1}{b} - \dfrac{6}{b^2}}$ 30. $\dfrac{\dfrac{18}{c^2} + \dfrac{11}{c} + 1}{1 - \dfrac{3}{c} - \dfrac{10}{c^2}}$

Simplify each complex fraction. See Example 5.

31. $\dfrac{x^{-2} - y^{-2}}{x^{-1} - y^{-1}}$ 32. $\dfrac{x^{-1} + y^{-1}}{x^{-1} - y^{-1}}$

33. $\dfrac{a - b^{-2}}{b - a^{-2}}$ 34. $\dfrac{m^{-1} + 2n}{m^{-1} - n}$

Simplify each complex fraction. See Example 6.

35. $\dfrac{\dfrac{3}{z-3} + \dfrac{2}{z-2}}{\dfrac{5z}{z^2 - 5z + 6}}$ 36. $\dfrac{\dfrac{h}{h^2 + 3h + 2}}{\dfrac{4}{h+2} - \dfrac{4}{h+1}}$

37. $\dfrac{\dfrac{2}{x+3} - \dfrac{1}{x-3}}{\dfrac{3}{x^2 - 9}}$ 38. $\dfrac{2 + \dfrac{1}{x^2 - 1}}{1 + \dfrac{1}{x-1}}$

TRY IT YOURSELF

Simplify each complex fraction.

39. $\dfrac{1 + \dfrac{x}{y}}{1 - \dfrac{x}{y}}$ 40. $\dfrac{\dfrac{x}{y} + 1}{1 - \dfrac{x}{y}}$

41. $\dfrac{\dfrac{x^2 + 5x + 6}{3xy}}{\dfrac{9 - x^2}{6xy}}$ 42. $\dfrac{\dfrac{x - y}{xy}}{\dfrac{y - x}{x}}$

43. $\dfrac{1 + \dfrac{6}{x} + \dfrac{8}{x^2}}{1 + \dfrac{1}{x} - \dfrac{12}{x^2}}$ 44. $\dfrac{1 - x - \dfrac{2}{x}}{\dfrac{6}{x^2} + \dfrac{1}{x} - 1}$

45. $\dfrac{\dfrac{ac - ad - c + d}{a^3 - 1}}{\dfrac{c^2 - 2cd + d^2}{a^2 + a + 1}}$ 46. $\dfrac{\dfrac{2x - tx + 2y - ty}{x^2 + 2xy + y^2}}{\dfrac{t^3 - 8}{15x + 15y}}$

47. $\dfrac{\dfrac{8}{x} + \dfrac{x}{8}}{\dfrac{x}{8} - \dfrac{8}{x}}$ 48. $\dfrac{\dfrac{x}{7} - \dfrac{7}{x}}{\dfrac{1}{7} + \dfrac{1}{x}}$

49. $-\dfrac{a}{\dfrac{1}{a} + \dfrac{1}{b} + \dfrac{1}{c}}$ 50. $-\dfrac{m}{\dfrac{1}{m} - \dfrac{1}{n} + \dfrac{1}{t}}$

51. $\dfrac{\dfrac{1}{a+1} + 1}{\dfrac{3}{a-1} + 1}$ 52. $\dfrac{2 + \dfrac{4}{y-7}}{\dfrac{4}{y-7}}$

53. $\dfrac{\dfrac{5ab^2}{}}{\dfrac{ab}{25}}$ 54. $\dfrac{\dfrac{6a^2b}{4t}}{3a^2b^2}$

55. $\dfrac{a - 4 + \dfrac{1}{a}}{\dfrac{1}{a} - a + 4}$ 56. $\dfrac{a + 1 + \dfrac{1}{a^2}}{\dfrac{1}{a^2} + a - 1}$

57. $\dfrac{\dfrac{2}{a+1} + \dfrac{1}{a+3}}{\dfrac{2a}{a^2 + 4a + 3}}$ 58. $\dfrac{\dfrac{2}{t-2} - \dfrac{1}{t^2 + t - 6}}{\dfrac{4}{t+3}}$

59. $\dfrac{y}{x^{-1} - y^{-1}}$ 60. $\dfrac{x^{-1} + y^{-1}}{(x+y)^{-1}}$

61. $\dfrac{\dfrac{1}{x^2} - \dfrac{3}{xy} + \dfrac{2}{y^2}}{\dfrac{2}{x^2} - \dfrac{1}{xy} - \dfrac{1}{y^2}}$ 62. $\dfrac{\dfrac{3}{s^2} + \dfrac{7}{st} + \dfrac{2}{t^2}}{\dfrac{2}{t^2} - \dfrac{5}{st} - \dfrac{3}{s^2}}$

63. $\dfrac{5xy}{1 + \dfrac{1}{xy}}$ 64. $\dfrac{3a}{a + \dfrac{1}{a}}$

65. $\dfrac{\dfrac{2}{y-1} - \dfrac{2}{y}}{\dfrac{3}{y-1} - \dfrac{1}{1-y}}$ 66. $\dfrac{\dfrac{1}{x} - \dfrac{4}{x-1}}{\dfrac{3}{x-1} + \dfrac{2}{x}}$

67. $\dfrac{\dfrac{t}{x^2 - y^2}}{\dfrac{t}{x+y}}$ 68. $\dfrac{\dfrac{7}{a-b}}{\dfrac{b}{a^3 - b^3}}$

69. $\dfrac{\dfrac{4}{cd}}{c^{-1} + d^{-1}}$ 70. $\dfrac{\dfrac{y}{x} + x^{-1}}{x^{-1} + \dfrac{2x}{y}}$

APPLICATIONS

71. Engineering. The stiffness k of the shaft shown is given by the formula

$$k = \dfrac{1}{\dfrac{1}{k_1} + \dfrac{1}{k_2}}$$

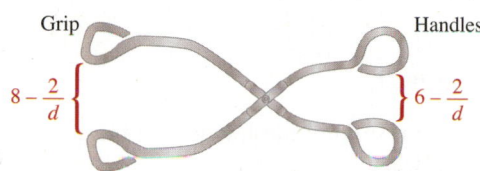

Section 1   Section 2

where k_1 and k_2 are the individual stiffnesses of each section. Simplify the complex fraction.

72. Transportation. If a bus travels a distance d_1 at a speed s_1, and then travels a distance d_2 at a speed s_2, the average (mean) speed $\bar{s}$ is given by the following formula. Simplify the complex fraction.

$$\bar{s} = \dfrac{d_1 + d_2}{\dfrac{d_1}{s_1} + \dfrac{d_2}{s_2}}$$

73. Kitchen Utensils. What is the ratio of the width of the grip of the ice tongs to the width of the opening of the handles? Express the result in simplest form.

Grip   Handles

$8 - \dfrac{2}{d}$ $6 - \dfrac{2}{d}$

74. Data Analysis. Use the data in the table to find the average measurement for the three-trial experiment. Express the answer as a simplified rational expression.

	Trial 1	Trial 2	Trial 3
Measurement	$\dfrac{k}{3}$	$\dfrac{k}{5}$	$\dfrac{k}{6}$

WRITING

75. What is a complex fraction?

76. Which of the two methods for simplifying a complex fraction do you think is simpler? Explain.

77. What does it mean when we say that

$$\dfrac{\dfrac{1}{x+2}}{1 + \dfrac{1}{x+2}} \quad \text{and} \quad \dfrac{1}{x+3}$$

are *equivalent expressions*?

78. To verify the simplification below, a student graphed the associated function for the expression on the left side of the equation, and the associated function for the expression on the right side. See figure (a) and figure (b) below. What conclusion can be made about the simplification from the graphs?

$$\dfrac{\dfrac{4}{3} + x}{\dfrac{x}{2} + \dfrac{2}{x}} = \dfrac{x^2 + 4}{x + 2}$$

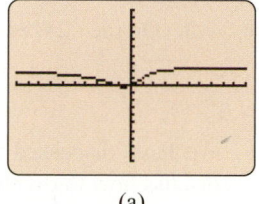   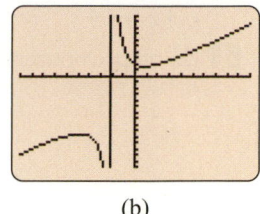

(a)   (b)

REVIEW

Solve each equation.

79. $\dfrac{8(a-5)}{3} = 2(a-4)$ **80.** $\dfrac{3t^2}{5} + \dfrac{7t}{10} = \dfrac{3t+6}{5}$

81. $a^4 - 13a^2 + 36 = 0$ **82.** $|2x - 1| = 9$

CHALLENGE PROBLEMS

Simplify each expression.

83. $a + \dfrac{a}{1 + \dfrac{a}{a+1}}$ **84.** $b + \dfrac{b}{1 - \dfrac{b+1}{b}}$

85. $\dfrac{x - \dfrac{1}{1 - \dfrac{x}{2}}}{\dfrac{3}{x + \dfrac{2}{3}} + x}$ **86.** $\dfrac{3x - \dfrac{1}{3 - \dfrac{x}{2}}}{\dfrac{3}{\dfrac{x}{2} - 3} + x}$

87. $(x^{-1}y^{-1})(x^{-1} + y^{-1})^{-1}$

88. $[(x^{-1} + 1)^{-1} + 1]^{-1}$

89. $\left(\dfrac{\dfrac{x}{y} - \dfrac{y}{x}}{\dfrac{x+y}{x}}\right) \div \left(\dfrac{\dfrac{x^2}{y} - y}{\dfrac{y^2}{x} - x}\right)$

90. $\left(\dfrac{\dfrac{2}{b} + \dfrac{1}{2b}}{b + \dfrac{b}{2}}\right) + \left(\dfrac{b - \dfrac{b-3}{3}}{\dfrac{4}{9} + \dfrac{2}{3b}}\right)$

SECTION 6.5

Dividing Polynomials

OBJECTIVES

1 Divide a monomial by a monomial.

2 Divide a polynomial by a monomial.

3 Divide a polynomial by a polynomial.

4 Divide polynomials with missing terms.

ARE YOU READY?

The following problems review some basic skills that are needed when dividing polynomials.

1. Simplify: **a.** $\dfrac{12}{30}$ **b.** $\dfrac{8a^8}{4a^3}$

2. Multiply:

$$4x^2y^3\left(2x^3y^2 - 5x^2y^4 + \dfrac{1}{2xy}\right)$$

3. Divide: $24\overline{)864}$

4. Subtract: $\begin{array}{r} 4x^3 + 7x^2 \\ -(4x^3 + 5x^2) \\ \hline \end{array}$

We have discussed addition, subtraction, and multiplication of polynomials. We will now discuss how to divide polynomials. This topic appears in a chapter about rational expressions because rational expressions indicate division of polynomials. For example,

$$\frac{x^2 - 3x + 7}{x + 1} = (x^2 - 3x + 7) \div (x + 1)$$

We begin with the simplest case, dividing a monomial by a monomial.

1 Divide a Monomial by a Monomial.

To divide monomials, we can use the method for simplifying fractions or the quotient rule for exponents.

EXAMPLE 1   Simplify. Write answers using positive exponents.   **a.** $\dfrac{21x^5}{7x^2}$ **b.** $\dfrac{10r^6s}{6rs^3}$

Strategy   First, we will simplify the rational expression by factoring and removing any factors that are common to the numerator and the denominator. Then we will simplify the rational expression using the rules for exponents.

Why   We want to show that the result is the same using either method.

Solution

By simplifying rational expressions

a. $\dfrac{21x^5}{7x^2} = \dfrac{3 \cdot \overset{1}{\cancel{7}} \cdot \overset{1}{\cancel{x}} \cdot \overset{1}{\cancel{x}} \cdot x \cdot x \cdot x}{\underset{1}{\cancel{7}} \cdot \underset{1}{\cancel{x}} \cdot \underset{1}{\cancel{x}}}$

$= 3x^3$

b. $\dfrac{10r^6s}{6rs^3} = \dfrac{\overset{1}{\cancel{2}} \cdot 5 \cdot \overset{1}{\cancel{r}} \cdot r \cdot r \cdot r \cdot r \cdot r \cdot \overset{1}{\cancel{s}}}{\underset{1}{\cancel{2}} \cdot 3 \cdot \underset{1}{\cancel{r}} \cdot \underset{1}{\cancel{s}} \cdot s \cdot s}$

$= \dfrac{5r^5}{3s^2}$

Using the rules for exponents

$\dfrac{21x^5}{7x^2} = 3x^{5-2}$ Divide the coefficients. Keep the common base x and subtract exponents.

$= 3x^3$

$\dfrac{10r^6s}{6rs^3} = \dfrac{5}{3}r^{6-1}s^{1-3}$ Simplify $\frac{10}{6}$. Keep each base and subtract exponents.

$= \dfrac{5}{3}r^5s^{-2}$

$= \dfrac{5r^5}{3s^2}$ Move s^{-2} to the denominator and change the sign of the exponent.

Success Tip

In this section, you will see that regardless of the number of terms involved, every polynomial division is a series of monomial divisions.

Self Check 1   Simplify: **a.** $\dfrac{30y^4}{5y^2}$ **b.** $\dfrac{8c^2d^6}{32c^5d^2}$

Now Try ▶   Problems 15 and 17

2 Divide a Polynomial by a Monomial.

Recall that to add two fractions with the same denominator, we add their numerators and keep the common denominator.

$$\frac{a}{d} + \frac{b}{d} = \frac{a + b}{d}$$

We can use this rule in *reverse* to divide polynomials by monomials.

Dividing a Polynomial by a Monomial	To divide a polynomial by a monomial, divide each term of the polynomial by the monomial. Let A, B, and D represent monomials, where D is not 0. $$\frac{A + B}{D} = \frac{A}{D} + \frac{B}{D}$$

EXAMPLE 2 Divide: **a.** $\dfrac{9x^2 + 3x}{3x}$ **b.** $\dfrac{12a^4b^3 - 18a^3b^2 + 2a^2}{6a^2b^2}$

Strategy We will divide each term of the polynomial in the numerator by the monomial in the denominator.

Why A fraction bar indicates division of the numerator by the denominator.

Solution

a. Here, we have a binomial divided by a monomial.

$$\frac{9x^2 + 3x}{3x} = \frac{9x^2}{3x} + \frac{3x}{3x}$$
Divide each term of the numerator, $9x^2 + 3x$, by the denominator, $3x$.

$$= 3x^{2-1} + 1x^{1-1}$$
Perform each monomial division. Divide the coefficients. Keep each base and subtract the exponents.

$$= 3x^1 + 1x^0$$
Do the subtraction.

$$= 3x + 1$$
Recall that $x^0 = 1$.

Check: We multiply the divisor, $3x$, and the quotient, $3x + 1$. The result should be the dividend, $9x^2 + 3x$.

$$3x(3x + 1) = 9x^2 + 3x$$ The answer checks.

b. Here we have a trinomial divided by a monomial.

$$\frac{12a^4b^3 - 18a^3b^2 + 2a^2}{6a^2b^2} = \frac{12a^4b^3}{6a^2b^2} - \frac{18a^3b^2}{6a^2b^2} + \frac{2a^2}{6a^2b^2}$$
Divide each term of the numerator by the denominator, $6a^2b^2$.

$$= 2a^{4-2}b^{3-2} - 3a^{3-2}b^{2-2} + \frac{a^{2-2}}{3b^2}$$
Perform each monomial division. Simplify: $\frac{2}{6} = \frac{1}{3}$.

$$= 2a^2b - 3a + \frac{1}{3b^2}$$
Do the subtraction.

Since the variables in a polynomial must have whole-number exponents, $2a^2b - 3a + \frac{1}{3b^2}$ is not a polynomial because the last term can be written as $\frac{1}{3}b^{-2}$. We can check the result using multiplication.

Check: $6a^2b^2\left(2a^2b - 3a + \dfrac{1}{3b^2}\right) = 12a^4b^3 - 18a^3b^2 + 2a^2$ The answer checks.

The Language of Algebra

The names of the parts of a division statement are

Dividend
$$\frac{9x^2 + 3x}{3x} = 3x + 1$$
Divisor   Quotient

Caution

Recall that only *factors* common to the numerator and denominator can be removed. In part a, $3x$ is not a factor of the numerator, it is a *term,* and, therefore, cannot be removed.

$$\frac{9x^2 + \overset{1}{\cancel{3x}}}{\underset{1}{\cancel{3x}}}$$

Success Tip

The sum, difference, and product of two polynomials are always polynomials. However, as seen in Example 2(b), the quotient of two polynomials is not always a polynomial.

Self Check 2   Divide: **a.** $\dfrac{50h^3 + 15h^2}{5h^2}$ **b.** $\dfrac{22s^5t^2 - s^4t^3 + 44s^2t}{11s^2t^2}$

Now Try ▶   Problems 21 and 25

---

3   Divide a Polynomial by a Polynomial.

To divide a polynomial by a polynomial (other than a monomial), we use a method similar to long division in arithmetic. To use long division to divide $x^2 + 7x + 12$ (the *dividend*) by $x + 4$ (the *divisor*), we proceed as follows:

Step 1:
$$\begin{array}{r} x \\ x + 4 \overline{)x^2 + 7x + 12} \end{array}$$

Divide the first term of the dividend by the first term of the divisor: $\frac{x^2}{x} = x$. Write the result, x, above the long division symbol.

Step 2:
$$\begin{array}{r} x \\ x + 4 \overline{)x^2 + 7x + 12} \\ x^2 + 4x \end{array}$$

Multiply each term of the divisor by x. Write the result, $x^2 + 4x$, under $x^2 + 7x$, and draw a line. Be sure to align the like terms.

Step 3:
$$\begin{array}{r} x \\ x + 4 \overline{)x^2 + 7x + 12} \\ -(x^2 + 4x) \\ 3x + 12 \end{array}$$

Write parentheses around $x^2 + 4x$ so that both its terms are subtracted.
Subtract $x^2 + 4x$ from $x^2 + 7x$. Work column by column: $x^2 - x^2 = 0$ and $7x - 4x = 3x$.
Bring down the next term, 12.

Step 4:
$$\begin{array}{r} x + 3 \\ x + 4 \overline{)x^2 + 7x + 12} \\ -(x^2 + 4x) \\ 3x + 12 \end{array}$$

Divide the first term of $3x + 12$ by the first term of the divisor: $\frac{3x}{x} = 3$. Write $+3$ above the long division symbol to form the second term of the quotient.

Step 5:
$$\begin{array}{r} x + 3 \\ x + 4 \overline{)x^2 + 7x + 12} \\ -(x^2 + 4x) \\ 3x + 12 \\ 3x + 12 \\ 0 \end{array}$$

Multiply each term of the divisor by 3. Write the result, $3x + 12$, under $3x + 12$ and draw a line. Be sure to align the like terms.

Step 6:
$$\begin{array}{r} x + 3 \\ x + 4 \overline{)x^2 + 7x + 12} \\ -(x^2 + 4x) \\ 3x + 12 \\ -(3x + 12) \\ 0 \end{array}$$

Write parentheses around $3x + 12$ so that both its terms are subtracted.
Subtract $3x + 12$ from $3x + 12$. Work vertically: $3x - 3x = 0$ and $12 - 12 = 0$.
There are no more terms to bring down. This is the remainder.

Step 7:   The division process stops when the result of the subtraction is a constant or a polynomial with degree less than the degree of the divisor. Here, the quotient is $x + 3$ and the remainder is 0.

We can check the answer using the fact that for any division:

$$\text{Divisor} \cdot \text{quotient} + \text{remainder} = \text{dividend}$$

$$\underbrace{\text{Divisor} \cdot \text{quotient}}_{} + \underbrace{\text{remainder}}_{} = \underbrace{\text{dividend}}_{}$$

Check: $(x + 4)(x + 3) +$ 0 $= x^2 + 7x + 12$ The answer checks.

Success Tip

Notice that this method is much like that used for division of whole numbers.

$$\begin{array}{r} 13 \\ 12 \overline{)156} \\ -12\downarrow \\ 36 \\ -36 \\ 0 \end{array}$$

Hundreds ⤶
Tens ⤶
Ones ⤶

Success Tip

The long division method aligns like terms vertically.

$$\begin{array}{r} x + 3 \\ x + 4 \overline{)x^2 + 7x + 12} \\ -(x^2 + 4x)) \\ 3x + 12 \\ -(3x + 12) \\ 0 \end{array}$$

x^2-terms ⤶
x-terms ⤶
Constants ⤶

EXAMPLE 3   Divide: $\dfrac{2a^3 + 13a^2 + 11a - 6}{2a + 3}$

Strategy   We will use the long division method. The dividend is $2a^3 + 13a^2 + 11a - 6$ and the divisor is $2a + 3$.

Why   Since the divisor has more than one term, we must use the long division method to divide the polynomials.

Solution

Success Tip

The long-division method is a series of four steps:
- Divide
- Multiply
- Subtract
- Bring down the next term

Use parentheses to avoid sign errors when subtracting.

$$\begin{array}{r} a^2 \\ 2a + 3 \overline{)\,2a^3 + 13a^2 + 11a - 6} \end{array}$$

Divide the first term of the dividend by the first term of the divisor: $\frac{2a^3}{2a} = a^2$. Write the result, a^2, above the long division symbol.

$$\begin{array}{r} a^2 \\ 2a + 3 \overline{)\,2a^3 + 13a^2 + 11a - 6} \\ -(2a^3 + 3a^2) \downarrow \\ \hline 10a^2 + 11a \end{array}$$

Multiply each term in the divisor by a^2 to get $2a^3 + 3a^2$. Subtract $2a^3 + 3a^2$ from $2a^3 + 13a^2$. Bring down the 11a.

$$\begin{array}{r} a^2 + 5a \\ 2a + 3 \overline{)\,2a^3 + 13a^2 + 11a - 6} \\ -(2a^3 + 3a^2) \\ \hline 10a^2 + 11a \end{array}$$

Divide the first term of $10a^2 + 11a$ by the first term of the divisor: $\frac{10a^2}{2a} = 5a$. Write $+5a$ above the long division symbol to form the second term of the quotient.

$$\begin{array}{r} a^2 + 5a \\ 2a + 3 \overline{)\,2a^3 + 13a^2 + 11a - 6} \\ -(2a^3 + 3a^2) \\ \hline 10a^2 + 11a \\ -(10a^2 + 15a)\downarrow \\ \hline -4a - 6 \end{array}$$

Multiply each term in the divisor by $5a$ to get $10a^2 + 15a$. Subtract $10a^2 + 15a$ from $10a^2 + 11a$. Bring down the -6.

$$\begin{array}{r} a^2 + 5a - 2 \\ 2a + 3 \overline{)\,2a^3 + 13a^2 + 11a - 6} \\ -(2a^3 + 3a^2) \\ \hline 10a^2 + 11a \\ -(10a^2 + 15a) \\ \hline -4a - 6 \end{array}$$

Divide the first term of $-4a - 6$ by the first term of the divisor: $\frac{-4a}{2a} = -2$. Write -2 above the long division symbol to form the third term of the quotient.

$$\begin{array}{r} a^2 + 5a - 2 \\ 2a + 3 \overline{)\,2a^3 + 13a^2 + 11a - 6} \\ \underline{2a^3 + 3a^2} \\ 10a^2 + 11a \\ \underline{-(10a^2 + 15a)} \\ -4a - 6 \\ \underline{-(-4a - 6)} \\ 0 \end{array}$$

Multiply each term in the divisor by -2 to get $-4a - 6$. Subtract $-4a - 6$ from $-4a - 6$ to get 0.

There are no more terms to bring down. This is the remainder.

Since the remainder is 0, the quotient is $a^2 + 5a - 2$. We can check the quotient by verifying that

$$\underbrace{\text{Divisor} \cdot \text{quotient}} \quad + \quad \text{remainder} \quad = \quad \underbrace{\text{dividend}}$$

Check: $(2a + 3)(a^2 + 5a - 2) \quad + \quad 0 \quad = 2a^3 + 13a^2 + 11a - 6$ The result checks.

Self Check 3   Divide: $\dfrac{4x^3 + 11x^2 + 2x - 3}{4x + 3}$

Now Try ▶ Problems 27 and 33

Dividing a Polynomial by a Polynomial	To divide a polynomial by a polynomial (other than a monomial) use long division.
	1. To begin, arrange the terms of the polynomials in descending powers of the variable. If a power of the variable is missing in either polynomial, insert a placeholder term by writing 0 times the variable raised to the missing power.
	2. Perform the long division process until the degree of the remainder is less than the degree of the divisor. If there is a remainder, write the result in the form: $$\text{quotient} + \frac{\text{remainder}}{\text{divisor}}$$

The long division method used in algebra can have a remainder just as long division in arithmetic does.

EXAMPLE 4 Divide: $\dfrac{3x^3 + 2x^2 - 3x - 28}{x - 2}$

Strategy   We will use the long division method. The dividend is $3x^3 + 2x^2 - 3x - 28$ and the divisor is $x - 2$.

Why   Since the divisor has more than one term, we must use the long division method to divide the polynomials.

Solution

$$
\begin{array}{r}
3x^2 + 8x + 13 \\
x - 2 \overline{\smash{)}\, 3x^3 + 2x^2 - 3x - 28} \\
-(3x^3 - 6x^2) \\
\hline
8x^2 - 3x \\
-(8x^2 - 16x) \\
\hline
13x - 28 \\
-(13x - 26) \\
\hline
-2
\end{array}
$$

The first division: $\frac{3x^3}{x} = 3x^2$.

The second division: $\frac{8x^2}{x} = 8x$.

The third division: $\frac{13x}{x} = 13$.

Subtract: $-28 - (-26) = -2$. There are no more terms to bring down. This is the remainder.

Success Tip

The long division method for polynomials continues until the degree of the remainder is less than the degree of the divisor. Here, the remainder, -2, has degree 0. The divisor, $x - 2$, has degree 1. Therefore, the division process ends.

This division gives a quotient of $3x^2 + 8x + 13$ and a remainder of -2. It is common to form a fraction with the remainder as the numerator and the divisor as the denominator and to write the result as

$$3x^2 + 8x + 13 + \frac{-2}{x - 2} \qquad \text{or} \qquad 3x^2 + 8x + 13 - \frac{2}{x - 2}$$

To check, we use multiplication to verify that

$$(x - 2)(3x^2 + 8x + 13) + (-2) = 3x^3 + 2x^2 - 3x - 28 .$$ The result checks.

Self Check 4   Divide: $\dfrac{2a^3 + 3a^2 - a - 85}{a - 3}$

Now Try ▶ Problem 37

The division method works best when the terms of the divisor and the dividend are written in descending powers of the variable. If the powers in the dividend or divisor are not in descending order, we use the commutative property of addition to write them that way.

EXAMPLE 5 Divide: $(-10x + 12x^3 + x^2 - 3) \div (1 + 3x)$

Strategy We will write the dividend and divisor in descending powers of x and use the long division method.

Why It is easier to align like terms in columns when the powers of the variable in the dividend and divisor are written in descending order.

Solution After writing the terms of the dividend and divisor in descending powers of x, we proceed as follows.

$$
\begin{array}{r}
4x^2 - x - 3 \\
3x + 1 \overline{) 12x^3 + x^2 - 10x - 3} \\
-(12x^3 + 4x^2) \\
-3x^2 - 10x \\
-(-3x^2 - x) \\
-9x - 3 \\
-(-9x - 3) \\
0
\end{array}
$$

The first division: $\frac{12x^3}{3x} = 4x^2$.

The second division: $\frac{-3x^2}{3x} = -x$.

The third division: $\frac{-9x}{3x} = -3$.

There are no more terms to bring down. This is the remainder.

Thus,

$$\frac{-10x + 12x^3 + x^2 - 3}{1 + 3x} = 4x^2 - x - 3 \qquad \text{The result checks.}$$

Self Check 5 Divide: $2 + 3a \overline{) -4a + 15a^2 + 18a^3 - 4}$

Now Try Problems 39 and 43

4 Divide Polynomials with Missing Terms.

If a power of the variable is missing in the dividend, we will insert **placeholder terms** or leave space. This keeps like terms in the same column, which is necessary when performing the subtraction in vertical form.

EXAMPLE 6 Divide $64x^3 + 1$ by $4x + 1$.

Strategy The dividend, $64x^3 + 1$, does not have an x^2-term or an x-term. We will insert a $0x^2$ term and a $0x$ term as placeholders and use the long division method.

Why We insert placeholder terms so that like terms will be aligned in the same column when we subtract.

Solution After inserting placeholder terms of $0x^2$ and $0x$ in the dividend, we proceed as follows.

$$
\begin{array}{r}
16x^2 - 4x + 1 \\
4x + 1 \overline{) 64x^3 + 0x^2 + 0x + 1} \\
-(64x^3 + 16x^2) \\
-16x^2 + 0x \\
-(-16x^2 - 4x) \\
4x + 1 \\
-(4x + 1) \\
0
\end{array}
$$

The first division: $\frac{64x^3}{4x} = 16x^2$.

The second division: $\frac{-16x^2}{4x} = -4x$.

The third division: $\frac{4x}{4x} = 1$.

There are no more terms to bring down. This is the remainder.

The Language of Algebra

Notice how the instruction to divide by $4x + 1$ translates:

Divide $64x^3 + 1$ by $4x + 1$.

$4x + 1 \overline{) 64x^3 + 1}$

Thus,

$$\frac{64x^3 + 1}{4x + 1} = 16x^2 - 4x + 1 \qquad \text{Use multiplication to check the result.}$$

Self Check 6 Divide $125a^3 - 1$ by $5a - 1$.

Now Try ▶ Problem 47

EXAMPLE 7 Divide: $\dfrac{-17x^2 + 5x + x^4 + 2}{x^2 - 1 + 4x}$

Strategy We will write the terms of the divisor and the dividend in descending powers of x. Then we will insert any needed placeholder terms and use the long division method.

Why We write the terms of the divisor and the dividend in descending powers of x (and insert necessary placeholder terms) so that like terms will be aligned in the same column when we subtract.

Solution After writing the divisor and the dividend in descending powers of x and inserting $0x^3$ for the missing term in the dividend, we proceed as follows:

$$
\begin{array}{r}
x^2 - 4x \\
x^2 + 4x - 1 \overline{\smash{\big)}\, x^4 + 0x^3 - 17x^2 + 5x + 2} \\
-(x^4 + 4x^3 - x^2) \\
\hline
-4x^3 - 16x^2 + 5x \\
-(-4x^3 - 16x^2 + 4x) \\
\hline
x + 2
\end{array}
$$

The first division: $\dfrac{x^4}{x^2} = x^2$.

The second division: $\dfrac{-4x^3}{x^2} = -4x$.

There are no more terms to bring down. This is the remainder.

> **Success Tip**
>
> If a power of a variable is missing in a *divisor*, it is also helpful to insert a placeholder, as shown below.
>
> $m^2 + 0m + 1\overline{\smash{\big)}\, m^3 - 4m^2 + 2m - 1}$

The degree of the remainder, $x + 2$, is 1 and the degree of the divisor, $x^2 + 4x - 1$, is 2. Since the degree of the remainder is less than the divisor, the division process stops. Thus,

$$\frac{-17x^2 + 5x + x^4 + 2}{x^2 - 1 + 4x} = x^2 - 4x + \frac{x + 2}{x^2 + 4x - 1}$$

Self Check 7 Divide: $\dfrac{2a^2 + 3a^3 + a^4 - 6 + a}{a^2 + 1 - 2a}$

Now Try ▶ Problem 51

SECTION 6.5 ▸ STUDY SET

VOCABULARY

Fill in the blanks.

1. The expression $\dfrac{18x^7}{9x^4}$ is a monomial divided by a _____.

   The expression $\dfrac{6x^3 - 4x^2 + 8x - 2}{2x^4}$ is a _____ divided by a monomial. The expression $\dfrac{x^2 - 8x + 12}{x - 6}$ is a trinomial divided by a _____.

2. The powers of x in $2x^4 + 3x^3 + 4x^2 - 7x - 8$ are written in _____ order.

3.
$$
\begin{array}{r}
x - 2 \\
x - 6 \overline{\smash{\big)}\, x^2 - 8x - 4} \\
-(x^2 - 6x) \\
\hline
-2x - 4 \\
-(-2x + 12) \\
\hline
-16
\end{array}
$$

4. Since $5x^2 + 6$ is missing an x-term, we insert a _____ $0x$ term in a division and write the polynomial as $5x^2 + 0x + 6$.

CONCEPTS

Fill in the blanks.

5. a. To divide a polynomial by a monomial, divide each _____ of the polynomial by the monomial.

   b. $\dfrac{18x + 9}{9} = \dfrac{18x}{} + \dfrac{9}{}$

   c. $\dfrac{30x^2 + 12x - 24}{6} = \dfrac{30x^2}{} + \dfrac{12x}{} - \dfrac{24}{}$

6. Divisor · _____ + remainder = dividend

7. Suppose that after dividing $2x^3 + 5x^2 - 11x + 4$ by $2x - 1$, you obtain $x^2 + 3x - 4$. Show how multiplication can be used to check the result.

8. Consider the first step of the division process for

$$2x^2 - 1\overline{)4x^4 + 0x^3 + 0x^2 + 0x - 1}$$

How many times does $2x^2$ divide $4x^4$?

NOTATION

Complete each solution.

9.
$$
\begin{array}{r}
x + 7 \\
x + 4{\overline{\smash{\big)}\,x^2 + 11x + 28}} \\
-(\;\;\;\; + 4x) \\
\hline
+ 28 \\
-(7x + \;\;\;\;) \\
\hline
0
\end{array}
$$

10.
$$
\begin{array}{r}
2x - 1 \\
3x + 4{\overline{\smash{\big)}\,6x^2 + 5x - 4}} \\
-(6x^2 + \;\;\;\;) \\
\hline
- 4 \\
-(-3x - \;\;\;\;) \\
\hline
\end{array}
$$

11. If a polynomial is divided by $3a - 2$ and the quotient is $3a^2 + 5$ with a remainder of 6, how do we write the result?

12. A polynomial is divided by $3a - 2$. The quotient is $3a^2 + 5$ with a remainder of -6. Write the answer to the division in two ways.

13. List three ways we can use symbols to write $x^2 - x - 12$ divided by $x - 4$.

14. Is the following statement true or false? Justify your answer.

$$2x^3 - 9 = 2x^3 + 0x^2 + 0x - 9$$

GUIDED PRACTICE

Simplify. Write answers using positive exponents. See Example 1.

15. $\dfrac{4x^2y^3}{8x^5y^2}$

16. $\dfrac{25x^4y^7}{5xy^9}$

17. $\dfrac{33a^2b^2}{44a^4b^2}$

18. $\dfrac{63a^4}{81a^6b^3}$

Perform each division. See Example 2.

19. $\dfrac{4x^4 + 6x}{2}$

20. $\dfrac{11a^3 - 99a^2}{11}$

21. $\dfrac{4x^2 - x^3}{6x}$

22. $\dfrac{5y^4 + 45y^3}{15y^2}$

23. $\dfrac{54a^3y^2 - 18a^4y^3}{27a^2y^2}$

24. $\dfrac{12x^2y^3 + x^3y^2}{6xy}$

25. $\dfrac{24x^6y^7 - 12x^5y^{12} + 36xy}{-48x^2y^3}$

26. $\dfrac{9x^4y^3 + 18x^2y - 27xy^4}{-9x^3y^3}$

Perform each division. See Objective 3 and Example 3.

27. $\dfrac{x^2 + 5x + 6}{x + 3}$

28. $\dfrac{x^2 + 10x + 21}{x + 7}$

29. $\dfrac{x^2 - 10x + 21}{x - 7}$

30. $\dfrac{x^2 - 5x + 6}{x - 3}$

31. $\dfrac{16x^2 - 16x - 5}{4x + 1}$

32. $\dfrac{6x^2 - x - 12}{2x - 3}$

33. $\dfrac{6x^3 - x^2 - 6x - 9}{2x - 3}$

34. $\dfrac{16x^3 + 16x^2 - 9x - 5}{4x + 5}$

Perform each division. See Example 4.

35. $\dfrac{t^3 + 8t^2 + 13t + 9}{t + 6}$

36. $\dfrac{s^3 + 10s^2 + 17s + 12}{s + 8}$

37. $\dfrac{6x^3 + 11x^2 - 19x - 2}{3x - 2}$

38. $\dfrac{6x^3 + 11x^2 - 9x - 20}{2x + 3}$

Perform each division. See Example 5.

39. $(2a + 1 + a^2) \div (a + 1)$

40. $(a - 15 + 6a^2) \div (2a - 3)$

41. $(6y - 4 + 10y^2) \div (5y - 2)$

42. $(-10x + x^2 + 16) \div (x - 2)$

43. $\dfrac{3x^2 + 9x^3 + 4x + 4}{2 + 3x}$

44. $\dfrac{3 + 5x + 6x^3 + 11x^2}{3 + 2x}$

45. $\dfrac{13x + 16x^4 + 3x^2 + 3}{3 + 4x}$

46. $\dfrac{4x^3 - 12x^2 + 17x - 12}{2x - 3}$

Perform each division. See Example 6.

47. Divide $8a^3 + 1$ by $2a + 1$.

48. Divide $27a^3 - 8$ by $3a - 2$.

49. Divide $15a^3 - 29a^2 + 16$ by $3a - 4$.

50. Divide $15c^3 - 19c^2 + 4$ by $5c + 2$.

Perform each division. See Example 7.

51. Divide $7x^2 - x + x^4 + 5x^3 - 12$ by $x^2 - 3 + 2x$.

52. Divide $x^4 + 2 + 4x^2 + 3x + 2x^3$ by $x^2 + 2 + x$.

53. $3x^2 - 7x + 4{\overline{\smash{\big)}\,7x - 1 + 6x^3 - 5x^2}}$

54. $3m^2 - m + 4{\overline{\smash{\big)}\,5m - 11 + 9m^3 - 6m^2}}$

TRY IT YOURSELF

Perform each division.

55. $y - 2 \overline{)-24y + 24 + 6y^2}$

56. $a - 3 \overline{)54 - 21a + a^2}$

57. $\dfrac{4a^3 + a^2 - 3a + 7}{a + 1}$

58. $\dfrac{3x^3 - 2x^2 + x - 6}{x - 1}$

59. $(x^6 - x^4 + 2x^2 - 8) \div (x^2 - 2)$

60. $(x^3 + 3x + 5x^2 + 6 + x^4) \div (x^2 + 3)$

61. $\dfrac{5a^5 - 10a}{25a^3}$ **62.** $\dfrac{24b^7 - 32b^2}{16b^5}$

63. Divide $2s^2 + 13s + 5$ by $2s + 3$.

64. Divide $4s^2 + 6s + 1$ by $2s - 1$.

65. $\dfrac{40m^{17}n^{20}}{35m^{15}n^{30}}$ **66.** $\dfrac{34s^{30}t^{15}}{14s^{40}t^{12}}$

67. Divide $m^3 - 4m^2 + 2m - 1$ by $m^2 + 1$.
(*Hint:* See the Success Tip on page 518.)

68. Divide $6m^3 + 2m^2 + m + 4$ by $2m^2 - 3$.
(*Hint:* See the Success Tip on page 518.)

69. $(y^3 - 64) \div (y - 4)$

70. $(8w^3 + 1) \div (2w + 1)$

71. $\dfrac{a^8 + a^6 - 4a^4 + 5a^2 - 3}{a^4 + 2a^2 - 3}$

72. $\dfrac{2x^4 + 3x^3 + 3x^2 - 5x - 3}{2x^2 - x - 1}$

73. $\dfrac{40x^3z^2 - 8x^2z - 4z}{4xz}$ **74.** $\dfrac{22a^2b^2 - 18a^2b - 52a}{2ab}$

75. $\dfrac{x^5 + 3x + 2}{x^3 + 1 + 2x}$

76. $\dfrac{9a^4 + 6a^3 + 55a^2 + 18a + 81}{3a^2 + a + 9}$

77. Divide $11x^2 - 4x + 8x^4 - 6x^3 + 3$ by $3 + 4x^2 - x$.

78. Divide $4x - 2 + x^4 - x^2 - x^3$ by $x - 1 + x^2$.

79. $\dfrac{15x^2 + 9x - 3}{27}$ **80.** $\dfrac{8x^2 + 12x + 9}{6}$

81. $x^2 + 3 \overline{)x^6 + 2x^4 - 6x^2 - 9}$

82. $m^2 - 2 \overline{)m^4 - 3m^2 + 10}$

83. Let $f(x) = 4x^4 + 20x^3 - x^2 - 2x + 15$ and $g(x) = x + 5$.
Find $\dfrac{f(x)}{g(x)}$ in simplified form.

84. Let $s(t) = t^5 - t^4 + 7t^2 - 27t + 10$ and $h(t) = t^2 - t + 5$.
Find $\dfrac{s(t)}{h(t)}$ in simplified form.

85. $\dfrac{4x^3 + 4x^2 + 7x - 5}{x - \dfrac{1}{2}}$

86. $\dfrac{3t^3 + 10t^2 + 21t + 6}{t + \dfrac{1}{3}}$

Look Alikes . . .

87. a. $\dfrac{16n^2 - 16n - 5}{4n}$

   b. $\dfrac{16n^2 - 16n - 5}{4n + 1}$

88. a. $\dfrac{9a^3 + 3a^2 + 4a + 4}{3a}$

   b. $\dfrac{9a^3 + 3a^2 + 4a + 4}{3a + 2}$

APPLICATIONS

89. Advertising. Find the length of one of the longer sides of the rectangular billboard if its area is represented by $x^3 - 4x^2 + x + 6$.

90. Masonry. The trowel shown is in the shape of an isosceles triangle. Find the height if its area is represented by $6 + 18t + t^2 + 3t^3$.

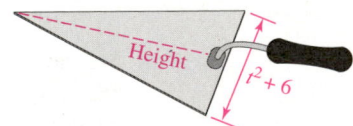

91. Winter Travel. Complete the following table, which lists the rate (mph), time traveled (hr), and distance traveled (mi) by an Alaskan trail guide using two different means of travel.

	r	$\cdot$	t	$=$	d
Dog sled			$4x + 7$		$12x^2 + 13x - 14$
Snowshoes	$3x + 4$				$3x^2 + 19x + 20$

92. Pricing. Complete the table for two items sold at a produce store.

	Price per lb $\cdot$ Number of lb $=$ Total Value		
Cashews	$x^2 + 2x + 4$		$x^4 + 4x^2 + 16$
Sunflower seeds		$x^2 + 6$	$x^4 - x^2 - 42$

WRITING

93. Explain how to check to determine whether
$(3x^2 - 15) \div (x + 3) = 3x - 9 + \dfrac{12}{x + 3}$.

94. Explain the error in the following long division. Use the word *degree* in your answer.

$$x + \cancel{\dfrac{9x + 3}{3x + 1}}$$

$$3x + 1 \overline{)3x^2 + 10x + 3}$$
$$\underline{-(3x^2 + x)}$$
$$9x + 3$$

REVIEW

Simplify each expression.

95. $2(x^2 + 4x - 1) + 3(2x^2 - 2x + 2)$

96. $3(2a^2 - 3a + 2) - 4(2a^2 + 4a - 7)$

97. $-2(3y^3 - 2y + 7) - (y^2 + 2y - 4) + 4(y^3 + 2y - 1)$

98. $3(4y^3 + 3y - 2) + 2(3y^2 - y + 3) - 5(2y^3 - y^2 - 2)$

CHALLENGE PROBLEMS

Perform each division.

99. $\left(3c^2 - \dfrac{7}{4}c - 3\right) \div (4c + 3)$

100. Divide $x^5 - 1$ by $x - 1$.

101. $\dfrac{c^4 - c^2d^2 + 10c^2 - 6d^2 + 23}{c^2 + 6}$

102. $\dfrac{0.03a^2 + 0.17a + 0.1}{0.03a + 0.02}$

(*Hint:* Think of a way to simplify the division.)

103. $x - 2 \overline{)9.8x^2 - 3.2x - 69.3}$

104. $2.5x - 3.7 \overline{)-22.25x^2 - 38.9x - 16.65}$

SECTION 6.6

Synthetic Division

OBJECTIVES

1 Perform synthetic division.

2 Use the remainder theorem to evaluate polynomials.

3 Use the factor theorem to factor polynomials.

ARE YOU READY?

The following problems review some basic skills that are needed when using synthetic division.

1. Divide: $x + 5 \overline{)3x^2 + 19x + 20}$

2. Divide: $\dfrac{2x^2 + x - 9}{x - 2}$

3. Let $f(x) = x^4 + 3x^3 - x - 12$. Find $f(2)$.

4. Subtract: $\begin{array}{r} 9x^3 + 0x^2 \\ -(9x^3 - 27x^2) \\ \hline \end{array}$

We have discussed how to divide polynomials by polynomials using a long division process. We will now discuss a shortcut method, called **synthetic division,** that we can use to divide a polynomial by a binomial of the form $x - k$.

1 Perform Synthetic Division.

To see how synthetic division works, we consider the division of $4x^3 - 5x^2 - 11x + 20$ by $x - 2$.

$$
\begin{array}{r}
4x^2 + 3x - 5 \\
x - 2 \overline{)4x^3 - 5x^2 - 11x + 20} \\
\underline{-(4x^3 - 8x^2)} \\
3x^2 - 11x \\
\underline{-(3x^2 - 6x)} \\
-5x + 20 \\
\underline{-(-5x + 10)} \\
10 \text{ (remainder)}
\end{array}
$$

$$
\begin{array}{r}
4 \quad 3 - 5 \\
1 - 2 \overline{)4 - 5 - 11 \quad 20} \\
\underline{-(4 - 8)} \\
3 - 11 \\
\underline{-(3 - 6)} \\
-5 \quad 20 \\
\underline{-(-5 \quad 10)} \\
10 \text{ (remainder)}
\end{array}
$$

On the left is the long division, and on the right is the same division with the variables and their exponents removed and the coefficients of the quotient moved to the left. We can remember the various powers of x without actually writing them, because the exponents of the terms in the divisor, dividend, and quotient were written in descending order.

We can further shorten the version on the right. The numbers printed in color need not be written, because they are duplicates of the numbers above them. If we remember to perform subtraction at the proper times, the minus symbols and the parentheses also can be dropped. Thus, we can write the division in the following form:

$$
\begin{array}{r}
4 \quad\; 3 \; - \; 5 \\
1-2\overline{)4 \; - \; 5 \; - \; 11 \quad\;\; 20} \\
\underline{-\;8} \\
3 \\
\underline{-\;6} \\
-\;5 \\
\underline{10} \\
10
\end{array}
$$

We can shorten the process further by compressing the work vertically and eliminating the 1 (the coefficient of x in the divisor):

$$
\begin{array}{r}
4 \quad\;\; 3 \quad\; -5 \\
-2\overline{)4 \quad -5 \quad -11 \quad\;\; 20} \\
\underline{-8 \quad -6 \quad 10} \\
3 \quad -5 \quad 10
\end{array}
$$

If we write the 4 in the quotient on the bottom line, the bottom line gives the coefficients of the quotient and the remainder. If we eliminate the top line, the division appears as follows:

$$
\begin{array}{r}
-2\underline{|} \quad 4 \quad -5 \quad -11 \quad\;\; 20 \\
\underline{-8 \quad -6 \quad 10} \\
4 \quad\;\; 3 \quad -5 \quad 10
\end{array}
$$

The bottom line is obtained by subtracting the middle line from the top line. If we replace the -2 in the divisor by 2, the division process will reverse the signs of every entry in the middle line, and then the bottom line can be obtained by addition. This gives the final form of the synthetic division.

$$
\begin{array}{r}
2\underline{|} \quad 4 \quad -5 \quad -11 \quad\;\; 20 \\
\underline{8 \quad\;\; 6 \quad -10} \\
4 \quad\;\; 3 \quad -5 \quad\;\; 10
\end{array}
$$ These are the coefficients of the dividend.

These are the coefficients of the quotient and the remainder.

$$
4x^2 + 3x - 5 + \frac{10}{x - 2}
$$ Read the result from the bottom row.

Thus,

$$
\frac{4x^3 - 5x^2 - 11x + 20}{x - 2} = 4x^2 + 3x - 5 + \frac{10}{x - 2}
$$

EXAMPLE 1   Divide: $(6x^2 - 29x - 5) \div (x - 5)$

Strategy   We will use synthetic division to divide.

Why   When a polynomial is divided by a binomial of the form $x - k$, synthetic division produces the quotient (and possible remainder) with less effort than long division.

Solution   We write the coefficients in the dividend and the 5 in the divisor in the following form:

Since we are dividing the polynomial by $x - 5$, the synthetic divisor is 5.

$$
5\underline{|} \quad 6 \quad -29 \quad -5
$$ ← This represents the dividend $6x^2 - 29x - 5$.

Then we follow these steps:

$\underline{5|}\ \ \ 6\ \ \ -29\ \ \ -5$ *Begin by bringing down the 6.*

6

$\underline{5|}\ \ \ 6\ \ \ -29\ \ \ -5$ *Multiply 5 by 6 to get 30.*
$\qquad\qquad 30$
$\boxed{6}$

$\underline{5|}\ \ \ 6\ \ \ -29\ \ \ -5$ *Add −29 and 30 to get 1.*
$\qquad\qquad 30$
$6\ \ \ .\ \ \ \boxed{1}$

$\underline{5|}\ \ \ 6\ \ \ -29\ \ \ -5$ *Multiply 1 by 5 to get 5.*
$\qquad\qquad 30\ \ \ 5$
$6\ \ \ \boxed{1}$

$\underline{5|}\ \ \ 6\ \ \ -29\ \ \ \boxed{-5}$ *Add −5 and 5 to get 0.*
$\qquad\qquad 30\ \ \ 5$
$6\ \ \ 1\ \ \ \boxed{0}$ *The remainder is 0.*

> Your solution should look like this:
>
> $\underline{5|}\ \ \ 6\ \ \ -29\ \ \ -5$
> $\qquad\qquad 30\ \ \ 5$
> $6\ \ \ 1\ \ \ 0$

The numbers 6 and 1 represent the quotient $6x + 1$, and 0 is the remainder. Thus,

$$\frac{6x^2 - 29x - 5}{x - 5} = 6x + 1$$

Self Check 1   Use synthetic division to divide: $5x^2 - 4x - 33$ by $x - 3$

Now Try ▶ Problem 15

Before using the synthetic division method to divide, write the terms of the dividend in descending powers of the variable. If a power of the variable is missing in the dividend, insert a placeholder term by writing 0 times the variable raised to the missing power.

EXAMPLE 2   Divide: $\dfrac{x^3 + x^2 - 1}{x - 3}$

Strategy   We will use synthetic division to divide.

Why   When a polynomial is divided by a binomial of the form $x - k$, synthetic division produces the quotient (and possible remainder) with less effort than long division.

Solution   We begin by writing

$\underline{3|}\ \ \ 1\ \ \ 1\ \ \ 0\ \ \ -1$ *Write 0 for the coefficient of x, the missing term.*

and complete the division as follows.

$\underline{3|}\ \ 1\ \ \boxed{1}\ \ 0\ \ -1$ $\qquad$ $\underline{3|}\ \ 1\ \ 1\ \ \boxed{0}\ \ -1$ $\qquad$ $\underline{3|}\ \ 1\ \ 1\ \ 0\ \ \boxed{-1}$
$\qquad\quad 3$ $\qquad\qquad\qquad\quad 3\ \ 12$ $\qquad\qquad\qquad\quad 3\ \ 12\ \ 36$
$1\ \ 4$ $\qquad\qquad\quad\ 1\ \ 4\ \ 12$ $\qquad\qquad\quad\ 1\ \ 4\ \ 12\ \ 35$

Multiply, then add. $\qquad$ *Multiply, then add.* $\qquad$ *Multiply, then add.*

The numbers 1, 4, and 12 represent the quotient $x^2 + 4x + 12$ and 35 is the remainder. Thus,

$$\frac{x^3 + x^2 - 1}{x - 3} = x^2 + 4x + 12 + \frac{35}{x - 3}$$

Self Check 2 Use synthetic division to divide: $\dfrac{x^3 + 3x - 62}{x - 4}$

Now Try ▶ Problems 21 and 25

EXAMPLE 3   Divide $5a^2 + 6a^3 + 2 - 4a$ by $a + 2$.

Strategy   Here, the dividend and the divisor are polynomials in the variable a. Since $a + 2$ can be written as $a - (-2)$, the divisor can be written in the form $a - k$ and we can use synthetic division.

Why   We will use synthetic division because it produces the quotient (and possible remainder) with less effort than long division.

Solution   First, we write the dividend with the powers of a in descending order.

$$6a^3 + 5a^2 - 4a + 2$$

Then we write the divisor in $a - k$ form: $a - (\mathbf{-2})$. Thus, $k = -2$. Using synthetic division, we begin by writing

This represents division by $a + 2$. → $\quad -2 \,|\ \ \ 6 \qquad 5 \qquad -4 \qquad 2$

and complete the division.

$$\begin{array}{r|rrrr} -2 & 6 & 5 & -4 & 2 \\ & & -12 & 14 & -20 \\ \hline & 6 & -7 & 10 & -18 \end{array}$$ The remainder is negative.

Notation

Because the remainder is negative, we can also write the result as

$$6a^2 - 7a + 10 - \dfrac{18}{a + 2}$$

Thus,

$$\dfrac{5a^2 + 6a^3 + 2 - 4a}{a + 2} = 6a^2 - 7a + 10 + \dfrac{-18}{a + 2}$$

Self Check 3 Use synthetic division to divide $2a - 4a^2 + 3a^3 - 3$ by $a + 1$.

Now Try ▶ Problems 29 and 33

---

2 Use the Remainder Theorem to Evaluate Polynomials.

Synthetic division is important because of the **remainder theorem**.

Remainder Theorem	If a polynomial $P(x)$ is divided by $x - k$, the remainder is $P(k)$.

It follows from the remainder theorem that we can evaluate polynomials using synthetic division. We illustrate this in the following example.

EXAMPLE 4   Let $P(x) = 2x^3 - 3x^2 - 2x + 1$. Find each value:
a. $P(3)$ **b.** the remainder when $P(x)$ is divided by $x - 3$

Strategy   To find $P(3)$, we will substitute 3 for x in $P(x)$ and simplify. To find the remainder when $P(x)$ is divided by $x - 3$, we will use synthetic division.

Why   After finding the remainder in two ways, we will see that the method using synthetic division is easier.

Solution To find $P(3)$ we evaluate the function for $x = 3$.

a. $P(x) = 2x^3 - 3x^2 - 2x + 1$

$P(3) = 2(3)^3 - 3(3)^2 - 2(3) + 1$ Substitute 3 for x.

$= 2(27) - 3(9) - 6 + 1$

$= 54 - 27 - 6 + 1$

$= 22$

Thus, $P(3) = 22$.

b. We use synthetic division to find the remainder when $2x^3 - 3x^2 - 2x + 1$ is divided by $x - 3$.

$$
\begin{array}{r|rrrr}
3 & 2 & -3 & -2 & 1 \\
  &   & 6 & 9 & 21 \\
\hline
  & 2 & 3 & 7 & \mathbf{22}
\end{array}
$$

Thus, the remainder is 22.

The same results in parts (a) and (b) show that instead of substituting 3 for x in $P(x) = 2x^3 - 3x^2 - 2x + 1$, we can divide the polynomial $2x^3 - 3x^2 - 2x + 1$ by $x - 3$ and note the remainder to find $P(3)$.

Self Check 4   Let $P(x) = 5x^3 - 3x^2 + x + 6$. Find each value: **a.** $P(1)$

b. use synthetic division to find the remainder when $P(x)$ is divided by $x - 1$

Now Try ▶ Problem 41

---

3 Use the Factor Theorem to Factor Polynomials.

If two quantities are multiplied, each is called a *factor* of the product. Thus, $x - 2$ is a factor of $6x - 12$, because $6(x - 2) = 6x - 12$. A theorem, called the *factor theorem,* tells us how to find one factor of a polynomial if the remainder of a certain division is 0.

Factor Theorem

If $P(x)$ is a polynomial in x, then

$$P(k) = 0 \text{ if and only if } x - k \text{ is a factor of } P(x)$$

If $P(x)$ is a polynomial in x and if $P(k) = 0$, k is called a **zero of the polynomial function**.

EXAMPLE 5   Let $P(x) = 3x^3 - 5x^2 + 3x - 10$. Show that:

a. $P(2) = 0$ **b.** $x - 2$ is a factor of $P(x)$

Strategy   We will substitute 2 for x in $P(x)$ to verify that $P(2) = 0$. We will then use synthetic division to divide $P(x)$ by $x - 2$.

Why   Since $P(2) = 0$, the division has a remainder of 0. This means that the divisor and the quotient are factors of the dividend.

Solution   **a.** Use the remainder theorem to evaluate $P(2)$ by dividing $P(x) = 3x^3 - 5x^2 + 3x - 10$ by $x - 2$.

$$
\begin{array}{r|rrrr}
2 & 3 & -5 & 3 & -10 \\
  &   & 6 & 2 & 10 \\
\hline
  & 3 & 1 & 5 & \mathbf{0}
\end{array}
$$

The remainder in this division is 0. By the remainder theorem, the remainder is $P(2)$. Thus, $P(2) = 0$, and 2 is a zero of the polynomial.

---

b. Because the remainder is 0, the numbers 3, 1, and 5 in the synthetic division in part (a) represent the quotient $3x^2 + x + 5$. Thus,

$$\underbrace{(x - 2)}_{} \cdot \underbrace{(3x^2 + x + 5)}_{} + \underbrace{0}_{} = \underbrace{3x^3 - 5x^2 + 3x - 10}_{}$$

Divisor · quotient + remainder = the dividend, $P(x)$

or

$$(x - 2)(3x^2 + x + 5) = 3x^3 - 5x^2 + 3x - 10$$

Thus, $x - 2$ is a factor of $3x^3 - 5x^2 + 3x - 10$.

Self Check 5 Let $P(x) = x^3 - 4x^2 + x + 6$. Show that $x + 1$ is a factor of $P(x)$ using synthetic division.

Now Try ▶ Problem 65

The result in Example 5 is true, because the remainder, $P(2)$, is 0. If the remainder had not been 0, then $x - 2$ would not have been a factor of $P(x)$.

Using Your Calculator ▶ **Approximating Zeros of Polynomials**

We can use a graphing calculator to approximate the real zeros of a polynomial function. For example, to find the real zeros of $f(x) = 2x^3 - 6x^2 + 7x - 21$, we graph the function as in the figure.

It is clear from the display that the function f has a zero at $x = 3$.

$$f(\mathbf{3}) = 2(\mathbf{3})^3 - 6(\mathbf{3})^2 + 7(\mathbf{3}) - 21 \qquad \text{Substitute 3 for x.}$$
$$= 2(27) - 6(9) + 21 - 21$$
$$= 0$$

From the factor theorem, we know that $x - 3$ is a factor of the polynomial. To find the other factor, we can synthetically divide by 3.

$$\begin{array}{r|rrrr} 3 & 2 & -6 & 7 & -21 \\ & & 6 & 0 & 21 \\ \hline & 2 & 0 & 7 & 0 \end{array}$$

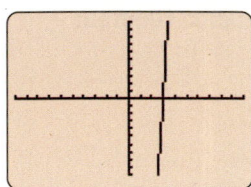

Thus, $f(x) = (x - 3)(2x^2 + 7)$. Since $2x^2 + 7$ cannot be factored over the real numbers, we can conclude that 3 is the only real zero of the polynomial function.

SECTION 6.6 ▶ **STUDY SET**

VOCABULARY

Fill in the blanks.

1. The method of dividing $x^2 + 2x - 9$ by $x - 4$ shown below is called _____ division.

$$\begin{array}{r|rrr} 4 & 1 & 2 & -9 \\ & & 4 & 24 \\ \hline & 1 & 6 & 15 \end{array}$$

2. Synthetic division is used to divide a polynomial by a _____ of the form $x - k$.

3. In Exercise 1, the synthetic _____ is 4.

4. By the _____ theorem, if a polynomial $P(x)$ is divided by $x - k$, the remainder is $P(k)$.

5. The factor _____ tells us how to find one factor of a polynomial if the remainder of a certain division is 0.

6. If $P(x)$ is a polynomial and if $P(k) = 0$, then k is called a _____ of the polynomial.

CONCEPTS

7. a. What division is represented below?

b. What is the answer?

$$
\begin{array}{r}
-2\,\big|\;\;5\quad\;\;0\quad\;\;1\quad\;-3 \\
\underline{\quad\;-10\quad20\quad-42} \\
5\quad-10\quad21\quad-45
\end{array}
$$

Fill in the blanks.

8. In the synthetic division process, numbers below the line are _____ by the synthetic divisor and that product is carried above the line to the next column. Numbers above the horizontal line are _____.

9. Rather than substituting 8 for x in $P(x) = 6x^3 - x^2 - 17x + 9$, we can divide the polynomial ▨▨▨▨▨ by ▨▨▨ to find $P(8)$.

10. For $P(x) = x^3 - 4x^2 + x + 6$, suppose we know that $P(3) = 0$. Then ▨▨▨ is a factor of $x^3 - 4x^2 + x + 6$.

NOTATION

Complete each synthetic division.

11. Divide $6x^3 + x^2 - 23x + 2$ by $x - 2$.

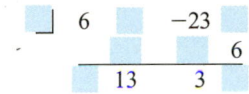

$$
\begin{array}{r}
\boxed{}\,\big|\;\;6\quad\boxed{}\quad-23\quad\boxed{} \\
\underline{\qquad\qquad\qquad 6} \\
\boxed{}\quad13\quad\;\;3\quad\boxed{}
\end{array}
$$

12. Divide $2x^3 - 4x^2 - 25x + 15$ by $x + 3$.

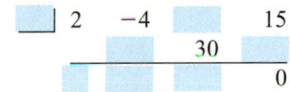

$$
\begin{array}{r}
\boxed{}\,\big|\;\;2\quad\;\;-4\quad\quad\quad 15 \\
\underline{\qquad\qquad\qquad 30\qquad} \\
\boxed{}\quad\boxed{}\quad\boxed{}\quad 0
\end{array}
$$

GUIDED PRACTICE

Use synthetic division to perform each division. See Example 1.

13. $(2x^2 + x - 3) \div (x - 1)$
14. $(4x^2 - 5x - 6) \div (x - 2)$
15. $(5x^2 - 27x + 10) \div (x - 5)$
16. $(6x^2 - 29x + 20) \div (x - 4)$
17. $(3x^2 - 13x + 12) \div (x - 3)$
18. $(2x^2 - 23x + 63) \div (x - 7)$
19. $(5x^2 - 24x - 36) \div (x - 6)$
20. $(3x^2 - 14x - 24) \div (x - 6)$

Use synthetic division to perform each division. See Example 2.

21. $\dfrac{a^3 - 3a^2 + 4}{a - 2}$

22. $\dfrac{a^3 - 2a^2 - 9}{a - 3}$

23. $\dfrac{3a^3 - 47a - 4}{a - 4}$

24. $\dfrac{2a^3 - 7a + 5}{a - 1}$

25. $\dfrac{3b^3 - 31b + 13}{b - 3}$

26. $\dfrac{4c^3 - 107c + 37}{c - 5}$

27. $\dfrac{4t^3 - t - 18}{t - 2}$

28. $\dfrac{m^3 + 2m + 5}{m - 2}$

Use synthetic division to perform each division. See Example 3.

29. Divide $x - 4x^2 + x^3 + 6$ by $x + 1$.
30. Divide $4x^2 - 10x + 12 + x^3$ by $x + 6$.
31. Divide $20x^2 - 36x - 42 + 3x^3$ by $x + 8$.

32. Divide $3x - 6x^2 + 5x^3 + 10$ by $x + 1$.

33. Divide $8 - 3x + 7x^2 + 2x^3$ by $x + 5$.

34. Divide $1 - 4x + 7x^2 + 3x^3$ by $x + 3$.

35. Divide $27 + x^3 - 17x + 8x^2$ by $x + 10$.

36. Divide $1 + x^3 - 23x + 5x^2$ by $x + 8$.

Use a calculator and synthetic division to perform each division. See Examples 1–3.

37. $\dfrac{7.2x^2 - 2.1x + 0.5}{x - 0.2}$

38. $\dfrac{2.7x^2 + x - 5.2}{x + 1.7}$

39. $\dfrac{9x^3 - 25}{x + 57}$

40. $\dfrac{0.5x^3 + x}{x - 2.3}$

Let $P(x) = 2x^3 - 4x^2 + 2x - 1$. Evaluate $P(x)$ by substituting the given value of x into the polynomial and simplifying. Then evaluate the polynomial by using the remainder theorem and synthetic division. See Example 4.

41. $P(1)$ **42.** $P(2)$
43. $P(-2)$ **44.** $P(-1)$
45. $P(3)$ **46.** $P(-4)$
47. $P(0)$ **48.** $P(4)$

Let $Q(x) = x^4 - 3x^3 + 2x^2 + x - 3$. Evaluate $Q(x)$ by substituting the given value of x into the polynomial and simplifying. Then evaluate the polynomial by using the remainder theorem and synthetic division. See Example 4.

49. $Q(-1)$ **50.** $Q(1)$
51. $Q(2)$ **52.** $Q(-2)$
53. $Q(3)$ **54.** $Q(0)$
55. $Q(-3)$ **56.** $Q(-4)$

Use the remainder theorem and synthetic division to find each function value. See Example 4.

57. $P(x) = x^3 - 4x^2 + x - 2$; find $P(2)$
58. $P(x) = x^3 - 3x^2 + x + 1$; find $P(1)$
59. $P(x) = 2x^3 + x + 2$; find $P(3)$
60. $P(x) = x^3 + x^2 + 1$; find $P(-2)$
61. $P(x) = x^4 - 2x^3 + x^2 - 3x + 2$; find $P(-2)$
62. $P(x) = x^5 + 3x^4 - x^2 + 1$; find $P(-1)$
63. $P(x) = 3x^5 + 1$; find $P\left(-\dfrac{1}{2}\right)$
64. $P(x) = 5x^7 - 7x^4 + x^2 + 1$; find $P(2)$

Use the factor theorem and determine whether the first expression is a factor of P(x). See Example 5.

65. $x - 3$; $P(x) = x^3 - 3x^2 + 5x - 15$

66. $x + 1$; $P(x) = x^3 + 2x^2 - 2x - 3$
 (*Hint:* Write $x + 1$ as $x - (-1)$.)

67. $x + 2$; $P(x) = 3x^2 - 7x + 4$
 (*Hint:* Write as $x - (-2)$.)

68. x; $P(x) = 7x^3 - 5x^2 - 8x$
 (*Hint:* $x = x - 0$.)

83. $(2x^3 - 50 - 16x^2 - 35x) \div (x - 10)$

84. $(m^3 - m^2 - m - 1) \div (m - 1)$

85. $(4x^3 - 1 + 5x^2) \div (x + 2)$

86. $(t^3 + t^2 + t + 2) \div (t + 1)$

87. Divide $8a^3 - 10a^2 - 32a - 15$ by $a + \dfrac{3}{4}$.

88. Divide $4a^3 - 2a^2 - 18a - 9$ by $a + \dfrac{3}{2}$.

TRY IT YOURSELF

Use synthetic division to perform each division.

69. $\dfrac{5x^2 + 4 + 6x^3}{x + 1}$

70. $\dfrac{-4 + 3x^2 - x}{x - 4}$

71. $(x^2 - 5x + 14) \div (x + 2)$

72. $(x^2 + 13x + 42) \div (x + 6)$

73. Divide $a^5 - 1$ by $a - 1$.

74. Divide $b^4 - 81$ by $b - 3$.

75. $\dfrac{-6c^5 + 14c^4 + 38c^3 + 4c^2 + 25c - 36}{c - 4}$

76. $\dfrac{-5x^5 + 4x^4 + 30x^3 + 2x^2 + 20x + 3}{x - 3}$

77. $\dfrac{9a^3 + 3a^2 - 21a - 7}{a + \dfrac{1}{3}}$

78. $\dfrac{8t^3 - 4t^2 + 2t - 1}{t - \dfrac{1}{2}}$

79. $\dfrac{4x^4 + 12x^3 - x^2 - x + 12}{x + 3}$

80. $\dfrac{x^4 - 9x^3 + x^2 - 7x - 20}{x - 9}$

81. $\dfrac{3x^3 - 25x^2 + 10x - 16}{x - 8}$

82. $\dfrac{2x^3 + 3x^2 - 8x + 3}{x + 3}$

WRITING

89. When dividing a polynomial by a binomial of the form $x - k$, synthetic division is considered to be faster than long division. Explain why.

90. Let $P(x) = x^3 - 6x^2 - 9x + 4$. You now know two ways to find $P(6)$. What are they? Which method do you prefer?

91. Explain the factor theorem.

92. This section includes a feature entitled *Using Your Calculator: Approximating Zeros of Polynomials*. What is a *zero* of a polynomial?

REVIEW

Solve each equation.

93. $|3x - 7| + 8 = 22$

94. $|75 - x| = -1$

95. $6 - 3|10x + 5| = 6$

96. $|2 - x| = |3x + 2|$

CHALLENGE PROBLEMS

Suppose that $P(x) = x^{100} - x^{99} + x^{98} - x^{97} + \cdots + x^2 - x + 1$.

97. Find the remainder when $P(x)$ is divided by $x - 1$.

98. Find the remainder when $P(x)$ is divided by $x + 1$.

99. Find 2^6 by using synthetic division to evaluate the polynomial $P(x) = x^6$ at $x = 2$.

100. Find $(-3)^5$ by using synthetic division to evaluate the polynomial $P(x) = x^5$ at $x = -3$.

SECTION 6.7

Solving Rational Equations

OBJECTIVES

1. Solve rational equations.
2. Solve rational equations with extraneous solutions.
3. Solve formulas for a specified variable.

ARE YOU READY?

The following problems review some basic skills that are needed when solving rational equations.

1. What is the LCD of $\dfrac{2}{3x}$, $\dfrac{1}{5x}$, and $\dfrac{11}{6x}$?

2. Multiply: $8(x - 2)\left(\dfrac{x}{x - 2}\right)$

3. Solve: $x^2 - 3x - 54 = 0$

4. Find all real numbers for which the rational expression $\dfrac{x + 9}{x - 4}$ is undefined.

In Chapter 1, we solved equations such as $\frac{1}{6}x + \frac{5}{2} = \frac{1}{3}$ by multiplying both sides by the LCD. With this approach, the equation that results is equivalent to the original equation, but easier to solve because it is cleared of fractions.

In this section, we will extend the fraction-clearing strategy to solve another type of equation, called a *rational equation*.

1 Solve Rational Equations.

If an equation contains one or more rational expressions, it is called a **rational equation.** Rational equations often have a variable in a denominator. Some examples are:

$$\frac{3}{5} + \frac{7}{x+2} = 2, \qquad \frac{x+3}{x-3} = \frac{2}{x^2-4}, \qquad \text{and} \qquad \frac{-x^2+10}{x^2-1} + \frac{3x}{x-1} = \frac{2x}{x+1}$$

To solve a rational equation, we find all the values of the variable that make the equation true. Any value of the variable that makes a denominator in a rational equation equal to 0 cannot be a solution of the equation. Such a number must be rejected, because division by 0 is undefined.

EXAMPLE 1   Solve: $\dfrac{3}{5} + \dfrac{7}{x+2} = 2$

Strategy This equation contains a rational expression that has a variable in the denominator. We begin by asking, "What value(s) of x make that denominator 0?"

Why   If a number makes the denominator of a rational expression 0, that number cannot be a solution of the equation because division by 0 is undefined.

Solution   We note that x cannot be -2, because this would produce a 0 in the denominator of $\dfrac{7}{x+2}$.

Since the denominators of the rational expressions in the equation are 5 and $x + 2$, we multiply both sides by the LCD, $5(x + 2)$, to clear the equation of fractions.

$$\frac{3}{5} + \frac{7}{x+2} = 2 \qquad \text{This is the equation to solve.}$$

$$5(x+2)\left(\frac{3}{5} + \frac{7}{x+2}\right) = 5(x+2)(2) \qquad \begin{array}{l}\text{Write each side of the equation}\\ \text{within parentheses and then}\\ \text{multiply both sides by the LCD.}\end{array}$$

$$5(x+2)\left(\frac{3}{5}\right) + 5(x+2)\left(\frac{7}{x+2}\right) = 5(x+2)(2) \qquad \begin{array}{l}\text{On the left side, distribute the}\\ \text{multiplication by } 5(x+2).\end{array}$$

$$\overset{1}{5}(x+2)\left(\frac{3}{\underset{1}{5}}\right) + 5(x+2)\left(\frac{7}{\underset{1}{x+2}}\right) = 5(x+2)(2) \qquad \begin{array}{l}\text{On the left side, simplify:}\\ \frac{5}{5} = 1 \text{ and } \frac{x+2}{x+2} = 1.\end{array}$$

$$3(x+2) + 5(7) = 10(x+2) \qquad \text{The fractions are cleared.}$$

The resulting linear equation does not contain any fractions. We now solve it for x.

$$3x + 6 + 35 = 10x + 20 \qquad \text{Use the distributive property and simplify.}$$
$$3x + 41 = 10x + 20 \qquad \text{Combine like terms: } 6 + 35 = 41.$$
$$-7x = -21 \qquad \text{Subtract } 10x \text{ and } 41 \text{ from both sides.}$$
$$x = 3 \qquad \text{To isolate } x, \text{ divide both sides by } -7.$$

The solution is 3 and the solution set is {3}. To check, we substitute 3 for x in the original equation and simplify:

$$\textit{Check:} \quad \frac{3}{5} + \frac{7}{x+2} = 2 \qquad \text{This is the original equation.}$$

$$\frac{3}{5} + \frac{7}{3+2} \overset{?}{=} 2 \qquad \text{Substitute.}$$

$$\frac{3}{5} + \frac{7}{5} \overset{?}{=} 2 \qquad \text{Do the addition.}$$

$$2 = 2 \qquad \text{True}$$

> **Self Check 1**   Solve: $\frac{2}{5} + \frac{8}{x-4} = 2$
>
> **Now Try ▶**   Problem 13

Using Your Calculator ▶ Solving Rational Equations Graphically

To use a graphing calculator to solve $\frac{3}{5} + \frac{7}{x+2} = 2$, we graph the functions $f(x) = \frac{3}{5} + \frac{7}{x+2}$ and $g(x) = 2$. If we trace and move the cursor closer to the intersection point of the two graphs, we will get the approximate value of x shown in figure (a) below. If we zoom twice and trace again, we get the results shown in figure (b). As we saw in Example 1, the exact solution is 3.

An alternate way of finding the point of intersection of the two graphs is to use the INTERSECT feature. In figure (c), the display shows that the graphs intersect at the point $(3, 2)$. This implies that the solution of the rational equation is 3.

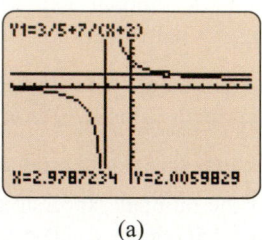 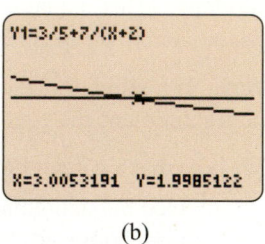 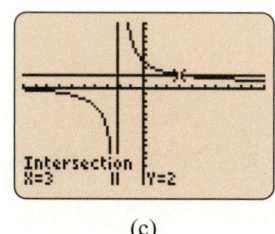

        (a)   (b)   (c)

EXAMPLE 2   Solve: $\dfrac{-x^2 + 10}{x^2 - 1} + \dfrac{3x}{x-1} = \dfrac{2x}{x+1}$

Strategy   We will begin by factoring the first denominator.

Why   To determine any restrictions on the variable and to find the LCD, we need to write $x^2 - 1$ in factored form.

Solution   Since $x^2 - 1$ factors as $(x+1)(x-1)$, we can write the given equation as:

$$\frac{-x^2 + 10}{(x+1)(x-1)} + \frac{3x}{x-1} = \frac{2x}{x+1} \qquad \text{\color{red}{Factor the denominator } } x^2 - 1.$$

> **Caution**
>
> After multiplying both sides by the LCD and simplifying, the equation should not contain any fractions. If it does, check for an algebraic error, or perhaps your LCD is incorrect.

We see that -1 and 1 cannot be solutions of the equation because they make rational expressions in the equation undefined.

We can clear the equation of fractions by multiplying both sides by $(x+1)(x-1)$, which is the LCD of the three rational expressions.

$$(x+1)(x-1)\left[\frac{-x^2 + 10}{(x+1)(x-1)} + \frac{3x}{x-1}\right] = (x+1)(x-1)\left(\frac{2x}{x+1}\right) \qquad \text{\color{red}{Multiply both sides by the LCD.}}$$

$$(x+1)(x-1)\left[\frac{-x^2 + 10}{(x+1)(x-1)}\right] + (x+1)(x-1)\left(\frac{3x}{x-1}\right) = (x+1)(x-1)\left(\frac{2x}{x+1}\right) \qquad \text{\color{red}{On the left side, distribute the multiplication by } } (x+1)(x-1).$$

$$\overset{1}{\cancel{(x+1)}}\overset{1}{\cancel{(x-1)}}\left[\frac{-x^2 + 10}{\underset{1}{\cancel{(x+1)}}\underset{1}{\cancel{(x-1)}}}\right] + (x+1)\overset{1}{\cancel{(x-1)}}\left(\frac{3x}{\underset{1}{\cancel{x-1}}}\right) = \overset{1}{\cancel{(x+1)}}(x-1)\left(\frac{2x}{\underset{1}{\cancel{x+1}}}\right) \qquad \text{\color{red}{Remove common factors of the numerator and denominator.}}$$

$$-x^2 + 10 + 3x(x + 1) = 2x(x - 1) \qquad \text{\color{red}{The fractions are cleared.}}$$
$$-x^2 + 10 + 3x^2 + 3x = 2x^2 - 2x \qquad \text{\color{red}{Use the distributive property twice.}}$$
$$2x^2 + 10 + 3x = 2x^2 - 2x \qquad \text{\color{red}{Combine like terms on each side.}}$$
$$10 + 3x = -2x \qquad \text{\color{red}{Subtract $2x^2$ from both sides. The result}}$$
$$\text{\color{red}{is a linear equation.}}$$
$$10 + 5x = 0 \qquad \text{\color{red}{Add $2x$ to both sides.}}$$
$$5x = -10 \qquad \text{\color{red}{Subtract 10 from both sides.}}$$
$$x = -2 \qquad \text{\color{red}{To isolate x, divide both sides by 5.}}$$

The solution is -2. Verify that it satisfies the original equation.

Self Check 2   Solve: $\dfrac{2}{x - 3} = \dfrac{-x}{x^2 - 9} + \dfrac{4}{x + 3}$

Now Try ▶ Problem 17

We can summarize the procedure used to solve rational equations.

Solving Rational Equations

1. Factor all denominators.
2. Determine which numbers cannot be solutions of the equation.
3. Multiply both sides of the equation by the LCD of all rational expressions in the equation.
4. Use the distributive property to remove parentheses, remove any factors equal to 1, and write the result in simplified form.
5. Solve the resulting equation.
6. Check all possible solutions in the original equation.

Using Your Calculator ▶ **Checking Apparent Solutions**

We can use a scientific calculator to check the solution -2 found in Example 2 by evaluating

$$\frac{-x^2 + 10}{x^2 - 1} + \frac{3x}{x - 1} \qquad \text{and} \qquad \frac{2x}{x + 1}$$

In each case, the result is 4. Since the results are the same, -2 is a solution of the equation.
We also can check by using a graphing calculator. One way of doing this is to enter

$$Y_1 = \frac{-x^2 + 10}{x^2 - 1} + \frac{3x}{x - 1} \qquad \text{and} \qquad Y_2 = \frac{2x}{x + 1}$$

and compare the values of the expressions when $x = -2$ in the table mode. See the figure.
We know that -2 is a solution of

$$\frac{-x^2 + 10}{x^2 - 1} + \frac{3x}{x - 1} = \frac{2x}{x + 1}$$

because the value of Y1 and Y2 are the same (namely, 4) for $x = -2$.

X	Y₁	Y₂
-2	4	4
-1	ERROR	ERROR
0	-10	0
1	ERROR	1
2	8	1.3333
3	4.625	1.5
4	3.6	1.6

X= -2

EXAMPLE 3   Solve: $\dfrac{a}{2} = \dfrac{a-6}{3a-9} - \dfrac{1}{3}$

Strategy   We will begin by factoring the second denominator.

Why   To determine any restrictions on the variable and to find the LCD, we need to write $3a - 9$ in factored form.

Solution   Since the binomial $3a - 9$ factors as $3(a - 3)$, we can write the given equation as:

$$\frac{a}{2} = \frac{a-6}{3(a-3)} - \frac{1}{3} \qquad \text{\color{red}Factor the denominator } 3a - 9 \text{ as } 3(a - 3).$$

> **Caution**
>
> When solving rational equations, each term on both sides must be multiplied by the LCD.

We see that 3 cannot be a solution of the equation, because it makes one of the rational expressions in the equation undefined.

We can clear the equation of fractions by multiplying both sides by $2 \cdot 3 \cdot (a - 3)$, which is the LCD of the three rational expressions.

$$2 \cdot 3 \cdot (a-3)\left(\frac{a}{2}\right) = 2 \cdot 3 \cdot (a-3)\left[\frac{a-6}{3(a-3)} - \frac{1}{3}\right] \qquad \text{\color{red}Multiply both sides by the LCD, } 2 \cdot 3 \cdot (a-3).$$

$$2 \cdot 3 \cdot (a-3)\left(\frac{a}{2}\right) = 2 \cdot 3 \cdot (a-3)\left[\frac{a-6}{3(a-3)}\right] - 2 \cdot 3 \cdot (a-3)\left(\frac{1}{3}\right) \qquad \text{\color{red}On the right side, distribute } 2 \cdot 3 \cdot (a-3).$$

$$\overset{1}{2} \cdot 3(a-3)\left(\frac{a}{\underset{1}{2}}\right) = 2 \cdot \overset{1}{3}(a-3)\left[\frac{a-6}{3(a-3)}\right] - 2 \cdot \overset{1}{3}(a-3)\left(\frac{1}{\underset{1}{3}}\right) \qquad \text{\color{red}Remove common factors of the numerator and denominator.}$$

$$3a(a-3) = 2(a-6) - 2(a-3) \qquad \text{\color{red}The fractions are cleared.}$$
$$3a^2 - 9a = 2a - 12 - 2a + 6 \qquad \text{\color{red}Use the distributive property three times.}$$
$$3a^2 - 9a = -6 \qquad \text{\color{red}Combine like terms: } 2a - 2a = 0 \text{ and } -12 + 6 = -6.$$

To use factoring to solve the resulting quadratic equation, we must write it in standard form $ax^2 + bx + c = 0$.

$$3a^2 - 9a + 6 = 0 \qquad \text{\color{red}To get 0 on the right side, add 6 to both sides.}$$
$$a^2 - 3a + 2 = 0 \qquad \text{\color{red}Divide both sides by 3.}$$
$$(a-1)(a-2) = 0 \qquad \text{\color{red}Factor the trinomial.}$$
$$a - 1 = 0 \quad \text{or} \quad a - 2 = 0 \qquad \text{\color{red}Set each factor equal to 0.}$$
$$a = 1 \qquad \qquad a = 2 \qquad \text{\color{red}Solve each linear equation.}$$

Verify that 1 and 2 both satisfy the original equation and that the solution set is $\{1, 2\}$.

> **Self Check 3**   Solve: $\dfrac{b}{5} = \dfrac{b-14}{2b-16} - \dfrac{1}{2}$
>
> **Now Try** ▶ Problem 25

Recall that the quotient of a polynomial and its opposite is -1. For example, $\dfrac{y-1}{1-y} = -1$. We can use this fact when solving rational equations whose denominators contain factors that are opposites.

EXAMPLE 4   Solve: $\dfrac{1}{6y-6} + \dfrac{1}{1-y} = \dfrac{1}{6}$

Strategy   We will begin by factoring the first denominator.

Why   To determine any restrictions on the variable and to find the LCD, we need to write $6y - 6$ in factored form.

Solution Since the binomial $6y - 6$ factors as $6(y - 1)$, we can write the given equation as:

$$\frac{1}{6(y-1)} + \frac{1}{1-y} = \frac{1}{6}$$

We see that 1 cannot be a solution of the equation because it makes two of the rational expressions in the equation undefined.

We note that $y - 1$ and $1 - y$ are opposites. We can clear the equation of fractions by multiplying both sides by $6(y - 1)$.

$$6(y-1)\left[\frac{1}{6(y-1)} + \frac{1}{1-y}\right] = 6(y-1)\left(\frac{1}{6}\right) \quad \text{Multiply both sides by the LCD, } 6(y-1).$$

$$6(y-1)\left[\frac{1}{6(y-1)}\right] + 6(y-1)\left(\frac{1}{1-y}\right) = 6(y-1)\left(\frac{1}{6}\right) \quad \text{On the left side, distribute } 6(y-1).$$

$$\overset{1}{6}(y-1)\left[\frac{1}{6(y-1)}\right] + 6(y-1)\overset{-1}{\left(\frac{1}{1-y}\right)} = \overset{1}{6}(y-1)\left(\frac{1}{6}\right) \quad \text{Simplify: } \frac{y-1}{1-y} = -1 \text{ (shown in blue).}$$

$$1 - 6 = y - 1 \quad \text{The fractions are cleared.}$$
$$-5 = y - 1 \quad \text{Combine like terms.}$$
$$-4 = y \quad \text{Add 1 to both sides.}$$

The solution is -4. Verify that it satisfies the original equation.

Self Check 4 Solve: $\frac{1}{2h-8} + \frac{11}{4-h} = \frac{3}{2}$

Now Try ▶ Problem 33

2 Solve Rational Equations with Extraneous Solutions.

When we multiply both sides of an equation by a quantity that contains a variable, we can get false solutions, called *extraneous solutions*. This happens when we multiply both sides of an equation by 0 and get a solution that gives a 0 in the denominator of a rational expression. Extraneous solutions must be discarded.

EXAMPLE 5 Solve: $3 - \frac{1-2t}{t+2} = \frac{t-3}{t+2}$

Strategy We will clear the equation of fractions by multiplying both sides by the LCD, $t + 2$.

Why Equations that contain only integers are usually easier to solve than equations that contain fractions.

Solution We note that t cannot be -2, because this would give a 0 in a denominator.

$$3 - \frac{1-2t}{t+2} = \frac{t-3}{t+2}$$

$$(t+2)\left(3 - \frac{1-2t}{t+2}\right) = (t+2)\left(\frac{t-3}{t+2}\right) \quad \text{Multiply both sides by the LCD.}$$

$$(t+2)(3) - (t+2)\left(\frac{1-2t}{t+2}\right) = (t+2)\left(\frac{t-3}{t+2}\right) \quad \text{On the left side, distribute the multiplication by } t+2.$$

$$(t+2)(3) - \overset{1}{(t+2)}\left(\frac{1-2t}{t+2}\right) = \overset{1}{(t+2)}\left(\frac{t-3}{t+2}\right) \quad \text{Remove common factors of the numerator and denominator.}$$

Success Tip

Even if you do not make an arithmetic or algebraic error when solving a rational equation, a possible solution may not check. When the possible solution -2 is substituted for t in the original equation, note that two undefined rational expressions appear in the check:

$$3 - \frac{5}{0} = \frac{-5}{0}$$

$$3t + 6 - (1 - 2t) = t - 3 \qquad \text{The fractions are cleared. Don't forget to write the parentheses shown in blue.}$$

$$3t + 6 - 1 + 2t = t - 3 \qquad \text{Change the sign of each term within the parentheses.}$$

$$5t + 5 = t - 3 \qquad \text{Combine like terms: } 3t + 2t = 5t \text{ and } 6 - 1 = 5.$$

$$4t = -8 \qquad \text{Subtract } t \text{ and 5 from both sides.}$$

$$t = -2 \qquad \text{To isolate } t, \text{ divide both sides by 4.}$$

Since t cannot be -2, it is an extraneous solution and must be discarded. This equation has *no solution* and the solution set is $\{\ \}$ or $\varnothing$.

Self Check 5   Solve: $2 - \dfrac{2a}{a-1} = \dfrac{3a-5}{a-1}$

Now Try ▶ Problem 37

3 Solve Formulas for a Specified Variable.

Many formulas involve rational expressions. We can use the fraction-clearing method of this section to solve such formulas for a specified variable.

EXAMPLE 6

Physics.   The *law of gravitation,* formulated by Sir Isaac Newton in 1684, states that if two masses, m_1 and m_2, are separated by a distance of r, the force F exerted by one mass on the other is given by the following equation. Solve for m_2.

$$F = \frac{Gm_1m_2}{r^2} \qquad G \text{ is the gravitational constant.}$$

Strategy   To solve for m_2, we will treat it as if it were the only variable in the equation. To isolate this variable, we will use the same strategy that we used in previous examples to solve rational equations in one variable.

Why   We can solve a formula as if it were an equation in one variable because all the other variables are treated as if they were numbers (constants).

Solution

$$F = \frac{Gm_1m_2}{r^2} \qquad \text{This is the given formula. Note that } m_1 \text{ and } m_2 \text{ are different variables.}$$

$$r^2(F) = r^2\left(\frac{Gm_1m_2}{r^2}\right) \qquad \text{Multiply both sides by the LCD, } r^2.$$

$$r^2(F) = \overset{1}{\cancel{r^2}}\left(\frac{Gm_1m_2}{\underset{1}{\cancel{r^2}}}\right) \qquad \text{Simplify the right side: } \frac{r^2}{r^2} = 1.$$

$$\frac{r^2F}{Gm_1} = \frac{Gm_1m_2}{Gm_1} \qquad \text{To isolate } m_2, \text{ divide both sides by } Gm_1.$$

$$\frac{r^2F}{Gm_1} = m_2 \qquad \text{Simplify the right side by removing the factors } G \text{ and } m_1, \text{ which are common to the numerator and denominator.}$$

$$m_2 = \frac{r^2F}{Gm_1} \qquad \text{Reverse the sides of the equation so that } m_2 \text{ is on the left.}$$

Self Check 6   Solve the law of gravitation formula for r^2.

Now Try ▶ Problem 41

EXAMPLE 7   **Electronics.**   In electronic circuits, resistors oppose the flow of an electric current. The total resistance R of a parallel combination of two resistors as shown is given by the following equation, where R_1 is the resistance of the first resistor and R_2 is the resistance of the second resistor. Solve for R.

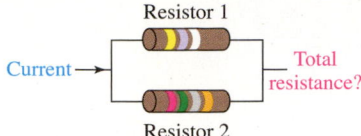

Resistor 1
Current →
Total resistance?
Resistor 2

$$\frac{1}{R} = \frac{1}{R_1} + \frac{1}{R_2}$$

Strategy   To solve for R, we will treat it as if it were the only variable in the equation. To isolate this variable, we will use the same strategy that we used in previous examples to solve rational equations in one variable; we will clear the equation of fractions.

Why   We can solve a formula as if it were an equation in one variable because all the other variables are treated as if they were numbers (constants).

Solution   We begin by clearing the equation of fractions by multiplying both sides by the LCD, which is RR_1R_2.

Success Tip
This formula contains three variables: R, R_1, and R_2. It is often helpful to circle the variable that you are solving for in a formula: $$\frac{1}{\widehat{R}} = \frac{1}{R_1} + \frac{1}{R_2}$$

$$\frac{1}{R} = \frac{1}{R_1} + \frac{1}{R_2}$$ This is the given formula.

$$RR_1R_2\left(\frac{1}{R}\right) = RR_1R_2\left(\frac{1}{R_1} + \frac{1}{R_2}\right)$$ Multiply both sides by the LCD, RR_1R_2.

$$RR_1R_2\left(\frac{1}{R}\right) = RR_1R_2\left(\frac{1}{R_1}\right) + RR_1R_2\left(\frac{1}{R_2}\right)$$ On the right side, distribute the multiplication by RR_1R_2.

$$\overset{1}{\cancel{R}}R_1R_2\left(\frac{1}{\cancel{R}}\right) = R\cancel{R_1}R_2\left(\frac{1}{\cancel{R_1}}\right) + RR_1\cancel{R_2}\left(\frac{1}{\cancel{R_2}}\right)$$ Remove common factors of the numerator and denominator.

$$R_1R_2 = RR_2 + RR_1$$ The fractions are cleared.

$$R_1R_2 = R(R_2 + R_1)$$ Factor out the common factor R on the right side.

$$\frac{R_1R_2}{R_2 + R_1} = R$$ To isolate R, divide both sides by $R_2 + R_1$.

$$R = \frac{R_1R_2}{R_2 + R_1}$$ Reverse the sides of the equation to write R on the left side.

Self Check 7   Solve $\frac{1}{x} - \frac{1}{y} = \frac{1}{z}$ for z.

Now Try ▶   Problem 49

SECTION 6.7   **STUDY SET**

VOCABULARY

Fill in the blanks.

1. Equations that contain one or more rational expressions, such as $\frac{x}{x+2} = 4 + \frac{10}{x+1}$, are called _____ equations.

2. When solving a rational equation, if we obtain a number that does not satisfy the original equation, the number is called an _____ solution.

CONCEPTS

3. Classify each of the following as an expression or an equation.

   a. $\dfrac{7}{5x} - \dfrac{1}{2} = \dfrac{5}{6x} + \dfrac{1}{3}$

   b. $\dfrac{4}{x^2-4} - \dfrac{5}{x-2}$

   c. $\dfrac{27p^4}{35q} \div \dfrac{9p}{21q}$

   d. $\dfrac{4}{t+3} + \dfrac{8}{t^2-9} = \dfrac{2}{t-3}$

   e. $\dfrac{\dfrac{y}{x} - \dfrac{x}{y}}{\dfrac{1}{y} - \dfrac{1}{x}}$

   f. $\dfrac{t^2+t-6}{t^2-6t+9} \cdot \dfrac{t^2-9}{t^2 4}$

4. Check to determine whether 2 is a solution of the following equations.

 a. $\dfrac{x}{2} + \dfrac{4}{x + 2} = x$

 b. $\dfrac{x + 2}{x - 2} + \dfrac{1}{x^2 - 4} = 1$

5. Consider the rational equation: $\dfrac{x}{x - 3} = \dfrac{1}{x} + \dfrac{2}{x - 3}$

 a. What values of x make a denominator 0?

 b. What values of x make a rational expression undefined?

 c. What numbers can't be solutions of the equation?

6. To clear the following equations of fractions, by what should both sides be multiplied?

 a. $\dfrac{1}{a} = \dfrac{1}{3} - \dfrac{2}{3a}$

 b. $\dfrac{2}{x - 2} + \dfrac{10}{x + 5} = \dfrac{2x}{x^2 + 3x - 10}$

NOTATION

Complete the solution.

7.
$$\dfrac{10}{3y} - \dfrac{7}{30} = \dfrac{9}{2y}$$

$$\blacksquare\left(\dfrac{10}{3y} - \dfrac{7}{30}\right) = \blacksquare\left(\dfrac{9}{2y}\right)$$

$$\blacksquare\left(\dfrac{10}{3y}\right) - \blacksquare\left(\dfrac{7}{30}\right) = \blacksquare\left(\dfrac{9}{2y}\right)$$

$$100 - \blacksquare = 135$$

$$-7y = \blacksquare$$

$$y = -5$$

8. Perform each multiplication.

 a. $4x\left(\dfrac{3}{4x}\right)$

 b. $(x + 6)(x - 2)\left(\dfrac{3}{x - 2}\right)$

 c. $8(x + 4)\left[\dfrac{7x}{2(x + 4)}\right]$

 d. $6(m - 5)\left(\dfrac{7}{5 - m}\right)$

GUIDED PRACTICE

Solve each equation. See Example 1.

9. $\dfrac{1}{4} + \dfrac{9}{x} = 1$

10. $\dfrac{1}{3} - \dfrac{10}{x} = -3$

11. $\dfrac{1}{a} = \dfrac{1}{3} - \dfrac{2}{3a}$

12. $\dfrac{1}{b} = \dfrac{1}{8} - \dfrac{3}{8b}$

13. $\dfrac{18}{y + 1} + \dfrac{2}{5} = 4$

14. $\dfrac{2}{3} + \dfrac{10}{a + 2} = 4$

15. $\dfrac{1}{2} + \dfrac{x}{x - 1} = 3$

16. $\dfrac{2}{3} + \dfrac{a}{a - 2} = 5$

Solve each equation. See Example 2.

17. $\dfrac{4}{t + 3} + \dfrac{8}{t^2 - 9} = \dfrac{2}{t - 3}$

18. $\dfrac{5}{x - 1} = \dfrac{1}{x^2 - 1} + \dfrac{1}{x - 1}$

19. $\dfrac{4}{x^2 - 4} - \dfrac{5}{x - 2} = \dfrac{1}{x + 2}$

20. $\dfrac{1}{m + 3} - \dfrac{m}{m^2 - 9} = \dfrac{-2}{m - 3}$

21. $\dfrac{2}{x - 2} + \dfrac{10}{x + 5} = \dfrac{2x}{x^2 + 3x - 10}$

22. $\dfrac{2}{a + 4} + \dfrac{2a - 1}{a^2 + 2a - 8} = \dfrac{1}{a - 2}$

23. $\dfrac{1}{n + 2} - \dfrac{2}{n - 3} = \dfrac{-2n}{n^2 - n - 6}$

24. $\dfrac{2x}{x^2 + 9x + 20} - \dfrac{3}{x + 4} = \dfrac{2}{x + 5}$

25. $\dfrac{x}{8} = \dfrac{x - 12}{3x - 27} - \dfrac{1}{3}$

26. $\dfrac{n}{7} = \dfrac{n - 19}{5n - 45} - \dfrac{1}{5}$

Solve each equation. See Example 3.

27. $\dfrac{p - 1}{2} + 1 = \dfrac{3}{p}$

28. $\dfrac{b + 1}{2} - \dfrac{3}{2} = \dfrac{4}{b}$

29. $\dfrac{16}{t + 3} + \dfrac{7}{t - 2} = 3$

30. $\dfrac{17}{s - 4} - \dfrac{10}{s + 2} = 2$

31. $\dfrac{2}{5x - 5} + \dfrac{x - 2}{15} = \dfrac{4}{5x - 5}$

32. $\dfrac{3}{2x + 4} = \dfrac{x - 2}{2} + \dfrac{x - 5}{2x + 4}$

Solve each equation. See Example 4.

33. $\dfrac{1}{3x - 18} + \dfrac{5}{6 - x} = \dfrac{1}{3}$

34. $\dfrac{1}{2x - 16} + \dfrac{14}{8 - x} = \dfrac{3}{2}$

35. $\dfrac{7}{3x - 9} + \dfrac{1}{3 - x} = \dfrac{4}{9}$

36. $\dfrac{1}{2d - 4} - \dfrac{1}{2 - d} = \dfrac{1}{4}$

Solve each equation. If a solution is extraneous, so indicate. See Example 5.

37. $4 - \dfrac{3x}{x - 9} = \dfrac{5x - 72}{x - 9}$

38. $2 - \dfrac{2x}{x - 10} = \dfrac{4x - 60}{x - 10}$

39. $\dfrac{6}{x + 3} + \dfrac{48}{x^2 - 2x - 15} - \dfrac{7}{x - 5} = 0$

40. $\dfrac{3}{x - 4} + \dfrac{2}{x + 5} + \dfrac{18}{x^2 + x - 20} = 0$

Solve each formula for the specified variable. See Examples 6 and 7.

41. $Q = \dfrac{A - I}{L}$ for A (from banking)

42. $z = \dfrac{x - \bar{x}}{s}$ for x (from statistics)

43. $I = \dfrac{E}{R_L + r}$ for r (from physics)

44. $P = \dfrac{R - C}{n}$ for C (from business)

45. $\mu_R = \dfrac{n_1(n_1 + n_2 + 1)}{2}$ for n_2 (from statistics)

46. $\dfrac{P_1 V_1}{T_1} = \dfrac{P_2 V_2}{T_2}$ for T_2 (from chemistry)

47. $P = \dfrac{Q_1}{Q_2 - Q_1}$ for Q_1 (from refrigeration/heating)

48. $S = \dfrac{a - \ell r}{1 - r}$ for r (from mathematics)

49. $\dfrac{1}{R} = \dfrac{1}{R_1} + \dfrac{1}{R_2} + \dfrac{1}{R_3}$ for R (from electronics)

50. $\dfrac{x}{a} + \dfrac{y}{b} = 1$ for a (from mathematics)

51. $\dfrac{E}{e} = \dfrac{R + r}{r}$ for r (from engineering)

52. $P + \dfrac{a}{V^2} = \dfrac{RT}{V - b}$ for b (from physics)

TRY IT YOURSELF

Solve each equation. If a solution is extraneous, so indicate.

53. $\dfrac{x + 2}{x + 3} - 1 = \dfrac{-1}{x^2 + 2x - 3}$

54. $\dfrac{m + 6}{3m - 12} + \dfrac{5}{4 - m} = \dfrac{2}{3}$

55. $\dfrac{3}{y} + \dfrac{7}{2y} = 13$

56. $\dfrac{2}{x} + \dfrac{1}{2} = \dfrac{7}{2x}$

57. $\dfrac{3}{r} + \dfrac{12}{r^2 - 4r} = -\dfrac{7}{r - 4}$

58. $\dfrac{4t^2 + 36}{t^2 - 9} - \dfrac{4t}{t + 3} = \dfrac{-12}{t - 3}$

59. $\dfrac{x + 4}{2x + 14} - \dfrac{x}{2x + 6} = \dfrac{3}{16}$

60. $\dfrac{30}{y - 2} + \dfrac{24}{y - 5} = 13$

61. $\dfrac{-10}{t + 3} = 1 - \dfrac{11}{t - 3}$

62. $\dfrac{3}{m} = 2 - \dfrac{m}{m - 2}$

63. $\dfrac{x + 2}{2x - 6} + \dfrac{3}{3 - x} = \dfrac{x}{2}$

64. $\dfrac{3}{4x - 8} = \dfrac{1}{36} - \dfrac{2}{6 - 3x}$

65. $\dfrac{2}{x} + \dfrac{1}{2} = \dfrac{9}{4x} - \dfrac{1}{2x}$

66. $\dfrac{7}{5x} - \dfrac{1}{2} = \dfrac{5}{6x} + \dfrac{1}{3}$

67. $\dfrac{3 - 5y}{2 + y} = \dfrac{-5y - 3}{y - 2}$

68. $\dfrac{a - 3}{a + 1} = \dfrac{a - 6}{a + 5}$

69. $\dfrac{21}{x^2 - 4} - \dfrac{14}{x + 2} = \dfrac{3}{2 - x}$

70. $\dfrac{-5}{c + 2} = \dfrac{3}{2 - c} + \dfrac{2c}{c^2 - 4}$

71. $\dfrac{x - 4}{x - 3} - \dfrac{x - 2}{3 - x} = x - 3$

72. $\dfrac{5}{x + 4} + \dfrac{1}{x + 4} = x - 1$

73. $\dfrac{a + 2}{a + 1} = \dfrac{a - 4}{a - 3}$

74. $\dfrac{z + 2}{z + 8} = \dfrac{z - 3}{z - 2}$

75. $x^{-1} - 3 = 4x^{-1}$

$\left(\text{Hint: Use } x^{-n} = \dfrac{1}{x^n}. \right)$

76. $x^{-1} + 2 = 3x^{-1}$

$\left(\text{Hint: Use } x^{-n} = \dfrac{1}{x^n}. \right)$

77. $\dfrac{5}{y - 1} + \dfrac{3}{y - 3} = \dfrac{8}{y - 2}$

78. $\dfrac{3 + 2a}{a^2 + 6 + 5a} + \dfrac{2 - 5a}{a^2 - 4} = \dfrac{2 - 3a}{a^2 - 6 + a}$

79. $\dfrac{3}{s - 2} + \dfrac{s - 14}{2s^2 - 3s - 2} - \dfrac{4}{2s + 1} = 0$

80. $\dfrac{1}{y^2 - 2y - 3} + \dfrac{1}{y^2 - 4y + 3} - \dfrac{1}{y^2 - 1} = 0$

81. $\dfrac{x}{x + 2} = 1 - \dfrac{3x + 2}{x^2 + 4x + 4}$

82. $\dfrac{a - 1}{a + 3} - \dfrac{1 - 2a}{3 - a} = \dfrac{2 - a}{a - 3}$

83. $3x^{-2} - 4x^{-1} + 1 = 0$

$\left(\text{Hint: Use } x^{-n} = \dfrac{1}{x^n}. \right)$

84. $3y^{-2} - y^{-1} - 2 = 0$

$\left(\text{Hint: Use } x^{-n} = \dfrac{1}{x^n}. \right)$

85. $\dfrac{5}{2z^2 + z - 3} - \dfrac{2}{2z + 3} = \dfrac{z + 1}{z - 1} - 1$

86. $\dfrac{x}{x - 5} + \dfrac{5}{x} = \dfrac{11}{6}$

87. $\dfrac{5}{3x + 12} - \dfrac{1}{9} = \dfrac{x - 1}{3x}$

88. $\dfrac{1}{y + 5} = \dfrac{1}{3y + 6} - \dfrac{y + 2}{y^2 + 7y + 10}$

Look Alikes . . .

For each expression in part (a), perform the indicated operations and then simplify, if possible. Solve each equation in part (b) and check the result.

89. a. $\dfrac{11}{12} - \dfrac{3}{2x} + \dfrac{4}{x}$

b. $\dfrac{11}{12} - \dfrac{3}{2x} = \dfrac{4}{x}$

90. a. $\dfrac{1}{6x} - \dfrac{2}{x - 6}$

b. $\dfrac{1}{6x} = \dfrac{2}{x - 6}$

91. a. $\dfrac{m}{m - 2} - \dfrac{1}{m - 3}$

b. $\dfrac{m}{m - 2} - \dfrac{1}{m - 3} = 1$

92. a. $\dfrac{a^2 + 1}{a^2 - a} - \dfrac{a}{a - 1}$

b. $\dfrac{a^2 + 1}{a^2 - a} - \dfrac{a}{a - 1} = \dfrac{1}{a}$

APPLICATIONS

93. Photography. The illustration below shows the relationship between distances when taking a photograph. The design of a camera lens uses the equation

$$\frac{1}{f} = \frac{1}{s_1} + \frac{1}{s_2}$$

which relates the focal length f of a lens to the image distance s_1 and the object distance s_2.

a. Solve the formula for f.

b. Find the focal length of the lens in the illustration. (*Hint:* Convert 5 feet to inches.)

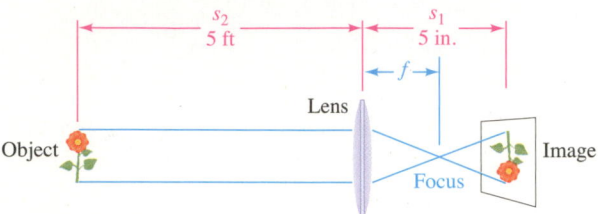

94. Optics. The focal length, f, of a lens is given by the lensmaker's formula,

$$\frac{1}{f} = 0.6\left(\frac{1}{r_1} + \frac{1}{r_2}\right)$$

where f is the focal length of the lens and r_1 and r_2 are the radii of the two circular surfaces. Solve the formula for f.

95. Accounting. As a piece of equipment gets older, its value usually lessens. One way to calculate *depreciation* is to use the formula

$$V = C - \left(\frac{C - S}{L}\right)N$$

where V denotes the value of the equipment at the end of year N, L is its useful lifetime (in years), C is its cost new, and S is its salvage value at the end of its useful life.

a. Solve the formula for L.

b. Determine what an accountant considered the useful lifetime of a forklift that cost $25,000 new, was worth $13,000 after 4 years, and has a salvage value of $1,000.

96. Engineering. The equation

$$a = \frac{9.8m_2 - f}{m_2 + m_1}$$

models the system shown, where a is the acceleration of the suspended block, m_1 and m_2 are the masses of the blocks, and f is the friction force. Solve for m_2.

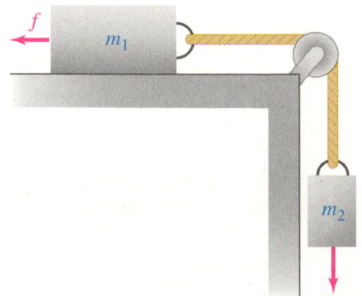

WRITING

97. Why is it necessary to check the solutions of a rational equation?

98. Explain what it means to *clear* a rational equation of fractions. Give an example.

99. Would you use the same approach to answer the following problems? Explain why or why not.

Simplify: $\dfrac{x^2 - 10}{x^2 - 1} - \dfrac{3x}{x - 1} - \dfrac{2x}{x + 1}$

Solve: $\dfrac{x^2 - 10}{x^2 - 1} - \dfrac{3x}{x - 1} = -\dfrac{2x}{x + 1}$

100. Explain how to solve the rational equation graphically:

$$\frac{3x}{x - 2} + \frac{1}{5} = 2$$

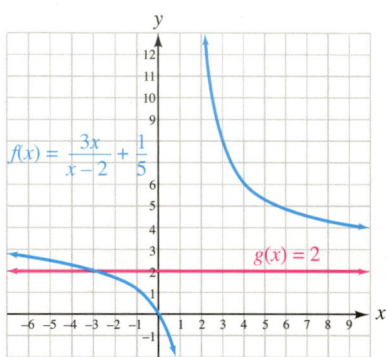

REVIEW

Write each italicized number in scientific notation.

101. Oil. The total cost of the Alaskan pipeline, running 800 miles from Prudhoe Bay to Valdez, was $*9,000,000,000*.

102. Natural Gas. The TransCanada Pipeline transported a record *2,352,000,000,000* cubic feet of gas in 1995.

103. Radioactivity. The least stable radioactive isotope is lithium 5, which decays in *0.00000000000000000000044* second.

104. Balances. The finest balances in the world are made in Germany. They can weigh objects to an accuracy of *35 × 10⁻¹¹* ounce.

CHALLENGE PROBLEMS

105. Solve: $\left(\dfrac{1}{2}\right)^{-1} = \dfrac{5b^{-1}}{2} + 2b(b + 1)^{-1}$

106. Write a rational equation that has an extraneous solution of 3.

107. Let $f(x) = \dfrac{x^3 - 3x^2 + 12}{x}$. For what values of x is $f(x) = 4$?

108. Let $f(x) = \dfrac{x^3 + 2x^2 - 32}{x}$. For what values of x is $f(x) = 16$?

109. Let $f(x) = \dfrac{2x^3 + x^2}{98x + 49}$. For what values of x is $f(x) = 1$?

110. Let $f(x) = \dfrac{x^3 + 4x^2}{25x + 100}$. For what values of x is $f(x) = 1$?

SECTION **6.8**

OBJECTIVES

1 Solve shared-work problems.

2 Solve uniform-motion problems.

Problem Solving Using Rational Equations

ARE YOU READY?

The following problems review some basic skills that are needed when using rational equations to solve application problems.

1. Solve the uniform motion formula $d = rt$ for t.

2. Multiply: $\dfrac{1}{5} \cdot x$

3. Multiply: $x(x - 30)\left(\dfrac{20}{x - 30}\right)$

4. What is the LCD for the fractions in the rational equation $\dfrac{x}{6} + \dfrac{x}{9} = 1$?

5. Write $\dfrac{40}{9}$ as a mixed number.

6. Solve: $x^2 - 29x + 100 = 0$

In this section, we will use rational equations to model situations involving work and uniform motion. As before, we will follow the six-step problem-solving strategy to organize our thinking.

1 Solve Shared-Work Problems.

Problems in which two or more people (or machines) work together to complete a job are called **shared-work problems.** To solve such problems, we must determine the **rate of work** for each person or machine involved. For example, suppose it takes you 4 hours to clean your house. Your rate of work can be expressed as $\frac{1}{4}$ of the job is completed per hour. If someone else takes 5 hours to clean the same house, they complete $\frac{1}{5}$ of the job per hour. In general, a rate of work can be determined in the following way.

Rate of Work	If a job can be completed in x hours, the rate of work can be expressed as:
	$\dfrac{1}{x}$ of the job is completed per hour
	If a job is completed in some other unit of time, such as x minutes or x days, the rate of work is expressed in that unit.

To solve shared-work problems, we must also determine the *amount of work* completed. To do this, we use a formula similar to the distance formula $d = rt$ used for motion problems.

Work completed = rate of work · time worked   or   **$W = rt$**

EXAMPLE 1

Home Construction. One crew can drywall a house in 4 days and another crew can drywall the same house in 5 days. If both crews work together, how long will it take to drywall the house?

Analyze Since the first crew can do the job alone in 4 days, and they are receiving help, we would expect the job to take less than 4 days if the crews work together.

Assign Let $x =$ the number of days it will take to drywall the house if both crews work together.

Form It is helpful to organize the facts of a shared-work problem in a table. Since the crews will be working for the same amount of time, enter x as the time worked for each crew.

If the first crew can drywall the house in 4 days, its rate working alone is $\frac{1}{4}$ of the job per day. If the second crew can drywall the house in 5 days, its rate working alone is $\frac{1}{5}$ of the job per day. To determine the work completed by each crew, multiply the rate by the time.

	Rate · Time = Work Completed		
1st crew	$\frac{1}{4}$	x	$\frac{x}{4}$
2nd crew	$\frac{1}{5}$	x	$\frac{x}{5}$

Think: $\frac{1}{4} \cdot \frac{x}{1} = \frac{x}{4}$.

Think: $\frac{1}{5} \cdot \frac{x}{1} = \frac{x}{5}$.

Enter this information first.

Multiply to get each of these entries; $W = rt$.

In shared-work problems, the number 1 represents one whole job completed. So we have

The part of the job done by 1st crew	plus	the part of the job done by 2nd crew	equals	1 job completed.
$\frac{x}{4}$	$+$	$\frac{x}{5}$	$=$	1

Solve

$$20\left(\frac{x}{4} + \frac{x}{5}\right) = 20(1)$$

Clear the rational equation of fractions by multiplying both sides by the LCD, 20.

$$20\left(\frac{x}{4}\right) + 20\left(\frac{x}{5}\right) = 20$$

On the left side, distribute the multiplication by 20.

$$\overset{1}{\cancel{4}} \cdot 5 \cdot \left(\frac{x}{\underset{1}{\cancel{4}}}\right) + 4 \cdot \overset{1}{\cancel{5}} \cdot \left(\frac{x}{\underset{1}{\cancel{5}}}\right) = 20$$

Factor 20 as $4 \cdot 5$, and remove common factors.

$$5x + 4x = 20$$

The fractions have been cleared.

$$9x = 20$$

Combine like terms: $5x + 4x = 9x$.

$$x = \frac{20}{9}$$

To isolate x, divide both sides by 9.

State If both crews work together, it will take $\frac{20}{9}$ or $2\frac{2}{9}$ days to drywall the house.

Check In $\frac{20}{9}$ days, the first crew drywalls $\frac{1}{4} \cdot \frac{20}{9} = \frac{5}{9}$ of the house and the second crew drywalls $\frac{1}{5} \cdot \frac{20}{9} = \frac{4}{9}$ of the house. The sum of these efforts, $\frac{4}{9} + \frac{5}{9}$, is $\frac{9}{9}$ or 1 house drywalled. The result checks.

> **Self Check 1**   **Painting a House.** One crew can paint a house in 3 days and another crew can paint the same house in 4 days. If both crews work together, how long will it take them to paint the house?
>
> **Now Try** ▶ Problem 11

Example 1 can be solved in a different way by considering the amount of work done by each crew in 1 day. As before, if we let $x =$ the number of days it will take to drywall the house if both crews work together, then together, in 1 day, they will complete $\frac{1}{x}$ of the job. If we add what the first crew can do in 1 day to what the second crew can do in 1 day, the sum is what they can do together in 1 day.

What 1st crew can do in 1 day	plus	what 2nd crew can do in 1 day	equals	what they can do together in 1 day.
$\dfrac{1}{4}$	$+$	$\dfrac{1}{5}$	$=$	$\dfrac{1}{x}$

To solve the equation, begin by clearing it of fractions.

$$20x\left(\frac{1}{4}+\frac{1}{5}\right) = 20x\left(\frac{1}{x}\right)$$ Multiply both sides by the LCD, 20x.

$$20x\cdot\frac{1}{4}+20x\cdot\frac{1}{5} = 20x\left(\frac{1}{x}\right)$$ Distribute the multiplication by 20x.

$$5x+4x = 20$$ Perform each multiplication by 20x.

$$9x = 20$$ Combine like terms: 5x + 4x = 9x.

$$x = \frac{20}{9}$$ To isolate x, divide both sides by 9.

This is the same as the solution obtained in Example 1.

EXAMPLE 2

Setting Up Seating.   It takes the head custodian at a school 30 minutes less time than his assistant to set up the chairs for a program in the auditorium. Working together, they can set up the chairs in 20 minutes. How long would it take each person working alone to set up the chairs?

Analyze   If they can do the job together in 20 minutes, we would expect it to take each of them longer than that if they did the job alone.

Assign   Let $x =$ the number of minutes that it takes the assistant to set up the chairs. Then $x - 30 =$ the number of minutes that it takes the head custodian to set up the chairs.

Form   We will organize the facts of this shared-work problem in a table. If the assistant can set up the chairs in x minutes, his rate working alone is $\frac{1}{x}$ of the job per minute. If the head custodian can set up the chairs in $(x - 30)$ minutes, his rate working alone is $\frac{1}{x-30}$ of the job per minute. Each rate is entered in the table below.

If they work together, it takes them 20 minutes to complete the job. We enter 20 minutes for the time worked for each person. To determine the work completed by each person, we multiply the rate by the time, as shown in the table.

	Rate	$\cdot$ Time	$=$ Work Completed	
Assistant	$\dfrac{1}{x}$	20	$\dfrac{20}{x}$	Think: $\frac{1}{x}\cdot\frac{20}{1}=\frac{20}{x}$.
Head custodian	$\dfrac{1}{x-30}$	20	$\dfrac{20}{x-30}$	Think: $\frac{1}{x-30}\cdot\frac{20}{1}=\frac{20}{x-30}$.

Enter this information first.   Multiply to get each of these entries: W = rt

In shared-work problems, the number 1 represents one whole job completed. So we have

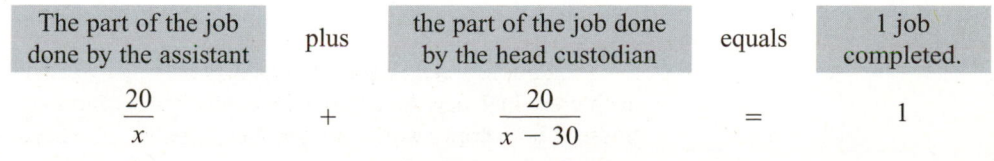

The part of the job done by the assistant	plus	the part of the job done by the head custodian	equals	1 job completed.
$\dfrac{20}{x}$	$+$	$\dfrac{20}{x-30}$	$=$	1

Solve

$$x(x - 30)\left(\frac{20}{x} + \frac{20}{x - 30}\right) = x(x - 30)(1)$$

Clear the equation of fractions by multiplying both sides by the LCD, $x(x - 30)$.

$$\overset{1}{\cancel{x}}(x - 30)\left(\frac{20}{\cancel{x}}\right) + x\overset{1}{\cancel{(x - 30)}}\left(\frac{20}{\cancel{x - 30}}\right) = x(x - 30)(1)$$

Distribute $x(x - 30)$ on the left side and then remove common factors.

$$(x - 30)(20) + x(20) = x(x - 30)(1)$$

The fractions are cleared.

$$20x - 600 + 20x = x^2 - 30x$$

Do the multiplication on each side. This is a quadratic equation.

$$40x - 600 = x^2 - 30x$$

Combine like terms.

$$0 = x^2 - 70x + 600$$

To get 0 on the left side, subtract $40x$ and add 600 to both sides.

$$0 = (x - 10)(x - 60)$$

Factor the trinomial.

$$x - 10 = 0 \quad \text{or} \quad x - 60 = 0$$

Set each factor equal to 0.

$$\cancel{x = 10} \qquad \qquad x = 60$$

Solve each equation.

State   The solutions of the equation are 10 and 60. If it takes them 20 minutes, working together, to set up the chairs, it does not make sense that it would take the assistant 10 minutes working alone. We reject that solution. Thus, we have found that, working alone, it takes the assistant 60 minutes and it takes the head custodian $60 - 30 = 30$ minutes to set up the chairs.

Check   In 20 minutes, the assistant completes $\frac{20}{60} = \frac{1}{3}$ of the job while the head custodian completes $\frac{20}{30} = \frac{2}{3}$ of the job. The sum of these efforts, $\frac{1}{3} + \frac{2}{3}$, is $\frac{3}{3}$ or 1 job completed. The results check.

Self Check 2   **Newspaper Routes.**   It takes a boy riding his bicycle 9 minutes longer to deliver the newspapers to the homes on his route than it does his father driving in his car. Working together, they can deliver the papers in 20 minutes. How long would it take each person working alone to deliver the papers?

Now Try ▶ Problem 21

---

2 Solve Uniform-Motion Problems.

In the next two examples, rational equations are used to model situations involving **uniform-motion.**

EXAMPLE 3   **Road Trips.**   A doctor drove 200 miles to attend a national convention. Because of poor weather, her average speed on the return trip was 10 mph less than her average speed going to the convention. If the return trip took 1 hour longer, how fast did she drive in each direction?

Analyze   The distance traveled was 200 miles each way. We need to find her rates of speed going to and returning from the convention.

Recall the distance formula that we have previously used to solve uniform-motion problems: $d = rt$. If we divide both sides of that equation by r, we obtain a way to describe the travel times:

$$t = \frac{d}{r}$$ The time traveled is equal to the distance traveled divided by the rate.

Assign Let r = the average rate of speed (in mph) going to the meeting. Then $r - 10$ = the average rate of speed on the return trip.

Form We can organize the facts of the problem in the table.

	Rate ·	Time	= Distance
Going	r	$\dfrac{200}{r}$	200
Returning	$r - 10$	$\dfrac{200}{r - 10}$	200

Enter this information first.

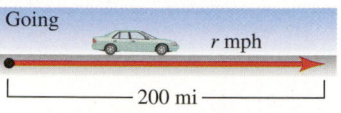

Going

r mph

200 mi

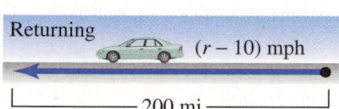

Returning

$(r - 10)$ mph

200 mi

We obtained these time entries by dividing the distance by the rate: $t = \dfrac{d}{r}$.

Because the return trip took 1 hour longer, we can form the following equation:

The time it took to travel to the convention	plus	1	equals	the time it took to return.
$\dfrac{200}{r}$	$+$	1	$=$	$\dfrac{200}{r - 10}$

Solve

$$r(r - 10)\left(\frac{200}{r} + 1\right) = r(r - 10)\left(\frac{200}{r - 10}\right)$$ Multiply both sides by the LCD, $r(r - 10)$.

$$r(r - 10)\frac{200}{r} + r(r - 10)1 = r(r - 10)\left(\frac{200}{r - 10}\right)$$ Distribute $r(r - 10)$ on the left side. Then simplify.

$$200(r - 10) + r(r - 10) = 200r$$ The fractions are cleared.

$$200r - 2{,}000 + r^2 - 10r = 200r$$ Distribute 200 and r. This is a quadratic equation.

$$190r - 2{,}000 + r^2 = 200r$$ Combine like terms.

$$r^2 - 10r - 2{,}000 = 0$$ To get 0 on the right side, subtract $200r$ from both sides.

$$(r - 50)(r + 40) = 0$$ Factor $r^2 - 10r - 2{,}000$.

$$r - 50 = 0 \quad \text{or} \quad r + 40 = 0$$ Set each factor equal to 0.

$$r = 50 \qquad \qquad r = -40$$ Solve each equation.

State We must exclude the solution of -40, because a speed cannot be negative. Thus, the doctor averaged 50 mph going to the convention, and she averaged $50 - 10$ or 40 mph returning.

Check At 50 mph, the 200-mile trip took 4 hours. At 40 mph, the return trip took 5 hours, which is 1 hour longer. The results check.

Self Check 3 **Road Trips.** A caravan of students traveled 150 miles to an academic competition. On the return trip, their average speed was 20 mph less than going, due to road construction. If the return trip took 2 hours longer, how fast did they drive in each direction?

Now Try ▶ Problem 31

EXAMPLE 4 **Riverboat Cruises.** The Forest City Queen can make a 9-mile trip down the Rock River and return in a total of 1.6 hours. If the riverboat travels 12 mph in still water, find the speed of the current in the Rock River.

Analyze Because of the current, the rate the boat travels down the river will be faster than on its return trip up the river. In each case, the distance traveled is 9 miles.

Assign Let c = the speed of the current (in mph). Since the boat travels 12 mph and a current of c mph pushes the boat while it is going downstream, the speed of the boat going downstream is $(12 + c)$ mph. On the return trip, the current pushes against the boat, and its speed is $(12 - c)$ mph. Since time = $\frac{\text{distance}}{\text{rate}}$, the time required for the downstream part of the trip is $\frac{9}{12 + c}$ hours, and the time required for the upstream part of the trip is $\frac{9}{12 - c}$ hours.

Form We can organize this information in the table.

The Language of Algebra

When a boat travels **downstream,** the speed of the boat is increased by the current. When a boat travels **upstream,** the speed of the boat is decreased by the current.

Caution

A common mistake is to represent the rate of the boat going upstream *incorrectly* as:

$c - 12$

	Rate ·	Time =	Distance
Going downstream	$12 + c$	$\frac{9}{12 + c}$	9
Going upstream	$12 - c$	$\frac{9}{12 - c}$	9

Enter this Information first. — Divide the distance by the rate.

Success Tip

Since we are using a rational equation to solve this application problem, it is best to write the decimal 1.6 as a fraction.

$1.6 = 1\frac{6}{10} = \frac{16}{10} = \frac{8}{5}$

We also know that the total time required for the round trip is 1.6 or $\frac{8}{5}$ hours.

The time it takes to travel downstream	plus	the time it takes to travel upstream	is	the total time for the round trip.
$\frac{9}{12 + c}$	$+$	$\frac{9}{12 - c}$	$=$	$\frac{8}{5}$

Solve

$$5(12 + c)(12 - c)\left(\frac{9}{12 + c} + \frac{9}{12 - c}\right) = 5(12 + c)(12 - c)\left(\frac{8}{5}\right)$$
Multiply both sides by the LCD, $5(12 + c)(12 - c)$.

$$5(12 + c)(12 - c)\left(\frac{9}{12 + c}\right) + 5(12 + c)(12 - c)\left(\frac{9}{12 - c}\right) = 5(12 + c)(12 - c)\left(\frac{8}{5}\right)$$
Distribute $5(12 + c)(12 - c)$ and remove common factors.

$$45(12 - c) + 45(12 + c) = 8(12 + c)(12 - c)$$ The fractions are cleared.

$$540 - 45c + 540 + 45c = 8(144 - c^2)$$ On the left side, distribute. On the right side, use the FOIL method.

$$1,080 = 1,152 - 8c^2$$ Combine like terms and multiply. This is a quadratic equation.

$$8c^2 - 72 = 0$$ To get 0 on the right side, add $8c^2$ and subtract 1,152 from both sides.

$$c^2 - 9 = 0$$ Divide both sides by 8.

$$(c + 3)(c - 3) = 0$$ Factor the difference of two squares.

$$c + 3 = 0 \quad \text{or} \quad c - 3 = 0$$ Set each factor equal to 0.

$$c = -3 \quad | \quad c = 3$$ Solve each linear equation.

State Since the current cannot be negative, the solution -3 must be discarded. The current in the Rock River is 3 mph.

Check The downstream trip is at $12 + 3 = 15$ mph for $\frac{9}{12+3} = \frac{3}{5}$ hr. Thus, the distance traveled is $15 \cdot \frac{3}{5} = 9$ miles. The upstream trip is at $12 - 3 = 9$ mph for $\frac{9}{12-3} = 1$ hr. Thus, the distance traveled is $9 \cdot 1 = 9$ miles. Since both distances are 9 miles, the result checks.

> **Self Check 4** **Water Travel.** A motorboat goes 5 miles upstream in the same time it requires to go 7 miles downstream. If the river flows at 2 miles per hour, find the speed of the boat in still water.
>
> **Now Try** ▶ Problem 39

SECTION 6.8 STUDY SET

VOCABULARY

Fill in the blanks.

1. In this section, we call problems that involve:
   - people or machines completing jobs, shared-_____ problems.
   - moving vehicles, uniform-_____ problems.
2. When a boat travels _____, the speed of the boat is increased by the current. When a boat travels _____, the speed of the boat is decreased by the current.

CONCEPTS

3. Fill in the blank: If a job can be completed in x hours, then the rate of work can be expressed as $\frac{1}{\boxed{}}$ of the job is completed per hour.
4. a. It takes a night security officer 35 minutes to check each of the doors in an office building to make sure they are locked. Fill in the blank: The officer's rate of work is $\frac{1}{\boxed{}}$ of the job per minute.
   b. It takes a high school mathematics teacher 4 hours to make out the semester report cards. What part of the job does she complete in x hours?
5. Complete the table.

	Rate · Time = Work completed	
1st crew	$\frac{1}{15}$	x
2nd crew	$\frac{1}{8}$	x

6. Solve $d = rt$ for t.

7. Complete the table.

	r	·	t	=	d
Running	x				12
Bicycling	$x + 15$				12

8. A boat can cruise at 30 mph in still water.
   a. What is its cruising speed upstream against a current of 4 mph?
   b. What is its cruising speed downstream with a current of 4 mph?

NOTATION

9. Write $\frac{41}{9}$ hours using a mixed number.
10. Fill in the blanks: In the formula $W = rt$, the variable W stands for the _____ completed, r is the _____, and t is the _____.

APPLICATIONS

11. **Roofing.** A homeowner estimates that it will take him 7 days to roof his house. A professional roofer estimates that he could roof the house in 4 days. How long will it take if the homeowner helps the roofer?
12. **Decorating.** One crew can put up holiday decorations in a department store in 12 hours. A second crew can put up the decorations in 15 hours. How long will it take if both crews work together to decorate the store?

13. **Housepainting.** The illustration shows two bids to paint a house.
    a. To get the job done quicker, the homeowner hired both the painters who submitted bids. How long will it take them to paint the house working together?
    b. What will the homeowner have to pay each painter?

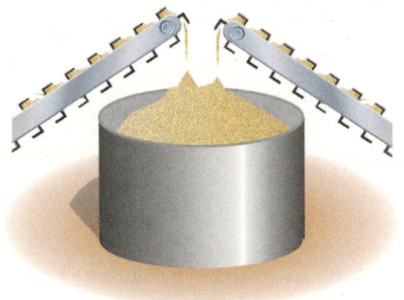

Santos Painting
Residential Bid:
3 days
@ $220 a day
Total: $660

Mays House Painting
Bid:
$200 per day
5 days work
Total: $1,000

14. **Groundskeeping.** It takes a groundskeeper 45 minutes to prepare a Little League baseball field for a game. It takes his assistant 55 minutes to prepare the same field. How long will it take if they work together to prepare the field?

15. **Farming.** In 10 minutes, a conveyor belt can move 1,000 bushels of corn into the storage bin shown. A smaller belt can move 1,000 bushels to the storage bin in 14 minutes. If both belts are used, how long will it take to move 1,000 bushels to the storage bin?

16. **Bottling.** At a packaging plant, the older of two machines can fill 5,000 bottles of shampoo in 6 hours. A newer machine can fill 5,000 bottles in 4 hours. If both machines are used, how long will it take to fill 5,000 bottles of shampoo?

17. **Thrill Rides.** At the end of an amusement park ride, a boat lands in a pool, splashing out a lot of water. Three inlet pipes, each working alone, can fill the pool in 10 seconds, 15 seconds, and 20 seconds, respectively. How long would it take to fill the pool if all three inlet pipes are used?

©Phil Degginger/Alamy

18. **Smoke Damage.** Three ventilation fans, each working alone, can clear the smoke out of a room in 12 hours, 16 hours, and 24 hours, respectively. How long would it take to clear out the smoke in the room if all three fans are used?

19. **Filling Ponds.** One pipe can fill a pond in 3 weeks, and a second pipe can fill it in 5 weeks. However, evaporation and seepage can empty the pond in 10 weeks. If both pipes are used, how long will it take to fill the pond?

20. **Housecleaning.** Sally can clean the house in 6 hours, her father can clean the house in 4 hours, and her younger brother, Dennis, can completely mess up the house in 8 hours. If Sally and her father clean and Dennis plays, how long will it take to clean the house?

21. **Fine Dining.** It takes a waiter 5 minutes less time than a busboy to fold the napkins used for the dinner seating in an upscale restaurant. Working together, they can fold the napkins in 6 minutes. How long would it take each person working alone to fold the napkins?

22. **Fire Drill.** If the east and west exit doors of a banquet hall are open, the occupants can clear out in 2 minutes. It takes 3 minutes longer to clear the hall if just the east door is open as it does if just the west door is open. How long does it take to clear the hall if just the west door is open?

23. **Fund-Raising Letters.** Working together, two secretaries can stuff the envelopes for a political fund-raising letter in 4 hours. Working alone, it takes the slower worker 6 hours longer to do the job than the faster worker. How long does it take each to do the job alone?

24. **Surveys.** It takes one team 9 days less than another to survey 1,000 people. If the teams work together, it takes them 20 days to complete such a survey. How long will it take each to do the survey alone?

25. **from Campus to Careers**
    Webmaster
    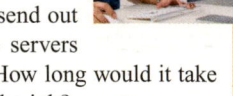
    It takes 1.5 hours less time for a Cisco Systems server to send out a set of email advertisements than it takes a Dell PowerEdge server to send out the same emails. Working together, the servers can complete the emailing in 1.8 hours. How long would it take each server, working alone, to complete the job?

    ©Ronen/Shutterstock.com

26. **Newsletters.** An elementary school teacher can assemble and staple the weekly newsletter three times faster than her student aide. Working together, they can assemble and staple the letters in 12 minutes. How long would it take each, working alone, to complete the job?

27. **Detailing a Car.** It takes a man 3 hours to wash and wax the family car. If his teenage son helps him, it only takes 1 hour. How long would it take the son, working alone, to wash and wax the car?

28. **Cleanup Crews.** It takes one crew 4 hours to clean an auditorium after an event. If a second crew helps, it only takes 1.5 hours. How long would it take the second crew, working alone, to clean the auditorium?

29. Oysters. According to the *Guinness Book of World Records,* the record for opening oysters is 100 in 140 seconds by Mike Racz in Invercargill, New Zealand, on July 16, 1990. If it would take a novice $8\frac{1}{2}$ minutes to perform the same task, how long would it take them working together to open 100 oysters? (*Hint:* Work in terms of seconds.)

30. End Zones. One groundskeeper can paint the end zone of a football field in 2 hours. Another can paint it in 1 hour 20 minutes. How many minutes will it take them working together to paint the end zone?

31. Truck Deliveries. A trucker drove 75 miles to make a delivery at a mountain lodge and returned home on the same route. Because of foggy conditions, his average speed on the return trip was 10 mph less than his average speed going. If the return trip took 2 hours longer, how fast did he drive in each direction?

32. Moving Houses. A house mover towed a historic Victorian home 45 miles to locate it on a new site. On his return, without the heavy house in tow, his average speed was 30 mph faster and the trip was 2 hours shorter. How fast did he drive in each direction?

33. Train Travel. An empty freight train traveled 60 miles from an auto assembly plant to an oil refinery. There, its tank cars were filled with petroleum products, and it returned on the same route to the plant. The total travel time for the train was $5\frac{1}{2}$ hours. If the train traveled 25 mph slower with the tank cars full, how fast did the train travel in each direction?

34. Boxing. For his morning workout, a boxer bicycles for 8 miles and then jogs back to camp along the same route. If he bicycles 6 mph faster than he jogs, and the entire workout lasts 2 hours, how fast does he jog?

35. Rates of Speed. Two trains made the same 315-mile run. Since one train traveled 10 mph faster than the other, it arrived 2 hours earlier. Find the speed of each train.

36. Deliveries. A FedEx delivery van traveled the 110 miles from Rockford to Chicago in 3 hours less time than it took a UPS van to travel the 275 miles from Rockford to St. Louis. If the vans traveled at the same average speed, for how long was the FedEx driver on the road?

37. Comparing Travel. A plane can fly 600 miles in the same time as it takes a car to go 240 miles. If the car travels 90 mph slower than the plane, find the speed of the plane.

38. Comparing Travel. A bicyclist can travel 40 miles in the same time that a motorcyclist can travel 60 miles. If the bicyclist travels 12 mph slower than the motorcyclist, find the speed of the motorcyclist.

39. Boating. It takes 6 hours for a boater to travel 16 miles upstream and 16 miles back. If the speed of the boat in still water is 6 mph, what is the speed of the current?

40. River Tours. A wave runner trip begins by going 60 miles upstream against a current. There, the driver turns around and returns with the current. If the still-water speed of the wave runner is set at 25 mph and the entire trip takes 5 hours, what is the speed of the current?

41. Boating. A man can drive a motorboat 45 miles down the Colorado River in the same amount of time that he can drive 27 miles upstream. Find the speed of the current if the speed of the boat is 12 mph in still water.

42. Crop Dusting. A helicopter spraying fertilizer over a field can fly 0.5 mile downwind in the same time as it can fly 0.4 mile upwind. Find the speed of the wind if the helicopter travels 45 mph in still air when dusting crops.

43. Catching a Plane. A walkway at an airport operates at a rate of 1.5 feet per second. Walking with the moving walkway, a man travels 65 feet in the same time that he could travel walking 35 feet in the opposite direction, against the walkway. What is the man's normal walking rate?

44. Kayaking. A kayaker can travel 1.2 miles downstream in the same time it takes him to go 0.4 miles upstream. If the river current flows at 2 mph, what is the kayaker's speed in still water?

WRITING

45. In Example 1, one crew could drywall a house in 4 days, and another crew could drywall the same house in 5 days. We were asked to find how long it would take them to drywall the house working together. Explain why each of the following approaches is incorrect.

The time it would take to drywall the house:

- is the *sum* of the lengths of time it takes each crew to drywall the house: 4 days + 5 days = 9 days.

- is the *difference* in lengths of time it takes each crew to drywall the house: 5 days − 4 days = 1 day.

- is the *average* of the lengths of time it takes each crew to drywall the house: $\frac{4 \text{ days} + 5 \text{ days}}{2} = \frac{9}{2}$ days $= 4\frac{1}{2}$ days.

46. Write a shared-work problem that can be modeled by the equation

$$\frac{x}{3} + \frac{x}{4} = 1$$

Simplify each expression. Write answers using positive exponents.

47. $\left(\dfrac{m^{10}}{n}\right)^{8}$ **48.** $\left(\dfrac{g^{20}}{t^{30}}\right)^{-4}$

49. $-w^{-2}$ **50.** $-3s^{0}t$

51. $-\dfrac{4x^{-9} \cdot x^{-3}}{x^{-12}}$ **52.** $\dfrac{y^{-3}y^{-4}y^{0}}{(2y^{-2})^{3}}$

53. $(-x^{2})^{5}y^{7}y^{3}x^{-2}y^{0}$ **54.** $5^{2}r^{-5}(r^{6})^{3}$

CHALLENGE PROBLEMS

55. Fireplaces. A mason and his assistant work together for 6 hours on a brick fireplace before the mason has to leave the job. The assistant finishes the job alone in 10 hours. If the mason can construct a fireplace in 18 hours working alone, how long does it take his assistant working alone to construct a fireplace?

56. Extended Vacation. Use the facts in the text message to determine how long the student had originally planned to stay in Europe. (*Hint:* Unit cost · number = total cost.)

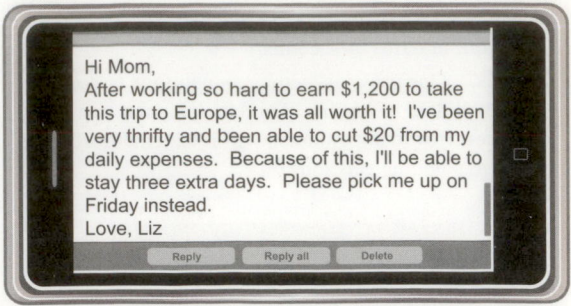

Hi Mom,
After working so hard to earn $1,200 to take this trip to Europe, it was all worth it! I've been very thrifty and been able to cut $20 from my daily expenses. Because of this, I'll be able to stay three extra days. Please pick me up on Friday instead.
Love, Liz

Reply   Reply all   Delete

Proportion and Variation

OBJECTIVES

1 Identify ratios, rates, and proportions.

2 Solve proportions.

3 Use proportions to solve problems.

4 Solve problems involving similar triangles.

5 Solve problems involving direct variation.

6 Solve problems involving inverse variation.

7 Solve problems involving joint variation.

8 Solve problems involving combined variation.

ARE YOU READY?

The following problems review some basic skills that are needed when working with proportions and variation.

1. Multiply: **a.** $bd \cdot \dfrac{a}{d}$ **b.** $bd \cdot \dfrac{c}{d}$ **2.** Solve: $2(x + 3) = 4(15)$

3. Solve: $x(x - 7) = 2(9)$ **4.** Find t if $t = \dfrac{90}{n}$ and $n = 18$.

5. Let $f = 0.0036Av^{2}$. Find f if $A = 500$ and $v = 30$.

6. What is the degree measure of a right angle?

In this section, we will discuss the *ratio-proportion model* and four *variation models*. They can be used to solve a variety of application problems.

1 Identify Ratios, Rates, and Proportions.

The quotient of two numbers or two quantities measured with the same units is often called a **ratio**. For example, $\frac{2}{3}$ can be read as "the ratio of 2 to 3." The notation 2:3 (read as "2 is to 3") is another way to denote a ratio. Some more examples of ratios are

Caution

A ratio that is the quotient of two quantities having the same units should be simplified so that no units appear in the final answer.

$\dfrac{4x}{7y}$ *The ratio of 4x to 7y*   and $\dfrac{x - 2}{3x}$ *The ratio of x − 2 to 3x*

When we compare two quantities having different units, we call the comparison a **rate**, and we can write it as a fraction. One example is an average rate of speed.

A distance traveled → $\dfrac{372 \text{ miles}}{6 \text{ hours}} = 62$ mph ← *The average rate of speed*
in a period of time →

Rates are often used to express **unit costs**, such as the cost per pound of ground beef.

The cost of a package of beef → $\dfrac{\$18.95}{5 \text{ lb}} = \$3.79$ per lb ← *The cost per pound*
The weight of the package →

The Language of Algebra

The word **proportion** implies a comparative relationship in size. For a picture to appear realistic, the artist must draw the shapes in the proper *proportion*. Sometimes the news media is accused of blowing things way out of *proportion*.

An equation indicating that two ratios or rates are equal is called a **proportion.** Two examples are

$$\frac{1}{4} = \frac{2}{8} \quad \text{and} \quad \frac{4}{7} = \frac{12}{21}$$

In the proportion $\frac{a}{b} = \frac{c}{d}$, the terms a and d are called the **extremes** of the proportion, and the terms b and c are called the **means.**

To develop a fundamental property of proportions, we suppose that

$$\frac{a}{b} = \frac{c}{d}$$

is a proportion and multiply both sides by bd to obtain

$$bd\left(\frac{a}{b}\right) = bd\left(\frac{c}{d}\right) \qquad \text{To clear the fractions, multiply both sides by the LCD, } bd.$$

$$\overset{1}{\cancel{bd}} \cdot \frac{a}{\underset{1}{\cancel{b}}} = \overset{1}{\cancel{bd}} \cdot \frac{c}{\underset{1}{\cancel{d}}} \qquad \text{Simplify by removing common factors of the numerator and denominator.}$$

$$ad = bc$$

Since $ad = bc$, the product of the extremes equals the product of the means.

The same products ad and bc can be found by multiplying diagonally in the proportion $\frac{a}{b} = \frac{c}{d}$. We call ad and bc **cross products**.

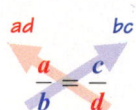

The Fundamental Property of Proportions	In a proportion, the product of the extremes is equal to the product of the means.
	If $\frac{a}{b} = \frac{c}{d}$, then $ad = bc$ and if $ad = bc$, then $\frac{a}{b} = \frac{c}{d}$.

---

2 Solve Proportions.

We have seen that a proportion contains four terms. If we know only three of the four terms of a proportion, we can use the fundamental property of proportions to find the value of the fourth term. This process is called **solving the proportion.**

EXAMPLE 1 Solve for x: $\dfrac{x + 3}{x} = \dfrac{x}{x + 6}$

Strategy To solve for x, we will set the cross products equal.

Why This equation is a proportion, and in a proportion the product of the means equals the product of the extremes.

Solution

$$\frac{x + 3}{x} = \frac{x}{x + 6} \qquad \text{This is the proportion to solve.}$$

$$(x + 3)(x + 6) = x \cdot x \qquad \text{Find each cross product and set them equal.}$$

$$x^2 + 9x + 18 = x^2 \qquad \text{Perform the multiplication on each side.}$$

$$9x + 18 = 0 \qquad \text{Subtract } x^2 \text{ from both sides.}$$

$$9x = -18 \qquad \text{Subtract 18 from both sides.}$$

$$x = -2 \qquad \text{To isolate } x, \text{ divide both sides by 9.}$$

Caution

The expression $\frac{x + 3}{x}$ is undefined if x is 0, because division by 0 would be indicated. Similarly, $\frac{x}{x + 6}$ is undefined if x is -6. Thus, we can rule out 0 and -6 as possible solutions of $\frac{x + 3}{x} = \frac{x}{x + 6}$.

The solution is -2 and the solution set is $\{-2\}$. To check, we substitute -2 for x in the proportion and simplify each side.

Check: $\dfrac{x+3}{x} = \dfrac{x}{x+6}$

$$\dfrac{-2+3}{-2} \overset{?}{=} \dfrac{-2}{-2+6}$$

$$\dfrac{1}{-2} \overset{?}{=} \dfrac{-2}{4}$$

$$-\dfrac{1}{2} = -\dfrac{1}{2}$$ True

Self Check 1   Solve for x: $\dfrac{x-1}{x} = \dfrac{x}{x+3}$

Now Try ▶ Problem 25

EXAMPLE 2   Solve: $\dfrac{5a+2}{2a} = \dfrac{18}{a+4}$

Strategy   To solve for a, we will set the cross products equal.

Why   This equation is a proportion, and in a proportion the product of the means equals the product of the extremes.

Solution   Because $\dfrac{5a+2}{2a}$ is undefined if $a = 0$, we have the restriction that $a \neq 0$. Because $\dfrac{18}{a+4}$ is undefined if $a = -4$, we also have the restriction that $a \neq -4$.

$$\dfrac{5a+2}{2a} = \dfrac{18}{a+4}$$ This is the proportion to solve.

$$(5a+2)(a+4) = 2a(18)$$ Find each cross product and set them equal.

$$5a^2 + 22a + 8 = 36a$$ Perform the multiplication on each side.

$$5a^2 - 14a + 8 = 0$$ To get 0 on the right side, subtract $36a$ from both sides.

$$(5a-4)(a-2) = 0$$ Factor to solve the quadratic equation.

$$5a - 4 = 0 \quad \text{or} \quad a - 2 = 0$$ Set each factor equal to 0.

$$5a = 4 \qquad\qquad a = 2$$ Solve each linear equation.

$$a = \dfrac{4}{5}$$

The solutions are $\dfrac{4}{5}$ and 2. Check each of them in the original equation.

> **Caution**
>
> Remember that a cross product is the product of the means and the extremes of a proportion. For example, it would be incorrect to try to compute cross products to solve
>
> $$\dfrac{5a+2}{2a} = \dfrac{18}{a+4} + \dfrac{1}{a}$$
>
> It is not a proportion. The right side is not a single ratio.

Self Check 2   Solve: $\dfrac{3x+1}{12} = \dfrac{x}{x+2}$

Now Try ▶ Problem 33

---

3 Use Proportions to Solve Problems.

We can use proportions to solve many application problems. If we are given a ratio (or rate) comparing two quantities, the words of the problem can be translated into a proportion and we can solve it to find the unknown.

EXAMPLE 3   **Gourmet Cooking.**   To make a dessert, a chef needs to purchase 14 pears. If they are on sale at 6 for $2.34, what will 14 cost?

Strategy   We will use the facts in the problem to set up a proportion.

Why   Three of the entries of the proportion (14, 6, and 2.34) are given. We can find the fourth entry, the unknown cost of 14 pears, by solving the proportion.

Solution   First, we let c = the cost of 14 pears in dollars. The price per pear when purchasing 6 pears is $\frac{\$2.34}{6}$, and the price per pear when purchasing 14 pears is $\frac{\$c}{14}$. Since these rates are equal, we have the following proportion:

$2.34 is **to** *6 pears* **as** *$c is* **to** *14 pears.*

Cost of 6 pears → $\dfrac{2.34}{6} = \dfrac{c}{14}$ ← Cost of 14 pears

6 pears →   ← 14 pears

The units can be written outside the proportion.

$$14(2.34) = 6c \qquad \text{Find each cross product and set them equal.}$$

$$32.76 = 6c \qquad \text{Multiply: } 14(2.34) = 32.76.$$

$$\frac{32.76}{6} = c \qquad \text{To isolate } c \text{, divide both sides by 6.}$$

$$c = 5.46 \qquad \text{Perform the division.}$$

Fourteen pears will cost $5.46.

We can use estimation to check the result. 14 pears are about 2 times as many as 6 pears, which cost $2.34. If we multiply $2.34 by 2, we get an estimate of the cost of 14 pears: $2.34 · 2 = $4.68. The result, $5.46, seems reasonable.

> ### Success Tip
>
> Since proportions are rational equations, they also can be solved by multiplying both sides by the LCD. For Example 3, an alternate approach is to multiply both sides by the LCD of 6 and 14, which is 42.
>
> $$42\left(\frac{2.34}{6}\right) = 42\left(\frac{c}{14}\right)$$

Self Check 3   **Movie Tickets.** Five adult admission tickets to a movie cost $42.50. What will 8 tickets cost?

Now Try ▶   Problem 69

4 Solve Problems Involving Similar Triangles.

If two angles of one triangle have the same measure as two angles of a second triangle, the triangles will have the same shape. In this case, we say that the triangles are **similar triangles.** Here are some facts about similar triangles.

Similar Triangles   If two triangles are similar, then

1. The three angles of the first triangle have the same measure, respectively, as the three angles of the second triangle.

2. The lengths of all corresponding sides are in proportion.

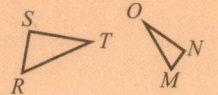
In the following figure, $\triangle ABC \sim \triangle DEF$. (Read the symbol $\sim$ as "is similar to.")

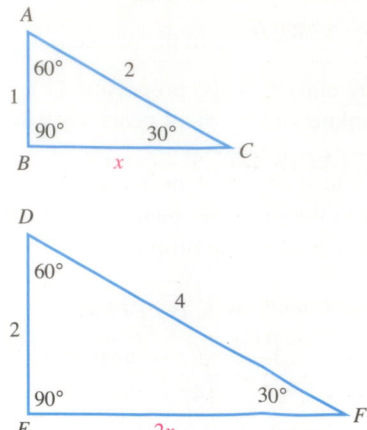

■ Corresponding angles have the same measure.

■ Since the corresponding sides are in proportion, we can write:

$$\frac{AB}{DE} = \frac{AC}{DF} \rightarrow \frac{1}{2} = \frac{2}{4}$$

$$\frac{AC}{DF} = \frac{BC}{EF} \rightarrow \frac{2}{4} = \frac{x}{2x}$$

$$\frac{BC}{EF} = \frac{AB}{DE} \rightarrow \frac{x}{2x} = \frac{1}{2}$$

The properties of similar triangles often enable us to find the lengths of the sides of triangles indirectly. For example, we can find the height of a tree and stay safely on the ground.

EXAMPLE 4 **Height of a Tree.** A tree casts a shadow of 29 feet at the same time as a vertical yardstick casts a shadow of 2.5 feet. Find the height of the tree.

Strategy We will use the facts in the problem to set up a proportion.

Why Three of the entries of the proportion are given. We can find the fourth entry, the unknown height of the tree, by solving the proportion.

Solution Refer to the figure, which shows the triangles determined by the tree and its shadow and the yardstick and its shadow.

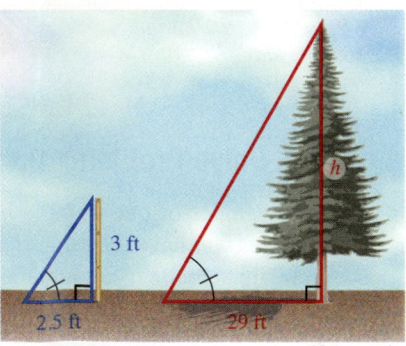

3 ft

2.5 ft   29 ft

Not to scale

Because the triangles have the same shape, they are similar, and the measures of their corresponding sides are in proportion. If we let $h =$ the height of the tree in feet, we can find h by setting up and solving the following proportion: h is to 3 as 29 is to 2.5.

Height of the tree $\rightarrow$ $\dfrac{h}{3} = \dfrac{29}{2.5}$ $\leftarrow$ Length of the tree's shadow
Height of the yardstick $\rightarrow$ $\phantom{\dfrac{h}{3}}$ $\leftarrow$ Length of the yardstick's shadow

$2.5h = 3(29)$ Find each cross product and set them equal.

$2.5h = 87$ Multiply: $3(29) = 87$.

$h = 34.8$ To isolate h, divide both sides by 2.5.

The tree is about 35 feet tall.

Self Check 4   **Height of a Tree.** Suppose the tree casts a shadow of 32 feet at the same time the yardstick casts a shadow of 4 feet. Find the height of the tree.

Now Try ▶   Problem 77

5 Solve Problems Involving Direct Variation.

To introduce direct variation, we consider the formula for the circumference of a circle

$$C = \pi D$$

where C is the circumference, D is the diameter, and $\pi \approx 3.14159$. If we double the diameter of a circle, we determine another circle with a larger circumference C_1 such that

$$C_1 = \pi(2D) = 2\pi D = 2C$$

Thus, doubling the diameter results in doubling the circumference. Likewise, if we triple the diameter, we will triple the circumference.

In the formula, $C = \pi D$, we say that the variables C and D *vary directly,* or that they are *directly proportional.* This is because C is always found by multiplying D by a constant. In this example, the constant π is called the *constant of variation* or the *constant of proportionality.*

Direct Variation	The words **y varies directly as x** or **y is directly proportional to x** means that $y = kx$ for some nonzero constant k. The constant k is called the **constant of variation** or the **constant of proportionality.**

Since the formula for direct variation ($y = kx$) defines a linear function, its graph is always a line with a y-intercept at the origin. The graph of $y = kx$ where $x \geq 0$ appears on the left for three positive values of k.

One example of direct variation is Hooke's law from physics. Hooke's law states that the distance a spring will stretch varies directly as the force that is applied to it.

If d represents a distance and f represents a force, this verbal model of Hooke's law can be expressed as

$$d = kf \quad \text{\color{red}{The direct variation model can be read as "d is directly proportional to f."}}$$

where k is the constant of variation. Suppose we know that a certain spring stretches 10 inches when a weight of 6 pounds is attached (see the figure). We can find k as follows:

$$\mathbf{d = kf}$$

$$\mathbf{10} = k(\mathbf{6}) \quad \text{\color{red}{Substitute 10 for d and 6 for f.}}$$

$$\frac{10}{6} = k \quad \text{\color{red}{To isolate k, divide both sides by 6.}}$$

$$\frac{5}{3} = k \quad \text{\color{red}{Simplify the fraction. This is the constant of variation.}}$$

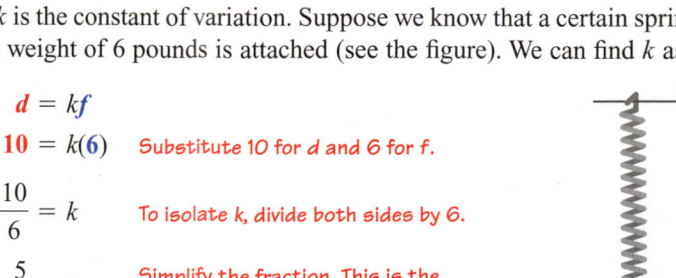

y = 6x *y = 2x* *y = 0.5x*

For any **direct variation** equation of the form $y = kx$, where $k > 0$: as x increases, y increases.

Success Tip

The value that we found for k is for this specific example. Another spring made out of a different type of steel will more than likely have a different value of k.

Unstretched length

10 in.

6 lb.

To find the force required to stretch the spring a distance of 35 inches, we can solve the equation $d = kf$ for f, with $d = 35$ and $k = \frac{5}{3}$.

$$\mathbf{d = kf} \quad \text{\color{red}{This is the direct variation model.}}$$

$$\mathbf{35} = \frac{\mathbf{5}}{\mathbf{3}}f \quad \text{\color{red}{Substitute 35 for d and $\frac{5}{3}$ for k.}}$$

$$105 = 5f \quad \text{\color{red}{Multiply both sides by 3.}}$$

$$21 = f \quad \text{\color{red}{To isolate f, divide both sides by 5.}}$$

Thus, the force required to stretch the spring a distance of 35 inches is 21 pounds.

We can use the following steps to solve variation problems.

Solving Variation Problems ▼	To solve a variation problem:
	1. Translate the verbal model into an equation.
	2. Substitute the first set of values into the equation from step 1 to determine the value of k.
	3. Substitute the value of k into the equation from step 1.
	4. Substitute the remaining set of values into the equation from step 3 and solve for the unknown.

EXAMPLE 5

Currency Exchange. The currency calculator shown below converts from U.S. dollars to Russian rubles. When exchanging these currencies, the number of rubles received is directly proportional to the number of dollars to be exchanged. How many rubles will an exchange of $1,200 bring?

Strategy   We will use a direct variation model to solve this problem.

Why   The words *the number of rubles received is directly proportional to the number of dollars to be exchanged* indicate that this type of model should be used.

Solution

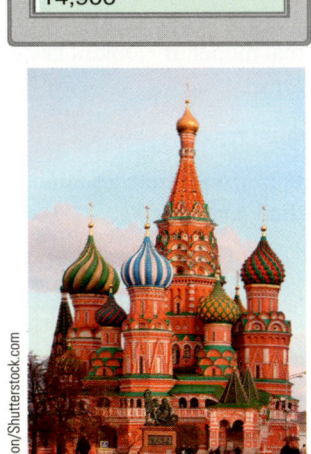

Step 1: The verbal model can be represented by the equation

$$r = kd \qquad \text{This is a direct variation model.}$$

where r is the number of rubles, k is the constant of variation, and d is the number of dollars.

Step 2: From the illustration, we see that an exchange of $500 brings 14,900 rubles. To find k, we substitute 500 for d and 14,900 for r, and then we solve for k.

$$r = kd$$
$$14{,}900 = k(500) \qquad \text{Substitute for } r \text{ and } d.$$
$$29.8 = k \qquad \text{To isolate } k, \text{ divide both sides by 500. This is the constant of variation.}$$

Step 3: Now we substitute the value of k, 29.8, into the equation $r = kd$, to get

$$r = 29.8d$$

Step 4: To find how many rubles an exchange of $1,200 will bring, we substitute 1,200 for d in the direct variation model, and then we evaluate the right side.

$$r = 29.8d \qquad \text{This is the ruble direct variation model.}$$
$$r = 29.8(1{,}200) \qquad \text{Substitute for } d.$$
$$r = 35{,}760 \qquad \text{Do the multiplication. The result is in rubles.}$$

An exchange of $1,200 will bring 35,760 rubles.

Self Check 5   **Currency Exchange.** When exchanging currencies, the number of British pounds received is directly proportional to the number of U.S. dollars to be exchanged. If $800 converts to 392 pounds, how many pounds will be received if $1,500 is exchanged?

Now Try 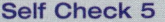 Problem 85

6 Solve Problems Involving Inverse Variation.

In the formula $w = \frac{12}{l}$, w gets smaller as l gets larger, and w gets larger as l gets smaller. Since these variables vary in opposite directions in a predictable way, we say that the variables *vary inversely*, or that they are *inversely proportional*. The constant 12 is the constant of variation.

Inverse Variation

The words **y varies inversely as x** or **y is inversely proportional to x** mean that $y = \dfrac{k}{x}$ for some nonzero constant k. The constant k is called the **constant of variation.**

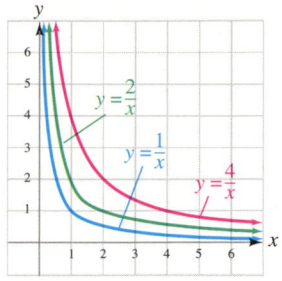

For any **inverse variation** equation of the form $y = \frac{k}{x}$, where $k > 0$: as x increases, y decreases.

The formula for inverse variation, $y = \frac{k}{x}$, defines a rational function whose graph will have the x- and y-axes as asymptotes. The graph of $y = \frac{k}{x}$ where $x > 0$ appears on the left for three positive values of k.

In an elevator, the amount of floor space per person varies inversely as the number of people in the elevator. If f represents the amount of floor space per person and n the number of people in the elevator, the relationship between f and n can be expressed by the equation:

$$f = \frac{k}{n}$$ This inverse variation model also can be read as "f is inversely proportional to n."

The figure on the right shows 6 people in an elevator; each has 8.25 square feet of floor space. To determine how much floor space each person would have if 15 people were in the elevator, we begin by determining k.

$$f = \frac{k}{n}$$ This is the inverse variation model.

$$8.25 = \frac{k}{6}$$ Substitute 8.25 for f and 6 for n.

$$k = 49.5$$ Multiply both sides by 6 to solve for k. This is the constant of variation.

To find the amount of floor space per person if 15 people are in the elevator, we proceed as follows:

$$f = \frac{k}{n}$$ This is the inverse variation model.

$$f = \frac{49.5}{15}$$ Substitute 49.5 for k and 15 for n.

$$f = 3.3$$ Do the division.

If 15 people were in the elevator, each would have 3.3 square feet of floor space.

EXAMPLE 6

©iStockphoto.com/kevinrusskevinruss

Photography. The intensity I of light received from a light source varies inversely as the square of the distance from the light source. If a photographer, 16 feet away from his subject, has a light meter reading of 4 foot-candles of luminance, what will the meter read if the photographer moves in for a close-up 4 feet away from the subject?

Strategy We will use the inverse variation model of the form $I = \frac{k}{d^2}$, where I represents the intensity and d^2 represents the square of the distance from the light source.

Why The words *intensity varies inversely as the square of the distance* indicate that this type of model should be used.

Solution

$$I = \frac{k}{d^2}$$ This inverse variation model also can be read as "I is inversely proportional to d^2."

$$4 = \frac{k}{16^2}$$ To find k, we substitute 4 for I and 16 for d and solve for k.

$$4 = \frac{k}{256}$$ Evaluate the denominator: $16^2 = 256$.

$$1,024 = k$$ To isolate k, multiply both sides by 256. This is the constant of variation.

Success Tip

The constant of variation is usually positive, because most real-life applications involve only positive quantities. However, the definitions of *direct, inverse, joint,* and *combined variation* allow for a negative constant of variation.

To find the intensity when the photographer is 4 feet away from the subject, we substitute 4 for *d* and 1,024 for *k* and simplify.

$$I = \frac{k}{d^2}$$ This is the inverse variation model.

$$I = \frac{1{,}024}{4^2}$$ Substitute for k and d.

$$I = 64$$ Evaluate the right side.

The intensity at 4 feet is 64 foot-candles.

Self Check 6 **Photography.** Find the intensity when the photographer is 8 feet away from the subject.

Now Try ▶ Problem 91

7 Solve Problems Involving Joint Variation.

There are times when one variable varies as the product of several variables. For example, the area of a triangle varies directly with the product of its base and height:

$$A = \frac{1}{2}bh$$

Such variation is called *joint variation*.

Joint Variation | If one variable varies directly as the product of two or more variables, the relationship is called **joint variation**. If *y* varies jointly with *x* and *z*, then *y = kxz*. The nonzero constant *k* is called the **constant of variation**.

EXAMPLE 7 **Force of the Wind.** The force of the wind on a billboard varies jointly as the area of the billboard and the square of the wind velocity. When the wind is blowing at 20 mph, the force on a billboard 30 feet wide and 18 feet high is 972 pounds. Find the force on a billboard having an area of 300 square feet caused by a 40-mph wind.

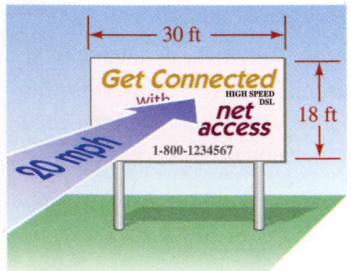

Strategy We will use the joint variation model $f = kAv^2$, where *f* represents the force of the wind, *A* represents the area of the billboard, and v^2 represents the square of the velocity of the wind.

Why The words *the force of the wind on a billboard varies jointly as the area of the billboard and the square of the wind velocity* indicate that this type of model should be used.

Solution $$f = kAv^2$$ The joint variation model also can be read as "f is directly proportional to the product of A and v^2."

Since the billboard is 30 feet wide and 18 feet high, it has an area of $30 \cdot 18 = 540$ square feet. We can find *k* by substituting 972 for *f*, 540 for *A*, and 20 for *v*.

$$f = kAv^2$$

$$972 = k(540)(20)^2$$

$$972 = k(216{,}000)$$ Evaluate: $(20)^2 = 400$. Then do the multiplication.

$$0.0045 = k$$ Divide both sides by 216,000 to solve for k.
This is the constant of variation.

To find the force exerted on a 300-square-foot billboard by a 40-mph wind, we use the formula $f = 0.0045Av^2$ and substitute 300 for A and 40 for v.

$$f = 0.0045Av^2 \quad \text{This is the joint variation model.}$$
$$f = 0.0045(300)(40)^2 \quad \text{Substitute for } A \text{ and } v.$$
$$f = 2{,}160 \quad \text{Evaluate the right side.}$$

The 40-mph wind exerts a force of 2,160 pounds on the billboard.

Self Check 7   **Force of the Wind.**   Refer to Example 7. Find the force of a 25-mph wind on a billboard having an area of 375 square feet.

Now Try ▶ Problem 95

8 Solve Problems Involving Combined Variation.

Many applied problems involve a combination of direct and inverse variation. Such variation is called **combined variation.**

EXAMPLE 8   **Highway Construction.**   The time it takes to build a highway varies directly as the length of the road, and inversely as the number of workers. If it takes 100 workers 4 weeks to build 2 miles of highway, how long will it take 80 workers to build 10 miles of highway?

Strategy   We will use the combined variation model $t = \frac{kl}{w}$, where t represents the time in days, l represents the length of road built in miles, and w represents the number of workers.

Why   The words *the time it takes to build a highway* **varies directly** *as the length of the road, and* **inversely** *as the number of workers* indicate that this type of model should be used.

Solution   The relationship between these variables can be expressed by the equation

$$t = \frac{kl}{w} \quad \text{This is a combined variation model.}$$

$$4 = \frac{k(2)}{100} \quad \text{Substitute 4 for } t, 100 \text{ for } w, \text{ and 2 for } l \text{ to find } k.$$

$$400 = 2k \quad \text{Multiply both sides by 100.}$$

$$200 = k \quad \text{To isolate } k, \text{ divide both sides by 2. This is the constant of variation.}$$

We now substitute 80 for w, 10 for l, and 200 for k in the equation $t = \frac{kl}{w}$ and simplify:

$$t = \frac{kl}{w} \quad \text{This is the combined variation model.}$$

$$t = \frac{200(10)}{80} \quad \text{Substitute for } k, l, \text{ and } w.$$

$$t = 25 \quad \text{Evaluate the right side.}$$

It will take 25 weeks for 80 workers to build 10 miles of highway.

Self Check 8   **Highway Construction.** How long will it take 60 workers to build 6 miles of highway?

Now Try ▶ Problem 101

SECTION 6.9 > STUDY SET

VOCABULARY

Fill in the blanks.

1. A _____ is the quotient of two numbers or two quantities with the same units.

2. An equation that states that two ratios are equal, such as $\frac{1}{2} = \frac{4}{8}$, is called a _____.

3. In $\frac{50}{3} = \frac{x}{9}$, the terms 50 and 9 are called the _____ and the terms 3 and x are called the _____ of the proportion. In a proportion, the product of the _____ is equal to the product of the _____.

4. The _____ products for the proportion $\frac{10}{3} = \frac{5}{x}$ are $10x$ and 15.

5. If two angles of one triangle have the same measure as two angles of a second triangle, the triangles are _____.

6. The equation $y = kx$ defines _____ variation: As x increases, y _____.

7. The equation $y = \frac{k}{x}$ defines _____ variation: As x increases, y _____.

8. The equation $y = kxz$ defines _____ variation, and $y = \frac{kz}{x}$ defines _____ variation.

CONCEPTS

Determine whether direct or inverse variation applies and sketch a possible graph for the situation. (Hint: Refer to the graphs on pages 553 and 555.)

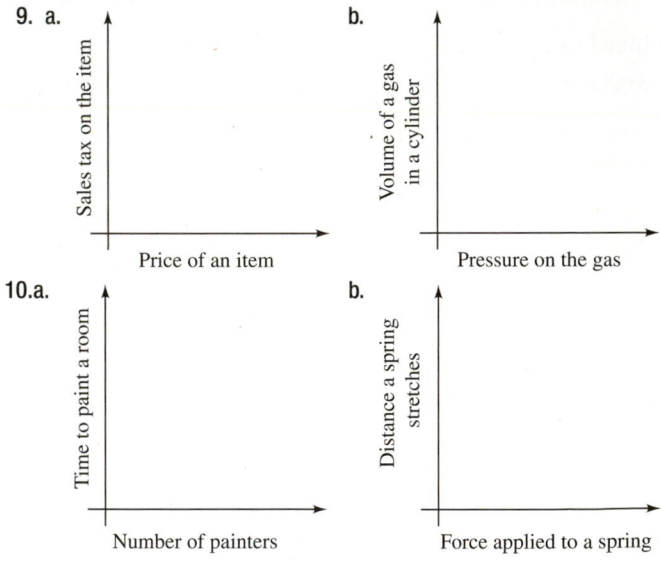

9. a.

b.

10. a.

b.

Tell whether each relationship suggests direct or inverse variation.

11. **Recycling.** The amount of money you receive and the number of aluminum cans you return

12. **Karate.** The force needed to break a board and the length of the board

13. **Tools.** The force you must exert on the handle of a wrench to loosen a bolt and the length of the handle

14. **Anatomy.** The volume of blood pumped from your heart each minute and your pulse rate

15. **Swimming Pools.** The amount of chlorine needed in a pool and the amount of water in the pool

16. **Lightning.** The time it takes you to hear the lightning after a strike and your distance from the strike

17. **Desserts.** The number of servings you can get from a wedding cake and the size of the piece that is served

18. **Remodeling.** The cost to remodel a house and the number of square feet to be added

NOTATION

Complete each solution.

19. Solve: $\dfrac{7}{6} = \dfrac{x+3}{12}$

$$\boxed{}(12) = \boxed{}(x+3)$$

$$84 = 6x + \boxed{}$$

$$\boxed{} = 6x$$

$$\boxed{} = x$$

20. Solve: $\dfrac{18}{2x+1} = \dfrac{3}{14}$

$$\boxed{}(14) = (\boxed{})3$$

$$252 = \boxed{} + 3$$

$$249 = \boxed{}$$

$$\boxed{} = x$$

GUIDED PRACTICE

Solve each proportion. See Example 1.

21. $\dfrac{x}{5} = \dfrac{15}{25}$

22. $\dfrac{4}{y} = \dfrac{6}{27}$

23. $\dfrac{r-2}{3} = \dfrac{r}{5}$

24. $\dfrac{x+1}{x-1} = \dfrac{6}{4}$

25. $\dfrac{5}{5z+3} = \dfrac{3}{2z+6}$

26. $\dfrac{9t+6}{t} = \dfrac{7}{3}$

27. $\dfrac{x-2}{x} = \dfrac{x+1}{x+2}$

28. $\dfrac{a}{a+1} = \dfrac{a+2}{a}$

Solve each proportion. See Example 2.

29. $\dfrac{2}{3x} = \dfrac{6x}{36}$

30. $\dfrac{y}{4} = \dfrac{4}{y}$

31. $\dfrac{2}{c} = \dfrac{c-3}{2}$

32. $\dfrac{2}{x+6} = \dfrac{-2x}{5}$

33. $\dfrac{1}{x+3} = \dfrac{-2x}{x+5}$

34. $\dfrac{x-1}{x+1} = \dfrac{2}{3x}$

35. $\dfrac{2b}{b+5} = \dfrac{-b}{3b+8}$

36. $\dfrac{-3c}{c-2} = \dfrac{c}{c+2}$

Express each verbal model in symbols. See Objectives 5 and 6.

37. A varies directly as the square of p.

38. t varies directly as s.

39. z varies inversely as the cube of t.

40. v varies inversely as the square of r.

Express each verbal model in symbols. See Objectives 7 and 8.

41. C varies jointly as x, y, and z.

42. d varies jointly as r and t.

43. P varies directly as the square of a and inversely as the cube of j.

44. M varies inversely as the cube of n and jointly as x and the square of z.

Express each variation model in words. In each equation,
k is the constant of variation. **See Objectives 5 and 6.**

45. $r = kt$

46. $A = kr^3$

47. $b = \dfrac{k}{h}$

48. $d = \dfrac{k}{W^4}$

Express each variation model in words. In each equation,
k is the constant of variation. **See Objectives 7 and 8.**

49. $U = krs^2t$

50. $L = kmn$

51. $P = \dfrac{km}{n}$

52. $R = \dfrac{kL}{d^2}$

TRY IT YOURSELF

Solve each proportion.

53. $\dfrac{b + 4}{5} = \dfrac{3b - 6}{3}$

54. $\dfrac{2y + 6}{3} = \dfrac{4y - 16}{5}$

55. $\dfrac{5}{b + 3} = \dfrac{b}{2}$

56. $\dfrac{p + 2}{p + 5} = \dfrac{p - 3}{p - 2}$

57. $\dfrac{9z + 6}{z^2 + 3z} = \dfrac{7}{z + 3}$

58. $\dfrac{3}{n^2 + 3n} = \dfrac{2}{n^2 + 4n + 3}$

59. $\dfrac{h^2}{5} = \dfrac{h}{2h - 9}$

60. $\dfrac{b^2}{5} = \dfrac{b}{6b - 13}$

61. $\dfrac{x}{x + 2} = \dfrac{6}{x + 2}$

62. $\dfrac{a}{a - 3} = \dfrac{5}{a - 3}$

63. $\dfrac{t^2 - 1}{5} = \dfrac{1 - t^2}{2t}$

64. $\dfrac{n^2}{6} = \dfrac{n}{n - 1}$

65. $\dfrac{2.5x + 1}{2} = \dfrac{4.5}{12}$

66. $\dfrac{2}{5} = \dfrac{1.5x - 2}{0.25}$

67. $\dfrac{t}{10} = \dfrac{10}{t}$

68. $\dfrac{-6}{m} = \dfrac{m}{-6}$

APPLICATIONS

Use a proportion to solve each problem.

69. **Caffeine.**   Many convenience stores sell super-size 44-ounce soft drinks in refillable cups. For each of the products listed in the table below, find the amount of caffeine contained in one of the large cups. Round to the nearest milligram.

Soft drink	Milligrams of caffeine in 12 ounces	Milligrams of caffeine in 44 ounces
Mountain Dew	55	?
Pepsi	38	?
Coca-Cola Classic	34	?

70. **Telephones.**   In 2009, the number of mobile telephone lines used by residents of the United Arab Emirates reached a record high of 191 lines per 100 people. If the Emirates' population was about 4,599,000 at that time, how many mobile telephone lines were being used? (Source: TheNationalBeta and World Bank)

71. **Wallpapering.**   Read the instructions on the label of wallpaper adhesive. Estimate the amount of adhesive needed to paper 500 square feet of kitchen walls if a heavy wallpaper will be used.

> COVERAGE: One-half gallon will hang approximately 4 single rolls (140 sq ft), depending on the weight of the wall covering and the condition of the wall.

72. **Recommended Dosages.**   The recommended child's dose of the sedative hydroxine is 0.006 gram per kilogram of body mass. Find the dosage for a 30-kg child in grams and in milligrams.

73. **Ergonomics.**   The science of ergonomics coordinates the design of working conditions with the requirements of the worker. The illustration gives guidelines for the dimensions (in inches) of a computer workstation to be used by a person whose height is 69 inches. Find a set of workstation dimensions for a person 5 feet 11 inches tall. Round to the nearest tenth.

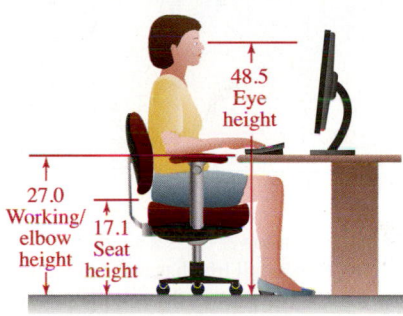

74. **Shopping.**   A recipe for guacamole dip calls for 5 avocados. If they are advertised at 3 for $1.98, what will 5 avocados cost?

75. **Drawing.**   See the illustration. To make an enlargement of the sailboat, an artist drew a grid over the smaller picture and transferred the contents of each small box to its corresponding larger box on another sheet of paper. If the smaller picture is 3 in. × 5 in. and if the width of the enlargement is 7.5 in., what is the length of the enlargement?

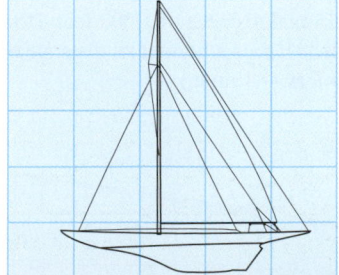

76. Drafting. In a scale drawing, a 280-foot antenna tower is drawn $7\frac{1}{2}$ inches high. The building next to it is drawn $2\frac{1}{4}$ inches high. How tall is the actual building?

Use similar triangles to solve each problem.

77. Flagpoles. A man places a mirror on the ground and sees the reflection of the top of a flagpole, as in the illustration. The two triangles in the illustration are similar. Find the height h of the flagpole.

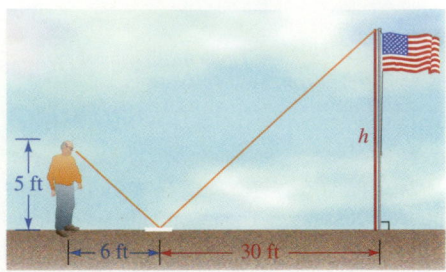

78. Washington, D.C. The Washington Monument casts a shadow of $166\frac{1}{2}$ feet at the same time as a 5-foot-tall tourist casts a shadow of $1\frac{1}{2}$ feet. Find the height of the monument.

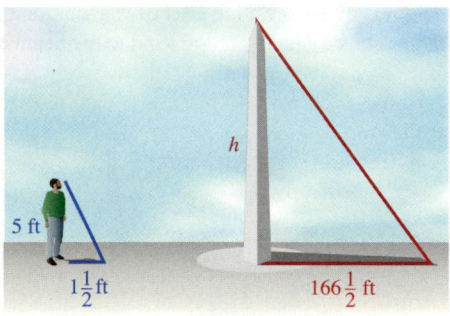

79. Towers. A cell phone tower casts a shadow of 75 feet at the same time that a 6-foot-tall tree has a shadow that is 3 feet 9 inches long. How tall is the cell phone tower?

80. Width of a River. Use the dimensions in the illustration to find w, the width of the river. The two triangles in the illustration are similar.

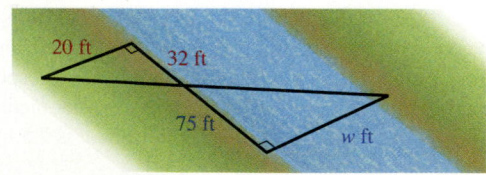

81. Flight Paths. An airplane ascends 150 feet as it flies a horizontal distance of 1,000 feet. How much altitude will it gain as it flies a horizontal distance of 1 mile? (*Hint:* 5,280 feet = 1 mile.)

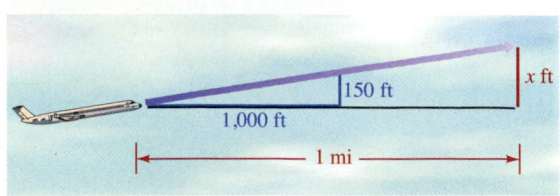

82. Ski Runs. A ski course with $\frac{1}{2}$ mile of horizontal run falls 100 feet in every 300 feet of run. Find the height of the hill.

83. ▶ from Campus to Careers

Webmaster

The language of variation is often used to describe various aspects of the Internet and websites. Determine whether each statement, generally speaking, is true or false.

a. The dollar amount of sales that an Internet website receives is inversely proportional to the amount of Internet traffic that visits the website.

b. The download time of an Internet website varies directly with the bandwidth being used.

c. Search engines like Google place a value on a website that is directly proportional to the number of sites that link to it.

84. Graphic Arts. The compass in the illustration is used to draw circles with different radii (plural for radius). For the setting shown, what radius will the resulting circle have?

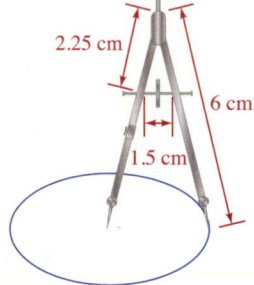

Solve each problem by writing a variation model.

85. Gravity. The force of gravity acting on an object varies directly as the mass of the object. The force on a mass of 5 kilograms is 49 newtons. What is the force acting on a mass of 12 kilograms?

86. Free Fall. An object in free fall travels a distance s that is directly proportional to the square of the time t. If an object falls 1,024 feet in 8 seconds, how far will it fall in 10 seconds?

87. Finding Distance. The distance that a car can go varies directly as the number of gallons of gasoline it consumes. If a car can go 288 miles on 12 gallons of gasoline, how far can it go on a full tank of 18 gallons?

88. Braking. Suppose the distance that a vehicle travels after its brakes have been applied varies directly as the square of the speed at which it was traveling. If the stopping distance for such a vehicle going 20 mph is 24 feet, what is the stopping distance for the vehicle traveling at 50 mph?

89. Farming. The number of days that a given number of bushels of corn will last when feeding cattle varies inversely as the number of animals. If x bushels will feed 25 cows for 10 days, how long will the feed last for 10 cows?

90. Organ Pipes. The frequency of vibration of air in an organ pipe is inversely proportional to the length of the pipe. If a pipe 2 feet long vibrates 256 times per second, how many times per second will a 6-foot pipe vibrate?

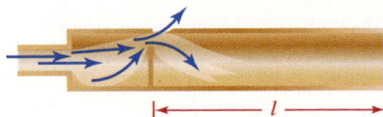

91. Gas Pressure. Under constant temperature, the volume occupied by a gas varies inversely to the pressure applied. If the gas occupies a volume of 20 cubic inches under a pressure of 6 pounds per square inch, find the volume when the gas is subjected to a pressure of 10 pounds per square inch.

92. Real Estate. The following table shows the listing price for three homes in the same general locality. Write the variation model (direct or inverse) that describes the relationship between the listing price and the number of square feet of a house in this area.

Number of square feet	Listing price
1,720	$180,600
1,205	$126,525
1,080	$113,400

93. Trucking Costs. The costs of a trucking company vary jointly as the number of trucks in service and the number of hours they are used. When 4 trucks are used for 6 hours each, the costs are $1,800. Find the costs of using 10 trucks, each for 12 hours.

94. Oil Storage. The number of gallons of oil that can be stored in a cylindrical tank varies jointly as the height of the tank and the square of the radius of its base. The constant of proportionality is 23.5. Find the number of gallons that can be stored in the cylindrical tank shown.

95. Electronics. The voltage (in volts) measured across a resistor is directly proportional to the current (in amperes) flowing through the resistor. The constant of variation is the **resistance** (in ohms). If 6 volts is measured across a resistor carrying a current of 2 amperes, find the resistance.

96. Electronics. The power (in watts) lost in a resistor (in the form of heat) varies directly as the square of the current (in amperes) passing through it. The constant of proportionality is the resistance (in ohms). What power is lost in a 5-ohm resistor carrying a 3-ampere current?

97. Structural Engineering. The deflection of a beam is inversely proportional to its width and the cube of its depth. If the deflection of a 4-inch-wide by 4-inch-deep beam is 1.1 inches, find the deflection of a 2-inch-wide by 8-inch-deep beam positioned as in figure (a) below.

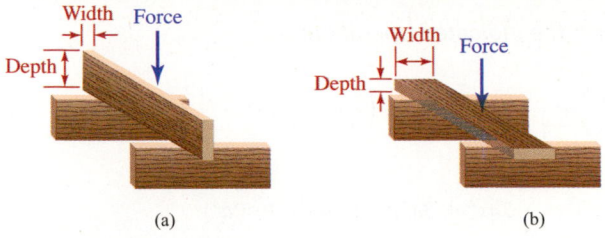

(a)   (b)

98. Structural Engineering. Find the deflection of the beam in Exercise 97 when the beam is positioned as in figure (b) above.

99. Electronics. The resistance of a wire is directly proportional to the length of the wire and inversely proportional to the square of the diameter of the wire. If the resistance is 11.2 ohms in a 80-foot-long wire with diameter 0.01 inch, what is the resistance in a 160-foot-long wire with diameter 0.04 inch?

100. Business Models. A businessman who sells widgets has found that the revenue from their sale varies directly as the advertising budget and inversely as the price. When $105,000 was spent on advertising and the widgets were priced at $19.95, the revenue from their sale was $200,000. How many widgets would he expect to sell at $17.50 each if $700,000 was spent on advertising?

101. Tension in a String. When playing with a Skip It toy, a child swings a weighted ball on the end of a string in a circular motion around one leg while jumping over the revolving string with the other leg. See the illustration. The tension T in the string is directly proportional to the square of the speed s of the ball and inversely proportional to the radius r of the circle. If the tension in the string is 6 pounds when the speed of the ball is 6 feet per second and the radius is 3 feet, find the tension when the speed is 8 feet per second and the radius is 2.5 feet.

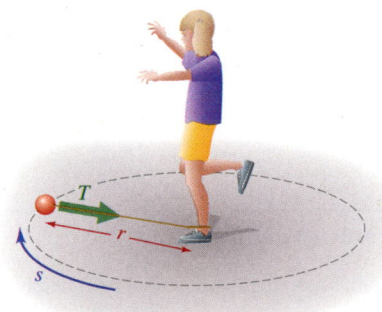

102. Gas Pressure. The pressure of a certain amount of gas is directly proportional to the temperature (measured on the Kelvin scale) and inversely proportional to the volume. A sample of gas at a pressure of 1 atmosphere occupies a volume of 1 cubic meter at a temperature of 273 Kelvin. When heated, the gas expands to twice its volume, but the pressure remains constant. To what temperature is it heated?

WRITING

103. Explain the difference between a *ratio* and a *proportion*.

104. Give examples of two quantities from everyday life that vary directly and two quantities that vary inversely.

REVIEW

Perform the indicated operations.

105. $\left(\dfrac{5}{2}w^3 + \dfrac{1}{4}w^2 + \dfrac{3}{5}\right) - \left(\dfrac{1}{3}w^3 + \dfrac{1}{2}w^2 - \dfrac{1}{5}\right)$

106. $(6a^2x^3 - 2ax^2 + 3a^3) + (-4a^2x^3 - 2a^3)$

107. $(3y + 1)(2y^2 + 3y + 2)$

108. $(5k - 6m^2)^2$

CHALLENGE PROBLEMS

109. As the cost of a purchase that is less than \$5 increases, the amount of change received from a five-dollar bill decreases. Is this inverse variation? Explain.

110. You've probably heard of Murphy's first law:

> If anything can go wrong, it will.

Another of Murphy's laws is:

> The chances of a piece of bread falling with the grape-jelly side down varies directly with the cost of the carpet.

Write one of your own witty sayings using the phrase *varies directly*.

6 ▶ Summary & Review

SECTION 6.1 ▶ Rational Functions and Simplifying Rational Expressions

DEFINITIONS AND CONCEPTS	EXAMPLES
A **rational expression** is an expression of the form $\dfrac{A}{B}$, where A and B are polynomials and B does not equal 0.	Rational expressions: $\dfrac{3x^2}{xy}$, $\dfrac{5b - 15}{b^2 - 25}$, and $\dfrac{a + 2}{a^2 - 3a - 4}$
A **rational function** is a function whose equation is defined by a rational expression in one variable.	Rational functions: $f(x) = \dfrac{6x}{x - 2}$ and $f(n) = \dfrac{n + 3}{n^3 + 2n - 9}$
Since division by 0 is undefined, any values that make the denominator 0 in a rational function must be excluded from the **domain** of the function. When finding the domain of a rational function, we *don't need to examine the numerator* of the expression; it can be any value, including 0.	Find the domain of the rational function: $f(x) = \dfrac{x + 3}{x^2 - 4}$ $x^2 - 4 = 0$ **Set the denominator equal to 0.** $(x + 2)(x - 2) = 0$ **Factor the difference of two squares.** $x + 2 = 0$ or $x - 2 = 0$ **Set each factor equal to 0.** $x = -2$ | $x = 2$ **Solve each equation.** The domain of the function is the set of all real numbers except -2 and 2. In interval notation, the domain is $(-\infty, -2) \cup (-2, 2) \cup (2, \infty)$.
To simplify a rational expression: **1.** Factor the numerator and denominator completely. **2.** Remove factors equal to 1 by replacing each pair of factors common to the numerator and denominator with the equivalent fraction $\dfrac{1}{1}$. **3.** Multiply the remaining factors in the numerator and in the denominator.	Simplify: $\dfrac{x^2 - 4}{2x + 4} = \dfrac{\overset{1}{\cancel{(x + 2)}}(x - 2)}{2\underset{1}{\cancel{(x + 2)}}} = \dfrac{x - 2}{2}$ Simplify: $\dfrac{2a^3 - 5a^2 - 12a}{2a^3 - 11a^2 + 12a} = \dfrac{\overset{1}{\cancel{a}}(2a + 3)\overset{1}{\cancel{(a - 4)}}}{\underset{1}{\cancel{a}}(2a - 3)\underset{1}{\cancel{(a - 4)}}} = \dfrac{2a + 3}{2a - 3}$

The quotient of any nonzero expression and its **opposite** is -1.	$\dfrac{7x - 6}{6 - 7x} = -1$ Because $7x - 6$ and $6 - 7x$ are opposites. Simplify: $\dfrac{10 - 2b}{b^2 - 5b} = \dfrac{2(5 - b)}{b(b - 5)} = \dfrac{\overset{-1}{2(5 - b)}}{b\underset{1}{(b - 5)}} = -\dfrac{2}{b}$

REVIEW EXERCISES

1. Use the graph of function f to find each of the following:

   a. $f(2)$

   b. $f(-1)$

   c. The value of x for which $f(x) = -1$

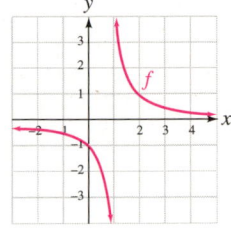

2. Find the domain of the rational function $f(x) = \dfrac{2x^2 + 8x}{x^2 + 2x - 24}$. Express your answer in words and using interval notation.

3. **Injections.** A hospital patient was given an injection for pain. The rational function

$$n(t) = \dfrac{28t}{t^2 + 1}$$

   gives the number of milligrams n of pain medication per liter in the patient's bloodstream, where t is the number of hours since the injection. Find $n(3)$ and explain its meaning.

4. Use a graphing calculator to graph the rational function $f(x) = \dfrac{3x + 2}{x}$. From the graph, determine the equations of the horizontal and vertical asymptotes and the domain and range.

Simplify each rational expression, if possible.

5. $\dfrac{48x^2 y}{76xy^8}$

6. $\dfrac{x^2 - 49}{x^2 + 14x + 49}$

7. $\dfrac{x^2 - 2x + 4}{2x^5 + 16x^2}$

8. $\dfrac{x^2 + 6x + 36}{x^7 - 216x^4}$

9. $\dfrac{5ac - 5ad + 5bc - 5bd}{5d^2 - 5c^2}$

10. $\dfrac{m^3 + m^2 n - 2mn^2}{2m^3 - mn^2 - m^2 n}$

11. $\dfrac{6x^2 - 5x - 4}{9x^2 - 24x + 16}$

12. $\dfrac{2m - 2n}{n - m}$

13. $\dfrac{s^2 + t^2}{s - t}$

14. $\dfrac{3m^2 - 10m + 8}{6 - m - m^2}$

SECTION 6.2 ▶ Multiplying and Dividing Rational Expressions

DEFINITIONS AND CONCEPTS	EXAMPLES
To **multiply rational expressions,** multiply the numerators and multiply the denominators. $\dfrac{A}{B} \cdot \dfrac{C}{D} = \dfrac{AC}{BD}$ Then simplify, if possible.	Multiply, and then simplify, if possible. $\dfrac{8z^2}{y^3} \cdot \dfrac{y}{4z} = \dfrac{8z^2 \cdot y}{y^3 \cdot 4z} = \dfrac{2 \cdot \overset{1}{4} \cdot \overset{1}{z} \cdot z \cdot \overset{1}{y}}{\underset{1}{y} \cdot y \cdot y \cdot \underset{1}{4} \cdot \underset{1}{z}} = \dfrac{2z}{y^2}$ Don't forget to simplify the result. Multiply, and then simplify, if possible. $\dfrac{x^2 - 4}{x + 3} \cdot \dfrac{3x + 9}{x + 2} = \dfrac{(x^2 - 4)(3x + 9)}{(x + 3)(x + 2)}$ Multiply the numerators. Multiply the denominators. $= \dfrac{\overset{1}{(x + 2)}(x - 2) \cdot 3 \cdot \overset{1}{(x + 3)}}{\underset{1}{(x + 3)}\underset{1}{(x + 2)}}$ Factor completely and then simplify. $= 3(x - 2)$ Multiply the remaining factors in the numerator. Multiply the remaining factors in the denominator.

To **divide rational expressions,** multiply the first by the reciprocal of the second.

$$\frac{A}{B} \div \frac{C}{D} = \frac{A}{B} \cdot \frac{D}{C} = \frac{AD}{BC}$$

Then simplify, if possible.

Divide, and then simplify, if possible.

$$\frac{x^2 + 4x + 3}{x^2 + 3x} \div \frac{3}{x} = \frac{x^2 + 4x + 3}{x^2 + 3x} \cdot \frac{x}{3}$$

Multiply the first rational expression by the reciprocal of the second.

$$= \frac{(x^2 + 4x + 3) \cdot x}{(x^2 + 3x) \cdot 3}$$

Multiply the numerators. Multiply the denominators.

$$= \frac{(x + 1)(\overset{1}{\cancel{x + 3}}) \cdot \overset{1}{\cancel{x}}}{\underset{1}{\cancel{x}}(\underset{1}{\cancel{x + 3}}) \cdot 3}$$

Factor completely and then simplify.

$$= \frac{x + 1}{3}$$

Multiply the remaining factors in the numerator. Multiply the remaining factors in the denominator.

REVIEW EXERCISES

Perform the operations and simplify when possible.

15. $\dfrac{3x^3 y^4}{35} \cdot \dfrac{10}{21x^5 y^4}$

16. Let $f(x) = \dfrac{x^3 + 4x^2 + 4x}{x^2 - x - 6}$ and $g(x) = \dfrac{9 - x^2}{x^2 + 5x + 6}$. Find $f(x) \cdot g(x)$.

17. $\dfrac{2a^2 - 5a - 3}{4a^3 - 36a} \div \dfrac{2a^2 + 5a + 2}{2a^2 + 5a - 3}$

18. $\dfrac{t^4 - 4t^2}{t} \div (t^3 + 2t^2)$

19. $\left(\dfrac{h - 2}{h^3 + 4} \right)^2$

20. $\dfrac{m^2 + 3m + 9}{m^2 + 4m + mr + 4r} \div \dfrac{m^3 - 27}{am + ar + 6m + 6r}$

21. $\dfrac{8m^2 + 6mn - 9n^2}{2m^2 + 5mn + 3n^2} \cdot \dfrac{6m^2 + 5mn - 4n^2}{12m^2 + 7mn - 12n^2}$

22. $\dfrac{x^3 + 3x^2 + 2x}{2x^2 - 2x - 12} \div \dfrac{x^3 - 3x^2 - 4x}{3x^2 - 3x} \cdot \dfrac{2x^2 - 4x - 16}{x^2 + 3x + 2}$

SECTION 6.3 ▶ Adding and Subtracting Rational Expressions

DEFINITIONS AND CONCEPTS	EXAMPLES

To **add (or subtract) two rational expressions with like denominators,** add (or subtract) the numerators and keep the common denominator.

$$\frac{A}{D} + \frac{B}{D} = \frac{A + B}{D}$$

$$\frac{A}{D} - \frac{B}{D} = \frac{A - B}{D}$$

Add:

$$\frac{x^2 - 26}{x - 5} + \frac{1}{x - 5} = \frac{x^2 - 26 + 1}{x - 5}$$

Add the numerators. Write the sum over the common denominator, x − 5.

$$= \frac{x^2 - 25}{x - 5}$$

Combine like terms.

$$= \frac{(x + 5)(\overset{1}{\cancel{x - 5}})}{\underset{1}{\cancel{x - 5}}}$$

Factor the numerator and simplify.

$$= x + 5$$

To find the **LCD** of several rational expressions, factor each denominator and use each factor the greatest number of times that it appears in any one denominator. The product of these factors is the LCD.

The denominators of two rational expressions are given. Find their LCD:

$$\left.\begin{array}{l} x^2 - 8x + 16 = (x - 4)(x - 4) \\ 9x - 36 = 3 \cdot 3 \cdot (x - 4) \end{array}\right\} \begin{array}{l} \text{LCD} = 3 \cdot 3 \cdot (x - 4)(x - 4) \\ = 9(x - 4)^2 \end{array}$$

To **add or subtract rational expressions with unlike denominators,** find the LCD and express each rational expression with a denominator that is the LCD. Add (or subtract) the resulting fractions and simplify the result, if possible. To **build a rational expression,** multiply it by 1 in the form of $\frac{c}{c}$, where c is any nonzero number or expression.	Subtract: $\dfrac{2x}{x+5} - \dfrac{1}{x} = \dfrac{2x}{x+5} \cdot \dfrac{x}{x} - \dfrac{1}{x} \cdot \dfrac{x+5}{x+5}$ Build each rational expression to have the LCD of x(x + 5). $= \dfrac{2x^2}{x(x+5)} - \dfrac{x+5}{x(x+5)}$ Multiply the numerators. Multiply the denominators. $= \dfrac{2x^2 - (x+5)}{x(x+5)}$ Subtract the numerators. Write the difference over the common denominator. Don't forget the parentheses. $= \dfrac{2x^2 - x - 5}{x(x+5)}$ The result does not simplify.

REVIEW EXERCISES

Perform the operations and simplify when possible.

23. $\dfrac{5y}{x-y} - \dfrac{3}{x-y}$

24. $\dfrac{d^2}{c^3 - d^3} + \dfrac{c^2 + cd}{c^3 - d^3}$

25. $\dfrac{4}{t-3} + \dfrac{6}{3-t}$

26. $\dfrac{p+3}{p^2 + 13p + 12} - \dfrac{2p+4}{p^2 + 13p + 12}$

The denominators of some rational expressions are given. Find the LCD.

27. $15a^2h,\ 20ah^3$

28. $ab^2 - ab,\ ab^2,\ b^2 - b$

29. $x^2 - 4x - 5,\ x^2 - 25$

30. $m^2 - 4m + 4,\ m^3 - 8$

Perform the operations and simplify when possible.

31. $9 - \dfrac{1}{a+1}$

32. $\dfrac{5x}{14z^2} + \dfrac{y^2}{16z}$

33. $\dfrac{4x}{x-4} - \dfrac{3}{x+3}$

34. $\dfrac{2a+4}{3} - \dfrac{9}{a+2}$

35. $\dfrac{6}{a^2 - 9} - \dfrac{5}{a^2 - a - 6}$

36. $\dfrac{4}{3xy - 6y} - \dfrac{4}{10 - 5x}$

37. $\dfrac{a}{a+2} - \dfrac{3}{a^2 + 2a} + \dfrac{1}{2a+4}$

38. Let $n(x) = \dfrac{x+7}{x+3}$ and $s(x) = \dfrac{x-3}{x+7}$. Find $n(x) - s(x)$.

SECTION 6.4 ▶ **Simplifying Complex Fractions**

DEFINITIONS AND CONCEPTS	EXAMPLES
Complex fractions contain fractions in their numerators and/or their denominators.	Complex fractions: $\dfrac{\frac{2}{t}}{\frac{5}{4t}}$, $\dfrac{\frac{3}{m} + \frac{m}{4}}{\frac{m}{2}}$, and $\dfrac{a^2 - b^2}{\frac{1}{a} + \frac{1}{b}}$

Two methods are used to simplify **complex fractions.**

Method 1: Write the numerator and denominator as single fractions. Then divide the fractions and simplify.

This method works well when a complex fraction is written, or can be easily written, as a quotient of two single rational expressions.

Simplify:

$$\frac{\dfrac{4x^2}{y^3}}{\dfrac{14x}{y}} = \frac{4x^2}{y^3} \div \frac{14x}{y}$$

The main fraction bar of the complex fraction indicates division.

$$= \frac{4x^2}{y^3} \cdot \frac{y}{14x}$$

To divide rational expressions, multiply the first by the reciprocal of the second.

$$= \frac{4x^2 \cdot y}{y^3 \cdot 14x}$$

Multiply the numerators.
Multiply the denominators.

$$= \frac{\overset{1}{2} \cdot \overset{1}{2} \cdot \overset{1}{\cancel{x}} \cdot x \cdot \overset{1}{\cancel{y}}}{\underset{1}{\cancel{y}} \cdot y \cdot y \cdot \underset{1}{2} \cdot 7 \cdot \underset{1}{\cancel{x}}}$$

Factor the numerator and denominator. Then simplify by removing common factors of the numerator and denominator.

$$= \frac{2x}{7y^2}$$

Multiply the remaining factors in the numerator. Multiply the remaining factors in the denominator.

Method 2: Determine the LCD of all the rational expressions in the complex fraction and multiply the complex fraction by 1, written in the form $\frac{\text{LCD}}{\text{LCD}}$.

This method works well when the complex fraction has sums and/or differences in the numerator or denominator.

Simplify:

$$\frac{\dfrac{1}{x} - y}{\dfrac{5}{2x}} = \frac{\dfrac{1}{x} - y}{\dfrac{5}{2x}} \cdot \frac{2x}{2x}$$

The LCD of all the rational expressions in the complex fraction is 2x. Multiply the complex fraction by 1 in the form $\frac{2x}{2x}$.

$$= \frac{\left(\dfrac{1}{x} - y\right)2x}{\left(\dfrac{5}{2x}\right)2x}$$

Multiply the numerators.
Multiply the denominators.

$$= \frac{\dfrac{1}{x} \cdot 2x - y \cdot 2x}{5}$$

In the numerator, distribute the multiplication by 2x. In the denominator, perform the multiplication by 2x.

$$= \frac{2 - 2xy}{5}$$

Perform each multiplication by 2x.

REVIEW EXERCISES

Simplify each complex fraction.

39. $\dfrac{\dfrac{4a^3b^2}{9c}}{\dfrac{14a^3b}{9c^4}}$

40. $\dfrac{\dfrac{p^2 - 9}{6pt}}{\dfrac{p^2 + 5p + 6}{3pt}}$

43. $\dfrac{(x-y)^{-2}}{x^{-2} - y^{-2}}$

44. $\dfrac{1 + \dfrac{1}{b+d}}{\dfrac{1}{b+d} - 1}$

41. $\dfrac{\dfrac{1}{a} + \dfrac{2}{b}}{\dfrac{2}{a} - \dfrac{1}{b}}$

42. $\dfrac{1 - \dfrac{1}{x} - \dfrac{2}{x^2}}{1 + \dfrac{4}{x} + \dfrac{3}{x^2}}$

45. $\dfrac{\dfrac{2b}{b-1} - \dfrac{3}{b}}{\dfrac{1}{b-1} + \dfrac{2}{b}}$

46. $\dfrac{\dfrac{8}{r+3}}{\dfrac{4}{r-2} - \dfrac{2}{r^2 + r - 6}}$

SECTION 6.5 ▶ **Dividing Polynomials**

DEFINITIONS AND CONCEPTS	EXAMPLES
To divide monomials, use the method for simplifying fractions or use the rules for exponents.	Divide the monomials: $$\frac{8p^2q}{20pq^3} = \frac{2 \cdot \overset{1}{\cancel{4}} \cdot \overset{1}{\cancel{p}} \cdot p \cdot \overset{1}{\cancel{q}}}{\underset{1}{\cancel{4}} \cdot 5 \cdot \underset{1}{\cancel{p}} \cdot \underset{1}{\cancel{q}} \cdot q \cdot q} \qquad \frac{8p^2q}{20pq^3} = \frac{2}{5}p^{2-1}q^{1-3}$ Keep each base and subtract the exponents.$$ $$= \frac{2p}{5q^2} \qquad\qquad\qquad = \frac{2p^1q^{-2}}{5}$ Do the subtraction.$$ $$= \frac{2p}{5q^2}$ $q^{-2} = \frac{1}{q^2}$$$
To divide a polynomial by a monomial, divide each term of the numerator by the denominator. $$\frac{A + B}{D} = \frac{A}{D} + \frac{B}{D}$$	Divide: $\dfrac{9c^9d^4 - 12c^3d^7}{27cd^5} = \dfrac{9c^9d^4}{27cd^5} - \dfrac{12c^3d^7}{27cd^5}$ $$= \frac{c^8}{3d} - \frac{4c^2d^2}{9}$ Do each monomial division.$$
Long division can be used to **divide a polynomial by a polynomial** (other than a monomial). The long division method is a series of four steps that are repeated: Divide, multiply, subtract, and bring down the next term. Don't forget to write the parentheses before performing the subtraction step. When the division has a remainder, write the answer in the form Quotient $+ \frac{\text{remainder}}{\text{divisor}}$.	Divide: $6x^3 - x^2 + 6x + 5$ by $2x + 1$ $$\begin{array}{r} 3x^2 - 2x + 4 \\ 2x+1{\overline{\smash{\big)}\,6x^3 - x^2 + 6x + 5}} \\ -(6x^3 + 3x^2) \\ -4x^2 + 6x \\ -(-4x^2 - 2x) \\ 8x + 5 \\ -(8x + 4) \\ 1 \end{array}$$ The first division: $\frac{6x^3}{2x} = 3x^2$. The second division: $\frac{-4x^2}{2x} = -2x$. The third division: $\frac{8x}{2x} = 4$. The remainder is 1. Thus, $\dfrac{6x^3 - x^2 + 6x + 5}{2x + 1} = 3x^2 - 2x + 4 + \dfrac{1}{2x + 1}$.
We can check the answer to any polynomial long division using: Divisor · quotient + remainder = dividend	To check the result of the division above, we note that: $(2x + 1)(3x^2 - 2x + 4) + 1 = 6x^3 - x^2 + 6x + 5$
The long division method works best when the terms of the divisor and the dividend are written in **descending powers of the variable.** When the dividend has **missing terms,** insert such terms with a coefficient of 0, or leave a blank space.	To divide $\dfrac{5x + x^3 + 3 + 3x^2}{x + 1}$, write: $x + 1{\overline{\smash{\big)}\,x^3 + 3x^2 + 5x + 3}}$ To divide $\dfrac{x^2 - 9}{x - 3}$, write: $x - 3{\overline{\smash{\big)}\,x^2 + 0x - 9}}$

REVIEW EXERCISES

Perform each division. Write answers using positive exponents.

47. $\dfrac{25h^4k^7}{55hk^9}$

48. $(5x^3y^3z^{10}) \div (10x^3y^6z^{20})$

Perform each division.

49. $\dfrac{36a + 32}{6}$

50. $\dfrac{30x^3y^2 - 15x^2y - 10xy^2}{-10xy}$

51. $b + 5{\overline{\smash{\big)}\,b^2 + 9b + 20}}$

52. $\dfrac{-33v - 8v^2 + 3v^3 - 10}{1 + 3v}$

53. Let $f(x) = x^3 + 8$ and $g(x) = x + 2$. Find $f(x) \div g(x)$.

54. Divide $8m^2 - 18m - 9$ by $4m + 1$.

55. $(3a^3 - 2a^2 - 8) \div (a^2 + 5)$

56. $\dfrac{m^8 + m^6 - 4m^4 + 5m^2 - 1}{m^4 + 2m^2 - 3}$

DEFINITIONS AND CONCEPTS	EXAMPLES
Synthetic division is used to divide a polynomial by a binomial of the form $x - k$.	To use synthetic division to divide $\dfrac{5x^3 - x^2 + 4x - 3}{x + 1}$, we write the divisor in $x - k$ form as $x - (-1)$ to determine that $k = -1$ and proceed as follows: $$\begin{array}{r} \text{This represents} \rightarrow \quad -1\rfloor \quad 5 \quad -1 \quad 4 \quad -3 \\ \text{division by } x + 1. \qquad\qquad -5 \quad 6 \quad -10 \\ \hline 5 \quad -6 \quad 10 \quad -13 \leftarrow \text{This is the remainder.} \end{array}$$ Thus, $\dfrac{5x^3 - x^2 + 4x - 3}{x + 1} = 5x^2 - 6x + 10 + \dfrac{-13}{x + 1}$.
Remainder theorem: If a polynomial $P(x)$ is divided by $x - k$, the remainder is $P(k)$. It follows from the remainder theorem that a polynomial can be evaluated using synthetic division.	Since the remainder in the previous division is -13, the remainder theorem guarantees that $P(-1) = -13$. We can check this by evaluating the function at $x = -1$: $$P(x) = 5x^3 - x^2 + 4x - 3$$ $$P(-1) = 5(-1)^3 - (-1)^2 + 4(-1) - 3 \qquad \text{Substitute } -1 \text{ for } x.$$ $$= -5 - 1 - 4 - 3$$ $$= -13$$
Factor theorem: If $P(x)$ is divided by $x - k$, then $P(k) = 0$, if and only if $x - k$ is a factor of $P(x)$.	Determine whether $x - 3$ is a factor of $2x^3 - 8x^2 + 9x - 9$. We can use synthetic division to find the remainder when the polynomial is divided by $x - 3$. If the remainder is 0, $x - 3$ is a factor. If the remainder is not 0, $x - 3$ is not a factor. $$\begin{array}{r} 3\rfloor \quad 2 \quad -8 \quad 9 \quad -9 \\ 6 \quad -6 \quad 9 \\ \hline 2 \quad -2 \quad 3 \quad 0 \end{array}$$ Since the remainder is **0**, $x - 3$ is a factor of $2x^3 - 8x^2 + 9x - 9$.

REVIEW EXERCISES

Use synthetic division to perform each division.

57. $(x^2 - 13x + 42) \div (x - 6)$

58. Divide $m^3 - 6m^2 + 11m - 6$ by $m - 3$.

59. $\dfrac{-3n^5 + 10n^4 + 7n^3 + 2n^2 + 9n - 4}{n - 4}$

60. $\dfrac{4x^3 + 5x^2 - 1}{x + 2}$

61. Divide $3a - a^2 + 3a^3 + 3a^4 + 10$ by $a + 1$.

62. $\dfrac{x^4 + 1}{x - 3}$

63. Let $P(x) = x^4 - 2x^3 + x^2 - 3x + 12$. Use the remainder theorem and synthetic division to find $P(-2)$.

64. Let $P(x) = x^3 - 13x^2 - 27$. Use the remainder theorem and synthetic division to find $P(5)$.

Use the factor theorem to determine whether the first expression is a factor of P(x).

65. $x - 5$; $P(x) = x^3 - 3x^2 - 8x - 10$

66. $x + 5$; $P(x) = x^3 + 4x^2 - 5x + 5$
(*Hint:* Write $x + 5$ as $x - (-5)$.)

SECTION 6.7 ▶ **Solving Rational Equations**

DEFINITIONS AND CONCEPTS	EXAMPLES
If an equation contains one or more rational expressions, it is called a **rational equation.**	Rational equations: $\dfrac{4}{5} = \dfrac{x}{x-1}$ and $\dfrac{2a}{a^2 + 9a + 20} = \dfrac{2}{a+5} + \dfrac{3}{a+4}$

To **solve a rational equation, we clear it of fractions:**

1. Factor all denominators.

2. Determine which numbers cannot be solutions of the equation.

3. Multiply both sides of the equation by the LCD of all rational expressions in the equation.

4. Use the distributive property to remove parentheses, remove any factors equal to 1, and write the result in simplified form.

5. Solve the resulting equation.

6. Check all possible solutions in the original equation.

It is important that you know the difference between **simplifying the expression**

$$\frac{3}{2} + \frac{1}{a-4} - \frac{5}{2a-8}$$ Build

and **solving the equation**

$$\frac{3}{2} + \frac{1}{a-4} = \frac{5}{2a-8}$$ Clear

Solve: $\dfrac{3}{2} + \dfrac{1}{a-4} = \dfrac{5}{2a-8}$

If we factor the last denominator, the equation can be written as:

$$\frac{3}{2} + \frac{1}{a-4} = \frac{5}{2(a-4)}$$

We see that 4 cannot be a solution of the equation, because it makes at least one of the rational expressions in the equation undefined.

We can clear the equation of fractions by multiplying both sides by $2(a-4)$, which is the LCD of the three rational expressions in the equation.

$$2(a-4)\left(\frac{3}{2} + \frac{1}{a-4}\right) = 2(a-4)\left[\frac{5}{2(a-4)}\right]$$ Multiply both sides by the LCD.

$$2(a-4)\left(\frac{3}{2}\right) + 2(a-4)\left(\frac{1}{a-4}\right) = 2(a-4)\left[\frac{5}{2(a-4)}\right]$$ Distribute.

$$2(a-4)\left(\frac{3}{2}\right) + 2(a-4)\left(\frac{1}{a-4}\right) = 2(a-4)\left[\frac{5}{2(a-4)}\right]$$ Remove common factors.

$$(a-4)3 + 2 = 5$$ Simplify.

$$3a - 12 + 2 = 5$$ Distribute on the left.

$$3a - 10 = 5$$ Combine like terms.

$$3a = 15$$

$$a = 5$$

The solution is 5. Use a check to verify that it satisfies the original equation.

Multiplying both sides of an equation by a quantity that contains a variable can lead to **extraneous solutions**.

All possible solutions of a rational equation must be checked.

Solve: $\dfrac{x+6}{x-1} = 2 + \dfrac{7}{x-1}$

$$(x-1)\left(\frac{x+6}{x-1}\right) = (x-1)\left(2 + \frac{7}{x-1}\right)$$ Multiply both sides by the LCD, $x-1$.

$$(x-1)\left(\frac{x+6}{x-1}\right) = (x-1)(2) + (x-1)\left(\frac{7}{x-1}\right)$$ Distribute on the right.

$$(x-1)\left(\frac{x+6}{x-1}\right) = (x-1)(2) + (x-1)\left(\frac{7}{x-1}\right)$$ Remove common factors.

$$x + 6 = (x-1)(2) + 7$$ Simplify.

$$x + 6 = 2x - 2 + 7$$ Distribute on the right.

$$x + 6 = 2x + 5$$ Combine like terms.

$$1 = x$$ Solve for x.

Since 1 makes the denominator of a rational expression 0, it does not check. It is an extraneous solution. The equation has no solution.

REVIEW EXERCISES

Solve each equation, if possible.

67. $\dfrac{4}{x} - \dfrac{1}{10} = \dfrac{7}{2x}$

68. $\dfrac{11}{t} = \dfrac{6}{t-7}$

69. $\dfrac{3}{y} - \dfrac{2}{y+1} = \dfrac{1}{2}$

70. $\dfrac{2}{3x+15} - \dfrac{1}{18} = \dfrac{1}{3x+12}$

71. $\dfrac{3}{x+2} = \dfrac{1}{2-x} + \dfrac{2}{x^2-4}$

72. $\dfrac{x+3}{x-5} + \dfrac{2x^2+6}{x^2-7x+10} = \dfrac{3x}{x-2}$

73. $\dfrac{5a}{a-3} - 7 = \dfrac{15}{a-3}$

74. a. Simplify: $\dfrac{10}{x^2-4x} - \dfrac{4}{x} + \dfrac{5}{x-4}$

 b. Solve: $\dfrac{10}{x^2-4x} - \dfrac{4}{x} = \dfrac{5}{x-4}$

Solve each formula for the indicated variable or expression.

75. $H = \dfrac{2ab}{a+b}$ for b

76. $\dfrac{x^2}{a^2} - \dfrac{y^2}{b^2} = 1$ for y^2

77. $\dfrac{1}{R} = \dfrac{1}{R_1} + \dfrac{1}{R_2}$ for R

78. $k = \dfrac{ma}{F}$ for F

SECTION 6.8 ▶ Problem Solving Using Rational Equations

DEFINITIONS AND CONCEPTS	EXAMPLES

Rate of Work: If a job can be completed in x units of time, the rate of work can be expressed as $\frac{1}{x}$ of the job is completed per unit of time.

Shared-work problems:

 Work completed = rate of work · time worked

Printers. Working alone, a 300-A model printer can print a company's payroll checks in 30 minutes. A 500-X model can print the same checks in 20 minutes. How long will it take if the printers work together to print the checks?

Analyze This is a shared-work problem.

Assign Let x = the number of minutes it will take the printers, working together, to print the checks.

Form Enter the data in a table.

	Rate ·	Time	=	Work Completed
Model 300-A	$\frac{1}{30}$	x		$\frac{x}{30}$
Model 500-X	$\frac{1}{20}$	x		$\frac{x}{20}$

The part of the job done by the 300-A model plus the part of the job done by the 500-X model equals 1 job completed.

$$\dfrac{x}{30} + \dfrac{x}{20} = 1$$

Solve

$$60\left(\dfrac{x}{30} + \dfrac{x}{20}\right) = 60(1) \qquad \text{Multiply both sides by the LCD, 60.}$$

$$60\left(\dfrac{x}{30}\right) + 60\left(\dfrac{x}{20}\right) = 60(1) \qquad \text{Distribute the multiplication by 60.}$$

$$2x + 3x = 60 \qquad \text{Perform each multiplication by 60.}$$

$$5x = 60 \qquad \text{Combine like terms.}$$

$$x = \dfrac{60}{5} = 12$$

State Working together, it will take the printers 12 minutes to print the checks.

Check In 12 minutes, the 300-A model will do $\frac{12}{30} = \frac{2}{5}$ of the job and the 500-X model will do $\frac{12}{20} = \frac{3}{5}$ of the job. Together they will do $\frac{2}{5} + \frac{3}{5} = 1$ whole job. The result checks.

To solve **uniform-motion problems,** the time traveled is given by the formula: Time $= \frac{\text{distance}}{\text{rate}}$. When a boat travels *downstream,* the boat's speed is increased by the current. When a boat travels *upstream,* the boat's speed is decreased by the current.	See pages 542–545 for two types of examples.

79. a. If a painter can complete a job in 10 hours, what is the painter's rate of work?

    b. If the painter works for x hours, how much of the job is completed?

80. Draining a Tank. If one outlet pipe can drain a tank in 24 hours and another pipe can drain the tank in 36 hours, how long will it take for both pipes to drain the tank?

81. Electricians. It takes an apprentice 5 days more than an experienced electrician to wire a 1,500-square foot house. Working together, they can wire such a house in 6 days. How long would it take each person working alone to wire the house?

82. Sinks. A faucet can fill a garage sink in 2 minutes. It takes 3 minutes for the drain to empty the sink when it is full. How long will it take to fill the sink if the drain is open and the faucet is on?

83. Advertising. A small plane pulling a banner can fly at a rate of 75 mph in calm air. Flying down the coast, with a tailwind, the plane flew 40 miles in the same time that it took to fly 35 miles up the coast, into a headwind. Find the rate of the wind.

84. Trip Length. Heavy traffic reduced a driver's usual speed by 10 mph, which lengthened her 200-mile trip by 1 hour. Find the driver's usual speed.

SECTION 6.9 ▶ **Proportion and Variation**

DEFINITIONS AND CONCEPTS	EXAMPLES
A **proportion** is a statement that two ratios are equal. In the proportion $\frac{a}{b} = \frac{c}{d}$, a and d are the **extremes** and b and c are the **means.**	Proportion: $\dfrac{4}{9} = \dfrac{28}{63}$ Extremes: 4 and 63 Means: 9 and 28
In any proportion, the product of the extremes is equal to the product of the means. (The **cross products** are equal.)	Proportion: $\dfrac{4}{9} \diagup\!\!\!\diagdown \dfrac{28}{63}$ Cross product: $9 \cdot 28 = 252$ Cross product: $4 \cdot 63 = 252$
To **solve a proportion,** set the product of the extremes equal to the product of the means and solve the resulting equation.	Solve the proportion: $\dfrac{x+2}{4} = \dfrac{6}{x}$ $(x+2)x = 4 \cdot 6$ In a proportion, the product of the extremes is equal to the product of the means. $x^2 + 2x = 24$ On the left side, distribute. $x^2 + 2x - 24 = 0$ Subtract 24 from both sides. $(x+6)(x-4) = 0$ Factor the trinomial. $x + 6 = 0$ or $x - 4 = 0$ Set each factor equal to 0. $x = -6$ $x = 4$ Solve each linear equation. The solutions are -6 and 4.

If two angles of one triangle have the same measure as two angles of a second triangle, the triangles are similar. The lengths of corresponding sides of **similar triangles** are proportional. Similar triangles have the same shape but not necessarily the same size.	In these similar triangles: $$\frac{a}{d} = \frac{b}{e} = \frac{c}{f}$$ 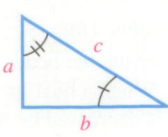 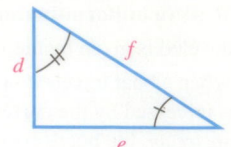 See page 552 for an example of how we can find the height of a tree and stay safely on the ground.
The words **y varies directly as x** or **y is directly proportional to x** mean that $y = kx$ for some nonzero constant k, called the **constant of variation**.	The distance d that a spring stretches *varies directly* as the force f attached to the spring: $d = kf$.
The words **y varies inversely as x** or **y is inversely proportional to x** mean that $$y = \frac{k}{x}$$ for some nonzero constant k.	If the voltage in an electric circuit is kept constant, the current I *varies inversely* as the resistance R: $I = \dfrac{k}{R}$.
Strategy for Solving Variation Problems 1. Translate the verbal model into an equation. 2. Substitute the first set of values into the equation from step 1 to determine the value of k. 3. Substitute the value of k into the equation from step 1. 4. Substitute the remaining set of values into the equation from step 3 and solve for the unknown.	Suppose d varies inversely as h. If $d = 5$ when $h = 4$, find d when $h = 10$. 1. The words d *varies inversely as h* translate to $d = \frac{k}{h}$. 2. If we substitute 5 for d and 4 for h in $d = \frac{k}{h}$, we have $$5 = \frac{k}{4}$$ $20 = k$ To isolate k, multiply both sides by 4. This is the constant of variation. 3. Since $k = 20$, the inverse variation equation is $d = \frac{20}{h}$. 4. To answer the final question, we substitute 10 for h in $d = \frac{20}{h}$. $$d = \frac{20}{10} = 2$$
Joint variation: One variable varies as the product of several variables. For example, $y = kxz$ (k is a constant). **Combined variation:** A combination of direct and inverse variation. For example, $$y = \frac{kx}{z}$$ (k is a constant)	The number of gallons g of oil that can be stored in a cylindrical tank *varies jointly* as the height h of the tank and the square of the radius r of its base: $g = khr^2$. The gravitational force F between two objects with masses m_1 and m_2 *varies directly* as the product of their masses and *inversely* as the square of the distance d between them: $F = \frac{km_1 m_2}{d^2}$.

REVIEW EXERCISES

Solve each proportion.

85. $\dfrac{x + 1}{8} = \dfrac{4x - 2}{24}$

86. $\dfrac{1}{x + 6} = \dfrac{x + 10}{12}$

87. $\dfrac{-r}{r + 2} = \dfrac{3r}{r - 2}$

88. $\dfrac{18}{t^2} = \dfrac{3t - 3}{t}$

89. **Similar Triangles.**   Find the height of a tree if it casts a 44-foot shadow when a 4-foot shrub casts a $2\frac{1}{2}$-foot shadow.

90. **Cooking.**   A recipe for spaghetti sauce requires four 16-ounce bottles of ketchup to make 2 gallons of sauce. How many bottles of ketchup are needed to make 10 gallons of sauce?

91. **Scale Models.**   A model of a playhouse was made using a 1/12th scale. If the scale model is 5.5 inches tall, how tall is the playhouse?

92. **Property Tax.**   The property tax in a certain county varies directly as assessed valuation. If a tax of $1,575 is charged on a single-family home assessed at $90,000, determine the property tax on an apartment complex assessed at $312,000.

93. **Elecricity.**   For a fixed voltage, the current in an electrical circuit varies inversely as the resistance in the circuit. If a certain circuit has a current of $2\frac{1}{2}$ amps when the resistance is 150 ohms, find the current in the circuit when the resistance is doubled.

94. Hurricane Winds. The wind force on a vertical surface varies jointly as the area of the surface and the square of the wind's velocity. If a 10-mph wind exerts a force of 1.98 pounds on the 3-foot-tall by 1.5-foot-wide sign shown above, find the force on the sign if the wind is blowing at 80 mph.

95. Does the graph show direct or inverse variation?

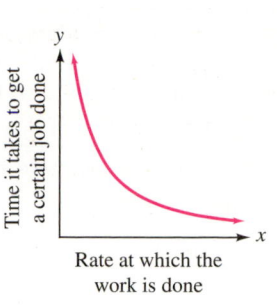

96. Assume that x_1 varies directly with the third power of t and inversely with x_2. Find the constant of variation if $x_1 = 1.6$ when $t = 8$ and $x_2 = 64$.

6 ▶ Chapter Test

1. Fill in the blanks.

a. A quotient of two polynomials, such as $\frac{x-8}{x^2-2x-3}$, is called a _____ expression.

b. The _____ of $\frac{x+1}{x-7}$ is $\frac{x-7}{x+1}$.

c. $\dfrac{\frac{x}{4}+\frac{1}{x}}{\frac{1}{8}+\frac{1}{x}}$ is an example of a _____ fraction.

d. When solving a rational equation, if we obtain a number that does not satisfy the original equation, the number is called an _____ solution.

e. An equation that states that two ratios are equal, such as $\frac{1}{2}=\frac{3}{6}$, is called a _____.

2. Explain the error that was made in the solution shown below.

$$\text{Simplify:}\quad \frac{2(x+2)+3(x-3)}{x+2}=\frac{\overset{1}{\cancel{2(x+2)}}+3(x-3)}{\underset{1}{\cancel{x+2}}}$$
$$=\frac{2+3x-9}{1}$$
$$=3x-7$$

Simplify each rational expression.

3. $\dfrac{12x^2y^3z^2}{18x^3y^6z^2}$

4. $\dfrac{2x+4}{x^2-4}$

5. $\dfrac{3y-6z}{2z-y}$

6. $\dfrac{2x^2+7xy+3y^2}{4xy+12y^2}$

7. Refer to the graph of function f on the right. Find

a. $f(5)$

b. $f(-1)$

c. $f(1)$

d. The value of x for which $f(x)=-2$

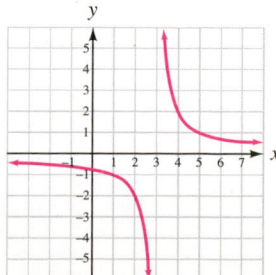

8. Wildlife. The number of prairie dogs p living in a given area of land is approximated by the rational function

$$p(t)=\frac{200t}{t+1}$$

where t is the time, in months, since the first of the year. Find the number of prairie dogs in that area by the end of July.

9. Are the expressions $\dfrac{x-4}{x^2+x-18}$ and $-\dfrac{x+4}{x^2+x-18}$ equivalent?

10. Find the domain of the rational function $f(x)=\dfrac{x^2+6x+5}{x-x^2}$. Express it using words and interval notation.

Perform the operations and simplify when necessary. Write all answers using positive exponents only.

11. $\dfrac{x^2}{x^3z^2y^2}\cdot\dfrac{x^2z^4}{y^2z}$

12. $\dfrac{a^2+12a+20}{a^2-4}\cdot\dfrac{a^2-12a+20}{a^2-100}$

13. $\dfrac{xu+2u+5x+10}{u^2+2u-15}\cdot\dfrac{13u-39}{x^2+3x+2}$

14. $\dfrac{x^3+y^3}{16x^2}\div\dfrac{x^2-xy+y^2}{8x^2+8xy}$

15. Let $f(a)=\dfrac{a^2+7a+12}{a+3}$ and $g(a)=\dfrac{16-a^2}{a-4}$. Find $f(a)\div g(a)$.

16. $\dfrac{(2x-3)^3}{x^2-2x+1}\div\dfrac{3x^2+7x+2}{3x^2-2x-1}\cdot\dfrac{x^2+x-2}{2x^7-3x^6}$

17. $\dfrac{-3t+4}{t^2+t-20}+\dfrac{6+5t}{t^2+t-20}$

18. $\dfrac{3wx}{wx-5}+\dfrac{wx+10}{5-wx}$

19. Let $f(x) = 8x - 5$ and $g(x) = \dfrac{5x + 4}{3x + 1}$.

Find $f(x) + g(x)$.

20. $\dfrac{a + 3}{a^2 - a - 2} - \dfrac{a - 4}{a^2 - 2a - 3}$

Simplify.

21. $\dfrac{\dfrac{2u^2w^3}{v^2}}{\dfrac{4uw^4}{uv}}$

22. $\dfrac{\dfrac{4}{3k} + \dfrac{k}{k + 1}}{\dfrac{k}{k + 1} - \dfrac{3}{k}}$

23. Divide: $\dfrac{18x^2y^3 - 12x^3y^2 + 9xy}{-3xy^4}$

24. Divide: $(y^3 - 48) \div (y + 2)$

25. Let $P(x) = 4x^3 + 3x^2 + 2x - 7$. Use synthetic division to find $P(2)$.

26. Use the factor theorem to decide whether $x + 3$ is a factor of $P(x) = x^4 + 3x^3 - 16x^2 - 27x + 63$.

Solve each equation.

27. $\dfrac{34}{x^2} + \dfrac{13}{20x} = \dfrac{3}{2x}$

28. $\dfrac{u - 2}{u - 3} + 3 = u + \dfrac{u - 4}{3 - u}$

29. $\dfrac{3}{x - 2} = \dfrac{x + 3}{2x}$

30. $\dfrac{4}{m^2 - 9} + \dfrac{5}{m^2 - m - 12} = \dfrac{7}{m^2 - 7m + 12}$

Solve each formula for the indicated variable.

31. $\dfrac{AB}{C} = \dfrac{DE}{F}$ for F

32. $\dfrac{1}{r} = \dfrac{1}{r_1} + \dfrac{1}{r_2}$ for r_2

33. Roofing.   One crew can finish a 2,800-square-foot roof in 12 hours, and another crew can do the job in 10 hours. If they work together, can they finish before a predicted rain in 5 hours? If not, how long will they have to work in the rain?

34. Hospitals.   It takes a technician 45 minutes longer than his supervisor to clean an outpatient surgery room. Working together, they can clean the room in 30 minutes. How long would it take each person working alone to clean the room?

35. Touring the Countryside.   A man bicycles 5 mph faster than he can walk. He bicycles a distance of 24 miles and then hikes back along the same route. If the entire trip takes 11 hours, how fast does he walk?

36. River Cruises.   A paddleboat can make an 8-mile trip up the Mississippi River and return in a total of 3 hours. If the boat travels 6 mph in still water, find the speed of the current.

37. Shadows.   Refer to the illustration below. Find the height of the tree.

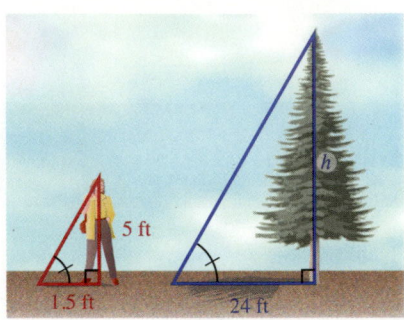

38. Anniversary Gifts.   A florist sells a dozen long-stemmed red roses for $57.99. In honor of their 16th wedding anniversary, a man wants to buy 16 roses for his wife. What will the roses cost?

39. Suppose r varies directly as x, and inversely as y and z. If $r = 1.5$ when $x = 1,000$, $y = 50$, and $z = 10$, find r when $x = 0.8$, $y = 0.02$, and $z = 0.1$.

40. Draw a possible graph showing that the weekly earnings of a person *varies directly* with the number of hours worked during the week. Label the axes.

41. Sound.   Sound intensity (loudness) varies inversely as the square of the distance from the source. If a rock band has a sound intensity of 100 decibels 30 feet away from the amplifier, find the sound intensity 60 feet away from the amplifier.

42. Graph the rational function $f(x) = \dfrac{2}{x}$. Label the horizontal and vertical asymptotes.

Group Project

Using Proportions When Cooking

Overview: In this activity, you will gain experience with proportions.

Instructions: Each student should bring to class the recipe for a favorite food written on the front of a 3 × 5 card. Form groups of 2 or 3 students. Working together, the group should write proportions to determine the amount of each ingredient needed to make the recipe for the exact number of people in the class. For instance, suppose there are 32 students in your class. If a recipe that serves 8 calls for 3 cups of flour, the amount of flour needed for the recipe to serve 32 is given by the proportion

$$\text{Cups of flour} \to \frac{3}{8} = \frac{x}{32} \gets \text{Cups of flour}$$
$$\text{People served} \to \qquad \qquad \gets \text{People served}$$

Write the new ingredient list on the back of each recipe card. Since the divisions involved might not be exact, be prepared to round when making the calculations. After all of the students' recipes have been adjusted, exchange recipe cards with someone in your class.

Continued Fractions

Overview: In this activity, as you gain experience simplifying complex fractions, you will make an interesting discovery about continued fractions.

Instructions: Form groups of 2 or 3 students. Working as a group, simplify each expression in the following list. Note that the third fraction, and all those that follow it, have a complex fraction in their denominator. These expressions are called *continued fractions*.

$$1 + \frac{1}{2}, \quad 1 + \cfrac{1}{1 + \frac{1}{2}}, \quad 1 + \cfrac{1}{1 + \cfrac{1}{1 + \frac{1}{2}}}, \quad 1 + \cfrac{1}{1 + \cfrac{1}{1 + \cfrac{1}{1 + \frac{1}{2}}}}, \quad 1 + \cfrac{1}{1 + \cfrac{1}{1 + \cfrac{1}{1 + \cfrac{1}{1 + \frac{1}{2}}}}}, \quad \dots$$

Each of these expressions can be simplified by using the value of the expression preceding it. For example, to simplify the second expression in the list, replace $1 + \frac{1}{2}$ with $\frac{3}{2}$. Show that the expressions in the list simplify to $\frac{3}{2}, \frac{5}{3}, \frac{8}{5}, \frac{13}{8}, \frac{21}{13}, \dots$. Then write the next 3 continued fractions in the list. From what you have learned, predict the answers if each of them were simplified.

CUMULATIVE REVIEW ▶▶ Chapters 1–6

1. Evaluate: $12 - 6[(130 - 4^3) - 2]$ [Section 1.3]

2. Simplify: $9(a^3 + 3a) - 5(3a - a^3) - 8(-a - a^3)$ [Section 1.4]

3. Solve: $\dfrac{3x - 4}{6} - \dfrac{x - 2}{2} = \dfrac{-2x - 3}{3}$ [Section 1.5]

4. Solve $l = a + (n - 1)d$ for n. [Section 1.6]

5. Find the area of a circle with diameter 25 inches. Round to one decimal place. [Section 1.6]

6. **Geometry.** In $\triangle ABC$, the measure of $\angle A$ is 10° larger than the measure of $\angle B$. The measure of $\angle C$ is 10° larger than the measure of $\angle A$. Find the measure of each angle. [Section 1.7]

7. **Amusement Parks.** The 2008 and 2009 attendance figures for Universal Studios in Orlando, Florida, are shown below. Find the percent decrease in attendance. Round to the nearest tenth of one percent. [Section 1.8]

   2008: 6,231,000   **2009:** 5,530,000

8. Find the midpoint of the line segment with endpoints $(-5, 2)$ and $(13, -3)$. [Section 2.1]

9. **Life Expectancy.** Determine the predicted average rate of change in the life expectancy of females during the years 2000–2050, as shown in the graph below. [Section 2.3]

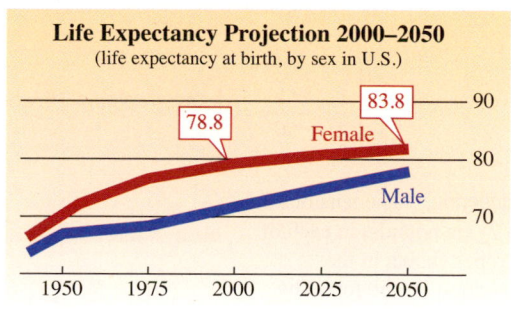

Life Expectancy Projection 2000–2050
(life expectancy at birth, by sex in U.S.)

Based on data from the Social Security Administration

10. Find the x- and y-intercepts of the graph of $7x - 10y = -7$. [Section 2.2]

Find an equation of the line with the given properties. Write the equation in slope–intercept form.

11. Passing through $(7, 5)$ and perpendicular to the line whose equation is $y = \frac{1}{7}x - 4$ [Section 2.4]

12. Passing through $(-4, 5)$ and $(2, -6)$ [Section 2.4]

13. The graph of a line is shown on the graphing calculator screen on the right. The axes are scaled in units of 1.

   a. What is the slope of the line? [Section 2.3]

   b. The line doesn't lie in one quadrant. Which quadrant is that? [Section 2.1]

   c. What is the equation of the line? Write your answer in slope–intercept form. [Section 2.4]

   d. Does the line pass through $(15, -42)$? [Section 2.2]

14. If $g(x) = -3x^3 + x - 4$, find $g(-2)$. [Section 2.5]

15. Determine whether each of the following defines y to be a function of x. If it does not, explain why. [Section 2.5]

   a. $x = |y|$

   b.

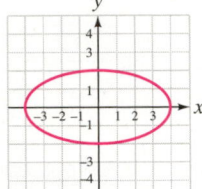

16. **The Atmosphere.**   The amount of carbon dioxide (CO_2) in our atmosphere can be modeled by a linear function. In 1998, there was approximately 365 parts per million. By the year 2008, that number had increased to about 384 parts per million. (Source: NOAA-ESRL) [Section 2.5]

   a. Let t be the number of years after 1990 and $C(t)$ be the parts per million of atmospheric carbon dioxide. Write a linear function to model the situation.

   b. Use your answer to part a to estimate how many parts per million of atmospheric carbon dioxide there will be in 2020.

17. First sketch the graph of the basic function associated with $g(x) = |x + 4|$. Then sketch the graph of g on the same coordinate system using a translation. Describe the domain and range of function g using interval notation. [Section 2.6]

18. Solve the system $\begin{cases} x = y + 3 \\ \frac{1}{4}x - \frac{1}{6}y = \frac{1}{3} \end{cases}$ by graphing. [Section 3.1]

19. **Engineering.**   The tensions T_1 and T_2 (in pounds) in each of the ropes shown in the illustration can be found by solving the following system. Find T_1 and T_2. [Section 3.2]

$$\begin{cases} 0.6T_1 - 0.8T_2 = 0 \\ 0.8T_1 + 0.6T_2 = 100 \end{cases}$$

20. Solve: $\begin{cases} x + 2y + 3z = 11 \\ 5x - y - z = 8 \\ 2x + y - 3z = -14 \end{cases}$ [Section 3.3]

21. Solve the following system using matrices or Cramer's Rule.
$$\begin{cases} 5x - 4y = 10 \\ x - 7y = 2 \end{cases}$$ [Section 3.4 or Section 3.5]

22. Evaluate: $\begin{vmatrix} 3 & -2 \\ -2 & 4 \end{vmatrix}$ [Section 3.5]

23. **Jewelry.**   A jeweler wants to make angel pins from an alloy that has a 40% silver content. The jeweler has on hand two alloys, one with 50% silver content and one with a 25% silver content. How many ounces of each alloy should be used to make 20 ounces of the 40% silver alloy? Use two variables to solve the problem. [Section 3.6]

24. **Pizza Sales.**   In 2009, Pizza Hut, Domino's, and Papa John's accounted for 28% of the market share of pizza sales in the United States. The combined market share of Domino's and Papa John's equaled that of Pizza Hut, and Papa John's market share was 2% less than Domino's. Find the market share for each company. (Source: franchisedirect.com) [Section 3.7]

Solve. Write the solution set in interval notation and then graph it.

25. $\frac{1}{2}x + 6 \geq 4 + 2x$ [Section 4.1]

26. $5(x + 2) \leq 4(x + 1)$ and $11 + x < 0$ [Section 4.2]

27. $-4(x + 2) \geq 12$ or $3x + 8 < 11$ [Section 4.2]

28. $\left| \frac{3a}{5} - 2 \right| + 1 \geq \frac{6}{5}$ [Section 4.3]

29. Graph: $y < 4 - x$ [Section 4.4]

30. Use a check to determine if $(4, -2)$ is a solution of the system
$$\begin{cases} 3x + 2y > 6 \\ x + y \leq 2 \end{cases}$$ [Section 4.5]

Simplify each expression. Write answers using positive exponents only. [Section 5.1]

31. $(3bb^2b^3c^0)^4$

32. $9^2d^{-8}(d^9)^2$

33. $-\dfrac{x^{-9}}{20k^{-7}}$

34. $\left(\dfrac{2x^{-2}y^3}{x^2x^3y^4} \right)^{-3}$

35. **Money.**   Express the *dollar value* of each type of U.S. coin and currency shown in the illustration below, using a power of 10. For example, one hundred dollars can be expressed as $\$10^2$. [Section 5.2]

Penny   Dime $1 bill $100,000 bill

36. Write each number in scientific notation and perform the operations. Give the answer in scientific notation and standard notation. [Section 5.2]

$$\frac{(0.00024)(96,000,000)}{640,000,000}$$

37. Refer to the graph of function f on the right. Find

    a. $f(-1)$

    b. $f(2)$

    c. The values of x for which $f(x) = 0$.

    d. Give the domain and range of f using interval notation.
    [Section 5.3]

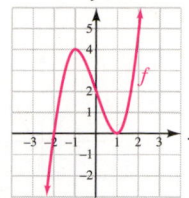

38. **Physics.**   An object shot from the ground straight up into the air with an initial velocity of 144 feet per second will reach a height of h feet after t seconds, according to the function $h(t) = -16t^2 + 144t$. Find the height of the object 3 seconds after being shot. [Section 5.3]

39. Find the degree of $17x^3y^4 + x^2y + 3$. [Section 5.3]

40. Simplify the polynomial: $\frac{9}{4}c^2 - \frac{5}{3}c - \frac{1}{2}c^2 + \frac{5}{6}c$

    [Section 5.3]

Perform the indicated operations.

41. Let $f(x) = x^3 + 3x^2 - 2x + 7$ and $g(x) = x^3 - 2x^2 + 2x + 5$. Find $f(x) + g(x)$.
    [Section 5.3]

42. $(2m^2n^2 + 2m - n) - (-2m^2n^2 - 2m + n)$
    [Section 5.3]

43. $\left(\frac{1}{16}r^9s^{10}\right)(32r^2s^{10})$ [Section 5.4]

44. $-3a^8(4a^4 + 3a^3 - 4a^2)$ [Section 5.4]

45. $(2x^3 - 1)^2$ [Section 5.4]

46. $(a + b + c)(2a - b - 2c)$ [Section 5.4]

Factor each expression.

47. $3r^2s^3 - 6rs^4$ [Section 5.5]

48. $5(x - y) - a(x - y)$ [Section 5.5]

49. $xu + yv + xv + yu$ [Section 5.5]

50. $3 - 10x^2 + 8x^4$ [Section 5.6]

51. $(x - y)^2 + 3(x - y) - 10$ [Section 5.6]

52. $81x^4 - 16y^4$ [Section 5.7]

53. $8x^3 - 27y^6$ [Section 5.7]

54. $x^2 + 10x + 25 - 16z^8$ [Section 5.8]

55. Solve $b^2x^2 + a^2y^2 = a^2b^2$ for b^2. [Section 5.5]

56. Solve: $6x^2 + 7 = -23x$ [Section 5.9]

57. Solve: $x^3 - 4x = 0$ [Section 5.9]

58. **Camping.**   The rectangular-shaped cooking surface of a small camping stove is 108 in.^{2}. If its length is 3 inches longer than its width, what are its dimensions? [Section 5.9]

59. Simplify: $\frac{2x^2y + xy - 6y}{3x^2y + 5xy - 2y}$ [Section 6.1]

60. Find the domain of the function $f(x) = \frac{2x + 1}{x^2 - 2x}$. Express your answer in words and using interval notation. [Section 6.1]

Perform the indicated operations and then simplify, if possible.

61. $(10n - n^2) \cdot \frac{n^6}{n^4 - 10n^3 - 2n^2 + 20n}$ [Section 6.2]

62. $\frac{p^3 - q^3}{q^2 - p^2} \div \frac{p^3 + p^2q + pq^2}{q^2 + pq}$ [Section 6.2]

63. $\frac{2}{x + y} + \frac{3}{x - y} - \frac{x - 3y}{x^2 - y^2}$ [Section 6.3]

64. $\dfrac{\dfrac{18}{c^2} + \dfrac{11}{c} + 1}{1 - \dfrac{3}{c} - \dfrac{10}{c^2}}$ [Section 6.4]

Perform the division.

65. $\frac{5y^4 + 45y^3}{15y^2}$ [Section 6.5]

66. $\frac{16x^3 + 16x^2 - 9x - 5}{4x + 5}$ [Section 6.5]

67. Solve: $\frac{3}{x - 2} + \frac{x^2}{x^2 + x - 6} = \frac{x + 4}{x + 3}$ [Section 6.7]

68. Solve: $\frac{5x - 3}{x + 2} = \frac{5x + 3}{x - 2}$ [Section 6.9]

69. Use synthetic division to find: $\frac{x^4 - 9x^3 + x^2 - 7x - 20}{x - 9}$

    [Section 6.6]

70. **Crowd Control.**   At the end of a sold-out concert, if just the east exits are opened, the concert hall can be cleared of people in 6 minutes. If just the west exits are opened, the hall can be cleared in 10 minutes. How long will it take to clear the hall of people if both sets of exits are opened? [Section 6.8]

71. **Economics.**   The controversial Phillips curve shown below depicts the trade-off between unemployment and inflation as seen by one school of economists. If unemployment drops to very low levels, what does the model predict will happen to the inflation rate? [Section 6.9]

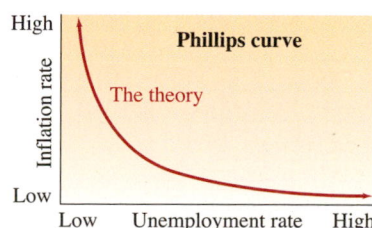

72. **Cooking.**   A recipe for brownies calls for 4 eggs and $1\frac{1}{2}$ cups of flour. If the recipe makes 15 brownies, how many cups of flour will be needed to make 130 brownies? [Section 6.9]

73. Farming. The number of days a given number of bushels of corn will last when feeding chickens varies inversely with the number of animals. If a certain number of bushels will feed 300 chickens for 4 days, how long will the feed last for 1,200 chickens? **[Section 6.9]**

74. Match each function with the correct graph shown below.

a. $f(x) = x$
   [Section 2.5]

b. $f(x) = x^2$
   [Section 2.6]

c. $f(x) = x^3$
   [Section 2.6]

d. $f(x) = \dfrac{1}{x}$
   [Section 6.1]

i.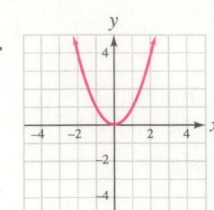

ii.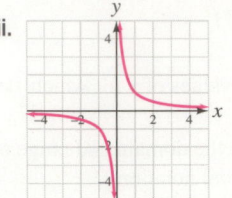

iii.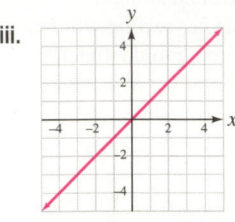

iv.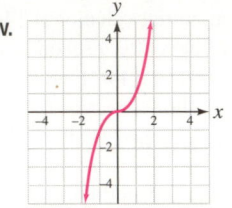

Radical Expressions and Equations

7

from Campus to Careers

General Contractor

The growing popularity of remodeling has created a boom for general contractors. If it's an additional bedroom you need or a makeover of a dated kitchen or bathroom, they can provide design and construction expertise, as well as knowledge of local building code requirements. From the planning stages of a project through its completion, general contractors use mathematics every step of the way.

Problem 141 in Study Set 7.2, problem 113 in Study Set 7.3, and problem 109 in Study Set 7.5 involve situations that a general contractor encounters on the job. The mathematical concepts in this chapter can be used to solve those problems.

JOB TITLE:
General Contractor

EDUCATION:
Courses in mathematics, science, drafting, business math, and English are important. Certificate programs are also available.

JOB OUTLOOK:
In general, employment is expected to increase by about 19% through the year 2018.

ANNUAL EARNINGS:
Mean annual salary $86,850

FOR MORE INFORMATION:
http://www.careers.stateuniversity.com

579

Study groups give students an opportunity to ask their classmates questions, share ideas, compare lecture notes, and review for tests. If something like this interests you, here are some suggestions for forming an effective study group.

GROUP SIZE: A study group should be small—from 3 to 6 people is best.

TIME AND PLACE: You should meet on a regular basis in a place where you can spread out and talk without disturbing others.

GROUND RULES: The study group will be more effective if, early on, you agree on some rules to follow.

Now Try This ▶

Would you like to begin a study group? If so, you need to answer the following questions. Who will be in your group? Where will your group meet? How often will it meet? For how long will each session last? Will you have a group leader? If so, what will be the leader's responsibilities? What will you try to accomplish each session? How will the members prepare for each meeting? Will you follow a set agenda each session? How will the members share contact information? When will you discuss ways to improve the study sessions?

SECTION 7.1

Radical Expressions and Radical Functions

OBJECTIVES

1 Find square roots.

2 Find square roots of expressions containing variables.

3 Graph the square root function.

4 Evaluate radical functions.

5 Find cube roots.

6 Graph the cube root function.

7 Find *n*th roots.

ARE YOU READY?

The following problems review some basic skills that are needed when working with radical expressions and radical functions.

1. Evaluate: **a.** 15^2 **b.** 4^3

2. Evaluate: **a.** $\left(\dfrac{7}{5}\right)^2$ **b.** $(0.2)^2$

3. Evaluate: **a.** $(-6)^3$ **b.** 3^4

4. Simplify: **a.** $a^4 \cdot a^4$ **b.** $x^3 \cdot x^3 \cdot x^3$

5. Multiply: $(x + 8)(x + 8)$

6. Let $f(x) = |2x - 3|$. Find $f(4)$.

In this section, we will reverse the squaring process and learn how to find *square roots* of numbers. Then we will generalize the concept of root and consider cube roots, fourth roots, and so on. We will also discuss a new family of functions, called *radical functions*.

1 Find Square Roots.

When we raise a number to the second power, we are squaring it, or finding its **square.**

- The square of 5 is 25 because $5^2 = 25$.
- The square of -5 is 25, because $(-5)^2 = 25$.

We can reverse the squaring process to find **square roots** of numbers. For example, to find the square roots of 25, we ask ourselves "What number, when squared, is equal to 25?" There are two possible answers.

- 5 is a square root of 25, because $5^2 = 25$.
- -5 is also a square root of 25, because $(-5)^2 = 25$.

In general, we have the following definition.

Square Root of *a* ▶ The number b is a **square root** of the number a if $b^2 = a$.

Every positive number has two square roots, one positive and one negative. For example, the two square roots of 9 are 3 and −3, and the two square roots of 144 are 12 and −12. The number 0 is the only real number with exactly one square root. In fact, it is its own square root, because $0^2 = 0$.

A **radical symbol** $\sqrt{}$ represents the **positive** or **principal square root** of a number. Since 3 is the positive square root of 9, we can write

$$\sqrt{9} = 3 \qquad \text{Read as "the square root of 9."}$$

The symbol $-\sqrt{}$ represents the **negative square root** of a number. It is the opposite of the principal square root. Since −12 is the negative square root of 144, we can write

$$-\sqrt{144} = -12 \qquad \text{Read as "the negative square root of 144 is −12" or "the opposite of the square root of 144 is −12."}$$

Square Root Notation

If a is a positive real number,

1. $\sqrt{a}$ represents the **positive** or **principal square root** of a. It is the positive number we square to get a.

2. $-\sqrt{a}$ represents the **negative square root** of a. It is the opposite of the principal square root of a: $-\sqrt{a} = -1 \cdot \sqrt{a}$.

3. The principal square root of 0 is 0: $\sqrt{0} = 0$.

The number or variable expression under a radical symbol is called the **radicand.** Together, the radical symbol and radicand are called a **radical.** An algebraic expression containing a radical is called a **radical expression.**

Radical symbol

$$\sqrt{81} \leftarrow \text{Radicand}$$

Radical

Read $\sqrt{81}$ as "the square root of 81" or as "radical 81."

Some more examples of radical expressions are:

$$\sqrt{100}, \qquad \sqrt{2} + 3, \qquad \sqrt{x^2}, \qquad \text{and} \qquad \sqrt{\dfrac{a-1}{49b^2}}$$

To evaluate (find the value of) square root radical expressions, you need to quickly recognize each of the natural-number **perfect squares** shown below in red.

$1 = 1^2$	$25 = 5^2$	$81 = 9^2$	$169 = 13^2$	$289 = 17^2$
$4 = 2^2$	$36 = 6^2$	$100 = 10^2$	$196 = 14^2$	$324 = 18^2$
$9 = 3^2$	$49 = 7^2$	$121 = 11^2$	$225 = 15^2$	$361 = 19^2$
$16 = 4^2$	$64 = 8^2$	$144 = 12^2$	$256 = 16^2$	$400 = 20^2$

EXAMPLE 1 Evaluate: **a.** $\sqrt{81}$ **b.** $-\sqrt{225}$ **c.** $\sqrt{\dfrac{49}{4}}$ **d.** $\sqrt{0.36}$

Strategy In each case, we will determine what positive number, when squared, produces the radicand.

Why The symbol $\sqrt{}$ indicates that the positive square root of the number written under it should be found.

Solution **a.** $\sqrt{81} = 9$ Because $9^2 = 81$ **b.** $-\sqrt{225} = -15$ Because $-\sqrt{225} = -1 \cdot \sqrt{225}$

c. $\sqrt{\dfrac{49}{4}} = \dfrac{7}{2}$ Because $\left(\dfrac{7}{2}\right)^2 = \dfrac{49}{4}$ d. $\sqrt{0.36} = 0.6$ Because $(0.6)^2 = 0.36$

Self Check 1 Evaluate: **a.** $\sqrt{64}$ **b.** $-\sqrt{1}$ **c.** $\sqrt{\dfrac{1}{16}}$

d. $\sqrt{0.09}$

Now Try Problems 23, 27, and 29

A number, such as 81, 225, $\dfrac{49}{4}$, and 0.36, that is the square of some rational number, is called a **perfect square**. In Example 1, we saw that the square root of a perfect square is a rational number.

If a positive number is not a perfect square, its square root is irrational. For example, $\sqrt{83}$ is an irrational number because 83 is not a perfect square. Since $\sqrt{83}$ is irrational, its decimal representation is nonterminating and nonrepeating. We can find a rational-number approximation of $\sqrt{83}$ using the square root key $\boxed{\sqrt{\ }}$ on a calculator or from the table of square roots found in Appendix I at the back of the book.

$$\sqrt{83} \approx 9.110433579 \qquad \text{or} \qquad \sqrt{83} \approx 9.11 \quad \text{Round to the nearest hundredth.}$$

CAUTION Square roots of negative numbers are not real numbers. For example, $\sqrt{-9}$ is not a real number, because no real number squared equals -9. Square roots of negative numbers come from a set called the **imaginary numbers**, which we will discuss later in this chapter. If we attempt to evaluate $\sqrt{-9}$ using a calculator, we will get an error message.

> **Success Tip**
>
> Estimation is helpful when approximating square roots. For example, $\sqrt{83}$ must be a number between 9 and 10, because $\sqrt{81} < \sqrt{83} < \sqrt{100}$.

> **Caution**
>
> Although they look similar, these radical expressions have very different meanings.
>
> $-\sqrt{9} = -3$
>
> $\sqrt{-9}$ is not a real number.

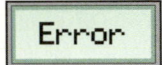

Error

```
ERR:NONREAL ANS
1▮Quit
2:Goto
```

Scientific calculator   Graphing calculator

We summarize three important facts about square roots as follows.

> **Square Roots**
>
> 1. If a is a perfect square, then $\sqrt{a}$ is rational.
>
> 2. If a is a positive number that is not a perfect square, then $\sqrt{a}$ is irrational.
>
> 3. If a is a negative number, then $\sqrt{a}$ is not a real number.

2 Find Square Roots of Expressions Containing Variables.

If $x \neq 0$, the positive number x^2 has x and $-x$ for its two square roots. To denote the positive square root of $\sqrt{x^2}$, we must know whether x is positive or negative.

If x is positive, we can write

$$\sqrt{x^2} = x \qquad \sqrt{x^2} \text{ represents the positive square root of } x^2, \text{ which is } x.$$

If x is negative, then $-x$ is positive and we can write

$$\sqrt{x^2} = -x \qquad \sqrt{x^2} \text{ represents the positive square root of } x^2, \text{ which is } -x.$$

If we don't know whether x is positive or negative, we can use absolute value symbols to ensure that $\sqrt{x^2}$ is not negative.

Simplifying $\sqrt{x^2}$	For any real number x,
	$\sqrt{x^2} = |x|$ The principal square root of x is equal to the absolute value of x.

We use this definition to *simplify* square root radical expressions.

EXAMPLE 2 Simplify. Assume all variables are unrestricted. **a.** $\sqrt{16x^2}$ **b.** $\sqrt{t^2 + 2t + 1}$
c. $\sqrt{m^6}$ **d.** $\sqrt{49r^8}$

Strategy In each case, we will determine what positive expression, when squared, produces the radicand.

Why The symbol $\sqrt{}$ indicates that the positive square root of the expression written under it should be found.

Solution If x, t, m, and r can be any real number, we have

The Language of Algebra

The phrase **"assume all variables are unrestricted"** in the instructions for Example 2 reminds us that negative numbers are permissible replacement values for the variables. Therefore, absolute value symbols may be necessary to ensure the answer is not negative.

a. $\sqrt{16x^2} = |4x|$ Because $(4x)^2 = 16x^2$. Since 4x could be negative (for example, when x = −5, the expression 4x is −20), absolute value symbols are needed to ensure that the result is not negative.

$\phantom{\sqrt{16x^2}} = 4|x|$ Since 4 is a positive constant in the product 4x, we can write it outside the absolute value symbols.

b. $\sqrt{t^2 + 2t + 1} = \sqrt{(t+1)^2}$ Factor the radicand: $t^2 + 2t + 1 = (t+1)^2$.

$\phantom{\sqrt{t^2+2t+1}} = |t+1|$ Since t + 1 can be negative (for example, when t = −5, the expression t + 1 is −4), absolute value symbols are needed to ensure that the result is not negative.

c. $\sqrt{m^6} = |m^3|$ Because $(m^3)^2 = m^6$. Since m^3 can be negative (for example, when m = −3, the expression m^3 is −27), absolute value symbols are needed to ensure that the result is not negative.

d. $\sqrt{49r^8} = 7r^4$ Because $(7r^4)^2 = 49r^8$. Since r^4 is never negative for any value of r, no absolute value symbols are needed.

Self Check 2 Simplify. Assume all variables are unrestricted. **a.** $\sqrt{25a^2}$
b. $\sqrt{b^{14}}$ **c.** $\sqrt{x^2 - 18x + 81}$
d. $\sqrt{100n^8}$

Now Try Problems 39, 41, 45, and 47

If we know that x, t, and m are positive in Example 2, we don't need to use absolute value symbols in the answers. For example,

The Language of Algebra

If we are told **"All variables represent positive real numbers,"** then absolute value symbols are not needed when simplifying square root radical expressions.

$\sqrt{16x^2} = 4x$ If x is positive, 4x is positive.

$\sqrt{m^6} = m^3$ If m is positive, m^3 is positive.

$\sqrt{t^2 + 2t + 1} = t + 1$ If t is positive, t + 1 is positive.

3 **Graph the Square Root Function.**

Since there is one principal square root for every nonnegative real number x, the equation $f(x) = \sqrt{x}$ determines a function, called a **square root function**. Square root functions belong to a larger family of functions known as **radical functions**.

EXAMPLE 3   Graph $f(x) = \sqrt{x}$ and find the domain and range of the function.

Strategy   We will graph the function by creating a table of function values and plotting the corresponding ordered pairs.

Why   After drawing a smooth curve through the plotted points, we will have the graph.

Solution   To graph the function, we select several values for x that are perfect squares, such as 0, 1, 4, and 9, and find the corresponding values of $f(x)$. We begin with $x = 0$, since 0 is the smallest input for which $\sqrt{x}$ is defined.

$f(x) = \sqrt{x}$	$f(x) = \sqrt{x}$	$f(x) = \sqrt{x}$	$f(x) = \sqrt{x}$
$f(0) = \sqrt{0}$	$f(1) = \sqrt{1}$	$f(4) = \sqrt{4}$	$f(9) = \sqrt{9}$
$= 0$	$= 1$	$= 2$	$= 3$

We also can select values of x that are not perfect squares when creating a table of function values. For example, if $x = 6$, then $f(6) = \sqrt{6}$, and it follows that the point $\left(2, \sqrt{6}\right)$ is on the graph. To help locate this point on the coordinate system, it is helpful to approximate: $\sqrt{6} \approx 2.45$.

We enter each value of x and its corresponding value of $f(x)$ in the table below. After plotting the ordered pairs, we draw a smooth curve through the points to get the graph shown in figure (a). Since the equation defines a function, its graph passes the vertical line test.

To find the domain of the function graphically, we project the graph onto the x-axis, as shown in figure (b). Because the graph begins at $(0, 0)$ and extends indefinitely to the right, the projection includes 0 and all positive real numbers. Thus, the domain of $f(x) = \sqrt{x}$ is the set of nonnegative real numbers, which can be written in interval notation as $[0, \infty)$.

To find the range of the function graphically, we project the graph onto the y-axis, as shown in figure (b). Because the graph of the function begins at $(0, 0)$ and extends indefinitely upward, the projection includes all nonnegative real numbers. Therefore, the range of the function is $[0, \infty)$.

The Language of Algebra

Together, 0 and the positive real numbers are called the **nonnegative real numbers.**

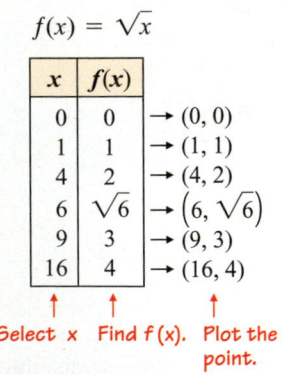

$f(x) = \sqrt{x}$

x	$f(x)$	
0	0	→ $(0, 0)$
1	1	→ $(1, 1)$
4	2	→ $(4, 2)$
6	$\sqrt{6}$	→ $\left(6, \sqrt{6}\right)$
9	3	→ $(9, 3)$
16	4	→ $(16, 4)$

↑ Select x ↑ Find $f(x)$.   ↑ Plot the point.

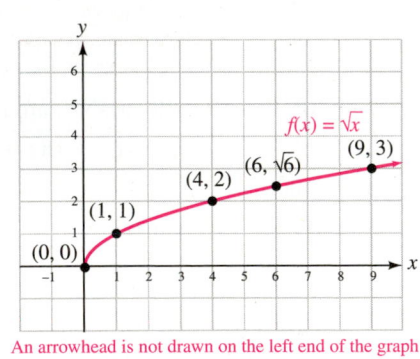
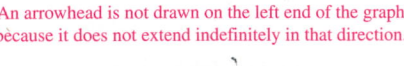

An arrowhead is not drawn on the left end of the graph because it does not extend indefinitely in that direction.

(a)

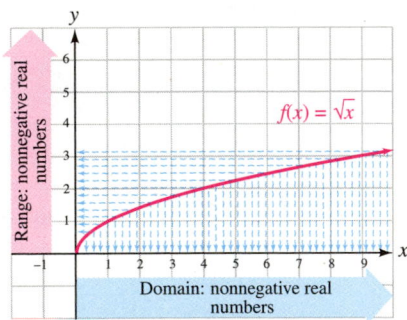

(b)

Self Check 3   Graph: $g(x) = \sqrt{x} + 2$. Then give its domain and range and compare it with the graph of $f(x) = \sqrt{x}$.

Now Try ▶ **Problem 51**

EXAMPLE 4   Let $g(x) = \sqrt{x + 3}$.   **a.** Find its domain algebraically.   **b.** Graph the function.
c. Find its range.

Strategy   We will determine the domain algebraically by finding all the values of x for which $x + 3 \geq 0$.

Why   Since the expression $\sqrt{x + 3}$ is not a real number when $x + 3$ is negative, we must require that $x + 3 \geq 0$.

Solution   **a.** To determine the domain of the function, we solve the following inequality:

$$x + 3 \geq 0 \quad \text{\textcolor{red}{Because we cannot find the square root of a negative number}}$$
$$x \geq -3 \quad \text{\textcolor{red}{To solve for x, subtract 3 from both sides.}}$$

The x-inputs must be real numbers greater than or equal to -3. Thus, the domain of $g(x) = \sqrt{x + 3}$ is the interval $[-3, \infty)$.

Success Tip

We solved linear inequalities in one variable, such as $x + 3 \geq 0$, in Section 4.1.

b. To graph the function, we construct a table of function values. We begin by selecting $x = -3$, since -3 is the smallest input for which $\sqrt{x + 3}$ is defined. Then we let $x = -2, 1,$ and 6 and list each corresponding function value $g(x)$ in the table.

$g(x) = \sqrt{x + 3}$	$g(x) = \sqrt{x + 3}$	$g(x) = \sqrt{x + 3}$	$g(x) = \sqrt{x + 3}$
$g(-3) = \sqrt{-3 + 3}$	$g(-2) = \sqrt{-2 + 3}$	$g(1) = \sqrt{1 + 3}$	$g(6) = \sqrt{6 + 3}$
$= \sqrt{0}$	$= \sqrt{1}$	$= \sqrt{4}$	$= \sqrt{9}$
$= 0$	$= 1$	$= 2$	$= 3$

After plotting the ordered pairs, we draw a smooth curve through the points to get the graph shown in figure (a). In figure (b), we see that the graph of $g(x) = \sqrt{x + 3}$ is the graph of $f(x) = \sqrt{x}$, translated 3 units to the left.

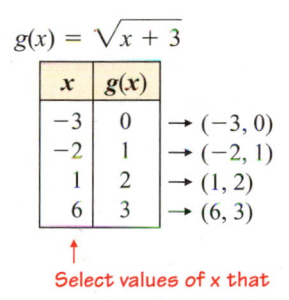

$g(x) = \sqrt{x + 3}$

x	$g(x)$	
-3	0	$\rightarrow (-3, 0)$
-2	1	$\rightarrow (-2, 1)$
1	2	$\rightarrow (1, 2)$
6	3	$\rightarrow (6, 3)$

↑
\textcolor{red}{Select values of x that make $x + 3$ a perfect square.}

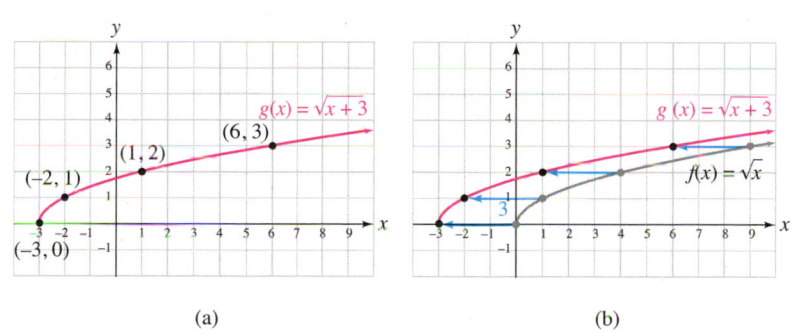

(a)   (b)

c. From the graph in figure (a), we see that the range of $g(x) = \sqrt{x + 3}$ is $[0, \infty)$.

Self Check 4   Let $h(x) = \sqrt{x - 2}$. **a.** Find its domain algebraically.
b. Graph the function.   **c.** Find its range.

Now Try ▶ Problems 53 and 55

4 Evaluate Radical Functions.

Radical functions can be used to model certain real-life situations that exhibit growth that eventually levels off. A calculator is often helpful when evaluating a radical function for a given value of the input variable.

EXAMPLE 5

Pendulums. The **period of a pendulum** is the time required for the pendulum to swing back and forth to complete one cycle. The period (in seconds) is a function of the pendulum's length L (in feet) and is given by $f(L) = 2\pi\sqrt{\frac{L}{32}}$. Find the period of the 5-foot-long pendulum of a clock. Round the result to the nearest tenth.

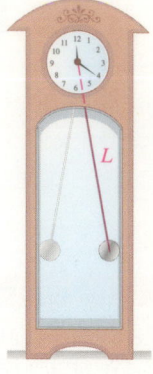

Strategy To find the period of the pendulum we will find $f(5)$.

Why The notation $f(5)$ represents the period (in seconds) of a pendulum whose length L is 5 feet.

Solution

$$f(L) = 2\pi\sqrt{\frac{L}{32}}$$

$$f(5) = 2\pi\sqrt{\frac{5}{32}}$$

Substitute 5 for L.

$$\approx 2.483647066$$

Evaluate the radical expression. Use a calculator to find an approximation. $2\pi\sqrt{\frac{5}{32}}$ means $2 \cdot \pi \cdot \sqrt{\frac{5}{32}}$.

The period is approximately 2.5 seconds.

Self Check 5 **Pendulums.** Find the period of a pendulum that is 3 feet long. Round the result to the nearest hundredth.

Now Try ▶ Problem 123

Using Your Calculator ▶ **Graphing and Evaluating a Square Root Function**

We can graph radical functions on a graphing calculator. To graph $f(x) = \sqrt{x}$ from Example 3, we press Y = and enter the right side of the equation. If the window settings for x and y are $[-1, 9]$, we get the graph shown in figure (a) when we press GRAPH. The function $g(x) = \sqrt{x} + 3$ from Example 4 is graphed in a similar way with window settings of $[-4, 10]$ for x and $[-1, 9]$ for y. See figure (b).

To answer Example 5 with a graphing calculator, we graph the function $f(x) = 2\pi\sqrt{\frac{x}{32}}$. We then trace and move the cursor toward an x-value of 5 until we see the coordinates shown in figure (c). The pendulum's period is given by the y-value shown on the screen. By zooming in, we can get better results.

After entering $Y_1 = 2\pi\sqrt{\frac{x}{32}}$, we can also use the TABLE mode to find $f(5)$. See figure (d).

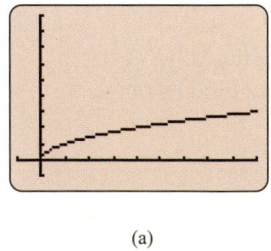

(a)

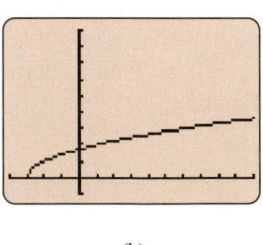

(b)

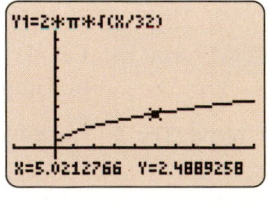

(c)

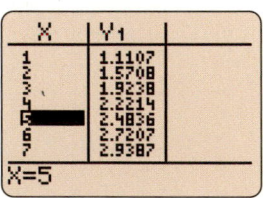

(d)

---

5 Find Cube Roots.

When we raise a number to the third power, we are cubing it, or finding its **cube.** We can reverse the cubing process to find **cube roots** of numbers. To find the cube root of 8, we ask "What number, when cubed, is equal to 8?" It follows that 2 is a cube root of 8, because $2^3 = 8$.

In general, we have this definition.

Cube Root of *a*	▼ The number b is a **cube root** of the real number a if $b^3 = a$.

All real numbers have one real cube root. A positive number has a positive cube root, a negative number has a negative cube root, and the cube root of 0 is 0.

Cube Root Notation	▼ The **cube root of *a*** is written as $\sqrt[3]{a}$. By definition, $\sqrt[3]{a} = b$ if $b^3 = a$

Earlier, we determined that the cube root of 8 is 2. In symbols, we can write: $\sqrt[3]{8} = 2$. The number 3 is called the **index**, 8 is called the **radicand**, and the entire expression is called a **radical.**

Index →
$$\overset{\text{Index}}{\sqrt[3]{8}} \leftarrow \text{Radicand}$$
Radical

Read as "the cube root of 8."

A number, such as 125, $\frac{1}{64}$, -27, and -8, that is the cube of some rational number is called a **perfect cube.**

To simplify cube root radical expressions, you need to quickly recognize each of the following natural-number perfect cubes shown in blue.

$$1 = 1^3 \qquad 27 = 3^3 \qquad 125 = 5^3 \qquad 343 = 7^3 \qquad 729 = 9^3$$
$$8 = 2^3 \qquad 64 = 4^3 \qquad 216 = 6^3 \qquad 512 = 8^3 \qquad 1{,}000 = 10^3$$

The following property is also used to simplify cube root radical expressions.

Simplifying $\sqrt[3]{x^3}$	▼ For any real number x, $$\sqrt[3]{x^3} = x$$

EXAMPLE 6   Simplify: **a.** $\sqrt[3]{125}$ **b.** $\sqrt[3]{-\dfrac{1}{64}}$ **c.** $\sqrt[3]{27x^3}$ **d.** $\sqrt[3]{-8a^9b^6}$

Strategy   In each case, we will determine what number or expression, when cubed, produces the radicand.

Why   The symbol $\sqrt[3]{\ }$ indicates that the cube root of the number written under it should be found.

Solution   **a.** $\sqrt[3]{125} = 5$ Because $5^3 = 5 \cdot 5 \cdot 5 = 125$

b. $\sqrt[3]{-\dfrac{1}{64}} = -\dfrac{1}{4}$ Because $\left(-\frac{1}{4}\right)^3 = \left(-\frac{1}{4}\right)\left(-\frac{1}{4}\right)\left(-\frac{1}{4}\right) = -\frac{1}{64}$

c. $\sqrt[3]{27x^3} = 3x$ Because $(3x)^3 = (3x)(3x)(3x) = 27x^3$. No absolute value symbols are needed.

d. $\sqrt[3]{-8a^9b^6} = -2a^3b^2$ Because $(-2a^3b^2)^3 = (-2a^3b^2)(-2a^3b^2)(-2a^3b^2) = -8a^9b^6$.

6 Graph the Cube Root Function.

Since there is one cube root for every real number x, the equation $f(x) = \sqrt[3]{x}$ defines a function, called the **cube root function.** Like square root functions, cube root functions belong to the family of radical functions.

EXAMPLE 7   Consider $f(x) = \sqrt[3]{x}$.   **a.** Graph the function.   **b.** Find its domain and range.
c. Graph: $g(x) = \sqrt[3]{x} - 2$

Strategy   We will graph the function by creating a table of function values and plotting the corresponding ordered pairs.

Why   After drawing a smooth curve through the plotted points, we will have the graph. The answers to parts (b) and (c) can then be determined from the graph.

Solution   **a.** To graph the function, we select several values for x, that are perfect cubes, such as $-8, -1, 0, 1,$ and 8, and find the corresponding values of $f(x)$. The results are entered in the table below.

$$
\begin{array}{c|c|c|c|c}
f(x) = \sqrt[3]{x} & f(x) = \sqrt[3]{x} & f(x) = \sqrt[3]{x} & f(x) = \sqrt[3]{x} & f(x) = \sqrt[3]{x} \\
f(-8) = \sqrt[3]{-8} & f(-1) = \sqrt[3]{-1} & f(0) = \sqrt[3]{0} & f(1) = \sqrt[3]{1} & f(8) = \sqrt[3]{8} \\
= -2 & = -1 & = 0 & = 1 & = 2
\end{array}
$$

After plotting the ordered pairs, we draw a smooth curve through the points to get the graph shown in figure (a).

$f(x) = \sqrt[3]{x}$

x	$f(x)$	
-8	-2	→ $(-8, -2)$
-1	-1	→ $(-1, -1)$
0	0	→ $(0, 0)$
1	1	→ $(1, 1)$
8	2	→ $(8, 2)$

↑
Select values of x that
are perfect cubes.

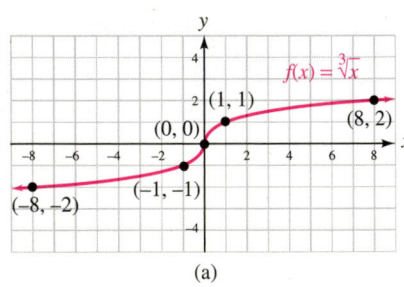

(a)

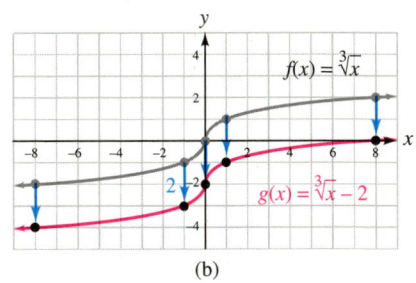

(b)

b. If we project the graph in figure (a) onto the x- and y-axes, we see that the domain and the range of the function are the set of real numbers. Thus, the domain is $(-\infty, \infty)$ and the range is $(-\infty, \infty)$.

c. Refer to figure (b). The graph of $g(x) = \sqrt[3]{x} - 2$ is the graph of $f(x) = \sqrt[3]{x}$, translated 2 units downward.

7 Find *n*th Roots.

Just as there are square roots and cube roots, there are fourth roots, fifth roots, sixth roots, and so on. In general, we have the following definition.

nth Roots of *a*	The **nth root** of *a* is written as $\sqrt[n]{a}$, and
	$$\sqrt[n]{a} = b \qquad \text{if} \qquad b^n = a$$
	The number *n* is called the **index** (or **order**) of the radical. If *n* is an even natural number, *a* must be positive or zero, and *b* must be positive.

When *n* is an odd natural number, the expression $\sqrt[n]{x}$, where $n > 1$, represents an **odd root**. Since every real number has just one real *n*th root when *n* is odd, we don't need absolute value symbols when finding odd roots. For example,

$$\sqrt[5]{243} = 3 \qquad \textcolor{red}{\text{Because } 3^5 = 243}$$

$$\sqrt[5]{-32x^5} = -2x \qquad \textcolor{red}{\text{Because } (-2x)^5 = -32x^5}$$

When *n* is an even natural number, the expression $\sqrt[n]{x}$, where $x > 0$, represents an **even root**. In this case, there will be one positive and one negative real *n*th root. For example, the real sixth roots of 729 are 3 and -3, because $3^6 = 729$ and $(-3)^6 = 729$. When finding even roots, we can use absolute value symbols to guarantee that the *n*th root is positive.

$$\sqrt[4]{(-3)^4} = |-3| = 3 \qquad \textcolor{red}{\text{We also could simplify this as follows: } \sqrt[4]{(-3)^4} = \sqrt[4]{81} = 3 \text{ .}}$$

$$\sqrt[6]{729x^6} = |3x| = 3|x| \qquad \textcolor{red}{\text{Because } (3x)^6 = 729x^6. \text{ The absolute value symbols ensure that the result is positive.}}$$

In general, we have the following rules.

Simplifying $\sqrt[n]{x^n}$	If *x* is a real number and $n > 1$, then
	If *n* is an odd natural number, $\sqrt[n]{x^n} = x$.
	If *n* is an even natural number, $\sqrt[n]{x^n} = |x|$.

EXAMPLE 8   Evaluate: **a.** $\sqrt[4]{625}$ **b.** $\sqrt[4]{-1}$ **c.** $\sqrt[5]{-32}$ **d.** $\sqrt[6]{\dfrac{1}{64}}$

Strategy   In each case, we will determine what number, when raised to the fourth, fifth, or sixth power, produces the radicand.

Why   The symbols $\sqrt[4]{}$, $\sqrt[5]{}$, and $\sqrt[6]{}$ indicate that the fourth, fifth, or sixth root of the number written under it should be found.

Solution   **a.** $\sqrt[4]{625} = 5$, because $5^4 = 625$. $\qquad$ \textcolor{red}{Read $\sqrt[4]{625}$ as "the fourth root of 625."}

b. $\sqrt[4]{-1}$ is not a real number. $\qquad$ \textcolor{red}{An even root of a negative number is not a real number.}

c. $\sqrt[5]{-32} = -2$, because $(-2)^5 = -32$.   \textcolor{red}{Read $\sqrt[5]{-32}$ as "the fifth root of -32."}

d. $\sqrt[6]{\dfrac{1}{64}} = \dfrac{1}{2}$, because $\left(\dfrac{1}{2}\right)^6 = \dfrac{1}{64}$.   \textcolor{red}{Read $\sqrt[6]{\frac{1}{64}}$ as "the sixth root of $\frac{1}{64}$."}

Self Check 8   Evaluate: **a.** $\sqrt[4]{\dfrac{1}{81}}$ **b.** $\sqrt[5]{10^5}$ **c.** $\sqrt[6]{-64}$

Now Try   Problems 87, 91, and 93

Using Your Calculator ▶ **Finding Roots**

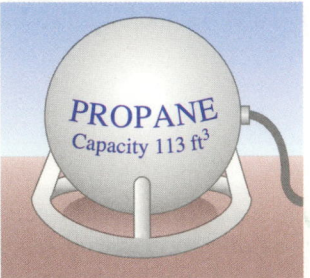

PROPANE
Capacity 113 ft³

The square root key $\boxed{\sqrt{}}$ on a reverse entry scientific calculator can be used to evaluate square roots. To evaluate roots with an index greater than 2, we can use the root key $\boxed{\sqrt[x]{y}}$. For example, the function $r(V) = \sqrt[3]{\dfrac{3V}{4\pi}}$ gives the radius of a sphere with volume V. To find the radius of the spherical propane tank shown on the left, we substitute 113 for V to get:

$$r(\mathbf{113}) = \sqrt[3]{\dfrac{3(\mathbf{113})}{4\pi}}$$

To evaluate a root, we enter the radicand and press the root key $\boxed{\sqrt[x]{y}}$ followed by the index of the radical, which in this case is 3.

$$\boxed{3}\ \boxed{\times}\ \boxed{113}\ \boxed{\div}\ \boxed{(}\ \boxed{4}\ \boxed{\times}\ \boxed{\pi}\ \boxed{)}\ \boxed{=}\ \boxed{2\text{nd}}\ \boxed{\sqrt[x]{y}}\ \boxed{3}\ \boxed{=}\qquad \boxed{2.999139118}$$

To evaluate the cube root of $\dfrac{3(113)}{4\pi}$ using a direct entry calculator, we enter:

$$\boxed{3}\ \boxed{2\text{nd}}\ \boxed{\sqrt[x]{y}}\ \boxed{(}\ \boxed{3}\ \boxed{\times}\ \boxed{113}\ \boxed{\div}\ \boxed{(}\ \boxed{4}\ \boxed{\times}\ \boxed{\pi}\ \boxed{)}\ \boxed{)}\ \boxed{\text{ENTER}}$$

To evaluate the cube root of $\dfrac{3(113)}{4\pi}$ with a graphing calculator, we enter:

$$\boxed{\text{MATH}}\ \boxed{4}\ \boxed{(}\ \boxed{3}\ \boxed{\times}\ \boxed{113}\ \boxed{)}\ \boxed{\div}\ \boxed{(}\ \boxed{4}\ \boxed{\times}\ \boxed{2\text{nd}}\ \boxed{\pi}\ \boxed{)}\ \boxed{)}\ \boxed{\text{ENTER}}$$

$\boxed{\begin{array}{l}{}^3\sqrt{((3*113)/(4*\pi)}\\ \\ \qquad\qquad 2.999139118\end{array}}$

If we round the result to the nearest foot, we see that the radius of the propane tank is about 3 feet.

EXAMPLE 9   Simplify each radical expression. Assume that x can be any real number.

a. $\sqrt[5]{x^5}$ **b.** $\sqrt[4]{16x^4}$ **c.** $\sqrt[6]{(x+4)^6}$ **d.** $\sqrt[4]{81x^8}$

Strategy   When the index n is odd, we will determine what expression, when raised to the nth power, produces the radicand. When the index n is even, we will determine what positive expression, when raised to the nth power, produces the radicand.

Why   This is the definition of nth root.

Solution

a. $\sqrt[5]{x^5} = x$ Since n is odd, absolute value symbols aren't needed.

The Language of Algebra

Another way to say that x can be any real number is to say that **the variable is unrestricted.**

b. $\sqrt[4]{16x^4} = |2x| = 2|x|$ Since n is even and x can be negative, absolute value symbols are needed to ensure that the result is positive.

c. $\sqrt[6]{(x+4)^6} = |x+4|$ Absolute value symbols are needed to ensure that the result is positive.

d. $\sqrt[4]{81x^8} = 3x^2$ Because $(3x^2)^4 = 81x^8$. Since $3x^2 \geq 0$ for any value of x, no absolute value symbols are needed.

Self Check 9   Simplify. Assume all variables are unrestricted.

a. $\sqrt[6]{x^6}$ **b.** $\sqrt[5]{(a+5)^5}$ **c.** $\sqrt[4]{16a^8}$

Now Try   Problems 95, 97, 99, and 101

If we know that x is positive in parts (b) and (c) of Example 9, we don't need to use absolute value symbols. For example, if $x > 0$, then

$$\sqrt[4]{16x^4} = 2x \qquad \text{If x is positive, 2x is positive.}$$

$$\sqrt[6]{(x + 4)^6} = x + 4 \qquad \text{If x is positive, x + 4 is positive.}$$

We now summarize the definitions concerning nth roots.

Summary of the Definitions of $\sqrt[n]{x}$	If n is a natural number greater than 1 and x is a real number,
	If $x > 0$, then $\sqrt[n]{x}$ is the positive number such that $\left(\sqrt[n]{x}\right)^n = x$.
	If $x = 0$, then $\sqrt[n]{x} = 0$.
	If $x < 0$ $\begin{cases} \text{and } n \text{ is odd, then } \sqrt[n]{x} \text{ is the negative number such that } \left(\sqrt[n]{x}\right)^n = x. \\ \text{and } n \text{ is even, then } \sqrt[n]{x} \text{ is not a real number.} \end{cases}$

SECTION 7.1   STUDY SET

VOCABULARY

Fill in the blanks.

1. $5x^2$ is the _____ root of $25x^4$ because $(5x^2)^2 = 25x^4$. The _____ root of 216 is 6 because $6^3 = 216$.

2. The symbol $\sqrt{}$ is called a _____ symbol or a _____ root symbol.

3. A radical symbol $\sqrt{}$ represents the _____ or principal square root of a number.

4. The number 4 has two square roots, -2 and 2. When we speak of *the* square root of 4, we mean only the _____ square root of 4, which is 2.

5. The number 100 has two square roots. The positive or _____ square root of 100 is 10.

6. In the expression $\sqrt[3]{27x^6}$, the _____ is 3 and $27x^6$ is the _____.

7. When we write $\sqrt{b^4} = b^2$, we say that we have _____ the radical expression.

8. When n is an odd number, $\sqrt[n]{x}$ represents an _____ root. When n is an _____ number, $\sqrt[n]{x}$ represents an even root.

9. $f(x) = \sqrt{x}$ and $g(x) = \sqrt[3]{x}$ are _____ functions.

10. Together, 0 and the positive real numbers are called the _____ real numbers.

CONCEPTS

Fill in the blanks.

11. b is a square root of a if $b^2 =$ ▢.

12. $\sqrt{0} =$ ▢ and $\sqrt[3]{0} =$ ▢.

13. $\sqrt{-4}$ is not a real number, because no real number _____ equals -4.

14. $\sqrt[3]{x} = y$ if $y^3 =$ ▢.

15. $\sqrt{x^2} =$ ▢ and $\sqrt[3]{x^3} =$ ▢.

16. **a.** The graph of $g(x) = \sqrt{x} + 3$ is the graph of $f(x) = \sqrt{x}$ translated ▢ units _____.

   b. The graph of $g(x) = \sqrt{x + 5}$ is the graph of $f(x) = \sqrt{x}$ translated ▢ units to the _____.

17. The graph of a square root function f is shown below. Find each of the following, if possible.

   a. $f(11)$ \qquad\qquad **b.** $f(2)$

   c. $f(-1)$

   d. The value of x for which $f(x) = 2$

   e. The value of x for which $f(x) = -1$

   f. The domain and range of f (Use interval notation.)

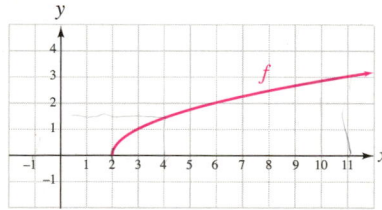

18. Refer to the graph in problem 17. Estimate each of the following function values.

   a. $f(4)$ \qquad\qquad **b.** $f(10)$

19. The graph of a cube root function f is shown below. Find each of the following.

a. $f(-8)$ b. $f(0)$

c. The value of x for which $f(x) = -2$

d. The domain and range of f
   (Use interval notation.)

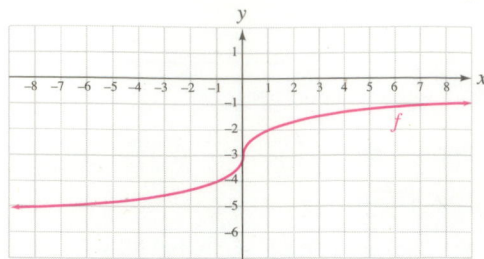

20. Match each function with the correct graph shown below.

a. $f(x) = x - 1$ b. $f(x) = \sqrt{x} - 1$

c. $f(x) = x^2 - 1$ d. $f(x) = \sqrt[3]{x} - 1$

e. $f(x) = |x| - 1$ f. $f(x) = \dfrac{1}{x - 1}$

i. 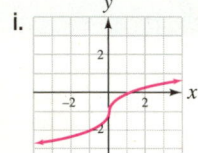   ii. 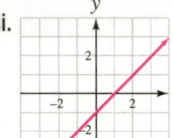   iii.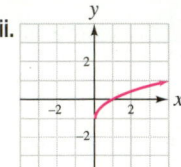

iv. 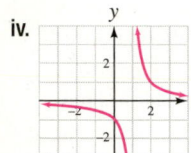   v. 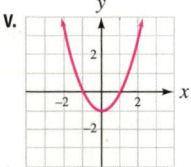   vi.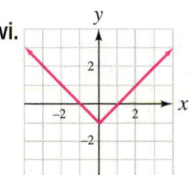

NOTATION

Translate each sentence into mathematical symbols.

21. a. The square root of x squared is the absolute value of x.

b. The cube root of x cubed is x.

c. The fifth root of negative thirty-two is negative two.

22. a. f of x equals the square root of the quantity x minus five.

b. g of x equals the cube root of x squared.

GUIDED PRACTICE

Evaluate each square root without using a calculator. See Objective 1 and Example 1.

23. $\sqrt{100}$ **24.** $\sqrt{49}$ **25.** $-\sqrt{64}$ **26.** $-\sqrt{1}$

27. $\sqrt{\dfrac{1}{9}}$ **28.** $\sqrt{\dfrac{4}{25}}$ **29.** $\sqrt{0.25}$ **30.** $\sqrt{0.16}$

31. $\sqrt{-81}$ **32.** $-\sqrt{49}$ **33.** $\sqrt{121}$ **34.** $\sqrt{144}$

Use a calculator to find each square root. Give each answer to four decimal places. See Objective 1.

35. $\sqrt{12}$ **36.** $\sqrt{340}$

37. $\sqrt{679.25}$ **38.** $\sqrt{0.0063}$

Simplify each expression. Assume that all variables are unrestricted and use absolute value symbols when necessary. See Example 2.

39. $\sqrt{4x^2}$ **40.** $\sqrt{64t^2}$

41. $\sqrt{81h^4}$ **42.** $\sqrt{36y^4}$

43. $\sqrt{36s^6}$ **44.** $\sqrt{9y^6}$

45. $\sqrt{144m^8}$ **46.** $\sqrt{4n^8}$

47. $\sqrt{y^2 - 2y + 1}$ **48.** $\sqrt{b^2 - 14b + 49}$

49. $\sqrt{a^4 + 6a^2 + 9}$ **50.** $\sqrt{x^4 + 10x^2 + 25}$

Complete each table and then graph the function. Give the domain and range. See Examples 3 and 4.

51. $f(x) = -\sqrt{x}$

x	y
0	
1	
4	
9	
16	

52. $f(x) = \sqrt{x} + 2$

x	y
0	
1	
4	
9	
16	

Graph each function. Give the domain and range. See Examples 3 and 4.

53. $f(x) = \sqrt{x + 4}$ **54.** $f(x) = \sqrt{x} - 1$

Find the domain of each function. See Example 4.

55. $f(x) = \sqrt{x + 6}$ **56.** $g(x) = \sqrt{x + 12}$

57. $g(x) = \sqrt{8 - 2x}$ **58.** $h(x) = \sqrt{35 - 5x}$

59. $s(t) = \sqrt{9t - 4}$ **60.** $T(a) = \sqrt{3a + 17}$

61. $c(x) = \sqrt{0.5x - 20}$ **62.** $H(b) = \sqrt{0.4b - 36}$

Find each function value, if possible. Do not use a calculator. See Example 5.

63. $f(x) = \sqrt{3x + 1}$

a. $f(8)$ b. $f(-2)$

64. $h(t) = \sqrt{t^2 + t - 3}$

a. $h(-4)$ b. $h(-1)$

65. $g(x) = \sqrt[3]{x - 4}$

a. $g(12)$ b. $g(-23)$

66. $s(a) = -\sqrt[3]{32a}$

a. $s(-2)$ b. $s(2)$

Use a calculator to find each function value. Round to the nearest ten-thousandth. See Example 5 and Using Your Calculator.

67. $f(x) = \sqrt{x^2 + 1}$

    a. $f(4)$ **b.** $f(2.35)$

68. $g(x) = \sqrt{7 - 4x}$

    a. $g(-\pi)$ **b.** $g(0.5)$

69. $g(x) = \sqrt[3]{x^2 + 1}$

    a. $g(6)$ **b.** $g(21.57)$

70. $h(t) = \sqrt[3]{2.1t + 11}$

    a. $h(-0.4)$ **b.** $h(15)$

Simplify each cube root. See Example 6.

71. $\sqrt[3]{1}$ **72.** $\sqrt[3]{8}$

73. $\sqrt[3]{-125}$ **74.** $\sqrt[3]{-27}$

75. $\sqrt[3]{\dfrac{8}{27}}$ **76.** $\sqrt[3]{\dfrac{125}{64}}$

77. $\sqrt[3]{64}$ **78.** $\sqrt[3]{1,000}$

79. $\sqrt[3]{-216a^3}$ **80.** $\sqrt[3]{-512x^3}$

81. $\sqrt[3]{-1,000p^6q^3}$ **82.** $\sqrt[3]{-343a^6b^3}$

Complete each table and then graph the function. Give the domain and range. See Example 7.

83. $f(x) = \sqrt[3]{x} - 3$ **84.** $f(x) = -\sqrt[3]{x}$

x	y
-8	
-1	
0	
1	
8	

x	y
-8	
-1	
0	
1	
8	

Graph each function. Give the domain and range. See Example 7.

85. $f(x) = \sqrt[3]{x} - 3$ **86.** $f(x) = \sqrt[3]{x} + 3$

Evaluate each radical expression, if possible, without using a calculator. See Example 8.

87. $\sqrt[4]{81}$ **88.** $\sqrt[6]{64}$

89. $-\sqrt[5]{243}$ **90.** $-\sqrt[4]{625}$

91. $\sqrt[6]{-256}$ **92.** $\sqrt[6]{-729}$

93. $\sqrt[5]{-\dfrac{1}{32}}$ **94.** $\sqrt[5]{-\dfrac{243}{32}}$

Simplify each radical expression. Assume all variables are unrestricted. See Example 9.

95. $\sqrt[5]{32a^5}$ **96.** $\sqrt[5]{-32x^5}$

97. $\sqrt[4]{81a^4}$ **98.** $\sqrt[8]{t^8}$

99. $\sqrt[6]{k^{12}}$ **100.** $\sqrt[6]{64b^6}$

101. $\sqrt[4]{(m + 4)^8}$ **102.** $\sqrt[4]{(x - 7)^8}$

TRY IT YOURSELF

Simplify each radical expression, if possible. Assume all variables are unrestricted.

103. $\sqrt[3]{64s^9t^6}$ **104.** $\sqrt[3]{1,000a^6b^6}$

105. $-\sqrt{49b^8}$ **106.** $-\sqrt{144t^4}$

107. $-\sqrt[5]{-\dfrac{1}{32}}$ **108.** $-\sqrt[5]{-243}$

109. $\sqrt[3]{-125m^6}$ **110.** $\sqrt[3]{-216z^9}$

111. $\sqrt{400m^{16}n^2}$ **112.** $\sqrt{169p^4q^2}$

113. $\sqrt[6]{64a^6b^6}$ **114** $\sqrt[6]{(x + 4)^6}$

115. $\sqrt[4]{-81}$ **116.** $\sqrt[6]{-1}$

117. $\sqrt{n^2 + 12n + 36}$ **118.** $\sqrt{s^2 - 20s + 100}$

Look Alikes . . .

119. a. $\sqrt{64}$ **b.** $\sqrt[3]{64}$ **120. a.** $\sqrt{-64}$ **b.** $\sqrt[3]{-64}$

121. a. $\sqrt{81}$ **b.** $\sqrt[4]{81}$ **122. a.** $\sqrt{16}$ **b.** $\sqrt[4]{16}$

APPLICATIONS

Use a calculator to solve each problem. Round answers to the nearest tenth.

123. Embroidery. The radius r of a circle is given by the formula $r = \sqrt{\dfrac{A}{\pi}}$, where A is its area. Find the diameter of the embroidery hoop if there are 38.5 in.^{2} of stretched fabric on which to embroider.

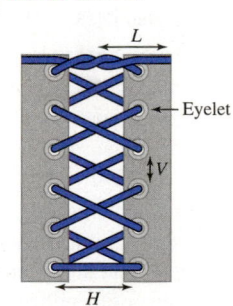

124. Pendulums. Find the period of a pendulum with length 1 foot. *See Example 5.*

125. Shoelaces. The formula

$$S = 2\left[H + L + (p - 1)\sqrt{H^2 + V^2}\right]$$ can be used to

calculate the correct shoelace length for the criss-cross lacing pattern shown in the illustration, where p represents the number of *pairs* of eyelets. Find the correct shoelace length if H(horizontal distance) = 50 mm, L (length of end) = 250 mm, and V(vertical distance) = 20 mm. Round to the nearest tenth. (*Source:* Ian's Shoelace Site at www.fieggen.com)

126. Baseball. The length of a diagonal of a square is given by the function $d(s) = \sqrt{2s^2}$, where s is the length of a side of the square. Find the distance from home plate to second base on a softball diamond and on a baseball diamond. Round to the nearest tenth. The illustration gives the dimensions of each type of infield.

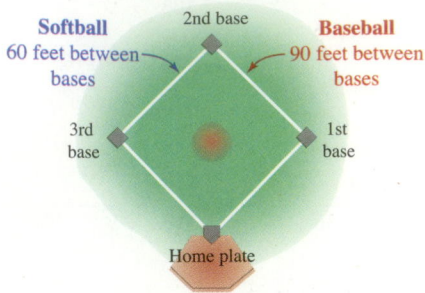

127. Pulse Rates. The approximate pulse rate (in beats per minute) of an adult who is t inches tall is given by the function $p(t) = \frac{590}{\sqrt{t}}$. The *Guinness Book of World Records 2008* lists Leonid Stadnyk of Ukraine as the tallest living man, at 8 feet, 5.5 inches. Find his approximate pulse rate as predicted by the function.

128. The Grand Canyon. The time t (in seconds) that it takes for an object to fall a distance of s feet is given by the formula $t = \frac{\sqrt{s}}{4}$. In some places, the Grand Canyon is one mile (5,280 feet) deep. How long would it take a stone dropped over the edge of the canyon to hit bottom?

129. Biology. Scientists will place five rats inside a clear plastic hemisphere and control the environment to study the rats' behavior. The function $d(V) = \sqrt[3]{12\left(\frac{V}{\pi}\right)}$ gives the diameter of a hemisphere with volume V. Use the function to determine the diameter of the base of the hemisphere, if each rat requires 125 cubic feet of living space.

130. Aquariums. The function $s(g) = \sqrt[3]{\frac{g}{7.5}}$ determines how long (in feet) an edge of a cube-shaped tank must be if it is to hold g gallons of water. What dimensions should a cube-shaped aquarium have if it is to hold 1,250 gallons of water?

131. Collectibles. The *effective rate of interest* r earned by an investment is given by the formula $r = \sqrt[n]{\frac{A}{P}} - 1$, where P is the initial investment that grows to value A after n years. Determine the effective rate of interest earned by a collector on a Lladró porcelain figurine purchased for $800 and sold for $950 five years later.

132. Law Enforcement. The graphs of the two radical functions shown in the illustration in the next column can be used to estimate the speed (in mph) of a car involved in an accident. Suppose a police accident report listed skid marks to be 220 feet long but failed to give the road conditions. Estimate the possible speeds the car was traveling prior to the brakes being applied.

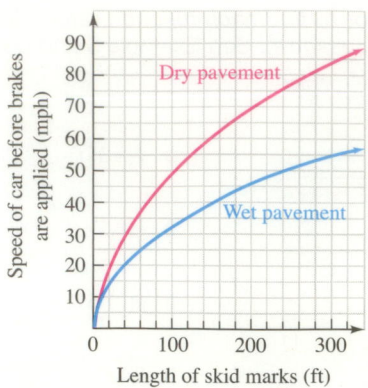

WRITING

133. If x is any real number, that is, if x is unrestricted, then $\sqrt{x^2} = x$ is not correct. Explain why.

134. Explain why $\sqrt{36}$ is just 6, and not also -6.

135. Explain what is wrong with the graph in the illustration if it is supposed to be the graph of $f(x) = \sqrt{x}$.

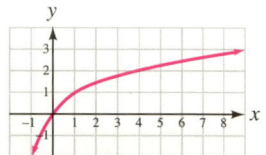

136. Explain how to estimate the domain and range of the radical function whose graph is shown here.

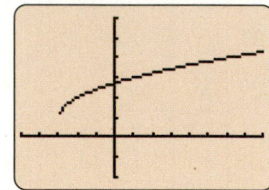

REVIEW

Perform the operations and simplify when possible.

137. $\dfrac{x^2 - 3xy - 4y^2}{x^2 + cx - 2yx - 2cy} \div \dfrac{x^2 - 2xy - 3y^2}{x^2 + cx - 4yx - 4cy}$

138. $\dfrac{2x + 3}{3x - 1} - \dfrac{x - 4}{2x + 1}$

CHALLENGE PROBLEMS

139. Graph $f(x) = -\sqrt{x - 2} + 3$ and find the domain and range.

140. Simplify $\sqrt{9a^{16} + 12a^8 b^{25} + 4b^{50}}$ and assume that $a > 0$ and $b > 0$.

SECTION 7.2

OBJECTIVES

1. Simplify expressions of the form $a^{1/n}$.
2. Simplify expressions of the form $a^{m/n}$.
3. Convert between radicals and rational exponents.
4. Simplify expressions with negative rational exponents.
5. Use rules for exponents to simplify expressions.
6. Simplify radical expressions.

Rational Exponents

ARE YOU READY?

The following problems review some basic skills that are needed when working with rational exponents.

1. Evaluate: **a.** $\sqrt{64}$ **b.** $\sqrt[3]{-64}$ 2. Evaluate: **a.** $\sqrt[4]{81}$ **b.** $\sqrt[5]{\dfrac{1}{32}}$

3. Simplify: **a.** $\sqrt{(x+5)^2}$ 4. Evaluate: $\left(\sqrt{36}\right)^3$
           b. $\sqrt[3]{27a^6}$

5. Simplify: 7^{-2} 6. Simplify: $\dfrac{x^5 x^7}{x^3}$

In this section, we will extend the definition of exponent to include rational (fractional) exponents. We will see how expressions such as $9^{1/2}$, $\left(\dfrac{1}{16}\right)^{3/4}$, and $(-32x^5)^{-2/5}$ can be simplified by writing them in an equivalent radical form using two new rules for exponents.

The Language of Algebra

Rational exponents are also called **fractional exponents**.

1 Simplify Expressions of the Form $a^{1/n}$.

It is possible to raise numbers to fractional powers. To give meaning to rational exponents, we first consider $\sqrt{7}$. Because $\sqrt{7}$ is the positive number whose square is 7, we have

$$\left(\sqrt{7}\right)^2 = 7$$

We now consider the notation $7^{1/2}$, which is read as "7 to the one-half power." If rational exponents are to follow the same rules as integer exponents, the square of $7^{1/2}$ must be 7, because

$$(7^{1/2})^2 = 7^{1/2 \cdot 2}$$ Keep the base and multiply the exponents.
$$= 7^1$$ Do the multiplication: $\frac{1}{2} \cdot 2 = 1$.
$$= 7$$

Since the square of $7^{1/2}$ and the square of $\sqrt{7}$ are both equal to 7, we define $7^{1/2}$ to be $\sqrt{7}$. Similarly,

$$7^{1/3} = \sqrt[3]{7}, \qquad 7^{1/4} = \sqrt[4]{7}, \qquad \text{and} \qquad 7^{1/5} = \sqrt[5]{7}$$

In general, we have the following definition.

The Definition of $x^{1/n}$	A rational exponent of $\dfrac{1}{n}$ indicates the nth root of its base. If n represents a positive integer greater than 1 and $\sqrt[n]{x}$ represents a real number,
	$$x^{1/n} = \sqrt[n]{x}$$ Read as "x to the $\frac{1}{n}$ power equals the nth root of x."

We can use this definition to simplify exponential expressions that have rational exponents with a numerator of 1. For example, to simplify $8^{1/3}$, we write it as an equivalent expression in radical form and proceed as follows:

Index

$$8^{1/3} = \sqrt[3]{8} = 2$$ The base of the exponential expression, 8, is the radicand of the radical expression. The denominator of the fractional exponent, 3, is the index of the radical.

Radicand

Thus, $8^{1/3} = 2$.

EXAMPLE 1 Evaluate: **a.** $9^{1/2}$ **b.** $(-64)^{1/3}$ **c.** $16^{1/4}$ **d.** $-\left(\dfrac{1}{32}\right)^{1/5}$

Strategy First, we will identify the base and the exponent of the exponential expression. Then we will write the expression in an equivalent radical form using the rule for rational exponents $x^{1/n} = \sqrt[n]{x}$.

Why We can then use the methods from Section 7.1 to evaluate the resulting square root, cube root, fourth root, and fifth root.

Solution

a. $9^{1/2} = \sqrt{9} = 3$

> The base is 9 and the exponent is $\frac{1}{2}$. Because the denominator of the exponent $\frac{1}{2}$ is 2, find the square root of the base, 9.

Success Tip

One key to successfully simplifying the exponential expressions in this section is this: When you see an exponent, always ask yourself, "What is the base?"

b. $(-64)^{1/3} = \sqrt[3]{-64} = -4$

> Read as " -64 to the one-third power." Because the denominator of the exponent $\frac{1}{3}$ is 3, find the cube root of the base, -64.

c. $16^{1/4} = \sqrt[4]{16} = 2$

> Because the denominator of the exponent $\frac{1}{4}$ is 4, find the fourth root of the base, 16.

d. $-\left(\dfrac{1}{32}\right)^{1/5} = -\sqrt[5]{\dfrac{1}{32}} = -\dfrac{1}{2}$

> Read as "the opposite of the one-fifth power of $\frac{1}{32}$." Because the denominator of the exponent $\frac{1}{5}$ is 5, find the fifth root of the base, $\frac{1}{32}$.

Self Check 1 Evaluate: **a.** $16^{1/2}$ **b.** $\left(-\dfrac{27}{8}\right)^{1/3}$ **c.** $-(81)^{1/4}$ **d.** $1^{1/5}$

Now Try ▶ Problems 19, 23, 25, and 27

As with radicals, when n is an *odd natural number* in the expression $x^{1/n}$, where $n > 1$, there is exactly one real nth root, and we don't need to use absolute value symbols.

When n is an *even natural number,* there are two nth roots. Since we want the expression $x^{1/n}$ to represent the positive nth root, we often must use absolute value symbols to ensure that the simplified result is positive. Thus, if n is even,

$$(x^n)^{1/n} = |x|$$

When n is even and x is negative, the expression $x^{1/n}$ is not a real number.

EXAMPLE 2 Simplify. Assume that the variables are unrestricted. **a.** $(-27x^3)^{1/3}$ **b.** $(256a^8)^{1/8}$
c. $[(y + 4)^2]^{1/2}$ **d.** $(25b^4)^{1/2}$ **e.** $(-256)^{1/4}$

Strategy We will write each exponential expression in an equivalent radical form using the rule for rational exponents $x^{1/n} = \sqrt[n]{x}$.

Why We can then use the methods of Section 7.1 to simplify the resulting radical expression.

Solution

a. $(-27x^3)^{1/3} = \sqrt[3]{-27x^3} = -3x$

> Read as "the quantity $-27x^3$ raised to the one-third power." Because $(-3x)^3 = -27x^3$. Since n is odd, no absolute value symbols are needed.

The Language of Algebra

The phrase **"assume that the variables are unrestricted"** in the instructions for Example 2 reminds us that negative numbers are permissible replacement values for the variables. Therefore, absolute value symbols may be necessary to ensure the answer is not negative.

b. $(256a^8)^{1/8} = \sqrt[8]{256a^8} = 2|a|$

> Because $(2a)^8 = 256a^8$. Since n is even and a can be any real number, $2a$ can be negative. Thus, absolute value symbols are needed.

c. $\left[(y + 4)^2\right]^{1/2} = \sqrt{(y + 4)^2} = |y + 4|$

> Because $|y + 4|^2 = (y + 4)^2$. Since n is even and y can be any real number, $y + 4$ can be negative. Thus, absolute value symbols are needed.

d. $(25b^4)^{1/2} = \sqrt{25b^4} = 5b^2$ Because $(5b^2)^2 = 25b^4$. Since $b^2 \geq 0$, no absolute value symbols are needed.

e. $(-256)^{1/4} = \sqrt[4]{-256}$, which is not a real number.   Because no real number raised to the 4th power is -256.

Self Check 2 Simplify. Assume that the variables are unrestricted.
a. $(-8n^3)^{1/3}$ **b.** $(625a^4)^{1/4}$ **c.** $(b^4)^{1/2}$ **d.** $(-64)^{1/6}$

Now Try ▶ Problems 29, 35, 37, and 39

If we were told that the variables represent positive real numbers in parts (b) and (c) of Example 2, the absolute value symbols in the answers would not be needed.

$(256a^8)^{1/8} = 2a$ If a represents a positive real number, then $2a$ is positive.

$[(y+4)^2]^{1/2} = y + 4$ If y represents a positive real number, then $y + 4$ is positive.

We summarize the cases as follows.

Summary of the Definitions of $x^{1/n}$

If n is a natural number greater than 1 and x is a real number,

If $x > 0$, then $x^{1/n}$ is the real number such that $(x^{1/n})^n = x$.

If $x = 0$, then $x^{1/n} = 0$.

If $x < 0$ $\begin{cases} \text{and } n \text{ is odd, then } x^{1/n} \text{ is the negative number such that } (x^{1/n})^n = x. \\ \text{and } n \text{ is even, then } x^{1/n} \text{ is not a real number.} \end{cases}$

② Simplify Expressions of the Form $a^{m/n}$.

We can extend the definition of $x^{1/n}$ to include fractional exponents with numerators other than 1. For example, since $8^{2/3}$ can be written as $(8^{1/3})^2$, we have

$8^{2/3} = (8^{1/3})^2$ Read $8^{2/3}$ as "8 to the two-thirds power."

$= (\sqrt[3]{8})^2$ Write $8^{1/3}$ in radical form.

$= 2^2$ Find the cube root first: $\sqrt[3]{8} = 2$.

$= 4$ Then find the power.

Thus, we can simplify $8^{2/3}$ by finding the second power of the cube root of 8.

The numerator of the rational exponent is the power.

$8^{2/3} = (\sqrt[3]{8})^2$ The base of the exponential expression is the radicand.

The denominator of the rational exponent is the index of the radical.

We also can simplify $8^{2/3}$ by taking the cube root of 8 squared.

$8^{2/3} = (8^2)^{1/3}$

$= 64^{1/3}$ Find the power first: $8^2 = 64$.

$= \sqrt[3]{64}$ Write $64^{1/3}$ in radical form.

$= 4$ Now find the cube root.

In general, we have the following definition.

The Definition of $x^{m/n}$	If m and n represent positive integers ($n \neq 1$) and $\sqrt[n]{x}$ represents a real number,
	$x^{m/n} = \left(\sqrt[n]{x}\right)^m$ and $x^{m/n} = \sqrt[n]{x^m}$

We read the first definition given above as "x to the m divided by n power equals the nth root of x, raised to the mth power."

Because of the previous definition, we can interpret $x^{m/n}$ in two ways:

1. $x^{m/n}$ means the nth root of the mth power of x.
2. $x^{m/n}$ means the mth power of the nth root of x.

We can use this definition to evaluate exponential expressions that have rational exponents with a numerator that is not 1. To avoid large numbers, we usually find the root of the base first and then calculate the power using the rule $x^{m/n} = \left(\sqrt[n]{x}\right)^m$.

EXAMPLE 3   Evaluate: **a.** $32^{2/5}$ **b.** $81^{3/4}$ **c.** $(-64)^{2/3}$ **d.** $-\left(\dfrac{1}{25}\right)^{3/2}$

Strategy   First, we will identify the base and the exponent of the exponential expression. Then we will write the expression in an equivalent radical form using the rule for rational exponents $x^{m/n} = \left(\sqrt[n]{x}\right)^m$.

Why   We know how to evaluate square roots, cube roots, fourth roots, and fifth roots.

Solution   **a.** To evaluate $32^{2/5}$, we write it in an equivalent radical form. The denominator of the rational exponent is the same as the index of the corresponding radical. The numerator of the rational exponent indicates the power to which the radical base is raised.

$$32^{2/5} = \left(\sqrt[5]{32}\right)^2 = (2)^2 = 4$$

Read as "32 to the two-fifths power." Because the exponent is 2/5, find the fifth root of the base, 32, to get 2. Then find the second power of 2.

b. $81^{3/4} = \left(\sqrt[4]{81}\right)^3 = (3)^3 = 27$

Read as "81 to the three-fourths power." Because the exponent is 3/4, find the fourth root of the base, 81, to get 3. Then find the third power of 3.

c. For $(-64)^{2/3}$, the base is -64.

$$(-64)^{2/3} = (\sqrt[3]{-64})^2 = (-4)^2 = 16$$

Read as "-64 to the two-thirds power." Because the exponent is 2/3, find the cube root of the base, -64, to get -4. Then find the second power of -4.

d. For $-\left(\dfrac{1}{25}\right)^{3/2}$, the base is $\dfrac{1}{25}$, not $-\dfrac{1}{25}$.

$$-\left(\frac{1}{25}\right)^{3/2} = -\left(\sqrt[2]{\frac{1}{25}}\right)^3 = -\left(\frac{1}{5}\right)^3 = -\frac{1}{125}$$

Read as "the opposite of the three-halves power of $\frac{1}{25}$." Because the exponent is 3/2, find the square root of the base, $\frac{1}{25}$, to get $\frac{1}{5}$. Then find the third power of $\frac{1}{5}$.

Caution

We also can evaluate $x^{m/n}$ using $\sqrt[n]{x^m}$. However, the resulting radicand is often extremely large. For example,

$$81^{3/4} = \sqrt[4]{81^3}$$
$$= \sqrt[4]{531,441}$$
$$= 27$$

Self Check 3   Evaluate: **a.** $16^{3/2}$ **b.** $125^{4/3}$ **c.** $(-216)^{2/3}$

d. $-\left(\dfrac{1}{32}\right)^{4/5}$

Now Try ▶ Problems 41, 45, and 47

EXAMPLE 4   Simplify. All variables represent positive real numbers.   **a.** $(36m^4)^{3/2}$ **b.** $(-8x^3)^{4/3}$
c. $-(x^5y^5)^{2/5}$

Strategy   We will write each exponential expression in an equivalent radical form using the rule for rational exponents $x^{m/n} = \left(\sqrt[n]{x}\right)^m$.

Why   We can then use the methods of Section 7.1 to simplify the resulting radical expression.

Solution   **a.**

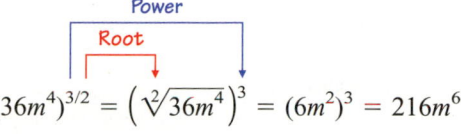

$(36m^4)^{3/2} = \left(\sqrt[2]{36m^4}\right)^3 = (6m^2)^3 = 216m^6$

Read as "the quantity of $36m^4$ raised to the three-halves power."
Because the exponent is 3/2, find the square root of the base, $36m^4$, to get $6m^2$. Then find the third power of $6m^2$.

b.

$(-8x^3)^{4/3} = \left(\sqrt[3]{-8x^3}\right)^4 = (-2x)^4 = 16x^4$

Read as "the quantity of $-8x^3$ raised to the four-thirds power."
Because the exponent is 4/3, find the cube root of the base, $-8x^3$, to get $-2x$. Then find the fourth power of $-2x$.

c.

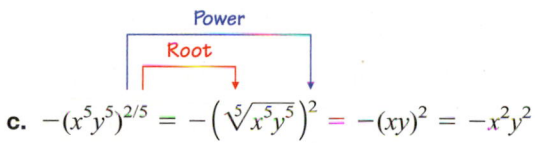

$-(x^5y^5)^{2/5} = -\left(\sqrt[5]{x^5y^5}\right)^2 = -(xy)^2 = -x^2y^2$

Read as "the opposite of the two-fifths power of the quantity x^5y^5."
Because the exponent is 2/5, find the fifth root of the base, x^5y^5, to get xy. Then find the second power of xy.

Self Check 4   Simplify. All variables represent positive real numbers. **a.** $(4c^4)^{3/2}$
b. $(-27m^3n^3)^{2/3}$ **c.** $-(32a^{10})^{3/5}$

Now Try ▶ Problems 49 and 51

Using Your Calculator ▶ **Rational Exponents**

We can evaluate expressions containing rational exponents using the exponential key $\boxed{y^x}$ or $\boxed{x^y}$ on a scientific calculator. For example, to evaluate $10^{2/3}$, we enter

$10\ \boxed{y^x}\ \boxed{(}\ 2\ \boxed{\div}\ 3\ \boxed{)}\ \boxed{=}$ $\boxed{4.641588834}$

Note that parentheses were used when entering the power. Without them, the calculator would interpret the entry as $10^2 \div 3$.

To evaluate the exponential expression using a direct entry or graphing calculator, we use the $\boxed{\wedge}$ key, which raises a base to a power. Again, we use parentheses when entering the power.

$10\ \boxed{\wedge}\ \boxed{(}\ 2\ \boxed{\div}\ 3\ \boxed{)}\ \boxed{\text{ENTER}}$ $\boxed{\begin{array}{l}\text{10\^{}(2/3)}\\ \qquad 4.641588834\end{array}}$

To the nearest hundredth, $10^{2/3} \approx 4.64$.

3 Convert Between Radicals and Rational Exponents.

We can use the rules for rational exponents to convert expressions from radical form to exponential form, and vice versa.

EXAMPLE 5   Write $\sqrt{5xyz}$ as an exponential expression with a rational exponent.

Strategy   We will use the first rule for rational exponents in reverse: $\sqrt[n]{x} = x^{1/n}$.

Why   We are given a radical expression and we want to write an equivalent exponential expression.

Solution   The radicand is $5xyz$, so the base of the exponential expression is $5xyz$. The index of the radical is an understood 2, so the denominator of the fractional exponent is 2.

$$\sqrt{5xyz} = (5xyz)^{1/2} \quad \text{Recall: } \sqrt[2]{5xyz} = \sqrt{5xyz}.$$

Self Check 5   Write $\sqrt[6]{7ab}$ as an exponential expression with a rational exponent.

Now Try ▶ Problem 57

Rational exponents appear in formulas used in many disciplines, such as science and engineering.

EXAMPLE 6   **Satellites.** The formula $r = \left(\dfrac{GMP^2}{4\pi^2}\right)^{1/3}$ gives the orbital radius (in meters) of a satellite circling Earth, where G and M are constants and P is the time in seconds for the satellite to make one complete revolution. Write the formula using a radical.

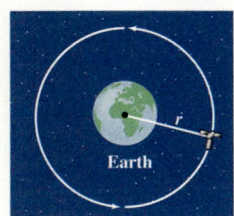
Earth

Strategy   We will use the first rule for rational exponents: $x^{1/n} = \sqrt[n]{x}$.

Why   We are given an exponential expression involving a rational exponent with a numerator of 1 and we want to write an equivalent radical expression.

Solution   The fractional exponent $\frac{1}{3}$, with a numerator of 1 and a denominator of 3, indicates that we are to find the cube root of the base of the exponential expression. So we have

$$r = \sqrt[3]{\frac{GMP^2}{4\pi^2}}$$

Self Check 6   **Statistics.**   The formula $\sigma = \left(\dfrac{\Sigma(x - \mu)^2}{N}\right)^{1/2}$ gives the population standard deviation. Write the formula using a radical.

Now Try ▶ Problem 61

4 Simplify Expressions with Negative Rational Exponents.

To be consistent with the definition of negative integer exponents, we define $x^{-m/n}$ as follows.

Definition of $x^{-m/n}$

If m and n are positive integers, $\frac{m}{n}$ is in simplified form, and $x^{1/n}$ is a real number, then

$$x^{-m/n} = \frac{1}{x^{m/n}} \quad \text{and} \quad \frac{1}{x^{-m/n}} = x^{m/n} \quad (x \neq 0)$$

From the definition, we see that another way to write $x^{-m/n}$ is to write its reciprocal and change the sign of the exponent.

EXAMPLE 7 Simplify. Assume that x can represent any nonzero real number.

a. $64^{-1/2}$ **b.** $(-16)^{-5/4}$ **c.** $-625^{-3/4}$ **d.** $(-32x^5)^{-2/5}$ **e.** $\dfrac{1}{25^{-3/2}}$

Strategy We will use one of the rules $x^{-m/n} = \dfrac{1}{x^{m/n}}$ or $\dfrac{1}{x^{-m/n}} = x^{m/n}$ to write the reciprocal of each exponential expression and change the exponent's sign to positive.

Why If we can produce an equivalent expression having a positive rational exponent, we can use the methods of this section to simplify it.

Solution

a. $64^{-1/2} = \dfrac{1}{64^{1/2}} = \dfrac{1}{\sqrt{64}} = \dfrac{1}{8}$ Read as "64 to the negative one-half power."
Because the exponent is negative, write the reciprocal of $64^{-1/2}$, and change the sign of the exponent.

Caution

A negative exponent does not, itself, indicate a negative number. For example,

$$64^{-1/2} = \frac{1}{8}$$

b. $(-16)^{-5/4}$ is read as "-16 to the negative five-fourths power." It is not a real number because $(-16)^{1/4}$ is not a real number.

c. In $-625^{-3/4}$, the base is 625.

$$-625^{-3/4} = -\frac{1}{625^{3/4}} = -\frac{1}{\left(\sqrt[4]{625}\right)^3} = -\frac{1}{(5)^3} = -\frac{1}{125}$$

Caution

A base of 0 raised to a negative power is undefined. For example, $0^{-2} = \frac{1}{0^2}$ is undefined because we cannot divide by 0.

d. This is read as "the quantity of $-32x^5$, raised to the negative two-fifths power."

$$(-32x^5)^{-2/5} = \frac{1}{(-32x^5)^{2/5}} = \frac{1}{\left(\sqrt[5]{-32x^5}\right)^2} = \frac{1}{(-2x)^2} = \frac{1}{4x^2}$$

e. This is read as "1 over 25 to the negative three-halves power."

$$\frac{1}{25^{-3/2}} = 25^{3/2} = \left(\sqrt{25}\right)^3 = (5)^3 = 125$$ Because the exponent is negative, write the reciprocal of $\frac{1}{25^{-3/2}}$, and change the sign of the exponent.

Self Check 7 Simplify. Assume that a can represent any nonzero real number.
a. $9^{-1/2}$ **b.** $(36)^{-3/2}$ **c.** $(-27a^3)^{-2/3}$ **d.** $-\dfrac{1}{81^{-3/4}}$

Now Try ▶ Problems 65, 69, and 75

---

5 Use Rules for Exponents to Simplify Expressions.

We can use the rules for exponents to simplify many expressions with fractional exponents. If all variables represent positive real numbers, absolute value symbols are not needed.

EXAMPLE 8 Simplify. All variables represent positive real numbers. Write all answers using positive exponents only. **a.** $5^{2/7} \cdot 5^{3/7}$ **b.** $(11^{2/7})^3$ **c.** $(a^{2/3}b^{1/2})^6$ **d.** $\dfrac{a^{8/3}a^{1/3}}{a^2}$

Strategy We will use the product, power, and quotient rules for exponents to simplify each expression.

Why The familiar rules for exponents discussed in Chapter 5 are valid for rational exponents.

Solution

a. $5^{2/7} \cdot 5^{3/7} = 5^{2/7+3/7}$ Use the rule $x^m x^n = x^{m+n}$. Do not multiply the bases.

$= 5^{5/7}$ Add the exponents: $\frac{2}{7} + \frac{3}{7} = \frac{5}{7}$.

b. $(11^{2/7})^3 = 11^{(2/7)(3)}$ Use the rule $(x^m)^n = x^{mn}$.

$= 11^{6/7}$ Multiply the exponents: $\frac{2}{7}(3) = \frac{6}{7}$.

c. $(a^{2/3} b^{1/2})^6 = (a^{2/3})^6 (b^{1/2})^6$ Use the rule $(xy)^n = x^n y^n$.

$= a^{12/3} b^{6/2}$ Use the rule $(x^m)^n = x^{mn}$ twice. Multiply the exponents.

$= a^4 b^3$ Simplify the exponents.

d. $\dfrac{a^{8/3} a^{1/3}}{a^2} = a^{8/3 + 1/3 - 2}$ Use the rules $x^m x^n = x^{m+n}$ and $\frac{x^m}{x^n} = x^{m-n}$.

$= a^{8/3 + 1/3 - 6/3}$ To establish an LCD, write -2 as $-\frac{6}{3}$.

$= a^{3/3}$ Simplify: $\frac{8}{3} + \frac{1}{3} - \frac{6}{3} = \frac{3}{3}$.

$= a$ Simplify: $\frac{3}{3} = 1$.

Success Tip

Now is a good time to review the rules for exponents on page 365 in Chapter 5. These rules are not only valid when m and n are integers, but when m and n are rational numbers, as well.

Self Check 8 Simplify. All variables represent positive real numbers.
a. $2^{1/5} \cdot 2^{2/5}$ **b.** $(12^{1/3})^4$ **c.** $(x^{1/3} y^{3/2})^6$ **d.** $\dfrac{x^{5/3} x^{2/3}}{x^{1/3}}$

Now Try ▶ Problems 77, 81, and 85

EXAMPLE 9 Perform each multiplication and simplify when possible. Assume all variables represent positive real numbers. Write all answers using positive exponents only.
a. $a^{4/5}(a^{1/5} + a^{3/5})$ **b.** $x^{1/2}(x^{-1/2} - x^{1/2})$

Strategy We will use the distributive property and multiply each term within the parentheses by the term outside the parentheses.

Why The first expression has the form $a(b + c)$ and the second has the form $a(b - c)$.

Solution

a. $a^{4/5}(a^{1/5} + a^{3/5}) = a^{4/5} a^{1/5} + a^{4/5} a^{3/5}$ Use the distributive property.

$= a^{4/5 + 1/5} + a^{4/5 + 3/5}$ Use the rule $x^m x^n = x^{m+n}$.

$= a^{5/5} + a^{7/5}$ Add the exponents.

$= a + a^{7/5}$ We cannot add these terms because they are not like terms.

Caution

A common error when distributing $a^{4/5}$ over the terms $a^{1/5}$ and $a^{3/5}$ is to multiply the exponents incorrectly :

$a^{4/5}(a^{1/5} + a^{3/5}) = a^{4/25} + a^{12/25}$

b. $x^{1/2}(x^{-1/2} - x^{1/2}) = x^{1/2} x^{-1/2} - x^{1/2} x^{1/2}$ Use the distributive property.

$= x^{1/2 + (-1/2)} - x^{1/2 + 1/2}$ Use the rule $x^m x^n = x^{m+n}$.

$= x^0 - x^1$ Add the exponents.

$= 1 - x$ Simplify: $x^0 = 1$.

Self Check 9 Simplify: $t^{5/8}(t^{3/8} - t^{-5/8})$. Assume t represents a positive real number.

Now Try ▶ Problem 91

---

6 Simplify Radical Expressions.

We can simplify many radical expressions by using the following steps.

Using Rational Exponents to Simplify Radicals

1. Change the radical expression into an exponential expression.
2. Simplify the rational exponents.
3. Change the exponential expression back into a radical.

EXAMPLE 10   Simplify: **a.** $\sqrt[4]{3^2}$ **b.** $\sqrt[8]{x^6}$ **c.** $\sqrt[9]{27x^6y^3}$ **d.** $\sqrt[5]{\sqrt[3]{t}}$

Strategy   We will write each radical expression as an equivalent exponential expression and use rules for exponents to simplify it. Then we will change that result back into a radical.

Why   When the given expression is written in an equivalent exponential form, we can use rules for exponents and our arithmetic skills with fractions to simplify the exponents.

Solution   **a.** $\sqrt[4]{3^2} = 3^{2/4}$ Use the rule $\sqrt[n]{x^m} = x^{m/n}$.

$= 3^{1/2}$ Simplify the fractional exponent: $\frac{2}{4} = \frac{1}{2}$.

$= \sqrt{3}$ Change back to radical form.

b. $\sqrt[8]{x^6} = x^{6/8}$ Use the rule $\sqrt[n]{x^m} = x^{m/n}$.

$= x^{3/4}$ Simplify the fractional exponent: $\frac{6}{8} = \frac{3}{4}$.

$= (x^3)^{1/4}$ Write $\frac{3}{4}$ as $3\left(\frac{1}{4}\right)$.

$= \sqrt[4]{x^3}$ Change back to radical form.

c. $\sqrt[9]{27x^6y^3} = (3^3x^6y^3)^{1/9}$ Write 27 as 3^3 and change the radical to an exponential expression.

$= 3^{3/9}x^{6/9}y^{3/9}$ Raise each factor to the $\frac{1}{9}$ power by multiplying the fractional exponents.

$= 3^{1/3}x^{2/3}y^{1/3}$ Simplify each fractional exponent.

$= (3x^2y)^{1/3}$ Use the rule $(xy)^n = x^ny^n$.

$= \sqrt[3]{3x^2y}$ Change back to radical form.

d. $\sqrt[5]{\sqrt[3]{t}} = \sqrt[5]{t^{1/3}}$ Change the radical $\sqrt[3]{t}$ to exponential notation.

$= (t^{1/3})^{1/5}$ Change the radical $\sqrt[5]{t^{1/3}}$ to exponential notation.

$= t^{1/15}$ Use the rule $(x^m)^n = x^{mn}$. Multiply: $\frac{1}{3} \cdot \frac{1}{5} = \frac{1}{15}$.

$= \sqrt[15]{t}$ Change back to radical form.

Self Check 10   Simplify: **a.** $\sqrt[6]{3^3}$ **b.** $\sqrt[4]{49x^2y^2}$ **c.** $\sqrt[3]{\sqrt[4]{m}}$

Now Try ▶   Problems 93, 99, and 101

SECTION 7.2   STUDY SET

VOCABULARY

Fill in the blanks.

1. The expressions $4^{1/2}$ and $(-8)^{-2/3}$ have _____ exponents.

2. We read $16^{3/2}$ as "16 to the three-_____ power."

3. We read $27^{-1/3}$ as "27 to the _____ one-third power."

4. We read $(-64a^5)^{4/5}$ as "the quantity of $-64a^5$, _____ to the four-fifths power."

5. In the radical expression $\sqrt[4]{16x^8}$, 4 is the _____, and $16x^8$ is the _____.

6. $32^{4/5}$ means the fourth _____ of the fifth _____ of 32.

CONCEPTS

7. Complete the table by writing the given expression in the alternate form. Also give the base and exponent for the exponential form.

Radical form	Exponential form	Base	Exponent
$\sqrt[5]{25}$			
	$(-27)^{2/3}$		
$\left(\sqrt[4]{16}\right)^{-3}$			
	$81^{3/2}$		
$-\sqrt{\frac{9}{64}}$			

8. In your own words, explain the three rules for rational exponents illustrated in the diagrams below.

a. $(-32)^{1/5} = \sqrt[5]{-32}$

b. $125^{4/3} = (\sqrt[3]{125})^4$

c. $8^{-1/3} = \dfrac{1}{8^{1/3}}$

9. Graph the following real numbers on a number line.

$$\left\{ 8^{2/3},\ (-125)^{1/3},\ -16^{-1/4},\ 4^{3/2},\ -\left(\frac{9}{100}\right)^{-1/2} \right\}$$

10. a. Evaluate $25^{3/2}$ by writing it in the form $(25^{1/2})^3$.
   b. Evaluate $25^{3/2}$ by writing it in the form $(25^3)^{1/2}$.
   c. Which way was easier?

Complete each rule for exponents.

11. $x^{1/n} = $ ▢

12. $x^{m/n} = $ ▢ $= \sqrt[n]{x^m}$

13. $x^{-m/n} = $ ▢

14. $\dfrac{1}{x^{-m/n}} = $ ▢

NOTATION

Complete each solution.

15. Simplify:

$(100a^4)^{3/2} = \left(\sqrt{}\right)^3$
$= \left(\right)^3$
$= 1,000a^6$

16. Simplify:

$(m^{1/3}n^{1/2})^6 = \left(\right)^6 (n^{1/2})^6$
$= m^{}n^{6/2}$
$= m^2 n^3$

GUIDED PRACTICE

Evaluate each expression. See Example 1.

17. $125^{1/3}$ **18.** $8^{1/3}$ **19.** $81^{1/4}$ **20.** $625^{1/4}$

21. $32^{1/5}$ **22.** $0^{1/5}$ **23.** $(-216)^{1/3}$ **24.** $(-1,000)^{1/3}$

25. $-16^{1/4}$ **26.** $-125^{1/3}$ **27.** $\left(\dfrac{1}{4}\right)^{1/2}$ **28.** $\left(\dfrac{1}{16}\right)^{1/2}$

Simplify each expression. Assume that the variables can be any real number, and use absolute value symbols when necessary. See Example 2.

29. $(4x^4)^{1/2}$ **30.** $(25a^8)^{1/2}$

31. $(x^2)^{1/2}$ **32.** $(x^3)^{1/3}$

33. $(-64p^8)^{1/2}$ **34.** $(-16q^4)^{1/2}$

35. $(-27n^9)^{1/3}$ **36.** $(-64t^9)^{1/3}$

37. $(-64x^8)^{1/8}$ **38.** $(243x^{10})^{1/5}$

39. $[(x+1)^6]^{1/6}$ **40.** $[(x+5)^8]^{1/8}$

Evaluate each expression. See Example 3.

41. $36^{3/2}$ **42.** $27^{2/3}$

43. $16^{3/4}$ **44.** $-100^{3/2}$

45. $\left(-\dfrac{1}{216}\right)^{2/3}$ **46.** $\left(\dfrac{4}{9}\right)^{3/2}$

47. $-4^{5/2}$ **48.** $(-125)^{4/3}$

Simplify each expression. All variables represent positive real numbers. See Example 4.

49. $(25x^4)^{3/2}$ **50.** $(27a^3b^3)^{2/3}$

51. $(-8x^6y^3)^{2/3}$ **52.** $(-32x^{10}y^5)^{4/5}$

53. $(81x^4y^8)^{3/4}$ **54.** $\left(\dfrac{1}{16}x^8y^4\right)^{3/4}$

55. $-\left(\dfrac{x^5}{32}\right)^{4/5}$ **56.** $-\left(\dfrac{27}{64y^6}\right)^{2/3}$

Change each radical to an exponential expression. See Example 5.

57. $\sqrt[5]{8abc}$ **58.** $\sqrt[7]{7p^2q}$

59. $\sqrt[3]{a^2-b^2}$ **60.** $\sqrt{x^2+y^2}$

Change each exponential expression to a radical. See Example 6.

61. $(6x^3y)^{1/4}$ **62.** $(7a^2b^2)^{1/5}$

63. $(2s^2-t^2)^{1/2}$ **64.** $(x^3+y^3)^{1/3}$

Simplify each expression. All variables represent positive real numbers. See Example 7.

65. $4^{-1/2}$ **66.** $49^{-1/2}$

67. $125^{-1/3}$ **68.** $8^{-1/3}$

69. $-(1,000y^3)^{-2/3}$ **70.** $-(81c^4)^{-3/2}$

71. $\left(-\dfrac{27}{8}\right)^{-4/3}$ **72.** $\left(\dfrac{25}{49}\right)^{-3/2}$

73. $\left(\dfrac{16}{81y^4}\right)^{-3/4}$ **74.** $\left(-\dfrac{8x^3}{27}\right)^{-1/3}$

75. $\dfrac{1}{9^{-5/2}}$ **76.** $\dfrac{1}{16^{-5/2}}$

Simplify each expression. Write the answers without negative exponents. All variables represent positive real numbers. See Example 8.

77. $9^{3/7} \cdot 9^{2/7}$ **78.** $4^{2/5} \cdot 4^{2/5}$

79. $6^{-2/3}6^{-4/3}$ **80.** $5^{1/3}5^{-5/3}$

81. $(m^{2/3}m^{1/3})^6$ **82.** $(b^{3/5}b^{2/5})^8$

83. $(a^{1/2}b^{1/3})^{3/2}$ **84.** $(mn^{-2/3})^{-3/5}$

85. $\dfrac{3^{4/3}3^{1/3}}{3^{2/3}}$ **86.** $\dfrac{2^{5/6}2^{1/3}}{2^{1/2}}$

87. $\dfrac{a^{3/4}a^{3/4}}{a^{1/2}}$ **88.** $\dfrac{b^{4/5}b^{4/5}}{b^{3/5}}$

Perform the multiplications. All variables represent positive real numbers. See Example 9.

89. $y^{1/3}(y^{2/3}+y^{5/3})$ **90.** $y^{2/5}(y^{-2/5}+y^{3/5})$

91. $x^{3/5}(x^{7/5}-x^{-3/5}+1)$ **92.** $x^{4/3}(x^{2/3}+3x^{5/3}-4)$

Use rational exponents to simplify each radical. All variables represent positive real numbers. See Example 10.

93. $\sqrt[4]{5^2}$ **94.** $\sqrt[6]{7^3}$

95. $\sqrt[9]{11^3}$ **96.** $\sqrt[12]{13^4}$

97. $\sqrt[6]{p^3}$ **98.** $\sqrt[8]{q^2}$

99. $\sqrt[10]{x^2y^2}$

100. $\sqrt[6]{x^2y^2}$

101. $\sqrt[9]{\sqrt{c}}$

102. $\sqrt[4]{\sqrt{x}}$

103. $\sqrt[5]{\sqrt[3]{7m}}$

104. $\sqrt[3]{\sqrt[4]{21x}}$

Use a calculator to evaluate each expression. Round to the nearest hundredth. See **Using Your Calculator: Rational Exponents.**

105. $15^{1/3}$

106. $(50.5)^{1/4}$

107. $(1.045)^{2/5}$

108. $(-1,000)^{3/5}$

TRY IT YOURSELF

Simplify each expression. All variables represent positive real numbers.

109. $(25y^2)^{1/2}$

110. $(-27x^3)^{1/3}$

111. $-\left(\dfrac{a^4}{81}\right)^{3/4}$

112. $-\left(\dfrac{b^8}{625}\right)^{3/4}$

113. $16^{-3/2}$

114. $(16)^{-5/4}$

115. $\dfrac{p^{8/5}p^{7/5}}{p^2}$

116. $\dfrac{c^{2/3}c^{2/3}}{c^{1/3}}$

117. $(-27x^6)^{-1/3}$

118. $(16a^4)^{-1/2}$

119. $\dfrac{1}{32^{-1/5}}$

120. $\dfrac{1}{64^{-1/6}}$

121. $n^{1/5}(n^{2/5} - n^{-1/5})$

122. $t^{4/3}(t^{5/3} + t^{-4/3})$

123. $\dfrac{1}{4^{-5/2}}$

124. $\dfrac{1}{100^{-5/2}}$

125. $(m^4)^{1/2}$

126. $(a^4)^{1/4}$

127. $\sqrt[4]{25b^2}$

128. $\sqrt[9]{8x^6}$

129. $(16x^4)^{1/4}$

130. $(-x^4)^{1/4}$

131. $-(8a^3b^6)^{-2/3}$

132. $-(25s^4t^6)^{-3/2}$

Look Alikes . . .

133. a. $-125^{2/3}$ b. $(-125)^{2/3}$

c. $-125^{-2/3}$ d. $\dfrac{1}{(-125)^{-2/3}}$

134. a. $81^{1/4}$ b. $81^{-1/4}$

c. $-81^{1/4}$ d. $\dfrac{1}{81^{-1/4}}$

135. a. $(64a^4)^{1/2}$ b. $(64a^4)^{-1/2}$

c. $-(64a^4)^{1/2}$ d. $\dfrac{1}{(64a^4)^{1/2}}$

136. a. $r^{1/3} \cdot r^{1/5}$ b. $(r^{1/3})^{1/5}$

c. $r^{1/3} \cdot r^{-1/5}$ d. $(r^{-1/3})^{-1/5}$

APPLICATIONS

137. **Ballistic Pendulums.** The formula $v = \dfrac{m+M}{m}(2gh)^{1/2}$ gives the velocity (in ft/sec) of a bullet with weight m fired into a block with weight M, that raises the height of the block h feet after the collision. See the illustration in the next column. The letter g represents a constant, 32. Find the velocity of the bullet to the nearest ft/sec.

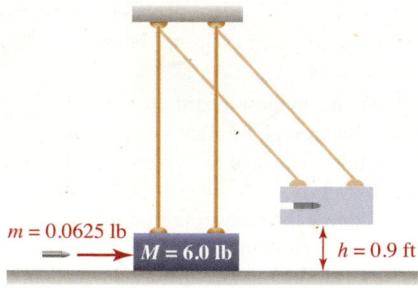

$m = 0.0625$ lb $M = 6.0$ lb $h = 0.9$ ft

138. **Geography.** The formula $A = [s(s-a)(s-b)(s-c)]^{1/2}$ gives the area of a triangle with sides of length a, b, and c, where s is one-half of the perimeter. Estimate the area of Virginia (to the nearest square mile) using the data given in the illustration.

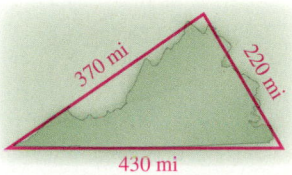

370 mi   220 mi   430 mi

139. **Relativity.** One concept of relativity theory is that an object moving past an observer at a speed near the speed of light appears to have a larger mass because of its motion. If the mass of the object is m_0 when the object is at rest relative to the observer, its mass m will be given by the formula $m = m_0\left(1 - \dfrac{v^2}{c^2}\right)^{-1/2}$ when it is moving with speed v (in miles per second) past the observer. The variable c is the speed of light, 186,000 mi/sec. If a proton with a rest mass of 1 unit is accelerated by a nuclear accelerator to a speed of 160,000 mi/sec, what mass will the technicians observe it to have? Round to the nearest hundredth.

140. **Logging.** The width w and height h of the strongest rectangular beam that can be cut from a cylindrical log of radius a are given by $w = \dfrac{2a}{3}\left(3^{1/2}\right)$ and $h = a\left(\dfrac{8}{3}\right)^{1/2}$. Find the width, height, and cross-sectional area of the strongest beam that can be cut from a log with *diameter* 4 feet. Round to the nearest hundredth.

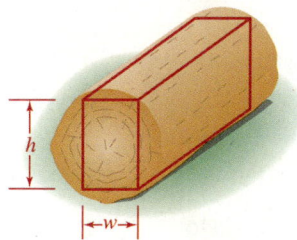

h w

141.

from **Campus to Careers**

General Contractor

©Goodluz/Shutterstock.com

The length L of the longest board that can be carried horizontally around the right-angle corner of two intersecting hallways is given by the formula $L = (a^{2/3} + b^{2/3})^{3/2}$, where a and b represent the widths of the hallways. Find the longest shelf that a carpenter can carry around the corner if $a = 40$ in. and $b = 64$ in. Give your result in inches and in feet. In each case, round to the nearest tenth.

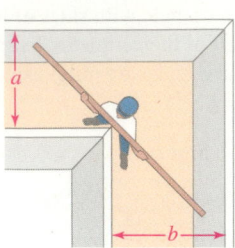

142. Cubicles. The area of the base of a cube is given by the function $A(V) = V^{2/3}$, where V is the volume of the cube. In a preschool room, 18 children's cubicles like the one shown are placed on the floor around the room. Estimate how much floor space is lost to the cubicles. Give your answer in square inches and in square feet.

Mary S.

Storage capacity 4,096 in.^{3}

WRITING

143. What is a rational exponent? Give some examples.

144. Explain how the root key $\boxed{\sqrt[x]{y}}$ on a scientific calculator can be used in combination with other keys to evaluate the expression $16^{3/4}$.

REVIEW

145. Commuting Time. The time it takes a car to travel a certain distance varies inversely with its rate of speed. If a certain trip takes 3 hours at 50 miles per hour, how long will the trip take at 60 miles per hour?

146. Bankruptcy. After filing for bankruptcy, a company was able to pay its creditors only 15 cents on the dollar. If the company owed a lumberyard $9,712, how much could the lumberyard expect to be paid?

CHALLENGE PROBLEMS

147. The fraction $\frac{2}{4}$ is equal to $\frac{1}{2}$. Is $16^{2/4}$ equal to $16^{1/2}$? Explain.

148. Explain how would you evaluate an expression with a mixed-number exponent. For example, what is $8^{1\frac{1}{3}}$? What is $25^{2\frac{1}{2}}$?

SECTION **7.3**	# Simplifying and Combining Radical Expressions

OBJECTIVES

1 Use the product rule to simplify radical expressions.

2 Use prime factorization to simplify radical expressions.

3 Use the quotient rule to simplify radical expressions.

4 Add and subtract radical expressions.

ARE YOU READY?

The following problems review some basic skills that are needed when adding and subtracting radical expressions.

1. Complete each factorization:
   a. $28 = \boxed{} \cdot 7$ **b.** $54 = \boxed{} \cdot 2$

2. Simplify:
   a. $\sqrt[3]{-8}$ **b.** $\sqrt[4]{16}$

3. Multiply: $3 \cdot 3 \cdot 3 \cdot a^4 \cdot b^8$

4. Which of the following are like terms?

   $7x \quad 3x^2 \quad 9x \quad 2x^3$

5. Combine like terms:
   $15m^2 + 5m - m^2 - 6m$

6. Simplify: $\dfrac{\sqrt{25}}{\sqrt{36}}$

In algebra, it is often helpful to replace an expression with a simpler equivalent expression. This is certainly true when working with radicals. In most cases, radical expressions should be written in simplified form. We use two rules for radicals to do this.

1 ## Use the Product Rule to Simplify Radical Expressions.

To introduce the product rule for radicals, we will find $\sqrt{4 \cdot 25}$ and $\sqrt{4}\sqrt{25}$, and compare the results.

Square root of a product	*Product of square roots*
$\sqrt{4 \cdot 25} = \sqrt{100}$	$\sqrt{4}\sqrt{25} = 2 \cdot 5$
$= 10$	$= 10$

In each case, the answer is 10. Thus, $\sqrt{4 \cdot 25} = \sqrt{4}\sqrt{25}$.

Similarly, we will find $\sqrt[3]{8 \cdot 27}$ and $\sqrt[3]{8}\sqrt[3]{27}$ and compare the results.

Cube root of a product	*Product of cube roots*
$\sqrt[3]{8 \cdot 27} = \sqrt[3]{216}$	$\sqrt[3]{8}\sqrt[3]{27} = 2 \cdot 3$
$= 6$	$= 6$

In each case, the answer is 6. Thus, $\sqrt[3]{8 \cdot 27} = \sqrt[3]{8}\sqrt[3]{27}$. These results illustrate the *product rule for radicals.*

Notation
The products $\sqrt{4}\sqrt{25}$ and $\sqrt[3]{8}\sqrt[3]{27}$ also can be written using a raised dot: $\sqrt{4} \cdot \sqrt{25}$ $\sqrt[3]{8} \cdot \sqrt[3]{27}$

The Product Rule for Radicals

The *n*th root of the product of two numbers is equal to the product of their *n*th roots. If $\sqrt[n]{a}$ and $\sqrt[n]{b}$ are real numbers,

$$\sqrt[n]{ab} = \sqrt[n]{a}\sqrt[n]{b}$$

Read as "the *n*th root of *a* times *b* equals the *n*th root of *a* times the *n*th root of *b*."

CAUTION The product rule for radicals applies to the *n*th root of a product. There is no such property for sums or differences. For example,

$\sqrt{9 + 4} \neq \sqrt{9} + \sqrt{4}$	$\sqrt{9 - 4} \neq \sqrt{9} - \sqrt{4}$
$\sqrt{13} \neq 3 + 2$	$\sqrt{5} \neq 3 - 2$
$\sqrt{13} \neq 5$	$\sqrt{5} \neq 1$

Thus, $\sqrt{a + b} \neq \sqrt{a} + \sqrt{b}$ and $\sqrt{a - b} \neq \sqrt{a} - \sqrt{b}$.

The product rule for radicals can be used to simplify radical expressions. When a radical expression is written in **simplified form,** each of the following is true.

Simplified Form of a Radical Expression

1. Each factor in the radicand is to a power that is less than the index of the radical.
2. The radicand contains no fractions or negative numbers.
3. No radicals appear in the denominator of a fraction.

To simplify radical expressions, we must often factor the radicand using two natural-number factors. Since one factor should be a perfect square, perfect cube, perfect-fourth power, and so/on, it is helpful to memorize the following lists.

Perfect squares: **1, 4, 9, 16, 25, 36, 49, 64, 81, 100, 121, 144, 169, 196, 225, . . .**

Perfect cubes: **1, 8, 27, 64, 125, 216, 343, 512, 729, 1,000, . . .**

Perfect-fourth powers: **1, 16, 81, 256, 625, . . .**

Perfect-fifth powers: **1, 32, 243, 1,024, . . .**

EXAMPLE 1 Simplify: **a.** $\sqrt{12}$ **b.** $\sqrt{98}$ **c.** $5\sqrt[3]{54}$ **d.** $-\sqrt[4]{48}$

Strategy We will factor each radicand into two factors, one of which is a perfect square, perfect cube, or perfect-fourth power, depending on the index of the radical. Then we can use the product rule for radicals to simplify the expression.

Why Factoring the radicand in this way leads to a square root, cube root, or fourth root of a perfect square, perfect cube, or perfect-fourth power that we can easily simplify.

Solution **a.** To simplify $\sqrt{12}$, we first factor 12 so that one factor is the largest perfect square that divides 12. Since 4 is the largest perfect-square factor of 12, we write 12 as $4 \cdot 3$, use the product rule for radicals, and simplify.

$$\sqrt{12} = \sqrt{4 \cdot 3} \quad \text{Write 12 as } 12 = 4 \cdot 3.$$

Write the perfect-square factor first.

$$= \sqrt{4}\sqrt{3} \quad \text{The square root of a product is equal to the product of the square roots.}$$

$$= 2\sqrt{3} \quad \text{Evaluate } \sqrt{4}. \text{ Read as "2 times the square root of 3" or as "2 radical 3."}$$

We say that $2\sqrt{3}$ is the simplified form of $\sqrt{12}$. Note that $\sqrt{12}$ and $2\sqrt{3}$ are just different representations of the same number. When we compare calculator approximations of each, that fact seems reasonable.

$$\sqrt{12} \approx 3.464101615 \qquad 2\sqrt{3} \approx 3.464101615$$

b. The largest perfect-square factor of 98 is 49. Thus,

$$\sqrt{98} = \sqrt{49 \cdot 2} \quad \text{Write 98 in factored form: } 98 = 49 \cdot 2.$$

$$= \sqrt{49}\sqrt{2} \quad \text{The square root of a product is equal to the product of the square roots: } \sqrt{49 \cdot 2} = \sqrt{49}\sqrt{2}.$$

$$= 7\sqrt{2} \quad \text{Evaluate } \sqrt{49}.$$

c. Since the largest perfect-cube factor of 54 is 27, we have

$$5\sqrt[3]{54} = 5\sqrt[3]{27 \cdot 2} \quad \text{Write 54 as } 27 \cdot 2.$$

$$= 5\sqrt[3]{27}\sqrt[3]{2} \quad \text{The cube root of a product is equal to the product of the cube roots: } \sqrt[3]{27 \cdot 2} = \sqrt[3]{27}\sqrt[3]{2}.$$

$$= 5 \cdot 3\sqrt[3]{2} \quad \text{Evaluate } \sqrt[3]{27}.$$

$$= 15\sqrt[3]{2} \quad \text{Multiply: } 5 \cdot 3 = 15.$$

d. The largest perfect-fourth power factor of 48 is 16. Thus,

$$-\sqrt[4]{48} = -\sqrt[4]{16 \cdot 3} \quad \text{Write 48 as } 16 \cdot 3.$$

$$= -\sqrt[4]{16}\sqrt[4]{3} \quad \text{The fourth root of a product is equal to the product of the fourth roots: } \sqrt[4]{16 \cdot 3} = \sqrt[4]{16} \cdot \sqrt[4]{3}.$$

$$= -2\sqrt[4]{3} \quad \text{Evaluate } \sqrt[4]{16}.$$

Self Check 1 Simplify: **a.** $\sqrt{18}$ **b.** $7\sqrt[3]{24}$ **c.** $\sqrt[4]{32}$
d. $\sqrt[5]{128}$

Now Try ▶ Problems 13, 17, and 19

The Language of Algebra

The instructions **simplify** and **approximate** do not mean the same thing.

Simplify: $\sqrt{12} = 2\sqrt{3}$ (exact)

Approximate:

$\sqrt{12} \approx 3.464$ (not exact)

When referrring to a square root, the instruction *simplify* means to remove any perfect-square factors from the radicand.

The Language of Algebra

In Example 1, **a radical of a product** is written as **a product of radicals**:

$$\sqrt[n]{ab} = \sqrt[n]{a}\sqrt[n]{b}$$

Variable expressions also can be perfect squares, perfect cubes, perfect-fourth powers, and so on. For example,

Perfect squares: $x^2, x^4, x^6, x^8, x^{10}, \ldots$

Perfect cubes: $x^3, x^6, x^9, x^{12}, x^{15}, \ldots$

Perfect-fourth powers: $x^4, x^8, x^{12}, x^{16}, x^{20}, \ldots$

Perfect-fifth powers: $x^5, x^{10}, x^{15}, x^{20}, x^{25}, \ldots$

EXAMPLE 2 Simplify. All variables represent positive real numbers.

  a. $\sqrt{m^9}$ **b.** $10\sqrt{128a^5}$ **c.** $\sqrt[3]{-24x^5}$ **d.** $\sqrt[5]{a^9b^5}$

Strategy We will factor each radicand into two factors, one of which is a perfect nth power.

Why We can then apply the rule *the nth root of a product is the product of the nth roots* to simplify the radical expression.

Solution **a.** The largest perfect-square factor of m^9 is m^8.

$$\sqrt{m^9} = \sqrt{m^8 \cdot m} \quad \text{Write } m^9 \text{ in factored form as } m^8 \cdot m.$$
$$= \sqrt{m^8}\sqrt{m} \quad \text{Use the product rule for radicals.}$$
$$= m^4\sqrt{m} \quad \text{Simplify } \sqrt{m^8}.$$

Caution

$\sqrt{m^9} \neq m^3$ because $(m^3)^2 \neq m^9$. By the power rule for exponents, $(m^3)^2 = m^6$. Similarly, $\sqrt{y^{25}} \neq y^5$ because $(y^5)^2 \neq y^{25}$.

b. Since the largest perfect-square factor of 128 is 64 and the largest perfect-square factor of a^5 is a^4, the largest perfect-square factor of $128a^5$ is $64a^4$. We write $128a^5$ as $64a^4 \cdot 2a$ and proceed as follows:

$$10\sqrt{128a^5} = 10\sqrt{64a^4 \cdot 2a} \quad \text{Write } 128a^5 \text{ in factored form as } 64a^4 \cdot 2a.$$
$$= 10\sqrt{64a^4}\sqrt{2a} \quad \text{Use the product rule for radicals.}$$
$$= 10 \cdot 8a^2\sqrt{2a} \quad \text{Simplify } \sqrt{64a^4}.$$
$$= 80a^2\sqrt{2a} \quad \text{Multiply: } 10 \cdot 8 = 80.$$

c. We write $-24x^5$ as $-8x^3 \cdot 3x^2$ and proceed as follows:

$$\sqrt[3]{-24x^5} = \sqrt[3]{-8x^3 \cdot 3x^2} \quad \text{$8x^3$ is the largest perfect-cube factor of $24x^5$. Since the radicand is negative, we factor it using $-8x^3$.}$$
$$= \sqrt[3]{-8x^3}\sqrt[3]{3x^2} \quad \text{Use the product rule for radicals.}$$
$$= -2x\sqrt[3]{3x^2} \quad \text{Simplify } \sqrt[3]{-8x^3}.$$

Caution

$\sqrt[3]{-24x^5} = -2x\sqrt[3]{3x^2}$

In simplified form, the radicand should not have any perfect-cube factors.

d. The largest perfect-fifth power factor of a^9 is a^5, and b^5 is a perfect-fifth power.

$$\sqrt[5]{a^9b^5} = \sqrt[5]{a^5b^5 \cdot a^4} \quad \text{a^5b^5 is the largest perfect-fifth power factor of a^9b^5.}$$
$$= \sqrt[5]{a^5b^5}\sqrt[5]{a^4} \quad \text{Use the product rule for radicals.}$$
$$= ab\sqrt[5]{a^4} \quad \text{Simplify } \sqrt[5]{a^5b^5}.$$

Self Check 2 Simplify. All variables represent positive real numbers.

  a. $6\sqrt{98b^3}$ **b.** $\sqrt[3]{-54y^5}$ **c.** $\sqrt[4]{t^8u^{15}}$

Now Try Problems 21, 25, and 27

2 Use Prime Factorization to Simplify Radical Expressions.

When simplifying radical expressions, prime factorization can be helpful in determining how to factor the radicand.

EXAMPLE 3 Simplify. All variables represent positive real numbers.

  a. $\sqrt{150}$ **b.** $\sqrt[3]{297b^4}$ **c.** $\sqrt[4]{224s^8t^7}$

Strategy In each case, the way to factor the radicand is not obvious. Another approach is to prime-factor the coefficient of the radicand and look for groups of like factors.

Why Identifying groups of like factors of the radicand leads to a factorization of the radicand that can be easily simplified.

Solution

a. $\sqrt{150} = \sqrt{2 \cdot 3 \cdot 5 \cdot 5}$ Write 150 in prime-factored form.

$= \sqrt{2 \cdot 3}\sqrt{5 \cdot 5}$ Group the pair of like factors together and use the product rule for radicals.

$= \sqrt{2 \cdot 3}\sqrt{5^2}$ Write $5 \cdot 5$ as 5^2.

$= \sqrt{6} \cdot 5$ Evaluate $\sqrt{5^2}$.

$= 5\sqrt{6}$ Write the factor 5 first.

$$\begin{array}{r|r} 2 & 150 \\ \hline 3 & 75 \\ \hline 5 & 25 \\ \hline & 5 \end{array}$$

> **Caution**
>
> To simplify $\sqrt{150}$, we could have factored 150 as $10 \cdot 15$. However, neither of the resulting radical expressions is the square root of a perfect square.
>
> $\sqrt{150} = \sqrt{10}\sqrt{15}$

b. $\sqrt[3]{297b^4} = \sqrt[3]{3 \cdot 3 \cdot 3 \cdot 11 \cdot b^3 \cdot b}$ Write 297 in prime-factored form. The largest perfect-cube factor of b^4 is b^3.

$= \sqrt[3]{3 \cdot 3 \cdot 3 \cdot b^3}\sqrt[3]{11b}$ Group the three like factors of 3 together and use the product rule for radicals.

$= \sqrt[3]{3^3 b^3}\sqrt[3]{11b}$ Write $3 \cdot 3 \cdot 3$ as 3^3.

$= 3b\sqrt[3]{11b}$ Simplify $\sqrt[3]{3^3 b^3}$.

$$\begin{array}{r|r} 3 & 297 \\ \hline 3 & 99 \\ \hline 3 & 33 \\ \hline & 11 \end{array}$$

c. $\sqrt[4]{224 s^8 t^7} = \sqrt[4]{2 \cdot 2 \cdot 2 \cdot 2 \cdot 2 \cdot 7 \cdot s^8 \cdot t^4 \cdot t^3}$ Write 224 in prime-factored form. The largest perfect-fourth power factor of t^7 is t^4.

$= \sqrt[4]{2 \cdot 2 \cdot 2 \cdot 2 \cdot s^8 \cdot t^4}\sqrt[4]{2 \cdot 7 \cdot t^3}$ Group the four like factors of 2 together and use the product rule for radicals.

$= \sqrt[4]{2^4 s^8 t^4}\sqrt[4]{2 \cdot 7 \cdot t^3}$ Write $2 \cdot 2 \cdot 2 \cdot 2$ as 2^4.

$= 2s^2 t\sqrt[4]{14 t^3}$ Simplify $\sqrt[4]{2^4 s^8 t^4}$.

$$\begin{array}{r|r} 2 & 224 \\ \hline 2 & 112 \\ \hline 2 & 56 \\ \hline 2 & 28 \\ \hline 2 & 14 \\ \hline & 7 \end{array}$$

Self Check 3 Simplify. All variables represent positive real numbers.

    a. $\sqrt{275}$ **b.** $\sqrt[3]{189 c^4 d^3}$ **c.** $\sqrt[4]{1{,}250 x^8 y^6}$

Now Try ▶ Problems 29 and 35

---

3   Use the Quotient Rule to Simplify Radical Expressions.

To introduce the quotient rule for radicals, we will find $\sqrt{\dfrac{100}{4}}$ and $\dfrac{\sqrt{100}}{\sqrt{4}}$ and compare the results.

Square root of a quotient	*Quotient of square roots*
$\sqrt{\dfrac{100}{4}} = \sqrt{25}$	$\dfrac{\sqrt{100}}{\sqrt{4}} = \dfrac{10}{2}$
$= 5$	$= 5$

Since the answer is 5 in each case, $\sqrt{\dfrac{100}{4}} = \dfrac{\sqrt{100}}{\sqrt{4}}$.

Similarly, we will find $\sqrt[3]{\dfrac{64}{8}}$ and $\dfrac{\sqrt[3]{64}}{\sqrt[3]{8}}$, and compare the results.

Cube root of a quotient	*Quotient of cube roots*
$\sqrt[3]{\dfrac{64}{8}} = \sqrt[3]{8}$	$\dfrac{\sqrt[3]{64}}{\sqrt[3]{8}} = \dfrac{4}{2}$
$= 2$	$= 2$

Since the answer is 2 in each case, $\sqrt[3]{\dfrac{64}{8}} = \dfrac{\sqrt[3]{64}}{\sqrt[3]{8}}$. These results illustrate the *quotient rule for radicals*.

| **The Quotient Rule for Radicals** | The *n*th root of the quotient of two numbers is equal to the quotient of their *n*th roots. If $\sqrt[n]{a}$ and $\sqrt[n]{b}$ are real numbers, then $$\sqrt[n]{\frac{a}{b}} = \frac{\sqrt[n]{a}}{\sqrt[n]{b}} \qquad (b \neq 0) \qquad \text{Read as "the nth root of a divided by b equals the nth root of a divided by the nth root of b."}$$ |

In the following examples, variables appear in the denominators of radical expressions. To avoid undefined situations, we will assume *all variables represent positive real numbers.*

EXAMPLE 4   Simplify each expression: **a.** $\sqrt{\dfrac{7}{64}}$ **b.** $\sqrt{\dfrac{15}{49x^2}}$ **c.** $\sqrt[3]{\dfrac{10x^2}{27y^6}}$

Strategy   In each case, the radical is not in simplified form because the radicand contains a fraction. To write each of these expressions in simplified form, we will use the quotient rule for radicals.

Why   Writing these expressions in $\dfrac{\sqrt[n]{a}}{\sqrt[n]{b}}$ form leads to square roots of perfect squares and cube roots of perfect cubes that we can easily simplify.

Solution   **a.** We can use the quotient rule for radicals to simplify each expression.

The Language of Algebra

In Example 4, a **radical of a quotient** is written as a **quotient of radicals:**
$$\sqrt[n]{\frac{a}{b}} = \frac{\sqrt[n]{a}}{\sqrt[n]{b}}$$

$$\sqrt{\frac{7}{64}} = \frac{\sqrt{7}}{\sqrt{64}} \qquad \text{The square root of a quotient is equal to the quotient of the square roots.}$$

$$= \frac{\sqrt{7}}{8} \qquad \text{Evaluate } \sqrt{64}.$$

b. $\sqrt{\dfrac{15}{49x^2}} = \dfrac{\sqrt{15}}{\sqrt{49x^2}} \qquad \text{The square root of a quotient is equal to the quotient of the square roots.}$

$$= \frac{\sqrt{15}}{7x} \qquad \text{Simplify the denominator: } \sqrt{49x^2} = 7x.$$

c. $\sqrt[3]{\dfrac{10x^2}{27y^6}} = \dfrac{\sqrt[3]{10x^2}}{\sqrt[3]{27y^6}} \qquad \text{The cube root of a quotient is equal to the quotient of the cube roots.}$

$$= \frac{\sqrt[3]{10x^2}}{3y^2} \qquad \text{Simplify the denominator.}$$

Self Check 4   Simplify: **a.** $\sqrt[3]{\dfrac{25}{27}}$ **b.** $\sqrt{\dfrac{11}{36a^2}}$ **c.** $\sqrt[4]{\dfrac{a^3}{625y^{12}}}$

Now Try ▶ Problems 37, 39, and 43

EXAMPLE 5   Simplify: **a.** $\dfrac{\sqrt{45xy^2}}{\sqrt{5x}}$ **b.** $\dfrac{\sqrt[3]{-432x^5}}{\sqrt[3]{8x}}$

Strategy   We will use the quotient rule for radicals in reverse: $\dfrac{\sqrt[n]{a}}{\sqrt[n]{b}} = \sqrt[n]{\dfrac{a}{b}}$

Why   When the radicands are written under a single radical symbol, the result is a rational expression. Our hope is that the rational expression can be simplified.

Solution **a.** We can write the quotient of the square roots as the square root of a quotient.

The Language of Algebra

In Example 5, a **quotient of radicals** is written as a **radical of a quotient**.

$$\dfrac{\sqrt[n]{a}}{\sqrt[n]{b}} = \sqrt[n]{\dfrac{a}{b}}$$

$$\dfrac{\sqrt{45xy^2}}{\sqrt{5x}} = \sqrt{\dfrac{45xy^2}{5x}}$$ 　Use the quotient rule for radicals. Note that the resulting radicand is a rational expression.

$$= \sqrt{9y^2}$$ 　Simplify the radicand: $\dfrac{45xy^2}{5x} = \dfrac{\overset{1}{\cancel{5}} \cdot 9 \cdot \overset{1}{\cancel{x}} \cdot y^2}{\underset{1}{\cancel{5}} \cdot \underset{1}{\cancel{x}}} = 9y^2.$

$$= 3y$$ 　Simplify the radical.

b. We can write the quotient of the cube roots as the cube root of a quotient.

$$\dfrac{\sqrt[3]{-432x^5}}{\sqrt[3]{8x}} = \sqrt[3]{\dfrac{-432x^5}{8x}}$$ 　Use the quotient rule for radicals. Note that the resulting radicand is a rational expression.

$$= \sqrt[3]{-54x^4}$$ 　Simplify the radicand: $-\dfrac{432x^5}{8x} = -54x^4.$

$$= \sqrt[3]{-27x^3 \cdot 2x}$$ 　$27x^3$ is the largest perfect cube that divides $54x^4$.

$$= \sqrt[3]{-27x^3}\sqrt[3]{2x}$$ 　Use the product rule for radicals.

$$= -3x\sqrt[3]{2x}$$ 　Simplify: $\sqrt[3]{-27x^3} = -3x.$

Success Tip

In parts a and b, we use skills learned in Chapter 6 to simplify the **rational expressions** in each radicand: $\dfrac{45xy^2}{5x}$ and $-\dfrac{432x^5}{8x}.$

Self Check 5 Simplify: **a.** $\dfrac{\sqrt{50ab^2}}{\sqrt{2a}}$ 　**b.** $\dfrac{\sqrt[3]{-2{,}000x^5v^3}}{\sqrt[3]{2x}}$

Now Try ▶ Problems 47 and 51

---

4 Add and Subtract Radical Expressions.

Radical expressions with the same index and the same radicand are called **like** or **similar radicals.** For example, $3\sqrt{2}$ and $2\sqrt{2}$ are like radicals. However,

- $3\sqrt{5}$ and $4\sqrt{2}$ are not like radicals, because the radicands are different.
- $3\sqrt[4]{5}$ and $2\sqrt[3]{5}$ are not like radicals, because the indices are different.

For an expression with two or more radical terms, we should attempt to combine like radicals, if possible. For example, to simplify the expression $3\sqrt{2} + 2\sqrt{2}$, we use the distributive property to factor out $\sqrt{2}$ and simplify.

Success Tip

Combining like radicals is similar to combining like terms.

$3\sqrt{2} + 2\sqrt{2} = 5\sqrt{2}$

$3x + 2x = 5x$

$$3\sqrt{2} + 2\sqrt{2} = (3 + 2)\sqrt{2}$$ 　Factor out $\sqrt{2}.$

$$= 5\sqrt{2}$$ 　Do the addition.

Radicals with the same index but different radicands often can be written as like radicals. For example, to simplify the expression $\sqrt{75} - \sqrt{27}$, we simplify both radicals first and then combine the like radicals.

$$\sqrt{75} - \sqrt{27} = \sqrt{25 \cdot 3} - \sqrt{9 \cdot 3}$$ 　Write 75 and 27 in factored form.

$$= \sqrt{25}\sqrt{3} - \sqrt{9}\sqrt{3}$$ 　Use the product rule for radicals.

$$= 5\sqrt{3} - 3\sqrt{3}$$ 　Evaluate $\sqrt{25}$ and $\sqrt{9}.$

$$= (5 - 3)\sqrt{3}$$ 　Factor out $\sqrt{3}.$

$$= 2\sqrt{3}$$ 　Do the subtraction.

As the previous examples suggest, we can add or subtract radicals as follows.

Adding and Subtracting Radicals 　To add or subtract radicals, simplify each radical, if possible, and combine like radicals.

EXAMPLE 6 Simplify: **a.** $18 + 2\sqrt{12} - 3\sqrt{48} + 9$ **b.** $\sqrt[3]{16} + \sqrt[3]{54} - \sqrt[3]{24}$

Strategy Since the radicals in each part are unlike radicals, we cannot add or subtract them in their current form. However, we will simplify the radicals and hope that like radicals result.

Why Like radicals can be combined.

Solution **a.** We begin by simplifying each radical expression:

$$18 + 2\sqrt{12} - 3\sqrt{48} + 9 = 18 + 2\sqrt{4 \cdot 3} - 3\sqrt{16 \cdot 3} + 9 \qquad \text{Factor the radicands, 12 and 48.}$$

$$= 18 + 2\sqrt{4}\sqrt{3} - 3\sqrt{16}\sqrt{3} + 9 \qquad \text{Use the product rule for radicals.}$$

$$= 18 + 2(2)\sqrt{3} - 3(4)\sqrt{3} + 9 \qquad \text{Evaluate } \sqrt{4} \text{ and } \sqrt{16}.$$

$$= 18 + \mathbf{4}\sqrt{3} - \mathbf{12}\sqrt{3} + 9 \qquad \text{Both radicals now have the same index, 2, and radicand, 3.}$$

$$= (\mathbf{4} - \mathbf{12})\sqrt{3} + 27 \qquad \text{Add the constants: } 18 + 9 = 27. \text{ Combine like radicals.}$$

$$= -8\sqrt{3} + 27 \qquad \text{Do the subtraction.}$$

b. We begin by simplifying each radical expression:

$$\sqrt[3]{16} + \sqrt[3]{54} - \sqrt[3]{24} = \sqrt[3]{8 \cdot 2} + \sqrt[3]{27 \cdot 2} - \sqrt[3]{8 \cdot 3} \qquad \text{Factor the radicands.}$$

$$= \sqrt[3]{8}\sqrt[3]{2} + \sqrt[3]{27}\sqrt[3]{2} - \sqrt[3]{8}\sqrt[3]{3} \qquad \text{Use the product rule.}$$

$$= \mathbf{2}\sqrt[3]{2} + \mathbf{3}\sqrt[3]{2} - 2\sqrt[3]{3} \qquad \text{Evaluate } \sqrt[3]{8} \text{ and } \sqrt[3]{27}.$$

$$= (\mathbf{2} + \mathbf{3})\sqrt[3]{2} - 2\sqrt[3]{3} \qquad \text{Combine the first two radical expressions because they have the same index and radicand.}$$

$$= 5\sqrt[3]{2} - 2\sqrt[3]{3} \qquad \text{Do the addition.}$$

CAUTION Even though the expressions $5\sqrt[3]{2}$ and $2\sqrt[3]{3}$ have the same index, we cannot combine them, because their radicands are different. Neither can we combine radical expressions having the same radicand but a different index. For example, the expression $\sqrt[3]{2} + \sqrt[4]{2}$ cannot be simplified.

Self Check 6 Simplify: **a.** $14 + 3\sqrt{75} - 2\sqrt{12} + 2\sqrt{48} - 8$
b. $\sqrt[3]{24} - \sqrt[3]{16} + \sqrt[3]{54}$

Now Try ▶ Problems 57 and 63

EXAMPLE 7 Simplify: $\sqrt[3]{16x^4} + 4\sqrt[3]{54x^4} - x\sqrt[3]{-128x}$

Strategy Since the radicals are unlike radicals, we cannot add or subtract them in their current form. However, we will simplify the radicals and hope that like radicals result.

Why Like radicals can be combined.

Solution We begin by simplifying each radical expression.

$$\sqrt[3]{16x^4} + 4\sqrt[3]{54x^4} - x\sqrt[3]{-128x}$$

$$= \sqrt[3]{8x^3 \cdot 2x} + 4\sqrt[3]{27x^3 \cdot 2x} - x\sqrt[3]{-64 \cdot 2x}$$

$$= \sqrt[3]{8x^3}\sqrt[3]{2x} + 4\sqrt[3]{27x^3}\sqrt[3]{2x} - x\sqrt[3]{-64}\sqrt[3]{2x}$$

$$= 2x\sqrt[3]{2x} + 4 \cdot 3x\sqrt[3]{2x} + x \cdot 4\sqrt[3]{2x} \qquad \text{All three radicals have the same index and radicand.}$$

$$= 2x\sqrt[3]{2x} + 12x\sqrt[3]{2x} + 4x\sqrt[3]{2x} \qquad \text{Combine like radicals.}$$

$$= (2x + 12x + 4x)\sqrt[3]{2x} \qquad \text{Multiply: } 4 \cdot 3x = 12x \text{ and } x \cdot 4 = 4x.$$

$$= 18x\sqrt[3]{2x} \qquad \text{Within the parentheses, combine like terms: } 2x + 12x + 4x = 18x.$$

Self Check 7 Simplify: $\sqrt{32x^3} + 4\sqrt{50x^3} - x\sqrt{18x}$

Now Try ▶ Problems 67 and 69

SECTION 7.3 STUDY SET

VOCABULARY

Fill in the blanks.

1. Radical expressions such as $\sqrt[3]{4}$ and $6\sqrt[3]{4}$ with the same index and the same radicand are called _____ radicals.

2. Numbers such as 1, 4, 9, 16, 25, and 36 are called perfect _____. Numbers such as 1, 8, 27, 64, and 125 are called perfect _____. Numbers such as 1, 16, 81, 256, and 625 are called perfect-fourth _____.

3. The largest perfect-square _____ of 27 is 9. The largest _____-cube factor of 16 is 8.

4. To _____ $\sqrt{24}$ means to write it as $2\sqrt{6}$.

CONCEPTS

Fill in the blanks.

5. The product rule for radicals: $\sqrt[n]{ab} =$ [ ]. In words, the nth root of the _____ of two numbers is equal to the product of their nth _____.

6. The quotient rule for radicals: $\sqrt[n]{\dfrac{a}{b}} =$ [ ]. In words, the nth root of the _____ of two numbers is equal to the quotient of their nth _____.

7. Consider the expressions $\sqrt{4 \cdot 5}$ and $\sqrt{4}\sqrt{5}$. Which expression is

   a. the square root of a product?

   b. the product of square roots?

   c. How are these two expressions related?

8. Consider $\dfrac{\sqrt[3]{a}}{\sqrt[3]{x^2}}$ and $\sqrt[3]{\dfrac{a}{x^2}}$. Which expression is

   a. the cube root of a quotient?

   b. the quotient of cube roots?

   c. How are these two expressions related?

9. a. Write two radical expressions that have the same radicand but a different index. Can the expressions be added?

   b. Write two radical expressions that have the same index but a different radicand. Can the expressions be added?

10. Fill in the blanks.

   a. $5\sqrt{6} + 3\sqrt{6} = \left(\boxed{} + \boxed{}\right)\sqrt{6} = \boxed{}\sqrt{6}$

   b. $9\sqrt[3]{n} - 2\sqrt[3]{n} = \left(\boxed{} - \boxed{}\right)\sqrt[3]{n} = 7\boxed{}$

NOTATION

Complete each solution.

11. Simplify:

$$\sqrt[3]{32k^4} = \sqrt[3]{\boxed{} \cdot 4k}$$

$$= \sqrt[3]{\boxed{}}\,\sqrt[3]{4k}$$

$$= \boxed{}\,\sqrt[3]{4k}$$

12. Simplify:

$$\frac{\sqrt{80s^2t^4}}{\sqrt{5s^2}} = \sqrt{\frac{80s^2t^4}{\boxed{}}}$$

$$= \sqrt{\boxed{}}$$

$$= \boxed{}$$

GUIDED PRACTICE

Simplify each expression. See Example 1.

13. $\sqrt{50}$

14. $\sqrt{28}$

15. $8\sqrt{45}$

16. $9\sqrt{54}$

17. $\sqrt[3]{32}$

18. $\sqrt[3]{40}$

19. $\sqrt[4]{48}$

20. $\sqrt[4]{32}$

Simplify each radical expression. All variables represent positive real numbers. See Example 2.

21. $\sqrt{75a^2}$

22. $\sqrt{50x^2}$

23. $\sqrt{128a^3b^5}$

24. $\sqrt{75b^8c}$

25. $2\sqrt[3]{-54x^6}$

26. $4\sqrt[3]{-81a^3}$

27. $\sqrt[4]{32x^{12}y^4}$

28. $\sqrt[5]{64x^{10}y^5}$

Simplify each radical expression. All variables represent positive real numbers. See Example 3.

29. $\sqrt{242}$

30. $\sqrt{363}$

31. $\sqrt{112a^3}$

32. $\sqrt{147a^5}$

33. $-\sqrt[3]{96a^4}$

34. $-\sqrt[7]{256t^6}$

35. $\sqrt[3]{405x^{12}y^4}$

36. $\sqrt[3]{280a^5b^6}$

Simplify each radical expression. All variables represent positive real numbers. See Example 4.

37. $\sqrt{\dfrac{11}{9}}$

38. $\sqrt{\dfrac{3}{4}}$

39. $\sqrt[4]{\dfrac{3}{625}}$

40. $\sqrt[5]{\dfrac{2}{243}}$

41. $\sqrt[5]{\dfrac{3x^{10}}{32}}$

42. $\sqrt[6]{\dfrac{5y^{12}}{64}}$

43. $\sqrt{\dfrac{z^2}{16x^2}}$

44. $\sqrt{\dfrac{b^4}{64a^8}}$

Simplify each expression. All variables represent positive real numbers. See Example 5.

45. $\dfrac{\sqrt{500}}{\sqrt{5}}$

46. $\dfrac{\sqrt{128}}{\sqrt{2}}$

47. $\dfrac{\sqrt{98x^3}}{\sqrt{2x}}$

48. $\dfrac{\sqrt{75y^5}}{\sqrt{3y}}$

49. $\dfrac{\sqrt[3]{48x^7}}{\sqrt[3]{6x}}$

50. $\dfrac{\sqrt[3]{64y^8}}{\sqrt[3]{8y^2}}$

51. $\dfrac{\sqrt[3]{189a^5}}{\sqrt[3]{7a}}$

52. $\dfrac{\sqrt[3]{243x^8}}{\sqrt[3]{9x}}$

Simplify by combining like radicals. See Objective 4 and Example 6.

53. $5\sqrt{7} + 3\sqrt{7}$

54. $11\sqrt{3} + 2\sqrt{3}$

55. $20\sqrt[3]{4} - 15\sqrt[3]{4}$

56. $30\sqrt[3]{6} - 10\sqrt[3]{6}$

57. $4 + \sqrt{8} + \sqrt{2} + 8$

58. $9 + \sqrt{45} + \sqrt{20} + 16$

59. $\sqrt{98} - \sqrt{50} - \sqrt{72}$

60. $\sqrt{20} + \sqrt{125} - \sqrt{80}$

61. $8 + \sqrt[3]{32} - \sqrt[3]{108} - 7$

62. $12 + \sqrt[3]{80} - \sqrt[3]{10{,}000} + 4$

63. $14\sqrt[4]{32} - 15\sqrt[4]{2}$

64. $23\sqrt[3]{3} + \sqrt[4]{48}$

Simplify by combining like radicals. All variables represent positive real numbers. See Example 7.

65. $4\sqrt{2x} + 6\sqrt{2x}$

66. $6\sqrt[3]{5y} + 3\sqrt[3]{5y}$

67. $\sqrt{18t} + \sqrt{300t} - \sqrt{243t}$

68. $\sqrt{80m} - \sqrt{128m} + \sqrt{288m}$

69. $2\sqrt[3]{16} - \sqrt[3]{54} - 3\sqrt[3]{128}$

70. $\sqrt[3]{250} - 4\sqrt[3]{5} + \sqrt[3]{16}$

71. $\sqrt[4]{32} + 5\sqrt[4]{2} - \sqrt[4]{162}$

72. $\sqrt[4]{48} - \sqrt[4]{243} - \sqrt[4]{768}$

TRY IT YOURSELF

Simplify each expression, if possible. All variables represent positive real numbers.

73. $\sqrt[6]{m^{11}}$

74. $\sqrt[6]{n^{13}}$

75. $2\sqrt[3]{64a} + 2\sqrt[3]{8a}$

76. $3\sqrt[4]{x^4y} - 2\sqrt[4]{x^4y}$

77. $\sqrt{8y^7} + \sqrt{32y^7} - \sqrt{2y^7}$

78. $\sqrt{y^5} - \sqrt{9y^5} - \sqrt{25y^5}$

79. $\sqrt{32b}$

80. $\sqrt{80c}$

81. $\sqrt{\dfrac{125n^5}{64n}}$

82. $\sqrt{\dfrac{72q^7}{25q^3}}$

83. $2\sqrt[3]{125} - 5\sqrt[3]{64}$

84. $3\sqrt[3]{27} + 12\sqrt[3]{216}$

85. $\sqrt{300xy}$

86. $\sqrt{200x^2y}$

87. $\sqrt[4]{\dfrac{5x}{16z^4}}$

88. $\sqrt[3]{\dfrac{11a^2}{125b^6}}$

89. $8\sqrt[5]{7a^2} - 7\sqrt[5]{7a^2}$

90. $10\sqrt[6]{12xy} - \sqrt[6]{12xy}$

91. $\sqrt[5]{x^6y^2} + \sqrt[5]{32x^6y^2} + \sqrt[5]{x^6y^2}$

92. $\sqrt[3]{xy^4} + \sqrt[3]{8xy^4} - \sqrt[3]{27xy^4}$

93. $\sqrt[4]{208m^4n}$

94. $\sqrt[4]{128p^8q^3}$

95. $\sqrt[3]{\dfrac{a^7}{64a}}$

96. $\sqrt[3]{\dfrac{b^3c^8}{125c^5}}$

97. $\sqrt[3]{\dfrac{7}{64}}$

98. $\sqrt[3]{\dfrac{4}{125}}$

99. $\sqrt{80} + \sqrt{45} - \sqrt{27}$

100. $\sqrt{63} + \sqrt{72} - \sqrt{28}$

101. $\sqrt[5]{64t^{11}}$

102. $\sqrt[5]{243r^{22}}$

103. $\sqrt[3]{24x} + \sqrt[3]{3x}$

104. $\sqrt[3]{16y} + \sqrt[3]{128y}$

Look Alikes . . .

105. a. $\sqrt{20} + \sqrt{20}$ **b.** $\sqrt{21} + \sqrt{21}$

106. a. $\sqrt{2} + \sqrt{18}$ **b.** $\sqrt{2} + \sqrt{19}$

107. a. $\sqrt{9x^2} - \sqrt{25x^2} + \sqrt{16x^2}$
    b. $\sqrt{9x^3} - \sqrt{25x^3} + \sqrt{16x^3}$

108. a. $2\sqrt{5} - \sqrt[3]{5} + 4\sqrt{5} - 6\sqrt[3]{5}$
    b. $2\sqrt[3]{5} - \sqrt[3]{5} + 4\sqrt[3]{5} - 6\sqrt[3]{5}$

109. a. $3\sqrt{16} + \sqrt{54}$ **b.** $3\sqrt[3]{16} + \sqrt[3]{54}$

110. a. $\sqrt[3]{27} - 5\sqrt[3]{8}$ **b.** $\sqrt{27} - 5\sqrt{8}$

111. a. $24\sqrt[5]{6x} + 16\sqrt[5]{6x}$ **b.** $24\sqrt[4]{6x} + 16\sqrt[4]{6x}$

112. a. $x\sqrt[3]{64x^6} - x\sqrt[3]{x^6}$ **b.** $x\sqrt{64x^6} - x\sqrt{x^6}$

APPLICATIONS

First give the exact answer, expressed as a simplified radical expression. Then give an approximation, rounded to the nearest tenth.

113.

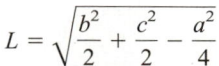

from **Campus to Careers**

General Contractor

Structural engineers have determined that two additional supports (shown in red) need to be added to strengthen the truss shown below. Find the length L of a support using the formula

$$L = \sqrt{\dfrac{b^2}{2} + \dfrac{c^2}{2} - \dfrac{a^2}{4}}$$

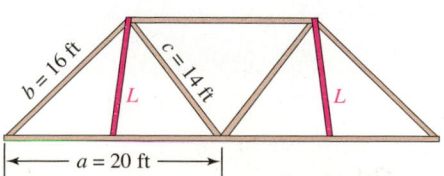

114. Unbrellas. The surface area of a cone is given by the formula $S = \pi r \sqrt{r^2 + h^2}$, where r is the radius of the base and h is its height. Use this formula to find the number of square feet of waterproof cloth used to make the umbrella shown.

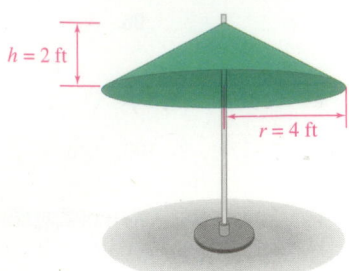

$h = 2$ ft

$r = 4$ ft

115. Blow Dryers. The current I (in amps), the power P (in watts), and the resistance R (in ohms) are related by the formula $I = \sqrt{\frac{P}{R}}$. What current is needed for a 1,200-watt hair dryer if the resistance is 16 ohms?

116. Communications Satellites. Engineers have determined that a spherical communications satellite needs to have a capacity of 565.2 cubic feet to house all of its operating systems. The volume V of a sphere is related to its radius r by the formula $r = \sqrt[3]{\frac{3V}{4\pi}}$. What radius must the satellite have to meet the engineer's specification? Use 3.14 as an approximation of π.

117. Ductwork. The following pattern is laid out on a sheet of galvanized tin. Then it is cut out and bent on the dashed lines to make an air conditioning duct connection. Find the total length of the cut that must be made. (All measurements are in inches.)

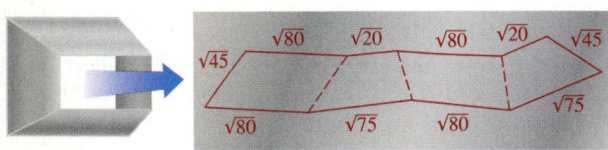

$\sqrt{45}$ $\sqrt{80}$ $\sqrt{20}$ $\sqrt{80}$ $\sqrt{20}$ $\sqrt{45}$

$\sqrt{80}$ $\sqrt{75}$ $\sqrt{80}$ $\sqrt{75}$

118. Outdoor Cooking. The diameter of a circle is given by the function $d(A) = 2\sqrt{\frac{A}{\pi}}$, where A is the area of the circle. Find the difference between the diameters of the barbecue grills.

Cooking area 147π in.^{3}

Cooking area 48π in.^{3}

WRITING

119. Explain why each expression is not in simplified form.

a. $\sqrt[3]{9x^4}$ b. $\sqrt{\frac{24m}{25}}$ c. $\frac{\sqrt[4]{c^3}}{\sqrt[4]{16}}$

120. How are the procedures used to simplify $3x + 4x$ and $3\sqrt{x} + 4\sqrt{x}$ similar?

121. Explain the mistake in the student's solution shown below.

Simplify: $\sqrt[3]{54}$

$\sqrt[3]{54} = \sqrt[3]{27 + 27}$

$= \sqrt[3]{27} + \sqrt[3]{27}$

$= 3 + 3$

$= 6$

122. Explain how the graphs of $Y_1 = 3\sqrt{24x} + \sqrt{54x}$ (on the left) and $Y_1 = 9\sqrt{6x}$ (on the right) can be used to verify the simplification $3\sqrt{24x} + \sqrt{54x} = 9\sqrt{6x}$. In each graph, settings of $[-5, 20]$ for x and $[-5, 100]$ for y were used.

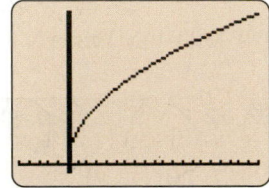   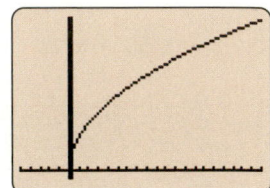

REVIEW

Perform each operation.

123. $3x^2y^3(-5x^3y^{-4})$

124. $(2x^2 - 9x - 5) \cdot \frac{x}{2x^2 + x}$

125. $2p - 5\overline{)6p^2 - 7p - 25}$

126. $\frac{xy}{\frac{1}{x} - \frac{1}{y}}$

CHALLENGE PROBLEMS

Simplify each expression. All variables represent positive real numbers.

127. $\frac{\sqrt{24}}{3} + \frac{\sqrt{6}}{5}$

128. $\sqrt{3} + \sqrt{3^2} + \sqrt{3^3} + \sqrt{3^4} + \sqrt{3^5}$

129. $\sqrt[3]{\frac{3b}{8}} - 9\sqrt[3]{3b}$

130. $\frac{\sqrt[4]{32}}{3y} - \frac{3\sqrt[4]{2}}{7y}$

131. $\sqrt{25x + 25} - \sqrt{x + 1}$

132. $\sqrt[3]{216m^4} + \sqrt[3]{125m}$

SECTION 7.4

OBJECTIVES

1 Multiply radical expressions.

2 Find powers of radical expressions.

3 Rationalize denominators.

4 Rationalize denominators that have two terms.

5 Rationalize numerators.

Multiplying and Dividing Radical Expressions

ARE YOU READY?

The following problems review some basic skills that are needed when multiplying and dividing radical expressions.

Perform the indicated operations and simplify, if possible.

1. $(5a^6)^2$

2. $9t^3(6t^3 + 2t^2)$

3. $(2x + 3)(x - 1)$

4. $7 - 3\sqrt{14} + \sqrt{14} - 6$

5. $(x + 10)(x - 10)$

6. Build an equivalent fraction for $\dfrac{2}{3a}$ with a denominator of $27a$.

In this section, we will discuss the methods we can use to multiply and divide radical expressions.

1 Multiply Radical Expressions.

We have used the *product rule for radicals* to write radical expressions in simplified form. We also can use this rule to multiply radical expressions that have the same index.

The Product Rule for Radicals

The product of the nth roots of two nonnegative numbers is equal to the nth root of the product of those numbers.

If $\sqrt[n]{a}$ and $\sqrt[n]{b}$ are real numbers,

$$\sqrt[n]{a} \cdot \sqrt[n]{b} = \sqrt[n]{a \cdot b}$$

EXAMPLE 1   Multiply and then simplify: **a.** $\sqrt{5}\sqrt{10}$ **b.** $3\sqrt{6}\left(2\sqrt{3}\right)$
c. $-2\sqrt[3]{7x} \cdot 6\sqrt[3]{49x^2}$

Strategy   In each expression, we will use the product rule for radicals to multiply factors of the form $\sqrt[n]{a}$ and $\sqrt[n]{b}$.

Why   The product rule for radicals is used to multiply radicals that have the same index.

Solution

a. $\sqrt{5}\sqrt{10} = \sqrt{5 \cdot 10}$ Use the product rule for radicals.

$\qquad\qquad = \sqrt{50}$ Multiply under the radical. Note that $\sqrt{50}$ can be simplified.

$\qquad\qquad = \sqrt{25 \cdot 2}$ Prepare to simplify: factor 50.

$\qquad\qquad = 5\sqrt{2}$ Simplify: $\sqrt{25 \cdot 2} = \sqrt{25}\sqrt{2} = 5\sqrt{2}$.

Success Tip

Here we use the product rule for radicals in two ways: to *multiply* and to *simplify* the result.

Notation

Multiplication of radicals can be shown in several ways:

$\sqrt{5} \cdot \sqrt{10} = \sqrt{5}\left(\sqrt{10}\right) = \sqrt{5}\sqrt{10}$

b. We use the commutative and associative properties of multiplication to multiply the integer factors and the radicals separately. Then we simplify any radicals in the product, if possible.

$3\sqrt{6}\left(2\sqrt{3}\right) = 3(2)\sqrt{6}\sqrt{3}$ Multiply the integer factors, 3 and 2, and multiply the radicals.

$\qquad\qquad = 6\sqrt{18}$ Use the product rule for radicals.

$\qquad\qquad = 6\sqrt{9}\sqrt{2}$ Simplify: $\sqrt{18} = \sqrt{9 \cdot 2} = \sqrt{9}\sqrt{2}$.

$\qquad\qquad = 6(3)\sqrt{2}$ Evaluate: $\sqrt{9} = 3$.

$\qquad\qquad = 18\sqrt{2}$ Multiply 6 and 3 to get 18.

c. $-2\sqrt[3]{7x} \cdot 6\sqrt[3]{49x^2} = -2(6)\sqrt[3]{7x}\sqrt[3]{49x^2}$ Write the integer factors together and the radicals together.

$= -12\sqrt[3]{7x \cdot 49x^2}$ Multiply the integer factors, -2 and 6, and multiply the radicals.

$= -12\sqrt[3]{7x \cdot 7^2 x^2}$ Write 49 as 7^2.

$= -12\sqrt[3]{7^3 x^3}$ Prepare to simplify: write $7x \cdot 7^2 x^2$ as $7^3 x^3$.

$= -12(7x)$ Simplify: $\sqrt[3]{7^3 x^3} = 7x$.

$= -84x$ Multiply.

> **Caution**
>
> Note that to multiply radical expressions, they must have the same index.
>
> $$\sqrt[n]{a} \cdot \sqrt[n]{b} = \sqrt[n]{a \cdot b}$$

Self Check 1   Multiply and then simplify: **a.** $\sqrt{7}\sqrt{14}$
b. $-2\sqrt[3]{2}\left(5\sqrt[3]{12}\right)$ **c.** $\sqrt[4]{4x^3} \cdot 9\sqrt[4]{8x^2}$

Now Try ▶ Problems 15, 23, and 25

Recall that to multiply a polynomial by a monomial, we use the distributive property. We use the same technique to multiply a radical expression that has two or more terms by a radical expression that has only one term.

EXAMPLE 2   Multiply and then simplify: $3\sqrt{3}\left(4\sqrt{8} - 5\sqrt{10}\right)$

Strategy   We will use the distributive property and multiply each term within the parentheses by the term outside the parentheses.

Why   The given expression has the form $a(b - c)$.

Solution $3\sqrt{3}\left(4\sqrt{8} - 5\sqrt{10}\right)$

$= 3\sqrt{3} \cdot 4\sqrt{8} - 3\sqrt{3} \cdot 5\sqrt{10}$ Distribute the multiplication by $3\sqrt{3}$.

$= 12\sqrt{24} - 15\sqrt{30}$ Multiply the integer factors and use the product rule to multiply the radicals.

$= 12\sqrt{4}\sqrt{6} - 15\sqrt{30}$ Simplify: $\sqrt{24} = \sqrt{4 \cdot 6} = \sqrt{4}\sqrt{6}$.

$= 12(2)\sqrt{6} - 15\sqrt{30}$ Evaluate: $\sqrt{4} = 2$.

$= 24\sqrt{6} - 15\sqrt{30}$ Multiply 12 and 2 to get 24.

Self Check 2   Multiply and then simplify: $4\sqrt{2}\left(3\sqrt{5} - 2\sqrt{8}\right)$

Now Try ▶ Problems 31 and 33

Recall that to multiply two binomials, we multiply each term of one binomial by each term of the other binomial and simplify. We multiply two radical expressions, each having two terms, in the same way.

EXAMPLE 3   Multiply and then simplify: **a.** $\left(\sqrt{7} + \sqrt{2}\right)\left(\sqrt{7} - 9\sqrt{2}\right)$
b. $\left(\sqrt[3]{x^2} - 4\sqrt[3]{5}\right)\left(\sqrt[3]{x} + \sqrt[3]{2}\right)$

Strategy   As with binomials, we will multiply each term within the first set of parentheses by each term within the second set of parentheses.

Why   This is an application of the FOIL method for multiplying binomials.

Solution   **a.** $\left(\sqrt{7} + \sqrt{2}\right)\left(\sqrt{7} - 9\sqrt{2}\right)$

$$
\begin{array}{ll}
\overset{\text{F}}{} \quad \overset{\text{O}}{} \quad \overset{\text{I}}{} \quad \overset{\text{L}}{} & \\
= \sqrt{7}\sqrt{7} - 9\sqrt{7}\sqrt{2} + \sqrt{2}\sqrt{7} - 9\sqrt{2}\sqrt{2} & \text{Use the FOIL method.} \\
= 7 - 9\sqrt{14} + \sqrt{14} - 9(2) & \text{Perform each multiplication.} \\
= 7 - 8\sqrt{14} - 18 & \text{Combine like radicals: } -9\sqrt{14} + \sqrt{14} = -8\sqrt{14}. \\
= -11 - 8\sqrt{14} & \text{Combine like terms: } 7 - 18 = -11.
\end{array}
$$

Caution

A common error is to "simplify" incorrectly by subtracting:

$-11 - 8\sqrt{14} \neq -19\sqrt{14}$

b. $\left(\sqrt[3]{x^2} - 4\sqrt[3]{5}\right)\left(\sqrt[3]{x} + \sqrt[3]{2}\right)$

$$
\begin{array}{ll}
= \sqrt[3]{x^2}\sqrt[3]{x} + \sqrt[3]{x^2}\sqrt[3]{2} - 4\sqrt[3]{5}\sqrt[3]{x} - 4\sqrt[3]{5}\sqrt[3]{2} & \text{Use the FOIL method.} \\
= \sqrt[3]{x^3} + \sqrt[3]{2x^2} - 4\sqrt[3]{5x} - 4\sqrt[3]{10} & \text{Perform each multiplication.} \\
= x + \sqrt[3]{2x^2} - 4\sqrt[3]{5x} - 4\sqrt[3]{10} & \text{Simplify the first term. There are no like} \\
& \text{radicals or terms to combine.}
\end{array}
$$

Self Check 3   Multiply and then simplify:
   a. $\left(\sqrt{5} + 2\sqrt{3}\right)\left(\sqrt{5} - \sqrt{3}\right)$
   b. $\left(\sqrt[3]{a} + 9\sqrt[3]{2}\right)\left(\sqrt[3]{a^2} - \sqrt[3]{3}\right)$

Now Try   Problems 37 and 41

2 Find Powers of Radical Expressions.

To find the power of a radical expression, such as $\left(\sqrt{5}\right)^2$ or $\left(\sqrt[3]{2}\right)^3$, we can use the definition of exponent and the product rule for radicals.

$$
\begin{array}{ll}
\left(\sqrt{5}\right)^2 = \sqrt{5}\sqrt{5} & \qquad \left(\sqrt[3]{2}\right)^3 = \sqrt[3]{2} \cdot \sqrt[3]{2} \cdot \sqrt[3]{2} \\
\qquad\;\; = \sqrt{25} & \qquad\qquad\quad = \sqrt[3]{8} \\
\qquad\;\; = 5 & \qquad\qquad\quad = 2
\end{array}
$$

These results illustrate the following property of radicals.

The *n*th Power of the *n*th Root	If $\sqrt[n]{a}$ is a real number, $\left(\sqrt[n]{a}\right)^n = a$

EXAMPLE 4   Find: **a.** $\left(\sqrt{5}\right)^2$ **b.** $\left(2\sqrt[3]{7x^2}\right)^3$ **c.** $\left(8 - \sqrt{3}\right)^2$ **d.** $\left(\sqrt{m+1} + 2\right)^2$

Strategy   In part (a), we will use the definition of square root. In part (b), we will use a power rule for exponents. In parts (c) and (d), we will use the FOIL method.

Why   Part (a) is the square of a square root, part (b) has the form $(xy)^n$, and part (c) has the form $(x + y)^2$.

Solution   **a.** $\left(\sqrt{5}\right)^2 = 5$ Because the square of the square root of 5 is 5

b. We can use the power of a product rule for exponents to find $\left(2\sqrt[3]{7x^2}\right)^3$.

$$
\begin{array}{ll}
\left(2\sqrt[3]{7x^2}\right)^3 = 2^3\left(\sqrt[3]{7x^2}\right)^3 & \text{Raise each factor of } 2\sqrt[3]{7x^2} \text{ to the 3rd power.} \\
\qquad\qquad\quad = 8(7x^2) & \text{Evaluate: } 2^3 = 8. \text{ Use } \left(\sqrt[n]{a}\right)^n = a. \\
\qquad\qquad\quad = 56x^2 & \text{Multiply: } 8 \cdot 7 = 56.
\end{array}
$$

c. $\left(8 - \sqrt{3}\right)^2 = \left(8 - \sqrt{3}\right)\left(8 - \sqrt{3}\right)$ Write the base $8 - \sqrt{3}$ as a factor twice.

$= 64 - 8\sqrt{3} - 8\sqrt{3} + \sqrt{3}\sqrt{3}$ Use the FOIL method.

$= 64 - 16\sqrt{3} + 3$ Combine like radicals: $-8\sqrt{3} - 8\sqrt{3} = -16\sqrt{3}$.

$= 67 - 16\sqrt{3}$ Combine like terms: $64 + 3 = 67$.

d. We can use the FOIL method to find the product.

$\left(\sqrt{m+1} + 2\right)^2 = \left(\sqrt{m+1} + 2\right)\left(\sqrt{m+1} + 2\right)$

$= \left(\sqrt{m+1}\right)^2 + 2\sqrt{m+1} + 2\sqrt{m+1} + 2 \cdot 2$

$= m + 1 + 2\sqrt{m+1} + 2\sqrt{m+1} + 4$ Use $\left(\sqrt[n]{a}\right)^n = a$.

$= m + 4\sqrt{m+1} + 5$ Combine like terms.

Self Check 4   Find: **a.** $\left(\sqrt{11}\right)^2$ **b.** $\left(3\sqrt[3]{4y}\right)^3$ **c.** $\left(\sqrt{3} + 4\right)^2$

               d. $\left(\sqrt{x-8} - 5\right)^2$

Now Try ▶   Problems 43, 49, and 53

3   Rationalize Denominators.

We have seen that when a radical expression is written in simplified form, each of the following statements is true.

<table>
<tr><td>**Simplified Form of a Radical Expression**</td><td>1. Each factor in the radicand is to a power that is less than the index of the radical.

2. The radicand contains no fractions or negative numbers.

3. No radicals appear in the denominator of a fraction.</td></tr>
</table>

We now consider radical expressions that do not satisfy requirements 2 or 3. We will introduce an algebraic technique, called **rationalizing the denominator,** that is used to write such expressions in an equivalent simplified form. In this process, we multiply the expression by a form of 1 and use the fact that $\sqrt[n]{a^n} = a$.

As an example, let's consider the following expression:

$\dfrac{\sqrt{5}}{\sqrt{3}}$ This radical expression is not in simplified form, because a radical appears in the denominator. It doesn't satisfy requirement 3 listed above.

We want to find a fraction equivalent to $\dfrac{\sqrt{5}}{\sqrt{3}}$ that does not have a radical in its denominator.

If we multiply $\dfrac{\sqrt{5}}{\sqrt{3}}$ by $\dfrac{\sqrt{3}}{\sqrt{3}}$, the denominator becomes $\sqrt{3} \cdot \sqrt{3} = 3$, a rational number.

$\dfrac{\sqrt{5}}{\sqrt{3}} = \dfrac{\sqrt{5}}{\sqrt{3}} \cdot \dfrac{\sqrt{3}}{\sqrt{3}}$ To build an equivalent fraction, multiply by a form of 1: $\dfrac{\sqrt{3}}{\sqrt{3}} = 1$.

$= \dfrac{\sqrt{15}}{3}$ Multiply the numerators: $\sqrt{5} \cdot \sqrt{3} = \sqrt{15}$. Multiply the denominators: $\sqrt{3} \cdot \sqrt{3} = \left(\sqrt{3}\right)^2 = 3$. The denominator is now a rational number, 3.

Thus, $\dfrac{\sqrt{5}}{\sqrt{3}} = \dfrac{\sqrt{15}}{3}$. These equivalent fractions represent the same number, but have different forms. Since there is no radical in the denominator, and $\sqrt{15}$ is in simplest form, the expression $\dfrac{\sqrt{15}}{3}$ is in simplified form. We say that we have *rationalized the denominator* of $\dfrac{\sqrt{5}}{\sqrt{3}}$.

EXAMPLE 5   Rationalize the denominator: **a.** $\sqrt{\dfrac{20}{7}}$ **b.** $\dfrac{4}{\sqrt[3]{2}}$

Strategy   We look at each denominator and ask, "By what must we multiply it to obtain a rational number?" Then we will multiply each expression by a carefully chosen form of 1.

Why   We want to produce an equivalent expression that does not have a radical in its denominator.

Solution   **a.** This radical expression is not in simplified form, because the radicand contains a fraction. (It doesn't satisfy requirement 2.) We begin by writing the square root of the quotient as the quotient of two square roots:

$$\sqrt{\dfrac{20}{7}} = \dfrac{\sqrt{20}}{\sqrt{7}} \qquad \text{Use the division property of radicals: } \sqrt[n]{\dfrac{a}{b}} = \dfrac{\sqrt[n]{a}}{\sqrt[n]{b}}.$$

To rationalize the denominator, we proceed as follows:

$$\dfrac{\sqrt{20}}{\sqrt{7}} = \dfrac{\sqrt{20}}{\sqrt{7}} \cdot \dfrac{\sqrt{7}}{\sqrt{7}} \qquad \text{To build an equivalent fraction, multiply by } \dfrac{\sqrt{7}}{\sqrt{7}} = 1.$$

$$= \dfrac{\sqrt{140}}{7} \qquad \text{Multiply the numerators. Multiply the denominators: } \sqrt{7} \cdot \sqrt{7} = \left(\sqrt{7}\right)^2 = 7. \text{ The denominator is now a rational number, 7.}$$

$$= \dfrac{2\sqrt{35}}{7} \qquad \text{Simplify: } \sqrt{140} = \sqrt{4 \cdot 35} = \sqrt{4}\sqrt{35} = 2\sqrt{35}.$$

> **Caution**
>
> Do not attempt to remove a common factor of 7 from the numerator and denominator of $\dfrac{2\sqrt{35}}{7}$. The numerator, $2\sqrt{35}$, does not have a factor of 7.
>
> $$\dfrac{2\sqrt{35}}{7} = \dfrac{2 \cdot \sqrt{5 \cdot 7}}{7}$$

b. This expression is not in simplified form because a radical appears in the denominator of a fraction. (It doesn't satisfy requirement 3.) Here, we must rationalize a denominator that is a cube root. We multiply the numerator and the denominator by a number that will give a perfect cube under the radical. Since $2 \cdot 4 = 8$ is a perfect cube, $\sqrt[3]{4}$ is such a number.

$$\dfrac{4}{\sqrt[3]{2}} = \dfrac{4}{\sqrt[3]{2}} \cdot \dfrac{\sqrt[3]{4}}{\sqrt[3]{4}} \qquad \text{To build an equivalent fraction, multiply by } \dfrac{\sqrt[3]{4}}{\sqrt[3]{4}} = 1.$$

$$= \dfrac{4\sqrt[3]{4}}{\sqrt[3]{8}} \qquad \text{Multiply the numerators. Multiply the denominators. This radicand is now a perfect cube.}$$

$$= \dfrac{4\sqrt[3]{4}}{2} \qquad \text{Evaluate the denominator: } \sqrt[3]{8} = 2. \text{ The denominator is now a rational number, 2.}$$

$$= 2\sqrt[3]{4} \qquad \text{Simplify the fraction: } \dfrac{4\sqrt[3]{4}}{2} = \dfrac{\overset{2}{\cancel{2}} \cdot 2\sqrt[3]{4}}{\underset{1}{\cancel{2}}} = 2\sqrt[3]{4}.$$

> **Caution**
>
> Multiplying $\dfrac{4}{\sqrt[3]{2}}$ by $\dfrac{\sqrt[3]{2}}{\sqrt[3]{2}}$ does not rationalize the denominator.
>
> $$\dfrac{4}{\sqrt[3]{2}} \cdot \dfrac{\sqrt[3]{2}}{\sqrt[3]{2}} = \dfrac{4\sqrt[3]{2}}{\sqrt[3]{4}}$$
> $$\qquad\qquad\qquad \uparrow$$
>
> Since 4 is not a perfect cube, this radical does not simplify.

An alternate way to rationalize the denominator is to use the equivalent form $\dfrac{\sqrt[3]{2^2}}{\sqrt[3]{2^2}}$ instead of $\dfrac{\sqrt[3]{4}}{\sqrt[3]{4}}$ to build an equivalent fraction:

$$\dfrac{4}{\sqrt[3]{2}} \cdot \dfrac{\sqrt[3]{2^2}}{\sqrt[3]{2^2}} = \dfrac{4\sqrt[3]{2^2}}{\sqrt[3]{2^3}} \qquad \text{Multiply the numerators. Multiply the denominators. In the denominator, the radicand is now a perfect cube because } 2 \cdot 2^2 = 2^3.$$

$$= \dfrac{4\sqrt[3]{4}}{2} \qquad \text{In the numerator, } 2^2 = 4. \text{ Simplify in the denominator, } \sqrt[3]{2^3} = 2.$$

$$= 2\sqrt[3]{4} \qquad \text{Simplify the fraction: } \dfrac{4\sqrt[3]{4}}{2} = \dfrac{\overset{2}{\cancel{2}} \cdot 2\sqrt[3]{4}}{\underset{1}{\cancel{2}}} = 2\sqrt[3]{4}.$$

Self Check 5   Rationalize the denominator: **a.** $\sqrt{\dfrac{8}{5}}$ **b.** $\dfrac{5}{\sqrt[3]{9}}$

Now Try ▶   Problems 57, 59, and 63

EXAMPLE 6   Rationalize the denominator: $\dfrac{\sqrt{5xy^2}}{\sqrt{xy^3}}$

Strategy   We will begin by using the quotient rule for radicals in reverse: $\dfrac{\sqrt[n]{a}}{\sqrt[n]{b}} = \sqrt[n]{\dfrac{a}{b}}$

Why   When the radicands are written under a single radical symbol, the result is a rational expression. Our hope is that the rational expression can be simplified, which could possibly make rationalizing the denominator easier.

Solution   There are two methods we can use to rationalize the denominator. In each method, we simplify the rational expression $\dfrac{5xy^2}{xy^3}$ that appears in the radicand first.

> **Caution**
>
> We will assume that all of the variables appearing in the following examples represent positive numbers.

Method 1

$$\dfrac{\sqrt{5xy^2}}{\sqrt{xy^3}} = \sqrt{\dfrac{5xy^2}{xy^3}}$$

$$= \sqrt{\dfrac{5}{y}} \qquad \text{Simplify the radicand.}$$

$$= \dfrac{\sqrt{5}}{\sqrt{y}} \qquad \text{Use the quotient rule.}$$

$$= \dfrac{\sqrt{5}}{\sqrt{y}} \cdot \dfrac{\sqrt{y}}{\sqrt{y}} \qquad \text{Multiply outside the radical.}$$

$$= \dfrac{\sqrt{5y}}{y} \qquad \text{Multiply the numerators and the denominators.}$$

Method 2

$$\dfrac{\sqrt{5xy^2}}{\sqrt{xy^3}} = \sqrt{\dfrac{5xy^2}{xy^3}}$$

$$= \sqrt{\dfrac{5}{y}} \qquad \text{Simplify the radicand.}$$

$$= \sqrt{\dfrac{5}{y} \cdot \dfrac{y}{y}} \qquad \text{Multiply within the radical.}$$

$$= \dfrac{\sqrt{5y}}{\sqrt{y^2}} \qquad \text{Use the quotient rule.}$$

$$= \dfrac{\sqrt{5y}}{y} \qquad \text{Simplify the denominator.}$$

Self Check 6   Rationalize the denominator: $\dfrac{\sqrt{4ab^3}}{\sqrt{2a^2b^2}}$

Now Try ▶   Problems 67 and 73

EXAMPLE 7   Rationalize the denominator: $\dfrac{11}{\sqrt{20q^5}}$

Strategy   We will simplify the radical expression in the denominator before rationalizing the denominator.

Why   We could begin by multiplying $\dfrac{11}{\sqrt{20q^5}}$ by $\dfrac{\sqrt{20q^5}}{\sqrt{20q^5}}$. However, to work with smaller numbers and simpler radical expressions, it is easier if we simplify $\sqrt{20q^5}$ first, and then rationalize the denominator.

Solution

$$\dfrac{11}{\sqrt{20q^5}} = \dfrac{11}{\sqrt{4q^4 \cdot 5q}} \qquad \text{To prepare to simplify } \sqrt{20q^5}, \text{ factor } 20q^5 \text{ as } 4q^4 \cdot 5q.$$

$$= \dfrac{11}{2q^2\sqrt{5q}} \qquad \text{Simplify: } \sqrt{4q^4 \cdot 5q} = \sqrt{4q^4}\sqrt{5q} = 2q^2\sqrt{5q}.$$

Success Tip

We usually simplify a radical expression before rationalizing the denominator.

$$= \frac{11}{2q^2\sqrt{5q}} \cdot \frac{\sqrt{5q}}{\sqrt{5q}}$$ To rationalize the denominator, multiply by $\frac{\sqrt{5q}}{\sqrt{5q}} = 1$.

$$= \frac{11\sqrt{5q}}{2q^2(5q)}$$ Multiply the numerators.
Multiply the denominators: $\sqrt{5q} \cdot \sqrt{5q} = \left(\sqrt{5q}\right)^2 = 5q$.

$$= \frac{11\sqrt{5q}}{10q^3}$$ Multiply in the denominator: $2 \cdot 5 = 10$ and $q^2 \cdot q = q^3$.

Self Check 7 Rationalize the denominator: $\frac{7}{\sqrt{18c^3}}$

Now Try ▶ Problems 75 and 81

EXAMPLE 8 Rationalize each denominator: **a.** $\dfrac{5}{\sqrt[3]{6n^2}}$ **b.** $\dfrac{\sqrt[4]{2}}{\sqrt[4]{3a}}$

Strategy In part (a), we will examine the radicand in the denominator and ask, "By what must we multiply it to obtain a perfect cube?" In part (b), we will examine the radicand in the denominator and ask, "By what must we multiply it to obtain a perfect-fourth power?"

Why The answers to those questions will determine what form of 1 we use to rationalize each denominator.

Solution **a.** To rationalize the denominator $\sqrt[3]{6n^2}$, we need the radicand to be a perfect cube. Since $6n^2 = 6 \cdot n \cdot n$, the radicand needs two more factors of 6 and one more factor of n.

It follows that we should multiply the given expression by $\dfrac{\sqrt[3]{6^2n}}{\sqrt[3]{6^2n}}$.

Success Tip

An alternate form of 1 that can be used to rationalize the denominator is $\dfrac{\sqrt[3]{36n}}{\sqrt[3]{36n}}$.

$$\frac{5}{\sqrt[3]{6n^2}} = \frac{5}{\sqrt[3]{6n^2}} \cdot \frac{\sqrt[3]{6^2n}}{\sqrt[3]{6^2n}}$$ Multiply by a form of 1 to rationalize the denominator.

$$= \frac{5\sqrt[3]{36n}}{\sqrt[3]{6^3n^3}}$$ Multiply the numerators. Multiply the denominators.
This radicand is now a perfect cube.

$$= \frac{5\sqrt[3]{36n}}{6n}$$ Simplify the denominator: $\sqrt[3]{6^3n^3} = 6n$.

b. To rationalize the denominator $\sqrt[4]{3a}$, we need the radicand to be a perfect-fourth power. Since $3a = 3 \cdot a$, the radicand needs three more factors of 3 and three more factors of a. It follows that we should multiply the given expression by $\dfrac{\sqrt[4]{3^3a^3}}{\sqrt[4]{3^3a^3}}$.

Success Tip

An alternate form of 1 that can be used to rationalize the denominator is $\dfrac{\sqrt[4]{27a^3}}{\sqrt[4]{27a^3}}$.

$$\frac{\sqrt[4]{2}}{\sqrt[4]{3a}} = \frac{\sqrt[4]{2}}{\sqrt[4]{3a}} \cdot \frac{\sqrt[4]{3^3a^3}}{\sqrt[4]{3^3a^3}}$$ Multiply by a form of 1 to rationalize the denominator.

$$= \frac{\sqrt[4]{54a^3}}{\sqrt[4]{3^4a^4}}$$ Multiply the numerators: $2 \cdot 27 = 54$. Multiply the denominators. This radicand is now a perfect-fourth power.

$$= \frac{\sqrt[4]{54a^3}}{3a}$$ Simplify the denominator: $\sqrt[4]{3^4a^4} = 3a$.

Self Check 8 Rationalize each denominator: **a.** $\dfrac{27}{\sqrt[3]{100a}}$

b. $\dfrac{\sqrt[4]{3}}{\sqrt[4]{4y^2}}$

Now Try ▶ Problems 83 and 87

4 Rationalize Denominators That Have Two Terms.

So far, we have rationalized denominators that have only one term. We will now discuss a method to rationalize denominators that have two terms.

One-termed denominators	*Two-termed denominators*

$$\frac{\sqrt{5}}{\sqrt{3}} \qquad \frac{11}{\sqrt{20q^5}} \qquad \frac{4}{\sqrt[3]{2}} \qquad\qquad \frac{1}{\sqrt{2}+1} \qquad \frac{\sqrt{x}+\sqrt{2}}{\sqrt{x}-\sqrt{2}}$$

To rationalize the denominator of $\dfrac{1}{\sqrt{2}+1}$, for example, we multiply the numerator and denominator by $\sqrt{2}-1$, because the product $\left(\sqrt{2}+1\right)\left(\sqrt{2}-1\right)$ contains no radicals.

$$\left(\sqrt{2}+1\right)\left(\sqrt{2}-1\right) = \left(\sqrt{2}\right)^2 - (1)^2 \qquad \text{\color{red}Use a special-product rule.}$$
$$= 2 - 1$$
$$= 1$$

Radical expressions that involve the sum and difference of the same two terms, such as $\sqrt{2}+1$ and $\sqrt{2}-1$, are called **conjugates.**

EXAMPLE 9   Rationalize the denominator: **a.** $\dfrac{1}{\sqrt{2}+1}$ **b.** $\dfrac{\sqrt{x}+\sqrt{2}}{\sqrt{x}-\sqrt{2}}$

Strategy   In each part, we will rationalize the denominator by multiplying the numerator and the denominator by the conjugate of the denominator.

Why   Multiplying each denominator by its conjugate will produce a new denominator that does not contain radicals.

Solution   **a.** To find a fraction equivalent to $\dfrac{1}{\sqrt{2}+1}$ that does not have a radical in its denominator, we multiply $\dfrac{1}{\sqrt{2}+1}$ by a form of 1 that uses the conjugate of $\sqrt{2}+1$.

$$\frac{1}{\sqrt{2}+1} = \frac{1}{\sqrt{2}+1} \cdot \frac{\color{red}\sqrt{2}-1}{\color{red}\sqrt{2}-1}$$

$$= \frac{\sqrt{2}-1}{\left(\sqrt{2}\right)^2 - (1)^2} \qquad \text{\color{red}Multiply the numerators. Multiply the denominators using a special-product rule.}$$

$$= \frac{\sqrt{2}-1}{2-1} \qquad \text{\color{red}In the denominator, } \left(\sqrt{2}\right)^2 = 2.$$

$$= \frac{\sqrt{2}-1}{1}$$

$$= \sqrt{2}-1$$

Success Tip

In the denominator, we can use the special-product rule $(A+B)(A-B) = A^2 - B^2$ to multiply $\sqrt{2}+1$ and $\sqrt{2}-1$ quickly.

b. We multiply the numerator and denominator by $\sqrt{x}+\sqrt{2}$, which is the conjugate of $\sqrt{x}-\sqrt{2}$, and simplify.

$$\frac{\sqrt{x}+\sqrt{2}}{\sqrt{x}-\sqrt{2}} = \frac{\sqrt{x}+\sqrt{2}}{\sqrt{x}-\sqrt{2}} \cdot \frac{\color{red}\sqrt{x}+\sqrt{2}}{\color{red}\sqrt{x}+\sqrt{2}}$$

$$= \frac{x + \sqrt{2x} + \sqrt{2x} + 2}{\left(\sqrt{x}\right)^2 - \left(\sqrt{2}\right)^2} \qquad \text{\color{red}Multiply the numerators. Multiply the denominators using a special-product formula.}$$

$$= \frac{x + \sqrt{2x} + \sqrt{2x} + 2}{x - 2} \qquad \text{\color{red}In the denominator, } \left(\sqrt{x}\right)^2 = x \text{ and } \left(\sqrt{2}\right)^2 = 2.$$

$$= \frac{x + 2\sqrt{2x} + 2}{x - 2} \qquad \text{\color{red}In the numerator, combine like radicals: } \sqrt{2x} + \sqrt{2x} = 2\sqrt{2x}.$$

Self Check 9 Rationalize the denominator: $\dfrac{\sqrt{x} - \sqrt{2}}{\sqrt{x} + \sqrt{2}}$

Now Try ▶ Problems 91 and 97

5 Rationalize Numerators.

In some advanced mathematics courses, like calculus, we sometimes have to rationalize a numerator by multiplying the numerator and denominator of the fraction by the conjugate of the numerator.

EXAMPLE 10 Rationalize the numerator: $\dfrac{\sqrt{x} - 3}{\sqrt{x}}$

Strategy To rationalize the numerator, we will multiply the numerator and the denominator by the conjugate of the numerator.

Why After rationalizing the numerator, we can simplify the expression. Although the result will not be in simplified form, this nonsimplified form is often desirable in calculus.

Solution We multiply the numerator and denominator by $\sqrt{x} + 3$, which is the conjugate of the numerator.

$$\dfrac{\sqrt{x} - 3}{\sqrt{x}} = \dfrac{\sqrt{x} - 3}{\sqrt{x}} \cdot \dfrac{\sqrt{x} + 3}{\sqrt{x} + 3}$$ Multiply by a form of 1 to rationalize the numerator.

$$= \dfrac{\left(\sqrt{x}\right)^2 - 3^2}{x + 3\sqrt{x}}$$ Multiply the numerators using a special-product rule. Multiply the denominators using the distributive property.

$$= \dfrac{x - 9}{x + 3\sqrt{x}}$$ In the numerator, $\left(\sqrt{x}\right)^2 = x$ and $3^2 = 9$.

Self Check 10 Rationalize the numerator: $\dfrac{\sqrt{x} + 3}{\sqrt{x}}$

Now Try ▶ Problem 101

SECTION 7.4 ▶ STUDY SET

VOCABULARY

Fill in the blanks.

1. In this section, we used the _____ rule for radicals in reverse: $\sqrt[n]{a} \cdot \sqrt[n]{b} = \sqrt[n]{ab}$.

2. To multiply $2\sqrt{5}\left(3\sqrt{8} + \sqrt{3}\right)$, use the _____ property.

3. To _____ the denominator of $\dfrac{4}{\sqrt{5}}$, we multiply the fraction by $\dfrac{\sqrt{5}}{\sqrt{5}}$.

4. The denominator of the fraction $\dfrac{4}{\sqrt{5}}$ is an _____ number.

5. To obtain a _____ -cube radicand in the denominator of $\dfrac{\sqrt[3]{7}}{\sqrt[3]{5n}}$, we multiply the fraction by $\dfrac{\sqrt[3]{25n^2}}{\sqrt[3]{25n^2}}$.

6. The _____ of $\sqrt{x} + 1$ is $\sqrt{x} - 1$.

CONCEPTS

7. Tell why each of the following expressions is not in simplified radical form. Then simplify it. Finally, use a calculator to approximate its value.

	Why isn't it in simplified form?	Simplified form	Approximation
$\dfrac{3}{\sqrt{2}}$			
$\dfrac{\sqrt{18}}{2}$			
$\sqrt{\dfrac{9}{2}}$			

8. Fill in the blank: To rationalize the denominator of $\dfrac{3}{\sqrt{2}}$, we multiply it by $\dfrac{\sqrt{2}}{\sqrt{2}}$, which is a form of ▢.

9. Fill in the blanks to complete this special product:

$$\left(5 - \sqrt{x}\right)^2 = (▢)^2 - ▢(5)\left(\sqrt{x}\right) + \left(▢\right)^2$$

$$= ▢ - 10\sqrt{x} + ▢$$

10. Fill in the blanks to complete this special product:

$$\left(\sqrt{7} + 2\right)\left(\sqrt{7} - 2\right) = \left(\sqrt{7}\right)^{▢} - (2)^{▢}$$

$$= ▢ - 4$$

$$= ▢$$

11. Perform each operation, if possible.

a. $4\sqrt{6} + 2\sqrt{6}$ b. $4\sqrt{6}\left(2\sqrt{6}\right)$

c. $3\sqrt{2} - 2\sqrt{3}$ d. $3\sqrt{2}\left(-2\sqrt{3}\right)$

12. Perform each operation, if possible.

a. $5 + 6\sqrt[3]{6}$ b. $5\left(6\sqrt[3]{6}\right)$

c. $\dfrac{30\sqrt[3]{15}}{5}$ d. $\dfrac{\sqrt[3]{15}}{5}$

NOTATION

Fill in the blanks.

13. Multiply:

$$5\sqrt{8} \cdot 7\sqrt{6} = 5(7)\sqrt{8\,▢}$$

$$= 35\sqrt{▢}$$

$$= 35\sqrt{▢ \cdot 3}$$

$$= 35\left(▢\right)\sqrt{3}$$

$$= 140\sqrt{3}$$

14. Rationalize the denominator:

$$\frac{9}{\sqrt[3]{4a^2}} = \frac{9}{\sqrt[3]{4a^2}} \cdot \frac{\sqrt[3]{2a}}{▢}$$

$$= \frac{9\sqrt[3]{2a}}{\sqrt[3]{▢}}$$

$$= \frac{9\sqrt[3]{2a}}{▢}$$

GUIDED PRACTICE

Multiply and simplify. All variables represent positive real numbers. See Example 1.

15. $\sqrt{3}\sqrt{15}$ **16.** $\sqrt{5}\sqrt{15}$

17. $2\sqrt{3}\sqrt{6}$ **18.** $-3\sqrt{11}\sqrt{33}$

19. $\left(3\sqrt[3]{9}\right)\left(2\sqrt[3]{3}\right)$ **20.** $\left(2\sqrt[3]{16}\right)\left(-\sqrt[3]{4}\right)$

21. $\sqrt[3]{2} \cdot \sqrt[3]{12}$ **22.** $\sqrt[3]{3} \cdot \sqrt[3]{18}$

23. $6\sqrt{ab^3}\left(8\sqrt{ab}\right)$ **24.** $3\sqrt{8x}\left(2\sqrt{2x^3y}\right)$

25. $\sqrt[4]{5a^3}\sqrt[4]{125a^2}$ **26.** $\sqrt[4]{2r^3}\sqrt[4]{8r^2}$

Multiply and simplify. All variables represent positive real numbers. See Example 2.

27. $3\sqrt{5}\left(4 - \sqrt{5}\right)$ **28.** $2\sqrt{7}\left(3 - \sqrt{7}\right)$

29. $\sqrt{2}\left(4\sqrt{6} + 2\sqrt{7}\right)$ **30.** $-\sqrt{3}\left(\sqrt{7} - \sqrt{15}\right)$

31. $-2\sqrt{5x}\left(4\sqrt{2x} - 3\sqrt{3}\right)$ **32.** $3\sqrt{7t}\left(2\sqrt{7t} + 3\sqrt{3t^2}\right)$

33. $\sqrt[3]{2}\left(4\sqrt[3]{4} + \sqrt[3]{12}\right)$ **34.** $\sqrt[3]{3}\left(2\sqrt[3]{9} + \sqrt[3]{18}\right)$

Multiply and simplify. All variables represent positive real numbers. See Example 3.

35. $\left(\sqrt{2} + 1\right)\left(\sqrt{2} - 3\right)$

36. $\left(2\sqrt{3} + 1\right)\left(\sqrt{3} - 1\right)$

37. $\left(\sqrt{3x} - \sqrt{2y}\right)\left(\sqrt{3x} + \sqrt{2y}\right)$

38. $\left(\sqrt{3m} + \sqrt{2n}\right)\left(\sqrt{3m} - \sqrt{2n}\right)$

39. $\left(2\sqrt[3]{4} - 3\sqrt[3]{2}\right)\left(3\sqrt[3]{4} + 2\sqrt[3]{10}\right)$

40. $\left(4\sqrt[3]{9} - 3\sqrt[3]{3}\right)\left(4\sqrt[3]{3} + 2\sqrt[3]{6}\right)$

41. $\left(\sqrt[3]{5z} + \sqrt[3]{3}\right)\left(\sqrt[3]{5z} + 2\sqrt[3]{3}\right)$

42. $\left(\sqrt[3]{3p} - 2\sqrt[3]{2}\right)\left(\sqrt[3]{3p} + \sqrt[3]{2}\right)$

Square or cube each quantity and simplify the result. See Example 4.

43. $\left(\sqrt{7}\right)^2$ **44.** $\left(\sqrt{11}\right)^2$

45. $\left(\sqrt[3]{12}\right)^3$ **46.** $\left(\sqrt[3]{9}\right)^3$

47. $\left(3\sqrt{2}\right)^2$ **48.** $\left(2\sqrt{5}\right)^2$

49. $\left(-2\sqrt[3]{2x^2}\right)^3$

50. $\left(-3\sqrt[3]{10y^3}\right)^3$

51. $\left(6 - \sqrt{3}\right)^2$

52. $\left(9 - \sqrt{11}\right)^2$

53. $\left(\sqrt{3x} + \sqrt{3}\right)^2$

54. $\left(\sqrt{5x} - \sqrt{3}\right)^2$

Rationalize each denominator. See Example 5.

55. $\sqrt{\dfrac{2}{7}}$

56. $\sqrt{\dfrac{5}{3}}$

57. $\sqrt{\dfrac{8}{3}}$

58. $\sqrt{\dfrac{8}{7}}$

59. $\dfrac{4}{\sqrt{6}}$

60. $\dfrac{8}{\sqrt{10}}$

61. $\dfrac{1}{\sqrt[3]{2}}$

62. $\dfrac{2}{\sqrt[3]{6}}$

63. $\dfrac{3}{\sqrt[3]{9}}$

64. $\dfrac{2}{\sqrt[3]{a}}$

65. $\dfrac{1}{\sqrt[4]{8}}$

66. $\dfrac{1}{\sqrt[5]{2}}$

Rationalize each denominator. All variables represent positive real numbers. See Example 6.

67. $\dfrac{\sqrt{10y^2}}{\sqrt{2y^3}}$

68. $\dfrac{\sqrt{15b^2}}{\sqrt{5b^3}}$

69. $\dfrac{\sqrt{48x^2}}{\sqrt{8x^2y}}$

70. $\dfrac{\sqrt{9xy}}{\sqrt{3x^2y}}$

71. $\dfrac{\sqrt[3]{12t^3}}{\sqrt[3]{54t^2}}$

72. $\dfrac{\sqrt[3]{15m^4}}{\sqrt[3]{12m^3}}$

73. $\dfrac{\sqrt[3]{4a^6}}{\sqrt[3]{2a^5b}}$

74. $\dfrac{\sqrt[3]{9x^5y^4}}{\sqrt[3]{3x^5y^5}}$

Rationalize each denominator. All variables represent positive real numbers. See Example 7.

75. $\dfrac{23}{\sqrt{50p^5}}$

76. $\dfrac{11}{\sqrt{75s^5}}$

77. $\dfrac{7}{\sqrt{24b^3}}$

78. $\dfrac{13}{\sqrt{32n^3}}$

Rationalize each denominator. All variables represent positive real numbers. See Example 8.

79. $\sqrt[3]{\dfrac{5}{16}}$

80. $\sqrt[3]{\dfrac{2}{81}}$

81. $\sqrt[3]{\dfrac{4}{81}}$

82. $\sqrt[3]{\dfrac{7}{16}}$

83. $\dfrac{19}{\sqrt[3]{5c^2}}$

84. $\dfrac{1}{\sqrt[3]{4m^2}}$

85. $\dfrac{\sqrt[3]{3}}{\sqrt[3]{2r}}$

86. $\dfrac{\sqrt[3]{7}}{\sqrt[3]{100s}}$

87. $\dfrac{\sqrt{2}}{\sqrt[4]{3t^2}}$

88. $\dfrac{\sqrt[4]{3}}{\sqrt[4]{5b^3}}$

89. $\dfrac{25}{\sqrt[4]{8a}}$

90. $\dfrac{4}{\sqrt[4]{9t}}$

Rationalize each denominator. All variables represent positive real numbers. See Example 9.

91. $\dfrac{\sqrt{2}}{\sqrt{5} + 3}$

92. $\dfrac{\sqrt{3}}{\sqrt{3} - 2}$

93. $\dfrac{2}{\sqrt{x} + 1}$

94. $\dfrac{3}{\sqrt{x} - 2}$

95. $\dfrac{\sqrt{7} - \sqrt{2}}{\sqrt{2} + \sqrt{7}}$

96. $\dfrac{\sqrt{3} + \sqrt{2}}{\sqrt{3} - \sqrt{2}}$

97. $\dfrac{\sqrt{x} - \sqrt{y}}{\sqrt{x} + \sqrt{y}}$

98. $\dfrac{\sqrt{x} + \sqrt{y}}{\sqrt{x} - \sqrt{y}}$

Rationalize each numerator. All variables represent positive real numbers. See Example 10.

99. $\dfrac{\sqrt{x} + 3}{x}$

100. $\dfrac{2 + \sqrt{x}}{5x}$

101. $\dfrac{\sqrt{x} + \sqrt{y}}{\sqrt{x}}$

102. $\dfrac{\sqrt{x} - \sqrt{y}}{\sqrt{x} + \sqrt{y}}$

TRY IT YOURSELF

The following problems involve addition, subtraction, and multiplication of radical expressions, as well as rationalizing the denominator. Perform the operations and simplify, if possible. All variables represent positive real numbers.

103. $\sqrt{x}\left(\sqrt{14x} + \sqrt{2}\right)$

104. $2\sqrt[3]{16} - 3\sqrt[3]{128} - \sqrt[3]{54}$

105. $\dfrac{3\sqrt{2} - 5\sqrt{3}}{2\sqrt{3} - 3\sqrt{2}}$

106. $\dfrac{3\sqrt{6} + 5\sqrt{5}}{2\sqrt{5} - 3\sqrt{6}}$

107. $\left(10\sqrt[3]{2x}\right)^3$

108. $\dfrac{\sqrt{3}}{\sqrt{98x^2}}$

109. $-4\sqrt[3]{5r^2s}\left(5\sqrt[3]{2r}\right)$

110. $-\sqrt[3]{3xy^2}\left(-\sqrt[3]{9x^3}\right)$

111. $\left(3p + \sqrt{5}\right)^2$

112. $\sqrt{288t} + \sqrt{80t} - \sqrt{128t}$

113. $\sqrt{\dfrac{72m^8}{25m^3}}$

114. $\left(\sqrt{14x} + \sqrt{3}\right)\left(\sqrt{14x} - \sqrt{3}\right)$

115. $\sqrt[4]{3n^2}\sqrt[4]{27n^3}$

116. $\dfrac{\sqrt{y} - 2}{\sqrt{y} + 3}$

117. $\dfrac{\sqrt[3]{x}}{\sqrt[3]{9}}$

118. $\sqrt[5]{\dfrac{2}{243}}$

119. $\left(3\sqrt{2r} - 2\right)^2$

120. $\left(2\sqrt{3t} + 5\right)^2$

121. $\sqrt{x(x + 3)}\sqrt{x^3(x + 3)}$ **122.** $\sqrt{y^2(x + y)}\sqrt{(x + y)^3}$

123. $\dfrac{2z - 1}{\sqrt{2z - 1}}$
(*Hint:* Do not perform the multiplication of the numerators.)

124. $\dfrac{3t - 1}{\sqrt{3t + 1}}$
(*Hint:* Do not perform the multiplication of the numerators.)

Look Alikes . . .

125. a. $\left(3\sqrt{a}\right)^2$ **b.** $\left(3 + \sqrt{a}\right)^2$

126. a. $\left(9\sqrt{x - 5}\right)^2$ **b.** $\left(9 + \sqrt{x - 5}\right)^2$

127. a. $\left(\sqrt{m - 6}\right)^2$ **b.** $\left(\sqrt{m} - 6\right)^2$

128. a. $\dfrac{1}{\sqrt{xy}}$ **b.** $\dfrac{1}{\sqrt{x} + \sqrt{y}}$

APPLICATIONS

129. Statistics. An example of a normal distribution curve, or *bell-shaped* curve, is shown. A fraction that is part of the equation that models this curve is $\dfrac{1}{\sigma\sqrt{2\pi}}$, where σ is a letter from the Greek alphabet. Rationalize the denominator of the fraction.

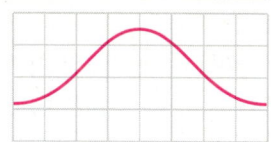

130. Analytical Geometry. The length of the perpendicular segment drawn from $(-2, 2)$ to the line with equation $2x - 4y = 4$ is given by

$$L = \frac{|2(-2) + (-4)(2) + (-4)|}{\sqrt{(2)^2 + (-4)^2}}$$

Find L. Express the result in simplified radical form. Then give an approximation to the nearest tenth.

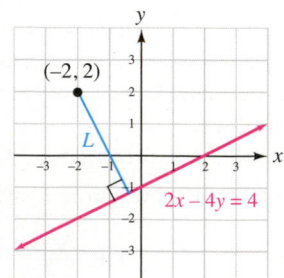

131. Trigonometry. In trigonometry, we often must find the ratio of the lengths of two sides of right triangles. Use the information in the illustration to find the ratio

$$\frac{\text{length of side } AC}{\text{length of side } AB}$$

Write the result in simplified radical form.

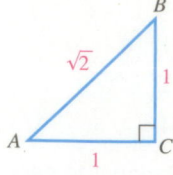

132. Engineering. Refer to the illustration below that shows a block connected to two walls by springs. A measure of how fast the block will oscillate when the spring system is set in motion is given by the formula $\omega = \sqrt{\dfrac{k_1 + k_2}{m}}$ where k_1 and k_2 indicate the stiffness of the springs and m is the mass of the block. Rationalize the right side and restate the formula.

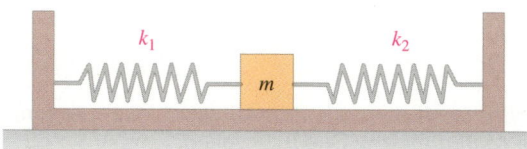

WRITING

133. Consider $\dfrac{\sqrt{3}}{\sqrt{7}} = \dfrac{\sqrt{3}}{\sqrt{7}} \cdot \dfrac{\sqrt{7}}{\sqrt{7}}$. Explain why the expressions on the left side and the right side of the equation are equal.

134. To rationalize the denominator of $\dfrac{\sqrt[4]{12}}{\sqrt[4]{3}}$, why wouldn't we multiply the numerator and denominator by $\dfrac{\sqrt[4]{3}}{\sqrt[4]{3}}$?

135. Explain why $\dfrac{\sqrt[3]{12}}{\sqrt[3]{5}}$ is not in simplified form.

136. Explain why $\sqrt{\dfrac{3a}{11k}}$ is not in simplified form.

137. Explain why $\sqrt{m} \cdot \sqrt{m} = m$ but $\sqrt[3]{m} \cdot \sqrt[3]{m} \neq m$. Assume that m represents a positive number.

138. Explain why the product of $\sqrt{m} + 3$ and $\sqrt{m} - 3$ does not contain a radical.

REVIEW

Solve each equation.

139. $\dfrac{8}{b - 2} + \dfrac{3}{2 - b} = -\dfrac{1}{b}$

140. $\dfrac{2}{x - 2} + \dfrac{1}{x + 1} = \dfrac{1}{(x + 1)(x - 2)}$

CHALLENGE PROBLEMS

141. Multiply: $\sqrt{2} \cdot \sqrt[3]{2}$. (*Hint:* Keep in mind two things. The indices (plural for *index*) must be the same to use the product rule for radicals, and radical expressions can be written using rational exponents.)

142. Show that $\dfrac{\sqrt[3]{a^2} + \sqrt[3]{a}\sqrt[3]{b} + \sqrt[3]{b^2}}{\sqrt[3]{a^2} + \sqrt[3]{a}\sqrt[3]{b} + \sqrt[3]{b^2}}$ can be used to rationalize the denominator of $\dfrac{1}{\sqrt[3]{a} - \sqrt[3]{b}}$.

SECTION 7.5

OBJECTIVES

1. Solve equations containing one radical.
2. Solve equations containing two radicals.
3. Solve formulas containing radicals.

Solving Radical Equations

ARE YOU READY?

The following problems review some basic skills that are needed when solving radical equations.

1. Simplify: **a.** $\left(\sqrt{x-1}\right)^2$

           b. $\left(\sqrt[3]{x^3+7}\right)^3$

2. Simplify: $\left(3\sqrt{2x+5}\right)^2$

3. Expand: $(x-4)^2$

4. Solve: $x^2 - 6x - 27 = 0$

5. Expand: $(x+2)^3$

6. Multiply: $\left(2-\sqrt{x}\right)^2$

When we solve equations containing fractions, we clear them of the fractions by multiplying both sides by the LCD. To solve equations containing radical expressions, we take a similar approach. The first step is to clear them of the radicals by raising both sides to a power.

1 Solve Equations Containing One Radical.

Radical equations contain at least one radical expression with a variable in the radicand. Some examples are

$$\sqrt{x+3} = 4 \qquad \sqrt[3]{x^3+7} = x+1 \qquad \sqrt{x} + \sqrt{x+2} = 2$$

To **solve a radical equation,** we find all the values of the variable that make the equation true. The goal when solving a radical equation is to use the following **power rule** to find an equivalent equation that we already know how to solve, such as a linear equation in one variable or a quadratic equation.

The Power Rule for Solving Radical Equations

If we raise two equal quantities to the same power, the results are equal quantities. If x, y, and n are real numbers and $x = y$, then

$$x^n = y^n \quad \text{for any exponent } n.$$

If both sides of an equation are raised to the same power, all solutions of the original equation are also solutions of the new equation. However, the resulting equation might not be equivalent to the original equation. For example, if we square both sides of the equation $x = 3$ with a solution set of $\{3\}$, we obtain the equation $x^2 = 9$ with a solution set of $\{3, -3\}$.

The equations $x = 3$ and $x^2 = 9$ are not equivalent, because they have different solution sets. The solution -3 of $x^2 = 9$ does not satisfy the equation $x = 3$. Since raising both sides of an equation to the same power can produce an equation with proposed solutions that don't satisfy the original equation, **we must always check each proposed solution in the original equation** and discard any **extraneous solutions.**

When we use the power rule to solve square root radical equations, it produces expressions of the form $\left(\sqrt{a}\right)^2$. We have seen that when this expression is simplified, the radical symbol is removed.

Caution

A similar warning about checking proposed solutions was given in Chapter 6 when we solved rational equations, such as:

$$\frac{11x}{x-5} = 6 + \frac{55}{x-5}$$

The Square of a Square Root

For any nonnegative real number a, $\left(\sqrt{a}\right)^2 = a$.

Here are some examples of the square of a square root. Notice how squaring such an expression *removes the square root symbol.*

$$\left(\sqrt{4n}\right)^2 = 4n, \qquad \left(\sqrt{x-3}\right)^2 = x-3, \qquad \text{and} \qquad \left(\sqrt{5a+8}\right)^2 = 5a+8$$

EXAMPLE 1   Solve: $\sqrt{x+3} = 4$

Strategy   We will use the power rule and square both sides of the equation.

Why   Squaring both sides will produce, on the left side, the expression $\left(\sqrt{x+3}\right)^2$ that simplifies to $x+3$. This step clears the equation of the radical.

Solution

$\sqrt{x+3} = 4$	This is the equation to solve.
$\left(\sqrt{x+3}\right)^2 = (4)^2$	To clear the equation of the square root, square both sides.
$x + 3 = 16$	Perform the operations on each side.
$x = 13$	Solve the resulting linear equation by subtracting 3 from both sides.

> **The Language of Algebra**
>
> When we square both sides of an equation, we are **raising both sides to the second power.**

We must check the proposed solution 13 to see whether it satisfies the original equation.

> **The Language of Algebra**
>
> **Proposed** solutions are also called **potential** or **possible** solutions.

Evaluate the left side. Do not square both sides when checking!

Check:

$$\sqrt{x+3} = 4 \qquad \text{This is the original equation.}$$
$$\sqrt{13+3} \overset{?}{=} 4 \qquad \text{Substitute 13 for } x.$$
$$\sqrt{16} \overset{?}{=} 4$$
$$4 = 4 \qquad \text{True}$$

Since 13 satisfies the original equation, it is the solution. The solution set is $\{13\}$.

Self Check 1   Solve: $\sqrt{a-2} = 3$

Now Try ▶ Problems 15 and 19

The method used in Example 1 to solve a radical equation containing a square root can be generalized, as follows.

> **Solving an Equation Containing Radicals**
>
> 1. Isolate a radical term on one side of the equation.
> 2. Raise both sides of the equation to the power that is the same as the index of the radical.
> 3. If it still contains a radical, go back to step 1. If it does not contain a radical, solve the resulting equation.
> 4. Check the proposed solutions in the original equation.

EXAMPLE 2   **Free Fall.**   The distance d in feet that an object will fall in t seconds is given by the formula $t = \sqrt{\frac{d}{16}}$. If the designers of the amusement park attraction want the riders to experience 3 seconds of vertical free fall, what length of vertical drop is needed?

Strategy   We will begin by substituting 3 for the time t in the formula.

Why   We can then solve the resulting radical equation in one variable to find the unknown distance d.

Solution

$$t = \sqrt{\dfrac{d}{16}}$$ This is the given formula.

$$3 = \sqrt{\dfrac{d}{16}}$$ Substitute 3 for t. Here the radical is isolated on the right side.

$$(3)^2 = \left(\sqrt{\dfrac{d}{16}}\right)^2$$ To clear the equation of the square root, square both sides.

$$9 = \dfrac{d}{16}$$ Perform the operations on each side.

$$144 = d$$ Solve the resulting equation by multiplying both sides by 16.

The amount of vertical drop needs to be 144 feet.

> **Caution**
>
> When using the power rule, don't forget to raise both sides to the same power. For this example, a common error would be to write
>
> $$3 = \left(\sqrt{\dfrac{d}{16}}\right)^2$$

Self Check 2 **Free Fall.** How long a vertical drop is needed if the riders are to free fall for 3.5 seconds?

Now Try ▶ Problem 105

When solving radical equations, always isolate a radical term before using the power rule.

EXAMPLE 3   Solve: $\sqrt{3x + 1} + 1 = x$

Strategy   Since 1 is outside the square root symbol, there are two terms on the left side of the equation. To isolate the radical term, we will subtract 1 from both sides.

Why   This will put the equation in a form in which we can square both sides to clear the radical.

Solution

> **Caution**
>
> If both sides of the equation in Example 3 are squared, as is, the radical is not cleared. In fact, the equation becomes more complicated.
>
> $$\left(\sqrt{3x + 1} + 1\right)^2 = (x)^2$$
> $$3x + 1 + 2\sqrt{3x + 1} + 1 = x^2$$

$$\sqrt{3x + 1} + 1 = x$$ This is the equation to solve.

$$\sqrt{3x + 1} = x - 1$$ To isolate the radical on the left side, subtract 1 from both sides.

$$\left(\sqrt{3x + 1}\right)^2 = (x - 1)^2$$ Square both sides to eliminate the square root. Don't forget the parentheses.

$$3x + 1 = x^2 - 2x + 1$$ On the right side, use a special-product formula to square the binomial: $(x - 1)^2 = x^2 - 2x + 1$. The resulting equation is quadratic.

$$0 = x^2 - 5x$$ To get 0 on the left side, subtract $3x$ and 1 from both sides.

$$0 = x(x - 5)$$ Factor out the GCF, x.

$$x = 0 \quad \text{or} \quad x - 5 = 0$$ Set each factor equal to 0.

$$x = 0 \quad | \quad \qquad x = 5$$

We must check each proposed solution to see whether it satisfies the original equation.

> **Success Tip**
>
> Even if you are certain that no algebraic mistakes were made when solving a radical equation, you must still check your solutions. Raising both sides to an even power can introduce extraneous solutions that must be discarded.

This is the check for **0**:

$$\sqrt{3x + 1} + 1 = x$$

$$\sqrt{3(0) + 1} + 1 \stackrel{?}{=} 0$$

$$\sqrt{1} + 1 \stackrel{?}{=} 0$$

$$2 = 0 \quad \text{False}$$

This is the check for **5**:

$$\sqrt{3x + 1} + 1 = x$$ This is the original equation.

$$\sqrt{3(5) + 1} + 1 \stackrel{?}{=} 5$$

$$\sqrt{16} + 1 \stackrel{?}{=} 5$$

$$5 = 5 \quad \text{True}$$

The proposed solution 0 does not check; it must be discarded. Since the only solution is 5, the solution set is $\{5\}$.

Self Check 3   Solve: $\sqrt{4x + 1} + 1 = x$

Now Try ▶   Problems 23 and 27

Using Your Calculator ▶ **Solving Radical Equations**

To find solutions for $\sqrt{3x + 1} + 1 = x$ with a graphing calculator, we graph the functions $f(x) = \sqrt{3x + 1} + 1$ and $g(x) = x$, as in figure (a). We then trace to find the approximate x-coordinate of their intersection point, as in figure (b). After repeated zooms, we will see that $x = 5$.

We also can use the INTERSECT feature to approximate the point of intersection of the graphs. See figure (c). The intersection point of (**5**, 5), with x-coordinate **5**, implies that 5 is a solution of the radical equation.

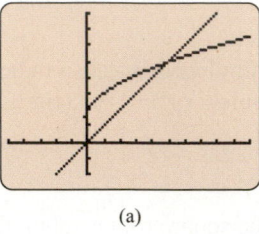   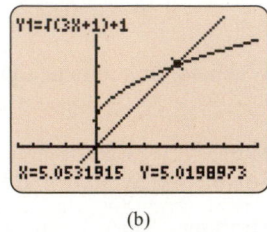   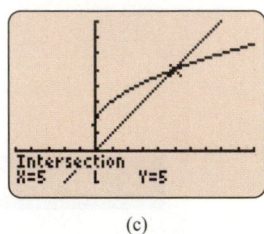

(a)   (b)   (c)

EXAMPLE 4   Solve: $\sqrt{3x} + 8 = 2$

Strategy   Since 8 is outside the square root symbol, there are two terms on the left side of the equation. To isolate the radical, we will subtract 8 from both sides.

Why   This will put the equation in a form in which we can square both sides to clear the radical.

Solution

$$\sqrt{3x} + 8 = 2 \qquad \text{\color{red}This is the equation to solve.}$$

$$\sqrt{3x} = -6 \qquad \text{\color{red}To isolate the radical on the left side, subtract 8 from both sides.}$$

$$\left(\sqrt{3x}\right)^2 = (-6)^2 \qquad \text{\color{red}Square both sides to eliminate the square root.}$$

$$3x = 36 \qquad \text{\color{red}Perform the operations on each side.}$$

$$x = 12 \qquad \text{\color{red}To solve the resulting linear equation, divide both sides by 3.}$$

We check the proposed solution 12 in the original equation.

$$\sqrt{3\color{red}{x}} + 8 = 2$$

$$\sqrt{3(\color{red}{12})} + 8 \stackrel{?}{=} 2 \qquad \text{\color{red}Substitute 12 for x.}$$

$$\sqrt{36} + 8 \stackrel{?}{=} 2$$

$$6 + 8 \stackrel{?}{=} 2$$

$$14 = 2 \qquad \text{\color{red}False}$$

Since 12 does not satisfy the original equation, it is extraneous. The equation $\sqrt{3x} + 8 = 2$ has no solution. The solution set is $\varnothing$.

Success Tip

After isolating the radical, we obtained the equation $\sqrt{3x} = -6$. Since the principal square root of a number cannot be negative, we immediately know that $\sqrt{3x} + 8 = 2$ has no solution.

Self Check 4   Solve: $\sqrt{a - 9} + 3 = 0$

Now Try ▶   Problem 31

Recall the following property of radicals from Section 7.4.

The *n*th Power of the *n*th Root	If $\sqrt[n]{a}$ is a real number, $$\left(\sqrt[n]{a}\right)^n = a$$

Here are some examples of the cube of a cube root, and the fourth power of a fourth root. Notice how raising each expression to the appropriate power *removes the radical symbol*.

$$\left(\sqrt[3]{5x}\right)^3 = 5x \qquad \left(\sqrt[3]{2a - 11}\right)^3 = 2a - 11 \qquad \left(\sqrt[4]{6 - y}\right)^4 = 6 - y$$

The power rule and the *n*th power of the *n*th root property can be used in combination to solve radical equations that involve cube roots, fourth roots, fifth roots, and so on.

EXAMPLE 5   Solve: $(x^3 + 7)^{1/3} = x + 1$

Strategy   We will use the definition of a fractional exponent with numerator 1 to rewrite the given equation as an equivalent radical equation.

Why   Then we can use the equation-solving strategy of this section to find the solution(s).

Solution   Recall from Section 7.2 that $x^{1/n} = \sqrt[n]{x}$. Thus, we can replace $(x^3 + 7)^{1/3}$ on the left side of the given equation with $\sqrt[3]{x^3 + 7}$.

Success Tip

Be careful cubing the binomial $x + 1$.
$$(x + 1)^3 \neq x^3 + 1^3$$

$$(x^3 + 7)^{1/3} = x + 1 \qquad \text{\color{red}{This is the equation to solve.}}$$
$$\sqrt[3]{x^3 + 7} = x + 1 \qquad \text{\color{red}{Write }}(x^3 + 7)^{1/3}\text{\color{red}{ using radical notation.}}$$
$$\left(\sqrt[3]{x^3 + 7}\right)^3 = (x + 1)^3 \qquad \begin{array}{l}\text{\color{red}{Cube both sides to eliminate the}}\\ \text{\color{red}{cube root. Don't forget the parentheses.}}\end{array}$$
$$x^3 + 7 = x^3 + 3x^2 + 3x + 1 \qquad \begin{array}{l}\text{\color{red}{Perform the operations on each side.}}\\ \text{\color{red}{On the right: }}(x + 1)^3 = (x + 1)(x + 1)^2.\end{array}$$
$$0 = 3x^2 + 3x - 6 \qquad \begin{array}{l}\text{\color{red}{To get 0 on the left side, subtract }}x^3\text{\color{red}{ and 7 from}}\\ \text{\color{red}{both sides. This is a quadratic equation.}}\end{array}$$
$$0 = x^2 + x - 2 \qquad \text{\color{red}{Divide both sides by 3.}}$$
$$0 = (x + 2)(x - 1) \qquad \text{\color{red}{Factor the trinomial.}}$$
$$x + 2 = 0 \quad \text{or} \quad x - 1 = 0 \qquad \text{\color{red}{Set each factor equal to 0.}}$$
$$x = -2 \qquad\qquad x = 1 \qquad \text{\color{red}{Solve each linear equation.}}$$

Success Tip

Raising both sides of an equation to an even power may not produce an equivalent equation. Raising both sides to an odd power does produce an equivalent equation and, in such cases, there will be no extraneous solutions. However, it is still recommended that all proposed solutions be checked regardless of the type of root involved.

We check each proposed solution, -2 and 1, to see whether they satisfy the original equation.

Check:

$$(x^3 + 7)^{1/3} = x + 1 \qquad\qquad (x^3 + 7)^{1/3} = x + 1$$
$$\sqrt[3]{x^3 + 7} = x + 1 \qquad\qquad \sqrt[3]{x^3 + 7} = x + 1$$
$$\sqrt[3]{(-2)^3 + 7} \stackrel{?}{=} -2 + 1 \qquad\qquad \sqrt[3]{1^3 + 7} \stackrel{?}{=} 1 + 1$$
$$\sqrt[3]{-8 + 7} \stackrel{?}{=} -1 \qquad\qquad\qquad \sqrt[3]{1 + 7} \stackrel{?}{=} 2$$
$$\sqrt[3]{-1} \stackrel{?}{=} -1 \qquad\qquad\qquad\quad \sqrt[3]{8} \stackrel{?}{=} 2$$
$$-1 = -1 \quad \text{\color{red}{True}} \qquad\qquad\qquad 2 = 2 \quad \text{\color{red}{True}}$$

Both -2 and 1 satisfy the original equation. Thus, the solution set is $\{-2, 1\}$.

Self Check 5   Solve: $(x^3 + 8)^{1/3} = x + 2$

Now Try ▶   Problems 35 and 39

EXAMPLE 6   Let $f(x) = \sqrt[4]{2x + 1}$. For what value(s) of x is $f(x) = 5$?

Strategy   We will substitute 5 for $f(x)$ and solve the equation $5 = \sqrt[4]{2x + 1}$. To do so, we will raise both sides of the equation to the fourth power.

Why   Raising both sides to the fourth power will produce, on the right side, the expression $\left(\sqrt[4]{2x + 1}\right)^4$ that simplifies to $2x + 1$. This step clears the equation of the radical.

Solution   To find the value(s) for which $f(x) = 5$, we substitute 5 for $f(x)$ and solve for x.

$$f(x) = \sqrt[4]{2x + 1}$$

$$5 = \sqrt[4]{2x + 1} \quad \text{\textcolor{red}{This is the equation to solve.}}$$

Since the equation contains a fourth root, we raise both sides to the fourth power to solve for x.

$$(5)^4 = \left(\sqrt[4]{2x + 1}\right)^4 \quad \text{\textcolor{red}{Use the power rule to clear the radical.}}$$

$$625 = 2x + 1 \quad \text{\textcolor{red}{Perform the operations on each side.}}$$

$$624 = 2x \quad \text{\textcolor{red}{To solve the resulting equation, subtract 1 from both sides.}}$$

$$312 = x \quad \text{\textcolor{red}{Divide both sides by 2.}}$$

If $x = 312$, then $f(x) = 5$. Verify this by evaluating $f(312)$ using a calculator, if necessary.

Self Check 6   Let $g(x) = \sqrt[5]{10x + 1}$. For what value(s) of x is $g(x) = 1$?

Now Try ▶   Problem 43

2 Solve Equations Containing Two Radicals.

To solve an equation containing two radicals, we want to have one radical on the left side and one radical on the right side.

EXAMPLE 7   Solve: $\sqrt{5x + 9} = 2\sqrt{3x + 4}$

Strategy   We will square both sides to clear the equation of both radicals.

Why   We can square both sides immediately since each radical is isolated on one side of the equation.

Solution

$$\sqrt{5x + 9} = 2\sqrt{3x + 4} \quad \text{\textcolor{red}{This is the equation to solve.}}$$

$$\left(\sqrt{5x + 9}\right)^2 = \left(2\sqrt{3x + 4}\right)^2 \quad \text{\textcolor{red}{Square both sides to eliminate the radicals.}}$$

$$5x + 9 = 2^2\left(\sqrt{3x + 4}\right)^2 \quad \text{\textcolor{red}{Simplify on the left. On the right, raise each factor of the product } 2\sqrt{3x + 4} \text{ to the second power.}}$$

$$5x + 9 = 4(3x + 4) \quad \text{\textcolor{red}{Perform the operations on the right.}}$$

$$5x + 9 = 12x + 16 \quad \text{\textcolor{red}{To solve the resulting linear equation, distribute the multiplication by 4.}}$$

$$-7 = 7x \quad \text{\textcolor{red}{Subtract 5x and 16 from both sides.}}$$

$$-1 = x \quad \text{\textcolor{red}{Divide both sides by 7.}}$$

Caution

When finding $\left(2\sqrt{3x + 4}\right)^2$, remember to square both of the factors, 2 and $\sqrt{3x + 4}$, to get:

$$2^2\left(\sqrt{3x + 4}\right)^2$$

This is an application of the power of a product rule for exponents:

$$(xy)^n = x^n y^n$$

We check the solution by substituting -1 for x in the original equation.

$$\sqrt{5\textcolor{red}{x} + 9} = 2\sqrt{3\textcolor{red}{x} + 4}$$

$$\sqrt{5(\textcolor{red}{-1}) + 9} \overset{?}{=} 2\sqrt{3(\textcolor{red}{-1}) + 4} \quad \text{\textcolor{red}{Substitute } -1 \text{ for x.}}$$

$$\sqrt{4} \overset{?}{=} 2\sqrt{1}$$

$$2 = 2 \quad \text{\textcolor{red}{True}}$$

The solution is -1 and the solution set is $\{-1\}$.

Self Check 7 Solve: $\sqrt{x-4} = 2\sqrt{x-16}$

Now Try ▶ Problem 49

When more than one radical appears in an equation, we often must use the power rule more than once.

EXAMPLE 8 Solve: $\sqrt{x} + \sqrt{x+2} = 2$

Strategy We will isolate $\sqrt{x+2}$ on the left side of the equation and square both sides to eliminate it. After simplifying the resulting equation, we will isolate the remaining radical term and square both sides a second time to eliminate it.

Why Each time that we square both sides, we are able to clear the equation of one radical.

Solution

$$\sqrt{x} + \sqrt{x+2} = 2 \qquad \text{This is the equation to solve.}$$

$$\sqrt{x+2} = 2 - \sqrt{x} \qquad \text{To isolate } \sqrt{x+2}, \text{ subtract } \sqrt{x} \text{ from both sides.}$$

$$\left(\sqrt{x+2}\right)^2 = \left(2 - \sqrt{x}\right)^2 \qquad \begin{array}{l}\text{Square both sides to eliminate} \\ \text{the square root on the left side.}\end{array}$$

$$x + 2 = \left(2 - \sqrt{x}\right)^2 \qquad \text{Perform the operation on the left side.}$$

Success Tip

Note in this example that we isolate the more complicated radical term, $\sqrt{x+2}$. As a result, we work with the less complicated radical term $\sqrt{x}$ when applying the special-product rule.

Caution

When finding $\left(2 - \sqrt{x}\right)^2$, it is like squaring a binomial. Remember to use a special-product rule (or FOIL). **Do not just square the first term and the last term.**

$$\left(2 - \sqrt{x}\right)^2 \neq 4 + x$$

To square the expression $2 - \sqrt{x}$ on the right side, we can use a special-product rule:

$$x + 2 = 2^2 - \underbrace{2(2)\left(\sqrt{x}\right)} + \left(\sqrt{x}\right)^2$$

Square	Twice the	Square the
the first	product of	last term,
term, 2.	both terms	$\sqrt{x}$.

$$x + 2 = 4 - 4\sqrt{x} + x$$

Since the equation still contains a radical, we need to square both sides again. Before doing that, we must isolate the radical on one side.

$$2 = 4 - 4\sqrt{x} \qquad \text{Subtract } x \text{ from both sides.}$$

$$-2 = -4\sqrt{x} \qquad \text{To isolate the radical term } -4\sqrt{x}, \text{ subtract 4 from both sides.}$$

$$\frac{1}{2} = \sqrt{x} \qquad \text{To isolate the radical, divide both sides by } -4.$$

$$\left(\frac{1}{2}\right)^2 = \left(\sqrt{x}\right)^2 \qquad \text{To eliminate the radical, square both sides again.}$$

$$\frac{1}{4} = x \qquad \text{Perform the operations on each side.}$$

Check: $\sqrt{x} + \sqrt{x+2} = 2 \qquad$ This is the original equation.

$$\sqrt{\frac{1}{4}} + \sqrt{\frac{1}{4} + 2} \stackrel{?}{=} 2 \qquad \text{Substitute } \tfrac{1}{4} \text{ for } x.$$

$$\frac{1}{2} + \sqrt{\frac{9}{4}} \stackrel{?}{=} 2 \qquad \text{Think of 2 as } \tfrac{8}{4} \text{ and add: } \tfrac{1}{4} + \tfrac{8}{4} = \tfrac{9}{4}.$$

$$\frac{1}{2} + \frac{3}{2} \stackrel{?}{=} 2 \qquad \text{Evaluate } \sqrt{\tfrac{9}{4}}.$$

$$2 = 2 \qquad \text{True}$$

The result $\frac{1}{4}$ checks. The solution set is $\left\{\frac{1}{4}\right\}$.

Self Check 8   Solve: $\sqrt{a} + \sqrt{a + 3} = 3$

Now Try ▶   Problems 55 and 59

Using Your Calculator ▶   **Solving Radical Equations**

To find solutions for $\sqrt{x} + \sqrt{x + 2} = 4$ (an equation similar to Example 8) with a graphing calculator, we graph the functions $f(x) = \sqrt{x}$ and $g(x) = \sqrt{x + 2}$ and $g(x) = 4$. We then trace to find an approximation of the x-coordinate of their intersection point, as in figure (a). From the figure, we can see that $x \approx 2.98$. We can zoom to get better results.

Figure (b) shows that the INTERSECT feature gives the approximate coordinates of the point of intersection of the two graphs as $(3.06, 4)$. Therefore, an approximate solution of the radical equation is 3.06. Check its reasonableness.

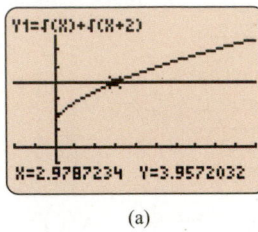   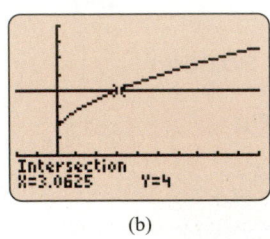

(a)   (b)

---

3 Solve Formulas Containing Radicals.

To *solve a formula for a variable* means to isolate that variable on one side of the equation, with all other quantities on the other side.

EXAMPLE 9   **Depreciation Rates.**   Some office equipment that is now worth V dollars originally cost C dollars 3 years ago. The rate r at which it has depreciated is given by $r = 1 - \sqrt[3]{\dfrac{V}{C}}$. Solve the formula for C.

Strategy   To isolate the radical, we will subtract 1 from both sides. We can then eliminate the radical by cubing both sides.

Why   Cubing both sides will produce, on the right, the expression $\left(\sqrt[3]{\dfrac{V}{C}}\right)^3$ that simplifies to $\dfrac{V}{C}$. This step clears the equation of the radical.

Solution   We begin by isolating the cube root on the right side of the equation.

$$r = 1 - \sqrt[3]{\frac{V}{C}} \qquad \text{\color{red}{This is the depreciation model.}}$$

$$r - 1 = -\sqrt[3]{\frac{V}{C}} \qquad \text{\color{red}{Subtract 1 from both sides to isolate the radical.}}$$

$$(r - 1)^3 = \left(-\sqrt[3]{\frac{V}{C}}\right)^3 \qquad \text{\color{red}{To eliminate the radical, cube both sides.}}$$

$$(r - 1)^3 = -\frac{V}{C} \qquad \text{\color{red}{Simplify the right side.}}$$

$$C(r - 1)^3 = -V \qquad \text{\color{red}{To clear the equation of the fraction, multiply both sides by C.}}$$

$$C = -\frac{V}{(r - 1)^3} \qquad \text{\color{red}{To isolate C, divide both sides by } (r - 1)^3.}$$

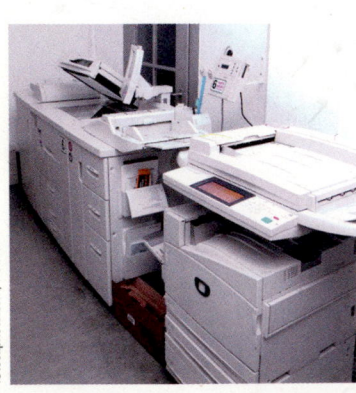

Self Check 9 **Statistics.** A formula used in statistics to determine the size of a sample to obtain a desired degree of accuracy is $E = z_0\sqrt{\frac{pq}{n}}$. Solve the formula for n.

Now Try ▶ Problem 67

SECTION 7.5 STUDY SET

VOCABULARY

Fill in the blanks.

1. Equations such as $\sqrt{x+4} - 4 = 5$ and $\sqrt[3]{x+1} = 12$ are called _____ equations.
2. To solve a radical equation, we find all the values of the variable that make the equation _____.
3. When we square both sides of a radical equation, we say we are _____ both sides to the second power.
4. When solving equations containing radicals, first we _____ one radical expression on one side of the equation.
5. Proposed solutions of a radical equation that don't satisfy it are called _____ solutions.
6. To _____ a proposed solution means to substitute it into the original equation and see whether a true statement results.

CONCEPTS

7. Fill in the blanks.
   a. The power rule for solving radical equations states that if x, y, and n are real numbers and $x = y$, then
   $$x^{\blacksquare} = y^{\blacksquare}$$
   b. If $\sqrt[n]{a}$ is a real number, then $\left(\sqrt[n]{a}\right)^n = \blacksquare$.
8. Determine whether 6 is a solution of each radical equation.
   a. $\sqrt{x+3} = x - 3$
   b. $\sqrt[3]{5x-3} + 9 = x$
9. What is the first step in solving each equation?
   a. $\sqrt{x+11} = 5$
   b. $\sqrt[3]{5x+4} + 3 = 30$
   c. $\sqrt{x+8} - \sqrt{2x} = 1$
10. Simplify each expression.
   a. $\left(\sqrt{x}\right)^2$
   b. $\left(\sqrt{x-5}\right)^2$
   c. $\left(\sqrt[3]{4x-8}\right)^3$
   d. $\left(\sqrt[4]{8x}\right)^4$
   e. $\left(4\sqrt{2x}\right)^2$
   f. $\left(3\sqrt[3]{x+1}\right)^3$
11. Find: $\left(\sqrt{x} - 3\right)^2$
12. Find: $\left(\sqrt{5x+2} - 4\right)^2$

NOTATION

Complete each solution.

13. Solve: $\sqrt{3x+3} - 1 = 5$
    $$\sqrt{3x+3} = \blacksquare$$
    $$\left(\sqrt{3x+3}\right)^{\blacksquare} = (6)^{\blacksquare}$$
    $$\blacksquare = 36$$
    $$3x = \blacksquare$$
    $$x = \blacksquare$$

    Does the proposed solution check?

14. Fill in the blanks. Write each radical equation using a rational exponent.
    a. $\sqrt{x+10} + 5 = 15$ can be written $(x+10)^{\blacksquare} + 5 = 15$
    b. $\sqrt[3]{2t+4} = t - 1$ can be written $(2t+4)^{\blacksquare} = t - 1$

GUIDED PRACTICE

Solve each equation. See Example 1.

15. $\sqrt{a-3} = 1$
16. $\sqrt{x-10} = 1$
17. $\sqrt{4x+5} = 5$
18. $\sqrt{5x-6} = 2$
19. $\sqrt{6x+13} = 7$
20. $\sqrt{6x+1} = 5$
21. $\sqrt{\frac{1}{3}x - 2} = 8$
22. $\sqrt{\frac{1}{2}x + 3} = 6$

Solve each equation. Write all proposed solutions. Cross out those that are extraneous. See Example 3.

23. $\sqrt{2x+11} + 2 = x$
24. $\sqrt{2a-3} + 3 = a$
25. $\sqrt{2r-3} + 9 = r$
26. $\sqrt{-x+2} + 2 = x$
27. $\sqrt{3t+7} - t = 1$
28. $\sqrt{t+3} - t = 1$
29. $\sqrt{9-a} - a = 3$
30. $\sqrt{4-a} - a = 2$

Solve each equation. Write all proposed solutions. Cross out those that are extraneous. See Example 4.

31. $\sqrt{5x+10} = 8$
32. $\sqrt[3]{3x+5} = 2$
33. $\sqrt{5-x} + 10 = 9$
34. $1 = 2 + \sqrt{4x+75}$

Solve each equation. See Example 5.

35. $\sqrt[3]{7n-1} = 3$
36. $\sqrt[3]{12m+4} = 4$
37. $\sqrt[3]{x^3-7} = x - 1$
38. $\sqrt[3]{b^3-63} = b - 3$

39. $(m^3 + 26)^{1/3} = m + 2$ **40.** $(x^3 + 56)^{1/3} = x + 2$

41. $(5r + 14)^{1/3} = 4$ **42.** $(2b + 29)^{1/3} = 3$

See Example 6.

43. Let $f(x) = \sqrt[4]{3x + 1}$. For what value(s) of x is $f(x) = 4$?

44. Let $f(x) = \sqrt{2x^2 - 7x}$. For what value(s) of x is $f(x) = 2$?

45. Let $f(x) = \sqrt[3]{3x - 6}$. For what value(s) of x is $f(x) = -3$?

46. Let $f(x) = \sqrt[5]{4x - 4}$. For what value(s) of x is $f(x) = -2$?

Solve each equation. See Example 7.

47. $\sqrt{3x + 12} = \sqrt{5x - 12}$ **48.** $\sqrt{m + 4} = \sqrt{2m - 5}$

49. $2\sqrt{4x + 1} = \sqrt{x + 4}$ **50.** $\sqrt{6 - 2x} = 4\sqrt{x - 3}$

51. $\sqrt{6t + 9} = 3\sqrt{t}$ **52.** $\sqrt{12x + 24} = 6\sqrt{x}$

53. $(34x + 26)^{1/3} = 4(x - 1)^{1/3}$ **54.** $(a^2 + 2a)^{1/3} = 2(a - 1)^{1/3}$

Solve each equation. Write all proposed solutions. Cross out those that are extraneous. See Example 8.

55. $\sqrt{x - 5} + \sqrt{x} = 5$ **56.** $\sqrt{x - 7} + \sqrt{x} = 7$

57. $\sqrt{z + 3} - \sqrt{z} = 1$ **58.** $\sqrt{x + 12} + \sqrt{x} = 6$

59. $3 = \sqrt{y + 4} - \sqrt{y + 7}$ **60.** $3 = \sqrt{u - 3} - \sqrt{u}$

61. $2 = \sqrt{2u + 7} - \sqrt{u}$ **62.** $1 = \sqrt{4s + 5} - \sqrt{2s + 2}$

Solve each equation for the specified variable or expression. See Example 9.

63. $v = \sqrt{2gh}$ for h **64.** $d = 1.4\sqrt{h}$ for h

65. $T = 2\pi\sqrt{\dfrac{l}{32}}$ for l **66.** $d = \sqrt[3]{\dfrac{12V}{\pi}}$ for V

67. $r = \sqrt[3]{\dfrac{A}{P}} - 1$ for A **68.** $r = \sqrt[3]{\dfrac{A}{P}} - 1$ for P

69. $L_A = L_B\sqrt{1 - \dfrac{v^2}{c^2}}$ for v^2 **70.** $R_1 = \sqrt{\dfrac{A}{\pi} - R_2{}^2}$ for A

TRY IT YOURSELF

Solve each equation. Write all proposed solutions. Cross out those that are extraneous.

71. $2\sqrt{x} = \sqrt{5x - 16}$ **72.** $3\sqrt{x} = \sqrt{3x + 54}$

73. $\sqrt{x + 5} + \sqrt{x - 3} = 4$ **74.** $\sqrt{b + 7} - \sqrt{b - 5} = 2$

75. $n = (n^3 + n^2 - 1)^{1/3}$ **76.** $(m^4 + m^2 - 25)^{1/4} = m$

77. $\sqrt{y + 2} + y = 4$ **78.** $\sqrt{22y + 86} - y = 9$

79. $\sqrt[3]{x + 8} = -2$ **80.** $\sqrt[3]{x + 4} = -1$

81. $2 = \sqrt{x + 5} - \sqrt{x + 1}$ **82.** $4 = \sqrt{x + 8} - \sqrt{x + 2}$

83. $x = \dfrac{\sqrt{12x - 5}}{2}$ **84.** $x = \dfrac{\sqrt{16x - 12}}{2}$

85. $(n^2 + 6n + 3)^{1/2} = (n^2 - 6n - 3)^{1/2}$

86. $(m^2 - 12m - 3)^{1/2} = (m^2 + 12m + 3)^{1/2}$

87. $\sqrt{x - 5} - \sqrt{x + 3} = 4$ **88.** $\sqrt{x + 8} - \sqrt{x - 4} = -2$

89. $\sqrt[4]{10y + 6} = 2\sqrt[4]{y}$ **90.** $\sqrt[4]{21a + 39} = 3\sqrt[4]{a - 1}$

91. $\sqrt{-5x + 24} = 6 - x$ **92.** $-s - 3 = 2\sqrt{5 - s}$

93. $\sqrt{2x} + 5 = 1$ **94.** $\sqrt{3x} + 10 = 1$

95. $\sqrt{6x + 2} - \sqrt{5x + 3} = 0$ **96.** $\sqrt{5x + 2} - \sqrt{x + 10} = 0$

97. Let $f(x) = \sqrt{x + 16}$ and $g(x) = 7 - \sqrt{x + 9}$. Find all values of x for which $f(x) = g(x)$.

98. Let $s(t) = \sqrt{t + 8}$ and $h(t) = 6 - \sqrt{t - 4}$. Find all values of t for which $s(t) = h(t)$.

99. Let $f(x) = \sqrt[4]{x + 8} - \sqrt[4]{2x}$. Find all values of x for which $f(x) = 0$.

100. Let $h(a) = \sqrt[4]{a + 11} - \sqrt[4]{2a + 6}$. Find all values of a for which $h(a) = 0$.

Look Alikes . . .

101. a. $3\sqrt{5n - 9} = \sqrt{5n}$ **b.** $3 + \sqrt{5n - 9} = \sqrt{5n}$

102. a. $\sqrt{8 + a} = 2\sqrt{a}$ **b.** $\sqrt{8 + a} = 2 + \sqrt{a}$

103. a. $\sqrt{2x} - 10 = 0$ **b.** $\sqrt{2x} + 10 = 0$

104. a. $x^{1/2} + 6 = 8$ **b.** $x^{1/4} + 6 = 8$

APPLICATIONS

105. Highway Design. A curved road will accommodate traffic traveling s mph if the radius of the curve is r feet, according to the formula $s = 3\sqrt{r}$. If engineers expect 40-mph traffic, what radius should they specify? Give the result to the nearest foot.

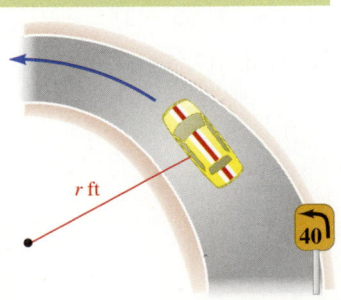

106. Forestry. The taller a lookout tower, the farther an observer can see. That distance d (called the *horizon distance,* measured in miles) is related to the height h of the observer (measured in feet) by the formula $d = 1.22\sqrt{h}$. How tall must a lookout tower be to see the edge of the forest, 25 miles away? (Round to the nearest foot.)

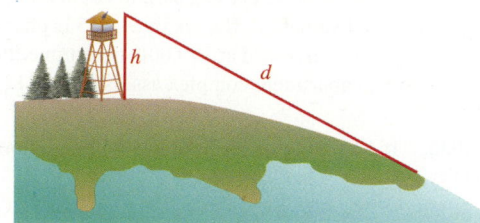

107. Wind Power. The power generated by a windmill is related to the velocity of the wind by the formula $v = \sqrt[3]{\dfrac{P}{0.02}}$ where P is the power (in watts) and v is the velocity of the wind (in mph). Find how much power the windmill is generating when the wind is 29 mph.

108. Diamonds. The *effective rate of interest r* earned by an investment is given by the formula $r = \sqrt[n]{\dfrac{A}{P}} - 1$ where P is the initial investment that grows to value A after n years. If a diamond buyer got \$4,000 for a 1.73-carat diamond that he had purchased 4 years earlier, and earned an annual rate of return of 6.5% on the investment, what did he originally pay for the diamond?

109.
from **Campus to Careers**

General Contractor

During construction, carpenters often brace walls as shown in the illustration, where the length L of the brace is given by the formula $L = \sqrt{f^2 + h^2}$. If a carpenter nails a 10-ft brace to the wall 6 feet above the floor, how far from the base of the wall should he nail the brace to the floor?

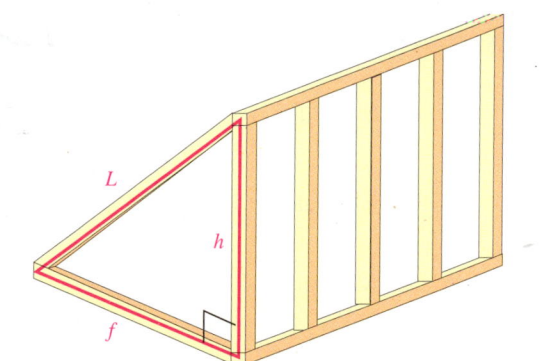

110. Theater Productions. The ropes, pulleys, and sandbags shown in the illustration are part of a mechanical system used to raise and lower scenery for a stage play. For the scenery to be in the proper position, the following formula must apply: $w_2 = \sqrt{w_1^2 + w_3^2}$. If $w_2 = 12.5$ lb and $w_3 = 7.5$ lb, find w_1.

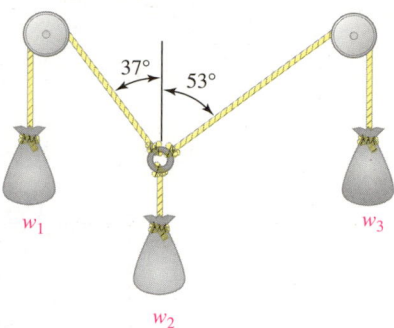

111. Supply and Demand. The number of wrenches that will be produced at a given price can be predicted by the formula $s = \sqrt{5x}$, where s is the supply (in thousands) and x is the price (in dollars). The demand d for wrenches can be predicted by the formula $d = \sqrt{100 - 3x^2}$. Find the equilibrium price—that is, find the price at which supply will equal demand.

112. Supply and Demand. The number of mirrors that will be produced at a given price can be predicted by the formula $s = \sqrt{23x}$, where s is the supply (in thousands) and x is the price (in dollars). The demand d for mirrors can be predicted by the formula $d = \sqrt{312 - 2x^2}$. Find the equilibrium price—that is, find the price at which supply will equal demand.

WRITING

113. What is wrong with the work shown below?

Solve: $\sqrt[3]{x+1} - 3 = 8$
$\sqrt[3]{x+1} = 11$
$\left(\sqrt[3]{x+1}\right)^3 = 11$
$x + 1 = 11$
$x = 10$

114. The first step of a student's solution is shown below. What is a better way to begin the solution?

Solve: $\sqrt{x} + \sqrt{x+22} = 12$
$\left(\sqrt{x} + \sqrt{x+22}\right)^2 = 12^2$

115. Explain the error in the following work.

Solve: $\sqrt{2y+1} = \sqrt{y+7} + 3$
$\left(\sqrt{2y+1}\right)^2 = \left(\sqrt{y+7} + 3\right)^2$
$2y + 1 = y + 7 + 9$

116. Explain why it is immediately apparent that $\sqrt{8x-7} = -2$ has no solution.

117. To solve the equation $\sqrt{2x+7} = \sqrt{x}$ we need only square both sides once. To solve the equation $\sqrt{2x+7} = \sqrt{x} + 2$ we have to square both sides twice. Why does the second equation require more work?

118. Explain how to solve $\sqrt{x-2} + 2 = 4$ using the graphs below.

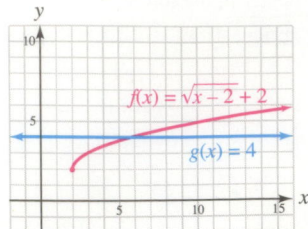

119. Explain how the table can be used to solve

$\sqrt{4x-3} - 2 = \sqrt{2x-5}$ if

$Y_1 = \sqrt{4x-3} - 2$ and

$Y_2 = \sqrt{2x-5}.$

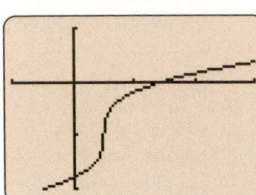

120. Explain how to use the graph of $f(x) = \sqrt[3]{x-0.5} - 1$, shown in the illustration, to approximate the solution of $\sqrt[3]{x-0.5} = 1.$

REVIEW

121. Lighting. The intensity of light from a lightbulb varies inversely as the square of the distance from the bulb. If you are 5 feet away from a bulb and the intensity is 40 foot-candles, what will the intensity be if you move 20 feet away from the bulb?

122. Property Tax. The property tax in a certain county varies directly as assessed valuation. If a tax of $1,575 is charged on a single-family home assessed at $90,000, determine the property tax on an apartment complex assessed at $312,000.

123. Typesetting. If 12-point type is 0.166044 inch tall, how tall is 30-point type?

124. Guitar Strings. The frequency of vibration of a string varies directly as the square root of the tension and inversely as the length of the string. Suppose a string 2.5 feet long, under a tension of 16 pounds, vibrates 25 times per second. Find k, the constant of proportionality.

CHALLENGE PROBLEMS

Solve each equation. Write all proposed solutions. Cross out those that are extraneous.

125. $\sqrt[4]{x} = \sqrt{\dfrac{x}{4}}$

126. $\sqrt[3]{2x} = \sqrt{x}$

127. $\sqrt{x+2} + \sqrt{2x} = \sqrt{18-x}$

128. $\sqrt{8-x} - \sqrt{3x-8} = \sqrt{x-4}$

129. $\sqrt{2\sqrt{x+1}} = \sqrt{16-4x}$

130. $(2x-1)^{2/3} = x^{1/3}$

SECTION 7.6

Geometric Applications of Radicals

OBJECTIVES

1 Use the Pythagorean theorem to solve problems.

2 Solve problems involving 45°–45°–90° triangles.

3 Solve problems involving 30°–60°–90° triangles.

4 Use the distance formula to solve problems.

ARE YOU READY?

The following problems review some basic skills that are needed when working with special triangles.

1. What is the sum of the measures of the angles of any triangle?

2. Simplify: $\sqrt{3a^2}$

3. Rationalize the denominator: $\dfrac{25}{\sqrt{3}}$

4. Evaluate: $\sqrt{(9-4)^2 + (-1-11)^2}$

5. Approximate to the nearest hundredth: $2\sqrt{3}$

6. Multiply: $2 \cdot \dfrac{7\sqrt{3}}{3}$

We will now consider applications of square roots in geometry. Then we will find the distance between two points on a rectangular coordinate system, using a formula that contains a square root. We begin by considering an important theorem about right triangles.

1 Use the Pythagorean Theorem to Solve Problems.

If we know the lengths of two legs of a right triangle, we can find the length of the **hypotenuse** (the side opposite the 90° angle) by using the **Pythagorean theorem.**

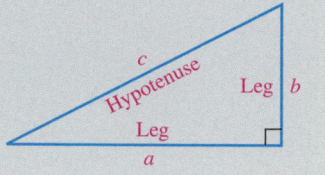

The Pythagorean Theorem	If a and b are the lengths of two legs of a right triangle and c is the length of the hypotenuse,
	$$a^2 + b^2 = c^2$$

In words, the Pythagorean theorem is expressed as follows:

In a right triangle, the sum of the squares of the lengths of two legs is equal to the square of the length of the hypotenuse.

Suppose the right triangle shown in the margin has legs of length 3 and 4 units. To find the length of the hypotenuse, we use the **Pythagorean equation.**

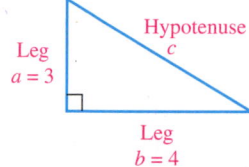

$$a^2 + b^2 = c^2$$
$$3^2 + 4^2 = c^2 \qquad \text{Substitute 3 for } a \text{ and 4 for } b.$$
$$9 + 16 = c^2$$
$$25 = c^2$$

To find c, we ask "What number, when squared, is equal to 25?" There are two such numbers: the positive square root of 25 and the negative square root of 25. Since c represents the length of the hypotenuse, and it cannot be negative, it follows that c is the positive square root of 25.

$$\sqrt{25} = c \qquad \text{Recall that a radical symbol } \sqrt{} \text{ is used to represent}$$
$$\text{the positive, or principal, square root of a number.}$$
$$5 = c$$

The length of the hypotenuse is 5 units.

The Language of Algebra

A **theorem** is a mathematical statement that can be proved. The Pythagorean theorem is named after Pythagoras, a Greek mathematician who lived about 2,500 years ago. He is thought to have been the first to prove the theorem.

EXAMPLE 1

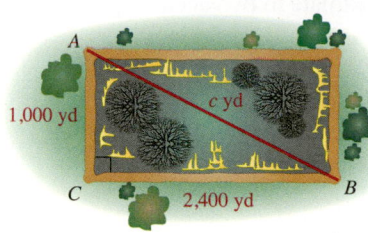

Firefighting. To fight a fire, the forestry department plans to clear a rectangular firebreak around the fire, as shown in the illustration on the left. Crews are equipped with mobile communications that have a 3,000-yard range. Can crews at points A and B remain in radio contact?

Strategy   We will use the Pythagorean theorem to find the distance between points A and B.

Why   If this distance is less than 3,000 yards, they can communicate. If it is greater than 3,000 yards, they cannot communicate.

Solution   The line segments connecting points A, B, and C form a right triangle. To find the distance c from point A to point B, we can use the Pythagorean theorem, substituting 2,400 for a and 1,000 for b and solving for c.

$$a^2 + b^2 = c^2 \qquad \text{This is the Pythagorean equation.}$$
$$2,400^2 + 1,000^2 = c^2 \qquad \text{Substitute for } a \text{ and } b.$$
$$5,760,000 + 1,000,000 = c^2 \qquad \text{Evaluate each exponential expression.}$$
$$6,760,000 = c^2 \qquad \text{Do the addition.}$$
$$\sqrt{6,760,000} = c \qquad \text{If } c^2 = 6,760,000, \text{ then } c \text{ must be a square root of}$$
$$6,760,000. \text{ Because } c \text{ represents a length, it must}$$
$$\text{be the positive square root of } 6,760,000.$$
$$2,600 = c \qquad \text{Use a calculator to find the square root.}$$

Caution

When using the Pythagorean equation $a^2 + b^2 = c^2$, we can let a represent the length of either leg of the right triangle. We then let b represent the length of the other leg. The variable c must always represent the length of the hypotenuse.

The two crews are 2,600 yards apart. Because this distance is less than the 3,000-yard range of the radios, they can communicate by radio.

2 Solve Problems Involving 45°–45°–90° Triangles.

An **isosceles right triangle** is a right triangle with two legs of equal length. Isosceles right triangles have angle measures of 45°, 45°, and 90°. If we know the length of one leg of an isosceles right triangle, we can use the Pythagorean theorem to find the length of the hypotenuse. Since the triangle shown in the margin is a right triangle, we have

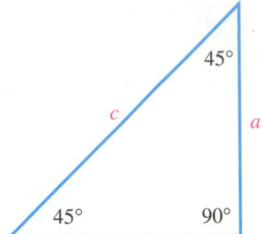

$$c^2 = a^2 + b^2 \qquad \text{This is the Pythagorean equation.}$$

$$c^2 = a^2 + a^2 \qquad \text{Both legs are } a \text{ units long, so replace } b \text{ with } a.$$

$$c^2 = 2a^2 \qquad \text{Combine like terms.}$$

$$c = \sqrt{2a^2} \qquad \text{If } c^2 = 2a^2, \text{ then } c \text{ must be a square root of } 2a^2. \text{ Because } c \text{ represents a length, it must be the positive square root of } 2a^2.$$

$$c = a\sqrt{2} \qquad \text{Simplify the radical: } \sqrt{2a^2} = \sqrt{2}\sqrt{a^2} = \sqrt{2}a = a\sqrt{2}.$$

Thus, **in an isosceles right triangle, the length of the hypotenuse is $\sqrt{2}$ times the length of one leg.**

EXAMPLE 2 If one leg of an isosceles right triangle is 10 feet long, find the exact length of the hypotenuse. Then approximate the length to two decimal places.

Strategy We will multiply the length of the known leg by $\sqrt{2}$.

Why The length of the hypotenuse of an isosceles right triangle is $\sqrt{2}$ times the length of one leg.

Solution Since the length of the hypotenuse is the length of a leg times $\sqrt{2}$, we have

$$c = 10\sqrt{2}$$

The exact length of the hypotenuse is $10\sqrt{2}$ feet. If we approximate to two decimal places, the length is 14.14 feet.

EXAMPLE 3 Find the exact length of each leg of the isosceles right triangle shown in the margin. Then approximate the lengths to two decimal places.

Strategy We will find the length of each leg of the triangle by substituting 25 for c in the formula $c = a\sqrt{2}$ and solving for a.

Why The variable a represents the unknown length of each leg of the triangle.

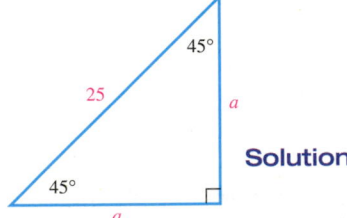

Solution

$$c = a\sqrt{2} \qquad \text{This is the formula for the length of the hypotenuse of an isosceles right triangle.}$$

$$25 = a\sqrt{2} \qquad \text{Substitute 25 for } c, \text{ the length of the hypotenuse.}$$

$$\frac{25}{\sqrt{2}} = \frac{a\sqrt{2}}{\sqrt{2}} \qquad \text{To isolate } a, \text{ undo the multiplication by } \sqrt{2} \text{ by dividing both sides by } \sqrt{2}.$$

$$\frac{25}{\sqrt{2}} = a \qquad \text{Simplify the right side.}$$

$$a = \frac{25}{\sqrt{2}} \qquad \text{Reverse the sides of the equation so that a is on the left.}$$

$$a = \frac{25}{\sqrt{2}} \cdot \frac{\sqrt{2}}{\sqrt{2}} \qquad \text{Rationalize the denominator.}$$

$$a = \frac{25\sqrt{2}}{2} \qquad \text{Simplify the denominator: } \sqrt{2} \cdot \sqrt{2} = 2.$$

The exact length of each leg is $\frac{25\sqrt{2}}{2}$ units. If we approximate to two decimal places, the length is 17.68 units.

Self Check 3   Find the exact length of each leg of an isosceles right triangle if the length of the hypotenuse is 9 inches. Then approximate the lengths to two decimal places.

Now Try ▶ Problem 27

3   Solve Problems Involving 30°–60°–90° Triangles.

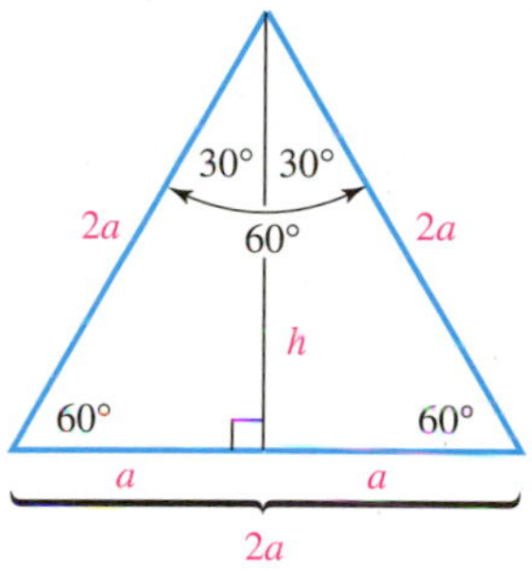

From geometry, we know that an **equilateral triangle** is a triangle with three sides of equal length and three 60° angles. Each side of the equilateral triangle shown in the margin is $2a$ units long. If an **altitude** (height) is drawn to its base, the altitude divides the base into two segments of equal length and divides the equilateral triangle into two 30°–60°–90° triangles. From the figure, we can see that the shorter leg of each 30°–60°–90° triangle (the side *opposite* the 30° angle) is a units long. Thus,

The length of the hypotenuse of a 30°–60°–90° triangle is twice as long as the shorter leg.

We can discover another relationship between the legs of a 30°–60°–90° triangle if we find the length of the altitude h in the figure. We begin by applying the Pythagorean theorem to one of the 30°–60°–90° triangles.

$$a^2 + b^2 = c^2 \qquad \text{This is the Pythagorean equation.}$$
$$a^2 + h^2 = (2a)^2 \qquad \text{The altitude is h units long, so replace b with h.}$$
$$\qquad\qquad\qquad \text{The hypotenuse is 2a units long, so replace c with 2a.}$$
$$a^2 + h^2 = 4a^2 \qquad (2a)^2 = (2a)(2a) = 4a^2.$$
$$h^2 = 3a^2 \qquad \text{Subtract } a^2 \text{ from both sides.}$$
$$h = \sqrt{3a^2} \qquad \text{If } h^2 = 3a^2, \text{ then h must be the positive square root of } 3a^2.$$
$$h = a\sqrt{3} \qquad \text{Simplify the radical: } \sqrt{3a^2} = \sqrt{3}\sqrt{a^2} = a\sqrt{3}.$$

We see that the altitude—the longer leg of the 30°–60°–90° triangle—is $\sqrt{3}$ times as long as the shorter leg. Thus,

The length of the longer leg of a 30°–60°–90° triangle is $\sqrt{3}$ times the length of the shorter leg.

EXAMPLE 4   Find the length of the hypotenuse and the length of the longer leg of the 30°–60°–90° triangle shown in the margin.

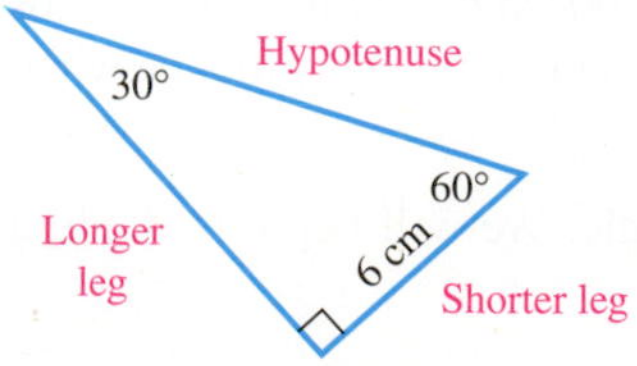

Strategy   To find the length of the hypotenuse, we will multiply the length of the shorter leg by 2. To find the length of the longer leg, we will multiply the length of the shorter leg by $\sqrt{3}$.

Why   These side-length relationships are true for any 30°–60°–90° triangle.

Solution   Since the length of the hypotenuse of a 30°–60°–90° triangle is twice as long as the shorter leg, and the length of the shorter leg is 6 cm, the hypotenuse is $2 \cdot 6 = 12$ cm.

Since the length of the longer leg is $\sqrt{3}$ times the length of the shorter leg, and the length of the shorter leg is 6 cm, the longer leg is $6\sqrt{3}$ cm (about 10.39 cm).

> **Self Check 4**   Find the length of the hypotenuse and the longer leg of a 30°–60°–90° triangle if the shorter leg is 8 centimeters long.
>
> **Now Try ▶**   Problem 31

EXAMPLE 5   Find the length of the hypotenuse and the length of the shorter leg of the 30°–60°–90° triangle shown in the margin. Then approximate the lengths to two decimal places.

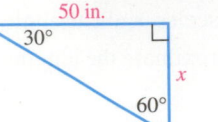

Strategy   We will find the length of the shorter leg first.

Why   Once we know the length of the shorter leg, we can multiply it by 2 to find the length of the hypotenuse.

Solution   If we let $x =$ the length in inches of the shorter leg of the triangle, we can form an equation by translating the following statement:

The length of the longer leg of a 30°–60°–90° triangle	is	$\sqrt{3}$	times	the length of the shorter leg.
50	=	$\sqrt{3}$	·	x

> **Success Tip**
>
> In a 30°–60°–90° triangle, the side opposite the 30° angle is the shorter leg, the side opposite the 60° angle is the longer leg, and the hypotenuse is opposite the 90° angle.

To find the length of the shorter leg, we solve the equation for x.

$$50 = \sqrt{3}x$$

$$\frac{50}{\sqrt{3}} = \frac{\sqrt{3}x}{\sqrt{3}} \qquad \text{To isolate } x \text{, divide both sides by } \sqrt{3}.$$

$$\frac{50}{\sqrt{3}} = x$$

The length of the shorter leg is exactly $\frac{50}{\sqrt{3}}$ inches. To write this number in simplified radical form, we rationalize the denominator.

$$\frac{50}{\sqrt{3}} \cdot \frac{\sqrt{3}}{\sqrt{3}} = \frac{50\sqrt{3}}{3}$$

Thus, the length of the shorter leg is exactly $\frac{50\sqrt{3}}{3}$ inches (about 28.87 inches).

Since the length of the hypotenuse of a 30°–60°–90° triangle is twice as long as the shorter leg, the hypotenuse is $2 \cdot \frac{50\sqrt{3}}{3} = \frac{100\sqrt{3}}{3}$ inches (about 57.74 inches).

> **Self Check 5**   Find the length of the hypotenuse and the shorter leg of a 30°–60°–90° triangle if the longer leg is 15 feet long. Then approximate the lengths to two decimal places.
>
> **Now Try ▶**   Problem 35

EXAMPLE 6   **Stretching Exercises.**   A doctor prescribed the exercise shown in figure (a) on the next page. The patient was instructed to raise his leg to an angle of 60° and hold the position for 10 seconds. If the patient's leg is 36 inches long, how high off the floor will his foot be when his leg is held at the proper angle?

Strategy   This situation is modeled by a 30°–60°–90° triangle. We will begin by finding the length of the shorter leg.

Why Once we know the length of the shorter leg, we can easily find the length of the longer leg, which represents the distance the patient's foot is off the ground.

Solution In figure (b), we see right triangle ABC, which models the situation. Since the length of the hypotenuse is twice as long as the side opposite the 30° angle, side AC is half as long as the hypotenuse. Since the hypotenuse is given to be 36 inches long, side AC must be 18 inches long.

Since the length of the longer leg (the leg opposite the 60° angle) is $\sqrt{3}$ times the length of the shorter leg (side AC), side BC is $18\sqrt{3}$, or about 31 inches long. So the patient's foot will be about 31 inches from the floor when his leg is in the proper position.

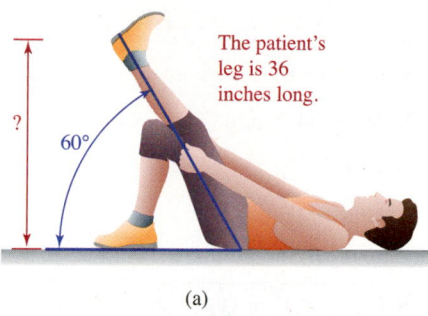

The patient's leg is 36 inches long.

60°

?

(a)

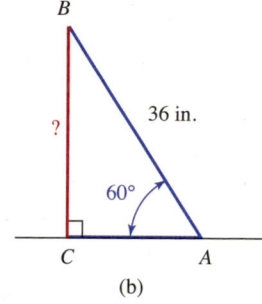

B

36 in.

?

60°

C A

(b)

Self Check 6   **Stretching Exercises.**   Refer to Example 6. The doctor prescribed the same exercise for a patient whose leg is 29 inches long. Find how high off the floor this patient's foot is when her leg is held at the proper angle.

Now Try ▶ Problems 39 and 61

4   Use the Distance Formula to Solve Problems.

With the **distance formula,** we can find the distance between any two points graphed on a rectangular coordinate system. To find the distance d between points $P(x_1, y_1)$ and $Q(x_2, y_2)$ shown in the figure on the right, we construct the right triangle PRQ. The distance between P and R is $|x_2 - x_1|$, and the distance between R and Q is $|y_2 - y_1|$. We apply the Pythagorean theorem to the right triangle PRQ to get

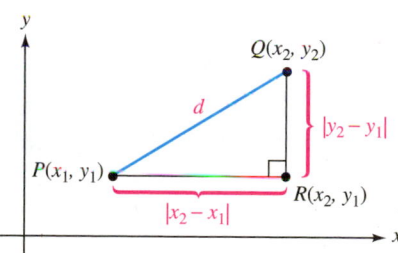

$$d^2 = |x_2 - x_1|^2 + |y_2 - y_1|^2$$
$$= (x_2 - x_1)^2 + (y_2 - y_1)^2 \quad \textcolor{red}{\text{Because } |x_2 - x_1|^2 = (x_2 - x_1)^2 \text{ and } |y_2 - y_1|^2 = (y_2 - y_1)^2}$$

Because d represents the distance between two points, it must be equal to the positive square root of $(x_2 - x_1)^2 + (y_2 - y_1)^2$.

$$d = \sqrt{(x_2 - x_1)^2 + (y_2 - y_1)^2}$$

We call this result the *distance formula.*

Distance Formula	The distance d between two points with coordinates (x_1, y_1) and (x_2, y_2) is given by $$d = \sqrt{(x_2 - x_1)^2 + (y_2 - y_1)^2}$$

EXAMPLE 7   Find the distance between the points: **a.** $(-2, 3)$ and $(4, -5)$ **b.** $(27, 46)$ and $(33, 50)$

Strategy   We will use the distance formula.

Why   We know the x- and y-coordinates of both points.

Solution   **a.** To find the distance, we can use the distance formula by substituting 4 for x_2, -2 for x_1, -5 for y_2, and 3 for y_1.

$$d = \sqrt{(x_2 - x_1)^2 + (y_2 - y_1)^2}$$ This is the distance formula.

$$= \sqrt{[4 - (-2)]^2 + (-5 - 3)^2}$$ Substitute for x_1, x_2, y_1, and y_2.

$$= \sqrt{(4 + 2)^2 + (-5 - 3)^2}$$ Simplify within the parentheses.

$$= \sqrt{6^2 + (-8)^2}$$ Do the addition and subtraction.

$$= \sqrt{36 + 64}$$ Evaluate each exponential expression.

$$= \sqrt{100}$$ Do the addition.

$$= 10$$ Evaluate the square root.

The distance between the points is 10 units.

b. $$d = \sqrt{(x_2 - x_1)^2 + (y_2 - y_1)^2}$$ This is the distance formula.

$$d = \sqrt{(33 - 27)^2 + (50 - 46)^2}$$ Substitute 33 for x_2, 27 for x_1, 50 for y_2, and 46 for y_1.

$$= \sqrt{6^2 + 4^2}$$ Do the subtraction.

$$= \sqrt{52}$$ Evaluate: $6^2 + 4^2 = 36 + 16 = 52$.

$$= 2\sqrt{13}$$ Simplify: $\sqrt{52} = \sqrt{4 \cdot 13} = 2\sqrt{13}$.

The distance between the points is exactly $2\sqrt{13}$ units, which is about 7.21 units.

Self Check 7   Find the distance between the points: **a.** $(-2, -2)$ and $(3, 10)$
b. $(55, 29)$ and $(61, 32)$
Now Try ▶ Problems 47 and 49

SECTION 7.6 ▶ STUDY SET

VOCABULARY

Fill in the blanks.

1. In a right triangle, the side opposite the 90° angle is called the
_____.

2. An _____ right triangle is a right triangle with two legs of equal length.

3. The _____ theorem states that in a right triangle, the sum of the squares of the lengths of the two legs is equal to the square of the hypotenuse.

4. An _____ triangle has three sides of equal length and three 60° angles.

CONCEPTS

Fill in the blanks.

5. If a and b are the lengths of the legs of a right triangle and c is the length of the hypotenuse, then ▢ + ▢ = ▢. This is called the Pythagorean _____.

6. In any right triangle, the square of the hypotenuse is equal to the _____ of the squares of the two _____.

7. In an isosceles right triangle, the length of the hypotenuse is ▢ times the length of one leg.

8. The shorter leg of a 30°–60°–90° triangle is _____ as long as the hypotenuse.

9. The length of the longer leg of a 30°–60°–90° triangle is ▢ times the length of the shorter leg.

10. In a 30°–60°–90° triangle, the shorter leg is opposite the ▢ angle, and the longer leg is opposite the ▢ angle.

11. The formula to find the distance between points (x_1, y_1) and (x_2, y_2) is $d = \sqrt{\rule{1.2cm}{0.4pt} + \rule{1.2cm}{0.4pt}}$.

12. Solve for c, where c represents the length of the hypotenuse of a right triangle. Simplify the result, if possible.

    a. $c^2 = 64$

    b. $c^2 = 15$

    c. $c^2 = 24$

NOTATION

Complete each solution.

13. Evaluate. Approximate to two decimal places.

$$\sqrt{(-1-3)^2 + [2-(-4)]^2} = \sqrt{(-4)^2 + []^2}$$
$$= \sqrt{}$$
$$= \sqrt{\cdot 13}$$
$$= \sqrt{13}$$
$$\approx $$

14. Solve $8^2 + 4^2 = c^2$ and assume $c > 0$. Approximate to two decimal places.

$$ + 16 = c^2$$
$$ = c^2$$
$$\sqrt{} = $$
$$\sqrt{\cdot 5} = c$$
$$\sqrt{5} = c$$
$$c 8.94$$

GUIDED PRACTICE

The lengths of two sides of the right triangle ABC are given. Find the length of the missing side. See Example 1.

15. $a = 6$ ft and $b = 8$ ft

16. $a = 5$ in. and $b = 12$ in.

17. $a = 8$ ft and $b = 15$ ft

18. $a = 24$ yd and $b = 7$ yd

19. $b = 9$ ft and $c = 41$ ft

20. $b = 18$ m and $c = 82$ m

21. $a = 10$ cm and $c = 26$ cm

22. $a = 14$ in. and $c = 50$ in.

Find the missing side lengths in each triangle. Give the exact answer and then an approximation to two decimal places when appropriate. See Example 2.

23.

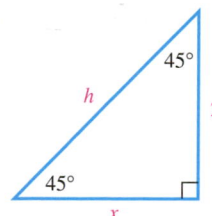

24.

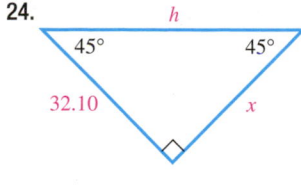

25. One leg of an isosceles right triangle is 3.2 feet long. Find the length of its hypotenuse. Give the exact answer and then an approximation to two decimal places.

26. One side of a square is $5\frac{1}{2}$ in. long. Find the length of its diagonal. Give the exact answer and then an approximation to two decimal places.

Find the missing side lengths in each triangle. Give the exact answer and then an approximation to two decimal places. See Example 3.

27.

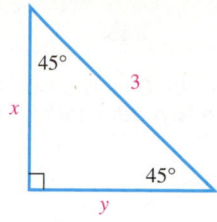

28.

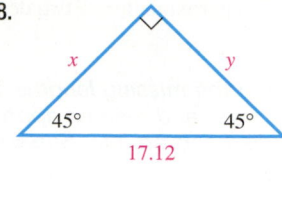

29. Photographs. The diagonal of a square photograph measures 10 inches. Find the length of one of its sides. Give the exact answer and then an approximation to two decimal places.

30. Parking Lots. The diagonal of a square parking lot is approximately 1,414 feet long.

  a. Find the length of one side of the parking lot. Round to the nearest foot.

  b. Find the approximate area of the parking lot.

Find the missing side lengths in each triangle. Give the exact answer and then an approximation to two decimal places, when appropriate. See Example 4.

31.

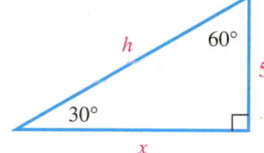

32.

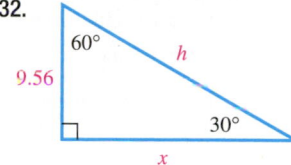

33. In a 30°–60°–90° triangle, the length of the leg opposite the 30° angle is 75 cm. Find the length of the leg opposite the 60° angle and the length of the hypotenuse. Give the exact answer and then an approximation to two decimal places, when appropriate.

34. In a 30°–60°–90° triangle, the length of the shorter leg is $5\sqrt{2}$ inches. Find the length of the hypotenuse and the length of the longer leg. Give the exact answer and then an approximation to two decimal places.

Find the missing lengths in each triangle. Give the exact answer and then an approximation to two decimal places. See Example 5.

35.

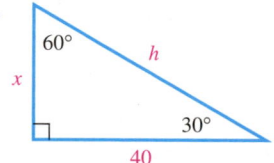

36.

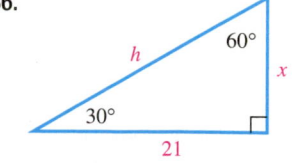

37. In a 30°–60°–90° right triangle, the length of the leg opposite the 60° angle is 55 millimeters. Find the length of the leg opposite the 30° angle and the length of the hypotenuse. Give the exact answer and then an approximation to two decimal places.

38. In a 30°–60°–90° right triangle, the length of the longer leg is 24 yards. Find the length of the hypotenuse and the length of the shorter leg. Give the exact answer and then an approximation to two decimal places.

Find the missing lengths in each triangle. Give the exact answer and then an approximation to two decimal places, when appropriate. See Example 6.

39.

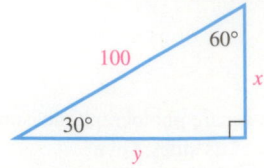

40.

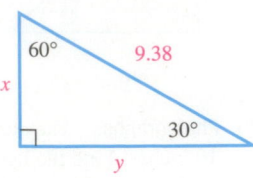

41. In a 30°–60°–90° right triangle, the length of the hypotenuse is 1.5 feet. To the nearest hundredth, find the length of the shorter leg and the length of the longer leg. Give the exact answer and then an approximation to two decimal places, when appropriate.

42. In a 30°–60°–90° right triangle, the length of the hypotenuse is $12\sqrt{3}$ inches. Find the length of the leg opposite the 30° angle and the length of the leg opposite the 60° angle. Give the exact answer and then an approximation to two decimal places, when appropriate.

Find the exact distance between each pair of points. See Example 7.

43. $(0, 0), (3, -4)$ **44.** $(0, 0), (-12, 16)$

45. $(-2, -8), (3, 4)$ **46.** $(-5, -2), (7, 3)$

47. $(6, 8), (12, 16)$ **48.** $(10, 4), (2, -2)$

49. $(-2, 1), (3, 4)$ **50.** $(2, -3), (4, -8)$

51. $(-1, -6), (3, -4)$ **52.** $(-3, 5), (-5, -5)$

53. $(-2, -1), (-5, 8)$ **54.** $(4, 7), (-4, -5)$

APPLICATIONS

55. Soccer. The allowable length of a rectangular soccer field used for international adult matches can be from 100 to 110 meters and the width can be from 64 to 75 meters.

  a. Find the length of the diagonal of the field that has the minimum allowable length and minimum allowable width. Give an approximation to two decimal places.

  b. Find the length of the diagonal of the field that has the maximum allowable length and maximum allowable width. Give the exact answer and an approximation to two decimal places.

56. Cubes. Find the exact length of the diagonal (in blue) of one of the *faces* of the cube shown here.

57. Cubes. Find the exact length of the diagonal (in green) of the cube shown here.

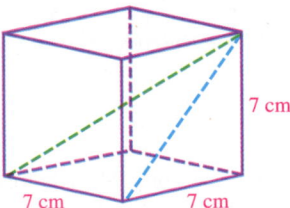

58. Geometry. Use the distance formula to show that a triangle with vertices $(-2, 4)$, $(2, 8)$, and $(6, 4)$ is isosceles.

59. Washington, D.C. The square in the map shows the 100-square-mile site selected by George Washington in 1790 to serve as a permanent capital for the United States. In 1847, the part of the district lying on the west bank of the Potomac was returned to Virginia. Find the exact coordinates of each corner of the original square that outlined the District of Columbia and approximations to two decimal places.

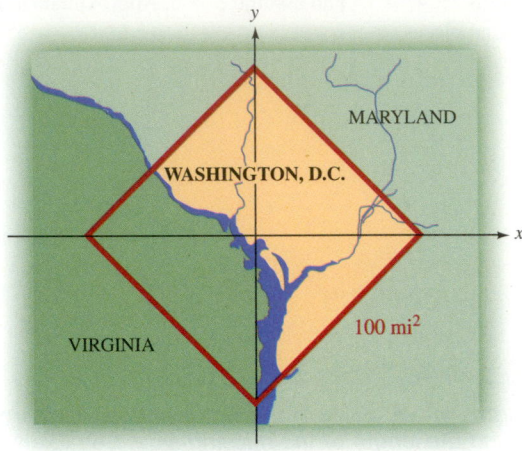

60. Paper Airplanes. The illustration gives the directions for making a paper airplane from a square piece of paper with sides 8 inches long. Find the length l of the plane when it is completed. Give the exact answer and an approximation to two decimal places.

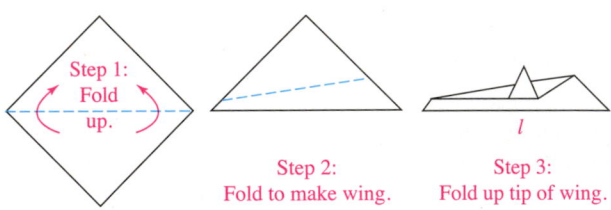

Step 1: Fold up.   Step 2: Fold to make wing.   Step 3: Fold up tip of wing.

61. Hardware. The sides of a regular hexagonal nut are 10 millimeters long. Find the height h of the nut. Give the exact answer and an approximation to two decimal places.

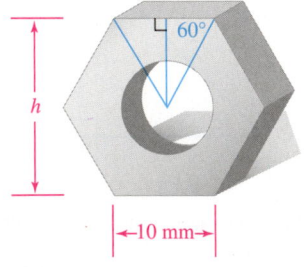

62. Ironing Boards.   Find the height *h* of the ironing board shown in the illustration. Give the exact answer and an approximation to two decimal places.

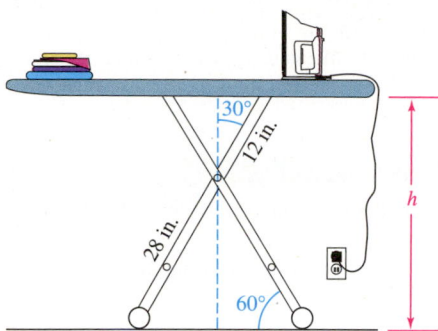

63. Baseball.   A baseball diamond is a square, 90 feet on a side. If the third baseman fields a ground ball 10 feet directly behind third base, how far must he throw the ball to throw a runner out at first base? Give the exact answer and an approximation to two decimal places.

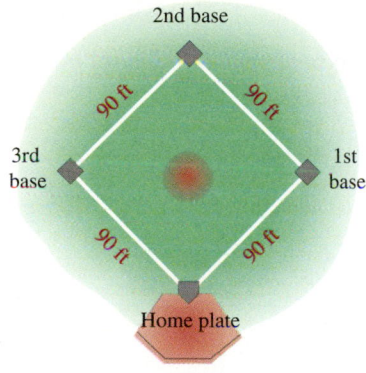

64. Baseball.   A shortstop fields a grounder at a point one-third of the way from second base to third base. How far will he have to throw the ball to make an out at first base? Give the exact answer and an approximation to two decimal places.

65. Clotheslines.   A pair of damp jeans are hung on a clothesline to dry. They pull the center down 1 foot. By how much is the line stretched? Give an approximation to two decimal places.

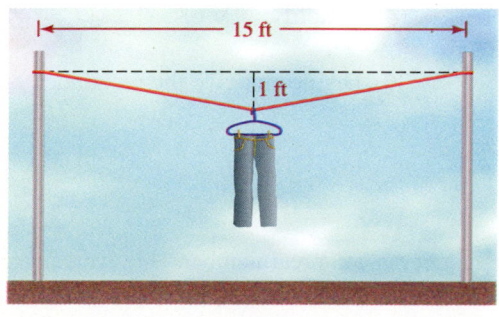

66. Firefighting.   The base of the 37-foot ladder is 9 feet from the wall. Will the top reach a window ledge that is 35 feet above the ground? Verify your result.

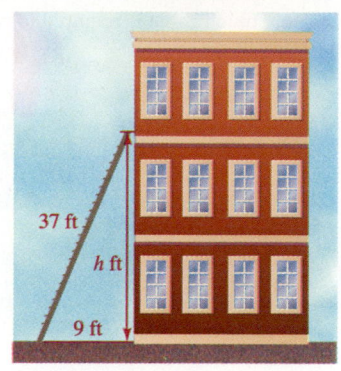

67. Art History.   A figure displaying some of the characteristics of Egyptian art is shown in the illustration. Use the distance formula to find the following dimensions of the drawing. Round your answers to two decimal places.

   a. From the foot to the eye

   b. From the belt to the hand holding the staff

   c. From the shoulder to the symbol held in the hand

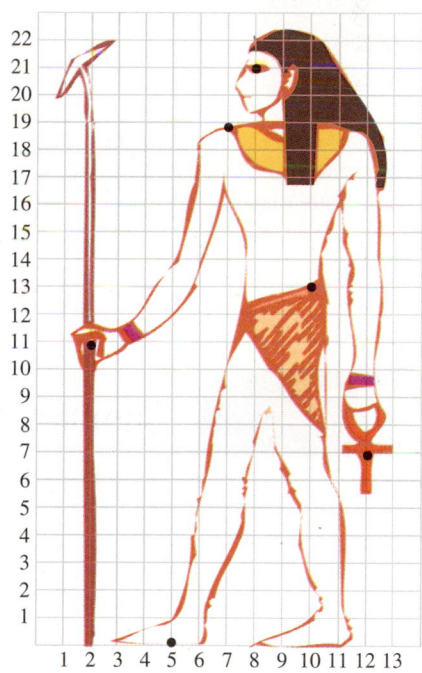

68. Packaging.   The diagonal *d* of a rectangular box with dimensions $a \times b \times c$ is given by $d = \sqrt{a^2 + b^2 + c^2}$. Will the umbrella fit in the shipping carton in the illustration? Verify your result.

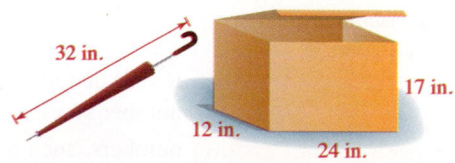

69. Packaging. An archaeologist wants to ship a 34-inch femur bone. Will it fit in a 4-inch-tall box that has a square base with sides 24 inches long? (See Exercise 68.) Verify your result.

70. Telephone. The telephone cable in the illustration runs from *A* to *B* to *C* to *D*. How much cable is required to run from *A* to *D* directly?

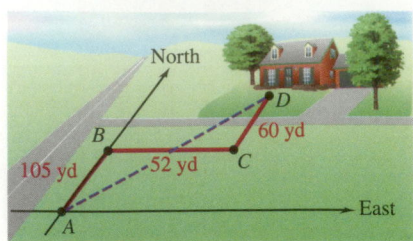

WRITING

71. State the Pythagorean theorem in words.

72. List the facts that you learned about special right triangles in this section.

73. When the lengths of the sides of a certain triangle are substituted into the equation of the Pythagorean theorem, the result is a false statement. Explain why.

$$a^2 + b^2 = c^2$$
$$2^2 + 4^2 = 5^2$$
$$4 + 16 = 25$$
$$20 = 25 \quad \text{False}$$

74. Explain how the distance formula and the Pythagorean theorem can be used to show that a triangle with vertices $(2, 3)$, $(-3, 4)$, and $(1, -2)$ is a right triangle.

REVIEW

75. Discount Buying. A repairman purchased some washing-machine motors for a total of $224. When the unit cost decreased by $4, he was able to buy one extra motor for the same total price. How many motors did he buy originally?

76. Aviation. An airplane can fly 650 miles with the wind in the same amount of time as it can fly 475 miles against the wind. If the wind speed is 40 mph, find the speed of the plane in still air.

CHALLENGE PROBLEMS

77. Find the length of the diagonal of the cube shown in figure (a) below.

78. Show that the length of the diagonal of the rectangular solid shown in figure (b) below is $\sqrt{a^2 + b^2 + c^2}$ cm.

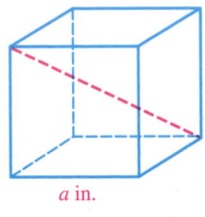
(a)

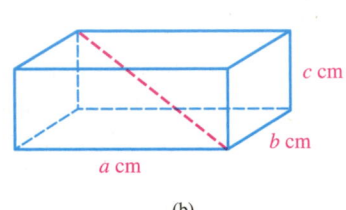
(b)

Find the distance between each pair of points.

79. $\left(\sqrt{48}, \sqrt{150} \right)$ and $\left(\sqrt{12}, \sqrt{24} \right)$

80. $\left(\sqrt{8}, -\sqrt{20} \right)$ and $\left(\sqrt{50}, -\sqrt{45} \right)$

SECTION 7.7

Complex Numbers

OBJECTIVES

1 Express square roots of negative numbers in terms of *i*.

2 Write complex numbers in the form $a + bi$.

3 Add and subtract complex numbers.

4 Multiply complex numbers.

5 Divide complex numbers.

6 Perform operations involving powers of *i*.

ARE YOU READY?

The following problems review some basic skills that are needed when working with complex numbers.

1. Explain why $\sqrt{-16}$ is not a real number.

2. Simplify: $(x^2 + 8x) + (5x^2 - 10x)$

3. Multiply: $(2n + 3)(7n - 1)$

4. Simplify: $\sqrt{63}$

5. Rationalize the denominator:
$$\frac{7}{\sqrt{x} + 4}$$

6. Divide and give the remainder: $4\overline{)87}$

Recall that the square root of a negative number is not a real number. However, an expanded number system, called the *complex number system,* gives meaning to square roots of negative numbers, such as $\sqrt{-9}$ and $\sqrt{-25}$. To define complex numbers, we use a number that is denoted by the letter *i*.

1 Express Square Roots of Negative Numbers in Terms of *i*.

Some equations do not have real-number solutions. For example, $x^2 = -1$ has no real-number solutions because the square of a real number is never negative. To provide a solution to this equation, mathematicians have defined the number i so that $i^2 = -1$.

The Number *i*	The **imaginary number *i*** is defined as $$i = \sqrt{-1}$$ From the definition, it follows that $i^2 = -1$.

This definition enables us to write the square root of any negative number in terms of i. We can use extensions of the product and quotient rules for radicals to write the square root of a negative number as the product of a real number and i.

EXAMPLE 1 Write each expression in terms of i: **a.** $\sqrt{-9}$ **b.** $\sqrt{-7}$ **c.** $-\sqrt{-18}$

d. $\sqrt{-\dfrac{24}{49}}$

Strategy We will write each radicand as the product of -1 and a positive number. Then we will apply the appropriate rules for radicals.

Why We want our work to produce a factor of $\sqrt{-1}$ so that we can replace it with i.

Solution After factoring the radicand, we use an extension of the product rule for radicals.

a. $\sqrt{-9} = \sqrt{-1 \cdot 9} = \sqrt{-1}\sqrt{9} = i \cdot 3 = 3i$ Replace $\sqrt{-1}$ with i.

b. $\sqrt{-7} = \sqrt{-1 \cdot 7} = \sqrt{-1}\sqrt{7} = i\sqrt{7}$ or $\sqrt{7}i$ Replace $\sqrt{-1}$ with i.

c. $-\sqrt{-18} = -\sqrt{-1 \cdot 9 \cdot 2} = -\sqrt{-1}\sqrt{9}\sqrt{2} = -i \cdot 3 \cdot \sqrt{2} = -3i\sqrt{2}$ or $-3\sqrt{2}i$

d. After factoring the radicand, use an extension of the product and quotient rules for radicals.

$$\sqrt{-\dfrac{24}{49}} = \sqrt{-1 \cdot \dfrac{24}{49}} = \dfrac{\sqrt{-1 \cdot 24}}{\sqrt{49}} = \dfrac{\sqrt{-1}\sqrt{4}\sqrt{6}}{\sqrt{49}} = \dfrac{2i\sqrt{6}}{7} \text{ or } \dfrac{2\sqrt{6}}{7}i$$

Notation

Since it is easy to confuse $\sqrt{b}i$ with $\sqrt{bi}$, we usually write i first so that it is clear that the i is not under the radical symbol. However, both $i\sqrt{b}$ and $\sqrt{b}i$ are correct. For example: $i\sqrt{7} = \sqrt{7}i$.

Self Check 1 Write each expression in terms of i: **a.** $\sqrt{-25}$
b. $-\sqrt{-19}$ **c.** $\sqrt{-45}$ **d.** $\sqrt{-\frac{50}{81}}$

Now Try Problems 19, 21, and 27

The results from Example 1 illustrate a rule for simplifying square roots of negative numbers.

Square Root of a Negative Number	For any positive real number b, $$\sqrt{-b} = i\sqrt{b}$$

To justify this rule, we use the fact that $\sqrt{-1} = i$.

$$\sqrt{-b} = \sqrt{-1 \cdot b}$$
$$= \sqrt{-1}\sqrt{b}$$
$$= i\sqrt{b}$$

2 Write Complex Numbers in the Form $a + bi$.

The imaginary number i is used to define *complex numbers*.

Complex Numbers	A **complex number** is any number that can be written in the form $a + bi$, where a and b are real numbers and $i = \sqrt{-1}$.
	Complex numbers of the form $a + bi$, where $b \neq 0$, are also called **imaginary numbers.***
	*Some textbooks define imaginary numbers as complex numbers with $a = 0$ and $b \neq 0$.

Notation

It is acceptable to use $a - bi$ as a substitute for the form $a + bi$. For example:

$$6 - 9i = 6 + (-9)i$$

For a complex number written in the **standard form** $a + bi$, we call a the **real part** and b the **imaginary part.** Some examples of complex numbers written in standard form are

$2 + 11i$	$6 - 9i$	$-\dfrac{1}{2} + 0i$	$0 + i\sqrt{3}$
$a = 2, b = 11$	$a = 6, b = -9$	$a = -\frac{1}{2}, b = 0$	$a = 0, b = \sqrt{3}$

Two complex numbers $a + bi$ and $c + di$ are equal if and only if $a = c$ and $b = d$. Thus, $0.5 + 0.9i = \frac{1}{2} + \frac{9}{10}i$ because $0.5 = \frac{1}{2}$ and $0.9 = \frac{9}{10}$.

EXAMPLE 2   Write each number in the form $a + bi$:   **a.** 6   **b.** $\sqrt{-64}$ **c.** $-2 + \sqrt{-63}$

Strategy   We will determine a, the real part, and we will simplify the radical (if necessary) to determine the bi part.

Why   We can put the two parts together to produce the desired $a + bi$ form.

Solution   **a.** $6 = 6 + 0i$ The real part a is 6. The imaginary part b is 0.

Caution

Use i only when working with the *square root* of a negative number. It does not apply to cube roots.

$\sqrt{-64} = 8i$ and $\sqrt[3]{-64} = -4$

b. $\sqrt{-64} = 0 + 8i$ The real part a is 0. Simplify: $\sqrt{-64} = \sqrt{-1}\sqrt{64} = 8i$. Thus, b is 8.

c. $-2 + \sqrt{-63} = -2 + 3i\sqrt{7}$ The real part a is -2.
Simplify: $\sqrt{-63} = \sqrt{-1}\sqrt{63} = \sqrt{-1}\sqrt{9}\sqrt{7} = 3i\sqrt{7}$.
Thus, $b = 3\sqrt{7}$.

Self Check 2   Write each number in the form $a + bi$:   **a.** -18
b. $\sqrt{-36}$ **c.** $1 + \sqrt{-24}$

Now Try   Problems 29 and 33

The following illustration shows the relationship between the real numbers, the imaginary numbers, and the complex numbers.

Success Tip

Just as real numbers are either rational or irrational, but not both, complex numbers are either real or imaginary, but not both.

The complex numbers: $a + bi$

The real numbers: $a + bi$ where $b = 0$				The imaginary numbers: $a + bi$ where $b \neq 0$		
-6	$\dfrac{5}{16}$	-1.75	π	$9 + 7i$	$-2i$	$\dfrac{1}{4} - \dfrac{3}{4}i$
$48 + 0i$	0	$-\sqrt{10}$	$-\dfrac{7}{2}$	$0.56i$	$\sqrt{-10}$	$6 + i\sqrt{3}$
$\sqrt[3]{-10}$	$4 + \pi$	$15\dfrac{5}{8}$	$\sqrt[4]{99}$	$25i\sqrt{5}$	$\dfrac{7}{8}i$	$\sqrt{3} + 3i\sqrt{2}$

3 Add and Subtract Complex Numbers.

Adding and subtracting complex numbers is similar to adding and subtracting polynomials.

Addition and Subtraction of Complex Numbers	1. To add complex numbers, add their real parts and add their imaginary parts.
	2. To subtract complex numbers, add the opposite of the complex number being subtracted.

EXAMPLE 3

Perform each operation. Write the answers in the form $a + bi$. **a.** $(8 + 4i) + (12 + 8i)$
b. $(-6 + 4i) - (3 + 2i)$ **c.** $\left(7 - \sqrt{-16}\right) + \left(9 + \sqrt{-4}\right)$

Strategy To add the complex numbers, we will add their real parts and add their imaginary parts. To subtract the complex numbers, we will add the opposite of the complex number to be subtracted.

Why We perform the indicated operations as if the complex numbers were polynomials with i as a variable.

Solution

a. $(\mathbf{8} + \mathbf{4}i) + (\mathbf{12} + \mathbf{8}i) = (\mathbf{8} + \mathbf{12}) + (\mathbf{4} + \mathbf{8})i$

The sum of the imaginary parts
The sum of the real parts

$$= 20 + 12i$$

Add

b. $(-6 + 4i) - (3 + 2i) = (-6 + 4i) + (-3 - 2i)$ To find the opposite, change the sign of each term of $3 + 2i$.

the opposite

$$= [-6 + (-3)] + [4 + (-2)]i$$ Add the real parts. Add the imaginary parts.

$$= -9 + 2i$$

c. $\left(7 - \sqrt{-16}\right) + \left(9 + \sqrt{-4}\right)$

$$= (7 - 4i) + (9 + 2i)$$ Write $\sqrt{-16}$ and $\sqrt{-4}$ in terms of i.

$$= (7 + 9) + (-4 + 2)i$$ Add the real parts. Add the imaginary parts.

$$= 16 - 2i$$ Write $16 + (-2i)$ in the form $16 - 2i$.

Self Check 3 Perform the operations. Write the answers in the form $a + bi$.
a. $(3 - 5i) + (-2 + 7i)$
b. $\left(3 - \sqrt{-25}\right) - \left(-2 + \sqrt{-49}\right)$

Now Try Problems 37 and 43

The Language of Algebra

For years, mathematicians thought numbers like $\sqrt{-4}$ and $\sqrt{-16}$ were useless. In the 17th century, French mathematician René Descartes (1596–1650) called them **imaginary numbers.** Today they have important uses such as describing alternating electric current.

Success Tip

Always express square roots of negative numbers in terms of i before performing any operations.

4 Multiply Complex Numbers.

Since imaginary numbers are not real numbers, some properties of real numbers do not apply to imaginary numbers. For example, we cannot use the product rule for radicals to multiply two imaginary numbers.

CAUTION If a and b are both negative, then $\sqrt{a}\sqrt{b} \neq \sqrt{ab}$. For example, if $a = -4$ and $b = -9$,

$$\sqrt{-4} \cdot \sqrt{-9} = \sqrt{-4(-9)} = \sqrt{36} = 6 \qquad \sqrt{-4}\sqrt{-9} = 2i(3i) = 6i^2 = 6(-1) = -6$$

EXAMPLE 4 Multiply: $\sqrt{-2} \cdot \sqrt{-20}$

Strategy To multiply the imaginary numbers, first we will write $\sqrt{-2}$ and $\sqrt{-20}$ in $i\sqrt{b}$ form. Then we will use the product rule for radicals.

Why We cannot use the product rule for radicals immediately because it does not apply when both radicands are negative.

Solution

$$\sqrt{-2} \cdot \sqrt{-20} = (i\sqrt{2})(2i\sqrt{5}) \quad \text{Simplify: } \sqrt{-20} = i\sqrt{20} = 2i\sqrt{5}.$$
$$= 2i^2\sqrt{2 \cdot 5} \quad \text{Multiply: } i \cdot 2i = 2i^2. \text{ Use the product rule for radicals.}$$
$$= 2i^2\sqrt{10}$$
$$= 2(-1)\sqrt{10} \quad \text{Replace } i^2 \text{ with } -1.$$
$$= -2\sqrt{10} \quad \text{Multiply.}$$

Success Tip

The algebraic methods we have used with polynomials and rational expressions (such as combining like terms, distributive property, FOIL, and building fractions) are also used with complex numbers. However, when you encounter i^2 in your work, remember to replace it with -1.

Self Check 4 Multiply: $\sqrt{-3} \cdot \sqrt{-32}$

Now Try Problem 47

Multiplying complex numbers is similar to multiplying polynomials.

EXAMPLE 5 Multiply. Write the answers in the form $a + bi$. **a.** $6(2 + 9i)$ **b.** $-5i(4 - 8i)$

Strategy We will use the distributive property to find the products.

Why We perform the indicated operations as if the complex numbers were polynomials with i as a variable.

Solution **a.** $6(2 + 9i) = 6(2) + 6(9i)$ Use the distributive property.
$$= 12 + 54i \quad \text{Perform each multiplication.}$$

Caution

A common mistake is to replace i with -1. Remember, $i \neq -1$. By definition, $i = \sqrt{-1}$ and $i^2 = -1$.

b. $-5i(4 - 8i) = -5i(4) - (-5i)8i$ Use the distributive property.
$$= -20i + 40i^2 \quad \text{Perform each multiplication.}$$
$$= -20i + 40(-1) \quad \text{Replace } i^2 \text{ with } -1.$$
$$= -20i - 40 \quad \text{Multiply.}$$
$$= -40 - 20i \quad \text{Write the real part, } -40, \text{ as the first term.}$$

Self Check 5 Multiply. Write the answers in the form $a + bi$.
a. $-2(-9 - i)$ **b.** $10i(7 + 4i)$

Now Try Problems 49 and 55

EXAMPLE 6 Multiply. Write the answers in the form $a + bi$. **a.** $(2 + 3i)(3 - 2i)$ **b.** $(-4 + 2i)(2 + i)$

Strategy We will use the FOIL method to multiply the two complex numbers.

Why We perform the indicated operations as if the complex numbers were binomials with i as a variable.

F   O   I   L

Solution

a. $(2 + 3i)(3 - 2i) = 6 - 4i + 9i - 6i^2$ Use the FOIL method.

$= 6 + 5i - 6(-1)$ Combine the imaginary terms: $-4i + 9i = 5i$.
Replace i^2 with -1.

$= 6 + 5i + 6$ Simplify the last term.

$= 12 + 5i$ Combine like terms.

b. $(-4 + 2i)(2 + i) = -8 - 4i + 4i + 2i^2$ Use the FOIL method.

$= -8 + 0i + 2(-1)$ Combine like terms: $-4i + 4i = 0i$.
Replace i^2 with -1.

$= -8 + 0i - 2$ Multiply.

$= -10 + 0i$ Combine like terms.

> ### Success Tip
>
> i is not a variable, but you can think of it as one when adding, subtracting, and multiplying. For example:
>
> $-4i + 9i = 5i$
> $6i - 2i = 4i$
> $i \cdot i = i^2$
>
> Remember that the expression i^2 simplifies to -1.

Self Check 6   Multiply. Write the answers in the form $a + bi$. $(-2 + 3i)(3 - 2i)$

Now Try ▶   Problem 59

5   Divide Complex Numbers.

Before we can discuss division of complex numbers, we must introduce an important fact about *complex conjugates*.

Complex Conjugates	The complex numbers $a + bi$ and $a - bi$ are called **complex conjugates**.

> ### The Language of Algebra
>
> Recall that the word **conjugate** was used earlier when we rationalized the denominators of radical expressions such as
>
> $$\dfrac{5}{\sqrt{6} - 1}$$

For example,

- $7 + 4i$ and $7 - 4i$ are complex conjugates.
- $5 - i$ and $5 + i$ are complex conjugates.
- $-6i$ and $6i$ are complex conjugates, because $-6i = 0 - 6i$ and $6i = 0 + 6i$.

In general, the product of the complex number $a + bi$ and its complex conjugate $a - bi$ is the real number $a^2 + b^2$, as the following work shows:

$(a + bi)(a - bi) = a^2 - abi + abi - b^2i^2$ Use the FOIL method.

$= a^2 - b^2(-1)$ $-abi + abi = 0$. Replace i^2 with -1.

$= a^2 + b^2$

EXAMPLE 7   Find the product of $3 + 5i$ and its complex conjugate.

Strategy   The complex conjugate of $3 + 5i$ is $3 - 5i$. We will find their product by using the FOIL method.

Why   We perform the indicated operations as if the complex numbers were binomials with i as a variable.

Solution We can find the product as follows:

$$(3 + 5i)(3 - 5i) = 9 - 15i + 15i - 25i^2 \qquad \text{Use the FOIL method.}$$

$$= 9 - 25i^2 \qquad \text{Combine like terms: } -15i + 15i = 0.$$

$$= 9 - 25(-1) \qquad \text{Replace } i^2 \text{ with } -1.$$

$$= 9 + 25$$

$$= 34$$

The product of $3 + 5i$ and its complex conjugate $3 - 5i$ is the real number 34.

Self Check 7 Multiply: $(2 + 3i)(2 - 3i)$

Now Try ▶ Problem 65

Recall that to divide *radical expressions,* we rationalized the denominator. We will use a similar approach to divide complex numbers. To divide two complex numbers when the divisor has two terms, we use the following strategy.

Division of Complex Numbers	To divide complex numbers, multiply the numerator and denominator by the complex conjugate of the denominator.

EXAMPLE 8 Divide. Write the answers in the form $a + bi$. **a.** $\dfrac{3}{6 + i}$ **b.** $\dfrac{1 + 2i}{3 - 4i}$

Strategy We will build each fraction by multiplying it by a form of 1 that uses the conjugate of the denominator.

Why This step produces a *real number* in the denominator so that the result then can be written in the form $a + bi$.

Solution **a.** We want to build a fraction equivalent to $\dfrac{3}{6 + i}$ that does not have i in the denominator. To make the denominator, $6 + i$, a real number, we need to multiply it by its complex conjugate, $6 - i$. It follows that $\dfrac{6 - i}{6 - i}$ should be the form of 1 that is used to build $\dfrac{3}{6 + i}$.

$$\frac{3}{6 + i} = \frac{3}{6 + i} \cdot \frac{6 - i}{6 - i} \qquad \text{To build an equivalent fraction, multiply by } \frac{6 - i}{6 - i} = 1.$$

$$= \frac{18 - 3i}{36 - 6i + 6i - i^2} \qquad \begin{array}{l}\text{To multiply the numerators, distribute the multiplication} \\ \text{by 3. Use the FOIL method to multiply the denominators.}\end{array}$$

$$= \frac{18 - 3i}{36 - (-1)} \qquad \begin{array}{l}\text{Combine like terms: } -6i + 6i = 0. \text{ Replace } i^2 \text{ with } -1. \\ \text{Note that the denominator no longer contains } i.\end{array}$$

$$= \frac{18 - 3i}{37} \qquad \begin{array}{l}\text{Simplify the denominator. This notation represents the} \\ \text{difference of two fractions that have the common} \\ \text{denominator 37: } \frac{18}{37} \text{ and } \frac{3i}{37}.\end{array}$$

$$= \frac{18}{37} - \frac{3}{37}i \qquad \text{Write the complex number in the form } a + bi.$$

b. We can make the denominator of $\dfrac{1 + 2i}{3 - 4i}$ a real number by multiplying it by the complex conjugate of $3 - 4i$, which is $3 + 4i$. It follows that $\dfrac{3 + 4i}{3 + 4i}$ should be the form of 1 that is used to build $\dfrac{1 + 2i}{3 - 4i}$.

$$\dfrac{1 + 2i}{3 - 4i} = \dfrac{1 + 2i}{3 - 4i} \cdot \dfrac{3 + 4i}{3 + 4i}$$

To build an equivalent fraction, multiply by $\dfrac{3 + 4i}{3 + 4i} = 1$.

$$= \dfrac{3 + 4i + 6i + 8i^2}{9 + 12i - 12i - 16i^2}$$

Use the FOIL method to multiply the numerators and the denominators.

$$= \dfrac{3 + 10i + 8(-1)}{9 - 16(-1)}$$

Combine like terms in the numerator and denominator. Replace i^2 with -1. The denominator is now a real number.

$$= \dfrac{3 + 10i - 8}{9 + 16}$$

Simplify the numerator and denominator.

$$= \dfrac{-5 + 10i}{25}$$

Combine like terms in the numerator and denominator.

$$= \dfrac{\overset{1}{\cancel{5}}(-1 + 2i)}{\underset{1}{\cancel{5} \cdot 5}}$$

Factor out 5 in the numerator and remove the common factor of 5.

$$= \dfrac{-1 + 2i}{5}$$

Simplify. This notation represents the *sum* of two fractions that have the common denominator 5.

$$= -\dfrac{1}{5} + \dfrac{2}{5}i$$

Write the complex number in the form $a + bi$.

Self Check 8   Divide. Write the answers in the form $a + bi$.   **a.** $\dfrac{6}{5 + 2i}$

b. $\dfrac{2 - 4i}{5 - 3i}$

Now Try Problems 71 and 77

EXAMPLE 9   Divide and write the answer in the form $a + bi$: $\dfrac{4 + \sqrt{-16}}{2 + \sqrt{-4}}$

Strategy   We will begin by writing $\sqrt{-16}$ and $\sqrt{-4}$ in $i\sqrt{b}$ form.

Why   To perform any operations, the numerator and denominator should be written in the form $a + bi$.

Solution

$$\dfrac{4 + \sqrt{-16}}{2 + \sqrt{-4}} = \dfrac{4 + 4i}{2 + 2i}$$

Simplify: $\sqrt{-16} = \sqrt{-1}\sqrt{16} = 4i$ and $\sqrt{-4} = \sqrt{-1}\sqrt{4} = 2i$.

$$= \dfrac{2(2 + 2i)}{2 + 2i}$$

Factor out 2 in the numerator and remove the common factor of $2 + 2i$.

$$= 2$$

$$= 2 + 0i$$

Write 2 in the form $a + bi$.

Self Check 9   Divide and write the answer in the form $a + bi$: $\dfrac{21 - \sqrt{-49}}{3 - \sqrt{-1}}$

Now Try Problem 81

EXAMPLE 10 Divide and write the result in the form $a + bi$: $\dfrac{7}{2i}$

Strategy We will use $\dfrac{-2i}{-2i}$ as the form of 1 to build $\dfrac{7}{2i}$.

Why Since the denominator $2i$ can be expressed as $0 + 2i$, its complex conjugate is $0 - 2i$. However, instead of building with $\dfrac{0 - 2i}{0 - 2i}$, we will drop the zeros and just use $\dfrac{-2i}{-2i}$.

Solution

$$\frac{7}{2i} = \frac{7}{2i} \cdot \frac{-2i}{-2i} \qquad \text{To build an equivalent fraction, multiply by } \frac{-2i}{-2i} = 1.$$

$$= \frac{-14i}{-4i^2} \qquad \text{Multiply the numerators and multiply the denominators.}$$

$$= \frac{-14i}{-4(-1)} \qquad \text{Replace } i^2 \text{ with } -1. \text{ The denominator is now a real number.}$$

$$= \frac{-14i}{4} \qquad \text{Simplify the denominator.}$$

$$= -\frac{7i}{2} \qquad \text{Simplify the fraction: } -\frac{\overset{1}{\cancel{2}} \cdot 7i}{\underset{1}{\cancel{2}} \cdot 2}.$$

$$= 0 - \frac{7}{2}i \qquad \text{Write in the form } a + bi.$$

Success Tip

In this example, the denominator of $\frac{7}{2i}$ is of the form bi. In such cases, we can eliminate i in the denominator by simply multiplying by $\frac{i}{i}$.

$$\frac{7}{2i} = \frac{7}{2i} \cdot \frac{i}{i} = \frac{7i}{2i^2} = -\frac{7}{2}i$$

Self Check 10 Divide and write the result in the form $a + bi$: $\dfrac{3}{4i}$

Now Try ▶ Problem 85

6 Perform Operations Involving Powers of i.

The powers of i produce an interesting pattern:

$$i = \sqrt{-1} = i \qquad\qquad i^5 = i^4 i = 1i = i$$

$$i^2 = \left(\sqrt{-1}\right)^2 = -1 \qquad i^6 = i^4 i^2 = 1(-1) = -1$$

$$i^3 = i^2 i = -1i = -i \qquad i^7 = i^4 i^3 = 1(-i) = -i$$

$$i^4 = i^2 i^2 = (-1)(-1) = 1 \qquad i^8 = i^4 i^4 = (1)(1) = 1$$

Success Tip

Note that the powers of i cycle through four possible outcomes:

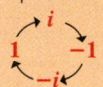

The pattern continues: $i, -1, -i, 1, \ldots$.

Larger powers of i can be simplified by using the fact that $i^4 = 1$. For example, to simplify i^{29}, we note that 29 divided by 4 gives a quotient of 7 and a remainder of 1. Thus, $29 = 4 \cdot 7 + 1$ and

$$i^{29} = i^{4 \cdot 7 + 1} \qquad 4 \cdot 7 = 28.$$

$$= (i^4)^7 \cdot i^1 \qquad \text{Use the rules for exponents } x^{m \cdot n} = (x^m)^n \text{ and } x^{m+n} = x^m \cdot x^n.$$

$$= 1^7 \cdot i \qquad \text{Simplify: } i^4 = 1.$$

$$= i \qquad \text{Simplify: } 1 \cdot i = i.$$

The result of this example illustrates the following fact.

Powers of i

If n is a natural number that has a remainder of R when divided by 4, then

$$i^n = i^R$$

EXAMPLE 11   Simplify: **a.** i^{55} **b.** i^{98}

Strategy   We will examine the remainder when we divide the exponents 55 and 98 by 4.

Why   The remainder determines the power to which i is raised in the simplified form.

Solution   **a.** We divide 55 by 4 and get a remainder of 3. Therefore,

$$i^{55} = i^3 = -i$$

$$\begin{array}{r} 13\text{R }\mathbf{3} \\ 4\overline{)55} \\ -4 \\ \hline 15 \\ -12 \\ \hline \mathbf{3} \end{array}$$

b. We divide 98 by 4 and get a remainder of 2. Therefore,

$$i^{98} = i^2 = -1$$

$$\begin{array}{r} 24\text{R }\mathbf{2} \\ 4\overline{)98} \\ -8 \\ \hline 18 \\ -16 \\ \hline \mathbf{2} \end{array}$$

Success Tip

If we divide the natural number exponent n of a power of i by 4, the remainder indicates the simplified form of i^n:

R = 1: i
R = 2: -1
R = 3: $-i$
R = 0:   1

Self Check 11   Simplify: **a.** i^{62} **b.** i^{105}

Now Try ▶ Problem 91

SECTION 7.7   STUDY SET

VOCABULARY

Fill in the blanks.

1. The _____ number i is defined as $i = \sqrt{-1}$. We call i^{25} a _____ of i.

2. A _____ number is any number that can be written in the form $a + bi$, where a and b are real numbers and $i = \sqrt{-1}$.

3. For the complex number $2 + 5i$, we call 2 the _____ part and 5 the _____ part.

4. $6 + 3i$ and $6 - 3i$ are called complex _____.

CONCEPTS

Fill in the blanks.

5. **a.** $i =$ ▢   **b.** $i^2 =$ ▢
   c. $i^3 =$ ▢   **d.** $i^4 =$ ▢
   e. In general, the powers of i cycle through _____ possible outcomes.

6. Simplify:
   $$\sqrt{-36} = \sqrt{ \cdot 36} = \sqrt{}\sqrt{36} = 6$$

7. **a.** To add (or subtract) complex numbers, add (or subtract) their _____ parts and add (or subtract) their _____ parts.
   b. To multiply two complex numbers, such as $(2 + 3i)(3 + 5i)$, we can use the _____ method.

8. To divide $6 + 7i$ by $1 - 8i$, we multiply $\frac{6 + 7i}{1 - 8i}$ by 1 in the form of ▢.

9. Give the complex conjugate of each number.
   a. $2 - 3i$ **b.** 2   **c.** $-3i$

10. Factor each numerator. Then remove the factor common to the numerator and denominator. Write the result in the form $a + bi$.
    a. $\dfrac{3 + 6i}{3}$ **b.** $\dfrac{15 + 25i}{10}$

11. Complete the illustration. Label the *real numbers,* the *imaginary numbers,* the *complex numbers,* the *rational numbers,* and the *irrational numbers.*

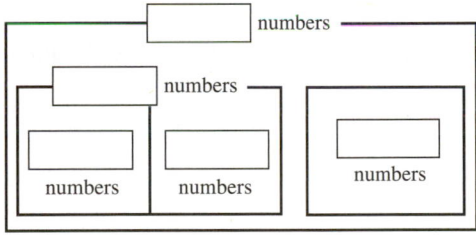

12. Determine whether each statement is true or false.
    a. Every complex number is a real number.
    b. Every real number is a complex number.
    c. i is a real number.
    d. The square root of a negative number is an imaginary number.
    e. The product of a complex number and its complex conjugate is always a real number.

NOTATION

Complete each solution.

13. $(3 + 2i)(3 - i) = \boxed{} - 3i + \boxed{} - 2i^2$

$= 9 + 3i + \boxed{}$

$= \boxed{} + 3i$

14. $\dfrac{3}{2 - i} = \dfrac{3}{2 - i} \cdot \dfrac{\boxed{}}{\boxed{}}$

$= \dfrac{6 + \boxed{}}{4 - \boxed{}}$

$= \dfrac{6 + 3i}{\boxed{}}$

$= \dfrac{\boxed{}}{\boxed{}} + \dfrac{3}{5}i$

15. Determine whether each statement is true or false.

a. $\sqrt{6}i = i\sqrt{6}$ b. $\sqrt{8}i = \sqrt{8i}$

c. $\sqrt{-25} = -\sqrt{25}$ d. $-i = i$

16. Write each number in the form $a + bi$.

a. $\dfrac{9 + 11i}{4}$ b. $\dfrac{1 - i}{18}$

GUIDED PRACTICE

Express each number in terms of i. See Example 1.

17. $\sqrt{-9}$ **18.** $\sqrt{-4}$

19. $\sqrt{-7}$ **20.** $\sqrt{-11}$

21. $\sqrt{-24}$ **22.** $\sqrt{-28}$

23. $-\sqrt{-72}$ **24.** $-\sqrt{-24}$

25. $5\sqrt{-81}$ **26.** $6\sqrt{-49}$

27. $\sqrt{-\dfrac{25}{9}}$ **28.** $-\sqrt{-\dfrac{121}{144}}$

Write each number in the form $a + bi$. See Example 2.

29. a. 5 **30.** a. -43

b. $\sqrt{-49}$ b. $\sqrt{-169}$

31. a. $1 + \sqrt{-25}$ **32.** a. $21 + \sqrt{-16}$

b. $-3 + \sqrt{-8}$ b. $-9 + \sqrt{-12}$

33. a. $76 - \sqrt{-54}$ **34.** a. $88 - \sqrt{-98}$

b. $-7 + \sqrt{-19}$ b. $-2 + \sqrt{-35}$

35. a. $-6 - \sqrt{-9}$ **36.** a. $-45 - \sqrt{-81}$

b. $3 + \sqrt{-6}$ b. $8 + \sqrt{-7}$

Perform the operations. Write all answers in the form $a + bi$. See Example 3.

37. $(3 + 4i) + (5 - 6i)$ **38.** $(8 + 3i) + (-7 - 2i)$

39. $(6 - i) + (9 + 3i)$ **40.** $(5 + 3i) - (6 - 9i)$

41. $(7 - 3i) - (4 + 2i)$ **42.** $(5 - 4i) - (3 + 2i)$

43. $\left(8 + \sqrt{-25}\right) - \left(7 + \sqrt{-4}\right)$

44. $\left(-7 + \sqrt{-81}\right) - \left(-2 - \sqrt{-64}\right)$

Multiply. See Example 4.

45. $\sqrt{-1} \cdot \sqrt{-36}$ **46.** $\sqrt{-9} \cdot \sqrt{-100}$

47. $\sqrt{-2}\sqrt{-12}$ **48.** $\sqrt{-3}\sqrt{-45}$

Multiply. Write all answers in the form $a + bi$. See Example 5.

49. $3(2 - 9i)$ **50.** $-4(3 + 4i)$

51. $7(5 - 4i)$ **52.** $-5(3 + 2i)$

53. $2i(7 - 3i)$ **54.** $i(8 + 2i)$

55. $-5i(5 - 5i)$ **56.** $2i(7 + 2i)$

Multiply. Write all answers in the form $a + bi$. See Example 6.

57. $(2 + i)(3 - i)$ **58.** $(4 - i)(2 + i)$

59. $(3 - 2i)(2 + 3i)$ **60.** $(3 - i)(2 + 3i)$

61. $\left(4 + \sqrt{-1}\right)\left(3 - \sqrt{-1}\right)$ **62.** $\left(1 - \sqrt{-25}\right)\left(1 - \sqrt{-16}\right)$

63. $(2 + i)^2$ **64.** $(3 - 2i)^2$

Find the product of the given complex number and its conjugate. See Example 7.

65. $2 + 6i$ **66.** $5 + 2i$

67. $-4 - 7i$ **68.** $-10 - 9i$

Divide. Write all answers in the form $a + bi$. See Example 8.

69. $\dfrac{9}{5 + i}$ **70.** $\dfrac{4}{2 - i}$

71. $\dfrac{11i}{4 - 7i}$ **72.** $\dfrac{2i}{3 + 8i}$

73. $\dfrac{3 - 2i}{4 - i}$ **74.** $\dfrac{6 - i}{2 + i}$

75. $\dfrac{7 + 4i}{2 - 5i}$ **76.** $\dfrac{2 + 3i}{2 - 3i}$

77. $\dfrac{7 + 3i}{4 - 2i}$ **78.** $\dfrac{5 - 3i}{4 + 2i}$

79. $\dfrac{1 - 3i}{3 + i}$ **80.** $\dfrac{3 + 5i}{1 - i}$

Divide. Write all answers in the form $a + bi$. See Example 9. (Hint: Factor the numerator.)

81. $\dfrac{8 + \sqrt{-144}}{2 + \sqrt{-9}}$ **82.** $\dfrac{3 + \sqrt{-36}}{1 + \sqrt{-4}}$

83. $\dfrac{-4 - \sqrt{-4}}{2 + \sqrt{-1}}$ **84.** $\dfrac{-5 - \sqrt{-25}}{1 + \sqrt{-1}}$

Divide. Write all answers in the form $a + bi$. See Example 10.

85. $\dfrac{5}{3i}$ **86.** $\dfrac{3}{8i}$

87. $-\dfrac{2}{7i}$ **88.** $-\dfrac{8}{5i}$

Simplify each expression. See Example 11.

89. i^{21} **90.** i^{19} **91.** i^{27} **92.** i^{22}

93. i^{100} **94.** i^{97} **95.** i^{42} **96.** i^{200}

TRY IT YOURSELF

Perform the operations. Write all answers in the form a + bi.

97. $(3 - i) - (-1 + 10i)$ **98.** $(14 + 4i) - (-9 - i)$

99. $\left(2 - \sqrt{-16}\right)\left(3 + \sqrt{-4}\right)$ **100.** $\left(3 - \sqrt{-4}\right)\left(4 - \sqrt{-9}\right)$

101. $(-6 - 9i) + (4 + 3i)$ **102.** $(-3 + 11i) + (-1 - 6i)$

103. $\dfrac{-2i}{3 + 2i}$ **104.** $\dfrac{-4i}{2 - 6i}$

105. $6i(2 - 3i)$ **106.** $-9i(4 - 6i)$

107. $\dfrac{4}{5i^{35}}$ **108.** $\dfrac{3}{2i^{17}}$

109. $\left(2 + i\sqrt{2}\right)\left(3 - i\sqrt{2}\right)$ **110.** $\left(5 + i\sqrt{3}\right)\left(2 - i\sqrt{3}\right)$

111. $\dfrac{5 + 9i}{1 - i}$ **112.** $\dfrac{5 - i}{3 + 2i}$

113. $(4 - 8i)^2$ **114.** $(7 - 3i)^2$

115. $\dfrac{\sqrt{5} - i\sqrt{3}}{\sqrt{5} + i\sqrt{3}}$ **116.** $\dfrac{\sqrt{3} + i\sqrt{2}}{\sqrt{3} - i\sqrt{2}}$

Look Alikes . . .

117. a. $\sqrt{-8}$ **b.** $\sqrt[3]{-8}$

118. a. $(3 + i) + (2 + 4i)$ **b.** $(3 + i)(2 + 4i)$

119. a. $(2i)^2$ **b.** $(2 + i)^2$

120. a. $\sqrt{-9}\sqrt{-16}$ **b.** $\sqrt{9}\sqrt{16}$

APPLICATIONS

121. Fractals. Complex numbers are fundamental in the creation of the intricate geometric shape shown below, called a *fractal*. The process of creating this image is based on the following sequence of steps, which begins by picking any complex number, which we will call z.

1. Square z, and then add that result to z.
2. Square the result from step 1, and then add it to z.
3. Square the result from step 2, and then add it to z.

If we begin with the complex number i, what is the result after performing steps 1, 2, and 3?

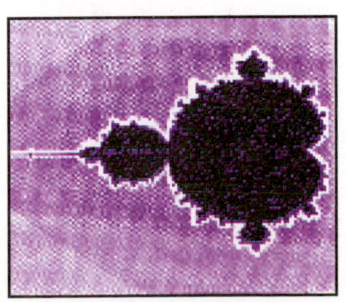

122. Electronics. The impedance Z in an AC (alternating current) circuit is a measure of how much the circuit impedes (hinders) the flow of current through it. The impedance is related to the voltage V and the current I by the formula $V = IZ$. If a circuit has a current of $(0.5 + 2.0i)$ amps and an impedance of $(0.4 - 3.0i)$ ohms, find the voltage.

WRITING

123. What is an imaginary number? What is a complex number?

124. The method used to divide complex numbers is similar to the method used to divide radical expressions. Explain why. Give an example.

125. Explain the error. Then find the correct result.

a. Add: ~~$\sqrt{-16} + \sqrt{-9} = \sqrt{-25}$~~

b. Multiply: ~~$\sqrt{-2}\sqrt{-3} = \sqrt{-2(-3)} = \sqrt{6}$~~

126. Determine whether the pair of complex numbers are equal. Explain your reasoning.

a. $4 - \dfrac{2}{5}i, \dfrac{8}{2} - 0.4i$ **b.** $0.25 + 0.7i, \dfrac{1}{4} + \dfrac{7}{10}i$

REVIEW

127. Wind Speeds. A plane that can fly 200 mph in still air makes a 330-mile flight with a tail wind and returns, flying into the same wind. Find the speed of the wind if the total flying time is $3\frac{1}{3}$ hours.

128. Finding Rates. A student drove a distance of 135 miles at an average speed of 50 mph. How much faster would she have to drive on the return trip to save 30 minutes of driving time?

CHALLENGE PROBLEMS

129. Simplify: $\left(i^{349}\right)^{-i^{456}}$

130. Simplify $(2 + 3i)^{-2}$ and write the result in the form $a + bi$.

7   Summary & Review

DEFINITIONS AND CONCEPTS	EXAMPLES
The number b is a **square root of a** if $b^2 = a$. Every positive real number has two square roots.	7 is a square root of 49 because $7^2 = 49$. -7 is also a square root of 49 because $(-7)^2 = 49$.
A **radical symbol** $\sqrt{}$ represents the **positive** or **principal square root** of a number. For any real number x, $\sqrt{x^2} = \|x\|$ The symbol $-\sqrt{}$ represents the **negative square root** of a number. Review the list of perfect squares on page 581.	$\sqrt{25} = 5$ because $5^2 = 25$. $\sqrt{36x^2} = \|6x\| = 6\|x\|$ because $(6x)^2 = 36x^2$. Since x could be negative, absolute value symbols ensure that the result is not negative. $\sqrt{\dfrac{r^8}{100}} = \dfrac{r^4}{10}$ because $\left(\dfrac{r^4}{10}\right)^2 = \dfrac{r^8}{100}$. Since $\frac{r^4}{10}$ cannot be negative, no absolute value symbols are needed. $-\sqrt{81} = -9$ and $\sqrt{-81}$ is not a real number.
A function of the form $f(x) = \sqrt{x}$ is called a **square root function.** Review its graph on page 584.	Let $f(x) = \sqrt{x - 2}$. Find $f(38)$. $f(38) = \sqrt{38 - 2} = \sqrt{36} = 6$
The **domain of a square root function** is the set of all real numbers for which the radicand is nonnegative. To find the domain, set the radicand greater than or equal to 0 and solve for the variable.	Find the domain of $g(x) = \sqrt{2x - 4}$. Since $\sqrt{2x - 4}$ is not a real number when $2x - 4$ is negative, we must require that $2x - 4 \geq 0$ Because we cannot find the square root of a negative number $2x \geq 4$ Solve for x. $x \geq 2$ The domain of g is $[2, \infty)$.
The **cube root** of x is denoted as $\sqrt[3]{x}$ and is defined as $\sqrt[3]{x} = y$ if $y^3 = x$ Review the list of perfect cubes on page 587.	$\sqrt[3]{8} = 2$ because $2^3 = 8$. $\sqrt[3]{-64} = -4$ because $(-4)^3 = -64$. $\sqrt[3]{-27a^6} = -3a^2$ because $(-3a^2)^3 = -27a^6$.
A function of the form $f(x) = \sqrt[3]{x}$ is called a **cube root function.** Review its graph on page 588.	Let $f(x) = \sqrt[3]{x + 4}$. Find $f(23)$. $f(23) = \sqrt[3]{23 + 4} = \sqrt[3]{27} = 3$
The ***n*th root of x** is denoted as $\sqrt[n]{x}$. Index $\sqrt[4]{16a^8}$ ← **Radicand** If x is a real number and $n > 1$, then: If n is an odd natural number, $\sqrt[n]{x^n} = x$. If n is an even natural number, $\sqrt[n]{x^n} = \|x\|$.	$\sqrt[4]{81x^4} = \|3x\| = 3\|x\|$ because $(3x)^4 = 81x^4$. $\sqrt[4]{-16}$ Not a real number $\sqrt[5]{32a^{10}} = 2a^2$ because $(2a^2)^5 = 32a^{10}$. No absolute value symbols are needed. $\sqrt[6]{(m - 4)^6} = \|m - 4\|$ because $(m - 4)^6 = (m - 4)^6$.

REVIEW EXERCISES

1. The graph of a square root function f is shown here. Find each of the following:

   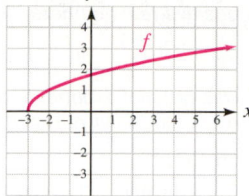

   a. $f(1)$ **b.** $f(-3)$

   c. The value of x for which $f(x) = 3$

   d. The domain and range of f

2. **a.** Simplify $\sqrt{100a^2}$. Assume that the variable a is unrestricted.

   b. Simplify $\sqrt{100a^2}$. Assume that the variable a is a positive real number.

Simplify each expression. Assume that all variables are unrestricted and use absolute value symbols when necessary.

3. $\sqrt{49}$

4. $-\sqrt{121}$

5. $\sqrt{\dfrac{225}{49}}$

6. $\sqrt{-4}$

7. $\sqrt{100a^{12}}$

8. $\sqrt{25x^2}$

9. $\sqrt{x^8}$

10. $\sqrt{x^2 + 4x + 4}$

11. $\sqrt[3]{-27}$

12. $-\sqrt[3]{216}$

13. $\sqrt[3]{64x^6y^3}$

14. $\sqrt[3]{\dfrac{x^9}{125}}$

15. $\sqrt[6]{64}$

16. $\sqrt[5]{-32}$

17. $\sqrt[4]{256x^8y^4}$

18. $\sqrt[5]{(x+1)^5}$

19. $-\sqrt[4]{\dfrac{1}{16}}$

20. $\sqrt[4]{-81}$

21. $\sqrt[6]{-1}$

22. $\sqrt[3]{0}$

23. Find the domain of $f(x) = \sqrt{3x + 15}$.

24. **Geometry.** The side of a square with area A square feet is given by the function $s(A) = \sqrt{A}$. Find the length of one side of a square that has an area of 169 ft².

25. **Cubes.** The total surface area of a cube is related to its volume V by the function $A(V) = 6\sqrt[3]{V^2}$. Find the surface area of a cube with a volume of 8 cm³.

26. Let $g(x) = \sqrt[3]{x^2 + 9}$. Use a calculator fo find $g(-1.9)$ to four decimal places.

Graph each function. Find the domain and range.

27. $f(x) = \sqrt{x}$

28. $f(x) = \sqrt[3]{x}$

29. $f(x) = \sqrt{x + 2}$

30. $f(x) = -\sqrt[3]{x} + 3$

SECTION 7.2 ▶ Rational Exponents

DEFINITIONS AND CONCEPTS	EXAMPLES
To simplify exponential expressions involving **rational (fractional) exponents,** use the following rules to write the expressions in an equivalent radical form. $$x^{1/n} = \sqrt[n]{x} \qquad x^{m/n} = \left(\sqrt[n]{x}\right)^m = \sqrt[n]{x^m}$$	Simplify. All variables represent positive real numbers. $$25^{1/2} = \sqrt{25} = 5 \qquad \left(\dfrac{256}{d^4}\right)^{1/4} = \sqrt[4]{\dfrac{256}{d^4}} = \dfrac{4}{d}$$ $$8^{2/3} = \left(\sqrt[3]{8}\right)^2 = (2)^2 = 4 \qquad (t^{10})^{6/5} = \left(\sqrt[5]{t^{10}}\right)^6 = (t^2)^6 = t^{12}$$
To be consistent with the definition of negative integer exponents, we define $x^{-m/n}$ as follows. $$x^{-m/n} = \dfrac{1}{x^{m/n}}$$ $$\dfrac{1}{x^{-m/n}} = x^{m/n}$$	Simplify. All variables represent positive real numbers. $$(125)^{-2/3} = \dfrac{1}{(125)^{2/3}} = \dfrac{1}{\left(\sqrt[3]{125}\right)^2} = \dfrac{1}{25}$$ $$\dfrac{1}{(-32x^5)^{-3/5}} = (-32x^5)^{3/5} = \left(\sqrt[5]{-32x^5}\right)^3 = -8x^3$$
The **rules for exponents** can be used to simplify expressions with rational (fractional) exponents.	Simplify: $$\dfrac{p^{5/3}p^{8/3}}{p^4} = p^{5/3 + 8/3 - 4} = p^{5/3 + 8/3 - 12/3} = p^{1/3}$$
We can write certain radical expressions as an equivalent exponential expression and use rules for exponents to simplify it. Then we can change that result back into a radical.	Simplify: $$\sqrt[4]{9} = \sqrt[4]{3^2} = 3^{2/4} = 3^{1/2} = \sqrt{3}$$

REVIEW EXERCISES

Write each expression in radical form.

31. $t^{1/2}$

32. $(5xy^3)^{1/4}$

Simplify each expression, if possible. Assume that all variables represent positive real numbers.

33. $25^{1/2}$

34. $-36^{1/2}$

35. $(-36)^{1/2}$

36. $1^{1/5}$

37. $\left(\dfrac{9}{x^2}\right)^{1/2}$

38. $(-8)^{1/3}$

39. $625^{1/4}$

40. $(81c^4d^4)^{1/4}$

41. $9^{3/2}$

42. $8^{-2/3}$

43. $-49^{5/2}$

44. $\dfrac{1}{100^{-1/2}}$

45. $\left(\dfrac{4}{9}\right)^{-3/2}$

46. $\dfrac{1}{25^{5/2}}$

47. $(25x^2y^4)^{3/2}$

48. $(8u^6v^3)^{-2/3}$

Perform the operations. Write answers without negative exponents. Assume that all variables represent positive real numbers.

49. $5^{1/4} \cdot 5^{1/2}$

50. $a^{3/7}a^{-2/7}$

51. $(k^{4/5})^{10}$

52. $\dfrac{3^{5/6}3^{1/3}}{3^{1/2}}$

Perform the multiplications. Assume all variables represent positive real numbers.

53. $u^{1/2}(u^{1/2} - u^{-1/2})$

54. $v^{2/3}(v^{1/3} + v^{4/3})$

Use rational exponents to simplify each radical. All variables represent positive real numbers.

55. $\sqrt[4]{a^2}$

56. $\sqrt[3]{\sqrt{c}}$

57. Visibility. The distance d in miles a person in an airplane can see to the horizon on a clear day is given by the formula $d = 1.22a^{1/2}$, where a is the altitude of the plane in feet. Find d.

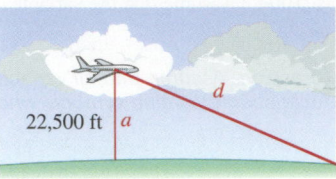

22,500 ft $\quad a$

58. Substitute the x- and y-coordinates of each point labeled in the graph into the equation

$$x^{2/3} + y^{2/3} = 32$$

Show that each one satisfies the equation.

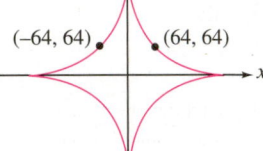

$(-64, 64)$ $(64, 64)$

SECTION 7.3 ▶ **Simplifying and Combining Radical Expressions**

DEFINITIONS AND CONCEPTS	EXAMPLES
Product rule for radicals: $$\sqrt[n]{ab} = \sqrt[n]{a}\sqrt[n]{b}$$ The product rule for radicals can be used to simplify radical expressions.	Simplify: $\sqrt{98} = \sqrt{49 \cdot 2}$ Write 98 as the product of its greatest perfect-square factor and one other factor. $\quad = \sqrt{49}\sqrt{2}$ The square root of a product is equal to the product of the square roots. $\quad = 7\sqrt{2}$ Evaluate $\sqrt{49}$.
Simplified form of a radical: 1. Except for 1, the radicand has no perfect-square factors. 2. No fraction appears in the radicand. 3. No radical appears in the denominator. Review the lists of perfect squares, cubes, fourth powers, and fifth powers on pages 607 and 608.	Simplify: $\sqrt[3]{16x^4} = \sqrt[3]{8x^3 \cdot 2x}$ Write $16x^4$ as the product of its greatest perfect-cube factor and one other factor. $\quad = \sqrt[3]{8x^3}\sqrt[3]{2x}$ The cube root of a product is equal to the product of the cube roots. $\quad = 2x\sqrt[3]{2x}$ Simplify $\sqrt[3]{8x^3}$.

Quotient rule for radicals: $$\sqrt[n]{\frac{a}{b}} = \frac{\sqrt[n]{a}}{\sqrt[n]{b}}$$	Simplify: $$\sqrt{\frac{10}{25x^4}} = \frac{\sqrt{10}}{\sqrt{25x^4}} = \frac{\sqrt{10}}{5x^2}$$ Simplify: $$\sqrt[3]{\frac{16y^3}{125a^3}} = \frac{\sqrt[3]{16y^3}}{\sqrt[3]{125a^3}} = \frac{\sqrt[3]{8y^3}\sqrt[3]{2}}{5a} = \frac{2y\sqrt[3]{2}}{5a}$$
Radical expressions with the same index and radicand are called **like radicals.** Like radicals can be combined by addition and subtraction. To **combine like radicals,** we use the distributive property in reverse.	Add: $3\sqrt{6} + 5\sqrt{6} = (3 + 5)\sqrt{6} = 8\sqrt{6}$ Subtract: $8\sqrt[4]{2y} - 9\sqrt[4]{2y} = (8 - 9)\sqrt[4]{2y} = -\sqrt[4]{2y}$
If a sum or difference involves unlike radicals, make sure that each one is written in simplified form. After doing so, like radicals may result that can be combined.	Simplify: $$\begin{aligned}\sqrt[3]{54a^4} - \sqrt[3]{16a^4} &= \sqrt[3]{27a^3 \cdot 2a} - \sqrt[3]{8a^3 \cdot 2a} \\ &= \sqrt[3]{27a^3}\sqrt[3]{2a} - \sqrt[3]{8a^3}\sqrt[3]{2a} \\ &= 3a\sqrt[3]{2a} - 2a\sqrt[3]{2a} \\ &= (3a - 2a)\sqrt[3]{2a} \\ &= a\sqrt[3]{2a}\end{aligned}$$

REVIEW EXERCISES

Simplify each expression. All variables represent positive real numbers.

59. $\sqrt{80}$ **60.** $\sqrt[3]{54}$

61. $\sqrt[4]{160}$ **62.** $\sqrt[5]{-96}$

63. $\sqrt{8x^5}$ **64.** $\sqrt[4]{r^{17}}$

65. $\sqrt[3]{-27j^7k}$ **66.** $\sqrt[3]{-16x^5y^4}$

67. $\sqrt{\dfrac{m}{144n^{12}}}$ **68.** $\sqrt{\dfrac{17xy}{64a^4}}$

69. $\dfrac{\sqrt[5]{64x^8}}{\sqrt[5]{2x^3}}$ **70.** $\dfrac{\sqrt[5]{243x^{16}}}{\sqrt[5]{x}}$

Simplify and combine like radicals. All variables represent positive real numbers.

71. $\sqrt{2} + 2\sqrt{2}$ **72.** $6\sqrt{20} - \sqrt{5}$

73. $2\sqrt[3]{3} - \sqrt[3]{24}$ **74.** $-\sqrt[4]{32a^5} - 2\sqrt[4]{162a^5}$

75. $2x\sqrt{8} + 2\sqrt{200x^2} + \sqrt{50x^2}$ **76.** $\sqrt[3]{54x^3} - 3\sqrt[3]{16x^3} + 4\sqrt[3]{128x^3}$

77. $2\sqrt[4]{32t^3} - 8\sqrt[4]{6t^3} + 5\sqrt[4]{2t^3}$ **78.** $10\sqrt[4]{16x^9} - 8x^2\sqrt[4]{x} + 5\sqrt[4]{x^5}$

79. Explain the error in each simplification.

  a. $2\sqrt{5x} + 3\sqrt{5x} = 5\sqrt{10x}$

  b. $30 + 30\sqrt[4]{2} = 60\sqrt[4]{2}$

  c. $7\sqrt[3]{y^2} - 5\sqrt[3]{y^2} = 2$

  d. $6\sqrt{11ab} - 3\sqrt{5ab} = 3\sqrt{6ab}$

80. **Sewing.** A corner of fabric is folded over to form a collar and stitched down as shown. From the dimensions given in the figure, determine the exact number of inches of stitching that must be made. Then give an approximation to one decimal place. (All measurements are in inches.)

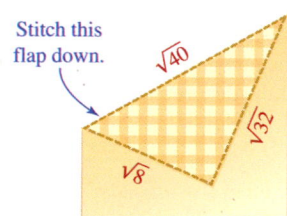

Stitch this flap down. $\sqrt{40}$ $\sqrt{32}$ $\sqrt{8}$

SECTION 7.4 ▶ **Multiplying and Dividing Radical Expressions**

DEFINITIONS AND CONCEPTS	EXAMPLES
We can use the product rule for radicals to **multiply radical expressions** that have the same index: $$\sqrt[n]{a}\,\sqrt[n]{b} = \sqrt[n]{ab}$$ provided $\sqrt[n]{a}$ and $\sqrt[n]{b}$ are real numbers.	Multiply and then simplify, if possible: $$\sqrt{6}\sqrt{8} = \sqrt{6\cdot 8} = \sqrt{48} = \sqrt{16\cdot 3} = \sqrt{16}\sqrt{3} = 4\sqrt{3}$$ $$\sqrt[3]{9x^4}\,\sqrt[3]{3x^2} = \sqrt[3]{9x^4\cdot 3x^2} = \sqrt[3]{27x^6} = 3x^2$$
We can use the **distributive property** to multiply a radical expression with two or more terms by a radical expression with one term.	Multiply and then simplify, if possible: $$2\sqrt{3}\left(4\sqrt{5} - 5\sqrt{2}\right) = 2\sqrt{3}\cdot 4\sqrt{5} - 2\sqrt{3}\cdot 5\sqrt{2}$$ $$= 2\cdot 4\sqrt{3\cdot 5} - 2\cdot 5\sqrt{3\cdot 2}$$ $$= 8\sqrt{15} - 10\sqrt{6}$$
We can use the **FOIL method** to multiply a radical expression with two terms by another radical expression with two terms.	Multiply and then simplify, if possible: $$\left(\sqrt[3]{x} - \sqrt[3]{3}\right)\left(\sqrt[3]{x} + \sqrt[3]{9}\right) = \sqrt[3]{x}\sqrt[3]{x} + \sqrt[3]{x}\sqrt[3]{9} - \sqrt[3]{3}\sqrt[3]{x} - \sqrt[3]{3}\sqrt[3]{9}$$ $$= \sqrt[3]{x^2} + \sqrt[3]{9x} - \sqrt[3]{3x} - \sqrt[3]{27}$$ $$= \sqrt[3]{x^2} + \sqrt[3]{9x} - \sqrt[3]{3x} - 3$$
If a radical appears in a denominator of a fraction, or if a radicand contains a fraction, we can write the radical in simplest form by **rationalizing the denominator.** To rationalize a denominator, we multiply the given expression by a carefully chosen form of 1.	Rationalize the denominator: $$\sqrt{\frac{10}{3}} = \frac{\sqrt{10}}{\sqrt{3}}\cdot\frac{\sqrt{3}}{\sqrt{3}} \qquad \sqrt[3]{\frac{5}{2p^2}} = \frac{\sqrt[3]{5}}{\sqrt[3]{2p^2}}\cdot\frac{\sqrt[3]{4p}}{\sqrt[3]{4p}}$$ $$= \frac{\sqrt{30}}{3} \qquad\qquad\qquad = \frac{\sqrt[3]{20p}}{\sqrt[3]{8p^3}}$$ $$\qquad\qquad\qquad\qquad = \frac{\sqrt[3]{20p}}{2p}$$
Radical expressions that involve the sum and difference of the same two terms are called **conjugates.**	Conjugates: $\sqrt{2x} + 3$ and $\sqrt{2x} - 3$
To **rationalize a two-termed denominator** of a fraction, multiply the numerator and the denominator by the **conjugate** of the denominator.	Rationalize the denominator: $$\frac{\sqrt{x} - 2}{\sqrt{x} + 2} = \frac{\sqrt{x} - 2}{\sqrt{x} + 2}\cdot\frac{\sqrt{x} - 2}{\sqrt{x} - 2}$$ $$= \frac{\sqrt{x}\sqrt{x} - 2\sqrt{x} - 2\sqrt{x} + 4}{\sqrt{x}\sqrt{x} - 2\sqrt{x} + 2\sqrt{x} - 4}$$ $$= \frac{x - 4\sqrt{x} + 4}{x - 4}$$

REVIEW EXERCISES

Simplify each expression. All variables represent positive real numbers.

81. $\sqrt{7}\sqrt{7}$

82. $\left(2\sqrt{5}\right)\left(3\sqrt{2}\right)$

83. $\left(-2\sqrt{8}\right)^2$

84. $2\sqrt{6}\sqrt{15}$

85. $\sqrt{9x}\sqrt{x}$

86. $\left(\sqrt[3]{x+1}\right)^3$

87. $-\sqrt[3]{2x^2}\,\sqrt[3]{4x^8}$

88. $\sqrt[5]{9}\cdot\sqrt[5]{27}$

89. $3\sqrt{7t}\left(2\sqrt{7t}+3\sqrt{3t^2}\right)$

90. $-\sqrt[4]{4x^5y^{11}}\,\sqrt[4]{8x^9y^3}$

91. $\left(\sqrt{3b}+\sqrt{3}\right)^2$

92. $\left(\sqrt[3]{3p}-2\sqrt[3]{2}\right)\left(\sqrt[3]{3p}+\sqrt[3]{2}\right)$

Rationalize each denominator. All variables represent positive real numbers.

93. $\dfrac{10}{\sqrt{3}}$

94. $\sqrt{\dfrac{3}{5xy}}$

95. $\dfrac{\sqrt[3]{6u}}{\sqrt[3]{u^5}}$

96. $\dfrac{\sqrt[4]{a}}{\sqrt[4]{3b^2}}$

97. $\dfrac{2}{\sqrt{2}-1}$

98. $\dfrac{4\sqrt{x}-2\sqrt{z}}{\sqrt{z}+4\sqrt{x}}$

99. Rationalize the numerator: $\dfrac{\sqrt{a}-\sqrt{b}}{\sqrt{a}}$

100. Volume. The formula relating the radius r of a sphere and its volume V is $r=\sqrt[3]{\dfrac{3V}{4\pi}}$. Write the radical in simplest form.

SECTION 7.5 ▶ Solving Radical Equations

DEFINITIONS AND CONCEPTS	EXAMPLES

We can use the **power rule** to solve radical equations.

If $x=y$, then $x^n=y^n$.

To **solve equations containing radicals:**

1. Isolate one radical expression on one side of the equation.
2. Raise both sides of the equation to the power that is the same as the index.
3. If it still contains a radical, go back to step 1. If it does not contain a radical, solve the resulting equation.
4. Check the proposed solutions in the original equation to eliminate **extraneous** solutions.

Solve each equation.

$$\sqrt{2x-2}+1=x$$
$$\sqrt{2x-2}=x-1$$
$$\left(\sqrt{2x-2}\right)^2=(x-1)^2$$
$$2x-2=x^2-2x+1$$
$$0=x^2-4x+3$$
$$0=(x-3)(x-1)$$
$$x-3=0 \quad\text{or}\quad x-1=0$$
$$x=3 \quad\quad\quad x=1$$

The solutions are 3 and 1. Verify that each satisfies the original equation.

$$\sqrt[3]{x+2}=3$$
$$\left(\sqrt[3]{x+2}\right)^3=3^3$$
$$x+2=27$$
$$x=25$$

The solution is 25. Verify that it satisfies the original equation.

When **more than one radical** appears in an equation, we often must use the power rule more than once to solve the equation.

The **special-product rules** are often helpful when squaring expressions containing two terms, such as $\left(5-\sqrt{x}\right)^2$.

Warning: The result is not $25+x$.

Solve:

$$\sqrt{x}+\sqrt{x+5}=5$$
$$\sqrt{x+5}=5-\sqrt{x} \qquad \text{To isolate one radical, subtract } \sqrt{x} \text{ from both sides.}$$
$$\left(\sqrt{x+5}\right)^2=\left(5-\sqrt{x}\right)^2 \qquad \text{To eliminate one radical, square both sides.}$$
$$x+5=25-10\sqrt{x}+x \qquad \text{Perform the operations on each side.}$$
$$-20=-10\sqrt{x} \qquad \text{To isolate the radical term, subtract 25 and x from both sides.}$$
$$2=\sqrt{x} \qquad \text{To isolate the radical, divide both sides by } -10.$$
$$(2)^2=\left(\sqrt{x}\right)^2 \qquad \text{To eliminate the radical, square both sides again.}$$
$$4=x$$

The solution is 4. Verify that it satisfies the original equation.

REVIEW EXERCISES

Solve each equation. Write all proposed solutions. Cross out those that are extraneous.

101. $\sqrt{7x - 10} - 1 = 11$ **102.** $u = \sqrt{25u - 144}$

103. $2\sqrt{y - 3} = \sqrt{2y + 1}$ **104.** $\sqrt{z + 1} + \sqrt{z} = 2$

105. $\sqrt[3]{x^3 + 56} - 2 = x$ **106.** $a = \sqrt{a^2 + 5a - 35}$

107. $(x + 2)^{1/2} - (4 - x)^{1/2} = 0$ **108.** $\sqrt{b^2 + b} = \sqrt{3 - b^2}$

109. $\sqrt[4]{8x - 8} + 2 = 0$

110. $\sqrt{2m + 4} - \sqrt{m + 3} = 1$

111. Let $f(x) = \sqrt{5x + 1}$ and $g(x) = x + 1$. For what values of x is $f(x) = g(x)$?

112. Let $f(x) = \sqrt{2x^2 - 7x}$. For what value(s) of x is $f(x) = 2$?

Solve each equation for the specified variable.

113. $r = \sqrt{\dfrac{A}{P} - 1}$ for P **114.** $h = \sqrt[3]{\dfrac{12I}{b}}$ for I

SECTION 7.6 ▶ **Geometric Applications of Radicals**

DEFINITIONS AND CONCEPTS	EXAMPLES
The Pythagorean theorem: If a and b are the lengths of the **legs** of a right triangle and c is the length of the **hypotenuse,** then $a^2 + b^2 = c^2$. 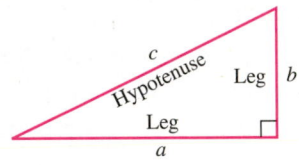	Find the length of the third side of the right triangle. $a^2 + b^2 = c^2$ This is the Pythagorean equation. $6^2 + b^2 = 10^2$ Substitute 6 for a and 10 for c. $36 + b^2 = 100$ $b^2 = 64$ To isolate b^2, subtract 36 from both sides. $b = \sqrt{64}$ Since b must be positive, find the positive square root of 64. $b = 8$ The length of the third side of the triangle is 8 ft.
In an **isosceles right triangle,** the length of the hypotenuse is $\sqrt{2}$ times the length of one leg.	If the length of one leg of an isosceles right triangle is 7 feet, the length of the hypotenuse is $7\sqrt{2}$ feet. 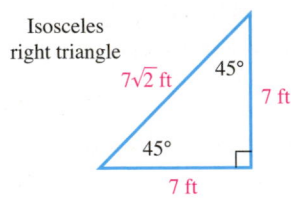
The hypotenuse of a **30°–60°–90° triangle** is twice as long as the shorter leg (the leg opposite the 30° angle.) The length of the longer leg (the leg opposite the 60° angle) is $\sqrt{3}$ times the length of the shorter leg.	If the shorter leg of a 30°–60°–90° triangle is 9 inches long: 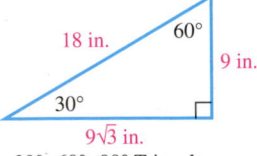 ■ The length of the hypotenuse is $2 \cdot 9 = 18$ inches. ■ The length of the longer leg is $\sqrt{3} \cdot 9 = 9\sqrt{3}$ inches.
The distance d between two points with coordinates (x_1, y_1) and (x_2, y_2) is given by **the distance formula:** $d = \sqrt{(x_2 - x_1)^2 + (y_2 - y_1)^2}$	The distance between points $(-2, 3)$ and $(1, 7)$ is: $d = \sqrt{(x_2 - x_1)^2 + (y_2 - y_1)^2}$ $= \sqrt{[1 - (-2)]^2 + (7 - 3)^2}$ $= \sqrt{3^2 + 4^2}$ $= \sqrt{9 + 16}$ $= \sqrt{25}$ $= 5$

REVIEW EXERCISES

115. Carpentry. The gable end of the roof shown below is divided in half by a vertical brace, 8 feet in height. Find the length of the roof line.

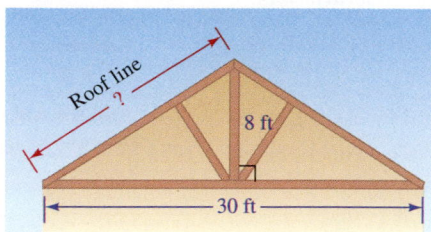

116. Sailing. A technique called *tacking* allows a sailboat to make progress into the wind. A sailboat follows the course shown below. Find *d*, the distance the boat advances into the wind after tacking.

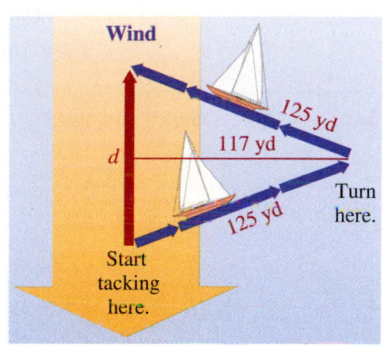

For problems 117–122, give the exact answer and then an approximation to two decimal places, when appropriate.

117. Find the length of the hypotenuse of an isosceles right triangle if the length of one leg is 7 meters.

118. The length of the hypotenuse of an isosceles right triangle is 15 yards. Find the length of one leg of the triangle.

119. The length of the hypotenuse of a 30°–60°–90° triangle is 12 centimeters. Find the length of each leg.

120. In a 30°–60°–90° triangle, the length of the longer leg is 60 feet. Find the length of the hypotenuse and the length of the shorter leg.

121. Find *x* and *y*.   **122.** Find *x* and *y*.

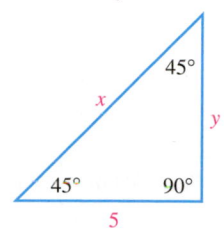   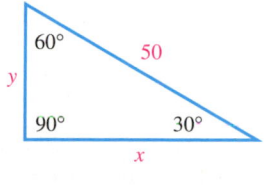

Find the distance between the points.

123. $(1, 3)$ and $(6, -9)$ **124.** $(-4, 6)$ and $(-2, 8)$

SECTION 7.7 ▶ Complex Numbers

DEFINITIONS AND CONCEPTS	EXAMPLES
The **imaginary number** i is defined as $$i = \sqrt{-1}$$ From the definition, it follows that $i^2 = -1$.	Write each expression in terms of i: $$\sqrt{-81} = \sqrt{-1 \cdot 81} \qquad \sqrt{-24} = \sqrt{-1 \cdot 24}$$ $$= \sqrt{-1}\sqrt{81} \qquad\qquad = \sqrt{-1}\sqrt{24}$$ $$= i \cdot 9 \qquad\qquad\qquad = i\sqrt{4}\sqrt{6}$$ $$= 9i \qquad\qquad\qquad = 2i\sqrt{6} \ \text{ or } \ 2\sqrt{6}i$$
A **complex number** is any number that can be written in the form $a + bi$, where a and b are real numbers and $i = \sqrt{-1}$. We call a the **real part** and b the **imaginary part**.	Complex numbers: $5 + 3i$ 5 is the real part *a* and 3 is the imaginary part *b*. $16 = 16 + 0i$ 16 is the real part *a* and 0 is the imaginary part *b*. $9i = 0 + 9i$ 0 is the real part *a* and 9 is the imaginary part *b*.

Adding and subtracting complex numbers is similar to adding and subtracting polynomials. To **add two complex numbers,** add their real parts and add their imaginary parts.

To **subtract two complex numbers,** add the opposite of the complex number being subtracted.

Add. Write the answer in the form $a + bi$.

$$(7 - 5i) + (3 + 9i) = (7 + 3) + (-5 + 9)i$$

Add the real parts. Add the imaginary parts.

$$= 10 + 4i$$

Subtract. Write the answer in the form $a + bi$.

$$(8 - i) - (-1 + 6i) = (8 - i) + (1 - 6i)$$

Add the opposite of $-1 + 6i$.

$$= (8 + 1) + [-1 + (-6)]i$$

Add the real parts. Add the imaginary parts.

$$= 9 - 7i$$

Multiplying complex numbers is similar to multiplying polynomials.

Multiply. Write the answers in the form $a + bi$.

$$3i(6 - 4i) = 18i - 12i^2$$
$$= 18i - 12(-1)$$
$$= 18i + 12$$
$$= 12 + 18i$$

$$(4 + 7i)(2 - i) = 8 - 4i + 14i - 7i^2$$
$$= 8 + 10i - 7(-1)$$
$$= 8 + 10i + 7$$
$$= 15 + 10i$$

The complex numbers $a + bi$ and $a - bi$ are called **complex conjugates.**

The complex numbers $7 - 2i$ and $7 + 2i$ are complex conjugates.

To **divide complex numbers,** multiply the numerator and denominator by the complex conjugate of the denominator. The process is similar to rationalizing denominators.

Divide. Write the answers in the form $a + bi$.

$$\frac{3}{1 + i} \cdot \frac{1 - i}{1 - i} = \frac{3(1 - i)}{1 - i + i - i^2}$$
$$= \frac{3(1 - i)}{1 - (-1)}$$
$$= \frac{3 - 3i}{2}$$
$$= \frac{3}{2} - \frac{3}{2}i$$

$$\frac{6 + i}{2 - i} \cdot \frac{2 + i}{2 + i} = \frac{12 + 6i + 2i + i^2}{4 + 2i - 2i - i^2}$$
$$= \frac{12 + 8i + (-1)}{4 - (-1)}$$
$$= \frac{11 + 8i}{5}$$
$$= \frac{11}{5} + \frac{8}{5}i$$

The **powers of i** cycle through four possible outcomes: i, -1, $-i$, and 1.

$$i^1 = i = i^5 = i^9 = \dots \quad R = 1$$
$$i^2 = -1 = i^6 = i^{10} = \dots \quad R = 2$$
$$i^3 = -i = i^7 = i^{11} = \dots \quad R = 3$$
$$i^4 = 1 = i^8 = i^{12} = \dots \quad R = 0$$

Simplify: i^{66}

We divide 66 by 4 to get a remainder of **2.** The remainder determines the power to which i is raised in the simplified form.

Thus, $i^{66} = i^2 = -1$.

$$\begin{array}{r} 16\ R2 \\ 4\overline{)66} \\ -4 \\ \hline 26 \\ -24 \\ \hline 2 \end{array}$$

Write each expression in terms of i.

125. $\sqrt{-25}$

126. $\sqrt{-18}$

127. $-\sqrt{-6}$

128. $\sqrt{-\dfrac{9}{64}}$

129. Complete the diagram.

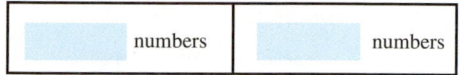

Complex numbers

___ numbers	___ numbers

130. Determine whether each statement is true or false.

a. Every real number is a complex number.

b. $3 - 4i$ is a complex number.

c. $\sqrt{-4}$ is a real number.

d. i is a real number.

Give the complex conjugate of each number.

131. a. $3 + 6i$

b. $19i$

132. a. $-1 - 7i$

b. $-i$

Perform the operations. Write all answers in the form a + bi.

133. $(3 + 4i) + (5 - 6i)$

134. $\left(7 - \sqrt{-9}\right) - \left(4 + \sqrt{-4}\right)$

135. $3i(2 - i)$

136. $(2 - 7i)(-3 + 4i)$

137. $\sqrt{-3} \cdot \sqrt{-9}$

138. $(9i)^2$

139. $\dfrac{5 + 14i}{2 + 3i}$

140. $\dfrac{3}{11i}$

Simplify each expression.

141. i^{42}

142. i^{97}

---

7 ▸ Chapter Test

1. Fill in the blanks.

a. The symbol $\sqrt{}$ is called a _____ symbol.

b. The _____ number i is defined as $i = \sqrt{-1}$.

c. Squaring both sides of an equation can introduce _____ solutions.

d. An _____ right triangle is a right triangle with two legs of equal length.

e. To _____ the denominator of $\dfrac{4}{\sqrt{5}}$, we multiply the fraction by $\dfrac{\sqrt{5}}{\sqrt{5}}$.

f. A _____ number is any number that can be written in the form $a + bi$, where a and b are real numbers and $i = \sqrt{-1}$.

2. a. State the product rule for radicals.

b. State the quotient rule for radicals.

c. Explain why $\sqrt[4]{-16}$ is not a real number.

3. Graph $f(x) = \sqrt{x - 1}$. Find the domain and range of the function.

4. **Diving.** Refer to the illustration in the next column. The velocity v of an object in feet per second after it has fallen a distance of d feet is approximated by the function $v(d) = \sqrt{64.4d}$. Olympic diving platforms are 10 meters tall (approximately 32.8 feet). Estimate the velocity at which a diver hits the water from this height. Round to the nearest foot per second.

32.8 feet

5. Use the graph of function f below to find each of the following.

a. $f(-1)$

b. $f(8)$

c. The value of x for which $f(x) = 1$

d. The domain and range of f

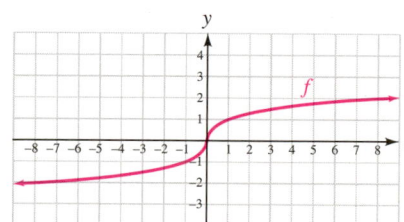

6. Find the domain of $f(x) = \sqrt{10x + 50}$.

Simplify each expression. All variables represent positive real numbers. Write answers without using negative exponents.

7. $(49x^4)^{1/2}$

8. $-27^{2/3}$

9. $36^{-3/2}$

10. $\left(-\dfrac{8}{125n^6}\right)^{-2/3}$

11. $\dfrac{2^{5/3}2^{1/6}}{2^{1/2}}$

12. $(a^{2/3})^{1/6}$

Simplify each expression. The variables are unrestricted.

13. $\sqrt{x^2}$

14. $\sqrt{y^2-10y+25}$

Simplify each expression. All variables represent positive real numbers.

15. $\sqrt[3]{-64x^3y^6}$

16. $\sqrt{\dfrac{4a^2}{9}}$

17. $\sqrt[5]{(t+8)^5}$

18. $\sqrt{540x^3y^5}$

19. $\dfrac{\sqrt[3]{24x^{15}y^4}}{\sqrt[3]{y}}$

20. $\sqrt[4]{32}$

Perform the operations and simplify. All variables represent positive real numbers.

21. $2\sqrt{48y^5}-3y\sqrt{12y^3}$

22. $2\sqrt[3]{40}-\sqrt[3]{5,000}+4\sqrt[3]{625}$

23. $\sqrt[4]{243z^{13}}+z\sqrt[4]{48z^9}$

24. $-2\sqrt{xy}\left(3\sqrt{x}+\sqrt{xy^3}\right)$

25. $\left(3\sqrt{2}+\sqrt{3}\right)\left(2\sqrt{2}-3\sqrt{3}\right)$

26. $\left(\sqrt[3]{2a}+9\right)^2$

27. $\dfrac{8}{\sqrt{10}}$

28. $\dfrac{\sqrt{x}+\sqrt{y}}{\sqrt{x}-\sqrt{y}}$

29. $\sqrt[3]{\dfrac{9}{4a}}$

30. Rationalize the numerator: $\dfrac{\sqrt{5}+3}{-4\sqrt{2}}$

Solve each equation. Write all proposed solutions. Cross out those that are extraneous.

31. $4\sqrt{x}=\sqrt{x+1}$

32. $\sqrt[3]{6n+4}-4=0$

33. $1=\sqrt{u-3}+\sqrt{u}$

34. $(2m^2-9)^{1/2}=m$

35. $\sqrt{t-2}-t+2=0$

36. $\sqrt{x-8}+10=0$

37. Let $f(x)=\sqrt[4]{15-x}$ and $g(x)=\sqrt[4]{13-2x}$. Find all values of x for which $f(x)=g(x)$.

38. Solve $r=\sqrt[3]{\dfrac{GMt^2}{4\pi^2}}$ for G.

Find the missing side lengths in each triangle. Give the exact answer and then an approximation to two decimal places, when appropriate.

39.

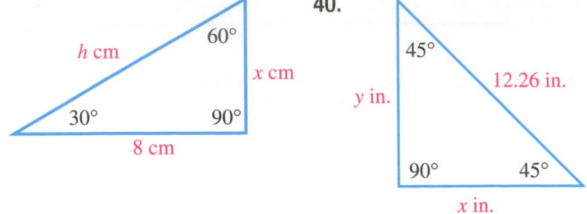

40.

41. Find the distance between $(-2,5)$ and $(22,12)$.

42. **Shipping Crates.**   The diagonal brace on the shipping crate in the illustration is 53 inches. Find the height h of the crate.

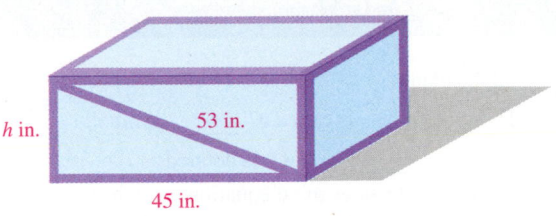

43. Express $\sqrt{-45}$ in terms of i.

44. Simplify: i^{106}

Perform the operations. Write all answers in the form **a + bi.**

45. $(9+4i)+(-13+7i)$

46. $\left(3-\sqrt{-9}\right)-\left(-1+\sqrt{-16}\right)$

47. $15i(3-5i)$

48. $(8+10i)(-7-i)$

49. $\dfrac{1}{i\sqrt{2}}$

50. $\dfrac{2+i}{3-i}$

Group Project

A Spiral of Roots

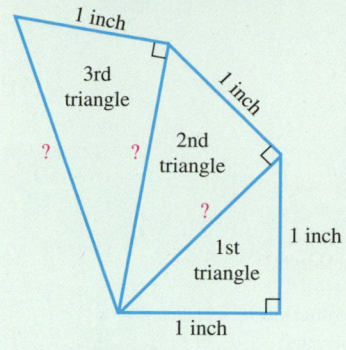

Overview: In this activity, you will create a visual representation of a collection of square roots.

Instructions: Form groups of 2 or 3 students. You will need a piece of unlined paper, a protractor, a ruler, and a pencil. Begin by drawing an isosceles right triangle with legs of length 1 inch in the middle of the paper. (See the illustration.) Use the Pythagorean theorem to determine the length of the hypotenuse. Draw a second right triangle using the hypotenuse of the first right triangle as one leg. Draw its second leg with length 1 inch. Find the length of the hypotenuse of the second triangle.

Continue creating right triangles, using the previous hypotenuse as one leg and drawing a new second leg of length 1 inch each time. Calculate the length of each resulting hypotenuse. When the figure begins to spiral onto itself, you may stop the process. Make a list of the lengths of each hypotenuse. What pattern do you see?

Graphing in Three Dimensions

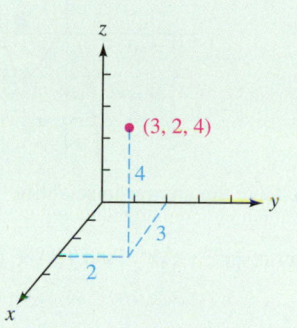

Overview: In this activity, you will find the distance between two points that lie in three-dimensional space.

Instructions: Form groups of 2 or 3 students. In a three-dimensional Cartesian coordinate system, the positive x-axis is horizontal and pointing toward the viewer (out of the page), the positive y-axis is also horizontal and pointing to the right, and the positive z-axis is vertical, pointing up. A point is located by plotting an ordered triple of numbers (x, y, z). In the illustration, the point $(3, 2, 4)$ is plotted.

In three dimensions, the distance formula is

$$d = \sqrt{(x_2 - x_1)^2 + (y_2 - y_1)^2 + (z_2 - z_1)^2}$$

1. Copy the illustration shown. Then plot the point $(1, 4, 3)$. Use the distance formula to find the distance between these two points.
2. Draw another three-dimensional coordinate system and plot the points $(-3, 3, -4)$ and $(2, -3, 2)$. Use the distance formula to find the distance between these two points.

CUMULATIVE REVIEW ▶▶ Chapters 1–7

1. Which of the following are natural numbers, whole numbers, integers, rational numbers, irrational numbers, and real numbers? [Section 1.2]

$$\left\{-\pi, \ -5, \ 3.4, \ \sqrt{19}, \ 1, \ \frac{16}{5}, \ 9.\overline{7}\right\}$$

2. Evaluate $\dfrac{|Ax_0 + By_0 + C|}{\sqrt{A^2 + B^2}}$ for $A = 6$, $B = -8$, $C = -5$, and $x_0 = y_0 = -2$. [Section 1.3]

3. Solve: $\dfrac{2}{3}(b + 3) = \dfrac{5}{4}b + \dfrac{17}{12}$ [Section 1.5]

4. Solve $S = \dfrac{a - \ell r}{1 - r}$ for ℓ. [Section 1.6]

5. **Electronics.** The illustration shows a closed circuit with two voltage sources and three resistors. The sum of the voltages in the loop must be 0. Find x. [Section 1.7]

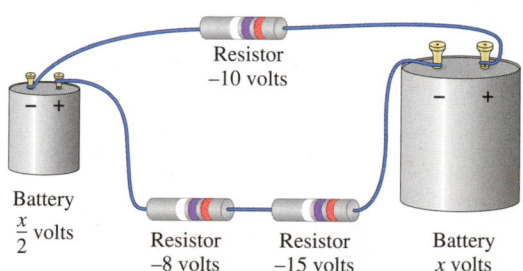

6. **Salad Dressing.** A caterer is going to combine 10% vinegar-oil dressing with 18% vinegar-oil dressing to make 10 cups of a 15% vinegar-oil dressing. How many cups of each type of dressing will she need? [Section 1.8]

7. Find the slope of the line that passes through $(3, -1)$ and $(-6, 2)$. [Section 2.2]

8. Graph: $x = 3$ [Section 2.2]

9. Refer to the illustration.
   [Section 2.4]

   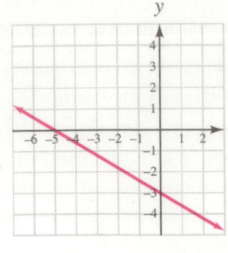

   a. What is the slope of the line?

   b. What is the y-intercept of the line?

   c. What is the equation of the line?

10. Refer to the illustration.
    [Section 2.5]

    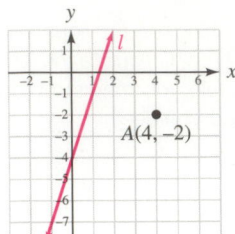

    a. What is the slope of line l?

    b. Write the equation of a linear function whose graph passes through point A and is perpendicular to line l.

11. Fill in the blanks: The graph of $g(x) = (x + 4)^2$ is the same as the graph of $f(x) = x^2$ except that it is shifted ▢ units to the ____. [Section 2.6]

12. Determine whether the relation $\{(-1, 6), (4, 7), (5, -3), (2, -3), (6, 1)\}$ defines a function. [Section 2.5]

13. Determine the domain and range of the function $f(x) = -|x| - 2$, which is graphed on the right. (The x- and y-axes are scaled in units of 1.) [Section 2.6]

    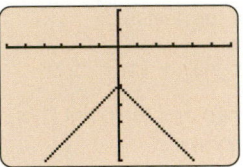

14. Graph: $g(x) = 1 + x^3$ [Section 2.6]

15. Solve by graphing: $\begin{cases} 3x - y = -3 \\ y = -2x - 7 \end{cases}$ [Section 3.1]

16. Solve: $\begin{cases} x = \frac{3}{2}y + 5 \\ 2x - 3y = 8 \end{cases}$ [Section 3.2]

17. Evaluate: $\begin{vmatrix} -9 & 7 \\ 4 & -2 \end{vmatrix}$ [Section 3.5]

18. **Budgets.** See the family budget guidelines below. It is recommended that housing, utilities, and transportation costs should be 60% of the budget; and food, clothing, and savings should be 25% of the budget. Find x, y, and z. (*Hint:* The sum of the percents from all categories in a circle graph is what number?) [Section 3.7]

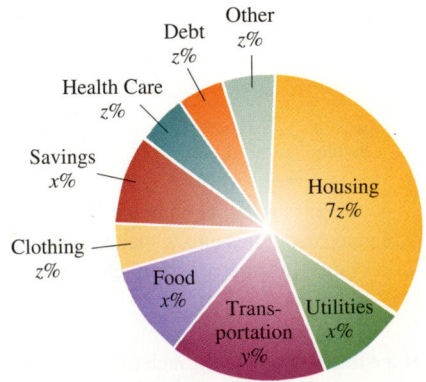

Solve each inequality or compound inequality and graph the solution set. Then express the solution set using interval notation.

19. $5(x + 1) \leq 4(x + 3)$ and $x + 12 < -3$ [Section 4.2]

20. $|-1 - 2x| > 5$ [Section 4.3]

21. Graph: $\begin{cases} x + 2y < 3 \\ 2x + 4y < 8 \end{cases}$ [Section 4.5]

22. Simplify: $\left(\dfrac{4a^{-2}b}{3ab^{-3}} \right)^3$ [Section 5.1]

23. The graph of a polynomial function f is shown on the right. Find each of the following. [Section 5.1]

    a. $f(0)$

    b. $f(4)$

    c. The values of x for which $f(x) = -4$

    d. The domain and range of f

    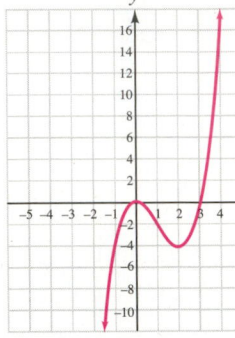

24. Find $(6.1 \times 10^8)(3.9 \times 10^5)$. Give the answer in scientific notation. [Section 5.2]

25. Let $f(x) = \frac{1}{5}x^6 + \frac{3}{4}x^2$ and $g(x) = \frac{2}{3}x^6 - \frac{1}{12}x^2$. Find $f(x) + g(x)$. [Section 5.3]

26. Subtract:
    $(-2x^2y^3 + 6xy + 5y^2) - (-4x^2y^3 - 7xy + 2y^2)$
    [Section 5.3]

27. Multiply: $(3y + 1)(2y^2 + 3y + 2)$
    [Section 5.4]

28. Simplify: $(x + 3)(x - 3) + (2x - 1)(x + 2)$
    [Section 5.4]

29. Expand: $(2y^5 + 5z)^2$ [Section 5.4]

30. Expand: $(5s - t^3)^3$ [Section 5.4]

Factor each polynomial completely.

31. $3c - cd + 3d - c^2$ [Section 5.5]

32. $x^3 - 8y^3$ [Section 5.6]

33. $(a + b)^2 - 2(a + b) + 1$ [Section 5.7]

34. $x^4 - 17x^2 + 16$ [Section 5.8]

Solve each equation and check the result.

35. $2z^3 - 200z = 0$ [Section 5.9]

36. $3m^2 + 10m = -3$ [Section 5.9]

37. **Framing Posters.** A cardboard mat of uniform width was used to frame a 12 inch by 16 inch poster. If 128 square inches of matting was used, what is the width of the border?
    [Section 5.9]

38. Simplify: $\dfrac{3x^2 - 10xy - 8y^2}{4y^2 - xy}$ [Section 6.1]

39. For what values of x is $\dfrac{2}{x^2 - x - 56}$ undefined?
[Section 6.1]

40. Graph: $f(x) = \dfrac{1}{x}$. Give the domain and range. [Section 6.1]

41. Clean Up.   Suppose the cost c (in dollars) of removing $p\%$ of the pollution on a beach is given by the rational function $c(p) = \dfrac{60{,}000p}{100 - p}$. Find the cost to remove 98% of the pollution. [Section 6.1]

42. Let $f(t) = 10t - t^2$ and $g(t) = \dfrac{t^4 - 10t^3 - 2t^2 + 20t}{t^6}$. Find $f(t) \div g(t)$. [Section 6.2]

Perform the operations.

43. $(2x^2 - 9x - 5) \cdot \dfrac{x}{2x^2 + x}$ [Section 6.2]

44. $\dfrac{2x}{x^2 - 4} - \dfrac{1}{x^2 - 3x + 2} + \dfrac{x + 1}{x^2 + x - 2}$
[Section 6.3]

45. Divide $15a^3 - 29a^2 + 16$ by $3a - 4$. [Section 6.5]

46. Solve: $\dfrac{2}{5x - 5} + \dfrac{x - 2}{15} = \dfrac{4}{5x - 5}$ [Section 6.7]

47. Shared Work.   One pipe fills a tank in 4 hours, and another fills it in 6 hours. How long will it take to fill the tank using both pipes? [Section 6.8]

48. Assume that r varies inversely with s. If $r = 40$ when $s = 10$, find r when $s = 15$. [Section 6.9]

49. Graph: $f(x) = \sqrt{x}$. Give the domain and range. [Section 7.1]

50. Let $f(x) = \sqrt[3]{-2x - 4}$. Find $f(30)$. [Section 7.1]

Simplify each expression. All variables represent positive real numbers.

51. $\left(-\dfrac{8x^3}{27} \right)^{-1/3}$ [Section 7.2]

52. $\sqrt{200x^4y^3z}$ [Section 7.3]

53. $\sqrt[3]{16} + \sqrt[3]{128}$ [Section 7.3]

54. $\left(\sqrt{5z} + \sqrt{3} \right)\left(\sqrt{5z} + \sqrt{3} \right)$ [Section 7.4]

55. Rationalize the denominator: $\dfrac{\sqrt{3}}{\sqrt{50}}$ [Section 7.4]

56. Solve: $\sqrt{-5x + 24} = 6 - x$ [Section 7.5]

57. Geometry.   Find the length of one leg of a right triangle if the other leg is 10 cm long and the hypotenuse is 26 cm long. [Section 7.6]

58. Divide: $\dfrac{5 - 3i}{4 + 2i}$. Write the answer in $a + bi$ form.
[Section 7.7]

Quadratic Equations, Functions, and Inequalities

8

© Robert E Daemmrich/Getty Images

> from **Campus to Careers**

Police Patrol Officer

The responsibilities of a police patrol officer are extremely broad. Quite often, he or she must make a split-second decision while under enormous pressure. One internet vocational website cautions anyone considering such a career to "pay attention in your mathematics and science classes. Those classes help sharpen your ability to think things through and solve problems—an important part of police work."

Problem 93 in **Study Set 8.2, problem 87** in **Study Set 8.3,** and **problem 89** in **Study Set 8.4** involve situations that a police patrol officer encounters on the job. The mathematical concepts discussed in this chapter can be used to solve those problems.

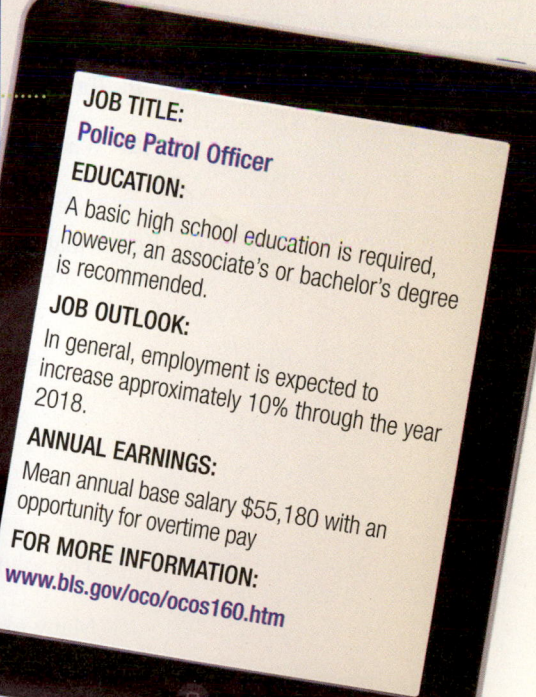

JOB TITLE:
Police Patrol Officer

EDUCATION:
A basic high school education is required, however, an associate's or bachelor's degree is recommended.

JOB OUTLOOK:
In general, employment is expected to increase approximately 10% through the year 2018.

ANNUAL EARNINGS:
Mean annual base salary $55,180 with an opportunity for overtime pay

FOR MORE INFORMATION:
www.bls.gov/oco/ocos160.htm

If you're like most students, your algebra notebook could probably use some attention at this stage of the course. You will definitely appreciate a well-organized notebook when it comes time to study for the final exam. Here are some suggestions to put it in tip-top shape.

ORGANIZE YOUR NOTEBOOK INTO SECTIONS: Create a separate section in the notebook for each chapter (or unit of study) that your class has covered this term.

ORGANIZE THE PAPERS WITHIN EACH SECTION: One recommended order is to begin each section with your class notes, followed by your completed homework assignments, then any study sheets or handouts, and, finally, all graded quizzes and tests.

Now Try This ▶

1. Organize your algebra notebook using the guidelines given above.
2. Write a Table of Contents to place at the beginning of your notebook. List each chapter (or unit of study) and include the dates over which the material was covered.
3. Compare your completed notebook with those of other students in your class. Have you overlooked any important items that would be useful when studying for the final exam?

SECTION 8.1

The Square Root Property and Completing the Square

OBJECTIVES

1. Use the square root property to solve quadratic equations.
2. Solve quadratic equations by completing the square.

ARE YOU READY?

The following problems review some basic skills that are needed when using the square root property and completing the square.

1. Simplify: **a.** $\sqrt{28}$
         b. $\sqrt{-36}$

2. Rationalize: $\sqrt{\dfrac{5}{6}}$

3. What is one-half of 9?

4. Find the square of $\dfrac{7}{2}$.

5. Fill in the blank: $1 + \dfrac{25}{144} = \dfrac{\boxed{}}{144}$

6. Factor: $x^2 - 8x + 16$

Recall that a *quadratic equation* is an equation of the form $ax^2 + bx + c = 0$, where a, b, and c are real numbers and $a \neq 0$. We have solved quadratic equations, such as $6x^2 - 7x - 3 = 0$, using factoring and the zero-factor property as shown below.

$$6x^2 - 7x - 3 = 0$$
$$(2x - 3)(3x + 1) = 0 \qquad \text{Factor the trinomial.}$$
$$2x - 3 = 0 \quad \text{or} \quad 3x + 1 = 0 \qquad \text{Set each factor equal to 0.}$$
$$x = \frac{3}{2} \qquad\qquad x = -\frac{1}{3} \qquad \text{Solve each linear equation.}$$

The solutions are $-\frac{1}{3}$ and $\frac{3}{2}$.

Many expressions do not factor as easily as $6x^2 - 7x - 3$. For example, it would be difficult to solve $2x^2 + 4x + 1 = 0$ by factoring, because $2x^2 + 4x + 1$ cannot be factored by using only integers. In this section and the next, we will develop more general methods that enable us to solve any quadratic equation. Those methods are based on the *square root property*.

1   Use the Square Root Property to Solve Quadratic Equations.

To develop general methods for solving quadratic equations, we first consider the equation $x^2 = c$. If $c \geq 0$, we can find the real solutions of $x^2 = c$ as follows:

$$x^2 = c$$

$$x^2 - c = 0 \qquad \text{Subtract } c \text{ from both sides.}$$

$$x^2 - \left(\sqrt{c}\right)^2 = 0 \qquad \text{Replace } c \text{ with } \left(\sqrt{c}\right)^2, \text{ since } c = \left(\sqrt{c}\right)^2.$$

$$\left(x + \sqrt{c}\right)\left(x - \sqrt{c}\right) = 0 \qquad \text{Factor the difference of two squares.}$$

$$x + \sqrt{c} = 0 \qquad \text{or} \qquad x - \sqrt{c} = 0 \qquad \text{Set each factor equal to 0.}$$

$$x = -\sqrt{c} \qquad\qquad x = \sqrt{c} \qquad \text{Solve each linear equation.}$$

The solutions of $x^2 = c$ are $\sqrt{c}$ and $-\sqrt{c}$.

The Square Root Property	For any nonnegative real number c, if $x^2 = c$, then $$x = \sqrt{c} \qquad \text{or} \qquad x = -\sqrt{c}$$

The Language of Algebra

The $\pm$ symbol is often seen in political polls. A candidate with 48% ($\pm 4\%$) support could be between $48 + 4 = 52\%$ and $48 - 4 = 44\%$.

We can write the conclusion of the square root property in a more compact form, called **double-sign notation.**

$$x = \pm\sqrt{c} \qquad \text{This is formally read as "x equals the positive or negative square root of c."}$$

However, it is often read more informally as "x equals plus or minus the square root of c."

EXAMPLE 1   Solve: $x^2 - 12 = 0$

Strategy   Since $x^2 - 12$ does not factor as a difference of two integer squares, we must take an alternate approach. We will add 12 to both sides of the equation and use the square root property to solve for x.

Why   After adding 12 to both sides, the resulting equivalent equation will have the desired form $x^2 = c$.

Solution

$$x^2 - 12 = 0 \qquad \text{This is the equation to solve. It is a quadratic equation that is missing an x-term.}$$

$$x^2 = 12 \qquad \text{To isolate } x^2 \text{ on the left side, add 12 to both sides.}$$

$$x = \sqrt{12} \quad \text{or} \quad x = -\sqrt{12} \qquad \text{Use the square root property.}$$

$$x = 2\sqrt{3} \qquad\qquad x = -2\sqrt{3} \qquad \text{Simplify: } \sqrt{12} = \sqrt{4}\sqrt{3} = 2\sqrt{3}.$$

Check:

$$x^2 - 12 = 0 \qquad\qquad\qquad x^2 - 12 = 0$$

$$\left(2\sqrt{3}\right)^2 - 12 \stackrel{?}{=} 0 \qquad\qquad \left(-2\sqrt{3}\right)^2 - 12 \stackrel{?}{=} 0$$

$$12 - 12 \stackrel{?}{=} 0 \qquad\qquad\qquad 12 - 12 \stackrel{?}{=} 0$$

$$0 = 0 \quad \text{True} \qquad\qquad\qquad 0 = 0 \quad \text{True}$$

The exact solutions are $2\sqrt{3}$ and $-2\sqrt{3}$, which can be written using double-sign notation as $\pm 2\sqrt{3}$. The solution set is $\left\{-2\sqrt{3}, 2\sqrt{3}\right\}$. We can use a calculator to approximate the solutions. To the nearest hundredth, they are ± 3.46.

Self Check 1   Solve: $x^2 - 18 = 0$

Now Try ▶ Problems 15 and 19

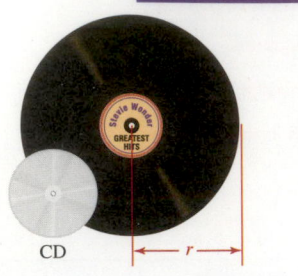

CD ←—r—→

EXAMPLE 2   **Phonograph Records.**   Before compact discs, music was recorded on thin vinyl discs. The discs used for long-playing records had a surface area of about 111 square inches per side. Find the radius of a long-playing record to the nearst tenth of an inch.

Strategy   The area A of a circle with radius r is given by the formula $A = \pi r^2$. We will find the radius of a record by substituting 111 for A and dividing both sides by π. Then we will use the square root property to solve for r.

Why   After substituting 111 for A and dividing both sides by π, the resulting equivalent equation will have the desired form $r^2 = c$.

Solution

$$A = \pi r^2 \qquad \text{\color{red}This is the formula for the area of a circle.}$$

$$111 = \pi r^2 \qquad \text{\color{red}Substitute 111 for A.}$$

$$\frac{111}{\pi} = r^2 \qquad \text{\color{red}To undo the multiplication by } \pi \text{, divide both sides by } \pi.$$

$$r = \sqrt{\frac{111}{\pi}} \quad \text{or} \quad r = -\sqrt{\frac{111}{\pi}} \qquad \text{\color{red}Use the square root property. Since the radius of the record cannot be negative, discard the second solution.}$$

The radius of a long-playing record is $\sqrt{\frac{111}{\pi}}$ inches—to the nearest tenth, 5.9 inches.

Self Check 2   **Restaurant Seating.**   A restaurant requires 14 square feet of floor area for one of their round tables. Find the radius of a table. Round to the nearest hundredth.

Now Try ▶ Problems 35 and 103

Some quadratic equations have solutions that are not real numbers.

EXAMPLE 3   Solve: $4x^2 + 25 = 0$

Strategy   We will subtract 25 from both sides of the equation and divide both sides by 4. Then we will use the square root property to solve for x.

Why   After subtracting 25 from both sides and dividing both sides by 4, the resulting equivalent equation will have the desired form $x^2 = $ a constant.

Solution

$$4x^2 + 25 = 0 \qquad \text{\color{red}This is the equation to solve.}$$

$$x^2 = -\frac{25}{4} \qquad \text{\color{red}To isolate } x^2 \text{, subtract 25 from both sides and divide both sides by 4.}$$

$$x = \pm\sqrt{-\frac{25}{4}} \qquad \text{\color{red}Use the square root property and write the result using double-sign notation.}$$

$$x = \pm\frac{5}{2}i \qquad \text{\color{red}Simplify the radical: } \sqrt{-\frac{25}{4}} = \sqrt{-1 \cdot \frac{25}{4}} = \sqrt{-1}\frac{\sqrt{25}}{\sqrt{4}} = \frac{5}{2}i.$$

Since the solutions are $\frac{5}{2}i$ and $-\frac{5}{2}i$, the solution set is $\left\{\frac{5}{2}i, -\frac{5}{2}i\right\}$.

Check:

$$4x^2 + 25 = 0 \qquad\qquad\qquad 4x^2 + 25 = 0$$

$$4\left(\frac{5}{2}i\right)^2 + 25 \stackrel{?}{=} 0 \qquad\qquad 4\left(-\frac{5}{2}i\right)^2 + 25 \stackrel{?}{=} 0$$

$$4\left(\frac{25}{4}\right)i^2 + 25 \stackrel{?}{=} 0 \qquad\qquad 4\left(\frac{25}{4}\right)i^2 + 25 \stackrel{?}{=} 0$$

$$25(-1) + 25 \stackrel{?}{=} 0 \qquad\qquad 25(-1) + 25 \stackrel{?}{=} 0$$

$$0 = 0 \quad \text{True} \qquad\qquad\qquad 0 = 0 \quad \text{True}$$

Self Check 3   Solve: $16x^2 + 49 = 0$

Now Try ▶ Problem 27

We can extend the square root property to solve equations that involve the square of a binomial and a constant.

EXAMPLE 4   Use the square root property to solve $(x - 1)^2 = 16$.

Strategy   Instead of a variable squared on the left side, we have a quantity squared. We still use the square root property to solve the equation.

Why   We want to eliminate the square on the binomial, so that we can eventually isolate the variable on one side of the equation.

Solution

$$(x - 1)^2 = 16 \qquad \text{This is the equation to solve.}$$

$$x - 1 = \pm\sqrt{16} \qquad \text{Use the square root property and write the result using double-sign notation.}$$

$$x - 1 = \pm 4 \qquad \text{Simplify: } \sqrt{16} = 4.$$

$$x = 1 \pm 4 \qquad \text{To isolate } x, \text{ add 1 to both sides. It is standard practice to write the addition of the 1 in front of the } \pm \text{ symbol. Read as "one plus or minus four."}$$

$$x = 1 + 4 \quad \text{or} \quad x = 1 - 4 \qquad \text{To find one solution, use +. To find the other, use −.}$$

$$x = 5 \qquad\qquad\quad x = -3 \qquad \text{Add (subtract).}$$

Verify that 5 and −3 satisfy the original equation.

> **Caution**
>
> It might be tempting to square the binomial on the left side of $(x - 1)^2 = 16$. However, that causes unnecessary, additional steps to solve the equation another way.
>
> $$(x - 1)^2 = 16$$
> $$\cancel{x^2 - 2x + 1 = 16}$$

Self Check 4   Use the square root property to solve $(x + 2)^2 = 9$.

Now Try ▶ Problems 31 and 33

EXAMPLE 5   Let $f(x) = (3x + 8)^2$. For what value(s) of x is $f(x) = 6$?

Strategy   We will substitute 6 for $f(x)$ and solve for x.

Why   In the equation, there are two unknowns, x and $f(x)$. If we replace $f(x)$ with 6, we can use equation-solving techniques to find x.

Solution

$$f(x) = (3x + 8)^2 \quad \text{This is the given function.}$$

$$6 = (3x + 8)^2 \quad \text{Substitute 6 for } f(x).$$

$$\pm \sqrt{6} = 3x + 8 \quad \text{Use the square root property and write}$$
$$\text{the result using double-sign notation.}$$

$$-8 \pm \sqrt{6} = 3x \quad \text{To isolate the variable term } 3x, \text{ subtract 8 from both sides. It is}$$
$$\text{standard practice to write the subtraction of } -8 \text{ in front of the } \pm.$$

$$\frac{-8 \pm \sqrt{6}}{3} = x \quad \text{To isolate } x, \text{ undo the multiplication by 3 by dividing both sides by 3.}$$

We have found that $f(x) = 6$ when $x = \frac{-8 - \sqrt{6}}{3}$ or $x = \frac{-8 + \sqrt{6}}{3}$. To check, verify that $f\left(\frac{-8 - \sqrt{6}}{3}\right) = 6$ and $f\left(\frac{-8 + \sqrt{6}}{3}\right) = 6$.

Self Check 5   Let $f(x) = (4x + 11)^2$. For what values of x is $f(x) = 3$?

Now Try ▶ Problem 35

---

2 Solve Quadratic Equations by Completing the Square.

When the polynomial in a quadratic equation doesn't factor easily, we can solve the equation by **completing the square.** This method is based on the following perfect-square trinomials (with leading coefficients of 1) and their factored forms:

$$x^2 + 2bx + b^2 = (x + b)^2 \qquad \text{and} \qquad x^2 - 2bx + b^2 = (x - b)^2$$

The Language of Algebra

Recall that trinomials that are the square of a binomial are called **perfect-square trinomials.**

In each of these perfect-square trinomials, the third term is the square of one-half of the coefficient of x.

- In $x^2 + 2bx + b^2$, the coefficient of x is $2b$. If we find $\frac{1}{2} \cdot 2b$, which is b, and square it, we get the third term, b^2.

- In $x^2 - 2bx + b^2$, the coefficient of x is $-2b$. If we find $\frac{1}{2}(-2b)$, which is $-b$, and square it, we get the third term: $(-b)^2 = b^2$.

We can use these observations to change certain binomials into perfect-square trinomials using a 3-step process. For example, to change $x^2 + 12x$ into a perfect-square trinomial, we find one-half of the coefficient of x, square the result, and add the square to $x^2 + 12x$.

$$x^2 + 12x + \boxed{}$$

Step 1: Find one-half of the coefficient of x.   Step 3: Add the square to the binomial.

$$\frac{1}{2} \cdot 12 = 6 \qquad 6^2 = 36$$

Step 2: Square the result.

We obtain the perfect-square trinomial $x^2 + 12x + 36$, which factors as $(x + 6)^2$. By adding 36 to $x^2 + 12x$, we say that we have *completed the square on $x^2 + 12x$.*

Completing the Square

To complete the square on $x^2 + bx$, add the square of one-half of the coefficient of x:

$$x^2 + bx + \left(\frac{1}{2}b\right)^2$$

EXAMPLE 6   Complete the square and factor the resulting perfect-square trinomial: **a.** $x^2 + 10x$

b. $x^2 - 11x$ **c.** $x^2 + \dfrac{7}{3}x$

Strategy   We will add the square of one-half of the coefficient of x to the given binomial.

Why   Adding such a term will change the binomial into a perfect-square trinomial that will factor.

Solution   **a.** To make $x^2 + 10x$ a perfect-square trinomial, we find one-half of 10, square it, and add the result to $x^2 + 10x$.

$$x^2 + 10x + \mathbf{25} \qquad \tfrac{1}{2} \cdot 10 = 5 \text{ and } 5^2 = 25. \text{ Add 25 to the given binomial.}$$

This trinomial factors as $(x + 5)^2$. To check, we square $x + 5$ and verify that the result is $x^2 + 10x + 25$.

b. To make $x^2 - 11x$ a perfect-square trinomial, we find one-half of -11, square it, and add the result to $x^2 - 11x$.

$$x^2 - 11x + \mathbf{\dfrac{121}{4}} \qquad \tfrac{1}{2}(-11) = -\tfrac{11}{2} \text{ and } \left(-\tfrac{11}{2}\right)^2 = \tfrac{121}{4}. \text{ Add } \tfrac{121}{4} \text{ to the given binomial.}$$

This trinomial factors as $\left(x - \dfrac{11}{2}\right)^2$. Check using multiplication or using a special-product rule.

c. To make $x^2 + \dfrac{7}{3}x$ a perfect-square trinomial, we find one-half of $\dfrac{7}{3}$, square it, and add the result to $x^2 + \dfrac{7}{3}x$.

$$x^2 + \dfrac{7}{3}x + \mathbf{\dfrac{49}{36}} \qquad \text{To complete the square: } \tfrac{1}{2}\left(\tfrac{7}{3}\right) = \tfrac{7}{6} \text{ and } \left(\tfrac{7}{6}\right)^2 = \tfrac{49}{36}.$$
$$\text{Add } \tfrac{49}{36} \text{ to the given binomial.}$$

The trinomial factors as $\left(x + \dfrac{7}{6}\right)^2$.

Self Check 6   Complete the square on $a^2 - 5a$ and factor the resulting trinomial.

Now Try ▶   Problems 39 and 41

Caution

Realize that when we complete the square on a binomial, we are not writing an equivalent trinomial expression. Since the result is a completely different polynomial, it would be incorrect to use an $=$ symbol between the two.

$x^2 + 10x = x^2 + 10x + 25$

The Language of Algebra

When we add $\frac{121}{4}$ to $x^2 - 11x$, we say we have **completed the square** on $x^2 - 11x$. For that reason, binomials such as $x^2 - 11x$ are often called **incomplete squares**.

To solve an equation of the form $ax^2 + bx + c = 0$ by completing the square, we use the following steps.

Completing the Square to Solve a Quadratic Equation in x

1. If the coefficient of x^2 is 1, go to step 2. If it is not, make it 1 by dividing both sides of the equation by the coefficient of x^2.

2. Get all variable terms on one side of the equation and constants on the other side.

3. Complete the square by finding one-half of the coefficient of x, squaring the result, and adding the square to both sides of the equation.

4. Factor the perfect-square trinomial as the square of a binomial.

5. Solve the resulting equation using the square root property.

6. Check your answers in the original equation.

EXAMPLE 7   Solve by completing the square: $x^2 - 8x - 5 = 0$. Approximate the solutions to the nearest hundredth.

Strategy   We will use the addition property of equality and add 5 to both sides. Then we will complete the square to solve for x.

Why   To prepare to complete the square, we need to isolate the variable terms, x^2 and $-8x$, on the left side of the equation and the constant term on the right side.

Solution

$$x^2 - 8x - 5 = 0 \quad \text{\color{red}This is the equation to solve.}$$
$$x^2 - 8x = 5 \quad \text{\color{red}Add 5 to both sides so that the constant term is on the right side.}$$

The Language of Algebra

For $x^2 - 8x - 5$, we call x^2 the **leading term**, the **first term**, or the **quadratic term**. We call $-8x$ the **second term** or the **linear term**. Together, x^2 and $-8x$ are called **variable terms**. We call -5 the **constant term** or the **third term**.

In $x^2 - 8x$, the coefficient of x is -8. One-half of -8 is -4, and $(-4)^2 = 16$. If we add 16 to $x^2 - 8x$, it becomes a perfect-square trinomial.

$$x^2 - 8x + 16 = 5 + 16 \qquad \text{\color{red}To complete the square on the}$$
$$ \qquad \text{\color{red}left side, add 16 to both sides.}$$

$$(x - 4)^2 = 21 \qquad \text{\color{red}Factor the perfect-square trinomial. Add on the right side.}$$

$$x - 4 = \pm\sqrt{21} \qquad \text{\color{red}By the square root property, } x - 4 = \sqrt{21} \text{ or}$$
$$\phantom{x - 4 = \pm\sqrt{21}} \qquad \text{\color{red}} x - 4 = -\sqrt{21}. \text{ Use double-sign notation to show this.}$$

$$x = 4 \pm \sqrt{21} \qquad \text{\color{red}To isolate x, add 4 to both sides. It is common practice to}$$
$$\phantom{x = 4 \pm \sqrt{21}} \qquad \text{\color{red}write the 4 in front of the } \pm \text{ symbol.}$$

The exact solutions of $x^2 - 8x - 5 = 0$ are two irrational numbers, $4 + \sqrt{21}$ and $4 - \sqrt{21}$.

     We also can approximate each solution by using the decimal approximation of $\sqrt{21}$, which is 4.582575695:

$$4 + \sqrt{21} \approx 4 + 4.582575695 \qquad\qquad 4 - \sqrt{21} \approx 4 - 4.582575695$$
$$\approx 8.58 \qquad\qquad\qquad\qquad\qquad\qquad \approx -0.58$$

To the nearest hundredth, the solutions are 8.58 and -0.58. These approximations can be used to check the exact solutions informally by substituting each of them into the original equation.

Self Check 7   Solve by completing the square: $x^2 - 10x - 4 = 0$. Approximate the solutions to the nearest hundredth.

Now Try ▶ Problem 47

If the coefficient of the squared variable (called the **leading coefficient**) of a quadratic equation is not 1, we must make it 1 before we can complete the square.

EXAMPLE 8   Solve $6x^2 + 5x - 6 = 0$ by completing the square.

Strategy   We will begin by dividing both sides of the equation by 6.

Why   This will create a leading coefficient that is 1 so that we can proceed to complete the square to solve the equation.

Solution   ***Step 1:*** To make the coefficient of x^2 equal to 1, we divide both sides of the equation by 6.

$$6x^2 + 5x - 6 = 0 \qquad \text{\color{red}This is the equation to solve.}$$

$$\frac{6x^2}{6} + \frac{5x}{6} - \frac{6}{6} = \frac{0}{6} \qquad \text{\color{red}Divide both sides by 6, term-by-term.}$$

$$x^2 + \frac{5}{6}x - 1 = 0 \qquad \text{\color{red}Simplify. The coefficient of } x^2 \text{ is now 1.}$$

Success Tip

When solving a quadratic equation using the factoring method, one side of the equation must be 0. When we complete the square, we are not concerned with that requirement.

Step 2: To have the constant term on one side of the equation and the variable terms on the other, add 1 to both sides.

$$x^2 + \frac{5}{6}x \qquad = 1$$

Some students find it helpful to leave additional space here to prepare for completing the square.

Step 3: The coefficient of x is $\frac{5}{6}$, one-half of $\frac{5}{6}$ is $\frac{5}{12}$, and $\left(\frac{5}{12}\right)^2 = \frac{25}{144}$. To complete the square, we add $\frac{25}{144}$ to both sides.

$$x^2 + \frac{5}{6}x + \mathbf{\frac{25}{144}} = 1 + \mathbf{\frac{25}{144}}$$

$$x^2 + \frac{5}{6}x + \frac{25}{144} = \frac{169}{144}$$

On the right side, add: $1 + \frac{25}{144} = \frac{144}{144} + \frac{25}{144} = \frac{169}{144}$.

Step 4: Factor the left side of the equation.

$$\left(x + \frac{5}{12}\right)^2 = \frac{169}{144}$$

$x^2 + \frac{5}{6}x + \frac{25}{144}$ is a perfect-square trinomial.

Caution

When using the square root property to solve an equation, always write the $\pm$ symbol. If you forget, you will lose one of the solutions.

Step 5: We can solve the resulting equation by using the square root property.

$$x + \frac{5}{12} = \pm\sqrt{\frac{169}{144}}$$

Don't forget to write the $\pm$ symbol.

$$x + \frac{5}{12} = \pm\frac{13}{12}$$

Simplify: $\pm\sqrt{\frac{169}{144}} = \pm\frac{13}{12}$.

$$x = -\frac{5}{12} \pm \frac{13}{12}$$

To isolate x, subtract $\frac{5}{12}$ from both sides. These are like terms that can be added and subtracted.

Success Tip

Since the solutions of $6x^2 + 5x - 6 = 0$ are rational numbers, that indicates it can also be solved by factoring:

$$(3x - 2)(2x + 3) = 0$$
$$x = \frac{2}{3} \quad \text{or} \quad x = -\frac{3}{2}$$

To find the first solution, we evaluate the expression using the $+$ symbol. To find the second solution, we evaluate the expression using the $-$ symbol.

$$x = -\frac{5}{12} + \frac{13}{12} \quad \text{or} \quad x = -\frac{5}{12} - \frac{13}{12}$$

$$x = \frac{8}{12} \qquad\qquad x = -\frac{18}{12}$$

Add (subtract) the fractions.

$$x = \frac{2}{3} \qquad\qquad x = -\frac{3}{2}$$

Simplify each fraction. The solutions are two rational numbers.

Step 6: Check each solution, $\frac{2}{3}$ and $-\frac{3}{2}$, in the original equation.

Self Check 8   Solve by completing the square: $3x^2 + 2x - 8 = 0$

Now Try ▶ Problem 55

EXAMPLE 9   Solve: $2x^2 + 4x + 1 = 0$. Approximate the solutions to the nearest hundredth.

Strategy   We will follow the steps for solving a quadratic equation by completing the square.

Why   Since the trinomial $2x^2 + 4x + 1$ cannot be factored using only integers, solving the equation by completing the square is our only option at this time.

Solution

$$2x^2 + 4x + 1 = 0$$ This is the equation to solve.

$$x^2 + 2x + \frac{1}{2} = 0$$ Divide both sides by 2 to make the coefficient of x^2 equal to 1.

$$x^2 + 2x \qquad = -\frac{1}{2}$$ Subtract $\frac{1}{2}$ from both sides so that the constant term is on the right side.

$$x^2 + 2x + 1 = -\frac{1}{2} + 1$$ To complete the square, square one-half of the coefficient of x and add it to both sides.

$$(x + 1)^2 = \frac{1}{2}$$ Factor on the left side and do the addition on the right side.

$$x + 1 = \pm\sqrt{\frac{1}{2}}$$ Use the square root property. Don't forget the ± symbol.

$$x = -1 \pm \sqrt{\frac{1}{2}}$$ To isolate x, subtract 1 from both sides.

To write $\sqrt{\frac{1}{2}}$ in simplified radical form, we use the quotient rule for radicals and then rationalize the denominator.

$$x = -1 + \frac{\sqrt{2}}{2} \quad \text{or} \quad x = -1 - \frac{\sqrt{2}}{2}$$ Rationalize: $\sqrt{\frac{1}{2}} = \frac{\sqrt{1}}{\sqrt{2}} = \frac{1 \cdot \sqrt{2}}{\sqrt{2}\sqrt{2}} = \frac{\sqrt{2}}{2}$.

We can express each solution in an alternate form if we write -1 as a fraction with a denominator of 2.

$$x = -\frac{2}{2} + \frac{\sqrt{2}}{2} \quad \text{or} \quad x = -\frac{2}{2} - \frac{\sqrt{2}}{2}$$ Write -1 as $-\frac{2}{2}$.

$$x = \frac{-2 + \sqrt{2}}{2} \quad \bigg| \quad x = \frac{-2 - \sqrt{2}}{2}$$ Add (subtract) the numerators and keep the common denominator 2.

The exact solutions are $\frac{-2 + \sqrt{2}}{2}$ and $\frac{-2 - \sqrt{2}}{2}$, or simply, $\frac{-2 \pm \sqrt{2}}{2}$. We can use a calculator to approximate them. To the nearest hundredth, they are -0.29 and -1.71.

> **Self Check 9**   Solve: $3x^2 + 6x + 1 = 0$. Approximate the solutions to the nearest hundredth.
>
> **Now Try** ▶ Problem 59

Using Your Calculator ▶ Checking Solutions of Quadratic Equations

We can use a graphing calculator to check the solutions of the quadratic equation $2x^2 + 4x + 1 = 0$ found in Example 9. After entering $Y_1 = 2x^2 + 4x + 1$, we call up the home screen by pressing ⎴2nd⎵ ⎴QUIT⎵. Then we press ⎴VARS⎵, arrow to Y-VARS, press ⎴ENTER⎵, and enter 1 to get the display shown in figure (a). We evaluate $2x^2 + 4x + 1$ for $x = \frac{-2 + \sqrt{2}}{2}$ by entering the solution using function notation, as shown in figure (b). When ⎴ENTER⎵ is pressed, the result of 0 is confirmation that $x = \frac{-2 + \sqrt{2}}{2}$ is a solution of the equation.

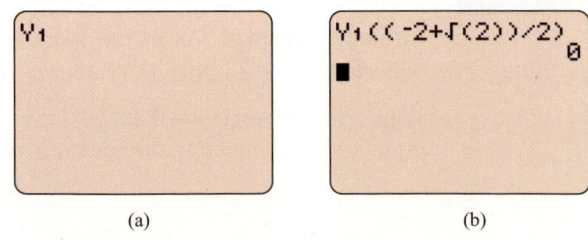

(a)   (b)

The solutions of some quadratic equations are complex numbers that contain i.

EXAMPLE 10

Solve: **a.** $x^2 - 4x + 16 = 0$ **b.** $x^2 + \frac{2}{3}x + 6 = 0$

Strategy   We will follow the steps for solving a quadratic equation by completing the square.

Why   Since the trinomials $x^2 - 4x + 16$ and $x^2 + \frac{2}{3}x + 6 = 0$ cannot be factored using only integers, solving the equation by completing the square is our only option.

Solution

a.

$x^2 - 4x + 16 = 0$	This is the equation to solve.
$x^2 - 4x \qquad = -16$	Subtract 16 from both sides so that the constant term is on the right side.
$x^2 - 4x + 4 = -16 + 4$	Complete the square: $\frac{1}{2}(-4) = -2$ and $(-2)^2 = 4$. Add 4 to both sides.
$(x - 2)^2 = -12$	Factor the left side. On the right side, add.
$x - 2 = \pm\sqrt{-12}$	Use the square root property. Don't forget the $\pm$ symbol.
$x = 2 \pm \sqrt{-12}$	To isolate x, add 2 to both sides.
$x = 2 \pm 2i\sqrt{3}$	Simplify the radical: $\sqrt{-12} = \sqrt{-1 \cdot 4 \cdot 3} = \sqrt{-1}\sqrt{4}\sqrt{3} = 2i\sqrt{3}$.

There are two complex number solutions involving i: $2 + 2i\sqrt{3}$ and $2 - 2i\sqrt{3}$. Verify this by checking them in the original equation.

b.

$x^2 + \frac{2}{3}x + 6 = 0$	This is the equation to solve.
$x^2 + \frac{2}{3}x \qquad = -6$	Subtract 6 from both sides so that the constant term is on the right side.
$x^2 + \frac{2}{3}x + \frac{1}{9} = \frac{1}{9} - 6$	Complete the square: $\frac{1}{2} \cdot \frac{2}{3} = \frac{1}{3}$ and $\left(\frac{1}{3}\right)^2 = \frac{1}{9}$. Add $\frac{1}{9}$ to both sides.
$\left(x + \frac{1}{3}\right)^2 = \frac{1}{9} - \frac{6}{1} \cdot \frac{9}{9}$	Factor the left side. On the right side, prepare to subtract. Write 6 as $\frac{6}{1}$ and multiply by $\frac{9}{9}$ to build an equivalent fraction with denominator 9.
$\left(x + \frac{1}{3}\right)^2 = \frac{1}{9} - \frac{54}{9}$	On the right side, multiply the numerators. Multiply the denominators.
$\left(x + \frac{1}{3}\right)^2 = -\frac{53}{9}$	On the right side, subtract the fractions.
$x + \frac{1}{3} = \pm\sqrt{-\frac{53}{9}}$	Use the square root property. Don't forget the $\pm$ symbol.
$x = -\frac{1}{3} \pm \sqrt{-\frac{53}{9}}$	To isolate x, subtract $\frac{1}{3}$ from both sides.
$x = -\frac{1}{3} \pm \frac{\sqrt{53}}{3}i$	Simplify the radical: $\sqrt{-\frac{53}{9}} = \sqrt{-1}\frac{\sqrt{53}}{\sqrt{9}} = \frac{\sqrt{53}}{3}i$.

The solutions are $-\frac{1}{3} + \frac{\sqrt{53}}{3}i$ and $-\frac{1}{3} - \frac{\sqrt{53}}{3}i$.

Self Check 10   Solve: **a.** $x^2 - 6x + 27 = 0$ **b.** $x^2 + \frac{2}{5}x + 3 = 0$

Now Try ▶ Problems 63 and 67

Success Tip

Example 8 can be solved by factoring. However, Examples 7, 9, and 10 cannot. This illustrates an important fact: completing the square can be used to solve any quadratic equation.

Success Tip

Notice how we multiply $\frac{6}{1}$ by a form of 1 to build an equivalent fraction with a denominator 9:

$$\frac{6}{1} \cdot \frac{9}{9}$$

Notation

The solutions are written in complex number form $a + bi$. They also could be written as

$$\frac{-1 \pm i\sqrt{53}}{3}$$

SECTION 8.1 ▶ STUDY SET

VOCABULARY

Fill in the blanks.

1. An equation of the form $ax^2 + bx + c = 0$, where $a \neq 0$, is called a _____ equation.

2. $x^2 + 6x + 9$ is called a _____-square trinomial because it factors as $(x + 3)^2$.

3. When we add 16 to $x^2 + 8x$, we say that we have completed the _____ on $x^2 + 8x$.

4. The _____ coefficient of $5x^2 - 2x + 7$ is 5 and the _____ term is 7.

CONCEPTS

Fill in the blanks.

5. For any nonnegative number c, if $x^2 = c$, then $x =$ ▭ or $x =$ ▭.

6. To complete the square on $x^2 + 10x$, add the square of _____ of the coefficient of x.

7. Find one-half of the given number and square the result.
   a. 12   b. -5

8. Fill in the blanks to factor the perfect-square trinomial.
   a. $x^2 + 8x + 16 = (x + \boxed{})^2$
   b. $x^2 - 9x + \dfrac{81}{4} = \left(x - \boxed{}\right)^2$

9. What is the first step to solve each equation by completing the square? **Do not solve.**
   a. $x^2 + 9x + 7 = 0$
   b. $4x^2 + 5x - 16 = 0$

10. Divide both sides of the equation by the proper number to make the coefficient of x^2 equal to 1. **Do not solve.**
   a. $2x^2 - 3x + 6 = 0$
   b. $3x^2 + 6x - 4 = 0$

11. Use a check to determine whether $-2 + \sqrt{2}$ is a solution of $x^2 + 4x + 2 = 0$.

12. Determine whether each statement is true or false.
   a. Any quadratic equation can be solved by the factoring method.
   b. Any quadratic equation can be solved by completing the square.

NOTATION

13. We read $8 \pm \sqrt{3}$ as "eight ____ ___ _____ the square root of 3."

14. When solving a quadratic equation, a student obtains $x = \dfrac{-5 \pm \sqrt{7}}{3}$.
   a. How many solutions are represented by this notation? List them.
   b. Approximate the solutions to the nearest hundredth.

GUIDED PRACTICE

Use the square root property to solve each equation. See Example 1.

15. $t^2 - 11 = 0$
16. $w^2 - 47 = 0$
17. $x^2 - 35 = 0$
18. $x^2 - 101 = 0$
19. $z^2 - 50 = 0$
20. $u^2 - 24 = 0$
21. $3x^2 - 16 = 0$
22. $5x^2 - 49 = 0$

Use the square root property to solve each equation. See Example 3.

23. $p^2 = -16$
24. $q^2 = -25$
25. $a^2 + 8 = 0$
26. $m^2 + 18 = 0$
27. $4m^2 + 81 = 0$
28. $9n^2 + 121 = 0$
29. $6b^2 + 144 = 0$
30. $9n^2 + 288 = 0$

Use the square root property to solve each equation. See Example 4.

31. $(x + 5)^2 = 9$
32. $(x - 1)^2 = 4$
33. $(t + 4)^2 = 16$
34. $(s - 7)^2 = 9$

See Example 5.

35. Let $f(x) = (x + 5)^2$. For what value(s) of x is $f(x) = 3$?

36. Let $f(x) = (x + 3)^2$. For what value(s) of x is $f(x) = 7$?

37. Let $g(a) = (7a - 2)^2$. For what value(s) of a is $g(a) = 8$?

38. Let $h(c) = (11c - 2)^2$. For what value(s) of c is $h(c) = 12$?

Complete the square and factor the resulting perfect-square trinomial. See Example 6.

39. $x^2 + 24x$
40. $y^2 - 18y$
41. $a^2 - 7a$
42. $b^2 + 11b$
43. $x^2 + \dfrac{2}{3}x$
44. $x^2 + \dfrac{2}{5}x$
45. $m^2 - \dfrac{5}{6}m$
46. $n^2 - \dfrac{5}{3}n$

Use completing the square to solve each equation. Approximate each solution to the nearest hundredth. See Example 7.

47. $x^2 - 4x - 2 = 0$
48. $x^2 - 6x - 4 = 0$
49. $x^2 - 12x + 1 = 0$
50. $x^2 - 18x + 2 = 0$
51. $t^2 + 20t + 25 = 0$
52. $t^2 + 22t + 93 = 0$

53. $t^2 + 16t - 16 = 0$ **54.** $t^2 + 14t - 7 = 0$

Use completing the square to solve each equation. See Example 8.

55. $2x^2 - x - 1 = 0$ **56.** $2x^2 - 5x + 2 = 0$

57. $12t^2 - 5t - 3 = 0$ **58.** $5m^2 + 13m - 6 = 0$

Use completing the square to solve each equation. Approximate each solution to the nearest hundredth. See Example 9.

59. $3x^2 - 12x + 1 = 0$ **60.** $6x^2 - 12x + 1 = 0$

61. $2x^2 + 5x - 2 = 0$ **62.** $2x^2 - 8x + 5 = 0$

Use completing the square to solve each equation. See Example 10.

63. $p^2 + 2p + 2 = 0$ **64.** $x^2 - 6x + 10 = 0$

65. $y^2 + 8y + 18 = 0$ **66.** $n^2 + 10n + 28 = 0$

67. $x^2 + \frac{2}{3}x + 7 = 0$ **68.** $x^2 + \frac{2}{5}x + 6 = 0$

69. $a^2 - \frac{1}{2}a + 1 = 0$ **70.** $b^2 - \frac{1}{4}b + 1 = 0$

TRY IT YOURSELF

Solve each equation. Approximate the solutions to the nearest hundredth when appropriate.

71. $(3x - 1)^2 = 25$ **72.** $(5x - 2)^2 = 64$

73. $3x^2 - 6x = 1$ **74.** $2x^2 - 6x = -3$

75. $x^2 + 8x + 6 = 0$ **76.** $x^2 + 6x + 4 = 0$

77. $6x^2 + 72 = 0$ **78.** $5x^2 + 40 = 0$

79. $x^2 - 2x = 17$ **80.** $x^2 + 10x = 7$

81. $m^2 - 7m + 3 = 0$ **82.** $m^2 - 5m + 3 = 0$

83. $7h^2 = 35$ **84.** $9n^2 = 99$

85. $\frac{7x + 1}{5} = -x^2$ **86.** $\frac{3}{8}x^2 = \frac{1}{8} - x$

87. $t^2 + t + 3 = 0$ **88.** $b^2 - b + 5 = 0$

89. $(8x + 5)^2 = 24$ **90.** $(3y - 2)^2 = 18$

91. $r^2 - 6r - 27 = 0$ **92.** $s^2 - 6s - 40 = 0$

93. $4p^2 + 2p + 3 = 0$ **94.** $3m^2 - 2m + 3 = 0$

Look Alikes . . .

95. a. $x^2 - 24 = 0$ **b.** $x^2 + 24 = 0$

96. a. $x^2 + 7x - 8 = 0$ **b.** $x^2 + 7x - 9 = 0$

97. a. $2m^2 - 8m = 0$ **b.** $2m^2 - 8m = 1$

98. a. $a^2 + a - 7 = 0$ **b.** $a^2 - a - 7 = 0$

99. a. $x^2 - 4x + 20 = 0$ **b.** $x^2 - 4x - 20 = 0$

100. a. $a^2 = 6a - 3$ **b.** $n^2 = 6n + 3$

101. a. $2r^2 - 4r + 3 = 0$ **b.** $2r^2 - 4r - 3 = 0$

102. a. $5y^2 + 15y + 12 = 0$ **b.** $5y^2 + 15y - 12 = 0$

APPLICATIONS

103. Movie Stunts. According to the *Guinness Book of World Records,* stuntman Dan Koko fell a distance of 312 feet into an airbag after jumping from the Vegas World Hotel and Casino. The distance d in feet traveled by a free-falling object in t seconds is given by the formula $d = 16t^2$. To the nearest tenth of a second, how long did the fall last?

104. Geography. The surface area S of a sphere is given by the formula $S = 4\pi r^2$, where r is the radius of the sphere. An almanac lists the surface area of the Earth as 196,938,800 square miles. Assuming the Earth to be spherical, what is its radius to the nearest mile?

105. Accidents. The height h (in feet) of an object that is dropped from a height of s feet is given by the formula $h = s - 16t^2$, where t is the time the object has been falling. A 5-foot-tall woman on a sidewalk looks directly overhead and sees a window washer drop a bottle from four stories up. How long does she have to get out of the way? Round to the nearest tenth. (A story is 12 feet.)

106. Flags. In 1912, an order by President Taft fixed the width and length of the U.S. flag in the ratio 1 to 1.9. If 100 square feet of cloth are to be used to make a U.S. flag, estimate its dimensions to the nearest $\frac{1}{4}$ foot.

107. Automobile Engines. As the piston shown moves upward, it pushes a cylinder of a gasoline/air mixture that is ignited by the spark plug. The formula that gives the volume of a cylinder is $V = \pi r^2 h$, where r is the radius and h is the height. Find the radius of the piston (to the nearest hundredth of an inch) if it displaces 47.75 cubic inches of gasoline/air mixture as it moves from its lowest to its highest point.

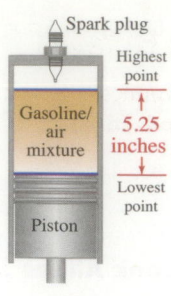

Spark plug

Highest point

Gasoline/air mixture

5.25 inches

Lowest point

Piston

108. Investments. If P dollars are deposited in an account that pays an annual rate of interest r, then in n years, the amount of money A in the account is given by the formula $A = P(1 + r)^n$. A savings account was opened on January 3, 2006, with a deposit of $10,000 and closed on January 2, 2008, with an ending balance of $11,772.25. Find the rate of interest.

109. Physics. Albert Einstein discovered a connection between energy and mass. This relationship (*energy equals mass times the velocity of light squared*) is expressed in the equation $E = mc^2$. Solve for c.

110. Right Triangles. The Pythagorean theorem relates the lengths of the sides in a right triangle: $a^2 + b^2 = c^2$, where a and b represent the lengths of the legs and c represents the length of the hypotenuse. Solve for b.

WRITING

111. Give an example of a perfect-square trinomial. Why do you think the word "perfect" is used to describe it?

112. Explain why completing the square on $x^2 + 5x$ is more difficult than completing the square on $x^2 + 4x$.

113. Explain the error in the work shown below.

a. $\dfrac{4 \pm \sqrt{3}}{8} = \dfrac{\overset{1}{\cancel{4}} \pm \sqrt{3}}{\underset{1}{\cancel{4}} \cdot 2} = \dfrac{1 \pm \sqrt{3}}{2}$

b. $\dfrac{1 \pm \sqrt{5}}{5} = \dfrac{1 \pm \sqrt{\overset{1}{\cancel{5}}}}{\cancel{5}} = \dfrac{1 \pm 1}{1}$

114. Explain the steps involved in expressing $8 \pm \dfrac{\sqrt{15}}{2}$ as a single fraction with denominator 2.

REVIEW

Simplify each expression. All variables represent positive real numbers.

115. $\sqrt[3]{40a^3b^6}$

116. $\sqrt[8]{x^{24}}$

117. $\sqrt[4]{\dfrac{16}{625}}$

118. $\sqrt{175a^2b^3}$

CHALLENGE PROBLEMS

119. What number must be added to $x^2 + \sqrt{3}x$ to make a perfect-square trinomial?

120. Solve $x^2 + \sqrt{3}x - \dfrac{1}{4} = 0$ by completing the square.

SECTION 8.2

The Quadratic Formula

OBJECTIVES

1. Derive the quadratic formula.

2. Solve quadratic equations using the quadratic formula.

3. Write equivalent equations to make quadratic formula calculations easier.

4. Use the quadratic formula to solve application problems.

ARE YOU READY?

The following problems review some basic skills that are needed when solving quadratic equations using the quadratic formula.

1. Evaluate: $\sqrt{5^2 - 4(4)(-6)}$

2. Simplify: $\sqrt{45}$

3. How many terms does $2x^2 - x + 7$ have? What is the coefficient of each term?

4. Evaluate: $\dfrac{-5 \pm 11}{8}$

5. Classify $\sqrt{-28}$ as a rational, irrational, or not a real number.

6. Approximate $\dfrac{6 + \sqrt{3}}{2}$ to the nearest hundredth.

We can solve quadratic equations by completing the square, but that method is often lengthy. In this section, we will develop a formula, called the *quadratic formula,* that enables us to solve quadratic equations with less effort.

1 Derive the Quadratic Formula.

To develop a formula that will produce the solutions of any given quadratic equation, we start with a quadratic equation in **standard form**, $ax^2 + bx + c = 0$, where $a > 0$. We can solve for x by completing the square.

$$ax^2 + bx + c = 0$$

$$\frac{ax^2}{a} + \frac{bx}{a} + \frac{c}{a} = \frac{0}{a}$$
Divide both sides by a so that the coefficient of x^2 is 1.

$$x^2 + \frac{b}{a}x + \frac{c}{a} = 0$$
Simplify: $\frac{ax^2}{a} = x^2$. Write $\frac{bx}{a}$ as $\frac{b}{a}x$.

$$x^2 + \frac{b}{a}x \qquad = -\frac{c}{a}$$
Subtract $\frac{c}{a}$ from both sides so that only the variable terms are on the left side of the equation and the constant is on the right side.

We can complete the square on $x^2 + \frac{b}{a}x$ by adding the square of one-half of the coefficient of x. Since the coefficient of x is $\frac{b}{a}$, we have $\frac{1}{2} \cdot \frac{b}{a} = \frac{b}{2a}$ and $\left(\frac{b}{2a}\right)^2 = \frac{b^2}{4a^2}$.

$$x^2 + \frac{b}{a}x + \frac{b^2}{4a^2} = -\frac{c}{a} + \frac{b^2}{4a^2}$$
To complete the square, add $\frac{b^2}{4a^2}$ to both sides.

$$x^2 + \frac{b}{a}x + \frac{b^2}{4a^2} = -\frac{4ac}{4aa} + \frac{b^2}{4a^2}$$
On the right side, build $-\frac{c}{a}$ by multiplying it by $\frac{4a}{4a}$. Now the fractions on that side have the common denominator $4a^2$.

$$\left(x + \frac{b}{2a}\right)^2 = \frac{b^2 - 4ac}{4a^2}$$
On the left side, factor the perfect-square trinomial. On the right side, add the fractions. In the numerator, write $-4ac + b^2$ as $b^2 - 4ac$.

$$x + \frac{b}{2a} = \pm\sqrt{\frac{b^2 - 4ac}{4a^2}}$$
Use the square root property.

$$x + \frac{b}{2a} = \pm\frac{\sqrt{b^2 - 4ac}}{\sqrt{4a^2}}$$
On the right side, the square root of a quotient is the quotient of square roots.

$$x + \frac{b}{2a} = \pm\frac{\sqrt{b^2 - 4ac}}{2a}$$
On the right side, simplify the denominator. Since $a > 0$, $\sqrt{4a^2} = 2a$.

$$x = -\frac{b}{2a} \pm \frac{\sqrt{b^2 - 4ac}}{2a}$$
To isolate x, subtract $\frac{b}{2a}$ from both sides.

$$x = \frac{-b \pm \sqrt{b^2 - 4ac}}{2a}$$
Combine the fractions. Write the sum (and difference) over the common denominator $2a$.

The Language of Algebra

To **derive** means to obtain by reasoning. To *derive* the quadratic formula means to solve $ax^2 + bx + c = 0$ for x, using the series of steps shown here, to obtain

$$x = \frac{-b \pm \sqrt{b^2 - 4ac}}{2a}$$

This result is called the **quadratic formula**. To develop this formula, we assumed that a was positive. If a is negative, similar steps are used, and we obtain the same result. This formula is very useful and should be memorized.

The Quadratic Formula

The solutions of $ax^2 + bx + c = 0$, with $a \neq 0$, are given by

$$x = \frac{-b \pm \sqrt{b^2 - 4ac}}{2a}$$
Read as "x equals the opposite of b plus or minus the square root of b squared minus $4ac$, all over $2a$."

The quadratic formula is a compact way of representing two solutions:

$$x = \frac{-b + \sqrt{b^2 - 4ac}}{2a} \qquad \text{or} \qquad x = \frac{-b - \sqrt{b^2 - 4ac}}{2a}$$

2 Solve Quadratic Equations Using the Quadratic Formula.

In the next example, we will use the quadratic formula to solve a quadratic equation.

EXAMPLE 1 Solve $2x^2 - 5x - 3 = 0$ by using the quadratic formula.

Strategy We will begin by comparing $2x^2 - 5x - 3 = 0$ to the standard form $ax^2 + bx + c = 0$.

Why To use the quadratic formula, we need to identify the values of a, b, and c.

Solution

$$2x^2 - 5x - 3 = 0 \quad \text{This is the equation to solve.}$$
$$ax^2 + bx + c = 0$$

> **Caution**
>
> Make sure to include the correct sign when determining a, b, and c. In this example, $b = -5$ and $c = -3$.

We see that $a = 2$, $b = -5$, and $c = -3$. To find the solutions of the equation, we substitute these values into the quadratic formula and evaluate the right side.

$$x = \frac{-b \pm \sqrt{b^2 - 4ac}}{2a} \quad \text{This is the quadratic formula.}$$

$$x = \frac{-(-5) \pm \sqrt{(-5)^2 - 4(2)(-3)}}{2(2)} \quad \text{Substitute 2 for } a, -5 \text{ for } b, \text{ and } -3 \text{ for } c.$$

$$x = \frac{5 \pm \sqrt{25 - (-24)}}{4} \quad \begin{array}{l}\text{Simplify: } -(-5) = 5. \text{ Evaluate the power and multiply} \\ \text{within the radical. Multiply in the denominator.}\end{array}$$

> **Caution**
>
> When writing the quadratic formula, be careful to draw the fraction bar so that it includes the entire numerator. Do not write
>
> $$x = -b \pm \frac{\sqrt{b^2 - 4ac}}{2a}$$
>
> or
>
> $$x = -b \pm \sqrt{\frac{b^2 - 4ac}{2a}}$$

$$x = \frac{5 \pm \sqrt{49}}{4} \quad \text{Simplify within the radical.}$$

$$x = \frac{5 \pm 7}{4} \quad \text{Evaluate the radical: } \sqrt{49} = 7.$$

To find the first solution, we evaluate the expression using the $+$ symbol. To find the second solution, we evaluate the expression using the $-$ symbol.

$$x = \frac{5 + 7}{4} \quad \text{or} \quad x = \frac{5 - 7}{4}$$
$$x = \frac{12}{4} \qquad\qquad x = \frac{-2}{4}$$
$$x = 3 \qquad\qquad x = -\frac{1}{2}$$

The solutions are 3 and $-\frac{1}{2}$ and the solution set is $\left\{3, -\frac{1}{2}\right\}$. Check each solution in the original equation.

Self Check 1 Solve $4x^2 - 7x - 2 = 0$ by using the quadratic formula.

Now Try ▶ Problem 13

To solve a quadratic equation in x using the quadratic formula, we follow these steps.

Solving a Quadratic Equation in *x* Using the Quadratic Formula	1. Write the equation in standard form: $ax^2 + bx + c = 0$.
	2. Identify a, b, and c.
	3. Substitute the values for a, b, and c in the quadratic formula and evaluate the right side to obtain the solutions.
	$$x = \dfrac{-b \pm \sqrt{b^2 - 4ac}}{2a}$$

EXAMPLE 2   Solve: $2x^2 = -4x - 1$

Strategy   We will write the equation in standard form $ax^2 + bx + c = 0$. Then we will identify the values of a, b, and c, and substitute these values into the quadratic formula.

Why   The quadratic equation must be in standard form to identify the values of a, b, and c.

Solution   To write the equation in standard form, we need to have all nonzero terms on the left side and 0 on the right side.

$$2x^2 = -4x - 1 \qquad \text{\color{red}This is the equation to solve.}$$

$$2x^2 + 4x + 1 = 0 \qquad \text{\color{red}To get 0 on the right side, add 4x and 1 to both sides.}$$

In the resulting equivalent equation, $a = 2$, $b = 4$, and $c = 1$.

$$x = \dfrac{-\boldsymbol{b} \pm \sqrt{\boldsymbol{b}^2 - 4\boldsymbol{ac}}}{2\boldsymbol{a}} \qquad \text{\color{red}This is the quadratic formula.}$$

$$x = \dfrac{-\boldsymbol{4} \pm \sqrt{\boldsymbol{4}^2 - 4(\boldsymbol{2})(\boldsymbol{1})}}{2(\boldsymbol{2})} \qquad \text{\color{red}Substitute 2 for a, 4 for b, and 1 for c.}$$

$$x = \dfrac{-4 \pm \sqrt{16 - 8}}{4} \qquad \text{\color{red}Evaluate the expression within the radical. Multiply in the denominator.}$$

$$x = \dfrac{-4 \pm \sqrt{8}}{4} \qquad \text{\color{red}Do the subtraction within the radical.}$$

$$x = \dfrac{-4 \pm 2\sqrt{2}}{4} \qquad \text{\color{red}Simplify the radical: } \sqrt{8} = \sqrt{4 \cdot 2} = 2\sqrt{2}.$$

Success Tip

Perhaps you noticed that Example 1 could be solved by factoring. However, that is not the case for Example 2. These observations illustrate an important fact: The quadratic formula can be used to solve **any** quadratic equation.

We can write the solutions in simpler form by factoring out 2 from the two terms in the numerator and removing the common factor of 2 in the numerator and denominator.

$$x = \dfrac{-4 \pm 2\sqrt{2}}{4} = \dfrac{2\left(-2 \pm \sqrt{2}\right)}{4} = \dfrac{\overset{1}{\cancel{2}}\left(-2 \pm \sqrt{2}\right)}{\underset{1}{\cancel{2} \cdot 2}} = \dfrac{-2 \pm \sqrt{2}}{2} \qquad \text{\color{red}Factor first.}$$

Notation

The solutions also can be written:

$$-\dfrac{2}{2} \pm \dfrac{\sqrt{2}}{2} = -1 \pm \dfrac{\sqrt{2}}{2}$$

The two irrational solutions are $\dfrac{-2 + \sqrt{2}}{2}$ and $\dfrac{-2 - \sqrt{2}}{2}$ and the solution set is $\left\{\dfrac{-2 + \sqrt{2}}{2}, \dfrac{-2 - \sqrt{2}}{2}\right\}$. We can approximate the solutions using a calculator. To the nearest hundredth, they are -0.29 and -1.71.

Self Check 2   Solve $3x^2 = 2x + 3$. Approximate the solutions to the nearest hundredth.

Now Try ▶   Problem 25

EXAMPLE 3   Solve: $m^2 + m = -1$

Strategy   We will write the equation in standard form $am^2 + bm + c = 0$. Then we will identify the values of a, b, and c, and substitute these values into the quadratic formula.

Why   The quadratic equation must be in standard form to identify the values of a, b, and c.

Solution   To write $m^2 + m = -1$ in standard form, we add 1 to both sides, to get

$$m^2 + m + 1 = 0 \qquad \text{This is the equation to solve.}$$

In the resulting equivalent equation, $a = 1$, $b = 1$, and $c = 1$:

$$m = \frac{-b \pm \sqrt{b^2 - 4ac}}{2a} \qquad \text{This is the quadratic formula.}$$

$$m = \frac{-1 \pm \sqrt{1^2 - 4(1)(1)}}{2(1)} \qquad \text{Substitute 1 for } a, \text{ 1 for } b, \text{ and 1 for } c.$$

$$m = \frac{-1 \pm \sqrt{1 - 4}}{2} \qquad \begin{array}{l}\text{Evaluate the expression within the radical.}\\ \text{Multiply in the denominator.}\end{array}$$

$$m = \frac{-1 \pm \sqrt{-3}}{2} \qquad \text{Do the subtraction within the radical.}$$

$$m = \frac{-1 \pm i\sqrt{3}}{2} \qquad \text{Simplify the radical: } \sqrt{-3} = \sqrt{-1 \cdot 3} = \sqrt{-1}\sqrt{3} = i\sqrt{3}.$$

The solutions are two complex numbers involving i: $-\frac{1}{2} + \frac{\sqrt{3}}{2}i$ and $-\frac{1}{2} - \frac{\sqrt{3}}{2}i$ and the solution set is $\left\{-\frac{1}{2} + \frac{\sqrt{3}}{2}i, -\frac{1}{2} - \frac{\sqrt{3}}{2}i\right\}$.

> **Caution**
>
> Since the variable in the equation is m, not x, we must change the variable in the quadratic formula to reflect this:
>
> $$m = \frac{-b \pm \sqrt{b^2 - 4ac}}{2a}$$

> **Notation**
>
> The solutions are written in complex number form $a + bi$. They also could be written as
>
> $$\frac{-1 \pm i\sqrt{3}}{2}$$

Self Check 3   Solve: $a^2 + 3a = -5$

Now Try ▶ Problem 29

3   Write Equivalent Equations to Make Quadratic Formula Calculations Easier.

When solving a quadratic equation by the quadratic formula, we often can simplify the calculations by solving a simpler, but equivalent equation.

EXAMPLE 4   For each equation below, write an equivalent equation so that the quadratic formula calculations will be simpler.

a. $-2x^2 + 4x - 1 = 0$ **b.** $x^2 + \frac{4}{5}x - \frac{1}{3} = 0$

c. $20x^2 - 60x - 40 = 0$ **d.** $0.03x^2 - 0.04x - 0.01 = 0$

Strategy   We will multiply both sides of each equation by a carefully chosen number.

Why   In each case, the objective is to find an equivalent equation whose values of a, b, and c are easier to work with than those of the given equation.

Solution   **a.** It is often easier to solve a quadratic equation using the quadratic formula if a is positive. If we multiply (or divide) both sides of $-2x^2 + 4x - 1 = 0$ by -1, we obtain an equivalent equation with $a > 0$.

$$-2x^2 + 4x - 1 = 0 \qquad \text{Here, } a = -2.$$

$$-1(-2x^2 + 4x - 1) = -1(0) \qquad \text{Don't forget to multiply each term by } -1.$$

$$2x^2 - 4x + 1 = 0 \qquad \text{Now } a = 2.$$

> **Success Tip**
>
> Unlike completing the square, the quadratic formula does not require the leading coefficient to be 1.

b. For $x^2 + \frac{4}{5}x - \frac{1}{3} = 0$, two coefficients are fractions: $b = \frac{4}{5}$ and $c = -\frac{1}{3}$. We can multiply both sides of the equation by their least common denominator, 15, to obtain an equivalent equation having coefficients that are integers.

$$x^2 + \frac{4}{5}x - \frac{1}{3} = 0 \qquad \textcolor{red}{\text{Here, } a = 1, b = \frac{4}{5}, \text{ and } c = -\frac{1}{3}.}$$

$$\textcolor{red}{15}\left(x^2 + \frac{4}{5}x - \frac{1}{3}\right) = \textcolor{red}{15}(0)$$

$$15x^2 + 12x - 5 = 0 \qquad \textcolor{red}{\text{On the left side, distribute the multiplication by 15.}}$$
$$\textcolor{red}{\text{Now } a = 15, b = 12, \text{ and } c = -5.}$$

The Language of Algebra

Recall that **equivalent equations** have the same solutions.

c. For $20x^2 - 60x - 40 = 0$, the coefficients 20, -60, and -40 have a common factor of 20. If we divide both sides of the equation by their GCF, we obtain an equivalent equation having smaller coefficients.

$$20x^2 - 60x - 40 = 0 \qquad \textcolor{red}{\text{Here, } a = 20, b = -60, \text{ and } c = -40.}$$

$$\frac{20x^2}{20} - \frac{60x}{20} - \frac{40}{20} = \frac{0}{20} \qquad \textcolor{red}{\text{The division by 20 is done term-by-term.}}$$

$$x^2 - 3x - 2 = 0 \qquad \textcolor{red}{\text{Now } a = 1, b = -3, \text{ and } c = -2.}$$

d. For $0.03x^2 - 0.04x - 0.01 = 0$, all three coefficients are decimals. We can multiply both sides of the equation by 100 to obtain an equivalent equation having coefficients that are integers.

$$0.03x^2 - 0.04x - 0.01 = 0 \qquad \textcolor{red}{\text{Here, } a = 0.03, b = -0.04, \text{ and } c = -0.01.}$$

$$\textcolor{red}{100}(0.03x^2 - 0.04x - 0.01) = \textcolor{red}{100}(0)$$

$$3x^2 - 4x - 1 = 0 \qquad \textcolor{red}{\text{On the left side, distribute 100.}}$$
$$\textcolor{red}{\text{Now } a = 3, b = -4, \text{ and } c = -1.}$$

Self Check 4   For each equation, write an equivalent equation so that the quadratic formula calculations will be simpler.

  a. $-6x^2 + 7x - 9 = 0$
  b. $\frac{1}{3}x^2 - \frac{2}{3}x - \frac{5}{6} = 0$
  c. $44x^2 + 66x - 99 = 0$
  d. $0.08x^2 - 0.07x - 0.02 = 0$

Now Try ▶   Problems 37 and 39

4   Use the Quadratic Formula to Solve Application Problems.

A variety of real-world applications can be modeled by quadratic equations. However, such equations are often difficult or even impossible to solve using the factoring method. In those cases, we can use the quadratic formula to solve the equation.

EXAMPLE 5   **Shortcuts.**   Instead of using the hallways, students are wearing a path through a planted quad area to walk 195 feet directly from the classrooms to the cafeteria. If the length of the hallway from the office to the cafeteria is 105 feet longer than the hallway from the office to the classrooms, how much walking are the students saving by taking the shortcut?

Analyze   The two hallways and the shortcut form a right triangle with a hypotenuse 195 feet long. We will use the Pythagorean theorem to solve this problem.

Assign   If we let $x =$ the length (in feet) of the hallway from the classrooms to the office, then the length of the hallway from the office to the cafeteria is $(x + 105)$ feet.

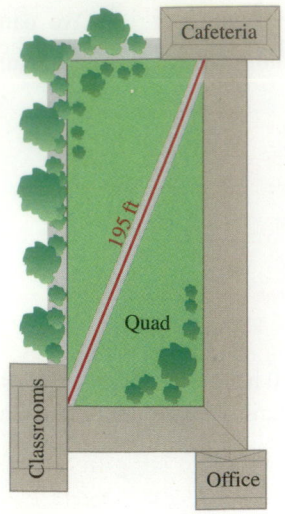

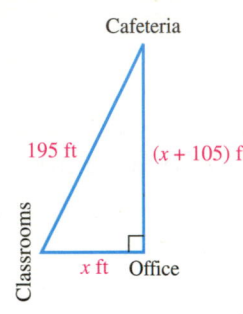

Form Substituting the lengths into the Pythagorean equation, we have

$$a^2 + b^2 = c^2$$ This is the Pythagorean equation.

$$x^2 + (x + 105)^2 = 195^2$$ Substitute x for a, $(x + 105)$ for b, and 195 for c.

$$x^2 + x^2 + 105x + 105x + 11{,}025 = 38{,}025$$ Find $(x + 105)^2$.

$$2x^2 + 210x + 11{,}025 = 38{,}025$$ Combine like terms.

$$2x^2 + 210x - 27{,}000 = 0$$ To get 0 on the right side, subtract 38,025 from both sides. This is a quadratic equation.

$$x^2 + 105x - 13{,}500 = 0$$ The coefficients have a common factor of 2: Divide both sides by 2.

Solve To solve $x^2 + 105x - 13{,}500 = 0$, we will use the quadratic formula with $a = 1$, $b = 105$, and $c = -13{,}500$.

$$x = \frac{-b \pm \sqrt{b^2 - 4ac}}{2a}$$ This is the quadratic formula.

$$x = \frac{-105 \pm \sqrt{105^2 - 4(1)(-13{,}500)}}{2(1)}$$ Substitute for a, b, and c.

$$x = \frac{-105 \pm \sqrt{65{,}025}}{2}$$ Evaluate within the radical: $105^2 - 4(1)(-13{,}500) =$ $11{,}025 + 54{,}000 = 65{,}025$. Multiply in the denominator.

$$x = \frac{-105 \pm 255}{2}$$ Use a calculator: $\sqrt{65{,}025} = 255$.

$$x = \frac{150}{2} \quad \text{or} \quad x = \frac{-360}{2}$$ Add: $-105 + 255 = 150$. Subtract: $-105 - 255 = -360$.

$$x = 75 \quad \bcancel{x = -180}$$ Do the division. Since the length of the hallway can't be negative, discard the solution -180.

State The length of the hallway from the classrooms to the office is 75 feet. The length of the hallway from the office to the cafeteria is $75 + 105 = 180$ feet. Instead of using the hallways, a distance of $75 + 180 = 255$ feet, the students are taking the 195-foot shortcut to the cafeteria, a savings of $(255 - 195)$, or 60 feet.

Check The length of the 180-foot hallway is 105 feet longer than the length of the 75-foot hallway. The sum of the squares of the lengths of the hallways is $75^2 + 180^2 = 38{,}025$. This equals the square of the length of the 195-foot shortcut: $195^2 = 38{,}025$. The result checks.

Self Check 5 | **Right Triangles.** The hypotenuse of a right triangle is 41 in. long. The longer leg is 31 inches longer than the shorter leg. Find the lengths of the legs of the triangle.

Now Try ▶ Problem 81

EXAMPLE 6 **Mass Transit.** A bus company has 4,000 passengers daily, each currently paying a 75¢ fare. For each 15¢ fare increase, the company estimates that it will lose 50 passengers. If the company needs to bring in $6,570 per day to stay in business, what fare must be charged to produce this amount of revenue?

©Krivosheev Vitaly/Shutterstock.com

Analyze To understand how a fare increase affects the number of passengers, let's consider what happens if there are two fare increases. We organize the data in a table. The fares are expressed in terms of dollars.

Number of increases	New fare	Number of passengers
One $0.15 increase	$0.75 + $0.15(1) = $0.90	4,000 − 50(1) = 3,950
Two $0.15 increases	$0.75 + $0.15(2) = $1.05	4,000 − 50(2) = 3,900

In general, the new fare will be the old fare ($0.75) plus the number of fare increases times $0.15. The number of passengers who will pay the new fare is 4,000 minus 50 times the number of $0.15 fare increases.

Assign If we let x = the number of $0.15 fare increases necessary to bring in $6,570 daily, then $(0.75 + 0.15x)$ is the fare that must be charged. The number of passengers who will pay this fare is $4,000 - 50x$.

Form We can now form an equation, using the words of the problem.

The bus fare	times	the number of passengers who will pay that fare	equals	$6,570.
$(0.75 + 0.15x)$	$\cdot$	$(4,000 - 50x)$	$=$	$6,570$

Solve

$$(0.75 + 0.15x)(4,000 - 50x) = 6,570$$

$$3,000 - 37.5x + 600x - 7.5x^2 = 6,570 \qquad \text{\color{red}Multiply the binomials on the left side.}$$

$$-7.5x^2 + 562.5x + 3,000 = 6,570 \qquad \text{\color{red}Combine like terms: } -37.5x + 600x = 562.5x.$$

$$-7.5x^2 + 562.5x - 3,570 = 0 \qquad \text{\color{red}To get 0 on the right side, subtract 6,570 from both sides. This is a quadratic equation.}$$

$$7.5x^2 - 562.5x + 3,570 = 0 \qquad \text{\color{red}Multiply both sides by } -1 \text{ so that the value of } a, \text{ 7.5, is positive.}$$

To solve this equation, we will use the quadratic formula.

$$x = \frac{-b \pm \sqrt{b^2 - 4ac}}{2a} \qquad \text{\color{red}This is the quadratic formula.}$$

$$x = \frac{-(-562.5) \pm \sqrt{(-562.5)^2 - 4(7.5)(3,570)}}{2(7.5)} \qquad \text{\color{red}Substitute 7.5 for } a, -562.5 \text{ for } b, \text{ and 3,570 for } c.$$

$$x = \frac{562.5 \pm \sqrt{209,306.25}}{15} \qquad \text{\color{red}Evaluate within the radical: } (-562.5)^2 - 4(7.5)(3,570) = 316,406.25 - 107,100 = 209,306.25. \text{ Multiply in the denominator.}$$

$$x = \frac{562.5 \pm 457.5}{15} \qquad \text{\color{red}Use a calculator: } \sqrt{209,306.25} = 457.5.$$

$$x = \frac{1,020}{15} \quad \text{or} \quad x = \frac{105}{15} \qquad \text{\color{red}Add: } 562.5 + 457.5 = 1,020. \text{ Subtract: } 562.5 - 457.5 = 105.$$

$$x = 68 \quad | \quad x = 7 \qquad \text{\color{red}Do the division.}$$

State If there are 7 fifteen-cent increases in the fare, the new fare will be $0.75 + $0.15(7) = $1.80. If there are 68 fifteen-cent increases in the fare, the new fare will be $0.75 + $0.15(68) = $10.95. Although this fare would bring in the necessary revenue, a $10.95 bus fare is unreasonable, so we discard it.

Check   A fare of $1.80 will be paid by $[4,000 - 50(7)] = 3,650$ bus riders. The amount of revenue brought in would be $1.80(3,650) = \$6,570$. The result checks.

> **Self Check 6**   **Airport Shuttles.**   A bus company shuttles 1,120 passengers daily between Rockford, Illinois, and O'Hare Airport. The current one-way fare is $10. For each 25¢ increase in the fare, the company predicts that it will lose 48 passengers. What increase in fare will produce daily revenue of $10,208$?
>
> **Now Try** ▶ Problem 91

EXAMPLE 7   **Lawyers.**   The number of lawyers in the United States is approximated by the function $N(x) = 45x^2 + 21,000x + 560,000$, where $N(x)$ is the number of lawyers and x is the number of years after 1980. In what year does this model indicate that the United States had one million lawyers? (Based on data from the American Bar Association)

Strategy   We will substitute 1,000,000 for $N(x)$ in the equation and solve for x.

Why   The value of x will give the number of years after 1980 that the United States had 1,000,000 lawyers.

Solution

$$N(x) = 45x^2 + 21,000x + 560,000 \qquad \text{This is the quadratic function model.}$$

$$1,000,000 = 45x^2 + 21,000x + 560,000 \qquad \text{Replace } N(x) \text{ with 1,000,000.}$$

$$0 = 45x^2 + 21,000x - 440,000 \qquad \begin{array}{l}\text{To get 0 on the left side, subtract}\\ \text{1,000,000 from both sides.}\end{array}$$

We can simplify the calculations by dividing both sides of the equation by 5, which is the greatest common factor of 45, 21,000, and 440,000.

$$9x^2 + 4,200x - 88,000 = 0 \qquad \text{Divide both sides by 5.}$$

We solve this equation using the quadratic formula.

$$x = \frac{-b \pm \sqrt{b^2 - 4ac}}{2a}$$

$$x = \frac{-4,200 \pm \sqrt{(4,200)^2 - 4(9)(-88,000)}}{2(9)} \qquad \begin{array}{l}\text{Substitute 9 for } a, 4,200\\ \text{for } b, \text{ and } -88,000 \text{ for } c.\end{array}$$

$$x = \frac{-4,200 \pm \sqrt{20,808,000}}{18} \qquad \begin{array}{l}\text{Evaluate the expression within the radical.}\\ \text{Multiply in the denominator.}\end{array}$$

$$x \approx \frac{362}{18} \quad \text{or} \quad x \approx \frac{-8,762}{18} \qquad \text{Use a calculator to evaluate each numerator.}$$

$$x \approx 20.1 \qquad \cancel{x \approx -486.8} \qquad \begin{array}{l}\text{Do the division. Since the model is defined for only}\\ \text{positive values of } x, \text{ we discard the second solution.}\end{array}$$

In 20.1 years after 1980, or in early 2000, the model predicts that the United States had approximately 1,000,000 lawyers.

> **Self Check 7**   **Lawyers.**   See Example 7. In what year does the model indicate that the United States had three-quarters of a million lawyers?
>
> **Now Try** ▶ Problem 95

EXAMPLE 8

Graduation Announcements. To create the announcement shown, a graphic artist must follow two design requirements.

- A border of uniform width should surround the text.
- Equal areas should be devoted to the text and to the border.

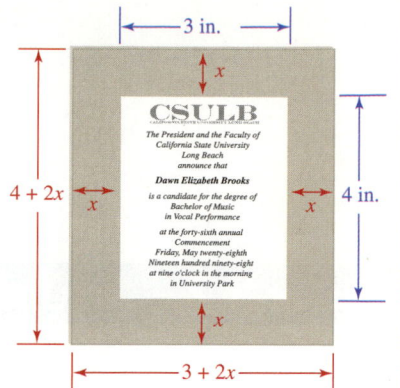

To meet these requirements, how wide should the border be?

Analyze The text occupies $4 \cdot 3 = 12$ in.^{2} of space. The border must also have an area of 12 in.^{2}.

Assign If we let $x =$ the width of the border in inches, the length of the announcement is $(4 + 2x)$ inches and the width is $(3 + 2x)$ inches.

Form We can now form an equation. Recall that the area of a rectangle is the product of its length and width.

The area of the announcement	minus	the area of the text	equals	the area of the border.
$(4 + 2x)(3 + 2x)$	$-$	12	$=$	12

Solve

$$(4 + 2x)(3 + 2x) - 12 = 12$$

$$12 + 8x + 6x + 4x^2 - 12 = 12 \qquad \text{On the left side, multiply the binomials.}$$

$$4x^2 + 14x = 12 \qquad \text{Combine like terms. This is a quadratic equation.}$$

$$4x^2 + 14x - 12 = 0 \qquad \text{To get 0 on the right side, subtract 12 from both sides.}$$

$$2x^2 + 7x - 6 = 0 \qquad \text{The coefficients have a common factor of 2. Divide both sides by 2.}$$

To solve this equation, we will use the quadratic formula with $a = 2$, $b = 7$, and $c = -6$.

$$x = \frac{-b \pm \sqrt{b^2 - 4ac}}{2a}$$

$$x = \frac{-7 \pm \sqrt{7^2 - 4(2)(-6)}}{2(2)} \qquad \text{Substitute for } a, b, \text{ and } c.$$

$$x = \frac{-7 \pm \sqrt{97}}{4} \qquad \begin{array}{l} \text{Evaluate within the radical:} \\ 7^2 - 4(2)(-6) = 49 + 48 = 97. \\ \text{Multiply in the denominator.} \end{array}$$

$$x \approx \frac{-7 + \sqrt{97}}{4} \qquad \text{or} \qquad x \approx \frac{-7 - \sqrt{97}}{4} \qquad \text{These are the two exact solutions.}$$

$$x \approx 0.71 \qquad \qquad \qquad x \approx -4.21$$

State The width of the border should be about 0.71 inch. (We discard the solution $\frac{-7 - \sqrt{97}}{4}$ since it is negative.)

Check If the border is 0.71 inch wide, the announcement has an area of about $5.42 \cdot 4.42 \approx 23.96$ in.^{2}. If we subtract the area of the text from the area of the announcement, we get $23.96 - 12 = 11.96$ in.^{2}. This represents the area of the border, which was to be 12 in.^{2}. The result seems reasonable.

> **Self Check 8**   **Graduation Announcements.**   See Example 8. Find the width of the border if the text occupies an area 3 inches by 5 inches.
>
> **Now Try** ▶ Problem 97

SECTION 8.2   STUDY SET

VOCABULARY

Fill in the blanks.

1. The standard form of a _____ equation is $ax^2 + bx + c = 0$.

2. $x = \dfrac{-b \pm \sqrt{b^2 - 4ac}}{2a}$ is called the _____ formula.

CONCEPTS

3. Write each quadratic equation in standard form.
   a. $x^2 + 2x = -5$ b. $3x^2 = -2x + 1$

4. For each quadratic equation, find the values of a, b, and c.
   a. $x^2 + 5x + 6 = 0$ b. $8x^2 - x = 10$

5. Determine whether each statement is true or false.
   a. Any quadratic equation can be solved by using the quadratic formula.
   b. Any quadratic equation can be solved by completing the square.
   c. Any quadratic equation can be solved by factoring using integers.

6. What is wrong with the beginning of the solution shown below?

   Solve: $x^2 - 3x = 2$
   $a = 1 \quad b = -3 \quad c = 2$

7. Evaluate each expression.
   a. $\dfrac{-2 \pm \sqrt{2^2 - 4(1)(-8)}}{2(1)}$
   b. $\dfrac{-(-1) \pm \sqrt{(-1)^2 - 4(2)(-4)}}{2(2)}$

8. A student used the quadratic formula to solve a quadratic equation and obtained $x = \dfrac{-2 \pm \sqrt{3}}{2}$.
   a. How many solutions does the equation have? What are they exactly?
   b. Graph the solutions on a number line.

9. Simplify each of the following.
   a. $\dfrac{3 \pm 6\sqrt{2}}{3}$ b. $\dfrac{-12 \pm 4\sqrt{7}}{8}$

10. a. Write an expression that represents the width of the larger rectangle shown in red.
    b. Write an expression that represents the length of the larger rectangle shown in red.

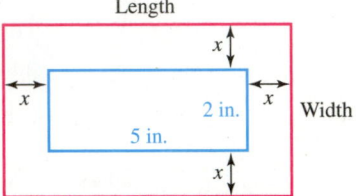

NOTATION

11. On a quiz, students were asked to write the quadratic formula. What is wrong with each answer shown below?
    a. $x = -b \pm \dfrac{\sqrt{b^2 - 4ac}}{2a}$
    b. $x = \dfrac{-b\sqrt{b^2 - 4ac}}{2a}$

12. When reading $\dfrac{-b \pm \sqrt{b^2 - 4ac}}{2a}$, we say, "The _____ of b, plus or _____ the square _____ of b _____ minus __ times a times c, all _____ $2a$."

GUIDED PRACTICE

Use the quadratic formula to solve each equation. See Example 1.

13. $x^2 - 3x + 2 = 0$ 14. $x^2 + 3x + 2 = 0$

15. $x^2 + 12x = -36$ 16. $x^2 - 18x + 81 = 0$

17. $2x^2 + x - 3 = 0$ 18. $6x^2 - x - 1 = 0$

19. $12t^2 - 5t - 2 = 0$ 20. $12z^2 + 5z - 3 = 0$

Solve each equation. Approximate the solutions to the nearest hundredth. See Example 2.

21. $x^2 = x + 7$ 22. $t^2 = t + 4$

23. $5x^2 + 5x = -1$ 24. $2x^2 + 7x = -1$

25. $3y^2 + 1 = -6y$

26. $4w^2 + 1 = -6w$

27. $4m^2 = 4m + 19$

28. $3y^2 = 12y - 4$

Solve each equation. See Example 3.

29. $2x^2 + x + 1 = 0$

30. $2x^2 + 3x + 5 = 0$

31. $3x^2 - 2x + 1 = 0$

32. $3x^2 - 2x + 5 = 0$

33. $x^2 - 2x + 2 = 0$

34. $x^2 - 4x + 8 = 0$

35. $4a^2 + 4a + 5 = 0$

36. $4b^2 + 4b + 17 = 0$

For each equation, write an equivalent quadratic equation that will be easier to solve. Do not solve the equation. See Example 4.

37. a. $-5x^2 + 9x - 2 = 0$

   b. $1.6t^2 + 2.4t - 0.9 = 0$

38. a. $\frac{1}{8}x^2 + \frac{1}{2}x - \frac{3}{4} = 0$

   b. $33y^2 + 99y - 66 = 0$

39. a. $45x^2 + 30x - 15 = 0$

   b. $\frac{1}{3}m^2 - \frac{1}{2}m - \frac{1}{3} = 0$

40. a. $0.6t^2 - 0.1t - 0.2 = 0$

   b. $-a^2 - 15a + 12 = 0$

TRY IT YOURSELF

Solve each equation. Approximate the solutions to the nearest hundredth when appropriate.

41. $x^2 - \frac{14}{15}x = \frac{8}{15}$

42. $x^2 = -\frac{5}{4}x + \frac{3}{2}$

43. $3x^2 - 4x = -2$

44. $2x^2 + 3x = -3$

45. $-16y^2 - 8y + 3 = 0$

46. $-16x^2 - 16x - 3 = 0$

47. $2x^2 - 3x - 1 = 0$

48. $3x^2 - 9x - 2 = 0$

49. $-x^2 + 10x = 18$

50. $-x^2 - 6x - 2 = 0$

51. $x(x - 6) = 391$

52. $x(x - 27) = 280$

53. $x^2 + 5x - 5 = 0$

54. $x^2 - 3x - 27 = 0$

55. $9h^2 - 6h + 7 = 0$

56. $5x^2 = 2x - 1$

57. $50x^2 + 30x - 10 = 0$

58. $120b^2 + 120b - 40 = 0$

59. $0.6x^2 + 0.03 - 0.4x = 0$

60. $2x^2 + 0.1x = 0.04$

61. $\frac{1}{8}x^2 - \frac{1}{2}x + 1 = 0$

62. $\frac{1}{2}x^2 + 3x + \frac{13}{2} = 0$

63. $\frac{a^2}{10} - \frac{3a}{5} + \frac{7}{5} = 0$

64. $\frac{c^2}{4} + c + \frac{11}{4} = 0$

65. $\frac{x^2}{2} + \frac{5}{2}x = -1$

66. $\frac{x^2}{8} - \frac{x}{4} = \frac{1}{2}$

67. $900x^2 - 8{,}100x = 1{,}800$

68. $14x^2 - 21x = 49$

69. $\frac{1}{4}x^2 - \frac{1}{6}x - \frac{1}{6} = 0$

70. $81x^2 + 12x - 80 = 0$

71. Let $f(x) = 0.7x^2 - 3.5x$. For what value(s) of x is $f(x) = 25$?

72. Let $g(x) = 4.5x^2 + 0.2x$. For what value(s) of x is $g(x) = 3.75$?

Look Alikes . . .

73. a. $a^2 + 4a - 7 = 0$

   b. $a^2 - 4a - 7 = 0$

74. a. $n^2 + 6n - 2 = 0$

   b. $n^2 + 6n + 2 = 0$

75. a. $(x + 2)(x - 4) = 16$

   b. $(x + 2)(x - 4) = -16$

76. a. $5n^2 - 14n - 3 = 0$

   b. $-5n^2 - 14n - 3 = 0$

77. a. $x^2 - 42x + 441 = 0$

   b. $x^2 + 42x + 441 = 0$

78. a. $0.3y^2 + 0.6y + 0.5 = 0$

   b. $0.003y^2 + 0.006y + 0.005 = 0$

APPLICATIONS

79. Crosswalks. Refer to the illustration below. Instead of using the Main Street and First Avenue crosswalks to get from Nordstrom to Best Buy, a shopper uses the diagonal crosswalk to walk 97 feet directly from one corner to the other. If the length of the Main Street crosswalk is 7 feet longer than the First Avenue crosswalk, how much walking does the shopper save by using the diagonal crosswalk?

80. Badminton. The person who wrote the instructions for setting up the badminton net shown below forgot to give the specific dimensions for securing the pole. How long is the support string?

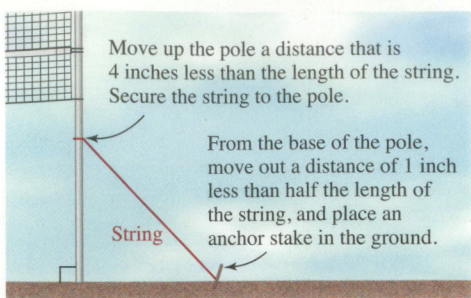

Move up the pole a distance that is 4 inches less than the length of the string. Secure the string to the pole.

From the base of the pole, move out a distance of 1 inch less than half the length of the string, and place an anchor stake in the ground.

String

81. Right Triangles. The hypotenuse of a right triangle is 2.5 units long. The longer leg is 1.7 units longer than the shorter leg. Find the lengths of the legs of the triangle.

82. Televisions. The screen size of a television is measured diagonally from one corner to the opposite corner. In 2007, Sharp developed the world's largest TV screen to date—a 108-inch flat-panel liquid crystal display (LCD). Find the width and the height of the rectangular screen if the width were 41 inches greater than the height. Round to the nearest inch.

83. IMAX Screens. The largest permanent movie screen is in the Panasonic Imax theater at Darling Harbor, Sydney, Australia. The rectangular screen has an area of 11,349 square feet. Find the dimensions of the screen if it is 20 feet longer than it is wide.

84. World's Largest LED Screen. A huge suspended LED screen is the centerpiece of The Place, a popular mall in Beijing, China. Find the length and width of the rectangular screen if the length is 10 meters more than 8 times its width, and the viewable area is 7,500 square meters.

©China Photos/Alamy

85. Parks. Central Park is one of New York's best-known landmarks. Rectangular in shape, its length is 5 times its width. When measured in miles, its perimeter numerically exceeds its area by 4.75. Find the dimensions of Central Park if we know that its width is less than 1 mile.

86. History. One of the important cities of the ancient world was Babylon. Greek historians wrote that the city was square. Measured in miles, its area numerically exceeded its perimeter by about 124. Find its dimensions. (Round to the nearest tenth.)

87. Polygons. A five-sided polygon, called a *pentagon*, has 5 diagonals. The number of diagonals d of a polygon of n sides is given by the formula $d = \frac{n(n-3)}{2}$. Find the number of sides of a polygon if it has 275 diagonals.

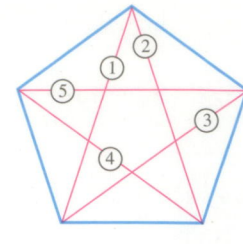

88. Metal Fabrication. A box with no top is to be made by cutting a 2-inch square from each corner of a square sheet of metal. After bending up the sides, the volume of the box is to be 220 cubic inches. Find the length of a side of the square sheet of metal that should be used in the construction of the box. Round to the nearest hundredth.

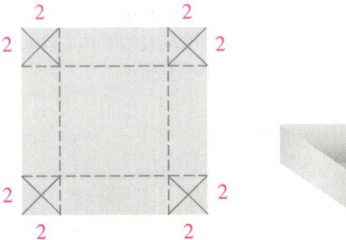
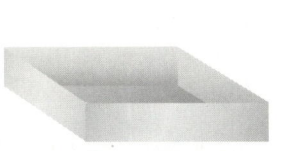

89. Dances. Tickets to a school dance cost $4 and the projected attendance is 300 people. It is further projected that for every 10¢ increase in ticket price, the average attendance will decrease by 5. At what ticket price will the receipts from the dance be $1,248?

90. Ticket Sales. A carnival usually sells three thousand 75¢ ride tickets on a Saturday. For each 15¢ increase in price, management estimates that 80 fewer tickets will be sold. What increase in ticket price will produce $2,982 of revenue on Saturday?

91. Magazine Sales. The *Gazette's* profit is $20 per year for each of its 3,000 subscribers. Management estimates that the profit per subscriber will increase by 1¢ for each additional subscriber over the current 3,000. How many subscribers will bring a total profit of $120,000?

92. Investment Rates. A woman invests $1,000 in a fund for which interest is compounded annually at a rate r. After one year, she deposits an additional $2,000. After two years, the balance in the account is $1,000(1 + r)^2 + $2,000(1 + r)$. If this amount is $3,368.10, find r.

93. **from Campus to Careers**

Police Patrol Officer

The number of female police officers (sworn) in the United States is approximated by the function $f(t) = -24t^2 + 1{,}534t + 72{,}065$, where t is the number of years after 2000. In what year does the model indicate that the number of female officers reached 75,000? (Source: *Crime in the United States, 2000–2010*)

94. Shopping Centers.   The number of shopping centers in the United States is approximated by the function $s(t) = 48t^2 + 581t + 77,383$, where t is the number of years after 1990. In what year does the model indicate that the number of shopping centers reached 100,000? (Source: U.S. Census Bureau)

95. Picture Framing.   The matting around the picture has a uniform width. How wide is the matting if its area equals the area of the picture? Round to the nearest hundredth of an inch.

96. Swimming Pools.   In the advertisement shown, how wide will the free concrete decking be if a uniform width is constructed around the perimeter of the pool? Round to the nearest hundredth of a yard. (*Hint:* Note the difference in units. Convert the dimensions of the pool to yards.)

SAHARA POOL & SPA
SUMMER SPECIAL
This 18 ft x 30 ft pool: only $28,500
Buy now and receive 28 square yards of concrete decking *FREE!*

97. Dimensions of Rectangle.   A rectangle is 4 feet longer than it is wide, and its area is 20 square feet. Find its dimensions to the nearest tenth of a foot.

98. Dimensions of a Triangle.   The height of a triangle is 4 meters longer than twice its base. Find the base and height if the area of the triangle is 10 square meters. Round to the nearest hundredth of a meter.

WRITING

99. Explain why the quadratic formula, in most cases, is easier to use to solve a quadratic equation than is the method of completing the square.

100. On an exam, a student was asked to solve the equation $-4w^2 - 6w - 1 = 0$. Her first step was to multiply both sides of the equation by -1. She then used the quadratic formula to solve $4w^2 + 6w + 1 = 0$ instead. Is this a valid approach? Explain.

REVIEW

Change each radical to an exponential expression.

101. $\sqrt{n}$

102. $\sqrt[7]{8r^2s}$

103. $\sqrt[4]{3b}$

104. $3\sqrt[3]{c^2 - d^2}$

Write each expression in radical form.

105. $t^{1/3}$

106. $(3m^2n^2)^{1/5}$

107. $(3t)^{1/4}$

108. $(c^2 + d^2)^{1/2}$

CHALLENGE PROBLEMS

All of the equations we have solved so far have had rational-number coefficients. However, the quadratic formula can be used to solve quadratic equations with irrational or even imaginary coefficients. Solve each equation.

109. $x^2 + 2\sqrt{2}x - 6 = 0$

110. $\sqrt{2}x^2 + x - \sqrt{2} = 0$

111. $x^2 - 3ix - 2 = 0$

112. $100ix^2 + 300x - 200i = 0$

SECTION 8.3

OBJECTIVES

1. Use the discriminant to determine number and type of solutions.

2. Solve equations that are quadratic in form.

3. Solve application problems using quadratic equations.

The Discriminant and Equations That Can Be Written in Quadratic Form

ARE YOU READY?

The following problems review some basic skills that are needed when working with discriminants and equations that can be written in quadratic form.

1. Find the value of the expression within the radical symbol only:

$$\frac{6 \pm \sqrt{(-6)^2 - 4(3)(-4)}}{6}$$

2. Fill in the blanks: **a.** $x^4 = (x^2)^{\square}$

   b. $(x^{1/3})^{\square} = x$

3. Fill in the blanks: **a.** $\left(\sqrt{x}\right)^2 = \square$

   b. $15a^{-2} = \dfrac{15}{\square}$

4. Solve: $x^2 = -4$

5. Solve: **a.** $\sqrt{x} = 3$

   b. $\sqrt[3]{x} = -\dfrac{1}{2}$

6. Solve: $\dfrac{1}{a} = \dfrac{1}{5}$

We have seen that solutions of the quadratic equation $ax^2 + bx + c = 0$ with $a \neq 0$ are given by the formula

$$x = \frac{-b \pm \sqrt{b^2 - 4ac}}{2a}$$

In this section, we will examine the radicand within the quadratic formula to distinguish or "*discriminate*" among the three types of solutions—rational, irrational, or imaginary.

1 Use the Discriminant to Determine Number and Type of Solutions.

The expression $b^2 - 4ac$ that appears under the radical symbol in the quadratic formula is called the **discriminant.** The discriminant can be used to predict what kind of solutions a quadratic equation has without solving it.

The Discriminant

For a quadratic equation of the form $ax^2 + bx + c = 0$ with rational-number coefficients and $a \neq 0$, the expression $b^2 - 4ac$ is called the **discriminant** and can be used to determine the number and type of the solutions of the equation.

Discriminant: $b^2 - 4ac$	*Number and type of solutions*
Positive	Two different real numbers
0	One repeated solution, a rational number
Negative	Two different imaginary numbers that are complex conjugates

Discriminant: $b^2 - 4ac$	*Number and type of solutions*
A perfect square	Two different rational numbers
Positive and not a perfect square	Two different irrational numbers

---

EXAMPLE 1 Determine the number and type of solutions for each equation:
a. $x^2 + x + 1 = 0$ **b.** $3x^2 + 5x + 2 = 0$

Strategy We will identify the values of a, b, and c in each equation. Then we will use those values to compute $b^2 - 4ac$, the discriminant.

Why Once we know whether the discriminant is positive, 0, or negative, and whether it is a perfect square, we can determine the number and type of the solutions of the equation.

Solution **a.** For $x^2 + x + 1 = 0$, the discriminant is:

$$b^2 - 4ac = 1^2 - 4(1)(1) \qquad \text{Substitute: } a = 1, b = 1, \text{ and } c = 1.$$
$$= -3 \qquad\qquad \text{The result is a negative number.}$$

Since $b^2 - 4ac < 0$, the solutions of $x^2 + x + 1 = 0$ are two different imaginary numbers that are complex conjugates.

b. For $3x^2 + 5x + 2 = 0$, the discriminant is:

$$b^2 - 4ac = 5^2 - 4(3)(2) \qquad \text{Substitute: } a = 3, b = 5, \text{ and } c = 2.$$
$$= 25 - 24$$
$$= 1 \qquad\qquad \text{The result is a positive number.}$$

Since $b^2 - 4ac > 0$ and $b^2 - 4ac$ is a perfect square, the solutions of $3x^2 + 5x + 2 = 0$ are two different rational numbers.

Success Tip

The discriminant also can be used to determine factorability. The trinomial $ax^2 + bx + c$ with integer coefficients and $a \neq 0$ will factor as the product of two binomials with integer coefficients if the value of $b^2 - 4ac$ is a perfect square. If $b^2 - 4ac = 0$, the factors will be the same.

Self Check 1 Determine the number and type of solutions for:
a. $x^2 + x - 1 = 0$
b. $3x^2 + 4x + 2 = 0$

Now Try ▶ Problems 11, 13, and 15

2 Solve Equations That Are Quadratic in Form.

We have discussed four methods that are used to solve quadratic equations. The table below shows some advantages and disadvantages of each method.

Method	Advantages	Disadvantages	Examples
Factoring and the zero-factor property	It can be very fast. When each factor is set equal to 0, the resulting equations are usually easy to solve.	Some polynomials may be difficult to factor and others impossible.	$x^2 - 2x - 24 = 0$ $4a^2 - a = 0$
Square root property	It is the fastest way to solve equations of the form $ax^2 = n$ (n = a number) or $(ax + b)^2 = n$.	It applies only to equations that are in these forms.	$x^2 = 27$ $(2y + 3)^2 = 25$
Completing the square*	It can be used to solve any quadratic equation. It works well with equations of the form $x^2 + bx = n$, where b is even.	It involves more steps than the other methods. The algebra can be cumbersome if the leading coefficient is not 1 or if b is odd or a fraction.	$x^2 + 4x = -1$ $t^2 - 14t - 9 = 0$
Quadratic formula	It can be used to solve any quadratic equation.	It involves several calculations in which sign errors can be made. Often the result must be simplified.	$x^2 + 3x - 33 = 0$ $4s^2 - 10s + 5 = 0$

> **Success Tip**
>
> The solutions of a quadratic equation can always be found using the quadratic formula. The solutions cannot always be found using factoring.

*The quadratic formula is just a condensed version of completing the square and is usually easier to use. However, you need to know how to complete the square because it is used in more advanced mathematics courses.

To determine the most efficient method for a given equation, we can use the following strategy.

> **Strategy for Solving Quadratic Equations**
>
> 1. See whether the equation is in a form such that the **square root method** is easily applied.
> 2. If step 1 does not apply, write the equation in $ax^2 + bx + c = 0$ form.
> 3. See whether the equation can be solved using the **factoring method**.
> 4. If you can't factor, solve the equation by the **quadratic formula. Completing the square** can also be used.

Many nonquadratic equations can be written in quadratic form ($ax^2 + bx + c = 0$) and solved using the techniques discussed in previous sections. For example, a careful inspection of the equation $x^4 - 5x^2 + 4 = 0$ leads to the following observations:

The leading term, x^4, is the square of the expression x^2 in the middle term: $x^4 = (x^2)^2$.

$$x^4 - 5x^2 + 4 = 0$$

The last term is a constant.

Equations that contain an expression, the same expression squared, and a constant term are said to be **quadratic in form**. One method used to solve such equations is to make a substitution.

EXAMPLE 2   Solve: $x^4 - 3x^2 - 4 = 0$

Strategy   Since the leading term, x^4, is the square of the expression x^2 in the middle term, we will substitute y for x^2.

Why   Our hope is that such a substitution will produce an equation that we can solve using one of the methods previously discussed.

Solution   If we write x^4 as $(x^2)^2$, the equation takes the form

$$(x^2)^2 - 3x^2 - 4 = 0$$

and it is said to be **quadratic in x^2**. We can solve this equation by letting $y = x^2$.

$$y^2 - 3y - 4 = 0 \qquad \text{Replace each } x^2 \text{ with } y.$$

> **Notation**
>
> The choice of the letter y for the substitution is arbitrary. We could just as well let $b = x^2$.

We can solve the resulting quadratic equation by factoring.

$$(y - 4)(y + 1) = 0 \qquad \text{Factor } y^2 - 3y - 4.$$

$$y - 4 = 0 \quad \text{or} \quad y + 1 = 0 \qquad \text{Set each factor equal to 0.}$$

$$y = 4 \qquad \qquad \qquad y = -1 \qquad \text{Solve for y.}$$

> **Caution**
>
> If you are solving an equation in x, you can't answer with values of y. Remember to reverse (undo) any substitutions, and solve for the variable in the original equation.

These are not the solutions for x. To find x, we reverse the substitution by replacing each y with x^2 and proceed as follows:

$$x^2 = 4 \qquad \text{or} \qquad x^2 = -1 \qquad \text{Undo the substitution. Substitute } x^2 \text{ for } y.$$

$$x = \pm\sqrt{4} \qquad \qquad x = \pm\sqrt{-1} \qquad \text{Use the square root property. Don't forget } \pm.$$

$$x = \pm 2 \qquad \qquad \quad x = \pm i \qquad \text{Simplify each radical.}$$

This equation has four solutions: 2, -2, i, and $-i$. Check each one in the original equation. In general, a fourth-degree polynomial equation, such as this, can have up to four distinct solutions.

Self Check 2   Solve: $x^4 - 5x^2 - 36 = 0$

Now Try ▶ Problem 23

EXAMPLE 3   Solve: $x - 7\sqrt{x} + 12 = 0$

Strategy   Since the leading term, x, is the square of the expression $\sqrt{x}$ in the middle term, we will substitute y for $\sqrt{x}$.

Why   Our hope is that such a substitution will produce an equation that we can solve using one of the methods previously discussed.

Solution   We examine the leading term and the middle term.

The leading term, x, is the square of the expression $\sqrt{x}$ in middle term: $x = \left(\sqrt{x}\right)^2$. $x - 7\sqrt{x} + 12 = 0$ The last term is a constant.

If we write x as $\left(\sqrt{x}\right)^2$, the equation takes the form

$$\left(\sqrt{x}\right)^2 - 7\sqrt{x} + 12 = 0$$

> **The Language of Algebra**
>
> Equations such as
>
> $x - 7\sqrt{x} + 12 = 0$ that are quadratic in form are also said to be **reducible** to a quadratic.

and it is said to be **quadratic in** $\sqrt{x}$. We can solve this equation by letting $y = \sqrt{x}$ and factoring.

$$y^2 - 7y + 12 = 0 \qquad \text{Replace each } \sqrt{x} \text{ with } y. \text{ This is a quadratic equation.}$$

$$(y - 3)(y - 4) = 0 \qquad \text{Factor the trinomial.}$$

$$y - 3 = 0 \quad \text{or} \quad y - 4 = 0 \qquad \text{Set each factor equal to 0.}$$

$$y = 3 \qquad \qquad \quad y = 4 \qquad \text{Solve for y.}$$

To find x, we reverse the substitution and replace each y with $\sqrt{x}$. Then we solve the resulting radical equations by squaring both sides.

$$\sqrt{x} = 3 \quad \text{or} \quad \sqrt{x} = 4 \qquad \text{Undo the substitution.}$$
$$x = 9 \qquad\qquad x = 16$$

The solutions are 9 and 16. Check each solution in the original equation.

Self Check 3 Solve: $x + \sqrt{x} - 6 = 0$

Now Try Problem 27

EXAMPLE 4 Solve: $2m^{2/3} - 2 = 3m^{1/3}$

Strategy We will write the equation in descending powers of m and look for a possible substitution to make.

Why Our hope is that a substitution will produce an equation that we can solve using one of the methods previously discussed.

Solution After writing the equation in descending powers of m, we see that

$$2m^{2/3} - 3m^{1/3} - 2 = 0$$

is **quadratic in $m^{1/3}$**, because $m^{2/3} = (m^{1/3})^2$. We will use the substitution $y = m^{1/3}$ to write this equation in quadratic form.

$$2m^{2/3} - 3m^{1/3} - 2 = 0 \qquad \text{This is the equation to solve.}$$
$$2(m^{1/3})^2 - 3m^{1/3} - 2 = 0 \qquad \text{Write } m^{2/3} \text{ as } (m^{1/3})^2.$$
$$2y^2 - 3y - 2 = 0 \qquad \text{Substitute: Replace each } m^{1/3} \text{ with } y.$$
$$\text{This is a quadratic equation.}$$
$$(2y + 1)(y - 2) = 0 \qquad \text{Factor } 2y^2 - 3y - 2.$$
$$2y + 1 = 0 \quad \text{or} \quad y - 2 = 0 \qquad \text{Set each factor equal to 0.}$$
$$y = -\frac{1}{2} \qquad\qquad y = 2 \qquad \text{Solve for } y.$$

The Language of Algebra

Recall that in $2m^{2/3}$, we call 2/3 a **rational** or **fractional exponent**.

To find m, we reverse the substitution and replace each y with $m^{1/3}$. Then we solve the resulting equations by cubing both sides.

$$m^{1/3} = -\frac{1}{2} \quad \text{or} \quad m^{1/3} = 2 \qquad \text{Undo the substitution.}$$
$$(m^{1/3})^3 = \left(-\frac{1}{2}\right)^3 \quad\quad (m^{1/3})^3 = (2)^3 \qquad \text{Recall that } m^{1/3} = \sqrt[3]{m}. \text{ To solve for } m, \text{ cube both sides.}$$
$$m = -\frac{1}{8} \qquad\qquad m = 8 \qquad \text{Find each power.}$$

Success Tip

We could write $m^{1/3} = -\frac{1}{2}$ in the equivalent radical form $\sqrt[3]{m} = -\frac{1}{2}$ and solve in the same way—by cubing both sides.

The solutions are $-\frac{1}{8}$ and 8. Check each solution in the original equation.

Self Check 4 Solve: $a^{2/3} = -3a^{1/3} + 10$

Now Try Problem 31

EXAMPLE 5 Solve: $(4t + 2)^2 - 30(4t + 2) + 224 = 0$

Strategy Since the leading term, $(4t + 2)^2$, is the square of the expression $4t + 2$ in the middle term, we will substitute y for $4t + 2$.

Why Our hope is that such a substitution will produce an equation that we can solve using one of the methods previously discussed.

Solution This equation is **quadratic in $4t + 2$**. If we make the substitution $y = 4t + 2$, we obtain

$$y^2 - 30y + 224 = 0$$

which can be solved by using the quadratic formula.

$$y = \frac{-b \pm \sqrt{b^2 - 4ac}}{2a}$$

$$y = \frac{-(-30) \pm \sqrt{(-30)^2 - 4(1)(224)}}{2(1)}$$ Substitute 1 for a, -30 for b, and 224 for c.

$$y = \frac{30 \pm \sqrt{4}}{2}$$ Evaluate within the radical.

$$y = \frac{30 \pm 2}{2}$$ Evaluate the radical: $\sqrt{4} = 2$.

$$y = 16 \quad \text{or} \quad y = 14$$ Evaluate: $\frac{30+2}{2} = 16$ and $\frac{30-2}{2} = 14$

To find t, we reverse the substitution and replace y with $4t + 2$. Then we solve for t.

$$4t + 2 = 16 \quad \text{or} \quad 4t + 2 = 14$$ Undo the substitution.
$$4t = 14 \qquad\qquad 4t = 12$$ Isolate the variable term, $4t$.
$$t = 3.5 \qquad\qquad t = 3$$ Solve for t.

Verify that 3.5 and 3 satisfy the original equation.

Self Check 5 Solve: $(n + 3)^2 - 6(n + 3) = -8$

Now Try Problem 35

EXAMPLE 6 Solve: $15a^{-2} - 8a^{-1} + 1 = 0$

Strategy We will write the equation with positive exponents and look for a possible substitution to make.

Why Our hope is that a substitution will produce an equation that we can solve using one of the methods previously discussed.

Solution When we write the terms $15a^{-2}$ and $-8a^{-1}$ using positive exponents, we see that this equation is **quadratic in $\frac{1}{a}$**.

Success Tip

We could also solve this equation by multiplying both sides by the LCD, a^2.

$$\frac{15}{a^2} - \frac{8}{a} + 1 = 0$$ Think of this equation as $15 \cdot \left(\frac{1}{a}\right)^2 - 8 \cdot \frac{1}{a} + 1 = 0$. Note that 0 is not a possible solution.

If we let $y = \frac{1}{a}$, the resulting quadratic equation can be solved by factoring.

$$15y^2 - 8y + 1 = 0$$ Substitute y^2 for $\frac{1}{a^2}$ and y for $\frac{1}{a}$. This is a quadratic equation.

$$(5y - 1)(3y - 1) = 0$$ Factor $15y^2 - 8y + 1$.
$$5y - 1 = 0 \quad \text{or} \quad 3y - 1 = 0$$ Set each factor equal to 0.
$$y = \frac{1}{5} \qquad\qquad y = \frac{1}{3}$$ Solve for y.

To find a, we reverse the substitution and replace each y with $\frac{1}{a}$. Then we proceed as follows:

$$\frac{1}{a} = \frac{1}{5} \quad \text{or} \quad \frac{1}{a} = \frac{1}{3} \qquad \textcolor{red}{\text{Undo the substitution.}}$$

$$5 = a \qquad\qquad\quad 3 = a \qquad \textcolor{red}{\text{Solve the proportions by finding the cross products.}}$$

The solutions are 5 and 3. Check each solution in the original equation.

Self Check 6   Solve: $28c^{-2} - 3c^{-1} - 1 = 0$

Now Try ▶ Problem 41

---

3 Solve Application Problems Using Quadratic Equations.

EXAMPLE 7

Electronic Temperature Control

Water Temp

Household Appliances. The illustration shows a temperature control on a washing machine. When the *warm* setting is selected, both the hot and cold water pipes open to fill the tub in 2 minutes 15 seconds. When the *cold* setting is chosen, the tub fills 45 seconds faster than when the *hot* setting is used. How long does it take to fill the washing machine if the *hot* setting is used?

Analyze   It is helpful to organize the facts of this shared-work problem in a table. We note that the hot and cold water inlets will be open for the same time: 2 minutes 15 seconds, or 135 seconds.

Assign   Let x = the number of seconds it takes to fill the tub with hot water. Since the cold water inlet fills the tub in 45 seconds less time, $x - 45$ = the number of seconds it takes to fill the tub with cold water.

Form   To determine the work completed by each inlet, multiply the rate by the time.

	Rate · Time = Work completed		
Hot water	$\frac{1}{x}$	135	$\frac{135}{x}$
Cold water	$\frac{1}{x - 45}$	135	$\frac{135}{x - 45}$

Enter this information first.   Multiply to get each of these entries: $W = rt$.

In shared-work problems, 1 represents one whole job completed. So we have

The fraction of tub filled with hot water	plus	the fraction of the tub filled with cold water	equals	1 tub filled.
$\frac{135}{x}$	$+$	$\frac{135}{x - 45}$	$=$	1

Solve

$$\frac{135}{x} + \frac{135}{x - 45} = 1$$

This is a rational equation.

$$x(x - 45)\left(\frac{135}{x} + \frac{135}{x - 45}\right) = x(x - 45)(1)$$

Multiply both sides by the LCD $x(x - 45)$ to clear the equation of fractions.

$$x(x - 45)\frac{135}{x} + x(x - 45)\frac{135}{x - 45} = x(x - 45)(1)$$

Distribute $x(x - 45)$.

$$135(x - 45) + 135x = x(x - 45)$$

Simplify each side.

$$135x - 6{,}075 + 135x = x^2 - 45x$$

Distribute 135 and x.

$$270x - 6{,}075 = x^2 - 45x$$

Combine like terms.

$$0 = x^2 - 315x + 6{,}075$$

Get 0 on the left side.

To solve for x, we will use the quadratic formula: $a = 1$, $b = -315$, and $c = 6{,}075$.

$$x = \frac{-b \pm \sqrt{b^2 - 4ac}}{2a}$$

$$x = \frac{-(-315) \pm \sqrt{(-315)^2 - 4(1)(6{,}075)}}{2(1)}$$

Substitute 1 for a, -315 for b, and 6,075 for c.

$$x = \frac{315 \pm \sqrt{74{,}925}}{2}$$

Evaluate the expression within the radical.

$$x \approx 294 \quad \text{or} \quad x \approx 21$$

Use a calculator to find each solution.

State We can discard the solution of 21 seconds, because this would imply that the cold water inlet fills the tub in a negative number of seconds ($21 - 45 = -24$). Therefore, the hot water inlet fills the washing machine tub in about 294 seconds, which is 4 minutes 54 seconds.

Check Use estimation to check the result. The work completed by the hot water inlet is $\frac{135}{294} \approx 0.46$ and the work completed by the cold water inlet is $\frac{135}{294 - 45} \approx 0.54$. Since $0.46 + 0.54 \approx 1$, the result seems reasonable.

Self Check 7 **Mowing Lawns.** Carly can mow a lawn in 1 hour less time than her friend Lindsey. Together they can finish the job in 5 hours. How long would it take Carly if she worked alone?

Now Try ▶ Problem 87

SECTION 8.3 ▶ **STUDY SET**

VOCABULARY

Fill in the blanks.

1. For the quadratic equation $ax^2 + bx + c = 0$, the _____ is $b^2 - 4ac$.

2. We can solve $x - 2\sqrt{x} - 8 = 0$ by making the _____ $y = \sqrt{x}$.

CONCEPTS

Consider the quadratic equation $ax^2 - bx + c = 0$, where a, b, and c represent rational numbers, and fill in the blanks.

3. If $b^2 - 4ac < 0$, the solutions of the equation are two different imaginary numbers that are complex _____.

4. If $b^2 - 4ac = $ ▨, the equation has one repeated rational-number solution.

5. If $b^2 - 4ac$ is a perfect square, the solutions of the equation are two different _____ numbers.

6. If $b^2 - 4ac$ is positive and not a perfect square, the solutions of the equation are two different _____ numbers.

7. For each equation, determine the substitution that should be made to write the equation in quadratic form.

    a. $x^4 - 12x^2 + 27 = 0$ Let $y =$ ▭

    b. $x - 13\sqrt{x} + 40 = 0$ Let $y =$ ▭

    c. $x^{2/3} + 2x^{1/3} - 3 = 0$ Let $y =$ ▭

    d. $x^{-2} - x^{-1} - 30 = 0$ Let $y =$ ▭

    e. $(x + 1)^2 - (x + 1) - 6 = 0$ Let $y =$ ▭

8. Fill in the blanks.

    a. $x^4 = \left(\text{▭}\right)^2$ **b.** $x = \left(\text{▭}\right)^2$

    c. $x^{2/3} = \left(\text{▭}\right)^2$ **d.** $\dfrac{1}{x^2} = \left(\text{▭}\right)^2$

NOTATION

Complete the solution.

9. To find the type of solutions for the equation $x^2 + 5x + 6 = 0$, we calculate the discriminant.

$$b^2 - \text{▭} = \text{▭}^2 - 4(1)\left(\text{▭}\right)$$
$$= 25 - \text{▭}$$
$$= 1$$

Since a, b, and c are rational numbers and the value of the discriminant is a perfect square, the solutions are two different _____ numbers.

10. Fill in the blanks to write each equation in quadratic form.

    a. $x^4 - 2x^2 - 15 = 0 \rightarrow \left(\text{▭}\right)^2 - 2\,\text{▭} - 15 = 0$

    b. $x - 2\sqrt{x} + 3 = 0 \rightarrow \left(\text{▭}\right)^2 - 2\,\text{▭} + 3 = 0$

    c. $8m^{2/3} - 10m^{1/3} - 3 = 0 \rightarrow 8\left(\text{▭}\right)^2 - 10\,\text{▭} - 3 = 0$

GUIDED PRACTICE

Use the discriminant to determine the number and type of solutions for each equation. Do not solve. See Example 1.

11. $4x^2 - 4x + 1 = 0$ **12.** $6x^2 - 5x - 6 = 0$

13. $5x^2 + x + 2 = 0$ **14.** $3x^2 + 10x - 2 = 0$

15. $2x^2 = 4x - 1$ **16.** $9x^2 = 12x - 4$

17. $x(2x - 3) = 20$ **18.** $x(x - 3) = -10$

19. $3x^2 - 10 = 0$ **20.** $5x^2 - 24 = 0$

21. $x^2 - \dfrac{14}{15}x = \dfrac{8}{15}$ **22.** $x^2 = -\dfrac{5}{4}x + \dfrac{3}{2}$

Solve each equation. See Example 2.

23. $x^4 - 17x^2 + 16 = 0$ **24.** $x^4 - 10x^2 + 9 = 0$

25. $x^4 + 5x^2 - 36 = 0$ **26.** $x^4 - 15x^2 - 16 = 0$

Solve each equation. See Example 3.

27. $x - 13\sqrt{x} + 40 = 0$ **28.** $x - 9\sqrt{x} + 18 = 0$

29. $2x + \sqrt{x} - 3 = 0$ **30.** $2x - \sqrt{x} - 1 = 0$

Solve each equation. See Example 4.

31. $a^{2/3} - 2a^{1/3} = 3$ **32.** $r^{2/3} + 4r^{1/3} = 5$

33. $x^{2/3} + 2x^{1/3} - 8 = 0$ **34.** $x^{2/3} - 7x^{1/3} + 12 = 0$

Solve each equation. See Example 5.

35. $(c + 1)^2 - 4(c + 1) + 3 = 0$

36. $(a - 5)^2 - 4(a - 5) - 21 = 0$

37. $2(2x + 1)^2 - 7(2x + 1) + 6 = 0$

38. $3(2 - x)^2 + 10(2 - x) - 8 = 0$

Solve each equation. See Example 6.

39. $m^{-2} + m^{-1} - 6 = 0$ **40.** $t^{-2} + t^{-1} - 42 = 0$

41. $8x^{-2} - 10x^{-1} - 3 = 0$ **42.** $2x^{-2} - 5x^{-1} - 3 = 0$

Solve each equation. See Example 7.

43. $1 - \dfrac{5}{x} = \dfrac{10}{x^2}$ **44.** $1 - \dfrac{3}{x} = \dfrac{5}{x^2}$

45. $\dfrac{1}{2} + \dfrac{1}{b} = \dfrac{1}{b - 7}$ **46.** $\dfrac{1}{4} - \dfrac{1}{n} = \dfrac{1}{n + 3}$

TRY IT YOURSELF

Solve each equation.

47. $2x - \sqrt{x} = 3$ **48.** $3x + 4\sqrt{x} = 4$

49. $x^{-2} + 2x^{-1} - 3 = 0$ **50.** $x^{-2} + 2x^{-1} - 8 = 0$

51. $x^4 + 19x^2 + 18 = 0$ **52.** $t^4 + 4t^2 - 5 = 0$

53. $(k - 7)^2 + 6(k - 7) + 10 = 0$

54. $(d + 9)^2 - 4(d + 9) + 8 = 0$

55. $\dfrac{2}{x - 1} + \dfrac{1}{x + 1} = 3$ **56.** $\dfrac{3}{x - 2} - \dfrac{1}{x + 2} = 5$

57. $x - 6x^{1/2} = -8$ **58.** $x - 5x^{1/2} + 4 = 0$

59. $(y^2 - 9)^2 + 2(y^2 - 9) - 99 = 0$

60. $(a^2 - 4)^2 - 4(a^2 - 4) - 32 = 0$

61. $x^{-4} - 2x^{-2} + 1 = 0$ **62.** $4x^{-4} + 1 = 5x^{-2}$

63. $t^4 + 3t^2 = 28$ **64.** $3h^4 + h^2 - 2 = 0$

65. $2x^{2/5} - 5x^{1/5} = -3$ **66.** $2x^{2/5} + 3x^{1/5} = -1$

67. $9\left(\dfrac{3m+2}{m}\right)^2 - 30\left(\dfrac{3m+2}{m}\right) + 25 = 0$

68. $4\left(\dfrac{c-7}{c}\right)^2 - 12\left(\dfrac{c-7}{c}\right) + 9 = 0$

69. $\dfrac{3}{a-1} = 1 - \dfrac{2}{a}$ **70.** $1 + \dfrac{4}{x} = \dfrac{3}{x^2}$

71. $\left(8 - \sqrt{a}\right)^2 + 6\left(8 - \sqrt{a}\right) - 7 = 0$

72. $\left(10 - \sqrt{t}\right)^2 - 4\left(10 - \sqrt{t}\right) - 45 = 0$

73. $x + \dfrac{2}{x-2} = 0$ **74.** $x + \dfrac{x+5}{x-3} = 0$

75. $3x + 5\sqrt{x} + 2 = 0$ **76.** $3x - 4\sqrt{x} + 1 = 0$

77. $x^4 - 6x^2 + 5 = 0$ **78.** $2x^4 - 26x^2 + 24 = 0$

79. $8(t+1)^{-2} - 30(t+1)^{-1} + 7 = 0$

80. $2(s-2)^{-2} + 3(s-2)^{-1} - 5 = 0$

81. $\dfrac{1}{x+2} + \dfrac{24}{x+3} = 13$ **82.** $\dfrac{3}{x} + \dfrac{4}{x+1} = 2$

Look Alikes . . .

83. a. $y^{2/3} + y^{1/3} - 20 = 0$ **b.** $y^{-2} + y^{-1} - 20 = 0$

84. a. $x + 6\sqrt{x} - 16 = 0$ **b.** $x^{2/3} + 6x^{1/3} - 16 = 0$

85. a. $\dfrac{1}{x} = \dfrac{2x}{x+1}$ **b.** $\dfrac{1}{x} + \dfrac{2x}{x+1} = 1$

86. a. $(m^2 - 1)^2 - (m^2 - 1) = 2$ **b.** $m^4 - m^2 = 2$

APPLICATIONS

87.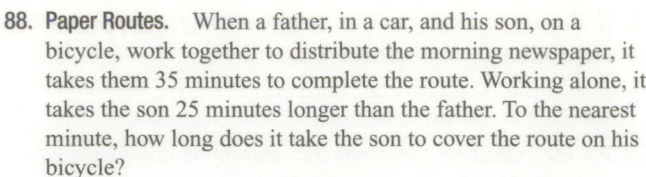

from Campus to Careers

Police Patrol Officer

Crowd Control. After a sporting event at a stadium, police have found that a public parking lot can be emptied in 60 minutes if both the east and west exits are opened. If just the east exit is used, it takes 40 minutes longer to clear the lot than it does if just the west exit is opened. How long does it take to clear the parking lot if every car must use the west exit? Round to the nearest minute.

88. Paper Routes. When a father, in a car, and his son, on a bicycle, work together to distribute the morning newspaper, it takes them 35 minutes to complete the route. Working alone, it takes the son 25 minutes longer than the father. To the nearest minute, how long does it take the son to cover the route on his bicycle?

89. Assembly Lines. A newly manufactured product traveled 300 feet on a high-speed conveyor belt at a rate of r feet per second. It could have traveled the 300 feet in 3 seconds less time if the speed of the conveyor belt was increased by 5 feet per second. Find r.

90. Bicycling. Tina bicycles 160 miles at the rate of r mph. The same trip would have taken 2 hours longer if she had decreased her speed by 4 mph. Find r.

91. Architecture. A **golden rectangle** is one of the most visually appealing of all geometric forms. The Parthenon, built by the Greeks in the 5th century B.C., fits into a golden rectangle if its ruined triangular pediment is included. See the illustration. In a golden rectangle, the length l and width w must satisfy the equation $\dfrac{l}{w} = \dfrac{w}{l-w}$. If a rectangular billboard is to have a width of 20 feet, what should its length be so that it is a golden rectangle? Round to the nearest tenth.

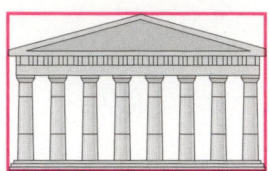

92. Door Designs. An architect needs to determine the height h of the window shown in the illustration. The radius r, the width w, and the height h of the circular-shaped window are related by the formula $r = \dfrac{4h^2 + w^2}{8h}$. If w is to be 34 inches and r is to be 18 inches, find h to the nearest tenth of an inch.

WRITING

93. Describe how to predict what type of solutions the equation $3x^2 - 4x + 5 = 0$ will have.

94. What **error** is made in the following solution?

Solve: $x^4 - 12x^2 + 27 = 0$

$y^2 - 12y + 27 = 0$ Let y = x².

$(y - 9)(y - 3) = 0$

$y - 9 = 0$ or $y - 3 = 0$

$y = 9$ | $y = 3$

The solutions of $x^4 - 12x^2 + 27 = 0$ are 9 and 3.

REVIEW

95. Write an equation of the vertical line that passes through (3, 4).

96. Find an equation for a linear function whose graph passes through $(-1, -6)$ and $(-2, -1)$.

97. Write an equation of the line with slope $\frac{2}{3}$ that passes through the origin.

98. Find an equation of the line that passes through $(2, -3)$ and is perpendicular to the line whose equation is $y = \frac{x}{5} + 6$. Write the equation in slope–intercept form.

100. Solve: $x^6 + 17x^3 + 16 = 0$

101. Solve: $x^3 - x^2 + 16x - 16 = 0$

102. Solve: $\sqrt{x^2 + 10} = 4\sqrt{x}$

CHALLENGE PROBLEMS

99. Find the real-number solutions of $x^4 - 3x^2 - 2 = 0$. Rationalize the denominators of the solutions.

SECTION 8.4

Quadratic Functions and Their Graphs

OBJECTIVES

1 Graph functions of the form $f(x) = ax^2$ and $f(x) = ax^2 + k$.

2 Graph functions of the form $f(x) = a(x - h)^2$ and $f(x) = a(x - h)^2 + k$.

3 Graph functions of the form $f(x) = ax^2 + bx + c$ by completing the square.

4 Find the vertex using $-\frac{b}{2a}$.

5 Determine minimum and maximum values.

6 Solve quadratic equations graphically.

ARE YOU READY?

The following problems review some basic skills that are needed when graphing quadratic functions.

1. Graph: $f(x) = x^2$

2. Let $f(x) = 2(x - 3)^2 - 4$. Find $f(4)$.

3. Complete the square on $x^2 + 8x$. Then factor the resulting trinomial.

4. Complete the square on $x^2 + x$. Then factor the resulting trinomial.

5. Solve: $x^2 + 6x + 9 = 0$

6. Solve: $-2x^2 - 8x - 8 = 0$

7. Evaluate $-\dfrac{b}{2a}$ for $a = -2$ and $b = -20$.

8. Is the graph of $x = -1$ a horizontal or a vertical line?

In this section, we will discuss methods for graphing *quadratic functions*.

Quadratic Functions

A **quadratic function** is a second-degree polynomial function that can be written in the form

$$f(x) = ax^2 + bx + c$$

where a, b, and c are real numbers and $a \neq 0$.

Notation

Since $y = f(x)$, quadratic functions can also be written as $y = ax^2 + bx + c$ and $y = a(x - h)^2 + k$.

Quadratic functions often are written in another form, called **standard form,**

$$f(x) = a(x - h)^2 + k$$

where a, h, and k are real numbers and $a \neq 0$. This form is useful because a, h, and k give us important information about the graph of the function. To develop a strategy for graphing quadratic functions written in standard form, we will begin by considering the simplest case, $f(x) = ax^2$.

---

1 Graph Functions of the Form $f(x) = ax^2$ and $f(x) = ax^2 + k$.

One way to graph quadratic functions is to plot points.

EXAMPLE 1   Graph: **a.** $f(x) = x^2$ **b.** $g(x) = 3x^2$ **c.** $s(x) = \dfrac{1}{3}x^2$

Strategy   We can make a table of values for each function, plot each point, and connect them with a smooth curve.

Why   At this time, this method is our only option.

Solution   After graphing each curve, we see that the graph of $g(x) = 3x^2$ is narrower than the graph of $f(x) = x^2$, and the graph of $s(x) = \dfrac{1}{3}x^2$ is wider than the graph of $f(x) = x^2$. For $f(x) = ax^2$, the smaller the value of $|a|$, the wider the graph.

$f(x) = x^2$

x	$f(x)$
-2	4
-1	1
0	0
1	1
2	4

$g(x) = 3x^2$

x	$g(x)$
-2	12
-1	3
0	0
1	3
2	12

$s(x) = \dfrac{1}{3}x^2$

x	$s(x)$
-2	$\dfrac{4}{3}$
-1	$\dfrac{1}{3}$
0	0
1	$\dfrac{1}{3}$
2	$\dfrac{4}{3}$

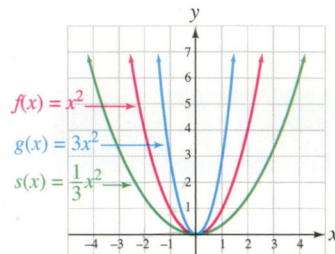

The values of g(x) increase faster than the values of f(x), making its graph steeper.

The values of s(x) increase more slowly than the values of f(x), making its graph flatter.

Self Check 1   Graph: $g(x) = \dfrac{2}{3}x^2$.

Now Try ▶ Problem 15

EXAMPLE 2   Graph: $f(x) = -3x^2$

Strategy   We make a table of values for the function, plot each point, and connect them with a smooth curve.

Why   At this time, this method is our only option.

Solution   After graphing the curve, we see that it opens downward and has the same shape as the graph of $g(x) = 3x^2$ that was graphed in Example 1.

$f(x) = -3x^2$

x	$f(x)$	
-2	-12	→ $(-2, -12)$
-1	-3	→ $(-1, -3)$
0	0	→ $(0, 0)$
1	-3	→ $(1, -3)$
2	-12	→ $(2, -12)$

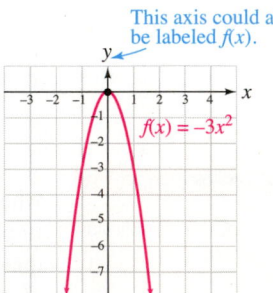

This axis could also be labeled $f(x)$.

Self Check 2   Graph: $f(x) = -\dfrac{1}{3}x^2$

Now Try ▶ Problem 17

The graphs of functions of the form $f(x) = ax^2$ are **parabolas.** The lowest point on a parabola that opens upward, or the highest point on a parabola that opens downward, is called the **vertex** of the parabola. The vertical line, called an **axis of symmetry,** that passes through the vertex divides the parabola into two congruent halves. If we fold the paper along the axis of symmetry, the two sides of the parabola will match.

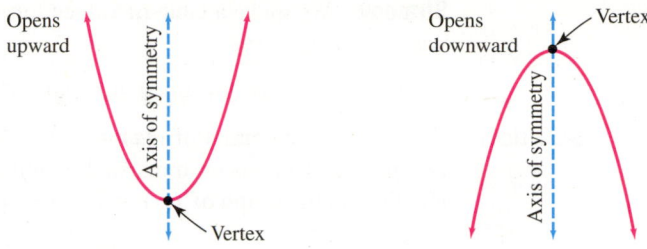

The results from Examples 1 and 2 confirm the following facts.

The Graph of $f(x) = ax^2$	The graph of $f(x) = ax^2$ is a parabola opening upward when $a > 0$ and downward when $a < 0$, with vertex at the point $(0, 0)$ and axis of symmetry the line $x = 0$.

EXAMPLE 3 Graph: **a.** $f(x) = 2x^2$ **b.** $g(x) = 2x^2 + 3$ **c.** $s(x) = 2x^2 - 3$

Strategy We make a table of values for each function, plot each point, and connect them with a smooth curve.

Why At this time, this method is our only option.

Solution After graphing the curves, we see that the graph of $g(x) = 2x^2 + 3$ is identical to the graph of $f(x) = 2x^2$, except that it has been translated 3 units upward. The graph of $s(x) = 2x^2 - 3$ is identical to the graph of $f(x) = 2x^2$, except that it has been translated 3 units downward. In each case, the axis of symmetry is the line $x = 0$.

$f(x) = 2x^2$

x	$f(x)$
-2	8
-1	2
0	0
1	2
2	8

$g(x) = 2x^2 + 3$

x	$g(x)$
-2	11
-1	5
0	3
1	5
2	11

$s(x) = 2x^2 - 3$

x	$s(x)$
-2	5
-1	-1
0	-3
1	-1
2	5

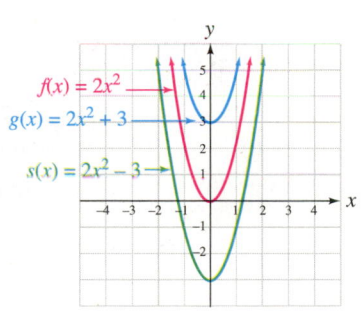

For each x-value, g(x) is 3 more than f(x).

For each x-value, s(x) is 3 less than f(x).

Self Check 3 Graph: $f(x) = 2x^2 + 1$

Now Try ▶ Problem 19

The results of Example 3 confirm the following facts.

The Graph of $f(x) = ax^2 + k$	The graph of $f(x) = ax^2 + k$ is a parabola having the same shape as $f(x) = ax^2$ but translated k units upward if k is positive and $|k|$ units downward if k is negative. The vertex is at the point $(0, k)$, and the axis of symmetry is the line $x = 0$.

2 Graph Functions of the Form $f(x) = a(x - h)^2$ and $f(x) = a(x - h)^2 + k$.

EXAMPLE 4   Graph: **a.** $f(x) = 2x^2$ **b.** $g(x) = 2(x - 3)^2$ **c.** $s(x) = 2(x + 3)^2$

Strategy   We make a table of values for each function, plot each point, and connect them with a smooth curve.

Why   At this time, this method is our only option.

Solution   We note that the graph of $g(x) = 2(x - 3)^2$ below is identical to the graph of $f(x) = 2x^2$, except that it has been translated 3 units to the right. The graph of $s(x) = 2(x + 3)^2$ is identical to the graph of $f(x) = 2x^2$, except that it has been translated 3 units to the left.

$f(x) = 2x^2$

x	$f(x)$
-2	8
-1	2
0	0
1	2
2	8

$g(x) = 2(x - 3)^2$

x	$g(x)$
1	8
2	2
3	0
4	2
5	8

$s(x) = 2(x + 3)^2$

x	$s(x)$
-5	8
-4	2
-3	0
-2	2
-1	8

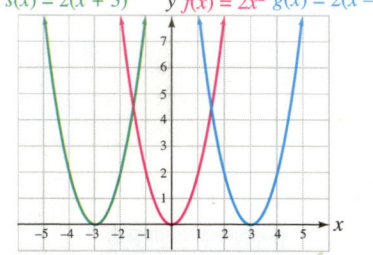

— When an x-value is increased by 3, the function's outputs are the same. —

— When an x-value is decreased by 3, the function's outputs are the same. —

Self Check 4   Graph: $g(x) = 2(x + 1)^2$

Now Try ▶ Problem 21

The results of Example 4 confirm the following facts.

The Graph of
$f(x) = a(x - h)^2$

The graph of $f(x) = a(x - h)^2$ is a parabola having the same shape as $f(x) = ax^2$ but translated h units to the right if h is positive and $|h|$ units to the left if h is negative. The vertex is at the point $(h, 0)$, and the axis of symmetry is the line $x = h$.

The results of Examples 1–4 suggest a general strategy for graphing quadratic functions that are written in the form $f(x) = a(x - h)^2 + k$.

Graphing a Quadratic
Function in Standard Form

The graph of the quadratic function $f(x) = a(x - h)^2 + k$, where $a \neq 0$, is a parabola with vertex at (h, k). The axis of symmetry is the line $x = h$. The parabola opens upward when $a > 0$ and downward when $a < 0$.

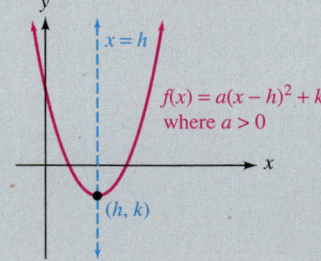

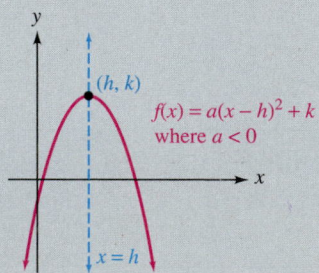

EXAMPLE 5   Graph: $f(x) = 2(x - 3)^2 - 4$. Label the vertex and draw the axis of symmetry.

Strategy   We will determine whether the graph opens upward or downward and find its vertex and axis of symmetry. Then we will plot some points and complete the graph.

Why   This method will be more efficient than plotting many points.

Solution   The graph of $f(x) = 2(x - 3)^2 - 4$ is identical to the graph of $g(x) = 2(x - 3)^2$, except that it has been translated 4 units downward. The graph of $g(x) = 2(x - 3)^2$ is identical to the graph of $s(x) = 2x^2$, except that it has been translated 3 units to the right.

We can learn more about the graph of $f(x) = 2(x - 3)^2 - 4$ by determining a, h, and k.

$$f(x) = \mathbf{2}(x - \mathbf{3})^2 - \mathbf{4}$$
$$f(x) = \mathbf{a}(x - \mathbf{h})^2 + \mathbf{k}$$

$a = 2$, $h = 3$, and $k = -4$

Success Tip

The most important point to find when graphing a quadratic function is the vertex.

Upward/downward: Since $a = 2$ and $2 > 0$, the parabola opens upward.

Vertex: The vertex of the parabola is $(h, k) = (3, -4)$, as shown below.

Axis of symmetry: Since $h = 3$, the axis of symmetry is the line $x = 3$, as shown below.

Plotting points: We can construct a table of values to determine several points on the parabola. Since the x-coordinate of the vertex is 3, we choose the x-values of 4 and 5, find $f(4)$ and $f(5)$, and record the results in a table. Then we plot $(4, -2)$ and $(5, 4)$, and use symmetry to locate two other points on the parabola: $(2, -2)$ and $(1, 4)$. Finally, we draw a smooth curve through the points to get the graph.

Success Tip

When graphing, remember that any point on a parabola to the right of the axis of symmetry yields a second point to the left of the axis of symmetry, and vice versa. Think of it as two-for-one.

$$f(x) = 2(x - 3)^2 - 4$$

x	$f(x)$
4	-2	→ $(4, -2)$
5	4	→ $(5, 4)$

↑

The x-coordinate of the vertex is 3. Choose values for x close to 3 and on the same side of the axis of symmetry.

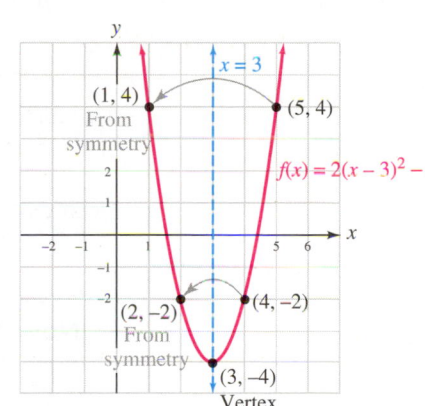

Self Check 5   Graph: $f(x) = 2(x - 1)^2 - 2$. Label the vertex and draw the axis of symmetry.

Now Try ▶   Problems 25 and 31

3 Graph Functions of the Form $f(x) = ax^2 + bx + c$ by Completing the Square.

To graph functions of the form $f(x) = ax^2 + bx + c$, we can complete the square to write the function in standard form $f(x) = a(x - h)^2 + k$.

EXAMPLE 6   Determine the vertex and the axis of symmetry of the graph of $f(x) = x^2 + 8x + 21$. Will the graph open upward or downward?

Strategy   To find the vertex and the axis of symmetry, we will complete the square on x and write the equation of the function in standard form.

Why   Once the equation is written in standard form, we can determine the values of a, h, and k. The coordinates of the vertex will be (h, k) and the equation of the axis of symmetry will be $x = h$. The graph will open upward if $a > 0$ or downward if $a < 0$.

Solution   To determine the vertex and the axis of symmetry of the graph, we complete the square on the right side so we can write the function in $f(x) = a(x - h)^2 + k$ form.

$$f(x) = x^2 + 8x + 21$$

$$f(x) = (x^2 + 8x \quad) + 21 \qquad \text{\color{red}{Prepare to complete the square on x}}$$
$$\text{\color{red}{by writing parentheses around } x^2 + 8x.}$$

To complete the square on $x^2 + 8x$, we note that one-half of the coefficient of x is $\frac{1}{2} \cdot 8 = 4$, and $4^2 = 16$. If we add 16 to $x^2 + 8x$, we obtain a perfect-square trinomial within the parentheses. Since this step adds 16 to the right side, we must also subtract 16 from the right side so that it remains in an equivalent form.

$$\text{\color{blue}{Add 16 to the}} \qquad\qquad \text{\color{blue}{Subtract 16 from the right side to}}$$
$$\text{\color{blue}{right side.}} \qquad\qquad\qquad \text{\color{blue}{counteract the addition of 16.}}$$

$$f(x) = (x^2 + 8x \mathbf{\color{red}{+ 16}}) + 21 \mathbf{\color{red}{- 16}}$$
$$f(x) = (x + 4)^2 + 5 \qquad \text{\color{red}{Factor } x^2 + 8x + 16 \text{ and combine like terms.}}$$

The function is now written in standard form, and we can determine a, h, and k.

$$\text{\color{red}{The standard form requires}}$$
$$\text{\color{red}{a minus symbol here.}}$$

$$f(x) = \left[(x \overset{\downarrow}{-} (\mathbf{-4}) \right]^2 + \mathbf{5} \qquad \text{\color{red}{Write } x + 4 \text{ as } x - (-4) \text{ to determine } h.}$$
$$\overset{\uparrow}{a = 1} \quad \overset{\uparrow}{h = -4} \quad \overset{\uparrow}{k = 5}$$

The vertex is $(h, k) = (-4, 5)$ and the axis of symmetry is the line $x = -4$. Since $a = 1$ and $1 > 0$, the parabola opens upward.

Self Check 6   Determine the vertex and the axis of symmetry of the graph of $f(x) = x^2 + 4x + 10$. Will the graph open upward or downward?

Now Try ▶ Problem 39

EXAMPLE 7   Graph: $f(x) = 2x^2 - 4x - 1$

Strategy   We will complete the square on x and write the equation of the function in standard form, $f(x) = a(x - h)^2 + k$.

Why   When the equation is in standard form, we can identify the values of a, h, and k from the equation. This information will help us sketch the graph.

Solution   Recall that to complete the square on $2x^2 - 4x$, the coefficient of x^2 must be equal to 1. Therefore, we factor 2 from $2x^2 - 4x$.

$$f(x) = 2x^2 - 4x - 1$$
$$f(x) = 2(x^2 - 2x \quad) - 1$$

To complete the square on $x^2 - 2x$, we note that one-half of the coefficient of x is $\frac{1}{2}(-2) = -1$, and $(-1)^2 = 1$. If we add 1 to $x^2 - 2x$, we obtain a perfect-square trinomial within the parentheses. Since this step adds 2 to the right side, we must also subtract 2 from the right side so that it remains in an equivalent form.

By the distributive property, when 1 is added to the expression within the parentheses, $2 \cdot 1 = 2$ is added to the right side.

Subtract 2 to counteract the addition of 2 shown in red.

$$f(x) = 2(x^2 - 2x + 1) - 1 - 2$$
$$f(x) = 2(x - 1)^2 - 3 \qquad \text{Factor } x^2 - 2x + 1 \text{ and combine like terms.}$$

We see that $a = 2$, $h = 1$, and $k = -3$. Thus, the vertex is at the point $(1, -3)$, and the axis of symmetry is $x = 1$. Since $a = 2$ and $2 > 0$, the parabola opens upward. We plot the vertex and axis of symmetry as shown below.

Finally, we construct a table of values, plot the points, use symmetry to plot the corresponding points, and then draw the graph.

$$f(x) = 2x^2 - 4x - 1$$
$$\text{or}$$
$$f(x) = 2(x - 1)^2 - 3$$

x	$f(x)$
2	-1	$\rightarrow (2, -1)$
3	5	$\rightarrow (3, 5)$

The x-coordinate of the vertex is 1. Choose values for x close to 1 and on the same side of the axis of symmetry.

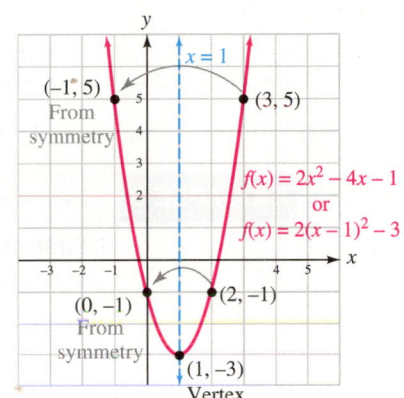

Success Tip

To find additional points on the graph, select values of x that are close to the x-coordinate of the vertex.

Self Check 7 Graph: $f(x) = 3x^2 - 12x + 8$.

Now Try ▶ Problem 45

4 Find the Vertex Using $-\dfrac{b}{2a}$.

Because of symmetry, if a parabola has two x-intercepts, the x-coordinate of the vertex is exactly midway between them. We can use this fact to derive a formula to find the vertex of a parabola.

In general, if a parabola has two x-intercepts, they can be found by solving $0 = ax^2 + bx + c$ for x. We can use the quadratic formula to find the solutions. They are

$$x = \frac{-b - \sqrt{b^2 - 4ac}}{2a} \qquad \text{and} \qquad x = \frac{-b + \sqrt{b^2 - 4ac}}{2a}$$

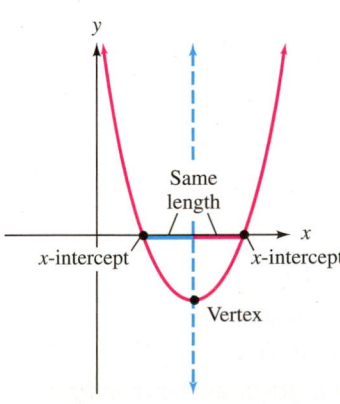

Thus, the parabola's x-intercepts are $\left(\dfrac{-b - \sqrt{b^2 - 4ac}}{2a}, 0\right)$ and $\left(\dfrac{-b + \sqrt{b^2 - 4ac}}{2a}, 0\right)$.

Since the x-value of the vertex of a parabola is halfway between the two x-intercepts, we can find this value by finding the average, or $\frac{1}{2}$ of the sum of the x-coordinates of the x-intercepts.

$$x = \frac{1}{2}\left(\frac{-b - \sqrt{b^2 - 4ac}}{2a} + \frac{-b + \sqrt{b^2 - 4ac}}{2a}\right)$$

$$x = \frac{1}{2}\left(\frac{-b - \sqrt{b^2 - 4ac} + (-b) + \sqrt{b^2 - 4ac}}{2a}\right)$$ Add the numerators and keep the common denominator.

$$x = \frac{1}{2}\left(\frac{-2b}{2a}\right)$$ Combine like terms: $-b + (-b) = -2b$ and $-\sqrt{b^2 - 4ac} + \sqrt{b^2 - 4ac} = 0$.

$$x = -\frac{b}{2a}$$ Remove the common factor of 2 in the numerator and denominator and simplify.

This result is true even if the graph has no x-intercepts.

Formula for the Vertex of a Parabola	The vertex of the graph of the quadratic function $f(x) = ax^2 + bx + c$ is $$\left(-\frac{b}{2a},\ f\left(-\frac{b}{2a}\right)\right)$$ and the axis of symmetry of the parabola is the line $x = -\frac{b}{2a}$.

EXAMPLE 8   Find the vertex of the graph of $f(x) = 2x^2 - 4x - 1$.

Strategy   We will determine the values of a and b and substitute into the formula for the vertex of a parabola.

Why   It is easier to find the coordinates of the vertex using the formula than it is to complete the square on $2x^2 - 4x - 1$.

Solution   The function is written in $f(x) = ax^2 + bx + c$ form, where $a = 2$ and $b = -4$. To find the vertex of its graph, we calculate

Success Tip

We can find the vertex of the graph of a quadratic function by completing the square or by using the formula.

$$-\frac{b}{2a} = -\frac{-4}{2(2)}$$
$$= -\frac{-4}{4}$$
$$= 1 \qquad \text{This is the } x\text{-coordinate of the vertex.}$$

$$f\left(-\frac{b}{2a}\right) = f(1)$$
$$= 2(1)^2 - 4(1) - 1$$
$$= -3 \qquad \text{This is the } y\text{-coordinate of the vertex.}$$

The vertex is the point $(1, -3)$. This agrees with the result we obtained in Example 7 by completing the square.

Self Check 8   Find the vertex of the graph of $f(x) = 3x^2 - 12x + 8$.

Now Try ▶   Problem 55

Using Your Calculator ▶   Finding the Vertex

We can use a graphing calculator to graph the function $f(x) = 2x^2 + 6x - 3$ and find the coordinates of the vertex and the axis of symmetry of the parabola. If we enter the function, we will obtain the graph shown in figure (a) on the next page.

We then trace to move the cursor to the lowest point on the graph, as shown in figure (b). By zooming in, we can see that the vertex is the point $(-1.5, -7.5)$, or $\left(-\frac{3}{2}, -\frac{15}{2}\right)$, and that the line $x = -\frac{3}{2}$ is the axis of symmetry.

Some calculators have an fmin or fmax feature that also can be used to find the vertex.

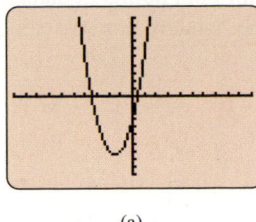 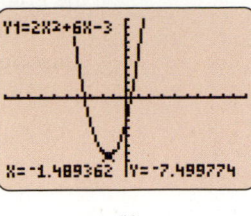

(a)   (b)

We can determine much about the graph of $f(x) = ax^2 + bx + c$ from the coefficients a, b, and c. This information is summarized as follows:

Graphing a Quadratic Function
$f(x) = ax^2 + bx + c$

- Determine whether the parabola opens upward or downward by finding the value of a.
- The x-coordinate of the vertex of the parabola is $x = -\dfrac{b}{2a}$.
- To find the y-coordinate of the vertex, substitute $-\dfrac{b}{2a}$ for x and find $f\left(-\dfrac{b}{2a}\right)$.
- The axis of symmetry is the vertical line passing through the vertex.
- The y-intercept is determined by the value of $f(x)$ when $x = 0$: the y-intercept is $(0, c)$.
- The x-intercepts (if any) are determined by the values of x that make $f(x) = 0$. To find them, solve the quadratic equation $ax^2 + bx + c = 0$.

EXAMPLE 9   Graph: $f(x) = -2x^2 - 8x - 8$

Strategy   We will follow the steps for graphing a quadratic function.

Why   This is the most efficient way to graph a general quadratic function.

Solution   **Step 1:   Determine whether the parabola opens upward or downward.** The function is in the form $f(x) = ax^2 + bx + c$, with $a = -2$, $b = -8$, and $c = -8$. Since $a < 0$, the parabola opens downward.

Step 2:   Find the vertex and draw the axis of symmetry. To find the coordinates of the vertex, we calculate

Success Tip

An easy way to remember the vertex formula is to note that $x = \frac{-b}{2a}$ is part of the quadratic formula:

$$x = \frac{-b \pm \sqrt{b^2 - 4ac}}{2a}$$

$$x = -\frac{b}{2a}$$

$$x = -\frac{-8}{2(-2)} \quad \text{Substitute } -2 \text{ for } a \text{ and } -8 \text{ for } b.$$

$$= -2 \quad \text{This is the } x\text{-coordinate of the vertex.}$$

$$f\left(-\frac{b}{2a}\right) = f(-2)$$

$$= -2(-2)^2 - 8(-2) - 8$$

$$= -8 + 16 - 8$$

$$= 0 \quad \text{This is the } y\text{-coordinate of the vertex.}$$

The vertex of the parabola is the point $(-2, 0)$. This point is in blue on the graph. The axis of symmetry is the line $x = -2$.

Step 3:   *Find the x- and y-intercepts.* Since $c = -8$, the y-intercept of the parabola is $(0, -8)$. The point $(-4, -8)$, two units to the left of the axis of symmetry, must also be on the graph. We plot both points in black on the graph.

To find the x-intercepts, we set $f(x)$ equal to 0 and solve the resulting quadratic equation.

$$f(x) = -2x^2 - 8x - 8 \qquad \text{This is the function to graph.}$$
$$0 = -2x^2 - 8x - 8 \qquad \text{Set } f(x) = 0.$$
$$0 = x^2 + 4x + 4 \qquad \text{Divide both sides by } -2.$$
$$0 = (x + 2)(x + 2) \qquad \text{Factor the trinomial.}$$
$$x + 2 = 0 \quad \text{or} \quad x + 2 = 0 \qquad \text{Set each factor equal to 0.}$$
$$x = -2 \qquad \qquad x = -2$$

Since the solutions are the same, the graph has only one x-intercept: $(-2, 0)$. This point is the vertex of the parabola and has already been plotted.

Step 4:   *Plot another point.* Finally, we find another point on the parabola. If $x = -3$, then $f(-3) = -2$. We plot $(-3, -2)$ and use symmetry to determine that $(-1, -2)$ is also on the graph. Both points are in green.

Step 5:   *Draw a smooth curve through the points.*

$$f(x) = -2x^2 - 8x - 8$$

x	$f(x)$
-3	-2	$\rightarrow (-3, -2)$

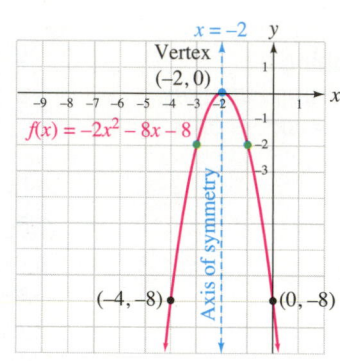

	Vertex $(-2, 0)$
$f(x) = -2x^2 - 8x - 8$	
$(-4, -8)$	$(0, -8)$

$x = -2$

Axis of symmetry

Self Check 9   Graph: $f(x) = -2x^2 + 12x - 16$

Now Try ▶   Problems 59 and 69

---

5 | Determine Minimum and Maximum Values.

It is often useful to know the smallest or largest possible value a quantity can assume. For example, companies try to minimize their costs and maximize their profits. If the quantity is expressed by a quadratic function, the y-coordinate of the vertex of the graph of the function gives its minimum or maximum value.

EXAMPLE 10

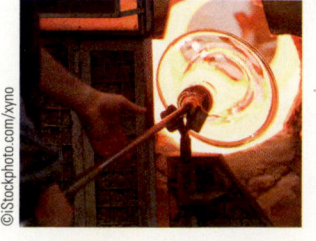

Minimizing Costs.   A glassworks that makes lead crystal vases has daily production costs given by the function $C(x) = 0.2x^2 - 10x + 650$, where x is the number of vases made each day. How many vases should be produced to minimize the per-day costs? What will the costs be?

Strategy   We will find the vertex of the graph of the quadratic function.

Why   The x-coordinate of the vertex indicates the number of vases to make to keep costs at a minimum, and the y-coordinate indicates the minimum cost.

Solution The graph of $C(x) = 0.2x^2 - 10x + 650$ is a parabola opening upward. The vertex is the lowest point on the graph. To find the vertex, we calculate

$$-\frac{b}{2a} = -\frac{-10}{2(0.2)} \qquad \text{$b = -10$ and $a = 0.2$.}$$

$$= -\frac{-10}{0.4}$$

$$= 25$$

$$f\left(-\frac{b}{2a}\right) = f(25)$$

$$= 0.2(25)^2 - 10(25) + 650$$

$$= 525$$

The vertex is $(25, 525)$, and it indicates that the costs are a minimum of \$525 when 25 vases are made daily.

    To solve this problem with a graphing calculator, we graph the function $C(x) = 0.2x^2 - 10x + 650$. By using TRACE and ZOOM, we can locate the vertex of the graph. The coordinates of the vertex indicate that the minimum cost is \$525 when the number of vases produced is 25.

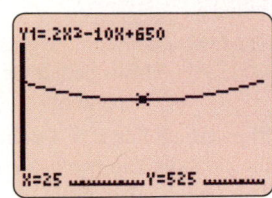

Self Check 10 **Minimizing Costs.** A manufacturing company has a daily production cost of $c(x) = 0.25x^2 - 10x + 800$, where x is the number of items produced and $c(x)$ is the cost. How many items should be produced to minimize the per-day cost? What is the minimum cost?

Now Try ▶ Problem 85

---

6 Solve Quadratic Equations Graphically.

When solving quadratic equations graphically, we must consider three possibilities. If the graph of the associated quadratic function has two x-intercepts, the quadratic equation has two real-number solutions. Figure (a) shows an example of this. If the graph has one x-intercept, as shown in figure (b), the equation has one repeated real-number solution. Finally, if the graph does not have an x-intercept, as shown in figure (c), the equation does not have any real-number solutions.

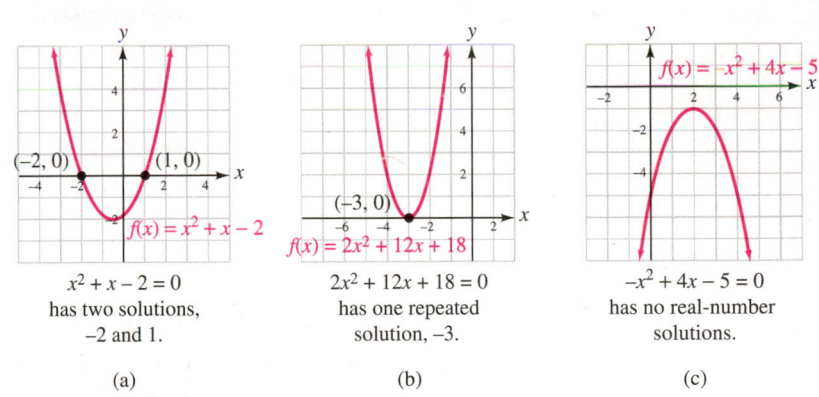

$x^2 + x - 2 = 0$ has two solutions, -2 and 1.	$2x^2 + 12x + 18 = 0$ has one repeated solution, -3.	$-x^2 + 4x - 5 = 0$ has no real-number solutions.
(a)	(b)	(c)

Using Your Calculator ▶ **Solving Quadratic Equations Graphically**

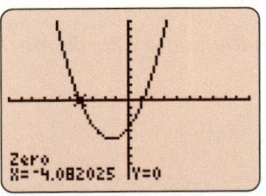

We can use a graphing calculator to find approximate solutions of quadratic equations. For example, the solutions of $0.7x^2 + 2x - 3.5 = 0$ are the numbers x that will make $y = 0$ in the quadratic function $f(x) = 0.7x^2 + 2x - 3.5$. To approximate these numbers, we graph the quadratic function and read the x-intercepts from the graph using the ZERO feature. (The ZERO feature can be found by pressing $\boxed{\text{2}^{\text{nd}}}$, CALC, and then 2.) In the figure, we see that the x-coordinate of the left-most x-intercept of the graph is given as -4.082025. This means that an approximate solution of the equation is -4.08. To find the positive x-intercept, we use similar steps.

SECTION 8.4 **STUDY SET**

VOCABULARY

Refer to the graph. Fill in the blanks.

1. $f(x) = 2x^2 - 4x + 1$ is called a _____ function. Its graph is a cup-shaped figure called a _____.

2. The lowest point on the graph is $(1, -1)$. This is called the _____ of the parabola.

3. The vertical line $x = 1$ divides the parabola into two halves. This line is called the _____ ___ _____.

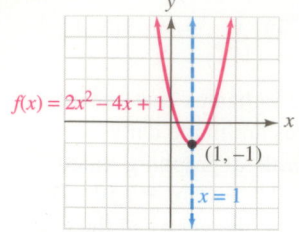

4. $f(x) = a(x - h)^2 + k$ is called the _____ form of the equation of a quadratic function.

CONCEPTS

5. Refer to the graph.
   a. What are the x-intercepts of the graph?
   b. What is the y-intercept of the graph?
   c. What is the vertex?
   d. What is the axis of symmetry?
   e. What are the domain and the range of the function?

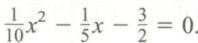

6. The vertex of a parabola is at $(1, -3)$, its y-intercept is $(0, -2)$, and it passes through the point $(3, 1)$, as shown in the illustration. Use the axis of symmetry shown in blue to help determine two other points on the parabola.

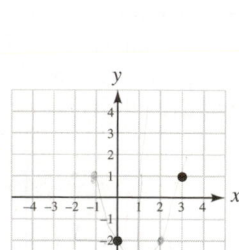

7. Draw the graph of a quadratic function using the given facts about its graph.
   - Opens upward
   - y-intercept: $(0, -3)$
   - Vertex: $(-1, -4)$
   - x-intercepts: $(-3, 0)$, $(1, 0)$

x	$f(x)$
2	5

8. For $f(x) = -x^2 + 6x - 7$, the value of $-\frac{b}{2a}$ is 3. Find the y-coordinate of the vertex of the graph of this function.

9. Fill in the blanks.
   a. To complete the square on the right side of $f(x) = 2x^2 + 12x + 11$, what should be factored from the first two terms?
   $$f(x) = \boxed{}(x^2 + 6x) + 11$$

   b. To complete the square on $x^2 + 6x$ shown below, what should be added within the parentheses and what should be subtracted outside the parentheses?
   $$f(x) = 2(x^2 + 6x + \boxed{}) + 11 - \boxed{}$$

10. Fill in the blanks. To complete the square on $x^2 + 4x$ shown below, what should be added within the parentheses and what should be added outside the parentheses?
   $$f(x) = -5(x^2 + 4x + \boxed{}) + 7 + \boxed{}$$

11. Use the graph of
   $$f(x) = \frac{1}{10}x^2 - \frac{1}{5}x - \frac{3}{2},$$
   shown here, to estimate the solutions of the equation
   $$\frac{1}{10}x^2 - \frac{1}{5}x - \frac{3}{2} = 0.$$

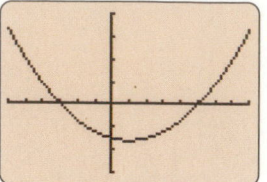

12. Three quadratic equations are to be solved graphically. The graphs of their associated quadratic functions are shown here. Determine which graph indicates that the equation has
   a. two real solutions.
   b. one repeated real solution.
   c. no real solutions.

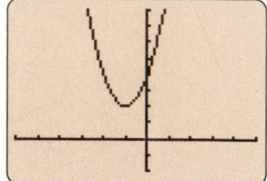

(i)

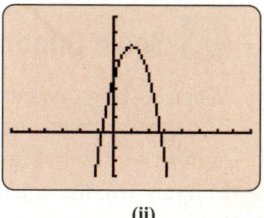

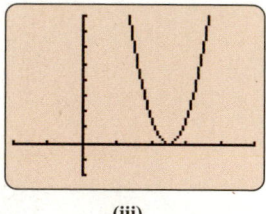

(ii)   (iii)

NOTATION

13. The function $f(x) = 2(x + 1)^2 + 6$ is written in the form $f(x) = a(x - h)^2 + k$. Is $h = -1$ or is $h = 1$? Explain.

14. Consider the function $f(x) = 2x^2 + 4x - 8$.
   a. What are a, b, and c?
   b. Find $-\frac{b}{2a}$.

GUIDED PRACTICE

Graph each group of functions on the same coordinate system. See Example 1.

15. $f(x) = x^2$, $g(x) = 2x^2$, $s(x) = \frac{1}{2}x^2$

16. $f(x) = x^2$, $g(x) = 4x^2$, $s(x) = \frac{1}{4}x^2$

Graph each pair of functions on the same coordinate system. See Example 2.

17. $f(x) = 2x^2$, $g(x) = -2x^2$

18. $f(x) = \frac{1}{2}x^2$, $g(x) = -\frac{1}{2}x^2$

Graph each group of functions on the same coordinate system. See Example 3.

19. $f(x) = 4x^2$, $g(x) = 4x^2 + 3$, $s(x) = 4x^2 - 2$

20. $f(x) = \dfrac{1}{3}x^2$, $g(x) = \dfrac{1}{3}x^2 + 4$, $s(x) = \dfrac{1}{3}x^2 - 3$

Graph each group of functions on the same coordinate system and describe how the graphs are similar and how they are different. See Example 4.

21. $f(x) = 3x^2$, $g(x) = 3(x + 2)^2$, $s(x) = 3(x - 3)^2$

22. $f(x) = \dfrac{1}{2}x^2$, $g(x) = \dfrac{1}{2}(x + 3)^2$, $s(x) = \dfrac{1}{2}(x - 2)^2$

Find the vertex and the axis of symmetry of the graph of each function. Do not graph the function, but determine whether the graph will open upward or downward. See Example 5.

23. $f(x) = (x - 1)^2 + 2$

24. $f(x) = 2(x - 2)^2 - 1$

25. $f(x) = -2(x + 3)^2 - 4$

26. $f(x) = -3(x + 1)^2 + 3$

27. $f(x) = -0.5(x - 7.5)^2 + 8.5$

28. $f(x) = -\dfrac{3}{2}\left(x + \dfrac{1}{4}\right)^2 + \dfrac{7}{8}$

29. $f(x) = 2x^2 - 4$

30. $f(x) = 3x^2 - 3$

Determine the vertex and the axis of symmetry of the graph of each function. Then plot several points and complete the graph. See Example 5.

31. $f(x) = (x - 3)^2 + 2$

32. $f(x) = (x + 1)^2 - 2$

33. $f(x) = -(x - 2)^2$

34. $f(x) = -(x + 2)^2$

35. $f(x) = -2(x + 3)^2 + 4$

36. $f(x) = -2(x - 2)^2 - 4$

37. $f(x) = \dfrac{1}{2}(x + 1)^2 - 3$

38. $f(x) = \dfrac{1}{3}(x - 1)^2 + 2$

Determine the vertex and the axis of symmetry of the graph of each function. Will the graph open upward or downward? See Example 6.

39. $f(x) = x^2 + 4x + 5$

40. $f(x) = x^2 - 4x - 1$

41. $f(x) = -x^2 + 6x - 15$

42. $f(x) = -x^2 - 6x + 3$

Complete the square to write each function in $f(x) = a(x - h)^2 + k$ form. Determine the vertex and the axis of symmetry of the graph of the function. Then plot several points and complete the graph. See Examples 6 and 7.

43. $f(x) = x^2 + 2x - 3$

44. $f(x) = x^2 + 6x + 5$

45. $f(x) = 4x^2 + 24x + 37$

46. $f(x) = 3x^2 - 12x + 10$

47. $f(x) = x^2 + x - 6$

48. $f(x) = x^2 - x - 6$

49. $f(x) = -4x^2 + 16x - 10$

50. $f(x) = -2x^2 + 4x + 3$

51. $f(x) = 2x^2 + 8x + 6$

52. $f(x) = 3x^2 - 12x + 9$

53. $f(x) = -x^2 - 8x - 17$

54. $f(x) = -x^2 + 6x - 8$

Use the vertex formula to find the vertex of the graph of each function. See Example 8.

55. $f(x) = x^2 + 2x - 5$ **56.** $f(x) = -x^2 + 4x - 5$

57. $f(x) = 2x^2 - 3x + 4$ **58.** $f(x) = 2x^2 - 7x - 4$

Find the x- and y-intercepts of the graph of the quadratic function. See Example 9.

59. $f(x) = x^2 - 2x - 35$ **60.** $f(x) = -x^2 - 10x - 21$

61. $f(x) = -2x^2 + 4x$ **62.** $f(x) = 3x^2 + 6x - 9$

Determine the coordinates of the vertex of the graph of each function using the vertex formula. Then determine the x- and y-intercepts of the graph. Finally, plot several points and complete the graph. See Example 9.

63. $f(x) = x^2 + 4x + 4$

64. $f(x) = x^2 - 6x + 9$

65. $f(x) = -x^2 + 2x - 1$

66. $f(x) = -x^2 - 2x - 1$

67. $f(x) = x^2 - 2x$

68. $f(x) = x^2 + x$

69. $f(x) = 2x^2 - 8x + 6$

70. $f(x) = 3x^2 - 12x + 12$

71. $f(x) = -6x^2 - 12x - 8$

72. $f(x) = -2x^2 + 8x - 10$

73. $f(x) = 4x^2 - 12x + 9$

74. $f(x) = 4x^2 + 4x - 3$

Use a graphing calculator to find the coordinates of the vertex of the graph of each quadratic function. Round to the nearest hundredth. See Using Your Calculator: Finding the Vertex.

75. $f(x) = 2x^2 - x + 1$ **76.** $f(x) = x^2 + 5x - 6$

77. $f(x) = -x^2 + x + 7$

78. $f(x) = 2x^2 - 3x + 2$

 Use a graphing calculator to solve each equation. If an answer is not exact, round to the nearest hundredth. See Using Your Calculator: Solving Quadratic Equations Graphically.

79. $x^2 + x - 6 = 0$

80. $2x^2 - 5x - 3 = 0$

81. $0.5x^2 - 0.7x - 3 = 0$

82. $2x^2 - 0.5x - 2 = 0$

APPLICATIONS

83. Crossword Puzzles. Darken the appropriate squares to the right of the dashed red line so that the puzzle has symmetry with respect to that line.

axis of symmetry

84. Graphic Arts. Draw an axis of symmetry over the letter shown here.

85. Operating Costs. The cost C in dollars of operating a certain concrete-cutting machine is related to the number of minutes n the machine is run by the function $C(n) = 2.2n^2 - 66n + 655$. For what number of minutes is the cost of running the machine a minimum? What is the minimum cost?

86. Water Usage. The height (in feet) of the water level in a reservoir over a 1-year period is modeled by the function $H(t) = 3.3(t - 9)^2 + 14$ where $t = 1$ represents January, $t = 2$ represents February, and so on. How low did the water level get that year, and when did it reach the low mark?

87. Fireworks. A fireworks shell is shot straight up with an initial velocity of 120 feet per second. Its height s in feet after t seconds is approximated by the equation $s = 120t - 16t^2$. If the shell is designed to explode when it reaches its maximum height, how long after being fired, and at what height, will the fireworks appear in the sky?

88. Projectiles. A ball is thrown straight upward from the top of a building with an initial velocity of 32 feet per second. The equation $s = -16t^2 + 32t + 48$ gives the height s of the ball in feet t seconds after it is thrown. Find the maximum height reached by the ball and the time it takes for the ball to hit the ground.

89.
Police Patrol Officer

Suppose you are a police patrol officer and you have a 300-foot-long roll of yellow "DO NOT CROSS" barricade tape to seal off an automobile accident, as shown in the illustration. What dimensions should you use to seal off the maximum rectangular area around the collision? What is the maximum area?

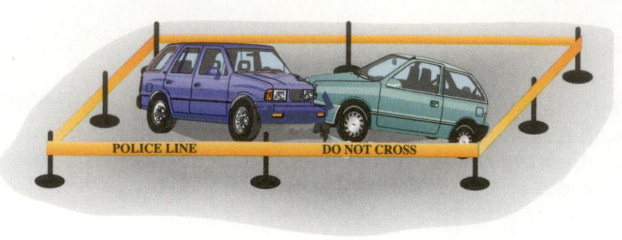

90. Ranching. See the illustration. A farmer wants to fence in three sides of a rectangular field with 1,000 feet of fencing. The other side of the rectangle will be a river. If the enclosed area is to be maximum, find the dimensions of the field.

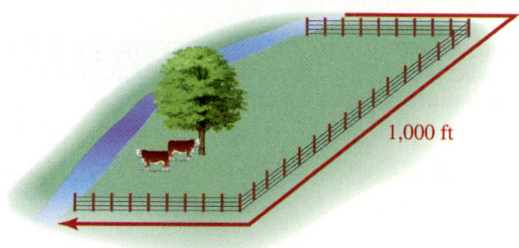

1,000 ft

91. Military History. The function $N(x) = -0.0534x^2 + 0.337x + 0.97$ gives the number of active-duty military personnel in the United States Army (in millions) for the years 1965–1972, where $x = 0$ corresponds to 1965, $x = 1$ corresponds to 1966, and so on. For this period, when was the army's personnel strength level at its highest, and what was it? Historically, can you explain why?

92. School Enrollment. After peaking in 1970, school enrollment in the United States fell during the 1970s and 1980s. The total annual enrollment (in millions) in U.S. elementary and secondary schools for the years 1975–1996 is given by the model $E(x) = 0.058x^2 - 1.162x + 50.604$ where $x = 0$ corresponds to 1975, $x = 1$ corresponds to 1976, and so on. For this period, when was enrollment the lowest? What was the enrollment?

93. Maximizing Revenue. The revenue R received for selling x stereos is given by the formula $R = -\dfrac{x^2}{5} + 80x - 1,000$. How many stereos must be sold to obtain the maximum revenue? Find the maximum revenue.

94. Maximizing Revenue.   When priced at \$30 each, a toy has annual sales of 4,000 units. The manufacturer estimates that each \$1 increase in price will decrease sales by 100 units. Find the unit price that will maximize total revenue. (*Hint:* Total revenue = price · the number of units sold.)

WRITING

95. Use the example of a stream of water from a drinking fountain to explain the concepts of the vertex and the axis of symmetry of a parabola. Draw a picture.

96. What are some quantities that are good to maximize? What are some quantities that are good to minimize?

97. A mirror is held against the y-axis of the graph of a quadratic function. What fact about parabolas does this illustrate?

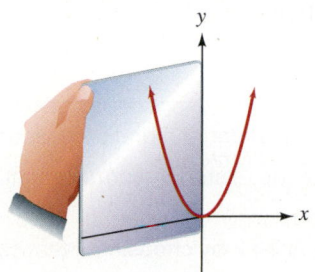

98. The vertex of a quadratic function $f(x) = ax^2 + bx + c$ is given by the formula $\left(-\frac{b}{2a}, \ f\left(-\frac{b}{2a}\right)\right)$. Explain what is meant by the notation $f\left(-\frac{b}{2a}\right)$.

99. A table of values for $f(x) = 2x^2 - 4x + 3$ is shown. Explain why it appears that the vertex of the graph of f is the point $(1, 1)$.

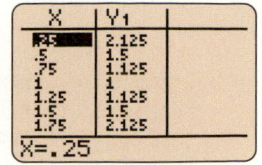

100. The illustration shows the graph of the quadratic function $f(x) = -4x^2 + 12x$ with domain $[0, 3]$. Explain how the value of $f(x)$ changes as the value of x increases from 0 to 3.

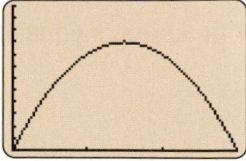

REVIEW

Simplify each expression. Assume all variables represent positive numbers.

101. $\dfrac{\sqrt{3}}{\sqrt{50}}$

102. $\dfrac{3}{\sqrt[3]{9}}$

103. $3\left(\sqrt{5b} - \sqrt{3}\right)^2$

104. $-2\sqrt{5b}\left(4\sqrt{2b} - 3\sqrt{3}\right)$

CHALLENGE PROBLEMS

105. Find a number between 0 and 1 such that the difference of the number and its square is a maximum.

106. Determine a quadratic function whose graph has x-intercepts of $(2, 0)$ and $(-4, 0)$.

SECTION 8.5

Quadratic and Other Nonlinear Inequalities

OBJECTIVES

1 Solve quadratic inequalities.

2 Solve rational inequalities.

3 Graph nonlinear inequalities in two variables.

ARE YOU READY?

The following problems review some basic skills that are needed when working with quadratic and nonlinear inequalities?

1. Does $x = -2$ satisfy $x^2 + x - 6 < 0$?

2. Solve: $x^2 - 5x - 50 = 0$

3. Graph the set of real numbers between -3 and 2 on a number line.

4. Graph the set of real numbers greater than 1 or less than or equal to -2 on a number line.

5. What values of x make the denominator of $\dfrac{x + 1}{x^2 - 9}$ equal to 0?

6. Graph: $y = -x^2 + 4$

We have previously solved *linear* inequalities in one variable such as $2x + 3 > 8$ and $6x - 7 < 4x - 9$. To find their solution sets, we used properties of inequalities to isolate the variable on one side of the inequality.

In this section, we will solve *quadratic* inequalities in one variable such as $x^2 + x - 6 < 0$ and $x^2 + 4x \geq 5$. We will use an interval testing method on the number line to determine their solution sets.

1 Solve Quadratic Inequalities.

Recall that a quadratic equation can be written in the form $ax^2 + bx + c = 0$. If we replace the $=$ symbol with an inequality symbol, we have a quadratic inequality.

Quadratic Inequalities	A **quadratic inequality** can be written in one of the standard forms $$ax^2 + bx + c < 0 \qquad ax^2 + bx + c > 0 \qquad ax^2 + bx + c \leq 0 \qquad ax^2 + bx + c \geq 0$$ where a, b, and c are real numbers and $a \neq 0$.

To solve a quadratic inequality in one variable, we will use the following steps to find the values of the variable that make the inequality true.

Solving Quadratic Inequalities	1. Write the inequality in standard form and solve its related quadratic equation. 2. Locate the solutions (called **critical numbers**) of the related quadratic equation on a number line. 3. Test each interval on the number line created in step 2 by choosing a test value from the interval and determining whether it satisfies the inequality. The solution set includes the interval(s) whose test value makes the inequality true. 4. Determine whether the endpoints of the intervals are included in the solution set.

EXAMPLE 1   Solve: $x^2 + x - 6 < 0$

Strategy   We will solve the related quadratic equation $x^2 + x - 6 = 0$ by factoring to determine the critical numbers. These critical numbers will separate the number line into intervals.

Why   We can test each interval to see whether numbers in the interval are in the solution set of the inequality.

Solution   The expression $x^2 + x - 6$ can be positive, negative, or 0, depending on what value is substituted for x. Solutions of the inequality are x-values that make $x^2 + x - 6$ less than 0. To find them, we will follow the steps for solving quadratic inequalities.

Step 1: Solve the related quadratic equation. For the quadratic inequality $x^2 + x - 6 < 0$, the related quadratic *equation* is $x^2 + x - 6 = 0$.

$$x^2 + x - 6 = 0$$
$$(x + 3)(x - 2) = 0 \qquad \text{\color{red}Factor the trinomial.}$$
$$x + 3 = 0 \quad \text{or} \quad x - 2 = 0 \qquad \text{\color{red}Set each factor equal to 0.}$$
$$x = -3 \quad \bigm| \quad x = 2 \qquad \text{\color{red}Solve each equation.}$$

The solutions of $x^2 + x - 6 = 0$ are -3 and 2. These solutions are the critical numbers.

Step 2: Locate the critical numbers on a number line. When we highlight -3 and 2 on a number line, they separate the number line into three intervals:

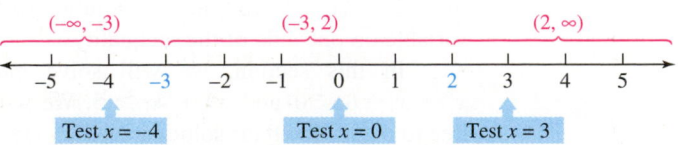

Step 3:   Test each interval. To determine whether the numbers in $(-\infty, -3)$ are solutions of the inequality, we choose a number from that interval, substitute it for x, and see whether it satisfies $x^2 + x - 6 < 0$. *If one number in that interval satisfies the inequality, all numbers in that interval will satisfy the inequality.*

If we choose -4 from $(-\infty, -3)$, we have:

$$x^2 + x - 6 < 0 \qquad \text{This is the original inequality.}$$
$$(-4)^2 + (-4) - 6 \overset{?}{<} 0 \qquad \text{Substitute } -4 \text{ for x.}$$
$$16 + (-4) - 6 \overset{?}{<} 0$$
$$6 < 0 \qquad \text{False}$$

Since -4 does not satisfy the inequality, none of the numbers in $(-\infty, -3)$ are solutions. To test the second interval, $(-3, 2)$, we choose $x = 0$.

$$x^2 + x - 6 < 0 \qquad \text{This is the original inequality.}$$
$$0^2 + 0 - 6 \overset{?}{<} 0 \qquad \text{Substitute 0 for x.}$$
$$-6 < 0 \qquad \text{True}$$

Since 0 satisfies the inequality, all of the numbers in $(-3, 2)$ are solutions. To test the third interval, $(2, \infty)$, we choose $x = 3$.

$$x^2 + x - 6 < 0 \qquad \text{This is the original inequality.}$$
$$3^2 + 3 - 6 \overset{?}{<} 0 \qquad \text{Substitute 3 for x.}$$
$$9 + 3 - 6 \overset{?}{<} 0$$
$$6 < 0 \qquad \text{False}$$

Since 3 does not satisfy the inequality, none of the numbers in $(2, \infty)$ are solutions.

Success Tip
If a quadratic inequality contains $\leq$ or $\geq$, the endpoints of the intervals are included in the solution set. If the inequality contains $<$ or $>$, they are not.

Step 4:   Are the endpoints included? From the interval testing, we see that only numbers from $(-3, 2)$ satisfy $x^2 + x - 6 < 0$. The endpoints -3 and 2 are not included in the solution set because they do not satisfy the inequality. (Recall that -3 and 2 make $x^2 + x - 6$ equal to 0.) The solution set is the interval $(-3, 2)$ as graphed on the right.

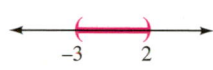

Self Check 1   Solve: $x^2 + x - 12 < 0$

Now Try ▶ Problem 15

EXAMPLE 2   Solve: $x^2 + 4x \geq 5$

Strategy   This inequality is not in standard form because it does not have 0 on the right side. We will write it in standard form and solve its related quadratic equation to find any critical numbers. These critical numbers will separate the number line into intervals.

Why   We can then test each interval to see whether numbers in the interval are in the solution set of the inequality.

Solution   To get 0 on the right side, we subtract 5 from both sides.

$$x^2 + 4x \geq 5 \qquad \text{This is the inequality to solve.}$$
$$x^2 + 4x - 5 \geq 0 \qquad \text{Write the inequality in the equivalent form } ax^2 + bx + c \geq 0.$$

We can solve the related quadratic *equation* $x^2 + 4x - 5 = 0$ by factoring.

$$x^2 + 4x - 5 = 0$$
$$(x + 5)(x - 1) = 0 \qquad\qquad\qquad \text{Factor the trinomial.}$$
$$x + 5 = 0 \quad \text{or} \quad x - 1 = 0 \qquad \text{Set each factor equal to 0.}$$
$$x = -5 \qquad\qquad x = 1$$

The critical numbers -5 and 1 separate the number line into three intervals. We pick a test value from each interval to see whether it satisfies $x^2 + 4x - 5 \geq 0$.

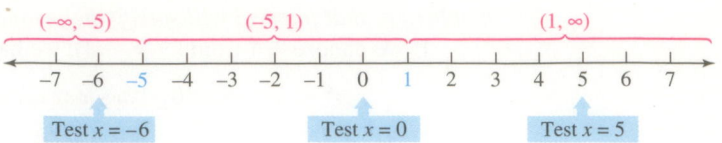

$$x^2 + 4x - 5 \geq 0$$
$$(-6)^2 + 4(-6) - 5 \stackrel{?}{\geq} 0$$
$$7 \geq 0 \quad \text{True}$$

$$x^2 + 4x - 5 \geq 0$$
$$0^2 + 4(0) - 5 \stackrel{?}{\geq} 0$$
$$-5 \geq 0 \quad \text{False}$$

$$x^2 + 4x - 5 \geq 0$$
$$5^2 + 4(5) - 5 \stackrel{?}{\geq} 0$$
$$40 \geq 0 \quad \text{True}$$

The numbers in the intervals $(-\infty, -5)$ and $(1, \infty)$ satisfy the inequality. Since the endpoints -5 and 1 also satisfy $x^2 + 4x - 5 \geq 0$, they are included in the solution set. (Recall that -5 and 1 make $x^2 + 4x - 5$ equal to 0.) Thus, the solution set is the union of two intervals: $(-\infty, -5] \cup [1, \infty)$. The graph of the solution set is shown on the right.

Self Check 2 Solve: $x^2 + 3x \geq 40$

Now Try ▶ Problem 19

2 Solve Rational Inequalities.

Rational inequalities in one variable such as $\frac{9}{x} < 8$ and $\frac{x^2 + x - 2}{x - 4} \geq 0$ can also be solved using the interval testing method.

<table>
<tr><td>**Solving Rational Inequalities**</td><td>

1. Write the inequality in standard form with a single quotient on the left side and 0 on the right side. Then solve its related rational equation.

2. Set the denominator equal to zero and solve that equation.

3. Locate the solutions (called **critical numbers**) found in steps 1 and 2 on a number line.

4. Test each interval on the number line created in step 3 by choosing a test value from the interval and determining whether it satisfies the inequality. The solution set includes the interval(s) whose test value makes the inequality true.

5. Determine whether the endpoints of the intervals are included in the solution set. Exclude any values that make the denominator 0.

</td></tr>
</table>

EXAMPLE 3   Solve: $\dfrac{9}{x} < 8$

Strategy This rational inequality is not in standard form because it does not have 0 on the right side. We will write it in standard form and solve its related rational equation to find any critical numbers. These critical numbers will separate the number line into intervals.

Why We can test each interval to see whether numbers in the interval are in the solution set of the inequality.

Solution   To get 0 on the right side, we subtract 8 from both sides. We then find a common denominator to write the left side as a single quotient.

$$\frac{9}{x} < 8 \qquad \text{This is the inequality to solve.}$$

$$\frac{9}{x} - 8 < 0 \qquad \text{Subtract 8 from both sides.}$$

$$\frac{9}{x} - 8 \cdot \frac{x}{x} < 0 \qquad \begin{array}{l}\text{To write the left side as a single quotient,}\\ \text{build 8 to a fraction with denominator } x.\end{array}$$

$$\frac{9}{x} - \frac{8x}{x} < 0$$

$$\frac{9 - 8x}{x} < 0 \qquad \text{Subtract the numerators and keep the common denominator, } x.$$

> **Caution**
>
> When solving rational inequalities such as $\frac{9}{x} < 8$, a common error is to multiply both sides by x to clear it of the fraction. However, we don't know whether x is positive or negative, so we don't know whether or not to reverse the inequality symbol.

Now we solve the related rational equation.

$$\frac{9 - 8x}{x} = 0$$

$$9 - 8x = 0 \qquad \text{If } x \neq 0, \text{ we can clear the equation of the fraction by multiplying both sides by } x.$$

$$-8x = -9 \qquad \text{To isolate the variable term } -8x, \text{ subtract 9 from both sides.}$$

$$x = \frac{9}{8} \qquad \text{Solve for } x. \text{ This is a critical number.}$$

> **Success Tip**
>
> Recall that if a fraction is equal to 0, its numerator must be 0.

If we set the denominator of $\frac{9 - 8x}{x}$ equal to 0, we obtain a second critical number, $x = 0$. When graphed, the critical numbers 0 and $\frac{9}{8}$ separate the number line into three intervals. We pick a test value from each interval to see whether it satisfies $\frac{9 - 8x}{x} < 0$.

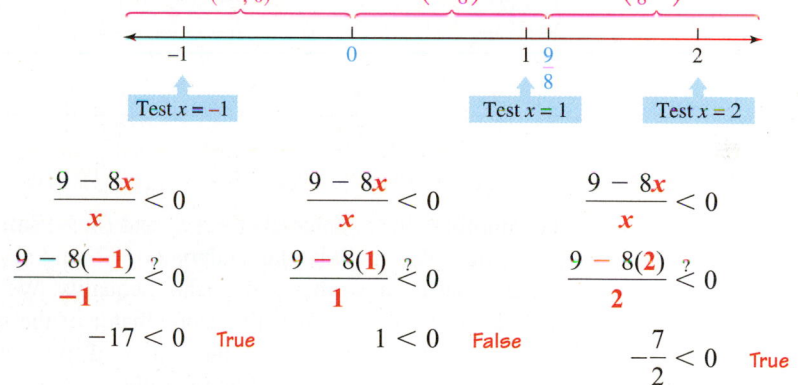

$$\frac{9 - 8x}{x} < 0 \qquad\qquad \frac{9 - 8x}{x} < 0 \qquad\qquad \frac{9 - 8x}{x} < 0$$

$$\frac{9 - 8(-1)}{-1} \overset{?}{<} 0 \qquad\qquad \frac{9 - 8(1)}{1} \overset{?}{<} 0 \qquad\qquad \frac{9 - 8(2)}{2} \overset{?}{<} 0$$

$$-17 < 0 \quad \text{True} \qquad\qquad 1 < 0 \quad \text{False} \qquad\qquad -\frac{7}{2} < 0 \quad \text{True}$$

The numbers in the intervals $(-\infty, 0)$ and $\left(\frac{9}{8}, \infty\right)$ satisfy the inequality. We do not include the endpoint 0 in the solution set, because it makes the denominator of the original inequality 0. Neither do we include $\frac{9}{8}$, because it does not satisfy $\frac{9 - 8x}{x} < 0$. (Recall that $\frac{9}{8}$ makes $\frac{9 - 8x}{x}$ equal to 0.) Thus, the solution set is the union of two intervals: $(-\infty, 0) \cup \left(\frac{9}{8}, \infty\right)$. Its graph is shown on the right.

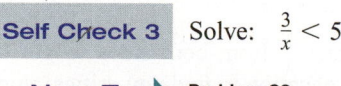

Self Check 3   Solve: $\frac{3}{x} < 5$

Now Try ▶ Problem 23

EXAMPLE 4   Solve: $\dfrac{x^2 + x - 2}{x - 4} \geq 0$

Strategy   This inequality is in standard form. We will solve its related rational equation to find any critical numbers. These critical numbers will separate the number line into intervals.

Why   We can test each interval to see whether numbers in the interval are in the solution set of the inequality.

Solution   To solve the related rational equation, we proceed as follows:

$$\frac{x^2 + x - 2}{x - 4} = 0$$

$$x^2 + x - 2 = 0 \qquad \textcolor{red}{\text{If } x \neq 4,\text{ we can clear the equation of the fraction by multiplying both sides by } x - 4.}$$

$$(x + 2)(x - 1) = 0 \qquad \textcolor{red}{\text{Factor the trinomial.}}$$

$$x + 2 = 0 \quad \text{or} \quad x - 1 = 0 \qquad \textcolor{red}{\text{Set each factor equal to 0.}}$$

$$x = -2 \quad \bigm| \quad x = 1 \qquad \textcolor{red}{\text{These are critical numbers.}}$$

Caution

We cannot multiply both sides of the inequality by $x - 4$ to clear it of fractions. We don't know if $x - 4$ is positive or negative, so we don't know whether or not to reverse the inequality symbol.

If we set the denominator of $\dfrac{x^2 + x - 2}{x - 4}$ equal to 0, we see that $x = 4$ is also a critical number. When graphed, the critical numbers, -2, 1, and 4, separate the number line into four intervals. We pick a test value from each interval to see whether it satisfies $\dfrac{x^2 + x - 2}{x - 4} \geq 0$.

$$\frac{(-3)^2 + (-3) - 2}{-3 - 4} \overset{?}{\geq} 0 \qquad \frac{0^2 + 0 - 2}{0 - 4} \overset{?}{\geq} 0 \qquad \frac{3^2 + 3 - 2}{3 - 4} \overset{?}{\geq} 0 \qquad \frac{6^2 + 6 - 2}{6 - 4} \overset{?}{\geq} 0$$

$$-\frac{4}{7} \geq 0 \qquad\qquad \frac{1}{2} \geq 0 \qquad\qquad -10 \geq 0 \qquad\qquad 20 \geq 0$$

$$\textbf{False} \qquad\qquad \textbf{True} \qquad\qquad \textbf{False} \qquad\qquad \textbf{True}$$

The numbers in the intervals $(-2, 1)$ and $(4, \infty)$ satisfy the inequality. We include the endpoints -2 and 1 in the solution set because they satisfy the inequality. We do not include 4 because it makes the denominator of the fraction 0. Thus, the solution set is the union of two intervals $[-2, 1] \cup (4, \infty)$, as graphed on the right.

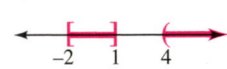

Self Check 4   Solve: $\dfrac{x + 2}{x^2 - 2x - 3} \geq 0$

Now Try ▶   Problem 27

EXAMPLE 5   Solve: $\dfrac{3}{x - 1} < \dfrac{2}{x}$

Strategy   We will subtract $\dfrac{2}{x}$ from both sides to get 0 on the right side and solve the resulting related rational equation to find any critical numbers. These critical numbers will separate the number line into intervals.

Why   We can test each interval to see whether numbers in the interval are in the solution set of the inequality.

Solution

$$\frac{3}{x-1} < \frac{2}{x}$$ This is the inequality to solve.

$$\frac{3}{x-1} - \frac{2}{x} < 0$$ Subtract $\frac{2}{x}$ from both sides.

$$\frac{3}{x-1} \cdot \frac{x}{x} - \frac{2}{x} \cdot \frac{x-1}{x-1} < 0$$ To get a single quotient on the left side, build each rational expression to have the common denominator $x(x-1)$.

$$\frac{3x - 2x + 2}{x(x-1)} < 0$$ Subtract the numerators and keep the common denominator.

$$\frac{x+2}{x(x-1)} < 0$$ Combine like terms.

The only solution of the related rational equation $\frac{x+2}{x(x-1)} = 0$ is -2. Thus, -2 is a critical number. When we set the denominator equal to 0 and solve $x(x-1) = 0$, we find two more critical numbers, 0 and 1. These three critical numbers create four intervals to test.

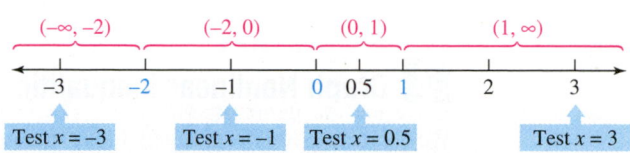

$$\frac{-3+2}{-3(-3-1)} \overset{?}{<} 0 \qquad \frac{-1+2}{-1(-1-1)} \overset{?}{<} 0 \qquad \frac{0.5+2}{0.5(0.5-1)} \overset{?}{<} 0 \qquad \frac{3+2}{3(3-1)} \overset{?}{<} 0$$

$$\frac{-1}{-3(-4)} \overset{?}{<} 0 \qquad \frac{1}{-1(-2)} \overset{?}{<} 0 \qquad \frac{2.5}{0.5(-0.5)} \overset{?}{<} 0 \qquad \frac{5}{3(2)} \overset{?}{<} 0$$

$$-\frac{1}{12} < 0 \qquad\qquad \frac{1}{2} < 0 \qquad\qquad -10 < 0 \qquad\qquad \frac{5}{6} < 0$$

True   False   True   False

The numbers 0 and 1 are not included in the solution set because they make the denominator 0, and the number -2 is not included because it does not satisfy the inequality. The solution set is the union of two intervals $(-\infty, -2) \cup (0, 1)$, as graphed on the right.

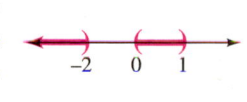

Self Check 5   Solve: $\frac{2}{x+1} > \frac{1}{x}$

Now Try ▶ Problem 31

Using Your Calculator ▶ **Solving Inequalities Graphically**

We can solve $x^2 + 4x \geq 5$ (Example 2) graphically by writing the inequality as $x^2 + 4x - 5 \geq 0$ and graphing the quadratic function $f(x) = x^2 + 4x - 5$, as shown in figure (a) on the next page. The solution set of the inequality will be those values of x for which the graph lies on or above the x-axis. We can trace to determine that this is the union of two intervals: $(-\infty, -5] \cup [1, \infty)$.

To solve $\frac{3}{x-1} < \frac{2}{x}$ (Example 5) graphically, we first write the inequality in the form $\frac{x+2}{x(x-1)} < 0$ and then graph the rational function $f(x) = \frac{x+2}{x(x-1)}$, as shown in figure (b) on the next page. The solution of the inequality will be those values of x for which the graph lies below the axis.

We can trace to see that the graph is below the x-axis when x is less than -2. Since we cannot see the graph in the interval $0 < x < 1$, we redraw the graph using window settings of $[-1, 2]$ for x and $[-25, 10]$ for y, as shown in figure (c).

Now we see that the graph is below the x-axis in the interval $(0, 1)$. Thus, the solution set of the inequality is the union of the two intervals: $(-\infty, -2) \cup (0, 1)$.

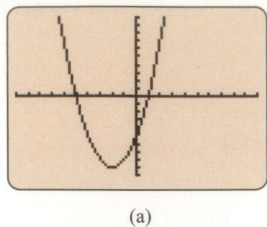   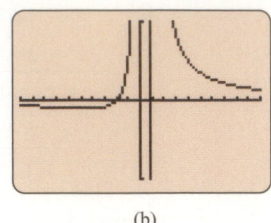   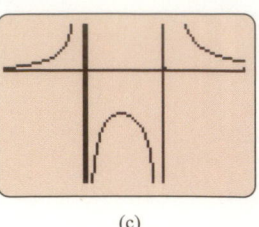

(a)   (b)   (c)

3 Graph Nonlinear Inequalities in Two Variables.

We have previously graphed linear inequalities in two variables such as $y > 3x + 2$ and $2x - 3y \le 6$ using the following steps.

Graphing Inequalities in Two Variables

1. Graph the related equation to find the boundary line of the region. If the inequality allows equality (the symbol is either $\le$ or $\ge$), draw the boundary as a solid line. If equality is not allowed ($<$ or $>$), draw the boundary as a dashed line.

2. Pick a test point that is on one side of the boundary line. (Use the origin if possible.) Replace x and y in the original inequality with the coordinates of that point. If the inequality is satisfied, shade the side that contains that point. If the inequality is not satisfied, shade the other side of the boundary.

We use the same procedure to graph **nonlinear inequalities** in two variables.

EXAMPLE 6   Graph: $y < -x^2 + 4$

Strategy   We will graph the related equation $y = -x^2 + 4$ to establish a boundary parabola. Then we will determine which side of the boundary parabola represents the solution set of the inequality.

Why   To *graph a nonlinear inequality* in two variables means to draw a "picture" of the ordered pairs (x, y) that make the inequality true.

Solution   The graph of the boundary $y = -x^2 + 4$ is a parabola opening downward, with vertex at $(0, 4)$ and axis of symmetry $x = 0$ (the y-axis). Since the inequality contains an $<$ symbol and equality is not allowed, we draw the parabola using a dashed curve.

To determine which region to shade, we pick the test point $(0, 0)$ and substitute its coordinates into the inequality. We shade the region containing $(0, 0)$ because its coordinates satisfy $y < -x^2 + 4$.

Graph the boundary

$$y = -x^2 + 4$$

Compare to $y = a(x - h)^2 + k$

$a = -1$: Opens downward

$h = 0$ and $k = 4$: Vertex $(0, 4)$

Axis of symmetry $x = 0$

x	y
1	3
2	0

Shading: Use the test point $(0, 0)$

$y < -x^2 + 4$

$0 \overset{?}{<} -0^2 + 4$

$0 < 4$ True

Since $0 < 4$ is true, $(0, 0)$ is a solution of $y < -x^2 + 4$.

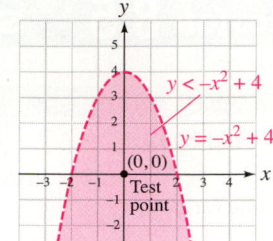

Self Check 6   Graph: $y \geq -x^2 + 4$

Now Try ▶ Problem 35

EXAMPLE 7   Graph: $x \leq |y|$

Strategy   We will graph the related equation $x = |y|$ to establish a boundary. Then we will determine which side of the boundary represents the solution set of the inequality.

Why   To *graph a nonlinear inequality* in two variables means to draw a "picture" of the ordered pairs (x, y) that make the inequality true.

Solution   To graph the boundary, $x = |y|$, we construct a table of solutions, as shown in figure (a). In figure (b), the boundary is a solid line because the inequality contains a $\leq$ symbol and equality is permitted. Since the origin is on the graph, we cannot use it as a test point. However, any other point, such as $(1, 0)$, will do. We substitute 1 for x and 0 for y into the inequality to get

$x \leq |y|$ This is the inequality to graph.

$1 \overset{?}{\leq} |0|$ Substitute.

$1 \leq 0$ False

Since $1 \leq 0$ is a false statement, the point $(1, 0)$ does not satisfy the inequality and is not part of the graph. Thus, the graph of $x \leq |y|$ is to the left of the boundary. The complete graph is shown in figure (c).

$x = |y|$

x	y
0	0
1	1
1	-1
2	2
2	-2

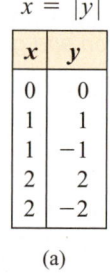

(a)

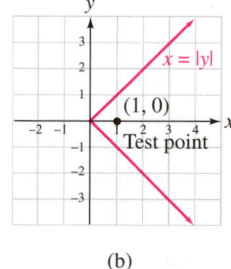

(b)

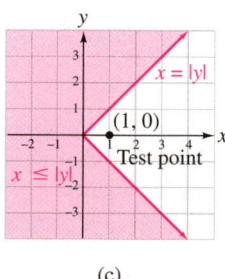

(c)

Self Check 7   Graph: $x \geq -|y|$

Now Try ▶ Problem 39

SECTION **8.5**   STUDY SET

VOCABULARY

Fill in the blanks.

1. $x^2 + 3x - 18 < 0$ is an example of a _____ inequality in one variable.

2. $\frac{x-1}{x^2 - x - 20} \leq 0$ is an example of a _____ inequality in one variable.

3. $y \leq x^2 - 4x + 3$ is an example of a nonlinear inequality in _____ variables.

4. The set of real numbers greater than 3 can be represented using the _____ notation $(3, \infty)$.

CONCEPTS

5. The critical numbers of a quadratic inequality are highlighted in red on the number line shown below. Use interval notation to represent each interval that must be tested to solve the inequality.

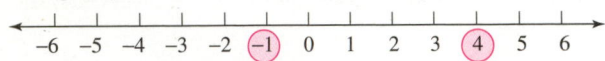

6. Graph each of the following solution sets.

   a. $(-2, 4)$ **b.** $(-\infty, -2) \cup (3, 5]$

7. The graph of the solution set of a rational inequality in one variable is shown. Determine whether each of the following numbers is a solution of the inequality.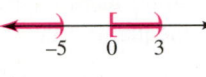

   a. -10 **b.** -5

   c. 0 **d.** 4

8. What are the critical numbers for each inequality?

   a. $x^2 - 2x - 48 \geq 0$ **b.** $\frac{x-3}{x(x+4)} > 0$

9. a. The results after interval testing for a quadratic inequality containing a $>$ symbol are shown below. (The critical numbers are highlighted in red.) What is the solution set?

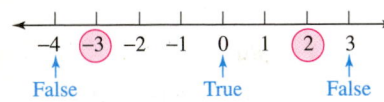

   b. The results after interval testing for a quadratic inequality containing a $\leq$ symbol are shown below. (The critical numbers are highlighted in red.) What is the solution set?

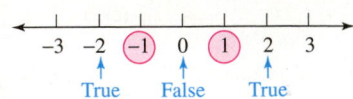

10. Fill in the blank to complete this important fact about the interval testing method discussed in this section: *If one number in an interval satisfies the inequality, _____ numbers in that interval will satisfy the inequality.*

11. a. When graphing the solution of $y \leq x^2 + 2x + 1$, should the boundary be solid or dashed?

   b. Does the test point $(0, 0)$ satisfy the inequality?

12. a. Estimate the solution of $x^2 - x - 6 > 0$ using the graph of $y = x^2 - x - 6$ shown in figure (a) below.

   b. Estimate the solution of $\frac{x-3}{x} \leq 0$ using the graph of $y = \frac{x-3}{x}$ shown in figure (b) below.

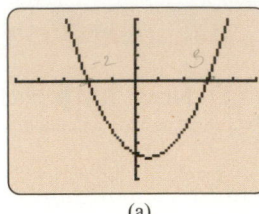 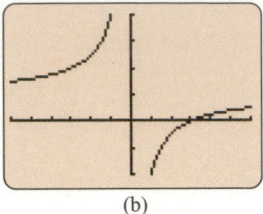

           (a)   (b)

NOTATION

13. Write the quadratic inequality $x^2 - 6x \geq 7$ in standard form.

14. The solution set of a rational inequality consists of the intervals $(-1, 4]$ and $(7, \infty)$. When writing the solution set, what symbol is used between the two intervals?

GUIDED PRACTICE

Solve each inequality. Write the solution set in interval notation and graph it. See Example 1.

15. $x^2 - 5x + 4 < 0$ **16.** $x^2 + 2x - 8 < 0$

17. $x^2 - 8x + 15 > 0$ **18.** $x^2 - 3x - 4 > 0$

Solve each inequality. Write the solution set in interval notation and graph it. See Example 2.

19. $x^2 - x \geq 42$ **20.** $x^2 - x \geq 72$

21. $x^2 + x \leq 12$ **22.** $x^2 - 8x \leq -15$

Solve each inequality. Write the solution set in interval notation and graph it. See Example 3.

23. $\frac{1}{x} < 2$ **24.** $\frac{1}{x} < 3$

25. $\frac{5}{x} \geq -3$ **26.** $\frac{4}{x} \geq 8$

Solve each inequality. Write the solution set in interval notation and graph it. See Example 4.

27. $\frac{x^2 - x - 12}{x - 1} < 0$ **28.** $\frac{x^2 + x - 6}{x - 4} \geq 0$

29. $\frac{6x^2 - 5x + 1}{2x + 1} \geq 0$ **30.** $\frac{6x^2 + 11x + 3}{3x - 1} < 0$

Solve each inequality. Write the solution set in interval notation and graph it. See Example 5.

31. $\dfrac{3}{x-2} < \dfrac{4}{x}$

32. $\dfrac{-6}{x+1} \geq \dfrac{1}{x}$

33. $\dfrac{7}{x-3} \geq \dfrac{2}{x+4}$

34. $\dfrac{-5}{x-4} < \dfrac{3}{x+1}$

Graph each inequality. See Example 6.

35. $y < x^2 + 1$

36. $y > x^2 - 3$

37. $y \leq x^2 + 5x + 6$

38. $y \geq x^2 + 5x + 4$

Graph each inequality. See Example 7.

39. $y < |x + 4|$

40. $y \leq |x - 3|$

41. $y \geq -|x| + 2$

42. $y > |x| - 2$

Use a graphing calculator to solve each inequality. Write the solution set in interval notation. See Using Your Calculator: Solving Inequalities Graphically.

43. $x^2 - 2x - 3 < 0$

44. $x^2 + x - 6 > 0$

45. $\dfrac{x+3}{x-2} > 0$

46. $\dfrac{3}{x} < 2$

TRY IT YOURSELF

Solve each inequality. Write the solution set in interval notation and graph it.

47. $\dfrac{x}{x+4} \leq \dfrac{1}{x+1}$

48. $\dfrac{x}{x+9} \geq \dfrac{1}{x+1}$

49. $x^2 \geq 9$

50. $x^2 \geq 16$

51. $x^2 + 6x \geq -9$

52. $x^2 + 8x < -16$

53. $\dfrac{x^2 + x - 2}{x - 3} > 0$

54. $\dfrac{x - 2}{x^2 - 1} > 0$

55. $2x^2 - 50 < 0$

56. $3x^2 - 243 < 0$

57. $\dfrac{2x - 3}{3x + 1} < 0$

58. $\dfrac{x - 5}{x + 1} < 0$

59. $x^2 - 6x + 9 < 0$

60. $x^2 + 4x + 4 > 0$

61. $\dfrac{5}{x + 1} > \dfrac{3}{x - 4}$

62. $\dfrac{3}{x - 2} \leq -\dfrac{2}{x + 3}$

APPLICATIONS

63. **Bridges.** If an x-axis is superimposed over the roadway of the Golden Gate Bridge, with the origin at the center of the bridge, the length L in feet of a vertical support cable can be approximated by the formula

$$L = \frac{1}{9{,}000}x^2 + 5$$

For the Golden Gate Bridge, $-2{,}100 < x < 2{,}100$. For what intervals along the x-axis are the vertical cables more than 95 feet long?

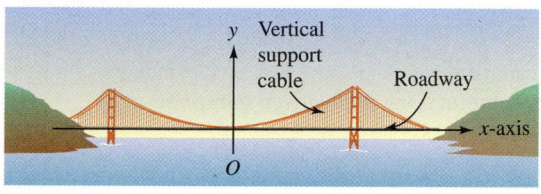

64. **Malls.** The number of people n in a mall is modeled by the formula

$$n = -100x^2 + 1{,}200x$$

where x is the number of hours since the mall opened. If the mall opened at 9 A.M., when were there 2,000 or more people in it?

WRITING

65. How are critical numbers used when solving a quadratic inequality in one variable?

66. Explain how to graph $y \geq x^2$.

67. The graph of $f(x) = x^2 - 3x + 4$ is shown here. Explain why the quadratic inequality $x^2 - 3x + 4 < 0$ has no solution.

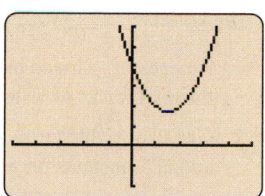

68. Describe the following solution set of a rational inequality in words: $(-\infty, 4] \cup (6, 7)$.

REVIEW

Translate each statement into an equation.

69. x varies directly as y.

70. y varies inversely as t.

71. t varies jointly as x and y.

72. d varies directly as t and inversely as u^2.

CHALLENGE PROBLEMS

73. **a.** Solve: $x^2 - x - 12 > 0$

b. Find a rational inequality in one variable that has the same solution set as the quadratic inequality in part (a).

74. **a.** Solve: $\dfrac{1}{x} < 1$

b. Now incorrectly "solve" $\dfrac{1}{x} < 1$ by multiplying both sides by x to clear it of the fraction. What part of the solution set is not obtained with this incorrect approach?

8 ▶ Summary & Review

SECTION 8.1 ▶ **The Square Root Property and Completing the Square**

DEFINITIONS AND CONCEPTS	EXAMPLES
We can use the **square root property** to solve equations of the form $x^2 = c$, where $c > 0$. The two solutions are $x = \sqrt{c}$ or $x = -\sqrt{c}$ We can write $x = \sqrt{c}$ or $x = -\sqrt{c}$ in more compact form using **double-sign notation**: $x = \pm\sqrt{c}$	Solve: $x^2 = 24$ $x = \sqrt{24}$ or $x = -\sqrt{24}$ Use the square root property. $x = \pm\sqrt{24}$ Use double-sign notation. $x = \pm 2\sqrt{6}$ Simplify $\sqrt{24}$. The solutions are $2\sqrt{6}$ and $-2\sqrt{6}$. Verify this using a check. Solve: $(x - 3)^2 = -81$ $x - 3 = \pm\sqrt{-81}$ Use the square root property and double-sign notation. $x = 3 \pm \sqrt{-81}$ To isolate x, add 3 to both sides. $x = 3 \pm 9i$ Simplify the radical expression. The solutions are $3 + 9i$ and $3 - 9i$. Verify this using a check.
To **complete the square** on $x^2 + bx$, add the square of one-half of the coefficient of x. $x^2 + bx + \left(\dfrac{1}{2}b\right)^2$	Complete the square on $x^2 + 8x$ and factor the resulting perfect-square trinomial. $x^2 + 8x \mathbf{+ 16}$ The coefficient of x is 8. To complete the square: $\frac{1}{2} \cdot 8 = 4$ and $4^2 = 16$. Add 16 to the binomial. Now we factor: $x^2 + 8x + 16 = (x + 4)^2$
To **solve a quadratic equation in x by completing the square:** 1. If necessary, divide both sides of the equation by the coefficient of x^2 to make its coefficient 1. 2. Get all variable terms on one side of the equation and all constants on the other side. 3. Complete the square. 4. Factor the perfect-square trinomial. 5. Solve the resulting equation by using the square root property. 6. Check your answers in the original equation.	Solve: $3x^2 - 12x + 6 = 0$ $\dfrac{3x^2}{3} - \dfrac{12x}{3} + \dfrac{6}{3} = \dfrac{0}{3}$ To make the leading coefficient 1, divide both sides by 3, term-by-term. $x^2 - 4x + 2 = 0$ Do the division. $x^2 - 4x = -2$ Subtract 2 from both sides so that the constant term, −2, is on the right side. $x^2 - 4x \mathbf{+ 4} = -2 \mathbf{+ 4}$ The coefficient of x is −4. To complete the square: $\frac{1}{2}(-4) = -2$ and $(-2)^2 = 4$. Add 4 to both sides. $(x - 2)^2 = 2$ Factor the perfect-square trinomial on the left side. Add on the right side. $x - 2 = \pm\sqrt{2}$ Use the square root property. $x = 2 \pm \sqrt{2}$ To isolate x, add 2 to both sides. The solutions are $2 + \sqrt{2}$ and $2 - \sqrt{2}$. Verify this using a check.

REVIEW EXERCISES

Solve each equation by factoring.

1. $x^2 + 9x + 20 = 0$ **2.** $6x^2 + 17x + 5 = 0$

Solve each equation using the square root property. Approximate the solutions to the nearest hundredth when appropriate.

3. $x^2 = 28$ **4.** $(t + 2)^2 = 36$

5. $a^2 + 25 = 0$ **6.** $5x^2 - 49 = 0$

7. Solve $A = \pi r^2$ for r. Assume all variables represent positive numbers. Express the result in simplified radical form.

8. Complete the square on $x^2 - x$ and then factor the resulting perfect-square trinomial.

Solve each equation by completing the square. Approximate the solutions to the nearest hundredth when appropriate.

9. $x^2 + 6x + 8 = 0$ **10.** $2x^2 - 6x + 3 = 0$

11. $6a^2 - 12a = -1$ **12.** $x^2 - 2x = -13$

13. Let $f(x) = x^2$. Find all values of x for which $f(x) = 32$.

14. Let $g(x) = (7x - 51)^2$. Find all values of x for which $g(x) = 11$.

15. Explain why completing the square on $x^2 + 7x$ is more difficult than completing the square on $x^2 + 6x$.

16. Happy New Year. As part of a New Year's Eve celebration, a huge ball is to be dropped from the top of a 605-foot-tall building at the proper moment so that it strikes the ground at exactly 12:00 midnight. The distance d in feet traveled by a free-falling object in t seconds is given by the formula $d = 16t^2$. To the nearest second, when should the ball be dropped from the building?

SECTION 8.2 ▶ **The Quadratic Formula**

DEFINITIONS AND CONCEPTS	EXAMPLES
To **solve a quadratic equation in x using the quadratic formula:** **1.** Write the equation in standard form: $$ax^2 + bx + c = 0$$ **2.** Identify a, b, and c. **3.** Substitute the values for a, b, and c in the quadratic formula $$x = \frac{-b \pm \sqrt{b^2 - 4ac}}{2a}$$ and evaluate the right side to obtain the solutions.	Solve: $3x^2 - 2x - 2 = 0$ Here, $a = 3$, $b = -2$, and $c = -2$. $x = \dfrac{-b \pm \sqrt{b^2 - 4ac}}{2a}$ This is the quadratic formula. $x = \dfrac{-(-2) \pm \sqrt{(-2)^2 - 4(3)(-2)}}{2(3)}$ Substitute 3 for a, -2 for b, and -2 for c. $x = \dfrac{2 \pm \sqrt{28}}{6}$ Evaluate the expression within the radical. $x = \dfrac{2 \pm 2\sqrt{7}}{6}$ Simplify the radical: $\sqrt{28} = \sqrt{4}\sqrt{7} = 2\sqrt{7}$. $x = \dfrac{\overset{1}{2}(1 \pm \sqrt{7})}{\underset{1}{2} \cdot 3}$ Factor out the GCF, 2, from the two terms in the numerator. In the denominator, factor 6 as $2 \cdot 3$. Then remove the common factor, 2. $x = \dfrac{1 \pm \sqrt{7}}{3}$ The exact solutions are $\dfrac{1 + \sqrt{7}}{3}$ and $\dfrac{1 - \sqrt{7}}{3}$. We can use a calculator to approximate them. To the nearest hundredth, they are 1.22 and -0.55.
When solving a quadratic equation using the quadratic formula, we can often **simplify the calculations** by solving an equivalent equation that does not involve fractions or decimals, and whose leading coefficient is positive.	**Before solving . . .** **do this . . .** **to get this** $-3x^2 + 5x - 1 = 0$ Multiply both sides by -1. $3x^2 - 5x + 1 = 0$ $x^2 + \frac{7}{8}x - \frac{1}{2} = 0$ Multiply both sides by 8. $8x^2 + 7x - 4 = 0$ $60x^2 - 40x + 90 = 0$ Divide both sides by 10. $6x^2 - 4x + 9 = 0$ $0.05x^2 + 0.16x + 0.71 = 0$ Multiply both sides by 100. $5x^2 + 16x + 71 = 0$

REVIEW EXERCISES

Solve each equation using the quadratic formula. Approximate the solutions to the nearest hundredth when appropriate.

17. $2x^2 + 13x = 7$

18. $-x^2 + 10x - 18 = 0$

19. $x^2 - 10x = 0$

20. $3y^2 = 26y - 2$

21. $\frac{1}{3}p^2 + \frac{1}{2}p + \frac{1}{2} = 0$

22. $3{,}000t^2 - 4{,}000t = -2{,}000$

23. $0.5x^2 + 0.3x - 0.1 = 0$

24. $x^2 - 3x - 27 = 0$

25. Let $h(x) = x^2 + 3x - 7$. Find all values of x for which $h(x) = 1$.

26. Let $T(x) = -4x^2 - 2x$. Find all values of x for which $T(x) = -3$.

27. Explain the error: $\dfrac{2 \pm \sqrt{7}}{2} = \dfrac{\overset{1}{2} \pm \sqrt{7}}{\underset{1}{2}}$

28. a. Write an expression that represents the width of the larger rectangle shown in red.

    b. Write an expression that represents the length of the larger rectangle shown in red.

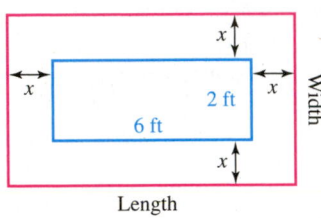

29. Posters. The specifications for a poster of Cesar Chavez call for a 615-square-inch photograph to be surrounded by a green border. The borders on the top and bottom of the poster are to be twice as wide as those on the sides. Find the width of each border.

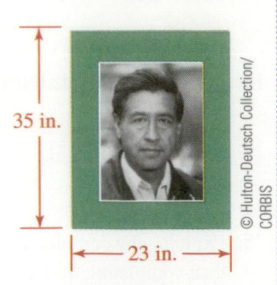

35 in.

23 in.

© Hulton-Deutsch Collection/CORBIS

30. Tutoring. A private tutoring company charges $20 for a 1-hour session. Currently, 300 students are tutored each week. Since the company is losing money, the owner has decided to increase the price. For each 50¢ increase, she estimates that 5 fewer students will participate. If the company needs to bring in $6,240 per week to stay in business, what price must be charged for a 1-hour tutoring session to produce this amount of revenue?

31. Acrobats. To begin his routine on a trapeze, an acrobat is catapulted upward as shown in the illustration. His distance d (in feet) from the arena floor during this maneuver is given by the function $d(t) = -16t^2 + 40t + 5$, where t is the time in seconds since being launched. If the trapeze bar is 25 feet in the air, at what two times will he be able to grab it? Round to the nearest tenth.

32. Triangles. The length of the longer leg of a right triangle exceeds the length of the shorter leg by 23 inches and the length of the hypotenuse is 65 inches. Find the length of each leg of the triangle.

SECTION 8.3 ▶ The Discriminant and Equations That Can Be Written in Quadratic Form

DEFINITIONS AND CONCEPTS	EXAMPLES
The **discriminant** predicts the type of solutions of $ax^2 + bx + c = 0$, where a, b, and c are rational numbers and $a \neq 0$. Review the table on page 704.	In the quadratic equation $2x^2 - 5x - 3 = 0$, we have $a = 2$, $b = -5$, and $c = -3$. So the value of the discriminant is $$b^2 - 4ac = (-5)^2 - 4(2)(-3) = 25 + 24 = 49$$ Since the value of the discriminant is positive and a perfect square, the equation $2x^2 - 5x - 3 = 0$ has two different rational-number solutions.

Equations that contain an expression, the same expression squared, and a constant term are said to be **quadratic in form.** One method used to solve such equations is to make a **substitution.**

The leading term, x, is the square of the expression $\sqrt{x}$ in the middle term: $x = \left(\sqrt{x}\right)^2$.

$$x - 9\sqrt{x} + 20 = 0$$

The last term is a constant.

Solve: $x^{2/3} - 6x^{1/3} + 5 = 0$

The equation can be written in quadratic form:

$$(x^{1/3})^2 - 6x^{1/3} + 5 = 0 \quad \text{Because } x^{2/3} = (x^{1/3})^2$$

We substitute y for $x^{1/3}$ and use factoring to solve the resulting quadratic equation $y^2 - 6y + 5 = 0$.

$$(y - 1)(y - 5) = 0 \quad \text{Let } y = x^{1/3}.$$
$$y = 1 \quad \text{or} \quad y = 5$$

Now we reverse (undo) the substitution $y = x^{1/3}$ and solve for x.

$$x^{1/3} = 1 \qquad \text{or} \qquad x^{1/3} = 5$$
$$(x^{1/3})^3 = (1)^3 \qquad (x^{1/3})^3 = (5)^3 \quad \text{Recall: } x^{1/3} = \sqrt[3]{x}.$$
$$x = 1 \qquad\qquad x = 125$$

The solutions are 1 and 125. Check both in the original equation.

REVIEW EXERCISES

Use the discriminant to determine the number and type of solutions for each equation.

33. $3x^2 + 4x - 3 = 0$

34. $4x^2 - 5x + 7 = 0$

35. $3x^2 - 4x + \dfrac{4}{3} = 0$

36. $m(2m - 3) = 20$

Solve each equation.

37. $x - 13\sqrt{x} + 12 = 0$

38. $a^{2/3} + a^{1/3} - 6 = 0$

39. $3x^4 + x^2 - 2 = 0$

40. $\dfrac{6}{x + 2} + \dfrac{6}{x + 1} = 5$

41. $(x - 7)^2 + 6(x - 7) + 10 = 0$

42. $m^{-4} - 2m^{-2} + 1 = 0$

43. $4\left(\dfrac{x + 1}{x}\right)^2 + 12\left(\dfrac{x + 1}{x}\right) + 9 = 0$

44. $2m^{2/5} - 5m^{1/5} + 2 = 0$

45. Weekly Chores.   Working together, two sisters can do the yard work at their house in 45 minutes. When the older girl does it all herself, she can complete the job in 20 minutes less time than it takes the younger girl working alone. How long does it take the older girl to do the yard work?

46. Road Trips.   A woman drives her automobile 150 miles at a rate of r mph. She could have gone the same distance in 2 hours less time if she had increased her speed by 20 mph. Find r.

SECTION 8.4 ▶ Quadratic Functions and Their Graphs

DEFINITIONS AND CONCEPTS	EXAMPLES
A **quadratic function** is a second-degree polynomial function of the form $$f(x) = ax^2 + bx + c$$	Quadratic functions: $$f(x) = 2x^2 - 3x + 5, \qquad g(x) = -x^2 + 4x, \qquad \text{and} \qquad s(x) = \frac{1}{4}x^2 - 10$$

The graph of the quadratic function $f(x) = a(x - h)^2 + k$ where $a \neq 0$ is a **parabola** with **vertex** at (h, k). The **axis of symmetry** is the line $x = h$. The parabola opens upward when $a > 0$ and downward when $a < 0$.

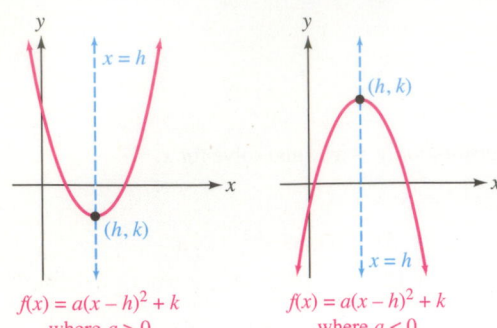

$f(x) = a(x - h)^2 + k$
where $a > 0$

$f(x) = a(x - h)^2 + k$
where $a < 0$

Graph: $f(x) = 2(x + 1)^2 - 8$

$$f(x) = 2[x - (-1)]^2 - 8$$
$$f(x) = a(x - h)^2 + k$$

We see that $a = 2$, $h = -1$, and $k = -8$. The graph is a parabola with vertex $(h, k) = (-1, -8)$ and axis of symmetry $x = -1$. Since a is positive, the parabola opens upward.

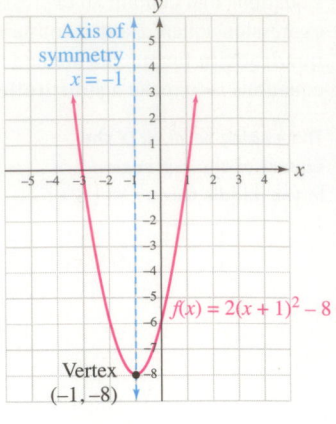

The **vertex** of the graph of $f(x) = ax^2 + bx + c$ is

$$\left(-\frac{b}{2a}, f\left(-\frac{b}{2a}\right) \right)$$

and the axis of symmetry is the line

$$x = -\frac{b}{2a}$$

The y-coordinate of the vertex of the graph of a quadratic function gives the **minimum or maximum value** of the function.

The **y-intercept** is determined by the value of $f(x)$ when $x = 0$: the y-intercept is $(0, c)$.

To find the **x-intercepts**, let $f(x) = 0$ and solve $ax^2 + bx + c = 0$.

Graph: $f(x) = -x^2 + 3x + 4$

Here, $a = -1$, $b = 3$, and $c = 4$.

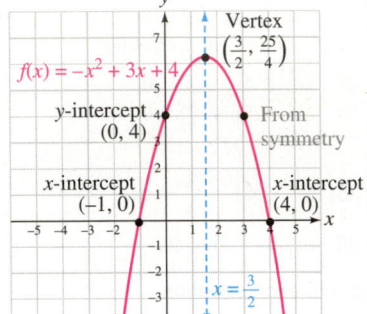

- Since $a < 0$, the graph opens downward.

- The x-coordinate of the vertex of the graph is:

$$-\frac{b}{2a} = -\frac{3}{2(-1)} = \frac{3}{2}$$

To find the y-coordinate of the vertex, we substitute $\frac{3}{2}$ for x in the function.

$$f\left(\frac{3}{2}\right) = -\left(\frac{3}{2}\right)^2 + 3\left(\frac{3}{2}\right) + 4 = \frac{25}{4}$$

The vertex of the parabola is $\left(\frac{3}{2}, \frac{25}{4}\right)$.

- The y-intercept is the value of the function when $x = 0$. Since $f(0) = 4$, the y-intercept is $(0, 4)$.

- To find the x-intercepts, we solve:

$$-x^2 + 3x + 4 = 0 \quad \text{Let } f(x) = 0.$$
$$x^2 - 3x - 4 = 0 \quad \text{Multiply both sides by } -1.$$
$$(x + 1)(x - 4) = 0 \quad \text{Factor.}$$
$$x = -1 \quad \text{or} \quad x = 4$$

The x-intercepts are $(-1, 0)$ and $(4, 0)$.

REVIEW EXERCISES

47. Hospitals.   The annual number of in-patient admissions to U.S. community hospitals for the years 1980–2008 can be modeled by the quadratic function $A(x) = 0.03x^2 - 0.88x + 37.3$, where $A(x)$ is the number of admissions in millions and x is the number of years after 1980. Use the function to estimate the number of in-patient admissions for the year 2005. Round to the nearest tenth of one million. (Source: American Hospital Association)

48. Fill in the blanks. The graph of the quadratic function $f(x) = a(x - h)^2 + k$ is a parabola with vertex at (___ , ___). The axis of symmetry is the line ___ $= h$. The parabola opens upward when $a > 0$ and downward when $a < 0$.

Graph each pair of functions on the same coordinate system.

49. $f(x) = 2x^2$, $g(x) = 2x^2 - 3$

50. $f(x) = -\dfrac{1}{4}x^2$, $g(x) = -\dfrac{1}{4}(x + 2)^2$

51. Find the vertex and the axis of symmetry of the graph of $f(x) = -2(x - 1)^2 + 4$. Then plot several points and complete the graph.

52. Complete the square to write $f(x) = 4x^2 + 16x + 9$ in the form $f(x) = a(x - h)^2 + k$. Determine the vertex and the axis of symmetry of the graph. Then plot several points and complete the graph.

53. Find the vertex of the graph of $f(x) = -2x^2 + 4x - 8$ using the vertex formula.

54. First determine the coordinates of the vertex and the axis of symmetry of the graph of $f(x) = x^2 + x - 2$ using the vertex formula. Then determine the x- and y-intercepts of the graph. Finally, plot several points and complete the graph.

55. Farming.   The number of farms in the United States for the years 1870–1970 is approximated by

$$N(x) = -1,526x^2 + 155,652x + 2,500,200$$

where $x = 0$ represents 1870, $x = 1$ represents 1871, and so on. For this period, when was the number of U.S. farms a maximum? How many farms were there?

56. Estimate the solutions of $-3x^2 - 5x + 2 = 0$ from the graph of $f(x) = -3x^2 - 5x + 2$, shown here.

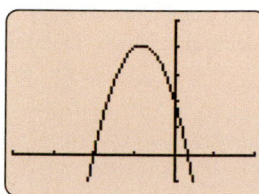

SECTION 8.5 ▶ **Quadratic and Other Nonlinear Inequalities**

DEFINITIONS AND CONCEPTS	EXAMPLES
To solve a quadratic inequality, get 0 on the right side and solve the related quadratic equation. Then locate the **critical numbers** on a number line, test each interval, and check the endpoints.	To solve $x^2 - x - 6 \geq 0$, we solve the related quadratic equation:

To solve $x^2 - x - 6 \geq 0$, we solve the related quadratic equation:

$$x^2 - x - 6 = 0$$

$$(x - 3)(x + 2) = 0 \qquad \text{Factor.}$$

$$x = 3 \quad \text{or} \quad x = -2 \qquad \text{These are the critical numbers that divide the number line into three intervals.}$$

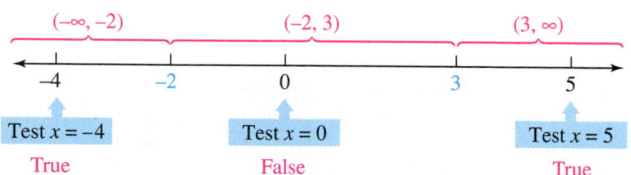

After testing each interval and noting that 3 and -2 satisfy the inequality, we see that the solution set is $(-\infty, -2] \cup [3, \infty)$.

To solve a rational inequality, get 0 on the right side and solve the related rational equation. Then locate the **critical numbers** (including any values that make the denominator 0) on a number line, test each interval, and check the endpoints.

To solve $\frac{x+1}{x-4} < 0$, we solve the related rational equation $\frac{x+1}{x-4} = 0$ to obtain the solution $x = -1$, which is a critical number. Another critical number is $x = 4$, the value that makes the denominator 0. These critical numbers divide the number line into three intervals.

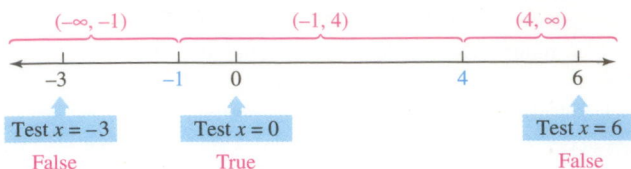

After testing each interval and noting that -1 and 4 do not satisfy the inequality, we see that the solution set is the interval $(-1, 4)$.

To graph a nonlinear inequality in two variables, first graph the boundary. Then use a test point to determine which side of the boundary to shade.

This is the graph of $y \le x^2 + 5x + 4$.

Since the inequality contains the symbol $\le$, and equality is allowed, we draw the parabola determined by $y = x^2 + 5x + 4$ using a **solid curve.**

We shade the region containing the test point $(0, 0)$ because its coordinates satisfy $y \le x^2 + 5x + 4$.

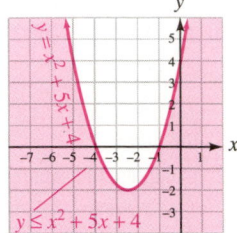

REVIEW EXERCISES

Solve each inequality. Write the solution set in interval notation and graph it.

57. $x^2 + 2x - 35 > 0$

58. $x^2 \le 81$

59. $\dfrac{3}{x} \le 5$

60. $\dfrac{2x^2 - x - 28}{x - 1} > 0$

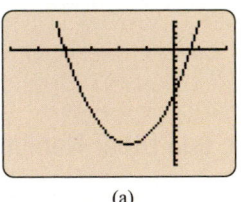
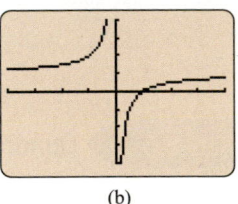

(a)   (b)

61. Estimate the solution set of $3x^2 + 10x - 8 \le 0$ from the graph of $f(x) = 3x^2 + 10x - 8$ shown in figure (a).

62. Estimate the solution set of $\frac{x-1}{x} > 0$ from the graph of $f(x) = \frac{x-1}{x}$ shown in figure (b).

Graph each inequality.

63. $y < \dfrac{1}{2}x^2 - 1$

64. $y \ge -|x|$

8 ▶ Chapter Test

1. Fill in the blanks.

   a. An equation of the form $ax^2 + bx + c = 0$, where $a \ne 0$, is called a _____ equation.

   b. When we add 81 to $x^2 + 18x$, we say that we have _____ the _____ on $x^2 + 18x$.

   c. The lowest point on a parabola that opens upward, or the highest point on a parabola that opens downward, is called the _____ of the parabola.

   d. $\frac{x-5}{x^2 - x - 56} > 0$ is an example of a _____ inequality in one variable.

   e. $y \le x^2 - 4x + 3$ is an example of a _____ inequality in two variables.

2. Solve $x^2 - 63 = 0$ using the square root property. Approximate the solutions to the nearest hundredth.

Solve each equation using the square root property.

3. $(a + 7)^2 = 50$

4. $m^2 + 4 = 0$

5. Add a number to make $x^2 + 11x$ a perfect-square trinomial. Then factor the result.

6. Solve $x^2 + 3x - 2 = 0$ by completing the square. Approximate the solutions to the nearest hundredth.

7. Solve $2x^2 + 8x + 12 = 0$ by completing the square.

8. Let $f(x) = (3x - 2)^2$. Find all values of x for which $f(x) = 18$.

Use the quadratic formula to solve each equation. Approximate each solution to the nearest hundredth, when appropriate.

9. $4x^2 + 4x - 1 = 0$

10. $\frac{1}{8}t^2 - \frac{1}{4}t = \frac{1}{2}$

11. $-t^2 + 4t - 13 = 0$ **12.** $0.01x^2 = -0.08x - 0.15$

13. $m^2 - 94m = -2{,}209$

14. Let $g(x) = 3x^2 - 20$. Find all values of x for which $g(x) = 10$.

Solve each equation by any method. Approximate each solution to the nearest hundredth, when appropriate.

15. $2y - 3\sqrt{y} + 1 = 0$ **16.** $3 = m^{-2} - 2m^{-1}$

17. $x^4 - x^2 - 12 = 0$

18. $4\left(\dfrac{x+2}{3x}\right)^2 - 4\left(\dfrac{x+2}{3x}\right) - 3 = 0$

19. $\dfrac{1}{n+2} = \dfrac{1}{3} - \dfrac{1}{n}$

20. $5a^{2/3} + 11a^{1/3} = -2$

21. $10x(x + 1) = -3$

22. $a^3 - 3a^2 + 8a - 24 = 0$

23. Solve $E = mc^2$ for c. Assume that all variables represent positive numbers. Express any radical in simplified form.

24. Use the discriminant to determine the number and type of solutions for each equation.

   a. $3x^2 + 5x + 17 = 0$

   b. $9m^2 - 12m = -4$

25. Tablecloths. In 1990, Sportex of Highland, Illinois, made what was at the time the world's longest tablecloth. Find the dimensions of the rectangular tablecloth if it covered an area of 6,759 square feet and its length was 8 feet more than 332 times its width.

26. Cooking. Working together, a chef and his assistant can make a pastry dessert in 25 minutes. When the chef makes it himself, it takes him 8 minutes less time than it takes his assistant working alone. How long does it take the chef to make the dessert?

27. Drawing. An artist uses four equal-sized right triangles to block out a perspective drawing of an old hotel. See the illustration in the next column. For each triangle, the leg on the horizontal line is 14 inches longer than the leg on the center line. The length of each hypotenuse is 26 inches. On the center line of the drawing, what is the length of the segment extending from the ground to the top of the building?

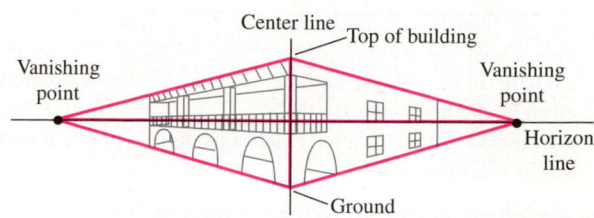

28. Home Decorating. A woman is going to put a cork border of uniform width around an 18-inch by 24-inch wall picture, as shown here. If the area covered by the cork is to be the same as the surface area of the picture, how wide should the border be? Round to the nearest tenth of an inch.

29. ER Rooms. The number of hospital emergency departments in the United States is approximated by the function $E(t) = 2.29t^2 - 75.72t + 5{,}206.95$, where t is the number of years after 1990. In what year does the model indicate that the number of hospital emergency departments was 5,000? (Source: American Hospital Association)

30. Anthropology. Anthropologists refer to the shape of the human jaw as a *parabolic dental arcade*. Which function is the best mathematical model of the parabola shown in the illustration?

   i. $f(x) = -\dfrac{3}{8}(x - 4)^2 + 6$ **ii.** $f(x) = -\dfrac{3}{8}(x - 6)^2 + 4$

   iii. $f(x) = -\dfrac{3}{8}x^2 + 6$ **iv.** $f(x) = \dfrac{3}{8}x^2 + 6$

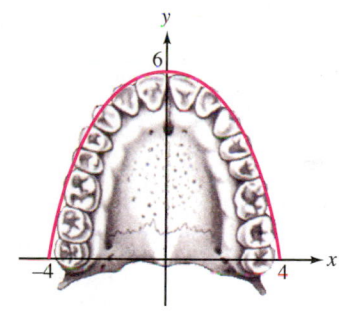

31. Find the vertex and the axis of symmetry of the graph of $f(x) = -3(x - 1)^2 + 2$. Then plot several points and complete the graph.

32. Complete the square to write the function $f(x) = 5x^2 + 10x - 1$ in the form $f(x) = a(x - h)^2 + k$. Determine the vertex and the axis of symmetry of the graph. Then plot several points and complete the graph.

33. First determine the coordinates of the vertex and the axis of symmetry of the graph of $f(x) = 2x^2 + x - 1$ using the vertex formula. Then determine the x- and y-intercepts of the graph. Finally, plot several points and complete the graph.

34. Distress Signals.   A flare is fired directly upward into the air from a boat that is experiencing engine problems. The height of the flare (in feet) above the water, t seconds after being fired, is given by the function $h(t) = -16t^2 + 112t + 15$. If the flare is designed to explode when it reaches its highest point, at what height will this occur?

Solve each inequality. Write the solution set in interval notation and then graph it.

35. $x^2 - 2x > 8$

36. $\dfrac{x - 2}{x + 3} \leq 0$

37. Climate.   The average monthly temperature in San Antonio, Texas, is approximated by the function $T(m) = -1.1m^2 + 15.3m + 29.5$, where m is the number of the month (January = 1, February = 2, and so on). Use the function to approximate the average monthly temperature in July. (Source: cityrating.com)

Average Temperature (°F) San Antonio, Texas

Jan. Feb. Mar. Apr. May. Jun. Jul. Aug. Sep. Oct. Nov. Dec.

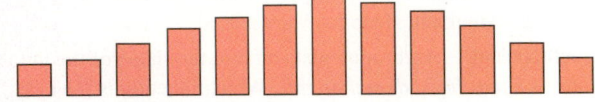

38. Graph: $y \leq -x^2 + 3$

39. The graph of a quadratic function of the form $f(x) = ax^2 + bx + c$ is shown. Estimate the solutions of the corresponding quadratic equation $ax^2 + bx + c = 0$.

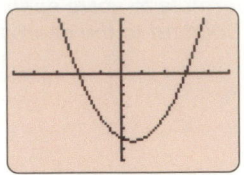

40. See Exercise 39. Estimate the solution of the quadratic inequality $ax^2 + bx + c \leq 0$.

Group Project

.... ▶

Picture Framing

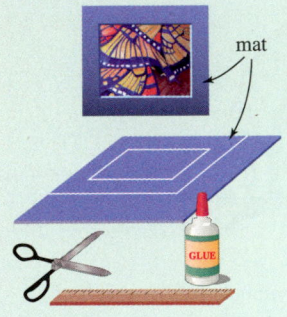

mat

Overview: When framing pictures, mats are often used to enhance the images and give them a sense of depth. In this activity, you will use the quadratic formula to design the matting for several pictures.

Instructions: Form groups of 3 students. Each person in your group is to bring a picture to class. You can use a picture from a magazine or newspaper, a picture postcard, or a photograph that is no larger than 5 in. × 7 in. You will also need a pair of scissors, a ruler, glue, and three pieces of construction paper (12 in. × 18 in.).

Select one of the pictures and find its area. A mat of *uniform* width is to be placed around the picture. The area of the mat should equal the area of the picture. To determine the proper width of the matting, follow the steps of Example 8 in Section 8.2. Once you have determined the proper width, cut out the mat from the construction paper and glue it to the picture.

Then, choose another picture and find its area. Determine the uniform width that a matting should have so that its area is double that of the picture. Cut out the proper-size matting from the construction paper and glue it to the second picture.

Finally, find the area of the third picture and determine the uniform width that a matting should have so that its area is one-half that of the picture. Cut out the proper-size matting from the construction paper and glue it to the third picture.

Is one size matting more visually appealing than another? Discuss this among the members of your group.

CUMULATIVE REVIEW ▶▶ Chapters 1–8

1. Solve: $3(x + 2) - 2 = -(5 + x) + x$ [Section 1.5]

2. **Pharmacists.** How many liters of a 1% glucose solution should a pharmacist mix with 2 liters of a 5% glucose solution to obtain a 2% glucose solution? [Section 1.8]

Find an equation of the line with the given properties. Write the equation in slope–intercept form.

3. Slope 3, passing through $(-2, -4)$ [Section 2.4]

4. Parallel to the graph of $2x + 3y = 6$ and passing through $(0, -2)$ [Section 2.4]

5. **Shortage of Nurses.** Use the data in the graph to find the projected rates of change in the supply and demand for registered nurses (RNs) in the United States for the years 2010–2020. [Section 2.3]

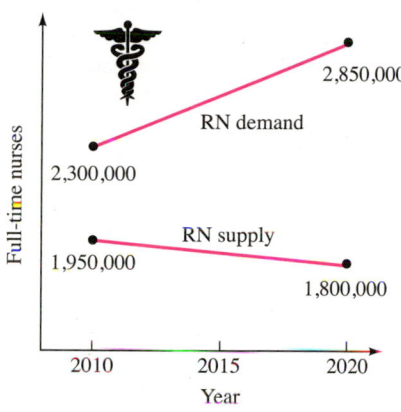

(Source: American Hospital Association)

6. **The Clothing Industry.** Since 1960, spending by Americans on clothing as a share of their disposable personal income has steadily declined. In 1970, for example, about 6.2% of disposable income was spent on clothing. By 1980, that percent dropped to 5.4%. The trend can be modeled by a linear function. (Source: fashion-incubator.com) [Section 2.5]

   a. Let t be the number of years after 1960 and P be the percent of personal disposable income used to buy clothing. Write a linear function $P(t)$.

   b. Use your answer to part a to estimate the percent of personal disposable income used to buy clothing in 2010.

7. Determine whether the graph on the right is the graph of a function. If it is not, find two ordered pairs where more than one value of y corresponds to a single value of x. [Section 2.5]

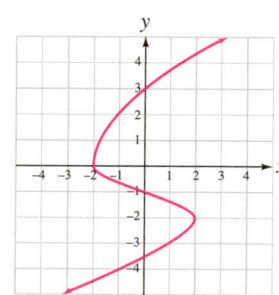

8. **Tides.** The illustration shows the graph of a function f, which gives the height of the tide for a 24-hour period in Seattle, Washington. (Note that military time is used on the x-axis: 3 A.M. = 3, noon = 12, 9 P.M. = 21, and so on.) [Section 2.6]

   a. Find the domain of the function.

   b. Find $f(6)$.

   c. What information does $f(12)$ give?

   d. Estimate the values of x for which $f(x) = 0$.

9. Solve the system by graphing:
$$\begin{cases} x + y = -1 \\ y = 2x - 4 \end{cases}$$ [Section 3.1]

10. Use substitution to solve the system:
$$\begin{cases} x - y = -5 \\ 3x - 2y = -7 \end{cases}$$ [Section 3.2]

11. Solve the system:
$$\begin{cases} x - y + z = 4 \\ x + 2y - z = -1 \\ x + y - 3z = -2 \end{cases}$$ [Section 3.3]

12. Evaluate the determinant: $\begin{vmatrix} -6 & -2 \\ 15 & 4 \end{vmatrix}$ [Section 3.5]

13. **TV History.** *Friends, The King of Queens,* and *That '70s Show* are three of the most popular television shows of all time. The total number of episodes of these three shows is 643. There were 36 more episodes of *Friends* than *That '70s Show,* and the difference between the number of episodes of *The King of Queens* and *That '70s Show,* was 7. Find the number of episodes of each show. (Source: angelfire.com) [Section 3.7]

14. **Job Offers.** A pharmaceutical company has offered a newly hired salesperson her choice of compensation packages:

   Package 1: $2,500 salary per month and a 5% sales commission

   Package 2: $3,500 salary per month and a 3% sales commission

   What amount of sales per month would the salesperson have to make so that Package 1 is more profitable for her? [Section 4.1]

Solve each inequality. Write the solution set in interval notation and then graph it.

15. Let $f(x) = 5(-2x + 2)$ and $g(x) = 20 - x$. Find all values of x for which $f(x) > g(x)$. [Section 4.1]

16. $5x - 3 \geq 2$ and $6 \geq 4x - 3$ [Section 4.2]

17. $|2x - 5| \geq 25$ [Section 4.3]

18. Graph the solution set of the system:

$$\begin{cases} 3x + 2y > 6 \\ x + 3y \leq 2 \end{cases}$$ [Section 4.5]

19. Simplify: $\left(\dfrac{2a^2 b^3 c^{-4}}{5a^{-2} b^{-1} c^3} \right)^{-3}$ [Section 5.1]

20. Write each number in scientific notation and perform the operations. Give the answer in scientific notation and in standard notation.   [Section 5.2]

$$\dfrac{(1{,}280{,}000{,}000)(2{,}700{,}000)}{240{,}000}$$

Perform the indicated operations.

21. Let $f(t) = -8.9t^3 - 2.4t$ and $g(t) = 2.1t^3 + 0.8t^2 - t$. Find $f(t) - g(t)$. [Section 5.3]

22. $(2a - b)(4a^2 + 2ab + b^2)$ [Section 5.4]

23. $(9mn^3 - 4)^2$ [Section 5.4]

24. $(2x^2 - y^2)(2x^2 + y^2)$ [Section 5.4]

Factor each expression.

25. $x^2 + 4y - xy - 4x$ [Section 5.5]

26. $30a^4 - 4a^3 - 16a^2$ [Section 5.6]

27. $49s^6 - 84s^3 n^2 + 36n^4$ [Section 5.6]

28. $x^2 + 10x + 25 - y^8$ [Section 5.7]

29. $x^4 - 16y^4$ [Section 5.7]

30. $8x^6 + 125y^3$ [Section 5.7]

Solve each equation.

31. $(m + 4)(2m + 3) - 22 = 10m$ [Section 5.9]

32. $6a^3 - 2a = a^2$ [Section 5.9]

33. **Projectiles.**   An object is shot vertically into the air. Its height h in feet t seconds after being shot is given by the function $h(t) = -16t^2 + 96t$. After how many seconds will the object strike the ground? [Section 5.9]

34. Find the domain of $f(x) = \dfrac{x^2 + 3x + 2}{x^2 - x - 56}$. [Section 6.1]

Graph each function and give its domain and range.

35. $f(x) = |x - 1| + 2$ [Section 2.6]

36. $f(x) = \dfrac{4}{x}$ for $x > 0$ [Section 6.1]

37. Simplify: $\dfrac{6x^2 - 7x - 5}{2x^2 + 5x + 2}$ [Section 6.1]

38. Divide: $\dfrac{x^3 + y^3}{x^3 - y^3} \div \dfrac{x^2 - xy + y^2}{x^2 + xy + y^2}$ [Section 6.2]

39. Perform the operations: $\dfrac{1}{x + y} - \dfrac{1}{x - y} + \dfrac{2y}{x^2 - y^2}$

[Section 6.3]

40. Simplify: $\dfrac{\dfrac{1}{r^2 + 4r + 4}}{\dfrac{r}{r + 2} + \dfrac{r}{r + 2}}$ [Section 6.4]

41. Divide: $\dfrac{24x^6 y^7 - 12x^5 y^{12} + 36xy}{48x^2 y^3}$ [Section 6.5]

42. Divide: $3a - 4 \overline{)15a^3 - 29a^2 + 16}$ [Section 6.5]

43. Solve: $\dfrac{x - 4}{x - 3} + \dfrac{x - 2}{x - 3} = x - 3$ [Section 6.7]

44. Solve for R: $\dfrac{1}{R} = \dfrac{1}{R_1} + \dfrac{1}{R_2} + \dfrac{1}{R_3}$ [Section 6.7]

45. **Sinks.**   A sink has two faucets, one for cold water and one for hot water. It can be filled by the cold-water faucet in 30 seconds and by the hot-water faucet in 45 seconds. How long will it take to fill the sink if both faucets are opened? [Section 6.8]

46. **Snow Removal.**   A state highway department uses a 7-to-2 sand-to-salt mix in the winter months for spreading across roadways covered with snow and ice. If they have 6 tons of salt in storage, how many tons of sand should be added to obtain the proper mix? [Section 6.9]

47. **Deliveries.**   The costs of a delivery company vary jointly with the number of trucks in service and the number of hours they are used. When 8 trucks are used for 12 hours each, the costs are $3,600. Find the costs of using 20 trucks, each for 12 hours. [Section 6.9]

48. Graph the function $f(x) = \sqrt{x - 2}$ and give its domain and range. [Section 7.1]

49. **Oceanography.**   The speed of a deep-water wave (in meters per second) is given by the function $C(L) = 1.25\sqrt{L}$, where L is the wave length (in meters). Find the speed of a wave if each adjacent crest is 6.5 meters apart. Round to the nearest tenth. [Section 7.1]

50. Let $f(t) = \sqrt[3]{t - 5}$ and $g(t) = 2\sqrt[3]{t + 9}$. Find all values of t for which $f(t) = g(t)$. [Section 7.5]

Simplify each expression. All variables represent positive real numbers.

51. $\sqrt[3]{-27x^3}$ [Section 7.1]   52. $64^{-2/3}$ [Section 7.2]

53. $\dfrac{x^{5/3} x^{1/2}}{x^{3/4}}$ [Section 7.2]   54. $\sqrt{48t^3}$ [Section 7.3]

55. $-3\sqrt[4]{32} - 2\sqrt[4]{162} + 5\sqrt[4]{48}$ [Section 7.3]

56. $3\sqrt{2} \left(2\sqrt{3} - 4\sqrt{12} \right)$ [Section 7.4]

Rationalize the denominator of each expression. All variables represent positive real numbers.

57. $\dfrac{\sqrt{x} + 2}{\sqrt{x} - 1}$ [Section 7.4]   58. $\dfrac{5}{\sqrt[3]{x}}$ [Section 7.4]

Solve each equation.

59. $5\sqrt{x + 2} = x + 8$ [Section 7.5]

60. $\sqrt{x} + \sqrt{x + 2} = 2$ [Section 7.5]

61. Find the length of the hypotenuse of the right triangle in figure (a). [Section 7.6]

62. Find the length of the hypotenuse of the right triangle in figure (b). [Section 7.6]

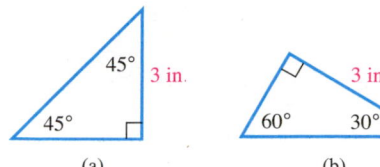

(a)   (b)

63. Find the distance between $(-2, 6)$ and $(4, 14)$. [Section 7.6]

64. Simplify: i^{43} [Section 7.7]

Perform the indicated operations. Write each result in a + bi form.

65. $\left(-7 + \sqrt{-81}\right) - \left(-2 - \sqrt{-64}\right)$ [Section 7.7]

66. $\dfrac{5}{3 - i}$ [Section 7.7]

67. $(2 + i)^2$ [Section 7.7]

68. $\dfrac{-4}{6i^7}$ [Section 7.7]

Solve each equation.

69. $x^2 = 28$ [Section 8.1]

70. $(x - 19)^2 = -5$ [Section 8.1]

71. Use the method of completing the square to solve $2x^2 - 6x + 3 = 0$. Approximate the solutions to the nearest hundredth. [Section 8.1]

72. Solve: $a^2 - \dfrac{2}{5}a = -\dfrac{1}{5}$

[Section 8.2]

73. Community Gardens. Residents of a community can work their own 16-ft $\times$ 24-ft plot of city-owned land if they agree to the following conditions:

- The area of the garden cannot exceed 180 square feet.
- A path of uniform width must be maintained around the garden.

Find the dimensions of the largest possible garden. [Section 8.2]

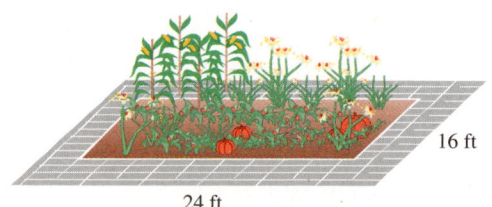

16 ft

24 ft

74. Sidewalks. A 170-meter-long sidewalk from the mathematics building M to the student center C is shown in red in the illustration. However, students prefer to walk directly from M to C, across a lawn. How long are the two segments of the existing sidewalk? [Section 8.2]

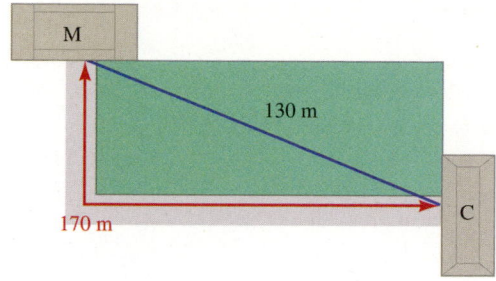

Solve each equation.

75. $t^{2/3} - t^{1/3} = 6$ [Section 8.3]

76. $x^{-4} - 2x^{-2} + 1 = 0$ [Section 8.3]

77. First determine the vertex and the axis of symmetry of the graph of $f(x) = -x^2 - 4x$ using the vertex formula. Then determine the x- and y-intercepts of the graph. Finally, plot several points and complete the graph. [Section 8.4]

Solve each inequality. Write the solution set in interval notation and then graph it.

78. $x^2 - 81 < 0$ [Section 8.5]

79. $\dfrac{1}{x + 1} \geq \dfrac{x}{x + 4}$ [Section 8.5]

80. a. The graph of $f(x) = 16x^2 + 24x + 9$ is shown below. Estimate the solution(s) of $16x^2 + 24x + 9 = 0$. [Section 8.5]

b. Use the graph to determine the solution of $16x^2 + 24x + 9 < 0$.

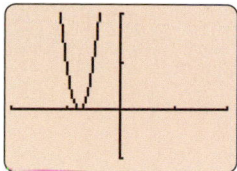

Exponential and Logarithmic Functions

9

©iStockphoto/Chris Schmidt

from Campus to Careers

Social Worker

For those with a desire to help improve other people's lives, social work is one career option to consider. Social workers offer guidance and counseling to people in crisis. They must be critical thinkers—able to use their logic and reasoning to brainstorm alternate solutions to problems faced by their clients. Social workers use their mathematical skills to construct family budgets, plan personnel schedules, gather and interpret data, and comprehend the statistical methods used in research studies.

Problem 49 in **Study Set 9.3, problem 115** in **Study Set 9.4,** and **problem 87** in **Study Set 9.5** involve situations that a social worker encounters on the job. The mathematical concepts discussed in this chapter can be used to solve those problems.

JOB TITLE:
Social Worker

EDUCATION:
The minimum requirements are a bachelor's degree in social work (BSW) and 2 to 4 years of experience in the field.

JOB OUTLOOK:
Employment is expected to increase between 14% and 19% through the year 2018.

ANNUAL EARNINGS:
Mean annual salary with a bachelor's degree is $44,869, and with a master's degree is $52,060.

FOR MORE INFORMATION:
www.bls.gov/oco/

751

Final exams can be stressful for many students because the number of topics to study can seem overwhelming. Here are some suggestions to help reduce the stress and prepare you for the test.

GET ORGANIZED: Gather all of your notes, study sheets, homework assignments, and especially all of your returned tests to review.

TALK WITH YOUR INSTRUCTOR: Ask your instructor to list the topics that may appear on the final and those that won't be covered.

MANAGE YOUR TIME: Adjust your daily schedule 1 week before the final so that it includes extended periods of study time.

Now Try This ▶

1. Review your old tests. Make a list of the test problems that you are still unsure about and see a tutor or your instructor to get help.
2. Make a practice final exam that includes one or more of each type of problem that may appear on the test.
3. Make a detailed study plan. Determine when, where, and what you will study each day for 1 week before the final.

SECTION 9.1

Algebra and Composition of Functions

OBJECTIVES

1 Add, subtract, multiply, and divide functions.

2 Find the composition of functions.

3 Use graphs to evaluate functions.

4 Use composite functions to solve problems.

ARE YOU READY?

The following problems review some basic skills that are needed when performing the algebra of functions.

1. Add:
$(7x^2 + 5x - 12) + (2x^2 - 3x + 4)$

2. Subtract:
$(8x^3 - 4x^2) - (6x^3 - 3x^2)$

3. Multiply: $(8x - 5)(9x^2 - 4)$

4. Divide: $(x^2 - x - 6) \div (x - 3)$

Just as it is possible to perform arithmetic operations on real numbers, it is possible to perform those operations on functions. We did exactly that in Chapters 5 and 6 when we added, subtracted, multiplied, and divided polynomial functions and rational functions. We call the process of adding, subtracting, multiplying, and dividing functions the **algebra of functions.**

1 Add, Subtract, Multiply, and Divide Functions.

We have seen that the sum, difference, product, and quotient of two functions are themselves functions. The new functions that result from such operations can be represented using the following notation.

Operations on Functions

Sum: $(f + g)(x) = f(x) + g(x)$ *Read as "f plus g of x equals f of x plus g of x."*

Difference: $(f - g)(x) = f(x) - g(x)$ *Read as "f minus g of x equals f of x minus g of x."*

Product: $(f \cdot g)(x) = f(x)g(x)$ *Read as "f times g of x equals f of x times g of x."*

Quotient: $(f/g)(x) = \dfrac{f(x)}{g(x)}$ where $g(x) \neq 0$ *Read as "f divided by g of x equals f of x divided by g of x."*

The domain of each sum, difference, and product function shown above is the set of real numbers x that are in the domains of both f and g. The domain of the quotient function is the set of real numbers x that are the domains of f and g, excluding any values of x where $g(x) = 0$.

EXAMPLE 1 Let $f(x) = 2x^2 + 1$ and $g(x) = 5x - 3$. Find each function and give its domain:
a. $f + g$ **b.** $f - g$ **c.** $f \cdot g$ **d.** f/g

Strategy We will add, subtract, multiply, and divide the functions as if they were binomials.

Why We add because of the plus symbol in $f + g$, we subtract because of the minus symbol in $f - g$, we multiply because of the raised dot in $f \cdot g$, and we divide because of the fraction bar in f/g.

Solution **a.** $(f + g)(x) = f(x) + g(x)$ This is the definition of a sum function.

$= (2x^2 + 1) + (5x - 3)$ Replace f(x) with $2x^2 + 1$ and g(x) with $5x - 3$.

$= 2x^2 + 5x - 2$ Drop the parentheses and combine like terms.

The domain of $f + g$ is the set of real numbers that are in the domain of both f and g. Since the domain of both f and g is $(-\infty, \infty)$, the domain of $f + g$ is the $(-\infty, \infty)$.

b. $(f - g)(x) = f(x) - g(x)$ This is the definition of a difference function.

$= (2x^2 + 1) - (5x - 3)$ Replace f(x) with $2x^2 + 1$ and g(x) with $5x - 3$.

$= 2x^2 + 1 - 5x + 3$ Change the sign of each term of $5x - 3$ and drop the parentheses.

$= 2x^2 - 5x + 4$ Combine like terms.

Since the domain of both f and g is $(-\infty, \infty)$, the domain of $f - g$ is $(-\infty, \infty)$.

Notation

The parentheses around $\frac{3}{5}$ in the notation $\left(-\infty, \frac{3}{5}\right) \cup \left(\frac{3}{5}, \infty\right)$ indicate that $\frac{3}{5}$ is not included in the domain of f/g. The domain could also be shown graphically:

c. $(f \cdot g)(x) = f(x) \cdot g(x)$ This is the definition of a product function.

$= (2x^2 + 1)(5x - 3)$ Replace f(x) with $2x^2 + 1$ and g(x) with $5x - 3$.

$= 10x^3 - 6x^2 + 5x - 3$ Multiply the binomials. There are no like terms.

The domain of $f \cdot g$ is the set of real numbers that are in the domain of both f and g. Since the domain of both f and g is $(-\infty, \infty)$, the domain of $f \cdot g$ is $(-\infty, \infty)$.

d. $(f/g)(x) = \dfrac{f(x)}{g(x)}$ This is the definition of a quotient function.

$= \dfrac{2x^2 + 1}{5x - 3}$ Replace f(x) with $2x^2 + 1$ and g(x) with $5x - 3$. The result does not simplify.

Since the denominator of the fraction cannot be 0, it follows that this function is undefined if $5x - 3 = 0$. If we solve for x, we see that x cannot be $\frac{3}{5}$. Thus, the domain of f/g is the union of two intervals: $\left(-\infty, \frac{3}{5}\right) \cup \left(\frac{3}{5}, \infty\right)$.

Self Check 1 Let $f(x) = 3x - 2$ and $g(x) = 2x^2 + 3x$. Find each function and give its domain:
a. $f + g$
b. $f - g$
c. $f \cdot g$
d. f/g

Now Try ▶ Problems 13, 15, 17, and 19

EXAMPLE 2 Use the results from Example 1 to find: **a.** $(f + g)(-3)$ **b.** $(f - g)(6)$
c. $(f \cdot g)(0)$ **d.** $(f/g)(10)$

Strategy We will substitute the given values within the second set of parentheses for each x in the sum, difference, product, and quotient functions found in Example 1. Then we will evaluate the right side of each equation.

Why The number that is within the second set of parentheses is the input of the function.

Solution   **a.** $(f + g)(x) = 2x^2 + 5x - 2$ This is the sum function found in Example 1, part a.

$(f + g)(-3) = 2(-3)^2 + 5(-3) - 2$ Substitute -3 for each x.

$= 18 + (-15) - 2$ Evaluate the right side.

$= 1$

b. $(f - g)(x) = 2x^2 - 5x + 4$ This is the difference function found in Example 1, part b.

$(f - g)(6) = 2(6)^2 - 5(6) + 4$ Substitute 6 for each x.

$= 72 - 30 + 4$ Evaluate the right side.

$= 46$

c. $(f \cdot g)(x) = 10x^3 - 6x^2 + 5x - 3$ This is the product function found in Example 1, part c.

$(f \cdot g)(0) = 10(0)^3 - 6(0)^2 + 5(0) - 3$ Substitute 0 for each x.

$= -3$ Evaluate the right side.

d. $(f/g)(x) = \dfrac{2x^2 + 1}{5x - 3}$ This is the quotient function found in Example 1, part d.

$(f/g)(10) = \dfrac{2(10)^2 + 1}{5(10) - 3}$ Substitute 10 for each x.

$= \dfrac{2(100) + 1}{50 - 3}$ Evaluate the right side.

$= \dfrac{201}{47}$

Self Check 2   Use the results from Self Check 1 to find: **a.** $(f + g)(5)$
b. $(f - g)(-9)$ **c.** $(f \cdot g)(2)$ **d.** $(f/g)(0)$

Now Try ▶ Problems 29 and 31

There is a relationship that can be seen between the graphs of two functions and the graph of their sum (or difference) function. For example, in the following illustration, the graph of $f + g$, which gives the total number of elementary and secondary students in the United States, can be found by adding the graph of f, which gives the number of secondary students, to the graph of g, which gives the number of elementary students. For any given x-value, we simply add the two corresponding y-values to get the graph of the sum function $f + g$.

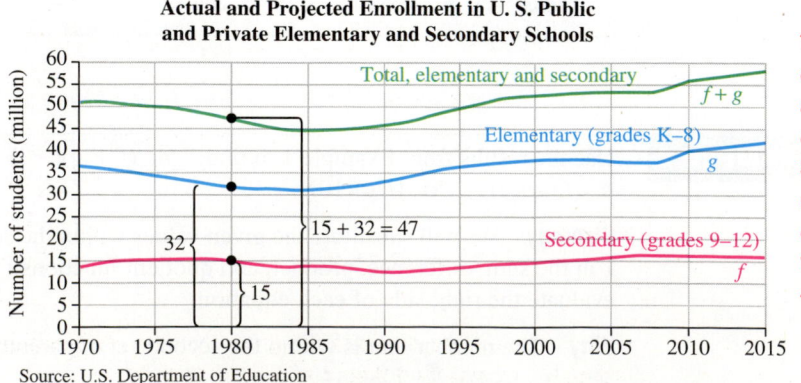

Actual and Projected Enrollment in U. S. Public and Private Elementary and Secondary Schools

To find the total enrollment in 1980, add the elementary enrollment of 32 million to the secondary enrollment of 15 million to get 47 million.

Source: U.S. Department of Education

2 Find the Composition of Functions.

We have seen that a function can be represented by a machine: We put in a number from the domain, and a number from the range comes out. For example, if we put the number 2 into the machine shown on the right, the number $f(2) = 8$ comes out. In general, if we put x into the machine, the value $f(x)$ comes out.

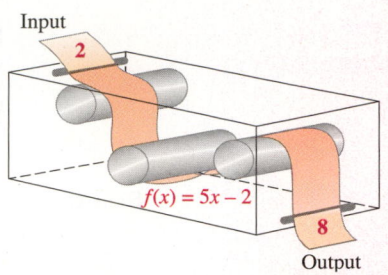

Often one quantity is a function of a second quantity that depends, in turn, on a third quantity. For example, the cost of a car trip is a function of the gasoline consumed. The amount of gasoline consumed, in turn, is a function of the number of miles driven. Such chains of dependence can be analyzed mathematically as **compositions of functions.**

Suppose that $y = f(x)$ and $y = g(x)$ define two functions. Any number x in the domain of g will produce the corresponding value $g(x)$ in the range of g. If $g(x)$ is in the domain of function f, then $g(x)$ can be substituted into f, and a corresponding value $f(g(x))$ will be determined. Because of the **nested parentheses** in $f(g(x))$, we read it as "f of g of x." This two-step process defines a new function, called a **composite function,** denoted by $f \circ g$. (This is read as "f composed with g" or "the composition of f and g" or "f circle g.")

The function machines shown below illustrate the composition $f \circ g$. When we put a number into the function g, a value $g(x)$ comes out. The value $g(x)$ then goes into function f, which transforms $g(x)$ into $f(g(x))$. If the function machines for g and f were connected to make a single machine, that machine would be named $f \circ g$.

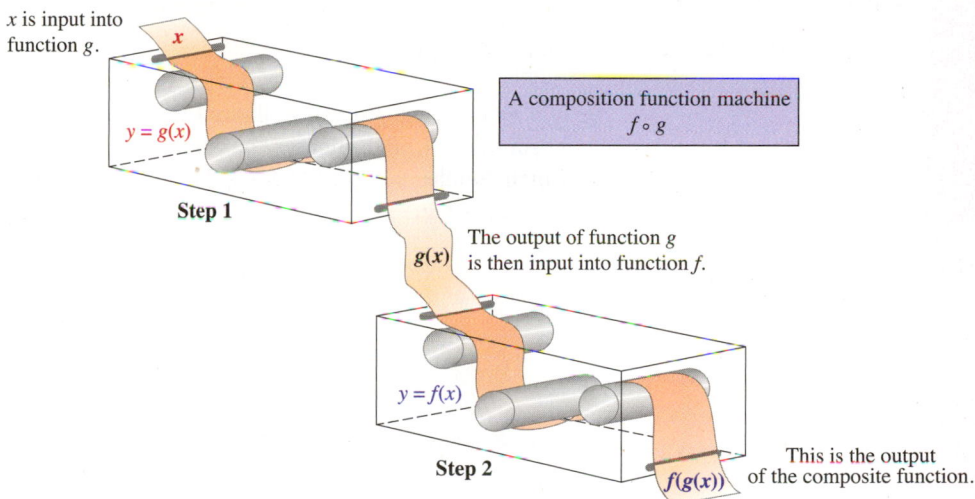

To be in the domain of the composite function $f \circ g$, a number x has to be in the domain of g and the output of g must be in the domain of f. Thus, the domain of $f \circ g$ consists of those numbers x that are in the domain of g, and for which $g(x)$ is in the domain of f.

Composite Functions

The **composite function** $f \circ g$ is defined by

$$(f \circ g)(x) = f(g(x))$$

If $f(x) = 4x$ and $g(x) = 3x + 2$, to find $f \circ g$ and $g \circ f$, we proceed as follows.

$$
\begin{aligned}
(f \circ g)(x) &= f(g(x)) & (g \circ f)(x) &= g(f(x)) \\
&= f(3x + 2) & &= g(4x) \\
&= 4(3x + 2) & &= 3(4x) + 2 \\
&= 12x + 8 & &= 12x + 2
\end{aligned}
$$

Different results

The different results illustrate that the composition of functions is not commutative. Usually, we will find that $(f \circ g)(x) \neq (g \circ f)(x)$.

EXAMPLE 3 Let $f(x) = 2x + 1$ and $g(x) = x - 4$. Find: **a.** $(f \circ g)(9)$ **b.** $(f \circ g)(x)$
c. $(g \circ f)(-2)$

Strategy In part (a), we will find $f(g(9))$. In part (b), we will find $f(g(x))$. In part (c), we will find $g(f(-2))$.

Why To evaluate a composite function written with the circle $\circ$ notation, we rewrite it using nested parentheses.

Solution **a.** $(f \circ g)(9)$ means $f(g(9))$, where the function g is applied first and function f is applied second.

$$(f \circ g)(9) = f(g(9))$$

Apply first

In figure (a) on the next page, we see that function g receives the number 9, subtracts 4, and releases the number $g(9) = 5$. Then 5 goes into the f function, which doubles 5 and adds 1. The final result, 11, is the output of the composite function $f \circ g$.

Read as "f composed   Read as "f
with g of 9."   of g of 9."

$$(f \circ g)(9) = f(g(9))$$ Change from $\circ$ notation to nested parentheses notation.
$$= f(5)$$ Evaluate: $g(9) = 9 - 4 = 5$.
$$= 2(5) + 1$$ Evaluate $f(5)$ using $f(x) = 2x + 1$.
$$= 11$$

Thus, $(f \circ g)(9) = 11$.

b. $(f \circ g)(x)$ means $f(g(x))$. In figure (a) on the next page, function g receives the number x, subtracts 4, and releases the number $x - 4$. Then $x - 4$ goes into the f function, which doubles $x - 4$ and adds 1. The final result, $2x - 7$, is the output of the composite function $f \circ g$.

Read as "f composed   Read as "f
with g of x."   of g of x."

$$(f \circ g)(x) = f(g(x))$$ Change from $\circ$ notation to nested parentheses notation.
$$= f(x - 4)$$ We are given $g(x) = x - 4$. Replace $g(x)$ with $x - 4$.
$$= 2(x - 4) + 1$$ Find $f(x - 4)$ using $f(x) = 2x + 1$.
$$= 2x - 8 + 1$$ Distribute the multiplication by 2.
$$= 2x - 7$$ Combine like terms.

Thus, $(f \circ g)(x) = 2x - 7$.

c. $(g \circ f)(-2)$ means $g(f(-2))$. In figure (b) on the next page, function f receives the number -2, doubles it and adds 1, and releases -3 into the g function. Function g subtracts 4 from -3 and outputs a final result of -7. Thus,

Read as "g composed   Read as "g
with f of −2."   of f of −2."

$$(g \circ f)(-2) = g(f(-2))$$ Change from $\circ$ notation to nested parentheses notation.
$$= g(-3)$$ Evaluate $f(-2)$ using $f(x) = 2x + 1$.
$$= -3 - 4$$ Evaluate $g(-3)$ using $g(x) = x - 4$.
$$= -7$$ Do the subtraction.

Thus, $(g \circ f)(-2) = -7$.

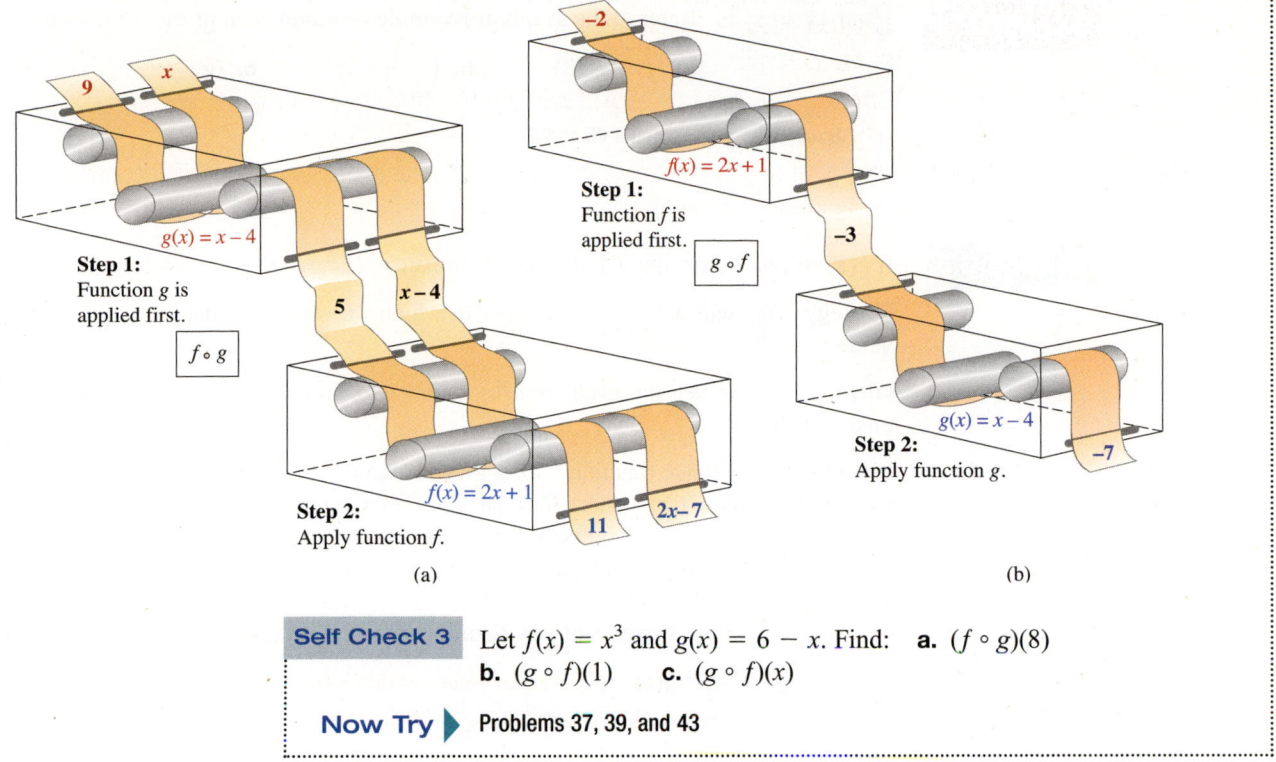

Step 1:
Function g is
applied first.

$f \circ g$

Step 2:
Apply function f.

(a)

Step 1:
Function f is
applied first.

$g \circ f$

Step 2:
Apply function g.

(b)

Self Check 3   Let $f(x) = x^3$ and $g(x) = 6 - x$. Find: **a.** $(f \circ g)(8)$
b. $(g \circ f)(1)$ **c.** $(g \circ f)(x)$

Now Try ▶ Problems 37, 39, and 43

3 Use Graphs to Evaluate Functions.

EXAMPLE 4   Refer to the graphs of functions f and g on the left to find each of the following.
a. $(f + g)(-4)$ **b.** $(f \cdot g)(2)$ **c.** $(f \circ g)(-3)$

Strategy   We will express the sum, product, and composite functions using the functions from which they are formed.

Why   We can evaluate sum, product, and composite functions at a given x-value by evaluating each function from which they are formed at that x-value.

Solution   **a.** The value of $f(-4)$ is found by looking in quadrant II of the
graph and the value of $g(-4)$ by looking in quadrant III.

$(f + g)(-4) = \mathbf{f(-4)} + \mathbf{g(-4)}$
$= \mathbf{3} + \mathbf{(-4)}$
$= -1$

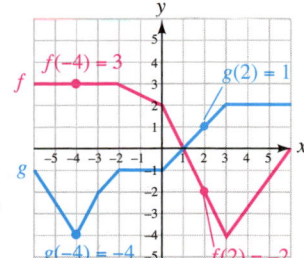

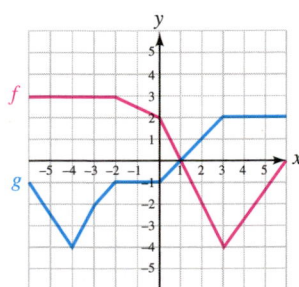

b. The value of $f(2)$ is found by looking in quadrant IV of the
graph and the value of $g(2)$ by looking in quadrant I.

$(f \cdot g)(2) = \mathbf{f(2)} \cdot \mathbf{g(2)}$
$= \mathbf{-2} \cdot \mathbf{1}$
$= -2$

c. The value of $g(-3)$ is found by looking in quadrant III of the graph and the value of
$f(-2)$ by looking in quadrant II.

$(f \circ g)(-3) = f(\mathbf{g(-3)})$ Change from ∘ notation to nested parentheses notation.
$= f(\mathbf{-2})$
$= 3$

Self Check 4   Refer to the graph in Example 4 to find each of the following.

    a. $(f - g)(3)$ **b.** $\left(\dfrac{f}{g}\right)(-2)$ **c.** $(g \circ f)(3)$

Now Try ▶ Problems 45 and 47

EXAMPLE 5   If $h(x) = (4x - 6)^2$, find f and g such that $h(x) = (f \circ g)(x)$.

Strategy   We will determine the order in which we would evaluate function h for a given value of x.

Why   If we can see the evaluation as a two-step process, it will help us determine the unknown functions f and g.

Solution   If we were to evaluate function h for a given value of x, we would first find $4x - 6$, and then we would square that result. This suggests that the first function that operates in the composition, function g, should receive an input x and produce $4x - 6$. Thus, $g(x) = 4x - 6$. The second function that operates in the composition, function f, should receive an input and square it. Thus, $f(x) = x^2$.

    We can check these results by forming the composition function.

$$(f \circ g)(x) = f(g(x)) \qquad \text{\color{red}Change from ∘ notation to nested parentheses.}$$
$$= f(4x - 6) \qquad \text{\color{red}Replace g(x) with 4x − 6.}$$
$$= (4x - 6)^2 \qquad \text{\color{red}Find f(4x − 6) using f(x) = x}^2.$$

The composition of functions $f(x) = x^2$ and $g(x) = 4x - 6$ does indeed produce $h(x) = (4x - 6)^2$. It is important to note that there are many other possibilities for f and g, but the ones that we found here are the most obvious.

Self Check 5   If $h(x) = \sqrt{x + 15}$, find f and g such that $h(x) = (f \circ g)(x)$.

Now Try ▶ Problems 49 and 51

---

4 Use Composite Functions to Solve Problems.

EXAMPLE 6   **Biological Research.**   A specimen is stored in refrigeration at a temperature of 15° Fahrenheit. Biologists remove the specimen and warm it at a controlled rate of 3°F per hour. Express its Celsius temperature as a function of the time t since it was removed from refrigeration.

Strategy   We will express the Fahrenheit temperature of the specimen as a function of the time t since it was removed from refrigeration. Then we will express the Celsius temperature of the specimen as a function of its Fahrenheit temperature and find the composition of the two functions.

Why   The Celsius temperature of the specimen is a function of its Fahrenheit temperature. Its Fahrenheit temperature is a function of the time since it was removed from refrigeration. This chain of dependence suggests that we write a composition of functions.

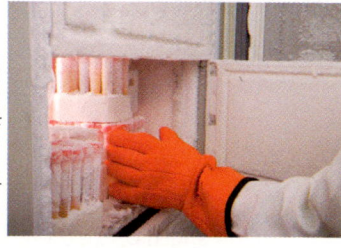

t $\longrightarrow$	$F(t)$ $\longrightarrow$	$C(F(t))$
Time in hours since specimen removed from refrigeration	Temperature of specimen in °F	Temperature of specimen in °C

Solution   The temperature of the specimen is 15°F when the time $t = 0$. Because it warms at a rate of 3°F per hour, its initial temperature of 15°F increases by $3t$°F in t hours. The Fahrenheit temperature at time t of the specimen is given by the function

$$F(t) = 3t + 15$$

The Celsius temperature C is a function of this Fahrenheit temperature F, given by the function

$$C(F) = \frac{5}{9}(F - 32)$$

To express the specimen's Celsius temperature as a function of *time,* we find the composite function $(C \circ F)(t)$.

$$(C \circ F)(t) = C(\mathbf{F(t)})$$ *Change from ∘ notation to nested parentheses notation.*

$$= C(\mathbf{3t + 15})$$ *Substitute 3t + 15 for F(t).*

$$= \frac{5}{9}\left(\mathbf{3t + 15} - 32\right)$$ *Find C(3t + 15) by substituting 3t + 15 for F in C(F) = $\frac{5}{9}$(F − 32).*

$$= \frac{5}{9}(3t - 17)$$ *Subtract within the parentheses: 15 − 32 = −17.*

$$= \frac{15}{9}t - \frac{85}{9}$$ *Distribute the multiplication by $\frac{5}{9}$.*

$$= \frac{5}{3}t - \frac{85}{9}$$ *Simplify $\frac{15}{9}$: $\dfrac{\overset{1}{\cancel{3}} \cdot 5}{\underset{}{\cancel{3}} \cdot 3} = \frac{5}{3}$.*

The composite function, $C(t) = \frac{5}{3}t - \frac{85}{9}$, gives the temperature of the specimen in degrees Celsius t hours after it is removed from refrigeration.

Self Check 6   **Weather Forecasting.**   A low-pressure area is bringing in colder weather for the next 12 hours. The temperature is now 86° Fahrenheit and is expected to fall 3° every 2 hours. Write a composite function that expresses the Celsius temperature as a function of the number of hours from now.

Now Try ▶ Problem 87

SECTION 9.1 ▶ STUDY SET

VOCABULARY

Fill in the blanks.

1. The _____ of f and g, denoted as $f + g$, is defined by $(f + g)(x) =$ _____ and the _____ of f and g, denoted as $f - g$, is defined by $(f - g)(x) =$ _____.

2. The _____ of f and g, denoted as $f \cdot g$, is defined by $(f \cdot g)(x) =$ _____ and the _____ of f and g, denoted as f/g, is defined by $(f/g)(x) =$ ___.

3. The _____ of the function $f + g$ is the set of real numbers x that are in the domain of both f and g.

4. The _____ function $f \circ g$ is defined by $(f \circ g)(x) =$ _____.

5. When we write $(f \circ g)(x)$ as $f(g(x))$, we have changed from ∘ notation to _____ parentheses notation.

6. When reading the notation $f(g(x))$, we say "f ___ g ___ x."

CONCEPTS

7. Fill in the blanks.
   a. $(f \circ g)(3) = f($ ___ $)$
   b. To find $f(g(3))$, we first find _____ and then substitute that value for x in $f(x)$.

8. a. If $f(x) = 3x + 1$ and $g(x) = 1 - 2x$, find $f(g(3))$ and $g(f(3))$.
   b. Is the composition of functions commutative?

9. Fill in the three blanks in the drawing of the function machines that show how to compute $g(f(-2))$.

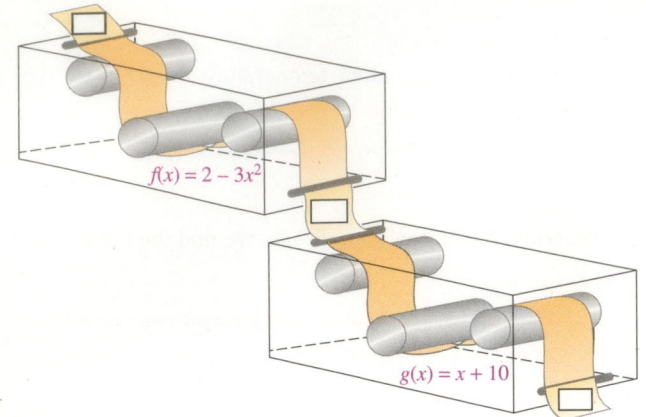

$f(x) = 2 - 3x^2$

$g(x) = x + 10$

10. If $f(x) = x^2 + 3$ and $g(x) = x - 4$, then $(f/g)(x) = \dfrac{x^2 + 3}{x - 4}$.

   a. What value of x makes $g(x) = 0$?

   b. Fill in the blank: The domain of f/g is $(-\infty, 4)$ ▮ $(4, \infty)$.

NOTATION

Complete each solution.

11. Let $f(x) = 3x - 1$ and $g(x) = 2x + 3$. Find $f \cdot g$.

$$(f \cdot g)(x) = f(x) \cdot \rule{1cm}{0.4pt}$$
$$= \rule{1cm}{0.4pt} (2x + 3)$$
$$= 6x^2 + \rule{0.6cm}{0.4pt} - \rule{0.6cm}{0.4pt} - 3$$
$$(f \cdot g)(x) = 6x^2 + 7x - 3$$

12. Let $f(x) = 3x - 1$ and $g(x) = 2x + 3$. Find $f \circ g$.

$$(f \circ g)(x) = f(\rule{0.8cm}{0.4pt})$$
$$= f(\rule{0.8cm}{0.4pt})$$
$$= 3(\rule{0.8cm}{0.4pt}) - 1$$
$$= \rule{0.6cm}{0.4pt} + \rule{0.6cm}{0.4pt} - 1$$
$$(f \circ g)(x) = 6x + 8$$

GUIDED PRACTICE

Let $f(x) = 2x + 1$ and $g(x) = x - 3$. Find each function and give its domain. See Example 1.

13. $f + g$
14. $f - g$
15. $g - f$
16. $g + f$
17. $f \cdot g$
18. f/g
19. g/f
20. $g \cdot f$

Let $f(x) = 3x$ and $g(x) = 4x$. Find each function and give its domain. See Example 1.

21. $f + g$
22. $f - g$

23. $g - f$
24. $g + f$
25. $f \cdot g$
26. f/g
27. g/f
28. $g \cdot f$

Let $f(x) = 2x - 5$ and $g(x) = x + 1$. Find each of the following function values. See Example 2.

29. $(f + g)(8)$
30. $(f - g)(-4)$
31. $(f \cdot g)(0)$
32. $(f/g)(2)$

Let $s(x) = 3 - x$ and $t(x) = x^2 - x - 6$. Find each function value. See Example 2.

33. $(s \cdot t)(-2)$
34. $(s + t)(3)$
35. $(s/t)(1)$
36. $(s - t)(12)$

Let $f(x) = 2x + 1$ and $g(x) = x^2 - 1$. Find each of the following. See Example 3.

37. $(f \circ g)(2)$
38. $(g \circ f)(2)$
39. $(g \circ f)(-3)$
40. $(f \circ g)(-3)$
41. $(f \circ g)\left(\dfrac{1}{2}\right)$
42. $(g \circ f)\left(\dfrac{1}{3}\right)$
43. $(g \circ f)(2x)$
44. $(f \circ g)(2x)$

Refer to graphs at the right. Find each function value. See Example 4.

45. a. $(f + g)(-5)$
   b. $(f - g)(3)$
   c. $(f \cdot g)(-3)$
46. a. $(f/g)(0)$
   b. $(f \circ g)(3)$
   c. $(g \circ f)(2)$

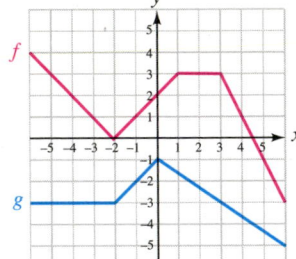

Refer to graphs at the right. Find each function value. See Example 4.

47. a. $(g + f)(2)$
   b. $(g - f)(-5)$
   c. $(g \cdot f)(1)$
48. a. $(g/f)(-6)$
   b. $(g \circ f)(4)$
   c. $(f \circ g)(6)$

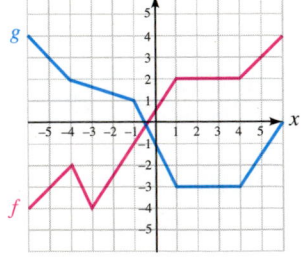

Find $f(x)$ and $g(x)$ such that $h(x) = (f \circ g)(x)$. Answers may vary. See Example 5.

49. $h(x) = (x + 15)^2$
50. $h(x) = (x - 9)^3$
51. $h(x) = x^5 + 9$
52. $h(x) = x^6 - 100$
53. $h(x) = \sqrt{16x - 1}$
54. $h(x) = \sqrt[3]{10 - x}$
55. $h(x) = \dfrac{1}{x - 4}$
56. $h(x) = \dfrac{1}{3x - 16}$

TRY IT YOURSELF

Let f(x) = 3x − 2 and g(x) = x² + x. Find each of the following.

57. $(f \circ g)(4)$ 58. $(g \circ f)(4)$

59. $(g \circ f)(-3)$ 60. $(f \circ g)(-3)$

61. $(g \circ f)(0)$ 62. $(f \circ g)(0)$

63. $(g \circ f)(x)$ 64. $(f \circ g)(x)$

Let f(x) = 3x − 2 and g(x) = 2x² + 1. Find each function and give its domain.

65. $f - g$

66. $f + g$

67. f/g

68. $f \cdot g$

Let $f(x) = \frac{1}{x}$ and $g(x) = \frac{1}{x^2}$. Find each of the following.

69. $(g \circ f)\left(\frac{1}{3}\right)$ 70. $(g \circ f)\left(\frac{1}{10}\right)$

71. $(g \circ f)(8x)$ 72. $(f \circ g)(5x)$

Let f(x) = x² − 1 and g(x) = x² − 4. Find each function and give its domain.

73. $f - g$

74. $f + g$

75. g/f

76. $g \cdot f$

Let $h(t) = \sqrt{t} + 3$ and k(t) = t − 5. Find each of the following.

77. $(h \circ k)(18)$ 78. $(h \circ k)(11)$

79. $(k \circ h)(22)$ 80. $(k \circ h)(-2)$

81. Use the tables of values for functions f and g to find each of the following.

    a. $(f + g)(1)$ b. $(f - g)(5)$

    c. $(f \cdot g)(1)$ d. $(g/f)(5)$

x	f(x)
1	3
5	8

x	g(x)
1	4
5	0

82. Use the table of values for functions f and g to find each of the following.

    a. $(f \circ g)(1)$ b. $(g \circ f)(2)$

x	f(x)
2	5
4	7

x	g(x)
1	2
5	−3

83. If $f(x) = x + 1$ and $g(x) = 2x - 5$, show that $(f \circ g)(x) \neq (g \circ f)(x)$.

84. If $f(x) = x^2 + 1$ and $g(x) = 3x^2 - 2$, show that $(f \circ g)(x) \neq (g \circ f)(x)$.

APPLICATIONS

85. **SAT Scores.** The graph of function m in the next column gives the average score on the mathematics portion of the SAT college entrance exam, the graph of function r gives the average score on the critical reading portion, and x represents the number of years since 2000.

    a. Find $(m + r)(4)$ and explain what information about SAT scores it gives.

    b. Find $(m - r)(4)$ and explain what information about SAT scores it gives.

    c. Find: $(m + r)(9)$

    d. Find: $(m - r)(9)$

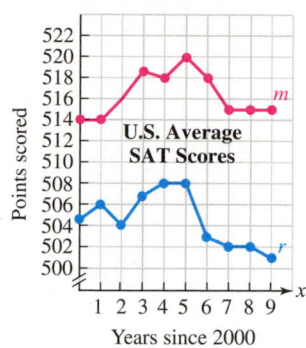

Source: National Center for Education Statistics

86. **Bachelor's Degrees.** The graph of function m below gives the number of bachelor's degrees awarded to men, and the graph of function w gives the number of bachelor's degrees awarded to women in the U. S. for the years 1990 through 2008.

    a. Estimate $(w + m)(2004)$ and explain what information about bachelor's degrees it gives.

    b. Estimate $(w - m)(2004)$ and explain what information about bachelor's degrees it gives.

    c. Estimate: $(w + m)(1994)$

    d. Estimate: $(w - m)(2000)$

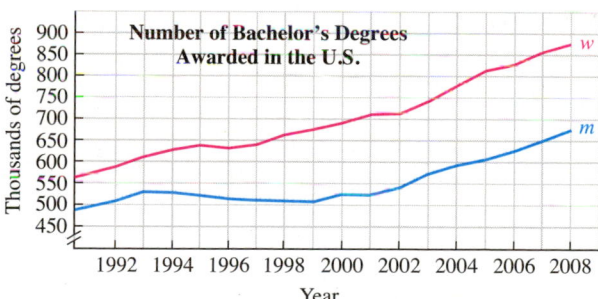

Source: National Center for Education Statistics

87. **Metallurgy.** A molten alloy must be cooled slowly to control crystallization. When removed from the furnace, its temperature is 2,700°F, and it will be cooled at 200° per hour. Write a composition function that expresses the Celsius temperature as a function of the number of hours t since cooling began. (*Hint*: $C(F) = \frac{5}{9}(F - 32)$.)

88. **Weather Forecasting.** A high-pressure area promises increasingly warmer weather for the next 48 hours. The temperature is now 34° Celsius and is expected to rise 1° every 6 hours. Write a composition function that expresses the Fahrenheit temperature as a function of the number of hours from now. (*Hint*: $F(C) = \frac{9}{5}C + 32$.)

89. Vacation Mileage Costs.

a. Use the following graphs to determine the cost of the gasoline consumed if a family drove 500 miles on a summer vacation.

b. Write a composition function that expresses the cost of the gasoline consumed on the vacation as a function of the miles driven.

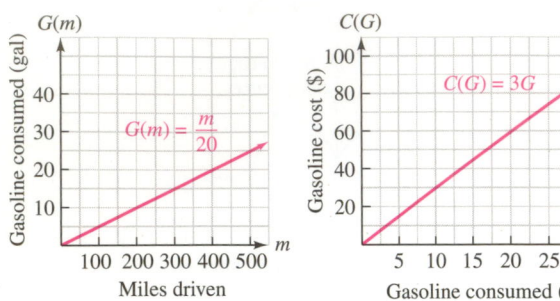

90. Halloween Costumes. The tables on the back of a pattern package can be used to determine the number of yards of material needed to make a rabbit costume for a child.

a. How many yards of material are needed if the child's chest measures 29 inches?

b. In this exercise, one quantity is a function of a second quantity that depends, in turn, on a third quantity. Explain this dependence.

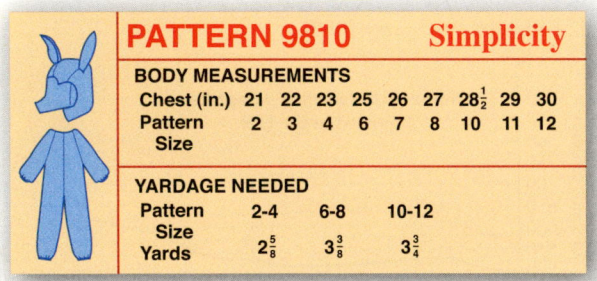

PATTERN 9810   **Simplicity**

BODY MEASUREMENTS

Chest (in.)	21	22	23	25	26	27	$28\frac{1}{2}$	29	30
Pattern Size	2	3	4	6	7	8	10	11	12

YARDAGE NEEDED

Pattern Size	2-4	6-8	10-12
Yards	$2\frac{5}{8}$	$3\frac{3}{8}$	$3\frac{3}{4}$

91. Exercise 89 illustrates a chain of dependence between the cost of the gasoline, the gasoline consumed, and the miles driven. Describe another chain of dependence that could be represented by a composition function.

92. In this section, what operations are performed on functions? Give an example of each.

93. Write out in words how to say each of the following:

$$(f \circ g)(2) \qquad g(f(-8))$$

94. If $Y_1 = f(x)$ and $Y_2 = g(x)$, explain how to use the following tables to find $g(f(2))$.

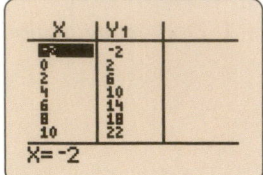

REVIEW

Simplify each complex fraction.

95. $\dfrac{\dfrac{ac - ad - c + d}{a^3 - 1}}{\dfrac{c^2 - 2cd + d^2}{a^2 + a + 1}}$

96. $\dfrac{2 + \dfrac{1}{x^2 - 1}}{1 + \dfrac{1}{x - 1}}$

CHALLENGE PROBLEMS

Fill in the blanks.

97. If $f(x) = x^2$ and $g(x) = $ _____ , then $(f \circ g)(x) = 4x^2 + 20x + 25$.

98. If $f(x) = \sqrt{3x}$ and $g(x) = $ _____ , then $(g \circ f)(x) = 9x^2 + 7$.

Refer to the following graphs of functions f and g.

99. Graph the sum function $f + g$ on the given coordinate system.

100. Graph the difference function $f - g$ on the given coordinate system.

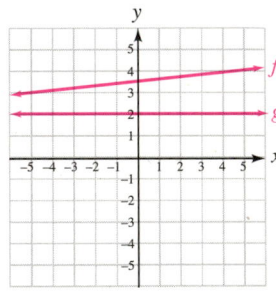

SECTION **9.2**

Inverse Functions

OBJECTIVES

1 Determine whether a function is a one-to-one function.

2 Use the horizontal line test to determine whether a function is one-to-one.

3 Find the equation of the inverse of a function.

4 Find the composition of a function and its inverse.

5 Graph a function and its inverse.

ARE YOU READY?

The following problems review some basic skills that are needed when working with inverse functions.

1. What are the domain and the range of the function $\{(-2, 8), (3, -3), (5, 10), (9, 1)\}$?

2. Is a parabola that opens upward the graph of a function?

3. Fill in the blank: If y is a _____ of x, the symbols y and $f(x)$ are interchangeable.

4. Let $f(x) = 2x + 6$. Find $f(8x)$.

In the previous section, we created new functions from given functions by using the operations of arithmetic and composition. Another way to create new functions is to find the *inverse of a function*.

1 Determine Whether a Function Is a One-to-One Function.

In figure (a) below, the arrow diagram defines a function f. If we reverse the arrows as shown in figure (b), we obtain a new correspondence where the range of f becomes the domain of the new correspondence, and the domain of f becomes the range. The new correspondence is a function because to each member of the domain, there corresponds exactly one member of the range. We call this new correspondence the **inverse** of f, or f inverse.

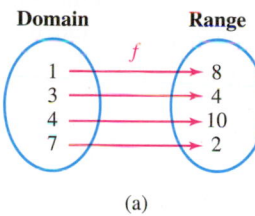
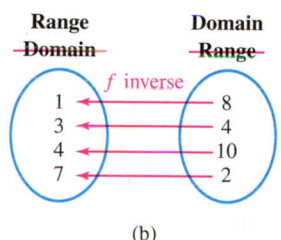

(a)   (b)

This reversing process does not always produce a function. For example, if we reverse the arrows in function g defined by the diagram in figure (a) below, the resulting correspondence shown in figure (b) is not a function. This is because to the number 2 in the domain, there corresponds two members of the range: 8 and 4.

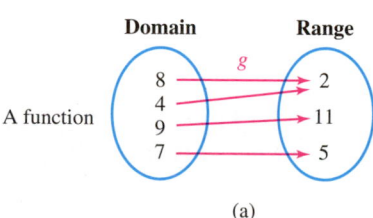
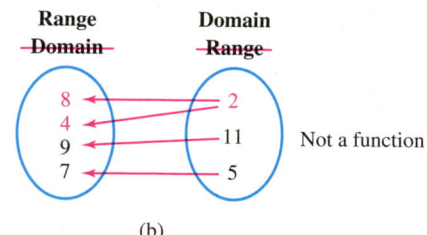

(a)   (b)

The question that arises is, "What must be true of an original function to guarantee that the reversing process produces a function?" The answer is: *The original function must be one-to-one.*

We have seen that in a function, each input determines exactly one output. For some functions, different inputs determine different outputs, as in figure (a) below. For other functions, different inputs might determine the *same* output, as in figure (b). When a function has the property that different inputs determine different outputs, as in figure (a), we say the function is *one-to-one*.

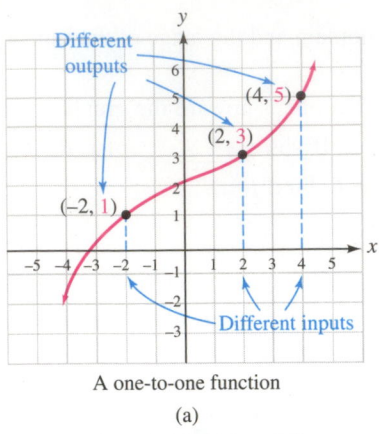

A one-to-one function

(a)

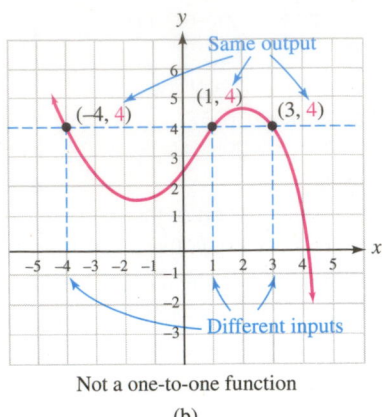

Not a one-to-one function

(b)

One-to-One Functions   A function is called a **one-to-one function** if different inputs determine different outputs.

EXAMPLE 1   Determine whether each function is one-to-one.   **a.** $f(x) = x^2$ **b.** $f(x) = x^3$

Strategy   We will determine whether different inputs have different outputs.

Why   If different inputs have different outputs, the function is one-to-one. If different inputs have the same output, the function is not one-to-one.

Solution   **a.** Since two different inputs, -3 and 3, have the same output 9, $f(x) = x^2$ is not a one-to-one function.

$$f(-3) = (-3)^2 = 9 \text{ and } f(3) = 3^2 = 9$$

x	$f(x)$
-3	9
3	9

The output 9 does not correspond to exactly one input.

Success Tip

Example 1 illustrates that not every function is one-to-one.

b. Since different numbers have different cubes, each input of $f(x) = x^3$ determines a different output. This function is one-to-one.

Self Check 1   Determine whether each function is one-to-one. If not, find an output that corresponds to more than one input.   **a.** $f(x) = 2x + 3$
b. $f(x) = x^4$

Now Try ▶   Problems 19 and 21

---

2 Use the Horizontal Line Test to Determine Whether a Function Is One-to-One.

To determine whether a function is one-to-one, it is often easier to view its graph rather than its defining equation. If two (or more) points on the graph of a function have the same y-coordinate, the function is not one-to-one. This observation suggests the following **horizontal line test.**

The Horizontal Line Test   A function is one-to-one if each horizontal line that intersects its graph does so exactly once.

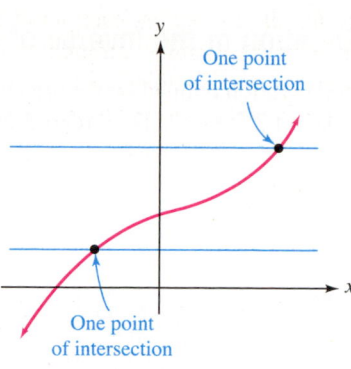

A one-to-one function

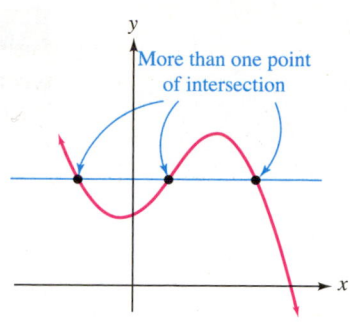

Not a one-to-one function

EXAMPLE 2 Use the horizontal line test to determine whether the following graphs of functions represent one-to-one functions.

Strategy We will draw horizontal lines through the graph of the function and see how many times each line intersects the graph.

Why If each horizontal line intersects the graph of the function exactly once, the graph represents a one-to-one function. If any horizontal line intersects the graph of the function more than once, the graph does not represent a one-to-one function.

Solution **a.** Because every horizontal line that intersects the graph of $f(x) = -\frac{3}{4}x - 2$ in figure (a) does so exactly once, the graph represents a one-to-one function. We simply say, the function $f(x) = -\frac{3}{4}x - 2$ is one-to-one.

b. Refer to figure (b). Because we can draw a horizontal line that intersects the graph of $f(x) = x^2 - 4$ twice, the graph does not represent a one-to-one function. We simply say, the function $f(x) = x^2 - 4$ is not one-to-one.

c. Because every horizontal line that intersects the graph of $f(x) = \sqrt{x}$ in figure (c) does so exactly once, the graph represents a one-to-one function. We simply say, the function $f(x) = \sqrt{x}$ is one-to-one.

> **Success Tip**
>
> Recall that we use the **vertical line test** to determine whether a graph represents a function. We use the **horizontal line test** to determine whether the function that is graphed is one-to-one.

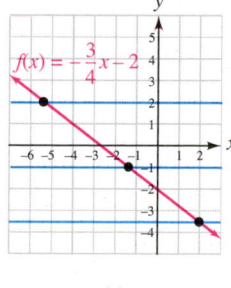

(a)

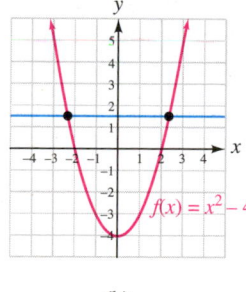

(b)

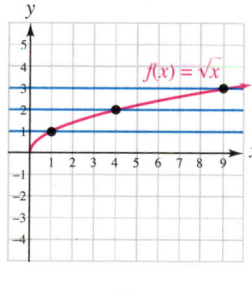

(c)

Self Check 2 Use the horizontal line test to determine whether the graph represents a one-to-one function.

a.

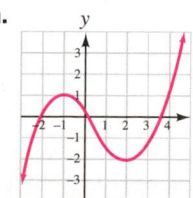

b.

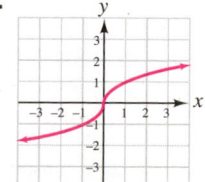

Now Try ▶ Problems 27 and 29

3 Find the Equation of the Inverse of a Function.

If f is the one-to-one function defined by the arrow diagram in figure (a), it turns the number 1 into 10, 2 into 20, and 3 into 30. The ordered pairs that define f can be listed in a table. Since the inverse of f must turn 10 back into 1, 20 back into 2, and 30 back into 3, it consists of the ordered pairs shown in the table in figure (b).

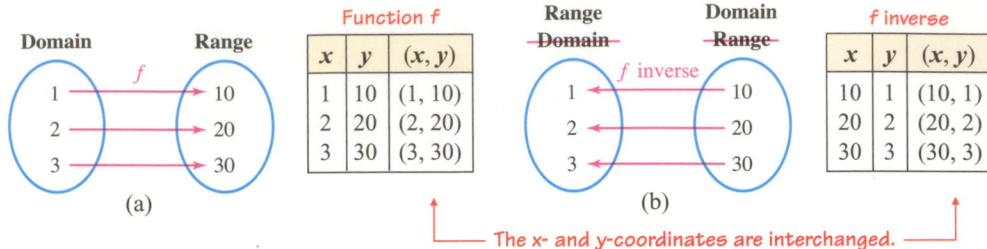

The x- and y-coordinates are interchanged.

We note that the domain of f and the range of its inverse is $\{1, 2, 3\}$. The range of f and the domain of its inverse is $\{10, 20, 30\}$.

This example suggests that to form the inverse of a function f, we simply interchange the coordinates of each ordered pair that determines f. When the inverse of a function is also a function, we call it **f inverse** and denote it with the symbol f^{-1}. The symbol $f^{-1}(x)$ is read as "the inverse of $f(x)$" or "f inverse of x."

The Inverse of a Function	If f is a one-to-one function consisting of ordered pairs of the form (x, y), **the inverse of f**, denoted f^{-1}, is the one-to-one function consisting of all ordered pairs of the form (y, x).

When a one-to-one function is defined by an equation, we use the following method to find the equation of its inverse.

Finding the Equation of the Inverse of a Function	If a function is one-to-one, we find its inverse as follows: 1. If the function is written using function notation, replace $f(x)$ with y. 2. Interchange the variables x and y. 3. Solve the resulting equation for y. 4. Substitute $f^{-1}(x)$ for y.

EXAMPLE 3 Determine whether each function is one-to-one. If so, find the equation of its inverse.
a. $f(x) = 4x + 2$ **b.** $f(x) = x^3$

Strategy We will determine whether each function is one-to-one. If it is, we can find the equation of its inverse by replacing $f(x)$ with y, interchanging x and y, and solving for y.

Why The reason for interchanging the variables is this: If a one-to-one function takes an input x into an output y, by definition, its inverse function has the reverse effect.

Solution **a.** We recognize $f(x) = 4x + 2$ as a linear function whose graph is a straight line with slope 4 and y-intercept $(0, 2)$. Since such a graph would pass the horizontal line test, we conclude that f is one-to-one.

To find the inverse of function f, we proceed as follows:

$f(x) = 4x + 2$ Function f multiplies all inputs by 4 and then adds 2.

$y = 4x + 2$ Replace $f(x)$ with y.

$x = 4y + 2$ Interchange the variables x and y.

$x - 2 = 4y$ To isolate the term $4y$, subtract 2 from both sides.

$\dfrac{x-2}{4} = y$ To solve for y, divide both sides by 4.

$y = \dfrac{x-2}{4}$ Write the equation with y on the left side.

Success Tip

Recall that $f(x) = 4x + 2$ is a linear function. Every linear function, except those of the form $f(x) = c$, where c is a constant, is one-to-one.

To denote that this equation is the inverse of function f, we replace y with $f^{-1}(x)$.

$f^{-1}(x) = \dfrac{x-2}{4}$ Function f^{-1} subtracts 2 from each input and then divides by 4.

As an informal check, we see below that if $x = 1$, function f produces an output of 6. And if $x = 6$, function f^{-1} produces an output of 1.

Success Tip

Notice that f^{-1} operates on an input using the inverse operations of f in the reverse order.

$$f(\mathbf{1}) = 4(\mathbf{1}) + 2$$
$$= 4 + 2$$
$$= \mathbf{6}$$
Ordered pair: $(\mathbf{1}, \mathbf{6})$

$$f^{-1}(\mathbf{6}) = \dfrac{\mathbf{6} - 2}{4}$$
$$= \dfrac{4}{4}$$
$$= \mathbf{1}$$
Ordered pair: $(\mathbf{6}, \mathbf{1})$

└── The coordinates are interchanged. ──┘

b. The graph of $f(x) = x^3$ is shown on the right. Since such a graph would pass the horizontal line test, we conclude that f is a one-to-one function.

To find its inverse, we proceed as follows:

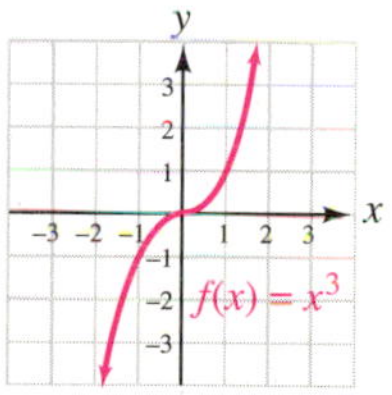

Caution

Only one-to-one functions have inverse functions.

$f(x) = x^3$ Function f cubes all inputs.

$y = x^3$ Replace $f(x)$ with y.

$x = y^3$ Interchange the variables x and y.

$\sqrt[3]{x} = y$ To solve for y, take the cube root of both sides.

$y = \sqrt[3]{x}$ Write the equation with y on the left side.

Replacing y with $f^{-1}(x)$, we have

$f^{-1}(x) = \sqrt[3]{x}$ Function f^{-1} finds the cube root of each input.

As an informal check, let $x = 4$ and determine whether f and f^{-1} produce ordered pairs whose coordinates are reversed.

Self Check 3   Determine whether each function is one-to-one. If it is, find the equation of its inverse.   **a.** $f(x) = -5x - 3$
   b. $f(x) = x^5$

Now Try ▶ Problems 35 and 45

---

4   Find the Composition of a Function and Its Inverse.

To emphasize a relationship between a function and its inverse, we substitute some number x, such as $x = 3$, into the function $f(x) = 4x + 2$ of Example 3(a). The corresponding value of y that is produced is

$f(\mathbf{3}) = 4(\mathbf{3}) + 2 = \mathbf{14}$ f determines the ordered pair $(3, 14)$.

If we substitute 14 into the inverse function, $f^{-1}(x) = \frac{x-2}{4}$, the corresponding value of y that is produced is

$$f^{-1}(14) = \frac{14-2}{4} = 3 \quad \textit{f}^{-1} \textit{ determines the ordered pair (14, 3).}$$

Thus, the function f turns 3 into 14, and the inverse function f^{-1} turns 14 back into 3.

In general, the composition of a function and its inverse function is the identity function, $f(x) = x$, such that any input x has the output x. This fact can be stated symbolically as follows.

The Composition of Inverse Functions	For any one-to-one function f and its inverse, f^{-1}, $$(f \circ f^{-1})(x) = x \quad \text{and} \quad (f^{-1} \circ f)(x) = x$$

We can use this property to determine whether two functions are inverses.

EXAMPLE 4   Show that $f(x) = 4x + 2$ and $f^{-1}(x) = \frac{x-2}{4}$ are inverses.

Strategy   We will find the composition of $f(x)$ and $f^{-1}(x)$ in both directions and show that the result is x.

Why   Only when the result of the composition is x in both directions are the functions inverses.

Solution   To show that $f(x) = 4x + 2$ and $f^{-1}(x) = \frac{x-2}{4}$ are inverses, we must show that for each composition, an input of x gives an output of x.

$$(f \circ f^{-1})(x) = f(\boldsymbol{f^{-1}(x)}) \qquad\qquad (f^{-1} \circ f)(x) = f^{-1}(\boldsymbol{f(x)})$$

$$= f\left(\frac{x-2}{4}\right) \qquad\qquad\qquad = f^{-1}(4x+2)$$

$$= 4\left(\frac{x-2}{4}\right) + 2 \qquad\qquad = \frac{4x+2-2}{4}$$

$$= x - 2 + 2 \qquad\qquad\qquad = \frac{4x}{4}$$

$$= x \qquad\qquad\qquad\qquad\quad = x$$

Because $(f \circ f^{-1})(x) = x$ and $(f^{-1} \circ f)(x) = x$, the functions are inverses.

Self Check 4   Show that $f(x) = x - 4$ and $g(x) = x + 4$ are inverses.

Now Try ▶ Problem 55

Success Tip

Recall that the line $y = x$ passes through points whose x- and y-coordinates are equal: $(-1, -1)$, $(0, 0)$, $(1, 1)$, $(2, 2)$, and so on.

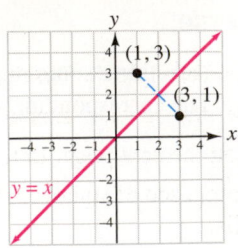

5 Graph a Function and Its Inverse.

If a point (a, b) is on the graph of function f, it follows that the point (b, a) is on the graph of f^{-1}, and vice versa. There is a geometric relationship between a pair of points whose coordinates are interchanged. For example, in the graph, we see that the line segment between $(1, 3)$ and $(3, 1)$ is perpendicular to and cut in half by the line $y = x$. We say that $(1, 3)$ and $(3, 1)$ are mirror images of each other with respect to $y = x$.

Since each point on the graph of f^{-1} is a mirror image of a point on the graph of f, and vice versa, the graphs of f and f^{-1} must be mirror images of each other with respect to $y = x$.

EXAMPLE 5   Find the equation of the inverse of $f(x) = -\frac{3}{2}x + 3$. Then graph f and its inverse on one coordinate system.

Strategy   We will determine whether the function has an inverse. If so, we will replace $f(x)$ with y, interchange x and y, and solve for y to obtain the equation of the inverse.

Why   The reason for interchanging the variables is this: If a one-to-one function takes an input x into an output y, by definition, its inverse function has the reverse effect.

Solution   Since $f(x) = -\frac{3}{2}x + 3$ is a linear function, it is one-to-one and has an inverse. To find the inverse function, we replace $f(x)$ with y, and interchange x and y to obtain

$$x = -\frac{3}{2}y + 3$$

Then we solve for y to get

$$x - 3 = -\frac{3}{2}y \qquad \text{Subtract 3 from both sides.}$$

$$-\frac{2}{3}x + 2 = y \qquad \text{To isolate } y, \text{ multiply both sides by } -\frac{2}{3}.$$

When we replace y with $f^{-1}(x)$, we have $f^{-1}(x) = -\frac{2}{3}x + 2$.

To graph f, we construct a table of values, plot points, and draw the graph in red as shown below. To graph f^{-1}, we don't need to do any calculations to construct a table of values. We can simply interchange the coordinates of the ordered pairs in the table for f and use them to graph f^{-1}. The result is the graph in blue shown below. Because the functions are inverses of each other, their graphs are **mirror images** about the line $y = x$.

$f(x) = -\frac{3}{2}x + 3$

x	$f(x)$	
0	3	$\rightarrow (0, 3)$
2	0	$\rightarrow (2, 0)$
4	-3	$\rightarrow (4, -3)$

$f^{-1}(x) = -\frac{2}{3}x + 2$

x	$f^{-1}(x)$	
3	0	$\rightarrow (3, 0)$
0	2	$\rightarrow (0, 2)$
-3	4	$\rightarrow (-3, 4)$

Simply interchange the coordinates to graph f^{-1}.

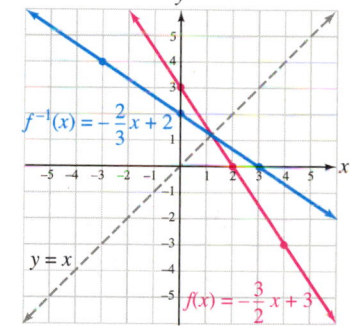

Self Check 5   Find the inverse of $f(x) = \frac{2}{3}x - 2$. Then graph the function and its inverse on one coordinate system.

Now Try   Problem 59

Using Your Calculator ▶ Graphing the Inverse of a Function

We can use a graphing calculator to check the result found in Example 5. First, we enter $f(x) = -\frac{3}{2}x + 3$ and then enter what we believe to be the inverse function, $f^{-1}(x) = -\frac{2}{3}x + 2$, as well as the equation $y = x$. See figure (a) on the next page. Before graphing, we adjust the display so that the graphing grid will be composed of squares. The line of symmetry $y = x$ is then at a 45° angle to the positive x-axis.

In figure (b), it appears that the two graphs are symmetric about the line $y = x$. Although it is not definitive, this visual check does help to validate the result of Example 5.

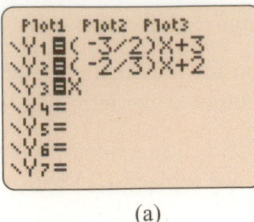

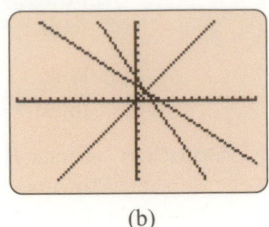

(a)

(b)

EXAMPLE 6 Graph the inverse of function f shown in figure (a).

Strategy We will find the coordinates of several points on the graph of f in figure (a). After interchanging the coordinates of these points, we will plot them as shown in figure (b).

Why The reason for interchanging the coordinates is this: If (a, b) is a point on the graph of a one-to-one function, then the point (b, a) is on the graph of its inverse.

Solution In figure (a), we see that the points $(-5, -3)$, $(-2, -1)$, $(0, 2)$, $(3, 3)$, $(5, 4)$, and $(7, 5)$ lie on the graph of function f. To graph the inverse, we interchange their coordinates, and plot them in blue, as shown in figure (b). Then we graph the line $y = x$ and use symmetry to draw a smooth curve through those points to get the graph of f^{-1}.

The Language of Algebra

We also can say that the graphs of f and f^{-1} are **reflections** of each other about the line $y = x$, or they are **symmetric about** $y = x$.

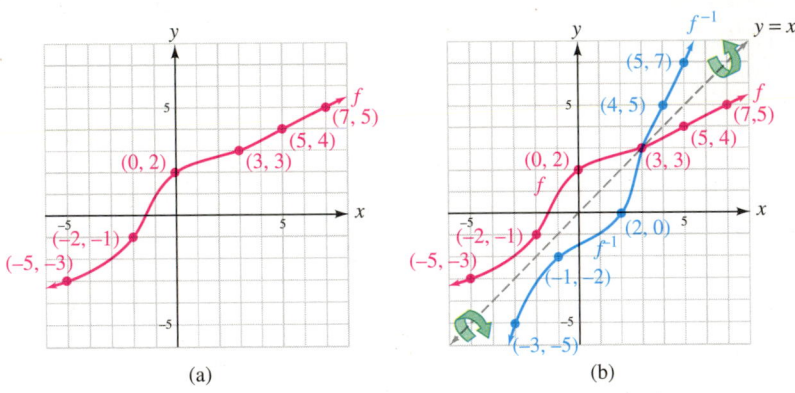

(a)   (b)

Now Try ▶ Problem 65

SECTION **9.2** **STUDY SET**

VOCABULARY

Fill in the blanks.

1. A function is called a _____ function if different inputs determine different outputs.

2. The _____ line test can be used to determine whether the graph of a function represents a one-to-one function.

3. The functions f and f^{-1} are _____.

4. The graphs of a function and its inverse are _____ images of each other with respect to $y = x$. We also say that their graphs are _____ with respect to the line $y = x$.

CONCEPTS

Fill in the blanks.

5. If any horizontal line that intersects the graph of a function does so more than once, the function is not _____.

6. To find the inverse of the function $f(x) = 2x - 3$, we begin by replacing $f(x)$ with y, and then we _____ x and y.

7. If f is a one-to-one function, the domain of f is the _____ of f^{-1}, and the range of f is the _____ of f^{-1}.

8. If a function turns an input of 2 into an output of 5, the inverse function will turn an input of 5 into the output ▢.

9. If f is a one-to-one function, and if $f(1) = 6$, then $f^{-1}(6) = ▢$.

10. If the point $(9, -4)$ is on the graph of the one-to-one function f, then the point (▢ , ▢) is on the graph of f^{-1}.

11. **a.** Is the correspondence defined by the arrow diagram in figure (a) below a one-to-one function?

    b. Is the correspondence defined by the table in figure (b) below a one-to-one function?

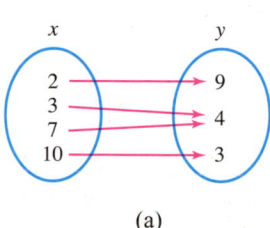

x	$f(x)$
-2	4
-1	1
0	0
2	4
3	9

(a)   (b)

12. Is the inverse of a one-to-one function always a function?

13. Use the table of values of the one-to-one function f to complete a table of values for f^{-1}.

x	$f(x)$
-4	-2
0	0
8	4

x	$f^{-1}(x)$
-2	
0	
4	

14. Redraw the graph of function f. Then graph f^{-1} and the axis of symmetry on the same coordinate system.

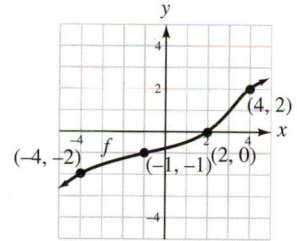

NOTATION

Complete each solution.

15. Find the inverse of $f(x) = 2x - 3$.

$$▢ = 2x - 3$$
$$x = ▢ - 3$$
$$x + ▢ = 2y$$
$$\frac{x + 3}{2} = ▢$$

The inverse of $f(x) = 2x - 3$ is $▢(x) = \dfrac{x + 3}{2}$.

16. Find the inverse of $f(x) = \sqrt[3]{x} + 2$.

$$▢ = \sqrt[3]{x} + 2$$
$$x = \sqrt[3]{▢} + 2$$
$$x - ▢ = \sqrt[3]{y}$$
$$(x - 2)^3 = ▢$$

The inverse of $f(x) = \sqrt[3]{x} + 2$ is $▢(x) = (x - 2)^3$.

17. The symbol f^{-1} is read as "the _____ of f" or "f _____."

18. Explain the difference in the meaning of the -1 in the notation $f^{-1}(x)$ as compared with a^{-1}.

GUIDED PRACTICE

Determine whether each function is one-to-one. See Example 1.

19. $f(x) = 2x$

20. $f(x) = |x|$

21. $f(x) = x^4$

22. $f(x) = x^3 + 1$

23. $f(x) = -x^2 + 3x$

24. $f(x) = \dfrac{2}{3}x + 8$

25. $\{(1, 1), (2, 1), (3, 1), (4, 1)\}$

26. $\{(3, 2), (2, 1), (1, 0)\}$

Each graph represents a function. Use the horizontal line test to determine whether the function is one-to-one. See Example 2.

27.
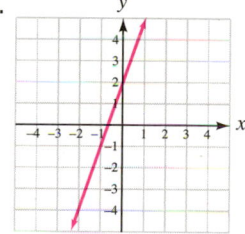

28.
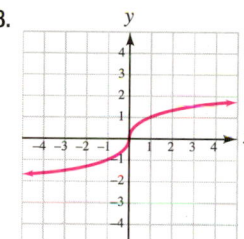

29.
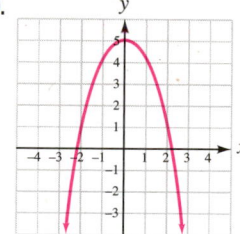

30.
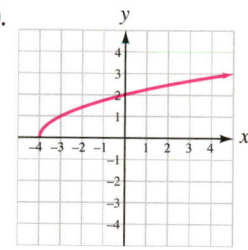

31.
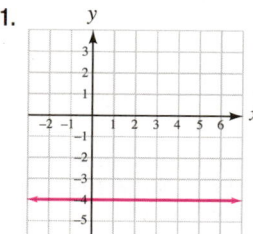

32.
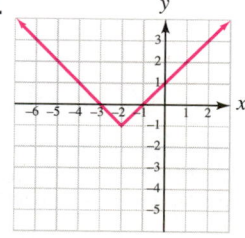

33.

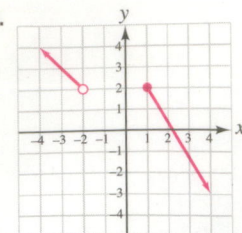

34.

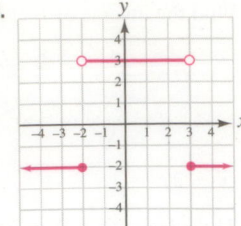

Each of the following functions is one-to-one. Find the inverse of each function and express it using $f^{-1}(x)$ notation. See Example 3.

35. $f(x) = 2x + 4$

36. $f(x) = 5x - 1$

37. $f(x) = \dfrac{x}{5} + \dfrac{4}{5}$

38. $f(x) = \dfrac{x}{3} - \dfrac{1}{3}$

39. $f(x) = \dfrac{x - 4}{5}$

40. $f(x) = \dfrac{2x + 6}{3}$

41. $f(x) = \dfrac{2}{x - 3}$

42. $f(x) = \dfrac{3}{x + 1}$

43. $f(x) = \dfrac{4}{x}$

44. $f(x) = \dfrac{1}{x}$

45. $f(x) = x^3 + 8$

46. $f(x) = x^3 - 4$

47. $f(x) = \sqrt[3]{x}$

48. $f(x) = \sqrt[3]{x - 5}$

49. $f(x) = (x + 10)^3$

50. $f(x) = (x - 9)^3$

51. $f(x) = 2x^3 - 3$

52. $f(x) = \dfrac{3}{x^3} - 1$

53. $f(x) = \dfrac{x^7}{2}$

54. $f(x) = \dfrac{x^9}{4}$

Show that each pair of functions are inverses. See Example 4.

55. $f(x) = 2x + 9, \ f^{-1}(x) = \dfrac{x - 9}{2}$

56. $f(x) = 5x - 1, \ f^{-1}(x) = \dfrac{x + 1}{5}$

57. $f(x) = \dfrac{2}{x - 3}, \ f^{-1}(x) = \dfrac{2}{x} + 3$

58. $f(x) = \sqrt[3]{x - 6}, \ f^{-1}(x) = x^3 + 6$

Find the inverse of each function. Then graph the function and its inverse on one coordinate system. Show the line of symmetry on the graph. See Examples 5 and 6.

59. $f(x) = 2x$

60. $f(x) = -3x$

61. $f(x) = 4x + 3$

62. $f(x) = \dfrac{x}{3} + \dfrac{1}{3}$

63. $f(x) = -\dfrac{2}{3}x + 3$

64. $f(x) = -\dfrac{1}{3}x + \dfrac{4}{3}$

65. $f(x) = x^3$

66. $f(x) = x^3 + 1$

67. $f(x) = x^2 - 1 \ (x \geq 0)$

68. $f(x) = x^2 + 1 \ (x \geq 0)$

APPLICATIONS

69. Interpersonal Relationships. Feelings of anxiety in a relationship can increase or decrease, depending on what is going on in the relationship. The graph shows how a person's anxiety might vary as a relationship develops over time.

a. Is this the graph of a function? Is its inverse a function?

b. Does each anxiety level correspond to exactly one point in time? Use the dashed lined labeled *Maximum threshold* to explain.

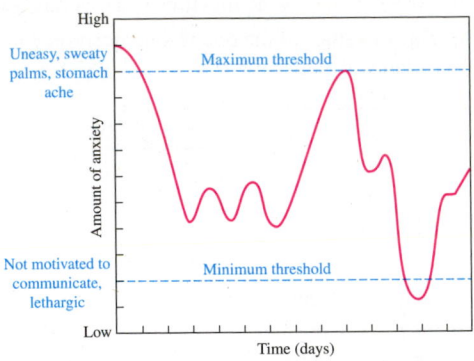

Source: Gudykunst, *Building Bridges: Interpersonal Skills for a Changing World* (Houghton Mifflin, 1994)

70. Lighting Levels. The ability of the eye to see detail increases as the level of illumination increases. This relationship can be modeled by a function E, whose graph is shown here.

a. From the graph, determine $E(240)$.

b. Is function E one-to-one? Does E have an inverse?

c. If the effectiveness of seeing in an office is 7, what is the illumination in the office? How can this question be asked using inverse function notation?

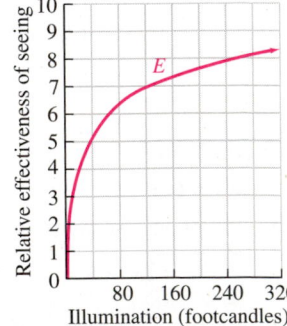

Based on information from *World Book Encyclopedia*

WRITING

71. In your own words, what is a one-to-one function?

72. Two functions are graphed on the square grid on the right along with the line $y = x$. Explain why the functions cannot be inverses of each other.

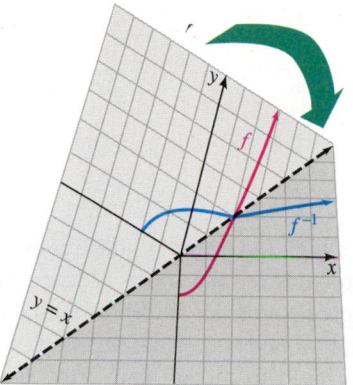

73. Explain how the graph of a one-to-one function can be used to draw the graph of its inverse function.

74. a. Explain the purpose of the vertical line test.

   b. Explain the purpose of the horizontal line test.

75. In the illustration, a function f and its inverse f^{-1} have been graphed on the same coordinate system. Explain what concept can be demonstrated by folding the graph paper on the dashed line.

76. Write in words how to read the notation.

   a. $f^{-1}(x) = \dfrac{1}{2}x - 3$

   b. $(f \circ f^{-1})(x) = x$

REVIEW

Simplify. Write the result in the form a + bi.

77. $3 - \sqrt{-64}$

78. $(2 - 3i) + (4 + 5i)$

79. $(3 + 4i)(2 - 3i)$

80. $\dfrac{6 + 7i}{3 - 4i}$

81. $(6 - 8i)^2$

82. i^{100}

CHALLENGE PROBLEMS

83. Find the inverse of $f(x) = \dfrac{x + 1}{x - 1}$.

84. Using the functions of Exercise 83, show that $(f \circ f^{-1})(x) = x$ and $(f^{-1} \circ f)(x) = x$.

85. A table of values for a function f is shown in figure (a). A table of values for f^{-1} is shown in figure (b). Use the tables to find $f^{-1}(f(4))$ and $f(f^{-1}(2))$.

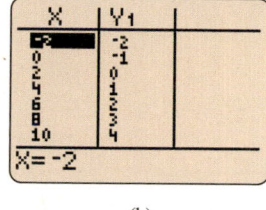

(a)   (b)

86. a. The graph of a one-to-one function lies entirely in quadrant I. In what quadrant does the graph of its inverse lie?

   b. The graph of a one-to-one function lies entirely in quadrant II. In what quadrant does the graph of its inverse lie?

   c. The graph of a one-to-one function lies entirely in quadrant III. In what quadrant does the graph of its inverse lie?

   d. The graph of a one-to-one function lies entirely in quadrant IV. In what quadrant does the graph of its inverse lie?

SECTION 9.3

Exponential Functions

OBJECTIVES

1 Define exponential functions.

2 Graph exponential functions.

3 Use exponential functions in applications involving growth or decay.

ARE YOU READY?

The following problems review some basic skills that are needed when working with exponential functions.

1. Simplify: **a.** $3^4 \cdot 3^8$ **b.** $(3^4)^8$

2. Evaluate: **a.** 2^3 **b.** 2^0 **c.** 2^{-2}

3. Evaluate: **a.** $\left(\dfrac{1}{3}\right)^2$ **b.** $\left(\dfrac{1}{3}\right)^0$ **c.** $\left(\dfrac{1}{3}\right)^{-3}$

4. Fill in the blanks: The graph of $f(x) = x$ is a _____ and the graph of $f(x) = x^2$ is a _____.

In previous chapters, we have discussed linear functions, polynomial functions, rational functions, and radical functions. We now begin a study of a new family of functions known as *exponential functions*.

As an example, consider the graph in figure (a) below, which models the soaring popularity of the social network website Twitter in recent years. The rapidly rising red curve is the graph of an exponential function.

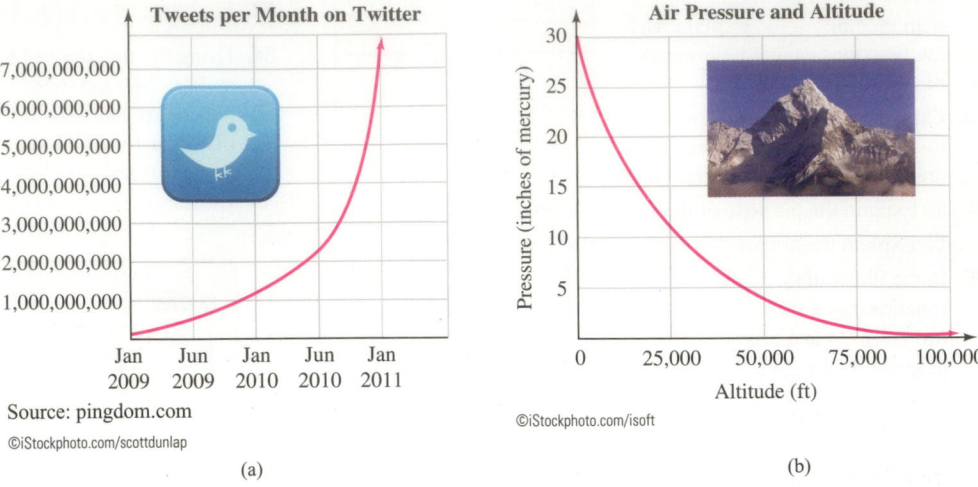

Source: pingdom.com
©iStockphoto.com/scottdunlap

©iStockphoto.com/isoft

(a)   (b)

If you have ever climbed a high mountain or gone up in an airplane that does not have a pressurized cabin, you have probably felt the effects of low air pressure. The graph in figure (b) above shows how the atmospheric pressure decreases with increasing altitude. The rapidly falling red curve is also the graph of an exponential function.

Exponential functions are used to model many other situations, such as population growth, the spread of an epidemic, the temperature of a heated object as it cools, and radioactive decay.

1   Define Exponential Functions.

In this course, we have evaluated exponential expressions with integer exponents, such as 2^5, 7^0, and 5^{-1}, and rational number exponents, such as $9^{1/2}$, $64^{3/4}$, and $8^{-2/3}$. To define exponential functions, we also must be able to evaluate exponential expressions that have irrational exponents. For example, let's consider the expression

$5^{\sqrt{2}}$ where $\sqrt{2}$ is the irrational number $1.414213562\ldots$

We can successively approximate $5^{\sqrt{2}}$ using the following *rational* powers:

$5^{1.4}$, $5^{1.41}$, $5^{1.414}$, $5^{1.4142}$, $5^{1.41421}$, $\ldots$

Using concepts from advanced mathematics, it can be shown that there is exactly one number that these powers approach. We define $5^{\sqrt{2}}$ to be that number. This process can be used to approximate $5^{\sqrt{2}}$ to as many decimal places as desired. Any other positive irrational exponent can be defined in the same manner, and negative irrational exponents can be defined using reciprocals.

This discussion leads us to the following conclusion: If b is positive, the exponential expression b^x has meaning and can be evaluated for any real number exponent x. Furthermore, it can be shown that all of the familiar rules of exponents are also true for irrational exponents.

Using Your Calculator ▶ Evaluating Exponential Expressions

We can use a calculator to obtain a very good approximation of an exponential expression with an irrational exponent. To find the value of $5^{\sqrt{2}}$ with a reverse entry scientific calculator, we enter:

5 $\boxed{y^x}$ 2 $\boxed{\sqrt{}}$ =   9.738517742

With a direct entry graphing calculator, we enter:

5 $\boxed{\wedge}$ $\boxed{\text{2nd}}$ $\boxed{\sqrt{}}$ 2 $\boxed{)}$ $\boxed{\text{ENTER}}$

```
5^√ (2)
          9.738517742
```

If $b > 0$ and $b \neq 1$, the function $f(x) = b^x$ is called an **exponential function.** Since x can be any real number, its domain is the set of real numbers, which can be written as $(-\infty, \infty)$.

Because b is positive, the value of $f(x)$ is positive, and the range is the set of positive numbers, which can be written as $(0, \infty)$.

Since $b \neq 1$, an exponential function cannot be the constant function $f(x) = 1^x$, in which $f(x) = 1$ for every real number x.

Exponential Functions

▼ An **exponential function with base b** is defined by the equations

$$f(x) = b^x \qquad \text{or} \qquad y = b^x$$

where $b > 0$, $b \neq 1$, and x is a real number. The domain of $f(x) = b^x$ is the interval $(-\infty, \infty)$, and the range is the interval $(0, \infty)$.

Exponential functions have a *constant base* and a *variable exponent*. Some examples of exponential functions are:

$$f(x) = 2^x \qquad g(x) = \left(\frac{1}{3}\right)^x \qquad s(x) = 4^{x+1} \qquad P(t) = (1.45)^{-3.04t}$$

The base is 2.　　　The base is $\frac{1}{3}$.　　　The base is 4.　　　The base is 1.45.

In the third example, we see that the exponent of an exponential function doesn't have to be just x. It can be a variable expression, such as $x + 1$. In the fourth example, we see that the base can be a positive decimal, and the input variable can be a letter other than x.

The following functions are not exponential functions.

$$f(x) = x^2 \qquad g(x) = x^{1/3} \qquad h(x) = (-4)^x \qquad f(x) = 1^x$$

These have a variable base　　The base cannot be negative.　　The base cannot be 1.
and a constant exponent.

2 Graph Exponential Functions.

Since the domain and range of $f(x) = b^x$ are sets of real numbers, we can graph exponential functions on a rectangular coordinate system. To do this, we will use the familiar **point-plotting method.**

EXAMPLE 1　Graph: $f(x) = 2^x$

Strategy　We will graph the function by creating a table of function values and plotting the corresponding ordered pairs.

Why　After drawing a smooth curve through the plotted points, we will have the graph.

Solution　To graph $f(x) = 2^x$, we select several values for x and find the corresponding values of $f(x)$. If x is -3, and if x is -2, we have:

Caution

We have previously graphed the linear function $f(x) = 2x$ and the squaring function $f(x) = x^2$. For the exponential function $f(x) = 2^x$, note that the variable is in the exponent.

$$f(x) = 2^x$$
$$f(-3) = 2^{-3} \quad \text{Substitute } -3 \text{ for } x.$$
$$= \frac{1}{2^3}$$
$$= \frac{1}{8}$$

$$f(x) = 2^x$$
$$f(-2) = 2^{-2} \quad \text{Substitute } -2 \text{ for } x.$$
$$= \frac{1}{2^2}$$
$$= \frac{1}{4}$$

The points $\left(-3, \frac{1}{8}\right)$ and $\left(-2, \frac{1}{4}\right)$ are on the graph of $f(x) = 2^x$. In a similar way, we find the corresponding values of $f(x)$ for x values of -1, 0, 1, 2, 3, and 4 and list them in a table. Then we plot the ordered pairs and draw a smooth curve through them, as shown below. Notice with the ordered pairs, as the value of x increases, the value of y also increases, but very rapidly as compared with x. This graph is an example of **exponential growth.**

$f(x) = 2^x$

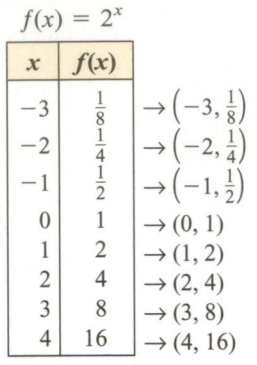

x	$f(x)$	
-3	$\frac{1}{8}$	$\rightarrow \left(-3, \frac{1}{8}\right)$
-2	$\frac{1}{4}$	$\rightarrow \left(-2, \frac{1}{4}\right)$
-1	$\frac{1}{2}$	$\rightarrow \left(-1, \frac{1}{2}\right)$
0	1	$\rightarrow (0, 1)$
1	2	$\rightarrow (1, 2)$
2	4	$\rightarrow (2, 4)$
3	8	$\rightarrow (3, 8)$
4	16	$\rightarrow (4, 16)$

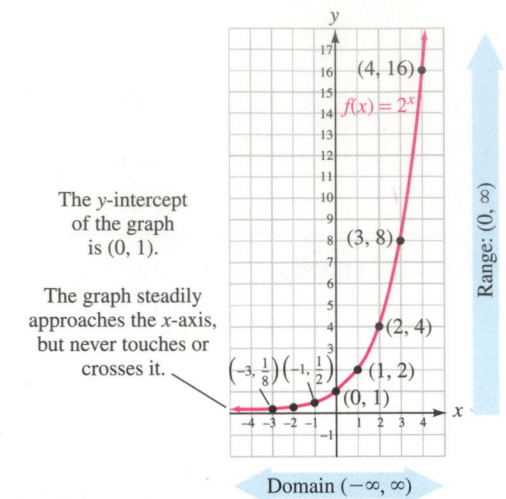

The y-intercept of the graph is $(0, 1)$.

The graph steadily approaches the x-axis, but never touches or crosses it.

Domain $(-\infty, \infty)$

Range: $(0, \infty)$

Because of the variable exponent in their equations, the graphs of exponential functions rise or fall sharply. When graphing them, make sure you plot enough points to show this.

By the vertical line test, we see that this is indeed the graph of a function. Because the graph extends indefinitely to the left and right, the projection of the graph onto the x-axis includes all real numbers. Thus, the domain of $f(x) = 2^x$ is $(-\infty, \infty)$.

Because the projection of the graph onto the y-axis covers only the positive portion of that axis, the range of the function is $(0, \infty)$. Since the graph passes the horizontal line test, the function is one-to-one.

Note that as x decreases, the values of $f(x)$ decrease and approach 0. Thus, the x-axis is a **horizontal asymptote** of the graph. The graph does not have an x-intercept, the y-intercept is $(0, 1)$, and the graph passes through the point $(1, 2)$.

Self Check 1   Graph: $g(x) = 4^x$

Now Try ▶   Problem 19

Exponential functions can have a base that is a real number between 0 and 1.

EXAMPLE 2   Graph: $f(x) = \left(\frac{1}{3}\right)^x$

Strategy   We will graph the function by creating a table of function values and plotting the corresponding ordered pairs.

Why   After drawing a smooth curve through the plotted points, we will have the graph.

Solution   If $x = -2$ and if $x = -1$, we have

$$f(x) = \left(\frac{1}{3}\right)^x$$

$$f(-2) = \left(\frac{1}{3}\right)^{-2}$$

$$= \left(\frac{3}{1}\right)^2 \quad \text{Recall: } \left(\frac{x}{y}\right)^{-n} = \left(\frac{y}{x}\right)^n.$$

$$= 9$$

$$f(x) = \left(\frac{1}{3}\right)^x$$

$$f(-1) = \left(\frac{1}{3}\right)^{-1}$$

$$= \left(\frac{3}{1}\right)^1$$

$$= 3$$

The points $(-2, 9)$ and $(-1, 3)$ are on the graph of $f(x) = \left(\frac{1}{3}\right)^x$. In a similar way, we find the corresponding values of $f(x)$ for $x = 0, 1$, and 2 and list them in a table. Then we plot the ordered pairs and draw a smooth curve through them, as shown below. Notice with the ordered pairs, as the value of x increases, the value of y decreases very rapidly as compared with x. This graph is an example of **exponential decay.**

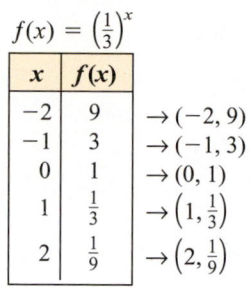

$$f(x) = \left(\frac{1}{3}\right)^x$$

x	$f(x)$	
-2	9	$\to (-2, 9)$
-1	3	$\to (-1, 3)$
0	1	$\to (0, 1)$
1	$\frac{1}{3}$	$\to \left(1, \frac{1}{3}\right)$
2	$\frac{1}{9}$	$\to \left(2, \frac{1}{9}\right)$

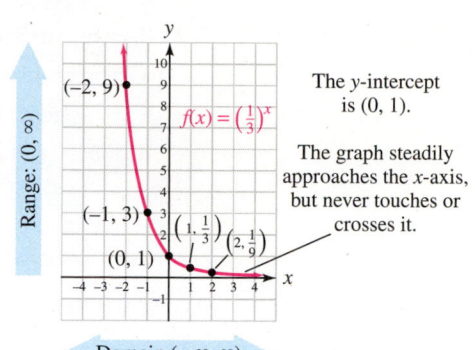

The graph passes the vertical line test, and so it is indeed the graph of a function. Because the graph extends indefinitely to the left and right, the projection of the graph onto the x-axis includes all real numbers. Thus, the domain of $f(x) = \left(\frac{1}{3}\right)^x$ is $(-\infty, \infty)$. Because the projection of the graph onto the y-axis covers only the positive portion of that axis, the range of the function is $(0, \infty)$.

Note that as x increases, the values of $f(x)$ decrease and approach 0. Thus, the x-axis is a horizontal asymptote of the graph. The graph does not have an x-intercept, the y-intercept is $(0, 1)$, and the graph passes through the point $\left(1, \frac{1}{3}\right)$.

Self Check 2 Graph: $g(x) = \left(\frac{1}{2}\right)^x$

Now Try ▶ Problem 23

In Example 1 (where $b = 2$), the values of y increase as the values of x increase. Since the graph rises as we move to the right, we call the function an *increasing function*. When $b > 1$, the larger the value of b, the steeper the curve, as shown in figure (a) below.

In Example 2 $\left(\text{where } b = \frac{1}{3}\right)$, the values of y decrease as the values of x increase. Since the graph drops as we move to the right, we call the function a *decreasing function*. When $0 < b < 1$, the smaller the value of b, the steeper the curve, as shown in figure (b) below.

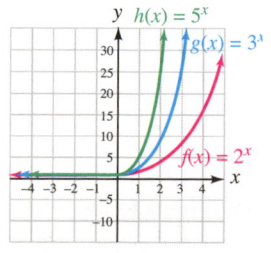

The bases of these exponential functions are 2, 3, and 5. Notice that the largest base, 5, has the steepest graph.

(a)

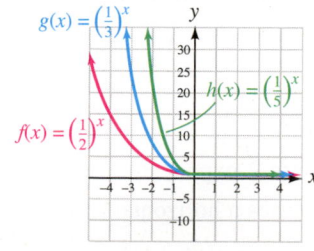

The bases of these exponential functions are $\frac{1}{2}$, $\frac{1}{3}$, and $\frac{1}{5}$, which are numbers between 0 and 1. Notice that the smallest base, $\frac{1}{5}$, has the steepest graph.

(b)

Examples 1 and 2 illustrate the following properties of exponential functions.

Properties of Exponential Functions

1. The domain of the exponential function $f(x) = b^x$ is the interval $(-\infty, \infty)$ and the range is the interval $(0, \infty)$.

2. The graph has a y-intercept of $(0, 1)$.

3. The x-axis is an asymptote of the graph.

4. The graph of $f(x) = b^x$ passes through the point $(1, b)$.

5. Exponential functions are one-to-one.

6. If $b > 1$, then $f(x) = b^x$ is an **increasing function**.

   If $0 < b < 1$, then $f(x) = b^x$ is a **decreasing function**.

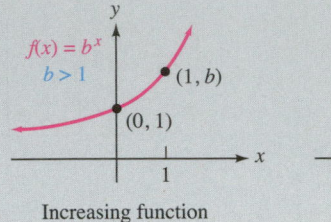

Increasing function

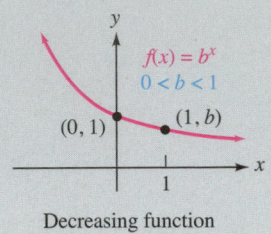

Decreasing function

Using Your Calculator ▶ Graphing Exponential Functions

To use a graphing calculator to graph $f(x) = \left(\frac{2}{3}\right)^x$ and $g(x) = \left(\frac{3}{2}\right)^x$, we enter the right sides of the equations after the symbols $Y_1 =$ and $Y_2 =$. The screen will show the following equations.

$$Y_1 = (2/3)^{\wedge}X$$
$$Y_2 = (3/2)^{\wedge}X$$

If we press $\boxed{\text{GRAPH}}$, we will obtain the display shown.

We note that the graph of $f(x) = \left(\frac{2}{3}\right)^x$ passes through $(0, 1)$. Since $\frac{2}{3} < 1$, the function is decreasing. The graph of $g(x) = \left(\frac{3}{2}\right)^x$ also passes through $(0, 1)$. Since $\frac{3}{2} > 1$, the function is increasing. Since both graphs pass the horizontal line test, each function is one-to-one.

The graphs of many exponential functions are horizontal and vertical translations of basic graphs.

EXAMPLE 3   Graph each function by using a translation:

a. $g(x) = 2^x - 4$ **b.** $g(x) = \left(\frac{1}{3}\right)^{x+3}$

Strategy   We will graph $g(x) = 2^x - 4$ by translating the graph of $f(x) = 2^x$ downward 4 units. We will graph $g(x) = \left(\frac{1}{3}\right)^{x+3}$ by translating the graph of $f(x) = \left(\frac{1}{3}\right)^x$ to the left 3 units.

Why   The subtraction of 4 in $g(x) = 2^x - 4$ causes a vertical shift of the graph of the base-2 exponential function 4 units downward. The addition of 3 to x in $g(x) = \left(\frac{1}{3}\right)^{x+3}$ causes a horizontal shift of the graph of the base-$\frac{1}{3}$ exponential function 3 units to the left.

Solution   **a.** The graph of $g(x) = 2^x - 4$ will be the same shape as the graph of $f(x) = 2^x$. We call this a **vertical translation.** To graph $g(x) = 2^x - 4$, simply translate each point on the graph of $f(x) = 2^x$ down 4 units. See figure (a) below.

b. The graph of $g(x) = \left(\frac{1}{3}\right)^{x+3}$ will be the same shape as the graph of $f(x) = \left(\frac{1}{3}\right)^x$. We call this a **horizontal translation.** To graph $g(x) = \left(\frac{1}{3}\right)^{x+3}$, simply translate each point on the graph of $f(x) = \left(\frac{1}{3}\right)^x$ to the left 3 units. See figure (b) below.

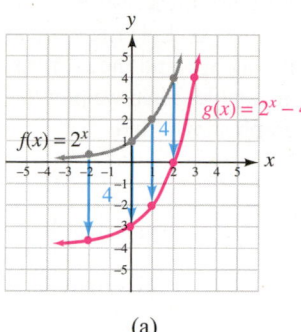   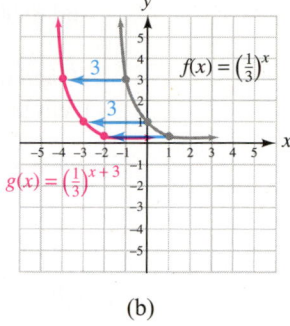

(a)   (b)

Self Check 3   Graph each function by using a translation:

a. $g(x) = \left(\frac{1}{4}\right)^x + 2$ **b.** $g(x) = 4^{x-3}$

Now Try ▶ Problems 27 and 29

3 Use Exponential Functions in Applications Involving Growth or Decay.

Many real-world situations can be modeled by exponential functions that describe how a quantity grows or decays over time. Some examples of this include the studies of populations, bacteria, heat transfer, radioactive substances, drug concentrations, and financial accounts. Two examples of such functions are:

Exponential growth: *Exponential decay:*

$$c(t) = 5(1.034)^t \qquad\qquad f(n) = 650(0.94)^n$$

A constant   A base greater   A constant   A base between
                  than 1   0 and 1

EXAMPLE 4

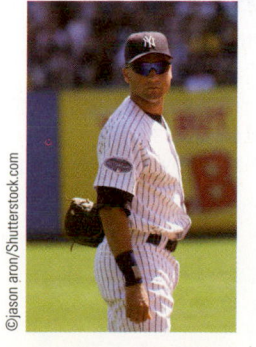

Professional Baseball Salaries. The exponential function $s(t) = 650{,}000(1.09)^t$ approximates the average annual salary of a major league baseball player, where t is the number of years after 1990. (Source: Baseball Almanac) **a.** Graph the function. **b.** Use the function to determine the average annual salary in 2020, if the current trend continues.

Strategy   For part a, we will graph the function by creating a table of function values and plotting the resulting ordered pairs. For part b, we will find $s(30)$.

Why   After drawing a smooth curve through the plotted points, we will have the graph. Since the year 2020 is 30 years after 1990, $t = 30$.

Solution **a.** The function values for $t = 0$ and $t = 5$ are calculated as follows:

$t = 0$ *(the year 1990)*	$t = 5$ *(the year 1995)*
$s(t) = 650{,}000(1.09)^t$	$s(t) = 650{,}000(1.09)^t$
$s(0) = 650{,}000(1.09)^0$	$s(5) = 650{,}000(1.09)^5$
$= 650{,}000(1)$	$\approx 1{,}000{,}106$ Use a calculator.
$= 650{,}000$	

To approximate $s(5)$, use the keystrokes $650000 \; \boxed{\times} \; 1.09 \; \boxed{y^x} \; 5 \; \boxed{=}$ on a scientific calculator and $650000 \; \boxed{\times} \; 1.09 \; \boxed{\wedge} \; 5 \; \boxed{\text{ENTER}}$ on a graphing calculator.

In a similar way, we find the corresponding values of $s(t)$ for t-values of 10, 15, and 20 and list them in a table. Then we plot the ordered pairs and draw a smooth curve through them to get the graph shown here.

t	$s(t)$
0	650,000
5	1,000,106
10	1,538,786
15	2,367,614
20	3,642,867

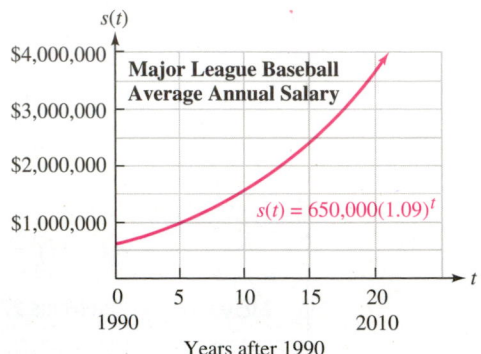

b. To estimate the average annual major league baseball salary in 2020, which is 30 years after 1990, we find $s(30)$.

$$s(t) = 650{,}000(1.09)^t \qquad \text{This is the exponential growth model.}$$
$$s(30) = 650{,}000(1.09)^{30} \qquad \text{Substitute 30 for } t.$$
$$\approx 8{,}623{,}991 \qquad \text{Use a calculator.}$$

If the trend continues, in 2020, the average annual salary will be approximately $8,623,991.

Self Check 4 **Salaries.** Use the function in Example 4 to determine the average annual salary in 2018, if the current trend continues.

Now Try ▶ Problem 43

Using Your Calculator ▶ **Graphing Exponential Functions**

To use a graphing calculator to graph the exponential function $s(t) = 170{,}000(1.12)^t$, we enter the right side of the equation after the symbol $Y_1 =$ and replace the variable t with x. The display will show the equation

$$Y_1 = 170000(1.12\text{^}X)$$

With window settings $[0, 30]$ for x and Xscale $= 5$ and $[0, 3000000]$ for y and Yscale $= 500000$, we obtain the display shown when we press $\boxed{\text{GRAPH}}$.

Up to this point, the financial application problems that we have solved involved **simple interest,** which is calculated using the formula $I = Prt$. However, most savings accounts and investments pay compound interest rather than simple interest. **Compound interest** is paid more than once a year on the principal and *previously earned interest*. The following compound interest formula is a useful application of exponential functions.

Formula for Compound Interest	If P is deposited in an account and interest is paid k times a year at an annual rate r, the amount A in the account after t years is given by $$A = P\left(1 + \frac{r}{k}\right)^{kt}$$

EXAMPLE 5

Educational Savings Plan. To save for college, parents of a newborn child invest $12,000 in a mutual fund at 10% interest, compounded quarterly.

a. Find a function for the amount in the account after t years.

b. If the quarterly interest paid is continually reinvested, how much money will be in the account when the child is 18 years old?

Strategy To write a function for the amount in the account after t years, we will substitute the given values for P, r, and k into the compound interest formula.

Why The resulting equation will involve only two variables, A and t. Then we can write that equation using function notation.

Solution

a. When we substitute 12,000 for P, 0.10 for r, and 4 for k in the formula for compound interest, the resulting formula involves only two variables, A and t.

$$A = P\left(1 + \frac{r}{k}\right)^{kt}$$ This is the compound interest model.

$$A = 12{,}000\left(1 + \frac{0.10}{4}\right)^{4t}$$ Since the interest is compounded quarterly, $k = 4$. Express $r = 10\%$ as a decimal.

Since the value of A depends on the value of t, we can express this relationship using function notation.

$$A(t) = 12{,}000\left(1 + \frac{0.10}{4}\right)^{4t}$$

$$A(t) = 12{,}000(1 + 0.025)^{4t}$$ Evaluate within the parentheses: $\frac{0.10}{4} = 0.025$.

$$A(t) = 12{,}000(1.025)^{4t}$$ The base of this exponential function is 1.025.

b. To find how much money will be in the account when the child is 18 years old, we need to find $A(18)$.

$$A(t) = 12{,}000(1.025)^{4(t)}$$ This is the exponential growth model.

$$A(18) = 12{,}000(1.025)^{4(18)}$$ Substitute 18 for t.

$$= 12{,}000(1.025)^{72}$$ Evaluate the exponent: $4(18) = 72$.

$$\approx 71{,}006.74$$ Use a scientific calculator and press these keys:
12000 × 1.025 y^x 72 = .

When the child is 18 years old, the account will contain $71,006.74.

Self Check 5   **Savings Plans.**   In Example 5, how much money would be in the account after 18 years if the parents initially invested $20,000?

Now Try ▶ Problem 53

Using Your **Calculator** ▶ Solving Investment Problems

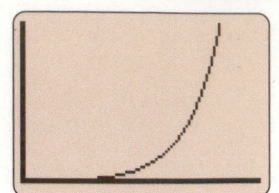

Suppose $1 is deposited in an account earning 6% annual interest, compounded monthly. To use a graphing calculator to estimate how much will be in the account in 100 years, we can substitute 1 for P, 0.06 for r, and 12 for k in the formula and simplify.

$$A = P\left(1 + \frac{r}{k}\right)^{kt} = 1\left(1 + \frac{0.06}{12}\right)^{12t} = (1.005)^{12t}$$

We now graph the function $A(t) = (1.005)^{12t}$ using window settings of [0, 120] and [0, 400] with Xscale = 1 and Yscale = 1 to obtain the graph shown. We can then trace and zoom to estimate that $1 grows to be approximately $397 in 100 years. From the graph, we can see that the money grows slowly in the early years and rapidly in the later years.

Examples 4 and 5 are applications illustrating exponential growth. In the next example, we see an application of exponential decay.

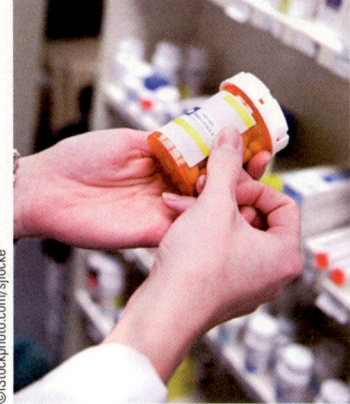

©iStockphoto.com/sjlocke

EXAMPLE 6   **Medications.**   The most common way people take medications is orally (by mouth). Medications, when swallowed, travel from the stomach or small intestine into the bloodstream, but they are eventually eliminated from the body by the kidneys and the liver. If a patient takes a 250-milligram dose of an antibiotic, the function $A(t) = 250(0.58)^t$ approximates the amount of medication (in milligrams) left in the patient's bloodstream t hours after it is taken.

a. Graph the function.

b. Use the function to determine the amount of medication in the patient's bloodstream 10 hours after taking the dose.

Strategy   For part a, we will graph the function by creating a table of function values and plotting the resulting ordered pairs. For part b, we will find $A(10)$.

Why   After drawing a smooth curve through the plotted points, we will have the graph. Since the variable t represents the time since taking the dose, $t = 10$.

Solution   **a.** The function values for $t = 0$ and $t = 2$ are calculated as follows:

$t = 0$:

$A(t) = 250(0.58)^t$
$A(0) = 250(0.58)^0$
$\quad = 250(1)$
$\quad = 250$

$t = 2$:

$A(t) = 250(0.58)^t$
$A(2) = 250(0.58)^2$
$\quad \approx 84.1$ Use a calculator.

Success Tip

$A(t) = 250(0.58)^t$ is a decreasing function because the base, 0.58, is such that $0 < 0.58 < 1$.

In a similar way, we find the corresponding values of $A(t)$ for t-values of 4 and 6, and list them in a table. Then we plot the ordered pairs and draw a smooth curve through them to get the graph shown below.

t	$A(t)$
0	250
2	84.1
4	28.3
6	9.5

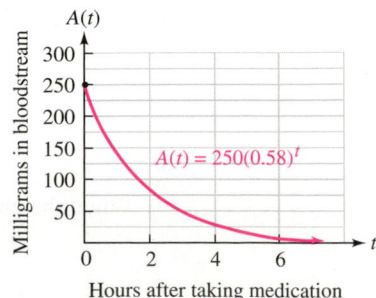

b. To estimate the amount of medication in the patient's bloodstream 10 hours after taking the dose, we find $A(10)$.

$$A(t) = 250(0.58)^t \qquad \text{This is the exponential decay model.}$$

$$A(\mathbf{10}) = 250(0.58)^{\mathbf{10}} \qquad \text{Substitute 10 for } t.$$

$$\approx 1.1 \qquad\qquad \text{Use a calculator.}$$

In 10 hours, there will be approximately 1.1 milligrams of medication in the patient's bloodstream.

Self Check 6 | **Medications.** Use the function in Example 6 to determine the amount of medication in the patient's bloodstream $8\frac{1}{2}$ hours after taking the dose.

Now Try ▶ Problem 44

SECTION 9.3 STUDY SET

VOCABULARY

Fill in the blanks.

1. $f(x) = 2^x$ and $f(x) = \left(\frac{1}{4}\right)^x$ are examples of _____ functions.

2. Exponential functions have a constant base and a variable _____ .

3. The graph of $f(x) = 3^x$ approaches, but never touches, the negative portion of the x-axis. Thus, the x-axis is an _____ of the graph.

4. _____ interest is paid on the principal and previously earned interest.

CONCEPTS

5. Refer to the graph shown at the right.
   a. What type of function is $f(x) = 3^x$?
   b. What is the domain of the function?
   c. What is the range of the function?
   d. What is the y-intercept of the graph? What is the x-intercept of the graph?
   e. Is the function one-to-one?
   f. What is an asymptote of the graph?
   g. Is f an increasing or a decreasing function?
   h. The graph passes through the point $(1, y)$. What is y?

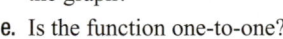

6. Which of the following functions are exponential functions?
   a. $f(x) = x^2$
   b. $g(x) = 4x$
   c. $h(x) = 8^x$
   d. $s(x) = \dfrac{1}{x}$
   e. $T(x) = (0.92)^{x+1}$
   f. $r(x) = x^3$
   g. $P(x) = \sqrt{x}$
   h. $d(x) = |x|$

7. Evaluate each expression without a calculator.
   a. 3^{-2}
   b. $\left(\frac{1}{2}\right)^4$
   c. $\left(\frac{1}{5}\right)^{-2}$

8. Evaluate each expression using a calculator. Round to the nearest tenth.
   a. $20{,}000(1.036)^{52}$
   b. $92(0.88)^6$

9. Match each function with its graph shown below.
   a. $f(x) = x^2$
   b. $f(x) = 2^x$
   c. $f(x) = 2$
   d. $f(x) = 2x$

   i.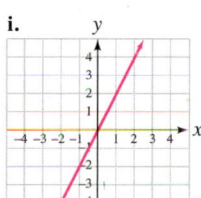
   ii.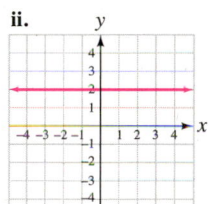
   iii.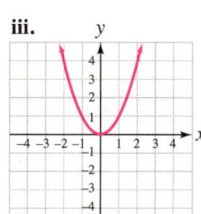
   iv.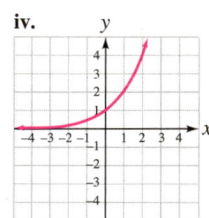

10. **a.** Two exponential functions of the form $f(x) = b^x$ are graphed in figure (a) below. Which function has the larger base b, the one graphed in red or the one graphed in blue?
    b. Two exponential functions of the form $f(x) = b^x$ are graphed in figure (b) below. Which function has the smaller base b?

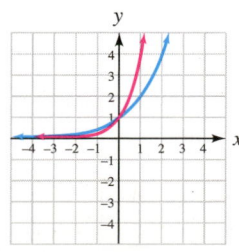
(a)

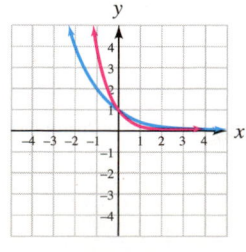
(b)

11. Determine the domain and range of each exponential function graphed below.

a.

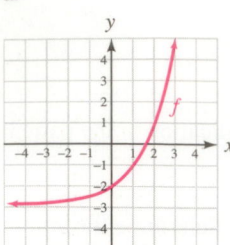

b.

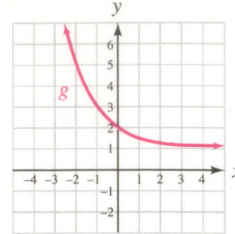

12. Fill in the blanks.

   a. The graph of $g(x) = 4^x + 3$ is similar to the graph of $f(x) = 4^x$, but it is translated 3 units _____.

   b. The graph of $g(x) = 4^{x-3}$ is similar to the graph of $f(x) = 4^x$, but it is translated 3 units to the ____.

13. Complete the table of function values shown here.

$$f(x) = 5^x$$

x	$f(x)$
-3	
-2	
-1	
0	
1	
2	
3	

14. Determine whether each of the following functions model exponential growth or exponential decay.

   a. $D(t) = 150(0.44)^t$ b. $H(t) = 15,000(1.03)^t$

15. Match each situation to the exponential graph that best models it.

   a. The number of cell phone subscribers in the world over the past 5 years

   b. The level of caffeine in the bloodstream after drinking a cup of coffee

   c. The amount of money in a bank account earning interest compounded quarterly

   d. The number of rabbits in a population with a high birth rate

   e. The amount of water in a shirt that was just washed and hung on a clothesline to dry

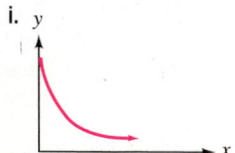

i.   ii.

16. What formula is used to determine the amount of money in a savings account earning compound interest?

NOTATION

17. For an exponential function of the form $f(x) = b^x$, what are the restrictions on b?

18. In $A(t) = 16,000\left(1 + \frac{0.05}{365}\right)^{365t}$, what is the base and what is the exponent?

GUIDED PRACTICE

Graph each function. See Examples 1 and 2.

19. $f(x) = 3^x$ 20. $f(x) = 6^x$

21. $f(x) = 5^x$ 22. $f(x) = 7^x$

23. $f(x) = \left(\frac{1}{4}\right)^x$ 24. $f(x) = \left(\frac{1}{5}\right)^x$

25. $f(x) = \left(\frac{1}{6}\right)^x$ 26. $f(x) = \left(\frac{1}{8}\right)^x$

Graph each function by plotting points or using a translation. See Example 4.

27. $g(x) = 3^x - 2$ 28. $g(x) = 2^x + 1$

29. $g(x) = 2^{x+1}$ 30. $g(x) = 3^{x-1}$

31. $g(x) = 4^{x-1} + 2$ 32. $g(x) = 4^{x+1} - 2$

33. $g(x) = -2^x$ 34. $g(x) = -3^x$

Use a graphing calculator to graph each function. Determine whether the function is an increasing or a decreasing function. See Using Your Calculator: Graphing Exponential Functions.

35. $f(x) = \frac{1}{2}(3^{x/2})$ 36. $f(x) = -3(2^{x/3})$

37. $f(x) = 2(3^{-x/2})$ 38. $f(x) = -\frac{1}{4}(2^{-x/2})$

APPLICATIONS

39. **CO_2 Concentration.** The exponential growth model below illustrates the rise in atmospheric carbon dioxide from 1744 to 2006. The historical data (shown with blue points) comes from ice cores, and modern data (shown with red points) was collected from the Mauna Loa Observatory in Hawaii.

   a. Estimate the atmospheric carbon dioxide concentrations in 1800, 1900, and 2000.

   b. In approximately what year did the concentration surpass 325 parts per million?

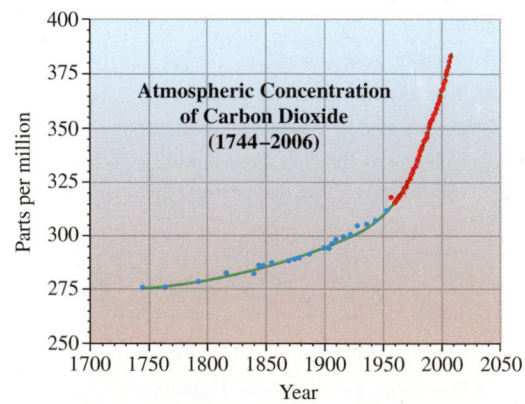

Source: www.eoearth.org

40. Global Warming. The following graph from the United States Environmental Protection Agency shows the projected sea level changes due to anticipated global warming.

a. What type of function does it appear could be used to model the sea level change?

b. When were the earliest instrumental records of sea level change made?

c. For the year 2100, what is the upper-end projection for sea level change? What is the lower-end projection?

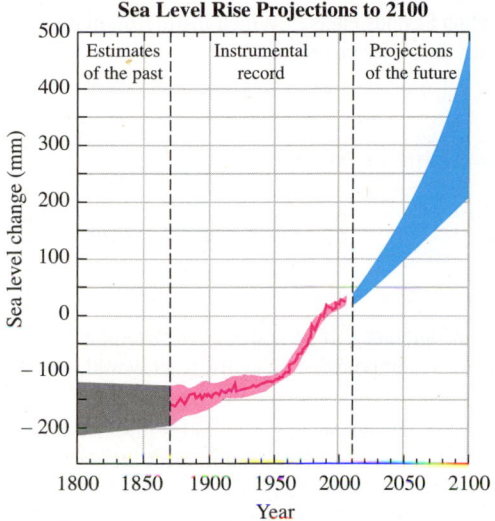

Sea Level Rise Projections to 2100

Source: United States Environmental Protection Agency

41. Value of a Car. The graph shows how the value of the average car depreciates as a percent of its original value over a 10-year period. It also shows the yearly maintenance costs as a percent of the car's value.

a. When is the car worth half of its purchase price?

b. When is the car worth a quarter of its purchase price?

c. When do the average yearly maintenance costs surpass the value of the car?

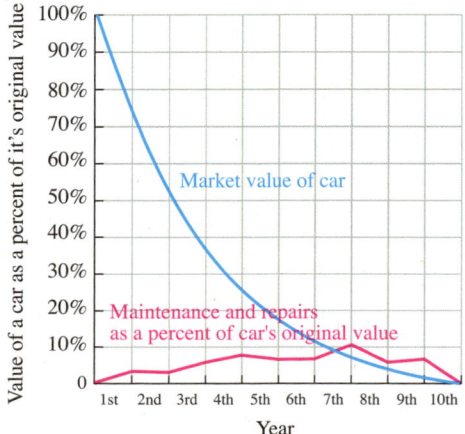

Source: U.S. Department of Transportation

42. Diving. *Bottom time* is the time a scuba diver spends descending plus the actual time spent at a certain depth. Graph the bottom time limits given in the table as ordered pairs of the form (depth, bottom time). Then draw a smooth curve through the points.

Depth (ft)	Bottom time limit (min)	Depth (ft)	Bottom time limit (min)
30	no limit	80	40
35	310	90	30
40	200	100	25
50	100	110	20
60	60	120	15
70	50	130	10

43. Computer Viruses. Suppose the number of computers infected by the spread of a virus through an e-mail is described by the exponential function $c(t) = 5(1.034)^t$, where t is the number of minutes since the first infected e-mail was opened.

a. Graph the function. Scale the t-axis from 0 to 400, in units of 50. Scale the $c(t)$-axis from 0 to 800,000 in units of 100,000.

b. Use the function to determine the number of infected computers in 8 hours, which is 480 minutes.

44. Salvage Value. A small business purchased a computer for $5,000. The value (in dollars) of the computer, t years after its purchase, is given by the exponential function $v(t) = 5,000(0.75)^t$.

a. Graph the function. Scale the t-axis from 0 to 10 in units of 2. Scale the $v(t)$-axis from 0 to 6,000 in units of 1,000.

b. Use the function to determine the value of the computer 12 years after it is purchased.

45. Guitars. The frets on the neck of a guitar are placed so that pressing a string against them determines the strings' vibrating length. The exponential function $f(n) = 650(0.94)^n$ gives the vibrating length (in millimeters) of a string on a certain guitar for the fret number n. Find the length of the vibrating string when a guitarist holds down a string at the 7th fret.

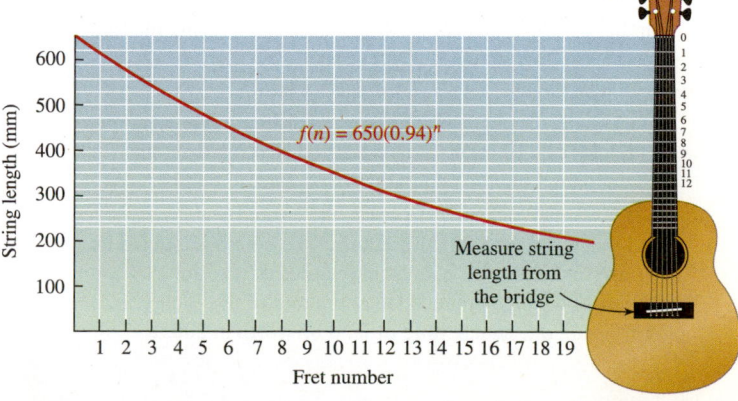

46. Bacterial Cultures. A colony of 6 million bacteria was determined to be growing in the culture medium shown in illustration (a). If the population P of bacteria after t hours is given by the function $P(t) = 6,000,000(2.3)^t$, find the population in the culture later in the day using the information given in illustration (b).

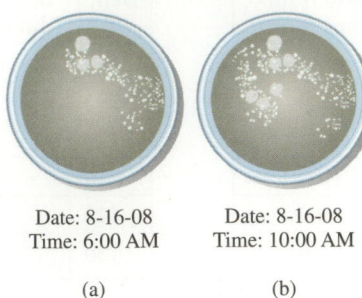

Date: 8-16-08　　Date: 8-16-08
Time: 6:00 AM　　Time: 10:00 AM

(a)　　　　　　　(b)

47. Radioactive Decay. Five hundred grams of a radioactive material decays according to the formula $A = 500\left(\frac{2}{3}\right)^t$, where t is measured in years. Find the amount present in 10 years. Round to the nearest one-tenth of a gram.

48. Discharging a Battery. The charge remaining in a battery decreases as the battery discharges. The charge C (in coulombs) after t days is given by the function $C(t) = 0.0003(0.7)^t$. Find the charge after 5 days.

---

49. from **Campus to Careers**

Social Worker

The function $P(t) = 35.8(1.06)^t$ approximates the number of people (in millions) in the United States living in poverty, where t is the number of years after 2006. Use the function to complete the table below. Round to the nearest tenth. (Source: U.S. Census Bureau)

Year	2006	2007	2008	2009
Number in poverty (in millions)				

50. Publishing Books. The function $W(f) = 4,066(0.8753)^f$ approximates the number of words that can be typeset on a standard page using the Times Roman font size f. Find the number of words that can be typeset on a page using the font size 12. (Source: writersservices.com)

51. Population Growth. The population of North Rivers is decreasing exponentially according to the formula $P = 3,745(0.93)^t$, where t is measured in years from the present date. Find the population in 6 years, 9 months.

52. The Louisiana Purchase. In 1803, the United States negotiated the Louisiana Purchase with France. The country doubled its territory by adding 827,000 square miles of land for $15 million. If the land appreciated at the rate of 6% each year, what would one square mile of land be worth in 2005?

In Exercises 53–58, assume that there are no deposits or withdrawals.

53. Compound Interest. An initial deposit of $10,000 earns 8% interest, compounded quarterly. How much will be in the account after 10 years?

54. Compound Interest. An initial deposit of $10,000 earns 8% interest, compounded monthly. How much will be in the account after 10 years?

55. Comparing Interest Rates. How much more interest could $1,000 earn in 5 years, compounded quarterly, if the annual interest rate were $5\frac{1}{2}\%$ instead of 5%?

56. Comparing Savings Plans. Which institution in the ads provides the better investment?

Fidelity Savings & Loan
Earn 5.25%
compounded monthly

Union Trust
Money Market Account
paying 5.35%
compounded annually

57. Compound Interest. If $1 had been invested on July 4, 1776, at 5% interest, compounded annually, what would it be worth on July 4, 2076?

58. Frequency of Compounding. $10,000 is invested in each of two accounts, both paying 6% annual interest. In the first account, interest compounds quarterly, and in the second account, interest compounds daily. Find the difference between the accounts after 20 years.

WRITING

59. If world population is increasing exponentially, why is there cause for concern?

60. How do the graphs of $f(x) = 3^x$ and $g(x) = \left(\frac{1}{3}\right)^x$ differ? How are they similar?

61. A snowball rolling downhill grows *exponentially* with time. Explain what this means. Sketch a simple graph that models the situation.

62. Explain why the change in temperature of a cup of hot coffee left unattended on a kitchen table is an example of exponential decay.

63. Let $f(x) = \left(\frac{1}{5}\right)^x$. Explain why we can rewrite the function equation as $f(x) = 5^{-x}$.

64. Explain why the graph of $f(x) = 3^x$ gets closer and closer to the x-axis as the values of x decrease. Does the graph ever cross the x-axis? Explain why or why not.

65. Describe the graphs of $f(x) = x^2$ and $g(x) = 2^x$ in words.

66. Write a paragraph explaining the concept that is illustrated in the graph.

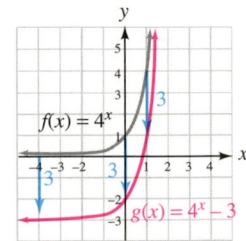

67. In the definition of the exponential function, b could not be negative. Why?

68. How does exponential growth differ from linear growth? Give an example.

In Exercises 69–72, refer to the illustration below in which lines r and s are parallel.

69. Find x.
70. Find the measure of $\angle 1$.
71. Find the measure of $\angle 2$.
72. Find the measure of $\angle 3$.

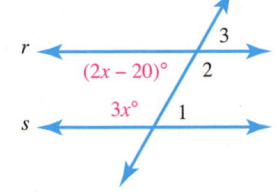

CHALLENGE PROBLEMS

73. Graph $f(x) = 3^x$. Then use the graph to estimate the value of $3^{1.5}$.

74. Graph $y = x^{1/2}$ and $y = \left(\frac{1}{2}\right)^x$ on the same set of coordinate axes. Estimate the coordinates of any point(s) that the graphs have in common.

75. Find the value of b that would cause the graph of $f(x) = b^x$ to look like the graph on the right.

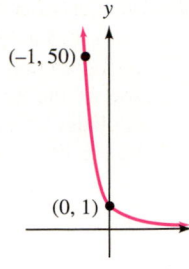

Simplify each expression. Write answers using positive exponents.

76. a. $\left(2^{\sqrt{3}}\right)^{\sqrt{3}}$ **b.** $7^{\sqrt{3}} 7^{\sqrt{12}}$
c. $\dfrac{5^{6\sqrt{2}}}{5^{4\sqrt{2}}}$ **d.** $5^{-\sqrt{5}}$

77. Graph: $f(x) = 2^{|x|}$
78. Graph the inverse of $f(x) = 3^x$.

SECTION 9.4

Logarithmic Functions

OBJECTIVES

1 Define logarithm.
2 Write logarithmic equations as exponential equations.
3 Write exponential equations as logarithmic equations.
4 Evaluate logarithmic expressions.
5 Graph logarithmic functions.
6 Use logarithmic formulas and functions in applications.

ARE YOU READY?

The following problems review some basic skills that are needed when working with logarithmic functions.

1. A table of values for a one-to-one function f is shown below. Complete the table of values for f^{-1}.

x	$f(x)$		x	$f^{-1}(x)$
0	1			
1	2			
2	4			

2. Fill in the blanks: **a.** $5^{-3} = \dfrac{1}{}$ **b.** $3^{} = 27$

3. Fill in the blanks: **a.** $4^{} = 4$ **b.** $7^{} = \sqrt{7}$

4. Evaluate: **a.** 10^3 **b.** 10^{-2}

In this section, we will discuss inverses of exponential functions. These functions are called *logarithmic functions,* and they can be used to solve problems from fields such as electronics, seismology (the study of earthquakes), and business.

1 Define Logarithm.

The graph of the exponential function $f(x) = 2^x$ is shown in red below. Since it passes the horizontal line test, it is a one-to-one function and has an inverse. To graph f^{-1}, we interchange the coordinates of the ordered pairs in the table, plot those points, and draw a smooth curve through them, as shown in blue. As expected, the graphs of f and f^{-1} are symmetric with respect to the line $y = x$.

To graph f^{-1}, interchange each pair of coordinates.

$f(x) = 2^x$

x	$f(x)$		
-3	$\frac{1}{8}$	$\rightarrow \left(-3, \frac{1}{8}\right)$	$\left(\frac{1}{8}, -3\right)$
-2	$\frac{1}{4}$	$\rightarrow \left(-2, \frac{1}{4}\right)$	$\left(\frac{1}{4}, -2\right)$
-1	$\frac{1}{2}$	$\rightarrow \left(-1, \frac{1}{2}\right)$	$\left(\frac{1}{2}, -1\right)$
0	1	$\rightarrow (0, 1)$	$(1, 0)$
1	2	$\rightarrow (1, 2)$	$(2, 1)$
2	4	$\rightarrow (2, 4)$	$(4, 2)$
3	8	$\rightarrow (3, 8)$	$(8, 3)$

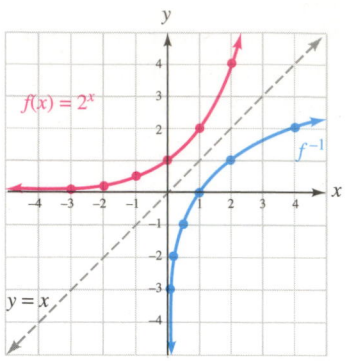

To write an equation for the inverse of $f(x) = 2^x$, we proceed as follows:

$$f(x) = 2^x$$
$$y = 2^x \qquad \text{Replace f(x) with y.}$$
$$x = 2^y \qquad \text{Interchange the variables x and y.}$$

We cannot solve the equation for y because we have not discussed methods for solving equations with a variable exponent. However, we can translate the relationship $x = 2^y$ into words:

$$y = \text{ the power to which we raise 2 to get } x$$

If we substitute the notation $f^{-1}(x)$ for y, we see that

$$f^{-1}(x) = \text{ the power to which we raise 2 to get } x$$

If we define the symbol $\log_2 x$ to mean *the power to which we raise 2 to get x*, we can write the equation for the inverse as

$$f^{-1}(x) = \log_2 x \qquad \text{Read } \log_2 x \text{ as "the logarithm, base 2, of x" or "log, base 2, of x."}$$

We have found that the inverse of the exponential function $f(x) = 2^x$ is $f^{-1}(x) = \log_2 x$. To find the inverse of exponential functions with other bases, such as $f(x) = 3^x$ and $f(x) = 10^x$, we define *logarithm* in the following way.

Definition of Logarithm

For all positive numbers b, where $b \neq 1$, and all positive numbers x,

$$y = \log_b x \quad \text{is equivalent to} \quad x = b^y$$

This definition guarantees that any pair (x, y) that satisfies the logarithmic equation $y = \log_b x$ also satisfies the exponential equation $x = b^y$. Because of this relationship, a statement written in logarithmic form can be written in an equivalent exponential form, and vice versa. They are just different ways of expressing the same thing. The following diagram will help you remember the respective positions of the exponent and base in each form.

Exponent

$$y = \log_b x \text{ means } x = b^y$$

Base

2 Write Logarithmic Equations as Exponential Equations.

The following table shows the relationship between logarithmic and exponential notation. We need to be able to work in both directions.

Logarithmic equation		*Exponential equation*
$\log_2 8 = 3$		$2^3 = 8$
$\log_3 81 = 4$		$3^4 = 81$
$\log_4 4 = 1$	means	$4^1 = 4$
$\log_5 \dfrac{1}{125} = -3$		$5^{-3} = \dfrac{1}{125}$

EXAMPLE 1

Write each logarithmic equation as an exponential equation:

a. $\log_4 64 = 3$ **b.** $\log_7 \sqrt{7} = \dfrac{1}{2}$ **c.** $\log_6 \dfrac{1}{36} = -2$

Strategy To write an equivalent exponential equation, we will determine which number will serve as the base and which will serve as the exponent.

Why We can then use the definition of logarithm to move from one form to the other: $\log_b x = y$ is equivalent to $x = b^y$.

Solution

a. 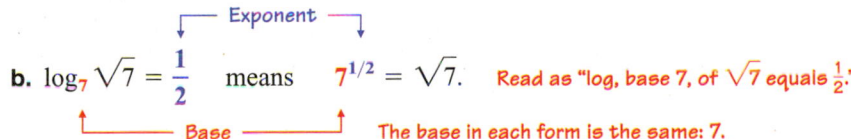 $\log_4 64 = 3$ means $4^3 = 64$.   Read as "log, base 4, of 64 equals 3."

The base in each form is the same: 4.

b. $\log_7 \sqrt{7} = \dfrac{1}{2}$ means $7^{1/2} = \sqrt{7}$.   Read as "log, base 7, of $\sqrt{7}$ equals $\frac{1}{2}$."

The base in each form is the same: 7.

c. 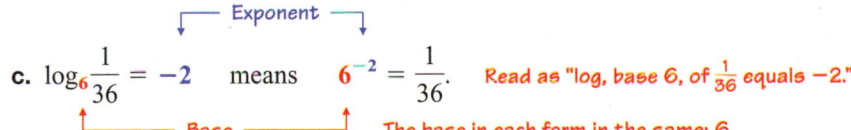 $\log_6 \dfrac{1}{36} = -2$ means $6^{-2} = \dfrac{1}{36}$.   Read as "log, base 6, of $\frac{1}{36}$ equals -2."

The base in each form in the same: 6.

Self Check 1 Write $\log_2 128 = 7$ as an exponential equation.

Now Try Problems 23, 27, and 35

---

3 Write Exponential Equations as Logarithmic Equations.

EXAMPLE 2

Write each exponential equation as a logarithmic equation:

a. $8^0 = 1$ **b.** $6^{1/3} = \sqrt[3]{6}$ **c.** $\left(\dfrac{1}{4}\right)^2 = \dfrac{1}{16}$

Strategy To write an equivalent logarithmic equation, we will determine which number will serve as the base and where we will place the exponent.

Why We can then use the definition of logarithm to move from one form to the other: $x = b^y$ is equivalent to $\log_b x = y$.

Solution

a. $8^0 = 1$ means $\log_8 1 = 0$ In each form, the base is 8.

(Exponent / Base labels pointing to the 0 exponent and 8 base)

b. $6^{1/3} = \sqrt[3]{6}$ means $\log_6 \sqrt[3]{6} = \dfrac{1}{3}$ In each form, the base is 6.

(Exponent / Base labels)

c. $\left(\dfrac{1}{4}\right)^2 = \dfrac{1}{16}$ means $\log_{1/4} \dfrac{1}{16} = 2$ In each form, the base is $\frac{1}{4}$.

(Exponent / Base labels)

Self Check 2 Write $9^{-1} = \frac{1}{9}$ as a logarithmic equation.

Now Try ▶ Problems 39, 43, and 49

Certain logarithmic equations can be solved by writing them as exponential equations.

EXAMPLE 3 Solve each equation for x: **a.** $\log_x 25 = 2$ **b.** $\log_3 x = -3$ **c.** $\log_{1/2} \dfrac{1}{16} = x$

Strategy To solve each logarithmic equation, we will instead write and solve an equivalent exponential equation.

Why The resulting exponential equation is easier to solve because the variable term is often isolated on one side.

Solution

a. Since $\log_x 25 = 2$ is equivalent to $x^2 = 25$, we can solve $x^2 = 25$ to find x.

(Exponent / Base labels)

$$x^2 = 25$$
$$x = \pm\sqrt{25}$$ Use the square root property.
$$x = \pm 5$$

In the expression $\log_x 25$, the base of the logarithm is x. Because the base must be positive, we discard -5 and we have

$$x = 5$$

To check the solution of 5, verify that $\log_5 25 = 2$. The solution set is written as $\{5\}$.

b. Since $\log_3 x = -3$ is equivalent to $3^{-3} = x$, we can instead solve $3^{-3} = x$ to find x.

(Exponent / Base labels)

$$3^{-3} = x$$
$$\frac{1}{3^3} = x$$
$$x = \frac{1}{27}$$

To check the solution of $\frac{1}{27}$, verify that $\log_3 \frac{1}{27} = -3$. The solution set is $\left\{\frac{1}{27}\right\}$.

c. Since $\log_{1/2}\frac{1}{16} = x$ is equivalent to $\left(\frac{1}{2}\right)^{x} = \frac{1}{16}$, we can instead solve $\left(\frac{1}{2}\right)^{x} = \frac{1}{16}$ to find x.

$$\left(\frac{1}{2}\right)^{x} = \frac{1}{16}$$

$$\left(\frac{1}{2}\right)^{x} = \left(\frac{1}{2}\right)^{4} \quad \text{Write } \frac{1}{16} \text{ as a power of } \frac{1}{2} \text{ to match the bases: } \frac{1}{2}\cdot\frac{1}{2}\cdot\frac{1}{2}\cdot\frac{1}{2} = \frac{1}{16}.$$

$$x = 4 \quad \text{Since the bases are the same, and since exponential functions are one-to-one, the exponents must be equal.}$$

To check the solution of 4, verify that $\log_{1/2}\frac{1}{16} = 4$. The solution set is $\{4\}$.

Self Check 3 Solve each equation for x: **a.** $\log_{x} 49 = 2$ **b.** $\log_{1/3} x = 2$ **c.** $\log_{6} 216 = x$

Now Try Problems 51, 53, and 55

4 Evaluate Logarithmic Expressions.

In the previous examples, we have seen that the logarithm of a number is an exponent. In fact,

$\log_{b} x$ is the exponent to which b is raised to get x.

Translating this statement into symbols, we have

$$b^{\log_{b} x} = x$$

EXAMPLE 4 Evaluate each expression: **a.** $\log_{8} 64$ **b.** $\log_{3}\frac{1}{3}$ **c.** $\log_{4} 2$

Strategy After identifying the base, we will ask "To what power must the base be raised to get the other number?"

Why That power is the value of the logarithmic expression.

Solution **a.** $\log_{8} 64 = 2$ Ask: "To what power must we raise 8 to get 64?" Since $8^{2} = 64$, the answer is the 2nd power.

An alternate approach for this evaluation problem is to let $\log_{8} 64 = x$. When we write the equivalent exponential equation $8^{x} = 64$, it is easy to see that $x = 2$.

b. $\log_{3}\frac{1}{3} = -1$ Ask: "To what power must we raise 3 to get $\frac{1}{3}$?" Since $3^{-1} = \frac{1}{3}$, the answer is the -1 power.

We could also let $\log_{3}\frac{1}{3} = x$. The equivalent exponential equation is $3^{x} = \frac{1}{3}$. Thus, x must be -1.

c. $\log_{4} 2 = \frac{1}{2}$ Ask: "To what power must we raise 4 to get 2?" Since $\sqrt{4} = 4^{1/2} = 2$, the answer is the $\frac{1}{2}$ power.

We could also let $\log_{4} 2 = x$. Then the equivalent exponential equation is $4^{x} = 2$. Thus, x must be $\frac{1}{2}$.

Self Check 4 Evaluate each expression: **a.** $\log_{9} 81$ **b.** $\log_{4}\frac{1}{16}$ **c.** $\log_{9} 3$

Now Try Problems 75 and 77

The Language of Algebra

London professor Henry Briggs (1561–1630) and Scottish lord John Napier (1550–1617) are credited with developing the concept of **common logarithms.** Their tables of logarithms were useful tools at that time for those performing large calculations.

In many applications, base-10 logarithms (also called **common logarithms**) are used. When the base b is not indicated in the notation log x, we assume that $b = 10$:

$$\log x \quad \text{means} \quad \log_{10} x$$

The table below shows the relationship between base-10 logarithmic notation and exponential notation.

Logarithmic form		**Exponential form**	
$\log 100 = 2$		$10^2 = 100$	Read log 100 as "log of 100."
$\log\dfrac{1}{10} = -1$	means	$10^{-1} = \dfrac{1}{10}$	
$\log 1 = 0$		$10^0 = 1$	

In general, we have

$$\log 10^x = x$$

EXAMPLE 5   Evaluate each expression: **a.** $\log 1{,}000$ **b.** $\log\dfrac{1}{100}$ **c.** $\log 10$ **d.** $\log(-10)$

Strategy   After identifying the base, we will ask "To what power must 10 be raised to get the other number?"

Why   That power is the value of the logarithmic expression.

Solution   **a.** $\log 1{,}000 = 3$ Ask: "To what power must we raise 10 to get 1,000?"
Since $10^3 = 1{,}000$, the answer is: the 3rd power.

b. $\log\dfrac{1}{100} = -2$ Ask: "To what power must we raise 10 to get $\frac{1}{100}$?"
Since $10^{-2} = \frac{1}{100}$, the answer is: the -2 power.

c. $\log 10 = 1$ Ask: "To what power must we raise 10 to get 10?"
Since $10^1 = 10$, the answer is: the 1st power.

d. To find $\log(-10)$, we must find a power of 10 such that $10^? = -10$. There is no such number. Thus, $\log(-10)$ is undefined.

Self Check 5   Evaluate each expression: **a.** $\log 10{,}000$ **b.** $\log\frac{1}{1{,}000}$
c. $\log 0$

Now Try   Problems 79 and 81

Many logarithmic expressions cannot be evaluated by inspection. For example, to find log 2.34, we ask, "To what power must we raise 10 to get 2.34?" This answer isn't obvious. In such cases, we use a calculator.

Using Your Calculator ▶ Evaluating Logarithms

To find log 2.34 with a scientific calculator we enter

2.34 |LOG| .369215857

On some calculators, the $\boxed{10^x}$ key also serves as the $\boxed{\text{LOG}}$ key when $\boxed{\text{2nd}}$ or $\boxed{\text{SHIFT}}$ is pressed. This is because $f(x) = 10^x$ and $f(x) = \log x$ are inverses.

To use a graphing calculator, we enter

$\boxed{\text{LOG}}$ 2.34 $\boxed{)}$ $\boxed{\text{ENTER}}$ log(2.34)
.369215857

To four decimal places, log 2.34 = 0.3692. This means, $10^{0.3692} \approx 2.34$.

If we attempt to evaluate logarithmic expressions such as log 0, or the logarithm of a negative number, such as $\log(-5)$, an error message like the following will be displayed.

| Error | ERR:DOMAIN
1:QUIT
2:Go to | ERR:NONREAL ANS
1:QUIT
2:Go to |

EXAMPLE 6 Solve $\log x = 0.3568$ and round to four decimal places.

Strategy   To solve this logarithmic equation, we will instead write and solve an equivalent exponential equation.

Why   The resulting exponential equation is easier to solve because the variable term is isolated on one side.

Solution   The equation $\log x = 0.3568$ means $\log_{10} x = 0.3568$, which is equivalent to $\mathbf{10}^{0.3568} = x$. Since we cannot determine $10^{0.3568}$ by inspection, we will use a calculator to find an approximate solution. We enter

$$10\ \boxed{y^x}\ .3568\ \boxed{=}$$

The display reads $\boxed{2.274049951}$. To four decimal places,

$$x = 2.2740$$

If your calculator has a $\boxed{10^x}$ key, enter .3568 and press it to get the same result. The solution is 2.2740. To check, use your calculator to verify that $\log 2.2740 \approx 0.3568$.

Self Check 6   Solve $\log x = 1.87737$ and round to four decimal places.

Now Try ▶   Problem 91

---

5 Graph Logarithmic Functions.

Because an exponential function defined by $f(x) = b^x$ is one-to-one, it has an inverse function that is defined by $x = b^y$. When we write $x = b^y$ in the equivalent form $y = \log_b x$, the result is called a *logarithmic function*.

Logarithmic Functions

If $b > 0$ and $b \neq 1$, the **logarithmic function with base b** is defined by the equations

$$f(x) = \log_b x \quad \text{or} \quad y = \log_b x$$

The domain of $f(x) = \log_b x$ is the interval $(0, \infty)$ and the range is the interval $(-\infty, \infty)$.

Caution

Since the domain of the logarithmic function is the set of positive real numbers, it is impossible to find the logarithm of 0 or the logarithm of a negative number. For example, $\log_2(-4)$ and $\log_2 0$ are undefined.

Since every logarithmic function is the inverse of a one-to-one exponential function, logarithmic functions are one-to-one.

We can plot points to graph logarithmic functions. For example, to graph $f(x) = \log_2 x$, we construct a table of function values, plot the resulting ordered pairs, and draw a smooth curve through the points to get the graph, as shown in figure (a) on the next page. To graph $f(x) = \log_{1/2} x$, we use the same method, as shown in figure (b).

By the vertical line test, we see that each graph is indeed the graph of a function. Because in each case the projection of the graph onto the x-axis covers only the positive portion of that axis, the domain of each function is $(0, \infty)$. Because the graphs extend indefinitely upward and downward, the projection of the graphs onto the y-axis includes all real numbers. Thus, the range of each function is $(-\infty, \infty)$.

$f(x) = \log_2 x$

x	$f(x)$	
$\frac{1}{4}$	-2	$\rightarrow \left(\frac{1}{4}, -2\right)$
$\frac{1}{2}$	-1	$\rightarrow \left(\frac{1}{2}, -1\right)$
1	0	$\rightarrow (1, 0)$
2	1	$\rightarrow (2, 1)$
4	2	$\rightarrow (4, 2)$
8	3	$\rightarrow (8, 3)$

↑

Because the base of the function is 2, choose values for x that are integer powers of 2.

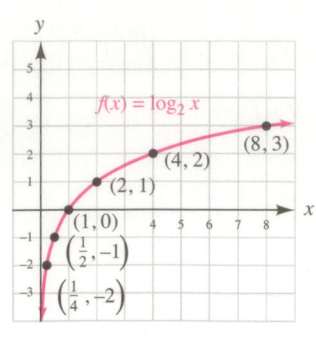

(a)

$f(x) = \log_{1/2} x$

x	$f(x)$	
$\frac{1}{4}$	2	$\rightarrow \left(\frac{1}{4}, 2\right)$
$\frac{1}{2}$	1	$\rightarrow \left(\frac{1}{2}, 1\right)$
1	0	$\rightarrow (1, 0)$
2	-1	$\rightarrow (2, -1)$
4	-2	$\rightarrow (4, -2)$
8	-3	$\rightarrow (8, -3)$

↑

Because the base of the function is $\frac{1}{2}$, choose values for x that are integer powers of $\frac{1}{2}$.

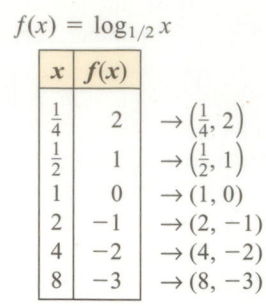
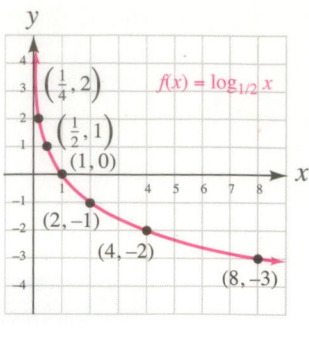

(b)

The graphs of all logarithmic functions are similar to those shown below. If $b > 1$, the logarithmic function is increasing, as in figure (a). If $0 < b < 1$, the logarithmic function is decreasing, as in figure (b).

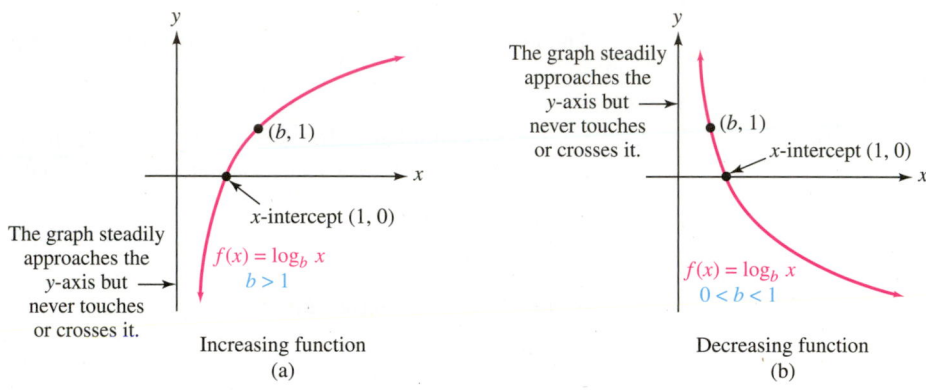

The graph steadily approaches the y-axis but never touches or crosses it.

$(b, 1)$

x-intercept $(1, 0)$

$f(x) = \log_b x$
$b > 1$

Increasing function
(a)

The graph steadily approaches the y-axis but never touches or crosses it.

$(b, 1)$

x-intercept $(1, 0)$

$f(x) = \log_b x$
$0 < b < 1$

Decreasing function
(b)

Properties of Logarithmic Functions	The graph of $f(x) = \log_b x$ (or $y = \log_b x$) has the following properties. 1. It passes through the point $(1, 0)$. 2. It passes through the point $(b, 1)$. 3. The y-axis (the line $x = 0$) is an asymptote. 4. The domain is the interval $(0, \infty)$ and the range is the interval $(-\infty, \infty)$.

The exponential and logarithmic functions are inverses of each other, so their graphs have symmetry about the line $y = x$. The graphs of $f(x) = \log_b x$ and $g(x) = b^x$ are shown in figure (a) when $b > 1$ and in figure (b) when $0 < b < 1$.

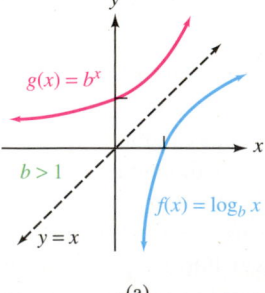

$g(x) = b^x$

$b > 1$

$f(x) = \log_b x$

$y = x$

(a)

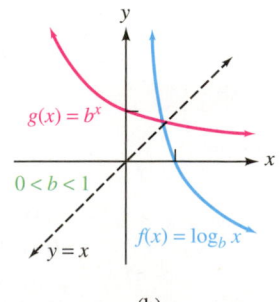

$g(x) = b^x$

$0 < b < 1$

$f(x) = \log_b x$

$y = x$

(b)

The graphs of many functions involving logarithms are translations of the basic logarithmic graphs.

EXAMPLE 7 Graph each function by using a translation: **a.** $g(x) = 3 + \log_2 x$
b. $g(x) = \log_{1/2}(x - 1)$

Strategy We will graph $g(x) = 3 + \log_2 x$ by translating the graph of $f(x) = \log_2 x$ upward 3 units. We will graph $g(x) = \log_{1/2}(x - 1)$ by translating the graph of $f(x) = \log_{1/2} x$ to the right 1 unit.

Why The addition of 3 in $g(x) = \mathbf{3 +} \log_2 x$ causes a vertical shift of the graph of the base-2 logarithmic function 3 units upward. The subtraction of 1 from x in $g(x) = \log_{1/2}(x \mathbf{- 1})$ causes a horizontal shift of the graph of the base-$\frac{1}{2}$ logarithmic function 1 unit to the right.

Solution **a.** The graph of $g(x) = 3 + \log_2 x$ will be the same shape as the graph of $f(x) = \log_2 x$. We simply translate each point on the graph of $f(x) = \log_2 x$ up 3 units. See figure (a) below.

b. The graph of $g(x) = \log_{1/2}(x - 1)$ will be the same shape as the graph of $f(x) = \log_{1/2} x$. We simply translate each point on the graph of $f(x) = \log_{1/2} x$ to the right 1 unit. See figure (b) below.

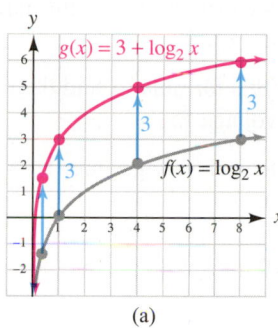

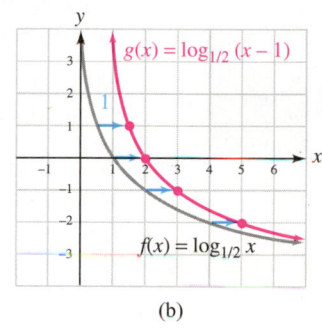

(a)   (b)

Self Check 7 Graph each function by using a translation: **a.** $g(x) = (\log_3 x) - 2$
b. $g(x) = \log_{1/3}(x + 2)$

Now Try ▶ Problems 103 and 105

To graph more complicated logarithmic functions, a graphing calculator is a useful tool.

Using Your Calculator ▶ **Graphing Logarithmic Functions**

To use a calculator to graph the logarithmic function $f(x) = -2 + \log_{10}\frac{x}{2}$, we enter the right side of the equation after the symbol $Y_1 = $. The display will show the equation

$$Y_1 = -2 + \log(X/2)$$

If we use window settings of $[-1, 5]$ for x and $[-4, 1]$ for y and press the $\boxed{\text{GRAPH}}$ key, we will obtain the graph shown.

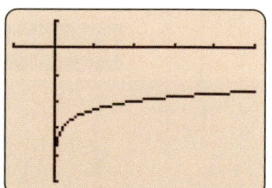

---

6 **Use Logarithmic Formulas and Functions in Applications.**

Logarithmic functions, like exponential functions, can be used to model certain types of growth and decay. Logarithms are especially useful when measuring a quantity that varies over a very large range of values, such as the intensity of earthquakes, the loudness of sounds, and the brightness of stars.

The following common logarithm formula is used in electrical engineering to express the gain (or loss) of an electronic device such as an amplifier as it takes an input signal and produces an output signal. The unit of gain (or loss) in such cases is called the **decibel**, which is abbreviated **dB.**

Decibel Voltage Gain

If E_O is the output voltage of a device and E_I is the input voltage, the decibel voltage gain of the device (dB gain) is given by

$$\text{dB gain} = 20 \log \frac{E_O}{E_I}$$

EXAMPLE 8

©iStockphoto.com/PeterAlbrektsen

dB Gain. If the input to an amplifier is 0.5 volt and the output is 40 volts, find the decibel voltage gain of the amplifier.

Strategy We will substitute into the formula for dB gain and evaluate the right side using a calculator.

Why We can use this formula to find the dB gain because we are given the input voltage E_I and the output voltage E_O.

Solution We can find the decibel voltage gain by substituting 0.5 for E_I and 40 for E_O into the formula for dB gain:

Notation

$20 \log \dfrac{E_O}{E_I}$ means $20 \cdot \log \dfrac{E_O}{E_I}$

$\text{dB gain} = 20 \log \dfrac{E_O}{E_I}$ Read as "20 times the log of E sub O divided by E sub I."

$\text{dB gain} = 20 \log \dfrac{40}{0.5}$ Substitute the input voltage 0.5 for E_I and the output voltage 40 for E_O.

$= 20 \log 80$ Divide: $\frac{40}{0.5} = 80$.

≈ 38 Use a scientific calculator and press: 20 ✕ 80 LOG =.

The amplifier provides a 38-decibel voltage gain.

Self Check 8   **dB Gain.** If the input to an amplifier is 0.6 volt and the output is 40 volts, find the decibel voltage gain of the amplifier.

Now Try ▶ Problem 111

EXAMPLE 9

©Coral Coolahan/Shutterstock.com

Stocking Lakes. To create the proper environmental balance, 250 hybrid bluegill were introduced into a lake by the local Fish and Game Department. Department biologists found that the number of bluegill in the lake could be approximated by the logarithmic function $f(t) = 250 + 400 \log (t + 1)$, where t is the number of years since the lake was stocked. Find the bluegill population in the lake after 5 years.

Strategy We will find $f(5)$.

Why Since the variable t represents the time since the lake was stocked, $t = 5$.

Solution

$f(t) = 250 + 400 \log (t + 1)$ This is the logarithmic growth model.

$f(5) = 250 + 400 \log (5 + 1)$ Substitute 5 for t.

$= 250 + 400 \log 6$ Do the addition within the parentheses.

≈ 561 Use a scientific calculator and press: 250 + 400 ✕ 6 LOG =.

There were approximately 561 bluegill in the lake 5 years after it was stocked.

Self Check 9   **Stocking Ponds.** A rancher stocked a pond on his property with 50 catfish. He was told that with the proper care, the catfish population could be approximated by the logarithmic function $f(t) = 50 + 22 \log (t + 1)$, where t is the number of years since the pond was stocked. Find the number of catfish he can expect in the pond in 3 years.

Now Try ▶ Problem 117

SECTION 9.4 ▶ STUDY SET

VOCABULARY

Fill in the blanks.

1. $f(x) = \log_2 x$ and $g(x) = \log x$ are examples of _____ functions.
2. Base-10 logarithms are called _____ logarithms.
3. The graph of $f(x) = \log_2 x$ approaches, but never touches, the negative portion of the y-axis. Thus the y-axis is an _____ of the graph.
4. $\log_x 81 = 4$ is _____ to $x^4 = 81$.

CONCEPTS

5. Refer to the graph on the right.

   a. What type of function is $f(x) = \log_4 x$?
   b. What is the domain of the function? What is the range of the function?
   c. What is the y-intercept of the graph? What is the x-intercept of the graph?
   d. Is f a one-to-one function?
   e. What is an asymptote of the graph?
   f. Is f an increasing or a decreasing function?
   g. The graph passes through the point $(4, y)$. What is y?

6. Determine the domain and range of each logarithmic function graphed below.

   a. 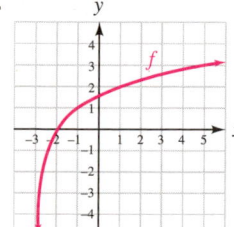   b.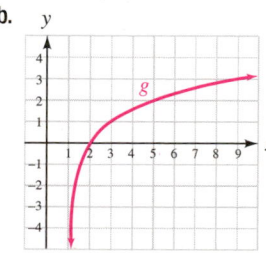

7. Match each function with its graph shown below.

   a. $f(x) = x^3$ b. $f(x) = \log_3 x$
   c. $f(x) = \sqrt[3]{x}$ d. $f(x) = 3x$

   i. 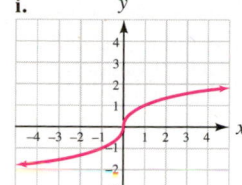   ii.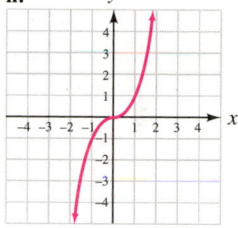

   iii. 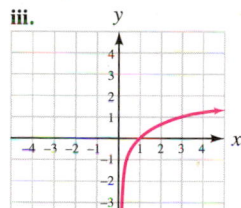   iv.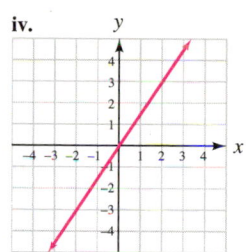

8. a. An exponential function is graphed below. Graph its inverse and the axis of symmetry on the same coordinate system.
   b. What type of function is the inverse function?

   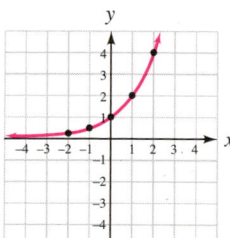

Fill in the blanks.

9. $\log_6 36 = 2$ means ▢ = ▢ .
10. $\log x = -2$ is equivalent to ▢ = ▢ .
11. $\log_b x$ is the _____ to which b is raised to get x.
12. The functions $f(x) = \log_{10} x$ and $f(x) = 10^x$ are _____ functions.
13. The inverse of an exponential function is called a _____ function.

14. Fill in the blanks.

    a. The graph of $g(x) = 4 + \log x$ is similar to the graph of $f(x) = \log x$, but it is translated 4 units _____.

    b. The graph of $g(x) = \log_4 (x + 2)$ is similar to the graph of $f(x) = \log_4 x$, but it is translated 2 units to the _____.

Complete the table of values.

15. $f(x) = \log x$

x	$f(x)$
100	
$\frac{1}{100}$	

16. $f(x) = \log_5 x$

x	$f(x)$
25	
$\frac{1}{25}$	

17. $f(x) = \log_6 x$

Input	Output
6	
-6	
0	

18. $f(x) = \log_8 x$

Input	Output
8	
-8	
0	

 19. a. Use a calculator to complete the table of values for $f(x) = \log x$. Round to the nearest hundredth.

    b. Graph $f(x) = \log x$. Note that the units on the x- and y-axes are different.

x	$f(x)$
0.5	
1	
2	
4	
6	
8	
10	

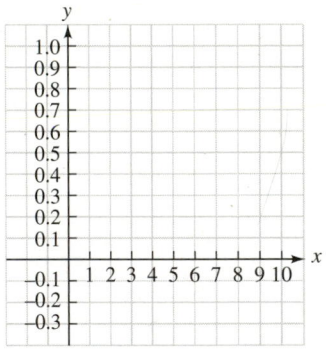

20. For each function, determine its inverse, $f^{-1}(x)$.

    a. $f(x) = 10^x$ **b.** $f(x) = 3^x$

    c. $f(x) = \log x$ **d.** $f(x) = \log_2 x$

Fill in the blanks.

21. a. $\log x = \log \ \_\_ \ x$ **b.** $\log_{10} 10^x = \ \_\_$

22. a. We read $\log_5 25$ as "log, ___ 5, ___ 25."

    b. We read $\log x$ as "___ of x."

Write each logarithmic equation as an exponential equation. **See Example 1. Do not solve.**

23. $\log_3 81 = 4$ **24.** $\log_7 7 = 1$

25. $\log_{10} 10 = 1$ **26.** $\log_{10} 100 = 2$

27. $\log_4 \frac{1}{64} = -3$ **28.** $\log_6 \frac{1}{36} = -2$

29. $\log_5 \sqrt{5} = \frac{1}{2}$ **30.** $\log_8 \sqrt[3]{8} = \frac{1}{3}$

31. $\log 0.1 = -1$ **32.** $\log 0.01 = -2$

33. $x = \log_8 64$ **34.** $x = \log_9 81$

35. $t = \log_b T_1$ **36.** $n = \log_b R_1$

37. $\log_n C = -42$ **38.** $\log_m P = 101$

Write each exponential equation as a logarithmic equation. **See Example 2.**

39. $8^2 = 64$ **40.** $10^3 = 1,000$

41. $4^{-2} = \frac{1}{16}$ **42.** $3^{-4} = \frac{1}{81}$

43. $\left(\frac{1}{2}\right)^{-5} = 32$ **44.** $\left(\frac{1}{3}\right)^{-3} = 27$

45. $x^y = z$ **46.** $m^n = p$

47. $y^t = 8.6$ **48.** $b^r = \frac{2}{3}$

49. $7^{4.3} = B + 1$ **50.** $12^{-2.6} = N + 1$

Solve for x. **See Example 3.**

51. $\log_x 81 = 2$ **52.** $\log_x 9 = 2$

53. $\log_8 x = 2$ **54.** $\log_7 x = 0$

55. $\log_5 125 = x$ **56.** $\log_4 16 = x$

57. $\log_5 x = -2$ **58.** $\log_3 x = -4$

59. $\log_{36} x = -\frac{1}{2}$ **60** $\log_{27} x = -\frac{1}{3}$

61. $\log_x 0.01 = -2$ **62.** $\log_x 0.001 = -3$

63. $\log_{27} 9 = x$ **64.** $\log_{12} x = 0$

65. $\log_x 5^3 = 3$ **66.** $\log_x 5 = 1$

67. $\log_{100} x = \frac{3}{2}$ **68.** $\log_x \frac{1}{1,000} = -\frac{3}{2}$

69. $\log_x \frac{1}{64} = -3$ **70.** $\log_x \frac{1}{100} = -2$

71. $\log_8 x = 0$ **72.** $\log_4 8 = x$

73. $\log_x \frac{\sqrt{3}}{3} = \frac{1}{2}$ **74.** $\log_x \frac{9}{4} = 2$

Evaluate each logarithmic expression. **See Examples 4 and 5.**

75. $\log_2 8$ **76.** $\log_3 9$

77. $\log_4 16$ **78.** $\log_6 216$

79. $\log 1,000,000$ **80** $\log 100,000$

81. $\log \frac{1}{10}$ **82.** $\log \frac{1}{10,000}$

83. $\log_{1/2} \dfrac{1}{32}$

84. $\log_{1/3} \dfrac{1}{81}$

85. $\log_9 3$

86. $\log_{125} 5$

 Use a calculator to find each value. Give answers to four decimal places. See Using Your Calculator: Evaluating Logarithms.

87. $\log 3.25$

88. $\log 0.57$

89. $\log 0.00467$

90. $\log 375.876$

 Use a calculator to solve each equation. Round answers to four decimal places. See Example 6.

91. $\log x = 3.7813$

92. $\log x = 2.8945$

93. $\log x = -0.7630$

94. $\log x = -1.3587$

95. $\log x = -0.5$

96. $\log x = -0.926$

97. $\log x = -1.71$

98. $\log x = 1.4023$

Graph each function. Determine whether each function is an increasing or a decreasing function. See Objective 5.

99. $f(x) = \log_3 x$

100. $f(x) = \log_{1/3} x$

101. $y = \log_{1/2} x$

102. $y = \log_4 x$

Graph each function by plotting points or by using a translation. (The basic logarithmic functions graphed in Exercises 99–102 will be helpful.) See Example 7.

103. $f(x) = 3 + \log_3 x$

104. $f(x) = (\log_{1/3} x) - 1$

105. $y = \log_{1/2} (x - 2)$

106. $y = \log_4 (x + 2)$

Graph each pair of inverse functions on the same coordinate system. Draw the axis of symmetry. See Objective 1.

107. $f(x) = 6^x$
$f^{-1}(x) = \log_6 x$

108. $f(x) = 3^x$
$f^{-1}(x) = \log_3 x$

109. $f(x) = 5^x$
$f^{-1}(x) = \log_5 x$

110. $f(x) = 8^x$
$f^{-1}(x) = \log_8 x$

APPLICATIONS

111. **dB Gain.** Find the dB gain of the amplifier shown below. Round to the nearest tenth.

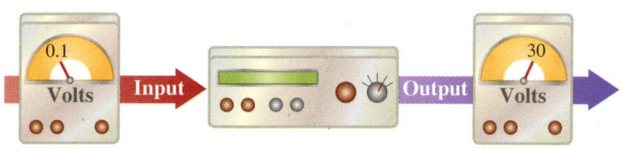

112. **Output Voltage.** Find the dB gain of an amplifier if the output voltage is 2.8 volts when the input voltage is 0.05 volt. Round to the nearest dB.

113. **Earthquakes.** Refer to the illustration in the next column. Common logarithms are used to measure the intensity of earthquakes. If R is the intensity of an earthquake on the **Richter scale**, A is the amplitude (measured in micrometers) of the ground motion and P is the period (the time of one oscillation of the Earth's surface measured in seconds), then $R = \log \dfrac{A}{P}$. If an earthquake has amplitude of 5,000 micrometers and a period of 0.2 second, what is its measure on the Richter scale? Round to the nearest tenth.

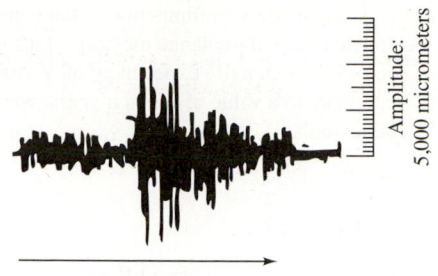

Time: 0.2 sec

114. **Earthquakes.** If an earthquake has amplitude of 95,000 micrometers and a period of $\frac{1}{4}$ second, what is its measure on the Richter scale? See problem 113. Round to the nearest tenth.

115.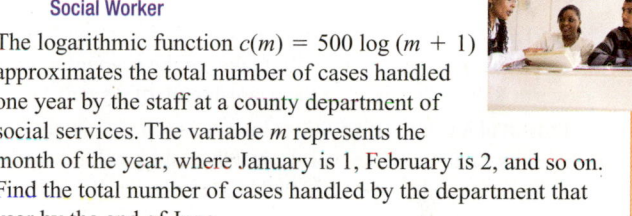

from Campus to Careers

Social Worker

The logarithmic function $c(m) = 500 \log (m + 1)$ approximates the total number of cases handled one year by the staff at a county department of social services. The variable m represents the month of the year, where January is 1, February is 2, and so on. Find the total number of cases handled by the department that year by the end of June.

116. **Advertising.** The dollar amount of sales of a certain new product is approximated by the logarithmic function $D(m) = 20{,}000 + 100{,}000 \log (15m + 1)$, where m is the number of minutes of advertising of the product that is shown on cable television. Find the total sales if 30 minutes of cable television advertising time is purchased.

117. **Stocking Lakes.** A farmer stocked a lake on her property with 75 sunfish. She was told that with the proper oversight, the sunfish population could be approximated by the logarithmic function $f(t) = 75 + 45 \log (t + 1)$, where t is the number of years since the lake was stocked. Find the number of sunfish she can expect in the lake in $2\frac{1}{2}$ years.

118. **Zoology.** A trap-and-release program run by zoologists found that the ground squirrel population in a wilderness area could be estimated by the logarithmic function $s(t) = 800 + 600 \log (50t + 1)$, where t is the number of months after the program started. Find the ground squirrel population 3 years after the program began. (*Hint:* Be careful to use the correct units in your solution.)

119. **Children's Height.** The logarithmic function $h(A) = 29 + 48.8 \log (A + 1)$ gives the percent of the adult height a male child A years old has attained. If a boy is 9 years old, what percent of his adult height will he have reached?

120. Depreciation. In business, equipment is often depreciated using the double declining-balance method. In this method, a piece of equipment with a life expectancy of N years, costing C, will depreciate to a value of V in n years, where n is given by the formula

$$n = \frac{\log V - \log C}{\log\left(1 - \frac{2}{N}\right)}$$

A computer that cost \$37,000 has a life expectancy of 5 years. If it has depreciated to a value of \$8,000, how old is it?

121. Investing. If P is invested at the end of each year in an annuity earning annual interest at a rate r, the amount in the account will be A after n years, where

$$n = \frac{\log\left(\frac{Ar}{P} + 1\right)}{\log(1 + r)}$$

If \$1,000 is invested each year in an annuity earning 12% annual interest, how long will it take for the account to be worth \$20,000? Round to the nearest tenth of a year.

122. Growth of Money. If \$5,000 is invested each year in an annuity earning 8% annual interest, how long will it take for the account to be worth \$50,000? (See Exercise 121.) Round to the nearest tenth of a year.

WRITING

123. Explain the mathematical relationship between $f(x) = \log x$ and $g(x) = 10^x$.

124. Explain why it is impossible to find the logarithm of a negative number.

125. A table of solutions for $f(x) = \log x$ is shown here. As x decreases and gets close to 0, what happens to the values of $f(x)$?

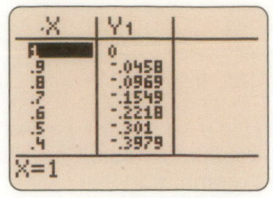

126. What question should be asked when evaluating the expression $\log_4 16$?

REVIEW

Solve each equation.

127. $\sqrt[3]{6x + 4} = 4$

128. $\sqrt{3x + 4} = \sqrt{7x + 2}$

129. $\sqrt{a + 1} - 1 = 3a$

130. $3 - \sqrt{t - 3} = \sqrt{t}$

CHALLENGE PROBLEMS

131. Without graphing, determine the domain of the function $f(x) = \log_5 (x^2 - 1)$. Express the result in interval notation.

132. Evaluate: $\log_6 (\log_5 (\log_4 1{,}024))$

133. Earthquakes. In 1985, Mexico City experienced an earthquake of magnitude 8.1 on the Richter scale. In 1989, the San Francisco Bay area was rocked by an earthquake measuring 7.1. By what factor must the amplitude of an earthquake change to increase its severity by 1 point on the Richter scale? (Assume that the period remains constant.)

134. Graph: $f(x) = \log_2 |x|$

SECTION 9.5

Base-*e* Exponential and Logarithmic Functions

OBJECTIVES

1. Define the natural exponential function.

2. Graph the natural exponential function.

3. Use base-*e* exponential formulas and functions in applications.

4. Define base-*e* logarithms.

5. Evaluate natural logarithmic expressions.

6. Graph the natural logarithmic function.

7. Use base-*e* logarithmic formulas and functions in applications.

ARE YOU READY?

The following problems review some basic skills that are needed when working with base-*e* exponential and logarithmic functions.

1. Evaluate $\left(1 + \frac{1}{n}\right)^n$ for $n = 2$.   **2.** Evaluate: $(2.718)^0$

3. Round 2.718281828459 to the nearest tenth.

4. Use a calculator to evaluate $55(2.718)^{0.4}$. Round to the nearest hundredth.

5. Fill in the blank: $\log_3 x = 6$ is equivalent to $\boxed{} = \boxed{}$.

6. Fill in the blank: If $(3, 8)$ is on the graph of a one-to-one function f, then the point $(\boxed{}, \boxed{})$ is on the graph of f^{-1}.

Any positive real number not equal to 1 can be used as a base of an exponential or a logarithmic function. However, some bases are used more often than others. Exponential and logarithmic functions that have many applications are ones whose base is an irrational number represented by the letter *e*.

1 Define the Natural Exponential Function.

n	$\left(1 + \frac{1}{n}\right)^n$
1	2
2	2.25
4	2.44140625 …
12	2.61303529 …
365	2.71456748 …
1,000	2.71692393 …
100,000	2.71826830 …
1,000,000	2.71828137 …

A scientific calculator was used to evaluate the expression.

The number called e is defined to be the value that $\left(1 + \frac{1}{n}\right)^n$ approaches as n gets larger and larger. The table in the margin shows the value of that expression as n increases from 1 to 1,000,000. It can be shown that as n approaches infinity, the value of $\left(1 + \frac{1}{n}\right)^n$ approaches:

$$e = 2.718281828459 \ldots$$

We give this number a letter-name because it makes communication easier. Reciting its first thirteen digits every time we refer to it would be overwhelming. So, we simply call it by the name "e."

Like π, the number e is an irrational number. That means its decimal representation is nonterminating and nonrepeating. Rounded to four decimal places, $e \approx 2.7183$.

Of all possible bases for an exponential function, e is the most convenient for problems involving growth or decay. Since these situations occur often in natural settings, we call $f(x) = e^x$ the *natural exponential function*.

The Natural Exponential Function	The function defined by $f(x) = e^x$ is the **natural exponential function** (or the **base-e exponential function**) where $e = 2.71828\ldots$. The domain of $f(x) = e^x$ is the interval $(-\infty, \infty)$. The range is the interval $(0, \infty)$.

The $\boxed{e^x}$ key on a calculator is used to find powers of e.

Using Your Calculator ▶ Finding powers of e

To find the value of $\boxed{e^5}$ with a reverse entry scientific calculator, we press:

5 $\boxed{e^x}$

```
148.4131591
```

On some calculators, the $\boxed{e^x}$ key also serves as the $\boxed{LN}$ key. (Later in this section we will see why.) To activate the $\boxed{e^x}$ key, we must begin by pressing $\boxed{2nd}$ or $\boxed{SHIFT}$.

With a direct entry graphing calculator, we press:

$\boxed{2nd}$ $\boxed{e^x}$ 5 $\boxed{)}$ $\boxed{ENTER}$

```
e^(5)
148.4131591
```

Since powers of e are irrational numbers (nonterminating, nonrepeating decimals), we often round such answers. For example, to the nearest hundredth, $e^5 \approx 148.41$

---

2 Graph the Natural Exponential Function.

To graph $f(x) = e^x$, we construct a table of function values by choosing several values for x and finding the corresponding values of $f(x)$. For example, if $x = -2$, we have

$$f(x) = e^x$$
$$f(-2) = e^{-2} \qquad \text{Substitute } -2 \text{ for each } x.$$
$$= 0.135335283 \ldots \qquad \text{Use a calculator. On a scientific calculator, press: 2 } \boxed{+/-} \text{ 2nd } e^x.$$
$$\approx 0.1 \qquad \text{Round to the nearest tenth.}$$

We enter $(-2, 0.1)$ in the table on the next page. Similarly, we find $f(-1)$, $f(0)$, $f(1)$, and $f(2)$, enter each result in the table, and plot the ordered pairs. We draw a smooth curve through the points to get the graph.

From the graph, we can verify that the domain of $f(x) = e^x$ is the interval $(-\infty, \infty)$ and the range is the interval $(0, \infty)$. Since the graph passes the horizontal line test, the function is one-to-one.

Note that as x decreases, the values of $f(x)$ decrease and approach 0. Thus, the x-axis is an asymptote of the graph. The graph does not have an x-intercept, the y-intercept is $(0, 1)$, and the graph passes through the point $(1, e)$.

$f(x) = e^x$

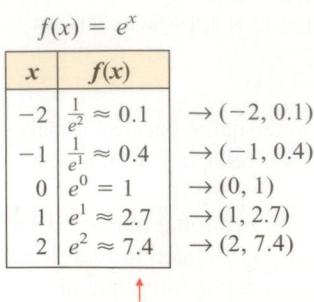

x	$f(x)$	
-2	$\frac{1}{e^2} \approx 0.1$	$\rightarrow (-2, 0.1)$
-1	$\frac{1}{e^1} \approx 0.4$	$\rightarrow (-1, 0.4)$
0	$e^0 = 1$	$\rightarrow (0, 1)$
1	$e^1 \approx 2.7$	$\rightarrow (1, 2.7)$
2	$e^2 \approx 7.4$	$\rightarrow (2, 7.4)$

The graph steadily approaches the x-axis, but never touches or crosses it.

The outputs can be found using the $\boxed{e^x}$ key on a calculator. In such cases, round to the nearest tenth to make point-plotting easier.

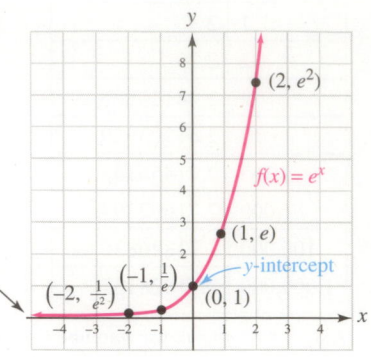

The graph of the natural exponential function can be translated horizontally and vertically, as shown below.

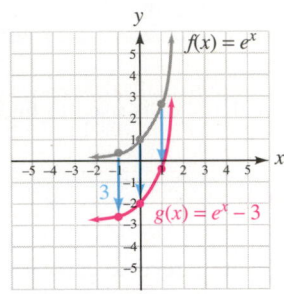

To graph $g(x) = e^x - 3$, translate each point on the graph of $f(x) = e^x$ down 3 units.

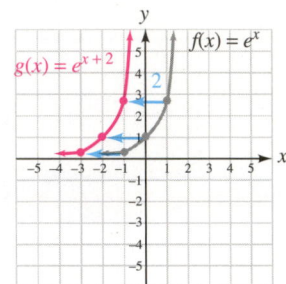

To graph $g(x) = e^{x+2}$, translate each point on the graph of $f(x) = e^x$ to the left 2 units.

We can illustrate the effects of vertical and horizontal translations of the natural exponential function by using a graphing calculator.

Using Your Calculator ▶ Graphing Base e (Natural) Exponential Functions

Figure (a) shows the calculator graphs of $f(x) = e^x$, $g(x) = e^x + 5$, and $h(x) = e^x - 3$. To graph them, we enter the right sides of the equations after $Y_1 =$, $Y_2 =$, and $Y_3 =$. The display will show:

$$Y_1 = e^\wedge(X) \qquad Y_2 = e^\wedge(X) + 5 \qquad Y_3 = e^\wedge(X) - 3$$

The graph of $g(x) = e^x + 5$ is 5 units above the graph of $f(x) = e^x$ and the graph of $h(x) = e^x - 3$ is 3 units below the graph of $f(x) = e^x$.

Figure (b) shows the calculator graphs of $f(x) = e^x$, $g(x) = e^{x+5}$, and $h(x) = e^{x-3}$. The graph of $g(x) = e^{x+5}$ is 5 units to the left of the graph of $f(x) = e^x$ and the graph of $h(x) = e^{x-3}$ is 3 units to the right of the graph of $f(x) = e^x$.

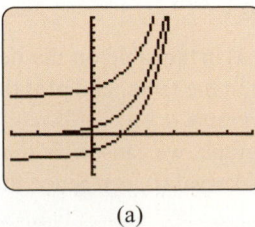

(a)

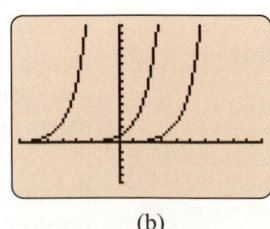

(b)

To graph complicated natural exponential functions, point-plotting can be tedious. In such cases, we will use a graphing calculator. For example, the figure shows the calculator graph of $f(x) = 3e^{-x/2}$. To graph this function, we enter the right side of the equation after the symbol $Y_1 =$. The display will show the equation

$$Y_1 = 3(e^{\wedge}(-X/2))$$

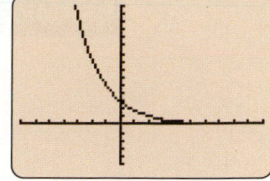

3 Use Base-*e* Exponential Formulas and Functions in Applications.

The following formula involving *e* provides a mathematical model for growth and decay applications. It is used in many areas, such as continuous compound interest, population trends, heat transfer, radioactivity, and learning retention.

Formula for Exponential Growth/Decay	If a quantity P increases or decreases at an annual rate r, compounded continuously, the amount A after t years is given by $A = Pe^{rt}$ Read as "A equals P times e to the rt power."

For the formula above, if the time is measured in years, r is called the **annual growth rate.** If r is negative, the growth represents a decrease.

For a given quantity P, say 10,000, and a given rate r, say 5%, we can write the formula for exponential growth using function notation:

$$A(t) = 10,000e^{0.05t}$$

Recall that if a bank pays interest twice a year, we say the interest is compounded *semiannually*. If it pays interest four times a year, we say the interest is paid *quarterly*. If it pays interest continuously (infinitely many times a year), we say that the interest is **compounded continuously.**

EXAMPLE 1   **Investing.**   If \$25,000 accumulates interest at an annual rate of 8%, compounded continuously, find the balance in the account in 50 years.

Strategy   We will substitute 25,000 for P, 0.08 for r, and 50 for t in the formula $A = Pe^{rt}$ and calculate the value of A.

Why   The words *compounded continuously* indicate that we should use the base-*e* exponential growth/decay formula.

Solution

$A = \boldsymbol{Pe^{rt}}$	This is the formula for continuous compound interest.
$A = \boldsymbol{25,000e^{0.08(50)}}$	Write 8% as 0.08. Substitute for *P*, *r*, and *t*.
$= 25,000e^4$	Evaluate the exponent: 0.08(50) = 4.
$\approx 1,364,953.751$	Use a calculator. On a scientific calculator, press: 25000 $\times$ 4 2nd e^x = .
$\approx 1,364,953.75$	Round to the nearest hundredth.

In 50 years, the balance will be \$1,364,953.75—more than a million dollars.

Self Check 1   **Investing.**   In Example 1, find the balance in 60 years.

Now Try   Problems 33 and 73

EXAMPLE 2

City Planning.   The population of a city is currently 15,000, but economic conditions are causing the population to decrease 3% each year. If this trend continues, find the population in 30 years.

Strategy   We will substitute 15,000 for P, -0.03 for r, and 30 for t in the formula $A = Pe^{rt}$ and calculate the value of A.

Why   Since the population is decreasing 3% each year, the annual growth rate is -3%, or -0.03.

Solution

Success Tip

For quantities that are decreasing, remember to enter a negative value for r, the annual rate, in the formula $A = Pe^{rt}$.

$$A = Pe^{rt}$$ This is the model for population growth/decay.

$$A = 15,000e^{-0.03(30)}$$ Substitute for P, r, and t.

$$= 15,000e^{-0.9}$$ Evaluate the exponent: $-0.03(30) = -0.9$.

$$\approx 6,098.544896$$ Use a calculator. On a scientific calculator, press: 15000 ✕ .9 +/− 2nd e^x = .

$$\approx 6,099$$ Round to the nearest whole number.

In 30 years, the expected population will be 6,099.

Self Check 2   **City Planning.**   In Example 2, find the population in 50 years.

Now Try ▶   Problems 37 and 81

EXAMPLE 3

©RoJo Images/Shutterstock.com

Baking.   A mother takes a cake out of the oven and sets it on a rack to cool. The function $T(t) = 68 + 220e^{-0.18t}$ gives the cake's temperature in degrees Fahrenheit after it has cooled for t minutes. If her children will be home from school in 20 minutes, will the cake have cooled enough for the children to eat it? (Assume that 80°F, or cooler, would be a comfortable eating temperature.)

Strategy   We will substitute 20 for t in the function $T(t) = 68 + 220e^{-0.18t}$.

Why   The variable t represents the number of minutes the cake has cooled.

Solution   When the children arrive home, the cake will have cooled for 20 minutes. To find the temperature of the cake at that time, we need to find $T(20)$.

$$T(t) = 68 + 220e^{-0.18t}$$ This is the cooling model.

$$T(20) = 68 + 220e^{-0.18(20)}$$ Substitute 20 for t.

$$= 68 + 220e^{-3.6}$$ Evaluate the exponent: $-0.18(20) = -3.6$.

$$\approx 74.01121894$$ Use a calculator. On a scientific calculator, press: 68 + 220 ✕ 3.6 +/− 2nd e^x = .

$$\approx 74.0$$ Round to the nearest tenth.

When the children return home, the temperature of the cake will be about 74°, and it can be eaten.

Self Check 3   **Baking.**   In Example 3, find the temperature of the cake 10 minutes after it is removed from the oven. Round to the nearest tenth.

Now Try ▶   Problem 91

4   **Define Base-e Logarithms.**

Of all possible bases for a logarithmic function, e is the most convenient for problems involving growth or decay. Since these situations occur often in natural settings, base-e

logarithms are called **natural logarithms** or **Napierian logarithms** after John Napier (1550–1617). They are usually written as $\ln x$ rather than $\log_e x$:

$$\ln x \quad \text{means} \quad \log_e x \qquad \textcolor{red}{\text{Read ln x letter-by-letter as "}\ell \dots n \dots \text{ of x."}}$$

In general, the logarithm of a number is an exponent. For natural logarithms,

$\ln x$ is the exponent to which e is raised to get x.

Translating this statement into symbols, we have

$$e^{\ln x} = x$$

<table>
<tr><td>

Caution

Because of the font used to print the natural log of x, some students initially misread the notation as $\ln x$. In handwriting, $\ln x$ should look like $\ell n \, x$.
</td></tr>
</table>

5 Evaluate Natural Logarithmic Expressions.

EXAMPLE 4 Evaluate each natural logarithmic expression:

a. $\ln e$ **b.** $\ln \dfrac{1}{e^2}$ **c.** $\ln 1$ **d.** $\ln \sqrt{e}$

Strategy Since the base is e in each case, we will ask "To what power must e be raised to get the given number?"

Why That power is the value of the logarithmic expression.

Solution

a. $\ln e = 1$ \textcolor{red}{Ask: "To what power must we raise e to get e?"}
\textcolor{red}{Since $e^1 = e$, the answer is: the 1st power.}

b. $\ln \dfrac{1}{e^2} = -2$ \textcolor{red}{Ask: "To what power must we raise e to get $\frac{1}{e^2}$?"}
\textcolor{red}{Since $e^{-2} = \frac{1}{e^2}$, the answer is: the -2 power.}

c. $\ln 1 = 0$ \textcolor{red}{Ask: "To what power must we raise e to get 1?"}
\textcolor{red}{Since $e^0 = 1$, the answer is: the 0 power.}

d. $\ln \sqrt{e} = \dfrac{1}{2}$ \textcolor{red}{Ask: "To what power must we raise e to get $\sqrt{e}$?"}
\textcolor{red}{Since $e^{1/2} = \sqrt{e}$, the answer is: the $\frac{1}{2}$ power.}

<table>
<tr><td>

Success Tip

Every natural logarithmic equation has a corresponding natural exponential equation. For example:

$\ln \dfrac{1}{e^2} = -2 \quad$ means $\quad e^{-2} = \dfrac{1}{e^2}$

and

$\ln \sqrt{e} = \dfrac{1}{2} \quad$ means $\quad e^{1/2} = \sqrt{e}$
</td></tr>
</table>

Self Check 4 Evaluate each expression: **a.** $\ln e^3$ **b.** $\ln \dfrac{1}{e}$ **c.** $\ln \sqrt[3]{e}$

Now Try ▶ Problems 41, 45, and 47

Many natural logarithmic expressions are not as easy to evaluate as those in the previous example. For example, to find $\ln 2.34$, we ask, "To what power must we raise e to get 2.34?" The answer isn't obvious. In such cases, we use a calculator.

Using Your Calculator ▶ **Evaluating Base-e (Natural) Logarithms**

To find $\ln 2.34$ with a reverse entry scientific calculator, we press:

2.34 $\boxed{\text{LN}}$ $\qquad\qquad\qquad\qquad\qquad\qquad$ `.850150929`

On some calculators, the $\boxed{e^x}$ key also serves as the $\boxed{\text{LN}}$ key when $\boxed{\text{2nd}}$ or $\boxed{\text{SHIFT}}$ is pressed. We will see why this is so later in this section.

To use a direct entry graphing calculator, we press:

$\boxed{\text{LN}}$ 2.34 $\boxed{)}$ $\boxed{\text{ENTER}}$ $\qquad\qquad\qquad\qquad$ `ln(2.34)`
$\qquad\qquad\qquad\qquad\qquad\qquad\qquad\qquad\qquad\qquad\qquad$ `.8501509294`

To four decimal places, $\ln 2.34 \approx 0.8502$. This means that $e^{0.8502} \approx 2.34$.

If we attempt to evaluate logarithmic expressions such as ln 0, or the logarithm of a negative number, such as ln (-5), then one of the following error statements will be displayed.

Error

```
ERR:DOMAIN
1:QUIT
2:Go to
```

```
ERR:NONREAL ANS
1:QUIT
2:Go to
```

Certain natural logarithmic equations can be solved by writing them as equivalent natural exponential equations.

EXAMPLE 5   Solve each equation: **a.** ln $x = 1.335$ and **b.** ln $x = -5.5$. Give each result to four decimal places.

Strategy   To solve these logarithmic equations, we will instead write and solve equivalent exponential equations.

Why   The resulting exponential equations are easier to solve because the variable term is isolated on one side.

Solution   **a.** Since the base of the natural logarithmic function is e:

Exponent

ln $x = 1.335$ is equivalent to $e^{1.335} = x$

Base

To use a reverse entry scientific calculator to find x, press:

1.335 $\boxed{\text{2nd}}$ $\boxed{e^x}$ $\boxed{3.799995946}$

To four decimal places, $x = 3.8000$. Thus, the solution is 3.8000. To check, use your calculator to verify that ln $3.8000 \approx 1.335$.

b. The logarithmic equation ln $x = -5.5$ is equivalent to the exponential equation $e^{-5.5} = x$. To use a reverse entry scientific calculator to find x, press:

5.5 $\boxed{+/-}$ $\boxed{\text{2nd}}$ $\boxed{e^x}$ $\boxed{0.004086771}$

To four decimal places, $x = 0.0041$. Thus, the solution is 0.0041. To check, use your calculator to verify that ln $0.0041 \approx -5.5$.

Self Check 5   Solve each equation. Give each result to four decimal places.
    a. ln $x = 1.9344$ **b.** $-3 = \ln x$

Now Try   Problems 61 and 65

6 Graph the Natural Logarithmic Function.

Because the natural exponential function defined by $f(x) = e^x$ is one-to-one, it has an inverse function that is defined by $x = e^y$. When we write $x = e^y$ in the equivalent form $y = \ln x$, the result is called the *natural logarithmic function*.

The Natural Logarithmic Function

The **natural logarithmic function** with base e is defined by the equations

$$f(x) = \ln x \text{ or } y = \ln x, \text{ where } \ln x = \log_e x$$

The domain of $f(x) = \ln x$ is the interval $(0, \infty)$, and the range is the interval $(-\infty, \infty)$.

Since the natural logarithmic function is the inverse of the one-to-one natural exponential function, the natural logarithmic function is one-to-one.

To graph $f(x) = \ln x$, we can construct a table of function values, plot the resulting ordered pairs, and draw a smooth curve through the points to get the graph shown in figure (a). Figure (b) shows the calculator graph of $f(x) = \ln x$. We see that the domain of the natural logarithmic function is $(0, \infty)$ and the range is $(-\infty, \infty)$.

$f(x) = \ln x$

To plot these ordered pairs, use a calculator to approximate x.

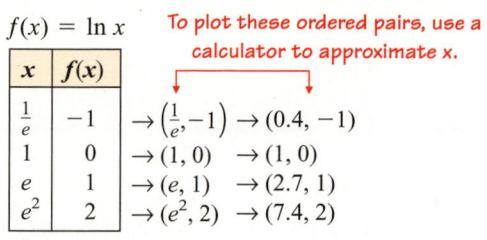

x	$f(x)$
$\frac{1}{e}$	-1
1	0
e	1
e^2	2

$\rightarrow \left(\frac{1}{e}, -1\right) \rightarrow (0.4, -1)$
$\rightarrow (1, 0) \quad \rightarrow (1, 0)$
$\rightarrow (e, 1) \quad \rightarrow (2.7, 1)$
$\rightarrow (e^2, 2) \rightarrow (7.4, 2)$

Since the base of the natural logarithmic function is e, choose x-values that are integer powers of e.

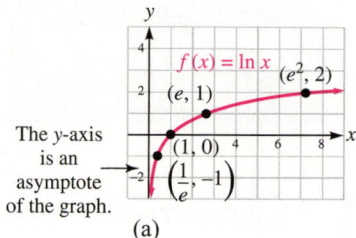

The y-axis is an asymptote of the graph.

(a)

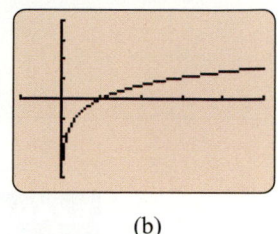

(b)

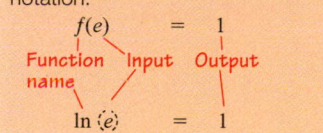

Notation

If it is helpful, imagine parentheses around the input value in the natural logarithmic function notation.

$f(e) \qquad = \quad 1$

Function Input Output
name

$\ln (e) \qquad = \quad 1$

The natural exponential function and the natural logarithm function are inverse functions. The figure shows that their graphs are symmetric to the line $y = x$.

You may want to confirm that these functions are indeed inverses using the composition of inverses test in Section 9.2.

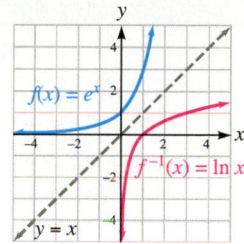

Using Your Calculator ▶ Graphing Base-e Logarithmic Functions

Many graphs of logarithmic functions involve translations of the graph of $f(x) = \ln x$. For example, the figure below shows calculator graphs of the functions $f(x) = \ln x$, $g(x) = (\ln x) + 2$, and $h(x) = (\ln x) - 3$.

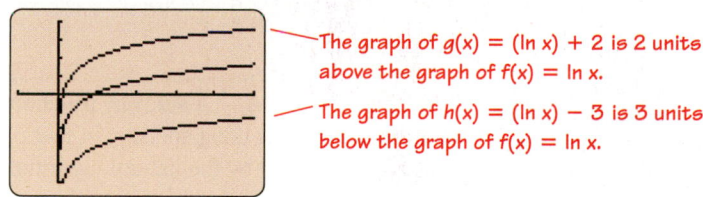

The graph of $g(x) = (\ln x) + 2$ is 2 units above the graph of $f(x) = \ln x$.

The graph of $h(x) = (\ln x) - 3$ is 3 units below the graph of $f(x) = \ln x$.

The next figure shows the calculator graph of the functions $f(x) = \ln x$, $g(x) = \ln (x - 2)$, and $h(x) = \ln (x + 3)$.

The graph of $h(x) = \ln (x + 3)$ is 3 units to the left of the graph of $f(x) = \ln x$.

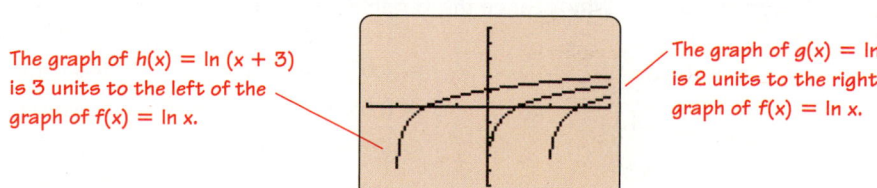

The graph of $g(x) = \ln (x - 2)$ is 2 units to the right of the graph of $f(x) = \ln x$.

7 Use Base-*e* Logarithmic Formulas and Functions in Applications.

If a population grows exponentially at a certain annual rate, the time required for the population to double is called the **doubling time.** It is given by the following formula.

Formula for Doubling Time	If r is the annual rate, compounded continuously, and t is the time required for a population to double, then $$t = \frac{\ln 2}{r}$$

EXAMPLE 6

Doubling Time. The population of the Earth is growing at the approximate rate of 1.133% per year. If this rate continues, how long will it take for the population to double? (Source: CIA World Fact Book, 2009 data)

Strategy We will substitute 1.133% expressed as a decimal for r in the formula for doubling time and evaluate the right side using a calculator.

Why We can use this formula because we are given the annual rate of continuous compounding.

Solution Since the population is growing at the rate of 1.133% per year, we substitute 0.0133 for r in the formula for doubling time and simplify.

$$t = \frac{\ln 2}{r}$$ Don't forget to substitute the decimal form of 1.133%, which is 0.0133, for r.

$$t = \frac{\ln 2}{0.0133}$$

≈ 52.11632937 Use a calculator. On a scientific calculator, press: 2 LN ÷ .0033 = .

≈ 52 Round to the nearest year.

At the current growth rate, the population of the Earth will double in about 52 years.

Self Check 6   **Doubling Time.**   If the annual growth rate of the Earth's population could be reduced to 1.1% per year, what would be the doubling time?

Now Try ▶   Problem 93

EXAMPLE 7

The Pace of Life.   A study by psychologists M. H. Bornstein and H. G. Bornstein found that the average walking speed s, in feet per second, of a pedestrian in a city of population p, is approximated by the natural logarithmic function $s(p) = 0.05 + 0.37 \ln p$. According to this study, what is the average walking speed of a pedestrian in Detroit if its population is 910,000? (Source: *Nature*, Volume 259, Feb. 19, 1976 and infoplease.com)

Strategy We will find $s(910,000)$.

Why Since the variable p represents the population of the city, $p = 910,000$.

Solution

$$s(p) = 0.05 + 0.37 \ln p$$ This is the natural logarithmic model.

$$s(910,000) = 0.05 + 0.37 \ln 910,000$$ Substitute 910,000 for p.

≈ 5.126843955 Use a calculator. On a scientific calculator, press: 0.05 + 0.37 × 910000 LN = .

≈ 5.1 Round to the nearest tenth.

The average walking speed of a pedestrian in Detroit is about 5.1 feet per second.

Self Check 7 **The Pace of Life.** Use the natural logarithmic function from Example 7 to find the average walking speed of a pedestrian in Dallas if its population is 1,300,000.

Now Try ▶ Problem 101

SECTION 9.5 STUDY SET

VOCABULARY

Fill in the blanks.

1. $f(x) = e^x$ is called the natural _____ function. The base is ☐.

2. $f(x) = \ln x$ is called the _____ logarithmic function. The base is ☐.

3. If a bank pays interest infinitely many times a year, we say that the interest is compounded _____.

4. Like π, the number e is an _____ number. Its decimal representation is nonterminating and _____.

CONCEPTS

5. Refer to the graph on the right.

   a. What is the name of the function $f(x) = e^x$?

   b. What is the domain of the function? What is the range of the function?

   c. What is the y-intercept of the graph? What is the x-intercept of the graph?

   d. Is the function one-to-one?

   e. What is an asymptote of the graph?

   f. Is f an increasing or a decreasing function?

   g. The graph passes through the point $(1, y)$. What is y?

6. a. Use a calculator to complete the table of values in the next column for $f(x) = \ln x$. Round to the nearest hundredth.

   b. Graph $f(x) = \ln x$. Note that the units on the x- and y-axes are different.

   c. What are the domain and range of the function?

   d. What is the x-intercept of the graph? What is the y-intercept?

   e. What is an asymptote of the graph?

f. Is f increasing or decreasing?

g. Is the function one-to-one?

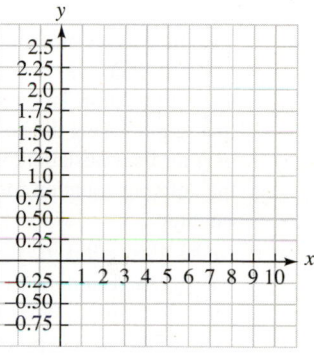

x	$f(x)$
0.5	
1	
2	
4	
6	
8	
10	

Fill in the blanks.

7. $e = $ ___ . ___ ___ ___ ___ ___ ___ ___ ___ ___ ___ ___ ...

8. To two decimal places, the value of e is ☐.

9. If n gets larger and larger, the value of $\left(1 + \frac{1}{n}\right)^n$ approaches the value of ☐.

10. The formula for exponential growth/decay is $A = $ ☐ e ☐.

11. To find $\ln e^2$, we ask, "To what power must we raise ☐ to get e^2?" Since the answer is the 2nd power, $\ln e^2 = $ ☐.

12. The logarithmic equation $\ln x = 1.5318$ is equivalent to the exponential equation ☐ $=$ ☐.

13. Graph each irrational number on the number line: $\{\pi, e, \sqrt{2}\}$.

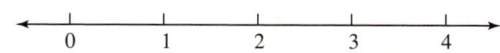

14. Complete the table of values. Use a calculator when necessary and round to the nearest hundredth.

x	-2	-1	0	1	2
e^x					

15. a. The function $f(x) = e^x$ is graphed in figure (a) below and the TRACE feature is used. What is the y-coordinate of the point on the graph having an x-coordinate of 1? What is the symbol that represents this number?

b. Figure (b) below shows a table of values for $f(x) = e^x$. As x decreases, what happens to the values of $f(x)$ listed in the Y_1 column? Will the value of $f(x)$ ever be 0 or negative?

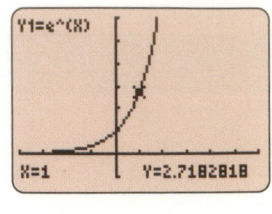 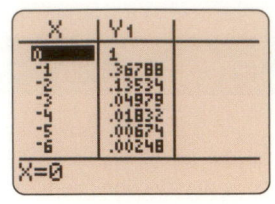

(a)   (b)

16. a. The illustration shows the graph of $f(x) = \ln x$, as well as a vertical translation of that graph. Using the notation $g(x)$ for the translation, write the defining equation for that function.

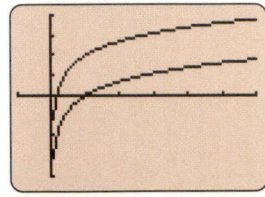

b. In the illustration, $f(x) = \ln x$ was graphed, and the TRACE feature was used. What is the x-coordinate of the point on the graph having a y-coordinate of 1? What is the name given this number?

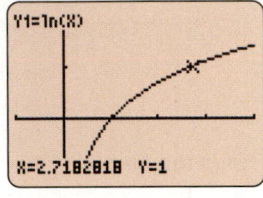

17. What is the inverse of the natural logarithmic function $f(x) = \ln x$?

18. Let $f(x) = 75 + 3,570 \ln x$. Find $f(28.1)$. Round to the nearest tenth.

NOTATION

Find A using the formula $A = Pe^{rt}$ given the following values of P, r, and t. Round to the nearest tenth.

19. $P = 1,000$, $r = 0.09$, and $t = 10$

$A = \boxed{} \, e^{(0.09)(\boxed{})}$

$= 1,000e^{\boxed{}}$

$\approx \boxed{}$ **Use a calculator.**

20. $P = 50,000$, $r = -0.12$, and $t = 50$

$A = 50,000e^{(\boxed{})(50)}$

$= 50,000e^{\boxed{}}$

$\approx \boxed{}$ **Use a calculator.**

Fill in the blanks.

21. We read $\ln x$ letter-by-letter as "$\boxed{}$ ... $\boxed{}$ ... of x."

22. a. $\ln 2$ means $\log_{\boxed{}} 2$.

b. $\log 2$ means $\log_{\boxed{}} 2$.

23. To evaluate a base-10 logarithm with a calculator, use the $\boxed{}$ key. To evaluate the base-e logarithm, use the $\boxed{}$ key.

24. If a population grows exponentially at a rate r, the time it will take the population to double is given by the formula $t = \boxed{}$.

Graph each function. See Objective 2.

25. $f(x) = e^x$

26. $f(x) = -e^x$

27. $f(x) = e^x + 1$

28. $f(x) = e^x - 2$

29. $y = e^{x+3}$

30. $y = e^{x-5}$

31. $f(x) = 2e^x$

32. $f(x) = \frac{1}{2}e^x$

Find A using the formula $A = Pe^{rt}$ given the following values of P, r, and t. Round to the nearest hundredth. See Example 1.

33. $P = 5,000$, $r = 8\%$, $t = 20$ years

34. $P = 15,000$, $r = 6\%$, $t = 40$ years

35. $P = 20,000$, $r = 10.5\%$, $t = 50$ years

36. $P = 25,000$, $r = 6.5\%$, $t = 100$ years

Find A using the formula $A = Pe^{rt}$ given the following values of P, r, and t. Round to the nearest hundredth. See Example 2.

37. $P = 15,895$, $r = -2\%$, $t = 16$ years

38. $P = 33,999$, $r = -4\%$, $t = 21$ years

39. $P = 565$, $r = -0.5\%$, $t = 8$ years

40. $P = 110$, $r = -0.25\%$, $t = 9$ years

Evaluate each expression without using a calculator. See Example 4.

41. $\ln e^5$

42. $\ln e^2$

43. $\ln e^6$

44. $\ln e^4$

45. $\ln \dfrac{1}{e}$

46. $\ln \dfrac{1}{e^3}$

47. $\ln \sqrt[4]{e}$

48. $\ln \sqrt[5]{e}$

49. $\ln \sqrt[3]{e^2}$

50. $\ln \sqrt[4]{e^3}$

51. $\ln e^{-7}$

52. $\ln e^{-10}$

Use a calculator to evaluate each expression, if possible. Express all answers to four decimal places. See Using Your Calculator: Evaluating Base-e (Natural) Logarithms.

53. $\ln 35.15$

54. $\ln 0.675$

55. $\ln 0.00465$

56. $\ln 378.96$

57. $\ln 1.72$

58. $\ln 2.7$

59. $\ln (-0.1)$

60. $\ln (-10)$

Solve each equation. Express all answers to four decimal places. See Example 5.

61. $\ln x = 1.4023$

62. $\ln x = 2.6490$

63. $\ln x = 4.24$

64. $\ln x = 0.926$

65. $\ln x = -3.71$

66. $\ln x = -0.28$

67. $\ln x = 1.001$

68. $\ln x = -0.001$

Use a graphing calculator to graph each function. See Objective 2. See Using Your Calculator: Graph Base-e Logarithmic Functions.

69. $f(x) = \ln \left(\dfrac{1}{2}x\right)$

70. $f(x) = \ln x^2$

71. $f(x) = \ln (-x)$

72. $f(x) = \ln (3x)$

APPLICATIONS

In Exercises 73–78, assume that there are no deposits or withdrawals.

73. Continuous Compound Interest. An initial investment of $5,000 earns 8.2% interest, compounded continuously. What will the investment be worth in 12 years?

74. Continuous Compound Interest. An initial investment of $2,000 earns 8% interest, compounded continuously. What will the investment be worth in 15 years?

75. Comparison of Compounding Methods. An initial deposit of $5,000 grows at an annual rate of 8.5% for 5 years. Compare the final balances resulting from annual compounding and continuous compounding.

76. Comparison of Compounding Methods. An initial deposit of $30,000 grows at an annual rate of 8% for 20 years. Compare the final balances resulting from annual compounding and continuous compounding.

77. Determining the Initial Deposit. An account now contains $11,180 and has been accumulating interest at 7% annual interest, compounded continuously, for 7 years. Find the initial deposit.

78. Determining the Previous Balance. An account now contains $3,610 and has been accumulating interest at 8% annual interest, compounded continuously. How much was in the account 4 years ago?

79. The 20th Century. The exponential function $A(t) = 123e^{0.0117t}$ approximates the population of the United States (in millions), where t is the number of years after 1930. Use the function to estimate the U.S. population for these important dates:

- 1937 The Golden Gate Bridge is completed
- 1941 The United States enters World War II
- 1955 Rosa Parks refuses to give up her seat on a Montgomery, Alabama, bus
- 1969 Astronaut Neil Armstrong walks on the moon
- 1974 President Nixon resigns
- 1986 The *Challenger* space shuttle explodes
- 1997 *The Simpsons* becomes the longest running cartoon television series in history

80. World Population Growth. The population of Earth is approximately 6.8 billion people and is growing at an annual rate of 1.133%. Use the exponential growth model to predict the world population in 30 years.

81. Highs and Lows. Kuwait, located at the head of the Persian Gulf, has one of the greatest population growth rates in the world. Bulgaria, in southeastern Europe, has one of the smallest. Use an exponential growth/decay model to complete the table.

Country	Population 2010	Annual growth rate	Estimated population 2025
Kuwait	2,789,132	3.501%	
Bulgaria	7,148,785	−0.768%	

Source: CIA World Factbook

82. Bed Bugs. If not checked, the population of a colony of bed bugs will grow exponentially at a rate of 65% per week. If a colony currently has 50 bedbugs, how many will there be in 6 weeks?

83. Epidemics. The spread of hoof-and-mouth disease through a herd of cattle can be modeled by the function $P(t) = 2e^{0.27t}$ (t is in days). If a rancher does not quickly treat the two cows that now have the disease, how many cattle will have the disease in 12 days?

84. Oceanography. The width w (in millimeters) of successive growth spirals of the sea shell *Catapulus voluto*, shown below, is given by the exponential function $w(n) = 1.54e^{0.503n}$ where n is the spiral number. Find the width, to the nearest tenth of a millimeter, of the sixth spiral.

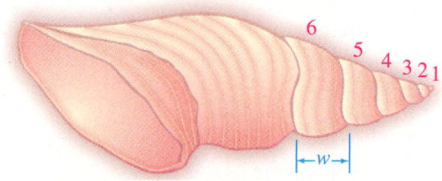

85. Ants. Shortly after an explorer ant discovers a food source, a recruitment process begins in which numerous additional ants travel to the source. The number of ants at the source grows exponentially according to the function $a(t) = 1.36\left(\frac{e}{2.5}\right)^t$, where t is the number of minutes since the explorer discovered the food. How many ants will be at the source in 40 minutes?

86. Half-Life of a Drug. The quantity of a prescription drug in the bloodstream of a patient t hours after it is administered can be modeled by an exponential function. (See the graph.) From the graph determine the time it takes to eliminate half of the initial dose from the body.

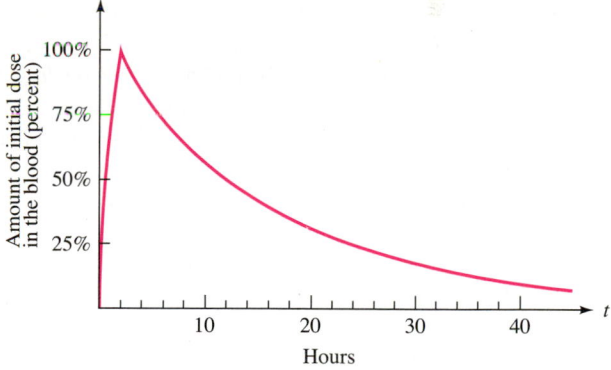

87.

from Campus to Careers

Social Worker

Social workers often use occupational test results when counseling their clients about employment options. The "learning curve" below shows that as a factory trainee assembled more chairs, the assembly time per chair generally decreased. If company standards required an average assembly time of 10 minutes or less, how many chairs did the trainee have to assemble before meeting company standards? (Notice that the graph is a model of exponential decay.)

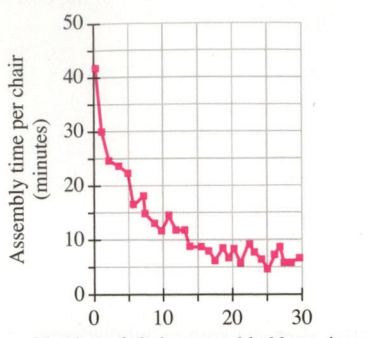

88. Ozone Concentrations.   A *Dobson* unit is the most basic measure used in ozone research. Roughly 300 Dobson units are equivalent to the height of 2 pennies stacked on top of each other. Suppose the ozone layer thickness (in Dobsons) over a certain city is modeled by the function $A(t) = 300e^{-0.0011t}$, where t is the number of years after 1990. Estimate how thick the ozone layer will be in 2015.

89. Disinfectants.   The exponential function $A(t) = 2,000,000e^{-0.588t}$ approximates the number of germs on a table top, t minutes after disinfectant was sprayed on it. Estimate the germ count on the table 5 minutes after it is sprayed.

90. Medicine.   The concentration of a certain prescription drug in an organ after t minutes is modeled by the function $f(t) = 0.08\left(1 - e^{-0.1t}\right)$ where $f(t)$ is the concentration at time t. Find the concentration of the drug at 30 minutes.

91. Sky Diving.   Before the parachute opens, a skydiver's velocity in meters per second is modeled by the function $f(t) = 50\left(1 - e^{-0.2t}\right)$ where $f(t)$ is the velocity at time t. Find the velocity after 20 seconds of free fall.

92. Free Fall.   After t seconds a certain falling object has a velocity in meters per second given by the function $f(t) = 50\left(1 - e^{-0.3t}\right)$. Which is falling faster after 2 seconds—the object or the skydiver in Exercise 91?

93. The Tarheel State.   The 4.3% annual population growth rate for the Raleigh-Cary metropolitan area in North Carolina is one of the largest of any metropolitan area in the United States. If its growth rate remains constant, how long will it take for its population to double? (Source: U.S. Bureau of the Census)

94. The Big Easy.   New Orleans has steadily won back some of the population it lost in the wake of Hurricane Katrina in 2005. If the current 8.2% annual increase in population remains constant, how long will it take for its population to double? (Source: money.cnn.com)

95. The Equality State.   In 2009, the state with the fastest annual population growth rate was Wyoming. If the 2.13% annual increase in population remains constant, what is the first full year that the population of Wyoming will be double what it was in 2009? (Source: U.S. Bureau of the Census)

96. Doubling Money.   How long will it take $1,000 to double if it is invested at an annual rate of 5% compounded continuously?

97. Population Growth.   A population growing continuously at an annual rate r will triple in a time t given by the formula $t = \dfrac{\ln 3}{r}$. How long will it take the population of a town to triple if it is growing at the rate of 12% per year?

98. Tripling Money.   Find the length of time for $25,000 to triple when it is invested at 6% annual interest, compounded continuously. See Exercise 97.

99. Forensic Medicine.   To estimate the number of hours t that a murder victim had been dead, a coroner used the formula $t = \dfrac{1}{0.25} \ln \dfrac{98.6 - T_s}{82 - T_s}$ where T_s is the temperature of the surroundings where the body was found. If the crime took place in an apartment where the thermostat was set at 70°F, approximately how long ago did the murder occur?

100. Making Jello.   After the contents of a package of JELL-O are combined with boiling water, the mixture is placed in a refrigerator whose temperature remains a constant 42°F. Estimate the number of hours t that it will take for the JELL-O to cool to 50°F using the formula $t = -\dfrac{1}{0.9} \ln \dfrac{50 - T_r}{200 - T_r}$ where T_r is the temperature of the refrigerator.

101. Cross Country Skiing.   The function $H(s) = -47.73 + 107.38 \ln s$ approximates the heart rate (in beats/minute) for an Olympic-class cross country skier traveling at s miles per hour, where $s > 5$ mph. Find the heart rate of a skier traveling at a rate of 7.5 miles per hour. (Source: btc.ontana.edu/Olympics/physiology)

102. Strength Loss.   After participating in an eight-week weight training program, a college student was selected to be part of a study to see how much strength he would lose if he discontinued working out. The function $M_B(w) = 225 - 14 \ln (4w + 1)$ approximates his maximum bench press (in pounds) w weeks after stopping weight training. What was his maximum bench press:

a. At the end of the eight-week training course?

b. 6 weeks after stopping the weight training?

103. The Pace of Life.   According to the study discussed in Example 7, how much faster does the average pedestrian in New York City (population 8,392,000) walk than the average pedestrian in Atlanta (population 541,000)? (Source: infoplease.com)

104. Maturity Levels. The function $P(a) = 41.0 + 20.4 \ln a$ approximates the percent of adult height attained by an early-maturing girl of age a years, for $1 \le a \le 18$. The function $P(a) = 37.5 + 20.2 \ln a$ does the same for a late-maturing girl. Find the difference in percent of their adult height for both maturity types on their 10th birthday. (Source: Growth, Maturation, and Physical Activity, Human Kinetic Books, Robert Malina)

WRITING

105. Explain why the graph of $y = e^x - 5$ is five units below the graph of $y = e^x$.

106. A feature article in a newspaper stated that the sport of snowboarding was growing *exponentially*. Explain what the author of the article meant by that.

107. As of 2007, the population growth rate for Russia was -0.37% annually. What are some of the consequences for a country that has a negative population growth?

108. What is e?

109. Explain the difference between the functions $f(x) = \log x$ and $g(x) = \ln x$.

110. How are the functions $f(x) = \ln x$ and $g(x) = e^x$ related?

111. Explain why $\ln e = 1$.

112. Why is $f(x) = \ln x$ called the natural logarithmic function?

113. A table of values for $f(x) = \ln x$ is shown in figure (a) below. Explain why ERROR appears in the Y_1 column for the first three entries.

114. The graphs of $f(x) = \ln x$, $g(x) = e^x$, and $y = x$ are shown in figure (b) below. Describe the relationship between the graphs in words.

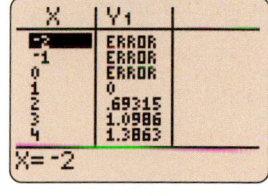

(a)

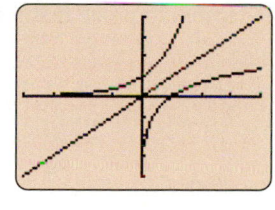

(b)

REVIEW

Simplify each expression. Assume that all variables represent positive numbers.

115. $\sqrt{240x^5}$

116. $\sqrt[3]{-125x^5y^4}$

117. $4\sqrt{48y^3} - 3y\sqrt{12y}$

118. $\sqrt[4]{48z^5} + \sqrt[4]{768z^5}$

CHALLENGE PROBLEMS

119. Without using a calculator, determine whether the statement $e^e > e^3$ is true or false. Explain your reasoning.

120. Graph the function defined by the equation $f(x) = \dfrac{e^x + e^{-x}}{2}$ from $x = -2$ to $x = 2$. The graph will look like a parabola, but it is not. The graph, called a **catenary,** is important in the design of power distribution networks, because it represents the shape of a uniform flexible cable whose ends are suspended from the same height.

121. If $e^{t+5} = ke^t$, find k.

122. If $e^{5t} = k^t$, find k.

123. Use the formula $P = P_0 e^{rt}$ to verify that P will be twice P_0 when $t = \dfrac{\ln 2}{r}$.

124. Use the formula $P = P_0 e^{rt}$ to verify that P will be three times as large as P_0 when $t = \dfrac{\ln 3}{r}$.

125. Use a graphing calculator to graph the function $f(x) = \dfrac{1}{1 + e^{-2x}}$. Describe its graph in words.

 126. Food Shortages. Suppose that a country with a population of 1,000 people is growing according to the formula $P = 1,000e^{0.02t}$ where t is in years. Furthermore, assume that the food supply F, measured in adequate food per day per person, is growing linearly according to the formula $F = 30.625t + 2,000$ (t is time in years). Use a graphing calculator to determine in how many years the population will outstrip the food supply.

Properties of Logarithms

ARE YOU READY?

The following problems review some basic skills that are needed when working with properties of logarithms.

1. Evaluate: $\log_2 8 + \log_2 1$

2. Evaluate: $\log 10{,}000 - \log 10$

3. Evaluate: $9 \log_3 \dfrac{1}{3}$

4. Evaluate: $\dfrac{\log_7 49}{\log_7 7}$

5. **a.** Write $\sqrt{x}$ using a fractional exponent.

   b. Write $(x - 2)^{1/2}$ using radical notation.

6. Use a calculator to find $\ln 5$. Round to four decimal places.

Since a logarithm is an exponent, we would expect there to be properties of logarithms just as there are properties of exponents. In this section, we will introduce seven properties of logarithms and use them to simplify and expand logarithmic expressions.

1 Use the Four Basic Properties of Logarithms.

The first four properties of logarithms follow directly from the definition of logarithm.

Properties of Logarithms

For all positive numbers b, where $b \neq 1$,

1. $\log_b 1 = 0$ 2. $\log_b b = 1$ 3. $\log_b b^x = x$ 4. $b^{\log_b x} = x \quad (x > 0)$

We can use the definition of logarithm to prove that these properties are true.

1. $\log_b 1 = 0$, because $b^0 = 1$.   Read as "the log base b of 1 equals 0."
2. $\log_b b = 1$, because $b^1 = b$.   Read as "the log base b of b equals 1."
3. $\log_b b^x = x$, because $b^x = b^x$.   Read as "the log base b of b to the x power equals x."
4. $b^{\log_b x} = x$, because $\log_b x$ is the exponent to which b is raised to get x.
   Read as "b raised to the log base b of x power equals x."

Properties 3 and 4 also indicate that the composition of the exponential and logarithmic functions (in both directions) is the identity function. This is expected, because the exponential and logarithmic functions are inverse functions.

EXAMPLE 1   Simplify: **a.** $\log_5 1$ **b.** $\log_3 3$ **c.** $\ln e^3$ **d.** $6^{\log_6 7}$

Strategy   We will compare each logarithmic expression to the left side of the previous four properties of logarithms.

Why   When we get a match, the property will provide the answer.

Solution   **a.** By property 1, $\log_5 1 = 0$, because $5^0 = 1$.

b. By property 2, $\log_3 3 = 1$, because $3^1 = 3$.

c. By property 3, $\ln e^3 = 3$, because $e^3 = e^3$.

d. By property 4, $6^{\log_6 7} = 7$, because $\log_6 7$ is the power to which 6 is raised to get 7.

Self Check 1   Simplify: **a.** $\log_4 1$ **b.** $\log_4 4$ **c.** $\log_2 2^4$ **d.** $5^{\log_5 2}$

Now Try   Problems 19, 21, 23, and 27

2 Use the Product Rule for Logarithms.

The next property of logarithms is related to the product rule for exponents: $x^m \cdot x^n = x^{m+n}$.

The Product Rule for Logarithms	The logarithm of a product equals the sum of the logarithms of the factors. For all positive real numbers M, N, and b, where $b \neq 1$, $$\log_b MN = \log_b M + \log_b N$$ Read as "the log base b of M times N equals the log base b of M plus the log base b of N."

As we apply properties of logarithms to rewrite expressions, we assume that all variables represent positive numbers.

EXAMPLE 2 Write each expression as a sum of logarithms. Then simplify, if possible. **a.** $\log_2(2 \cdot 7)$ **b.** $\log 100x$ **c.** $\log_5 125yz$

Strategy In each case, we will use the product rule for logarithms.

Why We use the product rule because each of the logarithmic expressions has the form $\log_b MN$.

Solution **a.** To avoid any confusion, the product $2 \cdot 7$ is written within parentheses.

$$\log_2(2 \cdot 7) = \log_2 2 + \log_2 7 \quad \text{Read as "the log base 2 of 2 times 7." The log of the product } 2 \cdot 7 \text{ is the sum of the logs of its two factors.}$$

$$= 1 + \log_2 7 \quad \text{Simplify: By property 2, } \log_2 2 = 1.$$

b. Recall that $100x$ means $100 \cdot x$.

$$\log 100x = \log 100 + \log x \quad \text{The log of the product } 100x \text{ is the sum of the logs of its two factors.}$$

$$= 2 + \log x \quad \text{Simplify: By property 3, } \log 100 = \log 10^2 = 2.$$

c. The product $125yz$ has three factors: $125 \cdot y \cdot z$.

$$\log_5 125yz = \log_5 125 + \log_5 y + \log_5 z \quad \text{The log of the product } 125yz \text{ is the sum of the logs of its three factors.}$$

$$= 3 + \log_5 y + \log_5 z \quad \text{Simplify: By property 3, } \log_5 125 = \log_5 5^3 = 3.$$

Self Check 2 Write each expression as the sum of logarithms. Then simplify, if possible. **a.** $\log_3(3 \cdot 4)$ **b.** $\log 1{,}000y$ **c.** $\log_5 25cd$

Now Try Problems 31 and 35

Success Tip

Your eyes should go immediately to the "input" of each log function, and then ask:

$\log 100x$

"What are the factors?"

$\log_5 125yz$

PROOF To prove the product rule for logarithms, we let $x = \log_b M$, $y = \log_b N$, and use the definition of logarithm to write each equation in exponential form.

$$M = b^x \text{ and } N = b^y$$

Then $MN = b^x b^y$, and a property of exponents gives

$$MN = b^{x+y} \quad \text{Keep the base and add the exponents: } b^x b^y = b^{x+y}.$$

We write this exponential equation in logarithmic form as

$$\log_b MN = x + y$$

Substituting the values of x and y completes the proof.

$$\log_b MN = \log_b M + \log_b N \quad \text{This is the product rule for logarithms.}$$

The log of a sum **does not equal** the sum of the logs. The log of a difference **does not equal** the difference of the logs.

CAUTION By the product rule, the logarithm of a *product* is equal to the *sum* of the logarithms. The logarithm of a sum or a difference usually does not simplify. **Do not incorrectly apply the distributive property in such cases.** In general,

$$\log_b (M + N) \neq \log_b M + \log_b N \quad \text{and} \quad \log_b (M - N) \neq \log_b M - \log_b N$$

For example,

$$\log_2 (2 + 7) \neq \log_2 2 + \log_2 7 \quad \text{and} \quad \log (100 - y) \neq \log 100 - \log y$$

Using Your Calculator ▶ Verifying Properties of Logarithms

We can use a calculator to illustrate the product rule for logarithms by showing that

$$\log (3.7 \cdot 15.9) = \log 3.7 + \log 15.9$$

We calculate the left and right sides of the equation separately and compare the results. To use a scientific calculator to find $\log (3.7 \cdot 15.9)$, we enter

3.7 $\boxed{\times}$ 15.9 $\boxed{=}$ $\boxed{\text{LOG}}$ `1.769598848`

To find $\log 3.7 + \log 15.9$, we enter

3.7 $\boxed{\text{LOG}}$ $\boxed{+}$ 15.9 $\boxed{\text{LOG}}$ $\boxed{=}$ `1.769598848`

Since the screen displays for the left and right sides are equal, that would suggest that the equation $\log (3.7 \cdot 15.9) = \log 3.7 + \log 15.9$ is true.

3 Use the Quotient Rule for Logarithms.

The next property of logarithms is related to the quotient rule for exponents: $\frac{x^m}{x^n} = x^{m-n}$.

The Quotient Rule for Logarithms

The logarithm of a quotient equals the difference of the logarithms of the numerator and denominator. For all positive real numbers M, N, and b, where $b \neq 1$,

$$\log_b \frac{M}{N} = \log_b M - \log_b N$$

Read as "the log base b of M divided by N equals the log base b of M minus the log base b of N."

The proof of the quotient rule for logarithms is similar to the proof for the product rule for logarithms.

EXAMPLE 3 Write each expression as a difference of logarithms. Then simplify, if possible.

a. $\ln \frac{10}{7}$ **b.** $\log_4 \frac{x}{64}$

Strategy In both cases, we will apply the quotient rule for logarithms.

Why We use the quotient rule because each of the logarithmic expressions has the form $\log_b \frac{M}{N}$.

Solution **a.** $\ln \frac{10}{7} = \ln 10 - \ln 7$ Recall that $\ln \frac{10}{7}$ is a natural logarithm and means $\log_e \frac{10}{7}$. The log of the quotient $\frac{10}{7}$ is the difference of the logs of its numerator and denominator.

b. $\log_4 \dfrac{x}{64} = \log_4 x - \log_4 64$ 〔The log of the quotient $\frac{x}{64}$ is the difference of the logs of its numerator and denominator.〕

$\qquad\qquad\qquad = \log_4 x - 3$ 〔Simplify: $\log_4 64 = \log_4 4^3 = 3$.〕

> **Self Check 3** Write each expression as a difference of logarithms. Then simplify, if possible. **a.** $\log_6 \dfrac{6}{5}$ **b.** $\ln \dfrac{y}{100}$
>
> **Now Try** ▶ Problem 39

CAUTION By the quotient rule, the logarithm of a *quotient* is equal to the *difference* of the logarithms. The logarithm of a quotient is not the quotient of the logarithms:

$$\log_b \frac{M}{N} \ne \frac{\log_b M}{\log_b N}$$

For example,

$$\ln \frac{10}{7} \ne \frac{\ln 10}{\ln 7} \quad \text{and} \quad \log_4 \frac{x}{64} \ne \frac{\log_4 x}{\log_4 64}$$

〔Using a calculator, we have:
$\ln \frac{10}{7} \approx 0.356674944$
$\frac{\ln 10}{\ln 7} \approx 1.183294662$〕

In the next example, the product and quotient rules for logarithms are used in combination to rewrite an expression.

EXAMPLE 4 Write $\log \dfrac{xy}{10z}$ as the sum and/or difference of logarithms of a single quantity. Then simplify, if possible.

Strategy We will use the quotient rule for logarithms and then the product rule.

Why We use the quotient rule because $\log \dfrac{xy}{10z}$ has the form $\log_b \dfrac{M}{N}$. We later use the product rule because the numerator and denominator of $\dfrac{xy}{10z}$ contain products.

Solution We begin by applying the quotient rule for logarithms.

$$\log \frac{xy}{10z} = \log xy - \log 10z$$ 〔The log of a quotient is the difference of the logs.〕

$$= \log x + \log y - (\log 10 + \log z)$$ 〔The log of a product is the sum of the logs.〕

〔Write parentheses here so that both terms of the sum $\log 10 + \log z$ are subtracted.〕

$$= \log x + \log y - \log 10 - \log z$$ 〔Change the sign of each term of $\log 10 + \log z$ and drop the parentheses.〕

$$= \log x + \log y - 1 - \log z$$ 〔Simplify: $\log 10 = 1$.〕

> **Success Tip**
>
> Your eyes should go immediately to the "input" of the log function, and identify any quotients and/or products.
>
> $\log \dfrac{xy}{10z}$
> ↑

> **Self Check 4** Write $\log_b \dfrac{x}{yz}$ as the sum and/or difference of logarithms of a single quantity. Then simplify, if possible.
>
> **Now Try** ▶ Problem 45

---

4 Use the Power Rule for Logarithms.

The next property of logarithms is related to the power rule for exponents: $(x^m)^n = x^{mn}$.

The Power Rule for Logarithms	The logarithm of a number raised to a power equals the power times the logarithm of the number.
	For all positive real numbers M and b, where $b \neq 1$, and any real number p,
	$$\log_b M^p = p \log_b M$$ Read as "the log base b of M to the p power equals p times the log base b of M."

EXAMPLE 5   Write each logarithm without an exponent or a square root: **a.** $\log_5 6^2$ **b.** $\log \sqrt{10}$

Strategy   In each case, we will use the power rule for logarithms.

Why   We use the power rule because $\log_5 6^2$ has the form $\log_b M^p$, as will $\log \sqrt{10}$ if we write $\sqrt{10}$ as $10^{1/2}$.

Solution   **a.** $\log_5 6^2 = \mathbf{2} \log_5 6$ The log of a power is equal to the power times the log. Write the exponent 2 in front of $\log_5 6$.

b. $\log \sqrt{10} = \log 10^{1/2}$ Write $\sqrt{10}$ using a fractional exponent: $\sqrt{10} = (10)^{1/2}$.

$\qquad\qquad = \dfrac{\mathbf{1}}{\mathbf{2}} \log 10$ The log of a power is equal to the power times the log. Write the exponent $\frac{1}{2}$ in front of log 10.

$\qquad\qquad = \dfrac{1}{2}$ Simplify: log 10 = 1.

Self Check 5   Write each logarithm without an exponent or a cube root:

a. $\ln x^4$ **b.** $\log_2 \sqrt[3]{3}$

Now Try ▶ Problems 51 and 53

PROOF   To prove the power rule, we let $x = \log_b M$, write the expression in exponential form, and raise both sides to the pth power:

$\qquad M = b^x$

$\qquad (M)^p = (b^x)^p$ Raise both sides to the pth power.

$\qquad M^p = b^{px}$ Keep the base and multiply the exponents.

Using the definition of logarithms gives

$\qquad \log_b M^p = p x$

Substituting $\log_b M$ for x completes the proof.

$\qquad \log_b M^p = p \log_b M$ This is the power rule.

It is often necessary to use more than one rule for logarithms to **expand a logarithmic expression.**

EXAMPLE 6   Write each expression as the sum and/or difference of logarithms of a single quantity:

a. $\log_b x^2 y^3 z$ **b.** $\ln \dfrac{y^3 \sqrt[4]{x}}{z}$

Strategy   In part (a), we will use the product rule and the power rules for logarithms. In part (b), we will use the quotient rule, the product rule, and the power rule for logarithms.

Why In part (a), we first use the product rule because the expression has the form $\log_b MN$. In part (b), we first use the quotient rule because the expression has the form $\log_b \frac{M}{N}$.

Solution

a. The expression $\log_b x^2 y^3 z$ is the logarithm of a product.

$$\log_b x^2 y^3 z = \log_b x^2 + \log_b y^3 + \log_b z \qquad \text{The log of a product is the sum of the logs.}$$
$$= \mathbf{2} \log_b x + \mathbf{3} \log_b y + \log_b z \qquad \text{The log of a power is the power times the log.}$$

b. The expression $\ln \dfrac{y^3 \sqrt[4]{x}}{z}$ is the logarithm of a quotient.

$$\ln \dfrac{y^3 \sqrt[4]{x}}{z} = \mathbf{\ln y^3 \sqrt[4]{x}} - \ln z \qquad \begin{array}{l}\text{This is a natural log expression.}\\ \text{The log of a quotient is the difference of the logs.}\end{array}$$

$$= \mathbf{\ln y^3 + \ln \sqrt[4]{x}} - \ln z \qquad \text{The log of a product is the sum of the logs.}$$
$$= \ln y^3 + \ln x^{1/4} - \ln z \qquad \text{Write } \sqrt[4]{x} \text{ as } x^{1/4}.$$
$$= \mathbf{3} \ln y + \frac{1}{4} \ln x - \ln z \qquad \text{The log of a power is the power times the log.}$$

Self Check 6 Expand: $\log \sqrt[4]{\dfrac{x^3 y}{z}}$

Now Try ▶ Problems 59 and 61

5 Write Logarithmic Expressions as a Single Logarithm.

We also can use properties of logarithms in reverse to **condense logarithmic expressions** having two or more terms.

EXAMPLE 7 Write each logarithmic expression as one logarithm:

a. $16 \log_8 x + \frac{1}{3} \log_8 y$ **b.** $\frac{1}{2} \log_b (x - 2) - \log_b y + 3 \log_b z$

Strategy In part (a), we will use the power rule and product rule for logarithms in reverse. In part (b), we will use the power rule, the quotient rule, and the product rule for logarithms in reverse.

Why We use the power rule because we see expressions of the form $p \log_b M$. The $+$ symbol between logarithmic terms suggests that we use the product rule and the $-$ symbol between such terms suggests that we use the quotient rule.

Solution

a. We begin by using the power rule on both terms of the expression.

$$16 \log_8 x + \frac{1}{3} \log_8 y = \mathbf{\log_8 x^{16} + \log_8 y^{1/3}} \qquad \text{A power times a log is the log of the power.}$$
$$= \mathbf{\log_8 (x^{16} \cdot y^{1/3})} \qquad \text{The sum of two logs is the log of the product.}$$
$$= \log_8 x^{16} y^{1/3}$$
$$= \log_8 x^{16} \sqrt[3]{y} \qquad \text{Write } y^{1/3} \text{ using radical notation: } \sqrt[3]{y}.$$

b. The first and third terms of this expression can be rewritten using the power rule of logarithms. Note that the base of each logarithm is b. We do not need to know the value of b to apply properties of logarithms.

$$\frac{1}{2}\log_b (x-2) - \log_b y + 3\log_b z$$

$$= \log_b (x-2)^{1/2} - \log_b y + \log_b z^3 \qquad \text{A power times a log is the log of the power.}$$

$$= \log_b \frac{(x-2)^{1/2}}{y} + \log_b z^3 \qquad \text{The difference of two logs is the log of the quotient.}$$

$$= \log_b \frac{\sqrt{x-2}}{y} + \log_b z^3 \qquad \text{Write } (x-2)^{1/2} \text{ using a radical notation: } \sqrt{x-2}.$$

$$= \log_b \left(\frac{\sqrt{x-2}}{y} \cdot z^3 \right) \qquad \text{The sum of two logs is the log of the product.}$$

$$= \log_b \frac{z^3 \sqrt{x-2}}{y}$$

Self Check 7   Write the expression as one logarithm:

$$2\log_a x + \frac{1}{2}\log_a y - 2\log_a (x-y)$$

Now Try ▶   Problems 75 and 79

The properties of logarithms can be used when working with numerical values.

EXAMPLE 8   If $\log 2 \approx 0.3010$ and $\log 3 \approx 0.4771$, use properties of logarithms to find approximations for: **a.** $\log 6$ **b.** $\log 18$

Strategy   We will express 6 and 18 using factors of 2 and 3 and then use properties of logarithms to simplify each resulting expression.

Why   We express 6 and 18 using factors of 2 and 3 because we are given values of $\log 2$ and $\log 3$.

Solution   **a.** $\log 6 = \log (2 \cdot 3)$ Write 6 using the factors 2 and 3.

$$= \log 2 + \log 3 \qquad \text{The log of a product is the sum of the logs.}$$

$$\approx 0.3010 + 0.4771 \qquad \begin{array}{l}\text{Substitute the given values for log 2 and log 3.}\\ \text{We must now use an } \approx \text{ symbol. Do the addition.}\end{array}$$

$$\approx 0.7781$$

b. $\log 18 = \log (2 \cdot 3^2)$ Write 18 using the factors 2 and 3.

$$= \log 2 + \log 3^2 \qquad \text{The log of a product is the sum of the logs.}$$

$$= \log 2 + 2\log 3 \qquad \begin{array}{l}\text{The log of a power is the power times the log.}\\ \text{Write the exponent 2 in front of log 3.}\end{array}$$

$$\approx 0.3010 + 2(0.4771) \qquad \begin{array}{l}\text{Substitute the given value for log 2 and log 3.}\\ \text{We must now use an } \approx \text{ symbol.}\end{array}$$

$$\approx 1.2552 \qquad \text{Evaluate the expression.}$$

Self Check 8   See Example 8. Find approximations for each logarithm.
a. log 1.5   **b.** log 0.75

Now Try ▶ Problems 87 and 89

We summarize the properties of logarithms as follows.

Properties of Logarithms	If b, M, and N are positive real numbers, $b \neq 1$, and p is any real number,

1. $\log_b 1 = 0$ 2. $\log_b b = 1$
3. $\log_b b^x = x$ 4. $b^{\log_b x} = x$

5. $\log_b MN = \log_b M + \log_b N$ 6. $\log_b \dfrac{M}{N} = \log_b M - \log_b N$

7. $\log_b M^p = p \log_b M$

6 Use the Change-of-Base Formula.

Most calculators can find common logarithms (base 10) and natural logarithms (base e). If we need to find a logarithm with some other base, we can use the following **change-of-base formula.**

Change-of-Base Formula	For any logarithmic bases a and b, and any positive real number x,

$$\log_b x = \frac{\log_a x}{\log_a b}$$ *This formula converts a logarithm of one base to a ratio of logarithms of a different base.*

We can use any positive number other than 1 for base a in the change-of-base formula. However, we usually use 10 or e because of the capabilities of a standard calculator.

EXAMPLE 9   Find: $\log_3 5$

Strategy   To evaluate this base-3 logarithm, we will substitute into the change-of-base formula.

Why   We assume that the reader does not have a calculator that evaluates base-3 logarithms (at least not directly). Thus, the only alternative is to change the base.

Solution   To find $\log_3 5$, we substitute 3 for b, 10 for a, and 5 for x in the change-of-base formula and simplify:

Caution

Don't misapply the quotient rule: $\frac{\log_{10} 5}{\log_{10} 3}$ means $\log_{10} 5 \div \log_{10} 3$.

It is the expression $\log_{10} \frac{5}{3}$ that means $\log_{10} 5 - \log_{10} 3$.

$$\log_b x = \frac{\log_a x}{\log_a b}$$ *This is the change-of-base formula.*

$$\log_3 5 = \frac{\log_{10} 5}{\log_{10} 3}$$ *The old base is 3. The new base we want to introduce is 10. Substitute: $b = 3$, $x = 5$, and $a = 10$.*

$$\approx 1.464973521$$ *Approximate. On a scientific calculator, enter:*
5 LOG ÷ 3 LOG = .

To four decimal places, $\log_3 5 = 1.4650$. To check this result, use a calculator to verify that $3^{1.4650} \approx 5$.

Caution

Wait until the final calculation has been made to round. Don't round any values when performing intermediate calculations. That could make the final result incorrect because of a build-up of rounding errors.

We also can use the natural logarithm function (base e) in the change-of-base formula to find a base-3 logarithm.

$$\log_b x = \frac{\log_a x}{\log_a b}$$ This is the change-of-base formula.

$$\log_3 5 = \frac{\log_e 5}{\log_e 3}$$ The old base is 3. The new base we want to introduce is e.
Substitute: $b = 3$, $x = 5$, and $a = e$.

$$\log_3 5 = \frac{\ln 5}{\ln 3}$$ Write $\log_e 5$ as ln 5 and $\log_e 3$ as ln 3.

$$\approx 1.464973521$$ Approximate. On a scientific calculator, enter: 5 LN ÷ 3 LN = .

We obtain the same result.

Self Check 9   Find $\log_5 3$ to four decimal places.

Now Try ▶ Problem 95

PROOF   To prove the change-of-base formula, we begin with the equation $\log_b x = y$.

$$y = \log_b x$$

$$x = b^y$$ Change the equation from logarithmic to exponential form.

$$\log_a x = \log_a b^y$$ Take the base-a logarithm of both sides.

$$\log_a x = y \log_a b$$ The log of a power is the power times the log.

$$y = \frac{\log_a x}{\log_a b}$$ Divide both sides by $\log_a b$.

$$\log_b x = \frac{\log_a x}{\log_a b}$$ Refer to the first equation, $y = \log_b x$ and substitute $\log_b x$ for y.
This is the change-of-base formula.

---

7 Use Properties of Logarithms to Solve Application Problems.

In chemistry, common logarithms are used to express how basic or acidic a solution is. The more acidic a solution, the greater the concentration of hydrogen ions. (A **hydrogen ion** is a positively charged hydrogen atom missing its electron.) The concentration of hydrogen ions in a solution is commonly measured using the **pH scale.** The pH of a solution is defined as follows.

pH of a Solution   If $[H^+]$ is the hydrogen ion concentration in gram-ions per liter, then
$$pH = -\log[H^+]$$

EXAMPLE 10   **pH Meters.**   One of the most accurate ways to measure pH is with a probe and meter. What reading should the meter give for lemon juice if it has a hydrogen ion concentration $[H^+]$ of approximately 6.2×10^{-3} gram-ions per liter?

Strategy   We will substitute into the formula for pH and use the power rule for logarithms to simplify the right side.

Why   After substituting 6.2×10^{-3} for $[H^+]$ in $-\log[H^+]$, the resulting expression will have the form $\log_b M^p$.

Solution   Since lemon juice has approximately 6.2×10^{-3} gram-ions per liter, its pH is

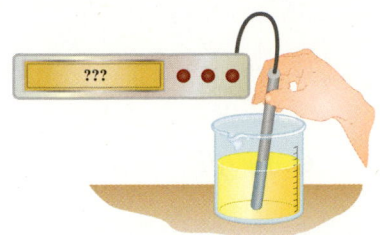

$$pH = -\log\,[\mathbf{H^+}]$$ This is the formula for pH. Read as "the opposite of the log of the hydrogen ion concentration."

$$pH = -\log(\mathbf{6.2 \times 10^{-3}})$$ Substitute 6.2×10^{-3} for $[H^+]$.

$$= -(\log 6.2 + \log 10^{-3})$$ The log of a product is the sum of the logs.

$$= -[\log 6.2 + (-3)\,\mathbf{\log 10}]$$ The log of a power is the power times the log.

$$= -[\log 6.2 + (-3) \cdot \mathbf{1}]$$ Evaluate: $\log 10 = 1$.

$$\approx 2.207608311$$ Use a calculator.

$$\approx 2.2$$ Round to the nearest tenth.

The meter should give a reading of approximately 2.2.

Self Check 10   **pH Meters.**   The hydrogen ion concentration range for a freshwater aquarium has a low-end value of 2.5×10^{-8}. Find the pH level that corresponds to this value.

Now Try   Problem 107

SECTION 9.6   **STUDY SET**

VOCABULARY

Fill in the blanks.

1. The logarithm of a _____, such as $\log_3 4x$, equals the sum of the logarithms of the factors.

2. The logarithm of a _____, such as $\log_2 \frac{5}{x}$, equals the difference of the logarithms of the numerator and denominator.

3. The logarithm of a number to a _____, such as $\log_4 5^3$, equals the power times the logarithm of the number.

4. The _____-of-base formula converts a logarithm of one base to a ratio of logarithms of a different base.

CONCEPTS

Fill in the blanks. In problem 6, also give the name of each rule.

5. a. $\log_b 1 =$ ▢   b. $\log_b b =$ ▢

   c. $\log_b b^x =$ ▢   d. $b^{\log_b x} =$ ▢

6. a. $\log_b MN = \log_b$ ▢ $+ \log_b$ ▢   _____ rule

   b. $\log_b \dfrac{M}{N} = \log_b M$ ▢ $\log_b N$ _____ rule

   c. $\log_b M^p = p \log_b$ ▢   _____ rule

   d. $\log_b x = \dfrac{\log_a ▢}{\log_▢ b}$ _____-of-_____ rule

Using Your Calculator: Verifying Properties of Logarithms.

Use a calculator to verify that each equation is true. See Using Your Calculator: Verifying Properties of Logarithms.

7. $\log(2.5 \cdot 3.7) = \log 2.5 + \log 3.7$

8. $\ln(2.25)^4 = 4 \ln 2.25$

9. $\ln \dfrac{11.3}{6.1} = \ln 11.3 - \ln 6.1$

10. $\log \sqrt{24.3} = \dfrac{1}{2} \log 24.3$

Match each expression with an equivalent expression from the list on the right.

11. $\log_3 10$ a. $\dfrac{\log 11}{\log 3}$

12. $\log_3 \dfrac{10}{11}$ b. $11 \log_3 10$

13. $\log_3 10^{11}$ c. $\log_3 5 + \log_3 2$

14. $\log_3 11$ d. $\log_3 10 - \log_3 11$

NOTATION

Complete each solution.

15. $\log_8 8a^3 = \log_8$ ▢ $+ \log_8$ ▢

    $= \log_8 8 +$ ▢ $\log_8 a$

    $=$ ▢ $+ 3 \log_8 a$

16. $\log \dfrac{r}{st} = \log r - \log($ ▢ $)$

    $= \log r - (\log$ ▢ $+ \log t)$

    $= \log r - \log s$ ▢ ▢

17. True or False?

$$\log 10{,}000x = \log (10{,}000x)$$

18. Fill in the blanks:

  a. $y^{1/3} = \sqrt[\Box]{}$

  b. $\sqrt[5]{x} = x^{\Box}$

GUIDED PRACTICE

In this Study Set, assume that all variables represent positive numbers and $b \neq 1$.

Evaluate each expression. See Example 1.

19. $\log_6 1$

20. $\log_9 9$

21. $\log_4 4^7$

22. $\ln e^8$

23. $5^{\log_5 10}$

24. $8^{\log_8 10}$

25. $\log_5 5^2$

26. $\log_4 4^2$

27. $\ln e$

28. $\log_7 1$

29. $\log_3 3^7$

30. $5^{\log_5 8}$

Write each logarithm as a sum. Then simplify, if possible. See Example 2.

31. $\log_2 (4 \cdot 5)$

32. $\log_3 (27 \cdot 5)$

33. $\log 25y$

34. $\log xy$

35. $\log 100pq$

36. $\log 1{,}000rs$

37. $\log 5xyz$

38. $\log 10abc$

Write each logarithm as a difference. Then simplify, if possible. See Example 3.

39. $\log \dfrac{100}{9}$

40. $\ln \dfrac{27}{e}$

41. $\log_6 \dfrac{x}{36}$

42. $\log_8 \dfrac{y}{8}$

Write each logarithm as the sum and/or difference of logarithms of a single quantity. Then simplify, if possible. See Example 4.

43. $\log \dfrac{7c}{2}$

44. $\log \dfrac{9t}{4}$

45. $\log \dfrac{10x}{y}$

46. $\log_2 \dfrac{ab}{4}$

47. $\ln \dfrac{exy}{z}$

48. $\ln \dfrac{5p}{e}$

49. $\log_8 \dfrac{1}{8m}$

50. $\log_6 \dfrac{1}{36r}$

Write each logarithm without an exponent or a radical symbol. Then simplify, if possible. See Example 5.

51. $\ln y^7$

52. $\ln z^9$

53. $\log \sqrt{5}$

54. $\log \sqrt[3]{7}$

55. $\log e^{-3}$

56. $\log e^{-1}$

57. $\log_7 \left(\sqrt[5]{100}\right)^3$

58. $\log_3 \left(\sqrt{10}\right)^5$

Write each logarithm as the sum and/or difference of logarithms of a single quantity. Then simplify, if possible. See Example 6.

59. $\log xyz^2$

60. $\log 4xz^2$

61. $\log_2 \dfrac{2\sqrt[3]{x}}{y}$

62. $\log_3 \dfrac{\sqrt[4]{x}}{yz}$

63. $\log x^3y^2$

64. $\log xy^2z^3$

65. $\log_b \sqrt{xy}$

66. $\log_b x^3\sqrt{y}$

67. $\log_a \dfrac{\sqrt[3]{x}}{\sqrt[4]{yz}}$

68. $\log_b \sqrt[4]{\dfrac{x^3y^2}{z^4}}$

69. $\ln x^{20}\sqrt{z}$

70. $\ln \sqrt{xy}$

71. $\log_5 \left(\dfrac{1}{t^3}\right)^d$

72. $\log_6 \left(\dfrac{1}{x^4}\right)^t$

73. $\ln \sqrt{ex}$

74. $\ln \sqrt[3]{e^2x}$

Write each logarithmic expression as one logarithm. See Example 7.

75. $\log_2 (x + 1) + 9 \log_2 x$

76. $2 \log x + \dfrac{1}{2} \log y$

77. $\log_3 x + \log_3 (x + 2) - \log_3 8$

78. $-2 \log x - 3 \log y + \log z$

79. $-3 \log_b x - 2 \log_b y + \dfrac{1}{2} \log_b z$

80. $3 \log_b (x + 1) - 2 \log_b (x + 2) + \log_b x$

81. $\dfrac{1}{3} \left[\log_b (M^2 - 9) - \log_b (M + 3)\right]$

82. $\dfrac{1}{4} \left[\log_r (n^2 - 16) - \log_r (n - 4)\right]$

83. $\ln \left(\dfrac{x}{z} + x\right) - \ln \left(\dfrac{y}{z} + y\right)$

84. $\ln (xy + y^2) - \ln (xz + yz) + \ln z$

85. $\dfrac{1}{2} \log_6 (x^2 + 1) - \log_6 (x^2 + 2)$

86. $\dfrac{1}{2} \log_8 (x^2 + 5) - \log_8 (x^2 + 5)$

Assume that $\log 4 \approx 0.6021$, $\log 7 \approx 0.8451$, and $\log 9 \approx 0.9542$. Use these values to evaluate each logarithm. See Example 8.

87. $\log_b 28$

88. $\log_b \dfrac{7}{4}$

89. $\log_b \dfrac{4}{63}$

90. $\log_b 36$

91. $\log_b \dfrac{63}{4}$

92. $\log_b 2.25$

93. $\log_b 64$

94. $\log_b 49$

Use the change-of-base formula to find each logarithm to four decimal places. See Example 9.

95. $\log_3 7$

96. $\log_7 3$

97. $\log_{1/3} 3$

98. $\log_{1/2} 6$

99. $\log_3 8$

100. $\log_5 10$

101. $\log_{\sqrt 2} \sqrt 5$

102. $\log_\pi e$

Look Alikes . . .
Which pair of expressions in each list are equivalent?

103. a. $\log(9 \cdot 3)$ **b.** $\log 9 \cdot \log 3$ **c.** $\log 9 + \log 3$

104. a. $\log_6 \dfrac{7}{9}$ **b.** $\dfrac{\log_6 7}{\log_6 9}$ **c.** $\log_6 7 - \log_6 9$

105. a. $\log_2 11^4$ **b.** $4\log_2 11$ **c.** $(\log_2 11)^4$

106. a. $\ln \sqrt t$ **b.** $\sqrt{\ln t}$ **c.** $\dfrac{1}{2}\ln t$

APPLICATIONS

107. pH of a Solution. Find the pH of a solution with a hydrogen ion concentration of 1.7×10^{-5} gram-ions per liter.

108. pH of Pickles. The hydrogen ion concentration of sour pickles is 6.31×10^{-4}. Find the pH.

109. Formulas. Use properties of logarithms to write the right side of each formula in an equivalent condensed form.
a. From sound engineering: $B = 10(\log I - \log I_0)$
b. From medicine: $T = \dfrac{1}{k}(\ln C_2 - \ln C_1)$

110. Doubling Time. The formula $t = \dfrac{\ln 2}{r}$ gives the time t for a population to double, where r is the annual rate of continuous compounding. Write the formula in an equivalent form so that it involves a common logarithm, not a natural logarithm.

WRITING

111. Explain the difference between a logarithm of a product and the product of logarithms.

112. How can the $\boxed{\text{LOG}}$ key on a calculator be used to find $\log_2 7$?

Explain why each statement is false.

113. $\log AB = (\log A)(\log B)$ **114.** $\log(A + B) = \log A + \log B$

115. $\log_b(A - B) = \dfrac{\log_b A}{\log_b B}$ **116.** $\dfrac{\log_b A}{\log_b B} = \log_b A - \log_b B$

117. Explain the meaning of the arrow:
$\log 7^{15}$

118. When is the change-of-base formula helpful?

REVIEW
Consider the line that passes through P(−2, 3) and Q(4, −4).

119. Find the slope of line PQ.

120. Find the distance between P and Q.

121. Find the midpoint of line segment PQ.

122. Write the equation in slope–intercept form of line PQ.

CHALLENGE PROBLEMS

123. Explain why $e^{\ln x} = x$.

124. If $\log_b 3x = 1 + \log_b x$, find b.

125. Show that $\log_{b^2} x = \dfrac{1}{2}\log_b x$.

126. Show that $e^{x \ln a} = a^x$.

SECTION 9.7

Exponential and Logarithmic Equations

OBJECTIVES

1 Solve exponential equations.

2 Solve logarithmic equations.

3 Use exponential and logarithmic equations to solve application problems.

ARE YOU READY?
The following problems review some basic skills that are needed when solving exponential and logarithmic equations.

1. Fill in the blanks: **a.** $16 = 2^{\boxed{}}$ **b.** $\dfrac{1}{125} = 5^{\boxed{}}$

2. Find $\dfrac{\log 12}{\log 9}$. Round to four decimal places.

3. Use the power rule for logarithms: $\log_3 8^x$

4. Evaluate: $\ln e$

5. Use a property of logarithms to write each expression as a single logarithm:
a. $\log_2 5 + \log_2 x$
b. $\ln 10 - \ln(2t + 1)$

6. Evaluate: $\log(-6)$

In earlier chapters, we have solved linear equations, absolute value equations, quadratic equations, rational equations, and radical equations. In this section, we will solve two new types of equations: exponential equations and logarithmic equations.

1 Solve Exponential Equations.

An **exponential equation** contains a variable in one of its exponents. Some examples of exponential equations are

$$3^{x+1} = 81, \qquad 6^{x-3} = 2^x, \qquad \text{and} \qquad e^{0.9t} = 8$$

If both sides of an exponential equation can be expressed as a power of the same base, we can use the following property to solve it.

Exponent Property of Equality	If two exponential expressions with the same base are equal, their exponents are equal. For any real number b, where $b \neq -1, 0,$ or 1, $$b^x = b^y \qquad \text{is equivalent to} \qquad x = y$$

EXAMPLE 1   Solve: $3^{x+1} = 81$

Strategy   We will express the right side of the equation as a power of 3.

Why   If each side of the equation is expressed as a power of the *same* base (in this case, 3), we can use the exponent property of equality to set the exponents equal and solve for x.

Solution

$$3^{x+1} = 81 \qquad \text{This is the equation to solve.}$$
$$3^{x+1} = 3^4 \qquad \text{Write 81 as a power of 3: } 81 = 3^4.$$
$$x + 1 = 4 \qquad \text{If two exponential expressions with the same base are equal, their exponents are equal.}$$
$$x = 3 \qquad \text{Solve for x by subtracting 1 from both sides.}$$

The solution is 3 and the solution set is $\{3\}$. To check this result, we substitute 3 for x in the original equation.

Check:
$$3^{x+1} = 81$$
$$3^{3+1} \stackrel{?}{=} 81$$
$$3^4 \stackrel{?}{=} 81$$
$$81 = 81 \qquad \text{True}$$

Self Check 1   Solve: $5^{3x-4} = 25$

Now Try ▶   Problem 21

EXAMPLE 2   Solve: $2^{x^2+2x} = \dfrac{1}{2}$

Strategy   We will express the right side of the equation as a power of 2.

Why   If each side of the equation is expressed as a power of the same base (in this case, 2), we can use the exponent property of equality to set the exponents equal and solve for x.

Solution

$$2^{x^2+2x} = \frac{1}{2}$$ This is the equation to solve.

$$2^{x^2+2x} = 2^{-1}$$ Write $\frac{1}{2}$ as a power of 2: $\frac{1}{2} = 2^{-1}$.

$$x^2 + 2x = -1$$ If two exponential expressions with the same base are equal, their exponents are equal.

$$x^2 + 2x + 1 = 0$$ Add 1 to both sides.

$$(x + 1)(x + 1) = 0$$ Factor the trinomial.

$$x + 1 = 0 \quad \text{or} \quad x + 1 = 0$$ Set each factor equal to 0.

$$x = -1 \quad | \quad x = -1$$ Solve each linear equation.

We see that the two solutions are the same. Thus, -1 is a repeated solution and the solution set is $\{-1\}$. Verify that -1 satisfies the original equation.

Self Check 2   Solve: $3^{x^2-2x} = \frac{1}{3}$

Now Try ▶ Problem 25

Using Your Calculator ▶ **Solving Exponential Equations Graphically**

To use a graphing calculator to approximate the solutions of $2^{x^2+2x} = \frac{1}{2}$ (see Example 2), we can subtract $\frac{1}{2}$ from both sides of the equation to get $2^{x^2+2x} - \frac{1}{2} = 0$ and graph the corresponding function $f(x) = 2^{x^2+2x} - \frac{1}{2}$ as shown in figure (a).

The solutions of $2^{x^2+2x} - \frac{1}{2} = 0$ are the x-coordinates of the x-intercepts of the graph of $f(x) = 2^{x^2+2x} - \frac{1}{2}$. Using the ZERO feature, we see in figure (a) that the graph has only one x-intercept, $(-1, 0)$. Therefore, -1 is the only solution of $2^{x^2+2x} - \frac{1}{2} = 0$.

We also can solve $2^{x^2+2x} = \frac{1}{2}$ using the INTERSECT feature found on most graphing calculators. After graphing $Y_1 = 2^{x^2+2x}$ and $Y_2 = \frac{1}{2}$, we select INTERSECT, which approximates the coordinates of the point of intersection of the two graphs. From the display shown in figure (b), we can conclude that the solution is -1. Verify this by checking.

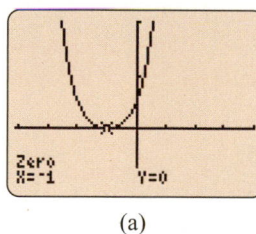   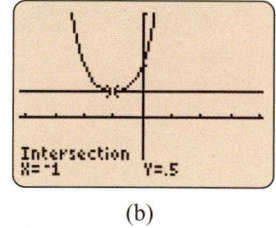

              (a)   (b)

When it is difficult or impossible to write each side of an exponential equation as a power of the same base, we often can use the following property of logarithms to solve the equation.

Logarithm Property of Equality

If two positive numbers are equal, the logarithms base-b of the numbers are equal. For any positive number b, where $b \neq 1$, and positive numbers x and y,

$$\log_b x = \log_b y \quad \text{is equivalent to} \quad x = y$$

EXAMPLE 3   Solve: $3^x = 5$

Strategy   We will take the base-10 logarithm on both sides of the equation.

Why   We can then use the power rule of logarithms to move the variable x from its current position as an exponent to a position as a factor.

Solution

Unlike Example 1, where we solved $3^{x+1} = 81$, it is not possible to write each side of $3^x = 5$ as an integer power of the same base 3. Instead, we use the logarithm property of equality and *take the logarithm on each side* to solve the equation. Although any base logarithm can be chosen, the calculations with a calculator are usually simplest if we use a common or natural logarithm.

$$3^x = 5$$ This is the equation to solve.

$$\log 3^x = \log 5$$ Take the common logarithm on each side.

$$x \log 3 = \log 5$$ The log of a power is the power times the log: $\log 3^x = x \log 3$. The variable x is now a factor of $x \log 3$ and not an exponent.

$$\frac{x \log 3}{\log 3} = \frac{\log 5}{\log 3}$$ $x \log 3$ means $x \cdot \log 3$. To isolate x, undo the multiplication by $\log 3$ by dividing both sides by $\log 3$.

$$\frac{x \overset{1}{\cancel{\log 3}}}{\underset{1}{\cancel{\log 3}}} = \frac{\log 5}{\log 3}$$ Simplify the left side by removing the common factor of $\log 3$ from the numerator and denominator.

$$x = \frac{\log 5}{\log 3}$$ This is the exact solution.

$$x \approx 1.464973521$$ Approximate. On a reverse-entry scientific calculator, press: 5 LOG ÷ 3 LOG = .

The exact solution is $\dfrac{\log 5}{\log 3}$ and the solution set is $\left\{ \dfrac{\log 5}{\log 3} \right\}$. Rounded to four decimal places, an approximate solution is 1.4650.

We also can take the natural logarithm on each side of the equation to solve for x.

$$3^x = 5$$

$$\ln 3^x = \ln 5$$ Take the natural logarithm on each side.

$$x \ln 3 = \ln 5$$ Use the power rule of logarithms: $\ln 3^x = x \ln 3$.

$$\frac{x \ln 3}{\ln 3} = \frac{\ln 5}{\ln 3}$$ $x \ln 3$ means $x \cdot \ln 3$. To isolate x, undo the multiplication by $\ln 3$ by dividing both sides by $\ln 3$.

$$\frac{x \overset{1}{\cancel{\ln 3}}}{\underset{1}{\cancel{\ln 3}}} = \frac{\ln 5}{\ln 3}$$ Simplify the left side by removing the common factor of $\ln 3$ from the numerator and denominator.

$$x = \frac{\ln 5}{\ln 3}$$ This is the exact solution.

$$x \approx 1.464973521$$ Approximate. On a scientific calculator, press: 5 LN ÷ 3 LN = .

The result is the same using the natural logarithm. To check the approximate solution, we substitute 1.4650 for x in 3^x and see if $3^{1.4650}$ is approximately 5.

Check:
$$3^x = 5$$
$$3^{1.4650} \overset{?}{=} 5$$
$$5.000145454 \approx 5$$ On a scientific calculator, press: 3 y^x 1.4650 = .

Self Check 3   Solve: $5^x = 4$

Now Try ▶   Problem 29

EXAMPLE 4   Solve: $6^{x-3} = 2^x$

Strategy   We will take the common logarithm on both sides of the equation.

Why   We can then use the power rule of logarithms to move the expressions $x - 3$ and x from their current positions as exponents to positions as factors.

Solution

$$6^{x-3} = 2^x$$ This is the equation to solve.

$$\log 6^{x-3} = \log 2^x$$ Take the common logarithm on each side.

$$(x - 3)\log 6 = x \log 2$$ The log of a power is the power times the log. The expression $x - 3$ is now a factor of $(x - 3)\log 6$ and not an exponent.

$$x \log 6 - 3 \log 6 = x \log 2$$ Distribute the multiplication by log 6.

$$x \log 6 - x \log 2 = 3 \log 6$$ To get the terms involving x on the left side, add 3 log 6 and subtract x log 2 on both sides.

$$x(\log 6 - \log 2) = 3 \log 6$$ Factor out x on the left side.

$$\frac{x \overset{1}{(\log 6 - \log 2)}}{\underset{1}{\log 6 - \log 2}} = \frac{3 \log 6}{\log 6 - \log 2}$$ To isolate x, undo the multiplication by log 6 − log 2 by dividing both side by log 6 − log 2. Then simplify the left side.

$$x = \frac{3 \log 6}{\log 6 - \log 2}$$ This is the exact solution.

$$x \approx 4.892789261$$ Approximate. On a reverse-entry scientific calculator, press:

3 ✕ 6 LOG ÷ (6 LOG − 2 LOG) = .

The solution is $\dfrac{3 \log 6}{\log 6 - \log 2}$ and the solution set is $\left\{ \dfrac{3 \log 6}{\log 6 - \log 2} \right\}$. To four decimal places, an approximate solution is 4.8928. To check, we substitute 4.8928 for each x in $6^{x-3} = 2^x$. The resulting values on the left and right sides of the equation should be approximately equal.

Self Check 4   Solve: $5^{x-2} = 3^x$

Now Try ▶   Problem 33

When an exponential equation involves an exponential expression with base e, it is easiest to take the natural logarithm on both sides to solve the equation.

EXAMPLE 5   Solve: $e^{0.9t} = 10$

Strategy   We will take the natural (base-e) logarithm on both sides of the equation.

Why   We can then use the power rule of logarithms to move the expression $0.9t$ from its current position as an exponent to a position as a factor.

Solution   The exponential expression on the left side has base e. In such cases, the calculations are easier when we take the natural logarithm of each side.

$$e^{0.9t} = 10$$ This is the equation to solve.

$$\ln e^{0.9t} = \ln 10$$ Take the natural logarithm on each side.

$$0.9t \ln e = \ln 10$$ Use the power rule of logarithms: $\ln e^{0.9t} = 0.9t \ln e$. The expression 0.9t is now a factor of 0.9t ln e and not an exponent.

$$0.9t \cdot 1 = \ln 10 \qquad \text{Simplify: } \ln e = 1.$$

$$0.9t = \ln 10 \qquad \text{Simplify the left side.}$$

$$t = \frac{\ln 10}{0.9} \qquad \text{To isolate } t, \text{ undo the multiplication by 0.9 by dividing both sides by 0.9.}$$

$$t \approx 2.558427881 \qquad \text{Approximate. On a reverse-entry scientific calculator, press:}$$
$$\text{10 } \boxed{\text{LN}} \div .9 \boxed{=}.$$

The exact solution is $\dfrac{\ln 10}{0.9}$. To four decimal places, an approximate solution is 2.5584. Verify this by using a calculator to show that $e^{0.9(\mathbf{2.5584})} \approx 10$.

Self Check 5   Solve: $e^{2.1t} = 35$

Now Try ▶   Problem 37

Strategy for Solving Exponential Equations	1. Isolate one of the exponential expressions in the equation.
	2. If both sides of the equation can be written as exponential expressions with the same base, do so. Then set the exponents equal and solve the resulting equation.
	3. If step 2 is difficult or impossible, take the common or natural logarithm on both sides. Use the power rule of logarithms to write the variable exponent as a factor, and then solve the resulting equation.
	4. Check the results in the original equation.

2   Solve Logarithmic Equations.

A **logarithmic equation** is an equation with a logarithmic expression that contains a variable. Some examples of logarithmic equations are

$$\log 5x = 3, \qquad \log (3x + 2) = \log (2x - 3), \qquad \text{and} \qquad \log_2 7 - \log_2 x = 5$$

Some logarithmic equations can be solved by rewriting them in equivalent exponential form.

EXAMPLE 6   Solve: $\log 5x = 3$

Strategy   Recall that $\log 5x = \log_{10} 5x$. To solve $\log 5x = 3$, we will instead write and solve an equivalent base-10 exponential equation.

Why   The resulting exponential equation is easier to solve because the variable term is isolated on one side.

Solution

$$\log 5x = 3 \qquad \text{This is the equation to solve.}$$

$$\log_{10} 5x = 3 \qquad \text{The base of the logarithm is 10.}$$

$$10^3 = 5x \qquad \text{Write the equivalent base-10 exponential equation.}$$

$$1{,}000 = 5x \qquad \text{Simplify: } 10^3 = 1{,}000.$$

$$200 = x \qquad \text{To isolate } x, \text{ divide both sides by 5.}$$

The solution is 200 and the solution set is $\{200\}$.

Caution
Always check your solutions to a logarithmic equation to identify any extraneous solutions.

Check:

$$\log 5x = 3 \qquad \text{This is the original equation.}$$

$$\log 5(\mathbf{200}) \overset{?}{=} 3 \qquad \text{Substitute 200 for } x.$$

$$\log 1{,}000 \overset{?}{=} 3 \qquad \text{Multiply } 5(200) = 1{,}000.$$

$$3 = 3 \qquad \text{Evaluate: } \log 1{,}000 = \log 10^3 = 3.$$

> **Self Check 6**   Solve: $\log_2 (x - 3) = -1$
>
> **Now Try** ▶ Problem 41

There is a possibility of obtaining **extraneous solutions** when solving logarithmic equations. Always discard any possible solutions that produce the logarithm of a negative number or the logarithm of 0 in the original equation.

EXAMPLE 7   Solve: $\log(3x + 2) = \log(2x - 3)$

Strategy   We will use the logarithmic property of equality to see that $3x + 2 = 2x - 3$.

Why   We can use the logarithm property of equality because the given equation, $\log(3x + 2) = \log(2x - 3)$, has the form $\log_b x = \log_b y$.

Solution

$\log(3x + 2) = \log(2x - 3)$ This is the equation to solve.

$3x + 2 = 2x - 3$ If the logarithms of two numbers are equal, the numbers are equal.

$x + 2 = -3$ Subtract 2x from both sides.

$x = -5$ To isolate x, subtract 2 from both sides.

Check:

$\log(3x + 2) = \log(2x - 3)$ This is the original equation.

$\log[3(-5) + 2] \stackrel{?}{=} \log[2(-5) - 3]$ Substitute −5 for x.

$\log(-13) \stackrel{?}{=} \log(-13)$ Evaluate within brackets.

Recall that $\log(-13)$ is undefined.

Caution

Don't make this error of trying to "distribute" log:

$\log(3x + 2)$

The notation log is not a number, it is the name of a function and cannot be distributed.

Since the logarithm of a negative number does not exist, the proposed solution of -5 must be discarded. This equation has no solution. Its solution set is $\varnothing$.

> **Self Check 7**   Solve: $\log(5x + 14) = \log(7x - 2)$
>
> **Now Try** ▶ Problem 49

In Examples 8 and 9, we will use the product and quotient rules of logarithms to "condense" one side of the equation first, before solving for the variable.

EXAMPLE 8   Solve: $\log x + \log(x - 3) = 1$

Strategy   We will use the product rule for logarithms in reverse: The sum of two logarithms is equal to the logarithm of a product. Then we will write and solve an equivalent exponential equation.

Why   We use the product rule of logarithms because the left side of the equation, $\log x + \log(x - 3)$, has the form $\log_b M + \log_b N$.

Solution

$\log x + \log(x - 3) = 1$ This is the equation to solve.

$\log x(x - 3) = 1$ On the left side, use the product rule of logarithms.

$\log_{10} x(x - 3) = 1$ The base of the logarithm is 10.

$x(x - 3) = 10^1$ Write the equivalent base-10 exponential equation.

$x^2 - 3x - 10 = 0$ Distribute the multiplication by x, and then subtract 10 from both sides.

$(x + 2)(x - 5) = 0$ Factor the trinomial.

$x + 2 = 0$ or $x - 5 = 0$ Set each factor equal to 0.

$x = -2$ $x = 5$

Success Tip

The objective is to use the product rule to "condense" the left side of the equation. We want to write an equivalent equation in which the variable x appears in only a single logarithmic expression.

Check: The number -2 is not a solution because it does not satisfy the equation (a negative number does not have a logarithm). We will check the other result, 5.

$$\log x + \log (x - 3) = 1 \qquad \text{This is the original equation.}$$

$$\log 5 + \log (5 - 3) \overset{?}{=} 1 \qquad \text{Substitute 5 for x.}$$

$$\log 5 + \log 2 \overset{?}{=} 1 \qquad \text{Do the subtraction within the parentheses.}$$

$$\log 10 \overset{?}{=} 1 \qquad \text{Use the product rule of logarithms:}$$
$$\log 5 + \log 2 = \log (5 \cdot 2) = \log 10.$$

$$1 = 1 \qquad \text{Evaluate: } \log 10 = 1.$$

Since 5 satisfies the equation, it is the solution.

Self Check 8   Solve: $\log x + \log (x + 3) = 1$

Now Try ▶ Problem 53

Using Your **Calculator** ▶ Solving Logarithmic Equations Graphically

To use a graphing calculator to approximate the solutions of the logarithmic equation $\log x + \log (x - 3) = 1$ (see Example 8), we can subtract 1 from both sides of the equation to get $\log x + \log (x - 3) - 1 = 0$ and graph the corresponding function $f(x) = \log x + \log (x - 3) - 1$ as shown in figure (a). Since the solution of the equation is the x-value that makes $f(x) = 0$, the solution is the x-coordinate of the x-intercept of the graph. We can use the ZERO feature to find that this x-value is 5.

We also can solve $\log x + \log (x - 3) = 1$ using the INTERSECT feature. After graphing $Y_1 = \log x + \log (x - 3)$ and $Y_2 = 1$, we select INTERSECT, which approximates the coordinates of the point of intersection of the two graphs. From the display shown in figure (b), we can conclude that the solution is 5.

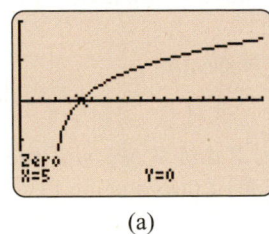
(a)

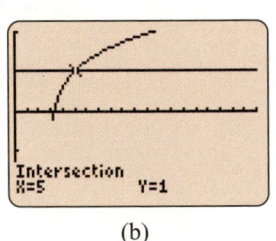
(b)

EXAMPLE 9   Solve: $\log_2 7 - \log_2 x = 5$

Strategy   We will use the quotient rule for logarithms in reverse: The difference of two logarithms is equal to the logarithm of a quotient. Then we will write and solve an equivalent exponential equation.

Why   We use the quotient rule for logarithms because the left side of the equation, $\log_2 7 - \log_2 x$, has the form $\log_b M - \log_b N$.

Solution

$$\log_2 7 - \log_2 x = 5 \qquad \text{This is the equation to solve.}$$

$$\log_2 \frac{7}{x} = 5 \qquad \text{On the left side, use the quotient rule for logarithms.}$$

$$\frac{7}{x} = 2^5 \qquad \text{Write the equivalent base-2 exponential equation.}$$

$$\frac{7}{x} = 32 \qquad \text{Evaluate: } 2^5 = 32.$$

$$7 = 32x \quad \text{To clear the equation of the fraction, multiply both sides by x.}$$

$$\frac{7}{32} = x \quad \text{To isolate x, divide both sides by 32.}$$

The solution is $\frac{7}{32}$. Verify that it satisfies the original equation.

Self Check 9　Solve:　$\log_2 9 - \log_2 x = 4$

Now Try　▶ Problem 57

---

3 Use Exponential and Logarithmic Equations to Solve Application Problems.

Recall from Section 9.6 that a **hydrogen ion** $[H^+]$ is the positively charged nucleus of a hydrogen atom, without its electron. The concentration of hydrogen ions in a solution is commonly measured using the **pH scale.** The pH of a solution is defined as follows.

---

pH of a Solution

If $[H^+]$ is the hydrogen ion concentration in gram-ions per liter, then

$$pH = -\log[H^+]$$

EXAMPLE 10

Hydrogen Ion Concentration.　Find the hydrogen ion concentration of seawater if its pH is 8.5.

Strategy　To find the hydrogen ion concentration, we will substitute 8.5 for pH in the formula $pH = -\log[H^+]$ and solve the resulting equation for $[H^+]$.

Why　After substituting for pH, the resulting logarithmic equation can be solved by solving an equivalent exponential equation.

Solution

$$\textbf{pH} = -\log[H^+] \quad \text{This is the formula for pH.}$$

$$\textbf{8.5} = -\log[H^+] \quad \text{Substitute 8.5 for pH.}$$

$$-8.5 = \log[H^+] \quad \text{Multiply both sides by } -1.$$

$$-8.5 = \log_{10}[H^+] \quad \text{The base of the logarithm is 10.}$$

$$[H^+] = \textbf{10}^{-8.5} \quad \text{Write the equivalent base-10 exponential equation.}$$

$$[H^+] \approx 0.000000003 \quad \text{Approximate. On a reverse-entry scientific calculator, press:}$$
$$10 \; \boxed{y^x} \; 8.5 \; \boxed{+/-} \; \boxed{=} \; .$$

Notation

In this logarithmic equation, the variable is the symbol $[H^+]$.

We can write the result using scientific notation:

$$[H^+] \approx 3.0 \times 10^{-9} \text{ gram-ions per liter}$$

Self Check 10　**Hydrogen Ion Concentration.**　Find the hydrogen ion concentration of a solution with a pH value of 4.8.

Now Try　▶ Problem 101

Experiments have determined the time it takes for half of a sample of a radioactive material to decompose. This time is a constant, called the material's **half-life.**

When living organisms die, the oxygen–carbon dioxide cycle common to all living things ceases, and carbon-14, a radioactive isotope with a half-life of 5,700 years, is no longer absorbed. By measuring the amount of carbon-14 present in an ancient object, archaeologists can estimate the object's age by using the radioactive decay formula.

| Radioactive Decay Formula | If A is the amount of radioactive material present at time t, A_0 was the amount present at $t = 0$, and h is the material's half-life, then $$A = A_0 2^{-t/h}$$ |

EXAMPLE 11   **Carbon-14 Dating.**   How old is a piece of wood that retains only one-third of its original carbon-14 content?

Strategy   If A_0 is the original carbon-14 content, then today's content $A = \frac{1}{3}A_0$. We will substitute $\frac{A_0}{3}$ for A and 5,700 for h in the radioactive decay formula and solve for t.

Why   The value of t is the estimated age of the piece of wood.

Solution   To find the time t when $A = \frac{1}{3}A_0$, we substitute $\frac{A_0}{3}$ for A and 5,700 for h in the radioactive decay formula and solve for t:

$A = A_0 2^{-t/h}$ This is the radioactive decay model.

$\dfrac{A_0}{3} = A_0 2^{-t/5,700}$ The half-life of carbon-14 is 5,700 years.

$1 = 3(2^{-t/5,700})$ Divide both sides by A_0 and multiply both sides by 3.

$\log 1 = \log 3(2^{-t/5,700})$ Take the common logarithm on both sides.

$0 = \log 3 + \log 2^{-t/5,700}$ $\log 1 = 0$, and use the product rule for logarithms.

$-\log 3 = -\dfrac{t}{5,700}\log 2$ Subtract $\log 3$ from both sides and use the power rule of logarithms.

$5,700\left(\dfrac{\log 3}{\log 2}\right) = t$ Multiply both sides by $-\frac{5,700}{\log 2}$.

$t \approx 9,034.286254$ Approximate. On a reverse-entry scientific calculator, press: 5700 ✕ 3 LOG ÷ 2 LOG = .

The piece of wood is approximately 9,000 years old.

Notation

The initial amount of radioactive material is represented by A_0, and it is read as "A sub 0."

Self Check 11   **Carbon-14 Dating.**   How old is a piece of wood that retains 25% of its original carbon-14 content?

Now Try ▶ Problem 103

When there is sufficient food and space available, populations of living organisms tend to increase exponentially according to the following growth model.

| Exponential Growth Model | If P is the population at some time t, P_0 is the initial population at $t = 0$, and k depends on the rate of growth, then $$P = P_0 e^{kt}$$ |

EXAMPLE 12   **Population Growth.**   The bacteria in a laboratory culture increased from an initial population of 500 to 1,500 in 3 hours. How long will it take for the population to reach 10,000?

Strategy   We will substitute 500 for P_0, 1,500 for P, and 3 for t in the exponential growth model and solve for k.

Why Once we know the value of k, we can substitute 10,000 for P, 500 for P_0, and the value of k in the exponential growth model and solve for the time t.

Solution

$$P = P_0 e^{kt} \qquad \text{This is the population growth formula.}$$
$$1{,}500 = 500(e^{k3}) \qquad \text{Substitute 1,500 for } P, 500 \text{ for } P_0, \text{ and 3 for } t.$$
$$3 = e^{3k} \qquad \text{Divide both sides by 500.}$$
$$3k = \ln 3 \qquad \text{Write the equivalent base-}e \text{ logarithmic equation.}$$
$$k = \frac{\ln 3}{3} \qquad \text{Divide both sides by 3.}$$

Notation

The initial population of bacteria is represented by P_0, and it is read as "P sub 0."

To find when the population will reach 10,000, we substitute 10,000 for P, 500 for P_0, and $\frac{\ln 3}{3}$ for k in the growth model and solve for t:

$$P = P_0 e^{kt}$$
$$10{,}000 = 500 e^{[(\ln 3)/3]t}$$
$$20 = e^{[(\ln 3)/3]t} \qquad \text{Divide both sides by 500.}$$
$$\left(\frac{\ln 3}{3}\right)t = \ln 20 \qquad \text{Write the equivalent base-}e \text{ logarithmic equation.}$$
$$t = \frac{3\ln 20}{\ln 3} \qquad \text{To isolate } t, \text{ multiply both sides by the reciprocal of } \tfrac{\ln 3}{3}, \text{ which is } \tfrac{3}{\ln 3}.$$
$$\approx 8.180499084 \qquad \text{Approximate. On a reverse-entry scientific calculator, press:}$$
$$3 \;\boxed{\times}\; 20 \;\boxed{\text{LN}}\; \boxed{\div}\; 3 \;\boxed{\text{LN}}\; \boxed{=}.$$

The culture will reach 10,000 bacteria in about 8 hours.

Self Check 12 **Population Growth.** In Example 12, how long will it take the population to reach 20,000?

Now Try ▶ Problem 115

SECTION 9.7 ▷ STUDY SET

VOCABULARY

Fill in the blanks.

1. An equation with a positive constant base and a variable in its exponent, such as $3^{2x} = 8$, is called an _____ equation.
2. An equation with a logarithmic expression that contains a variable, such as $\log_5(2x - 3) = \log_5(x + 4)$, is a _____ equation.

CONCEPTS

Fill in the blanks.

3. **a.** If two exponential expressions with the same base are equal, their exponents are _____.

   $b^x = b^y$ is equivalent to $\boxed{} = \boxed{}$.

   b. If the logarithms base-b of two numbers are equal, the numbers are _____.

   $\log_b x = \log_b y$ is equivalent to $\boxed{} = \boxed{}$.

4. The right side of the exponential equation $5^{x-3} = 125$ can be written as a power of $\boxed{}$.

5. If $6^{4x} = 6^{-2}$, then $4x = \boxed{}$.
6. **a.** Write the equivalent base-10 exponential equation for $\log(x + 1) = 2$.

   b. Write the equivalent base-e exponential equation for $\ln(x + 1) = 2$.

Fill in the blanks.

7. To solve $5^x = 2$, we can take the _____ of both sides of the equation to get $\log 5^x = \log 2$.
8. **a.** For $5^x = 2$, the power rule for logarithms provides a way of moving the variable x from its position as an _____ to a position as a factor.

   b. If the power rule for logarithms is used on the left side of the equation $\log 5^x = 2$, the resulting equation is $\boxed{} \log 5 = 2$.

9. If $e^{x+2} = 4$, then $\ln e^{x+2} = \boxed{}$.
10. Perform a check to determine whether -2 is a solution of $5^{2x+3} = \frac{1}{5}$.
11. Perform a check to determine whether 4 is a solution of $\log_5(x + 1) = 2$.

12. Use a calculator to determine whether 2.5646 is an approximate solution of $2^{2x+1} = 70$.

13. a. How do we solve $x \ln 3 = \ln 5$ for x?

    b. What is the exact solution?

    c. What is an approximate solution to four decimal places?

14. Use a property of logarithms to condense the left side of each equation to a single logarithm. **Do not solve.**

    a. $\log_5 x + \log_5 \cdot (4x - 1) = 1$

    b. $\log_3 4x - \log_3 7 = 2$

15. a. Find $\dfrac{\log 8}{\log 5}$. Round to four decimal places.

    b. Find $\dfrac{3 \ln 12}{\ln 4 - \ln 2}$. Round to four decimal places.

16. Does $\dfrac{\log 7}{\log 3} = \log 7 - \log 3$?

17. Complete each formula.

    a. pH $= -\ \boxed{}\ $ [H^+]

    b. Radioactive decay: $A = \boxed{}$

    c. Population growth: $P = \boxed{}$

18. Use the graphs below to estimate the solution of each equation.

    a. $2^x = 3^{-x+3}$

    b. $3 \log (x - 1) = 2 \log x$

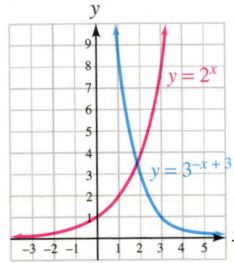

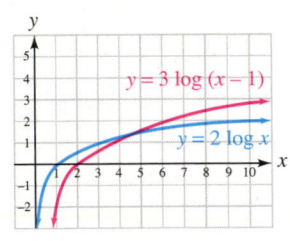

NOTATION

Complete each solution.

19. Solve: $2^x = 7$

$\boxed{}\, 2^x = \log 7$

$x\ \boxed{} = \log 7$

$x = \dfrac{\log 7}{\log 2}$

$x \approx \boxed{}$

20. Solve: $\log_2 (2x - 3) = \log_2 (x + 4)$

$\boxed{} = x + 4$

$x = \boxed{}$

GUIDED PRACTICE

Solve each equation. See Example 1.

21. $6^{x-2} = 36$

22. $3^{x+1} = 27$

23. $5^{4x} = \dfrac{1}{125}$

24. $8^{-2x+1} = \dfrac{1}{64}$

Solve each equation. See Example 2.

25. $2^{x^2 - 2x} = 8$

26. $3^{x^2 - 3x} = 81$

27. $3^{x^2 + 4x} = \dfrac{1}{81}$

28. $7^{x^2 + 3x} = \dfrac{1}{49}$

Solve each equation. Give the exact solution and an approximation to four decimal places. See Example 3.

29. $4^x = 5$

30. $7^x = 12$

31. $13^{x-1} = 2$

32. $5^{x+1} = 3$

Solve each equation. Give the exact solution and an approximation to four decimal places. See Example 4.

33. $2^{x+1} = 3^x$

34. $6^x = 7^{x-4}$

35. $5^{x-3} = 3^{2x}$

36. $8^{3x} = 9^{x+1}$

Solve each equation. Give the exact solution and an approximation to four decimal places. See Example 5.

37. $e^{2.9x} = 4.5$

38. $e^{3.3t} = 9.1$

39. $e^{-0.2t} = 14.2$

40. $e^{-0.7x} = 6.2$

Solve each equation. See Example 6.

41. $\log 2x = 4$

42. $\log 5x = 4$

43. $\log_3 (x - 3) = 2$

44. $\log_4 (2x - 1) = 3$

45. $\log (7 - x) = 2$

46. $\log (2 - x) = 3$

47. $\log \dfrac{1}{8} x = -2$

48. $\log \dfrac{1}{5} x = -3$

Solve each equation. See Example 7.

49. $\log (3 - 2x) = \log (x + 24)$

50. $\log (3x + 5) = \log (2x + 6)$

51. $\ln (3x + 1) = \ln (x + 7)$

52. $\ln (x^2 + 4x) = \ln (x^2 + 16)$

Solve each equation. See Example 8.

53. $\log x + \log (x - 48) = 2$

54. $\log x + \log (x + 9) = 1$

55. $\log_5 (4x - 1) + \log_5 x = 1$

56. $\log_2 (x - 7) + \log_2 x = 3$

Solve each equation. See Example 9.

57. $\log 5 - \log x = 1$

58. $\log 11 - \log x = 2$

59. $\log_3 4x - \log_3 7 = 2$

60. $\log_2 5x - \log_2 3 = 4$

TRY IT YOURSELF

Solve each equation. Give the exact solution and, when appropriate, an approximation to four decimal places.

61. $\log 2x = \log 4$

62. $\log 3x = \log 9$

63. $\ln x = 1$

64. $\ln x = 5$

65. $7^{x^2} = 10$

66. $8^{x^2} = 11$

67. $\log (x + 90) + \log x = 3$

68. $\log (x - 90) + \log x = 3$

69. $3^{x-6} = 81$

70. $5^{x+4} = 125$

71. $\log \dfrac{4x + 1}{2x + 9} = 0$

72. $\log \dfrac{2 - 5x}{2(x + 8)} = 0$

73. $15 = 9^{x+2}$

74. $29 = 5^{x-6}$

75. $\log x^2 = 2$

76. $\log x^3 = 3$

77. $\log (x - 6) - \log (x - 2) = \log \dfrac{5}{x}$

78. $\log (3 - 2x) - \log (x + 9) = 0$

79. $\log_3 x = \log_3 \left(\dfrac{1}{x}\right) + 4$

80. $\log_5 (7 + x) + \log_5 (8 - x) - \log_5 2 = 2$

81. $2 \log_2 x = 3 + \log_2 (x - 2)$

82. $2 \log_3 x - \log_3 (x - 4) = 2 + \log_3 2$

83. $\log (7y + 1) = 2 \log (y + 3) - \log 2$

84. $2 \log (y + 2) = \log (y + 2) - \log 12$

85. $e^{3x} = 9$

86. $e^{4x} = 60$

87. $\dfrac{\log (5x + 6)}{2} = \log x$

88. $\dfrac{1}{2} \log (4x + 5) = \log x$

Look Alikes . . .

89. a. $\log 5x = 1.7$ **b.** $\ln 5x = 1.7$

90. a. $\log_2 (x^2 - x) = 1$ **b.** $\log_6 (x^2 - x) = 1$

91. a. $4^{3x-5} = 90$ **b.** $e^{3x-5} = 90$

92. a. $\log x + 2 \log x = \log 8$ **b.** $\log x - 2 \log x = \log 8$

93. a. $\log_2 (x + 5) - \log_2 4x = \log_2 x$
   b. $\ln (x + 5) - \ln 4x = \ln x$

94. a. $5^{9x-1} = 125$ **b.** $5^{9x-1} = 124$

95. a. $\left(\dfrac{2}{3}\right)^{6-x} = \dfrac{8}{27}$ **b.** $\left(\dfrac{2}{3}\right)^{6-x} = \dfrac{16}{81}$

96. a. $\log x - \log (x + 7) = -1$ **b.** $\log x - \log (x + 7) = 1$

 Use a graphing calculator to solve each equation. If an answer is not exact, round to the nearest tenth. See Using Your Calculator: Solving Exponential Equations Graphically or Solving Logarithmic Equations Graphically.

97. $2^{x+1} = 7$

98. $3^x - 10 = 3^{-x}$

99. $\log x + \log (x - 15) = 2$

100. $\ln (2x + 5) - \ln 3 = \ln (x - 1)$

APPLICATIONS

101. Hydrogen Ion Concentration. Find the hydrogen ion concentration of a saturated solution of calcium hydroxide whose pH is 13.2.

102. Aquariums. The safe pH range for a freshwater aquarium is shown on the scale in the next column. Find the corresponding hydrogen ion concentration.

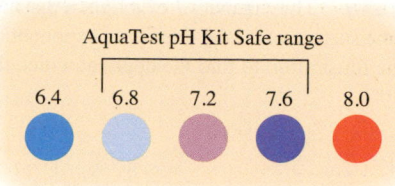

AquaTest pH Kit Safe range

6.4   6.8   7.2   7.6   8.0

103. Tritium Decay. The half-life of tritium is 12.4 years. How long will it take for 25% of a sample of tritium to decompose?

104. Radioactive Decay. In 2 years, 20% of a radioactive element decays. Find its half-life.

105. Thorium Decay. An isotope of thorium, written as ^{227}Th, has a half-life of 18.4 days. How long will it take for 80% of the sample to decompose?

106. Lead Decay. An isotope of lead, written as ^{201}Pb, has a half-life of 8.4 hours. How many hours ago was there 30% more of the substance?

107. Carbon-14 Dating. A bone fragment analyzed by archaeologists contains 60% of the carbon-14 that it is assumed to have had initially. How old is it?

108. Carbon-14 Dating. Only 10% of the carbon-14 in a small wooden bowl remains. How old is the bowl?

109. Compound Interest. If $500 is deposited in an account paying 8.5% annual interest, compounded semiannually, how long will it take for the account to increase to $800?

110. Continuous Compund Interest. In Exercise 109, how long will it take if the interest is compounded continuously?

111. Compound Interest. If $1,300 is deposited in a savings account paying 9% interest, compounded quarterly, how long will it take the account to increase to $2,100?

112. Compound Interest. A sum of $5,000 deposited in an account grows to $7,000 in 5 years. Assuming annual compounding, what interest rate is being paid?

113. Rule of Seventy. A rule of thumb for finding how long it takes an investment to double is called the **rule of seventy.** To apply the rule, divide 70 by the interest rate written as a percent. At 5%, an investment takes $\dfrac{70}{5} = 14$ years to double. At 7%, it takes $\dfrac{70}{7} = 10$ years. Explain why this formula works.

114. Bacterial Growth. A bacterial culture grows according to the function $P(t) = P_0 a^t$. If it takes 5 days for the culture to triple in size, how long will it take to double in size?

115. Rodent Control. The rodent population in a city is currently estimated at 30,000. If it is expected to double every 5 years, when will the population reach 1 million?

116. Population Growth. The population of a city is expected to triple every 15 years. When can the city planners expect the present population of 140 persons to double?

117. Bacterial Culture. A bacteria culture doubles in size every 24 hours. By how much will it have increased in 36 hours?

118. Oceanography.   The intensity I of a light a distance x meters beneath the surface of a lake decreases exponentially. Use the data in the illustration to find the depth at which the intensity will be 20%.

100%

6 m

70%

119. Newton's Law of Cooling.   Water initially at 100°C is left to cool in a room at temperature 60°C. After 3 minutes, the water temperature is 90°. The water temperature T is a function of time t given by the formula $T = 60 + 40e^{kt}$. Find k.

120. Newton's Law of Cooling.   Refer to Exercise 119 and find the time for the water temperature to reach 70°C.

WRITING

121. Explain how to solve the equation $2^{x+1} = 31$.

122. Explain how to solve the equation $2^{x+1} = 32$.

123. Write a justification for each step of the solution.

$15^x = 9$ This is the equation to solve.

$\log 15^x = \log 9$ _____.

$x \log 15 = \log 9$ _____.

$x = \dfrac{\log 9}{\log 15}$ _____.

124. What is meant by the term *half-life*?

REVIEW

125. Find the length of leg AC.

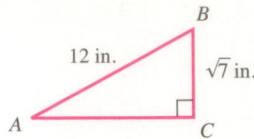

126. Dosages.   The amount of medicine a patient should take is often proportional to his or her weight. If a patient weighing 83 kilograms needs 150 milligrams of medicine, how much will be needed by a person weighing 99.6 kilograms?

CHALLENGE PROBLEMS

Solve each equation.

127. $\log_3 x + \log_3 (x + 2) = 2$

128. $x^{\log x} = 10{,}000$

129. $\dfrac{\log_2 (6x - 8)}{\log_2 x} = 2$

130. $\dfrac{\log (3x - 4)}{\log x} = 2$

---

9   Summary & Review

SECTION 9.1 ▶ Algebra and Composition of Functions

DEFINITIONS AND CONCEPTS	EXAMPLES
Just as it is possible to perform arithmetic operations on real numbers, it is possible to perform those operations on functions. The **sum, difference, product,** and **quotient functions** are defined as: $(f + g)(x) = f(x) + g(x)$ $(f - g)(x) = f(x) - g(x)$ $(f \cdot g)(x) = f(x)g(x)$ $(f/g)(x) = \dfrac{f(x)}{g(x)}, \quad$ with $g(x) \neq 0$	Let $f(x) = 2x + 1$ and $g(x) = x^2$. $\begin{aligned}(f + g)(x) &= f(x) + g(x) \\ &= 2x + 1 + x^2 \\ &= x^2 + 2x + 1\end{aligned}$ $\begin{aligned}(f - g)(x) &= f(x) - g(x) \\ &= 2x + 1 - x^2 \\ &= -x^2 + 2x + 1\end{aligned}$ $\begin{aligned}(f \cdot g)(x) &= f(x) \cdot g(x) \\ &= (2x + 1)x^2 \\ &= 2x^3 + x^2\end{aligned}$ $\begin{aligned}(f/g)(x) &= \dfrac{f(x)}{g(x)} \\ &= \dfrac{2x + 1}{x^2}\end{aligned}$

Often one quantity is a function of a second quantity that depends, in turn, on a third quantity. Such chains of dependence can be modeled by a **composition of functions.**

Read as "f of g of x."

$$(f \circ g)(x) = f(g(x))$$

Read as "f composed with g of x."

Let $f(x) = 4x - 9$ and $g(x) = x^3$. Find $(f \circ g)(2)$ and $(f \circ g)(x)$.

$(f \circ g)(2) = f(g(2))$ *Change to nested parentheses notation.*

$\quad\quad\quad\quad = f(8)$ *Evaluate: $g(2) = 2^3 = 8$.*

$\quad\quad\quad\quad = 4(8) - 9$ *Evaluate $f(8)$ using $f(x) = 4x - 9$.*

$\quad\quad\quad\quad = 23$

$(f \circ g)(x) = f(g(x)) = f(x^3) = 4x^3 - 9$

REVIEW EXERCISES

Let $f(x) = 2x$ and $g(x) = x + 1$. Find each function and give its domain.

1. $f + g$
2. $f - g$
3. $f \cdot g$
4. f/g

Let $f(x) = x^2 + 2$ and $g(x) = 2x + 1$. Find each of the following.

5. $(f \circ g)(-1)$
6. $(g \circ f)(0)$
7. $(f \circ g)(x)$
8. $(g \circ f)(x)$

9. Use the graphs of functions f and g to find each of the following.

   a. $(f + g)(2)$
   b. $(f \cdot g)(-4)$
   c. $(f \circ g)(4)$
   d. $(g \circ f)(6)$

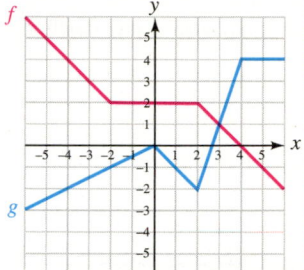

10. **Mileage Costs.** The function $f(m) = \frac{m}{8}$ gives the number of gallons of fuel consumed if a bus travels m miles. The function $C(f) = 3.25f$ gives the cost (in dollars) of f gallons of fuel. Write a composition function that expresses the cost of the fuel consumed as a function of the number of miles driven.

SECTION 9.2 ▶ Inverse Functions

DEFINITIONS AND CONCEPTS	EXAMPLES
A function is called a **one-to-one function** if different inputs determine different outputs.	The function $f(x) = 3x - 5$ is a one-to-one function because different inputs have different outputs. Since two different inputs, -2 and 2, have the same output 16, the function $f(x) = x^4$ is not one-to-one.
Horizontal line test: A function is one-to-one if every horizontal line intersects the graph of the function at most once.	The function $f(x) = |x| - 2$ is not a one-to-one function because we can draw a horizontal line that intersects its graph twice. 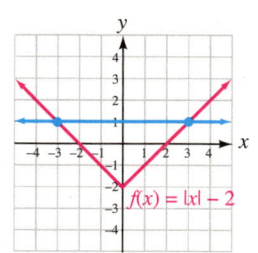
To find the inverse of a function, replace $f(x)$ with y, interchange the variables x and y, solve for y, and then replace y with $f^{-1}(x)$.	Find the inverse of the one-to-one function $f(x) = 2x + 1$. $f(x) = 2x + 1$ $y = 2x + 1$ *Replace $f(x)$ with y.* $x = 2y + 1$ *Interchange the variables x and y.* $\dfrac{x - 1}{2} = y$ *Solve for y.* $f^{-1}(x) = \dfrac{x - 1}{2}$ *Replace y with $f^{-1}(x)$.*

If a point (a, b) is on the graph of function f, it follows that the point (b, a) is on the graph of f^{-1}, and vice versa.

The graph of a function and its inverse are **symmetric about the line $y = x$**.

The graphs of $f(x) = 2x + 1$ and $f^{-1}(x) = \frac{x-1}{2}$ are symmetric about the line $y = x$ as shown in the illustration.

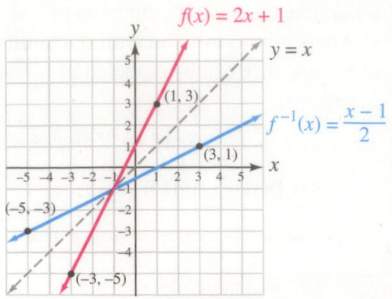

For any one-to-one function f and its inverse, f^{-1},

$$(f \circ f^{-1})(x) = x \quad \text{and} \quad (f^{-1} \circ f)(x) = x$$

The composition of $f(x) = 2x + 1$ and its inverse $f^{-1}(x) = \frac{x-1}{2}$ is the identity function $f(x) = x$.

$$(f \circ f^{-1})(x) = f(f^{-1}(x)) = f\left(\frac{x-1}{2}\right) = 2\left(\frac{x-1}{2}\right) + 1 = x - 1 + 1 = x$$

$$(f^{-1} \circ f)(x) = f^{-1}(f(x)) = f^{-1}(2x+1) = \frac{2x+1-1}{2} = \frac{2x}{2} = x$$

REVIEW EXERCISES

In Exercises 11–16, determine whether the function is one-to-one.

11. $f(x) = x^2 + 3$

12. $f(x) = \dfrac{1}{3}x - 8$

13. $\{(3, 4), (5, 10), (10, -1), (6, 6)\}$

14.

x	$f(x)$
0	-5
2	10
4	-5
6	15

15.

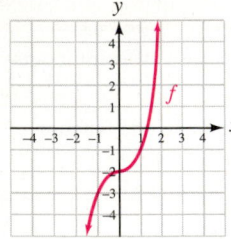

16.

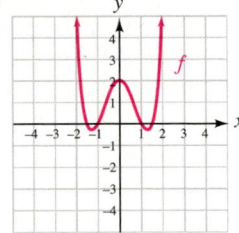

17. Use the table of values of the one-to-one function f to complete a table of values for f^{-1}.

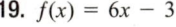

x	$f(x)$		x	$f^{-1}(x)$
-6	-6		-6	
-1	-3		-3	
7	12		12	
20	3		3	

18. Given the graph of function f, graph f^{-1} on the same coordinate axes. Label the axis of symmetry.

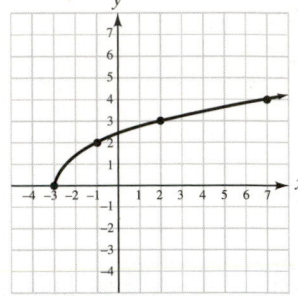

Find the inverse of each function.

19. $f(x) = 6x - 3$

20. $f(x) = \dfrac{4}{x - 1}$

21. $f(x) = (x + 2)^3$

22. $f(x) = \dfrac{x}{6} - \dfrac{1}{6}$

23. Find the inverse of $f(x) = \sqrt[3]{x - 1}$. Then graph the function and its inverse on one coordinate system. Show the axis of symmetry on the graph.

24. Use composition to show that $f(x) = 5 - 4x$ and $f^{-1}(x) = -\dfrac{x - 5}{4}$ are inverse functions.

SECTION 9.3 ▶ **Exponential Functions**

DEFINITIONS AND CONCEPTS	EXAMPLES

Exponential functions have a *constant base* and a *variable exponent* and are defined by the equation

$$f(x) = b^x, \quad \text{with } b > 0, \ b \neq 1$$

Properties of an exponential function $f(x) = b^x$:

The **domain** is the interval $(-\infty, \infty)$.

The **range** is the interval $(0, \infty)$.

Its graph has a **y-intercept** of $(0, 1)$.

The x-axis is an **asymptote** of its graph.

The graph **passes through** the point $(1, b)$.

If $b > 1$, then $f(x) = b^x$ is an **increasing function.**

If $0 < b < 1$, then $f(x) = b^x$ is a **decreasing function.**

The graphs of $f(x) = 2^x$ and $g(x) = \left(\frac{1}{2}\right)^x$ are shown below.

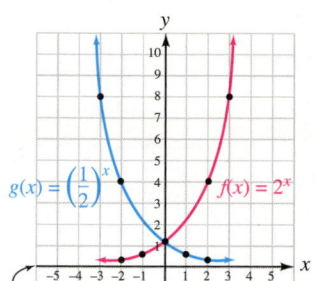
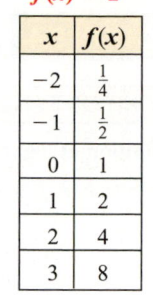

$g(x) = \left(\frac{1}{2}\right)^x$

x	$g(x)$
-3	8
-2	4
-1	2
0	1
1	$\frac{1}{2}$
2	$\frac{1}{4}$

$f(x) = 2^x$

x	$f(x)$
-2	$\frac{1}{4}$
-1	$\frac{1}{2}$
0	1
1	2
2	4
3	8

The x-axis is an asymptote of each graph.

Since the base 2 is greater than 1, the function $f(x) = 2^x$ is an increasing function.

Since the base $\frac{1}{2}$ is such that $0 < \frac{1}{2} < 1$, the function $g(x) = \left(\frac{1}{2}\right)^x$ is a decreasing function.

The graphs of exponential functions can be **translated** horizontally and vertically.

$h(x) = 2^{x+3}$ The graph of $f(x) = 2^x$ is moved 3 units to the left.

$s(x) = \left(\frac{1}{2}\right)^x - 4$ The graph of $g(x) = \left(\frac{1}{2}\right)^x$ is moved 4 units downward.

Exponential functions are used to model many situations, such as population **growth,** the spread of an epidemic, the temperature of a heated object as it cools, and radioactive **decay.**

Exponential functions are suitable models for describing **compound interest.**

If $\$P$ is the deposit, and interest is paid k times a year at an annual rate r, the amount A in the account after t years is given by

$$A = P\left(1 + \frac{r}{k}\right)^{kt}$$

If \$15,000 is deposited in an account paying an annual interest rate of 7.5%, compounded monthly, how much will be in the account in 60 years?

$$A(t) = 15{,}000\left(1 + \frac{0.075}{12}\right)^{12t}$$

To write the formula in function notation, substitute for P, r, and k.

$$A(60) = 15{,}000\left(1 + \frac{0.075}{12}\right)^{12(60)}$$

Substitute 60 for the time t.

$$= 15{,}000\left(1 + \frac{0.075}{12}\right)^{720}$$

Evaluate the exponent: 12(60) = 720.

$$\approx 1{,}331{,}479.52$$

Use a calculator with an exponential key: y^x or ^.

In 60 years, the account will contain about \$1,331,479.52.

REVIEW EXERCISES

25. a. Which of the following are exponential functions?

$$f(x) = 2x \qquad g(x) = x^2 \qquad h(a) = \sqrt{a}$$

$$n(x) = 2^x \qquad t(x) = \frac{1}{x} \qquad s(t) = 1.08^t$$

b. Use a calculator to find $0.9(1.42)^{14}$. Round to four decimal places.

26. Determine whether each application is an example of exponential growth or decay.

a.

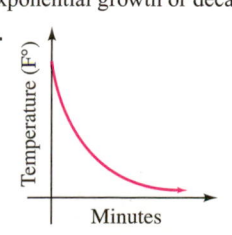

b.

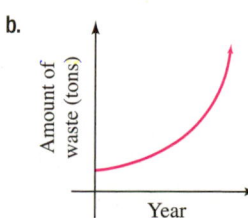

Graph each function and give the domain and the range. Label the y-intercept.

27. $f(x) = 3^x$

28. $f(x) = \left(\frac{1}{3}\right)^x$

29. $f(x) = \left(\frac{1}{2}\right)^x - 2$

30. $f(x) = 3^{x-1}$

31. In Exercise 30, what is the asymptote of the graph of $f(x) = 3^{x-1}$?

32. Coal Production. The function $c(t) = 128{,}000\,(1.08)^t$ approximates the number of tons of coal produced in the United States for the years 1800–1910, where t is the number of years after 1800. How many tons of coal did the U.S. produce in:

a. 1800?   **b.** 1900?

33. Compound Interest. How much will $10,500 become if it earns 9% annual interest, compounded quarterly, for 60 years?

34. Depreciation. The value (in dollars) of a certain model car is given by the function $V(t) = 12,000\left(10^{-0.155t}\right)$, where t is the number of years from the present. Find the value of the car in 5 years.

SECTION 9.4 ▶ **Logarithmic Functions**

DEFINITIONS AND CONCEPTS	EXAMPLES
Definition of logarithm: If $b > 0$, $b \neq 1$, and x is positive, then Exponent $y = \log_b x$ is equivalent to $x = b^y$ Base	*Logarithmic form* *Exponential form* Exponent $\log_5 125 = 3$ means $5^3 = 125$ Base Exponent $\log_2 \dfrac{1}{16} = -4$ means $2^{-4} = \dfrac{1}{16}$ Base
$\log_b x$ is the exponent to which b is raised to get x. $b^{\log_b x} = x$	To evaluate $\log_4 16$ we ask: "To what power must we raise 4 to get 16?" Since $4^2 = 16$, the answer is: the 2nd power. Thus, $\log_4 16 = 2$.
We cannot find the logarithm of 0 or a negative number.	$\log_4 0$ is undefined. $\log_8 (-64)$ is undefined.
For calculation purposes and in many applications, we use base-10 logarithms, called **common logarithms.** $\log x$ means $\log_{10} x$	$\log 1{,}000{,}000 = 6$ because $10^6 = 1{,}000{,}000$. $\log \dfrac{1}{1{,}000} = -3$ because $10^{-3} = \dfrac{1}{1{,}000}$.
If $b > 0$ and $b \neq 1$, the **logarithmic function with base b** is defined by $f(x) = \log_b x$. The domain is $(0, \infty)$ and the range is $(-\infty, \infty)$. If $b > 1$, then $f(x) = \log_b x$ is an increasing function. If $0 < b < 1$, then $f(x) = \log_b x$ is a decreasing function. The graphs of logarithmic functions can be **translated** horizontally and vertically.	The graph of the logarithmic function $f(x) = \log_2 x$ is shown at the right. From the graph, we see that $f(x) = \log_2 x$ is an increasing function. To graph $g(x) = \log_2(x - 3)$, move the graph of $f(x) = \log_2 x$ to the right 3 units. 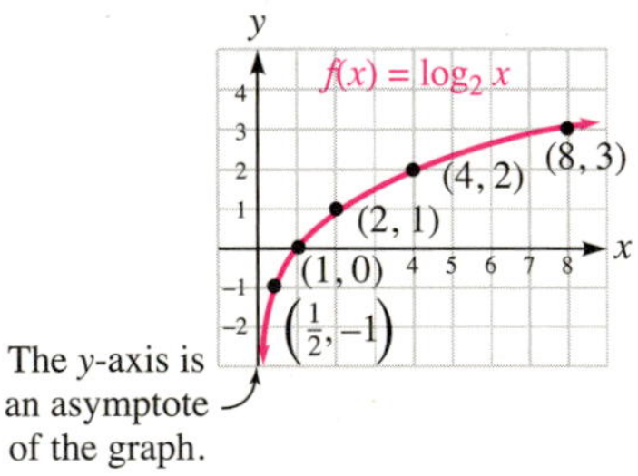 The y-axis is an asymptote of the graph.
The exponential function $f(x) = b^x$ and the logarithmic function $f(x) = \log_b x$ are inverses of each other.	$f(x) = 3^x$ and $f^{-1}(x) = \log_3 x$ are inverses of each other. Their graphs are symmetric about the line $y = x$. Similarly, $f(x) = 10^x$ and $f^{-1}(x) = \log x$ are inverses. 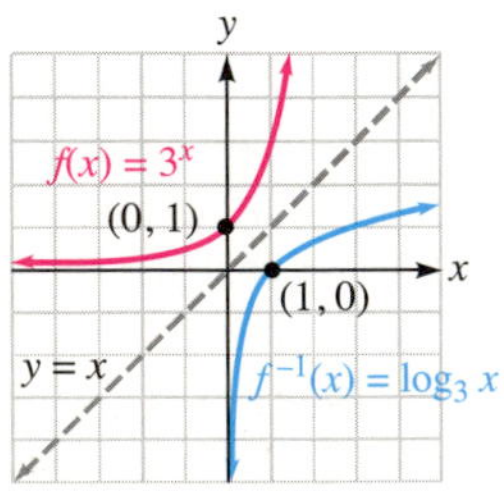

Logarithmic functions, like exponential functions, can be used to **model** certain types of growth and decay.

Decibel voltage gain: dB gain $= 20 \log \dfrac{E_O}{E_I}$

The Richter scale: $R = \log \dfrac{A}{P}$

The input to an amplifier is 0.4 volt and the output is 30 volts. Find the dB gain.

$$\text{dB gain} = 20 \log \dfrac{E_O}{E_I}$$

$$= 20 \log \dfrac{30}{0.4} \qquad \text{Substitute 30 for } E_O \text{ and 0.4 for } E_I.$$

$$\approx 37.50122527 \qquad \text{Use a calculator with a } \boxed{\text{LOG}} \text{ key.}$$

The dB gain is about 38 decibels.

REVIEW EXERCISES

35. Give the domain and range of $f(x) = \log x$.

36. Explain why a student got an error message when she used a calculator to evaluate log 0.

37. Write the statement $\log_4 64 = 3$ in exponential form.

38. Write the statement $7^{-1} = \frac{1}{7}$ in logarithmic form.

Evaluate.

39. $\log_3 9$

40 $\log_9 \frac{1}{81}$

41. $\log_{1/2} 1$

42. $\log_5 (-25)$

43. $\log_6 \sqrt{6}$

44. $\log 1,000$

Solve for x.

45. $\log_2 x = 5$

46. $\log_3 x = -4$

47. $\log_x 16 = 2$

48. $\log_x \frac{1}{100} = -2$

49. $\log_9 3 = x$

50. $\log_{27} 3 = x$

Use a calculator to find the value of x to four decimal places.

51. $\log 4.51 = x$

52. $\log x = 1.43$

Graph each function and its inverse on the same coordinate system. Draw the axis of symmetry.

53. $f(x) = \log_4 x$ and $g(x) = 4^x$

54. $f(x) = \log_{1/3} x$ and $g(x) = \left(\frac{1}{3}\right)^x$

Graph each function. Label the x-intercept.

55. $f(x) = \log (x - 2)$

56. $f(x) = 3 + \log x$

57. Electrical Engineering.   Find the dB gain of an amplifier with an output of 18 volts and an input of 0.04 volt.

58. Earthquakes.   An earthquake had a period of 0.3 second and an amplitude of 7,500 micrometers. Use the formula $R = \log \frac{A}{P}$ to find its measure on the Richter scale.

59. Organ Pipes.   The design for a set of brass pipes for a church organ is shown below. The function $h(n) = 52 + 25 \log n$ approximates the height (in centimeters) of the pipe with number n. Find the height of

a. Pipe 1

b. Pipe 8

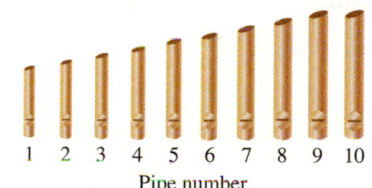

Pipe number

60. Girls' Heights.   The function $P(A) = 61.8 + 34.9 \log(A - 4)$ approximates the percent of the adult height a female child A years old has attained, where $5 \le A \le 15$. What percent of her adult height will a girl have reached the day she first becomes a teenager?

SECTION 9.5 ▶ **Base-*e* Exponential and Logarithmic Functions**

DEFINITIONS AND CONCEPTS	EXAMPLES
Of all possible bases for an exponential function, *e* is the most convenient for problems involving growth or decay. $e = 2.718281828459 \ldots$ As *n* approaches infinity, the value of $\left(1 + \frac{1}{n}\right)^n$ approaches *e*. The function defined by $f(x) = e^x$ is called the **natural exponential function.** It is called this because it can be applied to many natural settings.	From the graph, we see that the domain of the natural exponential function is $(-\infty, \infty)$ and the range is $(0, \infty)$. The graph of $f(x) = e^x$ can be **translated** horizontally and vertically. The *x*-axis is an asymptote of the graph. 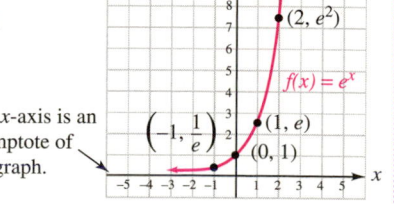

Exponential growth/decay: If a quantity increases or decreases at an annual rate r, **compounded continuously,** the amount A after t years is given by

$$A = Pe^{rt}$$

If r is negative, the amount decreases.

If interest is paid continuously (infinitely many times a year), we say that the interest is **compounded continuously.**

If \$30,000 accumulates interest at an annual rate of 9%, compounded continuously, find the amount in the account after 25 years.

$A = Pe^{rt}$	This is the formula for continuous compound interest.
$= 30{,}000e^{0.09 \cdot 25}$	Substitute 30,000 for P, 0.09 for r, and 25 for t.
$= 30{,}000e^{2.25}$	Evaluate the exponent: $0.09 \cdot 25 = 2.25$
$\approx 284{,}632.08$	Use a calculator with an e^x key.

In 30 years, the account will contain \$284,632.08.

Suppose the population of a city of 50,000 people is decreasing exponentially according to the function $P(t) = 50{,}000e^{-0.003t}$, where t is measured in years from the present date. Find the expected population of the city in 20 years.

$P(t) = 50{,}000e^{-0.003t}$	Since r is negative, this is the exponential decay model.
$P(20) = 50{,}000e^{-0.003(20)}$	Substitute 20 for t.
$= 50{,}000e^{-0.06}$	Evaluate the exponent: $-0.003\,(20) = -0.06$.
$\approx 47{,}088$	Use a calculator with an e^x key.

After 20 years, the expected population will be about 47,088 people.

Of all possible bases for a logarithmic function, e is the most convenient for problems involving growth or decay. Since these situations occur often in natural settings, base-e logarithms are called **natural logarithms:**

$$\ln x \quad \text{means} \quad \log_e x$$

To evaluate $\ln \frac{1}{e^4}$ we ask: "To what power must we raise e to get $\frac{1}{e^4}$?"

Since $e^{-4} = \frac{1}{e^4}$, the answer is: the -4th power. Thus,

$$\ln \frac{1}{e^4} = -4$$

$\ln 0$ and $\ln(-5)$ are undefined.

$\ln x$ is the exponent to which e is raised to get x.

$\ln 9 \approx 2.1972$ means $e^{2.1972} \approx 9$

The **natural logarithmic function** with base e is defined by

$$f(x) = \ln x$$

The domain is the interval $(0, \infty)$ and the range is the interval $(-\infty, \infty)$.

The graph of the natural logarithmic function $f(x) = \ln x$ is shown on the right.

From the graph, we see that $f(x) = \ln x$ is an increasing function.

The graph of $f(x) = \ln x$ can be translated horizontally and vertically.

The y-axis is an asymptote of the graph.

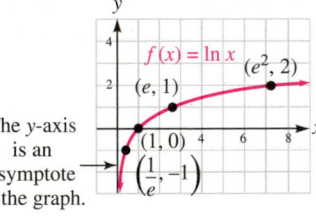

The natural exponential function $f(x) = e^x$ and the natural logarithmic function $f^{-1}(x) = \ln x$ are **inverses** of each other.

The graphs are symmetric about the line $y = x$.

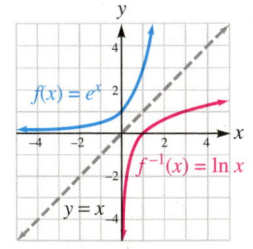

If a population grows exponentially at a certain annual rate r, the time required for the population to double is called the **doubling time.** It is given by the formula:

$$t = \frac{\ln 2}{r}$$

The population of a town is growing at a rate of 3% per year. If this rate continues, how long will it take the population to double?

We substitute 0.03 for r and use a calculator to perform the calculation.

$$t = \frac{\ln 2}{r} = \frac{\ln 2}{0.03} \approx 23.10490602 \quad \text{Use a calculator with an LN key.}$$

The population will double in about 23.1 years.

REVIEW EXERCISES

61. a. Approximate e to the nearest hundredth.

b. Fill in the blanks:

$\ln 15 \approx 2.7081$ means $e^{\boxed{}} \approx \boxed{}$

62. Use a calculator to find $16.4 + 252.7e^{-0.76(9)}$. Round to four decimal places.

Graph each function, and give the domain and the range.

63. $f(x) = e^x + 1$

64 $f(x) = e^{x-3}$

65. Interest Compounded Continuously. If \$10,500 accumulates interest at an annual rate of 9%, compounded continuously, how much will be in the account in 60 years?

66. The Sooner State. In 2009, Oklahoma had the largest *gross domestic product* growth rate of all fifty states: 6.6%. (The GDP, as it is called, is the value of all goods and services produced within a state.) If the 2009 GDP totaled \$142.5 billion, predict Oklahoma's GDP in 2015, assuming the growth rate remains the same. (Source: huffingtonpost.com)

67. Mortgage Rates. There was the housing boom in the 1980s as the baby boomers (those born from 1946 through 1964) bought their homes. The average annual interest rate in percent on a 30-year fixed-rate home mortgage for the years 1980–1996 can be approximated by the function $r(t) = 13.9e^{-0.035t}$, where t is the number of years since 1980. To the nearest hundredth of a percent, what does this model predict was the 30-year fixed rate in 1980? In 1985? In 1990?

68. Medical Tests. A radioactive dye is injected into a patient as part of a test to detect heart disease. The amount of dye remaining in his bloodstream t hours after the injection is given by the function $f(t) = 10e^{-0.27t}$. How can you determine from the function that the amount of dye in the bloodstream is decreasing?

Evaluate each expression. Do not use a calculator.

69. $\ln e$

70. $\ln e^2$

71. $\ln \frac{1}{e^5}$

72. $\ln \sqrt{e}$

73. $\ln (-e)$

74. $\ln 0$

75. $\ln 1$

76. $\ln e^{-7}$

Use a calculator to evaluate each expression. Express all answers to four decimal places.

77. $\ln 452$

78. $\ln 0.85$

Solve each equation. Express all answers to four decimal places.

79. $\ln x = 2.336$

80. $\ln x = -8.8$

81. Explain the difference between the functions $f(x) = \log x$ and $g(x) = \ln x$.

82. What function is the inverse of $f(x) = \ln x$?

Graph each function.

83. $f(x) = 1 + \ln x$

84. $f(x) = \ln (x + 1)$

85. Population Growth. How long will it take the population of Mexico to double if the growth rate is currently about 1.118%? (Source: CIA World Fact Book)

86. Botany. The height (in inches) of a certain plant is approximated by the function $H(a) = 13 + 20.03 \ln a$, where a is its age in years. How tall will it be when it is 19 years old?

SECTION 9.6 ▶ Properties of Logarithms

DEFINITIONS AND CONCEPTS	EXAMPLES
Properties of logarithms: If M, N, and b are positive real numbers, $b \neq 1$	Apply a property of logarithms and then simplify, if possible.

1. $\log_b 1 = 0$ **2.** $\log_b b = 1$

3. $\log_b b^x = x$ **4.** $b^{\log_b x} = x$

5. *Product rule for logarithms:*

$\log_b MN = \log_b M + \log_b N$

6. *Quotient rule for logarithms:*

$\log_b \frac{M}{N} = \log_b M - \log_b N$

7. *Power rule for logarithms:*

$\log_b M^p = p \log_b M$

1. $\log_3 1 = 0$ **2.** $\log_7 7 = 1$

3. $\log_5 5^3 = 3$ **4.** $9^{\log_9 10} = 10$

5. $\log_2 (6 \cdot 8) = \log_2 6 + \log_2 8$
$= \log_2 6 + 3$

6. $\log_3 \frac{81}{x} = \log_3 81 - \log_3 x$
$= 4 - \log_3 x$

7. $\log_8 7^3 = 3 \log_8 7$ and $\log \sqrt[4]{x} = \log x^{\frac{1}{4}} = \frac{1}{4} \log x$

Properties of logarithms can be used to **expand** logarithmic expressions.	Write $\log_3(x^2 y^3)$ as the sum and/or difference of logarithms of a single quantity. $\log_3(x^2 y^3) = \log_3 x^2 + \log_3 y^3$ The log of a product is the sum of the logs. $\qquad\qquad = \mathbf{2}\log_3 x + \mathbf{3}\log_3 y$ The log of a power is the power times the log.
Properties of logarithms can be used to **condense** certain logarithmic expressions. To condense, apply the power rule first to make the coefficients of logarithms 1. Then use the product and quotient rules.	Write $3\ln x - \frac{1}{2}\ln y$ as a single logarithm. $3\ln x - \frac{1}{2}\ln y = \ln x^3 - \ln y^{1/2}$ A power times a log is the log of the power. $\qquad\qquad = \ln\dfrac{x^3}{y^{1/2}}$ The difference of two logs is the log of the quotient. $\qquad\qquad = \ln\dfrac{x^3}{\sqrt{y}}$ Write $y^{1/2}$ as $\sqrt{y}$.
If we need to find a logarithm with some base other than 10 or e, we can use a conversion formula. **Change-of-base formula:** $$\log_b x = \frac{\log_a x}{\log_a b}$$	Find $\log_7 6$ to four decimal places. $$\log_7 \mathbf{6} = \frac{\log \mathbf{6}}{\log \mathbf{7}} \approx 0.920782221$$ Change to the ratio of base-10 logarithms. To four decimal places, $\log_7 6 = 0.9208$. To check, verify that $7^{0.9208}$ is approximately 6.
In chemistry, common logarithms are used to express the acidity of solutions using pH. **pH scale:** $\text{pH} = -\log[\text{H}^+]$ where the symbol $[\text{H}^+]$ represents the hydrogen ion concentration in gram-ions per liter.	Find the pH of a liquid with a hydrogen ion concentration of 10^{-8} gram-ions per liter. $\text{pH} = -\log[\mathbf{H^+}]$ This is the pH formula. $\text{pH} = -\log \mathbf{10^{-8}}$ Substitute 10^{-8} for $[\text{H}^+]$. $\qquad = -(-8)\log 10$ The log of a power is the power times the log. $\qquad = 8$ Simplify: $\log 10 = 1$.

REVIEW EXERCISES

Simplify each expression.

87. $\log_2 1$

88. $\log_9 9$

89. $\log 10^3$

90. $7^{\log_7 4}$

Write each logarithm as the sum and/or difference of logarithms of a single quantity. Then simplify, if possible.

91. $\log_3 27x$

92. $\log\dfrac{100}{x}$

93. $\log_5\sqrt{27}$

94. $\log_b 10ab$

Write each logarithm as the sum and/or difference of logarithms of a single quantity.

95. $\log_b\dfrac{x^2 y^3}{z}$

96. $\ln\sqrt{\dfrac{x}{yz^2}}$

Write each logarithmic expression as one logarithm.

97. $3\log_2 x - 5\log_2 y + 7\log_2 z$

98. $-3\log_b y - 7\log_b z + \frac{1}{2}\log_b(x+2)$

99. $\log_b(a^2 - 25) - \log_b(a+5)$

100. $3\log_8 x + 4\log_8 x$

Assume that $\log_b 5 = 1.1609$ and $\log_b 8 = 1.5000$ and find each value to four decimal places.

101. $\log_b 40$

102. $\log_b 64$

103. Find $\log_5 17$ to four decimal places.

104. pH of Grapefruit. Find the pH of grapefruit juice if its hydrogen ion concentration is 7.9×10^{-4} gram-ions per liter. Round to the nearest tenth.

SECTION 9.7 ▶ **Exponential and Logarithmic Equations**

DEFINITIONS AND CONCEPTS	EXAMPLES
An **exponential equation** contains a variable in one of its exponents. Two examples are: $6^{x-3} = 9$ and $e^{-2.5t} = 56$ If both sides of an exponential equation can be expressed as a power of the same base, we can use the following property to solve it: $b^x = b^y$ is equivalent to $x = y$	Solve: $3^{x+2} = 27$ $3^{x+2} = 3^3$ Express the right side of the equation as a power of 3: $27 = 3^3$. $x + 2 = 3$ If two exponential expressions with the same base are equal, their exponents are equal. $x = 1$ The solution is 1. Check it in the original equation.
When it is difficult to write each side of an exponential equation as a power of the same base, **take the logarithm on each side.** With this method, we often obtain **exact solutions** involving logarithms that we can **approximate.**	Solve: $4^x = 7$. Give an approximate solution to four decimal places. $\log 4^x = \log 7$ Take the base-10 logarithm on both sides of the equation. $x \log 4 = \log 7$ The log of a power is the power times the log. $x = \dfrac{\log 7}{\log 4}$ To isolate x, divide both sides by log 4. This is the exact solution. $x \approx 1.4037$ Use a calculator with a LOG key. To four decimal places, the approximate solution is 1.4037. To check it, we substitute 1.4037 for x in $4^x = 7$ and use a calculator to evaluate the left side: $4^{1.4037} \approx 7$
To solve an exponential equation that contains a base-e exponential expression, take the natural logarithm on both sides of the equation.	Solve: $e^{5t} = 11$ $\ln e^{5t} = \ln 11$ Take the base-e logarithm on both sides. $5t \ln e = \ln 11$ The log of a power is the power times the log. $5t \cdot 1 = \ln 11$ Simplify: $\ln e = 1$. $5t = \ln 11$ Simplify the left side. $t = \dfrac{\ln 11}{5}$ This is the exact solution. An approximate solution to four decimal places is 0.4796.
A **logarithmic equation** is an equation containing a variable in a logarithmic expression. Two examples are: $2 \log_3 x + \log_3 5x = 4$ and $\ln (x + 1) = \ln (3x - 4)$ Certain logarithmic equations can be solved using the following property: $\log_b x = \log_b y$ is equivalent to $x = y$	Solve: $\log(4x - 3) = \log(2x + 7)$ $4x - 3 = 2x + 7$ If the logarithms of two numbers are equal, the numbers are equal. $2x = 10$ $x = 5$ The solution is 5. Check it in the original equation.
To solve some logarithmic equations, we write and solve an equivalent exponential equation.	Solve: $\log_4(x + 1) = 2$ $x + 1 = 4^2$ Write the equivalent base-4 exponential equation. $x + 1 = 16$ $x = 15$ The solution is 15. Check it in the original equation.

To solve some logarithmic equations, we first apply properties of logarithms, such as:

- The product rule
- The quotient rule
- The power rule

Solve: $\log_2 (x + 14) - \log_2 x = 3$

$$\log_2 \frac{x + 14}{x} = 3$$ On the left side, use the quotient rule for logarithms.

$$\frac{x + 14}{x} = 2^3$$ Write the equivalent base-2 exponential equation.

$$\frac{x + 14}{x} = 8$$ Evaluate: $2^3 = 8$.

$$x + 14 = 8x$$ Multiply both sides by x.

$$14 = 7x$$ Subtract x from both sides.

$$2 = x$$ Divide both sides by 7.

The solution is 2. Check by substituting it into the original equation.

When there is sufficient food and space available, populations of living organisms tend to increase exponentially according to the following **growth model.**

Population growth: $P = P_0 e^{kt}$

Find the number of bacteria in a culture of 1,000 bacteria if they are allowed to reproduce for 5 hours. Assume $k = \frac{\ln 3}{3}$.

$$P = \boldsymbol{P_0 e^{kt}}$$ This is the population growth model.

$$= \boldsymbol{1{,}000 e^{\frac{\ln 3}{3} \cdot 5}}$$ Substitute for P_0, k, and t.

$$\approx 6{,}240$$ Use a calculator with LN and e^x keys.

In 5 hours, there will be approximately 6,240 bacteria.

REVIEW EXERCISES

Solve each equation. Give the exact solution and an approximate solution to four decimal places, when appropriate.

105. $5^{x+6} = 25$

106. $2^{x^2 + 4x} = \frac{1}{8}$

107. $3^x = 7$

108. $2^x = 3^{x-4}$

109. $e^x = 7$

110. $e^{-0.4t} = 25$

Solve each equation.

111. $\left(\frac{2}{5}\right)^{3x-4} = \frac{8}{125}$

112. $9^{x^2} = 33$

113. $\log (x - 4) = 2$

114. $\ln (2x - 3) = \ln 15$

115. $\log x + \log (29 - x) = 2$

116. $\log_2 x + \log_2 (x - 2) = 3$

117. $\dfrac{\log (7x - 12)}{\log x} = 2$

118. $\log_2 (x + 2) + \log_2 (x - 1) = 2$

119. $\log x + \log (x - 5) = \log 6$

120. $\log 3 - \log (x - 1) = -1$

121. Evaluate both sides of the statement $\frac{\log 8}{\log 15} \neq \log 8 - \log 15$ to show that the sides are indeed not equal.

122. Carbon-14 Dating.   A wooden statue found in Egypt has a carbon-14 content that is two-thirds of that found in living wood. If the half-life of carbon-14 is 5,700 years, how old is the statue?

123. Ants.   The number of ants in a colony is estimated to be 800. If the ant population is expected to triple every 14 days, how long will it take for the population to reach one million?

124. The approximate coordinates of the points of intersection of the graphs of $f(x) = \log x$ and $g(x) = 1 - \log (7 - x)$ are shown in parts (a) and (b) of the illustration. Use the graphs to estimate the solutions of the logarithmic equation $\log x = 1 - \log (7 - x)$. Then check your answers.

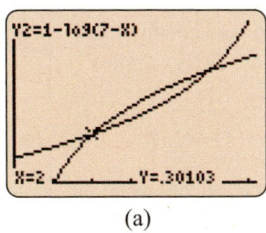   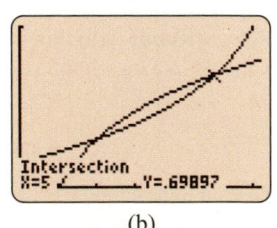

(a)   (b)

9 Chapter Test

1. Fill in the blanks.

   a. A _____ function is denoted by $f \circ g$.

   b. $f(x) = e^x$ is the _____ exponential function.

   c. In _____ compound interest, the number of compoundings is infinitely large.

   d. The functions $f(x) = \log_{10} x$ and $f(x) = 10^x$ are _____ functions.

   e. $f(x) = \log_4 x$ is a _____ function.

2. Write out in words how to say each of the following:

   a. $(f \circ g)(x)$

   b. $g(f(8))$

   c. $f^{-1}(x)$

Let $f(x) = x + 9$ and $g(x) = 4x^2 - 3x + 2$. Find each function and give its domain.

3. $f + g$

4. g/f

Let $f(x) = 2x^2 + 3$ and $g(x) = 4x - 8$. Find each composition.

5. $(g \circ f)(-3)$

6. $(f \circ g)(x)$

Use the tables of values for functions f and g to find each of the following.

7. a. $(f \cdot g)(9)$

   b. $(f \circ g)(-3)$

x	$f(x)$
9	-1
10	17

x	$g(x)$
-3	10
9	16

8. Refer to the graphs of functions f and g below to find each of the following.

   a. $(g/f)(-4)$

   b. $(f \circ g)(1)$

   c. $(f + g)(2)$

   d. $(f \cdot g)(0)$

   e. $(g - f)(1)$

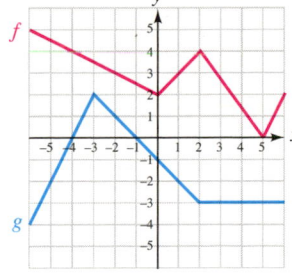

9. Determine whether each function is one-to-one.

   a. $f(x) = |x|$

   b. $\{(1,7), (7,1), (2,8), (8,2)\}$

   c.

   d.

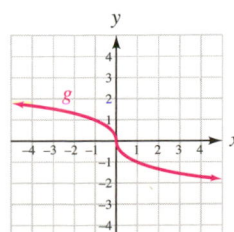

10. Find the inverse of $f(x) = -\frac{1}{3}x$ and then graph f and its inverse on the same coordinate axes. Label the axis of symmetry.

11. Determine whether $f(x) = \frac{1}{3}x + 2$ is a one-to-one function. If it is, find its inverse.

12. Find the inverse of $f(x) = (x - 15)^3$.

13. Use composition to show that $f(x) = 4x + 4$ and $f^{-1}(x) = \frac{x - 4}{4}$ are inverse functions.

14. Consider the following graph of the function f.

    a. Is f a one-to-one function?

    b. Is its inverse a function?

    c. What is $f^{-1}(260)$? What information does it give?

Relationship Between Car Speed and Tire Temperature

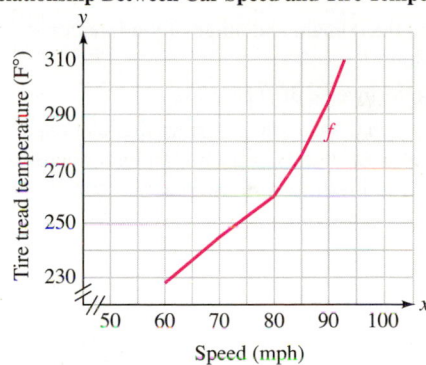

Graph each function and give the domain and the range.

15. $f(x) = 2^x + 1$

16. $f(x) = 3^{-x}$

17. **Radioactive Decay.**   A radioactive material decays according to the formula $A = A_0(2)^{-t}$. How much of a 3-gram sample will be left in 6 years?

18. **Compound Interest.**   An initial deposit of $1,000 earns 6% interest, compounded twice a year. How much will be in the account in one year?

19. a. Graph $f(x) = e^x$. Label the y-intercept and the asymptote of the graph.

    b. Give the domain and range.

    c. What is the inverse of $f(x) = e^x$?

20. **Population Growth.**   As of July 2010, the population of India was estimated to be 1,173,108,018, with an annual growth rate of 1.376%. If the growth rate remains the same, how large will the population be in July, 2020? Round to the nearest thousand. (Source: CIA World Fact Book)

21. Biology. Human growth hormone, known as HGH, is produced by the pituitary gland in the brain and released into the blood stream. It stimulates growth and cell production. After the age of 20, levels of HGH in the body decrease dramatically, as shown in the graph. Use the given function to approximate the amount of HGH produced per day by a person 55 years old.

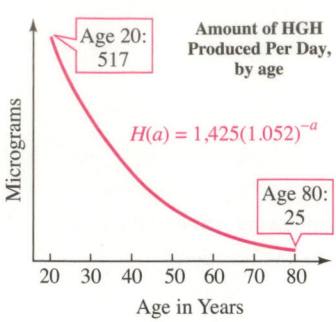

Source: Aging Well, James L.Holly, MD

22. Write the statement $\log_6 \frac{1}{36} = -2$ in exponential form.

23. a. What are the domain and range of the function $f(x) = \log x$?

   b. What is the inverse of $f(x) = \log x$?

24. Botany. Which phrase best describes the relationship between the rate of photosynthesis in plants and light intensity that is graphed below: linear growth, exponential growth, or logarithmic growth?

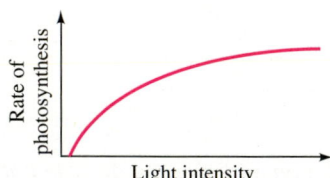

Evaluate each logarithmic expression, if possible.

25. $\log_5 25$

26. $\log_9 \frac{1}{81}$

27. $\log (-100)$

28. $\ln \frac{1}{e^6}$

29. $\log_4 2$

30. $\log_{1/3} 1$

Solve for x.

31. $\log_x 32 = 5$

32. $\log_8 x = \frac{4}{3}$

33. $\log_3 x = -3$

34. $\ln x = 1$

Graph each function.

35. $f(x) = -\log_3 x$

36. $f(x) = \ln x$

37. Chemistry. Find the pH of a solution with a hydrogen ion concentration of 3.7×10^{-7} gram-ions per liter. (*Hint:* pH $= -\log [H^+]$.)

38. Electronics. Find the dB gain of an amplifier when $E_O = 60$ volts and $E_I = 0.3$ volt. *Hint:* dB gain $= 20 \log \frac{E_O}{E_I}$.

39. Use a calculator to find x to four decimal places: $\log x = -1.06$

40. Use the change-of-base formula to find $\log_7 3$ to four decimal places.

41. Write the expression $\log_b a^2bc^3$ as the sum and/or difference of logarithms of a single quantity. Then simplify, if possible.

42. Write the expression $\frac{1}{2} \ln (a + 2) + \ln b - 3 \ln c$ as a logarithm of a single quantity.

Solve each equation. Give the exact solution and an approximate solution to four decimal places, when appropriate.

43. $5^x = 3$

44. $3^{x-1} = 27$

45. $\left(\frac{3}{2}\right)^{6x+2} = \frac{27}{8}$

46. $e^{0.08t} = 4$

47. $2 \log x = \log 25$

48. $\log_2 (x + 2) - \log_2 (x - 5) = 3$

49. $\ln (5x + 2) = \ln (2x + 5)$

50. $\log x + \log (x - 9) = 1$

51. The illustration shows the graphs of $y = \frac{1}{2} \ln (x - 1)$ and $y = \ln 2$ and the approximate coordinates of their point of intersection. Estimate the solution of the logarithmic equation $\frac{1}{2} \ln (x - 1) = \ln 2$. Then check the result.

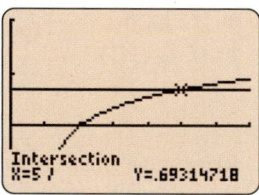

52. Insects. The number of insects attracted to a bright light is currently 5. If the number is expected to quadruple every 6 minutes, how long will it take for the number to reach 500?

Group Project

The Number *e*

Overview: In this activity, you will use a calculator to find progressively more accurate approximations of *e*.

Instructions: Form groups of two students. Each student will need a scientific calculator.

Begin by finding an approximation of *e* using the $\boxed{e^x}$ key on your calculator. Copy the table shown below, and write the number displayed on the calculator screen at the top of the table.

The value of *e* can be calculated to any degree of accuracy by adding the terms of the following pattern:

$$e = 1 + 1 + \frac{1}{2} + \frac{1}{2 \cdot 3} + \frac{1}{2 \cdot 3 \cdot 4} + \frac{1}{2 \cdot 3 \cdot 4 \cdot 5} + \cdots$$

The more terms that are added, the closer the sum will be to *e*.

You are to add as many terms as necessary until you obtain a sum that matches the value of *e*, given by the $\boxed{e^x}$ key on your calculator. Work together as a team. One member of the group should compute the fractional form of the term to be added. (See the middle column of the table.) The other member should take that information and calculate the cumulative sum. (See the right column of the table.)

How many terms must be added so that the cumulative sum approximation and the $\boxed{e^x}$ key approximation match in each decimal place?

Approximation of *e* found using the $\boxed{e^x}$ key: *e* ≈ _____

Number of terms in the sum	Term (expressed as a fraction)	Cumulative sum (an approximation of *e*)
1	1	1
2	1	2
3	$\frac{1}{2}$	2.5
4	$\frac{1}{2 \cdot 3} = \frac{1}{6}$	2.666666667
⋮	⋮	⋮

CUMULATIVE REVIEW ▶▶ Chapters 1–9

Write the formula associated with each concept.

1. Perimeter and area of a rectangle [Section 1.6]

2. Area and circumference of a circle [Section 1.6]

3. Area of a triangle [Section 1.6]

4. Volume of a rectangular solid and a cube [Section 1.6]

5. Percent formula [Section 1.8]

6. Total value [Section 1.8]

7. Simple interest [Section 1.8]

8. Distance (uniform motion) [Section 1.8]

9. Midpoint of a line segment $\left(, \right)$ [Section 2.1]

10. a. Formula for slope of a line [Section 2.3]

    b. Slope $= \dfrac{}{\text{run}}$

11. Slope–intercept form of the equation of a line [Section 2.4]

12. Point–slope form of the equation of a line [Section 2.4]

13. Equations of a horizontal line and a vertical line [Section 2.4]

14. Evaluate the determinant. [Section 3.5] $\begin{vmatrix} a & b \\ c & d \end{vmatrix} =$

15. a. Direct variation [Section 6.9]

    b. Inverse variation [Section 6.9]

16. Pythagorean theorem [Section 7.6]

17. Distance between two points [Section 7.6]

18. Quadratic formula [Section 8.2]

19. Standard form of a quadratic function [Section 8.4]

20. Composition of functions: $(f \circ g)(x) =$ [Section 9.1]

21. Exponential growth [Section 9.5]

22. Change-of-base formula for logarithms [Section 9.7]

23. Fill in the blanks. Parallel lines have the _____ slope. The slopes of perpendicular lines are _____ _____ . [Section 2.3]

24. Which method would be the most efficient to solve the following system? [Section 3.2]

$$\begin{cases} 3x - 2y = -10 \\ 6x + 5y = 25 \end{cases}$$

Fill in the blanks.

25. a. The solution set of a compound inequality containing the word *and* consists of all the numbers that make _____ inequalities true. [Section 4.2]

b. The solution set of a compound inequality containing the word *or* consists of all the numbers that make _____ or the other or _____ inequalities true. [Section 4.2]

26. Solve each absolute value equation and inequality. [Section 4.3]

a. $|X| = k$ is equivalent to _____ or _____ .

b. $|X| = |Y|$ is equivalent to _____ or _____ .

c. $|X| > k$ is equivalent to _____ or _____ .

d. $|X| < k$ is equivalent to _____ .

27. *Fill in the blanks to complete the rules for exponents.* [Section 5.1]

a. $x^1 =$ _____

b. $x^m x^n =$ _____

c. $(x^m)^n =$ _____

d. $(xy)^n =$ _____

e. $x^0 =$ _____

f. $\left(\dfrac{x}{y}\right)^n =$ _____

g. $\dfrac{x^m}{x^n} =$ _____

h. $x^{-n} =$ _____

i. $\dfrac{1}{x^{-n}} =$ _____

j. $\left(\dfrac{x}{y}\right)^{-n} =$ _____

k. $x^{1/n} =$ _____ [Section 7.2]   **l.** $x^{m/n} =$ _____ [Section 7.2]

28. To graph $y \geq x$, we first graph the boundary line $y = x$ as shown. Explain how we determine which side of the boundary to shade. [Section 4.4]

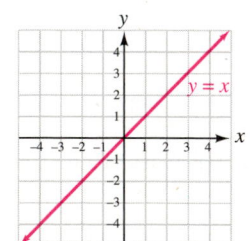

29. Complete each special product. [Section 5.4]

a. $(x + y)^2 =$ _____

b. $(x - y)^2 =$ _____

c. $(x + y)(x - y) =$ _____

30. Complete each factorization. [Section 5.6]

a. $x^2 - y^2 =$ _____

b. $x^3 - y^3 =$ _____

c. $x^3 + y^3 =$ _____

31. What is a quadratic equation? [Section 5.9]

32. Consider the equation $(x + 1)(x - 7) = 0$. How is the zero-factor property used to solve this equation? [Section 5.9]

33. Explain what the slashes and 1's mean. [Section 6.1]

$$\frac{6a^2 - 13a + 6}{3a^2 + a - 2} = \frac{\overset{1}{(3a - 2)}(2a - 3)}{\underset{1}{(3a - 2)}(a + 1)}$$

34. A fraction is undefined when the _____ is 0. [Section 6.1]

35. Perform the operations. [Section 6.2]

a. $\dfrac{A}{B} \cdot \dfrac{C}{D} =$ _____

b. $\dfrac{A}{B} \div \dfrac{C}{D} =$ _____

36. Perform the operations. [Section 6.3]

a. $\dfrac{A}{D} + \dfrac{B}{D} =$ _____

b. $\dfrac{A}{D} - \dfrac{B}{D} =$ _____

37. Explain the purpose of multiplying by a form of 1.

a. $\dfrac{5a}{24b} + \dfrac{11a}{18b^2} = \dfrac{5a}{24b} \cdot \dfrac{3b}{3b} + \dfrac{11a}{18b^2} \cdot \dfrac{4}{4}$ [Section 6.3]

b. $\dfrac{\frac{1}{x} + \frac{1}{y}}{\frac{1}{x} - \frac{1}{y}} = \dfrac{\left(\frac{1}{x} + \frac{1}{y}\right)}{\left(\frac{1}{x} - \frac{1}{y}\right)} \cdot \dfrac{xy}{xy}$ [Section 6.4]

c. $\dfrac{\sqrt{70}}{\sqrt{3}} = \dfrac{\sqrt{70}}{\sqrt{3}} \cdot \dfrac{\sqrt{3}}{\sqrt{3}}$ [Section 7.4]

38. Fill in the blanks to complete the fundamental property of proportions. [Section 6.9]

If $\dfrac{a}{b} = \dfrac{c}{d}$, then $ad =$ _____

The _____ _____ are equal.

39. What is the first step used to solve:

$$\frac{x - 3}{x - 2} - \frac{1}{x} = \frac{x - 3}{x}$$ [Section 6.7]

40. Complete each property of radicals. [Section 7.2]

a. $\sqrt[n]{ab} =$ _____

b. $\sqrt[n]{\dfrac{a}{b}} =$ _____ $(b \neq 0)$

41. What is the first step used to solve:

$$\sqrt{2x - 2} = x - 1$$ [Section 7.5]

42. Approximate each irrational number to the nearest hundredth. **[Section 1.2]**

 a. π

 b. $\sqrt{2}$

 c. e **[Section 9.5]**

43. Complete each property of logarithms. **[Section 9.7]**

 a. $\log_b 1 = $ ▨

 b. $\log_b b = $ ▨

 c. $\log_b b^x = $ ▨

 d. $b^{\log_b x} = $ ▨

 e. $\log_b MN = $ ▨

 f. $\log_b \dfrac{M}{N} = $ ▨

 g. $\log_b M^p = $ ▨

44. What is the first step used to solve: **[Section 9.8]**

 a. $4^{x+1} = 17$

 b. $\log x + \log(x - 3) = 1$

Fill in the blanks to complete each definition.

45. If a and b are real numbers, $a - b = a$ ▨ $(-b)$. **[Section 1.3]**

46. For any real number x,

$$\begin{cases} \text{if } x \geq 0, \text{ then } |x| = \text{▨} \\ \text{if } x < 0, \text{ then } |x| = \text{▨} \end{cases}$$

[Section 4.3]

47. a. The number b is a square root of a if ▨ $= a$. **[Section 7.1]**

 b. If x can be any real number, then $\sqrt{x^2} = $ ▨ . **[Section 7.1]**

 c. If x is any real number, then $\sqrt[3]{x^3} = $ ▨ . **[Section 7.1]**

48. a. $i = $ ▨ **[Section 7.7]**

 b. $i^2 = $ ▨

Conic Sections; More Graphing

10

©Andresr/Shutterstock.com

from Campus to Careers

Civil Engineer

Civil engineers design roads, buildings, airports, tunnels, dams, bridges, or water supply and sewage systems. They often must manage people as well as projects. A civil engineer may oversee a construction site or be a city engineer. Others may work in design, construction, research, and teaching. During the planning stages, they make detailed drawings and graphs of the project. Because construction projects are often publicly funded, they make budgets, submit bid proposals, and perform cost analysis studies to make sure that public funding is spent wisely.

Problem 89 in Study Set 10.1, problem 57 in Study Set 10.2, and problem 63 in Study Set 10.3 involve situations that a civil engineer might encounter on the job. The mathematical concepts discussed in this chapter can be used to solve those problems.

JOB TITLE:
Civil Engineer

EDUCATION:
A bachelor's degree in civil engineering is required.

JOB OUTLOOK:
Good; It is expected to increase 14% to 19% through 2018.

ANNUAL EARNINGS:
The median starting salary in 2009 was $52,048.

FOR MORE INFORMATION:
http://bls.gov/k12/build05.htm

Before moving on to a new mathematics course, it's worthwhile to take some time to reflect on your effort and performance in this course.

Now Try This ▶

As this course draws to a close, here are some questions to ask yourself.

1. How was my attendance?
2. Was I organized? Did I have the right materials?
3. Did I follow a regular schedule?
4. Did I pay attention in class and take good notes?
5. Did I spend the appropriate amount of time on homework?
6. How did I prepare for tests? Did I have a test-taking strategy?
7. Was I part of a study group? If not, why not? If so, was it worthwhile?
8. Did I ever seek extra help from a tutor or from my instructor?
9. In what topics was I the strongest? In what topics was I the weakest?
10. If I had it to do over, would I do anything differently?

SECTION 10.1

The Circle and the Parabola

OBJECTIVES

1 Identify conic sections and some of their applications.

2 Graph equations of circles written in standard form.

3 Write the equation of a circle, given its center and radius.

4 Convert the general form of the equation of a circle to standard form.

5 Solve application problems involving circles.

6 Convert the general form of the equation of a parabola to standard form to graph it.

ARE YOU READY?

The following problems review some basic skills that are needed when working with circles and parabolas.

1. Use a special-product formula to find $(x - 8)^2$.

2. Complete the square on $x^2 - 6x$ and factor the resulting trinomial.

3. Factor: $y^2 + 10y + 25$

4. Factor out -3 from the terms of the expression $-3y^2 - 12y$.

We have previously graphed first-degree equations in two variables such as $y = 3x + 8$ and $4x - 3y = 12$. Their graphs are lines. In this section, we will graph second-degree equations in two variables such as $x^2 + y^2 = 25$ and $x = -3y^2 - 12y - 13$. The graphs of these equations are *conic sections*.

1 Identify Conic Sections and Some of Their Applications.

The curves formed by the intersection of a plane with an infinite right-circular cone are called **conic sections.** Those curves have four basic shapes, called **circles, parabolas, ellipses,** and **hyperbolas,** as shown on the next page.

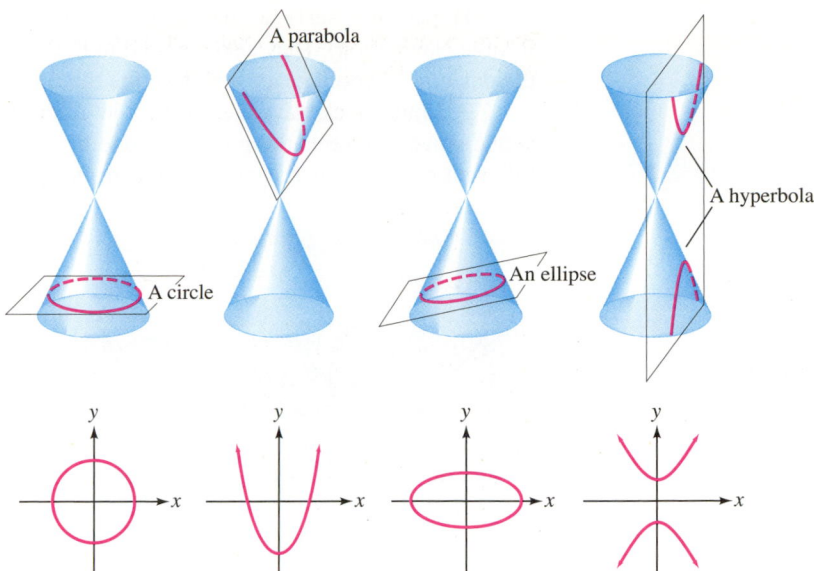

Conic sections have many applications. For example, everyone is familiar with circular wheels and gears, pizza cutters, and hula hoops.

Parabolas can be rotated to generate dish-shaped surfaces called **paraboloids.** Any light or sound placed at the **focus** of a paraboloid is reflected outward in parallel paths. This property makes parabolic surfaces ideal for flashlight and headlight reflectors. It also makes parabolic surfaces good antennas, because signals captured by such antennas are concentrated at the focus. Parabolic mirrors are capable of concentrating the rays of the sun at a single point, thereby generating tremendous heat. This property is used in the design of solar furnaces.

Any object thrown upward and outward travels in a parabolic path. An example of this is a stream of water flowing from a drinking fountain. In architecture, many arches are parabolic in shape, because this gives them strength. Cables that support suspension bridges hang in the shape of a parabola.

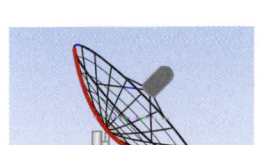

Radar dish

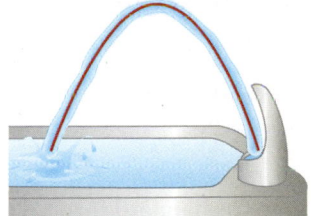

Stream of water

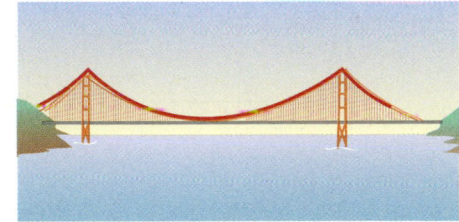

Support cables

Ellipses have optical and acoustical properties that are useful in architecture and engineering. Many arches are portions of an ellipse, because the shape is pleasing to the eye. The planets and many comets have elliptical orbits. Certain gears have elliptical shapes to provide nonuniform motion.

Arches

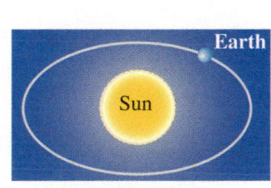

Earth's orbit

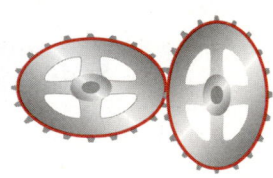

Gears

Hyperbolas serve as the basis of a navigational system known as LORAN (LOng RAnge Navigation). They also are used to find the source of a distress signal, are the basis for the design of hypoid gears, and describe the orbits of some comets.

A sonic shock wave created by a jet aircraft has the shape of a cone. In level flight, the sound wave intersects the ground as one branch of a hyperbola, as shown below. People in different places along the curve on the ground hear and feel the sonic boom at the same time.

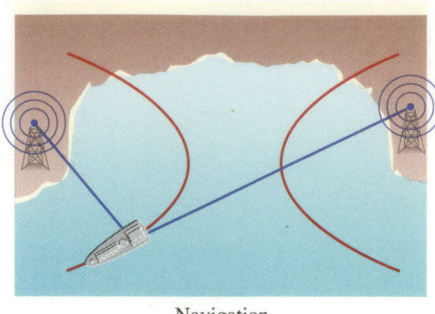

Navigation   Sonic boom

2 Graph Equations of Circles Written in Standard Form.

Every conic section can be represented by a second-degree equation in x and y. To find the equation of a circle, we use the following definition.

Definition of a Circle	A **circle** is the set of all points in a plane that are a fixed distance from a fixed point called its **center**. The fixed distance is called the **radius** of the circle.

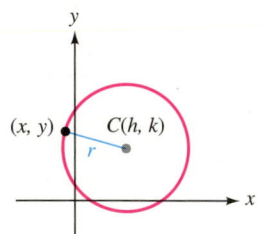

If we let (h, k) be the center of a circle and (x, y) be some point on a circle that is graphed on a rectangular coordinate system, the distance from (h, k) to (x, y) is the radius r of the circle. We can use the distance formula to find r.

$$r = \sqrt{(x - h)^2 + (y - k)^2}$$

We can square both sides to eliminate the radical and obtain

$$r^2 = (x - h)^2 + (y - k)^2$$

This result is called the **standard form of the equation of a circle** with radius r and center at (h, k).

Equation of a Circle	The **standard form of the equation of a circle** with radius r and center at (h, k) is $$(x - h)^2 + (y - k)^2 = r^2$$

EXAMPLE 1   Find the center and the radius of each circle and then graph it:
a. $(x - 4)^2 + (y - 1)^2 = 9$ **b.** $x^2 + y^2 = 25$ **c.** $(x + 3)^2 + y^2 = 12$

Strategy   We will compare each equation to the standard form of the equation of a circle, $(x - h)^2 + (y - k)^2 = r^2$, and identify h, k, and r.

Why   The center of the circle is the point with coordinates (h, k) and the radius of the circle is r.

Solution   **a.** The color highlighting shows how to compare the given equation to the standard form to find h, k, and r.

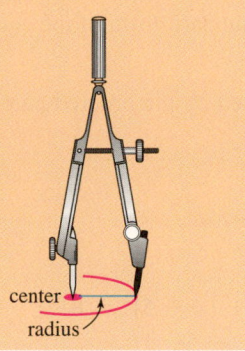
$$(x - \mathbf{4})^2 + (y - \mathbf{1})^2 = \mathbf{9}$$
$$\uparrow \qquad\qquad \uparrow \qquad\quad \uparrow$$
$$(x - \mathbf{h})^2 + (y - \mathbf{k})^2 = \mathbf{r}^2$$

$h = 4$, $k = 1$, and $r^2 = 9$. Since the radius of a circle must be positive, $r = 3$.

The center of the circle is $(h, k) = (4, 1)$ and the radius is 3.

To plot four points on the circle, we move up, down, left, and right 3 units from the center, as shown in figure (a). Then we draw a circle through the points to get the graph of $(x - 4)^2 + (y - 1)^2 = 9$, as shown in figure (b).

The center (4, 1) is not part of the graph of a circle; it only helps us sketch the graph.

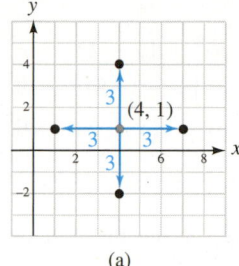

(a)

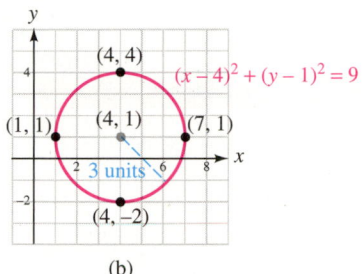

(b)

b. To find h and k, we will write $x^2 + y^2 = 25$ in the following way:

$$(x - \mathbf{0})^2 + (y - \mathbf{0})^2 = \mathbf{25}$$
$$\uparrow \qquad\qquad \uparrow \qquad\quad \uparrow$$
$$(x - \mathbf{h})^2 + (y - \mathbf{k})^2 = \mathbf{r}^2$$

$h = 0$, $k = 0$, and $r^2 = 25$. Since the radius must be positive, $r = 5$.

The center of the circle is at $(0, 0)$ and the radius is 5.

To plot four points on the circle, we move up, down, left, and right 5 units from the center. Then we draw a circle through the points to get the graph of $x^2 + y^2 = 25$, as shown.

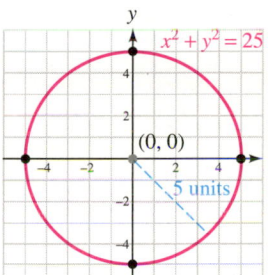

c. To find h, we will write $x + 3$ as $x - (-3)$.

Standard form requires a minus symbol here.
$$\downarrow$$
$$[x - (\mathbf{-3})]^2 + (y - \mathbf{0})^2 = \mathbf{12}$$
$$\uparrow \qquad\qquad\qquad \uparrow \qquad\quad \uparrow$$
$$(x - \mathbf{h})^2 + (y - \mathbf{k})^2 = \mathbf{r}^2$$

$h = -3$, $k = 0$, and $r^2 = 12$.

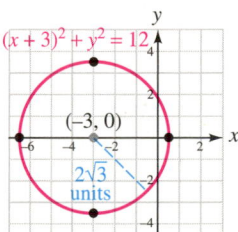

Since $r^2 = 12$, we have

$$r = \pm\sqrt{12} = \pm 2\sqrt{3}$$ Use the square root property.

Since the radius can't be negative, $r = 2\sqrt{3}$. The center of the circle is at $(-3, 0)$ and the radius is $2\sqrt{3}$.

To plot four points on the circle, we move up, down, left, and right $2\sqrt{3} \approx 3.5$ units from the center. We then draw a circle through the points to get the graph of $(x + 3)^2 + y^2 = 12$, as shown on the left.

Self Check 1   Find the center and the radius of each circle and then graph it:

a. $(x - 3)^2 + (y + 4)^2 = 4$ **b.** $x^2 + y^2 = 8$

Now Try ▶ Problems 15, 19, and 21

3 | Write the Equation of a Circle, Given Its Center and Radius.

Because a circle is determined by its center and radius, that information is all we need to know to write its equation.

EXAMPLE 2   Write the equation of the circle with radius 9 and center at $(6, -5)$.

Strategy   We substitute 9 for r, 6 for h, and -5 for k in the standard form of the equation of a circle, $(x - h)^2 + (y - k)^2 = r^2$.

Why   When writing the standard form, the center is represented by the ordered pair (h, k) and the radius as r.

Solution

$$\begin{aligned}
(x - \boldsymbol{h})^2 + (y - \boldsymbol{k})^2 &= \boldsymbol{r}^2 \qquad &&\text{This is the standard form.} \\
(x - \boldsymbol{6})^2 + [y - (\boldsymbol{-5})]^2 &= \boldsymbol{9}^2 \qquad &&\text{Substitute 6 for } h, -5 \text{ for } k, \text{ and 9 for } r. \\
(x - 6)^2 + (y + 5)^2 &= 9^2 \qquad &&\text{Write } y - (-5) \text{ as } y + 5.
\end{aligned}$$

> **Notation**
>
> Standard form can be written
> $$(x - 6)^2 + (y + 5)^2 = 9^2$$
> or
> $$(x - 6)^2 + (y + 5)^2 = 81$$

If we express 9^2 as 81, we can write the equation as $(x - 6)^2 + (y + 5)^2 = 81$.

Self Check 2   Write the equation of the circle with radius 10 and center at $(-7, 1)$.

Now Try ▶ Problems 23, 27, and 31

4 | Convert the General Form of the Equation of a Circle to Standard Form.

In Example 2, the result was written in standard form: $(x - 6)^2 + (y + 5)^2 = 81$. If we square $x - 6$ and $y + 5$, we obtain a different form for the equation of the circle.

$$\begin{aligned}
(\boldsymbol{x - 6})^2 + (\boldsymbol{y + 5})^2 &= \boldsymbol{9}^2 \\
x^2 - 12x + 36 + y^2 + 10y + 25 &= 81 \qquad &&\text{Square each binomial.} \\
x^2 - 12x + y^2 + 10y - 20 &= 0 \qquad &&\text{Subtract 81 from both sides. Combine like terms.} \\
x^2 + y^2 - 12x + 10y - 20 &= 0 \qquad &&\text{Rearrange the terms, writing} \\
& && \text{the squared terms first.}
\end{aligned}$$

> **Success Tip**
>
> This example illustrates an important fact: The equation of a circle contains both x^2 and y^2 terms on the same side of the equation with equal coefficients.

This result is written in the *general form of the equation of a circle*.

Equation of a Circle	The **general form of the equation of a circle** is $$x^2 + y^2 + Dx + Ey + F = 0$$

We can convert from the general form to the standard form of the equation of a circle by completing the square.

EXAMPLE 3   Write the equation $x^2 + y^2 - 4x + 2y - 11 = 0$ in standard form and graph it.

Strategy   We will rearrange the terms to write the equation in the form $x^2 - 4x + y^2 + 2y = 11$ and complete the square on x and y.

Why   Standard form contains the expressions $(x - h)^2$ and $(y - k)^2$. We can obtain a perfect-square trinomial that factors as $(x - 2)^2$ by completing the square on $x^2 - 4x$. We can complete the square on $y^2 + 2y$ to obtain an expression of the form $(y + 1)^2$.

Solution To write the equation in standard form, we complete the square twice.

$$x^2 + y^2 - 4x + 2y - 11 = 0$$

$$\boxed{x^2 - 4x} + \boxed{y^2 + 2y} = 11 \qquad \text{Write the x-terms together, the y-terms together, and add 11 to both sides.}$$

To complete the square on $x^2 - 4x$, we note that $\frac{1}{2}(-4) = -2$ and $(-2)^2 = \mathbf{4}$. To complete the square on $y^2 + 2y$, we note that $\frac{1}{2}(2) = 1$ and $1^2 = \mathbf{1}$. We add $\mathbf{4}$ and $\mathbf{1}$ to both sides of the equation.

$$x^2 - 4x + \mathbf{4} + y^2 + 2y + \mathbf{1} = 11 + \mathbf{4} + \mathbf{1}$$

$$(x - 2)^2 + (y + 1)^2 = 16 \qquad \text{Factor } x^2 - 4x + 4 \text{ and } y^2 + 2y + 1.$$

$x^2 + y^2 - 4x + 2y - 11 = 0$
or
$(x-2)^2 + (y+1)^2 = 16$

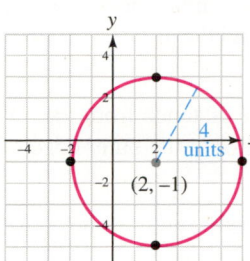

The equation also can be written as $(x - 2)^2 + (y + 1)^2 = 4^2$.

We can determine the circle's center and radius by comparing this equation to the standard form of the equation of a circle, $(x - h)^2 + (y - k)^2 = r^2$. We see that $h = 2$, $k = -1$, and $r = 4$. We can use the center, $(h, k) = (2, -1)$ and the radius $r = 4$, to graph the circle as shown on the left.

Self Check 3 Write the equation $x^2 + y^2 + 12x - 6y - 4 = 0$ in standard form and graph it.

Now Try ▶ Problem 35

Using Your Calculator ▶ **Graphing Circles**

Since the graphs of circles fail the vertical line test, their equations do not represent functions. It is more difficult to use a graphing calculator to graph equations that are not functions. For example, to graph the circle described by $(x - 1)^2 + (y - 2)^2 = 4$, we must split the equation into two functions and graph each one separately. We begin by solving the equation for y.

$$(x - 1)^2 + (y - 2)^2 = 4$$

$$(y - 2)^2 = 4 - (x - 1)^2 \qquad \text{Subtract } (x - 1)^2 \text{ from both sides.}$$

$$y - 2 = \pm\sqrt{4 - (x - 1)^2} \qquad \text{Use the square root property.}$$

$$y = 2 \pm \sqrt{4 - (x - 1)^2} \qquad \text{Add 2 to both sides.}$$

This equation defines two functions. If we graph

$$y = 2 + \sqrt{4 - (x - 1)^2} \quad \text{and} \quad y = 2 - \sqrt{4 - (x - 1)^2}$$

we get the distorted circle shown in figure (a). To get a better circle, we can use the graphing calculator's square window feature, which gives an equal unit distance on both the x- and y-axes. (Press ZOOM , 5, ENTER .) Using this feature, we get the circle shown in figure (b). Sometimes the two arcs will not connect because of approximations made by the calculator at each endpoint.

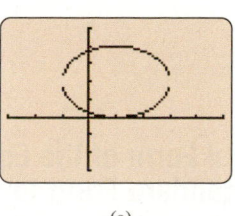

(a)

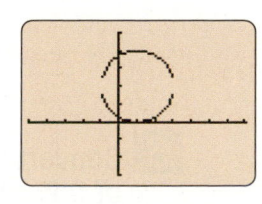

(b)

The graph of
$y = 2 + \sqrt{4 - (x - 1)^2}$
is the top half of the circle.

The graph of
$y = 2 - \sqrt{4 - (x - 1)^2}$
is the bottom half of the circle.

5 Solve Application Problems Involving Circles.

EXAMPLE 4   **Radio Translators.**   The broadcast area of a television station is bounded by the circle $x^2 + y^2 = 3{,}600$, where x and y are measured in miles. A translator station picks up the signal and retransmits it from the center of a circular area bounded by $(x + 30)^2 + (y - 40)^2 = 1{,}600$. Find the location of the translator and the greatest distance from the main transmitter that the signal can be received.

Strategy   Refer to the figure below. We will find two distances: the distance from the TV station transmitter to the translator and the distance from the translator to the outer edge of its coverage.

Why   The greatest distance of reception from the main transmitter is the sum of those two distances.

Solution   The coverage of the TV station is bounded by $x^2 + y^2 = 60^2$, a circle centered at the origin with a radius of 60 miles, as shown in yellow in the figure. Because the translator is at the center of the circle $(x + 30)^2 + (y - 40)^2 = 1{,}600$, it is located at $(-30, 40)$, a point 30 miles west and 40 miles north of the TV station. The radius of the translator's coverage is $\sqrt{1{,}600}$, or 40 miles.

As shown in the figure, the greatest distance of reception is the sum of d, the distance from the translator to the television station, and 40 miles, the radius of the translator's coverage.

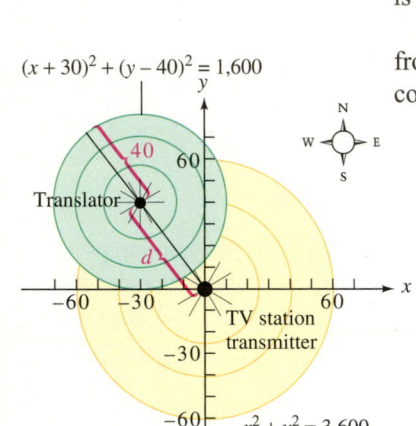

$(x + 30)^2 + (y - 40)^2 = 1{,}600$

$x^2 + y^2 = 3{,}600$

To find d, we use the distance formula to find the distance between the origin, $(x_1, y_1) = (0, 0)$, and $(x_2, y_2) = (-30, 40)$.

$$d = \sqrt{(x_2 - x_1)^2 + (y_2 - y_1)^2}$$ The distance formula was introduced in Section 7.6.

$$d = \sqrt{(-30 - 0)^2 + (40 - 0)^2}$$ Substitute for x_1, x_2, y_1, and y_2.

$$d = \sqrt{(-30)^2 + 40^2}$$ Simplify within the radical.

$$= \sqrt{2{,}500}$$ Evaluate the expression within the radical.

$$= 50$$ Find the square root.

The translator is located 50 miles from the television station, and it broadcasts the signal 40 miles. The greatest reception distance from the main transmitter signal is, therefore, $50 + 40$, or 90 miles.

Self Check 4   **Landscaping.** A landscape architect is designing a circular flower bed bounded by the circle $x^2 + y^2 = 36$ where x and y are measured in feet. Another circular flower bed in his design is bounded by $(x - 3)^2 + (y + 4)^2 = 16$. Find the length of the sidewalk from the center of the first circle to the furthest edge of the second circle.

Now Try ▸ Problem 87

6 Convert the General Form of the Equation of a Parabola to Standard Form to Graph It.

Another type of conic section is the parabola.

| **Definition of a Parabola** ▼ | A **parabola** is the set of all points in a plane that are equidistant from a fixed point, called the **focus,** and a fixed line, called the **directrix.** | 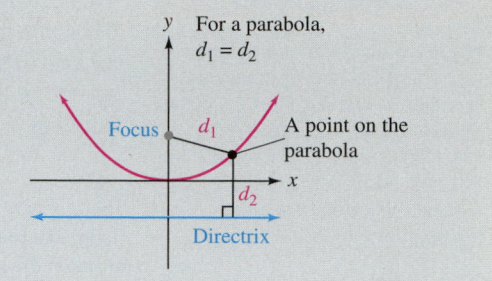 |

We have discussed parabolas whose graphs open upward or downward in Section 8.4: Quadratic Functions and Their Graphs.

| **Standard Form of the Equation of a Parabola** ▼ | The **graph of the quadratic function** $$y = a(x - h)^2 + k \quad \text{where } a \neq 0$$ is a parabola with vertex at (h, k). The axis of symmetry is the line $x = h$. The parabola opens upward when $a > 0$ and downward when $a < 0$. |

Parabolas can also open to the right and to the left, but they do not define functions because their graphs fail the vertical line test.

The two general forms of the equation of a parabola are similar.

| **Equation of a Parabola** ▼ | The **general forms of the equation of a parabola** are:

1. $y = ax^2 + bx + c$ The graph opens upward if $a > 0$ and downward if $a < 0$.
2. $x = ay^2 + by + c$ The graph opens to the right if $a > 0$ and to the left if $a < 0$. |

EXAMPLE 5   Write $y = -2x^2 + 12x - 13$ in standard form and graph it.

Strategy   We will complete the square on x to write the equation in standard form, $y = a(x - h)^2 + k$.

Why   Standard form contains the expression $(x - h)^2$. We can obtain a perfect-square trinomial that factors into that form by completing the square on x.

Solution   Because the equation is not in standard form, the coordinates of the vertex are not obvious. To write the equation in standard form, we complete the square on x.

$$y = -2x^2 + 12x - 13$$

$$y = -2(x^2 - 6x \qquad) - 13 \qquad \text{Factor out } -2 \text{ from } -2x^2 + 12x.$$

This step adds $-2 \cdot 9$ or -18 to this side.   Add 18 to counteract the addition of -18.

$$y = -2(x^2 - 6x + 9) - 13 + 18 \qquad \text{Complete the square on } x^2 - 6x.$$

$$y = -2(x - 3)^2 + 5 \qquad \text{Factor } x^2 - 6x + 9 \text{ and add: } -13 + 18 = 5.$$

$$y = \quad a(x - h)^2 + k$$

Success Tip

When the equation of a parabola that opens upward or downward is written in standard form, h is the number within the parentheses that *follows* the $-$ (subtraction symbol), and k is the number outside the parentheses that follows the $+$ (addition symbol).

This equation is written in the form $y = a(x - h)^2 + k$, where $a = -2$, $h = 3$, and $k = 5$. Thus, the graph of the equation is a parabola that opens downward with vertex at $(3, 5)$ and an axis of symmetry $x = 3$. We can construct a table of solutions and use symmetry to plot several points on the parabola. Then we draw a smooth curve through the points to get the graph of $y = -2x^2 + 12x - 13$, as shown below.

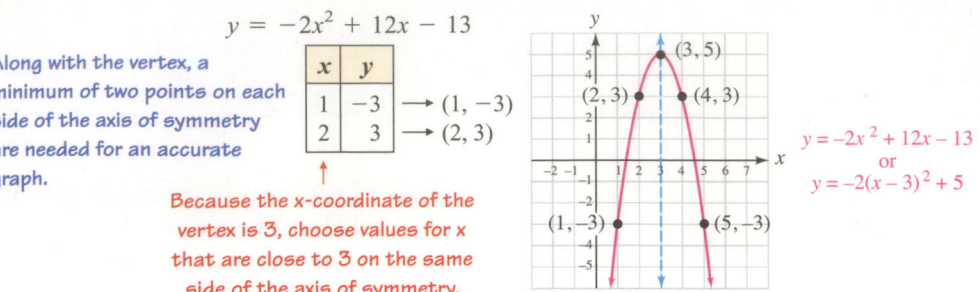

$$y = -2x^2 + 12x - 13$$

Along with the vertex, a minimum of two points on each side of the axis of symmetry are needed for an accurate graph.

x	y
1	-3	$\rightarrow (1, -3)$
2	3	$\rightarrow (2, 3)$

Because the x-coordinate of the vertex is 3, choose values for x that are close to 3 on the same side of the axis of symmetry.

$$y = -2x^2 + 12x - 13$$
or
$$y = -2(x - 3)^2 + 5$$

Self Check 5   Write $y = 2x^2 + 4x + 5$ in standard form and graph it.

Now Try ▶   Problem 39

The **standard form for the equation of a parabola** that opens to the right or left is similar to $y = a(x - h)^2 + k$, except that the variables, x and y, exchange positions as do the constants, h and k.

Standard Form of the Equation of a Parabola

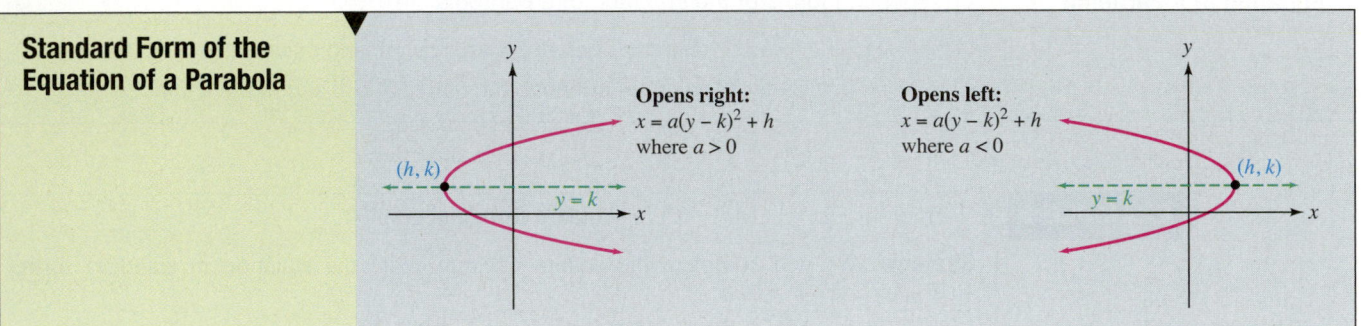

Opens right:
$x = a(y - k)^2 + h$
where $a > 0$

Opens left:
$x = a(y - k)^2 + h$
where $a < 0$

(h, k) $y = k$

(h, k) $y = k$

EXAMPLE 6   Graph: $x = \dfrac{1}{2}y^2$

Strategy   We will compare the equation to the standard form of the equation of a parabola to find a, h, and k.

Why   Once we know these values, we can locate the vertex of the graph. We also know whether the parabola will open to the left or to the right.

Solution   This equation is written in the form $x = a(y - k)^2 + h$, where $a = \frac{1}{2}$, $k = 0$, and $h = 0$. The graph of the equation is a parabola that opens to the right with vertex at $(0, 0)$ and an axis of symmetry $y = 0$.

To construct a table of solutions, we choose values of y and find their corresponding values of x. For example, if $y = 1$ and if $y = 2$, we have:

$$x = \frac{1}{2}y^2 \qquad\qquad x = \frac{1}{2}y^2$$

$$x = \frac{1}{2}(\mathbf{1})^2 \quad \text{Substitute 1 for } y. \qquad x = \frac{1}{2}(\mathbf{2})^2 \quad \text{Substitute 2 for } y.$$

$$x = \frac{1}{2} \qquad\qquad\qquad x = 2$$

The point $\left(\frac{1}{2}, 1\right)$ is on the parabola.   The point $(2, 2)$ is on the parabola.

We plot the ordered pairs from the table and use symmetry to plot three more points on the parabola. Then we draw a smooth curve through the points to get the graph of $x = \frac{1}{2}y^2$, as shown below.

$$x = \frac{1}{2}y^2$$

Along with the vertex, three points on each side of the axis of symmetry were plotted to obtain an accurate graph.

x	y	
$\frac{1}{2}$	1	$\rightarrow \left(\frac{1}{2}, 1\right)$
2	2	$\rightarrow (2, 2)$
8	4	$\rightarrow (8, 4)$

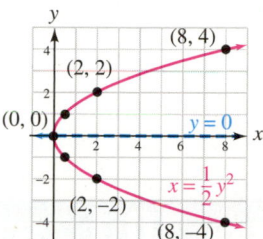

Because the y-coordinate of the vertex is 0, choose values for y that are close to 0 on the same side of the axis of symmetry.

Self Check 6 Graph: $x = -\frac{2}{3}y^2$

Now Try ▶ Problem 43

EXAMPLE 7 Write $x = -3y^2 - 12y - 13$ in standard form and graph it.

Strategy We will complete the square on y to write the equation in standard form, $x = a(y - k)^2 + h$.

Why Standard form contains the expression $(y - k)^2$. We can obtain a perfect-square trinomial that factors into that form by completing the square on y.

Solution To write the equation in standard form, we complete the square.

$$x = -3y^2 - 12y - 13$$

$$x = -3(y^2 + 4y \quad) - 13 \qquad \text{Factor out } -3 \text{ from } -3y^2 - 12y.$$

$$x = \mathbf{-3}(y^2 + 4y + \mathbf{4}) - 13 + \mathbf{12} \qquad \begin{array}{l}\text{Complete the square on } y^2 + 4y. \text{ Then add 12}\\\text{to the right side to counteract } -3 \cdot 4 = -12.\end{array}$$

$$x = -3(y + 2)^2 - 1 \qquad \text{Factor } y^2 + 4y + 4 \text{ and add: } -13 + 12 = 1.$$

$$x = -3[y - (-2)]^2 + (-1)$$

$$x = \quad a[y - \quad k]^2 + \quad h \qquad \text{Rewrite to help determine } h \text{ and } k.$$

This equation is in the standard form $x = a(y - k)^2 + h$, where $a = -3$, $k = -2$, and $h = -1$. The graph of the equation is a parabola that opens to the left with vertex at $(-1, -2)$ and an axis of symmetry $y = -2$.

We can construct a table of solutions and use symmetry to plot several points on the parabola. Then we draw a smooth curve through the points to get the graph of $x = -3y^2 - 12y - 13$, as shown below.

$$x = -3y^2 - 12y - 13$$
or
$$x = -3(y + 2)^2 - 1$$

x	y
-4	-1	→ $(-4, -1)$
-13	0	→ $(-13, 0)$

↑
Choose at least two values for y close to −2, and find the corresponding x-values.

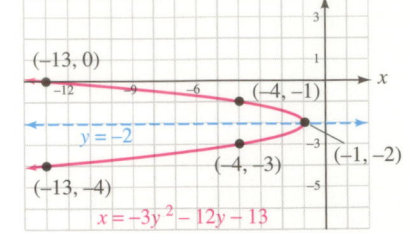

Success Tip

The equation of a circle contains an x^2 and a y^2 term. The equation of a parabola has either an x^2 term or a y^2 term, but not both.

Self Check 7 Write $x = 3y^2 - 6y - 1$ in standard form and graph it.

Now Try ▶ Problem 49

SECTION 10.1 ▶ STUDY SET

VOCABULARY

Fill in the blanks.

1. The curves formed by the intersection of a plane with an infinite right-circular cone are called _____ _____.

2. Give the name of each conic shown below.

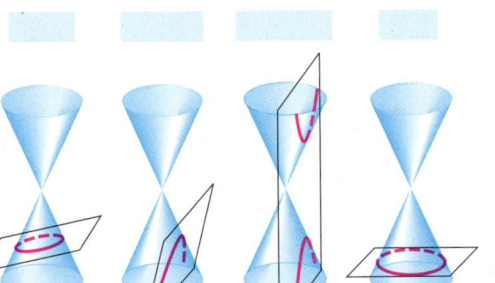

3. A _____ is the set of all points in a plane that are a fixed distance from a fixed point called its center. The fixed distance is called the _____.

4. A parabola is the set of all points in a plane that are equidistant from a fixed point and a fixed _____.

CONCEPTS

5. **a.** Write the standard form of the equation of a circle.

   b. Write the standard form of the equation of a circle with the center at the origin.

6. **a.** Find the center and the radius of the circle graphed on the right.

   b. Write the equation of the circle.

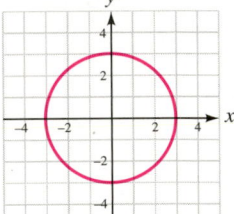

7. **a.** Find the center and the radius of the circle graphed on the right.

   b. Write the equation of the circle.

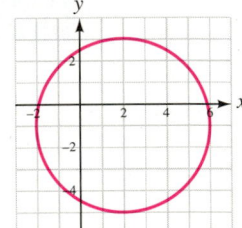

8. Fill in the blanks. To complete the square on $x^2 + 2x$ and on $y^2 - 6y$, what numbers must be added to each side of the equation?

$$x^2 + 2x + y^2 - 6y = 2$$
$$x^2 + 2x + \boxed{} + y^2 - 6y + \boxed{} = 2 + \boxed{} + \boxed{}$$

9. **a.** What is the standard form of the equation of a parabola opening upward or downward?

   b. What is the standard form of the equation of a parabola opening to the right or left?

10. Fill in the blanks.

   a. To complete the square on the right side, what should be factored from the first two terms?

$$x = 4y^2 + 16y + 9$$
$$x = \boxed{}\,(y^2 + 4y \quad\quad) + 9$$

b. To complete the square on $y^2 + 4y$, what should be added within the parentheses, and what should be subtracted outside the parentheses?

$$x = 4(y^2 + 4y + \boxed{}) + 9 - \boxed{}$$

11. Determine whether the graph of each equation is a circle or a parabola.

a. $x^2 + y^2 - 6x + 8y - 10 = 0$

b. $y^2 - 2x + 3y - 9 = 0$

c. $x^2 + 5x - y = 0$

d. $x^2 + 12x + y^2 = 0$

12. Draw a parabola using the given facts.

- Opens right
- Vertex $(-3, 2)$
- Passes through $(-2, 1)$
- x-intercept $(1, 0)$

NOTATION

13. Find h, k, and r: $(x - 6)^2 + (y + 2)^2 = 9$

14. a. Find a, h, and k: $y = 6(x - 5)^2 - 9$

b. Find a, h, and k: $x = -3(y + 2)^2 + 1$

GUIDED PRACTICE

Find the center and radius of each circle and graph it. See Example 1.

15. $x^2 + y^2 = 9$

16. $x^2 + y^2 = 16$

17. $x^2 + (y + 3)^2 = 1$

18. $(x + 4)^2 + y^2 = 1$

19. $(x + 3)^2 + (y - 1)^2 = 16$

20. $(x - 1)^2 + (y + 4)^2 = 9$

21. $x^2 + y^2 = 6$

22. $x^2 + y^2 = 10$

Write the equation of a circle in standard form with the following properties. See Example 2.

23. Center at the origin; radius 1

24. Center at the origin; radius 4

25. Center at $(6, 8)$; radius 5

26. Center at $(5, 3)$; radius 2

27. Center at $(-2, 6)$; radius 12

28. Center at $(5, -4)$; radius 6

29. Center at $(0, 0)$; radius $\dfrac{1}{4}$

30. Center at $(0, 0)$; radius $\dfrac{1}{3}$

31. Center at $\left(\dfrac{2}{3}, -\dfrac{7}{8}\right)$; radius $\sqrt{2}$

32. Center at $(-0.7, -0.2)$; radius $\sqrt{11}$

33. Center at the origin; diameter $4\sqrt{2}$

34. Center at the origin; diameter $8\sqrt{3}$

Write each equation of a circle in standard form and graph it. Give the coordinates of its center and give the radius. See Example 3.

35. $x^2 + y^2 - 2x + 4y = -1$

36. $x^2 + y^2 + 6x - 4y = -12$

37. $x^2 + y^2 + 4x + 2y = 4$

38. $x^2 + y^2 + 8x + 2y = -13$

Write each equation of a parabola in standard form and graph it. Give the coordinates of the vertex. See Example 5.

39. $y = 2x^2 - 4x + 5$

40. $y = x^2 + 4x + 5$

41. $y = -x^2 - 2x + 3$

42. $y = -2x^2 - 4x$

Graph each equation of a parabola. Give the coordinates of the vertex. See Example 6.

43. $x = y^2$

44. $x = 2y^2$

45. $x = 2(y + 1)^2 + 3$

46. $x = 3(y - 2)^2 - 1$

Write each equation of a parabola in standard form and graph it. Give the coordinates of the vertex. See Example 7.

47. $x = y^2 - 2y + 5$

48. $x = y^2 + 6y + 8$

49. $x = -3y^2 + 18y - 25$

50. $x = -2y^2 + 4y + 1$

Use a graphing calculator to graph each equation. (Hint: Solve for y and graph two functions.) See Using Your Calculator: Graphing Circles.

51. $x^2 + y^2 = 7$

52. $x^2 + y^2 = 5$

53. $(x + 1)^2 + y^2 = 16$

54. $x^2 + (y - 2)^2 = 4$

Use a graphing calculator to graph each equation. (Hint: Solve for y and graph two functions when necessary.)

55. $x = 2y^2$

56. $x = y^2 - 4$

57. $x^2 - 2x + y = 6$

58. $x = -2(y - 1)^2 + 2$

TRY IT YOURSELF

Write each equation in standard form, if it is not already so, and graph it. If the graph is a circle, give the coordinates of its center and its radius. If the graph is a parabola, give the coordinates of its vertex.

59. $x = y^2 - 6y + 4$

60. $x = y^2 - 8y + 13$

61. $(x - 2)^2 + y^2 = 25$

62. $x^2 + (y - 3)^2 = 25$

63. $x^2 + y^2 - 6x + 8y + 18 = 0$

64. $x^2 + y^2 - 4x + 4y - 3 = 0$

65. $y = 4x^2 - 16x + 17$ 66. $y = 4x^2 - 32x + 63$

67. $(x - 1)^2 + (y - 3)^2 = 15$ 68. $(x + 1)^2 + (y + 1)^2 = 8$

69. $x = -y^2 + 1$ 70. $x = -y^2 - 5$

71. $(x - 2)^2 + (y - 4)^2 = 36$ 72. $(x - 3)^2 + (y - 2)^2 = 36$

73. $x = -\dfrac{1}{4}y^2$ 74. $x = 4y^2$

75. $x = -6(y - 1)^2 + 3$ 76. $x = -6(y + 1)^2 - 4$

77. $x^2 + y^2 + 2x - 8 = 0$ 78. $x^2 + y^2 - 4y = 12$

79. $x = \dfrac{1}{2}y^2 + 2y$ 80. $x = -\dfrac{1}{3}y^2 - 2y$

81. $y = -4(x + 5)^2 + 5$ 82. $y = -4(x - 4)^2 - 4$

Look Alikes . . .

Find the center and radius of each circle.

83. **a.** $(x - 4)^2 + (y + 7)^2 = 28$ **b.** $(x + 4)^2 + (y - 7)^2 = 28$

84. **a.** $x^2 + y^2 + 10x - 14y - 7 = 0$
    b. $x^2 + y^2 - 10x + 14y - 7 = 0$

Find the coordinates of the vertex and the direction in which each parabola opens.

85. **a.** $y = 8(x - 3)^2 + 6$
    b. $x = 8(y - 3)^2 + 6$
86. **a.** $y = -x^2 + 4x + 1$
    b. $x = -y^2 + 4y + 1$

APPLICATIONS

87. **Broadcast Ranges.** Radio stations applying for licensing may not use the same frequency if their broadcast areas overlap. One station's coverage is bounded by $x^2 + y^2 - 8x - 20y + 16 = 0$, and the other's by $x^2 + y^2 + 2x + 4y - 11 = 0$. May they be licensed for the same frequency?

88. **Meshing Gears.** For design purposes, the large gear is described by the circle $x^2 + y^2 = 16$. The smaller gear is a circle centered at $(7, 0)$ and tangent to the larger circle. Find the equation of the smaller gear.

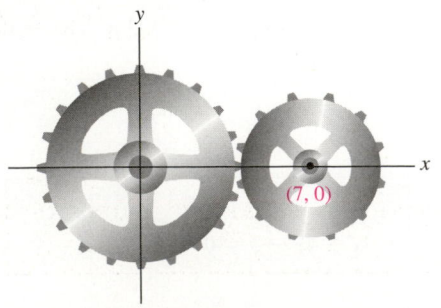

(7, 0)

89. from **Campus to Careers**
Civil Engineer

Two sections of a new freeway join with a curve that is one-quarter of a circle, as shown below. The equation of the circle is $x^2 + y^2 - 10x - 12y + 52 = 0$, where distances are measured in miles.

a. How far from City Hall will the new freeway intersect State Street?

b. How far from City Hall will the new freeway intersect Highway 60?

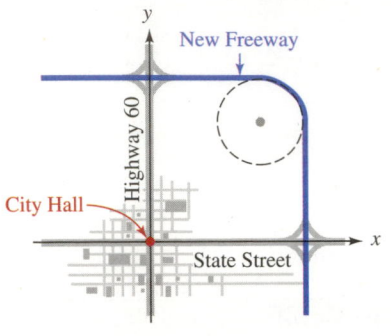

90. **Walkways.** The walkway shown is bounded by the two circles $x^2 + y^2 = 2,500$ and $(x - 10)^2 + y^2 = 900$, measured in feet. Find the largest and the smallest width of the walkway.

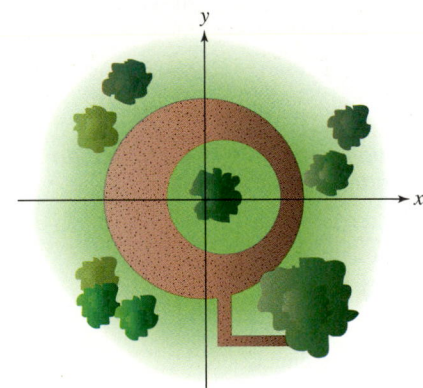

91. **Projectiles.** The cannonball in the illustration follows the parabolic path $y = 30x - x^2$. How far short of the castle does it land?

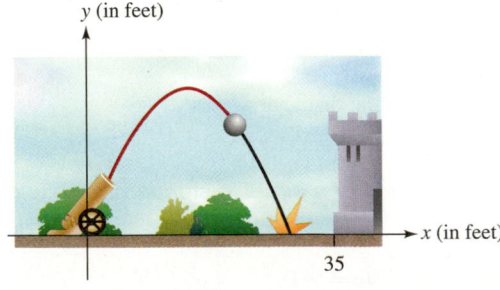

y (in feet)

x (in feet)

35

92. Projectiles. In Exercise 91, how high does the cannonball get?

93. Comets. If the orbit of the comet is approximated by the equation $2y^2 - 9x = 18$, how far is it from the sun at the vertex V of the orbit? Distances are measured in astronomical units (AU).

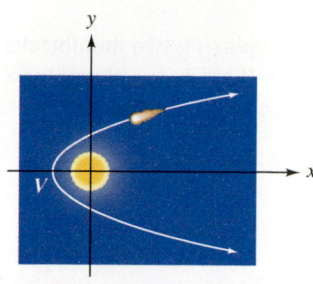

94. Satellite Antennas. The cross section of the satellite antenna in the illustration is a parabola given by the equation $y = \frac{1}{16}x^2$, with distances measured in feet. If the dish is 8 feet wide, how deep is it?

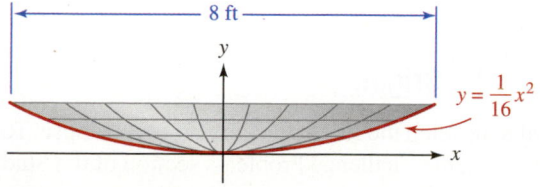

WRITING

95. Explain how to decide from its equation whether the graph of a parabola opens up, down, right, or left.

96. From the equation of a circle, explain how to determine the radius and the coordinates of the center.

97. On the day of an election, the warning *"No electioneering within a 1,000-foot radius of this polling place"* was posted in front of a school. Explain what it means.

98. What is meant by the *turning radius* of a truck?

REVIEW

Solve.

99. $|3x - 4| = 11$

100. $\left|\dfrac{4 - 3x}{5}\right| = 12$

101. $|3x + 4| = |5x - 2|$

102. $|6 - 4x| = |x + 2|$

CHALLENGE PROBLEMS

103. Could the intersection of a plane with an infinite right-circular cone as shown on page 857 be a single point? If so, draw a picture that illustrates this.

104. Under what conditions will the graph of $x = a(y - k)^2 + h$ have no y-intercepts?

105. Write the equation of a circle with a diameter whose endpoints are at $(-2, -6)$ and $(8, 10)$.

106. Write the equation of a circle with a diameter whose endpoints are at $(-5, 4)$ and $(7, -3)$.

SECTION 10.2

The Ellipse

OBJECTIVES

1 Define an ellipse.

2 Graph ellipses centered at the origin.

3 Graph ellipses centered at (h, k).

4 Solve application problems involving ellipses.

ARE YOU READY?

The following problems review some basic skills that are needed when working with ellipses.

1. Solve: $a^2 = 81$

2. Simplify: $y - (-4)$

3. Multiply: $16\left(\dfrac{x^2}{4} + \dfrac{y^2}{16}\right)$

4. Simplify: $\sqrt{8}$

A third conic section is an oval curve called an *ellipse*. Ellipses can be nearly round and look almost like a circle, or they can be long and narrow. In this section, we will learn how to construct ellipses and how to graph equations that represent ellipses.

1 Define an Ellipse.

To define a circle, we considered a fixed distance from a fixed point. The definition of an ellipse involves *two* distances from *two* fixed points.

Definition of an Ellipse	An **ellipse** is the set of all points in a plane for which the sum of the distances from two fixed points is a constant.

The figure below illustrates that any point on an ellipse is a constant distance $d_1 + d_2$ from two fixed points, each of which is called a **focus**. Midway between the **foci** is the **center** of the ellipse.

We can construct an ellipse by placing two thumbtacks fairly close together to serve as foci. We then tie each end of a piece of string to a thumbtack, catch the loop with the point of a pencil, and (keeping the string taut) draw the ellipse.

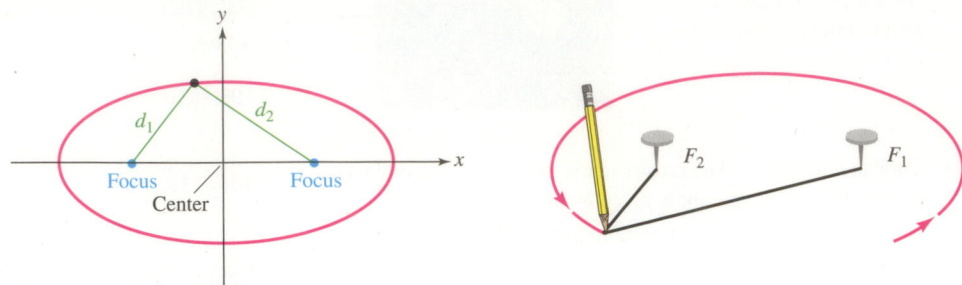

The Language of Algebra

The word **foci** (pronounced foe-sigh) is plural for the word *focus*. In the illustration on the right, the foci are labeled using subscript notation. One focus is F_1 and the other is F_2.

2 Graph Ellipses Centered at the Origin.

The definition of an ellipse can be used to develop the standard equation of an ellipse. To learn more about the derivation, see Problem 74 in the Challenge Problems section of the Study Set.

Equation of an Ellipse Centered at the Origin	The **standard form of the equation of an ellipse** that is symmetric with respect to both axes and centered at $(0, 0)$ is $$\frac{x^2}{a^2} + \frac{y^2}{b^2} = 1 \quad \text{where } a > 0 \text{ and } b > 0$$

To graph an ellipse centered at the origin, it is helpful to know the intercepts of the graph. To find the x-intercepts of the graph of $\frac{x^2}{a^2} + \frac{y^2}{b^2} = 1$ we let $y = 0$ and solve for x. To find the y-intercepts, we let $x = 0$ and solve for y.

x-intercepts:

$$\frac{x^2}{a^2} + \frac{\mathbf{0}^2}{b^2} = 1 \qquad \text{Substitute 0 for } y.$$

$$\frac{x^2}{a^2} + 0 = 1 \qquad \text{Simplify: } \frac{0^2}{b^2} = 0.$$

$$x^2 = a^2 \qquad \text{Multiply both sides by } a^2.$$

$$x = \pm a \qquad \text{Use the square root property.}$$

The x-intercepts are $(a, 0)$ and $(-a, 0)$.

y-intercepts:

$$\frac{\mathbf{0}^2}{a^2} + \frac{y^2}{b^2} = 1 \qquad \text{Substitute 0 for } x.$$

$$0 + \frac{y^2}{b^2} = 1 \qquad \text{Simplify: } \frac{0^2}{a^2} = 0.$$

$$y^2 = b^2 \qquad \text{Multiply both sides by } b^2.$$

$$y = \pm b \qquad \text{Use the square root property.}$$

The y-intercepts are $(0, b)$ and $(0, -b)$.

In general, we have the following results.

The Intercepts of an Ellipse	The graph of $\frac{x^2}{a^2} + \frac{y^2}{b^2} = 1$ is an ellipse, centered at the origin, with x-intercepts $(a, 0)$ and $(-a, 0)$ and y-intercepts $(0, b)$ and $(0, -b)$.

For $\frac{x^2}{a^2} + \frac{y^2}{b^2} = 1$, if $a > b$, the ellipse is horizontal, as shown on the next page, on the left. If $b > a$, the ellipse is vertical, as shown on the right. The points V_1 and V_2 are called the **vertices** of the ellipse. The line segment joining the vertices is called the **major axis**, and its midpoint is called the **center** of the ellipse. The line segment whose endpoints are on the ellipse and that is perpendicular to the major axis at the center is called the **minor axis** of the ellipse.

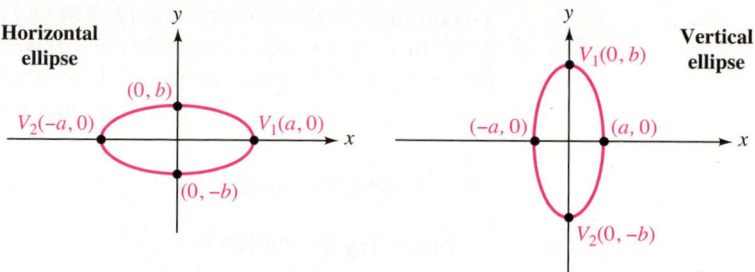

EXAMPLE 1   Graph: $\dfrac{x^2}{36} + \dfrac{y^2}{9} = 1$

Strategy   This equation is in standard $\dfrac{x^2}{a^2} + \dfrac{y^2}{b^2} = 1$ form. We will identify a and b.

Why   Once we know a and b, we can determine the intercepts of the graph of the ellipse.

Solution   The color highlighting shows how to compare the given equation to the standard form to find a and b.

$$\frac{x^2}{\mathbf{36}} + \frac{y^2}{\mathbf{9}} = 1 \qquad \frac{x^2}{\mathbf{a^2}} + \frac{y^2}{\mathbf{b^2}} = 1$$

Since $a^2 = 36$, it follows that $a = 6$.   Since $b^2 = 9$, it follows that $b = 3$.

The Language of Algebra

The word **vertices** is the plural form of the word *vertex*. From the graph, we see that one vertex of this horizontal ellipse is the point $(6, 0)$ and the other vertex is the point $(-6, 0)$.

The center of the ellipse is $(0, 0)$. The x-intercepts are $(a, 0)$ and $(-a, 0)$, or $(6, 0)$ and $(-6, 0)$. The y-intercepts are $(0, b)$ and $(0, -b)$, or $(0, 3)$ and $(0, -3)$. Using these four points as a guide, we draw an oval curve through them, as shown in figure (a). The result is a horizontal ellipse because $a > b$.

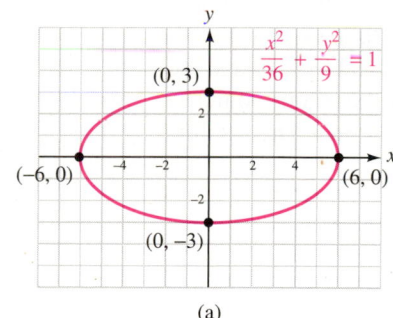

(a)

$$\frac{x^2}{36} + \frac{y^2}{9} = 1$$

x	y	
2	$\pm 2\sqrt{2}$	→ $(2, \pm 2.8)$
4	$\pm\sqrt{5}$	→ $(4, \pm 2.2)$

Approximate the radicals to plot these points.

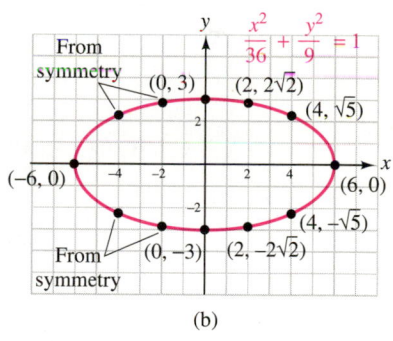

(b)

To increase the accuracy of the graph, we can find additional ordered pairs that satisfy the equation and plot them. For example, if $x = 2$, we have

$$\frac{2^2}{36} + \frac{y^2}{9} = 1$$ Substitute 2 for x in the equation of the ellipse.

$$36\left(\frac{4}{36} + \frac{y^2}{9}\right) = 36(1)$$ To clear the fractions, multiply both sides by the LCD, 36.

$$4 + 4y^2 = 36$$ Distribute the multiplication by 36 and simplify.

$$y^2 = 8$$ Subtract 4 from both sides and divide both sides by 4.

$$y = \pm\sqrt{8}$$ Use the square root property.

$$y = \pm 2\sqrt{2}$$ Simplify the radical.

Since two values of y, $2\sqrt{2}$ and $-2\sqrt{2}$, correspond to the x-value 2, we have found two points on the ellipse: $(2, 2\sqrt{2})$ and $(2, -2\sqrt{2})$. To plot them, approximate the y-coordinate to the nearest tenth: $(2, 2.8)$ and $(2, -2.8)$.

In a similar way, we can find the corresponding values of y for the x-value 4. In figure (b) on the previous page, we record these ordered pairs in a table, plot them, use symmetry with respect to the y-axis to plot four other points, and draw the graph of the ellipse.

Self Check 1 Graph: $\dfrac{x^2}{49} + \dfrac{y^2}{25} = 1$

Now Try ▶ Problem 17

EXAMPLE 2 Graph: $16x^2 + y^2 = 16$

Strategy We will write the equation in standard $\dfrac{x^2}{a^2} + \dfrac{y^2}{b^2} = 1$ form.

Why When the equation is in standard form, we will be able to identify the center and the intercepts of the graph of the ellipse.

Solution The given equation is not in standard form. To write it in standard form with 1 on the right side, we divide both sides by 16.

$$16x^2 + y^2 = 16$$

$$\frac{16x^2}{16} + \frac{y^2}{16} = \frac{16}{16} \quad \color{red}{\text{Divide both sides by 16, term-by-term.}}$$

$$\frac{x^2}{1} + \frac{y^2}{16} = 1 \quad \color{red}{\text{Simplify the fractions: } \frac{16x^2}{16} = x^2 = \frac{x^2}{1} \text{ and } \frac{16}{16} = 1.}$$

Success Tip

Although the term $\dfrac{16x^2}{16}$ simplifies to x^2, we write it as the fraction $\dfrac{x^2}{1}$ so that it has the form $\dfrac{x^2}{a^2}$.

The center of the ellipse is $(0, 0)$. To determine a and b, we can write the equation in the form

$$\frac{x^2}{1^2} + \frac{y^2}{4^2} = 1 \quad \color{red}{\text{To find } a, \text{ write 1 as } 1^2. \text{ To find } b, \text{ write 16 as } 4^2.}$$

Since a^2 (the denominator of x^2) is 1^2, it follows that $a = 1$, and since b^2 (the denominator of y^2) is 4^2, it follows that $b = 4$. Thus, the x-intercepts of the graph are $(1, 0)$ and $(-1, 0)$ and the y-intercepts are $(0, 4)$ and $(0, -4)$. We use these four points as guides to sketch the graph of the ellipse, as shown. The result is a vertical ellipse because $b > a$.

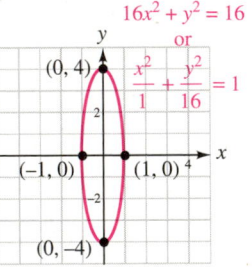

Self Check 2 Graph: $9x^2 + y^2 = 9$

Now Try ▶ Problem 21

---

3 Graph Ellipses Centered at (*h, k*).

Not all ellipses are centered at the origin. As with the graphs of circles and parabolas, the graph of an ellipse can be translated horizontally and vertically.

Equation of an Ellipse Centered at (*h, k*)

The standard form of the equation of a horizontal or vertical ellipse centered at (h, k) is

$$\frac{(x - h)^2}{a^2} + \frac{(y - k)^2}{b^2} = 1 \quad \text{where } a > 0 \text{ and } b > 0$$

For a horizontal ellipse, a is the distance from the center to a vertex. For a vertical ellipse, b is the distance from the center to a vertex.

EXAMPLE 3   Graph: $\dfrac{(x-2)^2}{16} + \dfrac{(y+3)^2}{25} = 1$

Strategy   The equation is in standard $\dfrac{(x-h)^2}{a^2} + \dfrac{(y-k)^2}{b^2} = 1$ form. We will identify h, k, a, and b.

Why   If we know h, k, a, and b, we can graph the ellipse.

Solution   To determine h, k, a, and b, we write the equation in the form

$$\dfrac{(x-2)^2}{4^2} + \dfrac{[y-(-3)]^2}{5^2} = 1$$

To find k, write $y + 3$ as $y - (-3)$.
To find a, write 16 as 4^2. To find b, write 25 as 5^2.

Success Tip

When the equation of an ellipse is written in standard form, h and k are the numbers within the parentheses that *follow* the $-$ (subtraction) symbols.

We find the center of the ellipse in the same way we would find the center of a circle, by examining $(x-2)^2$ and $(y+3)^2$. Since $h = 2$ and $k = -3$, this is the equation of an ellipse centered at $(h, k) = (2, -3)$. From the denominators, 4^2 and 5^2, we find that $a = 4$ and $b = 5$. Because $b > a$, it is a vertical ellipse.

We first plot the center, as shown below. Since b is the distance from the center to a vertex for a vertical ellipse, we can locate the vertices by counting 5 units above and 5 units below the center. The vertices are the points $(2, 2)$ and $(2, -8)$.

To locate two more points on the ellipse, we use the fact that a is 4 and count 4 units to the left and to the right of the center. We see that the points $(-2, -3)$ and $(6, -3)$ are also on the graph.

Using these four points as guides, we draw the graph shown below on the left. The illustration on the right shows how the graph of $\dfrac{(x-2)^2}{16} + \dfrac{(y+3)^2}{25} = 1$ can be obtained by translating the graph of $\dfrac{x^2}{16} + \dfrac{y^2}{25} = 1$ to the right 2 units, and then down 3 units.

The center $(2, -3)$ is not part of the graph of the ellipse; it only helps us sketch the graph.

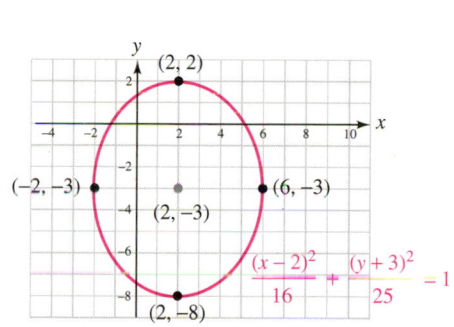

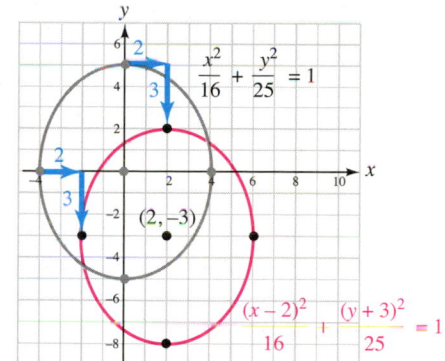

Self Check 3   Graph: $\dfrac{(x-1)^2}{9} + \dfrac{(y+2)^2}{16} = 1$

Now Try   ▶   Problem 25

Using Your Calculator ▶ **Graphing Ellipses**

To use a graphing calculator to graph the equation from Example 3, $\dfrac{(x-2)^2}{16} + \dfrac{(y+3)^2}{25} = 1$, we clear the equation of fractions and solve for y.

$$25(x-2)^2 + 16(y+3)^2 = 400$$ Multiply both sides by 400.

$$16(y+3)^2 = 400 - 25(x-2)^2$$ Subtract $25(x-2)^2$ from both sides.

$$(y+3)^2 = \dfrac{400 - 25(x-2)^2}{16}$$ Divide both sides by 16.

$$y + 3 = \pm\dfrac{\sqrt{400 - 25(x-2)^2}}{4}$$ Use the square root property.

$$y = -3 \pm \dfrac{\sqrt{400 - 25(x-2)^2}}{4}$$ Subtract 3 from both sides.

The previous equation represents two functions. On a calculator, we can graph them in a square window to get the ellipse shown here.

$$y = -3 + \frac{\sqrt{400 - 25(x-2)^2}}{4}$$

and

$$y = -3 - \frac{\sqrt{400 - 25(x-2)^2}}{4}$$

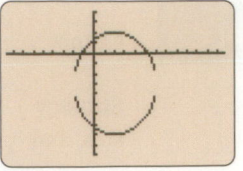

As we saw with circles, the two portions of the ellipse do not quite connect. This is because the graphs are nearly vertical there.

EXAMPLE 4   Graph: $4(x-2)^2 + 9(y-1)^2 = 36$

Strategy   We will write the equation in standard $\dfrac{(x-h)^2}{a^2} + \dfrac{(y-k)^2}{b^2} = 1$ form. Then we will identify h, k, a, and b.

Why   If we know h, k, a, and b, we can graph the ellipse.

Solution   This equation is not in standard form. To write it in standard form with 1 on the right side, we divide both sides by 36.

$$4(x-2)^2 + 9(y-1)^2 = 36$$

$$\frac{4(x-2)^2}{36} + \frac{9(y-1)^2}{36} = \frac{36}{36} \qquad \text{Divide both sides by 36.}$$

$$\frac{(x-2)^2}{9} + \frac{(y-1)^2}{4} = 1 \qquad \text{Simplify: } \tfrac{4}{36} = \tfrac{1}{9},\ \tfrac{9}{36} = \tfrac{1}{4},\ \text{and } \tfrac{36}{36} = 1.$$

This is the standard form of the equation of a horizontal ellipse, centered at $(2, 1)$, with $a = 3$ and $b = 2$. The graph of the ellipse is shown in the margin.

$4(x-2)^2 + 9(y-1)^2 = 36$
or
$\dfrac{(x-2)^2}{9} + \dfrac{(y-1)^2}{4} = 1$

$\bullet\,(2, 1)$

Self Check 4   Graph: $12(x-1)^2 + 3(y+1)^2 = 48$

Now Try ▶ Problem 29

---

4 Solve Application Problems Involving Ellipses.

EXAMPLE 5   **Landscape Design.**   A landscape architect is designing an elliptical pool that will fit in the center of a 20-by-30-foot rectangular garden, leaving 5 feet of clearance on all sides, as shown in the illustration on the next page. Find the equation of the ellipse.

Strategy   We will establish a coordinate system with its origin at the center of the garden. Then we will determine the x- and y-intercepts of the edge of the pool.

Why   If we know the x- and y-intercepts of the graph of the edge of the elliptical pool, we can use that information to write its equation.

Solution   We place the rectangular garden in the coordinate system shown below. To maintain 5 feet of clearance at the ends of the ellipse, the x-intercepts must be the points $(10, 0)$ and $(-10, 0)$. Similarly, the y-intercepts are the points $(0, 5)$ and $(0, -5)$.

Since the ellipse is centered at the origin, its equation has the form

$$\frac{x^2}{a^2} + \frac{y^2}{b^2} = 1$$

with $a = 10$ and $b = 5$. Thus, the equation of the boundary of the pool is

$$\frac{x^2}{100} + \frac{y^2}{25} = 1$$

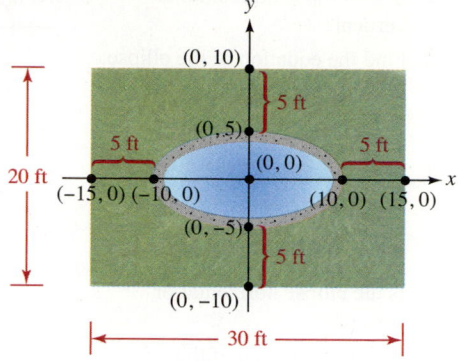

Self Check 5   **Decorating.** An interior decorator is designing an elliptical mirror that will fit in the center of a 52-in. tall and 40-in. wide rectangular panel, leaving 2 in. of clearance on all sides. Find the equation of the ellipse.

Now Try ▶   Problem 59

Ellipses, like parabolas, have **reflective properties** that are used in many practical applications. For example, any light or sound originating at one focus of an ellipse is reflected by the interior of the figure to the other focus.

Whispering Galleries

In an elliptical dome, even the slightest whisper made by a person standing at one focus can be heard by a person standing at the other focus.

Elliptical billiards tables

When a ball is shot from one focus, it will rebound off the side of the table into a pocket located at the other focus.

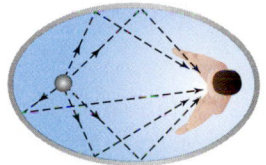

Treatment for kidney stones

The patient is positioned in an elliptical tank of water so that the kidney stone is at one focus. High-intensity sound waves generated at another focus are reflected to the stone to shatter it.

SECTION 10.2 ▶ STUDY SET

VOCABULARY

Fill in the blanks.

1. The curve graphed at the right is an _____.

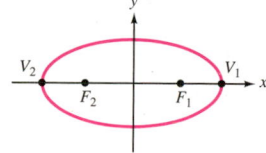

2. An _____ is the set of all points in a plane for which the sum of the distances from two fixed points is a constant.

3. In the graph above, F_1 and F_2 are the _____ of the ellipse. Each one is called a _____ of the ellipse.

4. In the graph above, V_1 and V_2 are the _____ of the ellipse. Each one is called a _____ of the ellipse.

5. The line segment joining the vertices of an ellipse is called the _____ axis of the ellipse.

6. The midpoint of the major axis of an ellipse is the _____ of the ellipse.

CONCEPTS

7. Write the standard form of the equation of an ellipse centered at the origin and symmetric to both axes.

8. Write the standard form of the equation of a horizontal or vertical ellipse centered at (h, k).

9. Find the x- and the y-intercepts of the graph of $\frac{x^2}{a^2} + \frac{y^2}{b^2} = 1$.

10. **a.** Find the center of the ellipse graphed on the right. What are a and b?

    b. Is the ellipse horizontal or vertical?

    c. Find the equation of the ellipse.

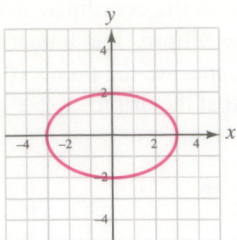

11. **a.** Find the center of the ellipse graphed on the right. What are a and b?

    b. Is the ellipse horizontal or vertical?

    c. Find the equation of the ellipse.

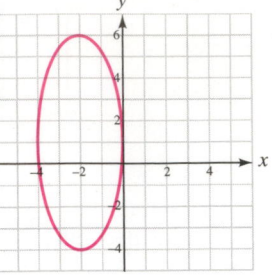

12. Find two points on the graph of $\frac{x^2}{16} + \frac{y^2}{4} = 1$ by letting $x = 2$ and finding the corresponding values of y.

13. Divide both sides of the equation by 64 and write the equation in standard form:

$$4(x-1)^2 + 64(y+5)^2 = 64$$

14. Determine whether the graph of each equation is a circle, a parabola, or an ellipse.

    a. $x = y^2 - 2y + 10$

    b. $\frac{x^2}{49} + \frac{y^2}{64} = 1$

    c. $(x-3)^2 + (y+4)^2 = 25$

    d. $2(x-1)^2 + 8(y+5)^2 = 32$

NOTATION

15. Find h, k, a, and b: $\dfrac{(x+8)^2}{100} + \dfrac{(y-6)^2}{144} = 1$

16. Write each denominator in the equation $\frac{x^2}{81} + \frac{y^2}{49} = 1$ as the square of a number.

GUIDED PRACTICE

Graph each equation. See Example 1.

17. $\dfrac{x^2}{25} + \dfrac{y^2}{4} = 1$ **18.** $\dfrac{x^2}{16} + \dfrac{y^2}{9} = 1$

19. $\dfrac{x^2}{4} + \dfrac{y^2}{9} = 1$ **20.** $\dfrac{x^2}{16} + \dfrac{y^2}{25} = 1$

Graph each equation. See Example 2.

21. $x^2 + 9y^2 = 9$ **22.** $25x^2 + 9y^2 = 225$

23. $16x^2 + 4y^2 = 64$ **24.** $4x^2 + 9y^2 = 36$

Graph each equation. See Example 3.

25. $\dfrac{(x-2)^2}{9} + \dfrac{(y-1)^2}{4} = 1$ **26.** $\dfrac{(x-1)^2}{9} + \dfrac{(y-3)^2}{4} = 1$

27. $\dfrac{(x+2)^2}{64} + \dfrac{(y-2)^2}{100} = 1$ **28.** $\dfrac{(x-6)^2}{36} + \dfrac{(y+6)^2}{144} = 1$

Graph each equation. See Example 4.

29. $(x+1)^2 + 4(y+2)^2 = 4$

30. $25(x+1)^2 + 9y^2 = 225$

31. $16(x-2)^2 + 4(y+4)^2 = 256$

32. $4(x-2)^2 + 9(y-4)^2 = 144$

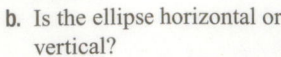 *Use a graphing calculator to graph each equation. See Using Your Calculator: Graphing Ellipses.*

33. $\dfrac{x^2}{9} + \dfrac{y^2}{4} = 1$ **34.** $x^2 + 16y^2 = 16$

35. $\dfrac{x^2}{4} + \dfrac{(y-1)^2}{9} = 1$ **36.** $\dfrac{(x+1)^2}{9} + \dfrac{(y-2)^2}{4} = 1$

TRY IT YOURSELF

Write each equation in standard form, if it is not already so, and graph it. The problems include equations that describe circles, parabolas, and ellipses.

37. $(x+1)^2 + (y-2)^2 = 16$ **38.** $(x-3)^2 + (y+1)^2 = 25$

39. $\dfrac{x^2}{16} + \dfrac{y^2}{1} = 1$ **40.** $\dfrac{x^2}{1} + \dfrac{y^2}{9} = 1$

41. $x = \dfrac{1}{2}(y-1)^2 - 2$ **42.** $x = -\dfrac{1}{2}(y+4)^2 + 5$

43. $x^2 + y^2 - 25 = 0$ **44.** $x^2 = 36 - y^2$

45. $x^2 = 100 - 4y^2$ **46.** $x^2 = 36 - 4y^2$

47. $y = -3x^2 - 24x - 43$ **48.** $y = 5x^2 - 60x + 173$

49. $x^2 + y^2 - 2x + 4y - 4 = 0$

50. $x^2 + y^2 + 4x + 6y + 9 = 0$

51. $9(x-1)^2 + 4(y+2)^2 = 36$

52. $16(x-5)^2 + 25(y-4)^2 = 400$

Look Alikes . . .

Graph the ellipses described by the equations in parts a and b on the same coordinate system.

53. **a.** $\dfrac{x^2}{9} + \dfrac{y^2}{25} = 1$ **b.** $\dfrac{x^2}{25} + \dfrac{y^2}{9} = 1$

54. **a.** $\dfrac{x^2}{169} + \dfrac{y^2}{25} = 1$ **b.** $\dfrac{x^2}{25} + \dfrac{y^2}{169} = 1$

55. **a.** $\dfrac{(x-3)^2}{100} + \dfrac{(y-2)^2}{36} = 1$ **b.** $\dfrac{(x+3)^2}{100} + \dfrac{(y+2)^2}{36} = 1$

56. **a.** $\dfrac{(x-4)^2}{9} + \dfrac{(y+5)^2}{4} = 1$ **b.** $\dfrac{(x+4)^2}{9} + \dfrac{(y-5)^2}{4} = 1$

APPLICATIONS

57. **from Campus to Careers**

Civil Engineer

The arch of an underpass shown below is part of an ellipse.

a. Find the equation of the ellipse.

b. Find the height of the arch at a point 10 feet to the right of the center line of the roadway.

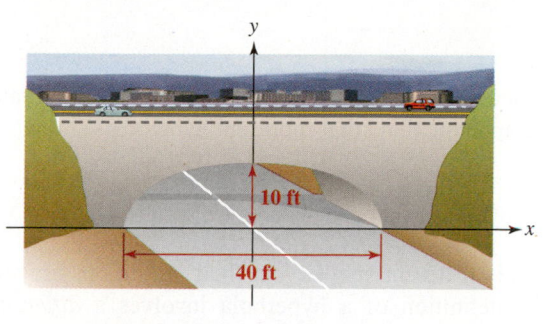

58. Fitness Equipment. With elliptical cross-training equipment, the feet move through the natural elliptical pattern that one experiences when walking, jogging, or running. Write the equation of the elliptical pattern shown below.

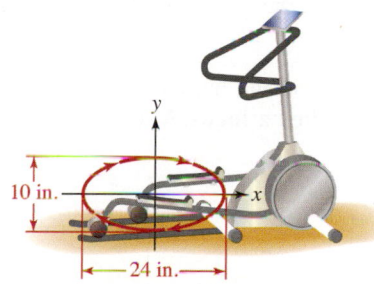

59. Koi Ponds. A landscape architect is designing an elliptical fish pond that will fit in the center of a 110-by-100-foot rectangular Japanese rock garden, leaving 15 feet of clearance on all sides. If she establishes a coordinate system with the 110-foot length along the x-axis, and the center of the pond at the origin, find the equation of the ellipse.

60. Pool Tables. Find the equation of the outer edge of the elliptical pool table shown in the next column.

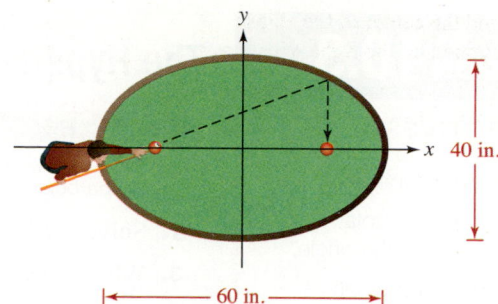

61. Area of an Ellipse. The area A bounded by an ellipse with the equation $\frac{x^2}{a^2} + \frac{y^2}{b^2} = 1$ is given by $A = \pi ab$. Find the area bounded by the ellipse described by $9x^2 + 16y^2 = 144$.

62. Area of a Track. The elliptical track shown in the figure is bounded by the ellipses $4x^2 + 9y^2 = 576$ and $9x^2 + 25y^2 = 900$. Find the area of the track. (See Exercise 61.)

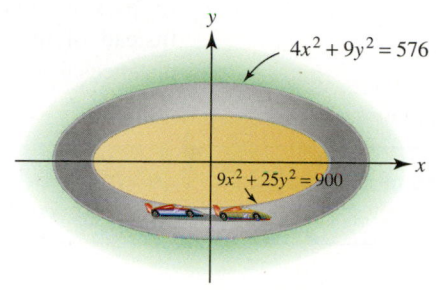

WRITING

63. What is an ellipse?

64. Explain the difference between the focus of an ellipse and the vertex of an ellipse.

65. Compare the graphs of $\frac{x^2}{81} + \frac{y^2}{64} = 1$ and $\frac{x^2}{64} + \frac{y^2}{81} = 1$. Do they have any similarities?

66. What are the reflective properties of an ellipse?

REVIEW

Find each product.

67. $3x^{-2}y^2(4x^2 + 3y^{-2})$

68. $(2a^{-2} - b^{-2})(2a^{-2} + b^{-2})$

Simplify each expression.

69. $\dfrac{x^{-2} + y^{-2}}{x^{-2} - y^{-2}}$

70. $\dfrac{2x^{-3} - 2y^{-3}}{4x^{-3} + 4y^{-3}}$

CHALLENGE PROBLEMS

71. What happens to the graph of the equation $\frac{x^2}{a^2} + \frac{y^2}{b^2} = 1$ when $a = b$?

72. Graph: $9x^2 + 4y^2 = 1$

73. Write the equation $9x^2 + 4y^2 - 18x + 16y = 11$ in the standard form of the equation of an ellipse.

74. Let the foci of an ellipse be $(c, 0)$ and $(-c, 0)$. Suppose that the sum of the distances from any point (x, y) on the ellipse to the two foci is the constant $2a$. Show that the equation for the ellipse is $\frac{x^2}{a^2} + \frac{y^2}{a^2 - c^2} = 1$. Then let $b^2 = a^2 - c^2$ to obtain the standard form of the equation of an ellipse.

SECTION 10.3

The Hyperbola

OBJECTIVES

1. Define a hyperbola.
2. Graph hyperbolas centered at the origin.
3. Graph hyperbolas centered at (h, k).
4. Graph equations of the form $xy = k$.
5. Solve application problems involving hyperbolas.
6. Identify conic sections by their equations.

ARE YOU READY?

The following problems review some basic skills that are needed when working with hyperbolas.

1. Solve: $b^2 = 16$
2. Simplify: $x - (-2)$
3. What is the slope of the line represented by $y = \dfrac{2}{3}x$?
4. If $xy = 20$, find y when $x = -2$.

The final conic section that we will discuss, the *hyperbola,* is a curve that has two branches. In this section, we will learn how to graph equations that represent hyperbolas.

1 Define a Hyperbola.

Ellipses and hyperbolas have completely different shapes, but their definitions are similar. Instead of the *sum* of distances, the definition of a hyperbola involves a *difference* of distances.

| **Definition of a Hyperbola** | A **hyperbola** is the set of all points in a plane for which the difference of the distances from two fixed points is a constant. |

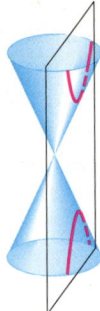

The figure below illustrates that any point P on the hyperbola is a constant distance $d_1 - d_2$ from two fixed points, each of which is called a **focus.** Midway between the **foci** is the **center** of the hyperbola.

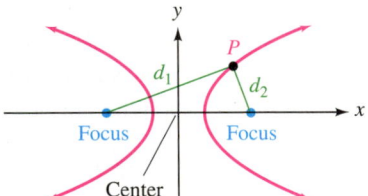   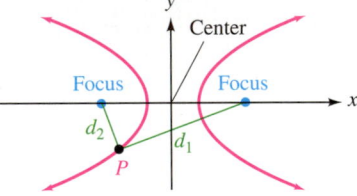

2 Graph Hyperbolas Centered at the Origin.

The graph of the equation

$$\frac{x^2}{25} - \frac{y^2}{9} = 1$$

is a hyperbola. To graph the equation, we make a table of solutions that satisfy the equation, plot each point, and join them with a smooth curve.

$$\frac{x^2}{25} - \frac{y^2}{9} = 1$$

x	y	
-7	± 2.9	$\longrightarrow (-7, \pm 2.9)$
-6	± 2.0	$\longrightarrow (-6, \pm 2.0)$
-5	0	$\longrightarrow (-5, 0)$
5	0	$\longrightarrow (5, 0)$
6	± 2.0	$\longrightarrow (6, \pm 2.0)$
7	± 2.9	$\longrightarrow (7, \pm 2.9)$

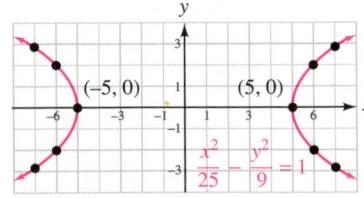

This graph is centered at the origin and intersects the x-axis at $(5, 0)$ and $(-5, 0)$. We also note that the graph does not intersect the y-axis.

It is possible to draw a hyperbola without plotting points. For example, if we want to graph the hyperbola with an equation of

$$\frac{x^2}{a^2} - \frac{y^2}{b^2} = 1$$

we first find the x- and y-intercepts. To find the x-intercepts, we let $y = 0$ and solve for x:

$$\frac{x^2}{a^2} - \frac{\mathbf{0}^2}{b^2} = 1$$
$$x^2 = a^2$$
$$x = \pm a \qquad \text{Use the square root property.}$$

The hyperbola crosses the x-axis at the points $V_1(a, 0)$ and $V_2(-a, 0)$, called the **vertices** of the hyperbola.

To attempt to find the y-intercepts, we let $x = 0$ and solve for y:

$$\frac{\mathbf{0}^2}{a^2} - \frac{y^2}{b^2} = 1$$
$$y^2 = -b^2$$
$$y = \pm\sqrt{-b^2}$$

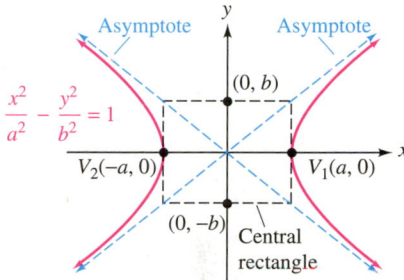

Since b^2 is always positive, $\sqrt{-b^2}$ is an imaginary number. This means that the hyperbola does not intersect the y-axis.

As shown in the illustration above, we can construct a rectangle, called the **central rectangle**, whose sides pass horizontally through $\pm b$ on the y-axis and vertically through $\pm a$ on the x-axis. The extended diagonals of the rectangle are a pair of intersecting straight lines called **asymptotes** of the hyperbola. As the hyperbola gets farther away from the origin, its branches get closer and closer to the asymptotes. Asymptotes should be drawn as dashed lines because they are not part of the hyperbola, but they do serve as a helpful guide when drawing its graph. Since the slopes of the diagonals are $\frac{b}{a}$ and $-\frac{b}{a}$, and since the diagonals pass through the origin, the equations of the asymptotes are

$$y = \frac{b}{a}x \quad \text{and} \quad y = -\frac{b}{a}x$$

The Language of Algebra

The central rectangle is also called the **fundamental rectangle**.

Standard Form of the Equation of a Horizontal Hyperbola Centered at the Origin

The equation $\frac{x^2}{a^2} - \frac{y^2}{b^2} = 1$ has a graph that is a hyperbola centered at the origin. The x-intercepts are the vertices $V_1(a, 0)$ and $V_2(-a, 0)$. There are no y-intercepts.

The asymptotes of the hyperbola are the extended diagonals of the central rectangle, and their equations are $y = \frac{b}{a}x$ and $y = -\frac{b}{a}x$.

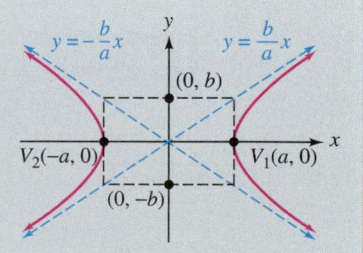

The branches of the hyperbola in previous discussions open to the left and to the right. It is possible for hyperbolas to have different orientations with respect to the x- and y-axes. For example, the branches of a hyperbola can open upward and downward. In that case, the following equation applies.

Standard Form of the Equation of a Vertical Hyperbola Centered at the Origin

The equation $\frac{y^2}{a^2} - \frac{x^2}{b^2} = 1$ has a graph that is a hyperbola centered at the origin. The y-intercepts are the vertices $V_1(0, a)$ and $V_2(0, -a)$. There are no x-intercepts.

The asymptotes of the hyperbola are the extended diagonals of the central rectangle, and their equations are $y = \frac{a}{b}x$ and $y = -\frac{a}{b}x$.

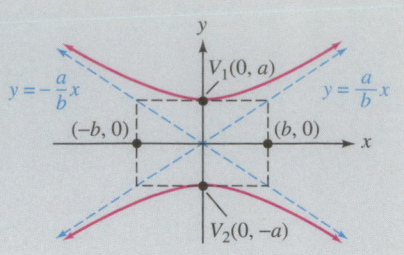

EXAMPLE 1   Graph: $\dfrac{x^2}{9} - \dfrac{y^2}{16} = 1$

Strategy   This equation is in standard $\frac{x^2}{a^2} - \frac{y^2}{b^2} = 1$ form. We will identify a and b.

Why   We can use a and b to find the vertices of the graph of the hyperbola and the location of the central rectangle.

Solution   The color highlighting shows how to compare the given equation with the standard form to find a and b.

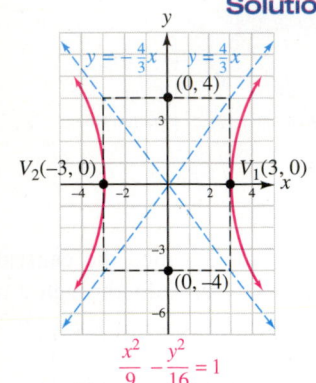

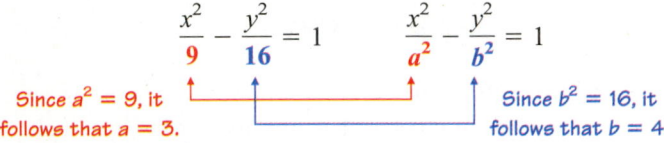

Since $a^2 = 9$, it follows that $a = 3$.   Since $b^2 = 16$, it follows that $b = 4$.

This is the standard form of the equation of a hyperbola, centered at the origin, that opens left and right. The x-intercepts are $(a, 0)$ and $(-a, 0)$, or $(3, 0)$ and $(-3, 0)$. They are also the vertices of the hyperbola.

To construct the central rectangle, we use the values of $a = 3$ and $b = 4$. The rectangle passes through $(3, 0)$ and $(-3, 0)$ on the x-axis, and $(0, 4)$ and $(0, -4)$ on the y-axis. We draw extended diagonal dashed lines through the rectangle to obtain the asymptotes and write their equations: $y = \frac{4}{3}x$ and $y = -\frac{4}{3}x$. Then we draw a smooth curve through each vertex that gets close to the asymptotes.

Self Check 1   Graph: $\dfrac{x^2}{25} - \dfrac{y^2}{4} = 1$

Now Try ▶ Problem 17

EXAMPLE 2   Graph: $9y^2 - 4x^2 = 36$

Strategy   We will write the equation in standard $\frac{y^2}{a^2} - \frac{x^2}{b^2} = 1$ form.

Why   When the equation is in standard form, we will be able to identify the center and the vertices of the graph of the hyperbola and the location of the central rectangle.

Solution   To write the equation in standard form, we divide both sides by 36.

$$9y^2 - 4x^2 = 36$$

$$\frac{9y^2}{36} - \frac{4x^2}{36} = \frac{36}{36}$$ To get a 1 on the right side, divide both sides by 36.

The positive variable term in the standard form equation determines whether a hyperbola is vertical or horizontal. In this example, the positive variable term involves y, so the hyperbola is vertical.

$$\frac{y^2}{4} - \frac{x^2}{9} = 1$$ Simplify each fraction.

Caution

Although the branches of a hyperbola look like parabolas, they are not two parabolas. The curvature of each branch of a hyperbola is different from that of a parabola. A parabola is not constrained by asymptotes as are the branches of a hyperbola.

This is the standard form of the equation of a hyperbola, centered at the origin, that opens up and down. The color highlighting shows how we compare the resulting equation to the standard form to find a and b.

$$\frac{y^2}{4} - \frac{x^2}{9} = 1 \qquad \frac{y^2}{a^2} - \frac{x^2}{b^2} = 1$$

Since $a^2 = 4$, it follows that $a = 2$.

Since $b^2 = 9$, it follows that $b = 3$.

The y-intercepts are $(0, a)$ and $(0, -a)$, or $(0, 2)$ and $(0, -2)$. They are also the vertices of the hyperbola.

Since $a = 2$ and $b = 3$, the central rectangle passes through $(0, 2)$ and $(0, -2)$, as well as $(3, 0)$ and $(-3, 0)$. We draw its extended diagonals and sketch the hyperbola.

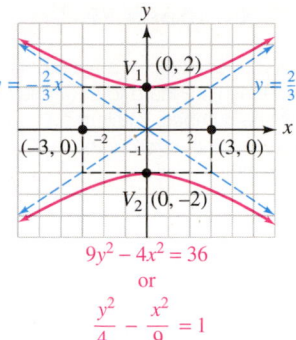

$$9y^2 - 4x^2 = 36$$
or
$$\frac{y^2}{4} - \frac{x^2}{9} = 1$$

Self Check 2   Graph: $16y^2 - x^2 = 16$

Now Try ▶ Problem 21

Using Your Calculator ▶ Graphing Hyperbolas

To graph $\frac{x^2}{9} - \frac{y^2}{16} = 1$ from Example 1 using a graphing calculator, we follow the same procedure that we used for circles and ellipses. To write the equation as two functions, we solve for y to get $y = \pm\frac{\sqrt{16x^2 - 144}}{3}$. Then we graph the following two functions in a square window setting to get the graph of the hyperbola shown at the right.

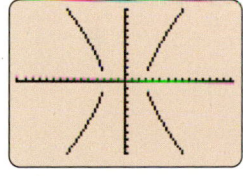

$$y = \frac{\sqrt{16x^2 - 144}}{3} \quad \text{and} \quad y = -\frac{\sqrt{16x^2 - 144}}{3}$$

---

3 **Graph Hyperbolas Centered at (h, k).**

If a hyperbola is centered at a point with coordinates (h, k), the following equations apply.

Standard Form of the Equation of a Hyperbola Centered at (h, k)

The equation $\frac{(x - h)^2}{a^2} - \frac{(y - k)^2}{b^2} = 1$ has a graph that is a hyperbola centered at (h, k) and that opens left and right.

The equation $\frac{(y - k)^2}{a^2} - \frac{(x - h)^2}{b^2} = 1$ has a graph that is a hyperbola centered at (h, k) and that opens up and down.

EXAMPLE 3   Graph: **a.** $\dfrac{(x-3)^2}{16} - \dfrac{(y+1)^2}{4} = 1$ **b.** $\dfrac{(y-2)^2}{9} - \dfrac{(x-1)^2}{9} = 1$

Strategy   We will write each equation in a form that makes it easy to identify h, k, a, and b.

Why   If we know h, k, a, and b, we can graph the hyperbola and the central rectangle.

Solution   **a.** We can write the given equation as

$$\dfrac{(x-\mathbf{3})^2}{\mathbf{4}^2} - \dfrac{[y-(\mathbf{-1})]^2}{\mathbf{2}^2} = 1$$

To find k, write $y + 1$ as $y - (-1)$.
To find a, write 16 as 4^2. To find b, write 4 as 2^2.

Because the term involving x is positive, the hyperbola opens left and right. We find the center by examining $(x-3)^2$ and $[y-(-1)]^2$. Since $h = 3$ and $k = -1$, the hyperbola is centered at $(h, k) = (3, -1)$. From the denominators, 4^2 and 2^2, we find that $a = 4$ and $b = 2$. Thus, its vertices are located 4 units to the right and left of the center, at $(7, -1)$ and $(-1, -1)$. Since $b = 2$, we can count 2 units

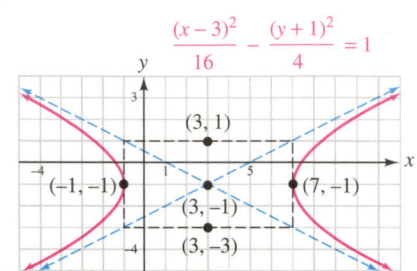

above and below the center to locate points $(3, 1)$ and $(3, -3)$. With these four points, we can draw the central rectangle along with its extended diagonals (the asymptotes). We then can sketch the hyperbola, as shown.

b. We can write the given equation as

$$\dfrac{(y-\mathbf{2})^2}{\mathbf{3}^2} - \dfrac{(x-\mathbf{1})^2}{\mathbf{3}^2} = 1$$

Because the term involving y is positive, the hyperbola opens up and down. We find its center by examining $(y-2)^2$ and $(x-1)^2$. Since $k = 2$ and $h = 1$, the hyperbola is centered at $(h, k) = (1, 2)$. From the denominators, 3^2 and 3^2, we find that $a = 3$ and $b = 3$, and we use that information to draw the central rectangle and its extended diagonals (the asymptotes), as shown.

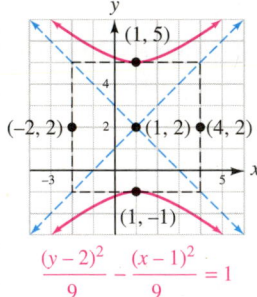

Self Check 3   Graph: **a.** $\dfrac{(x+2)^2}{9} - \dfrac{(y-1)^2}{4} = 1$ **b.** $\dfrac{(y+1)^2}{1} - \dfrac{(x+1)^2}{4} = 1$

Now Try ▶   Problems 25 and 27

---

4   Graph Equations of the Form *xy* = *k*.

There is a special type of hyperbola (also centered at the origin) that does not intersect either the x- or the y-axis. These hyperbolas have equations of the form $xy = k$, where $k \neq 0$.

EXAMPLE 4   Graph: $xy = -8$

Strategy   We will make a table of solutions, plot the points, and connect the points with a smooth curve.

Why   Since this equation cannot be written in standard form, we cannot use the methods used in the previous examples.

Solution   To make a table of solutions, we can solve the given equation for y to get $y = \dfrac{-8}{x}$. Then we choose several values for x, find the corresponding values of y, and record the results in the table below. We plot the ordered pairs and join them with a smooth curve to obtain the graph of the hyperbola.

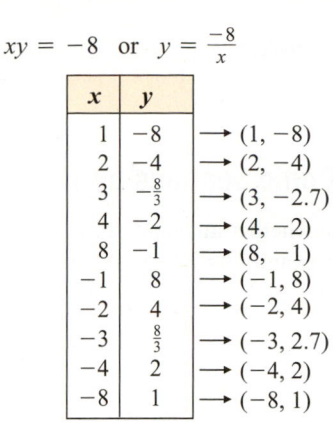

$$xy = -8 \quad \text{or} \quad y = \dfrac{-8}{x}$$

x	y	
1	-8	$\rightarrow (1, -8)$
2	-4	$\rightarrow (2, -4)$
3	$-\frac{8}{3}$	$\rightarrow (3, -2.7)$
4	-2	$\rightarrow (4, -2)$
8	-1	$\rightarrow (8, -1)$
-1	8	$\rightarrow (-1, 8)$
-2	4	$\rightarrow (-2, 4)$
-3	$\frac{8}{3}$	$\rightarrow (-3, 2.7)$
-4	2	$\rightarrow (-4, 2)$
-8	1	$\rightarrow (-8, 1)$

The y-values for x = 3 and x = −3 were approximated.

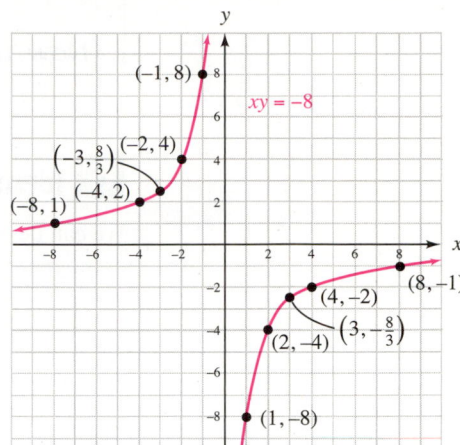

The Language of Algebra

The asymptotes of this hyperbola are the x- and y-axes. A hyperbola for which the aymptotes are perpendicular is called a **rectangular hyperbola**.

Self Check 4   Graph: $xy = 6$

Now Try ▶ Problem 33

The result in Example 4 illustrates the following general equation.

Equations of Hyperbolas of the Form $xy = k$

Any equation of the form $xy = k$, where $k \neq 0$, has a graph that is a **hyperbola**, which does not intersect either the x- or y-axis.

---

5 Solve Application Problems Involving Hyperbolas.

EXAMPLE 5   **Atomic Structure.**   In an experiment that led to the discovery of the atomic structure of matter, Lord Rutherford (1871–1937) shot high-energy alpha particles toward a thin sheet of gold. Many of them were reflected, and Rutherford showed the existence of the nucleus of a gold atom. An alpha particle is repelled by the nucleus at the origin; it travels along the hyperbolic path given by $4x^2 - y^2 = 16$. How close does the particle come to the nucleus?

Strategy   We will write the equation in standard form and find the coordinates of point V.

Why   The distance from the origin to point V is the closest the particle comes to the nucleus.

Solution   To find the distance from the nucleus at the origin, we must find the coordinates of the vertex V. To do so, we write the equation of the particle's path in standard form:

$$4x^2 - y^2 = 16 \qquad \text{This is the given equation.}$$

$$\dfrac{4x^2}{16} - \dfrac{y^2}{16} = \dfrac{16}{16} \qquad \text{Divide both sides by 16, term-by-term.}$$

$$\dfrac{x^2}{4} - \dfrac{y^2}{16} = 1 \qquad \text{Simplify the fractions.}$$

$$\dfrac{x^2}{2^2} - \dfrac{y^2}{4^2} = 1 \qquad \text{To determine } a \text{ and } b \text{, write 4 as } 2^2 \text{ and 16 as } 4^2.$$

This equation is in the form $\dfrac{x^2}{a^2} - \dfrac{y^2}{b^2} = 1$ with $a = 2$. Thus, the vertex of the path is $(2, 0)$. The particle is never closer than 2 units from the nucleus.

Self Check 5   **Astronomy.** Some comets have a hyperbolic orbit, with the sun as one focus and Earth at the center. For one such comet, the equation of its path is $\dfrac{x^2}{1 \times 10^{18}} - \dfrac{y^2}{2 \times 10^{18}} = 1$. The units are miles. How close does this comet come to Earth?

Now Try   ▶   Problem 61

6 Identify Conic Sections by Their Equations.

We can determine whether an equation, when graphed, will be a circle, a parabola, an ellipse, or a hyperbola by examining its variable terms.

$x^2 + y^2 = 16$ With the variable terms on the same side of the equation, we see that the coefficients of the squared terms are the same. The graph is a circle.

$4x^2 + 9y^2 = 144$ With the variable terms on the same side of the equation, we see that the coefficients of the squared terms are different, but have the same sign. The graph is an ellipse.

$4x^2 - 9y^2 = 144$ With the variable terms on the same side of the equation, we see that the coefficients of the squared terms have different signs. The graph is a hyperbola.

$x = y^2 + y - 16$ Since one variable is squared and the other is not, the graph is a parabola.

SECTION 10.3   STUDY SET

VOCABULARY

Fill in the blanks.

1. The two-branch curve graphed on the right is a _____.

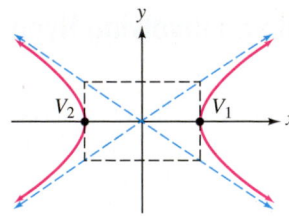

2. A _____ is the set of all points in a plane for which the difference of the distances from two fixed points is a constant.

3. In the graph above, V_1 and V_2 are the _____ of the hyperbola.

4. In the graph above, the figure drawn using dashed black lines is called the _____ _____.

5. The extended _____ of the central rectangle are asymptotes of the hyperbola.

6. To write $9x^2 - 4y^2 = 36$ in _____ form, we divide both sides by 36.

CONCEPTS

7. Write the standard form of the equation of a hyperbola centered at the origin that opens left and right.

8. Write the standard form of the equation of a hyperbola centered at (h, k) that opens up and down.

9. Write the standard form of the equation of a hyperbola centered at (h, k) that opens left and right.

10. **a.** Find the center of the hyperbola graphed on the right. What are a and b?

    b. Find the x-intercepts of the graph. What are the y-intercepts of the graph?

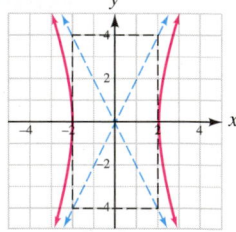

    c. Find the equation of the hyperbola.

    d. Find the equations of the asymptotes.

11. **a.** Find the center of the hyperbola graphed on the right. What are a and b?

    b. Find the equation of the hyperbola.

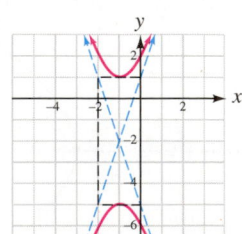

12. a. Fill in the blank: An equation of the form $xy = k$, where $k \neq 0$, has a graph that is a _____ that does not intersect either the x-axis or the y-axis.

b. Complete the table of solutions for $xy = 10$.

x	y
-2	
	5

13. Divide both sides of the equation by 100 and write the equation in standard form:

$$100(x + 1)^2 - 25(y - 5)^2 = 100$$

14. Determine whether the graph of the equation will be a circle, a parabola, an ellipse, or a hyperbola.

a. $x^2 + y^2 = 10$

b. $9y^2 - 16x^2 = 144$

c. $x = y^2 - 3y + 6$

d. $4x^2 + 25y^2 = 100$

NOTATION

15. Find h, k, a, and b: $\dfrac{(x-5)^2}{25} - \dfrac{(y+11)^2}{36} = 1$

16. Write each denominator in the equation $\dfrac{x^2}{36} - \dfrac{y^2}{81} = 1$ as the square of a number.

GUIDED PRACTICE

Graph each hyperbola. **See Example 1.**

17. $\dfrac{x^2}{9} - \dfrac{y^2}{4} = 1$

18. $\dfrac{x^2}{4} - \dfrac{y^2}{4} = 1$

19. $\dfrac{y^2}{4} - \dfrac{x^2}{9} = 1$

20. $\dfrac{y^2}{4} - \dfrac{x^2}{64} = 1$

Graph each hyperbola. **See Example 2.**

21. $y^2 - 4x^2 = 16$

22. $9y^2 - 25x^2 = 225$

23. $25x^2 - y^2 = 25$

24. $9x^2 - 4y^2 = 36$

Graph each hyperbola. **See Example 3.**

25. $\dfrac{(x-2)^2}{9} - \dfrac{y^2}{16} = 1$

26. $\dfrac{(x+2)^2}{16} - \dfrac{(y-3)^2}{25} = 1$

27. $\dfrac{(y+1)^2}{1} - \dfrac{(x-2)^2}{4} = 1$

28. $\dfrac{(y-2)^2}{4} - \dfrac{(x+1)^2}{1} = 1$

29. $\dfrac{(x+1)^2}{9} - \dfrac{(y+1)^2}{9} = 1$

30. $\dfrac{(x-2)^2}{16} - \dfrac{(y-1)^2}{16} = 1$

31. $\dfrac{(y-3)^2}{25} - \dfrac{x^2}{25} = 1$

32. $\dfrac{(y-1)^2}{9} - \dfrac{x^2}{9} = 1$

Graph each equation. **See Example 4.**

33. $xy = 8$

34. $xy = 4$

35. $xy = -10$

36. $xy = -12$

Use a graphing calculator to graph each equation. **See Using Your Calculator: Graphing Hyperbolas.**

37. $\dfrac{x^2}{9} - \dfrac{y^2}{4} = 1$

38. $y^2 - 16x^2 = 16$

39. $\dfrac{x^2}{4} - \dfrac{(y-1)^2}{9} = 1$

40. $\dfrac{(y+1)^2}{9} - \dfrac{(x-2)^2}{4} = 1$

TRY IT YOURSELF

Write each equation in standard form, if it is not already so, and graph it. The problems include equations that describe circles, parabolas, ellipses, and hyperbolas.

41. $(x + 1)^2 + (y - 2)^2 = 16$

42. $(x - 3)^2 + (y + 4)^2 = 1$

43. $9x^2 - 49y^2 = 441$

44. $25y^2 - 16x^2 = 400$

45. $4(x + 1)^2 + 9(y + 1)^2 = 36$

46. $16x^2 + 25(y - 3)^2 = 400$

47. $4(x + 3)^2 - (y - 1)^2 = 4$

48. $(x + 5)^2 - 16y^2 = 16$

49. $xy = -6$

50. $xy = 10$

51. $x = \dfrac{1}{2}(y - 1)^2 - 2$

52. $x = -\dfrac{1}{4}(y - 3)^2 + 2$

53. $\dfrac{y^2}{25} - \dfrac{(x-2)^2}{4} = 1$

54. $\dfrac{y^2}{36} - \dfrac{(x+2)^2}{4} = 1$

55. $y = -x^2 + 6x - 4$

56. $y = x^2 - 2x + 5$

57. $\dfrac{x^2}{1} + \dfrac{y^2}{36} = 1$

58. $\dfrac{x^2}{4} + \dfrac{y^2}{16} = 1$

59. $x^2 + y^2 + 4x - 6y - 23 = 0$

60. $x^2 + y^2 + 8x - 2y - 8 = 0$

APPLICATIONS

61. Alpha Particles. The particle in the illustration below approaches the nucleus at the origin along the path $9y^2 - x^2 = 81$ in the coordinate system shown. How close does the particle come to the nucleus?

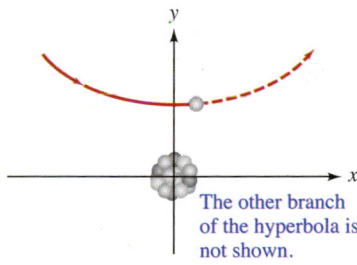

The other branch of the hyperbola is not shown.

62. LORAN. By determining the difference of the distances between the ship in the illustration and two radio transmitters, the LORAN navigation system places the ship on the hyperbola $x^2 - 4y^2 = 576$ in the coordinate system shown. If the ship is 5 miles out to sea, find its coordinates.

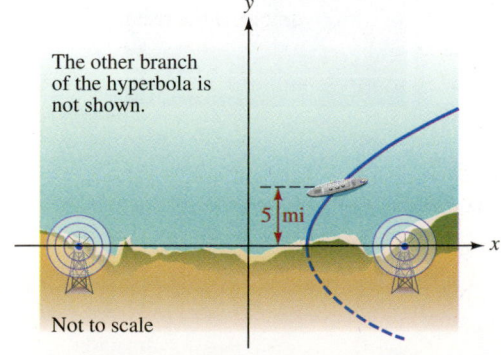

The other branch of the hyperbola is not shown.

5 mi

Not to scale

63. from **Campus to Careers**

Civil Engineer

A new subdivision of homes is planned for an area that military jets often fly over. On the coordinate system shown below, a sonic boom is heard by those on the ground within one branch of the hyperbola $y^2 - x^2 = 25$. How wide is the hyperbola 5 miles from its vertex?

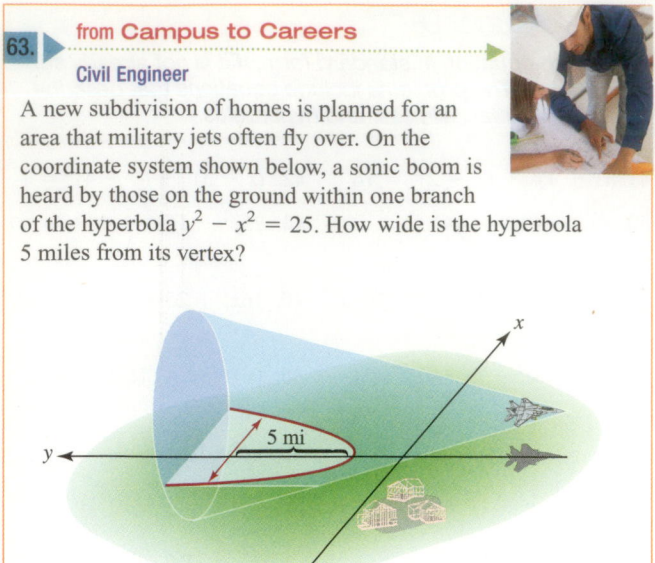

64. Lighting. Refer to the illustration. The "cones" of light emitted upward and downward by a lamp with a cylindrical shade cast an interesting pattern on the wall. What type of conic section can be seen?

65. Nuclear Power. The photograph below is an example of a standard cooling tower of a nuclear power plant. What type of conic section appears to be used in the design?

66. Fluids. See the illustration on the right. Two glass plates in contact at the left, and separated by about 5 millimeters on the right, are dipped in beet juice, which rises by capillary action to form a hyperbola. The hyperbola is modeled by an equation of the form $xy = k$. If the curve passes through the point $(12, 2)$, what is k?

WRITING

67. What is a hyperbola?

68. Compare the graphs of $\frac{x^2}{81} - \frac{y^2}{64} = 1$ and $\frac{y^2}{81} - \frac{x^2}{64} = 1$. Do they have any similarities?

69. Explain how to determine the dimensions of the central rectangle that is associated with the graph of $\frac{x^2}{36} - \frac{y^2}{25} = 1$.

70. Explain why the graph of $\frac{x^2}{a^2} - \frac{y^2}{b^2} = 1$ has no y-intercept.

REVIEW

Find each value of x.

71. $\log_8 x = 2$

72. $\log_{25} x = \frac{1}{2}$

73. $\log_{1/2} \frac{1}{8} = x$

74. $\log_{12} x = 0$

75. $\log_x \frac{9}{4} = 2$

76. $\log_6 216 = x$

77. $\log_x 1{,}000 = 3$

78. $\log_2 \sqrt{2} = x$

CHALLENGE PROBLEMS

79. Write the equation $x^2 - y^2 - 2x + 4y = 12$ in standard form to show that it describes a hyperbola.

80. Write the equation $x^2 - 4y^2 + 2x - 8y = 7$ in standard form to show that it describes a hyperbola.

81. Write the equation $36x^2 - 25y^2 - 72x - 100y = 964$ in standard form to show that it describes a hyperbola.

82. Write an equation of a hyperbola whose graph has the following characteristics:
- vertices $(\pm 1, 0)$
- equations of asymptotes: $y = \pm 5x$

83. Graph: $16x^2 - 25y^2 = 1$

84. Show that the equations of the extended diagonals of the fundamental rectangle of the hyperbola $\frac{x^2}{a^2} - \frac{y^2}{b^2} = 1$ are

$$y = \frac{b}{a}x \quad \text{and} \quad y = -\frac{b}{a}x$$

SECTION 10.4 · Solving Nonlinear Systems of Equations

OBJECTIVES

1 Solve systems by graphing.

2 Solve systems by substitution.

3 Solve systems by elimination (addition).

ARE YOU READY?

The following problems review some basic skills that are needed when solving nonlinear systems of equations.

1. Graph: $2x - 3y = 6$

2. Solve: $9y^2 + 4y - 5 = 0$

3. Solve: $\begin{cases} 3x + 2y = 36 \\ 4x - y = 4 \end{cases}$

4. Solve: **a.** $x^2 = \dfrac{5}{9}$ **b.** $y^2 = 18$

In Chapter 3, we discussed how to solve systems of linear equations by the graphing, substitution, and elimination methods. In this section, we will use these methods to solve systems in which at least one of the equations is nonlinear.

1 Solve Systems by Graphing.

A solution of a **nonlinear system of equations** is an ordered pair of real numbers that satisfies all of the equations in the system. The **solution set of a nonlinear system** is the set of all such ordered pairs. One way to solve a system of two equations in two variables is to graph the equations on the same rectangular coordinate system.

EXAMPLE 1

Solve $\begin{cases} x^2 + y^2 = 25 \\ 2x + y = 10 \end{cases}$ by graphing.

Strategy We will graph both equations on the same coordinate system.

Why If the equations are graphed on the same coordinate system, we can see whether they have any common solutions.

Solution

Success Tip

It is helpful to sketch the possibilities before solving the system:

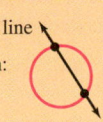

Secant line
2 points of intersection:
(2 real solutions)

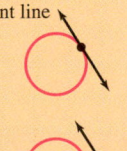

Tangent line
1 point of intersection:
(1 real solution)

No points of intersection:
(0 real solutions)

The graph of $x^2 + y^2 = 25$ is a circle with center at the origin and radius of 5. The graph of $2x + y = 10$ is a line with x-intercept $(5, 0)$ and y-intercept $(0, 10)$. Depending on whether the line is a **secant** (intersecting the circle at two points) or a **tangent** (intersecting the circle at one point) or does not intersect the circle at all, there are two, one, or no solutions to the system, respectively.

After graphing the circle and the line, it appears that the points of intersection are $(5, 0)$ and $(3, 4)$. To verify that they are solutions of the system, we need to check each one.

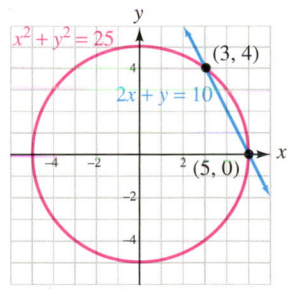

Check:

For (5, 0)

$2x + y = 10$	$x^2 + y^2 = 25$
$2(5) + 0 \overset{?}{=} 10$	$5^2 + 0^2 \overset{?}{=} 25$
$10 = 10$ True	$25 = 25$ True

For (3, 4)

$2x + y = 10$	$x^2 + y^2 = 25$
$2(3) + 4 \overset{?}{=} 10$	$3^2 + 4^2 \overset{?}{=} 25$
$10 = 10$ True	$25 = 25$ True

The ordered pair $(5, 0)$ satisfies both equations of the system, and so does $(3, 4)$. Thus, there are two solutions, $(5, 0)$ and $(3, 4)$, and the solution set is $\{(5, 0), (3, 4)\}$.

Self Check 1 Solve $\begin{cases} x^2 + y^2 = 25 \\ y = -2x - 5 \end{cases}$ by graphing.

Now Try ▶ Problem 15

To solve Example 1 with a graphing calculator, we graph the circle and the line on one set of coordinate axes. See figure (a). We then trace to find the coordinates of the intersection points of the graphs. See figures (b) and (c).

We can zoom for better results.

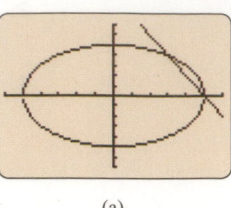　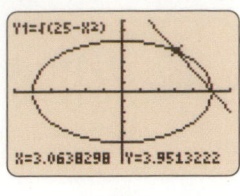　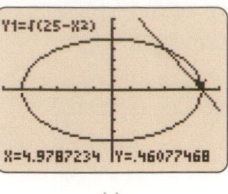

(a)　　　　　　　　　(b)　　　　　　　　　(c)

---

2 Solve Systems by Substitution.

When solving a system by graphing, it is often difficult to determine the coordinates of the intersection points. A more precise algebraic method called the **substitution method** can be used to solve certain systems involving nonlinear equations.

EXAMPLE 2

Solve $\begin{cases} x^2 + y^2 = 2 \\ 2x - y = 1 \end{cases}$ by substitution.

Strategy We will solve the second equation for y and substitute the result for y in the first equation.

Why We can solve the resulting equation for x and then back substitute to find y.

Solution This system has one second-degree equation and one first-degree equation. We can solve this type of system by substitution. Solving the linear equation for y gives:

$$2x - y = 1$$
$$-y = -2x + 1 \quad \text{Subtract 2x from both sides.}$$
$$y = 2x - 1 \quad \text{Multiply both sides by } -1. \text{ We call this the substitution equation.}$$

Because y and $2x - 1$ are equal, we can substitute $2x - 1$ for y in the first equation of the system.

Success Tip

With this method, the objective is to use an appropriate substitution to obtain *one* equation in *one* variable.

$$y = \boxed{2x - 1} \qquad\qquad x^2 + y^2 = 2$$

Then we solve the resulting quadratic equation for x.

$$x^2 + y^2 = 2$$
$$x^2 + (2x - 1)^2 = 2 \quad \text{Substitute } 2x - 1 \text{ for y. This equation is in one variable.}$$
$$x^2 + 4x^2 - 4x + 1 = 2 \quad \text{Use a special-product rule to find } (2x - 1)^2.$$
$$5x^2 - 4x - 1 = 0 \quad \text{To get 0 on the right side, subtract 2} $$
$$\text{from both sides and then combine like terms.}$$

$$(5x + 1)(x - 1) = 0 \quad \text{Factor.}$$
$$5x + 1 = 0 \quad \text{or} \quad x - 1 = 0 \quad \text{Set each factor equal to 0.}$$
$$x = -\frac{1}{5} \qquad\qquad x = 1 \quad \text{Solve each equation.}$$

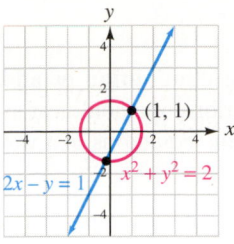

$2x - y = 1$ $x^2 + y^2 = 2$

If we substitute $-\frac{1}{5}$ for x in the equation $y = 2x - 1$, we get $y = -\frac{7}{5}$. If we substitute 1 for x in $y = 2x - 1$, we get $y = 1$. Thus, the system has two solutions, $\left(-\frac{1}{5}, -\frac{7}{5}\right)$ and $(1, 1)$. Verify that each ordered pair satisfies both equations of the original system.

The graph in the margin confirms that the system has two solutions, and that one of them is $(1, 1)$. However, it would be virtually impossible to determine from the graph that the coordinates of the second point of intersection are $\left(-\frac{1}{5}, -\frac{7}{5}\right)$.

Self Check 2 Solve $\begin{cases} x^2 + y^2 = 10 \\ y = x + 2 \end{cases}$ by substitution.

Now Try ▶ Problem 23

EXAMPLE 3 Solve: $\begin{cases} 4x^2 + 9y^2 = 5 \\ y = x^2 \end{cases}$

Strategy Since $y = x^2$, we will substitute y for x^2 in the first equation.

Why This will give an equation in one variable that we can solve for y. We can then find x by back substitution.

Solution We can solve this system by substitution.

$$4x^2 + 9y^2 = 5 \qquad y = x^2$$

When we substitute y for x^2 in the first equation, the result is a quadratic equation in y.

$$4x^2 + 9y^2 = 5$$
$$4y + 9y^2 = 5 \qquad \text{Substitute } y \text{ for } x^2. \text{ This equation is in one variable.}$$
$$9y^2 + 4y - 5 = 0 \qquad \text{To get 0 on the right side, subtract 5 from both sides.}$$
$$(9y - 5)(y + 1) = 0 \qquad \text{Factor } 9y^2 + 4y - 5.$$
$$9y - 5 = 0 \quad \text{or} \quad y + 1 = 0 \qquad \text{Set each factor equal to 0.}$$
$$y = \frac{5}{9} \qquad\qquad y = -1 \qquad \text{Solve each equation.}$$

Since $y = x^2$, the values of x are found by solving the equations

$$x^2 = \frac{5}{9} \quad \text{or} \quad x^2 = -1$$

Because $x^2 = -1$ has no real solutions, this possibility is discarded. The solutions of $x^2 = \frac{5}{9}$ are

$$x = \sqrt{\frac{5}{9}} = \frac{\sqrt{5}}{\sqrt{9}} = \frac{\sqrt{5}}{3} \quad \text{or} \quad x = -\sqrt{\frac{5}{9}} = -\frac{\sqrt{5}}{\sqrt{9}} = -\frac{\sqrt{5}}{3}$$

Thus, the solutions of the system are $\left(\frac{\sqrt{5}}{3}, \frac{5}{9}\right)$ and $\left(-\frac{\sqrt{5}}{3}, \frac{5}{9}\right)$. Verify this by using a check.

Self Check 3 Solve: $\begin{cases} x^2 + y^2 = 20 \\ y = x^2 \end{cases}$

Now Try ▶ Problem 27

Success Tip

$4x^2 + 9y^2 = 5$ is the equation of an ellipse centered at $(0, 0)$, and $y = x^2$ is the equation of a parabola with vertex at $(0, 0)$, opening upward. We would expect two solutions.

Caution

In this section, we are solving for only the real values of x and y.

3 Solve Systems by Elimination (Addition).

Another method for solving nonlinear system of equations is the **elimination** or **addition method.** The elimination method is most often used when the equations of a nonlinear system are both second-degree equations. With this method, we combine the equations in a way that will eliminate the terms of one of the variables.

EXAMPLE 4

Solve: $\begin{cases} 3x^2 + 2y^2 = 36 \\ 4x^2 - y^2 = 4 \end{cases}$

Strategy   We will multiply both sides of the second equation by 2 and add the result to the first equation.

Why   This will eliminate the y^2-terms and produce an equation that we can solve for x.

Solution

To solve this system of two second-degree equations, we can use either the substitution or the elimination method. We will use the elimination method because the y^2-terms can be eliminated by multiplying the second equation by 2 and adding it to the first equation.

$$\begin{cases} 3x^2 + 2y^2 = 36 \xrightarrow{\;\;\text{Unchanged}\;\;} \\ 4x^2 - y^2 = 4 \xrightarrow{\;\;\text{Multiply by 2}\;\;} \end{cases} \begin{cases} 3x^2 + 2y^2 = 36 \\ 8x^2 - 2y^2 = 8 \end{cases}$$

> **Success Tip**
>
> The elimination method is generally better than the substitution method when both equations of the system are of the form $Ax^2 + By^2 = C$.

We add the two equations on the right to eliminate y^2 and solve the resulting equation for x:

$$11x^2 = 44$$
$$x^2 = 4 \qquad \text{Divide both sides by 11.}$$
$$x = 2 \quad \text{or} \quad x = -2 \qquad \text{Use the square root property.}$$

> **Success Tip**
>
> $3x^2 + 2y^2 = 36$ is the equation of an ellipse, centered at $(0, 0)$, and $4x^2 - y^2 = 4$ is the equation of a hyperbola, centered at $(0, 0)$, opening left and right. It is possible to have four solutions.
>
>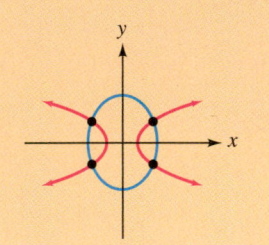

To find y, we can substitute 2 for x and then -2 for x into any equation containing both variables. It appears that the calculations will be simplest if we use $3x^2 + 2y^2 = 36$.

For x = 2:

$$3\textbf{\textit{x}}^2 + 2y^2 = 36$$
$$3(\textbf{2})^2 + 2y^2 = 36$$
$$12 + 2y^2 = 36$$
$$2y^2 = 24$$
$$y^2 = 12$$

For x = -2:

$$3\textbf{\textit{x}}^2 + 2y^2 = 36$$
$$3(\textbf{-2})^2 + 2y^2 = 36$$
$$12 + 2y^2 = 36$$
$$2y^2 = 24$$
$$y^2 = 12$$

Now we use the square root property to solve for y.

For x = 2:

$$y = \sqrt{12} \quad \text{or} \quad y = -\sqrt{12}$$
$$y = 2\sqrt{3} \qquad\qquad y = -2\sqrt{3}$$

For x = -2:

$$y = \sqrt{12} \quad \text{or} \quad y = -\sqrt{12}$$
$$y = 2\sqrt{3} \qquad\qquad y = -2\sqrt{3}$$

The four solutions of this system are $\left(2, 2\sqrt{3}\right)$, $\left(2, -2\sqrt{3}\right)$, $\left(-2, 2\sqrt{3}\right)$, and $\left(-2, -2\sqrt{3}\right)$. Verify this by using a check.

Self Check 4   Solve: $\begin{cases} x^2 + 4y^2 = 16 \\ x^2 - y^2 = 1 \end{cases}$

Now Try ▶ Problem 31

SECTION 10.4 ▶ STUDY SET

VOCABULARY

Fill in the blanks.

1. $\begin{cases} 4x^2 + 6y^2 = 24 \\ 9x^2 - y^2 = 9 \end{cases}$ is a _____ of two nonlinear equations.

2. The graph of $2x + y = 10$ is a _____ and the graph of $x^2 + y^2 = 25$ is a _____.

3. When solving a system by graphing, it is often difficult to determine the coordinates of the points of _____ of the graphs.

4. Two algebraic methods for solving systems of nonlinear equations are the _____ method and the _____ method.

5. A _____ is a line that intersects a circle at two points.

6. A _____ is a line that intersects a circle at one point.

CONCEPTS

7. **a.** A line can intersect an ellipse in at most _____ points.
   b. An ellipse can intersect a parabola in at most _____ points.
   c. An ellipse can intersect a circle in at most _____ points.
   d. A hyperbola can intersect a circle in at most _____ points.

8. Determine whether $(1, -1)$ is a solution of the system:
$$\begin{cases} 2x + y - 1 = 0 \\ x^2 - y^2 = 3 \end{cases}$$

9. Find the solutions of the system $\begin{cases} x^2 + 4y^2 = 25 \\ x^2 - 2y^2 = 1 \end{cases}$ that is graphed on the right.

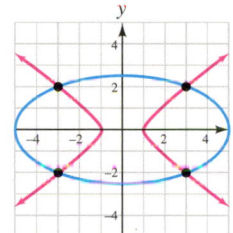

10. Find a substitution equation that can be used to solve the system: $\begin{cases} x^2 + y^2 = 9 \\ 2x - y = 3 \end{cases}$

11. Consider the system: $\begin{cases} 6x^2 + y^2 = 9 \\ 3x^2 + 4y^2 = 36 \end{cases}$

   a. If the y^2-terms are to be eliminated, by what should the first equation be multiplied?

   b. If the x^2-terms are to be eliminated, by what should the second equation be multiplied?

12. Suppose you begin to solve the system $\begin{cases} x^2 + y^2 = 10 \\ 4x^2 + y^2 = 13 \end{cases}$ and find that x is ± 1. Use the first equation to find the corresponding y-values for $x = 1$ and $x = -1$. State the solutions as ordered pairs.

NOTATION

Complete each solution to solve the system.

13. Solve: $\begin{cases} x^2 + y^2 = 5 \\ y = 2x \end{cases}$

$$x^2 + y^2 = 5 \qquad \text{This is the first equation.}$$
$$x^2 + \left(\boxed{} \right)^2 = 5$$
$$x^2 + 4x^2 = \boxed{}$$
$$\boxed{} x^2 = 5$$
$$x^2 = \boxed{}$$
$$x = 1 \quad \text{or} \quad x = -1$$

If $x = 1$, then $y = 2\left(\boxed{} \right) = 2.$ Use the second equation.

If $x = -1$, then $y = 2\left(\boxed{} \right) = -2.$

The solutions are $(1, 2)$ and $\left(-1, \boxed{} \right)$.

14. Solve: $\begin{cases} y = x^2 + 2 \\ y = -x^2 + 4 \end{cases}$

$$2y = \boxed{} \qquad \text{Add the equations.}$$
$$y = \boxed{}$$

If $y = 3$, then

$$\boxed{} = x^2 + 2 \qquad \text{This is the first equation.}$$
$$1 = x^2$$
$$\boxed{} 1 = x$$

The solutions are $\left(1, \boxed{} \right)$ and $\left(\boxed{}, 3 \right)$.

GUIDED PRACTICE

Solve each system of equations by graphing. See Example 1.

15. $\begin{cases} x^2 + y^2 = 9 \\ y - x = 3 \end{cases}$

16. $\begin{cases} x^2 + y^2 = 16 \\ y - x = -4 \end{cases}$

17. $\begin{cases} 9x^2 + 16y^2 = 144 \\ 9x^2 - 16y^2 = 144 \end{cases}$

18. $\begin{cases} x^2 + 9y^2 = 9 \\ 9y^2 - x^2 = 9 \end{cases}$

19. $\begin{cases} y = x^2 - 4x \\ x^2 + y = 0 \end{cases}$

20. $\begin{cases} x^2 - y = 0 \\ y = -x^2 + 4x \end{cases}$

21. $\begin{cases} x^2 + 4y^2 = 4 \\ x = 2y^2 - 2 \end{cases}$

22. $\begin{cases} 4x^2 + y^2 = 4 \\ y = 2x^2 - 2 \end{cases}$

Solve each system of equations by substitution for real values of x and y. See Examples 2 and 3.

23. $\begin{cases} x^2 + y^2 = 5 \\ x + y = 3 \end{cases}$

24. $\begin{cases} x^2 - x - y = 2 \\ 4x - 3y = 0 \end{cases}$

25. $\begin{cases} y = x^2 + 6x + 7 \\ 2x + y = -5 \end{cases}$

26. $\begin{cases} 2x + y = 1 \\ x^2 + y = 4 \end{cases}$

27. $\begin{cases} x^2 + y^2 = 13 \\ y = x^2 - 1 \end{cases}$

28. $\begin{cases} x^2 + y^2 = 10 \\ y = 3x^2 \end{cases}$

29. $\begin{cases} x^2 + y^2 = 30 \\ y = x^2 \end{cases}$

30. $\begin{cases} x^2 + y^2 = 20 \\ y = x^2 \end{cases}$

Solve each system of equations by elimination for real values of x and y. See Example 4.

31. $\begin{cases} x^2 + y^2 = 20 \\ x^2 - y^2 = -12 \end{cases}$

32. $\begin{cases} x^2 + y^2 = 13 \\ x^2 - y^2 = 5 \end{cases}$

33. $\begin{cases} 9x^2 - 7y^2 = 81 \\ x^2 + y^2 = 9 \end{cases}$

34. $\begin{cases} x^2 + y^2 = 25 \\ 2x^2 - 3y^2 = 5 \end{cases}$

35. $\begin{cases} 2x^2 + y^2 = 6 \\ x^2 - y^2 = 3 \end{cases}$

36. $\begin{cases} x^2 + y^2 = 36 \\ 49x^2 + 36y^2 = 1{,}764 \end{cases}$

37. $\begin{cases} x^2 - y^2 = -5 \\ 3x^2 + 2y^2 = 30 \end{cases}$

38. $\begin{cases} 6x^2 + 8y^2 = 182 \\ 8x^2 - 3y^2 = 24 \end{cases}$

 Solve each system. See Using Your Calculator: Solving Systems of Equations.

39. $\begin{cases} x^2 - 6x - y = -5 \\ x^2 - 6x + y = -5 \end{cases}$

40. $\begin{cases} x^2 - y^2 = -5 \\ 3x^2 + 2y^2 = 30 \end{cases}$

TRY IT YOURSELF

Solve each system of equations for real values of x and y.

41. $\begin{cases} 2x^2 - 3y^2 = 5 \\ 3x^2 + 4y^2 = 16 \end{cases}$

42. $\begin{cases} 2x^2 - y^2 + 2 = 0 \\ 3x^2 - 2y^2 + 5 = 0 \end{cases}$

43. $\begin{cases} y = x^2 - 4 \\ x^2 - y^2 = -16 \end{cases}$

44. $\begin{cases} y - x = 0 \\ 4x^2 + y^2 = 10 \end{cases}$

45. $\begin{cases} 3y^2 = xy \\ 2x^2 + xy - 84 = 0 \end{cases}$

46. $\begin{cases} x^2 + y^2 = 10 \\ 2x^2 - 3y^2 = 5 \end{cases}$

47. $\begin{cases} y^2 = 40 - x^2 \\ y = x^2 - 10 \end{cases}$

48. $\begin{cases} 25x^2 + 9y^2 = 225 \\ 5x + 3y = 15 \end{cases}$

49. $\begin{cases} 3x - y = -3 \\ 25y^2 - 9x^2 = 225 \end{cases}$

50. $\begin{cases} x - 2y = 2 \\ 9x^2 - 4y^2 = 36 \end{cases}$

51. $\begin{cases} x^2 - y = 0 \\ x^2 - 4x + y = 0 \end{cases}$

52. $\begin{cases} xy = -\dfrac{9}{2} \\ 3x + 2y = 6 \end{cases}$

53. $\begin{cases} x^2 - 2y^2 = 6 \\ x^2 + 2y^2 = 2 \end{cases}$

54. $\begin{cases} x^2 + 9y^2 = 1 \\ x^2 - 9y^2 = 3 \end{cases}$

55. $\begin{cases} y = x^2 - 4 \\ 6x - y = 13 \end{cases}$

56. $\begin{cases} y = x + 1 \\ x^2 - y^2 = 1 \end{cases}$

57. $\begin{cases} x^2 + y^2 = 4 \\ 9x^2 + y^2 = 9 \end{cases}$

58. $\begin{cases} 2x^2 - 6y^2 + 3 = 0 \\ 4x^2 + 3y^2 = 4 \end{cases}$

59. $\begin{cases} xy = \dfrac{1}{6} \\ y + x = 5xy \end{cases}$

60. $\begin{cases} xy = \dfrac{1}{12} \\ y + x = 7xy \end{cases}$

61. $\begin{cases} x^2 = 4 - y \\ y = x^2 + 2 \end{cases}$

62. $\begin{cases} 3x + 2y = 10 \\ y = x^2 - 5 \end{cases}$

63. $\begin{cases} x^2 - y^2 = 4 \\ x + y = 4 \end{cases}$

64. $\begin{cases} x - y = -1 \\ y^2 - 4x = 0 \end{cases}$

APPLICATIONS

Use a nonlinear system of equations to solve each problem.

65. **Integer Problem.**   The product of two integers is 32, and their sum is 12. Find the integers.

66. **Number Problem.**   The sum of the squares of two numbers is 221, and the sum of the numbers is 9. Find the numbers.

67. Archery.   See the illustration below. An arrow shot from the base of a hill follows the parabolic path $y = -\frac{1}{6}x^2 + 2x$, with distances measured in meters. The inclined hill has a slope of $\frac{1}{3}$ and can therefore be modeled by the equation $y = \frac{1}{3}x$. Find the coordinates of the point of impact of the arrow and then its distance from the archer.

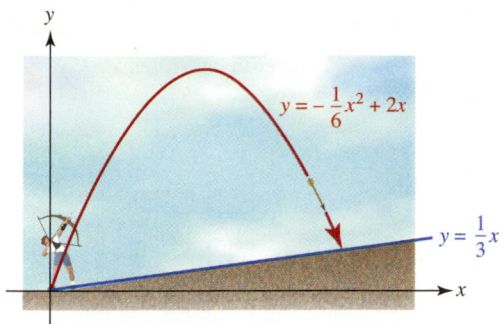

68. Geometry.   The area of a rectangle is 63 square centimeters, and its perimeter is 32 centimeters. Find the dimensions of the rectangle.

69. Fencing Pastures.   The rectangular pasture shown here is to be fenced in along a riverbank. If 260 feet of fencing is to enclose an area of 8,000 square feet, find the dimensions of the pasture.

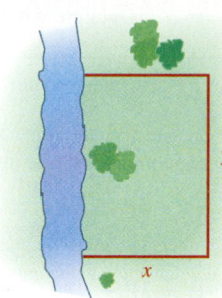

70. Driving Rates.   Jim drove 306 miles. Jim's brother made the same trip at a speed 17 mph slower than Jim did and required an extra $1\frac{1}{2}$ hours. What was Jim's rate and time?

71. Investing.   Grant receives $225 annual income from one investment. Jeff invested $500 more than Grant, but at an annual rate of 1% less. Jeff's annual income is $240. What are the amount and rate of Grant's investment?

72. Investing.   Carol receives $67.50 annual income from one investment. John invested $150 more than Carol at an annual rate of $1\frac{1}{2}$% more. John's annual income is $94.50. What are the amount and rate of Carol's investment? (*Hint:* There are two answers.)

WRITING

73. a. Describe the benefits of the graphical method for solving a system of nonlinear equations.

b. Describe the drawbacks of the graphical method.

74. Explain why the elimination method, not the substitution method, is the better method to solve the system:

$$\begin{cases} 4x^2 + 9y^2 = 52 \\ 9x^2 + 4y^2 = 52 \end{cases}$$

REVIEW

Solve each equation.

75. $\log 5x = 4$

76. $\log 3x = \log 9$

77. $\dfrac{\log(8x - 7)}{\log x} = 2$

78. $\log x + \log(x + 9) = 1$

CHALLENGE PROBLEMS

79. a. The graphs of the two independent equations of a system are parabolas. How many solutions might the system have?

b. The graphs of the two independent equations of a system are hyperbolas. How many solutions might the system have?

80. Solve the system for real solutions: $\begin{cases} \dfrac{1}{x} + \dfrac{2}{y} = 1 \\ \dfrac{2}{x} - \dfrac{1}{y} = \dfrac{1}{3} \end{cases}$

81. Solve the system for real solutions: $\begin{cases} \dfrac{1}{x} + \dfrac{3}{y} = 4 \\ \dfrac{2}{x} - \dfrac{1}{y} = 7 \end{cases}$

82. Solve the system $\begin{cases} x^2 - y^2 = 16 \\ x^2 + y^2 = 9 \end{cases}$ over the complex numbers.

10 Summary & Review

SECTION 10.1 ▶ The Circle and the Parabola

DEFINITIONS AND CONCEPTS	EXAMPLES
A **circle** is the set of all points in a plane that are a fixed distance from a fixed point called its **center.** The fixed distance is called the **radius** of the circle. **Standard forms of the equation of a circle:** $x^2 + y^2 = r^2$ Center $(0, 0)$, radius r $(x - h)^2 + (y - k)^2 = r^2$ Center (h, k), radius r	The graph of the equation $x^2 + y^2 = 16$, which can be written $x^2 + y^2 = 4^2$, is a circle with center at $(0, 0)$ and a radius of **4**. 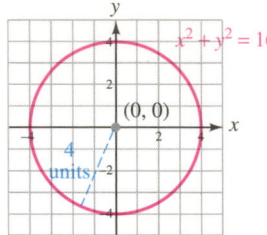 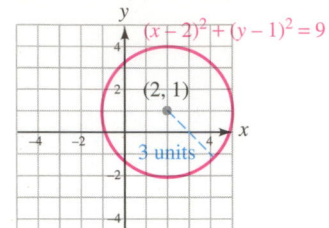 The graph of the equation $(x - 2)^2 + (y - 1)^2 = 9$, which can be written $(x - \textbf{2})^2 + (y - \textbf{1})^2 = 3^2$, is a circle with center at $(\textbf{2}, \textbf{1})$ and a radius of **3**.
Because a circle is determined by its center and radius, that information is all we need to know to write its equation.	Write the equation of a circle centered at $(4, -3)$ and with a radius of 5. In this problem, $h = 4$, $k = -3$, and $r = 5$. We substitute these values into the standard form of the equation of a circle and simplify. $(x - \textbf{h})^2 + (y - \textbf{k})^2 = \textbf{r}^2$ $(x - \textbf{4})^2 + [y - (\textbf{-3})]^2 = \textbf{5}^2$ $(x - 4)^2 + (y + 3)^2 = 25$
A **parabola** is the set of all points in a plane that are equidistant from a fixed point, called the **focus,** and a fixed line, called the **directrix.** **General forms of the equation of a parabola:** $y = ax^2 + bx + c$ $a > 0$: up; $a < 0$: down $x = ay^2 + by + c$ $a > 0$: right; $a < 0$: left **Standard forms of the equation of a parabola:** $y = a(x - h)^2 + k$ $a > 0$: up; $a < 0$: down Vertex at (h, k) Axis of symmetry is $x = h$ $x = a(y - k)^2 + h$ $a > 0$: right; $a < 0$: left Vertex at (h, k) Axis of symmetry is $y = k$	The equation $x = 2y^2 - 4y + 5$ is the equation of a parabola that opens to the right. To find its vertex and axis of symmetry, we complete the square on y and write the equation in standard form. $x = 2y^2 - 4y + 5$ $x = 2(y^2 - 2y\quad) + 5$ Factor out 2. $x = 2(y^2 - 2y \textbf{ + 1}) + 5 - 2$ Complete the square. $x = 2(y - 1)^2 + 3$ Factor and simplify. From the standard form, we see that $h = 3$ and $k = 1$. Thus, the vertex is at $(3, 1)$ and the axis of symmetry is $y = 1$. To construct a table of solutions, we choose values of y and find their corresponding values of x. $x = 2(y - 1)^2 + 3$ | x	y	
---	---	---
5	2	→ $(5, 2)$
11	3	→ $(11, 3)$	 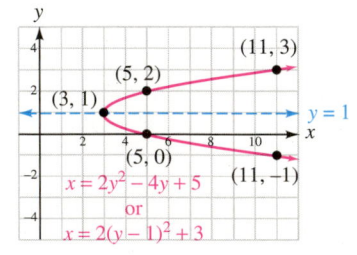

REVIEW EXERCISES

Graph each equation.

1. $x^2 + y^2 = 16$

2. $(x - 4)^2 + (y + 3)^2 = 4$

3. Write the equation $x^2 + y^2 + 4x - 2y = 4$ in standard form and graph it.

4. **Art History.** Leonardo da Vinci's *Vitruvian Man* (1492) is one of the most famous pen-and-ink drawings of all time. Use the coordinate system that is superimposed on the drawing to write the equation of the circle in standard form.

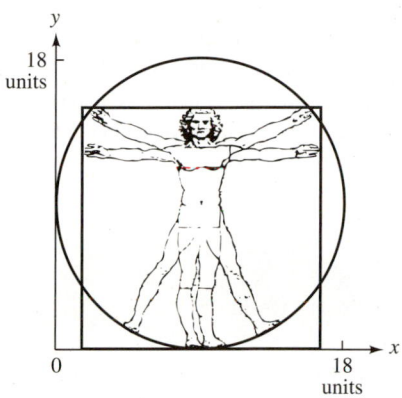

5. Find the center and the radius of the circle whose equation is $(x + 6)^2 + y^2 = 24$.

6. Fill in the blanks: A circle is the set of all points in a plane that are a fixed distance from a point called its _____ . The fixed distance is called the _____ of the circle.

Graph each parabola and give the coordinates of the vertex.

7. $x = y^2$

8. $x = 2(y + 1)^2 - 2$

Write each equation in standard form and graph it.

9. $x = -3y^2 + 12y - 7$

10. $y = x^2 + 8x + 11$

11. The axis of symmetry, the vertex, and two additional points on the graph of a parabola are shown. Find the coordinates of two other points on the parabola.

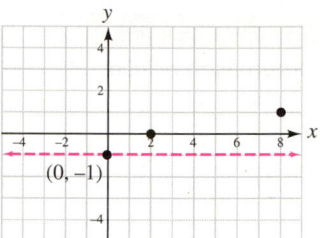

12. **Long Jump.** The equation describing the flight path of the long jumper is $y = -\frac{5}{121}(x - 11)^2 + 5$. Show that she will land at a point 22 feet away from the take-off board.

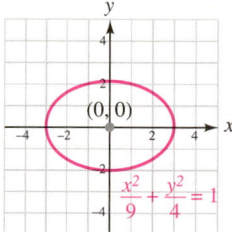

Take-off board ←——— 22 ft ———→ Landing

SECTION 10.2 ▶ The Ellipse

DEFINITIONS AND CONCEPTS	EXAMPLES

An **ellipse** is the set of all points in a plane for which the sum of the distances from two fixed points is a constant.

  Standard forms of the equation of an ellipse:

  $$\frac{x^2}{a^2} + \frac{y^2}{b^2} = 1 \quad \text{Center } (0, 0)$$

  $$\frac{(x - h)^2}{a^2} + \frac{(y - k)^2}{b^2} = 1 \quad \text{Center } (h, k)$$

The equation $\frac{x^2}{9} + \frac{y^2}{4} = 1$, which can be written $\frac{x^2}{3^2} + \frac{y^2}{2^2} = 1$, represents an ellipse that is centered at the origin. Here, $a = 3$ and $b = 2$.

$$\frac{x^2}{9} + \frac{y^2}{4} = 1$$

$$\frac{(x-1)^2}{4} + \frac{(y+2)^2}{16} = 1$$

The equation $\frac{(x - 1)^2}{4} + \frac{(y + 2)^2}{16} = 1$, which can be written $\frac{(x - 1)^2}{2^2} + \frac{(y + 2)^2}{4^2} = 1$, represents an ellipse that is centered at $(1, -2)$. Here, $a = 2$ and $b = 4$.

To write the equation $25x^2 + 16y^2 = 400$ in standard form, divide both sides by 400 and simplify.

$$25x^2 + 16y^2 = 400$$

$$\frac{25x^2}{400} + \frac{16y^2}{400} = \frac{400}{400}$$ To get 1 on the right side, divide both sides by 400.

$$\frac{x^2}{16} + \frac{y^2}{25} = 1$$ Simplify each fraction.

This result represents an ellipse that is centered at $(0, 0)$, with $a = 4$ and $b = 5$.

REVIEW EXERCISES

Graph each ellipse.

13. $\dfrac{x^2}{16} + \dfrac{y^2}{9} = 1$

14. $\dfrac{(x - 2)^2}{4} + \dfrac{(y - 1)^2}{25} = 1$

15. $4(x + 1)^2 + 9(y - 1)^2 = 36$

16. Consider the equation $\dfrac{x^2}{144} + \dfrac{y^2}{1} = 1$. Write each term on the left side with a denominator that is the square of a number.

17. Determine whether the graph of each equation is a circle, a parabola, or an ellipse.

   a. $(x - 1)^2 + (y + 9)^2 = 100$

   b. $\dfrac{x^2}{49} + \dfrac{y^2}{121} = 1$

   c. $x = y^2 - 2y + 6$

   d. $16(x - 4)^2 + 4(y + 8)^2 = 16$

18. Salami. When a delicatessen slices a cylindrical salami at an angle, the results are elliptical pieces that are larger than circular pieces. See the illustration. Write the equation of the shape of the slice of salami shown if it was centered at the origin of a coordinate system.

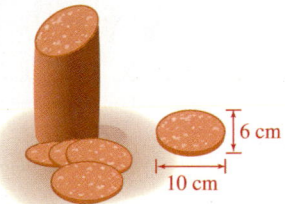

6 cm
10 cm

19. Fill in the blanks: An _____ is the set of all points in a plane for which the sum of the distances from two fixed points is a constant. Each of the fixed points is called a _____.

20. Construction. Sketch the path of the sound when a person, standing at one focus, whispers something in the whispering gallery dome shown below.

Focus   Focus

SECTION 10.3 ▶ **The Hyperbola**

DEFINITIONS AND CONCEPTS	EXAMPLES

A **hyperbola** is the set of all points in a plane for which the difference of the distances from two fixed points is a constant.

Standard forms of the equation of a hyperbola:

$\dfrac{x^2}{a^2} - \dfrac{y^2}{b^2} = 1$ Center $(0, 0)$, opens left and right

$\dfrac{y^2}{a^2} - \dfrac{x^2}{b^2} = 1$ Center $(0, 0)$, opens up and down

$\dfrac{(x - h)^2}{a^2} - \dfrac{(y - k)^2}{b^2} = 1$ Center (h, k), opens left and right

$\dfrac{(y - k)^2}{a^2} - \dfrac{(x - h)^2}{b^2} = 1$ Center (h, k), opens up and down

The equation $\dfrac{x^2}{4} - \dfrac{y^2}{9} = 1$, which can be written $\dfrac{x^2}{2^2} - \dfrac{y^2}{3^2} = 1$, represents a hyperbola, centered at $(0, 0)$, that opens left and right. Here, $a = 2$ and $b = 3$.

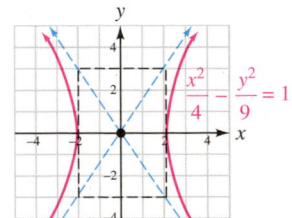

$\dfrac{x^2}{4} - \dfrac{y^2}{9} = 1$

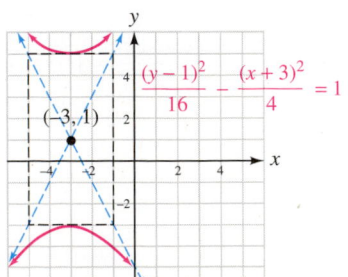

$\dfrac{(y - 1)^2}{16} - \dfrac{(x + 3)^2}{4} = 1$

$(-3, 1)$

The equation $\dfrac{(y - 1)^2}{16} - \dfrac{(x + 3)^2}{4} = 1$, which can be written $\dfrac{(y - 1)^2}{4^2} - \dfrac{(x + 3)^2}{2^2} = 1$, represents a hyperbola, centered at $(-3, 1)$, that opens up and down. Here, $a = 4$ and $b = 2$.

To write the equation $25y^2 - 9x^2 = 225$ in standard form, divide both sides by 225 and simplify.

$$25y^2 - 9x^2 = 225$$

$$\frac{25y^2}{225} - \frac{9x^2}{225} = \frac{225}{225} \qquad \text{To get 1 on the right side, divide both sides by 225.}$$

$$\frac{y^2}{9} - \frac{x^2}{25} = 1 \qquad \text{Simplify each fraction.}$$

This result represents a hyperbola centered at the origin that opens up and down. Here, $a = 3$ and $b = 5$.

REVIEW EXERCISES

Graph each hyperbola.

21. $\dfrac{y^2}{9} - \dfrac{x^2}{1} = 1$

22. $9(x - 1)^2 - 4(y + 1)^2 = 36$

23. $\dfrac{(y - 2)^2}{25} - \dfrac{(x + 1)^2}{25} = 1$

24. $xy = 9$

25. Electrostatic Repulsion. Two similarly charged particles are shot together for an almost head-on collision, as in the illustration. They repel each other and travel the two branches of the hyperbola given by $x^2 - 4y^2 = 4$ on the given coordinate system. How close do they get?

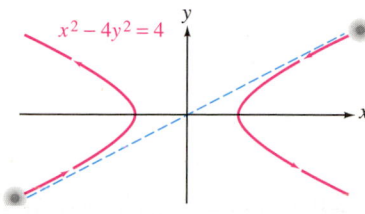

26. Determine whether the graph of each equation will be a circle, parabola, ellipse, or hyperbola.

a. $\dfrac{(x - 4)^2}{16} + \dfrac{y^2}{49} = 1$

b. $16(x + 3)^2 - 4(y - 1)^2 = 64$

c. $x = -4y^2 - y + 1$

d. $x^2 + 2x + y^2 - 4y = 40$

SECTION 10.4 ▶ Solving Nonlinear Systems of Equations

DEFINITIONS AND CONCEPTS	EXAMPLES
A **nonlinear system of equations** is a system that contains at least one nonlinear equation. Systems of nonlinear equations are solved by **graphing**, by **substitution**, or by **elimination (addition)**.	To solve the nonlinear system $\begin{cases} x^2 + y^2 = 20 \\ y = x^2 \end{cases}$ by graphing, we graph the equations on the same rectangular coordinate system, and determine the coordinates of the points of intersection of the graphs. Since the points of intersection of the graphs are $(-2, 4)$ and $(2, 4)$, the solutions of the system are $(-2, 4)$ and $(2, 4)$ and the solution set is $\{(-2, 4), (2, 4)\}$. 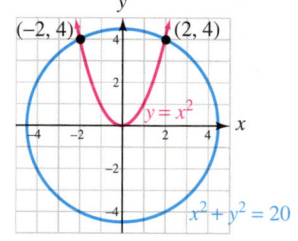

With the **substitution method,** the objective is to use an appropriate substitution to obtain *one* equation in *one* variable.

To use substitution to solve the nonlinear system $\begin{cases} y = 3x - 5 \\ x^2 + y^2 = 5 \end{cases}$, we substitute $3x - 5$ for y in the second equation and solve for x.

$$x^2 + y^2 = 5$$
$$x^2 + (3x - 5)^2 = 5 \quad \text{This is a quadratic equation in x.}$$
$$x^2 + 9x^2 - 30x + 25 = 5$$
$$10x^2 - 30x + 20 = 0 \quad \text{Combine terms and subtract 5 from both sides.}$$
$$x^2 - 3x + 2 = 0 \quad \text{Divide both sides by 10.}$$
$$(x - 2)(x - 1) = 0 \quad \text{Factor.}$$
$$x - 2 = 0 \quad \text{or} \quad x - 1 = 0 \quad \text{Set each factor equal to 0.}$$
$$x = 2 \quad | \quad x = 1$$

If $x = 2$, then $y = 3x - 5 = 3(2) - 5 = 1$.

If $x = 1$, then $y = 3x - 5 = 3(1) - 5 = -2$.

The two solutions of the system are $(2, 1)$ and $(1, -2)$.

With the **elimination (addition) method,** we combine the equations in a way that will eliminate the terms of one of the variables.

To use elimination to solve the nonlinear system $\begin{cases} x^2 - y = 0 \\ x + y = 0 \end{cases}$, we add the equations to get $x^2 + x = 0$. Then we factor this result to get $x = 0$ or $x = -1$. We can substitute these values into the second equation to find y.

If $x = 0$: $\quad x + y = 0 \quad \text{This is the second equation.}$
$$0 + y = 0 \quad \text{Substitute 0 for x.}$$
$$y = 0$$

If $x = -1$: $\quad x + y = 0 \quad \text{This is the second equation.}$
$$-1 + y = 0 \quad \text{Substitute −1 for x.}$$
$$y = 1$$

The two solutions of the system are $(0, 0)$ and $(-1, 1)$.

REVIEW EXERCISES

27. Determine whether $\left(-\sqrt{11}, -3\right)$ is a solution of the system:
$$\begin{cases} x^2 + y^2 = 20 \\ x^2 - y^2 = 2 \end{cases}$$

28. The graphs of $y^2 - x^2 = 9$ and $x^2 + y^2 = 9$ are shown. Estimate the solutions of the system
$$\begin{cases} y^2 - x^2 = 9 \\ x^2 + y^2 = 9 \end{cases}$$

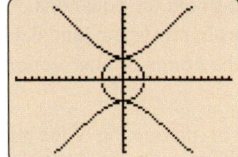

29. Solve the system $\begin{cases} xy = 4 \\ y = 2x - 2 \end{cases}$ by graphing.

30. Determine the maximum number of solutions there could be for a system of equations consisting of the given curves.

   a. A line and an ellipse   **b.** Two hyperbolas

   c. An ellipse and a circle   **d.** A parabola and a circle

31. Suppose the x-coordinate of both points of intersection of the circle, represented by $x^2 + y^2 = 1$, and the hyperbola, defined by $4y^2 - x^2 = 4$, is 0. Without graphing, determine the y-coordinates of both points of intersection. Express the answers as ordered-pair solutions.

32. Find a substitution equation that can be used to solve $\begin{cases} x^2 + y^2 = 16 \\ 3x - y = 1 \end{cases}$. Do not solve the system.

Solve each system for real values of x and y.

33. $\begin{cases} y^2 - x^2 = 16 \\ y + 4 = x^2 \end{cases}$

34. $\begin{cases} y = -x^2 + 2 \\ x^2 - y - 2 = 0 \end{cases}$

35. $\begin{cases} x^2 + 2y^2 = 12 \\ 2x - y = 2 \end{cases}$

36. $\begin{cases} 3x^2 + y^2 = 52 \\ x^2 - y^2 = 12 \end{cases}$

37. $\begin{cases} \dfrac{x^2}{16} + \dfrac{y^2}{12} = 1 \\ \dfrac{x^2}{1} - \dfrac{y^2}{3} = 1 \end{cases}$

38. $\begin{cases} xy = 4 \\ \dfrac{x^2}{1} + \dfrac{y^2}{2} = 9 \end{cases}$

39. $\begin{cases} y = -x^2 + 1 \\ x + y = 5 \end{cases}$

40. $\begin{cases} x = y^2 - 3 \\ x = y^2 - 3y \end{cases}$

10 Chapter Test

1. Fill in the blanks.
   a. The curves formed by the intersection of a plane with an infinite right-circular cone are called _____ sections.
   b. A circle is the set of all points in a plane that are a fixed distance from a point called its _____. The fixed distance is called the _____ of the circle.
   c. The standard form for the equation of a _____ centered at the origin that opens left and right is $\frac{x^2}{a^2} - \frac{y^2}{b^2} = 1$.
   d. $\begin{cases} y = x^2 + x - 4 \\ x^2 + y^2 = 36 \end{cases}$ is a(n) _____ system of equations.
   e. The standard form for the equation of an _____ centered at the origin is $\frac{x^2}{a^2} + \frac{y^2}{b^2} = 1$.

2. Find the center and the radius of the circle represented by the equation $x^2 + y^2 = 100$ and graph it.

3. Find the center and the radius of the circle represented by the equation $x^2 + y^2 + 4x - 6y = 5$.

4. **TV History.** In the early days of television, stations broadcast a black-and-white test pattern like that shown here during the early morning hours. Use the given coordinate system to write an equation of the large, bold circle in the center of the pattern.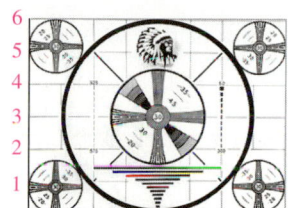

5. **Frisbee.** In the illustration, an Ultimate Frisbee is centered on a rectangular coordinate system. Determine the *diameter* of the circular disc from the given equation.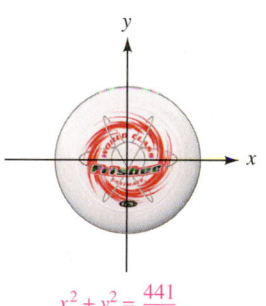

$$x^2 + y^2 = \frac{441}{16}$$

6. Fill in the blanks to complete the square on the right side of the following equation.
   $$x = y^2 + 8y + 10$$
   $$x = (y^2 + 8y + \boxed{}) + 10 - \boxed{}$$
   $$x = (y + \boxed{})^2 - \boxed{}$$

Graph each equation.

7. $(x + 2)^2 + (y - 1)^2 = 9$
8. $x = y^2 - 2y + 3$
9. $y = -2x^2 - 4x + 5$
10. $xy = -4$
11. $9x^2 + 4y^2 = 36$
12. $\frac{(x - 2)^2}{9} - \frac{y^2}{1} = 1$
13. $\frac{(x - 3)^2}{49} + \frac{(y + 2)^2}{16} = 1$
14. $x^2 + y^2 = 7$
15. $x = -\frac{1}{2}y^2$
16. $\frac{y^2}{25} - \frac{x^2}{9} = 1$

17. Write the equation in standard form of the ellipse graphed here.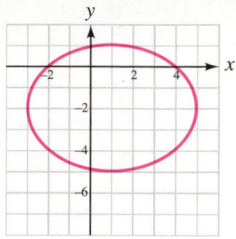

18. **Light.** The cross section of a parabolic mirror is given by the equation $x = \frac{1}{10}y^2$, with distances measured in inches. If the dish is 10 inches wide, how deep is it?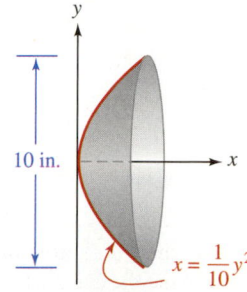

19. **Advertising.** An elliptical logo for Tom's Shoes is to be centered on a 36-by-60-inch rectangular background, leaving 2 inches of space on all sides. Find the equation of the ellipse.

20. Find the center and the length and width of the central rectangle of the graph of $(x + 1)^2 - (y - 1)^2 = 4$.

21. Find the equation in standard form of the hyperbola graphed here.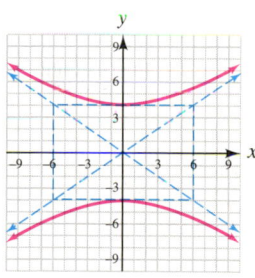

22. Determine whether the graph of each equation will be a circle, a parabola, an ellipse, or a hyperbola.
   a. $25x^2 + 100y^2 = 400$
   b. $9x^2 - y^2 = 9$
   c. $x^2 + 8x + y^2 - 16y - 1 = 0$
   d. $x = 8y^2 - 9y + 4$

23. Solve $\begin{cases} x^2 + y^2 = 25 \\ y - x = 1 \end{cases}$ graphically.

Solve each system for real values of x and y.

24. $\begin{cases} 2x - y = -2 \\ x^2 + y^2 = 16 + 4y \end{cases}$

25. $\begin{cases} 5x^2 - y^2 - 3 = 0 \\ x^2 + 2y^2 = 5 \end{cases}$

26. $\begin{cases} xy = -\dfrac{9}{2} \\ 3x + 2y = 6 \end{cases}$

27. $\begin{cases} y = x + 1 \\ x^2 - y^2 = 1 \end{cases}$

28. $\begin{cases} x^2 + 3y^2 = 6 \\ x^2 + y = 8 \end{cases}$

Group Project

Parabolas

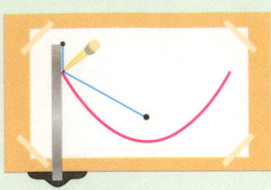

Overview: In this activity, you will construct several models of parabolas.

Instructions: Form groups of 2 or 3 students. You will need a T-square, string, paper, pencil, and a thumbtack. To construct a parabola, secure one end of a piece of string that is as long as the T-square to a large piece of paper using a brad or thumbtack, as shown at the left. Attach the other end of the string to the upper end of the T-square. Hold the string taut against the T-square with a pencil and slide the T-square along the edge of the table. As the T-square moves, the pencil will trace a parabola.

Each point on the parabola is the same distance away from a given point as it is from a given line. With this model, what is the given point, and what is the given line?

Make other models by moving the fixed point closer and farther away from the edge of the table. How is the shape of the parabola affected?

Ellipses

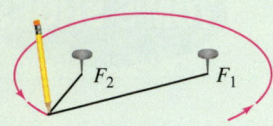

Overview: In this activity, you will construct several ellipses.

Instructions: Form groups of 2 or 3 students. You will need two thumbtacks, a pencil, and a length of string with a loop tied at one end. To construct an ellipse, place two thumbtacks (or brads) fairly close together, as shown in the illustration. Catch the loop of the string with the point of the pencil and, keeping the string taut, draw the ellipse.

Make several models by moving one of the thumbtacks farther away and then closer to the other thumbtack. How does the shape of the ellipse change?

For each point on the ellipse, the sum of the distances of the point from two given points is a constant. With this method of construction, what are the two points? What is the constant distance?

Miscellaneous Topics

from Campus to Careers

Real Estate Sales Agent

Buying a house is probably the biggest purchase that most people will make in their lives. The complex process of purchasing a home is much easier with the help of a real estate agent. Real estate agents use their mathematical skills in many ways. They calculate square footage, appraise property, calculate commissions, and write offer sheets. Technology is widely used in the real estate industry. Most sales agents use computers to locate and list available properties and identify sources of financing.

Problem 81 in **Study Set 11.3** involves a situation that real estate agents and their clients are always concerned with—changing property values.

JOB TITLE:
Real Estate Sales Agent

EDUCATION:
Must be a high school graduate, attend formal training classes, and pass a written licensing examination.

JOB OUTLOOK:
Good; it is expected to increase 14% through 2018.

ANNUAL EARNINGS:
The middle 50% of agents earned between $27,390 and $64,820 in salary and commissions in 2008.

FOR MORE INFORMATION:
www.bls.gov/oco/ocos120.htm

901

Study Skills Workshop ▶ Exploring Careers

Ultimately, your choice of career will determine the math course(s) that you need to take after Intermediate Algebra. Before the end of this term, it would be wise to have at least a general idea of your career goals.

HOW DO YOU DECIDE?: Seek the advice of a counselor, visit your school's career center, search the Internet, or read books that will help you discover your interests and possible related careers.

ONCE YOU'VE DECIDED: Talk to your counselor and consult the appropriate college catalogs to develop a long-term plan that will put you on the correct educational path.

Now Try This

1. Do you have a career goal in mind? If so, what is it?
2. Take at least two personality tests and two career-choice tests. A list of tests offered online can be found at www.cengage.com/math/tussy
3. Visit a counselor to discuss which classes you should take during your next term and beyond. Make a list of classes that your counselor suggests that you take.

SECTION 11.1

OBJECTIVES

1 Use Pascal's triangle to expand binomials.

2 Use factorial notation.

3 Use the binomial theorem to expand binomials.

4 Find a specific term of a binomial expansion.

The Binomial Theorem

ARE YOU READY?

The following problems review some basic skills that are needed when working with the binomial theorem.

1. Determine the coefficient of each term: $x^4 + 4x^3 + 6x^2 + 4x + 1$

2. Find the product: $(a + b)^2$

3. Evaluate: $5 \cdot 4 \cdot 3 \cdot 2 \cdot 1$

4. Simplify: $\dfrac{10 \cdot 9 \cdot 8}{2 \cdot 3 \cdot 4}$

In Chapter 5, we discussed how to raise binomials to positive-integer powers. For example, we have seen that

$$(a + b)^2 = a^2 + 2ab + b^2$$

The result, $a^2 + 2ab + b^2$, is called the **binomial expansion** of $(a + b)^2$.

To find the binomial expansion of $(a + b)^3$, we multiply $(a + b)^2$ by $(a + b)$. As we see below, this process involves several steps and many ways to make errors.

$$\begin{aligned}(a + b)^3 &= (a + b)(a + b)^2 \\ &= (a + b)(a^2 + 2ab + b^2) \\ &= a^3 + 2a^2b + ab^2 + a^2b + 2ab^2 + b^3 \\ &= a^3 + 3a^2b + 3ab^2 + b^3\end{aligned}$$

In this section, we will discuss a method for finding binomial expansions quickly, without having to perform all of the steps shown above.

The Language of Algebra

Recall that two-term polynomial expressions such as $a + b$ and $3u - 2v$ are called **binomials**.

1 Use Pascal's Triangle to Expand Binomials.

To **expand** a binomial of the form $(a + b)^n$, where n is a nonnegative integer, means to write it as a sum of terms. To develop a method for expanding such binomials, consider the following:

$(a + b)^0 =$	1	1 term
$(a + b)^1 =$	$a + b$	2 terms
$(a + b)^2 =$	$a^2 + 2ab + b^2$	3 terms
$(a + b)^3 =$	$a^3 + 3a^2b + 3ab^2 + b^3$	4 terms
$(a + b)^4 =$	$a^4 + 4a^3b + 6a^2b^2 + 4ab^3 + b^4$	5 terms
$(a + b)^5 =$	$a^5 + 5a^4b + 10a^3b^2 + 10a^2b^3 + 5ab^4 + b^5$	6 terms
$(a + b)^6 =$	$a^6 + 6a^5b + 15a^4b^2 + 20a^3b^3 + 15a^2b^4 + 6ab^5 + b^6$	7 terms

The Language of Algebra

When we **expand** a power of a binomial, the result is called a **binomial expansion.** For powers greater than or equal to 2, an expansion has more terms than the original binomial.

Several patterns appear in these expansions:

1. Each expansion has one more term than the power of the binomial.

2. For each term of an expansion, the sum of the exponents on a and b is equal to the exponent of the binomial being expanded. For example, in the expansion of $(a + b)^5$, the sum of the exponents in each term is 5:

$$4+1=5 \quad 3+2=5 \quad 2+3=5 \quad 1+4=5$$
$$(a + b)^5 = a^5 + 5a^4b + 10a^3b^2 + 10a^2b^3 + 5ab^4 + b^5$$

3. The first term in each expansion is a, raised to the power of the binomial, and the last term in each expansion is b, raised to the power of the binomial.

4. The exponents on a decrease by one in each successive term, ending with $a^0 = 1$ in the last term. The exponents on b, beginning with $b^0 = 1$ in the first term, increase by one in each successive term. For example, the expansion of $(a + b)^4$ could be written as

$$a^4b^0 + 4a^3b^1 + 6a^2b^2 + 4a^1b^3 + a^0b^4$$

Thus, the variables have the pattern

$$a^n, \quad a^{n-1}b, \quad a^{n-2}b^2, \quad \ldots, \quad ab^{n-1}, \quad b^n$$

5. The coefficients of each expansion begin with 1, increase through some values, and then decrease through those same values, back to 1.

To see another pattern, we write just the coefficients of each expansion of $a + b$ in a triangular array:

The Language of Algebra

This array of numbers is named **Pascal's triangle** in honor of the French mathematician Blaise Pascal (1623–1662).

```
              1                  Row for (a + b)⁰
           1     1               Row for (a + b)¹
        1     2     1            Row for (a + b)²
     1     3     3     1         Row for (a + b)³
  1     4     6     4     1      Row for (a + b)⁴
1    5    10    10    5    1     Row for (a + b)⁵
1  6   15   20   15   6   1      Row for (a + b)⁶
```

In this array, called **Pascal's triangle,** each entry between the 1's is the sum of the closest pair of numbers in the line immediately above it. For example, the first 15 in the bottom row is the sum of the 5 and 10 immediately above it. Pascal's triangle continues with the same pattern forever. The next two lines are

```
1   7   21   35   35   21   7   1      Row for (a + b)⁷
1  8   28   56   70   56   28  8   1    Row for (a + b)⁸
```

EXAMPLE 1 Expand: $(x + y)^5$

Strategy We will use the pattern shown on the previous page for raising binomials to powers and Pascal's triangle.

Why The pattern provides the variable expressions in the expansion and Pascal's triangle provides their coefficients.

Solution The first term in the expansion is x^5, and the exponents on x decrease by one in each successive term. A y first appears in the second term, and the exponents on y increase by one in each successive term, concluding when the term y^5 is reached. Thus, the variable expressions in the expansion are

$$x^5, \qquad x^4y, \qquad x^3y^2, \qquad x^2y^3, \qquad xy^4, \qquad y^5$$

Since the exponent of the binomial that is being expanded is 5, the coefficients of these variables are found in row 5 of Pascal's triangle.

```
                    1
                  1   1
                1   2   1
              1   3   3   1
            1   4   6   4   1
          1   5  10  10   5   1
        1   6  15  20  15   6   1
      1   7  21  35  35  21   7   1
    1   8  28  56  70  56  28   8   1
```

$$\quad 1 \quad 5 \quad 10 \quad 10 \quad 5 \quad 1 \quad$$ *Remember, the 1 at the top of Pascal's triangle is labeled row 0.*

Combining this information gives the following expansion:

$$(x + y)^5 = x^5 + 5x^4y + 10x^3y^2 + 10x^2y^3 + 5xy^4 + y^5$$

Self Check 1 Expand: $(x + y)^4$

Now Try ▶ Problem 25

EXAMPLE 2 Expand: $(u - v)^4$

Strategy We will use the pattern shown on the previous page for raising binomials to powers and Pascal's triangle.

Why The pattern provides the variable expressions in the expansion and Pascal's triangle provides their coefficients.

Solution We note that $(u - v)^4$ can be written as $[u + (-v)]^4$. The variable expressions in this expansion are

$$u^4, \qquad u^3(-v), \qquad u^2(-v)^2, \qquad u(-v)^3, \qquad (-v)^4$$

and the coefficients are given in row 4 of Pascal's triangle:

$$\qquad 1 \qquad 4 \qquad 6 \qquad 4 \qquad 1 \qquad$$ *Remember, the 1 at the top of Pascal's triangle is labeled row 0.*

Thus, the required expansion is

$$(u - v)^4 = u^4 + 4u^3(-v) + 6u^2(-v)^2 + 4u(-v)^3 + (-v)^4$$

The Language of Algebra

To **alternate** means to change back and forth. In this expansion, the signs + and − alternate.

Now we simplify each term. When $-v$ is raised to an even power, the sign is positive, and when $-v$ is raised to an odd power, the sign is negative. This causes the signs of the terms in the expansion to alternate between + and −.

$$(u - v)^4 = u^4 - 4u^3v + 6u^2v^2 - 4uv^3 + v^4$$

Self Check 2 Expand: $(x - y)^5$

Now Try ▶ Problem 27

2 Use Factorial Notation.

Although Pascal's triangle gives the coefficients of the terms in a binomial expansion, it is not the easiest way to expand a binomial. To develop a better way, we introduce **factorial notation.** The symbol $n!$ (read as "n **factorial**") is defined as follows.

Factorial Notation

$n!$ is the product of consecutively decreasing natural numbers from n to 1.
For any natural number n,

$$n! = n(n - 1)(n - 2)(n - 3) \cdot \cdots \cdot 3 \cdot 2 \cdot 1$$

Zero factorial is defined as $0! = 1$.

EXAMPLE 3

Evaluate each expression: **a.** 4! **b.** 6! **c.** $3! \cdot 2!$ **d.** $5! \cdot 0!$

Strategy We will use the definition of $n!$.

Why The definition explains how to evaluate factorials.

Solution
a. $4! = 4 \cdot 3 \cdot 2 \cdot 1 = 24$ Read as "4 factorial."
b. $6! = 6 \cdot 5 \cdot 4 \cdot 3 \cdot 2 \cdot 1 = 720$ Read as "6 factorial."
c. $\mathbf{3!} \cdot \mathbf{2!} = (\mathbf{3 \cdot 2 \cdot 1}) \cdot (\mathbf{2 \cdot 1}) = 6 \cdot 2 = 12$ Find each factorial and multiply the results.
d. $5! \cdot \mathbf{0!} = (5 \cdot 4 \cdot 3 \cdot 2 \cdot 1) \cdot \mathbf{1} = 120$ Simplify: 0! = 1.

Self Check 3 Evaluate each expression: **a.** 7! **b.** $4! \cdot 3!$ **c.** $1! \cdot 0!$

Now Try ▶ Problems 31 and 37

Using Your Calculator ▶ Factorials

We can find factorials using a calculator. For example, to find 12! with a scientific calculator, we enter

12 $\boxed{x!}$ (You may have to use a $\boxed{\text{2nd}}$ or $\boxed{\text{SHIFT}}$ key first.) $\boxed{479001600}$

To find 12! on a graphing calculator, we enter

12 $\boxed{\text{MATH}}$ $\boxed{\blacktriangleright}$ to PRB 4 $\boxed{\text{ENTER}}$

```
12!
   479001600
```

The following property follows from the definition of factorial.

Factorial Property

For any natural number n,

$$n(n - 1)! = n!$$

We can use this property to evaluate many expressions involving factorials.

EXAMPLE 4   Evaluate each expression: **a.** $\dfrac{6!}{5!}$ **b.** $\dfrac{10!}{8!(10-8)!}$

Strategy   We will use the factorial property to partially expand the factorial and then we will simplify the fraction.

Why   By using this approach, we can avoid difficult multiplications and divisions.

Solution   **a.** If we write 6! as $6 \cdot 5!$, we can simplify the fraction by removing the common factor 5! in the numerator and denominator.

$$\frac{6!}{5!} = \frac{6 \cdot 5!}{5!} = \frac{6 \cdot \overset{1}{\cancel{5!}}}{\underset{1}{\cancel{5!}}} = 6 \qquad \text{Simplify: } \tfrac{5!}{5!} = 1.$$

b. We subtract within the parentheses, write 10! as $10 \cdot 9 \cdot 8!$, and simplify.

$$\frac{10!}{8!(10-8)!} = \frac{10!}{8! \cdot 2!} = \frac{10 \cdot 9 \cdot \overset{1}{\cancel{8!}}}{\underset{1}{\cancel{8!}} \cdot 2!} = \frac{\overset{5}{\cancel{}} \cdot \overset{1}{\cancel{2}} \cdot 9}{\underset{1}{\cancel{2}} \cdot 1} = 45 \qquad \text{Simplify: } \tfrac{8!}{8!} = 1. \text{ Factor 10 as } 5 \cdot 2 \text{ and simplify: } \tfrac{2}{2} = 1.$$

Self Check 4   Evaluate each expression: **a.** $\dfrac{4!}{3!}$ **b.** $\dfrac{7!}{5!(7-5)!}$

Now Try ▶   Problems 39 and 47

3 Use the Binomial Theorem to Expand Binomials.

The following theorem summarizes our observations about binomial expansions and our work with factorials. Known as the **binomial theorem**, it is usually the best way to expand a binomial.

The Binomial Theorem	For any positive integer n, $$(a+b)^n = a^n + \frac{n!}{1!(n-1)!}a^{n-1}b + \frac{n!}{2!(n-2)!}a^{n-2}b^2 + \frac{n!}{3!(n-3)!}a^{n-3}b^3$$ $$+ \cdots + \frac{n!}{r!(n-r)!}a^{n-r}b^r + \cdots + b^n$$

In the binomial theorem, the exponents on the variables follow the familiar pattern:

- The sum of the exponents on a and b in each term is n.
- The exponents on a decrease by 1 in each successive term.
- The exponents on b increase by 1 in each successive term.

The method of finding the coefficients involves factorials. Except for the first and last terms, the numerator of each coefficient is $n!$. If the exponent on b in a particular term is r, the denominator of the coefficient of that term is $r!(n-r)!$.

EXAMPLE 5   Use the binomial theorem to expand $(a+b)^3$.

Strategy   We will substitute 3 for n in the binomial theorem and simplify.

Why   The binomial theorem is the fastest way to expand expressions of the form $(a+b)^n$.

Solution

$$(a + b)^3 = a^3 + \frac{3!}{1!(3-1)!}a^2b + \frac{3!}{2!(3-2)!}ab^2 + b^3$$

$$= a^3 + \frac{3!}{1! \cdot 2!}a^2b + \frac{3!}{2! \cdot 1!}ab^2 + b^3$$

$$= a^3 + \frac{3 \cdot 2!}{1! \cdot 2!}a^2b + \frac{3 \cdot 2!}{2! \cdot 1!}ab^2 + b^3$$

Write 3! as 3 · 2! to simplify the fractions.

$$= a^3 + 3a^2b + 3ab^2 + b^3$$

Self Check 5 Use the binomial theorem to expand $(a + b)^4$.

Now Try Problem 57

We can find expansions of binomials in variables other than a and b by making substitutions into the binomial theorem.

EXAMPLE 6 Use the binomial theorem to expand $(x - y)^4$.

Strategy First, we will write $(x - y)^4$ as $[x + (-y)]^4$. Then we will use the binomial theorem with $a = x$, $b = -y$, and $n = 4$.

Why To substitute directly into the binomial theorem, the difference within the parentheses, $x - y$, must be expressed as a sum.

Solution

$$(x - y)^4 = [x + (-y)]^4$$

$$= x^4 + \frac{4!}{1!(4-1)!}x^3(-y) + \frac{4!}{2!(4-2)!}x^2(-y)^2 + \frac{4!}{3!(4-3)!}x(-y)^3 + (-y)^4$$

$$= x^4 - \frac{4!}{1! \cdot 3!}x^3y + \frac{4!}{2! \cdot 2!}x^2y^2 - \frac{4!}{3! \cdot 1!}xy^3 + y^4$$

$$= x^4 - \frac{4 \cdot 3!}{1! \cdot 3!}x^3y + \frac{4 \cdot 3 \cdot 2!}{2! \cdot 2 \cdot 1}x^2y^2 - \frac{4 \cdot 3!}{3! \cdot 1!}xy^3 + y^4$$

Write 4! as 4 · 3! and as 4 · 3 · 2! to simplify the fractions.

$$= x^4 - 4x^3y + 6x^2y^2 - 4xy^3 + y^4$$

Note the alternating signs.

Self Check 6 Use the binomial theorem to expand $(x - y)^3$.

Now Try Problem 59

EXAMPLE 7 Use the binomial theorem to expand $(3u - 2v)^4$.

Strategy We will write the expansion of $(a + b)^4$. Then we will substitute for a and b to find the expansion of $(3u - 2v)^4$.

Why For binomials with more complicated terms, the calculations are often easier if the general expansion is written first, followed by the appropriate substitutions.

Solution   We can use the binomial theorem to expand $(a + b)^4$.

$$(a + b)^4 = a^4 + \frac{4!}{1!(4 - 1)!}a^3b + \frac{4!}{2!(4 - 2)!}a^2b^2 + \frac{4!}{3!(4 - 3)!}ab^3 + b^4$$

$$= a^4 + 4a^3b + 6a^2b^2 + 4ab^3 + b^4$$

If we write $(3u - 2v)^4$ as $[3u + (-2v)]^4$, we see that the expressions $3u$ and $-2v$ can be substituted for a and b respectively in the expansion of $(a + b)^4$.

$$(3u - 2v)^4 = (3u)^4 + 4(3u)^3(-2v) + 6(3u)^2(-2v)^2 + 4(3u)(-2v)^3 + (-2v)^4$$

$$= 81u^4 - 216u^3v + 216u^2v^2 - 96uv^3 + 16v^4$$

> **Self Check 7**   Use the binomial theorem to expand $(4a - 5b)^3$.
>
> **Now Try** ▶ Problem 69

4 Find a Specific Term of a Binomial Expansion.

To find a specific term of a binomial expansion, we don't need to write out the entire expansion. The binomial theorem and the pattern of the terms suggest the following method for finding a single term of an expansion.

Finding a Specific Term of a Binomial Expansion	The $(r + 1)$st term of the expansion of $(a + b)^n$ is $$\frac{n!}{r!(n - r)!}a^{n-r}b^r$$

EXAMPLE 8   Find the 4th term of the expansion of $(a + b)^9$.

Strategy   We will determine n and r and substitute into the formula for finding a specific term of a binomial expansion.

Why   We will use the formula because it enables us to find the fourth term of the expansion without us having to write out all the terms of the expansion.

Solution   To use the formula for finding a specific term of $(a + b)^9$, we must determine n and r. Since $r + 1 = 4$ in the fourth term, $r = 3$ and since this binomial is raised to the 9th power, $n = 9$. We substitute 3 for r and 9 for n into the formula to find the fourth term.

$$\frac{n!}{r!(n - r)!}a^{n-r}b^r = \frac{9!}{3!(9 - 3)!}a^{9-3}b^3$$

$$= \frac{9!}{3!6!}a^6b^3 \qquad \text{Evaluate: } \frac{9!}{3!6!} = \frac{9 \cdot 8 \cdot 7 \cdot \overset{1}{\cancel{6!}}}{3 \cdot 2 \cdot 1 \cdot \underset{1}{\cancel{6!}}} = 84.$$

$$= 84a^6b^3$$

> **Self Check 8**   Find the 3rd term of the expansion of $(a + b)^9$.
>
> **Now Try** ▶ Problem 77

EXAMPLE 9   Find the 6th term of the expansion of $\left(x^2 - \dfrac{y}{2}\right)^7$.

Strategy   We will determine n, r, a, and b and substitute these values into the formula for finding a specific term of a binomial expansion.

Why We will use the formula because it enables us to find the sixth term of the expansion without us having to write out all the terms of the expansion.

Solution To use the formula for finding a specific term of $\left(x^2 - \frac{y}{2}\right)^7$, we must determine n, r, a, and b. In the sixth term, $r + 1 = 6$. So $r = 5$. By comparing $\left(x^2 - \frac{y}{2}\right)^7$ to $(a + b)^n$, we see that $a = x^2$, $b = -\frac{y}{2}$, and $n = 7$. We substitute these values into the formula as follows.

$$\frac{n!}{r!(n-r)!}a^{n-r}b^r = \frac{7!}{5!(7-5)!}(x^2)^{7-5}\left(-\frac{y}{2}\right)^5$$

$$= \frac{7!}{5!2!}(x^2)^2\left(-\frac{y^5}{32}\right)$$

$$= -\frac{21}{32}x^4 y^5$$

Evaluate: $\frac{7!}{5!2!} = \frac{7 \cdot 6 \cdot 5!}{5! \cdot 2 \cdot 1} = 21.$

Self Check 9 Find the 5th term of the expansion of $\left(c^2 - \frac{d}{3}\right)^7$.

Now Try ▶ Problem 87

SECTION **11.1** ▶ STUDY SET

VOCABULARY

Fill in the blanks.

1. The two-term polynomial expression $a + b$ is called a _____.

2. $a^4 + 4a^3b + 6a^2b^2 + 4ab^3 + b^4$ is the binomial _____ of $(a + b)^4$. To _____ a binomial of the form $(a + b)^n$ means to write it as a _____ of terms.

3. We can use the _____ theorem to raise binomials to positive-integer powers without doing the actual multiplication.

4. The array of numbers that gives the coefficients of the terms of a binomial expansion is called _____ triangle.

5. $n!$ (read as "n _____") is the product of consecutively _____ natural numbers from n to 1.

6. In the expansion $a^3 - 3a^2b + 3ab^2 - b^3$, the signs _____ between + and −.

CONCEPTS

Fill in the blanks.

7. The binomial expansion of $(m + n)^6$ has _____ more term than the power of the binomial.

8. For each term of the expansion of $(a + b)^8$, the sum of the exponents of a and b is ▨.

9. The first term of the expansion of $(r + s)^{20}$ is $r^▨$ and the last term is $s^▨$.

10. In the expansion of $(m - n)^{15}$, the exponents on m _____ and the exponents on n _____.

11. The coefficients of the terms of the expansion of $(c + d)^{20}$ begin with ▨, increase through some values, and then decrease through those same values, back to ▨.

12. Complete Pascal's Triangle:

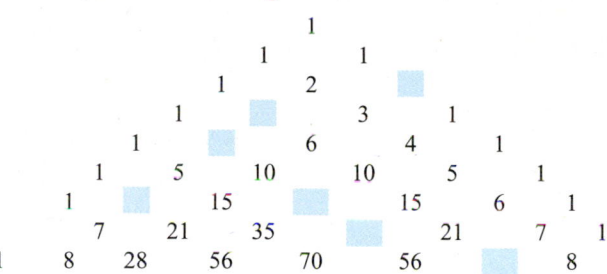

13. $n \cdot (▨ - ▨)! = n!$ 14. $8! = 8 \cdot ▨!$

15. $0! = ▨$

16. According to the binomial theorem, the third term of the expansion of $(a + b)^n$ is $\dfrac{▨!}{▨!(n-2)!} a^{n-2}b^▨$.

17. The coefficient of the fourth term of the expansion of $(a + b)^9$ is $9!$ divided by $3!(▨ - ▨)!$.

18. The exponent on a in the fourth term of the expansion of $(a + b)^6$ is ▨ and the exponent on b is ▨.

19. The exponent on a in the fifth term of the expansion of $(a + b)^6$ is ▨ and the exponent on b is ▨.

20. The expansion of $(a - b)^4$ is

$$a^4 \;▨\; 4a^3b \;▨\; 6a^2b^2 \;▨\; 4ab^3 \;▨\; b^4$$

21. $(x + y)^3$

$$= x^▨ + \frac{▨!}{1!(3-1)!}x^2 + \frac{3!}{▨!(3-2)!}xy^▨ + y^▨$$

22. Fill in the blanks.

a. The $(r + 1)$st term of the expansion of $(a + b)^n$ is
$$\frac{n!}{r!(n - \boxed{})!}a^{\boxed{}-r}b^{\boxed{}}.$$

b. To use the specific term formula to find the 6th term of the expansion of $\left(m + \frac{x}{2}\right)^8$, we note that $r = \boxed{}$, $n = \boxed{}$, $a = \boxed{}$, and $b = \boxed{}$.

NOTATION

Fill in the blanks.

23. $n! = n(\boxed{} - \boxed{})(n - 2) \cdot \ \cdots \ \cdot 3 \cdot 2 \cdot 1$

24. The symbol 5! is read as "_____ _____" and it means $5 \cdot \boxed{} \cdot \boxed{} \cdot \boxed{} \cdot \boxed{}$.

GUIDED PRACTICE

Use Pascal's triangle to expand each binomial. See Examples 1 and 2.

25. $(a + b)^3$

26. $(m + p)^4$

27. $(m - p)^5$

28. $(a - b)^3$

Evaluate each expression. See Examples 3 and 4.

29. $3!$ **30.** $7!$ **31.** $5!$ **32.** $6!$

33. $3! + 4!$ **34.** $4! + 4!$ **35.** $3!(4!)$ **36.** $2!(3!)$

37. $8(7!)$ **38.** $4!(5)$ **39.** $\dfrac{49!}{47!}$ **40.** $\dfrac{101!}{100!}$

41. $\dfrac{9!}{11!}$ **42.** $\dfrac{13!}{10!}$ **43.** $\dfrac{9!}{7!0!}$ **44.** $\dfrac{7!}{5!0!}$

45. $\dfrac{5!}{1!(5 - 1)!}$ **46.** $\dfrac{15!}{14!(15 - 14)!}$

47. $\dfrac{5!}{3!(5 - 3)!}$ **48.** $\dfrac{6!}{4!(6 - 4)!}$

49. $\dfrac{5!(8 - 5)!}{4! \cdot 7!}$ **50.** $\dfrac{6! \cdot 7!}{(8 - 3)!(7 - 4)!}$

51. $\dfrac{7!}{5!(7 - 5)!}$ **52.** $\dfrac{8!}{6!(8 - 6)!}$

 Use a calculator to evaluate each expression. See Using Your Calculator: Factorials.

53. $11!$ **54.** $13!$

55. $20!$ **56.** $55!$

Use the binomial theorem to expand each expression. See Examples 5 and 6.

57. $(m + n)^4$

58. $(a - b)^4$

59. $(c - d)^5$

60. $(c + d)^5$

61. $(a - b)^9$

62. $(a + b)^7$

63. $(s + t)^6$

64. $(s - t)^6$

Use the binomial theorem to expand each expression. See Example 7.

65. $(2x + y)^3$

66. $(x + 2y)^3$

67. $(2t - 3)^5$

68. $(2b + 1)^4$

69. $(5m - 2n)^4$

70. $(2m + 3n)^5$

71. $\left(\dfrac{x}{3} + \dfrac{y}{2}\right)^3$ **72.** $\left(\dfrac{x}{2} - \dfrac{y}{3}\right)^3$

73. $\left(\dfrac{x}{3} - \dfrac{y}{2}\right)^4$ **74.** $\left(\dfrac{x}{2} + \dfrac{y}{3}\right)^4$

75. $(c^2 - d^2)^5$

76. $(u^2 - v^3)^5$

Find the indicated term of each binomial expansion. See Examples 8 and 9.

77. $(x + y)^8$; 3rd term   **78.** $(x + y)^9$; 7th term

79. $(r + s)^6$; 5th term   **80.** $(r + s)^7$; 5th term

81. $(x - 1)^{13}$; 3rd term   **82.** $(x - 1)^{10}$; 5th term

83. $(x - 3y)^4$; 2nd term   **84.** $(3x - y)^5$; 3rd term

85. $(2x - 3y)^5$; 5th term   **86.** $(3x - 2y)^4$; 2nd term

87. $\left(\dfrac{c}{2} - \dfrac{d}{3}\right)^4$; 2nd term   **88.** $\left(\dfrac{c}{3} + \dfrac{d}{2}\right)^5$; 4th term

89. $(2t - 5)^7$; 4th term   **90.** $(2t - 3)^6$; 6th term

91. $(a^2 + b^2)^6$; 2nd term   **92.** $(a^2 + b^2)^7$; 6th term

WRITING

93. Describe how to construct Pascal's triangle.

94. Explain why the signs alternate in the expansion of $(x - y)^9$.

95. Explain why the third term of the expansion of $(m + 3n)^9$ could not be $324m^7n^3$.

96. Using your own words, write a definition of $n!$.

REVIEW

Assume that x, y, z, and b represent positive numbers. Use the properties of logarithms to write each expression as the logarithm of a single quantity.

97. $2 \log x + \dfrac{1}{2} \log y$

98. $-2\log x - 3\log y + \log z$

99. $\ln(xy + y^2) - \ln(xz + yz) + \ln z$

100. $\log_2(x + 1) - \log_2 x$

CHALLENGE PROBLEMS

101. Find the constant term in the expansion of $\left(x + \frac{1}{x}\right)^{10}$.

102. Find the coefficient of a^5 in the expansion of $\left(a - \frac{1}{a}\right)^9$.

103. a. If we applied the pattern of the coefficients to the coefficient of the first term in a binomial expansion, the coefficient would be $\frac{n!}{0!(n-0)!}$. Show that this expression is 1.

   b. If we applied the pattern of the coefficients to the coefficient of the last term in a binomial expansion, the coefficient would be $\frac{n!}{n!(n-n)!}$. Show that this expression is 1.

104. Expand $(i - 1)^7$, where $i = \sqrt{-1}$.

SECTION 11.2

Arithmetic Sequences and Series

OBJECTIVES

1 Find terms of a sequence given the general term.

2 Find terms of an arithmetic sequence by identifying the first term and the common difference.

3 Find arithmetic means.

4 Find the sum of the first n terms of an arithmetic sequence.

5 Solve application problems involving arithmetic sequences.

6 Use summation notation.

ARE YOU READY?

The following problems review some basic skills that are needed when working with arithmetic sequences and series.

1. Let $f(x) = 6x - 3$. Find $f(4)$.

2. Evaluate: **a.** $(-1)^4$ **b.** $(-1)^5$

3. Find the difference: $11 - 3$

4. Evaluate: $12 + (6 - 1)8$

The word *sequence* is used in everyday conversation when referring to an ordered list. For example, a history instructor might discuss the sequence of events that led up to the sinking of the *Titanic*. In mathematics, a **sequence** is a list of numbers written in a specific order.

1 Find Terms of a Sequence Given the General Term.

Each number in a sequence is called a **term** of the sequence. **Finite sequences** contain a finite number of terms and **infinite sequences** contain infinitely many terms. Two examples of sequences are:

> *Finite sequence:* 1, 5, 9, 13, 17, 21, 25
>
> *Infinite sequence:* 3, 6, 9, 12, 15, . . .   The . . . indicates that the sequence goes on forever.

Sequences are defined formally using the terminology of functions.

Finite and Infinite Sequences	A finite sequence is a function whose domain is the set of natural numbers $\{1, 2, 3, 4, \ldots, n\}$ for some natural number n. An **infinite sequence** is a function whose domain is the set of natural numbers: $\{1, 2, 3, 4, \ldots\}$.

Instead of using $f(x)$ notation, we use a_n (read as "*a* sub *n*") notation to write the value of a sequence at a number n. For the infinite sequence introduced earlier, we have:

1st term	2nd term	3rd term	4th term	5th term	
3, ↑ a_1	6, ↑ a_2	9, ↑ a_3	12, ↑ a_4	15, . . . ↑ a_5	Read a_1 as "a sub 1."

To describe all the terms of a sequence specifically, we can write a formula for a_n, called the **general term** of the sequence. For the sequence 3, 6, 9, 12, 15, . . . , we note that $a_1 = 3 \cdot 1$, $a_2 = 3 \cdot 2$, $a_3 = 3 \cdot 3$, and so on. In general, the nth term of the sequence is found by multiplying n by 3.

$$a_n = 3n$$

We can use this formula to find any term of the sequence. For example, to find the 12th term, we substitute 12 for n.

$$a_{12} = 3(12) = 36$$

EXAMPLE 1   Given an infinite sequence with $a_n = 2n - 3$, find each of the following:
a. the first four terms   **b.** a_{50}

Strategy   We will substitute 1, 2, 3, 4, and 50 for n in the formula that defines the sequence.

Why   To find the first term of the sequence, we let $n = 1$. To find the second term, let $n = 2$, and so on.

Solution   **a.** $a_1 = 2(1) - 3 = -1$ Substitute 1 for n. $a_2 = 2(2) - 3 = 1$ Substitute 2 for n.
$a_3 = 2(3) - 3 = 3$ Substitute 3 for n. $a_4 = 2(4) - 3 = 5$ Substitute 4 for n.

The first four terms of the sequence are $-1, 1, 3,$ and 5.

b. To find a_{50}, the 50th term of the sequence, we let $n = 50$ in the formula for the nth term:

$$a_{50} = 2(50) - 3 = 97$$

Notation

The variable n represents the position of the term in the sequence, while the notation a_n represents the value of the term in that position.

Self Check 1   Given an infinite sequence with $a_n = 3n + 5$, find each of the following:
a. the first three terms   **b.** a_{100}

Now Try ▶ Problem 17

EXAMPLE 2   Find the first four terms of the sequence whose general term is $a_n = \dfrac{(-1)^n}{2^n}$.

Strategy   We will substitute 1, 2, 3, and 4 for n in the formula that defines the sequence.

Why   To find the first term of the sequence, we let $n = 1$. To find the second term, let $n = 2$, and so on.

Solution $a_1 = \dfrac{(-1)^1}{2^1} = -\dfrac{1}{2}$ $(-1)^1 = -1$ $a_2 = \dfrac{(-1)^2}{2^2} = \dfrac{1}{4}$ $(-1)^2 = 1$

$a_3 = \dfrac{(-1)^3}{2^3} = \dfrac{-1}{8} = -\dfrac{1}{8}$ $(-1)^3 = -1$ $a_4 = \dfrac{(-1)^4}{2^4} = \dfrac{1}{16}$ $(-1)^4 = 1$

The first four terms of the sequence are $-\dfrac{1}{2}, \dfrac{1}{4}, -\dfrac{1}{8},$ and $\dfrac{1}{16}$.

Success Tip

The factor $(-1)^n$ in $a_n = \dfrac{(-1)^n}{2^n}$ causes the signs of the terms to alternate between positive (when n is even) and negative (when n is odd).

Self Check 2   Find the first four terms of the sequence whose general term is $a_n = \dfrac{(-1)^n}{n}$.

Now Try ▶ Problem 25

---

2   Find Terms of an Arithmetic Sequence by Identifying the First Term and the Common Difference.

A sequence in which each term is found by adding the same number to the previous term is called an **arithmetic sequence**. Two examples are

The Language of Algebra

We pronounce the adjective **arithmetic** in the phrase *arithmetic sequence* as:

air'-rith-met'-ic

$5, 12, 19, 26, 33, 40$ This is a finite arithmetic sequence in which each term is found by adding 7 to the previous term.

Add 7

$3, 1, -1, -3, -5, -7, \ldots$ This is an infinite arithmetic sequence in which each term is found by adding -2 to the previous term.

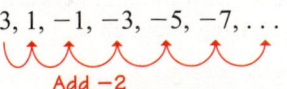

Add -2

Arithmetic Sequence	An **arithmetic sequence** is a sequence of the form

$$a_1, \quad a_1 + d, \quad a_1 + 2d, \quad a_1 + 3d, \quad \ldots, \quad a_1 + (n-1)d, \ldots$$

where a_1 is the **first term** and d is the **common difference**. The nth term is given by the **general term formula**

$$a_n = a_1 + (n-1)d \quad \text{Read } a_n \text{ as "a sub } n\text{."}$$

We note that the second term of an arithmetic sequence has an addend of $1d$, the third term has an addend of $2d$, the fourth term has an addend of $3d$, and the nth term has an addend of $(n-1)d$. We also note that the **difference between any two consecutive terms in an arithmetic sequence is d.**

EXAMPLE 3 An arithmetic sequence has a first term 5 and a common difference 4. Write the first five terms of the sequence and find the 25th term.

Strategy To find the first five terms, we will write the first term and add 4 to each successive term until we produce five terms. To find the 25th term, we will substitute 5 for a_1, 4 for d, and 25 for n in the formula $a_n = a_1 + (n-1)d$.

Why The same number is added to each term of an arithmetic sequence to get the next term. However, successively adding 4 to find the 25th term would be time consuming. Using the formula is faster.

Solution Since the first term is 5 and the common difference is 4, the first five terms are

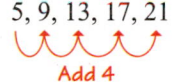

5, 9, 13, 17, 21

Add 4

Since the first term is $a_1 = 5$ and the common difference is $d = 4$, the arithmetic sequence is defined by the formula

$$a_n = 5 + (n-1)4 \quad \text{In } a_n = a_1 + (n-1)d, \text{ substitute 5 for } a_1 \text{ and 4 for } d.$$

To find the 25th term, we substitute 25 for n and simplify.

$$a_{25} = 5 + (\mathbf{25} - 1)4$$
$$= 5 + (24)4$$
$$= 101$$

The 25th term is 101.

Self Check 3 Write the first five terms of an arithmetic sequence with a first term 10 and a common difference of 8. Then find the 30th term.

Now Try ▶ Problem 29

EXAMPLE 4 The first three terms of an arithmetic sequence are 3, 8, and 13. Find the 100th term.

Strategy We can use the first three terms to find the common difference. Then we will know the first term of the sequence and the common difference.

Why Once we know the first term and the common difference, we can use the formula $a_n = a_1 + (n-1)d$ to find the 100th term by letting $n = 100$.

Solution The common difference d is the difference between any two successive terms. Since $a_1 = 3$ and $a_2 = 8$, we can find d using subtraction.

$$d = a_2 - a_1 = 8 - 3 = 5 \quad \text{Also note that } a_3 - a_2 = 13 - 8 = 5.$$

Success Tip

The common difference d of an arithmetic sequence is defined to be

$$d = a_{n+1} - a_n$$

To find the 100th term, we substitute 3 for a_1, 5 for d, and 100 for n in the formula for the nth term.

$$a_n = a_1 + (n-1)d \qquad \text{This is the general term formula.}$$
$$a_{100} = 3 + (100-1)5$$
$$= 3 + (99)5 \qquad \text{Evaluate the right side.}$$
$$= 498$$

Self Check 4   The first three terms of an arithmetic sequence are $-3, 6$, and 15. Find the 99th term.

Now Try ▶ Problem 37

EXAMPLE 5   The first term of an arithmetic sequence is 12 and the 50th term is 3,099. Write the first six terms of the sequence.

Strategy   We will find the common difference by substituting 3,099 for a_n, 12 for a_1, and 50 for n in the formula $a_n = a_1 + (n-1)d$.

Why   Once we know the first term and the common difference, we can successively add the common difference to each term to produce the first six terms.

Solution   Since the 50th term of the sequence is 3,099, we substitute 3,099 for a_{50}, 12 for a_1, and 50 for n in the formula $a_n = a_1 + (n-1)d$ and solve for d.

$$a_{50} = a_1 + (n-1)d$$
$$3{,}099 = 12 + (50-1)d \qquad \text{Substitute 3,099 for } a_{50}, \text{ 12 for } a_1, \text{ and 50 for } n.$$
$$3{,}099 = 12 + 49d \qquad \text{Simplify.}$$
$$3{,}087 = 49d \qquad \text{Subtract 12 from both sides.}$$
$$63 = d \qquad \text{To isolate } d, \text{ divide both sides by 49.}$$

Since the first term is 12 and the common difference is 63, the first six terms are

$$12, 75, 138, 201, 264, 327 \qquad \text{Add 63 to a term to get the next term.}$$

Self Check 5   The first term of an arithmetic sequence is 15 and the 12th term is 92. Write the first four terms of the sequence.

Now Try ▶ Problem 41

---

3 Find Arithmetic Means.

If numbers are inserted between two numbers a and b to form an arithmetic sequence, the inserted numbers are called **arithmetic means** between a and b. If a single number is inserted, it is called **the arithmetic mean** between a and b.

EXAMPLE 6   Insert two arithmetic means between 6 and 27.

Strategy   Because two arithmetic means are to be inserted between 6 and 27, we will consider a sequence of four terms, with a first term of 6 and a fourth term of 27. We will then use $a_n = a_1 + (n-1)d$ to find the common difference d.

Why   Once we know the first term and the common difference, we can add the common difference to find the two unknown terms.

Solution   The first term is $a_1 = 6$ and the fourth term is $a_4 = 27$. We must find the common difference so that the terms

$$6, \quad 6+d, \quad 6+2d, \quad 27$$
$$\uparrow \qquad \uparrow \qquad \quad \uparrow \qquad \quad \uparrow$$
$$a_1 \qquad a_2 \qquad \quad a_3 \qquad \quad a_4$$

form an arithmetic sequence. To find the common difference d, we substitute 6 for a_1, 4 for n, and 27 for a_4 in the formula for the 4th term:

$a_4 = a_1 + (n - 1)d$ This gives the 4th term of any arithmetic sequence.

$27 = 6 + (4 - 1)d$ Substitute.

$27 = 6 + 3d$ Subtract within the parentheses.

$21 = 3d$ Subtract 6 from both sides.

$7 = d$ To isolate d, divide both sides by 3.

To find the two arithmetic means between 6 and 27, we add the common difference 7, as shown:

$6 + d = 6 + 7$ or $6 + 2d = 6 + 2(7)$

$\quad\quad = 13$ This is a_2. $= 6 + 14$

$\quad\quad\quad\quad\quad\quad\quad\quad\quad\quad\quad\quad\quad = 20$ This is a_3.

Two arithmetic means between 6 and 27 are 13 and 20.

Self Check 6   Insert two arithmetic means between 8 and 44.

Now Try ▶ Problem 45

4   Find the Sum of the First n Terms of an Arithmetic Sequence.

To develop a formula for finding the sum of the first n terms of an arithmetic sequence, we let S_n represent the sum of the first n terms of an arithmetic sequence:

$$S_n = a_1 + [a_1 + d] + [a_1 + 2d] + \cdots + [a_1 + (n - 1)d]$$

We write the same sum again, but in reverse order:

$$S_n = [a_1 + (n - 1)d] + [a_1 + (n - 2)d] + [a_1 + (n - 3)d] + \cdots + a_1$$

Adding these equations together, term by term, we get

$$2S_n = [2a_1 + (n - 1)d] + [2a_1 + (n - 1)d] + [2a_1 + (n - 1)d] + \cdots + [2a_1 + (n - 1)d]$$

Because there are n equal terms on the right side of the preceding equation, we can write

(1) $2S_n = n[2a_1 + (n - 1)d]$

(2) $2S_n = n[a_1 + a_1 + (n - 1)d]$ Write $2a_1$ as $a_1 + a_1$.

$\quad\quad 2S_n = n(a_1 + a_n)$ Substitute a_n for $a_1 + (n-1)d$.

$\quad\quad S_n = \dfrac{n(a_1 + a_n)}{2}$ To isolate S_n, divide both sides by 2.

This reasoning establishes the following formula.

The Language of Algebra

The sum of a finite number of terms of a sequence is called a **partial sum**. The word *partial* means not complete, as in a *partial* eclipse of the moon.

Sum of the First n Terms of an Arithmetic Sequence

The sum of the first n terms of an arithmetic sequence is given by the formula

$$S_n = \frac{n(a_1 + a_n)}{2}$$

where a_1 is the first term, a_n is the nth (or last) term, and n is the number of terms in the sequence.

EXAMPLE 7   Find the sum of the first 40 terms of the arithmetic sequence: 4, 10, 16, ...

Strategy   We know the first term is 4 and we can find the common difference d. We will substitute these values into the formula $a_n = a_1 + (n - 1)d$ to find the last term to be added, a_{40}.

Why   To use the formula $S_n = \dfrac{n(a_1 + a_n)}{2}$ to find the sum of the first 40 terms, we need to know the first term, a_1, and the last term, a_{40}.

Solution   We can substitute 4 for a_1, 40 for n, and $10 - 4 = 6$ for d into $a_n = a_1 + (n - 1)d$ to get $a_{40} = 4 + (40 - 1)6 = 238$. We then substitute these values into the formula for S_{40}:

> ### Success Tip
>
> An alternate form of the summation formula for arithmetic sequences can be obtained from Equation (1) on the previous page by dividing both sides by 2.

$$S_n = \frac{n(a_1 + a_{40})}{2}$$ This is the formula for the sum of the terms of an artithmetic sequence.

$$S_{40} = \frac{40(4 + 238)}{2}$$ Substitute: $a_1 = 4$, $n = 40$, and $a_{40} = 238$.

$$= 20(242)$$ Add within the parentheses.

$$= 4,840$$ Multiply.

The sum of the first 40 terms is 4,840.

Self Check 7   Find the sum of the first 50 terms of the arithmetic sequence: 3, 8, 13, . . .

Now Try ▶ Problem 53

---

5 Solve Application Problems Involving Arithmetic Sequences.

EXAMPLE 8   **Halftime Performances.**   Each row of a formation formed by the members of a college marching band has one more person in it than the previous row. If 4 people are in the front row and 21 are in the 18th (and last) row, how many band members are there?

Strategy   To find the number of band members, we will write an arithmetic sequence to model the situation and find the sum of its terms.

Why   We can use an arithmetic sequence to model this situation because each row has one more person in it than the previous one. Thus, the common difference is 1.

Solution   When we list the number of band members in each row of the formation, we get the arithmetic sequence 4, 5, 6, . . . , 21, where $a_1 = 4$, $d = 1$, $n = 18$, and $a_{18} = 21$. We can use the formula $S_n = \dfrac{n(a_1 + a_n)}{2}$ to find the sum of the terms of the sequence.

$$S_{18} = \frac{18(4 + 21)}{2} = \frac{18(25)}{2} = 225$$ 4 is the first term and 21 is the last term.

There are 225 members of the marching band.

Self Check 8   **Marching Bands.**   How many band members would it take to form a 10-row formation, if the first row has 5 people in it, the second row has 7 people, the third row has 9 people, and so on?

Now Try ▶ Problem 95

6 Use Summation Notation.

When the commas between the terms of a sequence are replaced with + signs, we call the sum a **series**. The sum of the terms of an arithmetic sequence is called an **arithmetic series**. Some examples are

$$4 + 8 + 12 + 16 + 20 + 24$$

Since this series has a limited number of terms, it is a finite arithmetic series.

$$5 + 8 + 11 + 14 + 17 + \cdots$$

Since this series has infinitely many terms, it is an infinite arithmetic series.

When the general term of a sequence is known, we can use a special notation to write a series. This notation, called **summation notation,** involves the Greek letter Σ (sigma). The expression

$$\sum_{k=1}^{4} 3k$$

Read as "the summation of 3k as k runs from 1 to 4."

designates the sum of all terms obtained if we successively substitute the numbers 1, 2, 3, and 4 for k, called the **index of the summation.** Thus, we have

$$
\begin{array}{cccc}
k = 1 & k = 2 & k = 3 & k = 4 \\
\downarrow & \downarrow & \downarrow & \downarrow
\end{array}
$$

$$\sum_{k=1}^{4} 3k = 3(1) + 3(2) + 3(3) + 3(4)$$

$$= 3 + 6 + 9 + 12$$

$$= 30$$

EXAMPLE 9

Write the series associated with each summation and find the sum:

a. $\displaystyle\sum_{k=1}^{3} (2k + 1)$ **b.** $\displaystyle\sum_{k=2}^{8} k^2$

Strategy   In part (a), we will substitute 1, 2, and 3 for k and add the resulting numbers. In part (b), we will substitute 2, 3, 4, 5, 6, 7, and 8 for k and add the resulting numbers.

Why   Think of k as a counter that begins with the number written at the bottom of the notation and successively increases by 1 until it reaches the number written at the top.

Solution   **a.** We substitute the integers 1, 2, and 3 for k and find the sum.

The Language of Algebra

In these examples, the letter k has been used as the **index of summation.** However, other letters, such as i (not the complex number), j, and n are also often used.

$$\sum_{k=1}^{3} (2k + 1) = [2(1) + 1] + [2(2) + 1] + [2(3) + 1]$$

Read as "the summation of 2k + 1 as k runs from 1 to 3."

$$= \qquad 3 \quad + \quad 5 \quad + \quad 7$$

$$= 15$$

b. We substitute the integers from 2 to 8 for k and find the sum.

$$\sum_{k=2}^{8} k^2 = 2^2 + 3^2 + 4^2 + 5^2 + 6^2 + 7^2 + 8^2$$

Read as "the summation of k^2 as k runs from 2 to 8."

$$= 4 + 9 + 16 + 25 + 36 + 49 + 64$$

$$= 203$$

SECTION 11.2 STUDY SET

VOCABULARY

Fill in the blanks.

1. A _____ is a function whose domain is the set of natural numbers.

2. A sequence with an unlimited number of terms is called an _____ sequence. A sequence with a specific number of terms is called a _____ sequence.

3. Each term of an _____ sequence is found by adding the same number to the previous term.

4. 5, 15, 25, 35, 45, 55, . . . is an example of an _____ sequence. The first _____ is 5 and the common _____ is 10.

5. If a single number is inserted between a and b to form an arithmetic sequence, the number is called the arithmetic _____ between a and b.

6. The sum of the terms of an arithmetic sequence is called an arithmetic _____.

CONCEPTS

7. Write the first three terms of an arithmetic sequence if $a_1 = 1$ and $d = 6$.

8. Given the arithmetic sequence 4, 7, 10, 13, 16, 19, . . . , find a_5 and d.

9. **a.** Write the formula for a_n, the general term of an arithmetic sequence.

   b. Write the formula for S_n, the sum of the first n terms of an arithmetic sequence.

10. An infinite arithmetic sequence is of the form

$$a_1, \quad a_1+d, \quad \boxed{}, \quad a_1+3d, \quad \boxed{}, \ldots$$

NOTATION

Fill in the blanks.

11. The notation a_n represents the _____ term of a sequence.

12. To find the common difference of an arithmetic sequence, we use the formula $d = a_{\underline{}} - a_{\underline{}}$.

13. The symbol Σ is the Greek letter _____.

14. In the notation $\displaystyle\sum_{k=1}^{5}(2k-5)$, k is called the _____ of summation.

15. We read $\displaystyle\sum_{k=1}^{10}3k$ as "the _____ of $3k$ as k _____ from 1 to 10."

16. $\displaystyle\sum_{k=1}^{5}k = \boxed{} + \boxed{} + \boxed{} + \boxed{} + \boxed{}$

GUIDED PRACTICE

Write the first five terms of each sequence and then find the specified term. See Example 1.

17. $a_n = 4n - 1$, a_{40}

18. $a_n = 5n - 3$, a_{25}

19. $a_n = -3n + 1$, a_{30}

20. $a_n = -6n + 2$, a_{15}

21. $a_n = -n^2$, a_{20}

22. $a_n = -n^3$, a_{10}

23. $a_n = \dfrac{n-1}{n}$, a_{12} 24. $a_n = \dfrac{n+1}{2n}$, a_{100}

Write the first four terms of each sequence. See Example 2.

25. $a_n = \dfrac{(-1)^n}{3^n}$ 26. $a_n = \dfrac{(-1)^n}{4^n}$

27. $a_n = (-1)^n(n+6)$ 28. $a_n = (-1)^n(7n)$

Write the first five terms of each arithmetic sequence with the given properties and find the specified term. See Example 3.

29. First term: 3, common difference: 2; find the 10th term.

30. First term: -2, common difference: 3; find the 20th term.

31. First term: -5, common difference: -3; find the 15th term.

32. First term: 8, common difference: -5; find the 25th term.

33. First term: 7, common difference: 12; find the 30th term.

34. First term: -1, common difference: 4; find the 55th term.

35. First term: -7, common difference: -2; find the 15th term.

36. First term: 8, common difference: -3; find the 25th term.

The first three terms of an arithmetic sequence are shown below. Find the specified term. See Example 4.

37. 1, 4, 7, . . . ; 30th term

38. 2, 6, 10, . . . ; 28th term

39. $-5, -1, 3, \ldots$; 17th term

40. $-7, -1, 5, \ldots$; 15th term

Write the first five terms of the arithmetic sequence with the following properties. See Example 5.

41. The first term is 5 and the fifth term is 29.

42. The first term is 4 and the sixth term is 39.

43. The first term is −4 and the sixth term is −39.

44. The first term is −5 and the fifth term is −37.

Insert the given number of arithmetic means between the numbers. See Example 6.

45. Two arithmetic means between 2 and 11

46. Four arithmetic means between 5 and 25

47. Four arithmetic means between 10 and 20

48. Three arithmetic means between 20 and 80

49. Three arithmetic means between 20 and 30

50. Two arithmetic means between 10 and 19

51. One arithmetic mean between −4.5 and 7

52. One arithmetic mean between −6.5 and 8.5

For each arithmetic sequence, find the sum of the specified number of terms. See Example 7.

53. The first 35 terms of 5, 9, 13, . . .

54. The first 50 terms of 7, 12, 17, . . .

55. The first 40 terms of −5, −1, 3, . . .

56. The first 25 terms of 2, −3, −8, . . .

Write the series associated with each summation. See Example 9.

57. $\displaystyle\sum_{k=1}^{4} (3k)$

58. $\displaystyle\sum_{k=1}^{4} (k-9)$

59. $\displaystyle\sum_{k=2}^{4} k^2$

60. $\displaystyle\sum_{k=3}^{5} (-2k)$

Find each sum. See Example 9.

61. $\displaystyle\sum_{k=1}^{4} (6k)$

62. $\displaystyle\sum_{k=2}^{5} (3k)$

63. $\displaystyle\sum_{k=3}^{4} k^3$

64. $\displaystyle\sum_{k=2}^{4} (-k^2)$

65. $\displaystyle\sum_{k=3}^{4} (k^2 + 3)$

66. $\displaystyle\sum_{k=2}^{6} (k^2 + 1)$

67. $\displaystyle\sum_{k=4}^{4} (2k + 4)$

68. $\displaystyle\sum_{k=3}^{3} (3k^2 - 7)$

69. $\displaystyle\sum_{k=2}^{5} (5k)$

70. $\displaystyle\sum_{k=2}^{5} (3k - 5)$

71. $\displaystyle\sum_{k=4}^{6} (4k - 1)$

72. $\displaystyle\sum_{k=3}^{5} (k^3)$

TRY IT YOURSELF

73. Find the common difference of the arithmetic sequence with a first term of 40 if its 44th term is 556.

74. Find the first term of the arithmetic sequence with a common difference of −5 if its 23rd term is −625.

75. Find the sum of the first 12 terms of the arithmetic sequence if its second term is 7 and its third term is 12.

76. Find the sum of the first 16 terms of the arithmetic sequence if its second term is 5 and its fourth term is 9.

77. Find the first five terms of the arithmetic sequence if the common difference is 7 and the sixth term is −83.

78. Find the first five terms of the arithmetic sequence if the common difference is 3 and the seventh term is 12.

79. Find the first six terms of the arithmetic sequence if the common difference is −3 and the ninth term is 10.

80. Find the first six terms of the arithmetic sequence if the common difference is −5 and the tenth term is −27.

81. The first three terms of an arithmetic sequence are 5, 12, and 19. Find the 200th term.

82. The first three terms of an arithmetic sequence are 10, 14, and 18. Find the 500th term.

83. Find the sum of the first 50 natural numbers.

84. Find the sum of the first 100 natural numbers.

85. Find the 37th term of the arithmetic sequence with a second term of −4 and a third term of −9.

86. Find the 40th term of the arithmetic sequence with a second term of 6 and a fourth term of 16.

87. Find the first term of the arithmetic sequence with a common difference of 11 if its 27th term is 263.

88. Find the common difference of the arithmetic sequence with a first term of −164 if its 36th term is −24.

89. Find the 15th term of the arithmetic sequence: $\frac{1}{2}, \frac{1}{4}, 0, \ldots$

90. Find the 14th term of the arithmetic sequence: $\frac{2}{3}, \frac{1}{2}, \frac{1}{3}, \ldots$

91. Find the sum of the first 50 odd natural numbers.

92. Find the sum of the first 50 even natural numbers.

APPLICATIONS

93. Saving Money. Yasmeen puts $60 in a bank safety deposit box. She decides to begin a savings plan by putting $50 more in the box every month. Write the first six terms of an arithmetic sequence that gives the monthly amounts in her safety deposit box. Then find the amount of money that she will have in the box after 10 years of the deposits.

94. Installment Loans. Maria borrowed $10,000, interest-free, from her mother. She agreed to pay back the loan in monthly installments of $275. Write the first six terms of an arithmetic sequence that shows the balance due after each month, and find the balance due after 17 months.

95. Designing Patios. Refer to the illustration. Each row of bricks in a triangular patio floor is to have one more brick than the previous row, ending with the longest row of 150 bricks. How many bricks will be needed?

96. Logging. Logs are stacked so that the bottom row has 30 logs, the next row has 29 logs, the next row has 28 logs, and so on.

a. If there are 20 rows in the stack, how many logs are in the top row?

b. How many logs are in the stack?

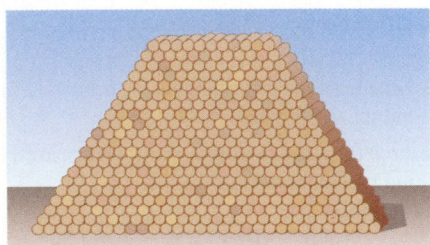

97. Holiday Songs. A popular song of European origin lists the gifts received from someone's "true love" over a 12-day span: a partridge in a pear tree, two turtle doves, three French hens, four calling birds, five gold rings, six geese a-laying, seven swans a-swimming, eight maids a-milking, nine ladies dancing, ten lords a-leaping, eleven pipers piping, and twelve drummers drumming. Use a formula from this section to determine the total number of gifts in this list.

98. Interior Angles. The sums of the angles of several polygons are given in the table below. Assuming that the pattern continues, complete the table.

Figure	Number of sides	Sum of angles
Triangle	3	180°
Quadrilateral	4	360°
Pentagon	5	540°
Hexagon	6	720°
Octagon	8	
Dodecagon	12	

WRITING

99. Explain why 1, 4, 8, 13, 19, 26, . . . is not an arithmetic sequence.

100. What is the difference between a sequence and a series?

101. What is the difference between a_n and S_n?

102. How is the symbol Σ used in this section?

REVIEW

Assume that x, y, z and b represent positive numbers. Use the properties of logarithms to write each expression in terms of the logarithms of x, y, and z.

103. $\log_2 \dfrac{2x}{y}$

104. $\ln x\sqrt{z}$

105. $\log x^3 y^2$

106. $\log x^3 y^{1/2}$

CHALLENGE PROBLEMS

107. Write the summation notation for

$$1 + 4 + 9 + 16 + 25$$

108. Write the summation notation for $3 + 4 + 5 + 6$ without using $k = 1$ in your answer.

109. For what value of x will $x - 2$, $2x + 4$, and $5x - 8$, in that order, form an arithmetic sequence?

110. For what value of x will the arithmetic mean of $x + 4$ and $x + 8$ be 5?

SECTION 11.3

Geometric Sequences and Series

OBJECTIVES

1 Find terms of a geometric sequence by identifying the first term and the common ratio.

2 Find geometric means.

3 Find the sum of the first n terms of a geometric sequence.

4 Define and use infinite geometric series.

5 Solve application problems involving geometric sequences.

ARE YOU READY?

The following problems review some basic skills that are needed when working with geometric sequences and series.

1. Let $f(x) = 6(2^x)$. Find $f(3)$.

2. Evaluate: **a.** $(5)^4$ **b.** $\left(\frac{1}{3}\right)^3$

3. Simplify: $\frac{6}{24}$

4. Solve: $r^2 = 144$

We have seen that the same number is added to each term of an arithmetic sequence to get the next term. In this section, we will consider another type of sequence in which we *multiply* each term by the same number to get the next term. This type of sequence is called a *geometric sequence*. Two examples are

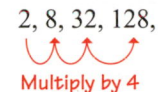

This is an infinite geometric sequence in which each term is found by multiplying the previous term by 4.

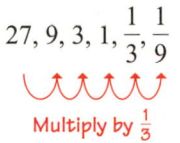

This is a finite geometric sequence in which each term is found by multiplying the previous term by $\frac{1}{3}$.

1 Find Terms of a Geometric Sequence by Identifying the First Term and the Common Ratio.

Each term of a geometric sequence is found by multiplying the previous term by the same number.

Geometric Sequence	A **geometric sequence** is a sequence of the form

$$a_1, \ a_1r, \ a_1r^2, \ a_1r^3, \ \ldots, \ a_1r^{n-1}, \ \ldots$$

where a_1 is the **first term** and r is the **common ratio**. The nth term is given by the **general form formula**

$$a_n = a_1r^{n-1}$$

We note that the second term of a geometric sequence has a factor of r^1, the third term has a factor of r^2, the fourth term has a factor of r^3, and the nth term has a factor of r^{n-1}. We also note that **r is the quotient obtained when any term is divided by the previous term**.

EXAMPLE 1 A geometric sequence has a first term 5 and a common ratio 3.
a. Write the first five terms of the sequence. **b.** Find the 9th term.

Strategy In part (a), we will write the first term and successively multiply each term by 3 until we produce five terms. In part (b), we will substitute 5 for a_1, 3 for r, and 9 for n in the formula for the nth term of a geometric sequence $a_n = a_1r^{n-1}$.

Why To find the terms of a geometric sequence, we multiply each term by the same number to get the next term. To answer part (b), successively multiplying by 3 to find the 9th term would be time consuming. Using the formula is faster.

Solution   **a.** Because the first term is $a_1 = 5$ and the common ratio is $r = 3$, the first five terms are

$$5, \quad 5(3), \quad 5(3^2), \quad 5(3^3), \quad 5(3^4) \qquad \text{Each term is found by multiplying}$$
$$\uparrow \qquad \uparrow \qquad \uparrow \qquad \uparrow \qquad \uparrow \qquad \text{the previous term by 3.}$$
$$a_1 \qquad a_2 \qquad a_3 \qquad a_4 \qquad a_5$$

or

$$5, 15, 45, 135, 405$$

b. We are given $a_1 = 5$ and $r = 3$. Because we want the ninth term, we let $n = 9$:

$$a_n = a_1 r^{n-1} \qquad \text{This is the general term formula.}$$
$$a_9 = 5(3)^{9-1} \qquad \text{Substitute.}$$
$$= 5(3)^8 \qquad \text{Subtract.}$$
$$= 5(6,561) \qquad \text{Evaluate } (3)^8.$$
$$= 32,805 \qquad \text{Multiply.}$$

Self Check 1   A geometric sequence has a first term 3 and a common ratio 4.
   a. Write the first four terms.   **b.** Find the 8th term.

Now Try ▶ Problem 15

EXAMPLE 2   The first three terms of a geometric sequence are 16, 4, and 1. Find the 7th term.

Strategy   We can use the first three terms to find the common ratio. Then we will know the first term and the common ratio.

Why   Once we know the first term, a_1, and the common ratio, r, we can use the formula $a_n = a_1 r^{n-1}$ to find the seventh term by letting $n = 7$.

Solution   The common ratio r is the ratio between any two successive terms. Since $a_1 = 16$ and $a_2 = 4$, we can find r as follows:

$$r = \frac{a_2}{a_1} = \frac{4}{16} = \frac{1}{4} \qquad \text{Also note that } \frac{a_3}{a_2} = \frac{1}{4}.$$

To find the seventh term, we substitute 16 for a_1, $\frac{1}{4}$ for r, and 7 for n in the formula for the nth term and simplify:

$$a_n = a_1 r^{n-1} \qquad \text{This is the general term formula.}$$
$$a_7 = 16\left(\frac{1}{4}\right)^{7-1} \qquad \text{Substitute 16 for } a_1, \text{ 7 for } n, \text{ and } \frac{1}{4} \text{ for } r.$$
$$= 16\left(\frac{1}{4}\right)^6 \qquad \text{Do the subtraction.}$$
$$= 16\left(\frac{1}{4,096}\right) \qquad \text{Use a calculator to evaluate } \left(\frac{1}{4}\right)^6.$$
$$= \frac{1}{256} \qquad \text{Simplify: } \frac{16}{4,096} = \frac{\overset{1}{\cancel{16}}}{\underset{1}{\cancel{16}} \cdot 256}.$$

Self Check 2   The first three terms of a geometric sequence are 25, 5, and 1. Find the 7th term.

Now Try ▶ Problem 19

EXAMPLE 3 Find the first five terms of the geometric sequence with a first term 2, a third term 32, and a common ratio that is positive.

Strategy We will substitute $a_1 = 2$, $a_3 = 32$, and $n = 3$ into the formula for the nth term of a geometric sequence $a_n = a_1 r^{n-1}$ and solve for r.

Why Once we know the common ratio, we can successively multiply each term by the common ratio to produce the first five terms.

Solution We will substitute 3 for n, 2 for a_1, 32 for a_3, and solve for r.

$$a_n = a_1 r^{n-1} \qquad \text{This is the formula for the nth term of a geometric sequence.}$$
$$a_3 = 2r^{3-1} \qquad \text{Substitute 3 for n and 2 for a_1.}$$
$$32 = 2r^2 \qquad \text{Substitute 32 for a_3 and simplify.}$$
$$16 = r^2 \qquad \text{Divide both sides by 2.}$$
$$\pm 4 = r \qquad \text{Use the square root property.}$$

> **Success Tip**
>
> The most important characteristics of a geometric sequence are:
>
> a_1: the first term
>
> r: the common ratio
>
> n: the number of terms

Since r is given to be positive, $r = 4$. The first five terms are produced by multiplying by the common ratio:

$$2,\ 2 \cdot 4,\ 2 \cdot 4^2,\ 2 \cdot 4^3,\ 2 \cdot 4^4$$

or

$$2,\ 8,\ 32,\ 128,\ 512$$

Self Check 3 Find the first five terms of the geometric sequence with a first term -2 and a fourth term -54.

Now Try ▶ Problem 23

2 Find Geometric Means.

If numbers are inserted between two numbers a and b to form a geometric sequence, the inserted numbers are called **geometric means** between a and b. If a single number is inserted, that number is called **the geometric mean** between a and b.

EXAMPLE 4 Insert two geometric means between 7 and 1,512.

Strategy Because two geometric means are to be inserted between 7 and 1,512, we will consider a sequence of four terms, with a first term of 7 and a fourth term of 1,512. We will then use $a_n = a_1 r^{n-1}$ to find the common ratio r.

Why Once we know the first term and the common ratio, we can multiply by the common ratio to find the two unknown terms.

Solution In this example, the first term is $a_1 = 7$, and the fourth term is $a_4 = 1,512$. To find the common ratio r so that the terms

$$\underset{a_1}{7},\quad \underset{a_2}{7r},\quad \underset{a_3}{7r^2},\quad \underset{a_4}{1,512}$$

form a geometric sequence, we substitute 4 for n and 7 for a_1 in the formula for the nth term of a geometric sequence and solve for r.

$$a_n = a_1 r^{n-1} \qquad \text{This is the general term formula.}$$
$$a_4 = 7r^{4-1} \qquad \text{Substitute 4 for } n \text{ and 7 for } a_1.$$
$$1{,}512 = 7r^3 \qquad \text{Substitute 1,512 for } a_4.$$
$$216 = r^3 \qquad \text{Divide both sides by 7.}$$
$$6 = r \qquad \text{Take the cube root of both sides.}$$

To find the two geometric means between 7 and 1,512, we multiply by the common ratio 6, as shown:

$$7r = 7(6) = \mathbf{42} \qquad \text{and} \qquad 7r^2 = 7(6)^2 = 7(36) = \mathbf{252}$$

The numbers 7, 42, 252, and 1,512 are the first four terms of a geometric sequence.

Self Check 4 Insert three positive geometric means between 1 and 16.

Now Try ▶ Problem 27

EXAMPLE 5 Find the geometric mean between 2 and 20.

Strategy Because one geometric mean is to be inserted between 2 and 20, we will consider a sequence of three terms, with a first term of 2 and a third term of 20. We will then use $a_n = a_1 r^{n-1}$ to find the common ratio r.

Why Once we know the first term and the common ratio, we can multiply by the common ratio to find the unknown term.

Solution We want to find the middle term of the three-termed geometric sequence

$$2, \quad 2r, \quad 20$$
$$\uparrow \qquad \uparrow \qquad \uparrow$$
$$a_1 \qquad a_2 \qquad a_3$$

with $a_1 = 2$, $a_3 = 20$, and $n = 3$. To find r, we substitute these values into the formula for the nth term of a geometric sequence:

$$a_n = a_1 r^{n-1} \qquad \text{This is the general term formula.}$$
$$a_3 = 2r^{3-1} \qquad \text{Substitute 3 for } n \text{ and 2 for } a_1.$$
$$20 = 2r^2 \qquad \text{Substitute 20 for } a_3.$$
$$10 = r^2 \qquad \text{Divide both sides by 2.}$$
$$\pm\sqrt{10} = r \qquad \text{Use the square root property.}$$

Because r can be either $\sqrt{10}$ or $-\sqrt{10}$, there are two values for the geometric mean. They are

$$2r = 2\sqrt{10} \qquad \text{and} \qquad 2r = -2\sqrt{10}$$

The sets of numbers 2, $2\sqrt{10}$, 20 and 2, $-2\sqrt{10}$, 20 both form geometric sequences. The common ratio of the first sequence is $\sqrt{10}$, and the common ratio of the second sequence is $-\sqrt{10}$.

Self Check 5 Find the positive geometric mean between 2 and 200.

Now Try ▶ Problem 31

3 Find the Sum of the First n Terms of a Geometric Sequence.

There is a formula that gives the sum of the first n terms of a geometric sequence. To develop this formula, we let S_n represent the sum of the first n terms of a geometric sequence.

(1) $S_n = a_1 + a_1 r + a_1 r^2 + a_1 r^3 + \cdots + a_1 r^{n-1}$

We multiply both sides of Equation 1 by r to get

(2) $S_n r = \qquad a_1 r + a_1 r^2 + a_1 r^3 + \cdots + a_1 r^{n-1} + a_1 r^n$

We now subtract Equation 2 from Equation 1 and solve for S_n:

$$S_n - S_n r = a_1 - a_1 r^n$$

$$S_n(1 - r) = a_1 - a_1 r^n \qquad \text{\color{red}Factor out } S_n \text{ \color{red}from the left side.}$$

$$S_n = \frac{a_1 - a_1 r^n}{1 - r} \qquad \text{\color{red}Divide both sides by } 1 - r.$$

This reasoning establishes the following formula.

Sum of the First n Terms of a Geometric Sequence

The sum of the first n terms of a geometric sequence is given by the formula

$$S_n = \frac{a_1 - a_1 r^n}{1 - r} \quad \text{or} \quad S_n = \frac{a_1(1 - r^n)}{1 - r} \quad \text{where } r \neq 1$$

where S_n is the sum, a_1 is the first term, r is the common ratio, and n is the number of terms.

EXAMPLE 6   Find the sum of the first six terms of the geometric sequence: $250, 50, 10, \ldots$

Strategy   We will find the common ratio r.

Why   If we know the first term and the common ratio, we can use the formula $S_n = \frac{a_1 - a_1 r^n}{1 - r}$ to find the sum of the first six terms by letting $n = 6$.

Solution   The common ratio r is the ratio between any two successive terms. Since $a_1 = 250$ and $a_2 = 50$, we can find r as follows:

$$r = \frac{a_2}{a_1} = \frac{50}{250} = \frac{1}{5} \qquad \text{\color{red}Also note that } \frac{a_3}{a_2} = \frac{10}{50} = \frac{1}{5}.$$

In this sequence, $a_1 = 250$, $r = \frac{1}{5}$, and $n = 6$. We substitute these values into the formula for the sum of the first n terms of a geometric sequence and simplify:

$$S_n = \frac{a_1 - a_1 r^n}{1 - r} \qquad \text{\color{red}This is the formula for the sum of terms of a geometric sequence.}$$

$$S_6 = \frac{250 - 250\left(\frac{1}{5}\right)^6}{1 - \frac{1}{5}} \qquad \text{\color{red}Substitute.}$$

$$= \frac{250 - 250\left(\frac{1}{15,625}\right)}{\frac{4}{5}} \qquad \text{\color{red}Use the calculator to evaluate: } \left(\frac{1}{5}\right)^6.$$

$$= \left(250 - \frac{250}{15,625}\right) \cdot \frac{5}{4} \qquad \text{\color{red}Multiply the numerator by the reciprocal of the denominator.}$$

$$= 312.48 \qquad \text{\color{red}Use a calculator.}$$

The sum of the first six terms is 312.48.

Self Check 6   Find the sum of the first five terms of the geometric sequence: 100, 20, 4, . . .

Now Try ▶ Problem 35

4 Define and Use Infinite Geometric Series.

When we add the terms of a geometric sequence, we form a **geometric series.** If we form the sum of the terms of an infinite geometric sequence, we get a series called an **infinite geometric series.** For example, if the common ratio r is 3, we have

Infinite geometric sequence   *Infinite geometric series*

2, 6, 18, 54, 162, . . . $2 + 6 + 18 + 54 + 162 + \cdots$

As the number of terms of this series gets larger, the value of the series gets larger. We can see that this is true by forming some **partial sums.**

The first partial sum of the series is $S_1 = 2$.

The second partial sum of the series is $S_2 = 2 + 6 = 8$.

The third partial sum of the series is $S_3 = 2 + 6 + 18 = 26$.

The fourth partial sum of the series is $S_4 = 2 + 6 + 18 + 54 = 80$.

> **The Language of Algebra**
>
> The word **partial** means only a part, not total. Have you ever seen a *partial* eclipse of the moon?

We can see that as the number of terms gets infinitely large, the value of this series gets infinitely large.

The values of some infinite geometric series get closer and closer to a specific number as the number of terms approaches infinity. One such series is

$$\frac{3}{2} + \frac{3}{4} + \frac{3}{8} + \frac{3}{16} + \frac{3}{32} + \frac{3}{64} + \cdots \quad \text{Here, } r = \tfrac{1}{2}.$$

To see that this is true, we form some partial sums.

The first partial sum is $S_1 = \dfrac{3}{2} = \mathbf{1.5}$

The second partial sum is $S_2 = \dfrac{3}{2} + \dfrac{3}{4} = \dfrac{9}{4} = \mathbf{2.25}$

The third partial sum is $S_3 = \dfrac{3}{2} + \dfrac{3}{4} + \dfrac{3}{8} = \dfrac{21}{8} = \mathbf{2.625}$

The fourth partial sum is $S_4 = \dfrac{3}{2} + \dfrac{3}{4} + \dfrac{3}{8} + \dfrac{3}{16} = \dfrac{45}{16} = \mathbf{2.8125}$

The fifth partial sum is $S_5 = \dfrac{3}{2} + \dfrac{3}{4} + \dfrac{3}{8} + \dfrac{3}{16} + \dfrac{3}{32} = \dfrac{93}{32} = \mathbf{2.90625}$

The sixth partial sum is $S_6 = \dfrac{3}{2} + \dfrac{3}{4} + \dfrac{3}{8} + \dfrac{3}{16} + \dfrac{3}{32} + \dfrac{3}{64} = \dfrac{189}{64} = \mathbf{2.953125}$

As the number of terms in this series gets larger, the values of the partial sums approach the number 3. We say that 3 is the **limit** of S_n as n approaches infinity, and we say that 3 is the **sum of the infinite geometric series.**

To develop a formula for finding the sum of an infinite geometric series, we consider the formula that gives the sum of the first n terms.

$$S_n = \frac{a_1 - \mathbf{a_1 r^n}}{1 - r} \quad \text{where } r \neq 1$$

If $|r| < 1$ and a_1 is constant, the term $a_1 r^n$ in the above formula approaches 0 as n becomes very large. For example,

$$a_1\left(\frac{1}{2}\right)^1 = \frac{1}{2}a_1, \qquad a_1\left(\frac{1}{2}\right)^2 = \frac{1}{4}a_1, \qquad a_1\left(\frac{1}{2}\right)^3 = \frac{1}{8}a_1$$

and so on. Thus, when n is very large, the value of $a_1 r^n$ is insignificant, and the term $a_1 r^n$ in the above formula can be ignored. This reasoning justifies the following formula.

Sum of the Terms of an Infinite Geometric Series	If a_1 is the first term and r is the common ratio of an infinite geometric sequence, and $|r| < 1$, the sum of the terms of the corresponding series is given by $$S = \frac{a_1}{1 - r}$$

EXAMPLE 7 Find the sum of the terms of the infinite geometric series: $125 + 25 + 5 + \cdots$

Strategy We will identify a_1 and find r.

Why To use the formula $S = \frac{a_1}{1-r}$ to find the sum, we need to know the first term, a_1, and the common ratio, r.

Solution In this geometric series, $a_1 = 125$ and $r = \frac{25}{125} = \frac{1}{5}$. Since $|r| = \left|\frac{1}{5}\right| = \frac{1}{5} < 1$, we can find the sum of the series. We do this by substituting 125 for a_1 and $\frac{1}{5}$ for r in the formula $S = \frac{a_1}{1-r}$ and simplifying:

Notation

The sum of the terms of an infinite geometric sequence also is written as S_∞.

$$S = \frac{a_1}{1 - r} = \frac{125}{1 - \frac{1}{5}} = \frac{125}{\frac{4}{5}} = 125 \cdot \frac{5}{4} = \frac{625}{4}$$

The sum of the series $125 + 25 + 5 + \cdots$ is $\frac{625}{4} = 156.25$.

Self Check 7 Find the sum of the infinite geometric series: $100 + 20 + 4 + \cdots$

Now Try ▶ Problem 43

EXAMPLE 8 Find the sum of the infinite geometric series: $64 + (-4) + \frac{1}{4} + \cdots$

Strategy We will identify a_1 and find r.

Why To use the formula $S = \frac{a_1}{1-r}$ to find the sum, we need to know the first term, a_1, and the common ratio, r.

Solution In this geometric series, $a_1 = 64$ and $r = \frac{-4}{64} = -\frac{1}{16}$. Since $|r| = \left|-\frac{1}{16}\right| = \frac{1}{16} < 1$, we can find the sum of the series. We substitute 64 for a_1 and $-\frac{1}{16}$ for r in the formula $S = \frac{a_1}{1-r}$ and simplify:

Notation

This infinite geometric series also could be written:

$$64 - 4 + \frac{1}{4} - \cdots$$

$$S = \frac{a_1}{1 - r} = \frac{64}{1 - \left(-\frac{1}{16}\right)} = \frac{64}{\frac{17}{16}} = 64 \cdot \frac{16}{17} = \frac{1{,}024}{17}$$

Caution

If $|r| \geq 1$ for an infinite geometric sequence, the sum of the terms of the sequence does not exist.

The sum of the geometric series $64 + (-4) + \frac{1}{4} + \cdots$ is $\frac{1{,}024}{17}$.

Self Check 8 Find the sum of the infinite geometric series: $81 + (-27) + 9 + \cdots$

Now Try ▶ Problem 51

EXAMPLE 9   Write $0.\overline{8}$ in fraction form.

Strategy   First, we will show that the decimal $0.888 \ldots$ can be represented by an infinite geometric series whose terms are fractions. Then we will identify a_1, find the common ratio r, and find the sum.

Why   When we use the formula $S = \dfrac{a_1}{1 - r}$ to find the sum, the result will be the required common fraction.

Solution   The decimal $0.\overline{8}$ can be written as the infinite geometric series

$$0.\overline{8} = 0.888 \ldots$$
$$= 0.8 + 0.08 + 0.008 + \cdots$$
$$= \frac{8}{10} + \frac{8}{100} + \frac{8}{1,000} + \cdots$$

Here, $a_1 = \frac{8}{10}$ and $r = \frac{1}{10}$. Because $|r| = \left|\frac{1}{10}\right| = \frac{1}{10} < 1$, we can find the sum as follows:

$$S = \frac{a_1}{1 - r} = \frac{\dfrac{8}{10}}{1 - \dfrac{1}{10}} = \frac{\dfrac{8}{10}}{\dfrac{9}{10}} = \frac{8}{10} \cdot \frac{10}{9} = \frac{8}{9}$$

Thus, $0.\overline{8} = \frac{8}{9}$. Long division will verify that $\frac{8}{9} = 0.888 \ldots$.

Self Check 9   Write $0.\overline{6}$ in fraction form.

Now Try ▶ Problem 55

---

5 Solve Application Problems Involving Geometric Sequences.

EXAMPLE 10   **Inheritances.**   A father decides to give his son part of his inheritance early. Each year, on the son's birthday, the father will pay the son 15% of what remains in the inheritance fund. If the fund initially begins with $100,000, how much money will be left in the fund after 20 years of payments?

Strategy   We will model the facts of the problem using a geometric sequence with a first term of 100,000 and a common ratio of 0.85.

Why   One of the terms of the sequence will represent the amount of money left in the inheritance fund after 20 years of payments.

Solution   If 15% of the money in the inheritance fund is given to the son each year, 85% of that amount remains after each payment. To find the amount of money that remains in the fund after a payment is made, we multiply the amount that was in the fund by 0.85. Over the years, the amounts of money that are left in the fund after a payment form a geometric sequence.

Amount of money remaining in the inheritance fund

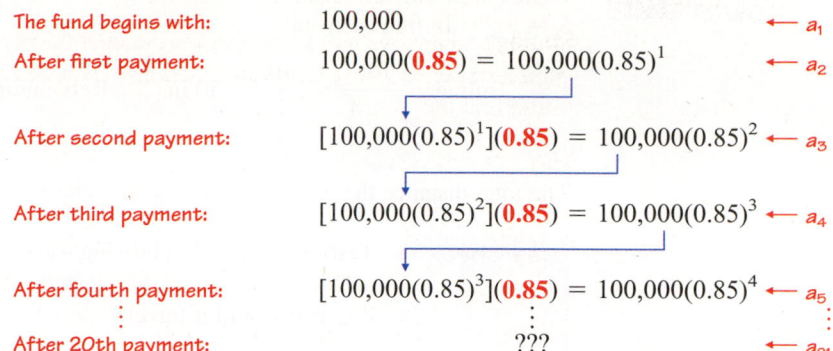

The fund begins with:	100,000	← a_1
After first payment:	$100{,}000(\mathbf{0.85}) = 100{,}000(0.85)^1$	← a_2
After second payment:	$[100{,}000(0.85)^1](\mathbf{0.85}) = 100{,}000(0.85)^2$	← a_3
After third payment:	$[100{,}000(0.85)^2](\mathbf{0.85}) = 100{,}000(0.85)^3$	← a_4
After fourth payment:	$[100{,}000(0.85)^3](\mathbf{0.85}) = 100{,}000(0.85)^4$	← a_5
After 20th payment:	???	← a_{21}

The amount of money remaining in the inheritance fund after 20 years is represented by the 21st term of a geometric sequence, where $a_1 = 100{,}000$, $r = 0.85$, and $n = 21$.

$$a_n = a_1 r^{n-1}$$ This is the formula for the *n*th term.

$$a_{21} = a_1 r^{21-1}$$ Substitute 21 for *n*.

$$= \mathbf{100{,}000(0.85)}^{21-1}$$ Substitute 100,000 for a_1 and 0.85 for *r*.

$$= 100{,}000(0.85)^{20}$$ Do the subtraction.

$$\approx 3{,}876$$ Use a calculator. Round to the nearest dollar.

In 20 years, approximately $3,876 of the inheritance fund will be left.

Self Check 10   **Inheritances.**   How much money will be left in the inheritance fund after 30 years of payments?

Now Try ▶ Problem 79

EXAMPLE 11

Testing Steel.   One way to measure the hardness of a steel anvil is to drop a ball bearing onto the face of the anvil. The bearing should rebound $\frac{4}{5}$ of the distance from which it was dropped. If a bearing is dropped from a height of 10 inches onto a hard forged steel anvil, and if it could bounce forever, what total distance would the bearing travel?

Strategy   We will show that the facts in the problem can be modeled by an infinite geometric sequence.

Why   The sum of the terms of the infinite geometric sequence will give the total distance the bearing will travel.

Solution   The total distance the ball bearing travels is the sum of two motions, falling and rebounding. The bearing falls 10 inches, then rebounds $\frac{4}{5} \cdot 10 = 8$ inches, and falls 8 inches, and rebounds $\frac{4}{5} \cdot 8 = \frac{32}{5}$ inches, and falls $\frac{32}{5}$ inches, and rebounds $\frac{4}{5} \cdot \frac{32}{5} = \frac{128}{25}$ inches, and so on.

10 in.

The distance the ball falls is given by the sum

$$10 + 8 + \frac{32}{5} + \frac{128}{25} + \cdots$$ This is an infinite geometric series with $a_1 = 10$ and $r = \frac{4}{5}$.

The distance the ball rebounds is given by the sum

$$8 + \frac{32}{5} + \frac{128}{25} + \cdots$$ This is an infinite geometric series with $a_1 = 8$ and $r = \frac{4}{5}$.

Since each of these is an infinite geometric series with $|r| < 1$, we can use the formula $S = \dfrac{a_1}{1-r}$ to find each sum.

Falling: $\dfrac{10}{1-\dfrac{4}{5}} = \dfrac{10}{\dfrac{1}{5}} = 50$ in.   Rebounding: $\dfrac{8}{1-\dfrac{4}{5}} = \dfrac{8}{\dfrac{1}{5}} = 40$ in.

The total distance the bearing travels is 50 inches + 40 inches $=$ 90 inches.

Self Check 11 **Testing Steel.** If a bearing was dropped from a height of 15 inches onto a hard forged steel anvil, and if it could bounce forever, what total distance would it travel?

Now Try ▶ Problem 85

SECTION 11.3 ▶ STUDY SET

VOCABULARY

Fill in the blanks.

1. Each term of a _____ sequence is found by multiplying the previous term by the same number.

2. 8, 16, 32, 64, 128, . . . is an example of a _____ sequence. The first _____ is 8 and the common _____ is 2.

3. If a single number is inserted between a and b to form a geometric sequence, the number is called the geometric _____ between a and b.

4. The sum of the terms of a geometric sequence is called a geometric _____. The sum of the terms of an infinite geometric sequence is called an _____ geometric series.

CONCEPTS

5. Write the first three terms of a geometric sequence if $a_1 = 16$ and $r = \frac{1}{4}$.

6. Given the geometric sequence 1, 2, 4, 8, 16, 32, . . . , find a_5 and r.

7. Write the formula for a_n, the general term of a geometric sequence.

8. **a.** Write the formula for S_n, the sum of the first n terms of a geometric sequence.

   b. Write the formula for S, the sum of the terms of an infinite geometric sequence, where $|r| < 1$.

9. Which of the following values of r satisfy $|r| < 1$?

   a. $r = \frac{2}{3}$ **b.** $r = -3$

   c. $r = 6$ **d.** $r = -\frac{1}{5}$

10. Write $0.\overline{7}$ as an infinite geometric series:

$$0.\overline{7} = 0.7 + 0.07 + 0.007 + \cdots$$

$$= \dfrac{7}{} + \dfrac{7}{} + \dfrac{7}{} + \cdots$$

NOTATION

Fill in the blanks.

11. An infinite geometric sequence is of the form

$$a_1, \quad a_1 r, \quad a_1 , \quad a_1 r^3, \quad , \ldots$$

12. The first four terms of the sequence defined by $a_n = 4(3)^{n-1}$ are , , , .

13. To find the common ratio of a geometric sequence, we use the formula: $r = \dfrac{a_{n+1}}{a_{}}$

14. S_8 represents the sum of the first _____ terms of a geometric sequence.

GUIDED PRACTICE

Write the first five terms of each geometric sequence with the given properties and then find the specified term. See Example 1.

15. First term: 3, common ratio: 2; find the 9th term

16. First term: -2, common ratio: 2; find the 8th term

17. First term: -5, common ratio: $\frac{1}{5}$; find the 8th term

18. First term: 8, common ratio: $\frac{1}{2}$; find the 10th term

Find the specified term of the geometric sequence with the following properties. See Example 2.

19. The first three terms are 2, 6, 18; find the 7th term

20. The first three terms are 50, 100, 200; find the 10th term

21. The first three terms are $\frac{1}{2}, -\frac{5}{2}, \frac{25}{2}$; find the 6th term

22. The first three terms are $\frac{1}{4}, -\frac{3}{4}, \frac{9}{4}$; find the 9th term

Write the first five terms of the geometric sequence with the following properties. See Example 3.

23. First term: 2, $r > 0$, third term: 18
24. First term: 2, $r < 0$, third term: 50
25. First term: 3, fourth term: 24
26. First term: -3, fourth term: -192

Find the geometric means to be inserted in each geometric sequence. See Example 4.

27. Insert three positive geometric means between 2 and 162.
28. Insert four geometric means between 3 and 96.
29. Insert four geometric means between -4 and $-12,500$.
30. Insert three geometric means (two positive and one negative) between -64 and $-1,024$.

Find a geometric mean to be inserted in each geometric sequence. See Example 5.

31. Find the geometric mean between 2 and 128.
32. Find the geometric mean between 3 and 243.
33. Find the geometric mean between 10 and 20.
34. Find the geometric mean between 5 and 15.

For each geometric sequence, find the sum of the specified number of terms. See Example 6.

35. The first 6 terms of 2, 6, 18, ...
36. The first 6 terms of 2, -6, 18, ...
37. The first 5 terms of 2, -6, 18, ...
38. The first 5 terms of 2, 6, 18, ...
39. The first 8 terms of 3, -6, 12, ...
40. The first 8 terms of 3, 6, 12, ...
41. The first 7 terms of 3, 6, 12, ...
42. The first 7 terms of 3, -6, 12, ...

Find the sum of each infinite geometric series, if possible. See Examples 7 and 8.

43. $8 + 4 + 2 + \cdots$
44. $12 + 6 + 3 + \cdots$
45. $54 + 18 + 6 + \cdots$
46. $45 + 15 + 5 + \cdots$
47. $-\frac{27}{2} + (-9) + (-6) + \cdots$
48. $-112 + (-28) + (-7) + \cdots$

49. $\frac{9}{2} + 6 + 8 + \cdots$
50. $\frac{18}{25} + \frac{6}{5} + 2 + \cdots$
51. $12 + (-6) + 3 + \cdots$
52. $8 + (-4) + 2 + \cdots$
53. $-45 + 15 + (-5) + \cdots$
54. $-54 + 18 + (-6) + \cdots$

Write each decimal in fraction form. Then check the answer by performing long division. See Example 9.

55. $0.\overline{1}$
56. $0.\overline{2}$
57. $0.\overline{3}$
58. $0.\overline{4}$
59. $0.\overline{12}$
60. $0.\overline{21}$
61. $0.\overline{75}$
62. $0.\overline{57}$

TRY IT YOURSELF

63. Find the common ratio of the geometric sequence with a first term -8 and a sixth term $-1,944$.
64. Find the common ratio of the geometric sequence with a first term 12 and a sixth term $\frac{3}{8}$.

65. Write the first five terms of the geometric sequence if its first term is -64, $r < 0$, and its fifth term is -4.
66. Write the first five terms of the geometric sequence if its first term is -64, $r > 0$, and its fifth term is -4.
67. Find a geometric mean, if possible, between -50 and 10.
68. Find a negative geometric mean, if possible, between -25 and -5.
69. Find the 10th term of the geometric sequence with $a_1 = 7$ and $r = 2$.
70. Find the 12th term of the geometric sequence with $a_1 = 64$ and $r = \frac{1}{2}$.
71. Write the first four terms of the geometric sequence if its first term is -64 and its sixth term is -2.
72. Write the first four terms of the geometric sequence if its first term is -81 and its sixth term is $\frac{1}{3}$.
73. Find the sum of the terms of the geometric sequence: $3, \frac{3}{4}, \frac{3}{16}, \frac{3}{64}, \cdots$
74. Find the sum of the terms of the geometric sequence: $1, -\frac{1}{2}, \frac{1}{4}, -\frac{1}{8}, \cdots$
75. Find the first term of the geometric sequence with a common ratio -3 and an eighth term -81.
76. Find the first term of the geometric sequence with a common ratio 2 and a tenth term 384.
77. Find the sum of the first five terms of the geometric sequence if its first term is 3 and the common ratio is 2.
78. Find the sum of the first five terms of the geometric sequence if its first term is 5 and the common ratio is -6.

APPLICATIONS

Use a calculator to help solve each problem.

79. **Declining Savings.** John has $10,000 in a safety deposit box. Each year, he spends 12% of what is left in the box. How much will be in the box after 15 years?
80. **Savings Growth.** Sally has $5,000 in a savings account earning 12% annual interest. How much will be in her account 10 years from now? (Assume that Sally makes no deposits or withdrawals.)

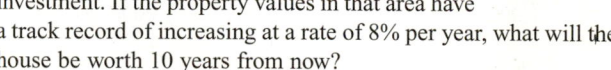

81. **from Campus to Careers**

Real Estate Sales Agent

Suppose you are a real estate sales agent and you are working with a client who is considering buying a $250,000 house in Seattle as an investment. If the property values in that area have a track record of increasing at a rate of 8% per year, what will the house be worth 10 years from now?

©Michael Pettigrew/Shutterstock.com

82. **Boat Depreciation.** A boat that cost $5,000 when new depreciates at a rate of 9% per year. How much will the boat be worth in 5 years?

83. Inscribed Squares. Each inscribed square in the illustration joins the midpoints of the next larger square. The area of the first square, the largest, is 1 square unit. Find the area of the 12th square.

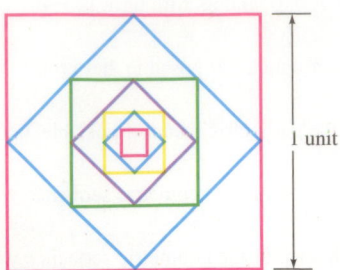

1 unit

84. Genealogy. The following family tree spans 3 generations and lists 7 people. How many names would be listed in a family tree that spans 10 generations?

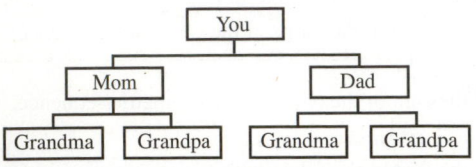

85. Bouncing Balls. On each bounce, the rubber ball in the illustration rebounds to a height one-half of that from which it fell. Find the total vertical distance the ball travels.

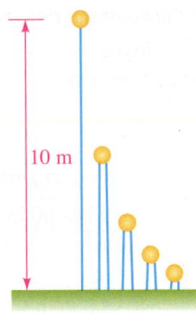

10 m

86. Bouncing Balls. A golf ball is dropped from a height of 12 feet. On each bounce, it returns to a height that is two-thirds of the distance it fell. Find the total vertical distance the ball travels.

87. Pest Control. To reduce the population of a destructive moth, biologists release 1,000 sterilized male moths each day into the environment. If 80% of these moths alive one day survive until the next, then after a long time the population of sterile males is the sum of the infinite geometric series

$$1,000 + 1,000(0.8) + 1,000(0.8)^2 + 1,000(0.8)^3 + \cdots$$

Find the long-term population.

88. Pendulums. On its first swing to the right, a pendulum swings through an arc of 96 inches. Each successive swing, the pendulum travels $\frac{99}{100}$ as far as on the previous swing. Determine the total distance that the pendulum will travel by the time it comes to rest.

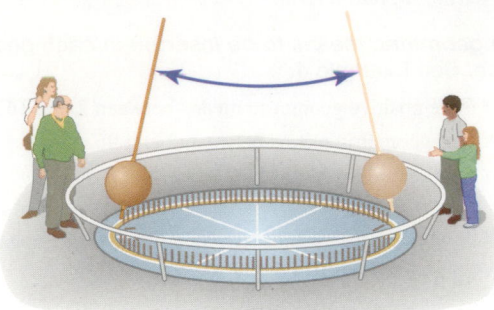

WRITING

89. Describe the real numbers that satisfy $|r| < 1$.

90. Why must the absolute value of the common ratio be less than 1 before an infinite geometric sequence can have a sum?

91. Explain the difference between an arithmetic sequence and a geometric sequence.

92. Why is $1 - \frac{1}{2} + \frac{1}{4} - \frac{1}{8} + \frac{1}{16} - \frac{1}{32} + \cdots$ called an alternating infinite geometric series?

REVIEW

Solve each inequality. Write the solution set using interval notation.

93. $x^2 - 5x - 6 \le 0$

94. $a^2 - 7a + 12 \ge 0$

95. $\dfrac{x - 4}{x + 3} > 0$

96. $\dfrac{t^2 + t - 20}{t + 2} < 0$

CHALLENGE PROBLEMS

97. If $f(x) = 1 + x + x^2 + x^3 + x^4 + \cdots$, find $f\left(\frac{1}{2}\right)$ and $f\left(-\frac{1}{2}\right)$.

98. Find the sum:

$$\frac{1}{\sqrt{3}} + \frac{1}{3} + \frac{1}{3\sqrt{3}} + \frac{1}{9} + \cdots$$

99. If $a > b > 0$, which is larger: the arithmetic mean between a and b or the geometric mean between a and b?

100. Is there a geometric mean between -5 and 5?

11 Summary & Review

SECTION 11.1 ▶ The Binomial Theorem

DEFINITIONS AND CONCEPTS	EXAMPLES
Pascal's triangle gives the coefficients of the terms of the expansion of $(a + b)^n$.	**Pascal's triangle:**

Pascal's triangle:

$$
\begin{array}{ccccccccccc}
 & & & & & 1 & & & & & \text{Row for } (a+b)^0 \\
 & & & & 1 & & 1 & & & & \text{Row for } (a+b)^1 \\
 & & & 1 & & 2 & & 1 & & & \text{Row for } (a+b)^2 \\
 & & 1 & & 3 & & 3 & & 1 & & \text{Row for } (a+b)^3 \\
 & 1 & & 4 & & 6 & & 4 & & 1 & \text{Row for } (a+b)^4 \\
1 & & 5 & & 10 & & 10 & & 5 & & 1 \quad \text{Row for } (a+b)^5
\end{array}
$$

$$(x + y)^5 = 1x^5 + 5x^4y + 10x^3y^2 + 10x^2y^3 + 5xy^4 + 1y^5$$

The symbol **$n!$** (**n factorial**) is the product of consecutively decreasing natural numbers from n to 1.

$$n! = n(n-1)(n-2) \cdot \cdots \cdot 2 \cdot 1$$

Evaluate each expression:

$$4! = 4 \cdot 3 \cdot 2 \cdot 1 = 24 \qquad 3! \cdot 2! = 3 \cdot 2 \cdot 1 \cdot 2 \cdot 1 = 12$$

$$\frac{6!}{5!} = \frac{6 \cdot \overset{1}{\cancel{5!}}}{\underset{1}{\cancel{5!}}} = 6 \qquad 1! = 1 \quad \text{and} \quad 0! = 1$$

The **binomial theorem** is usually the best way to expand a binomial.

$$(a + b)^n = a^n + \frac{n!}{1!(n-1)!}a^{n-1}b +$$

$$\frac{n!}{2!(n-2)!}a^{n-2}b^2 + \cdots + b^n$$

Use the binomial theorem to expand $(p + q)^4$, where $n = 4$.

$$(p + q)^4 = p^4 + \frac{4!}{1!(4-1)!}p^3q + \frac{4!}{2!(4-2)!}p^2q^2 + \frac{4!}{3!(4-3)!}pq^3 + q^4$$

$$= p^4 + 4p^3q + 6p^2q^2 + 4pq^3 + q^4$$

To find a specific term of an expansion:

The $(r + 1)$st term of the expansion of $(a + b)^n$ is

$$\frac{n!}{r!(n-r)!}a^{n-r}b^r$$

Remember that r is always 1 less than the number of the term that you are finding.

Find the third term of the expansion of $(a + b)^5$.

In the third term of $(a + b)^5$, $n = 5$ and $r = 2$.

$$\frac{5!}{2!(5-2)!}a^{5-2}b^2 = \frac{5 \cdot 4 \cdot \overset{1}{\cancel{3!}}}{2 \cdot 1 \cdot \underset{1}{\cancel{3!}}}a^3b^2 = 10a^3b^2$$

REVIEW EXERCISES

1. Complete Pascal's triangle. List the row that gives the coefficients for the expansion of $(a + b)^5$.

2. Consider the expansion of $(a + b)^{12}$.
   a. How many terms does the expansion have?
   b. For each term, what is the sum of the exponents on a and b?
   c. What is the first term? What is the last term?
   d. How do the exponents on a and b change from term to term?

Evaluate each expression.

3. $4! \cdot 3!$

4. $\dfrac{5!}{3!}$

5. $\dfrac{6!}{2!(6-2)!}$

6. $\dfrac{12!}{3!(12-3)!}$

7. $(n - n)!$

8. $\dfrac{8!}{7!}$

Use the binomial theorem to find each expansion.

9. $(x + y)^5$

10. $(x - y)^9$

11. $(4x - y)^3$

12. $\left(\dfrac{c}{2} + \dfrac{d}{3}\right)^4$

Find the specified term in each expansion.

13. $(x + y)^4$; third term

14. $(x - y)^6$; fourth term

15. $(3x - 4y)^3$; second term

16. $(u^2 - v^3)^5$; fifth term

SECTION 11.2 ▶ Arithmetic Sequences and Series

DEFINITIONS AND CONCEPTS	EXAMPLES
A **sequence** is a list of numbers written in a specific order.	**Finite sequence:** 1, 4, 7, 10, 13 **Infinite sequence:** 2, 6, 10, 14, 18, . . .
To describe all the terms of a sequence specifically, we can write a formula for a_n, called the **general term** of the sequence.	Find the first four terms of the sequence described by $a_n = 5n + 1$. We substitute 1, 2, 3, and 4, for n in the formula: $\quad a_1 = 5(1) + 1 = 6 \qquad a_2 = 5(2) + 1 = 11$ $\quad a_3 = 5(3) + 1 = 16 \qquad a_4 = 5(4) + 1 = 21$ The first four terms are 6, 11, 16, and 21.
A sequence in which each term is found by adding the same number to the previous term is called an **arithmetic sequence.** An arithmetic sequence has the form $a_1, a_1 + d, a_1 + 2d, \ldots, a_1 + (n - 1)d, \ldots$ where a_1 is the first term and d is the common difference. The ***n*th term of an arithmetic sequence** is given by $a_n = a_1 + (n - 1)d$. The **common difference d** of an arithmetic sequence is the difference between any two consecutive terms: $d = a_{n+1} - a_n$.	Find the 12th term of the arithmetic sequence $-4, -1, 2, 5, 8, \ldots$. First, we find the common difference d. It is the difference between any two successive terms: $d = a_2 - a_1 = -1 - (-4) = 3$. Then we substitute into the formula for the nth term: $\quad a_n = a_1 + (n - 1)d$ $\quad a_{12} = -4 + (12 - 1)3 \qquad$ In this sequence, $n = 12$, $a_1 = -4$, and $d = 3$. $\qquad\ = -4 + (11)3 \qquad$ Do the subtraction. $\qquad\ = -4 + 33 \qquad$ Evaluate the right side. $\qquad\ = 29$ The 12th term of the sequence is 29.
If numbers are inserted between two given numbers a and b to form an arithmetic sequence, the inserted numbers are **arithmetic means** between a and b.	Find three arithmetic means between -6 and 14. Since there will be five terms, $n = 5$. We also know that $a_1 = -6$ and $a_5 = 14$. We will substitute these values in the formula for the nth term and solve for d. $\quad a_5 = a_1 + (n - 1)d$ $\quad 14 = -6 + (5 - 1)d \qquad$ Substitute. $\quad 14 = -6 + 4d \qquad$ Do the subtraction. $\quad 20 = 4d \qquad$ Add 6 to both sides. $\quad\ \ 5 = d \qquad$ Solve for d. Since the common difference is 5, we successively add 5 to the terms to get the sequence $-6, -1, 4, 9, 14$. Thus, the three arithmetic means are -1, 4, and 9.

When the commas between the terms of a sequence are replaced with + signs, we call the sum a **series**.

The **sum of the first *n* terms of an arithmetic sequence** is given by

$$S_n = \frac{n(a_1 + a_n)}{2}$$

Find the sum of the first 12 terms of the arithmetic sequence $-4, -1, 2, 5, 8, \ldots$.

Here $a_1 = -4$ and $n = 12$. On the previous page, we found that for this sequence $a_{12} = 29$. We substitute these values into the formula for S_n and simplify.

$$S_n = \frac{\textcolor{red}{n(a_1 + a_n)}}{2}$$

$$S_{12} = \frac{\textcolor{red}{12(-4 + 29)}}{2} \quad \text{Substitute.}$$

$$= \frac{300}{2} \quad \text{Evaluate the numerator.}$$

$$= 150 \quad \text{Do the division.}$$

The sum of the first 12 terms is 150.

Summation notation involves the Greek letter sigma Σ. It designates the sum of terms called a **series**.

$$\sum_{n=1}^{4} (3k - 2) = [3(\textbf{1}) - 2] + [3(\textbf{2}) - 2] + [3(\textbf{3}) - 2] + [3(\textbf{4}) - 2]$$

$$= 1 + 4 + 7 + 10$$

$$= 22$$

REVIEW EXERCISES

17. Find the first four terms of the sequence defined by $a_n = 2n - 4$.

18. Find the first five terms of the sequence defined by $a_n = \frac{(-1)^n}{n + 1}$.

19. Find the 50th term of the sequence defined by $a_n = 100 - \frac{n}{2}$.

20. Find the eighth term of an arithmetic sequence whose first term is 7 and whose common difference is 5.

21. Write the first five terms of the arithmetic sequence whose ninth term is 242 and whose seventh term is 212.

22. The first three terms of an arithmetic sequence are 6, -6, and -18. Find the 101st term.

23. Find the common difference of an arithmetic sequence if its 1st term is -515 and the 23rd term is -625.

24. Find two arithmetic means between 8 and 25.

25. Find the sum of the first ten terms of the sequence $9, 6.5, 4, \ldots$.

26. Find the sum of the first 28 terms of an arithmetic sequence if the second term is 6 and the sixth term is 22.

Find each sum.

27. $\displaystyle\sum_{k=4}^{6} \frac{1}{2}k$

28. $\displaystyle\sum_{k=2}^{5} 7k^2$

29. $\displaystyle\sum_{k=1}^{4} (3k - 4)$

30. $\displaystyle\sum_{k=10}^{10} 36k$

31. What is the sum of the first 200 natural numbers?

32. **Seating.** The illustration shows the first 2 of a total of 30 rows of seats in an amphitheater. The number of seats in each row forms an arithmetic sequence. Find the total number of seats.

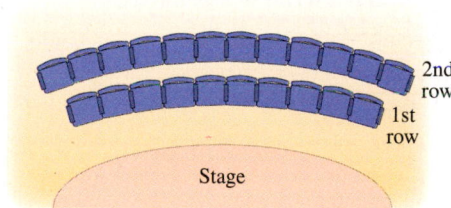

2nd row

1st row

Stage

SECTION 11.3 ▶ **Geometric Sequences and Series**

DEFINITIONS AND CONCEPTS	EXAMPLES
Each term of a **geometric sequence** is found by multiplying the previous term by the same number. A geometric sequence has the form: $$a_1, \quad a_1r, \quad a_1r^2, \quad a_1r^3, \quad \dots, a_1r^{n-1}, \dots$$ where a_1 is the first term and r is the common ratio. The ***n*th term of a geometric sequence** is given by: $$a_n = a_1r^{n-1}$$ The **common ratio r** of a geometric sequence is the quotient obtained when any term is divided by the previous term: $$r = \frac{a_{n+1}}{a_n}$$	Find the 8th term of the geometric sequence 8, 4, 2, ... First, we find the common ratio r. It is the quotient obtained when any term is divided by the previous term: $$r = \frac{a_2}{a_1} = \frac{4}{8} = \frac{1}{2}$$ Then we substitute into the formula for the *n*th term: $$a_n = a_1r^{n-1}$$ $$a_8 = 8\left(\frac{1}{2}\right)^{8-1} \qquad \text{Substitute: } n = 8, \ a_1 = 8, \text{ and } r = \tfrac{1}{2}.$$ $$= 8\left(\frac{1}{2}\right)^7 \qquad \text{Do the subtraction.}$$ $$= \frac{8}{128} \qquad \text{Evaluate: } \left(\tfrac{1}{2}\right)^7 = \tfrac{1}{128}.$$ $$= \frac{1}{16} \qquad \text{Simplify the fraction: Remove a common factor of 8 in the numerator and denominator.}$$
If numbers are inserted between a and b to form a geometric sequence, the inserted numbers are **geometric means** between a and b.	Find two geometric means between 6 and 162. Since there will be four terms, $n = 4$. We also know that $a_1 = 6$ and $a_4 = 162$. We will substitute these values in the formula for the *n*th term and solve for r. $$a_n = a_1r^{n-1}$$ $$162 = 6r^{4-1} \qquad \text{Substitute.}$$ $$27 = r^3 \qquad \text{Divide both sides by 6.}$$ $$3 = r \qquad \text{Take the cube root of both sides.}$$ Since the common ratio is 3, we successively multiply the terms by 3 to get the sequence 6, 18, 54, 162. Thus, the two geometric means are 18 and 54.
The **sum of the first *n* terms of a geometric sequence** is given by: $$S_n = \frac{a_1 - a_1r^n}{1 - r} \qquad r \neq 1$$	Find the sum of the first five terms of the geometric sequence 2, 6, 18, ... Here $a_1 = 2$, $r = 3$, and $n = 5$. We substitute these values into the formula for S_n and simplify. $$S_n = \frac{a_1 - a_1r^n}{1 - r}$$ $$S_5 = \frac{2 - (2)(3)^5}{1 - 3} \qquad \text{Substitute.}$$ $$= \frac{2 - 2(243)}{-2} \qquad \text{Evaluate: } (3)^5 = 243.$$ $$= \frac{2 - 486}{-2} \qquad \text{Do the multiplication.}$$ $$= \frac{-484}{-2} \qquad \text{Evaluate the numerator.}$$ $$= 242 \qquad \text{Do the division.}$$ The sum of the first five terms is 242.

If we form the sum of the terms of an infinite geometric sequence, we get a series called an **infinite geometric series**. The sum of an infinite geometric series is given by:

$$S = \frac{a_1}{1 - r} \quad \text{where } |r| < 1$$

Find the sum of the infinite geometric series: $12 + 8 + \frac{16}{3} + \cdots$

Here $a_1 = 12$ and $r = \frac{8}{12} = \frac{2}{3}$. (Note that $|r| < 1$.) We substitute these values into the formula for the sum of an infinite geometric series.

$$S = \frac{a_1}{1 - r} = \frac{12}{1 - \dfrac{2}{3}} = \frac{12}{\dfrac{1}{3}} = 12 \cdot \frac{3}{1} = 36$$

The sum is 36.

REVIEW EXERCISES

33. Find the sixth term of a geometric sequence with a first term of $\frac{1}{8}$ and a common ratio of 2.

34. Write the first five terms of the geometric sequence whose fourth term is 3 and whose fifth term is $\frac{3}{2}$.

35. Find the first term of a geometric sequence if it has a common ratio of -3 and the ninth term is 243.

36. Find two geometric means between -6 and 384.

37. Find the sum of the first seven terms of the sequence:

$$162, 54, 18, \ldots$$

38. Find the sum of the first eight terms of the sequence:

$$\frac{1}{8}, -\frac{1}{4}, \frac{1}{2}, \ldots$$

39. Feeding Birds. Tom has 50 pounds of birdseed stored in his garage. Each month, he uses 25% of what is left in the bag to feed the birds in his yard. How much birdseed will be left in 12 months?

40. Find the sum of the infinite series:

$$25 + 20 + 16 + \cdots$$

41. Change the decimal $0.\overline{05}$ to a common fraction.

42. Wham-O Toys. Tests have found that 1998 Superballs rebound $\frac{9}{10}$ of the distance from which they are dropped. If a Superball is dropped from a height of 10 feet, and if it could bounce forever, what total distance would it travel?

11 Chapter Test

1. Fill in the blanks.

  a. The array of numbers that gives the coefficients of the terms of a binomial expansion is called _____ triangle.

  b. In the expansion $a^3 - 3a^2b + 3ab^2 - b^3$, the signs _____ between $+$ and $-$.

  c. Each term of an _____ sequence is found by adding the same number to the previous term.

  d. The sum of the terms of an arithmetic sequence is called an arithmetic _____.

  e. Each term of a _____ sequence is found by multiplying the previous term by the same number.

2. Find the first 4 terms of the sequence defined by:
$$a_n = -6n + 8$$

3. Find the first 5 terms of the sequence defined by:
$$a_n = \frac{(-1)^n}{n^3}$$

4. Evaluate: $\dfrac{10!}{6!(10 - 6)!}$

5. Expand: $(a - b)^6$

6. Find the third term in the expansion of $(x^2 + 2y)^4$.

7. Find the tenth term of an arithmetic sequence whose first 3 terms are 3, 10, and 17.

8. Find the sum of the first 12 terms of the sequence:
$$-2, 3, 8, \ldots$$

9. Find two arithmetic means between 2 and 98.

10. Find the common difference of an arithmetic sequence if the second term is $\frac{5}{4}$ and the 17th term is 5.

11. Find the sum of the first 27 terms of an arithmetic sequence if the 4th term is -11 and the 20th term is -75.

12. Plumbing. Plastic pipe is stacked so that the bottom row has 25 pipes, the next row has 24 pipes, the next row has 23 pipes, and so on until there is 1 pipe at the top of the stack. If a worker removes the top 15 rows of pipe, how many pieces of pipe will be left in the stack?

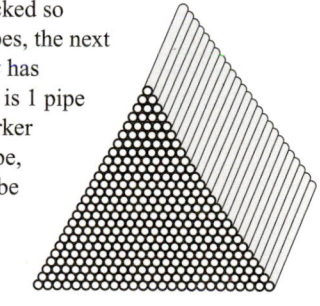

13. Falling Objects. If an object is in free fall, the sequence 16, 48, 80, . . . represents the distance in feet that object falls during the 1st second, during the 2nd second, during the 3rd second, and so on. How far will the object fall during the first 10 seconds?

14. Evaluate: $\displaystyle\sum_{k=1}^{3} (2k - 3)$

15. Find the seventh term of the geometric sequence whose first 3 terms are $-\frac{1}{9}, -\frac{1}{3}$, and -1.

16. Find the sum of the first 6 terms of the sequence:

$$\frac{1}{27}, \frac{1}{9}, \frac{1}{3}, \cdots$$

17. Find the first term of a geometric sequence if the common ratio is $-\frac{2}{3}$ and the fourth term is $-\frac{16}{9}$.

18. Find two geometric means between 3 and 648.

19. Find the sum of infinite geometric series: $9 + 3 + 1 + \cdots$

20. Depreciation. A yacht that cost $1,500,000 when new depreciates at a rate of 8% per year. How much will the yacht be worth in 10 years?

21. Pendulums. On its first swing to the right, a pendulum swings through an arc of 60 inches. Each successive swing, the pendulum travels $\frac{97}{100}$ as far as on the previous swing. Determine the total distance the pendulum will travel by the time it comes to rest.

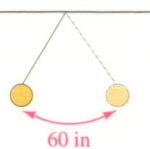

60 in

22. Change the decimal $0.\overline{7}$ to a common fraction.

Group Project

The Language of Algebra

Overview: This activity will help you review for the final exam.

Instructions: Form groups of two students. Match each instruction in column I with the most appropriate item in column II. Each letter in column II is used only once.

Column I

_____ 1. Use the FOIL method.

_____ 2. Apply a rule for exponents to simplify.

_____ 3. Add the rational expressions.

_____ 4. Rationalize the denominator.

_____ 5. Factor completely.

_____ 6. Evaluate the expression for $a = -1$ and $b = -6$.

_____ 7. Express in lowest terms.

_____ 8. Solve for t.

_____ 9. Combine like terms.

_____ 10. Remove parentheses.

_____ 11. Solve the system by graphing.

_____ 12. Find $f(g(x))$.

_____ 13. Solve using the quadratic formula.

_____ 14. Identify the base and the exponent.

_____ 15. Write without a radical symbol.

_____ 16. Write the equation of the line having the given slope and y-intercept.

_____ 17. Solve the inequality.

_____ 18. Complete the square to make a perfect-square trinomial.

_____ 19. Find the slope of the line passing through the given points.

_____ 20. Use a property of logarithms to simplify.

_____ 21. Set each factor equal to zero and solve for x.

_____ 22. State the solution of the compound inequality using interval notation.

_____ 23. Find the inverse function, $h^{-1}(x)$.

_____ 24. Write using scientific notation.

_____ 25. Write the logarithmic statement in exponential form.

_____ 26. Find the sum of the first 6 terms of the sequence.

Column II

a. $2,300,000,000$

b. e^3

c. $f(x) = x^2 + 1$ and $g(x) = 5 - 3x$

d. $-2x(3x^2 - 4x + 8)$

e. $4x - 7 > -3x - 7$

f. $\begin{cases} 2x = y - 5 \\ x + y = -1 \end{cases}$

g. $(x^2 - 5)(x^2 + 3)$

h. $(x + 2)(x - 10) = 0$

i. $\frac{x - 1}{2x^2} + \frac{x + 1}{8x}$

j. $\sqrt{4x^2}$

k. $\ln 6 + \ln x$

l. $\frac{10}{\sqrt{6} - \sqrt{2}}$

m. $2x - 8 + 6y - 14$

n. $(3, -2)$ and $(0, -5)$

o. $x^4 \cdot x^3$

p. $\frac{4x^2 y}{16xy}$

q. $h(x) = 10^x$

r. $\log_2 8 = 3$

s. $m = \frac{2}{3}$ and passes through $(0, 2)$

t. $2, 6, 18, \ldots$

u. $3y^3 - 243b^6$

v. $x + 7 \geq 0$ and $-x < -1$

w. $x^2 - 3x - 4 = 0$

x. $Rt = cd + 2t$

y. $-2\pi a^2 - 3b^3$

z. $x^2 + 4x$

CUMULATIVE REVIEW Chapters 1–11

1. Give the elements of the set $\left\{-\frac{4}{3}, \pi, 5.6, \sqrt{2}, 0, -23, e, 7i\right\}$ that belong to each of the following sets. [Section 1.2]

   a. Whole numbers

   b. Rational numbers

   c. Irrational numbers

   d. Real numbers

2. Solve: $6[x - (2 - x)] = -4(8x + 3)$ [Section 1.5]

3. Solve $A = \frac{1}{2}h(b_1 + b_2)$ for b_2. [Section 1.6]

4. **Martial Arts.** Find the measure of each angle of the triangle shown in the illustration. [Section 1.7]

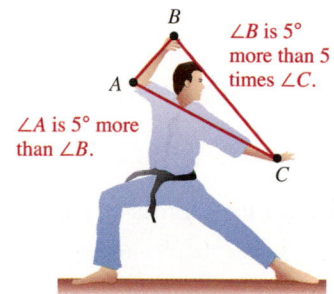

∠B is 5° more than 5 times ∠C.

∠A is 5° more than ∠B.

5. **Financial Planning.** Anna has some money to invest. Her financial planner tells her that if she can come up with $3,000 more, she will qualify for an 11% annual interest rate. Otherwise, she will have to invest the money at 7.5% annual interest. The financial planner urges her to invest the larger amount, because the 11% investment would yield twice as much annual income as the 7.5% investment. How much does she originally have on hand to invest? [Section 1.8]

6. **Boating.** Use the following graph to determine the average rate of change in the sound level of the engine of a boat in relation to the number of revolutions per minute (rpm) of the engine. [Section 2.3]

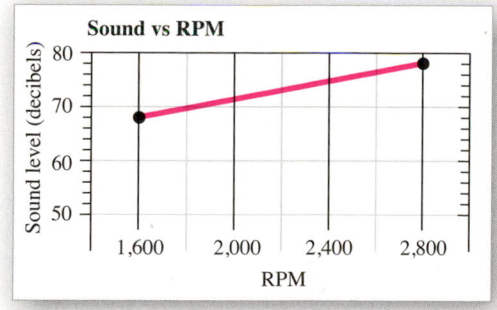

Sound vs RPM

7. Decide whether the graphs of the equations are parallel or perpendicular. [Section 2.3]

   a. $3x - 4y = 12,\ y = \frac{3}{4}x - 5$

   b. $y = 3x + 4,\ x = -3y + 4$

8. **Salvage Values.** A truck was purchased for $28,000. Its salvage value at the end of 6 years is expected to be $7,600. Find the straight-line depreciation equation. [Section 2.4]

Find an equation of the line with the given properties. Write the equation in slope–intercept form.

9. $m = -2$, passing through $(0, 5)$ [Section 2.4]

10. Passing through $(8, -5)$ and $(-5, 4)$ [Section 2.4]

11. If $f(x) = 3x^5 - 2x^2 + 1$, find $f(-1)$ and $f(a)$. [Section 2.5]

12. **Union Membership.** The percent of private-sector workers in the U.S. belonging to unions can be modeled by a linear function. In 1995, 10.3% of the private-sector workforce were union members. By the year 2000, that number had decreased to about 9%. (Source: unionstats.gsu.edu) [Section 2.5]

    a. Let t be the number of years after 1990 and P be the percent of the private-sector workforce that are union members. Write a linear function $P(t)$ to model the situation.

    b. Predict the percent of private-sector workers who will belong to a union in 2030, if the trend continues.

13. Explain why the graph does not represent a function. [Section 2.6]

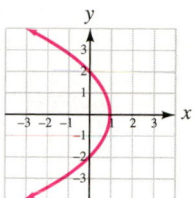

14. Use the graph of function h to find each of the following. [Section 2.6]

    a. $h(-3)$

    b. $h(4)$

    c. The value(s) of x for which $h(x) = 1$

    d. The value(s) of x for which $h(x) = 0$

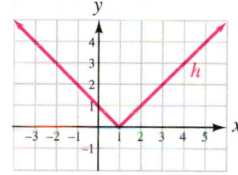

15. Consider the function $g(x) = (x - 6)^2$. First sketch the graph of its associated function. Then draw the graph of function g using a translation. Give the domain and range of g. [Section 2.6]

16. Solve the system $\begin{cases} 2x - y = 6 \\ y = -x \end{cases}$ by graphing. [Section 3.1]

17. Use substitution to solve: $\begin{cases} 2x - y = -21 \\ 4x + 5y = 7 \end{cases}$ [Section 3.2]

18. Use addition to solve: $\begin{cases} 4y + 5x - 7 = 0 \\ \dfrac{10}{7}x - \dfrac{4}{9}y = \dfrac{17}{21} \end{cases}$ [Section 3.2]

19. Solve: $\begin{cases} b + 2c = 7 - a \\ a + c = 8 - 2b \\ 2a + b + c = 9 \end{cases}$ [Section 3.3]

20. Use matrices to solve the system: $\begin{cases} 2x + y = 1 \\ x + 2y = -4 \end{cases}$ [Section 3.4]

21. Use Cramer's rule to solve: $\begin{cases} 2(x + y) + 1 = 0 \\ 3x + 4y = 0 \end{cases}$
[Section 3.5]

22. **Mixing Coffee.** How many pounds of regular coffee (selling for \$4 per pound) and how many pounds of Brazilian coffee (selling for \$11.50 per pound) must be combined to get 40 pounds of a mixture worth \$6 per pound? [Section 3.6]

23. **Aviation.** Flying with a tail wind, a passenger plane flew 1,000 miles in 2 hours. The return flight against the wind took 2.5 hours. Find the speed of the wind and the speed of the plane in still air. [Section 3.6]

24. **Sports Socks.** A company makes three types of high-end athletic socks: "ankle," "low cut," and "crew." The cost to make one pair of ankle socks is \$2 and they sell for \$3. The cost to make one pair of low cut socks is \$3 and they sell for \$5. The cost to make one pair of crew socks is \$4 and they sell for \$6. Each day, the cost of manufacturing 500 pairs of socks is \$1,650, and the daily revenue from their sale is \$2,550. How many pairs of each type of sock are manufactured daily? [Section 3.7]

Give the solution in interval notation and graph the solution set.

25. Solve: $4.5x - 1 < -10$ or $6 - 2x \geq 12$
[Section 4.2]

26. Solve: $|5 - 3x| - 14 \leq 0$ [Section 4.3]

27. Graph the solution set: $\begin{cases} 3x - 2y \leq 6 \\ y < -x + 2 \end{cases}$ [Section 4.5]

28. **Work Schedules.** A student works two part-time jobs. She earns \$12 an hour for working at the college book store and \$22.50 an hour working graveyard shift at an airport. To save time for study, she limits her work to 25 hours a week. If she enjoys the work at the bookstore more, how many hours can she work at the bookstore and still earn at least \$450 a week? [Section 4.1]

Simplify each expression. Write answers using positive exponents.

29. $(x^2)^5 y^7 y^3 x^{-2} y^0$

30. $\left(\dfrac{3x^5 y^2}{6x^5 y^{-2}} \right)^{-4}$

[Section 5.1]   [Section 5.1]

31. Write 173,000,000,000,000 and 0.000000046 in scientific notation. [Section 5.2]

32. Write each number in scientific notation and perform the indicated operations. Give the answer in scientific notation and in standard notation. [Section 5.2]

$$\frac{(0.00024)(96,000,000)}{(640,000,000)(0.025)}$$

33. **Produce.** The polynomial function $n(c) = \frac{1}{3}c^3 + \frac{1}{2}c^2 + \frac{1}{6}c$ gives the number of cantaloupes used in a display shaped like a square pyramid, having a base formed by c cantaloupes per side. Find the number of cantaloupes needed to make the display shown in the illustration in the next column. [Section 5.3]

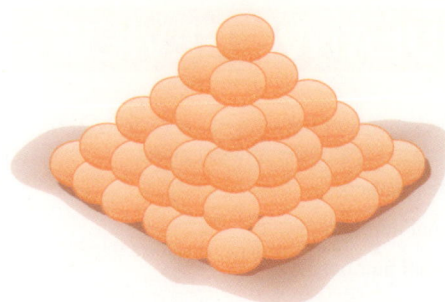

34. Simplify the polynomial: $\frac{9}{4}rt^2 - \frac{5}{3}rt - \frac{1}{2}rt^2 + \frac{5}{6}rt$
[Section 5.3]

Perform the indicated operations and simplify, if possible.

35. $(-2x^2y^3 + 6xy + 5y^2) - (-4x^2y^3 - 7xy + 2y^2)$
[Section 5.3]

36. $(x - 3y)(x^2 + 3xy + 9y^2)$ [Section 5.4]

37. $(2m^5 - 7)(3m^5 - 1)$ [Section 5.4]

38. $(9ab^2 - 4)^2$ [Section 5.4]

Factor the expression completely.

39. $3x^3y - 4x^2y^2 - 6x^2y + 8xy^2$ [Section 5.5]

40. $b^3 - 4b^2 - 3b + 12$ [Section 5.5]

41. $12y^2 + 23y + 10$ [Section 5.6]

42. $256x^4y^4 - z^8$ [Section 5.7]

43. $27t^3 + u^3$ [Section 5.7]

44. $a^4b^2 - 20a^2b^2 + 64b^2$ [Section 5.8]

45. Solve for λ: $\dfrac{A\lambda}{2} + 1 = 2d + 3\lambda$ [Section 5.5]

46. Use a check to determine whether $-\frac{5}{7}$ is a solution of $7t^2 - 2t - 5 = 0$. [Section 5.9]

47. Solve: $(x + 7)^2 = -2(x + 7) - 1$ [Section 5.9]

48. Solve: $x^3 + 8x^2 = 9x$ [Section 5.9]

49. **Painting.** When it is spread out, a rectangular painting tarp covers an area of 84 square feet. Its length is 1 foot longer than five times its width. Find its width and length. [Section 5.9]

50. **Projectiles.** The height (in feet) of an object thrown upward from the roof of an apartment building is given by the function $h(t) = -16t^2 + 64t + 80$, where t is the number of seconds since it was thrown. How long will it take for the object to hit the ground? [Section 5.9]

51. Find the domain of the rational function $f(x) = \dfrac{2x^2 - 3x - 2}{x^2 + 2x - 24}$.
Write the domain using interval notation. [Section 6.1]

52. Simplify: $\dfrac{6x^2 + 13x + 6}{6x^2 + 5x - 6}$ [Section 6.1]

Perform the indicated operations and simplify, if possible.

53. $\dfrac{p^3 - q^3}{q^2 - p^2} \cdot \dfrac{q^2 + pq}{p^3 + p^2q + pq^2}$ [Section 6.2]

54. $\dfrac{2}{a-2} + \dfrac{3}{a+2} - \dfrac{a-1}{a^2-4}$ [Section 6.3]

55. $\dfrac{\dfrac{y}{x} - \dfrac{x}{y}}{\dfrac{1}{x} + \dfrac{1}{y}}$ [Section 6.4]

56. $(16x^4 + 3x^2 + 13x + 3) \div (4x + 3)$ [Section 6.5]

57. Solve: $\dfrac{1}{a+5} = \dfrac{1}{3a+6} - \dfrac{a+2}{a^2+7a+10}$ [Section 6.7]

58. Solve $\dfrac{1}{R} = \dfrac{1}{R_1} + \dfrac{1}{R_2} + \dfrac{1}{R_3}$ for R. [Section 6.7]

59. Printing Paychecks. It takes a printer 6 hours to print the payroll checks for all of the employees of a large company. A faster printer can print the paychecks in 4 hours. How long will it take the two printers working together to print all of the paychecks? [Section 6.8]

60. Capture-Release Method. To estimate the ground squirrel population on his acreage, a farmer trapped, tagged, and then released two dozen squirrels. Two weeks later, the farmer trapped 31 squirrels and noted that 8 were tagged. Use this information to estimate the number of ground squirrels on his acreage. [Section 6.9]

61. Light. The intensity of a light source is inversely proportional to the square of the distance from the source. If the intensity is 18 lumens at a distance of 4 feet, what is the intensity when the distance is 12 feet? [Section 6.9]

62. Graph: $f(x) = \sqrt{x} + 2$. Give the domain and range of the function. [Section 7.1]

63. Find the domain of $f(x) = \sqrt{x+3}$. [Section 7.1]

64. Aquariums. The function $L(g) = \sqrt[3]{\dfrac{g}{7.5}}$ gives the length (in feet) of an edge of a cube-shaped tank that holds g gallons of water. What length of an edge should a cube-shaped aquarium have if it is to hold 480 gallons of water? [Section 7.1]

65. Evaluate: $\left(\dfrac{25}{49}\right)^{-3/2}$ [Section 7.2]

66. Simplify: $\sqrt{112a^3b^5}$ [Section 7.3]

Simplify each expression. All variables represent positive numbers.

67. $\sqrt{98} + \sqrt{8} - \sqrt{32}$ [Section 7.3]

68. $12\sqrt[3]{648x^4} + 3\sqrt[3]{81x^4}$ [Section 7.3]

69. $\left(2\sqrt{7} + 1\right)\left(\sqrt{7} - 1\right)$ [Section 7.4]

70. $3\left(\sqrt{5x} - \sqrt{3}\right)^2$ [Section 7.4]

Rationalize each denominator.

71. $\dfrac{\sqrt[3]{4}}{\sqrt[3]{b}}$ [Section 7.4]

72. $\dfrac{3t-1}{\sqrt{3t}+1}$ [Section 7.4]

Solve each equation and check the result.

73. $2x = \sqrt{16x - 12}$ [Section 7.5]

74. $\sqrt[3]{12m + 4} = 4$ [Section 7.5]

75. $\sqrt{x+3} - \sqrt{3} = \sqrt{x}$ [Section 7.5]

76. Accessories. A silk scarf has sides 16 inches long. When folded over itself, as shown below, the result is a triangular scarf that can be wrapped around one's neck. Find the length of the longest side. [Section 7.6]

77. Express $\sqrt{-25}$ in terms of i. [Section 7.7]

78. Simplify: i^{42} [Section 7.7]

Write each expression in a + bi form.

79. $(-7 + 9i) - (-2 - 8i)$ [Section 7.7]

80. $\dfrac{2-5i}{2+5i}$ [Section 7.7]

81. Solve: $t^2 = 24$ [Section 8.1]

82. Solve $m^2 + 10m - 7 = 0$ by completing the square. [Section 8.1]

Solve each equation.

83. $4w^2 + 6w + 1 = 0$ [Section 8.2]

84. $3x^2 - 4x = -2$ [Section 8.2]

85. $2(2x + 1)^2 - 7(2x + 1) + 6 = 0$ [Section 8.3]

86. $x^4 + 19x^2 + 18 = 0$ [Section 8.3]

87. Tire Wear. Refer to the graph below. [Section 8.4]

a. What type of function does it appear would model the relationship between the inflation of a tire and the percent of service it gives?

b. At what percent(s) of inflation will a tire offer only 90% of its possible service?

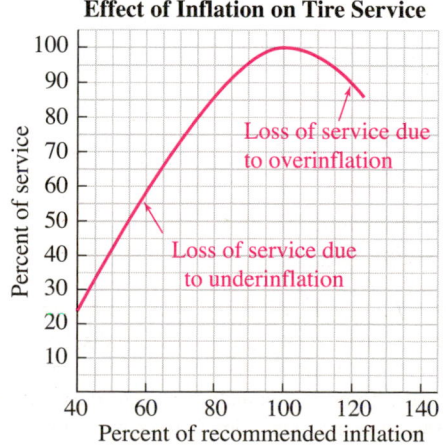

Effect of Inflation on Tire Service

88. Graph $f(x) = -6x^2 - 12x - 8$ using the vertex formula. Then determine the x- and y-intercepts of the graph. Finally, plot several points and complete the graph. [Section 8.4]

89. Solve $x^2 - 8x \leq -15$. Write the solution set in interval notation and graph it. [Section 8.5]

90. If $f(x) = x^2 - 2$ and $g(x) = 2x + 1$, find $(f \circ g)(x)$. [Section 9.1]

91. Find the inverse function of $f(x) = 2x^3 - 1$. [Section 9.2]

92. Graph $f(x) = \left(\frac{1}{2}\right)^x$ and give the domain and range of the function. [Section 9.3]

Find x.

93. $\log 1,000 = x$ [Section 9.4]   94. $\log_8 64 = x$ [Section 9.4]

95. $\log_3 x = -3$ [Section 9.4]   96. $\log_x 25 = 2$ [Section 9.4]

97. $\ln e = x$ [Section 9.5]   98. $\ln \frac{1}{e} = x$ [Section 9.5]

99. Graph $f(x) = e^x$ and its inverse on the same coordinate system. [Section 9.5]

100. **Population Growth.** As of 2010, the population of Mexico was about 114 million and the annual growth rate was 1.102%. If the growth rate remains the same, estimate the population of Mexico in 25 years. (Source: CIA World Factbook) [Section 9.5]

101. Find $\ln 0$, if possible. [Section 9.5]

102. Use the properties of logarithms to simplify $\log_6 \frac{36}{x^3}$. [Section 9.6]

103. Write the expression $\frac{1}{2} \ln x + \ln y - \ln z$ as a single logarithm. [Section 9.6]

104. **Bacteria Growth.** The bacteria in a laboratory culture increased from an initial population of 200 to 600 in 4 hours. How long will it take the population to reach 8,000? [Section 9.7]

Solve each equation. Round to four decimal places when necessary.

105. $5^{4x} = \frac{1}{125}$ [Section 9.7]

106. $2^{x+2} = 3^x$ [Section 9.7]

107. $\log x + \log(x + 9) = 1$ [Section 9.7]

108. $\log_3 x = \log_3 \left(\frac{1}{x}\right) + 4$ [Section 9.7]

109. Match each function to its graph shown below.

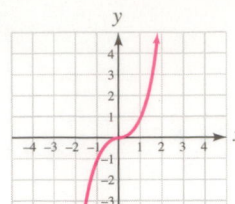

a. $f(x) = \frac{1}{x}$ [Section 6.1]   b. $f(x) = \log x$ [Section 9.4]

c. $f(x) = x^3$ [Section 2.6]   d. $f(x) = \sqrt[3]{x}$ [Section 7.1]

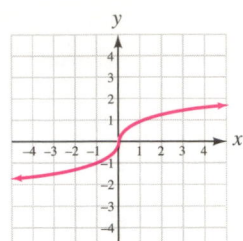
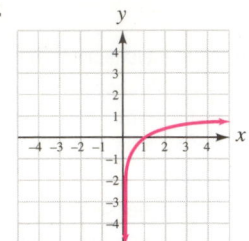

i.   ii.

iii.   iv.

110. Write the equation of the circle that has its center at $(1, -3)$ and a radius of 2. Graph the equation. [Section 10.1]

111. Complete the square to write the equation $y^2 + 4x - 6y = -1$ in $x = a(y - k)^2 + h$ form. Determine the vertex and the axis of symmetry of the graph. Then plot several points and complete the graph. [Section 10.1]

112. Graph: $\frac{(x + 1)^2}{4} + \frac{(y - 3)^2}{16} = 1$ [Section 10.2]

113. Write the equation in standard form and graph it: $(x - 2)^2 - 9y^2 = 9$ [Section 10.3]

114. Use the binomial theorem to expand $(3a - b)^4$. [Section 11.1]

115. Evaluate: $\frac{12!}{10!(12 - 10)!}$ [Section 11.1]

116. Find the 20th term of an arithmetic sequence with a first term -11 and a common difference 6. [Section 11.2]

117. Find the sum of the first 20 terms of an arithmetic sequence with a first term 6 and a common difference 3. [Section 11.2]

118. Evaluate: $\sum_{k=3}^{5} (2k + 1)$ [Section 11.2]

119. Find the seventh term of a geometric sequence with a first term $\frac{1}{27}$ and a common ratio 3. [Section 11.3]

120. **Boat Depreciation.** How much will a $9,000 boat be worth after 9 years if it depreciates 12% per year? [Section 11.3]

121. Find the sum of the first ten terms of the sequence: $\frac{1}{64}, \frac{1}{32}, \frac{1}{16}, \ldots$ [Section 11.3]

122. Find the sum of the infinite series: $9 + 3 + 1 + \cdots$ [Section 11.3]

n	n^2	$\sqrt{n}$	n^3	$\sqrt[3]{n}$	n	n^2	$\sqrt{n}$	n^3	$\sqrt[3]{n}$
1	1	1.000	1	1.000	51	2,601	7.141	132,651	3.708
2	4	1.414	8	1.260	52	2,704	7.211	140,608	3.733
3	9	1.732	27	1.442	53	2,809	7.280	148,877	3.756
4	16	2.000	64	1.587	54	2,916	7.348	157,464	3.780
5	25	2.236	125	1.710	55	3,025	7.416	166,375	3.803
6	36	2.449	216	1.817	56	3,136	7.483	175,616	3.826
7	49	2.646	343	1.913	57	3,249	7.550	185,193	3.849
8	64	2.828	512	2.000	58	3,364	7.616	195,112	3.871
9	81	3.000	729	2.080	59	3,481	7.681	205,379	3.893
10	100	3.162	1,000	2.154	60	3,600	7.746	216,000	3.915
11	121	3.317	1,331	2.224	61	3,721	7.810	226,981	3.936
12	144	3.464	1,728	2.289	62	3,844	7.874	238,328	3.958
13	169	3.606	2,197	2.351	63	3,969	7.937	250,047	3.979
14	196	3.742	2,744	2.410	64	4,096	8.000	262,144	4.000
15	225	3.873	3,375	2.466	65	4,225	8.062	274,625	4.021
16	256	4.000	4,096	2.520	66	4,356	8.124	287,496	4.041
17	289	4.123	4,913	2.571	67	4,489	8.185	300,763	4.062
18	324	4.243	5,832	2.621	68	4,624	8.246	314,432	4.082
19	361	4.359	6,859	2.668	69	4,761	8.307	328,509	4.102
20	400	4.472	8,000	2.714	70	4,900	8.367	343,000	4.121
21	441	4.583	9,261	2.759	71	5,041	8.426	357,911	4.141
22	484	4.690	10,648	2.802	72	5,184	8.485	373,248	4.160
23	529	4.796	12,167	2.844	73	5,329	8.544	389,017	4.179
24	576	4.899	13,824	2.884	74	5,476	8.602	405,224	4.198
25	625	5.000	15,625	2.924	75	5,625	8.660	421,875	4.217
26	676	5.099	17,576	2.962	76	5,776	8.718	438,976	4.236
27	729	5.196	19,683	3.000	77	5,929	8.775	456,533	4.254
28	784	5.292	21,952	3.037	78	6,084	8.832	474,552	4.273
29	841	5.385	24,389	3.072	79	6,241	8.888	493,039	4.291
30	900	5.477	27,000	3.107	80	6,400	8.944	512,000	4.309
31	961	5.568	29,791	3.141	81	6,561	9.000	531,441	4.327
32	1,024	5.657	32,768	3.175	82	6,724	9.055	551,368	4.344
33	1,089	5.745	35,937	3.208	83	6,889	9.110	571,787	4.362
34	1,156	5.831	39,304	3.240	84	7,056	9.165	592,704	4.380
35	1,225	5.916	42,875	3.271	85	7,225	9.220	614,125	4.397
36	1,296	6.000	46,656	3.302	86	7,396	9.274	636,056	4.414
37	1,369	6.083	50,653	3.332	87	7,569	9.327	658,503	4.431
38	1,444	6.164	54,872	3.362	88	7,744	9.381	681,472	4.448
39	1,521	6.245	59,319	3.391	89	7,921	9.434	704,969	4.465
40	1,600	6.325	64,000	3.420	90	8,100	9.487	729,000	4.481
41	1,681	6.403	68,921	3.448	91	8,281	9.539	753,571	4.498
42	1,764	6.481	74,088	3.476	92	8,464	9.592	778,688	4.514
43	1,849	6.557	79,507	3.503	93	8,649	9.644	804,357	4.531
44	1,936	6.633	85,184	3.530	94	8,836	9.695	830,584	4.547
45	2,025	6.708	91,125	3.557	95	9,025	9.747	857,375	4.563
46	2,116	6.782	97,336	3.583	96	9,216	9.798	884,736	4.579
47	2,209	6.856	103,823	3.609	97	9,409	9.849	912,673	4.595
48	2,304	6.928	110,592	3.634	98	9,604	9.899	941,192	4.610
49	2,401	7.000	117,649	3.659	99	9,801	9.950	970,299	4.626
50	2,500	7.071	125,000	3.684	100	10,000	10.000	1,000,000	4.642

ARE YOU READY? 1.1 (page 2)

1. a. Addition   **b.** Subtraction   **c.** Multiplication   **d.** Division
2. 20   **3.** 32   **4.** 11

SELF CHECKS 1.1

1. a. $d - 16$ **b.** $0.09p$ **c.** $3h + 44$ **d.** $\frac{1}{4}(x - 1)$
2. $I = prt$ **3.** $S = \frac{p}{3} - 2,000$ **4.**

h	c
6	800
7	900

STUDY SET SECTION 1.1 (page 7)

1. variable   **3.** equation   **5.** addition, subtraction
7. a. Expression   **b.** Equation   **c.** Expression   **d.** Expression
9. a. $7d = h$ (Answers may vary.)   **b.** $t = 2,500 - d$ (Answers may vary.)   **11.** $t - 7$ **13.** $0.54e$ **15.** $2p + 35$ **17.** $\frac{1}{2}(x + 4)$
19. $\frac{3}{4}p$ **21.** $\frac{w}{p}$ **23.** $950 + 0.1v$ **25.** $w - 500$ **27.** $\frac{150}{m}$
29. $7(77 + h + 88)$ **31.** $3w$ **33.** $4d - 15$ **35.** $0.95(200 + t)$
37. $0.01d$ or $\frac{1}{100}d$ **39.** $c = 13u + 24$ **41.** $w = \frac{c}{75}$
43. $A = t + 15$ **45.** $c = 12b$ **47.** $h = \frac{t}{2} - 75$
49. $b = 300 - 6s$ **51.** 2, 6, 15   **53.** 22.44, 21.43, 0
55. a. $S - s$ **b.** $s - S$ **57. a.** $|a - 2|$ **b.** $|a| - 2$
59. a. $0.155a + 6$ **b.** $0.155(a + 6)$ **61. a.** $(x - 14)^2$
b. $x^2 - 14$ **63.** 2, 4   **65.** $C = I - (m + a)$; answers may vary due to the variables chosen.   **67.** $b = t - 10$

ARE YOU READY? 1.2 (page 9)

1. 34   **2.** $-27°$ **3.** A decimal (3.6 GPA)   **4.** A fraction $\left(\frac{3}{4}\right.$ cup of flour$\left.\right)$

SELF CHECKS 1.2

1. a. True   **b.** False   **2.** Natural numbers: 1; whole numbers: 1; integers: -5, 1; rational numbers: -5, 3.4, 1, $\frac{16}{5}$, $9.\overline{7}$; irrational numbers: $-\pi$, $\sqrt{19}$; real numbers: all
3.

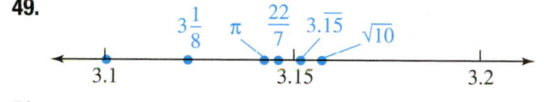

4. a. $\le$ **b.** $\ge$
5. a. 9.6   **b.** -12 **c.** $\frac{3}{2}$

STUDY SET SECTION 1.2 (page 17)

1. whole, natural, integers   **3.** prime, composite   **5.** Irrational
7. inequality   **9.** -8, 4   **11.** Nonrepeating, irrational
13. Repeating, rational

15.

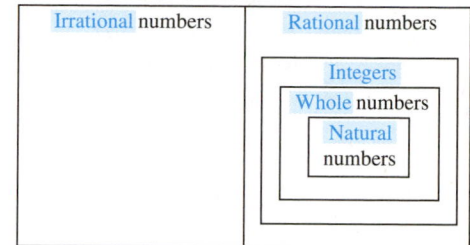

17. a. $12 < 19$ **b.** $-5 \ge -6$ **19.** is less than; is greater than or equal to   **21.** braces   **23.** $\left\{\frac{a}{b} \mid a \text{ and } b \text{ are integers, with } b \ne 0.\right\}$
25. a. Rational numbers   **b.** Irrational numbers
c. Real numbers   **27.** True   **29.** False   **31.** False   **33.** True
35. 1, 2, 9   **37.** -3, 0, 1, 2, 9   **39.** $\sqrt{3}$, π **41.** 2   **43.** 2   **45.** 9
47.

49.

51.

53.

55.

57.

59. $<$ **61.** $>$ **63.** $>$ **65.** $<$ **67.** $>$ **69.** $<$ **71.** $<$ **73.** $>$
75. 20   **77.** 5.9   **79.** -6 **81.** $-\frac{9}{4}$ **83.** $3\frac{2}{25} = 3.0800$
$\frac{77}{50} = 1.5400$ $\frac{15}{16} = 0.9375$ $2\frac{5}{8} = 2.6250$ $\frac{\pi}{4} \approx 0.7854$ $\sqrt{8} \approx 2.8284$
85. a. Georgia (1.3 m/yr)   **b.** Louisiana (-10.1 m/yr)
91. Expression   **93.** variable

ARE YOU READY? 1.3 (page 20)

1. 4, 6; -6 has the larger absolute value.   **2.** 12.17   **3.** 11
4. $\frac{5}{18}$ **5.** $\frac{4}{9}$ **6.** 8.5

SELF CHECKS 1.3

1. a. -9 **b.** -91.6 **c.** $\frac{1}{4}$ **d.** -4 **2. a.** -19 **b.** -4.5
c. $-\frac{2}{9}$ **d.** -6 **e.** 16   **3. a.** -30 **b.** 32.8   **c.** $-\frac{1}{6}$ **d.** -108
4. a. -11 **b.** 1.2   **5. a.** $-\frac{21}{16}$ **b.** $\frac{1}{50}$ **6. a.** -27 **b.** 0.64
c. 16   **d.** $\frac{49}{25}$ **7. a.** 6   **b.** -10 **c.** $\frac{2}{5}$ **d.** 1   **e.** 0.9   **f.** -20

8. a. 23   **b.** -46 **9. a.** -32 **b.** 719   **c.** 14   **10. a.** 125   **b.** $-\frac{5}{8}$

STUDY SET SECTION 1.3 (page 30)

1. sum, difference   **3.** reciprocal   **5.** squared, cubed
7. parentheses, outermost   **9. a.** multiplication   **b.** opposite
11. a. Negative   **b.** Negative   **c.** Positive   **d.** Negative
13. a. $(-4)^2 = 16$ **b.** $-4^2 = -16$ **15.** -8 **17.** -4.3
19. -13 **21.** $\frac{1}{6}$ **23.** -7 **25.** 0   **27.** $\frac{11}{10}$ **29.** -2
31. -12 **33.** -1.5 **35.** 60   **37.** $-\frac{6}{7}$ **39.** -2 **41.** -14
43. 9   **45.** $\frac{24}{25}$ **47.** 1,296   **49.** 62.41   **51.** -25 **53.** $-\frac{27}{125}$
55. 8   **57.** -9 **59.** $-\frac{3}{4}$ **61.** 0.2   **63.** -17 **65.** 32
67. -8 **69.** -32 **71.** 10,000   **73.** 64   **75.** -8 **77.** 13
79. -114 **81.** -32 **83.** $-\frac{1}{18}$ **85.** $\frac{1}{2}$ **87.** 5   **89.** $-\frac{1}{2}$
91. -24 **93.** -2 **95.** 61   **97.** 1   **99.** 10   **101.** 4
103. a. 85   **b.** 75   **c.** 25   **d.** 1   **105. a.** -7.3 **b.** 7.3
107. There was a net inflow of $328 billion.   **109.** New York, Atlanta, Boise, Omaha, Helena   **111.** 9   **115.** -7 and 3
117. $\{\ldots, -2, -1, 0, 1, 2, \ldots\}$ **119.** True

ARE YOU READY? 1.4 (page 32)

1. The factors are in a different order.   **2.** The position of the parentheses is different.   **3.** 42, 42; same result   **4.** Different variable factors (x and y); the same first factor (8)

SELF CHECKS 1.4

1. a. 23   **b.** $\left(-16 \cdot \frac{1}{2}\right) \cdot 7$ **2. a.** $42s$ **b.** $-4.8bt$ **c.** $6x$
3. a. $9r + 36$ **b.** $33x + 55$ **c.** $27k - 15$
4. a. $-4.8t + 2.4s - 8$ **b.** $6a + \frac{48}{5}b$ **5. a.** $13k$ **b.** $300a^2$
c. $-\frac{1}{12}xy$ **d.** $12d^2 - 18c$ **6. a.** $24a^3 + a^2$ **b.** $10r^2 + 11r$
7. $-48t + 4$

STUDY SET SECTION 1.4 (page 40)

1. term   **3.** constant   **5.** commutative, associative   **7.** Like
9. a. $(x + y) + z = x + (y + z)$ **b.** $xy = yx$
c. $r(s + t) = rs + rt$ **11. a.** a, Commutative Property of Multiplication   **b.** $a(bc)$, Associative Property of Multiplication
c. 0, Multiplicative Property of 0   **d.** a, Identity Property of Multiplication   **e.** 1, Multiplicative Inverse Property   **13. a.** 0
b. 1   **c.** $-x$ **d.** $\frac{1}{x}$ **15. a.** 5   **b.** $\frac{1}{5}$ **17.** Multiplication by -1
19. $3x^3$, $11x^2$, $-x$, 9; 3, 11, -1, 9   **21.** $\frac{11}{12}a^4$, $-\frac{3}{4}b^2$, $25b$; $\frac{11}{12}$, $-\frac{3}{4}$, 25
23. $7 + 3$ **25.** $3 \cdot 2 + 3d$ **27.** c **29.** 1   **31.** $(8 + 7) + a$
33. $2(x + y)$ **35.** $72m$ **37.** $-45q$ **39.** $-49x$ **41.** $64ry$
43. $81x + 18$ **45.** $12t - 12$ **47.** $-24 + d$ or $d - 24$
49. $2s^2 - 6$ **51.** $0.7m + 1.4n$ **53.** $9x + 2y$
55. $45t^2 - 60t - 15$ **57.** $4x - 5y + 1$ **59.** $2t + 3$
61. $-3y + 6$ **63.** $18x$ **65.** $-3.1h$ **67.** $0.8x^2$ **69.** $-2x$
71. $\frac{9}{10}ab$ **73.** $\frac{14}{15}t$ **75.** $12ad - 44a$ **77.** $6m + 2t$
79. $-2x^2 + 15x$ **81.** $-6p + 17$ **83.** $8x - 9$ **85.** $17y - 27$
87. $-4x + 87$ **89.** $56b + 6$ **91.** $-2a - A - 3$
93. $14cd + 62c$ **95.** $6.4a^2 + 2.6a + 5.7$ **97.** $-\frac{19}{16}x$
99. $14z - 5$ **101.** $18h^2 - 8h$ **103.** $-3.8y + 38.7$ **105.** 0
107. $26a + 64$ **109.** $-52a + 8$ **111. a.** $480n$ **b.** $96n + 60$
113. a. $18a$ **b.** Does not simplify   **115. a.** $20(x + 6)$ m^2
b. $(20x + 120)$ m^2 **c.** $20(x + 6) = 20x + 120$,
Distributive Property   **121.** $-\frac{7}{8}$ **123.** 988

ARE YOU READY? 1.5 (page 43)

1. -6 **2.** $10a$ **3.** $-11m + 8$ **4.** x **5.** $18n$ **6.** 27

SELF CHECKS 1.5

1. No   **2. a.** -5 **b.** -1.3 **3.** $-\frac{45}{2}$ **4.** -6 **5. a.** -5
b. $-\frac{4}{5}$ **6.** -6 **7.** -2 **8.** 6,000   **9.** All real numbers, $\mathbb{R}$
10. No solution, $\varnothing$

STUDY SET SECTION 1.5 (page 52)

1. equation   **3.** solution   **5.** identity   **7.** c, c, both
9. a. 3   **b.** 9   **c.** 2   **d.** 18   **11. a.** All real numbers; $\mathbb{R}$
b. No solution; $\varnothing$ **13.** $-2x$, 14, 14, -2, -2, -17, -10, $\overset{?}{=}$, 20, -17 **15.** Yes   **17.** No   **19.** 6   **21.** $\frac{15}{8}$ **23.** 28   **25.** 2.52
27. -0.26 **29.** 7   **31.** 15   **33.** $-\frac{5}{2}$ **35.** -16 **37.** 18
39. -11 **41.** 1   **43.** 1.7   **45.** $\frac{17}{4}$ **47.** -8 **49.** -11 **51.** 4
53. 2   **55.** 30   **57.** $-\frac{1}{2}$ **59.** 0   **61.** -2 **63.** 3   **65.** 24
67. All real numbers, $\mathbb{R}$; identity   **69.** All real numbers, $\mathbb{R}$; identity   **71.** No solution, $\varnothing$; contradiction   **73.** No solution, $\varnothing$;
contradiction   **75.** 63   **77.** 0   **79.** 29   **81.** 24   **83.** $\frac{21}{5}$
85. -0.5 **87.** 13   **89.** 6   **91.** -15 **93.** 1,000   **95.** $\frac{9}{13}$
97. No solution, $\varnothing$; contradiction   **99.** -1.2 **101.** -5 **103.** $\frac{28}{57}$
105. -1 **107. a.** $-x$ **b.** $\frac{12}{7}$ **109. a.** $102x - 12$ **b.** $\frac{6}{5}$
115. a. $a + b = b + a$ **b.** $(ab)c = a(bc)$
c. $a(b + c) = ab + ac$ **117. a.** $0 + a = a$ **b.** $1 \cdot a = a$

ARE YOU READY? 1.6 (page 55)

1. Volume   **2.** Circumference   **3.** Perimeter   **4.** Area

SELF CHECKS 1.6

1. 230 ft   **2.** 12 yd   **3.** 2.36 in.; 0.44 in.^{2}
4. $2,250\pi$ mm$^3 \approx 7,069$ mm^3 **5. a.** $r = \frac{I}{Pt}$ **b.** $f = \frac{2S - nl}{n}$
c. $T_c = T_h - ET_h$ **d.** $g = \frac{W}{h - 3t^2}$ **6.** $y = -\frac{6}{5}x + 7$ **7.** 100°C

STUDY SET SECTION 1.6 (page 61)

1. formula   **3.** volume   **5. a.** Area; ft^2 **b.** Volume; ft^3
c. Circumference; ft   **d.** Perimeter; ft   **7.** 2, 2, 4
9. ad, ad, bc, b, b, c, $\frac{t - ad}{b}$ **11.** 8 yd   **13.** 37 in.   **15.** 16.5 in.
17. 6.5 yd   **19.** 10.2 ft^2 **21.** 295.84 mi^2 **23.** 2.5 ft   **25.** 12 ft
27. 23.56 in.   **29** 15.71 ft   **31.** 102.1 in.^{2} **33.** 86.6 ft^2
35. 95.08 ft^3 **37.** 808.86 m^3 **39.** $t = \frac{d}{r}$ **41.** $h = \frac{V}{lw}$
43. $h = \frac{3V}{\pi r^2}$ **45.** $W = T - ma$ **47.** $a = \frac{2h - 96t}{t^2}$ or
$a = \frac{2(h - 48t)}{t^2}$ **49.** $b_2 = \frac{2A - b_1h}{h}$ or $b_2 = \frac{2A}{h} - b_1$
51. $n = \frac{l - a + d}{d}$ or $n = \frac{l - a}{d} + 1$ **53.** $w = \frac{P - 2h - 2l}{2}$ or
$w = \frac{P}{2} - h - l$ **55.** $A = \frac{\lambda}{x + B}$ **57.** $T_a = \frac{T_f}{1 - F}$ **59.** $d = \frac{l - a}{n - 1}$
61. $t = \frac{d_1 - d_2}{v}$ **63.** $y = \frac{2}{5}x - 4$ **65.** $y = -\frac{4}{3}x - 4$
67. $x = \frac{y - b}{m}$ **69.** $R = \frac{L - 2d - 3.25r}{3.25}$ **71.** $g = \frac{2(s - vt)}{t^2}$
73. $x = \frac{y - y_1 + mx_1}{m}$ **75.** $S = \frac{U + pV - G}{T}$ **77.** $r = \frac{PV}{nt}$
79. $R = \frac{E - Ir}{I}$ or $R = \frac{E}{I} - r$ **81.** $s_3 = 3A - s_1 - s_2$
83. $d = \frac{2S - 2an}{n(n - 1)}$ **85.** $h = \frac{3d}{4\pi}$ **87.** 216 in.   **89.** 1st term: area of
bottom flap; 2nd term: area of left and right flaps; 3rd term: area of top flap; 4th term: area of face; 42.5 in.^{2}

91. $C = \frac{5}{9}(F - 32)$ or $C = \frac{5(F-32)}{9}$; 432, −179; 58, −89; 17, −66   **93.** $d = \frac{360A}{\pi(r_1^2 - r_2^2)}$; 140, 160   **95.** $n = \frac{PV}{R(T + 273)}$; 0.008, 0.090   **97.** $n = \frac{C - 6.50}{0.07}$; 621, 1,000, about 1,692.9 kwh

99. Weight (lb) $= \frac{\text{BMI} \cdot \text{height}^2(\text{in.}^2)}{703}$ **105.** $12b + 10$

107. $-0.6pt + 11p$

ARE YOU READY? 1.7 (page 65)

1. $2x - 8$ **2.** 93   **3.** $59.50   **4.** $P = 2l + 2w$ **5.** 180°   **6.** 22

SELF CHECKS 1.7

1. FedEx: 254,142; UPS: 426,000   **2.** 3.5 mi, 2.5 mi, 11 mi
3. 30 people   **4.** iPads: 20; Kindles: 40; Nooks: 10
5. 32°, 52°, 96°   **6.** Width: 75 ft; length: 125 ft

STUDY SET SECTION 1.7 (page 71)

1. acute   **3.** complementary   **5.** right   **7.** angles
9. $d + 15, 2d - 10, \frac{d}{2} - 10$ **11.** B747: 403 passengers;
B777: 278 passengers   **13.** Base: 154 ft; statue: 151 ft   **15.** First
piece: 9 in.; middle piece: 14 in.; last piece: 6 in.   **17.** 6 in.
19. 20 students   **21.** 310 mi   **23.** 300 shares of BB,
150 shares of SS   **25.** 5,000 shares of stock funds,
7,000 shares of bond funds   **27.** 30°, 150°   **29. a.** 5.6°   **b.** 1.4°
31. $\angle 1$: 50°; $\angle 2$: 60°; $\angle 3$: 70°   **33.** 50°   **35.** 10°
37. Width: 3.6 in., height: 4.8 in.   **39.** 10 ft   **41.** 156 ft by 312 ft
43. 8 ft, 11 ft   **47.** Repeating   **49.** $\{\ldots, -2, -1, 0, 1, 2, \ldots\}$

ARE YOU READY? 1.8 (page 75)

1. a. 0.278   **b.** 83.56%   **2.** $300   **3.** 220 mi
4. 0.6 L of alcohol   **5.** $33.30   **6.** $(60,000 − x)$

SELF CHECKS 1.8

1. 16,750,000 racing video games   **2.** About 60%   **3. a.** 0.85%
decrease   **4.** $4,500 at 6%, $7,500 at 9%   **5.** 3 hr   **6.** 80 lb
7. 6 liters of 50%, 24 liters of 25%

STUDY SET SECTION 1.8 (page 83)

1. Percent   **3.** markdown   **5.** ☐ is ☐ % of ☐?
7. a. Prt **b.** rt **c.** price   **d.** strength   **9.** $x + 20$
11. $0.055x, 0.07(10,850 - x), 0.055x + 0.07(10,850 - x) = 1,205$
13. $1, 0.15x, x, 1, 0.18x$; $0.15x + 0.18x = 3,300$
15. $7.45p, 50 - p, 8.25(50 - p), 7.75(50)$
$7.75p + 8.25(50 - p) = 7.75(50)$ **17. a.** 0.025   **b.** 6%
19. $x = 0.05 \cdot 10.56$ **21.** $32.5 = 0.74x$ **23.** 511 quadrillion Btu
25. 307 accidents   **27.** 20%   **29.** $50   **31.** 9.3%   **33.** 0.3%
decrease, 1.0% decrease   **35.** CD: $10,000; Money market: $2,000
37. a. $15,000 at 7%; $30,000 at 10%   **b.** $45,000
39. $100,000   **41.** $\frac{1}{4}$ hr = 15 min   **43.** $\frac{2}{3}$ hr   **45.** 3:30 P.M.
47. $1\frac{1}{2}$ hr   **49.** 10 lb   **51.** 1.8 lb of each   **53.** 4,000 ft³ of the
premium mix; 2,000 ft³ of sawdust   **55.** 6 gal of Fruit Shake
57. 28 lb   **59.** 2 gal   **61.** 10 oz   **69.** 0   **71.** 8

CHAPTER 1 REVIEW (page 88)

1. a. $C = 2t + 15$ **b.** $l = \frac{25}{w}$ **c.** $P = u - 3$ **2.** 180, 195, 210, 225, 240   **3.** $w - 43$ **4.** $\frac{1}{3}(x - 10)$ **5. a.** 7   **b.** 0, 7

6. a. $-5, 0, 7$ **b.** $-5, 0, 2.4, 7, -\frac{2}{3}, -3.\overline{6}, \frac{15}{4}$

7. a. $-\sqrt{3}, \pi, 0.13242368\ldots$ **b.** all

8. a. $-5, -\sqrt{3}, -\frac{2}{3}, -3.\overline{6}$ **b.** $2.4, 7, \pi, \frac{15}{4}, 0.13242368\ldots$

9. a. 7   **b.** None   **10. a.** 0   **b.** −5, 7

11.

12.

13. a. False   **b.** True   **14. a.** False   **b.** True   **15. a.** $>$
b. $<$ **16. a.** False   **b.** True   **17.** 18   **18.** −6.26   **19.** −27
20. 10.1   **21.** $-\frac{3}{4}$ **22.** 2   **23.** 12.6   **24.** $-\frac{1}{32}$ **25.** 0.2
26. $-\frac{3}{56}$ **27.** −33   **28.** −5.7   **29.** −120   **30.** 1
31. −243   **32.** $\frac{4}{81}$ **33.** 0.064   **34.** −25   **35.** 2   **36.** −10
37. $\frac{3}{5}$ **38.** 0.8   **39.** 44   **40.** 1   **41.** −12   **42.** 58   **43.** 8
44. 3   **45.** 3,000   **46.** −16   **47.** 56   **48.** $-\frac{1}{2}$ **49.** $3x + 21$
50. $5t$ **51.** 0   **52.** $27 + (1 + 99)$ **53.** 1   **54.** m **55.** 1   **56.** 0
57. $-3(5 \cdot 2)$ **58.** $(z + t) \cdot t$ **59. a.** 1   **b.** −25   **60. a.** 0
b. Undefined   **61.** $72x + 48$ **62.** $-6y + 2$ **63.** $3.6x - 2.4y$
64. $6c^2 - 3c + \frac{3}{4}$ **65.** $48k$ **66.** $75xy$ **67.** $-189p$
68. $45a + 16$ **69.** 0   **70.** $3m - 40n$ **71.** $\frac{13}{20}x$ **72.** $-24.54l$
73. $40a^3 - 16a^2$ **74.** $\frac{1}{4}h + 8$ **75.** Yes   **76.** No   **77.** −225
78. 7.9   **79.** 0.014   **80.** −4   **81.** $\frac{12}{5}$ **82.** −9   **83.** −6   **84.** $\frac{11}{7}$
85. $\frac{88}{17}$ **86.** 12   **87.** 0.06   **88.** −8   **89.** 0   **90.** 3
91. No solution, $\varnothing$; contradiction   **92.** All real numbers, $\mathbb{R}$;
identity   **93.** 31 ft   **94.** 53.41 cm; 226.98 cm²   **95.** 1,767.15 m³
96. a. 100 in.²   **b.** $80\pi \approx 251.3$ in.³   **97.** $h = \frac{3V}{\pi r^2}$
98. $M = \frac{2K - Iw^2}{v_0^2}$ **99.** $d = \frac{l - a}{n - 1}$ **100.** $y = \frac{9}{5}x - 7$
101. Atlanta: 90.0 million; O'Hare: 69.3 million   **102.** $245 - 5c$
103. 600   **104.** 42 ft, 45 ft, 48 ft, 51 ft   **105.** 50°, 130°
106. 27 in. by 40 in.   **107.** 124 days   **108.** 32%   **109.** 18.7%
110. 2.3%   **111.** $18,000 at 10%; $7,000 at 9%
112. 2 min after the photographer leaves   **113.** $6\frac{2}{3}$ gal
114. Mild: 50 lb; robust: 40 lb

CHAPTER 1 TEST (page 98)

1. a. undefined   **b.** inequality   **c.** like terms   **d.** solve
e. addition, equality   **2. a.** $s = T + 10$ **b.** $A = \frac{1}{2}bh$
3. a. $0.64p$ **b.** $2p + 15$ **c.** $\frac{7}{8}x - 27$ **d.** $3(x - 1)$
4. a. $-2, 0, 5$ **b.** $-2, 0, -3\frac{3}{4}, 9.2, \frac{14}{5}, 5$ **c.** $\pi, -\sqrt{7}$ **d.** All
5. a. True   **b.** False   **c.** True   **d.** True
6.

7.

8. a. False   **b.** False   **9.** $\frac{4}{15}$ **10.** $\frac{8}{9}$ **11.** −209   **12.** −3
13. 100 mg   **14. a.** Commutative Property of Addition
b. Associative Property of Multiplication   **c.** Additive Inverse
Property   **d.** Multiplicative Identity Property
15. $11.1n^2 - 7.8n - 9.8$ **16.** $90st$ **17.** $-12c + 108$
18. $-\frac{1}{36}x + 16y$ **19.** 15   **20.** 6   **21.** No solution, $\varnothing$;
contradiction   **22.** 12   **23.** Yes   **24.** $i = \frac{f(P - L)}{s}$

25. $x_1 = \dfrac{y_1 + mx - y}{m}$ **26.** 1,018 m²   **27.** 8   **28.** 25
29. 85°, 85°, 10°   **30.** 4 cm by 9 cm   **31.** 8% decrease
32. 4 hr into the flights   **33.** $4,000   **34.** 400 mi   **35.** 10 oz
36. *Skin Soother:* 3 oz; *Cool Sport:* 5 oz

ARE YOU READY? 2.1 (page 102)

1.

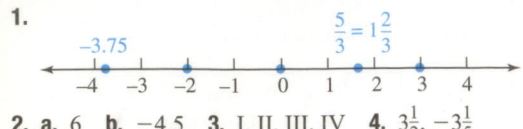

2. a. 6   **b.** −4.5   **3.** I, II, III, IV   **4.** $3\frac{1}{2}, -3\frac{1}{5}$

SELF CHECKS 2.1

1. a. 4 days before the storm began, 1 day and 9 days after the
storm began   **b.** 6 ft   **c.** 4 days   **2. a.** $140   **b.** $100
3. (2, 5)   **4.** (1, 13)

STUDY SET SECTION 2.1 (page 108)

1. ordered   **3.** origin   **5.** rectangular   **7.** midpoint
9. origin, right, down   **11.** II   **13.** Yes   **15.** A capital letter
17. $x^2 = x \cdot x$; x_2 represents the *x*-coordinate of a point.
19–26.

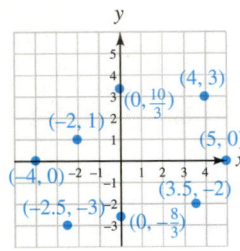

19. Quadrant I
21. Quadrant IV   **23.** *x*-axis
25. *y*-axis   **27.** (2, 4)
29. (−2.5, −1.5)   **31.** (3, 0)
33. (0, 0)

35.

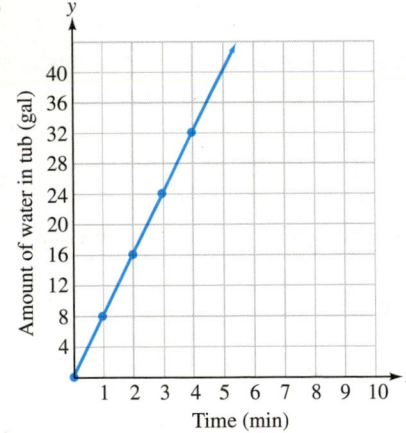

37. a. On the surface
b. Diving deeper
c. 3 hr
d. 500 ft
39. a. $2   **b.** $9
c. 3 days
41. (3, 4)
43. (9, 12)
45. $\left(\frac{1}{2}, -8\right)$
47. $\left(-\frac{3}{2}, \frac{5}{2}\right)$
49. (4, 1)
51. (−20, −3)

53. Jonesville (5, B), Easley (1, B), Hodges (2, E), Union (6, C)
55. a. (2, −1)   **b.** No   **c.** Yes   **57. a.** (50, 3,000), (128, 4,000),
(100, 11,000)   **b.** About 10,000 feet   **59. a.** iii   **b.** iv
c. v   **d.** ii   **e.** i   **61. a.** 78¢   **b.** 95¢   **c.** 3.5 oz   **63. a.** T
b. R   **71.** −5   **73.** 5   **75.** $w = \dfrac{P - 2l}{2}$

ARE YOU READY? 2.2 (page 113)

1. False   **2.** −3   **3.**

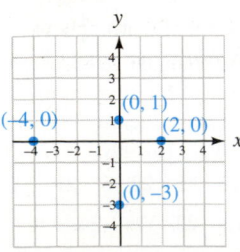

4. 0   **5.** −2

6.

SELF CHECKS 2.2

1. No   **2.**

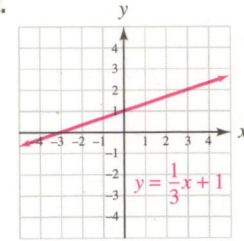

3.

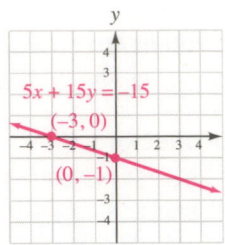

4.

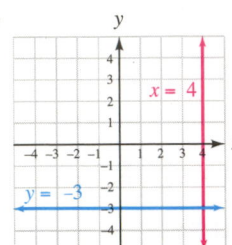

5. About 5,400 farmers markets
6. a. 3 yr   **b.** 5 yr

STUDY SET SECTION 2.2 (page 122)

1. ordered pair   **3.** linear   **5.** *y*-intercept, *x*-intercept
7. (0, 3), (3, 1), (6, −1)   **9.** 1, 1, 1   **11. a.** (−3, 0); (0, 4)
b. False   **13.** $y = -4x - 1$ **15. a.** The *y*-axis   **b.** The *x*-axis
17. a. Yes   **b.** No   **19. a.** Yes   **b.** No
21. 5, 4, 2   **23.** 0, −1, −2

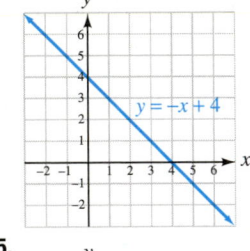
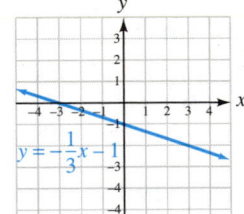

25.   **27.**

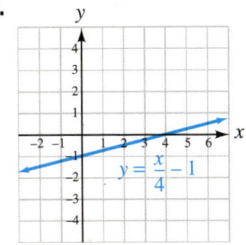
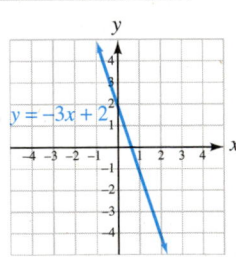

29.   **31.**

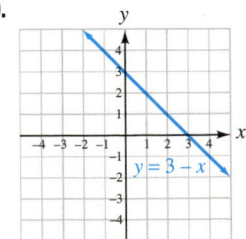
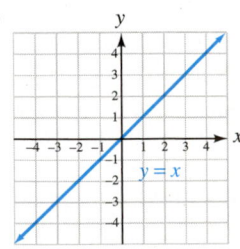

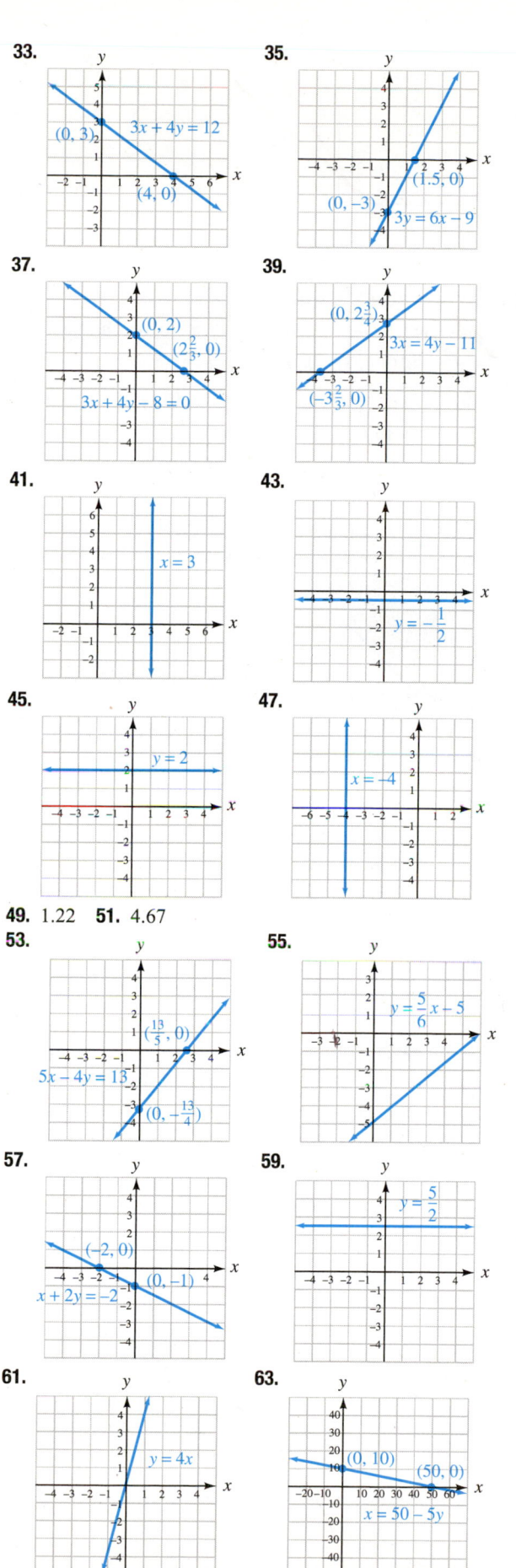

49. 1.22   **51.** 4.67

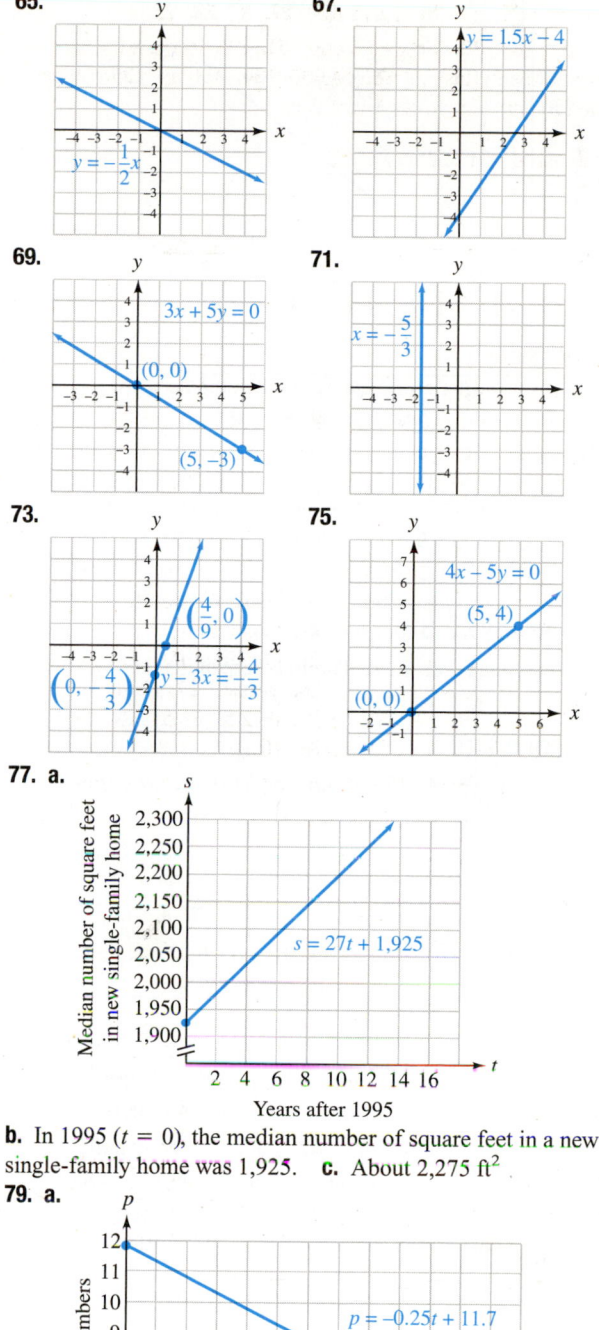

77. a.

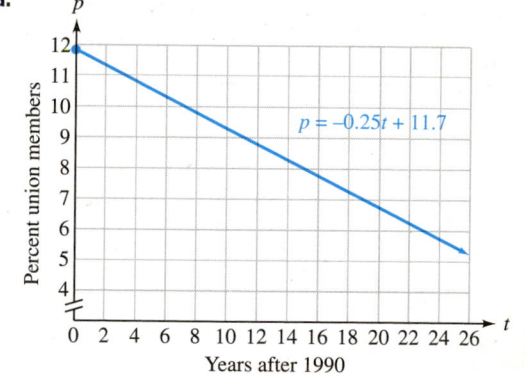

b. In 1995 ($t = 0$), the median number of square feet in a new single-family home was 1,925.   **c.** About 2,275 ft^2

79. a.

b. In 1990 ($t = 0$), 11.7% of the people working in the private sector were union members.   **c.** About 5.5%

81. a.

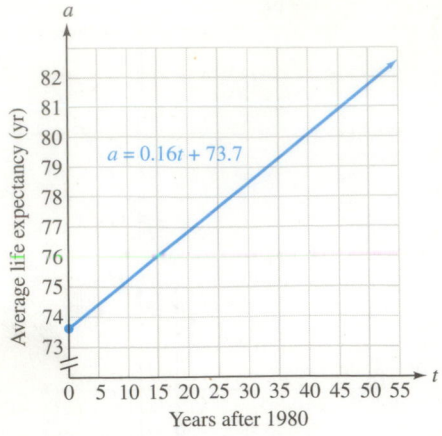

b. The average life expectancy in 1980 ($t = 0$) was 73.7 years.
c. About 81.7 yr   **83.** In 8 years ($x = 8$), the computer will have no value. When new ($x = 0$), the computer was worth $3,000.
85. 12.5 yr   **87.** 200   **93.** {11, 13, 17, 19, 23, 29}   **95.** III
97. $-480s$ **99.** $-24x - 64$

ARE YOU READY? 2.3 (page 126)

1. $\frac{3}{5}$ **2.** Undefined   **3.** $\frac{6}{7}$ **4.** Parallel, Perpendicular

SELF CHECKS 2.3

1. $\frac{1}{2}$ **2.** -2 **3.** $\frac{13}{24}$ **4.** 175 ft/min   **5.** They are not parallel.
6. Yes

STUDY SET SECTION 2.3 (page 133)

1. ratio, $\frac{a}{b}$ **3.** rate, change   **5.** change, rise   **7. a.** 200, 80
b. $-120, 20$ **c.** $-120, 20, -6$ **9. a.** -6 **b.** 2   **c.** -3
11. a. l_3; 0   **b.** l_2; undefined   **c.** l_1; 2   **d.** l_4; -3
13. a. $\frac{5}{7}, \frac{4}{7}$; no   **b.** $\frac{4}{7}, -\frac{7}{4}$; yes   **15.** per   **17.** $\frac{7}{5}$ **19.** $-\frac{8}{3}$
21. -3 **23.** $\frac{1}{2}$ **25.** 3   **27.** -1 **29.** $-\frac{1}{3}$ **31.** $\frac{11}{7}$ **33.** 0
35. Undefined   **37.** $\frac{9}{2}$ **39.** -0.5 or $-\frac{1}{2}$ **41.** Parallel
43. Perpendicular   **45.** Neither   **47.** Parallel   **49.** Parallel
51. Perpendicular   **53. a.** An increase of 65.5 million units/yr
b. A decrease of 70 million units/yr   **55.** 0.026°F/yr
57. $\frac{3}{140}, \frac{1}{15}, \frac{1}{20}$; part 2   **59.** $\frac{1}{25}$; 4%   **61.** $\frac{5}{6}$ **63.** 250 ft/min
65. No; they are equally steep. The steepness of each course would be found by finding the slope of the same line.   **71.** 40 lb licorice; 20 lb gumdrops

ARE YOU READY? 2.4 (page 138)

1. a. $5x, -8$ **b.** 5   **2.** $y = -\frac{4}{5}x + 4$ **3.** The y-axis   **4.** $\frac{2}{3}$
5. $y = 2x - 23$ **6.** $\frac{45}{8}$

SELF CHECKS 2.4

1. a. $y = x - 12$ **b.** $y = -\frac{1}{2}x + 1$ **2.** $y = -6x + 2$
3. a. $L = 0.12m + 0.75$ **b.** 2.19 in.   **4.** $y = \frac{5}{4}x - \frac{17}{2}$
5. $y = -\frac{4}{3}x + \frac{7}{3}$ **6.** $1,290

---

7. $m = \frac{3}{2}, (0, 2)$

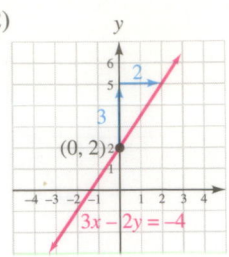

8. a. No   **b.** Yes
9. a. $y = 8x$
b. $y = \frac{3}{2}x - 14$

STUDY SET SECTION 2.4 (page 146)

1. slope–intercept   **3.** $m = -\frac{2}{3}, (0, 1)$ **5. a.** $(0, 0)$ **b.** None
7. a. $m = -\frac{2}{3}$ **b.** $(-1, 3)$ **9.** $m = -\frac{5}{2}, b = 75$
11. a. 8   **b.** $-\frac{4}{3}$ **13.** $\frac{1}{3}x, 1, 2, 2, 1, \frac{1}{3}, -1$ **15.** $y = 3x + 6$
17. $y = -\frac{2}{3}x - \frac{7}{3}$ **19.** $y = \frac{1}{2}x + 3$ **21.** $y = -\frac{7}{5}x - 1$
23. $y = 7x + 54$ **25.** $y = -9x + 14$ **27.** $y - 7 = 10(x - 1)$
29. $y + 4 = -\frac{2}{3}(x + 2)$ **31.** $y = 5x - 25$
33. $y = -9x - 28.8$ **35.** $y = -\frac{1}{2}x + 11$ **37.** $y = \frac{2}{3}x + \frac{11}{3}$
39. $1, (0, -1)$ **41.** $-\frac{5}{4}, (0, -3)$

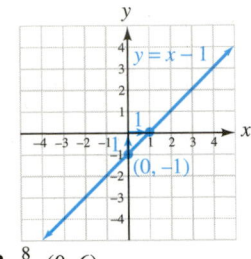   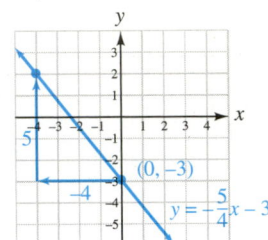

43. $\frac{8}{5}, (0, 6)$ **45.** $-\frac{3}{4}, (0, -2)$

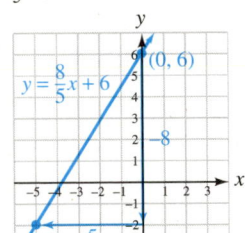   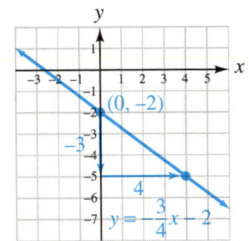

47. $\frac{3}{2}, (0, -4)$ **49.** $-\frac{1}{3}, \left(0, -\frac{5}{6}\right)$ **51.** Parallel
53. Perpendicular   **55.** Neither   **57.** Perpendicular
59. $y = 4x - 3$ **61.** $y = 3x - \frac{7}{4}$ **63.** $y = -\frac{1}{4}x$
65. $y = -\frac{9}{7}x + \frac{22}{7}$ **67.** $y = \frac{7}{3}x - 3$ **69.** $y = \frac{4}{3}x + \frac{7}{3}$
71. $y = -\frac{1}{4}x + \frac{11}{2}$ **73.** $y = \frac{7}{5}x - \frac{28}{5}$ **75.** $y = x$ **77.** $y = 4x$
79. $y = -3x + 6$ **81.** $y = -\frac{1}{3}x + \frac{17}{36}$ **83.** $y = \frac{4}{3}x + 4$
85. $y = \frac{4}{5}x - \frac{26}{5}$ **87. a.** $c = 7.8m + 220$ **b.** About $10\frac{1}{4}$ min
89. a. $b = -22,000,000t + 650,000,000$ **b.** b-intercept: The number of barrels produced in 1990 was about 650,000,000; Slope: The yearly decrease in production was about 22,000,000.
91. a. $B = \frac{1}{100}p - 195$ **b.** 905
93. a. $p = -505,000t + 148,525,000$ **b.** 138,425,000
95. a. $p = \frac{14.7}{33}d + 14.7$ or $p = \frac{147}{330}d + 14.7$ **b.** 126.1 psi
97. $y = -\frac{950}{3}x + 1,750$ **99. a.** $y = 12.5x + 100$
b. The year 2032   **101. a.** $w = 8h - 389$ **b.** 187 lb

103. a. $(0, 632)$; In 1980, about 632 billion cigarettes were consumed in the U.S.   **b.** -10; Since 1980, the number of cigarettes consumed in the U.S. has decreased by about 10 billion each year.   **111.** $29,100

ARE YOU READY? 2.5 (page 152)

1. $(3, 5), (3, 0)$ **2.** 7   **3.** $3, (0, -8)$ **4.** $-\frac{3}{2}$

SELF CHECKS 2.5

1. D: $\{-12, -6, 5, 8\}$; R: $\{-6, 4, 6\}$ **2. a.** No; $(0, 2), (0, 3)$
b. Yes   **c.** No; $(4, -1), (4, 4)$ **d.** Yes   **3. a.** Yes   **b.** No;
$(1, 2), (1, -2)$ **4. a.** -5 **b.** 5   **c.** $2t - 1$ **d.** 0.42   **5.** -27
6. a. The set of real numbers   **b.** The set of all real numbers
except -3 **7. a.** $f(x) = -\frac{1}{5}x + 9$ **b.** $f(x) = 2x - 13$
c. $f(x) = \frac{1}{10}x - 2$ **8. a.** $E(t) = 8t + 398$ **b.** 638 quadrillion Btu

STUDY SET SECTION 2.5 (page 162)

1. relation, domain, range   **3.** function, variable, dependent
5. linear   **7. a.** $\{(2000, 63), (2001, 56), (2002, 54), (2003, 50), (2004, 52), (2005, 51), (2006, 51)\}$
b. D: $\{2000, 2001, 2002, 2003, 2004, 2005, 2006\}$
R: $\{50, 51, 52, 54, 56, 63\}$

c.

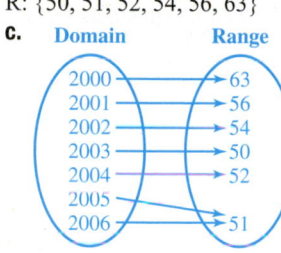

| Domain | Range |

9. If $x = -4$, the denominator of $\frac{1}{x + 4}$ is 0 and the fraction is undefined.
11. a. $(0, -4)$ **b.** $(-\frac{2}{3}, 0)$
13. a. of   **b.** of   **15.** $f(8)$
17. $f(x), y$ **19.** D: $\{-5, -1, 7, 8\}$;
R: $\{-11, -6, -1, 3\}$
21. D: $\{-23, 0, 7\}$; R: $\{1, 35\}$
23. Yes   **25.** No; $(4, 2), (4, 4), (4, 6)$ **27.** No; $(3, 4), (3, -4)$ or $(4, 3), (4, -3)$ **29.** Yes   **31.** Yes   **33.** No; $(-1, 0), (-1, 2)$
35. Yes   **37.** Yes   **39.** No; $(1, 1), (1, -1)$ **41.** Yes   **43.** Yes
45. No; $(1, 1), (1, -1)$ **47.** $9, -3$ **49.** $3, -5$ **51.** $-6, -1$
53. $-6, -24$ **55.** $9, 16$ **57.** $7, 16$ **59.** $7, 4$ **61.** $\frac{1}{8}, 1$
63. $\frac{5}{2}, \frac{2}{5}$ **65.** 0, undefined   **67.** 3.7, 1.1, 3.4
69. $-\frac{27}{64}, \frac{1}{216}, \frac{125}{8}$ **71.** $2w, 2w + 2$ **73.** $3w - 5, 3w - 2$
75. 0   **77.** 1   **79. a.** The set of real numbers   **b.** The set of all real numbers except 4   **81. a.** The set of real numbers   **b.** The set of all real numbers except $-\frac{1}{2}$

83.

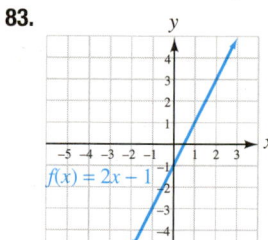

$f(x) = 2x + 1$

85.

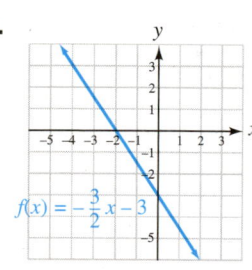

$f(x) = -\frac{3}{2}x - 3$

87.

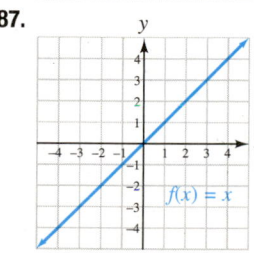

$f(x) = x$

89.

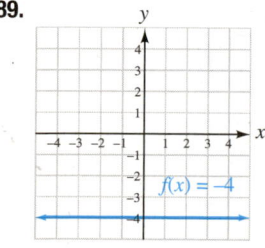

$f(x) = -4$

91.

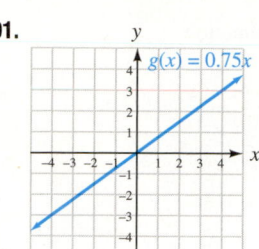

$g(x) = 0.75x$

93.

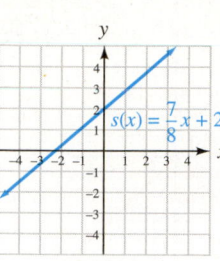

$s(x) = \frac{7}{8}x + 2$

95. $f(x) = 5x - 3$ **97.** $f(x) = \frac{1}{5}x - 1$ **99.** $f(x) = 2x + 5$
101. $f(x) = -\frac{2}{3}x + 2$ **103.** $f(x) = 6x - 4$ **105.** $f(x) = 12$
107. Between 20°C and 25°C   **109. a.** $p(b) = 4.75b - 125$
b. $397.50   **111. a.** $N(t) = 41,100t + 1,970,000$ **b.** 2,997,500
registered nurses   **113. a.** $L(a) = -1.2a + 132$ **b.** 36%
115. a. $3,331.25; the tax on an adjusted gross income of $25,000
is $ 3,331.25.   **b.** $T(a) = 4,681.25 + 0.25(a - 34,000)$
121. No solution, $\varnothing$; contradiction

ARE YOU READY? 2.6 (page 166)

1. $3, 9$ **2.** $1, 1$ **3.** $-8, 8$ **4.** $4, 4$

SELF CHECKS 2.6

1. a. -2 **b.** 2
2. D: the set of real numbers, R: the set of all real numbers greater than or equal to -2; the graph has the same shape, but is 2 units lower.

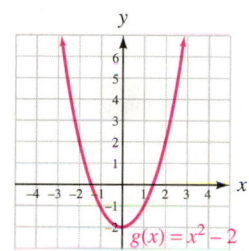

$g(x) = x^2 - 2$

3. D: the set of real numbers, R: the set of all real numbers; the graph has the same shape, but is 1 unit higher.

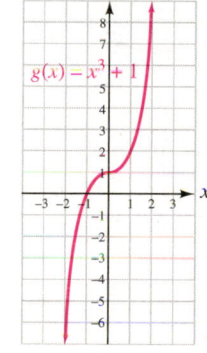

$g(x) = x^3 + 1$

4. D: the set of real numbers, R: the set of nonnegative real numbers; the graph has the same shape, but is 2 units to the right.

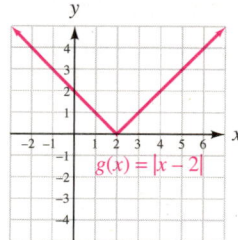

$g(x) = |x - 2|$

5.

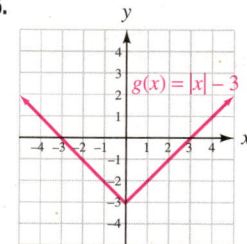

$g(x) = |x| - 3$

6.

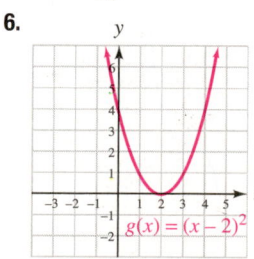

$g(x) = (x - 2)^2$

7.

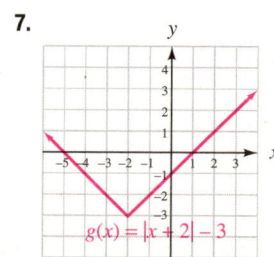

$g(x) = |x + 2| - 3$

8.

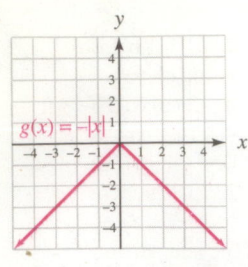

9. Not a function

33. D: the set of real numbers, R: the set of nonnegative real numbers

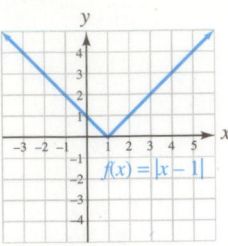

35. D: the set of real numbers, R: the set of nonnegative real numbers

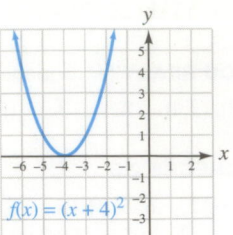

STUDY SET SECTION 2.6 (page 175)

1. nonlinear   **3.** nonnegative
5. a. The squaring function

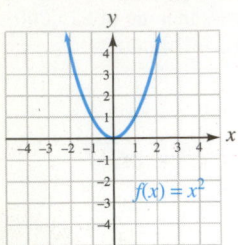

b. The cubing function

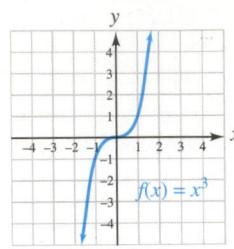

c. The absolute value function   **7.** $(5, 9)$

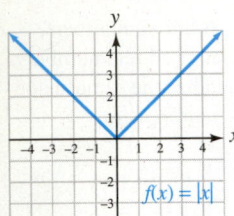

37. D: the set of real numbers, R: the set of real numbers greater than or equal to -2

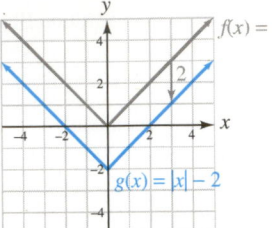

39. D: the set of real numbers, R: the set of real numbers

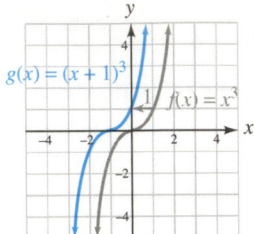

41. D: the set of real numbers, R: the set of all real numbers greater than or equal to -3

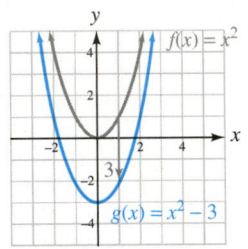

9. a.

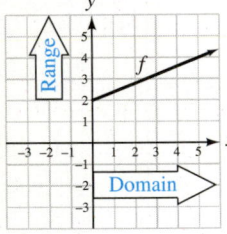

b. D: the set of nonnegative real numbers, R: the set of real numbers greater than or equal to 2
11. a. 4 **b.** 0 **c.** -2
13. a. $(-2, 4), (-2, -4)$ **b.** No; to the x-value -2, there corresponds more than one y-value (4 and -4). **15. a.** 4, left
b. 4, up **17. a.** -4 **b.** 0 **c.** 2
d. -1 **19. a.** 2 **b.** 4 **c.** $-1, 1, 5$ **d.** 2, 4 **21.** D: the set of real numbers, R: the set of real numbers **23.** D: the set of real numbers, R: the set of real numbers less than or equal to 5
25. D: the set of real numbers, R: the set of real numbers greater than or equal to -4 **27.** D: the set of nonnegative real numbers, R: the set of nonnegative real numbers
29. D: the set of real numbers, R: the set of real numbers greater than or equal to 2

31. D: the set of real numbers, R: the set of real numbers

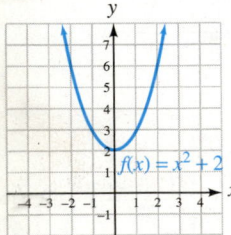

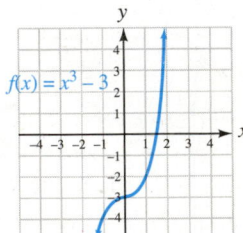

43. D: the set of real numbers, R: the set of real numbers

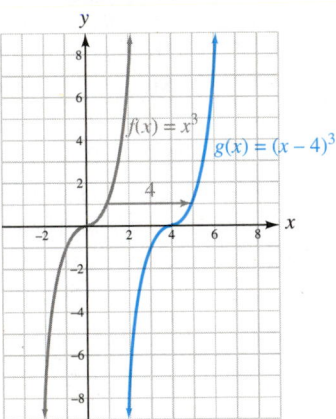

45. D: the set of real numbers, R: the set of real numbers

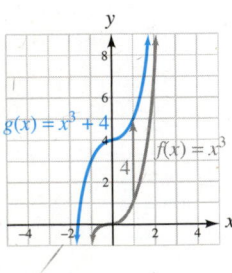

47. D: the set of real numbers, R: the set of nonnegative real numbers

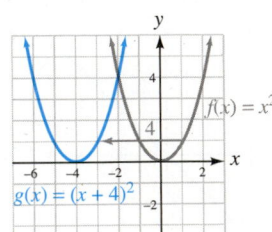

49.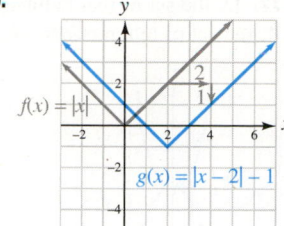

51.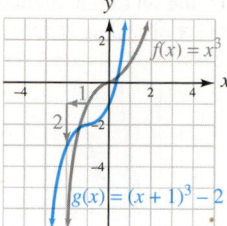

53.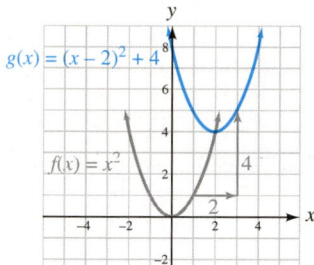

55.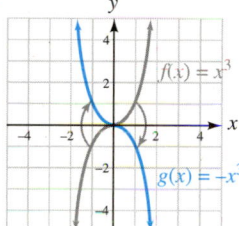

57. 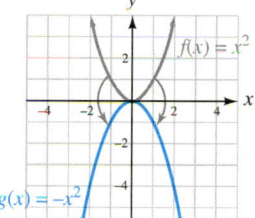 **59.**

61.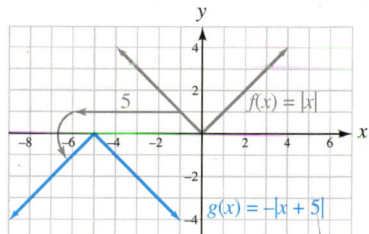

63.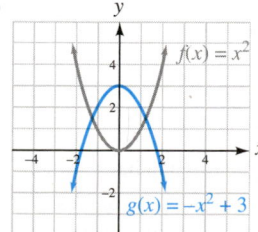

65. No, $(0, 2)$, $(0, -2)$ **67.** Yes **69.** Yes **71.** No, $(3, 0)$, $(3, 1)$

73.  **75.** **77.**

79. **81.** $f(x) = |x|$ **83.** A parabola

91. $W = T - ma$ **93.** $g = \frac{2(s - vt)}{t^2}$

CHAPTER 2 REVIEW (page 179)

1.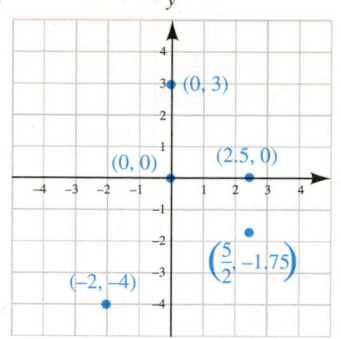

2. a. 1 ft below its normal level
b. Decreased by 3 ft
c. From day 3 to the beginning of day 4
3. a. $10 increments
b. $800 **4.** $(7, -3)$
5. Yes **6. a.** True
b. False **7.** 9, 0, -9
8. $-4, -\frac{5}{2}, -1$

9. 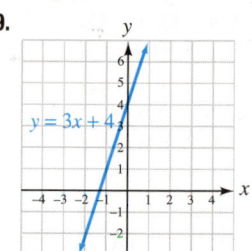 **10.**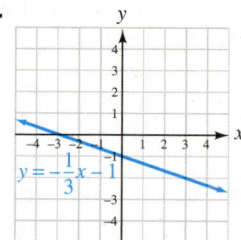

11. 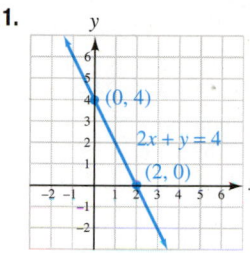 **12.**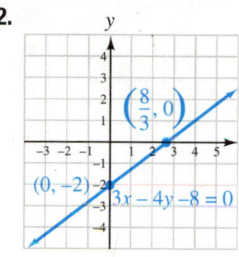

13. 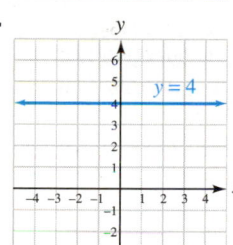 **14.**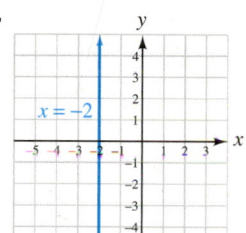

15. a. $55,200 **b.** 6 yr

16.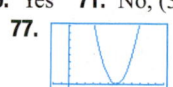

a. In 1980, it took about 25 cans to weigh one pound. **b.** 34 cans
17. Slope of $l_1 = \frac{4}{5}$; slope of $l_2 = -\frac{8}{5}$ **18.** 503 stores per yr
19. 1 **20.** $-\frac{14}{9}$ **21.** 0 **22.** Undefined **23.** Perpendicular
24. Parallel **25.** 6 ft **26.** 31.5% **27.** $y = 3x + 29$
28. $y = -\frac{13}{8}x + \frac{3}{4}$ **29.** $y = \frac{3}{2}x - \frac{1}{2}$ **30.** $y = -\frac{2}{3}x - 7$

31. $y = -\frac{3}{4}x - 3$; $m = -\frac{3}{4}$, $(0, -3)$

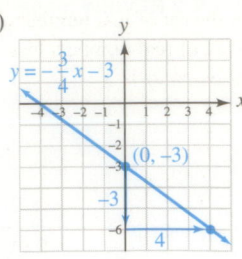

32. $m = -\frac{4}{5}$, $(0, 4)$; $y = -\frac{4}{5}x + 4$ **33. a.** $y = 0$ **b.** $x = 0$
34. Perpendicular **35.** $L = \frac{1}{150}p + 110$
36. a. $y = -1,720x + 8,700$ **b.** $(0, 8,700)$; it gives the value of
the saw blade when new: \$8,700 **37.** D: $\{-4, -1, 2, 5\}$,
R: $\{-2, 0, 16\}$ **38. a.** function **b.** function, variable, dependent
39. a. Yes **b.** No; $(-1, 8), (-1, 9)$ **c.** Yes **40.** 0 **41.** Yes
42. Yes **43.** No; $(25, 5), (25, -5)$ **44.** No; $(3, 4), (3, -4)$
45. -7 **46.** 18 **47.** 8 **48.** $3t + 8$ **49.** 3 **50.** $\frac{4}{3}$
51. The set of real numbers **52.** The set of real numbers
53. The set of all real numbers except 2 **54.** The set of all real
numbers except -5 **55.** -2, $(0, -16)$
56.

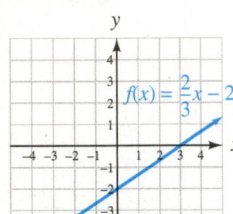

57. $f(x) = \frac{9}{10}x + \frac{7}{8}$
58. $f(x) = \frac{1}{5}x - 1$ **59.** $f(x) = 12$
60. $f(x) = 4x - 3$
61. $f(x) = \frac{1}{3}x + 5$
62. a. $R(t) = 0.02t + 5.05$
b. 7.05 milliohms
63. a. -4 **b.** 3 **c.** 1
64. a. 4 **b.** 1 **c.** $-4, 2$

65. D: the set of real numbers, R: the set of real numbers
66. D: the set of real numbers, R: the set of real numbers greater
than or equal to 1
67. D: the set of real numbers, R: the set of nonnegative real
numbers **68. a.** 6, up **b.** 6, left

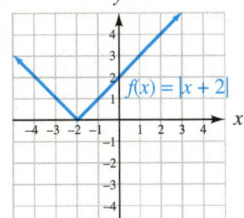

69. D: the set of real numbers, R: the set of real numbers greater
than or equal to -3

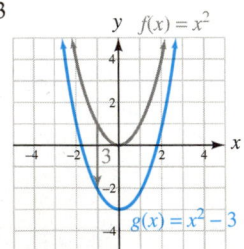

70. D: the set of real numbers, R: the set of nonnegative real
numbers

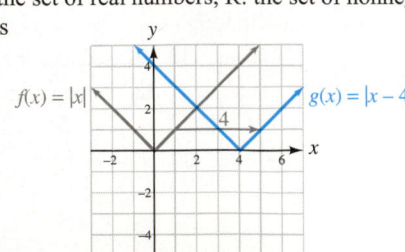

71. D: the set of real numbers,
R: the set of real numbers
72. D: the set of real numbers,
R: the set of real numbers

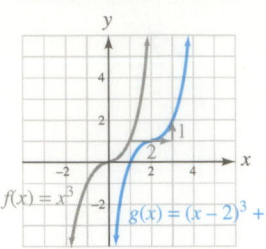
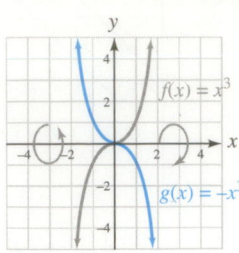

73. Function **74.** Not a function; $(0, 2), (0, 3)$

CHAPTER 2 TEST (page 188)

1. a. rectangular **b.** function **c.** function, variable, dependent
d. domain, range **e.** translation **2. a.** 11 P.M. **b.** 10 A.M.,
7 P.M. **c.** 4 hr **d.** 10 A.M. to 3 P.M.; 7 P.M. to 11 P.M. **e.** 3 P.M.
3. a. \$40 **b.** \$80 **c.** 3 days **4.** $\left(\frac{5}{2}, -8\right)$
5.

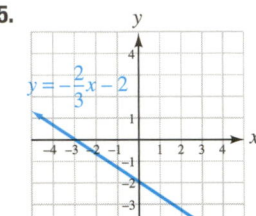

6. $(5, 0), \left(0, -\frac{10}{3}\right)$

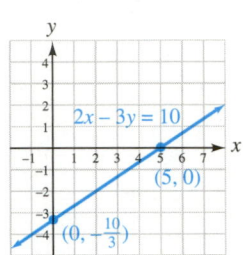

7.

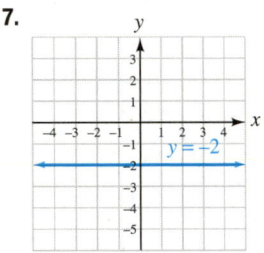

8. -1.5 degree/hr **9.** $\frac{1}{2}$ **10.** $\frac{2}{3}$ **11. a.** Undefined **b.** 0

12. $y = 3x + 1$ **13.** $y = 8x + 22$

14. $m = -\frac{4}{5}$, $(0, 5)$

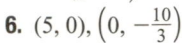

15. Perpendicular **16.** $y = -\frac{3}{2}x$ **17. a.** $v = -600x + 4,000$
b. $(0, 4,000)$; it gives the value of the copier when new: \$4,000
18. $n = \frac{11}{200}p + 1,400$ **19. a.** No; $(1, -8), (1, 6)$ **b.** Yes
c. Yes **d.** No; $(4, -4), (4, 4)$ **20.** The set of all real numbers
except 6 **21.** -20 **22.** 8, $(0, -9)$ **23.** $f(x) = -\frac{5}{4}x + \frac{5}{2}$
24. a. $T(m) = -0.16m + 331.3$ **b.** 8.9 sec **25.** 10 **26.** 49

27. $\frac{9}{16}$ **28.** $3r + 25$ **29.** -2 **30.** 2   **31.** Function   **32.** Not a function; $(2, 2), (2, -2)$ **33.** D: the set of real numbers, R: the set of real numbers greater than or equal to 3

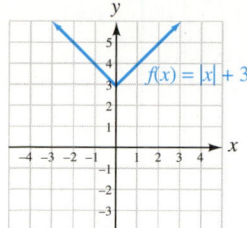

34.

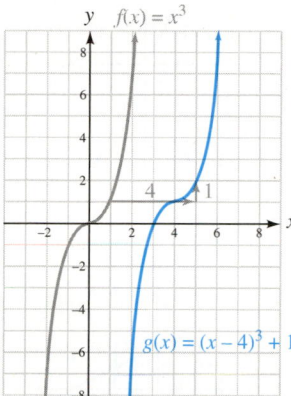

D: the set of real numbers, R: the set of real numbers

35. D: the set of real numbers, R: the set of nonnegative real numbers

36. $f(x) = -x^2 + 1$

CHAPTERS 1–2 CUMULATIVE REVIEW (page 190)

1. a. $1, 2, 6, 7$ **b.** $0, 1, 2, 6, 7$ **c.** $-2, 0, 1, 2, \frac{13}{12}, 6, 7$
d. $\sqrt{5}, \pi$ **e.** -2 **f.** $-2, 0, 1, 2, \frac{13}{12}, 6, 7, \sqrt{5}, \pi$ **g.** $2, 7$
h. $-2, 0, 2, 6$
2.

$$-\frac{35}{8} \quad -\pi \quad -1\frac{1}{2} \; -0.333... \; \sqrt{2} \qquad 3 \quad 4.25$$

number line from -5 to 5

3. -2 **4.** -149 **5.** $\frac{24}{25}$ **6.** 2   **7.** 33   **8.** -5
9. a. Associative property of addition   **b.** Distributive property
c. Commutative property of addition   **d.** Associative property of
multiplication   **10. a.** -6 **b.** -8 **11.** $-28st$ **12.** $15y + 22$
13. $-\frac{13}{12}s$ **14.** $-12x + 101$ **15.** $-\frac{8}{3}$ **16.** -1 **17.** 6
18. 24   **19.** No solution, $\varnothing$ **20.** $\frac{33}{5}$ **21.** $B = \frac{c + Tx}{3y}$
22. $h = \frac{2A}{b_1 + b_2}$ **23.** 164   **24.** $70°, 70°, 40°$ **25.** \$14,000
26. 39 mph going, 65 mph returning   **27.** $\left(1, \frac{1}{2}\right)$ **28.** Yes
29.

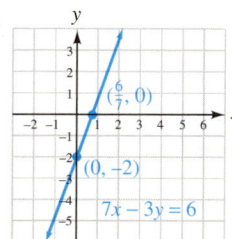

30. $-\frac{5}{6}$ **31.** $y = -\frac{7}{5}x + \frac{11}{5}$
32. $y = -3x - 3$
33. a. $T = \frac{1}{1,000}p + 25$ **b.** 60
34. function   **35. a.** 0   **b.** 2
36. D: The set of real numbers,
R: The set of real numbers
37. D: $\{-6, 0, 1, 5\}$;
R: $\{-12, 4, 7, 8\}$; no   **38.** No;
$(4, 2), (4, -2)$ **39.** -60 **40.** Yes   **41.** 5   **42.** -1 **43.** 3
44. $3r^2 + 2$ **45.** The set of all real numbers except -1
46. $6, (0, 15)$ **47.** $f(x) = -\frac{7}{8}$ **48. a.** $M(t) = 0.17t + 4.06$
b. 10.86 billion vehicle miles

49. D: the set of real numbers, R: the set of real numbers less than or equal to 1

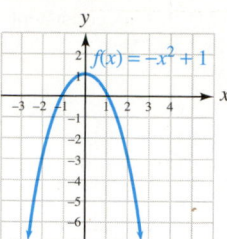

50. D: the set of real numbers, R: the set of real numbers greater than or equal to -4

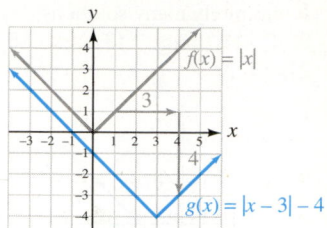

ARE YOU READY? 3.1 (page 194)

1. Not a solution

2.

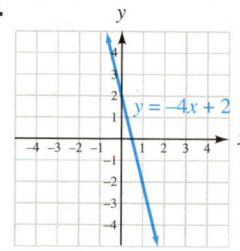

3.

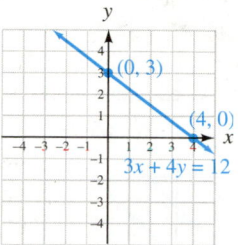

4. Parallel

SELF CHECKS 3.1

1. It is a solution.
2. $(1, 2)$ **3.** No solution; $\varnothing$

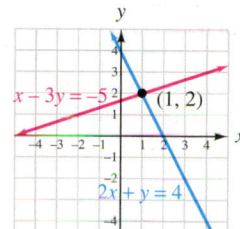

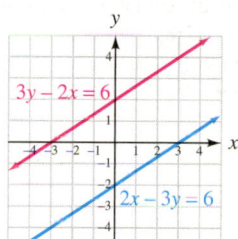

4. $\{(x, y) \mid 2x - y = 4\}$; infinitely many solutions

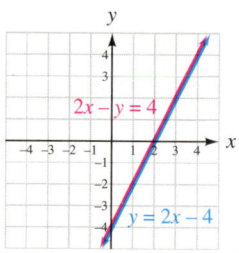

5. $(-6, 4)$ **6.** -1

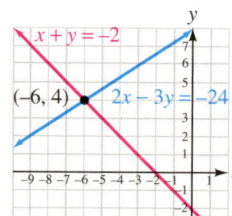

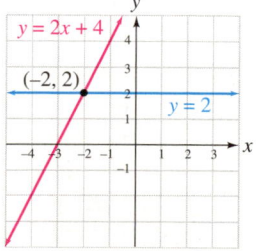

STUDY SET SECTION 3.1 (page 202)

1. system **3.** independent, dependent **5. a.** True **b.** False
c. True **d.** True **7. a.** No solution; independent
b. Infinitely many solutions; $(-3, 0), (-2, -2), (0, -6)$; consistent
9. 1 **11.** brace **13.** Yes **15.** No **17.** No **19.** Yes
21. $(4, 2)$ **23.** $(-1, 3)$

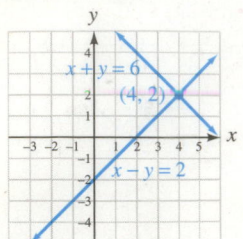 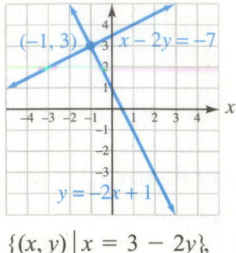

25. No solution, $\varnothing$; **27.** $\{(x, y) \mid x = 3 - 2y\}$,
inconsistent system   infinitely many solutions;
dependent equations

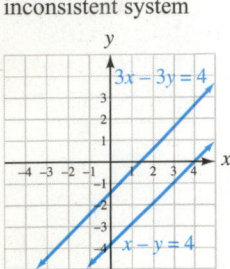 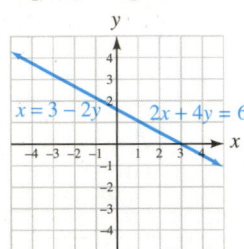

29. $(-3, -3)$ **31.** $(-2, -1)$

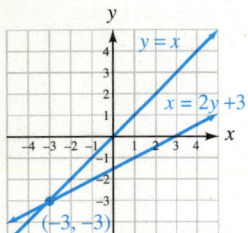 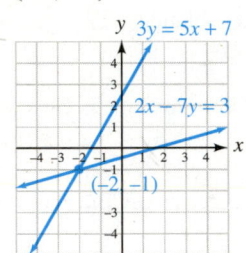

33. 2 **35.** 3

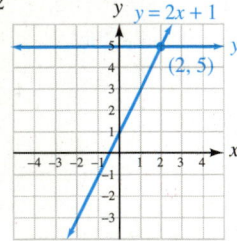 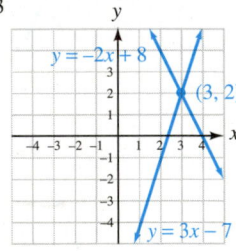

37. $(-0.37, -2.69)$ **39.** $(-7.64, 7.04)$ **41.** 4
43. -3 **45.** $(1, -2)$

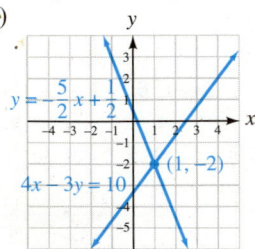

47. $(2, -2)$ **49.** $(2, 3)$

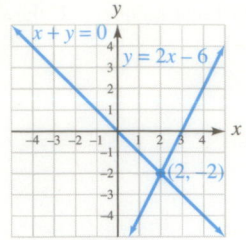 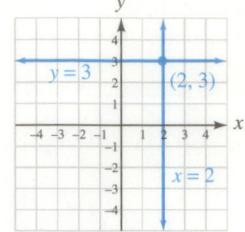

51. No solution, $\varnothing$; **53.** $\left(\frac{1}{2}, -1\right)$
inconsistent system

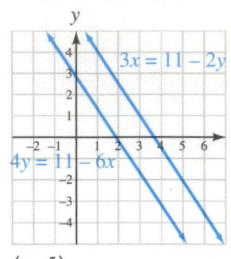 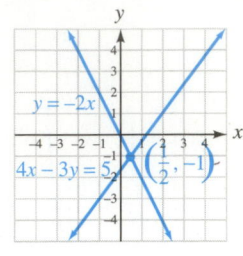

55. $\left(3, \frac{5}{2}\right)$ **57.** $(2, 1)$

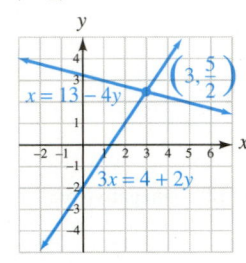 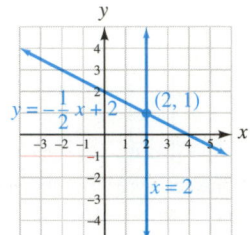

59. $(3, 2)$ **61.** $\{(x, y) \mid x + 3y = 6\}$,
infinitely many solutions;
dependent equations

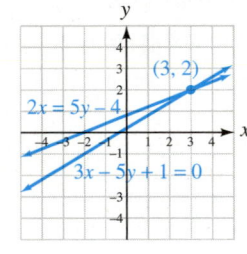 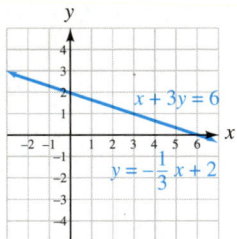

63. $\left(-1, \frac{2}{3}\right)$

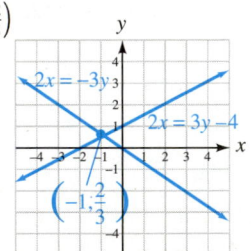

65. a. $(1978, 50\%)$ **b.** In 1978, the percent share of the U.S.
footwear market for shoes produced in the United States and
imports was the same: 50%. **67.** $(2,000, 50)$

69. a.

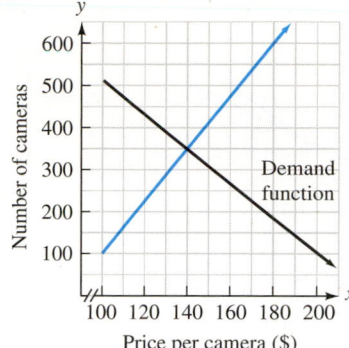

b. $140   **c.** Supply increases and demand decreases.

71. a. Yes

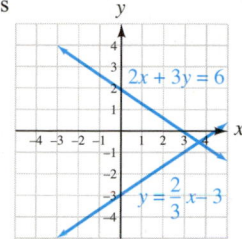

b. No. The ships would have to reach this point at the same time.
77. -3 **79.** 0 **81.** $-t^3 + 2t - 2$
83. D: the set of all real numbers except -2

ARE YOU READY? 3.2 (page 205)

1. 1 **2.** $y = 5x + 4$ **3.** -9 **4.** 3 **5.** 4
6. $-28x + 4y = -36$

SELF CHECKS 3.2

1. $(-3, 4)$ **2.** $\left(\frac{4}{3}, \frac{4}{3}\right)$ **3.** $(-1, -2)$ **4.** $\left(\frac{1}{5}, 6\right)$ **5.** $(1, -3)$
6. No solution, $\varnothing$; inconsistent system   **7.** $\{(x, y)\,|\,2x - 5y = 19\}$, infinitely many solutions; dependent equations

STUDY SET SECTION 3.2 (page 214)

1. standard   **3.** eliminated   **5.** y; second equation
7. a. $3; -4$ (answers may vary)   **b.** $2; -3$ (answers may vary)
9. Elimination method   **11. a.** $7x + 4y = 8, x - y = 8$
b. $2x + y = 12, 3x - 9y = 170$ **13.** $(2, 6)$ **15.** $(5, 3)$
17. $(-2, 4)$ **19.** $(4, -7)$ **21.** $(9, 2)$ **23.** $(-4, 0)$
25. $(3, -2)$ **27.** $(12, 4)$ **29.** $(22, 25)$ **31.** $\left(-2, \frac{3}{2}\right)$
33. No solution, $\varnothing$; inconsistent system
35. $\left\{(x, y)\,|\,4x - 5 = 2y\right\}$, infinitely many
solutions; dependent equations   **37.** $\left(5, \frac{3}{2}\right)$ **39.** $\left(\frac{1}{2}, \frac{2}{3}\right)$
41. $\{(x, y)\,|\,2x - 4 = y\}$, infinitely many solutions; dependent
equations   **43.** $(-20, -30)$ **45.** $(9, -1)$ **47.** $(4, 8)$
49. $(20, -12)$ **51.** No solution, $\varnothing$; inconsistent system
53. $\left(\frac{2}{3}, \frac{3}{2}\right)$ **55.** $\left(\frac{1}{2}, -3\right)$ **57.** $(2.75, -2.55)$ **59.** $(6, 2)$ **61.** $(2, 3)$
63. $\left(-\frac{1}{3}, 1\right)$ **65.** Alaska: 907; Hawaii: 808   **73.** $-\frac{5}{2}$ **75.** $-\frac{8}{5}$
77. $\frac{4}{3}$

ARE YOU READY? 3.3 (page 217)

1. $1, -4, 5$ **2.** 4 **3.** $7x + 5y - z = 9$ **4.** $(-5, 10)$

SELF CHECKS 3.3

1. Not a solution   **2.** $(2, 1, -4)$ **3.** $(1, 1, 2)$ **4.** $(40, -20, 5)$

5. $(-6, 0, 12)$ **6.** No solution, $\varnothing$; inconsistent system
7. Infinitely many solutions; dependent equations

STUDY SET SECTION 3.3 (PAGE 225)

1. system, standard   **3.** triples   **5.** dependent   **7. a.** No solution
b. No solution   **9.** $x + y + 4z = 3, 7x - 2y + 8z = 15,$
$3x + 2y - z = 4$ **11.** Yes   **13.** No   **15.** $(1, 1, 2)$
17. $(-1, 3, 0)$ **19.** $(3, -1, 3)$ **21.** $(-3, 0, 5)$ **23.** $(7, -6, 3)$
25. $\left(\frac{1}{2}, 2, -1\right)$ **27.** $(40, -20, 5)$ **29.** $\left(\frac{2}{3}, 2, -1\right)$ **31.** $(9, 0, -4)$
33. $(10, -5, 5)$ **35.** No solution, $\varnothing$; inconsistent system
37. Infinitely many solutions; dependent equations
39. $(8, 4, 5)$ **41.** $(12, -9, 4)$ **43.** $(3, 2, 1)$
45. No solution, $\varnothing$; inconsistent system   **47.** $\left(\frac{3}{4}, \frac{1}{2}, \frac{1}{3}\right)$
49. No solution, $\varnothing$; inconsistent system   **51.** $(2, 4, 8)$
53. Infinitely many solutions; dependent equations
55. $(2.5, 3, 3.5)$ **57.** $(2, 6, 9)$ **59. a.** Infinitely many solutions, all
lying on the line running down the binding   **b.** 3 parallel planes
(shelves); no solution   **c.** Each pair of planes (cards) intersect; no
solution   **d.** 3 planes (faces of die) intersect at a corner; 1 solution
61. $100, 81, 78$

67.

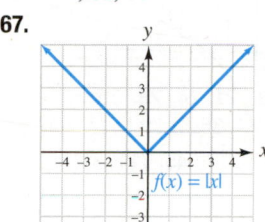

69.

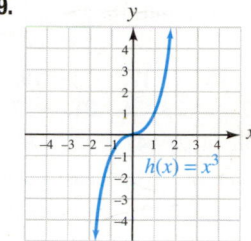

ARE YOU READY? 3.4 (page 228)

1. a. $3, -7$ **b.** $5, -1$ **2.** 1 **3.** 1 **4.** $-4x + 12y = -10$

SELF CHECKS 3.4

1. a. $\begin{bmatrix} 2 & -4 & | & 9 \\ 5 & -1 & | & -2 \end{bmatrix}$ **b.** $\begin{bmatrix} 1 & 1 & -1 & | & -4 \\ 0 & -2 & 7 & | & 0 \\ 10 & 8 & -4 & | & 5 \end{bmatrix}$

2. a. $\begin{bmatrix} 4 & -8 & | & 0 \\ 1 & -1 & | & 2 \end{bmatrix}$ **b.** $\begin{bmatrix} 1 & -1 & | & 2 \\ 0 & -4 & | & -8 \end{bmatrix}$ **c.** $\begin{bmatrix} 1 & \frac{1}{2} & -4 & | & 2 \\ 0 & 1 & 4 & | & -2 \\ 0 & 0 & -6 & | & 24 \end{bmatrix}$

3. $(-2, -5)$ **4.** $(1, -1, 2)$ **5. a.** No solution, $\varnothing$; inconsistent
system   **b.** $\{(x, y)\,|\,x - 3y = 6\}$, infinitely many solutions;
dependent equations

STUDY SET SECTION 3.4 (page 236)

1. matrix   **3.** rows, columns, by   **5.** row, Gauss, Jordan
7. a. 2×3 **b.** 3×4 **9. a.** $x, y, 10, 6$ **b.** $x, z, -16, 8, 4$
11. a. Interchange rows 1 and 2   **b.** Multiply row 1 by $\frac{1}{2}$
c. Add row 3 to 6 times row 2
13. $\begin{bmatrix} 1 & 2 & | & 6 \\ 3 & -1 & | & -10 \end{bmatrix}$ **15.** $\begin{cases} x + 6y = 7 \\ y = 4 \end{cases}$ **17.** $\begin{bmatrix} 1 & -4 & | & 4 \\ -3 & 1 & | & -6 \end{bmatrix}$
19. $\begin{bmatrix} 1 & -\frac{1}{3} & | & 2 \\ 1 & -4 & | & 4 \end{bmatrix}$ **21.** $\begin{bmatrix} 3 & 6 & -9 & | & 0 \\ -2 & 2 & -2 & | & 5 \\ 1 & 5 & -2 & | & 1 \end{bmatrix}$
23. $\begin{bmatrix} 3 & 6 & -9 & | & 0 \\ -2 & -1 & 7 & | & 1 \\ -2 & 2 & -2 & | & 5 \end{bmatrix}$ **25.** $(1, 1)$ **27.** $(2, -3)$

29. $(3, 2, 1)$ **31.** $(4, 5, 4)$ **33.** No solution, $\varnothing$; inconsistent system
35. $\{(x, y) \mid x + y = 3\}$, infinitely many solutions; dependent
equations   **37.** $(-2, -1, 1)$ **39.** $(-1, -1)$ **41.** $(0, -3)$
43. No solution, $\varnothing$; inconsistent system   **45.** $\{(x, y) \mid 4x - y = 2\}$,
infinitely many solutions; dependent equations
47. $\left(\frac{1}{2}, 1, -2\right)$ **49.** $(0, 1, 3)$ **51.** $(-4, 8, 5)$ **53.** $262{,}144$
55. $22°, 68°$ **57.** $\angle A: 40°, \angle B: 65°, \angle C: 75°$ **63.** $m = \frac{y_2 - y_1}{x_2 - x_1}$,
$(x_2 \neq x_1)$ **65.** $y - y_1 = m(x - x_1)$

ARE YOU READY? 3.5 (page 238)

1. 3 rows, 3 columns   **2.** $3, -1$ **3.** -54 **4.** -123

SELF CHECKS 3.5

1. 10   **2.** -44 **3.** -44 **4.** $(-2, 4)$ **5.** No solution, $\varnothing$;
inconsistent system   **6.** $(2, -2, 3)$

STUDY SET SECTION 3.5 (page 246)

1. determinant, diagonal   **3.** minor   **5.** ad, bc
7. The third column   **9.** $\begin{vmatrix} 1 & 2 & 0 \\ 3 & 1 & -1 \\ 8 & 4 & -1 \end{vmatrix}$ **11.** $(-2, -1, 1)$
13. $6, 30$ **15.** 4   **17.** -10 **19.** -170 **21.** 6   **23.** 35
25. 0   **27.** 4   **29.** 26   **31.** -37 **33.** 5   **35.** -79
37. 1   **39.** $(4, 2)$ **41.** $(-5, -8)$ **43.** $\left(5, \frac{3}{2}\right)$ **45.** $(11, 3)$
47. No solution, $\varnothing$; inconsistent system
49. $\{(x, y) \mid 5x = 12 - 6y\}$, infinitely many solutions;
dependent equations   **51.** $(1, 1, 2)$ **53.** $(-2, 0, 2)$
55. $(3, -2, 1)$ **57.** $(2, -1)$ **59.** $\left(-\frac{1}{2}, -1, -\frac{1}{2}\right)$
61. No solution, $\varnothing$; inconsistent system   **63.** No solution, $\varnothing$;
inconsistent system   **65.** $(-2, 3, 1)$ **67.** $\left(-\frac{1}{2}, \frac{1}{3}\right)$
69. Infinitely many solutions; dependent equations   **71.** $-46{,}811$
73. $-60{,}527{,}941$ **75.** $50°, 80°$ **83.** No   **85.** The graph of
function g is 2 units below the graph of function f.
87. y-intercept   **89.** $x; y$

ARE YOU READY? 3.6 (page 249)

1. $w - 150$ **2.** \$1,040   **3.** 72 mi   **4.** 32 lb

SELF CHECKS 3.6

1. Freshwater fish: 172 million; saltwater fish: 11 million
2. $27°, 63°$ **3.** Popcorn: \$5.50; medium drink: \$2.50
4. 5,000 plates   **5.** \$7,000 at 8%; \$8,000 at 3.5%   **6.** Boat:
15 mph; current: 6 mph   **7.** Peanuts: $5\frac{3}{4}$ lb; M&M's: $6\frac{1}{4}$ lb
8. Cream: 2 oz; 2% milk: 18 oz

STUDY SET SECTION 3.6 (page 259)

1. parallel   **3. a.** $70°, 75°$ **b.** alternate, interior
5. a. $x + c$ **b.** $x - c$ **7. a.** 0.06   **b.** 0.048   **c.** 0.135
9. $0.05x, 0.04y, 25{,}000, 1{,}050,$ $\begin{cases} x + y = 25{,}000 \\ 0.05x + 0.04y = 1{,}050 \end{cases}$
11. $13.90x, 5.10y, 6, 61.50,$ $\begin{cases} x + y = 6 \\ 13.90x + 5.10y = 61.50 \end{cases}$
13. 750 ohms, 625 ohms   **15.** Montana: 406; Idaho: 208
17. Gap: 3,094 stores; Aeropostalé, 952 stores   **19.** $75°, 25°$
21. $150°, 30°$ **23.** 16 m by 20 m   **25.** 15 sec: \$475; 30 sec: \$800

27. 85 racing bikes, 120 mountain bikes   **29.** Jonas Brothers,
\$123; Elton John, \$157   **31. a.** 400 tires
b.

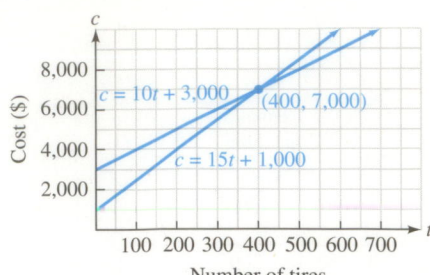

c. The second mold   **33. a.** $4{,}666\frac{2}{3}$ books   **b.** The newer press
35. a. 6,250 sets   **b.** At least 6,251 sets must be sold to make a
profit.   **37. a.** 590 units per month   **b.** 620 units per month
c. Process A (It results in a smaller loss)   **39.** \$3,000 at 10%,
\$5,000 at 12%   **41.** Credit union: \$8,000; money market: \$24,000
43. VISA: \$11,800; Robinsons-May: \$4,700   **45.** 525 mph,
75 mph   **47.** Walking: 6 ft per sec; moving walkway: 2 ft per sec
49. 25 mph, 5 mph   **51.** Gummy bears: 45 lb; jelly beans: 15 lb
53. Regular: $14\frac{2}{3}$ lb; Kona: $5\frac{1}{3}$ lb   **55.** Small flake: 900 lb;
mylar: 1,100 lb   **57.** 10%: 8 pints; 40%: 16 pints
59. 148 g of the 0.2%, 37 g of the 0.7%   **67.** rational
69. identity   **71.** isosceles

ARE YOU READY? 3.7 (page 265)

1. $n + 75$ **2.** \$150x   **3.** $5n¢$ **4.** $(-4, -4, 2)$

SELF CHECKS 3.7

1. 1-GB: 200 cards; 2-GB: 200 cards; 4-GB: 100 cards
2. Norway: 263 medals; United States: 193 medals;
Austria: 162 medals   **3.** $y = x^2 + 4x + 1$

STUDY SET SECTION 3.7 (page 270)

1. satisfy   **3.** $\begin{cases} x + y + z = 50 \\ 5x + 6y + 7z = 295 \\ 2x + 3y + 4z = 145 \end{cases}$ **5.** $-3 = 4a + 2b + c$
7. 30 large, 50 medium, 100 small
9. a. $2a, 2b, c, 13;$ $\begin{cases} 2a + 3b + c = 14 \\ 3a + 2b + c = 13 \\ 2a + b + 2c = 9 \end{cases}$
b. Food A: 2 oz, Food B: 3 oz, Food C: 1 oz   **11.** 120 coats, 200
shirts, 150 slacks   **13.** Young: 85 TD passes, Montana: 55 TD
passes, Gannon: 16 TD passes   **15.** Nitrogen: 78%, oxygen: 21%,
other: 1%   **17.** $\angle A: 40°, \angle B: 60°, \angle C: 80°$
19. *X-Files:* 201 episodes, *Will & Grace:* 194 episodes, *Seinfeld:*
180 episodes   **21.** r_1: 3 yd, r_2: 7 yd, r_3: 11 yd
23. 6 lb rose petals, 3 lb lavender, 1 lb buckwheat hulls
25. Nickels: 20, dimes: 40, quarters: 4   **27.** $y = \frac{1}{2}x^2 - 2x - 1$
29. $x^2 + y^2 - 2x - 2y - 2 = 0$ **33.** Yes   **35.** Yes
37. No; $(1, 2), (1, -2)$ **39.** No; $(4, 2), (4, -2)$

CHAPTER 3 REVIEW (page 273)

1. Yes   **2.** No   **3. a.** $(1, 3), (2, 1), (4, -3)$(Answers may vary)
b. $(0, -4), (2, -2), (4, 0)$(Answers may vary)   **c.** $(3, -1)$
4. 2019, about 15 hr

5. $(3, 5)$

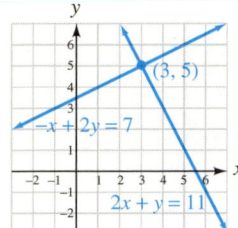

6. $(-2, 3)$

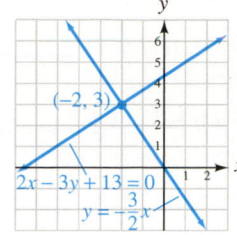

e. matrix   **2.** $(2, 1)$

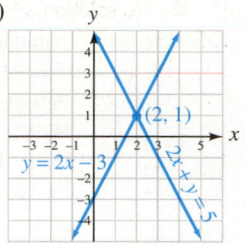

7. $\{(x, y) \mid 3x + 2y = 12\}$, infinitely many solutions; dependent equations

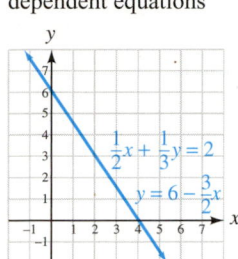

8. No solution, $\varnothing$; inconsistent system

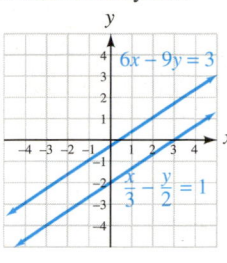

9. 2   **10.** -1 **11.** $(-1, 3)$ **12.** $(-3, -1)$ **13.** $(-5, -8)$
14. $(-1, -1)$ **15.** $(-6, 16)$ **16.** $\left(4, \frac{1}{2}\right)$ **17.** $\left(\frac{1}{4}, -\frac{1}{2}\right)$
18. $\{(x, y) \mid x + 3y = -2\}$, infinitely many solutions; dependent equations   **19.** No solution, $\varnothing$; inconsistent system   **20.** $(3, 4)$
21. $(-1, 0.7)$ (Answers may vary); $\left(-1, \frac{2}{3}\right)$ **23.** Not a solution
24. Yes; one solution   **25.** $(2, 0, -3)$ **26.** $(2, -1, 3)$
27. $(-1, 1, 3)$ **28.** $(-5, -14, -4)$ **29.** $(6, -2, 2)$ **30.** $(1, 4, 1)$
31. No solution, $\varnothing$; inconsistent system   **32.** Infinitely many solutions; dependent equations
33. $\begin{bmatrix} 5 & 4 & \vdots & 3 \\ 1 & -1 & \vdots & -3 \end{bmatrix}$ **34.** $\begin{bmatrix} 1 & 2 & 3 & \vdots & 6 \\ 1 & -3 & -1 & \vdots & 4 \\ 6 & 1 & -2 & \vdots & -1 \end{bmatrix}$
35. a. $\begin{bmatrix} 1 & 3 & \vdots & -2 \\ 6 & 12 & \vdots & -6 \end{bmatrix}$ **b.** $\begin{bmatrix} 1 & 2 & \vdots & -1 \\ 1 & 3 & \vdots & -2 \end{bmatrix}$ **c.** $\begin{bmatrix} 0 & -6 & \vdots & 6 \\ 1 & 3 & \vdots & -2 \end{bmatrix}$
36. a. $\begin{bmatrix} 1 & 1 & 0 & \vdots & -1 \\ 2 & -1 & 1 & \vdots & 3 \\ 3 & -1 & -2 & \vdots & 7 \end{bmatrix}$ **b.** $\begin{bmatrix} 2 & -1 & 1 & \vdots & 3 \\ 3 & 3 & 0 & \vdots & -3 \\ 3 & -1 & -2 & \vdots & 7 \end{bmatrix}$
c. $\begin{bmatrix} 0 & -3 & 1 & \vdots & 5 \\ 1 & 1 & 0 & \vdots & -1 \\ 3 & -1 & -2 & \vdots & 7 \end{bmatrix}$ **37.** $(1, -3)$ **38.** $(5, -3, -2)$
39. $\{(x, y) \mid -2x + y = -4\}$, infinitely many solutions; dependent equations   **40.** No solution, $\varnothing$; inconsistent system   **41.** 18
42. 38   **43.** -3 **44.** 28   **45.** $(2, 1)$ **46.** No solution, $\varnothing$; inconsistent system   **47.** $(1, -2, 3)$ **48.** $(-3, 2, 2)$
49. Austin–Houston: 162 mi; Austin–San Antonio: 83 mi
50. 8 mph, 2 mph   **51.** 500 oz of 6%, 250 oz of 18%
52. $4,000 at 6%, $6,000 at 12%   **53.** Teaspoon: 5 mL; tablespoon: 15 mL   **54.** 17,500 bottles   **55.** Small: 50; medium: 60; large: 40   **56.** $5,000 at 5%, $7,000 at 6%, and $10,000 at 7%
57. $a = -\frac{1}{16}, b = 2, c = 0$ **58.** 2 cups mix A, 1 cup mix B, 1 cup mix C

CHAPTER 3 TEST (page 282)

1. a. system   **b.** rows, columns   **c.** triples   **d.** plane

3. It is a solution.   **4.** The point of intersection is (6/06, 44%). Governor Schwarzenegger's job approval and disapproval ratings were the same in June of 2006, approximately 44%.
5. a. Inconsistent system; no solution, $\varnothing$ **b.** Dependent equations; infinitely many solutions   **6.** 3   **7.** $(7, 0)$ **8.** $(2, -3)$
9. $\{(x, y) \mid 2x + 3y = -3\}$, infinitely many solutions; dependent equations   **10.** $\left(\frac{1}{3}, 2\right)$ **11.** 55, 70   **12.** 15 gal of 40%, 5 gal of 80%
13. 1,375 impressions   **14.** $x = 68°; y = 40°$ **15.** 2.5% loan: $5,000; 4% loan: $3,500   **16.** Pear: $2.95; apple: $1.95
17. No   **18.** No solutions   **19.** $(3, 2, -1)$ **20.** $(-6, 2, 5)$
21. Children: 60, general admission: 30, seniors: 10
22. Bar: 45 lb; large plate: 45 lb; small plate: 10 lb
23. $\begin{bmatrix} 1 & 7 & \vdots & -3 \\ 0 & -22 & \vdots & 22 \end{bmatrix}$ **24.** $(2, 2)$ **25.** No solution, $\varnothing$; inconsistent system   **26.** $(-2, 4, 2)$ **27.** -2 **28.** 4   **29.** $(-3, 3)$ **30.** -1

CHAPTERS 1–3 CUMULATIVE REVIEW (page 284)

1.

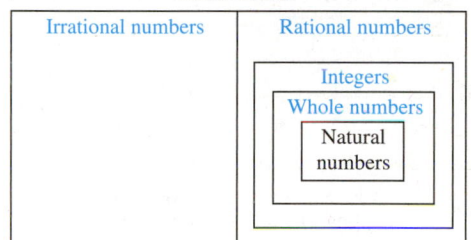

2. a. $<$
b. $>$ **3.** 70
4. 2
5. $15m + 22$
6. $-4a + 87$
7. -28
8. 2   **9.** -2
10. All real numbers, $\mathbb{R}$, identity
11. $B = \frac{\lambda - Ax}{A}$ **12.** $d_2 = d_1 - vt$ **13.** Los Angeles: 6 wk; Las Vegas: 4 wk; Dallas: 7 wk   **14.** $1,372 billion   **15.** 20 mph
16. 5 lb apple slices, 5 lb banana chips
17.

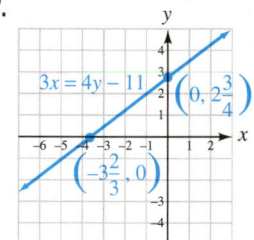

18.

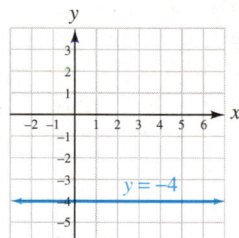

19. $-\frac{4}{5}$ **20.** Perpendicular   **21.** $y = -3x + 17$
22. $v = 17.5x + 300$ **23.** -105 **24.** $-r^2 - \frac{r}{2}$
25. $f(x) = x^3$ (Answers may vary)   **26.** Yes
27. a. $A(t) = -170t + 5,350$ **b.** 1,100 accidents   **28. a.** 1
b. 2   **29.** No. It does not pass the vertical line test.
30. D: the set of real numbers, R: the set of real numbers greater than or equal to 0   **31.** $g(x) = |x| + 3$

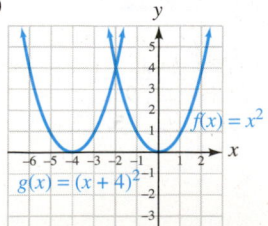

32. $(-2, -3)$

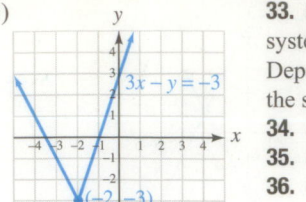

33. a. An inconsistent system has no solution. **b.** Dependent equations have the same graph.
34. $(2, -1)$
35. $(-1, 0, 2)$
36. $(8, 3)$ **37. a.** 40
b. 26   **38.** $\left(\frac{14}{3}, \frac{2}{3}\right)$

39. 200 of the $67 phones, 160 of the $100 phones
40. $2,500 in the 12-month CD, $15,000 in the 24-month CD, $12,500 in the 36-month CD

ARE YOU READY? 4.1 (page 288)

1. is greater than   **2.** False   **3.**

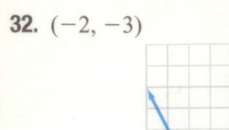

4. $0 < 6$

SELF CHECKS 4.1

1. a. $(1, \infty)$; $\{x \mid x > 1\}$

b. $(-\infty, -2]$; $\{x \mid x \le -2\}$

2. $(-\infty, -6)$; $\{x \mid x < -6\}$

3. $[3, \infty)$; $\{x \mid x \ge 3\}$

4. $\left(\frac{9}{2}, \infty\right)$; $\left\{s \mid s > \frac{9}{2}\right\}$

5. $\left(-\infty, -\frac{16}{3}\right]$; $\left\{x \mid x \le -\frac{16}{3}\right\}$

6. a. $(-\infty, \infty)$, $\mathbb{R}$ **b.** No solution, $\varnothing$

7. 4,600 rpm or less   **8.** Be Thrifty's plan is better if the car is going to be driven more than 100 miles that day.

STUDY SET SECTION 4.1 (page 298)

1. inequality   **3.** interval   **5.** builder, such that
7. $7t - 5 > 4$, $\frac{x}{2} \le -1$ **9. a.** Yes   **b.** No   **c.** Yes   **d.** No
11. a. $(-\infty, \infty)$ **b.** $\varnothing$

13. $-10, \le, -5, 2, -\infty, x \le 2$ **15.** $<$
17. $(-\infty, 14)$; $\{x \mid x < 14\}$

19. $[-2, \infty)$; $\{x \mid x \ge -2\}$

21. $(-\infty, 1)$; $\{x \mid x < 1\}$

23. $(-3, \infty)$; $\{x \mid x > -3\}$

25. $[-11, \infty)$; $\{x \mid x \ge -11\}$

27. $(-\infty, 2]$; $\{a \mid a \le 2\}$

29. $(0, \infty)$; $\{x \mid x > 0\}$

31. $(-\infty, 7]$; $\{t \mid t \le 7\}$

33. $\left[\frac{10}{3}, \infty\right)$; $\left\{x \mid x \ge \frac{10}{3}\right\}$

35. $\left(-\infty, -\frac{16}{7}\right)$; $\left\{x \mid x < -\frac{16}{7}\right\}$

37. $\left(-\infty, \frac{11}{5}\right)$; $\left\{x \mid x < \frac{11}{5}\right\}$

39. $\left[-\frac{18}{13}, \infty\right)$; $\left\{n \mid n \ge -\frac{18}{13}\right\}$

41. $(-\infty, \infty)$; $\mathbb{R}$ **43.** No solution; $\varnothing$

45. $\left[-\frac{2}{5}, \infty\right)$; $\left\{t \mid t \ge -\frac{2}{5}\right\}$

47. $\left(-\infty, \frac{1}{2}\right)$; $\left\{x \mid x < \frac{1}{2}\right\}$

49. $[60, \infty)$; $\{x \mid x \ge 60\}$

51. $(6, \infty)$; $\{a \mid a > 6\}$

53. $(-\infty, 3)$; $\{y \mid y < 3\}$

55. $(-\infty, -6]$; $\{d \mid d \le -6\}$

57. $(-\infty, 1.5]$; $\{x \mid x \le 1.5\}$

59. $(-\infty, 20]$; $\{z \mid z \le 20\}$

61. No solution; $\varnothing$ **63.** $[6, \infty)$; $\{x \mid x \ge 6\}$

65. $\left(-\infty, \frac{4}{3}\right]$; $\left\{x \mid x \le \frac{4}{3}\right\}$

67. $(-\infty, \infty)$; $\mathbb{R}$

69. $[-36, \infty)$; $\{y \mid y \ge -36\}$

71. $(-\infty, 10)$; $\{b \mid b < 10\}$

73. $\left(-\infty, \frac{45}{7}\right]$; $\left\{x \mid x \le \frac{45}{7}\right\}$

75. $(-\infty, 3]$; $\{t \mid t \le 3\}$

77. $(-4, \infty)$; $\{x \mid x > -4\}$

79. $(-1.5, \infty)$; $\{x \mid x > -1.5\}$

81. a. $(-\infty, 3.25]$; $\{x|x \le 3.25\}$

b. $(3.25, \infty)$; $\{x|x > 3.25\}$

83. a. $\left(-\infty, \frac{11}{2}\right)$; $\{x|x < \frac{11}{2}\}$

b. $\left(\frac{11}{2}, \infty\right)$; $\{x|x > \frac{11}{2}\}$

85. $(-\infty, 1)$ **87.** $(2, \infty)$ **89. a.** Midwest, South
b. Northeast, West   **91.** 50 min or longer   **93.** Years after 2002
95. More than 75 miles   **97.** They could have sold as many as 399
towels.   **99.** 88 or higher   **101.** 10 hr   **103.** 8 hr   **111.** 4, 5, 3

ARE YOU READY? 4.2 (page 302)

1. 3 and 4   **2.** $[-7, \infty)$ **3.** Yes

4.

SELF CHECKS 4.2

1. a. $\{9\}$ **b.** $\{3, 6, 8, 9, 10, 11, 12, 15\}$

2. $(-6, 10]$ **3.** $\left(\frac{1}{2}, \infty\right)$

4. No solution; $\varnothing$ **5.** $[1, 5]$

6. $(-\infty, 2) \cup (4, \infty)$

7. $(-\infty, \infty)$

STUDY SET SECTION 4.2 (page 310)

1. intersection, union   **3.** double   **5. a.** both   **b.** one, both
7. a. intersection   **b.** union   **9. a.** No   **b.** Yes
11. a. $[-2, 1)$ **b.** $[2, 2]$ **c.** $\varnothing$ **13.** union, intersection
15. All real numbers   **17.** $\{4, 6\}$ **19.** $\{-3, 1, 2\}$

21. $\{-3, -1, 0, 1, 2, 4, 6, 8, 10\}$ **23.** $\{-3, 0, 1, 2, 3, 4, 5, 6, 8\}$
25. $(-2, 5]$ **27.** $(2, 3]$

29. $(-\infty, -6]$

31. No solution; $\varnothing$
33. $[1, 4]$

35. $(0.8, 1.1)$

37. $(-\infty, -2] \cup (6, \infty)$

39. $(-\infty, -1) \cup (2, \infty)$

41. $(-\infty, 2) \cup (7, \infty)$

43. $(-\infty, \infty)$ **45.** $(-\infty, 1)$

47. $(-10, -9)$

49. $(-2, 5)$ **51.** $(-\infty, \infty)$

53. $[2, \infty)$ **55.** No solution; $\varnothing$

57. $[-4, 6)$ **59.** $[2, 2]$

61. $(-\infty, 4.8) \cup (6.5, \infty)$

63. $(-0.7, 0.2]$

65. $(-12, -6]$

67. $\left(-\infty, \frac{1}{6}\right] \cup (2, \infty)$

69. $[-2, 4]$ **71.** $[4, 4]$

73. $(1, 3]$ **75.** $(3, 6)$

77. a. $[6, \infty)$ **b.** $[2, \infty)$

79. a. No solution; $\varnothing$ **b.** $(-\infty, 3] \cup [6, \infty)$

81. a. 128, 192   **b.** $32 \le s \le 48$, $[32, 48]$ **83. a.** $(67, 77)$
b. $(62, 82)$ **85. a.** 2004   **b.** 2004, 2005, 2006, 2007   **c.** 2006,
2007   **d.** 2007   **87. a.**   **b.**
95. An 18% increase

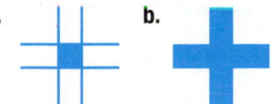

ARE YOU READY? 4.3 (page 314)

1. a. 12   **b.** 7.5   **2. a.** True   **b.** False
3. $(-\infty, -3] \cup [1, \infty)$

4. $(-8, 4)$

SELF CHECKS 4.3

1. a. $\frac{1}{2}, -\frac{1}{2}$ **b.** 5, -2 **c.** No solution, $\varnothing$ **2.** 7.5, 2.5   **3.** -6

4. 7, -15 **5.** $-1, -6$ **6.** $\left(-2, \frac{2}{3}\right)$

7. $[2.8985, 2.9015]$ **8.** No solution, $\varnothing$
9. $(-\infty, -2] \cup [6, \infty)$

10.
$(-\infty, -8) \cup \left(\frac{8}{3}, \infty\right)$

11. $(-\infty, \infty)$

STUDY SET SECTION 4.3 (page 323)

1. absolute value   **3.** isolate   **5.** compound   **7.** compound
9. a. $-2, 2$ **b.** $-1.99, -1, 0, 1, 1.99$ **c.** $-3, -2.01, 2.01, 3$

11. a. $8, -8$ **b.** $x - 3, -(x - 3)$ **13. a.** $x = 8$ or $x = -8$
b. $x \le -8$ or $x \ge 8$ **c.** $-8 \le x \le 8$ **d.** $5x - 1 = x + 3$ or
$5x - 1 = -(x + 3)$ **15. a.** No solution   **b.** No solution
c. All real numbers   **17. a.** ii   **b.** iii   **c.** i   **19.** $23, -23$
21. $13, -3$ **23.** $\frac{14}{3}, -6$ **25.** $50, -50$ **27.** $3.1, -6.7$
29. No solution; $\varnothing$ **31.** $25, -19$ **33.** $7, -\frac{7}{3}$ **35.** $2, -\frac{1}{2}$
37. $\frac{32}{7}, -16$ **39.** -10 **41.** -8 **43.** $\frac{9}{2}$ **45.** $-\frac{1}{2}$ **47.** $-4, \frac{28}{9}$
49. $-2, \frac{18}{11}$ **51.** $0, -2$ **53.** $11, \frac{1}{3}$
55. $(-4, 4)$
57. $[-21, 3]$
59. $\left(-\frac{8}{3}, 4\right)$ **61.** No solution; $\varnothing$
63. $(-\infty, -3) \cup (3, \infty)$
65. $(-\infty, -12) \cup (36, \infty)$
67. $\left(-\infty, -\frac{1}{5}\right] \cup \left[\frac{3}{5}, \infty\right)$
69. $(-\infty, \infty)$ **71.** $0, -6$ **73.** $(-3, 1)$
75. $\left(-\infty, -\frac{16}{3}\right) \cup (4, \infty)$
77. $[-10, 14]$ **79.** $40, -20$
81. $(-\infty, \infty)$ **83.** $-3, -\frac{3}{4}$
85. $\frac{2}{3}$ **87.** No solution; $\varnothing$ **89.** $\frac{12}{11}, -\frac{12}{11}$
91. $(-\infty, -2) \cup (5, \infty)$ **93.** No solution; $\varnothing$
95. a. 20 **b.** $20, 0$ **c.** $(20, \infty)$
d. $(-\infty, 0) \cup (20, \infty)$
97. a. -25 **b.** $-25, 31$ **c.** $(-\infty, -25)$
d. $(-\infty, -25) \cup (31, \infty)$
99. a. $[3, 7]$ **b.** $(-\infty, 3] \cup [7, \infty)$
101. a. $(-\infty, -8) \cup (10, \infty)$
b. $[-8, 10]$
103. $70° \le t \le 86°$ **105. a.** $|c - 0.6°| \le 0.5°$ **b.** $[0.1°, 1.1°]$
107. $26.45\%, 24.76\%$ **113.** $50°, 130°$

---

1. False   **2.** True   **3.**   **4. a.** Below
            b. Above   **c.** On

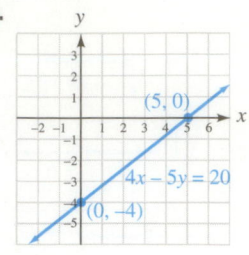

SELF CHECKS 4.4

1.
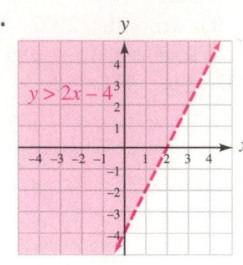

2.
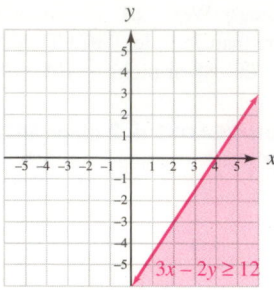

3.
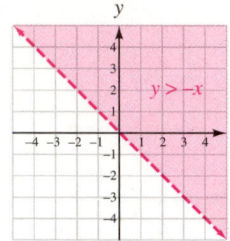

4.
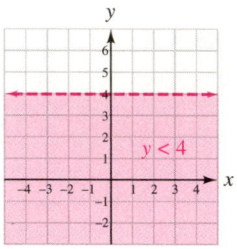

5.

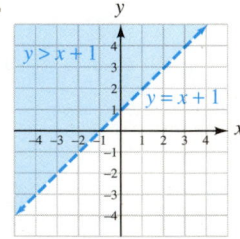

$(2, 12), (6, 10), (14, 2)$;
answers may vary.

---

1. linear, two   **3.** boundary   **5. a.** Yes   **b.** No   **c.** Yes   **d.** No
7. a. $m = 3; (0, -1)$ **b.** $(-3, 0), (0 -2)$ **9. a.** No; dashed
b. Yes; solid   **c.** Yes; solid   **d.** No; dashed
11.   **13.**

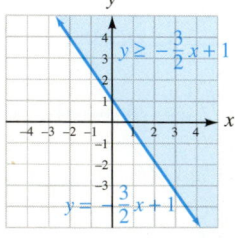

15.

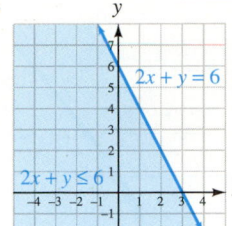

17.

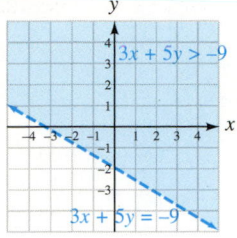

19.

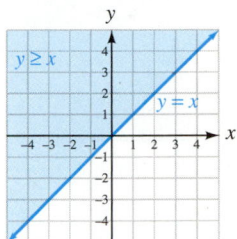

21.

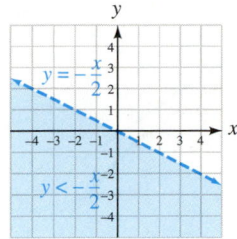

23.

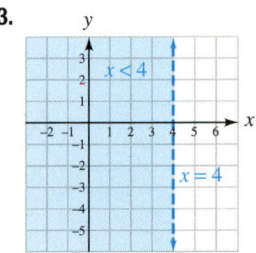

25.

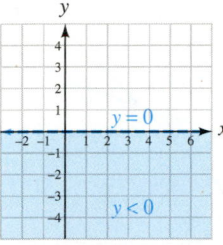

27.

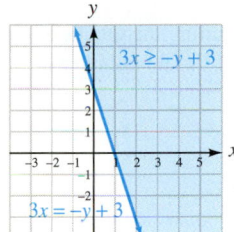

29.

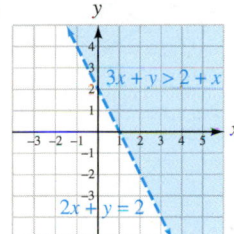

31.

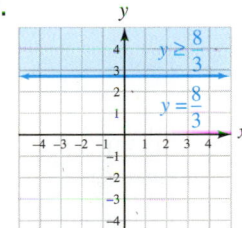

33.

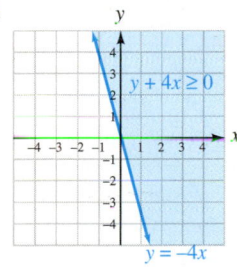

35.

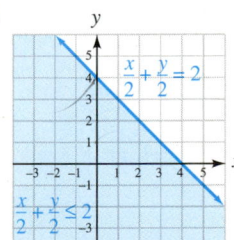

37.

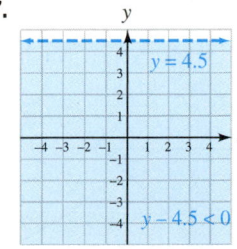

39.

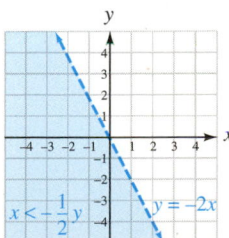

41.

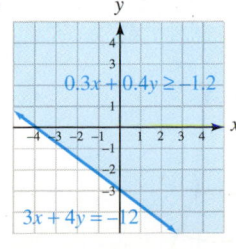

43. a.

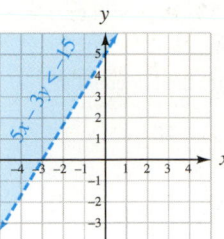

b.

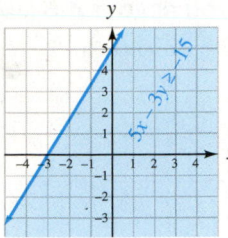

45. a.

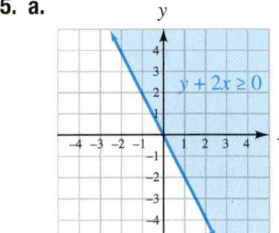

b.

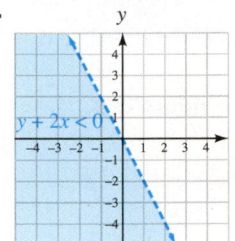

47.

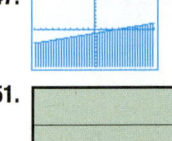

49.

51.

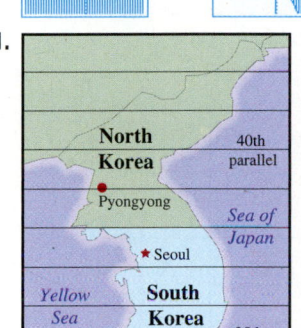

53. $4x + 6y \leq 120$; $(5, 15)$, $(15, 10)$, $(20, 5)$

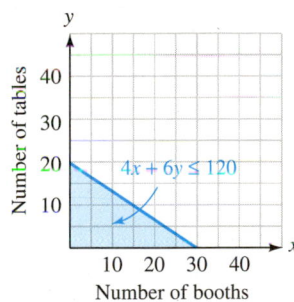

55. $10x + 15y \geq 1,200$; $(40, 80)$, $(80, 80)$, $(120, 40)$

61. Yes **63.** $(-2, -3)$

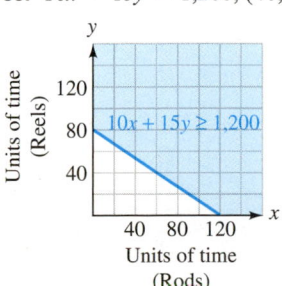

ARE YOU READY? 4.5 (page 335)

1. It is not a solution   **2.** A vertical line

3.

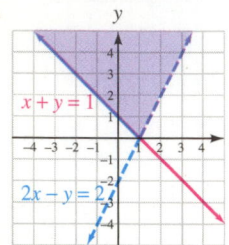

4. Greater than or equal to

SELF CHECKS 4.5

1.

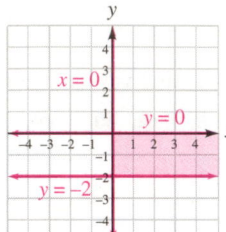

2.

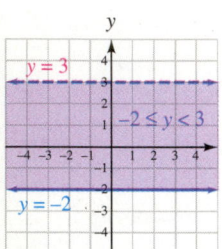

3.

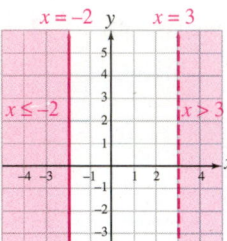

4.

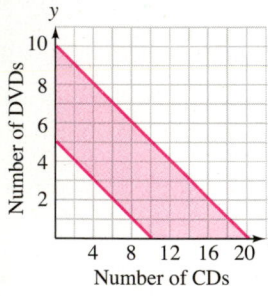

Some possible combinations are (2, 5), (4, 4), (14, 2); answers may vary.

5.

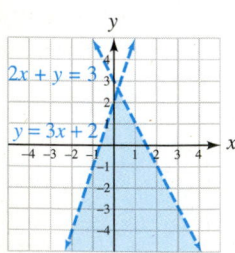

STUDY SET SECTION 4.5 (page 340)

1. inequalities   **3.** point   **5. a.** Yes   **b.** No   **c.** No   **d.** Yes
7. a. False   **b.** True   **c.** True   **d.** False   **e.** True   **f.** True

9.

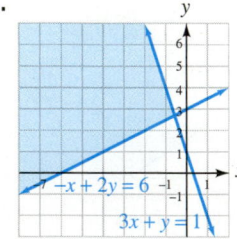

11.

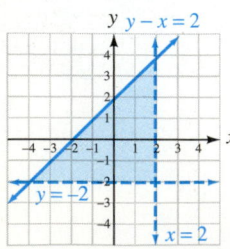

13.

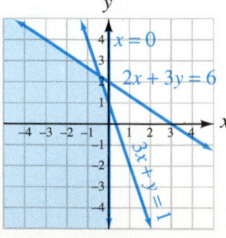

15.

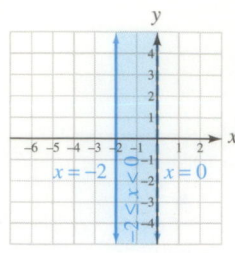

17.

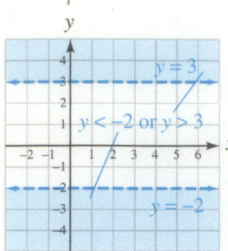

19.

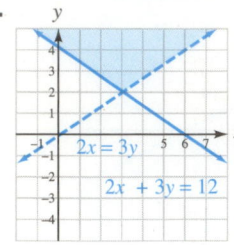

21.

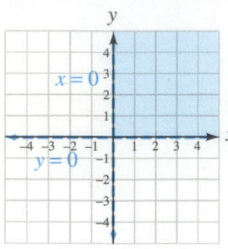

23.

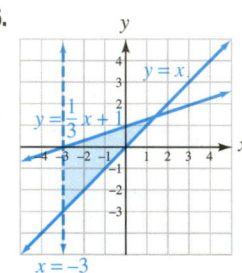

25.

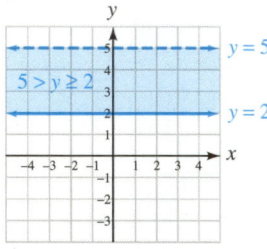

27.

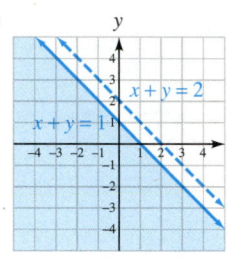

29.

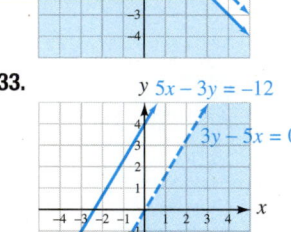

31. No solution; the graphs of the inequalities do not intersect.

33.

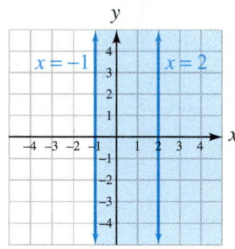

35.

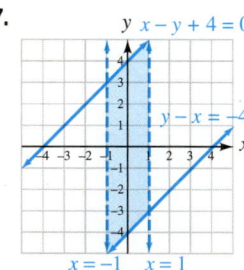

37.

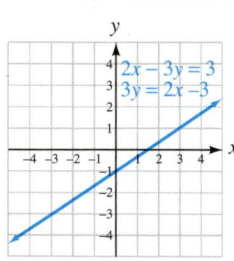

39.

41.

43.

45.

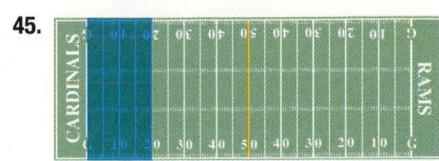

← Cardinals moving this direction

47.

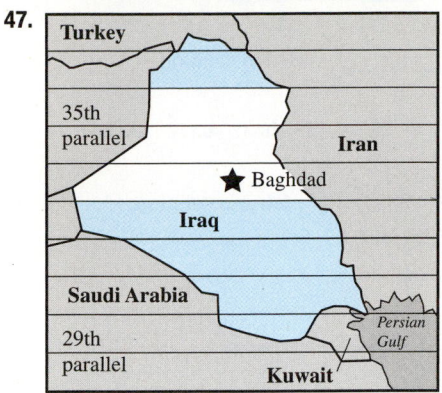

49. 1 $10 CD and 2 $15 CDs,
4 $10 CDs and 1 $15 CD

51. 1 air cleaner and 2 humidifiers;
2 air cleaners and 3 humidifiers

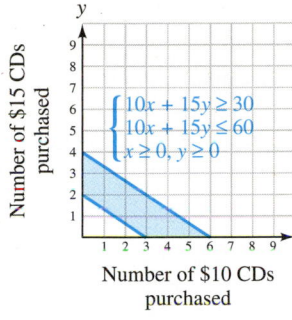

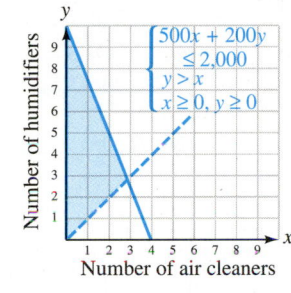

57. IV **59.** II

CHAPTER 4 REVIEW (page 343)

1. a. No **b.** Yes

2. $[-5, \infty)$, $\{x \mid x \geq -5\}$

3. $(-\infty, 3]$; $\{x \mid x \leq 3\}$

4. $[4, \infty)$; $\{x \mid x \geq 4\}$

5. $(-\infty, 20)$; $\{t \mid t < 20\}$

6. $\left(-\infty, -\frac{51}{11}\right)$; $\left\{x \mid x < -\frac{51}{11}\right\}$

7. $(-\infty, \infty)$, $\mathbb{R}$ **8.** No solution; $\varnothing$

9. $\left(-\frac{41}{12}, \infty\right)$ **10.** $\left(-\infty, \frac{1}{2}\right]$

11. More than 12.5 gallons of paint per week **12.** More than 2,100 business cards **13.** $20,000 or more **14.** She needs to receive a score that is greater than 5.2. **15.** 5 hr **17.** $\{-3, 3\}$
18. $\{-6, -5, -3, 0, 3, 6, 8\}$ **19.** Yes **20.** No

21. **22.**

23. $[-10, -4]$

24. $(-\infty, -11)$ **25.** No solution; $\varnothing$

26. $[0, 0]$ **27.** $\left(-\frac{1}{3}, 2\right)$

28. $[1, 9]$ **29.** Yes **30.** No

31. $(-\infty, -5) \cup (4, \infty)$

32. $(-\infty, \infty)$

33. $17 \leq 4x \leq 25$, 4.25 ft $\leq x \leq 6.25$ ft, $[4.25, 6.25]$
34. a. ii, iv **b.** i, iii **35.** $(-\infty, 7.36) \cup [137.6, \infty)$

36. $\left(-\frac{7}{3}, \frac{1}{3}\right]$ **37.** $2, -2$ **38.** $3, -\frac{11}{3}$ **39.** $\frac{26}{3}, -\frac{10}{3}$

40. No solution; $\varnothing$ **41.** 3 **42.** $\frac{1}{8}, -\frac{19}{8}$ **43.** $\frac{1}{5}, -5$ **44.** $\frac{13}{12}$

45. $[-3, 3]$

46. $(-5, -2)$

47. $\left[-3, \frac{19}{3}\right]$ **48.** No solution; $\varnothing$

49. $(-\infty, -1) \cup (1, \infty)$

50. $(-\infty, -4] \cup \left[\frac{22}{5}, \infty\right)$

51. $\left(-\infty, \frac{4}{3}\right) \cup (4, \infty)$

52. $(-\infty, \infty)$, $\mathbb{R}$

53. a. $8, 2$ **b.** $[6, 10]$ **54.** $3, -3$ **55.** $0, -8$
56. $(2, 12)$

57. Since $|0.04x - 8.8|$ is always greater than or equal to 0 for any real number x, this absolute value inequality has no solution.
58. Since $\left|\frac{3x}{50} + \frac{1}{45}\right|$ is always greater than or equal to 0 for any real number x, this absolute value inequality is true for all real numbers.
59. No **60.** Yes
61. **62.**

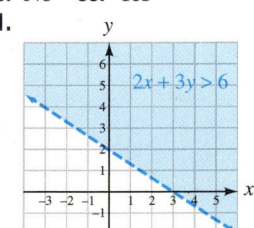

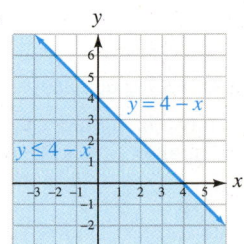

63.
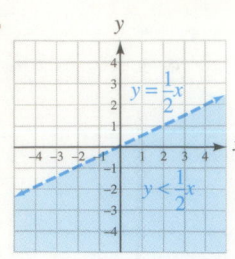

64.
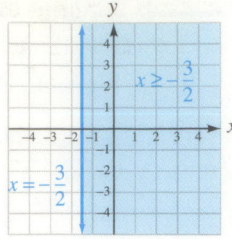

65. $3x - 4y > 12$

66. $6x + 4y \geq 10{,}200$; (1,800, 0), (1,000, 1,500), (2,000, 2,000)

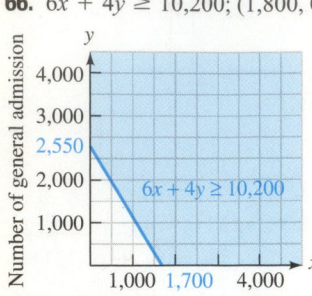

67. Yes   **68. a.** True
b. False   **c.** True   **d.** False
e. True   **f.** True

69.
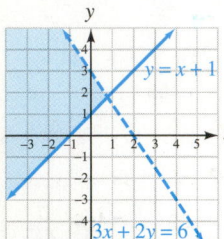

70.
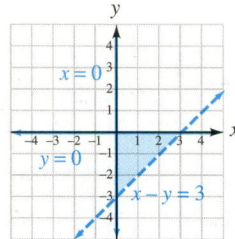

71.
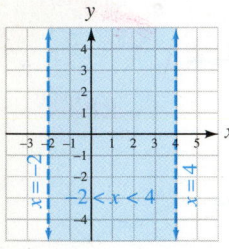

72.
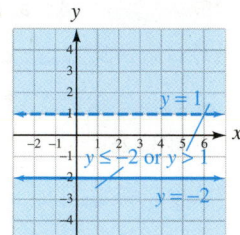

73. 5 $20 shirts and 15 $30 shirts, 15 $20 shirts and 10 $30 shirts
74. $\geq, \leq, \geq, \leq$

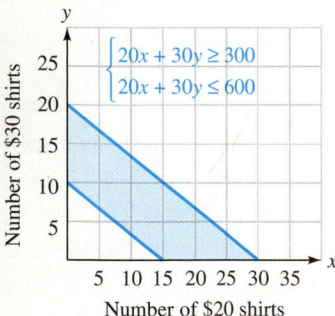

CHAPTER 4 TEST (page 350)

1. a. inequality   **b.** infinity   **c.** compound
d. union, intersection   **e.** system, two   **2.** Yes
3. $(12, \infty)$
4. $(-\infty, -5]$
5. $(-\infty, -14)$
6. $(-\infty, \infty)$
7. More than 78   **8.** 11 hr   **9.** The years from 1990 through 1999

10. 50 hours or more   **11.** No   **12.** $\{-4, 0, 11\}$
13. $\{-5, -4, 0, 7, 8, 9, 10, 11\}$ **14. a.**

b.   **15.** $\left[1, \dfrac{9}{4}\right]$

16. $(-\infty, -3) \cup (8, \infty)$

17. $(-2, 16)$ **18.** No solution; $\varnothing$

19. $(229, \infty)$ **20. a.** iii.   **b.** i.   **c.** iv.   **d.** ii.

21. $-5, \dfrac{23}{3}$ **22.** $4, -4$ **23.** $\dfrac{8}{9}, -\dfrac{8}{9}$ **24.** No solution; $\varnothing$
25. 10   **26.** $|x - 0.0625| \leq 0.0015$; [0.0610, 0.0640]
27. $[-7, 1]$

28. $(-\infty, -9) \cup (13, \infty)$

29. $(-\infty, 1) \cup (3, \infty)$

30. $\left[\dfrac{4}{3}, \dfrac{8}{3}\right]$

31. $(-\infty, \infty), \mathbb{R}$ **32.** No solution, $\varnothing$

33. $(-6, -3)$ **34.** $(5, 35)$

35.
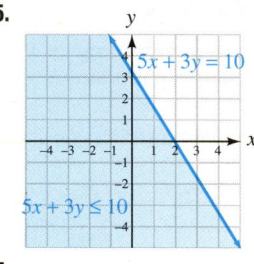

36.
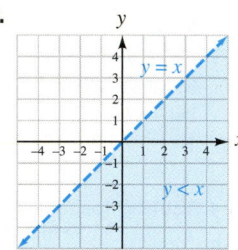

37.
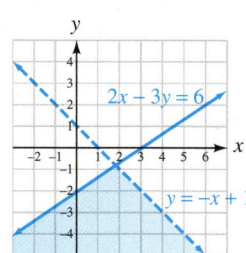

38.
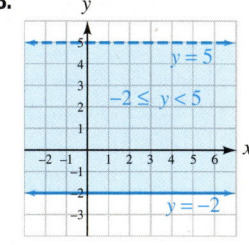

39.
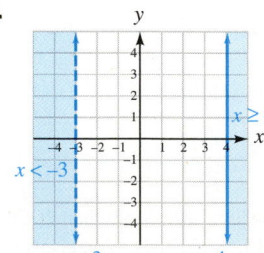

40. $(1, 1), (2, 1), (2, 2)$

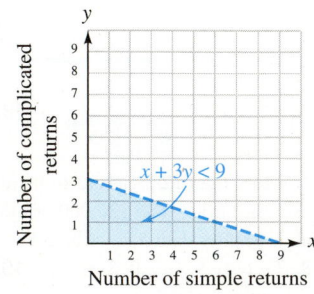

41. a. $(3, -4)$ is a solution of inequality 2.   **b.** No, it does not lie in the doubly shaded region.   **42.** $\leq, \geq, \geq, \leq$ **43. a.** ii.   **b.** i.
c. iv.   **d.** iii.   **44.** No, $10 \leq x < 3$ implies that $10 \leq 3$.

CHAPTERS 1–4 CUMULATIVE REVIEW (page 352)

1. rational, irrational

2. $\frac{p}{2} = 0.0625, \frac{p}{4} = 0.03125, .433p = 0.054125$

3. 10 **4.** -6 **5. a.** $x + 9$ **b.** $(6 \cdot 10)n$ **6. a.** $-x$ **b.** $\frac{1}{x}$

7. $-2a + b - 2$ **8.** $8t - 9$ **9.** 3 **10.** 6 **11.** No solution; $\varnothing$

12. -2 **13.** $d = \frac{l-a}{n-1}$ **14.** 201 ft^2 **15.** \$20,000 **16.** $\frac{1}{4}$ hr

17. $-\frac{8}{5}$ **18. a.** 26,400 prisoners/yr **b.** 1990–1995;

67,200 prisoners/yr **19.** Parallel **20.** $y = \frac{1}{3}x + \frac{11}{3}$

21. D: $\{-12, -6, 5, 8\}$, R: $\{-6, 4, 6\}$ **22.** No; $(1, 1), (1, -1)$

23. 14 **24.** $3t^2 - t$ **25. a.** $f(x) = -0.06x + 8.2$ **b.** 2.8 L/min

26. a. 1 **b.** $-2, 2$ **c.** D: $(-\infty, \infty)$, R: $[0, \infty)$

27. Not a function; $(3, 4), (3, 8)$ **28.** It defines a function.

29. D: all real numbers; R: all real numbers greater than or equal to -2

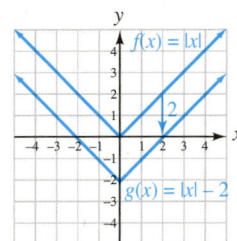

30. Yes **31.** $(2, 1)$

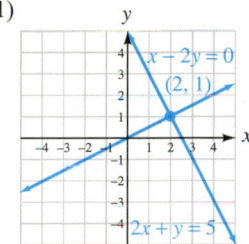

32. In 1982, 50% of the degrees awarded went to men and 50% went to women. **33.** $(3, 1)$ **34.** $(1, 1)$ **35.** $(-1, -1, 3)$

36. $(-1, -1)$ **37.** -10 **38.** $(3, -2)$ **39.** 750

40. 250 \$5 tickets, 375 \$3 tickets, 125 \$2 tickets

41. $(-\infty, 11]$

42. $(-3, 3)$

43. $(-\infty, 2) \cup (7, \infty)$

44. $\left[1, \frac{9}{4}\right]$ **45.** $3, -\frac{3}{2}$ **46.** $-5, -\frac{3}{5}$

47. $\left[-\frac{2}{3}, 2\right]$

48. $(-\infty, -4) \cup (1, \infty)$

49.

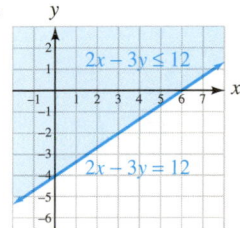

50.

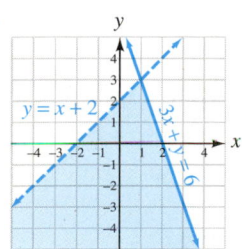

1. a. 18 **b.** 216 **2. a.** 32 **b.** 10 **3.** x^6 **4. a.** y^2 **b.** 7^3

5. $\frac{1}{8}$ **6.** •

SELF CHECKS 5.1

1. a. $2^8 = 256$ **b.** k^5 **c.** a^5b^7 **d.** $-8a^{64}$ **2. a.** a^{40}

b. 6^{15} **c.** a^{21} **d.** a^{15} **3. a.** a^8b^{10} **b.** $-\frac{216a^{15}}{b^{21}}$ **c.** $16d^{20}$

4. a. 1 **b.** $2x$ **c.** -1 **5. a.** $\frac{1}{64}$ **b.** $-\frac{1}{27}$ **c.** $\frac{12}{h^9}$ **d.** $-\frac{1}{c}$

6. a. $\frac{1}{a^4}$ **b.** a^{15} **7. a.** t^9 **b.** $\frac{64}{25}$ **c.** $-\frac{r^7}{8h^6}$ **8. a.** b^2 **b.** $\frac{3}{b^6}$

9. a. 1 **b.** $125a^6b^9$ **10.** $\frac{32}{243a^{15}b^{10}}$

STUDY SET SECTION 5.1 (page 366)

1. exponential **3.** factor **5.** power **7. a.** x^{m+n} **b.** x^{mn}

c. x^ny^n **d.** $\frac{x^n}{y^n}$ **e.** 1 **f.** $\frac{1}{x^n}$ **g.** x^{m-n} **h.** $\left(\frac{y}{x}\right), n$ **i.** $\frac{y^n}{x^m}$

9. multiply **11. a.** 1 **b.** reciprocal **13.** 9, (-2), 11

15. a. $6x$; 3 **b.** x; 5 **c.** b; 6 **17. a.** $\frac{n}{4}$; 3 **b.** $m - 8$; 6

c. 10; 0 **19.** x^5 **21.** y^{12} **23.** $-6t^{23}$ **25.** a^4b^5 **27.** 64

29. x^{28} **31.** r^{55} **33.** g^{32} **35.** $x^{20}y^4$ **37.** $64m^{21}$ **39.** $\frac{m^{80}}{n^8}$

41. $\frac{9a^{32}}{49n^{22}}$ **43.** 1 **45.** 1 **47.** 60 **49.** $-3t$ **51.** $\frac{1}{25}$

53. $-\frac{1}{27}$ **55.** $\frac{8}{x^9}$ **57.** $-\frac{1}{h}$ **59.** $\frac{1}{m^{10}}$ **61.** $\frac{1}{s^6}$ **63.** a^5 **65.** 49

67. $\frac{1}{16}$ **69.** $-\frac{p^8}{12t^6}$ **71.** m^{12} **73.** $\frac{33}{y^{12}}$ **75.** t^{14} **77.** $\frac{m^2}{6n^9}$

79. $\frac{9}{4}$ **81.** $\frac{t^{120}}{g^{80}}$ **83.** $16a^{20}$ **85.** $\frac{64b^{12}}{27a^9}$ **87.** -8 **89.** $x^{19}y^4$

91. x^{73} **93.** $25r^{13}$ **95.** $\frac{1}{m^{10}}$ **97.** $\frac{27z^{21}}{64a^{12}b^{12}}$ **99.** $\frac{1}{25}$ **101.** $\frac{1}{9x^3}$

103. $\frac{64d^6}{9}$ **105.** $-\frac{27}{c^{15}d^8}$ **107. a.** x^8 **b.** x^{16} **c.** $2x^4$

109. a. $-32t^{20}$ **b.** $-2t^8$ **c.** $\frac{1}{32t^{20}}$ **111. a.** $\frac{1}{6}$ **b.** $-\frac{1}{6}$ **c.** $\frac{1}{6}$

113. a. $\frac{8x}{y^2}$ **b.** $\frac{1}{64x^2y^2}$ **c.** $-\frac{xy}{64}$ **119.** $26^3 \cdot 10^4$; 175,760,000

121. $-6, -9$ **123.** $26x^{12}$ in.^{3}

129. $\left(-\infty, -\frac{10}{9}\right]$

ARE YOU READY? 5.2 (page 368)

1. 10,000 **2.** 9,630,000 **3.** $\frac{1}{1,000}$ **4.** 0.0431

SELF CHECKS 5.2

1. a. 1.101×10^{11} **b.** 2.0×10^{-9} **2. a.** 1.73×10^3

b. 4.5×10^{-6} **3. a.** 9,880,000,000 **b.** 0.0000003937

4. 5.88×10^{17} mi **5.** About 7,006 pounds of rice per acre

6. $2.0 \times 10^{11} = 200,000,000,000$

STUDY SET SECTION 5.2 (page 374)

1. scientific, standard **3.** 10^n, integer **5. a.** $N = 2.316; n = 54$

b. $N = 1.07; n = -21$ **7. a.** 60.22 is not between 1 and 10.

b. 0.6022 is not between 1 and 10. **9.** 3.9×10^3

11. 7.8×10^{-3} **13.** 1.73×10^{14} **15.** 9.6×10^{-6}

17. 2.03×10^{-9} **19.** 5.016×10^{16} **21.** 2.365×10^7

23. 9.009×10^{-10} **25.** 3.17×10^{-4} **27.** 5.27×10^3

29. 3.23×10^7 **31.** 6.0×10^{-4} **33.** 27,000 **35.** 0.00323

37. 796,000,000 **39.** 0.00000035 **41.** 5.23

43. 80,000,000,000,000 **45.** 2.6×10^9 **47.** 1.817×10^{12}

49. 5.005×10^8 **51.** 7.2×10^{-6} **53.** 4.3×10^9

55. 5.1×10^{11} **57.** 3.4×10^{16} **59.** 5.0×10^{-8}

61. 2.25×10^{16}; 22,500,000,000,000,000 **63.** 3.2×10^{-3}; 0.0032 **65.** 4.005×10^{20}; 400,500,000,000,000,000,000
67. 1.44×10^{7}; 14,400,000 **69.** 4.2×10^{9}; 4,200,000,000
71. 4.75×10^{-14}; 0.0000000000000475 **73.** 2,600,000 to 1
75. 1.0×10^{21} **77.** 4.5×10^{4} dollars; $45,000
79. 8.5×10^{-28} g **81.** About 9.5×10^{15} m
83. a. 2.5×10^{9} sec $= 2,500,000,000$ sec **b.** About 79 years
85. 1.209×10^{8} mi
93. $\left[1, \frac{9}{4}\right]$

95. $(-\infty, 2) \cup (7, \infty)$

ARE YOU READY? 5.3 (page 376)

1. a. 3 terms **b.** -8 **2.** function, domain **3.** -10 **4.** $10.2a^3$
5. $17t^2u$ **6.** $-3b^2 + b - 24$

SELF CHECKS 5.3

1. A binomial in two variables of degree 8 written in descending powers of p and ascending powers of q; terms: $-5.6p^6q$, p^5q^3; coefficients: $-5.6, 1$; degree: $7, 8$; leading term: $-5.6p^6q$; leading coefficient: -5.6 **2.** 256 ft **3. a.** -10 **b.** $-2, 0.2, 4.8$
4. a. $9m^4$ **b.** $11s^3t + 3s^2t$ **c.** $2x^2 + x$ **d.** $\frac{1}{8}rs + \frac{19}{9}r - 1$
5. a. $7a^2 + a + 3$ **b.** $-3a^3b^3 - 3a^3b^2 + ab^2$ **c.** $t^5 + \frac{19}{12}t^4$
6. a. $-3m^4 + 34m^2$ **b.** $8.9a^2b^3 - 10.8a^2b^2$ **7.** $2mn^3 + 5mn^2$

STUDY SET SECTION 5.3 (page 385)

1. polynomial **3.** leading, coefficient, constant **5.** degree
7. like **9. a.** No **b.** Yes **c.** No **d.** Yes **11. a.** 6 **b.** 5
13. a. 15 **b.** 5 **c.** $0, -2, -4$ **d.** D: $(-\infty, \infty)$; R: $(-\infty, \infty)$
15. a. 2 **b.** -5 **c.** $-2, 2$ **d.** D: $(-\infty, \infty)$; R: $(-\infty, 4]$
17. $3, 3, 3, -27, -29$ **19.** $25x^2, 25, 2$; $-x, -1, 1$; $4, 4, 0$;
Degree of the polynomial: 2; Type of polynomial: Trinomial
21. $5a^7b^4, 5, 11$; $-33a^2b, -33, 3$; Degree of the polynomial: 11;
Type of polynomial: Binomial
23. a. -1 **b.** 95 **25. a.** 6.5 **b.** 11
27. [graph] **29.** $20x^2 - x + 9$ **31.** $-10y^3 - 8y^2 - 4$

33. $5.1ab^2 + 0.6ab - 0.2a$ **35.** $\frac{7}{4}rst^2 - \frac{5}{6}rst$ **37.** $5x^2 - 5x + 6$
39. $8p^2q^2 - 2q + 8$ **41.** $\frac{13}{15}h^6 + \frac{2}{3}h^2$ **43.** $8a^4 - 4a^2 + 2a$
45. $4x^3 - 3x^2 + 3x - 7$ **47.** $4m^2n^2 + 4m - 2n$
49. $15.5y^3 + 4.9y^2 + 0.2y$ **51.** $\frac{13}{6}w^3 - \frac{1}{4}w^2 + \frac{4}{5}$
53. $x^2 - 8x + 22$ **55.** $2x^2 + 7a$ **57.** $-2xy^2 + 9x^2$
59. $-8x^2 - 2x + 2$ **61.** $7.9b^2 - 6.4bc + 0.3c^2$ **63.** $4x^2 - 11$
65. $6x^3 - 6x^2 + 14x - 17$ **67.** $ay^3 + 3ay^2 + 6ay - 5a$
69. $-0.3xy^7 + 1.4xy^5 - 0.2xy$ **71.** $-3y^3 + 18y^2 - 28y + 35$
73. $5x^3 - 6x^2 - x + 9$ **75.** $6k^4 - k^3 + 2k^2 + 5k - 16$
77. $-3x^2 + 4x + 1$ **79.** $\frac{1}{6}x^6 - \frac{2}{3}x^4 - \frac{1}{2}x^2$
81. a. $14d^3 - 4d^2 - 11d + 3$ **b.** $-8d^3 - 4d^2 + 5d + 7$
83. a. $-19m^2 + 3m + 10$ **b.** $3m^2 - 9m - 10$ **85.** 10 ft
87. 872 ft^3 **89.** 72 in.^{3} **91.** $V = \frac{4}{3}\pi r^2 h$ **93. a.** $J(9) \approx 12.2$;
in 2009, there were about 12.2 million manufacturing jobs in the U.S.
b. $x \approx 3$; in 2003, there were about 14.5 million maufacturing jobs
in the U.S. **95. a.** 5 **b.** 4 **c.** $\frac{1}{6}$ **d.** $\frac{65}{24} = 2\frac{17}{24}$ **103.** $[-5, 5]$
105. $(-1, 9)$

ARE YOU READY? 5.4 (page 390)

1. a. $90a$ **2.** $10x - 15$ **3.** $28y + 56$ **4.** x^{14} **5.** -30
6. $9x^2 + 3x + 2$

SELF CHECKS 5.4

1. a. $8y^5$ **b.** $4m^{12}n^7$ **2. a.** $2a^8 - 10a^5$
b. $-32c^6d^4 - 64c^5d^3 + 8cd^7$ **c.** $6m^{10}n^{10}p^7 - 54m^3n^9p^4$
3. a. $6a^2 - 19a - 36$ **b.** $12c^8 + 43c^4 + 35$
c. $50w^2z^4 + 2wz^2 - \frac{3}{50}$ **4.** $12y^4 - 2y^3 + 13y^2 + 5y$
5. $6x^4 + 4x^3 - 7x^2 + 2x - 5$ **6.** $(4x^2 - 120x + 864)$in.^{2}
7. $-10r^3 + 28r^2s - 16rs^2$ **8. a.** $64r^2 + 32rs + 4s^2$
b. $\frac{1}{9}a^6 - \frac{2}{3}a^3b^6 + b^{12}$ **c.** $0.16x^2 - 1.44y^8$
9. a. $s^2 + 8st + 16t^2 + 4s + 16t + 4$ **b.** $81 - m^2 + 2mn - n^2$
c. $64x^3 - 48x^2y + 12xy^2 - y^3$ **10.** $a^2 - 22a + 113$
11. $-15y^2 - 24y - 58$

STUDY SET SECTION 5.4 (page 398)

1. monomials, binomials **3.** first, outer, inner, last **5. a.** factors
b. term **c.** term **7.** square, square, $x^2 - y^2$ **9. a.** $6b - 2$
b. $2b$ **c.** $8b^2 - 6b + 1$ **11.** $6a^7$ **13.** $6g^{11}h^{16}$
15. $30x^{10} + 20x^7$ **17.** $-6d^7 + 6d^6 - 4d^5$
19. $7r^3s^3t + 7rs^5t - 7rs^3t^3$ **21.** $28x^7y^9z^4 - 16x^3y^7z^7$
23. $14t^2 + 17t - 6$ **25.** $6y^6 - 11y^3 + 4$ **27.** $32b^2 + b - \frac{3}{8}$
29. $27b^5c - 9b^4c^2 - 3b^2c + bc^2$ **31.** $6y^3 + 11y^2 + 9y + 2$
33. $x^3 - y^3$ **35.** $6a^3 + 17a^2 + 8a - 6$
37. $2x^4 - 6x^3 - 12x^2 + 58x - 42$ **39.** $18p^4 + 30p^3 - 72p^2$
41. $24m^3y - 20m^2y^2 + 4my^3$ **43.** $4a^2 + 4ab + b^2$
45. $25r^4 + 60r^2 + 36$ **47.** $\frac{1}{16}b^2 - b + 4$
49. $81a^2b^4 - 72ab^2 + 16$ **51.** $25y^2 - 5.76$ **53.** $x^4y^2 - \frac{36}{25}$
55. $36a^2 + 12ab + b^2 + 48a + 8b + 16$
57. $49 - 14x + 14y + x^2 - 2xy + y^2$
59. $25 - 4n^2 - 4np - p^2$ **61.** $27x^3 - 54x^2 + 36x - 8$
63. $b^2 - 6b - 5$ **65.** $a^2 - 8a + 3$ **67.** $48x^2 - 13x + 13$
69. $5x^2 - 36x + 7$ **71.** $9.2127x^2 - 7.7956x - 36.0315$
73. $299.29y^2 - 150.51y + 18.9225$ **75.** $8a^3 - b^3$
77. $9b^2 + 9b - 7$ **79.** $55m^3 + 22m^2n^2 + 15mn^3 + 6n^5$
81. $15s^6t^5u^2$ **83.** $2a^2 + ab - b^2 - 3bc - 2c^2$
85. $a^3 - 3a^2b - ab^2 + 3b^3$ **87.** $16k^2 - 104k + 169$
89. $\frac{1}{4}x^2 + 6x - 64$ **91.** $-15a^2b^3$ **93.** $y^6 - 4$
95. $9a^6 + 24a^3c + 16c^2$ **97.** $-6m^5n^2 - 6m^3n^5$
99. $-40x^3y^{14}z^{12}$ **101.** $125s^3 - 75s^2t^3 + 15st^6 - t^9$
103. $9d^2 + 6df + f^2 + 12d + 4f + 4$ **105.** $0.2t^2 - 2.7t + 9$
107. $25x^{10} - 36$ **109. a.** $-15a^2b^3$
b. $-15a^2 - 3ab + 5ab^2 + b^3$ **111. a.** $5x^2 + 2x + 6$
b. $6x^4 + 4x^3 + 17x^2 + 10x + 5$ **113. a.** $x^2 + 2x - 8$
b. $\frac{1}{2}b^2 + \frac{3}{2}b - 5$ **115. a.** $(x + y)(x - y)$ **b.** $x(x - y)$; $x^2 - xy$
c. $y(x - y)$; $xy - y^2$ **d.** They represent the same area.
$(x + y)(x - y) = x^2 - y^2$ **117.** $(4x^2 - 164x + 1,656)$ in.^{2}
123. **125.**

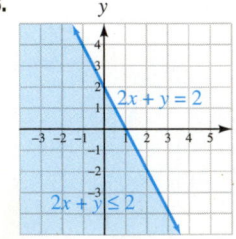

ARE YOU READY? 5.5 (page 401)

1. $2 \cdot 2 \cdot 3 \cdot 3 \cdot 3 = 2^2 \cdot 3^3$ **2.** t^5 **3.** $-h + 9$ **4.** 4
5. $6b^4 + 12b^2$ **6.** $m^2 - 3m - 10$

SELF CHECKS 5.5

1. $3x^2y$ **2. a.** $10t(3t + 2)$ **b.** $3x^3y^2(3x - 4y)$
c. $2a^2b(2a^2b + 3ab + 1)$ **3.** A prime polynomial
4. $-(b^4 + 3b^2 - 2)$ **5.** $-4ab^2(2a + 3b)$
6. a. $(y^2 + 1)(c + d)$ **b.** $(a + b - c)(x - y)$ **7.** $(r - s)(7 + s)$
8. a. $(x + 17)(x^4 + 1)$ **b.** $(a - b)(a - 1)$ **9.** $(y + 3)(y^2 - 2)$
10. $ab(3a - 2b)(a + 1)$ **11.** $p = \frac{A}{1 + rt}$

STUDY SET SECTION 5.5 (page 409)

1. factored **3.** greatest common factor **5.** binomial **7.** $6xy^2$
9. a. 4 **b.** No **c.** 5, s **11.** $2x$ **13.** $15a^4$ **15.** $8y^2$
17. $12r^3s^4$ **19.** $18xy^2z^6$ **21.** $c - d$ **23.** $8(3x + 2)$
25. $y(3y^2 + 5y - 9)$ **27.** $9a(5a - 1)$ **29.** $5x^2y(3 - 2y)$
31. Prime **33.** $3a^6b^4c^2(9a^3c^3 - 4a + 1)$ **35.** $-(a + b)$
37. $-(5xy - y + 4)$ **39.** $-8(a + 2)$ **41.** $-3x(2x + y)$
43. $-6ab(3a - 2b)$ **45.** $-4a^2c^8(2a^2 - 7a + 5c)$
47. $(x + y)(4 + t)$ **49.** $(a - b)(r - s + t)$
51. $(m + n + p)(3 + x)$ **53.** $(x + 7)^2(3x - 2)$
55. $(a + 8)(x + y)$ **57.** $(3 - c)(c + d)$ **59.** $(a + b)(a - 4)$
61. $(t - 3)(t^2 - 7)$ **63.** $(x + y)(x + 1)$ **65.** $(1 - n)(1 - m)$
67. $(y^2 + 3)(y - 4)$ **69.** $(t + v)(s + r)$
71. $2c(2b^3 + 1)(7a^3 - 1)$ **73.** $x(m + n)(p + q)$
75. $f = \frac{d_1 d_2}{d_2 + d_1}$ **77.** $a^2 = \frac{b^2x^2}{b^2 - y^2}$ **79.** $y = \frac{rx}{b + t}$ **81.** $r = \frac{S - a}{S - l}$
83. $-7u^2(9u - 4)$ **85.** $(a^3 - 3)(b^2 + 1)$
87. $9x^7y^3(5x^3 - 7y^4 + 9x^3y^7)$ **89.** $t(r^2 + s^2)(a - b)$ **91.** Prime
93. $\frac{1}{5}x^2(3ax^2 + b - 4ax)$ **95.** $6x^{30}(4x^{20} - 3x^{10} - 7)$

97. $(2x + y)(a - b + c)$ **99.** $5y^6(9y^6 + 6y^4 + 5y^2 - 1)$
101. $(a^2 + b)(x - 1)$ **103.** $(b + c)(a + b + c)$
105. a. $(5a + 6)(a^2 + 3)$ **b.** $3(a^3 + 2a^2 + 5a + 6)$
107. $\frac{1}{2}h(b_1 + b_2)$; the formula for the area of a trapezoid
109. $r^2(4 - \pi)$ **117.** \$4,900

ARE YOU READY? 5.6 (page 411)

1. 1 **2.** $x^2 + 2x - 48$ **3.** $4a^2 + 36a + 81$ **4.** 2 **5.** 2 and 5
6. -3 and 6

SELF CHECKS 5.6

1. a. $(m + 6)^2$ **b.** $(7b^2 - 2c^2)^2$ **2.** $(a + 5)(a + 8)$
3. $(m^2 - 3)(m^2 - 6)$ **4.** $3b^2(a - 5b)(a + 7b)$
5. $(5w + 3)(w + 1)$ **6.** $(2q + 1)(q - 9)$ **7.** A prime polynomial
8. $-3(x + 8y)(2x + 3y)$ **9.** $b(2a^2 + 1)(3a^2 + 4)$
10. $(a + b + 2)(a + b - 5)$ **11. a.** $(m + 7)(m + 6)$
b. $(3a + 4b)(5a - b)$ **12.** $a^2(7a - 2)(3a - 1)$

STUDY SET SECTION 5.6 (page 422)

1. perfect **3.** leading, coefficient, 2, descending **5. a.** Positive
b. Negative **c.** Positive **7.** 9, 6, -9, -6, product, sum **9.** No
11. $-$ **13.** $+$ **15.** 4, -4, 1 **17.** $(a + 9)^2$ **19.** $(2y + 7)^2$
21. $(y^2 - 5)^2$ **23.** $(3b^2 - 2c^2)^2$ **25.** $(x + 2)(x + 3)$
27. $(t + 6)(t + 8)$ **29.** $(x - 5)(x - 11)$ **31.** $(y - 8)(y - 9)$
33. $(y^2 - 3)(y^2 - 10)$ **35.** $(x^2 + 5)(x^2 + 10)$
37. $3(x + 7)(x - 3)$ **39.** $15(x + 2)(x + 1)$
41. $2y(x + y)(x - 7y)$ **43.** $3t^2(s - 2t)(s + 8t)$

45. $(5x + 3)(x + 2)$ **47.** $(7a + 5)(a + 1)$ **49.** $(3r - 2)(r + 5)$
51. $(11y - 1)(y + 3)$ **53.** Prime **55.** Prime
57. $(4a - 7y)(2a - y)$ **59.** $(7g + 2h)(g - 2h)$
61. $-2(9p + 2q)(p - q)$ **63.** $-5(3x - 4y)(2x + y)$
65. $a^2(3b^2 + 4)(3b^2 + 1)$ **67.** $c^8(4b^2 - 3)(2b^2 + 5)$
69. $b^2(2b^2 - 3)(5b^2 - 2)$ **71.** $m^3(3m^2 + 4)(3m^2 - 2)$
73. $(x + a + 1)^2$ **75.** $(a + b + 4)(a + b - 6)$
77. $(7q - 7r + 2)(2q - 2r - 3)$
79. $(8s + 8t + 9)(2s + 2t - 3)$ **81.** $-(a - 8)(a + 4)$
83. $(2z + 3)(3z + 4)$ **85.** $(5y - 1)^2$
87. $4h^4(8h - 1)(2h + 1)$ **89.** Prime **91.** $x^2(b^2 - 7)(b^2 - 5)$
93. $-(3a + 2b)(a - b)$ **95.** $14(4a^2 + 3a - 5)$
97. $(3t + 3w - 2)(2t + 2w + 5)$ **99.** $(3y^3 + 2)(4y^3 + 5)$
101. $(m + n)(6a - 5)(a + 3)$ **103.** $5(2a - 3b)^2$
105. $-3a^2(x - 3)(x - 2)$ **107.** $(5m^4 - 6n)^2$
109. a. $(x - 3)(x - 2)$ **b.** $(x^2 - 3)(x^2 - 2)$
c. $(x^3 - 3)(x^3 - 2)$ **111. a.** $(3x - 8)(x - 1)$
b. $(3x - 8y)(x - y)$ **113.** $5x - 4$ **119.** 5 **121.** 6

ARE YOU READY? 5.7 (page 425)

1. $n^2 - 81$ **2.** $64d^2$ **3. a.** 27 **b.** 216 **4.** $64a^3$ **5.** $b^3 + 64$
6. $8a^3 - 1$

SELF CHECKS 5.7

1. $(9p + 5)(9p - 5)$ **2.** $(6r^2 + s)(6r^2 - s)$
3. $(a^2 + 9)(a + 3)(a - 3)$
4. $(a^2 - 2ab + b^2 + c^2)(a - b + c)(a - b - c)$
5. $3(a^2 + 1)(a + 1)(a - 1)$ **6. a.** $(a + b)(a - b + 1)$
b. $(a + 2 + b)(a + 2 - b)$ **7.** $(p + 3)(p^2 - 3p + 9)$
8. $(2c^2 - 5d)(4c^4 + 10c^2d + 25d^2)$
9. $(p + q - r)(p^2 + 2pq + q^2 + pr + qr + r^2)$
10. $(1 + x)(1 - x + x^2)(1 - x)(1 + x + x^2)$
11. $3x^2(x + 2)(x^2 - 2x + 4)$

STUDY SET SECTION 5.7 (page 431)

1. squares **3. a.** 1, 4, 9, 16, 25, 36, 49, 64, 81, 100
b. 1, 8, 27, 64, 125, 216, 343, 512, 729, 1,000 **5. a.** $F - L$
b. $F^2 - FL + L^2$ **c.** $F^2 + FL + L^2$ **7. a.** $x^2 - 4$ (Answers
may vary) **b.** $(x - 4)^2$ (Answers may vary) **c.** $x^2 + 4$
(Answers may vary) **d.** $x^3 + 8$ (Answers may vary) **e.** $(x + 8)^3$
(Answers may vary) **9.** $(x + 4)(x - 4)$ **11.** $(3y + 8)(3y - 8)$
13. $(12 + c)(12 - c)$ **15.** $(10m + 1)(10m - 1)$
17. $(9a + 7b)(9a - 7b)$ **19.** Prime **21.** $(3r^2 + 11s)(3r^2 - 11s)$
23. $(4t + 5w^2)(4t - 5w^2)$ **25.** $(10rs^2 + t^2)(10rs^2 - t^2)$
27. $(6x^2y + 7z^3)(6x^2y - 7z^3)$ **29.** $(x^2 + y^2)(x + y)(x - y)$
31. $(4a^2 + 9b^2)(2a + 3b)(2a - 3b)$ **33.** $(x + y + z)(x + y - z)$
35. $(r - s + t^2)(r - s - t^2)$ **37.** $2(x + 12)(x - 12)$
39. $3x(x + 9)(x - 9)$ **41.** $5a(b^2 + 1)(b + 1)(b - 1)$
43. $4b(4 + b^2)(2 + b)(2 - b)$ **45.** $(c + d)(c - d + 1)$
47. $(a - b)(a + b + 2)$ **49.** $(x + 6 + y)(x + 6 - y)$
51. $(x - 1 + 3z)(x - 1 - 3z)$ **53.** $(a + 5)(a^2 - 5a + 25)$
55. $(2r + s)(4r^2 - 2rs + s^2)$
57. $(4t^2 - 3v)(16t^4 + 12t^2v + 9v^2)$
59. $(x - 6y^2)(x^2 + 6xy^2 + 36y^4)$
61. $(a - b + 3)(a^2 - 2ab + b^2 - 3a + 3b + 9)$
63. $(4 - a - b)(16 + 4a + 4b + a^2 + 2ab + b^2)$
65. $(x + 1)(x^2 - x + 1)(x - 1)(x^2 + x + 1)$
67. $(x^2 + y)(x^4 - x^2y + y^2)(x^2 - y)(x^4 + x^2y + y^2)$
69. $5(x + 5)(x^2 - 5x + 25)$ **71.** $4x^2(x - 4)(x^2 + 4x + 16)$

73. $(4a - 5b^2)(16a^2 + 20ab^2 + 25b^4)$
75. $2b^2(12 + b^2)(12 - b^2)$ **77.** $(x + y)(x - y + 8)$
79. $(x + y)(x^2 - xy + y^2)(x^6 - x^3y^3 + y^6)$
81. $(12at + 13b^3)(12at - 13b^3)$ **83.** Prime
85. $(9c^2d^2 + 4t^2)(3cd + 2t)(3cd - 2t)$
87. $2u^2(4v - t)(16v^2 + 4tv + t^2)$ **89.** $(y + 2x - t)(y - 2x + t)$
91. $(x + 10 + 3z)(x + 10 - 3z)$
93. $(c - d + 6)(c^2 - 2cd + d^2 - 6c + 6d + 36)$
95. $\left(\frac{1}{6} + y^2\right)\left(\frac{1}{6} - y^2\right)$
97. $(m + 2)(m^2 - 2m + 4)(m - 2)(m^2 + 2m + 4)$
99. $(a + b)(x + 3)(x^2 - 3x + 9)$
101. $(x^3 - y^4z^5)(x^6 + x^3y^4z^5 + y^8z^{10})$ **103. a.** $(q + 8)(q - 8)$
b. $(q - 4)(q^2 + 4q + 16)$ **105. a.** $(d + 5)(d - 5)$
b. $(d - 5)(d^2 + 5d + 25)$ **107. a.** $(a^2 - b)(a^4 + a^2b + b^2)$
b. $(a^2 + b)(a^4 - a^2b + b^2)$
109. a. $(5m + 2n)(25m^2 - 10mn + 4n^2)$
b. $(5m - 2n)(25m^2 + 10mn + 4n^2)$
111. $\frac{4}{3}\pi(r_1 - r_2)(r_1{}^2 + r_1r_2 + r_2{}^2)$
117.   **119.** $y = -4$

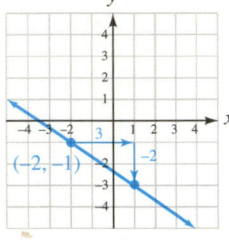

(−2,−1)

ARE YOU READY? 5.8 (page 433)

1. a. 3   **b.** 6   **2.** Yes, both do.   **3.** $3a^3 - 27a^2 - 210a$
4. $8mn^2$

SELF CHECKS 5.8

1. $3b^3c(5a - 2)(2a - 1)$ **2.** $3r^3(p^2 + q^2)(p + q)(p - q)$
3. $9(a + b)(a^2 - ab + b^2)(a + b)(a - b)$
4. $(a - 2)(a - 3)(a + b)$ **5.** $(a - 2)(a^3 + a^2 + 1)$

STUDY SET SECTION 5.8 (page 437)

1. prime   **3.** cubes, cubes   **5.** common   **7.** trinomial
9. Multiply the factors of $y^2z^3(x + 6)(x + 1)$ to see if the product
is $x^2y^2z^3 + 7xy^2z^3 + 6y^2z^3$.   **11.** $3ab, 2b, 2a$
13. $4bc(a - 5)(a + 6)$ **15.** $-3xy(x + y - 4)$
17. $(y + 1)(y - 1)(y - 3)(y^2 + 3y + 9)$
19. $36(x^2 + 1)(x + 1)(x - 1)$ **21.** Prime   **23.** $2(3x - 4)(x - 1)$
25. $y^2(2x + 1)^2$ **27.** $2x^2y^2z^2(2 - 13z)$ **29.** $a^2(3a^2 - 8)^2$
31. $6(a - b)(a^2 + ab + b^2)(a + b)(a - b)$
33. $(x - y + 5)(x^2 - 2xy + y^2 - 5x + 5y + 25)$
35. $(2a - 2b + 3)(a - b + 1)$ **37.** $(2x - 9)(3x + 7)$
39. $(x + 1)(x - 1)(x + 4)(x - 4)$ **41.** $(x + 5 + y^4)(x + 5 - y^4)$
43. $(3x - 1 - 5y)(3x - 1 + 5y)$ **45.** $(a + b - 2)(a - b + 2)$
47. $(x - y)^2(a - 1)$ **49.** $\left(\frac{9}{4}x^2 + y^{20}\right)\left(\frac{3}{2}x + y^{10}\right)\left(\frac{3}{2}x - y^{10}\right)$
51. $16(m^8 + 1)(m^4 + 1)(m^2 + 1)(m + 1)(m - 1)$
53. $(3y + 1)(3y^4 + y^3 + 1)$ **55.** $(x - 4)(x + y)(x - y)$
57. $c(c + 2a - b)(c - 2a + b)$ **59.** $(3x + 1)(3x^3 + x^2 + 1)$
61. $(2x + 1)^2$ **63.** Prime   **65.**
$1 - 16x^2 = (1 + 4x)(1 - 4x)$ in.^{2} **69.** 6   **71.** -13

ARE YOU READY? 5.9 (page 438)

1. 0   **2.** -4 **3.** 0   **4.** $(x - 3)(x + 2)$ **5.** $(3n + 2)(n - 1)$
6. $(x + 11)(x - 11)(x - 1)$

SELF CHECKS 5.9

1. $9, -5$ **2. a.** $0, 3$ **b.** $-\frac{7}{4}, \frac{7}{4}$ **3.** $\frac{3}{2}, -\frac{1}{3}$
4. A repeated solution of $-\frac{5}{2}$ **5.** $-11, 11$ **6.** $0, \frac{2}{5}, -3$
7. a. $-2, 2, -3, 3$ **b.** $-1, -10, 10$ **8.** 5 ft by 8 ft
9. 1 in. wide   **10.** 8 sec

STUDY SET SECTION 5.9 (page 447)

1. quadratic   **3.** standard   **5. a.** Yes   **b.** No   **c.** Yes
d. No   **7. a.** At least one is 0.   **b.** zero, 0, 0
9. 4 is a solution; -5 is not a solution   **11. a.** $10 + 2x$
b. $20 + 2x$ **13.** $y + 2, y - 4, -2$ **15.** $-3, -5$ **17.** $-2, -4$
19. $1, \frac{1}{2}$ **21.** $-\frac{1}{3}, -3$ **23.** $0, -1$ **25.** $0, 2$ **27.** $-4, 4$
29. $-\frac{3}{4}, \frac{3}{4}$ **31.** $\frac{1}{3}, -1$ **33.** $\frac{1}{5}, -2$ **35.** $\frac{1}{4}, -\frac{3}{2}$ **37.** $-3, 3$
39. A repeated solution of -8 **41.** $2, -\frac{5}{2}$ **43.** $1, 2$ **45.** $0, 2, 4$
47. $0, 7, -3$ **49.** $0, 7, -7$ **51.** $6, -6, 1, -1$ **53.** $3, -2, 2$
55. $2.78, 0.72$ **57.** 1   **59.** $2, -\frac{5}{6}$ **61.** $0, -1$ **63.** $1, -4$
65. $-4, 4, 2$ **67.** $0, \frac{25}{6}$ **69.** 0 and a repeated solution of 9
71. $1, -\frac{5}{7}$ **73.** $-1, -2, 2$ **75.** $-\frac{9}{8}, \frac{9}{8}$ **77.** $0, \frac{5}{6}, -7$
79. $11, -6$ **81.** $5, -\frac{7}{2}$ **83.** $5.5, -3$ **85.** $-5, 5, -3, 3$
87. $-1, \frac{2}{5}$ **89.** $-4, 7$ **91.** $16, 18$ **93.** Base: 8 ft; height: 4 ft
95. Length: 16 in.; width: 10 in.   **97.** 3 ft   **99.** 2.5 in. wide
101. Width: 20 ft; length: 40 ft   **103.** 4 sec   **105.** 2 sec
107. 6 m/sec   **109.** 4   **117.** 25 ft^2

CHAPTER 5 REVIEW (page 452)

1. 243   **2.** -32 **3.** -64 **4.** $\frac{4}{9}$ **5.** x^6 **6.** $\frac{m^3}{n^5}$ **7.** $64m^{16}$
8. t^{13} **9.** $9x^4y^6$ **10.** $\frac{x^{16}}{b^4}$ **11.** -10 **12.** $\frac{h^{12}}{5}$ **13.** $\frac{b}{2a}$
14. $\frac{x^5}{5}$ **15.** $\frac{x^6}{9}$ **16.** $\frac{2}{9x}$ **17.** $-c^{10}$ **18.** $\frac{25}{16}$ **19.** $\frac{16}{27}$ **20.** 64
21. s^{73} **22.** $-\frac{b^3c^3}{8a^{21}}$ **23.** 1.93×10^{10} **24.** 2.735×10^{-8}
25. 72,770,000   **26.** 0.0000000083   **27.** 7.6×10^2 sec
28. 1.67248×10^{-18} g   **29.** 8.4×10^6 **30.** 1.875×10^{-21}
31. No   **32.** Yes   **33.** Yes   **34.** No   **35.** Binomial, 2
36. Monomial, 4   **37.** None of these, 4   **38.** Trinomial, 8
39. -29 **40.** 134 in.^{3} **41.** $2t^3 - 2t^2 - 5t$ **42.** $\frac{3}{4}ab^2c - \frac{5}{6}abc$
43. $3x^2y^3 - 8x^2y + 8y$ **44.** $19m^2 - 13m - 9$ **45.** $\frac{1}{2}s^6 - \frac{5}{3}s^4$
46. $6k^4 - 6k^3 + 9k^2 - 2$ **47.** $4c^2d^2 + 4cd^2$
48. $5x^5 + 5x^2 - 3x$ **49.** $1.2t^3 + 0.4t^2 + 1.2t - 1.4$ **50. a.** -1
b. $-2, -1, 1$ **c.** D: $(-\infty, \infty)$; R: $[-1, \infty)$ **51.** $-4a^3$
52. $6x^3y^2z^5$ **53.** $2x^4y^3 - 8x^2y^7$ **54.** $a^4b + 2a^3b^2 - a^2b^3$
55. $6x^3 - 12x^2 + 4x - 8$ **56.** $25a^2t^2 - 60at + 36$
57. $49c^8d^6 - d^2$ **58.** $15x^4 - 22x^3 + 58x^2 - 40x$
59. $r^3 - 3r^2s - rs^2 + 3s^3$ **60.** $9c^2 - \frac{9}{2}c + \frac{9}{16}$
61. $25 - 10a + 10b + a^2 - 2ab + b^2$
62. $2x^2 + xy - y^2 - 3yz - 2z^2$ **63.** $41a^2 - 12a + 5$
64. $b^2 - b - 7$ **65.** $0.05x^2 + 1.01x - 0.84$ **66.** $x^3 + 27$
67. a. $(12x - 2)$ in.   **b.** $(8x^2 - 2x)$ in.^{2}
c. $(16x^3 + 12x^2 - 4x)$ in.^{3} **68. a.** $f(x) = x^3 + 3x^2 + 2x$
b. 210 in.^{3} **69.** 6   **70.** $3xy^3$ **71.** $4(x^4 + 2)$
72. $\frac{x}{5}(3x^2 - 6x + 1)$ **73.** Prime   **74.** $7a^3b(ab + 7)$
75. $5x^2(x + y)(1 - 3x)$ **76.** $9x^2y^3z^2(3xz + 9x^2y^2 - 10z^5)$
77. $-(x + 9)$ **78.** $-(-4r + 7)$ or $-(7 - 4r)$
79. $-7(b^3 - 2c)$ **80.** $-7a^2b^2(a - b)^3(7a^2 - 7ab - 9b^2)$
81. $(x + 2)(y + 4)$ **82.** $(ry - a + 1)(r - 1)$
83. $(t^2 + 1)(t - 9)$ **84.** $(1 - x)(1 - 3z)$ **85.** $m_1 = \frac{mm_2}{m_2 - m}$
86. $A = 2\pi r(r + h)$ **87.** $(x + 5)^2$ **88.** $(7a^3 + 6b^2)^2$

89. $(y - 20)(y - 1)$ **90.** $(z - 5)(z - 6)$ **91.** $-(x + 7)(x - 4)$
92. $(a - 8b)(a + 3b)$ **93.** $(4a - 1)(a - 1)$ **94.** Prime
95. $y^6(y + 2)(y - 1)$ **96.** $9st(3r - 2)(r + 4)$
97. $(r + s)(6t - 5)(t + 3)$ **98.** $(v^2 - 7)(v^2 - 6)$
99. $(w^4 - 10)(w^4 + 9)$ **100.** $(s + t - 1)^2$ **101.** $(z + 4)(z - 4)$
102. $(xy^2 + 8z^3)(xy^2 - 8z^3)$ **103.** Prime
104. $(c + a + b)(c - a - b)$
105. $10m^2(m^2 + 4)(m + 2)(m - 2)$ **106.** $(m + n)(m - n + 1)$
107. $2c(4a^2 + 9b^2)(2a + 3b)(2a - 3b)$
108. $(k + 1 + 3m)(k + 1 - 3m)$ **109.** $(t + 4)(t^2 - 4t + 16)$
110. $(2a - 5b^3)(4a^2 + 10ab^3 + 25b^6)$
111. $4d^4(d + 1)(d^2 - d + 1)$
112. $(b + c + 3)(b^2 + 2bc + c^2 - 3b - 3c + 9)$
113. $4rs(q - 5t)(q + 6t)$ **114.** $(2m + 2n + 3)(m + n - 1)$
115. $(z - 2)(z + x + 2)$ **116.** Prime
117. $(x + 2 + 2p^2)(x + 2 - 2p^2)$ **118.** $(y + 2)(y + 1 + x)$
119. $4c^2(ab + 4)(a^2b^2 - 4ab + 16)$
120. $(a + 3)(a - 3)(a + 2)(a - 2)$
121. $(2x + 3)(2x^3 + 3x^2 + 1)$ **122.** $\frac{\pi}{2}h(r_1 + r_2)(r_1 - r_2)$
123. $0, \frac{3}{4}$ **124.** $6, -6$ **125.** $\frac{1}{2}, -\frac{5}{6}$ **126.** $3, -3, 1, -1$
127. $0, -\frac{2}{3}, \frac{4}{5}$ **128.** $-\frac{2}{3}, 7, 0$ **129.** A repeated solution of -8
130. $-7, -1, 1$ **131.** $1, 3$ **132.** $-\frac{1}{3}, 2$ **133.** 7 sec
134. 7 m by 10 m **135.** 5 ft **136.** $-\frac{1}{2}, 1$

CHAPTER 5 TEST (page 461)

1. a. perfect **b.** product **c.** binomials **d.** quadratic
e. difference, cubes **f.** greatest common **2. a.** x^{10} ft^2
b. x^{15} ft^3 **3.** $\frac{a^5}{9m^5}$ **4.** $\frac{-8x^6y^9}{125}$ **5.** $-\frac{64}{b^8}$ **6.** $\frac{m^4}{9n^{10}}$
7. $2.9 \times 10^6 = 2,900,000$ **8.** 1.116×10^7 mi
9. a. $3, -4, -3, -\frac{5}{3}; 5$ **b.** $8, -1, 1, -6, 4; 13$ **10.** 110 ft
11. a. 0 **b.** 2, 6 **c.** D: $(-\infty, \infty)$; R: $(-\infty, 2]$
12. $1.15x^3 + 5.5x + 12$ **13.** $2x^2y^3 + 13xy + 3y^2$
14. $2x^5 + \frac{2}{15}x^2$ **15.** $-12a^4y^{12}z^{12}$ **16.** $-15a^3b^4 + 10a^3b^5$
17. $6y^7 + 9y^6 + 6y^5 + 2y^2 + 3y + 2$ **18.** $0.06d^2 + 1.6d - 6$
19. $36 + 12m - 12n + m^2 - 2mn + n^2$ **20.** $32s^3 - 50st^2$
21. $15t^2 - 21t + 13$ **22.** $c^2 - c + 4$ **23.** $x^3 - 343$
24. a. $8a + 2b$ **b.** $16a^2 + 8ab - 3b^2$
25. $3abc(4a^2b - abc + 2c^2)$ **26.** $(k + z)(h + b)$
27. $(x - 6)(x + 5)$ **28.** $(y^2 + 9)(y + 3)(y - 3)$
29. $-x^2(7x + 1)(x - 2)$ **30.** $(s + 3)(s - 3)(s + 2)(s - 2)$
31. $(5m - 4n)^2$ **32.** Prime **33.** $5(x + 5)(x^2 - 5x + 25)$
34. $(4a - 5b^2)(16a^2 + 20ab^2 + 25b^4)$
35. $(x - y + 5)(x - y - 2)$ **36.** $(3b + 2c)(2b - c)$
37. $(a + b)(a - b + 1)$ **38.** $(n - 3 + 2m)(n - 3 - 2m)$
39. $(a + 1)(a^2 - a + 1)(a - 1)(a^2 + a + 1)$ **40.** $v = \frac{v_1v_3}{v_3 + v_1}$
41. $0, 5$ **42.** $-5, \frac{7}{2}$ **43.** $\frac{3}{4}, -\frac{3}{2}$ **44.** $1, 0, -9$ **45.** $-1, -4, 4$
46. 4 m; 7 m **47.** 3 sec **48.** 1.5 ft

CHAPTERS 1–5 CUMULATIVE REVIEW (page 464)

1. a. True **b.** False **c.** False **d.** True **e.** True **2.** -27
3. $-\frac{19}{16}x$ **4.** 1 **5.** $C = \frac{5}{9}(F - 32)$ **6.** 1,687.22 cm^3 **7.** 3.5 hr
into the flights **8.** 12 oz water **9.** Yes **10.** x-intercept: $(5, 0)$;
y-intercept: $(0, -2)$ **11.** An increase of $47.5 billion per year
12. 0 **13.** $y = 8x + 21$ **14.** $y = -\frac{10}{13}x + \frac{2}{13}$ **15.** -3
16. $f(x) = 0.025x + 95$

17. D: the set of real numbers; R: the set of all real numbers greater than or equal to 2

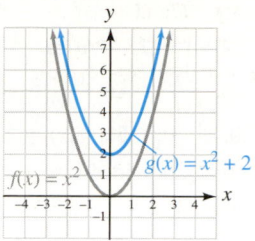

18. a. 4 **b.** 1 **c.** $-2, 4$ **19.** Not a function; $(1, 2)$, $(1, -2)$; answers may vary **20.** $(-1, 3)$

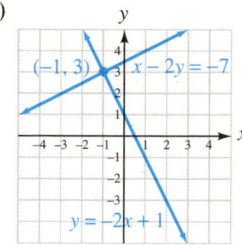

21. $\left(\frac{3}{4}, \frac{1}{3}\right)$ **22.** $(-2, 0, 2)$ **23.** $(2, -3)$ **24.** -23
25. 7.5%: $16,000; 6%: $34,000 **26.** $42°, 56°, 82°$
27. $(-\infty, 5]$ **28.** $[-21, -3)$
29. $(-\infty, -12] \cup [2, \infty)$
30. $(-\infty, -2)$ **31.** iii **32.**
33. $\frac{9a^2}{b^8}$ **34.** 0.000000090895
35. $0.7x^3 - 0.4x$
36. $36a^4b^2 - 35a^2bc^3 + 6c^6$
37. $8a^3 - b^3$ **38.** $13k^2 - 10k + 17$
39. $(x - y)(x - 4)$
40. $6s^2(s + 6)(s - 6)$
41. $(2x^2 + 5y)(4x^4 - 10x^2y + 25y^2)$
42. $-(3a + 2b)(a - b)$ **43.** $(a + 3)(a + 2)(a + b)$ **44.** $1, -\frac{1}{2}$
45. $m_1 = \frac{m_2g}{m_2 - g}$ **46.** 12 telephones

ARE YOU READY? 6.1 (page 468)

1. a. 0 **b.** Undefined **c.** -1 **2.** $\frac{9}{7}$ **3.** $4a^4(3a - 4)$
4. $(x + 10)(x - 10)$ **5.** $(5n - 2)(n + 5)$ **6.** $(2x + 1)(x^2 - 7)$

SELF CHECKS 6.1

1. $1.56 per hour **2.** The domain is the set of all real numbers except -7 and 7: $(-\infty, -7) \cup (-7, 7) \cup (7, \infty)$. **3.** $\frac{3a^3}{5b^2}$
4. a. $\frac{4}{x - 2}$ **b.** $\frac{x + 3}{5x}$ **5. a.** $\frac{x - 3}{6}$ **b.** $\frac{2b - 3}{2b + 3}$
6. $h(a) = \frac{a^2 + a + 1}{a^2 + 6}$, provided $a \neq 1$. **7.** $\frac{-2a + 3b}{b}$ or $\frac{-2a - 3b}{b}$

STUDY SET SECTION 6.1 (page 477)

1. rational **3.** factor **5.** undefined **7.** domain **9. a.** -1
b. 2 **c.** 2 **d.** 5 **11. a.** $\frac{3y}{7x^2}$ **b.** $\frac{x - 3}{x + 2}$ **c.** $-\frac{a^3}{a + 9}$
13. a. iii **b.** v. **c.** i. **d.** vi. **e.** iv. **f.** ii.
15. a. $-\frac{1}{4}$ **b.** $\frac{5}{6}$ **c.** Undefined **17.** Yes, Yes, Yes, No, Yes
19. All real numbers except 0; $(-\infty, 0) \cup (0, \infty)$
21. All real numbers except -2; $(-\infty, -2) \cup (-2, \infty)$

23. All real numbers except 0 and 1; $(-\infty, 0) \cup (0, 1) \cup (1, \infty)$
25. All real numbers except -7 and 8;
$(-\infty, -7) \cup (-7, 8) \cup (8, \infty)$ **27.** $\frac{2a^2}{3}$ **29.** $\frac{3}{5a^6}$ **31.** $\frac{3}{4t}$ **33.** $\frac{4y^7}{3x}$
35. $\frac{2}{x-6}$ **37.** $\frac{3n}{2n+3}$ **39.** $\frac{2}{x-9}$ **41.** $\frac{2a+5}{10}$ **43.** $\frac{5x}{x-2}$
45. $\frac{x+1}{x+3}$ **47.** $\frac{d+4}{d+2}$ **49.** $\frac{h+5}{2h-1}$ **51.** $f(x) = \frac{x+1}{x+6}$,
provided $x \neq 1$ **53.** $f(x) = \frac{9}{x}$, provided $x \neq -9, x \neq 0$
55. $g(x) = \frac{x^2 - 3x + 9}{x^2 + 4}$, provided $x \neq -3$ **57.** $s(a) = \frac{a^2 - 6}{a^2 + a + 1}$,
provided $a \neq 1$ **59.** $-\frac{3m+n}{m}$ or $\frac{-(3m+n)}{m}$ **61.** $-b - a$
63. $-\frac{x+2}{x+1}$ or $\frac{-(2+x)}{x+1}$ **65.** $-\frac{20x^3}{x-1}$ **67.** $x = 2$
69. $x = -2; x = 2$ **71.** $\frac{x+6}{3(x+4)}$ **73.** $\frac{a+2}{a^2 + 2a + 4}$
75. $-\frac{m+n}{n+2m}$ or $\frac{-m-n}{n+2m}$ **77.** $\frac{s-3}{s+6}$ **79.** Does not simplify
81. 3 **83.** $\frac{x^2}{x-3}$ **85.** $\frac{2}{x+2}$ **87.** $-\frac{2x+1}{x-2}$ or $\frac{2x+1}{2-x}$
89. $-\frac{x-3}{(9+x^2)(3+x)}$ or $-\frac{x-3}{(x^2+9)(x+3)}$ **91.** $\frac{2p^2}{3q^6}$
93. $-\frac{t-2}{3+t}$ or $-\frac{t-2}{t+3}$ **95. a.** $\$50,000$ **b.** $\$200,000$
97. a. $c(n) = 0.09n + 7.50$ **b.** $c(n) = \frac{0.09n + 7.50}{n}$
c. About 10¢ **99. a.** About 8 days **b.** About 9.5 days
105. $a^3 - 6a^2 + 5a + 6$ **107.** $-3m^4n^2 + 21m^2n^3 + 6m^3n^2$

ARE YOU READY? 6.2 (page 480)

1. $\frac{3}{56}$ **2.** $\frac{3}{2}$ **3.** $\frac{9}{5}$ **4.** $x^3(1 - x)$ **5.** $\frac{x+5}{4(x-5)}$ **6.** $\frac{x^2 + 5x + 25}{2}$

SELF CHECKS 6.2

1. $\frac{3x^4}{34}$ **2. a.** $\frac{a^2(a+3)}{42}$ **b.** 1 **3.** $\frac{a-3b}{a+b}$ **4.** $-\frac{x^2}{8(x^2+1)}$
5. $\frac{x^2 + 10x + 25}{x^4 - 12x^3 + 36x^2}$ **6.** $\frac{2a^6}{7t^3}$ **7.** $\frac{x(x-2)}{3}$ **8.** $\frac{m(m-3)}{a(a-3)}$ **9.** $\frac{1}{x(2x+3)}$

STUDY SET SECTION 6.2 (page 488)

1. rational **3.** invert **5.** numerators, denominators, $\frac{AC}{BD}$
7. $(x-5), (x+3), x, 5, x$ **9.** Yes, No, Yes **11.** $\frac{11}{4}$ **13.** $\frac{2}{5}$
15. $\frac{1}{25a^3}$ **17.** $\frac{3x^{34}}{14y^{14}}$ **19.** $\frac{y(y+3)}{10}$ **21.** $\frac{(x-2)^2}{x}$ **23.** $\frac{x+1}{9x^2}$
25. $\frac{x^2 - 3}{x^2}$ **27.** 1 **29.** $\frac{t-1}{t+1}$ **31.** $\frac{3}{2x}$ **33.** $\frac{a-b}{a}$
35. $x + 1$ **37.** $2(y + 8)$ or $2y + 16$ **39.** $-\frac{a^4}{2(a^2+3)}$
41. $-\frac{(x+1)^2(x+2)}{x+2c}$ **43.** $\frac{x^2 - 6x + 9}{x^4 + 8x^2 + 16}$
45. $\frac{4m^4 - 4m^3 - 11m^2 + 6m + 9}{x^4 - 2x^2 + 1}$ **47.** $\frac{5}{6}$ **49.** $\frac{9}{2}$ **51.** $\frac{2y^5}{3x^6}$
53. $\frac{3}{10p^9q^9}$ **55.** $\frac{x^{11}}{x^2 + 2x + 4}$ **57.** $\frac{1}{a^3(a-9)}$ **59.** $\frac{n+2}{n+1}$
61. $\frac{c+d}{d(25c^2 - 5cd + d^2)}$ **63.** $\frac{y(y+3)}{y+2}$ **65.** $\frac{2}{x+1}$ **67.** $\frac{a+1}{a-1}$
69. $\frac{3x}{2}$ **71.** $-\frac{(x-3)(x-6)}{(x+2)(x+3)}$ **73.** $\frac{2(x+1)}{x(x+3)}$ **75.** $\frac{q}{p}$ **77.** $\frac{5r^3}{2s^2}$
79. $-10(h - 3)$ or $-10h + 30$ **81.** $\frac{x+y}{x-y}$ **83.** $\frac{x+2}{x+1}$
85. $\frac{2x(x-5)}{x+5}$ **87.** $\frac{(x+3)^2}{(x-3)(x+2)}$ **89.** $\frac{2x(x-4)}{x+5}$ **91. a.** $\frac{(t+5)^2}{t+4}$
b. $\frac{t+4}{81}$ **93. a.** $\frac{16r}{25}$ **b.** $\frac{(r-2)^2}{r^3}$ **95.** $k_1(k_1 + 2), k_2 + 6$
101. a. x^{m+n} **b.** x^{mn} **c.** $x^n y^n$ **d.** $\frac{x^n}{y^n}$ **e.** 1 **f.** $\frac{1}{x^n}$

ARE YOU READY? 6.3 (page 491)

1. $\frac{8}{11}$ **2.** $\frac{7}{15}$ **3.** $x^2 + 3x + 4$ **4.** 1 **5.** $\frac{2}{a-2}$
6. $5(x+1)(x-1)$

SELF CHECKS 6.3

1. a. $\frac{15}{11t}$ **b.** $\frac{a+2}{a}$ **2. a.** $84z^3$ **b.** $(a-5)(a+5)(a+2)$

---

3. $\frac{15x + 4z^2}{84z^3}$ **4.** $-\frac{2a(a+12)}{(a+3)(a-3)}$ **5.** $\frac{-y+36}{6-y}$
6. $\frac{9a-2}{(a-2)(a-2)(a+2)}$ **7.** $\frac{2a-b}{a-b}$ **8.** $-\frac{45}{(a+5)(a-5)}$

STUDY SET SECTION 6.3 (page 498)

1. denominator **3.** build **5.** numerators, denominator,
$A + B, A - B$ **7.** 1 **9. a.** ii; adding or subtracting rational
expressions **b.** Simplifying a rational expression **11. a.** Once
b. Twice **c.** $3(x-2)^2$ **13. a.** $2 \cdot 2 \cdot 2 \cdot 5 \cdot x \cdot x$ **b.** $2x(x-3)$
c. $(n+8)(n-8)$ **15.** $3x - 2, 3, 3x - 1, 3$
17. $\frac{4v}{4v}, \frac{3}{3}, 12v^2, 32v, 12v^2, 12v^2$ **19.** $\frac{13}{3x}$ **21.** $\frac{t}{2r}$ **23.** 4 **25.** $\frac{3}{x+3}$
27. $36x^2y$ **29.** $x(x+3)(x-3)$ **31.** $(x+3)^2(x^2 - 3x + 9)$
33. $(2x+3)^2(x+1)^2$ **35.** $\frac{41}{30m}$ **37.** $\frac{3a - 10b}{4a^2b^2}$
39. $\frac{8x-2}{(x+2)(x-4)}$ **41.** $\frac{2x^2 - 30x}{(x+3)(x-3)}$ **43.** $\frac{2a^2 + a + 17}{(a+4)(a-5)}$
45. $\frac{5n+28}{(n-4)(n+4)}$ **47.** $\frac{4x-7}{x-2}$ **49.** $\frac{7x^2 + x}{7x-3}$ **51.** $\frac{1}{x+1}$
53. $\frac{m-5}{(m+3)(m+5)}$ **55.** $\frac{2x^2 + x}{(x+3)(x+2)(x-2)}$ **57.** $\frac{-x+28}{30(x-7)}$
59. $\frac{x}{x-3}$ **61.** $\frac{9m+2}{m-n}$ **63.** 3 **65.** 0 **67.** $\frac{11x^2 + 7x - 3}{(2x-1)(3x+2)}$
69. $\frac{-4x^2 + 14x + 54}{x(x+3)(x-3)}$ or $\frac{-2(2x^2 - 7x - 27)}{x(x+3)(x-3)}$ **71.** $\frac{21a - 22b}{6ab}$
73. $\frac{14s + 58}{(s+3)(s+7)}$ **75.** $\frac{5x-y}{6}$ **77.** $\frac{6}{x-3}$ **79.** $\frac{1}{a-b}$
81. $\frac{2x^2 + 5x + 4}{x+1}$ **83.** $\frac{d-6}{(d+4)(d+6)}$ **85.** $\frac{2}{x+1}$ **87.** $\frac{16y^2 + 3}{18y^4}$
89. $\frac{3}{(b+2)(b-3)}$ **91.** $\frac{a^2 - a}{a-5}$ **93.** $\frac{3a-3}{3a-2}$ **95.** $\frac{2x}{x+1}$ **97.** $\frac{z+1}{z}$
99. $\frac{9p^3}{5}$ **101.** $\frac{6-3d}{5d(d-1)}$ **103.** $\frac{2s-3t}{t(2s+3t)}$ **105.** $\frac{-x^2 + 11x + 8}{(3x+2)(x+1)(x-3)}$
107. $\frac{7x-20}{(x+5)(x-5)}$ **109. a.** $\frac{2x^2 + 2x + 5}{(x+2)(x-1)}$ **b.** 1
111. $\frac{10r + 20}{r}; \frac{3t+9}{t}$ **117.** A repeated solution of 3
119. $-1, 0$

ARE YOU READY? 6.4 (page 502)

1. Division **2.** $\frac{9}{28}$ **3.** $(a-2)(a+1)$ **4.** $12a^2 - 20$

SELF CHECKS 6.4

1. $\frac{5x}{6y^4}$ **2.** $\frac{5x}{6y^4}$ **3.** $\frac{3+2m}{m^2 + 2m}$ **4.** $\frac{5t-2s}{7t+3s}$ **5.** $\frac{y^2 + x^2}{xy^2 - x^2y}$ or $\frac{x^2 + y^2}{xy(y-x)}$
6. $\frac{b^2 + 5b + 8}{b}$

STUDY SET SECTION 6.4 (page 509)

1. rational, complex **3.** $\frac{t^2}{t^2}$ **5.** single, division, reciprocal
7. $\div, 3, 25m, m, 3, 3, m, 10$ **9. a.** $\div$ **b.** $6 - k - \frac{5}{k}; k^2 - 9$
11. $\frac{2a^5}{3}$ **13.** $\frac{5y}{9}$ **15.** $\frac{27y^3}{5x^5}$ **17.** $\frac{20}{49c^6d^4}$ **19.** $\frac{3-2a}{a^2 - 3a}$ **21.** $\frac{p^2 - 1}{3p-1}$
23. $y - x$ **25.** $-\frac{1}{a+b}$ **27.** $\frac{2+a}{2a+1}$ **29.** $\frac{b-1}{b+2}$ **31.** $\frac{y+x}{xy}$
33. $\frac{a^3b^2 - a^2}{a^2b^3 - b^2}$ or $\frac{a^2(ab^2 - 1)}{b^2(a^2b - 1)}$ **35.** $\frac{5z - 12}{5z}$ **37.** $\frac{x-9}{3}$ **39.** $\frac{y+x}{y-x}$
41. $\frac{2x+4}{3-x}$ **43.** $\frac{x+2}{x-3}$ **45.** $\frac{1}{c-d}$ **47.** $\frac{x^2 + 64}{(x-8)(x+8)}$
49. $-\frac{a^2bc}{bc + ac + ab}$ **51.** $\frac{a-1}{a+1}$ **53.** $125b$ **55.** -1 **57.** $\frac{3a+7}{2a}$
59. $\frac{xy}{y-x}$ **61.** $\frac{y-2x}{2y+x}$ **63.** $\frac{5x^2y^2}{xy+1}$ **65.** $\frac{1}{2y}$ **67.** $\frac{1}{x-y}$ **69.** $\frac{4}{c+d}$
71. $\frac{k_1k_2}{k_2 + k_1}$ **73.** $\frac{4d-1}{3d-1}$ **79.** 8 **81.** $2, -2, 3, -3$

ARE YOU READY? 6.5 (page 512)

1. a. $\frac{2}{5}$ **b.** $2a^5$ **2.** $8x^5y^5 - 20x^4y^7 + 2xy^2$ **3.** 36 **4.** $2x^2$

SELF CHECKS 6.5

1. a. $6y^2$ **b.** $\frac{d^4}{4c^3}$ **2. a.** $10h + 3$ **b.** $2s^3 - \frac{s^2t}{11} + \frac{4}{t}$
3. $x^2 + 2x - 1$ **4.** $2a^2 + 9a + 26 + \frac{-7}{a-3}$ or
$2a^2 + 9a + 26 - \frac{7}{a-3}$ **5.** $6a^2 + a - 2$ **6.** $25a^2 + 5a + 1$
7. $a^2 + 5a + 11 + \frac{18a - 17}{a^2 - 2a + 1}$

STUDY SET SECTION 6.5 (page 518)

1. monomial, polynomial, binomial **3.** Divisor, Quotient,
Dividend, Remainder **5. a.** term **b.** 9, 9 **c.** 6, 6, 6
7. $(2x - 1)(x^2 + 3x - 4) = 2x^3 + 5x^2 - 11x + 4$
9. $x^2, 7x, 28$ **11.** $3a^2 + 5 + \frac{6}{3a - 2}$
13. $\frac{x^2 - x - 12}{x - 4}, x - 4 \overline{)x^2 - x - 12}, (x^2 - x - 12) \div (x - 4)$
15. $\frac{y}{2x^3}$ **17.** $\frac{3}{4a^2}$ **19.** $2x^4 + 3x$ **21.** $\frac{2x}{3} - \frac{x^2}{6}$ **23.** $2a - \frac{2a^2y}{3}$
25. $-\frac{x^4y^4}{2} + \frac{x^3y^9}{4} - \frac{3}{4xy^2}$ **27.** $x + 2$ **29.** $x - 3$ **31.** $4x - 5$
33. $3x^2 + 4x + 3$ **35.** $t^2 + 2t + 1 + \frac{3}{t + 6}$
37. $2x^2 + 5x - 3 + \frac{-8}{3x - 2}$ **39.** $a + 1$ **41.** $2y + 2$
43. $3x^2 - x + 2$ **45.** $4x^3 - 3x^2 + 3x + 1$ **47.** $4a^2 - 2a + 1$
49. $5a^2 - 3a - 4$ **51.** $x^2 + 3x + 4$ **53.** $2x + 3 + \frac{20x - 13}{3x^2 - 7x + 4}$
55. $6y - 12$ **57.** $4a^2 - 3a + \frac{7}{a + 1}$ **59.** $x^4 + x^2 + 4$
61. $\frac{a^2}{5} - \frac{2}{5a^2}$ **63.** $s + 5 + \frac{-10}{2s + 3}$ **65.** $\frac{8m^2}{7n^{10}}$
67. $m - 4 + \frac{m + 3}{m^2 + 1}$ **69.** $y^2 + 4y + 16$ **71.** $a^4 - a^2 + 1$
73. $10x^2z - 2x - \frac{1}{x}$ **75.** $x^2 - 2 + \frac{-x^2 + 7x + 4}{x^3 + 2x + 1}$
77. $2x^2 - x + 1$ **79.** $\frac{5x^2}{9} + \frac{x}{3} - \frac{1}{9}$ **81.** $x^4 - x^2 - 3$
83. $4x^3 - x + 3$ **85.** $4x^2 + 6x + 10$
87. a. $4n - 4 - \frac{5}{4n}$ **b.** $4n - 5$ **89.** $x^2 - 5x + 6$
91. $3x - 2, x + 5$ **95.** $8x^2 + 2x + 4$
97. $-2y^3 - y^2 + 10y - 14$

ARE YOU READY? 6.6 (page 521)

1. $3x + 4$ **2.** $2x + 5 + \frac{1}{x - 2}$ **3.** 26 **4.** $27x^2$

SELF CHECKS 6.6

1. $5x + 11$ **2.** $x^2 + 4x + 19 + \frac{14}{x - 4}$
3. $3a^2 - 7a + 9 + \frac{-12}{a + 1}$ **4. a.** 9 **b.** 9
5. Since $P(-1) = 0$, $x + 1$ is a factor of $P(x)$.

STUDY SET SECTION 6.6 (page 526)

1. synthetic **3.** divisor **5.** theorem
7. a. $(5x^3 + x - 3) \div (x + 2)$ **b.** $5x^2 - 10x + 21 + \frac{-45}{x + 2}$
9. $6x^3 - x^2 - 17x + 9, x - 8$ **11.** 2, 6, 1, 12, 26, 2, 8
13. $2x + 3$ **15.** $5x - 2$ **17.** $3x - 4$ **19.** $5x + 6$
21. $a^2 - a - 2$ **23.** $3a^2 + 12a + 1$
25. $3b^2 + 9b - 4 + \frac{1}{b - 3}$ **27.** $4t^2 + 8t + 15 + \frac{12}{t - 2}$
29. $x^2 - 5x + 6$ **31.** $3x^2 - 4x - 4 + \frac{-10}{x + 8}$
33. $2x^2 - 3x + 12 + \frac{-52}{x + 5}$ **35.** $x^2 - 2x + 3 + \frac{-3}{x + 10}$
37. $7.2x - 0.66 + \frac{0.368}{x - 0.2}$ **39.** $9x^2 - 513x + 29{,}241 - \frac{1{,}666{,}762}{x + 57}$
41. -1 **43.** -37 **45.** 23 **47.** -1 **49.** 2 **51.** -1 **53.** 18
55. 174 **57.** -8 **59.** 59 **61.** 44 **63.** $\frac{29}{32}$ **65.** Yes **67.** No
69. $6x^2 - x + 1 + \frac{3}{x + 1}$ **71.** $x - 7 + \frac{28}{x + 2}$

73. $a^4 + a^3 + a^2 + a + 1$ **75.** $-6c^4 - 10c^3 - 2c^2 - 4c + 9$
77. $9a^2 - 21$ **79.** $4x^3 - x + 2 + \frac{6}{x + 3}$ **81.** $3x^2 - x + 2$
83. $2x^2 + 4x + 5$ **85.** $4x^2 - 3x + 6 + \frac{-13}{x + 2}$
87. $8a^2 - 16a - 20$ **93.** $7, -\frac{7}{3}$ **95.** $-\frac{1}{2}$

ARE YOU READY? 6.7 (page 528)

1. $30x$ **2.** $8x$ **3.** $-6, 9$ **4.** 4

SELF CHECKS 6.7

1. 9 **2.** 18 **3.** 3, 5 **4.** -3 **5.** No solution; 1 is extraneous
6. $r^2 = \frac{Gm_1m_2}{F}$ **7.** $z = \frac{xy}{y - x}$

STUDY SET SECTION 6.7 (page 535)

1. rational **3. a.** Equation **b.** Expression **c.** Expression
d. Equation **e.** Expression **f.** Expression **5. a.** 3, 0 **b.** 3, 0
c. 3, 0 **7.** $30y, 30y, 30y, 30y, 30y, 7y, 35$ **9.** 12 **11.** 5
13. 4 **15.** $\frac{5}{3}$ **17.** 5 **19.** $-\frac{2}{3}$ **21.** 1 **23.** 7 **25.** 1, 8
27. $2, -3$ **29.** $-\frac{1}{3}, 7$ **31.** $4, -1$ **33.** -8 **35.** 6
37. No solution; 9 is extraneous **39.** No solution; -3 is
extraneous **41.** $A = LQ + I$ **43.** $r = \frac{E - IR_L}{I}$
45. $n_2 = \frac{2\mu_R - n_1{}^2 - n_1}{n_1}$ **47.** $Q_1 = \frac{PQ_2}{1 + P}$
49. $R = \frac{R_1R_2R_3}{R_2R_3 + R_1R_3 + R_1R_2}$ **51.** $r = \frac{eR}{E - e}$ **53.** 2 **55.** $\frac{1}{2}$
57. No solution; 0 is extraneous **59.** $1, -11$ **61.** $-8, 9$
63. A repeated solution of 2 **65.** $-\frac{1}{2}$ **67.** 0 **69.** 5
71. 5; 3 is extraneous **73.** 1 **75.** -1 **77.** 6 **79.** 1 **81.** 2
83. 3, 1 **85.** $\frac{1}{6}$ **87.** $-\frac{3}{2}, 2$ **89. a.** $\frac{11x + 30}{12x}$ **b.** 6
91. a. $\frac{m^2 - 4m + 2}{(m - 2)(m - 3)}$ **b.** 4 **93. a.** $f = \frac{s_1s_2}{s_1 + s_2}$ **b.** $4\frac{8}{13}$ in.
95. a. $L = \frac{SN - CN}{V - C}$ **b.** 8 yr **101.** 9.0×10^9
103. 4.4×10^{-22}

ARE YOU READY? 6.8 (page 539)

1. $t = \frac{d}{r}$ **2.** $\frac{x}{5}$ **3.** $20x$ **4.** 18 **5.** $4\frac{4}{9}$ **6.** 25, 4

SELF CHECKS 6.8

1. If both crews work together, it will take $1\frac{5}{7}$ days to paint the
house. **2.** Father: 36 min; son: 45 min **3.** The caravan of
students averaged 50 mph going to the competition and 30 mph
returning. **4.** The speed of the boat in still water is 12 mph.

STUDY SET SECTION 6.8 (page 545)

1. work, motion **3.** x **5.** $\frac{x}{15}, \frac{x}{8}$ **7.** $\frac{12}{x}, \frac{12}{x + 15}$ **9.** $4\frac{5}{9}$ hr
11. $2\frac{6}{11}$ days **13. a.** $1\frac{7}{8}$ days **b.** Santos: $412.50, Mays: $375
15. $5\frac{5}{6}$ min **17.** $4\frac{8}{13}$ sec **19.** $2\frac{4}{13}$ weeks
21. Waiter: 10 min; busboy: 15 min
23. Faster worker: 6 hr; slower worker: 12 hr
25. Cisco: 3 hr; Dell: 4.5 hr **27.** $1\frac{1}{2}$ hours **29.** About 110 sec
31. Going: 25 mph; returning: 15 mph **33.** Going: 40 mph,
returning: 15 mph **35.** 35 mph and 45 mph **37.** 150 mph
39. 2 mph **41.** 3 mph **43.** 5 feet per sec **47.** $\frac{m^{80}}{n^8}$ **49.** $-\frac{1}{w^2}$
51. -4 **53.** $-x^8y^{10}$

ARE YOU READY? 6.9 (page 548)

1. **a.** ab **b.** bc **2.** 27 **3.** $-2, 9$ **4.** 5 **5.** 1,620 **6.** 90°

SELF CHECKS 6.9

1. $\frac{3}{2}$ **2.** $\frac{2}{3}, 1$ **3.** $68 **4.** 24 ft **5.** 735 British pounds
6. 16 foot-candles **7.** Approximately 1,055 lb **8.** 20 weeks

STUDY SET SECTION 6.9 (page 558)

1. ratio **3.** extremes, means, extremes, means **5.** similar
7. inverse, decreases **9. a.** Direct **b.** Inverse **11.** Direct
13. Inverse **15.** Direct **17.** Inverse **19.** 7, 6, 18, 66, 11
21. 3 **23.** 5 **25.** $\frac{21}{5}$ **27.** -4 **29.** $2, -2$ **31.** $4, -1$
33. $-\frac{5}{2}, -1$ **35.** $0, -3$ **37.** $A = kp^2$ **39.** $z = \frac{k}{t^3}$
41. $C = kxyz$ **43.** $P = \frac{ka^2}{j^3}$ **45.** r varies directly as t.
47. b varies inversely as h. **49.** U varies jointly as r, the square of s,
and t. **51.** P varies directly as m and inversely as n. **53.** $\frac{7}{2}$
55. $-5, 2$ **57.** No solution **59.** $-\frac{1}{2}, 0, 5$ **61.** 6
63. $-\frac{5}{2}, -1, 1$ **65.** -0.1 **67.** $-10, 10$
69. Mountain Dew: 202 mg, Pepsi: 139 mg, Coca-Cola Classic:
125 mg **71.** About 2 gal **73.** Eye: 49.9 in.; seat: 17.6 in.; elbow:
27.8 in. **75.** 12.5 in. **77.** 25 ft **79.** 120 ft **81.** 792 ft.
83. **a.** False **b.** False **c.** True **85.** 117.6 newtons
87. 432 mi **89.** 25 days **91.** 12 in.³ **93.** $9,000 **95.** 3 ohms
97. 0.275 in. **99.** 1.4 ohms **101.** 12.8 lb   **1**
05. $\frac{13}{6}w^3 - \frac{1}{4}w^2 + \frac{4}{5}$ **107.** $6y^3 + 11y^2 + 9y + 2$

CHAPTER 6 REVIEW (page 562)

1. **a.** 1 **b.** $-\frac{1}{2}$ **c.** 0 **2.** The domain is the set of all real
numbers except -6 and 4: $(-\infty, -6) \cup (-6, 4) \cup (4, \infty)$.
3. $n(3) = 8.4$; Three hours after the injection, the concentration of
pain medication in the patient's bloodstream was 8.4 milligrams per
liter. **4.** $y = 3, x = 0$; D: $(-\infty, 0) \cup (0, \infty)$; R: $(-\infty, 3) \cup (3, \infty)$
5. $\frac{12x}{19y^7}$ **6.** $\frac{x-7}{x+7}$ **7.** $\frac{1}{2x^2(x+2)}$ **8.** $\frac{1}{x^4(x-6)}$ **9.** $\frac{-a-b}{c+d}$
10. $\frac{m+2n}{2m+n}$ **11.** $\frac{2x+1}{3x-4}$ **12.** -2 **13.** Does not simplify
14. $-\frac{3m-4}{m+3}$ or $\frac{4-3m}{m+3}$ **15.** $\frac{2}{49x^2}$ **16.** $-x$ **17.** $\frac{2a-1}{4a(a+2)}$
18. $\frac{t-2}{t}$ **19.** $\frac{h^2 - 4h + 4}{h^6 + 8h^3 + 16}$ **20.** $\frac{a+6}{(m+4)(m-3)}$
21. $\frac{2m-n}{m+n}$ **22.** $\frac{3x(x-1)}{(x-3)(x+1)}$ **23.** $\frac{5y-3}{x-y}$ **24.** $\frac{1}{c-d}$
25. $-\frac{2}{t-3}$ **26.** $-\frac{1}{p+12}$ **27.** $60a^2h^3$ **28.** $ab^2(b-1)$
29. $(x-5)(x+5)(x+1)$ **30.** $(m^2 + 2m + 4)(m-2)^2$
31. $\frac{9a+8}{a+1}$ **32.** $\frac{40x + 7y^2z}{112z^2}$ **33.** $\frac{4x^2 + 9x + 12}{(x-4)(x+3)}$ **34.** $\frac{2a^2 + 8a - 19}{3(a+2)}$
35. $\frac{1}{(a+3)(a+2)}$ **36.** $\frac{12y + 20}{15y(x-2)}$ **37.** $\frac{2a-3}{2a}$ **38.** $\frac{14x + 58}{(x+3)(x+7)}$
39. $\frac{2bc^3}{7}$ **40.** $\frac{p-3}{2(p+2)}$ **41.** $\frac{b+2a}{2b-a}$ **42.** $\frac{x-2}{x+3}$
43. $\frac{x^2y^2}{(x-y)^2(y^2 - x^2)}$ or $\frac{x^2y^2}{(x-y)^3(x+y)}$ **44.** $\frac{1+b+d}{1-b-d}$
45. $\frac{2b^2 - 3b + 3}{3b-2}$ **46.** $\frac{4r-8}{2r+5}$ **47.** $\frac{5h^3}{11k^2}$ **48.** $\frac{1}{2y^3z^{10}}$ **49.** $6a + \frac{16}{3}$
50. $-3x^2y + \frac{3x}{2} + y$ **51.** $b + 4$ **52.** $v^2 - 3v - 10$
53. $x^2 - 2x + 4$ **54.** $2m - 5 + \frac{-4}{4m+1}$
55. $3a - 2 + \frac{-15a + 2}{a^2 + 5}$ **56.** $m^4 - m^2 + 1 + \frac{2}{m^4 + 2m^2 - 3}$
57. $x - 7$ **58.** $m^2 - 3m + 2$ **59.** $-3n^4 - 2n^3 - n^2 - 2n + 1$
60. $4x^2 - 3x + 6 + \frac{-13}{x+2}$ **61.** $3a^3 - a + 4 + \frac{6}{a+1}$
62. $x^3 + 3x^2 + 9x + 27 + \frac{82}{x-3}$ **63.** 54 **64.** -227 **65.** Yes

66. No **67.** 5 **68.** $\frac{77}{5}$ **69.** $-2, 3$ **70.** $-1, -2$ **71.** $\frac{3}{2}$ **72.** 0
73. No solution; 3 is extraneous **74. a.** $\frac{x+26}{x(x-4)}$ **b.** $\frac{26}{9}$
75. $b = \frac{Ha}{2a - H}$ **76.** $y^2 = \frac{x^2b^2 - a^2b^2}{a^2}$ **77.** $R = \frac{R_1 R_2}{R_2 + R_1}$
78. $F = \frac{ma}{k}$ **79. a.** $\frac{1}{10}$ of the job per hour
b. $\frac{x}{10}$ of the job is completed **80.** $14\frac{2}{5}$ hr
81. Experienced electrician: 10 days; apprentice: 15 days
82. 6 min **83.** 5 mph **84.** 50 mph **85.** 5 **86.** $-4, -12$
87. $0, -1$ **88.** $-2, 3$ **89.** 70.4 ft **90.** 20 **91.** 66 in.
92. $5,460 **93.** 1.25 amps **94.** 126.72 lb **95.** Inverse variation
96. 0.2

CHAPTER 6 TEST (page 573)

1. **a.** rational **b.** reciprocal **c.** complex **d.** extraneous
e. proportion **2.** $x + 2$ is not a factor of the entire numerator
and, therefore, cannot be removed. **3.** $\frac{2}{3xy^3}$ **4.** $\frac{2}{x-2}$ **5.** -3
6. $\frac{2x+y}{4y}$ **7. a.** 1 **b.** $-\frac{1}{2}$ **c.** -1 **d.** 2
8. $p(7) = 175$ **9.** No **10.** The set of all real numbers except 0
and 1; $(-\infty, 0) \cup (0, 1) \cup (1, \infty)$ **11.** $\frac{xz}{y^4}$ **12.** 1 **13.** $\frac{13}{x+1}$
14. $\frac{(x+y)^2}{2x}$ **15.** -1 **16.** $\frac{(2x-3)^2}{x^6}$ **17.** $\frac{2}{t-4}$ **18.** 2
19. $\frac{24x^2 - 2x - 1}{3x + 1}$ **20.** $\frac{6a - 17}{(a+1)(a-2)(a-3)}$ **21.** $\frac{u^2}{2vw}$
22. $\frac{3k^2 + 4k + 4}{3k^2 - 9k - 9}$ **23.** $-\frac{6x}{y} + \frac{4x^2}{y^2} - \frac{3}{y^3}$ **24.** $y^2 - 2y + 4 - \frac{56}{y+2}$
25. 41 **26.** $x + 3$ is a factor of $P(x)$. **27.** 40
28. 5; 3 is extraneous **29.** $6, -1$ **30.** 26 **31.** $F = \frac{CDE}{AB}$
32. $r_2 = \frac{rr_1}{r_1 - r}$ **33.** No, $\frac{5}{11}$ of an hour **34.** Supervisor: 45 min;
technician: 90 min **35.** 3 mph **36.** 2 mph **37.** 80 ft
38. $77.32 **39.** 300 **40.**
41. 25 decibels

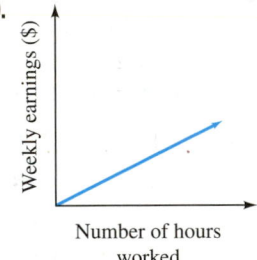

Weekly earnings ($) / Number of hours worked

42.

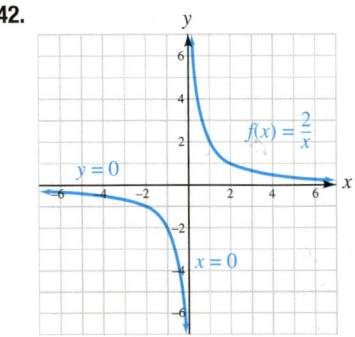

$f(x) = \frac{2}{x}$, $y = 0$, $x = 0$

CUMULATIVE REVIEW CHAPTERS 1–6 (page 575)

1. -372 **2.** $22a^3 + 20a$ **3.** -2 **4.** $n = \frac{l - a + d}{d}$
5. 490.9 in.² **6.** $\angle A$: 60°; $\angle B$: 50°; $\angle C$: 70°
7. 11.3% decrease **8.** $\left(4, -\frac{1}{2}\right)$ **9.** Life expectancy will increase
0.1 year each year during this period. **10.** x-intercept: $(-1, 0)$;
y-intercept: $\left(0, \frac{7}{10}\right)$ **11.** $y = -7x + 54$ **12.** $y = -\frac{11}{6}x - \frac{7}{3}$

13. a. $-\frac{7}{3}$ **b.** I **c.** $y = -\frac{7}{3}x - 7$ **d.** Yes **14.** 18

15. a. No; $(1, 1)$, $(1, -1)$; answers may vary **b.** No; $(0, 2)$, $(0, -2)$; answers may vary **16. a.** $C(t) = 1.9t + 349.8$

b. 406.8 parts per million

17. D: $(-\infty, \infty)$; R: $[0, \infty)$ **18.** $(-2, -5)$

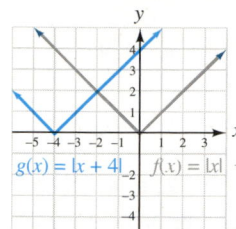

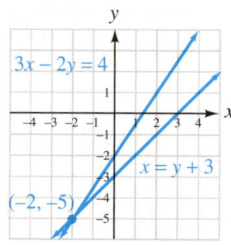

19. $T_1 = 80$, $T_2 = 60$ **20.** $(2, -3, 5)$ **21.** $(2, 0)$ **22.** 8
23. 12 oz 50% alloy, 8 oz 25% alloy **24.** Pizza Hut: 14%,
Domino's: 8%, and Papa John's: 6% **25.** $\left(-\infty, \frac{4}{3}\right]$

26. $(-\infty, -11)$ **27.** $(-\infty, 1)$

28. $(-\infty, 3] \cup \left[\frac{11}{3}, \infty\right)$

29.

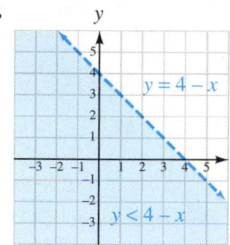

30. Yes **31.** $81b^{24}$ **32.** $81d^{10}$
33. $-\frac{k^7}{20x^9}$ **34.** $\frac{x^{21}y^3}{8}$
35. Penny: 10^{-2}, dime: 10^{-1},
one-dollar bill: 10^0,
one-hundred-thousand-dollar
bill: 10^5 **36.** 3.6×10^{-5};
0.000036 **37. a.** 4 **b.** 5 **c.** $-2, 1$
d. Domain: $(-\infty, \infty)$; range: $(-\infty, \infty)$

38. 288 ft **39.** 7 **40.** $\frac{7}{4}c^2 - \frac{5}{6}c$ **41.** $2x^3 + x^2 + 12$
42. $4m^2n^2 + 4m - 2n$ **43.** $2r^{11}s^{20}$ **44.** $-12a^{12} - 9a^{11} + 12a^{10}$
45. $4x^6 - 4x^3 + 1$ **46.** $2a^2 + ab - b^2 - 3bc - 2c^2$
47. $3rs^3(r - 2s)$ **48.** $(x - y)(5 - a)$ **49.** $(x + y)(u + v)$
50. $(4x^2 - 3)(2x^2 - 1)$ **51.** $(x - y + 5)(x - y - 2)$
52. $(9x^2 + 4y^2)(3x + 2y)(3x - 2y)$
53. $(2x - 3y^2)(4x^2 + 6xy^2 + 9y^4)$ **54.** $(x + 5 + 4z^4)(x + 5 - 4z^4)$
55. $b^2 = \frac{a^2y^2}{a^2 - x^2}$ **56.** $-\frac{1}{3}, -\frac{7}{2}$ **57.** $0, 2, -2$
58. 9 in. by 12 in. **59.** $\frac{2x - 3}{3x - 1}$ **60.** All real numbers except 0
and 2: $(-\infty, 0) \cup (0, 2) \cup (2, \infty)$ **61.** $-\frac{n^6}{n^2 - 2}$ **62.** $-\frac{q}{p}$
63. $\frac{4}{x - y}$ **64.** $\frac{c + 9}{c - 5}$ **65.** $\frac{y^2}{3} + 3y$ **66.** $4x^2 - x - 1$
67. -17 **68.** 0 **69.** $x^3 + x + 2 + \frac{-2}{x - 9}$ **70.** $3\frac{3}{4}$ min
71. It will rise sharply. **72.** 13 cups **73.** 1 day **74. a.** iii **b.** i
c. iv **d.** ii

ARE YOU READY? 7.1 (page 580)

1. a. 225 **b.** 64 **2. a.** $\frac{49}{25}$ **b.** 0.04 **3. a.** -216 **b.** 81
4. a. a^8 **b.** x^9 **5.** $x^2 + 16x + 64$ **6.** 5

SELF-CHECKS 7.1

1. a. 8 **b.** -1 **c.** $\frac{1}{4}$ **d.** 0.3 **2. a.** $5|a|$ **b.** $|b^7|$ **c.** $|x - 9|$

d. $10n^4$ **3.**

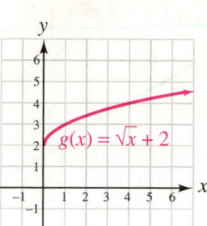

D: $[0, \infty)$; R: $[2, \infty)$; the graph
is 2 units higher

4. a. $[2, \infty)$ **b.**

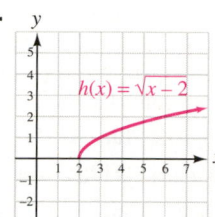

c. $[0, \infty)$

5. 1.92 sec **6. a.** 4 **b.** $-\frac{1}{10}$ **c.** $-5a$ **d.** $3m^2n^3$
7. a.

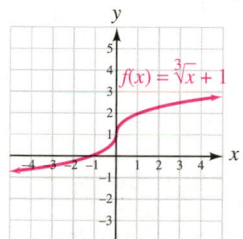

b. D: $(-\infty, \infty)$; R: $(-\infty, \infty)$

8. a. $\frac{1}{3}$ **b.** 10 **c.** Not a real number **9. a.** $|x|$ **b.** $a + 5$

c. $2a^2$

STUDY SET SECTION 7.1 (page 591)

1. square, cube **3.** positive **5.** principal **7.** simplified
9. radical **11.** a **13.** squared **15.** $|x|, x$ **17. a.** 3 **b.** 0
c. Undefined **d.** 6 **e.** None **f.** D: $[2, \infty)$; R: $[0, \infty)$ **19. a.** -5
b. -3 **c.** 1 **d.** D: $(-\infty, \infty)$, R: $(-\infty, \infty)$ **21. a.** $\sqrt{x^2} = |x|$,
b. $\sqrt[3]{x^3} = x$ **c.** $\sqrt[5]{-32} = -2$ **23.** 10 **25.** -8 **27.** $\frac{1}{3}$
29. 0.5 **31.** Not a real number **33.** 11 **35.** 3.4641
37. 26.0624 **39.** $2|x|$ **41.** $9h^2$ **43.** $6|s^3|$ **45.** $12m^4$
47. $|y - 1|$ **49.** $a^2 + 3$ **51.** $0, -1, -2, -3, -4$;
D: $[0, \infty)$; R: $(-\infty, 0]$ **53.** D: $[-4, \infty)$; R: $[0, \infty)$

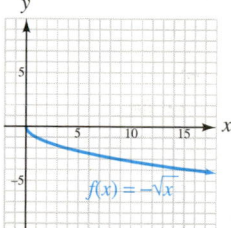

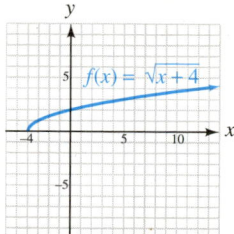

55. D: $[-6, \infty)$ **57.** D: $(-\infty, 4]$ **59.** D: $\left[\frac{4}{9}, \infty\right)$ **61.** D: $[40, \infty)$
63. a. 5 **b.** Undefined **65. a.** 2 **b.** -3 **67. a.** 4.1231
b. 2.5539 **69. a.** 3.3322 **b.** 7.7543 **71.** 1 **73.** -5 **75.** $\frac{2}{3}$
77. 4 **79.** $-6a$ **81.** $-10p^2q$

83. $-5, -4, -3, -2, -1;$ **85.** D: $(-\infty, \infty)$; R: $(-\infty, \infty)$
D: $(-\infty, \infty)$; R: $(-\infty, \infty)$

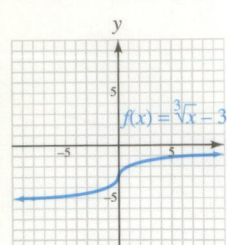
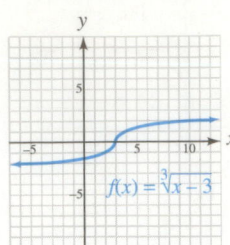

87. 3   **89.** -3 **91.** Not a real number   **93.** $-\frac{1}{2}$ **95.** $2a$
97. $3|a|$ **99.** k^2 **101.** $(m + 4)^2$ **103.** $4s^3t^2$ **105.** $-7b^4$
107. $\frac{1}{2}$ **109.** $-5m^2$ **111.** $20m^8|n|$ **113.** $2|ab|$ **115.** Not a real
number   **117.** $|n + 6|$ **119. a.** 8   **b.** 4   **121. a.** 9   **b.** 3
123. 7.0 in.   **125.** 1,138.5 mm   **127.** About 58.6 beats/min
129. 13.4 ft   **131.** 3.5%   **137.** $\frac{(x - 4y)^2}{(x - 2y)(x - 3y)}$

ARE YOU READY? 7.2 (page 595)

1. a. 8   **b.** -4 **2. a.** 3   **b.** $\frac{1}{2}$ **3. a.** $|x + 5|$ **b.** $3a^2$ **4.** 216
5. $\frac{1}{49}$ **6.** x^9

SELF-CHECKS 7.2

1. a. 4   **b.** $-\frac{3}{2}$ **c.** -3 **d.** 1   **2. a.** $-2n$ **b.** $5|a|$ **c.** b^2
d. Not a real number   **3. a.** 64   **b.** 625   **c.** 36   **d.** $-\frac{1}{16}$
4. a. $8c^6$ **b.** $9m^2n^2$ **c.** $-8a^6$ **5.** $(7ab)^{1/6}$ **6.** $\sigma = \sqrt{\frac{\Sigma(x - \mu)^2}{N}}$
7. a. $\frac{1}{3}$ **b.** $\frac{1}{216}$ **c.** $\frac{1}{9a^2}$ **d.** -27 **8. a.** $2^{3/5}$ **b.** $12^{4/3}$ **c.** x^2y^9
d. x^2 **9.** $t - 1$ **10. a.** $\sqrt{3}$ **b.** $\sqrt{7xy}$ **c.** $\sqrt[12]{m}$

STUDY SET SECTION 7.2 (page 603)

1. rational (or fractional)   **3.** negative   **5.** index, radicand
7.

Radical form	Exponential form	Base	Exponent
$\sqrt[5]{25}$	$25^{1/5}$	25	$\frac{1}{5}$
$\left(\sqrt[3]{-27}\right)^2$	$(-27)^{2/3}$	-27	$\frac{2}{3}$
$\left(\sqrt[4]{16}\right)^{-3}$	$16^{-3/4}$	16	$-\frac{3}{4}$
$\left(\sqrt{81}\right)^3$	$81^{3/2}$	81	$\frac{3}{2}$
$-\sqrt{\frac{9}{64}}$	$-\left(\frac{9}{64}\right)^{1/2}$	$\frac{9}{64}$	$\frac{1}{2}$

9.

$(-125)^{1/3}$ $-(9/100)^{-1/2}$ $-16^{-1/4}$ $8^{2/3}$ $4^{3/2}$
number line from -5 to 8 with points at -5, -3, -1, 4, 8

11. $\sqrt[n]{x}$ **13.** $\frac{1}{x^{m/n}}$ **15.** $100a^4, 10a^2$ **17.** 5   **19.** 3   **21.** 2
23. -6 **25.** -2 **27.** $\frac{1}{2}$ **29.** $2x^2$ **31.** $|x|$ **33.** Not a real
number   **35.** $-3n^3$ **37.** Not a real number   **39.** $|x + 1|$
41. 216   **43.** 8   **45.** $\frac{1}{36}$ **47.** -32 **49.** $125x^6$ **51.** $4x^4y^2$
53. $27x^3y^6$ **55.** $-\frac{x^4}{16}$ **57.** $(8abc)^{1/5}$ **59.** $(a^2 - b^2)^{1/3}$
61. $\sqrt[4]{6x^3y}$ **63.** $\sqrt{2s^2 - t^2}$ **65.** $\frac{1}{2}$ **67.** $\frac{1}{5}$ **69.** $-\frac{1}{100y^2}$ **71.** $\frac{16}{81}$
73. $\frac{27y^3}{8}$ **75.** 243   **77.** $9^{5/7}$ **79.** $\frac{1}{36}$ **81.** m^6 **83.** $a^{3/4}b^{1/2}$
85. 3   **87.** a **89.** $y + y^2$ **91.** $x^2 - 1 + x^{3/5}$ **93.** $\sqrt{5}$
95. $\sqrt[3]{11}$ **97.** $\sqrt{p}$ **99.** $\sqrt[5]{xy}$ **101.** $\sqrt[18]{c}$ **103.** $\sqrt[15]{7m}$
105. 2.47   **107.** 1.02   **109.** $5y$ **111.** $-\frac{a^3}{27}$ **113.** $\frac{1}{64}$ **115.** p

117. $-\frac{1}{3x^2}$ **119.** 2   **121.** $n^{3/5} - 1$ **123.** 32   **125.** m^2
127. $\sqrt{5b}$ **129.** $2x$ **131.** $-\frac{1}{4a^2b^4}$ **133. a.** -25 **b.** 25   **c.** $-\frac{1}{25}$
d. 25   **135. a.** $8a^2$ **b.** $\frac{1}{8a^2}$ **c.** $-8a^2$ **d.** $\frac{1}{8a^2}$ **137.** 736 ft/sec
139. 1.96 units   **141.** 145.8 in. or 12.1 ft   **145.** $2\frac{1}{2}$ hr

ARE YOU READY? 7.3 (page 606)

1. a. 4   **b.** 27   **2. a.** -2 **b.** 2   **3.** $27a^4b^8$ **4.** $7x, 9x$
5. $14m^2 - m$ **6.** $\frac{5}{6}$

SELF-CHECKS 7.3

1. a. $3\sqrt{2}$ **b.** $14\sqrt[3]{3}$ **c.** $2\sqrt[4]{2}$ **d.** $2\sqrt[5]{4}$ **2. a.** $42b\sqrt{2b}$
b. $-3y\sqrt[3]{2y^2}$ **c.** $t^2u^3\sqrt[4]{u^3}$ **3. a.** $5\sqrt{11}$ **b.** $3cd\sqrt[3]{7c}$
c. $5x^2y\sqrt[4]{2y^2}$ **4. a.** $\frac{\sqrt[3]{25}}{3}$ **b.** $\frac{\sqrt{11}}{6a}$ **c.** $\frac{\sqrt[4]{a^3}}{5y^3}$ **5. a.** $5b$
b. $-10xv\sqrt[3]{x}$ **6. a.** $19\sqrt{3} + 6$ **b.** $2\sqrt[3]{3} + \sqrt[3]{2}$ **7.** $21x\sqrt{2x}$

STUDY SET SECTION 7.3 (page 614)

1. like   **3.** factor, perfect   **5.** $\sqrt[n]{a}\sqrt[n]{b}$, product, roots
7. a. $\sqrt{4 \cdot 5}$ **b.** $\sqrt{4}\sqrt{5}$ **c.** $\sqrt{4 \cdot 5} = \sqrt{4}\sqrt{5}$ **9. a.** $\sqrt{5}$,
$\sqrt[3]{5}$ (Answers may vary); no   **b.** $\sqrt{5}$, $\sqrt{6}$ (Answers may vary); no
11. $8k^3, 8k^3, 2k$ **13.** $5\sqrt{2}$ **15.** $24\sqrt{5}$ **17.** $2\sqrt[3]{4}$ **19.** $2\sqrt[4]{3}$
21. $5a\sqrt{3}$ **23.** $8ab^2\sqrt{2ab}$ **25.** $-6x^2\sqrt[3]{2}$ **27.** $2x^3y\sqrt[4]{2}$
29. $11\sqrt{2}$ **31.** $4a\sqrt{7a}$ **33.** $-2\sqrt[5]{3a^4}$ **35.** $3x^4y\sqrt[3]{15y}$
37. $\frac{\sqrt{11}}{3}$ **39.** $\frac{\sqrt[4]{3}}{5}$ **41.** $\frac{x^2\sqrt[5]{3}}{2}$ **43.** $\frac{z}{4x}$ **45.** 10   **47.** $7x$
49. $2x^2$ **51.** $3a\sqrt[3]{a}$ **53.** $8\sqrt{7}$ **55.** $5\sqrt[3]{4}$ **57.** $12 + 3\sqrt{2}$
59. $-4\sqrt{2}$ **61.** $1 - \sqrt[3]{4}$ **63.** $13\sqrt[4]{2}$ **65.** $10\sqrt{2x}$
67. $3\sqrt{2t} + \sqrt{3t}$ **69.** $-11\sqrt[3]{2}$ **71.** $4\sqrt[4]{2}$ **73.** $m\sqrt[6]{m^5}$
75. $12\sqrt[3]{a}$ **77.** $5y^3\sqrt{2y}$ **79.** $4\sqrt{2b}$ **81.** $\frac{5n^2\sqrt{5}}{8}$ **83.** -10
85. $10\sqrt{3xy}$ **87.** $\frac{\sqrt[4]{5x}}{2z}$ **89.** $\sqrt[5]{7a^2}$ **91.** $4x\sqrt[5]{xy^2}$
93. $2m\sqrt[4]{13n}$ **95.** $\frac{a^2}{4}$ **97.** $\frac{\sqrt[3]{7}}{4}$ **99.** $7\sqrt{5} - 3\sqrt{3}$
101. $2t^2\sqrt[5]{2t}$ **103.** $3\sqrt[3]{3x}$ **105. a.** $4\sqrt{5}$ **b.** $2\sqrt{21}$ **107. a.** $2x$
b. $2x\sqrt{x}$ **109. a.** $12 + 3\sqrt{6}$ **b.** $9\sqrt[3]{2}$ **111. a.** $40\sqrt[5]{6x}$
b. $40\sqrt[4]{6x}$ **113.** $3\sqrt{14}$ ft, 11.2 ft   **115.** $5\sqrt{3}$ amps; 8.7 amps
117. $\left(26\sqrt{5} + 10\sqrt{3}\right)$ in.; 75.5 in.   **123.** $-\frac{15x^5}{y}$
125. $3p + 4 - \frac{5}{2p - 5}$

ARE YOU READY? 7.4 (page 617)

1. a. $25a^{12}$ **2.** $54t^6 + 18t^5$ **3.** $2x^2 + x - 3$ **4.** $1 - 2\sqrt{14}$
5. $x^2 - 100$ **6.** $\frac{18}{27a}$

SELF-CHECKS 7.4

1. a. $7\sqrt{2}$ **b.** $-20\sqrt[3]{3}$ **c.** $18x\sqrt[4]{2x}$ **2.** $12\sqrt{10} - 32$
3. a. $-1 + \sqrt{15}$ **b.** $a - \sqrt[3]{3a} + 9\sqrt[3]{2a^2} - 9\sqrt[3]{6}$ **4. a.** 11
b. $108y$ **c.** $19 + 8\sqrt{3}$ **d.** $x - 10\sqrt{x - 8} + 17$
5. a. $\frac{2\sqrt{10}}{5}$ **b.** $\frac{5\sqrt[3]{3}}{3}$ **6.** $\frac{\sqrt{2ab}}{a}$ **7.** $\frac{7\sqrt{2c}}{6c^2}$ **8. a.** $\frac{27\sqrt[3]{10a^2}}{10a}$
b. $\frac{\sqrt[4]{12y^2}}{2y}$ **9.** $\frac{x - 2\sqrt{2x} + 2}{x - 2}$ **10.** $\frac{x - 9}{x - 3\sqrt{x}}$

STUDY SET SECTION 7.4 (page 625)

1. product   **3.** rationalize   **5.** perfect

7.

	Why isn't it in simplified form?	Simplified form	Approximation
$\dfrac{3}{\sqrt{2}}$	A radical appears in the denominator.	$\dfrac{3\sqrt{2}}{2}$	2.121320344
$\dfrac{\sqrt{18}}{2}$	There is a perfect-square factor in the radicand: 9	$\dfrac{3\sqrt{2}}{2}$	2.121320344
$\sqrt{\dfrac{9}{2}}$	The radicand contains a fraction.	$\dfrac{3\sqrt{2}}{2}$	2.121320344

9. $5, 2, \sqrt{x}, 25, x$ **11. a.** $6\sqrt{6}$ **b.** 48 **c.** Can't be simplified
d. $-6\sqrt{6}$ **13.** $\sqrt{6}, 48, 16, 4$ **15.** $3\sqrt{5}$ **17.** $6\sqrt{2}$ **19.** 18
21. $2\sqrt[3]{3}$ **23.** $48ab^2$ **25.** $5a\sqrt[4]{a}$ **27.** $12\sqrt{5} - 15$
29. $8\sqrt{3} + 2\sqrt{14}$ **31.** $-8x\sqrt{10} + 6\sqrt{15x}$ **33.** $8 + 2\sqrt[3]{3}$
35. $-1 - 2\sqrt{2}$ **37.** $3x - 2y$
39. $12\sqrt[3]{2} + 8\sqrt[3]{5} - 18 - 6\sqrt[3]{20}$
41. $\sqrt[3]{25z^2} + 3\sqrt[3]{15z} + 2\sqrt[3]{9}$ **43.** 7 **45.** 12 **47.** 18
49. $-16x^2$ **51.** $39 - 12\sqrt{3}$ **53.** $3x + 6\sqrt{x} + 3$ **55.** $\dfrac{\sqrt{14}}{7}$
57. $\dfrac{2\sqrt{6}}{3}$ **59.** $\dfrac{2\sqrt{6}}{3}$ **61.** $\dfrac{\sqrt[3]{4}}{2}$ **63.** $\sqrt[3]{3}$ **65.** $\dfrac{\sqrt[4]{2}}{2}$ **67.** $\dfrac{\sqrt{5y}}{y}$
69. $\dfrac{\sqrt{6y}}{y}$ **71.** $\dfrac{\sqrt[3]{6t}}{3}$ **73.** $\dfrac{\sqrt[3]{2ab^2}}{b}$ **75.** $\dfrac{23\sqrt{2p}}{10p^3}$ **77.** $\dfrac{7\sqrt{6b}}{12b^2}$ **79.** $\dfrac{\sqrt[3]{20}}{4}$
81. $\dfrac{\sqrt[3]{36}}{9}$ **83.** $\dfrac{19\sqrt[3]{25c}}{5c}$ **85.** $\dfrac{\sqrt[3]{12r^2}}{2r}$ **87.** $\dfrac{\sqrt[4]{54t^2}}{3t}$ **89.** $\dfrac{25\sqrt[3]{2a^3}}{2a}$
91. $\dfrac{3\sqrt{2} - \sqrt{10}}{4}$ **93.** $\dfrac{2(\sqrt{x} - 1)}{x - 1}$ or $\dfrac{2\sqrt{x} - 2}{x - 1}$ **95.** $\dfrac{9 - 2\sqrt{14}}{5}$
97. $\dfrac{x - 2\sqrt{xy} + y}{x - y}$ **99.** $\dfrac{x - 9}{x(\sqrt{x} - 3)}$ **101.** $\dfrac{x - y}{\sqrt{x}(\sqrt{x} - \sqrt{y})}$
103. $x\sqrt{14} + \sqrt{2x}$ **105.** $\dfrac{3\sqrt{6} + 4}{2}$ **107.** $2{,}000x$
109. $-20r\sqrt[3]{10s}$ **111.** $9p^2 + 6p\sqrt{5} + 5$ **113.** $\dfrac{6m^2\sqrt{2m}}{5}$
115. $3n\sqrt[4]{n}$ **117.** $\dfrac{\sqrt[3]{3x}}{3}$ **119.** $18r - 12\sqrt{2r} + 4$
121. $x^2(x + 3)$ **123.** $\sqrt{2z} + 1$ **125. a.** $9a$ **b.** $9 + 6\sqrt{a} + a$
127. a. $m - 6$ **b.** $m - 12\sqrt{m} + 36$ **129.** $\dfrac{\sqrt{2\pi}}{2\pi\sigma}$ **131.** $\dfrac{\sqrt{2}}{2}$
139. $\dfrac{1}{3}$

ARE YOU READY? 7.5 (page 629)
1. a. $x - 1$ **b.** $x^3 + 7$ **2.** $18x + 45$ **3.** $x^2 - 8x + 16$
4. $-3, 9$ **5.** $x^3 + 6x^2 + 12x + 8$ **6.** $4 - 4\sqrt{x} + x$

SELF-CHECKS 7.5
1. 11 **2.** 196 ft **3.** 6, 0 is extraneous **4.** 18 is extraneous,
no solution, $\varnothing$ **5.** $0, -2$ **6.** 0 **7.** 20 **8.** 1 **9.** $n = \dfrac{z_0^2pq}{E^2}$

STUDY SET SECTION 7.5 (page 637)
1. radical **3.** raising **5.** extraneous **7. a.** n, n **b.** a
9. a. Square both sides. **b.** Subtract 3 from both sides.
c. Add $\sqrt{2x}$ to both sides. **11.** $x - 6\sqrt{x} + 9$
13. $6, 2, 2, 3x + 3, 33, 11,$ Yes **15.** 4 **17.** 5 **19.** 6 **21.** 198
23. $7, \not{1}$ **25.** $14, \not{6}$ **27.** $3, \not{2}$ **29.** $0, \not{7}$ **31.** $\frac{4}{5}$, no solution
33. $\not{4}$, no solution **35.** 4 **37.** $2, -1$ **39.** $1, -3$ **41.** 10
43. 85 **45.** -7 **47.** 12 **49.** 0 **51.** 3 **53.** 3 **55.** 9 **57.** 1

59. $\not{3}$, no solution **61.** 1, 9 **63.** $h = \dfrac{v^2}{2g}$ **65.** $l = \dfrac{8T^2}{\pi^2}$
67. $A = P(r + 1)^3$ **69.** $v^2 = c^2\left(1 - \dfrac{L_A^2}{L_B^2}\right)$ **71.** 16
73. 4 **75.** $-1, 1$ **77.** $2, \not{1}$ **79.** -16 **81.** 4 **83.** $\frac{5}{2}, \frac{1}{2}$ **85.** $-\frac{1}{2}$
87. $\not{6}$, no solution **89.** 1 **91.** 4, 3 **93.** $\not{8}$, no solution **95.** 1
97. 0 **99.** 8 **101. a.** $\frac{81}{40}$ **b.** $\frac{9}{5}$ **103. a.** 50 **b.** $\not{50}$, no
solution **105.** 178 ft **107.** About 488 watts **109.** 8 ft
111. \$5 **121.** 2.5 foot-candles **123.** 0.41511 in.

ARE YOU READY? 7.6 (page 640)
1. $180°$ **2.** $a\sqrt{3}$ **3.** $\dfrac{25\sqrt{3}}{3}$ **4.** 13 **5.** 3.46 **6.** $\dfrac{14\sqrt{3}}{3}$

SELF-CHECKS 7.6
1. Yes **2.** $12\sqrt{2}$ m ≈ 16.97 m **3.** $\dfrac{9\sqrt{2}}{2}$ in. ≈ 6.36 in.
4. 16 cm, $8\sqrt{3}$ cm **5.** $10\sqrt{3}$ in. ≈ 17.32 in., $5\sqrt{3}$ in. ≈ 8.66 in.
6. $\dfrac{29\sqrt{3}}{2}$ in. ≈ 25 in. **7. a.** 13 units **b.** $3\sqrt{5} \approx 6.71$ units

STUDY SET SECTION 7.6 (page 646)
1. hypotenuse **3.** Pythagorean **5.** a^2, b^2, c^2, equation **7.** $\sqrt{2}$
9. $\sqrt{3}$ **11.** $(x_2 - x_1)^2, (y_2 - y_1)^2$ **13.** 6, 52, 4, 2, 7.21
15. 10 ft **17.** 17 ft **19.** 40 ft **21.** 24 cm **23.** $x = 2$,
$h = 2\sqrt{2} \approx 2.83$ **25.** $3.2\sqrt{2}$ ft ≈ 4.53 ft
27. $x = \dfrac{3\sqrt{2}}{2} \approx 2.12$, $y = \dfrac{3\sqrt{2}}{2} \approx 2.12$
29. $5\sqrt{2}$ in. ≈ 7.07 in. **31.** $x = 5\sqrt{3} \approx 8.66$, $h = 10$
33. $75\sqrt{3}$ cm ≈ 129.90 cm, 150 cm **35.** $x = \dfrac{40\sqrt{3}}{3} \approx 23.09$,
$h = \dfrac{80\sqrt{3}}{3} \approx 46.19$ **37.** $\dfrac{55\sqrt{3}}{3}$ mm ≈ 31.75 mm,
$\dfrac{110\sqrt{3}}{3}$ mm ≈ 63.51 mm **39.** $x = 50, y = 50\sqrt{3} \approx 86.60$
41. 0.75 ft, $0.75\sqrt{3}$ ft ≈ 1.30 ft **43.** 5 **45.** 13 **47.** 10
49. $\sqrt{34}$ **51.** $2\sqrt{5}$ **53.** $3\sqrt{10}$ **55. a.** 118.73 m
b. 133.14 m **57.** $7\sqrt{3}$ cm **59.** $(5\sqrt{2}, 0), (0, 5\sqrt{2})$,
$(-5\sqrt{2}, 0), (0, -5\sqrt{2})$; $(7.07, 0), (0, 7.07), (-7.07, 0), (0, -7.07)$
61. $10\sqrt{3}$ mm ≈ 17.32 mm **63.** $10\sqrt{181}$ ft ≈ 134.54 ft
65. About 0.13 ft **67. a.** 21.21 units **b.** 8.25 units
c. 13.00 units **69.** Yes **75.** 7

ARE YOU READY? 7.7 (page 650)
1. No real number squared is equal to -16. **2.** $6x^2 - 2x$
3. $14n^2 + 19n - 3$ **4.** $3\sqrt{7}$ **5.** $\dfrac{7(\sqrt{x} - 4)}{x - 16}$ **6.** 21R3

SELF-CHECKS 7.7
1. a. $5i$ **b.** $-i\sqrt{19}$ **c.** $3i\sqrt{5}$ **d.** $\dfrac{5\sqrt{2}}{9}i$ **2. a.** $-18 + 0i$
b. $0 + 6i$ **c.** $1 + 2i\sqrt{6}$ **3. a.** $1 + 2i$ **b.** $5 - 12i$ **4.** $-4\sqrt{6}$
5. a. $18 + 2i$ **b.** $-40 + 70i$ **6.** $0 + 13i$ **7.** 13
8. a. $\dfrac{30}{29} - \dfrac{12}{29}i$ **b.** $\dfrac{11}{17} - \dfrac{7}{17}i$ **9.** $7 + 0i$ **10.** $0 - \dfrac{3}{4}i$
11. a. -1 **b.** i

STUDY SET SECTION 7.7 (page 659)
1. imaginary, power **3.** real, imaginary **5. a.** $\sqrt{-1}$ **b.** -1
c. $-i$ **d.** 1 **e.** four **7. a.** real, imaginary **b.** FOIL
9. a. $2 + 3i$ **b.** $2 - 0i$ **c.** $0 + 3i$

11.

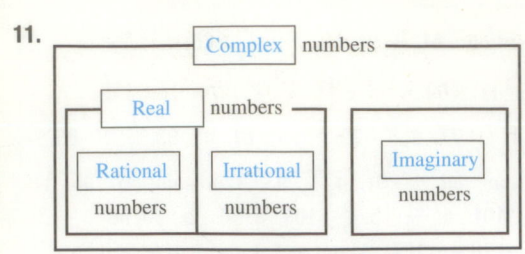

13. 9, 6i, 2, 11   **15. a.** True   **b.** False   **c.** False   **d.** False
17. 3i   **19.** $\sqrt{7}i$ or $i\sqrt{7}$ **21.** $2\sqrt{6}i$ or $2i\sqrt{6}$
23. $-6\sqrt{2}i$ or $-6i\sqrt{2}$ **25.** 45i   **27.** $\frac{5}{3}i$ **29. a.** $5 + 0i$
b. $0 + 7i$ **31. a.** $1 + 5i$ **b.** $-3 + 2i\sqrt{2}$ **33. a.** $76 - 3i\sqrt{6}$
b. $-7 + i\sqrt{19}$ **35. a.** $-6 - 3i$ **b.** $3 + i\sqrt{6}$ **37.** $8 - 2i$
39. $15 + 2i$ **41.** $3 - 5i$ **43.** $1 + 3i$ **45.** -6 **47.** $-2\sqrt{6}$
49. $6 - 27i$ **51.** $35 - 28i$ **53.** $6 + 14i$ **55.** $-25 - 25i$
57. $7 + i$ **59.** $12 + 5i$ **61.** $13 - i$ **63.** $3 + 4i$ **65.** 40
67. 65   **69.** $\frac{45}{26} - \frac{9}{26}i$ **71.** $-\frac{77}{65} + \frac{44}{65}i$ **73.** $\frac{14}{17} - \frac{5}{17}i$
75. $-\frac{6}{29} + \frac{43}{29}i$ **77.** $\frac{11}{10} + \frac{13}{10}i$ **79.** $0 - i$ **81.** $4 + 0i$
83. $-2 + 0i$ **85.** $0 - \frac{5}{3}i$ **87.** $0 + \frac{2}{7}i$ **89.** i **91.** $-i$
93. 1   **95.** -1 **97.** $4 - 11i$ **99.** $14 - 8i$ **101.** $-2 - 6i$
103. $-\frac{4}{13} - \frac{6}{13}i$ **105.** $18 + 12i$ **107.** $0 + \frac{4}{5}i$
109. $8 + \sqrt{2}i$ or $8 + i\sqrt{2}$ **111.** $-2 + 7i$ **113.** $-48 - 64i$
115. $\frac{1}{4} - \frac{\sqrt{15}}{4}i$ **117. a.** $2i\sqrt{2}$ **b.** -2 **119. a.** $-4 + 0i$
b. $3 + 4i$ **121.** $-1 + i$ **127.** 20 mph

CHAPTER 7 REVIEW (page 662)

1. a. 2   **b.** 0   **c.** 6   **d.** D: $[-3, \infty)$, R: $[0, \infty)$ **2. a.** $10|a|$
b. $10a$ **3.** 7   **4.** -11 **5.** $\frac{15}{7}$ **6.** Not a real number   **7.** $10a^6$
8. $5|x|$ **9.** x^4 **10.** $|x + 2|$ **11.** -3 **12.** -6 **13.** $4x^2y$
14. $\frac{x^3}{5}$ **15.** 2   **16.** -2 **17.** $4x^2|y|$ **18.** $x + 1$ **19.** $-\frac{1}{2}$
20. Not a real number   **21.** Not a real number   **22.** 0
23. $[-5, \infty)$ **24.** 13 ft   **25.** 24 cm²   **26.** 2.3276
27. D: $[0, \infty)$; R: $[0, \infty)$ **28.** D: $(-\infty, \infty)$; R: $(-\infty, \infty)$

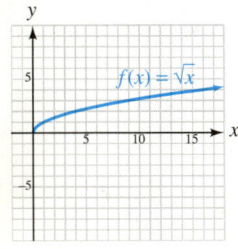   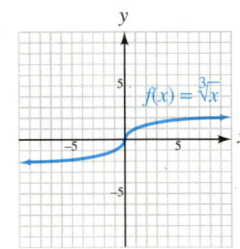

29. D: $[-2, \infty)$; R: $[0, \infty)$ **30.** D: $(-\infty, \infty)$; R: $(-\infty, \infty)$

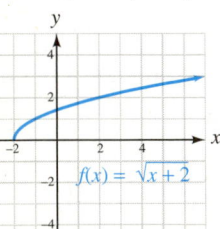   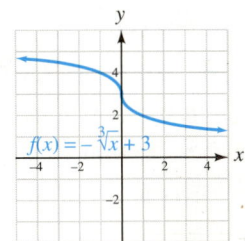

31. $\sqrt{t}$ **32.** $\sqrt[4]{5xy^3}$ **33.** 5   **34.** -6 **35.** Not a real number
36. 1   **37.** $\frac{3}{x}$ **38.** -2 **39.** 5   **40.** $3cd$ **41.** 27   **42.** $\frac{1}{4}$
43. $-16,807$ **44.** 10   **45.** $\frac{27}{8}$ **46.** $\frac{1}{3,125}$ **47.** $125x^3y^6$
48. $\frac{1}{4u^4v^2}$ **49.** $5^{3/4}$ **50.** $a^{1/7}$ **51.** k^8 **52.** $3^{2/3}$ **53.** $u - 1$

54. $v + v^2$ **55.** $\sqrt{a}$ **56.** $\sqrt[6]{c}$ **57.** 183 mi   **58.** Two true
statements result: $32 = 32$.   **59.** $4\sqrt{5}$ **60.** $3\sqrt[3]{2}$ **61.** $2\sqrt[4]{10}$
62. $-2\sqrt[5]{3}$ **63.** $2x^2\sqrt{2x}$ **64.** $r^4\sqrt[4]{r}$ **65.** $-3j^2\sqrt[3]{jk}$
66. $-2xy\sqrt[3]{2x^2y}$ **67.** $\frac{\sqrt{m}}{12n^6}$ **68.** $\frac{\sqrt{17xy}}{8a^2}$ **69.** $2x$ **70.** $3x^3$
71. $3\sqrt{2}$ **72.** $11\sqrt{5}$ **73.** 0   **74.** $-8a\sqrt[4]{2a}$ **75.** $29x\sqrt{2}$
76. $13x\sqrt[3]{2}$ **77.** $9\sqrt[4]{2t^3} - 8\sqrt[4]{6t^3}$ **78.** $12x^2\sqrt[4]{x} + 5x\sqrt[4]{x}$
80. $\left(6\sqrt{2} + 2\sqrt{10}\right)$ in., 14.8 in.   **81.** 7   **82.** $6\sqrt{10}$ **83.** 32
84. $6\sqrt{10}$ **85.** $3x$ **86.** $x + 1$ **87.** $-2x^3\sqrt[3]{x}$ **88.** 3
89. $42t + 9t\sqrt{21t}$ **90.** $-2x^3y^3\sqrt[4]{2x^2y^2}$ **91.** $3b + 6\sqrt{b} + 3$
92. $\sqrt[3]{9p^2} - \sqrt[3]{6p} - 2\sqrt[3]{4}$ **93.** $\frac{10\sqrt{3}}{3}$ **94.** $\frac{\sqrt{15xy}}{5xy}$ **95.** $\frac{\sqrt[3]{6u^2}}{u^2}$
96. $\frac{\sqrt[4]{27ab^2}}{3b}$ **97.** $2\left(\sqrt{2} + 1\right)$ or $2\sqrt{2} + 2$
98. $\frac{12\sqrt{xz} - 16x - 2z}{z - 16x}$ **99.** $\frac{a - b}{a + \sqrt{ab}}$ **100.** $r = \frac{\sqrt[3]{6\pi^2V}}{2\pi}$
101. 22   **102.** 16, 9   **103.** $\frac{13}{2}$ **104.** $\frac{9}{16}$ **105.** 2, -4 **106.** 7
107. 1   **108.** $-\frac{3}{2}$, 1   **109.** 3, no solution   **110.** 6, -2 **111.** 0, 3
112. $-\frac{1}{2}$, 4   **113.** $P = \frac{A}{(r + 1)^2}$ **114.** $I = \frac{h^3b}{12}$
115. 17 ft   **116.** 88 yd   **117.** $7\sqrt{2}$ m ≈ 9.90 m
118. $\frac{15\sqrt{2}}{2}$ yd ≈ 10.61 yd   **119.** Shorter leg: 6 cm, longer leg:
$6\sqrt{3}$ cm ≈ 10.39 cm   **120.** $40\sqrt{3}$ ft ≈ 69.28 ft,
$20\sqrt{3}$ ft ≈ 34.64 ft   **121.** $x = 5\sqrt{2} \approx 7.07$, $y = 5$
122. $x = 25\sqrt{3} \approx 43.30$, $y = 25$ **123.** 13   **124.** $2\sqrt{2}$
125. $5i$ **126.** $3i\sqrt{2}$ **127.** $-i\sqrt{6}$ **128.** $\frac{3}{8}i$
129. Real, Imaginary   **130. a.** True   **b.** True   **c.** False
d. False   **131. a.** $3 - 6i$ **b.** $0 - 19i$ **132. a.** $-1 + 7i$
b. $0 + i$ **133.** $8 - 2i$ **134.** $3 - 5i$ **135.** $3 + 6i$
136. $22 + 29i$ **137.** $-3\sqrt{3} + 0i$ **138.** $-81 + 0i$ **139.** $4 + i$
140. $0 - \frac{3}{11}i$ **141.** -1 **142.** i

CHAPTER 7 TEST (page 671)

1. a. radical   **b.** imaginary   **c.** extraneous   **d.** isosceles
e. rationalize   **f.** complex   **2. a.** If $\sqrt[n]{a}$ and $\sqrt[n]{b}$ are real
numbers, then $\sqrt[n]{ab} = \sqrt[n]{a}\sqrt[n]{b}$.   **b.** If $\sqrt[n]{a}$ and $\sqrt[n]{b}$ are real
numbers, then $\sqrt[n]{\frac{a}{b}} = \frac{\sqrt[n]{a}}{\sqrt[n]{b}}$, $(b \neq 0)$.   **c.** No real number raised to
the fourth power is -16.

3. D: $[1, \infty)$; R: $[0, \infty)$

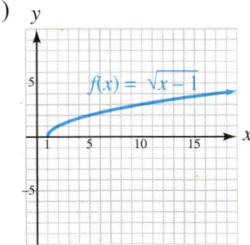

4. 46 ft/sec   **5. a.** -1 **b.** 2   **c.** 1   **d.** D: $(-\infty, \infty)$;
R: $(-\infty, \infty)$ **6.** $[-5, \infty)$ **7.** $7x^2$ **8.** -9 **9.** $\frac{1}{216}$ **10.** $\frac{25n^4}{4}$
11. $2^{4/3}$ **12.** $a^{1/9}$ **13.** $|x|$ **14.** $|y - 5|$ **15.** $-4xy^2$ **16.** $\frac{2}{3}a$
17. $t + 8$ **18.** $6xy^2\sqrt{15xy}$ **19.** $2x^5y\sqrt[3]{3}$ **20.** $2\sqrt[4]{2}$
21. $2y^2\sqrt{3y}$ **22.** $14\sqrt[3]{5}$ **23.** $5z^3\sqrt[4]{3z}$ **24.** $-6x\sqrt{y} - 2xy^2$
25. $3 - 7\sqrt{6}$ **26.** $\sqrt[3]{4a^2} + 18\sqrt[3]{2a} + 81$ **27.** $\frac{4\sqrt{10}}{5}$
28. $\frac{x + 2\sqrt{xy} + y}{x - y}$ **29.** $\frac{\sqrt[3]{18a^2}}{2a}$ **30.** $\frac{1}{\sqrt{2}(\sqrt{5} - 3)} = \frac{1}{\sqrt{10} - 3\sqrt{2}}$
31. $\frac{1}{15}$ **32.** 10   **33.** 4, no solution   **34.** 3, -3 **35.** 2, 3
36. 108, no solution   **37.** -2 **38.** $G = \frac{4\pi^2r^3}{Mt^2}$

39. $x = \frac{8\sqrt{3}}{3}$ cm ≈ 4.62 cm, $h = \frac{16\sqrt{3}}{3}$ cm ≈ 9.24 cm

40. $x = 6.13\sqrt{2}$ in. ≈ 8.67 in., $y = 6.13\sqrt{2}$ in. ≈ 8.67 in.

41. 25   **42.** 28 in.   **43.** $3i\sqrt{5}$ **44.** -1 **45.** $-4 + 11i$

46. $4 - 7i$ **47.** $75 + 45i$ **48.** $-46 - 78i$ **49.** $0 - \frac{\sqrt{2}}{2}i$

50. $\frac{1}{2} + \frac{1}{2}i$

CUMULATIVE REVIEW CHAPTERS 1–7 (page 673)

1. Natural: 1; whole: 1; integers: $-5, 1$; rational: $-5, 3.4, 1, \frac{16}{5}, 9.\overline{7}$; irrational: $-\pi, \sqrt{19}$; real: all   **2.** $\frac{1}{10}$ **3.** 1   **4.** $\ell = \frac{a - S + Sr}{r}$

5. 22   **6.** 10%: $3\frac{3}{4}$ cups; 18%: $6\frac{1}{4}$ cups   **7.** $-\frac{1}{3}$

8.

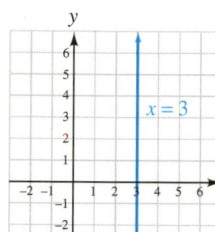

9. a. $-\frac{3}{5}$ **b.** $(0, -3)$

c. $y = -\frac{3}{5}x - 3$ **10. a.** 3

b. $f(x) = -\frac{1}{3}x - \frac{2}{3}$ **11.** 4, left

12. Yes

13. D: $(-\infty, \infty)$; R: $(-\infty, -2]$

14.

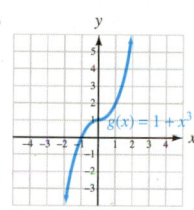

15. $(-2, -3)$

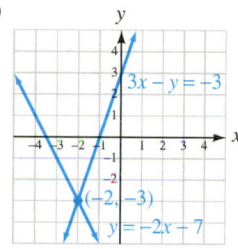

16. No solution   **17.** -10 **18.** 10, 15, 5

19. $(-\infty, -15)$

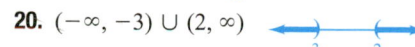

20. $(-\infty, -3) \cup (2, \infty)$

21.

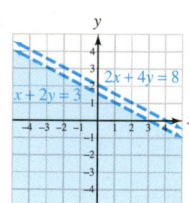

22. $\frac{64b^{12}}{27a^9}$ **23. a.** 0   **b.** 16   **c.** $-1, 2$

d. D: (∞, ∞); R: (∞, ∞)

24. 2.379×10^{14} **25.** $\frac{13}{15}x^6 + \frac{2}{3}x^2$

26. $2x^2y^3 + 13xy + 3y^2$

27. $6y^3 + 11y^2 + 9y + 2$

28. $3x^2 + 3x - 11$ **29.** $4y^{10} + 20y^5z + 25z^2$

30. $125s^3 - 75s^2t^3 + 15st^6 - t^9$ **31.** $(3 - c)(c + d)$

32. $(x - 2y)(x^2 + 2xy + 4y^2)$ **33.** $(a + b - 1)^2$

34. $(x + 1)(x - 1)(x + 4)(x - 4)$ **35.** 0, 10, -10 **36.** $-\frac{1}{3}, -3$

37. 2 in.   **38.** $\frac{-3x - 2y}{y}$ **39.** $-7, 8$

40. D: $(-\infty, 0) \cup (0, \infty)$; R: $(-\infty, 0) \cup (0, \infty)$

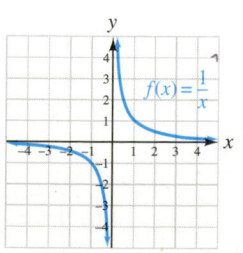

41. $2,940,000   **42.** $-\frac{t^6}{t^2 - 2}$ **43.** $x - 5$ **44.** $\frac{3x + 2}{(x + 2)(x - 1)}$

45. $5a^2 - 3a - 4$ **46.** $4, -1$ **47.** $2\frac{2}{5}$ hr   **48.** $\frac{80}{3}$ or $26\frac{2}{3}$

49. D: $[0, \infty)$; R: $[0, \infty)$

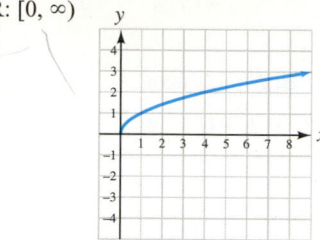

50. -4 **51.** $-\frac{3}{2x}$ **52.** $10x^2y\sqrt{2yz}$ **53.** $6\sqrt[3]{2}$

54. $5z + 2\sqrt{15z} + 3$ **55.** $\frac{\sqrt{6}}{10}$ **56.** 4, 3   **57.** 24 cm

58. $\frac{7}{10} - \frac{11}{10}i$

ARE YOU READY? 8.1 (page 678)

1. a. $2\sqrt{7}$ **b.** $6i$ **2.** $\frac{\sqrt{30}}{6}$ **3.** $\frac{9}{2}$ **4.** $\frac{49}{4}$ **5.** 169   **6.** $(x - 4)^2$

SELF CHECKS 8.1

1. $\pm 3\sqrt{2}$ **2.** $\sqrt{\frac{14}{\pi}}$ ft ≈ 2.11 ft   **3.** $\pm\frac{7}{4}i$ **4.** 1, -5

5. $\frac{-11 \pm \sqrt{3}}{4}$ **6.** $a^2 - 5a + \frac{25}{4} = \left(a - \frac{5}{2}\right)^2$

7. $5 \pm \sqrt{29}$; 10.39, -0.39 **8.** $\frac{4}{3}, -2$

9. $\frac{-3 \pm \sqrt{6}}{3}$; $-0.18, -1.82$ **10. a.** $3 \pm 3i\sqrt{2}$ **b.** $-\frac{1}{5} \pm \frac{i\sqrt{74}}{5}$

STUDY SET SECTION 8.1 (page 688)

1. quadratic   **3.** square   **5.** $\sqrt{c}, -\sqrt{c}$ **7. a.** 36   **b.** $\frac{25}{4}$

9. a. Subtract 7 from both sides   **b.** Divide both sides by 4

11. It is a solution.   **13.** plus or minus   **15.** $\pm\sqrt{11}$

17. $\pm\sqrt{35}$ **19.** $\pm 5\sqrt{2}$ **21.** $\pm\frac{4\sqrt{3}}{3}$ **23.** $\pm 4i$ **25.** $\pm 2i\sqrt{2}$

27. $\pm\frac{9}{2}i$ **29.** $\pm 2i\sqrt{6}$ **31.** $-8, -2$ **33.** 0, -8

35. $-5 \pm \sqrt{3}$ **37.** $\frac{2 \pm 2\sqrt{2}}{7}$ **39.** $x^2 + 24x + 144 = (x + 12)^2$

41. $a^2 - 7a + \frac{49}{4} = \left(a - \frac{7}{2}\right)^2$ **43.** $x^2 + \frac{2}{3}x + \frac{1}{9} = \left(x + \frac{1}{3}\right)^2$

45. $m^2 - \frac{5}{6}m + \frac{25}{144} = \left(m - \frac{5}{12}\right)^2$ **47.** $2 \pm \sqrt{6}$; 4.45, -0.45

49. $6 \pm \sqrt{35}$; 11.92, 0.08   **51.** $-10 \pm 5\sqrt{3}$; $-1.34, -18.66$

53. $-8 \pm 4\sqrt{5}$; 0.94, -16.94 **55.** $-\frac{1}{2}, 1$ **57.** $\frac{3}{4}, -\frac{1}{3}$

59. $\frac{6 \pm \sqrt{33}}{3}$; 3.91, 0.09   **61.** $\frac{-5 \pm \sqrt{41}}{4}$; 0.35, -2.85

63. $-1 \pm i$ **65.** $-4 \pm i\sqrt{2}$ **67.** $-\frac{1}{3} \pm \frac{i\sqrt{62}}{3}$ **69.** $\frac{1}{4} \pm \frac{i\sqrt{15}}{4}$

71. $2, -\frac{4}{3}$ **73.** $\frac{3 \pm 2\sqrt{3}}{3}$; 2.15, -0.15 **75.** $-4 \pm \sqrt{10}$;

$-7.16, -0.84$ **77.** $\pm 2i\sqrt{3}$ **79.** $1 \pm 3\sqrt{2}$; 5.24, -3.24

81. $\frac{7 \pm \sqrt{37}}{2}$; 6.54, 0.46   **83.** $\pm\sqrt{5}$; ± 2.24

85. $\frac{-7 \pm \sqrt{29}}{10}$; $-0.16, -1.24$ **87.** $-\frac{1}{2} \pm \frac{\sqrt{11}}{2}i$

89. $\frac{-5 \pm 2\sqrt{6}}{8}$; $-1.24, -0.01$ **91.** $-3, 9$ **93.** $-\frac{1}{4} \pm \frac{\sqrt{11}}{4}i$

95. a. $\pm 2\sqrt{6}$; ± 4.90 **b.** $\pm 2i\sqrt{6}$ **97. a.** 0, 4

b. $\frac{4 \pm 3\sqrt{2}}{2}$; 4.12, -0.12 **99. a.** $2 \pm 4i$

b. $2 \pm 2\sqrt{6}$; 6.90, -2.90

101. a. $\frac{2 \pm i\sqrt{2}}{2}$ **b.** $\frac{2 \pm \sqrt{10}}{2}$; 2.58, -0.58 **103.** 4.4 sec

105. 1.6 sec **107.** 1.70 in. **109.** $c = \frac{\sqrt{Em}}{m}$ **115.** $2ab^2\sqrt[3]{5}$

117. $\frac{2}{5}$

ARE YOU READY? 8.2 (page 690)

1. 11 **2.** $3\sqrt{5}$ **3.** 3; 2, -1, 7 **4.** $\frac{3}{4}$, -2 **5.** Not a real number **6.** 3.87

SELF CHECKS 8.2

1. 2, $-\frac{1}{4}$ **2.** $\frac{1 \pm \sqrt{10}}{3}$; -0.72, 1.39 **3.** $-\frac{3}{2} \pm \frac{\sqrt{11}}{2}i$

4. a. $6x^2 - 7x + 9 = 0$ **b.** $2x^2 - 4x - 5 = 0$

c. $4x^2 + 6x - 9 = 0$ **d.** $8x^2 - 7x - 2 = 0$ **5.** 9 in., 40 in.

6. \$1 **7.** Late 1988 **8.** $\frac{4 + \sqrt{31}}{2}$ in. ≈ 0.78 in.

STUDY SET SECTION 8.2 (page 700)

1. quadratic **3. a.** $x^2 + 2x + 5 = 0$ **b.** $3x^2 + 2x - 1 = 0$

5. a. True **b.** True **c.** False **7. a.** 2, -4 **b.** $\frac{1 \pm \sqrt{33}}{4}$

9. a. $1 \pm 2\sqrt{2}$ **b.** $\frac{-3 \pm \sqrt{7}}{2}$ **11. a.** The fraction bar wasn't drawn under both parts of the numerator.

b. A $\pm$ sign wasn't written between $-b$ and the radical. **13.** 1, 2

15. A repeated solution of -6 **17.** $-\frac{3}{2}$, 1 **19.** $\frac{2}{3}$, $-\frac{1}{4}$

21. $\frac{1 \pm \sqrt{29}}{2}$; 3.19, -2.19 **23.** $\frac{-5 \pm \sqrt{5}}{10}$; -0.28, -0.72

25. $\frac{-3 \pm \sqrt{6}}{3}$; -0.18, -1.82 **27.** $\frac{1 \pm 2\sqrt{5}}{2}$; 2.74, -1.74

29. $-\frac{1}{4} \pm \frac{\sqrt{7}}{4}i$ **31.** $\frac{1}{3} \pm \frac{\sqrt{2}}{3}i$ **33.** $1 \pm i$ **35.** $-\frac{1}{2} \pm i$

37. a. $5x^2 - 9x + 2 = 0$ **b.** $16t^2 + 24t - 9 = 0$

39. a. $3x^2 + 2x - 1 = 0$ **b.** $2m^2 - 3m - 2 = 0$ **41.** $\frac{4}{3}$, $-\frac{2}{5}$

43. $\frac{2}{3} \pm \frac{\sqrt{2}}{3}i$ **45.** $\frac{1}{4}$, $-\frac{3}{4}$ **47.** $\frac{3 \pm \sqrt{17}}{4}$; 1.78; -0.28

49. $5 \pm \sqrt{7}$; 2.35, 7.65 **51.** 23, -17

53. $\frac{-5 \pm 3\sqrt{5}}{2}$; 0.85, -5.85 **55.** $\frac{1}{3} \pm \frac{\sqrt{6}}{3}i$

57. $\frac{-3 \pm \sqrt{29}}{10}$; 0.24, -0.84 **59.** $\frac{10 \pm \sqrt{55}}{30}$; 0.58, 0.09

61. $2 \pm 2i$ **63.** $3 \pm i\sqrt{5}$ **65.** $\frac{-5 \pm \sqrt{17}}{2}$; -0.44, -4.56

67. $\frac{9 \pm \sqrt{89}}{2}$; 9.22, -0.22 **69.** $\frac{1 \pm \sqrt{7}}{3}$; 1.22, -0.55

71. $\frac{35 \pm 5\sqrt{329}}{14}$; 8.98, -3.98 **73. a.** $-2 \pm \sqrt{11}$; 1.32, -5.32

b. $2 \pm \sqrt{11}$; 5.32, -1.32 **75. a.** 6, -4 **b.** $1 \pm i\sqrt{7}$

77. a. A repeated solution of 21 **b.** A repeated solution of -21

79. 40 ft **81.** 0.7, 2.4 **83.** 97 ft by 117 ft

85. About 0.5 mi by 2.5 mi **87.** 25 sides **89.** \$4.80 or \$5.20

91. 4,000 **93.** 2002 **95.** 0.92 in. **97.** 2.9 ft, 6.9 ft **101.** $n^{1/2}$

103. $(3b)^{1/4}$ **105.** $\sqrt[3]{t}$ **107.** $\sqrt[4]{3t}$

ARE YOU READY? 8.3 (page 703)

1. 84 **2. a.** 2 **b.** 2/3 **3. a.** x **b.** a^2 **4.** $\pm 2i$ **5. a.** 9

b. $-\frac{1}{8}$ **6.** 5

SELF CHECKS 8.3

1. a. Two different irrational numbers **b.** Two different imaginary numbers that are complex conjugates

---

2. 3, -3, $2i$, $-2i$ **3.** 4, -9 does not check **4.** -125, 8

5. -1, 1 **6.** -7, 4 **7.** About 9.5 hours

STUDY SET SECTION 8.3 (page 710)

1. discriminant **3.** conjugates **5.** rational **7. a.** x^2 **b.** $\sqrt{x}$

c. $x^{1/3}$ **d.** $\frac{1}{x}$ **e.** $x + 1$ **9.** $4ac$, 5, 6, 24, rational

11. One repeated rational-number solution **13.** Two imaginary-number solutions (complex conjugates) **15.** Two different irrational-number solutions **17.** Two different rational-number solutions **19.** Two different irrational-number solutions **21.** Two different rational-number solutions **23.** -1, 1, -4, 4

25. 2, -2, $3i$, $-3i$ **27.** 25, 64 **29.** 1, $\frac{9}{4}$ is extraneous **31.** -1, 27

33. -64, 8 **35.** 0, 2 **37.** $\frac{1}{4}$, $\frac{1}{2}$ **39.** $-\frac{1}{3}$, $\frac{1}{2}$ **41.** -4, $\frac{2}{3}$

43. $\frac{5 \pm \sqrt{65}}{2}$ **45.** $\frac{7 \pm \sqrt{105}}{2}$ **47.** $\frac{9}{4}$, 1 is extraneous **49.** $-\frac{1}{3}$, 1

51. $-i$, i, $-3i\sqrt{2}$, $3i\sqrt{2}$ **53.** $4 \pm i$ **55.** $\frac{3 \pm \sqrt{57}}{6}$ **57.** 16, 4

59. $\pm i\sqrt{2}$, $\pm 3\sqrt{2}$ **61.** Repeated solutions of 1 and -1

63. 2, -2, $i\sqrt{7}$, $-i\sqrt{7}$ **65.** $\frac{243}{32}$, 1 **67.** A repeated solution of $-\frac{3}{2}$

69. $3 \pm \sqrt{7}$ **71.** 49, 225 **73.** $1 \pm i$ **75.** No solution

77. 1, -1, $\sqrt{5}$, $-\sqrt{5}$ **79.** $-\frac{5}{7}$, 3 **81.** -1, $-\frac{27}{13}$

83. a. 64, -125 **b.** $\frac{1}{4}$, $-\frac{1}{5}$ **85. a.** 1, $-\frac{1}{2}$ **b.** $\pm i$

87. 103 min **89.** 20 feet per second **91.** 32.4 ft **95.** $x = 3$

97. $y = \frac{2}{3}x$

ARE YOU READY? 8.4 (page 713)

1.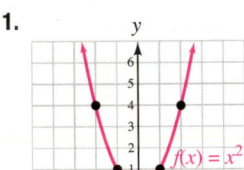

2. -2 **3.** $x^2 + 8x + 16 = (x + 4)^2$
4. $x^2 + x + \frac{1}{4} = \left(x + \frac{1}{2}\right)^2$
5. A repeated solution of -3
6. A repeated solution of -2
7. -5 **8.** A vertical line

SELF CHECKS 8.4

1.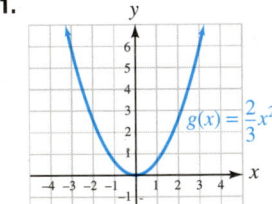

2.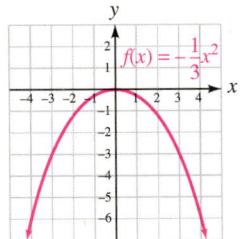

3.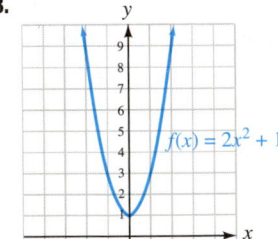

4.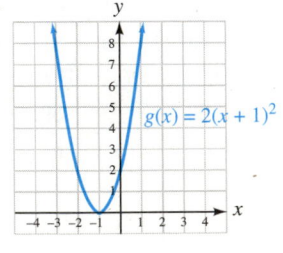

5.

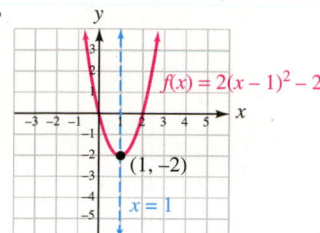

$f(x) = 2(x-1)^2 - 2$

$(1, -2)$

$x = 1$

6. $(-2, 6)$; $x = -2$; opens upward **7.**

8. $(2, -4)$

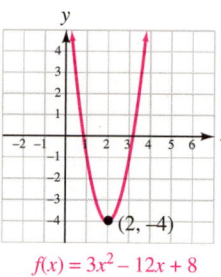

$(2, -4)$

$f(x) = 3x^2 - 12x + 8$

9.

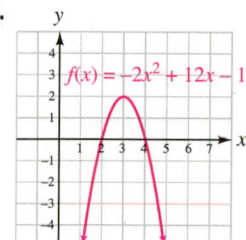

$f(x) = -2x^2 + 12x - 16$

10. 20, $700

STUDY SET SECTION **8.4** (page 724)

1. quadratic, parabola **3.** axis of symmetry
5. a. $(1, 0), (3, 0)$ **b.** $(0, -3)$ **c.** $(2, 1)$ **d.** $x = 2$
  e. Domain: $(-\infty, \infty)$; range: $(-\infty, 1]$
7.

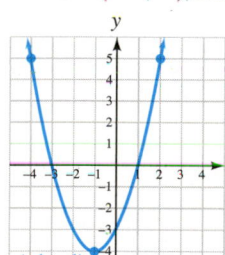

$(-1, -4)$

9. a. 2 **b.** 9, 18 **11.** $-3, 5$
13.
$h = -1$; $f(x) = 2[x - (-1)]^2 + 6$

15.

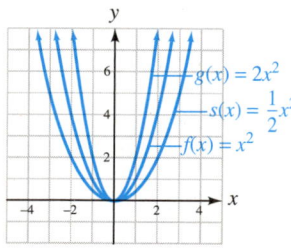

$g(x) = 2x^2$
$s(x) = \frac{1}{2}x^2$
$f(x) = x^2$

17.

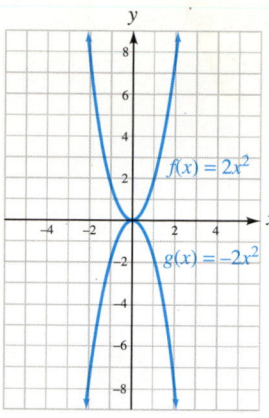

$f(x) = 2x^2$
$g(x) = -2x^2$

19.

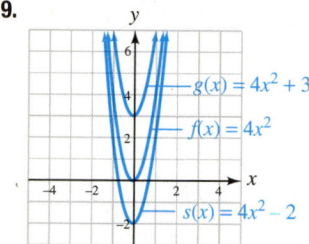

$g(x) = 4x^2 + 3$
$f(x) = 4x^2$
$s(x) = 4x^2 - 2$

21.

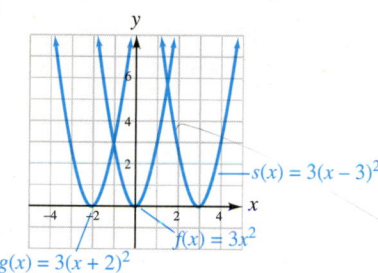

$s(x) = 3(x - 3)^2$
$f(x) = 3x^2$
$g(x) = 3(x + 2)^2$

23. $(1, 2)$; $x = 1$; upward **25.** $(-3, -4)$; $x = -3$; downward
27. $(7.5, 8.5)$; $x = 7.5$; downward **29.** $(0, -4)$; $x = 0$; upward
31. $(3, 2), x = 3$ **33.** $(2, 0), x = 2$

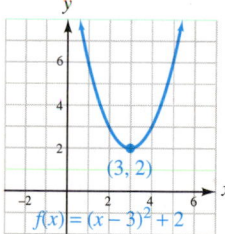

$(3, 2)$
$f(x) = (x - 3)^2 + 2$

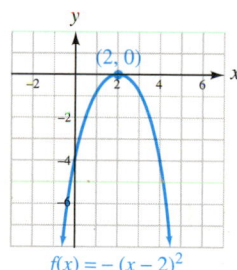

$(2, 0)$
$f(x) = -(x - 2)^2$

35. $(-3, 4), x = -3$ **37.** $(-1, -3), x = 1$

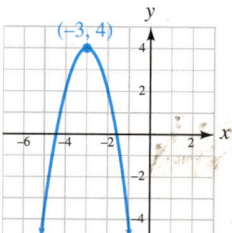

$(-3, 4)$
$f(x) = -2(x + 3)^2 + 4$

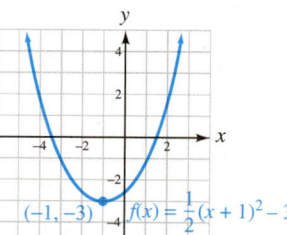

$(-1, -3)$ $f(x) = \frac{1}{2}(x + 1)^2 - 3$

39. $(-2, 1)$; $x = -2$; upward **41.** $(3, -6)$; $x = 3$; downward

43. $f(x) = (x + 1)^2 - 4$; $(-1, -4)$, $x = -1$

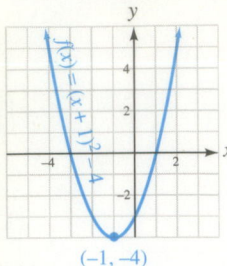

45. $f(x) = 4(x + 3)^2 + 1$; $(-3, 1)$, $x = -3$

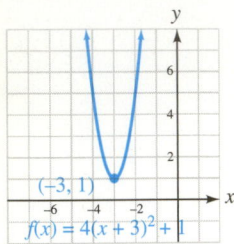

47. $f(x) = \left(x + \frac{1}{2}\right)^2 - \frac{25}{4}$; $\left(-\frac{1}{2}, -\frac{25}{4}\right)$; $x = -\frac{1}{2}$

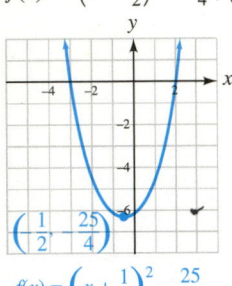

49. $f(x) = -4(x - 2)^2 + 6$; $(2, 6)$, $x = 2$

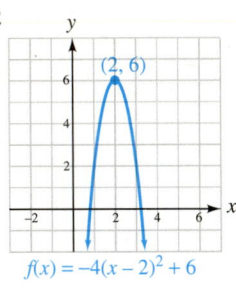

51. $f(x) = 2(x + 2)^2 - 2$; $(-2, -2)$, $x = -2$

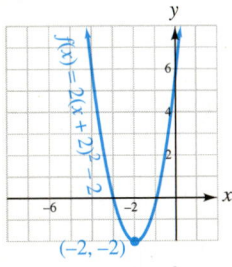

53. $f(x) = -(x + 4)^2 - 1$; $(-4, -1)$, $x = -4$

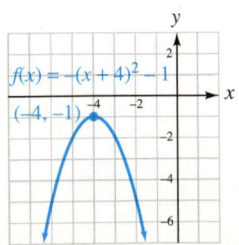

55. $(-1, -6)$ **57.** $\left(\frac{3}{4}, \frac{23}{8}\right)$ **59.** $(-5, 0), (7, 0); (0, -35)$

61. $(0, 0), (2, 0); (0, 0)$

63. $(-2, 0); (-2, 0); (0, 4)$

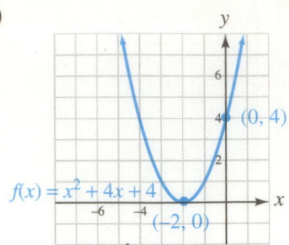

65. $(1, 0); (1, 0); (0, -1)$

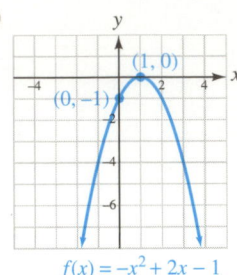

67. $(1, -1); (0, 0), (2, 0); (0, 0)$

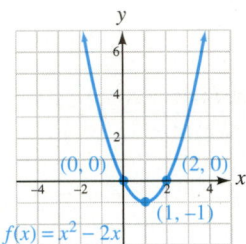

69. $(2, -2); (1, 0), (3, 0); (0, 6)$

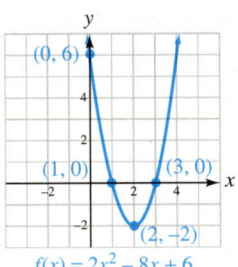

71. $(-1, -2)$; no x-intercept; $(0, -8)$

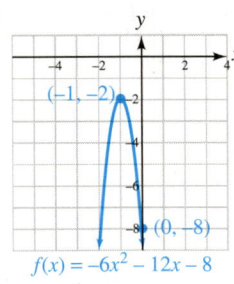

73. $\left(\frac{3}{2}, 0\right); \left(\frac{3}{2}, 0\right); (0, 9)$

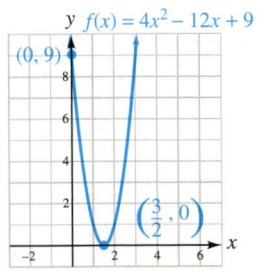

75. $(0.25, 0.88)$ **77.** $(0.50, 7.25)$ **79.** $2, -3$ **81.** $-1.85, 3.25$

83.

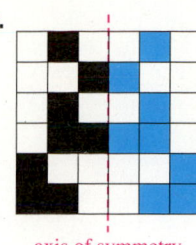

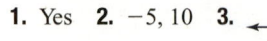
axis of symmetry

85. 15 min, $160
87. 3.75 sec, 225 ft
89. 75 ft by 75 ft, 5,625 ft²
91. 1968, about 1.5 million; the U.S. involvement in the war in Vietnam was at its peak
93. 200, $7,000 **101.** $\frac{\sqrt{6}}{10}$
103. $15b - 6\sqrt{15b} + 9$

ARE YOU READY? 8.5 (page 727)

1. Yes **2.** $-5, 10$ **3.**

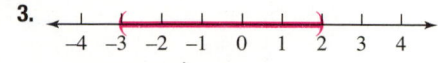

4.

5. $-3, 3$ **6.**

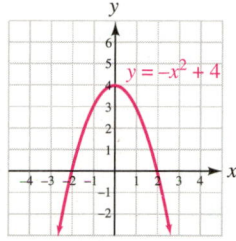

$y = -x^2 + 4$

SELF CHECKS 8.5

1. $(-4, 3)$

2. $(-\infty, -8] \cup [5, \infty)$

3. $(-\infty, 0) \cup \left(\frac{3}{5}, \infty\right)$

4. $[-2, -1) \cup (3, \infty)$

5. $(-1, 0) \cup (1, \infty)$

6.

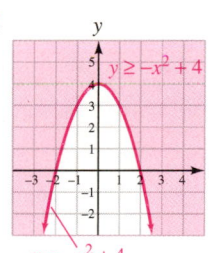

$y \geq -x^2 + 4$
$y = -x^2 + 4$

7.

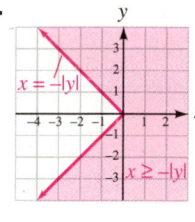

$x = -|y|$
$x \geq -|y|$

STUDY SET SECTION 8.5 (page 736)

1. quadratic **3.** two **5.** $(-\infty, -1), (-1, 4), (4, \infty)$
7. a. Yes **b.** No **c.** Yes **d.** No
9. a. $(-3, 2)$ **b.** $(-\infty, -1] \cup [1, \infty)$ **11. a.** Solid
b. Yes **13.** $x^2 - 6x - 7 \geq 0$ **15.** $(1, 4)$

17. $(-\infty, 3) \cup (5, \infty)$

19. $(-\infty, -6] \cup [7, \infty)$

21. $[-4, 3]$

23. $(-\infty, 0) \cup \left(\frac{1}{2}, \infty\right)$

25. $\left(-\infty, -\frac{5}{3}\right] \cup (0, \infty)$

27. $(-\infty, -3) \cup (1, 4)$

29. $\left(-\frac{1}{2}, \frac{1}{3}\right] \cup \left[\frac{1}{2}, \infty\right)$

31. $(0, 2) \cup (8, \infty)$

33. $\left[-\frac{34}{5}, -4\right) \cup (3, \infty)$

35.

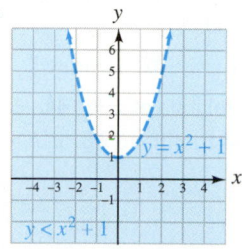

$y = x^2 + 1$
$y < x^2 + 1$

37.

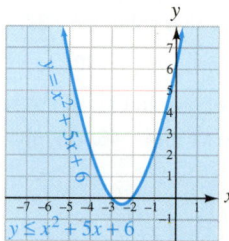

$y = x^2 + 5x + 6$
$y \leq x^2 + 5x + 6$

39.

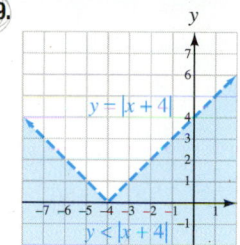

$y = |x + 4|$
$y < |x + 4|$

41.

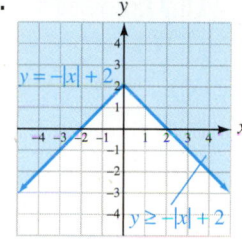

$y = -|x| + 2$
$y \geq -|x| + 2$

43. $(-1, 3)$ **45.** $(-\infty, -3) \cup (2, \infty)$
47. $(-4, -2] \cup (-1, 2]$

49. $(-\infty, -3] \cup [3, \infty)$

51. $(-\infty, \infty)$

53. $(-2, 1) \cup (3, \infty)$

55. $(-5, 5)$

57. $\left(-\frac{1}{3}, \frac{3}{2}\right)$ **59.** No solutions

61. $(-1, 4) \cup \left(\frac{23}{2}, \infty\right)$

63. $(-2,100, -900) \cup (900, 2,100)$ **69.** $x = ky$ **71.** $t = kxy$

CHAPTER 8 REVIEW (page 738)

1. $-5, -4$ **2.** $-\frac{1}{3}, -\frac{5}{2}$ **3.** $\pm 2\sqrt{7}; \pm 5.29$ **4.** $4, -8$
5. $\pm 5i$ **6.** $\pm \frac{7\sqrt{5}}{5}; \pm 3.13$ **7.** $r = \frac{\sqrt{\pi A}}{\pi}$
8. $x^2 - x + \frac{1}{4} = \left(x - \frac{1}{2}\right)^2$ **9.** $-4, -2$ **10.** $\frac{3 \pm \sqrt{3}}{2}; 2.37, 0.63$
11. $\frac{6 \pm \sqrt{30}}{6}; 1.91, 0.09$ **12.** $1 \pm 2i\sqrt{3}$ **13.** $\pm 4\sqrt{2}$

14. $\frac{51 \pm \sqrt{11}}{7}$ **15.** Because 7 is an odd number and not divisible by 2, the calculations involved in completing the square on $x^2 + 7x$ involve fractions. The calculations involved in completing the square on $x^2 + 6x$ do not.   **16.** 6 seconds before midnight

17. $\frac{1}{2}, -7$ **18.** $5 \pm \sqrt{7}$; 7.65, 2.35   **19.** 0, 10

20. $\frac{13 \pm \sqrt{163}}{3}$; 8.59, 0.08   **21.** $-\frac{3}{4} \pm \frac{\sqrt{15}}{4}i$ **22.** $\frac{2}{3} \pm \frac{\sqrt{2}}{3}i$

23. $\frac{-3 \pm \sqrt{29}}{10}$; 0.24, −0.84   **24.** $\frac{3 \pm 3\sqrt{13}}{2}$; 6.91, −3.91

25. $\frac{-3 \pm \sqrt{41}}{2}$ **26.** $\frac{-1 \pm \sqrt{13}}{4}$ **27.** 2 is not a factor of the numerator—it is a term. Only common factors of the numerator and denominator can be removed.   **28. a.** $(2 + 2x)$ ft   **b.** $(6 + 2x)$ ft
29. Sides: 1.25 in. wide; top/bottom: 2.5 in. wide
30. $24 or $26   **31.** 0.7 sec, 1.8 sec   **32.** 33 in., 56 in.
33. Two different irrational-number solutions
34. Two imaginary-number solutions that are complex conjugates
35. One repeated solution, a rational number   **36.** Two different rational-number solutions   **37.** 1, 144   **38.** 8, −27
39. $i, -i, \frac{\sqrt{6}}{3}, -\frac{\sqrt{6}}{3}$ **40.** $1, -\frac{8}{5}$ **41.** $4 \pm i$ **42.** Repeated solutions of −1 and 1   **43.** A repeated solution of $-\frac{2}{5}$ **44.** $\frac{1}{32}$, 32
45. About 81 min   **46.** 30 mph   **47.** 34.1 million   **48.** h, k, x
49.

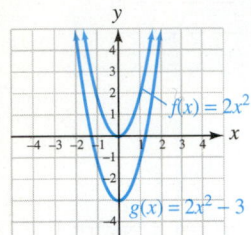

50.

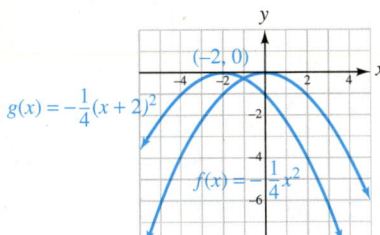

51. $(1, 4), x = 1$

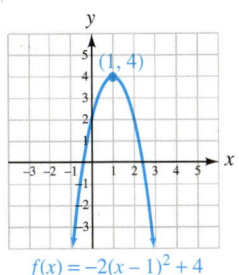

52. $f(x) = 4(x + 2)^2 - 7$; $(-2, -7), x = -2$

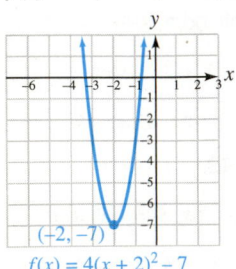

53. $(1, -6)$

54. $\left(-\frac{1}{2}, -\frac{9}{4}\right)$; $x = -\frac{1}{2}$; $(-2, 0), (1, 0)$; $(0, -2)$

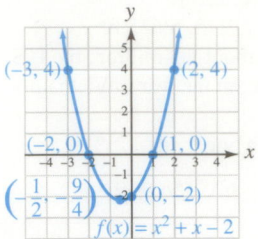

55. 1921; 6,469,326   **56.** $-2, \frac{1}{3}$
57. $(-\infty, -7) \cup (5, \infty)$

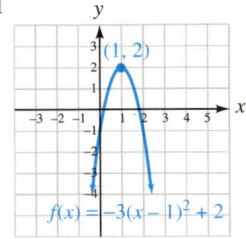

58. $[-9, 9]$
59. $(-\infty, 0) \cup \left[\frac{3}{5}, \infty\right)$
60. $\left(-\frac{7}{2}, 1\right) \cup (4, \infty)$
61. $\left[-4, \frac{2}{3}\right]$ **62.** $(-\infty, 0) \cup (1, \infty)$
63.

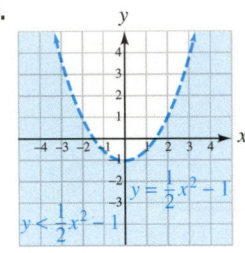

64.

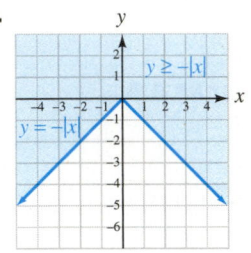

CHAPTER 8 TEST (page 744)

1. a. quadratic   **b.** completed, square   **c.** vertex   **d.** rational   **e.** nonlinear   **2.** $\pm 3\sqrt{7} \approx \pm 7.94$ **3.** $-7 \pm 5\sqrt{2}$ **4.** $\pm 2i$
5. $x^2 + 11x + \frac{121}{4} = \left(x + \frac{11}{2}\right)^2$ **6.** $\frac{-3 \pm \sqrt{17}}{2}$; −3.56; 0.56
7. $-2 \pm i\sqrt{2}$ **8.** $\frac{2 \pm 3\sqrt{2}}{3}$ **9.** $\frac{-1 \pm \sqrt{2}}{2}$; −1.21; 0.21
10. $1 \pm \sqrt{5}$; −1.24; 3.24   **11.** $2 \pm 3i$ **12.** $-5, -3$
13. A repeated solution of 47   **14.** $\pm\sqrt{10}$ **15.** $1, \frac{1}{4}$ **16.** $-1, \frac{1}{3}$
17. $2, -2, i\sqrt{3}, -i\sqrt{3}$ **18.** $-\frac{4}{5}, \frac{4}{7}$ **19.** $2 \pm \sqrt{10}$; 5.16, −1.16
20. $-8, -\frac{1}{125}$ **21.** $-\frac{1}{2} \pm \frac{\sqrt{5}}{10}i$ **22.** $3, \pm 2i\sqrt{2}$ **23.** $c = \frac{\sqrt{Em}}{m}$
24. a. Two different imaginary-number solutions that are complex conjugates   **b.** One repeated solution, a rational number
25. 4.5 ft by 1,502 ft   **26.** About 46 min   **27.** 20 in.   **28.** 4.3 in.
29. 1993   **30.** iii
31. $(1, 2), x = 1$

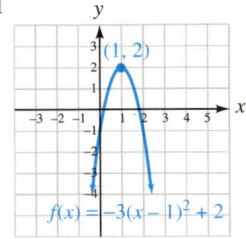

32. $f(x) = 5(x + 1)^2 - 6$; $(-1, -6)$, $x = -1$

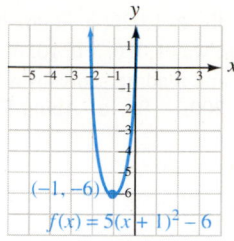

33. $\left(-\frac{1}{4}, -\frac{9}{8}\right)$, $x = -\frac{1}{4}$, $(-1, 0)$, $\left(\frac{1}{2}, 0\right)$; $(0, -1)$

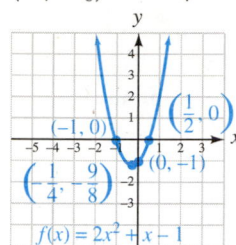

34. 211 ft   **35.** $(-\infty, -2) \cup (4, \infty)$

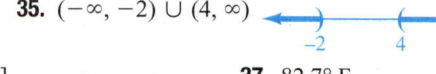

36. $(-3, 2]$ **37.** 82.7° F

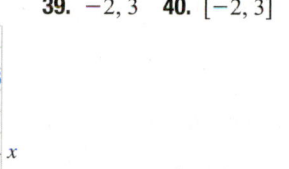

38.   **39.** $-2, 3$ **40.** $[-2, 3]$

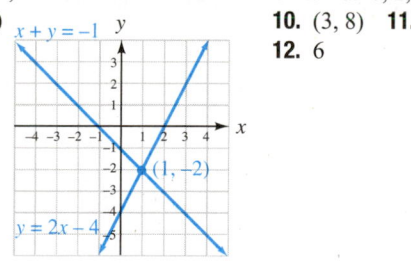

CUMULATIVE REVIEW CHAPTERS 1–8 (page 747)

1. -3 **2.** 6 L   **3.** $y = 3x + 2$ **4.** $y = -\frac{2}{3}x - 2$

5. Supply: a decrease of 15,000 nurses per year; demand: an increase of 55,000 nurses per year   **6. a.** $P(t) = -0.08t + 7$
b. 3%   **7.** Not a function; $(0, 3)$, $(0, -1)$; answers may vary.
8. a. Domain: $[0, 24]$ **b.** 1.5
c. At noon, the low tide mark was -2.5 m.   **d.** 0, 2, 9, 17
9. $(1, -2)$ **10.** $(3, 8)$ **11.** $(2, -1, 1)$
12. 6

13. *Friends:* 236 episodes; *The King of Queens:* 207 episodes, *That '70s Show:* 200 episodes   **14.** More than \$50,000 per month
15. $\left(-\infty, -\frac{10}{9}\right)$ **16.** $\left[1, \frac{9}{4}\right]$

17. $(-\infty, -10] \cup [15, \infty)$

18.

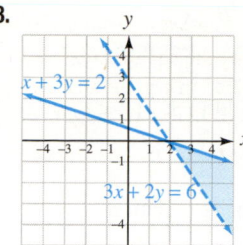

19. $\frac{125c^{21}}{8a^{12}b^{12}}$
20. 1.44×10^{10}; 14,400,000,000
21. $f(t) - g(t) = -11t^3 - 0.8t^2 - 1.4t$
22. $8a^3 - b^3$
23. $81m^2n^6 - 72mn^3 + 16$
24. $4x^4 - y^4$
25. $(x - y)(x - 4)$
26. $2a^2(3a + 2)(5a - 4)$
27. $(7s^3 - 6n^2)^2$ **28.** $(x + 5 + y^4)(x + 5 - y^4)$
29. $(x^2 + 4y^2)(x + 2y)(x - 2y)$
30. $(2x^2 + 5y)(4x^4 - 10x^2y + 25y^2)$
31. $2, -\frac{5}{2}$ **32.** $0, \frac{2}{3}, -\frac{1}{2}$ **33.** 6 sec
34. All real numbers except -7 and 8; $(-\infty, -7) \cup (-7, 8) \cup (8, \infty)$
35. D: $(-\infty, \infty)$; R: $[2, \infty)$

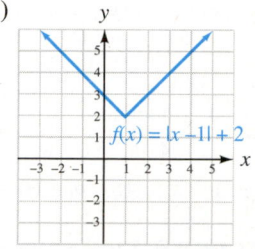

36. D: $(0, \infty)$; R: $(0, \infty)$

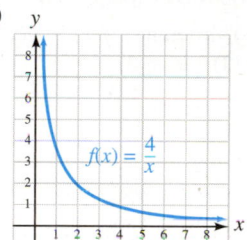

37. $\frac{3x - 5}{x + 2}$ **38.** $\frac{x + y}{x - y}$ **39.** 0   **40.** $\frac{1}{2r(r + 2)}$
41. $\frac{x^4y^4}{2} - \frac{x^3y^9}{4} + \frac{3}{4xy^2}$ **42.** $5a^2 - 3a - 4$ **43.** 5; 3 is extraneous
44. $R = \frac{R_1R_2R_3}{R_2R_3 + R_1R_3 + R_1R_2}$ **45.** 18 sec   **46.** 21 tons
47. \$9,000   **48.** D: $[2, \infty)$; R: $[0, \infty)$

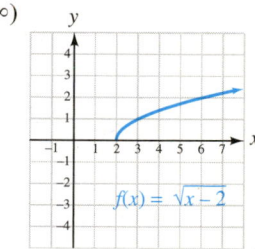

49. About 3.2 meters per sec   **50.** -11 **51.** $-3x$ **52.** $\frac{1}{16}$
53. $x^{17/12}$ **54.** $4t\sqrt{3t}$ **55.** $-12\sqrt[4]{2} + 10\sqrt[4]{3}$ **56.** $-18\sqrt{6}$
57. $\frac{x + 3\sqrt{x} + 2}{x - 1}$ **58.** $\frac{5\sqrt[3]{x^2}}{x}$ **59.** 2, 7   **60.** $\frac{1}{4}$ **61.** $3\sqrt{2}$ in.
62. $2\sqrt{3}$ in.   **63.** 10   **64.** $-i$ **65.** $-5 + 17i$ **66.** $\frac{3}{2} + \frac{1}{2}i$
67. $3 + 4i$ **68.** $0 - \frac{2}{3}i$ **69.** $\pm 2\sqrt{7}$ **70.** $19 \pm i\sqrt{5}$
71. $\frac{3 \pm \sqrt{3}}{2}$; 2.37, 0.63   **72.** $\frac{1}{5} \pm \frac{2}{5}i$ **73.** 10 ft by 18 ft
74. 50 m and 120 m   **75.** $-8, 27$ **76.** Repeated solutions of -1 and 1

77. $(-2, 4)$, $x = -2$; $(-4, 0)$, $(0, 0)$; $(0, 0)$

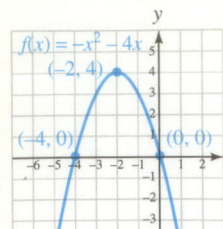

78. $(-9, 9)$

79. $(-4, -2] \cup (-1, 2]$

80. a. $-\frac{3}{4}$ **b.** No solution

75. $(g/f)(x) = \frac{x^2 - 4}{x^2 - 1}$, $(-\infty, -1) \cup (-1, 1) \cup (1, \infty)$ **77.** 4
79. 0 **81. a.** 7 **b.** 8 **c.** 12 **d.** 0 **85. a.** 1,026; in 2006, the average combined math and reading score was 1,026. **b.** 10; in 2006, the average difference in the math and reading scores was 10.
c. 1,016 **d.** 14 **87.** $C(t) = \frac{5}{9}(2,668 - 200t)$ **89. a.** About $75
b. $C(m) = \frac{3m}{20} = 0.15m$ **95.** $\frac{1}{c - d}$

ARE YOU READY? 9.1 (page 752)

1. $9x^2 + 2x - 8$ **2.** $2x^3 - x^2$ **3.** $72x^3 - 45x^2 - 32x + 20$
4. $x + 2$

SELF CHECKS 9.1

1. a. $(f + g)(x) = 2x^2 + 6x - 2$; D: $(-\infty, \infty)$
b. $(f - g)(x) = -2x^2 - 2$; D: $(-\infty, \infty)$
c. $(f \cdot g)(x) = 6x^3 + 5x^2 - 6x$; D: $(-\infty, \infty)$
d. $(f/g)(x) = \frac{3x - 2}{2x^2 + 3x}$; $\left(-\infty, -\frac{3}{2}\right) \cup \left(-\frac{3}{2}, 0\right) \cup (0, \infty)$ **2. a.** 78
b. -164 **c.** 56 **d.** Undefined **3. a.** -8 **b.** 5
c. $(g \circ f)(x) = 6 - x^3$ **4. a.** -6 **b.** -3 **c.** -4
5. $f(x) = \sqrt{x}$, $g(x) = x + 15$ **6.** $C(t) = -\frac{5}{6}t + 30$

STUDY SET SECTION 9.1 (page 759)

1. sum, $f(x) + g(x)$, difference, $f(x) - g(x)$ **3.** domain
5. nested **7. a.** $g(3)$ **b.** $g(3)$
9.

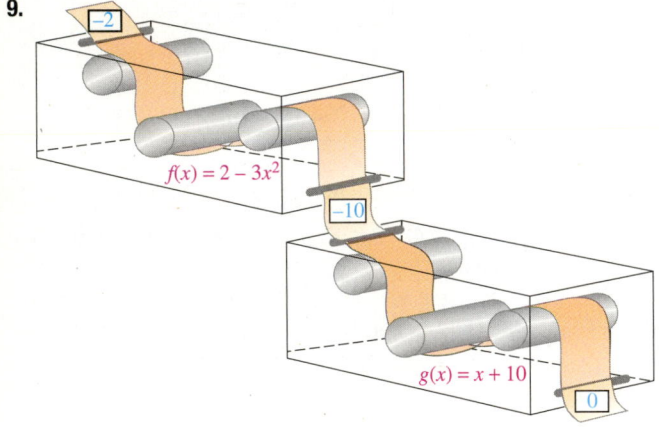

11. $g(x)$, $(3x - 1)$, $9x$, $2x$ **13.** $(f + g)(x) = 3x - 2$, $(-\infty, \infty)$
15. $(g - f)(x) = -x - 4$, $(-\infty, \infty)$
17. $(f \cdot g)(x) = 2x^2 - 5x - 3$, $(-\infty, \infty)$
19. $(g/f)(x) = \frac{x - 3}{2x + 1}$, $\left(-\infty, -\frac{1}{2}\right) \cup \left(-\frac{1}{2}, \infty\right)$
21. $(f + g)(x) = 7x$, $(-\infty, \infty)$ **23.** $(g - f)(x) = x$, $(-\infty, \infty)$
25. $(f \cdot g)(x) = 12x^2$, $(-\infty, \infty)$
27. $(g/f)(x) = \frac{4}{3}$, $(-\infty, 0) \cup (0, \infty)$ **29.** 20 **31.** -5 **33.** 0
35. $-\frac{1}{3}$ **37.** 7 **39.** 24 **41.** $-\frac{1}{2}$ **43.** $(g \circ f)(2x) = 16x^2 + 8x$
45. a. 0 **b.** 6 **c.** -3 **47. a.** -1 **b.** 6 **c.** -6
49. $f(x) = x^2$; $g(x) = x + 15$ **51.** $f(x) = x + 9$; $g(x) = x^5$
53. $f(x) = \sqrt{x}$; $g(x) = 16x - 1$ **55.** $f(x) = \frac{1}{x}$; $g(x) = x - 4$
57. 58 **59.** 110 **61.** 2 **63.** $(g \circ f)(x) = 9x^2 - 9x + 2$
65. $(f - g)(x) = -2x^2 + 3x - 3$, $(-\infty, \infty)$
67. $(f/g)(x) = \frac{3x - 2}{2x^2 + 1}$, $(-\infty, \infty)$ **69.** $\frac{1}{9}$
71. $(g \circ f)(8x) = 64x^2$ **73.** $(f - g)(x) = 3$, $(-\infty, \infty)$

ARE YOU READY? 9.2 (page 763)

1. D: $\{-2, 3, 5, 9\}$, R: $\{-3, 1, 8, 10\}$ **2.** Yes **3.** function
4. $16x + 6$

SELF CHECKS 9.2

1. a. Yes **b.** No, $(-1, 1)$, $(1, 1)$ **2. a.** No **b.** Yes
3. a. $f^{-1}(x) = \frac{-x - 3}{5}$ **b.** $f^{-1}(x) = \sqrt[5]{x}$
5.

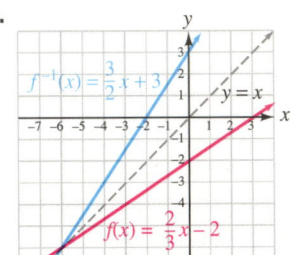

STUDY SET SECTION 9.2 (page 770)

1. one-to-one **3.** inverses **5.** one-to-one **7.** range, domain
9. 1 **11. a.** No **b.** No **13.** $-4, 0, 8$ **15.** y, $2y$, 3, y, f^{-1}
17. inverse, inverse **19.** Yes **21.** No **23.** No **25.** No
27. One-to-one **29.** Not one-to-one **31.** Not one-to-one
33. One-to-one **35.** $f^{-1}(x) = \frac{x - 4}{2}$ **37.** $f^{-1}(x) = 5x - 4$
39. $f^{-1}(x) = 5x + 4$ **41.** $f^{-1}(x) = \frac{2}{x} + 3$ **43.** $f^{-1}(x) = \frac{4}{x}$
45. $f^{-1}(x) = \sqrt[3]{x - 8}$ **47.** $f^{-1}(x) = x^3$
49. $f^{-1}(x) = \sqrt[3]{x} - 10$ **51.** $f^{-1}(x) = \sqrt[3]{\frac{x + 3}{2}}$
53. $f^{-1}(x) = \sqrt[7]{2x}$
59. $f^{-1}(x) = \frac{1}{2}x$ **61.** $f^{-1}(x) = \frac{x - 3}{4}$

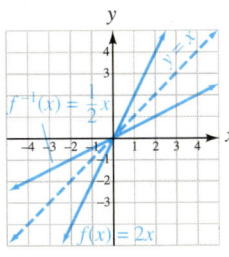

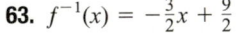

63. $f^{-1}(x) = -\frac{3}{2}x + \frac{9}{2}$ **65.** $f^{-1}(x) = \sqrt[3]{x}$

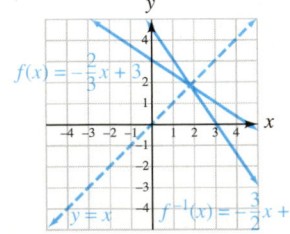

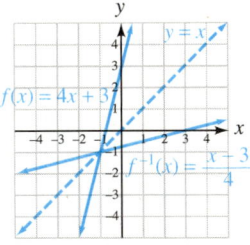

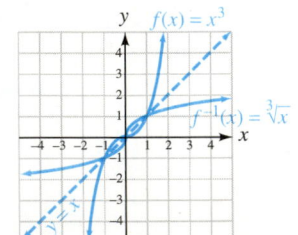

67. $f^{-1}(x) = \sqrt{x+1}$

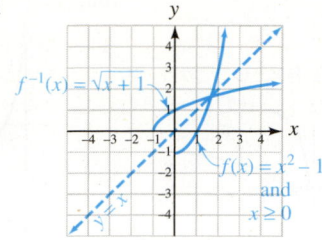

69. a. Yes; no   **b.** No. Twice during this period, the person's anxiety level was at the maximum threshold value.   **77.** $3 - 8i$
79. $18 - i$ **81.** $-28 - 96i$

ARE YOU READY? 9.3 (page 773)

1. a. 3^{12} **b.** 3^{32} **2. a.** 8   **b.** 1   **c.** $\frac{1}{4}$ **3. a.** $\frac{1}{9}$ **b.** 1   **c.** 27
4. line, parabola

SELF CHECKS 9.3

1.

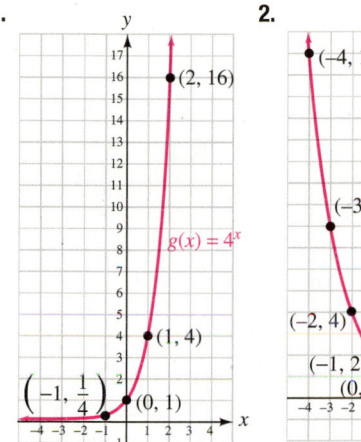

2.

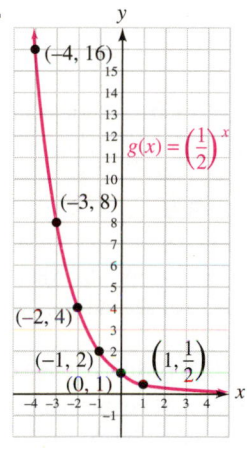

3. a.

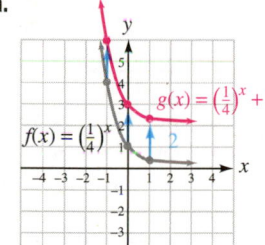

b.

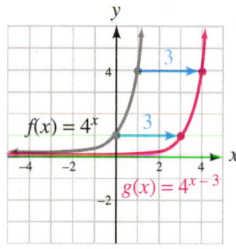

4. $7,258,641   **5.** $118,344.56   **6.** About 2.4 milligrams

STUDY SET SECTION 9.3 (page 783)

1. exponential   **3.** asymptote   **5. a.** Exponential   **b.** $(-\infty, \infty)$
c. $(0, \infty)$ **d.** $(0, 1)$, none   **e.** Yes   **f.** The x-axis ($y = 0$)
g. Increasing   **h.** 3   **7. a.** $\frac{1}{9}$ **b.** $\frac{1}{16}$ **c.** 25   **9. a.** iii   **b.** iv
c. ii   **d.** i   **11. a.** D: $(-\infty, \infty)$, R: $(-3, \infty)$
b. D: $(-\infty, \infty)$, R: $(1, \infty)$ **13.**
15. a. ii   **b.** i   **c.** ii   **d.** ii
e. i   **17.** $b > 0, b \neq 1$

$f(x) = 5^x$

x	$f(x)$
-3	$\frac{1}{125}$
-2	$\frac{1}{25}$
-1	$\frac{1}{5}$
0	1
1	5
2	25
3	125

19.

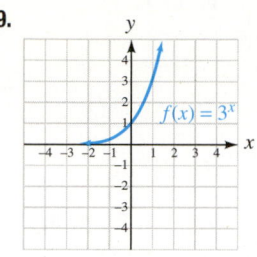

21.

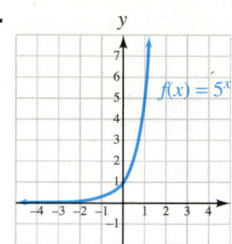

23.

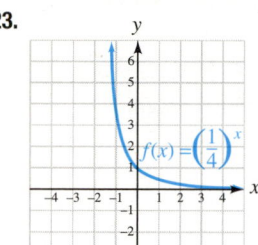

25.

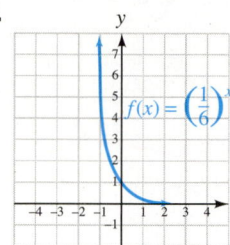

27.

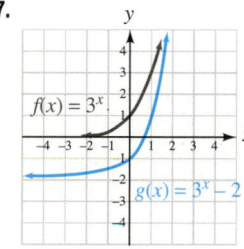

29.

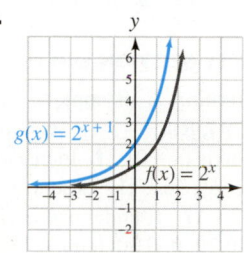

31.

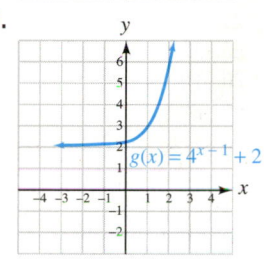

33.

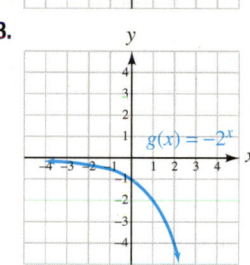

35. Increasing      **37.** Decreasing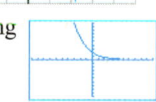

39. a. 280 ppm, 295 ppm, 370 ppm   **b.** About 1970
41. a. At the end of the 2nd year   **b.** At the end of the 4th year
c. During the 7th year
43. a.

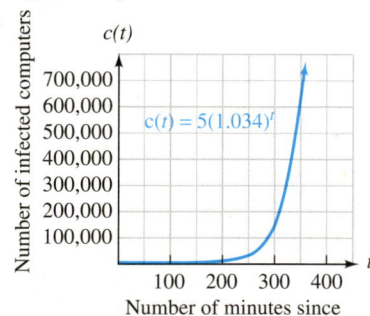

Number of minutes since
first infected e-mail opened

b. About 46,650,000 computers   **45.** About 422 mm   **47.** 8.7 gm

49.

Year	2006	2007	2008	2009
Number in poverty (in millions)	35.8	37.9	40.2	42.6

51. 2,295   **53.** $22,080.40   **55.** $32.03   **57.** $2,273,996.13
69. 40   **71.** 120°

ARE YOU READY? 9.4 (page 787)

1.

x	$f^{-1}(x)$
1	0
2	1
4	2

2. a. 125 **b.** 3 **3. a.** 1 **b.** 1/2
4. a. 1,000 **b.** $\frac{1}{100}$

SELF CHECKS 9.4

1. $2^7 = 128$ **2.** $\log_9 \frac{1}{9} = -1$ **3. a.** 7 **b.** $\frac{1}{9}$ **c.** 3 **4. a.** 2
b. -2 **c.** $\frac{1}{2}$ **5. a.** 4 **b.** -3 **c.** Undefined **6.** 75.3998
7. a. 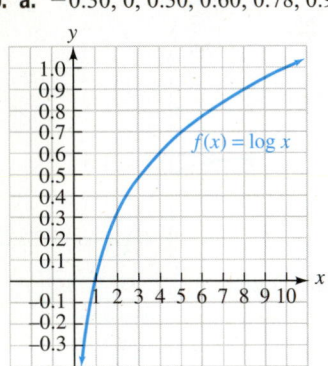   **b.**

8. About 36 dB **9.** About 63 catfish

STUDY SET SECTION 9.4 (page 797)

1. logarithmic **3.** asymptote **5. a.** Logarithmic
b. D: $(0, \infty)$; R: $(-\infty, \infty)$ **c.** None, $(1, 0)$ **d.** Yes
e. The y-axis $(x = 0)$ **f.** Increasing **g.** 1 **7. a.** ii **b.** iii **c.** i
d. iv **9.** $6^2, 36$ **11.** exponent **13.** logarithmic
15. $2, -2$ **17.** 1, undefined, undefined
19. a. $-0.30, 0, 0.30, 0.60, 0.78, 0.90, 1$
b.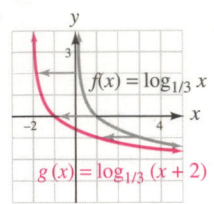

21. a. 10 **b.** x
23. $3^4 = 81$
25. $10^1 = 10$
27. $4^{-3} = \frac{1}{64}$
29. $5^{1/2} = \sqrt{5}$
31. $10^{-1} = 0.1$
33. $8^x = 64$
35. $b^t = T_1$
37. $n^{-42} = C$
39. $\log_8 64 = 2$
41. $\log_4 \frac{1}{16} = -2$
43. $\log_{1/2} 32 = -5$

45. $\log_x z = y$ **47.** $\log_y 8.6 = t$ **49.** $\log_7 (B + 1) = 4.3$
51. 9 **53.** 64 **55.** 3 **57.** $\frac{1}{25}$ **59.** $\frac{1}{6}$ **61.** 10 **63.** $\frac{2}{3}$
65. 5 **67.** 1,000 **69.** 4 **71.** 1 **73.** $\frac{1}{3}$ **75.** 3 **77.** 2
79. 6 **81.** -1 **83.** 5 **85.** $\frac{1}{2}$ **87.** 0.5119 **89.** -2.3307
91. 6,043.6597 **93.** 0.1726 **95.** 0.3162 **97.** 0.0195
99. Increasing **101.** Decreasing

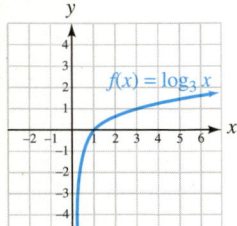 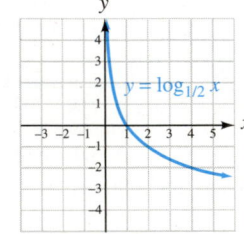

103. 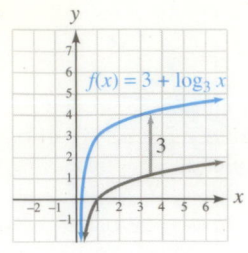 **105.**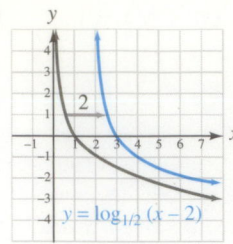

107. 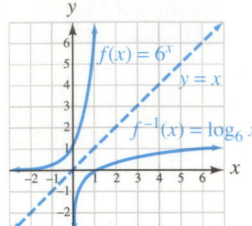 **109.**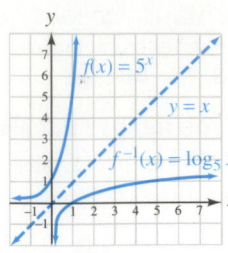

111. 49.5 dB **113.** 4.4 **115.** About 423 cases
117. About 99 sunfish **119.** 77.8% **121.** 10.8 yr **127.** 10
129. $0; -\frac{5}{9}$ is extraneous

ARE YOU READY? 9.5 (page 800)

1. 2.25 **2.** 1 **3.** 2.7 **4.** 82.05 **5.** $3^6, x$ **6.** 8, 3

SELF CHECKS 9.5

1. $3,037,760.44 **2.** 3,347 **3.** 104.4° **4. a.** 3 **b.** -1 **c.** $\frac{1}{3}$
5. a. 6.9199 **b.** 0.0498 **6.** About 63 years
7. About 5.3 feet per second

STUDY SET SECTION 9.5 (page 809)

1. exponential; e **3.** continuously **5. a.** The natural exponential
function **b.** D: $(-\infty, \infty)$; R: $(0, \infty)$ **c.** $(0, 1)$, none **d.** Yes
e. The x-axis $(y = 0)$ **f.** Increasing **g.** e **7.** 2.718281828459
9. e **11.** $e, 2$
13.

15. a. $2.7182818\ldots$; e **b.** They decrease, no **17.** $f^{-1}(x) = e^x$
19. 1,000, 10, 0.9, 2,459.6 **21.** l, n **23.** LOG, LN
25. 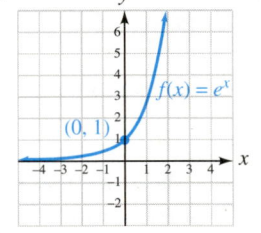 **27.**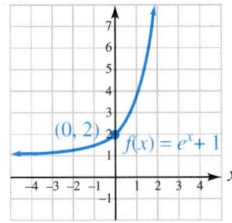

29. 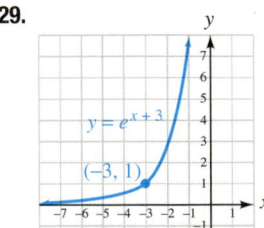 **31.**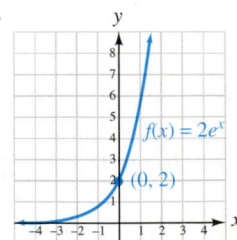

33. 24,765.16 **35.** 3,811,325.37 **37.** 11,542.14 **39.** 542.85
41. 5 **43.** 6 **45.** -1 **47.** $\frac{1}{4}$ **49.** $\frac{2}{3}$ **51.** -7 **53.** 3.5596
55. -5.3709 **57.** 0.5423 **59.** Undefined **61.** 4.0645
63. 69.4079 **65.** 0.0245 **67.** 2.7210

69.    **71.**

73. $13,375.68   **75.** $7,518.28 from annual compounding, $7,647.95 from continuous compounding   **77.** $6,849.16
79. 133 million, 140 million, 165 million, 194 million, 206 million, 237 million, 269 million   **81.** 4,715,620; 6,370,911
83. About 51   **85.** About 39   **87.** 14 chairs   **89.** 105,731 germs
91. About 49 meters per second   **93.** About 16 years
95. The year 2042   **97.** About 9.2 yr   **99.** About 3.5 hr
101. About 169 beats per minute   **103.** About 1 foot per second faster (5.9 ft/sec − 4.9 ft/sec)   **115.** $4x^2\sqrt{15x}$ **117.** $10y\sqrt{3y}$

ARE YOU READY? 9.6 (page 814)

1. 3   **2.** 3   **3.** −9   **4.** 2   **5. a.** $x^{1/2}$ **b.** $\sqrt{x-2}$ **6.** 1.6094

SELF CHECKS 9.6

1. a. 0   **b.** 1   **c.** 4   **d.** 2   **2. a.** $1 + \log_3 4$ **b.** $3 + \log y$
c. $2 + \log_5 c + \log_5 d$ **3. a.** $1 - \log_6 5$ **b.** $\ln y - \ln 100$
4. $\log_b x - \log_b y - \log_b z$ **5. a.** $4 \ln x$ **b.** $\frac{1}{3}\log_2 3$
6. $\frac{3}{4}\log x + \frac{1}{4}\log y - \frac{1}{4}\log z$ **7.** $\log_a \frac{x^2\sqrt{y}}{(x-y)^2}$ **8. a.** 0.1761
b. −0.1249   **9.** 0.6826   **10.** 7.6

STUDY SET SECTION 9.6 (page 823)

1. product   **3.** power   **5. a.** 0   **b.** 1   **c.** x **d.** x **11.** c   **13.** b
15. $8, a^3, 3, 1$ **17.** True   **19.** 0   **21.** 7   **23.** 10   **25.** 2   **27.** 1
29. 7   **31.** $2 + \log_2 5$ **33.** $\log 25 + \log y$
35. $2 + \log p + \log q$ **37.** $\log 5 + \log x + \log y + \log z$
39. $2 - \log 9$ **41.** $\log_6 x - 2$ **43.** $\log 7 + \log c - \log 2$
45. $1 + \log x - \log y$ **47.** $1 + \ln x + \ln y - \ln z$
49. $-1 - \log_8 m$ **51.** $7 \ln y$ **53.** $\frac{1}{2}\log 5$ **55.** $-3\log e$
57. $\frac{3}{5}\log_7 100$ **59.** $\log x + \log y + 2\log z$
61. $1 + \frac{1}{3}\log_2 x - \log_2 y$ **63.** $3\log x + 2\log y$
65. $\frac{1}{2}\log_b x + \frac{1}{2}\log_b y$ **67.** $\frac{1}{3}\log_a x - \frac{1}{4}\log_a y - \frac{1}{4}\log_a z$
69. $20\ln x + \frac{1}{2}\ln z$ **71.** $-3d\log_5 t$ **73.** $\frac{1}{2} + \frac{1}{2}\ln x$
75. $\log_2 x^9(x+1)$ **77.** $\log_3 \frac{x(x+2)}{8}$ **79.** $\log_b \frac{\sqrt{z}}{x^3 y^2}$
81. $\log_b \sqrt[3]{M-3}$ **83.** $\ln \frac{\frac{x}{z}+x}{\frac{y}{z}+y} = \ln \frac{x}{y}$ **85.** $\log_6 \frac{\sqrt{x^2+1}}{x^2+2}$
87. 1.4472   **89.** −1.1972   **91.** 1.1972   **93.** 1.8063   **95.** 1.7712
97. −1.0000   **99.** 1.8928   **101.** 2.3219   **103.** a, c   **105.** a, b
107. About 4.8   **109. a.** $B = \log\left(\frac{I}{I_0}\right)^{10}$ **b.** $T = \ln\sqrt{\frac{C_2}{C_1}}$
119. $-\frac{7}{6}$ **121.** $\left(1, -\frac{1}{2}\right)$

ARE YOU READY? 9.7 (page 825)

1. a. 4   **b.** −3   **2.** 1.1309   **3.** $x\log_3 8$ **4.** 1   **5. a.** $\log_2 5x$
b. $\ln \frac{10}{2t+1}$ **6.** Undefined

SELF CHECKS 9.7

1. 2   **2.** A repeated solution of 1   **3.** $\frac{\log 4}{\log 5} \approx 0.8614$
4. $\frac{2\log 5}{\log 5 - \log 3} \approx 6.3013$ **5.** $\frac{\ln 35}{2.1} \approx 1.6930$ **6.** $\frac{7}{2}$ **7.** 8   **8.** 2
9. $\frac{9}{16}$ **10.** About 1.58×10^{-5} gram-ions per liter
11. About 11,400 years   **12.** About 10 hours

STUDY SET SECTION 9.7 (page 835)

1. exponential   **3. a.** equal; x, y **b.** equal; x, y **5.** −2
7. logarithm   **9.** $\ln 4$ **11.** Not a solution   **13. a.** Divide both sides by $\ln 3$ **b.** $\frac{\ln 5}{\ln 3}$ **c.** 1.4650   **15. a.** 1.2920   **b.** 10.7549
17. a. log   **b.** $A_0 2^{-t/h}$ **c.** $P_0 e^{kt}$ **19.** log, log 2, 2.8074
21. 4   **23.** $-\frac{3}{4} = -0.75$ **25.** 3, −1
27. A repeated solution of −2   **29.** $\frac{\log 5}{\log 4} \approx 1.1610$
31. $\frac{\log 2 + \log 13}{\log 13} \approx 1.2702$ **33.** $\frac{\log 2}{\log 3 - \log 2} \approx 1.7095$
35. $\frac{3\log 5}{\log 5 - 2\log 3} \approx -8.2144$ **37.** $\frac{\ln 4.5}{2.9} \approx 0.5186$
39. $-\frac{\ln 14.2}{0.2} \approx -13.2662$ **41.** 5,000   **43.** 12   **45.** −93
47. 0.08   **49.** −7   **51.** 3   **53.** 50   **55.** $\frac{5}{4} = 1.25$ **57.** 0.5
59. 15.75   **61.** 2   **63.** $e \approx 2.7183$ **65.** $\pm\sqrt{\frac{\log 10}{\log 7}} \approx \pm 1.0878$
67. 10   **69.** 10   **71.** 4   **73.** $\frac{\log 15 - 2\log 9}{\log 9} \approx -0.7675$
75. 10, −10   **77.** 10   **79.** 9   **81.** A repeated solution of 4
83. 1, 7   **85.** $\frac{\ln 9}{3} \approx 0.7324$ **87.** 6   **89. a.** $\frac{10^{1.7}}{5} \approx 10.0237$
b. $\frac{e^{1.7}}{5} \approx 1.0948$ **91. a.** $\frac{\log 90 + 5\log 4}{3\log 4} \approx 2.7486$
b. $\frac{5 + \ln 90}{3} \approx 3.1666$ **93. a.** $\frac{5}{4}$ **b.** $\frac{5}{4}$ **95. a.** 3   **b.** 2   **97.** 1.8
99. 20   **101.** About 6.3×10^{-14} gram-ions per liter
103. About 5.1 yr   **105.** About 42.7 days   **107.** About 4,200 yr
109. About 5.6 yr   **111.** About 5.4 yr   **113.** Because $\ln 2 \approx 0.7$
115. About 25.3 yr   **117.** 2.828 times larger
119. $\frac{1}{3}\ln 0.75 \approx -0.0959$ **125.** $\sqrt{137}$ in.

CHAPTER 9 REVIEW (page 838)

1. $(f + g)(x) = 3x + 1, (-\infty, \infty)$
2. $(f - g)(x) = x - 1, (-\infty, \infty)$
3. $(f \cdot g)(x) = 2x^2 + 2x, (-\infty, \infty)$
4. $(f/g)(x) = \frac{2x}{x+1}, (-\infty, -1) \cup (-1, \infty)$ **5.** 3   **6.** 5
7. $(f \circ g)(x) = 4x^2 + 4x + 3$ **8.** $(g \circ f)(x) = 2x^2 + 5$
9. a. 0   **b.** −8   **c.** 0   **d.** −1   **10.** $C(m) = \frac{3.25m}{8}$
11. No   **12.** Yes   **13.** Yes   **14.** No   **15.** Yes   **16.** No
17. −6, −1, 7, 20   **18.**

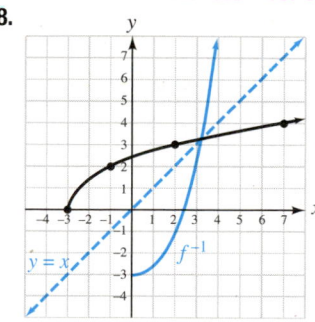

19. $f^{-1}(x) = \frac{x+3}{6}$ **20.** $f^{-1}(x) = \frac{4}{x} + 1$ **21.** $f^{-1}(x) = \sqrt[3]{x} - 2$
22. $f^{-1}(x) = 6x + 1$
23. $f^{-1}(x) = x^3 + 1$

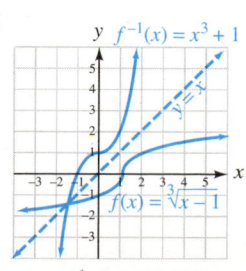

25. a. $n(x) = 2^x$, $s(t) = 1.08^t$ **b.** 121.9774

26. a. Exponential decay **b.** Exponential growth
27. D: $(-\infty, \infty)$; R: $(0, \infty)$ **28.** D: $(-\infty, \infty)$; R: $(0, \infty)$

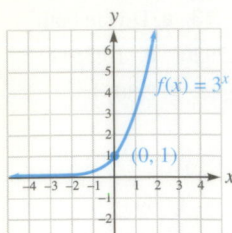 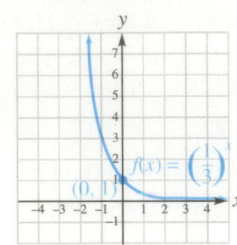

29. D: $(-\infty, \infty)$; R: $(-2, \infty)$ **30.** D: $(-\infty, \infty)$; R: $(0, \infty)$

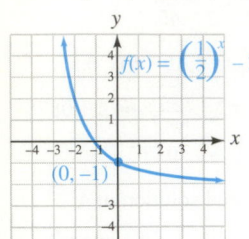 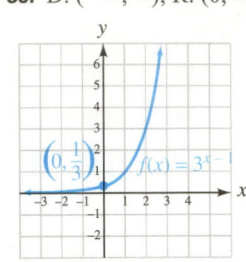

31. The x-axis ($y = 0$) **32. a.** 128,000 tons
b. About 281,569,441 tons **33.** About $2,189,703.45
34. About $2,015 **35.** D: $(0, \infty)$; R: $(-\infty, \infty)$ **36.** Since there is
no real number such that $10^? = 0$, log 0 is undefined. **37.** $4^3 = 64$
38. $\log_7 \frac{1}{7} = -1$ **39.** 2 **40.** -2 **41.** 0 **42.** Undefined
43. $\frac{1}{2}$ **44.** 3 **45.** 32 **46.** $\frac{1}{81}$ **47.** 4 **48.** 10 **49.** $\frac{1}{2}$
50. $\frac{1}{3}$ **51.** 0.6542 **52.** 26.9153
53.

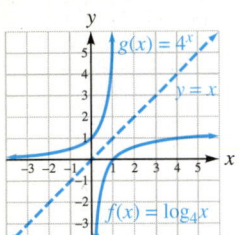

54. **55.**

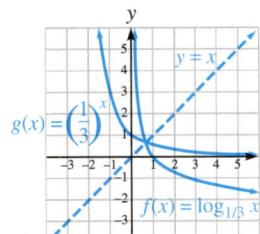 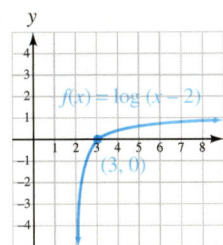

56.

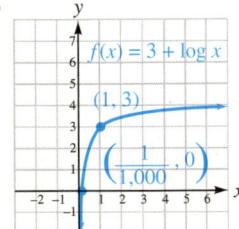

57. About 53 dB **58.** About 4.4 **59. a.** 52 cm **b.** About 74.6 cm
60. About 95% **61. a.** $e \approx 2.72$ **b.** 2.7081, 15 **62.** 16.6704

63. D: $(-\infty, \infty)$; R: $(1, \infty)$ **64.** D: $(-\infty, \infty)$; R: $(0, \infty)$

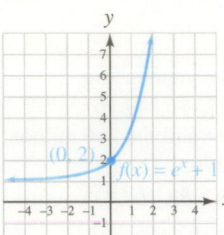 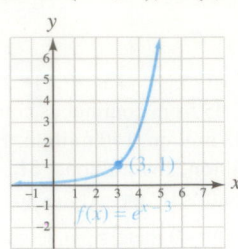

65. $2,324,767.37 **66.** About $211.7 billion **67.** 13.9%,
11.67%, 9.80% **68.** The exponent on the base e is negative.
69. 1 **70.** 2 **71.** -5 **72.** $\frac{1}{2}$ **73.** Undefined **74.** Undefined
75. 0 **76.** -7 **77.** 6.1137 **78.** -0.1625 **79.** 10.3398
80. 0.0002
81. They have different bases: $\log x = \log_{10} x$ and $\ln x = \log_e x$.
82. $f^{-1}(x) = e^x$
83. **84.**

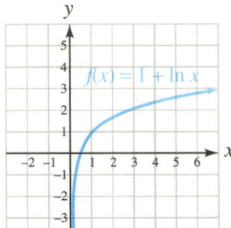 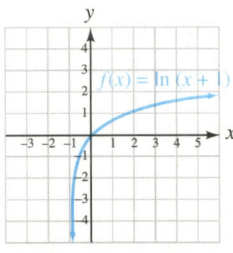

85. About 62 yr **86.** About 72 in. (6 ft) **87.** 0 **88.** 1 **89.** 3
90. 4 **91.** $3 + \log_3 x$ **92.** $2 - \log x$ **93.** $\frac{1}{2} \log_5 27$
94. $\log_b 10 + \log_b a + 1$ **95.** $2 \log_b x + 3 \log_b y - \log_b z$
96. $\frac{1}{2} \ln x - \frac{1}{2} \ln y - \ln z$ **97.** $\log_2 \frac{x^3 z^7}{y^5}$ **98.** $\log_b \frac{\sqrt{x+2}}{y^3 z^7}$
99. $\log_b(a - 5)$ **100.** $\log_8 x^7$ **101.** 2.6609 **102.** 3.0000
103. 1.7604 **104.** 3.1 **105.** -4 **106.** $-3, -1$
107. $\frac{\log 7}{\log 3} \approx 1.7712$ **108.** $\frac{4 \log 3}{\log 3 - \log 2} \approx 10.8380$
109. $\ln 7 \approx 1.9459$ **110.** $-\frac{\ln 25}{0.4} \approx -8.0472$ **111.** $\frac{7}{3}$
112. $\pm \sqrt{\frac{\log 33}{\log 9}} \approx \pm 1.2615$ **113.** 104 **114.** 9 **115.** 25, 4
116. 4 **117.** 4, 3 **118.** 2 **119.** 6 **120.** 31
121. $0.76787 \neq -0.27300$ **122.** About 3,300 yr
123. About 91 days **124.** 2, 5

CHAPTER 9 TEST (page 849)

1. a. composite **b.** natural **c.** continuous **d.** inverse
e. logarithmic **2. a.** f composed with g of x **b.** g of f of eight
c. f inverse of x **3.** $(f + g)(x) = 4x^2 - 2x + 11, (-\infty, \infty)$
4. $(g/f)(x) = \frac{4x^2 - 3x + 2}{x + 9}, (-\infty, -9) \cup (-9, \infty)$ **5.** 76
6. $32x^2 - 128x + 131$ **7. a.** -16 **b.** 17 **8. a.** 0 **b.** 3
c. 1 **d.** -2 **e.** -5 **9. a.** No **b.** Yes **c.** Yes **d.** No
10. $f^{-1}(x) = -3x$

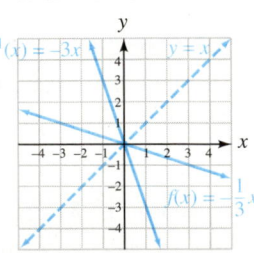

11. One-to-one function; $f^{-1}(x) = 3x - 6$
12. $f^{-1}(x) = \sqrt[3]{x} + 15$ **14. a.** Yes **b.** Yes
c. 80; when the temperature of the tire tread is 260°, the vehicle is
traveling 80 mph

15. D: $(-\infty, \infty)$; R: $(1, \infty)$ **16.** D: $(-\infty, \infty)$; R: $(0, \infty)$

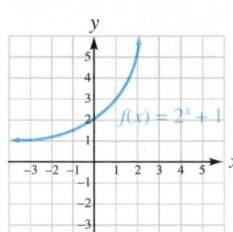   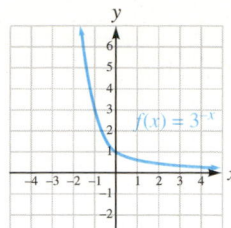

17. $\frac{3}{64}$ g = 0.046875 g   **18.** About $1,060.90

19. a.

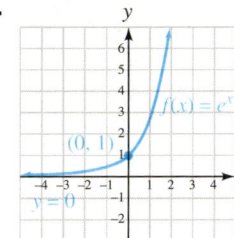

b. D: $(-\infty, \infty)$; R: $(0, \infty)$
c. $f^{-1}(x) = \ln x$
20. 1,346,161,000
21. About 88 micrograms
per day   **22.** $6^{-2} = \frac{1}{36}$
23. a. D: $(0, \infty)$; R: $(-\infty, \infty)$
b. $f^{-1}(x) = 10^x$
24. Logarithmic growth   **25.** 2

26. -2 **27.** Undefined   **28.** -6 **29.** $\frac{1}{2}$ **30.** 0   **31.** 2   **32.** 16
33. $\frac{1}{27}$ **34.** e

35.   **36.**

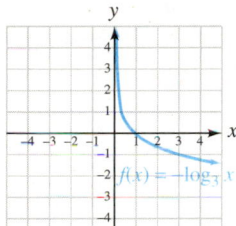   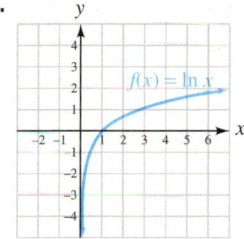

37. About 6.4   **38.** About 46 dB   **39.** 0.0871   **40.** 0.5646
41. $2\log_b a + 1 + 3\log_b c$ **42.** $\ln\frac{b\sqrt{a+2}}{c^3}$ **43.** $\frac{\log 3}{\log 5} \approx 0.6826$
44. 4   **45.** $\frac{1}{6}$ **46.** $\frac{\ln 4}{0.08} \approx 17.3287$ **47.** 5   **48.** 6   **49.** 1   **50.** 10
51. 5   **52.** About 20 min

CUMULATIVE REVIEW CHAPTERS 1–9 (page 851)

1. $P = 2l + 2w$; $A = lw$ **2.** $A = \pi r^2$; $c = 2\pi r = \pi D$
3. $A = \frac{1}{2}bh$ **4.** $V = lwh$; $V = s^3$ **5.** Amount = percent · base
6. Total value = amount · price   **7.** $I = Prt$ **8.** $d = rt$
9. $\frac{x_1 + x_2}{2}, \frac{y_1 + y_2}{2}$ **10. a.** $m = \frac{y_2 - y_1}{x_2 - x_1}$ **b.** rise   **11.** $y = mx + b$
12. $y - y_1 = m(x - x_1)$ **13.** $y = b, x = a$ **14.** $ad - bc$
15. a. $y = kx$ **b.** $y = \frac{k}{x}$ **16.** $a^2 + b^2 = c^2$
17. $d = \sqrt{(x_2 - x_1)^2 + (y_2 - y_1)^2}$ **18.** $x = \frac{-b \pm \sqrt{b^2 - 4ac}}{2a}$
19. $f(x) = a(x - h)^2 + k$ **20.** $f(g(x))$ **21.** $A = Pe^{rt}$
22. $\log_b x = \frac{\log_a x}{\log_a b}$ **23.** Same, negative reciprocals
24. The addition (elimination) method   **25. a.** both   **b.** one, both
26. a. $X = k$; $X = -k$ **b.** $X = Y$; $X = -Y$ **c.** $X > k$; $X < -k$
d. $-k < X < k$ **27. a.** x **b.** x^{m+n} **c.** x^{mn} **d.** $x^n y^n$ **e.** 1
f. $\frac{x^n}{y^n}$ **g.** x^{m-n} **h.** $\frac{1}{x^n}$ **i.** x^n **j.** $\left(\frac{y}{x}\right)^n$ **k.** $\sqrt[n]{x}$ **l.** $\left(\sqrt[n]{x}\right)^m$
28. Pick a test point on one side of the boundary line. In $y \geq x$,
replace x and y with the coordinates of that point. If the inequality
is satisfied, shade the side that contains that point. If the inequality
is not satisfied, shade the other side.
29. a. $x^2 + 2xy + y^2$ **b.** $x^2 - 2xy + y^2$ **c.** $x^2 - y^2$

30. a. $(x + y)(x - y)$ **b.** $(x - y)(x^2 + xy + y^2)$
c. $(x + y)(x^2 - xy + y^2)$
31. An equation of the form $ax^2 + bx + c = 0$, where $a \neq 0$
32. $x + 1 = 0$ or $x - 7 = 0$
33. A factor equal to 1 is being removed: $\frac{3a - 2}{3a - 2} = 1$
34. denominator   **35. a.** $\frac{AC}{BD}$ **b.** $\frac{AD}{BC}$ **36. a.** $\frac{A + B}{D}$ **b.** $\frac{A - B}{D}$
37. a. To build up the fractions so they have the same denominator
b. To simplify a complex fraction   **c.** To rationalize the
denominator   **38.** bc, cross products
39. Multiply both sides of the equation by the LCD, which is
$x(x - 2)$.   **40. a.** $\sqrt[n]{a}\sqrt[n]{b}$ **b.** $\frac{\sqrt[n]{a}}{\sqrt[n]{b}}$ **41.** Square both sides.
42. a. 3.14   **b.** 1.41   **c.** 2.72   **43. a.** 0   **b.** 1   **c.** x **d.** x
e. $\log_b M + \log_b N$ **f.** $\log_b M - \log_b N$ **g.** $p \log_b M$
44. a. Take the log on both sides.   **b.** Write the left side as a single
logarithm: $\log x(x - 3) = 1$.   **45.** $+$ **46.** $x, -x$
47. a. b^2 **b.** $|x|$ **c.** x **48. a.** $\sqrt{-1}$ **b.** -1

ARE YOU READY? 10.1 (page 856)

1. $x^2 - 16x + 64$ **2.** $x^2 - 6x + 9 = (x - 3)^2$ **3.** $(y + 5)^2$
4. $-3(y^2 + 4y)$

SELF CHECK 10.1

1. a.   **b.**

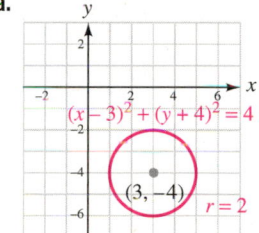   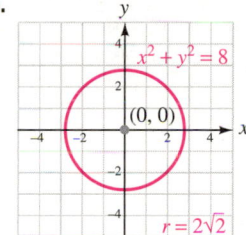

2. $(x + 7)^2 + (y - 1)^2 = 100$ **3.**

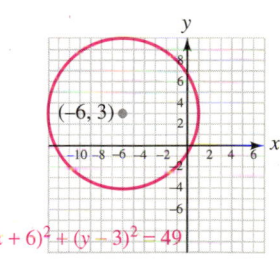

4. 9 ft   **5.**

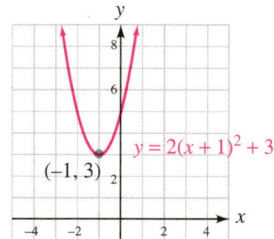

6.   **7.**

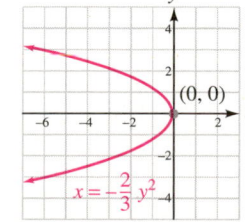   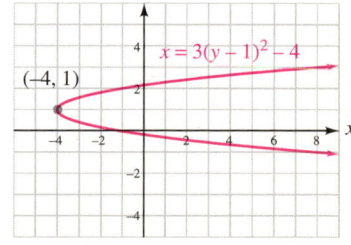

STUDY SET SECTION 10.1 (page 866)

1. conic sections **3.** circle, radius
5. a. $(x - h)^2 + (y - k)^2 = r^2$ **b.** $x^2 + y^2 = r^2$
7. a. $(2, -1); r = 4$ **b.** $(x - 2)^2 + (y + 1)^2 = 16$
9. a. $y = a(x - h)^2 + k$ **b.** $x = a(y - k)^2 + h$
11. a. Circle **b.** Parabola **c.** Parabola **d.** Circle
13. $6, -2, 3$ **15.** $(0, 0), r = 3$

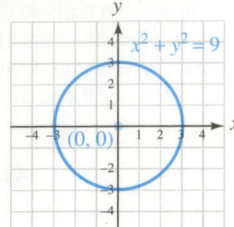

17. $(0, -3), r = 1$ **19.** $(-3, 1), r = 4$

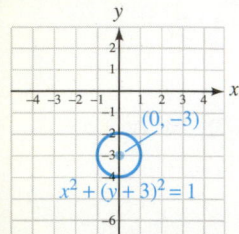   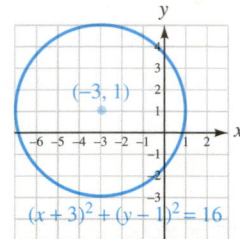

21. $(0, 0), r = \sqrt{6} \approx 2.4$

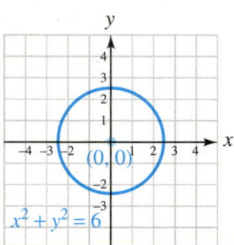

23. $x^2 + y^2 = 1$ **25.** $(x - 6)^2 + (y - 8)^2 = 25$
27. $(x + 2)^2 + (y - 6)^2 = 144$ **29.** $x^2 + y^2 = \frac{1}{16}$
31. $\left(x - \frac{2}{3}\right)^2 + \left(y + \frac{7}{8}\right)^2 = 2$ **33.** $x^2 + y^2 = 8$
35. $(x - 1)^2 + (y + 2)^2 = 4; (1, -2), r = 2$

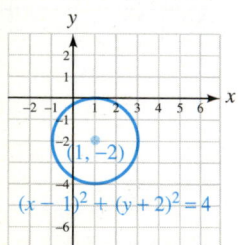

37. $(x + 2)^2 + (y + 1)^2 = 9; (-2, -1), r = 3$

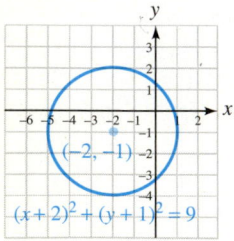

39. $y = 2(x - 1)^2 + 3$; vertex: $(1, 3)$

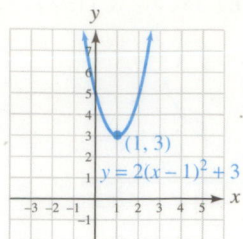

41. $y = -(x + 1)^2 + 4$; vertex: $(-1, 4)$

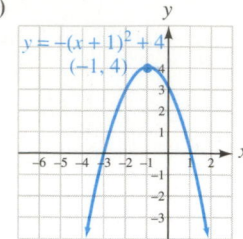

43. Vertex: $(0, 0)$ **45.** Vertex: $(3, -1)$

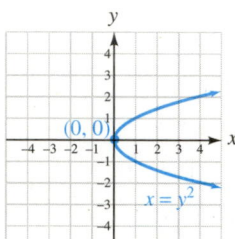   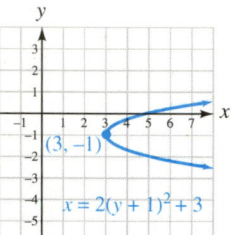

47. $x = (y - 1)^2 + 4$; vertex: $(4, 1)$

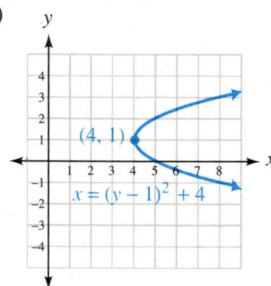

49. $x = -3(y - 3)^2 + 2$; vertex: $(2, 3)$

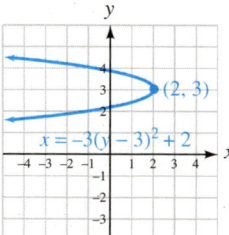

51.   **53.**

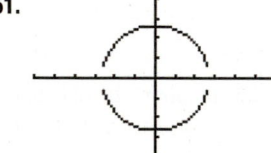   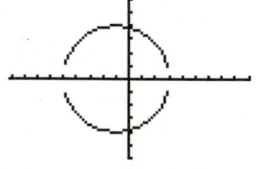

55.   **57.**

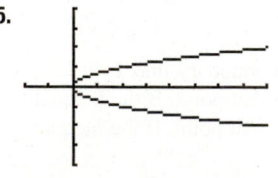   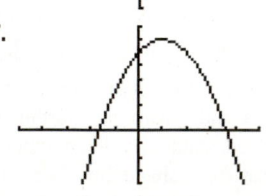

59. $x = (y - 3)^2 - 5$; vertex: $(-5, 3)$

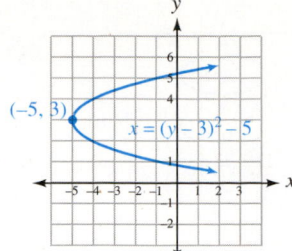

61. $(2, 0)$; $r = 5$

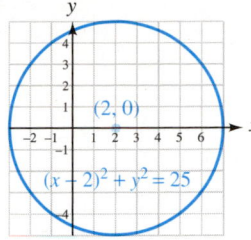

63. $(x - 3)^2 + (y + 4)^2 = 7$; $(3, -4)$; $r = \sqrt{7} \approx 2.6$

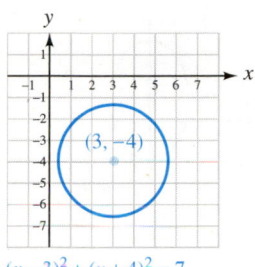

65. $y = 4(x - 2)^2 + 1$; vertex: $(2, 1)$

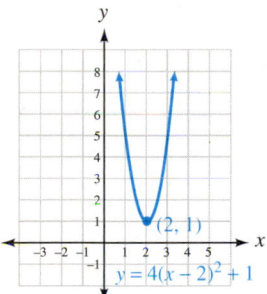

67. $(1, 3)$; $r = \sqrt{15} \approx 3.9$ **69.** Vertex: $(1, 0)$

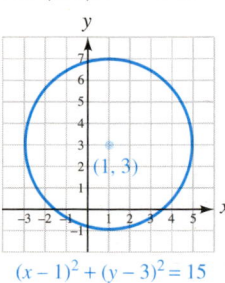 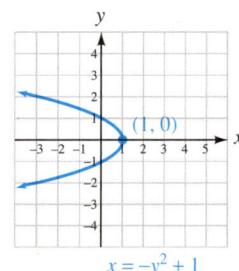

71. $(2, 4)$; $r = 6$

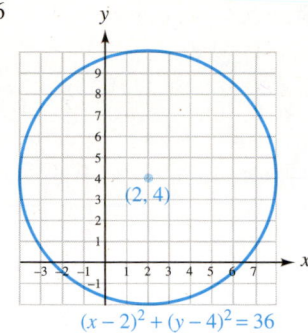

73. Vertex: $(0, 0)$ **75.** Vertex: $(3, 1)$

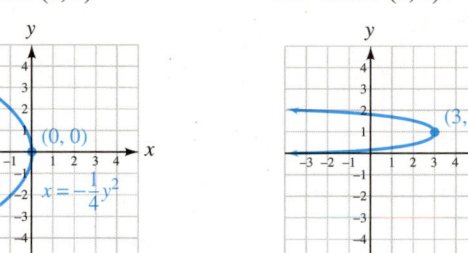

77. $(x + 1)^2 + y^2 = 9$; $(-1, 0)$; $r = 3$

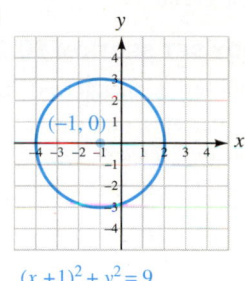

79. $x = \frac{1}{2}(y + 2)^2 - 2$; vertex: $(-2, -2)$

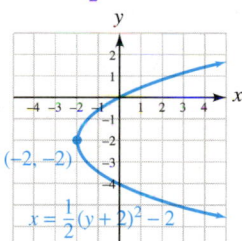

81. Vertex: $(-5, 5)$

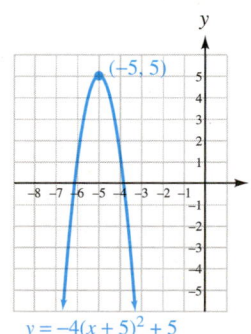

83. a. $(4, -7)$; $r = 2\sqrt{7} \approx 5.3$
b. $(-4, 7)$; $r = 2\sqrt{7} \approx 5.3$
85. a. $V(3, 6)$; opens upward
b. $V(6, 3)$; opens to the right
87. No   **89. a.** 8 mi
b. 9 mi   **91.** 5 ft
93. 2 AU   **99.** $5, -\frac{7}{3}$ **101.** $3, -\frac{1}{4}$

ARE YOU READY? 10.2 (page 869)

1. ± 9 **2.** $y + 4$ **3.** $4x^2 + y^2$ **4.** $2\sqrt{2}$

SELF CHECK 10.2

1.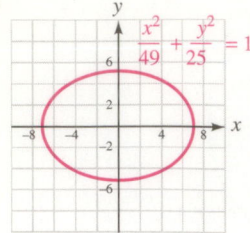
$\dfrac{x^2}{49} + \dfrac{y^2}{25} = 1$

2.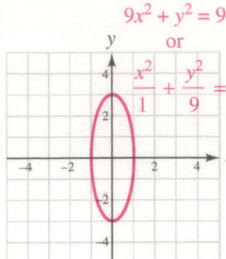
$9x^2 + y^2 = 9$ or $\dfrac{x^2}{1} + \dfrac{y^2}{9} = 1$

3.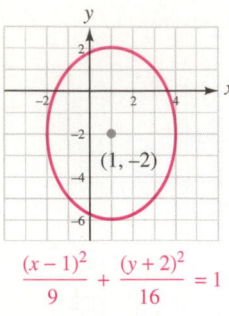
$(1, -2)$
$\dfrac{(x-1)^2}{9} + \dfrac{(y+2)^2}{16} = 1$

4.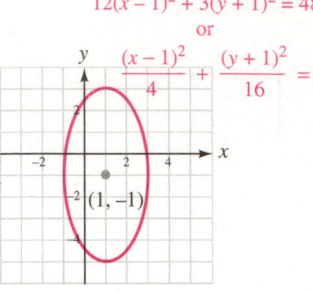
$12(x-1)^2 + 3(y+1)^2 = 48$ or $\dfrac{(x-1)^2}{4} + \dfrac{(y+1)^2}{16} = 1$
$(1, -1)$

5. $\dfrac{x^2}{324} + \dfrac{y^2}{576} = 1$

STUDY SET SECTION 10.2 (page 875)

1. ellipse **3.** foci, focus **5.** major **7.** $\dfrac{x^2}{a^2} + \dfrac{y^2}{b^2} = 1$
9. x-intercepts: $(a, 0), (-a, 0)$; y-intercepts: $(0, b), (0, -b)$
11. a. $(-2, 1)$; $a = 2, b = 5$ **b.** Vertical
c. $\dfrac{(x+2)^2}{4} + \dfrac{(y-1)^2}{25} = 1$ **13.** $\dfrac{(x-1)^2}{16} + \dfrac{(y+5)^2}{1} = 1$
15. $h = -8, k = 6, a = 10, b = 12$ **17.**
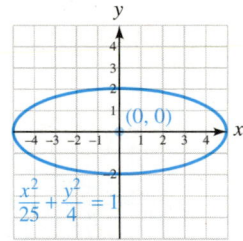
$(0, 0)$
$\dfrac{x^2}{25} + \dfrac{y^2}{4} = 1$

19.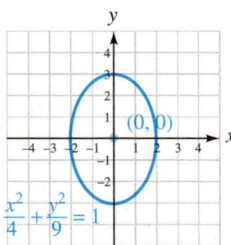
$(0, 0)$
$\dfrac{x^2}{4} + \dfrac{y^2}{9} = 1$

21.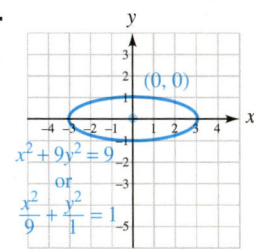
$(0, 0)$
$x^2 + 9y^2 = 9$ or $\dfrac{x^2}{9} + \dfrac{y^2}{1} = 1$

23.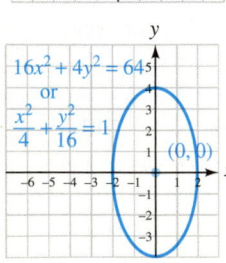
$16x^2 + 4y^2 = 64$ or $\dfrac{x^2}{4} + \dfrac{y^2}{16} = 1$
$(0, 0)$

25.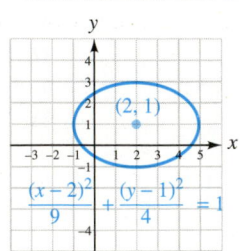
$(2, 1)$
$\dfrac{(x-2)^2}{9} + \dfrac{(y-1)^2}{4} = 1$

27.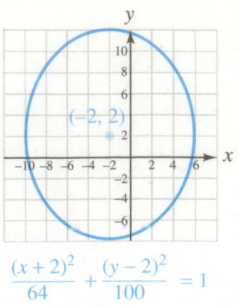
$(-2, 2)$
$\dfrac{(x+2)^2}{64} + \dfrac{(y-2)^2}{100} = 1$

29.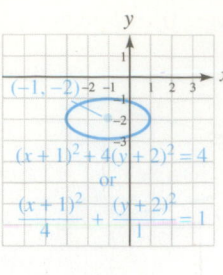
$(-1, -2)$
$(x+1)^2 + 4(y+2)^2 = 4$ or $\dfrac{(x+1)^2}{4} + \dfrac{(y+2)^2}{1} = 1$

31.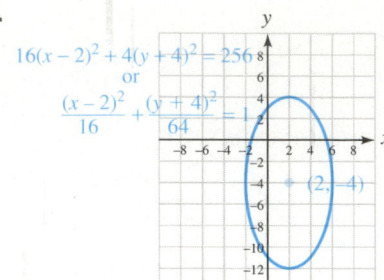
$16(x-2)^2 + 4(y+4)^2 = 256$ or $\dfrac{(x-2)^2}{16} + \dfrac{(y+4)^2}{64} = 1$
$(2, -4)$

33.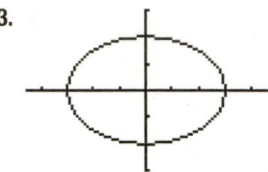

35.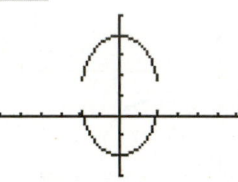

37.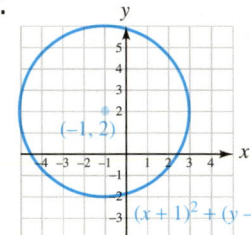
$(-1, 2)$
$(x+1)^2 + (y-2)^2 = 16$

39.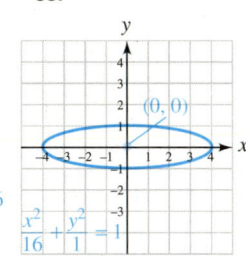
$(0, 0)$
$\dfrac{x^2}{16} + \dfrac{y^2}{1} = 1$

41.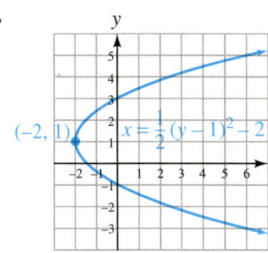
$(-2, 1)$
$x = \frac{1}{2}(y-1)^2 - 2$

43. $x^2 + y^2 = 25$
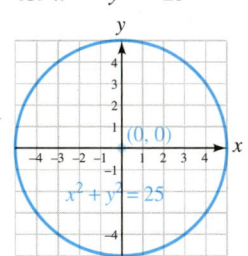
$(0, 0)$
$x^2 + y^2 = 25$

45.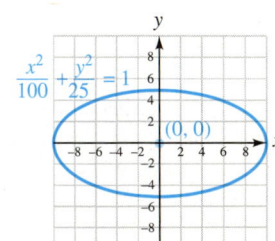
$\dfrac{x^2}{100} + \dfrac{y^2}{25} = 1$
$(0, 0)$

47.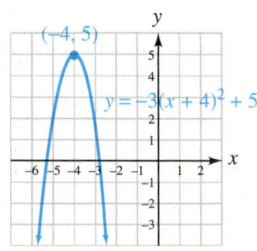
$(-4, 5)$
$y = -3(x+4)^2 + 5$

49.

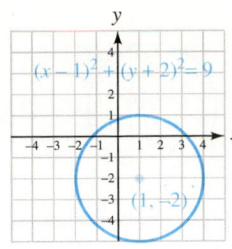

$(x-1)^2 + (y+2)^2 = 9$

51.

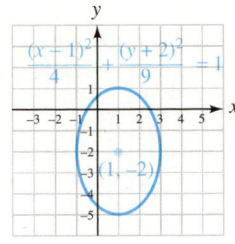

$\dfrac{(x-1)^2}{4} + \dfrac{(y+2)^2}{9} = 1$

4.

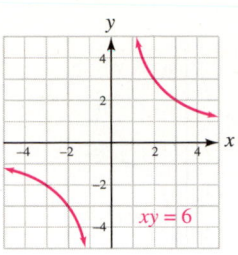

$xy = 6$

5. 1×10^9 mi

53.

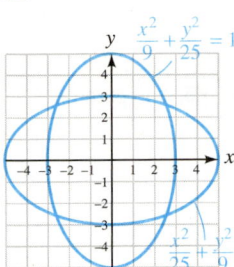

$\dfrac{x^2}{9} + \dfrac{y^2}{25} = 1$

$\dfrac{x^2}{25} + \dfrac{y^2}{9} = 1$

55.

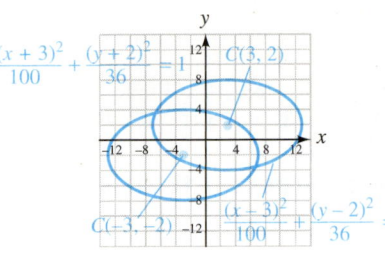

$\dfrac{(x+3)^2}{100} + \dfrac{(y+2)^2}{36} = 1 \quad C(3,2)$

$C(-3,-2) \quad \dfrac{(x-3)^2}{100} + \dfrac{(y-2)^2}{36} = 1$

57. a. $\dfrac{x^2}{400} + \dfrac{y^2}{100} = 1$ **b.** $5\sqrt{3}$ ft ≈ 8.7 ft

59. $\dfrac{x^2}{1,600} + \dfrac{y^2}{1,225} = 1$ **61.** 12π sq. units ≈ 37.7 sq. units

67. $12y^2 + \dfrac{9}{x^2}$ **69.** $\dfrac{y^2 + x^2}{y^2 - x^2}$

ARE YOU READY? 10.3 (page 878)

1. ± 4 **2.** $x + 2$ **3.** $\frac{2}{3}$ **4.** -10

SELF CHECKS 10.3

1.

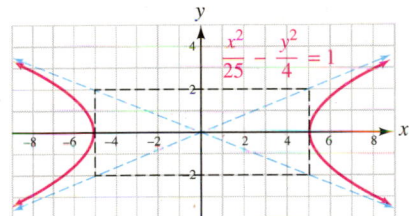

$\dfrac{x^2}{25} - \dfrac{y^2}{4} = 1$

2.

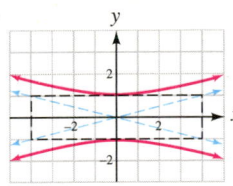

$16y^2 - x^2 = 16$
or
$\dfrac{y^2}{1} - \dfrac{x^2}{16} = 1$

3. a.

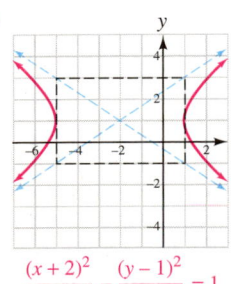

$\dfrac{(x+2)^2}{9} - \dfrac{(y-1)^2}{4} = 1$

b.

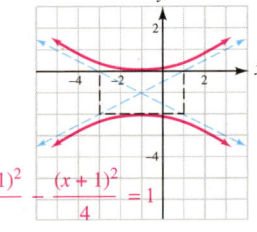

$\dfrac{(y+1)^2}{1} - \dfrac{(x+1)^2}{4} = 1$

STUDY SET SECTION 10.3 (page 884)

1. hyperbola **3.** vertices **5.** diagonals **7.** $\dfrac{x^2}{a^2} - \dfrac{y^2}{b^2} = 1$

9. $\dfrac{(x-h)^2}{a^2} - \dfrac{(y-k)^2}{b^2} = 1$ **11. a.** $(-1, -2)$; $a = 3, b = 1$

b. $\dfrac{(y+2)^2}{9} - \dfrac{(x+1)^2}{1} = 1$ **13.** $\dfrac{(x+1)^2}{1} - \dfrac{(y-5)^2}{4} = 1$

15. $h = 5, k = -11, a = 5, b = 6$

17.

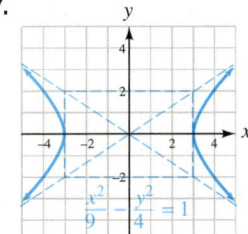

$\dfrac{x^2}{9} - \dfrac{y^2}{4} = 1$

19.

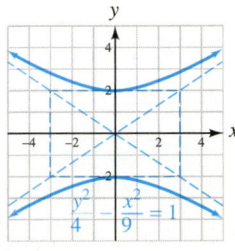

$\dfrac{y^2}{4} - \dfrac{x^2}{9} = 1$

21.

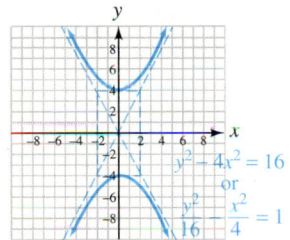

$y^2 - 4x^2 = 16$
or
$\dfrac{y^2}{16} - \dfrac{x^2}{4} = 1$

23.

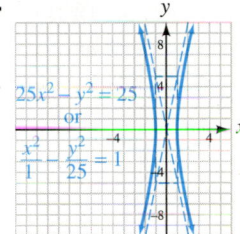

$25x^2 - y^2 = 25$
or
$\dfrac{x^2}{1} - \dfrac{y^2}{25} = 1$

25.

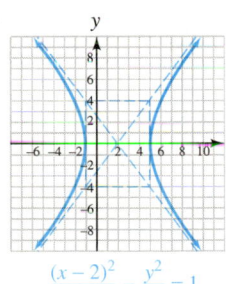

$\dfrac{(x-2)^2}{9} - \dfrac{y^2}{16} = 1$

27.

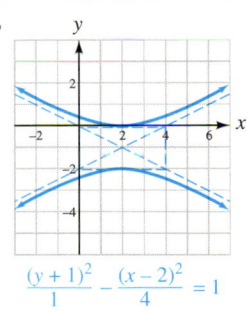

$\dfrac{(y+1)^2}{1} - \dfrac{(x-2)^2}{4} = 1$

29.

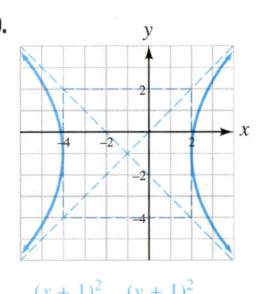

$\dfrac{(x+1)^2}{9} - \dfrac{(y+1)^2}{9} = 1$

31.

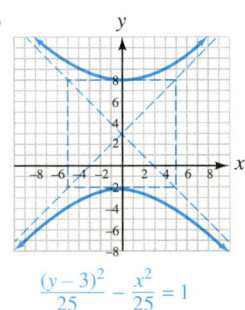

$\dfrac{(y-3)^2}{25} - \dfrac{x^2}{25} = 1$

33.
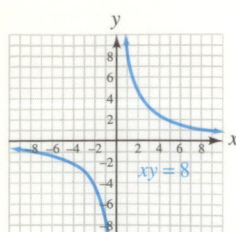

35.
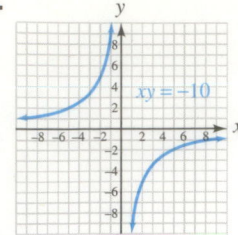

37.
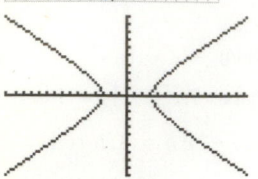

39.
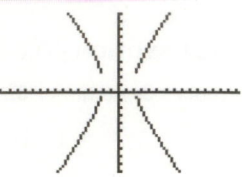

41.
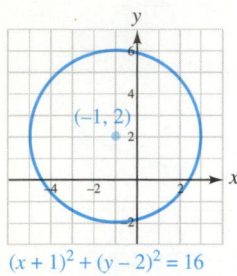

$(x + 1)^2 + (y - 2)^2 = 16$

43.
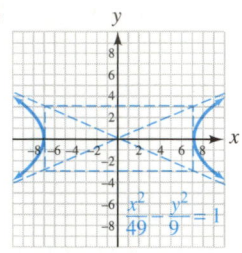

$\dfrac{x^2}{49} - \dfrac{y^2}{9} = 1$

45.
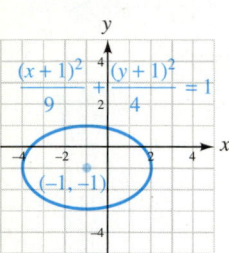

$\dfrac{(x+1)^2}{9} + \dfrac{(y+1)^2}{4} = 1$

47.
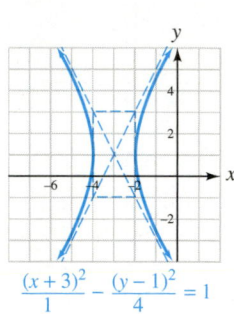

$\dfrac{(x+3)^2}{1} - \dfrac{(y-1)^2}{4} = 1$

49.
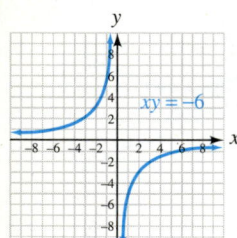

51.
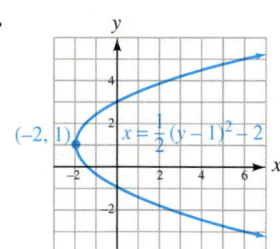

53.
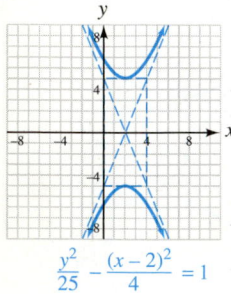

$\dfrac{y^2}{25} - \dfrac{(x-2)^2}{4} = 1$

55.
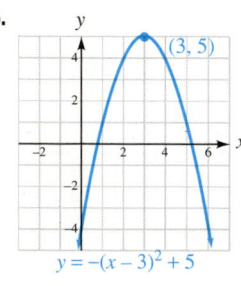

$y = -(x - 3)^2 + 5$

57.
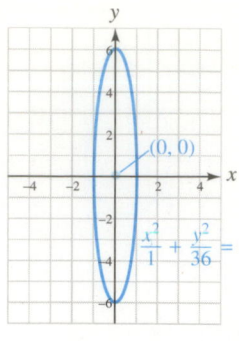

59.
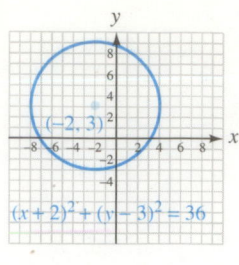

61. 3 units   **63.** $10\sqrt{3}$ miles ≈ 17.3 miles   **65.** A hyperbola
71. 64   **73.** 3   **75.** $\dfrac{3}{2}$ **77.** 10

ARE YOU READY? 10.4 (page 887)

1.
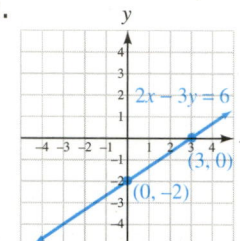

2. $-1, \dfrac{5}{9}$ **3.** $(4, 12)$

4. a. $\pm\dfrac{\sqrt{5}}{3}$ **b.** $\pm 3\sqrt{2}$

SELF CHECK 10.4

1. $(-4, 3), (0, -5)$

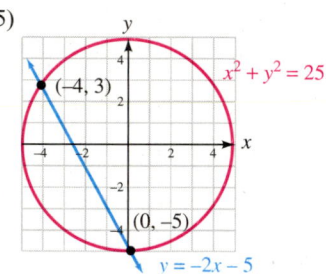

2. $(1, 3), (-3, -1)$ **3.** $(2, 4), (-2, 4)$

4. $\left(2, \sqrt{3}\right), \left(2, -\sqrt{3}\right), \left(-2, \sqrt{3}\right), \left(-2, -\sqrt{3}\right)$

STUDY SET SECTION 10.4 (page 891)

1. system   **3.** intersection   **5.** secant   **7. a.** two   **b.** four
c. four   **d.** four   **9.** $(-3, 2), (3, 2), (-3, -2), (3, -2)$
11. a. -4 **b.** -2 **13.** $2x, 5, 5, 1, 1, -1, -2$
15. $(0, 3), (-3, 0)$ **17.** $(-4, 0), (4, 0)$

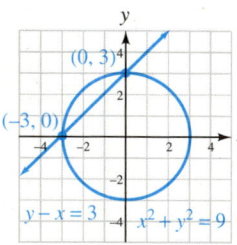

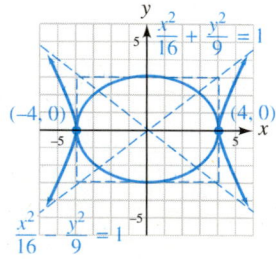

19. $(0, 0), (2, -4)$

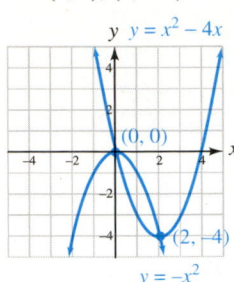

21. $(-2, 0), (0, -1), (0, 1)$

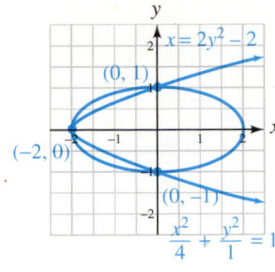

9.

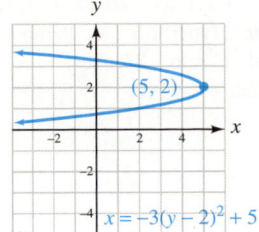

10.

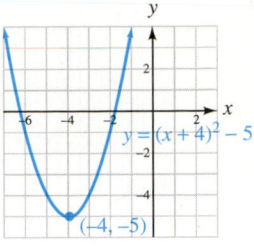

11. $(2, -2), (8, -3)$

12. When $x = 22$, $y = 0$: $-\frac{5}{121}(22 - 11)^2 + 5 = 0$

23. $(1, 2), (2, 1)$ **25.** $(-6, 7), (-2, -1)$ **27.** $(-2, 3), (2, 3)$
29. $(\sqrt{5}, 5), (-\sqrt{5}, 5)$ **31.** $(2, 4), (2, -4), (-2, 4), (-2, -4)$
33. $(3, 0), (-3, 0)$ **35.** $(\sqrt{3}, 0), (-\sqrt{3}, 0)$
37. $(-2, 3), (2, 3), (-2, -3), (2, -3)$
39. $(1, 0), (5, 0)$

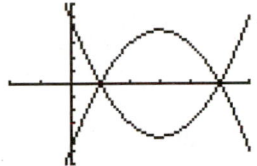

13.

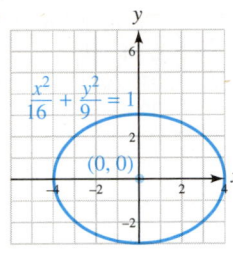

14.

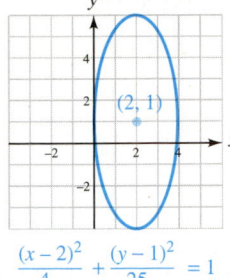

$$\frac{(x-2)^2}{4} + \frac{(y-1)^2}{25} = 1$$

41. $(2, 1), (-2, 1), (2, -1), (-2, -1)$ **43.** $(0, -4), (-3, 5), (3, 5)$
45. $(6, 2), (-6, -2), \left(-\sqrt{42}, 0\right), \left(\sqrt{42}, 0\right)$
47. $\left(-\sqrt{15}, 5\right), \left(\sqrt{15}, 5\right), (-2, -6), (2, -6)$
49. $(0, 3), \left(-\frac{25}{12}, -\frac{13}{4}\right)$ **51.** $(0, 0), (2, 4)$ **53.** No solution, $\varnothing$
55. $(3, 5)$
57. $\left(\frac{\sqrt{10}}{4}, \frac{3\sqrt{6}}{4}\right), \left(\frac{\sqrt{10}}{4}, -\frac{3\sqrt{6}}{4}\right), \left(-\frac{\sqrt{10}}{4}, \frac{3\sqrt{6}}{4}\right), \left(-\frac{\sqrt{10}}{4}, -\frac{3\sqrt{6}}{4}\right)$
59. $\left(\frac{1}{2}, \frac{1}{3}\right), \left(\frac{1}{3}, \frac{1}{2}\right)$ **61.** $(-1, 3), (1, 3)$ **63.** $\left(\frac{5}{2}, \frac{3}{2}\right)$ **65.** $4, 8$
67. $\left(10, \frac{10}{3}\right); \frac{10}{3}\sqrt{10}$ m   **69.** 80 ft by 100 ft or 50 ft by 160 ft
71. $2,500$ at 9%   **75.** $2,000$ **77.** 7

15.

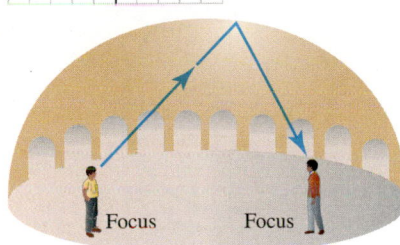

16. $\frac{x^2}{12^2} + \frac{y^2}{1^2} = 1$ **17. a.** Circle
b. Ellipse   **c.** Parabola
d. Ellipse   **18.** $\frac{x^2}{25} + \frac{y^2}{9} = 1$
19. ellipse, focus

20.

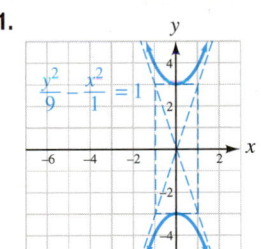

Answers may vary.

CHAPTER 10 REVIEW (page 894)

1.

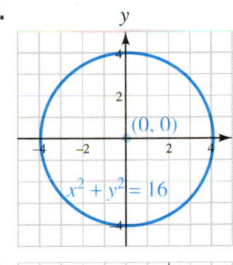

2.

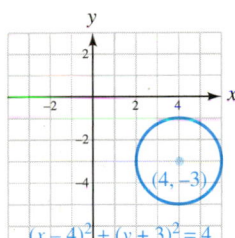

3.

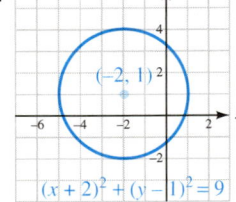

4. $(x - 9)^2 + (y - 9)^2 = 9^2$ or $(x - 9)^2 + (y - 9)^2 = 81$
5. $(-6, 0); r = 2\sqrt{6}$ **6.** center, radius
7. $(0, 0)$ **8.** $(-2, -1)$

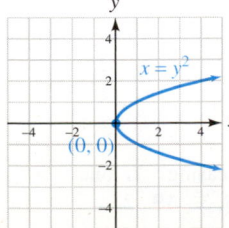

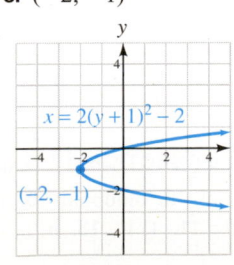

21.

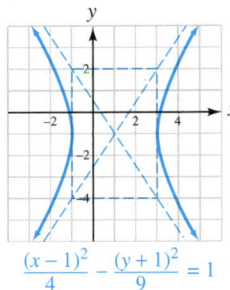

22.

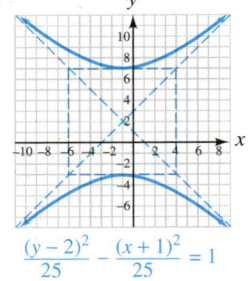

$$\frac{(x-1)^2}{4} - \frac{(y+1)^2}{9} = 1$$

23.

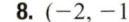

$$\frac{(y-2)^2}{25} - \frac{(x+1)^2}{25} = 1$$

24.

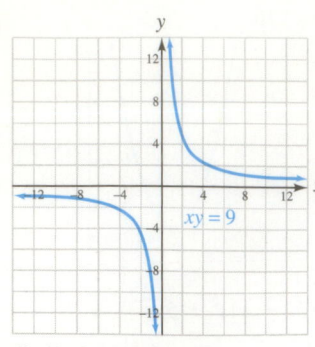

$xy = 9$

25. 4 units **26. a.** Ellipse
b. Hyperbola **c.** Parabola
d. Circle **27.** Yes
28. $(0, 3), (0, -3)$

11.

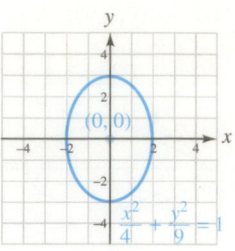

$(0,0)$
$\frac{x^2}{4} + \frac{y^2}{9} = 1$

12.

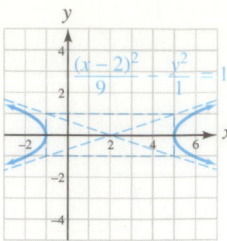

$\frac{(x-2)^2}{9} - \frac{y^2}{1} = 1$

29. $(2, 2), (-1, -4)$

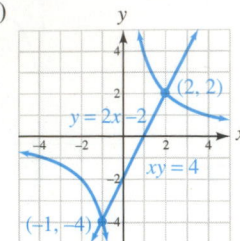

$y = 2x - 2$
$(2, 2)$
$xy = 4$
$(-1, -4)$

30. a. 2 **b.** 4
c. 4 **d.** 4
31. $(0, 1), (0, -1)$
32. $y = 3x - 1$

13.

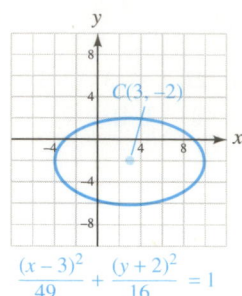

$C(3, -2)$

$\frac{(x-3)^2}{49} + \frac{(y+2)^2}{16} = 1$

14.

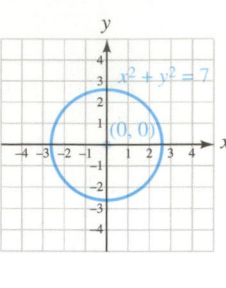

$x^2 + y^2 = 7$
$(0, 0)$

33. $(0, -4), (-3, 5), (3, 5)$ **34.** $\left(\sqrt{2}, 0\right), \left(-\sqrt{2}, 0\right)$
35. $(2, 2), \left(-\frac{2}{9}, -\frac{22}{9}\right)$ **36.** $(4, 2), (4, -2), (-4, 2), (-4, -2)$
37. $(2, 3), (2, -3), (-2, 3), (-2, -3)$
38. $\left(2\sqrt{2}, \sqrt{2}\right), \left(-2\sqrt{2}, -\sqrt{2}\right), (1, 4), (-1, -4)$
39. No solution, $\varnothing$ **40.** $(-2, 1)$

CHAPTER 10 TEST (page 899)

1. a. conic **b.** center, radius **c.** hyperbola **d.** nonlinear
e. ellipse **2.** $(0, 0); r = 10$

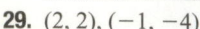

$(0, 0)$

$x^2 + y^2 = 100$

3. $(-2, 3), r = 3\sqrt{2}$ **4.** $(x - 4)^2 + (y - 3)^2 = 9$
5. $\frac{21}{2}$ in. = 10.5 in. **6.** 16, 16, 4, 6
7.

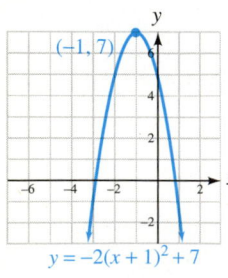

$(-2, 1)$
$(x + 2)^2 + (y - 1)^2 = 9$

8.

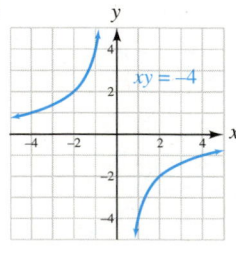

$(2, 1)$
$x = (y - 1)^2 + 2$

9.

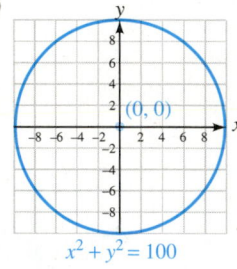

$(-1, 7)$
$y = -2(x + 1)^2 + 7$

10.

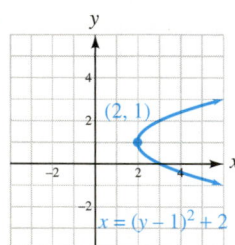

$xy = -4$

17. $\frac{(x - 1)^2}{16} + \frac{(y + 2)^2}{9} = 1$ **18.** 2.5 in.
19. $\frac{x^2}{784} + \frac{y^2}{256} = 1$
20. $(-1, 1)$; length: 4 units, width: 4 units **21.** $\frac{y^2}{16} - \frac{x^2}{36} = 1$
22. a. Ellipse **b.** Hyperbola **c.** Circle **d.** Parabola
23. $(-4, -3), (3, 4)$

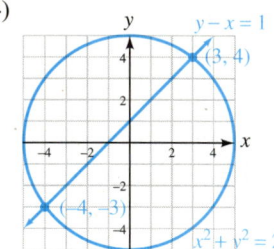

$y - x = 1$
$(3, 4)$
$(-4, -3)$
$x^2 + y^2 = 25$

24. $(2, 6), (-2, -2)$
25. $\left(1, \sqrt{2}\right), \left(1, -\sqrt{2}\right), \left(-1, \sqrt{2}\right), \left(-1, -\sqrt{2}\right)$
26. $\left(-1, \frac{9}{2}\right), \left(3, -\frac{3}{2}\right)$ **27.** $(-1, 0)$ **28.** No solution; $\varnothing$

ARE YOU READY? 11.1 (page 902)

1. 1, 4, 6, 4, 1 **2.** $a^2 + 2ab + b^2$ **3.** 120 **4.** 30

SELF CHECK 11.1

1. $x^4 + 4x^3y + 6x^2y^2 + 4xy^3 + y^4$
2. $x^5 - 5x^4y + 10x^3y^2 - 10x^2y^3 + 5xy^4 - y^5$ **3. a.** 5,040
b. 144 **c.** 1 **4. a.** 4 **b.** 21
5. $a^4 + 4a^3b + 6a^2b^2 + 4ab^3 + b^4$ **6.** $x^3 - 3x^2y + 3xy^2 - y^3$
7. $64a^3 - 240a^2b + 300ab^2 - 125b^3$ **8.** $36a^7b^2$ **9.** $\frac{35}{81}c^6d^4$

STUDY SET SECTION 11.1 (page 909)

1. binomial **3.** binomial **5.** factorial, decreasing **7.** one
9. 20, 20 **11.** 1, 1 **13.** $n, 1$ **15.** 1 **17.** 9, 3 **19.** 2, 4

21. $3, 3, y, 2, 2, 3$ **23.** $n, 1$ **25.** $a^3 + 3a^2b + 3ab^2 + b^3$
27. $m^5 - 5m^4p + 10m^3p^2 - 10m^2p^3 + 5mp^4 - p^5$ **29.** 6
31. 120   **33.** 30   **35.** 144   **37.** 40,320   **39.** 2,352
41. $\frac{1}{110}$ **43.** 72   **45.** 5   **47.** 10   **49.** $\frac{1}{168}$ **51.** 21
53. 39,916,800   **55.** $2.432902008 \times 10^{18}$
57. $m^4 + 4m^3n + 6m^2n^2 + 4mn^3 + n^4$
59. $c^5 - 5c^4d + 10c^3d^2 - 10c^2d^3 + 5cd^4 - d^5$
61. $a^9 - 9a^8b + 36a^7b^2 - 84a^6b^3 + 126a^5b^4 - 126a^4b^5 +$
$84a^3b^6 - 36a^2b^7 + 9ab^8 - b^9$
63. $s^6 + 6s^5t + 15s^4t^2 + 20s^3t^3 + 15s^2t^4 + 6st^5 + t^6$
65. $8x^3 + 12x^2y + 6xy^2 + y^3$
67. $32t^5 - 240t^4 + 720t^3 - 1,080t^2 + 810t - 243$
69. $625m^4 - 1,000m^3n + 600m^2n^2 - 160mn^3 + 16n^4$
71. $\frac{x^3}{27} + \frac{x^2y}{6} + \frac{xy^2}{4} + \frac{y^3}{8}$ **73.** $\frac{x^4}{81} - \frac{2x^3y}{27} + \frac{x^2y^2}{6} - \frac{xy^3}{6} + \frac{y^4}{16}$
75. $c^{10} - 5c^8d^2 + 10c^6d^4 - 10c^4d^6 + 5c^2d^8 - d^{10}$ **77.** $28x^6y^2$
79. $15r^2s^4$ **81.** $78x^{11}$ **83.** $-12x^3y$ **85.** $810xy^4$
87. $-\frac{1}{6}c^3d$ **89.** $-70,000t^4$ **91.** $6a^{10}b^2$ **97.** $\log x^2y^{1/2}$ or
$\log x^2\sqrt{y}$ **99.** $\ln y$

ARE YOU READY? 11.2 (page 911)

1. 21   **2. a.** 1   **b.** -1 **3.** 8   **4.** 52

SELF CHECK 11.2

1. a. $8, 11, 14$ **b.** 305   **2.** $-1, \frac{1}{2}, -\frac{1}{3}, \frac{1}{4}$
3. $10, 18, 26, 34, 42; 242$ **4.** 879   **5.** $15, 22, 29, 36$ **6.** $20, 32$
7. 6,275   **8.** 140 members   **9.** 52

STUDY SET SECTION 11.2 (page 918)

1. sequence   **3.** arithmetic   **5.** mean   **7.** $1, 7, 13$
9. a. $a_n = a_1 + (n-1)d$ **b.** $S_n = \frac{n(a_1 + a_n)}{2}$ **11.** nth
13. sigma   **15.** summation, runs   **17.** $3, 7, 11, 15, 19; 159$
19. $-2, -5, -8, -11, -14; -89$
21. $-1, -4, -9, -16, -25; -400$ **23.** $0, \frac{1}{2}, \frac{2}{3}, \frac{3}{4}, \frac{4}{5}; \frac{11}{12}$
25. $-\frac{1}{3}, \frac{1}{9}, -\frac{1}{27}, \frac{1}{81}$ **27.** $-7, 8, -9, 10$ **29.** $3, 5, 7, 9, 11; 21$
31. $-5, -8, -11, -14, -17; -47$ **33.** $7, 19, 31, 43, 55; 355$
35. $-7, -9, -11, -13, -15; -35$ **37.** 88   **39.** 59
41. $5, 11, 17, 23, 29$ **43.** $-4, -11, -18, -25, -32$ **45.** $5, 8$
47. $12, 14, 16, 18$ **49.** $\frac{45}{2}, 25, \frac{55}{2}$ **51.** $\frac{5}{4}$ **53.** 2,555
55. 2,920   **57.** $3 + 6 + 9 + 12$ **59.** $4 + 9 + 16$ **61.** 60
63. 91   **65.** 31   **67.** 12   **69.** 70   **71.** 57   **73.** 12   **75.** 354
77. $-118, -111, -104, -97, -90$ **79.** $34, 31, 28, 25, 22, 19$
81. 1,398   **83.** 1,275   **85.** -179 **87.** -23 **89.** -3
91. 2,500   **93.** $\$60, \$110, \$160, \$210, \$260, \$310; \$6,060$
95. 11,325   **97.** 78   **103.** $1 + \log_2 x - \log_2 y$
105. $3\log x + 2\log y$

ARE YOU READY? 11.3 (page 921)

1. 48   **2. a.** 625   **b.** $\frac{1}{27}$ **3.** $\frac{1}{4}$ **4.** ± 12

SELF CHECK 11.3

1. a. $3, 12, 48, 192$ **b.** 49,152   **2.** $\frac{1}{625}$
3. $-2, -6, -18, -54, -162$ **4.** $1, 2, 4, 8, 16$ **5.** 20
6. 124.96   **7.** 125   **8.** $\frac{243}{4}$ **9.** $\frac{2}{3}$ **10.** About $763
11. 135 in.

STUDY SET SECTION 11.3 (page 930)

1. geometric   **3.** mean   **5.** $16, 4, 1$ **7.** $a_n = a_1r^{n-1}$
9. a. Yes   **b.** No   **c.** No   **d.** Yes   **11.** r^2, a_1r^4 **13.** n
15. $3, 6, 12, 24, 48; 768$ **17.** $-5, -1, -\frac{1}{5}, -\frac{1}{25}, -\frac{1}{125}; -\frac{1}{15,625}$
19. 1,458   **21.** $-\frac{3,125}{2}$ **23.** $2, 6, 18, 54, 162$
25. $3, 6, 12, 24, 48$ **27.** $6, 18, 54$ **29.** $-20, -100, -500, -2,500$
31. -16 or 16   **33.** $-10\sqrt{2}$ or $10\sqrt{2}$ **35.** 728   **37.** 122
39. -255 **41.** 381   **43.** 16   **45.** 81   **47.** $-\frac{81}{2}$ **49.** No sum
51. 8   **53.** $-\frac{135}{4}$ **55.** $\frac{1}{9}$ **57.** $\frac{1}{3}$ **59.** $\frac{4}{33}$ **61.** $\frac{25}{33}$ **63.** 3
65. $-64, 32, -16, 8, -4$ **67.** No geometric mean exists.
69. 3,584   **71.** $-64, -32, -16, -8$ **73.** 4   **75.** $\frac{1}{27}$ **77.** 93
79. $\$1,469.74$ **81.** About $539,731
83. $\left(\frac{1}{2}\right)^{11} \approx 0.0005$ square unit   **85.** 30 m   **87.** 5,000
93. $[-1, 6]$ **95.** $(-\infty, -3) \cup (4, \infty)$

CHAPTER 11 REVIEW (page 933)

1. $1, 3, 1, 1, 10, 15; 1, 5, 10, 10, 5, 1$ **2. a.** 13   **b.** 12
c. a^{12}, b^{12} **d.** a: decrease; b: increase   **3.** 144   **4.** 20
5. 15   **6.** 220   **7.** 1   **8.** 8
9. $x^5 + 5x^4y + 10x^3y^2 + 10x^2y^3 + 5xy^4 + y^5$
10. $x^9 - 9x^8y + 36x^7y^2 - 84x^6y^3 + 126x^5y^4 - 126x^4y^5 + 84x^3y^6 - 36x^2y^7 + 9xy^8 - y^9$
11. $64x^3 - 48x^2y + 12xy^2 - y^3$
12. $\frac{c^4}{16} + \frac{c^3d}{6} + \frac{c^2d^2}{6} + \frac{2cd^3}{27} + \frac{d^4}{81}$ **13.** $6x^2y^2$ **14.** $-20x^3y^3$
15. $-108x^2y$ **16.** $5u^2v^{12}$ **17.** $-2, 0, 2, 4$
18. $-\frac{1}{2}, \frac{1}{3}, -\frac{1}{4}, \frac{1}{5}, -\frac{1}{6}$ **19.** 75   **20.** 42
21. $122, 137, 152, 167, 182$ **22.** $-1,194$ **23.** -5
24. $\frac{41}{3}, \frac{58}{3}$ **25.** $-\frac{45}{2}$ **26.** 1,568   **27.** $\frac{15}{2}$ **28.** 378
29. 14   **30.** 360   **31.** 20,100   **32.** 1,170   **33.** 4
34. $24, 12, 6, 3, \frac{3}{2}$ **35.** $\frac{1}{27}$ **36.** $24, -96$ **37.** $\frac{2,186}{9}$
38. $-\frac{85}{8}$ **39.** About 1.6 lb   **40.** 125   **41.** $\frac{5}{99}$ **42.** 190 ft

CHAPTER 11 TEST (page 937)

1. a. Pascal's   **b.** alternate   **c.** arithmetic   **d.** series
e. geometric   **2.** $2, -4, -10, -16$ **3.** $-1, \frac{1}{8}, -\frac{1}{27}, \frac{1}{64}, -\frac{1}{125}$
4. 210   **5.** $a^6 - 6a^5b + 15a^4b^2 - 20a^3b^3 + 15a^2b^4 - 6ab^5 + b^6$
6. $24x^4y^2$ **7.** 66   **8.** 306   **9.** $34, 66$ **10.** $\frac{1}{4}$ **11.** $-1,377$
12. 205   **13.** 1,600 ft   **14.** 3   **15.** -81 **16.** $\frac{364}{27}$ **17.** 6
18. $18, 108$ **19.** $\frac{27}{2}$ **20.** About $651,583   **21.** 2,000 in.   **22.** $\frac{7}{9}$

GROUP PROJECT (page 938)

1. g   **2.** o   **3.** i   **4.** l   **5.** u   **6.** y   **7.** p   **8.** x   **9.** m   **10.** d
11. f   **12.** c   **13.** w   **14.** b   **15.** j   **16.** s   **17.** e   **18.** z
19. n   **20.** k   **21.** h   **22.** v   **23.** q   **24.** a   **25.** r   **26.** t

CUMULATIVE REVIEW CHAPTERS 1–11 (PAGE 939)

1. a. 0   **b.** $-\frac{4}{3}, 5.6, 0, -23$ **c.** $\pi, \sqrt{2}, e$
d. $-\frac{4}{3}, \pi, 5.6, \sqrt{2}, 0, -23, e$ **2.** 0
3. $b_2 = \frac{2A - b_1h}{h}$ or $b_2 = \frac{2A}{h} - b_1$ **4.** $85°, 80°, 15°$ **5.** $\$8,250$
6. $\frac{1}{120}$ decibels/rpm   **7. a.** Parallel   **b.** Perpendicular
8. $y = -3,400x + 28,000$ **9.** $y = -2x + 5$
10. $y = -\frac{9}{13}x + \frac{7}{13}$ **11.** $-4; 3a^5 - 2a^2 + 1$

12. a. $P(t) = -0.26t + 11.6$ **b.** 1.2%   **13.** It doesn't pass the vertical line test. The graph passes through $(0, 2)$ and $(0, -2)$.

14. a. 4   **b.** 3   **c.** 0, 2   **d.** 1

15. D: the set of real numbers $(-\infty, \infty)$; R: the set of nonnegative real numbers $[0, \infty)$

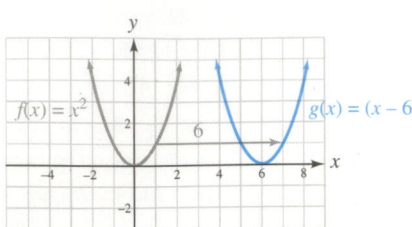

16. $(2, -2)$

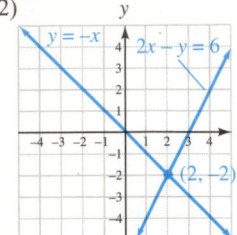

17. $(-7, 7)$

18. $\left(\frac{4}{5}, \frac{3}{4}\right)$ **19.** $(3, 2, 1)$

20. $(2, -3)$ **21.** $\left(-2, \frac{3}{2}\right)$

22. Regular: $29\frac{1}{3}$ lb;
Brazilian: $10\frac{2}{3}$ lb

23. Wind: 50 mph; plane: 450 mph

24. Ankle: 100 pair, low cut: 150 pair; crew: 250 pair

25. $(-\infty, -2)$ 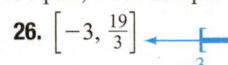   **26.** $\left[-3, \frac{19}{3}\right]$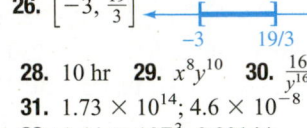

27.

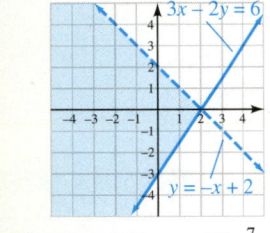

28. 10 hr   **29.** $x^8 y^{10}$ **30.** $\frac{16}{y^{16}}$

31. 1.73×10^{14}; 4.6×10^{-8}

32. 1.44×10^{-3}; 0.00144

33. 91 cantalopes   **34.** $\frac{7}{4}rt^2 - \frac{5}{6}rt$ **35.** $2x^2y^3 + 13xy + 3y^2$

36. $x^3 - 27y^3$ **37.** $6m^{10} - 23m^5 + 7$

38. $81a^2b^4 - 72ab^2 + 16$ **39.** $xy(3x - 4y)(x - 2)$

40. $(b - 4)(b^2 - 3)$ **41.** $(3y + 2)(4y + 5)$

42. $(16x^2y^2 + z^4)(4xy + z^2)(4xy - z^2)$

43. $(3t + u)(9t^2 - 3tu + u^2)$

44. $b^2(a + 4)(a - 4)(a + 2)(a - 2)$ **45.** $\lambda = \frac{4d - 2}{A - 6}$

46. It is a solution.   **47.** A repeated solution of -8 **48.** 0, 1, -9

49. Length: 21 ft; width: 4 ft   **50.** 5 sec

51. $(-\infty, -6) \cup (-6, 4) \cup (4, \infty)$ **52.** $\frac{3x + 2}{3x - 2}$ **53.** $-\frac{q}{p}$

54. $\frac{4a - 1}{(a + 2)(a - 2)}$ **55.** $y - x$ **56.** $4x^3 - 3x^2 + 3x + 1$

57. $-\frac{7}{5}$ **58.** $R = \frac{R_1R_2R_3}{R_2R_3 + R_1R_3 + R_1R_2}$

59. $2\frac{2}{5}$ hr $= 2.4$ hr   **60.** 93   **61.** 2 lumens

62. D: $[0, \infty)$; R: $[2, \infty)$

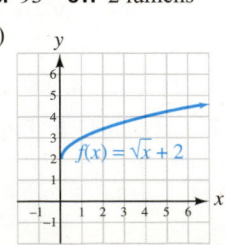

63. $[-3, \infty)$ **64.** 4 ft   **65.** $\frac{343}{125}$ **66.** $4ab^2\sqrt{7ab}$ **67.** $5\sqrt{2}$

68. $81x\sqrt[3]{3x}$ **69.** $13 - \sqrt{7}$ **70.** $15x - 6\sqrt{15x} + 9$

71. $\frac{\sqrt[3]{4b^2}}{b}$ **72.** $\sqrt{3t} - 1$ **73.** 1, 3   **74.** 5   **75.** 0

76. About $22\frac{1}{2}$ in.   **77.** $5i$ **78.** -1 **79.** $-5 + 17i$

80. $-\frac{21}{29} - \frac{20}{29}i$ **81.** $\pm 2\sqrt{6}$ **82.** $-5 \pm 4\sqrt{2}$ **83.** $\frac{-3 \pm \sqrt{5}}{4}$

84. $\frac{2}{3} \pm \frac{\sqrt{2}}{3}i$ **85.** $\frac{1}{4}, \frac{1}{2}$ **86.** $-i, i, -3i\sqrt{2}, 3i\sqrt{2}$

87. a. A quadratic function

b. At about 85% and 120% of the suggested inflation

88. $(-1, -2)$; $(0, -8)$; no x-intercepts

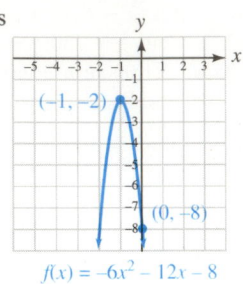

89. $[3, 5]$ **90.** $4x^2 + 4x - 1$

91. $f^{-1}(x) = \sqrt[3]{\frac{x + 1}{2}}$

92. D: $(-\infty, \infty)$; R: $(0, \infty)$

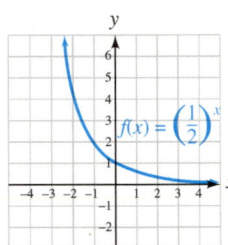

93. 3   **94.** 2   **95.** $\frac{1}{27}$ **96.** 5   **97.** 1   **98.** -1

99.

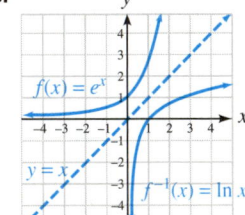

100. About 150 million

101. Undefined

102. $2 - 3\log_6 x$ **103.** $\ln \frac{y\sqrt{x}}{z}$

104. About 13.4 hr   **105.** $-\frac{3}{4}$

106. 3.4190

107. 1, -10 does not check

108. 9   **109. a.** ii   **b.** iv   **c.** i

d. iii   **110.**

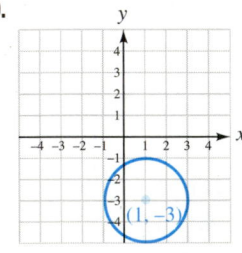

111. $y = 3$

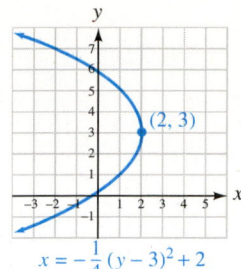

$x = -\frac{1}{4}(y-3)^2 + 2$

112.

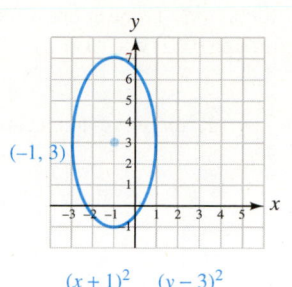

$\dfrac{(x+1)^2}{4} + \dfrac{(y-3)^2}{16} = 1$

113.

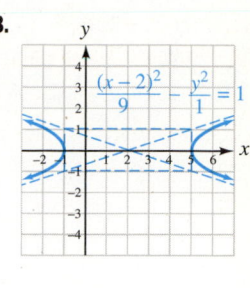

114. $81a^4 - 108a^3b + 54a^2b^2 - 12ab^3 + b^4$ **115.** 66
116. 103 **117.** 690 **118.** 27 **119.** 27 **120.** $2,848.31
121. $\frac{1,023}{64}$ **122.** $\frac{27}{2}$

INDEX

Minors of determinant, 240
Mirror images, 715, 769
Missing terms, 517–518
Mixed operations, 487, 497–498
Mixture problems, solving, 80–82,
    256–259
Monomials
    defined, 377
    dividing by monomials, 512–514
    multiplying, 390–391
Multiplication property
    of equality, 44
    of inequalities, 291–292
Multiplication property of zero, 34
Multiplicative identity, 34–35
Multiplicative inverses, 34–35
Multiplying
    binomials, 391–392
    complex numbers, 653–655
    distributive property and, 36–38
    monomials, 390
    polynomial by monomial, 390–391
    polynomial by polynomial, 393–394
    radical expressions, 617–619
    rational expressions, 480–484, 617–625
    real numbers, 22–24
    to simplify expressions, 398
    three polynomials, 394–395

N
Napierian logarithms, 805
Natural exponential function
    defined, 801
    graphing, 801–803
Natural logarithmic expressions,
    evaluating, 791–793, 805–806
Natural logarithmic function, graphing,
    806–807
Natural logarithms, 805
Natural-number exponents, 24, 357
Natural numbers, defined, 9–10
Negative exponents, 361–365
Negative fractions, 11–12
Negative infinity, 290
Negative integer exponent rules, 360–363
Negative numbers, 10, 651–652
Negative rational exponents, simplifying,
    600–601
Negative reciprocals, 132–133
Negative square roots, 581
Nested parentheses, 755
n factorial, 905
Nonlinear equations, solving in one
    variable, 44
Nonlinear functions, graphing, 168–171
Nonlinear inequalities, graphing in two
    variables, 734–735
Nonlinear systems of equations, solving,
    887–890
Nonnegative real numbers, 584
Notation
    cube root, 587
    decimal, 368
    double-sign, 679

factorial, 905–906
function, 155–157
interval, 289–291
scientific, 368, 369–374
set-builder, 12, 289–291
square roots, 581
standard, 368, 370–371
subscript, 58, 128
summation, 917–918
nth power of the nth root, 619, 633
nth roots, 589–591
Number i, 651–652
Number lines, 10
Numbers
    composite, 10
    critical, 728
    irrational, 12–13
    negative, 10, 651–652
    positive, 10
    prime, 10
    rational, 11–12
    real, 13–17, 20–29, 32–40, 548, 650
    whole, 10
Number-value problems
    solving, 68–69
    solving using systems of equations,
        251–252
Numerators, rationalizing, 625
Numerical coefficients, 33, 377

O
Obtuse angles, defined, 69
Odd integers, defined, 11
Odd root, 589
One-to-one functions, 763–765
Open intervals, 305
Operations
    on functions, 752–754
    involving powers of i, 658–659
    with real numbers, 20–29
Opposites
    additive inverses, 209
    defined, 10
    denominators as, 496–498
    quotient of, 476
    of real numbers, 16–17
Order, of matrices, 228
Ordered pairs
    graphs of, 102–103
    as solutions of equations, 113–114
    as solutions of systems of equations,
        195–196
Ordered triples, 217–218
Ordering, real numbers, 15–16
Order of operations, 26–28
Origin, 870–872, 878–881
Output value, 154

P
Paired data, graphing, 104
Parabolas
    as conic sections, 856–857
    defined, 863
    equations of, 862–866

graphing, 168–169, 862–866
    as quadratic functions, 715
    vertex of, 715, 719–722
Paraboloids, 857
Parallel lines
    recognizing, 144–146
    slopes of, 131–133
Parallelograms, 251
Parentheses, 289, 755
Partial sum, 915, 926
Partition, 728
Pascals (Pa), 136
Pascal's triangle, expanding binomials
    using, 903–904
Percent of increase or decrease, 77–78
Percent problems, solving, 75–78
Perfect-cube integers, 428
Perfect-square integers, 425
Perfect squares, 581–582
Perfect-square trinomials, 412, 682–683
Perimeters, finding, 55
Period of a pendulum, 586
Perpendicular lines
    recognizing, 144–146
    slopes of, 131–133
pH of a solution, 822–823
pH scale, 833
Plane, 218
Plotting a point, 103
Point-plotting method, 116, 158, 168–170,
    775
Point-slope form, writing equations of
    lines using, 141–142
Polynomial equations, 438, 443–444
Polynomial functions
    domain of, 380–381
    evaluating, 379–381
    quadratic functions as, 379, 446
Polynomials
    adding, 382–383
    classifying, 377–379
    defined, 376–377
    degree of, 378–379
    in descending powers of n, 377
    dividing by monomial, 512–514
    dividing by polynomial, 514–517
    dividing polynomial with missing terms,
        517–518
    evaluating using remainder theorem,
        524–525
    factoring by grouping, 405–408
    factoring random, 434–437
    factoring steps, 434
    factoring using factor theorem, 525–526
    greatest common factor from, 401–405
    multiplying by monomial, 390–391
    multiplying by polynomials, 393–394
    multiplying three, 394–395
    in one variable, 377
    prime, 404
    simplifying by combining like terms,
        381–382
    special products, 395–398
    subtracting, 383–385